渔药安全使用风险评估及其控制
（上册）

杨先乐　主　编

胡　鲲　副主编

海洋出版社

2020年·北京

图书在版编目（CIP）数据

渔药安全使用风险评估及其控制/杨先乐主编．—北京：海洋出版社，2016.12
ISBN 978-7-5027-9664-8

Ⅰ.①渔… Ⅱ.①杨… Ⅲ.①水产生物-药物学-安全评价②水产生物-药物学-安全管理
Ⅳ.①S948

中国版本图书馆 CIP 数据核字（2016）第 303306 号

责任编辑：高显刚 杨海萍
责任印制：赵麟苏

海洋出版社 出版发行

http://www.oceanpress.com.cn
北京市海淀区大慧寺路 8 号 邮编：100081
北京朝阳印刷厂有限责任公司印刷 新华书店北京发行所经销
2020 年 1 月第 1 版 2020 年 1 月第 1 次印刷
开本：889mm×1194mm 1/16 印张：66.75
字数：3075 千字 定价：368.00 元
发行部：62132549 邮购部：68038093 总编室：62114335
海洋版图书印、装错误可随时退换

《渔药安全使用风险评估及其控制》
编委会

主　编：杨先乐（上海海洋大学）

副主编：胡鲲（上海海洋大学）

编　委：艾晓辉（中国水产科学研究院长江水产研究所）

　　　　房文红（中国水产科学研究院东海水产研究所）

　　　　李忠琴（集美大学）

　　　　林茂（集美大学）

　　　　刘永涛（中国水产科学研究院长江水产研究所）

　　　　卢彤岩（中国水产科学研究院黑龙江水产研究所）

　　　　沈锦玉（浙江省淡水水产研究所）

　　　　汤菊芬（广东海洋大学）

　　　　王荻（中国水产科学研究院黑龙江水产研究所）

　　　　王伟利（中国水产科学研究院珠江水产研究所）

　　　　尹文林（浙江省淡水水产研究所）

序

 水产动物病害是制约水产养殖业健康、持续发展的关键因素之一。控制水产动物病害主要有三种方法：免疫防治、生态防治和药物防治。其中药物防治是一种简单、方便、廉价、效果易见的方法，容易被养殖者所接受。但药物防治却存在着很大的隐患，如果不正确使用甚至滥用，不仅会直接影响水产品的质量安全，而且会导致渔业水域环境受到相应的污染和破坏，进而对生态环境，乃至人类健康带来了潜在的、甚至持久严重的威胁。因此对渔药使用安全风险进行科学的评估，对渔药使用安全风险有效地控制是确保水域生态环境安全、水产品质量安全一个重要的技术措施和手段。

 在科技部、农业部以及上海市农委等部门的支持下，由上海海洋大学、中国水产科学研究院长江水产研究所、中国水产科学研究院东海水产研究所、中国水产科学研究院珠江水产研究所、中国水产科学研究院黑龙江水产研究所、浙江淡水水产研究所、集美大学、广东海洋大学等单位科研人员所组成的研究团队，经过十余载的潜心研究，在渔药安全使用风险评估及其控制的理论和技术研究方面取得了一批重大成果，编著了国内外第一部在渔药领域的著作。这本著作既是他们心血的结晶，又是我国科学研究者在渔药领域成果的一次重大展示。相信这本著作将会在很大程度上推动我国乃至全世界水产养殖病害药物防治领域理论与技术的发展，也将会为我国在渔药安全使用、水产品质量安全、生态环境安全的指导与监管上起到重要的作用。

张显良

农业部渔业渔政管理局

2016 年 11 月 10 日

前　言

我国是水产养殖大国，也是渔药生产和使用大国。但因渔药相关的理论与应用技术研究滞后，加上国外在该方面可资借鉴的资料匮乏，导致渔药的生产和使用缺乏科学的指导，存在着一定的安全隐患。因此，针对我国水产养殖的特点，加强渔药基础理论研究，科学评估和控制渔药使用风险，对保障公共卫生安全、促进水产养殖业持续、稳定、健康发展具有重要的意义。

十余年来，以上海海洋大学牵头，联合有关高校、科研院所组成的团队聚焦渔药基础研究，从渔药作用机理、药代动力学和药效学、新型渔药创制、药物残留检测、耐药性分析控制、渔药使用风险评估及其安全使用技术等方面开展了系统的研究，积累了一大批适合我国水产养殖特点的技术成果和推广应用经验。特别是"十二五"期间，研究团队在农业部公益性行业（农业）科研专项项目《渔药使用风险评估及其控制技术研究与示范》（项目编号201203085）等项目的支持下，集成了前期渔药安全使用风险评估及其控制技术体系的相关成果，在我国主要水产养殖区域建立了风险可控的渔药安全使用示范区，为从技术上解决长期困扰水产养殖产业发展的药物安全难题奠定了基础。

为了总结在渔药安全使用方面所积累的经验和成果，更好地促进我国在渔药领域的理论发展和技术进步，研究团队决定编写《渔药安全使用风险评估及其控制》一书。2015 年 10月成立了《渔药安全使用风险评估及其控制》编撰委员会，制定编撰宗旨。2015 年底，确定了编撰方案和大纲，并进行了编写分工。2016 年底 10 月完成了该书第一稿，11 月对本书进行了最后一次修改和审阅。

根据集成研究成果的特点，本书在理念、结构及撰写方式上，与传统的专著有较大的不同。第一，全书从已有的成果中进行提炼和加工，在编排体系上先进行概述，再根据逻辑顺序编排相关研究论文全文或摘要，为了体现最新进展，部分未见刊的研究论文也经过筛选一并收入。第二，内容上突出理论基础对实践应用的指导，突出了渔药的作用机理，如从渔药代谢酶 P450、渔药受体 GABA 和转运蛋白 P 糖蛋白等角度阐释了渔药代谢的机理。第三，突出了实践和创新，如孔雀石绿替代制剂——"复方甲霜灵粉"的创制研究从技术源头上解决了孔雀石绿禁用后留下的技术真空难题。

本书共分为七章，第一章渔药应用的基础理论，阐述了渔药作用的机理、代谢消除规律及残留研究模型，由杨先乐、胡鲲、房文红编写；第二章渔药代谢动力学参数及其使用技术，由杨先乐、艾晓辉、房文红、沈锦玉、姜兰、卢彤岩、胡鲲、林茂、汤菊芬编写；第三章新渔药及其制剂的研究，由杨先乐、胡鲲、沈锦玉、艾晓辉、汤菊芬编写；第四章渔药检测方法的研究和应用，由杨先乐、艾晓辉、房文红、沈锦玉、姜兰、卢彤岩、胡鲲、林茂、汤菊芬编写；第五章渔药的耐药性与控制，由房文红、姜兰、沈锦玉、卢彤岩、杨先乐编写；第六章 渔药使用的风险评估和第七章渔药的安全使用技术的应用、推广和示范由研究团队全体人员总结和编写。全书由杨先乐、胡鲲统稿。

本书的编写和整理得到了姜莺颖、杨柳、范宁宁、刘力硕、董龙香、张一柳、何山、刘付翠等人的协助，在此表示感谢。本书在编写过程中参阅了大量的国内外出版发行的（或即将出版发行的）文献、资料和书籍，限于篇幅的原因，未能一一列出，在此一并向原作者和出版单位致谢。由于现有的资料有限，加上编者水平和时间的限制，本书疏忽、错漏之处在所难免，敬请广大读者批评指正。

期盼《渔药安全使用风险评估及其控制》一书能为促进渔药基础理论和应用技术的发展，促进我国水产养殖的进步，保护水域生态环境，提高我国水产品质量，起到重要的推动作用。

编者

2016 年 12 月

目 录

第一章　渔药应用的基础理论

渔药的作用受到水生动物体内的酶、受体等因素的影响，而渔药的使用安全与其代谢、消除乃至残留等息息相关。研究渔药作用的酶、受体，研究渔药的药-动学模型及渔药在水产动物体内的检测方法，研究其代谢与消除规律，是保证渔药使用安全、合理使用的理论基础。

第一节　渔药作用的机理

在建立渔药代谢酶体外研究模型的基础上，对鱼类细胞色素 P450 系列酶的活性、表达及其诱导与抑制进行了研究，同时通过转运蛋白 P 糖蛋白和渔药作用受体 GABA 与药物作用的研究，探讨了渔药作用的机理。

一、渔药的代谢酶

细胞色素 P450（cytochromeP450 或 CYP450，简称 CYP450）在生物体内广泛分布，其主要功能是参与动物体内的药物代谢，有生理上功能的 CYP450 约有 50 种，包括：CYP1A、CYP3A、CYP2E1 等。如异育银鲫主要组织微粒体具有 CYP450 亚型活性，且它们在银鲫体内的分布和活性存在着组织与器官差异性，以肝组织中的含量和活性为最高（贾娴等，2011）。CYP3A 在异育银鲫组织中分布广泛，其中肝和肠组织中转录水平最高，在肾和鳃中次之（朱磊等，2011）。

CYP 数量和活性直接影响药物在体内的活化与代谢。CO 还原差示光谱法可以准确测定草鱼肝微粒体中 CYP 酶及细胞色素 b5 含量，其中细胞色素 b5 含量约为 P450 含量的一半（张宁等，2007）。红霉素-N-脱甲基酶（ERND）是细胞色素 P45023A（CYP3A）依赖的酶类，可通过测定该酶催化代谢产物甲醛的量来反映其活性（李聃等，2007）。

草鱼肝细胞系（GCL）是一种研究 CYP3A 的模型（Dan Li 等，2008）。草鱼前肠、肝脏和肾脏组织中 CYP3A 高度表达，而在其他组织低度表达（符贵红等，2012）。GCL 中，Log-Normal 模型能较好的拟合混 O-脱乙基酶（7-ethoxyresorufin-O-deethylase，EROD）的剂量-效应曲线，诱导剂 2，3，7，8-四氯二苯并（p）二（2，3，7，8-TCDD）对 GCL 细胞作用时间越长，EC100（诱导效应最大时的诱导剂浓度）越大，p100（诱导效应最大时的 EROD 活性）越小，但在作用 48 h 后即达到较大诱导效应，延长时间并无显著加强（林茂等，2006）。

许多药物能诱导 CYP 活性，草鱼肝脏中的 CYP3A 就能被利福平（rifampicin，RIF）诱导（符贵红等，2012）。乙醇也能诱导草鱼肝细胞中 CYP2E1，而加快底物氯唑沙宗（CZX）的代谢（王翔凌等，2008）。乙醇对 CHSE-1 细胞中的 CYP2E1 也具有诱导作用，在 Gentox 模型中，乙醇对 CYP2E1 有先诱导后抑制作用，CYP2E1 酶活性随着诱导时间的增加逐渐增强，24 h 内酶活性随时间的增加而增强（张宁等，2007）。40 μM 利福平诱导大口黑鲈（*Micropterus salmoides*）原代肝细胞 24 h 后，CYP450 酶系中的氨基比林-N-脱甲基酶（ANDM）活性达到最高（喻文娟等，2008）。在银鲫肝细胞中，双氟沙星可能是 CYP1A 的基质和抑制剂（Gui HongFu et al，2011）。恩诺沙星在鲫鱼肝微粒体中的代谢以 N-脱乙基反应为主，可能还通过其他途径生成另外代谢产物。恩诺沙星 N-脱乙基酶与底物结合力不强，酶促反应强度较弱（贾娴等，2009）。恩诺沙星也能抑制史氏鲟体内的细胞色素 P450 的酶活性（卢彤岩等，2011）。β-萘黄酮（BNF）能诱导草鱼肝、肾和卵巢细胞中 CYP1A 的表达（林茂等，2010）。恩诺沙星能以剂量依赖型抑制异育银鲫体内 CYP1A 表达，从而抑制其催化活性；CYP3A 的抑制是由于恩诺沙星引起的 CYP3A（蛋白和 mRNA）表达水平下降所致（X.HU et al.，2011）。氟甲喹对鲫鱼肝 CYP1A 的诱导是在翻译后水平，可能是加强蛋白的稳定性（胡晓等，2011）。

草鱼（*Ctenopharyngodon idellus*）肾细胞系（CIK）中利福平（rifampicin，RIF）能诱导 CYP3A 基因表达和红霉素-N-脱甲基酶（ERND）活性。RIF 还能促进恩诺沙星（ENR）的脱乙基代谢，且 RIF 诱导的酶与底物的亲合力较高，酶促反应的强度较大（林茂等，2007）。RIF 对 CYP3A 的诱导可能主要作用于转录水平而不直接作用于酶活（符贵红等，2011）。RIF 在浓度为 40μM 时对草鱼肝细胞系（GCL）中 CYP3A 活性具有显著诱导作用，而苯巴比妥（PB）和地塞米松（DEX）无显著诱导作用（P<0.05），RIF 诱导的 CYP3A 亲和力较强，催化效能较高（Dan Lia et al.，2008；李聃等，2008）。双氟沙星（DIF）与 β-萘黄酮（BNF）在草鱼肾细胞（CIK）中共孵育后，DIF 的代谢量增加了 1 倍，该结果证明 BNF 可能是 CYP1A 的特异性诱导剂，且 CYP1A 可能参与了 DIF 的代谢（于灵芝等，2010）。在草鱼肾细胞中，CYP1A 还能介导了双氟沙星的 N-脱甲基反应，即双氟沙星转换成沙拉沙星的过程（Ling Zhi Yu et al.，2010）。

不同种鲟体内的谷胱甘肽-S-转移酶（glutathione-S-transferases，GST）能与谷胱甘肽（glutathione，GSH）具有种差异

1

性，且其肝脏中 GST 活性和 GSH 含量均受恩诺沙星的影响（陈琛，2011）。诺氟沙星可以在鲟鱼体内诱导药物代谢相关酶如 ECOD 和 GST 等的酶活变化（S. W. Li，2013）。

某些中草药也能影响药物酶的表达。黄连素很可能是异育银鲫 CYP1A 的有效抑制剂，但却在较高的黄连素剂量下才对 CYP3A 起抑制作用（Chang Zhou，2011）。黄芩、柴胡、连翘、吴茱萸和野菊花 5 种中药提取物对异育银鲫肝微粒体中 CYP1A 的抑制均较显著，而对 CYP3A 的抑制则不明显（周常，2012）。黄芩苷和甘草酸很可能通过诱导 CYP1A 和 CYP3A 活性减少异育银鲫（*Carassius auratus gibelio*）对恩诺沙星的吸收，并促进其代谢产物 CF 的生成（房文红等，2012）。

1. P450 酶细胞诱导模型的建立

利用 GCL、大口黑鲈原代肝细胞等建立了 P450 酶的体外诱导模型，为鱼类药物代谢酶诱导、酶促动力学的研究创建了平台。

该文发表于《高技术通讯》【2007，17（3）：314-318】

草鱼肝细胞中恩诺沙星脱乙基代谢的酶动力学
Enzymatics on deethylation metabolism of enrofloxacin in grass carp hepatocyte

林 茂[1,2]，杨先乐[1]，王翔凌[1]，房文红[1]，喻文娟[1]

（上海水产大学农业部渔业动植物病原库，上海 200090；1. 上海高校水产养殖学 E-研究院，上海 200090；
2. 集美大学水产学院，福建 厦门 361021）

摘要：用 RP-HPLC 法检测了草鱼肝细胞系（GCL）细胞中恩诺沙星（ENR）的代谢情况，结果显示，药物利福平（RIF）和苯巴比妥（PB）的浓度分别在（12.6±0.9）μM 和（34.8±3.6）μM 时对 ENR 脱乙基酶具有最大的诱导倍数，分别为（6.0±0.4）和（2.3±0.3）；而红霉素（ERM）对 ENR 脱乙基酶具有抑制作用，其半抑制浓度（EC_{50}）为（20.3±1.8）μM。RIF 诱导组和对照组的 GCL 细胞中 ENR 代谢的药时曲线方程分别为 $Ct/Co = -0.0654Ln（t）+0.7342$ 和 $Ct/Co = -0.0280Ln（t）+0.9300$，酶动力学方程分别为 $1/V = 72.5530×1/[S]+1.4922$ 和 $1/V = 530.9800×1/[S]+3.5274$。ENR 消除率和环丙沙星（CIP）转化率的研究结果表明 RIF 对 ENR 的脱乙基代谢具有显著的促进作用，而酶动力学参数表明，RIF 诱导的酶与底物的亲合力较高，酶促反应的强度较大。

关键词：恩诺沙星，草鱼肝细胞，酶动力学，高效液相色谱，脱乙基

参与生物转化的主要酶类是依赖细胞色素 P450（cytochrome P450，CYP）的混合功能氧化酶系统，P450 酶系是一个超大家族（superfamily），种类众多而且催化药物代谢的反应非常多样化，包括脂肪族氧化、芳香族羟化、N-去烷基、O-去烷基、环氧化、氧化去胺、去硫、亚砜生成、N-氧化或 N-羟化、去卤等反应类型[1]。

恩诺沙星（enrofloxacin，ENR）是上市的第一种兽用氟喹诺酮类药物（fluoroquinolones，FQs），该药具较为广谱的抗菌性，可用于治疗多种感染性疾病。我国在 1994 年投入市场，并广泛应用于畜禽生产上，目前在水产养殖过程中的应用也相当广泛。ENR 在动物体内的主要代谢途径是通过氧化脱乙基反应生成环丙沙星（ciprofloxacin，CIP），这个反应属于典型的 N-去烷基反应[2]。

近年来，FQs 药物对 P450 药物代谢酶活性影响的研究甚为活跃，较多报道认为 FQs 药物如 ENR、CIP 等对 P450 中的某些药物代谢酶如红霉素脱甲基酶、氨基比林脱甲基酶、乙基吗啡脱甲基酶、乙氧异吩噁唑酮脱乙基酶、苯并芘羟化酶、戊巴比妥侧链羟化酶的活性有抑制作用[3-5]，也有少量研究表明了一些 FQs 药物如 ENR 在一定浓度时对 P450 酶也有明显诱导作用[6]。相对于 FQs 药物作为药物代谢酶的抑制剂或诱导剂，其作为酶底物的研究还很欠缺，对何种 P450 酶在相应的 FQs 药物生物转化中起关键作用尚不明确。

本文通过对 ENR-N-脱乙基酶的体外诱导，研究了 ENR 在草鱼（*Ctenopharyngodon idellus*）肝细胞系中的代谢时间曲线以及酶动力学参数，为进一步阐明 ENR 的代谢途径和机理提供借鉴，对指导临床合理用药具有重要的应用价值。

1 材料和方法

1.1 细胞

草鱼肝细胞系（GCL）由农业部渔业动植物病原库提供，以 M199 培养基置 28℃培养。

1.2　试剂

M199 基础培养基为 GIBCO-BRL 产品，红霉素（erythromycin，ERM）为 BIB 产品，利福平（rifampicin，RIF）、苯巴比妥（phenobarbital，PB）以及 ENR 和 CIP 标准品购自 sigma 公司，ENR 原粉（98% 纯度）购自上海三维制药厂。

1.3　色谱条件

Agilent 1100 系列高效液相色谱（HPLC）工作站。色谱柱：C18 柱（4.6 mm×150 mm）；流动相：甲醇/盐溶液 = 25/75（V/V）（盐溶液含 0.05 M 柠檬酸和 0.01 M 醋酸铵，用三乙胺调 pH 值到 4.5）；流速：1.0 mL·min^{-1}；检测波长：276 nm；柱温：40℃；进样量：10 μL[2]。

1.4　细胞样品前处理

细胞样品与甲醇 1∶1 混合，漩涡振荡 2 min 后离心（4 000 r·min^{-1}，20 min）。吸取上清液置广口小瓶中，在通风橱中 60℃ 水浴风干。加 1 mL 流动相盐溶液和 1 mL 正己烷溶解，取水相以 0.45 μm 滤膜过滤[2]。

1.5　细胞中 ENR 脱乙基酶活性的检测

GCL 细胞以 105 个·mL^{-1}的密度每孔 1 mL 接种于 24 孔板中预培养 24 h，然后加入药酶诱导剂或抑制剂作用 72 h，倒去培养基，以 PBS 缓冲液清洗。加入 1 mL 含 20 μM ENR 的无酚红无血清培养基，振荡后置培养箱中孵育 1 h。24 孔中样品每孔加 1 mL 甲醇，振荡 5 min，静置 10 min 后进行样品前处理，之后再进行 RP-HPLC 检测。ENR-N-脱乙基酶的活性以酶促反应速率 V，即每毫克蛋白单位时间内的 CIP 生成量表示（μmol·min^{-1}·mg^{-1}）。

1.6　诱导剂剂量效应曲线的拟合

通过 SigmaPlot 8.0 软件 Peak Log-Normal 模型对药酶诱导剂所形成的钟形效应曲线进行拟合[7]。该模型方程式如下：

$$F(x) = p_0 + (p_{100} - p_0) \exp\left[-0.5\left(\frac{\ln(x/EC_{100})}{b}\right)\right]$$

式中，p_0 表示最小药酶活性，p_{100} 表示最大药酶活性，b 为斜率系数，EC_{100} 表示诱导效应达 100%（值等于 p_{100}）时的诱导剂浓度。

1.7　抑制剂剂量效应曲线的拟合

利用 SigmaPlot 8.0 软件的 Logistic 模型对药酶抑制剂所形成的反 "S" 形剂量效应曲线进行拟合[8]。方程式如下：

$$F(x) = p_0 + \frac{p_{100} - p_0}{1 + \left(\dfrac{x}{EX_{50}}\right)^{nH}}$$

式中，nH 表示 Hill 常数，EC_{50} 表示抑制效应达 50% 时的抑制剂浓度。

2　结果

2.1　细胞中恩诺沙星（ENR）和环丙沙星（CIP）RP-HPLC 检测方法的建立

在上述色谱条件下，ENR 及其主要代谢产物 CIP 保留时间稳定，其出峰时间分别为 9.49 min 和 4.75 min，ENR 另有两个次要代谢产物出峰时间，分别在 5.74 min 和 5.92 min 附近（图 1-C）。分别以 ENR 和 CIP 的浓度（μM）为横坐标，峰面积为纵坐标作图，所得标准曲线的方程分别为 $Y=10.796X-4.2526$（$R^2=0.9988$）和 $Y=9.7234X-1.5593$（$R^2=0.9987$）。

此外，还检测了 ENR 和 CIP 在细胞悬液中的回收率和精密度，平均回收分别为（98.10±2.98）% 和（94.61±0.89）%，日内平均变异系数分别为（3.19±1.17）% 和（5.01±1.86）%，日间平均变异系数分别为（3.63±1.35）% 和（5.59±1.88）%。

2.2　药物对恩诺沙星脱乙基酶活性的影响

药物 RIF、PB 和 ERM 对 ENR 脱乙基酶活性作用的剂量-效应曲线如图 2 所示，RIF 和 PB 对 ENR 脱乙基酶具诱导作用，其 EC_{100} 分别为（12.6±0.9）μM 和（34.8±3.6）μM，最大诱导倍数（p_{100}/p_0）分别为 6.0±0.4 和 2.3±0.3。ERM 对 ENR 脱乙基酶具有抑制作用，其 EC_{50} 为（20.3±1.8）μM。

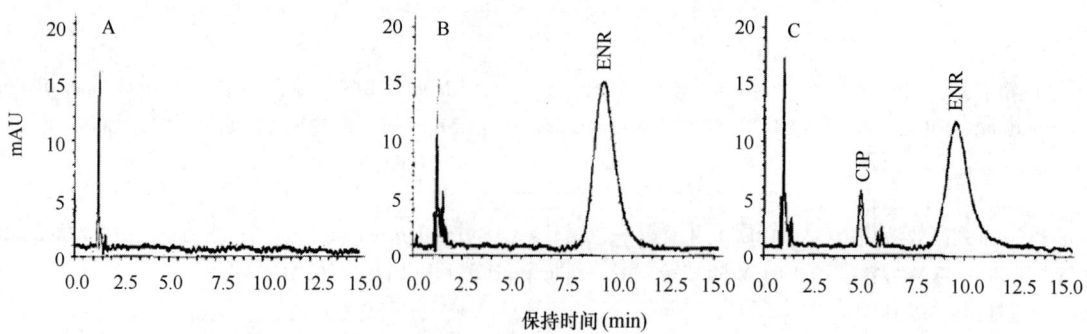

A: 空白细胞；B: ENR 与细胞悬液；C: ENR 与 RIF 诱导细胞，共孵育 12 h

图 1　ENR 及其代谢产物 CIP 的色谱图

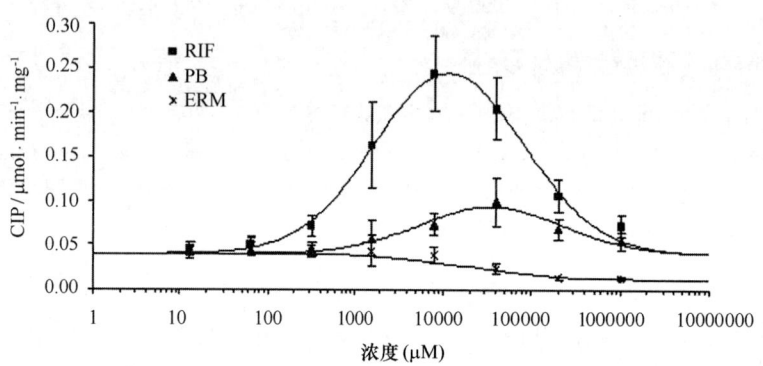

图 2　药物对 ENR 脱乙基酶活性的剂量效应曲线 （$n = 5$，$\bar{x} \pm s$）

2.3　恩诺沙星细胞体外代谢的药时曲线

GCL 细胞与初始浓度（C_o）为 20 μM 的 ENR 分别共孵育 0.25、0.5、1、3、6、12h 后检测 ENR 的剩余浓度（C_t）和 CIP 生成量（C_n）。经回归分析 RIF 诱导组和对照组的药时曲线方程分别为

$$C_t/C_o = -0.0654\mathrm{Ln}(t) + 0.7342 \quad (R^2 = 0.9744)$$
$$C_t/C_o = -0.0280\mathrm{Ln}(t) + 0.9300 \quad (R^2 = 0.9556)$$

其半衰期（T_1/T_2）分别为 35.9 h 和 4, 672, 210 h，结果显示对照组 ENR 的代谢较为缓慢，而 RIF 诱导组代谢较快（图 3）。

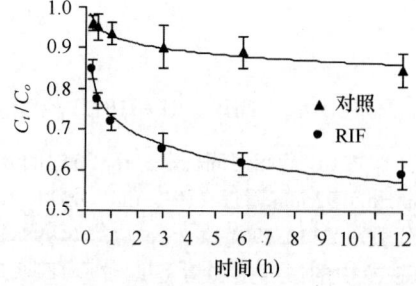

图 3　ENR 细胞体外代谢的药时曲线 （$n = 5$，$\bar{x} \pm s$）

CIP 转化率表明对照组脱乙基代谢途径平均占 ENR 全部代谢的大约 1/2，而诱导组却大约有 4/5，这说明 RIF 主要是通过对 ENR 的 N-脱乙基代谢途径的诱导作用促进 ENR 的消除（表 1）。

表 1　不同代谢时间内的 ENR 消除率和 CIP 转化率

时间（h）	对照组		RIF 诱导组	
	ENR 消除率（%）	CIP 转化率（%）	ENR 消除率（%）	CIP 转化率（%）
0.25	4.0 ±2.0	45.0 ±7.3	15.3 ±2.4**	75.7 ±3.6**
0.5	4.9 ±3.3	51.1 ±9.7	22.9 ±2.0**	66.9 ±5.6*
1	6.4 ±3.7	53.0 ±7.4	28.2 ±2.1**	71.5 ±8.7*
3	9.7 ±5.2	66.2 ±17.8	35.0 ±4.1**	87.0 ±8.4*
6	11.0 ±3.9	56.5 ±5.9	38.5 ±2.6**	86.3 ±6.5**
12	15.3 ±4.2	50.0 ±11.5	41.2 ±3.4**	86.4 ±4.1**

注：ENR 消除率（%）=（$1-C_t/C_o$）×100；CIP 转化率（%）= C_n×100/（C_t-C_o）。BIF 诱导组与对照组进行 T—Test，"*"表示显著差异（$P<0.05$）."**"表示极显著差异（$P<0.01$）

2.4　恩诺沙星在细胞中代谢的酶动力学研究

GCL 细胞分别与 6.25、12.5、25、50、100 μM 的底物 ENR 共孵育 1 h 后检测 CIP 的生成量。根据 Lineweaver-Burk 作图（图4）[9,10]，获得诱导组和对照组的酶促反应动力学方程分别为：

$$1/V = 72.5530×1/[S] +1.4922（R^2 = 0.9701）$$
$$1/V = 530.9800×1/[S] +3.5274（R^2 = 0.9946）$$

根据方程计算酶动力学参数，结果显示诱导组中的最大反应速率 V_{max} 和内在清除率 Cli 均极显著大于对照组，表明 RIF 诱导的酶催化效能较高。此外对照组米氏常数 Km 值是诱导组的 17.3 倍，表明 RIF 诱导的酶与底物的亲合力较高，酶促反应的强度较大（表2）。

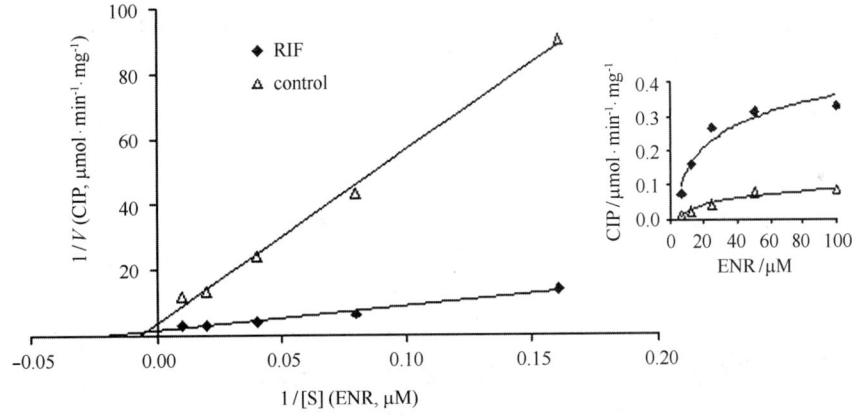

图 4　ENR-N-脱乙基代谢的 lineweaver-Burk 作图（$n=5$，$\bar{x}±s$）

表 2　细胞中 ENR-N-脱乙基代谢的酶动力学参数

实验组	Km（μmol · L^{-1}）	V_{max}（μmol · min^{-1} · mg^{-1}）	Cli（L · min^{-1} · g^{-1}）
RIF 诱导	108.26 ±9.61**	0.67 ±0.05**	6.19 ±0.57**
对照	1872.98 ±221.58	0.28 ±0.04	0.15 ±0.02

3　讨论

通过实验研究 RIF、PB、ERM 对 ENR-N-脱乙基酶活性的影响，可在一定程度上推测该酶可能依赖的 CYP 亚家族或个体。目前研究认为，RIF 可诱导 CYP 超家族中 2C8、2C9、2C18、2C19、3A4、3A5、4A 等酶的活性，PB 可诱导 2B1、2B4、2B6、2C6、3A4 等酶的活性，而 ERM 除了可以诱导 4A 的活性外，还能抑制 1A2、3A4、3A5 等酶的活性[11]。由实验结果可知 RIF、PB 对 ENR 脱乙基酶有诱导作用，而 ERM 有抑制作用，因此认为 CYP3A4 有可能介导 ENR 的 N-脱乙基反应。

不同的 FQs 药物在体内代谢过程中生物转化的程度不尽相同，洛美沙星（LOM）、氧氟沙星（OFL）、左氧沙星

（LEV）绝大部分以原药从肾脏排泄，几乎不经过肝药酶代谢消除，本实验中 ENR 的细胞体外消除率结果表明 ENR 的生物转化程度较高。ENR 的生物转化除了通过 I 相 N-脱乙基反应形成主要代谢产物 CIP 外，还可能通过直接与葡萄糖醛酸或硫酸酯等进行 II 相结合反应。由文中 ENR 代谢产物色谱图可见 CIP 并非唯一的代谢产物，而 CIP 转化率能较好地表示 ENR 转化为 CIP 的 N-脱乙基反应占整个代谢途径的比重，对于实验所显示出的 N-脱乙基反应之外的代谢途径我们将进行更为深入的研究。

4　结论

GCL 细胞中的药物剂量效应曲线表明 RIF 和 PB 对 ENR 脱乙基酶的活性具有诱导作用，而 ERM 具有抑制作用。此外，药时曲线表明 RIF 可显著缩短 ENR 在细胞中的半衰期，ENR 消除率和 CIP 转化率的数据也说明了 RIF 对 ENR 代谢的脱乙基途径具有显著的促进作用。在上述基础上，酶动力学方程对 ENR-N-脱乙基酶促反应的各种参数进行了更量化的研究，这些研究有助于进一步阐明 ENR 在鱼类细胞中的代谢途径和机理。

参考文献

［1］ Mansuy D., The great diversity of reactions catalyzed by cytochromes P450. Comp Biochem Physiol, 1998, 121C: 5-14.

［2］ Dunnett M., Richardson D. W., Lees P.. Detection of enrofloxacin and its metabolite ciprofloxacin in equine hair. Res Vet Sci, 2004, 77: 143-151.

［3］ Vaccaro E., Giorgi M. Longo V, et al. Inhibition of cytochrome P450 enzymes by enrofloxacin in the sea bass (Dicentrarchus labrax). Aquat Toxicol, 2003, 62: 27-33.

［4］ Ershov E., Bellaiche M., Hanji V., et al. Interaction of fluoroquinolones and certain ionophores in broilers: effect on blood levels and hepatic cytochrome P450 monooxygenase activity. Drug Metabol Drug Interact, 2001, 18 (3): 209-219.

［5］ Vancutsem P. M., Babish J. G.. In vitro and in vivo study of the effects of enrofloxacin on hepatic cytochrome P-450. Vet Hum Toxicol, 1996, 38 (4): 254-259.

［6］ Sakar D., Prukner E., Prevendar C. A., et al. Marek's disease vaccination, with turkey herpesvirus, and enrofloxacin modulate the activities of hepatic microsomal enzymes in broiler chickens. Acta Vet Hung, 2004, 52 (2): 211-217.

［7］ Fent K.. Fish cell lines as versatile tools in ecotoxicology: assessment of cytotoxicity, cytochrome P4501A induction potential and estrogenic activity of chemicals and environmental samples. Toxicol In Vitro, 2001, 15 (4): 477-488.

［8］ Tom D. J., Lee L. E., Lew J.. Induction of EROD activity by planar chlorinated hydrocarbons and polycyclic aromatic hydrocarbons in cell lines from the rainbow trout pituitary. Comp Biochem Physiol, 2001, 128A: 185-198.

［9］ Kudo S., Uchida M., Odomi M.. Metabolism of carteolol by cDNA-expressed human cytochrome P450. Eur J Clin Pharmacol, 1997, 52: 479-485.

［10］ 赵莉，楼雅卿. 中国人肝微粒体体外代谢奥美拉唑的酶促反应动力学. 药学学报, 1996, 31 (5): 352-357.

［11］ Anzenbacher P, Anzenbacherova E. Cytochromes P450 and metabolism of xenobiotics. Cell Mol Life Sci, 2001, 58: 737-747.

该文发表于《水产学报》【2006：(3)：311-315】

草鱼肝细胞中诱导剂对 EROD 作用的剂量效应研究

Research on dose-response of inducers to effect on EROD in grass carp hepatocyte

林茂[1,2]，杨先乐[1]，房文红[1]，张宁[1]，李聃[1]

（1. 上海高校水产养殖学 E-研究院，上海水产大学农业部渔业动植物病原库，上海 200090；
2. 集美大学水产学院，福建 厦门 361021）

摘要　在研究 GCL 细胞中 EROD 催化底物反应时间的基础上分别以三种数学模型拟合，分析比较了诱导剂 TCD-DPCB、PB、BaP、3-MC、BNF 对 GCL 细胞 EROD 酶活的剂量-效应曲线。结果表明 Log-Normal 模型能较好的拟合药酶诱导前升后降的钟形效应曲线，所得方程参数值 p_{100} 和 EC_{100} 可用于药物代谢酶的诱导研究；Gentox 模型拟合度相对较低且其 EC_{50} 值显著高于半效应浓度；Logistic 模型只能拟合曲线诱导上升部分而且 EC_{50} 和 p_{100} 与 Log-Normal 模型相比显著偏高。进一步研究结果显示诱导剂 TCDD 对 GCL 细胞作用时间越长，EC_{100} 越大，p_{100} 越小，但在作用 48 h 后即达到较大诱导效应，延长时间并无显著加强作用。

关键词 7-乙氧异吩唑酮-O-脱乙基酶（EROD）；诱导；草鱼肝细胞；药物代谢酶

P450 酶系是参与药物和环境化学物质体内合功能氧化酶（MFO）。鱼类中 7-乙氧异吩唑酮-生物灭活或激活反应的重要药物代谢酶，亦称混 O-脱乙基酶（7-ethoxyresorufin-O-deethylase，EROD）的活性反应是极具代表性的混合功能氧化酶的典型反应。在正常水体环境中，鱼体内 EROD 的活性是相对较低的，但在某些特定的外来化学物质（特别是各种多环芳香烃类化合物）的诱导下，它的活性异常增高，因此可将鱼体内 EROD 酶的活性作为水体环境中特定污染物的监测指标，目前 EROD 已作为生物标志物应用于生态毒理学研究领域[1,2]。

本文以 EROD 为模式药物代谢酶，多环芳香烃类化合物为诱导剂，数学模型为工具，研究草鱼肝细胞中 EROD 诱导剂的剂量效应曲线，从而为鱼类药物代谢酶体外诱导细胞模型的建立奠定基础，进而为研究水域污染对水产动物的生理影响以及渔药的代谢机理等提供有利条件。

1 材料和方法

1.1 细胞

草鱼肝细胞系（GCL）由农业部渔业动植物病原库提供，以 M199 细胞培养基置于 28℃下培养。

1.2 试剂

M199 基础培养基购自 GIBCO 公司，苯巴比妥（PB）、乙氧异吩唑酮（ERF）购于 SIGMA 公司，多氯联苯-1254（PCB-1254）、苯并（a）芘（BaP）、2, 3, 7, 8-四氯二苯并（p）二（2, 3, 7, 8-TCDD）为 SUPELCO 公司产品，异吩唑酮（RF）、β-萘黄酮（BNF）、3-甲基胆蒽（3-MC）为 FLUKA 公司产品。

1.3 细胞中 EROD 酶活诱导与检测

GCL 细胞以密度 105 cells·mL^{-1}、每孔 1 mL 接种于 24 孔板中预培养 24 h，然后加入药酶诱导剂作用 72 h，倒去培养基，以 PBS 缓冲液清洗。加入 1 mL 含有 2 μmol·L^{-1} ERF 的无酚红无血清培养基，置培养箱中孵育 60 min。以荧光分光光度计（SHIMADZURF-10A）测定上清液中 RF 的相对荧光强度，激发和发射波长分别为 560 nm 和 590 nm。根据标准曲线计算 RF 浓度，EROD 活性（pmol·min^{-1}·mg^{-1}）以每 mg 蛋白单位时间内的 RF 生成量表示[3]。

1.4 EROD 诱导剂剂量效应曲线的拟合

Excel MacroREGTOX_ EV7.0 程序中的 Gentox 模型以峰为界将 EROD 诱导剂的钟形剂量效应曲线分为上升和下降两个阶段，方程式如下，

$$F(x) = \frac{p_0 + kx}{1 + \left(\dfrac{xp}{EC_{50}}\right)^{nH}}$$

式中，$F(x)$ 表示当诱导剂浓度为 x 时的 EROD 活性，p_0 表示无诱导剂对照组的 EROD 活性，k 表示上升阶段 EROD 活性线性增加的斜率，nH 表示 Hill 常数，EC_{50} 表示概率函数 $f(x) = EC_{50}nH / (xnH + EC_{50}nH)$ 等于 0.5 时的诱导剂浓度[4]。

SigmaPlot8.0 软件的 Peak Log-Normal 模型方程式如下，

$$F(x) = p_0 + (p_{100} - p_0)\exp\left[-0.5\left(\frac{\ln(x/EC_{100})}{b}\right)\right]$$

式中，p_{100} 表示诱导效应最大时的 EROD 活性，EC_{100} 则表示此时的诱导剂浓度，b 为斜率系数。

SigmaPlot8.0 软件的 Logistic 模型只对 EROD 诱导剂在上升阶段的诱导效应进行拟合，方程式如下：

$$F(x) = p_0 + \frac{p_{100} - p_0}{1 + \left(\dfrac{x}{EC_{50}}\right)^{nH}}$$

式中，EC_{50} 表示 EROD 诱导效应达 50% 时的诱导剂浓度[5]。

1.5 诱导剂的相对效能因子

相对效能因子（relative potency factor，RPF）计算公式：RPF [test] = EC_{50} [min] / EC_{50} [test]。以诱导剂中对 EROD 诱导的 EC_{50} 最小的作为标准，计算各诱导剂的诱导效能。EC_{50} 越小，RPF 值越大，诱导能力越强[6]。

1.6 统计分析

通过 SPSS11.5 软件进行显著性分析，两组间比较采用 t 检验（t-Test），以" * "表示显著差异（$P<0.05$），" * * "表示极显著差异（$P<0.01$）；多重比较采用单向方差分析（One-WayANOVA）中的 Student-Neuman-Keuls（SNK）检验，样本组以英文字母进行标记，含相同字母表明组间无显著差异，反之则表明组间有显著差异（$P<0.05$）。

2 结果

2.1 底物反应时间

GCL 细胞以 $0.2\ \mathrm{nmol \cdot L^{-1}}$ TCDD 诱导 72 h 后，加底物 ERF 反应一段时间后检测 RF 生成量并以此计算 EROD 酶活性，实验结果表明，RF 生成量随反应时间延长而不断增加，但在 90 min 以后趋于平缓（图1）。进一步计算表示 RF 形成速率的 EROD 活性，结果表明反应时间在 60 min 时，EROD 值最高（30~90 min 之间无显著差异），随着时间的延长 EROD 值逐渐减小。

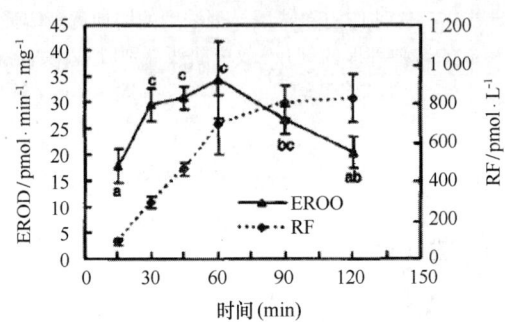

图1　RF 不同反应时间内的 RF 生成量及 EROD 酶活性（$n=3$，$\bar{x}\pm s$）

Fig. 1　RF concentration and EROD activity produced after a period of time（$n=3$，$\bar{x}\pm s$）

在诱导作用逐渐增强的上升阶段，采用 EROD 诱导剂研究中通用的 Logistic 模型对曲线进行拟合（图3），该模型的参数 EC_{50} 为生物意义上的半效应浓度，而 Gentox 模型的参数 EC_{50} 则是数理上的半概率浓度，后者所得结果极显著大于前者；由于 Logistic 模型只拟合了诱导递增阶段并对上升趋势进行了无限延伸，因此所得到的 EC_{50} 和 p_{100} 与 Log-Normal 模型相比显著偏高（表1）。

表1　EROD 诱导剂的剂量-效应曲线拟合模型的比较

Tab. 1　Comparison of different dose-response regression modules

比值 ratio	TCDD	PCB	PB	BaP	3-MC	BNF	$\bar{x}\pm s$
EC_{50}［Gentox］/EC_{50}［Logistic］	3.61	3.61	2.17	3.66	11.00	2.93	4.50±3.24 * *
EC_{50}■［Log-Normal］/EC_{50}［Logistic］	0.66	0.68	0.62	0.99	0.94	0.80	0.78±0.16 *
p_{100}［Log-Normal］/p_{100}［Logistic］	0.83	0.83	0.77	1.04	0.93	0.88	0.88±0.09 *
RPF［Gentox］/RPF［Logistic］	1.00	1.00	1.66	0.99	0.33	1.23	1.04±0.43
RPF［Log-Normal］/RPF［Logistic］	1.00	1.87	3.11	1.52	0.98	1.05	1.59±0.83

注：各比值以"1"为检验值进行单样本 T-Test；"■"所标 EC_{50} 是通过回归方程间接计算所得。

Notes：Each ratio was analyzed by one-sample t-Test with 1 as test value. ■The EC_{50} was deduced indirectly by the fitted equation

目前，许多研究通常是通过 Logistic 模型回归得到的 EC_{50} 计算 RPF，而实验结果表明通过 LogNormal 或 Genotox 模型计算得到的 RPF 与 Logistic 模型无显著差异（表1）。根据这3种方程所得参数计算的 RPF 的大小顺序均为：TCDD>PCB>3-MC>BaP>BNF>PB（表2）。在药酶诱导 GCL 细胞模型研究中，选择 60 min 为最佳药物代谢反应时间。

表2 各种多环芳香烃（PAHs）对 EROD 诱导的相对效能因子（RPFs）
Tab. 2 Relative potency factors（RPFs）of polyaromatic hydrocarbons（PAHs）

模型 model	TCDD	PCB	PB	BaP	3－MC	BNF
Gentox	1.000 0	0.161 8	0.000 2	0.002 0	0.002 7	0.000 4
Log－Normal	1.000 0	0.302 6	0.000 4	0.003 0	0.008 1	0.000 4
Logistic	1.000 0	0.161 7	0.000 1	0.002 0	0.008 2	0.000 3

注：Log－Normal 模型以 EC_{100} 参数的比值计算 RPFs。

Notes：The parameters EC_{100} was available to calculate RPFs in Log－Normal model

2.2 诱导剂对 EROD 酶活的剂量效应曲线

GCL 细胞以诱导剂 TCDD、PCB、PB、BaP、3－MC、BNF 分别作用 72 h 后检测 EROD 酶活性，实验结果表明，诱导剂可以在较低浓度下对 EROD 酶产生诱导作用，这种作用随着诱导剂浓度的增加逐渐增强，在 EC_{100} 处达到最强后又开始逐渐减弱至零（图2）。EROD 酶诱导剂所形成的这种钟形的剂量-效应曲线分别以 Log－Normal 和 Gentox 模型进行拟合，平均拟合度（R^2）分别为（0.9898±0.0051）和（0.9683±0.0280），Log－Normal 比 Gentox 的回归模型可获得更显著的相关性（P <0.05）。

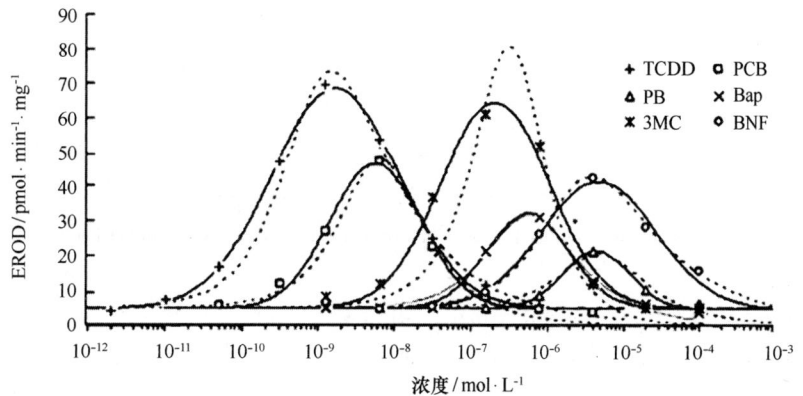

图2 诱导剂对 EROD 酶的剂量-效应完整曲线（$n=3$，$\bar{x}\pm s$）
（实线表示以 Log－Normal 模型拟合，虚线表示以 Gentox 模型拟合）

Fig. 2 Complete dose－response curves of drugs to induce EROD activity（Solid lines fitted by Log－Normal while broken lines fitted by Gentox）

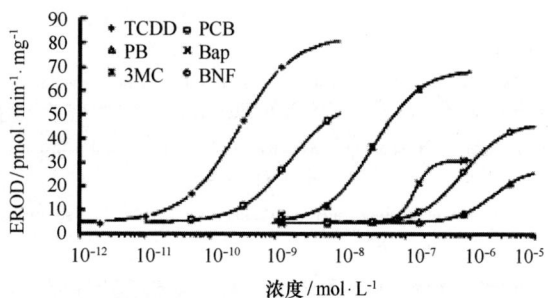

图3 Logistic 模型拟合诱导上升阶段的剂量-效应部分曲线（$n=3$，$\bar{x}\pm s$）

Fig. 3 Dose－response curves in ascending phase of EROD induction fitted by Logistic（$n=3$，$\bar{x}\pm s$）

2.3 诱导时间对 EROD 诱导剂剂量效应的影响

GCL 细胞以 TCDD 分别诱导作用 12、24、48 和 72 h 后检测 EROD 酶活性，并通过 Log－Normal 模型对诱导的剂量-效应进行回归分析，实验结果显示（图4），诱导剂作用时间越长，EC_{100} 越大，p_{100} 越小，不过 48 h 与 72 h 无显著差异（P <

0.05）。由此可见诱导剂 TCDD 对 GCL 细胞 EROD 酶的诱导在作用 48 h 后就已达到最大诱导效应，进一步延长时间并无显著加强。

3 讨论

利用体外培养细胞模型研究药物代谢酶的诱导，较之整体药物实验[7]，其优点是适合所有由 P450 代谢的化学物质的代谢研究，可排除机体和环境等其他因素的干扰。利用细胞进行酶学实验可不加入 NADPH 辅酶系统，反应在整个细胞水平实现，较之微粒体的亚细胞实验更能反映肝肾脏实际的药物代谢特征[8]。有许多报道利用原代培养细胞刚刚离体，生物性状尚未发生很大变化的特点进行药物代谢酶的诱导研究。由于原代细胞每次取组织消化细胞，易于受个体差异的影响而导致批次间的误差增大，因此不适于作为一个稳定的细胞研究模型。

国外有诸多的药理及毒理学家采用建系的细胞作为孵育系统，最常见的是鼠肝细胞培养，但药物代谢及 P450 系统存在着显著的种属差异，鱼类与鼠之间无论 P450 的含量和内部组成还是个别同工酶的结构、命名及构效关系均存在一定的差异[8]。因此，利用鱼类细胞作为研究体系才能更准确的研究鱼类药物代谢酶在渔药和水域污染物代谢中的作用，从而为渔药与环境污染物的代谢机理和毒理学，以及残留检测技术等研究提供良好的实验平台。

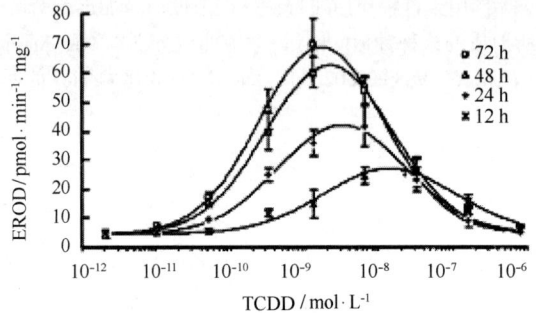

图 4 TCDD 作用时间对诱导 EROD 剂量-效应曲线的影响（$n=3$，$\bar{x}\pm s$）

Fig. 4 Effect of exposure time to TCDD on dose-response curves of EROD induction

目前文献报道对 EROD 诱导剂量效应的研究主要以 Logistic 模型进行回归分析[5]，但它只能拟合半边曲线，对处于诱导下降阶段的数据不能够利用，因此不能够很全面的考虑 EROD 诱导前升后降的这种趋势，所得到的 EC_{50} 和 p_{100} 一般会略高于实际值。故此本文提出以 Gentox 或 LogNormal 模型拟合完整的 EROD 诱导的剂量效应曲线，Gentox 方程既考虑了前阶段线性递增的趋势，又考虑了后阶段因为高浓度毒性而形成反 "S" 形递减的趋势，不过它提供的可用参数较少，唯一可用的 EC_{50} 还不是真正的半效应浓度；Log-Normal 模型也可以拟合出完整的钟形曲线，而且它给出的 EC_{100} 和 p_{100} 具有较大的研究价值，其缺陷是拟合小于 p_0 的数据点有偏差。总体而言，LogNormal 模型能较好的拟合药酶诱导前升后降的药效曲线，可用于药物代谢酶诱导的相关研究。

参考文献

[1] 霍传林，王菊英，韩庚辰，等. 鱼体内 EROD 活性对多氯联苯类的指示作用 [J]. 海洋环境科学，2002，21（1）：5-8.

[2] 黎雯，徐盈，吴文忠. 鱼肝 EROD 酶活力诱导作为二的水生态毒理学指标 [J]. 水生生物学报，2000，24：201-207.

[3] Pesonen M., Teivainen P., Lundstrom J., et al. Biochemical responses of fish sac fry and a primary cell culture of fish hepatocytes exposed to polychlorinated naphthalenes [J]. Arch Environ Contam Toxicol, 2000, 3: 52-58.

[4] Vindimian E., Robaut C., Fillion G. . A method for cooperative and noncooperative binding studies using non linear regression analysis on a microcomputer [J]. J ApplBiochem, 1983, 5: 261-268.

[5] Tom D. J., Lee L. E., Lew J. . Induction of EROD activity by planar chlorinated hydrocarbons and polycyclic aromatic hydrocarbons in cell lines from the rainbow trout pituitary [J]. Comp Biochem Physiol, 2001, 128 A: 185-198.

[6] Billiarda S. M., Bolsb N. C., Hodson P V. In vitro and in vivo comparisons of fish-specific CYP1A induction relative potency factorsfor selected polycyclic aromatic hydrocarbons [J]. Ecotoxicol and Environ Safe, 2004, 59: 292-299.

[7] 方展强，张凤君，郑文彪等. 多氯联苯对剑尾鱼 Na+/K+-ATPase 活性的影响 [J]. 水产学报，2004，28（1）：89-92.

[8] FentK. Fish cell lines as versatile tools in ecotoxicology: assessment of cytotoxicity, cytochrome P450A induction potential and estrogenic activity of chemicals and environmental samples [J]. Toxicol In Vitro, 2001, 15（4）：477-488.

该文发表于《高技术通讯》【2008,(8):863-867】

草鱼肝细胞中细胞色素 P450 3A 的酶促动力学

李聃*，杨先乐***，张书俊*，胡鲲*，林茂***，邱军强*

(*上海水产大学农业部渔业动植物病原库 上海 200090；**上海高校水产养殖学 E-研究院 上海 200090；
***集美大学渔业学院 厦门 361021)

摘要 用分光光度计法和反相高效液相色谱法研究了草鱼肝细胞系（GCL）中 CYP3A 探针酶类红霉素-N-脱甲基酶（ERND）和 6β-睾酮羟化酶（6β-TOH）的酶促反应。药物利福平（RIF）在浓度为 40μM 时对 GCL 细胞中 CYP3A 活性具有显著诱导作用，而苯巴比妥（PB）和地塞米松（DEX）无显著诱导作用（$P<0.05$）。在低浓度底物条件下，对照组和 40μMRIF 诱导组 GCL 细胞中 ERND 的酶促动力学方程分别为 $1/V=234.08×1/[S]+2.6544$ 和 $1/V=85.632×1/[S]+1.9458$，6β-TOH 的酶促动力学方程分别为 $1/V=529.91×1/[S]+14.615$ 和 $1/V=56.337×1/[S]+2.1017$，酶动力学参数表明，RIF 诱导的 CYP3A 亲和力较强，催化效能较高。当底物浓度进一步增高，CYP3A 活性反而降低，致使动力学曲线偏离经典的米氏动力学曲线，该结果表明 GCL 细胞中 CYP3A 存在特殊的负协同效应，这一现象是对鱼类细胞色素 P450（CYP）酶促动力学的有利补充。
关键词 细胞色素 P4503A，草鱼肝细胞，酶促动力学，协同效应，利福平

0 引言

细胞色素 P4503A（cytochrome P4503A，CYP3A）是药物代谢中最重要的 CYP 亚家族同工酶，广泛参与药物的生物转化，由于 CYP3A 酶类可被许多外源物诱导或抑制，导致体内药物相互作用，越来越受到人们的关注[1]。尽管 CYP3A 亚家族存在遗传多态性和显著的属差异，各成员的催化特性却较为相似[2]，其活性一般通过红霉素-N-脱甲基酶（erythromycin N-demethylase，ERND）、6β-睾酮羟化酶（6β-testosterone hydroxylase，6β-TOH）等特异性依赖 CYP3A 的探针酶类活性来反映[3-4]，但最为理想的单一探针酶类至今仍无定论。

近年来，许多鱼类 CYP3A 成员陆续被克隆[5-6]，CYP3A 在渔药代谢中的作用也得到了证实[7-8]，外源物对 CYP3A 的影响成为当前研究的热点。较多报道研究了环境污染物如内分泌干扰物（EDCs）和前致癌物对鱼类 CYP3A 的调控[9-11]，也有少量研究表明典型人类 CYP3A 诱导药物对鱼类 CYP3A 亦具有诱导作用[6]，但大多针对 CYP3A 的表达情况，而活性方面的研究还很欠缺，特别是淡水养殖鱼类 CYP3A 的酶促反应情况尚无相关报道。

选择合适的体外模型可以在一定程度上反映体内药物代谢的情况，相对于体内实验较为稳定与方便，在临床药理研究和药物开发中使用极为普遍[12]。如果能利用鱼类体外模型，研究清楚 CYP3A 的酶促反应机理，阐明药物相互作用的机理，就可能预测和防止渔药互作给水产品带来的安全隐患。有报道指出，草鱼（*Ctenopharyngodon idellus*）肝细胞系具有 CYP 代谢和诱导潜力，在药酶研究中具有一定的应用价值[8]。

本文通过药物对 CYP3A 的体外诱导，研究了 ERND 和 6β-TOH 在草鱼肝细胞系中的酶促动力学，首次发现了鱼类细胞模型中 CYP3A 的负协同效应，为进一步阐明 CYP3A 的酶促反应机理提供了基础，对新药开发和保障水产品安全具有一定的意义。

1 材料与方法

1.1 细胞

草鱼肝细胞系（GCL，编号 BYK-C12-02）来自农业部渔业动植物病原库，以 M199 培养液置 28℃培养。

1.2 试剂

M199 基础培养基为 Gibco-BRL 产品，利福平（rifampicin，RIF）和地塞米松（dexamethasone，DEX）为 Sigma-Aldrich 产品，苯巴比妥（phenobarbital，PB）购自上海新亚药业有限公司，红霉素（erythromycin，ERY）购自上海生工生物工程有限公司，二甲亚砜（dimethylsulfoxide，DMSO）、睾酮（testosterone，TT）、6β-羟化睾酮（6β-hydroxytestosterone，6β-HTT）和皮质酮（corticosterone，CCT）购自 Sigma 公司。

1.3 细胞样品处理

GCL 细胞以 $2×10^5$ 个·mL^{-1} 的密度接种于 50 mL 一次性培养瓶中预培养 24 h，然后加入 CYP3A 典型诱导剂或 DMSO（溶剂对照，终浓度小于 0.1%）作用 24 h，倒去培养基，以 PBS 缓冲液清洗。加入 5 mL 含 CYP3A 依赖性底物（400 μM ERY 或 100 μM TT）的无酚红无血清 M199 培养基，置培养箱中孵育 1 h 后制成细胞悬液用于酶活性检测，蛋白浓度利用 Lowry 比色法进行测定[13]。

1.4 ERND 活性的检测

细胞样品中加入 0.35 mL 15%的硫酸锌，冰浴 5 min，然后加入 0.35 mL 饱和氢氧化钡，静置 5 min 后离心（5 000 r·min^{-1}，60 min）。吸取上清液 2 mL，采用 Nash 比色法[14]于 413 nm 处紫外分光光度计检测代谢产物甲醛生成量。

1.5 6β-TOH 活性的检测

预先加入内标 CCT（20 μM）的细胞样品与二氯甲烷 1∶6 混合，漩涡振荡 2 min 后离心（7 000 r·min^{-1}，5 min），取有机相用液氮吹干仪于 65℃吹干。加 300 μL 甲醇溶解后利用 Agilent1100 系列高效液相色谱（HPLC）工作站检测代谢产物 6β-HTT 生成量。色谱柱：C18 柱（150 mm×4.6 mm）；流动相：20%甲醇（溶于 0.04%醋酸）A 与甲醇（溶于 0.04%醋酸）B，梯度洗脱（45 min 内 B 从 20%到 50%，50%B 再持续 10 min）；流速：0.8 mL·min^{-1}；检测波长：254 nm；柱温：40℃；进样量：50 μL[15]。

1.6 数据处理

CYP3A 依赖酶类的活性以酶促反应速率 V，即每毫克蛋白单位时间内的代谢产物生成量表示（pmol·min^{-1}·mg^{-1}）。利用 SigmaPlot10.0 软件 Log-Normal 模型拟合诱导剂的剂量效应曲线[16]，Single Rectangular 模型获得酶促动力学参数；采用 SPSS15.0 软件进行显著性分析。

2 结果

2.1 药物对体外 CYP3A 依赖酶活性的影响

CYP3A 典型诱导剂 PB、RIF 和 DEX 对 GCL 细胞中 ERND 活性的影响如图 1A 所示，PB 对 ERND 无显著诱导作用，而 40 μM 的 RIF 和 40~150 μM 的 DEX 对 ERND 具有显著诱导作用，其拟合的最大诱导浓度（EC_{100}）分别为（38.3±0.7）μM 和（123.4±46.9）μM，DEX 在 100 μM 时 ERND 活性出现突增导致拟合度较差。进一步研究 RIF 和 DEX 在 EC_{100} 时对 ERND 活性的影响以及 40 μM RIF 和 100 μM DEX 对 6β-TOH 活性的影响，结果表明两种药物对 ERND 最大诱导倍数（p_{100}/p_0）分别为 1.4±0.1 和 1.7±0.2，然而仅 RIF 对 6β-TOH 具诱导作用，在 40 μM 时诱导倍数为 8.1±0.4（图 1B）。

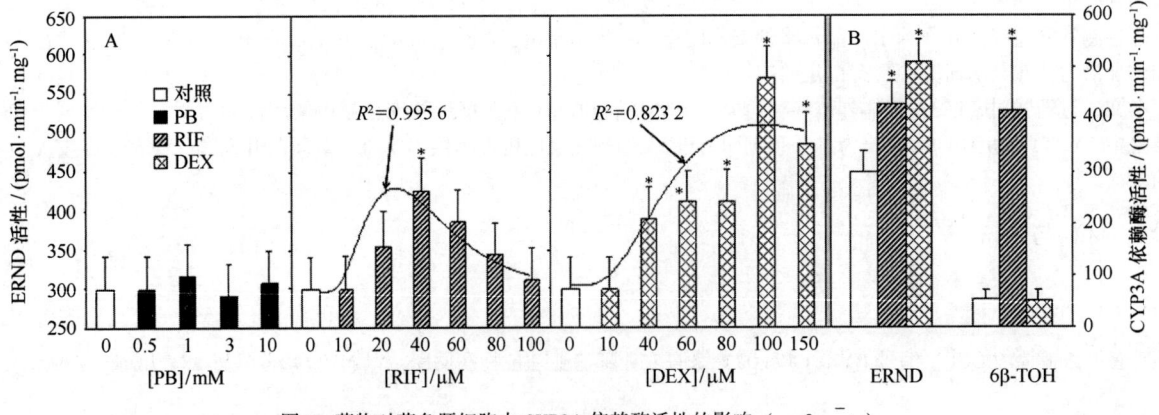

图 1　药物对草鱼肝细胞中 CYP3A 依赖酶活性的影响（$n=3$，$\bar{x}±s$）

Fig 1　Drug dependent on CYP3A enzyme activities of grass carp hepatocytes（$n=3$，$\bar{x}±s$）

注：A：三种 CYP3A 典型诱导剂对 ERND 活性的剂量效应；B：RIF 与 DEX 在拟合最大效应浓度（EC_{100}）时对 ERND 活性的影响以及 40 μM RIF 与 100 μM DEX 对 6β-TOH 活性的影响。其中利用 SigmaPlot 10.0 软件 Log-Normal 模型拟合两种影响显著药物对 ERND 活性的剂量效应曲线；与对照组比较，＊ $P<0.05$。

2.2　CYP3A 在细胞中的负协同效应

GCL 细胞分别与不同浓度的 ERY（10~800 μM）或 TT（10~400 μM）孵育 1 h 后检测 ERND 和 6β-TOH 活性（图 2）。对照组在两种底物浓度为 10 μM 时酶活性过低，接近检测限而无法准确测定。值得注意的是，两种 CYP3A 依赖酶类呈现非经典的酶促动力学行为，分别当底物浓度超过 400 μM 和 200 μM 时，ERND 和 6β-TOH 活性呈下降趋势而偏离经典的米氏动力学曲线，且 RIF 诱导组比对照组明显，ERND 比 6β-TOH 明显。这说明 CYP3A 存在底物分子之间的同源负协同效应和诱导剂底物之间的异源负协同效应，但在底物浓度较低时表现不明显。

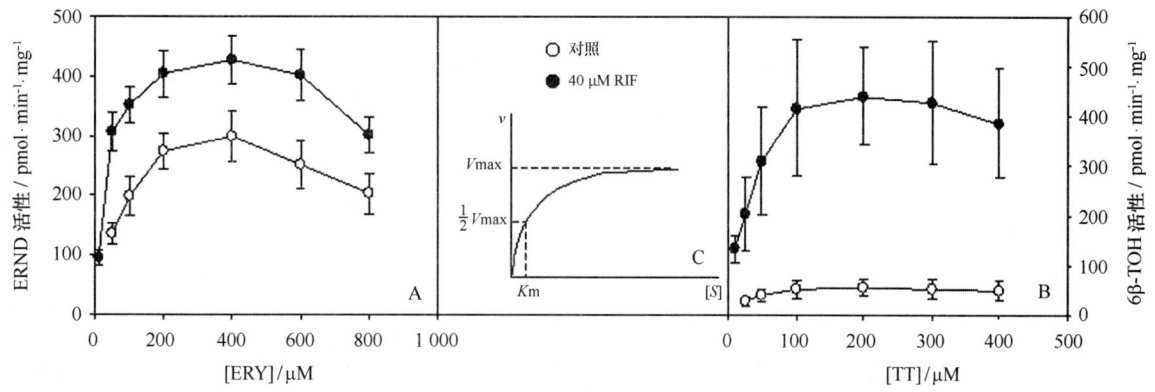

A：ERY-N-脱甲基代谢；B：6β-TT 羟化代谢；C：经典的米氏动力学曲线

图 2　草鱼肝细胞中 CYP3A 依赖性底物代谢的酶促动力学曲线（$n=3$，$\bar{x}\pm s$）

Fig2　Enzyme kinetics grass carp hepatocytes CYP3A substrate metabolism dependent curve（$n=3$，$\bar{x}\pm s$）

2.3　CYP3A 在细胞中的酶促动力学

GCL 细胞分别与 10、50、100、200、400 μM 的 ERY 或 10、25、50、100、200 μM 的 TT 孵育 1 h 后再次检测 ERND 和 6β-TOH 活性，其中对照组不检测底物浓度为 10 μM 时的 CYP3A 依赖酶活性。根据 Lineweaver-Burk 作图（图 3），结果表明在低浓度底物条件下，CYP3A 呈现经典的米氏动力学行为，对照组和 RIF 诱导组 GCL 细胞中 ERY 代谢的酶促反应动力学方程分别为：

$$1/V = 234.08 \times 1/[S] + 2.6544（R^2 = 0.9952）和 1/V = 85.632 \times 1/[S] + 1.9458（R^2 = 0.9958）$$

TT 代谢的酶促反应动力学方程分别为：

$$1/V = 529.91 \times 1/[S] + 14.615（R^2 = 0.9952）和 1/V = 56.337 \times 1/[S] + 2.1017（R^2 = 0.9766）$$

式中，[S] 为底物浓度。

利用 SigmaPlot 软件计算酶促动力学参数（表 1），结果显示诱导组米氏常数（K_m）值均显著小于对照组，而最大反应速率 V_{max} 与 K_m 的比值均显著大于对照组，进一步证实了 RIF 对 CYP3A 活性的诱导作用。

表 1　草鱼肝细胞中 CYP3A 依赖性底物低浓度代谢的酶促动力学主要参数

Tab 1　The main parameters of grass carp hepatocytes CYP3A-dependent low concentration of substrate metabolism enzyme kinetics

实验组	ERY-N-脱甲基			6β-TT 羟化		
	K_m（μmol·L^{-1}）	V_{max}（pmol·min^{-1}·mg^{-1}）	V_{max}（Km）	K_m（μmol·L^{-1}）	V_{max}（pmol·min^{-1}·mg^{-1}）	V_{max}（Km）
对照	88.2±11.2	376.7±16.6	4.3±0.5	36.3±3.6	68.4±2.2	1.9±0.2
诱导	44.0±4.2**	513.9±14.6**	11.7±0.8**	26.8±5.3*	475.8±26.4**	17.7±1.3**

3　讨论

通过实验研究了典型 CYP3A 诱导剂 PB、RIF 和 DEX 对鱼类细胞中 CYP3A 依赖酶促反应的影响，分析表明仅 RIF 对

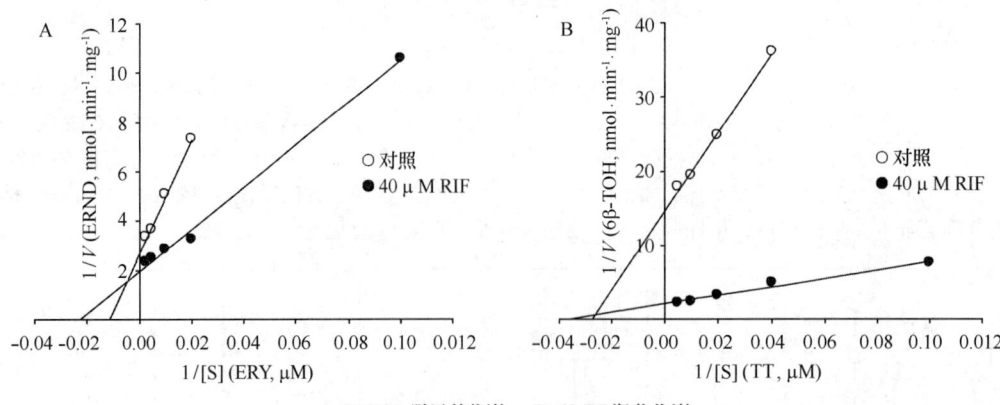

A: ERY-N- 脱甲基代谢；   B: 6β-TT 羟化代谢

图 3　草鱼肝细胞中 CYP3A 依赖性底物低浓度代谢的 Lineweaver-Burk 作图（$n=5$，$\bar{x}\pm s$）

Fig3　CYP3A substrates low concentration-dependent metabolism Lineweaver-Burk plot of grass carp hepatocytes（$n=5$，$\bar{x}\pm s$）

CYP3A 具有诱导作用，从而证实了 CYP3A 诱导存在种属差异，鉴于孕烷 X 受体（pregnaneX-receptor，PXR）在 CYP3A 调控中所起的关键作用[9]，可以推测这种差异的分子机理在于 CYP3A 本身和 PXR 的遗传多态性。此外，探针酶 ERND 和 6β-TOH 在考察 DEX 对 CYP3A 活性的影响时出现差异，这主要源于 CYP 复杂的介导机制及其多态性，有研究认为尽管 ERND 对人 CYP3A 依赖的专一性强，但在鱼类中可能依赖于多种 CYP 亚家族酶类[17]，而近年来鱼类中 6β-TOH 的专一性尚未遭到质疑，由此我们认为 6β-TOH 能较好反映鱼类 CYP3A 酶促反应的情况，为增强可靠性最好同时考察两种依赖酶类。

　　RIF 常被作为渔药使用，尽管人们已熟知临床上 RIF 的众多配伍禁忌[18]，但未认识到 RIF 对鱼类 CYP3A 的影响。本研究得到了低底物浓度条件下 CYP3A 的酶促动力学参数，结果表明经 RIF 诱导后 GCL 细胞中 Km 值显著降低，说明 CYP3A 与底物的亲和力增强，而 Vmax/Km 值显著增大，说明 CYP3A 催化效能升高，有力地证实了 RIF 对草鱼 CYP3A 活性的诱导作用。值得注意的是，对 CYP3A 的影响会间接改变鱼体内其他渔药的生物利用率和药动学参数[7]，从而可能导致渔药残留，也可能导致药物加速从体内清除，影响药物的疗效。利用 GCL 等鱼类体外模型研究渔药对 CYP3A 酶促动力学的影响，将有利于预测此类药物相互作用和指导合理用药。

　　本文首次利用鱼类细胞模型发现草鱼 CYP3A 的负协同效应，阐明了底物浓度对 CYP3A 酶促反应的影响，这种偏离经典米氏曲线的动力学行为是对鱼类 CYP 酶促动力学的有利补充，将为合理解释渔药浓度导致的代谢差异提供依据。此外，由于很多渔药和 EDCs 均为 CYP3A 的潜在底物和诱导剂，负协同效应极有可能在渔药互作及内分泌干扰中起着一定作用。推测造成负协同效应的主要原因在于：（1）CYP3A 具有独特的超大活性结构域可能是造成负协同效应的主要原因。由于存在多个底物结合位点，当底物浓度增大，单个或多个底物结合 CYP3A 后导致其蛋白空间结构发生改变，从而影响后续底物分子的结合，使酶活呈现相应变化。（2）负协同效应可能是 CYP3A 通过异构作用维持自身稳态的一种非遗传机制。RIF 是 PXR 的有效配体[19]，可激活 PXR 从而诱导 CYP3A 的表达和活性；与此同时，RIF 又可能结合到 CYP3A 的活性或异构位点，从而影响特异性底物的结合和转移，促使 CYP3A 的活性发生回降。对于负协同效应我们将进行更为深入的研究。

参考文献

[1]　Anzenbacher P, Anzenbacherová E. Cytochromes P450 and metabolism of xenobiotics. Cell Mol Life Sci, 2001, 58：737-747.

[2]　Bork R W, Muto T, Beaune P H, et al. Characterization of mRNA species related to human liver cytochrome P450 nifedipine oxidase and the regulation of catalytic activity. J Biol Chem, 1989, 264：910-919.

[3]　Vaccaro E, Giorgi M, Longo V, et al. Inhibition of cytochrome P450 enzymes by enrofloxacin in the sea bass (Dicentrarchus labrax). Aquat Toxicol, 2003, 62 (1)：27-33.

[4]　James M O, Lou Z, Rowland-Faux L, et al. Properties and regional expression of a CYP3A-like protein in channel catfish intestine. Aquat Toxicol, 2005, 72 (4)：361-371.

[5]　Barber D S, McNally A J, Garcia-Reyero N, et al. Exposure to p, p'-DDE or dieldrin during the reproductive season alters hepatic CYP expression in largemouth bass (Micropterus salmoides). Aquat Toxicol, 2007, 81 (1)：27-35.

[6]　Tseng H P, Hseu T H, Buhler D R, et al. Constitutive and xenobiotics-induced expression of a novel CYP3A gene from zebrafish larva. Toxicol Appl Pharmacol, 2005, 205 (3)：247-258.

[7]　Tang J, Yang X L, Zheng Z L, et al. Pharmacokinetics and its active metabolite of enrofloxacin in Chinese mitten-handed crab (Eriocheir sinensis). Aquaculture, 2006, 260：69-76.

[8]　林茂，杨先乐，王翔凌等．草鱼肝细胞中恩诺沙星脱乙基代谢的酶动力学．高技术通讯，2007，17（3）：314-318.

[9]　Bresolin T，Rebelo M D F，Bainy A C D. Expression of PXR，CYP3A and MDR1 genes in liver of zebrafish. Comp Biochem Physiol，2005，140C：403-407.

[10]　Gagné F，Blaise C，André C. Occurrence of pharmaceutical products in a municipal effluent and toxicity to rainbow trout（Oncorhynchus mykiss）hepatocytes. Ecotoxicol Environ Saf，2006，64（3）：329-336.

[11]　Kashiwada S，Kameshiro M，Tatsuta H，et al. Estrogenic modulation of CYP3A38，CYP3A40，and CYP19 in mature male medaka（Oryzias latipes）. Comp Biochem Physiol，2007，145C：370-378.

[12]　Brandon E F A，Raap C D，Meijerman I，et al. An update on in vitro test methods in human hepatic drug biotransformation research：pros and cons. Toxicol Appl Pharmacol，2003，189（3）：233-246.

[13]　Lowry O H，Rosebrough N J，Farr A L，et al. Protein measurement with the Folin phenol reagent. Biol Chem，1951，193：265-275.

[14]　Nash T. The colorimetric estimation of formaldehyde by means of the Hantzech reaction. Biochemistry，1953，55：416-426.

[15]　Reinerinck E J M，Doorn L，Jansen E H J M，et al. Measurement of enzyme activities of cytochrome P-450 isoenzymes by high-performance liquid chromatographic analysis of products. J Chromatogr，1991，553：233-241.

[16]　林茂，杨先乐，房文红等．草鱼肝细胞中诱导剂对 EROD 作用的剂量效应研究．水产学报，2006，30（3）：311-315.

[17]　Stegeman J J，Hahn M E. Current perspective on forms，functions，and regulation of cytochrome P450 in aquatic species. In：Malins D C，Ostrander G K（Eds.）. Biochemistry and molecular biology of monooxygenases. Boca Raton：CRC Press，1994. 87-206.

[18]　农业部《新编渔药手册》编撰委员会．新编渔药手册．北京：中国农业出版社，2005. 169-218.

[19]　Handschin C，Meyer U A. Regulatory network of lipid-sensing nuclear receptors：roles for CAR，PXR，LXR，and FXR. Arch Biochem Biophys，2005，433（2）：387-396.

该文发表于《水产学报》【2010，34（6）：685-690】

草鱼体外培养细胞中 β-萘黄酮对 CYP1A 基因表达的诱导

林茂[1,2]，杨先乐[2]，纪荣兴[1]

（1. 集美大学水产学院，福建 厦门 361021；2. 上海海洋大学农业部渔业动植物病原库，上海 200090）

摘要： 为了建立以 CYP1A cDNA 为探针的水环境毒理学细胞模型，本文以 β-萘黄酮（BNF）作为诱导剂，通过半定量 PCR 技术研究 CYP1A 在草鱼不同细胞系及其对应组织中的诱导表达情况。在半定量 PCR 反应参数研究中，对关键的退火温度和循环次数进行了优化，结果显示退火温度为 57℃，循环次数为 30 次较为合适，该条件下 Gauss 迹量的 CYP1A/ACT 比值能更准确的反映 CYP1A 的表达水平。对照组和诱导组草鱼细胞中 ACT 和 CYP1A cDNA 扩增结果表明，细胞中 CYP1A 的基础表达量较低，而 BNF 诱导使 GCL、CIK 和 CO 细胞中 CYP1A 的表达水平得到显著的提高。GCL、CIK 和 CO 细胞三者之间诱导后 CYP1A 的表达水平有显著差异（$P<0.05$），诱导表达量大小顺序为 GCL>CIK>CO。草鱼细胞系的相应组织中 ACT 和 CYP1A，cDNA 扩增以及电泳的结果与细胞中较为相似，组织中 CYP1A 诱导表达量大小顺序也与相应的细胞相同：肝>肾>卵巢。比较分析的结果表明，CYP1A 在体内与体外的诱导表达水平具有一定的相关性。

关键词： 草鱼；细胞；CYP1A；cDNA；诱导

　　细胞色素 P450 酶系（cytochrome P450，简称 CYP 或 P450）是参与药物与环境等化学物质体内生物转化的主要药物代谢酶，CYP1A 则属于 CYP 超家族中非常重要的一个亚家族，它已成为水环境毒理学研究领域的一个热点。CYP1A 主要参与多环芳香烃（PAHs）和芳香胺类化合物的代谢，在正常鱼体组织中的浓度很低，但在某些外来化学物质特别是 PAHs 的诱导下，它的活性异常增高。进一步研究发现，这主要是由于 CYP1A 基因的表达被 PAHs 等诱导剂高度诱导，其表达水平以几何级数增加。正因如此，鱼类 CYP1A 基因的表达水平已被作为水体环境中特定污染物的监测指标[1-4]。CYP1A 诱导情况的评价可以通过多种方法实现，在酶活水平可以检测乙氧异吩噁唑酮-O-脱乙基酶（EROD）活性的变化[3-5]，在蛋白水平可以利用免疫组化、ELISA 和 Western 印迹等手段[1,6]，在 mRNA 水平可以采用狭孔印迹、Northern 或者定量 PCR（quantitive PCR，Q-PCR）等技术[7-12]。在这些方法中定量 PCR 技术是最为敏感的，目前它已经被大量运用于鲑鳟等鱼类的药酶诱导研究中[8-11]。

　　定量 PCR 分外标法和内标法，内标法包括竞争性和非竞争性 PCR 等。非竞争性 PCR 也称为半定量 PCR（semiquantitive PCR），它是通过目的基因与内参物 PCR 产物的相对比值来定量的。本实验采用 β-肌动蛋白（β-actin，

ACT）基因作为定量 PCR 的内参物，在对草鱼 CYP1A 和 ACT cDNA 片段分别克隆和鉴定基础之上[12]，通过半定量 PCR 技术研究 CYP1A 基因在草鱼不同细胞中的基础和诱导表达情况，并与细胞相应的来源组织进行比较，从而为进一步建立以 CYP1A cDNA 为探针的水环境毒理学细胞模型奠定基础。

1 材料和方法

1.1 主要试剂

Trizol 试剂为 Invitrogen 公司产品，DNA Marker DL2000 为 TaKaRa 公司产品，Taq DNA 聚合酶、dNTPs、RNasin、DEPC 和 MOPS 等购自上海鼎国公司，First Strand cDNA Synthesis Kit（MMLV）为 Promega 公司产品，β-萘黄酮（BNF）为 Fluka 公司产品。

1.2 细胞系的诱导

本实验所用草鱼肝细胞 GCL、肾细胞 CIK、囊胚细胞 GCB、吻端成纤维细胞 PSF、卵巢细胞 CO 等均为农业部渔业动植物病原库（APCCMA）保藏物。培养基为 90%M199 + 10%NBS，培养条件为 28℃、5%CO_2。各细胞系和囊胚分别以 5 μM BNF（溶解于 0.01%DMSO）诱导 72 h 后[4]，检测 CYP1A 的表达水平，设 3 个重复，以溶剂处理为空白对照组。

1.3 鱼体内的诱导

草鱼连续 3 天，每天腹腔注射 5 mg·kg^{-1} 体重的 BNF（以玉米油溶解）后[13]，提取与上述细胞系相对应的来源组织的总 RNA，检测 CYP1A 的表达水平，设 3 个重复，以溶剂处理为空白对照组。

1.4 引物设计

以 Primer Premier 5.0 软件设计 CYP1A 和 ACT cDNA 片段的特异性扩增引物[12]，其中 CYP1A 的上游引物为 5′-ACAA-CATCCGAGACATCACA-3′，下游引物为 5′-TTCCACAGTTCTGGGTCA-3′；ACT 的上游引物为 5′-TCGGTATGGGACAGAAGG-3′，下游引物为 5′-GTCAGCAATGCCAGGGTA-3′。引物由上海华冠生物公司合成。

1.5 总 RNA 的提取

采用 Trizol 法[12]提取草鱼肝组织总 RNA，通过紫外分光光度计测 260 nm 和 280 nm 处的吸收值，检测 RNA 的产量和纯度。

1.6 cDNA 的扩增

根据 MMLV 产品说明书合成 cDNA 第一链，以此为模板进行 PCR 扩增。PCR 反应体系总体积 25 μL，包含 100 ng cDNA 模板，200 ng PCR 引物，1U Taq DNA 聚合酶，200 μmol·L^{-1} dNTPs 等。PCR 反应在 Eppendorf PCR 仪上进行，首先 94℃预变性 2 min，之后以 94℃变性 30 s、55℃退火 30 s、72℃延伸 1 min 的程序循环 35 次，最后 72℃再延伸 10 min。以上扩增完成后，取 5 μL 反应产物，在 1% 的琼脂糖凝胶上以 4 V·cm^{-1} 恒定电压进行电泳检测。

1.7 扩增产物电泳条带的定量分析

利用 Quantity One 4.5 软件对扩增产物电泳条带进行定量分析，以 Gauss 建模的光密度分布曲线下面积（Gauss 迹量）分别表示条带中 CYP1A 和 ACT cDNA 的相对含量，再通过 CYP1A/ACT 的比值对 CYP1A 的表达水平进行相对定量。不同细胞或组织中 CYP1A 相对表达量利用 SPSS 11.5 软件进行差异显著性分析。

2 结果

2.1 半定量 PCR 条件优化

分别以特异性引物扩增可得到预期大小的草鱼 CYP1A（439 bp）和 ACT（800 bp）cDNA 片断，这两个 cDNA 序列分别进行克隆和鉴定后[12]，已提交 GenBank 并获得登录号 DQ211095 和 DQ211096。在此基础上对半定量 PCR 条件进行优化，实验结果表明，退火温度为 57℃较为合适，该条件下扩增产物的电泳条带清晰、明亮（图 1）；PCR 循环次数则以 30 个循环为宜，此时 CYP1A 和 ACT 均处于指数扩增期，而 35 个循环时二者扩增产物已处于饱和状态，扩增平台的出现使目的序列对内参物的 PCR 产物的比值与初始数目不再成比例，CYP1A/ACT 的 Gauss 迹量比值也就不能准确反映 CYP1A 的表达水平（图 2）。

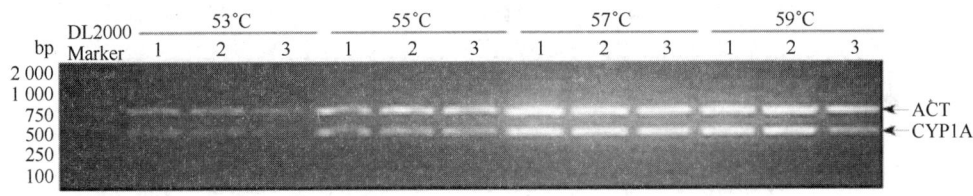

图 1 不同退火温度下草鱼 CYP1A 和 ACT cDNA 的扩增产物

Fig. 1 Amplification product of CYP1A and ACT cDNA annealing at different temperature

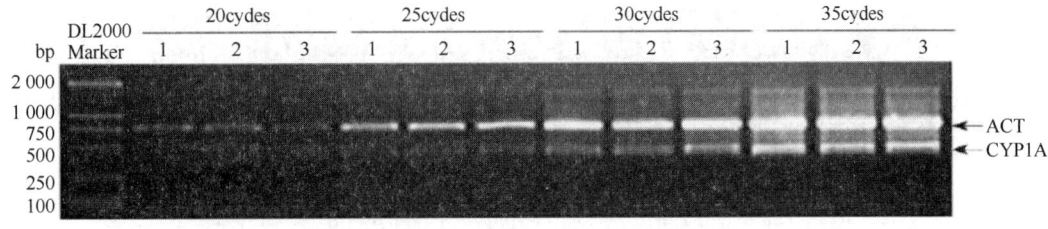

图 2 不同 PCR 循环次数扩增的草鱼 CYP1A 和 ACT cDNA 扩增产物

Fig. 2 Amplification product of CYP1A and ACT cDNA after different PCR cycles

2.2 草鱼不同细胞系中 CYP1A 基因的基础和诱导表达

草鱼各细胞系中 CYP1A 和 ACT cDNA 扩增结果表明，不论是对照组还是诱导组，ACT 在不同细胞中的表达均较为恒定，而 CYP1A 在不同细胞中基础和诱导表达的水平均有所不同（图 3）。CYP1A 在对照组 CIK、GCB、PSF 和 CO 细胞中的表达量太低以至于无法检测，但经过诱导后 CIK 和 CO 细胞扩增产物的电泳图谱均有明显的条带出现。GCL 细胞中 CYP1A 基础表达水平较高，而 BNF 诱导后表达量极显著提高（$P<0.01$），为基础水平的 4.17 倍。GCL、CIK 和 CO 细胞诱导后 CYP1A 的表达水平三者之间有显著差异（$P<0.05$），诱导表达量大小顺序为 GCL>CIK>CO。

2.3 草鱼不同组织中 CYP1A 基因的基础和诱导表达

选取上述草鱼细胞系的相应组织，比较研究 CYP1A 在体外培养细胞与组织中的表达差异，结果表明，各组织中 CYP1A 和 ACT cDNA 扩增以及电泳的结果与细胞中较为相似，对照组中各组织 CYP1A 的基础表达量也很低，而在诱导组肝、肾、卵巢等组织中 CYP1A 的表达量均得到了明显的提高，组织中 CYP1A 诱导表达量大小顺序与相应的细胞相同：肝>肾>卵巢（图 4）。定量分析结果表明，CYP1A 在体内与体外的表达水平具有一定的相关性。

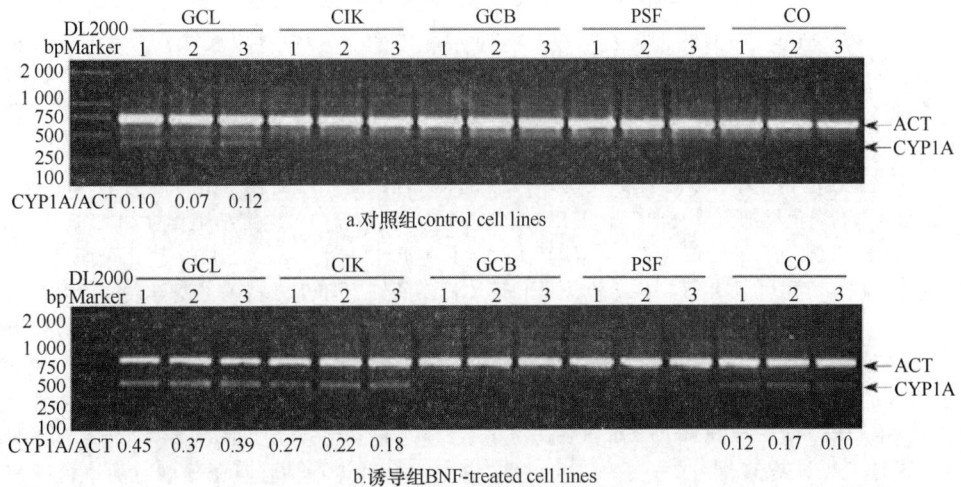

a.对照组 control cell lines

b.诱导组 BNF-treated cell lines

图 3 对照组和 BNF 诱导组草鱼细胞系中 CYP1A 和 ACT cDNA 的扩增产物

Fig. 3 Amplification product of CYP1A and ACT cDNA from control and BNF-treated grass carp cell lines

A：control cell lines；B：BNF-treated cell lines

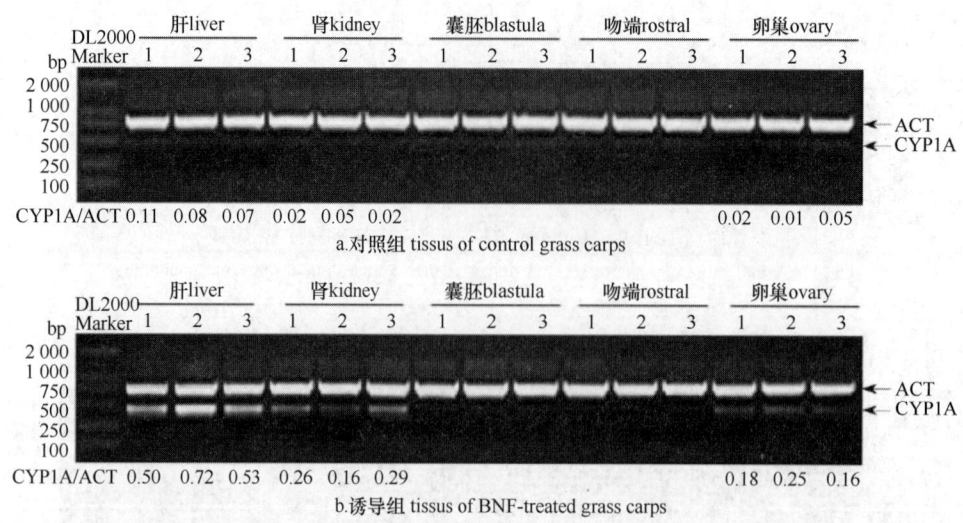

图 4　对照组和 BNF 诱导组草鱼组织中 CYP1A 和 ACT cDNA 的扩增产物

Fig. 4　Amplification product of CYP1A and ACT cDNA from control and BNF-treated grass carp tissues

A：tissus of control grass carps；B：tissues of BNF-treated grass carps

3　讨论

自从 Elcombe 等（1979）[14] 率先报道虹鳟肝 CYP 依赖型单加氧酶的诱导和鉴定，就有较多研究表明苯巴比妥、多氯联苯、苯并芘、萘黄酮、二噁英、多氯氧芴、3-甲基胆蒽等芳香烃化合物能诱导鱼类 CYP 相关的单加氧酶活性[1-20]。PAHs 化合物如 β-萘黄酮对 CYP1A 基因表达和酶活性的诱导能力极强，而研究表明化学物质对 CYP1A 的诱导能力越大，则往往意味着具有更大的生态毒性，这也使得 CYP1A 表达量可以作为评价环境污染物的生物标记[15]。CYP1A 被 PAHs 化合物诱导的机制主要是诱导剂与细胞内的芳香烃受体（AhR）结合，刺激 mRNA 的转录，从而使粗面内质网内 CYP 含量增加或活性增强[6]。笔者对草鱼 GCL、CIK 和 CO 细胞的 CYP 含量和 EROD 酶活检测的结果表明（文中未列）[12]，BNF 诱导 CYP1A 基因表达量升高的同时，也导致了 CYP 的相对含量以及 CYP1A 依赖的 EROD 酶活性在不同程度上相应的提高，这也侧面印证了 PAHs 等化合物诱导的机制。

许多文献报道表明鱼类肝、肾、鳃等组织中 CYP 含量或酶活较高，而另外一些证据表明来源于这些组织的原代或建系细胞的 CYP 含量和酶活与之有一定的相关性[15-17]，本文的研究结果也显示草鱼细胞系 CYP1A 基因的基础和诱导表达与来源组织均具有一定程度的相关性。国内外对鱼类 CYP 基因和酶活性研究以体内试验居多[1-3,7-10]，鱼类细胞也已开始作为毒理学研究和环境污染物评估的一种有效工具[4,5,15-20]。Sadar 等（1996）[11] 以虹鳟原代培养的肝细胞研究了苯巴比妥对 CYP1A 的诱导情况，Pesonen 等（2000）[17] 则进一步比较了幼鳟原代肝细胞和卵黄囊鱼苗中多氯萘对 CYP1A 酶活性的调控，国内学者万小琼等（2002）[18] 也利用原代培养的草鱼肝细胞对环境剧毒物质二噁英的毒性效应进行了评价。原代细胞容易获取，而且相对于传代细胞而言更多地保留了来源器官的生化反应特性，但原代细胞也有其欠缺性，如不能保证各批次细胞的均一性，不能稳定的传承下来，因此寻找诱导敏感的永久性细胞系作为技术平台成为了 CYP 研究的热点之一。Ackermann 等（1998）[19] 利用鱼类细胞系 PLHC-1 和 RTG-2 检测了不同培养基对测定细胞毒性和 CYP 活性的影响，Araujo 等（2000）[20] 利用虹鳟（RTG-2）和鲇鱼（BB）细胞系进行苯并芘毒性研究，Fent（2001）[16] 则全面性的评价了鱼类细胞（PLHC-1）在生态毒理学研究中的多用途——评价化学和环境样品的细胞毒性、CYP1A 诱导效能以及雌激素活性等。与体内研究相比，在药物代谢酶研究中运用鱼类细胞，具有许多优点：（1）细胞的均一性好，在遗传上极为相似；（2）药物与细胞直接接触，可较迅速获得结果；（3）可结合多种技术方法测知药物效应；（4）比用活鱼更经济、更方便。

草鱼是我国"四大家鱼"之一，是非常有代表性的鲤科鱼，目前国内已建成多株草鱼细胞系，仅农业部渔业动植物病原库保存的 GCB、PSF、CO、GCL、CIK 等细胞可供筛选。此外，笔者已利用 CYP1A mRNA 诱导前后的差异表达，对草鱼 CYP1A cDNA 片段进行了克隆和鉴定[12]。本文的研究结果表明，利用代谢活性强且诱导敏感的草鱼肝细胞系，以草鱼 CYP1A cDNA 片段为探针，可以快速、灵敏的检测外源化合物的诱导效能。下一步将以此建立水产动物细胞模型，开展渔药药理学和环境污染物生态毒理学等方面的研究工作。

参考文献

[1]　Parrott J L, Hodson P V, Servos M R, et al. Relative potency of polychlorinated dibenzo-p-dioxins and dibenzofurans for inducing mixed-func-

tion oxygenase activity in rainbow trout [J]. Environ Toxicol Chem, 1995, 14 (6): 1041-1050.

[2] Celander M, Broman D, Foerlin L. Effects of petroleum hydrocarbons on the hepatic cytochrome P450 1A1 system in rainbow trout [J]. Mar Environ Res, 1995, 39 (1): 61-65.

[3] Petrulis J R, Bunce N J. Competitive inhibition by inducer as a confounding factor in the use of the ethoxyresorufin-O-deethylase (EROD) assay to estimate exposure to dioxin-like compounds [J]. Toxicol Lett, 1999, 105 (3): 251-260.

[4] Tom D J, Lee L E, Lew J. Induction of 7-ethoxyresorufin-O-deethylase activity by planar chlorinated hydrocarbons and polycyclic aromatic hydrocarbons in cell lines from the rainbow trout pituitary [J]. Comp Biochem Physiol (Part A), 2001, 128 (2): 185-198.

[5] 林茂, 杨先乐, 房文红, 等. 草鱼肝细胞中诱导剂对 EROD 作用的剂量效应研究 [J]. 水产学报, 2006, 30 (3): 311-315.

[6] Fallone F, Villard P H, Decome L, et al. PPARα activation potentiates AhR-induced CYP1A1 expression. Toxicology, 2005, 216 (2): 122-128.

[7] Chaty S, Rodius F, Vasseur P. A comparative study of the expression of CYP1A and CYP4 genes in aquatic invertebrate (freshwater mussel, Unio tumidus) and vertebrate (rainbow trout, Oncorhynchus mykiss) [J]. Aquat Toxicol, 2004, 69 (1): 81-93.

[8] Campbell P M, Devlin R H. Expression of CYP1A1 in livers and gonads of Pacific salmon: quantitation of mRNA levels by RT-cPCR [J]. Aquat Toxicol, 1996, 34 (1): 47-69.

[9] Courtenay S C, Grunwald C M, Kreamer GL, et al. A comparison of the dose and time response of CYP1A1 mRNA induction in chemically treated Atlantic tomcod from two populations [J]. Aquat Toxicol, 1999, 47 (1): 43-69.

[10] Rees C B, McCormick S D, Heuvel JPV, et al. Quantitative PCR analysis of CYP1A induction in Atlantic salmon (Salmo salar) [J]. Aquat Toxicol, 2003, 62 (1): 67-78.

[11] Sadar M D, Ash R, Sundqvist J, et al. Phenobarbital induction of CYP1A1 gene expression in a primary culture of rainbow trout hepatocytes [J]. J Bio Chem, 1996, 271 (30): 17635-17643.

[12] 林茂. 鱼类细胞中药物代谢酶体外诱导的研究 [D]. 2006: 85-95.

[13] Augustine Arukwe. Complementary DNA cloning, sequence analysis and differential organ expression of b-naphthoflavone-inducible cytochrome P4501A in Atlantic salmon (Salmo salar) [J]. Comp Biochem Physiol (Part C), 2002, 133 (4): 613-624.

[14] Elcombe C R, Lech J J. Induction and characterization of hemoprotein (s) P450 and monooxygenation in rainbow trout (Salmo Gairdneri). Toxicol Appl Pharmacol, 1979, 49 (4): 437-450.

[15] Billiarda S M, Bolsb N C, Hodson P V. In vitro and in vivo comparisons of fish-specific CYP1A induction relative potency factorsfor selected polycyclic aromatic hydrocarbons [J]. Ecotoxicol and Environ Safe, 2004, 59 (3): 292-299.

[16] Fent K. Fish cell lines as versatile tools in ecotoxicology: assessment of cytotoxicity, cytochrome P4501A induction potential and estrogenic activity of chemicals and environmental samples [J]. Toxicol In Vitro, 2001, 15 (4): 477-488.

[17] Pesonen M, Teivainen P, Lundstrom J, et al. Biochemical responses of fish sac fry and a primary cell culture of fish hepatocytes exposed to polychlorinated naphthalenes [J]. Arch Environ Contam Toxicol, 2000, 38 (1): 52-58.

[18] 万小琼, 吴文忠, 贺纪正. 利用草鱼原代肝细胞培养评价二恶英毒性效应. 中国环境科学, 2002, 22 (2): 114-117.

[19] Ackermann G E, Fent K. The adaptation of the permanent fish cell lines PLHC-1 and RTG-2 to FCS-free media results in similar growth rates compared to FCS-containing conditions. Mar Environ Res, 1998, 46 (4): 363~367.

[20] Araujo C S A, Marques S A F, Carrondo M J T. In vitro response of the brown bullhead catfish (BB) and rainbow trout (RTG-2) cell lines to benzo [a] pyrene. Science of the Total Environment, 2000, 247 (2): 127-135.

该文发表于《海洋渔业》【2008, 30 (1): 31-36】

大口黑鲈原代肝细胞的培养及其应用于 CYP450 活性的诱导

Studies of Micropterus salmoides on the culture of primary hepatocytes and on the induction of CYP450 activity

喻文娟, 李聃, 王翔凌, 张宁, 邱军强, 杨先乐

(上海水产大学农业部渔业动植物病原库, 上海 200090)

摘要 体外培养大口黑鲈 (*Micropterus salmoides*) 原代肝细胞, 观察其在培养过程中的形态变化, 同时检测培养上清液中的乳酸脱氢酶 (LDH)、白蛋白 (ALb) 含量, 从而判定肝细胞的增殖和代谢能力, 进而应用于 CYP450 的诱导研究。结果表明: 培养 5 d 后, 肝细胞连接成片, 分裂增多; 培养至 7 d, 细胞渐至单层; 14 d 后, 细胞停止生长, 出现老化迹象。培养至第 3~4 天的细胞上清液中 LDH 的渗出量呈上升趋势, 在第 4 天时达到峰值, 随

着培养时间的延长，肝细胞的酶渗出量逐渐减少。ALb 的分泌量在第 3~5 天呈上升的趋势，第 4~8 天维持在较高水平，在第 5 天达到最高，此后随着培养时间的延长，ALb 的分泌量呈下降趋势，故选择培养至第 5 天的大口黑鲈肝细胞应用于 CYP450 酶系中的氨基比林-N-脱甲基酶（ANDM）的诱导研究。40 μM 利福平诱导 24 h 后，ANDM 活性最高。ANDM 的诱导研究为鱼类细胞药物代谢酶体外诱导奠定基础。

关键词 大口黑鲈；肝细胞；原代培养；乳酸脱氢酶；白蛋白；氨基比林-N-脱甲基酶

2. P450 酶的活性检测、表达、分布及其酶促动力学

红霉素-N-脱甲基酶（ERND）等可诱导草鱼 CYP3A 相关酶的活性，建立了草鱼肝细胞系中 CYP3A 的活性检测方法，建立了相应的诱导模型。

该文发表于《Comparative Biochemistry and Physiology Part C：Toxicology & Pharmacology》【2008，147（1）：17-29】

Effects of mammalian CYP3A inducers on CYP3A-related enzyme activities in grass carp（Ctenopharyngodon idellus）：Possible implications for the establishment of a fish CYP3A induction model

Dan Lia[a]，Xian-Le Yang[a]，Shu-Jun Zhanga[a]，Mao Lin[b]，Wen-Juan Yua[a]，Kun Hua[a]

（a. Aquatic Pathogen Collection Centre of Ministry of Agriculture，Shanghai Fisheries University，334 Jungong Road，Shanghai 200090，China
b. Fisheries College，Jimei University，Xiamen 361021，China）

Abstract：Unexpected drug-drug interactions in fish are generally associated with the induction of CYP3A activity and may lead to the formation of drug residues and thus threaten the safety of fishery products. However，little information is available about CYP3A induction in fish. In the present study，we determined the in vivo and in vitro effects of typical mammalian CYP3A inducers（rifampicin，phenobarbital and dexamethasone）on CYP3A-related enzyme activities in a freshwater teleost，the grass carp（Ctenopharyngodon idellus）. Our results showed that the response to rifampicin was similar for grass carp liver cell line（GCL），liver microsomes and the primary hepatocytes of grass carp，as indicated by the activity of aminopyrine N-demethylase（APND）. When erythromycin N-demethylase（ERND）and 6β-testosterone hydroxylase（6β-TOH）were taken into consideration，the GCL displayed a greater capacity for conducting CYP3A metabolism and induction than the C. idellus kidney cell line（CIK）. Using erythromycin and testosterone as substrates，we demonstrated that CYP3A catalysis exhibited non-Michaelis-Menten kinetics in GCL cells，and that V_{max}/K_m values were significantly increased due to rifampicin-treatment. Overall，this study may have implications for the use of GCL as a CYP3A induction model to identify physiological changes in fish as well as the similarities or differences between fish and mammals.

key words：Aminopyrine N-demethylase；CYP3A；Erythromycin N-demethylase；Grass carp；Induction；In vitro model；Rifampicin；Teleost；6β-Testosterone hydroxylase

哺乳动物 CYP3A 诱导物对草鱼 CYP3A 相关酶活性的影响：可建立鱼类 CYP3A 诱导模型

李聃[a]，杨先乐[a]，张书俊[a]，林茂[b]，喻文娟[a]，胡鲲[a]

（a. 上海水产大学农业部渔业动植物病原库，上海海洋大学，军工路 334 号 上海 中国 200090；
b. 集美大学水产学院，福建厦门 中国 361021）

摘要 渔药药物在联合使用时常发生不可预知的相互作用，这可能产生药残甚至威胁水产品的安全，通常这跟 CYP3A 活性有关。然而，对于 CYP3A 诱导物的了解很少。本研究确定了典型的哺乳动物 CYP3A 诱导物（利福

平、苯巴比妥米娜、地塞米松）对淡水硬骨鱼（以草鱼为例）CYP3A 相关酶活性的影响。结果显示，对利福平的反应跟草鱼肝细胞系、肝微粒体、原代肝细胞的反应相近，就像文中的通过氨基比林和 N-脱甲基酶的反应。考虑到对红霉素的 N-脱甲基酶和 6β-睾酮羟化酶的作用，草鱼肝细胞系比肾细胞系能更好地诱导 CYP3A 代谢和诱导物的产生。通过以红霉素和睾丸素为基质的研究，证明在肝细胞系的非米氏动力学；还有就是由于利福平的作用，V_{max}/Km 值明显提高。综上所述，研究证明，肝细胞系可以作为 CYP3A 诱导物模型来证明 CYP3A 在鱼类体内的作用，还有就是鱼类和哺乳动物的区别。

关键词　氨基比林的 N-去甲基；CYP3A；红霉素的 N-去甲基化；草鱼；诱导物；体外模型；利福平；硬骨鱼；6β-睾酮羟化酶

1 Introduction

The measurement of biomarkers in fish is crucial in providing instructive information on the metabolism of drugs in aquatic vertebrates as well as higher animals (Cravedi et al., 1998), and also in predicting the possible drug-drug interactions so as to ensure the safety of food products. Indeed, induction of drug metabolism by the hemoprotein family of cytochrome P450s (CYPs) in fish is well established as a useful tool for environmental contamination biomonitoring (Förlin and Celander, 1995; Fent et al., 1998; Malmström et al., 2004; Lee et al., 2006) and in ecotoxicology assays (Fent, 2001; Gravato and Santos, 2002; Billiard et al., 2004). However, little attention is paid to the use of biomarkers for predicting drug interactions in fish. This is surprising given that drug interactions may result in clinical failures and drug residues with serious consequences for human consumers.

In both mammalian and fish species the CYP3A family, considered to be the drug-metabolizing enzyme of principal importance, collectively comprises the largest portion of the liver and small intestinal CYPs (Hegelund and Celander, 2003; Lee and Buhler, 2003; Nallani et al., 2004). In fish, CYP3A is involved in the metabolism of an extensive range of drugs and can be strongly induced by a variety of structurally unrelated xenobiotics. Because of this, several drug-drug interactions are associated with induction of this enzyme (Tseng et al., 2005). Thus, the kinetic behavior and regulation of CYP3A enzymes in fish have attracted growing attention. Recently, a number of CYP3A genes have been isolated from several teleost species (McArthur et al., 2003; Kashiwada et al., 2005). Although CYP3A genes in fish contain multiple paralogs differing in gene expression pattern and tissue distribution, CYP3A enzymes exhibit similar catalytic properties because of structural similarity (Bork et al., 1989). This is of importance because it allows the assessment of overall CYP3A enzyme induction by determining activity of CYP3A-dependent enzymes. However, CYP3A, as the unique subfamily of CYPs thought to be able to bind multiple substrate molecules at any given time (Kashiwada et al., 2007), has very few specific probes that allow researchers to determine its activity. To date, 6β-testosterone hydroxylase (6β-TOH) is a well-known bioindicator widely used to determine the CYP3A activity in fish as in mammals (James et al., 2005), and erythromycin N-demethylase (ERND) is a marker for various fish CYPs especially CYP3A while it is also a strong marker for mammalian CYP3A (Vaccaro et al., 2003). In addition, aminopyrine N-demethylase (APND), which is a useful biomarker for evaluation of general toxic load of environmental contamination, is also known to be at least partially linked to CYP3A (Yawetz et al., 1998; Paolini et al., 1999). In brief, the enzymes are all CYP3A-related enzymes and can be used together to reflect CYP3A induction in fish more accurately by comparison and analysis.

Despite the primary focus of most CYP-induction studies on mammalian species, recent research has significantly advanced our understanding of CYP3A induction in fish. In general, activation of receptors and allosteric mechanisms are thought to serve as the key reasons explaining the response to inducers. Various authors have demonstrated that the induction of CYP3A expression in fish is probably regulated by the pregnane X receptor (PXR) (Bresolin et al., 2005; Meucci and Arukwe, 2006). In addition, fish CYP3A expression may be modulated by another closely related xeno sensor: constitutive androstane receptor (CAR), as it is encoded in some fish (Maglich et al., 2003). It is also worth noting that both the aryl hydrocarbon receptor 2 (AHR2) and AHR nuclear translocator (ARNT) play an important role in the up regulation of fish CYP3A expression (Tseng et al., 2005). A significant correlation between enzyme activity and mRNA expression for CYP3A has been demonstrated in fish as in mammals (Lange et al., 1999; Araújo et al., 2000; James et al., 2005), thus CYP3A activity is modulated via this molecular mechanism. In addition, allosteric effects also can alter fish CYP3A activity. Alterations in substrate recognition sites can modify binding specificity and orientation of ligand binding (Domanski et al., 2001). CYP3A protein is also capable of binding multiple substrates, which may result in cooperative activation of catalysis through allosteric mechanism (Kashiwada et al., 2005). Activation by endocrine disruptors may represent this non genomic mechanism in fish (Kullman et al., 2004). Regardless of the induction pathway, species-specific differences in the induction of CYP3A are generally observed (Roymans et al., 2004). For example, CYP3A in the PLHC[-1] cell line (Topminnow) was not induced by the prototypical mammalian CYP3A inducers (Celander et al., 1996). Therefore, it is sometimes difficult or even impossible to make the extrapolation from mammals to fish in this respect. Earlier studies have shown that CYP3A expression in some fish species could be enhanced by certain xenobiotics (Bainy et al. , 1999; Hasselberg et al. , 2004; Bresolin et al. , 2005; Meucci and Arukwe, 2006). There are, however, very few reports showing successful induction of the fish CYP3A-related enzymes mentioned above despite a number of reports on fish CYP3A catalytic activities. As we know, several CYP1A inducers exhibit this potentiality. Machala et al. (1997) demonstrated that APND activity was induced by PCBs in common carp. 6β-TOH activity was also induced by PCBs in rainbow trout according to Machala et al. (1998). Furthermore, APND as well as 6β-TOH were found to be induced by β-naphthoflavone (BNF) in blue striped grunt (Stegeman et al. , 1997). It was also reported that ERND activity was slightly enhanced by BNF in sea bass and gilthead seabream (Novi et al. , 1998; Pretti et al. , 2001). Nevertheless, the inducibility of CYP3A activity by rifampicin (RIF), phenobarbital (PB) and dexamethasone (DEX), which are typical mammalian CYP3A inducers (Sy et al. , 2002) and commonly used medicines, has not been previously reported in fish.

Grass carp (Ctenopharyngodon idellus) is an herbivorous species belonging to the family of Cyprinidae of the class Osteichthyes. It is an economically important freshwater fish in China and is considered to be the national fish. Grass carp is not only utilized as a food source but also used for biomonitoring of contaminated water owing to its widespread existence (Wan et al. , 2002; Wan et al. , 2004; Lin et al. , 2006). However, knowledge on the induction of CYP-related monooxygenases in grass carp is limited. Due to the various disadvantages of in vivo experiments, in vitro models with a strong predictive power have been developed to obtain early information about biotransformation pathways as well as induction possibilities and to forecast drug-drug interactions at the metabolic level (Brandon et al. , 2003). With regards to fish, besides some popular in vitro models such as liver microsomes and primary hepatocytes (Winzer et al. , 2002; Henczová et al. , 2006), established cell lines, including PLHC1 (Celander et al. , 1996), RTG-2, BB (Araújo et al. , 2000), RTL-W1 (Babín et al. , 2005) and so on, are a valuable substitute because they are generally easier to culture and have relatively stable enzyme co2ncentrations (Janošek et al. , 2006). However, to our knowledge, no information is available regarding the successful induction of CYP3A activity in these cell lines. We have recently demonstrated that GCL, a hepatic cellular model derived from grass carp, possesses drug metabolizing activity and CYP-related monooxygenase inducibility (Lin et al. , 2006; Lin et al. , 2007). Therefore, the aim of the present study was to further evaluate the usefulness of GCL as an in vitro model system for induction studies of CYP3A activity. The relationships between the different CYP3A-related enzymes in GCL cells were also examined. An additional objective was to investigate whether mammalian CYP3A inducers could enhance the CYP3A-related enzyme activities in grass carp in vitro models. Such information will facilitate the establishment of fish CYP3A induction models which could predict the unsuitable pharmacokinetic properties of drugs, thus leading to the appropriate use of fisheries drugs and further contributing to residue control.

2 Materials and methods

2.1 Chemicals

Testosterone, 6β-hydroxytestosterone and corticosterone were purchased from Sigma (St Louis, MO, USA; N99% purity as determined by HPLC). Rifampicin (RIF), dexamethasone (DEX) (99.2% purity as determined by HPLC), metyrapone (MET), pyridine (PD) and NADPH were obtained from Sigma-Aldrich (China). Erythromycin (ERY) was obtained from Sangon (Shanghai, China). Aminopyrine and phenobarbital (PB) were obtained from Xinya (Shanghai, China). Ketoconazole (KET), α-naphthoflavone (ANF), quinidine (QD), bovine serum albumin (BSA), trypsin, insulin and dimethyl sulphoxide (DMSO, 99.9%) were obtained from Sigma Chemicals. M199 and Liebovitz L15 medium, foetal bovine serum (FBS) and penicillin-streptomycin were obtained from Gibco BRL (UK). Methanol was of HPLC grade (Darmstadt, Germany). All other chemicals and solvents were of analytical grade obtained from common commercial sources.

2.2 Fish and treatments

Healthy immature grass carp (C. idellus) (weighing 150-250 g) and largemouth bass (Micropterus salmoides) (weighing 200-300 g) were obtained from a farm product market in Shanghai, China. The latter is a freshwater teleost possessing CYP3A inducibility (Barber et al. , 2007) and was used as a positive reference. Fish were fed with commercial fish food and housed in plastic tanks (500 L) supplied with a constant flow (1 L · h⁻¹) of aerated fresh water. The water quality was monitored daily, and adjusted as necessary. The pH was maintained between 6.8 and 7.5 and water temperature was adjusted to 23.0±1.0℃. Each species of fish was randomly divided into two groups and acclimated for two weeks before experimentation. The fish of group 1 received a single intraperitoneal (i. p.) injection or one injection daily for three consecutively days of corn oil (vehicle, 2 mL kg body weight per day) containing variable doses of rifampicin, while control fish were given vehicle alone (n=3～5). Fish were not anesthetized during the administration because of possible interference with the treatment (Marionnet et al., 1997). Twenty-four hours after the last injection fish were sacrificed by a blow to the head. The liver was then excised and microsomes prepared immediately. The fish of group 2 were used for cell culture without any treatment.

2.3　Microsomal preparation and measurement of hepatic aminopyrine N-demethylase activity

Liver microsomes from fish of group 1 were prepared as described by Pesonen et al. (2000b) except that the homogenization solution contained 1 mM EDTA, 1 mM phenylmethy l sulfonylfluoride (PMSF), 1 mM phenylthiourea (PTU), 0.1 mM dithiothreitol (DTT) and 10% glycerol. The microsomes were resuspended in 20% glycerol containing 1 mM EDTA, 1 mM PMSF, 0.1 mM DTT and stored at -70℃ in small polypropylene plastic bottles until use (b1 week, thawed once). All the procedures involved in the preparations of microsomes were performed at $0\sim4$℃. Microsomal aminopyrine N-demethylase (APND) activity was assayed as reported previously by Stegeman et al. (1979).

2.4　Cell culture and treatments

2.4.1　Primary culture cells

Hepatocytes were isolated from cultured grass carp (C. idellus) and largemouth bass (M. salmoides) in group 2 following the methods of Smeets et al. (1999) and Yu et al. (2006), respectively. Cell viability after isolation was over 90% in both cases as determined by the MTT reduction test (Vignati et al., 2005). The isolated hepatocytes from grass carp (0.5×10^6 cells mL^{-1} medium) and largemouth bass (1.0×10^6 cells mL medium) were cultured in sterile 50-cm polystyrene flasks (Corning Inc., Corning, NY, USA) in phenol red-free M199/L15 basal medium supplemented with 20% (v/v) FBS, 10 μg · mL^{-1} insulin and 1% (v/v) penicillin-streptomycin (10,000 IU mL^{+-1}-10,000 μg · mL^{-1}), and were incubated at 25℃ in 4.0% CO_2 (Thermo, USA). The hepatocytes were allowed to attach overnight after which the medium was changed every day until day 5 when the hepatocytes were exposed in a serum-free medium.

2.4.2　Subculture cell lines

Grass carp liver cell line (GCL, No. BYK-C12-02, passages 42 to 48 and 52 to 65), C. idellus kidney cell line (CIK, No. BYK-C06-02, passages 72 to 82) and Chinook salmon embryo cell line (CHSE, No. BYK-C16-01, passages 46 to 53) were obtained from the Aquatic Pathogen Collection Centre of Ministry of Agriculture (APCCMA, Shanghai, China). Cells were grown in M199 medium supplemented with 15% (v/v) FBS at 28℃ in 5.0% CO_2. One day prior to each experiment, the cells were subcultured into 50-cm^2 polystyrene flasks in M199 medium at a density of 0.2×10^6 cells · mL^{-1}, 2.0×10^6 cells · mL^{-1} and 1.0×10^7 cells · mL^{-1}, respectively. We used flasks rather than microplates because we found large size flasks contribute to better maintenance of the activity and inducibility of CYP3A-related enzymes in culture cells (data not shown). The serum medium was also replaced with a serum-free medium to better establish the influence of xenobiotics on the CYP3A activity as suggested previously (Celander et al., 1996).

2.4.3　Treatments for induction and inhibition studies in vitro

The drugs were dissolved in DMSO or ethanol and added to the medium. For the induction studies, all cells received fresh M199 medium and were then treated with solutions (final sublethal concentrations) of RIF (10~100 μM), PB (0.5~10 mM), or DEX (10~150 μM). During the inhibition studies, GCL cells received fresh M199 medium and were then treated with different CYP specific inhibitors, ANF (CYP1A), MET (CYP2B), QD (CYP2D), PD (CYP2E), KET (CYP3A) at concentrations ranging from 0.5 to 50 μM. Control cultures were treated with solvent only (n=3). Tightly capped flasks were placed in the 28℃ incubator for 24 h, at which time they received fresh phenol red-free M199 medium containing the substrates (10 mM aminopyrine, 400 μM erythromycin, or 100 μM testosterone) or vehicle alone. The final concentration of solvent in the medium never exceeded 0.1%. Flasks were then incubated for another 1 h before analysis. At that time, the medium was harvested from each sample group (10 flasks from each treatment) used for enzyme assay and cells were washed three times with 1 mL of 0.1 M phosphate buffer saline (PBS). After washing they were detached by flushing the flask with 0.1 M PBS (pH 7.4). Each cell suspension was divided into two parts. One part was stored at -40℃ and used for protein content measurement, the other part was homogenized by sonication with ultrasound for 2 min. The homogenates obtained were centrifuged at 3000×g for 30 min at 4℃ and the supernatants were immediately used in the activity assay.

2.4.4　Treatments for measurement of the enzyme kinetic parameters

The untreated and 40 μM RIF-treated GCL were incubated for 1 h at 28℃ with increasing substrate concentrations of erythromycin (from 10 μM to 800 μM) and testosterone (from 10 μM to 400 μM). Subsequently, medium and cells were harvested as described above.

2.5　Enzyme assays

APND and erythromycin N-demethylase (ERND) activities in all cell samples were determined by measuring the formation of formaldehyde according to the method of Nash (1953). 6βTestosterone hydroxylase (6β-TOH) was assayed in accordance with an HPLC method described by Hoen et al. (2000), with some modifications. The Agilent 1100 HPLC system (USA) consisted of a quaternary pump (Germany), vacuum degasser (Japan), and auto-injector (Germany). To correct for incomplete recovery, aliquots of 1 mL were supplemented with 20 nmol of corticosterone as an internal standard (eluted between 6β-hydroxytestosterone and testosterone, Fig. 1). Testosterone, 6β-hydroxytestosterone metabolite, and the internal standard were extracted by the addition of 6 vol. of dichloromethane, mixed vigorously for at least 1 min, and centrifuged for 10 min at 4, 000 ×g at 4℃ to separate the two phases. The aqueous phase was discarded by careful aspiration and the organic phase was transferred to clean tubes and evaporated under a nitrogen stream at 65℃. The residue was dissolved in 300 μL methanol and 50 μL of the solution was injected onto a reverse phase analytical column, ZORBAX SB-C18 (5 μm, 150 mm×4.6 mm, I.D., Agilent Technologies, USA). The mobile phase consisted of 20% methanol in 0.04% acetic acid (A) and methanol with 0.04% acetic acid (B), and the proportion of B was increased linearly from 20% to 50% over the initial 45 min, followed by a period of isocratic elution for a further 10 min. The flow rate was 0.8 mL min and the column eluate was monitored at 254 nm. For testosterone and 6β-hydroxytestosterone, average recovery in the medium was 100.26%±4.40% and 104.49%±4.14%. In each experiment, a calibration curve ranging from 1 to 100 nM 6β-hydroxytestosterone was constructed. Protein levels were measured by the method of Lowry et al. (1951), using BSA as standard. All water used in the experiment was formaldehyde-free thricedistilled water. Activities for each enzyme were presented as absolute values.

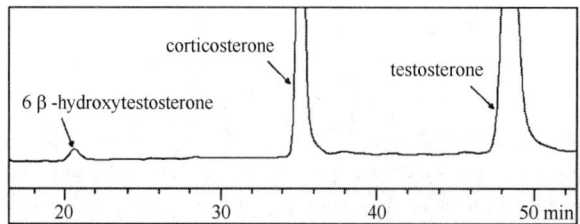

Fig. 1　Representative HPLC profile of medium samples in which testosterone (substrate), 6β-hydroxytestosterone (metabolite) and corticosterone (internal standard) are detectable

Table 1　Effect of RIF on liver microsomal aminopyrine N-demethylase (APND) activity in the grass carp (Ctenopharyngodon idellus) and largemouth bass (Micropterus salmoides)

Species	Control	RIF-treated	
		50	80
Grass carp	1143.55±237.70[a]	1353.96±231.58[ab]	1707.32±240.13[bc]
Largemouth bass	1093.89±231.56[a]	2070.73±241.32[cd]	2239.19±240.89[d]

[a] Fish were treated with i.p. injections of corn oil or RIF for three days, and RIF-1 content are given in mg RIF kg wet weight each day. Results represent the mean±S.D. of three experiments, each experiment used liver microsomes pooled from 3 to 5 control or RIF-treated fish 24 h after last injection. APND[-1], activities are expressed as pmol min · mgprotein. Values not sharing a common letter are significantly different, Pb0.05.

2.6　Data analysis and statistics

In vivo (n=3-5) and in vitro (n=3) measurements were performed at least three times. Statistical analyses were performed with SPSS[TM] software version 15.0 (SPSS Inc., Chicago, IL, USA) and experimental data were expressed as mean±standard deviation (S.D.). Significant differences between means of various treatment groups were determined by one-way analysis of variance (ANOVA) and means were contrasted using a one-tailed Dunnett's t-test. A probability level of Pb0.05 was considered significant. Apparent enzyme kinetic parameters (i.e., half-maximal substrate concentration, Km; maximal velocity, Vmax) were calculated from LineweaverBurk inverse plots. Enzyme induction analyses were calculated using SigmaPlot version 10.0 software (Jandel

Scientific).

3 Results

3.1 In vivo effect of RIF on APND activity in the liver microsomes of grass carp and largemouth bass

Preliminary experiments indicated that repeated injections of RIF over three consecutive days led to a greater increase of APND activity in grass carp liver than a single injection of RIF at a 3-fold higher concentration (data not shown). In the current study we examined the effect of RIF treatments (one injection of 50 mg · kg^{-1} or 80 mg · kg^{-1} daily for three consecutively days) on hepatic APND activity in grass carp as well as in largemouth bass.

APND activity of microsomes from vehicle-treated controls and RIF-treated fish is shown in Table 1. Basal level enzyme activity in the two species was within the range reported for other tropical and temperate fish (Stegeman et al., 1997). Treatment with RIF at 80 mg · kg^{-1}, but not 50 mg · kg^{-1}, significantly elevated APND activity compared to control values in grass carp. In contrast, APND activity in largemouth bass was elevated following treatment with both 50 and 80 mg · kg^{-1} RIF. Furthermore, largemouth bass displayed a greater increase in APND activity relative to grass carp when treated with RIF.

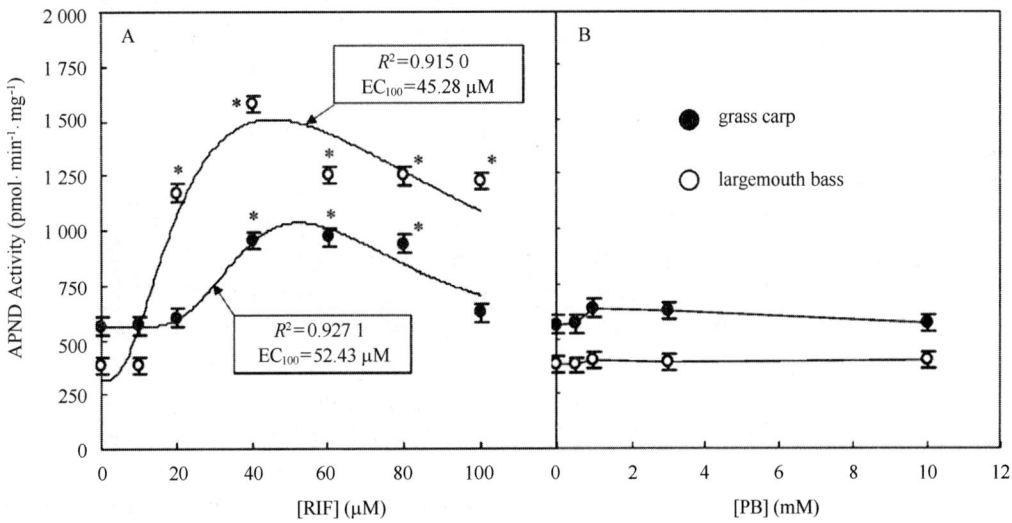

Fig. 2 Effect of RIF (A) or PB (B) on aminopyrine N-demethylase (APND) activity in the primary culture hepatocytes of grass carp (Ctenopharyngodon idellus) and largemouth bass (Micropterus salmoides). Cells were treated as described in Materials and methods section. Data are presented as the mean of four separate experiments±SD. (n=3). Points at y-axes indicate the APND activities of the controls treated with solvent instead of mammalian CYP3A inducer. Asterisk denotes values significantly different from control values at P<0.05. PB showed no induction potency. RIF induction values were fitted by non-linear regression analysis (Log-Normal model) as described earlier by Lin et al. (2006) using Sigma-Plot 10.0. Regressions for significant correlations (R^2) and optimal effect concentration (EC$_{100}$) are shown.

3.2 In vitro effects of RIF and PB on APND activity in grass carp and largemouth bass hepatocytes

Previously, we showed that APND activity in grass carp and largemouth bass liver could be induced by RIF in vivo, therefore we investigated the levels of this monooxygenase in control and treated primary culture hepatocytes of the two species. Measurements performed in control hepatocytes of grass carp showed that the metabolites were not detectable in cell suspension when the cell suspension and medium were analyzed separately, indicating that the metabolites were primarily secreted into the culture medium. Although this result was in agreement with many published reports (Kawanishi et al., 1981; Pesonen et al., 2000a), we cannot exclude the possibility that small amounts of metabolites were present in the cell suspensions under some conditions, thus cell suspension and medium were analyzed together in the subsequent experiments. The effects of RIF (10, 20, 40, 60, 80, and 100 μM) and PB (0.5, 1, 3, and 10 mM) on hepatic APND activity are illustrated in Fig. 2. There was no significant difference in APND activity in response to PB treatment. In contrast, both species hepatocytes reacted to RIF exposure by increasing the level of APND activity in a concentration dependent manner. Non-linear regression analysis indicates that the highest net increase in APND activity in grass carp and largemouth bass hepatocytes, compared with corresponding controls, was reached at 52.43 μM and 45.28 μM RIF respectively. Interestingly, despite the initial hepatic activity of APND being higher in grass carp than in largemouth bass, the former had lower maximal APND activity (approximately 1. 0 nmol min^{-1} · mg^{-1}) compared with the latter (approximately 1. 5 nmol min^{-1} · mg^{-1}).

Table 2　The concentrations of RIF and DEX used in the induction assay of CYP3Arelated enzymes in grass carp cell lines

Inducer	GCL			CIK		
	APND	ERND	6β-TOH	APND	ERND	6β-TOH
RIF (μM)	60	40	40	30	33. 3	40
DEX (μM)	90	120	100	80	100	100

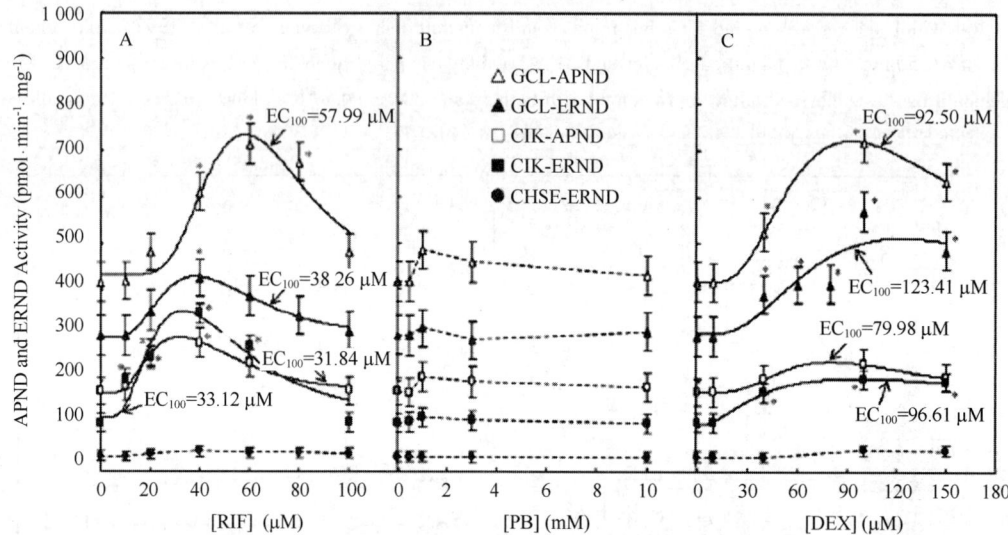

Fig. 3　Effect of RIF (A), PB (B), or DEX (C) on aminopyrine N-demethylase (APND, open points) and erythromycin N-demethylase (ERND, filled points) activities in GCL (triangle), CIK (square) and CHSE (circle) cells. Cells were treated as described in Materials and methods section. Each point corresponds to the mean±S. D. for three separate experiments (n = 3). Points at y-axes indicate enzyme activities of the controls treated with solvent instead of mammalian CYP3A inducer. Asterisk denotes values significantly different from control values at Pb0. 05. The solid lines are concentration-response curves fitted by non-linear regression analysis (LogNormal model) as described earlier by Lin et al. (2006) using SigmaPlot 10. 0 (R^{2}N0. 9, EC_{100} are shown). Note：APND activities in CHSE cells were not determined.

3. 3　In vitro effects of mammalian CYP3A inducers on the CYP3A-related enzyme activities in cell lines

3. 3. 1　Concentration-response assay

Since the APND activity in grass carp was shown to be inducible in vivo and in vitro, we further conducted experiments to determine the effects of RIF, PB and DEX on activities of APND and ERND in grass carp cell lines (GCL, CIK) as well as ERND in a salmon cell line (CHSE). The latter has been proven as a potential model for the study of APND induction (unpublished work). The time-course induction of CYP3A related enzymes and reaction of corresponding substrates has been determined previously by our laboratory (unpublished work). The half-inhibition concentration for cell viability (IC$_{50}$) was determined by MTT reduction test (data not shown) for all three drugs as a first step. Then the concentration-response experiments were conducted at the sub-lethal level induction concentrations with identical effective induction (24 h) and reaction (1 h) time (Fig. 3). As observed in fish hepatocytes, APND activity in grass carp cell lines responded to RIF but not to PB. Interestingly, all three drugs at the concentration employed seemed to exert no significant influence on ERND activity in CHSE cells. Moreover, PB was found to have no induction capacity for both CYP3A-related enzymes in the tested cells over the high mM concentration range. With regards to the GCL and CIK cells, RIF (10-100 μM) and DEX (10~150 μM) both induced APND and ERND activity in a concentration dependent manner. Notably, we did not observe a significant increase of APND in CIK cells exposed to DEX. Regarding RIF treatments, the highest significant increases compared to controls were obtained for APND and ERND activities at the 60 μM and 40 μM concentrations in GCL cells, and both 40 μM in CIK cells, resulting in an increase factor of 1. 72 (APND) and 1. 43 (ERND) in GCL cells, 1. 62 (APND) and 3. 23

（ERND）in CIK cells. As for DEX，we observed the largest significant increase of APND activity in GCL cells and ERND activity in both cell lines at a concentration of 100 μM，resulting in an increase factor of 1. 73 （APND）and 1. 91 （ERND）in GCL cells，and 1. 89 （ERND）in CIK cells. The concentration-response for RIF and DEX was fitted to a Log-Normal model using SigmaPlot 10. 0. The optimal effect concentration （EC_{100}），as shown in Fig. 3，in GCL cells was larger than the corresponding value in CIK cells，indicating that CIK cells are more sensitive to inducers.

3. 3. 2 Inducibility of enzyme activities：comparison of models and inducers

The concentration of inducers was optimized to favor the induction of CYP3A-related enzymes in GCL and CIK cells based on the above results. The concentrations of RIF and DEX used in the APND，ERND and 6β-TOH assays are presented in Table 2，while all other conditions were held constant in the assay. The results are shown in Fig. 4. With respect to APND and ERND，the increase factors were almost identical to those we calculated previously. The induction ability of RIF and DEX showed no significant difference although，the RIF-effect on ERND activity in CIK cells was more pronounced. Interestingly，RIF elicited a marked increase in the 6β-TOH activity of GCL cells （approximately 8 - fold），whereas DEX failed to produce any significant effect under the same conditions. Furthermore，the 6β-TOH activity in CIK cells was too low to be detected with reliable results. Notably，a batch-to-batch variation was observed in enzyme activity and inducibility of grass carp cell lines （GCLbCIK，data not shown）. Overall，the GCL cell line could maintain CYP3A-related enzyme inducibility and relatively high initial activity. And RIF seemed to be superior to DEX at inducing the CYP3A-related enzyme activities in fish cells. This would make RIF more suitable for fish CYP3A induction studies.

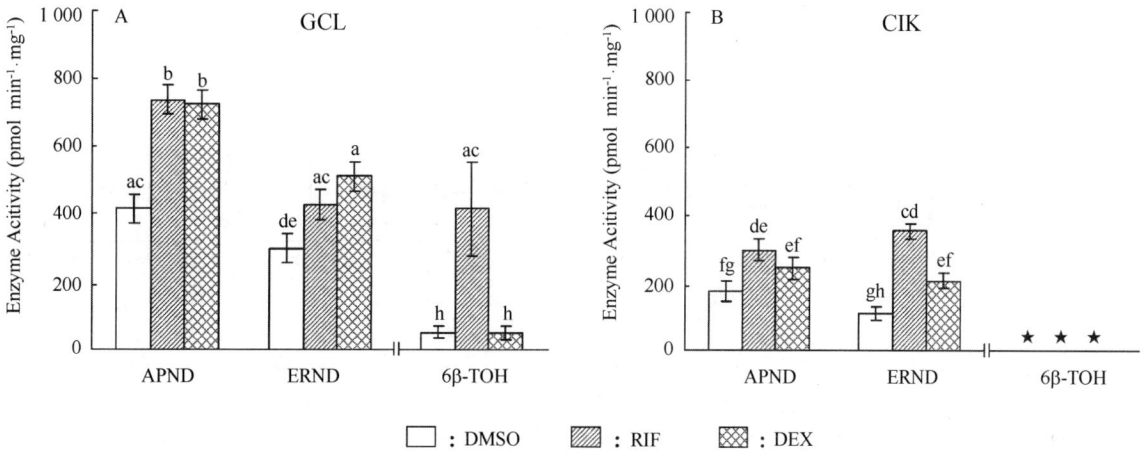

Fig. 4 Comparison of aminopyrine N-demethylase （APND），erythromycin N-demethylase （ERND）and 6β-testosterone hydroxylase （6β-TOH）activities in GCL （A）and CIK （B）cells from controls，RIF and DEX 24 h pre-treated cells. The concentration of inducers was optimized to favor the induction of CYP3A-related enzymes in grass carp cell lines. For those enzyme activities marked ' ★ ' the absorbances were below the lowest detection limit and were equal to zerotime blank values. Columns and error bars represent the mean± S. D. for three separate experiments （n=3）. Values not sharing a common letter are significantly different，P<0. 05.

3. 3. 3 Correlation analysis

The relationship between the different CYP3A-related enzyme activities in GCL cells as well as APND activity，measured in all cells at a relatively high level，in the different grass carp cells are presented （Fig. 5）. The linear regression of all values was extremely poor between 6β-TOH and ERND and APND （R^2<0. 20）. If we exclude the outlying values the correlation between ERND and 6β-TOH improved （R^2=0. 63）as did the correlation between ERND and APND （R^2=0. 53）. However，the correlation between APND and 6β-TOH remained low （R^2=0. 36）（Fig. 5A）. Since APND was distinct from other two enzymes，we further studied the effect of different CYP specific inhibitors on the APND activity in the GCL in order to determine other APND isozymes. The results indicated that only 10 μM KET （CYP3A）and 50 μM MET （CYP2B）significantly decreased APND activity （Fig. 6）. On the other hand，the linear regression reflected a good correlation between APND activities in primary culture hepatocytes and in GCL cells （R^2=0. 96）. In contrast，activities in GCL cells，as expected，did not correlate well with activities in CIK cells （R^2=0. 35）（Fig. 5B）.

3. 3. 4 Effect of substrate concentration on ERND and 6β-TOH activities：determination of kinetic properties

In order to confirm the rationale for the substrate concentrations we used in this study and further assess the induction of CYP3A by RIF in the GCL model，the kinetics of the ERND and 6β-TOH in the presence and absence of RIF were characterized by studying

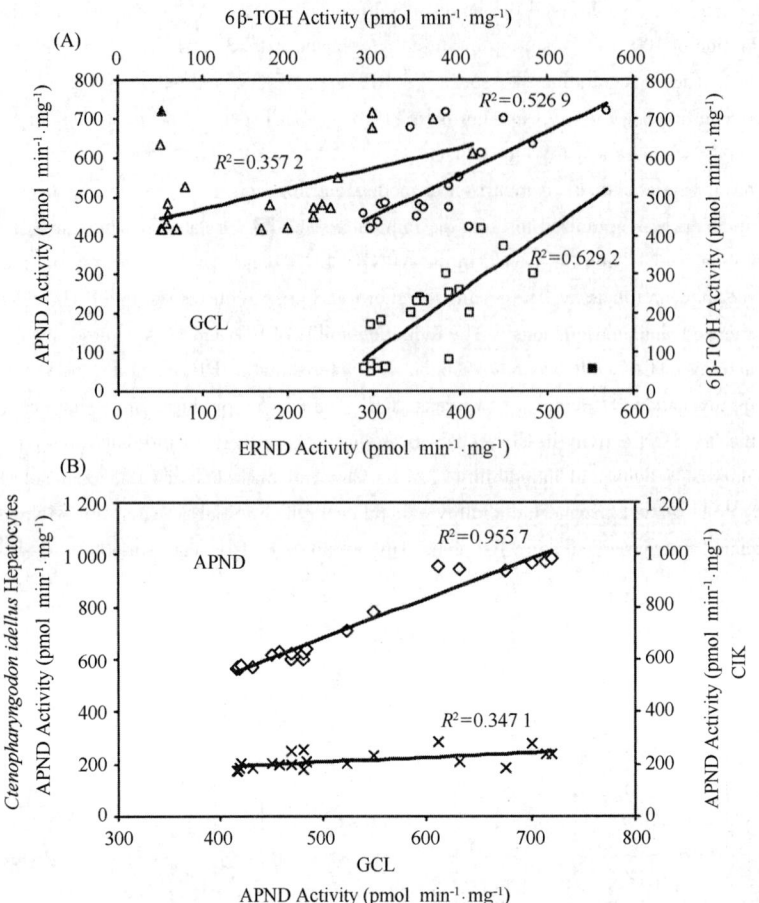

Fig. 5　Correlations and regressions between the CYP3A-related activities aminopyrine N-demethylase (APND), erythromycin N-demethylase (ERND) and 6β-testosterone hydroxylase (6β-TOH) in grass carp cells with various treatments as described in Materials and methods section. (A) APND vs ERND (○), GCL; ERND vs 6β-TOH (□), GCL; APND vs 6β-TOH (Δ), GCL. (B) primary culture hepatocytes vs GCL (◇), APND; GCL vs CIK (×), APND. Outlier values marked as filled points were not considered in calculations.

the effects of erythromycin (10～800 μM) and testosterone (10～400 μM) concentrations on corresponding enzyme activity. The values in control cells were too small to be obtained precisely when substrate concentration was set at 10 μM. In control and RIF-treated GCL cells, the saturation level of erythromycin available for ERND was around 200 μM and above (Fig. 7A), while that of testosterone available for 6β-TOH was around 50 μM and above (Fig. 7C). Accordingly, the substrate concentrations used in our study were not limiting. Kinetic analysis using erythromycin concentrations between 50 and 400 μM (Fig. 7B) as well as testosterone concentrations between 25 and 200 μM (Fig. 7D) showed that the enzymes exhibited classical Michaelis Menten (MM) kinetics. Addition of RIF to GCL cells resulted in a significant decrease in the Km value for ERND and a significant increase in the Vmax value for ERND. Although the Km value for 6β-TOH in control cells was only slightly larger than in RIFtreated cells, the Vmax/Km value for 6β-TOH was much higher in RIF-treated cells than in controls cells. Nevertheless, employing higher substrate concentrations (eg. 800 μM erythromycin or 400 μM testosterone) dramatically decreased reciprocal CYP3A-related enzyme activities, particularly in induced cells (more clearly observed with ERND), indicative of a deviation from the MM kinetic model over higher substrate concentration ranges.

4　Discussion

Fish CYP3A enzymes metabolize a wide range of substances (Tseng et al., 2005) and respond to various xenobiotics (Kashiwada et al., 2007), likely including new or existing fisheries drugs (Moutou et al., 1998; Vaccaro et al., 2003). This has led to demand for the development and application of in vitro models to follow the induction of CYP3A activity in fish. The present study demonstrated the potential effects of mammalian CYP3A inducers on fish CYP3A activity by assessing their capacity to induce CYP3A-re-

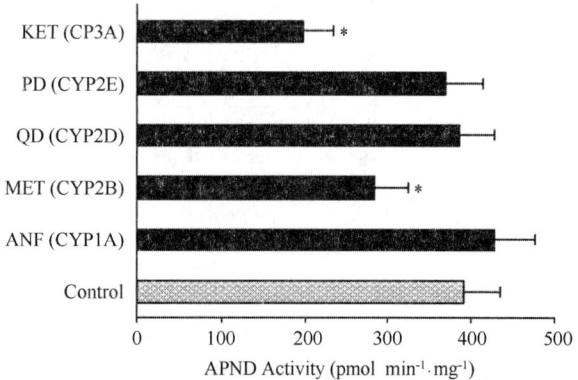

Fig. 6 Effects of CYP specific inhibitors on the APND activity in GCL cells. GCL cells were preincubated with 0. 5 μM α-naphthoflavone （ANF）, 50 μM metyrapone （MET）, 10 μM quinidine （QD）, 10 μM pyridine （PD）, or 10 μM ketoconazole （KET） for 24 h and then incubated with 10 mM aminopyrine for an additional 1 h. The experimental details are described in Materials and methods section. Columns and error bars represent the mean±S. D. for three separate experiments （n=3）. Asterisk denotes values significantly different from control values at P<0. 05.

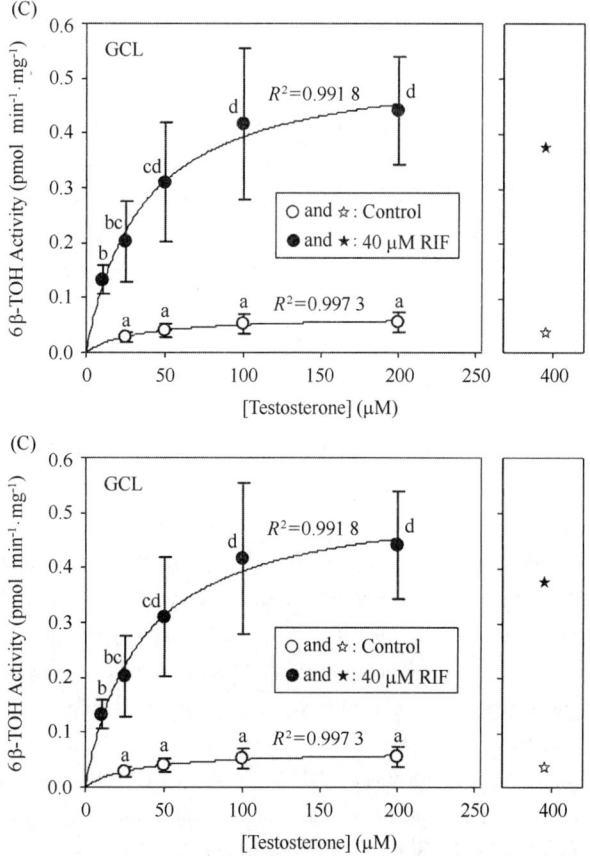

Fig. 7 Induction of the CYP3A-related activities erythromycin N-demethylase （ERND, A）, 6β-testosterone hydroxylase （6β-TOH, C） by RIF in GCL cells and the respective Lineweaver-Burk plots （ERND, B; 6β-TOH, D）. Assays were incubated for 24 h at 28℃ without （open points） or with （filled points） 40 μM RIF before reaction. Each value represents the mean of triplicate incubations. Enzyme induction values were calculated by non-linear regression analysis （Single-Rectangular model） using SigmaPlot 10. 0. Values marked as stars were not considered in calculations （notice the split x-axes）. Regressions for significant correlations （R2） and apparent enzyme kinetic parameters （Km, Vmax） ±standard error are shown. Values not sharing a common letter are significantly different, P<0. 05.

lated enzyme activities of grass carp in vitro models. We first determined the basal level activity and inducibility of APND in liver microsomes and primary hepatocytes of grass carp and largemouth bass (reference). Then we showed that RIF and DEX were able to increase both APND and ERND activities in two grass carp cell lines in a concentration-dependent manner. However, only RIF was able to induce 6β-TOH activity in the GCL, which is indicative of significant differences among inducers, enzymes as well as models. As ERND and 6β-TOH exhibited a better relationship in our study, we further characterized the kinetic properties of these two enzymes in GCL cells in the presence and absence of RIF to confirm the usefulness of GCL as a simple and reproducible in vitro model system for induction studies of CYP3A activity.

In the current study we used three indicator enzymes for CYP3A activity. Interestingly, although induction of all three enzymes is likely to be highly correlated with an increased activity of fish CYP3A following drug exposure, we did not observe this relationship in our study. This is may be due to complex roles of CYPs, inducers and multi-dependent patterns of CYP3A-related enzymes in fish. Generally, our data are in accordance with previous studies showing that the response of APND is somewhat distinct when compared to ERND and 6βTOH in fish (Vaccaro et al., 2003). Previously, Stegeman et al (1997) and Vaccaro et al. (2003) reported that APND might be related to CYP2B in fish. In this study, in addition to the CYP3A selective inhibitor KET, we found that MET, a typical inhibitor of CYP2B (Lin and Lin, 2006), inhibited APND activity in GCL cells. Therefore we presume APND is also dependent on CYP2B forms. Nonetheless, APND activity does partially reflect CYP3A activity and has the advantage of being easier to measure than 6β-TOH activity.

To our knowledge, this paper is the first to report the influences of three typical mammalian CYP3A inducers including RIF, PB and DEX on the CYP3A-related enzyme activities in grass carp cell lines. We demonstrated that none of the models tested in the current study were responsive to PB within the concentration range we tested. It is possible that PB-type responses do not occur in grass carp as previously suggested in some other fish species (Stegeman et al., 1997; Bainy et al., 1999). In addition, our data showed that none of the inducers we tested was able to elicit a response in the CHSE cell line. This suggests that the CHSE cell line no longer has the capacity to respond to CYP3A inducers, a phenomenon that has been described in several cell lines (Celander et al., 1996). Interestingly, we found that both RIF and DEX induced the CYP3A-related enzyme activities of both grass carp cell lines in a similar concentration-dependent manner although DEX did not affect 6β-TOH activity. We suggest that this may be due to species/substrate selectivity or that the indicators/inducers are not strictly CYP3A-specific. For example, ERND, as with APND, also depends on other CYP forms besides CYP3A in fish (Stegeman and Hahn, 1994). Given that 6β-TOH is the most specific indicator of CYP3A in fish (Kullman et al., 2000; Schlenk et al., 2000; Vaccaro et al., 2005), our results suggest that only RIF is able to induce CYP3A activity in the GCL. It is known that the mammalian CYP3A in cultured cells is induced by RIF via a PXR-mediated signaling pathway (Lehmann et al., 1998). However, the regulatory mechanism of teleost CYP3A is still not fully characterized. Indeed, PXR gene sequences have been isolated from zebrafish and puffer fish (Moore et al., 2002; Maglich et al., 2003). Meucci and Arukwe (2006) demonstrated that PXR plays an important role in the induction of CYP3A in Atlantic salmon. Nevertheless, Moore et al. (2002) indicated that RIF (10 μM) did not activate PXR strongly (below 2.5-fold) in zebrafish. Tseng et al. (2005) demonstrated that teleost RIF-induced CYP3A transcription might be mediated by PXR and AHR2 activation. Our study showed that PB and DEX, which are also ligands to PXR in fish (Moore et al., 2002; Corley-Smith et al., 2006), did not significantly induce CYP3A activity. Therefore, we hypothesize that the induction of CYP3A activity in grass carp is not solely dependent on PXR. Moreover, RIF dramatically changed the kinetic behavior of CYP3A as was previously observed using a number of pharmaceuticals (Stresser et al., 2000). We propose that inducers bind to an allosteric site in or near the active site of CYP3A, and thus influence the binding and turnover of the substrates. Small differences in the protein structure and the substrate binding at the proposed allosteric site may explain the observed differences in the effects of RIF and DEX on CYP3Arelated enzyme activities in different species. The response of CYP3A after treatment with inducers provides additional information on the fate and interaction of drugs in fish (Kashiwada et al., 2005). As a consequence, induction of CYP3A may represent a potential risk for fish populations and may also play a role in forming the drug residues in fish harvested for consumption. The data derived from in vitro models are valuable because they can be extrapolated to in vivo models. Furthermore, with appropriate correction for non specific binding they can result in extrapolations that have improved accuracy (Lin et al., 1980; Lave et al., 1999). In this respect, the degree of induction of CYP3A activity in fish in vitro models deserves more attention. The CYP3A-like protein level in herbivorous fish may be relatively high due to the effects of natural dietary products (Stegeman et al., 1997). Therefore, we determined the induction of CYP3A-related activities using four in vitro models from grass carp (an herbivorous species), namely liver microsomes, primary hepatocytes and two cell lines, GCL and CIK. The response of each of these models was similar for eachin ducer. However, basal level APND activity was significantly higher in the microsomes due to the differences in methodology.

Table 3　Summary of in vitro significant inductive effects of RIF, PB and DEX on CYP3A activity in fish and mammalian hepatic cell models

Species	Cell	Treatment			Bioindicator
		RIF (μM)	PB (mM)	DEX (μm)	
Grass carp	GCL cell[C]	40	–	–	6β-TOH ERND APND
Human	HeG2 cell[C]	–	–	[(i)]25	LBOD
	DPX-2 cell[D]	[(i)]10	–	–	LBOD
	HepaRG cell[D]	[(d)]50	–	–	6β-TOH
	BC2 cell[D]	–	–	[(c)]100	6β-TOH
	Hepatocytes[A]	[(g,j)]10	[(e)]0.75,[(h)]2	[(g)]100	6β-TOH
	Hepatocytes[B]	[(m)]50,[(n)]25	[(e)]0.75	[(g)]100	6β-TOH
Rat	Hepatocytes[A]	[(r)]50	[(q)]2	[(f,o)]10 [(g)]100	[(f,g,o)]6β-TOH [(q)]ECOD [(r)]ERND
	Hepatocytes[B]	–	–	[(o)]10	6β-TOH
Monkey	Hepatocytes[A]	–	[(k)]0.5	–	6β-TOH
Pig	Hepatocytes[A]	[(l,p)]50	[(b)]1.5,[(j)]0.05	[(j)]50	[(l,p)]6β-TOH [(b)]NIF
Minipig	Hepatocytes[A]	[(g)]10 (♂) [(g)]100 (♀)	–	–	6β-TOH
Dog	Hepatocytes[A]	[(g)]33 (♂) [(g)]100 (♀)	–	[(g)]33	6β-TOH
Rabbit	Hepatocytes[A]	[(a)]25	–	–	ERND

Abbreviations used: RIF, rifampicin; PB, phenobarbital; DEX, dexamethasone; APND, aminopyrine N-demethylase; ERND, erythromycin N-demethylase; 6βTOH, 6β-testosterone hydroxylase; ECOD, 7-ethoxycoumarin O-deethylase; NIF, nifedipine 6-hydroxylase; LBOD, luciferin 6′-benzylether O-debenzylase. Values concerned in treatment refer to optimal concentrations used in studies and showing significant inductions of CYP3A activities in vitro. Treatments only proved to induce CYP3A expression were not taken into consideration. "–" Reports not found.

A Freshly isolated primary culture cells, Bcryopreserved primary culture cells, Csubculture cell line, Dderivative cell line.

(a) Calleja et al., 1998; (b) Gillberg et al., 2006; (c) Gómez-Lechón et al., 2001; (d) Guillouzo et al., 2007; (e) Hengstler et al., 2000; (f) Hoen et al., 2000; (g) Lu and Li, 2001; (h) Martin et al., 2003; (i) Miranda and Meyer, 2007; (j) Nallani et al., 2004; (k) Nishibe and Hirata, 1995; (l) Olsen et al., 1997; (m) Reinach et al., 1999; (n) Roymans et al., 2004; (o) Silva et al., 1999; (p) Skaanild and Friis, 2000; (q) Wang et al., 2004; (r) Xia et al., 2004.

It should be noted that results obtained with microsomes cannot be used for quantitative estimations of in vivo biotransformation. This is due to the enrichment of CYPs in the microsomal fraction and the lack of competition with other enzymes (Brandon et al., 2003). However, microsomal data in the present study is useful for the qualitative determination of enzyme inducibility in grass carp as in largemouth bass. In addition, the linear regression reflected the correlation between primary hepatocytes and GCL but not between primary hepatocytes and CIK possibly owing to distinct organ sources. Even though enzyme activities occur to a lesser extent in cell lines, these results do not preclude their use in study. On the contrary, the decline in activity over time of enzymes after plating of primary hepatocytes is rapid (Butura et al., 2004) while cell lines are reproducible, generally easier to culture and have relatively stable enzyme activities. Such knowledge can be of value for developing cell lines for further study. When compared with GCL cells, CIK cells were more sensitive to drugs with relatively low EC100 values (Fig. 3). This is most likely owing to larger basal level activity and IC50 values of GCL (Yang et al., 2007). We also found that the enzyme activities in GCL cells remained comparatively constant during the culture period. Notably, it was difficult to detect testosterone metabolites in CIK cells, even in the enzyme-induced state. Bainy et al. (1999) pointed out that CYP3A induction varies by tissue, being high in the liver and nonexistent in the kidney of tilapia. Although the inducibility of CYP3A activity in CIK cells is unclear, it is most likely that CIK is not a suitable tool for assessing CYP3A induction. In order to confirm the usefulness of GCL we simultaneously determined, for the first time, Km and Vmax values for both ERND and 6β-TOH activities in control and RIF-treated GCL cells. Based on calculated data, Km values in the controls were very similar to the results recorded for untreated human hepatic models (Riley and Howbrook, 1998; Easterbrook et al., 2001; Grand et al., 2002). Moreover, the Vmax/Km value, an indicator of catalytic efficiency, for both enzymes was significantly higher in RIF treated cells than in control cells. Based on these results, we suggest that RIF is an effective inducer of CYP3A activity in grass carp. Interestingly, the effect of substrate on CYP3A activity varied with substrate concentration. At low concentrations there was an initial increase in activity followed by a decrease in activity with higher concentrations (Fig. 7). Deviation from a MM kinetic model at higher substrate concentrations is consistently observed with CYP3A enzymes in other fish (Kashiwada et al., 2007). This cooperative activity of the CYP3A-related enzymes, accounting for the non-Michael is Menten kinetics of these enzymes, probably results from the fact that the substrate-binding pocket of CYP3A is larger than that of other members of the P450 superfamily (Kullman et al., 2004).

At present, in addition to primary hepatocytes, various mammalian hepatic cell lines have been used as model systems to evaluate the CYP3A activity, especially for humans. These include: HepG2, DPX-2 (Miranda and Meyer, 2007), TONG (Schuetz et al., 1993), BC2 (Gómez-Lechón et al., 2001), FLC-5 (Iwahori et al., 2003), C3A (Brandon et al., 2003), B16A2 (Butura et al., 2004), HepaRG (Guillouzo et al., 2007) and so on. The in vitro findings have indicated some inductive effect ofRIF, PB and DEX on CYP3A indicators in mammalian hepatic cell models. However, data are scarce in fish cell models. As summarized in Table 3, differences are evident both among species and inducers. This species specificity of CYP3A limits the direct application of induction information from mammals to fish.

In conclusion, the experiments presented here show that the GCL cell line, preferably in the enzyme-induced state, holds promise as a model to examine physiological responses to xenobiotics in fish. It may also be useful for identifying the similarities or differences in CYP3A induction between fish and mammals. By using the stimulation of CYP3A-related enzyme activities, this study also demonstrates that in vivo and in vitro induction of CYP3A activity by RIF may be observed in in vitro hepatic models of grass carp. However, there might be other CYP isoform (s) relating to the CYP3A-related enzymes in grass carp. Since the structure-function relationships of grass carp CYP3A are unknown, further detailed studies are required to establish the CYP3A induction model using the GCL cell line for the purpose of predicting drug-drug interactions, and thus facilitating drug development and rational administration of fisheries drugs.

Acknowledgements

This study was financially supported by the National Natural Science Foundation of China (No. 30371109), Shanghai Leading Academic Discipline Project (No. Y1101) and E-institute of Shanghai Municipal Education Commission (No. E03009). We would like to express our appreciation to Dr. Shaun Clements at Oregon State University (Oregon, USA) for English corrections and anonymous reviewers for their helpful suggestions.

References

Araújo, C. S. A., Marques, S. A. F., Carrondo, M. J. T., Gonçalves, L. M. D., 2000. In vitro response of the brown bullhead catfish (BB) and rainbow trout (RTG-2) cell lines to benzo [a] pyrene. Sci. Total. Environ. 247, 127-135.

Babín, M., Casado, S., Chana, A., Herradón, B., Segner, H., Tarazona, J. V., Navas, J. M., 2005. Cytochrome P4501A induction caused by the imidazole derivative Prochloraz in a rainbow trout cell line. Toxicol. In Vitro 19, 899-902.

Bainy, A. C. D., Woodin, B. R., Stegeman, J. J., 1999. Elevated levels of multiple cytochrome P450 forms in tilapia from Billings Reservoir-São Paulo, Brazil. Aquat. Toxicol. 44, 289-305.

Barber, D. S., McNally, A. J., Garcia-Reyero, N., Denslow, N. D., 2007. Exposure to p, p'-DDE or dieldrin during the reproductive season alters hepatic CYP expression in largemouth bass (Micropterus salmoides). Aquat. Toxicol. 81, 27-35.

Billiard, S. M., Bols, N. C., Hodson, P. V., 2004. In vitro and in vivo comparisons of fish-specific CYP1A induction relative potency factors for selected polycyclic aromatic hydrocarbons. Ecotoxicol. Environ. Saf. 59, 292-299.

Bork, R. W., Muto, T., Beaune, P. H., Srivastava, P. K., Lloyd, R. S., Guengerich, F. P., 1989. Characterization of mRNA species related to human liver cytochrome P450 nifedipine oxidase and the regulation of catalytic activity. J. Biol. Chem. 264, 910-919.

Brandon, E. F. A., Raap, C. D., Meijerman, I., Beijnen, J. H., Schellens, J. H. M., 2003. An update on in vitro test methods in human hepatic drug biotransformation research: pros and cons. Toxicol. Appl. Pharmacol. 189, 233-246.

Bresolin, T., Rebelo, M. D. F., Bainy, A. C. D., 2005. ExpressionofPXR, CYP3Aand MDR1 genes in liver of zebrafish. Comp. Biochem. Physiol. C. 140, 403–407.

Butura, A., Johansson, I., Nilsson, K., Wärngård, L., Ingelman–Sundberg, M., Schuppe–Koistinen, I., 2004. Differentiation of human hepatoma cells during confluence as revealed by gene expression profiling. Biochem. Pharmacol. 67, 1249–1258.

Calleja, C., Eeckhoutte, C., Dacasto, M., Larrieu, G., Dupuy, J., Pineau, T., Galtier, P., 1998. Comparative effects of cytokines on constitutive and inducible expression of the gene encoding for the cytochrome P450 3A6 isoenzyme in cultured rabbit. Biochem. Pharmacol. 56, 1279–1285.

Celander, M., Hahn, M. E., Stegeman, J. J., 1996. Cytochromes P450 (CYP) in the Poeciliopsis lucida hepatocellular carcinoma cell line (PLHC–1): doseand time–dependent glucocorticoid potentiation of CYP1A induction without induction of CYP3A. Arch. Biochem. Biophys. 329, 113–122.

Corley – Smith, G. E., Su, H. T., Wang – Buhler, J. L., Tseng, H. P., Hu, C. H., Hoang, T., Chung, W. G., Buhler, D. R., 2006. CYP3C1, the first member of a new cytochrome P450 subfamily found in zebrafish (Danio rerio). Biochem. Biophys. Res. Commun. 340, 1039–1046.

Cravedi, J. P., Perdu – Durand, E., Paris, A., 1998. Cytochrome P450 – dependent metabolic pathways and glucuronidation in trout liver slices. Comp. Biochem. Physiol. C. 121, 267–275.

Domanski, T. L., He, Y. A., Khan, K. K., Roussel, F., Wang, Q., Halpert, J. R., 2001. Phenylalanine and tryptophan scanning mutagenesis of CYP3A4 substrate recognition site residues and effect on substrate oxidation and cooperativity. Biochemistry 40, 10150–10160.

Easterbrook, J., Fackett, D., Li, A. P., 2001. A comparison of aroclor 1254induced and uninduced rat liver microsomes to human liver microsomes in phenytoin O–deethylation, coumarin 7–hydroxylation, tolbutamide 4–hydroxylation, S–mephenytoin 4′–hydroxylation, chloroxazone 6–hydroxylationandtestosterone6b–hydroxylation. Chem. Biol. Interact. 134, 243–249.

Fent, K., Woodin, B. R., Stegeman, J. J., 1998. Effects of triphenyltin and other organotins on hepatic monooxygenase system in fish. Comp. Biochem. Physiol. C. 121, 277–288.

Fent, K., 2001. Fish cell lines as versatile tools in ecotoxicology: assessment ofcytotoxicity, cytochromeP4501Ainductionpotentialandestrogenicactivity of chemicals and environmental samples. Toxicol. In Vitro 15, 477–488.

Förlin, L., Celander, M., 1995. Studies of theinducibility of P4501A in perchfrom the PCB–contaminatedlake JärnsjöninSweden. Mar. Environ. Res. 39, 85–88.

Gillberg, M., Skaanild, M. T., Friis, C., 2006. Regulation of gender – dependent CYP2A expression in Pigs: involvement of androgens and CAR. Basic Clin. Pharmacol. Toxicol. 98, 480–487.

Gómez–Lechón, M. J., Donato, Y., Jover, R., Rodriguez, C., Ponsoda, X., Glaise, D., Castell, J. V., Guguen – Guillouzo, C., 2001. Expression and induction of a large set of drug – metabolizing enzymes by the highly differentiated human hepatoma cell line BC2. Eur. J. Biochem. 268, 1448–1459.

Grand, F., Kilinc, I., Sarkis, A., Guitton, J., 2002. Application of isotopic ratio mass spectrometry for the in vitro determination of demethylation activity in human liver microsomes using N–methyl–13C–labeled substrates. Anal. Biochem. 306, 181–187.

Gravato, C., Santos, M. A., 2002. β–Naphthoflavone liver EROD and erythrocytic nuclear abnormality induction in juvenile Dicentrarchus labrax L. Ecotoxicol. Environ. Saf. 52, 69–74.

Guillouzo, A., Corlu, A., Aninat, C., Glaise, D., Morel, F., Guguen–Guillouzo, C., 2007. The human hepatoma HepaRG cells: a highly differentiated model for studies of liver metabolism and toxicity of xenobiotics. Chem. Biol. Interact. 168, 66–73.

Hasselberg, L., Meier, S., Svardal, A., Hegelund, T., Celander, M. C., 2004. Effects of alkylphenols on CYP1A and CYP3A expression in first spawning Atlantic cod (Gadus morhua). Aquat. Toxicol. 67, 303–313.

Hegelund, T., Celander, M. C., 2003. Hepatic versus extrahepatic expression of CYP3A30 and CYP3A56 in adult killifish (Fundulus heteroclitus). Aquat. Toxicol. 64, 277–291.

Henczová, M., Deér, A. K., Komlósi, V., Mink, J., 2006. Detection of toxic effects of Cd2 + on different fish species via liver cytochrome P450dependent monooxygenase activities and FTIR spectroscopy. Anal. Bioanal. Chem. 385, 652–659.

Hengstler, J. G., Ringel, M., Biefang, K., Hammel, S., Milbert, U., Gerl, M., Klebach, M., Diener, B., Platt, K. L., Böttger, T., Steinberg, P., Oesch, F., 2000. Cultures with cryopreserved hepatocytes: applicability for studies of enzyme induction. Chem. Biol. Interact. 125, 51–73.

Hoen, P. A. C., Commandeur, J. N. M., Vermeulen, N. P. E., Berkel, T. J. C. V., Bijsterbosch, M. K., 2000. Selective induction of cytochrome P450 3A1 by dexamethasone in cultured rat hepatocytes. Biochem. Pharmacol. 60, 1509–1518.

Iwahori, T., Matsuura, T., Maehashi, H., Sugo, K., Saito, M., Hosokawa, M., Chiba, K., Masaki, T., Aizaki, H., Ohkawa, K., Suzuki, T., 2003. CYP3A4 inducible model for in vitro analysis of human drug metabolism using a bioartificial liver. Hepatology 37, 665–673.

James, M. O., Lou, Z., Rowland–Faux, L., Celander, M. C., 2005. Properties and regional expression of a CYP3A–like protein in channel catfish intestine. Aquat. Toxicol. 72, 361–371.

Janošek, J., Hilscherová, K., Bláha, L., Holoubek, I., 2006. Environmental xenobiotics and nuclear receptors–interactions, effects and in vitro assessment. Toxicol. In Vitro 20, 18–37.

Kashiwada, S., Hinton, D. E., Kullman, S. W., 2005. Functional characterization of medaka CYP3A38 and CYP3A40: kinetics and catalysis by expression in a recombinant baculovirus system. Comp. Biochem. Physiol. C. 141, 338-348.

Kashiwada, S., Kameshiro, M., Tatsuta, H., Sugaya, Y., Kullman, S. W., Hinton, D. E., Goka, K., 2007. Estrogenic modulation of CYP3A38, CYP3A40, and CYP19 in mature male medaka (Oryzias latipes). Comp. Biochem. Physiol. C. 145, 370-378.

Kawanishi, S., Seki, Y., Sano, S., 1981. Polychlorobiphenyls that induce δ-aminolevulinic acid synthetase inhibit uroporphyrinogen decarboxylase in cultured chick embryo liver cells. FEBS Lett. 129, 93-96.

Kullman, S. W., Hamm, J. T., Hinton, D. E., 2000. Identification and characterization of a cDNA encoding cytochrome P450 3A from the fresh water teleost medaka (Oryzias latipes). Arch. Biochem. Biophys. 380, 29-38.

Kullman, S. W., Kashiwada, S., Hinton, D. E., 2004. Analysis of medaka cytochrome P450 3A homotropic and heterotropic cooperativity. Mar. Environ. Res. 58, 469-473.

Lange, U., Goksøyr, A., Siebers, D., Karbe, L., 1999. Cytochrome P450 1Adependent enzyme activities in the liver of dab (Limanda limanda): kinetics, seasonal changes and detection limits. Comp. Biochem. Physiol. B. 123, 361-371.

Lave, T., Coassolo, P., Reigner, B., 1999. Prediction of hepatic metabolic clearance based on interspecies allometric scaling techniques and in vitro-in vivo correlations. Clin. Pharmacokinet. 36, 211-231.

Lee, S. J., Buhler, D. R., 2003. Cloning, tissue distribution, and functional studies of a new cytochrome P450 3A subfamily member, CYP3A45, from rainbow trout (Oncorhynchus mykiss) intestinal ceca. Arch. Biochem. Biophys. 412, 77-89.

Lee, Y. M., Seo, J. S., Kim, I. C., Yoon, Y. D., Lee, J. S., 2006. Endocrine disrupting chemicals (bisphenol A, 4-nonylphenol, 4-tert-octylphenol) modulate expression of two distinct cytochrome P450 aromatase genes differently in gender types of the hermaphroditic fish Rivulus marmoratus. Biochem. Biophys. Res. Commun. 345, 894-903.

Lehmann, J. M., McKee, D. D., Watson, M. A., Willson, T. M., Moore, J. T., Kliewer, S. A., 1998. The human orphan nuclear receptor PXR is activated by compounds that regulate CYP3A4 gene expression and cause drug interactions. J. Clin. Invest. 102, 1016-1023.

Lin, J. H., Sugiyama, Y., Awazu, S., Hanano, M., 1980. Kinetic studies on the deethylation of ethoxybenzamide: a comparative study with isolated hepatocytes and liver microsomes of rat. Biochem. Pharmacol. 29, 2825-2830.

Lin, C. H., Lin, P. H., 2006. Induction of ROS formation, poly (ADP-ribose) polymerase-1 activation, and cell death by PCB126 and PCB153 in human T47D and MDA-MB-231 breast cancer cells. Chem. Biol. Interact. 162, 181-194.

Lin, M., Yang, X. L., Fang, W. H., Zhang, N., Li, D., 2006. Dose-response of inducers to effect on EROD in grass carp hepatocyte. Shui Chan Xue Bao 30, 311-315.

Lin, M., Yang, X. L., Wang, X. L., Fang, W. H., Yu, W. J., 2007. Enzymatic on deethlation metabolism of enrofloxacin in grass carp hepatocyte. Gao Ji Shu Tong Xun 17, 314-318.

Lowry, O. H., Rosebrough, N. J., Farr, A. L., Randall, R. J., 1951. Protein measurement with the Folin phenol reagent. Biol. Chem. 193, 265-275.

Lu, C., Li, A. P., 2001. Species comparison in P450 induction: effects of dexamethasone, omeprazole, and rifampin on P450 isoforms 1A and 3A in primary cultured hepatocytes from man, Sprague-Dawley rat, minipig, and beagle dog. Chem. Biol. Interact. 134, 271-281.

Machala, M., Nezveda, K., Pet? ivalsk?, M., Jarošová, A. B., Pia? ka, V., Svobodová, Z., 1997. Monooxygenase activities in carp as biochemical markers of pollution by polycyclic and polyhalogenated aromatic hydrocarbons: choice of substrates and effects of temperature, gender and capture stress. Aquat. Toxicol. 37, 113-123.

Machala, M., Drábek, P., Ne? a, J., Kolá? ová, J., Svobodová, Z., 1998. Biochemical markers for differentiation of exposures to nonplanar polychlorinated biphenyls, organochlorine pesticides, or 2, 3, 7, 8-tetrachlorodibenzo-p-dioxin in trout liver. Ecotoxicol. Environ. Saf. 41, 107-111.

Maglich, J. M., Caravella, J. A., Lambert, M. H., Willson, T. M., Moore, J. T., Ramamurthy, L., 2003. The first completed genome sequence from a teleost fish (Fugu rubripes) adds significant diversity to the nuclear receptor superfamily. Nucleic Acids Res. 31, 4051-4058.

Malmström, C. M., Koponen, K., Lindström-Seppä, P., Bylund, G., 2004. Induction and localization of hepatic CYP4501A in flounder and rainbow trout exposed to benzo [a] pyrene. Ecotoxicol. Environ. Saf. 58, 365-372.

Marionnet, D., Taysse, L., Chambras, C., Deschaux, P., 1997. 3-Methylcholanthrene-induced EROD activity and cytochrome P450 in immune organs of carp (Cyprinus carpio). Comp. Biochem. Physiol. C. 118, 165-170.

Martin, H., Sarsat, J. P., Waziers, I., Housset, C., Balladur, P., Beaune, P., Albaladejo, V., Lerche-Langrand, C., 2003. Induction of cytochrome P450 2B6 and 3A4 expression by phenobarbital and cyclophosphamide in cultured human liver slices. Pharm. Res. 20, 557-568.

McArthur, A. G., Hegelund, T., Cox, R. L., Stegeman, J. J., Liljenberg, M., Olsson, U., Sundberg, P., Celander, M. C., 2003. Phylogenetic analysis of the cytochrome P450 3 (CYP3) gene family. J. Mol. Evol. 57, 200-211.

Meucci, V., Arukwe, A., 2006. The xenoestrogen 4-nonylphenol modulates hepatic gene expression of pregnane X receptor, aryl hydrocarbon receptor, CYP3A and CYP1A1 in juvenile Atlantic salmon (Salmo salar). Comp. Biochem. Physiol. C. 142, 142-150.

Miranda, S. R., Meyer, S. A., 2007. Cytotoxicity of chloroacetanilide herbicide alachlor in HepG2 cells independent of CYP3A4 and CYP3A7. Food Chem. Toxicol. 45, 871-877.

Moore, L. B., Maglich, J. M., McKee, D. D., Wisely, B., Willson, T. M., Kliewer, S. A., Lambert, M. H., Moore, J. T., 2002. Pregnane X receptor (PXR), constitutive androstane receptor (CAR), and benzoate X receptor (BXR) define three pharmacologically distinct classes of nuclear receptors. Mol. Endocrinol. 16, 977-986.

Moutou, K. A., Burke, M. D., Houlihan, D. F., 1998. Hepatic P450 monooxygenase response in rainbow trout administered aquaculture antibiotics. Fish Physiol. Biochem. 18, 97-106.

Nallani, S. C., Goodwin, B., Buckley, A. R., Buckley, D. J., Desai, P. B., 2004. Differences in the induction of cytochrome P450 3A4 by taxane anticancer drugs, docetaxel and paclitaxel, assessed employing primary human hepatocytes. Cancer Chemother. Pharmacol. 54, 219-229.

Nash, T., 1953. The colorimetric estimation of formaldehyde by means of the Hantzech reaction. Biochemistry 55, 416-426.

Nishibe, Y., Hirata, M., 1995. Induction of cytochrome P-450 isoenzymes in cultured monkey hepatocytes. Int. J. Biochem. Cell Biol. 27, 279-285.

Novi, S., Pretti, C., Cognetti, A. M., Longo, V., Marchetti, S., Gervasi, P. G., 1998. Biotransformation enzymes and their induction by β-naphtoflavone in adult sea bass (Dicentrarchus labrax). Aquat. Toxicol. 41, 63-81.

Olsen, A. K., Hansen, K. T., Friis, C., 1997. Pig hepatocytes as an in vitro model to study the regulation of human CYP3A4: prediction of drug-drug interactions with 17α-thynylestradiol. Chem. Biol. Interact. 107, 93-108.

Paolini, M., Barillari, J., Trespidi, S., Valgimigli, L., Pedulli, G. F., Cantelli-Forti, G., 1999. Captan impairs CYP-catalyzed drug metabolism in the mouse. Chem. Biol. Interact. 123, 149-170.

Pesonen, M., Korkalainen, M., Laitinen, J. T., Andersson, T. B., Vakkuri, O., 2000a. 2, 3, 7, 8-Tetrachlorodibenzo-p-dioxin alters melatonin metabolism in fish hepatocytes. Chem. Biol. Interact. 126, 227-240.

Pesonen, M., Teivainen, P., Lundström, J., Jakobsson, E., Norrgren, L., 2000b. Biochemical responses of fish sac fry and a primary cell culture of fish hepatocytes exposed to polychlorinated naphthalenes. Arch. Environ. Contam. Toxicol. 38, 52-58.

Pretti, C., Salvetti, A., Longo, V., Giorgi, M., Gervasi, P. G., 2001. Effects of β-naphthoflavone on the cytochrome P450 system, and phase II enzymes in giltheadseabream (Sparusaurata). Comp. Biochem. Physiol. C. 130, 133-144. Reinach, B., Sousa, G., Dostert, P., Ings, R., Gugenheim, J., Rahmani, R., 1999. Comparative effects of rifabutin and rifampicin on cytochromes P450 and UDP-glucuronosyl-transferases expression in fresh and cryopreserved human hepatocytes. Chem. Biol. Interact. 121, 37-48.

Riley, R. J., Howbrook, D., 1998. In vitro analysis of the activity of the major human hepatic CYP enzyme (CYP3A4) using [N-methyl-14C]-erythromycin. J. Pharmacol. Toxicol. Methods 38, 189-193.

Roymans, D., Looveren, C. V., Leone, A., Parker, J. B., McMillian, M., Johnson, M. D., Koganti, A., Gilissen, R., Silber, P., Mannens, G., Meuldermans, W., 2004. Determination of cytochrome P450 1A2 and cytochrome P450 3A4 induction in cryopreserved human hepatocytes. Biochem. Pharmacol. 67, 427-437.

Schlenk, D., DeBusk, B., Perkins, E. J., 2000. 2-Methylisoborneol disposition in three strains of catfish: absence of biotransformation. Fish Physiol. Biochem. 23, 225-232.

Schuetz, E. G., Schuetz, J. D., Strom, S. C., Thompson, M. T., Fisher, R. A., Molowa, D. T., Li, D., Guzelian, P. S., 1993. Regulation of human liver cytochromes P-450 in family 3A in primary and continuous culture of human hepatocytes. Hepatology 18, 1254-1262.

Silva, J. M., Day, S. H., Nicoll-Griffith, D. A., 1999. Induction of cytochromeP450 in cryopreserved rat and human hepatocytes. Chem. Biol. Interact. 121, 49-63.

Skaanild, M. T., Friis, C., 2000. Expression changes of CYP2A and CYP3A in microsomes from pig liver and cultured hepatocytes. Pharmacol. Toxicol. 87, 174-178.

Smeets, J. M. W., Rankouhi, T. R., Nichels, K. M., Komen, H., Kaminski, N. E., Giesy, J. P., Berg, M. V. D., 1999. In vitro vitellogenin production by carp (Cyprinus carpio) hepatocytes as a screening method for determining (anti) estrogenic activity of xenobiotics. Toxicol. Appl. Pharmacol. 157, 68-76.

Stegeman, J. J., Binder, R. L., Orren, A., 1979. Hepatic and extrahepatic microsomal electron transport components and mixed function oxygenases in the marine fish Stenotomus versicolor. Biochem. Pharmacol. 28, 3431-3439.

Stegeman, J. J., Hahn, M. E., 1994. Current perspective on forms, functions, and regulation of cytochrome P450 in aquatic species. In: Malins, D. C., Ostrander, G. K. (Eds.), Biochemistry and molecular biology of monooxygenases. Aquatic Toxicology, vol. 1. CRC Press, Boca Raton, FL, USA, pp. 87-206.

Stegeman, J. J., Woodin, B. R., Singh, H., Oleksiak, M. F., Celander, M., 1997. Cytochromes P450 (CYP) in tropical fishes: catalytic activities, expression of multiple CYP proteins and high levels of microsomal P450 in liver of fishes from Bermuda. Comp. Biochem. Physiol. C. 116, 61-75.

Stresser, D. M., Blanchard, A. P., Turner, S. D., Erve, J. C. L., Dandeneau, A. A., Miller, V. P., Crespi, C. L., 2000. Substrate-dependent modulation of CYP3A4 catalytic activity: analysis of 27 test compounds with four fluorometric substrates. Drug Metab. Dispos. 28, 1440-1448.

Sy, S. K. B., Ciaccia, A., Li, W., Roberts, E. A., Okey, A., Kalow, W., Tang, B. K., 2002. Modeling of human hepatic CYP3A4 enzyme kinetics, protein, and mRNA indicates deviation from log-normal distribution in CYP3A4 gene expression. Eur. J. Clin. Pharmacol. 58, 357-365.

Tseng, H. P., Hseu, T. H., Buhler, D. R., Wang, W. D., Hu, C. H., 2005. Constitutive and xenobiotics-induced expression of a novel CYP3A gene from zebrafish larva. Toxicol. Appl. Pharmacol. 205, 247-258.

Vaccaro, E., Giorgi, M., Longo, V., Mengozzi, G., Gervasi, P. G., 2003. Inhibition of cytochrome P450 enzymes by enrofloxacin in the sea bass（Dicentrarchus labrax）. Aquat. Toxicol. 62, 27–33.

Vaccaro, E., Meucci, V., Intorre, L., Soldani, G., Bello, D.D., Longo, V., Gervasi, P.G., Pretti, C., 2005. Effects of 17β-estradiol, 4-nonylphenol and PCB 126 on the estrogenic activity and phase 1 and 2 biotransformation enzymes in male sea bass（Dicentrarchus labrax）. Aquat. Toxicol. 75, 293–305.

Vignati, L., Turlizzi, E., Monaci, S., Grossi, P., Kanter, R., Monshouwer, M., 2005. An in vitro approach to detect metabolite toxicity due to CYP3A4dependent bioactivation of xenobiotics. Toxicology 216, 154–167.

Wan, X.Q., Wu, W.Z., He, J.Z., 2002. Evaluating toxic effects of PCDD/F using grass carp primary hepatocyte culture. China Environ. Sci. 22, 114–117.

Wan, X.Q., Ma, T.W., Wu, W.Z., Wang, Z.J., 2004. EROD activities in a primary cell culture of grass carp（Ctenopharyngodon idellus）hepatocytes exposed to polychlorinated aromatic hydrocarbonas. Ecotoxicol. Environ. Saf. 58, 84–89.

Wang, A.G., Deng, S.K., Xia, T., Yuan, J., Yu, R.A., Chen, X.M., Yang, K.D., 2004. Effect of phenobarbital on metabolism and toxicity of diclofenac sodium in vitro. Chin. J. Public Health 20, 662–663.

Winzer, K., Noorden, C.J.F.V., Köhler, A., 2002. Sex-specific biotransformation and detoxification after xenobiotic exposure of primary cultured hepatocytes of European flounder（Platichthys flesus L.）. Aquat. Toxicol. 59, 17–33.

Xia, X.Y., Peng, R.X., Wang, J., 2004. Culture of rat hepatocytes in collagen sandwich system and its cytochrome P450 activity measurement. Chin. Pharmacol. Bull. 20, 350–353.

Yang, X.L., Lin, M., Yu, W.J., Wang, X.L., Fang, W.H., 2007. MTT assay applied to detect the toxicity of drug on fish cell lines. Shang Hai Shui Chan Da Xue Xue Bao 16, 157–161.

Yawetz, A., Zilberman, B., Woodin, B., Stegeman, J.J., 1998. CytochromesP-4501A, P-4503A and P-4502B in liver and heart of Mugil capito treated with CYP1A inducers. Environ. Toxicol. Phar. 6, 13–25.

Yu, W.J., Yang, X.L., Tang, J., Zheng, Z.L., Zhang, N., 2006. Study on primary culture of hepatocytes from Micropterus salmoides. Shang Hai Shui Chan Da Xue Xue Bao 15, 430–435.

该文发表于《上海水产大学学报》【2007，（5）：495–499】

草鱼肝细胞系中 CYP3A 活性检测方法

Analysis of the activity of cytochrome P450 3A in grass carp liver cell line

李聃[1]，杨先乐[1,2]，张书俊[1]，喻文娟[1]，陈昱[1]

（1. 上海水产大学农业部渔业动植物病原库，上海 200090；2. 上海高校水产养殖学 E-研究院，上海 200090）

摘要 红霉素-N-脱甲基酶（ERND）是细胞色素 P450 3A（CYP3A）依赖的酶类，可通过测定该酶催化代谢产物甲醛的量来反映其活性，从而评价 CYP3A 活性的高低。以草鱼肝细胞系（GCL）为模型，红霉素为底物，采用 Nash 比色法检测 ERND 产物甲醛含量，Lowry 比色法测定细胞蛋白含量，建立鱼类细胞中 CYP3A 活性的简便检测方法。研究表明，底物以 0.4 mmol/L 作用 60 min 后进行检测较为合适。细胞样品中 ERND 产物甲醛平均回收率为 78.80%±4.37%，平均日内精密度为 2.80%±1.40%，平均日间精密度为 3.91%±1.45%，通过计算探针酶 ERND 活性可为鱼类细胞 CYP3A 活性的评价提供简便可靠的方法，从而为鱼类药物代谢酶体外诱导模型的研究奠定了基础。

关键词 细胞色素 P450 3A；红霉素-N-脱甲基酶；鱼类；细胞；甲醛

该文发表于《中国水产科学》【2011，18（4）：720-727】

草鱼肾细胞中细胞色素 P450 3A 基因诱导表达及其酶活性分析

Induction of CYP3A gene and enzyme activity expression in kidney cell of grass carp（*Ctenopharyngodon idellus*）

符贵红[1]，杨先乐[1]，喻文娟[1]，胡鲲[1]，章海鑫[1]，黄宣运[2]

（1. 上海海洋大学水产与生命学院，农业部渔业动植物病原库，上海 201306；
2. 中国水产科学研究院东海水产研究所，农业部水产品质量监督检验测试中心，上海 200090）

摘要　采用基因克隆技术获取草鱼（*Ctenopharyngodon idellus*）细胞色素 P450 3A（CYP3A）cDNA 序列长 1849bp，包含完整开放阅读框（openreadingframe，ORF）1542bp，1 个终止密码和含 polyA 信号 3′UTR；ORF 编码 514 个氨基酸，含信号肽（29aa），2 个跨膜螺旋区（23aa，23aa），血红素结合域（21aa）和 6 个底物识别位点（SRS1-6）；与其他脊椎动物 CYP3A 氨基酸序列相似度达 60%～92%。用实时定量 PCR 和分光光度计法分别研究了草鱼肾细胞系（Ctenopharyngodon idellus kidney cell line CIK）中利福平（rifampicin，RIF）诱导 CYP3A 基因表达和红霉素-N-脱甲基酶（ERND）活性，采用红霉素脱甲基酶活性法测定 CYP3A 的活性。结果显示，诱导组 CIK 细胞中 CYP3AmRNA 表达量在 8 h 高，而 CYP3A 酶活在 10 h 高；CYP3AmRNA 和酶活性与对照组均具显著差异性（$P<0.05$）。CYP3A 转录水平先于相应酶活变化表达，且转录水平表达显著高于酶活性，这很可能反映了 RIF 对 CYP3A 的诱导主要作用于转录水平而不直接作用于酶活，本研究旨为药物对 CYP3A 转录调控和活性调节提供科学依据，并为渔药代谢酶的研究提供理论依据。

关键词　细胞色素 P4503A；肾细胞；草鱼；CYP3A 基因和酶活性；诱导表达；利福平

细胞色素 P4503A（cytochromeP4503A，CYP3A）是药物代谢中重要且唯一具多个底物识别位点（substrate recognition sites，SRS）的 CYP 亚家族同工酶，其代谢底物种类繁多，60%临床药物通过 CYP3A 参与代谢[1]。鱼类 CYP3A 活性诱导会导致鱼体内药物相互作用，从而可能产生渔药残留或影响渔药疗效。药物对 CYP3A 诱导调控主要体现在转录和酶活性这 2 个水平进行[2]，然而，鱼类对此研究非常少，特别是缺乏 CYP3A 受到外源物诱导或抑制后其在基因转录和酶活性的相关性研究。

药物对 CYP3A 诱导或抑制效应不仅影响 CYP3A 基因转录而且对 CYP3A 酶催化活性起到作用。目前，CYP3A 亚型基因结构及功能研究已在虹鳟（Oncorhynchus mykiss）[3]、鳉（Fundulus heteroclitus）[4]和（Paraluteresprionurus）[5]等硬骨鱼类中展开。本实验室已建立鱼类 CYP3A 活性体外诱导细胞模型[6]，结果显示，RIF 对 CIK 中 CYP3A 依赖酶 ERND 活性具较好诱导作用，能建立理想的诱导条件，是一种较好的诱导剂。因此，本研究报道了草鱼 CYP3A 基因 cDNA 全长，并分析了序列主要功能位点；通过 RIF 对 CYP3A 体外诱导，研究了时间动态过程中 CYP3A 基因转录和酶活水平变化及相关性。本研究对草鱼（Ctenopharyngodon idellus）CYP3A 基因特征和酶活性进行研究，为渔药代谢酶的研究提供理论依据。

1　材料与方法

1.1　试剂

M199 基础培养基和胎牛血清（FBS），利福平（rifampicin，RIF），红霉素（erythromycin，ERY），二甲基亚砜（dimethylsulfoxide，DMSO），RNA 提取试剂 TrizolReagent，cDNA 逆转录试剂盒，pMD19-T 载体，PCR 产物纯化试剂盒，SYBR-greenPCRmix 试剂盒。

1.2　CYP3A 基因

cDNA 序列克隆总 RNA 提取按 Trizol 试剂盒（Invitrogen 公司）说明书方法进行，检测后-80℃保存备用。根据 GenBank 中鲦（Pimephalespromelas，ACA35027.1）和斑马鱼（NP_ 001032515.1）等 CYP3AcDNA 序列，应用 PrimerPrimer5.0 软件设计引物（表1）。按 Prime Script TMR Treagent Kit 试剂盒（TaKaRa 公司）说明书逆转录合成 cDNA 第一链。PCR 扩增，反应条件：94℃3 min；30 个循环 94℃30 s，52℃30 s，72℃35 s；72℃延伸 10 min。反应产物经 1.2%琼脂糖电泳并用试剂盒回收（TaKaRa 公司）纯化，经克隆检测后，样品委托上海生工生物工程技术有限公司双向测序，序列测定仪为美国 ABI 公司 377 型全自动序列分析仪。

1.3 序列分析及蛋白质结构预测将所测

cDNA 序列用 NCBI 网站中 Blastn 软件[7]进行同源基因搜索并分析；通过 Jellyfish 软件推导 ORF 氨基酸序列；使用 ClustalX2.0 软件进行氨基酸序列多重同源比较；用 MEGA4.0 软件包中系统发育分析程序以邻接法（neighbor-joining, NJ 法）构建系统发育树，用 Bootstrap 法进行 1000 次评估。将推导氨基酸序列利用 SignalP3.0 在线预测可能的信号肽及切割位点位置。对氨基酸一级和二级结构在 PredictProtein：http：//www.predictprotein.org/网站上在线分析，预测可能的跨膜结构、血红素结合位点和底物识别位点等功能区域分布。

1.4 细胞及样品处理

草鱼肾细胞系（CIK，编号 BYK-C06-02）来自农业部渔业动植物病原库，以含 10%（v/v）FBS 的 M199 培养基（GibcoBRL（UK）产品）置 28℃培养。48 h 后贴满壁，加入新鲜 M199 培养基，0.25%胰酶消化，悬浮液分别在 6 孔板中培养，至密度为 5×10^5 个/mL。用无血清 40 μmol/LRIF（Sigma-Akdrich 产品）培养基处理，对照组用不含药物无血清培养基处理，分别作用 8 个时间点（1 h、2 h、4 h、6 h、8 h、10 h、12 h、24 h），倒去培养基，用 D-hanks 缓冲液清洗，经处理后细胞分别用于 CYP3A 酶活性和基因表达检测。

1.5 CYP3A 基因表达

采用 Trizol 试剂盒（Invitrogen 公司）提取样品的 RNA，并用 PrimeScriptTMRTase（TaKaRa 公司）合成 cDNA 第一链，Primer5.0 和 BeaconDesigner7.0 软件设计 CYP3A 和内参 β-actin 基因的荧光定量引物，采用 iQ5RealTimePCR 检测仪（Bio-Rad），选用 SYBRgreenPCRmix（TaKaRa 公司）进行 Real-timePCR 扩增，扩增体系和扩增标准程序按说明书使用，用 2-△△Ct 法分析 CYP3AmRNA 在 CIK 中各时间点处理的相对表达量。

表 1 草鱼 CYP3A 基因 PCR 扩增引物序列
Tab. 1 Primer used to amplify the grass carp CYP3A gene

扩增 application	引物序列（5′-3′）primer sequence（5′-3′）	退火温度（℃）annealing temperature（℃）
cds 引物 CYP3A-F（1） CYP3A-R（1）	ATGAGCTACGACCTGTTCT GTCACAGATGTCGGGAAA	52.2
CYP3A-F（2） CYP3A-R（2）	CGTTCAGCGTGGACATCG CTTTGGGAGCCAGTAGACC	53.0
3′ RACE CYP3A-F（3） CYP3A-R（3）	GCCTCAATGGTCTACTGG TCATTTCCATATAATTTCTTAACAG	50.0
Real time PCR：CYP3A F Real time PCR：CYP3A R	GTCCCAACCTTCGTCCTCCA ATCTCATCCCGATGCAGTTCC	60.0
Real time PCR：β-actin-F Real time PCR：β-actin-R	CTGTATGCCTCTGGTCGT GCTGTAGCCTCTCTCGGT	60.0
cds 引物 CYP3A-F（1） CYP3A-R（1）	ATGAGCTACGACCTGTTCT GTCACAGATGTCGGGAAA	52.2
CYP3A-F（2） CYP3A-R（2）	CGTTCAGCGTGGACATCG CTTTGGGAGCCAGTAGACC	53.0
3′ RACE CYP3A-F（3） CYP3A-R（3）	GCCTCAATGGTCTACTGG TCATTTCCATATAATTTCTTAACAG	50.0
Real time PCR：CYP3A F Real time PCR：CYP3A R	GTCCCAACCTTCGTCCTCCA ATCTCATCCCGATGCAGTTCC	60.0
Real time PCR：β-actin-F Real time PCR：β-actin-R	CTGTATGCCTCTGGTCGT GCTGTAGCCTCTCTCGGT	60.0

1.6　CYP3A 活性检测

RIF 处理的 CIK 细胞样本，用 2 mL 含 CYP3A 依赖性底物 400 μmol/LERY（上海生工生物工程有限公司）的 TMS 缓冲液替代培养液，置 28℃培养箱中孵育 1 h 后制成细胞悬液用于酶活性检测。蛋白浓度利用 BCA 法进行测定；细胞样品中加入 0.35 mL15% 的硫酸锌，冰浴 5 min，然后加入 0.35 mL 饱和氢氧化钡，静置 5min 后离心（5 000 r/min，60 min），吸取上清液 2 mL 至反应管，采用 Nash 比色法[8] 于 413 nm 处紫外分光光度测定吸光值，根据标准曲线计算甲醛含量。细胞中 CYP3A 依赖酶类（ERND）活性（pmol·min^{-1}·mg^{-1}）以每毫克蛋白单位时间内的甲醛生成量表示，CYP3A 活性直接以依赖性酶 ERND 活性进行评价，每个样品设置 3 个平行。CYP3AmRNA 相对表达量和 CYP3A 酶活性结果采用 SPSS15.0 软件进行显著性分析，由 ANOVA（单因素方差分析）的 Duncan 氏进行多重比较，P<0.05 为差异显著；利用 SigmaPlot10.0 软件作图。

2　结果与分析

2.1　CYP3A 基因 cDNA 序列结构

已获取 CYP3A 基因 cDNA 序列 1849bp，（GenBank 登录号：HQ191278），经 BLAST 分析确定为 CYP3A 基因完整序列。包含完整 ORF 区 1542bp，编码 513 个氨基酸；3′UTR 区含一个保守的 mRNA 加尾信号序列 AATAAA 及含 23 个碱基的 PolyA 加尾信号。

2.2　氨基酸序列和二级结构预测

由 Jellyfish 软件推导的草鱼 CYP3A 氨基酸序列由 514aa 编码，经预测 CYP3A 蛋白量为 57.56kD，等电点为 6.71。结合 CYP3A 基因二级结构在线分析显示（图 1）：CYP3A 是一种膜蛋白，其蛋白质序列含有 1 个长 29 个 aa 的信号肽序列（1Met-Thr29），切割位点位于 29~30 之间（T-H）。两个跨膜螺旋区均由 23 个 aa 组成（5Leu-27Ser；214Asp-Met236）。血红素结合域由 21aa 组成，其中草鱼、鲦（Pimephalespromelas）和斑马鱼（Danio rerio）完全相同，而鲫（Carassius gibelio）与草鱼在 Ser 和 Ile 两位点不同；虹鳟和大西洋鲑（Salmo salar）两者序列完全相同，但与草鱼间保守性降至 76.1%（表 2）；哺乳类间仅存 1~2 个位点差异且与鱼类保守性达 76.1%。6 个底物识别位点（SRS1~6），其中，草鱼 CYP3A 基因血红素结合域和底物识别位点 SRS-1、-2、-4、-5 在硬骨鱼类中保守性高，而 SRS-3、-6 保守性较低（表 2）。而 CYP3A 蛋白中 SRS-3，-6 序列差异为明显，可进一步研究并证实其功能。2.3 CYP3A 氨基酸多重比对及系统发育树应用 Clustal2.0 软件对草鱼 CYP3A 基因（Ctenopharyngodonidellus HQ191278）与鳉（Fundulu sheteroclitus Q9PVE8.2）、鲦（Pimephalespromelas ACA35027.1）、鲫（Carassius gibelio ADF87312.1）、斑马鱼（Daniorerio NP_ 001032515.1）、大西洋鲑（Salmosalar ACI33861.1）、虹鳟（Oncorhynchus mykiss O42563.1）、人（Homosapiens NP_ 059488.2）、野猪（SusscrofaNP_ 999587）及小家鼠（MusmusculusNP_ 031844.1）CYP3A 基因进行多重比对（图 1），鱼类中各功能区域保守性较高，与哺乳类中功能区域保守性降低，物种间氨基酸序列差异可以从基因水平来解释功能的差异性。系统发育树结果显示（图 2），所有陆生动物共聚为一支，鱼类共聚为一支，而草鱼、鲦和鲫鱼则进一步聚在一起。

2.3　CIK 中 RIF 处理 CYP3A 基因表达

经 40μmol/LRIF 诱导的 CIK 细胞样本，提取总 RNA，经 Real-timeRT-PCR 扩增，选用管家基因（β-actin）对 CYP3A 基因 mRNA 进行相对表达量分析。实验结果表明（图 3-A），诱导组比对照组 CYP3AmRNA 表达量增高，经 SPSS15.0 统计软件中 Paired-SamplesT-Test 检验样本差异分析。诱导组 CYP3AmRNA 量在 8 h 升至高，结果显示其极显著高于其他时间点的表达量（P<0.01）。

2.4　CIK 中 RIF 处理 CYP3A 催化活性

（ERND）经 40 μmol/LRIF 诱导的 CIK 细胞样本均以 400 μmol/L 红霉素作用 60min 后检测 ERND 活性用于评价 CYP3A 活性，实验结果表明（图 3-B），诱导组 CYP3A 比对照组酶活性提高，经统计学分析，诱导组具有显著性诱导（P<0.05），CYP3A 酶活性在 10 h 升至高。

2.5　CIK 中 CYP3A 酶活性与基因表达量相关性分析

经 40 μmol/LRIF 诱导处理 CIK，分析 CYP3A 酶活性（ERND）和 mRNA 表达量显示 CYP3A 转录水平在 2 h 诱导量开始提高，而酶活性在 4 h 开始升高；当转录水平在 8 h 达高，而酶活性在 10 h 达高。CYP3A 转录水平先于相应酶活 2 h 表达，且转录水平的诱导表达量明显高于酶活性，这很可能反映了 RIF 诱导 CYP3A 表达是直接作用于基因的转录水平，而不直接作用于酶活性，且在诱导过程中具有一定时间依赖性。

3 讨论

3.1 CYP3A 基因结构及活性位点分析

SRS 位点被认为参与底物识别和酶催化活性[9-10]，而 SRS 位点发生氨基酸替代可使 CYP3A 底物特异性发生变化[11]，分析了硬骨鱼类 CYP3A 成员的催化活性不同于其他脊椎动物可能是受到 SRS 位点保守性影响。鱼类和哺乳类底物识别的功能差异首先体现在，硬骨鱼类 CYP3A 基因底物识别位点（SRS-1，-3，-6）与其他物种保守性较低，由此推断鱼类和哺乳类间功能差异的分子基础原因可能是序列保守性低；而 SRS-4 保守性高于 68%，可由 SRS-4 是反应机制中重要保守区且是与底物结合相连的部分来解释[12]。其次，SRS-5 为主要活性位点，在硬骨鱼类和哺乳类间保守性达 77.7%~100%。鱼类 CYP3A 可能由于这些高度保守位点使其保留了哺乳类 CYP3A 蛋白原始结构，终使鱼类和哺乳类 CYP3A 功能进化趋势大致相同。后 SRS-3 和 SRS-6 是 CYP3A 蛋白中在各物种间差异显著的区域，是影响各物种 CYP3A 底物特异性和酶活性差异的主要结构位点，因此其功能值得更深层次研究和探索。CYP450 为整合膜蛋白，分析了 CYP3A 两个主要跨膜螺旋结构域，其跨膜螺旋与 SRS 区域共同参与底物结合，与底物特异性密切相关。血红素结合位点经序列分析在各物种间保守性高，可解释为其功能差异不明显主要参与氧化还原作用（表 2，图 1）。

表 2 鱼类和脊椎动物 CYP3A 成员在底物识别位点和血红素结合域序列比对

Tab. 2 Sequence comparison between CYP3A in fish and other vertebrate CYP3A members at substrate recognition sites （SRS1-6）and heme binding domain （HBD）

区域 domain	草鱼 Ctenopharyngodon idellus	人 Homo sapiens	野猪 Sus scrofa	小鼠 Mus muscullus
	3A	3A4	3A29	3A11
SRS-1	100（102-123）[a]	52.1	43.4	47.8
SRS-2	100（201-211）[a]	50.0	50.0	37.5
SRS-3	100（238-243）[a]	28.5	57.1	28.5
SRS-4	100（300-318）[a]	73.6	68.4	78.9
SRS-5	100（347-382）[a]	77.7	77.7	77.7
SRS-6	100（482-491）[a]	60.0	40.0	50.0
HBD	100（441-461）[a]	80.9	80.9	76.1

注：每组数字表示与草鱼 CYP3A 序列比较后百分比；[a] 表示 CYP3A 氨基酸位点．

Note：Percentage of identigy comparing with sequence of Ctenopharyngodon idellus CYP3A；[a] Represent the residues in CYP3A.

3.2 CYP3A 基因诱导与酶活性分析

渔药代谢酶 CYP3A 代谢谱非常广[13]，大量已用或新用渔药由 CYP3A 参与代谢[14-15]，且 CYP3A 转录表达和催化活性易外源物诱导和抑制，导致 CYP3A 活性变化直接影响药物治疗效果和毒性反应（图 2）。CYP450 活性及表达受药物种类、研究物种和性别差异等因素影响，而外源物对 CYP450 影响体现在编码 CYP450 蛋白的 mRNA[2]、CYP450 蛋白浓度[16]和酶催化活性[17]。鱼类研究着重于诱导剂对 CYP3A 酶活性影响[18]和 CYP3A 活性诱导后导致的药物间相互作用，而药物诱导 CYP3A 不同水平的机理及相关性研究甚少。哺乳类中如仔鸡用探针法检测喹诺酮药物对 CYP450 抑制效应，表现为 CYP1A 和 CYP3A 蛋白表达下调，而 CYP1A4 和 CYP3A37mRNA 表达不受影响，显示喹诺酮药物在鸡体内不是通过 CYP450 酶转录水平，而是转录后翻译来影响 CYP450 表达[19]。人肝细胞通过 IL-6 处理检测是否能抑制 CYP450 酶表达，结果显示 IL-6 抑制 CYP3A4mRNA 和蛋白表达量来降低药物代谢，而酶活性影响小，分析可能是 CYP3AmRNA 转录受到抑制或 CYP3A4mRNA 降解性增强而导致 CYP3A4 蛋白表达受抑制[20]。而本研究结果显示 CIK 中 CYP3A 经 RIF 诱导，其转录和酶活性均被诱导，但 RIF 对 CYP3A 诱导主要体现在转录水平，即 CYP3A 基因水平的上调趋势明显高于酶活性变化，主要是通过诱导 CYP3A 转录水平变化达到酶活性增强的效果。

信号肽　signal peptide　　跨膜螺旋 1　transmembrane helices 1

```
C.idell3A        -MSYGLFFSAETWALLVLFGALLVIYGSWTHSIFKKLGIAGPKPIPFFGTMLRYREGFHNFDLECFKKYGRVWGIYDARQPVLCIMDQSIIKTILIKECY
P.promelas3A126  -...D........S.LV....L..H.P.RQ..................................I..............T........
C.gibelio3A37    -.........I..VT..F....P.GV.......P......L.......E..K...M....N..............L......V...
D.rerio3A65      -----.M......T..AFLA....P.RY.....P...A.........KD...V..MD...E....I.....G.......T....V...
O.mykiss3A27     M...FLPY.....T..A.LIT..I.V..Y.PYGV.T.M..P....L.Y......E..KK...T...T....Q.....I........K..M...V...
S.salar3A        M...FLPY.....T..A.LIT..I.V..Y.PYGV.T.M.VP....L.Y.....ME..KK...T...T....Q.....I........T..K.M...
F.heteroclitus3A30 -.G..F-YLT.....T..A.VT..LV.AY.PYGT..R...S....V.......H..R..FT..E...K......K......VT.PE...AV..V...L
H.sapiens3A4     -.ALIPDLAM....L..AVSLV..YL..THS..GL.....P..T.L..L.NI.S.HK..CM...M..H.....K....F...GQ...A.T.PDM...V.V...
S.scrofa3A29     -.DLIPG..T...V..ATSLV...YL..TYS..GL.....P..R.L.Y..NI.G..K.VDH...KK...QQ...KM..V..G....L.AVT.PNM...SV.V...
M.musculus3A11   -.DLVSAL.L....V..AISLV..YR...TRK..EL....Q..P...L..V.N.YK.LWK....M...Y.....KT...LF.GQT..L.AVT.PET...NV.V...F
```

SRS-1

```
C.idell3A        SLFTNRRNFRLNGPLYDAVSIVEDDDWRRIRSVLSPSFTSGRLKEMFGIMKAHSKILVENLGKSATRGENVDIKEFFGAYSMDVVTSTAFSVDIDSLNNP
P.promelas3A126  .................................................Q...Q.............V........
C.gibelio3A37    ...............................................T..HT..D....T....A.E..
D.rerio3A65      ...........A.......A............................K..H...DSM...T.K...SA..
O.mykiss3A27     NI..........H...E.F...L.VA.....T.............Q..ST.LSGMK..Q.DKDQTIEV....P..
S.salar3A        NI..........H...E.F....VA.....T.............Q..AN.LSGMK..Q.DKDQAIEL....P..
F.heteroclitus3A30 .F...............A...Q.K.........E....N..AN.IRSMK..K.DKD..PL..L....S...
H.sapiens3A4     ..V.........P.GPV.FMKS..I..A...EE..K.L...L....T...K....VP..IAQYGDV...R...RRE..ET..KP..TL..DV........I...S..G..N..
S.scrofa3A29     ..V.........S.GPL..AMRN..L..LA...EE..K...TL...L.........P..ISHYGDL...S..R..E..EK..KP..TM..DI........I.........G..N..
M.musculus3A11   ..V.........D.GPV..IMSK..I...SK...E.K.Y..AL....L.......PVIEQYGD....KY..RQK..KK..KP..TM..DVL........I...S..G..NV..
```

SRS-2　　跨膜螺旋 2　　　SRS-3

transmembrane helices 2

```
C.idell3A        KDPFVTNIKKMLKSDFLNPLFLISALFPFVIPVLEKMDFALFPTSVTDFFYAALQKIKSER----VANDHKKKRVDFLQLMVDSQTAGKTQHDEEHTEKG
P.promelas3A126  ..............F...S..V.......T....I...ELGF..I.....................-----..............................QG..
C.gibelio3A37    ...............F....V....V...IT......G..F.............K.......---.SSEQ...........GS---..
D.rerio3A65      ..............F.L....LI.F...MA.....................D.DTKT..K..NT............GV..HKS..G..
O.mykiss3A27     S.....S.V....F.LF.....LV.....TG.I....K.SF...A..........S.A....G.......-----DTGNSTN......I....KGSD..KTG...Q...
S.salar3A        S.....S.V....F.LF.....LV.....TG.I....K.SF....A........S.A....G.......-----DTGNSTS......I....KGND..KTG...Q...
F.heteroclitus3A30 S...........F......AV.F...LG.I....FELSF..K.........S.E....N.......EASQQ..S.......I....KNS---GAQQD..S
H.sapiens3A4     Q.....E.T..L.RF...D.F..SITV....L..I...VLNICV..RE...N.LRKSVKRM..ES.....-----LE..TQ..H....I....N-S.-----..TESH..A
S.scrofa3A29     Q.....E.S..L..FS.FD.FL.SLIF....LT..IF.VLNIT....K.SVN...TKSVKRM..ES.....-----LT..QQ..R....L...IN..N-S.-----..MDPH..S
M.musculus3A11   E.....EKA..L.RF...FD..LFSVV.....LT...Y.MLNICM..KDSIE...KKFVDRM..ES.....-----LDSKQ..H...MN..HNNS..-----DKVSH..A
```

SRS-4　　　　　　　　　　　　　　　　　　　SRS-5

```
C.idell3A        LSDHEILSQSMIFIFAGYETSSSTLMFFFYNLATNPETMKKLQEEIDETFPDKAPVDYEAVMNMDYLDAALNESLRLYPVAARLERVCKKTVDINGLIIP
P.promelas3A126  .............A..................S...........A......V...N..............................F.............E..V...
C.gibelio3A37    .............G..................S.......H..A.E....G...QS.SRED.....GI...E................F.IV...........LV..
D.rerio3A65      ....................S.L.............NQ.......TL.S...........S................F..............E..L...
O.mykiss3A27     .T.........A........MS.LA......HHV..T.......TV...N...IQ...L.Q......CV.......I.P.......A...E...IV...
S.salar3A        .T.........A........MS.LA......HV..A.......TV...N...IQ...L.Q......CV.......I.P.......A...E...IV...
F.heteroclitus3A30 .T..............S..T.LA...........A....N....H.QPL..E..E...CVI.......F.L........A..AA..E...VV...
H.sapiens3A4     ..L.LVA...I.......T...V.S..IM.E....H.DVQQ.......AVL..N......PT..DT..LQ..E....MVV...T...F..I.M.......D..E...MF...
S.scrofa3A29     ..NE.LVA..GI.........T...A..SLLA..E...H.DVQQ.......EA....N......PT..D..LAQ..E....MVV...T...I.......A...D..E..H..VFV...
M.musculus3A11   ..M...TA...I.........S..TLHS...H.DIQ........D....AL..N......PT..DT..E..E....MV...T...I.N.........D..EL...VY...
```

图 1　草鱼预测氨基酸序列与其他物种 CYP3A 基因用 CLUSTAL-W 法同源比较

"·"表示相同氨基酸，"～～～"表示预测的信号肽序列，斜体字母表示预测跨膜螺旋区域，阴影部分表示底物识别位点，"——"表示血红素结合域.

Fig. 1　Sequence alignment of CYP3A with other vertebrate CYP3A members by the CLUSTAL-Wmethod The consensus amino acids are indicated by

"·". The predicted signal peptide indicated as "～～～". The transmembrane helices region were indicated in italics The putative substrate recognition sites SRS1-6 were in shadows. The heme binding domain （HBD） were showed with "——".

	血红蛋白结合域 hemoglobin combining domain	SRS-6
*C.idell*3A	KDMVVMVPTFVLHRDPDYWSDPESFKPERFTKGNKELIDPYMYMP<u>FGLGPRNCIGMRFAQVTMKLA</u>IVEILQRFDVSVCEETQVPLELGLNGLLAPKDPI	
*P.promelas*3A126	.....I...A.............................A..............................A.......	K..KS...
*C.gibelio*3A37	..V...I...A......E.D..R.Q..S.D.R.S.....F.................SI......Q....	DTS.....S..
*D.rerio*3A65	..L......YA.....E...............S..............................D......	F....S...
*O.mykiss*3A27	..CI..L...WT.....EI.....E......S.E...S.....T......A........LIMI...M....S.TF...D..EI..	MDNQ...M..R..
*S.salar*3A	..CI..L...WT.....EI.....E......S.E...S.....T......A........LIMI...M....S.TF...D..EI..	MDNQ...M..R..
*F.heroclitus*3A30	.......I..WP.....EI.PE..A....S.K..DN....I.....S........L.LI...V......QYSF...K..E..F.	MDIQ......R..
*H.sapiens*3A4	.GV....I.SYA......K..TE..K.L....S.K..DN....I.T.....S........LMN......LIRV...N.SFKP..K....I...K.S.G...	Q.EK.V
*S.scrofa*3A29	.GT...V..V...........L.PE...E..R....S.KH.DT.N..T.L....T........LMN.......L.RV...N.SFKP..K....I...K.TTQ...	TQ.EK.V
*M.musculus*3A11	.GST...I.SYA...H...QH...E...E.Q.....S.E..GS....V.L......N.....LMN.......LTK.M..N.SFQP..K....I...K.SRQ...	Q.EK..

*C.idell*3A	KLRLKPRT---ASDICNNNKS---
*P.promelas*3A126	N.KFEH..---.A.V...ES----
*C.gibelio*3A37	..QF...KPSLSE......NTS--
*D.rerio*3A65	..K.Q..KLSQSP.V....QKS--
*O.mykiss*3A27	....EA.RNTPSNTTATTL..PTT
*S.salar*3A	....E..SNTPSNTTVTSL-----

续图 1 草鱼预测氨基酸序列与其他物种 CYP3A 基因用 CLUSTAL-W 法同源比较

"·"表示相同氨基酸,"~~~"表示预测的信号肽序列,斜体字母表示预测跨膜螺旋区域,阴影部分表示底物识别位点,"——"表示血红素结合域.

Fig. 1 Sequence alignment of CYP3A with other vertebrate CYP3A members by the CLUSTAL-Wmethod The consensus a-mino acids are indicated by

"·". The predicted signal peptide indicated as "~~~". The transmembrane helices region were indicated in italics The putative substrate recognition sites SRS1-6 were in shadows. The heme binding domain (HBD) were showed with "——".

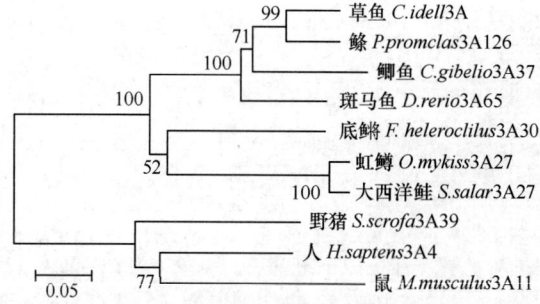

图2 草鱼和其他鱼类、物种的 CYP3A 系统发育树

Fig. 2 Construction of Phylogenetic tree based on CYP3A amino acids in grass carp and other species

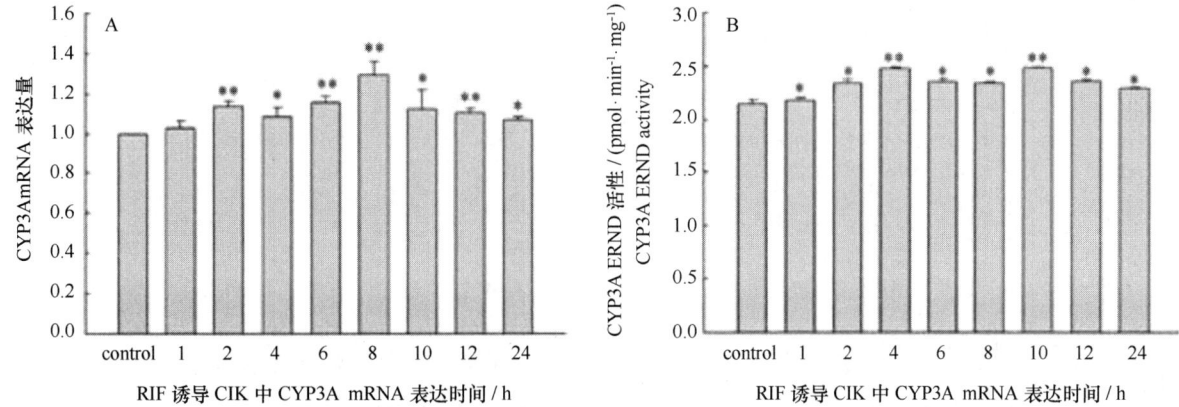

图 3　RIF 诱导 CIK 中 CYP3A mRNA 表达量和 CYP3A ERND 活性

3-A：RIF 诱导 CIK 中 8 个时间点 CYP3A mRNA 表达量；3-B：RIF 诱导 CIK 中 8 个时间点 CYP3A ERND 活性. 利用 SigmaPlot 10.0 软件作图分析，实验组与对照组相比具显著差异性（＊P<0.05；＊＊P<0.01）.

Fig. 3　CYP3A mRNA expression and CYP3A ERND activity induced by RIF in CIK

A：CYP3A mRNA expression induced by RIF with 8 time-point. B：The activity of CYP3A ERND induced by RIF in CIK with. Analysis using SigmaPlot 10.0 software, the test compared with the control indicated that significant differences（single asterisk：P<0.05；double asterisk：P<0.01）.

参考文献

［1］ Li A. P., Kaminski D. L., Rasmussen A. . Substrates of human hepatic cytochrome P450 3A4［J］. Toxicology, 1995, 104：1-8.

［2］ Courtenay S., Grunwald C., Kreamer G., et al. Induction and clearance of cytochrome P4501A mRNA in Atlantic tomcod caged in bleached kraft mill effluent in the Miramichi River［J］. Aquat Toxicol, 1993, 27：225-244.

［3］ Lee S. J., Buhler D. R. . Cloning, tissue distribution, and functional studies of a new cytochrome P450 3A subfamily member, CYP3A45, from rainbow trout（Oncorhynchus mykiss）intestinal ceca［J］. Arch Biochem Biophys, 2003, 412：77-89.

［4］ McArthur A. G., Hegelund T., Cox R. L., et al. Phylogenetic analysis of the cytochrome P4503（CYP3）gene family［J］. J Mol Evol, 2003, 57：200-211.

［5］ Nelson D. R. . Comparison of P450s from human and fugu：420 million years of vertebrate P450 evolution［J］. Arch Biochem Biophys, 2003, 409：18-24.

［6］ Li D., Yang X. L., Zhang S. J. . Effects of mammalian CYP3A inducers on CYP3A - related enzyme activities in grass carp（Ctenopharyngodon idellus）：Possible implications for the establishment of a fish CYP3A induction model［J］. Comp Biochem Physiol, 2008, 147 C：17-29.

［7］ Altschul S. F., Thomas L. M., Alejandro A., et al. Gapped BLAST and PSI-LAST：a new generation of protein database search program［J］. Nucl Acids Res, 1997, 25：3389-3402.

［8］ Nash. The colorimetric estimation of formaldehyde by means of the Hantzech reaction［J］. Biochemistry, 1953, 55：416-426.

［9］ Roussel F., Khan K. K., Halpert J. R., et al. The importance of SRS-1 residues in catalytic specificity of human cytochrome P450 3A4［J］. Arch Biochem Biophys, 2000, 374：269-278.

［10］ Xue L., Zgoda V. G., Arison B., et al. Structure-function relationships of rat liver CYP3A9 to its human liver orthologs：site-directed active site mutagenesis to a progesterone dihydroxylase［J］. Arch Biochem Biophys, 2003, 409：113-126.

［11］ Kullman S. W., Hamm J. T., Hinton D. E., et al. Identification and characterization of a cDNA encoding cytochrome P450 3A from the fresh water teleost medaka（Oryzias latipes）［J］. Arch Biochem Biophys, 2000, 380：29-38.

［12］ 冷欣夫, 邱星辉. 细胞色素 P450 酶系的结构、功能与应用前景［M］. 北京：科学出版社, 2001：11-13.

［13］ Tseng H. P., Hseu T. H., Buhler D. R., et al. Constitutive and xenobiotics-induced expression of a novel CYP3A gene from zebrafish larva. ［J］. Toxicol Appl Pharmacol, 2005, 205：247-258.

［14］ Kashiwada S., Kameshiro M., Tatsuta H., et al. Estrogenic modulation of CYP3A38, CYP3A40, and CYP19 in mature male medaka（Oryzias latipes）［J］. Comp Biochem Physiol, 2007, 145C：370-378.

［15］ Vaccaro E., Giorgi M., Longo V., et al. Inhibition of cytochrome P450 enzymes by enrofloxacin in the sea bass（Dicentrarchus labrax）［J］. Aquat Toxicol, 2003, 62：27-33.

［16］ Giorgi M., Marini S., Longo V., et al. Cytochrome P450 dependent monooxygenase activities and their inducibility by classic P450 inducers in the liver, kidney, and nasal mucosa of male adult ring-necked pheasants［J］. Toxicol Appl Pharmacol, 2000, 167：237-245.

[17] Martel P. H., KovacsT. G., O'Connor B. I., et al. A survey of pulp and paper efffluents for their potential to induce mixed function oxidase enzyme activity in fish [J]. Water Res, 1994, 28: 1835-1844.
[18] 李聃, 杨先乐, 张书俊, 等. 草鱼肝细胞系中 CYP3A 活性检测方法 [J]. 上海水产大学学报, 2007, 16 (5): 495-499.
[19] Zhang L. L., Zhang J. R., Guo K., et al. Effects of fluoroquinolones on CYP4501A and 3A in male broilers [J]. Res Vet Sci, 2010.
[20] Yang J., Hao D. F.. Pregnane X receptor is required for interleukin-6-mediated down-regulation of cytochrome P450 3A4 in human hepatocytes [J]. Toxicol Lett, 2010, 197: 219-226.

该文发表于《Environmental Toxicology and Pharmacology》【2011, 31: 307-313】

Effects of cytochrome P4501A substrate (difloxacin) on enzyme gene expression and pharmacokinetics in crucian cap (*hybridized Prussian carp*)

Gui Hong Fu, Xian Le Yang, Hai Xin Zhang, Wen Juan Yu, Kun Hu

(National Center for Aquatic Pathogen Collection, College of Fisheries and Life Science, Shanghai Ocean University, 999 Hucheng Huan Road, Shanghai 201306, China)

Abstract: Cytochrome P450s (CYPs) play a prominent role in drug metabolism and biotransformation which are distributed in liver of aquatic animals. However, limited information is available about CYP genes involved in drug metabolism in fish. In the present study, we explore CYP1A characterization for DIF metabolism. Firstly, we cloned and characterized the full length cDNA sequence of a CYP1A gene from crucian cap (hybridized Prussian carp), the predicted protein sequence for CYP1A comprise 496 amino acids. The heme-binding region of the CYP1A, encompassing the amino acid sequence GLGKRRCIG, which is identical to the same region of other homologues. Secondly, we studied the difloxacin (DIF) kinetics and the effects of DIF on their corresponding CYP1A mRNA levels in liver of crucian cap. CYP1A1 mRNA expression was analyzed by real-time PCR, and DIF concentration was determined by reversed-phase high-performance liquid chromatography (RP-HPLC). Results showed that the concentration of DIF in liver reached its peak (67.70 mgkg^{-1}) at 0.5 h, while the CYP1A1 gene expression was at the lowest point. CYP1A mRNA was down-regulated by 6.5 mgml^{-1} DIF in the liver of crucian cap. Thus, our work confirmed that DIF is both the substrate and inhibitor of CYP1A. The information provided a model for the potential utility of gene expression analysis and drug metabolization in fish.
key words: Cytochromes P450 1A; Difloxacin; Gene expression; Pharmacokinetics

双氟沙星对异育银鲫细胞色素 P4501A 的酶基因和药动学的影响

符贵红, 杨先乐, 章海鑫, 喻文娟, 胡鲲

(国家水生动物病原库, 上海海洋大学水产与生命学院, 沪城环路 999 号 上海 中国 201306)

摘要 分布于水产动物的肝脏 CPY450s 在药物代谢和生物转化具有重要作用。然而, 关于鱼类中 CYP 参加药物代谢基因信息了解很少。此研究中, CYP1A 在双氟沙星代谢的作用得到了证明。首先, 克隆和翻译银鲫的一条 CYP1A 基因中的全长 cDNA 序列, CYP1A 蛋白序列包含 496 个氨基酸。CYP1A 的血红素结合区域包含氨基酸 GLGKRRCIG 序列, 这与其他同系物的同一区域相似。第二, 本研究证明了双氟沙星的动力学和双氟沙星在银鲫肝细胞相应 CYP1A mRNA 的作用。通过实时 PCR 分析得到 CYP1A1 mRNA 的表达, 并且通过反相高效相色谱法确定双氟沙星的浓度。结果显示, 肝细胞中双氟沙星的浓度在 0.5 h 时达到峰值 (67.7 mg/kg), 而 CYP1A1 基因表达量最低。银鲫肝脏中双氟沙星降至 6.5 mg/kg。因此研究确认双氟沙星是 CYP1A 的基质和抑制剂。这些信息为鱼类基因表达分析的潜在利用和药物代谢提供了理论。
关键词 CYP1A; 双氟沙星; 基因表达; 药代动力学

1　Introduction

Cytochrome P450s (CYPs) are a large superfamily of heme-proteins that play key roles in biotransformation of xenobiotics and in synthesis and degradation of physiologically important endogenous substrates (Gigolos et al., 1996; Guengerich, 1999). Research on the CYP450 detoxification system has become an important model to reveal how an aquatic organism metabolizes to organic contaminants. There are also differences in the activity of enzymes that could influence individual species susceptibility to toxicity, and possibly influence clinical application. CYP450s are involved in the drugs, food additinves, and environmental pollutants. These compounds are known to stimulate expression of various members of the CYP450 family of genes, particularly those of the CYP1 family (Hahn and Stegeman, 1994; Stegeman et al., 2001). CYP1A1 is primarily involved in the biotransformation of drugs, and is often referred to as drug-metabolizing enzymes. Although studies have been published on the role of CYP1A1 in drug metabolism of mammals (Nemoto and Sakurai, 1992; Hanlona et al., 2008; Shinkyoa et al., 2003), very little data are available on the effect of drugs on CYP enzymes in fish (Li et al., 2008).

Fluorophenyl difloxacin is a new aryl-fluoroquinolone which is used only for animals. Fluoroquinones exhibit bactericidal action by targeting the bacterial DNA topoisomerases II (gyrase) and IV which is responsible for supercoiling of DNA around RNA core to provide a suitable spatial arrangement of DNA within the bacterial cell (Wolfson and Hooper, 1989; Drlica and Zhao, 1997). Difloxacin (DIF) is a broadspectrum antibacterial chemical agent of the fluoroquinone family for animals. DIF have been recently introduced as effective antibacterial agents in fish farming. Because of significant and progressive increase in the use of quinolones in animal production industries over the last decade, residue problems of these drugs have become serious (Gigolos et al., 2000). Pharmacokinetics of DIF as effective antibacterial agents have been investigated in fish farming, i. e. in grass carp (Yu and Yang, 2010), crucian cap (Carassius auraturs) (Ding et al., 2006).

CYP genes play an important role in the metabolism of xenobiotics and chemicals metabolism (Hu et al., 2004; Goldstone and Stegeman, 2006; Isin and Guengerich, 2007). The expression of CYP1A enzyme is a major factor influencing the role of CYPs in substrate oxidation (Willsa et al., 2009; Hawkins et al., 2002). CYP1A1 mRNA and protein expression levels were induced by BaP then decreased of estrogenic compound nonylphenol metabolism (Levine and Oris, 1999). It provided us with information to understand the relationship between CYP1A1 metabolic activity and drug metabolism. DIF is commonly used in fish, so it is crucial to understand the effect of the extensive utility of DIF on CYP enzymes in crucian cap. In this study, we first cloned a full-length cDNA sequence of CYP1A1, and investigated its tissue-specific distribution in crucian cap. After that, the DIF kinetics and CYP1A expression were analyzed in crucian cap liver. Our results may serve as reference for research on drug metabolism.

2　Materials and methods

2.1　Chemicals

DIF-HCl (>99.8%) was obtained from Sigma-Aldrich (China). Acetonitrile and methanol were HPLC grade (Darmstadt, Germany). All other chemicals and solvents were analytical grade obtained from local commercial sources.

2.2　Fish culture, chemical treatments and sampling

One hundred disease-free crucian caps (mean body weight, 70±10 g mean±S. D.) were obtained from a fish farm in Jiangsu province. All fish were never exposed to any drug in fish farm prior to sample collection. All fish were drug-free fed and acclimatized to laboratory condition for two weeks prior to experimentation. During acclimatization and experimental periods, fish were kept in plastic tanks (500l) supplied with a constant flow (1lh-1) in aerated freshwater (dissolved oxygen: 6±0.5 mg · L^{-1} mean±S. D.) at 20℃, with a pH 7±0.3 (mean±S. D.). Fish were neither fed under laboratory adaptation nor during the experimental procedure to avoid possible interference with the treatment during the period of experiment.

Fish were evenly divided into two large groups and cultured under the same condition, one group for test and the other for control. The water temperature was maintained at 20℃. An oxygen tank was used to ensure a dissolved oxygen concentration of 6±0.5mg · L^{-1}. Fish for the test group were treated with 6.5g · mL^{-1} of DIF, and each fish was given a single dose of 20mg · kg^{-1} b. w. by oral administration (Yang et al., 2005; Ding et al., 2006). Fish from the control group were not fed with any drug. They were used for expression pattern and pharmacokinetics analysis. The sampling times were 0.016, 0.25, 0.5, 1, 2, 3, 6, 12, 48, 96, 120 and 144 h after treatment. Three fish were sampled at each time-point (n=3). 2.0 g of liver was excised from each fish, and then it was divided into two parts: one part was stored at-80℃ prior to RNA isolation, the other part was preserved in-20℃ prior to RP-HPLC analysis. All fish were handled in accordance with "Regulation on Animal Experimentation" (State Scientific and Technological Commission, China).

2.3 Cloning of full-length crucian cap CYP1A1 cDNA by PCR amplification

Total RNAs from liver were extracted using Trizol reagent kit according to the manufacturer's instructions (Invitrogen, Shanghai Branch, China). The quality of purified RNA was checked by 1% agarose gel electrophoresis and the RNA concentration was determined by spectrophotometer. 2.0 g RNA was reverse-transcribed using Primer script First-strand cDNA Synthesis Kit (TAKARA, Japan). The first-strand of cDNA was subsequently used as the template for PCR with the CYP1A1 primers. Primers were designed according the homologous cDNA sequences of fish CYP1A1 from the GenBank (Table 1). The PCR cycling conditions were 94℃ for 3 min, followed by 30 cycles of 94℃ for 30 s, 55℃ for 30 s, and 72℃ for 30 s, 1 cycle of 72℃ for 8 min. PCR products were isolated from gel by using a Gel Extraction Kit (TaKaRa, Japan) and they were ligated with pGEM-T vector and transformed into DH5-competent cells and then sequenced.

2.4 Sequence analysis

The Blast alignment search tool from the National Center for Biotechnology Information (NCBI) was used to search homologous sequences in GenBank and protein predictions were performed using the DNAstar software. The putative amino acid sequences were analyzed for the presence of primary structure analysis using ProParam and Compute pI/Mw software (http://www.expasy.ch/tools/secondary) (Bendtsen et al., 2004). Multiple sequence alignments were performed using the CLUSTALW 1.8 program.

2.5 RT-PCR analysis

100 mg of tissues were extracted RNA from samples, cDNAs were synthesized as described in Section 2.3. The first-strand cDNA was subsequently used as the template for PCR with the CYP1A1 primers. -Actin was used as an internal control with specific primers. The reactions were completed in a thermocycler with following profile: one cycle of 2min at 94℃, 30 cycles of 30s at 94℃, 30s at 55℃, and 30s at 72℃, followed by 1 cycle of 8 min at 72℃. The RT-PCR products were analyzed by electrophoresis and verified by sequencing.

Table 1 CYP1A1 sequence used for similarity comparison

Species	Protein	Similarity (%)	GenBank accession no
Human (Homo sapiens)	CYP1A1	54.1	AF253322
Pig (Susscrofa)	CYP1A1	53.5	BAB85660
Mouse (Mus musculus)	CYP1A1	56.3	NM 009992
Xenopus laevis	CYP450	57.8	BAA37079
Tilapia nilotica (Oreochromis niloticus)	CYP1A	58.4	BAB39157
Common carp (Cyprinus carpio)	CYP1A	66.9	BAB39379
Grey mullet (Mugil cephalus)	CYP1A1	69.9	ABZ88706
Mugilidae (Liza aurata)	CYP1A1	71.4	AAB70307
Pagrus major (Red seabream)	CYP1A1	71.8	ABX10186
Atlantic salmon (Salmo salar)	CYP1A1	73.0	AAM00254
Rainbow trout (Oncorhynchus mykiss)	CYP1A3	73.1	NP 001118226
Crucian cap (Carassius auratus gibello)	CYP1A1	−	GQ353297.1

2.6 Real-time PCR analysis

100 mg of liver was extracted RNA from samples with DIF treated in crucian cap, all cDNA samples were synthesized as described in Section 2.3. Primer was designed using the primer 5.0 and Beacon Designer 7.0 software for amplification of CYP1A1 and β-actin cDNA fragments (Table 2). Optimal primer pairs were isolated based on their amplification efficiency and specificity. Plasmid DNAs containing the CYP1A1 and β-actin cDNA fragments were isolated using the Mini Plasmid Kit (TaKaRa, Japan). Serial 10-fold dilutions of plasmids were used as templates to derive standard curves for each PCR run (Changand Nie,

2007). Real-time PCR was performed using the iQ5 Real Time PCR Detection System (Bio-Rad) in 50l reactions containing the following components: 2l cDNA sample, 25l SYBR Premix Ex Taq (TaKaRa, Japan), 1l of each primer and 21l H_2O. The real-time PCR profile was: 10s at 95℃ 40 cycles consisting of 5 s at 95℃, 30 s at℃, followed by melting curve analysis. One-way ANOVA tests were performed using SPSS14.0 to determine whether there were any significant differences between the DIF-treated group and the control group. All statistical analysis was based on CYP1A1 expression levels normalized by-β-actin expression.

Table 2 cDNA target genes, GenBank identification and primers used for real-time PCR

Gene	GenBank ID	Forward primer (5' -3') Reverse primer (5' -3')	Amplicon size (nucleotides)
β-Actin	DQ835399.1	CTGTATGCCTCTGGTCGTGCTGTAGCCTCTCTCGGT	174
CYP1A1	GQ353297.1	CAAGGACAACATCCGAGATAGACGACACCCCAAGAC	174

2.7 RP-HPLC assay for difloxacin and pharmacokinetic analysis

The Aligen 1 100 HPLC system (USA) consisted of a quaternary pump (Germany), vacuum degasser (Japan) and auto-injector (Germany). DIF was extracted by addition of 6 volume of dichloromethane, mixed vigorously for at least 5 min, and centrifuged for 10min at 8 000×g at 4℃ to separate the two phases. The aqueous phase was discarded by careful aspiration and the organic phase was transferred to clean tubes and evaporated under a nitrogen stream at 45℃. The residue was dissolved in 1.0ml mobile phase and 10l of the solution was injected onto a reverse phase analytical column, ZORBAX SBC18 (5 m, 150 mm×4.6 mm, I.D., Agilent Technologies, USA). The mobile phase consisted of 5% acetonitrile and 95% tetrabutyl ammonium bromide. The flow rate was 1.5 mL·min^{-1} and the column eluate was monitored at 276 nm. Column temperature was 40℃.

All processes followed first-order kinetics. Pharmacokinetic analysis was performed using the computer program Drug and Statistics (DAS ver 1.0, Center for Drug Clinical Evaluation, Wannan Medical College, China). The program used a compartmental open model based on non-linear regression analysis and a non-compartmental analysis based on statistical moment theory to analyze concentration-time data for difloxacin.

2.8 Data analysis and statistics

Statistical analyses were performed with SPSS software version 16.0 (SPSS Inc., USA) and the data were used as mean±standard deviation (S.D.). Significant differences between means were determined using one-way analyses of variance (ANOVAs) and the Duncan's test for multiple range comparison with significance level established at $P<0.05$.

3 Results

3.1 Characterization of crucian cap CYP1A1

The full-length, 1 816 bp crucian cap CYP1A1 cDNA (GenBank accession no. GQ353297.1) contains a 14 91 bp open reading frame (ORF), a 63 bp 5-untranslated region (UTR), a 262 bp 3-UTR and its transcript product contains 496 amino acids. The sequence also possesses all major functional domains and characteristics. The heme-binding region of the CYP1A1, encompassing the amino acid sequence GLGKRRCIG (position 427-435), is identical to the same region of other homologues (Fig. 1). The protein was predicted at a putative molecular weight of 55.85 kDa and an isoelectric point of 7.18.3.2. Constitutive expression of the CYP1A1 gene in different tissues

The CYP1A1 gene displayed constitutive expression in most tissues examined, including liver, muscle, gill, kidney, intestine, heart, brain and gonad. Comparing with the relative expression rate of CYP1A1 gene in these organs requires a similar concentration of a reference transcript in all tissues. Among these tissues, liver, intestine and kidney exhibited the higher basal expression, and gonad and muscle were the lower expression of CYP1A1 gene (Fig.2).

3.2 Gene expression of liver CYP1A1 treated with DIF

Using a real-time PCR, we confirmed that CYP1A1 mRNA levels in liver demonstrated a time-dependent response to DIF metabolize (Fig.3). CYP1A1 mRNA level in DIF-treated group was significantly different compared with the control ($P<0.05$) (Fig.4). However, with the extension of treatment time, CYP1A1 mRNA expression was decreased, indicating that DIF is an inhibitor

```
GTGGCACTCTGTGTCTGAGAGCCTGGTGGCCATCATCACGATATGCGCCTCATGCGTACT    60
                                            M  R  L  M  R  T      6
AAGATCCCAGAAGGGCTCCAGAAGCTGCCGGGACCGAAGCCTCTTCCCATCATCGGAAAT   120
K  I  P  E  G  L  Q  K  L  P  G  P  K  P  L  P  I  I  G  N      26
GTCCTGGAAGTGGGAAACAACCCACATCTGAGCCTGACCGCTATGAGTAAGTTCTACGGC   180
V  L  E  V  G  N  N  P  H  L  S  L  T  A  M  S  K  F  Y  G      46
CCTGTCTTCCAGATCCAGATCGGGATGCGTCCCGTTGTAGTACTCAGCGGAAACGACGTG   240
P  V  F  Q  I  Q  I  G  M  R  P  V  V  V  L  S  G  N  D  V      66
ATCCGTCAAGCTCTCCTCAAACAAGGCGAGGAGTTCGCCGGACGTCCGGATCTCCACAGC   300
I  R  Q  A  L  L  K  Q  G  E  E  F  A  G  R  P  D  L  H  S      86
AGCAAGTTCATCAGCGACGGAAAGAGCCTGGCCTTTAGTACGGATCAGGTGGGTGTCTGG   360
S  K  F  I  S  D  G  K  S  L  A  F  S  T  D  Q  V  G  V  W     106
CGCGCCCGTCGCAAGCTGGCCCTCAGCGCCCTGCGCACATTTTCCACGGTGCAAGCCAAA   440
R  A  R  R  K  L  A  L  S  A  L  R  T  F  S  T  V  Q  A  K     126
AGCTCAGAGTATTCATGTGCCCTTGAGGAGCACATCAGCAAGGAAGGTCTCTATCTGATC   500
S  S  E  Y  S  C  A  L  E  E  H  I  S  K  E  G  L  Y  L  I     146
GAAAGGCTTCACAGCGTCATGAAGGCCGACGGGAGTTTCGATCCCTTCCGGCACATCGTG   560
E  R  L  H  S  V  M  K  A  D  G  S  F  D  P  F  R  H  I  V     166
GTGTCTGTGGCCAACGTGATCTGCGGGATCTGCTTCGGCCAGCGCTACAGCCACGACGAC   620
V  S  V  A  N  V  I  C  G  I  C  F  G  Q  R  Y  S  H  D  D     186
GACGAGCTGGTGAGCTTGGTCAATTTGAGCGACGAGTTCGGGAAGATCGTGGGAAGCGGC   680
D  E  L  V  S  L  V  N  L  S  D  E  F  G  K  I  V  G  S  G     206
AATCCTGCGGACTTCATCCCTTTCTTGCGTATCCTGCCCAGCACGATGATGAAGAAGTTC   740
N  P  A  D  F  I  P  F  L  R  I  L  P  S  T  M  M  K  K  F     226
GTGGCCATCAACGCTCGCTTCAGCAAGTTGATGAAGAAGATGGTCAATGACCATTACGAC   800
V  A  I  N  A  R  F  S  K  L  M  K  K  M  V  N  D  H  Y  D     246
ACTTTCAACAAGGACAACATCCGAGACATCACCGACTCGCTCATCAACCACTGCGAAGAC   860
T  F  N  K  D  N  I  R  D  I  T  D  S  L  I  N  H  C  E  D     266
CGGAAACTGGACGAGAACTCAAACGTGCAAGTGTCCGATGAGAAGATCGTCGGAATCGTC   920
R  K  L  D  E  N  S  N  V  Q  V  S  D  E  K  I  V  G  I  V     286
AATGATCTCTTCGGAGCTGGTTTCGACACTATCAGTACGGCTCTGTCTTGGGGTGTCGTC   980
N  D  L  F  G  A  G  F  D  T  I  S  T  A  L  S  W  G  V  V     306
TATCTAGTGGCCTACCCTGAGATCCAGGAGCGACTGCAAAGAGAGCTGAGAGAAAAGATC  1040
Y  L  V  A  Y  P  E  I  Q  E  R  L  Q  R  E  L  R  E  K  I     326
GGAATGGATCGAACGCCACGTTTGTCAGACAGAACGGATTTGCCACTTCTCGAGGCCTTC  1100
G  M  D  R  T  P  R  L  S  D  R  T  D  L  P  L  L  E  A  F     346
ATCCTGGAGATCTTCCGCCATTCGTCCTTCCTTTCACCATTCCTCACTGTACGTCT  1160
I  L  E  I  F  R  H  S  S  F  L  P  F  T  I  P  H  C  T  S     366
AAAGACACGTCGCTCAACGGATATTTCATTCCCAAAGACACCTGTGTGTTTGTAAACCAG  1220
K  D  T  S  L  N  G  Y  F  I  P  K  D  T  C  V  F  V  N  Q     386
TGGCAAGTCAACCATGACCCAGAACTGTGGAAGGATCCGTCAAGCTTCAATCTGGACCGA  1280
W  Q  V  N  H  D  P  E  L  W  K  D  P  S  S  F  N  L  D  R     406
TTCCTCACCGCAGACGGTACAGAGCTGAACAAGATTGAAGGAGAGAAGGTGTTGGTTTTT  1340
 F  L  T  A  D  G  T  E  L  N  K  I  E  G  E  K  V  L  V  F    426
GGGCTTGGGAAGCGGCGCTGCATTGGAGAGTCCATCGGACGTGCCGAGGTCTTCCTGTTC  1400
 G  L  G  K  R  R  C  I  G  E  S  I  G  R  A  E  V  F  L  F    446
ATGGCCATCCTTCTCCAGAGGTTAAAGTTCAGCGGGATGCCGGGAGAAATGCTGGATATG  1460
M  A  I  L  L  Q  R  L  K  F  S  G  M  P  G  E  M  L  D  M     466
ACGCCAGAGTACGGGCTGACCATGAAACACAAGCGCTGTATGCTGCGCGTCACGCCACAA  1520
T  P  E  Y  G  L  T  M  K  H  K  R  C  M  L  R  V  T  P  Q     486
CCCGGCTTCGAATTGCCCGGCACTGCCGGCCTGActctgacgaaaccgtcatcatagtttga  1580
P  G  F  E  L  P  G  T  A  A  *                                496
gctgaaatccacagcgctgttacttctgattctggttcggtgcagacctcgcagtcaactcacaaagcc
aaatatttgaaagatgtttgattgtgattcaggcaaggcatccggtgcattttactgcagttatttcat
aatgctggacgcaatgctcataaccagagtctgcatgtggctttatcatgtacatacatctgtgttttc
atatgtcatatgcggtgttgagatgggca
```

Fig. 1 Nucleotide and deduced amino acid residue sequence of Crucian cap CYP1A1. The characteristic heme-binding domain is indicated with a box. The start codon and stop codon are indicated in bold （GenBank accession no. GQ353297.1）.

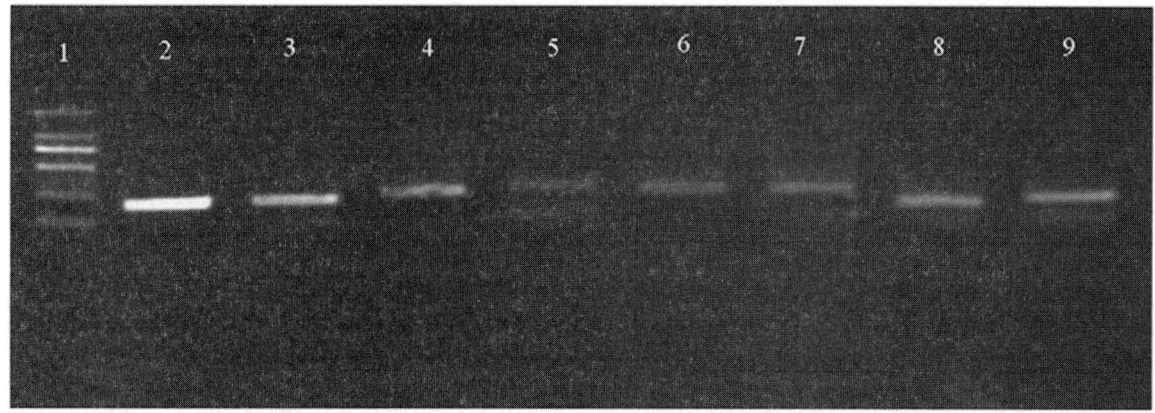

Fig. 2 RT-PCR analysis of CYP1A1 gene expression in various tissues. The products of RT-PCR were detected on agarose gel by electrophoresis. Lanes 1-9: (1) marker, (2) liver, (3) intestine, (4) kidney, (5) muscle, (6) gonad, (7) heart, (8) brain, and (9) gill.

of CYP1A1.

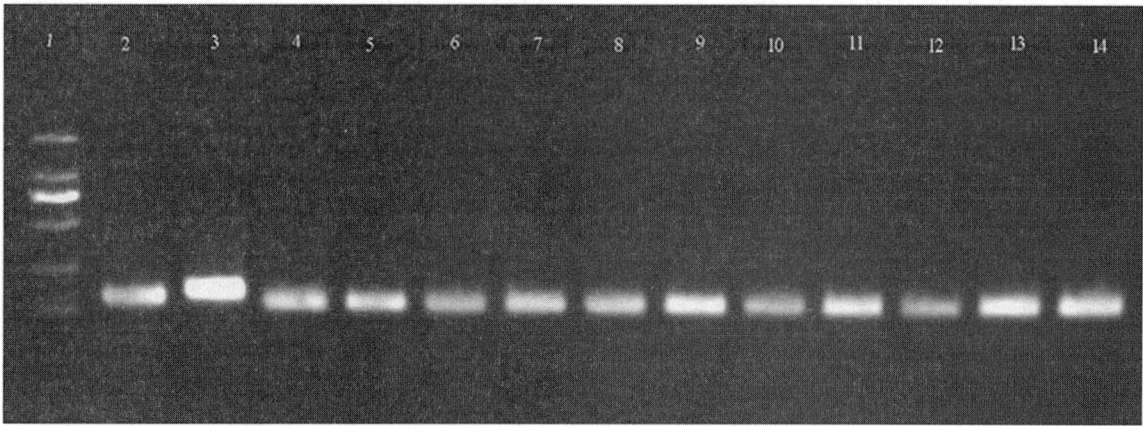

Fig. 3 Examination of crucian cap CYP1A1 mRNA in different time-point in liver after administration DIF, 2% gel analysis of the real-time PCR extraction and PCR conditions are described in Section 3. 3. Levels of CYP1A expression were normalized to-β-actin expression. Lane 1-14 were: (1) marker, (2) control, (3) 0.016 h, (4) 0.25 h, (5) 0.5 h, (6) 1 h, (7) 2 h, (8) 3 h, (9) 6 h, (10) 12 h, (11) 48 h, (12) 96 h, (13) 120 h, and (14) 144 h.

3. 3　Pharmacokinetics of DIF

DIF was separated at base line by using the chromatographic condition developed. There were no interfering peaks found at the same retention time of DIF in chromatogram. Retention time of DIF was 9. 25 min. Pharmacokinetic parameters for DIF in crucian cap after a single dose of 20 mgkg^{-1} administered orally are presented (Table 3) . The absorption half-time of the drug was found to be 0. 205 h, elimination half-life ($t_{1/2}$) of the drug was 26. 01 h, peak time (T_{max}) was 0. 5 h, The mean extent of absorption (AUC) was 294. 96 mg^{-1} kg^{-1} h^{-1} and the maximum plasma concentration (C_{max}) was 67. 7 mgkg^{-1}; total clearance (CL) was estimated to be 0. 07 mgkg^{-1} h^{-1} and the Vd/F was 0. 22 mgkg^{-1}.

4　Discussion

4. 1　Sequence comparison of crucian cap CYP1A1

In the present study, we first cloned and characterized cDNA sequences of CYP1A1 gene in Crucian cap. Extensive phylogenetic analyses of CYP1 gene families in vertebrates suggested that there were high similarities among the CYP1A1 family members. The deduced crucian cap CYP1A1 amino acid sequences were aligned with their orthologs. Based on the putative amino acid sequence,

CYP1A1 is also highly conservative with teleost CYP1A1 isozymes (Table 1). The apparently high degree of sequence similarity to these proteins in teleosts suggested that CYP1A1 isozymes were highly conservative during teleost evolution. Meanwhile, these lines of evidence indicate its physiological function in crucian cap.

4.2 Tissue-specific expression of CYP1A1 transcripts in crucian cap

The tissue distribution of CYP1A1 assayed by RT-PCR revealed that the gene was expressed in most tissues, including muscle, heart, liver, kidney, brain, gonad, gill and intestine, with the highest level of transcripts found in liver, and the weakest expression in muscle (Fig. 2). These results were consistent with those reported from yellow catfish and Takifugu obscures (Kima et al., 2008). It was also demonstrated that CYP1A is expressed in various tissues in zebrafish, Rivulus marmoratus, mouse and human, especially with the highest expression in liver (Jönsson et al., 2007; Lee et al., 2005; Dey et al., 1999; Kima et al., 2008). CYP1A gene expression in liver and intestine were significantly higher than those in other tissues (Arukwe, 2002). And the tissue-specific distribution of CYP1A may indicate its physiological function for detoxification in fish. CYP1A may play an important role in eliminating endogenous AHR ligands of endogenous metabolites and food-derived AHR agonists (Neber et al., 2000).

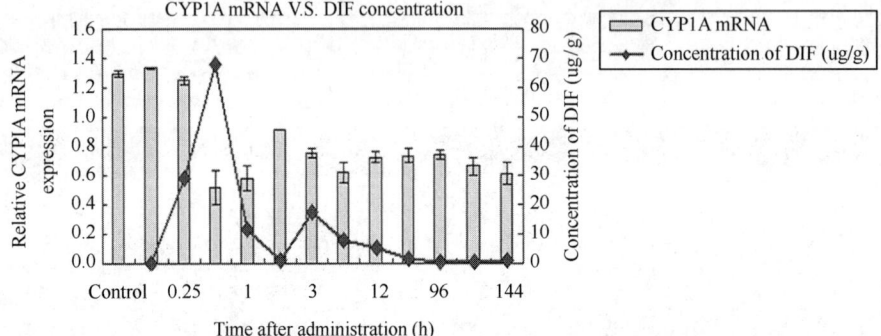

Fig. 4　Time after administration DIF, the CYP1A1 mRNA level and DIF concentration in liver. CYP1A1 mRNA expression was determined by real-time PCR using crucian cap. CYP1A1 and-β-actin primers developed in this study. Fold induction of CYP1A1 mRNA in each of time-point in treatment groups were determined compared with the expression of the control group. Levels of CYP1A1 expression were normalized to-β-actin expression. Bars indicated standard errors of the means. DIF concentrations and CYP1A1 mRNA were time-point responsively following by DIF treatment at 20 mg/kg b. w

Table 3　Pharmacokinetic parameters/profiles of difloxacin (DIF) after oral administration (20 mg/kg b.w.) to crucian cap.

Parameters	Value
α (h^{-1})	2.53
β (h^{-1})	0.03
K_a (h^{-1})	3.39
$t_{1/2Ka}$ (h)	0.20
$t_{1/2\alpha}$ (h)	0.27
$t_{1/2\beta}$ (h)	26.01
V_d/F (mg·kg^{-1})	0.22
AUC (mg·kg^{-1}·h)	294.96
CL_b (mg·h^{-1}·kg^{-1})	0.07
T_p (h)	0.36
T_{max} (h)	0.50
C_{max} (mg·kg^{-1})	67.70

Notes: Liver samples were collected at the scheduled time after oral administration. Each point represents the mean±S.D. (n=3). : distribution rate constant; : elimination rate constant; K_a: absorption rate constant; $t_{1/2}$: distribution half-life of the drug; $t_{1/2}$: elimination half-life of the drug; $t_{1/2Ka}$: absorption half-life of the drug; V_d/F: extensive apparent volume of the central compartment; AUC: area under the curve; CL_b: total body clearance of the drug; T_p: the time point of maximum plasma concentration of the drug; T_{max}: peak time; C_{max}: the maximum plasma concentration.

4.3 Correlation between CYP1A1 mRNA transcription level and DIF concentration in crucian cap liver

The concentration of DIF in liver reached its peak （67.70 mg·kg^{-1}） at 0.5 h, at the same moment, the CYP1A1 gene expression was at the lowest point. The pharmacokinetic characteristics of DIF and analysis of CYP1A1 gene expression showed that the metabolism of DIF in fish may be related to the activity of CYP1A （Fig. 4）. After DIF treatment, CYP1A expression gradually decreased and was significantly lower than the control. Thus, DIF seems to be an inhibitor of CYP1A in crucian cap.

In fish, CYP3A could be inhibited by enrofloxacin which is able to provoke the inactivation of the enzyme （Vaccaro et al., 2003）. In the present research, CYP1A is one of the major P450 enzymes involved in biotransformation of DIF. DIF can inhibit enzyme activity of CYP1A, while metabolism of DIF was affected by CYP1A activity. We further investigated the relationship between CYP1A1 transcription level and DIF concentration in crucian cap. The results suggested that mRNA expression of CYP1A1 gene is down-regulated by the DIF, and DIF is both the substrate and inhibitor of CYP1A. Moreover, other studies reported that CYP450 in liver plays a critical role in the overall elimination of many drugs （Pana et al., 2010; Levine and Oris, 1999; Woznya et al., 2010）. This study not only provides instructive information to ensure treatment success in fisheries medication with two or more drugs, but also reveals the relationship of pharmacokinetic characteristics and the expression of drug metabolizing enzymes.

5 Conclusion

In the study, we cloned the CYP1A1 gene cDNA sequence from crucian cap, analyzed its tissue-specific expression and its relation to DIF time-dependent manner in liver. We also confirmed that CYP1A1 transcript level was down-regulated by DIF, suggesting that DIF is both the substrate and inhibitor of CYP1A and the CYP1A is responsible for DIF metabolism in crucian cap liver. These results provided important information for illustrating the mechanism of CYP （s） in drug metabolism, which could contribute to drug application in fish farming.

Conflict of interest

None.

Acknowledgements

This work was financially supported by the earmarked fund for Modern Agro-industry Technology Research System and the Eleventh Five-Year Plan to support "Fisheries drugs safety using technology and the development of the new fishing drugs preparation" （2006BAD0304）.

References

A rukwe, A., 2002. Complementary DNA cloning, sequence analysis and differential organ expression of - naphthoflavone - inducible cytochrome P4501A in Atlantic salmon （Salmo salar）. Comp. Biochem. Physiol. 133C, 613-624.

Bendtsen, J. D., Nielsen, H., von Heijine, G., Brunak, S., 2004.

Improved prediction of signal peptides: Signal P3.0. J. Mol. Biol. 340, 783-795.

Chang, M. X., Nie, P., 2007. Intelectin gene from the grass carp Ctenopharyngodon idella: cDNA cloning, tissue expression, and immunohistochemical localization. Fish Shellfish Immunol. 23, 128-140.

Ding, F. K., Cao, J. Y., Ma, L. B., Pan, Q. S., Fang, Z. P., Lu, X. C., 2006. Pharmacokinetics and tissue residues of difloxacin in crucian cap （Carassius auratus） after oral administration. Aquaculture 256, 121-128.

Drlica, K., Zhao, X., 1997. DNA gyrase, topoisomerase IV, and the 4-quinolones. Microbiol. Mol. Biol. Rev. 61, 377-392.

Dey, A., Jones, J. E., Nebert, D. W., 1999. Tissue-and cell type-specific expression of cytochrome P450 1A1 and cytochrome P450 1A2 mRNA in the mouse localized in situ hybridization. Biochem. Pharmacol. 58, 525-537.

Gigolos, P. G., Revesado, P. R., Cadahia, O., Fente, C. A., Vazquez, Nelson, D. R., 1996. P450 superfamily: update on newsequences, genemapping, accession numbers and nomenclature. Pharmacogenetics 6, 1-42.

Gigolos, P. G., Revesado, P. R., Cadahia, O., Fente, C. A., Vazquez, B. I., Franco, C. M., Cepeda, A., 2000. Determination of quinolones in animal tissues and eggs by high-performance liquid chromatography with photodiode-array detection. J. Chromatogr. A 871, 31-36.

Guengerich, F. P., 1999. Cytochrome P-450 3A4: regulation and role in drug metabolism. Annu. Rev. Pharmacol. Toxicol. 39, 1-17.

Goldstone, H. M., Stegeman, J. J., 2006. A revised evolutionary history of the CYP1A subfamily: gene duplication, gene conversion, and positive selection. J. Mol. Evol. 62, 708-717.

Hahn, M. E., Stegeman, J. J., 1994. Regulation of cytochrome P450 1A1 in teleosts: sustained induction of CYP1A messenger-RNA, protein, and catalytic activity by 2, 3, 7, 8 tetrachlorodibenzofuran in the marine fish Stenotomus chrysops. Toxicol. Appl. Pharmacol. 127, 187-198.

Hawkins, S. A., Billiard, S. M., et al., 2002. Altering cytochrome P4501A activity affects polycyclic aromatic hydrocarbon metabolism and toxicity in rainbow trout (Oncorhynchus mykiss). Environ. Toxicol. Chem. 21, 1845-1853.

Hu, M.C., Hsu, H.J., Guo, I.C., Chung, B.-C., 2004. Function of Cyp11a1 in animal models. Mol. Cell. Endocrinol. 215, 95-100.

Hanlona, N., Coldhamb, N., Sauerb, M.J., Ioannides, C., 2008. Up-regulation of the CYP1 family in rat and human liver by the aliphatic iso-thiocyanates erucin and sulforaphane. Toxicology 252, 92-98.

Isin, E.M., Guengerich, F.P., 2007. Complex reactions catalyzed by cytochrome P450 enzymes. Biochim. Biophys. Acta 1770, 314-329.

Jönsson, M.E., Orrego, R., Woodin, B.R., 2007. Basal and, 5-pentachlorobiphenyl-induced expression of cytochrome P450 1A, 1B and 1C genes in zebrafish. Toxicol. Appl. Pharmacol. 221, 29-41.

Kima, J.-H., Raisuddin, S., et al., 2008. Molecular cloning and-naphthoflavone-induced expression of a cytochrome P450 1A (CYP1A) gene from an anadromous river pufferfish, Takifugu obscurus. Mar. Pollut. Bull. 57, 433-440.

Lee, Y.M., Williams, T.D., Jung, S.O., Lee, J.S., 2005. cDNA cloning and expression of a cytochrome P450 1A (CYP1A) gene from the hermaphroditic fish Rivulus marmoratus. Mar. Pollut. Bull. 51, 769-775.

Li, D., Yang, X.L., Zhang, S.J., Lin, M., Yu, W.J., Hu, K., 2008. Effects of mammalian CYP3A inducers on CYP3A-related enzyme activities in grass carp (Ctenopharyngodon idellus): possible implications for the establishment of a fish CYP3A induction model. Comp. Biochem. Physiol. C 147, 17-29.

Levine, S.L., Oris, J.T., 1999. CYP1A expression in liver and gill of rainbow trout following waterborne exposure: implications for biomarker deter-mination. Aquat. Toxicol. 46, 29-287.

Nemoto, N., Sakurai, J., 1992. Altered regulation of Cyp1a-1 gene expression during cultivation of mouse hepatocytes in primary culture. Biochem. Pharmacol. 44, 51-58.

Pana, J., Liua, G.-Y., Chenga, J., et al., 2010. CoMFA and molecular docking studies of benzoxazoles and benzothiazoles as CYP450 1A1 in-hibitors. Eur. J. Med. Chem. 45, 967-972.

Stegeman, J.J., Schlezinger, J.J., Craddock, K.E., Tillitt, D.E., 2001. Cytochrome P4501A expression in midwater fishes: potential effects of chemical contaminants in remote oceanic zones. Environ. Sci. Technol. 35, 54-62.

Shinkyoa, R., Sakakia, T., Ohtab, M., Inouye, K., 2003. Metabolic pathways of dioxin by CYP1A1: species difference between rat and human CYP1A subfamily in the metabolism of dioxins. Arch. Biochem. Biophys. 409, 180-187.

Vaccaro, E., Giorgi, M., Longo, V., Mengozzi, G., Gervasi, P.G., 2003. Inhibition of cytochrome P450 enzymes by enrofloxacin in the sea bass (Dicentrarchus labrax). Aquat. Toxicol. 62, 27-33.

Willsa, L.P., Zhub, S., Willettb, K.L., Di Giulioa, R.T., 2009. Effect of CYP1A inhibition on the biotransformation of benzo [a] pyrene in two populations of Fundulus heteroclitus with different exposure histories. Aquat. Toxicol. 92, 195-201.

Woznya, M., Brzuzana, P., Łuczy'nskib, M.K., et al., 2010. CYP1A' expression in liver and gills of rainbow trout (Oncorhynchus mykiss) after short-term exposure to dibenzothiophene (DBT). Chemosphere 79, 110-112.

Wolfson, J.S., Hooper, D.C., 1989. Fluoroquinolone antimicrobial agents. Clin. Microbiol. Rev. 2, 378-424. Yang, X.L., Lu, C.P., Zhan, W.B., et al., 2005. New Fishery Drugs Handbook 211-212, first ed (Beijing, China).

Yu, Ling Zhi, Yang, Xian Le, 2010. Effects of fish cytochromes P450 inducers and inhibitors on difloxacin N-demethylation in kidney of Chinese idle (Ctenopharyngodon idellus). Environ. Toxicol. Pharmacol. 29, 202-208.

该文发表于《水产学报》【2010，(3)：404-409】。

草鱼肾细胞中双氟沙星代谢酶的酶动学

Enzymaticson DIF metabolism in kidney cell of grass carp (*Ctenopharyngodon idellus*)

于灵芝[1]，杨先乐[1,2]*，王翔凌[3]，喻文娟[1]，胡鲲[1]

(1. 上海海洋大学农业部渔业动植物病原库，上海 201306; 2. 上海海洋大学水产与生命学院，
上海市高校水产养殖学 E-研究院，上海 201306; 3. 上海美迪西普亚医药科技有限公司，上海 201200)

摘要 细胞色素 P450s (CYPs) 主要参与动物体内药物代谢。水产养殖中不合理的联合用药常会导致治疗失败，这通常与 CYP 活性的诱导有关。然而，关于鱼类 CYP 的诱导却知之甚少。为获得有关 CYP 诱导的信息，实验采用反相高效液相色谱 (RP-HPLC) 的方法测定了草鱼肾细胞 (CIK) 中 CYPs 的特异性诱导剂对双氟沙星 (DIF) 的代谢作用及酶动学分析。对照组和 β-萘黄酮 (BNF) 诱导组的酶动学方程分别为 $1/V = 0.1375 \times 1/[S] + 0.003$ 和 $1/V = 0.0245 \times 1/[S] + 0.0013$。DIF 与经 BNF 处理的 CIK 共孵育后，其代谢量增加了 1 倍，酶动学参数 Clint

和 Vmax 值分别增加了 7 倍和 2 倍。BNF 是 CYP1A 的特异性诱导剂，因此 CYP1A 可能参与了 DIF 的代谢。

关键词　草鱼；细胞色素 P450；双氟沙星；酶动力学；反相高效液相色谱

细胞色素 P450s（CYPs）主要参与药物代谢，在水产动物的肝和肾中广泛分布[1-2]。CYPs 的活性能被诱导或抑制，因而一种药物可通过对酶的诱导或抑制改变其他药物的代谢[3]。因此，两种药物联用不当常会发生药物间的代谢性相互作用。酶动学参数 Vmax 和 Clint 表示了其产生代谢物的催化效率，鱼类 CYPs 的酶动学分析日渐引起人们的兴趣，在草鱼（Ctenopharyngodon idellus）和大口黑鲈（Micropterus salmoides）的体内外已经研究了喹诺酮类药物代谢酶的酶动力学，及 CYP2E1 及 CYP3A 活性的诱导和抑制作用[4-9]。同时证实，鱼类细胞模型适用于渔药的酶动力学分析[4-9]。CYP1A、CYP3A 和 CYP2E 是鱼类体内含量丰富的 CYPs，β-萘黄酮（BNF）、乙醇（EA）和利福平（RIF）分别是鱼类 CYP1A、2E 和 3A 的特异性诱导剂[4,10-11]。然而，关于鱼类 CYP 的诱导剂在草鱼肾细胞（CIK）中对喹诺酮类药物的作用，以及双氟沙星（DIF）代谢酶的酶动学分析，至今未见报道。

本文采用反相高效液相色谱（RP-HPLC）方法，研究了草鱼肾细胞中鱼类 CYP 诱导剂对 DIF 代谢的作用及 DIF 代谢酶的酶动学。这些信息为用鱼类细胞模型预测渔药联用的相互作用提供理论支持。

1　材料与方法

1.1　试剂

盐酸双氟沙星（>99.8%），RIF（>97%），MTT，BNF 和 DMSO（99.9%）购自 Sigma。Lowry 蛋白测定试剂盒购自上海尚谊化工科技有限公司。M199 和 FBS 购自 GibcoBRL（UK）。HPLC 级乙腈和甲醇购自德国 Darmstadt。其他化学药品和试剂是分析纯，购自国药集团上海化学试剂公司。

1.2　试剂处理

盐酸双氟沙星用氢氧化钠助溶。BNF 和 RIF 溶于 DMSO，DIF、BNF 和 RIF 溶解后加入到培养基中，EA 直接加入培养基。本实验所有用水都是三蒸水。

1.3　细胞培养

CIK 由农业部渔业动植物病原库提供，细胞培养用含 10%（v/v）FBS 的 M199 培养基，条件为 28℃，5.0%CO$_2$。48 h 后贴满壁，加入新鲜 M199 培养基，消化，悬浮液在 6 孔板中培养，至密度为每毫升 5×10^5 个细胞，供诱导处理用。

1.4　草鱼肾细胞的诱导处理

6 孔板中的细胞完全贴壁后，用无血清的含相应诱导浓度的 BNF（10 μmol/L）[11]，EA（100 μmol/L）[12] 和 RIF（40 μmol/L）[4] 的培养基处理，对照只用无血清的培养基处理，放置于 28℃培养箱，24 h 之后，弃去原来的培养基，用无血清的含非致死浓度的 DIF（62.5 μmol/L）的培养基孵育，于 28℃培养箱中孵育诱导 2 h。之后，每份细胞悬液分为两份，一份用于该细胞悬液蛋白含量测定，另一份用于鱼类诱导剂对 DIF 代谢的诱导分析。每一组做 3 个重复。

1.5　色谱条件

Agilent1100 型高效液相色谱系统包括四元泵（德国），真空脱气机（日本）和自动进样器（德国）。细胞样品中的 DIF 用 6 倍体积二氯甲烷萃取，充分混合至少 5 min，4℃8 000×g 离心 10 min 分离两相，小心弃去液相，将有机相转移到干净的管子里，45℃样品蒸干仪中恒温中蒸干。残留溶于 1.0 mL 流动相中，10 μL 溶解物转移进反相分析柱中进行 HPLC 测定（色谱柱类型：ZORBAXSB-C18，4.6 mm×150 mm，5 μm）。流动相包括 5%乙腈（A）和 95%四丁基溴化铵（B）。流速为 1.5 mL/min，紫外检测波长 276 nm。柱温是 40℃。

1.6　酶动力学分析

为确定 DIF 代谢的酶动力学，在 28℃培养箱中用浓度为 3.9，7.8，15.6，31.25，62.5 和 125 μmol/L 的 DIF 与细胞孵育 2 h。部分经诱导处理的细胞先进行样品前处理，之后再通过 RPHPLC 方法检测，DIF 代谢酶的活性以酶促反应速率，即每毫克蛋白单位时间内的 DIF 代谢量表示 [nmol/（min·mgprotein）]。K_m 和 V_{max} 值是由米氏方程决定的，即 $V=V_{max}[S]/(K_m+[S])$，其中，V 表示酶促反应速率，K_m 为米式常数，反映酶与底物亲和力的大小，V_{max} 指该酶促反应的最大速度。K_m 和 V_{max} 由 Lineweaver-Burk 方程式作图确定。细胞蛋白浓度测定参照商品说明书。

1.7 数据分析和统计

数据分析用软件 SPSS16.0 进行显著性分析，由 ANOVA（单因素方差分析）的 Duncan 氏进行多重比较，$P<0.05$ 为差异显著。

2 结果

2.1 DIF 检测方法的确定

DIF 在所用色谱条件下基线分离良好，空白组在滞留时间内无杂峰。DIF 的标准曲线是由细胞与一系列梯度浓度的 DIF 共孵育，然后用 RPHPLC 方法确定的。结果显示了良好的相关性，相关系数为 0.9992（$n=3$）。以峰面积为纵坐标，相应的浓度为横坐标（μmol/L），所得标准曲线的回归方程为 $y=5.8757x-1.02$。DIF 的检测限和定量限分别是 0.5 μmol/L 和 1 μmol/L（$S/N=3$）。回收率为 84.5%±0.85%。平均相对标准偏差（RSD.）都小于 1.9%。DIF 的检测方法可重复并且精确（表 1）。

2.2 CIK 中 DIF 的代谢

MTT 实验结果表明，浓度为 125 μmol/L 的 DIF 与 CIK 孵育 6 h 为 DIF 代谢研究的亚致死条件（数据未列出）。为了优化 DIF 代谢的最佳孵育浓度和时间，采用不同浓度（从 3.9 μmol/L 到 125 μmol/L）和不同时间（从 1 h 到 6 h）与 CIK 孵育，样品处理后，采用 RP-HPLC 检测 DIF 量，DIF 在孵育浓度为 62.5 μmol/L 和孵育时间 2 h 时有最大代谢量（图 1），因此，DIF 浓度 62.5 μmol/L 和孵育时间 2 h 作为诱导分析的最佳条件。CIK 与诱导剂 BNF，RIF 和 EA 诱导后 DIF 的代谢量与对照相比，BNF 处理组导致了 DIF 代谢水平增加了 1 倍，而 RIF 和 EA 处理组没有使 DIF 代谢发生明显改变（图 2）。

2.3 DIF 代谢的酶动学分析

DIF 浓度为 3.9~62.5 μmol/L，其代谢酶显示了典型的 MichaelisMenten 行为（图 3）。根据 Lineweaver-Burk 图，获得对照组和 BNF 诱导组的酶动学方程分别为 $1/V=0.1375×1/[S]+0.003$ 和 $1/V=0.0245×1/[S]+0.0013$（图 3）。蛋白浓度是（0.35±0.02）mg/mL 细胞。

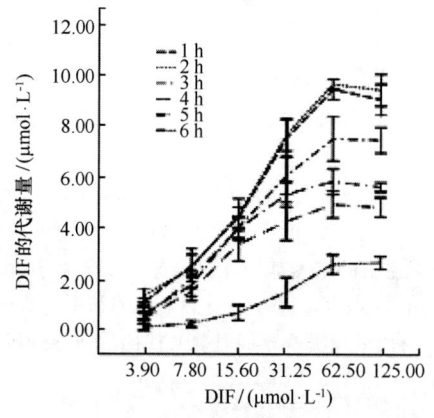

图 1 CIK 中 DIF 的代谢量（$n=3$，$\bar{x}±s$）

Fig. 1 Amounts of DIF metabolism in CIK（$n=3$，$\bar{x}±s$）

表 1 CIK 中 DIF 的回收率

Tab. 1 Recovery of DIF in CIK（$n=3$，$\bar{x}±s$）

空白浓度（μmol/L）Spiked amount	检测浓度（μmol/L）Measured amount	日内回收率（%）recovery intra-day	日间回收率（%）recovery inter-day	相对标准偏差（%）Relative standard deviation	
				intra-day	inter-day
1	0.98±0.07	98.33±6.80	81.49±1.00	0.12	1.73
10	10.50±0.64	105.03±6.37	82.38±1.06	1.11	1.84
100	99.26±0.93	99.26±9.30	85.49±0.60	1.61	0.60

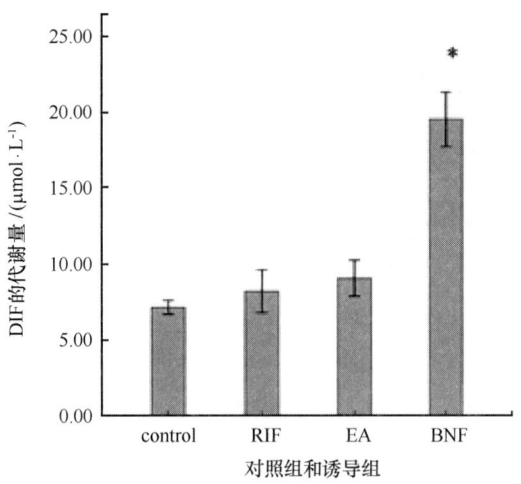

图 2　CYP 诱导剂处理后 DIF 的代谢量 （$n=3$）＊表示与对照差异显著 （$P<0.05$）

Fig. 2　Amounts of DIF metabolism in CIK pre-treated with fish CYP inducers

Asterisks indicate significant difference from the control （$P<0.05$）

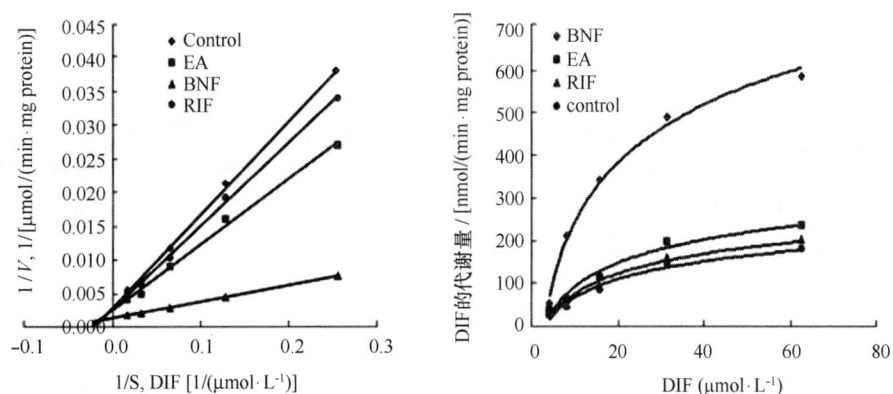

图 3　CIK 中由诱导剂处理的 DIF 代谢酶的 Lineweaver-Burk 图和 MichaelisMenten 图 （$n=3$）

Fig. 3　Line weaver-Burkplots and Michaelis Mentenplots of DIF metabolism by inducers in CIK （$n=3$）

3　讨论

3.1　体外模型在 CYPs 研究中的利用

体外模型可以得到生物转化途径的早期信息并可预测代谢水平的药物间相互作用[13]。目前已经建立的几种细胞系包括 PLHC-1[14]、RTG-2[15]、BB[15]、RTL-W1[16]、CIK[17] 和 GCL[4-6] 已被成功用于环境化合物的毒性检测，如有机锡、酚类和药物制剂[18]。结果表明，鱼类体外与体内的急性毒性有相似的趋势[19-22]。细胞培养系统适合研究化合物的毒性和酶动力学[5,6,20-22]，这还有助于减少水产药理学研究中活体动物的利用。细胞系还有助于减少水产药理学研究中活体动物的利用，并且自身有代谢外源物的优点。首先它们有研究外源物代谢的代谢活性，其次是细胞容易培养，最后是它们有稳定的诱导活性，这对外源物的代谢和解毒很重要[22]。在鱼类细胞系中研究 CYP 酶的酶动学及其调控引起了广泛关注，草鱼肝细胞（GCL）已经被用于研究渔药的 CYP 酶动学分析[4,6]。本研究证实，CIK 是适合研究 CYP 对 DIF 代谢作用的影响和渔药之间相互作用的体外模型。

3.2　鱼类 CYPs 诱导后的酶动学

与对照相比，RIF 处理的 CIK 中 DIF 代谢酶的酶动学参数 V_{max} 和 Clint 表明，DIF 代谢酶有较低的底物亲和力和催化 DIF 代谢的作用 （表 2）。DIF 的代谢量表明，RIF 对 DIF 的代谢没有诱导作用 （图 2）。本文结果与下述结论是一致的，即罗非鱼不同组织间 CYP3A 的诱导活性不同，肝和小肠中广泛存在，肾中不存在[23]。最近还有结果表明，RIF 在 GCL 中有增强恩诺沙星脱乙基转化为环丙沙星的作用[6]。因此可以推出，适合研究 CYP3A 诱导的体外模型是 GCL，而不是 CIK[4]。

表 2　CIK 中 DIF 代谢的酶动力学参数表

Tab. 2　Enzymatic parameters of DIF metabolism in CIK（$n=3$, $\bar{x}\pm s$）

处理组 treatments	米式常数 （μmol/L）K_m	最大反应速率 [nmol/（min·mg protein）]V_{max}	内在清除率 [L/（min·g protein）]Clint
对照	38.19±4.33	401.36±23.47	8.72±0.54
RIF	36.65±3.28	428.57±23.25	9.77±0.69
EA	31.03±3.98	461.45±32.34	12.39±1.14
BNF	16.5±2.39＊＊	1225±124.96＊＊	60.61±5.24＊＊

注：Clint=V_{max}/K_m；诱导组与对照进行 t-Test，"＊＊"表示差异极显著（$P<0.01$）。

Notes：Clint=V_{max}/K_m；＊＊ means significant difference（$P<0.01$）.

酶动力学参数表明，BNF 处理组有较高的底物亲和力和较强的代谢 DIF 的能力（表 2 和图 2）。BNF 是鱼类 CYP1A 的特异性诱导剂，因此 CYP1A 可能参与了上述反应（图 3）。为了研究 CIK 中参与 DIF 代谢的酶，还需要确定相应 CYP 抑制剂存在时的代谢信息。

3.3　与喹诺酮类药物代谢相关的 CYPs 的诱导及抑制作用

哺乳动物中有许多喹诺酮类药物都能对 CYP 酶系统产生抑制效应，从而导致某些药物代谢和清除率降低，但也并不是所有喹诺酮类药物有此效应[24-27]。此外，药物可同时作为 P450 底物、诱导剂和抑制剂，在某些情况下能调控自身的代谢[28]。Moutou 等[29]的研究结果表明，虹鳟线粒体中的喹诺酮类抗生素试剂噁喹酸和氟甲喹是 CYP1A 亚家族弱的剂量依赖性抑制剂。恩诺沙星在草鱼体内外显著抑制 CYP2E1 活性[7,12]，GCL 中 CYP3A4 参与了恩诺沙星脱乙基代谢[6]，其结果仍不明朗，因此喹诺酮类药物与 CYP（s）之间的相互作用需要深入研究。

总之，鱼类 CYP 诱导为预测可能的药物相互作用保证治疗效果提供了有利的信息。本研究表明，BNF 明显地增加了 DIF 的代谢作用，因此推测 CYP1A 参与了该反应。这些结果为鉴定代谢 DIF 的 CYP 酶和保证联合用药的疗效，提供了有利的信息。

参考文献

[1] Pesonen M, Celander M, Förlin L, etal. Comparison of xenobiotic biotrans formation enzymes in kidney and liver of rainbow trout（Salmogairdneri）[J]. Toxicol and Appl Pharmacol, 1987, 91：75-84.

[2] Lorenzana RM, Hedstrom OR, Buhler DR, Localization of cytochrome P450 in the head and trunk kidney of rainbow rout（Salmogairdneri）[J]. ToxicolApplPharmacol, 1988, 96：159-167.

[3] Khan MN, Reddy PK, Renaud RL, etal. Effect of cortisol on the metabolism of 17-hydroxyprogesterone by Arcticcharr and rainbow trout embryos [J]. Fish Physiol Biochem, 1997, 16：197-209. 408 水产学报 34 卷

[4] Li D, Yang XL, Zhang SJ, etal. Effects of mammalian CYP3A inducers on CYP3A-related enzyme activities in grasscarp（Ctenopharyngodon idellus）：Possible implications for the establishment of a fish CYP3A induction model [J]. Compa Biochem and Physiol, PartC, 2008, 147：17-29.

[5] 林茂, 杨先乐, 房文红, 等. 草鱼肝细胞中诱导剂对 EROD 作用的剂量效应研究 [J]. 水产学报, 2006, 30（3）：311-315.

[6] 林茂, 杨先乐, 王翔凌, 等. 草鱼肝细胞中恩诺沙星脱乙基代谢的酶动力学研究 [J]. 高技术通讯, 2007, 16（12）：38-42.

[7] 王翔凌, 杨先乐, 林茂, 等. 大口黑鲈肝细胞中恩诺沙星的消除及其代谢酶活性的测定 [J]. 中国水产科学, 2007, 14（6）：1004-1009.

[8] 李聘, 杨先乐, 张书俊, 等. 草鱼肝细胞系中 CYP3A 活性检测方法 [J]. 上海水产大学学报, 2007, 16（5）：495-499.

[9] 李聘, 杨先乐, 张书俊, 等. 草鱼肝细胞中细胞色素 P4503A 的酶促动力学 [J]. 高技术通讯, 2008, 18（8）：863-867.

[10] Kim JH, Raisuddin S, Ki JS, etal. Molecular cloning and beta-naphth of lavone-induced expression of a cytochromeP450 1A（CYP1A）gene from an anadromousriverpuffer fish, Takifuguobscurus [J]. MariPollutBull, 2008, 57：433-440.

[11] BhaktaKY, JiangW, CouroucliXI, etal. Regulation of cytochrome P4501A1 expression by hyperoxia in human lung celllines：Implications for hyperoxic lung in jury [J]. Toxicol and Appl Pharmacol, 2008, 233：169-178.

[12] 王翔凌, 杨先乐, 张宁, 等. 草鱼肝细胞中 CYP2E1 活性的诱导研究 [J]. 水生生物学报, 2008, 32（4）：469-474.

[13] Brandon EFA, Raap CD, Meijerman I, etal. An update on in vitro test methods in human hepatic drug biotransformation research：prosand-

cons [J]. Toxicol Appl Pharmacol, 2003, 189: 233-246.

[14] Celander M, HahnM E, Stegeman JJ. CytochromesP450 (CYP) in the Poeciliops is lucida hepatocellular carcinoma a cellline (PLHC-1): dose and time-dependent glucocorticoid potentiation of CYP1A induction without induction of CYP3A [J]. Arch Biochem Biophys, 1996, 329: 113-122.

[15] Araújo CSA, Marques SAF, Carrondo MJT, et al. In vitro response of the brown bullhead catfish (BB) and rainbow trout (RTG-2) cell-linesto benzo [a] pyrene [J]. Sci Total Environ, 2000, 247: 127-135.

[16] Babín M, Casado S, Chana A, etal. Cytochrome P4501A induction caused by the imidazole derivative Prochloraz ina rainbow trout cellline [J]. Toxicol In Vitro, 2005, 19: 899-902.

[17] Tan FX, Wang M, WangW M, etal. Comparative evaluation of the cytotoxicity sensitivity of six fish celllines to four heavy metals in vitro [J]. Toxicol In Vitro, 2008, 22: 164-170.

[18] Jano ekJ, Hilscherová K, BláhaL, etal. Environmental xenobiotics and nuclear receptors-inter actions, effects and in vitro assessment [J]. Toxicol InVitro, 2006, 20: 18-37.

[19] Caminada D, Escher C, Fent K, etal. Cytotoxicity of pharmaceuticals found in aquatic systems: comparison of PLHC-1 and RTG-2 fish celllines [J]. Aquat Toxicol, 2006, 79: 114-123.

[20] Ackermann GE and Fent K. The adaptation of the permanent fish celllines PLHC-1 and RTG-2 to FCS-freemedia results in similar growth rates compared to FCS-containing conditions [J]. Mar EnvironRes, 1998, 46: 363-367.

[21] Ackermann GE, Schwaiger J, Negele RD, etal. Effects of long-term nonylphenol exposure on gonadal development and biomarkers of estrogen exposure in juvenile rainbow trout (Oncorhynchus mykiss) [J]. Aquat Toxicol, 2002, 60: 203-221.

[22] Fent K. Fish celllines asversatile tools in ecotoxicology: Asssessment of cytotoxicity, cytochrome P4501A induction potential and estrogenic activity of chemicals and environmental samples [J]. Toxicol In Vitro, 2001, 15: 477-488.

[23] Bainy ACD, Woodin BR, Stegeman JJ. Elevated levels of multiple cytochromeP450 forms in tilapia from Billings Reservoir-Sao Paulo, Brazil [J]. Aquat Toxicol, 1999, 44: 289-305.

[24] Brouwers JR. Drug interactions with quinolone antibacterials [J]. Drug Saf, 1992, 7: 268-281.

[25] Davies BI, Maesen FP. Drug interactions with quinolones [J]. RevInfectDis, 1989, 11: 1083-1090.

[26] Harder S, Fuhr U, Staib AH, etal. Ciprofloxacin caffeine: a drug interaction established using in vivo and invitro investigations [J]. AmJMed, 1989, 87: 89-91.

[27] Xu X, Liu HY, Liu L, etal. The influence of a newly developed quinolone: antofloxacin, on CYP activity in rats [J]. Eur J Drug Metab Pharmokinet, 2008, 33: 1-7.

[28] Plant NJ, Ogg MS, Crowder M, etal. Control and statistical analysis of in vitro reporter gene assays [J]. Anal Biochem, 2000, 278: 170-174.

[29] Moutou KA, Burke MD, Houlihan DF. Hepatic P450 monooxygenase response in rainbow trout (Oncorhynchus mykiss Walbaum) administered aquaculture antibiotics [J]. Fish Physioland Biochem, 1998, 18: 97-106.

该文发表于《海洋渔业》【2012, 34 (1): 76-82】

草鱼孕烷 X 受体与细胞色素 P450 3AmRNA 表达相关性初步研究

The correlation of pregnane X receptor and cytochrome P450 3A in grass carp (*Ctenopharyngodon idellus*)

符贵红[1]，杨先乐[2]

(1. 中国水产科学研究院东海水产研究所，农业部海洋与河口渔业资源及生态重点开发实验室，上海 200090；

2. 上海海洋大学水产与生命学院，国家水生动物病原库，上海 201306)

摘要　根据 GenBank 中斑马鱼 (*Danio rerio*) 和虹鳟 (*Oncorhynchus mykiss*) 孕烷 X 受体 (pregnane Xreceptor, PXR) 基因序列设计保守区域简并引物，通过 PCR 扩增获得草鱼 (*Ctenopharyngodonidellus*) PXRcDNA 核苷酸序列片段为 509 bp，经推导获取 169aa 序列。对草鱼与其他物种 PXR 氨基酸序列进行同源性比较，用以鉴定其在物种中的进化。应用实时荧光定量 PCR (qRT-PCR) 检测草鱼 9 种组织 PXR 和细胞色素 P4503A (cytochrome P4503A, CYP3A) 基因 mRNA 相对含量，结果表明，其表达丰度依次为：前肠＞肝脏＞肾脏＞鳃＞肌肉＞后肠＞性腺＞心脏＞脾脏，CYP3A 和 PXR 基因在草鱼组织中表达分布基本一致，在前肠、肝脏和肾脏高度表达，而在其他组织低度表达。为了进一步研究 PXR 和 CYP3A 基因表达相关性，通过细胞，特选用诱导剂利福平

（rifampicin，RIF），最佳诱导浓度 40 μM，孵育 1、2、4、6、8、10、12、24 h 后，用 qRT-PCR 检测 CYP3A-PXR 基因动态表达过程。结果显示，诱导组 CYP3A 和 PXR 基因表达量均高于对照组，显示出显著诱导效应。

关键词 草鱼；孕烷 X 受体；细胞色素 P4503A；草鱼肾细胞系

该文发表于《海洋渔业》【2007，（2）：148-152】

草鱼肝微粒体的提取及 CYP 酶活性的测定

Cytochrome P450 contents and activity in hepaticmicrosome of grass carp（*Ctenopharyngodon idellus*）

张宁[1]，喻文娟[1]，王翔凌[1]，林茂[2]，邱军强[1]，杨先乐[1,3]

（1. 上海水产大学，农业部渔业动植物病原库，上海 200090；2. 集美大学水产学院，厦门 361021；
3. 上海高校水产养殖学 E-研究院，上海 200090）

摘要 CYP 酶代谢是药物生物转化的主要途径，其数量和活性大小直接影响药物在体内的活化与代谢。我们对草鱼肝微粒体 CYP 酶含量及其活性进行了初步研究，以差速离心法提取草鱼肝微粒体，以 CO 还原差示光谱法测得 CYP 酶及细胞色素 b5 含量分别为（0.619±0.102）nmol/mg、（0.264±0.042）nmol/mg。以 7-乙氧异吩噁唑酮-O-脱乙基反应、苯胺-4-羟化反应、氨基比林-N-脱甲基反应作为 CYP1A、CYP2E、CYP3A 的探针反应，测得 EROD 酶活为（0.043±0.004）nmol/mg·min，ANH 酶活为（0.028±0.002）nmolm/g·min，AMND 酶活为（0.207±0.035）nmol/mg·min。结果表明草鱼肝微粒体中 CYP 酶发育完好，并且具有参与药物代谢的 3 种主要亚型活性，其含量与活性大小与其他实验动物相差较大。本实验的方法与结果为草鱼 CYP 酶的系统研究提供可靠手段，最终为指导水产合理用药提供理论依据。

关键词 草鱼；肝微粒体；CYP 酶

3. P450 酶的诱导或抑制

该文发表于《水生生物学报》【2008，（4）：469-474】

草鱼肝细胞中 CYP2E1 活性的诱导研究

Induction of CYP2E1 Activity in Ctenopharyngodon Idellus Hepatocyte

王翔凌[1]，杨先乐[1,2]，张宁[1]，喻文娟[1]，胡鲲[1]

（1. 上海水产大学农业部渔业动植物病原库，上海 200090；2. 上海高校水产养殖学 E-研究院，上海 200090）

摘要 CYP2E1 为代谢大部分药物及环境中毒物的关键酶。以草鱼（*Ctenopharyngodon idellus*）肝细胞为反应体系，选取氯唑沙宗（CZX）为底物，采用反相高效液相色谱（RP-HPLC）法测定其产物 6-OH-氯唑沙宗（HCZX）的量，Lowry 法测定肝细胞中蛋白的浓度从而反映 CYP2E1 活性，并采用该酶特异性诱导剂乙醇对其进行诱导，观察其酶活变化及 CZX 在细胞中代谢情况。结果表明，CZX 在草鱼肝细胞中的基础代谢较低，经过对 CYP2E1 诱导条件的优化及筛选，得到最佳诱导剂剂量为 4 μg/mL、诱导时间为 24 h、底物浓度为 50 μg/mL 并且孵育时间为 1 h 时，其酶活达到最高，约为 0.47 μg/min·mg。对照组和诱导组的草鱼肝细胞中 CZX 的消除半衰期（t1/2）分别为 202.10 h 和 28.75 h，差异极显著，表明乙醇诱导的 CYP2E1 能够加快底物的代谢。酶促反应动力学参数表明乙醇诱导的 CYP2E1 与底物的亲和力较高，酶促反应强度较大。该结果能够为 CYP2E1 代谢的药物及环境毒物的研究提供理论依据。

关键词 反相高效液相色谱法；草鱼肝细胞；CYP2E1；氯唑沙宗；诱导

该文发表于《Fish Physiol Biochem》【2010，(29)：202-208】

Effects of fish CYP inducers on difloxacin N-demethylation in kidney cell of Chinese idle (*Ctenopharyngodon idellus*)

Lingzhi Yu, Xian Le Yang, Xiang Ling Wang, Wen Juan Yu, Kun Hu

Aquatic Pathogen Collection Centre of Ministry of Agriculture, Shanghai Ocean University, 999 Hucheng Ring Road, Shanghai 201306, China

Abstract：A drug-drug interaction occurs when the effect of one drug is altered by the presence of another drug which is generally associated with the induction of cytochrome P450s (CYPs) activity. Thus, unexpected treatment failures often happen resulting from inappropriate coadministration in fisheries. However, little information is available about CYP induction in fish. The reaction of difloxacin (DIF) biotransformation to sarafloxacin (SAR) belongs to N-demethylation catalyzed mainly by CYP (s). In order to supply useful information on CYP induction, the present study assessed the effects of fish-specific CYP inducers on DIF N-demethylation and enzyme kinetics in kidney cell of Chinese idle (CIK; grass carp (*Ctenopharyngodon idellus*)) by RP-HPLC. Results demonstrated that the amounts of SAR formation and enzymatic parameters Clint and Vmax were significantly increased due to b-naphthoflavone (BNF) pretreatment. Therefore, we suggest that CYP1A may be involved in DIF N-demethylation in CIK. This study provides instructive information to ensure treatment success via avoiding CYP induction in fisheries.

key words：CYP；Difloxacin；Grass carp；Kinetics；RP-HPLC；Sarafloxacin

鱼类 CYP 诱导物对草鱼肾细胞中双氟沙星去甲基化的影响

于灵芝，杨先乐，王翔凌，喻文娟，胡鲲

（上海海洋大学国家水生动物病原库，中国上海，沪城环路 999 号）

摘要 当联合用药时，经常会发生由于细胞色素 P450s 参与药物代谢而导致药效改变。因此，不恰当地联合用药物经常导致治疗失败。然而，对于 CYP 了解仍然很少。双氟沙星的反应（即双氟沙星在生物体内转换成沙氟沙星）主要由 CYP (s) 引导的 N-去甲基化催化反应。为了提供 CYP 感应的有用信息，研究利用反相液相色谱法评估草鱼肾细胞的 CYP 诱导物在双氟沙星中 N-去甲基化和酶动力学。实验结果证明，经过萘黄酮的预处理，沙氟沙星形成和酶克林特和最大值参数量显著增加了。因此，本实验得出 CYP1A 可能草鱼肾细胞双氟沙星的 N-脱甲基反应。此研究为使用药物时避免由于鱼类 CYP 反应而导致的药效下降提供理论依据。

关键词 细胞色素；双氟沙星；草鱼代谢；反相高效液相色谱；沙拉沙星

Abbreviations

BNF	b-Naphthoflavone
CIK	Kidney cell of Chinese idle
CYPs	Cytochrome P450s
DIF	Difloxacin
EA	Ethyl alcohol
GCL	Grass carp liver cell line
RIF	Rifampicin
RP-HPLC	Reversed-phase high-performance liquid chromatography
SAR	Sarafloxacin

Introduction

Various antimicrobial drugs are used extensively to control fish diseases in fish farming. Fluorophenyldifloxacin is a new aryl-fluoroquinolone which is used only for animals (Ding et al. 2006). Pharmacokinetics of difloxacin (DIF) and its main metabolite sarafloxacin (SAR) as effective antibacterial agents have been widely investigated in in vivo fish, i. e., in crucian cap (Ding et al. 2006), penaeid shrimp (Park et al. 1994), gilthead sea bream (Tyrpenou et al. 2003), juvenile channel catfish (Gingerich et al. 1995), eels (Ho et al. 1999), and Atlantic salmon (Martinsen and Horsberg1995). However, in vitro models also have a strong predictive power to obtain early information about biotransformation pathways and to forecast drug-drug interactions at the metabolic level (Brandon et al. 2003). Besides some popular in vitro models such as liver microsomes and primary hepatocytes (Winzer et al. 2002; Henczova' et al. 2006), several established fish cell lines, including PLHC-1 (Celander et al. 1996), RTG-2 (Arau'jo et al. 2000), BB (Arau'jo et al. 2000), RTL-W1 (Babı'n et al. 2005), CIK (Tan et al. 2008), and GCL (Li et al. 2008; Lin et al. 2006, 2007) have successfully been used for acute toxicity assessment of a variety of environmental chemicals such as organotins, substituted phenols, and pharmaceuticals (Janos˘ek et al. 2006). The in vitro results showed a trend similar to the in vivo acute toxicity in fish (Fent 2001; Caminada et al. 2006). These cells are therefore a promising tool in the toxicity screening and in the enzyme kinetics evaluation of pharmaceuticals. The reasons are as follows. Firstly, they have metabolic activities which are necessary to study the metabolism of xenobiotics action via metabolic activation. Secondly, cells are easy to cultivate. Thirdly, they possess inducible and stable cytochrome P450s (CYPs) enzymes, which are important for metabolism and detoxification of xenobiotics (Fent 2001). These cell culture systems are ideally suited to assess the toxicity and enzyme kinetics of chemicals (Ackermann and Fent 1998; Fent 2001; Ackermann et al. 2002; Lin et al. 2006, 2007). It will be helpful in a reduction of living animal testing in aquatic pharmacology.

CYPs are one of the major phase I-type classes of detoxification enzymes found in terrestrial and aquatic organisms ranging from bacteria to vertebrates which metabolize a wide variety of substrates including endogenous molecules (e. g., fatty acids, eicosenoids, steroids) and xenobiotics (e. g., hydrocarbons, pesticides, drugs) (reviewed by Snyder 2000). A drug-drug interaction occurs when the effect of one drug is altered by the presence of another drug with involvement of CYP induction (Muto et al. 1997). Indeed, induction of xenobiotics metabolism by CYPs in fish is well established as a useful tool for environmental contamination biomonitoring (Monod et al. 1994; Förlin and Celander 1995; Fent et al. 1998; Malmström et al. 2004) and in ecotoxicology assays (Seubert and Kennedy 1997; Fent2001; Gravato and Santos 2002; Billiard et al. 2004), which provide instructive information to predict the possible drug-drug interactions so as to ensure the suitable use of fisheries drugs.

The function of CYPs can also be inhibited or induced causing alterations in their ability to catalyse a reaction (Khan et al. 1997), which means induction and inhibition may contribute to less effective drug action when two or more drugs are concurrently administrated. CYP1A, CYP3A, and CYP2E are three important CYP enzymes responsible for drug metabolism in fish. b-naphthoflavone (BNF) is a specific inducer for CYP1A in fish (Celander and Förlin1991; Gooneratne et al. 1997; Ronisz and Förlin1998; Chung-Davidson et al. 2004; Kim et al. 2008). Results have shown that ethyl alcohol (EA) is the specific inducer of CYP2E1 in grass carp (Wang et al. 2008), while rifampicin (RIF) is an effective inducer for CYP3A activity in grass carp (Li et al. 2008). Nevertheless, the effects of fish CYP inducers on quinolones have not been previously reported in fish.

The N-demethylation reaction of DIF to SAR belongs to demethylation among the broad versatility of reactions catalyzed by CYPs, including carbon hydroxylation, heteroatom oxygenation, dealkylation, epoxidation, aromatic hydroxylation, reduction, and dehalogenation (Fig. 1; Sono et al. 1996; Werck Reichhart and Feyereisen 2000; Bernhardt 2004; Guengerich 2004).

Grass carp (Ctenopharyngodon idellus), or Chinese idle, is a herbivorous species belonging to the family of Cyprinidaeofthe class Osteichthyes which is viewed as the national fish in China (Li et al. 2008). It is not only utilized as a food source but also used for biomonitoring of contaminated water owing to its widespread existence (Wan et al. 2002, 2004; Lin et al. 2006). CYP activity is also quite high in the kidney in addition to the liver in fish (Pesonen et al. 1987; Lorenzana et al. 1988). The enzymatic parameters Vmax and Clint represent an index of catalytic efficiency in producing the metabolites. The kinetic behaviors of CYPs in fish have attracted growing attention. However, there are only few results for CYP enzyme kinetic behavior for fisheries drugs in grass carp liver cell line (GCL) (Lin et al. 2007; Li et al. 2008). The current study was to evaluate the effects of fish CYP inducers on DIF N-vdemethylation by reversed-phase high-performance liquid chromatography (RP-HPLC) in kidney cell of Chinese idle (CIK). Such information will facilitate the establishment of in vitro models in fish which could predict the unexpected drug-drug interactions, thus contributing to the appropriate use of fisheries drugs.

60

1. Materials and methods

Chemicals

DIF-HCL (>99. 8%), RIF (>97%), 3- (4, 5-dimethylthiazol-2-yl) -2, 5-diphenyltetrazolium bromide (MTT), BNF, and dimethyl sulphoxide (DMSO, 99.9%) were obtained from Sigma-Aldrich (China) . SAR hydrochloride (>99. 6%) was purchased from the China Institute of Veterinary Drug Control. Lowry Protein Assay kit was obtained from Sunny (Shanghai, China) . M199, foetal bovine serum (FBS) were from Gibco BRL (UK) . Acetonitrile and methanol were of HPLC grade (Darmstadt, Germany) . All other chemicals and solvents were of analytical grade obtained from common commercial sources.

Treatments for reagents

The DIF-HCl and SAR hydrochloride were dissolved with help of sodium hydroxide. Both BNF and RIF were dissolved in DM-SO. DIF, BNF, and RIF were added to medium after solution. EA was added in medium directly.

Cell culture

C. idellus kidney cell line (No. BYK-C06-02) was obtained from the Aquatic Pathogen Collection Centre of Ministry of Agriculture (APCCMA, Shanghai, China) . Cells were grown in M199 medium supplemented with 10% (v/v) FBS at 28℃ in 5. 0% CO_2. After 48 h for full adherence, the cells were subcultured into 96 and six-well plates with fresh M199 medium at a density of 5×10^5 cells mL^{-1} for following MTT test and induction treatments. All water used in the experiment was thrice distilled.

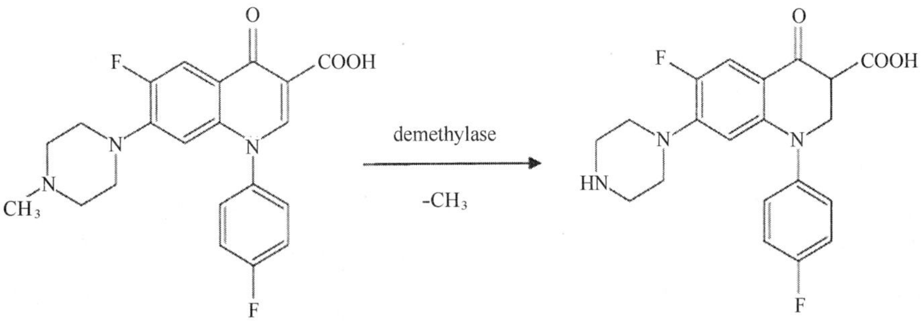

Fig. 1　The chemical structures of DIF and its main metabolite SAR

MTT assay

The viability of CIK treated and untreated DIF was studied using a slightly modified MTT assay (Mosmann1983) . After full adherence, the media in 96 well plates were replaced by fresh serum-free media containing DIF concentrations ranging from 3. 9 to 1, 000 l μM, and the cells at each concentration were incubated for 1, 2, 6, 12, 24, and 48 h, respectively (final volume, 200 μL) . Then, 20 μL MTT solution (5 mg mL^{-1} in PBS) were added to each well. After 4 h of incubation, the supernatant containing the unreacted dye was replaced with DMSO (150 μL $well^{-1}$) , then plates were vigorously shaken for 10 min, and absorbances at 490 nm were measured by BIO-TEK EL311 within 1 h. Within each experiment, determinations were performed in five replications. The percentages of cell survival were calculated from the absorbance values as follows (O. D. tested -O. D. blank) / (O. D. untreated control - O. D. blank) 9 100, with O. D. blank referring to the absorbance of wells that contained only medium and MTT.

Treatments for induction in CIK

After full adherence, the cells in six-well plates were then treated with serum-free media containing BNF, EA, and RIF at their respective referred inductive concentrations. Control cultures were treated with a serum-free medium. The six-well plates were placed in a 28℃ incubator for 24 h induction, at which time they were replaced by a serum-free medium containing DIF at the highest nonlethal concentrations or vehicle alone. The six-well plates were then incubated for another 2 h before analysis, at which time the media and cells were harvested from each sample group used for enzyme assay. Each cell suspension was divided into two parts. One part was used for protein content measurement, and the other part was treated for RP-HPLC. The measurements for the induction assays had three respective batches.

Equipment and chromatographic conditions

The Agilent 1100 HPLC system (USA) consisted of a quaternary pump (Germany), vacuum degasser (Japan), and auto-injector (Germany) . DIF and metabolite SAR were extracted by addition of 6 vols of dichloromethane, mixed vigorously for at least 5 min, and centrifuged for 10 min at 8, 000 g at 4℃ to separate the two phases. The aqueous phase was discarded by careful aspiration and the organic phase was transferred to clean tubes and evaporated under a nitrogen stream at 45℃. The residue was dissolved in 1. 0

mL mobile phase and 10 μL of the solution was injected onto a reverse phase analytical column, ZORBAX SB-C18 (5μm, 150 mm 9 4.6 mm id; Agilent Technologies, USA). The mobile phase consisted of 5% acetonitrile and 95% tetrabutyl ammonium bromide. The flow rate was 1.5 mL min^{-1} and the column eluate was monitored at 276 nm. Column temperature was 40℃.

Enzymatic analysis

DIF-demethylase activities in cell samples were determined by measuring enzymatic reaction rate V that is, the formation of SAR per minute per mg (μM min^{-1}mg^{-1}) according to the RP-HPLC method (Lin et al. 2007; Wang et al. 2007). To determine the enzyme kinetics for DIF N-demethylation metabolism, the enzyme assay was conducted with DIF concentrations at 3.9, 7.8, 15.6, 31.25, 62.5, and 125 μM. Apparent enzyme kinetic parameters (i.e., half-maximal substrate concentration, Km; maximal velocity, Vmax) were calculated from Lineweaver-Burk inverse plots [1/V against 1/ (S)]. Each cellular preparation for the protein concentration was provided by the manufacturer.

Data analysis and statistics

Statistical analyses were performed with SPSS software version 16.0 (SPSS, Germany) and the data were used as mean ± standard deviation (SD). Significant differences between means were determined using one-way analyses of variance (ANOVAs) and the Duncan's test for multiple range comparison with significance level established at P \ 0.01.

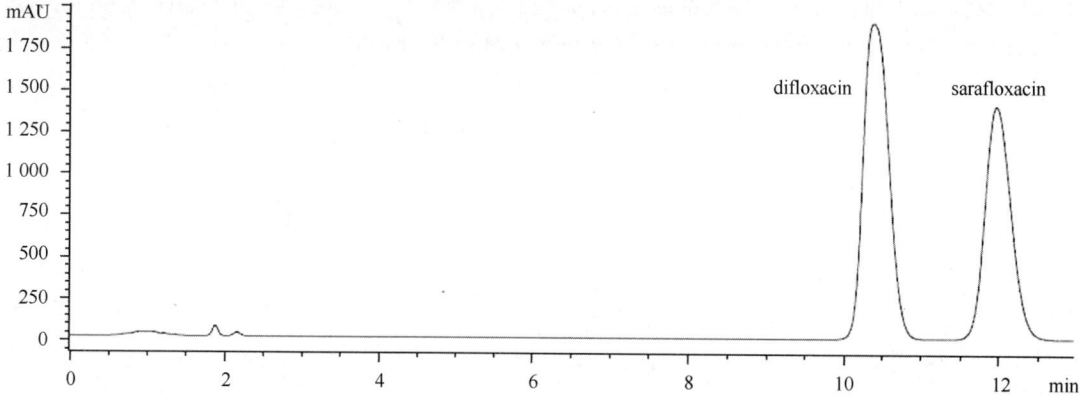

Fig. 2   Representative HPLC profile of CIK spiked with DIF and its metabolite SAR. Retention times of DIF and SAR were 10.37 min and 11.96 min, respectively

Table 1   Recoveries of SAR in CIK

Spiked amount (μM)	Measured amount (μM)	Recovery intra-day (%)	Recovery inter-day (%)	Relative standard deviation (%)	
				Intra-day	Inter-day
1	0.82 ±0.04	82.04 ±4.13	81.67 ±1.53	0.40	1.53
10	9.41 ±0.41	94.11 ±4.14	84.53 ±0.74	0.71	1.27
100	99.55 ±1.00	99.55 ±0.28	87.31 ±0.28	1.48	0.49

The values were the average of three batches of cells treated with SAR at concentrations of 1, 10 and 100μM (n=3, mean ±SD)

Results

Validation of SAR assay method

Difloxacin and its metabolite SAR were separated at baseline by using the chromatographic condition developed. There were no interfering peaks found at the same retention time of DIF and SAR in the chromatogram of blank cellular incubation (Fig. 2). Retention times of DIF and SAR were 10.37 min and 11.96 min, respectively. The SAR calibration curve was constructed by analyzing the cellular incubation spiked with a series concentrations of SAR. The results exhibited excellent linearity with correlations coefficient 0.9998 at SAR concentrations ranging from 5 to 100 μM (samples at each concentration n=3). The regression equation of the calibration curve based on SAR corresponding concentrations against its peak-areas spiked in the blank cellular incubation was y = 5.4826x + 0.7383. The limits of detection and of quantification for SAR were 1 and 0.5 μM (S/N =3). The average recovery was 83.27±0.89%. The average relative standard deviation was less than 1.6%. The assay method for SAR was reproducible and accurate (Table 1).

Nonlethal condition for CIK following DIF treatment

MTT reduction was used to optimize the nonlethal concentrations and time for induction test. The relative viability of CIK sharply decreased when DIF concentrations were set at 250 μM with incubation time for 12 h (Fig. 3). Therefore, incubation concentration at 125 μM and time for 6 h were used as the highest nonlethal condition to screen the optimal one for DIF N-demethylase induction studies.

DIF metabolic condition for induction in CIK

In order to optimize the concentration and time for DIF N-demethylation studies, different DIF concentrations ranging from 3. 9 to 125 μm were incubated with CIK for increasing time from 0. 25 to 6 h. SAR exhibited the largest amount under the condition of DIF concentration at 62. 5 μm and incubation time for 2 h by the RP-HPLC method (Fig. 4). Therefore, the condition described above was chosen for further induction analyses.

SAR formation in CIK pretreated with CYP inducers

CIK pretreated with BNF (10 μm) (Bhakta et al. 2008), RIF (40 μm) (Li et al. 2008) and EA (100 μm) (Wang et al. 2008) for 24 h were then incubated separately with 62. 5 μm DIF for 2 h to analyze the effects of fish CYP inducers on DIF demethylation. The concentrations in parentheses and the incubation time were optimized by our previous work according to the references with some modifications. The data are not shown. BNF led to an almost 3-fold increase in SAR formation level in the BNF-treated CIK incubations compared to control groups, while RIF-and EA-treated samples did not exhibit obvious differences (Fig. 5).

Kinetic analysis for DIF metabolism

The cells treated with BNF, RIF, and EA were selected to determine the kinetic parameters of DIF N-demethylation. The protein concentration is 0. 35±0. 02 mg mL^{-1} cell. The Km, Vmax and Clint values are summarized in Table 2. Addition of BNF to CIK resulted in an almost 2-fold increase in the Vmax value and a nearly 6-fold increase in the Clint value. While the Clint and Vmax values for EA and RIF pretreatments varied little compared to controls. In order to further assess the effects of BNF on DIF demethylation in the CIK, the kinetics in the presence and absence of BNF were characterized by studying the actions of demethylation at various DIF concentrations. Kinetic analysis using DIF concentrations from 3. 9 to 62. 5 μm showed that the enzyme action exhibited classical Michaelis-Menten kinetics (Fig. 6).

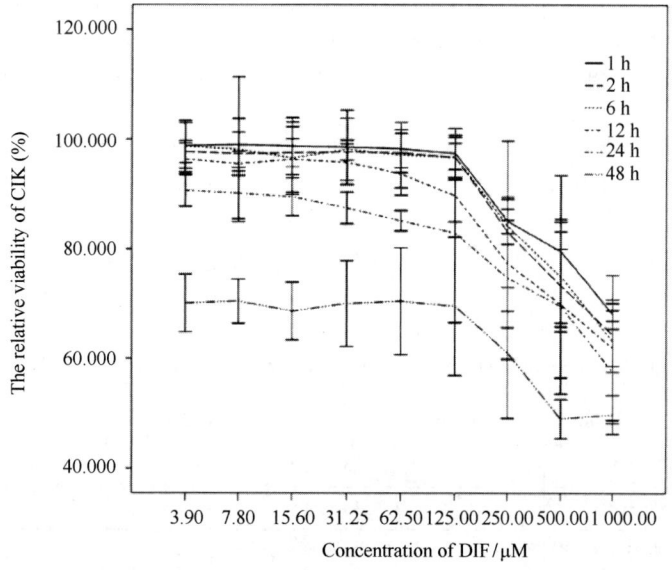

Fig. 3　The highest nonlethal condition for CIK following DIF treatment detected by MTT. DIF were incubated with CIK in 96 well plates at concentrations from 3. 9 to 1, 000 μM for different incubation times from 1 to 48 h at 28℃. The percentages of cell survival were calculated by formula (O. D. $_{tested}$ −O. D. $_{blank}$) / (O. D. $_{untreated\ control}$ −O. D. $_{blank}$) ×100. Data were the average values of in quintuplicate

Discussion

CYPs are responsible for the metabolism of an extensive range of fisheries drugs and can be strongly induced by a variety of structurally unrelated xenobiotics (Li et al. 2008). Because of this, drug-drug interactions are often associated with induction of these enzymes. The kinetic behavior and regulation of CYP enzymes in fish cell lines have been paid more attention (Li et al. 2008). GCL has been employed to research CYP kinetics for fisheries drugs (Lin et al. 2007; Li et al. 2008). In the present study, CIK appears to be a suitable in vitro model for evaluating effects of fish CYP inducers on DIF N-demethylation and for further studies of its potential

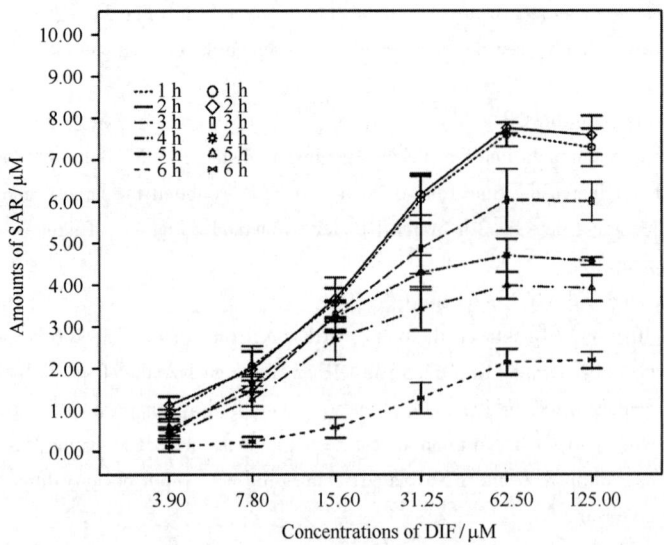

Fig. 4　Amounts of SAR formation from DIF in CIK. DIF were incubated with CIK at different concentrations
from 3. 9 to 125 μM for different times from 0. 25 to 6 h at 28℃. The RP-HPLC method was employed to test
SAR amounts. Each value represented the average of in triplicate incubations

for metabolic interactions with other fisheries drugs.

MTT reduction measures succinate dehydrogenase activity in mitochondria and was used to survey cell viability of CIK in the present study. MTT results suggested that incubation concentration at 125 μm and time for 6 h were the final nonlethal condition for DIF metabolic studies. However, SAR amounts determined by RP-HPLC method suggested that CIK occupy metabolic repression under this condition and CIK had the largest metabolic level for DIF at 62. 5 μm for 2 h incubation which were then chosen for further induction analyses.

The Km, Vmax, and Clint values in Table 2 revealed that DIF demethylase in CIK pretreated by RIF had a lower affinity with the substrate and slightly stronger potency to catalyze demethylation than the control. The amounts of SAR formation which indicated DIF Ndemethylation level suggested that RIF did not show inductive activity for DIF N-demethylation (Fig. 5). RIF is an effective inducer for CYP3 Aingrasscarp, thus CYP3A may not participate in DIF N-demethylation in CIK. Recent results have shown that CYP3A may be responsible for deethylation metabolism of enrofloxacin in GCL cells (Lin et al. 2007). These findings are consistent with the previous results that CYP3A inductive activity varies in different issues, being high in liver but non-existent in kidney of tilapia (Bainy et al. 1999). Therefore, it is most likely that GCL but not CIK is a suitable tool for assessing CYP3A induction (Li et al. 2008).

Table 2　Enzymatic parameters of DIF metabolism in CIK

Treatments	Km （μM）	Vmax （nmol min^{-1}mg^{-1}protein）	Clint （L min^{-1}g^{-1}protein）[a]
Control	45. 83 ±5. 03	333. 33 ±27. 91	7. 27 ±0. 69
RIF	43. 98 ±3. 79	357. 14 ±25. 88	8. 14 ±0. 73
EA	37. 23 ±4. 13	384. 62 ±34. 57	10. 33 ±1. 02
BNF	19. 8 ±2. 96 * *	1, 000 ±118. 73 * *	50. 51 ±4. 75 * *

The values are expressed as mean ±SD （n=3）

[a] Clint = V_{max}/K_m

* * $P<0.01$, compared with control

As shown in Fig. 5, amounts of SAR formation were significantly higher in BNF-treated samples than controls. Enzyme kinetic parameters showed higher substrate affinity with DIF N-demethylase and stronger potency for the demethylation in CIK pretreated by BNF than control (Table 2). BNF is the specific inducer of CYP1A in fish, thus CYP1A may be responsible for the reaction in CIK (Fig. 6). To determine CYP (s) responsible for N-demethylation in CIK, amounts of SAR need to be determined in the presence of specific inhibitors of the corresponding CYP (s). In addition to catalytical assays, immunochemical and molecular biology techniques

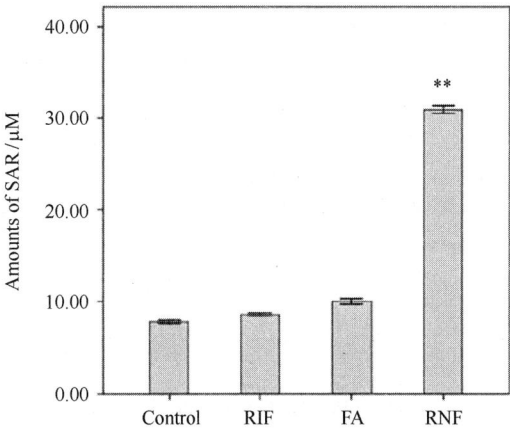

Fig. 5　Amounts of SAR formation from DIF in CIK pretreated with fish CYP inducers. DIF was incubated with CIK pretreated with RIF, EA and BNF at DIF concentration of 62. 5 μM for 2 h at 28℃. The controls were treated only by incubation with 62. 5 μM DIF for 2 h at 28℃. The RP−HPLC method was used to detect the amounts of SAR formation. Data were the means of in triplicate batches. Asterisks indicate significant difference from the control （$P < 0.01$）

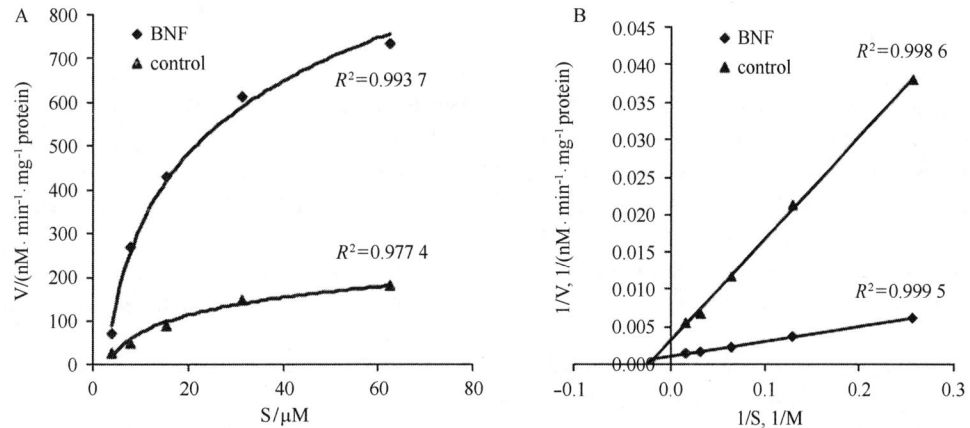

Fig. 6　Michaelis−Menten kinetics （a） and the Lineweaver−Burk plots （b） for induction of DIF N−demethylation by BNF in CIK. DIF were incubated with CIK for 2 h at 28℃ pretreated with or without 10 μM BNF before RP−HPLC. Each value represents the average of in triplicate incubations. Regressions for significant correlations （R^2） are shown.

will also be requisite.

Many, but not all, quinolones can have inhibitory effects on the CYP enzyme system, which will lead to reduced metabolism and clearance of certain other drugs （Davies and Maesen1989; Harder et al. 1989; reviewed by Brouwers1992; Xu et al. 2008）. It has been shown that two quinolone antibacterial agents, oxolinic acid and flumequine, were weak dose−independent inhibitors of CYP1A subfamily in rainbow trout microsomes （Moutou et al. 1998）. Recent results have shown that enrofloxacin markedly inhibited CYP2E1 activity in in vivo and in vitro grass carp （Wang et al. 2008）. Moreover, selected drugs can act simultaneously as CYP substrates, inducers, and inhibitors, and in some cases can regulate their own metabolism （Plant et al. 2000）. Thus, interactions between quinolones and CYP （s） need further investigation.

Conclusion

In brief, induction in fish is crucial in providing instructive information on the metabolism of drugs in aquatic vertebrates as well as higher animals, and also in predicting the possible drug−drug interactions so as to ensure therapeutic efficacy. This study shows that BNF obviously increases DIF N−demethylation to SAR and that CYP1A may be involved in the former reaction in CIK. Furthermore, these results provide instructive information for assessment of drug−drug interactions associated with CYP induction which can contribute to avoiding treatment failure in fisheries comedication.

References

1. Ackermann GE, Fent K (1998) The adaptation of the permanent fish cell lines PLHC-1 and RTG-2 to FCS-free media results in similar growth rates compared to FCS-containing conditions. Mar Environ Res 46: 363-367.

2. Ackermann GE, Schwaiger J, Negele RD, Fent K (2002) Effects of long-term nonylphenol exposure on gonadal development and biomarkers of estrogen exposure in juvenile rainbow trout (*Oncorhynchus mykiss*). Aquat Toxicol 60: 203-221.

3. Arau'jo CSA, Marques SAF, Carrondo MJT, Goncalves LMD (2000) In vitro response of the brown bullhead catfish (BB) and rainbow trout (RTG-2) cell lines to benzo [a] pyrene. Sci Total Environ 247: 127-135.

4. Babı'n M, Casado S, Chana A, Herrado'n B et al (2005) Cytochrome P4501A induction caused by the imidazole derivative Prochloraz in a rainbow trout cell line. ToxicolIn Vitro 19: 899-902.

5. Bainy ACD, Woodin BR, Stegeman JJ (1999) Elevated levels of multiple cytochrome P450 forms in tilapia from Billings Reservoir—Sao Paulo, Brazil. Aquat Toxicol 44: 289-305.

6. Bernhardt R (2004) Optimized chimeragenesis: creating diverse P450 functions. ChemBiol 11: 287-288.

7. Bhakta KY, Jiang W, Couroucli XI et al (2008) Regulation of cytochrome P4501A1 expression by hyperoxia in human lung cell lines: Implications for hyperoxic lung injury. Toxicol Appl Pharmacol 233: 169-178.

8. Billiard SM, Bols NC, Hodson PV (2004) In vitro and in vivo comparisons of fish-specific CYP1A induction relative potency factors for selected polycyclic aromatic hydrocarbons. Ecotoxicol Environ Saf 59: 292-299.

9. Brandon EFA, Raap CD, Meijerman I, Beijnen JH, Schellens JHM (2003) An update on in vitro test methods in human hepatic drug biotransformation research: pros and cons. Toxicol Appl Pharmacol 189: 233-246.

10. Brouwers JR (1992) Drug interactions with quinolone antibacterials. Drug Saf 7: 268-281.

11. Caminada D, Escher C, Fent K et al (2006) Cytotoxicity of pharmaceuticals found in aquatic systems: comparison of PLHC-1 and RTG-2 fish cell lines. AquatToxicol 79: 114-123.

12. Celander M, Förlin L (1991) Catalytic activity and immunochemical quantification of hepatic cytochrome P-450 in b-naphthoflavone and isosafrol treated rainbow trout (Oncorhynchus mykiss). Fish PhysiolBiochem 9: 189-197 Celander M, Hahn ME, Stegeman JJ (1996) Cytochromes P450 (CYP) in the Poeciliopsislucida hepatocellular carcinoma cell line (PLHC-1): dose and time-dependent glucocorticoid potentiation of CYP1A induction without induction of CYP3A. Arch Biochem Biophys 329: 113-122.

13. Chung-Davidson YW, Rees CB, Wu H, Yun SS, Li W (2004) beta-naphthoflavone induction of CYP1A in brain of juvenile lake trout (*Salvelinus namaycush* Walbaum). J Exp Biol 207: 1533-1542.

14. Davies BI, Maesen FP (1989) Drug interactions with quinolones. Rev Infect Dis 11: S1083-S1090.
Ding FK, Cao JY, Ma LB et al (2006) Pharmacokinetics and tissue residues of difloxacin in crucian cap (*Carassius auratus*) after oral administration. Aquaculture 256: 121-128.

15. Fent K (2001) Fish cell lines as versatile tools in ecotoxicology: asssessment of cytotoxicity, cytochrome P450 1A induction potential and estrogenic activity of chemicals and environmental samples. Toxicol In Vitro 15: 477-488.

16. Fent K, Woodin BR, Stegeman JJ (1998) Effects of triphenyltin and other organotins on hepatic monooxygenase system in fish. Comp BiochemPhysiol C 121: 277-288.

17. Förlin L, Celander M (1995) Studies of the inducibility of P450 1A in perch from the PCB-contaminated lake Järnsjön in Sweden. Mar Environ Res 39: 85-88.

18. Gingerich WH, Meinertz JR, Dawson VK et al (1995) Distribution and elimination of 14C sarafloxacin hydrochloride from tissues of juvenile channel catfish (*Ictalurus punctatus*). Aquaculture 131: 23-36.

19. Gooneratne R, Miranda CL, Henderson MC, Buhler DR (1997) Beta-naphthoflavone induced CYP1A1 and 1A3 proteins in the liver of rainbow trout (*Oncorhynchus mykiss*). Xenobiotica 27: 175-187.

20. Gravato C, Santos MA (2002) b-Naphthoflavone liver EROD and erythrocytic nuclear abnormality induction in juvenile (*Dicentrarchus labrax L.*). Ecotoxicol Environ Saf 52: 69-74.

21. Guengerich FP (2004) Cytochrome P450: what have we learned and what are the future issues? Drug Metab Rev 36: 159-197.

22. Harder S, Fuhr U, Staib AH, Wolff T (1989) Ciprofloxacincaffeine: a drug interaction established using in vivo and in vitro investigations. Am J Med 87: 89S-91S.

23. Henczova' M, Dee'r AK, Komlo'si V, Mink J (2006) Detection of toxic effects of Cd2? on different fish species via liver cytochrome P450-dependent monooxygenase activities and FTIR spectroscopy. Anal BioanalChem 385: 652-659.

24. Ho SH, Cheng CF, Wang WS (1999) Pharmacokinetics and depletion studies of sarafloxacin after oral administration to eel (*Anguilla anguilla*). J Vet Sci 61: 459-463.

25. Janos˘ek J, Hilscherova' K, Bla'ha L, Holoubek I (2006) Environmental xenobiotics and nuclear receptors-interactions, effects and in vitro assessment. ToxicolIn Vitro 20: 18-37.

26. Khan MN, Reddy PK, Renaud RL, Leatherland JF (1997) Effect of cortisol on the metabolism of 17-hydroxyprogesterone by Arctic charr and rainbow trout embryos. Fish PhysiolBiochem 16: 197-209.

66

27. Kim JH, Raisuddin S, Ki JS, Han KN （2008） Molecular cloning and beta－naphthoflavone－induced expression of a cytochrome P450 1A （CYP1A） gene from an anadromous river pufferfish. Takifugu obscurus. Mar Pollut Bull57：433－440.

28. Li D, Yang XL, Zhang SJ et al （2008） Effects of mammalian CYP3A inducers on CYP3A－related enzyme activities in grass carp （*Ctenopharyngodonidellus*）：Possible implications for the establishment of a fish CYP3A induction model. Comp BiochemPhysiol C 147：17－29.

29. Lin M, Yang XL, Fang WH et al （2006） Dose－response of inducers to effect on EROD in grass carp hepatocyte. J Fish China 30：311－315.

30. Lin M, Yang XL, Wang XL et al （2007） Enzymatic on deethlation metabolism of enrofloxacin in grass carp hepatocyte. High Technol Lett （Chinese version） 17：314－318.

31. Lorenzana RM, Hedstrom OR, Buhler DR （1988） Localization of cytochrome P-450 in the head and trunk kidney of rainbow trout （*Salmo gairdneri*）. Toxicol Appl Pharmacol 96：159－167.

32. Malmström CM, Koponen K, Lindström－Seppä P et al （2004） Induction and localization of hepatic CYP4501A in flounder and rainbow trout exposed to benzo ［a］ pyrene. Ecotoxicol Environ Saf 58：365－372.

33. Martinsen B, Horsberg TE （1995） Comparative single－dose pharmacokinetics of four quinolones, oxolinic acid, flumequine, sarafloxacin and enrofloxacin in Atlantic salmon （*Salmo salar*） held in seawater at 10C. Antimicrob Agents Chemother 39：1059－1064.

34. Monod G, Saucier D, Perdu－Durand E et al （1994） Biotransformation enzyme activities in the olfactory organ of rainbow trout （*Oncorhynchus mykiss*）. Immunocytochemical localization of cytochrome P4501A1 and its induction by b－naphthoflavone. Fish Physiol *Biochem* 13：433－444.

35. Mosmann T （1983） Rapid colorimetric assay for cellular growth and survival：application to proliferation and cytotoxicity assays. J Immunol Methods 65：55－63.

36. Moutou KA, Burke MD, Houlihan DF （1998） Hepatic P450 monooxygenase response in rainbow trout （*Oncorhynchus mykiss* （Walbaum） ） administered aquaculture antibiotics. Fish Physiol Biochem 18：97－106.

37. Muto N, Hirai H, Tanaka T, Itoh N, Tanaka K （1997） Induction and inhibition of cytochrome P450 isoforms by imazalil, a food contaminant, in mouse small intestine and liver. Xenobiotica 27：1215－1223.

38. Park ED, Lightner DV, Stamm JM, Bell TA （1994） Preliminary studies on the palatability, animal safety, and tissue residues of sarafloxacin－HCl in the penaeid shrimp, Penaeusvannamei. Aquaculture 126：231－241.

39. Pesonen M, Celander M, Förlin L, Andersson T （1987） Comparison of xenobiotic biotransformation enzymes in kidney and liver of rainbow trout （*Salmo gairdneri*）. Toxicol Appl Pharmacol 91：75－84.

40. Plant NJ, Ogg MS, Crowder M, Gibson GG （2000） Control and statistical analysis of in vitro reporter gene assays. Anal Biochem 278：170－174.

41. Ronisz D, Förlin L （1998） Interaction of isosafrole, betanaphthoflavone and other CYP1A inducers in liver of rainbow trout （*Oncorhynchus mykiss*） and eelpout （Zoarcesviviparus）. Comp Biochem Physiol C Pharmacol Toxicol Endocrinol 121：289－296.

42. Seubert JM, Kennedy CJ （1997） Thetoxicokinetics of benzo ［a］ pyrene in juvenile coho salmon, Oncorhynchuskisutch, during smoltification. Fish PhysiolBiochem16：437－447.

43. Snyder MJ （2000） Cytochrome P450 enzymes in aquatic invertebrates：recent advances and future directions. AquatToxicol 48：529－547.
    Sono M, Roach MP, Coulter ED, Dawson JH （1996） Hemecontainingoxygenases. Chem Rev 96：2841－2888.

44. Tan FX, Wang M, Wang WM, Lu YA （2008） Comparative evaluation of the cytotoxicity sensitivity of six fish cell lines to four heavy metals in vitro. ToxicolIn Vitro 22：164－170.

45. Tyrpenou AE, Iossifidou EG, Psomas IE et al （2003） Tissue distribution and depletion of sarafloxacin hydrochloride after in feed administration in gilthead seabream （Sparusaurata L.）. Aquaculture 215：291－300.

46. Wan XQ, Wu WZ, He JZ （2002） Evaluating toxic effects of PCDD/F using grass carp primary hepatocyte culture. China Environ Sci 22：114－117.

47. Wan XQ, Ma TW, Wu WZ, Wang ZJ （2004） EROD activities in a primary cell culture of grass carp （*Ctenopharyngodon idellus*） hepatocytes exposed to polychlorinated aromatic hydrocarbonas. Ecotoxicol Environ Saf 58：84－89.

48. Wang XL, Yang XL, Lin M, Yu WJ et al （2007） Elimination of enrofloxacin and detection of metabolic enzyme activity in Micropterussalmoides hepatocytes. J Fish Sci China 14：1004－1005.

49. Wang XL, Yang XL, Zhang N et al （2008） Induction of CYP2E1 activity in Ctenopharyngodonidellus hepatocyte. Acta Hydrobiol Sin 32：469－474.
    Werck－Reichhart D, Feyereisen R （2000） Cytochromes P450：a success story. Genome Biol 1：reviews 3003. 1－3003. 9.

50. Winzer K, Noorden CJFV, Köhler A （2002） Sex－specific biotransformation and detoxification after xenobiotic exposure of primary cultured hepatocytes of European flounder （Platichthysflesus L.）. Aquat Toxicol 59：17－33.

51. Xu X, Liu HY, Liu L, Xie L, Liu XD （2008） The influence of a newly developed quinolone：antofloxacin, on CYP activity in rats. Eur J Drug Metab Pharmacokinet 33：1－7.

该文发表于《中国水产科学》【2007，14（5）：856-859】

乙醇对 CHSE-1 细胞中细胞色素 P450 2E1 的诱导研究
Induced effect of ethanol on cytochrome P450 2E1 in CHSE-1 cell line

张宁[1]，杨先乐[1,2]，王翔凌[1]，林茂[3]，喻文娟[1]，张书俊[1]

(1. 上海水产大学农业部渔业动植物病原库，上海 200090；2. 上海高校水产养殖学 E-研究院，上海 200090；
3. 集美大学水产学院，福建 厦门 361021)

摘要 以乙醇为细胞色素 P450 2E1 的诱导剂，在 CHSE-1 细胞中研究了该诱导剂与酶活性之间的时间效应关系和剂量效应关系。细胞传代后培养 48 h，加入含有不同浓度乙醇的新鲜培养基，孵育诱导 24 h，之后加入底物苯胺反应 30 min。以苯胺的代谢产物 4-氨基酚的生成量来反映 CYP2E1 活性。苯胺浓度为 10～16 mmol/L 时 CYP2E1 活性最大，可达（0.152±0.095）nmol·min^{-1}·mg^{-1}，该浓度范围的苯胺适合用于指示 CYP2E1 的活性。以 Gentox 模型对诱导的剂量效应进行拟合，得到一个先升高后降低的曲线，说明乙醇对 CHSE-1 细胞中 CYP2E1 有先诱导后抑制作用。拟合方程 $F(x) = (p_0 + kx) / (1 + [x/EC_{50}] nH)$ 参数 p_0 为 0.140，nH 为 2.456，k 为 0.024，EC_{50} 为 34.938。CYP2E1 酶活性随着诱导时间的增加逐渐增强，24 h 内酶活性随时间的增加而增强，可达（0.446±0.092）nmol/（min·mg），呈典型的酶诱导现象，建立了乙醇对 CHSE-1 中 CYP2E1 的诱导模型。

关键词 细胞色素 P450 2E1；CHSE-1 细胞系；乙醇；诱导

细胞色素 P450 2E1（Cytochrome P450 2E1，CYP2E1）属于药物代谢酶 P450 家族，其代谢的药物占 P450 代谢的药物总量的 7%[1]，参与多种外来化学物的代谢活化，包括大量的强亲电子物和前致癌物，至少有 70 多种有机化合物（如乙醇、酮、二烷基亚硝胺、卤族溶剂等）在体内被 CYP2E1 代谢转化为有毒物质[2-4]。1998 年 Wall 等以氯唑沙宗为底物，首次报道硬骨鱼 winter flounder（*Pseudopleuronectes americanus*）肝脏中有 CYP 2E1 代谢活性酶，并发现 CYP 2E1 将前毒物代谢转化为毒物，造成肝损伤，鱼类的这种肝损伤与哺乳动物极为相似[5]，开创了以鱼肝脏中 P450 酶系的诱导作为评价环境污染状况的生物学反应的局面。这种生物学反应的实验材料包括鱼肝微粒体和细胞系。细胞系相比较肝微粒体，可不用加入 NADPH 辅酶系统，反应在整个细胞水平实现，可以更加准确地反映外源化合物的代谢、毒理和一般药理学特征[6-9]。而且细胞培养模型易控制外源化合物跟与细胞的接触浓度和接触时间，故成为药理毒理检测中的理想实验材料。但是细胞系在体外的培养过程中，细胞本身的酶活性会逐渐降低甚至消失，契诺克鲑鱼胚胎细胞系 CHSE-1（*Chinook Salinon embryo*）细胞形态呈上皮状，生长周期短，一般 1～2 d 长满单层，适合做毒理和药理方面检测用材料，但是在长期的传代培养中，细胞中蛋白酶活性会逐渐降低甚至消失，而酶的活性高低决定了外源性化合物在细胞内代谢为有毒物质的速度，与毒物对机体的最终毒性大小呈正相关[10]。以诱导剂诱导增加鱼类传代细胞中 CYP2E1 活性的研究国内外尚未见报道。本实验以苯胺为特异性底物，以苯胺羟化酶（Aniline hydroxylase，ANH）活性作为 CYP2E1 的活性指标，以乙醇为 CYP2E1 诱导剂，诱导增加 CHSE-1 细胞中 CYP2E1 活性，并以 Gentox 模型对诱导的剂量效应进行拟合，建立乙醇对 CYP2E1 酶的诱导模型，为研究环境污染物和毒素对水生生物的作用提供良好的技术平台。

1 材料与方法

1.1 材料与试剂

契诺克鲑鱼胚胎细胞系 CHSE-1（*Chinoo salinon* embryo）由农业部渔业动植物病原库提供，细胞编号为 BYK-C16-01。4-氨基酚为 Panya Chemical 公司产品，DMEM 无酚红培养基为 HyClone 公司产品，胎牛血清购自四季青生物公司，其他试剂均为国产分析纯。

1.2 细胞培养及 CYP2E1 酶的诱导

CHSE-1 细胞以 DMEM 培养基加 10% 胎牛血清培养，细胞传代后培养至对数生长期，分别换入含有不同浓度乙醇（0、4、40、50、100、150、200 mmol/L）的新鲜培养液，每个浓度设 5 个平行样，于 28℃ CO$_2$ 培养箱中培养 24 h 后，加入底物苯胺孵育 30 min。

1.3　CYP2E1 酶活性的测定及模型拟合

以苯胺为底物，以测定苯胺羟化酶（Aniline hydroxylase，ANH）活性作为 CYP2E1 的活性指标[11]。反应原理如下：

$$C_6H_5HN_2(苯胺)\xrightarrow[O_2]{CYP2E1}HO-C_6H_4-HN_2(4-氨基酚)$$

生成的 4-氨基酚代谢物可转变为一种酚-吲哚复合物，于 630 nm 处有最大吸收峰，4-氨基酚的测定依照 Schenkmen 法[12]进行，蛋白含量按 Lowry 法[13]进行。CYP2E1 活性以每毫克蛋白 1min 内的 4-氨基酚生成量（nmol）表示。

根据 Excel Macro REGTOX EV7.0 中的 Gentox 模型对 ANH 诱导剂的剂量效应曲线进行拟合，以峰为界可将此曲线分为上升和下降两个阶段。方程式如下：

$$F(x)=\cfrac{p_0+kx}{1+\left(\cfrac{x}{EC_{50}}\right)}$$

式中，$F(x)$ 表示当诱导剂浓度为 x 时的 ANH 活性，p_0 表示无诱导剂对照组的 ANH 活性，k 表示上升阶段 ANH 活性线性增加的斜率，nH 表示 Hill 常数，EC_{50} 表示概率函数 $f(x)=EC_{50}^{nH}/(x^{nH}+EC_{50}^{nH})$ 等于 0.5 时的诱导剂浓度，即诱导剂产生 50% 最大效应时的诱导剂浓度[14]。

1.4　数据处理

通过 SPSS13.0 软件进行显著性分析，采用 oneway ANOVA（单因素方差分析）中的 Duncan's 检验进行多重比较。数据以平均值±标准差（$\bar{X}\pm SD$）表示。样本组以英文字母进行标记，相同字母表明组间无显著差异，反之则表明组间有显著差异（$P<0.05$）。

2　结果与分析

2.1　CHSE-1 细胞中 CYP2E1 活性与底物浓度的关系

CYP2E1 与底物苯胺浓度的关系如图 1 所示。由图 1 可知，苯胺浓度在 2~16 mmol/L 时，产物 4-氨基酚的生成量回归方程为 $Y=0.0092X+0.018$（$R^2=0.91$），产物量呈线性增加，说明 CYP2E1 活性随苯胺浓度的升高而升高。苯胺浓度在 10~16 mmol/L 时，CYP2E1 活性最大，可达到 0.152 nmol/（min·mg），是苯胺浓度为 2.5 mmol/L 时的 6.57 倍。该浓度范围的苯胺适用于指示 CYP2E1 的活性。

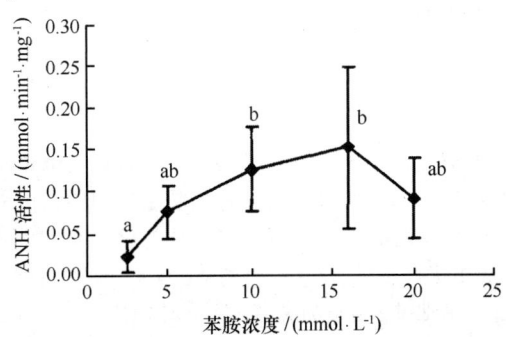

图 1　不同苯胺浓度下苯胺羟化酶（ANH）活性（$n=5$）

Fig. 1　Aniline hydroxylase activity（ANH）at different aniline concentration（$n=5$）

2.2　乙醇诱导的剂量效应

CHSE-1 细胞以不同浓度乙醇诱导 24 h 后分别检测 ANH 酶活性，以 Gentox 模型对诱导的剂量效应关系进行拟合，拟合模型曲线如图 2 所示，拟合参数 p_0 为 0.140，nH 为 2.456，k 为 0.024，EC_{50} 为 34.936。乙醇可以在较低浓度下对 CHSE-1 细胞中的 CYP2E1 产生诱导作用，这种作用随着诱导剂浓度的增加而增强，达到峰值后随着乙醇浓度的增加逐渐减弱至零。

2.3　乙醇诱导的时间效应

分别选取乙醇浓度为 4 mmol/L 和 40 mmol/L 的细胞样品，在 24 h 内连续检测诱导后活性，研究乙醇的诱导结果与诱导时间的关系。

由图3可见，即便是较低的乙醇浓度（4 mmol/L），在前6 h，苯胺羟化酶活性也没有显著变化，说明低浓度的乙醇对CHSE-1细胞的诱导需要一定时间才能发生。40 mmol/L乙醇的诱导也是在4 h后才会使苯胺羟化酶活性明显增加。0~24h时间内，4 mmol/L乙醇诱导的回归方程$Y=0.014X+0.152 1$（$R^2=0.816$），40 mmol/L乙醇诱导的回归$Y=0.007X+0.152 1$（$R^2=0.981$），说明在24 h内随着诱导的持续，酶的活性持续增加，呈典型的酶诱导现象。

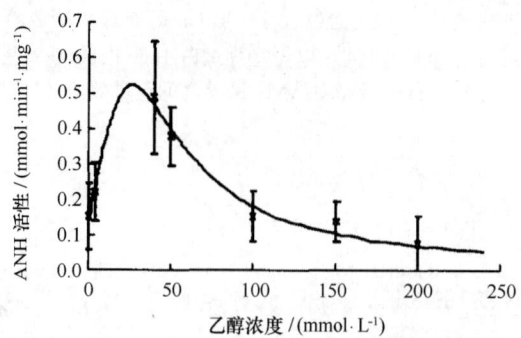

图2　不同乙醇浓度诱导的CHSE-1细胞苯胺羟化酶活性水平 （$n=5$）

Fig. 2　Induction of ANH in CHSE-1 following incubation with ethanol of different concentration （$n=5$）

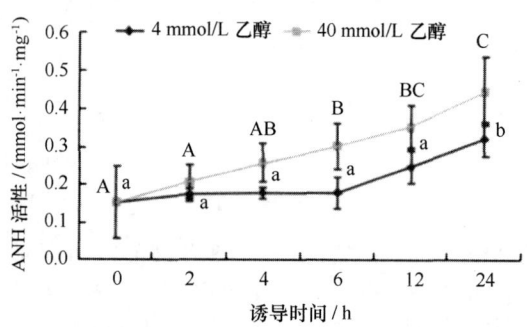

图3　4 mmol/L和40 mmol/L乙醇诱导24 h内CHSE-1细胞的ANH活性 （$n=5$）

Fig. 3　ANH activity in CHSE-1 cell following incubation with 4 mmol/L and 40 mmol/L ethanol within 24 h （$n=5$）

3　讨论

将细胞体系应用于药物代谢酶的诱导研究，可以排除体内因素干扰直接观察诱导剂对酶的调节以及酶对底物的选择代谢性，为整体试验提供可靠的理论依据[15]。本实验以契诺克鲑鱼胚胎细胞为研究对象，探讨了乙醇对细胞体系中对CYP2E1的影响，确定了乙醇对CHSE-1细胞的可诱导性，确定了底物剂量、诱导时间以及诱导剂剂量效应参数，建立了乙醇对CHSE-1细胞中CYP2E1的诱导模型。

以特异性底物代谢反应研究CYP2E1活性时，合适的底物浓度很重要[16]。本研究发现在CHSE-1细胞反应体系中，苯胺终浓度在2.5~16 mmol/L时苯胺的羟化代谢有较好的线性（$R^2=0.91$），随着苯胺浓度的增大CYP2E1活性增强，苯胺浓度超过16 mmol/L时，CYP2E1的活性有下降趋势，本实验以10~16 mmol/L的苯胺浓度作为底物浓度。

本研究发现，乙醇要与细胞共孵一段时间后才会诱导CYP2E1活性。不管是较低剂量还是较高诱导剂量，在与乙醇共孵2 h后都没有改变苯胺羟化酶的活性，4 mmol/L的乙醇在与细胞共孵6 h后都没有影响苯胺羟化酶活性。在24 h内4 mmol/L和40 mmol/L的乙醇对CHSE-1细胞中CYP2E1的诱导与时间相关系数分别为0.816和0.980，呈典型的酶诱导现象，说明CHSE-1细胞可作为理想的体外模型来研究乙醇的诱导作用。乙醇对CHSE-1细胞中CYP2E1的影响表现为复合效应，即低剂量的诱导作用和高剂量的抑制作用。与氯乙烯对大鼠染毒研究中大鼠肝微粒体CYP2E1活力增加现象相似[17]，都是低剂量组诱导剂会随着时间的增加CYP2E1活性明显增强，而中高剂量组CYP2E1活性有一个先升高后降低的过程，表明长期接触较高剂量的诱导剂不会一直引起该酶的诱导而表现为消耗和抑制。

CYP2E1在体内主要负责小分子量外源性药物或毒物的代谢[18]，其介导的反应既有利于机体健康的一方面，如加快二氯甲烷在体内的代谢消除，减少二氯甲烷对机体的毒性；也有对机体有害的一方面，如转化一些前毒物或前致癌物为毒物和致癌物[19]。本研究在CHSE-1中研究乙醇对CYP2E1的诱导效应，并以Gentox模型对乙醇诱导的剂量效应进行拟合，建立乙醇诱导的模型参数。得到的参数 p_0 可预测无诱导剂存在时的基础酶活，k 值反映了诱导剂与酶之间的反应性，EC_{50} 是诱导剂产生 50%最大效应时的诱导剂浓度。这些参数对评价诱导剂的效应有重要的意义，也为以鱼类体外细胞评价环境污染物和毒素对水生生物的影响提供新的思路。

参考文献

［1］　伍忠銮，谢红光，周宏灏，等. 细胞色素 P450 2E1 的研究进展 ［J］. 中国临床药理学杂志，1997，13（1）：57-62.

［2］　Coon M J，Ding X，Pemecky S J，et al. Cytochrome P450：Progress and prediction ［J］. FASEB J，1992，6：669-673.

［3］　Gonzalez F J. Human CytochromeP450：Problems and prospect ［J］. Trends Bio Sci，1992，13：346-352.

［4］　Guengerich P. Reactions and significance of Cytochrome P450 enzymes ［J］. J Biochem，1991，266：10019-10022.

［5］　Wall K L，Crivello J. Chlorzoxazone metabolism by winter flounder liver microsomes：evidence for existence of a CYP2E1-like isoform in teleosts ［J］. Toxicol and Appl Pharmacol. 1998，15：98-104.

［6］　Sirpa E H，Mark E H，Pirjo L S. A fish hepatoma cell line（PLHC-1）as a tool to study cytotoxicity and cyp1a induction properties of cellulose and wood chip extract ［J］. Chemosphere，1998，36（14）：2921-2932.

［7］　Courtney S，Sandra L T，Benita L M，et al. WIF-B cells as a model for alcohol-induced hepatocyte injury ［J］. Biochem Pharmacol，2004，67：2167-2174.

［8］　Babin M M，Tarazona J V. In vitro toxicity of selected pesticides on RTG-2 and RTL-W1 fish cell lines ［J］. Environ Poll，2005，135：267-274.

［9］　Daniel C，Claudia E，Karl F. Cytotoxicity of pharmaceuticals found in aquatic systems：Comparison of PLHC-1 and RTG-2 fish cell lines ［J］. Aqua Toxicol，2006，79：114-123.

［10］　王爱红，夏昭林. 细胞色素 P4502E1 活性的测定方法 ［J］. 中国工业医学杂志，2003，16（6）：354-356.

［11］　刘苏艳，刘毓谷，邱信芳，等. 诱导物对大鼠细胞色素 P450 2E1 基因表达和活化的影响 ［J］. J Fudan Univ：Natural Science，1991，30（2）：178-186.

［12］　徐叔云，卞如濂，陈修. 药理实验方法学：第 3 版 ［M］. 北京：人民卫生出版社，2002：513-519.

［13］　Lowry O H，Rosebrough N J，Farr A L，et al. Protein measurement with the Folin phenol reagent ［J］. Biol Chem，1951，193（1）：265-275.

［14］　Vindimian E，Robaut C，Fillion G. A method for cooperative and noncooperative binding studies using non linear regression analysis on a microcomputer ［J］. Appl Biochem，1983，5：261-268.

［15］　李文东，马辰. 药物体外肝代谢研究进展 ［J］. 中国药学杂志. 2003，38（10）：737-740.

［16］　Koop D R. Oxidative and reductive metabolism by cytochrome P450 2E1 ［J］. FASEB J，1992，6：724-730.

［17］　王爱红，朱守民，周元陵，等. 氯乙烯染毒对大鼠肝细胞色素 P450 2E1 活力和 mRNA 表达的影响 ［J］. 工业卫生与职业病，2005，31（3）：146-148.

［18］　Omiecinski C J，Remmel R P，Hosagrahara V P. Concise review of cytochrome P450s and their roles in toxicology ［J］. Toxicol Sci，1999，48（2）：151-156.

［19］　Dorothee M R，Thomas L，Donna B S. Expression of cytochrome P450 2E1 in normal human bronchial epithelial cells and activation by ethanol in culture ［J］. Arch Toxicol，2001，75：335-345.

该文发表于《水产学报》【2011，35（10）：1450-1457】

异育银鲫 P450 家族 CYP3A136 基因的克隆与表达分析

Cloning，sequencing and tissue expression of the cytochrome P450 3A136 in crucian cap（*Carassius auratus gibelio*）

朱磊[1,2]，胡晓[1,2]，房文红[1*]，胡琳琳[1]，李新苍[1]，李国烈[1,2]，写腊月[1]

（1. 中国水产科学研究院东海水产研究所，农业部海洋与河口渔业资源及生态重点开放实验室，上海 200090；
2. 四川农业大学动物医学院，雅安 625014）

摘要：根据 5 种硬骨鱼已知 CYP3A 序列保守区设计兼并引物，以异育银鲫 cDNA 为模板扩增得到 CYP3A 基因片段，根据得到片段序列设计特异引物并利用 cDNA 末端快速扩增技术（RACE）获得全长 cDNA。得到异育银鲫 CYP3A 基因全长为 1769 bp，开放阅读框为 1545 个核苷酸，编码 514 个氨基酸。其预测蛋白质分子量为 58.624 ku，理论等电点为 6.30。将异育银鲫 CYP3A 序列理论编码氨基酸序列提交细胞色素 P450 命名委员会

（Cytochrome P450 NomenclatureCommittee）并由其命名为 CYP3A136。氨基酸序列分析显示，其与稀有鲫、草鱼、鲮同源性较高，并且具有高度保守的血红素结合区域 FXXGXXXCXG。异育银鲫 CYP3A 基因的半定量 RTPCR 显示，在肝和肠组织中转录水平最高，在肾和鳃中次之，其余组织较低。

关键词：异育银鲫；CYP3A136；克隆；组织分布

细胞色素 P450（cytochrome P450，简称 CYPs）是一类含有亚铁血红素的超家族单加氧酶，它参与许多外来物质（包括药物，环境化合物等）以及内源性内固醇和胆汁酸的生物转化[1]。根据氨基酸序列的相似性，CYPs 可以划分到不同的家族和亚家族[2]。在人和哺乳动物中，CYP3A 同工酶类占肝脏和小肠 CYPs 蛋白的最大比重，且具有丰富的多样性，在哺乳动物使用的药物中有超过 50% 的药物是 CYP3A 的底物或者抑制剂[3]。这些酶参与了许多化学结构不同的化合物代谢，如毒素、杀虫剂、治疗药物、致癌物、饮食产品和激素等[4]。与哺乳动物相比，鱼类中有关 CYP3A 的研究较少，主要有舌齿鲈（*Dicentrarchus labrax*）[5]、斑马鱼（*Daniorerio*）[6]、稀有鲫（*Pimephales promelas*）[7]、鲮（*Gobiocypris rarus*）[8] 等，国内仅见草鱼（*Ctenopharyngodon idellus*）[9]。和哺乳动物一样，鱼类 CYP3A 同工酶具有多种亚型，主要存在于肝脏组织，在肠和鳃等其他组织中也有存在。药酶的诱导或抑制是研究药物相互作用的基础，分析 CYP3A 基因有助于揭示药物相互作用模式的分子机制。

异育银鲫（*Carassius auratus gibelio*）是我国重要的淡水养殖经济鱼类，在我国淡水水域广泛分布[10]，常用作环境污染评价。本研究使用同源序列法对异育银鲫 CYP3A 基因进行克隆，预测氨基酸序列、二级结构、功能区域、分析组织表达，旨在探明异育银鲫 CYP3A 基因分子背景，为后续设计定点突变实验，构建分子相互作用模型并进一步揭示其诱导和抑制等调控分子机制，以及蛋白分子和底物之间相互作用的关系等提供理论基础。

1 材料与方法

1.1 实验材料

实验试剂 3′-Full RACE Core Set Ver. 2.0、RNAiso Plus、5′-Full RACE Kit、PrimeScript RT-PCR Kit、TaKaRa LA Taq © Hot Start Version、pMD-18T 载体、DNase I 及胶回收试剂盒均为大连宝生物公司产品，所用引物为上海生工合成，其他试剂为国产分析纯。

实验动物 异育银鲫（C. auratus gibelio）采自上海市青浦淡水鱼养殖场，体重（160±10）g，暂于水循环、充氧玻璃水族箱，水温 25-27℃，投食饲养 4 周后开始实验。

1.2 实验方法

CYP3A 基因片段的获得 从异育银鲫肝脏中提取总 RNA，并用 DNase I 除去基因组 DNA。紫外分光光度计测定总 RNA 浓度，取 5 μg 总 RNA 合成 cDNA 第一链。

根据硬骨鱼类底鳉（*Fundulus heteroclitus*）GenBank：AF105068.2、日本青鳉（*Oryzias latipes*）GenBank：AF105018.1、虹鳟（*Oncorhynchus mykiss*）GenBank：U96077.1、鲮鱼（*Pimephales promelas*）GenBank：EU332794.1、红鲤（*Cyprinus carpio*）GenBank：GU046696.1 所公布完全编码区序列设计兼并引物 F-304、R-949 扩增 cDNA 并得到 CYP3A 中间片段。扩增条件为：94℃ 3 min（1 循环），94℃ 30 s，55℃ 30 s，70℃ 1 min（30 循环），72℃ 10 min（1 循环）。扩增片段在 1% 的琼脂糖凝胶中电泳并回收，将回收片段克隆到 18T 载体并送交大连宝生物公司完成测序。

CYP3A 基因全长的获得 根据测序验证后 CYP3A 的部分片段，设计 RACE 3′ 引物 F-567 和 5′ 引物 R-312，按照 RACE 试剂盒说明分别进行 3′ RACE 和 5′ RACE 扩增。反应条件：94℃ 3 min（1 循环），94℃ 30 s，60℃ 30 s，70℃ 2 min（30 循环），72℃ 10 min（1 循环）。PCR 产物分别克隆 18T 载体并由大连宝生物公司完成测序。拼接全序列后得到 CYP3A mRNA 完全编码序列，将预测的氨基酸序列提交细胞色素 P450 命名委员会（Cytochrome P450 Nomenclature Committee）命名。实验过程中所用引物设计软件为 Primer premier 5.0，引物序列见表 1。

表 1 异育银鲫 CYP3A 基因 RACE 和半定量 RT-PCR 相关引物列表

Tab. 1 universal and specific primers used for RACE and semi-quantitative RT-PCR

引物名称 Primer	序列（5′-3′）Sequences（5′-3′）
F-304	TCTTCACCAACMGMAGGAACT
R-949	CCGGCGAAGATGAAGATCATGG
F-567	ACCAGCACAGCCTTCAGCGTCG

续表

引物名称 Primer	序列（5'-3'）Sequences（5'-3'）
R-312	TGATGTTGGTCACAAAAGGGT
SQF	ATGGGCTCTGCTCATACTGTT
SQR	TCCTCCTCCAGTCGTCGTCTT
Beta-F	TGCCCTGGTCGTTGATAA
Beta-R	TGGCATACAGGTCCTTACGAA

序列生物信息学分析　使用 NCBI 数据库中 Blast 程序对序列进行比对分析。使用 DNAMAN 6.0 软件和 ORF Finder（http://www.ncbi.nlm.nih.gov/gorf/）预测氨基酸序列、组成和蛋白质基本理化性质。信号肽和结构域分析采用 SignalP 3.0 Server（http://www.cbs.dtu.dk/services/SignalP/）和 Motif Scan（http://hits.isb-sib.ch/cgi-bin/PFSCAN）使用 MEGA 4.1 软件构建系统发生树。用 predictprotein（http://www.predictprotein.org/）初步预测蛋白结构。

半定量检测 CYP3A mRNA 在异育银鲫组织分布　使用 RNAiso Plus 分别提取 3 尾异育银鲫肝、头肾、肾、前肠、中肠、后肠、脾、鳃和肌肉总 RNA，以 Oligo dT 和随机引物合成 cDNA 第一链。根据已获得的 CYP3A 全长序列设计引物 SQF 和 SQR，同时以异育银鲫 β-actin 基因序列（GenBank：AB039726）为内参设计引物 beta-F 和 beta-R，对各组织中 CYP3A 的表达进行半定量分析。扩增条件同为 94℃ 1 min（1 循环），94℃ 30 s，57℃ 30s，70℃ 1 min（28 循环），72℃ 10 min（1 循环）。产物经 1.5% 琼脂糖电泳，通过 Quantity One 4.6 软件对条带进行分析，以 Gauss 痕迹量的 CYP1A/β-actin 表示 DNA 的相对含量。

2　结果

2.1　异育银鲫 CYP3A136 基因的克隆和氨基酸序列分析

根据 5 种硬骨鱼类 CYP3A 同源性区域设计简并引物，获得了异育银鲫细胞色素 P450 3A136 的部分片段，在此基础上进行 RACE-PCR 得到了 2 个长度分别为 653 bp 和 1 135 bp 大小的 5' 和 3' 端序列片段，由此拼接成长 1 769 bp 的全长 cDNA 序列（GenBank：GU998964），polyA 加尾信号 AATAAA（1 740~1 745 bp），5' 端非编码区 30 bp，3' 端非编码区 194 bp，开放阅读框 1 545 bp，编码 514 个氨基酸（图 1）。提交氨基酸序列到 P450 命名委员会并将该基因命名为 CYP3A136。预测其蛋白质分子量为 58.624 KDa，理论等电点为 6.30。

通过异育银鲫 CYP3A136 亚铁血红素结合区域和其他物种 CYP3A 基因的亚铁血红素结合区域进行对比，发现血红素结合区域 FXXGXXXCXG 在不同物种间具有高度保守性，除第二位氨基酸有物种间差异，其余氨基酸基本相同（图 2）

利用蛋白预测软件 predictprotein 以氨基酸序列对蛋白结构进行预测，结果显示在 10-27 和 216-233 氨基酸序列段存在高疏水区域，在 264-268 位氨基酸存在蛋白结合信号。

2.2　CYP3A136 基因序列比对和系统进化分析

以异育银鲫 CYP3A136 开放阅读框序列通过 NCBI 网站 Blast x 同源对比发现：本实验所得异育银鲫 CYP3A136 基因与稀有鮈鲫（Gobiocypris rarus）CYP3A（Genebank：ABV01347.1）相似性为 83%，与草鱼（Ctenopharyngodon idella）CYP3A（Genebank：ADO19749.1）相似性为 82%，与鲦鱼（P. promelas）CYP3A126（ACA35026.1）相似性为 80%。基于不同物种的 CYP3A 氨基酸序列构建系统发生树、进化树显示，异育银鲫与硬骨鱼类同属一支，哺乳动物、爬行动物、鸟类、脊索动物属于另外 4 支。（图 3）

2.3　异育银鲫 CYP3A136 基因的各组织活性

使用半定量 RT-PCR 方法对异育银鲫 CYP3A136 基因在各组织中的转录情况进行研究，结果显示，CYP3A136 转录活性最强的组织是肝和后肠，在前肠和中肠也有活性，在其他组织中转录保持较低水平（图 4）。

3　讨论

首个 P450 蛋白于在大鼠肝微粒体中被发现[11]，后续从古细菌、植物到各类动物，细胞色素 P450 超家族几乎在所有生物中被发现[12]。1982 年，细胞色素 P450 首次被克隆、测序[13]，这标志着 P450 研究重点也慢慢由生物化学、生物物理学特征的鉴定及酶学功能的研究，转变为基因表达的调控机制和结构与功能的对应关系上来。截至 2009 年 8 月，已发布 997 个家族，2 519 个亚家族总计 11 294 个 P450 超家族基因。在众多 P450 酶系中，CYP3A 亚家族占主导地位，广泛参与外源

```
1    GTTTTAGTCTGAGCGTCCTGGAGCAGGAACCATGAGCTACGGTCTGTTCTTCTCTGCTGAAACATGGGCTCTGCTCATACTGTTTGTGAC
1                                        M  S  Y  G  L  F  F  S  A  E  T  W  A  L  L  I  L  F  V  T
91   ACTTCTGTTCATATATGGATCCTGGCCTCATGGTGTCTTCAAGAAGTTGGGGATCCCGGGGCCCAAACCTCTGCCGTTCTTCGGAACCAT
31   L  L  F  I  Y  G  S  W  P  H  G  V  F  K  K  L  G  I  P  G  P  K  P  L  P  F  F  G  T  M
181  GCTGGAATATAGAAAGGGGTTTCACAACTTCGATATGGAGTGTTTCAAGAAGAACGGACGAGTCTGGGGTATTTACGATGCGAGGCAGCC
61      L  E  Y  R  K  G  F  H  N  F  D  M  E  C  F  K  K  N  G  R  V  W  G  I  Y  D  A  R  Q  P
271  TGTTCTGTGCATCATGGACCTATCCATCATCAAAACCATCCTGGTTAAAGAATGCTACTCTCTCTTCACCAACAGAAGGAACTTCCGTCT
91      V  L  C  I  M  D  L  S  I  I  K  T  I  L  V  K  E  C  Y  S  L  F  T  N  R  R  N  F  R  L
361  GAACGGGCCGCTGTACGATGCCGTGTCCATCGTAGAAGACGACGGAGGAGGATCCGCAGCGTCCTCTCGCCCTCCTTCACCAGCGG
121     N  G  P  L  Y  D  A  V  S  I  V  E  D  D  D  W  R  R  I  R  S  V  L  S  P  S  F  T  S  G
451  GAGGTTAAAGGAGATGTTCGGTATCATGAAGACTCACTCTCACACTCTGGTTGATAATCTGGGGAAAACAGCAACGCGAGGAGAAGCGGT
151     R  L  K  E  M  F  G  I  M  K  T  H  S  H  T  L  V  D  N  L  G  K  T  A  T  R  G  E  A  V
541  GGAAATTAAAGAGTTCTTCGGGGCGTACGGTATGGATGTGGTTACCAGCACGTTCAGCGTCGACATCGACTCCCTCAACAACCCTAA
181     E  I  K  E  F  F  G  A  Y  G  M  D  V  V  T  S  T  A  F  S  V  D  I  D  S  L  N  N  P  K
631  AGACCCTTTTGTGACCAACATCAAGAAGATGCTGAAGTTTGACTTCCTGAACCCTGTGTTCCTGATCAGCGCTGTATTTCCTTTCATCAC
211     D  P  F  V  T  N  I  K  K  M  L  K  F  D  F  L  N  P  V  F  L  I  S  A  V  F  P  F  I  T
721  TCCTGTCCTGGAGAAAATGGGTTTCGCCTTCTTCCCCGACTTCTGTGACCGACTTCTTCTACGCTGCCTTGAAGAAGATCAAGTCTGAAAG
241     P  V  L  E  K  M  G  F  A  F  F  P  T  S  V  T  D  F  F  Y  A  A  L  K  K  I  K  S  E  R
811  AGTGTCCAGCGAGCAGAAGAAGAAGCGAGTGGACTTCCTGCAGCTGATGGTGGATTCTCAGACGGCTGGGGGATCTGAGGAGCACACTGA
271     V  S  S  E  Q  K  K  K  R  V  D  F  L  Q  L  M  V  D  S  Q  T  A  G  G  S  E  E  H  T  E
901  GAAAGGTCTGAGCGACCACGAGATCCTCTCTCAGTCCATGATCTTCATCTTCGGCGGGTACGAGACCAGCAGCAGCACGCTGTCCTTCTT
301     K  G  L  S  D  H  E  I  L  S  Q  S  M  I  F  I  F  G  G  Y  E  T  S  S  S  T  L  S  F  F
991  CTTCTACAATCTGGCCCACACACCCCGAGGCCATGGAGAAGCTGCAGGGGGAGATCGACCAGAGCTTCTCTAGAGAGGATCCGGTGGACTA
331     F  Y  N  L  A  T  H  P  E  A  M  E  K  L  Q  G  E  I  D  Q  S  F  S  R  E  D  P  V  D  Y
1081 TGAAGGCATCATGAACATGGAGTATCTGGACGCAGCGCTGAACGAGTCTCTGCGGCTCTTCCCCATCGTTGCTCGACTGGAGCGCGTCTG
361     E  G  I  M  N  M  E  Y  L  D  A  A  L  N  E  S  L  R  L  F  P  I  V  A  R  L  E  R  V  C
1171 TAAGAAAACGGTGGACATCAACGGCCTCCTGGTTCCTAAAGACGTGGTGGTGATGATCCCGACCTTCGCCCTCCACAGAGACCCGGACTA
391     K  K  T  V  D  I  N  G  L  L  V  P  K  D  V  V  V  M  I  P  T  F  A  L  H  R  D  P  D  Y
1261 CTGGAGCGAGCCGGACAGCTTCAGACCGCAGAGGTTCTCTAAAGACAACAGAGAGTCGATCGACCCCTACATGTTCATGCCCTTCGGTCT
421     W  S  E  P  D  S  F  R  P  Q  R  F  S  K  D  N  R  E  S  I  D  P  Y  M  F  M  P  F  G  L
1351 GGGGCCCAGGAACTGCATCGGGATGAGGTTTGCTCAGGTGAGCATCAAGCTGGCCATCGTGGAGATCCTGCAGCGCTTCGACGTCTCTGT
451     G  P  R  N  C  I  G  M  R  F  A  Q  V  S  I  K  L  A  I  V  E  I  L  Q  R  F  D  V  S  V
1441 GTGTGAGCAGACTCAGGTTCCTCTGGAGCTCGACACCAGCGGACTCCTGGCCCCCAAGAGCCCCATCAAACTCCAGTTCAAGCCTCGCAA
481     C  E  Q  T  Q  V  P  L  E  L  D  T  S  G  L  L  A  P  K  S  P  I  K  L  Q  F  K  P  R  K
1531 ACCTTCCCTCTCAGAGGACATCTGTAACAACAACAACACGTCGTGAAGCAGCTCTCAGGCCGACGAGAAGCTTGCCCTGCTCTTCATCTT
511     P  S  L  S  E  D  I  C  N  N  N  N  T  S  •
1621 TTCAGACTGAGAGCTTGAGTAACTTAGCTTCTGAGAGCACAGGTTCAGATTCAGAGGTTTATTCTGATTGAGAGAACTAAACATGAGATC
1711 AGCCTTTAAAGATGTTCTGTGGAAATATGAAATAAAATGTATGAAAAGTGAAAAAAAAAA
```

图 1　异育银鲫 CYP3A136 开放阅读框及其编码的氨基酸序列

Fig. 1　open reading frame and deduced amino acid sequence of CYP3A136

注：起始密码子用粗体标记，终止密码子用星号标出，信号肽用矩形框标出，血红素结合区用阴影标出。

Note：initiation codon is indicate in bold，terminator codon is indicate by sterisk，signal peptidase
is boxed by black，heme-binding region are shadowed.

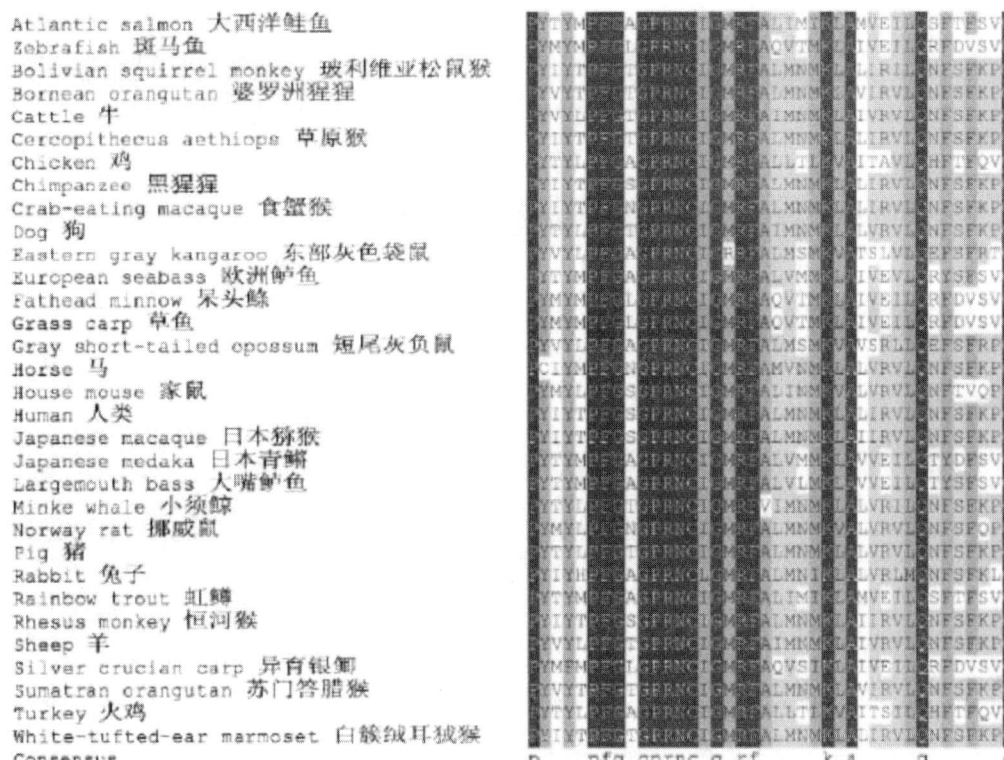

图2　不同物种间 CYP3A 氨基酸亚铁血红素结合区域序列对比

Fig. 2　CYP3A heme-binding region amino acids sequence alignment in different species

注：血红素结合区域高亮标出。

Note：Heme-binding region is highlight.

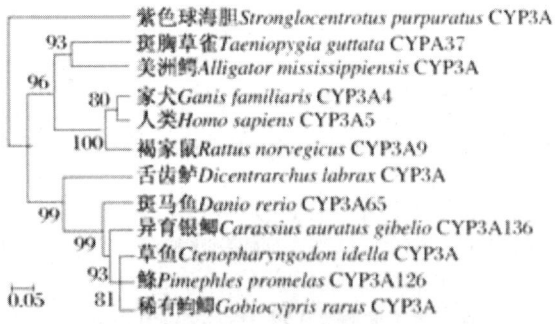

图3　基于 CYP3A 氨基酸序列构建的几种生物 NJ 系统发生树（1000 Bootstrap）

Fig. 3　phylogenitic tree constructed from the amino acid sequence by neighbor-joining（1000 Bootstrap）

性脂质有机物及内源性类固醇激素类代谢[14]。在遗传进化上，脊椎动物 CYP3A 基因主要包括 4 个分支，硬骨鱼类（Teleost）、爬行类（Diapsid）、鸟类（Aves）和哺乳动物（Mammalian），虽然 CYP3A 具有较大物种间差异，但整个硬骨鱼类 CYP3A 有着较为保守的进化趋势。细胞色素 P450 参与催化的底物种类和反应类型具有广泛性和多样性，各亚家族序列之间具有差异性，但所有的 P450 都具有高度保守的血红素结合区域（FXXGXXXCXG），其中绝对保守的半胱氨酸与催化活性中心亚铁血红素中铁元素形成硫醇盐离子键，从而成为铁的一个配体，这正是蛋白能够与氧结合的关键结构，同时也是作用于底物的酶活性中心[12]。P450 与底物的结合能诱导酶发生构象上的改变并引起氧化还原配体的相互作用，铁自旋态平衡，氧化还原电位和氧的结合、活化以及插入底物等等[15]。本研究表明，在 CYP3A136 中血红素结合区序列为 FGLGPRNCIG，其中 FG 组成一个转角（turn），RNC 组成另一个转角，在不同物种间存在差异的第三号氨基酸正好处于两个转角之间。这种关键结构域中间存在的非保守氨基酸可能通过周围氨基酸的位阻效应而屏蔽其对催化活性可能带来的负面影响。

75

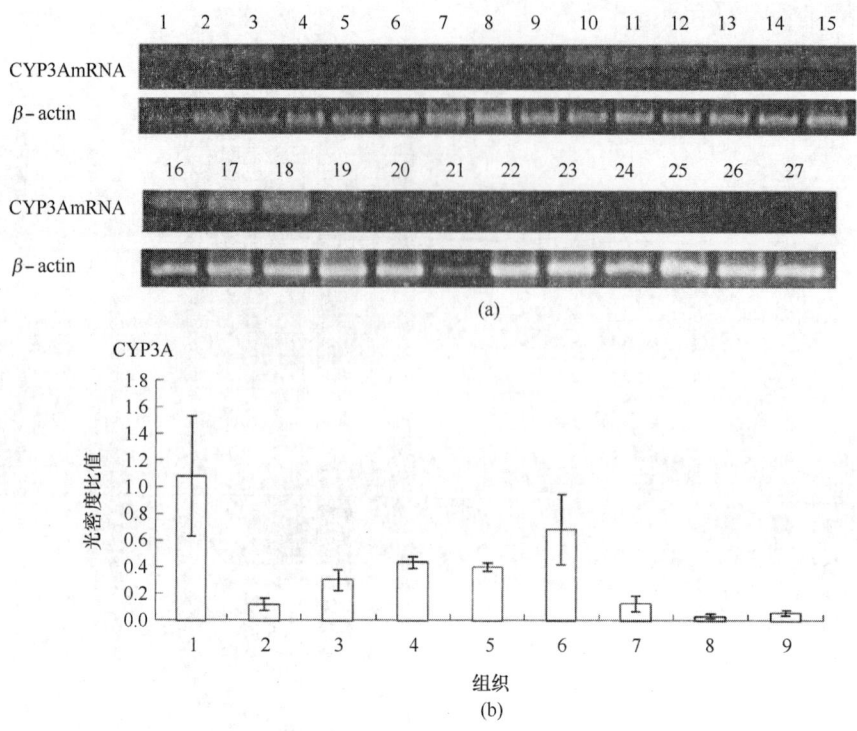

图 4 半定量 RT-PCR 检测 CYP3A136 基因 mRNA 在异育银鲫的组织分布

Fig. 4 mRNA distributing in different organs detect by semi-quantitative RT-PCR

注：（图 a）1-3 号泳道为肝组织，4-6 号泳道为头肾组织，7-9 号泳道为肾组织，10-12 泳道为前肠，13-15 泳道为中肠，16-18 泳道为后肠，19-21 泳道为鳃组织，22-24 号泳道为脾组织，25-27 为肌肉组织。（图 b）以 Gauss 痕迹量的 CYP3A/β-actin 表示 DNA 相对含量，异育银鲫各组织 CYP3A 分布柱状图（$n=3$）

Notes：（Fig a）Detect each organs repetitive group per every three lane, according the number, organs are liver, head-kidney, kidney, former/middle/latter intestine, gill, spleen and muscle in turn. （Fig b）semi-quantitative RT-PCR analysis of CYP3A mRNA distribution. The trace of the PCR products was analysed by using Gauss model and β-actin was used to normalize the data. Results are expressed as means±S. D.（bar）（$n=3$）

通过对 CYP3A136 蛋白二级结构的预测所发现的高疏水区域氨基酸序列段提示可能存在跨膜结构域，鉴于 10~17aa 区域是信号肽序列，故该蛋白可能为单次跨膜蛋白，此分析结果和 P450 超家族蛋白分布主要于内质网膜上的研究结果一致[16]。有假说认为，由于存在疏水性的膜结合结构域，其 F-G 环和 β-折叠能通过和膜结合的方式提高底物进出酶的能力[17]。脂质和 CYP450 的相互作用能增强磷脂横向扩散速率使底物接近膜结合的血红素部位[18]。CYP3A136 第 30-215aa 构成内质网膜结构域，而 234-514 构成其膜外结构域，血红素结合结构域和酶活性中心也存在细胞胞质溶胶侧。CYP3A 作用底物绝大多为脂溶性药物和通过转运蛋白进入细胞膜的药物分子这一现象从而可以得到解释[19-20]。蛋白质相互作用分析表明，264-268 位序列（EQKKK）有一个强有力的蛋白相互结合信号，提示 CYP3A136 可能通过此位点与内质网膜蛋白相互结合从而将其锚定在质膜或细胞骨架上。

通过组织分布半定量结果表明，CYP3A136 在各个组织上均有表达，表明其基因可以通过组成型表达模式进行表达，而肝肠等组织较高的基因转录水平的差异则在胚胎时期就已经建立起来[11]。

细胞色素 P450 作为一类异源物质代谢酶类，其家族成员在进化过程中不断增加，并在进化压力下获得更多不同的新功能，这些进化趋势主要体现在增加了单种酶类对多种底物的适应性[21-24]。部分氨基酸位点的突变导致酶对底物识别的特异性降低从而识别更多底物，但是仍然有大量的 P450 酶具有非常精确的底物专一性，对此有学者认为进化负压的存在加强了对有用表型的选择[25-26]。CYP3A 在人类基因组中共发现 4 个亚型和两个假基因，很难想象如此少量的分子竟然参与几乎 60% 的药物代谢。P450 分子对底物的识别能力和关键氨基酸残基的理化结构有着重要关系，在兔 P450s 2C1 和 2C3 中第 113 号和 365 号氨基酸残基决定了黄体酮水解酶活性的特异性[27-28]。鼠 P450 2B1 依赖 114，206，363 和 487 号氨基酸残基改变底物特异性[29]。在人类 P450 2C9 中第 359 位的亮氨酸残基突变为异亮氨酸残基会改变苯妥英水解酶的活性以及丙酮苄羟香豆素水解酶的活性[30-31]。如此众多的例子同时表明，氨基酸的改变，对整个 P450 家系的进化和演变来说都是十分关键的。包括 P450 家族基因的命名，主要是依靠其序列差异性进行划分的，同一家族的 P450 氨基酸同源性能达到 40%

以上，亚家族同源性则达到 55% 以上。本实验得到异育银鲫的 CYP3A136 的 cDNA 序列，为下一步对该基因实现定点突变以及用酶动力学方法研究其与底物和诱导剂抑制剂的结合提供了一个物质基础。

参考文献

[1] Guengrich F P. Reactions and significance of cytochrome P-450 enzymes [J]. Journal of Biological Chemistry, 1991, 266 (16): 10019-10022.

[2] Nelson D R. Comparison of P450s from human and fugu: 420 million years of vertebrate P450 evolution [J]. Archives of Biochemistry and Biophysics, 2003, 409 (1): 18-24.

[3] Lampen A, Christians U, Gonschior A K, et al. Metabolism of the macrolide immunosuppressant, tacrolimus, by the pig gut mucosa in the Ussing chamber [J]. British Journal of Pharmacology, 1996, 117 (8): 1730-1734.

[4] Nebert D W, Gonzalez F J. P450 genes: structure, evolution, and regulation [J]. Annual Review of Biochemistry, 1987, 56: 945-993.

[5] Vaccaro E. Cloning, tissue expression, and inducibility of CYP 3A79 from sea bass (*Dicentrarchus labrax*) [J]. Journal of Biochemical and Molecular Toxicology, 2007, 21 (1): 32-40.

[6] Tseng H P. Constitutive and xenobiotics-induced expression of a novel CYP3A gene from zebrafish larva [J]. Toxicology and Applied Pharmacologyl, 2005, 205 (3): 247-258.

[7] Christen V. Identification of a CYP3A form (CYP3A126) in fathead minnow (*Pimephales promelas*) and characterisation of putative CYP3A enzyme activity [J]. Analytical and Bioanalytical Chemistry, 2010, 396 (2): 585-595.

[8] Liu Y. Molecular characterization of cytochrome P450 1A and 3A and the effects of perfluorooctanoic acid on their mRNA levels in rare minnow (*Gobiocypris rarus*) gills [J]. Aquatic Toxicology, 2008, 88 (3): 183-190.

[9] Ii D. Effects of mammalian CYP3A inducers on CYP3Arelated enzyme activities in grass carp (*Ctenopharyngodon idellus*): Possible implications for the establishment of a fish CYP3A induction model [J]. Comparative Biochemistry and Physiology Part C: Toxicology & Pharmacology, 2008, 147 (1): 17-29.

[10] Labells F S, Stein D, Qqeen G. Occupation of the cytochrome P450 substrate pocket by diverse compounds at general anesthesia concentrations [J]. European Journal of Pharmacology, 1998, 358 (2): 177-185.

[11] Jensen S L, Cohen-bazire G, Nakyama T O, et al. The path of carotenoid synthesis in a photosynthetic bacterium [J]. Biochimica et Biophysica Acta, 1958, 29 (3): 477-498.

[12] Anzenbacher P, Anzenbacherova E. Cytochromes P450 and metabolism of xenobiotics [J]. Cellular and Molecular Life Sciences, 2001, 58 (5-6): 737-747.

[13] Chen Y T, Negishi M. Expression and subcellular distribution of mouse cytochrome P1-450 mRNA as determined by molecular hybridization with cloned P1-450 DNA [J]. Biochemical and Biophysical Research Communications, 1982, 104 (2): 641-648.

[14] Danielson P B. The cytochrome P450 superfamily: biochemistry, evolution and drug metabolism in humans [J]. Current Drug Metabolism, 2002, 37 (3): 561-597.

[15] Ravichandran K G. Crystal structure of hemoprotein domain of P450BM-3, a prototype for microsomal P450's [J]. Science, 1993, 261 (5122): 731-736.

[16] Olsen M J. Function-based isolation of novel enzymes from a large library [J]. Nature Biotechnology, 2000, 18 (10): 1071-1074.

[17] Amold G E, Ormstein R L. Molecular dynamics study of time-correlated protein domain motions and molecular flexibility: cytochrome P450BM-3 [J]. Biophysics Journal, 1997, 73 (3): 1147-1159.

[18] Lewis D F V. Cytochromes P450 structure, function and mechanism [M]. London: Taylor & Francis, 1996: 79-88.

[19] Prot J M, Videau O, Brochot C, et al. A cocktail of metabolic probes demonstrates the relevance of primary human hepatocyte cultures in a microfluidic biochip for pharmaceutical drug screening [J]. International Journal of Pharmaceutics, 2011, 408 (1-2): 67-75.

[20] Ekins S, Stresser D M, Andrew Williams J. In vitro and pharmacophore insights into CYP3A enzymes [J]. Trends in Pharmacological Sciences, 2003, 24 (4): 161-166.

[21] Matsumura I, Ellington A D. In vitro evolution of beta-glucuronidase into a beta-galactosidase proceeds through non-specific intermediates [J]. Journal of Molecular Biology, 2001, 305 (2): 331-339.

[22] Zaccolo M, Gherardi M. The effect of highfrequency random mutagenesis on in vitro protein evolution: a study on TEM-1 beta-lactamase [J]. Journal of Molecular Biology, 1999, 285 (2): 775-783.

[23] Yang J. Structure-based engineering of E. coli galactokinase as a first step toward in vivo glycorandomization [J]. Chemistry & Biologyl, 2005, 12 (6): 657-664.

[24] Varadarajan N. Highly active and selective endopeptidases with programmed substrate specificities [J]. Nature Chemical Biology, 2008, 4 (5): 290-294.

[25] O'loughlin T L, Greene D L, Matsumura I. Diversification and specialization of HIV protease function during in vitro evolution [J]. Molecular Biology and Evolution, 2006, 23 (4): 764-772.

[26] Krobach T, Kemper B, Johnson E F. A hypervariable region of P450IIC5 confers progesterone 21-hydroxylase activity to P450IIC1 [J]. Biochemistry, 1991, 30 (25): 6097-6102.

［27］ Hsu M H. A single amino acid substitution confers progesterone 6 beta-hydroxylase activity to rabbit cytochrome P450 2C3 ［J］. Journal of Biological Chemistry，1993，268（10）：6939-6944.

［28］ Keszie K M. Molecular basis for a functionally unique cytochrome P450IIB1 variant ［J］. Journal of Biological Chemistry，1991，266（33）：22515-22521.

［29］ Veronese M E. Site-directed mutation studies of human liver cytochrome P-450 isoenzymes in the CYP2C subfamily ［J］. Biochemical Journal，1993，289（Pt 2）：533-538.

［30］ Kaminsky L S. Correlation of human cytochrome P4502C substrate specificities with primary structure：warfarin as a probe ［J］. Molecular Pharmacology，1993，43（2）：234-239.

［31］ Szczesna-skorupa E，Chen C D，Kemoer B. Cytochromes P450 2C1/2 and P450 2E1 are retained in the endoplasmic reticulum membrane by different mechanisms ［J］. Archives of Biochemistry and Biophysics，2000，374（2）：128-136.

该文发表于《Journal of Veterinary Pharmacology & Therapeutics》【2011，35（3）：216-223】

Effects of enrofloxacin on cytochromes P4501A and P4503A in *Carassius auratus* gibelio（crucian cap）

恩诺沙星对异育银鲫细胞色素 P4501A 和 P4503A 的影响

X. HU，X. C. LI，B. B. SUN，W. H. FANG，S. ZHOU，L. L. HU & J. F. ZHOU

（East China Sea Fisheries Research Institute，Chinese Academy of Fishery Sciences；the Key Laboratory of Marine and Estuarine Fisheries Resources and Ecology，Ministry of Agriculture，Shanghai，China）

Abstract：Currently，although enrofloxacin（EF）as a widely used veterinary medicine has begun to apply to treating fish bacterial infections，the researches on the effects of EF on their main drug metabolic enzymes are limited. To investigate the effects of EF on fish cytochromes P450（CYPs）1A and 3A，the enzymatic activities and expressions（mRNA and protein）of crucian cap CYP1A and CYP3A after EF administration were examined. For CYP1A，in the in vivo experiments，EF exhibited potent inhibition on the CYP1A-related ethoxyres-orufin-O-deethylase（EROD）activity，as well as CYP1A expressions at both protein and mRNA levels，at 24 h after administration with different EF dosages（3，10，30，and 60 mg·kg）；Furthermore，CYP1A enzymatic activity and expressions at both protein and mRNA levels decreased more with increasing EF dosages. Additionally，the in vitro experimental results showed that，after incubated with microsomes，EF did not change the EROD activity through interacting directly with CYP1A. For CYP3A，the in vitro and in vivo experimental results demonstrated that EF could inhibit the CYP3A-related erythromycin N-demethylase activity in a time-and dose-dependent manner，while it did not suppress CYP3A expressions at both protein and mRNA levels after administration with EF for a short period（no more than 24 h）；however，after injection with EF at a high dose（10 mg/kg）for a long period，the CYP3A protein and mRNA reached their lowest levels at 96 and 48 h，respectively. These results indicate that EF can suppress CYP1A expressions in a dose-dependent manner，thereby inhibiting further its catalytic activity；meanwhile，both the interactions of EF with CYP3A and the expressions decrease（protein and mRNA）caused by EF contribute to the CYP3A inhibition.

key words：Carassius auratus gibelio；Enrofloxacin；P4501A；P4503A；Inhibition

摘要：尽管恩诺沙星作为兽药被用于防治细菌性感染，而有关恩诺沙星对主要药物代谢酶影响的研究很少。为了调查恩诺沙星对鱼的细胞色素 CYP1A 和 CYP3A 的影响，研究了腹腔注射恩诺沙星后对药酶活性及对 mRNA 表达、蛋白表达的影响。在体内实验中，给予不同剂量的恩诺沙星（3、10、30 和 60 mg/kg），恩诺沙星能有效抑制 CYP1A 相关的药酶 EROD 活性，CYP1A 在蛋白水平的表达和 mRNA 水平的表达。而且，随着恩诺沙星剂量升高，其对 CYP1A 活性及蛋白和 mRNA 水平的表达抑制作用增强。另外，在体外实验中，与微粒体共孵育后，恩诺沙星并没有通过直接与 CYP1A 作用而改变 EROD 活性。对于 CYP3A 而言，体内和体外实验显示，恩诺沙星能抑制 CYP3A 相关药酶 ERND 活性；在较短时间内（不超过 24 h），恩诺沙星不能抑制 CYP3A 在蛋白水平和 mRNA 水平的表达，但恩诺沙星在高剂量（10 mg/kg）下作用较长时间后，CYP3A 蛋白和 mRNA 表达分别在 96 h 和 48 h 达到最低。研究结果显示，恩诺沙星以剂量依赖型抑制 CYP1A 表达，从而抑制其催化活性；CYP3A 的抑制是由于恩诺沙星引起的 CYP3A（蛋白和 mRNA）表达水平下降所致。

关键词：异育银鲫；恩诺沙星；P4501A；P4503A；抑制

1 Introduction

Cytochrome P450s (CYPs) are well known for their participation in the oxidative and reductive metabolism of drugs, environmental chemicals, and endogenous compounds (Anzenbacher & Anzenbacherova, 2001). CYPs derived from families 1 to 3 are primarily involved in the biotransformation of most drugs and are also referred to as the drug metabolism enzymes (Snyder, 2000). Many research have illustrated that drug biotransforma-tion is frequently affected by inducers or inhibitors of CYPs, which are likely other drugs, chemical agents, and so on. Thus, when the inducer or inhibitor of CYPs was co-administrated with the substrate drugs (the substrates of CYPs), the adverse effects might come out (Hasselberg et al., 2005, 2008; Yamauchi et al., 2008). Currently, research involving the effects of drugs on CYPs, and CYPs-mediated drug-drug interaction received grow-ing attraction in various economical animals, including fishes (Yu et al., 2009; Yu & Yang, 2010).

Enrofloxacin (EF), as a chemotherapeutic agent, is widely used to control systemic infections in veterinary medicine because of its broad-spectrum antibacterial activity. And the metabolism of EF has been investigated in various animal species after oral or parenteral administration (Della Rocca et al., 2004; Kim et al., 2006; Fang et al., 2008) and it exhibits good absorption and bioavailability (Lewbart et al., 1997; Stoffregen et al., 1997; Intorre et al., 2000).

It was documented that EF could inhibit CYP isozymes (Shlosberg et al., 1995; Vancutsem & Babish, 1996; Mayeaux & Winston, 1998) and further affected the metabolism of their substrates (drugs). For example, it was reported that EF acted as a noncompetitive inhibitor of CYP1A and showed no effect on CYP3A in dogs (Regmi et al., 2005, 2007). By inhibiting enzymatic activity, EF could decrease the metabolism of theophylline in dogs (Intorre et al., 1995), reduce the elimina-tion of diclofenac in sheep (Rahal et al., 2008), and increase monensin toxicity and liver damage (Sureshkumar et al., 2004) in chickens when co-administered with EF. The latest report demonstrated that high doses of EF could depress the expres-sion of CYP1A and CYP3A proteins in male broilers but not their mRNA, thereby inhibiting their enzymatic activities (Zhang et al., 2011).

In fish, the CYP1A-linked ethoxyresorufin-O-deethylase (EROD) activity can be depressed by EF, and EF has also been demonstrated as a mechanism-based inhibitor of CYP3A (Vac-caro et al., 2003). However, the research on the effects of the varying dosages of EF administration on CYP1A and CYP3A at different time after injection is rare. Additionally, the mechanism underlying the effects of EF on these two drug metabolic enzymes is not very clear. Crucian cap, Carassius auratus gibelio, isan economically important freshwater fish belonging to the Family Cyprinidae of Class Osteichthyes. Currently, EF is frequently used to treat crucian cap bacterial infections through injection or oral administration with a dose of approximately 20 mg/kg. In this study, the effects of EF on crucian cap CYP1A and CYP3A were examined. In vivo experiments investigating the enzymatic activities and expressions of CYP1A and CYP3A at the different time after a single EF injection with varying dosages, as well as in vitro experiments evaluating the enzymatic activities of microsomes preincubated with EF were conducted to reveal the mechanisms underlying the effects of EF on CYP1A and CYP3A in crucian cap.

2 Materials and Methods

2.1 Chemicals

Enrofloxacin was supplied by Suzhou Hengyi Pharmaceutical Co., Ltd. (Suzhou, China) 7-Ethoxyresorufin and resorufin were purchased from Sigma-Aldrich (St. Louis, MO, USA). NADPH, as well as erythromycin was purchased from Sangon (Shanghai, China). The primary antibodies (rabbit anti-rat CYP1A1 and rabbit anti-human CYP3A4) and horseradish peroxidase-labeled goat anti-rabbit secondary antibody were bought from Millipore (Shanghai, China). The reagents used for RNA preparation and SYBR Premix Ex Taq were bought from Takara (Dalian, China). All other chemicals and solvents were of analytical grade and purchased from common commercial sources.

2.2 Fish treatment

Healthy crucian cap (160±10 g) were purchased from a farm product market in Shanghai, China. The fish were held indoors in tanks (180 L) with a continual flow of aerated fresh water at a mean temperature of 27.0±1.0℃. Experiments began after an acclimatization period of at least 4 weeks. Up to 60 fish were randomly assigned into ten groups (six fish in each group). Five groups were used for the dose-response experiment, and the rest were used for the time-course experiment. In the dose-response experiment, fish from each of the four experimental groups were injected once intraperitoneally (i. p.) with EF at a dose of 3, 10, 30, or 60 mg/kg, except for the control group, which received an equal amount of normal saline. All the fish were sacrificed via a sharp blow to the head at 24 h postinjection. In the time-course experiment, fish from the four experimental groups were injected i. p. with EF at a single dose of 10 mg/kg, whereas the control was treated with normal saline. Fish from these groups were killed at 1, 2, 4, or 8 days after injection. Subsequently, the liver tissue mixtures harvested from six fish of each group was used for total RNA isolation, microsomes preparation, and protein sample preparation, which was used for Western blotting.

2.3 Microsomal preparation and measurement of hepatic microsomal enzyme activities

Microsomes were prepared according to a previous study with slight modifications (Novi et al. , 1998). In brief, after perfusion in 1. 15% KCl solution, the liver tissue was homogenized and then centrifugated at 10 000 g for 20 min. Subsequently, the supernate collected was ultracentrifugated at 105 000 g for 60 min. Finally, the washed microsomal pellets were re-suspended in 100 mM potassium phosphate buffer (supple-mented with 1 mM EDTA, pH 7. 4) and used for enzymatic activity assays.

CYP1A-associated EROD and CYP3A-associated erythromycin N-demethylase (ERND) activities were assayed at 37℃ (the optimum temperature for this assay, unpublished data) with the hepatic microsomes of crucian cap to simulate CPY1A and CYP3A enzymatic activities, respectively. The ERND enzymatic activity was determined by measuring the formation of formaldehyde based on a previous report (Tu & Yang, 1983). To analyze EROD activity, 0. 6 mg microsomes (10 mg/mL), 0. 5 mM NADPH, and 0. 3 mM 7-ethoxyresorufin were incubated in 0. 05 M Tris-HCl (pH 7. 4) with a total volume of 1 mL for 5 min at 37℃. The reaction was stopped by the addition of 1 mL methanol. After centrifugation at 5 000 g for 10 min, the reaction product resorufin in the supernate was measured fluorometrically (Lange et al. , 1999).

2.4 Gel electrophoresis and immunoblotting

Total microsomal protein concentration was assayed using the Bradford method (Bradford, 1976). Each sample (10 lg protein) was separated on 10% polyacrylamide gel according to the Laemmli method (Laemmli, 1970), and then transferred onto polyvinylidene difluoride membrane (Millipore Company, Bill-erica, MA, USA), using a Bio-rad Semi-Dry Electrophoretic Transfer Cell (Hercules, CA, USA). Membranes were blocked in a Tris-HCl buffer (pH 8. 0) with 0. 1% Tween-20 and 3% nonfat dry milk, and then incubated with rabbit anti-rat CYP1A1 and rabbit anti-human CYP3A4 polyclonal antibodies (1/500 dilu-tion). Immune complexes were visualized using peroxidase-labeled secondary antibody (1/2 000 dilution) with a chemilu-minescent reagent. The content of each sample was calculated by densitometric analysis of the corresponding bands using a Digital Science Imaging System. The experiment was conducted three times to eliminate the slight variation derived from the loading of samples. In this analysis, no reference protein was included. Instead, the samples were normalized by virtue of the total protein content subjected to the gel.

2.5 Quantitative RT-PCR

Total RNA was isolated from the liver tissue using RNAiso Reagent Kit (Takara). Then, the cDNAs were synthesized as the templates with the total RNA using the PrimeScriptRT reagent kit (Takara) according the manufacturer's instructions. To detect the expression profiles of CYP1A and 3A at mRNA level, quantitative real-time polymerase chain reaction (qRT-PCR) was performed following the methods described in a previous report (Li et al. , 2010). The primers, F1 (5? -CAT CCC TTT CTT GCG TAT CCT-3?) and R1 (5? -CGT TTG AGT TCT CGT CCA GTT T-3?) for CYP1A (Accession no. DQ517445) whereas F2 (5? -CGA CCT TCG CCC TCC ACA G-3?) and R2 (5? -ACC TCA TCC CGA TGC AGT TCC-3?) for CYP3A (GU998964), were designed to generate the corresponding target fragments, respectively. In addition, the primers SF (5? -TCT TTT CCA GCC ATC CTT CCT A-3?) and SR (5? -GGT CAG CAA TGC CAG GGT A-3?) were synthesized to amplify b-actin (AB039726) as the reference. The qRT-PCR was programmed at 95℃ for 5 min, followed by 40 cycles of 95℃ for 10 sec, 58℃ for 30 sec, 72℃ for 20 sec, and then melting from 60 to 95℃. The data (CTs) collected in each group were subjected to the equation 2) 44CT and statistical analysis. Significant differences were considered if P<0. 05.

2.6 In vitro effect of EF on CYP1A and CYP3A

To examine the in vitro effects of EF on EROD and ERND activities, different concentrations of EF (0, 20, 50, and 100 lM) were preincubated at 37℃ at a volume of 3 mL with crucian cap microsomes (1 mg/mL) in the presence of an NADPH-generating system. Then, aliquots of the mixture were trans-ferred to reaction tubes containing the corresponding substrates, and then the remaining EROD and ERND activities at the different times (0, 2, 5, 10, and 20 min for EROD activity, and 0, 2, 5, 8, and 10 min for ERND activity) were determined based on the aforementioned methods. The experiment was repeated with the same batch of microsomes, and the data were expressed as the mean of these two experiments.

2.7 Statistical analysis

Statistical analyses were performed with SPSS Software Version 16.0 (SPSS Inc., Chicago, IL, USA), and the data were expressed as mean±SD. Significant differences between the means were determined using one-way analyses of variance (ANOVA) and the Duncan's test for multiple range comparison with significance level accepted at $P<0.05$.

3 Results

3.1 Dose-response effect of EF on CYP1A

To study the in vivo effect of EF on the hepatic CYP1A in crucian cap, EROD activity, as well as the expression profiles at both the translational (protein) and the transcriptional (mRNA) levels at 24 h postinjection with different doses of EF, was investigated. The behavior of the enzymatic activity and the expression profiles are summarized in Fig. 1a. The CYP1A-related EROD activity is apparently inhibited by EF at higher dosages (no < 10 mg/kg) in a dose-dependent manner. When the dosage of EF is above 10 mg/kg, the higher the dosage used, the lower the EROD activity becomes. This result is in agreement with the expression profile after administration with different doses of EF at the protein and mRNA level. In addition, the EROD activity and the expression remained unaffected at a low EF dosage (3 mg/kg), indicating that a high dosage of EF could inhibit the expression and the enzymatic activity of CYP1A.

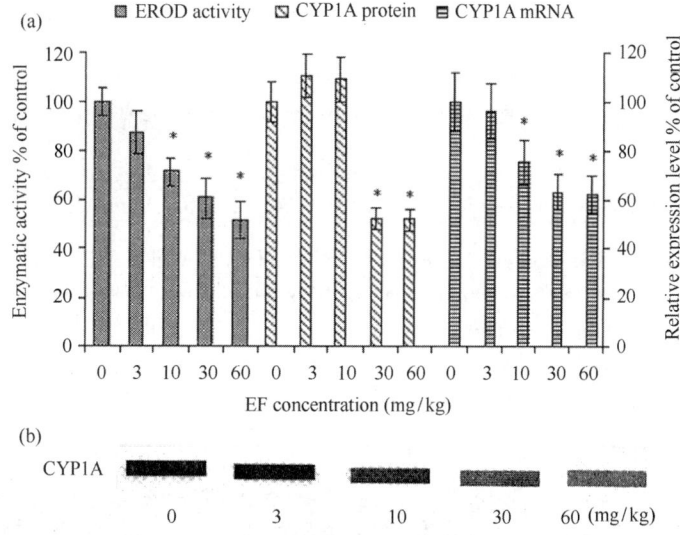

Fig. 1 Dose-response effects of enrofloxacin (EF) on CYP1A enzymatic activity, expression profiles at both protein and mRNA level after a single injection with different dosages (0, 3, 10, 30, and 60 mg/kg). (a) The five bars on the left show enzymatic activity% of control based on the mean of ethoxyresorufin-O-deethylase activity (36.7 pmol/Æmin/mg of protein) after injection with EF (0, 3, 10, 30, 60 mg/kg). The five bars in the middle and the five bars on the right indicate the expression pattern of CYP1A at protein and mRNA level, respectively. (b) Panel b was the result of Western blot. Contents of protein are calculated by densitometric analysis. Data represent% of the control. Expression profile of CYP1A is revealed by qRT-PCR. Asterisk ' * ' shows significant difference from the control ($P<0.05$).

3.2 Dose-response effect of EF on CYP3A

To elucidate further the in vivo effect of EF on CYP3A, the CYP3A-related ERND activity and the expression profiles of the CYP3A protein and mRNA at 24 h after treatment with different dosages of EF were examined. The ERND activity decreased significantly after injection with EF at a dose of no <3 mg/kg (Fig. 2). Furthermore, the ERND activity was further depressed with an increasing dosage of EF. Hence, EF collected its strongest inhibition (about 52%) at a dose of 60 mg/kg. Nevertheless, the expression profiles show that after treatments with different dosages of EF, the expressions of CYP3A both at the protein and mRNA levels do not exhibit evident changes (Fig. 2), implying that various dosages of EF can depress the enzymatic activity of CYP3A at 24 h postinjection, but not its expression.

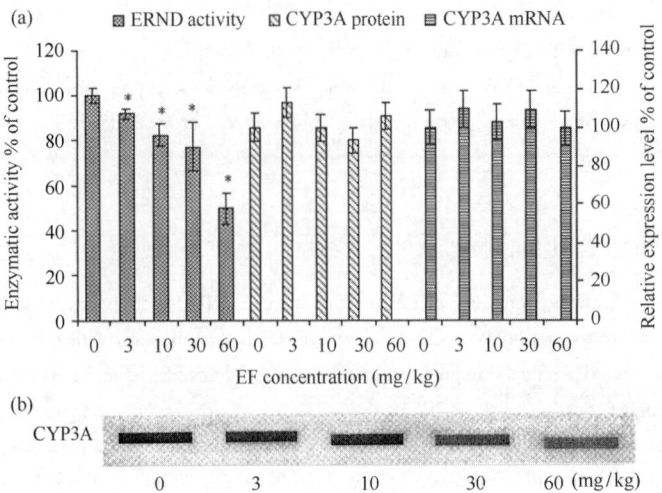

Fig. 2　Dose-response effects of enrofloxacin (EF) on CYP3A enzymatic activity, expression profiles at both protein and mRNA level after a single injection with different dosages (0, 3, 10, 30, and 60 mg/kg). (a) The five bars on the left show the enzymatic activity% of control based on the mean of erythromycin N-demethylase activity (717.3 nmol/min/mg of protein) after injection with EF. The five bars in the middle and on the right indicate the expression pattern of CYP3A at protein and mRNA levels, respectively. (b) Panel b was the result of Western blot. Contents of CYP3A protein were calculated by densitometric analysis. Data represent% of the control. Expression profile of CYP3A was revealed by qRT-PCR. Asterisk ' * ' shows significant difference from the control (P<0.05).

3.3　The temporal effects of EF on the enzymatic activities of P450 1A and 3A

To explore the in vivo temporal effects of EF on the enzymatic activity of CYP1A and CYP3A, the EROD and ERND activities of the microsomes from fish treated with a single injection of 10 mg/kg EF were examined. The results demonstrate that EF could inhibit both EROD and ERND activity at different time (Fig. 3). After injection with EF, ERND activity began to decrease and reached its minimum (about 60% remaining) at 48 h postinjection, and normalized by 4 days. In contrast, EROD activity decreased to its minimum (about 83% remaining) at 24 h, and then normalized by 2 days.

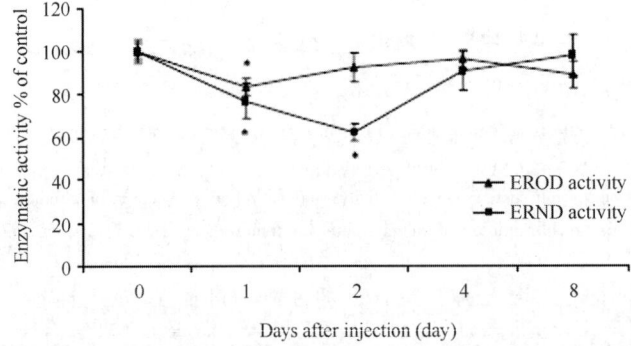

Fig. 3　The temporal effects of enrofloxacin on ethoxyresorufin-O-deethylase (EROD) (▲) and erythromycin N-demethylase (ERND) (■) activity at 0, 1, 2, 4 and 8 days after a single injection (i. p.) with a dose of 10 mg/kg. The means of EROD activity were 36, 30, 33.3, 34.6, and 32 pmol/min/mg of protein, and the means of ERND activity were 715.3, 548.5, 441.9, 648.6, and 700.5 nmol/min/mg of protein. Each point represents% of control. Asterisk ' * ' shows significant difference from the control (P< 0.05).

3.4　The temporal expression profiles of CYP1A and CYP3A at protein level

The expression profiles of CYP1A and CYP3A proteins after the treatments with EF were investigated using Western blotting.

The densitometric analysis showed that EF induces a slight increase in the CYP1A protein at 48 h postinjection, while the CYP3A expression was observed drop at 96 h after administra-tion with EF and then normalized by 8 days (Fig. 4a, b). However, the expression of CYP1A and CYP3A protein was not affected before 24 and 48 h postinjection, respectively. These results suggest that for a longer period after injection EF could inhibit CYP3A, but increase CYP1A expression at the protein level.

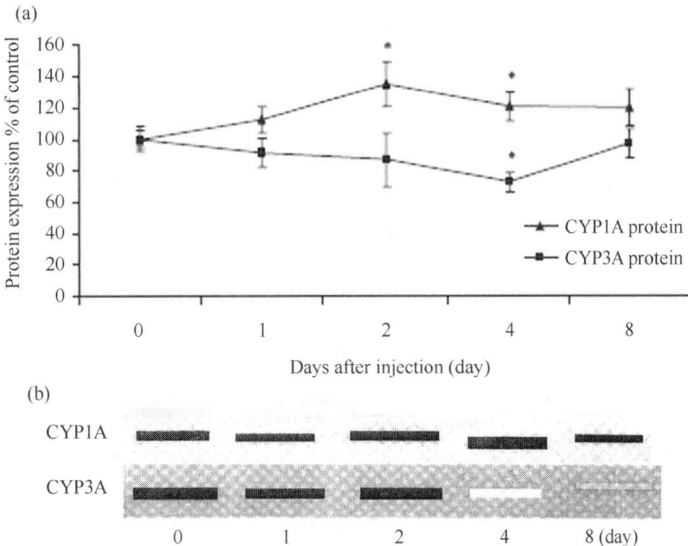

Fig. 4   The temporal effects of enrofloxacin on the protein expression of CYP1A (▲) and 3A (■) after a singleinjection (i. p.) with a dose of 10 mg/kg. (a) The data were calculated as the means of densitometric analysis results. Each point represents% of control. Asterisk ' * ' shows significant difference from the control (P<0.05). (b) The panels demonstrate the expression profiles detected by Western blot using rabbit anti-rat CYP1A1 and rabbit anti-human CYP3A4 polyclonal antibodies, respectively. The samples were normalized by virtue of the total protein contents quantified with the Bradford method.

3.5   Temporal expression profiles of CYP1A and CYP3A at mRNA level

Quantitative RT-PCR was conducted to study the effects of EF on the transcripts of CYP1A and CYP3A. The results demonstrate that the level of CYP1A transcripts decreased by approximately 40% at 24~48 h postadministration, and the CYP3A mRNA also declined after injection with EF and collected its maximum inhibition (decrease by approximately 36%) at 48 h (Fig. 5), and then the amounts of both the CYP1A and the CYP3A mRNA came back gradually.

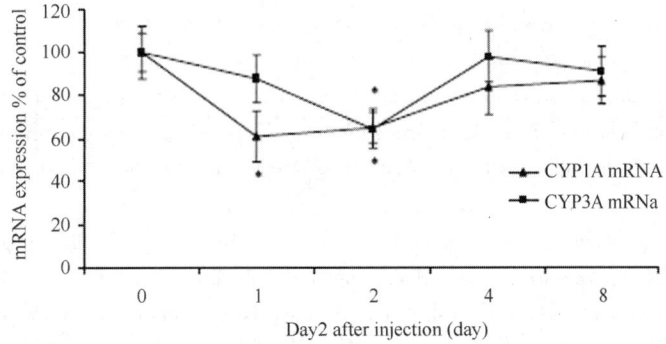

Fig. 5   Temporal effects of enrofloxacin on mRNA expression of CYP1A (▲) and 3A (■) after a single injection (i. p.) with a dose of 10 mg/kg. Temporal expression of CYP1A and 3A was revealed by qRT-PCR. b-actin was used as the reference. Asterisk ' * ' shows significant difference from the control (P< 0.05).

3.6   In vitro inhibitory activities of EF on CYP1A and CYP3A

To analyze further the in vitro effects of EF on CYP1A and CYP3A, the concentrations of EF ranging from 0 to 100 lM was added to the reactive mixture to assay the inactivation on EROD and ERND activity in microsomes. After incubation with EF, the EROD activity was hardly affected whereas the ERND activity appeared more susceptible (Fig. 6). When the liver microsomes were incubated with EF in the presence of NADPH, the activity of ERND was inhibited in a dose-and time-dependent manner (Fig. 6b). After incubation with 100 lM EF for 10 min, the ERND activity remained at only 27% that of the control. These results imply that EF can inactivate CYP3A but not CYP1A in a dose-and time-dependent manner.

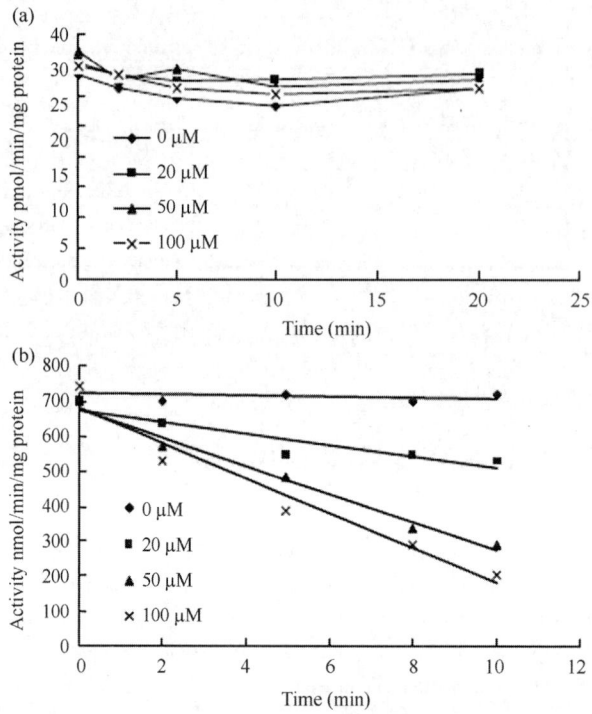

Fig. 6   The ethoxyresorufin-O-deethylase (EROD) and erythromycin N-demethylase (ERND) activities of microsomes preincubated with enrofloxacin (EF). Samples were collected from the primary reaction mixture at the indicated time points and assayed for EROD (a) and ERND (b) activity as described in experimental procedures. Each point represents the mean of duplicate values. The final concentrations of EF were 0 (◆), 20 (▲), 50 (■), and 100 lM (×).

4   Discussion

Until recently, researches on the effects of EF on crucian cap CYP1A and CYP3A are limited, although EF, as a widely used veterinary medicine, has been used in aquaculture. Currently, detecting microsomal EROD and ERND activity has become a popular method for studying the corresponding CYP1A and CYP3A catalytic functions in animals (Hasselberg et al., 2004; Li et al., 2008; Monari et al., 2009). When EF was administrated, assaying CYPs catalytic activities and their expressions at the same time would find out whether the variations of CYPs catalytic activities are because of the changes of their expres-sions. In addition, it will be validated whether EF interacts with CYPs according to the variations of microsomal EROD and ERND activities after preincubated with EF. In this study, we found that, similar to the previous reports in other animal species (Shlosberg et al., 1995; Vancutsem & Babish, 1996; Zhang et al., 2011; Vaccaro et al., 2003), EF could inhibit crucian cap CYP1A and CYP3A catalytic activities; furthermore, the CYP1A inhibition would be enhanced further as the administration dosage of EF increased; meanwhile, a longer period after injection with a higher dose of EF could also suppress CYP3A expressions (protein and mRNA), thereby inhibiting further its catalytic activity.

The current results demonstrate a complex mechanism of inhibition of the CYP1A in the crucian cap by EF. The in vitro experiments indicated that EF did not directly interact with CYP1A as the in vitro EROD activity was not affected after different concentrations of EF administration, and from the in vivo experiments, we speculated the inhibition of the CYP1A enzymatic activity might be through depressing mRNA expres-sion because both the catalytic activity and the mRNA expression were inhibited simultaneously after injection with EF. Furthermore, when the administration dosage of EF was increased, the catalytic activity of CYP1A was further inhibited with its transcripts decreasing. Interestingly, the CYP1A protein was up-regulated during the period between 48 and 96 h after administration. According to the latest research, in which EF can significantly depress the protein expression of CYP1A in chicken but it does not affect the mRNA content (Zhang et al., 2011), it was supposed that post-transcriptional modification might occur during this inhibitory process. Given that this kind of modifica-tion is widespread in eukaryotes, the CYP1A protein increase may be due to the protein modification at the post-transcrip-tional level, which causes inactivation of the enzyme, thereby making the protein hard to degrade. Thus, the amount of CYP1A increases at the protein level, suggesting that the mechanism of CYP1A inhibition by EF is much more complicated than that of an inhibitor.

In the present study, the in vitro experiment indicates that EF is a mechanism-based inhibitor of CYP3A in crucian cap, which is inhibited by EF in a time-and concentration-dependent manner, similar with the results in sea bass (Vaccaro et al., 2003). In addition, it was demonstrated that a longer period after injection with a high dose of EF could inhibit enzymatic activity by depressing the expression of CYP3A, as the ERND activity decreased as both the protein contents and transcripts of CYP3A were inhibited by EF. Based on the latest report, which indicates that EF can depress the CYP3A expression at protein level but not at the mRNA level in chickens (Zhang et al., 2011), and another previous research reporting that EF can depress the protein expression of CYP3A (Vaccaro et al., 2003), we conjec-tured that EF might inhibit enzymatic activity by affecting CYP3A expression given that the ex-pression (at protein and or mRNA level) after administration was decreased at different degrees. These results suggest that besides be-ing a mechanism-based inhibitor, the decrease in CYP3A expression also was responsible for the CYP3A inhibition of catalytic activi-ty.

From our results, the expression of CYP1A can be inhibited by the high dose of EF but not with a low dose, while the enzymatic activity of CYP3A can be inhibited by a low dose of EF (3 mg/kg), which is consistent with the result of a study in another fish (Vaccaro et al., 2003). In addition, the expression of CYP3A mRNA could be inhibited at 48 h after administration with a high dose of EF even though it was not depressed at 24 h. This may be because of the long depletion time of EF and the existence of its me-tabolites in body, because a former study in another fish (rainbow trout) demonstrates that after adminis-tration with EF for 4 days, the concentration of EF still remains rather high, and also ciprofloxacin (one metabolite of EF) (Lucchetti et al., 2004). Although the underlying mechanisms of inhibitions in fish need to further study, we think the variations in the expression profiles of CYP1A and CYP3A are related to the administration dosage of EF and the period after injection. Thus, if different dosages of EF were used to per-form the experiments and further examined the results at different time after injection, several incompatible results might be deter-mined, which may explain the inconsistencies of the expression profiles of CYP1A and CYP3A from different labora-tories.

CYP1A and CYP3A isoforms, especially CYP3A, have been documented to play a primary role in the metabolism of a variety of medications (Bibi, 2008). According to the results of this study, the administration of EF with a therapeutic dose (approximately 20 mg/kg) can inhibit microsomal CYP3A and CYP1A catalytic activities. Thus, injection with EF in fish may affect the metabolism of drugs mediated by these two enzymes. Although the research on the metabolism of drugs in crucian cap is limited, CYP1A and CYP3A may also play a similar role in the metabolism of drugs of this kind of fish according to the relevant studies in other ani-mals. This suggests that when the substrate drugs of CYP1A and CYP3A were co-administered with EF, the toxicity caused by the ap-plication of EF should be considered carefully, as the administration with EF at a single dose of approximately 20 mg/kg could lead to the CYP1A and CYP3A inhibitions at least for 1 and 2 days, respectively.

ACKNOWLEDGMENTS

This study was supported by a special research fund (Grant No. 2007M06) for the National Nonprofit Institutes (East China Sea Fisheries Research Institute).

References

Anzenbacher, P. & Anzenbacherova, E. (2001) Cytochromes P450 and metabolism of xenobiotics. Cellular and molecular life sciences, 58, 737-747.

Bibi, Z. (2008) Role of cytochrome P450 in drug interactions. Nutrition and Metabolism, 5, 27.

Bradford, M. M. (1976) A rapid and sensitive method for the quantitation of microgram quantities of protein utilizing the principle of protein-dye bind-ing. Analytical Biochemisty, 72, 248-254.

Della Rocca, G., Di Salvo, A., Malvisia, J. & Sello, M. (2004) The dispo-sition of enrofloxacin in sea bream (Sparus aurata) after single intra-venous injection or from medicated feed administration. Aquaculture, 232, 53-62.

Fang, W.H., Hu, L.L., Yang, X.L., Hu, K., Liang, S.C. & Zhou, S. (2008) Effect of temperature on pharmacokinetics of enrofloxacin in mud crab, Scylla serrata (Forsskal), following oral administration. Journal of Fish Diseases, 31, 171-176.

Hasselberg, L., Meier, S., Svardal, A., Hegelund, T. & Celander, M.C. (2004) Effects of alkylphenols on CYP1A and CYP3A expression in first spawning Atlantic cod (*Gadus morhua*). Aquatic Toxicology, 67, 303-313.

Hasselberg, L., Grosvik, B. E., Goksoyr, A. & Celander, M. C. (2005) Interactions between xenoestrogens and ketoconazole on hepatic CYP1A and CYP3A, in juvenile Atlantic cod (*Gadus morhua*). Com-parative Hepatology, 4, 2.

Hasselberg, L., Westerberg, S., Wassmur, B. & Celander, M. C. (2008) Ketoconazole, an antifungal imidazole, increases the sensitivity of rainbow trout to 17 alpha-ethynylestradiol exposure. Aquatic Toxicol-ogy, 86, 256-264.

Intorre, L., Mengozzi, G., Maccheroni, M., Bertini, S. & Soldani, G. (1995) Enrofloxacin-theophylline interaction: influence of enrofloxacin on theophylline steady-state pharmacokinetics in the beagle dog. Journal of Veterinary Pharmacology and Therapeutics, 18, 352-356.

Intorre, L., Cecchini, S., Bertini, S., Cognetti Varriale, A. M., Soldani, G. & Mengozzi, G. (2000) Pharmacokinetics of enrofloxacin in the sea bass (*Dicentrarchus labrax*). Aquaculture, 182, 49-59.

Kim, M. S., Lim, J. H., Park, B. K., Hwang, Y. H. & Yun, H. I. (2006) Pharmacokinetics of enrofloxacin in Korean catfish (*Silurus asotus*). Journal of Veterinary Pharmacology and Therapeutics, 29, 397-402.

Laemmli, U. K. (1970) Cleavageof structuralproteins during the assembly of the head of bacteriophage T4. Nature, 227, 680-685.

Lange, U., Goksøyr, A., Siebers, D. & Karbe, L. (1999) Cytochrome P4501A-dependent enzyme activities in the liver of dab (*Limanda limanda*): kinetics, seasonal changes and detection limits. Comparative Biochem-istry and Physiology, Biochemistry and Molecular Biology, 123, 361-371.

Lewbart, G., Vaden, S., Deen, J., Manaugh, C., Whitt, D., Doi, A., Smith, T. & Flammer, K. (1997) Pharmacokinetics of enrofloxacin in the red pacu (*Colossoma brachypomum*) after intramuscular, oral and bath administration. Journal of Veterinary Pharmacology and Therapeutics, 20, 124-128.

Li, D., Yang, X. L., Zhang, S. J., Lin, M., Yu, W. J. & Hu, K. (2008) Effects of mammalian CYP3A inducers on CYP3A-related enzyme activities in grass carp (*Ctenopharyngodon idellus*): possible implications for the establishment of a fish CYP3A induction model. Comparative Biochemistry and Physiology, Toxicology and Pharmacology, 147, 17-29.

Li, X. C., Zhang, R. R., Sun, R. R., Lan, J. F., Zhao, X. F. & Wang, J. X. (2010) Three Kazal-type serine proteinase inhibitors from the red swamp crayfish Procambarus clarkii and the characterization, function analysis of hcPcSPI2. Fish and Shellfish Immunology, 28, 942-951.

Lucchetti, D., Fabrizi, L., Guandalini, E., Podesta, E., Marvasi, L., Zaghini, A. & Coni, E. (2004) Long depletion time of enrofloxacin in rainbow trout (*Oncorhynchus mykiss*). Antimicrobial Agents and Chemotherapy, 48, 3912-3917.

Mayeaux, M. H. & Winston, G. W. (1998) Antibiotic effects on cyto-chromes P450 content and mixed-function oxygenase (MFO) activities in the American alligator, Alligator mississippiensis. Journal of Veterinary Pharmacology and Therapeutics, 21, 274-281.

Monari, M., Foschi, J., Matozzo, V., Marin, M. G., Fabbri, M., Rosmini, R. & Serrazanetti, G. P. (2009) Investigation of EROD, CYP1A immuno-positive proteins and SOD in haemocytes of Chamelea gallina and their role in response to B [a] P. Comparative Biochemistry and Physiology, Toxicology and Pharmacology, 149, 382-392.

Novi, S., Pretti, C. & Cognetti, A. M. (1998) Biotransformation enzymes and their induction by b-naphtoflavone in adult sea bass (*Dicentrarchus labrax*). Aquatic Toxicology, 41, 63-81.

Rahal, A., Kumar, A., Ahmad, A. H. & Malik, J. K. (2008) Pharmacokinetics of diclofenac and its interaction with enrofloxacin in sheep. Research in Veterinary Science, 84, 452-456.

Regmi, N. L., Abd El-Aty, A. M., Kuroha, M., Nakamura, M. & Shimoda, M. (2005) Inhibitory effect of several fluoroquinolones on hepatic microsomal cytochrome P-450 1A activities in dogs. Journal of Veter-inary Pharmacology and Therapeutics, 28, 553-557.

Regmi, N. L., Abd El-Aty, A. M., Kubota, R., Shah, S. S. & Shimoda, M. (2007) Lack of inhibitory effects of several fluoroquinolones on cyto-chrome P-450 3A activities at clinical dosage in dogs. Journal of Vet-erinary Pharmacology and Therapeutics, 30, 37-42.

Shlosberg, A., Ershov, E., Bellaiche, M., Hanji, V., Weisman, Y. & Soback, S. (1995) The effects of enrofloxacin on hepatic microsomal mixed function oxidases in broiler chickens. Journal of Veterinary Pharmacol-ogy and Therapeutics, 18, 311-313.

Snyder, M. J. (2000) Cytochrome P450 enzymes in aquatic invertebrates: recent advances and future directions. Aquatic Toxicology, 48, 529-547.

Stoffregen, D. A., Wooster, G. A., Bustos, P. S., Bowser, P. R. & Babish, J. G. (1997) Multiple route and dose pharmacokinetics of enrofloxa-cin in juvenile Atlantic salmon. Journal of Veterinary Pharmacology and Ther-apeutics, 20, 111-123.

Sureshkumar, V., Venkateswaran, K. V. & Jayasundar, S. (2004) Inter-action between enrofloxacin and monensin in broiler chickens. Vet-erinary and Human Toxicology, 46, 242-245.

Tu, Y. Y. & Yang, C. S. (1983) High-affinity nitrosamine dealkylase system in rat liver microsomes and its induction by fasting. Cancer Research, 43, 623-629.

Vaccaro, E., Giorgi, M., Longo, V., Mengozzi, G. & Gervasi, P. G. (2003) Inhibition of cytochrome p450 enzymes by enrofloxacin in the sea bass (*Dicentrarchus labrax*). Aquatic Toxicology, 62, 27-33.

Vancutsem, P. M. & Babish, J. G. (1996) In vitro and in vivo study of the effects of enrofloxacin on hepatic cytochrome P-450. Potential for drug in-teractions. Veterinary and Human Toxicology, 38, 254-259.

Yamauchi, R., Ishibashi, H., Hirano, M., Mori, T., Kim, J. W. & Arizono, K. (2008) Effects of synthetic polycyclic musks on estrogen re-ceptor, vitellogenin, pregnane X receptor, and cytochrome P450 3A gene expression in the livers of male medaka (*Oryzias latipes*). Aquatic Tox-icology, 90, 261-268.

Yu, L. Z. & Yang, X. L. (2010) Effects of fish cytochromes P450 inducers and inhibitors on difloxacin N-demethylation in kidney of Chinese idle (*Ctenopharyngodon idellus*). Environmental Toxicology and Pharmacology, 29, 202-208.

Yu, L. Z., Yang, X. L., Wang, X. L., Yu, W. J. & Hu, K. (2009) Effects of fish CYP inducers on difloxacin N-demethylation in kidney cell of Chinese idle (*Ctenopharyngodon idellus*). Fish Physiology and Biochemistry, 36, 677-686.

Zhang, L. L., Zhang, J. R., Guo, K., Ji, H., Zhang, Y. & Jiang, S. X. (2011) Effects of fluoroquinolones on CYP4501A and 3A in male broilers. Research in Veterinary Science, 90, 99-105.

该文发表于《Comparative Biochemistry & Physiology Part C: Toxicology & Pharmacology》【2011, 154 (4): 360-366】

Inhibition of CYP450 1A and 3A by berberine in crucian cap *Carassius auratus* gibelio

Chang Zhou[a,b], Xin-Cang Li[a], Wen-Hong Fang[a, *], Xian-Le Yang[b], Lin-Lin Hu[a], Shuai Zhou[a], Jun-Fang Zhou[a]

(a. East China Sea Fisheries Research Institute, Chinese Academy of Fishery Sciences; the Key and Open Laboratory of Marine and Estuarine Fisheries Resources and Ecology, Ministry of Agriculture, shanghai 200090, Chinab. School of Aquaculture and Life Science, Shanghai Ocean University, Shanghai 201306, China)

Abstract: Berberine has long been considered as an antibiotic candidate in aquaculture. However, studies regarding its effects on drug-metabolizing enzymes in fish are still limited. In the present study, the effects of berberine on cytochrome P4501A (CYP1A) and CYP3A in crucian cap were investigated. Injection of different concentrations of berberine (0, 5, 25, 50 and 100 mg/kg) inhibited the CYP1A mRNA expression, thereby inhibiting further the catalytic activity of CYP1A-related ethoxyresorufin-O-deethylase (EROD). Furthermore, both CYP1A expression and EROD activity were further inhibited with increasing berberine concentrations. In addition, the CYP3A expressions at both the mRNA and the protein levels were downregulated by higher berberine concentrations. The catalytic activity of CYP3A-related erythromycin N-demethylase (ERND) was also inhibited by berberine at a dose of no less than 25 mg/kg. Moreover, at the berberine concentration exceeding 25 mg/kg, the inhibition of CYP3A expression and ERND activity increased with increasing berberine concentrations. In vitro experiments were also performed. When berberine was pre-incubated with the crucian cap liver microsomes, it competitively inhibited the corresponding EROD activity with the IC50 of 11.7 μM. However, the ERND activity was slightly inhibited by berberine with the IC50 of 206.4 μM. These results suggest that, in crucian cap, berberine may be a potent inhibitor to CYP1A, whereas the CYP3A inhibition needs a higher concentration of berberine.

key words: Berberine; Crucian cap; CYP1A; CYP3A; Inhibition

黄连素对异育银鲫 CYP450 1A 和 3A 的抑制作用

摘要：黄连素是水产养殖用候选抗菌药物，然而其对鱼类药物代谢酶影响的研究十分有限，本文研究了黄连素对异育银鲫细胞色素 CYP1A 和 CYP3A 的影响。腹腔注射不同剂量黄连素（0、5、25、50 和 100 mg/kg）抑制异育银鲫 CYP1A mRNA 表达，进而抑制 EROD 的催化活性，而且随着黄连素浓度升高，对 CYP1A 表达和 EROD 活性的抑制越强。另外，在较高黄连素浓度下，CYP3A 在 mRNA 和蛋白水平的表达下调。在黄连素剂量不低于 25 mg/kg，与 CYP3A 相关的 ERND 的催化活性受到抑制；当黄连素剂量超过 25 mg/kg 时，黄连素剂量越高，CYP3A 表达和 ERND 活性受到抑制越强。在体外实验中，当黄连素与鲫肝微粒体预孵育时，EROD 活性受到抑制，抑制常数 IC_{50} 为 11.7 μM；ERND 活性轻微受到抑制，IC_{50} 为 206.4 μM。研究结果表明，黄连素很可能是异育银鲫 CYP1A 的有效抑制剂，而在较高的黄连素剂量下才对 CYP3A 起抑制作用。

关键词：盐酸小檗碱；异育银鲫；CYP1A；CYP3A；抑制

1 Introduction

Cytochrome P450 enzymes (CYPs) represent a superfamily of monooxygenases that play a pivotal role in drug metabolism

(Anzenbacher and Anzenbacherova, 2001). In humans, the most important monooxygenases are CYP3A4, CYP2C9, CYP2C19, CYP2D6, and CYP1A2, of which CYP3A4 and 1A2 participate in the metabolism of about 60% of the drugs (Uttamsingh et al., 2005). Several drugs and chemical reagents that are the inducers or inhibitors of the major CYPs associated with drug metabolism can change the expression of enzymes involved in drug metabolism in vivo. In addition, these substances may also interact with CYPs, thereby affecting the metabolism of corresponding substrate drugs. Therefore, the improper co-administration of the substrate drugs with inducers or inhibitors of CYPs can lead to unexpected results (Hasselberg et al., 2005, 2008; Yamauchi et al., 2008; Wills et al., 2009). To date, research regarding the effects of drugs upon CYPs and metabolism-based drug-drug interaction in a variety of animal species, including fishes, has received increasing attention (Li et al., 2008; Yu and Yang, 2009; Yu et al., 2010).

Berberine, a quaternary isoquinoline alkaloid, is an active ingredient used in traditional Chinese medicine and is known to treat human diseases. As demonstrated in modern pharmacology, it exhibits multiple pharmacological effects, such as an antitumor (Iizuka et al., 2000), antimicrobial (Hwang et al., 2003), and anti-inflammatory (Kuo et al., 2004). In aquatic animals, it has currently become an important antibiotic for preventing enteritis and other diseases through the co-administration with other clinical drugs (Guo and Chen, 2005). Therefore, further research regarding the effects of berberine on enzymes involved in drug metabolism is needed. Berberine reportedly induces or inhibits CYPs. For instance, berberine inhibits the enzymatic activity of rat CYP1A2 (Zhao et al., 2008) and shows bimodal effects on CYP1A1 in HepG2 cells, where berberine induces CYP1A1 expression but decreases its enzymatic activity (Vrzal et al., 2005). When berberine is co-administered with cyclosporin A, the adverse effects of CsA are reduced through the inhibition of CYP3A (Xin et al., 2004).

Crucian cap, Carassius auratus gibelio, is an economically significant freshwater fish belonging to the family Cyprinidae. Currently, research on the effects of berberine on CYP1A and CYP3A isoforms in crucian cap is limited. In a previous work, we cloned the full-length cDNA sequence of crucian cap CYP3A (GU998964). In addition, a partial sequence of crucian cap CYP1A (GQ353297) is available from the Genbank. In the present study, in vivo experiments were done to examine CYP1A and CYP3A expression and their enzymatic activity after berberine injection. Furthermore, in vitro experiments were conducted to evaluate the enzymatic activity of liver microsomes pre-incubated with berberine. Ultimately, the present study aims to reveal the mechanisms underlying the effects of berberine on CYP1A and CYP3A.

2 Materials and methods

2.1 Chemicals

Berberine chloride (C20H18ClNO4 MW: 371.84, purity N 90%), 7-ethoxyresorufin, and resorufin were obtained from Sigma-Aldrich (Shanghai, China). NADPH and erythromycin were purchased from Sangon (Shanghai, China). Rabbit anti-rat CYP1A1, rabbit anti-human CYP3A4, and mouse anti-GAPDH monoclonal antibody were obtained from Millipore (Shanghai, China). The molecular grade reagents used for RNA isolation and reverse transcription polymerase chain reaction were obtained from Takara (Japan). All other chemicals and solvents were analytical grade from common commercial companies.

2.2 Fish treatment

Crucian cap (Carassius auratus gibelio) (160±10 g, sexual maturity) were purchased from a farm produce market in Shanghai, China. The fish were held indoors in tanks (180 liters) with flowing aerated fresh water. After an acclimatization period of 4 weeks, experiments were carried out. Up to 30 fish were randomly assigned into 5 groups (6 fish each) and bred temporarily in 5 different tanks. In the first 4 groups, the fish from each experimental group received a single intraperitoneal injection with different berberine concentrations (5, 25, 50 and 100 mg/kg), respectively. Fish in the fifth group (the control) were injected with an equal amount of normal saline. The fish were then sacrificed at 24 h after injection. Subsequently, the liver tissues harvested from six fish in each group and used for microsomal preparation, total RNA isolation and protein sample preparation.

2.3 Liver microsome preparation and measurement of liver microsomal enzymatic activity

Liver microsomes were prepared according to the previous work with slight modifications (Novi et al., 1998). In brief, after perfusion in 1.15% KCl solution, the hepatic tissues were excised and homogenized, and then the homogenate was centrifuged at 10 000 g for 20 min. Subsequently, the supernatant was ultracentrifuged at 105,000 g for 60 min. After washing, the obtained liver microsomal pellets were resuspended in 100 mM potassium phosphate buffer (supplemented with 1 mM EDTA, pH 7.4) and used to assay enzymatic activity.

To simulate CPY1A and CYP3A enzymatic activities, CYP1A-related ethoxyresorufin-O-deethylase (EROD) and CYP3A-related erythromycin N-demethylase (ERND) activity were respectively assayed using crucian cap hepatic microsomes. To analyze EROD activity, 60 μL of 10 mg/mL microsomes, 50 μL of 0.5 mM NADPH, and 7 μL of 0.3 mM 7-ethoxyresorufin were incubated in 0.05 M Tris-HCl (pH 7.4) with a total volume of 1 mL for 5 min at 27℃. The reaction was terminated by adding 1 mL methanol. After centrifugation at 5000 g for 10 min, the reaction product resorufin in the supernatant was measured fluorometrically (Lange et al., 1999). In addition, to assay ERND activity, 100 μL of 10 mg/mL microsomes, 100 μL of 0.5 mM NADPH and 100 μL of 1 mM erythromycin were mixed in 70 mM Tris-HCl (supplemented with 0.07 mM EDTA, pH 7.4) with a total volume of 2 mL. After reacting for 30 min at 27℃ (the reaction is linear over 30 min, unpublished data), ERND activity was determined by testing the formation of formaldehyde at 412 nm based on a previous study (Tu and Yang, 1983).

2.4　Gel electrophoresis and immunoblotting

The total microsomal protein concentration was determined with the previous method (Bradford, 1976). Each sample (10 μg protein) was separated on 10% polyacrylamide gel (Laemmli, 1970) and then transferred onto a polyvinylidene difluoride membrane. Membranes were blocked in a Tris-HCl buffer (pH 8.0) with 0.1% Tween-20 and 3% nonfat dry milk, and then incubated with polyclonal antibodies of rabbit anti-rat CYP1A1 and anti-human CYP3A4, or monoclonal antibody of mouse anti-GAPDH. The immune complexes were visualized using peroxidase-labeled secondary antibody with a chemiluminescent reagent. The relative expression level of each sample was quantified through analyzing protein bands by densitometry using a Digital Science Imaging System. In these assays, GAPDH was used as the reference. The experiments were repeated twice to eliminate the slight variations derived from the loading of samples.

2.5　qRT-PCR analysis

To analyze CYP3A and CYP1A mRNA expression profiles after injection with berberine, quantitative real-time PCR (qRT-PCR) was performed based on a previous method (Li et al., 2010). First, according to the protocol of the RNAiso Reagent Kit (TakaRa), total RNA was isolated from the liver tissue mixture to synthesize cDNA as the templates. Next, the primers used to identify the corresponding crucian cap genes (F1 (5'-CAT CCC TTT CTT GCG TAT CCT-3') and R1 (5'-CGT TTG AGT TCT CGT CCA GTT T-3') for CYP1A, F3 (5'-CGA CCT TCG CCC TCC ACA G-3') and R3 (5'-ACC TCA TCC CGA TGC AGT TCC-3') for CYP3A, and F (5'-TCT TTT CCA GCC ATC CTT CCT A-3') and R (5'-GGT CAG CAA TGC CAG GGT A-3') for β-actin as the reference) were designed according to the sequences retrieved from GenBank and then were synthesized (Sangon, shanghai, China). Finally, the qRT-PCR was programmed at 95℃ for 5 min, followed by 40 cycles of 95℃ for 10 s, 58℃ for 30 s, 72℃ for 20 s, and then melted from 60 to 95℃ for analyzing the melting curve. The data (CTs) obtained in each group were subjected to the equation $2^{\triangle Ct}$ ($\triangle Ct=Ct$ reference gene-Ct target gene). The average of relative mRNA expression in each experimental group was compared to the mean of the control group. All other expressions were adjusted accordingly.

2.6　The inhibition of CYP1A and CYP3A by berberine in vitro and determination of median inhibitory concentration (IC_{50})

To study the effect of berberine on the catalytic activity of microsomal protein in vitro and to determine further the IC50sofberberine forthe CYP1A and CYP3A, the prepared liver microsomal protein was pre-incubated with different berberine concentrations (0, 5, 10, and 50 μM for EROD; 0, 50, 100, 200, and 300 μM for ERND) for 10 min to detect the remaining enzymatic activity. The values of enzymatic activities were fit into a curve versus the corresponding berberine concentrations. After the dose-response curves were analyzed, the IC50s were calculated using the SigmaPlot nonlinear curve-fitting module.

2.7　Kinetics of CYP1A inhibition

The type of inhibition and kinetic constants (Km and Vmax) were determined at various berberine concentrations using four different concentrations of substrate. In brief, this experiment consisted of four groups of reactions, where four different berberine concentrations (0, 0.5, 1.0 and 2.0 μM) were applied. Each group consisted of four different concentrations of substrate (0.6, 1.2, 1.8 and 2.4 μM), whereas the amount of microsomes and berberine were fixed. Reaction mixtures were incubated at 27℃ for 5 min. The values of Km and Vmax values were determined by fitting the data into a Lineweaver-Burk plot, while the value of Ki was obtained using a Dixon plot (Dixon, 1953).

2.8　Statistical analysis

Statistical analysis was performed with SPSS Software Version 16.0 (SPSS Inc.) and the data were expressed as mean±standard deviation (S. D.). Significant differences between the means were determined using one-way analyses of variance (ANOVA) and Duncan's test for multiple range comparison with significant level accepted at p b 0.05.

3 Results

3.1 Effects of berberine on CYP1A and CYP3A enzymatic activities

After treatment with different concentrations of berberine, the EROD and ERND activities were measured. The results were summarized in Fig. 1. After injection of berberine with a dose of more than 5 mg/kg, the EROD activity was inhibited and the inhibition became stronger with increasing berberine concentration (Fig. 1A). However, the ERND activity was inhibited by a higher dose of berberine (no less than 25 mg/kg) (p b 0.05), and the inhibition was shown in a dose-dependent manner (Fig. 1B). At the maximum dose of 100 mg/kg, both EROD and ERND activities decreased by 79.1% and 91.4%, respectively, compared with that of the control group (p b 0.05). These results indicate that the CYP1A and CYP3A enzymatic activities may be inhibited by berberine in a concentration-dependent manner, except for the inhibition on CYP3A, which requires higher berberine concentrations.

3.2 Effects of berberine on CYP1A and CYP3A protein expressions

The expression profiles of CYP1A and CYP3A proteins after treatment with berberine were revealed by western blot. The results are shown in Figs. 2 and 3. As observed in other fishes (Celander et al., 1996; Novi et al., 1998), these two polyclonal antibodies also recognize their respective CYP isoforms (CYP1A and CYP3A) of crucian cap. Through densitometric analysis in the experimental group injected with higher berberine concentrations (50 or 100 mg/kg) caused a decrease in CYP3A protein, and exhibited maximum inhibition (a decrease of 84.9%) at the dose of 100 mg/kg; meanwhile, the CYP1A protein did not exhibit an evident change (pN 0.05), which suggests that injection with berberine did not affect the CYP1A protein content but inhibited the CYP3A expression at higher concentrations.

23.3 Effects of berberine on CYP1A and CYP3A mRNA expressions

Next, qRT-PCR was performed to reveal the expression profiles of CYP1A and CYP3A after berberine injection. The statistical data are presented in Fig. 4, which shows that CYP1A and CYP3A expression were inhibited in a dose-dependent manner upon treatment with different concentrations of berberine. Even a low dose of berberine (5 mg/kg) can inhibit the expression of CYP1A (Fig. 4 A) whereas the CYP3A inhibition only occurs at higher doses (no less than 25 mg/kg) (Fig. 4 B). Thus, the CYP1A and CYP3A expressions dropped to their lowest levels (decrease by 85.5% and 90.0%) at a dose of 100 mg/kg. The trends of CYP1A and CYP3A variations are in agreement with those of their enzymatic activities respectively, suggesting that the decrease of CYP1A and CYP3A enzymatic activity might be due to inhibition of their expressions at the transcript level.

3.4 The correlations between CYP expressions and enzymatic activity

To elucidate further the effects of CYP1A and CYP3A expressions on their respective enzymatic activity, the data on the gene (protein) expression fold inhibition vs. the enzymatic activity fold inhibition were fitintoplots (Fig. 5), and the corresponding correlation coefficients were calculated from the automatically generated equations. From the plots, we can find that the correlation coefficient (R^2) of CYP1A mRNA expression vs. enzymatic activity is rather high (Fig. 5 C), but much lower for protein expression vs. enzymatic activity (Fig. 5 A) and mRNA expression vs. protein expression (Fig. 5 E), indicating that the inhibition of CYP1A enzymatic activity is closely related with the decrease of its mRNA expression.

In addition, it was shown that the CYP3A enzymatic activity correlated with its expressions, given that the correlation coefficients of CYP3A protein expression vs. enzymatic activity, mRNA expression vs. enzymatic activity, and mRNA expression vs. protein expression were 0.68, 0.76, and 0.50, respectively (Fig. 5 B, D and F), suggesting that the inhibition of CYP3A mRNA expression contributed mainly to the decrease in protein content, thereby inhibiting its enzymatic activity.

3.5 Determination of IC_{50}

To clarify further the in vitro effects of berberine on CYP1A and CYP3A, the EROD and ERND activities of the crucian cap hepatic microsome pre-incubated with berberine were measured. According to the results, berberine inhibited the microsomal EROD activity in a concentration-dependent manner, and the IC_{50} was 11.7 μM (Fig. 6 A). In addition, the ERND activity was slightly inhibited by berberine with the IC_{50} of 206.4 μM (Fig. 6 B). Hence, in vitro, these suggest that berberine might inhibit EROD activity through interaction with CYP1A protein directly, and thus it was partly responsible for CYP1A inhibition (including expression in-

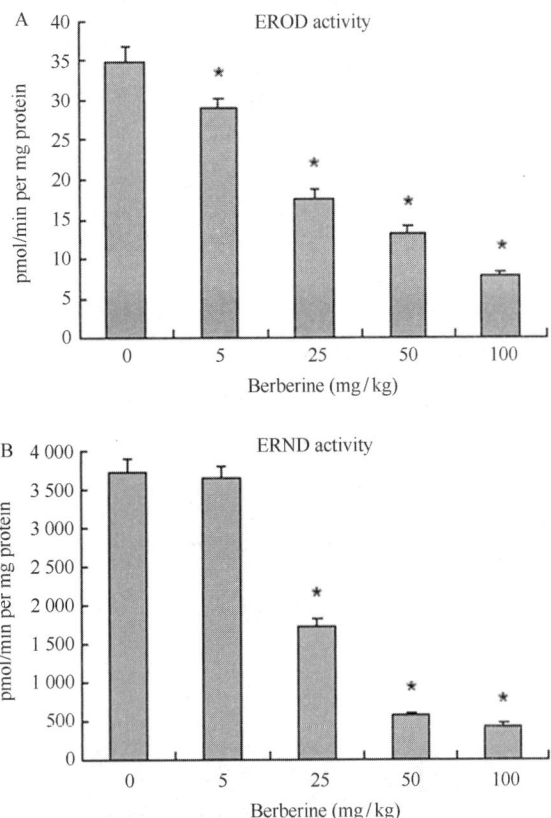

Fig. 1 The effects of different concentrations of berberine on CYP1A and CYP3A enzymatic activities. After injection with berberine (0, 5, 25, 50, and 100 mg/kg), the means of the EROD activity were 34.8, 29.1, 17.4, 13.2, and 7.8 pmol/min per mg protein (A), and the means of the ERND activity were 3729.1, 3653.8, 1717.6, 572.9, and 412.6 pmol/min per mg protein (B). 100% activity was measured in the absence of tested berberine. The asterisk " * " shows significant difference from the control (P<0.05).

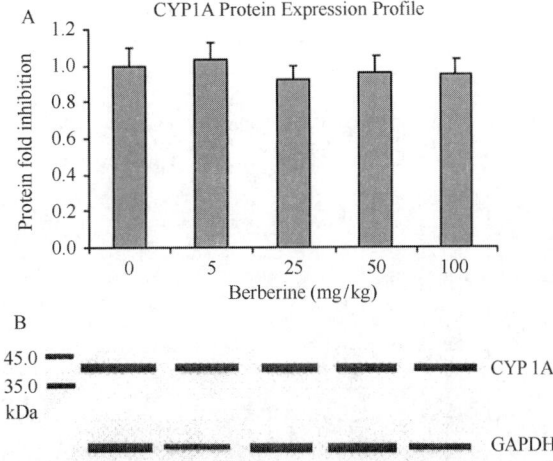

Fig. 2 The effects of varying concentrations of berberine on CYP1A protein expression. (A) The CYP1A expression pattern after injection with berberine (0, 5, 25, 50, 100 mg/kg) was investigated with western blot. Data represent percentages in terms of the control. (B) Panel B is the result of the western blot. Protein contents were calculated by densitometric analysis. The asterisk " * " indicates significant difference from the control (P<0.05).

hibition); however, berberine hardly contributed to CYP3A inhibition in vitro because the IC_{50} is quite high.

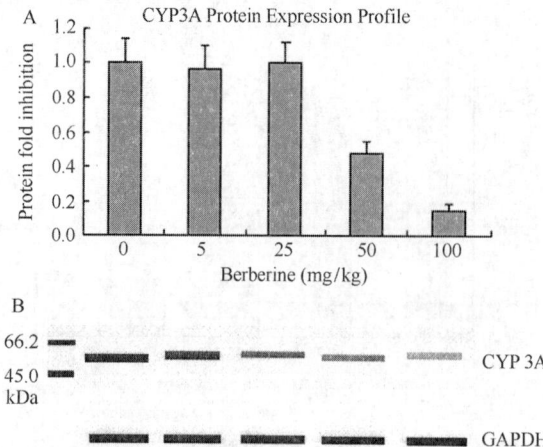

Fig. 3   The effects of varying concentrations of berberine on CYP3A protein expression. (A) The CYP3A expression profile after injection with berberine (0, 5, 25, 50, 100 mg/kg) was investigated with western blot. Data represent percentages in terms of the control. (B) Panel B is the result of the western blot. Contents of protein were calculated by densitometric analysis. The asterisk " * " shows significant difference from the control (P<0.05).

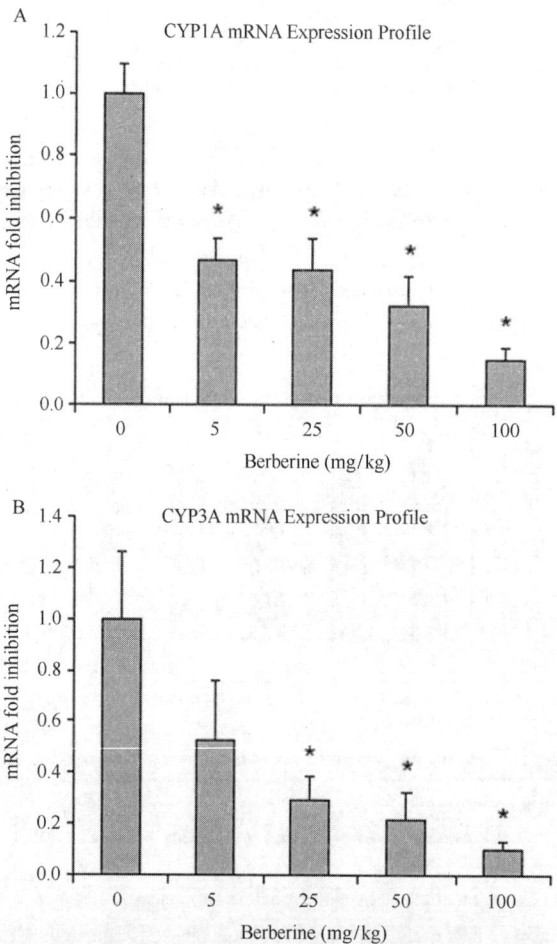

Fig. 4   The effects of different concentrations of berberine on CYP1A and CYP3A mRNA expressions. The CYP1A (A) and CYP3A (B) mRNA expressions after injection with berberine (0, 5, 25, 50, 100 mg/kg) were examined by qRT-PCR. The asterisk " * " shows significant difference from the control (P<0.05).

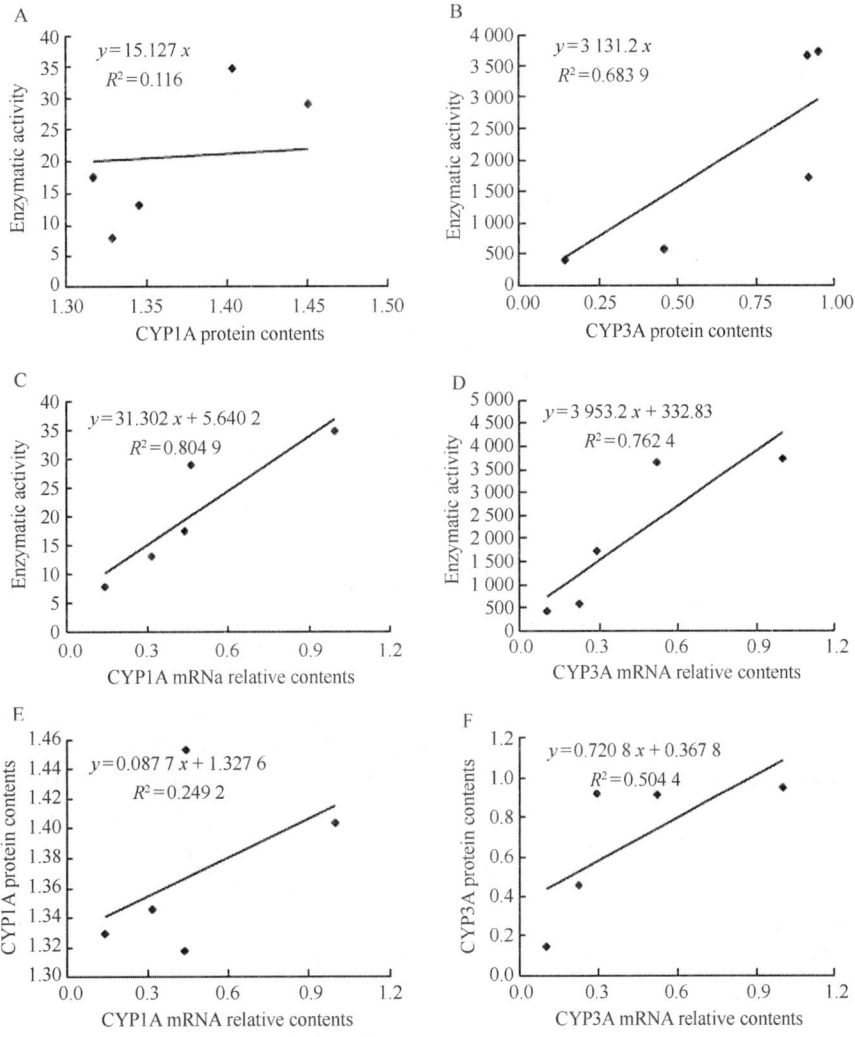

Fig. 5　Calculation of the correlation coefficients of the fold inhibition among the mRNA expressions, protein expressions, and enzymatic activities. The comparisons of CYP1A protein expression vs. enzymatic activity, mRNA expression vs. enzymatic activity, and mRNA expression vs. protein expression are shown in Panel A, C and E, respectively. The comparisons of CYP3A protein expression vs. enzymatic activity, mRNA expression vs. enzymatic activity, and mRNA expression vs. protein expression are also shown in Panel B, D and F, respectively.

3.6　Inhibition kinetics of berberine on CYP1A

The Lineweaver-Burk plots ($1/[V]$ vs. $1/[S]$) showed that the inhibition type was competitive because, along with the increasing concentration of berberine, the obtained lines had a common intercept in the 1/V axis although they had different slopes (Fig. 7A). Km and Vmax were 0.86 μM and 92.3 pmol/min per milligram of protein, respectively. The inhibition constant (Ki) was determined by replotting the data into a Dixon plot ($1/[V]$ vs $[I]$) (Fig. 7B), from which the Ki (0.83 μM) were calculated by virtue of the intersection of the regression lines. This indicates that berberine is a competitive inhibitor to CYP1A.

4　Discussion

Berberine has long been considered as an antibiotic candidate for controlling systemic bacterial infections in fish. Atpresent, reportson the effect of berberine on fish drug-metabolizing monooxygenases (CYPs) are limited. In this work, the inhibitions of berberine on crucian cap CYP1A and CYP3A were investigated, and the results indicated that berberine might be a potent inhibitor to CYP1A, which could not only reduce CYP1A mRNA expression in a dose-dependent manner but also directly inhibit this enzyme competitively; meanwhile, high berberine doses inhibit CYP3A through the downregulation of its expression at both the mRNA and the

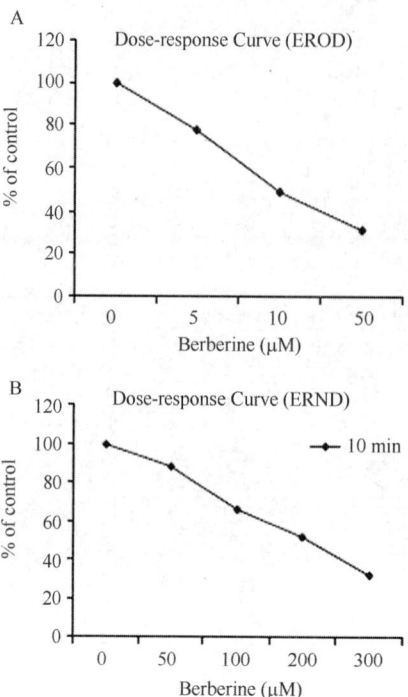

Fig. 6 The dose-response curves of berberine for EROD (A) and ERND (B) activities. The EROD catalytic activities were 34.9, 27.2, 17, 1 and 10.8 pmol/min per mg protein in the presence of varying concentrations of berberine (0, 5, 10 and 50 μM); the ERND catalytic activities were 3644.6, 3220.9, 2405.3, 1895.7 and 1182.2 pmol/min per mg protein after incubation with 0, 50, 100, 200 and 300 μM of berberine, respectively. Data represent percentages in terms of the control.

protein level.

The effects of xenobiotics on CYPs are sorted into two types, gene expression and substrate-like interaction regulation. Berberine has been documented to interact directly with CYP1A protein. For example, the administration of berberine decreased rat liver microsomal CYP1A2 activity in a concentration-dependent manner (Zhao et al., 2008). In another report, berberine was shown to be a potent inhibitor of CYP1A1 in HepG2 cell culture (Vrzal et al., 2005). In this work, it was further demonstrated that berberine directly interacted with crucian cap CYP1A protein and served as a competitive inhibitor with the IC50 of 11.7 μM. In addition, it was shown that berberine could inhibit the CYP1AmRNA expression. Given that the correlation coefficient of CYP1A mRNA expression vs. enzymatic activity was rather high, we predictedthat the decrease of CYP1A catalytic activity might be due to the inhibition of its mRNA expression by berberine. Our study suggested that both the interaction of CYP1A protein with berberine and the inhibition of CYP1A mRNA expression by berberine contributed to the decrease of CYP1A catalytic activity.

In the present study, berberine was proven to downregulate the CYP1A isoform mRNA expression in a dose-dependent manner. However, in a previous research, berberine reportedly augments CYP1A1 expression in HepG2 cell at both the mRNA and the protein levels, thereby increasing CYP1A1 catalytic activity (Vrzal et al., 2005). By comparing the identity of this crucian cap CYP1A isoform with those of human at the amino acid level, we find this crucian cap CYP1A isoform has a higher identity with human CYP1A2 than CYP1A1 (their identity at the amino acid level are 55.23% and 53.82%, respectively). This indicates that this crucian cap CYP1A isoform may be the homolog of human CYP1A2 instead of CYP1A1. Hence, the mechanisms underlying the effects by berberine on crucian cap CYP1A isoform and human CYP1A1 expressions may be quite different.

The effects of xenobiotics on CYPs enzymatic activity, mRNA and protein expression may be a comprehensive result of gene transcription and protein translation. During this process, the modifications, including the post-transcription (mRNA stabilization) (Coon et al., 1992) and post-translational levels (protein stability and modification) (White et al., 1997), are always involved. Thus, the content of the CYPs protein may sometimes disagree with its mRNA expression or enzymatic activity. For example, acetoacetate can inhibit CYP2E1 mRNA expression but promote the increase of its protein (Abdelmegeed et al., 2005). Also, lipopolysaccharide downregulated CYP2C11 mRNA expression and enzyme activity but not change its protein content (Shimamoto et al., 1999). In addition, enrofloxacin inhibited CYP1A and CYP3A activity and protein expression, whereas it had no effects on

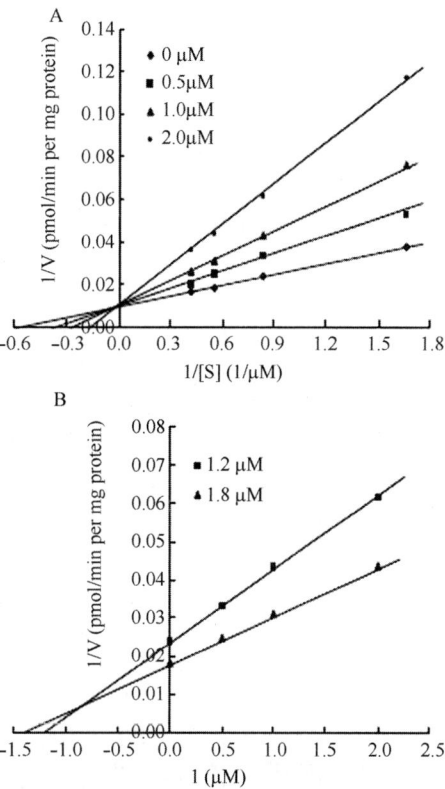

Fig. 7 Lineweaver-Burk and Dixon plots of EROD activity assays were performed at different concentrations of berberine (0, 0.5, 1.0 and 2.0 μM) in the presence of varying concentrations of substrates (0.6, 1.2, 1.8 and 2.4 μM). The concentrations of berberine or substrate are shown in the graphs. The values were the median of two independent experiments.

mRNA expression (Zhang et al., 2011). In the present study, berberine suppressed CYP1A expression but it did not affect the protein content, indicating that protein level modification may occur during this inhibition process. Thus, it is feasible for berberine to downregulate crucian cap CYP1A expression at the mRNA level but not at the protein level.

Berberine reportedly inhibits CYP3A4-related testosterone 6β-hydroxylation activity in human hepatic microsomes with the IC50 of about 400 μM (Chatterjee and Franklin, 2003). Furthermore, in our work the in vitro experiment further proved that berberine inhibits the catalytic activity of the crucian cap CYP3A isoform at an IC50 value of 206.4 μM. In an earlier research (Brahn et al., 1994), when the value of IC50 is higher than 50 μM, berberine was considered not to contribute to the inhibitory activity. Hence, the in vitro interaction of berberine with the CYP3A isozyme was hardly responsible for the inhibition of its catalytic activity. Compared with the in vitro study, the in vivo research on CYP3A expression by berberine is much less. In the present study, the results demonstrated that high berberine concentrations reduce CYP3A expression in a concentration-dependent manner at both the mRNA and the protein levels, thereby further inhibiting its catalytic activity. Although the CYP3A expression profiles at the mRNA and the protein levels are not completely consistent, in terms of the correlation coefficients of CYP3A mRNA expression vs. catalytic activity, protein expression vs. catalytic activity, and mRNA expression vs. protein expression, we can conclude that, to a great degree, the CYP3A catalytic activity decreasing is correlated with the inhibition of its expression, suggesting that the decrease of CYP3A catalytic activity in crucian cap may be due to the inhibition of its expression rather than its interaction with berberine.

We found that the inhibition of CYP1A and CYP3A by berberine were mainly due to the administration concentration. Both CYP1A mRNA expression and catalytic activity were inhibited in a concentration-dependent manner. In addition, after treatment with high berberine concentrations, the inhibition on CYP3A was also shown in a dose-dependent manner; more precisely, these inhibitory activities actually became stronger with increasing the concentration of berberine. Thus, when berberine was administrated to control the diseases of crucian cap or other fishes, careful attention must be given to its application concentration.

Berberine is extracted from plants and it possesses potent antibacterial activities. Thus, using berberine to control fish diseases could substitute for some antibiotics and reduce the pollution to the environment. From this work, it was further found that, to some degree, berberine inhibits the crucian cap CYP1A and CYP3A in a concentration-dependent manner. Given that CYP1A and CYP3A are the major drug-metabolizing enzymes in a variety of animal species, we assume that these two enzymes also play a major role in the drug metabolism of crucian cap and other fishes. This indicates that when berberine is co-administrated with other drugs metabolized by CYP1A and CYP3A, it will increase the efficacy of these drugs and reduce the amount of its application. Meanwhile, it may also lower the threshold of these drugs toxicity and thus increase their virulence. Hence, this study will help guide the coadministration of berberine with other drugs to fishes and further diminish the pollution caused by drugs.

Acknowledgements

This work was financed by a special research grant (No. 2007M06) for the National Nonprofit Institutes (East China Sea Fisheries Research Institute).

References

Abdelmegeed, M. A., Carruthers, N. J., Woodcroft, K. J., Kim, S. K., Novak, R. F., 2005. Acetoacetate induces CYP2E1 protein and suppresses CYP2E1 mRNA in primary cultured rat hepatocytes. J. Pharmacol. Exp. Ther. 315, 203-213.

Anzenbacher, P., Anzenbacherova, E., 2001. Cytochromes P450 and metabolism of xenobiotics. Cell Mol. Life Sci. 58, 737-747.

Bradford, M. M., 1976. A rapid and sensitive method for the quantitation of microgram quantities of protein utilizing the principle of protein-dye binding. Anal. Biochem. 72, 248-254.

Brahn, E., Tang, C., Banquerigo, M. L., 1994. Regression of collagen-induced arthritis with taxol, a microtubule stabilizer. Arthritis Rheum. 37, 839-845.

Celander, M., Buhler, D. R., Forlin, L., Goksoyr, A., Miranda, C. L., Woodin, B. R., Stegeman, J. J., 1996. Immunochemical relationships of cytochrome P4503A-like proteins in teleost fish. Fish Biochem. Physiol. 15, 323-332.

Chatterjee, P., Franklin, M. R., 2003. Human cytochrome p450 inhibition and metabolic-intermediate complex formation by goldenseal extract and its methylenedioxyphenyl components. Drug Metab. Dispos. 31, 1391-1397.

Coon, M. J., Ding, X. X., Pernecky, S. J., Vaz, A. D., 1992. Cytochrome P450: progress and predictions. FASEB J. 6, 669-673.

Dixon, M., 1953. The determination of enzyme inhibitor constants. Biochem. J. 55, 170-171.

Guo, Y. J., Chen, X. N., 2005. Current progress on utilization of Chinese herbal medicines in prevention and treatment of aquatic animal diseases in China. Fisher Sci. 2005 (24), 34-38.

Hasselberg, L., Grosvik, B. E., Goksoyr, A., Celander, M. C., 2005. Interactions between xenoestrogens and ketoconazole on hepatic CYP1A and CYP3A, in juvenile Atlantic cod (Gadus morhua). Comp. Hepatol. 4, 2.

Hasselberg, L., Westerberg, S., Wassmur, B., Celander, M. C., 2008. Ketoconazole, an antifungal imidazole, increases the sensitivity of rainbow trout to 17alpha-ethynylestradiol exposure. Aquat. Toxicol. 86, 256-264.

Hwang, B. Y., Roberts, S. K., Chadwick, L. R., Wu, C. D., Kinghorn, A. D., 2003. Antimicrobial constituents from goldenseal (the rhizomes of Hydrastis canadensis) against selected oral pathogens. Planta. Med. 69, 623-627.

Iizuka, N., Miyamoto, K., Okita, K., Tangoku, A., Hayashi, H., Yosino, S., Abe, T., Morioka, T., Hazama, S., Oka, M., 2000. Inhibitory effect of Coptidis rhizoma and berberine on the proliferation of human esophageal cancer cell lines. Cancer Lett. 148, 19-25.

Kuo, C. L., Chi, C. W., Liu, T. Y., 2004. The anti-inflammatory potential of berberine in vitro and in vivo. Cancer Lett. 203, 127-137.

LaemmLi, U. K., 1970. Cleavage of structural proteins during the assembly of the head of bacteriophage T4. Nature 227, 680-685.

Lange, U., Goksøyr, A., Siebers, D., Karbe, L., 1999. Cytochrome P450 1A-dependent enzyme activities in the liver of dab (Limanda limanda): kinetics, seasonal changes and detection limits. Comp. Biochem. Physiol. B Biochem. Mol. Biol. 123, 361-371.

Li, D., Yang, X. L., Zhang, S. J., Lin, M., Yu, W. J., Hu, K., 2008. Effects of mammalian CYP3A inducers on CYP3A-related enzyme activities in grass carp (Ctenopharyngodon idellus): Possible implications for the establishment of a fish CYP3A induction model. Comp. Biochem. Physiol. C Toxicol. Pharmacol. 147, 17-29.

Li, X. C., Zhang, R. R., Sun, R. R., Lan, J. F., Zhao, X. F., Wang, J. X., 2010. Three Kazal-type serine proteinase inhibitors from the red swamp crayfish Procambarus clarkii and the characterization, function analysis of hcPcSPI2. Fish Shellfish Immunol. 28, 942-951.

Novi, S., Pretti, C., Cognetti, A. M., 1998. Biotransformation enzymes and their induction by β-naphtoflavone in adult sea bass (Dicentrarchus labrax). Aquat. Toxicol. 41, 63-81. Shimamoto, Y., Tasaki, T., Kitamura, H., Hirose, K., Kazusaka, A., Fujita, S., 1999. Decrease in hepatic CYP2C11 mRNA and increase in heme oxygenase activity after intracerebroventricular injection of bacterial endotoxin. J. Vet. Med. Sci. 61, 609-613.

Tu, Y. Y., Yang, C. S., 1983. High-affinity nitrosamine dealkylase system in rat liver microsomes and its induction by fasting. Cancer Res. 43, 623-629.

Uttamsingh, V., Lu, C., Miwa, G., Gan, L. S., 2005. Relative contributions of the five major human cytochromes P450, 1A2, 2 C9, 2 C19, 2D6, and 3A4, to the hepatic metabolism of the proteasome inhibitor bortezomib. Drug Metab. Dispos. 33, 1723-1728.

Vrzal, R., Zdarilova, A., Ulrichova, J., Blaha, L., Giesy, J. P., Dvorak, Z., 2005. Activation of the aryl hydrocarbon receptor by berberine in HepG2 and H4IIE cells: Biphasic effect on CYP1A1. Biochem. Pharmacol. 70, 925-936.

White, R. D., Shea, D., Solow, A. R., Stegeman, J. J., 1997. Induction and post-transcriptional suppression of hepatic cytochrome P450 1A1 by

3, 3′, 4, 4′-tetrachlorobiphenyl. Biochem. Pharmacol. 53, 1029-1040.

Wills, L. P., Zhu, S., Willett, K. L., Di Giulio, R. T., 2009. Effect of CYP1A inhibition on the biotransformation of benzo [a] pyrene in two populations of Fundulus heteroclitus with different exposure histories. Aquat. Toxicol. 92, 195-201.

Xin, H. W., Wu, X. C., Li, Q., Yu, A. R., Zhoug, M. Y., Zhu, M., Liu, Y. Y., 2004. Effects of berberine chloride and co-administration with cyclosporin on CPY3A1 in rat liver and small intestine. Chin. J. Clin. Pharmacol. Ther. 5, 565-568.

Yamauchi, R., Ishibashi, H., Hirano, M., Mori, T., Kim, J. W., Arizono, K., 2008. Effects of synthetic polycyclic musks on estrogen receptor, vitellogenin, pregnane X receptor, and cytochrome P450 3A gene expression in the livers of male medaka (Oryzias latipes). Aquat. Toxicol. 90, 261-268.

Yu, L. Z., Yang, X. L., 2009. Effects of fish cytochromes P450 inducers and inhibitors on difloxacin N-demethylation in kidney of Chinese idle (Ctenopharyngodon idellus). Environ. Toxicol. Phar. 29, 202-208.

Yu, L. Z., Yang, X. L., Wang, X. L., Yu, W. J., Hu, K., 2010. Effects of fish CYP inducers on difloxacin N-demethylation in kidney cell of Chinese idle (Ctenopharyngodon idellus). Fish Physiol. Biochem. 36, 677-686.

Zhang, L. L., Zhang, J. R., Guo, K., Ji, H., Zhang, Y., Jiang, S. X., 2011. Effects of fluoroquinolones on CYP4501A and 3A in male broilers. Res. Vet. Sci. Sci. 90, 99-105.

Zhao, X., Zhang, J. J., Wang, X., Bu, X. Y., Lou, Y. Q., Zhang, G. L., 2008. Effect of berberine on hepatocyte proliferation, inducible nitric oxide synthase expression, cytochrome P450 2E1 and 1A2 activities in diethylnitrosamine - and phenobarbital - treated rats. Biomed. Pharmacol. Ther. 62, 567-572.

该文发表于《Journal of Applied Ichthyology》【2013, 29 (6): 1204-1207】

Effects of Norfloxacin on the drug metabolism enzymes of two species of sturgeons (*Acipenser schrencki* and *A. ruthenus*)

S. W. Li, D. Wang, H. B. Liu, T. Y. Lu *

(Heilongjiang River Fisheries Research Institute, Chinese Academy of Fishery Sciences, Harbin 150070, China)

SUMMARY: The enzyme activities of 7-ethoxycoumarin O-deethylase (ECOD) and glutathione S-transferase (GST), and on glutathione (GSH) content, which are involved in the metabolism of antibiotic Norfloxacin (NFLX), were investigated in Acipenserschrencki and Acipenserruthenus. Sturgeons weighing 45-55 g were kept in an aquarium (0.5 m×0.5 m×0.9 m) for 2 weeks under controlled conditions (fish density 88 individuals per m^3, 18℃) before the experiment. The two species of sturgeon were divided into 5 groups each (n=15 in each group), with each group subdivide into 3 replicates of 5 fish per tank. A control group, in which distilled water was administered orally, was also tested. NFLX was forced into the stomach of the fish at the concentration of 20, 40, 60, 80 and 100 mg · kg^{-1} body weight, respectively. ECOD activities in liver microsomes, and GST activityandGSH content in liver microsomes and blood plasma, were measured and compared. Our results indicate that ECOD activity is progressively inhibited with increasing NFLX concentrations. ECOD activity varied from 0.12 nmoL · mg^{-1} · min^{-1} to 0.07 nmoL · mg^{-1} · min^{-1}, demonstrating an inhibition rate of 60.83% in A. schrencki and 65.14% in A. ruthenus. In both species tested, GST and GSH levels exhibited a trend of first increasing, and then decreasing with increasing NFLX levels, reaching a peak value at 40 mg per kg^{-1} body weight. Thus, the results we present indicate that NFLX can induce a change in the activity of some drug metabolism related enzymes such as ECOD and GSTin vivo.

key words: Norfloxacin, 7-ethoxycoumarin O-deethylase, Glutathione S-transferase, Glutathione, Acipenser schrencki, Acipenser ruthenus

诺氟沙星对施氏鲟和小体鲟药物代谢相关酶的影响

李绍戊，王荻，刘红柏，卢彤岩

(中国水产科学研究院黑龙江水产研究所，哈尔滨，中国)

摘要： 本研究检测了施氏鲟和小体鲟体内参与诺氟沙星（NFLX）代谢的 7-乙氧基香豆素-O-脱乙基酶（ECOD）、谷胱甘肽巯基转移酶（GST）的活性和谷胱甘肽（GSH）的含量。实验前，选取体重在 45~55 g 之间的实验鲟于水族箱中（0.5 m×0.5 m×0.9 m）暂养 2 周，养殖密度为 88 尾/m³，养殖水温为 18℃。将两种实验鲟分别随机分为 5 组（每组 15 尾），分别按 NFLX 20、40、60、80 和 100 mg·kg⁻¹ 的浓度口灌给药，并设空白对照组（0 mg·kg⁻¹）。停药后，对肝脏微粒体中的 ECOD 酶活、肝微粒体和血浆中的 GST 酶活及 GSH 含量进行测定和比较。结果表明，随着 NFLX 浓度的提高，ECOD 酶活逐渐受到抑制，从 $0.12 \ nmoL \cdot mg^{-1} \cdot min^{-1}$ 的浓度降至 $0.07 \ nmoL \cdot mg^{-1} \cdot min^{-1}$；其中，ECOD 在施氏鲟中的抑制率为 60.83%，在小体鲟中的抑制率为 65.14%。随着 NFLX 浓度的升高，GST 和 GSH 水平在两种鲟体内均呈现先升高后降低的趋势，且在药物浓度为 40 mg·kg⁻¹ 时达到峰值。综上，本研究结果表明，诺氟沙星可以在鲟鱼体内诱导药物代谢相关酶如 ECOD 和 GST 等的酶活变化。

关键词： 诺氟沙星；ECOD；GST；GSH；施氏鲟；小体鲟

Introduction

Fluoroquinolones are important group of antibiotics that are used to control pulmonary, urinary and digestive-tract bacterial infections in farmed animals (Papich and Riviere, 2001). They inhibit bacterial DNA gyrase, a bacterial topoisomerase Ⅱ which is essential for DNA replication and transcription (Yorke and Froc, 2000). In contrast to third generation fluoroquinolones used previously, Norfloxacin (NFLX) is inexpensive, and has been widely applied for disease control in fish farming for several years. NFLX has strong and broad-spectrum antibacterial activity, especially against aerobic gram-negative and gram-positive bacteria. It also has few side-effects (Holmes et al., 1985). In China, NFLX is administrated to farmed fish as medicated feed at a rate of 30~50 mg per kg⁻¹ body weight per day for 3-5 days to treat bacterial infection (Yang et al., 2003; Fan et al., 2005; Shi et al., 2005).

Cytochrome P450 (CYP450) enzymes are known for their role in the metabolism of non-polar compounds (Slaughter and Edwards, 1995; Anzenbacher and Anzenbacherová, 2001). Within the complex group of cytochromes, the enzymes in CYP families 1-3 are primarily involved in the biotransformation of drugs, and are often referred to as drug-metabolizing enzymes (Fu et al., 2011; Gildayet al., 1996). In liver microsomes, this enzyme is involved in an NADPH-dependent electron transport pathway. It oxidizes a variety of structurally unrelated compounds, including steroids, fatty acids, and xenobiotics, and participates in the phase Ⅰ enzyme reactions which are important in drug metabolism. Therefore, the activities of CYP450 enzymes determine the metabolism rate of drugs and are related to the clearance rate of drugs (Oguand Maxa, 2000; Lynch and Price, 2007). It is estimated that the biotransformation of more than 60% of drugs is achieved through CYP450 enzymes (Venkatakrishnanet al., 2001). Glutathione S-transferase (GST) catalyze the formation of thioether conjugated between glutathione (GSH) and reactive xenobiotic compounds in phase Ⅱ reactions (Chooet al., 2004). Their major biological function is believed to be defense against electrophilic chemical species, many of which are formed by cellular oxidative reactions catalyzed by cytochromes P450 and other oxidases (Arriagaet al., 2000). GST are widely distributed in many tissues of the fish body, and play an important role in anti-oxidation fish, by acting in combined with GSH. The activities of GSTenzymes and the content of GSH are commonly used to evaluate the capability of organisms to metabolize drugs.

Acipenserschrencki and Acipenserruthenus are two of the main sturgeon species cultured in China. In recent years, bacterial outbreaks have occurred frequently in sturgeon fish farms. NFLX has been widely applied in the prevention and control of bacterial diseases. However, few studies investigated the use of drugs for treating diseases in sturgeons. Although it is known that CYP450 plays a role in NFLX metabolism of humans and several other animals, very little information is available on the activities of the biotransformation enzymes and phase Ⅱ conjugation enzymes in sturgeons exposed to NFLX, such as 7-ethoxycoumarin O-deethylase (ECOD), GST enzymes and GSH proteins. In this study, ECOD and GST enzyme activities andGSH content in sturgeons treated with varying concentrations of NFLX were determined in vivo, and the effects of NFLX on these enzymes were evaluated.

Materials and methods

Chemicals

NFLX was manufactured by Zhejiang Guobang Pharmaceutical Corporation (Zhejiang, China, approval number X087212). The liver microsomes were extracted using GENMED Animal Tissue Microsomes Isolation Kit (GMS18002, GENMED Scientifics Inc., USA) and total ECOD enzyme activity was measured using the GENMED Fluorescence Quantitative Diagnostic Kit (GMS18016.1, GENMED Scientifics Inc., USA). GST activity and GSH content were determined using kits from Nanjing Jiancheng Bioengineering Institute (Nanjing, China). All other organic solvents and chemicals used were analytical grade.

Fish and experimental treatments

Acipenserschrenckiand Acipenserruthenus were provided by the Technological and Engineering Center of Sturgeon's Reproduction, Chinese Academy of Fishery Sciences (Beijing, China). All fish were disease-free and healthy. The sturgeons weighed $45 \sim 55$ g and were kept in an oxygenated aquarium (0.5 m×0.5 m×0.9 m) for 2 weeks before the experiment began. The fish were maintained under the following conditions: fish density in the aquarium was 88 individuals per m^3. Water temperature was maintained at 18℃.

Fish of both species were divided at random into 5 groups per species, with 15 individuals in each group. Each group was then subdivided randomly into 3 groups of 5 individuals each, to allow 3 replicates per treatment. NFLX was dissolved in distilled water to make a suspension with a concentration of 1.25 mg · mL^{-1}. This suspension was stored in the dark at 4℃ until needed. NFLX was inserted into the stomach of the fish using a 1 mL plastic syringe. Five NFLX concentrations, 20, 40, 60, 80 and 100 mg kg^{-1} body weight, and one control treatment, were tested. To avoid the drug flowing out of the mouth and to ensure exact dosage, fish were medicated for five days with only minimal feeding, and then fed continuously for two daysso as to recover body condition. The control treatmentwas carried out by oral administration of distilled water. Before sampling, fish were briefly subjected to percussive stunningso as to reduce their discomfort. Blood was sampled from the caudal vein using heparin sodium (Wheat Warehouse Biological Technology Co. Ltd, Shanghai, China) treated syringe, and blood plasma was collected by centrifugation at 4500 rpm for 10 min. The fish then were killed and the livers collected and frozen at−80℃. Control fish (0 mg kg^{-1} body weight) were sampled before the initiation of the study.

Enzyme measurements

The activity unit of 7-ethoxycoumarin O-deethylase (CYP-ECOD) was defined as the enzyme amount need for 1 nmoL 7-ethoxycoumarin to be transformed into 7-hydroxycoumarin per minute at pH 7.5 and 37℃. The microsomes in livers were prepared through high-speed centrifugation of homogenate, and the protein content was measured following the Coomassie brilliant blue protocol described by Bradford (1976). CYP-ECOD was determined in vivo following the kit protocol. Fluorescent values were detected at an excitation wavelength of 368 nm and emission wavelength of 456 nm.

Glutathione-S-transferase activity (GST) was evaluated by absorbance at 340 nm due to the conjugation of GSH to 1-chloro-2, 4-dinitrobenzene (CDNB). The reaction mixture contained in final volume of 1.0 mL, including 765 μL sodium phosphate buffer (50 mM, pH 7.5), 25 μL GSH (40 mM), 10 μL CDNB (40 mM) and 200 μL enzyme extract, as described by Habiget al. (1974).

Statistical analysis

The results were reported as the mean±SD of the six fish in each treatment. Statistical analysis was carried out using Microsoft Excel 2003 and SPSS 17. Differences between the controls and the treatments were evaluated using a one-way analysis of variance (ANOVA) and P values<0.05 were considered significant and labeled with the (∗) symbol.

Results

Effect of NFLX on ECOD activity

Table 1 shows the effects of the five treatment strengths of NFLX on the activity of ECOD in liver microsomes of Acipenserschrenckiand Acipenserruthenus. NFLX strongly reduced ECOD activity in both A. schrencki and A. ruthenus, in which ECOD activity was reduced to 60.83% and 65.14% of the control value, respectively, at aconcentration of 100 mg kg^{-1} body weight of NFLX. As NFLX concentration was increased from 0 to 100 mg kg^{-1} body weight, the activity of ECOD reduced linearly, and was higher in A. schrenckithan in A. ruthenus. The ECOD activity in both the 80 and 100 mg kg^{-1} body weight groups were significantly lower than in the control group in A. schrencki, and significantly lower in the A. ruthenus100 mg kg^{-1} body weight group. The difference in ECOD activity between the two sturgeon species was highest at the 20 mg kg^{-1} body weight group, with a difference of 0.017 nmoL mg^{-1} min^{-1}. The regression equation between the CYP-ECOD inhibition rate and the naturallogarithm of NFLX concentration was y = $0.0021e^{2.0779x}$ ($R^2 = 0.949$) in A. schrencki and y = $0.0016e^{2.399x}$ ($R^2 = 0.970$) in A. ruthenus.

Table 1　The activities of 7-ethoxycoumarin O-deethylase（CYP-ECOD）in the liver microsome of Acipenserschrenckiand Acipenserruthenus

NFLX（mg · kg⁻¹ body weight）Fish species	0	20	40	60	80	100
	The activities of CYP-ECOD（nmoL mg⁻¹min⁻¹）					
Acipenserschrencki	0. 120±0. 010	0. 114±0. 005	0. 095±0. 003	0. 088±0. 011	0. 080±0. 012 *	0. 073±0. 015 *
Acipenserruthenus	0. 109±0. 004	0. 097±0. 016	0. 091±0. 013	0. 086±0. 007	0. 078±0. 002	0. 071±0. 008 * *
D-value	0. 011	0. 017	0. 006	0. 002	0. 002	0. 002

Activity levels of CYP – ECOD enzymes in the livermicrosomes of Acipenserschrencki and Acipenserruthenus in relation to theapplied dosage of NFLX. Data represent mean values and standard deviations（3 replicates with 5 fish each）. D-value＝the differences between the two sturgeon species; ∗P<0. 05, ∗∗P<0. 01

As shown in Table 2, significant positive correlations were found between hepatic CYP-ECOD and GST activity in both species of sturgeon when NFLX concentration was increased from 0 to 40 mg · kg⁻¹ body weight（0<Pearson's r <1）. However, we found negative correlations between CYP-ECOD and GST activity at the NFLX concentration of 60, 80 and 100 mg · kg⁻¹ body weight（−1< Pearson's r <0）.

Table 2　Correlation analysis of hepatic7-ethoxycoumarin O-deethylase（CYP-ECOD）with GST

NFLX（mg · kg⁻¹ body weight）	0	20	40	60	80	100
Pearson's r	Hepatic GST（mg g⁻¹）					
Acipenserschrencki Hepatic CYP-ECOD	0. 978	0. 477	0. 470	−0. 883	−0. 997	−0. 845
Acipenserruthenus Hepatic CYP-ECOD	0. 936	0. 442	0. 250	−0. 992	−0. 493	−0. 365

Pearson's r, correlation coefficient by Pearson's analysis

Effect of NFLX on GST activity and GSH content in the liver microsome and blood plasma

In the control groups the GST activity of A. ruthenus was found to be higher than that of A. schrencki in both the liver microsomes and blood plasma. Compared to the control group, the GST activity was observed to increase at the lower concentrations of NFLX, reaching a peak value at the 40 mg kg⁻¹ body weight concentration in both the liver microsomes and blood plasma of both species of sturgeon. At higher NFLX levels GST activity was gradually reduced. It is noteworthy that GST levels in the blood plasma of A. ruthenuswere found to be a little lower in the 100 mg · kg⁻¹ body weight treatment than in the control group. Overall, GST activity in the liver microsomes and the blood plasma exhibited a trend of first increasing and then decreasing with the increase of NFLX content, as shown in Figure 1.

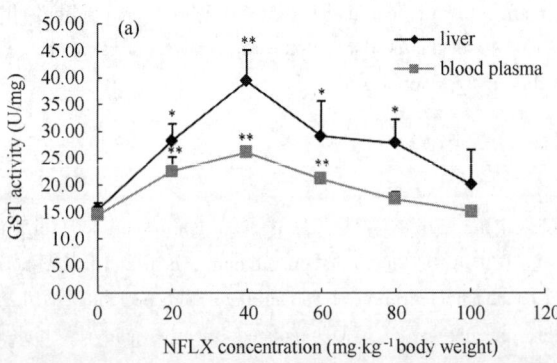 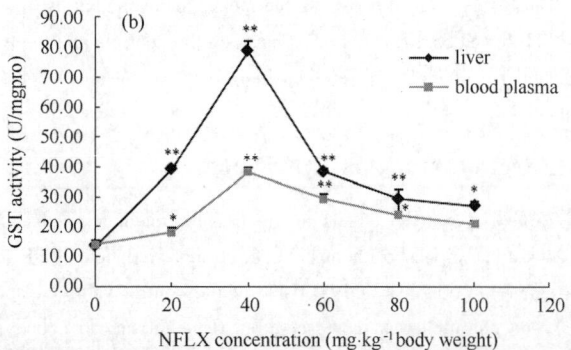

Fig. 1　The effects of different NFLX concentrations on the GST activity in liver microsome and blood plasma in A. ruthenus（a）and A. schrencki（b）. Samples were taken 48h after drug administration. Points are means and whiskers are half the standarddeviation（3 replicates with 5 fish each）.

Similarly to the change of GST activity, the GSH content also exhibited a trend of first increasing then decreasing with an increase in NFLX concentration in both species of sturgeon, with peak GSH content reached at 40 mg $\cdot$ kg^{-1} body weight concentration. GSH content was significantly higher in liver microsomes than in blood plasma in both fish species. As shown in Table 3, the maximum GSH content recorded was 20.746±1.486 mg $\cdot$ g^{-1} in the blood plasma and 23.684±1.060 mg $\cdot$ g^{-1} in liver microsomes of Acipenserruthenus, while for Acipenserschrenckithe GSH content was 23.684±1.060 mg $\cdot$ g^{-1} and 46.304±1.258 mg $\cdot$ g^{-1} in the blood plasma and liver microsomes, respectively.

Table 3　The GSH contents in blood plasma and liver microsomes in two sturgeons speciestreated with different NFLX concentrations

NFLX (mg $\cdot$ kg^{-1} body weight)	GSH content (mg $\cdot$ g^{-1})			
	Acipenserruthenus		Acipenserschrencki	
	Blood plasma	Liver microsome	Blood plasma	Liver microsome
0	11.400±1.400	16.374±2.423	9.320±1.630	10.943±0.749
20	16.585±2.077	24.161±3.319	11.593±0.205	21.717±0.589*
40	20.746±1.486**	38.192±4.274**	23.684±1.060**	46.304±1.258**
60	14.498±1.969**	22.665±1.349	16.168±1.585**	30.013±3.132**
80	12.180±1.797	18.851±2.288	9.799±1.259	17.795±1.130*
100	12.056±0.413	17.367±2.649	9.505±0.812	13.980±3.262

Note: values are mean±standard deviation; * $P<0.05$, ** $P<0.01$

Discussion

The metabolism of most xenobiotics, such as antibiotics, is always attributed to phase I enzymes, which are generally the first step in detoxification and excretion of toxins. In this study, ECOD activity was inhibited by treatment with NFLX in a dose-dependent manner. Vaccaroet al. (2003) reported that a fluoroquinolone, NOR, could induce the inhibition of cytochrome P450 monoxygenase activities in sea bass. This may help explain the dose-dependent decrease of ECOD during the metabolism of NFLX: ECOD may bind with NFLX and produce some intermediates with electrophilic properties responsible for suicidal enzymatic reactions. The results here also indicated that, in vivo, NFLX inhibits ECOD enzymes in an irreversible manner.

As a consequence, some intermediates products of phase I metabolism may induce the activity of some phase II enzymes, such as GST. In our experiment, GST activityand GSH contents first increased and then decreased with increasing NFLX treatment concentrations. The change of GST activity and GSH content make it clear that they also participate in the metabolism of NFLX, whether through phase I induced enzymes or not. The GST activity and GSH content reached a maximum value at 40 mg kg^{-1} body weight concentration of NFLX and then decrease gradually. Thus, high concentrations of NFLX may lead to some toxic reaction in vivo and may be harmful to treated fish. However, GST activity and GSH content were nonetheless higher in the 100 mg kg^{-1} body weight treatment group than in the control group, which demonstrates that NFLX can increase the activity of GST and content of GSH.

A. schrencki and A. ruthenus are both part of the genus Acipenser, in the family Acipenseridae, which display highly variable growth traits and disease-resistance. A. schrenckiexhibitscertaincharacteristics, such as rapid growth, largesize, late sexual maturity, and disease-resistance, in constrast to A. ruthenus, which aresmaller, reach sexual maturity earlier, and are less disease resistantthanA. schrencki (Zhang et al., 2012). Except for Acipenserbaeriand hybrids, these two species of sturgeon arethe main sturgeon species cultured in China (Sun et al., 2011). In addition, Liu et al. (2006/2007) found that there was a significant difference in lysozyme and phosphatase activities between these two species. Due to the recent widespread application of NFLX in prevention and control of bacterial diseases in sturgeon, the investigation of NFLX metabolism in vivo is of great importance, as these determine several pharmacological and toxicological properties of drugs. Studies on drug metabolism related enzymes arevaluable for understanding the mechanism of drug metabolism. However, no information was available on the difference in ECOD and GST enzyme activities in vivo betweenA. schrencki and A. ruthenus. Here, the activities of ECOD and GST, and the GSH content, were found to vary somewhat between A. schrenckiand A. ruthenus, which indicates that they may show different physiological responses to NFLX.

In summary, this study is the first to report that NFLX can affect the ECOD and GSTactivities of liver microsomes and blood plasma in A. schrenckiand A. ruthenus and thereby to examine the mechanism of NLFX metabolism in these two species. In future studies the expression level of CYP450 and its members such as CYP1A, CYP3A both at the molecular and protein levels, should be examined, so as to better clarify the metabolism of NFLX in sturgeons.

Acknowledgement

This project was supported by grants from the Special Fund for Agro – scientific Research in the Public Interest of China (No. 201003055) and the Central–Level Non–profit Scientific Research Institutes Special Funds (No. 201001).

References

A rriaga, M.; Rivera, S. R.; Parra, C. G.; Barron, M. F.; Flores, P. R.; Garcia, J. E., 2000: Antimutagenesis of b–carotene to mutations induced by quinolone on Salmonella typhimurium. Arch. Med. Res. 31, 156–161.

Anzenbacher, P.; Anzenbacherová, E., 2001: Cytochromes P450 and metabolism of xenobiotics, cellular and molecular. Life. Sci. 58 (5–6), 737–747.

Bradford, M. M., 1976: A rapid and sensitive method for the quantitation of microgram quantities of protein utilizing the principle of protein–dye binding. Anal. Biochem. 72, 248–254.

Choo, K. S.; Snoeijs, P.; Pedersen, M., 2004: Oxidative stress tolerance in the filamentous green algae Cladophoraglomerata andEnteromorphaahlneriana. J. Exp. Mar. Biol. Ecol. 298 (1), 111–123.

Fan, H. P.; Zeng, Z. Z.; Lin, Y., 2005: The vitro antibacterial activity of enrofloxacin with other four antimicrobial agents against aquatic pathogenic in bacterin. J. Fijian. Fish. 3 (1), 34–37.

Fu, G. H.; Yang, X. L.; Zhang, H. X.; Yu, W. J.; Hu, K., 2011: Effects of cytochrome P450 1A substrate (difloxacin) on enzyme gene expression and pharmacokinetics in crucian cap (*hybridized Prussian carp*). Environ. Toxicol. Pharmacol. 31 (2), 307–313.

Gilday, D.; Gannon, M.; Yutzey, K.; Bader, D.; Rifkind, A. B., 1996: Molecular cloning and expression of two novel avian cytochrome P450 1A enzymes induced by 2, 3, 7, 8-tetrachlorodibenzop-dioxin. J. Biol. Chem. 271, 33054–33059.

Habig, W. H.; Pabst, M. J.; Jacoby, W. B., 1974: Glutathiene S – transferases, the first enzymatic step in the mercapturic acid formation. J. Biol. Chem. 249, 7130–7139.

Holmes, B.; Brogden, R. N.; Richards, D. M., 1985: Norfloxacin. A review of its antibacterial activity, pharmacokinetic properties and therapeutic use. Drugs. 30 (6), 482–513.

Liu, H. B.; Zhang, Y.; Bai, X. J., 2006: Comparison of phosphatase content in different tissues of Amur sturgeon Acipenserschrencki Brandt and the starlet Acipenserruthenus Linnaeus. Heilongjiang. Anim. Sci. Vet. Med. 6, 92–93.

Liu, H. B.; Wei, W., 2007: Comparison of lysozyme content in different tissues of Amur sturgeon Acipenserschrencki Brandt and the starlet Acipenserruthenus Linnaeus. Chin. J. Fish. 20 (1), 57–60.

Lynch, T.; Price, A., 2007: The effect of cytochrome P450 metabolism on drug response, interactions, and adverse effects. Am. Fam. Physician. 76 (3), 391–396.

Ogu, C. C.; Maxa, J. L., 2000: Drug interactions due to cytochrome P450. Proc. (BaylUniv Med Cent) 13 (4), 421–423.

Papich, M. G.; Riviere, J. E., 2001: Fluoroquinolones antimicrobial drugs. Vet PharmacolTher. In: Adams, H. R. (Ed.), eighth ed. Iowa State University Press: Iowa, pp898–917.

Shi, Y. S.; Tong, G. Z.; Xue, C. B., 2005: Identification and drug sensitive test of Vibrio harveyi isolated from the tail–rotted Pseudosciaeracrocea in Zhoushan. Chin. J. Health. Lab. Tech. 15 (3), 267–269.

Slaughter, R. L.; Edwards, D. J., 1995: Recent advances: the cytochrome P450 enzymes. Ann. Pharmacother. 29 (6), 619–624.

Sun, D. J.; Qu, Q. Z.; Zhang, Y.; Ma, G. J.; Wang, N. M., 2011: Sturgeon aquaculture in China. Chin. J. Fish. 24 (4), 67–70.

Vaccaro, E.; Giorgi, M.; Longo, V.; Mengozzi, G.; Gervasi, P. G., 2003: Inhibition of cytochrome P450 enzymes by enrofloxacin in the sea bass (*Dicentrarchus labrax*). Aquat. Toxicol. 62 (1), 27–33.

Venkatakrishnan, K.; Von Moltke, L. L.; Greenblatt, D. J., 2001: Human drug metabolism and the cytochrome P450: application and relevance ofinvitromodels. J. Clin. Pharmacol. 41 (11), 1149–1179.

Yang, Y. H.; Tong, H. M.; Lu, T., 2003: Pharmacodynamics of several fluoroquinolones against A. hydrophila in vitro. J. Northeast. Agric. Univ. 34 (4), 368–371.

Yorke, J. C.; Froc, P., 2000: Quantitation of nine quinolones in chicken tissues by high performance liquid chromatography with fluorescence detection. J. Chromatogr. 882 (1–2), 63–77.

Zhang, Y.; Sun, H. W.; Liu, X. Y.; Qu, Q. Z.; Sun, D. J., 2012: Histological observation of gonadal differentiation and effect of rearing temperature on sex differentiation in Amur sturgeon Acipenserschrenckii. J. Fish. Sci. China. 19 (6): 1008–1017.

该文发表于《中国水产科学》【2011, 18 (2): 392–399】

氟甲喹对异育银鲫细胞色素 CYP450 主要药酶的影响

Effects of flumequine on cytochrome P450 enzymes in allogynogenetic silver crucian cap, *Carassius auratus gibelio*

胡晓[1,2]，房文红[1]，汪开毓[2]，孙贝贝[1]，胡琳琳[1]，周帅[1]，周俊芳[1]

(1. 中国水产科学研究院东海水产研究所 农业部海洋与河口渔业资源及生态重点开放实验室，上海 200090；

2. 四川农业大学，四川 雅安 625014)

摘要：本试验在分析了氟甲喹腹腔注射后对异育银鲫主要药酶活性影响的基础上，从蛋白表达和 mRNA 转录水平探究了氟甲喹对 CYP1A（EROD）的诱导机制。氟甲喹以剂量 35 mg/kg 一次性腹腔注射异育银鲫 24 h 后，肝微粒体的 7-乙氧基异吩唑酮-O-脱乙基酶（EROD）活性为 54.33 pmol/mg·min，显著高于对照组（34.00 pmol/mg·min）（$P < 0.01$），而红霉素-N-脱甲基酶（ERND）（177.98 pmol/mg·min）、氨基比林-N-脱甲基酶（APD）（934.40 pmol/mg·min）及 7-乙氧基香豆素-O-脱乙基酶（ECOD）（9.84 pmol/mg·min）与对照组（分别为 140.90 pmol/mg·min、850.71 pmol/mg·min 和 8.93 pmol/mg·min）相比无显著性差异。除肾组织中 ERND 活性高于肝脏外，其他 CYP 亚型药酶活性（APD、EROD 和 ECOD）均以肝组织中最高。Western-blotting 印记表明，试验组肝脏中 CYP1A 蛋白含量明显高于对照组，与酶活（EROD）相符合；而肾、肠中没有检测到特异性条带。半定量 RT-PCR 结果显示，CYP1A mRNA 在肝、肾和肠均有表达，但试验组与对照组并无明显差异。体外试验中，不同浓度的氟甲喹与微粒体共孵育，未见 EROD 活性与浓度、时间的依赖关系。综上推测，氟甲喹对鲫鱼肝 CYP1A 的诱导是在翻译后水平，可能是加强蛋白的稳定性。

关键词：细胞色素 P450；异育银鲫；氟甲喹；诱导

细胞色素 P450（CYP450）是一种多功能酶系，参与多种内源性和外源性化合物的氧化代谢。细胞色素 P450 家族包含许多同工酶，CYP1、CYP2 和 CYP3 三个家族在药物代谢中起着重要作用，在参与药物代谢的同时，还受着其他药物的影响，或被诱导或被抑制。黑鲈（*Dicentrarchus labrax*）腹腔注射 80 mg/kg 剂量的 β-萘黄酮后，其 CYP1A 酶活提高了 6 倍[1]，而注射恩诺沙星其酶活则降低了 85%[2]；多氯联苯和阿特拉津对异育银鲫 P450 酶活有显著的诱导作用[3]，而五倍子、复方抗菌剂则抑制其肝微粒体 CYP3A 活性[4]。氟甲喹（Flumequine，FLU）是一种专用于动物的抗菌药，由于其抗菌谱广、与其他抗菌药无交叉耐药性等特点，广泛应用于水产动物的细菌性疾病的预防和治疗[5]。药物的诱导或抑制作用，将会改变与之配伍的药物的代谢，影响着药物的毒性反应，以及其在体内的消除速率。因此，研究渔用药物对水产动物细胞色素 P450 活性及其诱导或抑制作用，是了解渔药代谢及药物间相互作用的基础，对临床合理用药和安全性评价具有理论意义和实用价值。本研究以异育银鲫（*Carassius auratus gibelio*）为研究对象，采用单次腹腔注射给药，比较分析了氟甲喹对肝、肠和肾组织药酶活性的影响，并在蛋白表达和转录水平进一步探讨氟甲喹诱导肝 CYP1A 的机制，为今后氟甲喹的临床合理应用提供科学依据。

1 材料方法

1.1 试剂及仪器

氟甲喹原药为苏州恒益医药原料有限公司产品，纯度>97%；7-乙氧基异吩唑酮和试卤灵购自 Sigma 公司；7-乙氧基香豆素和 7-羟基香豆素购自 Fluka 公司；还原性辅酶Ⅱ（NADPH）、红霉素和氨基比林购自上海生工生物技术服务有限公司；兔抗大鼠 CYP1A1 多克隆抗体和 PVDF 膜均购自 Millipore 公司；大鼠 GAPDH 单克隆抗体购自上海康成生物公司。其余试剂均为国药集团上海化学试剂公司产品——分析纯。而仪器及其生产厂家为：超速冷冻离心机（Beckman Optima L-100XP）；垂直凝胶电泳系统（Bio-Rad Min-PROTEAN Tetra Cell）；荧光分光光度计（上海三科仪器有限公司，970CRT）；紫外可见分光光度计（上海尤尼柯仪器有限公司，UV-2802S）。

1.2 试验动物

异育银鲫购自上海市郊青浦某养殖场，体重 70~80 g，水温 25~27℃ 下饲养 4 周后用于实验。挑选体表无损、健康的 12 尾异育银鲫，随机分成试验组和对照组，每组 6 尾。试验组腹腔注射氟甲喹，给药剂量为 35 mg/kg，同时设立对照组。24 h 后在冰上采集肝、肠和肾组织样品，保存于-80℃ 备用。

1.3 微粒体制备

参照 Novi 等[8]方法作了修改。肝、肠和肾组织用 4℃预冷的 1.15%KCl 漂洗后，按质量体积比 1∶4 加入预冷的的 1.15%KCl 匀浆，于 4℃、10 000×g 下离心 20 min，取出上清液，再于 4℃、105 000×g 下离心 60 min，沉淀即为微粒体。参照 Bradford 法[6]测定微粒体蛋白浓度后，用 20%甘油（v/v）、1 mM EDTA、100 mM PBS（pH 值为 7.4）稀释微粒体至蛋白浓度为 10 mg/mL，于-80℃保存备用。

1.4 P450 酶活测定

1.4.1 氨基比林-N-脱甲基酶（APD）和红霉素-N-脱甲基酶（ERND）

测定方法参照 Tu 和 Yang[7]。微粒体反应体系（2 mL）中含 70 mmol/L Tris-HCL、0.07 mM EDTA、5 mmol/L 氨基比林（或者 1 mmol/L 红霉素）、1 mg 微粒体蛋白，10 mmol/L NADPH（同时以蒸馏水代替作空白），于 37℃孵育 30 min 后，加入 0.2 mL、25%ZnSO$_4$终止反应。再加入 0.2 mL 饱和 Ba（OH）$_2$，3 000×g 离心 5 min，取 1.4 mL 上清液与 0.6 mLNash 试剂于 50℃下显色 30 min，紫外可见分光光度于 412 nm 测定吸光值。

1.4.2 7-乙氧基香豆素-O-脱乙基酶（ECOD）

除缓冲液为 0.3 MTris-HCl（pH 值为 7.4）外，其他均参照 Frank 等[11]方法，荧光分光光度计于 Ex368 nm、Em456 nm 测定其产物 7-羟基香豆素浓度。

1.4.3 7-乙氧基异吩唑酮-O-脱乙基酶（EROD）

在参照 Lange[8]方法上作了修改。反应体系（1 mL）中含 0.6 mg 微粒体蛋白、0.05M Tris-HCL（pH 值为 7.4）、6 μM 7-乙氧基异吩唑酮，10 mM 的 NADPH（同时以蒸馏水代替作空白），37℃孵育 5 min 后，加入等体积甲醇终止反应。5 000×g 离心 10 min，取上清液，荧光分光光度计于 Ex535 nm、Em586 nm 测定其产物试卤灵浓度。

1.5 CYP1A1 的 Western 免疫印迹

10 μg 微粒体蛋白经 5%、10%不连续聚丙烯酰胺凝胶电泳后[9]，半干法转至 PVDF 膜上，以兔抗大鼠 CYP1A1 为一抗，HRP 标记的羊抗兔 IgG 为二抗，大鼠 GAPDH 单克隆抗体为内参抗体，ECL 荧光检测对应 CYP 亚型条带。采用 Quantity One 4.6 软件对条带进行光密度分析，以 Gauss 痕迹量表示蛋白的相对含量。

1.6 CYP1A 的 mRNA 表达

采用 Trizol 试剂（TAKARA 公司产品）分别提取肝、肠和肾组织总 RNA。经 DNase I（TAKARA）处理后溶于 DEPC 处理水，取 1 μl 进行 3%琼脂糖凝胶电泳检测质量。根据已公布的异育银鲫 CYP1A、ß-actin cDNA 序列（GenBank 登录号分别为 DQ517445 和 AB039726）采用 Oligo 6.71 软件设计引物：CYP 1A：Primer（F）5′-ATCCCTTTCTTGCGTATCCT-3′，Primer（R）5′-CGATCCATTCCGATCTTTT-3′。ß-actin：Primer（F）5′-AAGGGCCAGACTCATCGT-3′，Primer（R）5′-CAATCCCAAAGCCAACAG-3′，上海生工合成。以 Prime Script RT-PCR 试剂盒进行 RT-PCR，产物经 1.5%琼脂糖电泳，采用 Quantity One 4.6 软件对条带进行光密度分析，以 Gauss 痕迹量的 CYP1A/ß-actin 表示 DNA 的相对含量。

1.7 微粒体体外孵育氟甲喹对 EROD 活性影响试验

在微粒体和 NADPH 的体外反应体系中，分别加入不同浓度氟甲喹，使其终浓度分别为 0 μM、20 μM、50 μM 和 100 μM，37℃下孵育，分别在孵育后 0 min、2 min、5 min、10 min 和 20 min 取样测定其 EROD 活性，每个时间点 2 个平行，反应体系及酶活测定方法同"1.4.3 7-乙氧基异吩唑酮-O-脱乙基酶（EROD）"。

1.8 数据分析

数据采用 SPSS 13.0 软件进行显著性分析（ANOVA），以平均数±标准差（Mean±SD）表示，组间显著性差异以字母 a 表示（$P<0.01$）。Microsoft Office Excel 2003 绘图。

2 结果

2.1 氟甲喹对异育银鲫组织中 P450 主要药酶活性的影响

采用 CYP 亚型酶的探针底物测定氟甲喹对异育银鲫肝、肠和肾等组织中 P450 药酶活性的影响（表 1）。除肾组织中 ERND 活性高于肝脏外，其他 CYP 亚型药酶活性（APD、EROD 和 ECOD）均以肝组织中最高，且远高于肠和肾脏。在肝组织中，给药组的 EROD 活性显著高于对照组（$p<0.01$），平均酶活升高 60%；而 ERND、APD 和 ECOD 活性没有显著性

差异（p> 0.01）。

表1　氟甲喹作用下异育银鲫肝、肠、肾细胞色素 P450 酶活性（n=6）

Table 1 CYP activities in liver, intestine and kidney microsomes from control and FLU-treated carp (*Carassius auratus*)

药酶 Enzyme	肝 Liver		肠 Intestine		肾 Kidney	
	氟甲喹 FLU	对照 Control	氟甲喹 FLU	对照 Control	氟甲喹 FLU	对照 Control
ERND	177.98±32.94	140.90±22.89	<50	<50	213.63	187.89
APD	934.40±173.29	850.71±229.68	<50	<50	<50	<50
EROD	54.33±5.42ᵃ	34.00±5.87	6.26±1.76	6.01±1.49	4.92	4.88
ECOD	9.84±3.29	8.93±2.77	0.61±0.02	0.62±0.40	0.79	0.71

注：酶活大小以平均数±标准差表示，单位：pmol/min per mg protein。"a" 表示与对照组相比差异极显著（$P<0.01$），t 检验。

Notes：Values are reported as means±S. D. （n=6） and are expressed as pmol/min per mg protein. ᵃ Significantly different from control microsomes：$P<0.01$ t-test.

2.2　氟甲喹对异育银鲫肝组织中 CYP1A 蛋白表达的影响

以给药组鲫肝 EROD 酶活最低和最高个体的微粒体样品、给药组6尾鲫肝微粒体等体积混合样品，以及对照组6尾鲫肝微粒体等体积混合样品上样电泳，每孔上样蛋白量均为 10 μg，其微粒体 CYP1A 蛋白免疫印迹见图1。结果表明，肝 EROD 酶活最高的个体和混合样品的 CYP1A 蛋白条带明显深于对照组等体积混合样品，即使是给药组 EROD 酶活最低的个体，其 CYP1A 蛋白条带也高于对照组等份混合样品。光密度分析显示（图2），试验组 CYP1A 蛋白平均表达量高于对照组 93%，说明氟甲喹能使 CYP1A 含量明显上升，与 EROD 酶活变化一致。采用同样的方法，对肠和肾组织中微粒体进行了 Western 免疫分析，结果并未检测到相应的 CYP1A 条带。

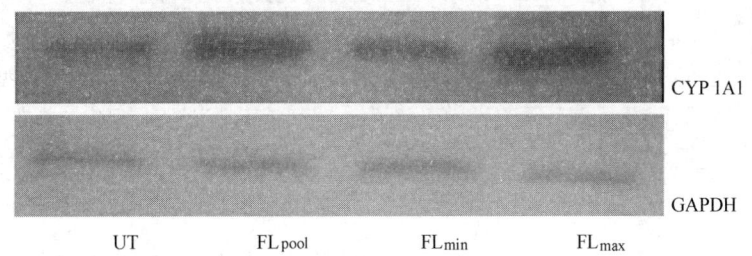

图1　氟甲喹作用下异育银鲫肝组织中 CYP1A 的 Western 免疫印迹图

Fig. 1 CYP1A Western blot from liver microsomal samples prepared from FLU-treated and control carps

注：UT：对照组6尾异育银鲫肝微粒体等体积混合样品；FLpool：给药组6尾异育银鲫肝微粒体等体积混合样品；FLmin 和 FLmax：分别表示给药组中 EROD 酶活最低个体和最高个体的肝微粒体样品。GAPDH 为内参。每孔上样蛋白量均为 10 μg。

Notes：UT：Pools of microsomes of 6 untreated fish. FLpool：Pools of microsomes of 6 flumequine treated fish. FLmin/FLmax：microsomes of flumequine treated individuals exhibiting the lowest/highest EROD activity. The protein content per well was 10 μg.

2.3　氟甲喹对异育银鲫肝、肠和肾组织 CYP1A 基因表达影响

异育银鲫腹腔注射氟甲喹后，其肝、肠和肾组织 CYP1A mRNA 的 RT-PCR 电泳图见图3，CYP1A mRNA 在肝、肠和肾组织中均有分布。采用 CYP1A mRNA 与 ß-actin 光密度比值进行半定量分析，对照组和实验组的肝、肠和肾组织 CYP1A mRNA 含量比值分别为：0.927/1.060、1.479/1.490 和 1.210/1.300。相比氟甲喹对蛋白含量变化而言，氟甲喹对 CYP1A mRNA 的影响不明显（图3）。

2.4　氟甲喹对异育银鲫肝微粒体 EROD 活性体外影响

在体外微粒体体系中，不同浓度氟甲喹与异育银鲫肝微粒体共同孵育后，其 EROD 活性见图4。由图可以看到，EROD 活性并没有随氟甲喹浓度与时间而发生变化，即在体外微粒体体系中，CYP1A 不受氟甲喹的诱导或抑制。

3　讨论

采用特异的探针药物测定 CYP 各亚型酶活是研究 P450 的重要方法。本试验依据哺乳动物和鱼类 CYP1A 依赖的7-乙氧

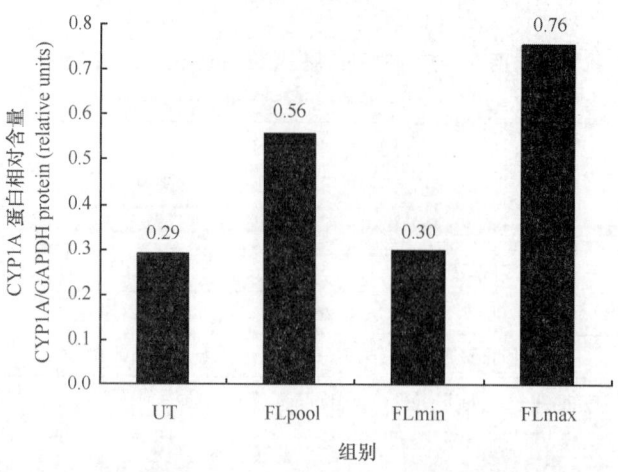

图2 氟甲喹作用下异育银鲫肝组织中 CYP1A 的 Western 免疫印迹的光密度分析图

Fig. 2 Densitometric analysis of CYP1A protein from liver microsomal samples prepared from FLU-treated and control carps

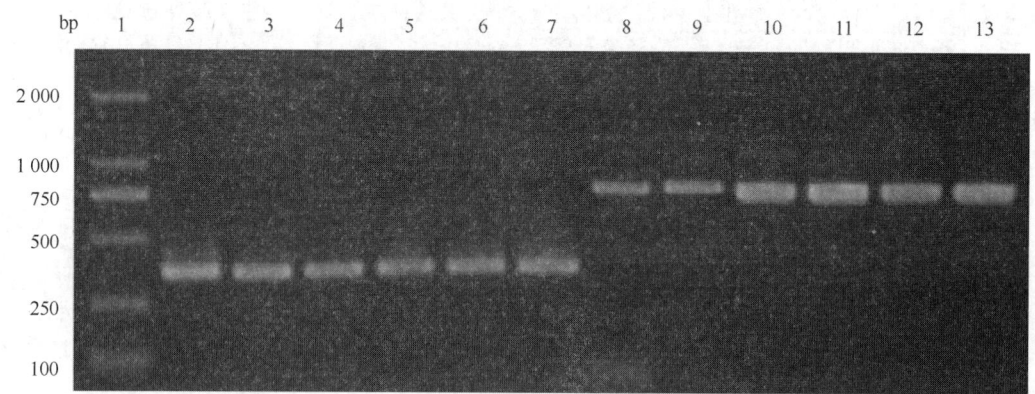

图3 氟甲喹作用下异育银鲫肝、肠和肾组织 CYP1A 的 mRNA 及 ß-actin RT-PCR 分析

Fig. 3 RT-PCR analysis of CYP1A mRNA and ß-actin in liver, intestine and kidney of untreated and

FLU-treated carps (*Carassius auratus*)

注：356bp 片段为 CYP1A mRNA；770bp 片段为 ß-actin。条带从左至右依次为：1：Marker 2000（Takara）；2、4 和 6：分别为对照组肝、肠和肾组织 CYP1A mRNA；3、5 和 7：分别为试验组肝、肠和肾组织 CYP1A mRNA；8、10 和 12：分别为对照组肝、肠和肾组织 ß-actin；9、11 和 13：分别为试验组肝、肠和肾组织 ß-actin。

Notes：The 356bp fragment corresponds to CYP1A mRNA and the 770bp fragment to constitutive gene （ß-actin）. Lane1：Marker 2000（Takara）；Lane 2-3：control/experiment group of liver CYP1A mRNA；Lane4-5：control/experiment group of intestine CYP1A mRNA；Lane 6-7：control/experiment group of kidney CYP1A mRNA；Lane 8-9：control/experiment group of liver ß-actin；Lane 10-11：control/experiment group of intestine ß-actin；Lane 12-13：control/experiment group of kidney ß-actin.

基异吩唑酮-O-脱乙基酶（EROD）[8,10,11]；CYP2B 依赖的氨基比林-N-脱甲基酶（APD）[12,13]；CYP1A、CYP2B 依赖的7-乙氧基香豆素-O-脱乙基酶（ECOD）[8,13]以及 CYP3A 依赖的红霉素-N-脱甲基酶（ERND）[14]研究了氟甲喹对异育银鲫P450 各亚型酶活的影响。本试验除肾组织中 ERND 活性高于肝脏外，APD、EROD 和 ECOD 活性均以肝组织中最高，这说明P450 各亚型药酶在鲫鱼体内的分布存在组织差异性。肝脏是药物生物转化的主要场所，通过细胞色素 P450 为主的 I 相氧化反应及谷胱甘肽转移酶等 II 相酶的结合反应，使药物活性降低、水溶性增加以利于从肾脏排泄[15]。在人和哺乳动物中，CYP3A 主要分布于肝脏和小肠[16]，参与代谢超过 50% 的药物[17]。Briubo 等[18]曾证明，鱼类的 CYP450 主要分布于肝、肾和鳃。本研究观察到的异育银鲫 P450 酶活分布与已有报道基本一致，肝脏和肾脏组织中 CYP3A 的集中分布可能与该组织参与较多外源化学物质的生物转化和解毒代谢等生理功能有关。

喹诺酮类药物对人类 CYP1A2 的抑制作用已有报道，恩诺沙星、诺氟沙星和氧氟沙星对人肝微粒体的抑制分别达到74.9%、55.7% 和 11.8%[19]。Moutou 等报道了氟甲喹和噁喹酸对虹鳟（*Oncorhynchus mykiss*）EROD 活性的影响，饲料中分别添加 1.2mg/g 和 1.0mg/g，连续投喂 10 天后，EROD 活性分别升高了 4.2 倍和 8 倍[10]。本研究报道了单剂量氟甲喹对异

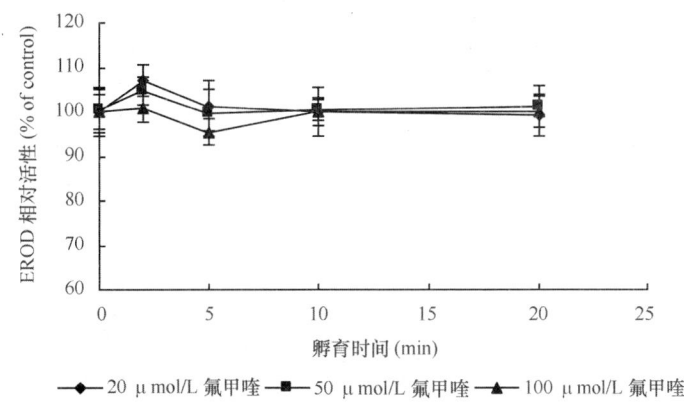

图4 不同浓度氟甲喹与异育银鲫肝微粒体体外孵育下 EROD 酶活变化曲线（$n=2$）

Fig. 4 *In vitro* EROD activities of liver microsomes from *Carassius auratus* incubated with flumequine of different concentration

育银鲫 CYP1A 的诱导作用，与人类中的结果相反，但与虹鳟一致。CYP1A 是一类古老的家族，在进化上，CYP1A 分为 3 个分支，哺乳动物 CYP1A1、CAP1A2 以及鱼类 CYP1A。CYP1A1、1A2 的分化已有 1.2 亿年的历史，而鱼类 CYP1A 的分化则在 3 亿年以前[20]。药物效应的不同或许与鱼类和哺乳动物 CYP1A 的差异有关。例如，哺乳动物 CYP2B 的典型诱导剂苯巴比妥不能诱导鱼类 CYP2B，相反却能够诱导鱼类的 CYP1A，这可能与 Ah 受体的激活有关，而 Ah 正是哺乳动物 CYP1A 的主要转录因子[21]。

药物对鱼类 CYP1A 的诱导程度受多种因素影响有关。一方面，不同种类的鱼对诱导剂的敏感程度不同。5mg/kg 的 ß-萘黄酮分别使鮈鱼（*Gobio gobio*）雄鱼和雌鱼 EROD 活性上升了 15 倍和 10 倍[22]，在黑鲈中上升了 4 倍[1]，而在虹鳟中，仅 50μg/kg 的 ß-萘黄酮则使其活性上升了 25 倍[23]。另一方面，诱导还受性别、温度等因素影响。40mg/kg 的 ß-萘黄酮诱导沟篮子鱼（*Siganus canaliculatus*）其 EROD 上升了约 2 倍，且雄鱼比雌鱼更易被诱导[24]，ß-萘黄酮对虹鳟 EROD 的诱导作用在 17℃ 较 5℃ 快，但低水温的诱导程度更高[25]。本研究氟甲喹单剂量注射给药后异育银鲫 EROD 活性仅升高了 60%，而虹鳟连续摄食 10 天含氟甲喹的饲料，其 EROD 活性升高了 4.2 倍[10]。由此看来，对 EROD 诱导作用还可能与动物种类、给药剂量与方式、药物作用时间等因素有关，今后尚需进一步研究。

鱼类和哺乳动物纯化的肝细胞色素 P450 还原酶具有相似的光谱、电泳及生物催化特性[28]。本研究中，采用 Western-blotting 验证了大鼠 CYP1A1 多克隆抗体与异育银鲫肝微粒体蛋白间的抗原抗体反应，鼠源 P450 抗体在黑鲈、变色窄牙鲷（*Stenotomus chrysops*）和虹鳟的研究中也有成功应用的报道[2,10,27]。Celander 等[29]报道了哺乳动物及人源的抗体（大鼠 CYP3A1、人类 CYP3A4）不仅能与纯化的虹鳟 LMC5 产生免疫反应，同时也能识别虹鳟、变色窄牙鲷和大西洋鳕（*Gadus morhua*）肝粒体中的 P450 蛋白。另外，本试验还尝试了人类 CYP3A4 抗体对异育银鲫肝微粒体的作用，但仅能检测到微弱的条带。这表明，鱼类与哺乳动物之间 CYP3A 亚型有一定相似性，同时鱼类种属之间也存在差异性。

外源物质对 CYP1A 的诱导发生在三个水平：转录水平（如转录激活）、转录后水平（如 mRNA 的稳定化或修饰）或翻译后水平（如蛋白的稳定化）[1,30]。本研究中，在体内氟甲喹对异育银鲫 CYP1A 活性和蛋白含量具有诱导作用，然而体外试验中并没有观察到氟甲喹对 CYP1A 活性的抑制作用，这与氟甲喹在虹鳟中的研究结果[10]一致。在此基础上，CYP1A mRNA 的半定量 RT-PCR 结果显示，试验组 CYP1A mRNA 表达与对照组无显著差异。综上结果，我们推测氟甲喹对 CYP1A 的诱导作用可能在翻译后，与蛋白修饰（如蛋白稳定化）有关。

由此可见，异育银鲫单剂量注射氟甲喹，对 CYP1A 具有诱导作用，经 CYP1A 代谢的药物，其组织分布、消除速率可能会受到影响[10]。目前联合用药已成为临床上重要的治疗手段，当几种药物合用时，药酶的诱导与抑制将改变药物的代谢，可能使药物的疗效降低或者产生严重的不良反应。研究 CYP450 的诱导或抑制机制对临床合理用药、提高药物疗效和降低药物的毒副作用具有重要意义。

参考文献

[1] Novi S, Pretti C, Cognetti A M, et al. Biotransformation enzymes and their induction by β-naphtoflavone in adult sea bass (*Dicentrarchus labrax*) [J]. Aquatic Toxicology, 1998, 41 (1): 63-81.

[2] Vaccaro E, Giorgi M, Longo V, et al. Inhibition of cytochrome P450 enzymes by enrofloxacin in the sea bass (*Dicentrarchus labrax*) [J]. Aquatic Toxicology, 2003, 62 (1): 27-33.

[3] 舒耀皋. 几种农药对鲫鱼肝脏微粒体 P450 酶系的影响 [D]. 乌鲁木齐：新疆农业大学，2008.

[4] 唐江芳，郑曙明. 复方抗菌剂对鲫鱼细胞色素 P4503A 活性的影响 [J]. 水利渔业，2007，27 (3): 105-107.

［5］ 杨先乐. 新编渔药手册［M］. 北京：中国农业出版社，2005：199-213.

［6］ Bradford M M. A rapid and sensitive method for the quantitation of microgram quantities utilizing the principle of protein dye binding［J］. Anal Biochem, 1976, 72（1-2）：248-254.

［7］ Tu Y and Yang C. High-affinity nitrosamine dealkylase system in rat liver microsomes and its induction by fasting［J］. Cancer Research, 1983, 43（2）：623.

［8］ Lange U, Goksøyr A, Siebers D, et al. Cytochrome P450 1A-dependent enzyme activities in the liver of dab (*Limanda limanda*)：kinetics, seasonal changes and detection limits［J］. Comparative Biochemistry and Physiology Part B, 1999, 123（4）：361-371.

［9］ Laemmli U K. Cleavage of structural proteins during the assembly of the head of bacteriophage T4［J］. Nature, 1970, 227（5259）：680-685.

［10］ Moutou K A, Burke M D and Houlihan D F. Hepatic P450 monooxygenase response in rainbow trout (*Oncorhynchus mykiss* (*Walbaum*)) administered aquaculture antibiotics［J］. Fish Physiology and Biochemistry, 1998, 18（1）：97-106.

［11］ Nannelli A, Chirulli V, Longo V, et al. Expression and induction by rifampicin of CAR and PXR regulated CYP2B and CYP3A in liver, kidney and airways of pig［J］. Toxicology, 2008, 252（1-3）：105-112.

［12］ HU Y Z and YAO T W. Induction of diphenytriazol on cytochrome CYP1A［J］. Acta Pharmacol Sin, 2004, 25（4）：528-533.

［13］ Machala M, Nezveda L, Petfivalskjrb M, et al. Monooxygenase activities in carp as biochemical markers of pollution by polycyclic and polyhalogenated aromatic hydrocarbons：choice of substrates and effects of temperature, gender and capture stress［J］. Aquatic Toxicology, 1997, 37（2-3）：113-123.

［14］ Li D, Yang X L, Zhang S J, et al. Effects of mammalian CYP3A inducers on CYP3A-related enzyme activities in grass carp (*Ctenopharyngodon idellus*)：Possible implications for the establishment of a fish CYP3A induction model［J］. Comparative Biochemistry and Physiology, Part C, 2008, 147（1）：17-29.

［15］ Benet L Z, Kroetz D L and Sheiner L B. Pharmacokinetics：the dynamics of drug absorption, distribution and elimination, in Goodman and Gilman's The Pharmacological Basis of Therapeutic (*Hardman GH and Limbird LE eds*)［M］. New York：McGraw-Hill Companies, 1996：3-28.

［16］ Tseng H P, Hseu T H, Buhler D R, et al. Constitutive and xenobiotics-induced expression of a novel CYP3A gene from zebrafish larva［J］. Toxicology and Applied Pharmacology, 2005, 205（3）：247-258.

［17］ Anzenbacher P and Anzenbacherová E. Cytochromes P450 and metabolism of xenobiotics［J］. Cell. Mol. Life Sci, 2001, 58（5-6）：737-747.

［18］ Briubo E B. Metabolism-dependent blinding of the beteroerelie ainine Trip-P. I in endothelial cells of choroids plexus and in large cerebral veins of cytochrome P450-induced mice［J］. Brain Res, 1994, 659（1-2）：91-98.

［19］ Fuhr U, Anders E M, Mahr G, et al. Inhibitory potency of quinolone antibacterial agents against cytochrome P4501A2 in vivo and in vitro［J］. Antimicrobial Agents Chemother, 1992, 36（5）：942-948.

［20］ Lewis D F V, Lake B G, George S G, et al. Molecular modelling of CYP1 family enzymes CYP1A1, CYP1A2, CYP1A6 and CYP1B1 based on sequence homology with CYP102［J］. Toxicology, 1999, 139（1-2）：53-79.

［21］ Williams D E, Lech J J and Buhler D R. Xenobiotics and xenoestrogens in fish：modulation of cytochrome P450 and carcinogenesis. Mutation Research, 1998, 399（2）：179-192.

［22］ Flammarion P, Migeon B, Urios S b, et al. Effect of methidathion on the cytochrome P-450 1A in the cyprinid fish gudgeon (*Gobio gobio*)［J］. Aquatic Toxicology, 1998, 42（2）：93-102.

［23］ Zhang Y, Andersson T and F rlin L. Induction of hepatic xenobiotic biotransformation enzymes in rainbow trout by［beta］-naphthoflavone. Time-course studies. Comparative Biochemistry and Physiology Part B：Biochemistry and Molecular Biology, 1990, 95（2）：247-253.

［24］ Raza H, Otaiba A and Montague W. ß-naphthoflavone-inducible cytochrome p450 1A1 activity in liver microsomes of the marine safi fish (*Siganus canaliculatus*)［J］. Biochem Pharmacol, 1995, 50（9）：1401-1406.

［25］ Andersson T and Koivusaan U. Influence of environmental temperature on the induction of xenobiotic metabolism by b-naphthoflavone in rainbow trout, Salmo gairdneri［J］. Toxicol Appl Pharmacol, 1985, 80（1）：45-50.

［26］ Snegaroff J, Bach J and Prevost V. Effects of chloramphenicol on hepatic cytochrome P-450 in rainbow trout. Comparative Biochemistry and Physiology Part C：Comparative Pharmacology, 1989, 94（1）：215-222.

［27］ Stegeman J J and Hahn E H. Biochemistry and molecular biology of monooxygenases：Current perspectives on forms, functions, and regulation of cytochrome P-450 in aquatic species, in Aquatic Toxicology, Molecular, Biochemical, and Cellular Perspectives［M］. Florida：Lewis Publishers, 1994：87-206.

［28］ Sen A and Arinc E. Purification and characterization of cytochrome P450 reductase from liver microsomes of feral leaping mullet (*Liza saliens*)［J］. J Biochem Mol Toxicol, 1998, 12（2）：103-113.

［29］ Celander M, Buhler D R, Forlin L, et al. Immunochemical relationships of cytochrome P450 3A-like proteins in teleost fish［J］. Fish Physiology and Biochemistry, 1996, 15（4）：323-332.

［30］ White R D, Shea D, Solow A R, et al. Induction and Post-Transcriptional Suppression of Hepatic Cytochrome P450 1Al by 3, 3', 4, 4'-Tetrachlorobiphenyl［J］. Biochemical Pharmacology, 1997, 53（7）：1029-1040.

该文发表于《中国水产科学》【2012，19（1）：154-160】

黄芩苷与甘草酸对恩诺沙星在异育银鲫体内代谢的影响

Effect of baicalin and glycyrrhizin on enrofloxacin *in vivo* in crucian cap（*Carassius auratus gibelio*）

房文红[1]，周常[1,2]，孙贝贝[1]，李国烈[1]，杨先乐[1]，李新苍[1]，胡琳琳[1]

(1. 中国水产科学研究院东海水产研究所农业部海洋与河口渔业资源及生态重点开放实验室，上海 200090；
2. 上海海洋大学国家水生动植物病原库，上海 201306)

摘要：中西药合用时，某些药物诱导或抑制细胞色素 P450 活性会导致合用药物的药动学发生改变，对指导临床用药具有重要意义，但在水产养殖中还未见相关报道。本实验将黄芩苷（Baicalin，BL）和甘草酸（Glycyrrhizin，GZ）口灌异育银鲫（*Carassius auratus gibelio*），探讨其对恩诺沙星（Enrofloxacin，EF）在体内代谢和肝微粒体 CYP1A、CYP3A 活性的影响。异育银鲫连续 7 d 分别口灌黄芩苷（100 mg/kg）和甘草酸（100 mg/kg），以口灌玉米油作为对照；末次给药 24 h 后每组随机取 10 尾腹腔注射恩诺沙星（10 mg/kg），采用单个动物连续采血，HPLC 测定血浆恩诺沙星和其代谢产物环丙沙星（Cipmfloxacin，CF）浓度，分析药动学及其参数；同时每组选取 6 尾检测肝微粒体 CYP1A 和 CYP3A 活性。结果表明：(1) 黄芩苷（BL）和甘草酸（GZ）对恩诺沙星的吸收有明显抑制作用，主要表现在恩诺沙星峰浓度（C_{max}）降低，曲线下面积（AUC）减少；(2) 口灌 BL 和 GZ 后，恩诺沙星及其代谢产物环丙沙星的消除半衰期（$t_{1/2z}$）明显小于对照组，而总体清除率（CLz/F）则增大，说明黄芩苷和甘草酸促进了恩诺沙星和环丙沙星的消除；(3) 口灌 BL 和 GZ 后，BL 组和 GZ 组的 $C_{max\,CF}/C_{max\,EF}$ 比值分别 1.48%、2.22%，对照为 0.95%；BL 组和 GZ 组的 $AUC_{0-t\,CF}/AUC_{0-t\,EF}$ 比值分别为 2.16%、1.76%，而对照组为 1.7%。综合分析代谢产物环丙沙星峰浓度、$C_{max\,CF}/C_{max\,EF}$ 和 AUC_{CF}/AUC_{EF} 比值可以得出，黄芩苷和甘草酸对恩诺沙星 N-脱乙基具有诱导作用；(4) 与对照组相比较，BL 组和 GZ 组的 7-乙氧基异吩唑酮-O-脱乙基酶（EROD，CYP1A 标志酶）和红霉素-N-脱甲基酶（ERND，CYP3A 标志酶）活性显著升高（$P<0.05$），说明黄芩苷和甘草酸对 CYP1A 和 CYP3A 都有诱导作用。结合以上结果，认为黄芩苷和甘草酸减少了恩诺沙星的吸收、加速了恩诺沙星的消除、促进了代谢产物 CF 的生成，很可能与诱导 CYP1A 和 CYP3A 活性有关。

关键词：异育银鲫；恩诺沙星；药动学；黄芩苷；甘草酸；CYP450

临床上经常见到中西药联合使用，如黄连、黄柏和四环素合用治疗痢疾，其疗效成倍提高[1]。可见联合用药会产生药物-药物相互作用（drug-drug interactions，DDIs），DIs 分为理化、药动学和药效学相互作用，其中药动学相互作用可能发生在吸收、分布、代谢和排泄中，代谢性药物相互作用的发生率最高，占药动学 DDIs 的 40%[2]，而 DDIs 主要由药物对细胞色素 P450 酶（CYP450）产生诱导或抑制作用所致。同样，中西药合用产生的相互作用中，由 CYP450 引起的占有很大比重[3]，如银杏叶提取物可诱导大鼠 CYP1A2，并显著降低苯巴比妥的血药浓度[4]；尼非地平与 CYP3A4 的抑制剂人参合用后，血药浓度明显上升[5]。

黄芩苷（Baicalin，BL）和甘草酸（Glycyrrhizin，GZ）分别是中药黄芩和甘草的主要有效成分之一，临床上主要用于抗菌消炎[6]和抗病毒感染[7]。中医上认为它们能够"调合百药、缓解药性"，说明它们对与之配伍的药物具有重要影响，或增加药性，或降低药效。胡椒碱能显著增大口服抗癫痫药苯妥英钠的吸收速率常数，减小其消除速率常数，提高其血药浓度[8]；甘草可影响氨茶碱的药动学，血药 C_{max} 和 AUC 减小，总体清除率增大[9]。在水产养殖中，黄芩苷和甘草酸等中药由于具有低毒、低污染和不易产生耐药性等优点，在水产动物疾病防治中应用日益受到重视，已逐渐应用于鱼类细菌和病毒性疾病防治，且常被推荐和化学渔药联合使用，如黄连和诺氟沙星合用防治细菌性烂鳃病，三黄散与恩诺沙星合用防治出血病[10]，但对药物代谢酶 P450 的影响及药物相互作用的研究几乎是空白。鉴于此，本研究选择我国常见淡水鱼异育银鲫（*Carassius auratus gibelio*），研究黄芩苷和甘草酸对恩诺沙星在其体内代谢的影响，为中西药联合用药的相互影响和作用机制提供理论基础，促使水产药物更加安全有效地使用。

1 材料与方法

1.1 实验药品和试剂

黄芩苷和甘草酸，纯度≥98%，购自鼎瑞化工（上海）有限公司；恩诺沙星（Enrofloxacin, EF）和环丙沙星（Cipofloxacin, CF）标准品，纯度≥99.5%，购自 Sigma 公司；恩诺沙星原料药，纯度≥98.0%，购于浙江新昌兽药厂。7-乙氧基异吩唑酮、试卤灵和红霉素购自 Sigma 公司，还原性辅酶Ⅱ（NADPH）购自上海生工生物工程技术服务有限公司，甲醇和乙腈均（HPLC 级）购自德国默克公司，四丁基溴化铵、磷酸等（分析纯）购自国药集团上海化学试剂公司。

1.2 仪器和设备

高效液相色谱仪，美国沃特斯公司 Waters 2695，色谱柱为 Zorbax SB-C$_{18}$，（5 μm, 150 mm×4.6 mm, I. D.），荧光检测器 Waters 2475。超速冷冻离心机（Beckman Optima L-100XP），荧光分光光度计（上海三科仪器有限公司，970CRT），紫外分光光度计（上海尤尼克仪器有限公司，UV-2802S），超低温冰箱（-80℃，美国 Thermo），超纯水仪（美国产，Millipore Advantage A10），电子分析天平（瑞士产，Mettler Toledoab 204，感量 0.000 1 g），高速冷冻离心机（日本产，HITACHI CF16RXⅡ），以及 0.22 μm 微孔滤膜等。

1.3 实验动物

实验用异育银鲫购自上海青浦某养殖场，体质量（200±13）g，暂养于 5 个循环过滤装置的体积为 35 cm×80 cm×45 cm 的玻璃缸中，循环水养殖，水温 26～27℃，投喂不含喹诺酮类药物的人工配合饲料并及时排污和残饵，暂养 4 周后挑选健康的个体用于试验，实验期间异育银鲫成活率为 100%。

1.4 口灌给药与样品采集

黄芩苷和甘草酸用玉米油作为溶剂溶解，恩诺沙星用 0.1 M 氢氧化钠助溶，再用无菌蒸馏水配成 10 mg/mL 的贮存液。实验鱼随机分成 3 组，每组 20 尾，分别口灌黄芩苷（BL 组）和甘草酸（GZ 组），给药剂量为 100 mg/（kg·d），口灌玉米油作为对照组。分别连续口灌 7 d，每天 1 次，观察 5 min 无饵料回吐用于试验。末次给药后 24 h，每组随机 6 尾取肝组织用于制备微粒体；另外，从 BL 组、GZ 组和对照组中分别随机选择 10 尾单次腹腔注射恩诺沙星，给药剂量为 10 mg/kg，给药后每尾鱼单独饲养于 20 cm×40 cm×30 cm 的玻璃缸中，于 5 min、30 min、1 h、2 h、4 h、8 h、12 h、1 d、2 d、3 d、4 d 下每尾鱼连续抽取 0.3 mL 血液，置于预先装有肝素钠抗凝剂的 1.5 mL 离心管中，全血于 8 000 rpm 下离心 10 min，取上清于-80℃保存备用。

1.5 血药样品处理

血浆样品于室温自然融解后，摇匀，吸取 200 μL 上清液置于 1.5 mL 离心管内，加入等体积甲醇，旋涡混合仪剧烈混合 2 min，于 10 000 rpm. 转速离心 5 min，吸取上清液经 0.22 μm 微孔滤膜过滤后，用于 HPLC 分析。

1.6 流动相和色谱条件

流动相为乙腈：0.01 mol/L 四丁基溴化铵溶液（磷酸调至 pH=2.8）= 5∶95（V/V），使用前用超声波脱气 10 min；柱温，40℃；流速，1.2 mL/min；进样量，10 μL；荧光检测波长，E_x=280 nm，E_m=450 nm。

1.7 微粒体制备及酶活性测定

异育银鲫肝微粒体制备参照 Novi 等[11]方法；7-乙氧基异吩唑酮-O-脱乙基酶（EROD）活性参照 Lange[12]方法；红霉素-N-脱甲基酶（ERND）活性参照 Tu 和 Yang[13]。

1.8 数据分析

腹腔注射给药下异育银鲫血药浓度-时间关系曲线药动学参数统计矩原理来推算，模型的拟合和参数的推理采用上海中医药大学药物临床研究中心的 DAS 2.1.1 药物与统计软件。

2 结果

2.1 异育银鲫血浆恩诺沙星浓度与时间关系

腹腔注射给药后，异育银鲫血药恩诺沙星浓度与时间关系曲线见图 1。BL 组 EF 浓度先上升后下降，达峰时间（T_{max}）

和峰浓度（C_{max}）分别为 0.5 h、（7.685±2.392）mg/L；GZ 组和对照组 EF 浓度均在第 1 个采样点（T_{max} = 0.083 h）最高，EF 峰浓度（C_{max}）分别为（5.769±1.164）mg/L 和（10.470±2.088）mg/L。BL 组和 GZ 组 EF 的 C_{max} 仅为对照组的 76% 和 55%。

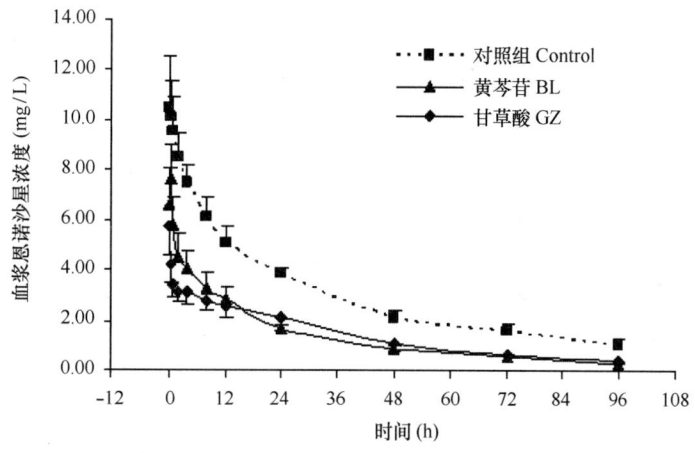

图 1 口服黄芩苷和甘草酸后异育银鲫血浆恩诺沙星浓度-时间关系（$n = 10$）

Fig. 1 Meanplasma EF concentration vs. time curves of crucian carp after coadministration of BL and GZ

2.2 异育银鲫血浆恩诺沙星药动学参数

腹腔注射给药后，血浆恩诺沙星药动学参数采用统计矩原理推算，参数见表 1。BL 组和 GZ 组血药曲线下面积（AUC_{0-t}）分别为 132.301 mg·h/L 和 136.358 mg·h/L，为对照组的 46.40% 和 47.81%。BL 组消除最快，$T_{1/2z}$ 为 25.90 h，其次是 GZ 组，为 30.21 h，明显低于对照组（37.38 h）；BL 组和 GZ 组总体清除率（$CL_{z/F}$）较对照均有明显提高，均达到对照组的 2 倍以上。

表 1 腹腔注射给药后恩诺沙星在异育银鲫血浆中药动学参数（统计矩原理）（$n = 10$）

Tab. 1 Pharmacokinetic parameters of enrofloxacin in plasma from crucian carp after intraperitoneal injection based on statistical moment theory

药动学参数 Pharmacokinetic parameters		对照组 Control	黄芩苷 Baicalin	甘草酸 Glycyrrhizin
C_{max}	（mg/L）	10.470±2.088	7.685±2.392	5.769±1.164
T_{max}	（h）	0.083	0.5	0.083
AUC_{0-t}	（mg·h/L）	285.142	132.301	136.358
$AUC_{0-\infty}$	（mg·h/L）	339.674	143.585	152.555
MRT_{0-t}	（h）	30.867	26.305	30.500
$MRT_{0-\infty}$	（h）	49.984	34.720	42.084
$V_{z/F}$	（L/kg）	1.588	2.603	2.858
$T_{1/2z}$	（h）	37.384	25.904	30.217
$CL_{z/F}$	（L/h/kg）	0.029	0.070	0.066

注：C_{max} 为药峰浓度；T_{max} 为达峰时间；AUC_{0-t}、$AUC_{0-\infty}$ 分别为 0-t 时间、0~∞ 时间的曲线下面积；MRT_{0-t}、$MRT_{0-\infty}$ 为 0-t 时间、0~∞ 时间的保留时间；$V_{z/F}$ 为表观分布容积；$T_{1/2z}$ 为消除半衰期；$CL_{z/F}$ 为总体清除率。

Note：C_{max}-peak OTC concentration；T_{max}-time to reach peak concentration；AUC_{0-t}-area under the concentration-time curve from zero to time；$AUC_{0-\infty}$-area under the concentration-time curve from zero to infinity；MRT_{0-t}-area residence time from zero to time；$MRT_{0-\infty}$-area residence time from zero to infinity；$V_{z/F}$-the apparent volume of distribution；$t_{1/2z}$-elimination half-life；$CL_{z/F}$-total body clearance.

2.3 异育银鲫血浆代谢产物环丙沙星浓度与时间关系

腹腔注射给药后，异育银鲫恩诺沙星的代谢产物环丙沙星血药浓度与时间关系曲线见图 2。三组的 CF 浓度变化趋势相同，都表现为先上升后下降，但达峰时间和峰浓度不同。对照组和 GZ 组的 T_{max} 相同，均为 8 h，但 C_{max} 不同，分别为（0.099±0.029）mg/L 和（0.128±0.024）mg/L；BL 组达峰较晚，T_{max} = 12 h，C_{max} 为（0.114±0.028）mg/L。BL 组和 GZ 组的 $C_{max\ CF}/C_{max\ EF}$ 比值分别 1.48%、2.22%，对照组为 0.95%。

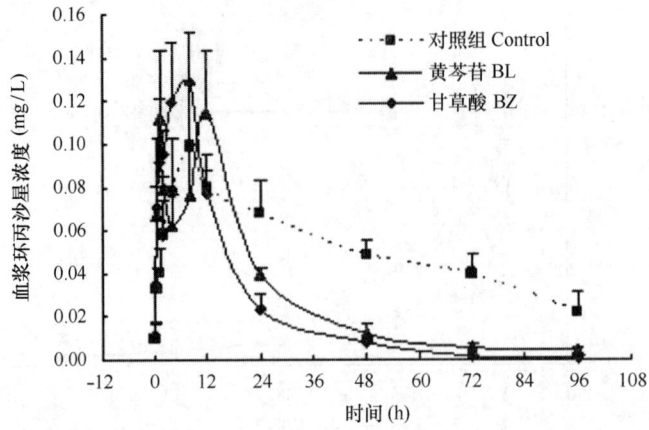

图 2　口服黄芩苷和甘草酸后异育银鲫血浆环丙沙星浓度-时间关系

Fig. 2　Meanplasma CF concentration vs. time curves of crucian carp after coadministration of BL and GZ

2.4　异育银鲫血浆代谢产物环丙沙星药动学参数

腹腔注射给药后，血浆中 EF 代谢产物 CF 处置过程药动学参数同样采用统计矩原理推算，参数见表 2。BL 组和 GZ 组的 MRT、$V_{z/F}$ 和 $T_{1/2z}$ 均明显低于对照组，而总体清除率（$CL_{z/F}$）却高于对照组。虽然 BL 组和 GZ 组血药 CF 曲线下面积（AUC）低于对照组，但 BL 组和 GZ 组的血药 $AUC_{0-t\ CF}/AUC_{0-t\ EF}$ 比值分别为 2.16%、1.76%，而对照组为 1.7%。

表 2　腹腔注射给药后代谢产物环丙沙星在异育银鲫血浆中药动学参数（统计矩原理）（$n=10$）

Tab. 2　**Pharmacokinetic parameters of ciprofloxacin in plasma from crucian carp after intraperitoneal injection based on statistical moment theory**

药动学参数 Pharmacokinetic parameters		对照组 Control	黄芩苷 Baicalin	甘草酸 Glycyrrhizin
C_{max}	（mg/L）	0.099±0.029	0.114±0.028	0.128±0.024
T_{max}	（h）	8	12	8
AUC_{0-t}	（mg·h/L）	5.022	2.859	2.400
$AUC_{0-\infty}$	（mg·h/L）	7.567	2.906	2.422
MRT_{0-t}	（h）	38.108	21.564	15.192
$MRT_{0-\infty}$	（h）	86.388	23.151	16.160
$V_{z/F}$	（L/kg）	113.179	83.359	93.906
$T_{1/2z}$	（h）	59.350	16.786	15.764
$CL_{z/F}$	（L/h/kg）	1.322	3.441	4.128

注：C_{max} 为药峰浓度；T_{max} 为达峰时间；AUC_{0-t}、$AUC_{0-\infty}$ 分别为 0-t 时间、0-∞ 时间的曲线下面积；MRT_{0-t}、$MRT_{0-\infty}$ 为 0-t 时间、0-∞ 时间的保留时间；$V_{z/F}$ 为表观分布容积；$T_{1/2z}$ 为消除半衰期；$CL_{z/F}$ 为总体清除率。

Note：C_{max}-peak OTC concentration；T_{max}-time to reach peak concentration；AUC_{0-t}-area under the concentration-time curve from zero to time；$AUC_{0-\infty}$-area under the concentration-time curve from zero to infinity；MRT_{0-t}-area residence time from zero to time；$MRT_{0-\infty}$-area residence time from zero to infinity；$V_{z/F}$-the apparent volume of distribution；$t_{1/2z}$-elimination half-life；$CL_{z/F}$-total body clearance.

2.5　黄芩苷和甘草酸对异育银鲫肝 P450 药酶活性的影响

BL 和 GZ 口灌异育银鲫后，其肝 CYP1A 和 CYP3A 活性见表 3。BL 组肝微粒体 EROD 和 ERND 活性显著高于对照组

（$P<0.05$），GZ 组肝微粒体 EROD 和 ERND 活性与对照组为极显著差异（$P<0.01$），两个药酶活性增加了 1 倍左右。

表 3　BL 和 GZ 对异育银鲫肝 CYP1A 和 CYP3A 活性的影响（$n=6$）

Tab. 3　The effects of BL and GZ on CYP1A and CYP3A activitiesof liver from crucian carp

组别 Group	剂量 Dose （mg/kg/d）	EROD 活性 EROD activity	ERND 活性 ERND activity
黄芩苷 Baicalin	100	52.65±2.64[a]	2185.72±207.11[b]
甘草酸 Glycyrrhizin	100	51.43±3.99[a]	2108.50±98.96[b]
对照组 Control	0	35.4±3.6	1166.47±125.71

注：酶活大小以平均数±标准差表示，单位：pmol/min per mg protein。"a" 表示与对照组差异显著（$P<0.05$），"b" 表示与对照组差异极显著（$P<0.01$）。

Note：Values are showed as means±S. D. and expressed as pmol/min per mg protein. "a" donates singnificant difference（$P<0.05$）. "b" donates extremely singnificant difference（$P<0.01$）.

3　讨论

许多中药对 P450 药酶具有诱导或抑制作用，明确中药及其有效成分对 P450 药酶的影响对临床用药具有指导意义。在哺乳动物中，甘草灌胃给药大鼠 7d 对 CYP1A 活性具有诱导作用[14]，人口服甘草酸 14d 能诱导 CYP3A 活性升高[15]，黄芩苷对人肠细胞 LS174T 的 CYP3A4 活性和蛋白表达都具有诱导作用[16]。韩华等研究了黄芩苷对牙鲆肝 CYP1A 活性和基因表达的影响，结果发现黄芩苷对 CYP1A 具有诱导作用，且与诱导时间和剂量呈正相关[17]。本实验中，采用 EROD 和 ERND 分别作为 CYP1A 和 CYP3A 的标志酶，黄芩苷和甘草酸连续用药 7d 后能显著提高两个药酶活性，由此可见黄芩苷和甘草酸对药酶诱导作用具有普遍性。

人们更为关注的是，药酶诱导或抑制后将影响其底物的代谢。中药对 P450 药酶活性的影响而产生的药物相互作用日益受到临床用药的重视[18]。本实验中，恩诺沙星进入机体内后，一方面通过原型药的方式消除，另一方面还可以通过代谢生成各种代谢产物消除。在多种动物体内环丙沙星是恩诺沙星的主要代谢产物[19]，口灌黄芩苷和甘草酸后，代谢产物环丙沙星血药 C_{max} 高于对照组，BL 组和 GZ 组的 $C_{max\ CF}/C_{max\ EF}$ 比值分别 1.48%、2.22%，对照为 0.95%；BL 组和 GZ 组的 $AUC_{0-t\ CF}/AUC_{0-t\ EF}$ 比值分别为 2.16%、1.76%，而对照组为 1.7%。综合恩诺沙星代谢生成环丙沙星的 C_{max}、$C_{max\ CF}/C_{max\ EF}$ 和 $AUC_{0-t\ CF}/AUC_{0-t\ EF}$ 比值可以得出，黄芩苷和甘草酸对恩诺沙星 N-脱乙基生成环丙沙星具有诱导作用。Vaccaro 等在舌齿鲈（Dicentrarchus labrax）肝体外微粒体恩诺沙星代谢实验中，发现恩诺沙星代谢生成环丙沙星是一个典型的 P450 反应，需要氧分子和 NADPH 体系，但未能说明参与该反应的 P450 药酶亚型。本实验中，黄芩苷和甘草酸促进了异育银鲫体内恩诺沙星 N-脱乙基反应，与 CYP1A 和 CYP3A 活性诱导呈正相关性。另有研究表明，甘草酸能活化孕烷 X 受体（PXR），从而诱导肝脏代谢酶 CYP3A 的表达，所以很可能是 CYP3A 参与了恩诺沙星的 N-脱乙基反应[15]，但仅此并不能排除 CYP1A 参与的可能。究竟是 P450 药酶哪个亚型参与恩诺沙星的 N-脱乙基化反应，还需要更多的实验来证实，如重组的 CYP 同工酶和免疫抑制分析[21]。

本实验中，异育银鲫连续口灌黄芩苷和甘草酸后，除影响 CYP1A 和 CYP3A 活性、恩诺沙星 N-脱乙基作用外，还表现为血浆恩诺沙星 C_{max} 和 AUC_{0-t} 下降、恩诺沙星和环丙沙星的消除半衰期（$T_{1/2z}$）减小和总体清除率（CLz）升高，说明黄芩苷和甘草酸作用后异育银鲫对恩诺沙星的吸收减少、恩诺沙星和环丙沙星的消除加快，后者现象与黄芩苷作用后诺氟沙星在中国对虾体内消除加快相一致。药物合用或前后使用时，一种药物引起另一种药物药动学改变的现象较为常见，其中中西药合用时，中药对西药药动学的影响研究较多，如银杏叶提取物可诱导大鼠 CYP1A2，并显著降低苯巴比妥的血药浓度[4]；胡椒碱能显著增大口服抗癫痫药苯妥英钠的吸收速率常数，减小其消除速率常数，提高其血药浓度[8]；盐酸黄连素与环孢素 A 合用可使后者 $T_{1/2}$ 延长，消除率下降，AUC 增大，生物利用度提高[21]；高剂量川芎嗪和维拉帕米预处理大鼠后，环孢素 A 的 AUC 和 C_{max} 均有显著的增加，且维拉帕米处理后的环孢素 A 的 $T_{1/2\beta}$ 显著延长，总体清除率显著降低[22]。两种药物合用时，一种药物导致另一种药物生物利用度减少或增加，不仅与 P450 药酶（如 CYP3A）的诱导或抑制有关，P-糖蛋白的诱导或抑制也是其重要原因。P-糖蛋白主要分布于细胞膜，它属于 ATP 结合盒（ABC）转运子超家族成员，可以将包括多种药物在内的生物毒性物质单向泵出细胞，从而使细胞内药物浓度下降[23]。Bauer 等[24]报道圣约翰草（St John' S Wort）通过诱导肠道 P-糖蛋白可明显降低肾移植受者环孢素 A 的血浓度。黄芩苷[25]和甘草酸[26]等可以增加 P-糖蛋白的转运能力，上调其表达。许多喹诺酮类药物是 P-糖蛋白的底物，如环丙沙星，培氟沙星（pefloxacin）[27]，达氟沙星（danofloxacin）等[28]；根据 Seelig 等[29]关于 P-糖蛋白底物提出的假设，即底物的唯一要求是具有一定空间排序的氢键键合基团和共平面的芳香环区，由此可以推测恩诺沙星也是 P-糖蛋白的底物。本实验中，黄芩苷和甘草酸有可能诱导了 P-

糖蛋白的表达，使 P-糖蛋白底物恩诺沙星外排加快，从而导致异育银鲫对 EF 的吸收减少、消除加快。

目前我国渔药中抗菌化学药物有 24 种，中药制剂有 67 种，在病害防治中常被推荐联合使用。黄芩苷和甘草酸分别是黄芩和甘草的主要有效成分，它们和恩诺沙星都是目前批准使用的渔药。本实验揭示了中药对恩诺沙星吸收、代谢和消除的影响，并探讨了中西药在水产动物体内的相互作用；研究结果提示，我们应注意药物的合理配伍和联合使用，以达到提高疗效、降低毒副作用的目的，该结果不仅为水产养殖生产中处方药的合理搭配提供了理论基础，同时还具有很高的实用价值。

参考文献

[1] 赵晓梅. 常见的中西药相互作用分析 [J]. 中华现代中西医杂志, 2003, 1 (4)：351-352.

[2] 刘彦卿, 洪燕君, 曾苏. 代谢性药物-药物相互作用的研究进展 [J]. 浙江大学学报（医学版）, 2009, 28 (2)：215-224.

[3] 艾常虹, 李桦, 董德利. 基于细胞色素 P450 酶诱导的药物相互作用及评价 [J]. 国际药学研究杂志, 2011, 38 (1)：52-57.

[4] Kubota Y, Kobayashi K, Tanaka N, et al. Pretreatment with Ginkgo biloba extract weakens the hypnosis action of Phenobarbital and its plasma concentration in rats [J]. Pharmacol, 2004, 56 (3)：401-405.

[5] Harkey M R, Henderson G L, Gershwin M E, et al. Variability in commercial ginseng products：An analysis of 25 preparations [J]. Am J Clin Nutr, 2001, 73：1101-1106.

[6] Isbrucker R A, Burdock G A. Risk and safety assessment on the consumption of Licorice root (Glycyrrhiza sp.), its extract and powder as a food ingredient, with emphasis on the pharmacology and toxicology of glycyrrhizin [J]. Regulatory Toxicology and Pharmacology, 2006, (46)：167-192.

[7] Changa H H, Yi P L, Cheng C H, et al. Biphasic effects of baicalin, an active constituent of Scutellaria baicalensis Georgi, in the spontaneous sleep-wake regulation [J]. Journal of Ethnopharmacology. www. elsevier. com/locate/jethpharm.

[8] Velpandian T, Jasuja R, Bhardwaj R K, et al. Piperine in food：interference in the pharmacokinetics of phenytoin [J]. Eur J Drug M etab Pharmacokinet, 2001, 26 (4)：241-247.

[9] 刘建群, 李青, 张锐, 等. LC-MS/MS 法研究甘草对雷公藤甲素药代动力学及组织分布与排泄的影响 [J]. 药物分析杂志, 2010, 30 (9)：1664-1671.

[10] 杨先乐. 水产养殖用药处方大全 [M]. 化学工业出版社, 2008：53, 173.

[11] Novi S, Pretti C, Cognetti A M, et al. Biotransformation enzymes and their induction by β-naphtoflavone in adult sea bass (Dicentrarchus labrax) [J]. Aquatic Toxicology, 1998, 41：63-81.

[12] Lange U, Goksøyr A, Siebers D, et al. Cytochrome P450 1A-dependent enzyme activities in the liver of dab (Limanda limanda)：kinetics, seasonal changes and detection limits [J]. Comparative Biochemistry and Physiology Part B, 1999 (123)：361-371.

[13] Tu Y Y, Yang C S. High-affinity nitrosamine dealkylase system in rat liver microsomes and its induction by fasting [J]. Cancer Res, 1983, 43：623-629.

[14] 何益军, 石苏英, 金科涛, 等. 甘草与大戟甘遂芫花配伍对大鼠肝脏细胞色素 P4501A2 酶活性的影响 [J]. 中国药物与临床, 2007, 7 (4)：278-280.

[15] 涂江华. 甘草酸对 CYP450 酶的影响及其机制研究 [D]. 中南大学. 2010.

[16] Li Y, Wang Q, Yao X M, et al. Induction of CYP3A4 and MDR1 gene expression by baicalin, baicalein, chlorogenic acid, and ginsenoside Rf through constitutive androstane receptor-and pregnane X receptor-mediated pathways [J]. European Journal of Pharmacology, 2010, 640 46-54.

[17] 韩华, 李健, 李吉涛, 等. 黄芩苷对牙鲆肝 CYP1A 酶活性及基因表达的影响 [J]. 中国水产科学, 2010, 17 (5)：1121-1127.

[18] 艾常虹, 孙汉雄, 李桦, 等. 中药有效成分对细胞色素 P450 酶的抑制活性评价. 中国药理学通报, 2011, 27 (4)：519-523.

[19] Vaccaro E, Giorgi M, Longo V, et al. Inhibition of cytochrome P450 enzymes by enrofloxacin in the sea bass (Dicentrarchus labrax) [J]. Aquatic Toxicology, 2003, 62：27-33.

[20] Yoo H H, Kim N S, Lee J, et al. Characterization of human cytochrome P450 enzymes involved in the biotransformation of eperisone [J]. Xenobiotica, 2009, 39：1-10.]

[21] 李馨, 吴笑春, 辛华雯, 等. 盐酸黄连素对环孢素增效作用的药物动力学研究 [J]. 华南国防医学杂志, 2002, 16 (1)：12-15.

[22] 刘晓磊, 唐靖, 宋娟, 等. 川芎嗪对大鼠灌服环孢素 A 药代动力学的影响 [J]. 药学学报, 2006, 41 (9)：882-887.

[23] Schinkel A H, Jonker J W. Mammalian drug efflux transporters of the ATP binding cassette (ABC) family：an overview [J]. Adv Drug Deliv Rev, 2003, 55 (1)：3-29.

[24] Bauer S, Stormer E, Johne A, et al. Alterations in cyclosporin A pharmacokinetics and metabolism during treatment with St John's wort in renal transplant patients [J]. Br J Clin Pharmacol, 2003, 55：203-211.

[25] 范岚. 黄芩苷对 CYP450 代谢酶和药物转运体影响的研究 [D]. 中南大学. 2008.

[26] 颜苗. 甘草酸 18 位差向异构体及其水解产物对 P-糖蛋白影响的研究 [D]. 中南大学. 2010.

[27] Seelig A, Landwojtowicz E. Structure-activity relationship of P-glycoprotein substrates and modifiers [J]. European Journal of Pharmaceutical Sciences, 2000, 12：31-40.

［28］ Schrickx J A, Gremmels J F. Danofloxacin-mesylate is a substrate for ATP-dependent efflux transporters ［J］. British Journal of Pharmacology, 2007, 150（4）: 463-469.

［29］ Seelig A, Landwojtowicz E. Structure-activity relationship of P-glycoprotein substrates and modifiers ［J］. European Journal of Pharmaceutical Sciences, 2000, 12: 31-40.

该文发表于《上海海洋大学学报》【2012, 21（3）: 389-395】

黄芩等中药提取物对异育银鲫肝微粒体 CYP1A 和 CYP3A 的抑制作用

Inhibition of Chinese herbal extracts on CYP1A and CYP3A of liver microsomes from crucian carp（*Carassius auratus gibelio*）

周常[1,2]，周帅[1]，房文红[1]，李国烈[1]，杨先乐[2]，李新苍[1]，胡琳琳[1]

（1. 中国水产科学研究院东海水产研究所 农业部海洋与河口渔业资源及生态重点开放实验室，上海 200090；

2. 上海海洋大学国家水生动植物病原库，上海 201306）

摘要：通过肝微粒体体外孵育实验，研究黄芩、柴胡、连翘、吴茱萸和野菊花 5 种中药提取物对异育银鲫肝微粒体 CYP450 的影响。体外测定 IC_{50} 以及相关酶动力学参数，并推测其可能的作用机制。结果显示，5 种中药对 7-乙氧基异吩唑酮-O-脱乙基酶（EROD）（CYP1A 标志酶）的抑制程度为黄芩>连翘>野菊花>柴胡≈吴茱萸，半抑制浓度（IC_{50}）依次为 0.27、2.88、7.62、16.20 和 16.42 mg/mL，酶促反应动力学常数 V_{max} 和 K_m 为 83.33±11.32 pmol/min/mg 和 3.05±0.89 μmol/L。据此计算黄芩、连翘、野菊花、柴胡和吴茱萸的抑制常数分别为 0.16、0.39、0.61、0.40 和 0.59 mg/mL。黄芩、连翘和吴茱萸对红霉素-N-脱甲基酶（ERND）（CYP3A 标志酶）抑制作用较弱，IC_{50} 分别为 90.9、70.8 和 43.5 mg/mL；未检测到柴胡和野菊花对 ERND 有抑制作用。结果表明，这 5 种中药对 CYP 酶的影响因亚型不同而差别较大，对 CYP1A 的抑制均较显著，而对 CYP3A 的抑制则不明显。进一步研究它们对 CYP1A 的抑制机制，推测黄芩、柴胡和吴茱萸对 EROD 的抑制为竞争性抑制，野菊花和连翘则是反竞争性抑制。笔者建议这 5 种中药与其他渔药合用时，要注意其引起其他渔药的药动学的变化，使疗效发生改变。

关键词：中药提取物；异育银鲫；肝微粒体；CYP1A；CYP3A；抑制

该文发表于《海洋渔业》【2009, 31（2）: 161-166】

异育银鲫肝微粒体体外恩诺沙星 N-脱乙基酶酶促反应动力学初步研究

Metabolism of enrofloxacin and detection of metabolic enzymeactivity in the liver microsome of Johnny carp（Carasius auratus gibelio）

贾娴[1,2]，房文红[2]，汪开毓[1]，胡晓[2]

（1. 四川农业大学动物医学学院，四川 雅安 625014；2. 中国水产科学研究院东海水产研究所

农业部海洋与河口渔业资源及生态重点开放实验室，上海 200090）

摘要：本研究建立了异育银鲫（*Carassius auratus*）肝微粒体体外孵育恩诺沙星（Enrofloxacin, EF）的酶促反应，应用 RP-HPLC 法检测孵育体系中恩诺沙星的代谢时间曲线和酶动力学，以其主要代谢产物环丙沙星（Ciprofloxacin, CF）来间接反映恩诺沙星 N-脱乙基酶的活性。肝微粒体体系中恩诺沙星和代谢产物环丙沙星采用二氯甲烷提取、正己烷去脂，样品回收率均大于 81.4%，日间变异系数小于 3.58%。恩诺沙星在鲫肝微粒体中代谢的消除方程为 $C = 140e^{-0.072t}$，消除速率常数（K）为 0.072/h，消除半衰期（$T_{1/2}$）为 9.627 h。酶动力学方程为 $1/V = 276.9215 × 1/[S] + 3.2458$，孵育过程中的酶活研究发现，以 140 μmol/L 恩诺沙星与 0.8 mg/mL 微粒体蛋白共孵育 45 min 时，反应体系达到最大反应速率（V_{max}）0.2342 nm/mg/min，米氏常数（K_m）为 85.3195 μmol/L，内在清除率（C_{lint}）为 0.0036 mL/（min·mg）。恩诺沙星消除率和环丙沙星转化率的研究结果表明，恩诺沙星在鲫鱼肝微粒体中的代谢以 N-脱乙基反应为主，提示可能还通过其他途径生成另外代谢产物。而酶动力学研

究数据表明，恩诺沙星 N-脱乙基酶与底物结合力不强，酶促反应强度较弱。据此，笔者推测恩诺沙星在鲫鱼肝微粒体中的代谢速度较为缓慢。

关键词：恩诺沙星；异育银鲫；肝微粒体；酶促反应动力学；恩诺沙星 N-脱乙基酶

该文发表于《大连海洋大学学报》【2011, 26 (1)：74-78】

异育银鲫组织中细胞色素 P450 酶系及主要药酶活性比较

Comparative Activity of microsomal cytochrome P450 in various tissues and organs in allogynogenesis silver crucian carp *Carassius auratus gibelio*

贾娴[1,2]，房文红[2]，胡琳琳[2]，汪开毓[1]，胡晓[1]

(1. 四川农业大学动物医学学院，四川 雅安 625014；2. 中国水产科学研究院东海水产研究所
农业部海洋与河口渔业资源及生态重点开放实验室，上海 200090)

摘要：建立了异育银鲫（*Carasius auratus gibelio*）主要组织中微粒体细胞色素 P450 （CYP450）及酶系中主要药酶活性的检测方法，并对其在组织中的分布进行了比较研究。结果显示：以 CO 还原差示光谱法测得异育银鲫肝、肾、鳃、肠、肌肉微拉体的细胞色素 P450 及 b5 含量均以肝微粒体最高，其次为肾、鳃、肠微粒体，肌肉最低。以氨基比林 N-脱甲基、红霉素 N-脱甲基、苯胺-4-羟化反应分别作为 CYP2B、CYP3A 和 CYP2E 的探针反应，测得氨基比林 N-脱甲基酶（APD）及红霉素 N-脱甲基（ERND）活性在上述组织中分布差异性类似，均表现为肝微粒体中最高，分别为 1.668 ± 0.104 nmol/min/mg 和 0.941 ± 0.061 nmol/min/mg，其次为肾、鳃、肠微粒体，肌肉微粒体中最低，两者活性分别为 0.245 ± 0.011 nmol/min/mg 和 0.078 ± 0.019 nmol/min/mg；苯胺-4-羟化酶（AH）活性以肝微粒体最高，为 0.052 ± 0.009 nmol/min/mg，其次为鳃微粒体，肌肉微粒体中 AHH 活性最低，不能检出。结果表明，异育银鲫主要组织微粒体具有 CYP450 亚型活性，且它们在银鲫体内的分布和活性存在着组织和器官差异性，以肝组织中的含量和活性为最高。

关键词：异育银鲫（*Carasius auratus gibelio*）；微粒体；细胞色素 P450；药酶；组织分布

该文发表于《江苏农业科学》【2011, 39 (5)：317-319】

恩诺沙星对 2 种鲟 GST 和 GSH 酶系活性影响的比较

Effects of Enrofloxacin on the activities of GST and GSH of Amur Sturgeon and Sterlet

陈琛[1,2]，李绍戊[2]，王荻[2]，马涛[2]，卢彤岩[2]

(1. 东北农业大学动物科学技术学院，黑龙江 哈尔滨 150030；2. 中国水产科学研究院黑龙江水产研究所，黑龙江 哈尔滨 150070)

摘要：将恩诺沙星（enrofloxacin）按照 0、20、40、60、80、100 mg/kg 的浓度，对小体鲟及史氏鲟口服给药 5 d，停药 2 d 后对其血浆及肝脏组织中 GST 活性及 GSH 含量进行测定。结果表明：2 种鲟血浆和肝脏组织中 GST 活性和 GSH 含量均随着给药浓度升高呈现先升高再降低的规律性变化，并在 40 mg/kg 给药浓度时达到最大值，小体鲟血浆及肝脏中 GST 活性分别为 23.235、68.670 U/（mg·pro），GSH 含量分别为 27.616、39.317 mg/（g·pro）；史氏鲟分别为 42.835、101.835 U/（mg·pro），27.852、51.192 mg/（g·pro）。2 种鲟肝脏组织中 GST 和 GSH 酶系的变化幅度明显大于血浆中，且史氏鲟血浆和肝脏组织在各给药浓度下 GST 活性和 GSH 含量均高于小体鲟。

关键词：恩诺沙星；小体鲟；史氏鲟；GST 活性；GSH 含量

该文发表于《生物技术通报》【2010，26（8）：199-203】

草鱼 CYP1A 和 ACT 基因 cDNA 片断的克隆与鉴定

林茂[1,2]，杨先乐[2]，纪荣兴[1]

（1. 集美大学水产学院，厦门 361021；2. 上海海洋大学农业部渔业动植物病原库，上海 200090）

摘要： 以 Trizol 法分别提取 BNF 诱导和对照处理草鱼的肝组织总 RNA 并合成 cDNA 第一链，以此为模板利用 1 对 ACT 特异性引物和 8 条 6 对 CYP1A 简并引物进行扩增，结果引物对 F0-R0 在对照和诱导草鱼中均扩增得到预期 ACT cDNA 片断，而引物对 F4-R4 在诱导草鱼中获得预期 CYP1A cDNA 产物。这两个 cDNA 片断分别进行克隆、测序和比对，BLAST 结果表明草鱼 ACT cDNA 片断（800 bp）与 Genbank 中 ACT 基因（登录号 M25013）同源性为 99.1%，推导氨基酸序列同源性为 99.2%；草鱼 CYP1A cDNA 片断（439 bp）与鲤鱼同源性最高，为 92.5%，推导氨基酸同源性为 96.6%。上述序列提交 GenBank，获得登录号分别为 DQ211096 和 DQ211095。通过 Mega 3.1 软件的 Neighbor-joining 程序对 CYP 基因的部分 cDNA 序列和氨基酸序列进行比对分析并绘制进化树，根据 CYP1A 部分蛋白的系统发育关系，在进化上可以将参与比对的真骨鱼划分为四个主要的分支。

关键词： 草鱼；CYP1A；β-肌动蛋白；cDNA；克隆

该文发表于《中国农学通报》【2011，27：（20）：87-91】

恩诺沙星对两种鲟细胞色素 P450 活性影响的比较研究

Comparison Study on the Effects of Enrofloxacin on the Cytochrome P450 Enzyme Activitiesin two Kinds of Sturgeons

卢彤岩[1]，陈琛[1,2]，李绍戊[1]，王荻[1]

（1. 中国水产科学研究院黑龙江水产研究所，哈尔滨 150070；2. 东北农业大学动物科学技术学院，哈尔滨 150030）

摘要： 将恩诺沙星按照 0、20、40、60、80 和 100 mg/kg 浓度，对小体鲟及史氏鲟口服给药 5 天，停药 2 天后对其肝脏组织中微粒体蛋白浓度及细胞色素 P450（cytochrome P450，CYP P450）酶活性进行测定。试验结果表明：2 种鲟体内均含有一定量的 P450 酶，且史氏鲟体内微粒体蛋白浓度 5.029 mg/g 及 P450 酶活性 0.102 nmol/（mg·min）均高于小体鲟相应的 4.833 mg/g 和 0.088 nmol/（mg·min）。而恩诺沙星对史氏鲟及小体鲟体内的细胞色素 P450 酶活性均有一定抑制作用，随着给药浓度从 0 到 100 mg/kg 的增加，2 种鲟体内的酶活性均逐渐降低，且史氏鲟体内 P450 酶活性降低幅度明显大于小体鲟，至 100 mg/kg 给药组，两者差值仅为 0.003 nmol/（mg·min）。上述结果对于探讨恩诺沙星在 2 种鲟体内的药物代谢及其与细胞色素 P450 酶之间相互作用的机制提供一定的理论依据。

关键词： 恩诺沙星；史氏鲟；小体鲟；肝微粒体蛋白浓度；CYP P450 活性

二、渔药作用的受体

目前，对于渔药作用的受体研究大多集中在 GABA、P 糖蛋白等方面。

GABA 受体（GABA γ2 亚基、GABA$_B$R1、GABA$_A$R 等）广泛分布于异育银鲫（*Carassius auratas gibelio*）的脑、肝、肾、心、肠和性腺等组织中，且有组织特异性：在脑组织中表达量最高（Yini Zhao, et al., 2015；Jiming Ruan, et al., 2014；A. L. Zhou1, et al., 2014；赵依妮等，2015；王祎等，2013；周爱玲等，2015）。在分类地位上，异育银鲫 GABA A

受体 γ2 亚基与斑马鱼亲缘关系最近（赵依妮等，2015），异育银鲫 GABA A α1 亚单位与斑马鱼的 GABA A 受体 α1 亚单位具有高度一致性（Yini Zhao, et al., 2015）。

作为受体，GABA 的表达收到药物的影响。杀虫剂阿维菌素（AVM）能提高 GABA$_A$R 的表达（Jiming Ruan, et al., 2014），且 AVM 对 GABA$_A$R 的表达具有一定程度的浓度依赖性（A. L. Zhou, et al., 2014）。氟喹诺酮类抗菌药双氟沙星（Difloxacin, DIF）能显著降低 GABA A 的表达量，且这种影响具有持久性（赵依妮等，2015）

P-糖蛋白（P-glycoprotein, P-gp）作为一种跨膜糖蛋白，是一种典型的药物外排泵，具有介导药物外排的功能。尼罗罗非鱼（Nile Tilapia）肝脏、肾脏组织中 P 糖蛋白（Permeability glycoprotein, P-gp）的表达参与了恩诺沙星（Enrofloxacin）在其体内的代谢过程（Kun Hu et al., 2014；胡鲲等，2013）壳聚糖（Chitosan）能通过抑制草鱼（Ctenopharyngodon idella）肠道中 P 糖蛋白（Permeability glycoprotein, P-gp）的表达而提高 NOR 在鱼体内的相对生物利用度（relative bioavailability）（Kun Hu et al., 2015）P-gp 的表达还与某些化合物（如异噻唑啉酮, isothiazolinone）对草鱼的毒性相关联，此外，温度升高可能会抑制 P-gp 的表达（Kun Hu et al., 2015）。

（一）GABA 的分布及其对渔药代谢的影响

该文发表于《Fish Physiol Biochem》【2014：DOI 10.1007/s10695-014-9925-8】

Distribution and quantitative detection of GABA$_A$ receptor in *Carassius auratus gibelio*

Jiming Ruan[1,2], Kun Hu[1], Haixin Zhang[1,3], Yi Wang[1],
Ailing Zhou[1], Yini Zhao[1], Xianle Yang[1]

(1. National Center for Aquatic Pathogen Collection, College of Fisheries and Life Science, Shanghai Ocean University, Shanghai 201306, People's Republic of China; 2. College of Sciences and Technology, Jiangxi Agricultural, Nanchang 330045, People's Republic of China; 3. Jiangxi Fisheries Research Institute, Nanchang 330039, People's Republic of China)

Abstract：Gamma - aminobutyric acid (GABA), a major inhibitory neurotransmitter in brain, is synthesized from glutamate and metabolized to succinic semialdehyde by glutamic acid decarboxylase (GAD) and GABA transaminase (GABA-T), respectively. The fast inhibitory effect of GABA is mediated by GABA type A (GABA$_A$) receptors that are associated with several neurological disorders, and GABA$_A$ receptors are targets of several therapeutic agents. To date, information on the distribution and quantity of GABA$_A$ receptors in *Carassius auratus gibelio* is still limited. We investigated for the first time, the tissue-specific distribution of GABA$_A$ Rb2a and GABA$_A$ Rb2b, the two subunits of the predominant GABA$_A$ receptor subtype (a1b2c2), and then, the expression of GABA$_A$ Rb2a, GABA$_A$ Rb2b, GAD, and quantified GABA-T genes in different tissues by quantitative real-time PCR method and compared different expressions between two developmental stages of C. *auratus gibelio*. Results showed that GABA$_A$ Rb2a and GABA$_A$ Rb2b genes expressed in both brain and peripheral organs using reverse transcription-polymerase chain reaction. In addition, the majority of GABA$_A$ Rb2a, GABA$_A$ Rb2b, GAD, and GABA-T were mainly synthesized in brain; however, a considerable amount of GABA-T was secreted from the peripheral tissues, especially in the liver. Moreover, the expression of GABA$_A$ Rb2a and GABA$_A$ Rb2b genes in different tissues varied with body weight change. This study provides a reference for further studies on GABA and GABA$_A$ receptors subunits and an insight on the possible pharmacological properties of the GABA$_A$ receptor in C. *auratus gibelio*.

Key words：GABA; GABA$_A$; receptor; GAD; GABA-T; *Carassius auratus gibelio*

异育银鲫的 GABA$_A$ 受体的分布和定量分析

阮记明[1,2]，胡鲲[1]，章海鑫[1,3]，王祎[1]，赵依妮[1]，杨先乐[1]

（1. 上海海洋大学水产与生命学院国家水生动物病原库，中国上海；2. 江西农业大学动物科学技术学院，中国南昌；

3. 江西水产研究所，中国南昌）

摘要： γ-氨基丁酸（GABA）是大脑的神经递质抑制物，GABA 由谷氨酸盐合成并且经分别经谷胺脱羧酶（GAD）和 GABA 转氨酶（GABA-T）代谢后形成琥珀酸半醛。GABA 的 A 类型具有快速抑制调解功能，GABAA 跟几种神经紊乱情况有关并且其受体也是几种治疗药物的靶物质。目前为止，对于异育银鲫 GABAA 受体的分布和数量了解很少。本研究首次做了关于特殊组织 GABA$_A$Rb2a 和 GABA$_A$Rb2b 的分布相关研究，还有 GABAA 的两个主要亚型子单位受体的研究（a1b2c2），然后是 GABA$_A$Rb2a，GABA$_A$Rb2b，GAD 的表达，利用定量实时 PCR 检测不同组织 GABA-T 基因的表达并比较异育银鲫两个不同发育阶段的表达情况。结果显示，经逆转录-聚合酶链式反应检测显示 GABA$_A$Rb2a 和 GABA$_A$Rb2b 基因在大脑和神经末梢器官两个部位有表达。此外，GABA$_A$Rb2a、GABA$_A$Rb2b、GAD 以及 GABA-T 物质大多数主要在脑中合成，也有不少 GABA-T 从神经末梢组织分泌出来，尤其是在肝脏。而且，不同组织的 GABA$_A$Rb2a 和 GABA$_A$Rb2b 基因表达情况随体重变化而不同。此研究为异育银鲫的 GABA 和 GABA$_A$ 受体子单位的深入研究和可能的药理学性能研究提供了理论依据。

关键词： γ-氨基丁酸；γ-氨基丁酸受体；谷氨酸脱羧酶；γ-氨基丁酸转氨酶；异育银鲫

Introduction

Gamma-aminobutyric acid (GABA) was first discovered in the brain in 1950 (Awapara et al. 1950; Roberts and Frankel 1950). GABA functions as the main inhibitory neurotransmitter in the central nervous system (CNS) of vertebrates (Farrant and Kaila 2007), which originates from the intestinal absorption of food or synthesis inside the body. The endogenous GABA is synthesized and metabolized in the GABA pathway, which involves two primary enzymes, namely glutamic acid decarboxylase (GAD) [EC4. 1. 1. 15] and GABA transaminase (GABA-T) [EC2. 6. 1. 19]. In the GABA pathway, glutamate can be catalyzed into GABA using GAD. GABA is then metabolized to succinic semialdehyde by GABA-T. As a GABA marker enzyme, GAD could exist in two isoforms, namely GAD$_{65}$ (65 kDa) and GAD$_{67}$ (67 kDa). GAD$_{65}$ has a restricted synaptic distribution, whereas GAD$_{67}$ exists in both dendrites and soma, which would be a more favorable position for the production of "chaperone" GABA (Eshaq et al. 2010). As a major inhibitory neurotransmitter in brain, GABA is released in approximately one third of all synapses (Nutt et al. 2006).

Gamma-aminobutyric acid is a multi-function substance that has different situational functions in the CNS, peripheral nervous system, and several nonneuronal tissues. GABA binds and activates GABA$_A$, GABA$_B$, and GABA$_C$ receptors (Brambilla et al. 2003). Previous studies have suggested that 30% or more of the central neurons use GABA as a neurotransmitter. GABA is a major neurotransmitter that may affect prefrontal cortex (PFC) activity in depression (Hashimoto 2011). GABA stimulates the acrosome reaction in bull spermatozoa acting through a classical GABA$_A$ receptor (Puente et al. 2011).

GABA$_A$ receptors are ligand-operated chloride channels that are assembled from five subunits, which belong to the cys-loop ligand-gated ion channel family, in a heteropentameric manner (Lester et al. 2004). GABA-binding pocket is formed at the a/b subunit interface in the subunit N-terminal regions, and each pentamer has two GABA-binding sites. Nineteen GABA$_A$ receptor subunit isoforms are cloned and grouped into eight subfamilies of mammals, namely a (1-6), b (1-3), c (1-3), d, e, h, p, and q (1-3) (Barnard 1998; Werner 2000). GABA$_A$ receptors are pentameric proteins with an integral chloride-permeable ion channel and normally gated by GABA (Lüddens and Korpi 1995; Rudolph and Möhler 2006). An integral chloride channel within GABA$_A$ receptor is closed or opened by binding with GABA, which allows chloride influx and leads to membrane hyperpolarization. GABA$_A$ receptors are ligand-gated chloride ion channels and have a number of binding sites for other ligands, such as benzodiazepines, barbiturates, convulsants (e.g., picrotoxinin), several general anesthetics, neurosteroids, ethanol, and fishery drugs.

Carassius auratus gibelio is one of the most important and widely bred fish species in China. However, reports on the distribution of GABA$_A$ receptors and the gene expression of GAD$_{65}$, GAD$_{67}$ and GABA-T in C. *auratus gibelio* are very limited. The predominant GABA$_A$ receptor subtype is composed of a1b2c2 subunits, which has two copies of a1 and b2 in each subunit (McKernan and Whiting 1996; Jackel et al. 1998; Pirker et al. 2000; Lüscher and Keller 2004). In this paper, the tissue specific distributions of GABA$_A$

Rb2a and $GABA_A$ Rb2b, two b2 subunits of $GABA_A$ receptor, were examined in C. *auratus gibelio*, by reverse transcription–polymerase chain reaction (RT-PCR), and the mRNA expressions of $GABA_A$ Rb2a, $GABA_A$ Rb2b, GAD_{65}, GAD_{67}, and GABA-T genes were quantified with quantitative real-time PCR (qPCR). Main aims of this study were to determine whether there has $GABA_A$ receptor exists in C. *auratus gibelio* and to explore their expression tissue specific distributions of $GABA_A$ Rb2a, $GABA_A$ Rb2b, GAD_{65}, GAD_{67}, and GABA-T genes.

Materials and methods

Fish culture and sampling

Normal C. *auratus gibelio* with body weights of 60.04±15.02 g (marketable group) and 4.67±0.93 g (fingerling group) were purchased from a farm in Jiangsu Province, East China. All fish were acclimatized to favorable laboratory condition for 2 weeks prior to experimentation. During acclimatization and experimental periods, the fish were maintained in aerated aquaria (3 000 L) that were constantly supplied with flowing (5 L/h) aerated freshwater (dissolved oxygen: 6.5±0.4 mg/L) at 16℃ and pH 7.5±0.3. The fish were fed twice a day. Six fish were utilized (n=6), and 2.0 g of each sample was excised from corresponding tissues. The samples were stored at -80℃ for RNA isolation. Different tissue samples (telencephalon, mesencephalon, cerebella, medulla oblongata, liver, kidney, heart, intestine, swim bladder, gill, muscle, and fin) were obtained for tissue distribution assay. All fish were handled in accordance with the "Regulation on Animal Experimentation" (State Scientific and Technological Commission, China).

Table 1 Information of primers used in the study

Genes	Primer sequence (5^0-3^0)	GenBank ID	Size (bp)	Temp. (C)
b-Actin	Forward TACGTTGCCATCCAGGCTGTG Reverse CATGGGGCAGGGCCGTAACC	M24113.1	124	55-60
$GABA_A$ Rb2a	Forward CCCGTGTGGCTTTAGGTATCA Reverse GACAAAGCAGCCCATCAGGTA	AM904760.1	125	58.9
$GABA_A$ Rb2b	Forward CCCGTGTGGCTTTAGGTATCA Reverse GACAAAGCAGCCCATCAGGTA	AM904761.1	125	57.4
GAD_{65}	Forward TTCTCTGTCGCTGCTCTGAT Reverse CTCTCGGCTGTAGACCCAT	AF149832.1	246	57.4
GAD_{67}	Forward GTTTTCTGATATCAAGCGTCTCAC Reverse TGGCAGGTTGTCGTAAATTAG	AF149833.1	209	56.1
GABA-T	Forward GCTGCCTGGCCACAACACA Reverse TCCCTCACAAACTCCTCCAGA	DQ287923.1	115	57.5

Total RNA extraction

Total RNA was extracted from the prepared samples using RNAiso Plus (Takara, Dalian, China) according to the manufacturer's instructions. RNA quality was detected by 1% agarose-gel electrophoresis. RNA concentration was quantified using a spectrophotometer (DU800, Beckman coulter, America). The absorption ratios (A260/A280 nm) were between 1.8 and 2.0. Reverse transcription

RT-PCR was performed with First Stand cDNA Synthesis Kit (TaKaRa code: D6110A). Total RNA (1μg) was used for reverse transcription in a reaction microtube containing 1 μL Oligo dT primer (50 lmol/L) and 1 μL dNTP mixture (10 mmol/L). The final volume was 10 μL with RNase-free dH_2O. The mixture was incubated at 65℃ for 5 min and then cooled immediately on ice. Then, 4 μL buffer, 0.5 μL RNase inhibitor (40 U/μL), 1 μL RTase (200 U/μL), and 4.5 μL RNase-free dH_2O were added into the reaction tube (20 μL final volume). The RT-PCR cycle parameters included one cycle at 30℃ for 10 min, 42℃ for 40 min, and 95℃ for 5 min. The RT products were stored at-20℃.

RT-PCR analysis

The primers (β-actin, $GABA_A$ Rb2a, $GABA_A$ Rb2b, GAD_{65}, GAD_{67}, and GABA-T) for RT-PCR and qPCR, which are listed in Table 1, were synthesized by Sangon (Shanghai, China), and b-actin was used as the reference gene. The reactions were performed in a thermocycler (Eppendorf, Germany) with the following profile: one cycle for 2 min at 94℃, followed by 35 cycles for 30 s at 94℃, 30 s at the corresponding temperature (Table 1), and 30 s at 72℃, and then 1 cycle for 8 min at 72℃.

qPCR analysis

Nowadays, qPCR has become an accurate technique for gene expression analysis. The resultant cDNA samples, which were used as templates to quantify the genes (GABA$_A$Rb2a, GABA$_A$Rb2b, GAD$_{65}$, GAD$_{67}$, and GABA-T) relative to b-actin, were adjusted to 100 ng/μL before qPCR test in this paper. In addition, qPCR was performed in the CFX_ 96 real-time PCR system (Bio-Rad, USA). The qPCR reactions with 20 μL total volume per reaction (10 μL iQ SYBR Green Supermix, 7.4 μL double distilled water, 0.3 μL of each primer, and 2 μL cDNA template) were set up with iQ^{TM}SYBR Green Supermix (Bio-Rad, USA) according to the manufacturer's protocol. The cycling procedure for the genes, as described in section 2.4, was 1 cycle at 95℃ for 30 s, followed by 39 cycles at 95℃ for 5 s and 57.5℃ for 30 s. Each sample was processed in triplicate. The comparative threshold method (2^{DDCt}) was employed to calculate relative expression level of the genes, and the specificity of each gene was determined using the melting curve analysis.

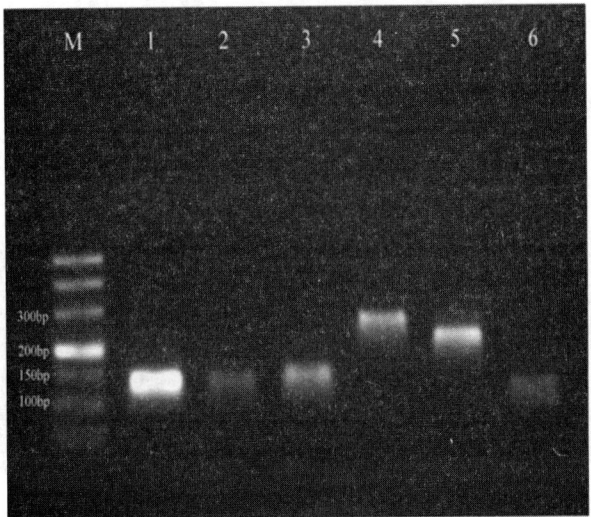

Fig. 1　RT-PCR analyses of the genes in C. auratus gibelio. With Lane M as the marker, Lanes 1-6 are (1) b-Actin, (2) GABA$_A$Rb2a, (3) GABA$_A$Rb2b, (4) GAD65, (5) GAD67, and (6) GABA-T. The RT-PCR products were tested by agarose-gel electrophoresis

The data were expressed as mean±standard deviation (SD), and statistical analyses were performed using SPSS 17.0 (Chicago, IL, USA). Significant differences between means were determined using one-way ANOVA. P \ 0.05 and P \ 0.01 were considered to indicate statistically significant and most significant difference, respectively.

Results

Verification of the RT-PCR products

The RT-PCRproducts of the genes (b-actin, GABA$_A$Rb2a, GABA$_A$Rb2b, GAD$_{65}$, GAD$_{67}$, and GABA-T) were analyzed by electrophoresis (Fig. 1). Then, the products were purified, ligated into the pMD18T vector (TaKaRa, Dalian, China) and then sequenced by Sangon (Shanghai, China). The retrieved sequences were aligned and confirmed by BLAST homology (http://www.ncbi.nlm.nih.gov/blast).

Tissue-specific distribution of GABAARb2a and GABAARb2b genes

The tissue-specific distribution of GABA$_A$Rb2a and GABA$_A$Rb2b was examined by RT-PCR. GABA$_A$Rb2a and GABA$_A$Rb2b mRNA expression was found both in brain regions, such as telencephalon, mesencephalon, cerebella and medulla oblongata, and in the peripheral tissues, such as liver, kidney, heart, intestine, swim bladder, gill, muscle, and fin (Figs. 2, 3). Relative expression of GABA$_A$Rb2a and GABA$_A$Rb2b genes.

Nowadays, qPCR has become an accurate technique for gene expression analysis. The fish in the marketable group were assigned as the experimental objects. The relative expression of $GABA_A Rb2a$ and $GABA_A Rb2b$ genes in different tissues was examined. As shown in Table 2, the mRNA expression of the $GABA_A Rb2a$ gene in telencephalon, mesencephalon, cerebella, medulla oblongata, heart, and swim bladder was significantly or most significantly higher than that of the liver, with the expression in the cerebella as the most significantly different. The $GABA_A Rb2a$ gene expression in the kidney, gill, muscle, and fin was adverse. The $GABA_A Rb2b$ gene expression in the telencephalon, mesencephalon, cerebella, medulla oblongata, and heart were higher than that in the liver, whereas the $GABA_A Rb2b$ gene expression in the kidney, intestine, swim bladder, gill, muscle, and fin was lower than that in the liver. In general, the $GABA_A Rb2a$ and $GABA_A Rb2b$ gene expression in the brain was much higher than that in the peripheral organs. Mesencephalon and medulla oblongata had higher $GABA_A Rb2a$ and $GABA_A Rb2b$ expression in the brain, respectively, than that in telencephalon (Fig. 4). The heart tissues exhibited the highest gene expression of $GABA_A Rb2a$ and $GABA_A Rb2b$ among the peripheral organs (Fig. 5).

The expression of the two receptors genes in the marketable and the fingerling groups was compared to understand the changes in the $GABA_A Rb2a$ and $GABA_A Rb2b$ gene expression during the developmental process. Upregulated expression of these genes was found in the telencephalon, mesencephalon, cerebella, medulla oblongata, liver, kidney, heart, and intestine in the marketable group. The liver and heart tissues exhibited the most up-regulated $GABA_A Rb2a$ and $GABA_A Rb2b$ gene expression, respectively, in contrast to the downregulated expression of these genes in the gill, muscle, and fin. Almost no change was observed in the swim bladder of the two fish groups (Fig. 6).

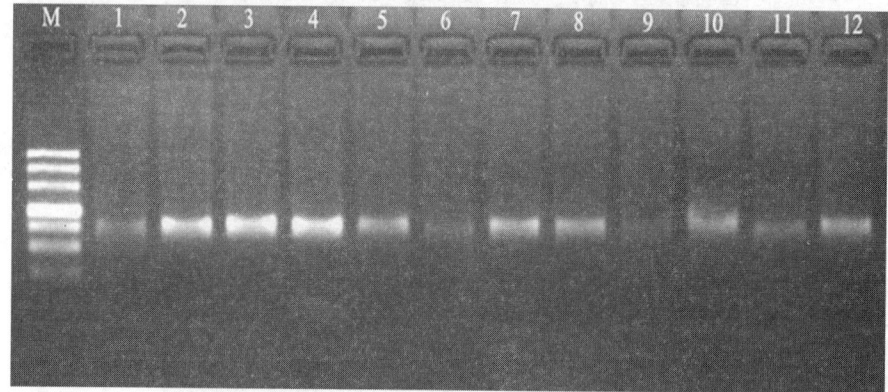

Fig. 2 RT-PCR analyses of $GABA_A Rb2a$ in the various tissues of marketable group. With lane M as marker, lanes 1-12 are 1 telencephalon, 2 mesencephalon, 3 cerebella, 4 medulla oblongata, 5 liver, 6 kidney, 7 heart, 8 intestine, 9 swim bladder, 10 gill, 11 muscle, and 12 fin. The RT-PCR products were tested by agarose-gel electrophoresis

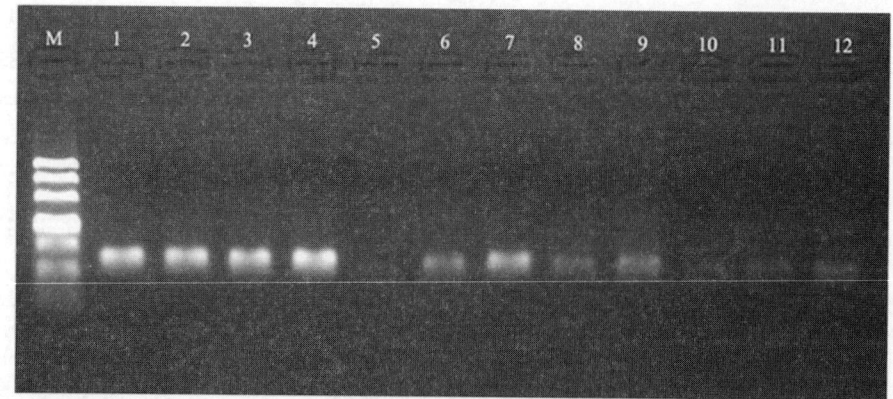

Fig. 3 RT-PCR analysis of $GABA_A Rb2b$ in the various tissues of marketable group. With lane M as marker, lanes 1-12 are 1 telencephalon, 2 mesencephalon, 3 cerebella, 4 medulla oblongata, 5 liver, 6 kidney, 7 heart, 8 intestine, 9 swim bladder, 10 gill, 11 muscle, and 12 fin. The RT-PCR products were tested by agarose-gel electrophoresis

Relative expression of GAD65, GAD67, and GABA-T

The two primary enzymes, GAD and GABA-T, are typically involved in the GABA pathway. GAD, the GABA marker enzyme, has two isoforms （GAD_{65} and GAD_{67}）. The relative expression proportions of GAD_{65} and GAD_{67} in the different tissues of C. auratus gibelio in the marketable group are shown in Table 3. The results showed that the GAD_{65} and GAD_{67} were synthesized mostly in the brain. Among the peripheral organs, the tissues in the swim bladder and liver had the highest expression of GAD_{65} and GAD_{67}, respectively, whereas the kidney has the least GAD expression. Furthermore, the amount of GAD_{65} in the three parts of the brain （telencephalon, mesencephalon, and cerebella）, with the exception of the medulla oblongata, was twice as much as that of GAD_{67} （Table 4）. GABA–T, an important enzyme in the GABA pathway, has synthesis characteristics that are different from that of GAD_{65}. GABA–T was synthesized in the brain and in the peripheral organs, especially in the liver （Table 5）.

Table 2　Relative proportions of $GABA_A$Rb2a and $GABA_A$Rb2b genes in different tissues of C. *auratus gibelio*

Tissues	Genes	
	$GABA_A$Rb2a	$GABA_A$Rb2b
Liver	0. 998±0. 002	1. 000±0. 002
Telencephalon	395. 071±88. 087 * *	107. 539±10. 822 * *
Mesencephalon	602. 443±119. 578 * *	130. 826±11. 078 * *
Cerebella	959. 243±153. 222 * *	154. 758±14. 506 * *
Medulla oblongata	944. 828±190. 099 * *	269. 841±18. 853 * *
Kidney	0. 286±0. 085[DD]	0. 520±0. 002[DD]
Heart	9. 139±0. 722 * *	2. 722±0. 118 * *
Intestine	1. 662±0. 327	0. 264±0. 034[DD]
Swim bladder	1. 749±0. 264 *	0. 308±0. 012[DD]
Gill	0. 368±0. 038[DD]	0. 034±0. 002[DD]
Muscle	0. 608±0. 090[D]	0. 046±0. 004[DD]
Fin	0. 339±0. 078[DD]	0. 025±0. 004[DD]

　＊ Abundant than that in the liver; D scarce than that in the liver. In the same column, ＊ and D significantly different from that in the liver （$P<$ 0. 05）, ＊＊ and DD most significantly different from that in the liver （$P<0. 01$）

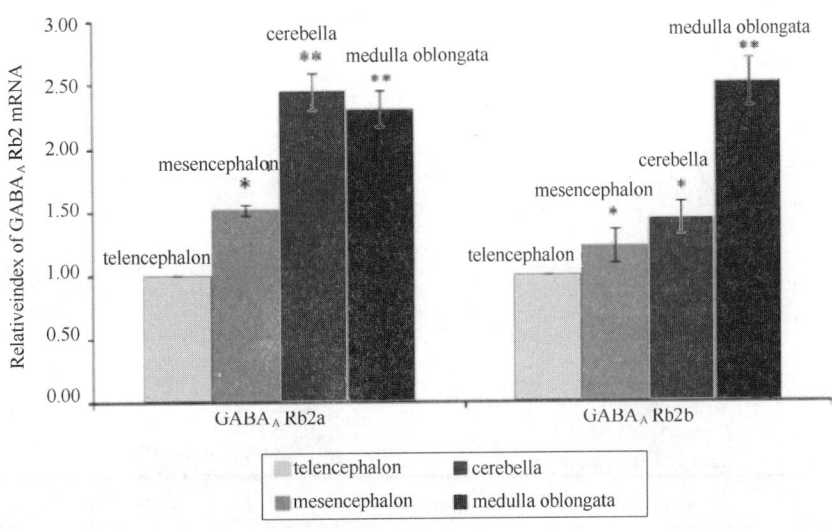

Fig. 4　Relativeexpressions of $GABA_A$Rb2a and $GABA_A$Rb2b genes in the brain of marketable group. Relative indices were normalized by the relative expression in the telencephalon respectively. Asterisk indicates significantly different from that in the telencephalon （$P<0.05$）, and double asterisk indicates most significantly different from that in the telencephalon （$P<0.01$）.

Discussion

Tissue-specific distribution of $GABA_A$ Rb2a and $GABA_A$ Rb2b genes in C. auratus gibelio

The $GABA_A$ receptors are some of the most extensively studied receptors in the GABAergic system. The $GABA_A$ receptors are also the most broadly distributed GABA subtype and responsible for diverse and important actions in the CNS (Wisden and Seeburg, 1992; Watanabe et al., 2002). The distribution of neurons that contain c_2-subunit and the expression of the mRNAs encoding a_3 and a_4 subunits of $GABA_A$ receptors were examined in the rat CNS with in situ hybridization histochemistry (Araki and Tohyama, 1992; Araki et al., 1992). All magnocellular neurons in the supraoptic and paraventricular nuclei of female rat were detected expressing the mRNA encoding of the c_2 subunit of the $GABA_A$ receptor, but not the c_1 or c_3 subunits with ^{35}S-labeled antisense oligonucleotides (Fenelon and Herbison 1995). A cDNA of $GABA_A$ receptor that encodes the fourth type of c subunit has been isolated from the chicken brain (Harvey et al., 1993). The $GABA_A$ and $GABA_B$ families are found in the avian brain (Glencorse et al. 1991). Studies on the GABA receptors in reptiles, such as turtles, have been performed (Chen and Chesler, 1990; Castro et al., 2011). GABA and GABA analog binding indicated that subtypes of $GABA_A$/benzodiazepine receptor were existed in the brain of Atlantic salmon (Anzelius et al., 1995). $GABA_A$ receptors were also found in bullfrogs (Hollis and Boyd, 2003; Asay and Boyd, 2006). In the bullfrog brain, a hypothesis that amphibian possesses a $GABA_A$-like receptor protein, which similar to the $GABA_A$ receptor characterized in mammals, was demonstrated by the binding characteristics and ligand specificity of the [3H] muscimol binding sites. Baclofen, a kind of mammalian $GABA_B$ receptor agonist, does not inhibit [3H] muscimol binding in bullfrogs, which suggested that [3H] muscimol binding would mediated by a $GABA_A$-like receptor instead of a $GABA_B$-like receptor (Hollis and Boyd, 2003). A novel class of GABA receptors exists in cockroach (Periplaneta Americana) (Lummis and Sattelle, 1985). The cloned Drosophila melanogaster GABA receptors elucidated the contribution of particular subunits to the pharmacological differences from their vertebrate counterparts (Hosie et al., 1997). Crustaceans, such as crayfish, have also been confirmed to possess $GABA_A$ receptors (Rashkovan et al., 1997; Adelsberger et al., 1998). Furthermore, $GABA_A$ receptors have been carefully studied outside the brain. The existences of GABA receptors were found in peripheral non-neural tissues, such as smooth muscle, endocrines, and female reproductive system, even in parts of the peripheral nervous system (Ong and Kerr, 1990). The GABA receptor subtypes have been verified in the rat stomach in vitro (Rotondo et al., 2010). $GABA_A$ receptor subunits exist in several peripheral blood mononuclear cells (Alam et al., 2006), blood leukocytes (Plummer et al., 2011), and T cells (Tian et al., 1999; Bjurstöm et al., 2008). Furthermore, the $GABA_A$ receptor subunits were expressed in multiple endocrine tissues, including the adrenal, ovary, testis, and small intestine of rats in a tissue-specific manner (Akinci and Schofield, 1999). $GABA_A$ receptors were demonstrated on the gastric membranes of the rat stomach (Erdö et al., 1989). The expression of different $GABA_A$ receptor subunits was also observed in the pancreas (Borboni et al., 1994; Yang et al., 1994).

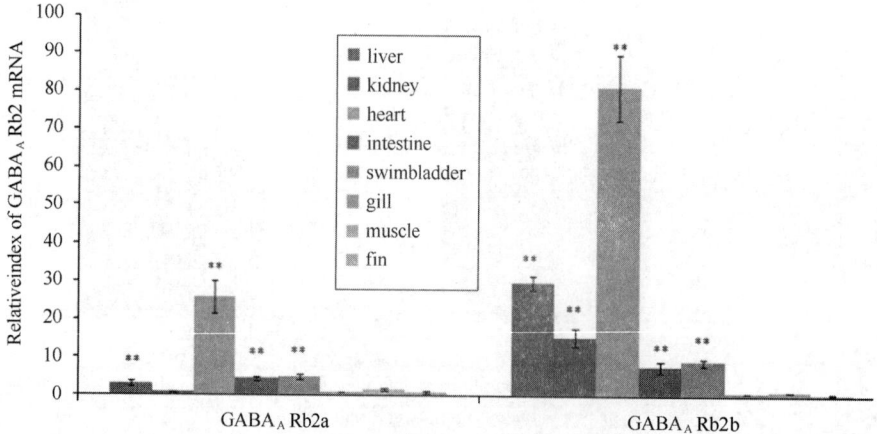

Fig. 5 Relative expressions of $GABA_A$ Rb2a and $GABA_A$ Rb2b in the peripheral tissues of marketable group. Relative indices were normalized by the relative expression in the gill respectively. Asterisk significantly different from that in the gill (P<0.05); Double asterisk most significantly different from that in the gill (P<0.01)

Ps: The order of the tissues from left to right is: liver-kidney-heart-intestine-swim bladder-gill-muscle-fin

Based on the results from the aforementioned species like mammals, birds, reptiles, amphibians, and crustaceans, it would hy-

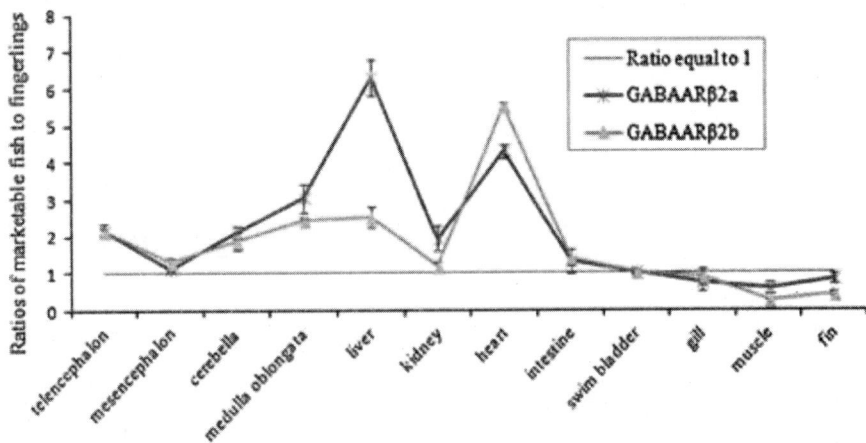

Fig. 6　Comparison of GABA$_A$Rb2a and GABA$_A$Rb2b mRNA expression between marketable group and fingerling group in different tissues of C. *auratus gibelio*. A ratio greater than 1 signifies up-regulated expression in marketable group, whereas a ratio less than 1 signifies down regulated expression

pothesize that GABA$_A$ receptors may exist in C. *auratus gibelio*. In this paper, GABA$_A$Rb2a and GABA$_A$Rb2b genes, the predominant subunits of GABA$_A$ receptors, were testified existing both in the brain and in the peripheral tissues of C. auratus gibelio by RT-PCR (Figs. 1, 2, 3), which showed our hypothesis was correct.

Table 3　Relative proportions of GAD$_{65}$ and GAD$_{67}$ that are expressed in different tissues of C. *auratus gibelio*

Tissues	Genes	
	GAD$_{65}$	GAD$_{67}$
Liver	1.00±0.00	1.00±0.00
Telencephalon	281.34±25.00**	238.59±29.96**
Mesencephalon	557.26±44.05**	509.44±54.17**
Cerebella	120.80±16.03**	111.59±18.27**
Medulla oblongata	351.96±28.93**	701.60±91.68**
Kidney	0.13±0.05^{DD}	0.09±0.03^{DD}
Heart	1.03±0.34	0.09±0.03^{DD}
Intestine	1.46±0.16	0.15±0.04^{DD}
Swim bladder	4.12±0.62**	0.42±0.08
Gill	1.05±0.14	0.16±0.06^{DD}
Muscle	1.40±0.24	0.27±0.07^{D}
Fin	2.69±0.36*	0.35±0.09

* Abundant than that in the liver; ^{D} scarce than that in the liver. In the same column; * and ^{D} significantly different from that in the liver (P<0.05); ** and ^{DD} most significantly different from that in the liver (P<0.01)

Table 4　Expression ratios of GAD$_{65}$/GAD$_{67}$ in the brain of C. *auratus gibelio*

Ratios	Tissues			
	Telencephalon	Mesencephalon	Cerebella	Medulla oblongata
GAD$_{65}$/GAD$_{67}$	2.12±0.38	2.04±0.33	2.01±0.23	0.91±0.22

Table 5　Relative proportions of GABA-T expressed in the different tissues of C. *auratus gibelio*

Tissues	GABA-T
Kidney	1.00 ± 0.00
Telencephalon	$4.44\pm0.65^{**}$
Mesencephalon	$10.08\pm0.98^{**}$
Cerebella	$4.41\pm0.31^{**}$
Medulla oblongata	$7.11\pm0.66^{**}$
Liver	$11.05\pm1.15^{**}$
Heart	0.75 ± 0.10^{D}
Intestine	1.03 ± 0.28
Swim bladder	0.13 ± 0.02^{DD}
Gill	0.24 ± 0.07^{DD}
Muscle	0.41 ± 0.06^{DD}
Fin	0.73 ± 0.09^{DD}

* Abundant than that in the kidney; [D] scarce than that in the kidney. In the same column, * and [D] significantly different from that in the kidney ($P<0.05$), ** and [DD] most significantly different from that in the kidney ($P<0.01$)

GABA$_A$Rb2a and GABA$_A$Rb2b relative gene expression

qPCR has recently become a powerful tool in gene expression analysis for its sensitivity, accuracy, and high throughput. The results of the quantitative relative expression of GABA$_A$Rb2a and GABA$_A$Rb2b genes of the marketable fish show a wide, but uneven, distribution in different tissues. As shown in Table 2, although GABA$_A$Rb2a and GABA$_A$Rb2b genes were found in the peripheral tissues, considerable amounts of the two subunits were synthesized mainly in the brain, in which the telencephalon, mesencephalon, cerebella, and medulla oblongata exhibited the strongest mRNA expression for b2a and b2b subunits. In addition, mesencephalon and medulla oblongata exhibited the highest expression of GABA$_A$ Rb2a and GABA$_A$ Rb2b in the brain, respectively, than the telencephalon (Fig. 4). The heart tissues had the highest GABA$_A$Rb2a and GABA$_A$Rb2b gene expression among the peripheral organs (Fig. 5).

Several factors, such as pathological state, stress, or age, would influence the expression of GABA$_A$ receptors. The transcript levels of GABA$_A$ receptor beta 2 were found to be significantly decreased in the human postmortem anterior cingulate cortex (ACC) in mood disorder (Zhao et al., 2012). The mRNA and protein expressions of GABARa1 subunit were significantly increased in hippocampus of tremor rat (Mao et al., 2010). GABR$_{A3}$ (a kind of GABA$_A$ receptor subunit) and GABA$_B$ receptor 2 were significantly down-regulated in peripheral blood leukocytes of migraineurs (Plummer et al., 2011). The level of the GABA$_A$ receptor a1 subunit was significantly decreased under acute restraint stress in rat hippocampus and PFC (Zheng et al., 2007). Several other reports have noted variations in the subunitlevels of GABA$_A$ receptors in brain according to acute or repeated stress (Gruen et al., 1995; Baulieu et al., 2001; Benavidez and Arce, 2002; Chadda and Devaud, 2005). Moreover, significant aging-related down-regulated mRNA expression of various GABA$_A$ receptor subunits (a1, c2, b2, b3, and d) was found in both cerebellum and cerebral cortex using quantitative dot blot and in situ hybridization techniques (Gutie rrez et al., 1997). In several brain areas of Sprague-Dawley and Fischer 344 rat, the mRNA expression level of c_{2s} and c_{2L} subunits of the GABA$_A$ receptors has been observed having aging-related alterations (Gutie rrez et al., 1996). The c_4-subunit was first detected at embryonic day 13 in the developing chicken brain (Harvey et al., 1993). In the grafted striatal neurons, the time course of the expression of the GABA$_A$ receptor subunits closely matched those of the development of functional synaptic activity and morphological maturation of the transplant in normal developing striatum (Liste et al., 1997). The expression in the two subunits between the marketable and fingerling groups of C. *auratus gibelio* was compared in this paper. The results showed up-regulated expressions of GABA$_A$Rb2a and GABA$_A$Rb2b in telencephalon, mesencephalon, cerebella, medulla oblongata, liver, kidney, heart, and intestine in the marketable group, relative to that in the fingerling group. The liver and heart had the most up-regulated GABA$_A$Rb2a and GABA$_A$Rb2b expression, respectively, whereas thegill, muscle, and fin exhibited down-regulated expression. While there almost has no changes of GABA$_A$Rb2a and GABA$_A$Rb2b expression was observed in the swim bladder of the two groups (Fig. 6).

Relative expression of GAD$_{65}$, GAD$_{67}$, and GABA-T

GABA is considered a multi-function substance that has different situational functions in CNS, peripheral nervous system, and several non-neuronal tissues. GABA is synthesized primarily from glutamate by GAD and metabolized to succinic semialdehyde by GABA-T. The distribution of GABA-containing neurons in brain of adult zebra fish was investigated using Nissl staining and immunohistochemistry (Kim et al., 2004). GABA functions are initiated by binding with its ionotropic receptors, $GABA_A$ and $GABA_C$, which are ligand-gated chloride channels, and its metabotropic receptor, $GABA_B$. An increase expression of $GABA_A$ receptors in HEK 293 cells was observed after GABA treatment (Eshaq et al., 2010) GABA concentrations in the cells of the pancreatic islet, oviduct, myenteric plexus of the gut, and adrenal chromaffin are comparable with that in the CNS (Tanaka, 1985). However, after comparing the proportions of GAD_{65}, GAD_{67}, and GABA-T (Tables 3, 5), the concentrations of GABA in the brain was far greater than the sum of the concentrations in the peripheral tissues in C. auratus gibelio. High GABA concentration was manifested in the synaptic terminals. However, the extracellular space in the brain only exhibited GABA concentrations from 0 to 1 $\mu mol/L$ (Lerma et al., 1986; Segovia et al., 1986; de Groote and Linthorst, 2007). This condition could explain the higher GAD expression in the brain than that in the peripheral tissues, which contain fewer nervous cells.

GAD exists in two isoforms, GAD_{65} and GAD_{67}, which are abundant in the nerve terminal and in the neuronal soma and dendrites respectively. Interestingly, the GAD_{65} contents in telencephalon, mesencephalon, and cerebella, with the exception of medulla oblongata, are approximately twice as much as that of GAD_{67} (Table 4). This result suggests that most of the endogenous GABA in C. auratus auratas is synthesized by GAD_{65} from glutamate. GAD_{65} and GAD_{67} are mostly synthesized in the brain, whereas a significant proportion of GABA-T is secreted in the peripheral tissues, especially in the liver, which suggests that liver is the main metabolic organ in the body.

In addition to its role in neural development, GABA has a wide variety of physiological functions in the tissues and organs outside the brain. Moreover, GABA acts as a hormone or trophic factor in non-neuronal peripheral tissue and as a neurotransmitter or neuromodulator in the autonomic nervous system (Ong and Kerr 1990; Gladkevich et al. 2006). However, GAD_{65} and GAD_{67} also exist in the peripheral organs, but not in the endocrine system, which would suggest that GABA could play the role of a trophic factor.

Conclusion

In summary, $GABA_A$Rb2a and $GABA_A$Rb2b genes are widely exist both in brain (telencephalon, mesencephalon, cerebella, and medulla oblongata) and in peripheral tissues (liver, kidney, heart, intestine, swim bladder, gill, muscle, and fin) of C. auratus gibelio, but the majority of the two genes was synthesized in the brain, as well as GAD_{65} and GAD_{67}. As for GABA-T, there has a considerable proportion secreted in the peripheral tissues, especially in the liver. The relative expressions of the $GABA_A$Rb2a and $GABA_A$Rb2b genes in the different tissues of C. auratus gibelio were varied in accordance with body weight. This phenomenon indicates that the mRNA expression of $GABA_A$Rb2a and $GABA_A$Rb2b genes is related to aging. This paper was the first to perform a panorama tracing of the distributions of $GABA_A$Rb2a and $GABA_A$Rb2b genes in C. auratus gibelio.

Acknowledgments

This work was financially supported by the National Natural Science Foundation of China (No. 31172430), the Special Fund for Agro-scientific Research in the Public Interest (No. 201203085), and grants from the National High Technology Research and Development Program of China (No. 2011AA10A216).

References

A delsberger H, Brunswieck S, Dudel J (1998) Modulatory effects of c-hydroxybutyric acid on a $GABA_A$ receptor from crayfish muscle. Eur J Pharmacol 350: 317-323.

Akinci MK, Schofield PR (1999) Widespread expression of $GABA_A$ receptor subunits in peripheral tissues. Neurosci Res 35: 145-153.

Alam S, Laughton DL, Walding A, Wolstenholme AJ (2006) Human peripheral blood mononuclear cells express $GABA_A$ receptor subunits. Mol Immunol 43: 1432-1442.

Anzelius M, Ekström P, Möhler H, Grayson Richards J (1995) Immunocytochemical localization of $GABA_A$ receptor b2b3-subunits in the brain of Atlantic salmon (Salmo salar L). J Chem Neuroanat 8: 207-221.

Araki T, Tohyama M (1992) Region-specific expression of $GABA_A$ receptor a3 and a4 subunits mRNAs in the rat brain. Mol Brain Res 12: 293-314.

Araki T, Sato M, Kiyama H, Manabe Y, Tohyama M (1992) Localization of $GABA_A$-receptor c2-subunit mRNAcontaining neurons in the rat central nervous system. Neuroscience 47: 45-61.

Asay MJ, Boyd SK (2006) Characterization of the binding of [3H] CGP54626 to $GABA_B$ receptors in the male bullfrog (Rana catesbeiana). Brain Res 1094: 76-85.

Awapara J, Landua AJ, Fuerst R, Seale B (1950) Free gammaaminobutyric acid in brain. J Biol Chem 187: 35-39.

Barnard EA (1998) Finding apaththrough the forestof the $GABA_A$ receptors. Eur Neuropsychopharmacol 8 (supplement2): S53.

Baulieu EE, Robel P, Schumacher M (2001) Neurosteroids: beginning of the story. Int Rev Neurobiol 46: 1-32.

Benavidez E, Arce A (2002) Effects of phosphorylation and cytoskeleton-affecting reagents on GABA$_A$ receptor recruitment into synaptosomes following acute stress. Pharmacol Biochem Behav 72: 497-506.

Bjurstöm H, Wang J, Ericsson I, Bengtsson M, Liu Y, KumarMendu S, Issazadeh-Navikas S, Birnir B (2008) GABA, a natural immunomodulator of T lymphocytes. J Neuroimmunol 205: 44-50.

Borboni P, Porzio O, Fusco A, Sesti G, Lauro R, Marlier LNJL (1994) Molecular and cellular characterization of the GABA$_A$ receptor in the rat pancreas. Mol Cell Endocrinol 103: 157-163.

Brambilla F, Biggio G, Pisu MG, Bellodi L, Perna G, Bogdanovich-Djukic V, Purdy RH, Serra M (2003) Neurosteroid secretion in panic disorder. Psychiatry Res 118: 107-116.

Castro A, Aguilar J, Andres C, Felix R, Delgado-Lezama R (2011) GABA$_A$ receptors mediate motoneuron tonic inhibition in the turtle spinal cord. Neuroscience 192: 74-80.

Chadda R, Devaud LL (2005) Differential effects of mild repeated restraint stress on behaviors and GABA$_A$ receptors in male and female rats. Pharmacol Biochem Behav 81: 854-863.

Chen JC, Chesler M (1990) A bicarbonate-dependent increase in extracellular pH mediated by GABA$_A$ receptors in turtle cerebellum. Neurosci Lett 116: 130-135.

De Groote L, Linthorst ACE (2007) Exposure to novelty and forced swimming evoke stressor-dependent changes in extracellular GABA in the rat hippocampus. Neuroscience 148: 794-805.

Erdö SL, Ezer E, Matuz J, Wolff JR, Amenta F (1989) GABA$_A$ receptors in the rat stomach may mediate mucoprotective effects. Eur J Pharmacol 165: 79-86.

Eshaq RS, Stahl LD, Stone R II, Smith SS, Robinson LC, Leidenheimer NJ (2010) GABA acts as a ligand chaperone in the early secretory pathway to promote cell surface expression of GABA$_A$ receptors. Brain Res 1346: 1-13.

Farrant M, Kaila K (2007) The cellular, molecular and ionic basis of GABA$_A$ receptor signaling. Prog Brain Res 160: 59-87.

Fenelon VS, Herbison AE (1995) Characterisation of GABA$_A$ receptor gamma subunit expression by magnocellular neurones in rat hypothalamus. Mol Brain Res 34: 45-56.

Gladkevich A, Korf J, Hakobyan VP, Melkonyan KV (2006) The peripheral GABA ergic system as a target in endocrine disorders. Auton Neurosci 124: 1-8.

Glencorse TA, Bateson AN, Hunt SP, Darlison MG (1991) Distribution of the GABA$_A$ receptor a1-and c2-subunit mRNAs in chick brain. Neurosci Lett 133: 45-48.

Gruen RJ, Wenberg K, Elahi R, Friedhoff AJ (1995) Alterations in GABA$_A$ receptor binding in the prefrontal cortex following exposure to chronic stress. Brain Res 684: 112-114.

Gutierrez A, Khan ZU, Miralles CP, De Blas AL (1996) Altered expression of c2L and c2S GABA$_A$ receptor subunits in the aging rat brain. Mol Brain Res 35: 91-102.

Gutierrez A, Khan ZU, Miralles CP, Mehta AK, Ruano D, Araujo F, Vitorica J, De Blas AL (1997) GABA$_A$ receptor subunit expression changes in the rat cerebellum and cerebral cortex during aging. Mol Brain Res 45: 59-70.

Harvey RJ, Kim HC, Darlison MG (1993) Molecular cloning reveals the existence of a fourth c subunit of the vertebrate brain GABA (A) receptor. FEBS Lett 331: 211-216.

Hashimoto K (2011) . The role of glutamate on the action of antidepressants. Prog Neuropsychopharmacol Biol Psychiatry 35: 1558-1568.

Hollis DM, Boyd SK (2003) Characterization of the GABA$_A$ receptor in the brain of the adult male bullfrog, Rana catesbeiana. Brain Res 992: 69-75.

Hosie A, Sattelle D, Aronstein K, Ffrench-Constant R (1997) Molecular biology of insect neuronal GABA receptors. Trends Neurosci 20: 578-583.

Jackel C, Kleinz R, Makela R, Hevers W, Jezequel S, Korpi ER, Lüddens H (1998) The main determinant of furosemide inhibition on GABA$_A$ receptors is located close to the first transmembrane domain. Eur J Pharmacol 357: 251-256.

Kim Y, Nam R, Yoo YM, Lee C (2004) Identification and functional evidence of GABA ergic neurons in parts of the brain of adult zebrafish (Danio rerio) . Neurosci Lett 355: 29-32.

Lerma J, Herranz AS, Herreras O, Abraira V, Del Rio RM (1986) In vivo determination of extracellular concentration of amino acids in the rat hippocampus. A method based on brain dialysis and computerized analysis. Brain Res 384: 145-155.

Lester HA, Dibas MI, Dahan DS, Leite JF, Dougherty DA (2004) Cys-loop receptors: new twists and turns. Trends Neurosci 27: 329-336.

Liste I, Caruncho HJ, Guerra MJ, Labandeira-Garcia JL (1997) GABAA receptor subunit expression in intrastriatal striatal grafts: comparison between normal developing striatum and developing striatal grafts. Dev Brain Res 103: 185-194.

Lüddens H, Korpi ER (1995) Biological function of GABA$_A$/benzodiazepine receptor heterogeneity. J Psychiatr Res 29: 77-94.

Lummis SCR, Sattelle DB (1985) Insect central nervous system c-aminobutyric acid. Neurosci Lett 60: 13-18.

Lüscher B, Keller CA (2004) Regulation of GABA$_A$ receptor trafficking, channel activity, and functional plasticity of inhibitory synapses. Pharmacol Therapeut 102: 195-221.

Mao X, Guo F, Yu J, Min D, Wang Z, Xie N, Chen T, Shaw C, Cai J (2010) Up-regulation of GABA transporters and GABA$_A$ receptor a1 subunit in tremor rat hippocampus. Neurosci Lett 486: 150-155.

McKernan RM, Whiting PJ (1996) Which GABA$_A$-receptor subtypes really occur in the brain? Trends Neurosci19: 139-143.

Nutt D, Argyropoulos S, Hood S, Potokar J (2006) Generalized anxiety disorder: a comorbid disease. Eur Neuropsychopharmacol 16 (Supplement 2): S109-S118.

Ong J, Kerr DIB (1990) GABA-receptors in peripheral tissues. Life Sci 46: 1489-1501.

Pirker S, Schwarzer C, Wieselthaler A, Sieghart W, Sperk G (2000) GABA_A receptors: immunocytochemical distribution of 13 subunits in the adult rat brain. Neuroscience 101: 815-850.

Plummer PN, Colson NJ, Lewohl JM, MacKay RK, Fernandez F, Haupt LM, Griffiths LR (2011) Significant differences in gene expression of GABA receptors in peripheral blood leukocytes of migraineurs. Gene 490: 32-36.

Puente MA, Tartaglione CM, Ritta MN (2011) Bull sperm acrosome reaction induced by gamma-aminobutyric acid (GABA) is mediated by GABAergic receptors type A. Anim Reprod Sci 127: 31-37.

Rashkovan G, Fisher K, Parnas I (1997) GABA_A receptors affect directly the release boutons in the neuromuscular junction of the crayfish opener muscle. Neurosci Lett 237 (Supplement 48): S40.

Roberts E, Frankel S (1950) Gamma-Aminobutyric acid in brain: its formation from glutamic acid. J Biol Chem 187: 55-63.

Rotondo A, Serio R, Mul F (2010) Functional evidence for different roles of GABA_A and GABA_B receptors in modulatingmousegastric tone. Neuropharmacology58: 1033-1037.

Rudolph U, Möhler H (2006) GABA-based therapeutic approaches: GABA_A receptor subtype functions. Curr Opin Pharmacol 6: 18-23.

Segovia J, Tossman U, Herrera-Marschitz M, Garcia-Munoz M, Ungerstedt U (1986) c-Aminobutyric acid release in the Globus pallidus in vivo after a 6-hydroxydopamine lesion in the substantia nigra of the rat. Neurosci Lett 70: 364-368.

Tanaka C (1985) c-aminobutyric acid in peripheral tissues. Life Sci 37: 2221-2235.

Tian J, Chau C, Hales TG, Kaufman DL (1999) GABA_A receptors mediate inhibition of T cell responses. J Neuroimmunol 96: 21-28.

Watanabe M, Maemura K, Kanbara K, Tamayama T, Hayasaki H (2002) GABA and GABA receptors in the central nervous system and other organs. Int Rev Cytol A Surv Cell Biol 213: 1-47.

Werner S (2000) Unraveling the function of GABA_A receptor subtypes. Trends Pharmacol Sci 21: 411-413.

Wisden W, Seeburg PH (1992) GABA_A receptor channels: from subunits to functional entities. Curr Opin Neurobiol 2: 263-269.

Yang W, Reyes AA, Lan NC (1994) Identification of the GABA_A receptor subtype mRNA in human pancreatic tissue. FEBS Lett 346: 257-262.

Zhao J, Bao AM, Qi XR, Kamphuis W, Luchetti S, Lou JS, Swaab DF (2012) Gene expression of GABA and glutamate pathway markers in the prefrontal cortex of nonsuicidal elderly depressed patients. J Affect Disord 138: 494-502.

Zheng G, Zhang X, Chen Y, Zhang Y, Luo W, Chen J (2007) Evidence for a role of GABA_A receptor in the acute restraint stress-induced enhancement of spatial memory. Brain Res 1181: 61-73.

该文发表于《Fish Physiology and Biochemistry》【2015: 29: 1-10】

Isolation, characterization, and tissue-specific expression of GABA_A receptor α1 subunit gene of Carassius auratus gibelio after avermectin treatment

Yini Zhao, Qi Sun, Kun Hu, Jiming Ruan, Xianle Yang

(National Pathogen Collection Center for Aquatic Animals, Shanghai Ocean University, No. 999 Hucheng Huan Road, Lingang New City, Pudong New District, Shanghai 201306, People's Republic of China)

Abstract: Carassius auratus gibelio has been widely cultivated in fish farms in China, with avermectin (AVM) being used to prevent parasite infection. Recently, AVM was found to pass through the Carassius auratus gibelio blood-brain barrier (BBB). Although AVM acts mainly through a GABA receptor and specifically the a1 subunit gene, the most common isoform of the GABA_A receptor, which is widely expressed in brain neurons and has been studied in other fish, Carassius auratus gibelio GABA_A receptor α1 subunit gene cloning, and whether AVM passes through the BBB to induce Carassius auratus gibelio GABA_A receptor a1 subunit gene expression have not been studied. The aim of this study was to clone, sequence, and phylogenetically analyze the GABA_A receptor a1 subunit gene and to investigate the correlation of its expression with neurotoxicity in brain, liver, and kidney after AVM treatment by quantitative real-time reverse transcription polymerase chain reaction. The a1 subunit gene was 1550 bp in length with an open reading frame of1380 bp encoding a predicted protein with 459 amino acid residues. The gene contained 128 bp of 5' terminal untranslated region (URT) and 72 bp of 30 terminal UTR. The a1 subunit structural features conformed to the Cys-loop ligand-gated ion channels family, which includes a signal peptide, an extracellular domain at the N-terminal, and four transmembrane domains. The established phylogenetic tree indicated that the a1 subunits of Carassius auratus gibelio and Danio rerio were the most closely related to each other. The a1 subunit was found to be highly expressed in brain and ovary, and the α1 mRNA transcription level increased significantly in brain. Moreover, the higher the concentration of AVM was, the higher the GABA$_A$ receptor expression was, indicating that AVM can induce significant neurotoxicity to Carassius auratus gibelio. Therefore, the α1 subunit mRNA expression was positively correlated with the neurotoxicity of AVM in Carassius auratus gibelio. Our findings suggest that AVM should be used carefully in Carassius auratus gibelio farming, and other alternate antibiotics with lower toxicity should be investigated with respect to toxicity via the induction of GABA$_A$ receptor expression for fish farming.

Key words: Carassius auratus gibelio; Avermectin; GABA$_A$ receptor α1 subunit; Sequence analysis; Phylogenetic tree; mRNA expression

异育银鲫 GABA$_A$ 受体 α1 基因的分离、鉴定及阿维菌素处理后的组织特异性表达

赵依妮，孙琪，胡鲲，阮记明，杨先乐

（上海海洋大学国家水生动物病原库，中国上海，浦东新区临港新城沪城环路 999 号）

摘要：异育银鲫是一种广泛养殖的水产动物，并且常用阿维菌素来预防寄生虫。最近发现，阿维菌素会通过异育银鲫的血脑屏障。虽然阿维菌素主要通过 GABA 受体和特殊 α1 基因表达，最常见的同种 GABA 受体在其他种类鱼脑神经末梢中发现并进行了研究，但关于异育银鲫 GABA$_A$ 受体 α1 的子基因克隆表达还有阿维菌素通过血脑屏障是否会诱导脑中 GABA$_A$ 受体 α1 子基因表达还不知道。本研究目的：1. 克隆、排序、并研究 GABA$_A$ 受体 α1 子基因的系统发生情况。2. 脑、肝脏、肾脏等经过阿维菌素处理，之后通过逆转录–聚合酶链反应研究阿维菌素的对其神经毒性的相关性。α1 基因长度为 1 550 bp，带有开放基因框架 1 380 bp，其编码的预测蛋白带有 459 氨基酸残基。该基因包含 128 bp 并且末端 5' 不进行编码，还有 72 bp 中 3' 不编码。α1 基因结构特征符合半胱氨酸环配体基门控离子通道家族特征，该家族特征是具有一个信号肽、细胞外 N 端区域和四个跨膜区。通过建立的系统发生树图谱说明异育银鲫和斑马鱼的 α1 基因结构特征最相近。α1 基因在脑和卵巢中表达量最高，脑中 α1mRNA 转录水平具有显著提高。此外，阿维菌素的浓度越高，GAMA$_A$ 受体的表达越强，这说明阿维菌素会诱导异育银鲫神经毒素的产生。因此，α1 亚基基因的表达跟阿维菌素导致神经毒性具有重要关系。研究表明，在养殖异育银鲫过程中要谨慎使用阿维菌素，应研究一下其他种类的具有低毒性的抗生素类药物，因为养鱼用药时需要考虑到毒素通过 GABA$_A$ 受体产生诱导物产生毒性的情况。

关键词：异育银鲫；阿维菌素；γ-氨基丁酸 A 受体 α1 亚基；序列分析；系统树；mRNA 表达

Introduction

Avermectin (AVM), a macrocyclic lactone isolated from *Strepomyces avermitilis* (Burg et al. 1979), is an effective pesticide usually used in agriculture against mites, true bugs, common pests (Novelli 2012), and aquatic animals to prevent parasitic diseases (Wang and Lu 2009). Because AVM has high lipid content and low water solubility, AVM is able to be accumulated through food chains polluted by AVM. In fact, it was found that AVM is highly toxic to *Cladocera crustacean*, *Gambusia affinis*, *Hypophthalmichthys molitrix*, *Macrobrachium nipponense*, *Cipangopaludin achincasis*, *Allogynogenetic crucian carp*, *Daphnia similis*, *Chironomus Xanthus*, *and Danio rerio* (Novelli 2012; Wang and Lu 2009). AVM is toxic to not only invertebrates but also vertebrates through passages of the blood−brain barrier (BBB), and more attention should be paid to the damage caused by AVM to vertebrates (Halley et al. 1993; Cheng et al. 2007).

AVM increases the permeability of chloride ion channels of the nervous system of organisms to cause a strong chloride influx, which results in disrupted neural signal transmission and obvious neurotoxicity through gamma−aminobutyric acid (GABA) ergic receptors and glutamate−gated chloride channels in invertebrates (Li et al. 2010a, b). In vertebrates, there are three classes of GABA receptors (GABAAR, GABABR, and GABACR), and GABA$_A$ receptors are widely expressed in brain neurons (Stewart et al. 2011). GABA$_A$ receptor is a hetero−oligomeric protein composed of five subunits (Nayeem et al. 1994; Tretter et al. 1997). So far, at least 19 GABA$_A$ receptor subunits belonging to several subunit classes (a1-6, b1-3, c1-3, d, e, p, h, and q1-3) have been identified in the mammalian nervous system (Bonnert et al. 1999; Barnard et al. 1998). Among all of the subunits, the α1 subunit of GABA$_A$ receptor is the most common a isoform (McKernan and Whiting 1996; Sieghart and Sperk 2002) and the dominant subunit important for the assembly and function of GABA$_A$ receptor (Sieghart and Sperk 2002; Mohler et al. 2002; Sieghart 2006). GABA$_A$ receptoral subunit genes of other fish such as *Danio rerio*, *Pundamilian yererei*, *Maylandia zabra*, *Oreochromisni loticus*, *Takifugu rubripes*, and *Oryzias latipes* have been cloned by others and submitted to GenBank.

Carassius auratus gibelio as an economic species has been widely cultivated in China. It was found that AVM penetrates the BBB into the *Carassius auratus gibelio* brain, and that the brain AVM content is positively correlated with initial AVM concentration (Li et al. 2010a, b; Ruan et al. 2013). However, *Carassius auratus gibelio* GABA$_A$ receptor a1 subunit gene cloning and whether AVM passes through the BBB to induce *Carassius auratus gibelio* GABA$_A$ receptor α1 subunit gene expression have not been studied. The aim of this study was to clone, sequence, and phylogenetically analyze the *GABA$_A$ receptor* α1 subunit gene and to examine its expression in association with neurotoxicity in brain, liver, and kidney after AVM treatment.

Materials and methods

Animals

Carassius auratus gibelio, weighing 60.04±5.02 g, were obtained from Nantong fish farm (Yancheng, Jiangsu, China). All fish were housed in flat-bottomed fiberglass tanks (500 L) supplied with a constant flow (1 L/h) of aerated fresh water for 2 weeks and fed daily with complete feed (meeting the nutrition requirements of the animal except water) that contained no medicine. The water was maintained at pH between 6.4 and 7.0 and adjusted to 18±1.0℃ by a cooling and heating device. The water quality was monitored daily, and the water pH adjusted when necessary.

AVM exposure

The AVM was provided by Keyang Biotechnology Company in Wuhan, China. Based on acute toxicity experiment, the LC$_{50}$ concentrations of AVM to *Carassius auratus gibelio* (60.04±5.02 g) at 24、48 and 96 h of AVM exposure were determined to be 0.127、0.071 and 0.039 mg/L, respectively (Ruan et al. 2013). The AVM dissolved in 95% ethanol was used as stock solution (10 mg·mL^{-1}). Dosing of the AVM stock solution was calculated according to the actual volume of water in the aquariums. Then, using medicated bath to administer the fish. Fish without AVM treatment were used as blank control. Throughout the entire exposure process, the fish were not fed with any food, and the water temperature was maintained at 18±1.0℃ and pH between 6.4 and 7.0. Tissue collection and preparation

After AVM treatment at 24 h LC$_{50}$, 48 h LC$_{50}$, and 96 h LC$_{50}$, brains, livers, and kidneys were collected from *Carassius auratus gibelio* (N=4) at 24 h, 48 h, and 96 h, respectively. Each concentration had three replicates. The control group was collected with their experimental group at the same time point. Skin, heart, muscle, kidney, intestinal, liver, ovary, and brain were collected from untreated fish (N=4), which were used for tissue-specific expression analysis. All the samples were stored at-80℃.

RNA extraction and cDNA synthesis

Total RNA was isolated from 2 g of each tissue/organ (skin, heart, muscle, kidney, intestinal, liver, ovary, and brain) using TRIzol reagent (Invitrogen, Carlsbad, CA, USA) according to the manufacturer's instructions. The total RNA concentration was determined by a Nanodrop ND2000 spectrophotometer (Thermo Electrom Corporation, Waltham, MA, USA), and the RNA (2 lg) according to the OD$_{260}$/OD$_{280}$ ratio between 1.8 and 2.0 was used for single-stranded cDNA synthesis using the PrimeScript 1st Strand cDNA Synthesis Kit (TaKaRa, Dalian, China) according to the manufacturer's instructions. The cDNA products were stored at-20℃ for later use.

Cloning and sequencing of GABA A receptor α1 subunit gene

The first strand of cDNA was used as a template to amplify the sequence of the α1 subunit gene using specific primers (F1 and F2, Table 1) designed according to the Danio rerio GABA A receptor α1 subunit gene sequence on NCBI (GI: 116268040). PCRs were performed in a total volume of 25 μL, including 2.5 μL 109 Ex Taq Buffer (Mg^{2+} Plus), 2.0 μL dNTP mixture, 0.5 μL of each primer, and 2.0 μL cDNA. The PCR conditions were set as follows: 3 min at 95 C for pre-denaturation, 35 cycles of 30 s at 94℃ for denaturation, 30 s at 55℃ for annealing, and 1 min at 72℃ for extension, and 10 min at 72℃ for final extension.

All PCR products were run on a 2.0% agarose gel, purified using a Fragment Purification kit, version 3.0 (TaKaRa) and ligated into pMD19-T vector (TaKaRa). Then, the pMD19-T vectors containing the cDNA were transformed into competent cells (DH5). The plasmid DNAs from each of the three bacterial colonies were prepared and confirmed by PCR and agarose gel electrophoresis before being sent to Sangon Biotech Company (Shanghai, China) for sequencing.

Sequencing and phylogenetic analysis

cDNA and deduced amino acid sequences of the α1 subunit gene were analyzed using DNAMAN 5.2 program. Sequence homology and conserved domain analyses were performed using the BLAST program on the National Center for Biotechnology Information (NCBI)

Web site. The signal peptide was analyzed using SignalP 3. 0 (http: //www. cbs. dtu. dk/services/SignalP/) (Bendtsen et al. 2004) . TMHMM 2. 0 (http: //www. cbs. dtu. dk/services/TMHMM‐2. 0/) was used to predict transmembrane domains. Online software programs found at http: //www. cbs. dtu. dk/services/NetNGlyc and http: //www. cbs. dtu. dk/services/NetPhos were used to analyze the N‐glycosylation sites. The software MEGA 4. 0 (http: //www. megasoftware. net/mega4/mega. html) was used to construct a phylogenetic tree based on the amino acid sequences of the α1 subunit in this study and those of the a1 subunit of other species using a bootstrapping N‐J tree method (algorithm: Poisson correction; bootstrap values: 1000 replicates) . The α1 subunit amino acid sequences of other species in the phylogenetic study included sequences of Takifugu rubripes (XP_ 003 970997. 1), Maylandia zebra (XP_ 004550437. 1), Danio rerio (NP_ 001070794. 1), Oryzias latipes (XP_ 00 4084249. 1), Oreochromisni loticus (XP_ 003446980. 1), Homo sapiens (NP _ 000797. 2), Gallusgallus (NP _ 98 9649. 1), Musmusculus (EDL32341. 1), Oryziaslatipes (XP_ 004084249. 1), and Pundamilian yererei (XP_ 005727812. 1).

Table 1　Primer sequences

Item	Primer	Sequence (5' ‐3')	Product size (bp)
α₁ (PCR)	F1	CTGTAGGACTGCGAATCTTCCT	1550
	R2	TGGGAACTTGACTATCGTTGAA	
α₂ (qPCR)	F2	ATCGGTTACTTCGTCATCCAGA	176
	R2	GCTACCTTGGGGAGAGAGTTTC	
β‐actin	F	CATGGGGCAGGGCGTAACC	125
	R	TACGTTGCCATCCAGGCTGTG	

Quantitative real‐time reverse transcription PCR (qRT‐PCR)

The tissue‐specific expression of α1 subunit mRNA was studied by qRT‐PCR. Total RNA was isolated from skin, heart, muscle, kidney, intestinal, liver, ovary, and brain using the TRIzol reagent as described above. The cDNA was synthesized using a First Strand cDNA Synthesis Kit (TaKaRa, Dalian, China) . The primers (F2 and R2, Table 1) were designed based on the cloned a1 subunit cDNA sequence using Premier 5 (Table 1) . Real‐time quantitative PCR was performed in quadruplicate on a CFX 96 Thermocycler (Bio‐Rad, Hercules, CA, USA) . The reaction mixture of the PCR consisted of 0. 5 μL of each primer (10 μL), 2. 0 μL cDNA template, and 10 μL iQ™SYBR GREEN.

Supermix (Bio‐Rad) in a total volume of 20 μL. The real‐time PCR was performed as follows: a cycle at 95℃ for 3 min, followed by 45 cycles at 95℃ for 10 s, 56. 5℃ for 30 s, and 72℃ for 20 s, and 10 min at 72℃ for final extension. The real‐time PCR data were analyzed with CFX Manager Software, version 1. 6 (Bio‐Rad) . The standard curves were obtained through a tenfold serial dilution (10^{-1}–10^{-6}) of brain cDNA. β‐actin cDNA was used as an internal control. In the tissue distribution test, the relative mRNA expression level of α1 subunit in skin was set as control group. Moreover, AVM‐treated brain, liver, and kidney tissues were compared to control brain, liver, and kidney tissues, respectively. The relative mRNA α1 subunit mRNA expression in each sample was calculated by the 2^{-DDCT} method using β‐actin as a reference gene.

Statistical analysis

The data are expressed as mean±standard deviation, and Microsoft Excel 2007 was used to analyze the data. The significance of intergroup differences was assessed by Student's t test using the SPSS statistical software for Windows (version 16. 0; SPSS, Chicago, IL, USA) . Differences were considered significant at a P value<0. 05.

Results

Sequence of GABA A receptor α1 subunit gene

The *GABA A receptor* α1 subunit gene cDNA was 1550 bp in length with an open reading frame (ORF) of 1380 bp encoding a predicted protein with 459 amino acid residues. The cDNA contained 128 bp of 50 terminal untranslated region (UTR) and 72 bp of 30 terminal UTR (Fig. 1) . The calculated molecular weight of the α1 subunit polypeptide with deduced 459 amino acid residues was 51. 8 kDa, and the theoretical isoelectric point (pI) was 9. 83. The signal peptide analysis indicated that the amino acid sequence included a putative signal peptide (amino acids [aa] 1‐24) (Fig. 1), followed by an ([aa] 25‐251) extracellular N‐terminal. Secondary structure analysis showed that the signal peptide was mainly composed of alpha helix, and most amino acids were hydrophobic amino acids. The structure prediction by SMART indicated that a1 subunit was a typical type I transmembrane protein that included four transmembrane domains (TM 1‐4; Fig. 1) in the form of alpha helices connecting each other by hydrophilic amino acids. The cysteine residues at the extracellular domain were predicted to play a vital role in the stabilization and formation of ligand binding sites in the three-dimensional structure of the activated $GABA_A$ receptor.

Amino acid sequence alignment and phylogenetic analysis

BLASTP in NCBI and multiple alignment analysis indicated that the α1 subunit shared different levels of homology with other fish α1 subunits (Fig. 2). The highest identity level was 97% with Danio rerio, followed by 93% with Pundamilia nyererei and Maylandia zebra, and 92% with Oreochromis niloticus, Takifugu rubripes, and Oryzias latipes. To assess the relationship between the α1 subunit and those of other species, the phylogenetic tree was constructed using the amino acid sequences of the other species retrieved from GenBank. It was found that the Carassius auratus gibelio α1 subunit was more closely related to the Danio rerio α1 subunit than to the α1 subunits of the other species (Fig. 3).

Tissue-specific distribution of $GABA_A$ receptor α1 subunit mRNA

To identify the tissue-specific expression of the α1 subunit for further analysis of $GABA_A$ receptor function, qRT-PCR was used to analyze the mRNA expression levels of α1 subunit in different tissues of healthy Carassius auratus gibelio. The α1 subunit was expressed ubiquitously in all the tested tissues, and the relative expression level of α1 subunit in the skin was the lowest among all the tested tissues. The relative expression levels of α1 subunit were 1.04-, 1.23-, 1.92-, 6.59-, 25.11-, and 79573.48-fold higher in the heart, muscle, kidney, intestinal, ovary, and brain than in the skin, respectively. The relative expression levels of α1 subunit in all the tested tissues were significantly different from each other ($P<0.05$, Fig. 4), indicating that α1 subunit was mainly distributed in the central nervous system.

$GABA_A$ receptor α1 subunit mRNA expression in brain, kidney, and liver after AVM treatment

To investigate the toxic effects of AVM on Carassius auratus gibelio, the mRNA expression levels of the α1 subunit in brain, liver, and kidney of surviving fish in an acute toxicity test were examined. As the results of each tissue's control group that collected at different time point did not have significant different, we merged them and only showed one control group. The results showed that AVM treatment caused a significant ($P<0.05$) increase in α1 subunit mRNA expression compared to levels in the untreated brain, liver, and kidney (Fig. 5). In AVM-treated brain, the α1 subunit mRNA expression was 2.82-fold higher with 24 h of exposure to the LC_{50} (0.127 mg/L), 2.68-fold higher exposure for 48 h to the LC_{50} (0.071 mg/L), and 2.16-fold higher after 96 h at the LC_{50} (0.039 mg/L) compared to that in the untreated brain. In liver, the increase in α1 subunit mRNA expression in each treated group was 8833.74-fold at the 24 h LC_{50}, 24.39-fold at the 48 h LC_{50}, and 10.45-fold at the 96 h LC_{50} compared to that in the untreated liver. In the treated kidney, the increase in the mRNA expression was 5400.46-fold at the 24 h LC_{50}, 12.96-fold at the 48 h LC_{50}, and 10.72-fold at the 96 h LC_{50} compared to that in the untreated kidney. The relative expression levels of α1 subunit in untreated brain were 3172.204-fold and 41436.083-fold higher than in untreated liver and kidney; the increases according to each AVM treatment concentration (0.127, 0.071 and 0.039 mg/L) in brain were 116849.754-, 111048.702-and 89501.939-fold compared to levels in the untreated kidney, and 8945.615-, 8501.507- and 6851.961-fold compared to levels in the untreated liver, respectively. Therefore, the most significant increase in expression was observed in the treated brains.

Discussion

In this study, the $GABA_A$ receptor α1 subunit of Carassius auratus gibelio gene was cloned, sequenced, and phylogenetically analyzed, and its expression associated with neurotoxicity in brain, liver, and kidney after AVM treatment was investigated.

The α1 subunit structural features identified conformed to the characteristics of the $GABA_A$ receptor subunit family, which included a signal peptide, an extracellular domain at N-terminal, and four transmembrane domains (Olsen 1995). The established phylogenetic tree indicated that the Carassius auratus gibelio and Danio rerio α1 subunits were most closely related to each other. This may be ascribed to the fact that crucian carp and zebrafish are both Cypriniformes, and Carassius auratus gibelio was bred in captivity with the natural gynogenetic Fangzheng crucian carp as the female parent and Xingguo red carp as the male parent and bred through artificial insemination (Gui et al. 2008).

The α1 subunit mRNA expression was found to be the highest in brain in this study. This is in line with the previous finding that the α1 subunit transcripts are most abundant in rat brain (Clement 1996). Therefore, $GABA_A$ receptor has been considered as the most important receptor for inhibitory modulation in the nervous system (Korpi and Sinkkonen 2006). In addition, a higher expression level of the α1 subunit was also found in ovary, suggesting that the $GABA_A$ receptor may have additional functions outside the nervous system and may play an important role in early Carassius auratus gibelio development.

After AVM treatment, all tested tissues experienced an apparent increase in the transcription level of α1 subunit mRNA. Comparing the mRNA expression increases for the α1 subunit in brain, kidney, and liver of each group (untreated, 24 h LC_{50},

```
-128                                                              GGGATCAG
-120 GTAGCTGAAAGAGGATCATATTGTGCGATATTCCAGGCGCTGGGTGAGACACTGCTACAC
-60  GGTGATTTTGGGAAAGATTCGAGCCATCGGGATTTTTCGTGTGGATTCATCCTTTGGAGA
1    ATGATGTGGGGTGGAAGAGGAGCAGCTTGGCTTTGGATTTGGGCCTGTTTACTGGTGTCC
1     M  M  W  G  G  R  G  A  A  W  L  W  I  W  A  C  L  L  V  S
61   AGTGTTCTGGCTGGGAAAAGTTCCAGTCAAAGCGCAAATGAGCAGAAAGACAACACCACA
21    S  V  L  A  G  K  S  S  S  Q  S  A  N  E  Q  K  D  N  T  T
121  GTTTTCACCAGGATCCTGGACAGCCTCCTCGATGGCTACAACAACCGTCTCAGGCCTGGG
41    V  F  T  R  I  L  D  S  L  L  D  G  Y  N  N  R  L  R  P  G
181  CTCGGAGAGCGTGTAACCGAAGTCAAGACTGACATCTTCGTGACGAGTATTGGGCCGGTC
61    L  G  E  R  V  T  E  V  K  T  D  I  F  V  T  S  I  G  P  V
241  TCAGACCATGACATGGAGTACACCATCGATGTGTTCTTCAGGCAAAGCTGGAAGGACGAG
81    S  D  H  D  M  E  Y  T  I  D  V  F  F  R  Q  S  W  K  D  E
301  AGGCTGAAGTTCAAAGGTCCCATGGCAGTGCTCCGTCTCAACAATCTCATGGCCAGCAAA
101   R  L  K  F  K  G  P  M  A  V  L  R  L  N  N  L  M  A  S  K
361  ATCTGGACGCCCGACACGTTTTTCCACAACGGAAAGAAGTCAGTCGCCCATAACATGACC
121   I  W  T  P  D  T  F  F  H  N  G  K  K  S  V  A  H  N  M  T
421  ATGCCTAACAAACTTCTGCGGATCAAAGAGGAAGGAACTCTGCTGTACACCATGAGGCTT
141   M  P  N  K  L  L  R  I  K  E  E  G  T  L  L  Y  T  M  R  L
481  ACAGTCAGAGCTGAATGTCCAATGCATTTGGAGGACTTCCCAATGGATGCCCATGCTTGC
161   T  V  R  A  E  C  P  M  H  L  E  D  F  P  M  D  A  H  A  C
541  CCTCTCAAATTCGGCAGCTATGCCTACACGAGGGCTGAGGTGGTGTACGTTTGGACACGA
181   P  L  K  F  G  S  Y  A  Y  T  R  A  E  V  V  Y  V  W  T  R
601  GGGGCGGCTCAGTCTGTGGTCGTGGCTGACGATGGCTCTAGGCTCAACCAGTATGACTTG
201   G  A  A  Q  S  V  V  V  A  D  D  G  S  R  L  N  Q  Y  D  L
661  ATGGGACAGACGGTGGACTCAGGTGTGGTGCAGTCCAGCACAGGAGAGTATGTTGTCATG
221   M  G  Q  T  V  D  S  G  V  V  Q  S  S  T  G  E  Y  V  V  M
721  AAAACACATTTCCACCTCAAGAGGAAGATCGGTTACTTTGTCATCCAGACATATTTGCCG
241   K  T  H  F  H  L  K  R  K  I  G  Y  F  V  I  Q  T  Y  L  P
781  TGCATCATGACCGTGATCCTGTCCCAGGTGTCCTTCTGGCTCAACCGGGAATCTGTCCCC
261   C  I  M  T  V  I  L  S  Q  V  S  F  W  L  N  R  E  S  V  P
841  GCCAGAACTGTGTTTGGAGTGACCACTGTCCTTACCATGACCACCCTGAGTATCAGCGCA
281   A  R  T  V  F  G  V  T  T  V  L  T  M  T  T  L  S  I  S  A
901  AGAAACCCTCTCCCCAAGGTGGCCTACGCCACAGCCATGGACTGGTTCATCGCCGTCTGC
301   R  N  P  L  P  K  V  A  Y  A  T  A  M  D  W  F  I  A  V  C
961  TACGCCTTTGTCTTCTCGGCTCTCATTGAGTTTGCCACTGTAAACTACTTCACCAAGAGG
321   Y  A  F  V  F  S  A  L  I  E  F  A  T  V  N  Y  F  T  K  R
1021 GGGTACGCCTGGGACGGGAAAAGTGTGGTGCCGGAAAAGCAAAAGAAGAAGAAGGAGTCA
341   G  Y  A  W  D  G  K  S  V  V  P  E  K  Q  K  K  K  K  E  S
1081 TTGCTGAAAAAGAACAACACCTACACCGCTAAAACGGCGACGACGTTTGCTCCGAACATT
361   L  L  K  K  N  N  T  Y  T  A  K  T  A  T  T  F  A  P  N  I
1141 GCCAGAGACCCTGGTTTGGCAACTATTGCTAAAAGTGCCCCCCCTCCACCAACCGAGCCC
381   A  R  D  P  G  L  A  T  I  A  K  S  A  P  P  P  P  T  E  P
1201 AAGGAAGAGCCGAAACCCAAAGCCCCCGAGGCCAAGAAGACCTTCAACAGCGTGAGCAAG
401   K  E  E  P  K  P  K  A  P  E  A  K  K  T  F  N  S  V  S  K
1261 ATCGACAGGATCGCCAGAATAGCTTTCCCGCTGCTCTTCGGAACCTTTAACTTGGTGTAT
421   I  D  R  I  A  R  I  A  F  P  L  L  F  G  T  F  N  L  V  Y
1321 TGGGCAACTTACTTAAATAAAAAACCCAAATTACAGGGTATGAACGTGCAGCCACACTAA
441   W  A  T  Y  L  N  K  K  P  K  L  Q  G  M  N  V  Q  P  H  *
1381 TTCTATCGCCCCGGTTTTCCCTTTTACTTTCTCTATGTTTTCAAGAAACTCTCTTCAAGT
1441 AGCTTCCAACGT
```

Fig. 1 cDNA and deduced amino acid sequences of *Carassiusc auratus gibelio* GABA A receptor α1 subunit. The transmembrane domains (TM 1-4) are shaded in gray. The putative sequence of signal peptide is underlined. The start codon (ATG) and the stop codon (TAG) are double underlined. Cysteine residues are italicized. Potential N-linked glycosylation sites are shown in bold

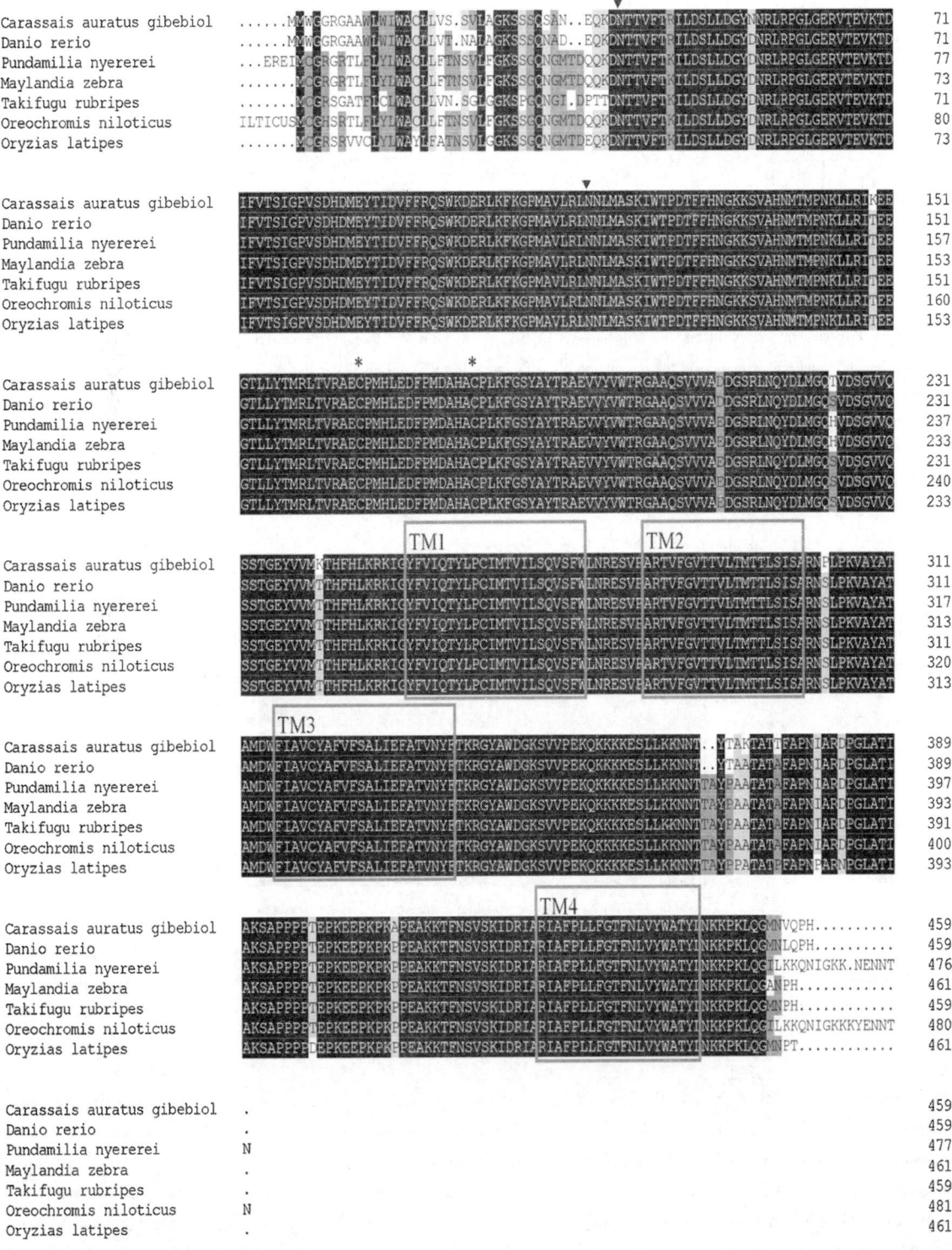

Fig. 2　Sequence alignment of *Carassius auratus gibelio* GABA$_A$ α1 subunit protein with those of other teleosts by ClustalW. The GenBank accession numbers are：Takifugu rubripes（gi. 410915044），Maylandia zebra（gi. 498982983），Danio rerio（gi. 116268041），Oryzias latipes（gi. 432949769），Oreochromis niloticus（gi. 348518922），and Pundamilian yererei（XP ＿ 005727812. 1）. The identical amino acid residues are shown in the black background；the gray background indicates similar amino acid residues；hyphens represent indels. Four transmembrane domains（TM 1-4）are shaded；the potential N-linked glycosylation sites are marked by black triangles；and the cysteine residues are marked by asterisks（＊）

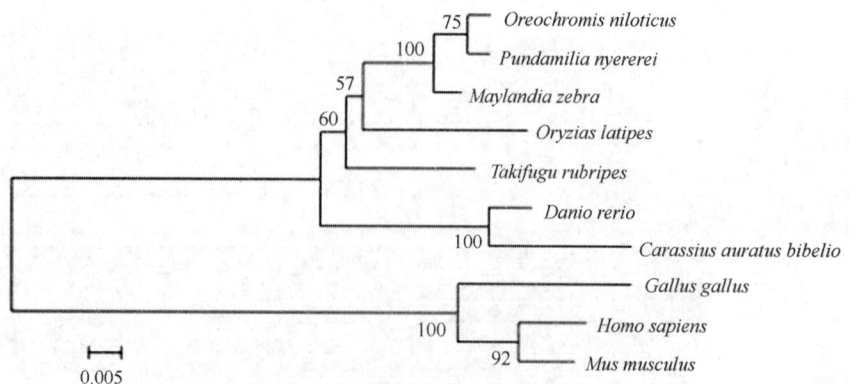

Fig. 3　Phylogenetic tree of amino acid sequences of GABA$_A$ α1 subunit of different species. Tree is based on the corresponding full-length amino acid sequences of the α1 subunit aligned using ClustalW and MEGA (version 5.05). The numbers at the branches denote the bootstrap majority consensus values for 1000 replicates and refer to percentage divergence. The gene accession numbers are: Takifugu rubripes (XP _ 003970997. 1), Maylandia zebra (XP _ 004550437. 1), Daniorerio (NP _ 001070794. 1), Oryzias latipes (XP_ 004084249. 1), Oreochromis niloticus (XP_ 003446980. 1), Homo sapiens (NP_ 000797. 2), Gallus gallus (NP_ 989649. 1), Musmusculus (EDL32341. 1), Oryzias latipes (XP_ 004084249. 1), and Pundamilia nyererei (XP_ 005727812. 1). The relative genetic distances are indicated by the scale bar and the branch lengths

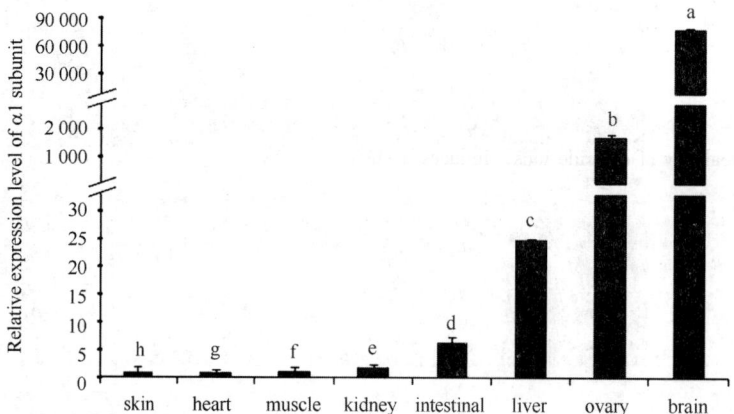

Fig. 4　Relative mRNA expression levels of GABA$_A$ α1 subunit in different tissues of *Carassius auratus gibelio*. The relative mRNA expression level of α1 subunit in skin was set at 1. The relative mRNA expression of each tissue was calculated by the 2-44CT method using β-actin as the reference gene, and each mRNA level was compared with that in skin. Values are expressed as the mean±SD. The differences between groups are indicated with different superscript letters (P<0. 05)

48 h LC$_{50}$, and 96 hLC$_{50}$), it seems that the liver experienced the most significant increase; however, considering the original expression level in normal brain was 3172. 204-fold and 41436. 083-fold higher than that in liver and kidney, respectively, the increase in brain was the greatest. AVM was found to be able to penetrate the BBB into the *Carassius auratus gibelio* brain when administered at different concentrations (0. 127, 0. 071, and 0. 039 mg/L), which is in accordance with a previous study showing that the brain AVM content is positively correlated with initial drug concentration (Ruan et al. 2013). In the present study, our results showed that AVM is able to induce a significant change in the GABA$_A$ receptor content in brain even at a relatively low concentration of 0. 039 mg/L. Therefore, as the balance of excitatory and inhibitory neurotransmission is the basic requirement for normal functioning ofthe central nervous system, we can draw a conclusion that the increase in GABA$_A$ receptor expression clearly reflects the fact that AVM can induce significant neurotoxicity in *Carassius auratus gibelio* and such damage cannot be ignored anymore. Additionally, α1 subunit mRNA was also found to be highly expressed in the kidney and liver in this study. The liver is the main metabolic and alexipharmic organ. In fact, the liver and fat are the main bioconversion organ/tissue of AVM and have the highest concentration of drug distribution and slowest elimination rate (Moline et al., 2000). The more highly increased expression level in kidney was in accordance with previous studies showing that AVM is excreted mainly through feces and urine: Up to 98% of AVM is excreted as non-metabolized drug, of which

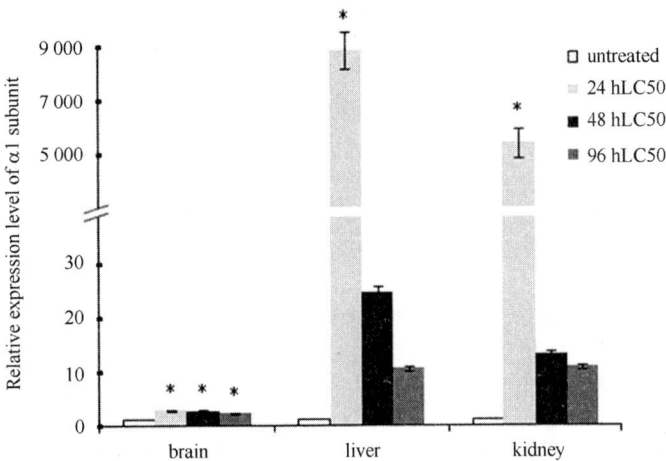

Fig. 5　Relative GABA$_A$ α1 subunit mRNA expression levels exposed to AVM after 24, 48, or 96 h of exposure at the LC$_{50}$ of 0.127, 0.071, or 0.039 mg/L, respectively, in brain, liver, and kidney. The relative mRNA α1 subunit mRNA expression in each tissue was calculated by the 2-44CT method using β-actin as the reference gene, and each mRNA level was compared with that in untreated tissue, respectively. Values are expressed as mean±SD. Significant differences over the untreated tissues (P<0.05) are indicated by *

2%～5% is excreted through the kidney (Moline et al., 2000; Molinari et al., 2009; Li et al., 2013).

The most significant increase in α1 subunit mRNA expression was at the 24 h LC$_{50}$, followed by the 48 h LC$_{50}$ and 96 h LC$_{50}$ compared to the control group, indicating that with a higher concentration of AVM, α1 subunit mRNA expression increased further, and thus, more GABA$_A$ receptor formation increased further. In fact, AVM stimulates GABA secretion, which excites GABA$_A$ receptor formation, increases the permeability of chloride ions, induces a strong chloride influx, results in disrupted neural signal transmission and defective cell functions, and eventually gives rise to neurotoxic symptoms, leading to death (Li et al. 2010a, b; McKellar and Benchaoui 1996; Lumaret and Errouissi 2002; Wu and Leng 2003; El-Shenawy 2010). Therefore, with greater GABA$_A$ receptor formation, AVM toxicity is greater. In all, the changes in α1 subunit mRNA expression may be used to represent AVM neurotoxicity in *Carassius auratus gibelio*.

Presently, AVMs are among the most highly used agents in veterinary medicine for the prevention of parasitic diseases in the world (Moline et al. 2000). AVM was considered as non-hazardous to vertebrates and mammals for a long time due to its physico-chemical properties according to previous studies (Halley et al. 1993). However, the notable increase in GABA$_A$ receptor expression induced by AVM, especially in the brain, suggests that AVM is neurotoxic to aquatic animals. These alterations may affect the composition, distribution, and pharmacology of GABA$_A$ receptor, which mediates both tonic presynaptic inhibition and phasic inhibitory synaptic transmission and thus affects the normal function of the animal body. Thus, our findings indicate that AVM should be used carefully, and it is necessary to develop other medicines with lower toxicity for protection of *Carassius auratus gibelio* from parasite infection.

Acknowledgments

This study was supported by the Chinese National Natural Science Foundation (Grant No. 31172430), the 863 Program (Grant No. 2011AA10A216), and the Special Fund for Agro-scientific Research in the Public Interest (Grant No. 201203085). There is no conflict of interest related to this manuscript.

References

Barnard E., Skolnick P., Olsen R. W., Mohler H, Sieghart W, Biggio G et al. (1998) International Union of Pharmacology. XV. Subtypes of c-aminobutyric acid A receptors: classification on the basis of subunit structure and receptor function. Pharmacol Rev 50 (2): 291-313.

Bendtsen JD, Nielsen H, von Heijne G, Brunak S (2004) Improved prediction of signal peptides: signalP 3.0. J MolBiol 340: 783-795.

Bonnert TP, McKernan RM, Farrar S, le Bourdelle's B, Heavens RP, Smith DW et al. (1999) Theta, a novel gammaaminobutyric acid type A receptor subunit. Proc Natl Acad Sci USA 96 (17): 9891-9896.

Burg RW, Miller BM, Baker EE, Birnbaum J, Currie SA, Hartman R et al. (1979) Avermectins, new family of potent anthelmintic agents: producing organism and fermentation. Antimicrob Agents Chemother 15 (3): 361-367.

Cheng GD, Xu SW, Li S (2007) Effect of avermectin on GABA$_A$ receptor in cerebrum of king pigeons. Vet Sci China11: 974-977.

Clement Y (1996) Structural and pharmacological aspects of the GABA receptor: involvement in behavioral pathogenesis. J Physiol Paris 90: 1-13.

El-Shenawy NS (2010) Effects of insecticides fenitrothion, endosulfan and abamectin on antioxidant parameters ofisolated rat hepatocytes. Toxicol In Vitro 24: 1148-1157.

Gui JF, Gong LU, Cheng XL, Li W (2008) Introduction of the new variety of *Carassius auratus gibelio* "zhongke 3". Fish Guide Rich 11: 51.

Halley BA, VandenHeuvel WJ, Wislocki PG (1993) Environmental effects of the usage of avermectins in livestock. Vet Parasitol 48: 109-125.

Korpi ER, Sinkkonen ST (2006) GABA$_A$ receptor subtypes as targets for neuropsychiatric drug development. Pharmacol Ther 109 (1): 12-32.

Li L, Xu Y, Yang WC, Zuo PX (2010a) Recent advances in toxicology of avermectins (in Chinese). J Med Pest Control 26: 608-610.

Li T, Long H, Wen H, Wei WX, Li KM, Liu ZJ et al. (2010b) Toxicological mechanisms and treatment of avermectins intoxication. Med Recapitul 16 (10): 1554-1556.

Li S, Li M, Cui LY, Wang XS (2013) Avermectin exposure induces apoptosis in king pigeon brain neurons. Pestic Biochem Phys 107 (2): 177.

Lumaret JP, Errouissi F (2002) Use of anthelmintics in herbivores and evaluation of risks for the non target fauna of pastures. Vet Res 33: 547-562.

McKellar QA, Benchaoui HA (1996) Avermectins and milbemycins. J Vet Pharmacol Ther 19: 331-351.

McKernan RM, Whiting PJ (1996) Which GABA$_A$-receptor subtypes really occur in the brain? Trends Neurosci 19: 139-143.

Mohler H, Fritschy JM, Rudolph U (2002) A new benzodiazepine pharmacology. J Pharmacol Exp Ther 300: 2-8.

Molinari G, Soloneski S, Reigosa MA, Larramendy ML (2009) In vitro genotoxic and cytotoxic effects of ivermectin and its formulation ivomec on Chinese hamster ovary (CHOK1) cells. J Hazard Mater 165: 1074-1082.

Moline JM, Golden AL, Bar-Chama N, Smith E, Rauch ME, Chapin RE et al (2000) Exposure to hazardous substances and male reproductive health: a research framework. Environ Health Perspect 108: 803-813.

Nayeem N, Green TP, Martin IL, Barnard EA (1994) Quarternary structure of the native GABA$_A$ receptor determined by electron microscopic image analysis. J Neurochem 62 (2): 815-818.

Novelli A (2012) Lethal effects of abamectin on the aquatic organisms Daphnia similis, Chironomus xanthus and Danio rerio. Chemosphere 86 (1): 36-40.

Olsen RW (1995) Functional domains of GABA$_A$ receptors. Trends Pharmacol Sci 16 (5): 162-168.

Ruan JM, Hu K, Yang XL, Zhang HX, Wang Y, Zhou AL et al. (2013) Blood-brain barrier permeability and residual tissue characteristics of Avermectin (AVM) in Carassius auratus gibelio. J Fish Sci China 20 (5): 1032-1038.

Sieghart W (2006) Structure, pharmacology, and function of GABA$_A$ receptor subtypes. Adv Pharmacol 54: 231-263.

Sieghart W, Sperk G (2002) Subunit composition, distribution and function of GABA (A) receptor subtypes. Top Med Chem 2: 795-816.

Stewart P, Williams EA, Stewart MJ, Soonklang N, Degnan SM, Cummins SF et al (2011) Characterization of a GABA$_A$ receptor b subunit in the abalone Haliotis asinina that is upregulated during larval development. J Exp Mar Biol Ecol 410: 53-60.

Tretter V, Ehya N, Fuchs K, Sieghart W (1997) Stoichiometry and assembly of a recombinant GABA$_A$ receptor subtype. J Neurosci 17 (8): 2728-2737.

Wang XZ, Lu HD (2009) Acute toxic effect of abamectin on fresh-water aquatic animals. J Environ Health 26: 593-597.

Wu YJ, Leng XF (2003) Recent advances in insecticide neurotoxicology. Acta Entmol Sin 46: 382-389.

该文发表于《Journal of Applied lchthyology》【2015. 31 (5): 862-869】

Effect of avermectin (AVM) on the expression of c-aminobutyric acid A receptor (GABA$_A$R) in Carassius auratas gibelio

A. L. Zhou[1], K. Hu[1], J. M. Ruan[2], H. P. Cao[1], Y. Wang[1], Y. N. Zhao[1] and X. L. Yang[1]

(1. National Pathogen Collection Center for Aquatic Animals, Shanghai Ocean University, Shanghai, China;

2. College of Animal Sciences and Technology, Jiangxi Agricultural University, Jiangxi, China)

Summary: This study describes the effect of avermectin (AVM) on the expression of c-aminobutyric acid A receptor (GABA$_A$R) in Carassius auratas gibelio. To assess the specific expression of GABA$_A$R in the brain, gonads, liver, kidneys, heart, muscles, and skin of C. auratas gibelio, the expression of GABA$_A$R a1 subunit (GABA$_A$Ra1) was measured by Western blotting. To study the effects of AVM on the expression of GABA$_A$R, the median lethal concentration (LC$_{50}$) at 24, 48, and 96 h of AVM was determined and the expression of GABA$_A$R in the brain, liver, and kidneys of the corresponding C. auratas gibelio evaluated by Western blotting and immunohistochemistry. The results show that GABA$_A$R was expressed in the brain, gonads, liver, kidneys, heart, intestines, muscles, and skin, while primarily distributed in the central nervous system and moderately distributed in peripheral tissues. The expression of GABA$_A$R in the brain, liver, and kidney tissues of C. auratas gibelio was increased with the treatment of AVM at 24 h LC_{50}, but attenuated by the treatment of AVM at 48 h LC_{50} and 96 h LC_{50}. This suggests a threshold effect of AVM.

阿维菌素对异育银鲫 c-氨基丁酸 A 受体表达的影响

周爱玲[1]，胡鲲[1]，阮记明[2]，曹海鹏[1]，王祎[1]，赵依妮[1]，杨先乐[1]

（1. 上海海洋大学国家水生动物病原库，中国上海；2. 江西农业大学动物科学技术学院，中国江西）

摘要：本实验研究了阿维菌素对异育银鲫 c-氨基丁酸 A 受体表达的影响。用蛋白免疫印迹法处理过的鲫鱼 $GABA_A R$ α1 亚基，以此来评价其对鲫鱼脑、生殖腺、肝脏、肾脏、心脏、皮肤等部位的 GABAA 表达情况影响。为了研究阿维菌素对 $GABA_A R$ 表达的影响，研究使用了阿维菌素的 24、48、96 h 半致死浓度数据，用蛋白免疫印迹法和免疫组织化学技术来检测脑、肝脏、肾脏中 $GABA_A R$ 的相关表达情况。结果显示，$GABA_A R$ 在脑、生殖腺、肝脏、肾脏、心脏、肠、肌肉、皮肤部位均有表达，但其首先分布在中枢神经系统并在边缘组织有一定分布。经阿维菌素处理后 GABAAR 在异育银鲫脑、肝、肾组织中的表达得到提升（24 h LC_{50}），但是在 48 h LC_{50}、94 h LC_{50} 的试验中相反。此研究为阿维菌素的临界效果提供了理论依据。

关键词：阿维菌素；异育银鲫；c-氨基丁酸,；A 受体

Introduction

Avermectin（AVM），also known as abamectin，is produced in Streptomyces avermitilis and features an antibiotic with antiparasitic activity. AVM is a new antibiotic for farm animals，and belongs to the insecticide-acaricide of macrolide antibiotics. AVM has a high biological activity against mites and pests on crops（Liu et al.，2006）. There are many forms and derivatives of AVM，such as ivermectin，doramectin，and emamectin. AVM and its derivatives are increasingly popular in the treatment of aquatic parasitic diseases and not only applicable in various types of freshwater aquaculture ponds，but also suitable in polyculture（ponds with fish，shrimp，and crab）for the prevention and treatment of Lernaeosis，Sinergasiliasis，and Balantidiasis in aquaculture.

The c-aminobutyric acid（GABA）is one of the most important inhibitory neurotransmitters in the central nervous systems of vertebrates and invertebrates（Watanabe et al.，2002）. As a member of the ligand ion channel receptor superfamily，c-aminobutyric acid A receptor（$GABA_A R$）mediates rapid inhibitory nerve conduction of chloride ions in the central nervous systems of vertebrates. Therefore，$GABA_A R$ is an active target of multiple drugs such as sedatives，inhibitory barbiturates，picrotoxins，alcohol，anesthetic nerve sterols，and phytochemical constituents. Research has demonstrated that AVM plays a major role in GABA-gated chloride ion channels（Lasota and Dybas，1991）. As a GABA agonist affecting $GABA_A R$，AVM can stimulate the presynaptic membrane，causing a massive release of GABA and promoting the opening of GABA-gated chloride ion channels. This causes a large number of chloride ions to localize in the cytomembranes，causing hyper-polarization of the postsynaptic membranes，and forming an inhibitory postsynaptic potential. These processes attenuate sensitivity to excitatory or inhibitory signal transfer reactions in insects，affect normal neural activities，and eventually cause paralysis and death（Wang，2012）. While the effects of AVM on small animals such as mice（Soderlund et al.，1987；Krflgek and Zemkovfi，1994），rats（Abalis et al.，1986），guinea pigs（Coccini et al.，1993），finfish（Deng et al.，1991；Lumaret et al.，2012），and mammals（Robertson，1989）and the distribution of $GABA_A R$ in C. auratas gibelio have been studied（Ruan et al.，2014），the effects of AVM on C. auratas gibelio remain unknown. Therefore，this paper investigated the effects of different medium lethal concentrations of AVM on the protein expression of $GABA_A R$ in C. auratas gibelio by Western blot and immunohistochemistry methods. This study may provide a theoretical basis for exploring the mechanism and safety evaluation of AVM in aquaculture.

Materials and methods

Fish

Purchased from a fishery in Jiangsu Province，East China，were 500 healthy C. auratas gibelio（60.04，5.02 g body weight）. All fish were acclimatized to favorable laboratory conditions for 2 weeks during which the fish were maintained in aerated aquaria（3 000 L）constantly supplied with flowing（5 L·h^{-1}）aerated freshwater（dissolved oxygen：6.5，0.4 mg·L^{-1}）at 18℃ and pH 7.5，0.3 prior to the begin of the experiment. The fish were fed twice daily.

Administration

The AVM was provided by the Keyang Biotechnology Company in Wuhan, China. Based on an acute toxicity experiment, 24 h, 48 h, and 96 h LC_{50} were determined to be 0.127, 0.071, 0.039 mg $\cdot$ L^{-1}, respectively. Each concentration had three replicates. Prior to dosage, the AVM was dissolved in 95% ethanol to a 10 mg $\cdot$ mL^{-1} stock solution. Dosage of the AVM stock solution was calculated according to the actual volume of water in the aquaria, then administered using a medicated bath. Water temperature was maintained at about 18°C.

Tissue collection and preparation

After AVM treatment at 24 h LC_{50}, 48 h LC_{50}, and 96 h LC_{50}, the brains, livers, kidneys, and muscles were collected from C. auratas gibelio (N=30) at 24, 48 and 96 h, respectively. Each concentration had three replicates. Some of the samples were stored at 80°C for Western blot, and the remainder fixed with Bouin's solution for 24 h and stored in 75% alcohol for paraffin embedding. Fish without AVM treatment were used as the control.

Extraction of total protein and Western blot assay

Western blot assay was used to detect the $GABA_A R$ $\alpha 1$ protein. Total protein was extracted with radio immunoprecipitation assay (RIPA) lysis buffer (Beyotime, China) according to the manufacturer's instructions. Protein content was measured using an enhanced BCA protein assay kit (Beyotime) on Powerwave XS2 (Biotek, VT). The supernatant was frozen at 80°C until Western blotting was performed. An aliquot containing 100 lg of protein was diluted in a loading buffer (1 : 5, loading buffer: sample, vol vol^1), heated to 98°C for 10 min and separated by 10% sodium dodecyl sulfate–polyacrylamide gel electrophoresis (SDS–PAGE). The protein was transferred onto a 0.45 lm–pore polyvinylidene fluoride membrane (PVDF; Bio-Rad, CA) at 100 V for 1 h. The membranes were blocked for 2 h with 5% skim milk in Tris–buffered saline containing 0.1% Tween–20 (TBST) and then incubated with rabbit anti–$GABA_A R\alpha1$ polyclonal antibody (diluted 1 : 200; Sigma–Aldrich, MO) at 4°C overnight. After washing with several changes of TBST (3, 9, 10 min), the membranes were incubated with horseradish peroxidase (HRP) conjugated goat anti–rabbit immunoglobulin G (IgG) (diluted 1 : 1 000; Jackson, MS) for 2 h at room temperature. After washing in TBST (3, 9, 10 min), bound conjugates were detected by Allergic ECL Chemiluminescence Substrate (Boster, China). Proteins were visualized by exposing the blot to an X–ray film, photographed with a digital camera, and then the net intensities of the individual bands were measured using (Bio–Rad). Rabbit antiGAPDH monoclonal antibodies (diluted 1: 2000; Boster) were used as a loading control, and $GABA_A R$ $\alpha 1$ protein expression was normalized to GAPDH. Quantity One V4.6.2 (Bio–Rad) was used to evaluate the gray values of $GABA_A R$ $\alpha 1$ and GADPH bands. The ratio of the gray values ($GABA_A R$ $\alpha 1$/GADPH) showed the relative expression intensity of $GABA_A R$ $\alpha 1$ protein in the different tissues. All four values in each tissue were used for statistical analyses.

Immunohistochemistry

The tissues used for immunohistochemistry were processed by paraffin embedding using a gradient of ethanol–xylene–paraffin. The samples were sectioned at 5 lm on a microtome (Leica, Germany). The sections were then spread on a 40°C Water Bath–Slide Drier (Leica) on poly–L–lysine slides and dried completely at 60°C for 2 h. Subsequently, the sections were dewaxed in two changes of fresh xylene before rehydrating in a descending ethanol series from 100% ethanol to sterile water and were used for immunohistochemistry. The sections were treated with 3% H_2O_2 to inactivate the endogenous peroxidase, and then exposed to boiling sodium citrate buffer (0.01 M, pH 6.0) for 30 min. After washing three times with 0.1 M phosphate–buffered saline (PBS), the sections were further processed with normal goat serum working liquid for 15 min to reduce non–specific background, and incubated overnight at 4°C with rabbit antiGABA$_A$R $\alpha 1$ polyclonal antibodies (diluted 1 : 200 by PBS; Sigma–Aldrich). After rinsing with PBS, the sections were incubated for 15 min at 25°C with biotinylated anti–rabbit IgG antibodies (SP–9001; Zsbio, China), followed by washing in PBS and incubation with an avidin–horseradish peroxidase complex for 15 min. After further rinsing with PBS, the sections were finally developed using 3, 3^0 – diaminobenzidine tetrahydrochloride (DAB; Tiangen, China), immersed in deionized water to stop the reaction, counterstained with haematoxylin, dehydrated, mounted, and checked with an electron microscope (Olympus, Japan). In each series of stained sections, positive and negative controls were included to assess the specificity of the assay. Negative control slides were sections in which the primary antibodies were replaced by PBS. All processes were conducted in a wet box. Nine different microscope fields were selected for three sections of the same tissue (9400). An Image–Pro plus 6.0 (Media Cybernetics) was used to evaluate the mean option density (MOD) of GABA$_A$R $\alpha 1$. All nine values in each tissue were used for statistical analyses.

Statistical analyses

Statistical analyses were performed using SPSS 13.0 (Chicago, IL). Least-significant difference (LSD) was adopted to determine the significant differences, in which the datum were expressed as means SD, and differences were considered significant at P<0.05 or P<0.01.

Results

Expression of $GABA_AR$ in different tissues of C. auratas gibelio

The distribution of the $GABA_AR$ protein was examined by Western blot analysis in different tissues. The results showed that $GABA_AR$ was primarily distributed in the central nervous system, but was also present in peripheral tissues. Highest relative expression intensity of $GABA_AR$ protein was recorded in the gonads, followed by intestines, kidneys, liver, heart, muscles, and skin in the peripheral tissues (Fig. 1A). Through statistical analysis, the relative expression intensity of $GABA_AR$ protein in the brain was significantly higher than in the peripheral tissues (P<0.05). The relative expression intensity of $GABA_AR$ protein in the gonads was significantly higher than others in the peripheral tissues (P<0.05). There was no significant difference in the relative expression intensity of $GABA_AR$ protein between the intestines and kidneys (P>0.05), but the expression in the intestines and kidneys was higher than in the liver, heart, muscles, and skin (P<0.05). There were no significant differences in the relative expression intensity of $GABA_AR$ in the liver, heart, or muscles (P>0.05). The relative expression intensity of $GABA_AR$ protein in the skin was significantly less than in other tissues (P<0.05) (Fig. 1B).

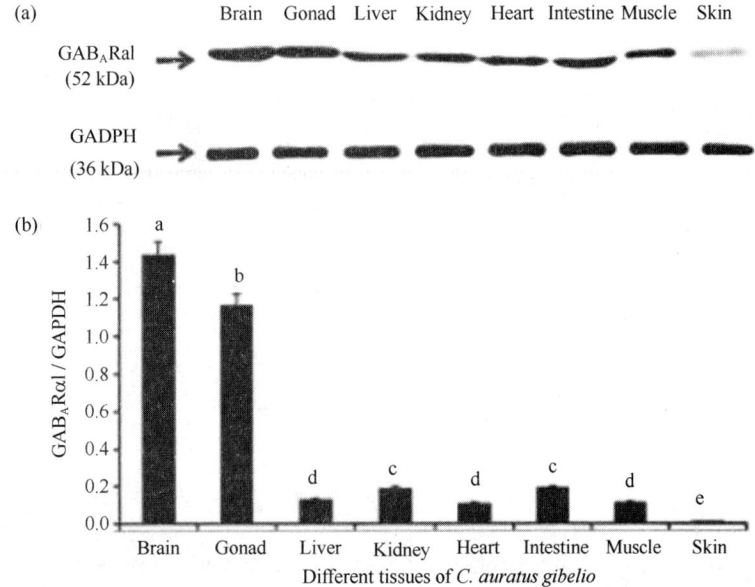

Fig. 1　Expression of $GABA_AR\alpha1$ protein in different C. gibilio tissues with no medication. Fish exposed formore than 2 weeksatroom temp. (A) Western blot analysis of $GABA_AR\alpha1$ (52 kDa, top) and GADPH (36 kDa, bottom); (B) Corresponding ratio analysis of gray values. Values are mean ± SD, n = 4 animals from each group. a - e: same letters = no signigicant difference (P>0.05); different letters = significant difference (P<0.05).

Effect of AVM on $GABA_AR$ protein in different tissues of C. auratas gibelio

Western blot identification of $GABA_AR$ in C. auratas gibelio

The results showed that AVM 24 h LC50 increased the relative expression intensity of $GABA_AR$ in the tissues of C. auratas gibelio. However, following treatment of AVM 48 h LC_{50} and AVM 96 h LC_{50}, the relative expression intensity of $GABA_AR$ was attenuated. With multiple comparison analyses, the relative expression intensity of $GABA_AR$ in the brain was reduced successively under AVM-24 h LC_{50}, AVM 48 h LC_{50}, and AVM-96 h LC_{50}, showing significant differences (P<0.05). The relative expression intensity of $GABA_AR$ in the liver and kidneys under AVM-24 h LC_{50} was significantly higher than AVM-48 h LC_{50} and AVM-96 h LC_{50} (P<0.05). However, there were no significant differences in the expressions of $GABA_AR$ in the liver and kidneys between AVM-48 h LC_{50} and AVM-96 h LC_{50} (P>0.05) (Fig. 2).

Immunohistochemical identification of $GABA_A R$ in the central nervous system

Immunoreactivity is presented as brown granules; the negative control is characterized by blue hema toxylin staining and an absence of brown granules (Fig. 3). Compared to normal tissues, $GABA_A R$-positive granules were more frequent and stained with greater intensity in the telencephalon, mesencephalon, cerebellum, and medulla oblongata following AVM–24 h LC_{50} treatment. These granules were primarily focused on the corpus striatum, optic tectum, Purkinje cells, and vagus lobe. However, the $GABA_A R$ positive granules were lower in the telencephalon, mesencephalon, cerebellum, and medulla oblongata after AVM48 h LC_{50} or AVM–96 h LC_{50} treatment.

With multiple comparison analyses of the mean optical density (Table 1), the densities of $GABA_A R$ in the telencephalon, mesencephalon, cerebellum, and medulla oblongata under AVM–24 h LC_{50} were higher than the corresponding normal tissues, of which the change of the mesencephalon was the most significant ($P<0.01$). The density of $GABA_A R$ in the telencephalon, mesencephalon, cerebellum, and medulla oblongata under AVM–48 h LC_{50} and AVM–96 h LC_{50} were lower than in the corresponding normal tissues. Compared to the normal cerebellum, the densities of $GABA_A R$ in the cerebellum under AVM–48 h LC_{50} and AVM–96 h LC_{50} had highly significant differences ($P<0.01$). There were no significant differences in the densities of $GABA_A R$ in the telencephalon among AVM–24 h LC_{50}, AVM–48 h LC_{50}, and AVM96 h LC_{50} treatments ($P>0.05$). The density of $GABA_A R$ in the mesencephalon under AVM–24 h LC_{50} was increased by 51.98 and 85.52% compared to AVM–48 h LC_{50} and AVM96 h LC_{50}, respectively ($P<0.05$). The density of $GABA_A R$ in the cerebellum under AVM–48 h LC_{50} and AVM–96 h LC_{50} was reduced by 77.08 and 65.61%, respectively, compared to AVM–24 h LC_{50} ($P<0.05$). The density of $GABA_A R$ in the medulla oblongata under AVM–48 h LC_{50} and AVM–96 h LC_{50} decreased by 57.89 and 59.86%, respectively, compared to AVM–24 h LC_{50} ($P<0.05$).

Table 1　Effects of AVM on the mean option density (MOD) of GABAAR in the brain of *C. auratus gibelio* (Mean±S. D.)

Group	Concentration ($mg \cdot L^{-1}$)	Mean option density (MOD)			
		Telencephalon	Mesencephalon	Cerebellum	Medulla oblongata
Blank Control	0	0.024±0.014	0.073±0.010[b]	0.078±0.009[a]	0.033±0.023[ab]
24 h LC_{50}	0.127	0.028±0.045	0.112±0.043[a]	0.111±0.076[a]	0.056±0.031[a]
48 h LC_{50}	0.071	0.023±0.011	0.074±0.032[b]	0.025±0.003[b]	0.024±0.011[b]
96 h LC_{50}	0.039	0.013±0.009	0.060±0.021[b]	0.038±0.010[b]	0.022±0.019[b]

Note: a, b: compared by multiple comparisons for same tissues at different AVM dosages, and groups with same letter showed no significant differences ($P>0.05$), while groups with different letters indicate a significant difference ($P<0.05$).

Immunohistochemical identification of $GABA_A R$ in the peripheral tissues

Compared to the normal liver, the $GABA_A R$ positive granules in the liver cell membranes did not change under AVM–24 h LC_{50} or AVM–48 h LC_{50}; however, there was a marked decrease in the $GABA_A R$-positive granules following AVM–96 h LC_{50} treatment. This paper also found that the hepaticlobules were loosely arranged and the nucleolus was dissolved. Compared to the normal kidney, the color of the $GABA_A R$-positive granules on renal tubule membranes was deepened after AVM–24 h LC_{50} treatment, but changed little under AVM–48 h LC_{50}. However, the $GABA_A R$-positive granules on the renal tubule membranes were reduced under AVM–96 h LC_{50} treatment (Fig. 4).

Table 2　Effects of AVM on the mean option density (MOD) of $GABA_A R$ in the liver and kidney of C. auratus gibelio (Mean±S. D.)

Group	Concentration ($mg \cdot L^{-1}$)	Mean option density (MOD)	
		Liver	Kidney
Blank Control	0	0.004±0.002[ab]	0.014±0.004[a]
24 h LC_{50}	0.127	0.005±0.002[a]	0.016±0.007[a]
48 h LC_{50}	0.071	0.003±0.01[b]	0.013±0.010[a]
96 h LC_{50}	0.039	0.001±0.001[c]	0.003±0.001[b]

With multiple comparison analyses of the mean optical density (Table 2), the density of $GABA_A R$ in the liver and kidney under

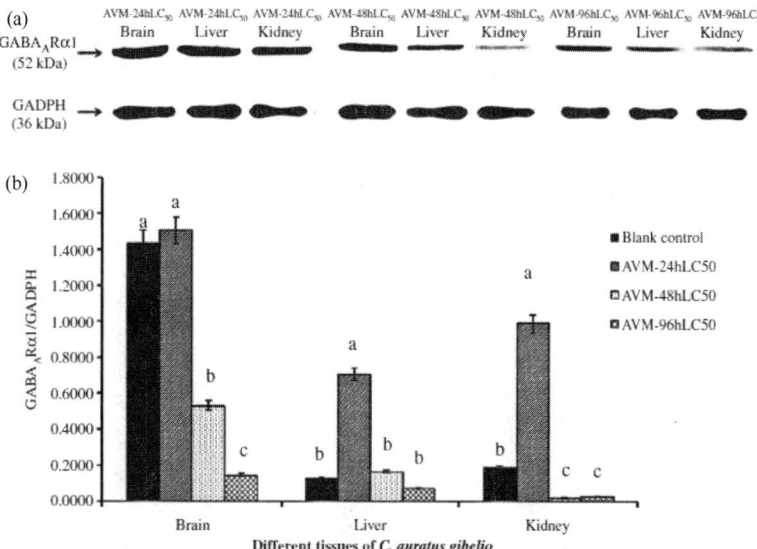

Fig. 2   To detect the effect of AVM on the expression of $GABA_A R$ in the tissues by Western blot. （A） Protein expression of $GABA_A R \alpha 1$ （52 kDa） and GADPH （36 kDa） in tissues. （B） Comparison of relative expression intensity of $GABA_A R \alpha 1$ in the treatment group and normal group. Values are mean±SD N=4. a，b，c：compared by multiple comparisons for same tissues at different AVM dosages；groups with same letter＝no significant differences （P>0.05）；groups with different letters＝significant differences （P<0.05）

AVM-24 h LC_{50} was increased compared to the normal tissues, but the differences were not significant （P>0.05）. There was no significant difference in the density of the $GABA_A R$ in the liver and kidney under AVM-48 h LC_{50} （P>0.05）. However, the density of the $GABA_A R$ in the liver and kidney under AVM-96 h LC_{50} was reduced by 72.09% in both （P<0.01）. Under AVM-24 h LC_{50}, AVM 48 h LC_{50}, and AVM-96 h LC_{50}, the density of the $GABA_A R$ in the liver was lower compared to normal liver tissue （P<0.05）. There were no significant differences between the densities of the $GABA_A R$ in the kidney under AVM-24 h LC_{50} or AVM-48 h LC_{50} treatment. The density of the $GABA_A R$ in the kidney following AVM-96 h LC_{50} treatment was significantly decreased by 82.17 and 78.46% compared to AVM-24 h LC_{50} and AVM-48 h LC_{50}, respectively （P< 0.05）.

Discussion

In other studies, research on GABA and $GABA_A R$ in fish concluded that elasmobranchii and osteichthyes have a high affinity of $GABA_A R$ （Betti et al., 2001）. These limited numbers of studies have primarily concentrated on saltwater fish such as Mugil cephalus and Salmon salar L. （Anzelius et al., 1995）, and have focused on gene cloning and pharmacology. The articles were often concerning portions of fish brain, such as the optic lobe, olfactory bulb, medulla oblongata and cerebellum, but rarely peripheral tissues. This experiment demonstrated that $GABA_A R$ is expressed in the brain, gonads, liver, kidneys, heart, intestines, muscles, and skin of C. auratus gibelio.

Expression of $GABA_A R$ in C. auratus gibelio

The predominant $GABA_A R$ subtype is composed of a1b2c2 subunits, thus the present experiment used $GABA_A R \alpha 1$ as the detection indicator to detect the distribution of $GABA_A R$ in the different tissues. The distribution of GABA-containing neurons in the olfactory bulb, telencephalon, tectum stratum, and hypothalamus of adult zebra fish was investigated using Nissl staining and immunohistochemistry （Yong-Jung Kim et al., 2004）. The expression and distribution of GABA, GAD65, $GABA_A R \alpha 1$, and $GABA_B R1$ existed in the cerebella of adult zebra fish （Delgado and Schmachtenberg, 2008）. Experiment results also showed that $GABA_A R$ exists in the central nervous system of C. auratus gibelio. It was hypothesized that $GABA_A R$ are critical to the function of the central nervous system.

This experiment found that the highest expression of $GABA_A R$ protein was recorded in the gonads, followed by intestines, kidneys, liver, heart, muscles, and skin in the peripheral tissues. Previous studies have shown that $GABA_A R$ subunits are expressed in multiple endocrine tissues of rats, such as in the placenta, ovaries and testes （Akinci and Schofield, 1999）. Some studies have also shown that $GABA_A R$ are widely distributed in the reproductive system （Geigerseder et al., 2003） because $GABA_A R$ has an important

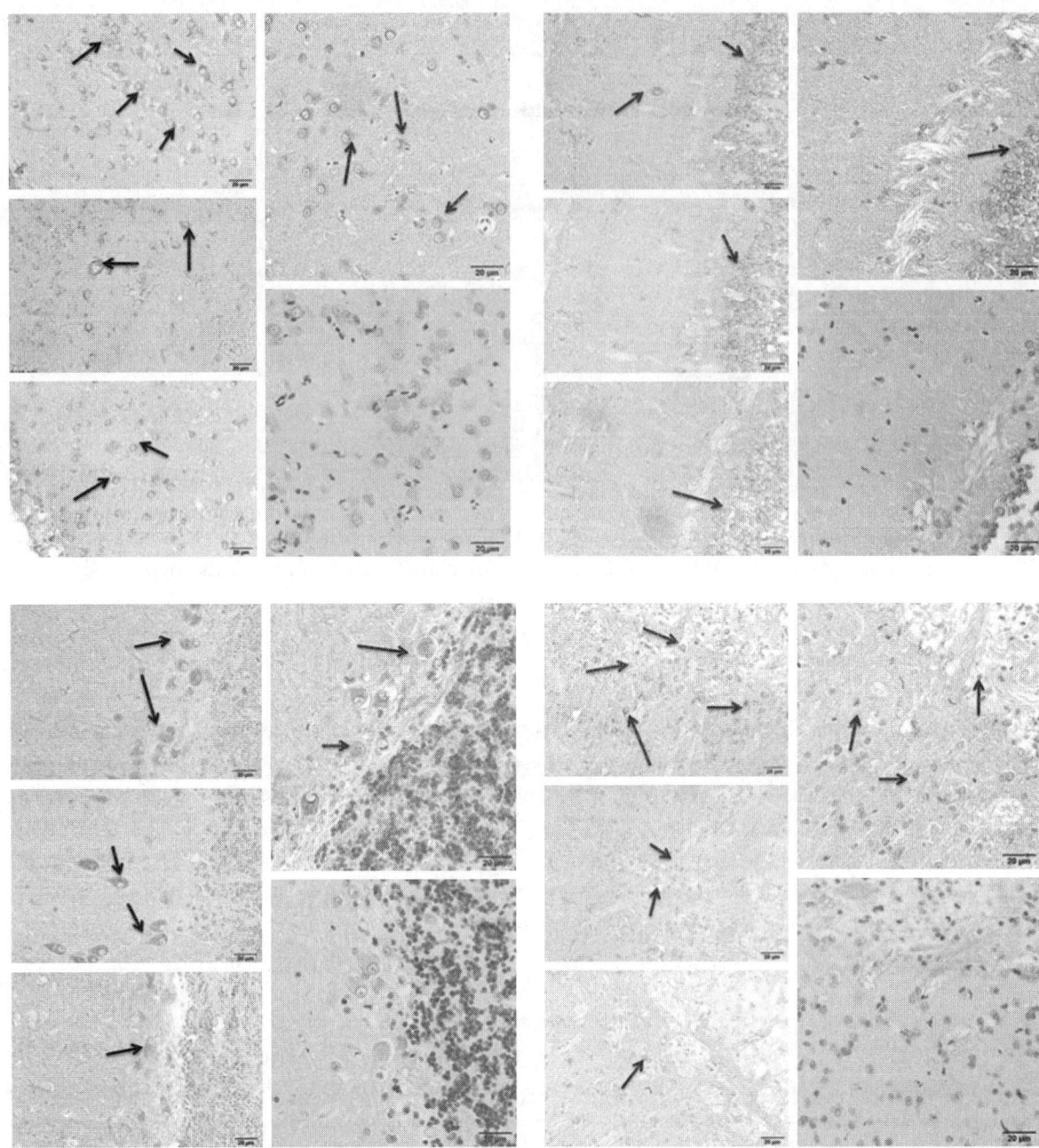

Fig. 3　Effects of AVM on $GABA_A R$ distribution in the telencephalon, mesencephalon, cerebellum, and medulla oblongata of C. auratus gibelio in (A) telencephalon; (B) mesencephalon; (C) cerebellum; and (D) medulla oblongata. a−e = treatment at 24 h LC_{50}, 48 h LC_{50}, 96 h LC_{50}, No AVM, No AVM−no antibody, respectively. Arrows = $GABA_A R$−positive granules. Scale = 20 lm

effect on reproduction. Furthermore, there is a functional GABA−ergic system in the livers of rats, which may regulate the liver and bile duct function through autocrine and paracrine mechanisms (Wang, 2011). Expressions of $GABA_A R$ subunits and transporter have also been seen in the livers of fathead minnows (Pimephales promelas) (Biggs et al., 2013). Present results demonstrate that $GABA_A R$ also exists in the livers of C. auratus gibelio. The distribution of GABAAR in fish kidneys has been rarely reported in the current literature, as it is primarily concentrated in mammals. The a1, a2, a6, b1, b2, b3, c1, c2 and c3 subunits of the $GABA_A R$ have been shown to be abundant in the kidneys of mice compared to other non−neural tissues (Tyagi et al., 2007). GABA might have a beneficial effect on renal function in nephrectomized rats by inhibiting fibrosis and atrophy primarily in tubuli and tubular interstitium (Liang et al., 2009). The experiment found that $GABA_A R$ was also expressed in kidney tubules of C. auratus gibelio. It suggested that $GABA_A R$ might regulate fish kidney functions to some extent. Several studies have discovered that $GABA_A R$ are expressed in the heart and intestinal tissues. Some researchers suggest that $GABA_A R$ plays a certain function in the cardiovascular and intestinal endocrine systems (Bentzen and Grunnet, 2011; Vigliano et al., 2011). These findings were consistent with prior $GABA_A R$ expression profiles.

Effects of AVM on $GABA_A R$ of C. auratus gibelio

Blood-brain barrier (BBB) (Deeken and Loscher, 2007) is a control interface between the capillaries in the brain and spinal cord and nerve tissues, which can prevent harmful substances from entering the brain, thereby protecting the cerebral environment. The methods of xenobiotics passing the blood-brain barrier include passive transport, receptormediated transport, carriermediated transport, absorptivemediated transport and active efflux transport (Pardridge, 2007). Recent studies have shown that AVM and its derivatives can cross the blood-brain barrier into the central nervous system. Research has revealed that 100 mg $\cdot$ kg^{-1} of AVM shows toxicity to the brain cells of rats through the blood-brain barrier (Hu et al., 2010). AVM has been shown to induce apoptosis from oxidative stress in nervous tissues of king pigeons, affecting normal physiological functions (You, 2011). Another study reported the inhibition of P-glycoprotein in the blood-brain barrier altered the AVM neurotoxicity and swimming performance of rainbow trout (Kennedy et al., 2014). Ruan et al., 2013, demonstrated that AVM could pass the blood-brain barrier of C. auratus gibelio using high performance liquid chromatography (HPLC).

GABA and GABAR mainly exist in the central nervous system. Experiment results also showed that AVM could transit through the blood-brain barrier into the brain tissue.

The 24 h LC_{50} (0.127 mg $\cdot$ kg^{-1}) of AVM increased the expression of $GABA_A R$ in the telencephalon, mesencephalon, cerebellum, and medulla oblongata, but $48hLC_{50}$ (0.071 mg $\cdot$ kg^{-1}) and 96 h LC_{50} (0.039 mg $\cdot$ kg^{-1}) of AVM decreased the expression of $GABA_A R$ in those tissues. At the same time, the Ruan et al. (2013) study found that the residual AVM in the brain was dependently reduced following treatment of 24 h LC_{50} (0.127 mg $\cdot$ kg^{-1}), 48 h LC_{50} (0.071 mg $\cdot$ kg^{-1}), and 96 h LC_{50} (0.039 mg $\cdot$ kg^{-1}). These studies demonstrate that AVM can reduce the relative expression of $GABA_A R$ in the brain, even in the presence of the blood-brain barrier. However, above a certain concentration, AVM might increase the relative expression of GABAAR in the brain. By cloning the GABAR gene from a Bactroceradorsalis strain resistant to AVM, research verified that the resistance of AVM-selected strains might be due to the changes of the GABAR number or metabolic enzyme activity, instead of the mutation of conserved region locus of the GABAR gene (Zhang et al., 2014). The study also demonstrated that the mRNA expression level of GABAR in Tetranychus cinnabarinus, induced by 0.018 mg $\cdot$ L^{-1} of AVM, was significantly reduced (Lu, 2009). Rat recombinant $GABA_A R$ and ivermectin, an AVM derivative, should activate the channel independently, resulting in different kinetic properties (Adelsberger et al., 2000).

The liver is the largest digestive gland and an important detoxification organ in fish. The kidney is also a major urinary organ and immune organs of fish. The results indicate that 24 h LC_{50} (0.127 mg $\cdot$ kg^{-1}) of AVM could stimulate $GABA_A R$ expression in the liver and kidneys, but that 48 h LC_{50} (0.071 mg $\cdot$ kg^{-1}) and 96 h LC_{50} (0.039 mg $\cdot$ kg^{-1}) of AVM decrease the expression of $GABA_A R$ in the liver and kidneys. The studies of Ruan et al. (2013, 2014) also found that the residual AVM in the liver and kidneys were reduced successively under the conditions of 24 h LC_{50} (0.127 mg $\cdot$ kg^{-1}), 48 h LC_{50} (0.071 mg $\cdot$ kg^{-1}), and 96 h LC_{50} (0.039 mg $\cdot$ kg^{-1}). One report revealed that AVM can affect the function of a cytochrome P450 enzyme system and oxidation stress, which cause tissue morphology, and induce apoptosis by expression of caspase-3, caspase-8, and fas mRNA in the liver and kidneys of king pigeons (Zhu, 2013). Another study demonstrated that AVM could decrease transcript abundance of GABA-gated chloride channels and neuronal acetylcholine receptor subunits in salmon lice (*Lepeophtheirus salmonis*), suggesting their involvement in AVM-induced toxicity in caligids (Carmichael et al., 2013).

Conclusion

$GABA_A R$ is mainly distributed in the central nervous systems of C. gibelio, but also has a moderate distribution in peripheral tissues. There was a concentration-dependent effect of AVM on the density of $GABA_A R$ in C. gibelio. After reaching a certain concentration, AVM can raise the density of $GABA_A R$ to a threshold where the body produces negative feedback, reducing the expression of $GABA_A R$.

Acknowledgements

This work was supported financially by the National Natural Science Foundation of China (No. 31172430) and the Special Fund for Agro-scientific Research in the Public Interest (No. 201203085).

References

A balis, I. M.; Eldefrawi, A. T.; Eldefrawi, M. E., 1986: Actions of avermectin B1a on the c-aminobutyric acid A receptor and chloride channels in rat brain. J. Biochem. Toxicol. 1, 69-82.

Adelsberger, H.; Lepier, A.; Dudel, J., 2000: Activation of rat recombinant a1b2c2 GABA$_A$ receptor by the insecticide ivermectin. Eur. J. Pharmacol. 394, 163-170.

Akinci, M. K.; Schofield, P. R., 1999: Widespread expression of GABAA receptor subunits in peripheral tissues. Neurosci. Res. 35, 145-153.

Anzelius, M.; Ekstrom, P.; M ohler, H.; Richards, J. G., 1995: Immu- nocytochemical localization of GABAA receptor b2b3-subunits in the brain of Atlantic salmon (Salmo salar L.). J. Chem. Neuroanat. 8, 207-221.

Bentzen, B. H.; Grunnet, M., 2011: Central and peripheral GABAA receptor regulation of the heart rate depends on the conscious state of the animal. Adv. Pharmacol. Sci. 2011, 1-10.

Betti, L.; Giannaccini, G.; Gori, M.; Bistocchi, M.; Lucacchini, A., 2001: [3H] Ro 15-1788 binding sites to brain membrane of the saltwater Mugil cephalus. Comp. Biochem. Physiol. C Toxicol. Pharmacol. 128, 291-297.

Biggs, K.; Seidel, J. S.; Wilson, A.; Martyniuk, C. J., 2013: c-Aminobutyric acid (GABA) receptor subunit and transporter expression in the gonad and liver of the fathead minnow (Pimephales promelas). Comp. Biochem. Physiol. A Mol. Integr. Physiol. 166, 119-127.

Carmichael, S. N.; Bron, J. E.; Taggart, J. B.; Ireland, J. H.; Bekaert, M.; Burgess, S. T. G.; Skuce, P. J.; Nisbet, A. J.; Gharbi, K.; Sturm, A., 2013: Salmon lice (Lepeophtheirus salmonis) showing varying emamectin benzoate susceptibilities differ in neuronal acetylcholine receptor and GABA-gated chloride channel mRNA expression. BMC Genom. 14, 1-16.

Coccini, T.; Candura, S. M.; Manzo, L.; Costa, L. G.; Tonini, M., 1993: Interaction of the neurotoxic pesticides ivermectin and lindane with the enteric GABA$_A$ receptor-ionophore complex in the guinea-pig. Eur. J. Pharmacol. 248, 1-6.

Deeken, J. F.; Loscher, W., 2007: The blood-brain barrier and can- cer: transporters, treatment, and Trojan horses. Clin. Cancer Res. 13, 1663-1674.

Delgado, L.; Schmachtenberg, O., 2008: Immunohistochemical localization of GABA, GAD65, and the receptor subunits GABA$_A$ α1 and GABAB1 in the zebrafish cerebellum. Cerebellum 3, 444-450.

Deng, L.; Nielsen, M.; Olsen, R. W., 1991: Pharmacological and biochemical properties of the c-aminobutyric acid-benzodiazepine receptor protein from codfish brain. J. Neurochem. 56, 968-977.

Geigerseder, C.; Doepner, R.; Thalhammer, A.; Frungieri, M. B.; Gamel-Didelon, K.; Calandra, R. S.; Kohn, F. M.; Mayerhofer, A., 2003: Evidence for a GABA ergic system in rodent and human testis: local GABA production and GABA receptors. Neuroendocrinology 77, 314-323.

Hu, Y. W.; Xu, Y.; Chen, S. F., 2010: Preliminary study of abamectin toxicity to the brain cells of rats. J. Toxicol. 3, 226-228. (In Chinese).

Kennedy, C. J.; Tierney, K. B.; Mittelstadt, M., 2014: Inhibition of P-glycoprotein in the blood-brain barrier alters avermectin neurotoxicity and swimming performance in rainbow trout. Aquat. Toxicol. 146, 176-185.

Krflgek, J.; Zemkovfi, H., 1994: Effect of ivermectin on y-aminobutyric acid-induced chloride currents in mouse hippocampal embryonic neurones. Eur. J. Pharmacol. 259, 121-128.

Lasota, J. A.; Dybas, R. A., 1991: Avermectins, a novel class of compounds: implications for use in arthropod pest control. Annu. Rev. Entomol. 36, 91-117.

Liang, P. P.; Wei, H. L.; Li, D. X.; Liu, S. W., 2009: c-aminobutyric acid inhibits progression of tubular interstitial fibrosis in rats with chronic renal failure. Herald Med. 28, 1120-1123. (In Chinese).

Liu, L. J.; He, H. J.; Qiu, L. N., 2006: Study on the development of avermectins. Biotechnology 84, 84-85. (In Chinese).

Lu, W. C., 2009: Effects of abamectin and heat stresses on the expression of GABA and GABA receptor in Tetranychus cinnabarinus. Southwest University, Chongqing.

Lumaret, J. P.; Errouissi, F.; Floate, K.; Rombke, J.; Wardhaugh, K., 2012: A review on the toxicity and non-target effects of macrocyclic lactones in terrestrial and aquatic environments. Curr. Pharm. Biotechnol. 13, 1004.

Pardridge, W. M., 2007: Blood-brain barrier delivery. Drug Discov. Today 12, 54-61.

Robertson, B., 1989: Actions of anaesthetics and avermectin on GABA$_A$ chloride channels in mammalian dorsal root ganglion neurones. Br. J. Pharmacol. 98, 167-176.

Ruan, J.; Hu, K.; Zhang, H.; Wang, Y.; Zhou, A.; Zhao, Y.; Yang, X., 2014: Distribution and quantitative detection of GABA$_A$ receptor in Carassius auratus gibelio. Fish Physiol. Biochem. 40, 1-11.

Ruan, J. M.; Hu, K.; Yang, X. L.; Zhang, H. X.; Wang, Y., 2013: Blood-brain barrier permeability and tissues residuals of avermectin (AVM) in Carassius auratus gibelio. J. Fish. Sci. China 5, 1032-1038. (In Chinese).

Soderlund, D. M.; Adams, P. M.; Bloomquist, J. R., 1987: Differences in the action of avermectin B1a on the GABA$_A$ receptor complex of mouse and rat. Biochem. Biophys. Res. Commun. 146, 692-698.

Tyagi, N.; Steed, M.; Gillespie, W.; Rosenberger, D. S.; Lominadze, D.; Sen, U.; Moshal, K. S.; Henderson, B. C.; Tyagi, S. C.,

2007：Differential expression of the GABA$_A$ receptor subunits in the kidney and cardiovascular system. FASEB J. 21, 594-598.

Vigliano, F. A.; Mu Oz, L.; Hernandez, D.; Cerutti, P.; Bermudez, R.; Quiroga, M. I., 2011：An immunohistochemical study of the gut neuroendocrine system in juvenile pejerrey Odontesthes bonariensis (Valenciennes). J. Fish Biol. 78, 901-911.

Wang, Q., 2012：Molecular characterization of AVM and IVM that bind to GABA receptor. Qingdao University of Science and Technology, Qingdao, Shandong.

Wang, S. L., 2011：A hepatic c-aminobutyric acid (GABA) -ergic system and its role in experimental acute liver failure in rats. Shandong University, Shandong.

Watanabe, M.; Maemura, K.; Kanbara, K.; Tamayama, T.; Hayasaki, H., 2002：GABA and GABA receptors in the central nervous system and other organs. Int. Rev. Cytol. 213, 1-47.

Kim, Y. J.; Nam, R.; Yoo, Y. M.; Lee, C., 2004：Identification and functional evidence of GABA ergic neurons in parts of the brain of adult zebrafish (Danio rerio). Neurosci. Lett. 355, 29-32.

You, Y. Z., 2011：The studies on apoptotic mechanism of subchronic avermectin in King Pigeon's brain. Northeast Agricultural University, Harbin.

Zhang, Y. P.; Zeng, L.; Lu, Y. Y., 2014：Clone and sequence analysis of GABA receptor 5' fragment from Bactrocera dorsalis strain resistant to avermectin. J. Environ. Entomol. 36, 51-57. (In Chinese).

Zhu, W. J., 2013：The studies on the effect of sub-chronic avermectin on apoptosis and CYP450 enzyme system in King Pigeon's liver and kidney. Northeast Agricultural University, Harbin.

Author's address：Ailing L. Zhou, National Pathogen Collection Center for Aquatic Animals, Shanghai Ocean University, 201306 Shanghai, China. E-mails：zhouailing1989@ aliyun. com and X. L. Yang (Corresponding author), National Pathogen Collection Center for Aquatic Animals, Shanghai Ocean University, 201306 Shanghai, China. Tel.：+86-021-61900451；E-mail：xlyang@ shou. edu. cn.

该文发表于《水生生物学报》【2015, 3：598-603】

基于GABA$_A$受体评估双氟沙星对异育银鲫的安全性

Safety evalution of difloxacin on *carassais auratus gibebiol based* on GABA$_A$ receptor

赵依妮[1]，孙琪[1,2]，胡鲲[1]，杨先乐[1]，阮记明[1]，周爱玲[1]

(1. 上海海洋大学国家水生动物病原库，上海 201306；2. 上海黛龙生物科技工程有限公司，上海 201306)

关键词：异育银鲫；GABA A 受体 γ2 亚基；克隆；组织分布；双氟沙星；安全性

神经传递的兴奋和抑制的相互平衡是维持中枢神经系统正常功能的必要条件，其中抑制性则是由最重要的神经递质 γ-氨基丁酸（GABA）调节的，而 GABA 则主要通过 GABA$_A$ 受体 GABA$_A$R 起作用。A 型受体是一种跨膜门控氯离子通道，可介导阶段性的抑制突触传递，也可强直性抑制外周突触传递，因此在神经性疾病如焦虑和癫痫的病理生理学中起着至关重要的作用[1]。刺激 GABA 的分泌，增加其与 GABA$_A$ 受体的结合将抑制中枢神经系统兴奋，导致抑郁等消极情绪的产生；相反，抑制 GABA 的分泌，减少其与 GABA$_A$ 受体的结合可刺激中枢神经兴奋，导致癫痫等疾病的产生[2]。喹诺酮类药物是人工合成的含 4-喹诺酮基本结构的抗菌药物，其抑菌机理主要是对细菌 DNA 回旋酶（DNA gyrase）具有选择性抑制作用[3]。喹诺酮类药物分为四代，目前临床应用较多的为第三代喹诺酮类药物，水产养殖业中常用的有双氟沙星、诺氟沙星、左氧氟沙星等。其中，双氟沙星（Difloxacin, DIF）抗菌活性高，对革兰氏阳性和阴性菌及多种厌氧菌都有较强的抗菌作用[4]，且双氟沙星质量稳定、口服吸收性好、体内浓度高、半衰期长且与其他抗菌药物无交叉耐药性，因而在水产养殖细菌性疾病的防治中得到广泛使用[5]。而在水产养殖过程中，对细菌、寄生虫等靶病原用药时，药物给养殖动物带来副作用是不可避免的。随着 DIF 应用的越来越广泛，导致水产动物不良反应的发生也越来越频繁[6]。目前在脊椎动物体内的研究表明，喹诺酮类药物产生的不良反应主要发生在消化系统，而神经系统是即消化系统后第二大不良反应作用部位[7]。研究发现大部分含有亲脂性氟原子的氟喹诺酮类药物可透过血脑屏障进入脑组织，通过抑制 GABA 与 GABA$_A$ 受体的结合，从而使得中枢神经兴奋性增加[7]。大鼠（Rattus norvegicus）动脉灌注培氟沙星、诺氟沙星、环丙沙星、氟罗沙星和司帕沙星等氟喹诺酮类药物后，在其大脑组织中可以检测到相应药物的存在[8]。最近，阮记明等[9]的研究表明，DIF 可透过异育银鲫（Carassais auratus gibebiol）脑血屏障，且 DIF 在大脑内的消除缓慢，对中枢神经系统造成的影响具有持久性。DIF 等喹诺酮类药物所造成的不良反应主要包括头晕、焦虑、恐惧感等，而鱼类、虾蟹等中枢神经系统的过度兴奋将导致鱼类急游、耗氧量增加，死亡率也随之增加。目前，有关 GABA 受体在水产动物领域的研究较少，其研究仍主要集中在癫痫病治疗药物、抗抑郁等神经药物和农用杀虫剂的创新与研制等方面[10—13]。

国内有关药物分子毒理学的研究工作刚刚起步，且以药物受体研究药物安全性具有针对性、特异性的优势，因此本研究以 $GABA_A$ 受体作为检测指标，从分子水平反映 DIF 对水产养殖动物的毒性，从而评估其药物安全性，促进 GABA 受体在药物安全性检测中的应用。$GABA_A$ 受体由五个亚基组装而成，目前已发现的亚基共有 19 种（α1-6、β1-3、γ1-3、δ、ε、π、θ、ρ1-3），其中，γ2 亚基是 $GABA_A$ 受体主要的功能性亚基之一[14]，且 84% 的 $GABA_A$ 受体包含 γ2 亚基[15,16]。同时免疫沉淀实验表明 γ2 亚基可以同 α1、α2、α3、α5、α6、β2、β3、γ3 和 δ 等亚基共同组装；此外，γ2 亚基也为氯离子通道正常导电所必需[17]。因此，我们选取异育银鲫 $GABA_A$ 受体的 γ2 亚基作为检测指标，为 $GABA_A$ 受体作为药物安全性评估指标提供一种新思路。

1 材料与方法

1.1 实验动物与试剂

实验动物及 DIF 处理健康的异育银鲫购买于江苏省南通某国营农场，体重（61±5.5）g。用 2% NaCl 浸泡 10 min，然后放置于高锰酸钾消毒后的水族箱（100 cm × 80 cm × 80 cm）内暂养 2 周，饲喂普通颗粒饲料。试验期间使用自动控温系统控制水温为（20±2）℃，24 h 充气，2 d 换一次水。以 4 尾未灌药的异育银鲫作为空白对照组，其组织也用于 GABA A 受体 γ2 亚基 mRNA 及蛋白组织特异性表达分析，所取组织包括脑、肝、肾、心、肌肉、肠、皮、性腺。按 20 mg/kg 鱼体重标准口灌盐酸双氟沙星原料药溶解液，分别于给药后 1 h、3 h、6 h、12 h、24 h、48 h、96 h 各取异育银鲫 4 尾，分别取脑、肝、肾、肌、性腺。组织样品用液氮迅速冷却，储存在-70℃条件下，用于转录水平和蛋白水平的检测。

主要试剂和仪器 Trizol 试剂（美国 Invitrogen 公司），反转录酶、ExTaq DNA 聚合酶（日本 TaKara 公司），iQ™ SYBR GREEN Supermix（美国 Bio-Rad 公司），引物、AXYGEN 柱式凝胶回收试剂盒、AXYGEN 小量质粒提取试剂盒（上海生工生物有限公司），BCA 蛋白浓度测定试剂盒、RIPA 强裂解液、6×SDS-PAGE 蛋白上样缓冲液、彩色预染蛋白质分子量、考马斯亮蓝染液、显影液和定影液购自碧云天生物科技有限公司；兔抗 GABAAR γ2 多克隆抗体（一抗）（美国 Sigma-Aldrich 公司）；兔抗 β-actin 多克隆抗体（内参一抗）（武汉博士德生物有限公司）；ECL 超敏化学发光试剂盒（上海威奥生物有限公司）；PVDF 膜（聚偏二氟乙烯膜）（美国 Bio-Rad 公司），滤纸（北京索莱宝生物有限公司）。PCR 仪为 eppend of Mastercycle Gradient，核酸测定仪为 eppend of Biophotomter，实时定量 PCR 仪为 BIO-RAD Mini Optica，165—8001 小型垂直电泳仪（美国 Bio-Rad 公司）；半干转膜仪（美国 Bio-Rad 公司）；温控摇床、脱色摇床（美国 Thermo 公司）。

1.2 实验方法

总 RNA 的提取与反转录分离异育银鲫脑、肝、肾、心脏、前肠、表皮、肌肉和性腺，按 Trizol 法提取总 RNA。采用紫外分光光度计与 1.0% 琼脂糖凝胶电泳检测 RNA 的浓度及完整性，-80℃保存备用。使用 M-MLV 反转录 cDNA。反转录体系及反应条件按照试剂盒推荐体系进行。

引物设计、基因片段的克隆及测序应用 Primer Premier 5.0 软件，根据 Gene Bank 发表的斑马鱼（NM_ 001256250.1）GABA A 受体的 γ2 亚基序列设计一对特异性引物 F1&R1（表1），引物由上海生工生物有限公司合成。PCR 反应体系及反应条件按照试剂盒推荐体系进行。1.0% 的琼脂糖凝胶电泳检验扩增产物，按琼脂糖凝胶 DNA 回收试剂盒操作说明纯化回收特异性扩增产物，并连接至 pMD19-T vector 载体，将链接好的质粒转化感受态 DH5，通过蓝白斑筛选和菌落 PCR 初步鉴定后，阳性菌落于 37℃震荡培养过夜。提取质粒后将鉴定为阳性的克隆送上海生工生物科技有限公司进行测序。

表 1 本文所用引物
Tab. 1 Sequences of primers used in the study

引物 Primer		序列 Sequence（5′-3′）
Part 1	F1	GGCGTCTGTACCAGTTCTCTTT
	R1	GGGTCTTGTTGGAATCTGTAAC
Q-PCR γ2	F2	TATGGACCTCTTCGTGTCTGTG
	R2	ATTCGTAGCCGTATTCTTCGTC
Q-PCR β-actin	F3	CATGGGGCAGGGCGTAACC
	R3	TACGTTGCCATCCAGGCTGTG

GABA A 受体 g2 亚基 mRNA 表达分析以获得的基因序列为模板，利用 Primer Primer 5.0 软件设计一对 γ2 亚基特异引物 F2 & R2（表 1），对不同组织（脑、肝、肾、心、肌肉、肠、皮、性腺）及双氟沙星处理后脑、肝、肾、肌肉和性腺中 γ2 亚基 mRNA 的表达进行检测，并以 β-actin 为内参基因，所用特异性引物 F3 & R3 见表 1。采用 iQ™ SYBR GREEN Supermix（Bio-Rad）进行实时定量 RCR，反应体系及反应条件按照试剂盒推荐体系进行。用内参 β-actin 对各样品的 C_t 值进行均一化处理，采用 $2^{-\triangle\triangle C_t}$ 确定不同样品 mRNA 的相对含量[18]。

GABA A 受体 g2 亚基蛋白表达分析采用 Western blot 法对不同组织（脑、肝、肾、心、肌肉、肠、皮、性腺）及双氟沙星处理后脑、肝、肾、肌肉和性腺中 γ2 亚基蛋白的表达量进行检测，并以 β-actin 为内参基因。具体方法为：按照每 20 mg 组织加入 200 μL 裂解液的比例加入裂解液。在匀浆机中研磨，直至裂解液中组织块消失。放在 4℃ 静置 30 min 充分裂解，并每隔 10 min 强烈振荡一次。充分裂解后，4℃、12 000 r/min 离心 20 min，取上清。提取组织总蛋白后，采用 Brandford 法（考马斯亮蓝结合显色法）蛋白定量试剂盒定量蛋白浓度。将上样液于 100℃ 的 PCR 仪中煮沸 10 min，使蛋白变性，并制备 11% 分离胶和 4% 积层胶，加电泳缓冲液后上样电泳。电泳完毕后将蛋白转印于硝酸纤维素膜上。脱脂奶粉封闭膜后分别加入兔抗 GABA$_A$R γ2 多克隆抗体（1：200）和兔抗 β-actin（1：2000）孵育 2 h，洗膜，再加入用封闭液稀释的辣根过氧化物酶（HRP）标记的羊抗兔二抗（1：2000），37℃ 摇床上孵育 2 h。洗膜后膜转入 ECL 超敏化学发光试剂（试剂 A：试剂 B＝1：1）显色液中反应 2 min，取出膜，甩去多余的液体，用保鲜膜包好 PVDF 膜，暗室中用 X 胶片感光、显影、定影。将 Western blot 的胶片置于凝胶成像仪中用化学发光法进行检测、照相，显影结果用 Quantity One V4.6.2 软件（美国 Bio-rad 公司）对蛋白质条带灰度进行半定量分析处理，结果用 GABA$_A$R γ2 蛋白目的条带与 β-actin 条带的灰度值比值表示 GABA$_A$R 蛋白相对表达强度（GABA$_A$R γ2/β-actin）。

1.3 统计方法

用 Excel 统计数据，SPSS 13.0 统计软件进行单因素方差分析，以 P<0.01（差异极显著），0.01<P<0.05（差异显著），P>0.05（差异不显著）作为差异显著性判断标准，差异显著时采用 LSD 法对各组平均样之间进行多重比较。结果用平均值±标准差（Mean±SD）表示，当 P<0.05 时，差异显著。

2 结果

2.1 GABA A 受体 γ2 亚基 mRNA 的组织表达分析

荧光定量 PCR 结果显示，异育银鲫 GABA$_A$ 受体 γ2 亚基和内参的实时荧光定量 PCR 扩增曲线指数增长明显，溶解曲线均为单一峰，扩增效率 E 值均接近 1.0，分别为 97.3 和 98.3，表明此条件下目的基因与内参基因的扩增效率一致，适合作为标准曲线。此外，4 个平行加样孔之间扩增结果偏差不显著，说明此条件下扩增体系和反应条件良好。荧光定量检测 γ2 亚基在异育银鲫各组织中的表达情况，以 β-actin 为内参基因，结果显示，GABA$_A$ 受体 γ2 亚基在脑、肝、肾、心、肌肉、肠、皮、性腺中均有表达。其中尤以脑组织中含量最高，性腺和肝脏次之，前肠中表达水平相对较低，而在肾脏、肌肉、心脏、表皮中的表达量最低。统计学分析显示 GABA$_A$ 受体 γ2 亚基 mRNA 在各组织间表达量差异显著（P<0.05）（图 1）。

2.2 GABA$_A$ 受体 γ2 亚基蛋白的组织表达分析

Western blot 结果显示（图 2），β-actin 蛋白在各组织中的表达无明显差异；以 β-actin 为内参分析 GABA$_A$ 受体 γ2 亚基蛋白的表达，可见 GABA$_A$ 受体 γ2 亚基蛋白在各组织中均有表达，其中尤以脑中的含量最多，性腺次之，肝脏、肾脏、前肠、肌肉、心脏表达水平依次降低，在皮中表达量最低。统计学分析显示 GABA$_A$ 受体 γ2 亚基蛋白在各组织间表达量差异显著（P<0.05）（表 2）。

表 2 各组织 GABA$_A$ 受体 γ2 亚基蛋白相对表达（光密度比值）
Tab. 2 The relative protein expression of GABA$_A$ receptor γ2 subunit（n=4, x±s）

组织 Tissue	GABARr2/β-actin
脑 Brain	3.2901±0.0397[a]
性腺 Gonad	1.5967±0.0282[b]
肝脏 Liver	1.2939±0.0429[c]
肾脏 Kidney	0.7692±0.0156[d]
心脏 Heart	0.2808±0.0091[f]

组织 Tissue	GABA_A Rr2/β-actin
前肠 Intestine	0.4473 ± 0.0022^{e}
肌肉 Muscle	0.3169 ± 0.0062^{f}
皮肤 Skin	0.0193 ± 0.0018^{g}

注：处理组之间的差异显著性以不同字母在平均数后用上标表示，上标不同表示差异显著（P<0.05）

Note：The difference between groups were showed with different superscript letters after each average numbers，different superscript showed significant difference（P<0.05）

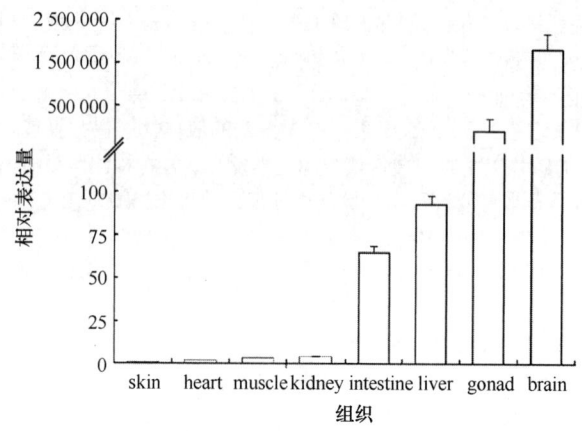

图 1　异育银鲫 GABA_A 受体 γ2 亚基基因在各组织中的表达情况

Fig. 1　Expression of GABA_A receptor γ2 subunit gene in different Carassaisauratus gibebiol tissues

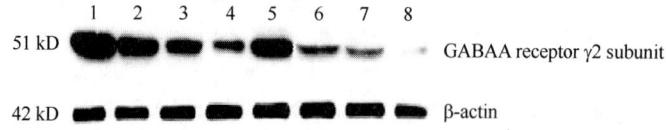

图 2　异育银鲫各组织 GABA_A 受体 γ2 亚基蛋白表达情况

Fig. 2　The protein expression of GABA_A receptor γ2 subunit in different tissues of Carassaisauratusgibebiol

1. 脑 brain；2. 肝 liver；3. 肾 kidney；4. 肌肉 muscle；5. 性腺 gonad；
6. 前肠 intestine；7. 心 heart；8. 皮 skin

2.3　DIF 对 GABA_A 受体 γ2 亚基 mRNA 和蛋白表达的影响

实时荧光定量 PCR 结果（图3）和 Western blot 结果（图5）显示，DIF 可明显抑制异育银鲫 GABA_A 受体 γ2 亚基 mRNA 和蛋白质的表达，与各组织的空白对照组比较差异有统计学意义（$P<0.05$）。相比于转录水平的迅速降低，蛋白表达水平的降低具有一定的滞后性，这主要是因为基因到蛋白表达需通过一系列复杂的反应才可以实现的缘故。

3　讨论

3.1　GABA_A 受体 γ2 亚基的组织表达特征

运用实时定量 PCR 和 Western Blot 检测 GABA_A 受体 γ2 亚基 mRNA 和蛋白在异育银鲫脑、肝、肾、心脏、肌肉、前肠、表皮和性腺 8 个组织中的表达，结果发现 GABA_A 受体 γ2 亚基在各组织中均有表达，且在脑组织中其基因水平和蛋白水平的表达量显著高于其他组织（$P<0.05$），这与 GABA_A 受体是中枢神经系统主要的抑制性神经调节递质 γ-氨基丁酸（GABA）的主要受体相符合[13,15]。Pirker 等[19]，Hörtnagl 等[20]的研究发现 GABA_A 受体 γ2 亚基在老鼠和家鼠体内也主要分布于脑组织，且主要集中于海马区、皮层和小脑等区域。此外，GABA_A 受体 γ2 亚基的基因和蛋白在性腺中也具有一个较高表达水平，Praphaporn Stewart 等[21]发现 GABA_A 受体在耳鲍幼体发育过程中的含量明显增加，因此，我们推测 GABA_A

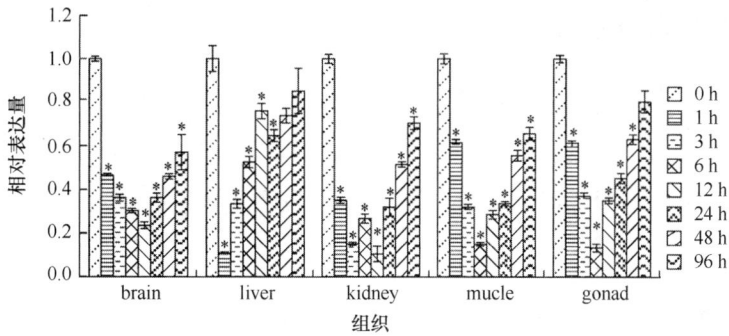

图 3　异育银鲫 GABA$_A$ 受体 γ2 亚基在 DIF 作用下 mRNA 表达情况

Fig. 3　The Expression of GABA$_A$ receptor γ2 subunit gene in Carassaisauratus gibebiol after DIF challenge

各组织均以空白样为对照，P<0.05 差异显著，以 ＊ 标示

Significant differences with control were indicated with ＊ P<0.05

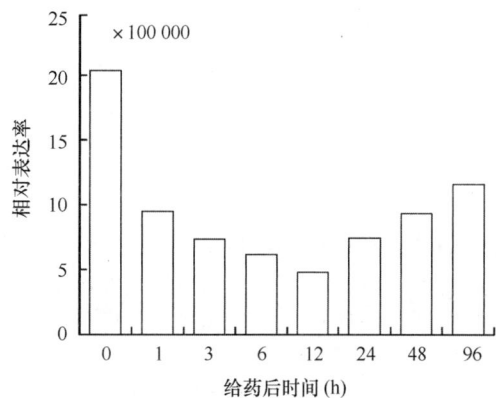

图 4　异育银鲫 GABA$_A$ 受体 γ2 亚基在 DIF 作用下脑组织内 mRNA 表达情况

Fig. 4　The Expression of GABA$_A$ receptor γ2 subunit gene in brain of Carassaisauratus gibebiol after DIF challenge

受体除了神经调节功能外，还将参与异育银鲫生殖及幼体发育过程。

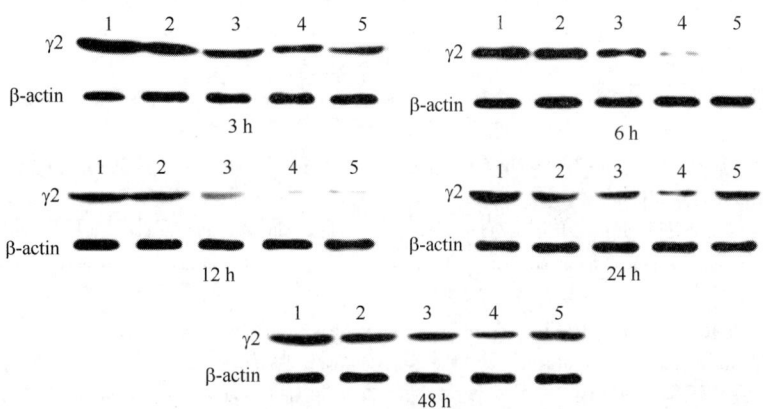

图 5　异育银鲫 GABA$_A$ 受体 γ2 亚基在 DIF 作用下蛋白表达情况

Fig. 5　The protein expression of GABA$_A$ receptor γ2 subunit in Carassaisauratus gibebiol after DIF challenge

1. 脑 brain；2. 肝 liver；3. 肾 kidney；4. 肌肉 muscle；5. 性腺 gonad

3.2 DIF 对异育银鲫 GABA$_A$ 受体表达量的影响

与各组织空白对照相比，DIF 可明显抑制 GABA$_A$ 受体 mRNA 和蛋白的表达，对异育银鲫各组织存在一定的神经毒性（图 3、图 5）。另外，在 DIF 用药组实验中，我们选取了 7 个时间点，分析药物代谢过程中 GABA$_A$ 受体在各组织中的变化情况，结果表明药物在体内浓度越高 GABA$_A$ 受体的含量越低，随着 DIF 被鱼体不断代谢，GABA$_A$ 受体的含量逐渐回升趋于正常值（图 3、图 5），这说明 GABA$_A$ 受体表达量与 DIF 在鱼体内的含量成负相关，可利用 GABA$_A$ 受体在组织中的表达量来推测 DIF 在鱼体不同组织中的含量。

肾脏中 GABA$_A$ 受体 mRNA 含量在第 12 小进出现二次回落，肝脏在第 24 小时出现二次回落，说明 DIF 在肾脏、肝脏中的含量出现再次回升，可能是因为 DIF 在异育银鲫体内存在肝肠循环所致。这与章海鑫等[5]、阮记明等[22]的研究结果 DIF 在异育银鲫体内的药时曲线均形成"多峰现象"相一致。

图 3 中脑组织受 DIF 影响后 GABA$_A$ 受体表达模式的变化同其他组织基本一致，然由于 GABA$_A$ 受体 mRNA 在脑组织中的初始表达量是肝脏、肾脏等外周组织的 2 万至 20 万倍，因此我们以脑组织空白对照组 GABA$_A$ 受体原始表达量为基准，单独构建了脑组织用药后 mRNA 表达量变化图（图 4），结果表明用 DIF 处理后，脑组织中 GABA$_A$ 受体 mRNA 表达量降低程度相对更大，因此 DIF 对脑组织中 GABA$_A$ 受体含量的抑制效果最显著，说明 DIF 对水产动物所造成的神经毒性作用在脑组织中最显著，与前人研究结果相符[3,6,7]。同时，相关研究显示，γ2 亚基表达量的降低将导致生物体神经性死亡[23,24]，因此 DIF 对水产动物所造成的神经毒性是不可忽视的。

本研究还发现，DIF 对脑组织中 GABA$_A$ 受体含量的影响具有持久性，恢复到原有水平需经过一段较长的代谢周期（图 3、图 4），阮记明等[9]的研究表明 DIF 可透过异育银鲫脑血屏障，且正是由于血脑屏障的存在，导致 DIF 在脑中的消除半衰期最长，到试验第 960 小时，大脑组织中 DIF 含量较肝脏、肾脏和肌肉中最高。

4 展望

异育银鲫经 DIF 处理后，所有检测组织中 GABA$_A$ 受体 mRNA 和蛋白表达量的影响均有明显降低，表明 DIF 对异育银鲫各组织器官均有一定的神经毒性作用，然尤以脑组织中含量变化最明显即神经毒性最显著。同时，进一步的实验表明，随着 DIF 被鱼体不断代谢，GABA$_A$ 受体的含量逐渐回升，并趋于正常值，即鱼体内 DIF 含量与 GABA$_A$ 受体含量呈负相关关系。本实验说明 GABA$_A$ 受体作为双氟沙星神经毒性的评估指标是可行的，并为后期 GABA$_A$ 受体在氟-喹诺酮类药物对水产动物神经毒性的评估提供理论及实验基础，从而在分子水平上体现药物对水生动物神经系统的影响。

参考文献

[1] Gangisetty O, Reddy D S. The optimization of TaqMan real-time RT-PCR assay for transcriptional profiling of GABA-A receptor subunit plasticity [J]. Journal of Neuroscience Methods, 2009, 181 (1): 58-66.

[2] Zhu W P. Analysis on adverse effects of quinolone drugs [J]. China Modern Medicine, 2010, 17 (22): 82-83.

[3] Sun H P, Cai L L, Yan F Q, et al. Mechanism and adverse effect of quinolones [J]. Chinese Journal of Nosocomiology, 2008, 18 (7): 1014-1016.

[4] Shen C, Shen J Z, Xiao X L, et al. Study on dominant lethal assay of difloxacin [J]. ChineseJournal of Veterinary Drug, 2000, (5): 16-18.

[5] Zhang H X, Ruan J M, Hu K, et al. Researches about pharmacokinetics of DIF in Carassais auratus gibebio [J]. Periodical of Ocean University of China, 2012, 42 (11): 81-86.

[6] Yang P. Analysis of mechanism and adverse effect of Quinolones [J]. China Journal of Pharmaceutical Economics, 2013, (4): 43-44.

[7] PAN F C. Discuss the importance of reasonable application of Quinolones [J]. China Journal of Pharmaceutical Economics, 2013, (2): 72-73, 77.

[8] De Lange E C M, Marchand S, van den Berg D, et al. In vitro and in vivo investigations on fluoroquinolones: effects of the P-glycoprotein efflux transporter on brain distribution of sparfloxacin [J]. European Journal of Pharmaceutical scie-nces, 2000, 12 (2): 85-93.

[9] Ruan J M, Hu K, Yang X L, et al. BLOOD-brain barrier permeability of dif and its elimination comparative study between brain and peripheral tissues in carassius auratus gibelio [J]. Acta Hydrobiologica Sinica, 2014, 38 (2): 272-278.

[10] Whiting P J. GABA-A receptor subtypes in the brain: a paradigm for CNS drug discovery [J]? Drug Discovery Today, 2003, 8 (10): 445-450.

[11] Whiting P J. The GABA$_A$ receptor gene family: new opportunities for drug development [J]. Current Opinion in Drug Discovery & Development, 2003, 6 (5): 648-657.

[12] Kinirons P, Cavalleri G L, Shahwan A, et al. Examining the role of common genetic variation in the γ2 subunit of the GABA$_A$ receptor in epilepsy using tagging SNPs [J]. Epilepsy Research, 2006, 70 (2-3): 229-238.

[13] Korpi E R, Sinkkonen S T. GABA (A) receptor subtypes as targets for neuropsychiatric drug development [J]. Pharmacology & Therapeu-

tics, 2006, 109 (1-2): 12-32.

[14] Hancili S, Önal Z E, Ata P, et al. The GABA_A Receptor γ2 Subunit (R43Q) Mutation in Febrile Seizures [J]. Pediatric Neurology, 2014, 50 (4): 353-356.

[15] Mascia M P, Biggio F, Mancuso L, et al. Changes in GABA (A) receptor gene expression induced by withdrawal of, but not by long-term exposure to, ganaxolone in cultured rat cerebellar granule cells [J]. Journal of Pharmacology and Experimental Therapeutics, 2002, 303 (3): 1014-1020.

[16] Brandon N J, Delmas P, Hill J, et al. Constitutive tyrosine phosphorylation of the GABA_A receptor γ2 subunit in rat brain [J]. Neuropharmacology, 2001, 41 (6): 745-752.

[17] Schweizer C, Balsiger S, Bluethmann H, et al. The γ2 subunit of GABA_A receptors is required for maintenance of receptors at mature synapses [J]. Molecular and Cellular Neuroscience, 2003, 24 (2): 442-450.

[18] Livak K J, Schmittgen T D. Analysis of relative gene expression data using real-time quantitative PCR and the 2 (-Delta Delta C (T)) method [J]. Methods, 2001, 25 (4): 402-408.

[19] Pirker S, Schwarzer C, Wieselthaler A, et al. GABAA receptors: immunocytochemical distribution of 13 subunits in the adult rat brain [J]. Neuroscience, 2000, 101 (4): 815-850.

[20] Hörtnagl H, Tasan R O, Wieselthaler A, et al. Patterns of mRNA and protein expression for 12 GABA_A receptor subunits in the mouse brain [J]. Neuroscience, 2013, 236: 345-372.

[21] Stewart P, Williams E A, Stewart M J, et al. Characterization of a GABAA receptor β subunit in the abalone Haliotisasinina that is upregulated during larval development [J]. Journal of Experimental Marine Biology and Ecology, 2011, 410: 53-60.

[22] Ruan J M, Hu K, Zhang H X, et al. Pharmacokinetics comparisons of difloxacin in crucian carp (Carassaisauratusgibebio) at two different water temperatures [J]. Journal of Shanghai Ocean University, 2011, 20 (6): 858-865.

[23] Jesper Karle, Michael Robin Witt, Mogens Nielsen. Diazepam protects against rat hippocampal neuronal cell death induced by antisense oligodeoxynucleotide to GABA receptor γ2 subunit [J]. Brain Research, 1997, 765 (1): 21-29.

[24] Karle J, Laudrup P, Sams-Dodd F, et al. Differential changes in induced seizures after hippocampal treatment of rats with an antisense oligodeoxynucleotide to the GABA_A receptor γ2 subunit [J]. European Journal of Pharmacology, 1997, 340 (2-3): 153-160.

该文发表于《生物技术通报》【2015, 31 (3): 191-198】

异育银鲫 GABA_A 受体 γ2 亚基全长克隆及生物信息学分析

Cloning and Bioinformatics Analysis of GABA_A R eceptor γ2 Subunit Gene in *Carassais auratus gibebiol*

赵依妮[1]，胡鲲[1]，孙琪[1,2]，杨先乐[1]，阮记明[1]，周爱玲[1]

(1. 上海海洋大学国家水生动物病原库，上海 201306；2. 上海黛龙生物科技工程有限公司，上海 201414)

摘要： 近年来 GABA_A 受体亚基特异性在药物筛选、研发过程中的应用得到广泛地关注，其中有关 α1、β2 和 γ2 三种功能性亚基的研究最为深入。异育银鲫因其良好的生长、繁殖优势，在国内得到广泛养殖。采用 RACE 法克隆得到了异育银鲫 GABA_A 受体 γ2 亚基基因全长 cDNA，并进行了生物信息学分析。该基因长 2 763 bp，其中 CDS 区长 1 437 bp，可编码 477 个氨基酸的前体蛋白。预测蛋白分子量 55.3 kD，理论等电点 9.13。异育银鲫体内 GABA_A 受体 γ2 亚基氨基酸序列 N 端存在 1 个长度为 35 个氨基酸的信号肽，4 个长度分别为 23、20、23 和 23 个氨基酸的跨膜区，3 个 N-糖基结合位点和 2 个 O-糖基化位点，1 个特异性结构域，其氨基酸序列具有明显的氯离子门控通道家族特征。氨基酸序列与其他物种氨基酸序列的同源性都在 89% 以上，表明该蛋白属于 GABA_A 受体亚基家族。系统进化树表明异育银鲫与斑马鱼聚为一支，亲缘关系最近。

关键词： 异育银鲫；GABA_A 受体 γ2 亚基；克隆；生物信息学分析

该文发表于《动物学杂志》【2013, 48 (6): 905-911】

异育银鲫 $GABA_B$ 受体 1 亚基 cDNA 部分序列的克隆及表达分析

Cloning and Expression of Partial cDNA Encoding $GABA_B$ Receptor Subunit 1 in *Carassius auratus gibelio*

王祎[1], 阮记明[1,2], 周爱玲[1], 赵依妮[1], 曹海鹏[1], 胡鲲[1], 杨先乐[1*]

(1. 上海海洋大学国家水生动物病原库, 上海 201306; 2. 江西农业大学动物科学技术学院)

摘要: 为了确定 γ-氨基丁酸 B 受体 (gamma-aminobutyric acid B receptor, $GABA_B$R) 基因在异育银鲫 (*Carassius auratus gibelio*) 不同组织中的表达, 本实验分别对异育银鲫不同组织中 $GABA_B$R1 基因进行 RT-PCR 扩增, 并进行了克隆和测序, 在与 GenBank 基因库中已知 $GABA_B$R1 序列进行同源性比对的基础上采用邻接法构建系统发育树, 并进一步分析其在异育银鲫不同组织内的表达水平。经克隆获得异育银鲫 $GABA_B$R1 基因 CDS 区序列383 bp, 编码 127 个氨基酸。荧光定量 PCR 结果显示, $GABA_B$R1 基因在异育银鲫脑、肝、肾、心、肠、鳔、鳃、肌、尾鳍、脾、卵巢、精巢组织中均有表达, 且在不同组织中的表达水平由高到低依次是: 脑>尾鳍>精巢>心、肠、鳔>卵巢、脾、鳃、肌>肝、肾。本研究证实了 $GABA_B$R1 基因在异育银鲫各组织中表达的广泛性, 且有明显的组织特异性。

关键词: 异育银鲫; $GABA_B$R1 基因; 克隆; 组织表达

该文发表于《中国农学通报》【2014, 30 (14): 33-38】

异育银鲫组织 γ-氨基丁酸 A 受体 ($GABA_A$R) 的免疫组织化学定位

Immunohistochemical Localization of γ-aminobutyric Acid A Receptor ($GABA_A$R) in the Tissues of *Carassius auratus gibelio*

周爱玲[1], 阮记明[1,2], 曹海鹏[1], 胡鲲[1], 王祎[1], 赵依妮[1], 杨先乐[1]

(1. 上海海洋大学国家水生动物病原库, 上海 201306; 2. 江西农业大学动物科学技术学院, 南昌 330045)

摘要: 观察 γ-氨基丁酸 A 受体 ($GABA_A$R) 异育银鲫中枢神经系统和外周组织中的特异性分布情况, 探讨其在不同组织中的作用。本试验以 $GABA_A$R 的 α1 亚基 ($GABA_A$Rα1) 作为 $GABA_A$R 的检测指标, 采用免疫组织化学方法对异育银鲫端脑、中脑、小脑、延脑、肝脏、肾脏组织中的 $GABA_A$R 进行定位研究。结果表明: 在异育银鲫端脑、中脑、小脑、延脑、肝脏、肾脏中均有 $GABA_A$R 存在, 各组织中的 $GABA_A$R 密度从高到低依次为: 小脑、中脑、延脑、端脑、肾脏、肝脏。其中, 小脑的分子层和蒲氏细胞层、中脑的视盖、延脑的迷走叶、端脑的纹状体、肾脏的肾小管以及肝细胞中 $GABA_A$R 的密度较高。研究显示, $GABA_A$R 主要分布于异育银鲫的中枢神经系统, 外周组织中也有少量分布。本研究旨在为异育银鲫 $GABA_A$R 分布的定位研究奠定一定的科学基础。

关键词: $GABA_A$R; 异育银鲫; 免疫组织化学; 定位

（二）P-gp 对渔药代谢的影响

该文发表于《Journal of Aquatic Animal Health》【2015，27：104-111】

Chitosan influences the expression of P-gp and metabolism of norfloxacin in *Ctenopharyngodon Idellus*

Kun Hu[1], Xinyan Xie[1], Yi-Ni Zhao[1], Yi Li[1], Jiming Ruan[1],

Hao-Ran Li[1], Tianyi Jin[2] and Xian-Le Yang[1]*

（1. National Pathogen Collection Center for Aquatic Animals, Shanghai Ocean University, 999 Hucheng Huan Road, Shanghai, 201306, China；
2. Southwest Weiyu Middle School, 671 Yishan Road, Shanghai, 200233, China）

Abstract：The aim of the study is to investigate the relationship between administration of chitosan（CTS）, expression of permeability glycoprotein（P-gp）, and the metabolism of norfloxacin（NOR）in *Ctenopharyngodon idellus*. Fishes were administrated with a single dose of either NOR, CTS, 1/5 NOR-CTS or 1/10 NOR-CTS. The P-gp expression was analyzed by immunohistochemistry and RT-PCR. NOR drug concentration was determined using HPLC. The mRNA and protein expression of P-gp in the fish intestine was significantly enhanced following a single dosage of 40 mg/kg NOR, with peak expression at 3 h after drug administration（$P<0.05$）. A single dosage of both 1/5 NOR-CTS and 1/10 NOR-CTS reduced the intestinal P-gp expression to levels significantly lower than NOR alone（$P<0.05$）, but significantly higher than the control（$P<0.05$）. Interestingly, CTS alone also led to a slight decrease in P-gp expression. In addition, pharmacokinetic assays revealed a marked increase in area under the curve（AUC）of NOR with 1/5 and 1/10 NOR-CTS, by approx. 1.5-fold and 3-fold, respectively. Finally, the relative bioavailability of NOR after a single oral dosage of 1/5 and 1/10 NOR-CTS was enhanced to 148.02% and 304.98%, respectively. In this study, we demonstrate that the transmembrane glycoprotein P-gp regulates NOR metabolism in the intestine of *Ctenopharyngodon idellus*, suggesting that NOR may be a direct substrate of P-gp. More importantly, we show that chitosan can inhibit P-gp expression in a dose-dependent manner and improve the relative bioavailability of NOR in these fishes.

Key word：chitosan；P-gp expression；norfloxacin；*Ctenopharyngodon idellus*；drug metabolism

壳聚糖对草鱼体内 P 糖蛋白表达及诺氟沙星代谢的影响

胡鲲[1]，谢欣燕[2]，赵依妮[1]，李怡[1]，阮记明[1]，李浩然[1]，金天逸[1]，杨先乐[1]

（1. 上海海洋大学国家水生动物病原库，中国上海沪城环路999号；2.）

摘要：本研究的目的是弄清可溶性甲壳素（CTS）的注射，蛋白的渗透率（P-gp）的表达与诺氟沙星（NOR）在草鱼体内代谢的关系。鱼类通常只注射 NOR，CTS，1/5 NOR-CTS 或者 1/10 NOR-CTS 中的一种。P-gp 的表达是通过免疫组织技术和 RT-PCR 来分析的。NOR 的浓度则通过 HPLC 来测定。鱼类肠道内的 P-gp 的 mRNA 和蛋白质的表达明显随着每多 40 mg 的 NOR 而增强。表达的峰值是在注射药物过后的三小时（$P<0.05$），每单剂量的 1/5 NOR-CTS 和 1/10 NOR-CTS 将会影响肠道 P-gp 的表达水平，表达水平明显的比使用 NOR 低（$P<0.05$），但却明显的比空白对照组高。有意思的是，可溶性甲壳素也会轻微的抑制 P-gp 的表达。除此之外，药代动力学分析曲线在注射 1/5 NOR 和 1/10 NOR-CTS 后分别都明显的增长了 1.5～3 倍。最后，口服单剂量的 1/5 NOR 和 1/10 NOR-CTS 后。NOR 的利用率将会分别增长到 148.02% 和 304.98%。这个研究证明了细胞膜上糖蛋白的渗透率会调控草鱼肠道内的 NOR 的代谢，显示 NOR 可能会是蛋白质渗透的直接物质。更重要的是我们证明了可溶性甲壳素会以一种与剂量相关的机制抑制鱼类 P-gp 的表达，提高 NOR 的利用率。

Introduction

Permeability glycoprotein（P-gp）is widely expressed in the kidney, intestine and liver, where it can regulate drug metabolism

(Hemmer et al. 1995; Hochman et al. 2001; Thomas et al. 2007; Lee et al. 2013; Sousa et al. 2013; Wessler et al. 2013). It also acts as a transmembrane efflux pump, and is a member of the adenosine triphosphate (ATP) –binding cassette transporter family (Rosenberg et al. 2001). Owing to a broad substrate specificity, that includes various structurally divergent drugs in clinical use today, it has the potential to reduce the levels of drug residues in tissues (Marzolini et al. 2004). In addition, since various aquatic animals also express P-gp, it has been proposed as a biomarker to measure exposure to pollution (Kleinow et al. 2000; Doi et al. 2001, Keppler & Ringwood 2001; Bard et al. 2002; Bresolin et al. 2005; Thomas et al. 2007; MA 2008).

Time-controlled oral drug delivery systems offer several advantages to those that facilitate immediate drug release (Streubel et al. 2006). The choice of suitable pharmaceutical excipients in order to achieve controlled drug release is of paramount importance in the clinical implementation of several drugs currently available in the market. A number of pharmaceutical excipients inhibit P-gp, and potentially enhance drug absorption (Cornaire et al. 2004; Shono et al. 2004; Yan et al. 2008; Iangcharoen et al. 2011). Chitosan (CTS) is a linear copolymer of β- (1-4) linked 2-acetamido-2-deoxy-β-d-glucopyranose and 2-amino-2-deoxy-β-d-glycopyranose obtained by deacetylation of chitin, and is known to be an excellent material for drug delivery (Sinha et al. 2004; Sezer et al. 2008; de la Fuente et al. 2010; Park et al. 2010; Saber et al. 2010). It has several beneficial features, including biodegradability, biocompatibility, low toxicity, mucoadhesiveness, easy absorption and antimicrobial properties (Rinaudo 2006). The intestinal absorption of various therapeutic compounds known to be metabolized by P-gp, is significantly improved when administered using a CTS-based drug delivery system (Foger et al. 2006; Foger et al. 2007; Mo et al. 2011; Jin et al. 2014).

Quinolones are synthetic antibacterial agents widely used for the treatment of various infections, especially against gram-negative bacteria. Their rampant use in aquaculture has led to the pollution of different aquatic compartments with these drugs (T et al. 2000). Many modern techniques have been used to estimate the level of quinolone residues in marine aquaculture environment, in order to obtain information that could help avoid potential health risks in consumers (Zhang et al. 2010; Silva et al. 2012; Quesada et al. 2013; Tao et al. 2013). In China, quinolones are also widely used in the treatment of animal diseases. Norfloxacin (NOR), ciprofloxacin and ofloxacin account for approximately 98% of the commercially produced quinolones.

Cyprinidae belongs to the class of osteichthyes or bony fishes, and is the most important fresh water aquatic product in China. To ensure the safety of certain food products, extensive research has been done on the metabolism of NOR in fishes. However, the relationship between pharmaceutical excipients like CTS and NOR metabolism in aquatic animals is still poorly understood. In the present study, we investigated the effect of CTS on the expression of P-gp and metabolism of NOR in *Ctenopharyngodon idellus*.

Methods

Reagents

NOR (content ≥98.5%) was purchased from Zhejiang Guobang Pharmaceutical, Zhejiang, China. Low molecular weight CTS (approx. 5-10 kDa), acetonitrile (HPLC grade), hydrochloric acid (HCL) and n-hexane (chemically pure) were obtained from Shanghai Anpel Scientific Instrument, Shanghai, China.

The NOR-CTS drug delivery system was prepared using a freeze-drying method. NOR and CTS were carefully mixed and dissolved first in 2% HCL, then water, to generate a 40% solution, and stored frozen. Vacuumizing, freezing and drying of this solution was performed sequentially at-50℃ to obtain 1/5 NOR-CTS (1 : 5 w/w ratio of NOR to CTS) and 1/10 NOR-CTS (1 : 10 w/w ratio of NOR to CTS).

Fishes

Ctenopharyngodon idellus, weighing 50 ±10. 6 g, were provided by a farm in Nantong, Jiangsu, China. All fishes were housed in two fiberglass tanks (approx. 250 fishes per 500 L tank) supplied with a constant flow (1 L/h) of aerated fresh water for 7 days, and fed a drug-free commercial diet. Water quality was monitored daily, and adjusted when necessary. The pH value was maintained between 6. 4 and 7. 0, and temperature was set to 25. 0±1. 0℃ using a combination of cooling and heating devices.

Drug administration and sampling

Fishes were divided into 4 experimental groups of 100 fishes each, that were then administered a single oral dosage of CTS, NOR (40 mg/kg), 1/5 NOR-CTS (40 mg/kg) or 1/10 NOR-CTS (40 mg/kg). Those not treated with any drug served as the control. At each assayed time point (0. 25, 0. 5, 1, 3, 6, 12, 24, 48 and 192 or 196 hours), ten fishes were sacrificed. Next, blood samples drawn from tail sinus and samples of the intestines were collected and frozen immediately and stored at −20℃ until analysis. All animal experiments were performed in accordance with the guidelines for the Regulation of Animal Experimentation, and were approved by the State Scientific and Technological Commission.

In vitro release assay of NOR-CTS drug delivery system

To estimate the release of NOR-CTS in vitro, 4 mL NOR, 1/5 NOR-CTS or 1/10 NOR-CTS solution was placed into separate dialysis bags, sealed at both ends with clips, immersed in 20 mL PBS (0.2 mol/L, pH 7.4) and stirred on a magnetic stirrer at 37℃. During a period of 24 h, release of freeze-dried NOR-CTS was tested three times by intermittently sampling the contents (1 mL) of the outer media. The buffer was replaced immediately after sampling. The drug release rate was calculated according to the formula: Release Rate= (NOR concentration in the outer media/total NOR concentration of the freeze-dried drug delivery system) ×100%.

The concentration of NOR was determined by high performance liquid chromatography (HPLC). The HPLC (Agilent, Santa Clara, CA, USA) instrument was equipped with a fluorescence detector and a reversed-phase C18 column (150 mm×4.6 mm) with a mobile phase comprised of 5 : 95 v/v ratio of acetonitrile: 0.01 M tetrabutyl ammonium bromide, pH 3.1. 10 μL sample solution was loaded, the flow rate was set at 1.5 mL/min and the column temperature was set at 40℃. A detection wavelength of 280 nm was used. Concentration of the drug was determined by the external standard method, whereby standard solutions of NOR (concentration range=0.02-10 mg/mL) were prepared and standard curves were prepared, associating drug concentration (x-axis) with peak area (y-axis).

RNA isolation and RT-qPCR

Total RNA was extracted from intestinal tissues using a TRIzol RNA extraction kit (Invitrogen, Carlsbad, CA, USA) according to the manufacturer's protocol. To ensure sufficient integrity of extracted RNA for subsequent experiments, the OD_{260}/OD_{280} ratio was measured on a spectrophotometer (DU800, Beckmen, USA). cDNA was synthesized from 1 μg total RNA using a reverse transcription kit (ReverTra Ace-α-©, Toyobo Co., Ltd., Japan) and a mixture of oligo (dT) and random primers. The primers and probes were synthesized by designed using Primer Premier 5.0 software (PREMIER Biosoft, Palo Alto, California); are listed in Table 1. Quantitative RT-PCR was carried out using SYBR Green PCR Master Mix (QPK-201, Toyobo Co., Ltd., Japan), in a Bio-Rad iQ5 real-time PCR instrument (Hercules, CA, USA). The RT-qPCR reaction for P-gp was performed in a final reaction volume of 10 μL under the following conditions: a preheating cycle at 94℃ for 3 min, followed by 35 cycles of amplification-94℃ for 30 sec, 51.5℃ for 30 sec, 72℃ for 30 sec, and finally an elongation reaction at 72℃ for 10 min. Amplification of β-actin RNA, a relatively invariant internal reference gene was performed in parallel, and cDNA amounts were normalized to equivalent β-actin mRNA levels (Michael, 2001). Sterile water was used as negative control.

Table 1 Primer sequences for RT-qPCR analysis.

	Primers	Accession number cDNA[b]	Annealing temperature (℃)	Length (bp)[a]
P-gp	Forward: 5' -CACCTGGACGTTACCAAAGAAGATATA-3' Reverse: 5' -TCACCAACCAGCGTCTCATATTT-3'	AB699096	52.3	127
β-actin	Forward: 5' -CCTCACCCTCAAGTACCCCAT-3' Reverse: 5' -TTGGCCTTTGGGTTGAGTG-3'	EU887951	54.5	153

[a] Amplicon length in base pairs.

[b] GenBank accession number, available at http: //www.ncbi.nlm.nih.gov/.

Immunohistochemistry

The level of P-gp protein expression was estimated using immunohistochemical staining (Kleinow et al. 2000). Intestinal samples were fixed with 4% buffered paraformaldehyde (PFA) at 4℃ for 24 h and washed with PBS (pH 7.4). Next, they were sequentially dehydrated in 75%, 85%, 95%, and 100% ethanol, permeabilized in xylene and embedded in paraffin. 5 μm thick tissue sections were sliced, mounted on slides coated with 3-aminopropyl-triethoxysilane, and placed in a 60℃ oven for 2 h. Sections were first blocked for 45 min with normal horse serum, followed by overnight incubation with primary anti-P-gp mouse monoclonal antibody C-219 (Gene Tex Inc, Irvine, California, USA) at 4℃. Incubation in PBS for 1.5 h was used as the negative control. The following day, labeled sections were washed three times with PBS, incubated for 30 min with biotinylated horse anti-mouse secondary antibody, washed in PBS, and further incubated with avidin-conjugated horse raddish peroxidase (HRP) complexes for 45 min. Diaminobenzidine was used as the substrate chromagen for HRP, to obtain a colored product corresponding to localization of the antigen protein. Next, the slides were washed in PBS and distilled water and counter stained in Mayer's hematoxylin. Images were procured on a fluorescence microscope (ECLIPSE TI-SR, NIKON Co., Ltd., Japan) and image analysis was performed on Image pro -plus 6.0. Two sections were prepared from each fish in the four experimental groups (CTS, NOR, 1/5 NOR-CTS and 1/10 NOR-CTS). In each slide, 10 randomly selected fields of view were imaged and analyzed at magnifications of 200X and 400X. The average cumulative integrated option density value of positive staining for each group is expressed as mean±standard error (SE).

Pharmacokinetic study

To perform pharmacokinetic analysis, we collected blood samples from all experimental groups at each time point (described earlier) following drug administration. The concentration of NOR in plasma was determined using the same HPLC procedure as described earlier in section 2.3. Plasma was extracted in acetonitrile, and centrifuged at 4500 rpm for 5 min. The supernatant was carefully collected and mixed with n-hexane. The aqueous phase of this mixture was discarded, while the organic phase was collected and concentrated in a rotary evaporator (Heidolph Instruments, Schwabach, Germany). The resulting residues were dissolved in the mobile phase and filtered through a 0.45 μm filter. In addition, to validate the HPLC method used here, a standard solution of NOR (5 mg/ L) was added to the blank samples. Next, we determined the intra-day and inter-day recoveries and relative standard deviations. Pharmacokinetic parameters such as peak concentration (C_{max}) and the time taken by the drug to reach peak levels (T_{max}), were estimated. Finally, the area under the curve (AUC_{0-t}) was calculated using the software Drug and Statistics (DAS version 3.0, Center for Drug Clinical Evaluation, Wannan Medical College, China).

Statistical analysis

Statistical analyses were performed with SPSS 16.0 software (SPSS Inc., Chicago, IL, USA) and data were presented as the mean±standard deviation (SD). Normal distribution and homogeneity of the variances was tested by Kolmogorov-Smirnov test and F-test, respectively. The analysis of variance (ANOVAs) and the Duncan's test for multiple range comparison with significant level established at $P < 0.05$.

Results

Dynamics of release of freeze-dried NOR-CTS in vitro

We first analyzed the dynamics of release of freeze dried NOR-CTS conjugates into aqueous solution by placing the complexes into dialysis bags and measuring the concentration of released NOR in the outer media. We found a distinct peak of NOR concentration, that was accompanied by a stable baseline. Specifically, we estimated the retention time of NOR to be 4.80 min, and obtained the following standard curve equation-$y = 61.803x - 0.5808$ ($R^2 = 0.996$), within the dose range of 0.02-5 mg/mL. In the external standard method, the NOR detection limit was 5 mg/mL.

The encapsulation efficiency of the freeze-drying procedure used to prepare NOR-CTS was more than 97.8% with good stability (data not shown). From the in vitro release assay (Fig 1), we found that complete release of NOR was achieved in approx. 0.25 h. In contrast, the release rate of 1/5 NOR-CTS and 1/10 NOR-CTS at 0.25 h was only 44% and 39%, respectively. Complete release of 1/5 NOR-CTS and 1/10 NOR-CTS was achieved in 12 h and 24 h, respectively.

NOR-CTS administration leads to reduced P-gp mRNA levels in fish intestine

Next, we examined the effect of NOR, alone and in combination with CTS, on the expression of P-gp in the intestine of Ctenopharyngodon idellus. We first quantified P-gp transcript levels using specific primers and RT-qPCR. In comparison to the control group, intestinal P-gp mRNA expression was significantly higher following the administration of a single dose of NOR, 1/5 NOR-CTS and 1/10 NOR-CTS ($P < 0.05$) (Fig 2). Peak expression was noted at 3 h after ingestion of the drug. Interestingly, a single dosage of CTS alone led to a slight decrease in P-gp expression. More importantly however, P-gp mRNA level after a single dosage of 1/5 NOR-CTS or 1/10 NOR-CTS was significantly lower than after a single dosage of NOR alone ($P < 0.05$). Finally, a single dosage of 1/10 NOR-CTS reduced P-gp mRNA levels in the intestine significantly below that recorded after administration of a single dosage of 1/5 NOR-CTS ($P < 0.05$).

NOR-CTS administration leads to reduced P-gp protein levels in fish intestine

In order to corroborate above results, we assayed P-gp protein levels in all experimental groups using the C-219 monoclonal antibody, which was found to be immunoreactive to the intestine of Ctenopharyngodon idellus. Compared with the control group, P-gp protein expression was significantly higher in response to a single dosage of NOR, 1/5 NOR-CTS or 1/10 NOR-CTS ($P < 0.05$) (Fig 3), with peak expression at 3 h after drug administration. As seen before, a single dosage of CTS alone led to a small decrease in P-

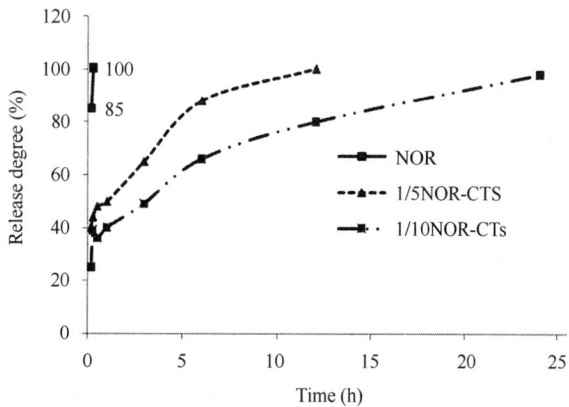

Fig. 1　Cumulative release of freeze-dried NOR-CTS at 37℃ in PBS（0.1 mol/L, pH 7.4）. Black squares on continuous line, NOR; Black triangles on broken line, 1/5 NOR-CTS; Black squares on broken line, 1/10 NOR-CTS.（NOR: norfloxacin; CTS, chitosan）

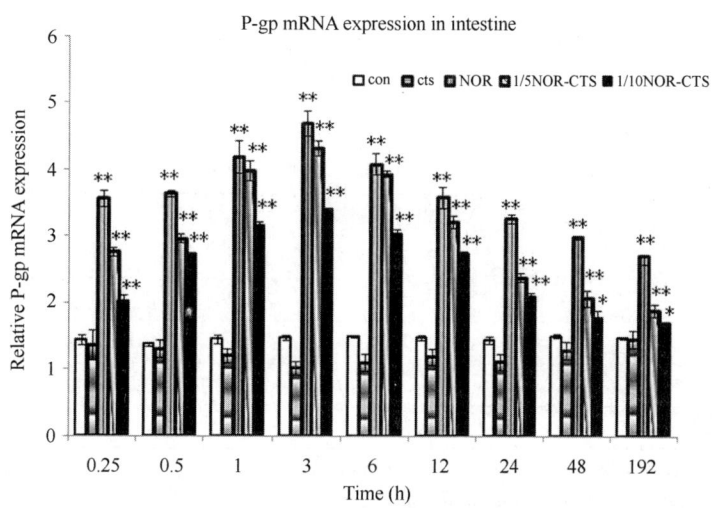

Fig. 2　Quantification of intestinal P-gp mRNA expression in *Ctenopharyngodon idellus* at indicated time points（0 min［control］, 0.25 h, 0.5 h, 1 h, 3 h, 6 h, 12 h, 24 h, 48 h and 192 h）after a single oral administration of NOR, 1/5 NOR-CTS or 1/10 NOR-CTS. ∗P<0.05, compare with the control, ∗∗P<0.01, compare with the control.（CON: control; NOR: norfloxacin; CTS, chitosan）. β-actin was used as an reference gene.

gp protein expression, in comparison to the control group. More significantly however, intestinal P-gp protein levels declined markedly after a single dosage of 1/5 NOR-CTS or 1/10 NOR-CTS, compared to the levels observed after a single dosage of NOR alone（P<0.05）. Furthermore, a single dosage of 1/10 NOR-CTS led to a decline in P-gp protein expression significantly below that observed after administration of a single dosage of 1/5 NOR-CTS（P<0.05）. These findings are consistent with the results obtained from P-gp RT-qPCR assay.

　　To further elucidate the differences between the experimental groups, we compared the intestinal mRNA and protein expression levels at 3 h after drug administration, since both parameters reached peak values at this time point（Table 2）. In contrast to the control group（1.47±0.03）, intestinal P-gp mRNA levels were significantly higher after a single dosage of NOR（4.70±0.68）, 1/5 NOR-CTS（4.32±0.10）or 1/10 NOR-CTS（3.39±0.01）（P<0.01）. Similarly, compared with the control group（3232.12± 1021.2）, intestinal P-gp protein expression was significantly elevated following a single dosage of NOR（12344.21±1031.31）, 1/5 NOR-CTS（9821.3±1031.31）or 1/10 NOR-CTS（7631.14±983.31）（P<0.01）. Furthermore, P-gp expression was also significantly different between the NOR group and the 1/5 and 1/10 NOR-CTS groups（P<0.05）, as well as between 1/5 NOR-CTS and 1/10 NOR-CTS groups（P<0.05）. Finally, both the mRNA and protein levels of P-gp in the CTS group did not vary significantly from the control group.

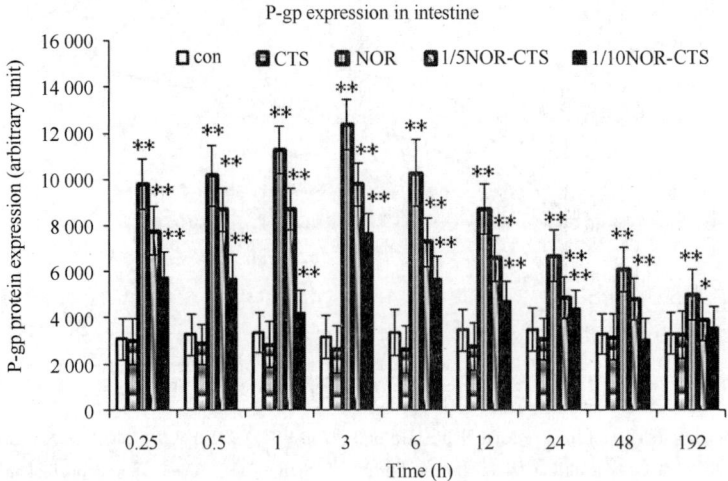

Fig. 3　Quantification of intestinal P-gp protein expression in *Ctenopharyngodon idellus* at indicated time points （0min ［control］，0.25 h，0.5 h，1 h，3 h，6 h，12 h，24 h，48 h and 192 h） after a single oral administration of NOR，1/5 NOR-CTS or 1/10 NOR-CTS. ＊P<0.05，compare with the control，＊＊P<0.01，compare with the control. （CON：control；NOR：norfloxacin；CTS，chitosan）

Table 2　Comparisons of the relative P-gp mRNA and protein expression in all experimental groups in *Ctenopharyngodon idellus* at 3 h after drug administration. （CON：control；NOR：norfloxacin；CTS，chitosan）

	mRNA expression	protein expression
CON	1.47±0.03 Aa	3232.12±1021.2 Aa
CTS	1.02±0.11 Bb	2663.11±1031.31 Aa
NOR	4.70±0.68 Ca	12344.21±1031.31 Bb
1/5 NOR-CTS	4.32±0.10 Cc	9821.3±1031.31 Ba
1/10 NOR-CTS	3.39±0.01 Dd	7631.14±983.31 Db

Data are presented as mean±standard error （SE）. Comparisons were made by one-way ANOVA or Duncan's multiple range test. Within a column，values with different uppercase letters indicate very significant difference （P<0.01）；values with the same uppercase letter and different lowercase letter represent significant difference （0.01<P<0.05）；values with the same uppercase and lowercase letters indicate no significant different （P>0.05）. Relative mRNA expression was determined by normalising to β-actin. Relative protein expression was calculated as the average cumulative integrated optical density of positive staining in each group.

Pharmacokinetic studies of NOR in *Ctenopharyngodon idellus*

Since，the conjugation of NOR to CTS seems to reduce the rate of release of NOR both in vitro as well as in vivo，we next asked whether CTS also influenced the metabolism of NOR in fishes. To determine this，we assayed for certain pharmacokinetic parameters. Using HPLC，we determined NOR concentration in the plasma，and found that the average recoveries spiked in the blank plasma at 5，500 and 1000 mg/mL ranged from 81.01 to 93.76%，while the relative standard deviation of NOR concentration ranged from 3.82 to 7.63% （Table 3）.

Table 3　Recovery of NOR in the plasma of *Ctenopharyngodon Idellus* （n=5）. （NOR：norfloxacin）

Spiked Concentration in blank plasma （mg/mL）	Intra-day		Inter-day	
	Average Recovery （%）	Relative standard deviation （%）	Average Recovery （%）	Relative standard deviation （%）
5	82.76	7.63	83.01	5.32
500	81.01	3.82	93.76	4.38
10000	93.73	3.83	92.42	4.73

After orally administering a single dosage of NOR, 1/5 NOR-CTS or 1/10 NOR-CTS, we measured the concentration of NOR in the plasma using HPLC. The resulting profile of NOR concentration is shown in Fig. 4. The residues in the intestine were fitted to the open two-compartment model using DAS pharmacokinetic software. The pharmacokinetic parameters derived from this compartment model are shown in Table 4. The Tmax and Cmax of NOR in *Ctenopharyngodon idellus* after a single oral dose of NOR, 1/5 NOR-CTS and 1/10 NOR-CTS (40 mg/kg each) were 0.5 h and 9.09 mg/L, 1 h and 7.54 mg/L, and 3 h and 4.27 mg/L, respectively. The elimination time of NOR in after a single oral dose of NOR, 1/5 NOR-CTS and 1/10 NOR-CTS (40 mg/L each) were 4.64 h, 9.14 h and 37.87 h, respectively. In comparison to NOR administration alone, the area under the curve (AUC) of NOR in Ctenopharyngodon idellus after a single oral dose of 1/5 NOR-CTS and 1/10 NOR-CTS, was significantly increased by approx. 1.5-fold and 3-fold, respectively (P<0.05). The relative bioavailability of NOR after ingestion of 1/5 NOR-CTS and 1/10 NOR-CTS was 148.02% and 304.98%, respectively.

Table 4　Pharmacokinetic properties of NOR in *Ctenopharyngodon Idellus* after administration of a single oral dose. (NOR: norfloxacin; CTS, chitosan).

	NOR	1/5 NOR-CTS	1/10 NOR-CTS
C_{max} (mg/L)	9.09	7.54	4.27
T_{max} (h)	0.5	1	3
$T_{1/2}$ (h)	4.64	9.14	37.87
AUC_{0-196} (μg/h/mL)	42.32	62.64	129.07
F_r (%)	–	148.02	304.98
Cl (kg/h/L)	470.66	319.08	151.85

Cmax: maximum concentration; Tmax: peak time; $T_{1/2}$: half-life of the drug; Fr: relative bioavailability; AUC: area under the curve; Cl: total body clearance of the drug

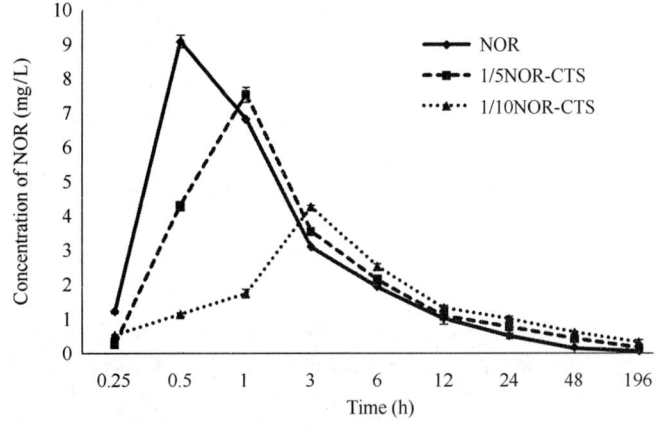

Fig. 4　Time-concentration profile of NOR after a single dose of NOR, 1/5 NOR-CTS or 1/10 NOR-CTS in *Ctenopharyngodon idellus*. Black circles on continuous line, NOR; Black squares on broken line, 1/5 NOR-CTS; Black triangles on broken line, 1/10 NOR-CTS. (NOR: norfloxacin; CTS, chitosan)

Discussion

Time-controlled oral drug delivery systems have several advantages over immediate drug release. The desired rate of drug release can be obtained using appropriate excipients and delivery methods. CTS is a biocompatible and biodegradable polymer (Rinaudo 2006), used effectively for drug delivery as hydrogels, drug conjugates, biodegradable release systems, and polyelectrolyte complexes. CTS-based delivery systems have been used for administration of proteins/peptides, growth factors, anti-inflammatory drugs and antibiotics. It is also being used in gene therapy and bio-imaging applications (Sinha et al. 2004; Sezer et al. 2008; de la Fuente et al. 2010; Park et al. 2010; Saber et al. 2010). In the present study, we demonstrate that delivery of NOR through a low molecular weight freeze-dried CTS-based system extends the time taken for complete drug release in vivo.

Permeability glycoprotein has been shown to regulate drug metabolism in mammals by mediating the efflux of therapeutic compounds and thus affecting their bioavailability in the liver, kidney and intestines (Marzolini et al. 2004; Thomas et al. 2007). The intestinal absorption of the drugs metabolized by P-gp is significantly improved when administered using a CTS-based system (Foger et al. 2006; Foger et al. 2007; Mo et al. 2011; Jin et al. 2014). For example, CTS-4-thiobutylamidine was shown to significantly enhance AUC and Cmax of saquinavir, a substrate of P-gp (Foger et al. 2007). In addition to mammals, P-gp is also expressed in a variety of aquatic animals (Kleinow et al. 2000; Doi et al. 2001; Keppler & Ringwood 2001; Bresolin et al. 2005; Hu et al. 2014), and is associated with chemical contamination of water bodies (Bard et al. 2002; Thomas et al. 2007). By using immunohistochemical analysis with a mammalian C-219 monoclonal antibody, Kleinow and colleagues detected Pgp protein expression in the liver and intestines of a catfish (Doi et al. 2001; Keppler & Ringwood 2001). In addition, P-gp protein levels were measured seasonally in the gills of southeastern oysters Crassostrea virginica (Keppler & Ringwood 2001). However, P-gp in Zebrafish shares only 49.4% and 49.7% nucleotide homology with humans and mice, respectively (MA 2008).

Quinolone antibacterial agents like NOR are widely used for the treatment of various infections in aquatic animals, and were found to be metabolized by P-gp. In this study, we show that both the mRNA and protein level of P-gp in the fish intestine is significantly enhanced after a single dosage of 40 mg/kg NOR, with peak expression at 3 h after drug administration ($P<0.05$). This finding was consistent with our previous study on enrofloxacin (Hu et al. 2014), and suggested that NOR metabolism in the intestine of Ctenopharyngodon idellus might be associated with P-gp activity. However, thus far, no studies have reported the effect of CTS on P-gp expression in aquatic animals. In this study, we show that both mRNA and protein levels of intestinal P-gp are downregulated after a single dosage of either 1/5 NOR-CTS or 1/10 NOR-CTS, when compared with administration of NOR alone ($P<0.05$), but are still significantly higher than the control ($P<0.05$) (Fig 3). Differences in P-gp protein level between NOR and control groups suggests that P-gp expression might be regulated by NOR. The effect of CTS on NOR metabolism is further demonstrated by the finding that P-gp expression, following ingestion of 1/10 NOR-CTS, is significantly lower than after a single dosage of 1/5 NOR-CTS ($P<0.05$). When we administered a single oral dosage of NOR (40 mg/kg), 1/5 NOR-CTS (40 mg/kg) and 1/10 NOR-CTS (40 mg/kg), We recorded Cmax values of 9.09 mg/L, 7.54 mg/L and 4.27 mg/L following a single oral dosage of NOR (40 mg/kg), 1/5 NOR-CTS (40 mg/kg) and 1/10 NOR-CTS (40 mg/kg), respectively (Table 4). In addition, both 1/5 NOR-CTS and 1/10 NOR-CTS markedly increased the AUC of NOR by approx. 1.5-fold and 3-fold, respectively ($P<0.05$), and its relative bioavailability in Ctenopharyngodon idellus to 148.02% and 304.98%, respectively. Altogether, our findings suggest that CTS might inhibit the expression of P-gp in a dose-dependent manner, leading to enhanced NOR absorption in the intestine of Ctenopharyngodon idellus.

In conclusion, our study demonstrates that norfloxacin is a substrate of P-gp which regulates drug metabolism in the intestine of Ctenopharyngodon idellus. Furthermore CTS inhibits P-gp expression in a dose-dependent manner and improve the relative bioavailability of NOR. Our findings suggest a novel mechanism that can improve drug bioavailability and reduce the risk of drug residues in animals.

Acknowledgement

This study was supported by the 863 Program (Grant 2011AA10A216), the Special Fund for Agro-scientific Research in the Public Interest (Grant 201203085) and the Key Project of Science & Technology of Shanghai Agriculture Committee. There is no conflict of interest related to this manuscript.

References

Backhaus, T., Scholze, M., and Grimme, L. H. 2000. The single substance and mixture toxicity of quinolones to the bioluminescent bacterium Vibrio fischeri. Aquatic Toxicology 49: 49-61.

Bard, S. M., Woodin, B. R., and Stegeman, J. J. 2002. Expression of P-glycoprotein and cytochrome p450 1A in intertidal fish (Anoplarchus purpurescens) exposed to environmental contaminants. Aquatic Toxicology 60: 17-32.

Bresolin, T., de Freitas Rebelo, M., and Celso Dias Bainy, A. 2005. Expression of PXR, CYP3A and MDR1 genes in liver of zebrafish. Comparative Biochemistry and Physiology Toxicology & Pharmacology: CBP 140: 403-407.

Cornaire, G., Woodley, J., Hermann, P., Cloarec, A., Arellano, C., and Houin, G. 2004. Impact of excipients on the absorption of P-glycoprotein substrates in vitro and in vivo. International journal of pharmaceutics 278: 119-131.

de la Fuente, M., Ravina, M., Paolicelli, P., Sanchez, A., Seijo, B., and Alonso, M. J. 2010. Chitosan-based nanostructures: a delivery platform for ocular therapeutics. Advanced drug delivery reviews 62: 100-117.

Doi, A. M., Holmes, E., and Kleinow, K. M. 2001. P-glycoprotein in the catfish intestine: inducibility by xenobiotics and functional properties. Aquatic toxicology 55: 157-170.

Foger, F., Hoyer, H., Kafedjiiski, K., Thaurer, M., and Bernkop-Schnurch, A. 2006. In vivo comparison of various polymeric and low molecular mass inhibitors of intestinal P-glycoprotein. Biomaterials 27: 5855-5860.

Foger, F., Kafedjiiski, K., Hoyer, H., Loretz, B., and Bernkop-Schnurch, A. 2007. Enhanced transport of P-glycoprotein substrate saquinavir in presence of thiolated chitosan. Journal of Drug Targeting 15: 132-139.

Hemmer, M. J., Courtney, L. A., and Ortego, L. S. 1995. Immunohistochemical detection of P-glycoprotein in teleost tissues using mammalian polyclonal and monoclonal antibodies. The Journal of Experimental Zoology 272: 69-77.

Hochman, J. H., Chiba, M., Yamazaki, M., Tang, C., and Lin, J. H. 2001. P-glycoprotein-mediated efflux of indinavir metabolites in Caco-2 cells expressing cytochrome P450 3A4. The Journal of Pharmacology and Experimental Therapeutics 298: 323-330.

Hu, K., Cheng, G., Zhang, H., Wang, H., Ruan, J., Chen, L., Fang, W., and Yang, X. 2014. Relationship between Permeability Glycoprotein (P-gp) Gene Expression and Enrofloxacin Metabolism in Nile Tilapia. Journal of Aquatic Animal Health 26: 59-65.

Iangcharoen, P., Punfa, W., Yodkeeree, S., Kasinrerk, W., Ampasavate, C., Anuchapreeda, S., and Limtrakul, P. 2011. Anti-P-glycoprotein conjugated nanoparticles for targeting drug delivery in cancer treatment. Archives of Pharmacal Research 34: 1679-1689.

Jin, X., Mo, R., Ding, Y., Zheng, W., and Zhang, C. 2014. Paclitaxel-loaded N-octyl-O-sulfate chitosan micelles for superior cancer therapeutic efficacy and overcoming drug resistance. Molecular Pharmaceutics 11: 145-157.

Keppler, C., and Ringwood, A. H. 2001. Expression of P-glycoprotein in the gills of oysters, Crassostrea virginica: seasonal and pollutant related effects. Aquatic Toxicology 54: 195-204.

Kleinow, K. M., Doi, A. M., and Smith, A. A. 2000. Distribution and inducibility of P-glycoprotein in the catfish: immunohistochemical detection using the mammalian C-219 monoclonal. Marine Environmental Research 50: 313-317.

Lee, S. D., Osei-Twum, J. A., and Wasan, K. M. 2013. Dose-dependent targeted suppression of P-glycoprotein expression and function in Caco-2 cells. Molecular Pharmaceutics 10: 2323-2330.

Ma, Y. F. 2008. cDNA cloning and expression analysis of P-glycoprotein in Zebrafish. Master, Southwest University, Chongqing, China.

Marzolini, C., Paus, E., Buclin, T., and Kim, R. B. 2004. Polymorphisms in human MDR1 (P-glycoprotein): recent advances and clinical relevance. International Journal of Clinical Pharmacology and Therapeutics 75: 13-33.

Michael W. Pfaffl. 2001. A new mathematical model for relative quantification in real-time RT-PCR. Nucleic acids research. 29 (9): e45.

Park, J. H., Saravanakumar, G., Kim, K., Kwon, I. C. 2010. Targeted delivery of low molecular drugs using chitosan and its derivatives. Advanced Drug Delivery 62: 28-41.

Quesada, S. P., Paschoal, J. A., and Reyes, F. G. 2013. A simple method for the determination of fluoroquinolone residues in tilapia (Oreochromis niloticus) and pacu (Piaractus mesopotamicus) employing LC-MS/MS QToF. Food Additives & Contaminants Part A, Chemistry, Analysis, Control, Exposure & Risk Assessment 30: 813-825.

Rinaudo, M. 2006. Chitin and chitosan: properties and applications. Progress in Polymer Science 31: 603-632

Rosenberg, M. F., Velarde, G., Ford, R. C., Martin, C., Berridge, G., Kerr, I. D., Callaghan, R., Schmidlin, A., Wooding, C., Linton, K. J., and Higgins, C. F. 2001. Repacking of the transmembrane domains of P-glycoprotein during the transport ATPase cycle. The EMBO Journal 20: 5615-5625.

Saber, A., Strand, S. P., and Ulfendahl, M. 2010. Use of the biodegradable polymer chitosan as a vehicle for applying drugs to the inner ear. European Journal of Pharmaceutical Sciences 39: 110-115.

Sezer, A. D., Cevher, E., Hatipoglu, F., Ogurtan, Z., Bas, A. L., and Akbuga, J. 2008. Preparation of fucoidan-chitosan hydrogel and its application as burn healing accelerator on rabbits. Biological & Pharmaceutical Bulletin 31: 2326-2333.

Shono, Y., Nishihara, H., Matsuda, Y., Furukawa, S., Okada, N., Fujita, T., and Yamamoto, A. 2004. Modulation of intestinal P-glycoprotein function by cremophor EL and other surfactants by an in vitro diffusion chamber method using the isolated rat intestinal membranes. Journal of Pharmaceutical Sciences 93: 877-885.

Silva, T. I., Moreira, F. T., Truta, L. A., and Sales, M. G. 2012. Novel optical PVC probes for on-site detection/determination of fluoroquinolones in a solid/liquid interface: application to the determination of Norfloxacin in aquaculture water. Biosensors & Bioelectronics 36: 199-206.

Sinha, V. R., Singla, A. K., Wadhawan, S., Kaushik, R., Kumria, R., Bansal, K., and Dhawan, S. 2004. Chitosan microspheres as a potential carrier for drugs. International Journal of Pharmaceutic 274: 1-33.

Sousa, E., Palmeira, A., Cordeiro, A. S., Sarmento, B., Ferreira, D., Lima, R. T., Vasconcelos, M. H., and Pinto, M. 2013. Bioactive xanthones with effect on P-glycoprotein and prediction of intestinal absorption. Medicinal Chemistry Research 22: 2115-2123.

Streubel, A., Siepmann, J., and Bodmeier, R. 2006. Gastroretentive drug delivery systems. Expert Opinion on Drug Delivery 3: 217-233.

Tao, X., Chen, M., Jiang, H., Shen, J., Wang, Z., Wang, X., Wu, X., and Wen, K. 2013. Chemiluminescence competitive indirect enzyme immunoassay for 20 fluoroquinolone residues in fish and shrimp based on a single-chain variable fragment. Analytical and Bioanalytical Chemistry 405: 7477-7484.

Thomas, V., Ramasamy, K., Sundaram, R., and Chandrasekaran, A. 2007. Effect of honey on CYP3A4 enzyme and P-glycoprotein activity in healthy human volunteers. Iranian Journal of Pharmacology & Therapeutics (IJPT) 6: 171-176.

Wessler, J. D., Grip, L. T., Mendell, J., and Giugliano, R. P. 2013. The P-glycoprotein transport system and cardiovascular drugs. Journal of the American College of Cardiology 61: 2495-2502.

Yan, F., Si, L. Q., Huang, J. G., and Li, G. 2008. [Advances in the study of excipient inhibitors of intestinal P-glycoprotein]. Acta Pharmaceutica Sinica 43: 1071-1076.

163

Zhang, H., Chen, S., Lu, Y., and Dai, Z. 2010. Simultaneous determination of quinolones in fish by liquid chromatography coupled with fluorescence detection: comparison of sub-2 microm particles and conventional C18 columns. Journal of the American College of Cardiology 33: 1959-1967.

该文发表于《Journal of Aquatic Animal Health》【2014. 26: 59-65】

Relationship between permeability glycoprotein gene expression and enrofloxacin metabolism in *Oreochomis niloticus* Linn (Nile tilapia)

Kun Hu[1], Gang Cheng[2], Haixin Zhang[1], HuicongWang[1], Jiming Ruan[1],
Li Chen[1], Wenhong Fang[3], Xianle Yang[1*]

(1. National Pathogen Collection Center for Aquatic Animals, Shanghai Ocean University, 999 HuchengHuan Road, Shanghai 201306, China;

2. South-Central College for Nationalities, 708 Nationalities Road, Wuhan, Hubei 430074, China;

3. East China Sea Fisheries Research Institute, Chinese Academy of Fishery Sciences, 300 Jungong Road, Shanghai 200090, China)

Abstract: The aim of this study was to analyze the influence of permeability glycoprotein (P-gp) gene expression on enrofloxacin (ENR) metabolism in aquatic animals. Nile Tilapia Oreochomisniloticuswere fed different doses of ENR ranging from 0 to 80 mg/kg. The P-gp gene expression levels were determined by quantitative real-time PCR (qRTPCR) at indicated time points after drug administration. Drug metabolism was determined by HPLC. The P-gp gene expression in liver and kidney was greatly enhanced 30 min after ENR administration at 40 mg/kg, peaked 3 h after drug administration, and then gradually decreased. Thirty minutes after a single oral administration of ENR (0, 20, 40, or 80 mg/kg), the P-gp gene expression increased in a dose-dependent manner. The P-gp gene expression levels in the kidney were significantly higher than those in the liver. Additionally, the metabolic rate of ENR in kidney was more rapid than that in liver. Furthermore, a close correlation was found between P-gp gene expression and ENR concentrations. These results suggest that P-gp may be involved in the ENR metabolism process in Nile Tilapia, providing a novel model for the potential utility of gene expression and drug metabolism studies in aquatic animals.

尼罗罗非鱼体内 P 糖蛋白的表达与恩诺沙星代谢的关联性分析

胡鲲[1]，程钢[2]，章海鑫[1]，王会聪[1]，阮记明[1]，陈力[1]，房文红[3]，杨先乐[1*]

(1. 上海海洋大学国家水生动物病原库，中国上海沪城环路 999 号；2. 中南民族学院，中国武汉民族路 708 号；

3. 中国水产科学研究院东海水产研究所，中国上海军工路 300 号)

摘要：本研究主要分析了水产动物服用恩诺沙星后对糖蛋白渗透基因表达的影响。实验过程中，以尼罗罗非鱼为研究对象，恩诺沙星浓度范围为 0~80 mg/kg。糖蛋白渗透功能基因表达在药物处理之后在特定时间点进行实时定量 PCR 测定。药物代谢经 HPLC 检测。用药物浓度为 40 mg/kg 处理 30 min 后肝脏和肾脏内的糖蛋白渗透功能基因表达得到明显提升，在 3 h 时达到峰值，之后逐渐减低。口服单剂量恩诺沙星 30 min 后（浓度分别为 0，20，40，80 mg/kg），在剂量依赖性方式处理的糖蛋白渗透功能基因表达得到增加。肾脏中表达情况明显高于肝脏。此外，肾脏中恩诺沙星的代谢速率高于肝脏。研究还发现，P-糖蛋白功能基因表达跟恩诺沙星浓度有关。这些结果说明 P-糖蛋白功能基因可能参与尼罗罗非鱼中恩诺沙星代谢，同时也为研究水产动物基因表达和药物代谢研究的潜在效用提供了新的实验动物。

Permeability glycoprotein (P-gp), also known as transmem-brane glycoprotein, is widely expressed in normal cells, including cells that constitute the barrier, and is involved in metabolism in the liver, kidney, intestines, pancreas, and brain microvascular endothelia (Saekietal. 1993; Hemmeretal. 1995; Thomasetal. 2007; Lee et al. 2013; Sousa et al. 2013; Wessler et al. 2013).

This protein is present in certain tissue compartments, such as the gastrointestinal tract, and plays an essential role in preventing absorption and eliminating xenobiotics or preventing exposure of sensitive tissues to xenobiotic agents (Hochman et al. 2001). Moreover, P-gp acts as a transmembrane efflux pump and is a member of the adenosine triphosphate (ATP) -binding cassette (ABC) transporter family (Hughes 1994; Rosenberg et al. 2001). This efflux transporter has broad substrate specificity, including various structurally divergent drugs in clinical use today, and may reduce drug residues in tissues (Marzolini et al. 2004). In addition, P-gp can mediate renal tubular secretion of drugs (Ito et al. 1993).

The expression of P-gp has been detected in a variety of species, including Zebrafish Danio rerio (Bresolin et al. 2005), a killi-fish (Miller 1993), and a catfish (Kleinow et al. 2000), and has been proposed as a biomarker of pollution exposure. By using immunohistochemical analysis with a mammalian C219 monoclonal antibody, Kleinow and colleagues detected Pgp protein expression in the liver and intestines of a catfish (Kleinow et al. 2000; Doi et al. 2001). In addition, P-gp protein levels were measured seasonally in the gills of southeastern oysters Crassostrea virginica (Keppler and Ringwood 2001). Furthermore, P-gp in Zebrafish shares 49.4% and 49.7% nucleotide homology with that in human and mouse, respectively (Ma 2008). Similarly, a highly conserved sequence of the Pgp gene (over 78%) was found in major fish groups such as Chondrichthyes and Osteichthyes (Ma 2008). Nile tilapia Oreochomis niloticus, which belongs to Osteichthyes, is an important Chinese aquatic product export.

Enrofloxacin (ENR) is a fluoroquinolone antibiotic that has been widely applied to treat fish commercially (Dalsgaard and Bjerregaard 1991). However, use of the primary metabolite of ENR, ciprofloxacin (CIP), may increase the risk of developing drug resistance and chronic diseases (Hernandez et al. 2002). The European Community has established a maximum residue limit (MRL) of 100 μg/kg for both ENR and CIP in the edible tissues of fish (DOCE 1990). Although the metabolism of ENR and CIP in fish has been given sufficient attention to ensure the safety of human food production (Nouws et al. 1988; Bowser etal. 1992; Stoffregenetal. 1997; Intorreetal. 2000; dellaRocca et al. 2004; Fang et al. 2012; Liang et al. 2012; Wack et al. 2012; Zhang et al. 2012), the in vivo metabolism of ENR and its potential correlation with P-gp expression is still poorly understood.

In the present study, we investigated P-gp gene expression levels in the liver and kidney of Nile Tilapia after ENR administration. In addition, the potential relationship between residue amounts of ENR and P-gp gene expression levels in these tissues was explored. These results may help to better understand the mechanism of ENR disposition in aquatic animals.

Methods

Reagents. —The antibiotics ENR and CIP (content ≥ 98.5%) were purchased from Zhejiang Guobang Pharmaceutical, Zhejiang, China. Acetonitrile (HPLC grade), sodium sulfate (chemically pure), hydrochloric acid (chemically pure), n-hexane (chemically pure), and tetrabutylammonium bromide (chemically pure) were obtained from Shanghai Anpel Scientific Instrument, Shanghai, China.

Fish. —Nile Tilapia, weighing 50 ± 10.4 g (mean±SD), were provided by a farm in Nantong, Jiangsu, China. All fish were housed in fiberglass tanks (500 L) supplied with a constant flow (1L/h) of aerated freshwater for 7 d before the experiments. The water quality was monitored daily and adjusted as necessary; the pH was maintained between 6.4 and 7.0, while water temperature was adjusted to 25.0 ± 1.0 ℃ by means of cooling and heating devices. Fish (240 in total) were divided into four groups of 60 fish each. Enrofloxacin (10 mg) was treated with 5 mL of acetic acid and then diluted with 95 mL of distilled water to give a stock solution of 0.1 mg ENR/mL. Flexible tubes of approximately 15 cm in length were connected with the injectors (1 mL). Based on the weights of the individual fish, the appropriate amount of ENR solution was loaded into the injectors. By inserting the flexible tube through the esophagus into the fish's stomach carefully, each group of fish was treated with a different dose of ENR (0, 20, 40, or 80 mg/kg body weight) by oral gavage. The procedure was finished within a few seconds. The fish not treated with ENR served as the control group. Ten fish were sacrificed at each time point (1 and 30 min, and 3, 6, 12, and 48 h) after drug administration. All experiments were performed in accordance with the guidelines of the Regulation on Animal Experimentation and were approved by the State Scientific and Technological Commission.

Isolation of RNA and qRT-PCR. —At indicated time points after drug administration, fish were sacrificed, and the total RNA was extracted from liver or kidney tissues with a TRIzol RNA extraction kit (Invitrogen, Carlsbad, California) according to the manufacturer's protocol. To ensure the extracted RNA was integrated enough for the following experiments, the optical density (OD) ratios, OD_{260}/OD_{280}, of extracted total RNA were measured by a spectrophotometer (model DU800, Beckman Coulter, Indianapolis, Indiana.). The RNA purity was evaluated by 1% gel electrophoresis. Single-stranded complementary DNA (cDNA) was generated by reverse transcription using ReverTra Ace-α-TM (Toyobo, Osaka, Japan). Quantitative real-time PCR (qRT-PCR) was performed with 0.4 μL of cDNA, 0.2 μL of forward primer (10 μmol/L), 5 μL of reverse primer (10 μmol/L), and 5 μL of SYBR-Green PCR Master Mix (QPK-201; Toyobo) in a Bio-Rad iQ5 real-time PCR instrument (Hercules, California). The PCR primers were designed by Primer Premier 5.0 software (PREMIER Biosoft, Palo Alto, California) as follows: for the P-gp gene, P-gp forward (F) (5CACCTGGACGTTACCAAAGAAGATATA-3) and P-gp reverse (R) (5-TCACCAACCAGCGTCTCATATTT-3); for the β-actin gene, β-actin F (5-CCTCACCCTCAAGTACCCCAT3) and β-actin R (5-TTGGCCTTTGGGTTGAGTG-3).

The β-actin gene was used as a control for the analysis. The PCR for P-gp was performed in a final reaction volume of 10 μL using the following conditions: a preheating cycle at 94℃ for 3 min, followed by 35 cycles at 94℃ for 30 s, 51.5℃ for 30 s, and 72℃ for 30 s, and finally elongation at 72℃ for 10 min. The PCR conditions for β-actin were similar, except that the annealing temperature was 49.6℃. The amplified products were identified by agarose gel electrophoresis. The amount of P-gp messenger RNA (mRNA) was normalized to β-actin expression. Data were quantified from three independent experiments for each group.

The qRT-PCR assays were run in triplicate for each cDNA sample, and melting curves were drawn to monitor the quality of amplicons and reactions. Amplification products were quantified by comparison to experimental threshold cycle (Ct) levels (Ct is defined as the PCR cycle where an increase in fluorescence over the background level first occurs). A standard curve was generated for each primer pair based on known quantities of cDNA (10-fold serial dilutions corresponding to cDNA transcribed from 100 to 0.01 ng of total RNA). The relative mRNA expression was calculated using the 2^{Ct} equation and β-actin was used as an internal control for the PCRs.

Evaluation of tissue drug concentrations. —Tissue drug concentrations were determined by HPLC. Tissue samples were extracted with acetonitrile. After centrifugation at 4,500 rpm for 5 min, the supernatants were carefully collected and completely mixed with n-hexane. The aqueous phase of the mixture was removed and the organic phase was evaporated using a rotary evaporator (Heidolph Instruments, Schwabach, Germany). The residues were dissolved in the mobile phase (acetonitrile : 0.01 M tetrabutyl ammonium bromide, pH 3.1 [5:95 v/v]), and the solution was then filtered through a 0.45-μm filter. An HPLC (Agilent, Santa Clara, California) instrument equipped with a fluorescence detector and a reversed-phase C18 column (150 × 4.6 mm) was used at a flow rate of 1.5 mL/min and a column temperature of 40℃. The detection wavelength was set at 280 nm, and 10 μL of sample solution was loaded. Standard solutions of ENR and CIP (0.2~10 μg/mL) were prepared, and standard curves were drawn with the drug concentration (μg/mL) on the x-axis and peak area on the y-axis. Drug concentrations in tissue samples were determined by the external standard method. Data were calculated from five independent experiments for each group. To certify the validation of the HPLC method, ENR and CIP standard solutions (5 mg/L) were eachaddedintoblanktissuesamples. Therecoveriesandrelative SD-values were determined in the intraday (five replicates in 1 d) and the interday (5 d). The main pharmacokinetic parameters such as the peak concentration (C_{max}) and the time for the drug to reach the peak level (T_{max}) were examined. In addition, the area under the curve (AUC_{0-t}) was calculated by the computer program Drug and Statistics (DAS version 3.0; Center for Drug Clinical Evaluation, Wannan Medical College, China).

Statistical analysis. —Statistical analyses were performed with SPSS version 16.0 software (SPSS, Chicago, Illinois) and data were presented as the mean±SD. Significant differences between means were determined using one-way ANOVA and the Duncan's multiple range test for comparison with significance level established at $P < 0.05$.

Results

P-gp Gene Expression Levels in Liver and Kidney Tissues after ENR Administration

In the present study, P-gp fragments (127 bp) were successfully amplified in both liver and kidney of Nile Tilapia by using specific primers (data not shown). The baseline P-gp gene expression levels in liver or kidney of Nile Tilapia were relatively low. However, administration of 40 mg/kg ENR significantly enhanced the P-gp gene expression levels 30 min after drug administration ($P < 0.05$) (Figure 1a, b). In addition, the P-gp gene expression levels in liver and kidney both peaked 3 h after drug administration and then gradually decreased (Figure 1a, b). Furthermore, 3 h after a single oral administration of ENR (0, 20, 40, or 80 mg/kg), the P-gp gene expression levels increased in a dose-dependent manner (Figure 2a, b). Moreover, the P-gp gene expression levels in the kidney were significantly higher than those in the liver ($P < 0.05$).

Drug Concentration in Liver and Kidney Tissues

Tissue concentrations of ENR and CIP were determined by HPLC analysis. The peaks of ENR and CIP could be separated properly with a stable baseline. The retention times for ENR and CIP were 5.50 min and 6.38 min, respectively, and the standard curves were $y = 5.524x - 0.135$ ($r^2 = 0.998$) and $y = 3.550x - 0.226$ ($r^2 = 0.999$), respectively, within a concentration range of 0.2~10 μg/mL. By using the external standard method, ENR and CIP residue concentrations were determined with a limit of detection of 1 μg/kg. The average recoveries were 70.01%~80.52% and the relative SD-values were 1.20%~7.92% for ENR, while the recoveries were 65.01%~83.43% and SDs were 2.63%~5.91% for CIP (Table 1). After a single oral administration of ENR at different dosages (20, 40, or 80 mg/kg), the concentrations of ENR and CIP in liver and kidney were examined. The residues in the two tissues were fitted to the open two-compartment model using DAS pharmacokinetic software. The pharmacokinetic parameters derived from the compartment model are shown in Tables 2 and 3. Thedistribution and elimination half-lives ($T_{1/2\alpha}$ and $T_{1/2\beta}$), total body clearance (Cl), and apparent volume of distribution (V_d) of ENR in Nile Tilapia after a single oral dose (40 mg/kg) were 0.14 h, 19.19 h, 636.44 kg·h^{-1}·kg^{-1}, and 4,012.21 kg/kg in liver, while the values were 0.36 h, 40.12 h, 213.82 kg·h^{-1}·kg^{-1}, and

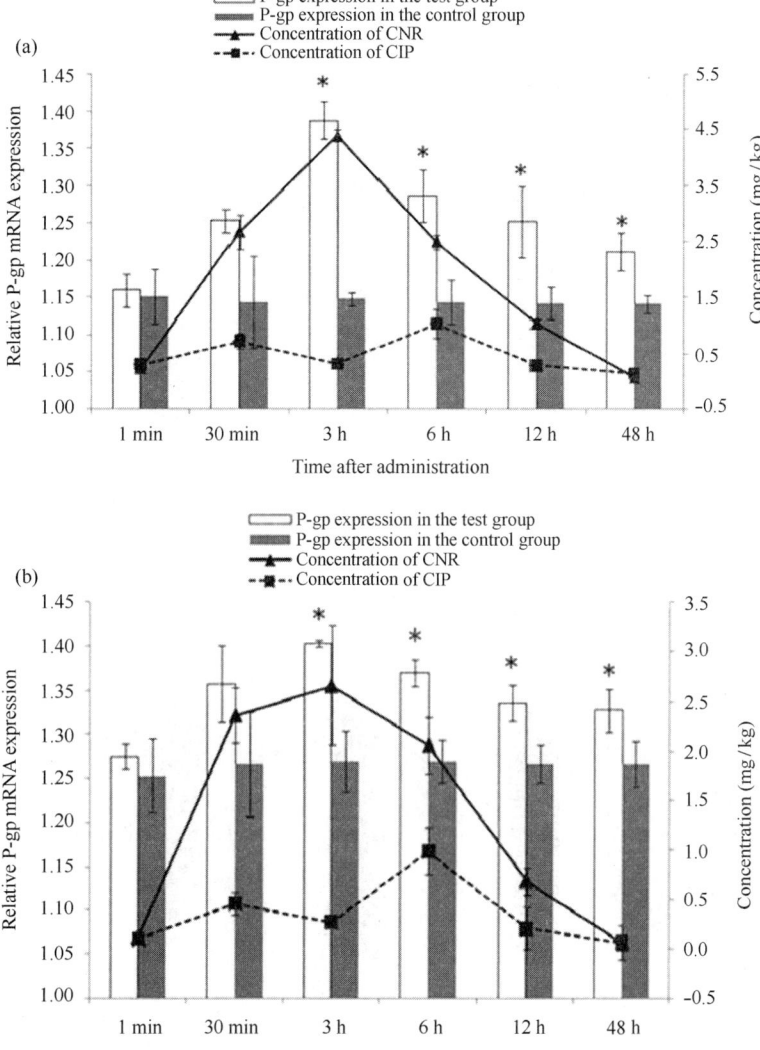

Fig. 1　Quantification of P-gp gene expression and ENR residue amounts in (upper) liver and (lower) kidney of Nile Tilapia at indicated time points (0 min [control], 30 min, 3 h, 6 h, 12 h, or 48 h) after a single oral administration of ENR (40 mg/kg). An asterisk indicates a significant difference (P<0.05) compared with control.

3, 372. 12 kg/kg in kidney. The distribution and elimination halflives, total body clearance, and apparent volume of distribution of CIP showed the similar tendency in liver and kidney. The metabolic rate of ENR and CIP at different concentrations in liver was more rapid than that in kidney. In addition, the C_{max} of ENR in Nile Tilapia after single oral doses of 20, 40, or 80 mg/kg were 1. 97, 4. 47, and 39. 44 mg/kg, respectively, in liver, and the respective values in kidney were 0. 93, 2. 56, and 28. 38 mg/kg. The C_{max} of CIP residues in liver were 0. 84, 1. 33, and 2. 80 mg/kg and in kidney were 0. 63, 0. 88, and 2. 38 mg/kg. The metabolism of ENR and CIP in the liver and kidney gradually increased in a dose-dependent pattern.

　Relationship between P-gp Expression and ENR Metabolism

　　As mentioned above, the P-gp gene expression levels in liver and kidney were significantly increased 30 min after ENR administration of 40 mg/kg; they peaked at 3 h and then gradually decreased. Normalized to β-actin, the P-gp gene expression maximum could reach 1. 38±0. 02 in liver and 1. 40±0. 01 in kidney (Figure 1a, b). A similar trend was detected for ENR residues in the kidney, whereas ENR residues in the liver peaked 3 h after drug application, which was slightly earlier than in the kidney (Figure 1a, b). At 3 h after drug application, the maximum concentration of ENR reached 4. 39±0. 10 mg/kg in liver and 2. 65±0. 61 mg/kg in kidney, while the concentrations of CIP in liver and kidney were 0. 31±0. 08 mg/kg and 0. 27±0. 04 mg/kg, respectively.

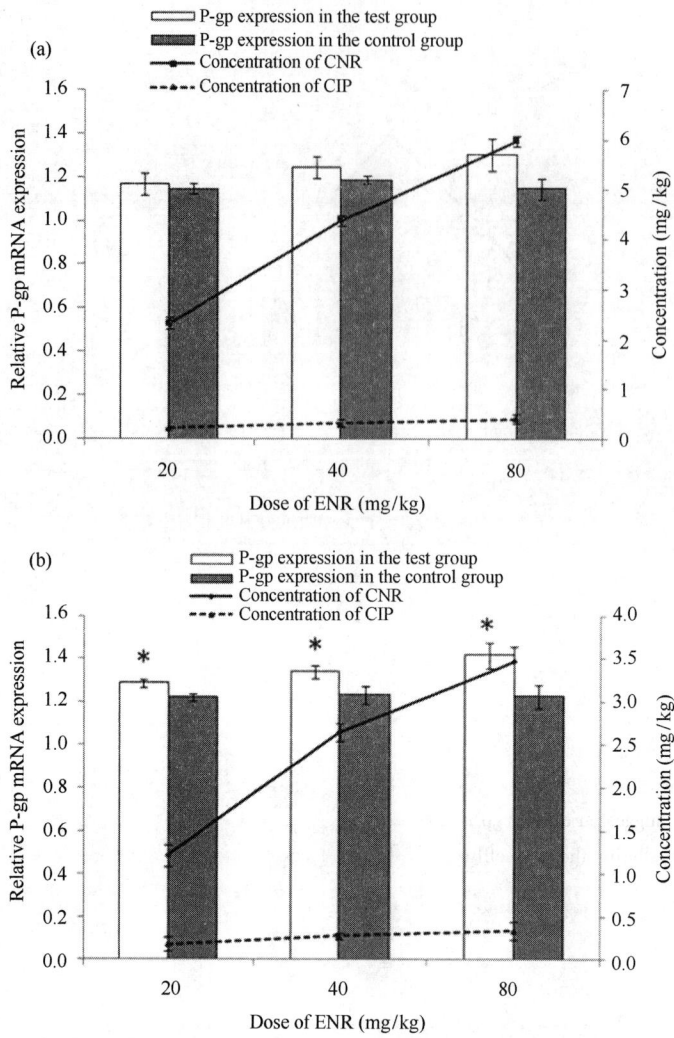

Fig. 2   Quantification of P-gp gene expression in (upper) liver and (lower) kidney of Nile Tilapia 3 h after a single oral administration of ENR (0 [control], 20, 40, or 80 mg/kg). An asterisk (∗) indicates a significant difference (P<0. 05) compared with control.

Table 1   Recoveries (mean±SD) of ENR and CIP from the liver and kidney of Nile Tilapia. The values were the average of five batches of samples (n=5) treated with ENR or CIP at concentrations of 5 mg/L.

Tissue		Recovery (%)		Relative SD (%)	
		Intraday	Interday	Intraday	Interday
Liver	ENR	72. 76±1. 63	70. 01±1. 43	5. 67	7. 92
	CIP	65. 01±4. 93	70. 27±1. 79	4. 52	5. 91
Kidney	ENR	80. 52±1. 03	78. 92±0. 73	1. 20	4. 93
	CIP	83. 43±3. 27	78. 23±2. 03	2. 63	3. 73

Table 2　Main pharmacokinetic parameters of ENR in Nile Tilapia after a single oral dose（0，20，40，or 80 mg/kg）.

Pharmacokinetic Parameter	0 mg/kg		20 mg/kg		40 mg/kg		80 mg/kg	
	Liver	Kidney	Liver	Kidney	Liver	Kidney	Liver	Kidney
C_{max}（mg/kg）	ND	ND	1.97	0.93	4.47	2.56	39.44	28.38
T_{max}（h）	ND	ND	2.11	2.89	2.00	3.02	2.33	3.12
$T_{1/2\alpha}$（h）	–	–	0.10	0.25	0.14	0.36	0.45	0.61
$T_{1/2\beta}$（h）	–	–	8.22	19.82	19.19	40.12	39.03	54.33
AUC_{0-240}（mg/h/kg）	–	–	30.32	49.34	70.33	163.22	180.32	282.98
$AUC_{0-\infty}$（mg/h/kg）	–	–	31.11	52.34	77.85	186.56	193.32	302.31
Cl（kg/h/kg）	–	–	1014.89	462.32	636.44	213.82	129.83	107.62
V_d（g/g）	–	–	5680.82	4972.12	4012.21	3372.12	2687.21	1883.76

C_{max}: maximum concentration；T_{max}: peak time；$T_{1/2\alpha}$: distribution half-life of the drug；$T_{1/2\beta}$: elimination half-life of the drug；AUC: area under the curve；Cl: total body clearance of the drug；V_d: extensive apparent volume of the central compartment；ND: not detected.

Discussion

Permeability glycoprotein, a plasma membrane ATP-binding cassette transporter, is responsible for the elimination of drug components in vivo, and the P-gp level in aquatic animals is associated with the extent of chemical contamination（Bard et al. 2002；Thomas et al. 2007）. Using immunohistochemical analysis, Kleinow et al.（2000）investigated the involvement of the P-gp in the liver and intestines during xenobiotic absorption and excretion in catfish. In addition, Bard et al.（2002）demonstrated that exposure to anthropogenic drugs and naturally occurring toxins led to enhanced P-gp gene expression in the intertidal fish, *Anoplarchuspurpurescens*. Hochman et al.（2001）suggested that P-gp may share similar functions with the cytochrome P450 enzyme. In mammals, P-gp regulates drug metabolism by mediating the drug efflux process and thus influencing in vivo drug bioavailablility（Lownetal. 1997；Marzolinietal. 2004；Thomas et al. 2007）. By comparing the main pharmacokinetic parameters, such as C_{max}, T_{max}, and $AUC_{(0-t)}$, Thomas et al.（2007）indicated that P-gp activity in human blood could be modulated by grapefruit juice and herbal drugs. In addition, Lown et al.（1997）measured the pharmacokinetics of orally administered cyclosporine on stable kidney transplant recipients and found that intestinal P-gp played a significant role in the firstpass elimination of cyclosporine. With regard to aquatic animals such as fish, P-gp was activated along with cytochrome P450 when animals were exposed to environmental contaminants, and their activities were related to drug resistance in fish（Bard et al. 2002）. A recent study reported that cytochrome P4501A participates in the metabolism of difloxacin in hybridized Crucian Carp

Carassius carassius（Fu et al. 2010）. Nonetheless, the effect of P-gp in the drug metabolism of aquatic animals is still poorly understood. The liver and kidneys are vital organs that serve critical roles for drug accumulation and efflux. The drug residue in liver and kidney could give rise to the changes of some protein involved in the metabolism. In the present study, we found that P-gp gene expression in liver and kidney was greatly enhanced 30 min after 40 mg/kg ENR administration, peaked 3 h after drug feeding, and then gradually decreased, a trend that was positively correlated with the amount of ENR and CIP residue in these organs（Figure 1）. Moreover, the upregulation of P-gp gene expression levels in the liver and kidneys depended on the dose of ENR（Figure 2）. The bioavailability（C_{max}, T_{max}, and AUC）of ENR and the residue of CIP seemed relative with the P-gpgene expression. Furthermore theoral dosage ofENR has a strong influence on the elimination efficiency（$T_{1/2\beta}$, Cl, and V_d）and the gene expression of P-gp. These observation ssuggest that P-gp gene expression may correlate with the ENR metabolism process in the liver and kidneys of Nile Tilapia. It is possible that P-gp alters the distribution of ENR in organs（Kemper et al. 2004）. As a similar example, difloxacin（a chemical agent of the fluoroquinone family）is confirmed as the substrate of CYP1A in hybridized Prussian carp（Fu et al. 2010）. On the other hand, P-gp may accelerate the efflux process of drugs（Pavek et al. 2003）. Permeability glycoprotein-modulated ENR elimination could serve two functions：（1）direct clearance of ENR so that systemic exposure to ENR is minimized, and（2）removal of ENR from the cytosol, preventing the accumulation of high concentrations of ENR within cells（Hochman et al. 2001）.

In summary, our study demonstrates that P-gp gene may be involved in the ENR metabolism process in Nile Tilapia while ENR may be a substrate for P-gp. Furthermore, CIP residue may be indicated by P-gp gene expression. It offers a novel strategy for promoting drug bioavailability and reducing the risks of drug residues in animals.

Acknowlededgments

This study was supported by the 863 Program (Grant 2011AA10A216), the Special Fund for Agro-scientific Research in the Public Interest (Grant 201203085), Shanghai University Knowledge Service Platform, and the Open Fund of the Key Laboratory of Marine and Estuarine Fisheries Resources and Ecology, Ministry of Agriculture (Grant Kai-09-06). There is no conflict of interest related to this manuscript.

Reference

B ard, S. M., B. R. Woodin, and J. J. Stegeman. 2002. Expression of Pglycoprotein and cytochrome p450 1A in intertidal fish (*Anoplarchus purpurescens*) exposed to environmental contaminants. Aquatic Toxicology 60: 17-32.

Bowser, P. R., G. A. Wooster, J. St Leger, and J. G. Babish. 1992. Pharmacokinetics of enrofloxacin in fingerling rainbow trout (*Oncorhynchus mykiss*).

Journal of Veterinary Pharmacology and Therapeutics 15: 62-71.

Bresolin, T., M. de Freitas Rebelo, and A. Celso Dias Bainy. 2005. Expression of PXR, CYP3A and MDR1 genes in liver of Zebrafish. Comparative Biochemistry and Physiology C 140: 403-407.

Dalsgaard, I., and J. Bjerregaard. 1991. Enrofloxacin as an antibiotic in fish. Acta Veterinaria Scandinavica Supplementum 87: 300-302.

della Rocca, G., A. Di Salvo, J. Malvisi, and M. Sello. 2004. The disposition of enrofloxacinin Seabream (SparusaurataL.) after single intravenousinjection or from medicated feed administration. Aquaculture 232: 53-62.

DOCE (DiarioOficial de las ComunidadesEuropeas). 1990. Por el que se estable ceunproce dimientocomunitario de fijacion de loslimites m aximos de residuos de *medicamentosveterinariosenlosalimentos* de origen animal. [Laying down a community procedure for the establishment of maximum residue limits of veterinary products in feed of animal origin.] DOCE Regulaion 2377/90 (18 August 1990): L224.

Doi, A. M., E. Holmes, and K. M. Kleinow. 2001. P-glycoprotein in the catfish intestine: inducibility by xenobiotics and functional properties. Aquatic Toxicology 55: 157-170.

Fang, X., X. Liu, W. Liu, and C. Lu. 2012. Pharmacokinetics of enrofloxacin in allogynogenetic Silver Crucian Carp, Carassius auratus gibelio. Journal of Veterinary Pharmacology and Therapeutics 35: 397-401.

Fu, G. H., X. L. Yang, H. X. Zhang, W. J. Yu, and K. Hu. 2010. Effects of cytochrome P450 1A substrate (difloxacin) on enzyme gene expression and pharmacokinetics in Crucian Carp (hybridized Prussian Carp). Environmental Toxicology and Pharmacology 31: 307-313.

Hemmer, M. J., L. A. Courtney, and L. S. Ortego. 1995. Immunohistochemical detection of P-glycoprotein in teleost tissues using mammalian polyclonal and monoclonal antibodies. Journal of Experimental Zoology 272: 69-77.

Hernandez, M., C. Aguilar, F. Borrull, and M. Calull. 2002. Determination of ciprofloxacin, enrofloxacinandflumequineinpigplasmasamplesbycapillary isotachophoresis-capillary zone electrophoresis. Journal of Chromatography B 772: 163-172.

Hochman, J. H., M. Chiba, M. Yamazaki, C. Tang, and J. H. Lin. 2001. Pglycoprotein-mediated efflux of indinavir metabolites in Caco-2 cells expressing cytochrome P450 3A4. Journal of Pharmacology and Experimental Therapeutics 298: 323-330.

Hughes, A. L. 1994. Evolution of the ATP-binding-cassette transmembrane transporters of vertebrates. Molecular Biology and Evolution 11: 899-910.

Intorre, L., S. Cecchini, S. Bertini, A. M. CognettiVarriale, G. Soldani, and G. Mengozzi. 2000. Pharmacokinetics of enrofloxacin in the seabass (Dicentrarchuslabrax). Aquaculture 182: 49-59.

Ito, S., C. Woodland, P. A. Harper, and G. Koren. 1993. P-glycoprotein-mediated renal tubular secretion of digoxin: the toxicological significance of the urineblood barrier model. Life Sciences 53: 25-31.

Kemper, E. M., W. Boogerd, I. Thuis, J. H. Beijnen, and O. van Tellingen. 2004. Modulation of the blood-brain barrier in oncology: therapeutic opportunities for the treatment of brain tumours? Cancer Treatment Reviews 30: 415-423.

Keppler, C., and A. H. Ringwood. 2001. Expression of P-glycoprotein in the gills of oysters, Crassostrea vir ginica: seasonal and pollutant related effects. Aquatic Toxicology 54: 195-204.

Kleinow, K. M., A. M. Doi, and A. A. Smith. 2000. Distribution and inducibility of P-glycoprotein in the catfish: immunohistochemical detection using the mammalian C-219 monoclonal. Marine Environmental Research 50: 313-317.

Lee, S. D., T-Osei, A. -Jo, and K. M. Wasan. 2013. Dose-dependent targeted suppression of P-glyco protein expression and function in Caco-2cells. Molecular Pharmaceutics 10: 2323-2330.

Liang, J. P., J. Li, F. Z. Zhao, P. Liu, and Z. Q. Chang. 2012. Pharmacokinetics and tissue behavior of enrofloxacin and its metabolite ciprofloxacin in Turbot Scophthalmus maximus at two water temperatures. Chinese Journal of Oceanology and Limnology 30: 644-653.

Lown, K. S., R. R. Mayo, A. B. Leichtman, H. L. Hsiao, D. K. Turgeon, P. Schmiedlin-Ren, M. B. Brown, W. Guo, S. J. Rossi, L. Z. Benet, and P. B. Watkins. 1997. Role of intestinal P-glycoprotein (mdr1) in interpatient variation in the oral bioavailability of cyclosporine. Clinical Pharmacology and Therapeutics 62: 248-260.

Ma, Y. F. 2008. cDNA cloning and expression analysis of P-glycoprotein in Zebrafish. Master's thesis. Southwest University, Chongqing, China.

Marzolini, C., E. Paus, T. Buclin, and R. B. Kim. 2004. Polymorphisms in human MDR1 (P-glycoprotein): recent advances and clinical relevance. Clinical Pharmacology and Therapeutics 75: 13-33.

Miller, D. S. 1993. Daunomycin secretion by killifish renal proximal tubules. American Journal of Physiology 269: R370-R379.

Nouws, J. F. , J. L. Grondel, A. R. Schutte, and J. Laurensen. 1988. Pharmacokinetics of ciprofloxacin in carp, African catfish and Rainbow Trout. Veterinary Quarterly 10: 211-216.

Pavek, P. , F. Staud, Z. Fendrich, H. Sklenarova, A. Libra, M. Novotna, M. Kopecky, M. Nobilis, and V. Semecky. 2003. Examination of the functional activity of P-glycoprotein in the rat placental barrier using rhodamine 123. Journal of Pharmacology and Experimental Therapeutics 305: 1239-1250.

Rosenberg, M. F. , G. Velarde, R. C. Ford, C. Martin, G. Berridge, I. D. Kerr, R. Callaghan, A. Schmidlin, C. Wooding, K. J. Linton, and C. F. Higgins. 2001. Repacking of the transmembrane domains of P-glycoprotein during the transport ATPase cycle. EMBO Journal 20: 5615-5625.

Saeki, T. , K. Ueda, Y. Tanigawara, R. Hori, and T. Komano. 1993. Human P-glycoprotein transports cyclosporin A and FK506. Journal of Biological Chemistry 268: 6077-6080.

Sousa, E. , A. Palmeira, A. S. Cordeiro, B. Sarmento, D. Ferreira, R. T. Lima, M. H. Vasconcelos, and M. Pinto. 2013. Bioactive xanthones with effect on P-glycoprotein and prediction of intestinal absorption. Medicinal Chemistry Research 22: 2115-2123.

Stoffregen, D. A. , G. A. Wooster, P. S. Bustos, P. R. Bowser, and J. G. Babish. 1997. Multiple route and dose pharmacokinetics of enrofloxacin in juvenile Atlantic Salmon. Journal of Veterinary Pharmacology and Therapeutics 20: 111-123.

Thomas, V. , K. Ramasamy, R. Sundaram, and A. Chandrasekaran. 2007. Effect of honey on CYP3A4 enzyme and P-glycoprotein activity in healthy human volunteers. Iranian Journal of Pharmacology and Therapeutics 6: 171-176.

Wack, A. N. , B. KuKanich, E. Bronson, M. Mary Denver. 2012. Pharmacokinetics of enrofloxacin after single dose oral and intravenous administration in the African penguin (Spheniscusdemersus) . Journal of Zoo and Wildlife Medicine 43: 309-316.

Wessler, J. D. , L. T. Grip, J. Mendell, and R. P. Giugliano. 2013. The Pglycoprotein transport system and cardiovascular drugs. Journal of the American College of Cardiology 61: 2495-502.

Zhang, Y. Y. , Y. Q. Huang, F. L. Zhai, R. Du, Y. D. Liu, and K. Q. Lai. 2012. Analyses of enrofloxacin, furazolidone and malachite green in fish products with surface-enhanced Raman spectroscopy. Food Chemistry 135: 845-850.

该文发表于《Environmental Toxicology and Pharmacology》【2014，37（2）：529-535】

Tissue accumulation and toxicity of isothiazolinone in *Ctenopharyngodon idellus* (grass carp): association with P-glycoprotein expression and location within tissues

Kun Hu[a,b,c], Hao-Ran Li[a], Ren-Jian Ou[a], Chun-Zeng Li[a], Xian-Le Yang[a,*]

(National Pathogen Collection Center for Aquatic Animals, Shanghai Ocean University, Shanghai 201306, China)

Abstract: Isothiazolinone is widely used as a broad-spectrum fungicide in various industries, such as oil, paper, pesticide, dyes, tanning and cosmetics. There is an increasing concern over protection of the aquatic environment due to its large-scale use. The acute toxicity (LC_{50}) of isothiazolinone in *Ctenopharyngodon idellus* was investigated. The residual time and accumulation in tissues, P-glycoprotein mRNA level and localization of P-glycoprotein in the liver and kidney were also analyzed. The LC_{50} (48 h) values of isothiazolinone to C. idelluswere 0.53 ± 0.17 mg/L and 0.41 ± 0.08 mg/L at 15℃ and 25℃, respectively. The LC_{50} values decreased as the temperature increased. The accumulation of isothiazolinone in livers and kidneys in the high temperature group (25℃) was significantly greater than that of the low temperature group (15℃). Prolonged tissue residual time of isothiazolinone was seen in all the groups. There were significant differences in P-glycoprotein mRNA expression between isothiazolinone-treated groups and control samples ($P<0.05-0.01$). Temperature affected accumulation and toxicity of isothiazolinone.

Key words: Isothiazolinone; *Ctenopharyngodon idellus*; Toxicity; P-glycoprotein; Expression; Location

草鱼组织内异噻唑啉毒素积累与 P-糖蛋白表达和 其组织内分布位置关系

胡鲲，李浩然，欧仁建，李春震，杨先乐

（上海海洋大学水产与生命学院国家水生动物病原库；中国上海）

摘要：在不同行业中异噻唑啉是经常使用的广谱杀菌剂，比如石油、造纸、杀虫剂、染料、制革、化妆品等行业。随着其在水产业中的大范围使用，人们越来越关注异噻唑啉水产环境的危害。试验中调查了草鱼的急性中毒半致死浓度。实验研究了药物在组织中的残留时间和积累，以及 p-糖蛋白在肝脏和肾脏中 mRNA 表达水平和位置。48 小时半致死浓度在 15℃和 25℃时分别为 0.53±0.17 mg/L 和 0.41±0.08 mg/L。半致死浓度值随着温度升高而降低。异噻唑啉在高温（25℃）组中的积累明显高于低温（15℃）组。在所有组中均存在延长组织残留时间。在实验组和对照组的样品对比中 P-糖蛋白 mRNA 表达方面具有显著差异（P< 0.05-0.01）。温度对异噻唑啉的积累和毒性具有影响。

关键词：异噻唑啉酮；草鱼；毒性；P-糖蛋白；表达；定位

1 Introduction

Isothiazolinone, a broad-spectrum industrial fungicide, has significant killing and peeling effects against fungi, bacteria, algae, mollusks, and slimes, and is widely used in industries such as oil, paper, pesticide, leather, ink, dyes, tanning and cosmetics (Hunziker, 1992). There is increasing concern over its potential effects on health and the environment, due to its wide use. For example, isothiazolinone is used as an agent for washing dandruff and is contained in many daily consumer products (such as shampoo and shower gel). Isothiazolinone has long been used for biological control of algae and sludge attached to sea cages or shipping vessels.

As isothiazolinone contained industrial and domestic wastewaters are discharged into the natural environments, the potential toxicity of isothiazolinone to fishes and other living organisms in the aquatic environment is of concern posing a major problem for environment safety. To date, the safety evaluation studies of isothiazolinone are mostly focused on mammals (Pazzaglia et al., 1996, Stejskal et al., 1990 and Hunziker et al., 1992), but its toxicity on fish and other aquatic organisms is largely unknown (Pereira et al., 2001).

Ctenopharyngodon idellus is one of the most common freshwater fish species worldwide. In China, C. *idellus* has the widest distribution among the farmed aquaculture species, including a variety of farming climates (Kucharczyk et al., 2008). Therefore, C. *idellus* is often used as a model for studies of water pollution. The present study was designed to determine the toxic effects of isothiazolinone on C. *idellus*, the underlying mechanisms of toxicity, the effects of environmental factors (i.e. water temperature) toxicity, and the value of using C. *idellus* as an environmental monitoring system.

P-glycoprotein (P-gp), a family member of ATP-binding cassette (ABC) membrane transporter family, is a transmembrane protein present in the liver and other organs (Agarwal et al., 2013). P-gp can pump out exogenous compounds, reducing their intracellular accumulation and thus protecting the cell from damage. Therefore, P-gp genetic/genomic variations may result in increased toxicity of exogenous compounds. P-gp is expressed in zebrafishes, killifishes, catfishes, and other aquatic organisms (He et al., 2012, Bard and Gadbois, 2007, Tan et al., 2010 and Bard et al., 2002) and can be used as biomarker for the toxicity of environmental pollutants. In the present study, C. *idellus* was used to determine the acute toxicity of isothiazolinone and the mechanism of accumulation in the liver and kidney at different temperatures. The mRNA and protein levels of P-gp were measured in order to evaluate the linkage between P-gp and the accumulation of isothiazolinone. It would be applied to assess the safety of isothiazolinone in the aquatic environment.

2 Materials and methods

2.1 Fish, chemicals and reagents

This study was approved by the institutional review board or ethics committee of Shanghai Ocean University (Permit Number: 2013022). All experiments were conducted following guidelines approved by the Shanghai Ocean University Committee on the Use and Care of Animals.

C. idellus, weighing 50 ± 10.4 g, were purchased from Nantong Farm Fisheries Management, Jiangsu, China. All fish were housed in a pond (10 mus) and were healthy, pathogen-free and never treated with any chemicals or drugs. The TRIzol RNA extraction kit was purchased from Invitrogen (Carlsbad, CA, USA). ReverTra Ace-α-TM was purchased from Toyobo Co., Ltd. (Osaka, Japan).

2.2　Acute toxicity

The *C. idellus* were cultivated in tanks (4 m $\times 2$ m $\times 1$ m) containing fresh water for three days and fasted for four days before testing. Water quality was monitored daily and the pH was maintained between 6.4 and 7.0. The *C. idellus* were divided two groups and cultivated at two different temperatures (15℃ and 25℃). Five concentrations of isothiazolinone were used in the cultivation water, including the minimum concentration that resulted in 100% deaths (0.90 mg/L), the highest concentration that resulted no death (0.05 mg/L), and three concentrations (0.15 mg/L, 0.30 mg/L, 0.60 mg/L) between the two (Steinbach et al., 2013). There were three groups (8 fish/group) for every concentration at 15℃ (low temperature) and 25℃ (high temperature). The blank control group (no isothiazolinone treatment) was included. Within 8 h of exposure to isothiazolinone, the fish were observed continuously to record their reactions to the toxin. After a 48-h exposure to isothiazolinone, the numbers of survival and deaths were recorded in each group to calculate the median lethal concentration (LC_{50}). If a dead fish was found, it was removed in a timely manner. The disposal of all experimental fishes followed the "Regulation on Animal Experimentation" (State Science and Technology Commission of China).

2.3　Residual time and accumulation in tissues

Isothiazolinone was dissolved in the cultivating water at a final concentration of 0.2 mg/L. After being exposed to isothiazolinone for 48 h or 96 h, the fish were sacrificed and samples of liver and kidneys were taken. To extract isothiazolinone from the tissue samples, 0.5 g of each tissue was homogenized and mixed with 5 mL of methanol. The mixture was centrifuged at 4500 rpm for 5 min; the supernatant was collected. The extraction procedure was repeated once and the supernatants were combined. The resultant supernatants were then mixed with n-hexane and thoroughly shaken. After sitting and the two layers of solvents were clearly separated, the upper portion was discarded and the lower portion was evaporated in a rotary evaporator until dry.

Drug analysis was accomplished using high performance liquid chromatography (HPLC). The above residues were dissolved in the mobile phase and filtered through a 0.45-μm microporous membrane before being injected onto the HPLC column. Chromatographic conditions were as follows: reversed-phase C18 column (4.6 mm × 150 mm); mobile phase, methanol/water = 25/75; flow rate, 1.0 mL/min; UV detection wavelength, 276 nm; column temperature, 30℃; injection volume, 10 μL. The standard solutions of isothiazolinone (0.5~10 mg/L) were prepared, and the standard curves were drawn with the drug concentrations (mg/L) on the x-axis and the peak areas on the y-axis. The contents of isothiazolinone in livers and kidneys were calculated using the external standard curves.

2.4　Quantitative PCR

At the indicated time points after drug administration, fishes were sacrificed, and the total RNA was extracted from liver and kidney samples with a TRIzol RNA extraction kit according to the manufacturer's protocol. The quality and quantity of the extracted RNA were measured by the OD_{260}/OD_{280} ratios using a spectrophotometer (DU800, BECKMEN, USA). The RNA purity was evaluated by 1% gel electrophoresis. Single-stranded cDNA was generated by reverse transcription using ReverTra Ace-α-TM. Real-time polymerase chain reaction (RT-PCR) was performed with a mixture of cDNA (0.4 μL), forward primer (0.2 μL; 10 μmol/L), reverse primer (5 μL; 10 μmol/L), and SYBR© Green PCR Master Mix (5 μL; QPK-201, Toyobo Co., Ltd., Japan), in a Bio-Rad iQ5 real-time PCR instrument (Hercules, CA, USA). β-actin was used as an internal control for the RT-PCRs. The PCR primers were designed using Primer Premier 5.0 software. The primers, NCBI accession numbers of sequences for *C. idellus*, and annealing temperatures are listed in Table 1. The PCR for P-gp was performed in a final reaction volume of 10 μL using the following conditions: a preheating cycle at 94℃ for 3 min, followed by 35 cycles at 94℃ for 30 s, 51.5℃ for 30 s, and 72℃ for 30 s, and finally elongation at 72℃ for 10 min. The PCR conditions for β-actin were similar, except that the annealing temperature was 49.6℃. The amplified products were identified by 1% agarose gel electrophoresis and were subsequently purified by a Midi Purification Kit (Biomiga Diagnostics, Agoura, CA, USA). The amount of P-gp mRNA was normalized to β-actin mRNA expression. Three independent experiments were carried out for this assay.

Table 1   RT-PCR primers used for assessing gene expression in different tissues.

Gene	Primers	Accession number	Annealing temperature（℃）
P-gp	Forward：5′-CACCTGGACGTTACCAAAGAAGATATA-3′ Reverse：5′-TCACCAACCAGCGTCTCATATTT-3′	AB699096	52.3
β-Actin	Forward：5′-CCTCACCCTCAAGTACCCCAT-3′ Reverse：5′-TTGGCCTTTGGGTTGAGTG-3′	EU887951	54.5

A standard curve was generated for each primer pair based on known quantities of cDNA（10-fold serial dilutions corresponding to cDNA transcribed from 100 to 0.01 ng of total RNA）. The relative P-gp mRNA expression was calculated using the $2^{\Delta Ct}$ equation.

2.5   Immunohistochemical staining

The liver and kidney samples were fixed with 4% buffered paraformaldehyde at 4℃ for 24 h and washed with PBS（pH 7.4）. The tissues were sequentially dehydrated in 75%，85%，95%，and 100% ethanol, permeabilized by xylene and then embedded in paraffin. Tissue sections（5 μm）were attached to APES sticky tablet handled slides and placed in a 60℃ oven for 2 h.

Mouse monoclonal antibody C219 to P-gp（Gene Tex Inc, Irvine, California, USA）was applied for immunohistochemistry（Kleinow et al., 2000）. Image pro-plus 6.0 software was used for the analysis of immunohistochemistry images. For each slide, 10 randomly selected viewing fields were reviewed at magnifications of 200× and 400×. The average cumulative optical density（IOD value）of positive staining for each group was expressed as mean±standard error（SE）.

2.6   Statistical analysis

Statistical analyses were performed with SPSS 16.0 software（SPSS Inc., Chicago, IL, USA）and data were presented as the mean±standard deviation（SD）. Significant differences between means were determined using one-way analyses of variance（ANOVAs）and the Duncan's test for multiple range comparison with significant level established at $P<0.05$.

3   Results

3.1   Acute toxicity of isothiazolinone

The LC_{50} values for isothiazolinone were shown in Table 2. The results indicated that there was the potential acute toxicity of isothiazolinone to *C. idellus*. Furthermore, it suggested that temperature was involved in the acute toxicity of isothiazolinone to *C. idellus*. At 15℃ the LC_{50}（48 h）was 0.53±0.17 mg/L while the one at 25℃ was 0.41±0.08 mg/L. The higher water temperature resulted in the lower LC_{50} value, which indicated that the increased temperature could increase the toxicity of isothiazolinone to *C. idellus*.

Table 2   The LC_{50} values of isothiazolinone on *C. idellus* at different temperatures.

Temperature（℃）	LC_{50}（95% confidence interval）/（mg/L）	
	48 h	96 h
15	0.53±0.17（0.45-0.57）	0.31±0.11（0.24-0.36）
25	0.41±0.08（0.36-0.45）	0.22±0.09（0.17-0.25）

3.2. Tissue accumulation of isothiazolinone

The accumulation of isothiazolinone in livers and kidneys were shown in Fig. 1. The group of *C. idellus* cultured in the temperature of 25℃ accumulated more isothiazolinone than those in 15℃. The concentrations of isothiazolinone in livers were higher than in kidneys at the same temperatures. These results indicated that there was a high binding affinity between isothiazolinone and *C. idellus*, which may be linked to its aquatic toxicity.

3.3. P-gp mRNA expression

Gene expression of P-gp could be detected in the kidneys and livers of *C. idellus*（Fig. 2）. The specific amplified fragment of the P-gp gene（127 bp）appeared as desired.

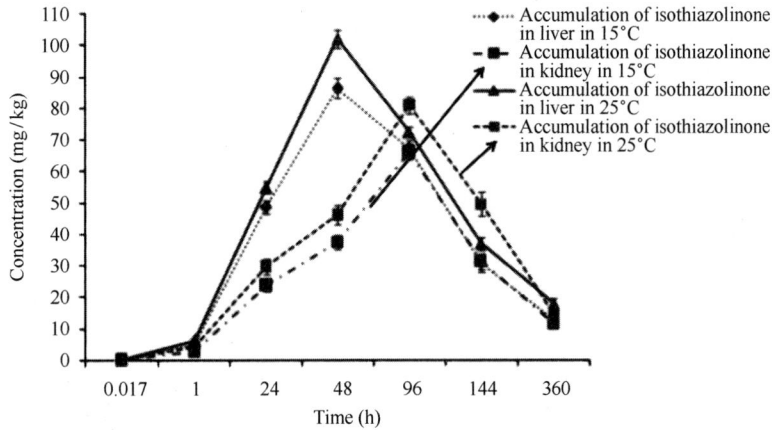

Fig. 1　The residual time and accumulation of isothiazolinone in tissues.

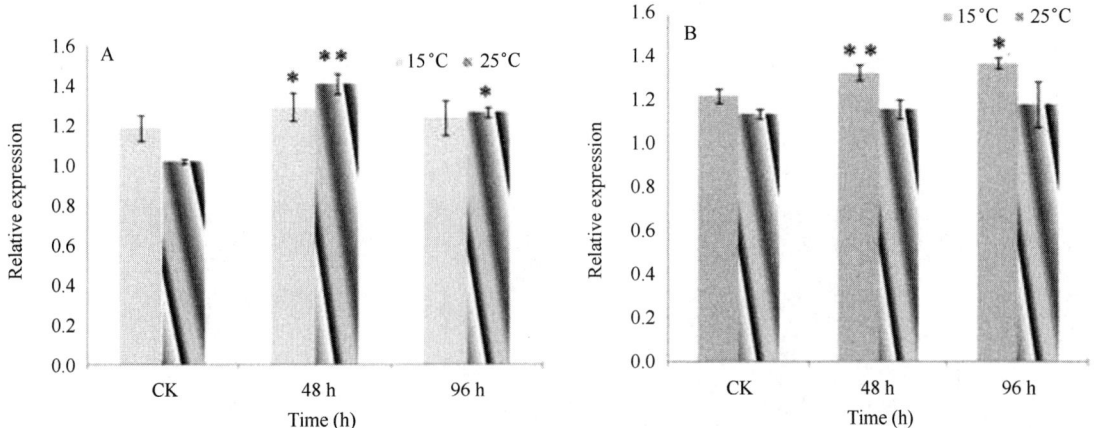

Fig. 2　The amounts of P-gp mRNA in livers and kidneys measured by PCR（β-actin used as the internal control）.
1：P-gp mRNA in livers；2：β-actin in livers；3：P-gp mRNA in kidneys；4：β-actin in kidneys；M：marker.

The P-gp mRNA expression levels in livers and kidneys at different temperatures were shown in Fig. 3. For the blank controls, the mean mRNA level at 25℃ was significantly lower than at 15℃. For livers, the mRNA levels at 25℃ were significantly higher than at 15℃. But for the kidneys, the opposite findings were noted. For livers, there were significant increases in P-gp mRNA expression after 48 h at 15℃ and after 48 h or 96 h at 25℃, compared with the P-gp mRNA expression of the corresponding control groups（P<0.05-0.01）. For kidneys, no differences were seen between the blank controls and other groups at all the time points at 25℃. P-gp mRNA expression in kidneys at 48 h or 96 h at 15℃ was significantly higher than in blank controls（P<0.05-0.01）.

3.4. P-gp protein expression detected by immunohistochemistry

P-gp protein was detected in livers（Fig. 4A and B）and kidneys（Fig. 4C and D）of *C. idellus*. The effects of isothiazolinone on the P-gp levels in liver and kidneys were shown in Fig. 5. For the blank controls, the P-gp level at 25℃ was much lower than at 15℃, which was consistent with the mRNA result. The protein levels in livers and kidneys at both time points（48 and 96 h）were significantly higher than those of the corresponding controls（P<0.01）.

4　Discussion

Isothiazolinones were used to control bacteria, fungi, and algae in cooling water systems, fuel storage tanks, water systems of pulp and paper mill, oil extraction systems, and wood preservation, and used as antifouling agents（Hunziker, 1992）. They were also frequently used in personal care products, such as shampoos and other hair care products, and in certain paint formulations（Smaoui and Hlima, 2012）.

It was estimated that a total of about 6000 registered chemical products on the Swiss market in 2001 were preserved with isothiazo-

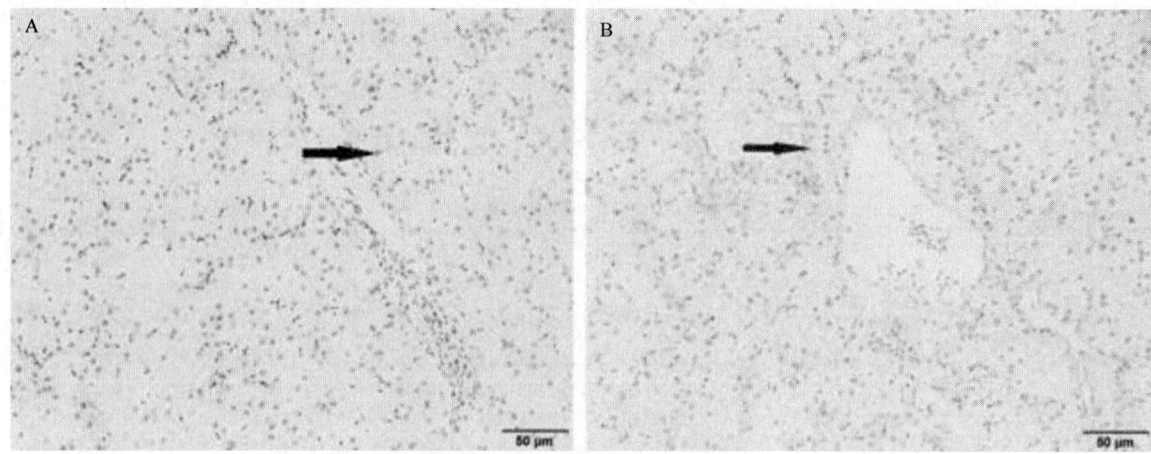

Fig. 3　（A）Effects of temperature on P-gp mRNA expression in livers；
（B）Effects of temperature on P-gp mRNA expression in kidneys.

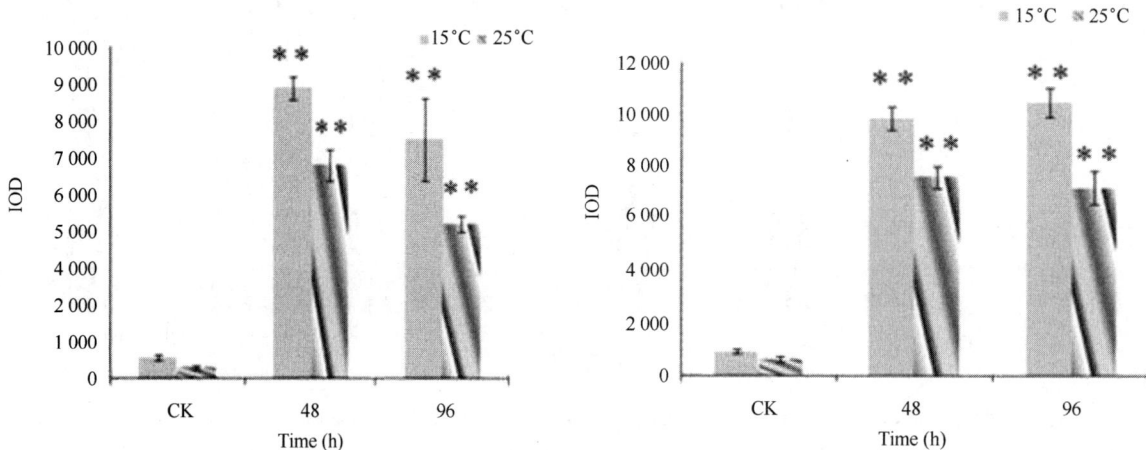

Fig. 4　（A）C219 immunohistological staining for liver samples from the blank control group （black arrow indicate P-gp protein）；（B） C219 immunohistological staining for liver samples from the test group after 48 h exposure to iosthiazolinone （black arrow indicate P-gp protein）；（C）C219 immunohistological staining for kidney samples from the blank control group （black arrow indicate P-gp protein）；（D） C219 immunohistological staining for kidney samples from the test group after 48 h exposure to iosthiazolinone （black arrow indicate P-gp protein）. The brown blot （indicated by black arrow） referred to the expression of P-gp and the nucleus of the cell was stained purple by hematoxylin.

linones （MCL/MI）（Reinhard et al., 2001）. Isothiazolinones could cause undesirable or toxic reactions （Morren et al., 1992, Burden et al., 1994, Bourke et al., 1997 and Basketter et al., 1999）. The MCI/MI concentrations in air samples of freshly painted rooms a few days after painting are high enough to elicit airborne reactions in already-sensitized patients （Reinhard et al., 2001）. The most frequently identified allergens in makeup and moisturizers were isothiazolinones and fragrance ingredients （Reinhard et al., 2001）. Exposure of HL60 cells to a mixture of MCI/MI induced the cells to a state of oxidative stress, inducing apoptosis and necrosis （Frosali et al., 2009）.

To reduce undesirable effects of traditional isothiazolinones, several novel antifoulants were discovered and characterized with low bioaccumulation and rapid biodegradation, such as 4, 5-dichloro-2-（n-octyl）-3（2H）-isothiazolone （DCOI, Sea-Nine 211™）. The short-term in vitro exposure of haemocytes to high concentrations of Sea-Nine 211™ provoked a marked reduction in haemocyte functionality （Cima et al., 2008）. It was suggested that Sea-Nine 211 was a P-gp substrate, which provided new insights into the interactions of P-gp with the antifouling contaminants in aquatic organisms （Xu et al., 2011）. In the present study, the mRNA and protein expression of P-gp were measured to evaluate the effects of isothiazolinone in the water environment. The P-gp protein levels in livers and kidneys of isothiazolinone-treated groups were significantly higher than in corresponding controls. For livers,

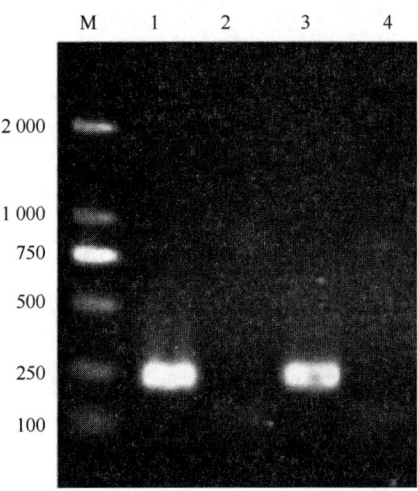

Fig. 5　（A）Effects of temperature on P-gp expression in livers；
（B）Effects of temperature on P-gp expression in kidneys.

there were significant increases in the P-gp mRNA expression after 48 h at 15℃ and after 48 h or 96 h at 25℃ compared with the corresponding control groups （P<0.05-0.01）. For kidneys, no differences were seen between blank controls and other time points at 25℃. Compared with the blank control group, the P-gp mRNA expression levels in kidneys at 48 h or 96 h at 15℃ were significantly higher （P<0.05-0.01）. During our research, isothiazolinone heavily affected P-gp mRNA expression, which was closely related to the toxicity of isothiazolinone.

Considering that C. idellus had the greatest numbers and widest distributions among the farmed aquaculture species in China, we investigated whether water temperature could affect the toxicity of isothiazolinone. In the acute toxicity experiments, the LC_{50} value of isothiazolinone on C. idellus at higher temperature was less than at lower temperature. Consistent with these results, the accumulation of isothiazolinone in livers and kidneys of the higher temperature group was greater than that of the lower temperature group. There was extensive residual time of isothiazolinone in livers and kidneys at both high and low temperatures. This observation indicated that there was high binding affinity between isothiazolinone and organs of C. idellus. The effects of temperature on isothiazolinone toxicity in aquatic environment should be considered in risk assessment. We inferred that it was easier for the efficient binding of isothiazolinone and P-gp at higher temperature.

Concentrations of isothiazolinone in water should be detected using high-performance liquid chromatography-tandem mass spectrometry （Lin et al. , 2010）. With the help of this analytical method, half-life of Sea-Nine 211™ was calculated as 2.5 days for nominal concentrations of 32 and 100 nM, as reported. The results showed that these concentrations of Sea-Nine 211™ significantly increased the community tolerance after only two days of exposure; the tolerance was maintained for a period of 16 days even when Sea-Nine 211™ was degraded during this period （Larsen et al. , 2003）. In our current study, the LC_{50} of isothiazolinone was 0.22 mg/L at 25℃ after a 96-h exposure to isothiazolinone in cultivating water. What is more, the P-gp protein levels in livers and kidneys at both time points （48 and 96 h） were significantly higher than that of the corresponding controls. These results indicated that isothiazolinone played an important role in the P-gp protein expression. In order to investigate the mechanism of isothiazolinone toxicity related to the expression of P-gp protein and mRNA, the intracellular changes and reactions should be investigated. Cellular mechanisms for tolerance to pollution were keys in risk assessment for aquatic environments. The ATP-binding cassette （ABC） transport proteins related with the multidrug/multixenobiotic resistance （MDR/MXR） phenomenon seemed to have an important role in the elimination of xenobiotics in aquatic organisms （Caminada et al. , 2008）. Multixenobiotic resistance in marine and freshwater organisms had been detected in some tissues by immunoreactivity with mammalian P-gp antibodies （Bard, 2000）. The biochemical mechanism underlying such "multixenobiotic" resistance in marine mussels and freshwater fishes was similar to the mechanism of multidrug resistance （MDR） found in tumor cells （Kurelec, 1992）. Some researchers analyzed a series of pharmaceuticals for their potential to modulate the activity of xenobiotic efflux transporters from the ABC sub-family in the Poeciliopsis lucida hepatoma cell line （PLHC-1/wt） and a doxorubicin （DOX） resistant subclone （PLHC-1/DOX） characterized by an elevated expression of the P-gp （ABCB1） （Caminada et al. , 2008）. The results demonstrated that 18 out of 33 tested pharmaceuticals showed MXR inhibitory activity with IC_{50} values being in the lower micromolar to millimolar range. The study revealed significant inhibitory effects of environmentally relevant pharmaceuticals on P-gp1 and MRP-like transporters in fish. In the present study, the mechanism of toxicity of isothiazolinone in aquatic environments might be related with the multixenobiotic resistance mediated by P-gp. Future toxicity bioassays with isothiazolinones could be performed to elucidate the tolerance role of the P-gp1 efflux transporter in the mechanism of multixenobiotic resistance. And as isothiazolinone could increase the expression of P-gp mRNA and protein, investigating the relationship of multixenobiotic resistance and isothiazolinone could be an interesting topic.

Conflict of interest statement

All authors declared there was no conflict interest involved.

Acknowledgements

This study was supported by the 863 Program (Grant No. 2011AA10A216), the Special Fund for Agro-scientific Research in the Public Interest (Grant No. 201203085), the Key Project of Science & Technology of Shanghai Agriculture Committee, Shanghai University Knowledge Service Platform, and the Open Fund of Key Laboratory of Fishery Drug Development.

References

S. Agarwal, V. Arya, L. Zhang; Review of P-gp inhibition data in recently approved new drug applications: utility of the proposed [I1] /IC50 and [I2] /IC50 criteria in the P-gp decision tree; J. Clin. Pharmacol. , 53 (2013), pp. 228-233.

S. M. Bard; Multixenobiotic resistance as a cellular defense mechanism in aquatic organisms; Aquat. Toxicol. , 48 (2000), pp. 357-389.

S. M. Bard, S. Gadbois; Assessing neuroprotective P-glycoprotein activity at the blood-brain barrier in killifish (*Fundulus heteroclitus*) using behavioural profiles; Mar. Environ. Res. , 64 (2007), pp. 679-682.

S. M. Bard, B. R. Woodin, J. J. Stegeman; Expression of P-glycoprotein and cytochrome p450 1A in intertidal fish (*Anoplarchus purpurescens*) exposed to environmental contaminants; Aquat. Toxicol. , 60 (2002), pp. 17-32.

D. A. Basketter, R. Rodford, I. Kimber, I. Smith, J. E. Wahlberg; Skin sensitization risk assessment: a comparative evaluation of 3 isothiazolinone biocides; Contact Dermat, 40 (1999), pp. 150-154.

S. J. Bourke, R. P. Convery, S. C. Stenton, R. M. Malcolm, D. J. Hendrick; Occupational asthma in an isothiazolinone manufacturing plant; Thorax, 52 (1997), pp. 746-748.

A. D. Burden, J. B. O' Driscoll, F. C. Page, M. H. Beck; Contact hypersensitivity to a new isothiazolinone; Contact Dermat, 30 (1994), pp. 179-180.

D. Caminada, R. Zaja, T. Smital, K. Fent; Human pharmaceuticals modulate P-gp1 (ABCB1) transport activity in the fish cell line PLHC-1; Aquat. Toxicol. , 90 (2008), pp. 214-222.

F. Cima, M. Bragadin, L. Ballarin; Toxic effects of new antifouling compounds on tunicate haemocytes I. Sea-nine 211 and chlorothalonil; Aquat. Toxicol. , 86 (2008), pp. 299-312.

S. Frosali, A. Leonini, A. Ettorre, G. Di Maio, S. Nuti, S. Tavarini, P. Di Simplicio, A. Di Stefano; Role of intracellular calcium and S-glutathionylation in cell death induced by a mixture of isothiazolinones in HL60 cells; Biochim. Biophys. Acta, 1793 (2009), pp. 572-583.

Q. He, K. Liu, S. Wang, H. Hou, Y. Yuan, X. Wang; Toxicity induced by emodin on zebrafish embryos; Drug Chem. Toxicol. , 35 (2012), pp. 149-154.

N. Hunziker The ' isothiazolinone story' Dermatology, 184 (1992), pp. 85-86.

N. Hunziker, F. Pasche, A. Bircher, L. Bruckner-Tuderman, T. Hunziker, P. Schmid, A. Suard, H. Suter, D. Perrenoud, J. Stäger, et al. Sensitization to the isothiazolinone biocide. Report of the Swiss Contact Dermatitis Research Group 1988-1990Dermatology, 184 (1992), pp. 94-97.

K. M. Kleinow, A. M. Doi, A. A. SmithDistribution and inducibility of P-glycoprotein in the catfish: immunohistochemical detection using the mammalian C-219 monoclonal Mar. Environ. Res. , 50 (2000), pp. 313-317.

D. Kucharczyk, K. Targońska, P. Hliwa, P. Gomułka, M. Kwiatkowski, S. Krejszeff, J. Perkowski Reproductive parameters of common carp (*Cyprinus carpio L.*) spawners during natural season and out-of-season spawning Reprod. Biol. , 8 (2008), pp. 285-289.

B. Kurelec; The multixenobiotic resistance mechanism in aquatic organisms Crit. Rev. Toxicol. , 22 (1992), pp. 23-43.

D. K. Larsen, I. Wagner, K. Gustavson, V. E. Forbes, T. Lund Long-term effect of Sea-Nine on natural coastal phytoplankton communities assessed by pollution induced community tolerance Aquat. Toxicol. , 62 (2003), pp. 35-44.

Q. B. Lin, T. J. Wang, H. Song, B. Li Analysis of isothiazolinone biocides in paper for food packaging by ultra-high-performance liquid chromatography-tandem mass spectrometry Food Addit. Contam. A: Chem. Anal. Control Expo. Risk Assess. , 27 (2010), pp. 1775-1781.

M. A. Morren, A. Dooms-Goossens, J. Delabie, C. De Wolf-Peeters, K. Marien, et al. Contact allergy to isothiazolinone derivatives: unusual clinical presentations Dermatology, 184 (1992), pp. 260-264.

M. Pazzaglia, C. Vincenzi, F. Gasparri, A. Tosti Occupational hypersensitivity to isothiazolinone derivatives in a radiology technician Contact Dermat, 34 (1996), pp. 143-144.

M. O. Pereira, M. J. Vieira, V. M. Beleza, L. F. Melo; Comparison of two biocides-carbamate and glutaraldehyde-in the control of fouling in pulp and paper industry; Environ. Technol. , 22 (2001), pp. 781-790.

E. Reinhard, R. Waeber, M. Niederer, T. Maurer, P. Maly, S. Scherer; Preservation of products with MCI/MI in Switzerland; Contact Dermat, 45 (2001), pp. 257-264.

S. Smaoui, H. B. Hlima; Effects of parabens and isothiazolinone on the microbiological quality of baby shampoo: the challenge test; Biocontrol Sci., 17 (2012), pp. 135-142.

C. Steinbach, G. Fedorova, M. Prokes, K. Grabicova, J. Machova, R. Grabic, O. Valentova, H. K. Kroupova; Toxic effects, bioconcentration and depuration of verapamil in the early life stages of common carp (*Cyprinus carpio L.*); Sci. Total Environ., 461-462 (2013), pp. 198-206.

V. D. Stejskal, M. Forsbeck, R. Nilsson; Lymphocyte transformation test for diagnosis of isothiazolinone allergy in man; J. Invest. Dermatol., 94 (1990), pp. 798-802.

X. Tan, S. Y. Yim, P. Uppu, K. M. Kleinow; Enhanced bioaccumulation of dietary contaminants in catfish with exposure to the waterborne surfactant linear alkylbenzene sulfonate; Aquat. Toxicol., 99 (2010), pp. 300-308.

X. Xu, J. Fu, H. Wang, B. Zhang, X. Wang, Y. Wang Influence of P-glycoprotein on embryotoxicity of the antifouling biocides to sea urchin (*Strongylocentrotus intermedius*) Ecotoxicology, 20 (2011), pp. 419-428.

该文发表于《Fish Physiology and Biochemistry》【2016】

Toxicity and accumulation of zinc pyrithione in the liver and kidneys of *Carassius auratus gibelio*: association with P-glycoprotein expression

Tao Ren[a,#], Gui-Hong Fu[b,#], Teng-Fei Liu[a], Kun Hu[a*],
Hao-Ran Li[a], Wen-Hong Fang[b], Xian-Le Yang[a]

(a. National Pathogen Collection Center for Aquatic Animals, Shanghai Ocean University, Shanghai 201306, China;

b. East China Sea Fisheries Research Institute Chinese Academy of Fishery Sciences, Shanghai 201306, PR China)

Abstract: Zinc pyrithione (ZPT) is a broad-spectrum antibacterial and antifungal agent; therefore, it is widely used in industry and civilian life. It is discharged into the aquatic environment with industrial and civilian waste water. *Carassius sp.* is one of the most widely distributed and farmed fish in China. The effects of aquatic ZPT on *Carassius sp.* remain unknown. In this study, we determined the acute toxicity of ZPT on *Carassius sp.*. The results showed that the median lethal concentration (LC_{50} 96 h) of ZPT on *Carassius sp.* cultivated in fresh water or water with 1.5‰ or 3‰ salinity was 0.163, 0.126, and 0.113 mg/L, respectively. ZPT has a higher affinity to the liver than the kidney, with a prolonged tissue residual time. P-glycoprotein (P-gp), an ATP-binding cassette (ABC) transporter, was found to be induced in the liver and kidney tissues of these *Carassius sp.* after ZPT treatment, based on the determination of its mRNA and protein levels by quantitative real-time reverse transcription polymerase chain reaction (qRT-PCR) and immunohistochemistry, respectively. The ZPT accumulation and magnitude of P-gp induction was also affected by the salinity of the cultivation water. These results suggest that aquatic ZPT is potentially toxic to *Carassius sp.*. We speculate that P-gp induction may play a protective role for *Carassius sp.*. Our findings provide a basis for assessing the potential risk of ZPT to aquatic animals including *Carassius sp.*.

Keywords: Zinc pyrithione; *Carassius auratus gibelio*; acute toxicity; tissue accumulation; P-glycoprotein

ZPT 对于鲫鱼的毒性及 P 糖蛋白的肝脏肾脏中的表达分析

任涛[a,#]，符贵红[b,#]，刘腾飞[a]，胡鲲[a*]，李浩然[a]，房文红[b]，杨先乐[a]

(a. 上海海洋大学国家水生动物病原库，中国上海；b. 中国水产科学研究院东海水产研究所，中国上海)

摘要：ZPT 是一种广泛应用的抗细菌，真菌的抗体，并且在工业和居民生活中也使用的很广泛。通过工业废水的和生活废水，ZPT 被释放到养殖水环境中。鲫鱼是中国养殖量最大也最常见的鱼。养殖水体中的 ZPT 对鲫鱼的影响尚未可知。本研究发现 ZPT 对鲫鱼有严重的毒害作用。实验结果显示：淡水和水中的鲫鱼对 ZPT 的半致死浓度（LC50 96 h）仅为 1.5‰ 和 3‰，此时盐度分别为 0.163，0.126，ZPT 的浓度为 0.113 mg/L。ZPT 相对于肾来说，在肝脏的残留时间更长。通过 qRT-PCR 的方法测定 mRNA 和蛋白质的水平，发现在 ZPT 处理之后，鲫鱼肾脏和肝脏中的一种转运蛋白（ABC）糖蛋白（P-gp）分别的将会减少。ZPT 加快加大 P-gp 的生产同样也影响了养殖水体的盐度。这些结果显示：ZPT 对鲫鱼有潜在的毒性，我们推测，P-gp 的产生可能对鲫鱼起保护作用，我们的发现为进一步研究 ZPT 对包括鲫鱼在内的养殖动物有潜在的毒害作用提供了基础。

关键词：砒啶硫酮锌；鲫鱼；急性毒性；组织蓄积

Introduction

Zinc pyrithione（ZPT），also known as zinc omadine，is a coordination complex that kills a broad-spectrum of fungi and bacteria. For this reason，it is widely used in industry and civilian life. For example，ZPT is used in many well-known brands of shampoo to clear dandruff（Dawson et al. 2004；Guthery et al. 2005）. As one of the most effective antidandruff ingredients，ZPT is used in cosmetic products with a maximum concentration of 0. 5%~1. 0%（Wang 2000）. In addition，it is used in the treatment of seborrheic dermatitis（Faergemann 2000；Brayfield 2011）. ZPT is also widely used in coatings，adhesives，and marine antifouling paint（Thomas 1999；Kobayashi and Okamura 2002；Voulvoulis et al. 2002；Doose et al. 2004；Wallströma et al. 2011）. Studies have shown that a high dose of ZPT（5~20 mg/kg）is toxic to rats，rabbits，dogs，cats，and some marine copepod（Nolen and Dierckman 1979；Konstantinou and Albanis 2004；Sanchez-Bayo and Goka 2006；Bao et al. 2014）.

Unsurprisingly，ZPT is found in the aquatic environment to which industrial waste water is discharged. The concentration of ZPT in the aquatic environment ranges from 0. 04 μg/L to 0. 32 μg/L（Bones et al. 2006）. Regarding its toxicity to aquatic animals，the toxicity of ZPT to both ostracod and cladoceran species appears to be similar，with 48-h median lethal concentrations（LC_{50} 96 h）in the range of 137~524 μg/L and 75~197 μg/L for ostracods and cladocerans，respectively，and similar values for the median effective concentrations（Sanchez-Bayo and Goka 2006）. Knowledge on whether ZPT in the aquatic environment is toxic to *Carassius sp.* and other aquatic animals remains very limited. *Carassius sp.* is one of the most widely distributed and farmed fish in China（Kalous et al. 2004；Gui and Zhou 2010；Wu et al. 2013）. *Carassius sp.* are healthy when the breeding water salinity is <3‰. The hemoglobin levels of *Carassius auratus gibelio* rise significantly when water salinity is increased from 1. 5‰ to 3‰（Lv et al. ，2007）. *Carassius sp.* are usually farmed in water with a low salinity（<3‰）（in Jiangsu Province，China）or fresh water（in Hubei and Hunan Provinces China）. We hypothesized that ZPT is toxic to *Carassius sp.* ，depending on its concentration in the water. Therefore，the goal of this study was to characterize the toxicity of ZPT to *Carassius sp.* .

P-glycoprotein（P-gp）is an ATP-binding cassette（ABC）transporter. It can efflux exogenous compounds from cells；therefore，it protects cells and the body from harm by hazardous chemicals. It has been shown that P-gp can be downregulated by exogenous compounds（Agarwal et al. 2013）. The presence of P-gp has been demonstrated in many fishes such as zebrafish，killifish，and catfish（Bard et al. 2002；Bard and Gadbois 2007；Tan et al. 2010；He et al. 2012）. The expression of P-gp and its location within tissues have been associated with the accumulation and toxicity of ZPT in *Ctenopharyngodon idellus*（Hu et al. 2014）. We hypothesized that P-gp expression and its location within tissues are also associated with the accumulation and toxicity of ZPT in *Carassius sp.* .

The aims of this study were to examine the acute toxicity of ZPT exposure to *Carassius sp.* as well as to determine the ZPT concentrations and P-gp expression levels in the liver and kidney tissues of *Carassius sp.* . Our findings may provide a reference for assessing the potential risk of ZPT to Carassius sp. .

Material and Methods

Maintenance and treatment of *Carassius sp.*

Experiments were conducted at the National Pathogen Collection Center for Aquatic Animals of Shanghai Ocean University（Shanghai，China）. All experiments were conducted following the guidelines approved by the Shanghai Ocean University Committee on the Use and Care of Animals.

Carassius sp. （100±20. 6 g，16±0. 8cm）were provided by Nantong Farm Fisheries Management，Jiangsu Province. These Carassius sp. are considered "adult" at this stage of development. All *Carassius sp.* were healthy，pathogen-free，and not treated with any drugs or chemicals. All *Carassius sp.* were randomly divided into three groups and were acclimatized in experimental tanks（4 m×2 m×1 m）filled with saline water（the salinity values were 0‰，1. 5‰，and 3. 0‰，respectively）at 26±1℃ for 7 days，and then they were fasted for 1 day before testing. The volume of water in the acclimation tanks was 2000 L. All *Carassius sp.* were fed a drug-free commercial diet during the 7-day exposure period. All fish were housed in a pond and were healthy，pathogen-free. Local tap water（Shanghai）was used as "fresh" water. The salinity of the water was adjusted using NaCl（Chemical grade，Shanghai Guoyao Chemical Reagent Co. ，Shanghai）. The pH of the water was between 7. 2 and 7. 4. The water was aerated with 6. 5 mg of O_2/L. The tanks were placed in a dark room because ZPT has a low photostability（Thomas 1999）. Fasted *Carassius sp.* were divided into groups with

10 fish/group. A total of 180 fasted *Carassius sp.* (6×3×10) were used in the formal experiment. The volume of water in the experimental tanks (80 cm× 40 cm× 45 cm) was 64 L. The density (grams of *Carassius sp* per liter of water) used during the LC50 experiments was 15. 6. During the experiments, the water was not changed, and all *Carassius sp.* were fasted.

Determination of the LC_{50} 96 h values of ZPT on *Carassius sp.*

The LC_{50} 96 h values of ZPT (Chemical grade, Hongsheng Technology Co. Ltd) were determined according to a published method (Lu and Song 2002) with modification. In brief, fasted Carassius sp. were treated with one of five concentrations of ZPT for 96 h: 0. 115 mg/L, 0. 138 mg/L, 0. 166 mg/L, 0. 199 mg/L, or 0. 238 mg/L. Fish were regarded as being dead if their swimming became slow, they were unresponsive, or they were lying down at the bottom of the cylinder. For each ZPT concentration, three parallel groups were treated. The survival or death of the *Carassius sp.* in each group was observed and recorded every 8 h. Dead *Carassius sp.* were promptly removed from the tank during the experiment and were disposed according to the "Regulation on Animal Experimentation" (State Science and Technology Commission of China). The 95% confidence limits of the LC50 96 h values were calculated according to the Karber method using SPSS16. 0 (SPSS Inc. , Chicago, IL, USA). The experiment was repeated three times.

Determination of tissue accumulation of ZPT in *Carassius sp.*

Fasted *Carassius sp.* were treated with ZPT (0. 1 mg/L) for 168 h, and samples were collected at 0, 0. 017, 0. 5, 1, 2, 4, 8, 12, 16, 24, 48, 96, 120, 144, and 168 h. Then, the liver and kidney tissues from the *Carassius sp.* were collected. Methanol (5 mL) was added to the minced liver or kidney sample (0. 5 g), and the mixture was shaken to extract the ZPT. After centrifugation at 2263. 95g (the relative centrifugal force) for 5 min, the supernatant was collected. The precipitate was extracted with another 5 mL of methanol. After combining the two supernatants, n-hexane was added, and the mixture was shaken vigorously. After the layers were separated, the substratum was collected and evaporated to dryness on a rotary evaporator. The residue was dissolved in 80% methanol and filtered through a 0. 45-μm Millipore filter. The concentration of ZPT was determined using a high-performance liquid chromatographic (HPLC) method, according to a published procedure (Hu et al. 2012). The chromatographic conditions were as follows: reversed-phase C18 column chromatography (4. 6 mm ×150 mm); mobile phase: methanol/water=80/20; flow velocity: 1. 0 mL/min; detection wavelength: 265 nm; column temperature: 25℃; and sample size: 10 μL. Standard ZPT concentrations were prepared with 80% methanol.

Determination of P-gp mRNA expression using quantitative real-time reverse transcription polymerase chain reaction (qRT-PCR)

Total RNA was extracted from the liver and kidney tissues using a TRIzol RNA extraction kit (Invitrogen, Carlsbad, CA, USA), according to the manufacturer's protocol. The integrity of the extracted RNA was confirmed with the OD_{260}/OD_{280} ratio and gel electrophoresis. Complementary DNA was synthesized from 1 μg of total RNA using a reverse transcription kit (ReverTra Ace-α-©, Toyobo Co. , Ltd. , Japan) and a mixture of oligo (dT) and random primers. The primers (Table 1) were designed using Primer Premier 5. 0 software (PREMIER Biosoft, Palo Alto, CA, USA) and synthesized by Shanghai Generay Biotech Co. , Ltd. (Shanghai, China). The GAPDH gene was used as an internal reference for normalization of the levels of target gene expression (Pfaffl, 2001). Sterile water was used as a negative control for sample amplification. qRT-PCR was carried out using the SYBR Green PCR Master Mix (QPK-201, Toyobo Co. , Ltd. , Japan) in a Bio-Rad iQ5 cycler (Hercules, CA, USA). The qRT-PCR for P-gp was performed in a final reaction volume of 10 μL under the following conditions: a preheating cycle at 95℃ for 3 min, followed by 40 cycles of amplification at 95℃ for 5 s and 60℃ for 30 s, and finally cooling to 4~10℃. A standard curve was generated for each primer pair based on known quantities of cDNA (10-fold serial dilutions corresponding to cDNA transcribed from 100 to 0. 01 ng of total RNA). The GAPDH gene was used as an internal standard reference for the relative levels of P-gp mRNA. The relative level of P-gp mRNA expression induced by ZPT was calculated using the $2^{-\Delta\Delta Ct}$ method.

Table 1 Primers for P-gp and GAPDH

Gene	Primers	GenBank Accession No.	Annealing temperature (℃)	Amplicon Length (bp)
P-gp	Forward 5′-GTGGTGCTGGATGACAATGATG-3′ Reverse 5′-TCTCCGTAGGCAATGTTCTCTG-3′	DQ059072	52. 3	125
GAPDH	Forward 5′-CCACAGTCCATGCCTACACC-3′ Reverse 5′-CACCAGTTGAAGCAGGGATGA-3′	AY641443	54. 5	111

Immunohistochemistry and determination of integrated option density

The level and localization of P-gp protein expression in the liver and kidney tissues were determined using immunohistochemistry staining, according to a published procedure (Kleinow et al. 2000). The liver and kidney tissue samples were fixed with 4% buffered paraformaldehyde at 4℃ for 24 h and washed with phosphate-buffered saline (PBS; pH 7.4). Next, they were sequentially dehydrated in 75%, 85%, 95%, and 100% ethanol, permeabilized in xylene, and embedded in paraffin. Tissue sections (5-μm thick) were sliced, mounted on slides coated with 3-aminopropyl-triethoxysilane, and placed in a 60℃ oven for 2 h. Sections were first blocked for 45 min with normal horse serum (Shanghai Jingquan Biological Technology Co., Ltd., Shanghai, China), followed by an overnight incubation with primary anti-P-gp mouse monoclonal antibody C-219 (Gene Tex Inc., Irvine, CA, USA) at 4℃. Incubation in PBS without anti-P-gp antibody for 1.5 h was used as the negative control. After washing with PBS, biotinylated horse anti-mouse secondary antibody (Shanghai Jingquan Biological Technology Co., Ltd., Shanghai, China) was added, and the sections were incubated for 30 min. After washing with PBS, avidin-conjugated horseradish peroxidase (HRP) complex was added, and the sections were incubated for an additional 45 min. Diaminobenzidine was applied as the substrate chromagen for HRP to visualize the expression and localization of P-gp. Subsequently, the slides were washed in PBS and distilled water and counter-stained in Mayer's hematoxylin. Images were procured on a light microscope (Eclipse Ti-SR, Nikon Co., Ltd., Japan), and image analysis was performed using Image Pro-plus 6.0.

Two sections were prepared from each *Carassius sp.* in the three experimental groups. On each slide, 10 randomly selected fields of view were imaged and analyzed for integrated optical density (IOD) at magnifications of 200× and 400×. To measure IOD, we first randomly selected three fields for each magnification of 200× and 400× under a microscope and took pictures in the stable conditions. Pro-plus 6.0 Image software was used to analyze the IOD of each picture. In each group, the mean IOD represents the IOD value for that group. The average cumulative IOD value of positive staining for each group was expressed as the mean±standard deviation (SD).

Statistical analyses

Statistical analyses were performed with SPSS 16.0 software (SPSS Inc., Chicago, IL, USA). Data were presented as the mean±SD. Normal distribution and homogeneity of the variances was analyzed by the Kolmogorov-Smirnov test and the F-test, respectively. Comparisons among multiple groups were analyzed using analysis of variance and Duncan's test. $P<0.05$ was considered significant.

Results

Acute toxicity of ZPT

To investigate the acute toxicity of ZPT on *Carassius sp.*, we determined the LC_{50} 96 h values for ZPT on *Carassius sp.* cultivated under various salinities. The results showed that the LC_{50} 96 h value was 0.163 mg/L for *Carassius sp.* cultivated in fresh water containing ZPT (0.115 mg/L, 0.138 mg/L, 0.166 mg/L, 0.199 mg/L, or 0.238 mg/L) for 96 h, and it was 0.126 and 0.113 mg/L, respectively, in water with salinities of 1.5‰ and 3.0‰ (Table 2).

Table 2 The LC_{50} 96 h values of ZPT on *Carassius sp.* at different salinities (96 h)

Salinity (‰)	Dose (mg/L) /Dead fish (n)	LC_{50} 96 h (95% confidence interval) (mg/L)
0	0/0, 0.115/0, 0.138/3, 0.166/6, 0.199/8, 0.238/10	0.163 (0.147 9~0.180)
1.5	0/1, 0.096/1, 0.115/4, 0.138/7, 0.166/9, 0.199/10	0.126 (0.111~0.139)
3.0	0/0, 0.080/0, 0.096/2, 0.115/6, 0.138/9, 0.166/10	0.113 (0.103~0.123)

Tissue accumulation of ZPT

To examine whether ZPT accumulates in the tissues of *Carassius sp.*, we determined the ZPT concentrations in the liver and kidney tissues of these *Carassius sp.* exposed to ZPT. The results showed that the ZPT concentrations in both the liver and kidney tissues of *Carassius sp.* cultivated in fresh water were increased after ZPT exposure. The ZPT concentrations in the liver and kidney tissues persisted and reached a peak at 16 h after ZPT exposure (Fig. 1). The increases were persistent up to 16 h after ZPT exposure (Fig. 1). The liver ZPT concentration was 25 μg/kg at 16 h after ZPT exposure in fresh cultivation water. It was 22 and 18 μg/kg, respectively, in water with salinities of 1.5‰ and 3.0‰. The kidney ZPT concentration was 13 μg/kg at 16 h after ZPT exposure in fresh cultivation water. It was 13 and 12 μg/kg, respectively, in cultivation water with salinities of 1.5‰ and 3.0‰. The ZPT concentrations in the liver and kidney tissues decreased to approximately 5 μg/kg at 168 h after ZPT exposure.

ZPT induces P-gp mRNA expression in the liver and kidneys of *Carassius sp.* after exposure

To examine the effect of ZPT on P-gp expression in the liver and kidney tissues of *Carassius sp.*, we determined the relative P-gp

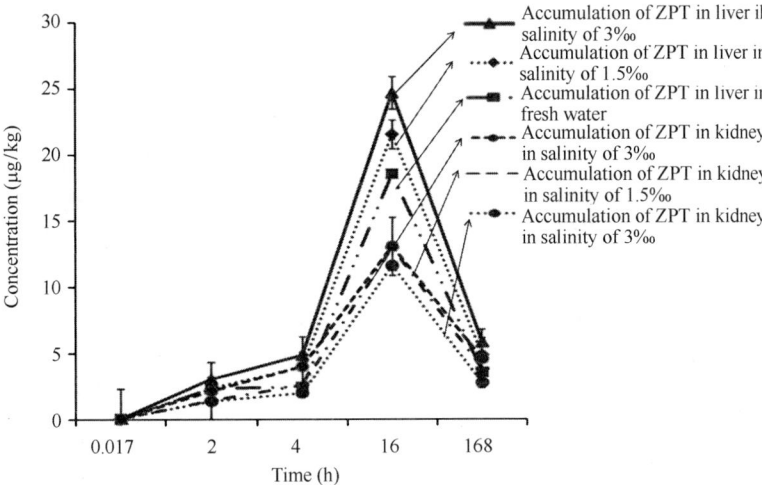

Fig. 1 ZPT concentrations in the liver and kidney tissues of *Carassius sp.* treated with ZPT （0. 1 mg/L）. Liver and kidney samples were collected from *Carassius sp.* exposed to ZPT in cultivation water with a designated salinity （0‰, 1. 50‰, and 3. 0‰）. The results are mean±SD of 3 fish/group.

mRNA levels in the liver and kidneys of *Carassius sp.* exposed to ZPT （0. 1 mg/L） for 16 h. The qRT-PCR results showed that P-gp mRNA was normally detected in the livers and kidneys of *Carassius sp.* （Fig. 2A and 2B, 0 h）. Its levels were significantly increased in the liver and kidney tissues at 2 h after ZPT treatment. At 16 h after ZPT treatment, P-gp expression was increased only in the liver of *Carassius sp.* exposed to ZPT in 1. 5‰ salinity （Fig. 2A and 2B）. This finding was salinity-dependent, and the effect was also observed at 2 h after ZPT exposure, but it was not obvious at 16 h after ZPT exposure （Fig. 2A and 2B）. These results suggest that ZPT caused increases in the P-gp mRNA levels in the liver and kidney tissues of Carassius sp. exposed to ZPT. In addition, the effect of long-term ZPT exposure on P-gp expression is not affected by salinity.

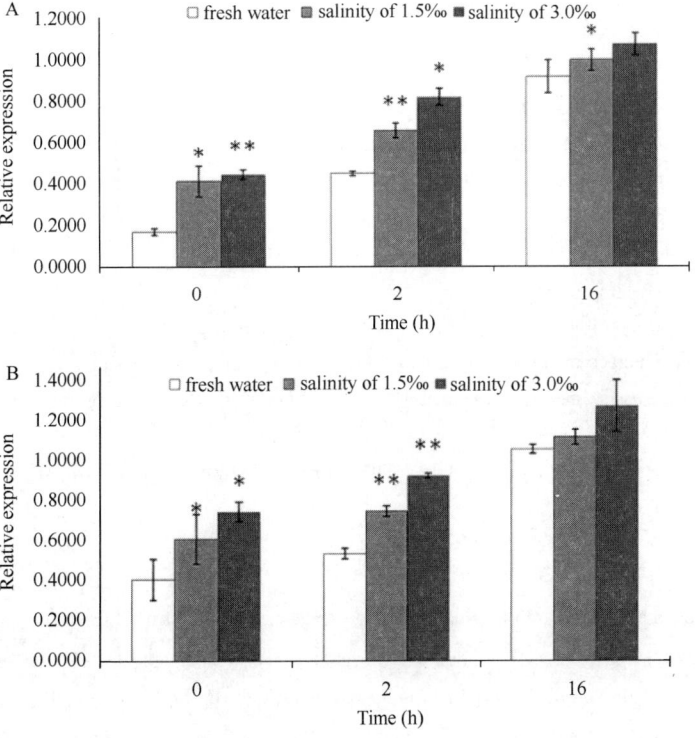

Fig. 2 Relative P-gp mRNA levels in the liver （A） and kidney （B） tissues of *Carassius sp.* exposed to ZPT （0. 1 mg/L） in cultivation water with a designated salinity （0‰, 1. 50‰ and 3. 0‰）. All "freshwater" groups considered the control group. The results are mean±SD of 3 fish/group. *, compared to their corresponding controls, P < 0.05; * *, compared to their corresponding controls, P < 0. 01. Representative data from three replicates are shown.

ZPT induces P-gp protein expression in the liver and kidney tissues of Carassius sp. exposed to ZPT in high salinity water

To further examine the effect of ZPT on P-gp expression in the liver and kidneys of *Carassius sp.*, we determined the P-gp protein levels and its cellular localization in the liver and kidney tissues of Carassius sp. exposed to ZPT. The results showed that the P-gp protein was localized in the liver of Carassius sp. both before and after exposure to ZPT in fresh water (Fig. 3A and 3B). There were no significant changes in the P-gp protein levels at 2 h after ZPT exposure, but an increasing trend was observed at 16 h after ZPT exposure in fresh water as well as in water with 1.5‰ salinity (Fig. 3C). The P-gp protein levels were significantly increased at 2 h and 16 h after ZPT exposure in water with 3‰ salinity (Fig. 3C).

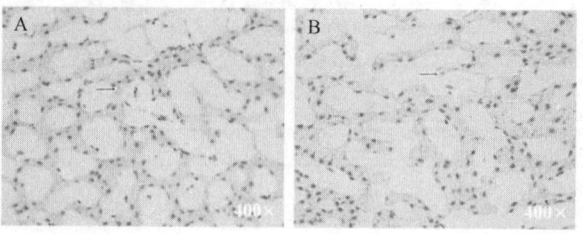

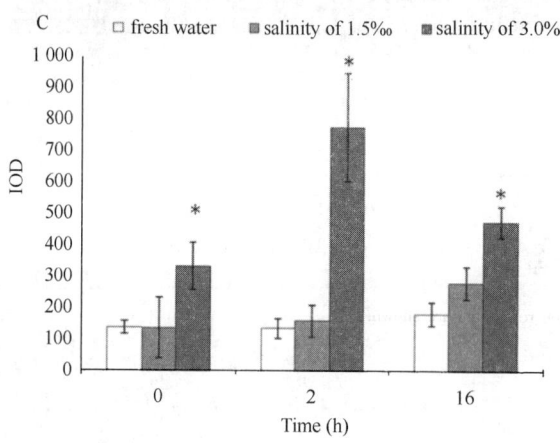

Fig. 3　Effects of ZPT on P-gp protein expression in the liver tissue of Carassius sp. . Liver samples were collected from Carassius sp. at 16 h after treatment with (B) or without (A) ZPT (0.126 mg/L) and subjected to immunohistological staining using the C219 monoclonal antibody for P-gp protein. P-gp protein stains brown as indicated by the black arrows. The cell nuclei were stained purple by hematoxylin. (C) Relative P-gp protein levels were quantitated using an IOD assay. All the "freshwater" groups were as control group. The results are mean ± SD of 3 fish/group. ∗, compared to control, P < 0.05; ∗∗, compared to control, P < 0.01. Representative data from three replicates are shown.

The P-gp protein was localized in the kidneys of *Carassius sp.* both before and after exposure to ZPT in fresh water (Fig. 4A and 4B). There were no significant changes in the P-gp protein levels at 2 h after ZPT exposure, but an increasing trend at 16 h after ZPT exposure was observed in *Carassius sp.* cultivated in fresh water (Fig. 4C). The P-gp protein levels were significantly increased at 2 h and 16 h after *Carassius sp.* were exposed to ZPT in water with 1.5‰ salinity (Fig. 4C). There was no significant change in the P-gp protein levels at 2 h, but a significant increase at 16 h after *Carassius sp.* were exposed to ZPT in water with 3‰ salinity was observed (Fig. 4C).

Discussion

In the current study, we tested the LC_{50} 96 h value of ZPT. After determination of the LC_{50} 96 h, a concentration close to the LC_{50} 96 h was used to assess the tissue uptake of ZPT, induction of a gene that is responsible for cellular export of ZPT (and other compounds), and the histological distribution of this export protein. We found that the LC_{50} 96 h of ZPT to *Carassius sp.* is between 0.163 mg/L in fresh water and 0.113 mg/L in water with 3.0‰ salinity at 96 h after exposure. The results indicated that ZPT was toxic to *Carassius sp.* and that the toxicity of ZPT decreased with increasing salinity of the cultivation water *Carassius sp.* is a freshwater *Carassius sp.*. Moreover, research shows that when euryhaline fish move from fresh water into the sea, in order to adapt to the environment, its gill structure and physiological function will adapt to the change by getting rid of Na^+ and Cl^- in the plasma. During this process, the

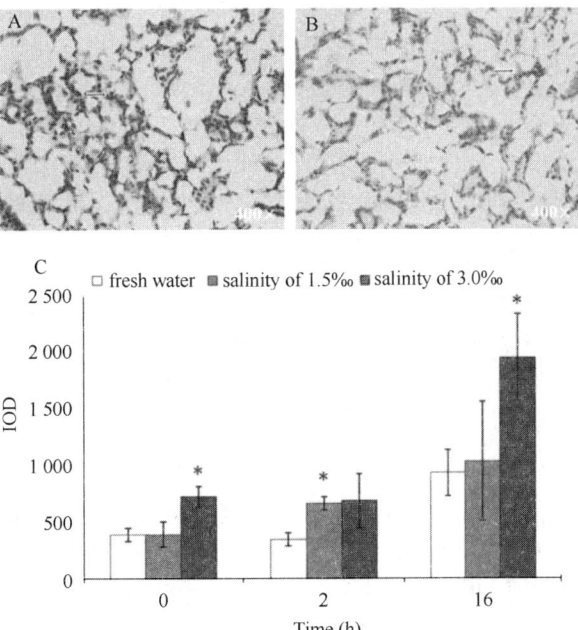

Fig. 4   Effects of ZPT on P-gp protein expression in the kidneys of *Carassius sp.*. Kidney samples were collected from *Carassius sp.* at 16 h after treatment with (B) or without (A) ZPT (0. 163 mg/L) and subjected to immunohistological staining using the C219 monoclonal antibody for P-gp protein. P-gp protein stains brown as indicated by the black arrows. The cell nuclei were stained purple by hematoxylin. (C) Relative P-gp protein levels were quantitated using an IOD assay. All "freshwater" groups were designated as controls. The results are mean ± SD of 3 fish/group. *, compared to control, $P < 0.05$; * *, compared to control, $P < 0.01$. Representative data from three replicates are shown.

chloride concentration and distribution as well as the Na^+/K^+-ATPase activity of the epithelial cells in the gill filaments will change significantly. Among them, the change in the Na^+/K^+-ATPase activity will directly affect the ability of ions to cross the plasma membrane and energy metabolism, which can lead to abnormal cell morphology, structure and function as well as apoptosis (Daborn et al. 2001; Martinez-Alvarez et al. 2005). So, it is possible that ZPT is more toxic to Carassius sp. in water with a higher salinity.

It has been reported that the LC_{50} 96 h values of ZPT for some freshwater fish (*Pimephales promelas* and *Oncorhynchus mykiss*) species range from 2. 6 to 400 μg/L (Madsen et al. 2000). In addition, the LC_{50} 96 h value of ZPT for freshwater crustaceans (Daphnia magna) is 6. 3 μg/L, and The LC_{50} 96 h value of ZPT for Oyster is 22 μg/L (Madsen et al. 2000). Besides, research indicates that the LC_{50} 120 h value of ZPT for the green alga (Selenastrum capricornutum) is 28 μg/L (Madsen et al. 2000). Since ZPT is toxic to these organisms, so we believe that ZPT may also pose acute toxicity to *Carassius sp.*.

It is known that the concentration of ZPT in the aquatic environment ranges from 0. 04 μg/L to 0. 32 μg/L (Bones et al. 2006), which is much lower than the LC_{50} 96 h value of ZPT to *Carassius sp.*. In the current study, we found that ZPT can accumulate in both the liver and kidney tissues of *Carassius sp.* treated with 0. 113-0. 163 mg/L ZPT. These results of tissue accumulation of ZPT indicate that ZPT is distributed to and accumulates in the liver and kidneys of these *Carassius sp.*. The amount of ZPT accumulation decreased with increasing salinity. Moreover, ZPT had a greater affinity to the liver than to the kidneys of *Carassius sp.*. In addition, the histological patterns between the liver and kidneys were different. Therefore, the affinity was different. However, whether ZPT in the aquatic environment is potentially toxic to *Carassius sp.* and poses a risk for *Carassius sp.* farming remains to be investigated in the future.

The results that ZPT induces P-gp mRNA expression suggest that ZPT may induce P-gp protein expression in the liver and kidney tissues of *Carassius sp.* exposed to ZPT. In addition, a higher P-gp protein expression was associated with water containing a higher salinity content.

It is well known that P-gp is an ABC efflux transporter. Its substrates include toxicants such as 4, 5-dichloro-2- (n-octyl) -3 (2H) -isothiazolone (DCOI, Sea-Nine 211TM) (Xu et al. 2011). It also has been shown that P-gp expression is associated with multi-xenobiotic resistance in marine and fresh water organisms (Bard 2000). Therefore, it protects aquatic organisms by facilitating the elimination of xenobiotics from the body (Caminada et al. 2008). In the current study, we found that the mRNA and protein levels of P-gp were increased in the liver and kidney tissues of *Carassius sp.* exposed to ZPT. This result was associated with the toxicity profile of ZPT. We speculate that P-gp may also protect Carassius sp. from toxicity by ZPT. However, whether ZPT is a substrate of P-gp and how P-gp effluxes ZPT remain unknown and deserve further investigation.

Carassius sp. is one of the most widely distributed and farmed fish in China (Kalous et al. 2004; Gui and Zhou 2010; Wu et al. 2013). It is usually farmed in low-salinity water (such as Jiangsu Province) or fresh water (Hubei, Hunan, etc.). In the current study, we found that the toxicity and tissue accumulation of ZPT as well as P-gp expression were affected by the salinity of the cultivation water. Therefore, the magnitude of ZPT toxicity on Carassius sp. should be considered in a region-dependent manner.

ZPT is commonly used in cosmetic products. Therefore, humans also have indirect contact with ZPT. Since a small amount of ZPT is harmful to *Carassius sp.*, is ZPT also harmful to humans? To answer this question, further research is needed to explore the toxicity of substances commonly employed as cosmetics.

In conclusion, exposure of *Carassius sp.* to the LC_{50} 96 h of ZPT causes its accumulation in the liver and kidney tissues as well as the upregulation of P-gp expression in both studied tissues. In addition, these ZPT-induced effects are affected by the salinity of the cultivation water. Our findings provide a basis for assessing the potential risk of ZPT to aquatic animals.

Acknowledgments

This study was supported by the Special Fund for Agro-scientific Research in the Public Interest (Grant 201203085), the 863 Program (Grant 2011AA10A216), the National Science and Technology Resources Platform, Key Laboratory of Genetic Resources for Freshwater Aquaculture and Fisheries and the Shanghai University Knowledge Service Platform.

References

Agarwal S, Arya V, Zhang L (2013) Review of P-gp inhibition data in recently approved new drug applications: utility of the proposed [I (1)] / IC (50) and [I (2)] /IC (50) criteria in the P-gp decision tree. J Clin Pharmacol 53 (2): 228-233.

Bao VW, Lui GC, Leung KM (2014) Acute and chronic toxicities of zinc pyrithione alone and in combination with copper to the marine copepod Tigriopus japonicus. Aquat Toxicol 157: 81-93.

Bard SM (2000) Multixenobiotic resistance as a cellular defense mechanism in aquatic organisms. Aquat Toxicol 48 (4): 357-389.

Bard SM, Gadbois S (2007) Assessing neuroprotective P-glycoprotein activity at the blood-brain barrier in killifish (*Fundulus heteroclitus*) using behavioural profiles. Mar Environ Res 64 (5): 679-682.

Bard SM, Woodin BR, Stegeman JJ (2002) Expression of P-glycoprotein and cytochrome p450 1A in intertidal fish (*Anoplarchus purpurescens*) exposed to environmental contaminants. Aquat Toxicol 60 (1-2): 17-32.

Bones J, Thomas KV, Paull B (2006) Improved method for the determination of zinc pyrithione in environmental water samples incorporating on-line extraction and preconcentration coupled with liquid chromatography atmospheric pressure chemical ionisation mass spectrometry. J Chromatogr A 1132 (1-2): 157-164.

Brayfield A (2011) " Pyrithione Zinc". Pharmaceutical Press, Martindale: The Complete Drug Reference

Caminada D, Zaja R, Smital T, Fent K (2008) Human pharmaceuticals modulate P-gp1 (ABCB1) transport activity in the fish cell line PLHC-1. Aquat Toxicol 90 (3): 214-222.

Daborn K, Cozzi RR, Marshall WS (2001) Dynamics of pavement cell-chloride cell interactions during abrupt salinity change in Fundulus heteroclitus. J Exp Biol 204 (Pt 11): 1889-1899.

Dawson TL, DeAngelis Y, Kaczvinsky J, Schwartz J (2004) Broad-spectrum anti-fungal activity of pyrithione zinc and its effect on the causes of dandruff and associated itch. Journal of the American Academy of Dermatology 3 (50): 36.

Doose CA, Ranke J, Stock F, Bottin-Weber U, Jastorff B (2004) Structure-activity relationships of pyrithiones-IPC-81 toxicity tests with the antifouling biocide zinc pyrithione and structural analogs. Green Chem 6: 259-266.

Faergemann J (2000) Management of seborrheic dermatitis and pityriasis versicolor. Am J Clin Dermatol 1 (2): 75-80.

Gui J, Zhou L (2010) Genetic basis and breeding application of clonal diversity and dual reproduction modes in polyploid *Carassius auratus gibelio*. Sci China Life Sci 53 (4): 409-415.

Guthery E, Seal LA, Anderson EL (2005) Zinc pyrithione in alcohol-based products for skin antisepsis: persistence of antimicrobial effects. Am J Infect Control 33 (1): 15-22.

He Q, Liu K, Wang S, Hou H, Yuan Y, Wang X (2012) Toxicity induced by emodin on zebrafish embryos. Drug Chem Toxicol 35 (2): 149-154.

Hu K, Li HR, Ou RJ, Li CZ, Yang XL (2014) Tissue accumulation and toxicity of isothiazolinone in *Ctenopharyngodon idellus* (grass carp): association with P-glycoprotein expression and location within tissues. Environ Toxicol Pharmacol 37 (2): 529-535.

Hu Q, Zhong JQ, Zheng R, Ji MS, Wang K (2012) Determination of zinc pyrithione and impurities test in cosmetic material by HPLC. Chinese Journal of Health Laboratory Technology 22 (9): 2037-2040.

Kalous L, Memis D, Bohlen J (2004) Finding of Triploid *Carassius ar gibelio* (Bloch, 1780) (Cypriniformes, Cyprinidae) in Turkey. Cybium 28

（1）：77-79.

Kleinow KM, Doi AM, Smith AA（2000）Distribution and inducibility of P-glycoprotein in the catfish：immunohistochemical detection using the mammalian C-219 monoclonal. Mar Environ Res 50（1-5）：313-317.

Kobayashi N, Okamura H（2002）Effects of new antifouling compounds on the development of sea urchin. Mar Pollut Bull 44（8）：748-751.

Konstantinou IK, Albanis TA（2004）Worldwide occurrence and effects of antifouling paint booster biocides in the aquatic environment：a review. Environ Int 30（2）：235-248.

Lu L, Song F（2002）Fish acute toxicity experiment. Bulletin of Biology 37（7）：50-52.

Lv F, Hu Y, Huang JT, Wang AM（2007）Effects of salinity on hemoglobin and protein of muscle of *Allogynogemetic Crucian* Carp. Freshwater Fisheries37（5）：56-59.

Madsen T, Samsøe-Petersen L, Gustavson K, Rasmussen D（2000）Ecotoxicological assessment of antifouling biocides and nonbiocidal antifouling paints. Environmental Project Report 531. Danish Environmental Protection Agency, Copenhagen, Denmark

Martinez-Alvarez RM, Sanz A, Garcia-Gallego M, Domezain A, Domezain J, Carmona R, del Valle Ostos-Garrido M, Morales AE（2005）Adaptive branchial mechanisms in the sturgeon Acipenser naccarii during acclimation to saltwater. Comp Biochem Physiol A Mol Integr Physiol 141（2）：183-190.

Nolen GA, Dierckman TA（1979）Reproduction and teratology studies of zinc pyrithione administered orally or topically to rats and rabbits. Food Cosmet Toxicol 17（6）：639-649.

Pfaffl MW（2001）A new mathematical model for relative quantification in real-time RT-PCR. Nucleic Acids Res 29：e45.

Sanchez-Bayo F, Goka K（2006）Influence of light in acute toxicity bioassays of imidacloprid and zinc pyrithione to zooplankton crustaceans. Aquat Toxicol 78（3）：262-271.

Tan X, Yim SY, Uppu P, Kleinow KM（2010）Enhanced bioaccumulation of dietary contaminants in catfish with exposure to the waterborne surfactant linear alkylbenzene sulfonate. Aquat Toxicol 99（2）：300-308.

Thomas KV（1999）Determination of the antifouling agent zinc pyrithione in water samples by copper chelate formation and high-performance liquid chromatography-atmospheric pressure chemical ionisation mass spectrometry. J Chromatogr A 833（1）：105-109.

Voulvoulis N, Scrimshaw MD, Lester JN（2002）Comparative environmental assessment of biocides used in antifouling paints. Chemosphere 47（7）：789-795.

Wallströma E, Jespersenb HT, Schaumburg K（2011）A new concept for anti-fouling paint for Yachts. Progress in Organic Coatings 72：109-114.

Wang LH（2000）Determination of Zinc Pyrithione in Hair Care Products on Metal Oxides Modified Carbon Electrodes. Electroanalysis 12（3）：227-232.

Wu T, Ding Z, Ren M, An L, Xiao Z, Liu P, Gu W, Meng Q, Wang W（2013）The histo-and ultra-pathological studies on a fatal disease of Prussian carp（Carassiusaryngodon gibelio）in mainland China associated with cyprinid herpesvirus 2（CyHV-2）. Aquaculture 412-413：8-13.

Xu X, Fu J, Wang H, Zhang B, Wang X, Wang Y（2011）Influence of P-glycoprotein on embryotoxicity of the antifouling biocides to sea urchin（*Strongylocentrotus intermedius*）. Ecotoxicology 20（2）：419-428.

该文发表于《中国水产科学》【2013, 20（2）：411-418】

基于 P-glycoprotein 基因表达评价尼罗罗非鱼体内恩诺沙星代谢"首过效应"

Association between permeability glycoprotein expression and enrofloxacin metabolism to evaluate the first-pass effct in *Oreochomis niloticus* Linn

胡鲲[1]，程钢[2]，吕利群[1]，章海鑫[1]，王会聪[1]，阮记明[1]，陈力[1]，房文红[3]，杨先乐[1]

（1. 上海海洋大学国家水生动物病原库，上海 201306；2. 中南民族学院，湖北 武汉 430074；
3. 中国水产科学研究院东海水产研究所，上海，200090）

摘要：为从药物酶的角度建立一种客观评价鱼类"首过效应"的方法，利用荧光定量 PCR 法测定尼罗罗非鱼（*Oreochomis niloticus* Linn）肝脏、肾脏组织中 P-glycoprotein（P-gp）基因表达量，分析了单剂量（40 mg·kg^{-1}）口服给药恩诺沙星（enrofloxacin, ENR）后，尼罗罗非鱼肠道、肝脏组织中 mRNA 水平的相对表达量与 ENR 血药浓度的时实相关性。实验结果显示：在尼罗罗非鱼肠道、肝脏组织中，P-gp 基因在分子量 127bp 处出现了与预期大小相待的特异性扩增片段。对尼罗罗非鱼口灌给药 ENR 后，ENR 能迅速通过肠道进入血浆，其在肠道、肝脏和血浆中的消除速度较快，其药物时量曲线关系符合一级吸收的二室开放动力学模型。当血浆中 ENR 浓度达到最高达峰时（1 h），实验组肠道和肝脏中 P-gp 基因的相对表达量相对于对照组均表现出显著性差异（$P<0.05$）；当肠道中 ENR 浓度达到最高峰时（2 h），实验组肠道 P-gp 基因的相对表达量则表现出极显著差异（$P<0.01$）；当肝脏中 ENR 浓度达到最高峰时（2 h），实验组肝脏 P-gp 基因的相对表达量与对照组相比则表现出显著性差异（$P<0.05$）。该结果证实了鱼类 P-gp 基因参与药物代谢过程，提供了一种从分子水平揭示水产动物体内药物代谢规律的思路。

关键词： P 糖蛋白；尼罗罗非鱼；恩诺沙星；首过效应；药物代谢

药物在动物体内的代谢过程中，存在着首过效应（first-pass effct）[1]：药物通过肠道、肝脏等器官时被部分分解或外排，致使进入血液中的药量减少（血药浓度降低），分布到各个靶组织的药物剂量随之减少，药效降低。首过效应具有剂量依赖性。在制定药物安全使用技术规范时，往往由于忽略了首过效应等因素，导致实践生产中药效不能达到预期效果（C_{max}/MIC 保持适当的倍数）。其后果是药物往往被过量使用，威胁水产品质量安全。建立恰当的方法评价药物在水产动物体内的"首过效应"可为药物安全使用技术的制定提供依据，对控制水产养殖品中药物残留风险具有较大的现实意义。

P 糖蛋白（P-glycoprotein，P-gp）是一种广泛地存在于动物肝肾、肠道等组织器官中的跨膜糖蛋白[2-4]，也被证实存在于鱼类中[5,6]等动物组织中。P-gp 是一种典型的药物外排泵，具有介导药物外排的功能。由于 P-gp 的对药物的外排作用，其表达是否与药物在水产动物肝脏、肠道等组织中的首过效应具有某种联系，成为一个令人感兴趣的问题。以往关于水产动物体内 P-gp 的研究大都集中于监测环境污染物方面[7-9]，尚未有将鱼类组织中 P-gp 与其药物代谢过程相联系的研究报道。

本文以恩诺沙星（enrofloxacin，ENR）为模式药物，分析了尼罗罗非鱼（*Oreochomis niloticus* Linn）肠道、肝脏组织中 P-gp 基因在 mRNA 水平的表达，评价了其相对表达量与 ENR 血药浓度的相关性，尝试建立一种利用 P-gp 基因评价水产动物体内首过效应的方法，为完善水产养殖品安全用药技术提供新的思路。

1 材料与方法

1.1 仪器设备

Agilent-1100 型高效液相色谱仪，配荧光检测器；Heidolph 4000 型旋转蒸发仪；KINEMTICA AG 型高速组织捣碎机；HitachiCR21G 型冷冻高速离心机；BIO-RAD iQ5 Real-time PCR 仪，BECKMEN DU800 型紫外分光光度计；IKA-C-MAG MS7 型振荡器。

1.2 试剂与耗材

ENR 标准品（含量≥99%）购自 Sigma 公司。总 RNA 提取试剂 Trizol 购自 Invitrogen 公司。反转录试剂盒（ReverTra Ace-α-TM）和荧光定量试剂盒（SYBR Green Realtime PCR Master Mix QPK-201）购自 TOYOBO 公司。乙腈为色谱级。无水硫酸钠、盐酸、正己烷四丁基溴化铵为化学纯，购自上海安谱科学仪器有限公司。

1.3 动物饲养、给药及取样

尼罗罗非鱼购于江苏省南通农场，健康无病，未用过任何药物。鱼平均体重（50 ±10.4）g。试验前于 4 m×2 m×1 m 的网箱中暂养 7 d，并在给药前 3 d 停止投饲。以 40 mg·kg⁻¹ 的剂量，用套有橡皮软管的注射器对尼罗罗非鱼口灌 ENR。给药后尼罗罗非鱼在 25℃ 水温下饲养。在给药后不同时间点随机取试验尼罗罗非鱼 3 组，每组 5 尾。采取血浆、肠道和肝脏组织，放入-80℃超低温冰箱保存备用。设立未给药的尼罗罗非鱼 3 组（每组 5 尾）为空白对照。所以实验用鱼的处置均遵循《动物实验规范（Regulation on Animal Experimentation）》（中国国家科学技术委员会，State Scientific and Technological Commission）

1.4 利用 P-gp mRNA 水平的表达分析首过效应

1.4.1 P-gp 基因的表达分析
1.4.1.1 通过 PCR 扩增克隆尼罗罗非鱼 P-gp 基因的 cDNA
采用 Trizol 试剂盒提取尼罗罗非鱼组织总 RNA。利用 1% 琼脂糖凝胶电泳检查纯化后的 RNA 质量。使用 Rerver Tra Ace-α-TM 试剂盒将 RNA 反转录得到 cDNA。
选取 β-actin 基因作为内参基因。β-actin 和 P-gp 的引物利用 Primer Premier 5.0 软件进行设计。引物序列如下：

β-actin　　Forward primer：CCTCACCCTCAAGTACCCCAT

　　　　　　Reverse primer：TTGGCCTTTGGGTTGAGTG

p-gp　　　　Forward primer：CACCTGGACGTTACCAAAGAAGATATA

Reverse primer：TCACCAACCAGCGTCTCATATTT

PCR 反应步骤为：94℃3 min；94℃30 s、N℃（β-actin 基因和 P-gp 基因分别为 49.6℃和 51.5℃）30 s、72℃30s，35 个循环；72℃10 min。利用电泳检测 PCR 产物。采取 Midi Purification Kit 回收经琼脂糖凝胶电泳分离的 PCR 产物。

1.4.1.2　P-gp 基因的实时 PCR 分析

称取 100 mg 动物组织提取 RNA，按照 1.4.1 的方法合成 cDNA。以 β-actin 做内参基因，对样品进行 real-time PCR 扩增。10 μL 的定量反应体系组成如下：cDNA 样品 0.4 μL、SYBR©Green PCR Master Mix 5 μL、Primer R（10 μmol/L）5μL 和 Primer F（10 μmol/L）0.2 μL。PCR 反应结束后设置 57~99℃的熔解温度，并每隔1℃采集 1 次信号。每 1 个反应重复 3 次。所有的统计数据由 P-gp 基因与 β-actin 基因表达量的比值得到。

1.4.2　组织中 ENR 及其代谢产物 CIP 残留量的分析

利用高效液相色谱法（high performance liquid chromatography，HPLC）分析不同给药时间点的组织中 ENR 的浓度。样品用 6 倍体积的乙腈振荡提取，4 500 r/min 离心 5 min，取上清。对残渣重复提取一次，并合并上清。上清液与正己烷混合，充分振荡，静置，抛弃水相。将收集的有机相在旋转蒸发器中蒸发至干，流动相溶解残渣并经微孔滤膜过滤后上 HPLC。色谱条件：反相 C_{18} 柱色谱柱（4.6 mm ×150 mm）；流动相为甲醇/柠檬酸-醋酸铵混合盐溶液 = 25/75（mL/mL）；流速 1.5 mL/min；激发波长 280 nm，发射波长 450 nm；柱温 40℃；进样量 10 μL。将 ENR、CIP 标准品配制成梯度浓度的标准溶液。以浓度（μg/mL）为横坐标，药物峰面积为纵坐标绘制标准曲线。利用外标法测定样品中药物浓度。

1.5　ENR 代谢过程中 P-gp 基因的表达分析

分析 ENR 代谢过程中的特征性时间点肠道、肝脏中 P-gp 基因的在 mRNA 水平的相对表达量。

1.6　数据处理

根据血药浓度-时间的半对数关系初步确定药物的处置过程模型。使用 DAS 3.0 药代动力学软件对药动学模型进行拟合并计算相关参数。实验数据由 SPSS16.0 统计软件进行处理和分析。对实验组与对照组的测定结果进行显著性分析，$P < 0.05$ 认定为两者具有显著性差异，$P < 0.01$ 认定为两者具有极显著差异。

2　结果

2.1　尼罗罗非鱼肠道组织中 P-gp 基因表达分析

对于提取的 RNA 样品，经紫外分析 OD_{260}/OD_{280} 值在 1.90~1.95，表明总 RAN 样品完整性较好，没有降解，可以作为反转录模板。总 RNA 经反转录为 cDNA 第一链后，以其为模板进行 RT-PAR 对肠道组织中的 β-actin 和 P-gp 进行扩增。PCR 产物经 1.0%琼脂凝胶电泳，结果如图 1 所示。未给药的对照组和给药 ENR（40 mg·kg^{-1}）的实验组肠道、肝脏组织中均能检测出 P-gp 基因的表达。P-gp 基因在 127bp 处出现了与预期大小小付的特异性扩增片段。

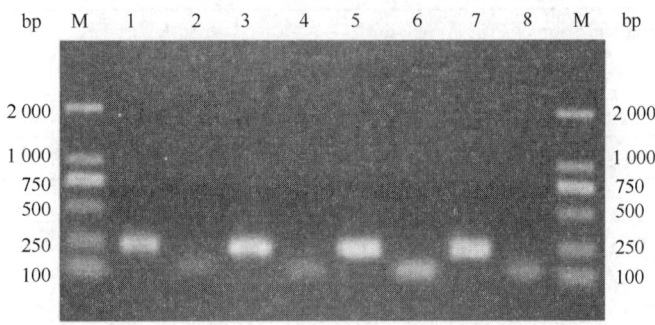

图 1　β-action 和 P-gp 基因 RT-PCR 扩增产物电泳图

Fig. 1　Gel electrophoresis of the RT-PCR amplification

M：DNA Marker；泳道 1、3、5、7 分别为空白对照组肠道、给药实验组肠道、空白对照组肝脏和给药实验组肝脏组织中 β-action 基因 PCR 扩增产物；泳道 2、4、6、8 分别为空白对照组肠道、给药实验组肠道、空白对照组肝脏和给药实验组肝脏组织中 P-gp 基因 PCR 扩增产物；

M：DNA Marker. Lane 1, 5 were the control group of the intestinal and liver tissues for β-action respectively. Lane 3, 7 were the experimental group of the intestinal and liver tissues for β-action respectively. Lane 2, 6 were the control group of the intestinal and liver tissues for P-gp respectively. Lane 4, 8 were the control group of the intestinal and liver tissues for P-gp respectively.

2.2 血浆中 ENR 代谢过程分析

利用本文的 HPLC 法测定 ENR，ENR 药物峰分离良好，基线平稳，其保留时间分别为 6.38 min。在标准品 0.2~10 μg·mL^{-1} 的浓度范围内 ENR 的标准曲线分别为 $Y=5.524X-0.135$，$R^2=0.998$。利用外标法检测尼罗罗非鱼肠道、肝脏和血浆中的 ENR 最低检测限可达 5 μg·kg^{-1}，平均回收率为 70.0%~80.5%，相对标准偏差分别为 1.2%~7.9%。

25℃水温条件下，尼罗罗非鱼单次口灌 ENR 药液后，利用 HPLC 法测定不同时间点血浆中 ENR 残留量，ENR 在肠道、肝脏和血液中的代谢规律见图 2。ENR 在血液中的最大达峰时间最早，为 1 h；肠道、肝脏的达峰时间均为 2 h。ENR 在肠道中最大达峰浓度最高，达 34.98 μg·g^{-1}；在肝脏和血浆中的最大达峰浓度分别为 4.84 μg·g^{-1} 和 3.47 μg·mL^{-1}。应用 DAS 软件分析药物水平−时间关系进行二室模型数据拟合，ENR 在肠道、肝脏和血液中药物时量曲线方程分别为：$C=227.34 \times e^{-0.54t}+185.73e^{-0.33t}-43.79\ e^{-0.77t}$、$C=28.31 \times e^{-0.90(t-0.07)}+2.27e^{-0.03(t-0.07)}-32\ e^{-1.22(t-0.07)}$ 和 $C=4.66 \times e^{-0.81t}+1.55e^{-0.07t}-151.13\ e^{-7.54t}$；拟合度（$R^2$）分别为 0.948、0.966 和 0.979。利用房室模型推算所得的 ENR 在血浆中主要药动学参数见表 1。

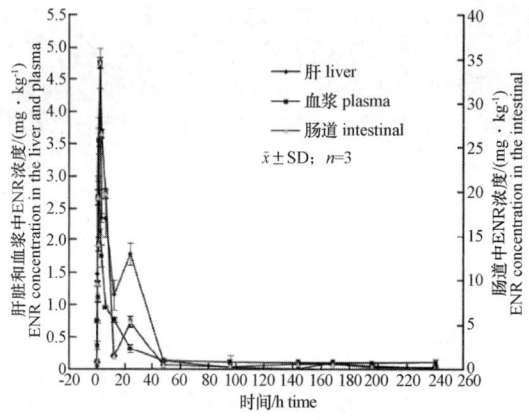

图 2　ENR 在尼罗罗非鱼肝脏、血浆和肠道组织中的代谢的过程

ENR 浓度为 HPLC 方法的实测值，以"平均值±标准差"（$n=3$）表示

Fig. 2　Pharmacokinetic of ENR after a single oral administrationin in intestinal, liver and plasma of *Oreochomis niloticus Linn* The concentration of ENR were assayed by HPLC method and the dates were showed as $\overline{x}\pm s$ （$n=3$）

表 1　ENR 在尼罗罗非鱼组织中药物代谢的主要参数（$\overline{x}\pm s$, $n=3$）

Tab 1　Main pharmacokinetic parameters of ENR after a single oral administration （40 mg mg·kg^{-1}） to *Oreochomis niloticus* Linn （$\overline{x}\pm s$, $n=3$）

代谢参数/Pharmacokinetic Parameters	肠道/Intestinal	肝脏/Liver	血浆/Plasma
C_{max}/μg·g^{-1}、μg·mL^{-1}	34.98	4.84	3.47
T_{max}/h	2	2	1
$t_{1/2\alpha}$/h	1.277	0.773	0.854
$t_{1/2\beta}$/h	2.124	21.19	10.328
V_1/L·kg^{-1}	229.459	4371.67	7024.08
CLs/L·h^{-1}·kg^{-1}	178.044	506.231	1447.51
AUC_{0-240}/μg·h^{-1}·mL^{-1}	224.657	78.987	27.631
AUC0-∞/μg·h^{-1}·mL^{-1}	224.664	79.015	27.634

V_1：中央室表观容积；AUC：血药浓度时间曲线下面积；$t_{1/2\alpha}$：吸收半衰期；$t_{1/2\beta}$：消除半衰期；CLs：清除率；C_{max}: the maximum concentration；T_{max}: peak time；$T_{1/2\alpha}$: distribution half−life of the drug；$T_{1/2\beta}$: elimination half−life of the drug；AUC: area under the curve；CL_b: total body clearance of the drugs.

2.3 ENR 最大达峰时间尼罗罗非鱼肠道、肝脏组织中的 P−gp 基因表达的差异性分析

利用 PCR 的方法，在对照组和给药实验组尼罗罗非鱼肠道、肝脏组织中均检测出 P−gp 基因的表达。选取 β−actin 基因

为内参基因，利用荧光定量 PCR 的方法测定 ENR 代谢过程中不同时间点尼罗罗非鱼肠道、肝脏组织中 P-gp 基因相对表达量，实验结果见图 3。ENR 代谢的初始时间点（0.017 h）和残留消除末期时间点（240 h）的实验组肠道组织中 P-gp 基因的相对表达量与对照组无显著性差异。当血浆中 ENR 浓度达到最高达峰时（1 h），实验组肠道 P-gp 基因的相对表达量相对于对照组表现出显著性差异（P<0.05）；当肠道中 ENR 浓度达到最高峰时（2 h），实验组肠道 P-gp 基因的相对表达量则表现出极显著差异（P<0.01）。在肝脏组织中，在 ENR 代谢的初始时间点（0.017 h）和残留消除末期时间点（240 h）P-gp 基因的相对表达量与对照组无显著性差异。在血浆中 ENR 的最大达峰时间（1 h）和肝脏中 ENR 的最大达峰时间（2 h），实验组肝脏 P-gp 基因的相对表达量与对照组相比均表现出显著性差异（P<0.05）。以上结果证明，P-gp 在尼罗罗非鱼肠道、肝脏组织中广泛；且其在肠道和肝脏组织中的存在通过对药物的外排作用，其表达可能会影响尼罗罗非鱼体内"首过效应"，其在 mRNA 水平的表达可以进一步作为评价"首过效应"提供了参考依据。

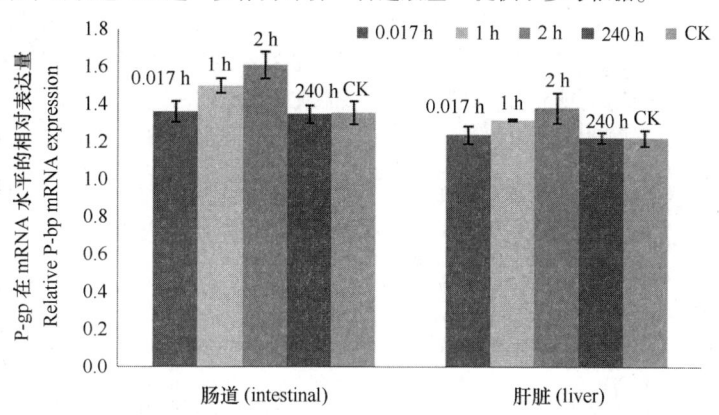

图 3　ENR 代谢过程中尼罗罗非鱼肠道、肝脏组织中的 P-gp 基因表达的差异性分析

Fig. 3　Quantification of P-gp mRNA expression in intestinal and liver tissues

3　讨论

3.1　鱼类组织中 P-gp 基因

P-gp 是一种的跨膜糖蛋白，属于 ATP 结合盒（ATP-binding cassette，ABC）转运蛋白超家族成员之一[3,10]。马云芳（2008）报道了斑马鱼（*Danio rerio*）肠道组织中克隆了 P-gp 基因，获得了其 cDNA 和保守区的氨基酸序列。经同源比较证实：斑马鱼 P-糖蛋白与人和小鼠的 P-糖蛋白同源性分别为 49.4% 和 49.7%；P-gp 基因在软骨鱼纲、硬骨鱼纲等水产动物中的序列保守度可达到 78% 以上[11]。尼罗罗非鱼分类地位上属于硬骨鱼纲，也是中国出口的重要水产经济养殖动物，这是本文选取尼罗罗非鱼作为研究动物的原因。

P-gp 作为一种能量依赖性药物外排泵，能降低细胞内药物浓度，并在内源性、外源性药物化合物的排序中起重要作用[3,10]。Kleinow（2000）通过检测免疫组化（Immunohistochemistry）的方法比较了 catfish 肝脏和肠组织中的 P-gp 在 xenobioti 等环境污染物吸收和外排过程中的作用[6]。以往一般将水生动物体内 P-gp 作为环境污染物的生物标记。P-gp 与药物代谢酶 P450 具有相似的功能（Hochman, et. al., 2001），但在调控机制上又不尽相同。在畜牧动物和人体内，P-gp 被证明能通过介导药物的外排过程而影响药物的代谢过程，是决定药物在动物体内生物利用度的重要因素之一[3,12,13]。但其对水生动物药物代谢过程影响的报道几乎为空白。本文在给药组和对照组的尼罗罗非鱼肝脏和肾脏中均扩增出与预期大小相符的 P-gp 特异性片段（图 1），证实了 P-gp 在尼罗罗非鱼肝脏和肾脏组织中的存在，并为随后分析其在 ENR 代谢过程中的作用提供了方法。

3.2　尼罗罗非鱼肠道、肝脏中 P-gp 的对 ENR 的"首过效应"的关联系分析

ENR 等喹诺酮类药物属于典型的浓度依赖性抗菌药，抗菌谱广，杀菌作用强。ENR 这类浓度依赖性抗菌药通常具有首过效应[1]。尽管关于有大量的关于 ENR 在水产动物组织中代谢规律的研究报道[14~20]，但这些研究报道均未从药效学的角度分析水产动物的"首过效应"问题。肠道和肝脏是鱼类口服给药后药物首先经过的器官，也是 P-gp 的主要分布器官。药物通过肠道和肝脏后，随血液的体内循环分布到全身各个组织。"首过效应"在很大程度上最终体现为 ENR 在鱼类血浆中浓度的变化。由图 2 和表 1 可知，血浆中药物达峰时间最早（1 h），说明口灌给药的 ENR 通过肠道、肝脏进入血浆的速度较快，该结果与张雅斌等（2004）测定的同为鲤科鱼类的鲤的血浆中 ENR 消除规律相似[20]。ENR 口灌给药后首先进入尼罗罗非鱼肠道，实验结果显示 ENR 大部分为在其肠道所吸收，肠道中 ENR 的最高残留量可达 34.98 $\mu g \cdot g^{-1}$，远远高于

肝脏。另一方面，血浆中的 V_1 值高于肠道和肝脏组织（表 1），说明 ENR 在尼罗罗非鱼血浆中具有较大的分布容积。表 1 中 $t_{1/2\alpha}$、$t_{1/2\beta}$ 和 CLs 等参数说明 ENR 在尼罗罗非鱼肠道、肝脏和血浆中的消除速度十分迅速。以上结果证实了 ENR 适于在尼罗罗非鱼养殖过程中作为抗菌药使用而残留的风险很小。

作为药物泵，P-gp 的表达能影响动物内药物代谢过程[21]。本文选取 ENR 代谢过程中不同的时间点为横坐标，评价了肠道和肝脏组织中 P-gp 表达的差异性。实验结果显示：在 ENR 低残留阶段（代谢的初始阶段和代谢末期），实验组肠道和肝脏中 P-gp 在 mRNA 水平的相对表达量与对照组无显著性差异；而在 ENR 高残留阶段（在血浆、肠道/肝脏中的最大达峰时间），实验组肠道和肝脏中 P-gp 在 mRNA 水平的相对表达量显著升高，且肠道组织有极显著性差异。

以上结果显示：①"P-gp 基因的表达与外源性药物 ENR 的摄入及其代谢过程紧密相关"。李静（2009）曾经报道过"环丙沙星（ENR 在动物体内的代谢产物）可能是 P-gp 的底物"[22]。出现这种结果的原因可能是由于 ENR 也是一种 P-gp 的诱导剂。②ENR 的残留量与 P-gp 在 mRNA 水平的相对表达量也存在着相关性。ENR 残留量越高，P-gp 的表达量越高，P-gp 介导的对于药物的外排作用越强。③肠道和肝脏中 P-gp 表达也存在差异；特别是在血浆中 ENR 达到峰值时，肠道中 P-gp 的表达显著增强。该结果一方面说明相对于肝脏，肠道可能对"首过效应"的影响更大，另一方面，肠道中的 ENR 更高的残留量也可能是引起该现象的原因。以上结果也与前面关于 ENR 在尼罗罗非鱼体组织器官水平消除迅速的结论相符。令人感兴趣的是，在某些特定的代谢时间点，P-gp 基因的相对表达量与血药浓度的关联性为从药物酶的角度评价药物"首过效应"提供了一个新的技术方法。

参考文献

［1］ 刘克辛，韩国柱．临床药物代谢动力学［M］．科学出版社，2009，北京．

［2］ Saeki T, Ueda K, Tanigawara Y, et a1. Human P-glycoprotein transports cyctosporin A and FK506［J］．Biol Chem, 1993, 268（9）：6077-6080.

［3］ Vinod Thomas, Kesavan Ramasamy, Rajan Sundaram, Adithan Chandrasekaran. Effect of honey on CYP3A4 Enzyme and P-Glycoprotein activity in healthy human volunteers［J］．Iranian Journal of Pharmacology & Therapeutics, 2007, 6：171-176.

［4］ Hemmer M J, Courtney L A, Ortego L S. Immunohistochemical detection of P-glycoprotein in teleost tissues using mammalian polyclonal and monoclonal antibodies［J］．J Exp Zool. 1995. 272：69-77.

［5］ Miller D S. Daunomycin secretion by killifish renal proximal tubules. Am J Physiol.（Regul Integrat Comp Physiol 38）. 1993. 269：370-379.

［6］ Kleinow K, Doi A, Smith A. Distribution and inducibility of P-glycoprotein in the catfish：immunohistochemical detection using the mammalian C-219 monoclonal. Mar Environ Res. 2000. 50, 313-318.

［7］ Kurelec B, Pivcevic B. Evidence for a multixenobiotic resistance mechanism in the mussel *Mytilus galloprovincialis*［J］．Aquat Toxicol. 1991. 19, 291-302.

［8］ Toomey B H, Epel D. Multixenobiotic resistance in Urechis caupo embryos：protection from environmental toxins［J］．Biol Bull. 1993. 185：355-364.

［9］ Liu J, Chen H, Miller DS, et a1. Overexpression of glutathione S-transferase II and multidrug resistance transport proteins is associated with acquired tolerance to inorganic arsenic［J］．Mol Pharmacol, 200l, 60（2）：302-309.

［10］ Shannon mala bard, Bruce R. woodin, John J. stegeman. Expression of P-glycoprotein and cytochrome P450 1A in intertidal fish（*Anoplarchus purpurescens*）exposed to environmental contaminants［J］．Aquatic Toxicology, 2002, 60：17-32.

［11］ 马云芳．斑马鱼 P-糖蛋白 cDNA 的克隆及表达分析．西南大学硕士论文，2008.

［12］ Lown KS, Mayo RR, Leichtman AB, et a1. Role of intestinal P-glycoprotein（mdrl）in interpatient variation in the oral bioavailability of cyclosporine［J］．Clin Pharmacol Ther, 1997, 62（3）：248-260.

［13］ Marzolini C, Paus E, Buclin T, Kim RB. Polymorphisms in human MDRl（P-glycoprotein）：recent advances and clinical relevance［J］．Clin Pharmacol Ther, 2004, 75：13-33.

［14］ 梁俊平，李健，张茹，王群，刘德月，王吉桥．肌注和口服恩诺沙星在大菱鲆体内的药代动力学比较［J］．水生生物学报，2010，34（6）：1122-1129.

［15］ 房文红，于慧娟，蔡友琼，周凯，黄冬梅．恩诺沙星及其代谢物环丙沙星在欧洲鳗鲡体内的代谢动力学［J］．中国水产科学，2007，14（4）：622-629.

［16］ 郑宗林，唐俊，喻文娟，胡鲲，叶金明，杨先乐．RP-HPLC 法测定中华绒螯蟹主要组织中的恩诺沙星及其代谢产物［J］．上海水产大学学报，2006，15（2）：156-162.

［17］ 余开，陈寅儿，赵青松，金珊，陆彤霞．恩诺沙星在人工感染溶藻弧菌的三疣梭子蟹体内的代谢动力学［J］．台湾海峡，2011，30（2）：257-263.

［18］ 郭娇娇，潘红艳，杨虎，廖鑫，宫智勇，李谷．恩诺沙星在杂交鲟体内的药物代谢动力学［J］．大连海洋大学学报，2011，26（4）：262-266.

［19］ 余培建．药浴给药恩诺沙星及其代谢产物在欧洲鳗鲡体内的药代动力学研究［J］．福建水产，2007，4：38-43.

［20］ 张雅斌，刘艳辉，张祚新，祖岫杰，唐虹，刘铁钢．恩诺沙星在鲤体内的药效学及药动力学研究［J］．大连水产学院学报，

2004, 19（4）：239-242.

[21]　Hochman JH, Chiba M, Yamazaki M, et a1. P-glycoprotein-mediated efflux of indinavir metabolites in Caco-2 cells expressing cytochrome P450 3A4 [J]. J Pharmacol Exp Ther, 2001, 298（1）：323-330.

[22]　李静. P-糖蛋白与药物相互作用模型的建立及其在药物评价中的应用. 军事医学科学院博士论文,

第二节　渔药在鱼体内代谢的模型、消除规律及其残留

一、渔药在鱼体内的药动学模型

药效与药物消除过程存在某种联系，采用体内药动学和体外药效学相结合的方法，建立相应的药动学模型，是研究渔药分布和代谢的一个重要手段。通过抑制效应的 Sigmoid Emax 模型方程得到了治疗斑点叉尾鮰细菌性败血症的最佳临床给药方案（艾晓辉等，2011）。强力霉素的药物动力学和抑制嗜水气单胞菌（Aeromonas hydrophila）的药效学之间关系密切，当 Cmax/MIC>3.96 时，强力霉素能够对嗜水气单胞菌起到持续的抑制作用（丁运敏等，2011）。

该文发表于《水生生物学报》【2011, 35（6）：893-899】

在斑点叉尾鮰血清中强力霉素对嗜水气单胞菌药动-药效模型研究

Study on Pharmacokinetics/Pharmacodynamics Model of Doxycycline Against Aeromonas hydrophila in Serum of Channel catfish ex vivo

艾晓辉[1,2]，丁运敏[2,3]，汪开毓[1]，刘永涛[2]，沈丹怡[2,3]

（1. 四川农业大学, 动物医学院, 雅安 625014；2. 中国水产科学研究院长江水产研究所, 荆州 434000；
3. 华中农业大学水产学院, 武汉 430070）

摘要：为合理应用强力霉素治疗斑点叉尾鮰（Ictalurus punctatus）细菌性败血症, 采用体内药动学和体外药效学相结合的方法, 研究了斑点叉尾鮰血清中强力霉素抗嗜水气单胞菌（A. hydropila）的活性, 数据使用 3p97 和 kinetica4.4 软件分析。结果表明：强力霉素在普通肉汤培养基和血清中对嗜水气单胞菌的最小抑菌浓度（MIC）均为 2.0 μg/mL。斑点叉尾鮰按 20 mg/kg 体重的剂量口灌强力霉素后, 药物吸收迅速、达峰快、消除缓慢, 血浆药物达峰时间（T_{max}）为 2.57 h, 峰浓度（C_{max}）为 1.72 μg/mL, 消除半衰期（$T_{(1/2)\beta}$）为 38.63 h。在半效应室内, 半效浓度参数（EC_{50}）为 16.95 h, 即血清药物浓度为 1.41 μg/mL 时可产生 50% 最大效应。PK-PD 同步模型参数 $C_{max}/MIC_{血清}$ 为 0.86, $AUC_{0 \to 24h}/MIC_{血清}$ 为 20.57 h。通过抑制效应的 Sigmoid E_{max} 模型方程可得到临床起到抑菌效果的最佳给药剂量范围为 10.68-41.42 mg/kg 体重。结论：临床上发生斑点叉尾鮰细菌性败血症时的最佳给药方案为：对出现临床病状的斑点叉尾鮰以 41.42 mg/kg 体重的剂量进行拌饲投喂进行治疗, 1 次/d。临床上预防斑点叉尾鮰细菌性败血症时以 10.68 mg/kg 体重的剂量进行拌饲投喂, 1 次/d。

关键词：强力霉素；嗜水气单胞菌；药动-药效同步模型；斑点叉尾鮰；半体内

强力霉素（doxycycline）又名多西环素、脱氧土霉素, 是半合成的四环素类抗菌药, 具有强效、长效的特点, 无明显肾毒性, 在人类和动物体内均具有广谱抗菌活性、还具有抗原虫能力, 现广泛应用于兽医临床和水产上[1-4]。强力霉素在兽医临床上主要用于防治畜禽的支原体病、大肠杆菌病、沙门氏病、巴氏杆菌病、布氏杆菌病、鹦鹉热及急性呼吸道感染, 也常被用来作为饲料添加剂, 用于防治肠道感染、促生长和提高奶产量[5,6]。在水产上, 常用于防治由嗜水气单胞菌引起的细菌性败血症[7]。

抗菌药物在临床应用过程中出现的耐药性, 已影响到对多种疾病的经验治疗, 由于新药开发费用巨大, 因此只有充分利用现有的抗菌药物, 合理的制定给药方案, 避免病原菌暴露在亚致死药物浓度水平, 来减少耐药性, 才能更好的延长现有抗菌药物的使用寿命[8]。传统给药方案的确定一般是综合考虑抗菌药物体外 MIC 和体内药动学特征, 根据体内药物浓度超过 MIC 的时间来确定给药间隔, 这样的方法在大多数情况下是有效的, 但由于未将血清、机体免疫和药物在机体内浓度的动态过程等因素考虑在内, 因此结果往往与药物在机体内的实际抗菌效果存在差异[9]。研究表明, PK-PD 同步信息确定

抗菌药物的疗效是最合理的，PK-PD 同步信息可以优化抗菌药物的给药方案，减少病原菌对这些抗菌药物耐药性的产生。为了获得抗菌药物的最佳疗效，减少毒副作用、耐药性和降低药物残留，应用抗菌药物的 PK-PD 同步信息确定最佳给药方案正成为现代药理学发展的一个新方向[10-12]。

目前，关于抗菌药物半体内的试验主要集中于陆地哺乳动物，如牛、羊、猪等，药物主要涉及麻保沙星、达氟沙星、阿莫西林、氟苯尼考等[13-15]，并且主要研究血清和组织液中药物的 PK-PD 同步模型。抗菌药物在水产动物半体内 PK-PD 同步模型在国内外均未见报道。本研究探讨了强力霉素口灌给药后在斑点叉尾鮰血清中的 PK 和体外 PD，并将这些结果进行 PK-PD 同步模型分析，为合理使用强力霉素提供理论指导，并为该类药物的使用和研究提供可借鉴的思路。

1 材料和方法

1.1 药品及培养基

强力霉素原粉（含量 ≥98.0%，东北制药总厂）；强力霉素标准品（含量 ≥99.0，德国 Dr. Ehrenstorfer 公司）。乙腈（色谱纯，美国 J. T. Baker 公司），磷酸二氢钠、高氯酸均为国产分析纯。液体培养基为普通肉汤培养基，固体培养基为普通琼脂培养基，均为本实验室配制。

1.2 实验动物及菌株

健康斑点叉尾鮰（*Ictalurus punctatus*）由中国水产科学研究院长江水产研究所窑墩试验场提供，体重为（179.1±8.58）g。实验前在 120 cm×60 cm×100 cm 水族箱中暂养 7 d，每天换水一次。实验用水为曝气 48 h 自来水，用氧气头鼓泡连续充气，保持水中溶氧大于 5.0 mg/L，维持 12 h 光照-12 h 黑暗的光周期。实验期间，水温为 28±1℃，pH 值为 7.5-8.0，实验过程中停饲。

嗜水气单胞菌（*A. hydrophila*）CY200908，2009 年 8 月分离于湖北长阳患暴发性出血病斑点叉尾鮰，经鉴定为嗜水气单胞菌。

1.3 仪器

ACQUITY UPLC 超高效液相色谱（美国 Waters 公司）配 TUV 双通道紫外检测器及 Empower 2 色谱工作站；20PR-520 型自动高速冷冻离心机（日本日立）；Mettler-TOLEDO AE-240 型精密电子天平（梅特勒-托利多公司）；调速混匀器（上海康华生化仪器制造厂）；Startorius PB-10 酸度计（德国赛多利斯公司）；YX400Z 全自动不锈钢双层立式电热蒸汽消毒器（上海三申医疗器械有限公司）；全自动新型生化培养箱、恒温培养振荡器（上海智城分析仪器制造有限公司）；净化工作台（上海新苗医疗器械制造有限公司）。

1.4 给药方案及试样采集

按 20 mg/kg 体重的剂量给斑点叉尾鮰进行单次灌胃，给药前采取空白血样，在给药后血样的采集按照 0.083、0.167、0.25、0.5、1、2、4、6、8、12、24、48、72、120、168 h 的时间进行。每个时间点，取 5 尾实验鱼，尾静脉采血 3 mL，加入盛有肝素钠的 5 mL 离心管中，4 000 r/min 离心 10 min 分离血浆，置-20℃冰箱，待进行血药浓度的测定；另取 5 尾实验鱼，尾静脉采血 3 mL 加入灭菌的 5 mL 离心管，常温放置，待血清析出后，4 000 r/min 离心 15 min，将血清转移至另一无菌的离心管，放入-20℃冰箱中冻存，待进行半体内药效试验。

1.5 普通肉汤和血清中强力霉素对嗜水气单胞菌的 MIC 测定

采用微量稀释法测定强力霉素在普通肉汤培养基和血清中对嗜水气单胞菌 CY200908 的最小抑菌浓度（MIC）。本研究中使用的菌液浓度为 10^6 CFU/mL。相同过程重复 3 次。

1.6 血清中强力霉素对嗜水气单胞菌抗菌活性的测定

首先将无药对照血清，给药后 0.083、0.5、1、2、4、6、8、12、24、48、72、120 h 的血清样品解冻，使用孔径为 0.22 μm 的针式滤器过滤除菌，取 1 mL 除菌血清加入 10 μL 对数期的嗜水气单胞菌 CY200908 培养物，使血清中嗜水气单胞菌的数量大约为 $3×10^7$ CFU/mL，将加入嗜水气单胞菌的血清放入 28℃恒温培养箱中，并于培养后 0、3、6、9、12、24 h 通过平板菌落计数确定血清中活菌数的变化。

1.7 血浆中强力霉素的超高压液相色谱测定

将血浆样品解冻，取 1.0 mL 血浆加入到装有 1.0 mL 5% 高氯酸的 5 mL 离心管中，振动 2 min，然后以 10 000 r/min 离心 5 min。所有上清液转移到干净的 5 mL 离心管中，它们经亲水性滤头（0.22 μm）过滤后进入超高压高效液相色谱系统。

其中一部分提取物（5 μL）用高效液相色谱仪进行分析。斑点叉尾鮰血浆药动学数据的分析使用3p97软件。

色谱条件：ACQUITY UPLC BEH C18（1.7 μm，2.1×50 mm）反相色谱柱；流动相：A相为0.01 mol/L磷酸二氢钠（含10%乙腈），B相为乙腈（体积比4∶1）；柱温：45℃；流速：0.300 mL/min；进样量：5 μL；紫外检测方式采用单波长运行模式，波长为350 nm。

1.8 强力霉素半体内药效数据及药动-药效学同步数据的分析

半体内药效数据使用kinetica4.4软件进行处理，选择抑制效应的Sigmoid Emax模型（Inhibitory Effect Sigmoid E_{max}，$C=0$ at E_{max}，$C=$infinity at E_0）。该模型可用以下公式描述：$E=E_{max}-C_e^N \times (E_{max}-E_0) / (EC_{50}^N+C_e^N)$。这里，$C_e$为半体内效应室药物的$AUC_{0 \to 24h}/MIC_{血清}$，$E$为抗菌效应，即含有强力霉素的血清加入嗜水气单胞菌孵育24 h，孵育前后血清中嗜水气单胞菌数量lgCFU的变化。E_0为包含强力霉素的血清样品加入嗜水气单胞菌孵育24 h后，嗜水气单胞菌达到检测限（10 CFU/mL）时，孵育前后血清中嗜水气单胞菌数量lgCFU的变化。E_{max}为空白对照血清样品中加入嗜水气单胞菌孵育24 h后，孵育前后血清中嗜水气单胞菌数量lgCFU的变化。N为Hill系数，它是描述$AUC_{0 \to 24h}/MIC_{血清}$与效应E直线化后的斜率。当$E=0$时没有抑菌效果，当$E=-3$时可以杀灭99.9%的细菌。

使用体外$MIC_{血清}$数据和斑点叉尾鮰体内的药动学数据计算$AUC_{0 \to 24h}/MIC_{血清}$、$C_{max}/MIC_{血清}$、$T>MIC_{血清}$等药动-药效学同步关系数据。

1.9 给药方案的预测

通过抑制效应的Sigmoid E_{max}模型方程可计算出血清刚开始起到抑菌效果时的ex vivo $AUC_{0 \to 24h}/MIC_{血清}$值和彻底清除病原菌的ex vivo $AUC_{0 \to 24h}/MIC_{血清}$值，代入公式，即可得到临床起到抑菌效果的最佳给药剂量范围。公式为：

$$X=d \times [AUC_{0 \to 24h}/MIC_{血清}(ex\ vivo)] / [AUC_{0 \to 24h}/MIC_{血清}(in\ vivo)]$$

式中，X为药物用量（mg/kg）；d为强力霉素药物动力学用药剂量，即可得到临床起到抑菌效果的最佳给药剂量范围。

2 结果

2.1 体外药效结果

强力霉素在普通肉汤培养基和血清中对嗜水气单胞菌CY200908的MIC均为2.0 μg/mL。

2.2 强力霉素在斑点叉尾鮰体内的药动学数据、参数及药效学数据

斑点叉尾鮰按照20 mg/kg体重的剂量口灌后，血浆中的药物浓度和血清中嗜水气单胞菌数量对数值减少量见表1，使用3p97软件分析药物浓度-时间数据，符合开放性二室模型，药动学参数见表2。强力霉素经消化道吸收进入血液，血药达峰时间（T_{max}）为2.57 h，峰浓度（C_{max}）为1.72 μg/mL，吸收速率（K_a）为0.53/h，分布半衰期（$T_{(1/2)\alpha}$）为1.76 h，消除半衰期（$T_{(1/2)\beta}$）为38.63 h，滞后时间（Lag time）为0.006 h，说明口灌给药方式吸收迅速，达峰快，消除缓慢。

表1　血浆中强力霉素浓度和血清中嗜水气单胞菌对数值数量减少量

Tab. 1　The concentrations of doxycycline in plasma and the decreased value of A. h in serum $n=5$；$\bar{x} \pm SE$

血样采集时间（h） Sampling time of blood	血浆强力霉素浓度（μg/mL） Doxycycline concentration in plasma	$AUC_{0 \to 24h}/MIC_{血清}$（h）	血清中细菌对数值减少量（lgCFU/mL） The decreased value of lgCFU in serum
0.083	0.40±0.12	4.78±1.43	1.02±0.71
0.167	0.46±0.14	5.57±1.65	
0.25	0.53±0.32	6.40±3.83	
0.5	1.28±0.20	15.32±2.39	-0.77±0.31
1	0.86±0.56	10.26±6.71	0.25±0.14
2	1.32±0.19	15.88±2.28	-0.86±0.92
4	3.75±3.03	45.06±36.37	-3.02±1.46
6	1.55±0.88	18.64±10.50	-1.63±0.80
8	0.99±0.30	11.84±3.63	0.10±0.24
12	0.61±0.23	7.26±2.78	0.69±0.53

血样采集时间（h） Sampling time of blood	血浆强力霉素浓度（μg/mL） Doxycycline concentration in plasma	$AUC_{0\to24h}/MIC_{血清}$（h）	血清中细菌对数值减少量（lgCFU/mL） The decreased value of lgCFU in serum
24	0.36±0.19	4.37±2.24	1.11±0.78
48	0.29±0.03	3.43±0.33	1.37±1.61
72	0.20±0.08	2.38±0.93	1.66±1.18
120	0.12±0.05	1.42±0.55	1.79±0.59
168	0.04±0.02	0.49±0.30	

表 2　斑点叉尾鮰单剂量口灌强力霉素（20mg/kg BW）后的血浆药物动力学参数

Table. 2　Pharmacokinetic parameters for doxycycline in channel catfish after single oral administration at 20 mg/kg（BW）

参数 Parameters	单位 Unit	数值 Value
d	μg/kg	20
K_a	h^{-1}	0.53
Lag time	h	0.006
$T_{(1/2)}\alpha$	h	1.76
$T_{(1/2)}ka$	h	1.31
$T_{(1/2)}\beta$	h	38.63
$AUC_{0\to\infty}$	μg·h/mL	47.37
$AUC_{0\to24}$	μg·h/mL	41.14
T_{max}	h	2.57
C_{max}	μg/mL	1.72

注：d 为口灌给药剂量；K_a 为一级吸收速率常数；Lag time 为滞后时间；$T_{1/2Ka}$ 为药物在中央室的吸收半衰期；$T_{(1/2)a}$、$T_{(1/2)\beta}$ 分别为总的吸收和消除半衰期；$AUC_{0\to\infty}$ $AUC_{0\to24}$ 为药-时曲线下面积；T_{max} 为出现最高血药质量浓度的时间；C_{max} 为最高血药质量浓度

Note：d-dose；Ka-First-order rate constant for drug absorption；$T_{(1/2)Ka}$-Absorption half-life in central compartment；$T_{(1/2)a}$，$T_{(1/2)\beta}$-The overall absorption half-life and Elimination half-life；$AUC_{0\to\infty}$，$AUC_{0\to24}$-Area under concentration-time curve；T_{max}-The time at highest peak of concentration；C_{max}-The biggest concentration of the drug in blood

2.3　强力霉素半体内药效学及药动-药效学同步关系结果

含有强力霉素的血清样品体外培养 24 h 的抗菌曲线见图 1。结果可见，空白血清中添加 10^7 左右嗜水气单胞菌培养 24 h 后，数量持续增长到 10^{11} 左右，而用药后的血清中添加同样数量的嗜水气单胞菌培养 24 h 后，细菌数量先不同程度的减少，然后重新开始增长，但增长程度不同。1~8 h 采集血清中，培养 3 h 后细菌数量减少明显，6 h 开始回升，24 h 后增长至 10^6~10^8 右，其中第 4 小时采集血清中，嗜水气单胞菌数量在 3 h 减少到 10^3，一直抑制到 12 h 后才开始继续生长；12 h 后采集血清中，培养 3 h 细菌减少量较小，6~24 h 持续增长至 10^8~10^{10}。

半效应室内 $AUC_{0\to24h}/MIC_{血清}$ 与 E 的药效学参数见表 3，半效应室内 $AUC_{0\to24h}/MIC_{血清}$ 为 0 时不产生抑制作用，E_{max} 为 1.68。半效应室内产生最大抑制作用为 -3.82。半效浓度（EC_{50}）为 16.95 h，即血清药物浓度为 1.41 μg/mL 时可产生 50% 最大效应。

药动-药效同步模型参数见表 4，将斑点叉尾鮰体内的药动学数据与体外测得的药效数据相结合，获得一组新的参数来分析药动学-药效学之间的关系。$C_{max}/MIC_{血清}$ 为 0.86，用药后药物浓度大于 $MIC_{血清}$ 的时间为 5.46 h。$AUC_{0\to24h}/MIC_{血清}$ 是体内 24 h 药时曲线下面积与 $MIC_{血清}$ 比值，本研究中这一比值为 20.57 h。

2.4　给药方案预测结果

从表 3 中可知，当 $AUC_{0\to24h}/MIC_{血清}$ 为 10.98 h 时，开始有抑菌作用；$AUC_{0\to24h}/MIC_{血清}$ 为 42.60 h 时能杀灭 99.9% 的细菌，并抑制剩余细菌生长。从表 4 可知，体内 $AUC_{0\to24h}/MIC_{血清}$ 为 20.57 h。代入公式 $X = d \times$ [targetted $AUC_{0\to24h}/MIC_{血清}$

（exvivo）］／［$AUC_{0\rightarrow24h}$/$MIC_{血清}$（in vivo）］，可得 $X_1 = 10.68$ mg/kg 即为临床最低给药剂量，$X_2 = 41.42$ mg/kg 即为临床最高给药剂量。根据 PK-PD 模型可以求得任意时刻抑菌效果所需的给药剂量，从而为临床更好的用药提供参考，避免滥用药物造成药物残留和细菌耐药。

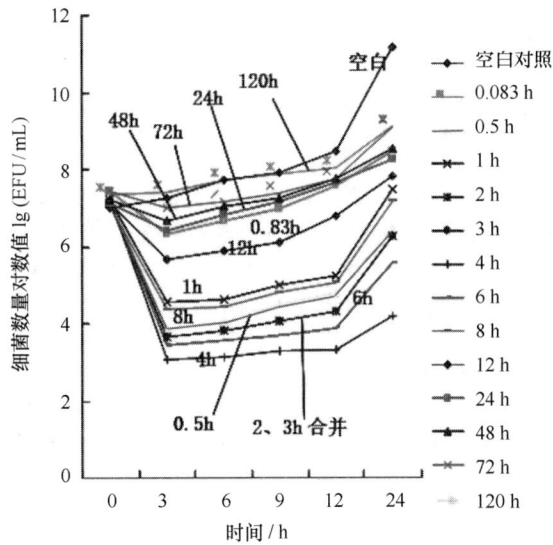

图1 含强力霉素的斑点叉尾鲴血清半体内抗菌曲线

Fig. 1 Ex vivo antibacterial activity curve of doxycycline against A. h in channel catfish serum

表3 AUC0-24h/MIC 血清与 E 的关系参数

Tab. 3 The relationship parameter of AUC0-24h/MIC

参数 Parameter	数值 Value
E_0／（lgCFU/mL）	−3.82
E_{max}／（lgCFU/mL）	1.68
$E_{max} - E_0$／（lgCFU/mL）	5.50
EC_{50}/h	16.95
$AUC_{0\rightarrow24h}$/$MIC_{血清}$ for bacteriostatic action/h	10.98
$AUC_{0\rightarrow24h}$/$MIC_{血清}$ for bacterial action/h	42.60
N	1.89

表4 PD-PK 同步关系的一些参数

Tab. 4 The parameters of PK-PD integration

$AUC_{0\rightarrow24h}$/$MIC_{血清}$（h）	C_{max}/$MIC_{血清}$	$C>MIC_{血清}$（h）
20.57	0.86	5.46

3 讨论

3.1 强力霉素在斑点叉尾鲴血液中的药动学特征

关于鱼类口灌强力霉素后的血浆药动学特征在国外未见报道，国内仅见陈红等[16]等报道了健康奥尼罗非鱼单剂量口灌多西环素 20 mg/kg，水温在（24±0.5）℃，药动学特征是吸收快，消除缓慢，用非房室模型的统计矩分析血药浓度-时间数据，可知达峰时间（T_{max}）为 24 h，最大血药浓度（C_{max}）为 2.27 μg/mL，消除半衰期（$T_{(1/2)\beta}$）为 22.69 h，滞后时间（Lag time）为 35.00 h。本实验得到的出现最高血药质量浓度的时间（T_{max}）为 2.57 h，C_{max} 为 1.72 μg/mL，K_a 为 0.53/h，$T_{(1/2)ka}$ 为 1.76 h，$T_{(1/2)\beta}$ 为 38.63 h，Lag time 为 0.006 h，可知强力霉素在斑点叉尾鲴体内吸收迅速，达峰快，消除缓慢。

这些数据的差异可能跟实验对象的种属以及实验条件（例如温度）的差异相关。本实验斑点叉尾鮰血浆中强力霉素药-时数据符合二室模型，这与杨海峰等[17]报道的健康猪肌注强力霉素（10 mg/kg），R. Bocker[18]等报道的给老年病人一次性静注200 mg强力霉素和V. K. Jha[19]等报道的健康雌性山羊静注5 mg/kg强力霉素后得到的开放性二室模型一致。王金花等[20]研究发现，强力霉素以2.5 mg/kg单剂量肌内注射给猪，药动学模型符合有吸收一室模型。这些模型的差异可能与试验对象的种属、品种差异相关。

3.2 斑点叉尾鮰血清中强力霉素对嗜水气单胞菌的杀菌情况

本研究使用微量稀释法测得强力霉素在普通肉汤培养基和血清中对嗜水气单胞菌CY200908的MIC均为2.0 μg/mL，结果表明在普通肉汤培养基和血清中强力霉素对此株嗜水气单胞菌的抑菌能力较好，此结果与刘金玉[21]的研究一致。斑点叉尾鮰口灌强力霉素后，体外测定血清中的强力霉素对嗜水气单胞菌的药效发现，给药后0.5~8 h的血清中添加的嗜水气单胞菌在3 h时被杀灭程度较高，主要是因为血清中药物浓度达到或者接近有效浓度所致，其中4 h的血清中药物浓度接近2 MIC，在3 h杀灭细菌程度达到99.99%。12~120 h的血清中添加的嗜水气单胞菌在3 h时被杀灭程度不高，说明血清中药物浓度没有达到有效浓度，并且细菌数目从6 h开始迅速增加，但不如空白血清中细菌的增加程度高。这些结果说明，强力霉素是浓度依赖性抑菌药物，药物浓度时决定临床疗效的因素，即药物浓度升高，杀菌活性增强。此类药物剂量对细菌的清除起着非常重要的作用，而药物暴露的时间就不那么重要[22]。其对致病菌的杀菌作用主要取决于峰浓度，可以通过提高C_{max}来提高临床疗效，但不能超过最低毒性剂量[23]。临床给药时，在日剂量不变的情况下，单次给药可以获得较大的峰浓度，从而明显提高抗菌活性和临床疗效。

3.3 斑点叉尾鮰血清中强力霉素对嗜水气单胞菌的PK-PD特征

本研究中，由于强力霉素对细菌发挥的是抑制作用，所以选择抑制效应的Sigmoid Emax模型描述强力霉素对嗜水气单胞菌的浓度-效应关系，每个时间点的血清都可看做一个效应室，对于细菌药物浓度与效应的关系中使用的浓度是$AUC_{0\rightarrow24h}/MIC_{血清}$，效应室室内细菌培养24 h后细菌对数值的减少量。通过计算，半效浓度（EC_{50}）为16.95 h，即血清药物浓度为1.41 μg/mL时可产生50%最大效应。EC_{50}处于S曲线斜度最大处，剂量稍有变化，此处的效应就出现很大的变动，因此敏感度高，重复定性好，是衡量反应强度的重要指标[24]。该结果说明当血浆中药物浓度高于1.41 μg/mL时，抗菌效果才能显著发挥。C_{max}/MIC和AUC/MIC是评价浓度依赖性药物抗菌活性的最重要PK-PD指标[22]。Meagher等[25]的研究表明，对于同为四环素类的替加环素，MIC高达16 mg/L，给药方案为人体腹部注射，首剂量100 mg，间隔12 h给药50 mg，如果$AUC_{0\rightarrow24}/MIC>30$，则治愈率>0.9。本研究中MIC为2.0 μg/mL，体内$AUC_{0\rightarrow24h}/MIC_{血清}$为20.57，从表3可知当$AUC_{0\rightarrow24h}/MIC_{血清}$为42.60 h时能杀灭99.9%的细菌，并抑制剩余细菌生长，说明$AUC_{0\rightarrow24h}/MIC_{血清}$为42.60 h时，治愈率可达99.9%。

4 结论

综合分析认为，强力霉素的抗菌效果与病原菌对强力霉素的敏感性和药物浓度等有关。血清中强力霉素对嗜水气单胞菌有较强的抑菌活性，所以可根据血清中强力霉素间接杀菌活性制定临床治疗斑点叉尾鮰细菌性败血症的给药剂量为10.68~41.42 mg/kg。根据药物动力学结果，血清中强力霉素超过最低抑菌浓度的时间为5.46 h，但由于强力霉素是浓度依赖性药物，暴露时间不是很重要，因此治疗斑点叉尾鮰细菌性败血症时的给药间隔可设置为24 h。同时，根据PK-PD模型可知血清中开始抑菌和彻底清除病原菌时的给药剂量分别是10.68 mg/kg和41.42 mg/kg。因此，临床上治疗斑点叉尾鮰细菌性败血症时的给药方案建议以41.42 mg/kg体重的剂量进行拌饲投喂，1次/d。

参考文献

[1] Ding J R, Wang K Y, Ai X H, et al. Studies on elimination regular ity of doxycycline residues in Ictalurus punctatus at different temperatures [J]. Journal of Fisheries of China, 2009, 33 (4): 672-678 [丁俊仁, 汪开毓, 艾晓辉, 等. 不同水温下强力霉素在斑点叉尾鮰体内的残留消除规律. 水产学报, 2009, 33 (4): 672-678].

[2] Shan Q, Yang F. Research progress on clinical application of doxycycline [J]. Veterinary Orientation, 2009, 148 (12): 38-39 [单奇, 杨帆. 多西环素在兽药临床上的应用. 兽医导刊, 2009, 148 (12): 38-39].

[3] Aronson A L, Pharmacotherapeutics of the newer tetracycline. Journal of the American Veterinary Medical Association [J]. 1980 (176): 1061-1068.

[4] Ole-Mapenay, Mitema E S. Some pharmacokinetic Parameters of Doxycycline in East African Goats after Intramuscular Administration of a Long-acting Formulation [J]. Veterinary Research Communication, 1995 (15): 425-432.

[5] Liu Y P, Leng K L, Wang Q Y, et al. Determination of four tetracyclines residues in aquatic products by High Performance Liquid Chromatography-Tandem Mass Spectrometry [J]. Marine Sciences, 2009, 33 (4): 34-39 [刘艳萍, 冷凯良, 王清印, 等. 高效液相色谱-串联

质谱法测定水产品中的 4 中四环素类药物残留量. 海洋科学, 2009, 33 (4): 34-39].

[6] Liu H. W., Liu Y. M.. Advances in doxycycline research [J]. Veterinary Orientation, 2008, 125 (1): 27-29 [刘辉旺, 刘义明. 多西环素研究进展. 兽医导刊, 2008, 125 (1): 27-29].

[7] Meng X. L., Chen C. F., Wu Z. X., et al. Development of Drug-resistance and Disappear rate of doxycycline against Aeromonas hydrophila [J]. Journal of Yangtze University (Natural Science Edition), 2009, 6 (1): 42-44, 58 [孟小亮, 陈昌福, 吴志新, 等. 嗜水气单胞菌对盐酸多西环素的耐药性获得与消失速率研究. 长江大学学报 (自然科学版), 2009, 6 (1): 42-44, 58].

[8] Durgess D. S., Phamarcodynamics Princinples of Antimicrobial Therapy in the Prevention of Resistance [J]. Chest, 1999, 115 (3): 19-23.

[9] Yang Y. H., Yang D., Ding H. Z., et al. Pharmacokinetics/Pharmacodynamics Integration of Florfenicol against E. coli in Pigs ex vivo [J]. Acta Veterinaria et Zootechnica Sinica, 2009, 40 (2): 243-247 [杨雨辉, 杨东, 丁焕中, 等. 猪半体内氟苯尼考对大肠杆菌的药动学-药效学同步关系研究. 畜牧兽医学报, 2009, 40 (2): 243-247].

[10] Wang R.. The significance of the pharmacokinetics and pharmacodynamics parameters of antimicrobials on dosage regimen rational design [J]. China Pharmacist, 2003, 6 (12): 806-809 [王睿. 抗菌药物 PK/PD 参数对合理设计给药方案的意义. 中国药师, 2003, 6 (12): 806-809].

[11] Gunderson B. W., Ross G. H., Ibrahim K. H., et al. What Do We Really Know About Antibiotic Pharmaco-dynamics [J]. Pharmacotherapy, 2001 (21): 302-318.

[12] Su L. Q.. The elucidation of pharmacokinetics and pharmacodynamics [M]. Beijing: Chemical Industry Press, 2007, 11 [苏乐群. 药效学与药动学诠释. 北京: 化学工业出版社, 2007, 11].

[13] Shojaee A. F., Lees P.. Pharmacokenetics and pharmacokinetic/pharmacodymic integration of marbofloxacin in calf serum, exudates and transudate. Journal of Vererinary Pharmacology and Therapeutics, 2002, 25: 161-174.

[14] Aliabadi F. S., Ali B. H., Landoni M. F., et al. Phramacokenetics and PK-PD modeling of danofloxacin in camel serum and tissue cage fluids. The Veterinary Journal, 2003, 165: 104-118.

[15] Shojaee A. F., Landoni M. F., Lee P.. Pharmacokinetics (PK), Pharmacodynamics (PD), and PK-PD Integration of danofloxacin in sheep biological fluids. Antimicrobial Agents and Chemotheragy, 2003, 47 (2): 626-635.

[16] Chen H., Li Z. L.. The Pharmacokinetics of doxycycline on the health Tilapiagalilaea [C]. Papers (Abstracts) of the 2008 academic conference of China Fisheries [陈红, 李智丽, 曾振灵. 多西环素在健康奥尼罗非鱼的药物代谢动力学研究. 2008 年中国水产学会年会论文摘要集].

[17] Yang H. F., Tan S. H., Wang J. C., et al. Comparison of pharmacokinetics of two different doxycycline injections in pigs [J]. Journal of Huazhong Agricultural University, 2007, 26 (1): 76-79 [杨海峰, 覃少华, 王加才, 等. 2 种多西环素注射液在健康猪体内的药动学比较. 华中农业大学学报, 2007, 26 (1): 76-79].

[18] Bocker R., Mühlberg W., Platt D., et al. Serum Level, Half-life and Apparent Volume of Distribution of Doxycycline in Geriatric Patients [J]. European Journal of Clinical Pharmacology, 1986 (30): 105-108.

[19] Jha V. K., Jayachandran C., Sinagh M. K., et al. Pharmacokinetic Data on Doxycycline and Its Distribution in Different Biological Fluids in Female Goats [J]. Veterinary Research Communications, 1989 (13): 11-16.

[20] Wang J. H., Cao J. Y., Zai K. Y., et al. Pharmacokinetics and residues of doxycycline hydrochloride in pigs [J]. Cinese Journal of Veterinary Science, 2008, 28 (11): 1313-1316, 1320 [王金花, 操继跃, 翟克影, 等. 盐酸多西环素在猪体内的药物动力学及残留. 中国兽医学报, 2008, 28 (11): 1313-1316, 1320].

[21] Liu J. Y., Yang W. M., Li A. H., et al. Preliminary Study on the Etiology of Channel Catfish Intussusception Disease. Acta Hydrobiologica Sinica, 2008, 32 (6): 824-831 [刘金玉, 杨五名, 李爱华, 等. 斑点叉尾鮰套肠症的病原学初步研究. 水生生物学报, 2008, 32 (6): 824-831].

[22] Liu X. H., Guo R. C., Huang M. H.. The significance of the pharmacokinetics and pharmacodynamics parameters of antimicrobials on clinical application [J]. Chinese Journal of Hospital Pharmacy, 2005, 25 (2): 154-155 [刘学红, 郭瑞臣, 黄明慧. 抗菌药物的药动学和药效学参数对临床用药的意义. 中国医院药学杂志, 2005, 25 (2): 154-155].

[23] Ma H. P.. The pharmacokineties and pharmaeodynam ics parameters of antimicrobials and optimized antibosis treatment [J]. Qilu Pharmaceutical Affairs, 2005, 29 (7): 422-425 [马慧萍. 抗菌药物的 PK/PK 参数与优化抗菌治疗. 齐鲁药事, 2005, 29 (7): 422-425].

[24] Sun R. Y., Zheng Q. S.. Fresh studies on serum pharmacology [M]. Beijing: People's Publishing House, 2004, 201 [孙瑞元, 郑青山. 数学药理学新论. 北京: 人民出版社, 2004, 201].

[25] Meagher A K, Passarell J A, Cirincione B B, et al. Exposure-response analysis of the efficacy of tigecycline in patients with complicated skin and skin structure infections [J]. Clin Microbiol Infect, 2007, 51 (6): 1939-1945.

该文发表于《淡水渔业》【2011. 41（4）：75-79】

体外药动模型中强力霉素对嗜水气单胞菌的药效研究

Pharmacodynamic effect of doxycycline on *Aeromonas hydrophila in vitro* pharmacokinetic model

丁运敏[1,2]，刘永涛[2]，沈丹怡[1,2]，余少梅[1,2]，索纹纹[1,2]，艾晓辉[2]

（1. 华中农业大学水产学院，武汉 430070；2. 中国水产科学研究院长江水产研究所，武汉 430223）

摘要： 为保证强力霉素（doxycyline）在防治水产动物疾病上的合理应用，通过建立强力霉素体外药动-药效同步（PK-PD）模型的方法，研究了强力霉素的药物动力学和抑制嗜水气单胞菌（*Aeromonas hydrophila*）的药效学之间的关系。结果显示，强力霉素对嗜水气单胞菌的最小抑菌浓度（MIC）和最小杀菌浓度（MBC）分别为 2.0 μg/mL 和 8.0 μg/mL。在消除半衰期为 38.61 h 的模型内，2 MIC 的强力霉素对嗜水气单胞菌仅能抑制 3 h，模型运行 3 h 后细菌出现再生长。4、8 和 16 MIC 的强力霉素对嗜水气单胞菌能够起到持续的抑制作用。结果表明，强力霉素对嗜水气单胞菌的药效与药物浓度关系密切。当 C_{max}/MIC>3.96 时，强力霉素能够对嗜水气单胞菌起到持续的抑制作用。

关键词： 体外；药动-药效同步模型；强力霉素；嗜水气单胞菌

该文发表于《中国渔业质量与标准》【2011，1（1）：46-53】

长期、低浓度暴露甲苯咪唑在银鲫体内的动态过程研究

Study on the Dynamic Process in Johnny Carp（*Carassius auratus* gibelio）after Long-term，Low Dose Exposure Mebendazole

刘永涛[1]，郭东方[2]，艾晓辉[1]，杨红[1]

（1. 中国水产科学研究院 长江水产研究所，湖北 武汉 430223；2. 湖北省荆州市荆州区水产局，湖北 荆州 434020）

摘要： 在（25±1）℃条件下，将鲫鱼长期暴露于 15 μg·L^{-1} 甲苯咪唑溶液中，研究了甲苯咪唑及其代谢物羟基甲苯咪唑和氨基甲苯咪唑在鲫鱼体内的动态分布及消除规律，同时还对水体中的甲苯咪唑消除规律进行了研究。结果表明：暴露 1 h 后在鲫鱼的皮和肌肉中甲苯咪唑浓度分别为（10.15±5.78）μg·kg^{-1} 和（3.21±1.23）μg·kg^{-1}，并逐渐上升，在暴露 12 h 和 48 h 后分别达到峰值，其浓度分别为（24.98±3.54）μg·kg^{-1} 和（23.97±9.87）μg·kg^{-1}；暴露 2 h 后可在鲫鱼肝脏和血浆中检测到甲苯咪唑浓度分别为（7.87±1.23）μg·kg^{-1} 和（12.22±7.77）μg·L^{-1}，并分别在暴露 72 h 和 48 h 后分别达到峰值，浓度分别为（131.31±4.32）μg·kg^{-1} 和（40.45±9.05）μg·L^{-1}；肾脏组织中甲苯咪唑在暴露 4 h 后才检测到甲苯咪唑，浓度分别为（6.56±1.56）μg·kg^{-1}，并于 72 h 达到峰值，浓度为（50.4±3.56）μg·kg^{-1}。各组织中甲苯咪唑在鲫鱼肝组织中浓度较高且易蓄积，其次为肾脏组织。甲苯咪唑在鲫鱼体内的主要代谢物为氨基甲苯咪唑（MBZ-NH$_2$）和羟基甲苯咪唑（MBZ-OH），在鲫鱼各组织中均有检出，其中鲫鱼肝脏组织中氨基甲苯咪唑和羟基甲苯咪唑的浓度最高，消除最慢，其最高浓度分别为（143.67±10.98）μg·kg^{-1} 和（522.17±8.25）μg·kg^{-1}。用药 1 h 后，水体中甲苯咪唑浓度为 14.56 μg·L^{-1}，288 h 时水体中甲苯咪唑浓度为 0.54 μg·L^{-1}，384 h 后在水体中检测不到甲苯咪唑。水体中并未检测到 MBZ-NH$_2$ 和 MBZ-OH。

关键词： 甲苯咪唑；氨基甲苯咪唑；羟基甲苯咪唑；代谢物；鲫鱼

二、渔药在鱼体内的分布和富集

药物在养殖水环境中的不同分配相和不同生物组织的富集作用差异大。阿维菌素在水体中消解较快，随后由水体向底

泥、伊乐藻和水产动物迁移。生物富集系数FBC值显示对阿维菌素的富集浓度由高到低依次为：鲫鱼>伊乐藻（*Elodea nuttallii*）>中华绒螯蟹>底泥（张卫卫等，2016）。养殖环境底泥中同时存在孔雀石绿和隐性孔雀石绿，以隐性孔雀石绿为主，隐性孔雀石绿呈现蓄积的趋势；而水体中几乎不存在隐性孔雀石绿（杨秋红等，2013）。泼洒呋喃西林后，水体中和斑点叉尾鮰内呋喃西林代谢物氨基脲（Semicarbazide，SEM）降至低于检出限（0.5 μg·kg⁻¹），底泥中氨基脲仍高于检出限；斑点叉尾鮰体内氨基脲含量与水环境密切相关，但鱼体中 SEM 并未出现明显富集（索纹纹等，2013）。喹烯酮（Quinocetone，QCT）在草鱼（*Ctenopharyngodon idellus*）与血浆蛋白结合强度较高，且无浓度依赖性（赵凤等，2014）。

该文发表于《中国水产科学》【2016，23（1）：225-232】

阿维菌素在模拟水产养殖生态系统中的分布、蓄积与消除规律

The distribution，accumulation and elimination of avermectin under a simulated aquaculture ecosystem

张卫卫¹，符贵红¹，王元¹，湛嘉²，房文红¹，沈锦玉³，周俊芳¹，姚嘉赟³

（1. 中国水产科学研究院东海水产研究所农业部东海与远洋渔业资源开发利用重点实验室，上海 200090；
2. 宁波市检验检疫科学技术研究院，浙江 宁波 315012；3. 浙江省淡水水产研究所，浙江 湖州 313001）

摘要：本文采用 UPLC-MS/MS 法，研究了阿维菌素泼洒用药后在模拟水产养殖池塘生态系统中的行为，探明了阿维菌素在水体中的消解规律，在底泥、伊乐藻（*Elodea nuttallii*）和鱼蟹水产动物体内的蓄积与消除规律。结果显示，以 6 μg/L 剂量单次泼洒用药后，水体中阿维菌素消解较快，其半衰期为 63.8 h。阿维菌素在养殖水环境中消减的同时，逐渐由水体向底泥、伊乐藻和水产动物迁移。底泥中阿维菌素峰浓度、AUC 和半衰期分别为 1.25 μg/kg、469.2 μg/kg·h 和 115.5 h，伊乐藻中的相应值分别为 8.75 μg/kg、2 521.7 μg/kg·h 和 315.0 h，伊乐藻对阿维菌素有明显的吸收和富集作用。该模拟系统中的异育银鲫（*Carassius auratus* gibelio）对阿维菌素具有明显的吸收，其血液、肾脏、鳃、肝脏和肌肉组织阿维菌素的最高浓度（C_{max}）依次为 50.9、45.37、21.25、15.47 和 11.9 μg/kg；而该系统中的中华绒螯蟹（*Eriocheir sinensis*）仅鳃组织检出阿维菌素，其 C_{max} 在 12 h 为 8.08 μg/kg，血淋巴、肌肉和肝胰腺等组织均未检出阿维菌素。生物富集系数 F_{BC} 值显示对阿维菌素的富集浓度由高到低依次为鲫鱼>伊乐藻>中华绒螯蟹>底泥，显示阿维菌素在不同分配相和不同生物组织的富集作用差异大。

关键词：阿维菌素；异育银鲫；中华绒螯蟹；伊乐藻；底泥；蓄积与消除

阿维菌素（Avermectin，AVM）属于大环内酯类化合物，主要破坏虫体细胞膜上氯离子通道而使其发生肌肉麻痹死亡[1]，对线虫类、节肢动物幼体以及蠕虫等具有良好杀灭和防治效果，被广泛应用于农业和畜牧业生产[2-4]，也被用于水产养殖防治鱼类中华鳋、锚头鳋[5]、虾蟹寄生虫[6]等疾病。在水产养殖中阿维菌素施用多采用全池泼洒方式，药物进入池塘后在驱虫杀虫的同时，将在池塘不同生态位中进行代谢和迁移，因此为了正确评价其对环境和水产动物源食品的安全性，研究其水产养殖池塘生态系统中的行为和归宿是十分必要的。关于阿维菌素在环境中行为变化的研究主要集中在光降解、水降解及在土壤中的迁移[7-9]，在水产养殖中主要是研究其在单一动物体内的蓄积和消除[10]。本文通过研究阿维菌素在实验室模拟水产养殖池塘生态系统中的环境行为变化，可以认识到阿维菌素在水体、沉积物和生物体不同生态位中蓄积与消除，及在水产动物不同组织、器官中的分布和迁移等规律，为水产养殖中阿维菌素合理应用提供科学依据。

1　材料与方法

1.1　仪器设备

UPLCTM 超高效液相色谱串联 XEVO 系统组成的三重四极杆串联质谱联用仪（Waters，美国）；3-18K 高速冷冻离心机（Sigma 公司，德国）；PT-MR 2100 高速匀浆机（KINEMATICA AG 公司，瑞士）；N-EVAPTM 氮吹仪（Organomation 公司，美国）；真空旋转蒸发仪（Heidolph，德国）；SK3300LHC 超声波清洗仪（科导，中国）；QL-901 旋涡混合器（其林贝尔，中国）；PL2002 电子天平（梅特勒-托利多，瑞士）；AC/KDF150-1-300 净水器（开能，中国）。

1.2 药品与试剂

阿维菌素标准品（92%阿维菌素 B1a，8%阿维菌素 B1b，Dr. Ehrenstorfer，德国）；阿维菌素原粉（B1a 纯度大于 90%，潍坊三江医药集团），1%阿维菌素微乳剂由实验室制备；乙酸乙酯、乙腈和甲醇，均为 HPLC 级（Tedia 公司）；超纯水由 Millipore 公司 Milli-Q Advantage A10 制备；D-SPE 净化填料（含 200mg Al_2O_3、100 mg PSA 和 150 mg C_{18}），博纳艾杰尔公司。

1.3 试验设计

中华绒螯蟹（*Eriocheir sinensis*）（100±20 g），购自江苏省兴化市某养殖场；异育银鲫（*Carassius auratus gibelio*）（165±20）g，购自上海市青浦某养殖场。实验前暂养 3 周，挑选健康鲫鱼 120 尾，蟹 150 只用于实验。伊乐藻（*Elodea nuttallii*）（长度 35±5 cm）和底泥分别采自上海市浦东新区河道和闲置地。上述试验材料在试验前均未检出阿维菌素。

试验在模拟水产养殖生态系统中进行。设置 6 个平行组，均为 1 m×1 m×1 m 水泥池，池底铺 5 cm 厚底泥，植入 40 棵伊乐藻，每两天补水一次保持水深 80 cm，水为经净水器处理的自来水。连续充气平衡 7 d，每个水泥池放入 20 尾异育银鲫和 25 只中华绒螯蟹，为了避免鱼和蟹活动对系统的影响，将其放在网箱中，见图 1。

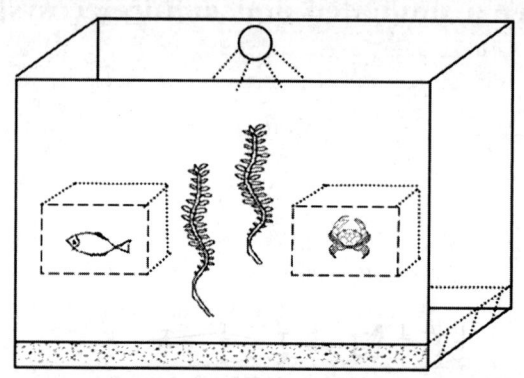

图 1 模拟水产养殖池塘生态系统

Fig. 1 Aquaculture pond ecosystem model

实验期间水温 26~28℃，连续充气，投喂适量颗粒饲料。采用白炽灯照射（12 h 照明、12 h 黑暗），水表面光照强度为 300 Lux。模拟系统稳定 7 d 后泼洒阿维菌素，使其水体浓度达 6 μg/L。在药物泼洒后 0.5 h、2 h、6 h、12 h、1 d、2 d、3 d、5 d、7 d、9 d、12 d、15 d、20 d、25 d、30 d、35 d、40 d 和 45 d 从每个水泥池中随机取 1 条异育银鲫（采集血液、肌肉、肝脏、鳃和肾脏）和 1 只中华绒螯蟹（采集血淋巴、肌肉、肝胰腺和鳃），2 棵伊乐藻，20 g 底泥和 5 mL 水样。血液或血淋巴加入含 2 mg 草酸铵抗凝剂的离心管混匀并离心 5 min（1 000 rpm，20℃），取上清至 2 mL 离心管中保存；水样放入 10 mL 离心管；其他样品直接放入密封袋，所有样品置超低温冰箱冷冻保存。

1.4 样品预处理

水样前处理：取 1 mL 水样，加入 1 mL 甲醇混匀，过 0.22 μm 聚四氟乙烯（PTFE）滤膜，采用 UPLC-MS/MS 分析。

血浆和组织样品前处理：鲫和蟹血浆、肌肉等组织样品的前处理同张卫卫等方法[11]。

底泥样品前处理：称取 10 g 底泥样品至 50 mL 离心管，加入 10 mL 乙腈、10 mL 乙酸乙酯和 5 g 氯化钠，旋涡震荡 1 min，于 4 700 rpm、15℃下离心 5 min，转移上层液至 15 mL 离心管，再用 5 mL 乙腈和 10 mL 乙酸乙酯重复提取一次，合并上层液，氮气吹干，用乙腈定容至 1 mL，加入 0.5 mL 去离子水，旋涡震荡 1 min，超声 2 min，加入 0.45 g D-SPE 和 0.25 g NaCl，旋涡震荡 2 min，离心 8min（10 000 rpm，-4℃），-20℃冷冻 2 h，取上清液过 0.22 μm 滤膜，UPLC-MS/MS 分析，样品处理综合参考邓立刚[12]。

伊乐藻样品前处理：称取剪碎伊乐藻 10 g 至 50 mL 离心管，参考底泥样品前处理提取上层液，合并上层液后，加入 100 mg C18，100 mg 石磨炭黑，漩涡 1 min，离心 5 min（4 700 rpm，15℃），转移上清液至 15 mL 离心管，定容与后续操作同底泥样品前处理，取上清液过 0.22 μm 滤膜，UPLC-MS/MS 分析。

1.5 色谱条件

采用 Waters 超高效液相色谱串联三重四级杆质谱（UPLC-MS/MS）法检测样品中阿维菌素含量，色谱条件参考张卫卫等方法[11]，所有样品的数据均为阿维菌素 B1a 组分的数据。

1.6　数据处理

水样、底泥和伊乐藻中阿维菌素消解方程为 $C_t = C_0 e^{-kt}$，式中：C_0 和 C_t 分别为起始浓度和即时浓度，k 为消解常数，t 为时间；方程拟合采用 Microsoft Excel 2010 计算。异育银鲫和中华绒螯蟹组织中阿维菌素药动学参数采用统计矩原理推算，由药物与统计软件（DAS3.1）求得。水样、底泥和伊乐藻中阿维菌素浓度–时间关系的曲线下面积（AUC）采用梯形法计算。生物富集系数 $F_{BC} = C_f / C_w$，式中：C_f 和 C_w 分别为阿维菌素在生物体内或组织内浓度和水中的浓度。

2　结果

2.1　模拟养殖池塘水和底泥中阿维菌素浓度变化

施药后池塘水体和底泥中的阿维菌素浓度随时间变化见图 2 施药后池塘水体中阿维菌素浓度呈快速下降，在 168 h 时降至 0.62 μg/L，随后下降缓慢，360h 时水体中阿维菌素浓度降至 0.1 μg/L。底泥中阿维菌素含量逐渐升高，在 120h 时达最高值 1.25 μg/kg，随后下降，至 960 h 时底泥中阿维菌素含量为 0.23 μg/kg。运用梯形法求得图 2 中水和底泥阿维菌素浓度–时间关系的曲线下面积（AUC）分别为 322.3 μg/kg·h 和 469.2 μg/kg·h。

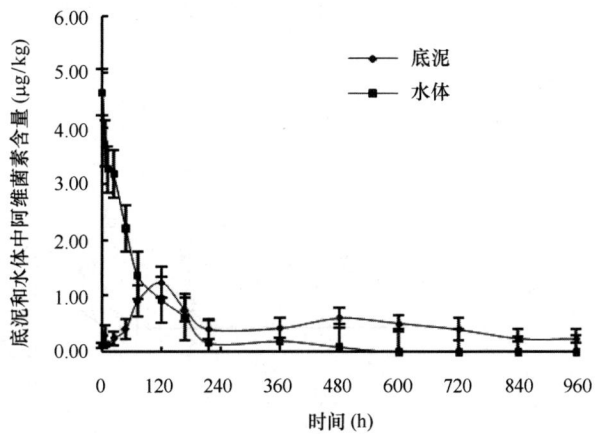

图2　模拟养殖池塘水体和底泥中阿维菌素浓度随时间的变化（$n=6$）

Fig. 2　The change of avermectin concentrations with time in aquatic water and sediments

施药后伊乐藻中阿维菌素含量变化呈现快速升高再下降的趋势。伊乐藻中阿维菌素蓄积在 12 h 时达最高值 8.75 μg/kg，后续阶段呈缓慢下降趋势，720 h 时阿维菌素浓度将至 1.14 μg/kg（图 3）。伊乐藻中阿维菌素浓度–时间关系的曲线下面积（AUC）为 2521.7 μg/kg·h。

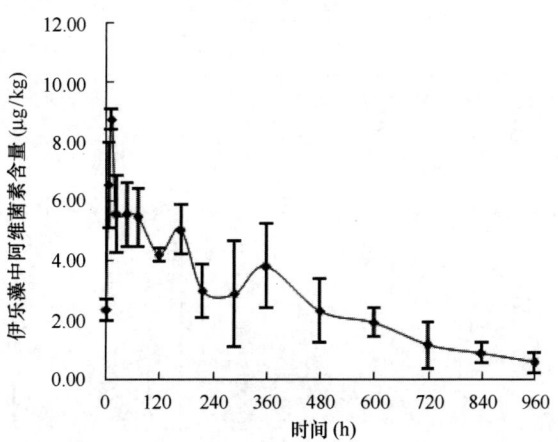

图3　伊乐藻中阿维菌素含量随时间的变化（$n=6$）

Fig. 3　The change of the concentration of AVM with time in *Elodea nuttallii*

阿维菌素在水体、底泥和伊乐藻中消解动态呈指数型消解，其消除曲线方程、相关系数及消除半衰期见表 1。

表1　阿维菌素在水体、底泥和伊乐藻中的消解方程及半衰期

Table 1　Elimination equation and half-life of AVM in aquatic water, sediment and Elodea nuttallii

介质 Media	消解方程 Elimination equation	相关系数（R^2） Correlation coefficient	半衰期 Elimination half-life time
养殖水 Aquatic water	$C_t = 3.780e^{-0.0114t}$	0.986	63.8 h
底泥 Sediment	$C_t = 1.642e^{-0.006t}$	0.907	115.5 h
伊乐藻 Elodea nuttallii	$C_t = 6.475e^{-0.0022t}$	0.943	315.0 h

2.2　阿维菌素在水产动物组织中蓄积与消除

泼洒用药后，异育银鲫和中华绒螯蟹体内阿维菌素浓度随时间变化见图4。异育银鲫血浆、肌肉、肝脏、肾脏和鳃中的阿维菌素浓度总体上呈现先升高再下降趋势，峰浓度依次为50.90 μg/L、11.91 μg/kg、15.47 μg/kg、45.37 μg/kg和21.25 μg/kg，除肾脏达峰时间为120 h，其他组织均为72 h。中华绒螯蟹仅在鳃组织中检测到阿维菌素，其变化趋势与异育银鲫相似，峰浓度为8.08 μg/kg，达峰时间为12 h；而在中华绒螯蟹血淋巴、肌肉和肝胰腺组织中均未检出阿维菌素。

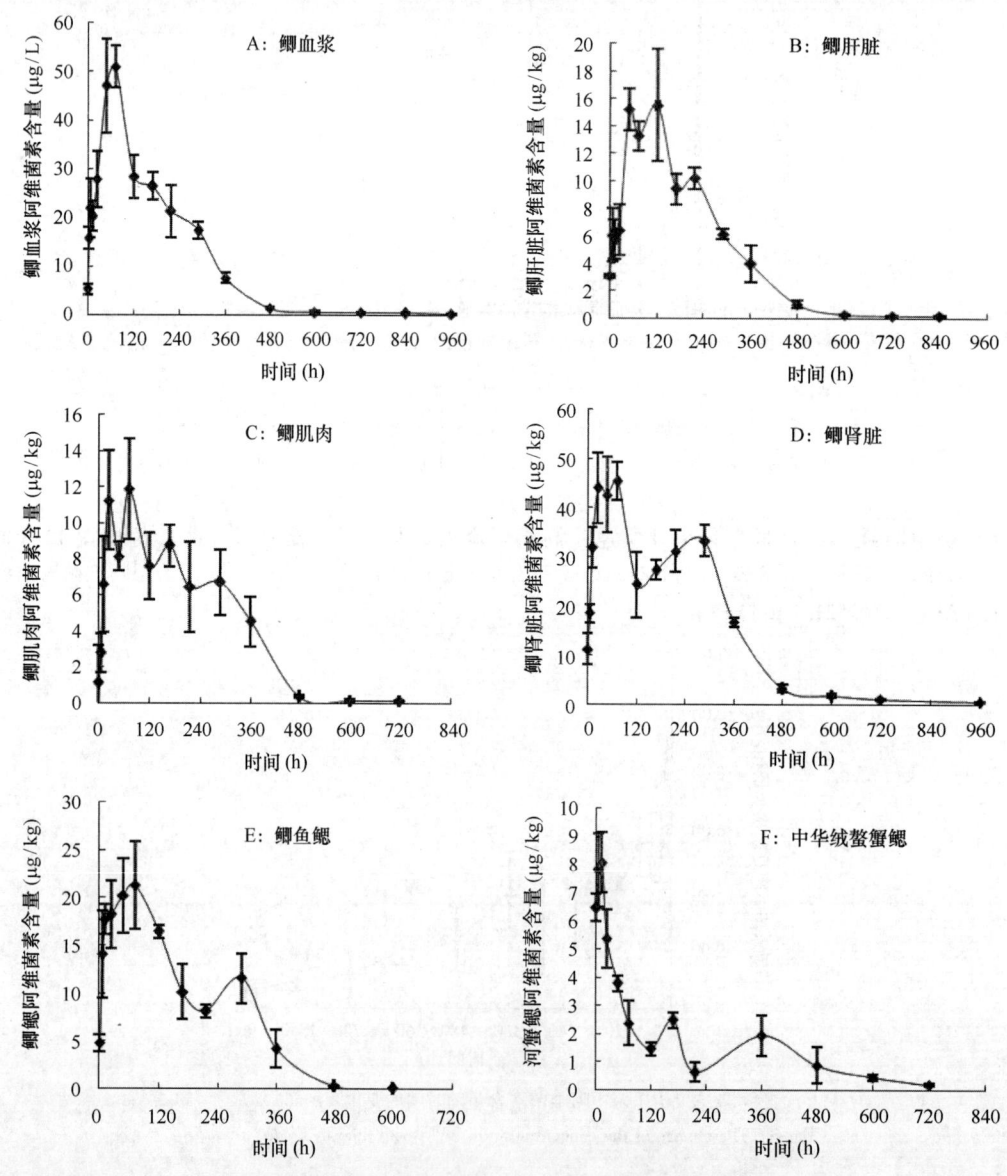

图4　药浴暴露鲫和中华绒螯蟹组织中阿维菌素浓度-时间曲线关系

Fig. 4　AVM concentration-time curves in different tissues of Carassius auratus qibelio and Eriocheir siensises

阿维菌素在异育银鲫和中华绒螯蟹体内蓄积和消除过程的药动学参数采用统计矩原理计算，见表2。阿维菌素在异育银鲫血浆和组织的 AUC_{0-t} 大小顺序依次为肾脏、血浆、鳃、肝脏和肌肉，中华绒螯蟹鳃 AUC_{0-t} 远低于异育银鲫的各个组织。

表2 基于统计矩原理的药动学参数

Table 2 Pharmacokinetic parameters based on statistical moment theoryfollowing ivermectin exposure

药动学参数 Pharmacokinetic parameters		异育银鲫 Carassius auratus					中华绒螯蟹 Eriocheir siensises
		血浆 plasm	肌肉 muscle	肝脏 liver	肾脏 kidney	鳃 gill	鳃 gill
C_{max}	μg/L（kg）*	50.90	11.91	15.47	45.37	21.25	8.08
T_{max}	h	72	72	120	72	72	12
AUC_{0-t}	μg/L（kg）·h*	10031.2	3070.5	3857.5	12892.0	4886.4	1103.7
$AUC_{0-\infty}$	μg/L（kg）·h*	10035.1	3085.0	3865.2	12950.6	4894.6	1139.0
MRT_{0-t}	h	164.5	188.6	183.6	200.3	160.0	222.9
$MRT_{0-\infty}$	h	166.0	191.6	184.8	203.1	160.8	243.4
$t_{1/2}$	h	69.17	74.25	64.69	73.20	47.05	113.55

＊注：血浆 C_{max} 和 AUC 的单位分别为 μg/L 和 μg/L·h，其他组织的单位分别为 μg/kg 和 μg/kg·h。

＊Note：The plasma concentration of C_{max} and AUC units is μg/L and μg/L·h，respectively，other tissue units were μg/kg and μg/kg·h，respectively。

2.3 底泥、伊乐藻和水产动物的富集作用

图5为底泥、伊乐藻和水产动物不同组织对阿维菌素的富集作用。阿维菌素的生物富集系数由高至低依次为鲫鱼、伊乐藻、中华绒螯蟹和底泥，鲫鱼体内组织以血浆和肾脏富集倍数最高，其次为鳃和肝脏，肌肉富集量最低；中华绒螯蟹仅在鳃中富集。结果显示，阿维菌素在水产动物体内的富集存在着种属差异，同一生物体内不同组织器官对阿维菌素的富集能力也存在着差异。

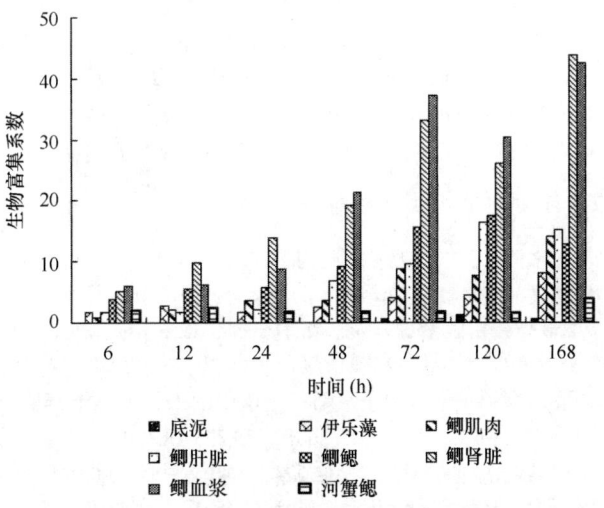

图5 底泥、伊乐藻和水产动物对阿维菌素的富集

Fig.5 The bioconcentration factors of AVM in sediments，between different compartments and water borne

注：图中生物富集排列顺序从左至右为：6-48 h 为伊乐藻-鲫肌肉-鲫肝脏-鲫血浆-中华绒螯蟹，
72-168 h 为底泥-伊乐藻-鲫肌肉-鲫肝脏-鲫血浆-中华绒螯蟹。

3 讨论

3.1 阿维菌素养殖池水中的光解和水解

阿维菌素作为一种大环内酯类抗生素类驱虫药，应用于水产养殖防治中华鳋、锚头鳋等寄生虫[5]。全池泼洒进入池塘生态系统后，可经光、水、微生物和动植物等多种途径降解[7,8,13]。有研究报道，阿维菌素在中性条件下比较稳定，避光条件下水解速率较慢，水温25℃、pH值为7.0水中降解半衰期为78.8 d，在杭州华家池池水（pH 7.21）中降解半衰期为88.9 d[14]。在不同光照强度下，阿维菌素降解速率随着光照强度增强而提高，在1 800 lx时半衰期为19.64 h，而在11 000 lx时半衰期仅为3.85 h[15]。本研究的光照强度设为300 lx，主要是考虑到水产养殖池塘池水透明度很低，试验水体中阿维菌素降解半衰期虽大于1 800 lx光照强度下的光解半衰期[15]，可从光照强度差异得到解释，但远远小于阿维菌素水解半衰期[8]。通过上述比较分析，作者认为在模拟池塘系统下阿维菌素的降解主要是通过光解作用降解。

3.2 水生植物对池水阿维菌素的蓄积和降解

国外研究报道了水生植物对有机磷农药的吸附、蓄积和代谢。Brock等在淡水生态系统模型中发现，水中大部分毒死蜱被伊乐藻吸附，只有很小比例沉积到底泥中[16]，毒死蜱能在沉水水生植物中蓄积主要是通过从水中吸收和从污染的底泥中吸收[17]，然而毒死蜱在植物中代谢的速率很慢[18]。本研究结果与上述毒死蜱的结果比较一致，在伊乐藻中阿维菌素的AUC和生物浓缩因子F_{BC}值高于底泥，说明伊乐藻能很好地蓄积阿维菌素，这是因为阿维菌素具有良好的层移活性，阿维菌素能渗透入叶片组织，并在表皮薄壁细胞内形成药囊并长期储存[19]，因此阿维菌素在伊乐藻中的消除半衰期很长。鉴于水生植物对环境污染物的蓄积和代谢作用，很多研究报道了利用水生植物来净化环境水质，如农药、抗生素和重金属，并且在实践中发挥了重要作用[20]。

3.3 阿维菌素在水产动物体内吸收和蓄积存在着种属差异

阿维菌素泼洒到养殖水环境中，除向底泥、伊乐藻等环境介质中迁移外，也可在水产动物体内富集。从C_{max}、AUC和F_{BC}值等参数来看，异育银鲫对阿维菌素具有很强的吸收和富集作用，在鲫肾脏、鳃、肝脏和肌肉组织中有明显的蓄积；而中华绒螯蟹仅在鳃组织中检出阿维菌素，其他组织均未检出。结果表明，不同水产动物对阿维菌素的吸收和蓄积存在着种属的差异。造成阿维菌素在鲫和蟹之间吸收和蓄积的差异很可能与它们的体表结构不同有关，异育银鲫可以通过皮肤从水环境中直接吸收[21]，而中华绒螯蟹体表是甲壳，甲壳的隔离作用导致其不能由体表直接吸收药物[22]。中华绒螯蟹只有通过鳃吸收水体中的阿维菌素，然而，在同一水环境中，中华绒螯蟹鳃中阿维菌素峰浓度和AUC（分别为8.08 μg/kg和1 103.7 μg/kg·h）均小于异育银鲫鳃（分别为21.25 μg/kg和4 886.4 μg/kg·h），说明蟹的鳃对阿维菌素的亲和性低于鲫鱼，由此可以解释中华绒螯蟹对阿维菌素吸收少于异育银鲫。另外，阿维菌素对异育银鲫的96 h LC_{50}为0.98 mg/L～2.15 mg/L[23]，远低于阿维菌素对中华绒螯蟹幼蟹的96 h LC_{50}（为73.44 mg/L）[24]，也就是说阿维菌素对中华绒螯蟹的毒性低于异育银鲫，这可以从中华绒螯蟹对阿维菌素的吸收大大少于异育银鲫得到解释。

3.4 阿维菌素在水产养殖环境中消解快、残留低

阿维菌素作为杀虫剂泼洒到养殖水体中，不仅要考虑到药物在水产动物体内的残留消除，还要考虑在养殖水和底泥环境中的残留消解。我国规定了动物源食品（MRL为20～100 μg·kg⁻¹）[25]和农产品（MRL为10～100 μg·kg⁻¹）中阿维菌素的最高残留限量[26]，没有土壤中最高残留限量的规定。在本研究中以6 μg/L剂量施用阿维菌素后，底泥中阿维菌素C_{max}为1.25 μg/kg，均低于动物源食品和农产品的最高残留限量。再者，阿维菌素在粮食、蔬菜、水果等农作物和土壤中消解规律研究报道较多。阿维菌素在湿润稻田土壤和淹水稻田土壤中的平均消解半衰期分别为17.20 d和13.51 d，在黄瓜、柑桔和油菜等土壤中半衰期分别为7.9～18.7 d[27]、7.86 d[28]和3.5～3.9 d[9]。本研究的模拟池塘底泥中阿维菌素消解半衰期除略大于油菜土壤外，均低于其他作物的土壤。阿维菌素在稻田水中平均消解半衰期为2.52 d[29]，与本研究模拟养殖池水消解半衰期接近。根据《化学农药环境安全评价试验准则》规定，土壤中农药的消解半衰期<30 d时，则将该农药归为易消解农药，本研究中在模拟养殖水环境中的阿维菌素属于易消解型农药。

参考文献

[1] Omura S. Ivermectin: 25 years and still going strong [J]. International journal of antimicrobial agents, 2008, 31 (2): 91-98.

[2] Halley B A, VandenHeuvel W J, Wislocki P G. Environmental effects of the usage of avermectins in livestock [J]. Veterinary Parasitology, 1993, 48 (1): 109-125.

[3] Chung K, Yang C, Wu M, et al. Agricultural avermectins: an uncommon but potentially fatal cause of pesticide poisoning [J]. Annals of emergency medicine, 1999, 34 (1): 51-57.

[4]　Lasota J A, Dybas R A. Avermectins, a novel class of compounds: implications for use in arthropod pest control [J]. Annual review of entomology, 1991, 36 (1): 91-117.

[5]　E Horsberg T. Avermectin use in aquaculture [J]. Current pharmaceutical biotechnology, 2012, 13 (6): 1095-1102.

[6]　尹敬敬. 阿维菌素对日本沼虾的毒性作用及其药物代谢动力学研究 [D]. 上海海洋大学, 2011.

[7]　刘卫国, 朱欣妍, 尹明明, 等. 阿维菌素 3 种剂型的光解研究 [J]. 农业环境科学学报, 2012, 31 (10): 1906-1912.

[8]　张卫, 虞云龙, 谭成侠, 等. 阿维菌素水解动力学的研究 [J]. 农业环境科学学报, 2004, 23 (1): 174-176.

[9]　袁蕾, 王会利, 李建中. 阿维菌素在油菜和土壤中残留及降解行为研究 [J]. 2011.

[10]　王锡珍, 陆宏达. 关于阿维菌素对异育银鲫的急性毒性和组织病理研究 [J]. 大连水产学院学报, 2010, 25 (1): 66-70.

[11]　张卫卫, 湛嘉, 王元, 等. 药浴暴露下阿维菌素在异育银鲫体内蓄积和消除规律 [J]. 海洋渔业, 2014, 36 (5): 461-468.

[12]　邓立刚. 阿维菌素 (abamectin) 在甘蓝和土壤中的残留检测技术及降解规律研究 [D]. 山东师范大学, 2009.

[13]　张卫, 林匡飞, 蔡兰坤, 等. 阿维菌素的微生物降解及其机理研究 [J]. 生态环境, 2007, 16 (2): 421-424.

[14]　张卫, 虞云龙, 林匡飞, 等. 阿维菌素在土壤中的微生物降解研究 [J]. 应用生态学报, 2005, 15 (11): 2175-2178.

[15]　张卫, 林匡飞, 虞云龙, 等. 农药阿维菌素在水中的光解动态及机理 [J]. 生态环境学报, 2009, 18 (5): 1679-1682.

[16]　Brock T C, Crum S, Van Wijngaarden R, et al. Fate and effects of the insecticide Dursban © 4E in indoor Elodea-dominated and macrophyte-free freshwater model ecosystems: I. Fate and primary effects of the active ingredient chlorpyrifos [J]. Archives of Environmental Contamination and Toxicology, 1992, 23 (1): 69-84.

[17]　Chambers J E, Levi P E. Organophosphates: chemistry, fate, and effects (ed) [M]. Academic Press, 1992. 47-48.

[18]　Smith G N, Watson B S, Fischer F S. Investigations on dursban insecticide. Uptake and translocation of [36cl] o, o-diethyl o-3, 5, 6-trichloro-2-pyridyl phosphorothioate and [14c] o, o-diethyl o-3, 5, 6-trichloro-2-pyridyl phosphorothioate by beans and corn [J]. Journal of Agricultural and Food Chemistry, 1967, 15 (1): 127.

[19]　杨会荣. 阿维菌素和高氯在甘蓝和土壤中的残留消解动态及阿维菌素的光解研究 [D]. 安徽农业大学, 2009.

[20]　阿丹. 人工湿地对 14 种常用抗生素的去除效果及影响因素研究 [D]. 暨南大学, 2012.

[21]　McKim J M, Nichols J W, Lien G J, et al. Dermal absorption of three waterborne chloroethanes in rainbow trout (Oncorhynchus mykiss) and channel catfish (Ictalurus punctatus) [J]. Toxicological Sciences, 1996, 31 (2): 218-228.

[22]　佟蕊, 成永旭, 吴旭干, 等. 3 种不同栖息环境下蟹鳃的超微结构, 脂类组成及含量的比较 [J]. 水产学报, 2011, 35 (9): 1426-1435.

[23]　周帅, 房文红, 吴淑勤. 渔用阿维菌素水乳剂的安全性和药效评价 [J]. 上海海洋大学学报, 2009 (3): 327-331.

[24]　李赫, 宋文华, 李文宽, 等. 三种常用农药对中华绒螯蟹幼蟹的急性毒性研究 [J]. 水产学杂志, 2013, 26 (6): 44-47.

[25]　中华人民共和国农业部第 235 号公告. 动物源性食品中兽药最高残留限量 [S]. 北京: 2002.

[26]　GB 2763-2012 食品中农药最大残留限量 [S]. [C] //北京: 2012.

[27]　金芬, 王, 魏闪闪, 等. 阿维菌素在黄瓜和土壤中的残留及其消解动态 [J]. 中国农业科学, 2014, 47 (18): 3684-3690.

[28]　傅强, 杨仁斌, 廖海玉, 等. 阿维菌素在柑桔和土壤中的消解动态研究 [J]. 分析科学学报, 2011, 27 (1): 85-88.

[29]　张杰. 阿维菌素在稻田中的生态环境行为与效应研究 [D]. 湖南农业大学, 2011.

该文发表于《淡水渔业》【2013, 43 (5): 43-49】

孔雀石绿及其代谢物在斑点叉尾鮰体内及养殖环境中的消解规律

Elimination rules of malachite green in aquaculture environment and assessment of the residual in the Channel catfish (*Ietalurus punetaus*) tissue

杨秋红[1], 刘永涛[1], 艾晓辉[1], 王群, 索纹纹[1,3], 吕思阳[1,2]

(1. 中国水产科学研究院长江水产研究所, 农业部淡水鱼类种质监督检验测试中心, 武汉 430223;
2. 中国水产科学研究院, 北京 100141; 3. 华中农业大学水产学院, 武汉 430070)

摘要: 研究了以全池泼洒的投药方式, 孔雀石绿 (池塘中孔雀石绿的理论浓度为 1 mg/L) 及其主要代谢物隐性孔雀石绿在斑点叉尾鮰肌肉和皮肤, 以及养殖水体和底泥中的残留消除规律。采用高效液相色谱串联质谱法 (HPLC-MS/MS) 分析孔雀石绿及其代谢物隐性孔雀石绿在斑点叉尾鮰体内及环境中的浓度水平。结果显示: 肌肉、皮肤中孔雀石绿用药后第 1 天最高浓度分别为: (42.77±5.26) μg/kg 和 (6.36±0.11) μg/kg, 肌肉、皮肤消除半衰期 $T_{1/2}$ 分别为 57.76 d、31.51 d; 皮肤和肌肉中隐性孔雀石绿分别在用药后第 3 d 和第 1 d 达到最高 (502.27±

20.43）μg/kg 和（125.26±12.76）μg/kg，肌肉、皮肤消除半衰期 $T_{1/2}$ 分别为 33.01 d、38.51 d。这表明孔雀石绿在斑点叉尾鮰体内会迅速转化为隐性孔雀石绿，且隐性孔雀石绿残留在皮肤中的浓度大于肌肉中的浓度。养殖环境底泥中同时存在孔雀石绿和隐性孔雀石绿，以隐性孔雀石绿为主，并且隐性孔雀石绿呈现蓄积的趋势，在第 360 d 出现最高浓度（5.92±1.23）μg/kg；水体中孔雀石绿最高浓度在第 1 d 的（46.44±7.39）μg/L，随后急剧降至 1 μg/L 左右，水体中几乎不存在隐性孔雀石绿。

关键词：斑点叉尾鮰（*Ietalurus punetaus*）；孔雀石绿；养殖环境；消除；残留

该文发表于《农业环境科学学报》【2013，32（4）：681-688】

环境中氨基脲消解规律及对斑点叉尾鮰残留评估

Elimination Rules of the Semicar bazide in Environ mentand Assessment of Semicar bazide in the Channel Catfish（*Letalurus Punetaus*）Tissue

索纹纹[1,2]，刘永涛[2]，艾晓辉[2*]，杨秋红[2]，吕思阳[1,2]，沈丹怡[3]

（1. 华中农业大学水产学院，武汉 430070；2. 中国水产科学研究院长江水产研究所农业部淡水鱼类种质监督检验测试中心，武汉 430223；3. 浙江慈溪市水产技术推广中心，浙江 慈溪 315300）

摘要：为研究环境中呋喃西林代谢物氨基脲（Semicarbazide，SEM）在池塘和鱼体中的蓄积消解规律，采用全池泼洒呋喃西林的给药方式，采集环境（水样和底泥）和斑点叉尾鮰组织（皮肤和肌肉）样品，采样工作持续 1 a，并采用高效液相色谱串联质谱（HPLC-ESI/MS/MS）法对样品中 SEM 含量进行了分析。结果表明：水体中 SEM 初始阶段具有较高的消除速率，随后消除速率趋于平缓，并维持较长时间，底泥中 SEM 浓度随时间呈非线性下降；实验前 5 d，鱼皮中 SEM 出现了短暂蓄积后随时间呈线性下降，肌肉中 SEM 消除规律与水体中相似；水体、肌肉和鱼皮的消除半衰期 $T_{1/2}$ 分别为 1.7、2.09、2.42 d，水环境与斑点叉尾鮰组织中 SEM 浓度值存在显著相关性（$P<0.01$）；至实验结束时，水体中和鱼体内氨基脲最终降至低于检出限（0.5 μg·kg^{-1}），底泥中氨基脲仍高于检出限；斑点叉尾鮰体内氨基脲含量与水环境密切相关，但鱼体中 SEM 并未出现明显富集。

关键词：养殖环境；斑点叉尾鮰；呋喃西林代谢物；氨基脲；消除；蓄积

呋喃西林（Nitrofurazone，NFZ）是一种化学合成的硝基呋喃类广谱抗菌药，在生物体内的半衰期短，进入动物体内数小时即代谢为氨基脲（Semicarbazide，SEM）[1]。大量研究表明，SEM 能够与组织蛋白质紧密结合，长时间残留于体内，在酸性条件下可从蛋白质中释放而被机体吸收[2]，并具有致癌、致突变和生殖毒性[3-5]。由于硝基呋喃类药物残留事件严重威胁消费者的身体健康，因此在欧盟[6]、日本[7]和我国[8]呋喃西林原形药在食用畜禽及水产动物中均为禁用，我国也已制定相应法律法规来管制此类事件的发生，但目前未使用呋喃类药物而实际检测中呋喃西林代谢物含量超标的现象时有发生，其原因尚无定论。有人指出，在水产品未经过加工包装上市之前呋喃西林代谢物的检出可能是由于环境中用药残留引起的蓄积所致，但目前国内外均尚未对此可能性进行相关的研究。本研究以斑点叉尾鮰为受试对象，通过对养殖水体进行直接泼洒呋喃西林的方式，首次对环境中呋喃西林代谢物 SEM 在水体、底泥、鱼肉和鱼皮中的蓄积和消解规律进行初步研究，评估环境中使用呋喃西林对斑点叉尾鮰可食组织残留的风险，为确定斑点叉尾鮰中氨基脲残留的渊源提供参考，并为进一步规范药物使用管理以及加强水产品中硝基呋喃类药物的监控提供科学依据。

1 材料与方法

1.1 材料

实验动物：健康斑点叉尾鮰由湖北省仙桃市水产局提供，平均质量为（124.13±25.2）g。试验前在池塘的网箱（1 m×2 m×1.5 m）内暂养 1 周。试验期间每日上午 8 点和下午 6 点投喂斑点叉尾鮰饲料，并记录水温。

药品及试剂：呋喃西林代谢物（SEM）标准品（纯度>93.5%，Dr. EhrenstorferGmbH）；SEM-^{13}C-^{15}N2 标准品（纯度>99%，德国 Wegita）；呋喃西林原粉（纯度>90.1%，济南金达药化有限公司）；甲醇（色谱纯，美国 J. T. Baker）；LC-MS-water（美国 CNW）；乙酸铵（分析纯，美国 J. T. Baker）；二硝基苯甲醛（分析纯，北京恒业中远化工有限公司）。

仪器：高效液相色谱-串联质谱（Suryeyor ms pumpplus，suryeyorautos am plerplus，Thermo tsq quantumacessmax）及

ThermoLCquan2.6 数据采集处理软件；自动高速冷冻离心机（日本 Hitachi20PR-520 型）；Mettler-toledoAE-240 型精密电子天平（梅特勒-托利多公司）；FS-1 高速匀浆机（华普达教学仪器有限公司）；调速混匀器（上海康华生化仪器制造厂）；氮吹仪（Aosheng，杭州奥盛仪器有限公司）。

1.2　方法

1.2.1　试验设计与采样

实验设计：试验池为长 35.8 m，宽 20 m，平均水深 0.9 m，水体积为 644.0 m³ 的土池。两个同样大小网箱（1 m×2 m×1.5 m）与养鱼池垂直放置，平行排列于池中，离池长边的距离均为 4 m，距池宽边的距离为 17 m（图 1）。称取 644 g 呋喃西林原粉，于 10 L 水中进行一定的溶解后，将呋喃西林溶液进行全池泼洒，使池中呋喃西林溶液理论浓度为 1 mg·L⁻¹。试验期间（2011 年 07 月 11 日至 2012 年 06 月 05 日）每日上午 8 点和下午 6 点投喂斑点叉尾鮰饲料，并记录水温。于给药后 1、3、5、10、15、30、45、60、90 d……采集肌肉、皮肤，同时分别随机采集池边及池中水样（水面下约 20 cm 处）及底泥（约 5 cm 深度）样品，置-20℃冰箱避光保存，待测定。

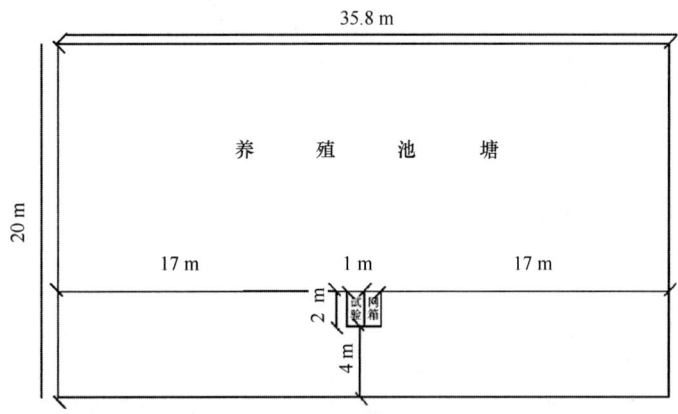

图 1　试验网箱在养殖池塘中的位置

Fig. 1　The position of net cages in pond

样品采集：每一时间点各取 5 尾鱼和 5 份环境样品，作为 5 个平行样品分别处理测定。

1.2.2　样品检测

按照中华人民共和国农业部 783 号公告-1-2006《水产品中硝基呋喃类代谢物残留量的测定液相色谱-串联质谱法》进行测定，略有改动。

水解和衍生化：将冷冻保存的肌肉、皮肤、水样和底泥样品室温下自然解冻，肌肉称取 2.0 g（精确到 0.01 g）、皮肤根据样品量称取并记录称取量和底泥称取 10.0 g（精确到 0.01 g）均于 50 mL 离心管中，水样取 50 mL 于 120 mL 塑料管中，分别加入 0.05 mL 100 ng·mL⁻¹ 内标工作液（SEM-¹³C-¹⁵N₂）涡旋 50 s，再加入 100 μL 浓盐酸和 0.15 mL 0.05 mol·L⁻¹2-硝基苯甲醛溶液，涡旋 50 s 后，37℃避光振荡 16 h。

提取净化：取塑料管和离心管冷却至室温，各加入 1.0 mol·L⁻¹ 磷酸氢二钾溶液，调节 pH 至 7.0~7.5，水样、底泥、肌肉和皮肤各加入 30、20、10、10 mL 乙酸乙酯剧烈振荡混匀 5 min，静置 30 min，取上层清液于鸡心瓶中；再重复上述操作，合并上清于鸡心瓶中，于 40℃旋转蒸发至近干；肌肉和鱼皮样品采用 1 mL 初始流动复溶，过 0.2 μm 滤膜，待测；水样和底泥样品采用 2 mL 乙酸乙酯反萃，重复 3 次，萃取液于 10 mL 离心管中，于 40℃氮吹至近干，1 mL 初始流动复溶，过 0.2 μm 滤膜，待测。

（1）色谱条件

Hypersil GoldC18（100 mm×2.1 mm×3 μm）反相色谱柱；柱温 30℃；流速 0.2 mL·min⁻¹；进样量 20.0 μL。梯度洗脱条件（表 1）：A 相为甲醇，B 相为 5 mmol·L⁻¹ 乙酸铵溶液。

表 1　流动相梯度洗脱程序

Table 1　The mobile phase gradient program

时间（min）	A（%）	B（%）	流速（mL·min⁻¹）
0	20	80	0.2
3.00	40	60	0.2

时间（min）	A（%）	B（%）	流速（mL·min⁻¹）
8.00	40	60	0.2
8.01	20	80	0.2
11.00	20	80	0.2

（2）质谱分析条件

加热大气压电喷雾离子源（HESI），正离子模式；采用选择反应监测（SRM）模式；喷雾电压 3 000 V；源内解离电压 0V；鞘气压力 40 arb；辅助气压力 20 arb；碰撞气（氩气）压力，1.5 mTorr；离子传输毛细管温度 350℃；Q1PW0.7，Q3PW0.7。母离子、定性离子对、定量离子对和碰撞能量见表 2。

表 2　呋喃西林代谢物及其同位素内标质谱条件
Table 2　Spectometric parameters of SEM and SEM-^{13}C-^{15}N$_2$

分析物 Analyst	母离子 Parention	子离子 Pruduction	碰撞能量 Collision energy/V
SEM	209	166*	7
		192	11
SEM-^{13}C-^{15}N$_2$	212	168	9

注：离子对为定量离子对。

（3）标准曲线的制备及回收率与精密度测定

标准曲线的制备：配制含 SEM 浓度为 1、2、5、10、20、50 ng·mL⁻¹，含 5 ng·mL⁻¹内标的低浓度标准溶液系列及含 SEM 浓度为 50、100、200、500、1 000 ng·mL⁻¹，含 5 ng·mL⁻¹内标的高浓度标准溶液系列，作 HPLC/MS/MS 分析，以测得的 SEM 与 SEM-^{13}C-^{15}N$_2$面积的比值 X 为横坐标，SEM 浓度 Y 为纵坐标，绘制标准曲线，求出回归方程和相关系数。用空白组织、空白水样和空白底泥制成低质量浓度药物的含药组织、含药水样和含药底泥，经预处理后测定，将引起 3 倍基线噪声的药物质量浓度定义为仪器最低检测限。

回收率与精密度测定：回收率＝C_r/C_0×100%（其中 C_r为空白水体、空白底泥、空白肌肉和空白皮肤中各加入一定量的 SEM 已知标准溶液，再按样品预处理方法进样后，测定 SEM 的质量浓度；C_0为加入一定量的 SEM 已知标准溶液）。在空白肌肉、空白皮肤、空白水样和空白底泥中分别添加 5 个浓度水平的标准溶液，使组织和底泥中质量浓度分别为 1.0、10.0、50.0、100.0、200.0 μg·kg⁻¹，水中为 1.0、10.0、50.0、100.0、200.0 μg·L⁻¹。每个浓度的样品，日内做 5 个重复，1 周内重复做 5 次，计算日内及日间精密度。

（4）样品中 SEM 含量计算

经过前处理所得样品按照上述仪器方法检测，将得到的样品中 SEM 所对应的峰面积值代入标准曲线，即得到 SEM 的浓度值。样品中呋喃西林代谢物残留量按式（1）计算（计算结果需扣除空白值）。

$$X = \frac{C_i \times V}{m} \tag{1}$$

式中：X 为样品中呋喃西林代谢物的含量，μg·kg⁻¹或 ng·mL⁻¹；C_i为样品制备液中 SEM 的浓度，ng·mL⁻¹；V 为最终定容体积，mL；m 为样品质量。

1.2.3　数据处理

标准曲线、药物经时曲线图、消除方程及休药期计算和回归图，采用 Micros of tExcel 2007 进行计算和绘制。消除方程采用 $C=C_0e^{-kt}$，其中 C 表示药物浓度，C_0为残留消除对数曲线的截距（μg·kg⁻¹或 ng·mL⁻¹），k 表示消除速率常数。

2　结果

2.1　标准曲线方程与相关系数

在 1~50.0 ng·mL⁻¹和 50.0~1 000.0 ng·mL⁻¹范围内 SEM 线性关系良好，低浓度线性方程为 $Y=0.0234054+0.130627X$，高浓度线性方程为 $Y=0.0801631+1.3252417X$，其相关性指数 R^2均大于 0.998 8。

2.2　方法检测限、回收率与精密度

本试验条件下水样 SEM 方法检测限为 0.02 ng·mL^{-1}，肌肉、皮肤组织和底泥中 SEM 方法检测限均为 0.5 μg·kg^{-1}。肌肉、鱼皮、水样和底泥中 5 个质量浓度中加标水平的 SEM 平均回收率为 96.32%~108.65%，测得的日内精密度与日间精密度均小于 10%。

2.3　SEM 在斑点叉尾鲴肌肉和鱼皮以及环境中的消除规律

将测得的 SEM 的峰面积分别与其氘代同位素内标峰面积的比值，代入标准曲线方程，可以求出 SEM 在斑点叉尾鲴肌肉、鱼皮以及养殖环境中的浓度值（表 3）。如图 2、图 3 和图 4 所示：肌肉、水体中 SEM，在实验前 30 d 呈现快速下降的趋势，后进入缓慢下降阶段，肌肉中 270 d 未检出，水体中 270 d 后低于检测限；鱼皮中在第 3 d 和第 5 d 呈现短期的蓄积现象，随后快速下降，30 d 后进入缓慢下降阶段，第 300 d 仍高于检测限；底泥中 SEM 整体呈逐渐降低的趋势，但并不呈一定的线性关系，至第 330 d，底泥中 SEM 的含量仍然高于检测限。

表 3　1 mg/L^{-1}呋喃西林泼洒于池塘后 SEM 在养殖环境和斑点叉尾鲴组织中的浓度（$n=5$）

Table 3　The concentrations of SEM in aquaculture environment and channel catfish (*Ietalurus Punetaus*) tissue after 1 mg/L^{-1}nitrofuranzone splashed into the pond (*n=5*)

时间 Time (d)	呋喃西林代谢物 SEM (x±SD)					
	肌肉 Muscle (μg·kg^{-1})	皮肤 Skin (μg·kg^{-1})	中间水样 Intermediate water (ng·mL^{-1})	边缘水样 Edge water (ng·mL^{-1})	中间底泥 Intermediate mud (μg·kg^{-1})	边缘底泥 Edge mud (g·kg^{-1})
1	28.95±1.54	75.26±10.03	133.53±18.85	165.48±34.76	92.85±5.62	101.77±17.98
3	27.68±1.12	86.20±0.02	77.63±7.52	91.15±12.05	37.70±3.79	144.15±6.34
5	23.06±1.08	84.87±17.72	50.15±13.23	56.66±6.25	25.08±1.73	37.47±0.86
10	19.88±0.45	57.11±4.87	33.10±0.42	30.94±2.35	21.52±0.31	60.21±5.22
15	10.53±0.97	38.16±12.42	26.83±3.31	26.83±1.77	70.09±5.33	98.24±8.45
30	5.27±0.70	20.97±5.64	16.29±0.24	15.15±0.07	78.16±9.48	83.64±3.40
45	4.65±0.88	15.52±6.13	14.71±0.19	12.05±0.74	81.63±0.61	67.75±8.54
60	2.84±0.87	13.42±4.85	9.66±0.57	6.90±0.18	74.50±4.33	64.03±1.64
90	2.01±0.58	11.31±6.15	7.89±0.33	5.90±0.23	8.31±0.61	47.52±0.01
120	1.67±0.26	8.86±3.61	5.23±0.17	5.30±0.20	14.91±1.55	8.48±1.12
180	1.26±0.52	8.28±5.44	4.46±0.12	5.06±0.12	19.54±2.09	66.02±6.58
210	0.99±0.20	6.68±0.29	2.21±0.07	2.50±0.07	18.89±1.17	34.71±0.23
240	0.93±0.23	6.30±0.57	0.94±0.03	1.13±0.03	5.66±0.47	22.88±2.23
270	ND	2.66±0.22	0.56±0.14	0.75±0.02	4.78±0.30	13.22±1.08
300	ND	0.86±0.24	0.26±0.03	0.30±0.02	8.29±0.21	9.83±0.39
330	ND	ND	0.24±0.03	0.21±0.02	1.20±0.11	4.74±0.56

注：ND-未检出。Notes：ND-no detection.

2.4　肌肉、皮肤和水体中 SEM 消除参数比较

数据经回归处理得到组织及水体中药物浓度（C）与时间（t）关系的消除曲线方程、相关性指数（R^2）及消除半衰期（$T_{1/2}$）（表 4）。SEM 在肌肉、皮肤和水样中的消除半衰期 $T_{1/2}$（d）为皮肤>肌肉>边缘水体=中间水体。

2.5　斑点叉尾鲴组织与环境中 SEM 含量的相关性分析

数据经 SPSS16.0 进行相关性分析，所得相关系数见表 5。

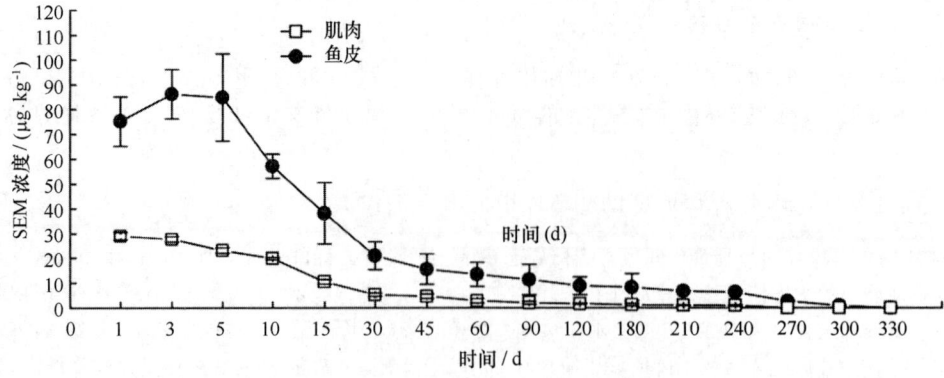

图 2　斑点叉尾鮰组织中 SEM 药时曲线图

Fig. 2　The curve of SEM concentrations-time in channel catfish（Ietalurus Punetaus）tissue

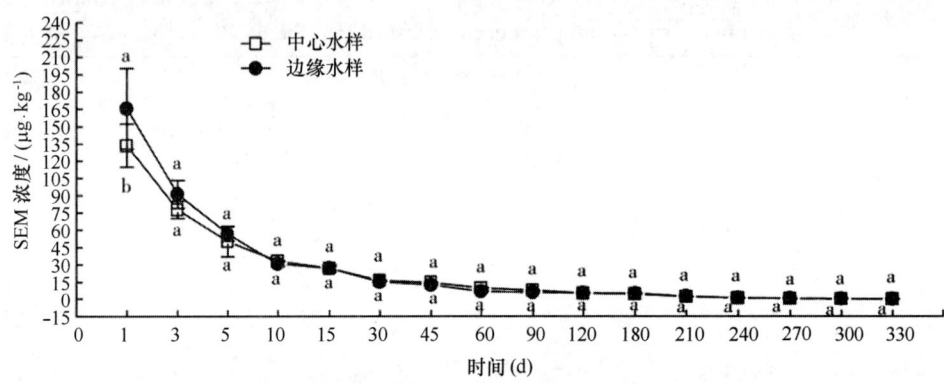

不同小写字母代表不同时间点所对应试验组差异显著（*P*<0.05）。下同

图 3　养殖水体中 SEM 药时曲线图

Fig. 3　The curve of SEMconcentrations-time in aquaculture water

注：不同小写字母代表不同时间点所对应试验组差异显著（P<0.05）。下同

Note ：Different small letter shows significant diference （P<0.05） between groups. The same beolow

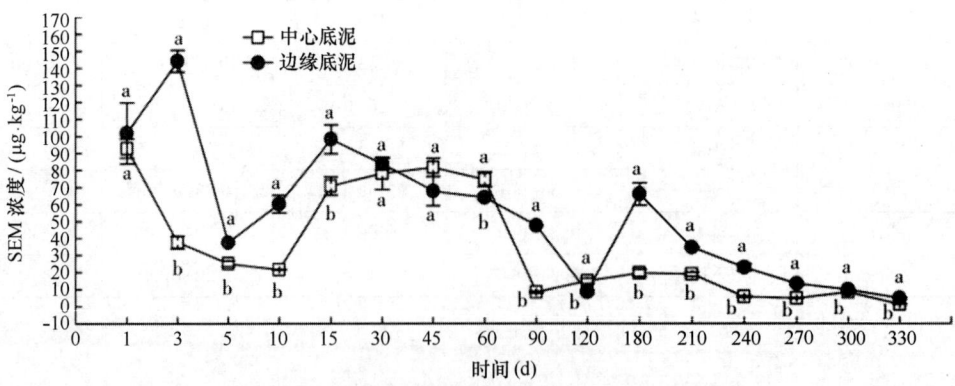

图 4　底泥中 SEM 药时曲线图

Fig. 4　The curve of SEM concentrations-time in mod

表4 SEM 在斑点叉尾鮰组织及环境中消除曲线方程及相关指数

Table 4 The equation of elimination curve and correlation index in channel catfish (*Ietalurus Punetaus*) tissue and aquaculture environment

采样位置 Sampling location	方程 Equation	相关指数 R^2 correlation index	消除半衰期 $T_{1/2}$/d
肌肉 Muscle	$C = 50.193e^{-0.331t}$	0.9727	2.09
皮肤 Skin	$C = 148.66e^{-0.287t}$	0.9331	2.42
中间水体 Intermediate water	$C = 213.55e^{-0.408t}$	0.9746	1.70
边缘水体 Edge water	$C = 214.79e^{-0.407t}$	0.9746	1.70

表5 斑点叉尾鮰组织与环境中 SEM 含量的相关性分析

Table 5 The correlation coefficient of SEM in channel catfish (*Ietalurus Punetaus*) and aquaculture environment

	肌肉 Muscle	鱼皮 Skin	中心水样 Intermediate water	边缘水样 Edge water	中心底泥 Intermediate mud	边缘底泥 Edge mud	时间 Time
肌肉 Muscle	1						
鱼皮 Skin	0.985**	1					
中心水样 Intermediate water	0.913**	0.854**	1				
边缘水样 Edge water	0.890**	0.826**	0.668**	1			
中心底泥 Intermediate mud	0.392	0.359	0.520*	0.497	1		
边缘底泥 Edge mud	0.674**	0.660**	0.668**	0.639**	0.686**	1	
时间 Time	-0.718**	-0.744**	-0.619**	-0.571*	-0.569*	-0.631**	1

注: *表示相关性显著 ($P<0.05$); **表示相关性极显著 ($P<0.01$)。

Note: * donates significant correlation ($P<0.05$); ** donates extremely significant correlation ($P<0.01$).

3 讨论

3.1 受试动物及实验环境的选择

氨基脲（SEM）一直被认为是呋喃西林的特征代谢物，并作为监测水产品非法使用呋喃西林的标志，研究表明，对于鲜活水产品而言，除了甲壳类动物甲壳和与甲壳相连的上皮层中自然产生的氨基脲[9]外，其他水动物尚未有自身能产生氨基脲的报道。本实验以斑点叉尾鮰为受试对象，排除了受试动物自身产生氨基脲的干扰情况；对实验环境（水体和底泥）以及所用斑点叉尾鮰进行了本底筛查，结果均不含 SEM，确保了实验的可行性。同时，选择偏离市区而毗邻武汉梁子湖的江夏养殖基地，完全在实际养殖条件下进行实验，使得实验结果更加符合实际。

3.2 SEM 在环境及斑点叉尾鮰组织中残留消除结果分析

第 1 d 的数据显示肌肉、鱼皮、底泥和水体中均出现了较高含量的 SEM，同时环境中明显高于鱼体。第 1 d 水体和底泥中具有较高含量 SEM，可能是由于环境中部分原药受到温度、光照和微生物等作用，被代谢为 SEM。对此虽然目前尚未有实验证明，但是李东燕等[11]通过实验证明随着光照时间的延长、反应温度的升高，呋喃西林分解速度加快，而鱼体中具有较高含量的 SEM，本文推断 SEM 可能来自鱼体自身通过体表、口腔摄入呋喃西林后机体代谢转化为代谢物 SEM，以及直接从环境中摄入 SEM。研究已表明，呋喃西林可以在水产品生物中富集，并在数小时之内被分解成代谢物 SEM[10]，而 SEM 与细胞膜蛋白稳定结合不易消除[2]，在鱼池面积相对于所放鱼数量较大的情况下，环境中第一天的 SEM 含量显著高于肌肉和鱼皮中的含量。由此推测，呋喃西林泼洒到池塘后，SEM 主要来源于环境中呋喃西林的降解。

本实验理论泼洒浓度为 1 mg·L⁻¹，而检测到的肌肉、鱼皮、底泥和水样中 SEM 最高浓度分别为 28.95、75.26、144.15 μg·kg⁻¹ 和 165.48 ng·mL⁻¹，最高浓度仅为理论泼洒浓度的 2.90%、7.53%、14.42% 和 16.55%，可见本实验条件下该原药在肌肉、鱼皮和养殖环境中代谢为 SEM 的转化率区别较大，肌肉、鱼皮中浓度明显低于环境可能与鱼体本身对药物吸收率有关。但是本实验得到的肌肉中 SEM 转化率与其他研究结果有较大区别：黄玉英等[12]通过单次口灌 98 mg·kg⁻¹ 的呋喃唑酮，鲤鱼肌肉中检测 AOZ 最高转化率仅约 0.25%；徐维海等[13]给罗非鱼投喂 30 mg·kg⁻¹·d⁻¹ 的呋喃唑酮，随后检测肌肉中 AOZ 最高转化率约为呋喃唑酮最高含量的 1/13；谭志军等[7]通过 20 mg·kg^{-1}呋喃西林浸泡大菱鲆后检测到肌肉中 SEM 最高转化率为 0.56%，同时又远高于与呋喃唑酮共同给药下的 SEM 转化率。因此，不同的给药方式下，组织中 SEM 累积能力并不相同，同时说明直接对鱼池进行泼洒的方式可能更容易导致鱼体内 SEM 的累积。

本实验前期，斑点叉尾鮰肌肉中呋喃西林代谢物 SEM 浓度消除规律与谭志军等[7]对大菱鲆肌肉中 SEM 消除规律相似，都表现出前期急剧下降，30 d 后进入缓慢消除阶段并持续较长时间，至 180 d 仍高于检测限。由于本实验前期（前 30 d）正处于 9 月至 10 月，水温相对较高，鱼体摄食与自身代谢速率较快，可能是导致前期 SEM 快速下降的重要原因，进入 11 月后，水温逐渐降低，SEM 浓度即呈现缓慢下降的趋势。斑点叉尾鮰皮肤中呋喃西林代谢物 SEM 呈现的规律与肌肉有所不同，起始浓度显著高于肌肉，第 3 天浓度累积至最高峰值为（86.20±10.02）μg·kg^{-1}，随后与肌肉呈现相同规律。这可能是由于药浴开始阶段，吸附在斑点叉尾鮰体表（皮）的呋喃西林原药被迅速代谢为 SEM，而经口进入体内的药物通过鱼体自身代谢后部分不断积累到鱼皮上，同时水体中的 SEM 也源源不断富集所致。但从 SEM 整个消除过程来看，养殖环境中直接泼洒呋喃西林，与一般药物通过饲料投喂后或者通过浸泡后放养水产品，组织中 SEM 呈现相似的消除规律：在实验前阶段浓度急剧下降，然后消除速率逐渐平缓并持续较长时间。另外本实验条件下，肌肉及鱼皮中 SEM 半衰期（$T_{1/2}$）显著小于其在大菱鲆中[7]的半衰期，可能与给药方法、给药剂量、给药环境和鱼类品种的差异有关。已有研究表明[14]，在太湖开放性水域条件下中华绒螯蟹体内的 SEM 的衰减速率要快于室内条件。这些实验结果都说明温度、光强、开放性水域的水交换和底泥吸附以及生物降解和腐植酸作用都可能是影响 SEM 消除的因素。

为避免采样期间风向变化造成采样不均匀等问题，本实验将每个时间点的采样点设置为：水池周边随机采取 5 次水样和底泥，水池中间采集 5 次水样和底泥，尽可能的确保采样的均匀性，使所得的实验结果更具有科学性。每个时间点，中间水体和边缘水体中 SEM 浓度的差异性结果由图 3 所示。边缘水体中呋喃西林代谢物 SEM 的起始浓度为（165.48±18.85）μg·kg^{-1}，显著高于（$P<0.01$）中间水样（133.53±34.76）μg·kg^{-1}，第 3 d 至采样结束，中间水样与边缘水样中 SEM 浓度值无差异性（$P>0.05$），这可能是由于在药物泼洒后，未溶解的原药受到风向作用向池边累积后快速代谢，从而导致首次结果显示出显著差异。实验至第 3 d，水中 SEM 经过水体自身的对流以及药物自身的扩散作用，逐步分布均匀。对每个时间点，中间底泥和边缘底泥中 SEM 浓度的差异性如图 4 所示，底泥中呋喃西林代谢物 SEM 的浓度整体呈现逐渐降低的趋势，但是各时间点之间并没有呈现一定的线性关系，可能是由于底泥采样不可重复性，以及底泥生物群落差异及腐植酸等因素的不均，导致吸附药物具有较大差异，其具体原因还有待研究。分析边缘与中间底泥中 SEM 浓度值的差异性，结果显示第 1 d、3 d 和第 45 d 差异不显著（$P>0.05$），剩余多个时间点边缘底泥中 SEM 浓度显著高于（$P<0.05$）中间底泥，可能是由于施药初期呋喃西林原药不均匀沉降，底泥吸附后不可移动性所致，与水体中前期结果（边缘水体高于中间水体）保持一致。到采样结束时，斑点叉尾鮰组织及养殖水体中 SEM 都已经完全消除，但底泥中 SEM 的浓度仍然高于检测限，可见 SEM 在底泥中的消除更加缓慢。

3.3 对斑点叉尾鮰组织 SEM 残留风险评估

由表 3 可以看出，呋喃西林代谢物 SEM 在斑点叉尾鮰组织及环境中的起始浓度水平为边缘水样>中心水样>边缘底泥>中间底泥>皮肤>肌肉，虽然前期鱼皮中 SEM 出现了短暂的蓄积现象，但在 5 d 后便与肌肉中 SEM 同步快速降低，最终皮肤于 330 天完全消除，肌肉于 270 d 完全消除，而水样和底泥至实验第 330 d 仍未完全消除。SEM 在肌肉、鱼皮和水环境中的消除半衰期之间无明显差别。肌肉、鱼皮与水体和边缘底泥中 SEM 值呈显著正相关，与时间呈显著性负相关，与 SEM 溶于水[15]，同时易与蛋白质稳定结合，只有在酸性条件下才被释放出来[2]等特性保持一致。徐英江等[16]通过实验说明文蛤能吸收水体中的氨基脲，说明环境中 SEM 与鱼体肌肉和鱼皮中 SEM 含量密切相关。但环境中的药物并没有导致鱼肉和鱼皮中 SEM 的明显蓄积。

4 结论

本实验首次报道了环境中呋喃西林代谢物 SEM 在池塘和斑点叉尾鮰中的消除规律，通过 1 年的研究发现，在养殖水体直接泼洒呋喃西林药物后的初期，环境中 SEM 浓度较高，更容易导致鱼体内 SEM 累积，环境和鱼体中 SEM 的消除均需要较长时间，但肌肉和鱼皮中 SEM 并未出现明显蓄积的现象。

参考文献

[1] Cooper K. M., Mulder P. P., et al. Depletion of four nitrofuran antibiotics and their tissue-bound metabolites in porcine tissues and determination using LC-MS/MS and HPLC-UC [J]. Food Addit Contam, 2005, 22 (5)：406.

[2] 何方洋, 沈建忠, 万宇平, 等. 呋喃西林及其代谢物残留检测研究进展 [J]. 中国兽药杂志, 2009, 43 (4)：51-54.
HE Fang-yang, SHEN Jian-zhong, WAN Yu-ping, et al. Research progress on determination of nitrofurazone and its metabolite residue [J]. Chinese Journal of Veterinary Drug, 2009, 43 (4)：51-54.

[3] Kelly B. D., Heneghan M. A., Bennani F., et al. Nitrofurantoin-induced hepatotoxicity mediated by CD8+Tcells [J]. American Journal of

Gastroenterology, 1998, 93（5）：819-821.

［4］ Bock C., Gowik P., Stachel C.. Matrix-comprehensive in-house validation and robustness check of a confirmatory method for the determination of four nitrofuran metabolites in poultry muscle and shrimp by LC-MS/MS［J］. Journal of Chromatography B, 2007, 856（1/2）：178-189.

［5］ Barbosa J., Moura S., Barbosa R., et al. Determination of nitrofurans in animal feeds by liquid chromatography-UV photodiode array detection and liquid chromatography-ionspray tandem mass spectrometry［J］. Analytica Chimica Acta, 2007, 586：359-365.

［6］ Council Regulation（EEC）2377/90, Off. J. Eur. Commun. No. L224（1990）1.

［7］ 谭志军，翟毓秀，冷凯良，等. 呋喃西林和呋喃唑酮代谢物在大菱鲆组织中的消除规律［J］. 中山大学学报（自然科学版），2008，47（增刊）：63-69.
TAN Zhi-jun, ZHAI Yu-xiu, LENG Kai-liang, et al. The depuration rules of the metabolites of furazolidone and nitrofurazone inTurbot（Scophthalmusmaximus）［J］. Journal of Sun Yatsen University（Social Science Edition），2008，47（Suppl）：63-69.

［8］ 食品动物禁用的兽药及其他化合物清单，中华人民共和国农业部文件农牧发（2002）1号.
The list of animal medicine and other compound forbidden using in food animals, Announcement No.（2002）1 of the Ministry of Agriculture of the People's Republic of China, published by Animal Husbandry Bureau.

［9］ Mccracken R., Hanna B., Ennis D., et al. The occurrence of semicarbazide in the meat and shell of Bangladeshi fresh-water shrimp［J］. Food Chem-istry, 2011, 11：088.

［10］ Xu W. H., Zhu J. B., Wang X. T., etal. Residues of enrofloxacin, furazoli-done and their metabolites in Nile tilapia（Oreochromis niloticus）［J］. Aquaculture, 2006, 254：1-8.

［11］ 李东燕，徐睿来，李光汉，等. 台阶恒温加速法对复方呋喃西林滴鼻液的稳定性观察［J］. 江西医学院学报，2003，43（6）：123-124.
LI Dong-yan, XU Rui-lai, LI Guang-han, et al. The stability observation of compound nitrofurazone nasal drops by the method of step constant temperature accelerated［J］. Acta Academiae Medicinae Jiangxi, 2003, 43（6）：123-124.

［12］ 黄玉英，彭爱红，黄志勇，等. 呋喃唑酮及代谢产物3-氨基-2-恶唑烷酮在鲤鱼体内的残留规律［J］. 福建农林大学学报（自然科学版），2009，38（2）：181-185.
HUANG Yu-ying, PENG Ai-hong, HUANG Zhi-yong, et al. Pharmacokinetics of furazolidone and itsmetabolite 3-amina-2-oxazolldinone in Cyprinus carpio［J］. Journal of Fujian Agriculture and Forestry University（Natural Science Edition），2009，38（2）：181-185.

［13］ 徐维海，林黎明，朱校斌，等. HLPC/MS法对呋喃唑酮及代谢物AOZ在罗非鱼体内残留研究［J］. 上海水产大学学报，2005，14（1）：35-39.
XU Wei-hai, LIN Li-ming, ZHU Xiao-bin, et al. The research of residues of furazolidone and its metabolite in tilapias by HPLC/MS［J］. Journal of Shanghai Fisherise University, 2005, 14（1）：35-39.

［14］ 樊新华，郑浩，钱伟，等. 呋喃西林代谢物氨基脲在中华绒螯蟹体内的衰减研究［J］. 江苏农业科学，2010（6）：238-370.
FAN Xin-hua, ZHEN Hao, Qian wei, et al. The reduction rules of the metabolites of nitrofurazone in Eriocheir sinensis［J］. Jiang su Agricultural Science, 2010（6）：238-370.

［15］ 吴晓君，施高茂. 硝基呋喃西林代谢物—氨基脲的应对设想［J］. 海洋与渔业，2007（4）：25-26.
WU Xiao-jun, SHI Gao-mao. The conceive of replying the metabolites of nitrofurazone-semicarbazide［J］. Ocean and Fishery, 2007（4）：25-26.

［16］ 徐英江，田秀慧，张秀珍，等. 文蛤（Meretrixmeretrix）体内氨基脲含量与环境相关性研究［J］. 海洋与湖沼，2011，42（4）：587-591.
XU Ying-jiang, TIAN Xiu-hui, ZHANG Xiu-zhen, et al. Study on the correlation between semicarbazide in Clam Meretrix meretrix and environment plloution［J］. Oceanologia Et Limnologia Sinica, 2011, 42（4）：587-591.

该文发表于《水生生物学报》【2013，37（2）：269-280】

浸泡条件下孔雀石绿及其代谢物隐色孔雀石绿在斑点叉尾鮰组织中分布及消除规律研究

Tissue Distribution and Elimination of Malachite Creen and Its Metabolite Leucomalahite Green From Channel Catfish （Ietaluruspunetaus）After Bath Treatment

刘永涛[1,2]，艾晓辉[1,2]，索纹纹[1,3]，杨秋红[1,2]

（1. 中国水产科学研究院长江水产研究所，农业部淡水鱼类种质监督检验测试中心，武汉 430223；
2. 中国水产科学研究院淡水研究中心，无锡 214081；3. 华中农业大学水产学院，武汉 430070）

摘要：以 7 mg/L 的孔雀石绿浸泡斑点叉尾鮰苗种 5 min 后将其饲养于池塘的网箱中，研究了在养殖模式下孔雀石绿及其代谢物隐色孔雀石绿在斑点叉尾鮰苗种各组织中的分布及消除规律。采用高效液相色谱串联质谱法（HPLC-MS/MS）分析孔雀石绿及其代谢物隐色孔雀石绿在斑点叉尾鮰血液、肌肉、皮肤、肝脏、肾脏组织中的浓度水平。采用药代动力学分析软件3p97对血药浓度时间数据进行分析。结果表明，孔雀石绿和隐色孔雀石绿血药浓度时间曲线符合有吸收二室模型，动力学方程分别为：$C_{孔雀石绿} = 683.063\ e^{-0.248t} + 11.176\ e^{-0.006t} - 694.239e^{-0.333t}$，$C_{隐色孔雀石绿} = 757.240\ e^{-0.222t} + 14.474\ e^{-0.007t} - 771.714\ e^{-0.382t}$。血液中孔雀石绿和隐色孔雀石绿达峰时间$T_{peak}$分别为3.480 h和3.623 h，峰浓度值$C_{max}$分别为81.560 ng/mL和159.619 ng/mL，表观分布容积V_d/F分别为37.689 L/kg和21.125 L/kg，分布相的一级速率常数α分别为0.248/h和0.222/h，消除相的一级速率常数β分别为0.006/h和0.007/h，吸收半衰期$T_{(1/2)a}$分别为2.794 h和3.124 h，消除半衰期$T_{(1/2)\beta}$分别为113.068 h和105.841 h，中央室向周边室转运的一级速率常数K_{12}分别为0.020/h和0.015/h，周边室向中央室转运的一级速率常数K_{21}分别为0.159/h和0.121/h，药-时曲线下面积AUC分别为2493.944 ng·h/mL和3601.863 ng·h/mL。肌肉、皮肤、肝脏和肾脏组织中孔雀石绿和隐色孔雀石绿浓度水平的结果表明，孔雀石绿在斑点叉尾鮰4种组织中浓度由高到低的顺序是皮肤>肌肉>肾脏>肝脏，其中斑点叉尾鮰皮肤组织易蓄积孔雀石绿，其残留时间最长，肝脏组织由于对孔雀石绿有极强的代谢转化功能而浓度较低。孔雀石绿在肌肉、皮肤、肝脏和肾脏组织中的消除方程分别为$C = 5.570\ e^{-0.009t}$、$C = 6.302\ e^{-0.007t}$、$C = 4.791\ e^{-0.006t}$和$C = 4.591\ e^{-0.002t}$，相关系数$r^2 \geq 0.773$，消除半衰期$T_{1/2}$肌肉、皮肤、肝脏和肾脏分别为3.2 d、4.1 d、4.8 d和14.4 d。肌肉、皮肤、肝脏和肾脏组织中孔雀石绿分别在45 d、60 d、30 d和60 d才未被检测到；隐色孔雀石绿在斑点叉尾鮰4种组织中浓度由高到低的顺序是肝脏>皮肤>肌肉>肾脏，残留时间最长的组织也是皮肤组织。隐色孔雀石绿在肌肉、皮肤、肝脏和肾脏组织中的消除方程分别为$C = 6.491\ e^{-0.004t}$、$C = 6.958\ e^{-0.003t}$、$C = 6.722\ e^{-0.007t}$和$C = 6.162\ e^{-0.002t}$，相关系数$r^2 \geq 0.673$，消除半衰期$T_{1/2}$肌肉、皮肤、肝脏和肾脏分别为7.2 d、9.6 d、4.1 d和14.4 d。肌肉、皮肤、肝脏和肾脏组织中隐色孔雀石绿分别在90 d、90 d、60 d和90 d才未被检出。试验期间（2011年5月17日至2011年7月15日）平均水温为26.4℃，孔雀石绿和隐色孔雀石绿90 d后在各组织中才未检测到，因此，使用7 mg/L孔雀石绿浸泡2龄斑点叉尾鮰苗种孔雀石绿及其代谢物隐色孔雀石绿至少应经过2 376℃·d后才能消除。

关键词：孔雀石绿；隐色孔雀石绿；斑点叉尾鮰；组织分布；消除

孔雀石绿是一种N-甲基三苯甲烷类染料，由于其抗菌效果好，价格便宜，容易获得，曾被广泛用于防治水产动物的真菌病、细菌病和寄生虫病等[1]。孔雀石绿多用于水产动物的繁殖、消毒和运输等环节。由于孔雀石绿及其代谢物隐色孔雀石绿对人体存在潜在致畸、致癌和致突变的"三致"作用[2-5]。许多国家都将孔雀石绿列为水产养殖禁用药物，我国也于2002年5月将孔雀石绿列入《食品动物禁用的兽药及其化合物清单》中，禁止用于所有食品动物中。欧盟从来没有允许孔雀石绿作为兽药，美国食品和药品管理局也没有批准使用这种染料[6,7]。从目前国内对于鲜活水产品的监测情况来看孔雀石绿仍有检出，孔雀石绿之所以屡禁不止，除孔雀石绿抗菌效果好，价格便宜外，没有效果相当的替代药物出现也是一个重要的原因。目前，孔雀石绿在水产动物体内代谢的报道多是关于其在水产动物肌肉组织中残留消除规律的报道[8-15]，涉及的水产动物主要有鳗鲡、罗非鱼、鲫鱼、中华绒螯蟹等，国内尚未见关于孔雀石绿及其代谢物隐色孔雀石绿在斑点叉尾鮰体内药代动力学、组织分布与消除规律的报道，国外Plakas, et al.[13]采用静注和浸泡方式给予斑点叉尾鮰（0.5-0.7 kg）^{14}C标记的孔雀石绿0.8 mg/L研究其代谢规律。由于用药浓度、用药时间、鱼体规格及养殖方式的不同孔雀石绿及其代谢物隐色孔雀石绿在鱼体内代谢规律会有较大差异。本实验根据我国斑点叉尾鮰养殖过程中实际用药浓度的高限和用药时间研究了养殖模式下孔雀石绿及其代谢物隐色孔雀石绿在斑点叉尾鮰苗种体内的药代动力学、组织分布与消除规律。为孔雀石绿及其代谢物隐色孔雀石绿在斑点叉尾鮰苗种中的代谢规律提供了理论数据，并为加强该药物的管理和监督非法使用提供理论依据。

1 材料与方法

1.1 试验动物

健康斑点叉尾鮰由湖北省仙桃市水产局提供，平均质量为（87.0±22.7）g。试验前在池塘的网箱（1.0 m×2.0 m×1.5 m）内暂养一周。实验池长37.1 m，宽18.0 m，平均深度1.1 m，水体积734.6 m³。试验期间（2011年5月17日至2011年7月15日）每天上午8点和下午6点投喂斑点叉尾鮰饲料，并记录水温，试验期间平均水温为26.4℃，池水的pH值为7.5-8.0。

1.2 仪器与色谱条件

高效液相色谱-串联质谱（Suryeyor Ms Pump Plus, Suryeyor Autosampler Plus, Thermo TSQ Quantum Access Max）及

216

Thermo LCquan 2.6 数据采集处理软件；自动高速冷冻离心机（日本 Hitachi 20PR-520 型）；Mettler-Toledo Ae-240 型精密电子天平（梅特勒-托利多公司）；FS-1 高速匀浆机（华普达教学仪器有限公司）；调速混匀器（上海康华生化仪器制造厂）；氮吹仪（Aosheng，杭州奥盛仪器有限公司）；固相萃取装置（美国，CNW）。

色谱柱：Hypersil Gold C18（150.0 mm×2.1 mm×5.0 μm）反相色谱柱；柱温：35℃；流速 0.20 mL/min；进样量 10.0 μL。梯度洗脱条件：A 相为乙腈，B 相为 5 mmol/L 乙酸铵溶液用乙酸调 pH4.5，0~3.00 min，75% A；3.00~3.01 min，75% A~90% A；3.01~8.00 min，90% A；8.00~8.01 min，75% A；8.01~10.00 min 75% A。

1.3　质谱分析条件

离子源：加热大气压电喷雾离子源（HESI）；离子化模式：正离子模式；检测方式采用选择反应监测（SRM）模式；喷雾电压：3 000 V；源内解离电压 8 V；鞘气压力 20 arb；辅助气压力 5 arb；碰撞气及压力：氩气，1.5 mTorr；离子传输毛细管温度：350℃；Q1 PW 0.7，Q3 PW 0.7。母离子、定性离子对、定量离子对、碰撞能量和管透镜补偿电压（表1）。

表1　孔雀石绿、隐色孔雀石绿、氘代孔雀石绿和氘代隐色孔雀石绿质谱条件

Tab. 1　Spectometric parameters of MG，MG-D$_5$，LMG and LMG-D$_6$

药物 Drug	母离子 Parent ion	子离子 Product ion	碰撞能量 Collision energy（V）	管透镜补偿电压 Tube lens offset（V）
MG	329.0	208.1*	33	104
		313.1	36	104
MG-D$_5$	334.0	318.2	37	104
LMG	331.0	239.0*	30	104
		316.1	20	104
LMG-D$_6$	337.0	322.2	20	94

注：* 离子对为定量离子对。

Note：* Quantitative ion pair.

1.4　药品及试剂

孔雀石绿标准品（纯度>93.5%，Dr. Ehrenstorfer GmbH）；隐色孔雀石绿标准品（纯度>92.5%，Dr. Ehrenstorfer GmbH）；氘代孔雀石绿标准品（纯度>99.0%，WiTEGA）；氘代隐色孔雀石绿标准品（纯度>99.0%，WiTEGA）；试剂级孔雀石绿（分析纯，天津市天力化学试剂有限公司）；乙腈、乙酸（色谱纯，美国 J. T. Baker）；质谱用水（美国，CNW）；乙酸铵（分析纯，美国 J. T. Baker），中性氧化铝柱（美国，CNW）。

1.5　试验设计与采样

药液的配制：称取 0.63 g 孔雀石绿试剂，溶于 90 L 水中配制成 7 mg/L 的孔雀石绿溶液中置于 125 L 塑料箱中。

样品采集：斑点叉尾鮰苗种于下塘前将斑点叉尾鮰置于上述含有 7 mg/L 的孔雀石绿溶液的塑料箱中浸泡 5 min，下塘后于 0.25 h、0.5 h、1 h、2 h、4 h、6 h、8 h、12 h、24 h、48 h、96 h、192 h、288 h、384 h、480 h、720 h、1 080 h、1 440 h 每一时间点尾静脉取血液，并取肌肉、皮肤、肝脏、肾脏组织置于-20℃冰箱中保存，直至样品处理。

每一时间点各取 5 尾斑点叉尾鮰，作为 5 个平行样品分别处理测定。

另设一对照组，定期采集样品进行分析。

1.6　样品预处理

血液、肝脏、肾脏、皮肤样品的处理参照中华人民共和国国家标准 GB/T 19857-2005《水产品中孔雀石绿和结晶紫残留量的测定》进行测定，略有改动。将冷冻保存的血液、肝脏、肾脏和皮肤样品室温下自然解冻，将每个时间点的肾脏样品合并为一个样品取样量约为 0.5 g，取 1.0 mL 血液，1.0 g 肝脏、皮肤于 10 mL 塑料离心管中，加入 80 μL，100 μg/L 混合内标溶液，加入 4 mL 乙腈，振荡 30 s，8 000 r/min 离心 5 min，转移到 10 mL 具塞刻度离心管再加入 4 mL 乙腈，振荡 30 s，离心后，合并提取液于 10 mL 具塞刻度离心管，用乙腈定溶至 10 mL。取 5 mL 提取液过中性氧化铝柱，用 4 mL 乙腈洗中性氧化铝柱。将其置于 45℃氮吹仪上吹至 1 mL，在加入 1 mL pH4.5 乙酸铵溶液，涡旋振荡 30 s，超声提取 1 min，取 1 mL 过 0.22 μm 滤头过滤后，上 HPLC/MS/MS 分析。

肌肉样品的处理按照中华人民共和国国家标准 GB/T 19857-2005《水产品中孔雀石绿和结晶紫残留量的测定》进行测定。

1.7 标准曲线的制备及回收率与精密度测定

标准曲线的制备：配制含氘代孔雀石绿和氘代隐色孔雀石绿为内标，孔雀石绿和隐色孔雀石绿浓度为 0.25、0.5、1、2、10、20 ng/mL 的低浓度标准曲线；配制含氘代孔雀石绿内标和氘代隐色孔雀石绿内标，孔雀石绿和隐色孔雀石绿浓度为 50、100、200、500、1 000 ng/mL 的高浓度标准曲线。作 HPLC/MS/MS 分析，以测得的孔雀石绿与氘代孔雀石绿和隐色孔雀石绿与氘代隐色孔雀石绿面积的比值 X 为横坐标和孔雀石绿和隐色孔雀石绿浓度 Y 为纵坐标，绘制标准曲线，求出回归方程和相关系数。用空白组织制成低质量浓度药物的含药组织，经预处理后测定，将引起 3 倍基线噪音的药物的质量浓度定义为最低检测限。

回收率与精密度测定：回收率=C_r/C_0×100%，其中 C_r 为用空白血液、肌肉肝脏、皮肤、肾脏组织中加入一定量的孔雀石绿和隐色孔雀石绿已知标准溶液，再按样品预处理方法进样后，测定孔雀石绿和隐色孔雀石绿的质量浓度；C_0 为加入一定量的孔雀石绿和隐色孔雀石绿已知标准溶液。在 5 种空白组织中分别添加 4 个浓度水平的标准溶液，使组织中质量浓度分别为 1.0、20.0、50.0、500 和 1 000 μg/kg，血液为 1.0、10.00、50.0、200 和 500 μg/L。每个浓度的样品，日内做 5 个重复，一周内重复做 5 次，计算日内及日间精密度。

1.8 数据处理

药物动力学模型拟合及参数计算采用中国药理学会数学专业委员会编制的 3P97 药动软件分析；标准曲线，药物经时曲线图，消除方程计算和回归图，采用 Microsoft Excel 2003，进行计算和绘制。消除方程采用 $C=C_0e^{-kt}$，C 表示药物浓度，C_0 为残留消除对数曲线的截距（μg/kg），k 表示消除速率常数[17]。

2 结果

2.1 标准曲线方程与相关系数

在 0.25~20.0 ng/mL 和 50.0~1000.0 ng/mL 范围内孔雀石绿和隐色孔雀石绿线性关系良好，低浓度线性方程分别为 $Y=0.07759+0.5499X$，$Y=0.1696+1.777X$；高浓度线性方程分别为 $Y=0.6803+7.108X$，$Y=1.230+11.652X$；相关指数均大于 0.9988。

2.2 方法检测限、回收率与精密度

在本实验条件下，血液、肌肉、肝脏、肾脏和皮肤组织中孔雀石绿和隐色孔雀石绿方法检测限均为 0.5 μg/L、0.5 μg/kg。各组织中 5 个质量浓度加标水平的孔雀石绿和隐色孔雀石绿平均回收率为 92.27%~110.67%，测得的日内精密度与日间精密度均小于 10%。

2.3 孔雀石绿及其代谢物隐色孔雀石绿在斑点叉尾鮰血液中的浓度

将测到的孔雀石绿及其代谢物隐色孔雀石绿的峰面积分别与其氘代同位素内标峰面的比值，代入标准曲线方程，可以求出孔雀石绿及其代谢物隐色孔雀石绿在斑点叉尾鮰血液中的浓度值（表2）。由表2可以看出，孔雀石绿和隐色孔雀石绿峰浓度分别出现在 4 h 和 6 h，浓度分别为（118.16±30.95）μg/L 和（203.20±24.25）μg/L，0.5 h 后孔雀石绿的代谢物隐色孔雀石绿浓度高于孔雀石绿的浓度，在 1 080 h 后孔雀石绿及其代谢物隐色孔雀石绿才检测不到。

表 2　7 mg/L 孔雀石绿浸泡后孔雀石绿及其代谢物在斑点叉尾鮰血液中的浓度（ng/mL）n=5
Tab. 2　The concentrations of GM and its metabolite LMG in channel catfish blood after 7 mg/L GM bath n=5

时间 Time（h）	孔雀石绿 MG	隐色孔雀石绿 LMG
0.25	88.10±26.99	18.75±10.89
0.5	46.05±15.59	81.79±14.23
1	39.58±13.65	78.33±28.45
2	88.22±6.34	118.31±49.02
4	118.16±30.95	176.06±51.77
6	88.18±13.60	203.20±24.25

时间 Time（h）	孔雀石绿 MG	隐色孔雀石绿 LMG
8	61.26±10.16	140.41±24.67
12	18.87±7.44	34.84±20.37
24	15.44±7.89	54.89±15.25
48	10.41±7.39	9.00±6.52
96	5.38±0.84	12.77±10.84
192	4.87±1.40	2.83±3.51
288	2.78±2.78	5.78±3.04
384	2.91±1.72	0.70±0.58
480	5.45±1.23	8.86±7.61
720	1.20±0.99	2.92±1.21
1080	ND	ND
1440	ND	ND

2.4　孔雀石绿及隐色孔雀石绿在斑点叉尾鮰体内的药代动力学特征

孔雀石绿以 7 mg/L 的浓度浸泡斑点叉尾鮰后，孔雀石绿及其代谢物隐色孔雀石绿药时数据经 3p97 实用药代动力学分析软件分析，血液中药物浓度与时间关系符合有吸收二室模型，孔雀石绿和隐色孔雀石绿动力学方程分别为：$C_{孔雀石绿} = 683.063\,e^{-0.248t}+11.176\,e^{-0.006t}-694.239e^{-0.333t}$，$C_{隐色孔雀石绿}=757.240\,e^{-0.222t}+14.474\,e^{-0.007t}-771.714\,e^{-0.382t}$。孔雀石绿通过浸泡方式进入斑点叉尾鮰体内进而进入血液循环，药动学软件模拟的孔雀石绿和隐色孔雀石绿的达峰时间 T_{peak} 分别为 3.480 h 和 3.623 h，峰浓度 C_{max} 分别为 81.506 ng/mL 和 159.619 ng/mL，吸收速率 K_a 分别为 0.333/h 和 0.382/h，分布半衰期 $T_{(1/2)a}$ 分别为 2.794 h 和 3.124 h，消除半衰期 $T_{(1/2)\beta}$ 较长 113.068 h 和 105.841 h。具体药动学参数（表3）。

表3　7 mg/L 的孔雀石绿浸泡斑点叉尾鮰孔雀石绿及隐色孔雀绿在其体内的药代动力学参数

Tab. 3　Pharmacokinetic parameters for MG and LMG in channel catfish after water bath administration of 7 mg/L

参数 Parameters		单位孔雀石绿 UnitMG	隐色孔雀石绿 LMG
A	ng/mL	683.063	757.240
a	1/h	0.248	0.222
B	ng/mL	11.176	14.474
β	1/h	0.006	0.007
K_a	1/h	0.333	0.382
Lag time	h	0.017	0.080
V_d/F	L/Kg	37.689	21.125
$T_{(1/2)a}$	h	2.794	3.124
$T_{(1/2)\beta}$	h	113.068	105.841
$T_{(1/2)Ka}$	h	2.079	1.816
K_{21}	1/h	0.020	0.015
K_{10}	1/h	0.074	0.092
K_{12}	1/h	0.159	0.121
AUC	ng·h/mL	2493.944	3601.863
CL（s）	(L/h)/kg	0.002	0.003

续表

参数 Parameters		单位孔雀石绿 UnitMG	隐色孔雀石绿 LMG
$T_{(peak)}$	h	3.480	3.623
$C_{(max)}$	ng/mL	81.506	159.619

注：A，B为药时曲线对数图上曲线在横轴和纵轴上的截距 Drug concentration-time curve on logarithmic graph curve on the horizontal axis and vertical axis intercept；a，β分别为分布相、消除相的一级速率常数 First-order rate constant for drug distribution and elimination phase；K_{21}由周边室向中央室转运的一级速率常数 First-order rate constant for drug distribution from peripheral compartment to central compartment；K_{10}由中央室消除的一级速率常数 First-order rate constant for drug elimination from central compartment；K_{12}由中央室向周边室转运的一级速率常数 First-order rate constant for drug distribution from central compartment to peripheral compartment；V_d/F表观分布容积 Apparent volume of distribution；AUC药时曲线下面积 Area under concentration-time curve；Lag time 滞后时间；Ka 为一级吸收速率常数 First-order rate constant for drug absorption；$T_{(1/2)Ka}$为药物在中央室的吸收半衰期 Absorption half-life in central compartment；$T_{(1/2)a}$、$T_{(1/2)β}$分别为总的吸收和消除半衰期 The overall absorption half-life and Elimination half-life；T_{peak}出现最高血药质量浓度的时间 The time at highest peak of concentration；C_{max}最高血药质量浓度 The biggest concentration of the drug in blood；CL（s 为总体清除率 The overall clearance rate

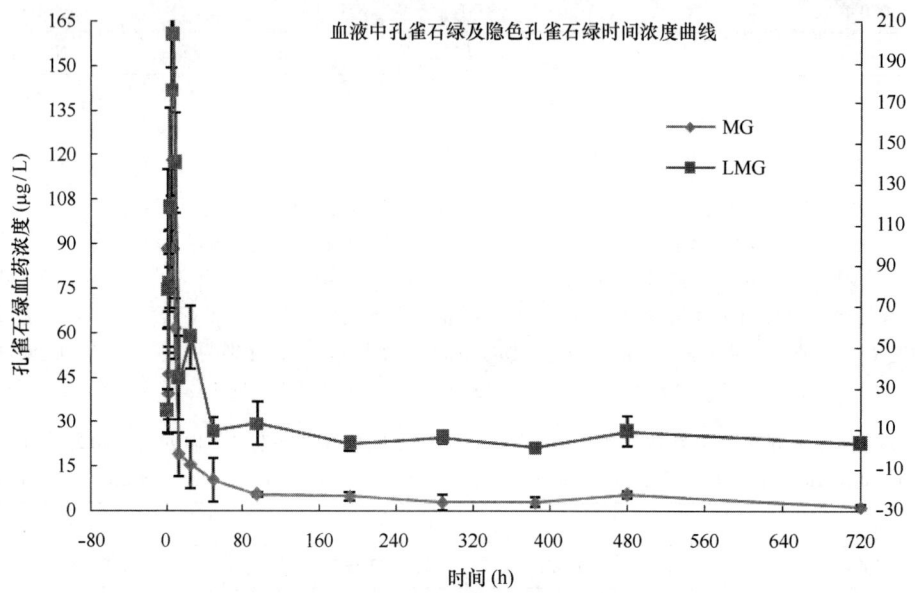

图1　孔雀石绿及其代谢物隐色孔雀石绿在斑点叉尾鲫血液中的代谢规律

Fig.1　The metabolic rule of MG and LMG in blood of channel catfish

2.5　孔雀石绿及其代谢物隐色孔雀石绿在斑点叉尾鲫组织中分布和消除规律

孔雀石绿及其代谢物隐色孔雀石绿在斑点叉尾鲫组织中的浓度（表4和表5）。孔雀石绿在斑点叉尾鲫肌肉、皮肤、肝脏和肾脏组织中分别在1 h、6 h、0.5 h，0.5 h出现最大浓度值分别为（487.7±142.5）μg/kg、（2 208.407±1 223.962）μg/kg、（242.183±73.521）μg/kg和270.13 μg/kg。4种组织中孔雀石绿分别在45 d、60 d、30 d和60 d才未被检测到；隐色孔雀石绿在斑点叉尾鲫肌肉、皮肤、肝脏和肾脏组织中分别在8 h、8 h、0.5 h，48 h时出现最大浓度值分别为（868.2±199.730）μg/kg、（1 206.096±217.526 1）μg/kg、（5 583.84±1 544.98）μg/kg和362.69 μg/kg。4种组织中隐色孔雀石绿分别在90 d、90 d、60 d和90 d才未被检出。

表4　7 mg/L孔雀石绿浸泡斑点叉尾鲫孔雀石绿及其代谢物隐色孔雀石绿在各组织中的浓度

Tab.4　The concentrations（μg/kg）of MG and LMG residues in all kinds of tissues of channel catfish（$n=5$）

时间 Time（h）	孔雀石绿 MG			隐色孔雀石绿 LMG		
	肌肉 Muscle	皮肤 Skin	肝脏 Liver	肌肉 Muscle	皮肤 Skin	肝脏 Liver
0.25	413.8±119.5	509.3±292.0	154.8±55.4	709.4±209.5	709.0±165.5	5448.4±1721.6

220

续表

时间 Time（h）	孔雀石绿 MG			隐色孔雀石绿 LMG		
	肌肉 Muscle	皮肤 Skin	肝脏 Liver	肌肉 Muscle	皮肤 Skin	肝脏 Liver
0.5	441.0±52.8	692.1±283.6	242.2±73.5	837.0±64.5	669.9±218.3	5583.8±1545.0
1	487.7±142.5	707.1±362.9	120.9±34.6	805.1±126.7	590.5±189.6	5109.4±1107.0
2	379.6±170.1	1459.9±1161.6	103.0±56.6	672.9±144.9	750.4±189.6	3010.1±1622.9
4	305.6±88.7	2032.3±932.7	139.4±61.0	759.8±58.6	882.4±766.6	2326.2±892.9
6	402.0±116.2	2208.4±1224.0	63.5±26.9	663.7±27.1	839.1±360.5	1117.4±422.4
8	411.7±110.2	1396.5±791.1	110.5±29.8	868.2±199.7	1206.1±217.5	988.3±5.83
12	307.8±174.5	1378.1±798.0	110.9±61.7	481.2±227.8	934.2±181.3	347.7±20.0
24	122.8±37.8	334.8±54.6	85.3±31.0	422.1±54.4	1078.5±502.1	205.0±6.53
48	116.7±99.1	220.8±55.2	93.8±17.8	420.6±42.7	914.6±148.3	105.0±22.3
96	35.5±28.5	73.3±22.5	71.1±59.7	357.8±196.3	1086.9±5.54	69.9±17.4
192	9.35±0.18	13.84±1.80	30.1±14.0	191.1±49.0	864.2±34.3	74.5±18.0
288	7.03±5.35	9.27±3.30	16.7±0.31	185.9±116.3	720.4±27.8	48.8±32.8
384	7.76±2.00	6.82±2.22	11.6±8.36	233.1±108.0	687.4±152.0	24.0±2.30
480	5.35±0.53	7.60±0.89	8.36±2.31	159.4±60.0	454.4±15.7	35.8±12.0
720	0.72±0.24	2.13±0.15	ND	17.5±5.12	97.7±53.5	7.12±3.54
1 080	ND	0.60±0.072	ND	8.01±6.01	15.28±8.61	1.762±0.32
1 440	ND	ND	ND	1.04±0.63	4.32±1.97	ND
2 160	ND	ND	ND	ND	ND	ND

表5　7 mg/L 孔雀石绿浸泡后孔雀石绿及其代谢物在斑点叉尾鮰肾脏中的浓度

Tab. 5　The concentrations（μg/kg）of GM and its metabolite LMG in channel catfish kidney after 7 mg/L GM bath

时间 Time（h）	孔雀石绿 MG	隐色孔雀石绿 LMG
0.25	255.45	84.90
0.5	270.13	101.24
1	117.91	99.31
2	118.25	156.23
4	102.26	174.43
6	65.91	208.36
8	117.79	301.81
12	93.86	331.50
24	60.00	279.53
48	76.89	362.69
96	89.08	338.71
192	41.07	326.33
288	27.73	222.06
384	35.71	324.28
480	29.82	138.42
720	31.30	41.88
1080	7.88	11.43
1440	ND	2.14
2160	ND	ND

数据经回归处理得到肌肉组织中药物浓度（C）与时间（t）关系的消除曲线方程、相关系数（R^2）及消除半衰期（$T_{1/2}$），（表6和表7）。孔雀石绿在4种组织中的消除半衰期 $T_{1/2(d)}$ 肾脏>肝脏>皮肤>肌肉；隐色孔雀石绿在4种组织中的消除半衰期 $T_{1/2(d)}$ 肾脏>皮肤>肌肉>肝脏。

表6 7 mg/L 孔雀石绿浸泡斑点叉尾鮰孔雀石绿在各组织中消除曲线方程及相关系数
Tab. 6 The equation of elimination curve and correlation index of MG in channel catfish after water bath of 7 mg/L MG

组织 Tissue	方程 Equation	相关系数 R^2 Correlation coefficient	消除半衰期 $T_{1/2}$（d）
肌肉 Muscle	$C=5.570e^{-0.009t}$	0.887	3.2
皮肤 Skin	$C=6.302e^{-0.007t}$	0.811	4.1
肝脏 Liver	$C=4.791e^{-0.006t}$	0.916	4.8
肾脏 Kidney	$C=4.591e^{-0.002t}$	0.773	14.4

表7 7 mg/L 孔雀石绿浸泡斑点叉尾鮰隐色孔雀石绿在各组织中消除曲线方程及相关系数
Tab. 7 The equation of elimination curve and correlation index of LMG in channel catfish after water bath of 7 mg/L MG

组织 Tissue	方程 Equation	相关系数 R^2 Correlation coefficient	消除半衰期 $T_{1/2}$（d）
肌肉 Muscle	$C=6.491e^{-0.004t}$	0.936	7.2
皮肤 Skin	$C=6.958e^{-0.003t}$	0.828	9.6
肝脏 Liver	$C=6.722e^{-0.007t}$	0.673	4.1
肾脏 Kidney	$C=6.162e^{-0.002t}$	0.891	14.4

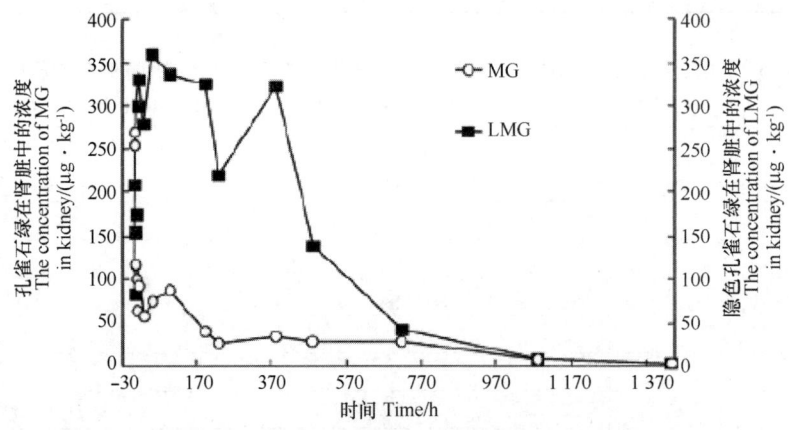

图2 孔雀石绿及其代谢物在斑点叉尾鮰肾脏中的代谢规律
Fig. 2 The metabolic rule of MG and LMG in kidney of channel catfish

3 讨论

3.1 孔雀石绿用药浓度及时间的确定及对斑点叉尾鮰代谢的影响

影响孔雀石绿在鱼体内残留的主要因素有用药时间、用药浓度、水产动物种类、水温、pH 等。彭景书等[15]采用3种浓度（0.1、0.2、0.15 mg/L）的孔雀石绿浸泡青石斑鱼5 h 和连续浸泡3次，1次/d，15 min/次，结果表明，在浸泡浓度相同的条件下连续浸泡3次，15 min/次组中孔雀石绿在青石斑鱼肌肉组织中的残留浓度和残留时间均小于浸泡5 h组；当采用不同浓度（0.15 mg/L）时连续浸泡3次，1次/d，15 min/次（0.15 mg/L组）孔雀石绿残留浓度和残留时间仍小于浸泡5h组（0.15 mg/L）可以看出浸泡时间对孔雀石绿在鱼体内残留量及残留时间影响较大，高露娇等[11]采用0.1 mg/L和0.2 mg/L的孔雀石绿药浴欧洲鳗鲡24 h后转移至清水中孔雀石绿在鳗鲡肌肉中的残留时间>120 d。曲志娜等[16]采用0.8

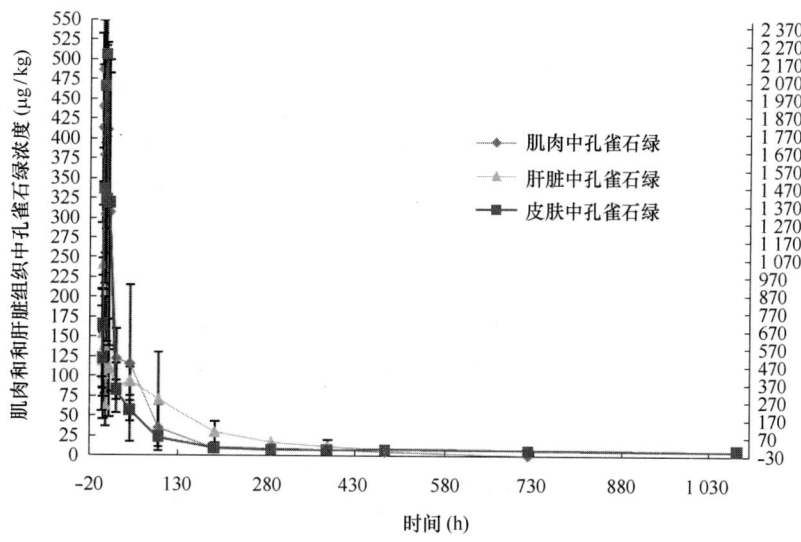

图 3 孔雀石绿在斑点叉尾鮰各组织中的分布与消除规律
Fig. 3 The distribution and deleption of MG in tissues of channel catfish

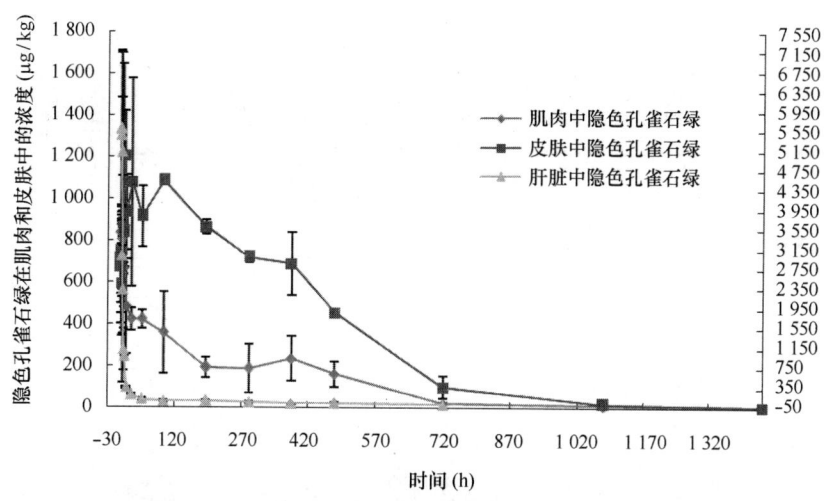

图 4 隐色孔雀石绿在斑点叉尾鮰各组织中的分布与消除规律
Fig. 4 The distribution and deleption of LMG in tissues of channel catfish

mg/L 的孔雀石绿浸泡大菱鲆 1 h 后采集大菱鲆的肌肉组织进行分析，结果表明，孔雀石绿降至检测限以下需要 20.3 d，而其代谢物隐色孔雀石绿消除则要 281.9~356.7 d。由以上报道可以看出除不同水产动物以外，浸泡用药的时间对于孔雀石绿残留的影响较大。本实验按照水产养殖过程中苗种下塘前孔雀石绿作为消毒剂对其进行浸泡消毒的方式给药，采用实际应用浓度的上限 7 mg/L 的浓度，浸泡 5 min 后将斑点叉尾鮰鱼苗养殖在池塘的网箱中。

3.2 孔雀石绿及隐色孔雀石绿在斑点叉尾鮰血液中的动力学规律

国内外关于孔雀石绿在水产动物体内药代动力学特征的报道较少，在本实验条件下孔雀石绿以 7 mg/L 的浓度浸泡斑点叉尾鮰后，孔雀石绿及其代谢物隐色孔雀石绿药时数据经 3p97 实用药代动力学分析软件分析，血液中药物浓度与时间关系符合有吸收二室模型，孔雀石绿和隐色孔雀石绿动力学方程分别为：$C_{孔雀石绿} = 683.063 \, e^{-0.248t} + 11.176 \, e^{-0.006t} - 694.239 e^{-0.333t}$，$C_{隐色孔雀石绿} = 757.240 \, e^{-0.222t} + 14.474 \, e^{-0.007t} - 771.714 \, e^{-0.382t}$，孔雀石绿和隐色孔雀石绿的表观分布容积（$V_d/F$）分别为 37.689 L/kg 和 21.125 L/kg，两者均大于 1 L/kg 说明孔雀石绿及其代谢物主要分布在血液以外的组织这与 Plaska, et al.[13] 报道的孔雀石绿有较强的血管外分布的能力相符。孔雀石绿通过浸泡方式进入斑点叉尾鮰体内进而进入血液循环，孔雀石绿的达峰时间是 3.480 h，浓度为 81.506 ng/mL，吸收速率为 0.333/h，分布半衰期为 2.794 h，消除半衰期较长

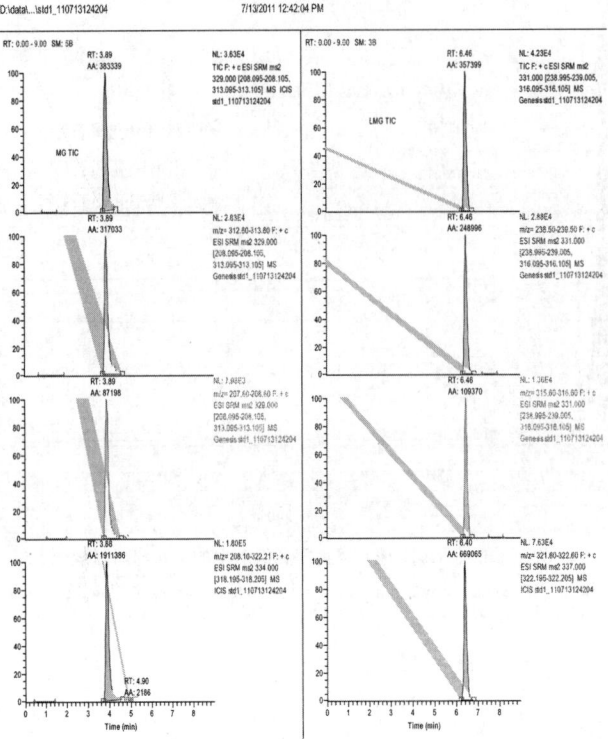

图 5 0.25 ng/mLMG 和 LMG 标准溶液选择离子监测色谱图

Fig. 5 SRM chromatograms of 0.25 ng/mL MG and LMG mixed standard

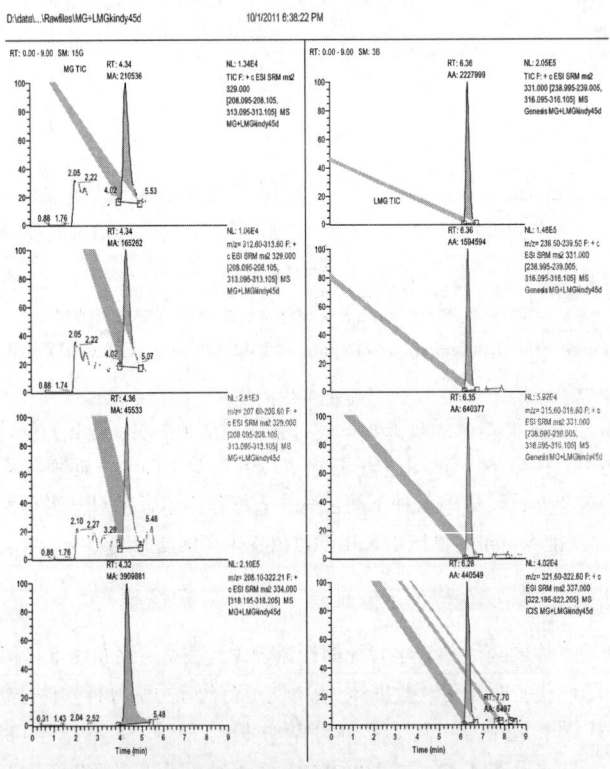

图 6 45 d 肾脏中 LMG 和 LMG 选择离子监测色谱图

Fig. 6 SRM chromatograms of MG and LMG in 45 d kidney

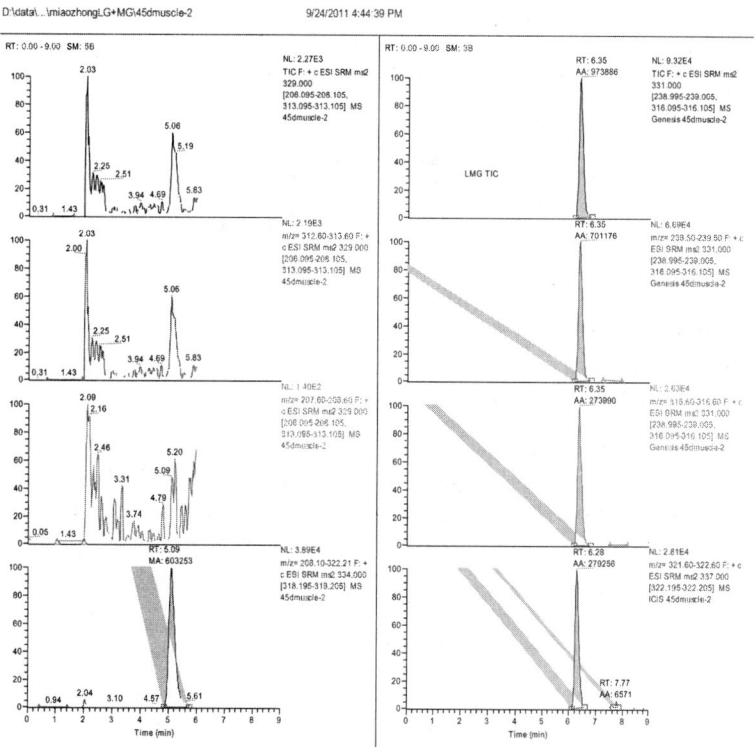

图 7 45 d 肌肉中 MG 和 LMG 选择离子监测色谱图

Fig. 7 SRM of MG and LMG chromatograms in 45 d muscle

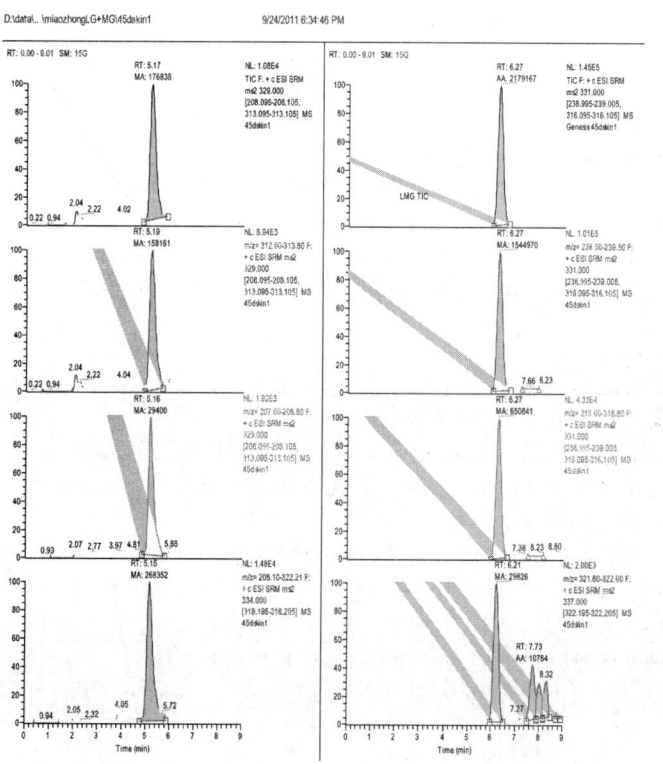

图 8 45 d 皮肤中 LMG 和 LMG 选择离子监测色谱图

Fig. 8 SRM of MG and LMG chromatograms in 45 d skin

（113.068 h）与药物分布量的曲线下面积（AUC）较大（2493.944 ng·h/mL）两者是相关的。$K_{21}<K_{12}$，说明药物从中央室进入外周室的速率比外周室回到中央室的速率大。孔雀石绿进入斑点叉尾鮰体内后主要代谢为隐色孔雀石绿，隐色孔雀石绿的达峰时间为 3.623 h，浓度为 159.619 ng/mL，吸收速率为 0.382/h，分布半衰期为 3.124 h，消除半衰期较长（105.841 h）与药物分布量的曲线下面积（AUC）较大（3601.863 ng·h/mL）。

3.3 孔雀石绿及隐色孔雀石绿在斑点叉尾鮰组织中的分布规律

目前，关于孔雀石绿及其代谢物在水产动物体内各组织中分布与消除规律的报道较少，多数仅研究了水产动物肌肉组织中孔雀石绿和隐色孔雀石绿消除规律[8-11,14-16]。本实验研究了孔雀石绿在斑点叉尾鮰皮肤、肝脏、肾脏和肌肉 4 种组织中的分布与消除过程，孔雀石绿在斑点叉尾鮰 4 种组织中浓度由高到低的顺序是皮肤>肌肉>肾脏>肝脏，其中斑点叉尾鮰皮肤组织易蓄积孔雀石绿，其残留时间最长，肝脏组织由于对孔雀石绿有极强的代谢转化功能而浓度较低；隐色孔雀石绿在斑点叉尾鮰 4 种组织中浓度由高到低的顺序是肝脏>皮肤>肌肉>肾脏，残留时间最长的组织也是皮肤组织。从孔雀石绿及其代谢物隐色孔雀石绿在各组织的消除半衰期来看皮肤组织消除半衰期 $T_{1/2}$ 较长分别为 4.1 d 和 9.6 d，而孔雀石绿和隐色孔雀石绿在肌肉组织中的消除半衰期 $T_{1/2}$ 3.2 d 和 7.2 d，这可能与斑点叉尾鮰皮肤组织中含有较多脂肪，而孔雀石绿具有亲脂性有关。目前，我国对于水产品中孔雀石绿和隐色孔雀石绿的监控主要是鱼体的可食组织-肌肉组织，从实验结果可以看出孔雀石绿和隐色孔雀石绿易在皮肤组织中蓄积，而且消除时间也相对较长，建议相关部门将水产动物的可食组织皮肤和肌肉一起列为检测靶组织。试验期间（2011 年 5 月 17 日至 2011 年 7 月 15 日）平均水温为 26.4℃，因此，使用 7 mg/L孔雀石绿浸泡 2 龄斑点叉尾鮰苗种孔雀石绿及其代谢物隐色孔雀石绿至少应经过 2 376℃·d 后才能消除。

参考文献

［1］ Wang S Y. How to effectively use chlorine-containing disinfectants and malachite green ［J］. Journal of Beijing Fisheries, 1999, （3）: 16［王声瑜. 怎样有效选用含氯消毒剂和孔雀石绿. 北京水产, 1999, （3）: 16］.

［2］ Meyer F P, JorgensenT A. Teratological and othereffects of malachite green on the development of rainbow trout and rabbits ［J］. Transaction of the American Fisheries Society, 1983, 112 （6）: 818-824.

［3］ Rao K V K. Inhibition of DNA synthesis in primary rat hepatocyte cultures by malachite green: a new liver tumor promoter ［J］. Toxicology Letters, 1995, 81 （2-3）: 107-113.

［4］ Gouranchat C. Malachite green in fish culture （state of the art and perspectives） ［R］. Veterinaire ENVT Nante France: Ecole Nat. 2000, 142.

［5］ Fernandes C, LalithaV S, Rao K V K. Enhancing effects of malachite green on development of hepatic preneoplastic lesions induced by N-nitrosodiethylamine in rats ［J］. Carcinogenesis, 1991, 12 （5）: 839-845.

［6］ Culp SJ, Blankenship LR, Kusewitt DF, et al. Toxicity and metabolism of malachite green and leucomalachite green during short-term feeding to fisher 344 rats and B6C3F1 mice ［J］. Chemico-Biological Interactions, 1999, 122 （3）: 153-170.

［7］ Srivastava S, Sinha R, Roy D. Toxicological effects of malachite green ［J］. Aquatic Toxicology, 2004, 66 （3）: 319-329.

［8］ Yang X Q, Sun M Y, Cen J W, et al. Elimination of malachite green and its metabolite in tilapia muscle ［J］. Journal of Tropical Oceanography, 2010, 29 （4）: 107-111 ［杨贤庆, 孙满义, 岑剑伟, 等. 孔雀石绿及其代谢产物在罗非鱼肌肉中残留规律的研究. 热带海洋学报, 2010, 29 （4）: 107-111］.

［9］ Qiu X J, Lin H, Wang L Z, et al. Study on Metabolism of Malachite Green in Crucian （Carassiusauratus） muscle ［J］. Periodical of Ocean University of China, 2006, 36 （5）: 745-748 ［邱绪建, 林洪, 王联珠, 等. 孔雀石绿及其代谢物物色孔雀石绿在鲫鱼肌肉中的代谢规律. 中国海洋大学学报, 2006, 36 （5）: 745-748］.

［10］ Huang J T, Lü F, Zheng J, et al. Elimination of malachite green and its metabolite Leucomalachite green in Eriocheirsinensis ［J］, Jiangsu Agricultural Sciences, 2010, （2）: 320-328 ［黄金田, 吕富, 郑浩, 等. 孔雀石绿及其代谢物隐性孔雀石绿在中华绒螯蟹体中的消除规律. 江苏农业科学, 2010, （2）: 320-328］.

［11］ Gao L J, Cai Y Q, Jiang C J, et al. Accumulation and elimination of malachite green and its primary metabolite in juvenile European eels ［J］. Journal of Fisheries of China, 2007, 31 （Suppl）: 104-108 ［高露娇, 蔡友琼, 姜朝军, 等. 孔雀石绿及其主要代谢产物在欧洲鳗鲡肌肉中的蓄积及消除规律. 水产学报, 2007, 31 （增刊）: 104-108］.

［12］ Li X A. Study on metabolism and residue of malachite green in grass carp ［D］. Thesis for Master of Science. Southwest University, Chongqing. 2008 ［李爱心. 孔雀石绿在草鱼体内的代谢及残留研究. 硕士学位论文, 西南大学, 重庆. 2008］.

［13］ PlakasSM, EL Said KR, Stehly G R, et al. Uptake, tissue distribution, and metabolism of malachite green in the channel catfish （Ictaluruspunctatus）［J］. Canadian Journal of Fisheries Aquatic Sciences, 1996, 53 （6）: 1427-1433.

［14］ Alborali L, Sangiorgi E, Leali M, et al. The persistence of malachite green in edible tissue of rainbow trout （Oncorhynchusmykiss）［J］. Rivista. Italiana Di Acquacoltura. Verona, 1997, 32 （2）: 45-60.

［15］ Peng J S, Huang M H, She Z M, et al. Accumulation and Elimination of Malachite Green in Muscle of Epinephelusawoara ［J］. Journal of Guangdong Ocean University, 2011, 31 （1）: 84-87 ［彭景书, 黄敏红, 佘忠明, 等. 孔雀石绿在清石斑鱼肌肉组织中的残留与消

除规律. 广东海洋大学学报，2011，31（1）：84-87].

［16］ Qu Z N, Li C J, Zhao S J, et al. Study on eliminate of malachite green and its metabolite in turbot muscle ［J］. Chinese Agricultural Science Bulletin, 2008, 24（6）：491-496 [曲志娜, 李存金, 赵思俊, 等. 孔雀石绿及其代谢物在大菱鲆肌肉中的消除规律. 中国农学通报，2008，24（6）：491-496].

［17］ Ding J R, Ai X H, Wang K Y, et al. Srudies on tissue distribution and the elimination regularity of Doxycycline residues in Ictaluruspunctatus ［J］. ActaHydrobiologicaSinica, 2012, 36（1）：126-132 [丁俊仁, 艾晓辉, 汪开毓, 等. 强力霉素在斑点叉尾鮰体内药物动力学及残留消除规律研究. 水生生物学报，2012，36（1）：126-132].

该文发表于《淡水渔业》【2014，44（2）：71-76】

喹烯酮在草鱼血浆中蛋白结合率的测定

Determination of protein binding of Quinocetone with grass carp（*Ctenopharyngodon idelus*）plasma

赵凤，刘永涛，胥宁，杨秋红，吕思阳，艾晓辉*

（1. 中国水产科学研究院长江水产研究所，农业部淡水鱼类种质监督检验测试中心，武汉 430223；2. 南京农业大学渔业学院，无锡 214000；3. 淡水水产健康养殖湖北省协同创新中心，武汉 430223；4. 华中农业大学水产学院，武汉 430070）

摘要：为了研究喹烯酮（Quinocetone，QCT）与草鱼（*Ctenopharyngodon idellus*）血浆蛋白的结合情况，实验采用超高效液相色谱仪（UPLC）法测定透析袋两侧溶液中喹烯酮的质量浓度，计算 QCT 的血浆蛋白结合率。血浆样品采用乙腈提取，正己烷脱脂净化。采用 ACQUITY UPLC BHE C_{18} 分离柱，柱温为 30℃，流动相为乙腈-水（30：70），流速为 0.3 mL/min，紫外检测波长为 320 nm。本方法的线性范围为 0.025~3.0 mL，相关指数 r > 0.998。添加水平为 0.05、0.1、0.2 g/mL 时，平均回收率 75%~90%，相对标准偏差均低于 15%。结果显示：QCT 在质量浓度为 0.05、0.2、1 mL 时与血浆蛋白结合率分别为（81.40~40.80）%、（79.60±0.80）%、（80.54±0.30）%，达到平衡的时间为 36 h。喹烯酮的血浆蛋白结合无浓度依赖性，且具有高强度的血浆蛋白结合率。

关键词：超高效液相色谱法；喹烯酮；血浆蛋白结合率；草鱼（*Ctenopharyngodon idellus*）

三、渔药在鱼体内的残留及其检测与消除规律

不同药物在不同养殖鱼类中会产生不同的代谢产物，因此决定其残留标示物是进行残留检测的一个重要手段。此外，不同药物在不同养殖鱼类中残留靶组织也各不相同，其残留消除规律也有较大的区别。呋喃唑酮的主要代谢产物 3-氨基-2-噁唑烷酮（3-amina-2-oxazolidinone，AOZ）在斑点叉尾鮰体皮肤中的浓度高于肌肉，而且残留时间长（刘永涛等，2012）。呋喃西林的主要代谢物氨基脲（SEM）在斑点叉尾鮰体内的消除较慢（刘永涛等，2013）。氟苯尼考（Florfenicol，FF）及其代谢物氟苯尼考胺（Florfenicol Amine，FFA）在斑点叉尾鮰肾脏组织中消除最慢，因此，建议将肾脏作为氟苯尼考在斑点叉尾鮰体内残留检测的靶组织（刘永涛等，2007）。喹烯酮及其代谢物在鱼体内残留消除较快，按照规定剂量（75mg/kg）对建鲤（*Cyprinus carpio* var. jian）和斑点叉尾鮰（*Ictalurus punctatus*）拌饲投喂喹烯酮，停药 1 d 后鱼体肌肉组织中喹烯酮及其代谢物浓度均低于计算所得的安全浓度（刘永涛等，2009）。养殖水体中低剂量的杀虫剂（0.5 mg/L 的敌百虫），养殖鱼体内该药的残留量均符合安全标准（吴京蔚等，2007）。

值得注意的是，某些渔药的代谢产物所占比较较大，应列入残留监控目标。甲苯咪唑银鲫（*Carassius auratus* gibelio）体内吸收较慢（郭东方等，2009），而甲苯咪唑在团头鲂体内的代谢产物羟基甲苯咪唑（MBZ-OH）和氨基甲苯咪唑（MBZ-NH4）逐渐升高，二者百分比都占到总残留的 10% 以上，建议将二者作为残留检测的主要监控目标（胥宁等，2013）。呋喃西林代谢物氨基脲（SEM）在斑点叉尾鮰皮肤中浓度高于肌肉组织，而且残留时间长（刘永涛等，2013）。呋喃唑酮（罗玉双等，2006）及其代谢产物 3-氨基-2-噁唑烷酮（3-amina-2-oxazolidinone，AOZ）（刘永涛等，2012）在斑点叉尾鮰（*Ietalurus Punetaus*）组织中均且残留时间较长。

该文发表于《淡水渔业》【2012，42（5）：38-44】

呋喃唑酮代谢物 AOZ 在斑点叉尾鮰体内组织分布与消除规律研究

Tissue distribution and elimination rules of the metabolite of Furazolidone（AOZ）residues in *Ietalurus Punetaus*

刘永涛[1,2]，艾晓辉[1,2]，索纹纹[1,3]，余少梅[1,3]，陈建武[1,2]

（1. 中国水产科学研究院淡水研究中心，江苏 无锡 214081；2. 中国水产科学研究院长江水产研究所
农业部淡水鱼类种质监督检验测试中心，武汉 430223；3. 华中农业大学 水产学院，武汉 430070）

摘要：研究了在池塘网箱养殖模式下，平均水温为 30.6℃，以 100 mg/kg 鱼体重的剂量投饲斑点叉尾鮰（*Ietalurus Punetaus*）苗种（95±20.9）g 含呋喃唑酮的药饵，连续 5 d，每天 2 次。采用高效液相色谱串联质谱（HPLC-ESI/MS/MS）法分析了呋喃唑酮主要代谢产物 3-氨基-2-噁唑烷酮（3-amina-2-oxazolidinone，AOZ）在斑点叉尾鮰体内的组织分布与消除规律。结果表明：血液、肌肉、皮肤、肝脏和肾脏中 AOZ 在停药后在 4 h 时的浓度分别为（2 649.0±463.1）ng/mL，（1 169.4±194.5）、（2 308.5±100.5）、（7 816.2±568.5）和 15 938.1 μg/kg，血液、肌肉、皮肤、肝脏和肾脏中 AOZ 在 60 d 时平均浓度分别为 6.66 ng/mL、4.50、33.7、17.1 和 21.3 μg/kg 分别是 AOZ 判定限 1 μg/kg 的 6.66、4.5、33.7、17.1 和 21.3 倍。90 d 时 AOZ 在血液和各组织中均检测不到。AOZ 在血液和各组织中的浓度水平肾脏>肝脏>皮肤>血液>肌肉，消除半衰期 $T_{1/2}$（d）皮肤（11.1）>肌肉（7.8）>血液（7.2）>肝脏（6.9）>肾脏（6.7）。斑点叉尾鮰皮肤中 AOZ 的浓度高于肌肉组织，而且残留时间长，在此养殖条件下，AOZ 在斑点叉尾鮰各组织中消除至少需要 27℃、54 d。

关键词：斑点叉尾鮰；呋喃唑酮；3-氨基-2-噁唑烷酮；组织分布；消除

该文发表于《中国水产科学》【2007，14（6）：1010-1016】

氟苯尼考及其代谢物氟苯尼考胺在斑点叉尾鮰体内的残留消除规律

Florfenicoland its metaboliteflorfenicol amineresiduedepletion in *Ictalurus punctatus*

刘永涛，艾晓辉，杨红

（中国水产科学研究院 长江水产研究所，农业部淡水鱼类种质监督检验测试中心，湖北 荆州 434000）

摘要：研究不同水温（18℃和28℃）下，氟苯尼考（Florfenicol，FF）及其代谢物氟苯尼考胺（Florfenicol Amine，FFA）在斑点叉尾鮰体内残留消除规律。以含氟苯尼考 4 g/kg 的饲料按 10 mg/kg 鱼体质量连续强饲斑点叉尾鮰（*Ictalurus punctatus*）3 d，于 1、2、3、5、7、9 d 分别将斑点叉尾鮰处死后取肌肉、肝脏、皮肤、肾脏 4 种组织。采用反相高效液相色谱紫外检测法测定斑点叉尾鮰组织中氟苯尼考及其代谢物氟苯尼考胺。结果表明，不同水温下同种组织，相同水温下不同组织中氟苯尼考及其代谢物氟苯尼考胺的消除速率快慢不一（差异显著）。氟苯尼考和氟苯尼考胺的和作为标示残留物，高水温时标示残留物（总量）在斑点叉尾鮰体内消除的更快。与其他组织相比标示残留物（总量）在肾脏中的消除最慢，因此，建议将肾脏作为标示残留物（总量）在斑点叉尾鮰体内残留的靶组织。在18℃和28℃时，标示残留物（总量）在肾脏中以 300 μg/kg 为最高残留限量（maximum residue limit，MRL）建议休药期分别为234℃·d和224℃·d。本研究为不同水温条件下制定该药的休药期提供了可靠的理论依据。

关键词：氟苯尼考；氟苯尼考胺；代谢物；斑点叉尾鮰；标示残留物；残留；消除规律

氟苯尼考（Florfenicol，FF）又称氟甲砜霉素是甲砜霉素的单氟衍生物。氟苯尼考以其广谱、高效、吸收迅速、分布广泛、安全等特点，在国内外畜牧和水产动物上迅速获得广泛应用。氟苯尼考不引起再生障碍性贫血，已知的副作用主要为

对哺乳动物具有胚胎毒性，随着对该药毒理学的深入研究，其他毒副作用可能进一步被发现[1]。氟苯尼考胺（Florfenicol Amine，FFA）是氟苯尼考的主要代谢物。欧盟将动物性食品中的氟苯尼考及其谢物氟苯尼考胺的和作为标示残留物[2]，中国以氟苯尼考的代谢物氟苯尼考胺作为标示残留物[3]。两者均规定标示残留物在鱼类带皮肌肉中的最高残留限量（MRL）为 1 000 μg/kg。加拿大规定鲑科鱼类肌肉中氟苯尼考的最大残留限量为 0.8 μg/kg[4]。美国 FDA 规定鲶肌肉中标示残留物氟苯尼考胺的最高残留限量为 1 000 μg/kg[5]。目前，国外关于鱼体中氟苯尼考及其代谢物氟苯尼考胺同时检测的方法有高效液相色谱紫外检测法[6]、液相色谱质谱检测法[7]和气相色谱电子捕获检测法[8]；而对氟苯尼考及其代谢物氟苯尼考胺在鱼体内的残留消除规律的研究相对较少，仅见在大西洋鲑（Salmo salar）肌肉与肝脏组织[9]和斑点叉尾鲴（Ictalurus punctatus）肌肉组织[10]中进行过相关研究。国内仅见鱼肌肉中氟苯尼考和氟苯尼考胺残留的高效液相色谱紫外检测法[11]和虾肉中氟苯尼考和氟苯尼考胺残留气相色谱检测法[12]的报道，关于氟苯尼考及其代谢物氟苯尼考胺在鱼体内的消除规律的报道尚未见。本研究采用同时检测氟苯尼考及其代谢物氟苯尼考胺的高效液相色谱紫外检测法同国内外相关报道有较大改进（另文报道），研究不同水温下氟苯尼考及其代谢物氟苯尼考胺在斑点叉尾鲴体内残留消除规律，旨在为水产养殖业中制定该药的休药期提供理论依据，从而为保证水产品安全提供技术保障。

1　材料与方法

1.1　实验鱼

斑点叉尾鲴体质量（128.40±10.51）g，由中国水产科学研究院长江水产研究所窑湾试验场提供，运至实验室后，用 2% 的食盐水消毒 10 min，然后放置于已清洗并用高锰酸钾消毒的水族箱（122 cm×80 cm×80 cm）中暂养 1 周，实验用水为经活性炭脱氯的自来水，水交换速率 20~25 L/h。采用连续充氧，保持水中溶氧大于 8.0 mg/L，暂养期间按鱼体重的 3% 投喂斑点叉尾鲴鱼配合饲料（通威 171）。每日 9：00—10：00 和 16：00—17：00 各投喂 1 次，每天换水 1 次，暂养结束后，选择无病无伤个体，规格均匀的鱼进行试验。试验期间停止饲喂并用加热棒控制水温，低温组水温为（18±1）℃，高温组水温为（28±1）℃。

1.2　试剂与溶液配制

氟苯尼考标准品（德国 Dr. Eherestorfer 公司，纯度 99.0%）；氟苯尼考胺标准品（纯度 ≥95%，sigma-aldrich 公司）；氟苯尼考原料药（批号 200504，含量 99.0%）由武汉九州神农有限公司提供；乙腈，HPLC 级，上海国药集团提供；丙酮、二氯甲烷、乙酸乙酯、十二烷基硫酸钠（SDS）、三乙胺、磷酸二氢钠、磷酸氢二钠、磷酸、正己烷均为国产分析纯。

1.2.1　磷酸缓冲液的配制

磷酸缓冲液（pH 7.0）配制：准确称取磷酸氢二钠（$Na_2HPO_4 \cdot 12H_2O$）71.6 g 和磷酸二氢钠（$NaH_2PO_4 \cdot 2H_2O$）27.6 g，分别溶于 1 L 蒸馏水中配成 0.2 mol/L 的溶液，然后按 61：39 的体积比混合即成。

1.2.2　定容溶液的配制

A 液：称取 0.895 g 磷酸氢二钠溶解在 250 mL 双蒸水中配制成 0.01 mol/L 的磷酸氢二钠溶液，用磷酸将溶液 pH 调至 2.8，经抽滤装置抽滤倒入具塞玻璃瓶中置 4℃ 冰箱中备用；B 液：乙腈。A 液与 B 液体积比 3：2。

1.3　仪器与设备

高效液相色谱仪为美国 Waters 公司的 515 泵、717 自动进样器、2487 双通道紫外检测器；Empower 液相色谱工作站。自动高速冷冻离心机（日本 HITACHI 20PR-520 型）；Mettler-TOLEDO AE-240 型精密电子天平（梅特勒-托利多公司）；80-2B 型台式离心机，FS-1 高速匀浆机，调速混匀器（上海康华生化仪器制造厂），HGC-12 氮吹仪（HENGAO T&D 公司）。

1.4　色谱条件

色谱柱：Waters symmetry C_{18} 反相色谱柱（250 mm×4.6 mm，5 μm）。流动相：乙腈-磷酸二氢钠溶液（0.01 mol/L，含 0.05 mol/LSDS 和 0.1% 三乙胺）体积比 2：3，用磷酸调 pH 值至 4.5。流速：0.6 mL·min⁻¹。柱温：室温。紫外检测波长 225 nm。进样量 20 μL。

1.5　实验设计与方法

1.5.1　药饵制备、给药剂量及取样

将购买的斑点叉尾鲴饲料（通威 171）用粉碎机粉碎，将氟苯尼考加到粉碎后的饲料中重新制成氟苯尼考的含量为 4 g/kg 颗粒饲料；按 10 mg/kg 鱼体质量进行强饲。先将投胃管装好加药饲料然后慢慢将投胃管插入鲴的胃内，将一根实心

管子从投胃管内径将其中的药饵慢慢推到鲫的胃内。给药后保定 1 min，然后将鱼单独放入一个水盆中观察 5 min，将返胃回吐药饵的鱼弃去，无回吐的鱼放入水族箱中。给药前同批鱼采 1 次空白组织，1 d 给药 1 次，连续给药 3 d，于停药后 1、2、3、5、7、9 d 将鲫鱼钝击头部致死取皮、背脊两侧肌肉、肝脏、肾脏，置塑料袋中 −20℃ 冰箱中保存，每个采样点取 5 尾鱼。

1.5.2　样品处理

将冷冻保存的组织样室温下自然解冻。剪取适量的肌肉组织，置高速匀浆机中匀浆呈糜状，准确称取 1.0 g；鱼皮先剪碎。准确称取 1.0 g，肾脏取 0.5 g，肝脏 0.5 g，分别放于 10 mL 的具塞离心管中，加入 1 mL 的 pH7.0 的磷酸盐缓冲液，置调速混匀器上涡旋振荡混匀 1 min，加入 3 mL 丙酮涡旋振荡混匀 1 min，以 10 000 r/min 离心 5 min，将溶液转移到另一支 10 mL 离心管中，在向溶液中加入 2 mL 二氯甲烷，涡旋振荡混匀 1 min，5 000 r/min 离心 5 min，弃去上层水相，置 50℃ 氮吹仪上吹干，用 1 mL 定容溶液溶解残渣，涡旋振荡混匀 1 min，加入 3 mL 正己烷涡旋振荡混匀 1 min，10 000 r/min 离心 10 min，弃去正己烷，重复去脂 1 遍，水相经 0.22 μm 滤头过滤置自动进样瓶，HPLC 测定。

1.5.3　标准工作曲线的制备及最低检测限（LOD）

在斑点叉尾鲫 4 种空白组织（肌肉、皮肤、肝脏、肾脏各 1 g）中加入已知浓度的标准溶液，制成含氟苯尼考和氟苯尼考胺质量分数分别为 0.05 μg/g、0.1 μg/g、0.25 μg/g、0.5 μg/g、1.0 μg/g、2.5 μg/g、5.0 μg/g、10.0 μg/g 组织样品，分别以氟苯尼考和氟苯尼考胺的峰面积（A_i）为纵坐标，质量浓度（C）为横坐标绘制标准工作曲线，求出回归方程和相关系数。按上述的色谱条件测定加标样品，以 3 倍信噪比（3S/N）计算氟苯尼考及氟苯尼考胺在血浆、肌肉、皮肤、肝脏、肾脏组织中的最低检测限。

1.5.4　回收率及方法精密度测定

在斑点叉尾鲫空白肌肉、皮肤、肝脏、肾脏组织样品中分别添加 5 个水平的 FF 和 FFA 混合标准溶液，使样品质量浓度分别为 0.1 μg/g、0.2 μg/g、0.5 μg/g、1.0 μg/g、2.0 μg/g，每个质量浓度做 3 个平行，并设一个空白对照，按样品前处理过程处理后测定回收率。在 4 种空白组织中分别添加 2 个浓度水平的 FF 和 FFA 混合标准溶液，使质量浓度分别为 0.1 μg/g、1.0 μg/g。每个浓度的样品，日内做 5 个重复，1 周内重复做 5 次，计算日内及日间精密度。

1.6　数据处理

消除方程及休药期计算和回归图，采用 Microsoft Excel 2003，SPSS（13.0），Statistic（6.0）进行计算和绘制。消除方程采用 $C=C_0 e^{-kt}$。C 表示药物质量分数（μg/kg），C_0 为残留消除对数曲线的截距（μg/kg），k 表示消除速率常数。根据休药期（WDT）计算各组织药物浓度降至规定水平（MRL）所需的时间：

$$WDT = \frac{\ln(C_0/MRL)}{k}$$

式中，WDT 为休药期，MRL 为最高残留限量（μg/kg），C_0 为残留消除对数曲线的截距（μg/kg），k 为残留消除曲线速率常数[13]。

2　结果与分析

2.1　标准工作曲线及最低检测限（LOD）

4 种组织中氟苯尼考和氟苯尼考胺标准工作曲线方程及相关系数如下，肌肉中 FF $A_{i肌肉}=77764C_{肌肉}-1908.2$，$R=0.9984$，FFA $A_{i肌肉}=79617C_{肌肉}+4059.8$，$R=0.9993$；肝脏中 FF $A_{i肝脏}=76475C_{肝脏}-3535.4$，$R=0.9991$，FFA $A_{i肝脏}=90704C_{肝脏}+1884.8$，$R=0.9994$；肾脏中 FF $A_{i肾脏}=76581C_{肾脏}+773.29$，$R=0.9984$，FFA $A_{i肾脏}=88623C_{肾脏}-10845$，$R=0.9980$；皮肤中 FF $A_{i皮肤}=73690C_{皮肤}-2316.8$，$R=0.9979$，FFA $A_{i皮肤}=81851C_{皮肤}-9453.5$，$R=0.9962$。肌肉、肝脏、肾脏、皮肤中 FF 和 FFA 的最低检测限分别为 20 μg/kg 和 15 μg/kg，50 μg/kg 和 50 μg/kg，50 μg/kg 和 50 μg/kg，50 μg/kg 和 50 μg/kg。

2.2　回收率及方法精密度

4 种组织中 FF 平均回收率为 75.14%～92.65%，FFA 平均回收率为 76.00%～91.50%，日内变异系数不大于 8.68%，日间变异系数不大于 9.22%。

2.3　不同水温下 FF 和 FFA 在斑点叉尾鲫体内残留及消除

氟苯尼考在动物体内的主要代谢产物是氟苯尼考胺，同时还有氟苯尼考醇、氟苯尼考草氨酸等代谢物，这些代谢物有极低的抗菌活性，对环境生物影响较小。目前，欧盟和中国以氟苯尼考和氟苯尼考胺作为残留检测的标示检出物[1]。

2.3.1　不同水温下 FF 和 FFA 在斑点叉尾鮰组织中的残留

不同水温下斑点叉尾鮰经强饲含有氟苯尼考的饲料 3 d 后氟苯尼考及其代谢物氟苯尼考胺在 4 种组织中的残留（表 1 和表 2）。

表 1　18℃斑点叉尾鮰连续 3 d 强饲氟苯尼考各组织中 FF 和 FFA 的残留量

Tab. 1　FF and FFAlevels in channel catfish tissues after oral gavages administration at

does of 10 mg/kg bw for 3 consecutive days at 18℃ n=5；$\overline{X}$±SD；μg/g

时间/d Time	肌肉 Muscule		皮肤 Skin		肝脏 Liver		肾脏 Kindey	
	FF	FFA	FF	FFA	FF	FFA	FF	FFA
1	4.43±0.36	0.10±0.02	3.69±0.44	0.27±0.02	5.44±0.71	0.51±0.25	6.34±0.60	0.54±0.16
2	1.46±0.26	0.07±0.01	1.14±0.20	0.26±0.02	1.35±0.49	0.37±0.12	2.86±0.59	0.51±0.11
3	0.83±0.20	0.03±0.02	0.58±0.13	0.25±0.07	0.89±0.16	0.24±0.06	1.59±0.37	0.35±0.07
5	0.14±0.02	ND	0.29±0.04	0.24±0.03	0.18±0.01	0.10±0.03	0.90±0.34	0.26±0.07
7	ND	ND	0.21±0.06	0.20±0.06	ND	ND	0.89±0.12	0.24±0.01
9	ND	ND	0.15±0.04	ND	ND	ND	0.71±0.27	0.19±0.02

注：ND-未检测到.

Note：ND-Undetectable.

表 2　28℃斑点叉尾鮰连续 3 d 强饲氟苯尼考后各组织中 FF 和 FFA 的残留量

Tab. 2　FF and FFA levels in channel catfish tissues after oral gavages administration at

does of 10 mg/kg bw for 3 consecutive days at 28℃ n=5；$\overline{X}$±SD；μg/g

时间/d Time	肌肉 Muscule		皮肤 Skin		肝脏 Liver		肾脏 Kindey	
	FF	FFA	FF	FFA	FF	FFA	FF	FFA
1	1.92±0.66	0.11±0.07	1.39±0.52	0.25±0.09	2.51±0.97	0.40±0.17	2.82±0.82	0.47±0.22
2	0.26±0.08	ND	0.54±0.09	0.19±0.03	0.60±0.20	0.27±0.09	1.36±0.28	0.30±0.04
3	0.08±0.02	ND	0.16±0.04	0.16±0.02	0.29±0.09	0.08±0.03	0.45±0.12	0.22±0.01
5	ND	ND	0.08±0.03	0.14±0.01	0.09±0.01	ND	0.29±0.09	0.19±0.02
7	ND	ND	ND	ND	ND	ND	0.19±0.14	0.17±0.03
9	ND	ND	ND	ND	ND	ND	0.17±0.07	ND

注：ND-未检测到.

Note：ND-Undetectable.

由表 1 和表 2 结果表明，不同水温（18℃和 28℃）下，同一采样点，相同组织或不同组织中，氟苯尼考及其代谢物氟苯尼考胺的残留量低水温时均比高水温时高，而且相差较大；相同水温下，不同组织中，氟苯尼考及其代谢物氟苯尼考胺的残留量也有较大差别。

2.3.2　不同水温下标示残留物（总量）在斑点叉尾鮰组织中残留消除规律

氟苯尼考及其代谢物氟苯尼考胺浓度的和作为动物性食品中氟苯尼考残留的标示残留物[2]的浓度见表 3 和表 4。标示残留物（总量）在斑点叉尾鮰各组织中的残留浓度随时间变化的规律见图 1 和图 2。数据经回归处理得到 4 种组织中药物浓度（C）与时间（t）关系的消除曲线方程、相关指数（R^2）及消除半衰期（$T_{1/2}$）见表 5 和表 6。

表3 18℃氟苯尼考及其代谢物氟苯尼考胺浓度的和作为标示残留物在斑点叉尾鮰各组织中的残留量

Tab. 3 The sum of FF and FFA levelsas the marker residue in channel catfish tissues at 18℃ μg/g

时间/d Time	肌肉 Muscule	皮肤 Skin	肝脏 Liver	肾脏 Kindey
1	4.53	3.96	5.95	6.88
2	1.53	1.40	1.72	3.37
3	1.03	0.83	1.13	1.94
5	0.16	0.53	0.28	1.16
7	ND	0.41	ND	1.13
9	ND	0.15	ND	0.90

注：ND-未检测到.

Note：ND-Undetectable.

表4 28℃氟苯尼考及其代谢物氟苯尼考胺浓度的和作为标示残留物在斑点叉尾鮰各组织中的残留量

Tab. 4 The sum of FF and FFA levels as the marker residue in channel catfish tissues at 28℃μg/g

时间（d） Time	肌肉 Muscule	皮肤 Skin	肝脏 Liver	肾脏 Kindey
1	2.03	1.64	2.91	3.29
2	0.26	0.73	0.87	1.66
3	0.08	0.32	0.37	0.67
5	ND	0.22	0.09	0.48
7	ND	ND	ND	0.36
9	ND	ND	ND	0.17

注：ND-未检测到.

Note：ND-Undetectable.

由表3和表4结果表明，不同水温（18℃和28℃）下，同一采样点，氟苯尼考的标示残留物在相同组织或不同组织中的残留量低水温时均比高水温时高，而且相差较大；相同水温下，不同组织中，氟苯尼考的标示残留物的残留量也有较大差别。

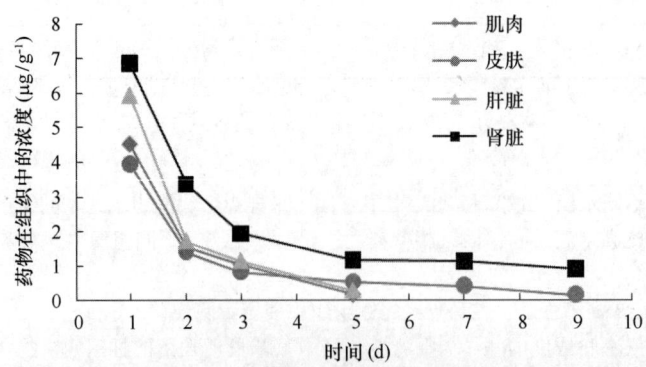

图1 18℃斑点叉尾鮰连续3d强饲氟苯尼考，标示残留物在组织中的消除曲线

Fig. 1 Elimination curve of the marker residue in channel catfish after oral gavages for 3 consecutive days at 18℃

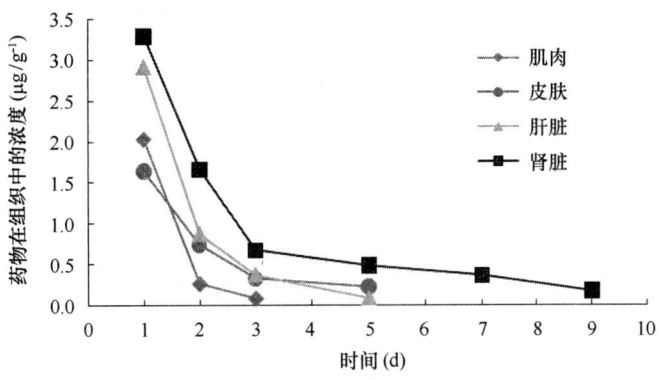

图 2　28℃斑点叉尾鮰连续 3 d 强饲氟苯尼考，标示残留物在组织中的消除曲线

Fig. 2　Elimination curve of the marker residue in channel catfish after oral gavages for 3 consecutive days at 28℃

表 5　18℃标示残留物在斑点叉尾鮰组织中的消除曲线方程及参数

Tab. 5　Equation of elimination curve and parameters of the marker residue in channel catfish tussues after oral gavages for 3 successive days at 18℃

组织 Tissue	方程 Equation	相关指数 r^2	消除半衰期/h $T_{1/2}$
肌肉 Muscule	$C_{肌肉} = 9.564e^{-0.809t}$	0.984	20.558
皮肤 Skin	$C_{皮肤} = 3.467e^{-0.347t}$	0.914	47.931
肝脏 Liver	$C_{肝脏} = 9.905e^{-0.727t}$	0.971	22.878
肾脏 Kindey	$C_{肾脏} = 5.432e^{-0.229t}$	0.815	72.629

表 6　28℃ 标示残留物在斑点叉尾鮰组织中的消除曲线方程及参数

Tab. 6　The equation of elimination curve and parameters of the marker residue in channel catfish tussues after oral gavages for 3 successive days at 28℃

组织 Tissue	方程 Equation	相关指数 R^2	消除半衰期 $T_{1/2}$（h）
肌肉 Muscule	$C_{肌肉} = 8.836e^{-1.617t}$	0.976	10.286
皮肤 Skin	$C_{皮肤} = 2.095e^{-0.494t}$	0.887	33.688
肝脏 Liver	$C_{肝脏} = 5.569e^{-0.849t}$	0.983	19.590
肾脏 Kindey	$C_{肾脏} = 3.060e^{-0.331t}$	0.909	50.248

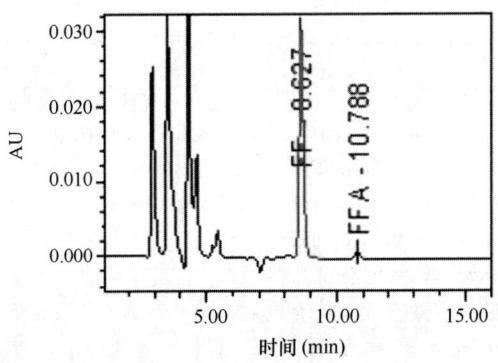

图 3　18℃多次口灌 FF 后 1 d 肌肉色谱图

Fig. 3　Chromatogram of muscule at 1 day after multidoes oral adiministratin of FF at 18℃

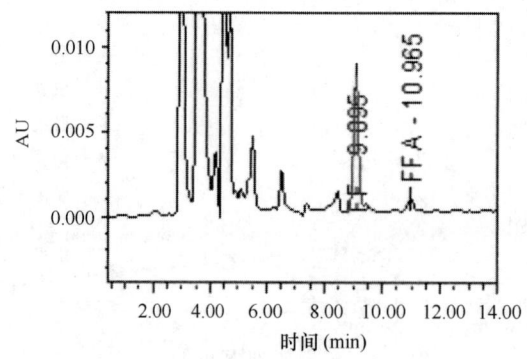

图 4 28℃多次口灌 FF 后 1 d 肌肉色谱图

Fig. 4 Chromatogram of muscule at 1 day after multidoes oral adiministratin of FF at 28℃

2.4 色谱图

图 3 和图 4 是 18℃和 28℃斑点叉鲴经多次强饲氟苯尼考后 1 d 时肌肉样品中氟苯尼考及其主要代谢物氟苯尼考胺的液相色谱图，由色谱图可以看出 28℃时氟苯尼考和氟苯尼考胺的浓度要比 18℃低；另外，在同一组织中同一采样点氟苯尼考和氟苯尼考胺的浓度高水温时要比低水温时低。本实验在检测鲴组织中氟苯尼考胺的过程中发现其浓度较低，与 Horsberg 等[9]报道的氟苯尼考胺在大西洋鲑肌肉和肝脏中的残留相差较大。

3 讨论

3.1 氟苯尼考及其代谢物氟苯尼考胺在鲴鱼组织中的分布规律

由表 1 和表 2 可以看出，氟苯尼考的代谢物氟苯尼考胺在鲴各组织中的残留均较低，氟苯尼考和氟苯尼考胺在组织中的分布规律相似，残留水平由高到低大致为肾脏、肝脏、肌肉、皮肤，与余培建等[14]报道的氟苯尼考在欧洲鳗鲴组织中的分布规律相似，Horsberg 等[9]也报道大西洋鲑肝脏中的氟苯尼考和氟苯尼考胺的残留要比肌肉中的高。而且 18℃水温下要比 28℃时同一采样点组织中氟苯尼考和氟苯尼考胺的残留要高。18℃时氟苯尼考在肌肉中于 7 d 后检测不到，而 28℃时氟苯尼考于第 5 天时已检测不到；皮肤中的氟苯尼考在 18℃ 9 d 时仍可检测到其质量分数为（0.15±0.04）μg/g，而 28℃ 5 d 时其残留仅为（0.08±0.03）μg/g，7 d 时已检测不到。在 18℃和 28℃下 9 d 时均可在肾脏组织检测到氟苯尼考，其残留分别为（0.71±0.27）μg/g 和（0.17±0.07）μg/g。在 18℃下 9 d 时氟苯尼考胺在肾脏中可被检测到其残留为（0.19±0.02）μg/g，而在 28℃下 9 d 时氟苯尼考胺不能被检测到。28℃下肌肉中氟苯尼考胺在 2 d 时已检测不到，18℃ 5 d 时肌肉中氟苯尼考胺才检测不到，而与 Horsberg 等[9]报道的 10℃下 20 d 时在大西洋鲑的肌肉和肝脏中仍可以检测到氟苯尼考胺的结果相差较大，而 Samuelsen 等[15]报道在大西洋鳕的血浆和组织中没有检测到氟苯尼考胺。这种差异作者认为是由水生动物种属以及水温差异造成的。

3.2 标示残留物氟苯尼考胺（总量）在鲴组织中残留特征

采用 HPLC 法可在斑点叉尾鲴的肌肉、皮肤、肝脏、肾脏组织中可同时检测到氟苯尼考及其代谢物氟苯尼考胺（图 3 和图 4）。欧盟规定将氟苯尼考及其代谢物氟苯尼考胺的和作为氟苯尼考在动物组织中残留的标示残留物。目前，欧盟和中国以氟苯尼考和氟苯尼考胺作为动物性食品中的检出物[1]，但国内尚未见有以氟苯尼考及其代谢物氟苯尼考胺的和作为标示残留物进行残留研究的文献。在 18℃和 28℃水温条件下标示残留物（总量）在斑点叉尾鲴组织中的分布及残留消除规律是一致的，标示残留物（总量）在组织中的残留由高到低依次均为肾脏、肝脏、肌肉、皮肤，消除速率由快到慢依次均为肌肉、肝脏、皮肤、肾脏，可以看出标示残留物（总量）在肝脏、肾脏中的浓度最高，这说明标示残留物（总量）在肝脏代谢和肾脏富集、排泄为主，与余培建等[14]报道的氟苯尼考在欧洲鳗鲴肾脏中残留最高结果相似，并在肾脏中富集和排泄的结果相似。从消除规律上看 18℃和 28℃水温下标示残留物（总量）在肌肉中均消除最快，消除半衰期分别为 20.558 h 和 10.268 h，而在肾脏中消除最慢，消除半衰期分别为 47.931 h 和 33.688 h。从两种水温条件下标示残留物（总量）在斑点叉尾鲴体内各组织中的浓度及消除规律来看，同一采样时间时高水温条件下标示残留物（总量）在各组织中的浓度明显低于低水温时，而消除速率比低水温时快。

国外关于氟苯尼考及其代谢物氟苯尼考的残留消除研究较少而且只报道了肌肉和肝脏中消除规律。以 10 mg/（kg·bw）的剂量连续 10 d 饲喂体质量为（2.784±922）g 的大西洋鲑含有氟苯尼考的饲料，于第 1 天、第 15 天、第 20 天、第

26 天、第 30 天、第 35 天、第 40 天和第 49 天分别取 10 尾大西洋鲑进行氟苯尼考及其代谢物氟苯尼考胺的消除规律进行研究。结果表明氟苯尼考在 1 d 时可以检测到其余时间均未检测到（此实验设计的时间间隔太长，不利于氟苯尼考消除规的研究），而氟苯尼考胺在大西洋鲑肝脏中的消除速率要慢于在肌肉中的消除速率[9]（与本实验的结果相似）。但在 20 d 时还可以在大西洋鲑的肌肉和肝脏组织中检测到氟苯尼考胺，而本实验 5 d 以后在肌肉和肝脏中就检测不到氟苯尼考胺了。出现这样的差异可能与种属、水温、盐度、投喂次数等因素的差异有关，本实验考虑到氟苯尼考在防治鱼类疾病时一般连用 3~5 d[16]，以及本实验采用强饲方式给药的工作量及对鱼体的影响，本实验采用连续给药 3 d 的试验方案。水温在 17.9~20.7℃斑点叉尾鮰连续饲喂 10 d 含氟苯尼考的饲料，并将肌肉中氟苯尼考及其代谢物转化为氟苯尼考胺进行测定，4 d 后标示残留物（总量）低于 1 000 μg/kg 与本实验 18℃测得的结果相似。本实验在不同水温条件下分别对肌肉、皮肤、肝脏、肾脏组织中标示残留物（总量）消除规律进行了研究。研究发现肾脏是残留量最高消除最慢的组织，因此，可将肾脏作为本试验中残留分析的靶组织。

3.3　休药期

欧盟[2]对氟苯尼考的标示残留物（总量）在牛、羊、家禽、猪可食性组织（牛和羊：肌肉、肝脏肾脏；家禽和猪：肌肉、皮肤+脂肪、肝脏、肾脏）中的最高残留限量（MRL）均做了详细的规定，而在水产品中只对有鳍鱼类可食组织中自然比例的肌肉和皮肤中标示残留物（总量）的最高残留限量（MRL）作了规定。本实验对不同水温条件下斑点叉尾鮰肌肉、皮肤、肝脏、肾脏组织中标示残留物（总量）的消除规律作了研究，进一步丰富了氟苯尼考在有鳍鱼类各组织中分布及消除资料，也为制定最高残留限量及休药期提供了参考资料。根据欧盟规定氟苯尼考在有鳍鱼类肌肉和皮肤组织及所在有动物源性食品肝脏和肾脏组织中规定的 MRL：分别为 1 000 μg/kg、1 000 μg/kg、2 000 μg/kg、300 μg/kg。本实验以此作为参考，通过 WDT 公式计算可知斑点叉尾鮰各组织的理论 WDT 为 18℃休药期肌肉 2.79 d、皮肤 3.58 d、肝脏 2.20 d、肾脏 12.65 d；28℃休药期肌肉 1.35 d、皮肤 1.50 d、肝脏 1.21 d、肾脏 7.02d。通常将残留量高、消除最慢的组织可看成是残留分析的靶组织。实验结果表明，肾脏组织残留量高、消除最慢，因此，可将肾脏作为残留分析的靶组织则休药期分别为 18℃时 234℃·d 和 28℃时 224℃·d。

参考文献

[1] 徐力文，廖昌容，刘广锋. 氟苯尼考用于水产养殖的安全性 [J]. 中国水产科学，2005，12 (4)：512-518.

[2] E. U. Commission Regulation（EC）No. 508/1999 of 4 March 1999. amending Annexes I to IV to Council Regulation（EEC）No. 2377/90 laying down a Community procedure for the setablishment of maximum residue limits of veterinary medicinal products in foodstuffs of animal origin. Official Journal of the European Communities，1999.

[3] 中华人民共和国农业部公告第 235 号.《动物性食品中兽药最高残留限量》，2002.

[4] Health Canada. Maximum Residue Limits（MRLs）set by Canada [Z]. http：//www. hc-sc. gc. ca/dhp-mps/vet/mrl-lmr/mrl-lmr_versus_ new-nouveau-e. html.

[5] FDA. Title 21-Food and Drugs，PART 556 Tolerances for residues of new animal drugs in food，Subpart B. 556. 283-Florfenicol [Z]. http：//www. accessdata. fda. gov/scripts/cdrh/cfcfr/CFRSearch. cfm？CFrpart=556. 283.

[6] Hormazabal V.，Steffenak I and Yndestad M. Simultaneous determination of residues of florfenicol and the metabolite florfenicol amine in fish tissues by high performance liquid chromatography [J]. Chromatogra，1993，616：161-165.

[7] Jeffery M. R.，Ross A. P.，Melissa C. F.，et al. Simultaneous determination of residues of chloramphenicol，thiamphenicol，florfenicol，and florfenicol amine in farmed aquatic species by liquid chromatography/mass spectrometry [J]. AOAC International，2003，86 (3)：510-514.

[8] Pfenning A. P.，Roybal J. E.，Rupp H. S.，et al. Simultaneous determination of residues of chloramphenicol，florfenicol，florfenicol amine，and thiamphenicol in shrimp tissue by gas chromatography with electron capture detection [J]. AOAC International，2000，83 (1)：26-30.

[9] Horsberg T. E.，Hoff K. A. and Nordmo R.. Pharmacokinetics of florfenicol and its metabolite florfenicol amine in Atlantic salmon [J]. Aqu Anim Health，1996，8：292-301.

[10] Wrzesinski C.，Crouch L.，Gaunt P.，et al. Florfenicol residue depletion in channel catfish，Ictalurus punctatus（Rafinesque）[J]. Aquaculture，2006，253 (1-4)：309-316.

[11] 郭霞，张素霞，沈建忠，等. 鱼肌肉中氟苯尼考和氟苯尼考胺残留的高效液相色谱检测 [J]. 中国兽医科学，2006，36 (09)：743-747.

[12] 孙丰云，张素霞，沈建忠，李建成. 虾肉中氯霉素 甲砜霉素 氟苯尼考及氟苯尼考胺残留气相色谱-微电子捕获检测法 [J]. 中国兽医杂志，2006，42 (10)：66-67.

[13] Riviere J. E.. Comparative Pharmacolinetics：Principle，Techniques，and Application，Ames：Iowa State University Press，1999：308.

[14] 余培建，翁祖桐，樊海平，等. 氟苯尼考在欧洲鳗鲡体内的药物代谢动力学的研究 [J]. 福建水产，2005，4：52-57.

[15] Samuelsen O. B.，Bergh O.，Ervik A.. Pharmacokinetics of florfenicol in cod Gadus morhua and in vitro antibacterial activity against Vibrio anguillarum [J]. Dis Aquat Organ，2003，56：127-133.

[16] 杨先乐，陆承平，战文斌等. 新编渔药手册 [J]. 北京：中国农业出版社，2005：189.

该文发表于《水生态学杂志》【2009，2（5）：95-98】

喹烯酮在建鲤和斑点叉尾鮰体内的残留消除规律研究

Study on Residues of Quinocetone in Jian Carp（*Cyprinus carpio* var. jian） and Channel Catfish（*Ictalurus punctatus*）

刘永涛[1]，郭东方[2]，杨莉[3]，杨红[1]，袁科平[1]，艾晓辉[1]

（1. 中国水产科学研究院长江水产研究所农业部淡水鱼类种质监督检验测试中心，湖北 荆州 434000；2. 华中农业大学水产学院，湖北 武汉 430070；3. 四川农业大学动物医学院，四川 雅安 625014）

摘要： 以含喹烯酮75 mg/kg的饲料，在池塘中饲喂建鲤（*Cyprinuscarpio* var. jian）和斑点叉尾鮰（*Ictalurus punctatus*）60 d，研究饲养条件下喹烯酮及其代谢物脱二氧喹烯酮在这2种鱼体内的残留。结果表明，喹烯酮平均回收率为75.4%~80.5%，脱二氧喹烯酮平均回收率为74.5%~79.1%，喹烯酮和脱二氧喹烯酮的检测限分别为7.35和16.23 µg/kg。在饲喂结束后1~15 d采集的建鲤和斑点叉尾鮰肌肉样品中均未检测到喹烯酮及脱二氧喹烯酮。因停药1 d后鱼体肌肉中喹烯酮及其代谢物浓度均低于计算所得的安全浓度，故喹烯酮用在水产养殖上是安全的。

关键词： 喹烯酮；脱二氧喹烯酮；建鲤；斑点叉尾鮰；残留

该文发表于《福建农林大学学报（自然科学版）》【2013，42（1）：72-76】

网箱养殖条件下呋喃西林代谢物SEM在斑点叉尾鮰体内组织分布及消除规律研究

Tissue distribution and elimination rules of the metabolite of Nitrofuranzone（SEM）residues channel catfish（*Ietalurus punetaus*）cultured in cage

刘永涛[1,2]，艾晓辉*[1,2]，索纹纹[1,3]，余少梅[1,3]，陈建武[1,2]，

（1. 中国水产科学研究院长江水产研究所 农业部淡水鱼类种质监督检验测试中心，湖北 武汉 430223；2. 中国水产科学研究院淡水研究中心，江苏 无锡 214081；3. 华中农业大学 水产学院，武汉 430070）

摘要： 采用高效液相色谱串联质谱（HPLC-ESI/MS/MS）法分析呋喃西林主要代谢物氨基脲（SEM）在斑点叉尾鮰体内的组织分布与消除规律。结果表明：在停药1 h后，血液和肾脏中的SEM浓度达最大值，分别为（37.5±7.45）ng·mL^{-1}和（383.3±89.2）µg·kg^{-1}；在停药2 h后，肌肉、皮肤、肝脏中的SEM达到最大值，其浓度分别为（33.3±8.25）、（168.2±43.47）和（105.0±48.4）µg·kg^{-1}；在480 h时，肌肉中的SEM浓度是其残留判定值（1 µg·kg^{-1}）的1.1倍；在1080 h时，血液、皮肤、肝脏和肾脏中的SEM分别是其残留判定值（1 µg·kg^{-1}）的1.17倍、8.75倍、1.0倍和5.47倍；在1440 h（60 d）时，血液、肌肉和皮肤中的SEM平均浓度分别为（0.57±0.25）ng·mL^{-1}、（0.28±0.03）和（0.83±0.24）µg·kg^{-1}，而在肝脏和肾脏中，未检测出SEM；在2 160 h（90 d）时，血液和各组织中均未检测SEM。SEM在血液和各组织中消除半衰期$T_{1/2}$为血液（11.11 d）>皮肤（9.02 d）>肌肉（8.25 d）>肾脏（7.40 d）>肝脏（6.71 d）。皮肤中SEM浓度高于肌肉组织，而且残留时间长。整个实验期间平均水温为28℃，在此养殖条件下SEM在斑点叉尾鮰苗种各组织中消除至少需要2 520℃·d。

关键词： 斑点叉尾鮰；呋喃西林代谢物；氨基脲；组织分布；消除

该文发表于《水产学杂志》【2007，20（2）：54-57】

养殖水体中敌百虫残留限量的确定——
敌百虫在鲤鱼及鲫鱼体内的积累实验

Determination of trichlorfon residue limits in the aquaculture water-trichlorfon accumulation in carp and *crucian carp*

吴京蔚[1]，卢形岩[2]，王荻[2]，刘红柏[2]

（1. 东北农业大学动物科学技术学院，黑龙江 哈尔滨 150030；2. 中国水产科学研究院黑龙江水产研究所，黑龙江哈尔滨市 150070）

摘要： 鲤鱼及鲫鱼种分别置于敌百虫浓度为 0.2、0.3、0.4、0.5、1 ppm 的水体中，并于实验开始后的 5、10、15、20、30、40 d 及 60 d 取实验鱼肌肉进行高效气相色谱分析，检测鱼肉中敌百虫含量，结果表明：此检测方法下线性范围为 0.02~5.00 μg/m L，测定限量为 0.02 mg/kg，且回收率和精密度均符合检测要求，从实验结果中可看出，养殖水体中敌百虫低于 0.5 ppm 时，鱼体中的敌百虫残留量均符合卫生标准，可安全食用。

关键词： 鱼体；敌百虫；积累

该文发表于《中国水产科学》【2009，16（3）：434-441】

甲苯咪唑在银鲫体内的药代动力学研究
Pharmacokinetics of mebendazole in Johnny carp（*Carassius auratus gibelio*）

郭东方[1,2]，刘永涛[1]，艾晓辉[1]，袁科平[1]，王淼[1,2]，秦改晓[1,2]

（1. 中国水产科学研究院长江水产研究所，湖北 荆州 434000；2. 华中农业大学水产学院，湖北 武汉 430070）

摘要： 采用高效液相色谱法检测甲苯咪唑（MBZ）及其代谢物氨基甲苯咪唑（MBZ-NH2）和羟基甲苯咪唑（MBZ-OH）在银鲫（*Carassius auratus gibelio*）血浆及组织中的浓度，数据经 3P97 药代动力学程序分析。结果表明：$(25±1)$℃的水温条件下，银鲫单剂量口灌 MBZ 20 mg·kg^{-1}，血药经时过程符合二室开放式模型。主要药代动力学参数为：吸收速率常数 Ka 0.235 h^{-1}；消除半衰 $T_{(1/2)\beta}$52.26 h；药时曲线下面积 AUC 180.07μg·h·mL^{-1}；表观容积分布 V_d/F 2.465 L·kg^{-1}；清除率 CL（s）0.111 mL·h-1·kg^{-1}；达峰时间 $T_{(peak)}$6.20 h；质量浓度 C_{max} 4.14 μg·mL^{-1}。与哺乳类相比，MBZ 在银鲫体内吸收较慢，消除半衰期明显延长，药代动力学参数也有较大差异。建议在做到合理用药的同时加强药物检测力度。

关键词： 银鲫；甲苯咪唑；氨基甲苯咪唑；羟基甲苯咪唑；药代动力学；高效液相色谱

甲苯咪唑（MBZ）属于苯并咪唑类药物，也称甲苯哒唑、二苯酮咪胺酯、安乐士。化学名为（5-苯甲酰基-1H-苯并咪唑-2-基）氨基甲酸甲酯，是一种常见的广谱驱虫类药物。分子式是：$C_{16}H_{13}N_3O_3$，1971 年由 Van Geldev 首次合成，并获美国专利，比利时杨森公司同年投入临床使用。MBZ 作为一种广谱杀虫剂可用来治疗蛔虫、线虫、钩虫、鞭虫的感染[1]，在哺乳动物中用来预防山羊肠胃和肺中的线虫和蠕虫[2]，对肥育犬消化道寄生虫病具有较好的治疗效果[3]，可治疗小鼠脊形管圆线虫病。在水产上有报道甲苯咪唑可治疗欧洲鳗鲡指环虫病[4-6]。口服甲苯咪唑可治疗石斑鱼鳃的单殖吸虫病[8]。采用药溶可治疗红鲷多微小吸盘虫的感染[9]。药物的药代动力学研究是临床设计合理给药方案、发挥药物最佳疗效和减少副作用的理论基础。目前国内还未见甲苯咪唑在水产动物体内的药代学研究报道。本研究初步探讨该药在银鲫体内的药代学特征，为在鱼病害防治中科学合理地用药提供参考。

1　材料与方法

1.1　实验动物

健康银鲫（*Carassius auratus gibelio*）由中国水产科学研究院长江水产研究所窑湾试验场提供，体质量为（200±10）g。

实验前在 1 m³ 水族箱内暂养 1 周,每天换水 1 次。实验用水为曝气 48 h 自来水,连续充氧,保持水中溶氧大于 5.0 mg · L⁻¹,实验水温为(25±1)℃,pH 值为 7.5~8.0,实验过程中停饲。

1.2 药品及试剂

MBZ 标准品(纯度≥99%,美国 Sigma 公司),羟基甲苯咪唑(MBZ-OH)和氨基甲苯咪唑(MBZ-NH₂)(纯度≥99%,Witega 实验室,Berlin-Adlershof GmbH);MBZ 原粉(纯度≥99%,九州神龙有限公司);乙腈、乙酸乙酯、正己烷等(均为色谱纯,Fisher ChemAlert 公司),二甲基亚砜和二甲基甲酰胺(分析纯,国药集团化学试剂有限公司);三乙胺、磷酸二氢铵(分析纯,天津福晨化学试剂公司),磷酸氢二钠、磷酸二氢钠(分析纯,天津天大化学试剂公司)。

磷酸盐缓冲液:准确称取磷酸氢二钠(Na₂HPO₄ · 12H₂O)17.9 g 和磷酸二氢钠(NaH₂PO₄ · 2H₂O)4.6 g,分别溶于 1 L 蒸馏水中配成 0.05 mol · L⁻¹ 的溶液,然后按 9∶1 的体积比混合用氨水调节 pH 值到 9.5。

定容液的配制:A 液,称取 1.438 g 磷酸二氢铵溶解在 250 mL 双蒸水中配制成 0.05 mol · L⁻¹ 的磷酸二氢铵溶液。B 液,二甲基甲酰胺。定容溶液为 A 液+B 液(体积比 7∶3),用磷酸将溶液 pH 调至 2.0,经抽滤倒入具塞玻璃瓶中置 4℃ 冰箱中备用。

标准混合液配制:取 MBZ 标准品 0.10 g 用二甲基亚砜溶解并定容到 100 mL,得到质量浓度 1 000 µg · mL⁻¹ 的标准储备液。取 MBZ-OH、MBZ-NH₂ 两种药物标准品各 0.10 g 用二甲基甲酰胺溶解定容到 100 mL,得到质量浓度为 1 000 µg · mL⁻¹ 的标准储备液。分别从 3 种标准储备液中各取 10 mL 混合后用流动相稀释至 100 mL,得到质量浓度 100.0 µg · mL⁻¹ 标准混合液。

1.3 仪器与设备

高效液相色谱仪(Waters 515 泵,717 自动进样器,2487 双通道紫外检测器及 Empower 色谱工作站);自动高速冷冻离心机(日本 HITACHI 20PR-520 型);Mettler-TOLEDO AE-240 型精密电子天平(梅特勒-托利多公司);FS-1 高速匀浆机(华普达教学仪器有限公司);调速混匀器(上海康华生化仪器制造厂);恒温烘箱(上海浦东荣丰科学仪器有限公司);HGC-12 氮吹仪(HENGAO T&D 公司);Sartorius PB-10 型酸度计(德国赛多利斯公司)。

1.4 实验设计与采样

口灌给药:MBZ 原粉 200 mg 用少许二甲基亚砜溶解再用 100 mL 蒸馏水定容,此时药物质量浓度为 2 mg/mL。按 20 mg · kg⁻¹ 单次给药,将药物用 2 mL 注射器接硅胶软管从银鲫口中插入,注意深度,无回吐鱼可用于实验。

采样:给药后的 0.25 h、0.5 h、1 h、2 h、4 h、8 h、12 h、24 h、48 h、72 h、120 h、168 h 尾静脉取血液,1% 肝素钠抗凝后血液以 3 000 r · min⁻¹ 离心 10 min,取上清液,同时解剖分离出肌肉、皮、肝脏、肾脏;将血清和其他组织于 -20℃ 保存备用。

每一时间点各取 5 尾鱼,作为 5 个平行样品分别处理测定。另取 5 尾未给药鱼作空白对照。

1.5 样品预处理

将冷冻保存的肌肉、皮、肝脏、肾脏和血液样品室温下自然解冻,剪取适量的肌肉组织,置高速匀浆机中匀浆至糜状,准确称取 5.0 g 于 50 mL 塑料离心管中,加 1 mL 的磷酸盐缓冲液和 20 mL 乙酸乙酯,置漩涡混合器上振荡 5 min,5 000 r/min 离心 10 min,取出提取液,再加入 15 mL 乙酸乙酯,重复操作一次,合并提取液于鸡心瓶中,置 40℃ 水温旋转蒸发至干。鱼皮先剪碎,准确称取 1.0 g;肾脏和肝脏都各取 0.5 g;血浆 1.0 mL。分别加入 15 mL 具塞离心管中,加入 0.5 mL 的磷酸盐缓冲液,然后加 4 mL 乙酸乙酯,置调速混匀器上涡旋振荡混匀 1 min,以 4 000 r · min⁻¹ 离心 10 min,吸出上清液于另一支 15 mL 具塞离心管中,重复提取一次,合并上清液氮吹至干。旋转蒸发和氮吹至干的剩余物分别用 1 mL 定溶液定容,再加入 1 mL 的正己烷,置漩涡混合器上振荡 1 min,混合溶液于 4 000 r · min⁻¹ 离心 10 min,去除上层正己烷液,下层液经 0.22 µm 滤头过滤,用于 HPLC 测定。

1.6 HPLC 分析方法

1.6.1 色谱条件 Waters symmetry C₁₈ 反相色谱柱(250 mm×4.6 mm,5 µm);流动相:乙腈+磷酸二氢铵溶液(0.05 mol · L⁻¹,0.1% 三乙胺)(33∶67,V/V);流速:0.8 mL · min⁻¹;柱温:室温;紫外检测波长:298 nm;进样量:50 µL。

1.6.2 血液与组织标准曲线的制备 空白血液、组织(肌肉、肝脏、肾脏、皮)中加入 MBZ、MBZ-OH 和 MBZ-NH₂ 已知标准混合液,使其药物质量浓度范围分别为 0.01~100.00 µg · mL⁻¹ 和 0.01~100.00 µg · g⁻¹。按样品处理方法处理,作 HPLC 分析,以测得的各平均峰面积 y 为纵坐标,相应的质量浓度 x 为横坐标绘制标准工作曲线,求出回归方程和相关系数。用空白组织制成低质量浓度药物的含药组织，经预处理后测定，将引起三倍基线噪声的药物的质量浓度定义为最低检测限。

1.6.3　回收率与精密度测定　回收率＝$C_r / C_0 \times 100\%$，其中 C_r 为用空白样品（血液或组织）加入一定量的 MBZ、MBZ-OH 和 MBZ-NH$_2$ 已知标准混合液，再按样品预处理方法进样后，测定 3 种药物的质量浓度；C_0 为加入一定量的 MBZ、MBZ-OH 和 MBZ-NH2 已知标准混合液，测得 3 种药物的质量浓度。在 4 种空白组织和血液中分别添加 4 个质量浓度水平的混合标准液，使组织中质量浓度分别为 0.01、0.10、1.00 和 10.0 μg·g^{-1}，血液为 0.01、0.10、1.00 和 10.0 μg·mL^{-1}。每个浓度的样品，日内做 5 个重复，1 周内重复做 5 次，计算日内及日间精密度。

1.7　数据处理

药物动力学模型拟合及参数计算采用中国药理学会数学专业委员会编制的 3P97 药动软件分析；标准曲线，药物经时曲线图，消除方程及休药期计算和回归图，采用 Microsoft Excel 2003，SPSS（13.0）进行计算和绘制。

2　结果与分析

2.1　标准曲线方程与相关系数

空白血液、组织（肌肉、肝脏、肾脏、皮）中加入 MBZ、MBZ-OH 和 MBZ-NH$_2$ 已知标准混合液，使其药物在组织中的水平为 0.01～100.00 μg·g^{-1}，血液中为 0.01～100.00 μg·mL^{-1}。按样品处理方法处理，在 0.01～100.00 μg·g^{-1} 和 0.01～100.00 μg·mL^{-1} 范围内线性关系良好。以测得的各平均峰面积对相应质量浓度作线性回归，并制作标准曲线。5 种组织中 MBZ、MBZ-OH 和 MBZ-NH2 的标准曲线及相关系数分别见表 1。

表 1　MBZ 及其代谢物在银鲫血浆和组织中标准曲线和相关系数

Table. 1　Standard curve and correlation coefficient of MBZ and its metabolites in Johny carp plasma and tissues

标准曲线名称 Standard curve	分析物 Analyte	线性方程 Linear equation	相关系数 Correlation coefficient
血浆标准曲线 Standard stand of plasama	MBZ	y＝14231x－10985	0.9993
	MBZ-OH	y＝6392.2x－4556	0.9978
	MBZ-NH$_2$	y＝17363x－15116	0.9864
肌肉标准曲线 Standard curve of muscle	MBZ	y＝25574x－8920	0.9233
	MBZ-OH	y＝17221x－6756.3	0.9550
	MBZ-NH$_2$	y＝26269x－18943	0.9540
肝脏标准曲线 Standard curve of liver	MBZ	y＝74170x－33536	0.9363
	MBZ-OH	y＝52504x－30581	0.9464
	MBZ-NH$_2$	y＝52163x－21133	0.9880
肾脏标准曲线 Standard curve of kindey	MBZ	y＝7683.5x－2385	0.9943
	MBZ-OH	y＝22116x－20349	0.9991
	MBZ-NH$_2$	y＝5348x－3281.3	0.9525
皮肤标准曲线 Standard curve of skin	MBZ	y＝29389x－10828	0.9940
	MBZ-OH	y＝28312x－17262	0.9786
	MBZ-NH$_2$	y＝23094x－8939	0.9698

2.2　回收率与精密度

本实验条件下，各组织和血液中 4 个质量浓度水平的 MBZ 回收率为 81.2%～86.3%，MBZ-OH 回收率为 86.5%～93%，MBZ-NH 2 回收率为 70.6%～75.0%。测得的日内精密度与日间精密度均小于 10%。

2.3　MBZ 在各组织中的浓度

将测到的 MBZ 的峰面积代入各自的标准曲线所绘制的回归方程（表 1），就可以求出药物在各组织中的浓度值

（表2）。

表2 单剂量 20 mg·kg⁻¹（BW）口灌 MBZ 在银鲫血液和组织中的残留

Table. 2 MBZ concentration in plasma and different organized samples in Johnny carp after oral administration at 20mg·kg⁻¹（BW） $n=5$：$x-\pm SE$

时间 Time	MBZ 残留水平 MBZresidual level				
	肌肉（μg·g⁻¹）Muscle	血液（μg·g⁻¹）Blood	肝脏（μg·g⁻¹）Licer	皮（μg·g⁻¹）Skin	肾脏（μg·g⁻¹）Kidney
0.25	0.136±0.003	0.193±0.026	0.873±0.012	0.087±0.112	0.017±0.006
0.5	0.309±0.045	0.427±0.016	1.01±0.164	0.179±0.203	0.121±0.008
1	0.543±0.206	1.18±0.248	1.289±0.277	0.540±0.231	0.340±0.011
2	0.881±0.220	3.16±1.394	2.03±0.182	1.039±0.634	0.734±0.204
4	1.305±0.413	4.04±1.573	4.59±0.335	1.45±1.082	0.898±0.082
8	1.755±0.536	4.73±0.461	5.789±0.803	2.369±0.528	1.369±0.228
12	3.027±0.253	3.72±0.547	7.880±1.235	3.232±1.225	1.533±0.225
24	2.801±0.665	1.51±0.714	9.546±2.121	5.173±1.413	1.673±0.213
48	1.434±0.135	1.22±0.898	6.907±2.378	3.298±1.657	0.898±1.657
72	0.663±0.048	0.89±0.467	4.101±1.053	0.529±0.081	0.529±0.781
120	0.242±0.023	0.534±0.233	1.786±0.034	0.381±0.102	0.381±0.362
168	0.012±0.004	0.141±0.163	0.887±0.136	0.095±0.066	0.165±0.066

2.4 MBZ 在银鲫体内的药代动力学参数（表3）

MBZ 以 20 mg·kg⁻¹ 剂量单次口灌银鲫后，有关动力学数据经 3P97 实用药代动力学软件分析，血液中药物浓度与时间关系符合开放性二室模型，动力学方程为：$C=17.38e^{-0.151t}+1.944e^{-0.0133t}$。$K_{21}<K_{12}$，说明药物从中央室进入外周室的速率比外周室回到中央室的速率大。

2.5 色谱图

图 1（A-F）是标准液及银鲫经单次口灌 MBZ 后血浆和各个组织样品中 MBZ 水平达到峰值时的色谱图。由图 1 可见，甲苯咪唑经鱼体吸收后在其各个组织中出现了 MBZ-NH₂ 和 MBZ-OH 两种代谢物。原药和代谢物消除规律的同时研究，可更加合理地制定休药期。

表3 单剂量 2020 mg·kg⁻¹（BW）口灌 MBZ 在银鲫体内的药代动力学参数

Table. 3 Pharmacokinetic parameters for MBZ in Johnny carp after single oral administration at 20 mg·kg⁻¹（BW）

参数 Parameter	单位 Unit	血液 Blood
A	μg·mL⁻¹	17.38
a	h⁻¹	0.151
B	μg·mL⁻¹	1.944
β	h⁻¹	0.0133
K_a	h⁻¹	0.235
Lag time	h	0.199
V_d/F	L·kg⁻¹	2.465
$T_{(1/2)a}$	h	4.612
$T_{(1/2)\beta}$	h	52.26
$T_{(1/2)Ka}$	h	2.947

续表

参数 Parameter	单位 Unit	血液 Blood
A	μg. mL^{-1}	17. 38
K_{21}	h^{-1}	0. 044
K_{10}	h^{-1}	0. 045
K_{12}	h^{-1}	0. 074
AUC	μg. h. mL^{-1}	180. 07
CL（s）	mL. h. kg^{-1}	0. 111
T_{peak}	h	6. 20
C_{max}	μg. mL^{-1}	4. 14

注：A，B为药时曲线对数图上曲线在横轴和纵轴上的截距；α、β分别为分布相、消除相的一级速率常数；K21由周边室向中央室转运的一级速率常数；K10由中央室消除的一级速率常数；K12由中央室向周边室转运的一级速率常数；Vd/F表观分布容积；AUC药-时曲线下面积；Lag time滞后时间；Ka为一级吸收速率常数；T（1/2）Ka为药物在中央室的吸收半衰期；T（1/2）α、T（1/2）β分别为总的吸收和消除半衰期；Tpeak为单剂量给药后出现最高血药质量浓度的时间；Cmax为单剂量给药后的最高血药质量浓度；CL（s）为总体清除率。

Note：A，B-Drug concentration-time curve on logarithmic graph curve on the horizontal axis and vertical axis intercept；α，β-First-order rate constant for drug distribution phase and elimination phase；K21-First-order rate constant for drug distribution from peripheral compartment to central compartment；K10-First-order rate constant for drug elimination from central compartment；K12-First-order rate constant for drug distribution from central compartment to peripheral compartmenti；Vd/F-Apparent volume of distribution；AUC-Area under concentration-time curve；Ka-First-order rate constant for drug absorption；T（1/2）Ka-Absorption half-life in central compartment；T（1/2）α，T（1/2）β-The overall absorption half-life and Elimination half-life；Tpeak-The time at highest peak of concentration；Cmax-The biggest concentration of the drug in blood；CL（s）-The overall clearance rate.

3　讨论

3.1　MBZ在银鲫体内的药动学特征

银鲫血浆中的MBZ药时数据符合二室开放模型。这与Pandey等[2]报道的静脉注射山羊得到的药代动力学房室模型一致。王长虹等[10]、罗兰等[11]、李岩等[12]报道了MBZ在兔血中的药代动力学模型为单室模型，KitzmanI等[13]报道了MBZ的同系物氟苯哒唑在斑点叉尾鮰体内药代动力学模型为二室模型。这些模型的差异可能与实验对象种属差异有关。T_{max}、K_a、$T_{(1/2)a}$是反应药物在体内吸收速率的重要指标。新西兰兔口服MBZ 40 mg/kg[11]，其血药浓度达峰时间T_{max}为4.48 h，吸收半衰期$T_{(1/2)a}$为1.74 h，吸收速率常数K_a为0.514 h^{-1}。Pandey等[2]报道的山羊单剂量口服MBZ 40 mg/kg，其血浆中的药物吸收参数T_{max}=（5.8±0.88）h，K_a=（0.34±0.06）h^{-1}，$T_{(1/2)a}$=（3.2±0.06）h。本实验条件下得到MBZ在银鲫体内血浆药物浓度达峰时间T_{max}为6.2 h，吸收速率常数K_a0.235 h^{-1}；吸收半衰期$T_{(1/2)a}$4.612 h。MBZ在银鲫体内的达峰时间及吸收半衰期要慢于前两者，说明药物在银鲫体内的吸收速率低于哺乳动物。$T_{(1/2)\beta}$是反应药物在体内消除速率的重要参数。本实验得到的消除半衰期$T_{(1/2)\beta}$52.26 h要大于Krishnaiah等[14]报道的2种不同包衣的MBZ在人体内的消除半衰期，分别为（13.2±5.6）h和（8.2±2.9）h。王长虹等[10]报道MBZ乳剂和片剂在兔体内的消除半衰期分别为（4.5±0.7）h和（6.3±2.6）h，消除时间都比在鱼体内短。与哺乳动物和人类相比，MBZ在银鲫体内吸收较慢，消除半衰期明显延长，药代动力学参数也有较大的差异。

动物体内的组织蛋白结合能力和少量的细胞内渗透是造成个体间药代动力学参数差异的原因，药物制剂也是导致差异的因素之一。另外，药物也会因为在血液中的循环途径不同，影响到以后的分布和消除。

单次口服剂量为20 mg·kg^{-1}的MBZ，药物在血浆和组织内的药-时浓度如表2所示：达峰浓度C_{max}从高到低依次为肝脏、皮、血液、肌肉、肾脏。MBZ在肝脏中峰浓度最高，这与Braithwaite等[15]报道的人口服MBZ主要被肝门静脉吸收，肝脏药物含量最高相符合。

3.2　MBZ在银鲫体内代谢物

银鲫口服MBZ后，在其体内产生MBZ-NH^{2}和MBZ-OH两种代谢物（图1）。Iosifidou等[16]报道了用1×10^{-6}的MBZ浸

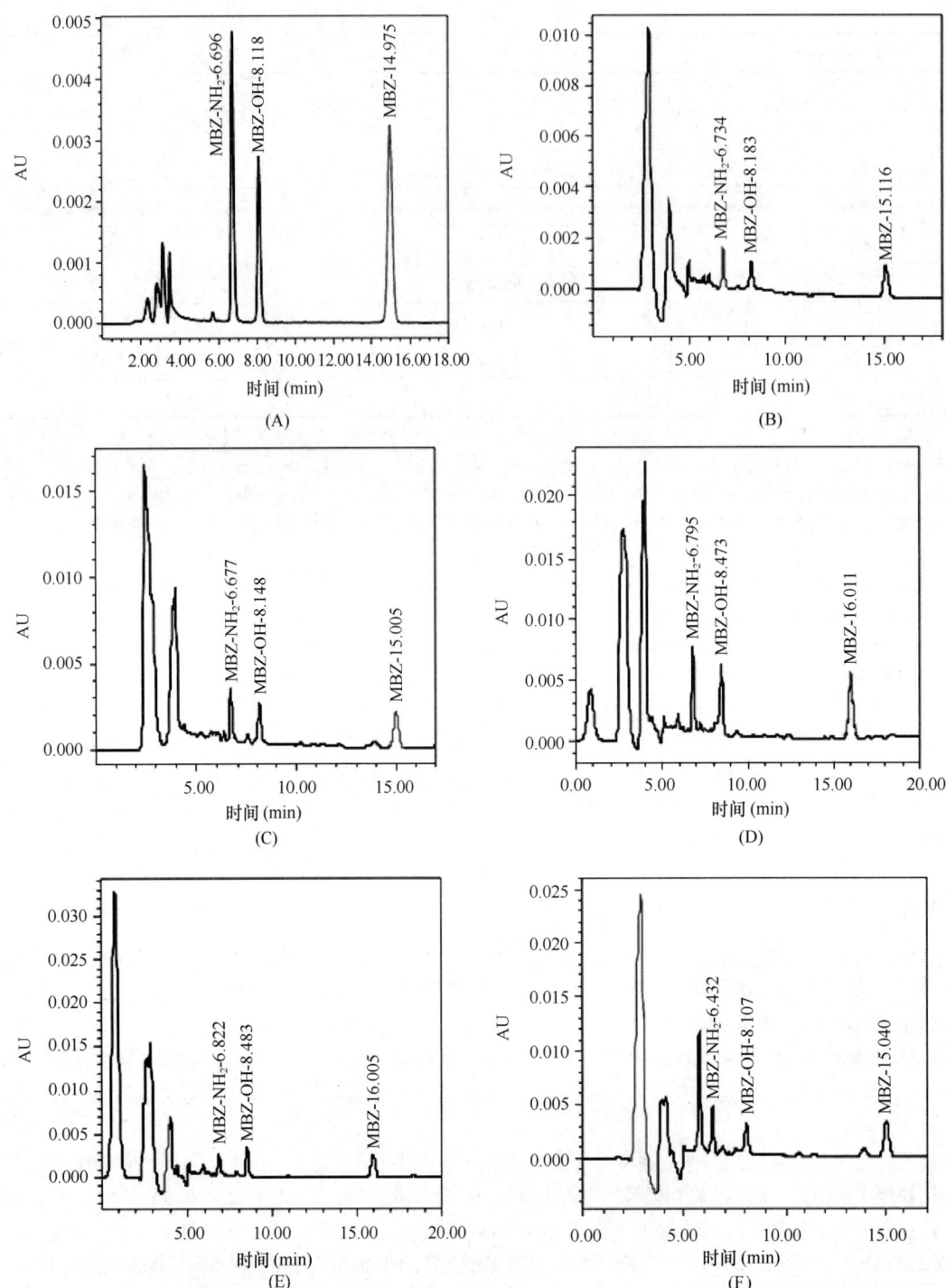

图1　1 μg·mL⁻¹标准混合溶液（A）、给药后8 h银鲫血浆（B）、给药后12 h银鲫肌肉（C）、给药后24 h银鲫肝脏（D）、给药后24 h银鲫肾脏（E）、给药后24 h银鲫皮肤（F）的色谱图

Fig. 1　Chromatogram of 1 μg·mL⁻¹ mixed standard solution （A）, Johnny carp plasama sample at 8 h after oral MBZ （B）, Johnny carp muscle sample at 12 h after oral MBZ （C）, Johnny carp liver sample at 24 h after oral MBZ （D）, Johnny carp kindey sample at 24 h after oral MBZ （E） and Johnny carp skin sample at 24 h after oral MBZ （F）

泡欧洲鳗鲡24 h，测得有MBZ-OH和MBZ-NH₂两种代谢物，其中MBZ-NH₂在体内的消除时间最长，需要14 d。欧盟规定MBZ及其代谢物残留（总量）在羊、马可食性组织（肌肉、脂肪、肝脏、肾脏）中规定的最高残留限量（MRL）分别为60 μg·kg⁻¹、60 μg·kg⁻¹、400 μg·kg⁻¹、60 μg·kg⁻¹ [17]，而未规定水产品各组织中限量。本实验对（25±1）℃水温条件

下银鲫血浆、肌肉、皮肤、肝脏、肾脏组织研究中发现了 MBZ 及其 2 种代谢物，进一步丰富了 MBZ 的在鱼类各组织中分布及消除资料，也为制定最高残留限量（MRL）及休药期（WDT）提供了参考资料。

3.3　MBZ 的用药方式及残留限量

MBZ 主要用作人和兽用驱虫药，中国近年在水产养殖中开始运用。李志青等[18]报道了养殖欧洲鳗鲡在水温低于 28℃时，建议药浴水平为 0.6~1.0 mg/L，药浴时间不超过 24 h。水温超过 28℃，建议使用药浴水平为 0.2~0.3 mg/L，可长时间药浴。李海燕等[19]研究建议生产上使用 0.8~1.0 mg/L 的 MBZ 治疗鳗的伪指环虫病。伪指环虫从产卵到孵化出幼虫一般需要 3~6 d，因此，建议使用 MBZ 后 7~8 d 再重复处理 1 次。此外，MBZ 还可用来治疗其他鱼类的单殖吸虫病，如鳜鳃上常寄生河鲈锚首虫。该药现用于水生动物多采用浸泡方式，药物直接进入水体对水环境会造成何种影响，国内外在这方面尚无研究报道，药物在水体中的代谢情况还需进一步探讨。本实验采用灌胃方式给药以研究药物在银鲫体内的动力学，减少药物对水环境的影响，为药物拌饲投喂提供科学数据。

MBZ 的允许最高残留限量（MRL）在不同国家和地区是不同的。欧盟对羊和马属动物规定的 MRL 为 0.04 mg/kg[20]；美国 FDA 只批准用于马和狗[21]；日本肯定列表中暂时规定在水产品中 MBZ 的 MRL 为 0.02 mg/kg。由于 MBZ 可引起人全身不适、搔痒、皮疹、视神经视网膜病变，出现血尿现象等症状[22-24]，因此建议在做到合理用药的同时也要加大药物检测力度。

参考文献

[1]　Swanepoela E，Liebenberga W，Melgardt M. Quality evaluation of generic drugs by dissolution test：changing the USP dissolution medium to distinguish between active and non-active mebendazole polymorphs ［J］. Eur J Pharmaceutics Biopharm，2003，55：345-349.

[2]　Pandey S N，Roy B K. Disposition kinetics of mebendazole in plasma，milk and ruminal fluid of goats ［J］. Small Ruminant Res，1998，27：111-117.

[3]　吴忠，贺永健，周鸿海. 复方甲苯咪唑对肥育犬的影响 ［J］. 动保药品，2006，31（2）：31-32.

[4]　Terada M，Kino H，Akyol CV，Sano M. Effects of mebendazole on Angiostrongylus costaricensis in mice，with special reference to the timing of treatment ［J］. Parasitnlngy Research，1993，79：441-443.

[5]　Buchmann K，Bjerregaard J. Mebendazole treatment of Pseudodactylogyrosis in an intensive eel-culture system ［J］. Aquaculture，1990，86：139-153.

[6]　Buchmann K，Bjerregaard J. Comparative efficacies of commercially available benzimidazoles against Pseudodactylogyrus infestations in eels ［J］. Dis Aquat Org，1990，9：117-120.

[7]　Mellergaard S. Mebendazole treatment against Pseudodactylogyrus infections in eel（Anguilla anguilla）［J］. Aquaculture，1990，91：15-21.

[8]　Kim K H，Choi E S. Treatment of Microcotyle sebastis（Monogenea）on the gills of cultured rockfish（Sebastes schelegeli）with oral administration of mebendazole and bithionol ［J］. Aquaculture，1998，167：115-121.

[9]　Pantelis K，Nikos P，Pascal D. Treatment of Microcotyle sp.（Monogenea）on the gills of cage-cultured red porgy，Pagrus pagrus following baths with formalin and mebendazole ［J］. Aquaculture，2006，252（2-4）：167-171.

[10]　王长虹，孙殿甲. O/W 型甲苯咪唑口服乳剂兔体内药物动力学研究 ［J］. 中国医院药学杂志，1999，19（10）：591-594.

[11]　罗兰，孙殿甲，苗爱东，等. 甲苯咪唑 β-CD 饱和物兔体内药物动力学研究 ［J］. 西北药学杂志，2000，15（3）：118-119.

[12]　李岩，孙殿甲. 口服甲苯咪唑微丸的药物动力学及生物利用度 ［J］. 中国医院药学杂志，1999，19（11）：649-651.

[13]　KitzmanI J V，Holley J H，Huberi W G. Pharmacokinetics and metabolism of fenbendazole in channel catfish ［J］. Veterin Res Commun，1990，14：217-226.

[14]　Krishnaiah Y S R，Rajua P V. Pharmacokinetic evaluation of guar gum-based colon-targeted drug delivery systems of mebendazole in healthy volunteers ［J］. Controlled Release，2003，88：95-103.

[15]　Braithwaite P A，Roberts M S，Allan R J，et al. Clinical pharmacokinetics of high dose mebendazole in patients treated for cystic hydatid disease ［J］. Eur J Clin Pharmacol，1982，22：161-169.

[16]　Iosifidou EG，Haagsma N，Olling M，et al. Residue study of mebendazole and its metabolites hydroxy-mebendazole and amino-mebendazole in eel（Anguilla anguilla）after bath treatment ［J］. Drug Metabolism and Disposition，1997，25（3）：317-320.

[17]　EEC Council Regulation No. 2377/1990. Community procedure for the establishment of maximum residue limits of veterinary medicinal products in foodstuffs of animal origin ［S］.

[18]　李志青，樊海平，曾占状，等. 复方甲苯咪唑对指环虫的杀灭效果和欧洲鳗的毒性 ［J］. 浙江农学院学报，2000，12（4）：221-223.

[19]　李海燕，李桂峰. 甲苯咪唑对鱼的急性毒性试验及驱虫效果研究 ［J］. 广州大学学报，2005，4（1）：33-35.

[20]　Commission regulation（EC）No 1680/2001 of 22 August 2001 amending Annexes I and II to Council Regulation（EEC）No 2377/90 laying down a Community ［S］.

[21]　Code of Federal Regulations Title 21，Volume 6 Revised as of April 1，2008，CITE：21CFR520. 132 ［S］.

[22] 席成亮. 复方甲苯咪唑致药物性荨麻疹 2 例报告 [J]. 中原医刊, 1997, 24 (4): 30.

[23] 白玲, 赵丕烈, 陈东平. 甲苯咪唑的不良反应 [J]. 医药导报, 1999, 18 (2): 128.

[24] 景宝洁. 甲苯咪唑引起肝损害 [J]. 药物不良反应杂志, 2007, 9 (6): 433.

该文发表于《淡水渔业》【2013, 43 (6): 39-44】

甲苯咪唑在团头鲂体内主要代谢物及其变化规律研究

The study of main metabolites about mebendazole and its variation in *Megalobrama amblycephala*

胥宁[1,2], 刘永涛[1,2], 艾晓辉[1,2], 赵凤[1,4], 杨秋红[1], 吕思阳[1,3]

(中国水产科学研究院长江水产研究所, 武汉 430223; 2. 淡水水产健康养殖湖北省协同创新中心 武汉 430223; 3. 华中农业大学水产学院, 武汉 430070; 4. 南京农业大学渔业学院, 无锡 214000)

摘要: 为了确定甲苯咪唑在团头鲂 (*Megalobrama amblycephala*) 体内主要代谢产物, 实验用 1 mg/L 甲苯咪唑水溶液浸泡团头鲂, 分别在浸泡后 1、2、4、8、12 和 24 h, 分别采集肝脏、肌肉、胆汁和血液, 采用高效液相串联质谱法 (HPLC-MS-MS) 定量和标准品比对法进行鉴定。结果显示, 甲苯咪唑 (MBZ) 在团头鲂体内所占的比例随着时间的延长逐渐降低, 羟基甲苯咪唑 (MBZ-OH) 和氨基甲苯咪唑 (MBZ-NH$_4$) 逐渐升高。MBZ-OH 和 MBZ-NH$_4$ 百分比都占到总残留的 10% 以上, 二者为主要代谢产物, 应作为残留检测的主要监控目标。

关键词: 团头鲂; 甲苯咪唑; 主要代谢物; 变化规律

该文发表于《水产科学》【2006, 25 (2): 75-78】

多次口灌呋喃唑酮在草鱼体内残留研究

Residues of furazolidon multiple-orally administrated in grass carp *ctenopharyngodon idellus*

罗玉双[1], 艾晓辉[2], 刘长征[2]

(1. 湖南文理学院生命科学系, 湖南 常德 415000; 2. 中国水产科学院长江水产研究所, 湖北 荆州 434000)

摘要: 报道了多次口灌呋喃唑酮 (60 mg/kg 连续 3 d, 每天 1 次) 在草鱼体内血浆、肌肉、肝脏和肾脏中的消除与残留研究。用高效液相色谱法 (HPLC) 检测用药后不同时间各组织中药物浓度, MILLEMNIUM[32] 工作站处理原始数据, MCPKP 药代动力学软件处理药时数据。结果表明: 多次口灌呋喃唑酮在草鱼体内的药时数据符合一级速率吸收一室开放模型, 其主要药动学参数为: AUC 2.3182 μg·h/mL, C_{max} 0.850 μg/mL, Ka 0.551 5 h^{-1}, K 0.093 6 h^{-1}, $T_{1/2ka}$ 1.256 5 h, $T_{1/2k}$ 7.430 8 h, T_p 2.873 2 h, 药物在血浆、肌肉、肝脏、肾脏中消除时间分别为: 72 h、15 d、20 d、20 d。

关键词: 呋喃唑酮; 草鱼; 组织残留; 消除

第三节 渔药药动学模型及影响渔药的代谢和消除的因素

渔药在水产动物机体内的分布、代谢等受很多因素的影响, 环境、动物性别、健康状态及药物的拓扑结构都能影响药物代谢过程, 尤其对处于环境条件比较复杂的水产动物来说, 除了遗传种属、性别等生理因素和病理、药理因素外, 环境因素比较重要, 其中水温能显著影响药其代动力学过程。研究不同条件情况下的药动学规律, 对于合理、正确地使用渔药有重要的指导作用。

一、渔药在水产动物体内的药代动力学研究概况

房室模型、非房室模型和生理学模型是水产动物药动学几种不同研究方法。不同种类的水产动物及其健康状况、不同药物、不同的给药方法以及水环境因素均会对渔药在水产动物体内的代谢产生一定影响。

该文发表于《中国水产科学》【2004，11（4）：379-384】

水产动物药物代谢动力学研究概况

Advances in the study on pharmacokinetics in the aquatic animals

房文红，郑国兴

（农业部海洋与河口渔业重点开放实验室，中国水产科学研究院东海水产研究所，上海 200090）

摘要： 本文总结了水产动物药物代谢动力学研究所涉及的受试动物、原形药物以及部分药物的代谢产物，比较了药浴法、肌肉注射法、口服法、血管注射法、心包内或血窦内注射给药等不同给药途径在水产动物药动学研究中的优缺点，探讨了种属差异、性别和健康状况等水产动物生理因素以及温度和盐度等环境因素对药动学的影响，分析了水产动物药动学研究模型如房室模型、非房室模型和生理学模型的特点，并提出了我国水产动物药动学的今后研究方向。

关键词： 药物代谢动力学；水产动物；生理因素；分析模型

自 Snieszko 等[1]首次报道了磺胺类药物在虹鳟体内药物浓度变化以来，水产动物药动学的研究内容和理论有了长足的发展。药动学研究由最初的药物浓度变化分析[1]，发展到目前广为采用的房室模型[2-38]，而基于生理学的药动学模型[39-41]，虽然方法复杂但已开始在水产动物药动学中应用，这将是描述水产动物药动学最为理想的模型。下面将从水产动物药动学研究对象、给药途径、影响药动学的水产动物生理、环境因素以及药动学分析模型等几方面来概述水产动物药动学研究状况，以推动我国始受重视的水产动物药动学研究的发展。

1　研究对象

1.1　受试药物

1.1.1　原形药物

国外水产动物药动学研究涉及的药物种类比较全面，主要有抗生素类、磺胺类、呋喃类、喹诺酮类、杀虫驱虫药类和激素类等。如抗生素类药物有土霉素（oxytetracycline，OTC）[2-6]、羟氨苄青霉素（amoxicillin）[7]、氟甲砜霉素（florfenicol）[8]等；磺胺类药物较多，主要有磺胺间二甲氧嘧啶（sulfadimethoxine，SDM）[9-12]、磺胺二甲基嘧啶（sulfadimidine，SDD）[12-13]、磺胺嘧啶（sulfadiazine，SD）[11]、磺胺甲基异噁唑（sulfamethoxazole，SMZ）[12-13]、磺胺间甲氧嘧啶（sulphamonomethoxine，SMM）[12,14]、磺胺胍（sulphaguanidine，SGD）[12]及磺胺增效剂嘧啶二胺（ormetoprim，OMP）[10-11]和甲氧苄氨嘧啶（trimethoprim，TMP）[13]；呋喃类药物有呋喃唑酮（furazolidone）[15]、呋喃妥因（nitrofurantoin）[16]和苯并呋喃（tetrachlorodibenzofuran）[17]等；喹诺酮类药物有萘啶酸（nalidixic acid）[18-19]、噁喹酸（oxolinic acid）[20-21]、氟甲喹（flumequine）[22]、恩诺沙星（enrofloxacin）[23]、沙拉沙星（sarafloxacin）[24]、环丙沙星（ciprofloxacin）[25]和米诺沙星（miloxacin）[26]等。涉及的杀虫驱虫药主要有对硫磷（parathion）和对氧磷（paraoxon）[27]、苯硫哒唑（fenbendazole）[28]、毒死蜱（chlorpyrifos）[29]、吖啶黄（acriflavine）[30]、三丁锡（tributyltin）和甲基汞（methylmercury）[31]等；激素类药物如甲基睾丸素（methyltestosterone）[32]、壬基酚（nonylphenol）[33]等。此外，还有一些药物，如9-羟基-苯甲芘（9-hydroxybenzo［α］pyrene，9-OH-BaP）[34]、萘酚（naphthol）[35]、苯佐卡因（benzocaine）[36]、甲基冰片（methylisoborneol）[37]、安息香酸（Benzoic acid）[38]等。

国内水产动物药动学研究由于起步晚，目前受试的药物还较少，研究主要集中在抗微生物药物上，如土霉素[42]、氯霉素（chloromycetin）[43-44]、磺胺二甲基嘧啶[45]、复方新诺明（SMZ 和 TMP）[46]、呋喃唑酮[47]、噁喹酸[48]、诺氟沙星（norfloxacin）[49-51]、环丙沙星[52]、喹乙醇（Olaquindox）[53]等。值得一提的是，为保障水产品安全，上述药物中的氯霉素、呋喃唑酮、环丙沙星和喹乙醇等药物，现已被包括中国在内的多数国家列为禁用药，但对它们药动学研究的结果和方法学，

为我国渔药的监管提供了科学依据，并对其他药物的研究提供了有价值的参考。

1.1.2 代谢产物

大多数药物在机体作用下都会发生生物转化，形成极性较强、水溶性较大的代谢产物。而目前对水产动物药动学的研究，多针对原形药物，较少涉及代谢产物，然而它们在体内的存在形式、消除速率及检测也是药动学研究的重点，尤其那些对水产品安全构成危害的代谢产物。国内尚未见药物代谢产物方面的报道，国外学者作了一些研究，如 Touraki 和 Niopas[13]研究了鲈幼鱼体内磺胺甲基异噁唑的代谢产物乙酰-磺胺甲基异噁唑的残留动力学。Uno 等[54]发现虹鳟口服 SMM、SDM 后，其各自主要代谢产物乙酰-SMM、乙酰-SDM 在虹鳟体内的半衰期和消除时间均长于相应的原药。Li 等[34]美洲鳌龙虾（Homarus americanus）心包内注射 9-OH-BaP 后，给药后 8 h 发现主要代谢产物苯甲芘硫酸盐（BaP 9-sulfate），给药后 1 d 达到最高浓度，且在血淋巴中的消除速率快于原药和另一代谢产物苯甲芘-葡萄糖甙（BaP 9-beta-D-glucoside）。

1.2 受试水产动物

国外研究涉及的水产动物种类范围十分广泛，且以经济价值高的海水鱼类为主，主要是斑点叉尾鮰（Ictalurus punctatus）[15,16,22,28-30,37,38,41]和鲑鳟类。如鲑鳟类有虹鳟（Oncorhynchus mykiss）[14,17,32,33,36,54]、银大麻哈鱼（O. kisutch）[55]、大西洋鲑（Salmo salar）[6,8,24,59]、北极红点鲑（Salvelinus alpinus）[1]、玫瑰大麻哈鱼（O. rhodurus）[2]和大鳞大麻哈鱼（O. tshawytscha）[6]等。此外，还有其他一些海水鱼类如五条鰤（Seriola quinqueradiata）[14]、美洲拟庸鲽（Hippoglossoides platessoides）[31]、庸鲽（H. hippoglossus）[12]、黄鲈（Perca flavescens）[41]、大菱鲆（Scophthalmus maximus）[21]等，以及淡水鱼类如香鱼（Plecoglossus altivelis）[4]、鲤（Cyprinus carpio）[5]、尼罗罗非鱼（Oreochromis niloticus）[20]、日本鳗鲡（Anguilla japonica）[26]等。对甲壳动物的研究，较为有限，仅收集到美洲鳌龙虾（Homarus americanus）[9,34-35]、南美白对虾（Penaeus vannamei）[10]、克氏原鳌虾（Procambarus clarki）[56]和雪蟹（Chionoecetes opilio）[57]等。

国内受试的水产动物在种类上与国外明显不同，主要是一些在我国养殖规模较大的经济鱼类，如鱼类的鲤（C. carpio）[50]、草鱼（Ctenopharyngodon idellus）[44]、日本鳗鲡（A. japonica）[42]、异育银鲫（Carassius auratus auratus）[44-45]、鲈（Lateolabrax janopicus）[46]、黑鲷（Sparus macrocephalus）[47]等；甲壳动物主要有中国对虾（P. chinensis）[43]、斑节对虾（P. monodon）[51]和中华绒鳌蟹（Eriocheir sinensis）[52]等；爬行动物仅中华鳖（Trioryx sinensis）[49]。由此可见，研究的种类较少，与我国现有的养殖规模还远远不相适应，但可以相信今后的研究将会更深更广。

2 给药途径

2.1 药浴法

药浴法又称为浸洗法、暴露法。不同药物即使药浴浓度相同但在组织中的浓度却相差很大。Samuelsen 等[12]将庸鲽在浓度为 200 μg/mL 的五种磺胺类药物（TMP、SDD、SGD、SDM、SMZ）中药浴 72 h，肌肉和腹部器官组织中的 TMP 浓度显著高于其他磺胺类药物，这可能是由于 TMP 具有比其他磺胺药更好的亲脂性。这也提示在药浴时，应考虑水产动物的不同组织对药物的吸收率及在体内的残留情况。

2.2 肌肉注射法

肌肉注射给药（intramuscular injection）是通过毛细血管壁被吸收，具有给药量准确、吸收迅速、疗效可靠等特点，如鲤和虹鳟[25]肌肉注射环丙沙星（15 mg/kg）吸收快，肌注后 1 h 血药浓度即达到最高，分别为 3.49 和 2.37 μg/mL；对虾肌肉注射诺氟沙星（10 mg/kg）后 2 min 血药浓度即达到最高[51]。此外，药物的水溶性、注射部位毛细血管丰富程度及血液循环速度也会影响吸收速率。

2.3 口服法

口服法分为投喂法与口灌法。投喂法即不对水产动物采取任何外力帮助而让其自然摄食药饵，多用于预防水产动物疾病，投喂法给药最贴近养殖生产中用药，能真实反映实际生产中药物在水产动物体内吸收、分布、代谢和排泄过程。如 Park 等[10]研究 SDM 和 OMP 在南美白对虾体内生物利用度和药动学时，采用单剂量投喂药饵法。Touraki 等[13]采用卤虫幼体作为 TMP、SMX 的载体，以鲜活饵料投喂给鲈鱼苗，研究鲈鱼幼体的药动学。

口灌法实际上是一种强制性口服法，其操作方法通常是用 MS-222[3,58]或苯佐卡因[22]等麻醉剂将鱼麻醉（避免伤害给药对象），然后用橡胶导管将调制好的药液或药饵灌入鱼的胃或肠。研究连续给药的药动学时常采用此法，可以保证定时给药及给药量。如 Bent-Samuelsen 等[59]对大西洋鲑连续 5 d 口灌药饵，研究 OMP 和 SDM 的组织分布和消除。但此法给药的缺点一是有时会出现药物回吐现象，二是麻醉剂的使用会对给药对象产生胁迫作用，影响药动学，如 MS-222 能降低药物的吸收[58]。

需指出的是，纯药液与药饵给药其药动学并不相同[50]，甚至药物采用不同的载体进行给药，其药动学也不相同。据 Sijm 等[58]报道，6 种不同的亲脂性化合物分别以油或明胶作为载体通过胃管灌服给虹鳟（O. mykiss），发现它们的吸收和药动学有明显差异。用明胶作载体时，这些化合物的吸收量、吸收速率均明显高于用油作载体时。这一结果对今后采用药饵载体给药研究药动学时应予以重视。

2.4　血管注射法

血管注射法是国外研究药动学用得比较多的方法，常用于评价一种药物剂型的生物利用度，即设定本法给药后获得的曲线下面积（AUC）为 100%，待测剂型的其他给药方法获得的 AUC 与之比较，从而作出评价。水产动物血管注射法包括动脉注射（intraarterial administration）[28,32,37]、静脉注射（intravenous administration）[24,25,29]、血管内给药（intravascular administration）[4,22,30,38]等。

2.5　心包内或血窦内注射给药

评价药物剂型在甲壳动物体内的生物利用度时，常采用心包内（intrapericardial administration）或血窦内注射给药（intra-sinus injection）。如 James 和 Barron[9]研究 SDM 在美洲螯龙虾（H. americanus）体内处置过程时，采用心包内注射和口服两种给药方法进行比较研究。Park 等（1995）[10]研究南美白对虾给予 SDM 和 OMP 后的药动学和生物利用度时，采用血窦注射给药。

总之，给药途径的不同，主要影响药物的吸收速度、吸收量，因而也影响药物作用的快慢与强弱。血管内注射、心包内或血窦内注射给药方法仅用于研究药物剂型的生物利用度研究。这类给药途径没有吸收过程，直接通过血液循环进入分布相。而以药浴法、肌肉注射法、口服法等方式给药，药物均需通过吸收过程才能进入血液循环。某些通过胃肠道吸收的药物，还会发生"第一关卡效应"（first-pass effect）。至于临床应采取哪一种给药途径则应根据具体情况和需要来定，探索出适合水产动物的最为有效的给药方式定会推进水产动物疾病学的发展。

3　影响药动学的水产动物生理因素

3.1　种属差异

水产养殖所涉及的对象十分广泛，从甲壳动物虾、蟹类，到鱼类（包括鲤科鱼类、鲈形目、鲷科等），再到爬行类等。关于种属差异对药动学影响，不同的药物其结果不一，但多数研究表明，同一药物在不同属的水产动物体内的代谢差异比较显著。如养殖的虹鳟、玫瑰大麻哈鱼口服土霉素（100 mg/kg）后[2]，AUC 分别为 32.1 μg·h/mL 和 58.7 μg·h/mL，平均残留时间分别为 50.3 h 和 24.6 h，半衰期分别为 23 h 和 16 h；草鱼和复合四倍体鲫单次腹腔注射氯霉素 100 mg/kg 后[44]，达峰时间分别为 0.92 h 和 0.63 h，体清除率分别为 70.067 mL/（kg·h）和 126.673 mL/（kg·h），表观分布容积分别为 2.07 L/kg 和 1.20 L/kg。斑节对虾[51]和南美白对虾[1]肌注诺氟沙星给药后，$T_{\beta1/2}$、AUC、CLs 和 V_d 等药动学参数也存在一定差别。有研究者认为不同种类水产动物之间的药动学及参数差异可能是由于解剖学上的体积差异以及药物与血浆蛋白、组织结合的差异所致[60]。种属间的药动学差异提示我们，不可轻易将某一药物在一个动物的药动学结果应用于其他动物。

3.2　性别

性别差异对药动学影响的研究还很少，目前发现性别对某些甲壳动物药动学有影响，如对美洲龙虾进行心包内给药，血淋巴中萘酚的处置过程用二室模型描述[35]，雄虾的 β 相半衰期明显长于雌虾，分别为（63.9±30.9）h 和（30.6±6.8）h；雌虾总体清除率为（26.4±6.5）mL/（kg·h），高于雄虾的（11.1±5.9）mL/（kg·h）。但 Poher 和 Blanc[21]在研究大菱鲆单剂量口服噁喹酸的药动学时发现，性别对药动学过程没有显著影响。这些可能与不同的药物、不同的水产动物有一定的关系，有待深入研究，尤其对全雄鱼或全雌鱼药动学的研究。

3.3　健康状况

研究水产动物药动学时，绝大多数实验是选用健康的动物进行，然而在养殖中药物使用对象多为发病的个体，理论上研究感染疾病个体的药动学最为真实，但实际操作难度很大。Uno[4]研究了健康的香鱼和弧菌感染的香鱼的 OTC 药动学和生物利用度。口服给药试验推算所得的生物利用度，健康鱼为 9.3%，而感染鱼为 3.8%；健康鱼血液、肌肉、肝和肾的消除半衰期分别为 53.1 h、106 h、125 h 和 117 h，而感染鱼分别为 63.2 h、92.9 h、107 h 和 123 h。两组鱼的消除是相近的，而生物利用度有明显差别，这有可能水产动物在健康状况较差时，影响了药物的分布、与血浆蛋白的结合率及药物代谢酶的活性。

4 影响药动学的环境因素

4.1 温度

一般说来，在一定温度范围内，药物的代谢强度与水温成正比。Karara 等[61]将杂色鳉（*Cyprinodon variegatus*）在 10、16、23、29 和 35℃下暴露于含 60 ng/mL 邻苯二甲酸二辛酯（di-2-ethylhexyl phthalate，DEHP）的水中，随温度升高，吸收清除速率呈线性上升，这可能是由于温度诱导鳃血流量增加。Namdari 等[6]研究大鳞大麻哈鱼组织中 OTC 的分布和消除时测得，水温 9℃、12℃下大鳞大麻哈鱼肌肉中 OTC 消除半衰期（$T_{1/2}$）分别为 13.59 d 和 10.34 d，差异显著，因此水温较低时停药期应适当延长。

4.2 盐度

盐度对药动学的影响在不同的水产动物中其结果不一。房文红和郑国兴[1]研究盐度 1 和 15 下对南美白对虾肌注诺氟沙星（10 mg/kg）药动学时，在给药后 2 min 对虾血药浓度达到最高峰，且峰浓度相近，但两盐度下部分药动学参数差别较大，如消除半衰期分别为 4.208 h 和 1.140 h。Ishida[62]在对海水虹鳟和淡水虹鳟口服噁喹酸（40 mg/kg）试验中得出，在 24 h 内两组鱼的各组织内药物浓度差别不大，但 24 h 后噁喹酸在海水虹鳟组织中的消除速率明显大于淡水虹鳟。两者结果相近。而 Abedini 等[3]在比较研究海水大鳞大麻哈鱼和淡水虹鳟动脉注射或口服 OTC 药动学时，在相同给药方式下，它们的 OTC 血药浓度－时间曲线相似，大鳞大麻哈鱼的 OTC 消除半衰期、分布容积和口服生物利用度分别为 88.29 h、0.89 L/kg 和 24.84%；虹鳟的相应值分别为 94.22 h、0.87 L/kg 和 30.30%，其结果表明盐度和种间差异对鲑鳟鱼的 OTC 吸收与消除并不起重要作用。

5 水产动物药动学分析模型

目前研究水产动物药动学普遍应用的模型是"房室模型"，而基于生理学的模型和非房室模型在哺乳动物药理学和毒理学应用越来越多，具有在水产动物中应用的潜力。

房室模型是把药物在体内的动力学状况按转运速度不同分成若干个隔室，用非线性的最小二乘法所建立的模型方程进行描述，从而得到药物在动物体内的分布和处置的量变规律，得出较充全的的参数。水产动物除测定血药浓度外，还常测定某些特定组织的药物浓度。房室模型中，通常认为吸收和消除在中央室进行，虽然有学者为水生动物建立了更为复杂的模型[63]，但一室或二室模型仍适用于描述大多数药物；同时房室模型已用来评定环境因子（如 Barron 等，1987）[64]和个体大小[65]对水产动物药动学的影响。尽管如此，但房室模型并不能描述组织间浓度差异较大的生理系统，且分析结果依赖于房室模型的选择，而房室模型的选择带有一定的不确定性。

非房室模型分析较为简便，不需要进行隔室设定、模型假设和对数据的模型拟合，同样可以描述药物在生物体内的吸收、分布、消除等，但提供的药动学参数较为有限，目前在水生动物毒理学中应用较少。

而基于生理学的药代动力学（physiologically-based pharmacokinetics，PBPK）模型将生理学过程和参与药物处置的重要组织一体化，可以用来推测其他环境和其他的动物，是描述动物药动学最为理想的模型。但模型的建立需要关于动物和药物性质的较为具体的生理学资料，如组织容积、药物在血液和组织间的分配系数和扩散系数等。PBPK 模型最大的优点是，一旦确立了主导的转运机制，它可以在超出原始的数据基础上将该模型外推到其他条件或种类上。Bungay 等[39]首先完成了水生动物的 PBPK 模型化；Abbas 和 Hayton[40]研究虹鳟体内对氧磷药动学时，采用 PBPK 模型描述水中对氧磷浓度与靶器官、对氧磷的乙酰胆碱酯酶抑制和羟酸酯酶去毒化之间的关系。该模型结构包括脑、心脏、肝、肾及其他与血液循环紧密相连的部分；Clark[41]采用 PBPK 模型比较研究了斑点叉尾鲖和黄鲈的种间药动学差异及温度对药动学的影响。由此可见，PBPK 模型有望在水产动物中得到越来越多的运用。

6 展望

国外学者在水产动物药动学方面已作了较广泛、深入的研究，而国内则刚刚起步，在广度和深度上均远落后于国外。借鉴国外水产动物药动学研究方法与经验，对提高我国水产动物药动学研究水平十分必要。药动学的特征参数，如吸收速率、达峰时间、峰浓度、分布和消除半衰期、曲线下面积和总体清除率等是选择药物种类、制定用药剂型、剂量和用药周期以及确定休药期的重要依据，为药物的科学使用和水产品安全监控与管理提供理论指导，确保我国水产养殖业持续、健康、稳定发展。今后我国水产动物药动学研究的发展方向可从以下几方面着手：（1）同一药物在不同的水产动物体内代谢动力学种属差异性研究；（2）研究环境因子（如温度、盐度、溶解氧等）对水产动物药动学影响；（3）开展中草药在水产动物体内的代谢、转化和消除规律研究；（4）比较研究水产动物健康与非健康水平时的药动学，建立人工诱发疾病的水产动物药动学模型；（5）进一步研究药物在水产动物体内生物转化及代谢产物在水产动物体内的代谢和消除规律；（6）深入开展水产动物药动学与药效学、毒理学的同步研究，全面评价药物在水产动物体内的效用。

参考文献

［1］ Snieszko S F, Friddle S O, Griffin P J. Successful treatment of ulcer disease in Brook Trout （Salvelinus fontinalis） with Terramycin ［J］. Science, 1951, 113: 717-718.

［2］ Uno K, Aoki T, Ueno R. Pharmacokinetic study of oxytetracycline in cultured rainbow trout, amago salmon, and yellowtail ［J］. Nippon Suisan Gakkaishi Bull, 1992, 58 （6）: 1151-1156.

［3］ Abedini S, Namdari R, Law F C P. Comparative pharmacokinetics and bioavailability of oxytetracycline in rainbow trout and chinook salmon ［J］. Aquacult, 1998, 162 （1）: 23-32.

［4］ Uno K. Pharmacokinetic study of oxytetracycline in healthy and vibriosis-infected ayu （Plecoglossus altivelis） ［J］. Aquacult, 1996, 143 （1）: 33-42.

［5］ Grondel J L, Nouws J F M, de-Jong M, et al. Pharmacokinetics and tissue distribution of oxytetracycline in carp, Cyprinus carpio L., following different routes of administration ［J］. J Fish Dis, 1987, 10 （3）: 153-163.

［6］ Namdari R, Abedini S, Albright L, et al. Tissue distribution and elimination of oxytetracycline in sea-pen cultured chinook salmon, Oncorhynchus tshawytscha, and Atlantic salmon, Salmo salar, following medicated-feed treatment ［J］. J Appl Aquacult, 1998, 8 （1）: 39-52.

［7］ Della-Rocca G, Zaghini A, Magni A. Pharmacokinetics of amoxicillin in sea bream （Sparus aurata） after oral and intravenous administrations. 1: comparison of serum concentrations evaluated with two different analytical methods ［J］. Boll Soc Ital Patol Ittica, 1998, 10 （24）: 45-50.

［8］ Martinsen B, Horsberg T E, Varma K J, et al. Single dose pharmacokinetic study of florfenicol in Atlantic salmon （Salmo salar） in seawater at 11 degree C ［J］. Aquacult, 1993, 112 （1）: 1-11.

［9］ James M O, Barron M G. Disposition of sulfadimethoxine in the lobster （Homarus americanus） ［J］. Vet Hum Toxicol, 1988, 30 （1 suppl.）: 36-40.

［10］ E D, Lightner D V, Milner N, et al. Exploratory bioavailability and pharmacokinetic studies of sulphadimethoxine and ormetoprim in the penaeid shrimp, Penaeus vannamei ［J］. Aquacult, 1995, 130 （2, 3）: 113-128.

［11］ Horsberg T E, Martinsen B, Sandersen K, et al. Potentiated sulfonamides: In vitro inhibitory effect and pharmacokinetic properties in Atlantic salmon in seawater ［J］. J Aquat Anim Health, 1997, 9 （3）: 203-210.

［12］ Samuelsen O B, Lunestad B T, Jelmert A. Pharmacokinetic and efficacy studies on bath-administering potentiated sulphonamides in Atlantic halibut, Hippoglossus hippoglossus L ［J］. J Fish Dis, 1997, 20 （4）, 287-296.

［13］ Touraki M, Niopas I, Kastritsis C. Bioaccumulation of trimethoprim, sulfamethoxazole and N-acetyl-sulfamethoxazole in Artemia nauplii and residual kinetics in seabass larvae after repeated oral dosing of medicated nauplii ［J］. Aquacult, 1999, 175 （1）: 15-30.

［14］ Uno K, Aoki T, Ueno R, et al. Pharmacokinetics and metabolism of sulphamonomethoxine in rainbow trout （Oncorhynchus mykiss） and yellowtail （Seriola quinqueradiata） following bolus intravascular administration ［J］. Aquacult, 1997, 153 （1）: 1-8.

［15］ Plakas S M, El-Said K R, Stehly G R. Furazolidone disposition after intravascular and oral dosing in the channel catfish ［J］. Xenobiotica, 1994, 24 （11）: 1095-1105.

［16］ Stehly G R, Plakas S M. Pharmacokinetics, tissue distribution, and metabolism of nitrofurantoin in the channel catfish （Ictalurus punctatus） ［J］. Aquacult, 1993, 113 （1-2）: 1-10.

［17］ Steward A R, Maslanka R, Pangrekar J, et al. Disposition and metabolism of 2, 3, 7, 8-tetrachlorodibenzofuran by rainbow trout （Oncorhynchus mykiss） ［J］. Toxicol Appl Pharmacol, 1996, 139 （2）: 418-429.

［18］ Uno K, Aoki T, Ueno R, et al. Pharmacokinetics of nalidixic acid and sodium nifurstyrenate in cultured fish following bolus intravascular administration ［J］. Fish Pathol, 1996, 31 （4）: 191-196.

［19］ Uno K, Kato M, Aoki T, et al. Pharmacokinetics of nalidixic acid in cultured rainbow trout and amago salmon ［J］. Aquacult, 1992, 102 （4）: 297-307.

［20］ Kim M S, Park S W, Huh M D, et al. Effect of temperature and bacterial infection on the absorption and elimination of oxolinic acid in Nile tilapia （Oreochromis niloticus） ［J］. J Korean Fish Soc, 1998, 31 （5）: 677-684.

［21］ Poher I, Blanc G. Pharmacokinetics of a discontinuous absorption process of oxolinic acid in turbot, Scophthalmus maximus, after a single oral administration ［J］. Xenobiotica, 1998, 28 （11）: 1061-1073.

［22］ Plakas S M, El-Said K R, Musser S M. Pharmacokinetics, tissue distribution, and metabolism of flumequine in channel catfish （Ictalurus punctatus） ［J］. Aquacult, 2000, 187: 1-14.

［23］ Intorre L, Cecchini S, Bertini S, et al. Pharmacokinetics of enrofloxacin in the seabass （Dicentrarchus labrax） ［J］. Aquacult, 2000, 182 （1-2）: 49-59.

［24］ Martinsen B, Horsberg T E, Sohlberg S, et al. Single dose kinetic study of sarafloxacin after intravenous and oral administration of different formulations to Atlantic salmon （Salmo salar） held in sea water at 8.5 degree C ［J］. Aquacult 1993, 118: 37-47.

［25］ Nouws J F M, Grondel J L, Schutte A R, et al. Pharmacokinetics of ciprofloxacin in carp, African catfish and rainbow trout ［J］. Vet Quart,

1988，（10）：211-216.

[26] Ueno R, Okada Y, Tatsuno T. Pharmacokinetics and metabolism of miloxacin in cultured eel [J]. Aquacult, 2001, 193 (1-2)：11-24.

[27] Abbas R, Hayton W. Gas chromatographic determination of parathion and paraoxon in fish plasma and tissues [J]. J Anal Toxicol, 1996, 20 (3)：151-154.

[28] Kitzman J V, Holley J H, Huber W G, et al. Pharmacokinetics and metabolism of fenbendazole in channel catfish [J]. Vet Res Commun, 1990, 14 (3)：217-226.

[29] Barron M G, Plakas S M, Wilga P C. Chlorpyrifos pharmacokinetics and metabolism following intravascular and dietary administration in channel catfish [J]. Toxicol Appl Pharmacol, 1991, 108 (3)：474-482.

[30] Plakas S M, El-Said K R, Bencsath F A. Pharmacokinetics, tissue distribution and metabolism of acriflavine and proflavine in the channel catfish (Ictalurus punctatus) [J]. Xenobiotica, 1998, 28 (6)：605-616.

[31] Rouleau C, Gobeil C, Tjaelve H. Pharmacokinetics and distribution of dietary tributyltin compared to those of methylmercury in the American plaice Hippoglossoides platessoides [J]. Mar Ecol Prog Ser, 1998, 171：275-284.

[32] Vick A M, Hayton W L. Methyltestosterone pharmacokinetics and oral bioavailability in rainbow trout (Oncorhynchus mykiss) [J]. Aquat Toxicol, 2001, 52 (3-4)：177-188.

[33] Thibaut R, Debrauwer L, Rao D, et al. Characterization of biliary metabolites of 4-n-nonylphenol in rainbow trout (Oncorhynchus mykiss) [J]. Xenobiotica, 1998, 28 (8)：745-757.

[34] Li C L J, James M O. The oral bioavailability, pharmacokinetics and biotransformation of 9-Hydroxybenzo [a] pyrene in the American lobster, Homarus americanus [J]. Mar Environ Res, 1998, 46 (1-5)：505-508.

[35] Li C L J, James M O. Pharmacokinetics of 2-naphthol following intrapericardial administration, and formation of 2-naphthyl-beta-D-glucoside and 2-naphthyl sulphate in the American lobster, Homarus americanus [J]. Xenobiotica, 1997, 27 (6)：609-626.

[36] Meinertz J R, Stehly G R, Gingerich W H. Pharmacokinetics of benzocaine in rainbow trout (Oncorhynchus mykiss) after intraarterial dosing [J]. Aquacult, 1996, 148 (1)：39-48.

[37] Martin J F, Plakas S M, Holley J H, et al. Pharmacokinetics and tissue disposition of the off-flavor compound 2-methylisoborneol in the channel catfish (Ictalurus punctatus) [J]. Can J Aquat Sci, 1990, 47 (3)：544-547.

[38] Plakas S M, James M O. Bioavailability, metabolism, and renal excretion of benzoic acid in the channel catfish (Ictalurus punctatus) [J]. Drug Metab Disposition, 1990, 18 (5)：552-556.

[39] Bungay P M, Dedrick R L, Guarino A M. Pharmacokinetic modeling of the dogfish shark (Squalus acanthias)：distribution and urinary and biliary excretion of phenol red and its glucuronide [J]. J Pharmcok Biopharm, 1976, 4：377-388.

[40] Abbas R, Hayton W L. A physiologically based pharmacokinetic and pharmacodynamic model for paraoxon in rainbow trout [J]. Toxicol Appl Pharmacol, 1997, 145 (1)：192-201.

[41] Clark, K. J. Temperature and species comparisons of benzocaine pharmacokinetics, metabolism and physiologically based pharmacokinetic model within Channel catfish, Ictalurus punctatus, and Yellow Perch, Perca flavescens [J]. Dissertation Abstracts International Part B：Science and Engineering. 2000, 60 (8)：3880.

[42] 李美同，郭文林，仲峰等. 土霉素在鳗鲡组织中残留的消除规律 [J]. 水产学报，1997，21 (1)：39-42.

[43] 李兰生，王勇强. 对虾体内氯霉素含量测试方法的研究 [J]. 青岛海洋大学学报，1995，25 (3)：400-406.

[44] 李爱华. 氯霉素在草鱼和复合四倍体异育银鲫体内的比较药代动力学 [J]. 中国兽医学报，1998，18 (4)：372-374.

[45] 艾晓辉，陈正望. 磺胺二甲嘧啶在银鲫体内的药动学及组织残留研究. 淡水渔业，2001，31 (6)：52-54.

[46] 王群，孙修涛，刘德用等. 复方新诺明在鲈鱼体内的药物代谢动力学研究. 海洋科学，2001，25 (2)：35-38.

[47] 王群，李健. 孙修涛. 呋喃唑酮在黑鲷体内的代谢动力学和残留研究. 青岛海洋大学学报，2002，32 (增)：45-49.

[48] 李健，王群，孙修涛等. 噁喹酸对水生生物细菌病的防治效果及残留研究 [J]. 中国水产科学，2001，8 (3)：45-49.

[49] 陈文银，印春华. 诺氟沙星在中华鳖体内的药代动力学研究 [J]. 水产学报，1997，21 (4). -434-437.

[50] 张祚新，张雅斌，杨永胜，等. 诺氟沙星在鲤鱼体内的药代动力学 [J]. 中国兽医学报，2000，20 (1)：66-69.

[51] 房文红，邵锦华，施兆鸿，等. 斑节对虾血淋巴中诺氟沙星含量测定及药代动力学 [J]. 水生生物学报，2003，27 (1)：13-17.

[52] 杨先乐，刘至治，孙文钦，等. 中华绒螯蟹血淋巴内环丙沙星反相高效液相色谱测定法的建立 [J]. 水产学报，2001，25 (4)：348-354.

[53] 曾子建，李逐波，吴绪田，等. 喹乙醇在鲤鱼体内的药代动力学研究 [J]. 四川农业大学学报，1993，11 (1)：109-112.

[54] Uno K, Aoki T, Ueno R. Pharmacokinetics of sulphamonomethoxine and sulphadimethoxine following oral administration to cultured rainbow trout (Oncorhynchus mykiss) [J]. Aquacult, 1993, 115 (3/4)：209-219.

[55] Barron M G. Mayes M A, Murphy P G. Pharmacokinetics and metabolism of triclopyr butoxyethyl ester in coho salmon [J]. Aquat Toxicol, 1990, 16 (1)：19-32.

[56] Barron M G, Hansen S C, Ball T. Pharmacokinetics and metabolism of triclopyr in the crayfish (Procambarus clarki) [J]. Drug Metab Disposition, 1991, 19 (1)：163-167.

[57] Rouleau C, Gobeil C, Tjaelve H. Pharmacokinetics and Distribution of Dietary Tributyltin and Methylmercury in the Snow Crab (Chionoecetes opilio) [J]. Environ Sci Tech, 1999, 33 (19)：3451-3457.

［58］　Sijm D T H M, Bol J, Seinen W, et al. Ethyl m-aminobenzoate methanesulfonate dependent and carrier dependent pharmacokinetics of extremely lipophilic compounds in rainbow trout［J］. Arch Environ Contam Toxicol, 1993, 25（1）：102-109.

［59］　Bent-Samuelsen O, Pursell L, Smith P, et al. Multiple-dose pharmacokinetic study of Romet super（30）in Atlantic salmon（Salmo salar）and in vitro antibacterial activity against Aeromonas salmonicida［J］. Aquacult, 1997 vol. 152, no. 1-4, pp. 13-24.

［60］　Barron M G, Gedutis C and James M O. Pharmacokinetics of sulphadimethoxine in the lobster. Homerus americannus, following intrapericardial administration［J］. Xenobiotica, 1988, 18（3）：269-277.

［61］　Krara A H, Hayton W L A. pharmacokinetic analysis of the effect of temperature on the accumulation of di-2-ethylhexyl phthalate（DEHP）in sheepshead minnow［J］. Aquat Toxicol, 1989, 15（1）：27-36.

［62］　Ishida N. Tissue levels of oxolinic acid after oral or intravascular administration to freshwater and seawater rainbow trout［J］. Aquacult, 1992, 102：9-15.

［63］　Barron M G, Schultz M A, Hayton W L. Presystemic branchial metabolism limits di-2-ethylhexyl phthalate accumulation in fish［J］. Toxicol Appl Pharmacol, 1989, 98：49-57.

［64］　Barron M G, Tarr B D, Hayton W L. Temperature dependence of di-2-ethylhexyl phthalate pharmacokinetics in rainbow trout［J］. Toxicol Appl Pharmacol, 1987, 88：305-312.

［65］　Tarr B D, Barron M G, Hayton W L. Effect of body size on the uptake and bioconcentration of di-2-ethylhexyl phthalate in rainbow trout［J］. Environ Toxicol Chem, 1990, 9：989-995.

该文发表于《中国兽药杂志》【2003：37（12）：38-41】

影响水产动物药代动力学的因素

湛嘉[1]、李佐卿[1]、康继韬[1]、俞雪均[1]、谢东华[1]、孙大为[1]、杨先乐[2]

（1. 宁波出入境检验检疫局，浙江 宁波 315012；2. 农业部上海水产大学渔业动植物病原库，上海 200090）

[摘要]：对影响水产动物药代动力学的机体因素、药理因素和环境因素进行了综述。影响药动学的因素很多，尤其对于所处环境因素复杂的水产动物来说，除了遗传种属、性别等生理因素和病理、药理等因素外，环境因子尤其是水温能显著影响药代动力学过程。研究不同条件下的药动学，对于合理、正确地使用药物有重要的指导作用。

[关键词]：水产动物；药代动力学；药物

随着水产养殖业的迅速发展，病害的流行和危害趋于严重，渔用药物的应用对确保养殖渔业的稳定和持续发展起着重要作用。但是，历来的药物防治研究大多集中于针对病原的药物筛选和使用浓度及使用方法等药效学方面，有关药物动力学、毒理学的基础研究工作都十分薄弱，滞后于水产养殖业的发展，致使水产动物的用药基本上借鉴于兽医甚至人医的用药经验[1,2]。

药物代谢动力学（pharmacokinetics，简称药代动力学或药动学）是研究机体对药物的吸收、分布、代谢与排泄的过程和血药浓度随时间变化规律的科学[3]，其本身又受药理、机体条件及环境等因素的影响，是指导正确、合理用药以及保障水产品食用安全的应用学科。本文就影响水产动物药代动力学因素的研究现状做一概述，旨在为药物防治水产动物病害及水产动物组织中药物残留监测等提供一定的参考。

1　水产动物的机体因素

1.1　生理性差异

1.1.1　种属差异

水产动物生物体是药物应用的对象，不同的生物体对药物的代谢类型和强度一般具有种的特异性。Uno 等[4]指出了土霉素（OTC）在虹鳟、大马哈鱼和五条鰤的曲线下面积、平均残留时间（MRT）和消除半衰期（hq）的差异性。磺胺间甲氧嘧啶（SMM）在虹鳟体内的 $t_{1/2\beta}$ 和总表观分布容积（V_d）明显较大，而在五条鰤的总清除率（CZ_B）和酰基化率分别较高，表明虹鳟对 SMM 的亲组织性显著高于五条鰤[5]。环丙沙星在鲤鱼、虹蹲及非洲鲶体内的的差别并不显著，但 V_d 值的差异显著，分别为（0.89±0.20）、（1.13±0.23）和（2.44±0.63）L·kg^{-1}[6]。Heijden 等[7]在同水温下比较了氟甲喹在鲤鱼、非洲鲶和欧洲鳗鲡体内的药学学规律，发现其平均最高血药浓度（C_{max}）、分布半衰期（$t_{1/2\alpha}$）和 $t_{1/2\beta}$ 差异显著，并指出为了有效利用药物和尽量减少药残，应依据不同的种属来确定相应药物的剂量范围和休药期。

水产动物的代谢类型和强度又存在种间共性。李爱华[8]指出,除了CZ_B和K差异显著外,氯霉素从腹腔均被迅速吸收,在草鱼和异育银鲫两种鱼体内的动力学过程相似,均可用带一级吸收的二室开放模型进行描述。OTC在淡水虹鳟与大鳞大马哈鱼体内的组织分布情况都非常相似,药动学参数可以互用[9]。金头鲷和海鲈对噁喹酸的代谢也非常类似[10]。药动学特征的相似性可能跟这些鱼类之间的亲缘关系有关。

Van等[11]认为鱼类药动学参数的种属差异可能由于其在肾功能、肌肉生化组成、活动性等生理上的不同而引起。由于同温下种属本身的代谢率也可能不同,因此,在各自最适水温条件下,水产动物的药动学差异或许要小些[7]。

1.1.2 年龄和性别因素的影响

年龄对药物的吸收、分布和消除具有明显影响,年龄不同,许多生理机能也不同,对药物的反应也不一致。Michel等[12]采用不同给药方式比较了未成熟与成熟的沟鲶,发现成熟鱼在多次给药时的药动学参数与单次给药时的一致,而幼鱼却不一致,表现为总消除时间减少。

雌雄水产动物生理机能的差异可能会对药物的体内分布、吸收过程产生影响。从已有报道来看,性别对甲壳动物药动学有影响,而对鱼类的影响至今尚缺乏证据。对美洲龙虾进行心包内给药,其血淋巴中萘酚的处置过程可以用二室模型描述[13],雄、雌虾的$T_{1/2\alpha}$分别为(26±19)min、(29±15)min,雄虾的$T_{1/2\beta}$明显长于雌虾,分别为(63.9±30.9)h和(30.6±6.8)h;雌虾的Cl_β为(26.4±6.5)mL·h^{-1}·kg^{-1},高于雌虾的(11.1±5.9)mL·h^{-1}·kg^{-1}。而Poher和Blanc[14]在研究大菱鲆单剂量口服噁喹酸的药动学时认为,性别对药动学过程没有显著影响。

1.1.3 病理因素

疾病及饥饿引起的营养不良均可改变药物在体内的代谢过程,从而影响药动学参数。Uno[15]研究指出OTC在健康的香鱼体内口服的生物利用度明显高于患弧菌病鱼的口服生物利用度(健康鱼为9.3%,而感染鱼为3.8%),消除规律和血浆蛋白结合率却非常相似,健康鱼血液、肌肉、肝和肾的$T_{1/2\beta}$分别为53.1 h、106 h、125 h和117 h,而感染鱼分别为63.2 h、92.9 h、107 h和123 h。Hustvedt等[16]研究了噁喹酸在患弧菌病大西洋鲑体内的药动学,指出在治疗期间连续10 d平均血药浓度高于1.2 μg/mL,对敏感致病菌具有较好的抑制作用。

由于水产动物疾病模型的建立难度较大,目前绝大部分抗菌素在水产动物体内的药代动力学研究尚局限于基础药代动力学范畴。

2 药理因素

2.1 药物的种类

不同的药物由于理化性质不同,在水产动物的代谢差异很大,即使是同一类药物,在同种水产动物体内的药动学特征及生物利用度差别也可能是明显的。Bjorklund等[17]指出在同样条件下,噁喹酸和OTC在虹鳟体内的$T_{1/2}$,分别为0.31 h和1.53 h,V_d分别为1.94 L/kg和1.34 L/kg,相比之下,噁喹酸的血浆结合率明显较低,而生物利用率高出一半。比较了混饲给药噁喹酸、氟甲喹、盐酸沙拉沙星和恩诺沙星等同一类药物在大西洋鲑体内的药动学特征,发现其生物利用率依次为30.1%、44.7%、2.2%和55.5%,恩诺沙星的$t_{1/2\beta}$接近噁喹酸的2倍。Regostad等[19]也指出大西洋鲑对氟甲喹比对噁喹酸生物利用度大,尤其在较高剂量使用时氟甲喹吸收更快,生物利用率更具优势,从而认为氟甲喹是一种比噁喹酸更好的治疗药物。

2.2 药物剂量及给药期

用药剂量不同导致药理效应差异。在安全范围内,药物剂量越大,药理效应越强。因此,要发挥药物的有效作用,同时限制其不良反应,必须严格掌握药物的剂量范围。加大剂量可能引起生物利用率的下降,大西洋鲑服用25 mg/kg体重的噁喹酸比50 mg/kg体重剂量的口服生物利用度高出15%[19]。Cravedi等[20]指出虹鳟对20 mg/kg体重剂量的噁喹酸口服生物利用度为40%,而采用100 mg/kg体重剂量时的生物利用度仅为12%。Guy等[21]以1 mg/kg和10 mg/kg体重的不同剂量导管注射研究呋喃妥因在沟鲶体内的药动学,结果表明:低剂量的是高剂量的一半,分别适合二室和三室药动学模型。

Michel等[12]指出成熟的斑点叉尾鮰在多次和单次给药下的药动学参数一致。Martinsen等[22]证实,将总剂量同为100 mg/kg体重的沙拉沙星以5 d和10 d的不同给药期分别投喂给大西洋鲑,结果显示前者的血浆、肌肉、肝脏中的最高平均药物浓度显著高于后者。

2.3 给药途径

药物的吸收速度受给药途径的影响,从快到慢一般依次为:静脉注射、肌肉注射、皮下注射、口服、药浴。不同给药途径的药动学参数和生物利用度差别显著。Grondel等[23]采用静脉注射、肌肉注射和口服法研究OTC在鲤鱼体内的药动学及组织分布情况,结果发现,肌肉注射与口服给药后,鲤鱼对OTC的生物利用度差别极其显著,分别为48%和0.6%。郭锦朱等[24]比较了在不同的给药方式下带点石斑鱼对氟甲喹的吸收、分布与消除，结果表明，经口投方式给药生物利用度相当高（达 44%），而以药浴方式投药则药物利用率非常低（仅 9%），不符合经济效率，而 Samuelsen 等[25]则肯定了药浴法的治疗潜力。张雅斌等[26]研究指出诺氟沙星在鲤鱼体内采用口灌服方式比肌注和混饲口服给药方式吸收程度都好，混饲口服给药吸收速度较慢且生物利用率最低。

2.4　其他

关于联合用药的药动学研究为数很少，磺胺与其增效剂体外药敏试验表明，二者联用可大大增加药物的杀菌效果。Droy 等[27]曾报道 [14]C 标记的 SDM 与虹鳟的血浆蛋白结合率没有受到未标记的奥美普林（Ormetoprim，OMP）的浓度增加的影响，同样未标记的 SDM 的浓度增加也不影响血浆蛋白对 [14]C 标记 OMP 的结合率，提出 SDM 的药动学和分布不受 OMP 加入的影响，从而推断出它们在体内也能同样起到相加作用。目前，Romet[3]°（OMP：SDM = 1：5）已成为美国和加拿大允许应用于控制水产动物疾病的一种商品药物。Samuelsen 等[28]的研究也表明，在多次给药下，Rom et[3]° 的两种组分能按不同的固定比例持续分布在大西洋鲑的不同组织中一段时期。

不同的剂型对药动学的影响也是显著的。Sijm 等[29]将 6 种不同的亲脂性化合物分别以油和明胶作为载体通过胃管灌服给虹鳟，发现它们的吸收和药动学有明显差异。用明胶作载体时，虹鳟对这些化合物的吸收百分率、吸收速率及浓度均比以油做载体高。

3　环境因素

水产动物是生活在水环境中的变温动物，水环境的变化对其活动影响很大，从而影响机体对药物的代谢过程。环境因素包括以下几方面：

3.1　温度

一般说来，在一定温度范围内，药物的代谢强度与水温成正比，水温越高，代谢速度越快，通常水温每升高 1℃，鱼类的代谢和消除速度将提高 10%[30]。虹鳟在注射牛磺胆酸 1 h 后，处在 14℃ 和 18℃ 水温下的胆汁排泄明显快于 10℃ 组，血浆 $t_{1/2\beta}$ 明显较短[31]；在 14℃ 和 24℃ 条件下口服噁喹酸的生物利用率分别为 56% 和 90.7%，分别为 54.3 和 33.1 h[32]，可见随着水温升高生物利用率提高，而消除也随之加快。Bjorklund 等[30]的研究结果表明：5、10 和 15℃ 条件下 OTC 在虹鳟血清中达到峰值的时间分别为 24、12 和 1 h，消除时间最多相差 108 d。总之，使用药物时应充分考虑水温，水温低时应适当延长停药期。

3.2　盐度和 pH 值

关于盐度对药动学的影响存在不同的看法。Noriko 等[33]证明虹鳟在海水中比在淡水中的代谢排除率高，海水中 72 h 后组织中噁喹酸已检测不到，而在淡水中 10 d 后仍能检测到该药物，因此使用噁喹酸防治海水鳟疾病时，必须考虑其代谢排除快的药动学特征，缩小投药间隔。Sohlber 等[34]证明大西洋鲑在海水中比在淡水中的代谢排除率高，在海水中血液和肌肉中的氟甲喹分别于 4 d 和 14 d 时无法检出，而在淡水中 64 d 时仍能检出。Hustvedt 和 Salte 等[16]也指出虹鳟在海水中比在淡水中对噁喹酸的吸收和分布更快，在淡水中的 V_d 则比在海水中略大。而 Namdari 等[35]证明盐度相差 24% 并未对 OTC 的吸收和消除起很大作用，而且提出淡水中的虹鳟可作为一个模型来研究海水中的虹鳟对 OTC 的药动学。由此可推断，盐度对药动学的影响程度很可能与药物结构不同有关。

pH 值对药动学影响的研究报道并不多。Hansen 等[36]通过药浴给药比较了不同的 pH 值对欧洲鳗鲡的氟甲喹药动学影响，发现 pH 值由 7.15 降低至 6.07 时，生物利用率由 19.8% 上升至 41%。还发现呼吸产生的 CO_2 使 pH 值降低，也可使生物利用率上升至 35%。

参考文献

[1] 王群，马向东，李绍伟. HPLC 法研究磺胺类药物在水产动物体内的代谢和残留 [J]. 海洋科学，1999，6：33-36.

[2] 李兰生，王勇强. 对虾体内氯霉素含量测试方法的研究 [J]. 青海海洋大学学报，1995，25（3）：400-406.

[3] 李剑勇，赵荣材. 我国兽医药物代谢动力学研究进展 [J]. 中国兽药杂志，1999，33（4）：44-47.

[4] Uno K, Aoki T, Ueno R. Pharmacokinetic study of oxytetracycline in cultured rainbow trout araago salmon and yellowtail [J]. Nippon Suisan-gakkaishi Bull Jap Soc Sci Fish, 1992, 58（6）：1151-1156.

[5] Undo Kazuaki, Aoki Takahiko, Ueno Ryuji, et cd. Pharmacokinetics and metabolism of sulphamonomethoxine in rainbow trout（Oncorhynchus mykiss）and yeilowtail（Seriola quinqueradiata）following bolus intravascular administration [J]. Aquaculture, 1997, 153（1-2）：1-8.

[6] Nouws JFM, Grondel JL, Schutte AR, et al. Phannacokinetics of ciprofloxacin in carp, African catfish and rainbow trout [J]. Vetquart, 1988, 10：211-216.

［7］ MHT van der，Heijden HJ，Keukens WHFX，et al. Plasma disposition of fluraequine in common carp，African catfish and European eel after a single peroral adminitstration ［J］. Aquaculture，1994，123：21-30.

［8］ 李爱华. 氯霉素在草鱼和复合四倍体异育银鲫体内的比较药代动力学 ［J］. 中国兽医学报，1998，18（4）：372-374.

［9］ Sid9el Sohlberg，Krystyna Czerwinska，Kynut Rasmussen，et al. Plasma concentrations of flumequine after intra-arterial and oral administralion to rainbow trout exposed to low water ［J］. Aquaculture，1990，84：355-36L

［10］ Christofilogianiiis P，Alexi MN，Richards RH. Antibiotics use in Greek maricultures：study of the oxolinic acid retention tirae in giltheadsea bream（Sparus aurata）and sea bass（DicerUrarchus labrax），at high temperatures after oral treatment ［S］. Hellenic Symposium Oceanography and Fisheries，Kavala（Greece），1997，（4）：15-18.

［11］ Grondel JL，Nouws JFM，De-jong M，et al. Pharmacokinetics and tissue distribution of oxytetracycline in carp，cyprinus carpio L.，follows different routes of administration ［J］. Journal of Fish Diseases，1987，10（3）：153-163.

［12］ Michel CMF，Hiazraacokinetics and n^tabolism of suHadimeihoxine in channel catfish（Ictalurus punctatus）［J］. Diss Abst Int Pt B Sci Eng，1990，51（1）：230-238.

［13］ Li C IJ，James MO，Pharmacokinetics of 2-naphthol following in-irapericardial administration，and formation of 2-naphthyl-beta-D-glucoside and 2-naphthyl in the American lobster，Homarus ameri-canus ［J］. Xenobiotica，1997，27（6）：609-626.

［14］ Poher I，Blanc G. Pharmacokinetics of a discontinuous absorption process of oxolinic acid in turbot Scophthalmus Macimus，after a single oral administration ［J］. Xenobiotica，1998，28（11）：1061-1073.

［15］ Uno K. Pharmacokinetic study of oxytetracycline in healthy and vibriosis-infected ayu（Plecoglossus altivelis）［J］. Aquaculture，1996，143（1）：33-42.

［16］ Svein Olaf Hustvedt，Aagnar Salte，Vidar Vassvik. Combatting cold-waler vibriosis in Atlantic Salomon（Salmo salar L）with oxolinic acid：a case report ［J］. Aquaculture，1992，103：213-219.

［17］ BjOrklund HV，Bylund G，Comparative pharmacokinetics and bio-availaJbility of oxolinic acid and oxytertracyccline in rainbow trout（Oncorhynchus mykiss）［J］. Xenobiolica，1991，21（11）：1511-1520.

［18］ Martinsen B，Horsberg TE. Comparative single-dose pharmacokinetics of four quinolones，oxolinic acid，fluraequine，sarafloxacin，and enrofloxacin，in Atlantic salmon held in seawater at 10 degree C ［J］. Antimicrob Agents Chemother，1995，39（5）：1059-1064.

［19］ Rogslad A，Ellingsen OF，Syverlsen C，Pharmacokinetics and bioavailability of flumequine and oxolinic acid after various routes of administration to Atlantic salmon in 9eawater ［J］. Aquaculture，1993，110（34）：207-220.

［20］ Cravedi JP，Choubert G，Delous G. Digestibility of chloramphenicol，oxolinic acid and oxytetracycline in rainbow trout and influence of these antibiotics on lipid digestibility ［J］. Aquaculture，1987，60：133-141.

［21］ Guy R，Stehly，steven M Plakas. Pharmacokinetics，tissue distribution t and metabolism of nitrofurantoin in the channel catfish（Ictalurus punctutus）［J］. Aquaculture，1993，113：1-10.

［22］ Martinsen B，Horskeng TE，sohbei^ S，et al. Single dose kinetic study of sarafloxacin after intravenous and oral administration of different formulations to Atlantic salmon（Salmo salar）held in sea water at 8.5 degree C ［J］. Aquaculture，1993，118（1-2）：37-47.

［23］ Grondel JL，Nouws JFM，De-jong M，et al. Pharmacokinetics and tissue distribution of oxytetracycline in carp，cyprinus carpio L，follows different routes of administration ［J］，Journal of Fish Diseases，1987，10（3）：153-163.

［24］ 郭锦朱，廖一久. 抗菌剂在养殖鱼类的药动学研究（1）在不同的给药方式下带点石斑鱼对 Flumequiiie 之吸收、分布与排除 ［J］. 水产研究，1996，4（2）：117-126，

［25］ Samuelsen OB，Lunestad BT，Jelment A. Pharmacokinetic and efficacy studies on bath-administering potentiated sulphonamides in Atlantic halibut，Hippoglossus hippoglossus L ［J］. J Fish Dis，1997，20（4）：287-296.

［26］ 张雅斌，张柞新，郑伟，等. 不同给药方式下鲤对诺氟沙星的药代动力学研究 ［J］. 水产学报，2000，24（6）：559-563.

［27］ Dnoy BF，Tate，T，Lech JJ，et al. Influence of ormetoprim on the bioavailability，distribution，and pharmacokinetics of sulfadime-thoxine in rainbow trout（Oncorhynchus mykiss）［J］. Comp Biochem Physiol，C，1989，94（1）：303-307.

［28］ Bent Samuelsen O，Pursell L，Smith P，et al. Multiple-dose pharmacokinetic study of Romet super 30 in Atlantic Salmon（Salmo salar）and in vitro antibacterial activity against Aeromnas salmonicida ［J］. Aquaculture，1997，152（1-4）：13-24.

［29］ Sijm DTHM，Bol J f Seinen W tet al. Ethyl m-aminobenzoate methanesulfonate dependent and carrier dependent pharmacokinetics of extremely lipophilic compounds in rainbow trout ［J］. Arch Environ Contam Toxicol，1993，25（1）：102-109.

［30］ Bjorklund H，Bylund G. Temperature-related absorption and excretion of oxytetracycline in rainbow trout（Salmon gairdneri R）［J］. Aquaculture，1990，84：363-372.

［31］ Curtis LR，Kemp CJ，Svec AV. Biliary excretion of（super（14）C）taurocholate by rainbow trout（5a/mo gairdneri）is stimulated at warmer acclimation temperature ［J］. Comp Biochem Physiol，C，1986，84（1）：87-90.

［32］ Kleinow KM，HH Jarboe，KE Shoemaker. Comparative pharmacokinetics and bioavailability of oxolinic acid in channel catfish（Ictalurus punctatus）and rainbow trout（oncorhynchus mykiss）［J］. Gan J Fish Aquaculture Sci，1994，51：1205-1211.

［33］ Noriko Ishida. Tissue levels of oxolinic acid after oral or intravascular administration to freshwater and seawater rainbow trout ［J］. Aquaculture，1993，102：9-15.

［34］ Sohlber S，Ingebrigtsen K，Hansen MKtet al. Flumequine in Atlanticsalmon（Salmo salar）：disposition in fish held in sea water versus fresh water［J］. Dis Aquat Organ, 2002, 49（1）: 39-44.

［35］ Namdari R, Abedini S, Law FCP. A comparative tissue distribution study of oxytetracycline in raibow trout, Oncorhynchus mykiss（Walbaum）, and Chinook salmon, oncorhynchus tshawytscha（Walbaum）［J］. Aquaculture Research, 1999, 30（4）: 279-286.

［36］ Hansen MK, Horsberg TE. Single-dose pharmacokinetics of flumequine in the eel（Anguilla anguilla）afterintxavascular, oral and bath administration［J］. J Vet Pharmacol Ther, 2000, 23（3）: 169-174.

二、影响抗生素类渔药代谢的因素

水温对氟甲砜霉素在斑点叉尾鮰血浆中的药代动力学有一定影响，随着水温升高，氟甲砜霉素在斑点叉尾鮰血浆中吸收、分布和消除明显，且在斑点叉尾鮰血浆样品中未检测到氟甲砜霉素的主要代谢物氟甲砜霉素胺（刘永涛，2009）。腹注给药方式下甲砜霉素在红笛鲷体内的吸收快于口灌给药（秦青英，2013）。美国红鱼体内注射给药氟苯尼考的吸收快于口灌给药（黄月雄，2014）。甲砜霉素在鲫鱼和鲤鱼体内的吸收分布迅速，但在组织中的消除半衰期有明显差异（杨洪波，2013；2013；2013；2013）。

该文发表于《水生生物学报》【2009，33（1）：1-6】

不同水温下氟甲砜霉素在斑点叉尾鮰体内的药代动力学研究
Pharmacokinetics of Florfenicol in Channel Catfish（*Ictaluruspunc Tatus*）at Different Watertemperatures

刘永涛，艾晓辉，杨红

（中国水产科学研究院长江水产研究所农业部淡水鱼类种质监督检验测试中心，荆州 434000）

摘要： 研究不同水温（18℃和28℃）条件下，单剂量（10mg/kg·bw）强饲氟甲砜霉素，在斑点叉尾鮰（*Ictalurus punctatus*）体内药代动力学特征。采用高效液相色谱紫外检测法可以同时检测血浆中氟甲砜霉素及其代谢物氟甲砜霉素的浓度。用3p97药代动力学软件处理药时数据。结果表明：在不同水温条件下氟甲砜霉素在斑点叉尾鮰体内的药时数据均符合一室开放式模型。药时规律符合理论方程 $C_{血浆} = 7.921（e^{-0.036t} - e^{-0.18t}）$ 和 $C_{血浆} = 9.061（e^{-0.081t} - e^{-0.301t}）$。18℃和28℃的条件下，主要药代动力学参数：吸收半衰期 $T_{1/2(ka)}$ 分别为 3.845 h 和 2.301 h，消除半衰期 $T_{1/2(ke)}$ 分别为 19.118 h 和 8.519 h，达峰时间 $T_{(peak)}$ 分别为 11.136 h 和 5.953 h，最大血药浓度 C_{max} 分别为 4.074 μg/mL 和 4.226 μg/mL，曲线下面积 AUC 分别为 174.547（μg/mL）/h 和 81.279（μg/mL）/h，平均驻留时间 MRT 分别为 27.581h 和 12.290 h，相对表观分布容积 V/F（c）分别为 1.580 L/kg 和 1.512 1 L/kg。采用氟甲砜霉素防治斑点叉尾鮰细菌性疾病，建议在 18℃左右口服 10 mg/kg 体重剂量的氟甲砜霉素，2 d 给药 1 次；在 28℃左右口服 10 mg/kg 体重剂量的氟甲砜霉素，1 d 给药 1 次。试验过程中在斑点叉尾鮰血浆样品中未检测到氟甲砜霉素的主要代谢物氟甲砜霉素胺。

关键词： 氟甲砜霉素；氟甲砜霉素胺；斑点叉尾鮰；药代动力学；水温

氟甲砜霉素（*Florfenicol*，FF）是氯霉素类广谱抗生素，其抗菌活性和抗菌谱优于氯霉素和甲砜霉素。氟甲砜霉素对杀鲑气单胞菌（*Aeromonas salmonicida*），嗜水气单胞菌（*A. hydropila*），迟缓爱德华氏（*Edwardsiella tarda*），鳗弧菌（*Vibrio anguillarum*），杀鱼巴斯德菌（*Pasteurella piscida*）等水产致病菌有较好的抗菌效果。同时氟甲砜霉素人工诱导导的黄鲫鱼（*Seriola quiqueradiata*）假结核病，迟缓爱德华氏菌诱导的鳗鲡（*Anguilla japonica*）迟缓爱德华氏菌病，鳗弧菌诱导的金鱼（*Carassius auratus*）弧菌病，杀鲑气单胞菌诱导的大西洋鲑（*Salmo salar*）疖疮病都有较好的效果[1]。因此，氟甲砜霉素被广泛应用在水产养殖业上。

目前，氟甲砜霉素在斑点叉尾鮰体内的药代动力学研究尚未见报道，本文拟研究不同水温条件下氟甲砜霉素在斑点叉尾鮰体内的药代动力学特征，为制定不同水温条件下氟甲砜霉素的给药方案提供理论依据。

1 材料和方法

1.1 试验鱼

斑点叉尾鮰，体重（128.40±10.51）g，由中国水产科学研究院长江水产研究所窑湾试验场提供，运输至实验室后，用2%的食盐水消毒10 min，然后放置于已清洗并用高锰酸钾消毒水族箱（122 cm×80 cm×80 cm）中暂养1周，试验用水为经活性炭脱氯的自来水，水交换速率20~25 L/h。采用连续充氧，保持水中溶氧大于8.0 mg/L，暂养期间按鱼体重的3%投喂斑点叉尾鮰鱼配合饲料。每日9：00—10：00和16：00—17：00各投喂1次，每天换水1次，暂养结束后，选择无病无伤个体，规格均匀的鮰鱼进行试验。试验期间用加热棒控制水温，低温组水温为（18±1）℃，高温组为（28±1）℃。

1.2 药品与试剂及配制方法

氟甲砜霉素标准品（纯度99.0%，德国Dr. Eherestorfer公司）；氟甲砜霉素胺标准品（FFA，纯度≥95%，sigma-aldrich公司），氟甲砜霉素原料药（批号20050816，纯度99%，武汉九州神农有限公司提供），乙腈、二氯甲烷、磷酸氢二钠（$Na_2HPO_4 \cdot 12H_2O$）、磷酸二氢钠（$NaH_2PO_4 \cdot 2H_2O$）、正己烷（色谱纯或分析纯，上海国药集团）、十二烷基硫酸钠（SDS）、丙酮（分析纯，武汉市江北化学试剂有限责任公司），三乙胺（分析纯，上海建北有机化工有限公司），肝素钠（生物级，上海生化试剂有限公司）。

1.2.1 磷酸缓冲液的配制

磷酸缓冲液（pH 7.0）配制：分别配制0.2 mol/L的磷酸氢二钠和磷酸二氢钠的溶液，然后按61：39的体积比混合即成。

1.2.2 定容溶液的配制

A液：0.01 mol/L的磷酸氢二钠溶液，用磷酸将溶液pH值调至2.8；B液：乙腈。定容溶液为A液：B液（3：2，V/V）。

1.2.3 肝素钠离心管的制备

将肝素钠溶解在0.85%生理盐水中配成1%的肝素钠溶液，然后，向每只塑料离心管加入0.2 mL，振荡10 s，住恒温干燥箱中50℃烘干。

1.3 仪器与设备

高效液相色谱仪（Waters 515泵，717自动进样器，2 475多通道荧光检测器及Empower色谱工作站）；自动高速冷冻离心机（日本HITACHI 20PR-520型）；Mettler-TOLEDO AE-240型精密电子天平（梅特勒-托利多公司）；FS-1高速匀浆机（华普达教学仪器有限公司）；调速混匀器（上海康华生化仪器制造厂）；HGC-12氮吹仪（HENGAO T&D公司）；恒温烘箱（上海浦东荣丰科学仪器有限公司）。

1.4 药饵制备、给药方式及取样

以商用鮰鱼饲料制备含4 g/kg氟甲砜霉素的颗粒饲料，并按10 mg/kg鱼体重进行强饲。先将投胃管装好加药饲料然后慢慢将投胃管插入鮰鱼的胃内，将一根实心管子从投胃管内径将其中的药饵慢慢推到鮰鱼的胃内。给药后保定1 min，然后将鮰鱼单独放入一个水盆中观察5 min，将反胃回吐药饵的鱼弃去，无回吐的鮰鱼放入水族箱中。试验前停喂24 h在进行给药操作。给药前同批鱼采1次空白血浆，以给药后于0.5、1、2、3、4、5、6、8、10、12、24、48、72、96、120、168 h自尾静脉采血样，每个时间点采样5尾，以肝素化注射器从尾静脉采血，制备血浆，−20℃冰箱内保存备用。

1.5 色谱条件

色谱柱：Waters symmetry C18反相色谱柱（250 mm×4.6 mm，5 μm），流动相：乙腈-磷酸二氢钠溶液（0.01 mol/L，含0.005 mol/LSDS和0.10%三乙胺）（2：3，V/V）用磷酸调pH值至4.5；流速：0.6 mL·min^{-1}；柱温：室温；检测波长为225 nm。

1.6 样品处理

将冷冻保存的血浆室温下自然解冻，摇匀后吸取1.0 mL于10 mL具塞离心管中，加入pH值为7.0磷酸盐缓冲液1 mL置调速混匀器上涡旋振荡混匀1 min，加入3 mL丙酮涡旋振荡混匀1 min，以10 000 r/min离心5 min，将溶液转移到另一支10 mL离心管中，再向溶液中加入2 mL二氯甲烷，涡旋振荡混匀1 min，5 000 r/min离心5 min，弃去上层水相，置45℃氮吹仪上吹干，用1 mL定容溶液溶解残渣，涡旋振荡混匀1 min，加入3 mL正己烷涡旋振荡混匀1 min，10 000 r/min离心10 min，弃去正己烷，重复去脂一遍，水相经0.22 μm滤头过滤，HPLC测定。

1.7　标准曲线制备及方法确证

分别准确称取（精确至 0.000 1 g）氟甲砜霉素及氟甲砜霉素胺标准品，用乙腈配制成 100 μg/mL 标准储备液，－20℃冰箱中冷藏保存。测定时以储备液为基础，分别从两种储备液中取一定体积的标准溶液混合后稀释成含氟甲砜霉素和氟甲砜霉素胺浓度均为 10、25、50、100、250、500、1 000、2 500、5 000、10 000 ng/mL 的混合标准液，进样量为 20 μL；分别以氟甲砜霉素和氟甲砜霉素胺的峰面积（A_i）为纵坐标，质量浓度（C）为横坐标绘制标准曲线，求出回归方程和相关系数。在 1 mL 空白血浆中加入已知浓度的系列标准溶液，以类似的方法绘制 FF 和 FFA 的峰面积（A_i）对质量浓度（C）的标准曲线，并求出回归方程和相关系数。在斑点叉尾鮰空白血浆分别添加 5 个浓度水平的 FF 和 FFA 混合标准溶液：使样品浓度为 0.10、0.20、0.50、1.00、2.00 mg/kg 每个浓度做 3 个平行，每个平行设一个空白对照，按样品前处理过程处理后测定回收率。方法精密度测定：在空白血浆中分别添加 2 个浓度水平的 FF 和 FFA 混合标准溶液，使浓度为 0.1、1 mg/kg。每个浓度的样品，在日内每个浓度做 5 个重复，在 1 周内每个浓度做重复做 3 次，计算日内及日间精密度。

2　数据处理

标准曲线，药物经时曲线图采用 Microsoft Excel 2003 软件绘制；药物动力学模型拟合及参数计算采用中国药理学会数学专业委员会编制的 3P97 药代动力学程序软件处理。

3　结果

3.1　色谱特征

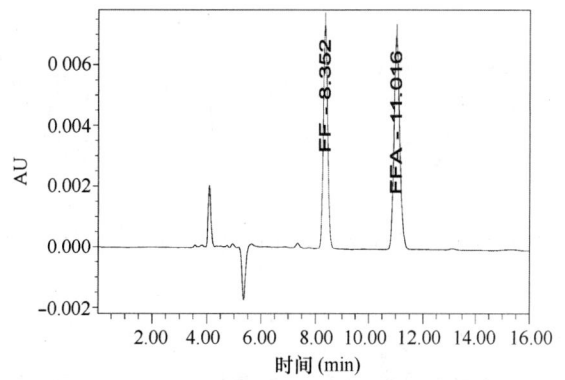

图 1　1 μg/mL FF，FFA 混合标准溶液色谱图

Fig. 1　Chromatography of 1 μg/mL

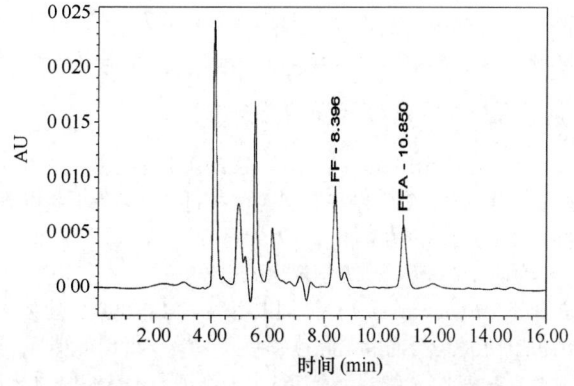

图 2　1 μg/mL 斑点叉尾鮰血浆加标样品色谱图

Fig. 2　Chromatography of 1 μg/mLFF，FFA spiked mixed standard solutionchannel catfish plasma sample

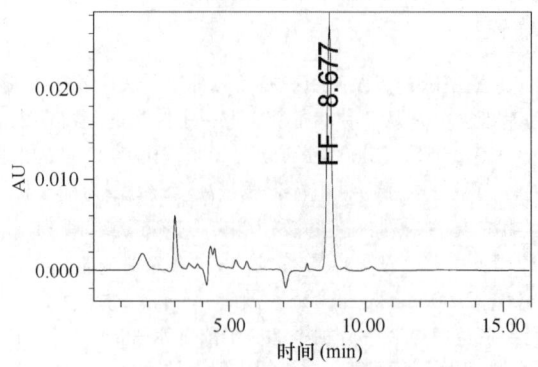

图 3 18℃ 24 h 时斑点叉尾鮰血浆样品色谱图

Fig. 3 Chromatogram of channel catfish plasma sample at 24 hoursat 18℃

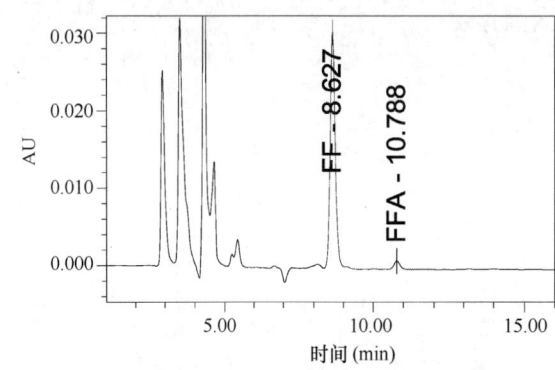

图 4 18℃ 24 h 时斑点叉尾鮰肌肉样品色谱图

Fig. 4 Chromatogram of channel catfish muscule sample at 24 hoursat 28℃

3.2 方法线性范围、回收率、检测限

氟甲砜霉素和氟甲砜霉素胺的标准溶液在 10~10 000 ng/mL 浓度范围内呈线性相关，回归方程及相关系数分别为 $Y=0.0118X+32.205$，$R=0.9997$；$Y=0.0093X+34.601$，$R=0.9997$；血浆工作曲线及相关系数分别为 $Y=0.0135X+35.138$ $R=0.9999$，$Y=0.0103X+72.368$，$R=0.9997$；以 0.10 μg/mL、0.20 μg/mL 和 1.0 μg/mL 3 个添加浓度测定血浆样品中氟甲砜霉素和氟甲砜素胺的回收率分别为（78.81±0.047932）%，（84.25±0.043183）%，（91.14±0.013939）% 和（78.29±0.019238）%，（80.59±0.033924）%，（82.48±0.010125）%；日内变异系数为 3.31%、4.30% 和 4.08%、9.04%；日间变异系数 3.14%、8.02% 和 8.68%、9.22%；按 3 倍信噪比（3S/N）计算血浆中氟甲砜霉素最低检测限为 25 μg/kg、20 μg/kg。

3.3 不同水温下 FF 在斑点叉尾鮰体内的药代动力学特征

不同水温（18℃ 和 28℃）条件下，单剂量（10 mg/kg·bw）强饲氟甲砜霉素在斑点叉尾鮰体内药代动力学特征。经药动学分析软件 3p97 进行动力学模型拟合和参数计算。结果表明，在不同水温条件下氟甲砜霉素在斑点叉尾鮰体内的药时数据均符合一室开放式模型，药时规律分别符合理论方程 $C_{血浆}=7.921$（$e^{-0.036t}-e^{-0.180t}$）和 $C_{血浆}=9.061$（$e^{-0.081t}-e^{-0.301t}$）。

3.3.1 不同水温下 FF 在斑点叉尾鮰血浆中药代动力学规律的比较

在 18℃ 和 28℃ 水温条件下，单剂量（10 mg/kg·bw）强饲氟甲砜霉素在斑点叉尾鮰体内的药代动力学结果表明，水温升高，氟甲砜霉素在斑点叉尾鮰血浆中的吸收、分布和消除速率明显加快；28℃ 条件下，给药 0.5 h 后，FF 的血药浓度为（0.23±0.028）μg/mL，3 h 时血药浓度迅速升至（3.42±0.763）μg/mL，5.95 h 时血药浓度达到峰浓度为 4.23 μg/mL，24 h 时迅速下降到（1.53±0.515）μg/mL，96 h 仅为（0.05±0.003）μg/mL，120 h 已经检测不到；18℃ 条件下，给药 0.5 h 和 3 h FF 的血药浓度仅为（0.17±0.013）μg/mL 和（1.59±1.090）μg/mL，11.14 h 时血药浓度达到峰值 4.07 μg/mL，24 h 和 96 h 血药浓度分别为（3.64±0.263）μg/mL 和（0.31±0.086）μg/mL，144 h 仍能检测到且为（0.23±0.0437）μg/mL。血药浓度（见表 1），药时曲线（见图 5）。

表1　在18℃和28℃水温条件下氟甲砜霉素在斑点叉尾鮰血浆中的浓度

Tab. 1　FF concentration in plasma after oral administration of 10mg/kg b·w at 18℃ and 28℃（n=5）

时间（h） Time	18℃ 血浆药物浓度（μg/mL）Plasma concentration	28℃ 血浆药物浓度（μg/mL）Plasma concentration
0.5	0.17±0.013	0.23±0.028
1	0.19±0.032	0.52±0.154
2	0.60±0.300	0.90±0.591
3	1.59±1.090	3.42±0.763
4	2.86±0.639	4.15±1.572
5	3.49±0.619	4.42±0.983
6	4.33±1.155	4.12±1.330
8	4.04±0.307	3.53±1.278
10	3.91±0.412	3.46±1.487
12	3.81±1.437	3.32±0.734
24	3.64±0.263	1.53±0.515
48	1.37±0.391	0.15±0.091
72	0.39±0.128	0.07±0.015
96	0.31±0.086	0.05±0.003
120	0.26±0.073	ND
144	0.23±0.0437	ND
168	ND	ND

注：ND-未检出

Notes：ND-no detection

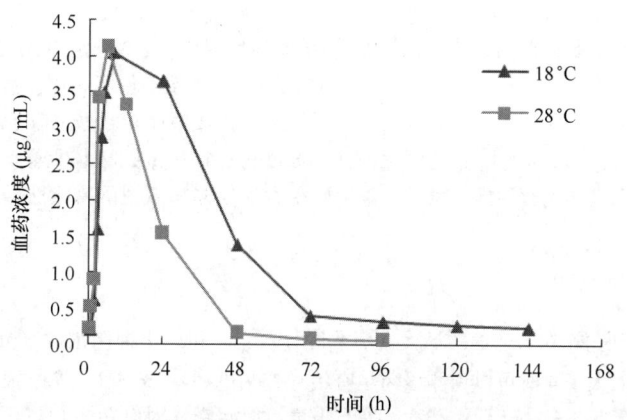

图5　18℃和28℃水温条件下氟甲砜霉素在斑点叉尾鮰体内的药时曲线

Fig. 5　Concentration-time curve offlorfenicol in channel catfish plasma at 18℃（▲）and 28℃（■）

3.3.2　不同水温下FF的药代动学参数

在18℃和28℃水温条件下，水温较高时吸收速率（K_a）和消除速率（K_e）均较快，吸收半衰期（$T_{1/2Ka}$），消除半衰期（$T_{1/2Ke}$）及达峰时间（T_p）较短。而平均驻留时间（MRT）和血药浓度-时间下面积（AUC）仅为18℃的1/2。药代动力学参数（见表2）。

表2　在18℃和28℃条件下单剂量10 mg/kg·bw强饲氟甲砜霉素在斑点叉尾鮰体内的药代动力学参数

Tab. 2　Pharmacokinetic parameters for FF following single oral administration of 10mg/kg·bw at 18℃ and 28℃

参数 Parameters	单位 Unit	18℃	28℃
A	μg/mL	7.921	9.061
Ke	1/h	0.036	0.081
Ka	1/h	0.180	0.301
$T_{1/2(Ka)}$	h	3.845	2.301
$T_{1/2(Ke)}$	h	19.118	8.519
$T_{(peak)}$	h	11.136	5.953
C_{max}	μg/mL	4.074	4.226
AUC	(μg/mL)/h	174.547	81.279
CL/F (s)	L/(h·kg)	0.057	0.123
V/F (c)	L/kg	1.580	1.512
MRT	h	27.581	12.290

4　讨论

4.1　氟甲砜霉素在斑点叉尾鮰体内的药动学特征

在18℃和28℃水温条件下，氟甲砜霉素在斑点叉尾鮰体内药时数据都符合一室开放式模型。与Martinsen等[2]报道的在水温11℃的海水里以10 mg/kg体重单剂量口灌氟甲砜霉素在大西洋鲑体内，朱丽敏等[3]报道的口服氟甲砜霉素在中华鳖体内和Samuelsen等[4]报道以10 mg/kg体重单剂量管饲口服氟甲砜霉素在大西洋鳕体内的药代动力学房室模型一致。28℃和18℃水温条件下，相对表观分布容积$V/F_{(c)}$分别为1.512 L/kg和1.580 L/kg表明氟甲砜霉素在斑点叉尾鮰体内分布广泛，Martinsen等[2]和Horsberg等[5]分别报道氟甲砜霉素在大西洋鲑体内的表观分布容积V_d为1.122 L/kg和1.32 L/kg。28℃水温条件下吸收半衰期$T_{1/2\alpha}$小于18℃水温条件下的吸收半衰期说明水温升高氟甲砜霉素从血液中分布到组织中的速率加快。消除半衰期$T_{1/2\beta}$表明28℃水温条件下氟甲砜霉素在斑点叉尾鮰血浆中消除的速率也比18℃水温条件下要快。达峰时间T_p受温度的影响较大28℃和18℃水温条件下分别为5.953 h和11.136 h，而最高血药浓度C_{max}受温度的影响不大28℃和18℃水温分别为4.226 μg/mL和4.074 μg/mL；在水温18℃时血药曲线下面积AUC和平均驻留时间MRT分别为174.547（μg/mL）/h和27.581 h，而28℃时血药曲线下面积AUC和平均驻留时间MRT分别为81.279（μg/mL）/h和12.290 h，可以看出水温对血药曲线下面积和平均驻留时间影响较大。18℃与28℃水温条件下相比，氟甲砜霉素吸收慢在体内的消除也慢。本试验在斑点叉尾鮰血浆中未检测到氟甲砜霉素的主要代谢产物氟甲砜霉素胺而在肌肉等组织中可检测到氟甲砜霉素胺（色谱图见图3和图4）。Samuelsen等[4]报道在大西洋鳕8℃单剂量10 mg/kg体重静脉注射及单剂量管饲口服氟甲砜霉素，用HPLC检测在血浆也未检测到氟甲砜霉素胺。氟甲砜霉素的代谢物氟甲砜霉素胺的产生可能与鱼的组织和鱼的种类有关。

4.2　给药剂量和给药间隔

制定给药方案时主要考虑到药物在鱼体内的最大血药浓度（C_{max}）和最小抑菌浓度（MIC），当最大血药浓度（C_{max}）与最小抑菌浓度（MIC）的比值大于4时可以预测该药是有治疗作用的[6]。Ho S. P. 等[7]报道氟甲砜霉素对嗜水气单胞菌（A. hydrophila）最小抑菌浓度为0.4 mg/mL；McGinnis等[8]报道氟甲砜霉素对鲇鱼爱德华氏菌（E. ictaluri）最小抑菌浓度为0.25 mg/mL；氟甲砜霉素对鳗弧菌（Vibrio anguillarum）最小抑菌浓度为0.2~0.8 mg/mL[9,10]，对杀鱼巴斯德菌（Pasteurella piscicida）菌的最小抑菌浓度为0.004~0.6 mg/L[4,9]。本试验在18℃和28℃的水温下以10 mg/kg体重的剂量给斑点叉尾鮰强饲氟甲砜霉素。经药代动力学软件3p97模拟，斑点叉尾鮰血浆中氟甲砜霉素的最高血药浓度分别为4.074 μg/mL和4.226 μg/mL与上述氟甲砜霉素对水产常见致病菌的（MIC）的比值大于4，说明该给药剂量是可行的。18℃水温下24 h和48 h时血浆中的氟甲砜霉素浓度分别为（3.64±0.263）μg/mL和（1.37±0.391）μg/mL而28℃下24 h和48 h时血浆中的氟甲砜霉素浓度分别为（1.53±0.515）μg/mL和（0.15±0.091）μg/mL。一般来说，在一定温度范围内，药物的代谢强度与水温成正比，水温越高，代谢速度加快。因此给药时水温是一个应予以考虑的重要因素，尤其是水温较低时停药期应予适当延长[11]。综合考虑以上因素，采用氟甲砜霉素防治斑点叉尾鮰细菌性疾病时，建议口灌剂量为 10 mg/kg 体重水温较低时给药的间隔为 2 d，水温较高时给药间隔 1 d。

5　小结

水温对氟甲砜霉素在斑点叉尾鮰血浆中的药代动力学有一定影响，随着水温升高，氟甲砜霉素在斑点叉尾鮰血浆中吸收、分布和消除明显加快。目前，国外对于氟甲砜霉素在动物体内的研究已不局限于氟甲砜霉素本身而是对氟甲砜霉素及其代谢物进行研究，欧盟和中国将氟甲砜霉素和氟甲砜霉素胺作为水产品中的检出物[1]。国内尚未见有关于动物体中氟甲砜霉素及其代谢物研究的相关报道。本试验在设计时本想采用建立的氟甲砜霉素及其代谢物氟甲砜霉素胺同时检测的高效液相色谱法（另文报道），对氟甲砜霉素及其代谢物氟甲砜霉素胺在血浆中的代谢规律进行研究，但在实际血浆样品的检测中并未检测到氟甲砜霉素的代谢物氟甲砜霉素胺（与 Samuelsen 等[4]的报道相似），但作者在检测斑点叉尾鮰肌肉、肝脏、肾脏、皮肤时均可检测到氟甲砜霉素及其代谢物氟甲砜霉素胺。对于氟甲砜霉素的代谢物在血浆中的代谢机理还需进一步研究。

参考文献

［1］ Xu L W, LIAO C R and LIU G F. Security about florfenicol used in aquaculture ［J］. Journal of Fishery Sciences of China, 2005, 12（4）：512-518 ［徐力文，廖昌容，刘广锋. 氟苯尼考用于水产养殖的安全性. 中国水产科学, 2005, 12（4）：512-518］.

［2］ Martinsen B, Horsberg T E, Varma K J and Sams R. Single-dose pharmacokinetic study of florfenicol in Atlantic salmon（Salmo salar）in sea water at 11℃ ［J］. Aquaculture, 1993, 112：1-1.

［3］ Zhu L M, Yang X L, Lin Q C, et al. The residues and pharmacokinetics of florfenicol in Trionyx sinensis following intramascular injection and oral administration ［J］. 2006, 30（4）：515-519 ［朱丽敏，杨先乐，林启存，等. 肌肉和口服氟苯尼考在中华鳖体内残留分析及药代动力学. 水产学报, 2006, 30（4）：515-519］.

［4］ Samuelsen O B, Bergh O, Ervik A. Pharmacokinetics of florfenicol in cod Gadus morhua and in vitro antibacterial activity against Vibrio anguillarum. Dis Aquat Organ, 2003, 56：127-133.

［5］ Horsberg T E, Hoff K A and Nordmo R. Pharmacokinetics of florfenicol and its metabolite florfenicol amine in Atlantic salmon ［J］. Journal of Aquatic Animal Health, 1996, 8：292-301.

［6］ BSAC, 1991. A guide to sensitivity testing. Report of the working party on antibiotic sensitivity testing of theBritish Society for Antimicrobial Chemotherapy. J. Antimicrob. Chemother 23, 1-47.

［7］ Ho S P, Hsu T Y, Chen M H, et al. Antibacterial effect of chloramphenicol, thiamphenicol and florfenicol against aquatic animalbacteria ［J］. The Journal of Veterinary Medical Science 2000, 62：479-485.

［8］ McGinnis A, Gaunt P, Santucci T, et al. In vitro evaluation ofthe susceptibility ofEdwardsiella ictaluri, etiological agent ofenteric septicemia in channel catfish, Ictalurus punctatus（Rafinesque）, to florfenicol ［J］. J Vet Diagn Invest, 2003, 15：576-579.

［9］ Fukui H, Fujihara Y, Kano T. In vitro and in vivo antibacterial activities of florfenicol, a new fluorinated analog of thiamphenicol against fish pathogens ［J］. Fish Pathology, 1987, 22：201-207.

［10］ Zhao J, Kim E-H, Kobayashi T, et al. Drug resistance of Vibrio angullarum isolated form ayu between 1989 and 1991 ［J］. Nippon Suisan Gakkishi, 1992, 58：1523-1527.

［11］ AI X H, LIU C Zand ZHOU Y T. A study on pharmacokinetic of sulphamethoxazole in grass carp at different temperatures and administration regimes ［J］. Acta Hydrobiologica Sinica, 2005, 29（2）：210-214 ［艾晓辉，刘长征，周运涛. 不同水温和给药方式下磺胺甲噁唑在草鱼体内的药学研究. 水生生物学报, 2005, 29（2）：210-214］.

该文发表于《中国兽药杂志》【2013，47（11）：31-36】

甲砜霉素在红笛鲷体内的组织分布和药代动力学研究

The Distribution and Pharmacokinetics of Thiamphenicolin Tissues of *Luthjanus sancguineus*

秦青英[1,2]，汤菊芬[1,2]，黄郁葱[1,2]，简纪常[1,2*]，黄月雄[1,2]，廖建萌[3]

（1. 广东海洋大学水产学院，广东 湛江 524088；2. 广东省水产经济动物病原生物学及流行病学重点实验室，广东 湛江 524088；3. 广东省湛江市质量计量监督检测所，广东 湛江 524022）

摘要：研究腹腔注射和口灌给药方式下甲砜霉素在红笛鲷体内的组织分布和药代动力学特征。甲砜霉素单剂量

10 mg/（kg b.w）分别腹注和口灌红笛鲷，给药后取血浆、肌肉、肝脏和肾脏，各组织中药物浓度用 HPLC-MS/MS 测定，所得药时数据用 DAS3.0 软件分析。结果显示：两种给药方式下红笛鲷血浆药时数据均符合一级吸收二室模型；血药达峰时间（t_p）分别为 2.39 h 和 4.51 h，血药浓度峰值（C_{max}）分别为 5.80 μg/mL 和 5.06 μg/mL，消除半衰期（$T_{1/2\beta}$）分别为 7.25 h 和 50.09 h。腹注和口灌给药肌肉、肝脏和肾脏的 C_{max} 分别为 2.504 μg/g 和 2.678 μg/g、4.612 μg/g 和 4.393 μg/g、12.464 μg/g 和 7.509 μg/g，T_p 分别为 6.0 h 和 4.0 h、2.0 h 和 4.0 h、1.5 h 和 4.0 h；甲砜霉素在红笛鲷体内消除速度较慢，肌肉、肝脏和肾脏的 $T_{1/2\beta}$ 分别为 37.168 h 和 33.519 h、22.499 h 和 33.649 h、19.672 h 和 8.673 h。腹注给药方式下甲砜霉素在红笛鲷体内的吸收快于口灌给药，在血浆和肝脏中的消除快于口灌给药，在肌肉和肾脏中的消除则慢于口灌给药。

关键词：甲砜霉素；红笛鲷；药代动力学；组织分布

该文发表于《广东海洋大学学报》【2014，34（3）：58-64】

氟苯尼考在美国红鱼体内的药代动力学和组织分布

Tissues Distribution and Pharmacokinetics of Florfenicol in *Sciaenop socellatus*

黄月雄[1,2]，汤菊芬[1,2]，简纪常[1,2]，秦青英[1,2]，廖建萌[3]，蔡佳[1,2]

（1. 广东海洋大学水产学院；2. 广东省水产经济动物病原生物学及流行病学重点实验室，广东 湛江 524088；
3. 广东省湛江市质量计量监督检测所，广东 湛江 524022）

摘要：在水温（20±2）℃的条件下，按 10 mg/kg 的剂量通过腹腔注射和口灌方式，以液-质联用法研究氟苯尼考在美国红鱼（*Sciaenop socellatus*）体内的药代动力学和组织分布，所得药时数据用 DAS3.0 软件分析。结果显示，美国红鱼腹注氟苯尼考后血药经时符合一级吸收二室模型，而口灌氟苯尼考则符合一级吸收一室模型；腹腔注射组，美国红鱼血浆、肝脏、肾脏、肌肉的氟苯尼考出峰时间（t_p）分别为 0.5、4、0.5、4 h，峰浓度（C_{max}）分别为 15.47、9.90、6.83、2.73 μg/g，药时曲线下面积（AUC）分别为 394.10、105.66、157.37、120.01 mg/（L·h），消除半衰期分别为 45.39、7.42、148.25、39.22 h；口灌组，美国红鱼血浆、肝脏、肾脏、肌肉的氟苯尼考出峰时间（t_p）分别为 2、1.5、4、4 h，峰浓度（C_{max}）分别为 6.31、5.44、9.10、5.12 μg/g，药时曲线下面积（AUC）分别为 136.21、162.32、213.32、157.37 mg/（L·h），消除半衰期 $t_{1/2\beta}$ 分别为 12.08、27.9、14.85、38.53 h。结果表明，氟苯尼考注射给药在美国红鱼体内的吸收快于口灌给药，腹注给药氟苯尼考除在肝脏中的消除快于口灌给药外，在血浆、肌肉和肾脏中的消除均慢于口灌给药。

关键词：氟苯尼考；美国红鱼；药代动力学；组织分布

该文发表于《淡水渔业》【2013，43（3）：72-76】

甲砜霉素在鲫鱼体内的药物代谢动力学研究

Pharmacokinetics of Thiamphenicol in *Carassius auratus*

杨洪波[1,2]，王荻[1]，卢彤岩[1]

（1. 中国水产科学研究院 黑龙江水产研究所，黑龙江 150070；2. 上海海洋大学 水产与生命学院，上海 201306）

摘要：采用液相色谱串联质谱法，研究通过单剂量口灌给药，对甲砜霉素在鲫（*Carassius auratus*）体内的药代动力学进行研究，为甲砜霉素在鲫疾病预防和治疗方面提供理论基础。给药剂量为 30 mg/kg 体重，实验水温（20±0.2）℃，鲫平均体重（40.70±7.87）g。取给药后 0.25、0.5、1、1.5、2、4、6、8、12、24、36、48、72 h 鲫的肌肉、血浆、肝脏、肾脏，测定各组织中甲砜霉素的浓度，用药动学 3p97 软件进行数据的处理和分析。结果表明：甲砜霉素在鲫体内吸收分布迅速，符合一级吸收二室开放模型，但消除缓慢。甲砜霉素在鲫血浆、肝脏、肾脏和肌肉中的主要药代动力学参数如下：分布半衰期（$T_{1/2\alpha}$）分别为 1.446 h、1.958 h、7.410 h 和 1.376 h；消除半衰期（$T_{1/2\beta}$）分别为 16.712 h、21.267 h、79.970 h 和 25.600 h；药时曲线下总面积（AUC）分别为 669.073μg/（mL·h）、271.260 μg/（g·h）、3616.060 μg/（g·h）和 158.634 μg/（g.h）。甲砜霉素在水产动物体内吸收快，肌肉和肾脏消除半衰期长，消除缓慢，因此，在该类药物使用时，应相对延长给药间隔时间，避免耐药性的产生。

关键词： 鲫鱼；甲砜霉素；药物代谢动力学

该文发表于《江苏农业科学》【2013，41（6）：192-195】

鲫鱼血清中甲砜霉素对嗜水气单胞菌的体外药效学研究

Pharmacodynamics of Thiamphenicol in Carp serum on *Aeromonas Hydrophila* in *Vitro*

杨洪波[1,2]，王荻[1]，卢彤岩[1]

（1. 中国水科学研究院黑龙江水产研究所，黑龙江 150070；2. 上海海洋大学水产与生命学院，上海 201306）

摘要： 本文采用体内药动学和体外药效学相结合的方法，应用 Excel 2007、药动学 3p97 软件和 kinetica4.4 软件进行数据处理和分析，对鲫鱼血清中甲砜霉素抗嗜水气单胞菌的活性进行研究，为甲砜霉素在水产动物疾病预防和治疗细菌性败血症提供重要的理论依据。结果表明：以 30 mg·kg^{-1} 的剂量对鲫鱼进行单剂量口灌甲砜霉素后，药物在鲫鱼体内吸收迅速，达峰快，消除缓慢。血药达峰时间（T_{peak}）为 1.5h，峰浓度（C_{max}）为 37.172 μg·mL^{-1}，吸收速率（Ka）为 1.523 h，分布半衰期 $T_{1/2(ka)}$ 为 0.455 h，滞后时间（TL）为 0.02 h，消除半衰期 $T_{1/2(ke)}$ 为 16.712 h。在半效应室内，EC_{50} 为 14.28 h。K-PD 同步模型参数 AUC/MIC 血清为 32.41 h，C_{max}/MIC 血清为 23.23。通过抑制效应 Sigmoid E_{max} 模型得到 8.61~46.20 mg·kg^{-1} 为临床使用 TAP 防治鲫鱼细菌性败血症的给药剂量。建议在水产动物中，用甲砜霉素预防和治疗细菌性败血症的最佳给药方案为：以 46.20 mg·kg^{-1} 剂量对发病鲫鱼进行拌饵投喂或口灌给药进行治疗，以 8.61 mg·kg^{-1} 剂量对鲫鱼进行拌饵投喂来预防细菌性败血症的发生。

关键词： 鲫鱼血清；甲砜霉素；嗜水气单胞菌；体外药效学

该文发表于《水产学杂志》【2013，26（2）：49-54】

松浦镜鲤口灌甲砜霉素对嗜水气单胞菌的 PK-PD 模型分析

Pharmacodynamic Effect of Thiamphenicol in Serum of Songpu Mirror Carp on *Aeromonas hydropila in vitro* Pharmacokinetic Model

杨洪波[1,2]，王荻[1]，卢彤岩[1]

（1. 中国水科学研究院黑龙江水产研究所，黑龙江 150070；2. 上海海洋大学水产与生命学院，上海 201306）

摘要： 本文采用体内药动学和体外药效学相结合的方法，应用 Excel 2007、药动学 3p97 软件和 kinetica4.4 软件进行数据处理和分析，研究了鲤鱼血清中甲砜霉素抗嗜水气单胞菌 Aeromonas hydropila 的活性，为甲砜霉素在水产动物疾病预防和治疗细菌性败血症提供理论依据。研究结果表明：给鲤鱼单剂量口灌 30 mg·kg^{-1} 剂量的甲砜霉素后，药物在鲤体内吸收迅速，达峰快，消除缓慢。1.361 h 血药达峰（T_{peak}），峰浓度（C_{max}）为 39.825 μg·mL^{-1}，吸收速率（Ka）为 2.58 h，分布半衰期 $T_{1/2(ka)}$ 为 7.317 h，滞后时间（TL）为 0.234 h，消除半衰期 $T_{1/2(ke)}$ 为 77.292 h。在半效应室内，EC_{50} 为 15.62 h。PK-PD 同步模型参数 AUC/MIC 血清为 35.37 h，C_{max}/MIC 血清为 24.89。通过抑制效应 Sigmoid E_{max} 模型得到 8.52~47.94 mg·kg^{-1} 为临床使用甲砜霉素（TAP）防治鲤鱼细菌性败血症的给药剂量。建议甲砜霉素在预防和治疗细菌性败血症的最佳给药方案为：药饵的治疗剂量为

47. 94 mg·kg^{-1}，预防剂量为 8. 52 mg·kg^{-1}。

关键词：鲤鱼血清；甲砜霉素；嗜水气单胞菌；体外药效学

该文发表于《大连海洋大学学报》【2013，28（3）：298-302】

甲砜霉素在松浦镜鲤体内的药物代谢动力学研究

Pharmacokinetics of Thiamphenicol in Carassius auratus

杨洪波[1,2]，王荻[1]，卢彤岩[1]

（1. 中国水科学研究院 黑龙江水产研究所，黑龙江 150070；2. 上海海洋大学 水产与生命学院，上海 201306）

摘要：采用液相色谱与质谱连用技术（LC-MS），研究了以 30 mg/kg（体质量）的药量在单次口灌给药条件下，甲砜霉素在松浦镜鲤 Cynipus carpio L. 体内的药物代谢动力学。试验水温为（20±0.2）℃，试验鲤（1+龄）体质量为（84.84±17.45）g。在给药后 0.25、0.50、0.75、1.00、1.50、2.00、4.00、6.00、8.00、12.00、24.00、36.00、48.00、72.00 h 采集鲤的肌肉、血浆、肝胰脏和肾脏，测定各组织中甲砜霉素的浓度，用药动学 3P97 软件进行数据分析和处理。结果表明，甲砜霉素在鲤体内吸收分布迅速，符合一级吸收二室开放模型，但消除缓慢。甲砜霉素在鲤血浆、肝胰脏、肾脏和肌肉中的主要药代动力学参数如下：消除半衰期（$T_{1/2\beta}$）分别为 77.292、24.625、79.966、25.600 h；分布半衰期（$T_{1/2\alpha}$）分别为 7.317、0.454、7.409、1.376 h；药时曲线下总面积（AUC）分别为 782.641 μg/（mL·h）、544.756 μg/（g·h）、3616.060 μg/（g·h）和 158.634 μg/（g·h）。鉴于甲砜霉素在鲤血浆和肾脏中的消除半衰期长，消除缓慢，建议使用该类药物时应相对延长给药间隔时间。

关键词：鲤鱼；甲砜霉素；药物代谢动力学

三、影响氟喹诺酮类渔药物代谢的因素

水温对恩诺沙星相对生物利用度有显著影响（W . H. Fang et al.，2008）。低温时双氟沙星能较好地分布于异育银鲫体内，而具有更高的药效（阮记明等，2011；汪文选等，2012）。中华绒螯蟹的性别对盐酸环丙沙星在其体内的代谢有较大的影响，但对盐酸沙拉沙星的影响不大（杨先乐等，2003；彭家红等，2013）。嗜水气单胞菌感染会使药物在动物体内的吸收、分布、代谢和消除受到不同程度的影响（章海鑫等，2012），并且不同的给药方式下药物学特征存在明显差异（李春雨等，2009；汤菊芬等，2012；简纪常等，2005；Ning Xu et al.，2016），史氏鲟感染嗜水气单胞菌后对口服达氟沙星给药方式的影响更为显著（卢彤岩等，2006）。相同注射给药剂量，相比氧氟沙星，左氧氟沙星在中华绒螯蟹体内浓度更高，分布更广（吴冰醒等，2015）。单次口服氧氟沙星后，罗非鱼组织中肝胰脏的药物浓度最高，肾脏组织中药物消除速度最快（王贤玉等，2011）；

该文发表于《Journal of Fish Diseases》【2008，31（3）：171-176】

Effect of temperature on pharmacokinetics of enrofloxacin in mud crab, Scylla serrata（Forsskål），following oral administration

W H Fang[1,2]，L L Hu[1]，X L Yang[2]，K Hu[2]，S C Liang[1] and S Zhou[1]

（1. East China Sea Fisheries Research Institute，Chinese Academy of Fisheries Science，Key and Open Laboratory of Marine and Estuarine Fisheries，Ministry of Agriculture，Shanghai，China；2. Shanghai Fisheries University，Shanghai，China）

Abstract：The study was conducted to evaluate the pharmacokinetics of enrofloxacin following a single oral gavage（10 mg kg^{-1}）in mud crab, Scylla serrata, at water temperatures of 19℃ and 26℃. Enrofloxacin concentration in haemolymph was determined using high-performance liquid chromatography（HPLC）. A multiple and repeated haemolymph sampling from the articular cavity of crab periopods was developed. The haemolymph of an individual crab was successfully sampled up to 11 times from the articular cavity. The profile of haemolymph enrofloxacin concentration of an individual crab versus time was thus achieved. The mean haemolymph enrofloxacin concentration versus time was described by a two-compartment model with first-order absorption at two water temperatures. The peak concentrations of haemolymph enrofloxacin at 19℃ and 26℃ were 7. 26 $\mu g \cdot mL^{-1}$ and 11. 03 $\mu g \cdot mL^{-1}$, at 6 h and 2 h, respectively. The absorption and distribution half-life time ($t_{1/2ka}$ and $t_{1/2\alpha}$) at 19℃ were 3. 7 h and 4. 5 h, respectively, which were markedly larger than the corresponding values (1. 1 h and 1. 5 h) at 26℃; the elimination half-life time ($t_{1/2\beta}$) was 79. 1 h and 56. 5 h at 19℃ and 26℃, respectively. The area under curve (AUC), total body clearance (CLs) and mean residence time ($MRT_{0-\infty}$) at 19℃ were 636. 0 $mg \cdot L^{-1}$ h, 0. 016 L $h^{-1}kg^{-1}$ and 102. 5 h, respectively; the corresponding values at 26℃ were 583. 4 $mg \cdot L^{-1}$h, 0. 018 L $h^{-1}kg^{-1}$ and 63. 7 h. These results indicate that enrofloxacin is absorbed and eliminated more rapidly in mud crab at 26℃ than at 19℃.

Key words: Scylla serrata; enrofloxacin; pharmacokinetics; temperature; haemolymph

温度对恩诺沙星口灌给药下在锯缘青蟹体内药动学的影响

房文红[1,2]，胡琳琳[1]，杨先乐[2]，胡鲲[2]，梁思成[1]，周帅[1]

（1. 中国水产科学研究院东海水产研究所，中国上海；2. 上海海洋大学，中国上海）

摘要：该论文研究了水温 19℃ 和 26℃ 下，恩诺沙星以 10 mg/kg 剂量口灌给药下在锯缘青蟹体内药动学，血淋巴中恩诺沙星浓度采用 HPLC 分析。建立了从蟹步足关节腔重复多次采血的方法，可从单一个体的血腔中连续采血 11 次，获得了个体水平的血药浓度-时间关系曲线。两个水温下，青蟹平均血药浓度-时间关系曲线均适合采用一级吸收二室模型来描述。水温 19℃ 和 26℃ 下，血淋巴恩诺沙星峰浓度分别是 7. 26 $\mu g \cdot mL^{-1}$ 和 11. 03 $\mu g \cdot mL^{-1}$，达峰时间分别为 6 h 和 2 h。19℃ 下，吸收和分布半衰期（$T_{1/2ka}$ and $T_{1/2\alpha}$）分别是 3. 7 h 和 4. 5 h，明显大于 26℃ 时的相应值（分别为 1. 1 h 和 1. 5 h）；19℃ 和 26℃ 的消除半衰期分别是 79. 1 h 和 56. 5 h；19℃ 时的曲线下面积（AUC）、总体清除率（CLs）和平均驻留时间（$MRT_{0-\infty}$）分别是 636. 0 $mg \cdot L^{-1} \cdot h$、0. 016 L $\cdot h^{-1} \cdot kg^{-1}$ 和 102. 5 h，26℃ 时的相应值分别为 583. 4 $mg \cdot L^{-1} \cdot h$、0. 018 L $h^{-1} \cdot kg^{-1}$ 和 63. 7 h。研究结果表明，26℃ 下恩诺沙星吸收和消除均快于 19℃。

关键词：锯缘青蟹；恩诺沙星；药动学；温度；血淋巴

1　Introduction

The mud crab, Scylla serrata (Forsskål), is widely distributed throughout the Indo-Pacific region from Hawaii, southern Japan, southern China, Philippines, to Australia, the Red Sea and East and South Africa (Motoh 1979; Dai & Yang 1991). Because of its economic value, the mud crab farming industry is now expanding rapidly in China, and it has been accompanied by recurrent problems with bacterial infections. The main pathogens isolated from diseased mud crabs were Vibrio cincinnatiensis, V. alginolyticus and V. parahemolyticus (Mao, Zhuo, Yang & Wu 2001), which were susceptive to fluoroquinolone antimicrobials, with in vitro MICs for enrofloxacin of approximately 0. 10-0. 50 $\mu g \cdot mL^{-1}$. Utilizing antimicrobials to reduce such problems has become an important part of mud crab farming.

Enrofloxacin is a quinolone carboxylic acid derivative with antimicrobial action, which is bactericidal through inhibition of DNA gyrase (Vancutsem, Babish & Schwark 1990) and is effective against Gram-negative and Gram-positive bacteria (Hooper & Wolfson 1985). Enrofloxacin is registered for use in aquaculture in China. It has received growing attention because of its potential efficacy for the treatment of diseases in fish (Dalsgaard & Bjerregaard 1991; Bowser, Wooster, St. Leger & Babish 1992; Brown 1996; Burka, Hammell, Horsberg, Johnson, Rainnie & Speare 1997; Lewbart, Vaden, Deen, Manaugh, Whitt, Doi, Smith & Flammer 1997; Stoffregen, Wooster, Bustos, Bowser & Babish 1997). Previous enrofloxacin pharmacokinetic studies were performed mainly with finfish. These included red pacu, Colossoma brachypomum (Cuvier) (Lewbart et al. 1997), rainbow trout, Oncorhynchus mykiss (Walbaum) (Bowser et al. 1992), seabream, Sparus aurata L. (Rocca, Salvo, Malvisi & Sello 2004), seabass, Dicentrarchus labrax (L.) (Intorre, Cecchini, Bertini, Varriale, Soldani & Mengozzi 2000), and Atlantic salmon, Salmo salar L. (Stoffregen et al. 1997). It has also been shown to be beneficial in crustacean aquaculture (Fang, Wang & Li 2004; Tang, Yang, Zheng, Yu, Hu & Yu 2006; Wu, Yong, Zhu & Cheng 2006).

Most invertebrates, including crab and shrimp as well as most fish, are ectothermic animals and studies on the effect of temperature on pharmacokinetics in rainbow trout (Björklund, Eriksson & Bylund 1992; Sohlberg, Aulie & Soli 1994), channel catfish, Ictalurus punctatus (Rafinesque) (Kleinow, Jarboe & Shoemaker 1994), sea bass (Rigos, Alexis, Andriopoulou & Nengas 2002a, b) and crucian carp, Carassius auratus (L.) (Ding, Cao, Ma, Pan, Fang & Lu 2006) revealed that temperature affects the absorption, distribution or elimination of drugs.

The present study was carried out in order to obtain information on the absorption and elimination of enrofloxacin in mud crab following a single oral gavage at different water temperatures.

2 Materials and methods

2.1 Experimental animals and chemicals

Two temperature experiments were performed on the pharmacokinetics of enrofloxacin in mud crab. Healthy subadult mud crabs (weight 120−150 g) were captured from Hangzhou Bay, Shanghai, China. The crabs were divided into two experimental groups and reared in twelve 40−L fibreglass tanks with 1 animal per tank. Six tanks were kept at 19.0±0.5℃, and the others at 26.0±0.5℃, supplied with recirculated and filtered brackish water of 8‰ salinity. The water renewal rate was 30%~50% a day with constant aeration. The crabs were fed twice a day with artificial feed at 2%~3% of their average body weight. After a 2−week acclimatization period, the crabs were administered enrofloxacin by oral gavage.

2.2 Chemicals

Standard enrofloxacin was purchased from Sigma (St Louis, MO, USA). Prototype enrofloxacin was supplied by Shanghai Sunve Pharmaceutical Co., Ltd. The active ingredient was determined by high performance liquid chromatography (HPLC) and found to exceed 98.5%. The drug was diluted in 1% acetic acid prior to use and was stored in a refrigerator in a dark bottle. The organic solvent acetonitrile was of HPLC grade (Sigma−Aldrich, Germany). All other chemicals were of at least analytical grade.

2.3 Oral administration

The crabs were given a single dose of 0.5 mL · kg^{-1} body weight of an aqueous solution containing enrofloxacin at 20 g · L^{-1}, corresponding to a final dose of 10 mg · kg^{-1} body weight. Enrofloxacin solution was forced into the stomach of the crab with a 1 mL syringe fitted with a 7$^{\#}$ gauge needle. After oral gavage, crabs which did not blow out foam were considered not to have regurgitated and only these animals were used in the experiment. Crabs were returned back to their tanks 5 min after administration.

Prior to administration of enrofloxacin, crabs were not fed for a 48−h period. Maintenance feeding at a rate of 2%~3% crab body weight per day, was restarted 24 h post−dosing and continued twice a day for the duration of the study.

2.4 Sample collection

Multiple haemolymph samples were obtained from individual crabs. Haemolymph samples (0.3 mL) were removed repeatedly from the articular cavity between the merus and carpopodium of the second periopod with a 1−mL syringe (5$^{\#}$ gauge needle). Samples were immediately placed into vials containing the same volume of acid citrate dextrose anticoagulant (ACD) consisting of 45 mM citrate acid, 20 mM trisodium citrate and 82 mM glucose (Cai, Li & Li 1983) and vortexed. All samples were frozen and maintained at −70℃ until extraction and HPLC analysis.

The time intervals for haemolymph sampling after administration of enrofloxacin dose were 0.5, 1, 2, 4, 6, 12 h and 1, 2, 4, 6 and 8 days post−dosing to ensure an adequate number of data points during the elimination phase. At each time six crabs were sampled.

2.5 Analytical procedures

Enrofloxacin concentrations in the haemolymph of mud crab were determined by HPLC analysis performed with an Agilent 1100 system. The column was a Zorbax SB C−18, 4.6×150 mm.

The mobile phase consisted of acetonitrile：phosphate buffer (14.5：85.5, v/v). The phosphate buffer contained 10mmol L^{-1} NaH$_2$PO$_4$ and 5mmol L^{-1} tetrabutylammonium bromide; the pH was adjusted to 2.0 with 25% phosphate acid. The phosphate buffer was filtered through a 0.45 μm Millipore filter and sonicated for 5 min before use. The flow rate was 1.0 mL · min^{-1}. The column temperature was maintained at 40℃. The injection volume was 10 μL and a fluorescence detector with Ex 280 nm and Em 418nm was used.

Haemolymph was thawed at room temperature and vortexed for 2 min prior to ensure homogeneity and then centrifuged for 5 min at 5000 rpm. Enrofloxacin was extracted from 0. 5 mL haemolymph supernatant after the addition of 5 mL acetonitrile. The mixture was vortexed for 2 min and centrifuged for 10 min at 5000 rpm at 4℃. The clear supernatant was dried at 40℃ using a vortex evaporator (Botongc Co. Ltd. , Shanghai, China) and reconstituted in 1. 0 mL of mobile phase. The samples were filtered through 0. 45 μm disposable syringe filters prior to analysis.

The standard calibration curves, recovery and precision of the method for enrofloxacin were similar to those of Tang et al. (2006) . The standard curve for enrofloxacin was linear in the range 0. 01-10. 00 μg · mL^{-1}. The limit of detection and limit of quantitation for enrofloxacin were 0. 005 μg · mL^{-1}and 0. 01 μg · mL^{-1}, respectively. The recovery obtained from haemolymph ranged from 92. 3% to 98. 1%, and the percentage relative standard deviations for inter-and intra-day determinations were 3. 63% and 4. 76%, respectively.

2. 6　Pharmacokinetic analysis

Pharmacokinetic non-compartmental modelling of the drug concentration-time data was performed using DAS Version 2. 0 (Drug and Statistics Software were developed by Mathematical Pharmacology Professional Committee of China and authorized and supported by China State Drug Administration) .

3　Results

The best curve fitting of the data set was described by a two-compartment open pharmacokinetic model with a first-order absorption. The mean haemolymph concentration-time curve was calculated from the following equation:

$$C_t = A \cdot e^{-\alpha(t-tlag)} + B \cdot e^{-\beta(t-tlag)} - K \cdot e^{-Ka(t-tlag)}$$

where $C_t$is the haemolymph mean concentration of enrofloxacin at time t; A, B and K are the coefficients of the compartments; α, β and Ka are the constants of the compartments, and t_{lag}represents the lag time. The mean haemolymph concentrations of enrofloxacin in six individual animals over time at 19℃ and 26℃ are given in Table 1. The survival rate was 100% during the experiment. After oral gavage, the enrofloxacin mean haemolymph concentration-time curve showed typical absorption and elimination phases for the two temperature groups. At 19℃, the mean peak concentration was 7. 26 μg · mL^{-1}, occurring at 6 h; the corresponding values at 26℃ were 11. 03 μg · mL^{-1} and 2 h (Table 1) . The biexponential curve indicated a two-compartment pharmacokinetic model with a fast distribution phase followed by a slow elimination phase. At 19℃ there was a lag time of 0. 026 h.

Table 1　Mean enrofloxacin concentration (mean±SD) in hemolymph (μg · mL^{-1}) following oral gavage of 10 mg · kg^{-1} of enrofloxacin at two water temperatures (n=6)

Time (h)	19 ℃	26 ℃
0. 5	1. 42±0. 40	8. 59±2. 23
1	1. 81±0. 74	9. 96±2. 08
2	2. 25±0. 36	11. 03±3. 50
4	4. 41±0. 62	9. 83±1. 58
6	7. 26±0. 93	7. 79±0. 92
12	5. 67±0. 88	5. 89±1. 41
24	4. 88±0. 66	5. 00±1. 23
48	3. 85±0. 95	4. 48±0. 96
96	2. 37±0. 69	1. 87±0. 55
144	1. 52±0. 55	0. 98±0. 36
192	0. 99±0. 39	0. 46±0. 14

The pharmacokinetic parameters derived from the compartment model are shown in Table 2. The absorption rate constant (Ka) was 0. 19 and 0. 64 h^{-1} at 19℃ and 26℃, respectively. At 19℃, the absorption, distribution and elimination half-life ($t_{1/2Ka}$, $t_{1/2\alpha}$ and $t_{1/2\beta}$) were 3. 7 h, 5. 0 h and 79. 1 h, respectively; the corresponding values at 26℃ were 1. 1 h, 1. 5 h and 56. 5 h, respectively. The rate of distribution of drug from compartment 1-2 at the higher temperature was four times that at the lower temperature, while the rate of backtransfer (K_{21}) was 2. 6 times higher. There was a minor difference observed in the AUC and CLs between the two temperature groups.

Table 2 Pharmacokinetic parameters of enrofloxacin derived from compartment model in S. serrata following in oral administration of 10 mg · kg⁻¹ at 19℃ and 26℃

Pharmacokinetic parameters	19 ℃	26 ℃	Pharmacokinetic parameters	19 ℃	26 ℃
α (h⁻¹)	0.140	0.453	CLs (L h⁻¹ kg⁻¹)	0.016	0.018
β (h⁻¹)	0.009	0.012	K_{10} (h⁻¹)	0.017	0.030
Ka (h⁻¹)	0.188	0.644	K_{12} (h⁻¹)	0.060	0.253
$T_{1/2\alpha}$ (h)	4.957	1.531	K_{21} (h⁻¹)	0.071	0.182
$T_{1/2\beta}$ (h)	79.083	56.521	Tlag (h)	0.026	0
$T_{1/2ka}$ (h)	3.682	1.076	C_{max} (μg · mL⁻¹)	7.26	11.03
$V_{d/F}$ (L kg⁻¹)	1.637	1.111	T_{max} (h)	6.0	2.0
AUC 0-∞ (mg · L⁻¹ h)	636.0	583.4			

α: distribution rate constant; β: elimination rate constant; Ka: the first-order absorption rate constant; $t_{1/2\alpha}$: distribution half-life of the drug; $t_{1/2\beta}$: elimination half-life of the drug; $t_{1/2Ka}$: absorption half-life of the drug; $V_{d/F}$: apparent volume of distribution; AUC: area under curve; CLs: total body clearance of the drug; K_{12}, K_{21}: first-order rate constants for drug distribution between the central and peripheral compartments; K_{10}: elimination rate constant from the central compartment; Tlag: lag time before absorption; C_{max}: the maximum hemolymph concentration; T_{max}: the time point of maximum enrofloxacin concentration.

Discussion

This is the first study on the pharmacokinetics of enrofloxacin in mud crab following oral gavage. The enrofloxacin concentration-time curve was well described by a two-compartment open model with first-order absorption following oral gavage at both water temperatures, which was in accord with the results for enrofloxacin in other crustaceans, such as Chinese mitten crab (Tang et al. 2006; Wu et al. 2006) and shrimp (Fang et al. 2004) following intramuscular injection. With regard to oral administration, the plasma concentration-time data best fitted an open two-compartmental model only in seabass (Intorre et al. 2000). In several other studies performed on finfish, such as Atlantic salmon (Stoffregen et al. 1997), red pacu (Lewbart et al. 1997) and seabream (Rocca et al. 2004), the plasma concentration-time curve could not be described by a compartment model. Whether a compartment or non-compartment model was used to describe the experimental data in previous work may be due to the different species involved and the different route of administration.

The absorption half-life ($t_{1/2Ka}$) and distribution half-life ($t_{1/2\alpha}$) in mud crab were markedly affected by water temperature. At 19℃, the $t_{1/2Ka}$ and $t_{1/2\alpha}$ of enrofloxacin were three times longer than at 26℃. In Chinese mitten crab following intramuscular injection, the $t_{1/2\alpha}$ at a dose of 10 mg · kg⁻¹ at 25℃ (Tang et al. 2006) was about three times shorter than that at a dose of 5 mg · kg⁻¹ at 17℃ (Wu et al. 2006). A shorter $t_{1/2\alpha}$ or $t_{1/2Ka}$ at higher temperature has also been reported for oxolinic acid in rainbow trout (Björklund et al. 1992), channel catfish (Kleinow et al. 1994) and sea bass (Rigos et al. 2002b), for oxytetracycline in sea bass (Rigos et al. 2002a) and for difloxacin in crucian carp (Ding et al. 2006). In the present study, the effect of temperature was also apparent in the relative magnitudes of the distribution rate constant from the central to the peripheral compartment (K_{12}) and from the peripheral to the central compartment (K_{21}), where the value at a high temperature is increased by almost 4-fold and 2.6-fold, respectively. A great K_{12} and K_{21} at high temperature has also been reported in sea bass (Rigos et al. 2002b).

There was a difference in the elimination half-life ($t_{1/2\beta}$) of enrofloxacin in mud crab with temperature. The $t_{1/2\beta}$ at 19℃ (79.1 h) was estimated to be longer than at 26℃ (56.5 h). The present study indicated that water temperature had less effect on elimination than on absorption and distribution. This finding is in agreement with the results for oxolinic acid in catfish (Kleinow et al. 1994), but contrasts with the results of Rigos et al. (2002a, b). Kleinow et al. (1994) and Rigos et al. (2002) suggested that a variety of factors may account for this temperature-dependent relationship such as cardiac output (Randall 1970), membrane lipid composition (Hazel 1984) and xenobiotic biotransformation (Kleinow, Melancon & Lech 1987).

A minor difference was observed in the AUC and CLs between the two temperature groups, similar to the result for oxytetracycline in sea bass (Rigos et al. 2002a), but in disagreement with the findings for difloxacin in crucian carp (Ding et al. 2006) and for flumequine in rainbow trout (Sohlberg et al. 1994).

In the present study, the haemolymph kinetic profile of enrofloxacin after gavage of 10 mg $\cdot$ kg^{-1} at both temperatures differed from that observed in Atlantic salmon (Stoffregen et al. 1997) or seabream (Rocca et al. 2004) given the same dose of the drug. In particular, oral gavage with enrofloxacin in mud crab resulted in much higher haemolymph concentrations than in Atlantic salmon (Stoffregen et al. 1997) and seabream (Rocca et al. 2004), while it showed a shorter T_{max} and higher AUC values in these fish species. The C_{max} after oral gavage in mud crab was close to that in Chinese mitten crab following intramuscular injection with the same drug (Tang et al. 2006; Wu et al. 2006). Enrofloxacin was thus rapidly absorbed in mud crab following oral administration.

Some researchers have suggested that differences in certain pharmacokinetic parameters such as C_{max}, T_{max} and AUC among species might be explained by differences in anatomical volumes and plasma protein and tissue binding of a drug (Oie & Tozer 1979; Barron, Gedutis & James 1988). We suggest that differences in certain pharmacokinetic parameters between crustaceans and finfish might also be due to differences in circulatory systems. Crabs, as other crustaceans, have an open circulatory system, as opposed to closed systems in finfish. In an open circulatory system, the blood or haemolymph flows freely throughout the haemocoelic cavity. Circulation of the haemolymph may result from body movements, muscular and gut contractions, or through the pumping action of the heart. The haemolymph is pumped into an interconnected system of sinuses, surrounding the organs (Bliss 1983).

For fluoroquinolones, which have a concentration-dependent bactericidal mechanism, the peak concentration/MIC ratio is considered one of the best indicators of efficacy (Intorre et al. 2000). Mean MIC values of enrofloxacin against mud crab pathogens ranged from 0.10 μg $\cdot$ mL^{-1} for Vibrio cincinnatiensis and 0.50 μg $\cdot$ mL^{-1} for V. alginolyticus to 0.25 μg $\cdot$ mL^{-1} for V. parahemolyticus. Using an MIC value of 0.50 μg $\cdot$ mL^{-1}, a single oral dose of enrofloxacin yielded a Cmax/MIC of 14.5 at 19℃ and 22 at 26℃. According to other studies (Drusano, Johnson, Rosen & Standiford 1993; Haritova, Lashev & Pashov 2003), a ratio of C_{max}/AUC = 8-10 is a breakpoint for clinical efficacy of fluoroquinolones. Values of enrofloxacin obtained after oral administration in this study may well be efficacious against vibrios.

In conclusion, the results of this study indicated that the absorption and distribution of enrofloxacin in mud crab were largely influenced by water temperature, but the effect of temperature on elimination was less. However, there was a minor effect of water temperature on CLs and AUC, which remains to be further studied. This study suggests a single dose of 10 mg $\cdot$ kg^{-1} body weight of enrofloxacin after oral administration might be sufficient for a therapeutic concentration in the haemolymph, regardless of the water temperature.

Acknowledgement

This work was supported financially by the Natural Science Foundation of Chinese Academy of Fisheries Science (2004-4-3).

References

B arron M. G., Gedutis C. & James M. O. (1988) Pharmacokinetics of sulphadimethoxine in the lobster, Homarus americanus, following intrapericardial administration. Xenobiotica18, 269-277.

Björklund H. V., Eriksson A. & Bylund G. (1992) Temperature-related absorption and excretion of oxolinic acid in rainbow trout (Oncorhynchus mykiss). Aquaculture102, 17-27.

Bliss D. E. (1983) The Biology of Crustacea, Volume 5, Internal Anatomy and Physiological Regulation (ed. by L. H. Mantel), pp. 3-19. Academic Press, London.

Bowser P. R., Wooster G. A., St. Leger J. & Babish J. G. (1992) Pharmacokinetics of enrofloxacin in fingerling rainbow trout Oncorhynchus mykiss. Journal of Veterinary Pharmacology and Therapeutics15, 62-71.

Brown S. A. (1996) Fluoroquinolones in animal health. Journal of Veterinary Pharmacology and Therapeutics19, 1-14.

Burka J. F., Hammell K. L., Horsberg T. E., Johnson G. R., Rainnie D. J. & Speare D. J. (1997) Drugs in salmonid aquaculture-a review. Journal of Veterinary Pharmacology and Therapeutics 20, 333-349.

Cai W. C., Li B. Y. & Li Y. M. (1983) The Technique Guide for Biochemical Experiment. Fudan University Press, Shanghai, China.

Dai A. Y. & Yang S. L. (1991) Crabs of the China Seas. China Ocean Press, Beijing, China.

Dalsgaard I. & Bjerregaard J. (1991) Enrofloxacin as an antibiotic in fish. ActaVeterinaria Scandinavica Suppl. 87, 300-302.

Ding F. K., Cao J. Y., Ma L. B., Pan Q. S., Fang Z. P. & Lu X. C. (2006) Pharmacokinetics and tissue residues of difloxacin in crucian carp (Carassius auratus) after oral administration. Aquaculture256, 121-128.

Drusano G. L., Johnson D. E., Rosen M. & Standiford H. C. (1993) Pharmacodynamics of fluoroquinolone antimicrobial agent in neutropenic rat model of Pseudomonas sepsis. Antimicrobial Agents and Chemotherapy37, 483-490.

Fang X. X., Wang Q. & Li J. (2004) Pharmacokinetics of enrofloxacin and its metabolite ciprofloxacin in Fenneropenaeus chinensis. Journal of Fisheries of China28 (suppl.), 35-41.

Haritova A., Lashev L. & Pashov D. (2003) Pharmacokinetics of enrofloxacin in lactating sheep. Research in Veterinary Science74, 241-245.

Hazel J. R.（1984）Effects of temperature on the structure and metabolism of cell membranes in fish. American Journal of Physiology246，460-470.

Hooper D. & Wolfson J.（1985）The fluoroquinolones：structures，mechanisms of action and resistance and spectra of activity in vitro. Antimicrobial Agents and Chemotherapy28，581-586.

Intorre L.，Cecchini S.，Bertini S.，Varriale A. M. C.，Soldani G. & Mengozzi G.（2000）Pharmacokinetics of enrofloxacin in the seabass（Dicentrarchus labrax）. Aquaculture182，49-59.

Kleinow K. M.，Melancon M. J. & Lech J. J.（1987）Biotransformation and induction：implications for toxicity，bioaccumulation and monitoring of environmental xenbiotics in fish. Environmental Health Perspectives71，105-119.

Kleinow K. M.，Jarboe H. H. & Shoemaker K. E.（1994）Comparative pharmacokinetics and bioavailability of oxolinic acid in channel catfish（Ictalurus punctatus）and rainbow trout. Canadian Journal of Fisheries and Aquatic Sciences 51，1205-1211.

Lewbart G.，Vaden S.，Deen J.，Manaugh C.，Whitt D.，Doi A.，Smith T. & Flammer K.（1997）Pharmacokinetics of enrofloxacin in the red pacu Colossoma brachypomum after intramuscular，oral and bath administration. Journal of Veterinary Pharmacology and Therapeutics 20，124-128.

Mao Z. J.，Zhuo H. L，Yang J. F. & Wu X. F.（2001）Studies on pathogen of bacteria epidemic against mud crab Scylla serrata. Journal of Oceanography in Taiwan Strait20，187-192.

Motoh H.（1979）Edible crustaceans in the Philippines，11th in a series. Asian Aquaculture2，5pp.

Oie S. & Tozer T. N.（1979）Effect of altered plasma protein binding on the apparent volume of distribution. Journal of Pharmacological Science 68，1203-1210.

Randall D. J.（1970）The circulatory system. In：Fish Physiology：The Nervous System，Circulation and Respiration，Vol. IV（ed. by W. S. Hoar & D. J. Randall），pp. 133-172. Academic Press，New York.

Rigos G.，Alexis M.，Andriopoulou A. & Nengas I.（2002a）Pharmacokinetics and tissue distribution of oxytetracycline in sea bass，Dicentrarchus labrax，at two water temperatures. Aquaculture210，59-67.

Rigos G.，Alexis M.，Andriopoulou A. & Nengas I.（2002b）Temperature-dependent pharmacokinetics and tissue distribution of oxolinic acid in sea bass，Dicentrarchus labrax L.，after a single intravascular injection. Aquaculture Research33，1175-1181.

Rocca G. D.，Salvo A. D.，Malvisi J. & Sello M.（2004）The disposition of enrofloxacin in seabream（Sparus aurata L.）after single intravenous injection or from medicated feed administration. Aquaculture232，53-62.

Sohlberg S.，Aulie A. & Soli N. E.（1994）Temperature-dependent absorption and elimination of flumequine in rainbow trout（Oncorhynchus mykiss Walbaum）in fresh water. Aquaculture119，1-10.

Stoffregen D. A.，Wooster G. A.，Bustos P. S.，Bowser P. R. & Babish J. G.（1997）Multiple route and dose pharmacokinetics of enrofloxacin in juvenile Atlantic salmon. Journal of Veterinary Pharmacology and Therapeutics 20，111-123.

Tang J.，Yang X. L.，Zheng Z. L.，Yu W. J.，Hu K. & Yu H. J.（2006）Pharmacokinetics and the active metabolite of enrofloxacin in Chinese mitten-handed crab（Eriocheir sinensis）. Aquaculture260，69-76.

Vancutsem P. M.，Babish J. G. & Schwark W. S.（1990）The fluoroquinolone antimicrobials：structure，antimicrobial activity，pharmacokinetics，clinical use in domestic animals and toxicity. Cornell Veterinarian 80，173-186.

Wu G..H.，Yong M.，Zhu X. H. & Cheng H.（2006）Pharmacokinetics and tissue distribution of enrofloxacin and its metabolite ciprofloxacin in the Chinese mitten-handed crab，Eriocheir sinensis. Analytical Biochemistry358，25-30.

该文发表于《上海海洋大学学报》【2011，20（6）：858-865】

两种水温条件下异育银鲫体内双氟沙星药代动力学比较

Pharmacokinetics comparisons of difloxacin in crucian carp (*Carassaisauratus gibebio*) at two different water temperatures

阮记明[1,2]，胡鲲[1]，章海鑫[1]，王会聪[1]，付乔芳[1]，杨先乐[1]

（1. 上海海洋大学国家水生动物病原库，上海 201306；2. 江西农业大学动物科学技术学院，江西南昌 330045）

摘要：运用反相高效液相色谱法（RP-HPLC），对比了单次口灌 20 mg/kg 鱼体重剂量条件下，16℃和25℃两种水温时双氟沙星（DIF）在异育银鲫体内代谢动力学差异。数据经过药代动力学软件（3p97）处理，结果显示：两种温度条件下，双氟沙星药时规律均符合开放性二室模型且存在明显的差异。其中，吸收半衰期（$T_{1/2Ka}$）与水温呈正相关，在两种温度条件下分别可达 0.010 h 和 0.122 h；消除半衰期（$T_{1/2\beta}$）、表观分布容积（V_d/F）、血浆和组织中药时曲线下总面积（AUC）等参数则与水温呈负相关，16℃条件下血浆 $T_{1/2\beta}$、V_d/F 和 AUC 分别为 70.968 h、1.875/kg 和 763.761 μg/（mL·h），25℃则为 18.322 h、0.676/kg 和 243.244 μg/（mL·h）；16℃条件下肝脏组织中 AUC 为 746.622 μg/（mL·h），肾脏组织中 AUC 为 1 095.711 μg/（mL·h），肌肉组织中 AUC 为 1 222.750 μg/（mL·h）；而 25℃ 条件下肝脏组织中 AUC 则为 294 857 μg/（mL·h），肾脏组织中 AUC 为 258.587 μg/（mL·h），肌肉组织中 AUC 为 344.630 μg/（mL·h）。另外，在两种温度条件下，DIF 代谢产物沙拉沙星在血浆及肝脏、肾脏和肌肉组织中均能被检出，且呈多峰现象。结果说明，相对于 25℃ 的水温条件，16℃ 水温时双氟沙星能较好地分布于异育银鲫体内。另外，16℃ 时 DIF $C_{max}/MIC_{90} \geqslant 10$ 的保持时间可达 72 h，而 25℃ 时 $C_{max}/MIC_{90} \geqslant 10$ 的保持时间仅为 4~6 h；因此，双氟沙星在较低水温时（16℃）具有更高的药效。研究亮点：（1）比较了不同水温条件下 DIF 在异育银鲫体内的药动学差异。在我国，不同地域、不同季节的水温变化和差异均很大，因此有必要对不同水温下的药动学进行比较研究，以期为不同水温下的用药提供理论参考；（2）从药物本身及其代谢产物两个层面整体考虑药效及药残等。

关键词：水温；异育银鲫；双氟沙星；沙拉沙星；药代动力学

该文发表于《黑龙江畜牧兽医》【2012，（2）：129-132】

两种水温条件下恩诺沙星在鲫鱼体内的药动学比较

Pharmacokinetics of enrofloxacin in crucian carp at 20℃ and 25℃ temperature

汪文选[1,2]，卢彤岩[2]，王荻[2]，陈俭清[1,2]，刘红柏[2]

（1. 东北农业大学动物科技学院，黑龙江 哈尔滨 150030；2. 中国水产科学研究院黑龙江水产研究所，黑龙江 哈尔滨 150070）

摘要：为了研究水温对鲫鱼体内恩诺沙星药动学规律的影响，试验在 20℃ 和 25℃ 水温条件下将恩诺沙星灌入鲫鱼的前肠，于给药后 0.25、0.5、0.75、1、1.5、2、4、6、8、12、24、36、48、72 h 分别采集血浆、肾脏和肝脏样品，以高效液相色谱法（HPLC）测定药物浓度，用药动学软件 3P97 处理药时数据。结果表明：在鲫鱼血浆中，两种水温条件下恩诺沙星的达峰时间和分布半衰期基本相同；20℃ 药时曲线下面积、消除半衰期明显大于 25℃，而清除率明显小于 25℃。在肾脏和肝脏中，20℃ 药物的达峰时间、分布半衰期、消除半衰期、药时曲线下面积均明显大于 25℃，而清除率明显小于 25℃。

关键词：恩诺沙星；鲫鱼；药动学；水温

该文发表于《水生生物学学报》【2003，27（1）：18-22】

盐酸环丙沙星在中华绒螯蟹体内药物代谢动力学研究

Pharmacokinetics of Ciprofloxacinum in Chinese Mitten-Handed Crab，Eriocheir Sinensis

杨先乐[1]，刘至治[1]，横山雅仁[2]

（1. 上海水产大学农业部水产增养殖生态、生理重点开放实验室，上海 200090；2. 日本国际农林水产研究中心，筑波 305-8686）

摘要：应用反向高效液相色谱法对盐酸环丙沙星在中华绒螯蟹雄、雌蟹血淋巴与肌肉内的代谢规律进行了研究。肌肉注射给药后血淋巴中盐酸环丙沙星的浓度立即达到峰值，药物开始向肌肉等组织中运转，血药浓度在 1 h 内迅速下降，而肌肉中的浓度达到峰值；此后无论是血淋巴还是肌肉中的盐酸环丙沙星浓度均缓慢下降，到第 120 h 雌、雄蟹中盐酸环丙沙星的浓度均在 1 μg/（$mL^{-1} \cdot g^{-1}$）以下，到 216 h 均不能检出。研究表明：中华绒螯蟹血淋巴中盐酸环丙沙星的药物浓度可用一级吸收二室开放模型描述，雄、雌蟹的消除半衰期（$T_{1/2\beta}$）分别为（40.34±1.48）h 和（22.07±0.31）h；中华绒螯蟹的性别对盐酸环丙沙星在其体内的代谢有较大的影响。

关键词：中华绒螯蟹；盐酸环丙沙星；血淋巴；肌肉；药物代谢动力学

盐酸环丙沙星（Ciprofloxacinum）是新一代的氟喹诺酮类合成抗菌药，由于它具有抗菌活性高、抗菌谱广、口服吸收完全、体内分布广泛且不易产生耐药性等优点，已广泛应用于水生动物细菌性疾病的防治[1-3]，对由弧菌、气单胞菌引起的中华绒螯蟹（*Eriocheirsinensis Milne-Edwards*）细菌性疾病也是首选药物之一[4]。由于对该药在中华绒螯蟹体内的代谢规律缺乏研究，在该药的使用上还存在着较大的盲目性，这不仅影响了药物的使用效果，而且也对药物在中华绒螯蟹体内的残留缺乏了解，影响着中华绒螯蟹的品质。因此在建立盐酸环丙沙星对中华绒螯蟹血淋巴内反向高效液相色谱测定法的基础上[5]，对盐酸环丙沙星在中华绒螯蟹血淋巴与肌肉内的代谢状况进行了探讨，为制定合理的给药方法提供依据。

1 材料与方法

1.1 实验动物

由上海前卫特种水产养殖公司一场提供，体重（61.5±4.11）g。试验前，在试验水族箱（1.0 m×0.5 m×0.4 m）中驯养一周。试验水温（20.0±0.5）℃。所有试验蟹经抽查血淋巴和肌肉均未检测到盐酸环丙沙星。

1.2 实验药品

甲醇，HPLC级，上海吴径化工厂生产；盐酸环丙沙星标准品（纯度为99.54%），盐酸环丙沙星原粉（纯度为98%），均由上海三维制药公司提供；抗凝剂ACD参照蔡武城等的方法配制[6]。

1.3 高效液相色谱仪及色谱条件

高效液相色谱仪采用岛津（SHIMADZU）LC-6A型系列，色谱条件参照杨先乐等方法进行[5]。检测灵敏度按0.01AUFS计算。

1.4 给药、样品的采集与处理

选择中华绒螯蟹第四步足与体关节膜处肌肉注射给药，给药剂量第一次雄蟹为（8.17±0.56）mg·kg^{-1}，雌蟹为（6.25±0.85）mg·kg^{-1}；第二次雄蟹为（5.99±0.27）mg·kg^{-1}，雌蟹为（6.05±0.32）mg·kg^{-1}。每个时间点平行设立2组，每组5只蟹。血淋巴从蟹背部心区采取，用ACD作抗凝剂分离血浆[5]，此后取体壁肌肉；合并5只蟹的血淋巴和肌肉样品，于-60℃保存备用。检测前，用0.5 mL的血淋巴与1 mL甲醇漩涡混合2 min，再加入0.5 mL甲醇，同法混合2 min，3 000 r/min离心10 min，取上清液1.5 mL于样品瓶中，自动进样器2 μL进样[5]。肌肉样品按1∶1的比例，加入双蒸水在研钵中反复研磨10 min，匀浆后取1.0 g于离心管，加入1 mL甲醇，漩涡混合2 min，再加1 mL甲醇，同法混合2 min，4 000 r/min离心10 min，上清液10 000 r/min离心10 min，再经0.45 μm微孔滤膜过滤器过滤后上液相。

1.5 血淋巴内盐酸环丙沙星的代谢规律

根据杨先乐等[5]所建立的方法，测定肌肉注射给药后雄蟹从0.017 h到48 h内的12个时间点、雌蟹从0.17 h到48 h内的9个时间点血淋巴内盐酸环丙沙星的浓度（以下简称血药浓度），以探讨其变化规律，并将所得的各时间点血药浓度数据根据药动学程序软件p56（第二医科大学瑞金医院临床药理研究室提供）的静脉注射给药模型，进行模型嵌合和参数计算。在此基础上为了获得盐酸环丙沙星在血淋巴中的消除规律，再次分别对雌、雄蟹在给药后24、48、74、96、120、144、168、192（仅雌蟹）、216 h后检测其血药浓度。

1.6 肌肉中盐酸环丙沙星的代谢规律

建立检测方法（另文详细报道）并以此分别测定雌、雄蟹给药0.5、1、3（仅雄蟹）、6（仅雌蟹）、12、24、48、72、96、120、144、168 h后肌肉中盐酸环丙沙星的浓度，以确定其在肌肉中的代谢与消除规律。肌肉样品的提取回收率和提取精密度试验均采用加入法。取离心管5支，各加入肌肉匀浆样品1.0 g，然后分别加入浓度为116.855 μg·mL^{-1}的标准盐酸环丙沙星溶液25、50、60、75、85 μL，旋涡混合2 min，其余同肌肉样品处理。回收率（%）=样品实测浓度/样品理论浓度×100%。在另5支离心管中各加入肌肉匀浆样品1.0 g和标准盐酸环丙沙星溶液50 μL，其余同回收率样品处理，并根据各次所测的峰面积计算变异系数（RSD）。日间精密度测定方法为：将浓度为0.011 69、0.116 9、1.168 6、2.337 1、5.842 8 μg·mL^{-1}的标准盐酸环丙沙星溶液各1 mL加入样品瓶，HPLC测定，连续测3次，每隔1 d测一次，计算平均峰面积及RSD。

2 结果

2.1 盐酸环丙沙星在中华绒螯蟹血淋巴内的代谢

第一次二组平行测定的结果表明，肌肉注射给药后，血药物浓度立即达到峰值，此后迅速下降：雄蟹由0.017 h（即

1 min）的（37.13±0.18）μg·mL⁻¹降至0.17 h（10 min）的（20.27±0.59）μg·mL⁻¹，1 h 为（15.16±0.78）μg·mL⁻¹，到 2 h 时只达（11.66±0.09）μg·mL⁻¹，仅为 0.17 h 的 57.5%，此后下降速度减缓，到 48 h 为（5.96±0.18）μg·mL⁻¹，是 0.17 h 时 29.4%（图1）；雌蟹的血药浓度下降的速度比雄蟹稍快，0.17 h（10 min）为（16.47±0.29）μg·mL⁻¹，2 h 降（8.18±0.02）μg·mL⁻¹，仅是前者的 48.8%，此后与雄蟹一样下降速度减缓，但比雄蟹稍快，48 h 仅为（1.98±0.46）μg·mL⁻¹，为 0.17 h 的 12.0%（图2）。第二次二组的平行测定的结果进一步说明，从 24 h 120 h，盐酸环丙沙星血药浓度缓慢、波浪式下降，但雌蟹仍较雄蟹稍快；由图1、图2可以看出雄蟹的血药浓度由 24 h 的（3.03±0.37）μg·mL⁻¹降至 48 h 的（1.43±0.04）μg·mL⁻¹，120 h 的（0.55±0.41）μg·mL⁻¹，分别只下降了 52.9% 与 81.9%，而雌蟹则由 24 h 的（2.40±0.20）μg·mL⁻¹降至 48 h 的（1.06±0.17）μg·mL⁻¹，120 h 的（0.37±0.19）μg·mL⁻¹，却分别下降了 55.0% 与 84.0%。第 144 小时雄、雌蟹血药浓度分别仅有（0.26±0.04）μg·mL⁻¹，（0.67±0.61）μg·mL⁻¹，第 168 小时血药浓度稍有回升，分别为（0.47±0.04）μg·mL⁻¹与（0.67±0.61）μg·mL⁻¹，但紧接着进一步下降，到 216 h 已不能检出（<0.01 μg·mL⁻¹）。对第一次测定的数据通过药动学软件分别进行开放型单室、二室及三室模型嵌合，结果表明，肌肉注射给药后中华绒螯蟹的药动学过程较符合一级吸收二室开放模型（$C = Ae^{-\alpha t} + Be^{-\beta t}$），它们的血药浓度–时间过程分别符合下列方程：

$$C = 11.74e^{-2.17t} + 7.67e^{-0.031t}（雌）\quad 和 \quad C = 28.75e^{-8.33t} + 13.39e^{-0.017t}（雄）$$

经检验理论值与实测值相关系数分别为 $r = 0.999\,2$ 和 $r = 0.996\,7$，曲线拟合良好。其药动学参数如表1所示。

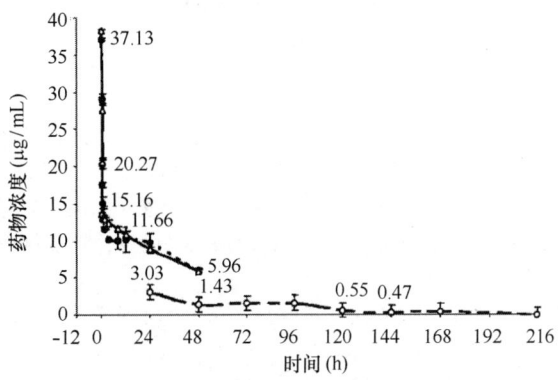

图1　盐酸环丙沙星在中华绒螯蟹雄蟹血淋巴内的浓度变化

Fig. 1　The concentration of Ciprofloxacinum in the hemolymph of maleE. sinensis

——实测值一 The values of the first experiment；

——实测值二 The values of the second experiment；

—△—理论值 The theorotic values

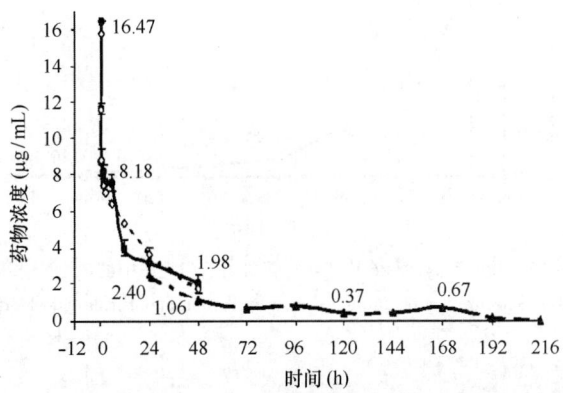

图2　盐酸环丙沙星在中华绒螯蟹雌蟹血淋巴内的浓度变化

Fig. 2　The concentration of Ciprofloxacinum in the hemolymph offemale E. sinensis

——实测值一 The values of the first experiment；

——实测值二 The values of the second experiment；

——理论值 The theorotic values

表1 盐酸环丙沙星在中华绒螯蟹体内的药动学参数

Tab. 1 Pharmacokinetic parameters of Ciprofloxacinum in E. sinensis

	药动学参数	Parameter	雄蟹 Male	雌蟹 Female
Dose	给药剂量	$mg \cdot kg^{-1}$	8.17±0.56	6.25±0.85
A	分布相的零时截距	$\mu g \cdot mL^{-1}$	28.75±0.15	11.74±0.84
B	消除相的零时截距	$\mu g \cdot mL^{-1}$	13.39±0.36	7.67±0.29
$T_{1/2\alpha}$	药物的分布相半衰期	h	0.084±0.008	0.39±0.23
$T_{1/2\beta}$	药物的消除相半衰期	h	40.34±1.48	22.07±0.31
K_{10}	药物自中央室的消除速率常数	h^{-1}	0.054±0.003	0.080±0.003
K_{12}	药物自中央室到周边室的一级转运速率常数	h^{-1}	5.65±0.47	1.27±0.81
K_{21}	药物自周边室中央室到中央室的一级转运速率常数	h^{-1}	2.64±0.30	0.85±0.50
V_d	表观分布容积	L/kg	0.015±0.005	0.080±0.002
AUC	血药浓度-时间曲线下面积	$\mu g \cdot h \cdot mL^{-1}$	783.24±49.40	250.58±2.46

2.2 盐酸环丙沙星在肌肉组织中的代谢

建立了盐酸环丙沙星在肌肉组织中的检测方法，结果表明，在试验所设计的色谱条件下，HPLC 基线走动平稳，盐酸环丙沙星的平均保留时间为（8.26±0.10）min，虽有杂峰出现，但无干扰峰，与药峰分离良好；标准盐酸环丙沙星浓度在 0.01~5.0 $\mu g \cdot mL^{-1}$ 的范围内，标准曲线 $Y = 1.64618 \times 10^{-5} X$（$Y$-盐酸环丙沙星浓度，$X$-峰面积）的相关系数为 $R = 0.9998$，相关性良好；检测灵敏度为 0.01 $\mu g \cdot mL^{-1}$，平均回收率为（80.37±16.36）%，提取精密度为 1.84%，平均日间精密度为（4.98±5.18）%。

用此方法对肌肉组织中盐酸环丙沙星的代谢规律进行了探讨，结果揭示肌肉中药峰出现在给药后 1 h 前，雌蟹的药峰时间要比雄蟹稍快，雄蟹 1 h 后的药物浓度达（6.21±0.14）$\mu g \cdot mL^{-1}$，是 0.5 h 113%，而雌蟹为（4.1±0.62）$\mu g \cdot mL^{-1}$，仅为 0.5 h 的 84%。此后开始下降，到 48 h 时雄蟹仅为（1.27±0.11）$\mu g \cdot mL^{-1}$，雌蟹为（1.24±1.03）$\mu g \cdot mL^{-1}$，分别比 1 h 降低了 79.0%和 69.0%；此后下降速度进一步加快，72、96、120、144 h 雌、雄蟹分别为（0.97±0.87）、（0.17±0.11）、（0.05±0.07）、（0.11±0.01）$\mu g \cdot mL^{-1}$ 和（0.49±0.09）、（0.31±0.07）、（0.18±0.02）、（0.10±0.05）$\mu g \cdot mL^{-1}$，到 168 h 时，雌、雄蟹肌肉中盐酸环丙沙星含量小于 0.01 $\mu g \cdot mL^{-1}$，不能检出（图3和图4）。

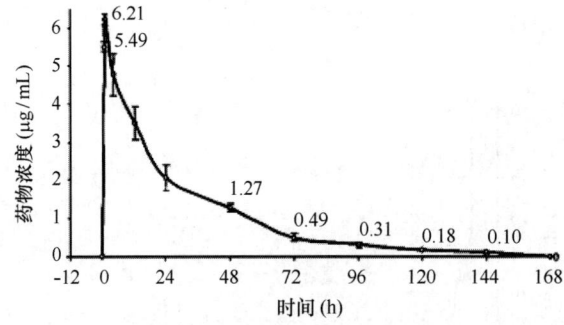

图3 盐酸环丙沙星在中华绒螯蟹雄蟹肌肉组织中的浓度变化

Fig. 3 The concentration of Ciprofloxacinum in the muscle of male Eriocheirsinensis

3 讨论

3.1 盐酸环丙沙星在中华绒螯蟹体内代谢的特点

注射给药后，中华绒螯蟹血药浓度在短时间（1 min）内就达到峰值，这是因为中华绒螯蟹属甲壳动物，循环系统为开放式，注射给药后药物直接进入血液循环，较相似于哺乳动物静脉给药。此后血药溶度迅速下降，尤其是在 1 h 内，这说明血淋巴内的药物迅速向体内的其他组织分布或消除，1 h 后下降速度明显减缓。出现这种现象，也许是因为药物对中华绒螯蟹来说，是一种外来的刺激物，进入血淋巴循环后，由于机体的保护性机能，通过触角腺、鳃等排泄器官将其排出体

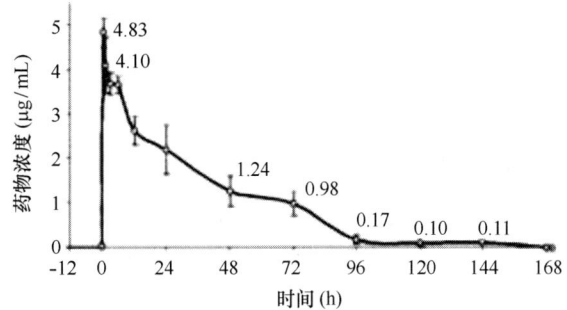

图 4　盐酸环丙沙星在中华绒螯蟹雌蟹肌肉组织中的浓度变化

Fig. 4　The concentration of Ciprofloxacinum in the musclue of female Eriocheirsinensis

外或通过肝胰脏等器官将其降解，以减轻对机体的刺激，所以在第 1 h 内血药浓度下降较快；随着时间的推移，一方面机体对其产生了一定适应性，另一方面药物处于分布与消除过程，消除与降解速度减慢，因而血药浓度的下降速度就减缓。

消除相半衰期（$T_{1/2\beta}$），反应了药物在机体内的消除速度，是药动学的一个重要参数。从本试验来看，无论是雌蟹还是雄蟹，盐酸环丙沙星在其体内半衰期均较长，与其他水生动物相比，如盐酸环丙沙星在鲤、非洲鲇、虹鳟体内的 $T_{1/2\beta}$ 分别为（14.5±0.5）h、（14.2±1.3）h、（11.2±1.6）h[7]，诺氟沙星在鲤和中华鳖体内的 $T_{1/2\beta}$ 分别为（3.403 2±0.587 3）h[8] 和 4.24 h，嘧啶酸和恩诺沙星在虹鳟体内的 $T_{1/2\beta}$ 分别为 13 h 和 2.52 h，均比中华绒螯蟹稍短。对于这一现象，作者认为中华绒螯蟹是较低等的水生动物，它们主要通过触角腺和肝胰腺对药物进行排泄或降解，其对药物的降解与排除的能力均较其他水生动物差，而鱼类等则可通过肾脏和呼吸器官（如鳃、鳃上腺、肺）等进行扩散消除，哺乳动物还可通过肾脏的主动运转予以消除（如猪、牛等对环丙沙星的 $T_{1/2\beta}$ 仅为 2.5 h），因而中华绒螯蟹的 $T_{1/2\beta}$ 就比其他鱼类和哺乳动物要长。Jame 和 Barron 通过心包膜注射给药研究磺胺间二甲氧嘧啶（SDM）在美洲螯虾体内的分布规律时也有同样的发现，其 $T_{1/2\beta}$ 达 77 h，比哺乳类要长得多。

盐酸环丙沙星在中华绒螯蟹血淋巴中的代谢与性别有较大的关系，其运转速度雌蟹要比雄蟹稍快。这主要表现在以下二方面：①消除相半衰期（$T_{1/2\beta}$）。雌、雄蟹差别明显，雄蟹差不多是雌蟹的二倍。这一现象与 Li 等对 2-萘酚在美洲螯虾体内代谢的研究有些相似，他们发现 $T_{1/2\beta}$ 雌虾远小于雄虾，分别为（7.70±4.19）h 和（41.63±21.35）h，二者差异非常显著（P<0.05）[11]；②药时曲线下面积（AUC）。虽然给药剂量雌、雄蟹稍有差别，但二者的药时曲线下面积却差别显著，雄蟹是雌蟹的 3.1 倍。Li 等对 9-羟基一吡喃酮在美洲螯虾体内的药动学研究也获得了相似的结果。药时曲线下面积是反应药物进入循环药量的多少，试验的结果说明，肌肉注射给药后，盐酸环丙沙星进入循环系统的量雄蟹比雌蟹多，这也就导致了雄蟹的 $T_{1/2\beta}$ 长于雌蟹。

3.2　盐酸环丙沙星在中华绒螯蟹血淋巴与肌肉组织中的消除

表观分布容积（V_d）大的药物与组织蛋白结合多；主要分布于细胞内液与组织间液；V_d 小的药物与血浆蛋白结合小；较集中于血浆。从雌、雄蟹的表观分布容积（V_d）非常小可以看出；盐酸环丙沙星进入中华绒螯蟹体内后；主要与血浆蛋白相结合，因此它较多地集中于血淋巴。试验证明；注射给药 1 h 内血淋巴中的药物向肌肉等组织中转移；但总的看来；肌肉的药物浓度要比血淋巴低；而且下降速度也较快；这说明肌肉组织不是药物在中华绒螯蟹体内的主要富集组织（器官）。120 h 以后，无论是血淋巴还是肌肉组织，无论是雌蟹还是雄蟹，盐酸环丙沙星的浓度均在 1 μg/（mL·g）以下，216 h 后均不能检出。由此，如果仅从血淋巴与肌肉中的残留考虑，当水温在（20.0±0.5）℃下中华绒螯蟹对盐酸环丙沙星的休药期可定为 9 d。当然，药物还有可能在蟹的其他可食部分（如性腺、肝胰腺等）残留，休药期还应根据这些组织药物中的消除状况最后确定，对此将进行深入研究。

3.3　盐酸环丙沙星的最适给药方法

最适给药方法就是确定药物的给药剂量和给药的间隔时间，一般由最低抑菌浓度（MIC）和血药浓度来决定。以往研究表明，盐酸环丙沙星对导致水生动物致病的革兰氏阴性菌的 MIC 都较低，如对河流弧菌-Ⅱ（Vibrio fluvialis-Ⅱ）为 0.32 μg·mL⁻¹，对气单胞菌（Aeromonas spp.）小于 0.01/0.02 μg·mL⁻¹，对假单胞菌（Pseudomonas spp.）为 0.32 μg·mL⁻¹ 等。根据试验，在注射给药剂量分别为（6.05±0.32）mg·kg⁻¹、（5.99±0.27）mg·kg⁻¹ 时，雌、雄蟹经 5 d 后血药浓度分别为（0.37±0.19）μg·mL⁻¹、0.55 μg·mL⁻¹，均在 0.3 μg·mL⁻¹ 以上（图 1 和图 2），还远高于盐酸环丙沙星对水生动物大部分致病菌的最低抑菌浓度。由此认为，若口服给药的生物利用度是肌肉注射给药 30%（实践经验），那么给药剂量则以 20 mg·kg⁻¹ 为宜，给药的间隔期不应超过 2 d，这一给药方法与防病实践中盐酸环丙沙星的常用的方法基本相仿[1]。

当然盐酸环丙沙星对中华绒螯蟹的口服给药与肌肉注射给药的药动学规律可能有一定的区别，对此将进行更深入的比较与研究。

参考文献

[1] Yang J. Fisheries medicine manual [M]. Beijing：Science and Technology Press of China. 1998，179—180.［杨坚主编，渔药手册. 北京：中国科学技术出版社. 1998，179—180］.

[2] Li T W，Ding M J，Song X M，et al. Preliminary studies on the mechanism of Vibrio fluvialis- Ⅱ resistance to antibiotics. Oceanologia ET LimnoligiaSinica，1996，27（6）：637—645［李太武，丁明进，宋协民，等. 皱纹盘鲍脓胞病病原菌———河流弧菌-Ⅱ的抗药机制的初步研究. 海洋与湖沼，1996，27（6）：637—645］.

[3] Lin T R，Yu N F. The synthetical treatment for Trionyxsinensis with coexistence symptom of acute gastroenteritis and branchial inflammation [J]. Aquaculture of interior Land. 1997，7：25.［林天然，余乃凤. 甲鱼急性肠胃炎鳃腺炎并发症的综合治疗. 内陆水产，1997，7：25］.

[4] Meng Q X，Yu K K. The diagnosis，prevention and cure for the disease of fish，shrimp，crab and shellfish [M]. Bingjing：Agriculture Press of China. 1996，255—281.［孟庆显，余开康. 鱼虾蟹贝疾病诊治和防治. 北京：中国农业出版社. 1996，255—281］.

[5] Yang X L，Liu Z Z，Sun W Q，et al. The method of RP-HPLC ofciprofloxacinum in hemolymph of Eriocheirsin enses [J]. Journal ofFisheries of China. 2000，25（4）：348—355.［杨先乐，刘至治，孙文钦，等. 中华绒螯蟹血淋巴内盐酸环丙沙星的反相高效液相色谱法的建立. 水产学报，2001，25（4）：348—355］.

[6] Cai W C，Li B Y，Li Y M. The experimental technique tutorial of Bio-chemical [M]. Shanghai：Fudan University Press. 1983：47.［蔡武城，李碧羽，李玉民. 生化实验技术教程［M］. 上海：复旦大学出版社. 1983：47］.

[7] Nouws J F M，Grondel J L，Schutte A R，et al.，Pharmacokinetics ofciprofloxacin in carp，African catfish and rainbow. trout [J]. Vet-Quart，1988，10：2-216.

[8] Zhang Z X，Zhang Y B，Yang R S. The pharmacokinetic of Nor-floxacin studies on common crap [J]. Journal of veterinary of China. 2000，20（1）：66-69.［张作新，张雅斌，杨永胜. 诺氟沙星在鲤鱼体内的药代动力学. 中国兽医学报，2000，20（1）：66-69］.

该文发表于《海洋渔业》【2013，35（3）：331-336】

盐酸沙拉沙星在中华绒螯蟹体内药动学及药效学研究

Pharmacokinetics and pharmacodynamics of sarafloxacin hydrochloride in *Eriocheir sinensis*

彭家红[1,2]，王元[1,2]，房文红[1]，姚小娟[1,2]，符贵红[1]

(1. 中国水产科学研究院东海水产研究所 农业部东海与远洋渔业资源开发利用重点实验室，上海 200090；
2. 上海海洋大学水产与生命学院，上海 201306)

摘要： 本文研究了盐酸沙拉沙星（Sarafloxacin hydrochloride）在中华绒螯蟹（Eriocheir sinensis）雄蟹和雌蟹体内药动学，检测了盐酸沙拉沙星对致病性气单胞菌的最小抑菌浓度（MIC），建立了药动/药效（PK/PD）关系。盐酸沙拉沙星以 30 mg·kg^{-1} 单剂量口灌给药中华绒螯蟹，雌蟹和雄蟹血浆沙拉沙星浓度-时间关系曲线均符合一级吸收的二室开放动力学模型。雌蟹和雄蟹血药达峰时间（T_{max}）为 4 h，达峰浓度（C_{max}）分别为（5.619±1.192）mg·L^{-1} 和（7.775±1.426）mg·L^{-1}。沙拉沙星在雌蟹和雄蟹体内分布广泛，其表观分布容积（V_d）分别为 3.801 L·kg^{-1} 和 2.871 L·kg^{-1}；消除半衰期（$T_{1/2\beta}$）分别为 51.982 h 和 55.956 h，总体清除率（CL_s）为 0.138 L/（kg·h）$^{-1}$ 和 0.087 L/（kg·h）$^{-1}$，曲线下面积（AUC）分别为 209.1 mg/（L·h）$^{-1}$ 和 325.5 mg/（L·h）$^{-1}$。中华绒螯蟹扣蟹致病菌嗜水气单胞菌 HX1b1、维氏气单胞菌 CMX313 和 CMX4 对盐酸沙拉沙星的最小抑菌浓度分别为 0.5、0.25 和 0.25μg·mL^{-1}。以 30 mg·kg^{-1} 单剂量给药，嗜水气单胞菌 HX1B1 的 C_{max}/MIC=15.55（雄蟹）和 11.24（雌蟹）、AUC$_{0-24}$/MIC=197（雄蟹）和 145.4（雌蟹），维氏气单胞菌 CMX313 和 CMX4 的 C_{max}/MIC= 31.1（雄蟹）和 22.48（雌蟹）、AUC$_{0-24}$/MIC=394（雄蟹）和 290.8（雌蟹）。由此可见，盐酸沙拉沙星在雌雄蟹体内的药动学数据有一定差异，但不是很大。以 30 mg·kg^{-1} 剂量拌料给药，即可以有效防治上述气单胞菌引起的细菌性疾病。

关键词： 中华绒螯蟹；盐酸沙拉沙星；药动学；气单胞菌；药效学

276

该文发表于《动物学杂志》【2012，47（6）：72~77】

双氟沙星在人工感染嗜水气单胞菌的异育银鲫体内的药代动力学

Pharmacokinetics of Difloxacin in the Aeromonas hydrophila–infected Carassiusaurutus gibelio

章海鑫[1,2]，曹海鹏[1]，阮记明[1]，胡鲲[1]，杨先乐[1]

（1. 上海海洋大学国家水生动物病原库上海 201306；2. 江西省水产科学研究所南昌 330039）

摘要：为了阐明双氟沙星在健康与处于患病状态下的异育银鲫（Carassiusaurutus gibelio）体内的药代动力学特征差异，为双氟沙星的正确合理用药提供参考，本研究通过人工创伤感染的方式采用最佳浓度的嗜水气单胞菌（Aeromonashydrophila）感染异育银鲫，在此基础上进一步以双氟沙星在健康异育银鲫体内的药代动力学特征为对照，采用反相高效液相色谱法测定双氟沙星在感染嗜水气单胞菌的异育银鲫体内的药代动力学特征。实验结果表明，以鱼体重的 20 mg/kg 口灌给药后，双氟沙星在人工感染嗜水气单胞菌的异育银鲫与健康异育银鲫体内的总药时曲线均符合一级吸收开放性二室模型，其药动学方程分别为 $C=6.227e^{-0.109t}-8.074e^{-2.752t}+1.847e^{-0.006t}$ 和 $C=110.295e^{-0.331t}+1.533e^{-0.01t}-111.828e^{-0.412t}$，但与双氟沙星在健康异育银鲫体内的药代动力学参数相比，双氟沙星在人工感染嗜水气单胞菌的异育银鲫体内的吸收、分布、消除速度减慢，其在人工感染嗜水气单胞菌的异育银鲫体内的分布半衰期、消除半衰期、吸收速率常数、曲线下面积分别增加了 4.25 h、36.17 h、2.34 h 和 74.52 mg·h/L，达峰时间延长了 5.75 h，峰浓度降低了 61.16%，且未出现重吸收现象。本研究证实嗜水气单胞菌感染能够导致异育银鲫肝肾功能损伤，因而双氟沙星在人工感染嗜水气单胞菌的异育银鲫体内的吸收、分布、代谢和消除均会减慢。

关键词：双氟沙星；嗜水气单胞菌；异育银鲫；药代动力学

该文发表于《集美大学学报》【2009，14（3）：8-13】

不同给药方式下恩诺沙星在鲤体内的药动学研究

Pharmacokinetic of Enrofloxacin in Carp Following Different Ways of Administration

李春雨[1]，李继昌[1]，卢彤岩[2]，王荻[2]，侯晓亮[3]

（1. 东北农业大学动物医学院，黑龙江 哈尔滨 150030；
2. 中国水产科学研究院黑龙江水产研究所，黑龙江 哈尔滨 150070；3. 黑龙江民族职业学院，黑龙江 哈尔滨 150081）

摘要：对两组鲤分别进行腹腔注射、口灌恩诺沙星后，应用高效液相色谱法测定了其组织中的药物浓度，研究了恩诺沙星在鲤（Cyprinus carpio）体内的吸收、分布与消除等药代动力学参数。结果表明：两种给药方式下，鲤血浆、肝脏、肾脏和肌肉组织的药时曲线符合一级消除二室模型。腹腔注射给药血浆动力学参数：AUC 为 59.185 6 $\mu g\cdot h\cdot mL^{-1}$、$K_a$ 为 75.762 7 h^{-1}、$t_{1/2\beta}$ 为 96.545 6 h、T_{peak} 为 0.073 0 h、C_{max} 为 3.297 0 $\mu g\cdot mL^{-1}$；灌服给药血浆动力学参数：AUC 为 600.296 1 $\mu g\cdot h\cdot mL^{-1}$、$K_a$ 为 0.169 3 h^{-1}、$t_{1/2\beta}$ 为 168.287 1 h、T_{peak} 为 3.665 5 h、C_{max} 为 3.266 1 $\mu g\cdot mL^{-1}$这说明腹腔注射给药比口灌给药吸收快，血药达峰时间短，达峰浓度高。

关键词：恩诺沙星；鲤；药代动力学；高效液相色谱

该文发表于《广东海洋大学学报》【2012, 32（4）：28-33】

恩诺沙星在红笛鲷体内的药代动力学

Pharmacokinetics of Enrofloxacin in *Lutjanussan guineus*

汤菊芬，简纪常，鲁义善，黄郁葱

（广东海洋大学水产学院，广东省水产经济动物病原生物学及流行病学重点实验室，广东 湛江 524088）

摘要：在水温 26±2℃、盐度 28 条件下，恩诺沙星（enrofloxacin, EF）按 5 mg/kg 的剂量腹腔注射和口灌红笛鲷（*Lutjanussan guineus*），用反相高效液相色谱法研究恩诺沙星在红笛鲷体内的药代动力学和组织分布。结果显示：注射和口灌后红笛鲷血浆的药时数据均符合一级吸收二室模型；血药达峰时间（T_p）分别为 0.75 和 1.5 h，血药质量浓度峰值（cmax）分别为 5.22 μg/mL 和 3.94 μg/mL，药时曲线下面积（AUC）分别 36.98 和 17.24mg/（L·h），消除半衰期（$t_{1/2β}$）分别为 13.28 μg/mL 和 9.18 h；恩诺沙星在组织中分布较广，注射和灌服给药肌肉、肝脏和肾脏的 C_{max} 分别为 3.53 μg/g 和 3.40 μg/g、4.08 μg/mL 和 3.98 μg/g、11.23 μg/mL 和 5.40 μg/g，T_p 分别为 1.5 和 2.0 h、1.0 和 2.0 h、0.75 和 1.5 h，AUC 分别 33.17 和 23.12 mg/（L·h）、39.36 和 18.20 mg/（L·h）、98.97 和 56.69 mg/（L·h）；EF 在红笛鲷体内消除速度较慢，注射和灌服给药肌肉、肝脏和肾脏的 $T_{1/2β}$ 分别为 22.08 和 83.32 h、41.52 h 和 94.47 h、20.94 和 21.81 h。表明恩诺沙星注射给药在红笛鲷体内的吸收和消除均快于口灌给药。

关键词：红笛鲷；恩诺沙星；药物代谢动力学

该文发表于《中国兽医学报》【2005, 25（2）：195-197】

恩诺沙星在眼斑拟石首鱼体内的药物代谢动力学

Pharmacokinetics of Enrofloxacin in *Sciaenop socellatus*

简纪常，吴灶和，陈刚

（湛江海洋大学 水产学院，广东 湛江 524025）

摘要：应用反相高效液相色谱法（RP-HPLC）研究了恩诺沙星在眼斑拟石首鱼（*Sciaenop socellatus*）体内的药物代谢动力学。实验数据经 DAS 药代动力学分析软件分析后得出：腹腔注射组血浆的药时数据符合一级吸收二室模型，动力学方程为：$C = 4.925e^{-1.452t} + 2.730e^{-0.075t}$，其主要药代动力学参数：AUC 37.533 mg/（L.h）、C_{max} 4.747 mg/L、T_{max} 0.750 h、$T_{1/2α}$ 0.477 h、$T_{1/2β}$ 9.292 h、$V_{d/F}$ 1.676 L/kg、$T_{1/2Ka}$ 0.170 h、K_a 4.081/h、K 6.714/h、K_{10} 0.191/h、K_{12} 0.769/h、K_{21} 0.566/h；灌服组血浆的药时数据符合一级吸收一室模型，动力学方程为：$C = 6.482$ $(e^{-7.092t} - e^{-10.356t})$，其主要药代动力学参数：AUC 15.805（mg/L.h）、C_{max} 2.770 mg/L、T_{max} 1.500 h、$T_{1/2α}$ 4.989 h、$V_{d/F}$ 2.072 L/kg、K_a 7.092/h、K 10.356/h。结果表明，腹腔注射给药比灌服给药吸收快，血药浓度达峰时间短于灌服给药，但血药浓度峰值明显高于灌服给药（$P < 0.05$），血浆中恩诺沙星的回收率为 94.579%。

关键词：眼斑拟石首鱼；恩诺沙星；药物代谢动力学

该文发表于《Journal of Veterinary Pharmacology and Therapeutics》【2016，39（2）：191-195】

Pharmacokinetics and bioavailability of flumequine in blunt snout bream（*Megalobrama amblycephala*） after intravascular and oral administrations

Ning Xu[*†], Jing Dong[*], Yibin Yang[*], Qiuhong Yang[*], Yongtao Liu[*†], Xiaohui Ai[*†]

（ * Freshwater Fish Germplasm Quality Supervision and Testing Center, Ministry of Agriculture, Yangtze River Fisheries Research Institute, Chinese Academy of Fishery Sciences, Wuhan, China; † Hu Bei Freshwater Aquaculture Collaborative Innovation Center, Wuhan, China）

Abstract：In present study, the pharmacokinetic profile of flumequine（FMQ）was investigated in blunt snout bream（Megalobrama amblycephala）after intravascular（3 mg/kg body weight（b. w.））and oral（50 mg/kg b. w.）administrations. The plasma samples were determined byultra-performance liquid chromatography（UPLC）with fluorescence detection. After intravascular administration, plasma concentration-time curves were best described by a two-compartment open model. The distribution half-life（$t_{1/2\alpha}$）, elimination half-life（$t_{1/2\beta}$）, and area under the concentration-time curve（AUC）of blunt snout bream were 0.6 h, 25.0 h, and 10612.7 h. μg/L, respectively. After oral administration, a two-compartment open model with first-order absorption was also best fit the data of plasma. The $t_{1/2\alpha}$, $t_{1/2\beta}$, peak concentration（C_{max}）, time-to-peak concentration（T_{max}）, and AUC of blunt snout bream were estimated to be 2.5 h, 19.7 h, 3946.5 μg/L, 1.4 h, and 56618.1 h. μg/L, respectively. The oral bioavailability（F）was 32.0%. The pharmacokinetics of FMQ in blunt snout bream displayed low bioavailability, rapid absorption, and rapid elimination.

Key words：pharmacokinetics；bioavailability；Flumequine；*Megalobrama amblycephala*

经静脉注射和口服给药后氟甲喹在团头鲂体内药动学和生物利用度研究

胥宁[*†]，董靖[*]，杨仪斌[*]，杨秋红[*]，刘永涛[*†]，艾晓辉[*†]

（* 农业部淡水鱼类种质资源监督检测中心，湖北 武汉 430223；* 中国水产科学研究院长江水产研究所 湖北 武汉 430223；† 淡水水产健康养殖湖北省协同创新中心 湖北 武汉 430223）

摘要：本研究主要考察了经口服和静脉给药后氟甲喹在团头鲂体内药动学特性。样品采用 UPLC 荧光检测器进行检测。经静脉注射之后，血浆药时曲线能够被二房室模型进行很好的拟合。分布半衰期、消除半衰期和药时曲线下面积分别为 0.6 h、25 h 和 10 612.7 h·μg/L。经口服之后，一级吸收二房室模型也很好的拟合了血浆药时曲线。分布半衰期、消除半衰期、峰浓度、达峰时间和药时曲线下面积分别为 2.5 h、19.7 h、3 946.5 μg/L、1.4 h 和 10 612.7 h·μg/L。口服生物利用度为 32.0%。氟甲喹在团头鲂体内呈现出了低生物利用度、快速吸收和快速消除的特性。

关键词：药动学，生物利用度，氟甲喹，团头鲂

INTRODUCTION

Flumequine（FMQ）is a broad spectrum synthetic antimicrobial agent belonging to the second-generation quinolone group. It is used for treating certain bacterial diseases in aquatic animals, such as enteric infections, hemorrhage and gill rot disease. Nowadays, it is wildly used in aquaculture of China due to good antibacterial effect. There is extensive information on the pharmacokinetics of FMQ in various aquatic species including Japanese sea perch（Lateolabras janopicus）（Hu et al., 2011）, Muraenesox cinereus（Liang et al., 2006）, European eel（Anguilla anguilla）（Hansen & Horsberg, 2000）, rainbow trout（Oncorhynchus mykiss）（Sohlberg et al., 1990）, Atlantic halibut（Hippoglossus hippoglossus）, turbot（Schophalmus maximus）（Hansen & Horsberg, 1999）, corkwing wrasse（Symphodus melops）（Samuelsen et al., 2000）, Atlantic salmon（Salmo salar）（Elema et al., 1994）, channel catfish（Ictalurus punctatus）（Plakas et al., 2000）.

However, little information is available on the pharmacokinetic profile of FMQ in freshwater fish in China. Therefore, the aim of the present study is to investigate the pharmacokinetics and bioavailability of FMQ in blunt snout bream（Megalobrama amblycephala），a freshwater fish cultured widely in Southern China, after intravascular and oral administrations.

MATERIALS AND METHODS

Chemicals and regents

Analytical standard FMQ（purity grade 99.0%）was purchased from Dr. Ehrenstorfer GmbH. , Augsburg, Germany FMQ powder（purity grade 98.0%）used for oral and intravenous administration was supplied by Zhongbo aquaculture Biotechnology Co. Ltd. , Wuhan, China. The anesthetic of MS222 was purchased from Aibo Biotechnology Co. Ltd. , Wuhan, China. The purity of other chemicals and reagents was analytical grade.

The FMQ solution（15 mg/mL）for oral administration was prepared by dissolving FMQ power in 0.1 mol/L NaOH solution. The solution was adjusted to pH 10.5 with HCl.

The FMQ solution（5 mg/mL）for intravenous administration was prepared by dissolving FMQ power in 0.1 mol/L NaOH solution. The solution was adjusted to pH 10.5 with HCl.

Animals

Blunt snout bream（300±10 g）were obtained from breeding base of Yangtze River Fisheries Research Institute. The fish were held in 30 plastic tanks（100×60×80 cm, length×width×depth）receiving flowing well water（26 L/min）with six individuals per tank. The water quality parameters were measured once daily in each tank, and maintained at the following: total ammonia nitrogen levels $\leqslant$ 0.73 mg/L; nitrite nitrogen concentrations at less than 0.05 mg/L; dissolved oxygen levels at 6.4~8.2 mg/L; temperature at 25.0±1.2℃ and pH 7.5±0.2. All fish were fed a drug-free feed, which was detected by UPLC, and were acclimatized for 14 days prior to commencing the experiment. A total of 128 healthy fish were selected and divided into two groups, one group（48 fish for formal experiment with six individuals per tank and 10 fish for controlholding in one tank）for intravenous administration and one group（60 fish for formal experiment with six individuals per tank and 10 fish for control holding in one tank）for oral administration. Prior to drug administration, the fish fasted for 24 h. All the experimental procedures involving animals in the study were approved by the animal care center, Hubei Academy of Medical Sciences.

Intravascular administration and sample collection

Prior to intravascular administration, fish were anaesthetized with MS222（50 mg/L）and weighted. The dose of FMQ was 3 mg/kg body weight（b.w.）for intravascular treatment. The control fish received corresponding volume of blank solution. The position of the needle in the caudal vein was confirmed by aspirating blood into the syringe prior to injection. If the fish was heavily bleeding after withdrawal of the needle or the needle had translocated during injection, the fish was excluded from the study and replaced. After intravascular treatment, six fishwere taken from tank for each sampling point at time intervals 0.083, 0.167, 0.5, 1, 2, 4, 8, 16h. At the end of the experiment, the control fish were allsampled. For each fish, approximately2mL of blood was drawn from the caudal vein away from the injection site. Blood samples were centrifuged at 4000 r/min for 10 min. The plasma was collected and stored at−20℃ until assayed. Following sampling, the fish were euthanized by an overdose in a solution of 300 mg/L MS 222.

Oral administration and sample collection

Prior to oral administration, fish were anaesthetized with MS222（50 mg/L）and weighted. The fish were administered orally at a dose of 50 mg/kg b.w. by inserting a plastic hose with 1−mL micro−injector. The control fish received corresponding volume of blank solution. After treatment, each fish was observed in tank for regurgitation of FMQ. If FMQ solution was regurgitated, the fish was removed from tank and replaced.

After oral dose, approximately 2 mL blood was taken from each of 6 fish at each time point（0.083, 0.167, 0.5, 1, 2, 4, 8, 16, 24, 48 h）. At the end of the experiment, the control fish were allsampled. Following sampling, the fish were euthanized by an overdose in a solution of 300 mg/L MS 222. Sample preparation and storage conditions were consistent with the protocol followed for the intravascular administration.

Analytical method

Protein of 0.5−mL plasma was precipitated by the same volume of acetonitrile. It was vortexed and centrifuged at 7 000 r/min for 5 min. The supernatant was filtered through a 0.22 μm disposable syringe filter for UPLC analysis（Hu et al, 2011）. The samples were analyzed by a Waters ACQUITY™UPLC with a fluorescence detector. The excitation was at 325 nm and emission was at 369nm. An ACQUITY UPLC BEH C18 column（1.7 μm, 2.1×100 mm）was used for separation. The mobile phase was tetrabutylammonium bromide buffer andacetonitrile（70:30, v/v）. The buffer contained 0.01 moL tetrabutylammonium bromide, and the pH was adjusted to 2.0 with phosphate acid. The column temperature was set to 30℃, and flow rate was 0.3 mL/min.

The calibration curves of FMQ in plasma were prepared separately with five different concentrations between 10 and 8000 μg/L using blank plasma. Precision and accuracy weredetermined by analyzing five replicates at a low, medium, andhigh concentration within the concentration range for the study.

Pharmacokinetic and statistical analyses

Pharmacokinetic parameters were calculated using Practical Pharmacokinetic Program 3P97 (Mathpharmacology Committee, Chinese Academy of Pharmacology, Beijing, China). The appropriate pharmacokinetic model was selected by the visual examination of concentration-time curve and by application of Akaike's Information Criterion (AIC) (Yamaokaet al., 1978). The oral bioavailability for FMQ was determined by calculating the area under concentration-time curve after oral and intravascular administration and using the following equation:

$$F = \frac{(AUCoraladministration) \times (doseintravascularinjection)}{(AUCintravascularinjection) \times (doseoraladministration)} \times 100\%$$

RESULTS

Method validation

The limit of detection for FMQ was 10 μg/L in blunt snout bream plasma by UPLC analysis. The good linearity of curves over range of 10 and 8000 μg/L was exhibited by the coefficient of correlation $R^2 = 0.999$. The mean±SD recovery rates of FMQ in blunt snout bream plasma were 88.2±3.5% and its percentage relative standard deviations for inter-day and intra-day precision were 3.3±1.0% and 4.1±0.9%.

Pharmacokinetics of FMQ in blunt snout bream

The plasma concentration-time curve after intravascular and oral treatment was shown in Fig. 1 and Fig. 2. The pharmacokinetic parameters were listed in Table1. After oral dosing, the data were best described by a two-compartment open model with first-order absorption. FMQ was rapidly absorbed by gastro - intestinal tract of blunt snout bream. The drug concentration already reached 1 924.8 μg/L at first sampling time. Until 1.4 h after dosing, it increased to the peak plasma concentration (3 946.5 μg/L), and then the plasma concentration began to decline. The half-life of distribution ($t_{1/2\alpha}$), half-life elimination ($t_{1/2\beta}$), absorption rate constant (K_a), area under concentration - time curve (AUC) and bioavailability were estimated to 2.5 h, 19.7 h, 1.81/h, 56618.1 h.μg/L and 32.0%, respectively. After intravascular treatment, the data best fit a two-compartment open model. The $t_{1/2\alpha}$, $t_{1/2\beta}$, and AUC were calculated to 0.6h, 25.0h, and 10612.7 h · μg/L, respectively.

Table 1 The pharmacokinetic parameters of flumequine for intravascular (3 mg/kg b. w.) andoral (50 mg/kg b. w.) administration in blunt snout bream (n=6)

Parameters	Units	i. v.	p. o.
A	μg/L	1 745.0	4 332.7
α	1/h	1.24	0.28
B	μg/L	255.7	1 566.0
β	1/h	0.028	0.035
K_a	1/h	–	1.81
$t_{1/2\alpha}$	h	0.6	2.5
$t_{1/2\beta}$	h	25.0	19.7
$t_{1/2 Ka}$	h	–	0.4
k_{10}	1/h	0.19	0.091
k_{12}	1/h	0.90	0.12
k_{21}	1/h	0.18	0.11
AUC_{0-t}	h.μg/L	10 612.7	56 618.1
T_{max}	h	–	1.4
C_{max}	μg/L	–	3 946.5
V_d	L/kg	0.45	2.89

A: Zero-time blood drug concentration intercept of distribution phase; B: Zero-time blood drug concentration intercept of elimination phase; α: Distribution rate constant; β: Elimination rate constant; Ka: Absorption rate constant; $t_{1/2\alpha}$: Distribution half-life; $t_{1/2\beta}$: Elimination half-life; $t_{1/2 Ka}$: Absorption half-life; k_{10}: Drug elimination rate constant from central compartment; k_{12}: First-order transport rate constant from central compartment to peripheral compartment; k_{21}: First-order transport rate constant from peripheral compartment to central compartment; AUC_{0-t}: Area under concentration-time curve; T_{max}: Time to peak concentration; C_{max}: Peak concentration; V_d: Apparent distribution volume.

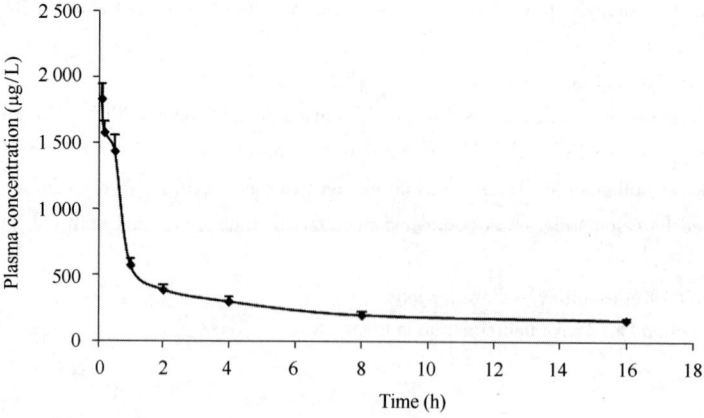

Fig. 1　Mean plasma concentration profiles of flumequine inblunt snout bream following intravascularadministration （3mg/kg b. w. ）（n=6）

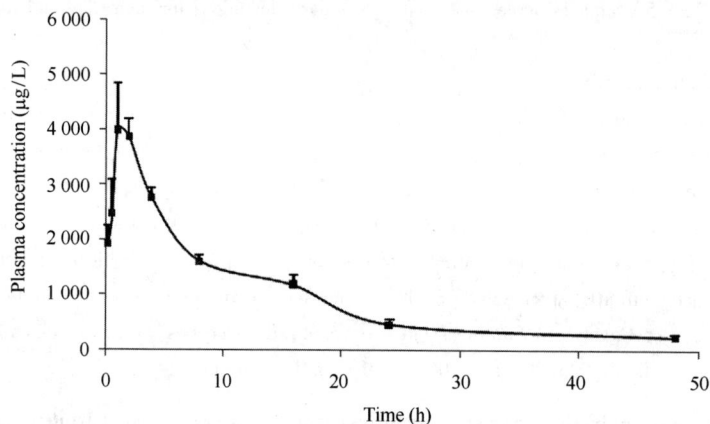

Fig. 2　Mean plasma concentration profiles of flumequine inblunt snout bream following oral administration （50 mg/kg b. w. ）（n=6）

DISCUSSION

The present study firstly investigated the pharmacokinetics and oral bioavailability of FMQ in blunt snout bream. These data would be helpful to the dosage design of FMQ in clinical treatment.

A two-compartment open model proved to best fit the data of the two different administrations. After intravascular administration, the distribution half-life （$t_{1/2\alpha}$）and the terminal elimination half-life （$t_{1/2\beta}$）werecalculated to be 0. 6 h and 25. 0 h, respectively. Following identical drug administration, the $t_{1/2\alpha}$ was longer than that in gilthead sea bream （0. 2 h）（Rigos et al. , 2003）but less than that in sea bass （1. 05 h）（Rigos et al. , 2002）. The $t_{1/2\beta}$ was similar to that in sea bass （30 h）, catfish （25 h）（Plakas et al. , 2000）, halibut （32 h）, turbot （34 h）（Hansen & Horsberg, 1999）, wrasse （31 h）（Hansen & Horsberg, 2000）, Atlantic salmon （30-40 h）（Elema et al. , 1995）, but far less than that in eel, Anguillaanguilla （314 h）（Hansen & Horsberg, 2000）. The volume of apparent distribution （0. 45 L/kg）in this study was less than above-mentioned fish, such as gilthead sea bream （0. 57 L/kg）, sea bass （1. 51 L/kg）, halibut （2. 99 L/kg）, turbot （3. 75 L/kg）, cod （2. 41 L/kg）, wrasse （2. 15 L/kg）, eel （3. 4 L/kg）, which suggested that tissue distribution of FMQ was low in blunt snout bream. The corresponding values of total body clearance （Cl）in halibut, turbot, eel, and wrasse were 0. 12, 0. 17, 0. 012, and 0. 14 L/h. kg, respectively. From those data, only value of eel was less than that of blunt snout bream. The reason for the slow elimination of FMQ in eels might be the lack of biotransformation and other possible physiological inter-species differences （Byczkowska-Smyk, 1958）. Furthermore, only traces of 7-OH FMQ （one main metabolite of FMQ）and no glucuronide metabolites were found in eel （Van der Heijden et al. , 1993）.

After oral administration, the Cmax was 3946. 5μg/kg at 1. 4 h, which was equal to or slightly higher than that in previous stud-

y. The elimination half-life was estimated to be 19. 7 h, which was less than that reported from above-mentioned study in halibut (43 h), turbot (42 h), cod (74 h), wrasse (41 h) after oral dose. Although many factors differed between our study and others, it might be accounted for by the different temperatures in studies. Our studies were performed at a temperature of 25. 0±1. 2℃. However, the temperatures in literatures were ranged from 8 to18℃. Previous studies demonstrated that a temperature increase of 1℃ corresponded to a 10% increase in metabolic and excretory rates of fish (Ellis, 1978). In addition, those references were obtained from sea water fish. Sohlberg, et al. (2002) reported that FMQ was eliminated at a substantially higher rate from fish in sea water than fresh water. This phenomenon might be explained by the influence of chelation on the compound's pharmacokinetic properties (Stein, 1996). Sea water fish drink actively, and the chelation of FMQ in the gut contents by Mg^{2+} and Ca^{2+} ions from sea water might have reduced the primary absorption and entero-hepatic cycling in seawater compared with freshwater conditions (Sohlberg et al., 2002). Rigos et al. (2003) also reported that the addition of 10mM/L Ca^{2+} and 55 mM/L Mg^{2+} to the medium resulted in an 8-to > 120-fold reduction in FMQ activity. Moreover, the $t_{1/2\beta}$ values of FMQ in fish were longer than that in poultry and mammals. Previous study indicated that $t_{1/2\beta}$ value in chicken was 6. 91 h (Anadon et al., 2008). A shorter $t_{1/2\beta}$ value was present in pigeons, only 1. 57 h (Dorrestein et al., 1983). In mammals, $t_{1/2\beta}$ values were 6. 35 h in pig (Villa et al., 2005), 11. 5 h in sheep (Delmas et al., 1997), 3 h in calves (Ziv et al., 1986), respectively. The differences might be due to the metabolic capacity of species or different temperature.

The oral bioavailability (F) of FMQ was 32. 0% in blunt snout bream after oral gavage of liquid dose, which was close to that in gilthead seabream (29. 0%) following same oral dose with the present study but less than that reported from above-mentioned studies in halibut (56. 0%), turbot (59. 0%), cod (65. 0%), wrasse (41. 0%) following oral gavage of FMQ mixed in a fish-feed emulsion, Atlantic salmon (40. 0%) following oral gavage of liquid dose. Elema et al. (1995) suggested that malabsorption, first-pass metabolism of FMQ might be possible causes. FMQ is a weak acid with a pKa value of 6. 0, and its absorption might be reduced by the alkaline intestinal environment of fish. Otherwise, first-pass metabolism of drug might occur in intestinal mucosa and intestinal flora (Sousa et al., 2008), which needed further studies.

For fluoroquinolone antimicrobials, both C_{max}-to-MIC ratio and AUC-to-MIC ratio were used to predict antibacterial success. Most experts agreed that a C_{max} to MIC of 8-10or an AUC-to-MIC ratio greater than 100-125 prevented the overgrowth of resistant bacteria (Wright et al., 2000; McKellar et al., 2004). Minimal inhibition concentrations of FMQ against Vibrio anguillarum Serotype 1b, Photobacterium damsel ssp. piscicida, V. alginolyticus, V. damsel and V. fluvialis were 0. 15, 0. 3, 1. 2, 0. 019 and 0. 15 μg/mL (Rigos et al., 2003). From our resultsafter given FMQ by oral gavage of liquid dose, the ratios of C_{max}/MIC were 26. 09, 13. 04, 3. 26, 205. 95 and 26. 09 in this study, respectively, and that of AUC/MIC were 340. 33, 170. 17, 42. 54, 2686. 84 and 340. 33, respectively. In contrast, ratios of C_{max}/MIC and AUC/MIC for V. alginolyticus were the lowest.
The pharmacokinetics of FMQ in blunt snout bream displayed low bioavailability, rapid absorption and rapid elimination. We showed that FMQ administered at a dose of 50 mg/kg b. w. by oral gavage of liquid dose offered an option for treatment of some diseases.

ACKNOWLEDGMENTS

This work is financially supported by the Special Fund for Agro-Scientific Research in the Public Interest of China (201203085) and National Nonprofit Institute Research Grant of Freshwater Fisheries Research Center, CAFS (2015JBFM29).

References

A nadon, A., Martinez, M. A., Martinez, M., De La Cruz, C., Diaz, M. J. & Martinez-Larranaga, M. R. (2008) Oral bioavailability, tissue distribution and depletion of flumequine in the food producing animal, chicken for fattening. Food and Chemical Toxicology, 46 (2), 662-670.

Byczkowska-smyk, W. (1958) The respiratory surface of the gills in teleosts. 2. Acta Biologica Cracoviensia (Serie: Zoologie), 1, 831-837

Delmas, J. M., Chapel, A. M., Gaudin, V., &Sanders, P. (1997) Pharmacokinetics of flumequine in sheep after intraveous and intramuscular administration: bioavailability and tissue residue studies. Journal of Veterinary Pharmacology and Therapeutics, 20 (4): 249-257

Dorrestein, G. M., van Gogh, H., Buitelaar, M. N. & Nouws, J. F. (1983) Clinical pharmacology and pharmacokinetics of flumequine after intravenous, intramuscular and oral administration in pigeons (Columba livia). Journal of Veterinary Pharmacology and Therapeutics, 6 (4), 281-292.

Ellis, A. E. (1978) The anatomy and physiology of teleost. In Fish Pathology, 3 Edn. Eds Roberts, R., PP. 13-54. Bailliere Tindall, London.

Elema, M. O., Hoff, K. A. & Kristensen, H. G. (1994) Multiple-dose pharmacokinetic study of flumequine in Atlantic salmon (Salmo salar L.). Aquaculture, 128 (1-2), 1-11.

Elema, M. O., Hoff, K. A. & Kristensen, H. G. (1995) Bioavailability of flumequine after oral administration to Atlantic salmon (Salmo salar L.). Aquaculture, 136 (3-4), 209-219.

Hansen, M. K. & Horsberg, T. E. (1999) Single-dose pharmacokinetics of flumequine in halibut (Hippoglossus hippoglossus) and turbot (Scophthalmus maximus）. Journal of Veterinary Pharmacology and Therapeutics, 22 (2), 122–126.

Hansen, M. K. & Horsberg, T. E. (2000) Single-dose pharmacokinetics of flumequine in cod (Gadus morhua) and goldsinny wrasse (Ctenolabrus rupestris). Journal of Veterinary Pharmacology and Therapeutics, 23 (3), 163–168.

Hansen, M. K. & Horsberg, T. E. (2000) Single-dose pharmacokinetics of flumequine in the eel (Anguilla anguilla) after intravascular, oral and bath administration. Journal of Veterinary Pharmacology and Therapeutics, 23 (3), 169–174.

Hu, L. L., Fang, W. H., Zhou, K., Xia, L. J., Zhou, S., Lou, B. & Shi, H. L. (2011) Pharmacokinetics, Tissue Distribution, and Elimination of Flumequine in Japanese Seabass (Lateolabrax japonicus) following Intraperitoneal Injection and Oral Gavage. Israeli Journal of Aquaculture-Bamidgeh, 63, 586.

Liang, Z. H., Lin, L. M., Liu, J. J., Jiang, Z. G. & Gong, Q. L. (2006) Pharmacokinetic study of oxolinic acid and flumequine in Muraenesox cinereus. Marine Fisheries Research, 27 (3), 6.

McKellar, Q. A., Sanchez Bruni, S. F. & Jones, D. G. (2004) Pharmacokinetic/pharmacodynamic relationships of antimicrobial drugs used in veterinary medicine. Journal of Veterinary Pharmacology and Therapeutics, 27 (6), 503–514.

Plakas, S. M., El Said, K. R. & Musser, S. M. (2000) Pharmacokinetics, tissue distribution, and metabolism of flumequine in channel catfish (Ictalurus punctatus). Aquaculture, 187 (1–2), 1–14.

Rigos, G., Tyrpenou, A., Nengas, I. & Alexis, M. (2002) A pharmacokinetic study of flumequine in sea bass, Dicentrarchus labrax (L.), after a single intravascular injection. Journal of Fish Diseases, 25 (2), 101–105.

Rigos, G., Tyrpenou, A. E., Nengas, I., Yiagnisis, M., Koutsodimou, M., Alexis, M. & Troisi, G. M. (2003) Pharmacokinetics of flumequine and in vitro activity against bacterial pathogens of gilthead sea bream Sparus aurata. Diseases of Aquatic Organisms, 54 (1), 35–41.

Samuelsen, O. B., Husgard, S., Torkildsen, L. & Bergh, O. (2000) The efficacy of a single intraperitoneal injection of flumequine in the treatment of systemic vibriosis in corkwing wrasse Symphodus melops. Journal of Aquatic AnimalHealth, 12 (4), 324–328.

Sohlberg, S., Czerwinska, K., Rasmussen, K. & Søli, N. E. (1990) Plasma concentrations of flumequine after intra–arterial and oral administration to rainbow trout (Salmo gairdneri) exposed to low water temperatures. Aquaculture, 84 (3–4), 355–361.

Sohlberg, S., Ingebrigtsen, K., Hansen, M. K., Hayton, W. L. & Horsberg, T. E. (2002) Flumequine in Atlantic salmon Salmo salar: disposition in fish held in sea water versus fresh water. Diseases of Aquatic Organisms, 49 (1), 39–44.

Sousa, T., Paterson, R., Moore, V., Carlsson, A., Abrahamsson, B. & Basit, A. W. (2008) The gastrointestinal microbiota as a site for the biotransformation of drugs. International Journal of Pharmaceutics, 363 (1–2), 1–25.

Stein, G. E. (1996) Pharmacokinetics and pharmacodynamics of newer fluoroquinolones. Clinical Infectious Diseases, 23 Suppl 1, S19–24.

Van der Heijden, M. H. T., Boon, J. H., Nouws, J. F. M., & Mengelers, M. J. B. (1993) residue depletion of flumequine in European eel. In Residues of Veterinary Drugs in Food. Proceedings of the EuroResidue II Conference, Eds Haagsma, N., Ruiter, A. Czedik-Eysenbergeds, P. B., Veldhoven, The Netherlands, May 3–5, 1993. pp. 357–361

Villa, R., Cagnardi, P., Acocella, F., Massi, P., Anfossi, P., Asta, F. & Carli, S. (2005) Pharmacodynamics and pharmacokinetics of flumequine in pigs after single intravenous and intramuscular administration. TheVeterinary Journal, 170 (1), 101–107.

Wright, D. H., Brown, G. H., Peterson, M. L. & Rotschafer, J. C. (2000) Application of fluoroquinolone pharmacodynamics. Journal of Antimicrobial Chemotherapy, 46 (5), 669–683.

Yamaoka, K., Nakagawa, T. & Uno, T. (1978) Statistical moment in pharmacokinetics. Journal of Pharmacokinetic and Biopharmaceutics, 6, 547–558.

Ziv, G., Soback, S., Bor, A. & Kurtz, B. (1986) Clinical pharmacokinetics of flumequine in calves. Journal of Veterinary Pharmacology and Therapeutics, 9 (2), 171–182.

该文发表于《水生生物学报》【2006, 3 (2): 54-57】

达氟沙星在史氏鲟体内药物代谢动力学比较研究

The Pharmacokinetic of Danofloxacin in Healthy and Diseased Acipenser Schrenckii Infectde By Aeromonas Hydrophila

卢彤岩，杨雨辉，徐连伟，孙大江

（中国水产科学研究院黑龙江水产研究所，哈尔滨 150070）

摘要：采用高效液相色谱法测定以 10 mg/kg 体重剂量静脉注射和口服给药后史氏鲟血浆中达氟沙星的浓度。该法采用 C18 色谱柱，流动相为乙腈-水相（15：85），荧光激发波长和发射波长分别为 280 nm 和 450 nm，样品用甲醇沉淀蛋白，离心取上清液进样。达氟沙星在 $0.005 \sim 1.0$ μg/mL 范围内线性关系良好，本方法的最低检测限为 0.005 μg/mL。健康鱼单剂量静注达氟沙星（10 mg/kg），其药时数据符合无吸收的三室开放模型，方程为 $C = 5.830 - 5.582t + 4.162 - 1.157t + 0.852 - 0.029t$，主要动力学参数如下：$t_{1/2\alpha}$ 0.552h；$t_{1/2\beta}$ 22.186 h；AUC 34.226 mg/ (L·h)；V 10.922L/kg；Vb 10.144 L/kg；Ke 10.317 h。Ah 感染组的 $V1$ 减小至 0.290 L/kg，静注感染组鱼体内达氟沙星的消除没有显著的改变。健康口服组数据结果符合一级吸收二室开放模型，血药浓度和时间方程为 $C = 1.278e - 0.073t + 0.177e - 0.089t - 1.455e - 0.329t$。药动学常数分别为：$t_{1/2ka}$ 9.491h，$t_{1/2\beta}$ 78.267h，T_{max} 6.284 h，C_{max} 0.791 mg/mL；α 0.073 h。但 Ah 感染改变达氟沙星口服给药后在史氏鲟体内的吸收、分布和消除。分布速率常数降低为 0.050/h。消除减慢，消除半衰期延长为 93.988 h，达峰时间延长至 9.060 h，峰浓度降低为 0.585 mg/mL。口服达氟沙星水溶液，健康及感染组史氏鲟对达氟沙星生物利用度分别为 96.503% 和 94.435%。本实验结果表明达氟沙星在健康史氏鲟体内分布广泛、吸收较完全。感染 Ah 对达氟沙星在史氏鲟体内的吸收、分布及消除规律均有不同程度的影响，其中口服给药的影响更为显著。达氟沙星可用于史氏鲟感染 Ah 的治疗。

关键词：达氟沙星；史氏鲟；药物代谢动力学

达氟沙星（Danofloxin）是继恩诺沙星后又一动物专用氟喹诺酮类药物，由于具有水溶性好、可多种途径给药、抗菌谱广、抗菌活性强、药动学特性优良、不良反应小及与其他抗菌药无交叉耐药性等优点，目前已被广泛用于兽医临床[1]。

嗜水气单胞菌（Aeromonas hydrophila，简称 Ah）是一种重要的致病菌，可以感染多种鱼类并造成大量死亡，随着鲟鱼养殖业的兴起，由 Ah 引起养殖鲟鱼的病害日趋严重[2]。在水产养殖生产中，抗菌药物的实际给药方案主要是根据健康鱼类的实验数据提出的，由此提出的方案假定药量和药效变化在患病和健康鱼类体内的反应是相同的，而未考虑疾病对药物量效关系的影响。事实上各种疾病均会导致动物机体的生理状况如器官组织的血流量和通透性、血液蛋白含量、药物的蛋白结合率、代谢能力等的变化，这些变化可能会影响药物在体内的吸收、分布、代谢和排泄，从而影响药物的效应[3]。

本研究探讨达氟沙星通过静注和口服在健康及实验室嗜水气单胞菌感染的史氏鲟体内代谢规律的影响。实验把疾病模型与药物动力学研究结合起来，为阐明达氟沙星在史氏鲟体内的吸收分布、消除规律和生物利用度以及 Ah 对达氟沙星在史氏鲟体内药物代谢动力学特征的影响提供重要的理论依据，从而为临床使用该药物提供参考。

1　材料与方法

1.1　试验动物

史氏鲟（Acipenser schrenckii）购自中国水产科学研究院鲟鱼繁育基地，体重范围 75~95 g，暂养于室内玻璃水族箱中 7 d 后进行实验。水族箱有效水体积为 180 L，试验期间以经过充分曝气的自来水作为水源，水温在（23±1）℃，持续以曝气的形式进行充氧。暂养期间每天投喂两次颗粒饵料。

1.2　药品和试剂

达氟沙星（甲磺酸达氟沙星）为河南郑州兽药厂生产，批号 990415，含量为 99.5%。乙腈为国产 HPLC 试剂，四丁基溴化铵、甲醇、磷酸、柠檬酸等均为国产分析纯试剂。

1.3　菌种

嗜水气单胞菌菌株由珠江水产研究所鱼病研究室馈赠，并经鉴定和保存。

1.4　仪器

分析仪器为自动进样的 Waters 2695 型配有 474 型荧光检测器的高效液相色谱仪，分离柱为 Waters C_{18} 反相分离柱。BF2000 氮气吹干仪（BFC 八方世纪公司）。

1.5　疾病模型的复制

经过预实验确定嗜水气单胞菌对史氏鲟的半数致死剂量（LD50），从暂养池中选择健康的史氏鲟，腹腔注射经培养 24 h 的嗜水气单胞菌培养液，注射体积为 2 mL/kg（约 10^6 CFU/mL），选择具有明显发病特征的史氏鲟作为实验用感染鱼。

1.6　实验分组及给药

实验分为健康口服组、健康静脉注射组、感染口服组和感染静脉注射组。取 150 尾健康鲟用于健康口服和静注达氟沙星药动学研究。选择感染症状明显的史氏鲟用于疾病模型组口服和静注药动学试验。配制 5 mg/mL 达氟沙星灭菌水溶液供静注和口服给药，单剂量给药剂量均为 10 mg/kg。口服给药时用医用静滴软管一端固定于 1 mL 注射器上，一端轻插入实验鱼胃部，并按体重大小推入相应体积的药液。静注时注射入史氏鲟尾静脉，注射成功与否通过观察注射前回抽尾静脉血的方法进行确定。

1.7 样品采集

实验用史氏鲟，分别于静注给药后 0.017、0.125、0.25、0.5、0.75、1、2、4、8、12、24、48、72 和 120 h 及口服给药后 0.125、0.25、0.5、0.75、1、2、4、8、12、24、48、72 和 120 h 各时间点于尾部采血，1% 肝素抗凝后离心吸取血浆保存于−24℃备用，每一时间点采集 5 尾史氏鲟血液用于药物浓度分析。

1.8 样品处理

取出血浆样品，于室温下自然解冻摇匀后取出 0.5 mL 血浆，加入 0.5 mL 甲醇，旋涡快速混匀器上混合 1 min，于 4 000 r/min转速离心 8 min。取全部上清液在 BF2000 氮气吹干仪 70℃下吹干，再加入 0.5 mL 流动相混匀后经 0.45 μm 微孔滤膜过滤后，用于 HPLC 分析。

1.9 色谱条件

流动相中乙腈：水相为 15∶85（V/V），每 1 000 mL 水相中含 12.4 mL 0.2 mol/L 磷酸氢二钠和 187.6 mL 0.1 mol/L 柠檬酸，每 1 000 mL 流动相中含 2.7 g 溴化四丁基铵，流动相 pH 为 3。柱温为 18℃，进样体积为 10 μL，荧光检测器激发和发射波长分别为 280 nm 和 450 nm。流速为 0.8 mL/min。

1.10 达氟沙星标准曲线的制备

采用外标法进行试验样品的测定，在 100 mL 棕色容量瓶中配制成 100 μg/mL 的达氟沙星标准品母液，在 4℃冰箱中避光保存备用，用流动相分别稀释成 0.005、0.01、0.025、0.05、0.1、0.25、0.5 和 1 μg/mL 达氟沙星水溶液。以峰面积为横坐标，药物浓度为纵坐标做标准曲线，用最小二乘法求出回归方程和相关系数。

1.11 回收率的测定

将 0.1 μg/mL、1 μg/mL、10 μg/mL 标准品分别加入 0.5 mL 的血浆中，使其终浓度分别为 0.01 μg/mL、0.1 μg/mL 和 1 μg/mL，按 1.8 方法进行处理后进行测定，各浓度重复测定 5 次，获得各浓度的峰面积，再按标准曲线回归方程计算达氟沙星的浓度，并与实验前加入量进行比较，回收率＝实际测定量/初始加入量×100%。

1.12 数据处理

药时数据用 MCPKP 药物代谢动力学参数程序计算，由程序自动拟合最佳模型。最后得出两种给药途径下健康及疾病模型史氏鲟体内药物代谢动力学参数、模型指数和系数，确定理论方程。

2 结果

2.1 感染嗜水气单胞菌疾病模型的复制

史氏鲟感染嗜水气单胞菌 8 h 后，出现典型的感染症状。感染鱼腹部、嘴四周、眼睛、腹部出血，肛门红肿，随机选择的感染鱼有淡红色腹水，无菌操作挑取患病鲟鱼肝脏，进行细菌分离，28℃培养 24 h 挑取单一菌落，在斜面上划线纯化，根据《发酵性革兰氏阴性杆菌分类鉴定编码手册》细菌鉴定方法，证实为嗜水气单胞菌。同时取未感染鱼作对照细菌培养，结果未感染鱼腹腔内无腹水，肝脏未分离培养出任何细菌。感染实验表明已成功复制出感染嗜水气单胞菌的疾病模型。

2.2 标准曲线及变异系数

标准曲线为 $y=6521.906+59617810x$（$R=0.9999$），达氟沙星在 0.005~1.0 μg/mL 范围内线性关系良好，当所测定的血浆的药物浓度大于线性范围的上限，则用流动相稀释后再进行测定，本方法的最低检测限为 0.005 μg/mL。高、中、低 3 个浓度的回收率均在 84% 以上，批内、批间、日内、日间变异系数均在 10% 以内，可满足达氟沙星药物代谢动力学的研究要求。

2.3 史氏鲟体内的血药浓度

史氏鲟健康及疾病状态下静注和口服达氟沙星水溶液后，不同时间理论及实际测定值见图 1 至图至图 4。

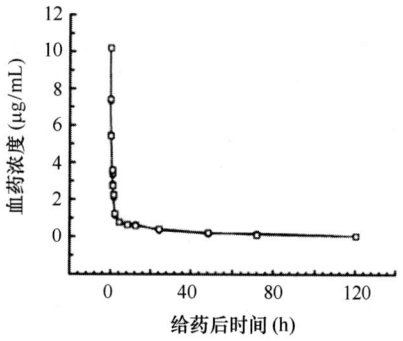

图 1 健康静注达氟沙星后血药浓度的实测值和理论值

Fig. 1 Theoretic and observed concentration of danof loxacin in plasma of healthy *Amur sturgeon* after IV administration

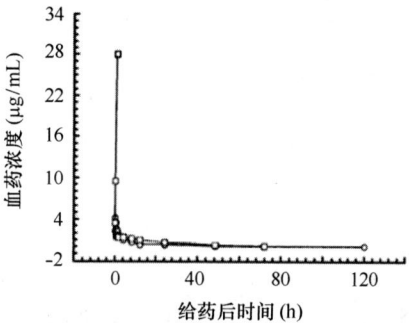

图 2 感染后史氏鲟静注达氟沙星后血药浓度的实测值和理论值

Fig. 2 Theoretic and observed concentration of danofloxacin in plasma of infected *Amur sturgeon* after IV administration

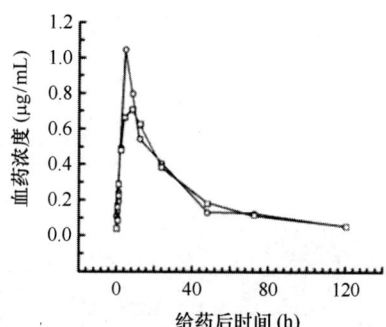

图 3 康口服达氟沙星后血药浓度的实测值和理论值

Fig. 3 Theoretic and observed concentration of danofloxacin in plasma of healthy after oraladministration

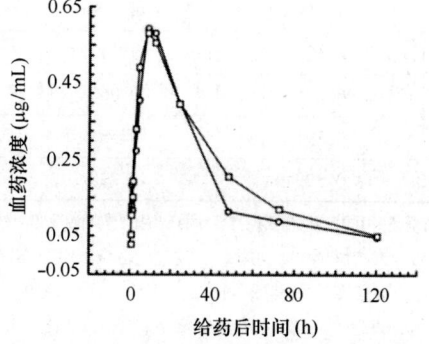

图 4 感染 Ah 后口服达氟沙星后血药浓度的实侧值和理论值

Fig 4 Theoretic and observed concentration of danofloxacin in plasma of infected fish after oral administration

通过处理血药浓度-时间数据表明，史氏鲟口服达氟沙星的药时数据结果符合一级吸收二室开放模型，健康组的血药浓度和时间方程为 $C=1.278e^{-0.073t}+0.177e^{-0.089t}-1.455e^{-0.329t}$。感染组口服给药的血药浓度和时间方程为 $C=1.038e^{-0.050t}+0.079e^{-0.074t}-1.117e^{-0.221t}$，相关系数分别为 0.842 及 0.877。

史氏鲟静注达氟沙星药时数据结果符合无吸收三室开放模型，健康组的方程为 $C=5.830^{-5.582t}+4.162^{-1.157t}+0.852^{-0.0289t}$，疾病组的方程为 $C=31.035e^{-13.4548t}+2.830e^{-0.433t}+0.648e^{-0.0336t}$，相关系数分别为 0.999 和 0.991。

2.4 生物利用度

达氟沙星在健康及感染 Ah 史氏鲟体内药物代谢动力学参数见表 1。口服达氟沙星水溶液，健康组生物利用度 96.5027%，感染组生物利用度为 94.4348%。

表 1 达氟沙星在健康和感染嗜水气单胞菌史氏鲟体内的药动学参数

Tab. 1 The pharmacokinetics parameters of danofloxacin lactate in *Amur sturgeon* by intravenous and oral admistration

参数 Parameters		口服 oral administration		静注 intravenous administration	
名称	单位	健康组 Control	感染组 Infected group	健康组 Cntrol	感染组 Infected group
D	mg/kg	10.000	10.000	10.000	10.000
Co	mg/mL	1.167	0.881	10.844	34.512
A	mg/mL	1.278	1.038	4.162	2.830
B	mg/mL	0.177	0.079	0.852	0.648
P	mg/mL	—	—	5.830	31.035
α	h^{-1}	0.073	0.050	1.157	0.433
β	h^{-1}	0.089	0.074	0.029	0.034
π	h^{-1}	—	—	5.582	13.455
$t_{1/2\pi}$	h			0.114	0.049
$t_{1/2\alpha}$	h	9.491	13.959	0.552	1.479
$t_{1/2\beta}$	h	78.267	93.987	22.186	19.004
K_{12}	h^{-1}	0.028	0.013	1.628	8.826
K_{21}	h^{-1}	0.183	0.110	3.133	1.662
K_{13}	h^{-1}	—	—	1.502	1.814
K_{31}	h^{-1}	—	—	0.187	0.094
AUC	mg/L/h	33.029	26.591	34.226	28.157
$V1$	L/kg			0.922	0.290
Vb	L/kg			10.144	10.562
CI_b	L/kg/h			0.292	0.355
TCP	h	142.490	62.554	98.452	76.176
$t_{1/2ka}$	h	2.104	3.142		
t_{max}	h	6.284	9.060		
C_{max}	mg/mL	0.791	0.584		
K_a	h^{-1}	0.329	0.220		

注：D 为给药剂量；Co 为初始浓度；A 为分布相的零时截距；B 为消除相的零时截距；P 为深室分布相的零时截距；α 为分布速率常数；β 为消除速率常数，π 为深室分布速度常数；Ka 为药物吸收速率常数；$t_{1/2\pi}$ 为深室分布相半衰期；$t_{1/2\alpha}$ 为浅室分布相半衰期；$t_{1/2\beta}$ 为消除相半衰期；K_{12} 为药物由中央室到周边室内的一级运转速率常数；K_{21} 为药物自周边室到中央室的一级运转速率；K_{13} 为药物由中央室到周边室的一级运转速率常数；K_{31} 为药物由深室到中央室的一级运转速率常数；AUC 为血药浓度-时间曲线下面积；$V1$ 为中央室表观分布容积；Vb 为表观分布容积；Clb 为清除率；$t_{1/2ka}$ 为吸收相半衰期；t_{max} 为达峰时间；C_{max} 为峰浓度；Ka 为药物吸收速率常数。

3 讨论

3.1 静注给药的药动学特征

大量的药物代谢动力学试验表明，由于动物种属间的生理代谢特点不同，不同动物及同种动物不同年龄之间达氟沙星的药物代谢动力学特点均有所不同[3-7]。但有关达氟沙星在鱼类体内的药物代谢动力学规律尚未见报道。Shojaee Aliabadi F 试验结果证明小牛静注达氟沙星有较高的分布容积和相对较快的清除率，半衰期为 2.65 h，清除半衰期分别为 10.2 h。山羊静注达氟沙星，血浆清除半衰期分别为 4.67 h，表观分布容积较高[4]。Atef M[5] 等人试验证实以 1.25 mg/kg 达氟沙星静注和健康的成年雌性山羊，血浆药物时间曲线符合二室开放模型，药物迅速分布和消除，半衰期分别为 17.71 h。药物在中央室分布率高，K_{12}/K_{21} 比率为 0.67，表观分布容积为 1.42 L/kg。研究证实以 1.25 mg/kg 在绵羊静注达氟沙星，表观分布容积为 2.766 L/kg，消除半衰期（$t_{1/2\beta}$）为 3.32 h，清除速率为 0.63 L/（kg·h）[6]。也有试验表明雏鸡静脉注射达氟沙星后，达氟沙星在鸡体内分布广泛、迅速，且消除较快，因此达氟沙星适合于临床上治疗全身感染和深部组织感染[7]。本试验结果表明，健康及感染组史氏鲟以 10 mg/kg 静脉注射达氟沙星后，最佳药物代谢动力学模型符合无吸收三室开放模型，达氟沙星在浅室分布较快，在深室分布较慢，中央室表观分布较小，表观分布较大，表明达氟沙星在健康史氏鲟体内分布广泛，组织中的药物浓度较高。这与达氟沙星在其他动物体内的药物代谢动力学特征基本一致。

3.2 口服给药的药动学特征

史氏鲟口服达氟沙星的药时数据结果符合二室开放模型，健康史氏鲟以 10 mg/kg 剂量口服达氟沙星水溶液后表现为吸收速度较慢，吸收半衰期为 9.49 h，消除半衰期为 78.27 h，达峰时间为 6.28 h，峰浓度为 0.79 μg/mL。实验表明口服噁喹酸、氟甲喹、恩诺沙星和沙拉沙星的大西洋鲑，这 4 种药物的达峰时间均高于 6 h[8]，而口服诺氟沙星和环丙沙星的鲤鱼达峰时间分别为 0.44 h 和 0.73 h[8,9]。笔者认为由于大西洋鲑属于冷水性鱼类（适温 15~18℃），而鲤鱼属温水性鱼类（适温 23~28℃），二者的基础体温和环境温度差异很大，由于温度的作用而药物的代谢速度有明显的差别，因而药物在冷水性鱼体内的达峰时间均长于温水性鱼类。史氏鲟的生长适温为 21℃ 左右，介于温水性鱼类和冷水性鱼类之间[11]，本实验的试验水温为（23±1）℃，达峰时间也介于温水性鱼和冷水性鱼类之间。

3.3 Ah 感染对静注给药的药动学的药动学的影响

动物机体的健康状况有时会对药物在体内的分布、代谢及吸收有一定的影响。有文献报道支原体与大肠杆菌合并感染能使司帕沙星在鸡体内的消除半衰期略有延长，机体消除率减少，药时曲线下面积增大，但影响不如口服给药明显[10]。认为是由于静注给药时药物直接入血，不需要吸收过程有关。研究表明猪大肠杆菌感染可使恩诺沙星的清除率显著下降、消除半衰落期延长，从而导致曲线下面积显著增大，表明大肠杆菌感染能显著改变恩诺沙星在猪体内的分布和消除，使中央室分布容积显著减小。本实验 Ah 感染明显改变达氟沙星在史氏鲟体内的中央室分布容积。但染组鱼体内达氟沙星的消除规律没有显著的改变。由于嗜水气单胞菌感染属全身性感染，会导致包括动物体肝脏、肾脏及其他器官组织生理状况及代谢功能的改变。而郑伟文等研究表明，嗜水气单胞菌感染后的中华鳖血清中肌酐的含量也由健康组的 21.8 显著升高至 39 μm/L。本试验结果表明 Ah 感染对达氟沙星在史氏鲟体内的分布规律有一定的影响。

3.4 Ah 感染对口服给药的药动学的影响

研究结果表明，由于不同药物的结构不同，药物代谢情况受实验动物感染情况的影响是不同的。研究表明雏鸡口服达氟沙星水溶液，支原体与大肠杆菌合并感染能改变药物在体内的动力学特征，使达氟沙星在感染鸡体内的吸收、分布和消除均减慢，达峰时间、有效浓度维持时间延长，药物在中央室与周边室的转运速度也减慢。刘雅红[11] 则证明感染霉形体与大肠杆菌感染鸡内服给药氧氟沙星药动学参数没有显著改变。也有报道霉形体与大肠杆菌合并感染鸡内服司帕沙星，吸收减慢，达峰时间推迟，峰浓度增加，消除减慢。本实验中嗜水气单胞菌感染明显改变达氟沙星口服给药后在史氏鲟体内的吸收、分布和消除。分布速率常数降低，分布减慢。同时药物的消除也减慢，消除半衰期延长。达峰时间延长，峰浓度降低。可见在口服给药方式下嗜水气单胞菌感染对达氟沙星在史氏鲟体内的代谢规律有较大的影响。

3.5 达氟沙星在史氏鲟体内生物利用度的评价

生物利用度是确定药物剂量与作用强度之间关系的重要因素，也是进行药物动力学评价时的主要参数之一[14-17]。本实验结果表明，健康及感染组鱼生物利用度均较高，说明该药在史氏鲟体内吸收较完全。张秀英等证明达氟沙星内服给药的生物利用度为 90.08%。Aliabadi 证明山羊肌注达氟沙星的生物利用度为 100%。El-Gendi[12] 实验表明火鸡肌注和口服达氟沙星的生物利用度分别为 96.56% 和 81.4%。Atef[5] 则发现健康山羊肌注达氟沙星生物利用度为（65.70±10.28）%。

Shem-Tov[13]证明哺乳奶牛静注达氟沙星生物利用度大于100%。McKellar[6]等发现奶牛肌注达氟沙星的生物利用度为95.71%。表明达氟沙星在动物体内的吸收过程虽然存在着种属的差异，但生物利用度均较高。同时嗜水气单胞菌感染并未使达氟沙星在史氏鲟体内的生物利用度发生明显的改变。

体外抑菌实验结果表明达氟沙星对嗜水气单胞菌的最小抑菌浓度（MIC）为0.05 μg/mL，结合本实验结果达氟沙星可用于史氏鲟嗜水气单胞菌感染的治疗，建议给药方式以间隔2 d、给药剂量为10 mg/kg为宜。

参考文献

[1] 王建元，刘玉年．氟喹诺酮类抗菌药在动物中的应用．中国兽医学报，1996，16（3）：304-309.
[2] 储卫华，于勇．鲟鱼嗜水气单胞菌的分离与鉴定．淡水渔业，2003，33（2）：16-17.
[3] Shojaee Aliabadi F, Less P. Pharmacokinetic-pharmacodynamic integration of danof loxacin in the calf [J]. Res Vet Sci, 2003, 74 (3): 247-259.
[4] Aliabadi FS, Lees P. Pharmacokinetics and pharmacodynamics of danof loxacin in serum and tissue fluids of goats f ollowing intravenous and intramuscular administration [J]. Am J Vet, 62 (12): 1979-1989.
[5] Atef M, El-Gendi A Y, Aziza, et al. Some pharmacokineti c data for danof loxacin in healthy goats [J]. Vet Res Commun, 2001, 25 (5): 367-377.
[6] McKellar Q A, Gibson I F, McCormack R Z. Pharmacokinetics and pharmacodynamics of danofloxacin in serum and tissue fluids of goats following intravenous and intramuscular administration [J]. Biopharm Drug Dipos, 1998, 19 (2): 123-129.
[7] Martinsen B, Horsberg T E, Varma K J, et al. Single dose pharmacokinetic study of f lorfenicol in At lanti c salmon in seawater at 11℃ [J]. Aquculture, 1993, 112: 1-11.
[8] 张雅斌，张祚新，郑伟，等．不同给药方式下鲤对诺氟沙星的药代动力学研究，水产学报，2000，24（6）：559-563.
[9] 杨雨辉，佟恒敏，卢彤岩，等．环丙沙星在鲤鱼体内吸收、代谢和生物利用度的研究，水产学报，2003，27（6）：582-589.
[10] 孙大江，曲秋芝，马国军，等．史氏鲟人工养殖研究现状与展望，中国水产科学，1998，5（3）：108-111.
[11] 刘雅红，冯淇辉．氧氟沙星在健康和霉形体与大肠杆菌感染鸡的药动学研究，畜牧兽医学报，1998，29（6）：536-540.
[11] el-Gendi AY, el-Banna HA, Abo Norag M, et al. Disposition kinetics of danof loxacin and ciprofloxacin in broi ler chi ckens [J]. Dtsch Tierarztl Wochenschr. 2001, 108 (10): 429-434.
[12] Shem-Tov M, Rav-Hon O, Ziv G, et al. Pharmacokineti cs and penetration of danofloxacin from the blood into the milk of cows [J]. J Vet Pharmacol Ther, 1998, 21 (3): 209-213.
[13] 艾晓辉，陈正望．喹乙醇在鲤鱼体内的药物代谢动力学及组织浓度．水生生物学报，2003，27（3）：273-277.
[14] 杨先乐，刘至治．盐酸环丙沙星在中华绒螯蟹体内药物代谢动力学研究．水生生物学报，2003，27（1）：18-22.
[15] 王翔凌，方之平，操继跃，等．盐酸沙拉沙星在鲫鱼体内的残留及消除规律研究．水生生物学报，2006，30（2）：196-201.

该文发表于《淡水渔业》【2015，45（2）：72-78】

氧氟沙星与左氧氟沙星在中华绒螯蟹体内的药代动力学和残留消除规律研究

Pharmacokinetics and elimination regularity of ofloxacin and levofloxacin in *Eriocheir sinensis*

吴冰醒，曹海鹏，阮记明，马荣荣，胡鲲，杨先乐

（上海海洋大学，国家水生动物病原库，上海 201306）

摘要：以15 mg/kg的剂量分别对中华绒螯蟹（*Eriocheir sinensis*）单次肌注氧氟沙星和左氧氟沙星，运用反相高效液相色谱法，测定了中华绒螯蟹血淋巴和肌肉、肝胰腺、精巢、卵巢中氧氟沙星和左氧氟沙星的浓度，分别阐明了它们在中华绒螯蟹各组织中的残留消除规律。结果显示，氧氟沙星和左氧氟沙星在中华绒螯蟹血淋巴内的药物-曲线时间关系符合开放性二室模型，其药代动力学方程分别为：$C_t = 21.31e^{-0.497t} + 9.779e^{-0.035t}$ 和 $C_t = 74.938e^{-1.822t} + 20.811e^{-0.038t}$。氧氟沙星在肌肉、肝胰腺、精巢、卵巢的消除半衰期分别为88.89、65.02、83.26、114.76 h。左氧氟沙星在肌肉、肝胰腺、精巢、卵巢的消除半衰期分别为86.03、344.83、106.19、154.73 h。结果表明，相同给药剂量下，左氧氟沙星在体内浓度更高，分布更广，对于中华绒螯蟹的细菌性疾病有着更好的预

防与治疗作用。

关键词：氧氟沙星；左氧氟沙星；中华绒螯蟹（*Eriocheir sinens*）；药代动力学；残留消除

该文发表于《大连海洋大学学报》【2011，26（2）：144-148】

氧氟沙星在吉富罗非鱼体内的药代动力学及残留的研究

Pharmacokinetics and residues of ofloxacin in GIFT tilapia strain *Oreoch romisniloticus*

王贤玉[1]，宋洁[1,2]，王伟利[1]，姜兰[1]，罗理[1]，柯剑[1,2]

（1. 中国水产科学研究院珠江水产研究所，广东 广州 510380；2. 上海海洋大学水产与生命学院，上海 201306）

摘要：在水温为（25±2）℃下，按 10 mg/kg 的剂量给吉富罗非鱼（*Oreoch romisniloticus*）单次口服氧氟沙星，用高效液相色谱法测定鱼血浆和组织中的药物浓度，研究氧氟沙星在吉富罗非鱼体内的代谢及消除规律。结果表明：血浆及组织药时数据均符合一级吸收二室开放模型，吸收分布迅速，但消除较为缓慢，血浆、肌肉、肝胰脏、肾脏中的达峰时间（T_{max}）分别为 0.41、3.19、0.18、0.59 h；最大血药浓度分别为 7.98 μg/mL、17.24、36.10、46.65 μg/g，组织中肝胰脏的药物浓度最高，在测定时间内各组织的药物浓度均高于血浆；药物消除速度依次为肾脏>肌肉>肝胰脏，消除半衰期（$T_{1/2\beta}$）分别为 12.90、19.45、28.27 h。以 10 μg/kg 为最高残留限量，在本试验条件下，建议休药期不低于 8 d。在治疗罗非鱼疾病时，氧氟沙星的给药剂量为 10 mg/kg，每天两次，连续使用 2~3 d。

关键词：吉富罗非鱼；药代动力学；氧氟沙星；残留

四、影响磺胺类渔药代谢的因素

温度对磺胺类药物在罗非鱼体内的吸收及消除影响显著，可提高药物的最大血药浓度与消除速率，但对复方磺胺嘧啶在血浆中的比值影响不显著（肖贺等，2014）。高温能降低磺胺甲噁唑（SMZ）在草鱼体内的半衰期和达峰时间，并提高其达峰浓度（艾晓辉等，2005）。

该文发表于《华南农业大学学报》【2014，35（6）：13-18】

不同温度下复方磺胺嘧啶在罗非鱼体内的药代动力学比较

Pharmacokinetic of compound sulfadiazine in *Oreochromis niloticus* at different temperature

肖贺[1,2]，王伟利[1]，姜兰[1]，鞠晶[1,2]，罗理[1]，邓玉婷[1]，谭爱萍[1]

（1. 中国水产科学研究院 珠江水产研究所/农业部渔用药物创制重点实验室/广东省免疫技术重点实验室，
广东 广州 510380；2. 上海海洋大学，上海，201306）

摘要：【目的】探讨不同温度条件下复方磺胺嘧啶在罗非鱼血液中的药代动力学（简称药动学）特点及其变化。【方法】以罗非鱼为研究对象，在不同水温（18、23、28 和 33℃）饲养条件下按复方磺胺嘧啶（磺胺嘧啶：甲氧苄啶=5：1 {g/g}）120 mg/kg 的剂量单次饲喂给药，分别于给药后 0.5、1、2、4、6、8、10、24、48、72 h 采集血液样品，使用 HPLC 方法检测罗非鱼血浆中的药物浓度，研究复方磺胺嘧啶在罗非鱼血液中的吸收和消除变化规律。【结果与讨论】18、23、28 和 33℃时，磺胺嘧啶在血浆中的峰浓度分别为 12.40、19.60、22.48 和 30.75 μg/mL，甲氧苄啶在血浆中的峰浓度分别为 1.22、2.06、2.44 和 2.70 μg/mL，2 个药物的峰浓度均随着温度的升高而增大。磺胺嘧啶在血浆中的消除半衰期（$T_{1/2\,ke}$）分别为 18.22、17.89、16.90 和 12.99 h，甲氧苄啶在血浆中的消除半衰期（$T_{1/2\,ke}$）分别为 16.39、7.08、5.99 和 4.04 h，药物的消除随温度的升高而加快；各温度下药物在罗非鱼血液中的药动学均为 1 级动力学过程。给药后 10 h 内血浆中磺胺嘧啶和甲氧苄啶的比例分别为 9.57：1～11.01：1、6.30：1～8.40：1、5.4：1～9.15：1 和 4.2：1～20.6：1，均维持在 1：1～40：1 的理想抑菌配比范围内。研究结果表明温度对药物在罗非鱼体内的吸收及消除影响显著，可提高药物的最大血药浓度与消除速率，但对复方磺胺嘧啶在血浆中的比值影响不显著。

关键词：温度；血药浓度；复方磺胺嘧啶；药代动力学；罗非鱼

该文发表于《水生生物学报》【2005，29（2）：210-214】

不同水温和给药方式下磺胺甲噁唑在草鱼体内的药动学研究

A study on pharmacokinetic of sulphamethoxazole in grass carp at different temperatures and administration regimes

艾晓辉，刘长征，周运涛

（农业部淡水鱼类种质资源与生物技术重点开放实验室，中国水产科学研究院长江水产研究所，荆州 434000；
中国水产科学研究院淡水渔业研究中心，无锡 214081）

摘要：在 18℃和 28℃两个不同水温条件下，分别采用口灌和腹腔注射的给药方式，给予草鱼 100 mg/kg 体重单剂量的磺胺甲噁唑（SMZ），以 HPLC 法测定草鱼血浆和肌肉中药物浓度，用 MCPKP 药代动力学软件处理药时数据，结果表明，18℃条件下，草鱼口灌 SMZ 的分布半衰期 $T_{1/2\alpha}$、消除半衰期 $T_{1/2\beta}$、达峰时间 T_p 均显著长于 28℃（$P<$ 0.01），其峰浓度 C_{max} 显著低于 28℃下的值（$P<0.01$）；腹腔注射给药时，SMZ 在鱼体血浆和肌肉中的吸收与分布较口服快，消除较口服也快。在水温 18℃条件下，口灌 SMZ 在草鱼血浆中的药时数据符合带时滞的二室开放模型，腹腔注射的符合无时滞二室开放模型；水温 28℃条件下，口灌 SMZ 在草鱼血浆中的药时数据符合可忽略时滞的二室开放模型。根据 SMZ 的药代动力学规律、最低有效血药浓度和抗菌药物应用的一般原则，制定了 SMZ 的用药方案：治疗温水性鱼类细菌性疾病，在水温 18℃左右，口服或注射 100 mg/kg 剂量的 SMZ，以每天给药两次，3 d 一个疗程为宜；在水温 28℃左右，以每天给药 3～4 次为宜。该项研究全面了解了在不同水温和不同给药方式下 SMZ 在鱼体内的药动学规律，为确定合理的临床用药方案以及无公害水产品中药物残留监测提供了可靠的理论依据。

关键词：磺胺甲噁唑；草鱼；药代动力学；水温；给药方式

磺胺类药物是水产养殖生产上广泛应用的抗微生物药，对鱼类多种革兰氏阴性细菌和一些革兰氏阳性细菌有较显著的抑菌效果，同时对某些真菌、衣原体等亦有抑制作用，是一种广谱的抑菌药物[1]。

近年来的研究资料表明，磺胺类药物在人体及不同动物体内的代谢情况存在明显的种属差异性[2-3]，同一种药物在同一种水生动物体内，即使性别不同，其代谢规律也存在差异性[4]。因此，只根据某一种实验动物或人的实验数据，为磺胺药在水产上的临床应用制订给药方案，很不合理。

磺胺甲噁唑（SMZ）在草鱼（*Cenopharyngodon idellus*）体内的药代动力学研究尚未见报道，本文就磺胺甲噁唑在不同水温和不同给药方式的条件下，在草鱼体内的药代动力学进行了研究，为制订磺胺甲噁唑在水产临床应用上的合理给药方案提供理论依据。

1　材料和方法

1.1　药品与试剂　磺胺甲噁唑（SMZ）粉，含量 99.7%，由昆山双鹤药业有限责任公司提供，批号 020709。二氯甲烷、无水醋酸钠为分析纯，85% 的磷酸为优级纯，配成 0.017 mol/L 磷酸缓冲液及 1 mol/L 的醋酸钠溶液备用；乙腈为色谱纯；蒸馏水为三蒸水。

1.2　仪器与色谱条件高效液相色谱仪（waters 600 型，由 717 自动进样器、515 高压泵、2487 双通道紫外检测器及 MILLEMNIUM32 组成）；WH-1 微型漩涡混合仪；TDL-16B 高速离心机；电热恒温水浴锅；SCQ-250 超声波清洗器；IKA ALLbasic 组织捣碎机；0.45 um 过滤器。ODS-C$_{18}$柱，150 mm×3.9 mm（id），粒度 5 μm；流动相：乙腈：0.017 mol/L 磷酸缓冲液（40∶60/V∶V）；流速：1.0 mL/min；检测波长：287 nm；柱温：室温；进样量：20 μL。

1.3　试验鱼饲养及试验分组　草鱼，体重（76.14±5.27）g，健康无伤，由长江水产研究所鱼类育种试验场提供，试验前在 100 cm×68 cm×80 cm 的两个室内小水泥池中暂养 5～7 d，不投食，连续充氧，水温一个池控制在 18℃，另一池控制在 28℃，pH6.8～7.0。试验共分 3 组，每组鱼数 80 尾。（1）水温 18℃单次口腔灌药组；（2）水温 28℃单次口腔灌药组；（3）水温 18℃单次腹腔注射组。

1.4　给药方法　（1）口腔灌药组　称取 7.614 g SMZ 粉于 1 000 mL 烧杯中，加水约 800 mL，同时加入少量淀粉加热搅拌均匀至乳浊状，冷却后定容至 1 000 mL，每尾鱼灌 1 mL，即灌药剂量为 100 mg/kg 鱼体重，灌药采用文献[5]的方法进行。（2）注射给药组用 0.2 mol/L 氢氧化钠溶液溶解 SMZ，使药液浓度为 0.7614 g/100 mL 水，过滤、灭菌，进行鱼体安全试验后，每尾鱼腹腔注射该药液 1 mL，即注射剂量为 100 mg/kg 鱼体重，其余操作同单次灌药。

1.5　采样每组分 16 个时间段采样，每次随机取草鱼 5 尾，从每尾鱼的尾动脉抽取血液 1.0 mL，分别置预先涂有肝素钠的塑料离心管中，做 5 个平行样，4 000 r/min 离心 5 min，分离出血浆；随后将取过血样的草鱼剖杀，迅速取出背脊两侧肌肉，分别装于白色塑料袋中，按样时间顺序编号，保存于−20℃冰箱。单次给药的采样时间为给药前取一次空白血样及肌肉样，给药后参照文献[6]的方法进行，分别于 0.25、0.5、1、2、3、4、6、8、12、24、48、72、96、120、144、168 h 采血样及组织样。

1.6　药物测定　鱼血浆及肌肉中 SMZ 的测定参照文献[3]的方法进行。该法的标准曲线的线性范围是在 0.02～10.0 μg/mL（r=1.00），回收率稳定在 75%～85%，日内精密度与日间精密度均小于 5%。

1.7　数据处理　采用 MCPKP 药物代谢动力学参数计算程序[7]，在微机上处理各给药组血浆、肌肉的血药浓度-时间数据，算出有关药动学参数。

2　结果

2.1　在不同水温条件和不同给药方式下，SMZ 在草鱼血浆中的药动学特征

以 100 mg/kg 鱼体重的单剂量给草鱼口灌 SMZ 后，在水温 18℃条件下，药物在（3.09±0.13）h 达到峰浓度，峰浓度均值为（152.67±3.21）μg/mL。在水温 28℃条件下，药物在（1.93±0.08）h 达到峰浓度，峰浓度均值为（414.05±7.31）μg/mL。SMZ 在两种不同水温条件下的血浆药动学最佳数学模型皆为带时滞的二室开放模型，28℃条件下的时滞可忽略不计。

以 100 mg/kg 鱼体重的单剂量给草鱼腹腔注射 SMZ 后，在水温 18℃条件下，药物在（0.888±0.251）h 达到峰浓度，峰浓度均值为（195.28±6.53）μg/mL。其血浆药动学最佳数学模型符合二室开放模型。

在 18℃和 28℃两种不同水温条件下给草鱼口灌 SMZ 以及在水温 18℃条件下给草鱼腹腔注射 SMZ，SMZ 在草鱼血浆中的主要药动学参数见表 1，给药后的时量曲线见图 1。

表 1　腹腔注射和口服 SMZ 在草鱼血浆中的主要药动学参数（n=5）

Tab. 1　Plasma pharmacokinetic parameters of SMZ after intraperitoneal and oral administration in grass carps（n=5）

参数		18℃（口灌） （Oral administration）	28℃（口灌） （Oral administration）	18℃（注射） （Intraperitoneal）
名称 Parameter	单位 Unit	参数值（X±SD） Value	参数值（X±SD） Value	参数值（X±SD） Value
D	mg/kg	100	100	100
Co	μg/mL	225.34±0.19	978.10±0.03	351.37±5.89
A	μg/mL	187.41±0.30	4669.78±3.24	608.64±3.24
B	μg/mL	97.45±0.72	80.83±0.62	124.97±1.35
α	/h	0.203±0.013	0.488±0.051	1.16±0.05
β	/h	0.052±0.045	0.085±0.064	0.071±0.012
K_A	/h	0.724±0.131	0.605±0.028	1.87±0.20
K_{12}	/h	0.049±0.046	0.095±0.026	0.604±0.141
K_{21}	/h	0.112±0.112	0.113±0.028	0.444±0.12
K_{el}	/h	0.094±0.015	0.364±0.024	0.184±0.013
LAGTIME	h	0.210±0.042	0.0002±0.0007	0.00

续表

名称 Parameter	单位 Unit	18℃（口灌） （Oral administration） 参数值（$X\pm$SD） Value	28℃（口灌） （Oral administration） 参数值（$X\pm$SD） Value	18℃（注射） （Intraperitoneal） 参数值（$X\pm$SD） Value
$T_{1/2\alpha}$	h	3.41±0.15	1.14±0.13	0.596±0.117
$T_{1/2\beta}$	h	13.31±0.61	8.19±0.47	9.85±0.64
AUC	mg/L.h	2401.3±12.7	2684.7±11.5	1908±14.6
T_{p}	h	3.09±0.13	1.93±0.08	0.888±0.251
C_{max}	μg/mL	152.67±3.21	414.05±7.31	195.28±6.53
TCP	h	12.82±0.43	5.04±0.25	13.02±0.13

注：D 为给药剂量，Co 为初始药物浓度，A 为分布相的零时截距，B 为消除相的零时截距，α 为分布相消除速率常数，β 为消除相消除速率常数，$T_{1/2\alpha}$ 为药物的分布相半衰期，$T_{1/2\beta}$ 为药物的消除相半衰期，K_{10} 为药物自中央室的消除速率常数，K_{12} 为药物自中央室到周边室的一级转运速率常数，K_{21} 为药物自周边室中央室到中央室的一级转运速率常数，AUC 为血药浓度-时间曲线下面积，T_{p} 为达峰时间，C_{max} 为达峰浓度，TCP 为有效血药浓度维持时间。

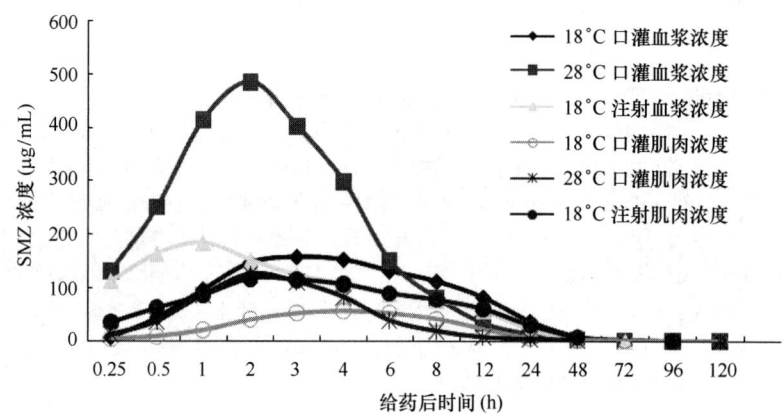

图 1　不同温度和给药方式下 SMZ 在草鱼血浆和肌肉中的浓度变化

Fig. 1　Variation of plasma and muscle SMZ concentration in grass carp after different administration at different temperatures

2.2　在不同水温条件和不同给药方式下，SMZ 在草鱼肌肉中的药动学特征

以 100 mg/kg 鱼体重的单剂量给草鱼口灌 SMZ 后，在水温 18℃ 条件下，草鱼肌肉中药物在 4.01±0.22 h 达到峰浓度，峰浓度均值为（57.22±2.27）μg/g。在水温 28℃ 条件下，药物在（2.06±0.12）h 达到峰浓度，峰浓度均值为（102.37±3.11）μg/g。在水温 18℃ 条件下，SMZ 在草鱼肌肉中药动学最佳数学模型皆为带时滞的二室开放模型，在水温 28℃ 条件下，SMZ 在草鱼肌肉中药动学最佳数学模型皆为无时滞的二室开放模型。

以 100 mg/kg 鱼体重的单剂量给草鱼腹腔注射 SMZ 后，在水温 18℃ 条件下，草鱼肌肉中药物在（2.37±0.36）h 达到峰浓度，峰浓度均值为（116.37±5.74）μg/g。其药动学最佳数学模型符合一室开放模型。

在 18℃ 和 28℃ 两种不同水温条件下给草鱼口灌 SMZ 以及在水温 18℃ 条件下给草鱼腹腔注射 SMZ，SMZ 在草鱼肌肉中的主要药动学参数见表 2，给药后的时量曲线见图 1。

由图 1 可见，在同等的温度条件下，以 100 mg/kg 鱼体重的单剂量给草鱼无论是口灌还是腹腔注射 SMZ 后，SMZ 在草鱼肌肉中的出峰时间（T_{p}）皆比血浆中长，而峰浓度则皆比血浆中低。SMZ 口灌给药后在草鱼肌肉中的消除速度比血浆中慢，而腹腔注射给药后，SMZ 在草鱼肌肉中的消除速度与血浆中差异不明显。

表2　腹腔注射和口服 SMZ 在草鱼肌肉中的主要药动学参数（n＝5）

Tab. 2　Muscles pharmacokinetic parameters of SMZ after intraperitoneal and oral administration of grass carp（n＝5）

参数		18℃（口灌） （Oral administration）	28℃（口灌） （Oral administration）	18℃（注射） （Intraperitoneal）
名称 Parameter	单位 Unit	参数值（$X\pm SD$） Value	参数值（$X\pm SD$） Value	参数值（$X\pm SD$） Value
D	mg/kg	100	100	100
C_0	μg/mL	151. 54±4. 46	258. 13±8. 13	137. 76±11. 48
A	μg/mL	8168. 78±12. 14	2173. 33±16. 14	–
B	μg/mL	4. 24±0. 34	6. 21±0. 31	–
α	/h	0. 263±0. 021	0. 465±0. 071	–
β	/h	0. 014±0. 009	0. 022±0. 006	–
K_A	/h	0. 267±0. 059	0. 526±0. 052	–
K_{12}	/h	0. 076±0. 058	0. 139±0. 079	–
K_{21}	/h	0. 021±0. 037	0. 032±0. 037	–
K_{el}	/h	0. 179±0. 056	0. 316±0. 077	–
LAGTIME	h	0. 143±0. 093	0. 00	0. 00
$T_{1/2\alpha}$（$T_{1/2ka}$）	h	2. 64±0. 33	1. 49±0. 11	0. 535±0. 421
$T_{1/2\beta}$（$T_{1/2k}$）	h	48. 59±1. 14	31. 94±0. 65	9. 73±0. 58
AUC	mg/L·h	843. 66±6. 9	816. 27±4. 82	1934. 5±17. 9
T_p	h	4. 01±0. 22	2. 06±0. 12	2. 37±0. 36
C_{max}	μg/mL	57. 22±2. 27	102. 37±3. 11	116. 37±5. 74

3　讨论

SMZ 在草鱼体内的药动学实验结果表明，在水温18℃条件下，SMZ 在鱼体血浆和肌肉中的分布较慢，而消除也慢；在水温28℃条件下，SMZ 在鱼体血浆和肌肉中的分布较快，而消除也相对较前者快。低温18℃条件下，草鱼口服 SMZ 的分布半衰期 $T_{1/2\alpha}$、消除半衰期 $T_{1/2\beta}$、达峰时间 T_p 均显著长于28℃高温（$P<0.01$），其峰浓度 C_{max} 显著低于28℃高温（$P<0.01$）。该结论与 Herman and Degurse（1967）[8] 报道的在7.7℃与14℃两个不同水温条件下研究磺胺甲基嘧啶在虹鳟体内的代谢规律以及 Lee and Kou（1978）[9] 在7℃与14℃两个不同水温条件下研究4-磺胺-6-甲氧嘧啶（SMM）在鳗鱼体内的代谢规律结论都是一致的。据 BjÖrklund 等（1990）[10] 报道，5℃、10℃、16℃下土霉素在虹鳟血清中达到峰值的时间分别为 24 h、12 h 和1 h，消除时间高水温与低水温相差 108 d 之久。BjÖrklund 等（1990）[10] 认为，通常水温每升高1℃，鱼类的代谢和消除速度将提高 10% 。以上研究均表明，水环境的变化对水生动物药动学的规律影响很大。一般来说，在一定温度范围内，药物的代谢强度与水温成正比，水温越高，代谢速度加快。因此给药时水温是一个应予以考虑的重要因素，尤其是水温较低时停药期应予适当延长。

在不同的给药方式同等水温条件下，SMZ 在草鱼体内进行的药动学研究结果表明，腹腔注射给药时，SMZ 在鱼体血浆和肌肉中的吸收与分布较口服快，消除较口服也快，血浆中 SMZ 的分布半衰期 $T_{1/2\alpha}$ 为（0.596±0.117）h，消除半衰期 $T_{1/2\beta}$ 为（9.85±0.64）h，肌肉中 SMZ 的分布半衰期 $T_{1/2\alpha}$ 为（0.535±0.421）h，消除半衰期 $T_{1/2\beta}$ 为（9.73±0.58）h。注射给药吸收和消除速度比口灌给药快，这与张雅斌和张祚新（2000）[11] 报道的在不同给药方式下诺氟沙星在鲤体内的代谢规律是一致的。

SMZ 的消除相半衰期在不同的对象体内也有差异，其消除半衰期在人体内为 11~12 h，泰和鸡为 5.369 h，绿头鸭为 5.7 h，兔体内为 2.89 h[2]，在草鱼体内为 8.19~13.31 h（水温18℃~28℃），比较而言，SMZ 在鱼体内为长效磺胺。

磺胺类药具有广谱抗菌作用，按磺胺类药在血中最低有效血药浓度为 50 μg/mL 计算，水温18℃口灌给药、水温28℃口灌给药及水温18℃腹腔注射给药的有效血药浓度维持时间（TCP）分别为（12.82±0.43）h、（5.04±0.25）h、（13.02±0.13）h，再根据不同条件下 SMZ 的消除半衰期和抗菌药物应用的一般原则，治疗温水性鱼类细菌性疾病，在水温18℃左右，口服或注射 100 mg/kg 剂量的 SMZ，以每天给药两次，3 d 一个疗程为宜；在水温 28℃左右，以每天给药 3~4 次为宜，考虑到水产养殖上投饲习惯一般是一天两次，投药时间为了与投饲习惯吻合，可将用药剂量增至 200 mg/kg。因而养殖者应根据实际情况调整用药剂量和次数。

关于 SMZ 的休药期，我们已对 SMZ 多次给药及与甲氧苄啶（TMP）混合给药的代谢动力学进行了研究，将结合本文研究结果另文报道。

参考文献

[1] Editorial board of "A handbook of medicines for aquatic animals" of Ministry of Agriculture. A Handbook of Medicines for Aquatic Animals [M]. Beijing: China Science and Technology Publishing House, 1998, 195~196. [农业部"渔药手册"编撰委员会. 渔药手册 [M]. 北京：中国科技出版社, 1998：195~196.].

[2] Ge Y H, Zhou J L. Studies on pharmacokinetics of SMZ and SMZ-TMP in domestic rabbits [J]. Journal of Hebei Agricultural University, 1989, 12 (1)：123~125. [葛印怀, 周建玲. 磺胺甲基异恶唑（SMZ）与复方磺胺甲基异恶唑（SMZ-TMP）在家兔体内代谢动力学的研究 [J]. 河北农业大学学报, 1989, 12 (1)：123~125.].

[3] Liu C Z, Ai X H. A method for the detection of sulphamethoxazole and trimethoprim content in aquatic products with high performance liquid chromatography [J]. Freshwater Fisheries, 2004, 34 (2)：23~24. [刘长征, 艾晓辉. 水产品中磺胺甲嗯唑与甲氧苄啶含量的高效液相色谱测定法 [J]. 淡水渔业, 2004, 34 (2)：23~24.].

[4] Yang X L, Liu Z Z, Masahito YOKOYAMA. Pharmacokinetics of ciprofloxacinum in chinese mitten-handed crab, Eriocheir sinensis [J]. Acta Hydrobiologica Sinica, 2003, 27 (1)：18~22. [杨先乐, 刘至治, 横山雅仁. 盐酸环丙沙星在中华绒螯蟹体内药物代谢动力学研究 [J]. 水生生物学报, 2003, 27 (1)：18~22.].

[5] AiX H, ChenZ W, Zhang C W, et al. Studies on pharmacokinetics and tissue concentration of olaquindox in carps [J]. Acta Hydrobiologica Sinica, 2003, 27 (3)：56~60. [艾晓辉, 陈正望, 张春光等. 喹乙醇在鲤体内的药物代谢动力学及组织浓度研究 [J]. 水生生物学报, 2003, 27 (3)：56~60.].

[6] Fang W H, Shao J H, Shi Z H. Analytical method of norfloxacin in the giant tiger shrimp (Penaeus monodon) hemolymph and brief study on pharmacokinetics [J]. Acta Hydrobiologica Sinica, 2003, 27 (1)：13~17. [房文红, 邵锦华, 施兆鸿等. 斑节对虾血淋巴中诺氟沙星含量测定及药代动力学 [J]. 水生生物学报, 2003, 27 (1)：13~17.].

[7] Xia W J, Cheng Z R. MCPKP-Amicrocomputer programme for kinetic analysis of drugs [J]. Journal of Pharmacology of China, 1988, 9：(2) 188~192. [夏文江, 成章瑞. MCPKP—药物动力学分析的一种微机程序 [J]. 中国药理学报, 1988, 9 (2)：188~192.].

[8] Herman, R. L. and Degurse, P. E. Sulfamerazine residues in trout tissues [J]. Ichthyologica, 1967, 39：73~9.

[9] Lee. M. J. and Kou. G. H. Absorption of sulfamonomethoxine by eels in medicated bath. JCRR [J] Fish. Series, 1978, 34：69~75.

[10] BjÖrklund, H and Bylund, G. Temperature-related absorption and excretion of oxytetracycline in rainbow trout [J]. Aquaculture, 1990, (84)：363~372.

[11] Zhang Y B, Zhang Z X, Zheng W, et al. Study on pharmacokinetics of norfloxacin in carp following different forms administration [J]. Journal of Fisheries of China, 2000, 24 (6)：559~563. [张雅斌, 张祚新, 郑伟等. 不同给药方式鲤鱼对诺氟沙星的药代动力学研究 [J]. 水产学报, 2000, 24 (6)：559~563.]

五、影响其他渔药代谢的因素

盐度会影响杀虫剂吡喹酮的代谢过程：咸水中草鱼体内吡喹酮的代谢比淡水中迅速，咸水中鱼体组织内吡喹酮残留的浓度也更低（Xinyan Xie et al., 2015；谢欣燕等, 2015）。

动物的年龄也会影响对药物残留的消除。氯霉素在一龄罗非鱼血浆中的吸收速度比二龄罗非鱼快，达峰时间短，消除速度明显减慢，但分布半衰期和曲线下面积（AUC）大小二者比较接近（杨先乐等, 2005）。

该文发表于《BMC Veterinary Research》【2015, 11：84. doi：10.1186/s12917-015-0400-2】

Comparison of praziquantel pharmacokinetics and tissue distribution in fresh and brackish water cultured grass carp (*Ctenopharyngodon idellus*) after oral administration of single bolus

XinyanXie, Yini Zhao, Xianle Yang and Kun Hu

(Shanghai Fisheries University, 999Huchenghuan Road, Shanghai 201306, PR China)

Abstract：Background：Praziquantel (PZQ) is an effective pesticide against monogeneans. Its pharmacokinetics in fish may be affected by water environment and temperature. The present study was designed to compare the pharmacokinetics, tissue distribution, and elimination of PZQ in freshwater-acclimated grass carp and brackish water cultured grass carp. Plasma and tissue PZQ concentrations were determined after a single 10 mg/kg oral PZQ dose. Results: The datas of plasma and tissues drug concentration was calculated by the software SPSS 13.0. According to the One-Way ANOVA, the results showed that the salinity had a significant effect on the drug concentration of plasma ($p<0.01$), muscle ($p<0.01$), liver ($p<0.01$) and kidney ($p<0.01$) in the all sampling time points between the brackish water grass carps and the freshwater grass carps, wherein, PZQ plasma and tissue concentrations in the brackish water group were constantly lower than that in the freshwater group. The peak PZQ levels of plasma, muscle, liver, and kidneys in the brackish water group were 0.76 μg/mL, 0.51 μg/g, 2.7 μg/g, and 2.99 μg/g, respectively; and that in the freshwater group were 0.91 μg/mL, 0.62 μg/g, 3.87 μg/g, and 3.39 μg/g, respectively. The elimination half-lives ($t_{1/2\beta}$) in plasma and all tissues of the freshwater group were significantly longer than that in the brackish water group. The elimination half-lives ($t_{1/2\beta}$) of plasma, muscle, liver and kidneys in brackish water grass carps were 56.46, 36.17, 15.31, and 132.64 h, respectively; and that in the freshwater grass carps were 71.15, 44.88, 23.86, and 150.23 h, respectively. Conclusion: These findings indicate that water environment affects the tissue distribution and elimination of PZQ in grass carps, the elimination in brackish water grass carps is more rapid than that in fresh water grass carps and tissue concentrations of brackish water grass carps are lower than that in freshwater grass carps after orally administrating the same dosage at the same water temperature. We speculate that the main excretion pathway of the drug is through renal elimination, and the decreased kidney function in brackish water grass carps is likely responsible for the considerable difference in pharmacokinetics between the two groups of grass carps.

Key words: Praziquantel, Pharmacokinetics, Grass carp, Salinity, Tissue distribution, Elimination

不同盐度条件下吡喹酮在草鱼体内的代谢和组织残留分布

谢欣燕，赵依妮，杨先乐，胡鲲

（上海海洋大学，中国上海沪城环路 999 号，201306）

摘要： 吡喹酮（PZQ）是针对单殖吸虫的一种有效的药物，其在鱼类体内的药代动力学可能受水环境和温度的影响。本研究的目的是比较淡水和咸水养殖条件下吡喹酮在草鱼体内药代动力学、组织分布。10 mg/kg 口服吡喹酮剂量后测定血浆和组织药物浓度。结果：血浆和组织中的数据通过 SPSS 13 软件计算药物浓度。根据单因素方差分析，结果表明，盐度对血浆药物浓度的影响显著（$P<0.01$），肌肉（$P<0.01$），肝（$P<0.01$）和肾脏（$P<0.01$）之间的咸水和淡水草鱼草鱼，其中所有采样的时间点，表明血浆和组织浓度盐水组不断低于淡水组。血浆、肌肉、肝脏中药物峰值水平，咸水组肾脏 0.76 μg/mL，0.51 μg/g，2.7 μg/g 和 2.99 μg/g；而在淡水组 0.91 μg/mL，0.62 μg/g，3.87 μg/g，3.39 μg/g。消除半衰期（$T_{1/2\beta}$）血浆和组织所有的淡水组明显高于盐水组延长。消除半衰期（$T_{1/2\beta}$）血浆、肌肉、肝脏和肾脏在咸水草鱼为 56.46 h，36.17 h，15.31 h 和 132.64 h，分别；并且在淡水草鱼为 71.15 h，44.88 h，23.86 h 和 150.23 h。结论：这些结果表明，水环境影响草鱼的组织分布和消除的药物，在苦咸水草鱼的消除比淡水草鱼和组织咸水草浓度低于淡水鲤鱼草鱼后口服相同剂量在相同的水的温度更迅速。我们推测该药的主要排泄途径是通过肾脏排泄。

关键词： 吡喹酮，代谢动力学，草鱼，盐度，组织分布，消除

Background

Praziquantel (PZQ) is widely used as a chemotherapeutic agent for treating helminths in captive fish. It is effective against monogeneans that infect the gills, skin, and branchial cavities[1-13], larval and encysted digeneans infecting the eyes[14] and skin[15], as well as intestinal cestodes[16] of teleosts and elasmobranchs.

Grass carp (*Ctenopharyngodon idellus*) is an economically important species of farmed fish. In China, grass carp has a widespread breeding distribution scope, covering different salinity areas in the eastern and northwestern regions, and freshwater areas of the central and southern regions. For example, the water salinity of LianYungang in Jiangsu province is 3.6‰ and that of Tachen Island in Zhejiang province is 5.8‰[17]. Several studies have indicated that environmental water salinity can influence the accumulation and elimination of chemicals in fish[18-22]. The difference in pharmacokinetics influences the effectiveness of PZQ in aquaculture, especially for treating monogenean infection in brackish water cultured grass carps.

Chemical/drug pharmacokinetics in aquatic animals can be affected by various environmental factors, such as water temperature, PH, and salinity[23]. However, the effect of salinity is often overlooked, especially in freshwater fish. Moreover, in order to improve grass carp muscle quality, the fish are often raised in brackish water or placed into brackish water for a period of time before listing[24]. Considering that grass carp are cultured in both freshwater and brackish water, it is necessary to investigate possible tissue distribution and praziquantel elimination differences in grass carp under different water environments. We believe that this study would help better formulate effective PZQ regimens in treating monogenean infections of fish.

Methods

Chemicals and reagents

Praziquantel (2-cyclohexyl-carbonyl-4-oxo-1, 2, 3, 6, 7, 11bhexahydro-4Hpyrazino [2, 1-a] isoquinoline) (99.9% purity) is used as an analytical standard and was purchased from Sigma-Aldrich (St. Louis, MO, USA); PZQ (purity 98%) used in the pharmacokinetic study was purchased from Feng Hua Pharmaceutical Co., Ltd., Hebei. HPLCgrade acetonitrile was purchased from Merck (KGaA, Germany); and all the other chemicals were of analytical grade.

Animals

All the experiment fish precedures were reviewed and approved by the Institutional Animal Care and Use Committee at Shanghai Ocean University at Shanghai Ocean University. According to the Shanghai ocean university animal health guidelines for animal care and experimentation. About 120 six month age Grass carps (Ctenopharyngodonidellus, weighing 80±5.6 g) were obtained from Nantong Fisheries Farm (Yancheng, China) and were randomly divided into two groups: freshwater group and brackish water group and number of fish in each group was 60. The two groups of fish were brought in the laboratory and placed in two ponds (1 m^3). The brackish water group was gradually acclimated to brackish water with 3.5‰ salinity for five days; circulating water (the salinity for the freshwater and brackish water groups were 0 and 3.5‰, respectively) and oxygen were continuously supplied using an inflation pump. Before the experiment, the two groups were acclimated in rearing tanks for 10 days at 22±1℃. During acclimation, they were fed twice daily with a drug-free commercial diet to apparent satiation, then starved for 24 h before drug administration.

Drug administration

To prepare the PZQ dosing solution, PZQ was dissolved in 5 mL of ethanol and mixed with water to achieve a final concentration of 2 g/L. The fish were given 5 μL/g of the PZQ solution via gavage using a stomach tube-with the final PZQ dosage being 10 mg/kg. After oral drug administration, each fish was placed in an observation tank for five minutes for possible drug regurgitation; regurgitated fish were excluded from the analysis.

Sample collection

Blood sampleswere collected from the tail sinus of five fish at various time points (before dosing; and 0.25, 0.5, 1, 3, 6, 12, 24, 48, and 96 h after dosing); then, five fish in each sampling time point were killed by breaking their spine. Muscle, liver, and kidney samples were taken, immediately frozen, and stored at −20℃ until analysis.

Sample preparation

The analytical procedure for PZQ analysis was modified according to the report of Jing Yao[25]. In brief, the plasma (1 mL) or ground tissue samples (1 g) were separately placed into 50-1mL centrifuge tubes, and mixed with 3 mL of ethylacetate using a Vortex vibration meter (Thermo Fisher Scientific, USA) for three minutes. After 10 minutes of centrifugation at 4, 500 rpm, the supernatant was removed and transferred to a 15 mL centrifuge tube. The extraction step was repeated once. The combined ethyl acetate extracts were evaporated at 45℃ to dryness using the rotary evaporation apparatus (Eppendorf, Germany). Residues were reconstituted with 1 mL of HPLC mobile phase, in which 2 mL of hexane was added. The mixture was vortexed for three minutes and centrifuged at 8, 000 rpm for two minutes. The supernatant was removed and transferred to a 2-mL centrifuge tube; then 20 μL was injected onto the HPLC system, as described below.

Chromatographic conditions

We used an Agilent (HP1100) HPLC system with a fluorescence detector. The separation was performed on a Zorbax XDB-C18 column (4.6 × 150 mm internal diameter, 5-μm particle size; Agilent Technologies, USA) using an isocratic mixture of acetonitrile: water (50: 50, v/v) as the mobile phase at a constant flow rate of 1 mL/min. Injection volume was 20 μL, optical maser wavelength was 265 nm, emission wavelength was 280 nm, and column temperature was 25℃.

Pharmacokinetic analysis

Pharmacokinetic was performed using the DAS 3.0 program; and pharmacokinetic parameters were analyzed based on the classical compartmental analysis for plasma and tissue concentration-time data. The following pharmacokinetic parameters were estimated: Cmax (peak concentration), $t_{1/2\alpha}$ (half-life of the absorption rate constant), $t_{1/2\beta}$ (half-life of the elimination rate constant), AUC (0-t) (area under the concentration-time curve from time zero to t), and CL/F (total body clearance).

Statistical analysis

Statistical analysis between the two groups was completed using the SPSS 13. 0 program.

Results

Analytical method validation

The analytical procedure for PZQ analysis was modified according to the report of Jing Yao[25]. A linear calibration curve over the 0. 05~20 μg/mL PZQ range was established, yielding a correlation coefficient exceeding 0. 9994. The drug concentration of the samples above the upper calibration limit was diluted with blank plasma and supernatants of the tissue samples. The results showed that all inter-day and intra-day coefficients of variation were below 3. 34%; while tissue sample accuracy ranged from 74% to 85%, and mean recovery was above 76%. Based on a signal-to-noise ratio>3, the lower limits of detection (LOD) and quantitation (LOQ) for this assay were 0. 05 μg/mL.

Pharmacokinetic analysis

The PZQ concentration-time curves are shown in Figure 1. The pharmacokinetic parameters are shown in Table 1. The plasma, muscle and liver of freshwater group grass carp and the plasma, muscle and kidney of brackish water group PZQ concentration-time profiles were bestdescribed by a two-compartmental open pharmacokineticmodel with first-order absorption. The kidney of freshwatergroup and liver of brackish water group PZQconcentration-time profiles were best described by a threecompartmentalopen pharmacokinetic model. The datas ofplasma and tissues drug concentration was calculated bythe software SPSS 13. 0. According to the One-Way ANOVA, the results showed that the salinity had a significant effect on the drug concentration of plasma (p<0.01), musle (p<0.01), liver (p<0.01) and kidney (p<0.01) in the all sampling time points between the brackish water grass carps and the freshwater grass carps.

PZQ distribution and elimination

Tissue concentrations are shown in Figure 1. The absorption of PZQ in freshwater grass carps was essentially the same as the brackish water carps. There were no significant differences in T_{max} of plasma, liver, and kidneys between the two groups.

The mean peak levels in plasma, muscle, liver and kidneys of the brackish water grass carps were 0. 76, 0. 51, 2. 7, and 2. 99 μg/g, respectively; which were lower than that in the freshwater grass carps (0. 91, 0. 62, 3. 87, and 3. 39 μg/g, respectively). In most sampling time points, the plasma and tissue PZQ levels in brackish water grass carps were lower than that in the freshwater grass carps. The plasma and tissue PZQ elimination half-lives (t1/2β) in brackish water grass carps were shorter than that in the freshwater grass carps (Table 1). The PZQ elimination half-lives (t1/2β) in plasma, muscle, liver and kidneys of the brackish water grass carps were 56. 46, 36. 17, 15. 31, and 132. 64 h, respectively; and the figures for freshwater grass carps were 71. 15, 44. 88, 23. 86, and 150. 23 h, respectively. These findings indicated that PZQ elimination in brackish water grass carps was more rapid than that in freshwater grass carps.

Table 1 Pharmacokinetic parameters of Praziquantel in grass carp (n=5)

Parameter	Unit	plasma		muscle		liver		kidney	
		Fresh water	Brackish water	Fresh water	Brackish water	Fresh water	Brackish water	Fresh water	Brackish water
C_{max}	ug/ (g or ml)	0. 91	0. 76	0. 62	0. 51	3. 87	2. 70	3. 39	2. 99
T_{max}	h	0. 5	0. 5	0. 5	1	0. 5	0. 5	1	1
$t_{1/2\alpha}$	h	3. 25	1. 85	0. 31	1. 12	0. 35	2. 69	0. 76	2. 87
$t_{1/2\beta}$	h	71. 15	56. 46	44. 88	36. 17	23. 86	15. 31	150. 23	132. 64
$t_{1/2Ka}$	h	0. 16	0. 028	0. 058	0. 57	0. 15	0. 031	0. 307	0. 64
V1/F	L/kg	23. 57	30. 15	21. 53	22. 22	28. 38	97. 33	34. 30	87. 17
CL/F	L/h/kg	0. 755	1. 042	1. 961	0. 974	0. 481	0. 308	1. 878	0. 350
$AUC_{(0-t)}$	μg/L · h	17. 951	14. 614	8. 096	6. 865	34. 984	44. 047	10. 411	38. 663
$AUC_{(0-\infty)}$	μg/L · h	26. 464	19. 189	10. 196	20. 515	41. 547	64. 823	10. 645	57. 132
K_{10}	1/h	0. 064	0. 035	0. 091	0. 044	0. 17	0. 048	0. 34	0. 16

续表

Parameter	Unit	plasma		muscle		liver		kidney	
		Fresh water	Brackish water	Fresh water	Brackish water	Fresh water	Brackish water	Fresh water	Brackish water
K_{12}	1/h	0.235	0.212	1.574	0.474	1.469	0.113	2.034	0.754
K_{21}	1/h	0.104	0.14	0.548	0.103	0.355	0.036	0.245	0.042
Ka	1/h	4.201	25.183	11.951	1.203	4.537	22.106	2.26	1.082

Cmax, the peak concentration in plasma; Tmax, the time point of the drug's maximum plasma concentration; $t_{1/2\alpha}$, distribution half-life of the drug; $t_{1/2\beta}$, eliminationhalf-life of the drug; $t_{1/2Ka}$, absorption half-life of the drug; V1/F, extensive apparent volume of the central compartment; CL/F, total body clearance of the drug; $AUC_{(0-t)}$, area under the concentration-time curve from 0 h to t; $AUC_{(0-\infty)}$, area under the concentration-time curve from 0 h to ∞; K_{12}, K_{21}, first-order rate constants for drug distribution between the central and peripheral compartments; K_{10}, elimination rate constant from the central compartment; Ka, absorption rate constant.

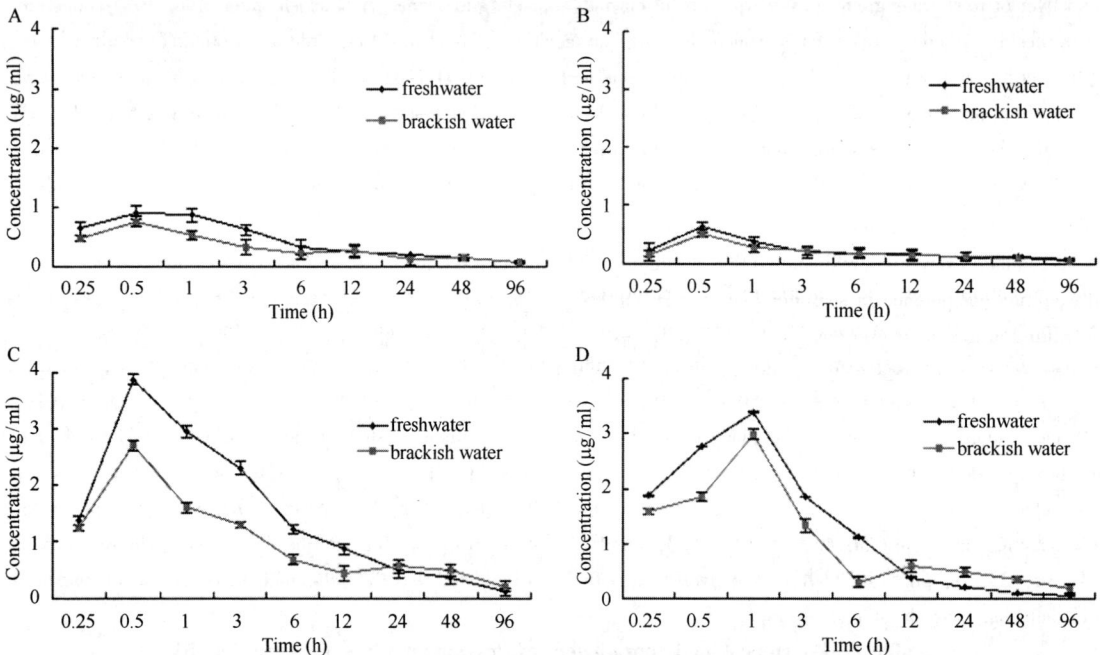

Fig. 1　Praziquantel in plasma and tissues of freshwater and brackish water grass carps after oral administration. Standard deviation is given vertical bars. (n=5, The mean±SD). (A: plasma; B: muscle; C: liver; D: kidney)

Discussion

Monogeneans can rapidly multiply in intensive farming facilities[26], necessitating early diagnosis, as well as rapid and effective treatment, to prevent benign infections from becoming pathogenic. PZQ is a useful chemotherapeutic against helminths of captive fish. There is a need for examining factors that affect its pharmacokinetics in order to minimize the potential development of anthelmintic resistance caused by extended exposure of parasites to subcurative doses[27].

Salinity conditions play an important role in pharmacokinetics and fish tissue residues. For example, oxolinic acid presents in lower concentrations for a longer period of time in seawater teleosts than freshwater teleosts[17]. Ishida[19] reported that oxolinic acid is excreted more slowly in freshwater trout than in seawater trout. Abedini et al.[20] suggested that freshwater trout may be used as a salmonid model to study oxytetracycline pharmacokinetics in seawater salmon; however, drug elimination and clearance rates in freshwater trout are remarkably lower than that in seawater salmon. Feng et al.[21] demonstrated that tissue drug concentrations of seawater tilapia are lower than that in freshwater tilapia; and the elimination of florfenicol in seawater tilapia is more rapid than that in freshwater tilapia. Even though grass carps could live in both fresh and brackish waters, possible PZQ pharmacokinetic differences between freshwater

-acclimated grass carps and brackish water grass carps have never been reported.

In the present study, the datas of plasma and tissues drug concentration was showed that the salinity had a significant effect on the drug concentration of plasma ($p < 0.01$), muscle ($p < 0.01$), liver ($p < 0.01$) and kidney ($p < 0.01$) in the all sampling time points between the brackish water grass carps and the freshwater grass carps. The plasma PZQ concentration-time curve patterns were similar in the two groups, which could be best described by a two-compartmental open pharmacokinetic model with first-order absorption.

The CL/F is the total body clearance of the drug and the $t_{1/2\beta}$ is the elimination half-life of the drug. The two pharmacokinetics parameters can describe how quickly the drug is eliminated from the fish. The CL/F of the freshwater group and brackish water grass carp plasma were 0.755 and 1.042 L/h/kg respectively. The elimination half-lives ($t_{1/2\beta}$) of plasma in freshwater water grass carps and brackish grass carp were 71.15 and 56.46 h, respectively. Compare to the freshwater grass carps, the brackish water grass carps have more rapidly drug clearance and shorter elimination half-lives when the grass carp cultured in the low salinity water. The water salinity change will influence the regulation of osmotic pressure in the grass carps. Even though seawater salinity is greater than brackish water, the results indicated that osmoregulation still existed in brackish water fish, and that their drug excretion pathway may be altered. The results were consistent with previous similar studies [18,20,21].

K_{12} and K_{21} were the first-order rate constants for drug distribution between the central and peripheral compartments. The datas display that the K_{12} and K_{21} of the brackish water group grass carp were lower than that of freshwater group in most tissues such as liver and kidney. This phenomenon indicated that the salinity water could decrease the drug distribution rate between the central and peripheral compartments.

In the most sampling time points, the liver drug concentrations of the brackish water group are lower than that of freshwater group which water has no salinity; indicating that water salinity can affect drug metabolism in the liver, and that the drug may be eliminated through other drug excretion pathways, such as branchial excretion. Sohlberg et al. [28] reported that gill excretion occurs in seawater fish; Atlantic salmon excrete approximately 60% of administered flumequine through the gills in seawater. Feng et al. [21] reported that branchial excretion competence in seawater fish may be attributed to active drinking for osmoregulatory purposes of seawater tilapia; possibly resulting to the rapid excretion of the drug and its metabolites or their ionized complexes via the branchial chloride cells.

Kidneys are the main organs that excrete drugs and its metabolites. When the water has no salinity, the PZQ concentration in kidney was gently rising in the previous time points. Strangely, the PZQ concentration of brackish water grass carp kidneys slowly increased between 0.25 and 0.5 h after dosing, which peaked between 0.5 and 1 h; indicating that water salinity can also affect brackish water fish kidneys due to osmoregulation result in kidney weakening function. For grass carps in brackish water, the drug and its metabolites are likely to stay in the body and be reabsorbed by entero-hepatic cycling to some extent, retarding drug elimination due to the decrease in urine output through osmoregulation. When the renal drug concentration of brackish water grass carps slowly increased to its peak concentration between 0.25 and 0.5 h after dosing, the drug concentrations of the other tissues decreased. This phenomenon illustrated that the drug in the surrounding tissues entered the kidneys at that time, increasing the kidney drug concentration to its peak concentration; which again rapidly declined after micturition.

Conclusion

In summary, our results indicate that salinity level greatly impacts the accumulation and tissue distribution of PZQ in grass carps, and that its elimination in brackish water grass carps is more rapid than that in fresh water grass carps. Further, tissue concentrations of brackish water grass carps are lower than that in freshwater grass carps, after orally administrating the same dosage at the same water temperature. We speculate that the main excretion pathway of the drug is through renal elimination, and the decreased kidney function in brackish water grass carps is likely responsible for the considerable difference in pharmacokinetics between the two groups of grass carps.

References

1. Schmahl G, Mehlhorn H. Treatment of fish parasites. 1. Praziquantel effective against monogenea (Dactylogyrusvastator, Dactylogyrusextensus, Diplozoonparadoxum). Zoo Para. 1985; 71: 727-37.

2. Schmahl G, Taraschewski H. Treatment of fish parasites. 2. Effects of praziquantel, niclosamide, levamisoleHCl, and metrifonate on monogenea (Gyrodactylusaculeati, Diplozoonparadoxum). Parasitol Res. 1987; 73: 341-51.

3. Thoney DA. The effects of trichlorfon, praziquantel and copper sulphate on various stages of the monogenean Benedeniellaposterocolpa, a skin parasite of the cownose ray, Rhinopterabonasus (Mitchill). J Fish Dis. 1990; 13: 385-9.

4. Kim KH, Park SI, Jee BY. Efficacy of oral administration of praziquantel and mebendazole against Microcotylesebastis (Monogenea) infestations of

cultured rockfish（Sebastes schlegeli）. Fish Pathol. 1998；33：467-71.

5. Hirazawa N，Ohtaka T，Hata K. Challenge trials on the anthelmintic effect of drugs and natural agents against the monogenean Heterobothriumokamotoiin the tiger puffer Takifugurubripes. Aquaculture. 2000；188：1-13.

6. Chisholm LA，Whittington ID. Efficacy of praziquantel bath treatments for monogenenean infections of the Rhinobatostypus. J AquatAnim Health. 2002；14：230-4.

7. Stephens FJ，Cleary JJ，Jenkins G，Jones JB，Raidal SR，Thomas JB. Treatments to control Haliotremaabaddon in the West Australian dhufish. Glaucosomahebraicum Aquaculture. 2003；215：1-10.

8. Janse M，Borgsteede FHM. Praziquantel treatment of captive white-spotted eagle rays（Aetobatusnarinari）infested with monogenean trematodes. Bull EurAssoc Fish Pathol. 2003；23：152-6.

9. Sharp N，Diggles BK，Poortenaar CW，Willis TJ. Efficacy of Aqui-S，formalin and praziquantel against the monogeneans，BenedeniaseriolaeandZeuxaptaseriolae，infecting yellowtail kingfish Seriolalalandilalandiin New Zealand. Aquaculture. 2004；236：67-83.

10. Hirazawa N，Mitsuboshi T，Hirata T，Shirasu K. Susceptibility of spotted halibut Verasπervariegatus（Pleuronectidae）to infection by the monogenean Neobenedeniagirellae（Capsalidae）and oral therapy trials using praziquantel. Aquaculture. 2004；234：83-95.

11. Hayward CJ，Bott NJ，Itoh N，Iwashita M，Okihiro M，Nowak BF. Three species of parasites emerging on the gills of mulloway. Argyrosomus japonicas Aquaculture. 2007；1-4：27-40.

12. Buchmann K，Kania PW，Neumann L，de'Besi G. Pseudodactylogyrosis in Anguilla anguilla（Actinopterygii：Anguilliformes：Anguillidae）：change of control strategies due to occurrence of anthelmintic resistance. ActaIchthyologicaEtPiscatoria. 2011；2：105-8.

13. Hirazawa N，Akiyama K，Umeda N. Differences in sensitivity to the anthelmintic praziquantel by the skin-parasitic monogeneans Benedeniaseriolae and Neobenedeniagirellae. Aquaculture. 2013；404-405：59-64.

14. Bylund G，Sumari O. Laboratory tests with Droncit against diplostomiasis in rainbow trout，Salmo gairdneri Richardson. J Fish Dis. 1981；4：259-264.

15. Mitchell AJ. Importance of treatment duration for praziquantel used against larval digenetic trematodes in sunshine bass. J AquatAnim Health. 1995；7：327-30.

16. Sanmartin Duran ML，Caamano-Garcia F，Fernandez Casal J，Leiro J，Ubeira FM. Anthelminthic activity of praziquantel，niclosamide，netobimin and mebendazole against Bothriocephalusscorpiinaturally infecting turbot（Scophthalmus maximus）. Aquaculture. 1989；76：199-201.

17. Zuo J，Yu Y，Chen Z. The analysis of sea level variation factor along China coast. Advance Earth Sci. 1994；5：48-53.

18. Ishida N. Comparison of tissue-level of oxolinic acid in fresh and sea-water fishes after the oral-administration. Nippon Suisan Gakkaishi. 1990；2：281-6.

19. Tachikawa M，Sawamura R，Okada S，Hamada A. Differences between freshwater and seawater killifish（Oryziaslatipes）in the accumulation and elimination of pentachlorophenol. Arch Environ ContamToxicol. 1991；21：51-146.

20. Ishida N. Tissue levels of oxolincic acid after oral or intravascular administration to freshwater and seawater rainbow trout. Aquaculture. 1992；102：9-15.

21. Abedini S，Namdari R，Law FCP. Comparative pharmacokinetics and bioavailability of oxytetracycline in rainbow trout and chinook salmon. Aquaculture. 1998；1：23-32.

22. Jing-Bin F，Xiao-Ping J，Liu-Dong L. Tissue distribution and elimination of florfenicol in tilapia（Oreochromisniloticus&×&O. caureus）after a single oral administration in freshwater and seawater at 28℃. Aquaculture. 2008；1-4：29-35.

23. Zhang H，Zhao M，Zhuang Z. The research status of domestic fishery drug metabolism dynamics. Progress Veterinary Med. 2010；31：138-143.

24. Li X，Li X，Leng X，Liu X，Wang X，Li J. The effect of salinity on growth of grass carp and muscle quality. Journal Fisheries China. 2007；31：343-8.

25. Jing Yao. The Toxicities and pharmacokineties Residues of Praziquantel on Goldfish（Catassiusauratus）by oral and bath treatment. Master thesis. Sichuan Agricultural University，College of Veterinary Medicine；2007.

26. Ernst I，Whittington I，Corneillie S，Talbot C. Monogenean parasites in sea-cage aquaculture. Austasia Aquaculture. 2002；1：46-8.

27. Tubbs LA，Tingle MD. Bioavailability and pharmacokinetics of a praziquantel bolus in kingfish Seriolalalandi. Dis Aquat Organ. 2006；2-3：233-8.

28. Sohlberg S，Martinsen B，Horsberg TE，Søli NE. Excretion of flumequine in free-swimming Atlantic salmonSalmosalar，determined by cannulation of the dorsal aorta，gall bladder and urethra. J Vet PharmacolTher. 1999；22：72-5.

该文发表于《华中农业大学学报》【2015，4：102-107】

盐度对吡喹酮预混剂在草鱼体内吸收及其残留消除规律的影响

Effect of salinity on tissue absorption and elimination of praziquantel premix in grass carp（*Ctenopharyngodon idellus*）

谢欣燕，赵依妮，杨先乐，胡鲲

（上海海洋大学国家水生动物病原库，上海 201306）

摘要： 运用高效液相色谱法（HPLC），在以每千克鱼体质量单次口灌 500 mg 剂量条件下，对比吡喹酮预混剂中有效成分吡喹酮在淡水及半咸水草鱼体内的药动学差异。结果显示，在不同的试验组间，血浆、肌肉、肝脏和肾脏在各个时间点的药物浓度存在极显著性差异（$P<0.01$）；淡水试验组的血浆、肌肉、肝脏和肾脏的药物峰质量浓度（C_{max}）分别为1.13、1.43、5.55和3.72 μg/mL，达峰时间均为1 h；半咸水试验组的血浆、肌肉、肝脏和肾脏的药物峰质量浓度（C_{max}）则分别为0.98、1.1、3.99和3.2 μg/mL，达峰时间均为1 h，表明淡水条件下草鱼体内的吡喹酮质量浓度高于在半咸水条件，而对吸收没有造成显著影响；淡水试验组的血浆、肌肉、肝脏和肾脏的消除半衰期分别为14.82、5.99、17.51和5.36 h，半咸水试验组的血浆、肌肉、肝脏和肾脏的消除半衰期分别为10.77、5.57、3.49和3.37 h，明显比淡水试验组的消除半衰期短，这可能说明了在水体中，渗透压会影响吡喹酮在草鱼体内的吸收和消除，建议在实际生产中吡喹酮预混剂的用药方案要根据养殖水体情况做适当的调整。

关键词：半咸水；吡喹酮预混剂；草鱼；吸收；残留；消除

该文发表于《水产学报》【2005，29（1）：60-65】

氯霉素在尼罗罗非鱼血浆中的药代动力学研究

The comparative pharmacokinetics of chloramphenicol in *Oreochromis niloticus*

杨先乐，湛嘉，胡鲲

（上海高校水产养殖 E-研究院，上海水产大学农业部水产种质资源与养殖生态重点开放实验室，
农业部渔业动植物病原库，上海 200090）

摘要：建立了尼罗罗非鱼血浆内氯霉素的反相液相色谱测定法（RT-HPLC），并研究了以 50 mg·kg^{-1} 剂量的氯霉素单次给药后，不同水温、不同年龄尼罗罗非鱼血浆中药代动力学的差异。结果表明，RT-HPLC 法检测血浆中的氯霉素基线走势平稳，药峰与杂质峰分离良好，在空白血浆样品中添加0.25，2.5 和12.5 μg·mL^{-1}标准品，3个水平的平均回收率可达（90.47 ± 3.42）%，且变异系数小于10%；日内精密度和日间精密度均较高，平均变异系数分别为1.199%和1.770%。研究揭示，水温和年龄对氯霉素在尼罗罗非鱼血浆中的药代动力学有一定的影响。水温升高，氯霉素在罗非鱼血浆中吸收、分布和消除速度则明显加快，26℃下 $T_{1/2 ka}$ 仅为 18℃的1/3，$T_{1/2\alpha}$ 和 $T_{1/2\beta}$ 也明显缩短，T_p 不到18℃的一半，CLs 是18℃的1.74 倍，氯霉素在一龄罗非鱼血浆中的吸收速度比二龄罗非鱼快，达峰时间短，消除速度明显减慢，但分布半衰期和曲线下面积（AUC）大小二者比较接近。鉴于 CAP 毒性强，在对水产品中的 CAP 残留进行监控时，应以养殖水温较低、年龄较小者作为监控的重点，以确保水产品的安全。

关键词：尼罗罗非鱼；氯霉素；水温；年龄；药物代谢动力学

氯霉素（chloramphenicol，CAP）由于对水产动物病原菌的抑菌作用强、价格便宜，曾广泛地应用于水产养殖中各种细菌病的预防与治疗上。但是因 CAP 较大的副作用和边缘效应，如导致"灰婴综合征"和"再生障碍性贫血"等[1]，世界卫生组织（WHO）和美国食品及药物管理局（FDA）规定禁止 CAP 用于所有食用动物的养殖中，世界各国对 CAP 的残留监控的标准也越来越高，欧美一些国家要求 CAP 的最高残留限量（maximum residue level，MRL）为零（检测限量为0.1 μg·kg^{-1}），我国也禁止在水产养殖中使用 CAP[2]。为了作好 CAP 的监控，了解它在水产动物体内药代动力学的规律，了解各种因素对 CAP 代谢影响显得十分重要。然而关于 CAP 在水产动物体内药代动力学的研究却鲜有报道。目前仅见 CAP 在虹鳟（Salmo gairdneri R.）[3]、对虾（Penaeus）[4]、草鱼（Ctenopharyngodon idellus）和异育银鲫（Carassius auratus gibelio）[5]等水产动物体内药动学的研究，这些研究除了李爱华[5]比较了 CAP 在两种不同种鱼体内药代动力学差异外，大多均为单因子试验。罗非鱼是世界性养殖的主要鱼类[6]，也是我国出口创汇的主要鱼类之一，探讨 CAP 在罗非鱼体内药动学规律，比较不同水温、年龄对它在罗非鱼体内的代谢差异，对加强 CAP 的管理和监控，保证食品安全等均具有重要的意义。

1 材料和方法

1.1 实验动物和试剂

吉富尼罗罗非鱼（Oreochomisniloticus，GIFT）购于上海宝山盛桥水产技术推广站，健康无病，未用过任何药物。1龄鱼平均体重为（65.02±7.70）g，2龄鱼平均体重（241.40±28.42）g；试验前于4 m×2 m×1 m 的网箱中暂养7d，并在给药前3 d 停止投饲。试验期间投喂罗非鱼专用饲料，饲料由上海大江水产饲料公司惠赠。

CAP 标准品（纯度为99.4%）由上海药品检验所提供，CAP 原粉（纯度为98.5%）购于上海第六制药厂。甲醇为HPLC 级，高氯酸、正己烷均为分析纯。

1.2 给药、取样与样品前处理

所有试验鱼均以 50 mg·kg^{-1} 鱼体重的剂量口灌给药，给药后 0.5 h，1 h，2 h，4 h，8 h，12 h，24 h，48 h，96 h，192 h，288 h 围心腔取血，肝素钠抗凝；取4尾鱼的血浆合并成1个样品后于-80℃保存备用。每组均设1个重复。给药前取空白的血浆样品作对照。样品检测前，准确吸取 300 μL 样品于 2 mL 离心管中，加入甲醇 600 μL，旋涡混合 10 min，10 000 r·min^{-1} 离心 10 min 取上清液，经 0.45 μm 微膜过滤后作自动进样分析。

1.3 RT-HPLC 建立及药代动力学参数计算

样品用 Agilent 1100 高效液相色谱仪分析，分析柱为 RP-ODS C$_{18}$柱（15 cm×4.6 mm，5 μm）。流动相：甲醇：水 = 35：65（V/V），波长：278 nm，流速：1.0 mL·min^{-1}，柱温：40℃，进样量：10 μL。以常规方法建立标准曲线。在空白血浆中分别加入 2.5，25 和 125 μg·mL^{-1} 的 CAP 系列标准溶液各 100 μL，每个浓度重复3次，测定回收率；随机取4个样品，每个样品每隔 2 h 平行测定一次，计算平均峰面积及 RSD，确定日内精密度；将4种不同浓度的标准 CAP 溶液测定2次，每隔3 d 一次，计算平均峰面积、标准差及相对标准偏差（RSD），确定日间精密度。

用药动学程序软件 MCPKP 口服给药模型对各组试验数据进行模型嵌合和参数计算。

1.4 温度和年龄对 CAP 在罗非鱼血浆中药代动力学影响

比较水温为 26 和 18℃时 2 龄罗非鱼以及在同一水温（18℃）下 1 龄罗非鱼与 2 龄罗非鱼药代动力学规律，以探讨它们之间的差异。

2 结果

2.1 检测方法的可靠性

用 RT-HPLC 法检测血浆中 CAP，基线走势平稳，CAP 保留时间约为 7.1 min，药峰与杂质峰分离良好（图1），其标准曲线方程为 $y = 16.814x$（$R = 0.9999$），线性良好。以两倍噪声为最低检测限，此法灵敏度为 0.05 μg·mL^{-1}。平均回收率达（90.47±3.42）%，相对标准偏差为（8.00±0.73）%（表1）。日内精密度和日间精密度均较高，相对标准偏差分别为 1.199% 和 1.770%（表2）

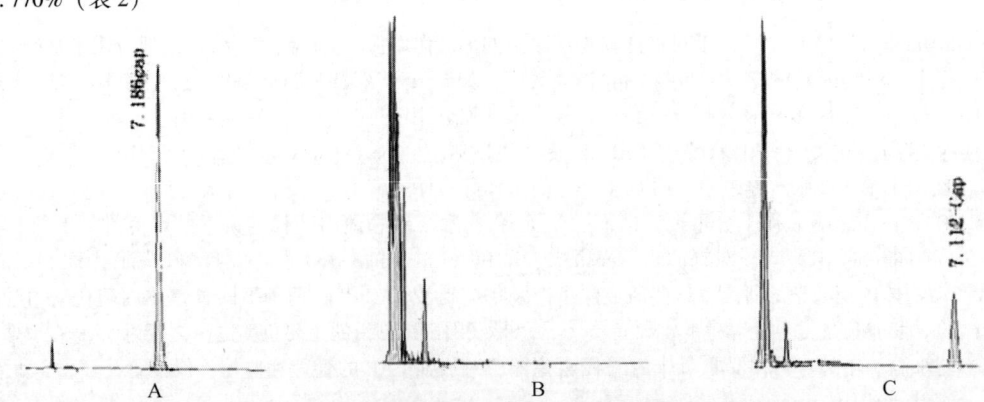

图1　尼罗罗非鱼血浆内 CAP 的色谱行为

A. CAP 标准品；B. 空白血浆；C. 在空白血浆中添加了 CAP

Fig. 1　The liquid chromatogram of CAP in the plasma of O. niloticus

A. the standard of CAP；B. the blank plasma；C. the blank plasma added CAP

表1　尼罗罗非鱼血浆内 CAP 反相高效液相色谱测定法的回收率

Tab. 1　The recoveries of RT-HPLC method for CAP in plasma of O. niloticus

样品的理论浓度 （μg·mL⁻¹） experiment concentration	实测浓度 （μg·mL⁻¹） measured concentration	回收率（%） recovery	平均回收率（%） mean±SEM	相对标准偏差 （%）RSD
0．25	0．248/0．257/0．202/0．232	93.93±7．12		
2．5	2．481/2.117/2．201/2．291/2.641	93.86±7．32	90．47±3．42	8．00±0．73
12．5	9．155/10．226/11．882	83.63±7．61		

表2　尼罗罗非鱼血浆内 CAP 反相高效液相色谱法测定的精密度

Tab. 2　The precision of RP-HPLC method of CAP in the plasma of O. niloticus

项目 item	测定样品 sample	日内精密度 intra-assay precision	日间精密度 inter-day precision
峰面积均值±标准差（n） mean±SD	1	39．430±0.561（4）	44．313±1．591（2）
	2	70．843±1.110（4）	83．575±0．494（2）
	3	107．496±1．012（4）	207．557±1.971（2）
	4	136．004±2．750（4）	423.445±11.270（2）
变异系数（%） CV	1	1．410	3．590
	2	1．570	0.591
	3	0．992	0.238
	4	2．024	2.661
平均变异系数（%） mean CV		1．199	1.770

2.2　不同水温条件下 CAP 在尼罗罗非鱼血浆中药代动力学规律比较

水温26℃和18℃时2龄尼罗罗非鱼药代动力学的结果表明，水温较高时，药物在血浆中的吸收、分布以及消除速度均较快；在水温26℃时，CAP 在第2 h 即达到峰值22.246 μg·mL⁻¹，随后开始下降，第4、8、12小进分别下降到20.513、14.289、13.742 μg·mL⁻¹，第24小时仅约为第2小时的1/8，为2.816 μg·mL⁻¹，第96小时降至0.210 μg·mL⁻¹，第192小时后则未能检出（检测限0.05 μg·mL⁻¹）；而水温18℃下，血浆中 CAP 在12 h 才到达最大值22.792 μg·mL⁻¹，此后下降幅度虽然较快，但在同一采样点均高于26℃，第24、48、96 h 分别为5.948、3.939、0.736 μg·mL⁻¹，第192小时仍高于26℃第24 h 的水平，为0.297 μg·mL⁻¹，第288小时仍达0.148 μg·mL⁻¹的水平（图2）。

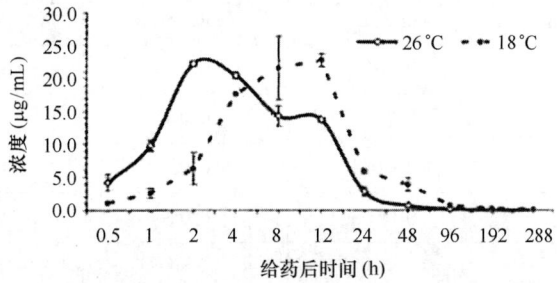

图2　水温对尼罗罗非鱼血浆中氯霉素药代动力学的影响

Fig. 2　The effect of water temperature on the pharmacokinetics of chloramphenicol in plasma of O. niloticus

2.3 CAP 在不同年龄尼罗罗非鱼血浆中药代动力学规律的比较

图 3 比较了水温18℃时 2 龄和 1 龄尼罗罗非鱼药代动力学的规律。结果表明 CAP 在 1 龄罗非鱼血浆中比 2 龄尼罗罗非鱼的分布要稍快；1 龄尼罗罗非鱼 4 h 时到达峰值 26.504 μg·mL^{-1}而 2 龄达到峰值却在 12 h 时，而且峰浓度略比 1 龄低；24 h 时二者基本上处于同一水平（5.557 μg·mL^{-1}和 5.948 μg·mL^{-1}），以后二者在同一水平上波动下降，至 288 h 时 1 龄和 2 龄罗非鱼降至 0.247 μg·mL^{-1}和 0.148 μg·mL^{-1}。

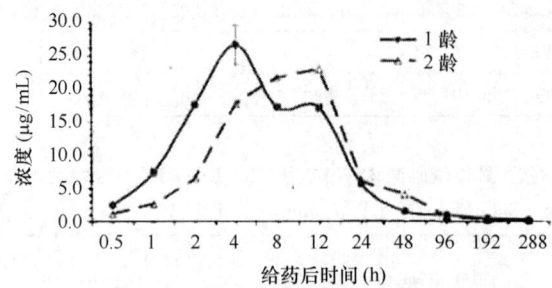

图 3　不同年龄的尼罗罗非鱼血浆氯霉素药代动力学的比较

Fig. 3　The comparison of pharmacokinetics of chloramphenicol in plasma between different ages of O. niloticus

2.4 不同水温和年龄的 CAP 药代动力学参数

MCPKP 药动学程序软件处理的结果表明，无论何种情况，CAP 在尼罗罗非鱼血浆中的药代动力学特征均适合用一次式二室开放模型来描述。根据此模型，可获得一些药代动力学参数（表 3）。从表 3 可以看出，水温较高时，吸收相半衰期（$T_{1/2ka}$）、分布相半衰期（$T_{1/2\alpha}$）和消除相半衰期（$T_{1/2\beta}$）消除率均较高，达峰时间（T_p）短，峰浓度高（C_{max}）、（CLs）快；而血液浓度-时间曲线下面积（AUC）药却较小，26℃仅为 18℃的 57%。1 龄和 2 龄罗非鱼相比，CAP 在 1 龄罗非鱼血浆中的吸收相半衰期（$T_{1/2ka}$）、分布相半衰期（$T_{1/2\alpha}$）都短，而消除相半衰期（$T_{1/2\beta}$）却较长，达峰时间短（T_p），峰浓度（C_{max}）高，二者的清除速率（CLs）和血液浓度-时间曲线下面积（AUC）则基本相同。

表 3　水温和年龄对 CAP 在罗非鱼血浆中的药物动力学参数比较

Tab. 3　The comparison of pharmacokinetic parameters of CAP in plasma of O. niloticusbetween different water temperatures and ages

药动学参数 pharmacokinetic parameters	单位 unit	水温 water temp（℃）（年龄 age：2 龄 2 years old）		年龄 age（水温 water temp. 18.0±0.5 ℃）	
		26.0±0.5	18.0±0.5	2 龄 2 years old	1 龄 1 year old
dose	mg·kg^{-1}	50.000	50.000	50.000	50.000
Cor.	–	0.980	0.977	0.977	0.988
C_0	μg·mL^{-1}	30.351	34.556	34.556	32.979
K_a	h^{-1}	0.704	0.247	0.247	0.655
K_{12}	h^{-1}	0.013	0.016	0.016	0.031
K_{21}	h^{-1}	0.016	0.014	0.014	0.009
KeL	h^{-1}	0.091	0.059	0.059	0.055
$T_{1/2ka}$	h	0.985	2.809	2.809	1.059
$T_{1/2\alpha}$	h	6.521	8.865	8.865	7.731
$T_{1/2\beta}$	h	50.721	65.402	65.402	118.477
AUC	μg·h^{-1}·mL^{-1}	334.300	583.410	583.410	599.820

续表

药动学参数 pharmacokinetic parameters	单位 unit	水温 water temp（℃） （年龄 age：2 龄 2 years old）		年龄 age （水温 water temp. 18.0±0.5 ℃）	
		26.0±0.5	18.0±0.5	2 龄 2 years old	1 龄 1 year old
T_p	h	3.494	7.756	7.756	4.196
C_{max}	μg·mL^{-1}	21.736	20.513	20.513	24.070
CLs	L·kg^{-1}·h	0.150	0.860	0.860	0.830

注：$C_{or.}$ 为相关系数；Co 为初始药物浓度；K_{10} 为药物自中央室的消除速率常数；K_{12} 为药物自中央室到周边室的一级转运速率常数；K_{21} 为药物自周边室中央室到中央室的一级转运速率常数；$T_{1/2ka}$ 为吸收半衰期；$T_{1/2\alpha}$ 为药物的分布相半衰期；$T_{1/2\beta}$ 为药物的消除相半衰期；Tp 为达峰时间；C_{max} 为峰浓度；AUC 为血药浓度-时间曲线下面积；CLs 为清除率。

Notes：Cor.：relation coefficient；Co：initial concentration；K$_{10}$：drug elimination constants from mid-room；K$_{12}$：first-order transport constant from mid-room to peripheral-room；K$_{21}$：first-order transport constant from peripheral-room to mid-room；T$_{1/2ka}$：half-life of absorption；T$_{1/2\alpha}$：half-life of distribution；T$_{1/2\beta}$：half-life of elimination；T$_p$：time to peak concentration；Cmax：peak concentration；AUC：area under concentration-time curve；CLs：the body clearance

3　讨论

HPLC 方法是药物检测的一种常用方法，具有分离效果好、特异性强、测定精密度高、重现性好和使用较方便等特点。在研究 CAP 在对虾[4]、草鱼和高倍体银鲫[5] 体内的药代动力学时，均用 HPLC 方法。本试验对 CAP 在罗非鱼血浆内的检测方法的研究进一步证实，该方法不仅基线走动平稳，无干扰峰，而且有较高和稳定的回收率、较高的精密度。虽然它的灵敏度相对较低，不能达到对 CAP 最高残留限量的要求，在对 CAP 进行精密的残留检测和监控上有一定的缺陷，但由于它前处理方法简单，操作方便，对于样品量大的药动学研究，以及较大规模的 CAP 残留普查，仍具有较强的使用意义。

鱼类是水生变温动物，水温的高低会影响其生命活动，也会影响机体对药物作用。一般说来，在一定温度范围内，药物的代谢强度与水温成正比，水温越高，代谢速度加快。有人认为水温每升高 1℃，鱼类的代谢和消除速度将提高 10%[7]。Curtis 等[8] 在研究不同水温下虹鳟胆汁对药物的排泄时发现，在水温 18℃ 和 14℃ 下排泄速度要明显快于 10℃，血浆消除半衰期也明显较短。Kleinow 等[9] 发现，14℃ 和 24℃ 的口服生物利用率分别为 56%，30.7%，$T_{1/2\beta}$ 分别为 54.3 h 和 33.1 h。Björklund 等[7] 认为，5℃、10℃、16℃ 下土霉素在虹鳟血清中达峰时间分别为 24 h、12 h 和 1 h，消除时间最长相差可达 108 d 之久。Karara 等[10] 研究温度对 2-乙基己基邻苯二甲酸盐（DEHP）在红鲈（Cyprinodonvariegatlls）体内药动学规律时也发现，在 10、16、23、29 和 35℃ 等不同温度下，药物的吸收和清除随温度升高呈线性上升，他认为温度升高诱导了鳃血流量的增加。Namdari 等[11] 在研究水温对大鳞大麻哈鱼组织中土霉素（OTC）分布和消除的影响时也得到了类似结果，当水温由 9℃ 上升到 15℃，OTC 在大鳞大麻哈鱼肌肉中的 $T_{1/2\beta}$ 则由 13.59 d 缩短到 7.14 d。本文的研究结果与上述报道基本相同。

我们认为温度对 CAP 在尼罗罗非鱼血浆内的吸收、分布与消除有较大的影响，水温较高时 CAP 在罗非鱼体内的运转速度快，表现在 $T_{1/2ka}$、$T_{1/2\alpha}$ 和 $T_{1/2\beta}$ 增高，达峰时间（T_p）缩短，峰浓度高（C_{max}）、消除率（CLs）加快；另一方面，当水温升高后 AUC 却明显降低，表明水温也较大地影响着生物利用率。因此在对 CAP 进行监控时应要特别注意水温对 CAP 代谢的影响，特别是对于曾用过或曾被 CAP 污染过水体，在水温较低时，应要特别注意 CAP 在其水生动物中的残留，以免造成较大的危害。

Michel 等[12] 在研究磺胺在斑点叉尾（Ictalurus punctatus）体内的药代动力学的规律时发现，未成熟幼鱼比成鱼的对药物的清除速度要快，总消除的时间明显减少。而我们的试验结果与其有所区别，1 龄的罗非鱼比 2 龄罗非鱼对 CAP 的吸收速度快，达峰时间短，但消除速度明显缓慢。我们认为造成不同年龄罗非鱼对 CAP 吸收与消除速度差异的原因可能是 1 龄鱼对药物吸收的表面积相对比 2 龄鱼大，而呈现出较快的吸收速度，从而使达峰时间缩短；另一方面由于 1 龄鱼的肾脏排泄功能可能尚未发育完全，或者容易受 CAP 的影响而使其功能下降，从而影响了药物的排泄，而导致消除速度较 2 龄鱼慢。

CAP 是酰胺醇类的一种抗菌药物，它主要通过阻碍细菌 70S 核糖体与 50S 亚基的结合，干扰细菌蛋白质合成而产生抑菌作用。因为它对较多的水产动物致病菌的抑菌效果好，使用价格较便宜，曾广泛地用于防治水产动物的疾病上。但由于哺乳动物和人骨髓细胞等组织的线粒体核糖体与细菌的相同，CAP 能抑制多种动物（包括人）的骨髓造血功能，使红细胞、白细胞和血小板减少，它还能导致一种与剂量无关的不可逆的再生性贫血，严重地危害着动物和人的生命[13]；因此对

307

CAP 的监控已引起我国和世界上大部分国家的高度重视。由于 CAP 在水产养殖疾病防治上惯性作用，以及在水产养殖品的生产和加工过程中可能出现的 CAP 污染，CAP 在水产养殖动物中的残留仍不可忽视，因此加强对 CAP 的监控显得十分必要。本研究所建立的 RT-HPLC 法检测 CAP 因操作较简单可作为监控的一种普查方法。此外根据本研究，在较低温度下 CAP 的代谢速度较慢，CAP 在年龄较小的罗非鱼中消除要比较大年龄的罗非鱼要慢，因此我们在进行监控时，要十分注意将可能受到 CAP 污染、养殖水温较低、年龄较小者作为监控的重点，以确保水产品的安全。

参考文献

［1］ Kijak P J. Confirmation of chloramphenicol residues in bowine milk by gas chromatography/mass spectrometry ［J］. J AOACInt, 1994, 77 (1): 34-45.

［2］ 农业部渔业局全国水产标准化技术委员会. 中国水产标准化汇编-无公害食品卷 ［S］. 北京: 中国标准出版社, 2002. 147-154.

［3］ Cravedi J P, Heuillet G, Peleran J C, et al. Disposition and metabolism of chloramphenicol in trout ［J］. Xenobiotica, 1985, 15 (2): 115-121.

［4］ 李兰生, 王勇强. 对虾体内氯霉素含量测试方法的研究 ［J］. 青岛海洋大学学报, 1995, 25 (3): 400-406.

［5］ 李爱华. CAP 在草鱼和复合四倍体异育银鲫体内的比较药代动力学 ［J］. 中国兽医学报, 1998, 18 (4): 372-374.

［6］ Kim M S, Park S W, Huh M D, et al. Effect of temperature and bacterial infection on the absorption and elimination of oxolinic acid in Nile tilapia (Oreochromisniloticus) ［J］. J Korean Fish Soc, 1998, 31 (5): 677-684.

［7］ Björklund H, Bylund G. Temperature-related absorption and excretion of oxytetracycline in rainbow trout (SalmogairdneriR.) ［J］. Aquac, 1990, 84: 363-372.

［8］ Curtis L R, Kemp C J, Svec AV. Biliary excretion of ［super (14) C］ taurocholate by rainbow trout (Salmo gairdneri) is stimulated at warmer acclimation temperature ［J］. Comp BlochemPhysiol, 1986, 84 (1): 87-90.

［9］ Kleinow K M, Jarboe H H, Shoemaker K E. Comparative pharmacokinetics and bioavailability of oxolinic acid in channel catfish (Ictalurus punctatus) and rainbow trout (Oncorhynchusmykiss) ［J］. Can J Fish AquacSci, 1994, 51: 1205-1211.

［10］ Krara A H, Hayton W L. A pharmacokinetic analysis of the effect of temperature on the accumulation of di-2-ethylhexyl phthalate (DEHP) in sheepshead minnow ［J］. AquatToxicol, 1989, 15 (1): 27-36.

［11］ Namdari R, Abedini S, Law F C P. Tissue distribution and elimination of oxytetracyline in seawater chinook and coho salmon following medicated-feed treatment ［J］. Aquac, 1996 (144): 27-38.

［12］ Michel C M F. Pharmacokinetics and metabolism of sulfadimethoxine in channel catfish (Ictalurus punctatus) ［J］. Xenobiotica, 1990, 51 (1): 230-238.

［13］ 《医用药理学》编写组. 医用药理学 (第二版) ［M］. 北京: 人民卫生出版社, 1984. 750-752.

第二章　主要渔药的药代动力学及其合理使用

渔药在水产动物体内的变化可从体内过程和速率过程来描述。体内过程指渔药在水产动物体内的吸收、分布、转化、排泄（ADMA）规律及其影响这个过程的因素，由此可以了解渔药的起效时间，效应强度和持续时间，了解渔药在体内的变化规律；速率过程是一个动力学过程，定量地描述渔药在水产动物体内的动态变化，由此可绘制相应的曲线，选取适当的模型，以及建立准确的数学方程，获得一些重要的药动学参数，为制定和调整用药方案、合理使用渔药提供重要依据。

本章主要阐述了几类常用渔药在水产动物体内的药动学规律及其相应的参数。

第一节　抗生素类渔药在水产动物体内的药代动力学

抗生素（antibiotics）主要来源于细菌、放线菌和丝状真菌等微生物，但现在已有较多的抗生素能人工合成或半合成。目前用于防治水产动物疾病的抗生素主要有氨基糖苷类、四环素类和酰胺醇类等。为了确保水产品的质量安全和生态环境的安全，防止耐药性的产生和传播，对抗生素的使用进行严格限制，并根据其药理学特性合理使用已是一个不容忽视的重要问题。

一、四环素类

四环素类的抗生素是一类具有共同多环并四苯羧基酰胺母核的衍生物，它们仅在5、6、7位的取代基有所不同。四环素类可分为天然品和半合成两类，前者是由不同链霉菌的培养液中提取获得的，包括四环素、土霉素、金霉素等，具有性质稳定的特点，后者为半合成衍生物，如强力霉素（多西环素）、美他霉素、米诺霉素（二甲胺四环素）等。水产常用的有土霉素、四环素、金霉素和强力霉素等。研究发现，土霉素在牙鲆血液中的浓度明显低于土霉素在斑节对虾中的血液浓度（纪荣兴等，2008；孙贝贝等，2011）；强力霉素在斑点叉尾鮰（Ictalurus punctatus）体内吸收迅速、达峰快、消除缓慢（丁俊仁等，2012）。用剂量20 mg/kg的强力霉素连续对斑点叉尾鮰口服5 d，其在肝脏中浓度最高，肌肉+皮中浓度最低，在不同组织中强力霉素的消除速率明显不同（丁俊仁等，2012）。

药效与药物消除过程存在某种联系，采用体内药动学和体外药效学相结合的方法，通过抑制效应的 Sigmoid Emax 模型方程可得到治疗斑点叉尾鮰细菌性败血症的最佳临床给药方案：认为用 10.68 mg/kg 体重的剂量强力霉素拌饲投喂，每 1 天 1 次，可预防斑点叉尾鮰由嗜水气单胞菌（Aeromonas hydrophila）引起的细菌性败血症（艾晓辉等，2011）。

该文发表于《水生生物学报》【2012, 36 (1)：126-132】

强力霉素在斑点叉尾鮰体内药物动力学及残留消除规律研究

STUDIES ON TISSUE DISTRIBUTION AND THE ELIMINATION REGULARITY OF DOXYCYCLINE RESIDUES IN *ICTALURUS PUNCTATUS*

丁俊仁[1]，艾晓辉[2]，汪开毓[3]，刘永涛[2]，袁科平[2]，袁丹宁[4]

（1. 四川省畜牧科学研究院，成都 610066；2. 中国水产科学研究院长江水产研究所，农业部淡水鱼类种质监督检验测试中心，武汉 430223；3. 四川农业大学动物医学院，雅安 625014；4. 华中农业大学水产学院，武汉 430070）

摘要： 研究利用高效液相色谱法研究了强力霉素在斑点叉尾鮰（Ictalurus punctatus）体内的药物动力学与消除规律，有助于制定合理用药方案和休药期，为水产品质量安全提供理论依据。（1）单次口服剂量 20 mg/kg 强力霉素在斑点叉尾鮰体内的药时数据符合二室开放式模型。药-时曲线呈明显双峰现象：第一次达峰时，强力霉素在肾、血和肌肉中浓度迅速上升，达峰时间 $T_{max(1)}$ 出现在 30 min，强力霉素在肝脏中浓度上升缓慢，出现在 1 h；肝、肾、血和肌肉第二次达峰的时间 $T_{max(2)}$ 出现在 8 h，第二次达峰浓度 $C_{max(2)}$ 大于第一次的浓度 $C_{max(1)}$。药-时曲线下面积（AUC）：肾、肝、血和肌肉分别为 63.242、1282.077、142.379、62.348 μg·h/mL。消除半衰期 $[T_{1/2b}]$：肾、肝、血和肌肉分别为 40.668、48.767、36.527、31.091 h，平均滞留时间（MRT）：肾、肝、血和肌肉分别为 46.585、56.989、48.859、42.428 h；（2）连续口服剂量 20 mg/kg 的强力霉素 5 d，停药后强力霉素在斑点叉尾鮰尾肝脏中浓度最高，肌肉+皮中浓度最低。在不同组织中强力霉素的消除速率不同（$P<0.05$），药物消除速度由高到低依次为肌肉+皮、肾脏、肝脏。若以肝脏为靶组织，最高残留限量 300 μg/kg，休药期不低于 30 d；若以可食组织肌肉+皮为靶组织，最高残留限量 300 μg/kg，休药期不低于 19 d。

关键词：斑点叉尾鮰；强力霉素；组织分布；残留消除规律

强力霉素（Doxycycline，DOTC）又名多西环素、脱氧土霉素，是半合成的四环素类抗菌药，该药于 1963 年由 Stophens，et al. 合成，1971 年在我国开始投入生产，具有抗菌活性和组织穿透性强、体内分布广、生物利用度高、半衰期较长等优点，现广泛应用在兽医临床[1-3]和水产上。目前，四环素类抗生素是我国畜禽饲养业中生产量和临床使用量最大的抗生素，强力霉素是四环素类抗生素，其抗菌活性较四环素提高了一倍，尤其在大肠杆菌、巴氏杆菌、支原体感染等水产疾病上效果显著，美国和日本仍在使用其作为饲料药物添加剂[4]。由于人体长期摄入含抗生素残留的水产品，抗生素会不断在体内蓄积，当达到一定量时，就会对人体产生毒性作用。我国水产品出口也因抗菌素药物超标问题而多次遭到欧盟、美国、日本等国家"绿色壁垒"的抵制，给国家造成了巨大经济损失，因而在水产养殖中如何安全使用强力霉素是个重要问题。目前，国内外对强力霉素在鱼体内的组织分布和残留消除规律的研究未见报道，仅在羊、鸡、牛、马、猪等畜禽中进行过相关研究[5-7]。据实践证明，斑点叉尾有明显的经济及产业发展优势，农业部已将其列为三个淡水鱼类出口优势产品之一。本实验首次利用高效液相色谱法研究了强力霉素在斑点叉尾鮰体内的分布、残留与消除规律，有助于制定合理用药方案和休药期，从而为水产品质量安全提供理论依据。

1 材料与方法

1.1 试验动物

健康斑点叉尾鮰由中国水产科学研究院长江水产研究所窑湾试验场提供，平均质量为（100±6）g。实验前在 1 m³ 水族箱内暂养一周，每天换水一次以上，饲喂不加抗菌药物的斑点叉尾鮰全价配合饲料。试验用水为曝气 48 h 的自来水，连续充氧，保持水中溶氧大于 8.0 mg/L，试验水温为（28±1）℃，pH 为 7.5~8.0，试验过程中停饲。

1.2 药品及试剂

强力霉素原粉（纯度≥98.0%，为东北制药总厂生产）；强力霉素标准品（纯度≥99.0%，购自德国 Dr. Eherestorfer 公司）；化学试剂：乙腈（HPLC 级，上海国药集团产品），甲醇（HPLC 级，美国 Merck 公司产品），肝素钠（生物级，上海惠兴生化试剂有限公司），磷酸氢二钠（$Na_2HPO_4 \cdot 12H_2O$），柠檬酸（$C_6H_8O_7 \cdot H_2O$），乙二胺四乙酸二钠（$Na_2EDTA \cdot 2H_2O$），草酸，高氯酸均为国产分析纯。

1.3 仪器与设备

高效液相色谱仪（Waters 515 泵，717 自动进样器，2487 双通道紫外检测器及 Empower 色谱工作站）；自动高速冷冻离心机（日本 HITACHI 20PR-520 型）；Mettler-TOLEDO AE-240 型精密电子天平（梅特勒-托利多公司）；FS-1 高速匀浆机（华普达教学仪器有限公司）；调速混匀器（上海康华生化仪器制造厂）；恒温烘箱（上海浦东荣丰科学仪器有限公司）；Sartorius PB-10 型酸度计（德国赛多利斯公司）。

1.4 试剂

0.1 mol/L Na_2EDTA-麦氏缓冲液，10% 高氯酸提取液。

1.5 试验设计与采样

试验分代谢、残留两组，DOTC 原粉用蒸馏水配成混悬液，口服灌胃给药。代谢组按 20 mg/kg 剂量单次给药，残留组按 20 mg/kg 剂量连续 5 d 给药。

药代组：给药后的 0.083、0.167、0.25、0.5、1、2、4、8、12、16、24、48、72、96、120、144 h 尾静脉取血液，1% 肝素钠抗凝后血液以 3 000 r/min 离心 10 min，取上清液，同时解剖分离出肌肉、肝脏、肾脏；将血清和其他组织于 -20℃

保存，直至样品处理。

残留组：连续灌胃 5 d 后第 1、3、5、7、9、12、15、24、30、40 天各取样一次，在每一时间点取肌肉+皮、肝脏、肾脏等，至于-20℃冰箱，避光保存，待测定。

每一时间点各取 5 尾鱼，作为 5 个平行样品分别处理测定。另取 5 尾未给药的鱼作空白对照。

1.6　样品预处理

血液样品处理　将冷冻保存的血浆室温下自然解冻，摇匀后吸取 1.0 mL 于 5 mL 具塞离心管中，加入 1 mL 的 10%高氯酸提取液置调速混匀器上涡旋振荡混匀 2 min，以 3 000 r/min 离心 10 min，吸出上清液于另一支 5 mL 具塞离心管中，经 0.45 μL 滤头过滤，HPLC 测定。

肝脏、肾脏样品处理　将冷冻保存的肝脏、肾脏样品室温下自然解冻，各称取 0.5 g 于 5 mL 具塞离心管中，加入 2.5 mL 的 10%高氯酸提取液置调速混匀器上涡旋振荡混匀 2 min，以 3 000 r/min 离心 10 min，吸出上清液于另一支 5 mL 具塞离心管中，经 0.45 μL 滤头过滤置自动进样瓶中，HPLC 测定。

肌肉（加皮）样品处理　将冷冻保存的肌肉或肌肉+皮样品室温下自然解冻，置高速匀浆机中匀浆至组织呈糜状，放入小塑料袋中。准确称取 5 g 放入 50 mL 具塞离心管中，加入 20 mL pH = 4.0 的 Na₂EDTA-麦氏缓冲液，置于振荡器上振荡 3 min，以 5 000 r/min 离心 10 min；残渣中加入 15 mL pH=4.0 的 Na₂EDTA-麦氏缓冲液，重复提取一次，合并上清液。上清液用滤纸过滤后倒入另一支 50 mL 离心管中。用 5 mL 甲醇和 5 mL 蒸馏水活化 Oasis HLB 固相萃取柱，待上清液完全流出后，用 5 mL 20%甲醇洗柱，弃去全部流出液，最后用 2 mL 甲醇洗脱，收集洗脱液于 5 mL 离心管中，将洗脱液经 0.45 μL 滤头过滤置自动进样瓶中，HPLC 测定。

1.7　HPLC 分析方法

色谱条件色谱柱：Waters symmetry C18（250 mm ×4.6 mm，5 μm）；流动相：乙腈：甲醇：0.01 mol/L 草酸溶液（2：1：7，$V/V/V$）；流速：0.7 mL/min；柱温：室温；紫外检测波长：350 nm；进样量：20 μL。

标准工作曲线的制备与最低检测限（LOD）

空白组织［肌肉（加皮）、肝脏、肾脏］中加入 DOTC 已知标准液，使其药物浓度范围组织中为 0.05～10.00 μg/g。按样品处理方法处理，作 HPLC 分析，以测得的各平均峰面积 A 为横坐标，相应的质量浓度 Ci 为纵坐标绘制标准工作曲线，求出回归方程和相关系数。用空白组织制成低质量浓度药物的含药组织，经预处理后测定，将引起三倍基线噪音的药物的质量浓度定义为最低检测限。

回收率与精密度测定回收率= $C_r/C_0 \times 100\%$，其中 C_r 为用空白样品加入一定量的 DOTC 标准品，再按样品预处理方法进样后，测定的 DOTC 的质量浓度；C_0 为加入一定的 DOTC 标准品，测得 DOTC 的质量浓度。在四种空白组织中分别添加 2 个浓度水平的 DOTC 标准液，使组织中质量浓度分别为 0.1 μg/g 和 1 μg/g。每个浓度的样品，日内做 5 个重复，一周内重复做 5 次，计算日内及日间精密度。

1.8　休药期（WDT）的确定

强力霉素是按一级动力学过程从体内消除的，即在消除后期服从指数消除：$C=C_0 e^{-kt}$，可以根据消除后期测定的组织药物浓度及规定的 MRL，计算各组织药物浓度降至规定水平（MRL）所需的时间（WDT）：

$$WDT = \frac{\ln(C_0/MRL)}{k}$$

式中，WDT 为休药期，MRL 为最高残留限量（μg/kg），C_0 为残留消除对数曲线的截距（μg/kg），k 为残留消除曲线速率常数。

1.9　数据处理

药物动力学模型拟合及参数计算采用中国药理学会数学专业委员会编制的 3p97 药动软件分析；标准曲线、药物经时曲线图、消除方程及休药期计算和回归图，采用 Microsoft Excel 2003 进行计算和绘制。消除方程采用 $C=C_0 e^{-kt}$，C 表示药物浓度，C_0 为残留消除对数曲线的截距（μg/kg），k 表示消除速率常数。

2　结果

2.1　标准曲线与最低检测限

DOTC 在 0.05～10.00 μg/g 和 0.05～10.00 μg/mL 范围内线性关系良好。以测得的各平均峰面积对相应质量浓度作线性回归，并制作标准曲线。5 种组织中 DOTC 的标准曲线及相关系数（表 1）。

表 1　强力霉素在斑点叉尾鮰血液和组织中标准曲线和相关系数

Tab. 1　The calibration curve and correlation coefficient of DOTC in channal catfish plasma and tissues

标准曲线名称 Calibration curve	线性方程 Linear equation	相关系数 Correlation coefficient
血液工作曲线 Calibration curve of blood	$Y=27557x+179.79$	0.9998
肝脏工作曲线 Calibration curve of liver	$Y=17335x+43.796$	0.9998
肌肉工作曲线 Calibration curve of muscle	$Y=22773x+81.214$	0.9999
肾脏工作曲线 Calibration curve of kidney	$Y=26362x+161.67$	0.9998

2.2　回收率与精密度

在本实验条件下，各组织和血液中 4 个质量浓度水平的 DOTC 回收率为 75.13%~92.05%。测得的日内精密度为（1.75±0.072）%，日间精密度为（2.83±0.89）%。

2.3　DOTC 在斑点叉尾鮰体内的药动学

DOTC 以 20 mg/kg 剂量单次口灌斑点叉尾鮰后，各组织的药-时曲线均出现明显的双峰现象，药物的质量浓度在 30 min 和 1 h 达峰值后即迅速下降，在 8 h 和 12 h 又明显升高形成第二峰，其中在肝脏中的浓度最高。药-时曲线用 3p97 软件处理分析，结果表明 DOTC 经斑点叉尾鮰单次口服后在血液、肝脏、肾脏和肌肉中药物经时过程均符合二室开放模型。所得药动学参数（表 2），药-时曲线图（图 1）。

表 2　单剂量 20 mg/kg b.w 口服 DOTC 在斑点叉尾鮰体内的药代动力学参数

Tab. 2　Pharmacokinetic parameters for DOTC in channel catfish after single oral administration of 20 mg/kg b.w

参数 Parameters	单位 Unit	肝脏 Liver	肾脏 Kidney	肌肉 Muscle	血液 Blood
$T_{1/2(Ka)}$[1]	h	5.340	0.207	1.274	2.986
$T_{1/2(alpa)}$[2]	h	9.356	0.290	23.946	4.459
$T_{1/2(b)}$[3]	h	48.767	40.668	31.091	36.527
K_{21}[4]	h^{-1}	0.035	1.031	0.027	0.021
K_{10}[5]	h^{-1}	0.030	0.020	0.023	0.039
K_{12}[6]	h^{-1}	0.023	1.345	0.0002	0.099
AUC[7]	μg/（mL·h）	1282.0769	61.242	62.348	142.379
Tpeak[8]	h	12.394	0.507	5.994	5.654
Cmax[9]	μg/mL	21.454	0.742	1.269	2.647
MRT[10]	h	56.989	46.585	42.428	48.859

注：1）吸收半衰期 Absorption half-life；2）分布半衰期 Distribution half-life；3）消除半衰期 Elimination half-life；4）由周边室向中央室转运的一级速率常数 First-order rate constant for drug distribution from peripheral compartment to central compartment；5）由中央室消除的一级速率常数 First-order rate constant for drug elimination from central compartment；6）由中央室向周边室转运的一级速率常数 First-order rate constant for drug distribution from central compartment to peripheral compartment；7）药-时曲线下面积 Area under con-centration-time curve；8）峰时 Peak time；9）峰浓度 Peak concentration；10）平均驻留时间 Mean retention time

2.4　多次口服给药后药物残留及消除

斑点叉尾鮰连续 5 d 口服强力霉素 20 mg/kg，DOTC 在斑点叉尾鮰体内的药物消除曲线（图 2）。在停药 1 d 后，药物在肝脏中的浓度最高，为（101.016±0.010 3）μg/g；其次是肾脏，肌肉加皮中最低，为（2.758 0±0.004 2）μg/g。40 d 时，肌肉中还能检测到 0.066 μg/g，肝脏中为 0.092 μg/g。消除速度为：肝<肾<肌肉。所得的药-时数据经回归处理得到各组织中药物的质量浓度（C）与时间（t）关系的消除曲线方程（表 3）。

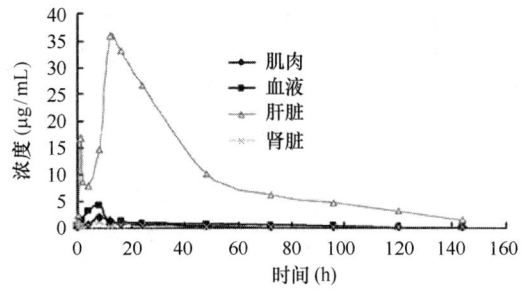

图 1　单剂量 20 mg/kg·b·w 口服 DOTC 在斑点叉尾鮰体内的药-时曲线

Fig. 1　Concentration-time curve of DOTC in channel catfish after single oral administration of 20 mg/kg b·w

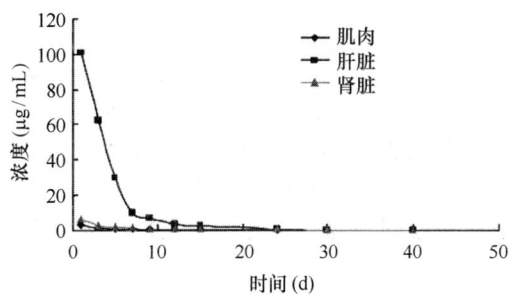

图 2　斑点叉尾鮰连续 5 d 口服 20 mg/kg b·w DOTC，第 5 次给药后的药物消除曲线

Fig. 2　Elimination curve of DOTC in channel catfish after con-tinuous oral administration of 20 mg/kg b·w for five days

表 3　连续 5 d 口服 20 mg/kg b·w DOTC 在斑点叉尾鮰中消除曲线方程及相关指数

Tab. 3　The equation of elimination curve and correlation index in channel catfish after continuous oral administration of 20 mg/kg b·w for five days

组织 Tissue	方程 Equation	相关指数 R^2 Correlation index
肝脏 Liver	$C = 27.319e^{-0.9992t}$	0.3906
肌肉+皮 Muscle and skin	$C = 3.232e^{-0.3813t}$	0.7258
肾脏 Kidney	$C = 6.405e^{-0.7895t}$	0.6394

3　讨论

3.1　DOTC 在斑点叉尾鮰中的药动学

给斑点叉尾鮰单次口服 20 mg/kg 强力霉素后，吸收半衰期（$T_{1/2ka}$）为 2.98 h，与同类药物土霉素在鲫鱼体内[8]的吸收半衰期（3.83 h）相似；表观分布容积（Vd）为 3.564 L/kg；消除半衰期（$T_{1/2b}$）为 36.527 h，两次达峰时间（T_{peak}）为 30 min 和 8 h，说明药物在体内吸收较缓慢，分布广泛，消除也较慢。EI-Aty, et al.[9] 对山羊单次静脉注射 5 mg/kg 强力霉素时，药物在体内分布半衰期为 0.52 h，消除半衰期 4.62 h，达峰时间为 0.86 h。I. M. Ole-Mapenay, et al.[10] 报道矮山羊肌内注射 DOTC，消除半衰期为 24.51 h，达峰浓度为 5.5 μg/mL。与绵羊、山羊相比较，DOTC 在斑点叉尾鮰体内吸收较慢，消除半衰期明显延长，药动参数也有较大的差异。Riond & Riviere[11] 认为 DOTC 在动物体内的组织蛋白结合能力和少量的细胞内渗透是造成个体间药动学参数差异的原因。Jha et al.[12] 认为脂肪沉积的差异也可能是导致 DOTC 在动物体内分布容积和消除半衰期差异的原因。本作者也认为种属间的差异对 DOTC 的药物动力学参数影响较大。另外，药物往往也会因为在血液中的循环途径不同，影响到以后的分布和消除。从研究结果可以看出，给药后药物一方面向血液中转移，使血药浓度迅速上升，以至 30 min 就达到峰值；另一方面药物也向肝脏转移，肝脏中的药物浓度 1 h 也达到峰值，而且进入肝脏中的药物要比血液中多。这一现象，杨先乐等[13] 认为一部分药物经吸收后可能直接进入肝脏，另一部分也因首过效应的原因流经血液的药物也被送到肝脏中，从而导致肝脏中的药物浓度一直处在比血浆中高的水平，最高浓度可达到（35.907±0.378）μg/g。

单次口服剂量为 20 mg/kg 的强力霉素，药物在斑点叉尾鮰体内的药-时曲线呈明显双峰现象：第一次达峰时间 $T_{max(1)}$ 肝脏出现在 1 h，肾、血和肌肉出现在 30 min；第一次达峰浓度 $C_{max(1)}$ 从高到低依次为肝脏>肾脏>肌肉>血液；第二次达峰的时间 $T_{max(2)}$ 出现在 8 h，肝脏为 12 h，第二次达峰浓度 $C_{max(2)}$ 大于第一次的浓度 $C_{max(1)}$。关于药物吸收多峰现象及其机制的探讨多见在人体和哺乳动物中的报道[14]，水产动物体内未见深入研究。文献报道中把此现象出现的原因多解释为肠肝循环、胃肠循环、多部位吸收等，而肠—肝循环（EHC）被认为是产生吸收多峰现象最可能的一种机制。如果重吸收的药量足够大，导致血药质量的浓度一次、再次的升高，药-时曲线便会出现多峰[15]。本实验的结果表明强力霉素在斑点叉尾鮰体内出现双峰现象可能与肠肝循环有关，与同类药物土霉素在鳗鲡中的规律相似[16]。

3.2 三种组织中强力霉素代谢情况的比较

强力霉素在各组织中分布并不均匀。本实验测得强力霉素在斑点叉尾鮰血液、肌肉、肝和肾中的代谢动力学均符合二室开放模型，药物进入体内后在三种组织中的吸收都较动物缓慢[9,10]，从药时曲线上可以看出，在给药后 8 h 肌肉和肾脏达到最高；肝脏在 10 min—1 h 曲线呈上升趋势，之后迅速下降，在 12 h 达到最高点；与肌肉和肾脏相比，DOTC 在肝脏中的吸收稍慢。药物在肌肉、肝脏和肾脏三种组织中的表观分布容积（Vd）分别为 13.650 5、0.512 7、15.601 3 L/kg，说明 DOTC 在各种组织中的分布广泛，其中肝脏中的分布更多一些；比较药时曲线下面积 AUC，在肝、肾和肌肉三种组织中分别为 1282.076 9、61.242、62.348 μg·h/mL，AUC 表示组织内循环药量的多少，可以看出进入肝脏中的药量最多，这表明了肝脏是药物的重要吸收代谢器官。由药-时曲线上来看三种组织中的药物浓度的变化在开始时肝脏与肾脏、肌肉就有明显差异，比较药物在三种组织中的消除半衰期 $T_{1/2b}$，肝脏中的消除时间最长，为 48.767 h，说明药物在肝脏中驻留时间较长，消除过程比其他两种组织缓慢，残留很严重。此与同类药物土霉素在草鱼组织中残留[17]结果相似。

3.3 多次给药后的残留消除规律

从图 2 可以看出在 (28±1)℃水温条件下，斑点叉尾鮰以 20 mg/kg 剂量连续经口灌强力霉素 5d 后，停药后第 1 天，肝脏中浓度最高 [（101.016±0.0103）μg/g]，肌肉+皮中浓度最低 [（2.758 0±0.004 2）μg/g]。在停药初期，强力霉素消除速度最快，随体内强力霉素含量的降低，药物消除速度减慢。徐维海等[18]报道给吉富罗非鱼一次灌服 50 mg/kg 剂量的四环素和土霉素，12 h 药物含量达最高，TC 在肌肉、肝脏和血液中分别为 0.92、2.703、1.225 mg/kg；OTC 分别为 0.860、2.188、1.075 mg/kg，在肝脏中药物浓度达最高，肌肉中最低。强力霉素与此结果相似。

对强力霉素在斑点叉尾鮰体内的残留消除研究表明，DOTC 在斑点叉尾鮰肌肉中第 40 天时仍然能检测到 0.066 μg/g。人们食用鱼常常是肌肉和皮一起食用的，所以如果研究结果用于制定停药期，研究时就不能只研究肌肉，而应该连皮肤一起研究。

3.4 休药期

欧盟对强力霉素在牛、家禽和猪的可食性组织（牛：肌肉、肝脏、肾脏；家禽和猪：肌肉、皮肤+脂肪、肝脏、肾脏）中的最高残留限量（MRL）均做了详细规定[19]，而在水产品各组织中未见规定。本实验首次对强力霉素在斑点叉尾鮰肌肉+皮、肝脏和肾脏组织中的消除规律作了研究，提供了强力霉素在斑点叉尾鮰组织中分布及消除资料，也为制定强力霉素在水产品中的最高残留限量及休药期提供了理论基础。通常将残留量高、消除最慢的组织看成是残留分析的靶组织，本实验结果表明，肝脏组织残留量最高、消除最慢，因此，将肝脏作为强力霉素残留分析的靶组织。欧盟规定，强力霉素在动物肝脏内最高残留限量为 300 μg/kg，土霉素在水产中肌肉加皮最高残留限量为 300 μg/kg。本实验以此作为参考来计算 WDT，通过计算可知：强力霉素在肝脏中的理论 WDT 为 29.53 d。但人们常常是去掉内脏，食用的是肌肉和皮，若根据肌肉加皮中的残留浓度来分析，休药期为 18.92 d。

参考文献

[1] Liu T Z, Zhang Z B. Doxycycline cure suspected infectious laryngotracheitis of chickens [J]. Livestock and Poultry Ind-vstry, 1999, 7 (7): 43-44 [刘同忠, 张泽波. 盐酸多西环素防治疑似鸡传染性喉气管炎. 畜禽业, 1999, 7 (7): 43-44].

[2] Chen Z H. Drugs of prevention and curing avian colibacillo-sis [J]. China Poultry, 2004, 26 (12): 27-29 [陈志华. 防治禽大肠杆菌病常用药物. 中国家禽, 2004, 26 (12): 27-29].

[3] Cao J Y. Drug therapy of respiratory infectious diseases of pigs [J]. Animals Breeding and Feed, 2005, 4: 39-43 [操继跃. 猪呼吸系统感染性疾病药物治疗. 养殖与饲料, 2005, 4: 39-43].

[4] Chen Y Z, Zhang Y Y, Yuan X P, et al. Present research status and development prospect of Tetracycline [J]. Veteri-nary Pharmaceuticals & Feed Additives, 2006, 11 (3): 16-17 [陈育枝, 张元元, 袁希平, 等. 动物四环素类抗生素现状及前景. 兽药与饲料添加剂, 2006, 11 (3): 16-17].

[5] Sárközy G, Semjén G, Laczay P. Disposition of norfloxacin in broiler chickens and turkeys after different methods of oral administration [J].

The Veterinary Journal, 2004, 168: 312-316.

[6] Riond J L, Tyczkowska K, Riviere J E. Pharmacokinetics and metabolic inertness of Doxycycline in calves with ma-ture and immature rumen function [J]. American Journal of Veterinary Research, 1989, 50: 1329-1333.

[7] Yang H F, Qin S H, Wang J C, et al. Comparison of pharma-cokinetics of two different doxycycline injections in pigs [J]. Journal of Hua-zhong Agricultural University, 2007, 26 (1): 76-79 [杨海峰, 覃少华, 王加才, 等. 2种多西环素注射液在健康猪体内的药动学比较. 华中农业大学学报, 2007, 26 (1): 76-79].

[8] Zhang Q Z, Li X M. Comparison of Oxytetracycline phar-makokinetic parameters of two different administration mod-els in crucian [J]. Freshwater Fisheries, 2006, 36 (2): 47-48 [张其中, 李雪梅. 两种给药方式下土霉素在鲫鱼体内药物动力学参数比较. 淡水渔业, 2006, 36 (2): 47-48].

[9] Abd El-Aty A M, Goudah A, Zhou H H. Pharmacokinetics of Doxycycline after administration as a single intravenous bo-lus and intramuscular doses to non-lactating Egyptian goats [J]. Pharmacological Research, 2004, 49: 487-491.

[10] Ole-Mapenay I M, Mitema E S, Maitho T E. Aspects of the pharmacokinetics of Doxycycline given to healthy and pneumonic East African dwarf goats by intramuscular injec-tion [J]. Veterinary Research Communications, 1997, 21 (6): 453-462.

[11] Riond J, L&Riviere J E. Pharmacokinetics and metabolic inertness of Doxycycline in young pigs [J]. American Jour-nal of Veterinary Research, 1998, 51: 1271-1275.

[12] Jha V K, Singh M K, Singh S D, et al. Pharmacokinetic data on Doxycycline and its distribution in different biological fluids in female goats [J]. Veterinary Research Communica-tions, 1989, 13: 11-16.

[13] Yang X L, Zhan J. Metabolism and elimination of chloram-phenicol in tissues of Nile [J]. Acta Hydrobiologica Sinica, 2005, 29 (3): 266-271 [杨先乐, 湛嘉. 氯霉素在罗非鱼体内的代谢和消除规律. 水生生物学报, 2005, 29 (3): 266-271].

[14] Zhu M Z. Veterinary Drug Handbook [M]. Beijing: Chemi-cal Industry Press. 2002, 73-76 [朱模忠. 兽药手册. 北京: 化学工业出版社. 2002, 73-76].

[15] Zhou H W. Studies on dynamics of multimodal phenomenon of drug [J]. Chinese Journal of Modern Applied Pharmacy, 1989, 6 (2): 37-40 [周怀梧. 药物吸收多峰现象的动力学研究. 中国现代应用药学, 1989, 6 (2): 37-40].

[16] Li M T, Guo W L. Study on the depletion of Oxytetracycline residues on eel tissues [J]. Journal of Fisheries of China, 1997, 21 (1): 39-43 [李美同, 郭文林. 土霉素在鳗鲡组织中残留的消除规律. 水产学报, 1997, 21 (1): 39-43].

[17] Xiao L, Wang J, Wang C M, et al. Studies on tissue residue of Oxytetracycline in grass carp [J]. Hubei Agricultural Sci-ences, 2007, 46 (3): 466-488 [肖亮, 王婧, 王承明, 等. 土霉素在草鱼体内的组织残留研究. 湖北农业科学, 2007, 46 (3): 466-488].

[18] Xu W H, Lin L M, Zhu X B, et al. Pharmacokinetic and residues study of Tetracycline in Tilapias [J]. Fisheries Science, 2004, 23 (3): 1-3 [徐维海, 林黎明, 朱校斌, 等. 四环素类抗菌药物在吉富罗非鱼体内的代谢动力学研究. 水产科学, 2004, 23 (3): 1-3].

[19] EU. Commission Regulation (EC) No. 508/1999 of 4 March 1999. amending Annexes I to IV to Council Regulation (EEC) No. 2377/90 lay-ing down a Community procedure for the es-tablishment of maximum residue limits of veterinary me-dicinal products in foodstuffs of animal origin. Official Journal of the European Communities, 1999.

该文发表于《中国兽医学报》【2011, 31 (11): 1653-1659】

土霉素在斑节对虾体内药代动力学和生物利用度

Pharmacokinetics and bioavailaility of Oxytetracycline in black tiger shrimp, Penaeus monodon

孙贝贝[1,2], 胡琳琳[1], 房文红[1], 周俊芳[1], 周帅[1], 李国烈[1]

(1. 中国水产科学研究院东海水产研究所业部海洋与河口渔业资源及生态重点开放实验室, 上海 200090;

2. 上海理工大学医疗器械与食品学院, 上海 200093)

摘要: 本实验在自然海水 (盐度 33), 水温为 (28.0±1.0)℃ 养殖条件下, 采用反相高效液相色谱法 (RP-HPLC), 研究口灌 (100 mg/kg) 和围心腔注射 (20 mg/kg) 两种给药途径下, 土霉素在斑节对虾 (Penaeus monodon) 体内的药代动力学和生物利用度。围心腔注射和口灌给药下, 血药药时曲线均适合采用二室模型拟合。围心腔注射下血药达峰浓度 (C_{max})、药时曲线下面积 (AUC_{0-t})、消除半衰期 ($T_{1/2\beta}$) 分别为 (80.71±13.12) μg/mL、378.25 mg·h/L、17.398 h; 口灌给药下的相应值分别为 (21.98±3.32) μg/mL、324.52 mg·h/L、23.372 h, 土霉素在斑节对虾体内的生物利用度 (F) 为 17.16%。口灌土霉素后, 肝胰腺 C_{max} 为 (138.655±21.375) μg/g, 是血药的 6.3 倍、肌肉峰浓度的 130.2 倍, 药物在肝胰腺中含量最高; 然而, 肌肉和肝胰腺中土霉素消除较快，消除半衰期（$T_{1/2z}$）分别为28.18 h和19.311 h。根据我国水产品中药物残留限量规定，水产品中土霉素的最高残留限量（NY5070-2002）为0.1 mg/kg，结合本试验研究结果，斑节对虾使用土霉素后的休药期为5 d，肌肉可食组织即符合无公害食品标准要求。

关键词：土霉素；斑节对虾；药代动力学；生物利用度；围心腔注射；口灌

斑节对虾（*Penaeus monodon*）是我国南方沿海主要养殖的对虾品种之一，但在精养模式下常常发生以弧菌为主的细菌性疾病，危害着斑节对虾的养殖与发展[1]。使用抗菌素预防和控制细菌性疾病成为养殖必不可少的一部分。土霉素（Oxytetracycline，OTC）属于四环素类抗菌药，具光谱抗菌作用，美国FDA第一个批准允许在鱼类养殖上使用的抗菌素，对因气单胞菌、弧菌和曲挠杆菌等引发的细菌性疾病有较好的疗效，在防治对虾弧菌病和肝胰腺坏死感染中具有较大潜力[2]。研究药动学可以掌握药物在水产动物体内代谢和消除规律，有助于合理用药方案和休药期的制定。国内外有关鱼类土霉素药动学研究报道很多[3-11]；在甲壳动物方面有斑节对虾（*Penaeus monodon*）、日本囊对虾（*Penaeus japonicus*）、凡纳滨对虾（*Litopenaeus vannamei*）、白滨对虾（*Litopenaeus setiferus*）、南美蓝对虾（*Penaeus stylirostris*）等[12-18]，且多为国外的研究，在国内仅见锯缘青蟹（*Scylla serrata*）[19]。本文采用反相高效液相色谱法，研究在盐度33海水中，口灌和围心腔注射两种方式给药下，土霉素在斑节对虾体内的药动学和生物利用度，探明其在斑节对虾体内吸收、分布和消除规律，为指导生产合理用药提供理论依据。

1 材料与方法

1.1 实验药品和试剂

土霉素盐酸盐（Oxytetracycline hydrochloride，OTC）标准品（纯度≥98.5%），购于Dr. Ehrenstorfer公司；土霉素盐酸盐原料药（纯度≥98%，以下简称土霉素），购于潍坊医药集团。

甲醇和乙腈均为HPLC级，购自德国默克公司；高氯酸，优级纯，购自国药集团上海化学试剂公司；磷酸氢二钠，磷酸，分析纯，购自国药集团上海化学试剂公司。

1.2 仪器和设备

高效液相色谱仪，美国沃特斯公司Waters 2695，色谱柱为Zorbax SB-C$_{18}$，（5 μm，150 mm×4.6 mm，I. D.），紫外检测器。超低温冰箱（-80℃，美国产，Thermo）；超纯水仪（德国产，Millipore）；电子分析天平（瑞士产，Mettler Toledoab 204，感量0.000 1 g）；高速冷冻离心机（日本产，HITACHI CF16RXⅡ）；以及0.22 μm微孔滤膜等。

1.3 实验动物

实验用斑节对虾购买自海南省琼海市某对虾养殖场，体重9.5~12.8 g。暂养于3.5 m×4 m×2 m水泥池中，暂养和实验期间用水为自然海水（盐度33），水温为（28.0±1.0）℃，保持24 h充气，每天换水1/3体积。暂养10 d后挑选体表无损伤、活力强的斑节对虾用于实验。实验期间斑节对虾成活率为100%。

1.4 给药与采样

1.4.1 口灌给药与样品采集

糊状药饵配制：准确称取1.00 g盐酸土霉素，用少许蒸馏水溶解后，加入20 g 100目过筛的鱼粉，加蒸馏水搅拌混合，最终定容至50 mL，即为20 mg盐酸土霉素/毫升糊状饵料。

给药方式：单剂量混饲口灌给药，给药剂量为100 mg/kg。采用1 mL注射器（13#针头磨圆）从对虾口器将糊状药饵直接送到对虾胃内，观察10 min无饵料回吐方可用于试验。

样品采集：于给药后10 min、30 min、1 h、2 h、6 h、12 h、1 d、2 d、3 d、4 d和5 d抽取0.3 mL血淋巴，置于预先装有2 mg草酸铵抗凝剂的1.5 mL离心管中，同时取肌肉、肝胰腺组织，每个时间点采样6尾对虾。所有样品均于-80℃保存备用。

1.4.2 围心腔注射给药与样品采集

注射液配制：准确称取0.50 g盐酸土霉素，用少许蒸馏水溶解后，生理盐水定容至50 mL，配置成10 mg/mL注射用药液。

给药方式：单剂量围心腔注射给药，给药剂量为20 mg/kg。采用0.5 mL胰岛素注射器将药液从对虾头胸甲底部直接注入围心腔内，给药时应注意针头的位置，避免对心脏造成损伤。

样品采集：于给药后 1 min、5 min、10 min、30 min、1 h、2 h、6 h、12 h、1 d、2 d、3 d 抽取 0.3 mL 血淋巴，置于预先装有 2mg 草酸铵抗凝剂的 1.5 mL 离心管中，同时取肌肉、肝胰腺组织，每个时间点采样 6 尾对虾。所有样品均于−80℃保存备用。

1.5　样品处理

1.5.1　血淋巴样品处理

血淋巴于室温自然溶解后，取 200 μL 血淋巴，加入等体积的 0.5%高氯酸，旋涡混匀器上混合 2 min，于 10 000 rpm. 转速离心 10 min，吸取上清液经 0.22 μm 微孔滤膜过滤后，用 HPLC 分析。

1.5.2　肌肉、肝胰腺样品处理

参照陈玉露等[19]方法做了修改。样品于室温自然解冻后，将所取组织样品置于研钵中研磨达到均质后，准确称取（1.00±0.01）g 样品，置于 10 mL 离心管中，加入 2 倍体积的 4%高氯酸，振荡 5 min，8 000 rpm. 转速离心 5 min，收集上清液于 50 mL 离心管中。残渣中再加入 1 倍体积的 4%高氯酸，振荡 5 min，8 000 rpm. 转速离心 5 min，收集上清液，重复操作一次。合并三次上清液，缓慢过 C$_{18}$ 固相萃取柱，用 5 mL 超纯水洗柱，以 5 mL 甲醇洗脱，收集洗脱液，40℃下氮气吹干仪吹干，加入 1 mL 流动相溶解，1 mL 正己烷去脂，置于 20℃摇床中震荡至残留物完全溶解，8 000 rpm. 转速离心 5 min，下清液经 0.22 μm 微孔滤膜过滤，用 HPLC 分析。

1.6　流动相和色谱条件

流动相：乙腈：0.01 mol/L 磷酸二氢钠溶液（pH 值为 2.5）= 26：74（V/V），使用前用超声波脱气 10 min；柱温，37℃；流速，1.0 m L/min；进样量，10 μL；紫外检测波长，355 nm。

1.7　数据分析

口灌和围心腔注射给药下斑节对虾血药浓度−时间关系曲线采用房室模型拟合分析，其药动学参数同时采用房室模型和统计矩原理来推算，肌肉和肝胰腺组织中药物处置规律的药动学参数采用统计矩原理来推算。模型的拟合和参数的推理采用上海中医药大学药物临床研究中心的 DAS 2.1.1 药物与统计软件。生物利用度计算公式为：

$$F（\%）=（AUC_{oral} \times Dose_{IS}）/（AUC_{IS} \times Dose_{oral}）\times 100$$

式中，Dose$_{IS}$、Dose$_{oral}$ 分别为围心腔注射和口灌给药剂量，AUC$_{IS}$、AUC$_{oral}$ 分别为围心腔注射和口灌给药（时间 0～120 h）曲线下面积。

2　结果

2.1　围心腔注射和口灌给药后斑节对虾血淋巴土霉素浓度−时间关系曲线及药动学参数

以 20 mg/kg 剂量围心腔注射给药下，斑节对虾血淋巴土霉素浓度随时间的变化曲线见图 1。血药浓度与时间关系曲线最适合采用二室模型拟合，拟合所得药动学方程为：

$$C_t = 49.872 \cdot e^{-0.254t} + 5.011 \cdot e^{-0.040t}（R^2 = 0.9997，AIC = 3.252）$$

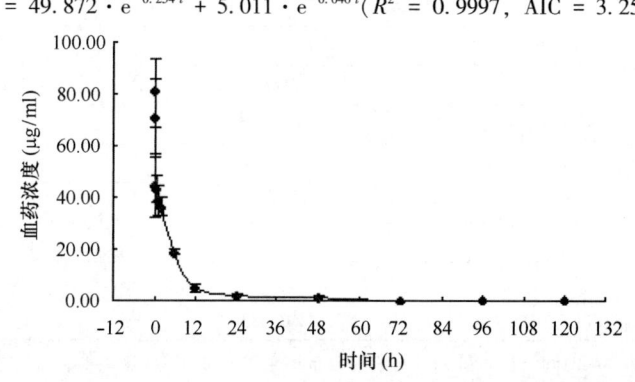

图 1　斑节对虾围心腔注射土霉素后血药浓度−时间曲线（n=6）

Fig. 1　Time course of OTC concentration in the hemolymph of black tiger shrimp following intra−sinus injection（n=6）

以 100 mg/kg 剂量口灌给药下，斑节对虾血淋巴土霉素浓度随时间的变化曲线见图 2。血药浓度与时间关系曲线最适合采用具一级吸收的二室模型拟合，拟合所得药动学方程为：

$$C_t = 77.431 \cdot e^{-0.152(t-0.014)} + 2.234 \cdot e^{-0.029(t-0.014)} - 79.74 \cdot e^{-0.223(t-0.014)} (R^2 = 0.998, \text{AIC} = 16.067)$$

同时采用房室模型和统计矩原理推算药动学参数，见表1。以统计矩原理推算围心腔注射和口灌给药下的血药曲线下面积（AUC_{0-t}）分别为378.25 mg·h/L和324.52 mg·h/L，计算所得生物利用度（F）为17.16%。

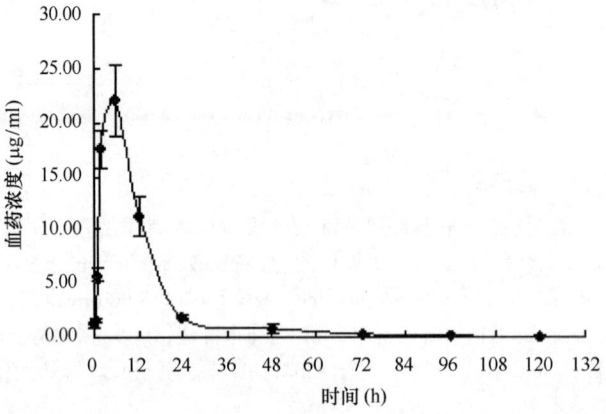

图2　斑节对虾口灌土霉素后血药浓度-时间曲线（$n=6$）

Fig. 2　Time course of OTC concentration in the hemolymph of black tiger shrimp following oral administration（$n=6$）

表1　围心腔注射和口灌给药后土霉素在斑节对虾血淋巴中药动学参数（n=6）

Tab. 1　Pharmacokinetic parameters of oxytetracycline in hemolymph from black tiger shrimp（Penaeus monodon）after an intra-sinus injection and oral administration

药动学参数 Pharmacokinetic parameters		围心腔注射 intra-sinus injection	口灌 Oral administration
房室模型 compartmental model			
$T_{1/2\alpha}$	（h）	2.731	4.355
$T_{1/2\beta}$	（h）	17.398	23.372
$T_{1/2ka}$	（h）	—	3.109
AUC_{0-t}	（mg·h/L）	346.16	283.65
$AUC_{0-\infty}$	（mg·h/L）	349.00	283.80
V_d	（L/kg）	0.364	4.153
CL_s	L/（kg·h）	0.057	0.468
Tlag	（h）	—	0.014
统计矩原理药动学参数 statistical moment theory			
C_{max}（C_0）	（μg/mL）	80.71±13.12	21.98±3.32
T_{max}	（h）	0.0167	6
AUC_{0-t}	（mg·h/L）	378.25	324.52
$AUC_{0-\infty}$	（mg·h/L）	380.00	325.51
MRT	H	12.47	14.091
MRT	H	13.121	14.489
$T_{1/2z}$	H	13.615	17.223
CLz	L/（kg·h）	0.053	0.307

注：C_{max}为血药峰浓度；T_{max}为达峰时间；$T_{1/2\alpha}$和$T_{1/2\beta}$分别为分布相半衰期和消除相半衰期，$T_{1/2ka}$分别为吸收相半衰期；CL_s（CLz）为总体清除率；AUC_{0-t}、$AUC_{0-\infty}$为0-t时间、0-∞时间的曲线下面积；V_d（Vz）为表观分布容积；Tlag为时滞；MRT_{0-t}、$MRT_{0-\infty}$为0-t时间、0-∞时间的保留时间。

Note: C_{max}-peak OTC concentration in hemolymph; T_{max}-time to reach peak concentration; $t_{1/2\alpha}$-distribution half-life; $t_{1/2\beta}$-elimination half-life; $t_{1/2ka}$-absorption half-life; CL_s（CLz）-total body clearance; AUC_{0-t}-area under the concentration-time curve from zero to time; $AUC_{0-\infty}$-area under the concentration-time curve from zero to infinity; V_d（Vz）-the apparent volume of distribution; Tlag-Lag Time; MRT_{0-t}-area residence time from zero to time; $MRT_{0-\infty}$-area residence time from zero to infinity.

2.2　两种方式给药后土霉素在斑节对虾肌肉和肝胰腺中的分布与消除

围心腔注射给药后，斑节对虾肌肉和肝胰腺组织中土霉素浓度随时间变化曲线见图3，两组织中土霉素浓度变化趋势相同，二者都表现为先上升后下降，但达峰时间和峰浓度不同。肌肉中分别为 1 h 和 (1.385 ± 0.303) μg/g，肝胰腺分别为 6 h 和 (4.181 ± 1.186) μg/g；两组织中土霉素消除较快，给药后第 4 d 肌肉和肝胰腺药物浓度均小于 0.1 μg/g，以统计矩原理推算的消除半衰期 $(T_{1/2z})$ 分别为 20.313 h 和 23.28 h，其他药动学参数见表 3。

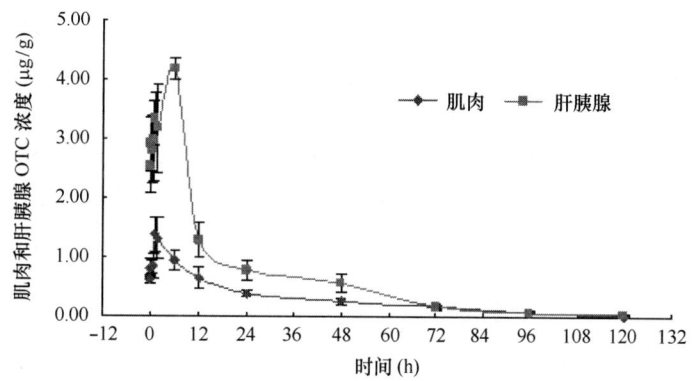

图 3　斑节对虾围心腔注射土霉素后组织中药物浓度-时间曲线 ($n=6$)

Fig. 3 Tissue OTC concentration vs. time following intra-sinus injection

口灌给药后，斑节对虾肌肉和肝胰腺组织中土霉素浓度随时间变化曲线见图4，两组织中土霉素浓度变化趋势虽相同，表现为先上升后下降，但两组织中 OTC 峰浓度相差甚大，肌肉峰浓度仅为 (1.065 ± 0.159) μg/g，而肝胰腺峰浓度高达 (138.655 ± 21.375) μg/g。口灌给药下，肌肉和肝胰腺中药物消除同样较快，肌肉中第 4 天的 OTC 浓度为 (0.082 ± 0.027) μg/g，肝胰腺第 5 天的 OTC 浓度为 (0.079 ± 0.008) μg/g，肌肉和肝胰腺的 OTC 消除半衰期 $(T_{1/2z})$ 分别为 28.18 h 和 19.311 h，其他药动学参数见表 2。

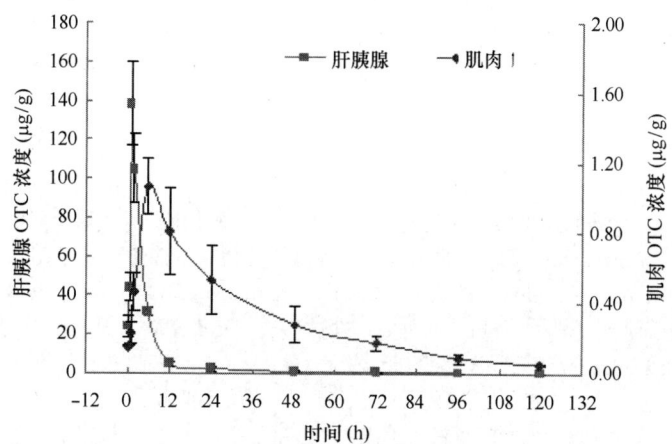

图 4　斑节对虾口灌土霉素后肌肉和肝胰腺中药物浓度-时间曲线 ($n=6$)

Fig. 4　Muscle and hepatopancreas OTC concentration vs. time following oral administration

表 2　围心腔注射和口灌给药后土霉素在斑节对虾肌肉和肝胰腺中药动学参数（统计矩原理）（n=6）

Tab. 2　Pharmacokinetic parameters of oxytetracycline in muscle and hepatopancreas from black tiger shrimp after a single intra-sinus injection and oral administration based on statistical moment theory

药动学参数 Pharmacokinetic parameters		围心腔注射 Intrasinus injection		口灌 Oral administration	
		肌肉 Muscle	肝胰腺 Hepatopancteas	肌肉 Muscle	肝胰腺 Hepatopa-ncteas
C_{max}	（μg/g）	1.385±0.303	4.181±0.186	1.065±0.159	138.655±21.375
T_{max}	（h）	1	6	6	1
AUC_{0-t}	（mg·h/kg）	35.55	80.094	36.759	668.02
$AUC_{0-\infty}$	（mg·h/kg）	36.574	81.60	38.818	670.27
MRT_{0-t}	（h）	31.521	23.322	33.167	8.167
$MRT_{0-\infty}$	（h）	34.82	25.727	39.929	8.637
$T_{1/2z}$	（h）	20.313	23.28	28.18	19.311
CL_z	［kg/（kg·h）］	0.547	0.245	2.576	0.149

注：C_{max} 为药峰浓度；T_{max} 为达峰时间；AUC_{0-t}、$AUC_{0-\infty}$ 分别为 0~t 时间、0~∞ 时间的曲线下面积；MRT 为平均滞留时间；$T_{1/2z}$ 为消除半衰期；CL_z 为总体清除率

Note：C_{max}-peak OTC concentration；T_{max}-time to reach peak concentration；AUC_{0-t}-area under the concentration-time curve from zero to time；$AUC_{0-\infty}$-area under the concentration-time curve from zero to infinity；MRT-residence time；$T_{1/2z}$-elimination half-life；CL_z-total body clearance

3　讨论

本试验中，盐酸土霉素围心腔注射和口灌给药下，斑节对虾血淋巴药时曲线关系均可采用二室模型拟合，且拟合度很高。盐酸土霉素在围心腔注射日本囊对虾[13]、白滨对虾（*P. setisferse*），以及盐度 5 下的斑节对虾[12]，它们的血药药时曲线均可采用二室模型拟合，这与本试验结果相一致。但在口灌给药下，日本囊对虾不能用任何房室模型拟合；即使同是斑节对虾，在盐度 5 下所得到的血药药时曲线也不能用房室模型拟合。为了便于与其他文献比较分析，本试验同时采用经典房室模型和非房室模型的统计矩原理来描述土霉素在斑节对虾体内的动态变化规律。

3.1　口灌给药下对虾对土霉素的生物利用度明显大于鱼类

日本囊对虾口灌 50 mg/kg 土霉素后其血药 C_{max} 为 24.3 μg/mL，生物利用度为 43.2%；斑节对虾在盐度 5 下口灌 50 mg/kg土霉素后其血药 C_{max} 为 21.1 μg/mL，生物利用度为 35.6%，而本试验口灌给药剂量是它们的 2 倍，其血药峰浓度仅与它们相近（21.98 μg/mL），生物利用度明显低（F 为 17.16%），这说明本试验中斑节对虾对土霉素的吸收减少。造成土霉素吸收差除与动物种类有关外，还可能与盐度有关。四环素类药物具有多个与离子结合高亲和力的官能团，如亲水的土霉素，具有很强的与阳离子结合形成复合物的能力（Goodman & Gilman，1985；Lunestad & Goksovr，1990）[20,21]，土霉素与阳离子结合后将导致跨膜通透性下降[22]。斑节对虾是广盐性渗透调变类动物，血淋巴中离子随环境的变化而变化，盐度 32 时血淋巴 Ca^{2+} 浓度为 10.2 mM，高于盐度 8 时的 Ca^{2+} 浓度（6 mM）[23]，因此在高盐度海水中，土霉素可能与血淋巴中二价阳离子（Ca^{2+}、Mg^{2+}）结合形成更多的复合物，从而阻碍了土霉素的跨膜运动。

对于鱼类而言，口服给药后对土霉素的吸收更低。黑鲷（*Sparus macrocephlus*）口服 75 mg/kg 剂量土霉素其血药 C_{max} 为 1.398 μg/mL[6]；牙鲆（*Paralichthys olivaceus*）口灌 200 mg/kg 土霉素，其血药 C_{max} 仅为 0.54 μg/mL[7]；虹鳟（*Oncorhynchus mykiss*）口服 50 mg/kg 土霉素其血药峰浓度 6.72 μg/mL，生物利用度为 30.3%[24]；大西洋鲑（*Salmo salar* L.）口灌 50 mg/kg 土霉素，其血药浓度为 0.42 μg/mL，生物利用度为 2%[25]；金头鲷鱼（*Sparus aurata* L.）口灌 75 mg/kg 剂量土霉素，其血药峰浓度为 2.5%，生物利用度为 9%[26]。同是口服给药条件下，甲壳类和鱼类血药达峰浓度和生物利用度差别很大，这很可能与甲壳动物属于开放式循环系统有关，血淋巴由心脏通过血管驱动输送到胃肠道，组织直接浸浴在血淋巴中[27]。

3.2　口灌给药下斑节对虾肝胰腺的"首过效应"

组织中的 C_{max} 和 AUC 是衡量药物在分布的主要药动学参数。本试验中，口灌土霉素后，斑节对虾肝胰腺药物峰浓度显著高于血药和肌肉峰浓度，分别是它们的 6.3 倍和 130.2 倍，肝胰腺药物曲线下面积（AUC_{0-t}）是它们的 2.1 倍和 18.9

倍，药物在肝胰腺中含量最高，这于 Rigos 等[28]（2004）对黑鲈（*Dicentrarchus labrax L.*）和 Bjorklund 等[29]（1990）对虹鳟的研究类似，这一现象称之为"肝首过效应"[28]。斑节对虾口灌给药时，肝胰腺药物达峰时间为 1h，而血淋巴和肌肉的达峰时间为 6h，由此可见，药物是首先进入肝胰腺，然后再向其他组织分布。土霉素在组织中分布及达峰时间均与凡纳滨对虾[14]口灌二水土霉素的结果相一致。许多证据表明肝胰腺对吸收、代谢、消除都起着重要作用[30]。Faroongsarng 等[14]（2007）在深入分析二水土霉素在凡纳滨对虾体内吸收和药动学后，提出口灌给药下肝胰腺在药物的吸收起着主要作用，同时也是药物消除的主要场所。尽管斑节对虾（本试验）和凡纳滨对虾[14]肝胰腺土霉素峰浓度很高，但消除都比较快，且斑节对虾的消除半衰期（19.31 h）小于凡纳滨对虾（51.3 h）。

3.3　临床给药方案的制定

药动学参数的实际应用是为设计临床剂量方案服务的，多次重复给药后要维持血药浓度高于最低抑菌浓度（MIC）和低于中毒浓度。Aliabadi F S and Lees[31]（2000）认为，对于四环素类、苯丙醇类（phenicols）、大环内酯类和林可酰胺类（lincosamides），维持感染部位的药物浓度超过 MIC_{90} 方可达到药效作用。Ruangpan and Kitao[32]（1992）对 205 株分离自发病斑节对虾的弧菌进行了体外药效实验，这些弧菌对土霉素的 MIC_{90} 为 37.5 μg/mL。因此，当所期望的稳态血药浓度（C_{ss}）为 37.5 μg/mL、给药剂量（Dose）为 100 mg/kg 时，推算给药时间间隔（τ）的公式如下：

$$\tau = F \cdot Dose/C_{ss} \cdot CLs$$

式中，CLs 为围心腔注射给药下的总体清除率，F 为生物利用度。Park 等[33]（1995）和 Uno 等[12,13]均采用此方法推算临床给药方案。按照此公式，本试验推算所得到的给药时间间隔为 8.03 h，也就是说，每天给药 3 次，每次给药剂量为 100 mg/kg，即可达到治疗所需稳态血药浓度。当然，本推算所选细菌的 MIC 偏高，如果实际的 MIC 较低，可以通过延长给药时间间隔或降低用药剂量，从而可减少药物使用总量，达到药物治疗效果。

3.4　土霉素在斑节对虾体内消除与休药期确定

消除半衰期（$T_{1/2\beta}$ 或 $T_{1/2}$）、总体清除率（CLs 或 CLz）是评价药物消除的重要参数。土霉素在水产动物中消除的研究报道较多，如血管给药金头鲷[34]消除半衰期（$t_{1/2\beta}$）为 53 h，CLz 为 50 mL · kg^{-1} · h^{-1}；血管注射和口灌给药下北极红点鲑（*Salvelinus alpinus*）[35]的 $t_{1/2\beta}$ 分别为 266.3~326.9 h 和 367.0~468.5 h，血管注射的 CLs 为 6.27~6.54 mL · kg^{-1} · h^{-1}；血管注射和口灌给药下，香鱼（*Plecoglossus altivelis*）[36]的 $t_{1/2\beta}$ 分别为 52.1 h 和 53.1 h，血管注射的 CLs 为 17.4 mL · kg^{-1} · h^{-1}。本试验中，在围心腔注射和口灌两种给药方式下，斑节对虾血药消除半衰期（分别为 17.398 h 和 17.233 h）与盐度 5 下斑节对虾[12]（分别为 16.2 h 和 17.233 h）比较接近，略快于日本囊对虾[13]（分别为 24.7 h 和 33.6 h），显著快于上述几种鱼类。

虽然本试验口灌给药下斑节对虾肌肉 $t_{1/2z}$（28.18 h）高于盐度 5 下的斑节对虾（$T_{1/2}=22.0$ h），但本试验肌肉中土霉素残留消除要快些，盐度 5 下的斑节对虾第 5 d 肌肉中土霉素残留低于 0.2 μg/g（日本囊对虾也是如此），而本试验第 5 天肌肉中的残留则低于 0.1 μg/g，这可能是因为本试验中肌肉峰浓度低于其他两种对虾的原因。根据我国水产品中药物残留限量规定，水产品中土霉素的最高残留限量为 0.1 mg/kg[37]，结合本试验研究结果，斑节对虾使用土霉素后的休药期为 5 d，肌肉可食组织即符合无公害食品标准要求。

参考文献

[1] 陶保华，胡超群，吴蔚. 斑节对虾弧菌病的病原生物学研究 [J]. 热带海洋学报. 2001, 20（2）：80-87.

[2] Reed L A, Siewicki T C, Shah J C. Pharmacokinetics of oxytetracycline in the white shrimp, Litopenaeus setiferus [J]. Aquaculture. 2004, 232 (1-4): 11-28.

[3] 郑增忍，赵思俊，邹明，等. 土霉素在大菱鲆体内药代动力学及残留消除规律研究 [J]. 中国农学通报. 2008, 24（10）：559-563.

[4] Meinertz J R, Gaikowski M P, Stehly G R, et al. Oxytetracycline depletion from skin-on fillet tissue of coho salmon fed oxytetracycline medicated feed in freshwater at temperatures less than 9℃ [J]. Aquaculture. 2001, 198: 29-39.

[5] 孙绪文，王群，颜显辉，等. 土霉素在红鳍东方鲀体内的残留及休药期研究 [J]. 渔业科学进展. 2009, 30（6）：75-80.

[6] 王群，孙修涛，刘德月，等. 土霉素在黑鲷体内的药物代谢动力学研究 [J]. 海洋水产研究. 2001, 22（1）：42-47.

[7] 纪荣兴，刘爱原，张春丽，等. 土霉素在牙鲆体内的药代动力学研究 [J]. 海洋科学. 2008, 32（7）：38-41, 45.

[8] Abedini S, Namdari R, Law F C P. Comparative pharmacokinetics and bioavailability of Oxytetracycline in rainbow trout and chinook salmon [J]. Aquaculture. 1998, 162: 23-32.

[9] Rigos G, Alexis M, Andriopoulou A, et al. Pharmacokinetics and tissue distribution of oxytetracycline in sea bass, Dicentrarchus labrax, at two water temperatures [J]. Aquaculture. 2002, 210 (1-4): 59-67.

[10] Rigos G, Nengas I, Tyrpenou A E, et al. Pharmacokinetics and bioavailability of oxytetracycline in gilthead sea bream (Sparus aurata) after a

single dose〔J〕. Aquaculture. 2003, 221（1-4）: 75-83.

[11] 于相满, 张其中, 叶柏喜, 等. 土霉素在奥尼罗非鱼体内的药动学研究〔J〕. 淡水渔业. 2009, 39（1）: 63-67, 75.

[12] Uno K, Aoki T, Kleechaya W, et al. Pharmacokinetics of oxytetracycline in black tiger shrimp, Penaeus monodon, and the effect of cooking on the residues〔J〕. Aquaculture. 2006, 254（1-4）: 24-31.

[13] Uno K. Pharmacokinetics of oxolinic acid and oxytetracycline in kuruma shrimp, Penaeus japonicus〔J〕. Aquaculture. 2004, 230（1-4）: 1-11.

[14] Faroongsarng D, Chandumpai A, Chiayvareesajja S, et al. Bioavailability and absorption analysis of oxytetracycline orally administered to the standardized moulting farmed Pacific white shrimps（Penaeus vannamei）〔J〕. Aquaculture. 2007, 269（1-4）: 89-97.

[15] Gmez-Jimenez S, Espinosa-Plascencia A, Valenzuela-Villa F, et al. Oxytetracycline（OTC）accumulation and elimination in hemolymph, muscle and hepatopancreas of white shrimp Litopenaeus vannamei following an OTC-feed therapeutic treatment〔J〕. Aquaculture. 2008, 274（1）: 24-29.

[16] Reed L A, Siewicki T C, Shah J C. The biopharmaceutics and oral bioavailability of two forms of oxytetracycline to the white shrimp, Litopenaeus setiferus〔J〕. Aquaculture. 2006, 258（1-4）: 42-54.

[17] Nogueira-Lima A C, Gesteira T C V, Mafezoli J. Oxytetracycline residues in cultivated marine shrimp（Litopenaeus vannamei Boone, 1931）（Crustacea, Decapoda）submitted to antibiotic treatment〔J〕. Aquaculture. 2006, 254（1-4）: 748-757.

[18] Mohney L L, Williams R R, Bell T A, et al. Residues of oxytetracycline in cultured juvenile blue shrimp, Penaeus stylirostris（Crustacea: Decapod）, fed medicated feed for 14 days〔J〕. Aquaculture. 1997, 149（3-4）: 193-202.

[19] 陈玉露, 房文红, 周凯, 等. 土霉素在锯缘青蟹体内的药物代谢和消除规律〔J〕. 海洋渔业. 2009, 31（2）: 154-160.

[20] Lunestad B T, Goks Yr J. Reduction in the antibacterial effect of oxytetracycline in sea water by complex formation with magnesium and calcium. 〔J〕. Diseases of aquatic organisms. 1990, 9（1）: 67-72.

[21] Goodman A G, Gilman L S, Rall W, et al. The chemotherapy of Tuberculosis. The Pharmacological Basis of Therapeutics〔J〕. 1985.

[22] Clive D. Chemistry of tetracyclines〔J〕. Quarterly Reviews, Chemical Society. 1968, 22（4）: 435-456.

[23] Ferraris R P, Parado-Estepa F D, Ladja J M, et al. Effect of salinity on the osmotic, chloride, total protein and calcium concentrations in the hemolymph of the prawn Peneaus monodon（fabricius）〔J〕. Comparative Biochemistry and Physiology Part A: Physiology. 1986, 83（4）: 701-708.

[24] Abedini S, Namdari R, Law F C P. Comparative pharmacokinetics and bioavailability of oxytetracycline in rainbow trout and chinook salmon〔J〕. Aquaculture. 1998, 162（1-2）: 23-32.

[25] Elema M O, Hoff K A, Kristensen H G. Bioavailability of oxytetracycline from medicated feed administered to Atlantic salmon（Salmo salar L.）in seawater〔J〕. Aquaculture. 1996, 143（1）: 7-14.

[26] Fikri B, Haşmet Ç. Oxytetracycline residues in cultured gilthead sea bream（Sparus aurata L. 1758）tissues〔J〕. African Journal of Biotechnology. 2010, 9（42）: 7192-7196.

[27] Goy M F. Nitric oxide: an inhibitory retrograde modulator in the crustacean heart〔J〕. Comp Biochem Physiol A Mol Integr Physiol. 2005, 142（2）: 151-163.

[28] Rigos G, Nengas I, Alexis M, et al. Bioavailability of oxytetracycline in sea bass, Dicentrarchus labrax（L.）〔J〕. Journal of Fish Diseases. 2004, 27（2）: 119-122.

[29] Bj Rklund H, Bylund G. Temperature-related absorption and excretion of oxytetracycline in rainbow trout（Salmo gairdneri R.）〔J〕. Aquaculture. 1990, 84（3-4）: 363-372.

[30] Verri T, Mandal A, Zilli L, et al. -Glucose transport in decapod crustacean hepatopancreas * 1〔J〕. Comparative Biochemistry and Physiology-Part A: Molecular & Integrative Physiology. 2001, 130（3）: 585-606.

[31] Aliabadi F S, Lees P. Antibiotic treatment for animals: effect on bacterial population and dosage regimen optimisation〔J〕. Int J Antimicrob Agents. 2000, 14（4）: 307-313.

[32] Ruangpan L, Kitao T. Minimal inhibitory concentration of 19 chemotherapeutants against Vibrio bacteria of shrimp, Penaeus monodon〔J〕. Diseases in Asian Aquaculture. 1992, 1: 135-142.

[33] Park E D, Lightner D V, Milner N, et al. Exploratory bioavailability and pharmacokinetic studies of sulphadimethoxine and ormetoprim in the penaeid shrimp, Penaeus vannamei〔J〕. Aquaculture. 1995, 130（2-3）: 113-128.

[34] Rigos G, Nengas I, Tyrpenou A E, et al. Pharmacokinetics and bioavailability of oxytetracycline in gilthead sea bream（Sparus aurata）after a single dose〔J〕. Aquaculture. 2003, 221（1-4）: 75-83.

[35] Hansen M K, Horsberg T E. Single - dose pharmacokinetics of flumequine in cod（Gadus morhua）and goldsinny wrasse（Ctenolabrus rupestris）〔J〕. Journal of Veterinary Pharmacology and Therapeutics. 2000, 23（3）: 163-168.

[36] Uno K. Pharmacokinetic study of oxytetracycline in healthy and vibriosis-infected ayu（Plecoglossus altivelis）〔J〕. Aquaculture. 1996, 143（1）: 33-42.

[37] 中华人民共和国农业部. NY 5070-2002 中华人民共和国农业行业标准无公害食品水产品中渔药残留限量〔S〕. 北京: 中国标准出版社, 2002.

该文发表于《海洋科学》【2008，32（7）：38-45】

土霉素在牙鲆体内的药代动力学研究

The characteristic analysis and application of Babin model

纪荣兴，刘爱原，张春丽，邹文政

（集美大学水产学院，福建省高校水产科学技术与食品安全重点实验室（集美大学），福建厦门 361021）

摘要：本文采用高效液相色谱法为定性、定量手段，研究土霉素在牙鲆体内的药代动力学过程，采用 DAS 药代动力学程序对数据进行分析。结果表明，牙鲆单剂量口服土霉素后（200 mg·kg^{-1}），血药经时过程符合二室模型，主要动力学参数如下：$T_{1/2a}$ 为 10.043 h，T_{max} 为 4.000 h，C_{max} 为 0.54 mg·L^{-1}，AUC$_{0-72}$ 为 17.15 mg·L^{-1}·h^{-1}，Ka 为 0.223，k 为 0.476 h^{-1}。牙鲆肌肉中土霉素的经时过程符合一级吸收一室模型，主要动力学参数：$T_{1/2a}$ 为 74.893 h，T_{max} 为 4.000 h，C_{max} 为 3.58 mg·L^{-1}，AUC$_{0-72}$ 为 148.56 mg·L^{-1}·h^{-1}，Ka 为 0.731，k 为 2.991。牙鲆肝脏中土霉素的经时过程符合一级吸收一室模型，主要动力学参数：$T_{1/2a}$ 为 31.376 h，T_{max} 为 4.000 h，C_{max} 为 13.78 mg·L^{-1}，AUC$_{0-72}$ 为 494.14 mg·L^{-1}·h^{-1}，Ka 为 0.876，k 为 4.940 h^{-1}。

关键词：土霉素；牙鲆；药代动力学；高效液相色谱

二、氨基糖苷类

氨基糖苷类（aminoglycosides）是一种有机碱的抗生素，水溶性好，其化学结构中均含有一个氨基醇环和氨基糖分子结合而成的苷，制剂多为性质稳定的硫酸盐，有天然和人工半合成的两类。天然氨基糖苷类抗生素有人工半合成的有阿米卡星、奈替米星等。该类抗生素抗菌谱较广，对革兰氏阴性杆菌作用强，对革兰氏阳性菌的作用较弱。链霉素、卡那霉素、新霉素、庆大霉素常用于水产养殖上。研究表明，硫酸链霉素在鳗鲡体内消除较快（HE Ping et al.，2012），硫酸新霉素在吉富罗非鱼（Genetically improved farmed tilapia）肌肉组织中残留消除最快（宋洁等，2010）。

该文发表于《pharamacology&pharmacy》【2012，3：195-200】

Pharmacokinetic Disposition of Streptomycin Sulfate in Japanese Eel（Anguilla japonica）after Oral and Intramuscular Administrations

HE Ping[1,2]，SHEN Jin-Yu[1]，YIN Wen-Lin[1]，YAO Jia-Yun[1]，

XU Yang[1]，PAN Xiao-Yi[1]，HAO Gui-Jie[1]

（1. Zhejiang Institute of Freshwater Fisheries, Huzhou, Zhejiang, 313001, China；

2. Zhejiang San Men food inspection center, San Men, Zhejiang, 317100, China）

Abstract：Pharmacokinetics and residue elimination of streptomycin sulfate（STR）are important in the determination of optimal dosage regimens and in establishing safe withdrawal periods in farmed fishes. The pharmacokinetics of STR was studied after a single dose（50 mg/kg）of intramuscular（i. m.）or oral gavage（p. o.）administration to Japanese eel（Anguilla japonica）in freshwater at 25℃. Eight fish per sampling point were examined after treatment. Plasma and muscle were collected and analyzed by high-performance liquid chromatography（HPLC）method with 0.05 μg/mL detection limit. The data of pharmacokinetics conformed to the two-compartment open model for intramuscular and one-compartment open model for oral administrations. After intramuscular administration, the elimination half-life（t$_{1/2\beta}$）was calculated to be 11.346 h, the maximum plasma concentration（Cmax）to be 29.524 μg/mL, the time to peak plasma streptomycin concentration（Tmax）to be 0.218 h, and the area under the plasma concentration-time curve（AUC）to be 90.206 μg/mL·h. Following p. o. administration, the corresponding estimates were 13.239 h, 0.346 μg/mL, 11.960 h, and 12.356 μg/mL·h. After intramuscular administration, a therapeutic concentration of the drug was maintained for 12 hours in the plasma, however, a therapeutic level could not be achieved after oral administration, and the results suggest that the drug can be used clinically by intramuscular administration against streptomycin susceptible systemic infections in Japanese eel.

Keyword：Pharmacokinetic；streptomycin sulfate；Japanese eel；intramuscular；oral gavage

硫酸链霉素在鳗鲡体内的药物代谢动力学及残留的研究

何平[1,2]，沈锦玉[1]，尹文林[1]，姚嘉赟[1]，徐洋[1]，潘晓艺[1]，郝贵杰[1]

(1. 浙江省淡水水产研究所 浙江 湖州 313001；2. 浙江三门食品检验中心，浙江 三门 317100)

摘要：固相萃取，氢氧化钠、硫酸铁铵衍生后，高效液相紫外检测法测定了鳗鲡组织中的硫酸链霉素的药代动力学。链霉素的最佳衍生条件为：0.05 M 氢氧化钠 80 μL，70℃水浴 1 h，冷却后，加 0.2% 硫酸铁铵 40 μL。净化的最佳条件为：在组织提取液中加 0.2 M 庚烷磺酸钠 5 mL，C_{18} 固相萃取柱净化，10 mL 甲醇洗脱，血液和肌肉中的最低检测限为 0.05 μg/mL（μg/g），血液中的平均回收率为 80.9±4.7%，肌肉中的平均回收率为（77.6±2.9）%。鳗鲡在（26±1）℃水温下，以 50 mg/kg 单剂量肌肉注射硫酸链霉素后，鳗鲡血液中的药-时数据用一级吸收二室开放模型来描述。主要的药动学参数为：达峰时间（T_{peak}）0.218 h，吸收相半衰期（$T_{1/2\alpha}$）为 0.196 h，消除相半衰期（$T_{1/2\beta}$）为 11.346 h，峰浓度（C_{max}）为 29.524 μg/mL。结果表明鳗鲡肌肉注射硫酸链霉素后，药物在体内吸收迅速，达峰短，消除速度快。鳗鲡在（26±1）℃时，以 50 mg/kg 单剂量肌肉注射后，4 d 后血液中未检出硫酸链霉素，5 d 后肌肉中未出检出。在（26±1）℃水温下，以 50 mg/kg 拌饵口灌硫酸链霉素后，最大的血液浓度也仅为 0.4888 μg/mL，药-时数据用开放的一室模型描述，主要的药动学参数为：达峰时间（T_{peak}）为 11.960 h，峰浓度（C_{max}）为 0.346 μg·mL^{-1}，吸收半衰期（$T_{1/2ka}$）为 5.529 h，消除半衰期（$T_{1/2ke}$）为 13.239 h。拌饵口灌后，血液和肌肉 3 d 后未检出。肌肉作为鱼类最主要的可食性组织，将其作为该药残留监控的靶组织，相应的休药期至少为 5 d。

关键词：鳗鲡；药物代谢动力学；固相萃取；衍生；残留

1 Introduction

Streptomycin (STR) is an aminoglycoside antibiotic widely used in medical and veterinary practices to treat infections from gram-negative bacteria and is now used in export eel farming as a chemotherapeutic agent in China. However, incorrect use of this drug can lead to the presence of residues of this drug and may lead to direct toxic effects on consumers, allergic reactions of those sensitive to the antibiotics or may indirectly cause problems through the induction of resistant strains of bacteria[1].

Japanese eel (Anguilla japonica) is one kind of the most economically important freshwater fishes in China and have become a consumer and farmers' favorite for its nutritional value, taste, strong adaptability, and efficient growth. However, bacterial disease resulted in huge losses in recent years; drug residue problems have now become serious owing to the increasing and unreasonable use of the drug. Therefore, pharmacokinetics (PK) and residue elimination studies are important in determination of optimal dosage regimens and in establishing a safe withdrawal time. The PK of STR has been studied extensively in many animal species, including camels[2], buffaloes[3], dog[4], horse[5], goat[6], and human being[7]. To the best of our knowledge, no data on STR PK and residue elimination in Japanese eel are available. The objective of the present study was to obtain information on the PK and residue elimination of STR for Japanese eel, and to determine reasonable dosage regimens of the drug in Japanese eel.

2 Materials and methods

2.1 Chemicals

Streptomycin sulfate (712 U/mg, NO. k0051201) were purchased from China Institute of Veterinary Drug Control, acetonitrile, methanol, perchloric acid, ammonium ferric sulfate, sodium 1-heptanesulfonate were of HPLC or analytical grade.

2.2 Fish

Clinically healthy Japanese eel, weighing 95.4±5.2 g, were obtained from aquatic fry farm of Zhejiang Institute of Freshwater Fisheries in China and maintained in ponds (1 m³), supplied with circulating water and oxygen continuously by an inflation pump. The water temperature was maintained by circulating water system (Shanghai Haisheng Company) at 25±1℃. The fish were allowed to acclimate for three days, fed a drug-free commercial diet. The fish had no previous exposure to any antibiotic, and no drugs were given to the animals during the acclimation or study periods. To rule out the influence of food content on the absorption of STR, the fish were starved for 24 h before drug administration.

2. 3　Standard calibration curve

Working calibration curves were prepared at concentrations of 0.25, 1.25, 2.5, 6.25, 12.5, 25, 125and250 μg/mL was drawn by plotting the known STR concentrations against the average peak area. The STR concentrations in unknown samples were calculated from the standard curves.

2. 4　Dosing and sampling

Japanese eelwere divided randomlyinto two groups for i. m. and oral administrations, respectively. The STR solution for i. m. (50 mg/mL) and oral (50 mg/mL) was prepared by dissolving the pure STR powder in distilled water. Individual fish was injected with STR solution at the dorsal fin (i. m.) or drenched with STR solution with gavage needle (p. o.) at the dose of 50 mg/kg (b. w.). The blood samples were taken from tail sinus ofeight fish at 0.25, 0.5, 1, 3, 6, 9, 12, 18, 24, 48, 72, 120, 168 h, 5 d, 7 d, 9 d, 12 d, 15 d after i. m. and p. o treatment, respectively. All samples were immediately frozen and stored at−20℃until analysis.

2. 5　STR analysis protocol

To 15 mL graduated plastic-stoppered centrifuge tubes was added 1 mL plasma (1 g of ground tissue) sample and 5 mL of perchloric acid. Each sample was whirlmixed for 5 min and then centrifuged for 15 min at 5000 rpm. The remaining tissue pellet in the centrifuge tube was extracted again for another two times, the supernatant were combined and transferred to another plastic-stoppered centrifuge tube and adjusted to pH 2.4 with perchloric acid, 5mL of 0.2 M Sodium 1-heptanesulfonate was added to each tube, then were transferred to Bond Elute C−18 solid-phase extraction columns pre-washed with methanol (10 mL) and distilled water (30 mL), and allowed to run dry for at least 2 min. STR was eluted from C−18 column with 10 mL methanol. The methanol eluate was collected and evaporated under reduced pressure in avacuum rotary evaporator for the complete removal ofsolvents. The dried eluate were dissolved in 200 μL distilled water, 80μl of 0.05 M NaOH were added and mixed for 1 min, then standing at 65℃for 1 h, 40 μl of 0.2% ammonium ferric sulfate were added when the solution temperature were 25℃, and diluted to 500 μl with distilled water. An aliquot (20 μL) of the diluted sample was analyzed with HPLC (Agilent XDB−C$_{18}$, 5μm, 4.6mm×250mm). The mobile phase of acetonitrile−0.2% sodium 1−heptanesulfonate (20:80, v/v) was filtered through a 0.45 μm millipore filter and degassed using sonication (5 min). The column was operated at 30℃. Flow rate was adjusted to 0.3 mL/min and the eluate was monitored at 330 nm.

2. 6　Detection limit and recovery percentage：

The detection limit is defined as the concentration of STR in Japanese eel tissues, which can produce a HPLC−UV signal that is three times greater than the signal-to-noise ratio.

2. 7　Method validation

The method was validated for plasma and muscle. The linearity, recovery, accuracy and precision were evaluated by analysis of the samples spiked with 50, 10, 5 and 1 μg/mL STR. Recovery was calculated by comparing the peak area of STR from processed samples with that from the STR standard in the mobile phase. The accuracy was determined by comparing the measured concentration with its true value. The limit of detection (LOD) of STR was defined as the STR concentration resulting in a peak height three times the signal noise.

2. 8　Pharmacokinetics analysis

Pharmacokinetic analysis was performed by the computer program 3P97 (version 1.0, edited by The Chinese Society of Mathematical Pharmacology, P. R. China). The model was selected on the basis of the residual sum of squares and the minimum Akaike's information criterion (AIC)[8]. The area under the concentration-time curve (AUC) was calculated using the trapezoidal rule and was extrapolated to infinity[9]. The elimination characteristics of the drug from each tissue were estimated using linear regression analysis of the terminal phase of the elimination time curve[10]. The elimination rate constant (β) was the slope of linear regression equation on logtransferred STR concentration (ln C) against time, and the elimination half-life (t$_{1/2β}$) was calculated from the equation $t_{1/2\beta} = 0.693/\beta$ for each tissue.

3　Results

3.1　Method validation

Calibration curves showed a good linearity ($r^2 = 0.999$) in the range of 0.1 to 100 μg/mL. The average inter-and intra-day precision (RSDs) were 0.9%~2% and 1.1%~3.1% for plasma and 0.5%~3.9% and 0.7%~4.0% for muscle. The STR accuracy for plasma ranged from 75% to 86% and mean recovery was 81%. The STR accuracy for muscle ranged from 73% to 81% and mean recovery was 77%. The limit of detection (LOD) of STR was determined as 0.05 μg/mL.

3.2　Pharmacokinetics analysis in serum

After i. m. administration, serum streptomycin concentration as a function of time is best described by an open two-compartment pharmacokinetic model (Fig. 1) . Pharmacokinetic data for streptomycin concentration in serum after i. m. administration of a single dose (50 mg/kg body weight) are presented in Table 1. STR was rapidly absorbed following administration. Plasma concentration reached 10.8 μg/mL at 0.167 h and the maximum plasma concentration (29.524 μg/mL) was reached at 0.218 h. After this the drug level declined rapidly and reached 0.24 μg/mlat 48 h. Correspondingly, the elimination half-life was 11.346 h. Other pharmacokinetic parameters were shown in Table 1.

The plasma concentration-time curve of STR after oral administration to fish at a single dose of 50 mg/kg (b. w.) is shown in Fig. 2. The data of pharmacokinetics fitted to one - compartment open pharmacokinetic model, the pharmacokinetic parameters calculated from this model are given in Table 2. STR levels were recorded at the first sample point (0.167 h) after oral dosing, followed by an increasing profile and the drug reached the peak plasma concentration (0.35 μg/mL) at 11.96 h (Tmax), then it slowly decreased. The absorption half-life ($T_{1/2ka}$) and elimination half-life ($T_{1/2b}$) of STR were 5.529h and 1

3.3　39 h respectively.

3.4　Residues of STR in Japanese eel

The STR residue concentrations in muscle after intramuscular administration are listed in Table 3. The elimination of STR in muscle was rapid, the drug concentrations declined to 0.18 μg/gin muscle on 4 d and were undetectable in muscle on 5 d, the corresponding date after oral administration were 0.12 μg/mlin muscle on 3 d and were undetectable in muscle on 4 d.

4　Discussion

Drug residue problems have now become serious owing to the increasing and unreasonable use ofantibiotic. Therefore, pharmacokinetics and residue elimination studies are important for establishing correct dosage regimes, promoting optimal use of the drug in question, and minimizing environmental impact. In the present study we determined the pharmacokinetics of STR in Japanese eel (Anguilla japonica) after a single dose (50 mg/kg) of intramuscular or oral administration in freshwater at 25℃. To our best knowledge, this is the first study on the pharmacokinetics and tissue disposition of streptomycin in Japanese eel.

The V/F (c) is an accurate indication of the diffusion of the drug in the body, and AUC represents the extent of drug absorption and determines bioavailability. In the present study, the V/F (c) and AUC were about 11 and 7 times larger when administered i. m. (0.847 L/kg and 90.206 μg/mL · h) than administered orally (0.077 L/kg and 12.356 μg/mL · h) . The time of plasma concentration above 2 μg/mL was approximately 12 h and 0 h following i. m. and oral administration (Table 1) . Based on MIC data studied on bacteria from fish, 2 μg/mL STR showed high efficacy against most bacteria[11,12] . All these results indicated that i. m. administration is more pharmacologically advantageous than p. o. administration, which was in consistent with the results in other species of other studies[2,3] .

STR was rapidly obsorbed following i. m. injection. The absorption half-life was 6 min and a mean peak serum concentration of 29.52 μg/mL was attained 30 min after drug administration. The values were considerably less than values previously reported for canels, 86 min and 16.8 min, respectively[13], and other animal species, including cattle (peak serum concentration 60 min:[14,15]), swine (peak serum concentration 60 min and absorption[16]), and dog (peak serum concentration 60 min,[17]) . The distribution half life, $T_{1/2\alpha}$, estimated in the present study for streptomycin following i. m. administrationwas about 0.2 hours. This shows that the drug is rapidly distributed from the central compartment into the body tissues.

The elimination half life, $T_{1/2\beta}$, was found to be 11.346 h. The reported tip of streptomycin was about 3.4 hours in horse[5], and $2 \sim 3$ h for normal adult human beings[7]. The above findings indicate that in fish and horse, the drug is eliminated comparatively more slowly than in man. The reason for the differencemay be attributableto differences in species and pharmaceutical factors for administration.

In summary, the pharmacokinetic profile of STR shown by the present study demonstrates rapid absorption and slow elimination, and i. m. injection may be better than oral administration for therapy of common bacterial infections. Further work is needed to understand the pharmacokinetics of STR more clearly in Japanese eel and other species, and to obtain accurate and reasonable dosing regimen.

5　Acknowledgements

This work was supported by a grant from the National High Technology Research and Development Program of China (863 Program) (No. 2011AA10A216), as well as Special Fund for Agro-scientific Research in the Public Interest (No. 201203085), China.

Reference

K rieger R., in: Handbook of Pesticide Toxicology, Principles, Part 1, Academic Press, San Diego, CA, 2001.

Hadi, A. A. A., Wasfi, I. A., Bsahir, A. K, 1998. Pharmacokinetics of streptomycin in camels. J Vet Pharmacol Ther. 21, 494-496.

Jayachandran, C., Singh, M. K., Singh, S. D., Banerjee, N. C., 1987. Pharmacokinetics of streptomycin with particular reference to its distribution in plasma milk and uterine fluid of shebuffaloes. Vet Res Commun. 71, 353-358.

Venzke, W. G., Smith, C. R, 1949. Streptomycin in small animal medicine. In: S. A. Waksrnan (ed), Streptomycin: nature and practical application, (Willim & Wilkins Baltimore). 580.

Baggot, J. D., Lone, D. N., Rose, R. J., Raus, J, 1981. Pharmacokinetics of some aminoglycoside antibiotics in the horse. J Vet Pharmacol Ther. 4, 277-284.

Wang, J. Y., Hu, W. J., Liu, Q. Y, 1981. Studies on Pharmacokinetics of Streptomycin in Milk Goats. J Northwest Sci - tech Agric - Forestry Univ. 4, 52-56.

Jackson, E. A. and McLeod, D, 1974. Pharmacokinetics and dosing of anti-microbial agents in renal impairment, American Journal of Hospital Pharmacy 31, 137-146

Yamaoka, K., Nakagawa, Y. and Uno, T.: Application of Akaike's information criterion (AIC) in the evaluation of linear pharmacokinetic equations. J. Pharmacokinet. Biopharm. 6, 165-175 (1978).

Guo, T.: Modern pharmacokinetics. China science and technology press, pp: 1-26 (2005).

Campbell, D. A., Pantazis, P. and Kelly, M. S.: Impact and residence time of oxytetracycline in the sea urchin, Psammechinus miliaris, a potential aquaculture species. Aquacult. 202, 73-87 (2001).

Zhou, Y. C., Fan, H. P., Liao, B. C. and Zeng, Z. Z.: An Anti-Bacterial Formula Study for Aeromonas hydrophila in Vitro. J Guangdong Ocean Univ. 30 (6), 35-39 (2010).

Song, L. G., Guang, N., Yi, B., Yang, C. H., Wan, R. L., Zhang X. Y., Shi, Q. and Zhang D.: Analysis for antimicrobial susceptibility of Streptococcus suis isolated from Sichuan Province. Chin J Vet Drug. 39 (10), 1-5 (2005).

EI-Gendi, A. Y. I., EI-Sayed, M. G. A., Atef, M. and Zaki Hussin, A.: Pharmacolinetic interpretation of some antibiotics in camels. Arc Int Phar, 261, 186-195 (1983).

Hannond, P. B.: Dihydrostreptomycin dose-serum level relationships in cattle. JAm Vet Med Assoc, 122, 203-206 (1953).

Teske, R. H., Rolins, L. D. and Carter, G. G.: Penicillin and dihydrostreptomycin serum concentrations after single and repeated doses to feeder steers. JAm Vet Med Assoc, 160, 873-878 (1972).

Mercer, H. D., Righter, H. F., Gordon, G and Carter, A. B.: Serum concentration of penicillin and dihydrostreptomycin after their parenteral administration in swine. JAm Vet Med Assoc, 160, 61-65 (1971).

Huber, W. G: Streptomycin, chloramphenicol, and other antibacterial agents. In veterinary Pharmacology and Therapeutics, 4th edn. Eds Jones, L. M., Boots, N. H. McDonald, L, E. Iowa State University Press, 1977, pp. 940-946.

该文发表于《中国水产科学》【2010，17（6）：1358-1363】

硫酸新霉素在吉富罗非鱼体内的药代动力学及休药期

Pharmacokinetics and withdrawal period of neomycin sulfate intilapia（GIFT）

宋洁[1,2]，王贤玉[1]，王伟利[1]，姜兰[1]，罗理[1]

（1. 中国水产科学研究院珠江水产研究所，广东 广州 510380；2. 上海海洋大学，上海 201306）

摘要：在实验水温（28±2）℃条件下，按 25 mg·kg^{-1} 的剂量对吉富罗非鱼（*Genetically improved farmed tilapia*，GIFT）单次口灌给药后，采用 UPLC-MS/MS 法测定吉富罗非鱼组织中的药物水平，研究硫酸新霉素在吉富罗非鱼体内的药代动力学及消除规律。结果表明，血药浓度时间数据符合一级吸收二室开放模型，药物在血浆中达峰时间 T_{max}、血药浓度高峰 C_{max} 和消除半衰期 $T_{1/2\beta}$ 分别为 1.299 h、16.138 μg·mL^{-1} 和 25.776 4 h。药物消除速度由快到慢依次为：肌肉、肝脏、肾脏，消除半衰期 $T_{1/2\beta}$ 分别为 31.802 h、34.917 h、45.175 h。选取吉富罗非鱼可食性肌肉组织作为残留检测靶组织，参考中华人民共和国第 235 号公告中对禽肌肉 MRL 规定，以 0.5 mg·kg^{-1} 为残留限量，建议休药期不低于 6 d。

关键词：吉富罗非鱼；硫酸新霉素；药代动力学；休药期

硫酸新霉素（Neomycinsulfate）为氨基糖苷类抗生素，可抑制细菌蛋白质合成，对葡萄球菌属、需氧革兰氏阴性杆菌、部分结核分支杆菌有较好的抗菌活性，常用于防治畜禽、养殖鱼类细菌性病害[1]。由于该药物具有较强的耳毒性和肾毒性[2]，残留在食品中的硫酸新霉素是潜在的安全隐患，因此，该药物的安全使用和残留检测技术就显得十分重要。常用的检测方法有微生物法[3]、旋光法[4]、分光光度法[5]和高效液相色谱法[6-8]，这些方法样品前处理较繁琐，通常需要对样品衍生后才能进行检测，且组织中检测限一般在 0.1 mg·kg^{-1} 左右，无法满足药代动力学研究的要求。本实验建立的硫酸新霉素 UPLC-MS/MS 检测法，样品前处理简单，且无需衍生，可直接上机检测，检测限达到 0.005 mg·kg^{-1}，可满足兽残和药代动力学研究的要求。

目前国内外未见有关硫酸新霉素在罗非鱼体内药代动力学和残留情况的研究报道和 MRL 的规定，对于硫酸新霉素在动物体内的药动和残留研究主要集中在牛、羊等畜禽种类，由于硫酸新霉素在不同动物种类间的代谢速率是有一定差别的[9]，因此本研究模拟水产实际养殖条件，研究了硫酸新霉素在吉富罗非鱼体内的药代动力学及残留情况，为其在罗非鱼养殖中的科学合理使用提供参考依据。

1 材料与方法

1.1 实验动物

吉富罗非鱼（*Genetically improved farmed tilapia*，GIFT）购自珠江水产研究所水产养殖基地，体质量（200±10）g，健康无病。实验前在实验池内暂养 2 周，水温为（28±2）℃。饲喂不含任何药物的全价饲料。实验前对罗非鱼进行抽检表明组织中均不含硫酸新霉素。

1.2 药品与试剂

硫酸新霉素标准品（每毫克相当于 709 单位）：购自中国药品生物制品检定所，批号 099310；三氯乙酸（分析纯）、磷酸二氢钠（分析纯）：购自广东光华化学厂；磷酸氢二钠、氯化钠（分析纯）：购自广州化学试剂厂；三氟乙酸（色谱纯）：购自杭州宝凯生物化学品有限公司；固相萃取柱：WatersOASIS HLB（500 mg/6mL）；磷酸盐缓冲液（pH 6.5）：取磷酸氢二钠（Na$_2$HPO$_4$·12H$_2$O）9.702g 和磷酸二氢钠（NaH$_2$PO$_4$·2H$_2$O）10.686 g 溶解于 500 mL 双蒸水。

1.3 实验仪器与设备

超高效液相色谱-串联四极杆质谱联用仪：美国 Waters 公司，配有电喷雾离子源（ESI）；固相萃取仪：美国 Suppcol 公司；小流量多通道泵：美国 Waters 公司；水浴氮吹干仪：美国 OrganomationAssociate 公司；破碎乳化机：日本 ACE 公司；台式高速冷冻离心机：德国 Sigma 公司。

1.4 色谱及质谱条件

1.4.1 色谱条件

色谱柱：UPLC Acquity BEH（50 mm×2.1 mm×1.7 μm）；进样量：10 μL；柱温：40℃；流动相 A 为乙腈，流动相 B 为 20 mmol 三氟乙酸水溶液（A、B 体积比 1∶19，V∶V）；等度洗脱，流速为 0.25 mL·min^{-1}。

1.4.2 质谱条件

电喷雾 ESI 正离子电离模式，多反应监测离子采集数据，毛细管电压 3.5 kV，萃取电压 4.0 V，透镜电压 0.5 V，离子源温度 110℃，脱溶剂气速 800 L·h^{-1}，锥孔反吹气速 50 L·h^{-1}，硫酸新霉素检测条件如表 1 所示。

表 1　硫酸新霉素质谱检测条件

Tab. 1　Mass spectrometry detection condition of neomycin sulfate

药物名称 Drug	保留时间（min） Retention time	定性离子对（m·z^{-1}） Qualitative ion	定量离子对（m·z^{-1}） Quantitative ion	锥孔电压/V Cone voltages	碰撞能量/eV Collision enery
硫酸新霉素 Neomycin sulfate	0.96	615.30> 160.80 615.30> 293.10	615.30> 160.80	70.00 70.00	30.00 25.00

1.5 实验设计及采样

实验用鱼共分 18 组，每组 4 尾，另取 4 尾未给药吉富罗非鱼作空白对照，采取口灌方式。将药品摇匀后，用长 10 cm 软胶管从鱼口中插入胃中，用注射器将药物灌入鱼胃，无回吐者作为实验对象保留，按照 25 mg·kg^{-1} 的剂量单次给药。给药前停饲 1 d，给药后 4 h 自由采食，于给药后 0.25、0.5、1、1.5、2、4、6、8、12、24、48、72、96、168、240、312、360、432 h 分别采集血液、肌肉、肝脏和肾脏，将样品放入−20℃冰箱保存用于实验分析。

1.6 样品处理

萃取：准确称取匀浆肌肉、肝脏、肾脏 2.0 g，血浆 2.0 mL，移至 50 mL 离心管，加入 20 mL 提取液（pH6.5 的磷酸盐缓冲液+5%三氯乙酸；V∶V=1∶1），水平摇床室温振荡 1h，离心 15 min（3 500 g，4℃），取上清液，上清液中加入 2 mL 0.2 mol·L^{-1} 三氟乙酸水溶液。

净化：用 5 mL 甲醇、5 mL 三氟乙酸水溶液（20 mmol·L^{-1}）活化平衡 HLB 柱，取全部上清液过柱，以 1.0 mL·min^{-1} 流速过柱，5 mL 纯水淋洗，抽干柱子，用 6 mL10%乙酸甲醇溶液以 1 mL·min^{-1} 洗脱至试管中，40℃水浴下氮气吹干。用 1.0 mL 三氟乙酸水溶液（20 mmol·L^{-1}）将其溶解，过 0.22 μm 针孔滤膜后上机检测。

1.7 回收率

准确取 1.0 mL 血浆及 2.0 g 鱼组织样品（包括肌肉、肝脏、肾脏），分别添加硫酸新霉素标样制得 0.1、0.5、1.0 mg·L^{-1}（mg·kg^{-1}）添加样品，静置 30 min，按照方法 1.6 处理，各浓度进行 3 批次实验，每批次进行 5 个样品的平行实验，计算回收率结果。

1.8 线性关系及检测限

1.8.1 线性关系

准确取 1.0 mL 血浆移至 50 mL 塑料离心管中，依次加入系列浓度的硫酸新霉素标准液，制得 0.05、0.1、0.2、0.5、1.0、5.0 mg·L^{-1} 的加标样品（n=4）；准确取 2.0 g 组织样品（包括肌肉、肝脏、肾脏）移至 50 mL 塑料离心管中，依次加入系列浓度的硫酸新霉素标准液，制得 0.05、0.1、0.2、0.5、1.0、5.0 mg·kg^{-1} 的加标样品（n=4），室温静置 30 min。按照方法 1.6 处理。将样品药物浓度作为横坐标（x），定量离子对色谱峰面积为纵坐标（y），求得回归方程和相关系数（r^2）。

1.8.2 检测限

准确取 1.0 mL 血浆及 2.0 g 组织样品（包括肌肉、肝脏、肾脏），分别添加标样制得 0.005、0.01、0.02、0.05、0.075、0.1 mg·L^{-1}（mg·kg^{-1}）添加样品，每个浓度做 4 个重复。按照 1.6 方法进行处理，以 3 倍信噪比为检测限，以 10 倍信噪比为定量限。

1.9 数据处理及休药期的计算

血浆中药代动力学模型拟合及参数计算采用 3P97 药动学软件分析；药物在各组织中的浓度曲线图和药物消除的药时曲线图采用 Microsoft Excel 2000 软件绘制。硫酸新霉素在吉富罗非鱼体内消除后期服从指数消除：$C_i = C_0 e^{-kt}$，这一过程与静脉注射一室模型类似，因此用 3P97 软件把消除相数据拟合成静脉注射一室模型，得出血液及各组织中的消除方程及相关参数。各组织药物浓度降至规定水平所需时间采用休药期软件 WT1.4 计算。

2 结果与分析

2.1 线性关系与检测限

硫酸新霉素标准曲线制备采用外标法定量，各组织中药物浓度相关系数 r^2 均大于 0.99，该药物在吉富罗非鱼血浆、肌肉、肝脏及肾脏中的标准曲线、平均回收率、检测限及定量限见表 2。

表 2　硫酸新霉素在吉富罗非鱼血浆、肌肉、肝脏及肾脏中的标准曲线、检测限及定量限
Tab. 2　Calibration, average recovery, LOD and LOQ of neomycin sulfate in plasma, muscle, liver and kidney of tilapia (GIFT)

样品 Sample	标准曲线 (n=4) calibration	线性范围 range of linearity (mg·kg⁻¹)	相关系数 (R^2) Relavent coefficient	平均回收率/% Average recovery	检测限 LOD (mg·kg⁻¹)	定量限 LOQ (mg·kg⁻¹)
血浆 plasma	$Y=126.68x-16581$	0.05~5.0	0.9928	88.16±3.95	0.002	0.005
肌肉 muscle	$Y=25.418x+116.6$	0.05~5.0	0.9969	87.63±3.42	0.005	0.01
肝脏 liver	$Y=1.2149x+2753.9$	0.05~5.0	0.9924	72.37±2.63	0.005	0.01
肾脏 kidney	$Y=7.0216x+3609.3$	0.05~5.0	0.9956	80.26±3.29	0.005	0.01

2.2 药物在鱼体内的药代动力学

在实验水温 (28±2)℃ 条件下，按 25 mg·kg⁻¹ 的剂量对吉富罗非鱼单次口灌给药后，药物在血浆中的药动学参数见表 3。经房室模型分析，其血药浓度时间数据符合一级吸收二室开放模型：$C = 0.0009e^{-7.7735t} + 13.9957e^{-0.0269t} - 13.9966e^{-0.2179t}$。药物在血浆中分布较快，其分布半衰期 $T_{1/2\alpha}$ 为 0.0892，且消除较为缓慢，消除半衰期 $T_{1/2\beta}$ 为 25.776 4 h，在给药后 312 h 未检测到药物。

表 3　吉富罗非鱼单剂量口灌硫酸新霉素 (25 mg·kg⁻¹ b·w) 的血浆药代动力学参数
Tab. 3　Pharmacokinetics in plasma following single oral administration of 25 mg·kg⁻¹ b·w of neomycin sulfate to tilapia (GIFT)

参数 Parameter	单位 Unit	参数值 Value	参数 Parameter	单位 Unit	参数值 Value
A	µg·mL⁻¹	0.001	AUC	µg·h·mL⁻¹	456.225
α	h⁻¹	7.774	$T_{1/2K\alpha}$	h	3.181
B	µg·mL⁻¹	13.996	K_{21}	h⁻¹	7.794
β	h⁻¹	0.027	K_{10}	h⁻¹	0.027
Ka	h⁻¹	0.218	K_{12}	h⁻¹	0.020
$T_{1/2\alpha}$	h	0.089	C_{max}	µg·mL⁻¹	16.138
$T_{1/2\beta}$	h	25.776	T_{max}	h	1.299

2.3 药物在各组织中的分布及消除规律

按照 25 mg·kg⁻¹ 单剂量口灌给药后，吉富罗非鱼血浆及各组织的药物残留浓度随时间变化的规律见图 1 和图 2。由图中可看出药物广泛分布于各组织中，给药后第 240 h 肌肉中药物水平为 0.18 mg·kg⁻¹；给药后第 312 h 肝脏中药物残留为

0.39 mg·kg^{-1}；其中肾脏中药物消除最慢，给药后第 360 h 肾脏中药物残留为 0.32 mg·kg^{-1}；第 432 h 各组织中均未检测到药物。数据经回归处理得到血浆和各种组织中药物残留水平（C）与时间（T）关系的消除曲线方程、相关指数（r^2）及消除半衰期（$T_{1/2}$）见表 4。

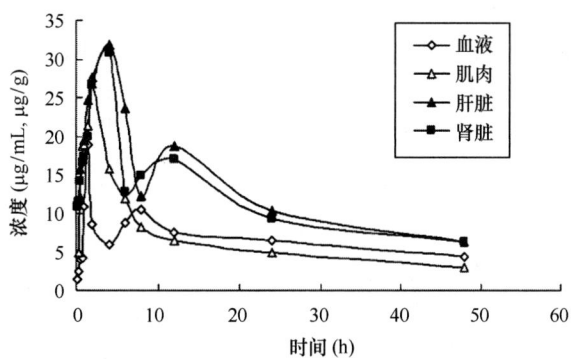

图 1　吉富罗非鱼单剂量口灌硫酸新霉素后各组织中药物残留

Fig. 1　Residue of neomycin sulfate in tissue of tilapia（GIFT）after oral administration

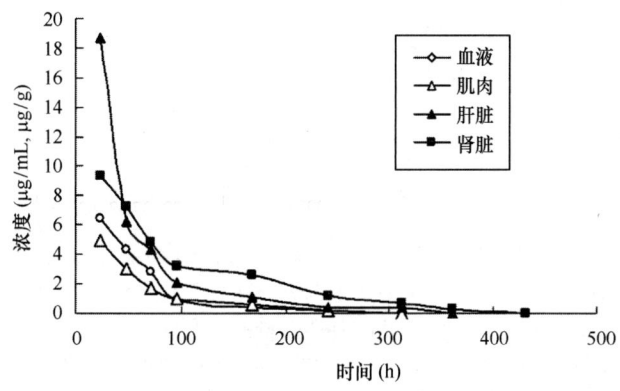

图 2　吉富罗非鱼口灌硫酸新霉素后药物消除曲线

Fig. 2　Elimination curve of neomycin sulfate in tilapia（GIFT）after single oral administration

表 4　吉富罗非鱼口灌新霉素后血浆和组织消除曲线方程及参数

Tab. 4　Equation of elimination curve and parameters of neomycin sulfate in plasma and

tissues after oral administration in tilapia（GIFT）

组织 tissue	方程 equation	相关指数 R^2	消除半衰期 $T_{1/2}$（h）
血浆 plasma	$C_{血} = 13.736e^{-0.027t}$	0.992	25.935
肌肉 muscle	$C_{肌} = 8.371e^{-0.022t}$	0.999	31.802
肝脏 liver	$C_{肝} = 25.52e^{-0.02t}$	0.999	34.917
肾脏 kidney	$C_{肾} = 17.096e^{-0.015t}$	0.999	45.175

2.4　休药期

联合国食品法典委员会（CAC）及世界各国均未制定鱼组织中新霉素的残留限量（MRL），鉴于中国的水产养殖量较大，在保障食品安全方面可参考农业部第 235 号公告，以禽肌肉组织的标准 0.5 mg·kg^{-1}作为残留限量（MRL），由于水产药物中大部分使用硫酸新霉素，根据以上标准对硫酸新霉素残留限量（MRL）计算得 0.74 mg·kg^{-1}。通过休药期软件 WT1.4 计算可知各组织的理论休药期为：肌肉 5.34 d；肝脏 8.19 d；肾脏 9.81 d。选取吉富罗非鱼可食性肌肉组织作为残留检测靶组织，建议休药期不低于 6 d。

3 讨论

3.1 硫酸新霉素在罗非鱼体内的药代动力学特征

目前，国内外对硫酸新霉素在动物体内的药代动力学研究的相关报道主要集中在畜牧及家禽方面，未见水产动物体内的药代动力学研究报道。在实验水温（28±2）℃条件下，药物在血浆中的消除半衰期 $T_{1/2\beta}$ 为 25.776 4 h，表明药物在罗非鱼体内消除速度较为缓慢，但硫酸新霉素在牛体内的消除速度较快，Ziv 等[10]按 20 mg·kg^{-1} 体质量给牛静脉注射，其消除半衰期 $T_{1/2\beta}$ 为（3.25±0.86）h；Black 等[11]按 12 mg·kg^{-1} 体质量给牛静脉注射，其消除半衰期 $T_{1/2\beta}$ 为（2.776±0.792）h；Shaikh 等[12]按 22 mg·kg^{-1} 体质量给牛肌肉注射，其消除半衰期 $T_{1/2\beta}$ 为（6.61±3.14）h。从以上比较可以看出该药物在罗非鱼和牛体内药动学特征存在显著差异，此差异可能是由于罗非鱼属于水生变温动物，在实验水温（28±2）℃条件下，体温维持在 28℃左右，而牛属于恒温动物，体温一般维持在 37.5℃。一般来说，在一定温度范围内，药物代谢强度与体温成正比，且牛具有较罗非鱼更为完善的代谢器官，可以通过肝脏和肾脏进行主动代谢，因此硫酸新霉素在罗非鱼体内代谢较慢。

硫酸新霉素在单次口灌给药后血液和肝脏中出现了双峰现象。目前关于口服给药后药物 C-T 曲线的双峰现象研究报道较多，认为引起该现象的主要原因是由于药物的肠肝循环和非齐性吸收所致，两峰间隔时间由 2 h 到 20 h 不等[13-14]。通过作者的实验发现，对吉富罗非鱼单次口灌给药后，12 h 肝脏中出现了第 2 个药物浓度高峰，说明药物排泄进入十二指肠后，可在小肠重吸收返回肝脏，形成肠肝循环，引起双峰现象。非齐性吸收：这种吸收现象主要是由于在胃肠道的不同部位和管壁对某些药物的通透性不同，当口服药物后，不同部位的吸收时间和吸收速率并不一致，一般有 2 个吸收部位，一个吸收部位在胃肠道上部，另一个在下部[15]。由于硫酸新霉素口服后大部分以原形排出，因此当药物在不同部位被吸收时，药物的形态主要以原形为主，所以在检测组织中药物浓度时，出现了 2 个药物浓度峰；吉富罗非鱼单次口灌给药后血液和肝脏中出现的双峰现象可能是由肠肝循环和非齐性吸收所引起的。

3.2 硫酸新霉素在罗非鱼组织中消除规律

在实验水温（28±2）℃条件下，按 25 mg·kg^{-1} 的剂量对吉富罗非鱼单次口灌给药后，药物在各组织中消除速度由快到慢依次为：肌肉、肝脏、肾脏，消除半衰期 $T_{1/2\beta}$ 分别为 31.802 h、34.917 h、45.175 h。Ronning 等[16]对新霉素在牛、羊、山羊和猪体内消除规律做了研究报道，他们按 22 mg·kg^{-1} 体质量自由饮水给药，采用微生物杯碟法进行残留分析，结果新霉素在牛肾脏中残留时间为 14 d，羊肾脏中残留为 7 d，山羊中肾脏残留为 4 d，猪肾脏中残留为 14 d，其他组织未检出有药物残留；Black 等[11]在同一时间检测牛的肾皮质和肾髓质组织发现肾皮质中的新霉素含量要远远高于髓质中药物的含量；沈川等[9]认为新霉素在动物体内与肾皮质以较强的离子键结合，在肾中蓄积时间很长，通常需要 28~30 d 的休药期。以上研究结果表明，虽然罗非鱼与牛、羊、猪等哺乳类动物种属间的消除半衰期相差较远，但消除规律一致，在肾脏中消除速率最慢，残留时间最长。

参考文献

[1] 常津，刘海峰，姚康德. 实用药物指南：[M]. 第 1 版. 北京：人民军医出版社，2000：264-285.

[2] Isoherranen N, Soback S. Determination of Gentamicins C1, C1a, and C2 in Plasma and Urine by HPLC [J]. J AOAC Int, 1999, 82：1017-1045.

[3] 国家药典委员会编. 中华人民共和国药典：二部 [S]. 北京：化学工业出版社，2000：866-898.

[4] 刘明忠. 旋光法测定硫酸新霉素溶液的含量 [J]. 药学实践杂志，1997，15（1）：38.

[5] 陈壬荃，陈桂良. 新霉素及其他氨基糖苷类抗生素的分光光度快速测定 [J]. 中国医药工业杂志，1994，25（2）：79.

[6] Stead D A, Richards R M E. Sensitive fluorimetric determination of gentamicin sulfate in biological matrices using solid-phase extraction, pre-column derivatization with 9-fluorenylmethyl chloroformate and reversed-phase highperformance liquid chromatography [J]. J Chromatogr B, 1996, 675：295-302.

[7] Stead D A, Richards R ME. Sensitive high-performance liquid chromatographic assay for aminoglycosides in biological matrices enables the direct estimation of bacterial drug uptake [J]. J Chromatogr B, 1997, 693：415-421.

[8] Andrzej P, Jan Z, Jolanta N, et al. Sample preparation for residue determination of gentamicin and neomycin by liquid chromatography [J]. J Chromatogr A, 2001, 914：59-66.

[9] 沈川，肖希龙. 新霉素在动物机体中的残留及其测定方法 [J]. 中国兽药杂志，1998，32（2）：53-56.

[10] Ziv G and Sulman FG. Distribution of aminoglycoside antibiotics in the blood and milk [J]. Res Veterin Sci, 1974, 17：68-74.

[11] Black W D, Holt J D, Gentry R D. Pharmacokinetic study of neomycin in calves following intravenous and intramuscular administration [J]. Can J Compar Med, 1983, 47：433-435.

[12] Shaikh B, Jackson J, Guyer G, et al. Determination of neomycin in plasma and urine by high-performance liquid chromatography. Application

to a preliminary pharmacokinetic study［J］. J Chromatogr Biomed Appl, 1991, 571: 189-198.

［13］　Kroboth P D, Smith R B, Rault R, et al. Effects of end-stage renal disease and aluminum hydroxide on temazepam kinetics［J］. Clin Pharmacol Therap, 1985, 37: 453-459.

［14］　Pentikainen P J, Neuvonen P J, Penttila A. Pharmacokinetics and pharmacodynamics of glipizide in healthy volunteers［J］. Int J Clin Pharmacol Ther Toxicol, 1983, 21: 98-107.

［15］　Staveris S, Houin G, Tillement JP, et al. Primary dosedependent pharmacokinetic study of veralipride［J］.J Pharmac Sci, 1985, 74: 94-96.

［16］　FAO. Redisues of Some Veterinary Drugs in Animds and Food［Z］, FAO Food and Nuition Paper41/7, 1994, 56-57.

三、酰胺醇类

酰胺醇类是一类广谱性的抗生素，可有效地抑制各种细菌。它主要作用于细菌70S核糖体的50S亚基，通过与rRNA分子可逆性结合，抑制由rRNA直接介导的转肽酶反应而阻断肽链延长，从而抑制细菌蛋白质合成，该类药物主要有甲砜霉素、氟苯尼考等。酰胺醇类抗生素在不同鱼类体内的吸收、分布、代谢和排泄过程不尽相同，即使是同一种鱼类不同的给药途径也有较大的区别。研究结果表明，淡水养殖凡纳滨对虾口灌给药甲砜霉素和氟苯尼考后，甲砜霉素吸收和消除快于氟苯尼考，可能是由于甲砜霉素的体外蛋白结合率（28.38%）低于氟苯尼考（37.91%）（Wenhong Fang et al, 2013）。氟苯尼考在花鲈（Lateolabrax japonicus）体内消除较快（黄聚杰等，2016）。氟苯尼考在日本鳗鲡（Anguilla japonica）和欧洲鳗鲡（A. Anguilla）体内的药代规律不同（林茂等，2011）。水温对氟苯尼考在日本鳗鲡体内的相对生物利用度有显著影响（林茂等，2013）。氟苯尼考在鱼体血液中的吸收和消除速率均快于肌肉（朱丽敏等，2006；王伟利等，2010），氟苯尼考和氟苯尼考胺在斑点叉尾鮰的肾脏中消除最慢，可以将肾脏作为其残留的靶组织（刘永涛等，2007）。氟苯尼考主要以原形药物形式代谢消除（王瑞雪等，2012）。用30 mg·kg⁻¹剂量拌饲口服氟苯尼考，它在中华鳖体内吸收快，血药浓度高，维持时间长，生物利用度高，而在肌肉中消除缓慢（朱丽敏等，2006）。

该文发表于《Journal of Aquatic Animal Health》【2013, 25（2）: 83-89】

Pharmacokinetics and tissue distribution of thiamphenicol and florfenicol in Pacific white shrimp（*Litopenaeus vannamei*）in freshwater following oral administration

Wenhong Fang[a], Guolie Li[a,b], Shuai Zhou[a], Xincang Li[a], Linlin Hu[a] & Junfang Zhou[a]

(a. East China Sea Fisheries Research Institute, Chinese Academy of Fishery Sciences, Key Laboratory of Marine and Estuarine Fisheries Resources and Ecology, Ministry of Agriculture, 300 Jungong Road, Shanghai, 200090, China; b. Nanchong Supervision and Test Center for Agricultural Products Quality, 137 Nongke Alley, Nanchong, Sichuan Province, 637000, China)

Abstract: This study evaluated the pharmacokinetic disposition of thiamphenicol (THA) and florfenicol (FLR) after oral administration at a single dose of 10 mg/kg body weight in Pacific white shrimp (Litopenaeus vannamei) held in freshwater at 25.0±1.0℃. The THA and FLR concentrations in the hemolymph, muscle and hepatopancreas were determined by high-performance liquid chromatography (HPLC). The profiles of hemolymph THA and FLR concentrations vs. time were best described by a two-compartment open pharmacokinetic model with first-order absorption. The peak concentration (C_{max}), peak time (T_{max}), absorption half-life ($t_{1/2ka}$) and elimination half-life ($t_{1/2\beta}$) of THA in hemolymph were 7.96 μg/mL, 2 h, 0.666 h and 10.659 h, respectively. The corresponding values for FLR were 5.53 μg/mL, 2 h, 1.069 h and 17.360 h, respectively. Following oral administration, THA and FLR were rapidly absorbed in white shrimp and THA in hemolymph was absorbed and eliminated more quickly than FLR. The parameters in muscle and hepatopancreas were calculated by non-compartment model based on statistical moment theory. The peak concentration (C_{max}), area under the concentration-time curve (AUC_{0-t}), mean residue time (MRT_{0-t}) and half-life ($t_{1/2z}$) in muscle were 2.98 μg/g, 29.10 mg/kg·h, 9.77 h and 6.84 h, respectively. The corresponding values for FLR were 1.91 μg/g, 15.97 mg/kg·h, 19.40 h and 18.32 h, respectively. THA in muscle was eliminated more quickly than FLR. The peak concentrations of THA and FLR in the hepatopancreas were 204.25 μg/g and 164.22 μg/g and the area under the concentration-time curve (AUC_{0-t}) were 1337.74 mg/kg·h and 871.73 mg/kg·h, respectively, which were much higher than those in hemolymph and muscle. The in vitro protein binding value of THA (28.38%) was lower than that of FLR (37.91%), which might be related to that THA in shrimp was absorbed and eliminated more quickly than FLR.

Keywords：Pharmacokinetics；Tissue distribution；Thiamphenicol；Florfenicol；Litopenaeus vannamei；Oral administration

口灌给药下甲砜霉素和氟苯尼考在淡水养殖
凡纳滨对虾体内药动学和组织分布

房文红[a]，李国烈[a,b]，周帅[a]，李新苍[a]，胡琳琳[a]，周俊芳[a]

（a. 中国水产科学研究院东海水产研究所，中国上海，军工路 300 号，200090；
b. 南充农产品监督测试中心，中国四川南充市，农科路 137 号，，637000）

摘要：本文研究了淡水养殖凡纳滨对虾（25.0±1.0℃）口灌给药甲砜霉素（THA）和氟苯尼考（FLR）后的药动学和组织分布。采用 HPLC 测定血淋巴、肌肉和肝胰腺中 THA 和 FLR 含量。血淋巴中 THA 和 FLR 浓度与时间关系曲线适合采用一级吸收二室开放模型描述。血淋巴 THA 峰浓度（C_{max}）、达峰时间（T_{max}）、吸收半衰期（$T_{1/2ka}$）和消除半衰期（$T_{1/2\beta}$）分别为 7.96 μg/mL、2 h、0.666 h 和 10.659 h；FLR 的相应值分别为 5.53 μg/mL、2 h、1.069 h 和 17.360 h。口灌给药后，THA 和 FLR 吸收快，且血淋巴 THA 吸收和消除快于 FLR。肌肉和肝胰腺药动学参数采用统计矩原理推算。肌肉中 THA 的 C_{max}、AUC_{0-t}、MRT_{0-t} 和 $T_{1/2z}$ 分别为 2.98 μg/g、29.10 mg/kg·h、9.77 h 和 6.84 h，FLR 的 C_{max}、AUC_{0-t}、MRT_{0-t} 和 $T_{1/2z}$ 分别为 1.91 μg/g、15.97 mg/kg·h、19.40 h 和 18.32 h。肌肉中 THA 消除快于 FLR。肝胰腺 THA 和 FLR 峰浓度分别是 204.25 μg/g 和 164.22 μg/g，AUC_{0-t} 分别是 1337.74 mg/kg·h 和 871.73 mg/kg·h，显著高于肌肉和血淋巴的。THA 的体外蛋白结合率（28.38%）低于 FLR（37.91%），这可能是 THA 吸收和消除快于 FLR 的解释。

关键词：药动学；组织分布；甲砜霉素；氟苯尼考；凡纳滨对虾；口灌给药

1 Introduction

The Pacific white shrimp （*Litopenaeus vannamei*）has become the most important cultivated shrimp in China due to its fast growth rate and strong adaptability. This shrimp is popularly farmed not only in seawater but also in freshwater （BOF 2011）. Given its euryhaline characteristics, L. vannamei has become the main aquatic animal cultured in brackish water and freshwater regions in recent years. In 2010, there were 615 010 tons of shrimp produced in freshwater ponds in China. Recently, diseases caused by Vibrio bacteria have become a serious problem causing economic losses for shrimp farming （Wang and Yang 2005；Zhou et al. 2012）.

Thiamphenicol （THA）and its fluorinated derivative, florfenicol （FLR）, are broad-spectrumantibiotics belonging to a part of the chloramphenicol family of drugs. Although THA and FLR have similar mechanisms of action toward chloramphenicol, they do not exert the potential fatal side effect of dose-unrelated aplastic anemia in humans （Dowling 2006）. After the prohibition of chloramphenicol in aquaculture, THA and FLR are expected to be used as alternatives.

Owing to their broad antibacterial spectrum and high potency, THA and FLR are commonly used to control susceptible bacterial diseases in poultry （Switala et al. 2007）and fish （Van de Riet et al. 2003；Burridge et al. 2010）farming. They have been demonstrated to be effective in the treatment of pseudotuberculosis in yellowtail fish （*Seriola quinqueradiata*）（Yasunaga and Yasumoto 1988）, edwardsiellosis in channel catfish （*Ictalurus punctatus*）（McGinnis et al. 2003）, vibriosis in goldfish （*Carassius auratus*）（Fukui et al. 1987）, and furunculosis in Atlantic salmon （*Salmo salar*）（Inglis et al. 1991）.

They also have potential use in the treatment of vibriosis and necrotizing hepatopancreatitis infections in farm-raised shrimps. The pharmacokinetic characteristics of THA and FLR in shrimp needs to be understood to determine the optimal dose regimens for achieving and maintaining therapeutic drug levels as well as predict the residue withdrawal time in edible tissues of treated shrimp.

Previous THA and/or FLR pharmacokinetic studies have been performed mainly on finfish, including seabass （*Dicentrarchus labrax* L.）（Castells et al. 2000；Malvisi et al. 2002）, seabream （*Sparus aurata* L.）（Malvisi et al. 2002）, olive flounder （Paralichthys olivaceus）（Lim et al. 2010；Lim et al. 2010）, red pacu （*Piaractus brachypomus*）（Lewbart et al. 2005）, Korean catfish （*Silurus asotus*）（Park et al. 2006）, and crucian carp （*Carassius auratus cuvieri*）（Zhao et al. 2011）. The results show that THA and FLR are absorbed rapidly, distributed extensively, and eliminated rapidly. Although the characteristics of THA and FLR have been studied in a variety of vertebrate species, little is known about the pharmacokinetics of THA and FLR in crustaceans, particularly in shrimp.

The objective of the present study was to investigate the pharmacokinetics and tissue distribution of THA and FLR in Pacific white shrimp held in freshwater after a single oral administration at 25.0±1.0℃.

2 Materials and methods

2.1 Chemicals

THA and FLR standards were purchased from Sigma (St. Louis, MO, USA). THA and FLR raw materials were supplied by Zhangjiagang Hengsheng Pharmaceutical Co., Ltd. The active ingredients were determined by high-performance liquid chromatography (HPLC) and found to exceed 98.0%. The organic solvents used, including methanol, ethyl acetate, acetonitrile, and n-hexane, were of HPLC grade (Tedia Company Inc., USA). All other chemicals were of analytical grade.

2.2 Experimental animals and aquarium conditions

Healthy white shrimps (Litopenaeus vannamei, 9.1~12.3 g) were obtained from a shrimp farm in Suburb Fengxian, Shanghai, China. They were kept in 120 L fiberglass tanks (6 individuals per tank) with freshwater. Tap water was dechlorinated by a purifier (AC/KDF-150BSE, Shanghai Canature Environmental Products Co., Ltd.). Water was recirculated, aeration was kept constant, and the temperature was maintained at 25.0±1.0℃. The shrimps were fed daily with pellet feed 2%~3% of their average body weight. The remnants, excretions and moults were promptly removed. The shrimps were acclimated for 10 d and analyzed to confirm the absence of THA and FLR before experiments. Prior to drug administration, the shrimps were not fed for 48 h.

2.3 Oral administration

THA (or FLR) was mixed in a slurry of food. The slurry contained 2 mg of THA (or FLR) per mL. The single oral dose was 10 mg THA (or FLR)/kg body weight. Thus, the dosing feed was gavaged at a single dose of 5 mL/kg body weight, corresponding to a final dose of 10 mg/kg body weight. The slurry was forced into the stomach of shrimps using a 1 mL syringe fitted with a blunt 12# gauge needle. Only those individuals that did not regurgitate were used in the experiment.

2.4 Sample collection

The sampling times were 0.25 h, 0.5 h, 1 h, 2 h, 4 h, 6 h, 9 h, 12 h, 24 h, 48 h, 5 d, and 8 d post-dosing. Six shrimps were sacrificed each time and tissues were sampled.

Hemolymph was collected from the pericardial cavity using 1 mL syringes, placed in vials containing ammonium oxalate (0.001 g) as an anticoagulant, and mixed. The collected hemolymph was centrifuged at 850×g (Hitachi CF16RXⅡ with type 49 rotor) for 10 min at room temperature. Plasma was removed and placed in 1.5 mL Eppendorf vials. Muscle and hepatopancreas were also collected and placed in individually marked plastic bags. The samples were frozen and stored at-80℃ prior to extraction.

2.5 Sample analysis

THA and FLR in plasma and tissues were measured by high-performance liquid chromatography (HPLC) method (Switala et al. 2007) with modification. The HPLC system used was an Agilent 1100 series consisting of a double pump, an auto-injector, a column temperature tank, and a UV detector. The column was a Zorbax SB C-18 (4.6 mm × 150 mm). The mobile phase consisted of acetonitrile: deionized water (16：84, v/v). The mobile phase was filtered through a 0.45 μm Millipore filter and sonicated for 10 min before use. The injection volume was 10 μL and the flow rate was 1.0 mL/min. The column temperature was controlled at 40℃ and the wavelength of the detector was set at 225 nm.

Plasma (500 μL) was placed in a 2 mL centrifuge tube and ethyl acetate (1 mL) was added as an extractant. The mixture was then vortexed for 5 min and centrifuged at 9500×g (Hitachi CF16RXⅡ with type 49 rotor) for 5 min. The supernatant was transferred to a clean centrifuge tube and the residue was re-extracted twice as the above procedure. The combined supernatants were evaporated to dryness using a vortex evaporator. The resulting residue was dissolved in 1.0 mL of mobile phase, and then filtered through 0.45 μm disposable syringe filters prior to analysis by the HPLC system.

Muscle samples (1.0 g) were thawed and triturated using a mortar and pestle. Hepatopancreas samples (0.7~1.0 g) were weighed after thawing and triturated with a glass rod in a centrifuge tube. About 1.0 mL of 0.02 M H_3PO_4 were added to the muscle and hepatopancreas samples, which were vortexed for 2 min. After adding 4 mL of ethyl acetate, the sample was vortexed for 5 min and centrifuged at 8300×g (Hitachi CF16RXⅡ with type 44 rotor) for 5 min. The supernatant was transferred to a clean centrifuge tube and the residue was re-extracted twice as the above procedure. The combined supernatants were evaporated to dryness using a vortex e-vaporator. The resulting residue was dissolved in 1. 0 mL of mobile phase and 2 mL of n-hexane. The mixture was shaken vigorously and then transferred into a 5 mL centrifuge tube. After centrifugation at 5000 rpm for 5 min, the bottom-layer liquid was filtered through a 0. 45 μm disposable syringe filters prior to analysis by the HPLC system.

The calibration standards for THA and FLR were in the range of 0. 05~10 μg/mL. The limits of quantification (LOQs) of THA and FLR were 0. 05 μg/mL for plasma, 0. 05 μg/g for hepatopancreas, and 0. 05 μg/g for muscle. The recoveries for the various THA concentrations assayed were 86%~97% in plasma and 80%~92% in tissues. The recoveries for the various FLR concentrations assayed were 88%~95% in plasma and 78%~93% in tissues.

2. 6 Plasma protein binding

The ultrafiltration method was used to determine the plasma protein binding values. Untreated hemolymph from 30 shrimps was collected and mixed by vortexing. Plasma was obtained after centrifugation at 2, 400×g (Hitachi CF16RX II with type 49 rotor) for 5min. About 500 μL of plasma was placed in a 2 mL centrifuge tube, and THA or FLR was added to a final concentration of 0. 5 μg/mL. Mixing and incubation for 4 h at 28℃ was subsequently performed. About 300 μL was added to each centrifugal ultrafiltration tube (Microncon YM-10, 10 kDa), and the sample was centrifuged at 13, 700×g (Hitachi CF16RX II with type 49 rotor) and room temperature for 60 min. The ultrafiltration filtrate from the plasma sample preparation was determined by HPLC.

2. 7 Pharmacokinetic analysis

The concentration-time profiles of the plasma and tissues were analyzed according to compartmental and non-compartmental modeling analyses, respectively. All compartmental and non-compartmental modelings were performed using Drug and Statistics Software Version 2. 1. 1 (Mathematical Pharmacology Professional Committee of China).

3 Results

3. 1 Pharmacokinetics of THA and FLR after oral administration

The plasma concentrations of THA and FLR afteroral administration at a dose of 10 mg/kg were shown in Fig. 1. The peak concentrations of THA and FLR in plasma were 7. 96 μg/mL and 5. 53 μg/mL, respectively, both occurred 2 h after administration, and followed by a rapid decrease until the 12 h post-gavage. The plasma concentrations of FLR and THA could be best described by a two-compartment open model with first-order absorption. The pharmacokinetic parameters derived from the compartment model were shown in Table 1. The distribution and elimination half-lives ($t_{1/2\alpha}$ and $t_{1/2\beta}$), total body clearance (CL_s), and apparent volume of distribution (V_d) of THA were found to be 2. 488 h, 10. 659 h, 227. 1 mL/kg/h, and 0. 942 L/kg, respectively. The corresponding FLR values were 1. 364 h, 17. 360 h, 242. 0 mL/kg/h, and 0. 887 L/kg, respectively. The plasma protein binding values of THA and FLR were 28. 38%±1. 26% and 37. 91%±7. 61%, respectively. These values were not high, and the value of THA was lower than that of FLR.

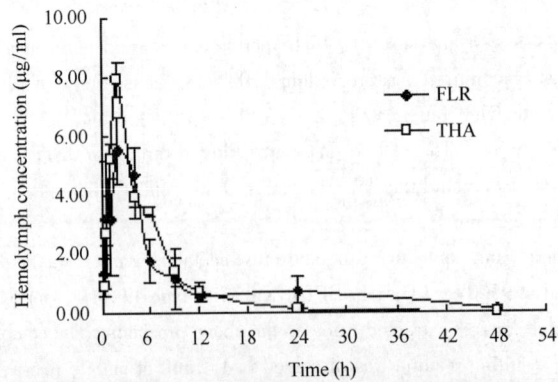

Fig. 1 Profiles of hemolymph thiamphenicol and florfenicol concentrations vs. time of Litopenaeus vannamei following oral gavage (n=6).

Table 1 Pharmacokinetic parameters of thiamphenicol and florfenicol in hemolymph from Litopenaeus vannamei based on the compartmental model (n=6).

Parameters	Unit	Thiamphenicol	Florfenicol
C_{max}	μg/mL	7.96±1.61	5.53±1.14
T_{max}	h	2.0	2.0
$t_{1/2\alpha}$	h	2.488	1.364
$t_{1/2\beta}$	h	10.659	17.360
$t_{1/2ka}$	h	0.666	1.069
CL_s	mL/kg/h	227.1	242.0
AUC_{0-t}	mg/L·h	43.70	38.94
$AUC_{0-\infty}$	mg/L·h	43.96	41.27
V_d	L/kg	0.942	0.887
V_{ss}	L/kg	1.547	4.381
t_{lag}	h	0.161	0.033

Note: C_{max}, maximum plasma concentration; T_{max}, time when the maximum concentration was reached; $t_{1/2\alpha}$, distribution half-life of the drug; $t_{1/2\beta}$, elimination half-life of the drug; $t_{1/2ka}$, absorption half-life of the drug; CL_s, total body clearance of the drug; AUC_{0-t}, area under the concentration-time curve from zero to time; $AUC_{0-\infty}$, area under the concentration-time curve from zero to infinity; V_d, apparent volume of distribution; V_{ss}, apparent volume of distribution at steady-state; and t_{lag}, lag time before absorption.

3.2 Tissue distribution and elimination

The tissue concentration-time curves of THA and FLR after oral administration were shown in Fig. 2 and Fig. 3. The peak concentrations of THA and FLR in muscle were 2.98 μg/g and 1.91 μg/g, respectively. Both occurred 2 h after administration followed by a rapid decrease until the 12th hour. THA and FLR were quickly cleared from muscle, and 0.023 μg/g and 0.040 μg/g levels were measured at 2 d, respectively.

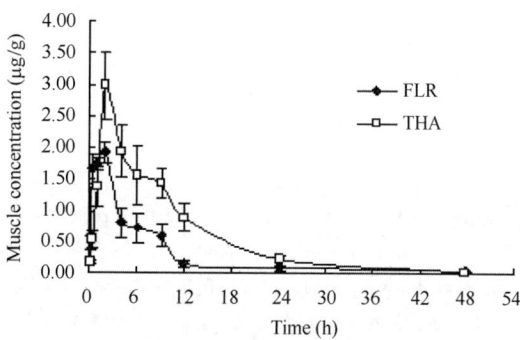

Fig. 2 Profiles of muscle thiamphenicol and florfenicol concentrations vs. time of Litopenaeus vannamei following oral gavage (n=6).

Although the trends of the concentration-time curve were similar, the concentrations of THA and FLR in the hepatopancreas were much higher than those in muscle and hemolymph. The THA and FLR peak concentrations were 204.25 μg/g and 164.22 μg/g, which occurred 1 h and 0.5 h post-dosing, respectively.

The behaviors of THA and FLR in tissues after oral administration were analyzed according to the statistical moment theory. The pharmacokinetic parameters were calculated and listed in Table 2. For muscle, the estimated AUC_{0-t} values were 29.10 mg/kg·h and 15.97 mg/kg·h for THA and FLR, respectively. The elimination half-lives ($t_{1/2z}$) were 6.84 h and 18.32 h for THA and FLR, respectively. For hepatopancreas, the estimated AUC_{0-t} values were 1 337.74 mg/kg·h and 871.73 mg/kg·h for THA and FLR, respectively. The elimination half-lives ($t_{1/2z}$) were 31.29 h and 41.38 h for THA and FLR, respectively.

337

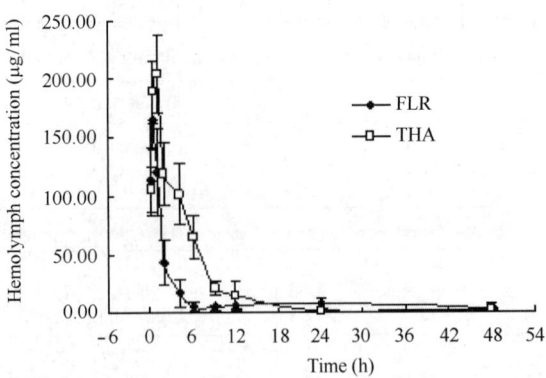

Fig. 3   Profiles of hepatopancreas thiamphenicol and florfenicol concentrations vs. time of Litopenaeus vannamei following oral gavage （n=6）.

Table 2   Pharmacokinetic parameters based on the statistical moment theory of thiamphenicol and florfenicol in muscle and hepatopancreas fromLitopenaeus vannameiafter a single oral administration （n=6）

Parameters	Unit	Muscle		Hepatopancreas	
		Thiamphenicol	Florfenicol	Thiamphenicol	Florfenicol
C_{max}	mg/kg	2.98±0.53	1.91±0.18	204.25±33.41	164.22±25.64
T_{max}	h	2.0	2.0	1.0	0.5
AUC_{0-t}	mg/kg · h	29.10	15.97	1337.74	871.73
$AUC_{0-\infty}$	mg/kg · h	29.33	16.04	1381.60	902.24
MRT_{0-t}	h	9.77	19.40	27.71	38.43
$MRT_{0-\infty}$	h	10.14	19.94	34.82	45.64
$t_{1/2z}$	h	6.84	18.32	31.29	41.38
CL_z	kg/h/kg	0.341	0.624	0.007	0.011

Note: C_{max}, maximum plasma concentration; T_{max}, time when the maximum concentration was reached; AUC_{0-t}, area under the concentration-time curve from zero to time; $AUC_{0-\infty}$, area under the concentration-time curve from zero to infinity; MRT_{0-t}, mean residue time of drug in body from zero to time; $MRT_{0-\infty}$, mean residue time of drug in body from zero to infinity; $t_{1/2z}$, half-life of the drug; and CL_z, total body clearance of the drug.

4   Discussion

Overall, the pharmacokinetic properties of THA and FLR in Pacific white shrimp were similar, and their plasma concentration-time curves could both be well described by a two-compartment open model with first-order absorption after oral administration.

Following oral administration, THA and FLR were rapidly absorbed in white shrimp. The mean C_{max} values （7.96 µg/mL and 5.53 µg/mL for THA and FLR, respectively） peaked at 2.0 h. The T_{max} value for THA in white shrimp was lower than that in seabass （Castells et al. 2000）. The T_{max} value for FLR in white shrimp was lower than those reported in Atlantic salmon （Horsberg et al. 1996）, cod （*Gadus morhua* L.） （Samuelsen et al. 2003）, tilapia （*Oreochromis niloticus*） （Feng et al. 2008）, and Korean catfish （*Silurus asotus*） （Park et al. 2006）, except for crucian carp （*Carassius carassius* L.） （Zhao et al. 2011）. In the present study, the rapid oral absorption in white shrimp could be attributed to its open circulatory system and active excretion of excess water for osmosis regulatory purposes in freshwater.

The volume of distribution at steady state （V_{ss}） is an accurate indicator of drug diffusion in body tissues. The V_{ss} of 1.547 L/kg for THA in white shrimp was lower than that for FLR （V_{ss} of 4.381 L/kg）. However, they were higher than the values for FLR reported in Korean catfish （Park et al. 2006）, Atlantic salmon （Horsberg et al. 1996）, and cod （Samuelsen et al. 2003）. The V_{ss} values of THA and FLR in the present paper revealed that the drugs were well distributed in the body organs and tissues of white shrimp.

The $t_{1/2z}$ and CL_z are important pharmacokinetic parameters that describe how quickly a drug is eliminated from the body. After oral administration, the $t_{1/2z}$ values for THA and FLR were similar to those reported in Korean catfish （Park et al. 2006）, tilapia （Feng and Jia 2009）, and Atlantic salmon （Horsberg et al. 1996）, but were lower than those in cod （Samuelsen et al. 2003） and olive flounder（*Paralichthys olivaceus*）（Lim et al. 2010）. However, the CL$_z$ values for THA and FLR in white shrimp were much larger than those reported in Korean catfish（Park et al. 2006）, tilapia（Feng and Jia 2009）, and Atlantic salmon（Horsberg et al. 1996）. In addition to differences in temperature and salinity affecting the drug elimination（Samuelsen et al. 2003; Park et al. 2006）, this result can be well explained by the differences in anatomical volumes, plasma proteins, and tissue binding of the drug between crustaceans and fishes（Oie and Tozer, 1979; Barron et al. 1988）.

Plasma protein binding determines the amount of drug in free form and that available for tissue distribution as well as therapeutic action. In the present study, the plasma protein binding values of THA and FLR were found to be 28.38%±1.26% and 37.91%±7.61%, respectively. This low protein binding suggested that almost 70% of THA and 60% of FLR in hemolymph were free and available for treating systemic infections. The plasma protein binding of THA was lower than FLR, which can perhaps explain the wider distribution of THA in the plasma, muscle, and hepatopancreas of shrimp than FLR, as well as the faster elimination of THA than FLR from the shrimp tissues.

In the present study, the drug concentrations in hepatopancreas were significantly higher than those in hemolymph and muscle. The highest mean concentration ratios of hepatopancreas/hemolymph and hepatopancreas/muscle were 25.6 and 68.5 for THA as well as 29.7 and 86.0 for FLR, respectively. The AUC$_{0-t}$ ratios of hepatopancreas/hemolymph and hepatopancreas/muscle were 30.57 and 45.97 for THA as well as 22.39 and 54.58 for FLR, respectively. These results indicated that only a small amount of the drugs penetrated the systemic circulation and other tissues before leaving the hepatopancreas. A large amount of the drugs were greatly reduced in hepatopancreas after oral administration, which is called the "first-pass effect." This phenomenon was also observed in other crustaceans, such as norfloxacin in L. vannamei（Fang et al. 2004）, sulfadimethoxine in American lobster（*Homarus americanus*）（James and Barron, 1988）, ormetoprim in shrimp（*Penaeus vannamei*）（Park et al. 1995）, triclopyr in crayfish（*Orconectes propinquus*）（Barron et al. 1991）, and oxytetracycline in Pacific white shrimp（Faroongsarng et al. 2007）. These studies showed that the drug concentration in hepatopancreas was 10^1-10^2 times higher than that in other tissues. In crustaceans, there is substantial evidence that the hepatopancreas plays a role not only in metabolism/elimination but also in absorption（Verri et al. 2001; Faroongsarng et al. 2007）. After oral administration, the drug is delivered from the shrimp stomach to the digestive gland, where both the drug uptake into and elimination out of the body competitively take place. In the present study, the peak times of THA and FLR in hepatopancreas（1 h and 0.5 h, respectively）were faster than those in hemolymph and muscle（2.0 h）, further illustrating that the drugs were absorbed from the stomach to the hepatopancreas of the shrimp.

Vibriosis is an important disease affecting the shrimp industry in China and other countries（Wang and Yang 2005; Zhou et al. 2012; Roque et al. 2001）. Despite many reports about the shrimp vibriosis, pharmacodynamic data about vibrios was very lack. Roque et al.（2001）determined the in vitro susceptibility to florfenicol of 98 stains of Vibrio spp isolated from shrimp in Northwestern Mexico. It was found the Mean minimum inhibitory concentration was 1.79 μg/mL with range of 0.25~8.0 μg/mL. Zhou et al.（2012）reported a specific bacterial pathogen, Vibrio harveyi HLB0905 from "white tail" shrimp, which generally accompanied by mass mortalities. Later we examined the sensitivity of this microorganism to THA and FLR, and found MIC values of 0.25 μg/mL for THA and 1.0 μg/mL for FLR, respectively. Shojaee AliAbadi and Lees（2000）suggested that for a bacteriostatic drug like florfenicol, an optional dosage regimen should maintain concentrations at the size of infection in excess of MIC90 for the entire medication period. On the basis of the results of this study, it is reasonable to assume that a dose of 10 mg/kg given orally at 12 hour intervals for THA and 8 hour intervals for FLR should be appropriate for control of "Bacterial White Tail Disease" of Litopenaeus vannamei. For the treatment of infected shrimp by other bacterial species, speculation about effective dose regimens for THA or FLR in shrimp requires the combination of pharmacokinetics and pathogen sensitivity. The dose size and dosing interval could be accordingly adapted to achieve better therapeutic efficacy.

In conclusion, the pharmacokinetic profiles of THA and FLR were shown. The rapid absorption, extensive distribution, and quick elimination of the drugs were also demonstrated. THA was distributed more extensively and eliminated more quickly than FLR in shrimp. The pharmacokinetic data revealed that THA and FLR could be used to treat bacterial infections in white shrimp.

Acknowledgement

This work was supported financially by the Special Fund for Agro-scientific Research in the Public Interest（200803012 and 201203085）.

References

Barron, M.G., Gedutis, C. and James, M.O. 1988. Pharmacokinetics of sulphadimethoxine in the lobster, Homarus americanus, following intrapericardial administration. Xenobiotica 18: 269-277.

Barron, M. G., Hensen, S. C. and Ball, T. 1991. Pharmacokinetics and metabolism of triclopyr in the crayfish (Procambarus clarki). Drug Metabolism and Disposition 19: 163-167.

BOF. 2011. China Fisheries Yearbook. Eds Bureau of Fisheries, Ministry of Agriculture of China. pp. 175-195. China Agriculture Press, Beijing.

Burridge, L., Weis, J. S., Cabello, F., Pizarro J. and Bostick, K. 2010. Chemical use in salmon aquaculture: A review of current practices and possible environmental effects. Aquaculture 306: 7-23.

Castells, G., Intorre, L., Bertini, S., Cristòfol, C., Soldani, G. and Arboix, M. 2000. Oral single-dose pharmacokinetics of thiamphenicol in the sea-bass (Dicentrarchus labrax). Journal of Veterinary Pharmacology and Therapeutics 23: 53-54.

Dowling, P. M. 2006. Chloramphenicol, thiamphenicol and florfenicol. In Antimicrobial Therapy in Veterinary Medicine. 4th edn. Eds Giguère, S., Prescott, J. F., Baggot, J. D., Walker, R. D. & Dowling, P. M. pp. 241-248. Blackwell Publishing, Ames, Iowa.

Fang, W. H., Yang, X. L. and Zhou, K. 2004. Disposition and elimination of norfloxacin in tissues from the whiteleg shrimp, Litopenaeus vannamei. Journal of Fisheries of China 28 (suppl.): 19-24.

Faroongsarng, D., Chandumpai, A., Chiayvareesajja, S. and Theapparat Y. 2007. Bioavailability and absorption analysis of oxytetracycline orally administered to the standardized moulting farmed Pacific white shrimps (Penaeus vannamei). Aquaculture 269: 89-97.

Feng, J. B. and Jia, X. P. 2009. Single dose pharmacokinetic study of flofenicol in tilapia (Oreochromis niloticus × O. aureus) held in freshwater at 22℃. Aquaculture 289: 129-133.

Feng, J. B., Jia, X. P. and Li, L. D. 2008. Tissue distribution and elimination of florfenicol in tilapia (Oreochromis niloticus× O. caureus) after a single oral administration in freshwater and seawater at 28℃. Aquaculture 276: 29-35.

Fukui, H., Fujihara, Y. and Kano, T. 1987. In vitro and in vivo antibacterial activities of florfenicol, a new fluorinated analog of thiamphenicol, against fish pathogens. Fish Pathology 22: 201-207.

Horsberg, T. E., Hoff, K. A. and Nordmo, R. 1996. Pharmacokinetics of florfenicol and its metabolite florfenicol amine in Atlantic salmon. Journal of Aquatic Animal Health 86: 292-301.

Inglis, V., Richards, R. H., Varma, K. J., Sutherland, I. and Brokken, E. S. 1991. Florfenicol in Atlantic salmon, Salmo salar L., parr: tolerance and assessment of efficacy against furunculosis. Journal of Fish Diseases 14: 343-351.

James, M. O. and Barron, M. G. 1988. Disposition of sulfadimethoxine in the lobster (Homarus americanus). Veterinary and Human Toxicology 30 (Suppl.): 36-43.

Lewbart, G. A., Papich, M. G. and Whitt-Smith, D. 2005. Pharmacokinetics of florfenicol in the red pacu (Piaractus brachypomum) after single-dose intramuscular administration. Journal of Veterinary Pharmacology and Therapeutics 28: 317-319.

Lim, J. H., Kim, M. S., Hwang, Y. H., Song, I. B., Park, B. K. and Yun, H. I. 2010. Pharmacokinetics of florfenicol following intramuscular and intravenous administration in olive flounder (Paralichthys olivaceus). Journal of Veterinary Pharmacology and Therapeutics 34: 206-208.

Malvisi, J., Della Rocca, G., Anfossi, P., Tomasi, L., Di Salvo, A., Zanchetta, S., Magni, A., Sello, M. and Giorgetti, G. 2002. Tissue distribution and residue depletion of thiamphenicol after multiple oral dosing in seabass (Dicentrarchus labrax L.) and seabream (Sparus aurata L.). Journal of Applied Ichthyology 18: 35-39.

McGinnis, A., Gaunt, P., Santucci, T., Simmons, R. and Endris, R. 2003. In vitro evaluation of the susceptibility of Edwardsiella ictaluri, etiological agent of enteric septicemia in channel catfish, Ictalurus punctatus (Rafinesque), to florfenicol. Journal of Veterinary Diagnostic Investigation 15: 576-579.

Oie, S. and Tozer, T. N. 1979. Effect of altered plasma protein binding on the apparent volume of distribution. Journal of Pharmacological Sciences 68: 1203-1210.

Park, B. K., Lim, J. H., Kim, M. S. and Yun, H. I. 2006. Pharmacokinetics of florfenicol and its metabolite, florfenicol amine, in the Korean catfish (Silurus asotus). Journal of Veterinary Pharmacology and Therapeutics 29: 37-40.

Park, E. D., Lightner, D. V., Milner, N., Mayersohn, M., Park, D. L., Gifford, J. M. and Bell, T. A. 1995. Exploratory bioavailability and pharmacokinetic studies of sulphadimethoxine and ormetoprim in the penaeid shrimp. Penaeus vannamei. Aquaculture 130: 113-128.

Roque A, Molina-Aja A, Bolán-Mejía C, Gomez-Gil B. 2001. In vitro susceptibility to 15 antibiotics of vibrios isolated from penaeid shrimps in Northwestern Mexico. Int J Antimicrob Agents 17 (5): 383-387.

Samuelsen, O. B., Bergh, Ø. and Ervik, A. 2003. Pharmacokinetics of florfenicol in cod Gadus morhua and in vitro antibacterial activity against Vibrio anguillarum. Diseases of Aquatic Organisms 56: 127-133.

Switala, M., Hrynyk, R., Smutkiewicz, A., Jaworski, K., Pawlowski, P., Okoniewski, P., Grabowski, T. and Debowy, J. 2007. Pharmacokinetics of florfenicol, thiamphenicol, and chloramphenicol in turkeys. Journal of Veterinary Pharmacology and Therapeutics 30: 145-150.

Van de Riet, J. M., Potter, R. A., Christie-Fougere, M. and Garth Burns, B. 2003. Simultaneous determination of residues of chloramphenicol, thiamphenicol, florfenicol, and florfenicol amine in farmed aquatic species by liquid chromatography/mass spectrometry. Journal of AOAC International 86: 510-514.

Verri, T., Mandal, A., Zilli, L., Bossa, D., Mandal, P. K., Ingrosso, L., Zonno, V., Vilella, S., Ahearn, G. A. and Storelli, C. 2001. D-glucose transport in decapod crustacean hepatopancreas. Comparative Biochemistry and Physiology. Molecular and Integrative Physiology 130: 585-606.

Wang, Q. Y. and Yang, C. H. 2005. Development and prospect of healthy culture of shrimp farming in China. China Fisheries 1：21-24.

Yasunaga, N. and Yasumoto, S. 1988. Therapeutic effect of florfenicol on experimentally induced pseudotuberculosis in yellowtail. Fish Pathology 23：1-5.

Zhao, H. Y., Zhang, G. H., Bai, L., Zhu, S., Shan, Q., Zeng, D. P. and Sun Y. X. 2011. Pharmacokinetics of florfenicol in crucian carp (Carassius auratus cuvieri) after a single intramuscular or oral administration. Journal of Veterinary Pharmacology and Therapeutics 34：460-463.

Zhou, J. F., Fang, W. H., Yang, X. L., Zhou, S., Hu, L. L., Li, X. C., Qi, X. Y., Su, H. and Xie, L. Y. 2012. A nonluminescent and highly virulent Vibrio harveyi strain is associated with 'Bacterial White Tail Disease' of Litopenaeus vannamei shrimp. PLoS ONE 7 (2), e29961. doi：10.1371/journal. pone. 0029961.

该文发表于《水产学报》【2006，30（4）：515-519】

肌注和口服氟苯尼考在中华鳖体内残留分析及药代动力学

The residues and pharmacokinetics of florphenicol inTrionyx sinensis Following intramascular injection and oral administration

朱丽敏[1]，杨先乐[2]，林启存[1]，蔡丽娟[1]，许宝青[1]，张惠宏[1]

(1. 杭州市农业科学研究院，浙江 杭州 310024；2. 上海水产大学生物科学与技术学院，上海 200090)

摘要：研究不同给药条件下，氟苯尼考在中华鳖体内的残留及药代动力学特征。健康中华鳖160只，随机分为2组，按30 mg·kg^{-1}剂量分别单次肌注或口服氟苯尼考，高效液相色谱法测定中华鳖血浆和肌肉药物残留浓度，利用3P87药代动力学软件分析数据。肌注和口服时均符合一室开放模型；肌注给药的动力学方程$C=16.72$（e$^{-0.15t}$－e$^{0.52t}$），主要药代动力学参数：AUC＝76.45 μg·mL^{-1}·h^{-1}，吸收半衰期（$T_{1/2Ka}$）= 1.31 h，半衰期（$T_{1/2Ke}$）= 4.48 h，最高血药浓度C_{max}＝7.09 μg·L^{-1}；口服给药的动力学方程为$C=39.99$（e$^{-0.19t}$－e$^{0.4t}$），主要药代动力学参数：AUC＝109.42 μg·mL^{-1}·h^{-1}，吸收半衰期（$T_{1/2Ka}$）= 1.73 h，半衰期（$T_{1/2Ke}$）= 3.63 h，最高血药浓度C_{max}＝10.64 μg·L^{-1}。实验结果表明，氟苯尼考口服情况下，在中华鳖体内吸收快，血药浓度高，维持时间长，生物利用度高；药物在肌肉中消除缓慢。

关键词：氟苯尼考；中华鳖；药代动力学；残留分析；高效液相色谱法

氟苯尼考（florphenicol）是一种合成的新型专用兽药，为氯霉素和甲砜霉素的替代产品，具有较高的抗菌活性，抑制细菌蛋白质合成，适用于由链球菌、巴氏杆菌、弧菌及其他革兰氏阴性菌引起的鱼病，该药使用后不产生再生障碍性贫血，在水产动物疾病防治上有着广阔的应用前景。本研究旨在探讨氟苯尼考在中华鳖体内的药代动力学特征和药物残留消除规律，为指导临床科学合理用药提供理论依据。

1 材料与方法

1.1 实验试剂

氟苯尼考原料药（含量99%，浙江海翔医药化工有限公司），氟苯尼考标准品（99.76%，Schering-Plough 公司）；氯霉素标准品（99.9%，Sigma 公司），乙腈（色谱纯，Sigma 公司）；乙酸乙脂（分析纯，上海五联化工厂公司）；正己烷（色谱纯，上海化学试剂研究所）。

1.2 实验仪器

Varian 高效液相色谱仪，Prostar330 二极管阵列紫外检测器，色谱工作条件：紫外检测波长223 nm，流动相乙腈-水（27/135：73/365，V/V），流速1.0 mL·min^{-1}，反相色谱柱 ODS-C18（5 μm）4.6 mm×250 mm 进样量为20 μL。氮吹仪 N-EVAPTM112，0.45 μm 微孔滤膜等。

1.3 实验动物

中华鳖（Trionyx sinensis）采用水族箱模拟生态控温培育，水族箱外围挂黑布，箱内挂网片，实验期间充气增氧，水温控制在（30±1）℃。中华鳖由平均规格每只（4.5±0.5）g 培养到每只（100±10）g 后进入正式实验。

1.4 给药方法和样品采集

取实验条件下饲养的健康中华鳖 160 只，体重（100±10）g，随机分组，每组 80 只，分养于 2 只缸，以 30 mg·kg^{-1} 剂量分别一次性肌注、口灌给药；分别给药后在 5、15、30、60、120、240、360、480、600、1 440 min 各点采集血液、肌肉，每一时间点采集 4 只鳖，断头取血 5~6 mL，置于含肝素的离心管，以 3 000 r·min^{-1} 离心 15 min，吸取上层血浆于塑料离心管中，备用；采集的组织样置于密封塑料袋，于-20℃冰箱保存。

1.5 样品前处理

血浆样品：取 2.00 mL 血浆于 10 mL 离心管中，加入 4.00 µg 氯霉素（100 µg·mL^{-1}）作为内标，加 2 mL 磷酸缓冲液后振荡 5 min，再加入 8 mL 乙酸乙酯振荡 25 min，静置 10 min 后吸取上层乙酸乙酯层于干净离心管中，同法重复提取一次，合并提取液，氮气吹干。加 2 mL 流动相，涡合混匀。吸取 5 mL 溶解物于离心管 3 000 r·min 离心 15 min，吸取上清液经过滤后待上机。

肌肉样品：称取 5.00 g 肌肉等样品于 50 mL 离心管中，加入 4.00 µg 氯霉素（100 µg·mL^{-1}）作为内标，并加 5 mL PBS（pH=7.0）振荡 5 min，加 20 mL 乙酸乙酯振荡 30 min（1 800 r·min^{-1}）离心 10 min，吸上层于离心管，重复提取一次合并提取液氮气吹干，加入 2 mL 正己烷和 3 mL 流动相振荡溶解后，离心弃去正己烷，重复 3 次，留下层液体过滤小瓶中待上机测定。

1.6 标准曲线的制备

氟苯尼考精制原药用流动相溶解配制成 100 µg·mL^{-1} 标准储备液，临用前用流动相稀释成 2.5 µg·mL^{-1}。称取 6 份空白样品（肌肉或血浆）分别加入 2.5 µg·mL^{-1} 标准溶液 100 µg，400 µL，800 µL，2 mL，4 mL 和 4 µg 氯霉素内标物。然后按照"1.5"方法进行提取和测定。

1.7 回收率的测定

采用加标回收法。取氟苯尼考标准储备液 1、5、20 µL，分别加入到 3 个不同体积的血浆、肌肉混匀后按样品前处理方法处理样品，HPLC 测定氟苯尼考浓度，每个浓度重复 3 次。根据氟苯尼考标准曲线计算出组织样氟苯尼考实测浓度。回收率（%）= 实际测定浓度/理论浓度值×100。

1.8 精密度测定

日内精密度：在同日内的试验初、中、末，分别取氟苯尼考标准储备液 5 µL，加入中体积的组织中混匀后按 1.5 处理，进样测定其血药和肌肉药物浓度，每次重复 3 次，记录氟苯尼考峰面积，计算日内相对标准偏差（RSD）。日间精密度：一周内重复测定 2 次，每次重复 3 次，计算日间相对标准偏差（RSD）。

1.9 数据处理

药代动力学模型拟合采用 3P87 软件处理数据；采用 Marquardt 法对一、二、三房室分别以权重 1，1/C、1/C2 3 种情况进行拟合，根据 WSS 和 AIC 值来判断最适合的药代动力学模型。

2 结果

2.1 氟苯尼考的色谱行为特征

氟苯尼考和氯霉素的色谱峰尖锐且对称，血浆和肌肉中的杂质与药物和内标分离良好（图1）。

2.2 氟苯尼考标准曲线和灵敏度

以测得氟苯尼考平均峰面积与氯霉素平均峰面积之比为纵坐标，对相应质量浓度作线性回归，并制作标准曲线（图2）；得回归方程 $y = 0.78981x - 0.01513$，在 0.25~15 µg·mL^{-1} 范围内具有很好的相关性（$P < 0.01$），按信噪比 $S/N = 3$ 计算，最低检测浓度为 0.05 µg·mL^{-1}。

2.3 回收率和精密度

本实验条件下氟苯尼考标准溶液在低、中、高 3 个浓度下平均回收率为 89.9%±1.2%；测得日内相对标准差（RSD）

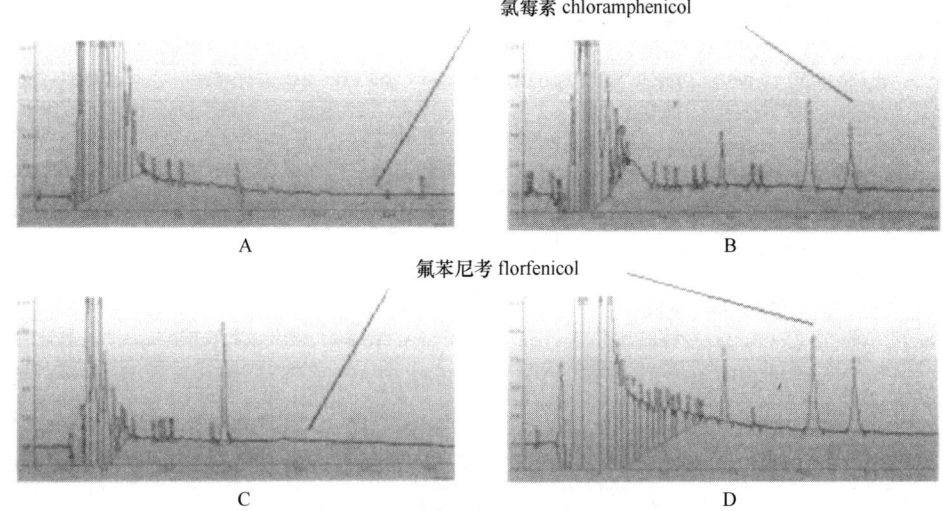

图 1　氟苯尼考色谱图

Fig. 1　Chromatograms of florphenicol

A：空白血浆色谱图；B：血浆加氟苯尼考标准品色谱图；C：空白肌肉色谱图；D：肌肉加氟苯尼考标准品色谱图。

A：chromatogram of plasma control；B：chromatogram of plasma treated with standard florphenicol；

C：chromatogram of musculaturecontrol；D：chromatogram of musculature treated with florphenicol

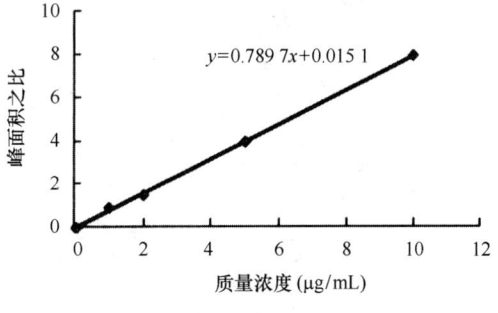

图 2　氟苯尼考标准曲线

Fig. 2　Relationship between the ratio of the peak area of florphenicol to that of chloramphenicol and the concentration

为 2.51%±0.89%；日间相对标准差为 3.36%±0.91%。

2.4　一次性肌注和口服给药后氟苯尼考药代动力学特征

中华鳖肌注和口服氟苯尼考后，不同时间的血浆样品经色谱仪检测所得药物平均浓度及中华鳖口服氟苯尼考药后不同时间段肌肉内残留量见表 1，肌注氟苯尼考药时曲线见图 3，口服氟苯尼考药时曲线见图 4，用 3P87 软件处理药时数据所得药物动力学参数列于表 2。

表 1　一次性肌注和口服给药后氟苯尼考药物浓度

Tab. 1　Residue level of florphenicol in different tissues following a single dose of
intramascular injection and oral adminstration

时间（h） time	肌注 intramascular injection	口服（$\mu g \cdot mL^{-1}$）oral administration	
	血浆（$\mu g \cdot mL^{-1}$）plasma	血浆 plasma	肌肉 muscle
0.083	1.23±0.64	0.98±0.71	1.30±1.02
0.5	2.30±1.06	3.75±1.28	4.84±1.79
1.5	5.16±1.10	6.08±1.60	7.89±2.43

时间（h）time	肌注 intramuscular injection	口服（μg·mL⁻¹）oral administration	
	血浆（μg·mL⁻¹）plasma	血浆 plasma	肌肉 muscle
2	6.11±0.11	8.75±1.81	9.54±2.67
3	8.02±1.47	11.62±2.45	13.04±3.58
4	7.46±0.32	12.91±3.36	14.53±3.94
6	5.35±0.31	10.56±2.35	11.52±2.87
9	3.52±1.05	7.29±1.02	7.38±2.39
12	2.71±0.52	2.26±0.34	5.39±2.18
24	1.28±0.37	1.03±0.43	1.22±1.05
48	1.05±0.29	0.89±0.23	1.01±0.81
72	0.27±0.20	0.098±0.21	0.13±0.12
96	0.17±0.12	ND	0.1±0.15
192	ND	ND	0.09±0.04
264	ND	ND	0.07±0.02

注：ND 未检出

Notes：ND means not detected

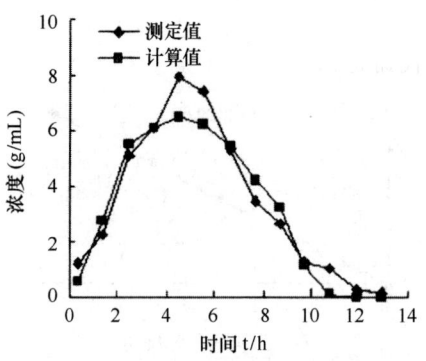

图 3　肌注氟苯尼考（30 mg·kg⁻¹）药时曲线

Fig. 3　Concentration-time curve of florphenicol after intramascular injection

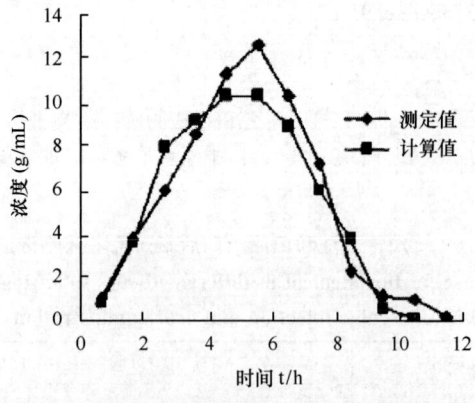

图 4　口灌氟苯尼考（30 mg·kg⁻¹）药时曲线

Fig. 4　Concentration-time curve of florphenicol after oral administration

表2　一次性肌注和口服氟苯尼考后药代动力学参数

Tab. 2　Pharmacokinetic parameters of florphenicol following a single dose of oral administration and intramuscular injection

参数 parameters	单位 unit	肌注 intramascular injection 血浆（μg·mL⁻¹）plasma	口服（μg·mL⁻¹）oral administration 血浆 plasma	口服（μg·mL⁻¹）oral administration 肌肉 muscle
A	$\mu g \cdot mL^{-1}$	16.72	39.99	27.81
K_e	h^{-1}	0.154	0.19	0.13
K_a	h^{-1}	0.52	0.4	0.47
$T_{1/2(Ka)}$	h	1.31	1.73	1.46
$T_{1/2(Ke)}$	h	4.48	3.63	5.28
$T_{(peak)}$	h	3.29	3.54	3.74
C_{max}	$\mu g \cdot mL^{-1}$	7.09	10.63	12.33
AUC	$(\mu g \cdot mL^{-1}) \cdot h^{-1}$	76.45	109.42	153.41
CL（s）	$L \cdot h^{-1} \cdot kg^{-1}$	0.32	0.27	0.19
MRT	h	24.61	29.01	128.17

3　讨论

3.1　氟苯尼考在中华鳖体内的代谢特征

中华鳖在肌注和口服 30 mg·kg⁻¹ 氟苯尼考药物后，药时数据均符合一级吸收一室开放模型，动力学方程分别为 $C=16.72\ (e^{-0.15t}-e^{-0.52t})$ 和 $C=3.99\ (e^{-0.19t}-e^{0.4t})$，两种给药方式其吸收半衰期（$T_{1/2Ka}$）分别为 1.31 h 与 1.73 h，消除半衰期（$T_{1/2Ke}$）分别 4.48 h 与 3.63 h，二者差异似乎不明显；但最高血药浓度（Cmax）差异显著，分别为 7.09 与 10.64 μg·mL⁻¹，血药浓度时间曲线下面积（AUC）分别为 76.45 与 109.42 μg·mL⁻¹·h⁻¹，说明口服吸收较肌注吸收完全，AUC 值大，表示血浆组织内药物含量多；口服药物平均滞留时间 MRT 为 29.01 h，维持血药浓度时间也较长，上述结果提示，理论上中华鳖的疾病防治口服氟苯尼考将优于肌肉注射。

口服试验中，氟苯尼考在中华鳖体内的消除半衰期（$T_{1/2Ke}$）为 3.63 h，与大西洋鲑（$T_{1/2Ke}$）的 12.2~14.7h[1,2] 和中国对虾（$T_{1/2Ke}$）的 9.07 h[3] 相比，氟苯尼考药物在中华鳖体内的消除速度较快，其差异可能来自于以下几方面：（1）与水生动物种属差异有关[4,5]，分类地位高其降解排泄的途多消除快[3]；（2）与养殖水温有关，水生动物在适温范围内水温增高，体内各类酶活性增强、代谢加快。

3.2　氟苯尼考给药方案与安全性探讨

确定给药方案时重点考虑血药浓度和药物最小抑菌浓度（MIC）。氟苯尼考抗菌活性高抗菌谱广，对多数病原菌显示高度敏感，据郭闯[6]等报道14株患病水生动物气单胞菌属细菌对氟苯尼考高度敏感；廖昌容等[7]对6种海洋致病弧菌进行体外抗菌试验，得到氟苯尼考对溶藻弧菌（*V. alginolyticus*）、副溶血弧菌（*V. parahaemolyticus*）、哈氏弧菌（*V. harveyi*）的 MIC 值为 2 μg·mL⁻¹，对漂浮弧菌（*V. natrigen*）、沙蚕弧菌（*V. nereis*）、鲨鱼弧菌（*V. carchariae*）的 MIC 为 4 μg·mL⁻¹；国外报道[8]氟苯尼考对鳗弧菌的 MIC 为 0.2~0.8 μg·mL⁻¹。根据一次性口服给药试验，检测结果显示48 h内血药浓度保持在（0.89±0.23）μg·mL⁻¹，能有效地杀灭病原菌，48 h后，迅速降至（0.098±0.21）μg·mL⁻¹，建议口服药剂量为 30 mg·kg⁻¹ 时，给药的间隔期不应超过 2 d，与实际应用相符。国内对氟苯尼考的药效和安全性研究尚未见相关报道。制定渔药休药期是为了保障动物食品的安全性，通常根据国家规定的最大残留限量（MRL）标准和动物对药物的残留消除规律来确定。欧盟规定鱼类肌肉中氟苯尼考最大残留限量（MRL）为 1 mg·kg⁻¹[9]，而加拿大规定 MRL 0.8 mg·kg⁻¹[10]，我国尚未对水产动物食品中氟苯尼考制定最大残留限量标准，本试验中华鳖口服给药后肌肉中药物浓度经 3.74 h 达高峰，72 h后肌肉内实测药物浓度已达欧盟残留限量 1 mg·kg⁻¹，若以此为标准，按本试验数据推算，中华鳖口服氟苯尼考的休药期为 ≥3 d，鉴于中华鳖为变温动物，其代谢速率与水温有关，建议休药期以温度时间为标准单位，中华鳖口服氟苯尼考的休药期不低于 90℃·d⁻¹。

参考文献

［1］　Martinsen B, Horsberg T E, Varma K J, et al. Single dose pharmacokinetic study of florfenicol in Atlantic salmon in seawater at 11℃ [J]. Aquaculture, 1993, 112: 1–11.

[2] Horsberg TE, Hoff K A, Nordmo R. Pharmacokinetics offlorfenicol and its metabolite florfenicol a mine in Atlantic salmon [J]. Journal of Aquatic Animal Health, 1996, 8: 292–301.

[3] 李静云、王群、李健、等. 氟苯尼考在中国对虾组织内药物代谢动力学及残留消除规律 [J]. 水产学报, 2004, 28 (12): 63-68.

[4] Uno K, Aoki T, Ueno R. Pharmacokinetic study of oxytetracycline in cultured rainbow trout amago salmon and yellowtail [J]. Bull Jap SocSci Fish, 1992, 58 (6): 1151 1156.

[5] 湛嘉、李佐卿、康继韬、等. 影响水产动物药代动力学的因素 [J]. 中国兽药杂志, 2003, 37 (12): 38-4.

[6] 郭闯、朱国强、王永坤、等. 14 株患病水生动物气单胞菌属细菌的分离鉴定与最佳治疗药物筛选试验 [J] 水产科学, 2003, 22 (4): 14-17.

[7] 廖昌容、徐力文、陈毕. 氟苯尼考对六种海洋致病弧菌的体外抗菌活性研究 [J]. 水产养殖, 2005, 26 (4): 1-4.

[8] Samuelsen O B, Bergh O, Ervik A. Pharmacokinetics of florfenicol in cod Gadusmorhua and in vitro antibacteral activity against Vibrio anguilla- rum [J]. Dis Aquat Organ, 2003, 56: 127–133.

[9] EMEA-E U. Committee for veterinary medicinal products: Florfenicol (extension to all food producing species), summary report (6) [Z]. London, UK. http://www.emea.eu.int/pdfs/vet/mrls/082202en.pdf. January 2002.

[10] HealthCanada. Maximum Residue Limits (MRLs) set by Canada [Z]. http://www.hc-sc.gc.ca/vetdrugs-medsvet/mrl_e.html. 2004.

该文发表于《中国水产科学》【2007, 14 (6): 1010-1016】

氟苯尼考及其代谢物氟苯尼考胺在斑点叉尾鮰体内的残留消除规律

Florfenicoland its metaboliteflorfenicol amineresiduedepletion in Ictalurus punctatus

刘永涛，艾晓辉，杨红

(中国水产科学研究院 长江水产研究所，农业部淡水鱼类种质监督检验测试中心，湖北 荆州 434000)

摘要： 研究不同水温（18℃和28℃）下，氟苯尼考（Florfenicol, FF）及其代谢物氟苯尼考胺（Florfenicol Amine, FFA）在斑点叉尾鮰体内残留消除规律。以含氟苯尼考 4 g/kg 的饲料按 10 mg/kg 鱼体质量连续强饲斑点叉尾鮰（Ictalurus punctatus） 3 d，于 1、2、3、5、7、9 d 分别将斑点叉尾鮰处死后取肌肉、肝脏、皮肤、肾脏 4 种组织。采用反相高效液相色谱紫外检测法测定斑点叉尾鮰组织中氟苯尼考及其代谢物氟苯尼考胺。结果表明，不同水温下同种组织，相同水温下不同组织中氟苯尼考及其代谢物氟苯尼考胺的消除速率快慢不一（差异显著）。氟苯尼考和氟苯尼考胺的和作为标示残留物，高水温时标示残留物（总量）在斑点叉尾鮰体内消除的更快。与其他组织相比标示残留物（总量）在肾脏中的消除最慢，因此，建议将肾脏作为标示残留物（总量）在斑点叉尾鮰体内残留的靶组织。在 18℃和 28℃时，标示残留物（总量）在肾脏中以 300 μg/kg 为最高残留限量（maximum residue limit, MRL）建议休药期分别为 234℃·d 和 224℃·d。本研究为不同水温条件下制定该药的休药期提供了可靠的理论依据。

关键词： 氟苯尼考；氟苯尼考胺；代谢物；斑点叉尾鮰；标示残留物；残留；消除规律

氟苯尼考（Florfenicol, FF）又称氟甲砜霉素是甲砜霉素的单氟衍生物。氟苯尼考以其广谱、高效、吸收迅速、分布广泛、安全等特点，在国内外畜牧和水产动物上迅速获得广泛应用。氟苯尼考不引起再生障碍性贫血，已知的副作用主要为对哺乳动物具有胚胎毒性，随着对该药毒理学的深入研究，其他毒副作用可能进一步被发现[1]。氟苯尼考胺（Florfenicol Amine, FFA）是氟苯尼考的主要代谢物。欧盟将动物性食品中的氟苯尼考及其代谢物氟苯尼考胺的和作为标示残留物[2]，中国以氟苯尼考的代谢物氟苯尼考胺作为标示残留物[3]。两者均规定标示残留物在鱼类带皮肌肉中的最高残留限量（MRL）为 1 000 μg/kg。加拿大规定鲑科鱼类肌肉中氟苯尼考的最大残留限量为 0.8 μg/kg[4]。美国 FDA 规定鲶肌肉中标示残留物氟苯尼考胺的最高残留限量为 1 000 μg/kg[5]。目前，国外关于鱼体中氟苯尼考及其代谢物氟苯尼考胺同时检测的方法有高效液相色谱紫外检测法[6]、液相色谱质谱检测法[7]和气相色谱电子捕获检测法[8]；而对氟苯尼考及其代谢物氟苯尼考胺在鱼体内的残留消除规律的研究相对较少，仅见在大西洋鲑（Salmo salar）肌肉与肝脏组织[9]和斑点叉尾鮰（Ictalurus punctatus）肌肉组织[10]中进行过相关研究。国内仅见鱼肌肉中氟苯尼考和氟苯尼考胺残留的高效液相色谱紫外检测法[11]和虾肉中氟苯尼考和氟苯尼考胺残留气相色谱检测法[12]的报道，关于氟苯尼考及其代谢物氟苯尼考胺在鱼体内的消除规律的报道尚未见。本研究采用同时检测氟苯尼考及其代谢物氟苯尼考胺的高效液相色谱紫外检测法同国内外相关报道有较大改进（另文报道），研究不同水温下氟苯尼考及其代谢物氟苯尼考胺在斑点叉尾鮰体内残留消除规律，旨在为水产养殖业中制定该药的休药期提供理论依据，从而为保证水产品安全提供技术保障。

1　材料与方法

1.1　实验鱼

斑点叉尾鮰（*Ictalurus punctatus*）体质量（128.40±10.51）g，由中国水产科学研究院长江水产研究所窑湾试验场提供，运至实验室后，用2%的食盐水消毒10 min，然后放置于已清洗并用高锰酸钾消毒的水族箱（122 cm×80 cm×80 cm）中暂养1周，实验用水为经活性炭脱氯的自来水，水交换速率20~25 L/h。采用连续充氧，保持水中溶氧大于8.0 mg/L，暂养期间按鱼体重的3%投喂斑点叉尾鮰鱼配合饲料（通威171）。每日9：00—10：00和16：00—17：00各投喂1次，每天换水1次，暂养结束后，选择无病无伤个体，规格均匀的鱼进行试验。试验期间停止饲喂并用加热棒控制水温，低温组水温为（18±1）℃，高温组水温为（28±1）℃。

1.2　试剂与溶液配制

氟苯尼考标准品（德国 Dr. Eherestorfer 公司，纯度99.0%）；氟苯尼考胺标准品（纯度≥95%，sigma-aldrich 公司）；氟苯尼考原料药（批号200504，含量99.0%）由武汉九州神农有限公司提供；乙腈，HPLC级，上海国药集团提供；丙酮、二氯甲烷、乙酸乙酯、十二烷基硫酸钠（SDS）、三乙胺、磷酸二氢钠、磷酸氢二钠、磷酸、正己烷均为国产分析纯。

1.2.1　磷酸缓冲液的配制

磷酸缓冲液（pH 7.0）配制：准确称取磷酸氢二钠（$Na_2HPO_4 \cdot 12H_2O$）71.6 g 和磷酸二氢钠（$NaH_2PO_4 \cdot 2H_2O$）27.6 g，分别溶于1 L蒸馏水中配成0.2 mol/L的溶液，然后按61：39的体积比混合即成。

1.2.2　定容溶液的配制

A液：称取0.895 g磷酸氢二钠溶解在250 mL双蒸水中配制成0.01 mol/L的磷酸氢二钠溶液，用磷酸将溶液pH调至2.8，经抽虑装置抽虑倒入具塞玻璃瓶中置4℃冰箱中备用；B液：乙腈。A液与B液体积比3：2。

1.3　仪器与设备

高效液相色谱仪为美国 Waters 公司的515泵、717自动进样器、2487双通道紫外检测器；Empower液相色谱工作站。自动高速冷冻离心机（日本 HITACHI 20PR-520 型）；Mettler-TOLEDO AE-240 型精密电子天平（梅特勒-托利多公司）；80-2B型台式离心机，FS-1高速匀浆机，调速混匀器（上海康华生化仪器制造厂），HGC-12氮吹仪（HENGAO T&D 公司）。

1.4　色谱条件

色谱柱：Waters symmetry C_{18}反相色谱柱（250 mm×4.6 mm，5 μm）。流动相：乙腈-磷酸二氢钠溶液（0.01 mol/L，含0.05 mol/LSDS和0.1%三乙胺）体积比2：3，用磷酸调pH值至4.5。流速：$0.6 \text{ mL} \cdot \text{min}^{-1}$。柱温：室温。紫外检测波长225 nm。进样量20 μL。

1.5　实验设计与方法

1.5.1　药饵制备、给药剂量及取样

将购买的斑点叉尾鮰饲料（通威171）用粉碎机粉碎，将氟苯尼考加到粉碎后的饲料中重新制成氟苯尼考的含量为4 g/kg颗粒饲料；按10 mg/kg鱼体质量进行强饲。先将投胃管装好加药饲料然后慢慢将投胃管插入鮰的胃内，将一根实心管子从投胃管内径将其中的药饵慢慢推到鮰的胃内。给药后保定1 min，然后将鱼单独放入一个水盆中观察5 min，将返胃回吐药饵的鱼弃去，无回吐的鱼放入水族箱中。给药前同批鱼取1次空白组织，1 d给药1次，连续给药3 d，于停药后1、2、3、5、7、9 d将鮰鱼钝击头部致死取皮、背脊两侧肌肉、肝脏、肾脏，置塑料袋中-20℃冰箱中保存，每个采样点取5尾鱼。

1.5.2　样品处理

将冷冻保存的组织样室温下自然解冻。剪取适量的肌肉组织，置高速匀浆机中匀浆呈糜状，准确称取1.0 g；鱼皮先剪碎。准确称取1.0 g，肾脏取0.5 g，肝脏0.5 g，分别放于10 mL的具塞离心管中，加入1 mL调pH值为7.0的磷酸盐缓冲液，置调速混匀器上涡旋振荡混匀1 min，加入3 mL丙酮涡旋振荡混匀1 min，以10 000 r/min离心5 min，将溶液转移到另一支10 mL离心管中，在向溶液中加入2 mL二氯甲烷，涡旋振荡混匀1 min，5 000 r/min离心5 min，弃去上层水相，置50℃氮吹仪上吹干，用1 mL定容溶液溶解残渣，涡旋振荡混匀1 min，加入3 mL正己烷涡旋振荡混匀1 min，10 000 r/min

离心 10 min，弃去正己烷，重复去脂 1 遍，水相经 0.22 μm 滤头过滤置自动进样瓶，HPLC 测定。

1.5.3 标准工作曲线的制备及最低检测限（LOD）

在斑点叉尾鮰 4 种空白组织（肌肉、皮肤、肝脏、肾脏各 1 g）中加入已知浓度的标准溶液，制成含氟苯尼考和氟苯尼考胺质量分数分别为 0.05、0.1、0.25、0.5、1.0、2.5、5.0、10.0 μg/g 组织样品，分别以氟苯尼考和氟苯尼考胺的峰面积（A_i）为纵坐标，质量浓度（C）为横坐标绘制标准工作曲线，求出回归方程和相关系数。按上述的色谱条件测定加标样品，以 3 倍信噪比（3S/N）计算氟苯尼考及氟苯尼考胺在血浆、肌肉、皮肤、肝脏、肾脏组织中的最低检测限。

1.5.4 回收率及方法精密度测定

在斑点叉尾鮰空白肌肉、皮肤、肝脏、肾脏组织样品中分别添加 5 个水平的 FF 和 FFA 混合标准溶液：使样品质量浓度分别为 0.1、0.2、0.5、1.0、2.0 μg/g，每个质量浓度做 3 个平行，并设一个空白对照，按样品前处理过程处理后测定回收率。在 4 种空白组织中分别添加 2 个浓度水平的 FF 和 FFA 混合标准溶液，使质量浓度分别为 0.1 μg/g、1.0 μg/g。每个浓度的样品，日内做 5 个重复，1 周内重复做 5 次，计算日内及日间精密度。

1.6 数据处理

消除方程及休药期计算和回归图，采用 Microsoft Excel 2003，SPSS（13.0），Statistic（6.0）进行计算和绘制。消除方程采用 $C=C_0 e^{-kt}$。C 表示药物质量分数（μg/kg），C_0 为残留消除对数曲线的截距（μg/kg），k 表示消除速率常数。根据休药期（WDT）计算各组织药物浓度降至规定水平（MRL）所需的时间：

$$WDT = \frac{\ln(C_0/MRL)}{k}$$

其中 WDT 为休药期，MRL 为最高残留限量（μg/kg），C_0 为残留消除对数曲线的截距（μg/kg），k 为残留消除曲线速率常数[13]。

2 结果与分析

2.1 标准工作曲线及最低检测限（LOD）

4 种组织中氟苯尼考和氟苯尼考胺标准工作曲线方程及相关系数如下，肌肉中 FF $A_{肌肉}=77764C_{肌肉}-1908.2$，$r=0.9984$，FFA $A_{肌肉}=79617C_{肌肉}+4059.8$，$r=0.9993$；肝脏中 FF $A_{肝脏}=76475C_{肝脏}-3535.4$，$r=0.9991$，FFA $A_{肝脏}=90704C_{肝脏}+1884.8$，$r=0.9994$；肾脏中 FF $A_{肾脏}=76581C_{肾脏}+773.29$，$r=0.9984$，FFA $A_{肾脏}=88623C_{肾脏}-10845$，$r=0.998\,0$；皮肤中 FF $A_{皮肤}=73690C_{皮肤}-2316.8$，$r=0.9979$，FFA $A_{皮肤}=81851C_{皮肤}-9453.5$，$r=0.9962$。肌肉、肝脏、肾脏、皮肤中 FF 和 FFA 的最低检测限分别为 20 μg/kg 和 15 μg/kg，50 μg/kg 和 50 μg/kg，50 μg/kg 和 50 μg/kg，50 μg/kg 和 50 μg/kg。

2.2 回收率及方法精密度

4 种组织中 FF 平均回收率为 75.14%～92.65%，FFA 平均回收率为 76.00%～91.50%，日内变异系数不大于 8.68%，日间变异系数不大于 9.22%。

2.3 不同水温下 FF 和 FFA 在斑点叉尾鮰体内残留及消除

氟苯尼考在动物体内的主要代谢产物是氟苯尼考胺，同时还有氟苯尼考醇、氟苯尼考草氨酸等代谢物，这些代谢物有极低的抗菌活性，对环境生物影响较小。目前，欧盟和中国以氟苯尼考和氟苯尼考胺作为残留检测的标示检出物[1]。

2.3.1 不同水温下 FF 和 FFA 在斑点叉尾鮰组织中的残留

不同水温下斑点叉尾鮰经强饲含有氟苯尼考的饲料 3 d 后氟苯尼考及其代谢物氟苯尼考胺在 4 种组织中的残留（表 1、表 2）。

表 1 18℃斑点叉尾鮰连续 3 d 强饲氟苯尼考各组织中 FF 和 FFA 的残留量

Tab. 1 FF and FFA levels in channel catfish tissues after oral gavages administration at does of

10 mg/kg bw for 3 consecutive days at 18℃ n=5；$\overline{X}\pm SD$；μg/g

时间/d Time	肌肉 Muscle		皮肤 Skin		肝脏 Liver		肾脏 Kindey	
	FF	FFA	FF	FFA	FF	FFA	FF	FFA
1	4.43±0.36	0.10±0.02	3.69±0.44	0.27±0.02	5.44±0.71	0.51±0.25	6.34±0.60	0.54±0.16

续表

时间/d Time	肌肉 Muscle		皮肤 Skin		肝脏 Liver		肾脏 Kindey	
	FF	FFA	FF	FFA	FF	FFA	FF	FFA
2	1.46±0.26	0.07±0.01	1.14±0.20	0.26±0.02	1.35±0.49	0.37±0.12	2.86±0.59	0.51±0.11
3	0.83±0.20	0.03±0.02	0.58±0.13	0.25±0.07	0.89±0.16	0.24±0.06	1.59±0.37	0.35±0.07
5	0.14±0.02	ND	0.29±0.04	0.24±0.03	0.18±0.01	0.10±0.03	0.90±0.34	0.26±0.07
7	ND	ND	0.21±0.06	0.20±0.06	ND	ND	0.89±0.12	0.24±0.01
9	ND	ND	0.15±0.04	ND	ND	ND	0.71±0.27	0.19±0.02

注：ND-未检测到。

Note：ND-Undetectable.

表 2　28℃斑点叉尾鮰连续 3 d 强饲氟苯尼考后各组织中 FF 和 FFA 的残留量

Tab. 2　FF and FFA levels in channel catfish tissues after oral gavages administration at does of

10 mg/kg bw for 3 consecutive days at 28℃ n=5；$\bar{X}$±SD；μg/g

时间/d Time	肌肉 Muscle		皮肤 Skin		肝脏 Liver		肾脏 Kindey	
	FF	FFA	FF	FFA	FF	FFA	FF	FFA
1	1.92±0.66	0.11±0.07	1.39±0.52	0.25±0.09	2.51±0.97	0.40±0.17	2.82±0.82	0.47±0.22
2	0.26±0.08	ND	0.54±0.09	0.19±0.03	0.60±0.20	0.27±0.09	1.36±0.28	0.30±0.04
3	0.08±0.02	ND	0.16±0.04	0.16±0.02	0.29±0.09	0.08±0.03	0.45±0.12	0.22±0.01
5	ND	ND	0.08±0.03	0.14±0.01	0.09±0.01	ND	0.29±0.09	0.19±0.02
7	ND	ND	ND	ND	ND	ND	0.19±0.14	0.17±0.03
9	ND	ND	ND	ND	ND	ND	0.17±0.07	ND

注：ND-未检测到.

Note：ND-Undetectable.

　　由表 1 和表 2 结果表明，不同水温（18℃和 28℃）下，同一采样点，相同组织或不同组织中，氟苯尼考及其代谢物氟苯尼考胺的残留量低水温时均比高水温时高，而且相差较大；相同水温下，不同组织中，氟苯尼考及其代谢物氟苯尼考胺的残留量也有较大差别。

2.3.2　不同水温下标示残留物（总量）在斑点叉尾鮰组织中残留消除规律

　　氟苯尼考及其代谢物氟苯尼考胺浓度的和作为动物性食品中氟苯尼考残留的标示残留物[2]的浓度见表 3 和表 4。标示残留物（总量）在斑点叉尾鮰各组织中的残留浓度随时间变化的规律见图 1 和图 2。数据经回归处理得到 4 种组织中药物浓度（c）与时间（t）关系的消除曲线方程、相关指数（r^2）及消除半衰期（$T_{1/2}$）见表 5 和表 6。

表 3　18℃氟苯尼考及其代谢物氟苯尼考胺浓度的和作为标示残留物在斑点叉尾鮰各组织中的残留量

Tab. 3　The sum of FF and FFAlevelsas the marker residue in channel catfish tissues at 18℃　μg/g

时间/d Time	肌肉 Muscle	皮肤 Skin	肝脏 Liver	肾脏 Kindey
1	4.53	3.96	5.95	6.88
2	1.53	1.40	1.72	3.37
3	1.03	0.83	1.13	1.94
5	0.16	0.53	0.28	1.16
7	ND	0.41	ND	1.13
9	ND	0.15	ND	0.90

注：ND-未检测到。

Note：ND-Undetectable.

表4 28℃氟苯尼考及其代谢物氟苯尼考胺浓度的和作为标示残留物在斑点叉尾鲴各组织中的残留量

Tab. 4 The sum of FF and FFA levels as the marker residue in channel catfish tissues at 28℃ μg/g

时间（d） Time	肌肉 Muscle	皮肤 Skin	肝脏 Liver	肾脏 Kindey
1	2.03	1.64	2.91	3.29
2	0.26	0.73	0.87	1.66
3	0.08	0.32	0.37	0.67
5	ND	0.22	0.09	0.48
7	ND	ND	ND	0.36
9	ND	ND	ND	0.17

注：ND-未检测到。

Note：ND-Undetectable.

由表3和表4结果表明，不同水温（18℃和28℃）下，同一采样点，氟苯尼考的标示残留物在相同组织或不同组织中的残留量低水温时均比高水温时高，而且相差较大；相同水温下，不同组织中，氟苯尼考的标示残留物的残留量也有较大差别。

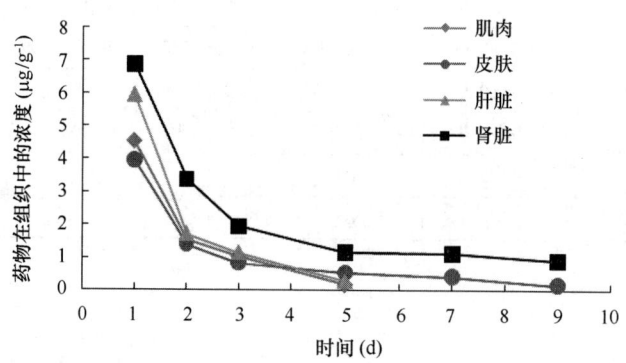

图1 18℃斑点叉尾鲴连续3 d强饲氟苯尼考，标示残留物在组织中的消除曲线

Fig. 1 Elimination curve of the marker residue in channel catfish after oral gavages for 3 consecutive days at 18℃

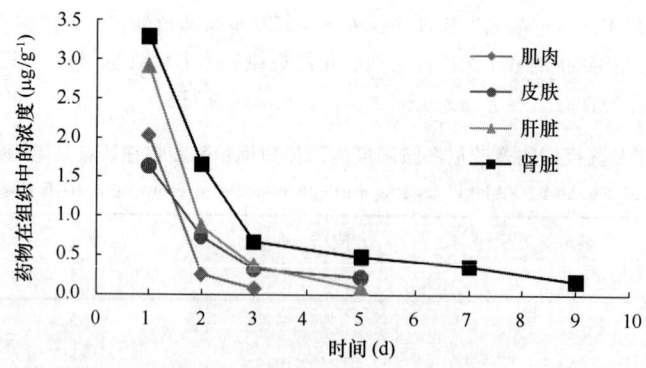

图2 28℃斑点叉尾鲴连续3 d强饲氟苯尼考，标示残留物在组织中的消除曲线

Fig. 2 Elimination curve of the marker residue in channel catfish after oral gavages for 3 consecutive days at 28℃

表5　18℃标示残留物在斑点叉尾鮰组织中的消除曲线方程及参数
Tab. 5　Equation of elimination curve and parameters of the marker residue in channel catfish tussues after oral gavages for 3 successive days at 18℃

组织 Tissue	方程 Equation	相关指数 r^2	消除半衰期 $T_{1/2}$（h）
肌肉 Muscle	$C_{肌肉} = 9.564e^{-0.809t}$	0.984	20.558
皮肤 Skin	$C_{皮肤} = 3.467e^{-0.347t}$	0.914	47.931
肝脏 Liver	$C_{肝脏} = 9.905e^{-0.727t}$	0.971	22.878
肾脏 Kindey	$C_{肾脏} = 5.432e^{-0.229t}$	0.815	72.629

表6　28℃ 标示残留物在斑点叉尾鮰组织中的消除曲线方程及参数
Tab. 6　The equation of elimination curve and parameters of the marker residue in channel catfish tussues after oral gavages for 3 successive days at 28℃

组织 Tissue	方程 Equation	相关指数 r^2	消除半衰期 $T_{1/2}$（h）
肌肉 Muscle	$C_{肌肉} = 8.836e^{-1.617t}$	0.976	10.286
皮肤 Skin	$C_{皮肤} = 2.095e^{-0.494t}$	0.887	33.688
肝脏 Liver	$C_{肝脏} = 5.569e^{-0.849t}$	0.983	19.590
肾脏 Kindey	$C_{肾脏} = 3.060e^{-0.331t}$	0.909	50.248

2.4　色谱图

图3和图4是18℃和28℃斑点叉尾鮰经多次强饲氟苯尼考后1 d时肌肉样品中氟苯尼考及其主要代谢物氟苯尼考胺的液相色谱图，由色谱图可以看出28℃时氟苯尼考和氟苯尼考胺的浓度要比18℃低；另外在同一组织中同一采样点氟苯尼考和氟苯尼考胺的浓度高水温时要比低水温时低。本实验在检测鮰组织中氟苯尼考胺的过程中发现其浓度较低，与 Horsberg 等[9]报道的氟苯尼考胺在大西洋鲑肌肉和肝脏中的残留相差较大。

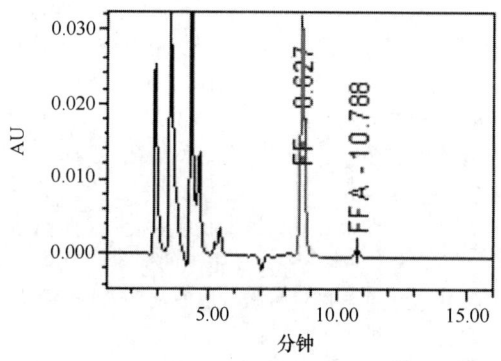

图3　18℃多次口灌 FF 后1 d 肌肉色谱图
Fig. 3　Chromatogram of muscule at 1 day after multidoes oral adiministratin of FF at 18℃

3　讨论

3.1　氟苯尼考及其代谢物氟苯尼考胺在鮰鱼组织中的分布规律

由表1和表2可以看出，氟苯尼考的代谢物氟苯尼考胺在鮰各组织中的残留均较低，氟苯尼考和氟苯尼考胺在组织中的分布规律相似，残留水平由高到低大致为肾脏、肝脏、肌肉、皮肤，与余培建等[14]报道的氟苯尼考在欧洲鳗鲡组织中的分布规律相似，Horsberg 等[9]也报道大西洋鲑肝脏中的氟苯尼考和氟苯尼考胺的残留要比肌肉中的高。而且18℃水温下要

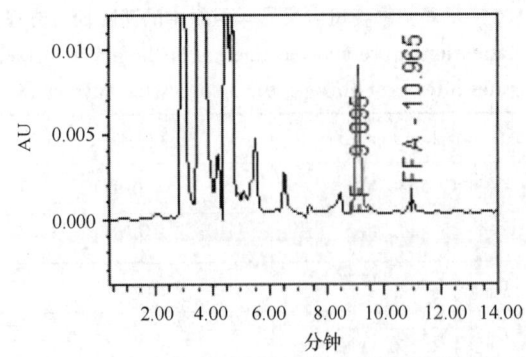

图 4 28℃多次口灌 FF 后 1 d 肌肉色谱图

Fig. 4 Chromatogram of muscule at 1 day after multidoes oral adiministratin of FF at 28℃

比 28℃时同一采样点组织中氟苯尼考和氟苯尼考胺的残留要高。18℃时氟苯尼考在肌肉中于 7 d 后检测不到，而 28℃时氟苯尼考于第 5 天时已检测不到；皮肤中的氟苯尼考在 18℃ 9 d 时仍可检测到其质量分数为（0.15±0.04）μg/g，而 28℃ 5 d 时其残留仅为（0.08±0.03）μg/g，7 d 时已检测不到。在 18℃和 28℃下 9 d 时均可在肾脏组织检测到氟苯尼考，其残留分别为（0.71±0.27）μg/g 和（0.17±0.07）μg/g。在 18℃下 9 d 时氟苯尼考胺在肾脏中可被检测到其残留为（0.19±0.02）μg/g，而在 28℃下 9 d 时氟苯尼考胺不能被检测到。28℃下肌肉中氟苯尼考胺在 2 d 时已检测不到，18℃下 5 d 时肌肉中氟苯尼考胺才检测不到，而与 Horsberg 等[9]报道的 10℃下 20 d 时在大西洋鲑的肌肉和肝脏中仍可以检测到氟苯尼考胺的结果相差较大，而 Samuelsen 等[15]报道在大西洋鳕的血浆和组织中没有检测到氟苯尼考胺。这种差异作者认为是由水生动物种属以及水温差异造成的。

3.2 标示残留物氟苯尼考胺（总量）在鮰组织中残留特征

采用 HPLC 法可在斑点叉尾鮰的肌肉、皮肤、肝脏、肾脏组织中可同时检测到氟苯尼考及其代谢物氟苯尼考胺（图 3 和图 4）。欧盟规定将氟苯尼考及其代谢物氟苯尼考胺的和作为氟苯尼考在动物组织中残留的标示残留物。目前，欧盟和中国以氟苯尼考和氟苯尼考胺作为动物性食品中的检出物[1]，但国内尚未见有以氟苯尼考及其代谢物氟苯尼考胺的和作为标示残留物进行残留研究的文献。在 18℃和 28℃水温条件下标示残留物（总量）在斑点叉尾鮰组织中的分布及残留消除规律是一致的，标示残留物（总量）在组织中的残留由高到低依次均为肾脏、肝脏、肌肉、皮肤，消除速率由快到慢依次均为肌肉、肝脏、皮肤、肾脏，可以看出标示残留物（总量）在肝脏、肾脏中的浓度最高，这说明标示残留物（总量）在肝脏代谢和肾脏富集、排泄为主，与余培建等[14]报道的氟苯尼考在欧洲鳗鲡肾脏中残留最高结果相似，并在肾脏中富集和排泄的结果相似。从消除规律上看 18℃和 28℃水温下标示残留物（总量）在肌肉中均消除最快，消除半衰期分别为 20.558 h 和 10.268 h，而在肾脏中消除最慢，消除半衰期分别为 47.931 h 和 33.688 h。从两种水温条件下标示残留物（总量）在斑点叉尾鮰体内各组织中的浓度及消除规律来看，同一采样时间时高水温条件下标示残留物（总量）在各组织中的浓度明显低于低水温时，而消除速率比低水温时快。

国外关于氟苯尼考及其代谢物氟苯尼考的残留消除研究较少而且只报道了肌肉和肝脏中消除规律。以 10 mg/kgbw 的剂量连续 10 d 饲喂体质量为（2.784±922）g 的大西洋鲑含有氟苯尼考的饲料，于第 1 天、第 15 天、第 20 天、第 26 天、第 30 天、第 35 天、第 40 天和 49 天分别取 10 尾大西洋鲑进行氟苯尼考及其代谢物氟苯尼考胺的消除规律进行研究。结果表明氟苯尼考在 1 d 时可以检测到其余时间均未检测到（此实验设计的时间间隔太长，不利于氟苯尼考消除规的研究），而氟苯尼考胺在大西洋鲑肝脏中的消除速率要慢于在肌肉中的消除速率[9]（与本实验的结果相似）。但在 20 d 时还可以在大西洋鲑的肌肉和肝脏组织中检测到氟苯尼考胺，而本实验 5 d 以后在肌肉和肝脏中就检测不到氟苯尼考胺了。出现这样的差异可能与种属、水温、盐度、投喂次数等因素的差异有关，本实验考虑到氟苯尼考在防治鱼类疾病时一般连用 3~5 d[16]，以及本实验采用强饲方式给药的工作量及对鱼体的影响，本实验采用连续给药 3 d 的试验方案。水温在 17.9~20.7℃斑点叉尾鮰连续饲喂 10 d 含氟苯尼考的饲料，并将肌肉中氟苯尼考及其代谢物转化为氟苯尼考胺进行测定，4 d 后标示残留物（总量）低于 1 000 μg/kg 与本实验 18℃测得的结果相似。本实验在不同水温条件下分别对肌肉、皮肤、肝脏、肾脏组织中标示残留物（总量）消除规律进行了研究。研究发现肾脏是残留量最高消除最慢的组织，因此，可将肾脏作为本试验中残留分析的靶组织。

3.3 休药期

欧盟[2]对氟苯尼考的标示残留物（总量）在牛、羊、家禽、猪可食性组织（牛和羊：肌肉、肝脏肾脏；家禽和猪：肌肉、皮肤+脂肪、肝脏、肾脏）中的最高残留限量（MRL）均做了详细的规定，而在水产品中只对有鳍鱼类可食组织中自然比例的肌肉和皮肤中标示残留物（总量）的最高残留限量（MRL）作了规定。本实验对不同水温条件下斑点叉尾鮰肌肉、皮肤、肝脏、肾脏组织中标示残留物（总量）的消除规律作了研究，进一步丰富了氟苯尼考在有鳍鱼类各组织中分布及消除资料，也为制定最高残留限量及休药期提供了参考资料。根据欧盟规定氟苯尼考在有鳍鱼类肌肉和皮肤组织及所在有动物源性食品肝脏和肾脏组织中规定的 MRL：分别为 1 000 μg/kg、1 000 μg/kg、2 000 μg/kg、300 μg/kg。本实验以此作为参考，通过 WDT 公式计算可知斑点叉尾鮰各组织的理论 WDT 为 18℃ 休药期肌肉 2.79 d、皮肤 3.58 d、肝脏 2.20 d、肾脏 12.65 d；28℃ 休药期肌肉 1.35 d、皮肤 1.50 d、肝脏 1.21 d、肾脏 7.02 d。通常将残留量高、消除最慢的组织可看成是残留分析的靶组织。实验结果表明，肾脏组织残留量高、消除最慢，因此，可将肾脏作为残留分析的靶组织则休药期分别为 18℃ 时 234℃·d 和 28℃ 时 224℃·d。

参考文献

［1］　徐力文，廖昌容，刘广锋．氟苯尼考用于水产养殖的安全性［J］．中国水产科学，2005，12（4）：512-518．

［2］　EU. Commission Regulation（EC）No. 508/1999 of 4 March 1999. amending Annexes I to IV to Council Regulation（EEC）No. 2377/90 laying down a Community procedure for the setablishment of maximum residue limits of veterinary medicinal products in foodstuffs of animal origin. Official Journal of the European Communities，1999．

［3］　中华人民共和国农业部公告第 235 号《动物性食品中兽药最高残留限量》，2002．

［4］　Health Canada. Maximum Residue Limits（MRLs）set by Canada［Z］. http：//www. hc-sc. gc. ca/dhp-mps/vet/mrl-lmr/mrl-lmr_ versus_ new-nouveau-e. html．

［5］　FDA. Title 21-Food and Drugs，PART 556 Tolerances for residues of new animal drugs in food，Subpart B. 556. 283-Florfenicol［Z］. http：//www. accessdata. fda. gov/scripts/cdrh/cfcfr/CFRSearch. cfm？CFrpart=556. 283．

［6］　Hormazabal V，Steffenak I and Yndestad M. Simultaneous determination of residues of florfenicol and the metabolite florfenicol amine in fish tissues by high performance liquid chromatography［J］. Chromatogra，1993，616：161-165．

［7］　Jeffery M R，Ross A P，Melissa C F，et al. Simultaneous determination of residues of chloramphenicol，thiamphenicol，florfenicol，and florfenicol amine in farmed aquatic species by liquid chromatography/mass spectrometry［J］. AOAC International，2003，86（3）：510-514．

［8］　Pfenning A P，Roybal J E，Rupp H S，et al. Simultaneous determination of residues of chloramphenicol，florfenicol，florfenicol amine，and thiamphenicol in shrimp tissue by gas chromatography with electron capture detection［J］. AOAC International，2000，83（1）：26-30．

［9］　Horsberg T E，Hoff K A and Nordmo R. Pharmacokinetics of florfenicol and its metabolite florfenicol amine in Atlantic salmon［J］. Aqu Anim Health，1996，8：292-301．

［10］　Wrzesinski C，Crouch L，Gaunt P，et al. Florfenicol residue depletion in channel catfish，Ictalurus punctatus（Rafinesque）［J］. Aquaculture，2006，253（1-4）：309-316．

［11］　郭霞，张素霞，沈建忠，等．鱼肌肉中氟苯尼考和氟苯尼考胺残留的高效液相色谱检测［J］．中国兽医科学，2006，36（09）：743-747．

［12］　孙丰云，张素霞，沈建忠，李建成．虾肉中氯霉素　甲砜霉素　氟苯尼考及氟苯尼考胺残留气相色谱-微电子捕获检测法［J］．中国兽医杂志，2006，42（10）：66-67．

［13］　Riviere J E. Comparative Pharmacolinetics：Principle，Techniques，and Application，Ames：Iowa State University Press，1999：308．

［14］　余培建，翁祖桐，樊海平，等．氟苯尼考在欧洲鳗鲡体内的药物代谢动力学的研究［J］．福建水产，2005，4：52-57．

［15］　Samuelsen O B，Bergh O，Ervik A. Pharmacokinetics of florfenicol in cod Gadus morhua and in vitro antibacterial activity against Vibrio anguillarum［J］. Dis Aquat Organ，2003，56：127-133．

［16］　杨先乐，陆承平，战文斌等．新编渔药手册［J］．北京：中国农业出版社，2005：189．

该文发表于《大连海洋大学学报》【2010，25（4）：285-288】

适温条件下氟苯尼考在罗非鱼体内的药物动力学

Pharmacokinetics of florfenicol in tilapia fed diet containing the drug

王伟利，罗理，姜兰，谭爱萍，邹为民，卢迈新

（中国水产科学研究院珠江水产研究所，广州 510380）

摘要：采用 HPLC-MS-MS 方法，研究了给罗非鱼 Oreochromis niloticus×O. caureus 单次口服（12 mg/kg）氟苯尼考后其体内的药代谢规律。结果表明：氟苯尼考在罗非鱼体内吸收迅速，1 h 后血液和肌肉中的药浓度均超过 1.0 μg/mL（mg/kg），能够有效杀灭绝大多数水产病菌，T_{max} 和 C_{max} 分别为 5.05 h、6.67 μg/mL 和 6.80 h、8.49

mg/kg，维持有效药物浓度（以 MIC=1.0 μg/mL 计）以上的时间大于 50 h，组织滞留时间较长；血液和肌肉中的 $T_{1/2\beta}$ 分别为 11.18 h 和 12.67 h，药时曲线下面积分别为 148.8 μg/（mL·h）和 225.2 mg/（kg·h）；氟苯尼考在血液中的吸收和消除速率均快于肌肉，但最高药物浓度较肌肉中低。在本试验条件下，给药 59 h 后罗非鱼各组织中的药物含量均低于美国 FDA（1 mg/kg）及加拿大标准（0.8 mg/kg），说明氟苯尼考具有良好的应用价值。

关键词：氟苯尼考；罗非鱼；药物动力学

该文发表于《上海海洋大学学报》【2012，21（4）：568-574】

氟苯尼考及氟苯尼考胺在西伯利亚鲟体内的药动学及组织分布研究

Pharmacokinetics and Tissue Distribution of Florfenicol and Florfenicol Amine in Acipenser baeri

王瑞雪[1,2]，李绍戊[1]，王荻[1]，卢彤岩[1]

(1. 中国水产科学研究院 黑龙江水产研究所，哈尔滨 150070；2. 东北农业大学 动物医学学院，哈尔滨 150070)

摘要：采用液质联用法（HPLC-MS/MS）建立了氟苯尼考和氟苯尼考胺同时检测的方法，研究了氟苯尼考口灌给药西伯利亚鲟后，氟苯尼考及其代谢物氟苯尼考胺在西伯利亚鲟体内的药动学和组织分布。水温 22℃ 下，氟苯尼考以 15 mg/kg 剂量单次口灌给药西伯利亚鲟，检测血浆、肝脏、肾脏和肌肉等组织中氟苯尼考及其代谢产物氟苯尼考胺的浓度，结果显示：氟苯尼考及其代谢产物氟苯尼考胺在西伯利亚鲟体内的药时数据均符合一级吸收二室开放模型，氟苯尼考在血浆中的达峰浓度（C_{max}）为 3.4 μg/mL，达峰时间（T_{peak}）为 2.943 h，表观分布容积（V/F）为 3.267 L/kg，消除半衰期（$T_{1/2\beta}$）为 31.21 h，药时曲线下总面积（AUC）为 76.51 μg·h/mL，C_{max}（FFA）/C_{max}（FF）和 AUC_{FFA}/AUC_{FF} 仅为 5.44% 和 20.73%；氟苯尼考在各组织中分布广泛，分布规律相近，肝脏、肾脏中药物浓度较高。结果表明：氟苯尼考在西伯利亚鲟体内具有吸收迅速、达峰浓度高、消除相对缓慢及组织中分布广泛的特征且氟苯尼考主要以原形药物形式代谢消除。

关键词：西伯利亚鲟；氟苯尼考；氟苯尼考胺；药动学；组织分布

该文发表于《中国渔业质量与标准》【2016，6（3）：6-13】

氟苯尼考在花鲈体内的代谢及残留消除规律

Pharmacokinetics and rules of residue elimination of florfenicol in Lateolabrax janponicus

黄聚杰[1]，林茂[1,2*]，鄢庆枇[1]，李忠琴[1,2]，李江森[1]

(1. 集美大学水产学院，福建厦门 361021；2. 鳗鲡现代产业技术教育部工程研究中心，福建厦门 361021)

摘要：为了解氟苯尼考在花鲈（Lateolabrax japonicus）体内的代谢动力学特征和残留消除规律，利用高效液相色谱法检测氟苯尼考混饲口灌给药后在花鲈血浆、肌肉、肝和肾等样品中的时间-浓度变化。在代谢动力学研究中，将 20 mg/kg 氟苯尼考单次混饲口灌给药于花鲈后，获得 48 h 内的药时数据，利用 DAS 和 WinNonlin 软件进行比较分析。结果表明，不同药代动力学分析软件或者不同权重系数所获得的房室参数值有较大差异，特别是消除相半衰期（$T_{1/2\beta}$）；而非房室参数值则比较接近。其中 DAS 软件非房室模型分析氟苯尼考在花鲈血浆中的药代动力学参数显示，药时曲线下面积（$AUC_{0-\infty}$）为 257.591 mg/（L·h），表观分布容积（Vz/F）为 1.401 L/kg，平均滞留时间（$MRT_{0-\infty}$）和消除半衰期（$T_{1/2z}$）分别为 18.505 h 和 12.508 h，达峰浓度（C_{max}）和达峰时间（T_{max}）分别为 18.356 μg/mL 和 3 h。在残留消除研究中，氟苯尼考以 60 mg/kg 的高剂量单次给药后，采集 30 d 内的药时数据，利用 WT 程序计算的结果显示，氟苯尼考在花鲈肌肉、肝、肾和血浆中的理论休药期分别为 6.54、8.69、8.30 和 5.89 d。研究结果为氟苯尼考在花鲈养殖中的用药方案和休药期的制定提供理论依据。

关键字：氟苯尼考；花鲈；药代动力学；残留；休药期

该文发表于《安徽农业科学》【2011，39（36）：341-343】

氟苯尼考在日本鳗鲡和欧洲鳗鲡体内的药代动力学

Pharmacokinetics of Florfenicol in A. japonica and A. Anguilla

林茂，纪荣兴，陈政强，范红照，谢吉林

（集美大学水产学院，福建 厦门 361021）

摘要： 本文旨在研究氟苯尼考在日本鳗鲡和欧洲鳗鲡体内的药代动力学特征。氟苯尼考以混饲口灌方式给药，剂量为 30 mg/kg，药时数据利用 DAS 软件进行药动学分析。结果显示，日本鳗鲡和欧洲鳗鲡体内药动学的房室参数：吸收速率常数（K_a）分别为 0.329 和 0.4491/h，消除相半衰期（$T_{1/2\beta}$）为 44.266 和 12.690 h。非房室参数：药时曲线下面积（$AUC_{0-\infty}$）分别为 257.099 和 285.945 mg·h/L，平均滞留时间（$MRT_{0-\infty}$）为 18.370 和 14.227 h，半衰期（$T_{1/2}$）为 12.341 和 9.919 h，达峰浓度（C_{max}）为 15.92 和 20.39 μg/mL，达峰时间（T_{max}）均为 4 h。[结论] 建议在鳗鲡中使用氟苯尼考进行治疗时可采用 30 mg/kg 体重的剂量，给药间隔为 12 h，即每天两次。
关键词： 氟苯尼考；日本鳗鲡；欧洲鳗鲡；药代动力学

该文发表于《集美大学学报（自然科学版）》【2011，16（2）：92-96】

氟苯尼考在两种鳗鲡体内残留及消除规律的研究

Residue and Elimination of Florfenicol in Eels Anguilla japonica and A. anguilla

林茂[1,2]，王雪虹[1,2]，姚志贤[1]

（1. 集美大学水产学院，福建 厦门 361021；2. 福建省高校水产科学与食品安全重点实验室，福建 厦门 361021）

摘要： 本文研究了氟苯尼考在日本鳗鲡和欧洲鳗鲡体内的残留消除规律。在 25℃ 下，以 30 mg/kg 体重的剂量多次口灌给药后 1 d、2 d、3 d、5 d、8 d、12 d、20 d、30 d，取鳗鲡的血浆和肌肉、肝脏、肾脏等样品，加入内标氯霉素混合，经萃取过滤后采用反相高效液相色谱法检测，测定的平均回收率在 97.8%~101.3% 之间，日内变异系数为（3.23±0.49）%，日间变异系数为（4.08±0.85）%。氟苯尼考在鳗鲡体内消除较慢，在日本鳗鲡和欧洲鳗鲡血浆中的半衰期（$T_{1/2}$）达 7.8 d 和 8.3 d。氟苯尼考的最高残留限量如规定为 0.2 μg/g，在日本鳗鲡和欧洲鳗鲡中用药后的最大休药期分别为 38.7 d 和 28.5 d；如规定为 1 μg/g，最大休药期则分别是 19.0 d 和 11.2 d。
关键词： 氟苯尼考；鳗鲡；残留；消除；休药期

该文发表于《上海海洋大学学报》【2013，22（2）：225-231】

不同温度下氟苯尼考在鳗鲡体内药代动力学的比较

Comparative pharmacokinetics of florfenicol in Japanese eels at different temperature

林茂[1,2]，陈政强[1]，纪荣兴[1]，杨先乐[3]，王见[1]

（1. 集美大学水产学院，福建 厦门 361021；2. 农业部东海海水健康养殖重点实验室，福建 厦门 361021；

3. 上海海洋大学国家水生动植物病原库，上海 201306）

摘要： 采用对个体连续采血的方法，研究了不同水温条件下氟苯尼考以 30 mg/kg 的单剂量混饲口灌给药后在日本鳗鲡（Anguilla japonica）体内的药代动力学特征。利用 DAS 软件的统计矩原理计算每个个体的药时曲线关系，获得药动学参数，单因素方差分析结果表明，不同温度实验组间多个药动学参数存在显著性差异（$P<0.05$）。20℃、24℃和28℃实验组药物峰浓度（C_{max}）分别为（7.839±1.125）、（13.010±2.334）和（18.267±3.717）μg/mL，达峰时间（T_{max}）分别为（6.500±2.070）、（4.500±1.414）和（3.429±0.926）h，这表明温度越高吸收越多越快。表观分布容积（$V_{z/F}$）分别为（3.964±0.594）、（2.466±0.672）和（1.841±0.485）L/kg，表明温度较高时氟苯尼考与血浆蛋白的结合更多。平均滞留时间（$MRT_{0-\infty}$）分别为（31.503±7.117）、（22.881±4.940）和（22.134±6.204）h，消除半衰期（$T_{1/2z}$）分别为（21.243±5.166）、（14.994±4.293）和（14.656±5.061）h，24℃和28℃水温下的消除速率显著快于20℃实验组。药时曲线下面积（$AUC_{0-\infty}$）分别为（235.580±62.013）、（271.983±75.023）和（353.192±92.491）μg·h/mL，表明水温对相对生物利用度有显著影响。

关键词：氟苯尼考；鳗鲡；药代动力学；温度

第二节　喹诺酮类渔药在水产动物体内的药代动力学

氟喹诺酮类（Qunolones）是一类人工合成的含有 4-喹酮母核的一类抗菌药物。由于该类药物具有抗菌谱广、抗菌活性强、给药方便、与常用抗菌药物无交叉耐药性、不需要发酵生产、价格低廉、疗效好的特点，被广泛用于水产动物疾病的控制上，是水产养殖上较常应用的抗菌药物之一。氟喹诺酮类药物在不同的水产动物体内代谢存在差异性。

一、恩诺沙星和环丙沙星

恩诺沙星是人工合成的第三代氟喹诺酮类抗菌药物，为畜禽和水产专用氟喹诺酮类抗菌药物。它的主要代谢产物是环丙沙星。恩诺沙星在罗氏沼虾血液中即刻达到峰值，并迅速向组织中分布（钱云云等，2007）。恩诺沙星在中华绒螯蟹体内只有极少部分代谢为环丙沙星（郑宗林等，2011）。中华绒螯蟹肝胰脏为盐酸环丙沙星的主要残留组织（Jun Tang et al.，2006，李正等，2004）。恩诺沙星与其代谢物环丙沙星在鳗鲡血浆、肌肉和肝脏中药物浓度的变化趋势基本相似（房文红等，2007），起药效作用仍是以恩诺沙星为主（周帅等，2011；何平，待发表）。

该文发表于《水产学报》【2011，35（8）：1182-1190】

恩诺沙星及其代谢产物环丙沙星在拟穴青蟹体内药代动力学研究

Pharmacokinetics Regularity of enrofloxacin and its Metabolite ciprofloxacin in Scylla paramamosain

周帅，胡琳琳，房文红，周凯，于慧娟

（中国水产科学研究院东海水产研究所农业部海洋与河口渔业资源及生态重点开放实验室，上海 200090）

摘要：采用高效液相色谱法，研究盐度 33 下恩诺沙星口灌和肌肉注射给药（剂量 10 mg/kg）后，恩诺沙星及其代谢物环丙沙星在拟穴青蟹（Scylla paramamosain）体内的药代动力学和组织分布。血淋巴和组织中药动学参数采用基于统计矩原理的非房室模型进行计算。恩诺沙星口灌和肌肉注射拟穴青蟹给药后，血药达峰快，分别为 0.5 h 和 1 min，达峰浓度分别为 12.90 μg/mL 和 31.86 μg/mL，曲线下面积（AUC）分别为 216.1 μg/mL·h 和 816.8 μg/mL·h。恩诺沙星在拟穴青蟹组织中分布较广，口灌给药下肌肉和肝胰腺 AUC 分别为 445.9 μg/g·h 和 817.6 μg/g·h，肌肉注射给药下的 AUC 分别为 554.7 μg/g·h 和 2573.7 μg/g·h。与其他水产动物相比，恩诺沙星在拟穴青蟹体内消除速度为中等水平，口灌和肌肉注射恩诺沙星后血药消除半衰期（$t_{1/2z}$）分别为 26.45 h 和 57.02 h，总体清除率（CLz）分别为 0.054 L/h/kg 和 0.012 L/h/kg。恩诺沙星在拟穴青蟹体内代谢生成环丙沙星的量较少，口灌给药下血淋巴、肌肉和肝胰腺的 AUC_{CIP}/AUC_{ENR} 百分比分别为 6.68%、3.60% 和 4.78%，肌肉注射给药下，其相应百分比分别为 4.16%、7.24% 和 1.48%，在拟穴青蟹体内起药效作用仍是以恩诺沙星为主。以 C_{max}/MIC 比值和 AUC_{0-24}/MIC 比值评价恩诺沙星在青蟹体内的药效作用，建议给拟穴青蟹以 10 mg/kg 剂量每隔 24 h 投喂一次恩诺沙星，对弧菌引起的细菌性疾病具有较好的防治效果。

关键词：拟穴青蟹；恩诺沙星；药动学；代谢产物环丙沙星；口灌给药；肌肉注射给药

356

拟穴青蟹（*Scylla paramamosain*）是我国主要海水养殖蟹类之一，近几年在浙江、福建、广东、广西和海南等沿海地区养殖面积已达 50 万亩。然而随着拟穴青蟹养殖的规模化发展，其病害发生情况日趋严重，弧菌病、黄水病和黑斑病等细菌性疾病现已危害到该品种的健康养殖与发展[1-2]。恩诺沙星（Enrofloxacin，ENR）作为动物专用的喹诺酮类抗菌药物，已被广泛用于水产养殖动物感染性疾病的预防与治疗[3]，不过有关恩诺沙星的药物学研究还是主要集中在鱼类[4-6]，在甲壳动物中仅收集到肌肉注射下恩诺沙星在中华绒螯蟹[7]、日本囊对虾[8]、罗氏沼虾[9]和中国明对虾[10]的药动学报道。恩诺沙星在许多动物体内发生脱乙基反应代谢生成具有活性作用的环丙沙星（Ciprofloxacin，CIP）[11]，其代谢程度存在着明显的种属差异[12]，因此了解代谢产物环丙沙星的代谢和消除规律同样重要。本文主要研究口灌和肌肉注射给药后，恩诺沙星及其代谢产物环丙沙星在拟穴青蟹体内药动学，有助于了解它们在拟穴青蟹组织中的代谢和消除规律，为科学、合理使用恩诺沙星防治青蟹细菌性疾病提供理论依据。

1 材料与方法

1.1 试验试剂

恩诺沙星原料药（含量 98%以上）由浙江新昌制药有限公司提供；恩诺沙星标准品（含量≥99.9%，批号 H040798），由中国兽药监察所提供。甲醇、乙腈和正己烷均为 HPLC 级，德国 SIGMA-ALDRICH 生产；盐酸、磷酸、无水硫酸钠等试剂为上海化学试剂公司产品，分析纯。

1.2 试验动物及试验设计

试验用拟穴青蟹（简称青蟹）购自海南琼海市某养殖场，挑选体表无伤、附肢齐全、体质健康的青蟹用于口灌和肌肉注射给药试验，体重 160~210 g，试验用海水为经过沙滤的天然海水，盐度为 33，水温控制在（26.0±1.0）℃，24 h 不间断充氧。暂养 10 d 后用于试验，试验期间投喂适量牡蛎肉，并及时排出残饵和污物。试验期间青蟹成活率为 100%。

1.3 试验仪器

高效液相色谱仪采用美国安捷伦公司的 Agilent 1100 系列，包括四元泵、自动进样器、柱温箱、荧光检测器及 HP 化学工作站，色谱柱为 ZORBAX SB-C$_{18}$，4.6 mm×250 mm。-80℃低温冰箱（SANYO，日本），微量移液器（Eppendorf，德国），电子天平（Mettler Toledoab 204，瑞士），超纯水仪（Milli-Q Advantage，Millipore 公司），高速冷冻离心机（TGL-16G，上海安亭科学仪器厂），高速组织捣碎机（IKA 18，德国），0.45 μm 微孔滤膜（Millipore，美国）等。

1.4 口灌和肌注给药

口灌和肌注给药剂量均为 10 mg/kg，所使用的恩诺沙星药液浓度为 2%，根据蟹体重计算给予药液的体积，如蟹体重 200 g，计算所得的给药量为 0.10 mL。口灌给药，使用 0.5 mL 注射器（7#针头磨成钝圆）从蟹的口器插入约 1.5 cm 深将药液注入，插入深度事先经解剖试验确证达到胃部，口灌给药后不漏液的蟹用于样品采集试验。肌肉注射给药，使用 0.5 mL 注射器插入到青蟹第三步足基部的腹部肌肉中，插入深度 1 cm 左右。

1.5 样品采集

口灌给药组采样时间点为给药后 0.25、0.5、1、2、4、6、12 h 和 1、2、4、8、12、16、20、25 d，肌肉注射组采样时间点为给药后 1、2、5、10 min、0.5、1、2、6、12 h，1、2、4、6、8、10 和 15 d，每个时间点采样 5 只。

血淋巴采样先用纱布擦干蟹体，用 1 mL 注射器插入第二步足的长节和腕节之间的关节膜，抽取血淋巴 0.8 mL 以上，转移到离心管中，然后加入等体积的 ACD 抗凝剂[13]，在震荡器上震荡 1 min，于 5 000 r/min 转速下离心 5 min，取出上层液贮存于-80℃低温冰箱中，直至药物浓度分析。

肝胰脏采样蟹血淋巴样品采集后立即将头胸甲和腹部分开，取出肝胰脏装入塑料离心管中密封，于-80℃保存直至使用。

肌肉采样用镊子和剪刀剔除甲壳，采集肌肉装入塑料密封袋，于-80℃保存直至使用。

1.6 样品前处理

拟穴青蟹血淋巴、肌肉、肝胰腺和性腺样品前处理参照房文红等[6]方法进行。

1.7 样品中恩诺沙星和环丙沙星分析

拟穴青蟹血淋巴、肌肉和肝胰脏样品中恩诺沙星和环丙沙星浓度分析采用反相高效液相色谱法。流动相为：乙腈：甲醇：磷酸盐缓冲液（含 0.03 mmol/L 乙酸铵和 0.02 mmol/L 柠檬酸，0.1 mol/L 磷酸用 1.0 mol/L NaOH 调节 pH = 12.0）= 10：20：70（$V/V/V$）。柱温：30℃。流速：1.0 mL/min。荧光检测器，检测波长：E_x = 280 nm，E_m = 450 nm。

在该条件下，恩诺沙星和环丙沙星保留时间分别为 9.26 min 和 7.13 min。血淋巴中恩诺沙星回收率为 94%，其他组织样品中的回收率为 82%~86%，日内精密度和日间精密度相对标准偏差小于 5%；血淋巴中环丙沙星回收率为 88%，其他组织中的回收率为 80%~83%，日内精密度和日间精密度相对标准偏差小于 7%。血淋巴恩诺沙星和环丙沙星检测限均为 0.02 μg/mL，肌肉、肝胰脏和性腺样品中的检测限为 0.02 μg/g。

1.8　数据处理

青蟹血淋巴、肌肉和肝胰腺等组织中药物浓度与时间关系药动学参数采用统计矩原理来推算，分析软件为上海中医药大学药物临床研究中心的 DAS2.0 药物与统计软件。

2　结果

2.1　恩诺沙星在拟穴青蟹体内药动学

恩诺沙星以 10 mg/kg 剂量口灌和肌肉注射给药后，拟穴青蟹血淋巴中恩诺沙星浓度随时间的变化曲线分别见图 1 和图 2。口灌给药后，青蟹血淋巴中的恩诺沙星浓度迅速上升，仅 0.5 h 即达最高峰（C_{max} 为 12.90 μg/mL），随后立即下降，4~6 h 下降速度较快，6 h 后下降速度明显变慢。肌肉注射给药后，第一个采样时间点（T_{max} = 1 min）即为药峰浓度，C_{max} 为 31.86 μg/mL；给药后 2 h 内血药浓度下降速度很快，2 h 后下降速度明显变缓，6 h 后进入缓慢消除阶段。

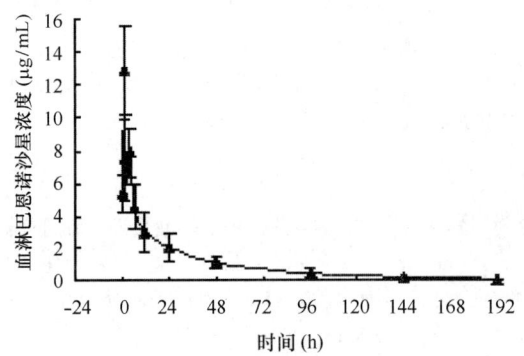

图 1　口灌给药下拟穴青蟹血淋巴恩诺沙星浓度与时间关系

Fig. 1　Hemolymph enrofloxacin concentration-time profile following oral administration

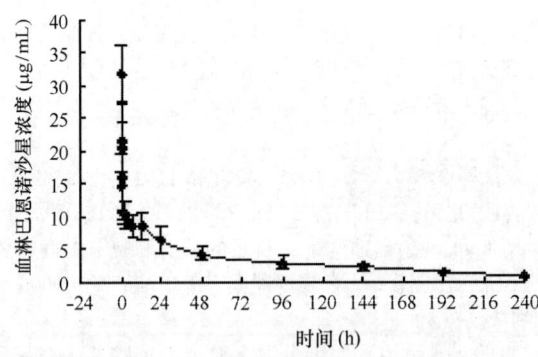

图 2　肌肉注射下拟穴青蟹血淋巴恩诺沙星浓度与时间关系

Fig. 2　Hemolymph enrofloxacin concentration-time profile following intramuscular injection

以 10 mg/kg 剂量口灌给药后，青蟹肌肉和肝胰腺中恩诺沙星浓度随时间的变化关系见图 3。给药后，肌肉中恩诺沙星表现为迅速上升，第 0.5 小时即达最高峰（C_{max} 为 6.05 μg/g），随后立即下降，6 h 后下降速度明显变慢。给药后肝胰腺中药物浓度也表现为迅速上升趋势，1 h 即达到药峰（C_{max} 为 22.30 μg/g）随后即以较快的速度下降，至 24 h 后肝胰腺中恩诺沙星浓度呈现出缓慢下降趋势。

以 10 mg/kg 剂量肌肉注射给药后，青蟹肌肉和肝胰腺中恩诺沙星浓度随时间的变化关系分别见图 4 和图 5。在第一个

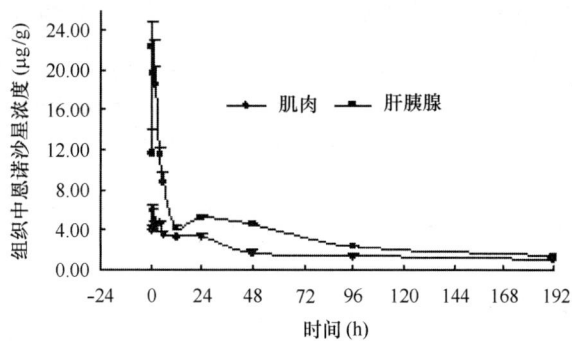

图3　口灌给药下拟穴青蟹组织中恩诺沙星浓度与时间关系

Fig. 3　Tissue enrofloxacin concentration-time profile of mud crabfollowing oral administration

采样时间点（1 min），肌肉中恩诺沙星浓度最高为（13.14±1.50）μg/g，随即几乎成垂直下降，1 h后下降速率明显趋于平缓（图4）。肝胰脏中的恩诺沙星浓度则首先表现出上升趋势，2 h到达最高药峰为（18.09±0.97）μg/g，随后开始下降，但下降过程十分缓慢，给药后96 h肝胰脏中恩诺沙星浓度仍高达（13.77±0.42）μg/g，远高于同采样时间点的肌肉中药物浓度（2.89±1.03）μg/g，96 h后肝胰腺中恩诺沙星下降速率开始增大（图5）。

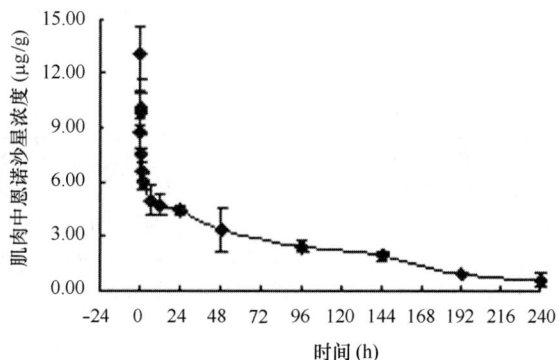

图4　肌注给药下拟穴青蟹肌肉中恩诺沙星浓度与时间关系

Fig. 4　Muscle enrofloxacin concentration-time profile of mud crab following intramuscular injection

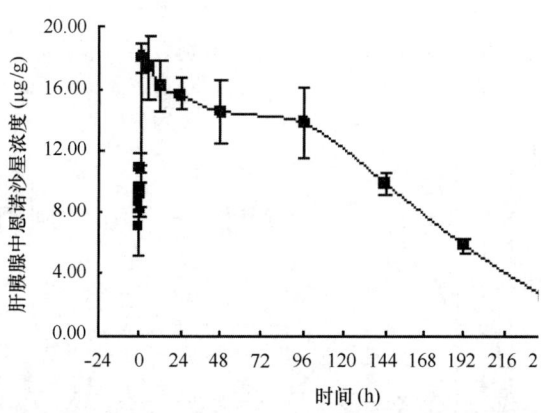

图5　肌注给药下拟穴青蟹肌肉中恩诺沙星浓度与时间关系

Fig. 5　Hepatopancreas enrofloxacin concentration-time profile of mud crab following intramuscular injection

采用统计矩原理推算所得的药动学参数见表1。从同一给药途径的不同组织来看，恩诺沙星在肝胰腺中的达峰浓度、曲线下面积（AUC）高于肌肉，在肝胰腺中的平均驻留时间（MRT）长于肌肉，总体清除率（CLz）小于肌肉，但相反地恩诺沙星在肌肉中的消除半衰期（$T_{1/2z}$）却长于肝胰腺。从不同给药途径来看，肌注给药下青蟹血淋巴中恩诺沙星的AUC值是口灌给药的近3倍，但消除比口灌给药慢，总体清除率低于口灌给药。肌注给药下肌肉和肝胰腺AUC均高于口灌给药，药物在组织中的平均驻留时间（MRT）长于口灌给药，且肌肉注射的总体清除率亦低于口灌给药。

表 1   口灌和肌肉注射下拟穴青蟹组织中恩诺沙星药动学参数

Tab. 1   Pharmacokinetic parameters of enrofloxacin in tissues from mud crabfollowing oral gavage and intramuscular injection

药动学参数 pharmacokinetic parameters	口灌给药 oral gavage			肌肉注射 intramuscular injection		
	血淋巴 hemolymph	肌肉 muscle	肝胰脏 hepatopancreas	血淋巴 hemolymph	肌肉 muscle	肝胰脏 hepatopancreas
C_{max}（µg/mL；µg/g）	12.90±2.65	6.05±0.45	22.30±2.51	31.86±4.27	13.14±1.50	18.09±0.97
T_{max}/（h）	0.50	0.50	0.50	0.017	0.017	2.00
AUC_{0-t} （µg/mL·h；µg/g·h）	216.1	445.9	817.6	816.8	554.7	2573.7
$AUC_{0-\infty}$ （µg/mL·h；µg/g·h）	255.7	448.1	835.0	830.8	606.9	2769.2
MRT_{0-t}（h）	25.87	36.36	53.29	86.98	79.08	91.16
$MRT_{0-\infty}$（h）	33.11	93.28	96.16	92.96	101.16	106.74
$T_{1/2z}$（h）	26.45	65.50	47.74	57.02	66.53	49.79
V_d（L/kg）	2.062	–	–	0.990	–	–
Clz（L/h/kg；kg/h/kg）	0.054	0.029	0.016	0.012	0.016	0.004

注：C_{max}表示组织中药物峰浓度；T_{max}表示药物达峰时间；AUC_{0-t}、$AUC_{0-\infty}$分别表示曲线下面积（0-t）和曲线下面积（0-∞）；MRT_{0-t}和$MRT_{0-\infty}$分别表示平均驻留时间；$T_{1/2z}$表示消除半衰期；V_d表示表观分布容积；CLz 表示总体清除率。

Notes：C_{max}--Maximum concentration in tissues；T_{max}--Time when maximum concentration was obtained；AUC_{0-t}（$AUC_{0-\infty}$）--Area under the drug concentration-time curve from the time zero to 240 h（from the time zero to infinity）；MRT_{0-t}（$MRT_{0-\infty}$）--Mean residue time of drug in body from zero to 240 h（from zero to infinity）；$T_{1/2z}$--Elimination half-life of drug；V_d--Apparent volume of distribution；CL_z--Total body clearance.

2.2   代谢产物环丙沙星在拟穴青蟹组织中的代谢和消除规律

口灌给药下，拟穴青蟹不同组织中的恩诺沙星主要代谢产物环丙沙星浓度-时间关系曲线见图 6。从图中可以看出，血淋巴、肌肉和肝胰脏中代谢产物环丙沙星浓度变化比较复杂，呈现多峰现象，但总的趋势是先上升后下降。这三组织中代谢产物环丙沙星峰浓度出现时间分别为 48 h、48 h 和 96 h，峰浓度分别为 0.07 µg/mL、0.10 µg/g 和 0.18 µg/g，远低于相应组织中恩诺沙星的达峰浓度。

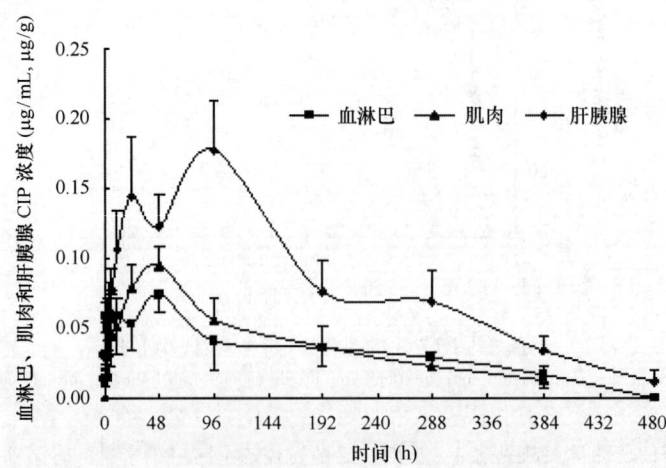

图 6   口灌给药拟穴青蟹组织中代谢产物环丙星浓度时间关系曲线

Fig. 6   The bomolymph muscle and hepatopancross of motabolite ciprofloxzchin in mud crab followingoral gavageo of onrofloxacin

　　肌肉注射给药下，拟穴青蟹不同组织中的恩诺沙星主要代谢产物环丙沙星浓度–时间关系曲线见图7。从图中可以看出，血淋巴、肌肉和肝胰脏中代谢产物环丙沙星浓度变化比较复杂，呈现多峰现象；在3种组织中药峰浓度出现时间并不一致，分别为48 h、96 h和96 h，峰浓度分别为0.23 μg/mL、0.19 μg/g和0.18 μg/g，同样远低于相应组织中恩诺沙星的达峰浓度。采用统计矩原理推算锯缘青蟹组织中代谢物环丙沙星的药动学参数（见表2）。

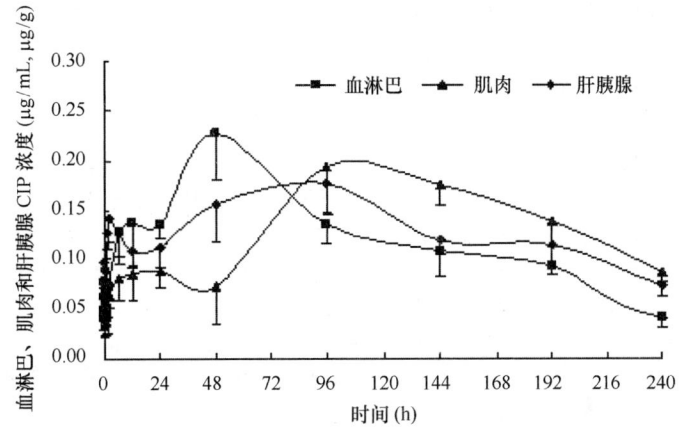

图7　肌注给药下拟穴青蟹组织中代谢物环丙沙星浓度–时间关系

Fig. 7　The hemolymoph, muslcle and hepatopancress concentrations of metabolite ciprofloxacin in mud crab following intramus cular injection of enrofloxacin

表2　拟穴青蟹口灌和肌肉注射恩诺沙星后代谢物环丙沙星的药动学参数
Tab. 2　Pharmacokinetic parameters of metabolite ciprofloxacin in tissues from mud crabfollowing oral gavage and intramuscular injection

药动学数 pharmacokinetic parameter	口灌给药 oral gavage			肌肉注射 intramuscular injection		
	血淋巴 hemolymph	肌肉 muscle	肝胰脏 hepatopancreas	血淋巴 hemolymph	肌肉 muscle	肝胰脏 hepatopancreas
C_{max}（μg/mL；μg/g）	0.07±0.02	0.10±0.01	0.18±0.04	0.23±0.04	0.19±0.05	0.18±0.03
T_{max}（h）	48	48	96	48	96	96
CLz（L/h/kg）	0.503	0.540	0.245	0.259	0.205	0.214
$AUC_{(0-t)}$（μg/mL·h；μg/g·h）	14.4	16.3	39.1	34.0	40.2	38.1
$AUC_{(0-\infty)}$（μg/mL·h；μg/g·h）	19.9	18.5	40.8	38.7	48.8	46.7
$MRT_{(0-t)}$（h）	148.60	125.38	162.57	121.38	157.66	141.83
$MRT_{(0-\infty)}$（h）	299.86	176.20	181.82	169.66	227.50	221.60

　　注：C_{max}表示组织中药物峰浓度；T_{max}表示药物达峰时间；CLz表示总体清除率；AUC_{0-t}和$AUC_{0-\infty}$分别表示曲线下面积（0-t）和曲线下面积（0-∞）；MRT_{0-t}和$MRT_{0-\infty}$分别表示平均驻留时间。

　　Notes：C_{max}—Maximum concentration in tissues；T_{max}—Time when maximum concentration was obtained；CL_z—Total body clearance；AUC_{0-t}（$AUC_{0-\infty}$）—Area under the drug concentration–time curve from the time zero to 240 h（from the time zero to infinity）；MRT_{0-t}（$MRT_{0-\infty}$）—Mean residue time of drug in body from zero to 240 h（from zero to infinity）.

3　讨论

3.1　恩诺沙星在拟穴青蟹体内药动学特征

尽管采用房室模型可以推算得到较多的药动学参数，但房室模型的选择也往往会影响药动学参数[14]。因此本研究中，采用统计矩原理推算了口灌和肌肉注射给药下恩诺沙星在拟穴青蟹血淋巴、肌肉和肝胰腺组织中的药动学参数，可以不受模型选择的限制，分析药物的吸收、分布和消除等处置过程。

血药 C_{max}、T_{max} 和 AUC 是反映药物在体内吸收快慢和程度的重要参数[15]。恩诺沙星以 10 mg/kg 剂量混饲口灌鲤（*Cyprinus carpio L.*）[16]，其血药峰浓度达 2.147 μg/mL，达峰时间为 1h，AUC 为 45.055 μg·h/mL。舌齿鲈（*Dicentrarchus labrax*）口服 5 mg/kg 恩诺沙星，吸收较慢，T_{max} 为 8 h，C_{max} 为 1.39 μg/mL，AUC 为 65.93 μg·h/mL[17]。短盖巨脂鲤（*Colossoma brachypomum*）口服 5 mg/kg 剂量的恩诺沙星，其吸收尤为缓慢，t_{max} 为 36 h，C_{max} 为 0.80 μg/mL，AUC 仅为 26.5 μg·h/mL[18]。本研究中，拟穴青蟹口灌恩诺沙星的血药达峰快（T_{max} 为 0.5 h），C_{max} 和 AUC 均明显高于上述几种水生脊椎动物。在肌肉注射给药下，青蟹对恩诺沙星的吸收更为迅速，在给药后的 1 min（第 1 个采样时间点）即达到最高值，且吸收效果好，这一结果与氟喹诺酮类药物在其他甲壳动物中相一致。中华绒螯蟹（*Eriocheir sinensis*）肌肉注射环丙沙星（雄蟹给药剂量为 8.17 μg/kg，雌蟹 6.25 μg/kg）时，血药达峰时间均为 1 min，峰浓度分别为 37.13 μg/mL，16.47 μg/mL[19]；南美白对虾（*Penaeus vannamei Boone*）肌肉注射 10 mg/kg 诺氟沙星，达峰时间为 2 min，血药峰浓度为 29.33-30.27 μg/mL[20]。大西洋鲑（*Salmo salar*）幼鱼肌注 10 mg/kg 恩诺沙星，0.41 h 才达到峰浓度[21]。猪和鸡肌肉注射恩诺沙星，其达峰时间更长，分别为 0.75 h，峰浓度分别仅为 0.77-0.79 μg/mL[12]和 1.33 μg/mL[22]。由此可见，恩诺沙星在甲壳类动物体内不仅吸收快，而且吸收程度高。其原因可能主要是：虾蟹等甲壳动物血液循环系统为开放式循环系统，血管末端开放，没有微血管相连，血淋巴由身体各部分的组织间隙逐渐汇集到混合体腔，肌肉注射给药类似于血窦内或血管注射给药[23]。

组织中 C_{max} 和 AUC 是反映药物在组织中分布的主要药动学参数。恩诺沙星以 10 mg/kg 剂量肌肉注射后，罗氏沼虾（*Macrobrachium rosenbergii*）[9]肌肉和肝胰腺的 C_{max} 分别为 6.635 μg/g 和 9.465 μg/g，AUC 分别为 207.04 μg/mL·h 和 308.07 μg/mL·h；中国明对虾（*Fenneropenaeus chinensis*）肌肉和肝胰腺的 C_{max} 分别为 6.29 μg/g 和 11.82 μg/g，AUC 分别为 73 μg/mL·h 和 320 μg/mL·h[10]。本研究中，同是肌肉注射相同给药剂量下，青蟹肌肉和肝胰腺的 C_{max} 和 AUC 均远大于上述的对虾和沼虾，这说明恩诺沙星在青蟹组织中分布广泛。

$T_{1/2z}$、CLz 和 MRT 等参数是反映药物在体内消除的主要参数，恩诺沙星在不同水产动物体内消除差别较大，在此将恩诺沙星在水产动物体内消除速度分为快、中、慢。眼斑拟石首鱼（*Sciaenps ocellatus*）灌服 5 mg/kg 恩诺沙星，其血浆药物 $T_{1/2z}$ 为 4.989 h，CLz 为 0.288 L/h/kg[5]；中国明对虾血淋巴 $T_{1/2z}$ 为 7.03 h，CLs 为 0.25 L/h/kg[10]，这两种动物体内恩诺沙星消除较快。本研究的恩诺沙星在青蟹体内的消除速度属于中等速度，青蟹口灌和肌肉注射恩诺沙星后，血药消除半衰期（$T_{1/2z}$）分别为 26.45 h 和 57.02 h，总体清除率（CL_z）分别为 0.054 L/h/kg 和 0.012 L/h/kg；与虹鳟（*Oncorhynchus mykiss*）幼鱼（$T_{1/2z}$ 为 29.5 h）[24]和罗氏沼虾（$T_{1/2z}$ 为 69.315 h，CLs 为 0.035 L/h/kg）[9]接近，青蟹口灌给药的 MRT 与虹鳟幼鱼（10 mg/kg 剂量下 MRT 为 35.3 h）相近。而恩诺沙星在欧洲鳗鲡（*Anguilla anguilla*）和大西洋鲑体内消除较慢，欧洲鳗鲡口灌给药恩诺沙星的血浆 $T_{1/2z}$ 为 161.10 h，MRT 为 267.3 h[6]；大西洋鲑口灌 10 mg/kg 的血浆消除半衰期为 105.1 h，MRT 为 151.7 h[21]。众所周知，水产动物种类多，形态和解剖学结构差别大，还受着生存环境的影响。药物在不同种类之间的消除上的差异可能是由于解剖学上的体积差异、药物与血浆蛋白和组织结合的差异[25-26]，以及生存环境温度的差异所致[27]。由此可见，药物在水产动物体内代谢与消除规律十分复杂，针对特定药物开展其在不同养殖动物体内的药动学研究，制定科学、合理的用药方案和休药期显得尤为重要。

3.2　拟穴青蟹体内恩诺沙星及其代谢物环丙沙星的关系

恩诺沙星在动物体内脱乙基代谢成为具有活性作用的环丙沙星，在陆生哺乳动物体内较多地代谢为环丙沙星。山羊肌肉注射恩诺沙星后，其血浆中代谢产物环丙沙星与恩诺沙星的曲线下面积之百分比（AUC_{CIP}/AUC_{ENR}）为 34%[11]，绵羊、猪、犬和马的 AUC_{CIP}/AUC_{ENR} 分别为 55%[28]、51.5%[29]、43%[30]和 20%~35%[31]。由此可见，在陆生哺乳动物体内，代谢产物环丙沙星和原形药物恩诺沙星共同起着药效作用。

而在水产动物中，恩诺沙星代谢生成环丙沙星的量差异较大。乌贼（*Sepia officinalis*）血管注射和药浴浸泡恩诺沙星时，所有实验样品中均未检测到代谢产物环丙沙星（检测限为 0.02 μg/mL 或者 μg/g）[32]；舌齿鲈口灌和药浴给药恩诺沙星时，在肝脏中经常检测到环丙沙星，血液中偶尔检测到，而在皮肤和肌肉中则从未检测到（检测限为 0.01 μg/mL 或者 μg/g）[17]；罗非鱼（*Tilapia*）投喂 50 mg/kg 剂量恩诺沙星药饵，肌肉中环丙沙星代谢产物最高含量为 0.22 μg/g[4]；欧洲鳗鲡口灌给药恩诺沙星，其血浆、肌肉和肝脏等组织中均可检测到环丙沙星，且最高浓度分别为 0.123 mg/L、0.388 mg/kg

和 1.212 mg/kg 和 1.212 mg/kg 和 1.212 mg/kg，但与相应组织中的恩诺沙星最高浓度相比，仅分别约为 1/40、1/15 和 1/7.5，血浆中 AUC_{CIP}/AUC_{ENR} 百分比为 6.24%[6]。在甲壳动物中，罗氏沼虾肌注恩诺沙星后，在血淋巴、肌肉和肝胰脏三组织中均可检测到环丙沙星，但含量处于较低水平，肝胰脏内浓度最高，为 0.198 $\mu g/g$[9]；中国明对虾肌肉内注射剂量 10mg/kg 恩诺沙星，其血药环丙沙星代谢物最高浓度为 0.024 $\mu g/mL$，其血药 AUC_{CIP}/AUC_{ENR} 百分比仅为 1.2%[10]。本研究中，肌肉注射给药下，青蟹血淋巴、肌肉和肝胰脏三组织 AUC_{CIP}/AUC_{ENR} 百分比分别为 4.16%、7.24% 和 1.48%；口灌给药下三组织 AUC_{CIP}/AUC_{ENR} 百分比分别为 6.68%、3.60% 和 4.78%。由此可见，无论是水生脊椎动物，还是甲壳类无脊椎动物，恩诺沙星代谢生成环丙沙星量都明显少于陆生哺乳动物，且在水生动物中起抑（杀）菌活性作用以恩诺沙星为主，代谢产物环丙沙星所起的药效作用甚微。

3.3　恩诺沙星的疗效及给药方案

Roque 等（2001）从墨西哥养殖的南美白对虾群体分离到 144 株弧菌，并进行了 15 种抗生素对所分离弧菌的敏感性试验。结果表明，所分离的 144 株弧菌中只有 3 株对恩诺沙星不敏感，恩诺沙星对余下 141 株菌的平均最小抑菌浓度（MIC）为 0.45 $\mu g/mL$，而氟甲砜霉素的平均 MIC 为 1.79 $\mu g/mL$，土霉素平均 MIC 为 304.0 $\mu g/mL$，说明恩诺沙星对弧菌具有较强的抗菌能力，且效果优于其他抗生素[33]。氟喹诺酮类药物属于典型的浓度依赖性药物，其临床药效既受最高血药浓度的影响还受曲线下面积（AUC）的影响，当 C_{max}/MIC 比值为 8~10 和 AUC_{0-24}/MIC 比值为 30~55 时，氟喹诺酮类药物显现出很好的临床药效[34]。弧菌是海水养殖动物的主要病原菌，毛芝娟等（2001）从养殖发病青蟹体内分离到具有致病性的辛辛那提弧菌、溶藻弧菌和副溶血弧菌，虽然未对这些病原菌作药敏测试，但我们可以参考上述的 MIC 结果，假设上述几株青蟹病原弧菌的 MIC 为 1.0 $\mu g/mL$，那么其 C_{max}/MIC 比值为 12.9、AUC_{0-24}/MIC 比值为 95.3，均达到氟喹诺酮类药物临床治疗要求，这说明给拟穴青蟹以 10 mg/kg 剂量每隔 24 h 口服投喂恩诺沙星，能够较好的防治弧菌引起的青蟹细菌性疾病。

参考文献

[1]　毛芝娟. 养殖青蟹的主要病害与防治 [J]. 科学养鱼, 2000, 7: 30-30.

[2]　冯振飞, 王国良, 倪海儿. 养殖锯缘青蟹黄水病流行病学及其预报模型 [J]. 水产科学, 2009, 28 (12): 713-716.

[3]　刘开永, 汪开毓. 恩诺沙星在水产中的应用与研究 [J]. 中国兽药杂志, 2004, 38 (10): 32-34.

[4]　徐维海, 林黎明, 朱校斌, 等. 恩诺沙星及其代谢产物在吉富罗非鱼、中国对虾体内的残留规律研究 [J]. 水产科学, 2004, 23 (7): 5-8.

[5]　简纪常, 吴灶和, 陈刚. 恩诺沙星在眼斑拟石首鱼体内的药物代谢动力学 [J]. 中国兽医学报, 2005, 25 (2): 195-197.

[6]　房文红, 于慧娟, 蔡友琼, 等. 恩诺沙星及其代谢物环丙沙星在欧洲鳗鲡体内的代谢动力学 [J]. 中国水产科学, 2007, 14 (4): 622-629.

[7]　Pharmacokinetics and the active metabolite of enrofloxacin in Chinese mitten-hanged crab (Eriocheir sinensis) [J]. Aquaculture, 2006, 260: 69-76.

[8]　宋维彦, 苏永全, 潘滢, 等. 磺胺甲基恶唑和恩诺沙星在日本囊对虾体内的药代动力学研究 [J]. 海洋科学, 2010, 34 (7): 22-27.

[9]　钱云云. 恩诺沙星在罗氏沼虾体内的药物代谢动力学 [J]. 动物学杂志, 2007, 42 (5): 62-69.

[10]　方星星, 王群, 李健. 恩诺沙星及其代谢物环丙沙星在中国对虾体内的药代动力学 [J]. 水产学报, 2004, 28 (增刊): 35-41.

[11]　TYCZKOWSKA K, HEDEEN K M, AUCOIN D P, et al. High-performance liquid chromatographic method for the simultaneous determination of enrofloxacin and its primary metabolite ciprofloxacin in canine serum and prostatic tissue [J]. Journal of Chromatograhpy, 1989, 493: 337-346.

[12]　曾振灵, 冯淇辉. 恩诺沙星在猪体内的生物利用度及药物动力学研究 [J]. 中国兽医学报, 1996, 16 (6): 605-612.

[13]　杨先乐, 刘至治, 孙文钦, 等. 中华绒螯蟹血淋巴内盐酸环丙沙星的反相高效液相色谱法的建立 [J]. 水产学报, 2001, 25 (4): 348-355.

[14]　王广基. 药物代谢动力学 [M]. 北京: 化学工业出版社, 2005: 315-318.

[15]　王广基. 药物代谢动力学 [M]. 北京: 化学工业出版社, 2005: 64-85.

[16]　张雅斌, 刘艳辉, 张祚新, 等. 恩诺沙星在鲤体内的药效学及药动学研究 [J]. 大连水产学院学报, 2004, 19 (4): 239-242.

[17]　INTORRE L S, Cecchini, S. Bertini, A. M, et al. Pharmacokinetics of enrofloxacin in the seabass (Dicentrarchus labrax) [J]. Aquaculture, 2000, 182: 49-59.

[18]　Lewbart G, Vaden S, Deen J, et al. Pharmacokinetics of enrofloxacin in the red pacu Colossoma brachypomum after intramuscular, oral and bath administration [J]. J. Vet. Pharmacol. Ther., 1997, 20: 124-128.

[19]　杨先乐, 刘至治, 横山雅仁. 盐酸环丙沙星在中华绒螯蟹体内药物代谢动力学研究 [J]. 水生生物学报, 2003, 27 (1): 18-22.

[20]　房文红, 郑国兴. 肌注和药饵给药下诺氟沙星在南美白对虾血淋巴中药代动力学 [J]. 水生生物学报, 2006, 30 (5): 541-546.

[21]　Stoffregen D A, Wooster G A, Bustos P S. Multiple route and dose pharmacokinetics of enrofloxacin in juvenile Atlantic salmon [J]. Journal of Veterinary Pharmacology and Therapeutics, 1997, 20 (2): 111-123.

[22] 应翔宇，李冬郊，田惠英，等．恩诺沙星在健康及巴氏杆菌感染鸡体内的药物代谢动力学研究［J］．中国兽医杂志，2002，36（1）：3-6.

[23] Bliss D E. The biology of crustacea［J］. In: Internal Anatomy and Physiological Regulation, 1983, 5: 3-19.

[24] Bowser P R, Wooster G A, StLeger J, et al. Pharmacokinetics of enrofloxacin in fingerling, rainbow trout Oncorhynhus mykiss［J］. Journal of Veterinary Pharmacology and Therapeutics, 1992, 15: 62-71.

[25] Barron M G, Gedutis C, James M O. Pharmacokinetics of sulphadimethoxine in the lobster, Homerus americannus, following intrapericardial administration［J］. Xenobiotica, 1988, 18 (3): 269-277.

[26] Oie S, Tozer T N. Effect of altered plasma protein binding on the apparent volume of distribution［J］. J. Pharm. Sci. 1979, 68: 1203-1208.

[27] Rigos G, Alexis M, Andriopoulou A, et al. Temperature-dependent pharmacokinetics and tissue distribution of oxolinic acid in sea bass, Dicentrarchus labrax L., after a single intravascular injection［J］. Aquaculture Research, 2002, 33: 1175-1181.

[28] Mengozzi G, Intorre L, Bertini S, et al. Pharmacokinetics of enrofloxacin and its metabolite, ciprofloxacin after intravenous and intramuscular administration in sheep［J］. Am. J. Vet. Res. 1996, 57: 1040-1043.

[29] Anadon A, Martinez-Larranaga M R, Diaz M J, et al. Pharmacokinetic variables and tissue residues of enrofloxacin and ciprofloxacin in healthy pigs［J］. Am. J. Vet. Res., 1999, 60: 1377-1382.

[30] Kung K L, Riond J L, Wanner M. Pharmacokinetics of enrofloxacin and its metabolite, ciprofloxacin after intravenous and oral administration of enrofloxacin in dogs［J］. J. Vet. Pharmacol. Ther., 1993, 16: 462-468.

[31] Kaartinen L, Pam S, Pyörälä S. Pharmacokinetics of enrofloxacin in horses after single intravenous and intramuscular administration［J］. Equine Vet. J., 1997, 29: 378-381.

[32] Gore R S, Harms C A, Kukanich B, et al. Enrofloxacin pharmacokinetics in the European cuttlefish, Sepia officinalis, after a single i. v. injection and bath administration［J］. J. Vet. Pharmacol. Ther., 2005, 28 (5): 433-439.

[33] Roque A, Gomez-Gil B. Therapeutic effects of enrofloxacin in anexperimental infection with a luminescent Vibrio harveyi in Artemia franciscana Kellog 1906［J］. Aquaculture, 2003, 220: 37-42

[34] Lacy M K, Lu W, Xu X W, et al. Pharmcodynamic comparisons of levofloxacin, ciprofloxacin and Ampicillin against Streptococcus pneumoniae in an in vitro model of infection［J］. Antimicrobial Agents and Chemotherapy, 1999, 43 (3): 672-677.

该文发表于《水产学报》【2004，28（12）：25-29】

盐酸环丙沙星药液口灌在中华绒螯蟹体内的代谢动力学研究

The pharmacokinetics of ciprofloxacinum in *Eriocheir sinensis*

李正[1]，杨勇[1]，杨先乐[1]，胡鲲[1]，房文红[2]，陆平[1]

(1. 上海水产大学生命科学与技术学院，农业部动植物病原库，上海 200090；2. 中国水产科学研究院东海水产研究所，上海 200090)

摘要： 盐酸环丙沙星药液口灌中华绒螯蟹，给药剂量（6.00±0.13）$\mu g \cdot g^{-1}$。给药后，盐酸环丙沙里在中华绒螯蟹血淋巴中处置过程为三室模型。其主要药动学参数为：雄蟹血淋巴中药物的分布和消除半衰期分别为 2.47 h 和 313.63 h，雌蟹分别为 1.24 h 和 84.14 h，雄蟹和雌蟹血淋巴中药物曲线下面积（AUC）分别为 582.00 $\mu g \cdot h^{-1} \cdot mL^{-1}$ 和 476.37 $\mu g \cdot h^{-1} \cdot mL^{-1}$，总体清除率分别为 10.31 $mL \cdot kg^{-1} \cdot h^{-1}$ 和 12.60 $mL \cdot kg^{-1} \cdot h^{-1}$。检测结果表明，肝胰脏组织为盐酸环丙沙里的主要残留组织，其最高残留 M 达 10.07 $\mu g \cdot g^{-1}$，36 d 后不能检出；肌肉组织中药物残留量较少，最高为 0.51 $\mu g \cdot g^{-1}$，多少天后不能检出。

关键词： 盐酸环丙沙星；中华绒螯蟹；药物代谢动力学

盐酸环丙沙星（ciprofloxacinum）是第 3 代氟喹诺酮类抗菌药，于 1983 年由西德 Bayer 公司研制成功，1986 年上市。由于环丙基的引入，盐酸环丙沙星改变了与细菌 DNA 旋转酶的结合，加上氟原子，使其成为喹诺酮类药物中抗菌活性最强，毒副作用最低的药物，对需氧性、厌氧性球菌和杆菌，革兰氏阴性菌及革兰氏阳性菌均具有强大的抗菌效果。因其疗效好副作用小，已风行于世界。国内外已广泛用于泌尿生殖系统、消化系统、呼吸系统、皮肤软组织感染及骨髓炎等。在抗感染药物中，其销售总额名列世界第二。目前，该药仍为人用药，但在我国有一段时间盐酸环丙沙星曾广泛应用于水产动物相关疾病的预防和治疗，尤其在经济价值较高的中华绒螯蟹养殖上。为了解它在中华绒螯蟹血淋巴内的处置过程和在组织中残留、消除过程，并为其禁用提供相应理论依据，试验进行了以下研究。

1　材料与方法

1.1　试验材料

中华绒螯蟹由上海前卫特种水产养殖公司提供。雌蟹体重 61.1±13.8 g，雄蟹体重 52.2±9.0 g。试验动物在水族箱（1.0 m×0.5 m×0.4 m）中驯养 1 周后，挑选健康活泼的开始正式试验，试验水温 20.0±0.5℃。盐酸环丙沙星标准品（纯度为 99.54%）和原粉（纯度为 98%）均由上海三维制药公司提供；抗凝剂参照蔡武城等[1]的方法配制；甲醇，HPLC 级，其他所用试剂分析纯。

1.2　试验方法

给药和取样：用注射器将 3 mg·mL⁻¹ 浓度的药物水溶液灌人中华绒螯蟹胃内，按每 10 g 蟹给药 0.02 mL（每只蟹注射 0.1 mL 左右），算得给药剂量为（6.00±0.13）µg·g⁻¹。给药后将蟹在空气中放置 15～20 min，不出现回吐的进行正式试验。给药后在试验设定时间点分别取血淋巴和组织样品，每 5 只蟹作为一组，每组两个平行。

样品制备：取 1 mL 血淋巴加入甲醇 2 mL，旋涡混合 2 min，4 000 r·min⁻¹ 离心 10 min，0.45 µm 滤膜过滤后自动进样 10 µL。其他组织样品加入相应蛋白沉淀剂后匀浆机捣碎漩涡混合 2 min，3 500 r·min⁻¹ 离心 10 min，10 000 r·min⁻¹ 离心 10 min，0.45 µm 滤膜过滤后自动进样 10 µL。

色谱条件：高效液相色谱仪采用 Agilent 1100 系列，包括四元泵、自动进样器、柱温箱、DAD 检测器。色谱柱为 C₁₈ 分析柱（15 cm×4.6 mm，5 µm），流动相为甲醇：磷酸缓冲液（0.05 mol·L⁻¹）：四丁基溴化铵（0.05 mol·L⁻¹）= 24.5：73.5：2（V/V/V），检测波长 280 nm，流速 1.0 mL·min⁻¹，柱温 40℃，自动进样 10 µL。

数据处理：根据血药浓度-时间的半对数关系初步确定药物的处置过程模型，药动学模型拟合及相关系数计算采用 SAS^R Proprietary Software Release 6.12（SAS Institute Inc., Cary, NC, USA.）软件中非线性回归分析部分 Marquardt 方法进行模型嵌和参数计算。给药后的肌肉、肝胰脏和性腺样品中药物的消除半衰期按 Bjorklund 等[2]使用的方法，以时间为横坐标，浓度的常用对数为纵坐标，作一元线性回归，计算斜率（β）和相关系数，消除半衰期为 $T_{1/2\beta}$ = ln2/β，预计停药期按 Salte 等[3]所使用的方法得出。

2　结果

2.1　标准曲线

试验设定条件下，HPLC 基线平稳，环丙沙星保留时间稳定，药物峰与杂质峰分离良好。标准品浓度为 0.01～5.0 µg·mL⁻¹ 时，标准曲线为 $y = 1.64618×10^{-5}x$，相关系数 $r = 0.9998$，相关性良好。

2.2　血淋巴中盐酸环丙沙星的处置过程

口灌给药，盐酸环丙沙星在中华绒螯蟹雌雄蟹血淋巴中浓度测定结果见表 1。给药后盐酸环丙沙星在雌蟹血淋巴中 8 h 达到最高值，雄蟹中 4 h 达到最高值，之后进入分布期和消除期，血药浓度开始下降，336 h 均不能检出。盐酸环丙沙星在中华绒螯蟹血淋巴中的处置过程为三室模型，雌蟹药-时关系方程为 $C_0 = -12.2116×e^{-0.551t} + 10.4641×e^{-4150t} + 3.8798×e^{-0.00821}$，拟合曲线相关指数 $R^2 = 0.963$，雄蟹药-时关系方程为 $C_0 = -5.0147×e^{-0.2789t} + 6.6690×e^{-0.0209t} + 0.6213×e^{-0.0022t}$，相关指数 $R^2 = 0.927$，以动力学模型方程计算药动学参数见表 2。盐酸环丙沙星在雄蟹血淋巴中的分布和消除半衰期分别为 2.47 h 和 313.63 h，雌蟹血淋巴中分别为 1.24 h 和 84.14 h。曲线下面积（AUC）分别为 582.00 µg·h⁻¹ 和 476.37 µg·h⁻¹，总体清除率分别为 10.31 mL·kg⁻¹·h⁻¹ 和 12.60 mL·kg⁻¹·h⁻¹。

2.3　盐酸环丙沙星在中华绒螯蟹组织中的消除

盐酸环丙沙星 6.00±0.13 µg·g⁻¹ 剂量口灌给予中华绒螯蟹，肌肉中药物消除速度开始较快，4 d 后消除速度变慢，药物浓度维持在一定水平，从 4 d 到 11 d 雌雄蟹肌肉中残留量分别从 0.51 µg·g⁻¹ 和 0.45 µg·g⁻¹ 降低至 0.37 µg·g⁻¹ 和 0.35 µg·g⁻¹，14 d 后不能检出。比较而言肝胰脏中残留量较高。最高可达 10.07 µg·g⁻¹，高于同时间点血淋巴和肌肉中药物浓度。随着时间的延长，残留量逐渐降低，29 d 后达到较低水平的 0.05 µg·g⁻¹，但此时雄蟹中仍有较高残留量，36 d 后不能检出。精巢中盐酸环丙沙星在 2 d 达到最大值 2.54 µg·g⁻¹，其后有所降低并一直维持在 1.0 µg·g⁻¹ 水平波动，而卵巢中在 4 d 达到最大值 3.57 µg·g⁻¹，其后逐渐降低，36 d 后不能检出。按 Bjorklund 等使用的方法算得盐酸环丙沙星在中华绒螯蟹肌肉、肝胰脏和性腺组织中的消除半衰期在 6～10 d。按 Salte 等的方法预计试验条件下以 0.001 µg·g⁻¹ 为最低残留限，停药期为 11 d（表 4）。

表1 中华绒螯蟹血淋巴中盐酸环丙沙星浓度

Tab. 1 Ciprofloxacinum concentrations in hemolymph from Eriocheir sinensis

时间（h）time	雌蟹 female		雄蟹 male	
	平均浓度（μg·mL⁻¹）averse values	理论浓度（μg·mL⁻¹）theory values	平均浓度（μg·mL⁻¹）averse values	理论浓度（μg·mL⁻¹）theory values
0.5	2.737±0,071	2.858	3.270±0.012	3.189
1	3.390±0.031	3.357	3.512±0.089	3.697
2	4.675±0.097	4.144	–	–
3	–	–	4.366±0.051	4.155
4	4.137±0.086	5.107	4.858±0.269	4.147
8	6.754±1.235	5.714	3.404±0.026	4.001
12	5.162±0.431	5.618	3.568±0.029	3.849
24	4.391±0.189	4.622	3.376±0.085	3.426
48	3.307±0.511	3.005	2.730±0.107	2.714
72	2.150±0.031	2.011	2.122±0.110	2.151
96	0.950±0.083	1.400	1.858±0.084	1.704
144	1.014±0.217	0.781	0.421±0,006	0.671
264	0,340±0.044	0.374	0.661±0.023	0.334
336	ND	–	ND	–

表2 盐酸环丙沙星口灌给药的药代动力学参数

Tab. 2 Pharmacokinetic parameters of ciprofloxacinum oral administration

dose	参数 parameters	单位 units	雄蟹 male	雌蟹 female
	给药剂量	μg·mL⁻¹	6.0	6.0
A	分布相的零时截距	μg·mL⁻¹	−5.014	−12.211
B	吸收相的零时截距	μg·mL⁻¹	6.669	10.464
P	消除相的零时截距	μg·mL	0.621	3.879
a	吸收相一级速率常数	h⁻¹	−0.278	−0.555
β	分布相一级速率常数	h⁻¹	−0.020	−0.415
p_i	消除相一级速韦常数	h⁻¹	−0.0022	−0.0082
$T_{1/2a}$	药物的吸收相半衰期	h	2.47	1.24
$T_{1/2\beta}$	药物的分布相半衰期	h	23.79	1.66
$T_{1/2p}$	药物的消除相半衰期	h	313.63	84.14
AUC	血药浓度-时N曲线下面积	μg·h⁻¹·mL⁻¹	582.00	476.37
CLs	总体清除率	mL·kg⁻¹·h⁻¹	10.31	12.60

表3 中华绒螯蟹不同组织中盐酸环丙沙星浓度的变化

Tab. 3 The concentrations of ciprofloxacinum in the tissues of Eriocheir sinensis

时间（d）time	肌肉（μg·g⁻¹）muscle		肝胰脏（μg·g⁻¹）liver		性腺（μg·g⁻¹）gonad	
	平均浓度♀ average values	平均浓度♂ averse values	平均浓度♀ average values	平均浓度♂ averse values	平均浓度♀ average values	平均浓度♂ averse values
1	1.24	1.39	8.55	8.19	0.97	1.30
2	1.19	0.68	7.55	10.07	1.55	2.54
3	0.95	0.95	4.26	6.66	2.76	1.94
4	0.51	0.45	–	5.63	3.57	1.23
6	0.51	0.52	7.26	5.74	2.78	1.42
8	0.49	0.44	5.15	4.18	3.23	1.15
11	0.37	0.35	3.06	3.23	2.41	1.26
14	ND	ND	1.91	2.54	1.93	1.01
22			1.02	1.16	0.97	0.44
29			0.05	0.33	0.19	ND
36			ND	ND	ND	

注：ND 表示未检出。

Notes：ND means no detect.

表4 盐酸环丙沙星在中华绒螯蟹组织中的消除半衰期

Tab. 4 Elimination half-life of ciprofloxacinum in tissues of Eriocheir sinensis

组织 tissue		浓度常用对数-时间消除曲线 log concentration-time curve	时间范围（d）time	相关系数 relativity	消除半衰期（d）$T_{1/2β}$	停药期（d）Pre
肌肉 muscle	♀	$y=-0.004\,7x+0.080$	1~11	0.8350	6	62
	♂	$y=-0.005\,1x+0.227$	1~11	0.8998	6	58
肝胰脏 liver	♀	$y=-0.002\,9x+0.840\,1$	6~22	0.9484	10	111
	♂	$y=-0.004\,6x+1.938\,1$	4~29	0.9522	6	80
性腺 gonad	♀	$y=-0.002\,9x+0.831\,4$	2~22	0.9254	10	111
	♂	$y=-0.004\,6x+1.938\,1$	4~29	0.9522	6	80

3 讨论

3.1 口灌给药盐酸环丙沙星在中华绒螯蟹体内代谢的特点

不同给药方式，盐酸环丙沙星在中华绒螯蟹体内代谢过程不同。口灌给药雌雄蟹血淋巴中药物达峰时间分别为8 h和4 h，相同条件下肌肉注射给药后血药浓度在短时间（1 min）内就达到峰值[1]。这是因为中华绒螯蟹的循环系统为开放式循环，注射给药后药物直接进入血淋巴循环，药物达峰时间短，峰浓度高。

与其他动物比较，盐酸环丙沙星在中华绒螯蟹血淋巴中消除更为缓慢。肌肉注射给药盐酸环丙沙星在鲤、非洲鲶和虹鳟体内的$T_{1/2β}$分别为（14.5±0.5）h、（14.2±1.3）h和（11.2±1.6）h[5]，而同为第3代氟喹诺酮类抗菌药的诺氟沙星在鲤体内的$T_{1/2β}$为（3.403 2±0.587 3）h[6]，在中华鳖体内的$T_{1/2β}$为4.24 h[7]，这些均比口灌给药中华绒螯蟹血淋巴中盐酸环丙沙星消除半衰期短。试验认为中华绒螯蟹是低等的水生动物，它们主要通过触角腺和肝胰腺对药物进行排泄或降解，其对药物的降解与排除的能力均较其他水生动物差，而鱼类等则可通过肾脏和呼吸器官（如鳃、鳃上腺等）等进行扩散消除，哺乳动物还可通过肾脏的主动运转予以消除（如猪、牛等对环丙沙星的$T_{1/2β}$仅为2.5 h）[8]，因而中华绒螯蟹的$T_{1/2β}$就比其他鱼类和哺乳动物要长。Jame和Barron通过心包膜注射给药研究磺胺间二甲氧嘧啶（SDM）在美洲鳌虾体内的分布规律时也有同样的发现，其$T_{1/2β}$达77 h，比哺乳类要长得多[9]。

盐酸环丙沙星对鱼体的各组织具有良好的渗透性，在鱼体内分布较广。蒋志伟等报道鲫口灌 10 mg·kg^{-1} 环丙沙星后，肝胰脏和肾脏中的药物浓度超过同期血药浓度[10]。试验结果表明，口灌给药后，肝胰脏也为盐酸环丙沙星的主要残留组织，其中最高残留量可达 10.07 μg·g^{-1}，约为同时间点血淋巴、肌肉中药物含量的 2 倍和 7 倍。比较而言肌肉组织内残留时间短，残留量少，14 d 后不能检出。

3.2 盐酸环丙沙星用药建议

试验结果表明，盐酸环丙沙星在中华绒螯蟹肝胰腺和性腺中残留量较高，而肝胰腺和性腺又是人们主要的取食部分，因此应严格控制这些组织中的药物残留，而且盐酸环丙沙星目前仍为人用药，为保证食品卫生，保障人们的身体健康，水产动物应禁止使用盐酸环丙沙星。

参考文献

[1] 蔡武城，李碧别，李玉民. 生化实验技术教程 [M]. 上海：复旦大学出版社，1983，47.

[2] Bjorklund H, Bylund G. Temperature-related absorption and excretion of oxytetracycline in rainbow trout [J]. Aquae, 1990 (84)：363-372.

[3] Salte R, Liestol K. Drug withdrawal from farmed fish. Deple-tion of oxytetracycline, sulfadiazime and trimethoprim from musclar tissue of rainbow trout (Salmo gairdneri) [J]. Acra Vet Scand, 1983, 24：418-430.

[4] 杨先乐，刘至治，横山雅仁. 盐酸环丙沙星在中华绒螯蟹体内药物代谢动力学研究 [J]. 水生生物学报，2003，27 (1)：18-22.

[5] Nouws J F M, Grondel J L, Schutte A R, et al. Pharmac-okinetics of ciprofloxacin in carp African catfish and rainbow trout [J]. Vet-Quart, 1988, 10：2-216.

[6] 张作新，张雅斌，杨永胜，等. 诺氟沙甩在鲤鱼体内的药代动力学 [J]. 中国兽医学报，2000，20 (1)：66-69.

[7] 陈文银，印春华. 诺氟沙星在中华鳖体内的药代动力学研究 [J]. 水产学报，1997，21 (4)：434-437.

[8] Kleibow K M, Jarboe H H, Shoemaker K E, et al. Compa-rative pharmacokineticx; s and bioavailability of oxolinic acid in channel catfish (Ictalurus punciaius) and rainbow trout (Oru'orhynchiLs mykiss) [J]. Can J Fish Aquat Sci, 1994, 51 (5)：1205-1211.

[9] Uno K, AokiT, Ueno R, Pharmacokinetions of sulphamono-methoxine and sulphadimetho-xine following oral administration to cultured rainbow trout (Oncorhynchus mykiss) [J]. Aquae, 1993, 115 (3-4)：209-219.

[10] 蒋志伟，王宝安，朱国强，等. 盐酸环丙沙星在鲫鱼体内的药代动力学及组织分布研究 [J]. 江苏农业研究，1999，20 (2)：14-16.

该文发表于《水生生物学学报》【2003，27 (1)：18-22】

盐酸环丙沙星在中华绒螯蟹体内药物代谢动力学研究

PHARMACOKINETICS OF CIPROFLOXACINUM IN CHINESE MITTEN-HANDED CRAB, ERIOCHEIR SINENSIS

杨先乐[1]，刘至治[1]，横山雅仁[2]

(1. 上海水产大学农业部水产增养殖生态、生理重点开放实验室，上海 200090；2. 日本国国际农林水产研究中心，筑波 305-8686)

摘要： 应用反向高效液相色谱法对盐酸环丙沙星在中华绒螯蟹雄、雌蟹血淋巴与肌肉内的代谢规律进行了研究。肌肉注射给药后血淋巴中盐酸环丙沙星的浓度立即达到峰值，药物开始向肌肉等组织中运转，血药浓度在 1 h 内迅速下降，而肌肉中的浓度达到峰值；此后无论是血淋巴还是肌肉中的盐酸环丙沙星浓度均缓慢下降，到第 120 h 雌、雄蟹中盐酸环丙沙星的浓度均在 1 μg·mL^{-1} (g^{-1}) 以下，到 216 h 均不能检出。研究表明：中华绒螯蟹血淋巴中盐酸环丙沙星的药物浓度可用一级吸收二室开放模型描述，雄、雌蟹的消除半衰期 ($T_{1/2\beta}$) 分别为 (40.34± 1.48) h 和 (22.07±0.31) h；中华绒螯蟹的性别对盐酸环丙沙星在其体内的代谢有较大的影响。

关键词： 中华绒螯蟹；盐酸环丙沙星；血淋巴；肌肉；药物代谢动力学

盐酸环丙沙星 (Ciprofloxacinum) 是新一代的氟喹诺酮类合成抗菌药，由于它具有抗菌活性高、抗菌谱广、口服吸收完全、体内分布广泛且不易产生耐药性等优点，已广泛应用于水生动物细菌性疾病的防治[1-3]，对由弧菌和气单胞菌引起的中华绒螯蟹 (Eriocheirsinensis Milne-Edwards) 细菌性疾病也是首选药物之一[4]。由于对该药在中华绒螯蟹体内的代谢规律缺乏研究，在该药的使用上还存在着较大的盲目性，这不仅影响了药物的使用效果，而且也对药物在中华绒螯蟹体内的残留缺乏了解，影响着中华绒螯蟹的品质。因此在建立盐酸环丙沙星对中华绒螯蟹血淋巴内反向高效液相色谱测定法的基础上[5]，对盐酸环丙沙星在中华绒螯蟹血淋巴与肌肉内的代谢状况进行了探讨，为制定合理的给药方法提供依据。

1 材料与方法

1.1 实验动物

由上海前卫特种水产养殖公司一场提供，体重61.5±4.11 g。试验前，在试验水族箱（1.0 m×0.5 m×0.4 m）中驯养一周。试验水温（20.0±0.5）℃。所有试验蟹经抽查血淋巴和肌肉均未检测到盐酸环丙沙星。

1.2 实验药品

甲醇，HPLC级，上海吴径化工厂生产；盐酸环丙沙星标准品（纯度为99.54%），盐酸环丙沙星原粉（纯度为98%），均由上海三维制药公司提供；抗凝剂ACD参照蔡武城等的方法配制[6]。

1.3 高效液相色谱仪及色谱条件

高效液相色谱仪采用岛津（SHIMADZU）LC-6A型系列，色谱条件参照杨先乐等方法进行[5]。检测灵敏度按0.01AUFS计算。

1.4 给药、样品的采集与处理

选择中华绒螯蟹第四步足与体关节膜处肌肉注射给药，给药剂量第一次雄蟹为（8.17±0.56）mg·kg^{-1}，雌蟹为（6.25±0.85）mg·kg^{-1}；第二次雄蟹为（5.99±0.27）mg·kg^{-1}，雌蟹为（6.05±0.32）mg·kg^{-1}。每个时间点平行设立2组，每组5只蟹。血淋巴从蟹背部心区采取，用ACD作抗凝剂分离血浆[5]，此后取体壁肌肉；合并5只蟹的血淋巴和肌肉样品，于-60℃保存备用。检测前，用0.5 mL的血淋巴与1 mL甲醇漩涡混合2 min，再加入0.5 mL甲醇，同法混合2 min，3 000 r/min离心10 min，取上清液1.5 mL于样品瓶中，自动进样器2 μL进样[5]。肌肉样品按1：1的比例，加入双蒸水在研钵中反复研磨10 min，匀浆后取1.0 g于离心管，加入1 mL甲醇，漩涡混合2 min，再加1 mL甲醇，同法混合2 min，4 000 r/min离心10 min，上清液10 000 r/min离心10 min，再经0.45 μm微孔滤膜过滤器过滤后上液相。

1.5 血淋巴内盐酸环丙沙星的代谢规律

根据杨先乐等[5]所建立的方法，测定肌肉注射给药后雄蟹从0.017 h到48 h内的12个时间点、雌蟹从0.17 h到48 h内的9个时间点血淋巴内盐酸环丙沙星的浓度（以下简称血药浓度），以探讨其变化规律，并将所得的各时间点血药浓度数据根据药动学程序软件p56（第二医科大学瑞金医院临床药理研究室提供）的静脉注射给药模型，进行模型嵌合和参数计算。在此基础上为了获得盐酸环丙沙星在血淋巴中的消除规律，再次分别对雌、雄蟹在给药后24、48、74、96、120、144、168、192（仅雌蟹）、216 h后检测其血药浓度。

1.6 肌肉中盐酸环丙沙星的代谢规律

建立检测方法（另文详细报道）并以此分别测定雌、雄蟹给药0.5、1、3（仅雄蟹）、6（仅雌蟹）、12、24、48、72、96、120、144、168 h后肌肉中盐酸环丙沙星的浓度，以确定其在肌肉中的代谢与消除规律。肌肉样品的提取回收率和提取精密度试验均采用加入法。取离心管5支，各加入肌肉匀浆样品1.0 g，然后分别加入浓度为116.855 μg·mL^{-1}的标准盐酸环丙沙星溶液25、50、60、75、85 μL，旋涡混合2 min，其余同肌肉样品处理。回收率（%）=样品实测浓度/样品理论浓度×100%。在另5支离心管中各加入肌肉匀浆样品1.0 g和标准盐酸环丙沙星溶液50 μL，其余同回收率样品处理，并根据各次所测的峰面积计算变异系数（RSD）。日间精密度测定方法为：将浓度为0.011 69、0.116 9、1.168 6、2.337 1、5.842 8 μg·mL^{-1}的标准盐酸环丙沙星溶液各1 mL加入样品瓶，HPLC测定，连续测3次，每隔1 d测一次，计算平均峰面积及RSD。

2 结果

2.1 盐酸环丙沙星在中华绒螯蟹血淋巴内的代谢

第一次二组平行测定的结果表明，肌肉注射给药后，血药物浓度立即达到峰值，此后迅速下降：雄蟹由0.017 h（即1 min）的（37.13±0.18）μg·mL^{-1}降至0.17 h（10 min）的（20.27±0.59）μg·mL^{-1}，1 h为（15.16±0.78）μg·mL^{-1}，到2 h时只达（11.66±0.09）μg·mL^{-1}，仅为0.17 h的57.5%，此后下降速度减缓，到48 h为（5.96±0.18）μg·mL^{-1}，

是 0. 17 h 时 29.4%（图 1）；雌蟹的血药浓度下降的速度比雄蟹稍快，0. 17 h（10 min）为（16.47±0.29）μg·mL⁻¹，2 h 降（8.18±0.02）μg·mL⁻¹，仅是前者的 48.8%，此后与雄蟹一样下降速度减缓，但比雄蟹稍快，48 h 仅为（1.98±0.46）μg·mL⁻¹，为 0.17 h 的 12.0%（图 2）。第二次二组的平行测定的结果进一步说明，从 24 h 至 120 h，盐酸环丙沙星血药浓度缓慢、波浪式下降，但雌蟹仍较雄蟹稍快；由图 1 和图 2 可以看出雄蟹的血药浓度由 24 h 的（3.03±0.37）μg·mL⁻¹ 降至 48 h 的（1.43±0.04）μg·mL⁻¹，120 h 的（0.55±0.41）μg·mL⁻¹，分别只下降了 52.9% 与 81.9%，而雌蟹则由 24 h 的（2.40±0.20）μg·mL⁻¹ 降至 48 h 的（1.06±0.17）μg·mL⁻¹，120 h 的（0.37±0.19）μg·mL⁻¹，却分别下降了 55.0% 与 84.0%。第 144 h 雄、雌蟹血药浓度分别仅有（0.26±0.04）μg·mL⁻¹、（0.67±0.61）μg·mL⁻¹，第 168 h 血药浓度稍有回升，分别为（0.47±0.04）μg·mL⁻¹ 与（0.67±0.61）μg·mL⁻¹，但紧接着进一步下降，到 216 h 已不能检出（<0.01 μg·mL⁻¹）。对第一次测定的数据通过药动学软件分别进行开放型单室、二室及三室模型嵌合，结果表明，肌肉注射给药后中华绒螯蟹的药动学过程较符合一级吸收二室开放模型（$C = Ae^{-\alpha t} + Be^{-\beta t}$），它们的血药浓度–时间过程分别符合下列方程：

$$C = 11.74e^{-2.17t} + 7.67e^{-0.031t} （雌） 和 C = 28.75e^{-8.33t} + 13.39e^{-0.017t} （雄）$$

经检验理论值与实测值相关系数分别为 $r = 0.999\ 2$ 和 $r = 0.996\ 7$，曲线拟合良好。其药动学参数如表 1 所示。

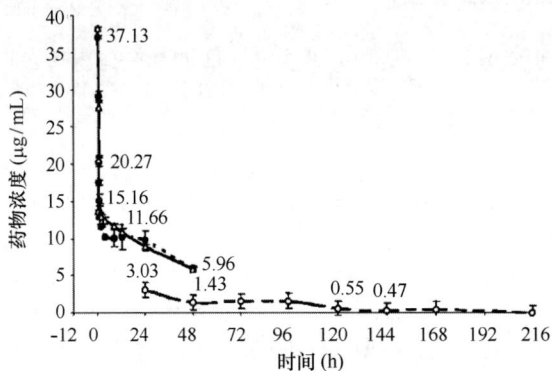

图 1　盐酸环丙沙星在中华绒螯蟹雄蟹血淋巴内的浓度变化

Fig. 1　The concentration of Ciprofloxacinum in the hemolymph of maleE. sinensis

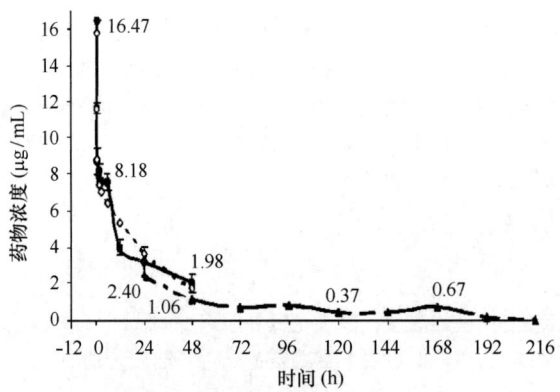

图 2　盐酸环丙沙星在中华绒螯蟹雌蟹血淋巴内的浓度变化

Fig. 2　The concentration of Ciprofloxacinum in the hemolymph offemale E. sinensis

表 1　盐酸环丙沙星在中华绒螯蟹体内的药动学参数

Tab. 1　Pharmacokinetic parameters of Ciprofloxacinum in E. sinensis

	药动学参数	Parameter	雄蟹 Male	雌蟹 Female
Dose	给药剂量	（mg·kg⁻¹）	8.17±0.56	6.25±0.85
A	分布相的零时截距	（μg·mL⁻¹）	28.75±0.15	11.74±0.84
B	消除相的零时截距	（μg·mL⁻¹）	13.39±0.36	7.67±0.29
$T_{1/2\alpha}$	药物的分布相半衰期	（h）	0.084±0.008	0.39±0.23

	药动学参数	Parameter	雄蟹 Male	雌蟹 Female
$T_{1/2\beta}$	药物的消除相半衰期	（h）	40.34±1.48	22.07±0.31
K_{10}	药物自中央室的消除速率常数	（h^{-1}）	0.054±0.003	0.080±0.003
K_{12}	药物自中央室到周边室的一级转运速率常数	（h^{-1}）	5.65±0.47	1.27±0.81
K_{21}	药物自周边室中央室到中央室的一级转运速率常数	（h^{-1}）	2.64±0.30	0.85±0.50
V_d	表观分布容积	（L/kg）	0.015±0.005	0.080±0.002
AUC	血药浓度–时间曲线下面积	（μg·h·mL^{-1}）	783.24±49.40	250.58±2.46

2.2　盐酸环丙沙星在肌肉组织中的代谢

建立了盐酸环丙沙星在肌肉组织中的检测方法，结果表明，在试验所设计的色谱条件下，HPLC 基线走动平稳，盐酸环丙沙星的平均保留时间为 （8.26±0.10） min，虽有杂峰出现，但无干扰峰，与药峰分离良好；标准盐酸环丙沙星浓度在 0.01~5.0 μg·mL^{-1}的范围内，标准曲线 $y = 1.646\,18×10^{-5}x$ （y-盐酸环丙沙星浓度，x-峰面积）的相关系数为 $r = 0.999\,8$，相关性良好；检测灵敏度为 0.01 μg·mL^{-1}，平均回收率为 80.37%±16.36%，提取精密度为 1.84%，平均日间精密度为 4.98%±5.18%。

用此方法对肌肉组织中盐酸环丙沙星的代谢规律进行了探讨，结果揭示肌肉中药峰出现在给药后 1 h 前，雌蟹的药峰时间要比雄蟹稍快，雄蟹 1 h 后的药物浓度达 （6.21±0.14） μg·mL^{-1}，是 0.5 h 113%，而雌蟹为 （4.1±0.62） μg·mL^{-1}，仅为 0.5 h 的 84%。此后开始下降，到 48 h 时雄蟹仅为 （1.27±0.11） μg·mL^{-1}，雌蟹为 （1.24±1.03） μg·mL^{-1}，分别比 1 h 降低了 79.0%和 69.0%；此后下降速度进一步加快，72、96、120、144 h 雌、雄蟹分别为 （0.97±0.87）、（0.17±0.11）、（0.05±0.07）、（0.11±0.01） μg·mL^{-1}和 （0.49±0.09）、（0.31±0.07）、（0.18±0.02）、（0.10±0.05） μg·mL^{-1}，到 168 h 时，雌、雄蟹肌肉中盐酸环丙沙星含量小于 0.01 μg·mL^{-1}，不能检出 （图 3 和图 4）。

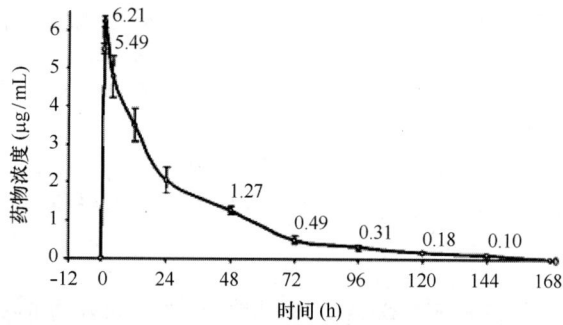

图 3　盐酸环丙沙星在中华绒螯蟹雄蟹肌肉组织中的浓度变化

Fig. 3　The concentration of Ciprofloxacinum in the muscle of male Eriocheirsinensis

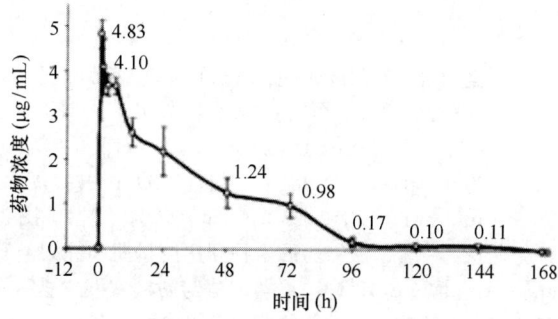

图 4　盐酸环丙沙星在中华绒螯蟹雌蟹肌肉组织中的浓度变化

Fig. 4　The concentration of Ciprofloxacinum in the musclue of female Eriocheirsinensis

3 讨论

3.1 盐酸环丙沙星在中华绒螯蟹体内代谢的特点

注射给药后，中华绒螯蟹血药浓度在短时间（1 min）内就达到峰值，这是因为中华绒螯蟹属甲壳动物，循环系统为开放式，注射给药后药物直接进入血液循环，较相似于哺乳动物静脉给药。此后血药浓度迅速下降，尤其是在 1 h 内，这说明血淋巴内的药物迅速向体内的其他组织分布或消除，1 h 后下降速度明显减缓。出现这种现象，也许是因为药物对中华绒螯蟹来说是一种外来的刺激物，进入血淋巴循环后，由于机体的保护性机能，通过触角腺和鳃等排泄器官将其排出体外或通过肝胰脏等器官将其降解，以减轻对机体的刺激，所以在第 1 h 内血药浓度下降较快；随着时间的推移，一方面机体对其产生了一定适应性，另一方面药物处于分布与消除过程，消除与降解速度减慢，因而血药浓度的下降速度就减缓。

消除相半衰期（$T_{1/2\beta}$），反应了药物在机体内的消除速度，是药动学的一个重要参数。从本试验来看，无论是雌蟹还是雄蟹，盐酸环丙沙星在其体内半衰期均较长，与其他水生动物相比，如盐酸环丙沙星在鲤、非洲鲇和虹鳟体内的 $T_{1/2\beta}$ 分别为（14.5±0.5）h、（14.2±1.3）h 和（11.2±1.6）h[7]，诺氟沙星在鲤和中华鳖体内的 $T_{1/2\beta}$ 分别为（3.403 2±0.587 3）h[8] 和 4.24 h，嘧啶酸和恩诺沙星在虹鳟体内的 $T_{1/2\beta}$ 分别为 13 h 和 2.52 h，均比中华绒螯蟹稍短。对于这一现象，作者认为中华绒螯蟹是较低等的水生动物，它们主要通过触角腺和肝胰腺对药物进行排泄或降解，其对药物的降解与排除的能力均较其他水生动物差，而鱼类等则可通过肾脏和呼吸器官（如鳃、鳃上腺等）等进行扩散消除，哺乳动物还可通过肾脏的主动运转予以消除（如猪、牛等对环丙沙星的 $T_{1/2\beta}$ 仅为 2.5 h），因而中华绒螯蟹的 $T_{1/2\beta}$ 就比其他鱼类和哺乳动物要长。Jame 和 Barron 通过心包膜注射给药研究磺胺间二甲氧嘧啶（SDM）在美洲螯虾体内的分布规律时也有同样的发现，其 $T_{1/2\beta}$ 达 77 h，比哺乳类要长得多。

盐酸环丙沙星在中华绒螯蟹血淋巴中的代谢与性别有较大的关系，其运转速度雌蟹要比雄蟹稍快。这主要表现在以下两方面：①消除相半衰期（$T_{1/2\beta}$）。雌、雄蟹差别明显，雄蟹差不多是雌蟹的二倍。这一现象与 Li 等对 2-萘酚在美洲螯虾体内代谢的研究有些相似，他们发现 $T_{1/2\beta}$ 雌虾远小于雄虾，分别为（7.70±4.19）h 和（41.63±21.35）h，二者差异显著（$P<0.05$）[11]；②药时曲线下面积（AUC）。虽然给药剂量雌、雄蟹稍有差别，但二者的药时曲线下面积却差别显著，雄蟹是雌蟹的 3.1 倍。Li 等对 9-羟基一吡喃酮在美洲螯虾体内的药动学研究也获得了相似的结果。药时曲线下面积是反应药物进入循环药量的多少，试验结果说明，肌肉注射给药后，盐酸环丙沙星进入循环系统的量雄蟹要比雌蟹多，这也就导致了雄蟹的 $T_{1/2\beta}$ 长于雌蟹。

3.2 盐酸环丙沙星在中华绒螯蟹血淋巴与肌肉组织中的消除

表观分布容积（V_d）大的药物与组织蛋白结合多，主要分布于细胞内液与组织间液，V_d 小的药物与血浆蛋白结合小，较集中于血浆。从雌、雄蟹的表观分布容积（V_d）非常小可以看出，盐酸环丙沙星进入中华绒螯蟹体内后，主要与血浆蛋白相结合，因此它较多地集中于血淋巴。试验证明，注射给药 1 h 内血淋巴中的药物向肌肉等组织中转移，但总的看来，肌肉的药物浓度要比血淋巴低，而且下降速度也较快，这说明肌肉组织不是药物在中华绒螯蟹体内的主要富集组织（器官）。120 h 以后，无论是血淋巴还是肌肉组织，无论是雌蟹还是雄蟹，盐酸环丙沙星的浓度均在 1 μg·mL⁻¹（g⁻¹）以下，216 h 后均不能检出。由此，如果仅从血淋巴与肌肉中的残留考虑，当水温在（20.0±0.5）℃下中华绒螯蟹对盐酸环丙沙星的休药期可定为 9 d。当然，药物还有可能在蟹的其他可食部分（如性腺和肝胰腺等）残留，休药期还应根据这些组织药物中的消除状况最后确定，对此将进行深入研究。

3.3 盐酸环丙沙星的最适给药方法

最适给药方法就是确定药物的给药剂量和给药的间隔时间，一般由最低抑菌浓度（MIC）和血药浓度来决定。以往研究表明，盐酸环丙沙星对导致水生动物致病的革兰氏阴性菌的 MIC 都较低，如对河流弧菌－Ⅱ（Vibrio fluvialis－Ⅱ）为 0.32 μg·mL⁻¹，对气单胞菌（Aeromonas spp.）小于 0.01/0.02 μg·mL⁻¹，对假单胞菌（Pseudomonas spp.）为 0.32 μg·mL⁻¹ 等。根据试验，在注射给药剂量分别为（6.05±0.32）mg·kg⁻¹、（5.99±0.27）mg·kg⁻¹ 时，雌、雄蟹经 5 d 后血药浓度分别为（0.37±0.19）μg·mL⁻¹、0.55 μg·mL⁻¹，均在 0.3 μg·mL⁻¹ 以上（图1，图2），还远高于盐酸环丙沙星对水生动物大部分致病菌的最低抑菌浓度。由此认为，若口服给药的生物利用度是肌肉注射给药 30%（实践经验），那么给药剂量则以 20 mg·kg⁻¹ 为宜，给药的间隔期不应超过 2 d，这一给药方法与防病实践中盐环丙沙星的常用的方法基本相仿[1]。当然盐酸环丙沙星对中华绒螯蟹的口服给药与肌肉注射给药的药动学规律可能有一定的区别，对此将进行更深入的比较与研究。

参考文献

[1] Yang J. Fisheries medicine manual [M]. Beijing: Science and Technology Press of China. 1998, 179—180. [杨坚主编, 渔药手册. 北京：

中国科学技术出版社.1998,179—180].

[2] Li T W, Ding M J, Song X M, et al. Preliminary studies on the mechanism of Vibrio fluvialis-Ⅱ resistance to antibiotics. Oceanologia ET Lim-noligiaSinica, 1996, 27（6）：637—645 [李太武, 丁明进, 宋协民, 等. 皱纹盘鲍脓胞病病原菌———河流弧菌-Ⅱ的抗药机制的初步研究. 海洋与湖沼, 1996, 27（6）：637—645].

[3] Lin T R, Yu N F. The synthetical treatment for Trionyxsinensis with coexistence symptom of acute gastroenteritis and branchial inflammation [J]. Aquaculture of interior Land. 1997, 7：25. [林天然, 余乃凤. 甲鱼急性肠胃炎鳃腺炎并发症的综合治疗. 内陆水产, 1997, 7：25].

[4] Meng Q X, Yu K K. The diagnosis, prevention and cure for the disease of fish, shrimp, crab and shellfish [M]. Bingjing：Agriculture Press of China. 1996, 255—281. [孟庆显, 余开康. 鱼虾蟹贝疾病诊治和防治. 北京：中国农业出版社.1996, 255—281].

[5] Yang X L, Liu Z Z, Sun W Q, et al. The method of RP-HPLC of ciprofloxacinum in hemolymph of Eriocheirsin enses [J]. Journal of Fisher-ies of China. 2000, 25（4）：348—355. [杨先乐, 刘至治, 孙文钦, 等. 中华绒螯蟹血淋巴内盐酸环丙沙星的反相高效液相色谱法的建立. 水产学报, 2001, 25（4）：348—355].

[6] Cai W C, Li B Y, Li Y M. The experimental technique tutorial of Bio-chemical [M]. Shanghai：Fudan University Press. 1983：47. [蔡武城, 李碧羽, 李玉民. 生化实验技术教程 [M]. 上海：复旦大学出版社.1983：47].

[7] Nouws J F M, Grondel J L, Schutte A R, et al., Pharmacokinetics ofciprofloxacin in carp, African catfish and rainbow. trout [J]. Vet-Quart, 1988, 10：2—216.

[8] Zhang Z X, Zhang Y B, Yang R S. The pharmacokinetic of Nor-floxacin studies on common crap [J]. Journal of veterinary of China. 2000, 20（1）：66—69. [张作新, 张雅斌, 杨永胜. 诺氟沙星在鲤鱼体内的药代动力学. 中国兽医学报, 2000, 20（1）：66—69].

该文发表于《集美大学学报》【2009, 14（3）：8-13】

不同给药方式下恩诺沙星在鲤体内的药动学研究

Pharmacokinetic of Enrofloxacin in Carp Following Different Ways of Administration

李春雨[1], 李继昌[1], 卢彤岩[2], 王荻[2], 侯晓亮[3]

（1. 东北农业大学动物医学院, 黑龙江 哈尔滨 150030; 2. 中国水产科学研究院黑龙江水产研究所, 黑龙江 哈尔滨 150070;
3. 黑龙江民族职业学院, 黑龙江 哈尔滨 150081）

摘要： 对两组鲤分别进行腹腔注射、口灌恩诺沙星后, 应用高效液相色谱法测定了其组织中的药物浓度, 研究了恩诺沙星在鲤 (*Cyprinus carpio*) 体内的吸收、分布与消除等药代动力学参数。结果表明：两种给药方式下, 鲤血浆、肝脏、肾脏和肌肉组织的药时曲线符合一级消除二室模型。腹腔注射给药血浆动力学参数：AUC 为 59.185 6 $\mu g \cdot h \cdot mL^{-1}$、Ka 为 75.762 7 h^{-1}、$t_{1/2\beta}$ 为 96.545 6 h、$T_{(peak)}$ 为 0.073 0 h、$C_{(max)}$ 为 3.297 0 $\mu g \cdot mL^{-1}$; 灌服给药血浆动力学参数：AUC 为 600.296 1 $\mu g \cdot h \cdot mL^{-1}$、Ka 为 0.169 3 h^{-1}、$t_{1/2\beta}$ 为 168.287 1 h、$T_{(peak)}$ 为 3.665 5 h、$C_{(max)}$ 为 3.266 1 $\mu g \cdot mL^{-1}$。这说明腹腔注射给药比口灌给药吸收快, 血药达峰时间短, 达峰浓度高。

关键词： 恩诺沙星; 鲤; 药代动力学; 高效液相色谱

该文发表于《广东海洋大学学报》【2012, 32（4）：28-33】

恩诺沙星在红笛鲷体内的药代动力学

Pharmacokinetics of Enrofloxacin in Lutjanussan guineus

汤菊芬, 简纪常, 鲁义善, 黄郁葱

（广东海洋大学水产学院, 广东省水产经济动物病原生物学及流行病学重点实验室, 广东 湛江 524088）

摘要： 在水温（26±2）℃、盐度 28 条件下, 恩诺沙星（enrofloxacin, EF）按 5 mg/kg 的剂量腹腔注射和口灌红笛鲷（*Lutjanussanguineus*）, 用反相高效液相色谱法研究恩诺沙星在红笛鲷体内的药代动力学和组织分布。结果显示：注射和口灌后红笛鲷血浆的药时数据均符合一级吸收二室模型; 血药达峰时间（T_p）分别为 0.75 和 1.5 h,

血药质量浓度峰值（C_{max}）分别为5.22和3.94 μg/mL，药时曲线下面积（AUC）分别36.98和17.24 mg/（L·h），消除半衰期（$T_{1/2\beta}$）分别为13.28和9.18 h；恩诺沙星在组织中分布较广，注射和灌服给药肌肉、肝脏和肾脏的C_{max}分别为3.53和3.40 μg/g、4.08和3.98 μg/g、11.23和5.40 μg/g，T_p分别为1.5和2.0 h、1.0和2.0 h、0.75和1.5 h，AUC分别33.17和23.12 mg/（L·h）、39.36和18.20 mg/（L·h）、98.97和56.69 mg/（L·h）；EF在红笛鲷体内消除速度较慢，注射和灌服给药肌肉、肝脏和肾脏的$T_{1/2\beta}$分别为22.08和83.32 h、41.52和94.47 h、20.94和21.81 h。表明恩诺沙星注射给药在红笛鲷体内的吸收和消除均快于口灌给药。

关键词：红笛鲷；恩诺沙星；药物代谢动力学

该文发表于《中国兽医学报》【2005, 25（2）：195-197】

恩诺沙星在眼斑拟石首鱼体内的药物代谢动力学

Pharmacokinetics of Enrofloxacin in *Sciaenop socellatus*

简纪常，吴灶和，陈刚

（湛江海洋大学 水产学院，广东 湛江 524025）

摘要：应用反相高效液相色谱法（RP-HPLC）研究了恩诺沙星在眼斑拟石首鱼（*Sciaenop socellatus*）体内的药物代谢动力学。实验数据经DAS药代动力学分析软件分析后得出：腹腔注射组血浆的药时数据符合一级吸收二室模型，动力学方程为：$C = 4.925e^{-1.452t} + 2.730e^{-0.075t}$，其主要药代动力学参数：AUC 37.533 mg/（L·h）、C_{max} 4.747 mg/L、T_{max} 0.750 h、$T_{1/2\alpha}$ 0.477 h、$T_{1/2\beta}$ 9.292 h、$V_{d/F}$ 1.676 L/kg、$T_{1/2Ka}$ 0.170 h、K_a 4.081/h、K 6.714/h、K_{10} 0.191/h、K_{12} 0.769/h、K_{21} 0.566/h；灌服组血浆的药时数据符合一级吸收一室模型，动力学方程为：$C=6.482（e^{-7.092t}-e^{-10.356t}）$，其主要药代动力学参数：AUC 15.805（mg/L.h）、C_{max} 2.770 mg/L、T_{max} 1.500 h、$T_{1/2\alpha}$ 4.989 h、$V_{d/F}$ 2.072 L/kg、Ka 7.092/h、K 10.356/h。结果表明，腹腔注射给药比灌服给药吸收快，血药浓度达峰时间短于灌服给药，但血药浓度峰值显著高于灌服给药（$P<0.05$），血浆中恩诺沙星的回收率为94.579%。

关键词：眼斑拟石首鱼；恩诺沙星；药物代谢动力学

该文发表于《黑龙江畜牧兽医》【2012，（2）：129-132】

两种水温条件下恩诺沙星在鲫鱼体内的药动学比较

Pharmacokinetics of enrofloxacin in crucian carp at 20℃ and 25℃ temperature

汪文选[1,2]，卢彤岩[2]，王荻[2]，陈俭清[1,2]，刘红柏[2]

（1. 东北农业大学动物科技学院，黑龙江 哈尔滨 150030；2. 中国水产科学研究院黑龙江水产研究所，黑龙江 哈尔滨 150070）

摘要：为了研究水温对鲫鱼体内恩诺沙星药动学规律的影响，试验在20℃和25℃水温条件下将恩诺沙星灌入鲫鱼的前肠，于给药后0.25、0.5、0.75、1、1.5、2、4、6、8、12、24、36、48、72 h分别采集血浆、肾脏和肝脏样品，以高效液相色谱法（HPLC）测定药物浓度，用药动学软件3P97处理药时数据。结果表明：在鲫鱼血浆中，两种水温条件下恩诺沙星的达峰时间和分布半衰期基本相同；20℃药时曲线下面积、消除半衰期明显大于25℃，而清除率明显小于25℃。在肾脏和肝脏中，20℃药物的达峰时间、分布半衰期、消除半衰期、药时曲线下面积均明显大于25℃，而清除率明显小于25℃。

关键词：恩诺沙星；鲫鱼；药动学；水温

该文发表于《动物学杂志》【2007，42（5）：62-69】

恩诺沙星在罗氏沼虾体内的药物代谢动力学

Pharmacokinetics of Enrofloxacin in *Macrobrachium rosenbergii*

钱云云，唐俊，郑宗林，杨先乐

（农业部渔业动植物病原库上海水产大学 上海 200090）

摘要： 应用反相高效液相色谱法（RP-HPLC）研究了恩诺沙星在罗氏沼虾（*Macrobra chiumrosenbergii*）体内的药物代谢动力学。实验结果表明，恩诺沙星在血淋巴、肝胰腺和肌肉中的平均回收率分别为（86.54±2.39）%、（85.43±2.75）%和（95.01±1.99）%，其代谢产物环丙沙星在血淋巴、肝胰腺和肌肉中的平均回收率分别为（94.34±8.30）%、（75.17±5.42）%和（80.42±1.67）%；恩诺沙星及其代谢产物环丙沙星在三种组织中的平均日内精密度分别为（3.39±0.53）%和（3.92±1.24）%，而日间精密度分别为（5.11±1.73）%和（5.28±2.10）%。恩诺沙星和环丙沙星的最低检测限分别为 0.02 μg/mL 和 0.01 μg/mL。罗氏沼虾以 10 mg/kg 虾体重剂量单次肌肉注射给药后，血液中药物浓度即刻达到峰值，并迅速向组织中分布。实验数据经 MCPKP 药动学软件分析，恩诺沙星在血淋巴中的主要药物代谢动力学参数为：$T_{1/2\alpha}$ 为 0.581 h、$T_{1/2\beta}$ 为 69.315 h、$V_{d/F}$ 为 7.230 L/kg、CL/F 为 0.035 L/h·kg、K_{12} 为 0.01 h、K_{21} 为 0.005 h、AUC 为 291.898 μg/mL·h、T_{max} 为 0.083 h、C_{max} 为 6.293 μg/mL；恩诺沙星在肝胰腺和肌肉组织中主要药动学参数：$T_{1/2\alpha}$ 为 1.941 h、0.000 h；$T_{1/2\beta}$ 为 70.732 h、59.456 h；AUC 为 308.07 μg/mL·h、217.039 μg/mL·h。三种组织中均能检测到恩诺沙星的活性代谢产物环丙沙星，但含量均处于较低水平，药物浓度-时间数据经 MCPKP 药动学软件处理后，不能用开放性一室模型或二室模型拟合。

关键词： 恩诺沙星；罗氏沼虾；药物代谢动力学

该文发表于《海洋渔业》【2011，33（1）：74-82】

恩诺沙星及其代谢产物在中华绒螯蟹血淋巴中的比较药代动力学

On comparative pharmacokinetic of Enrofloxacinin Hemolymph of Chinese mitten-handed crab，Eriocheir sinensis

郑宗林[1]，叶金明[2]，李代金[1]，杨先乐[3]，胡鲲[3]，唐俊[3]

（1. 西南大学水产系，重庆 402460；2. 江苏省扬州市水产生产技术指导站，扬州 225000；

3. 农业部渔业动植物病原库，上海海洋大学，上海 201306）

摘要： 研究了不同水温（16℃、25℃）、不同给药剂量（10 mg/kg、20 mg/kg）和不同给药方式（肌注、口灌）等条件下，恩诺沙星及其代谢产物环丙沙星在中华绒螯蟹体内的药代动力学，比较了不同条件下药物在蟹血淋巴中的吸收、分布和代谢的差异。结果表明，恩诺沙星在蟹血淋巴中的药-时数据符合开放式二室模型。16℃时以 10 mg/kg 剂量肌注给药后，恩诺沙星在蟹血淋巴的主要药代动力学参数为：AUC 96.818 mg/（L·h），C_{max} 6.54 μg/mL，$T_{1/2\alpha}$ 0.851 h，$T_{1/2\beta}$ 95.415 h；25℃时以 10 mg/kg 剂量肌注给药后，恩诺沙星的主要药代动力学参数为：AUC168.457 mg/（L·h），C_{max}7.12 μg/mL，$T_{1/2\alpha}$ 0.58 h，$T_{1/2\beta}$ 88.833 h；25℃时以 20 mg/kg 剂量肌注给药后，恩诺沙星的主要药代动力学参数为：AUC 155.612 mg/（L·h），C_{max}11.045 μg/mL，$T_{1/2\alpha}$5.239 h，$T_{1/2\beta}$ 88.378 h；25℃时以 10 mg/kg 剂量口灌给药后，恩诺沙星的主要药代动力学参数为：AUC 86.525 mg/（L·h），C_{max}3.469 μg/mL，$T_{1/2\alpha}$8.071 h，$T_{1/2\beta}$ 1.842 h。不同条件下，恩诺沙星在蟹血淋巴中的主要药代动力学参数差异较大。恩诺沙星的活性代谢产物环丙沙星在各种给药条件下的蟹血淋巴中均能检出，但含量均处于较低值，且药-时数据不能用房室模型拟合，表明恩诺沙星在蟹体内只有极少部分代谢为环丙沙星。

关键词： 恩诺沙星；环丙沙星；中华绒螯蟹；高效液相色谱；药代动力学

恩诺沙星及其代谢产物环丙沙星在青鱼体内的药代动力学和残留研究

Pharmacokinetics and residues of Enrofloxacin and its metabolite ciprofloxacin in Mylopharyn godon piceus

何平[1,2]，尹文林[2]，沈锦玉[2]

（1. 宁波大学，浙江 宁波 315000；2. 浙江省淡水水产研究所，浙江 湖州 313001）

摘要： 应用反相高效液相色谱法（RP-HPLC）研究了恩诺沙星及其代谢产物环丙沙星在青鱼（*Mylopharyngodon piceus*）体内的药物代谢动力学和残留。在（26±1）℃水温条件下，青鱼以 20 mg/kg 鱼体重混饲口灌后，恩诺沙星及其代谢产物环丙沙星的药代动力学模型为二室开放模型，恩诺沙星血液分布相半衰期 $T_{1/2\alpha}$ 为 13.588 h，消除相半衰期 $T_{1/2\beta}$ 为 88.384 h，吸收相半衰期 $T_{1/2Ka}$ 为 0.548 h，达峰时间 T_{Peak} 为 2.87 h，峰浓度 C_{max} 为 4.36 μg/mL，曲下面积 245.402 μg/mL·h，清除率 0.004 mg/kg·h。而环丙沙星血液分布相半衰期（$T_{1/2\alpha}$）为 4.106 h，消除相半衰期（$T_{1/2\beta}$）为 61.801 h，吸收相半衰期（$T_{1/2Ka}$）为 1.387 h，达峰时间 T_{Peak} 为 4.34 h，峰浓度 C_{max} 为 1.24 μg/mL，曲下面积为 75.223 μg/mL·h，清除率为 0.013 mg/kg·h。结果表明恩诺沙星在组织中分布迅速，恩诺沙星和环丙沙星在青鱼体内的消除半衰期较长，残留时间也较长。我国规定恩诺沙星的最大残留限量为恩诺沙星+环丙沙星≤100 μg，在本实验条件下，建议休药期为 10 d。

关键词： 恩诺沙星；青鱼；药代动力学；残留

二、诺氟沙星

诺氟沙星（Norfloxacin）又称氟哌酸，化学名称是（l-乙基-6-氟-4-氧代-1，4-二氢-7-（1-哌嗪基）-3-喹啉羧酸），它是第一个上市的第三代含氟喹诺酮类抗菌药物，在 C_6 位上引入氟原子。

诺氟沙星在罗氏沼虾（左梦丽等，2014）、青虾（何平等，2008）、斑节对虾（房文红等，2003）、凡纳滨对虾（房文红等，2004）等多个水产动物中的药代谢动力学特征表明渔药具有明显的种属差异性，烟酸诺氟沙星在鲤和鲫鱼体内也都呈现出了显著的差异性（Ning Xu et al.，2015）。大黄鱼口服诺氟沙星后，诺氟沙星在肝胰组织中药物浓度最高，肾脏组织中消除速度最快，肌肉中消除速度最慢（刘玉林等，2007）。随着给药剂量的增加，诺氟沙星的吸收和消除速率均加快（韩冰等，2015）。我国已经禁止在水产动物疾病防治上使用诺氟沙星。

该文发表于《JOURNAL OF VETERINARY PHARMACOLOGY AND THERAPEUTICS》【2015，38（3）：309-312】

Comparative pharmacokinetics of norfloxacin nicotinate in common carp（Cyprinus carpio）and crucian carp（Carassius auratus）after oral administration

Ning Xu[*†]，Xiaohui Ai[*†]，Yongtao Liu[*†]，Qiuhong Yang[*]

（ * Freshwater Fish Germplasm Quality Supervision and Testing Center, Ministry of Agriculture, Yangtze River Fisheries Research Institute, Chinese Academy of Fishery Sciences, Wuhan, China；† Hu Bei Freshwater Aquaculture Collaborative Innovation Center, Wuhan, China）

Abstract： Comparative pharmacokinetics of norfloxacin nicotinate（NFXNT）was investigated in common carp（Cyprinus carpio）and crucian carp（Carassius auratus）after a single oral administration of 10 mg/kg body weight（b.w.）. Analyses of plasma samples were performed usingultra-performance liquid chromatography（UPLC）with fluorescence detection. After oral dose，plasma concentration time curves of common carp and crucian carp were best described by a two-compartment open model with first-order absorption. The pharmacokinetic parameters of common carp were similar to those of crucian carp. The distribution half-life ($T_{1/2\alpha}$), elimination half-life ($T_{1/2\beta}$), peak concentration (C_{max}), time to peak concentration (T_{max}), and area under the concentration-time curve (AUC) of common carp were 1.58 h, 26.33 h, 6 069.79 μg/L, 1.08 h, and 103 072.36 h·μg/L, respectively, and those corresponding to crucian carp were 1.36 h, 26.55 h, 9 586.06 μg/L, 0.84 h, and 126 604.4 h·μg/L, respectively. These studies demonstrated that 10mg NFXNT/kg body weightin common carp and crucian carp following oraldosepresented good pharmacokinetic characteristics.

Key words: omparative pharmacokinetics, Norfloxacin nicotinate, Cyprinus carpio, Carassius auratus

口服烟酸诺氟沙星在鲤鱼和鲫鱼体内比较药动学研究

胥宁*†，艾晓辉*†，刘永涛*†，杨秋红*

（＊中国水产科学研究院长江水产研究所 中国 武汉 430223；＊农业部淡水鱼类种质资源监督检测中心 中国 武汉 430223；
†淡水水产健康养殖湖北省协同创新中心 湖北 武汉 430223）

摘要：本文研究了经单次口灌 10 mg/kg（b.w.）烟酸诺氟沙星在鲤鱼和鲫鱼体内比较药动学。样品采用 UPLC 荧光检测器进行检测。经口服之后，诺氟沙星在鲤鱼和鲫鱼血液中浓度时间曲线都能被一级吸收二房室模型进行拟合，二者参数较为相似。在鲤鱼体内分布半衰期、消除半衰期、峰浓度、达峰时间和药时曲线面积分别为 1.58 h、26.33 h、6 069.79 μg/L、1.08 h 和 103 072.36 h·μg/L。相对应在鲫鱼体内分别为 1.36 h、26.55 h、9586.06 μg/L、0.84 h 和 126 604.4 h.μg/L。这些数据表明经口服 10 mg/kg（b.w.）烟酸诺氟沙星在鲤鱼和鲫鱼体内都呈现出了良好的药动学特性。

关键词：比较药动学，烟酸诺氟沙星，鲤鱼，鲫鱼

Norfloxacin (NFX) is one of the first modern fluoroquinlolones, which has been widely used in human and veterinary medicine due to its broad spectrum antimicrobial properties, good absorption, wide distribution, and no cross-resistance with other antimicrobials (Vancutsem et al., 1990). Due to these advantages, NFX is approved for use in freshwater aquaculture in China. However, NFX is slightly soluble in ethanol or water resulting in its low bioavailability. In order to change its water solubility and increase its bioavailability, norfloxacin nicotinate (NFXNT, an adduct of NFX and nicotinic acid) has been developed and used commercially.

The pharmacokinetics of NFX has been studied in humans (Helmy, 2013), dogs (Brown, et al., 1990), laboratory animals (Gilfillan, et al., 1984) and other species (Anadon, et al., 1992; Anadon, et al., 1995; Gips & Soback, 1996; Gonzalez, et al., 2001). However, the pharmacokinetic study of NFXNT was not conducted in common carp (Cyprinus carpio) and crucian carp (Carassius auratus). Therefore, the present study investigates the comparative pharmacokinetics of NFXNT in the two species of fish for designing optimal regimen and residual control.

Analytical standard NFX (purity grade 99.5%) was purchased from Dr. Ehrenstorfer GmbH., Augsburg, Germany. The NFXNT powder (purity grade 98.0%) used for treatment was supplied by Zhongbo aquaculture Biotechnology Co. Ltd., Wuhan, China. NFXNT is a preparation of NFX, which is an adduct of nicotinic acid and NFX. After entering the blood circulation of animals, NFXNT becomes NFX and nicotinic acid by digestion and absorption in the gastrointestinal tract. Therefore, the target compound of detection was NFX in the blood and tissues of animals. In addition, a pharmacokinetic paper of NFXNT and NFX-glycine acetate (another preparation of NFX) had been published (Park et al., 1998) whose compound of determination was also NFX. In this study, NFXconcentrations in samples were analyzed by a Waters ACQUITY™ ultra-performance liquid chromatography (UPLC) with a fluorescence detector. The excitation was at 278 nm, and emission was at 450 nm. An ACQUITY UPLC BEH C18 column (1.7 μm, 2.1×100 mm) was used for separation. The mobile phase was tetrabutylammonium bromide buffer (pH 3.1) andacetonitrile (95:5, v/v), 0.3 mL/min flux, at 30℃. 0.5 mL plasma sample was precipitated protein by same volume of 6% three chloroacetic acid. It was vortexed and centrifuged at 8 000 r/min for 10 min. The supernatant was filtered through a 0.22 μm disposable syringe filter for UPLC analysis.

The UPLC analysis of NFX was rapidly performed with high reproducibility. The limit of quantification for NFX was 10 μg/L. A good linearity of curves over range of 10 and 20 000 μg/L was exhibited by the coefficient of correlation $R^2 = 0.997$. The mean±SD recovery rates of NFX in common carp plasma and crucian carp plasma were 82.02±4.42% and 83.14±3.21%, respectively. The percentage relative standard deviations for inter-day and intra-day precision of common carp plasma were 2.77±0.99% and 3.92±0.72%, respectively, and those of crucian carp plasma were 3.02±0.75% and 3.59±0.69%, respectively.

A total of 66 common carp (300±10g) or 66 crucian carp (300±10g) were randomly assigned to groups of 6 fishfor each tank

（100 cm×60 cm×80 cm, length×width×depth）receiving flowing well water（26 L/min）. The water quality parameters were measured once daily in each tankand maintained at the following values: total ammonia nitrogen levels at ≤0.75 mg/L; nitrite nitrogen concentrations at less than 0.04 mg/L; dissolved oxygen levels at 6.3–8.1 mg/L; temperature at 20.0±1.1℃ and pH 7.4±0.3. Prior to drug administration, all fish were fed a drug-free feed, which was detected by UPLC, and acclimatized for 14 days. Afterward, common carp and crucian carp were administered orally at a dose of 10mg NFXNT/kg body weight（b.w.）by inserting a plastic hose with 1mL micro-injector after fasting for 24 h. Then six fish（common carp or crucian carp）were taken from the tank for each sampling point and anesthetized with MS222（50 mg/L）at time intervals 0.083, 0.167, 0.5, 1, 2, 4, 8, 16, 24, 48, and 72 h. For each fish, 2mL of blood was collected from the caudal vein. Following sampling, the fish were euthanized by an overdose in a solution of 300 mg/L MS 222. Blood samples were centrifuged at 4 000 r/min for 10min and stored at −20℃ until assayed. All the experimental procedures involving animals in the study were approved by the animal care center, Hubei Academy of Medical Sciences.

Pharmacokinetic parameters following single oral dose were calculated using the Practical Pharmacokinetic Program 3P97（Math-pharmacology Committee, Chinese Academy of Pharmacology, Beijing, China）. The appropriate pharmacokinetic model was selected by the visual examination of concentration-time curve and by the application of Akaike's Information Criterion（AIC）（Yamaoka et al., 1978）.

The plasma concentration-time curve after oral treatment is shown in Fig. 1. NFXNT was rapidly absorbed by the gastro-intestinal tract of common carp and crucian carp from the results, and its trends of plasma concentration in the two species were remarkably similar. The pharmacokinetic parameters are listed in Table1.

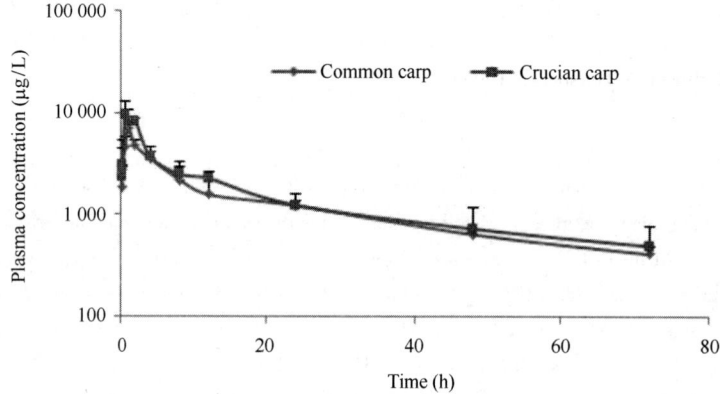

Fig. 1　Semilogarithmic plot of norfloxacin plasma concentration vs. time in common carp and crucian carp following a single oral dose of 10 mg norfloxacin nicotinate/kg b. w. （$n=6$）

Table 1　The pharmacokinetic parameters for oral administration of 10 mg norfloxacin nicotinate/kg b. w. in common carp and crucian carp（$n=6$）

Paramenters	Units	Common carp	Crucian carp
A	μg/L	7 587.79	12 904.06
a	1/h	0.44	0.51
B	μg/L	2 378.55	2 793.08
β	1/h	0.026	0.026
K_a	1/h	2.16	2.75
$t_{1/2a}$	h	1.58	1.36
$t_{1/2\beta}$	h	26.33	26.55
$t_{1/2ka}$	h	0.32	0.25
K_{10}	1/h	0.081	0.10
K_{12}	1/h	0.24	0.30
K_{21}	1/h	0.14	0.13

Paramenters	Units	Common carp	Crucian carp
AUC_{0-t}	h. μg/L	103 072. 36	126 604. 4
T_{max}	h	1. 08	0. 84
C_{max}	μg/L	6 069. 79	9 586. 06
V_d	L/kg	1. 19	0. 23

A: Zero-time blood drug concentration intercept of distribution phase; B: Zero-time blood drug concentration intercept of elimination phase; α: Distribution rate constant; β: Elimination rate constant; Ka: Absorption rate constant; $t_{1/2\alpha}$: Distribution half-life; $t_{1/2\beta}$: Elimination half-life; $t_{1/2\,Ka}$: Absorption half-life; k_{10}: Drug elimination rate constant from central compartment; k_{12}: First-order transport rate constant from central compartment to peripheral compartment; k_{21}: First-order transport rate constant from peripheral compartment to central compartment; AUC_{0-t}: Area under concentration-time curve; T_{max}: Time to peak concentration; C_{max}: Peak concentration; V_d: Apparent distribution volume.

The pharmacokinetic properties of NFXNT in common carp were similar to those in crucian carp. The peak concentration of crucian carp was higher than that in common carp, and the time to peak concentration of crucian carp (1. 08 h) was shorter than that of common carp (0. 84 h). A two-compartment open model with first-order absorption proved to best fit the data of the two species. The same class drugs with NFX presented a part of analogous pharmacokinetic characteristic with this study in aquatic animals. The peak concentration of difloxacinwas 6. 14 μg/kg at 9 h in olive flounder after oral administration, and the two-compartment open model was also best fit for data of plasma (Sun, et al., 2014). The time to peak concentrationwas 6 h for flumequine (1. 42 μg/kg) and 12 h for sarafloxacin (0. 08 μg/kg) in Atlantic salmon following oral administration, whose data in plasma were still described by the two-compartment open model (Martinsen & Horsberg, 1995).

In the present study, thedistribution half-life ($t_{1/2\alpha}$) and elimination half-life ($t_{1/2\beta}$) were 1. 58h and 26. 33h, respectively, in plasma of common carpand were1. 36 h and 26. 55 h, respectively, in plasma of crucian carp. From these results, the values of $t_{1/2\beta}$in the two species were relatively close. Compared to these values of NFX, enrofloxacin had a longer $t_{1/2\beta}$ (62. 7 h) in plasma of allogynogenetic silver crucian carpgiven the same dose at analogous temperature (Fang, et al., 2012). The $t_{1/2\beta}$ values weredifferent in different species of animalsand the values of NFX in plasma of common carp and crucian carp were longer than those in poultry and mammals. In poultry, $t_{1/2\beta}$ values were 11. 11 h in chicken, 9. 07 h in turkey, and 10. 65 h in goose (Laczay, et al., 1998), respectively; in mammals, they were 7. 13 h in pig (Chang, et al., 2007), 5. 44 h in horse (Park & Yun, 2003), and 3. 51 h in donkey (Lavy, et al., 1995), respectively. The observation was consistent with the depletion of melamine and cyanuric acid in fish compared with mammalian data] (Stine, et al., 2012). The differences might be due to the metabolic capacity of species.

The apparent distribution volume (V_d) of NFXNTwas 1. 19 L/kg in common carp, which was 5. 17 times that of crucian carp (0. 23 L/kg) butless than the previous value of NFX-glycine acetate in flounder after dipping administration (Park, et al., 1996). However, $V_d$values of broilers (Abu-Basha, et al., 2008), donkeys (Lavy, et al., 1995), goats (Wajeeha, et al., 2006), calves (Gips & Soback, 1996), and lactating cows (Gips & Soback, 1999) were higher than that of fish, which suggested that a better distribution of NFX occurred in birds and mammals. The ratio of k_{12} and k_{21} was 1. 71 in common carp and 2. 31 in crucian carp. This indicated that NFX was retained in the tissues.

For fluoroquinolone antimicrobials, both Cmax to MIC ratio and AUC to MIC ratio were used to predict antibacterial success. Most experts agreed that a Cmax toMIC of 8-10or an AUC to MIC ratio greater than 100-125 prevented the overgrowth of resistant bacteria (Wright, et al., 2000; McKellar, et al., 2004; Schentag, 1999). Minimal inhibition concentrations of NFX were 0. 2-0. 8 μg/mL for P. anguilliseptica, Edwardsiella tarda, Aeromonas hydrophila, Vibrio sp. and Pasteurella fluorescens (Park, et al., 1996). From our results, the ratios of Cmax/MIC and AUC/MIC were 7. 5-30. 3 and 128. 8-515. 4 in common carp and 12. 0-47. 9, 158. 3-633. 0 in crucian carp, respectively.

The pharmacokinetics of NFXNT in common carp and crucian carp after oral administration indicated good absorption and distribution. We showed that NFXNT administered orally at a dose of 10 mg/kg b. w. offered anoption for treatment of some diseases.

ACKNOWLEDGMENTS

This work is financially supported by the Special Fund for Agro-Scientific Research in the Public Interest of China (201203085) and Fundamental Research of Chinese Academy of Fishery Sciences (2014A09XK04).

REFERENCES

A bu-Basha, E. A., Gharaibeh, S. M., &Abudabos, A. M. (2008) Pharmacokinetics and bioequivalence of two norfloxacin oral dosage forms in healthy broiler chickens. International Journal of Poultry Science, 7 (3): 289-293.

Anadon, A., Martinez-Larranaga, M. R., Diaz, M. J., Fernandez, R., Martinez, M. A. & Fernandez, M. C. (1995) Pharmacokinetics and tissue residues of norfloxacin and its N-desethyl-and oxo-metabolites in healthy pigs. Journal of Veterinary Pharmacology and Therapeutics, 18 (3), 220-225.

Anadon, A., Martinez-Larranaga, M. R., Velez, C., Diaz, M. J. & Bringas, P. (1992) Pharmacokinetics of norfloxacin and its N-desethyl-and oxo-metabolites in broiler chickens. American Journal of Veterinary Research, 53 (11), 2084-2089.

Brown, S. A., Cooper, J., Gauze, J. J., Greco, D. S., Weise, D. W. & Buck, J. M. (1990) Pharmacokinetics of norfloxacin in dogs after single intravenous and single and multiple oral administrations of the drug. American Journal of Veterinary Research, 51 (7), 1065-1070.

Chang, Z. Q., Oh, B. C., Kim, J. C., Jeong, K. S., Lee, M. H., Yun, H. I., Hwang, M. H. & Park, S. C. (2007) Clinical pharmacokinetics of norfloxacin-glycine acetate after intravenous and oral administration in pigs. Journal of Veterinary Science, 8 (4), 353-356.

Fang, X., Liu, X., Liu, W. & Lu, C. (2012) Pharmacokinetics of enrofloxacin in allogynogenetic silver crucian carp, Carassius auratus gibelio. Journal of Veterinary Pharmacology and Therapeutics, 35 (4), 397-401.

Gilfillan, E. C., Pelak, B. A., Bland, J. A., Malatesta, P. F. & Gadebusch, H. H. (1984) Pharmacokinetic studies of norfloxacin in laboratory animals. Chemotherapy, 30 (5), 288-296.

Gips, M. & Soback, S. (1996) Norfloxacin nicotinate pharmacokinetics in unweaned and weaned calves. Journal of Veterinary Pharmacology and Therapeutics, 19 (2), 130-134.

Gips, M. & Soback, S. (1999) Norfloxacin pharmacokinetics in lactating cows with sub-clinical and clinical mastitis. Journal of Veterinary Pharmacology and Therapeutics, 22, 202-208.

Gonzalez, F., San Andres, M. I., Nieto, J., San Andres, M. D., Waxman, S., Vicente, M. L., Lucas, J. J. &Rodriguez, C. (2001) Influence of ruminal distribution on norfloxacin pharmacokinetics in adult sheep. Journal of Veterinary Pharmacology and Therapeutics, 24 (4), 241-245.

Helmy, S. A. (2013) Simultaneous quantification of linezolid, tinidazole, norfloxacin, moxifloxacin, levofloxacin, and gatifloxacin in human plasma for therapeutic drug monitoring and pharmacokinetic studies in human volunteers. Therapeutic Drug Monitoring, 35 (6), 770-777.

Laczay, P., Semjen, G., Nagy, G. & Lehel, J. (1998) Comparative studies on the pharmacokinetics of norfloxacin in chickens, turkeys and geese after a single oral administration. Journal of Veterinary Pharmacology and Therapeutics, 21 (2), 161-164.

Lavy, E., Ziv, G. & Glickman, A. (1995) Intravenous disposition kinetics, oral and intramuscular bioavailability and urinary excretion of norfloxacin nicotinate in donkeys. Journal of Veterinary Pharmacology and Therapeutics, 18 (2), 101-107.

Martinsen, B. & Horsberg, T. E. (1995) Comparative single-dose pharmacokinetics of four quinolones, oxolinic acid, flumequine, sarafloxacin, and enrofloxacin, in Atlantic salmon (Salmo salar) held in seawater at 10 degrees C. Antimicrobial Agents Chemothermotherapy, 39 (5), 1059-1064.

McKellar, Q. A., Sanchez Bruni, S. F. & Jones, D. G. (2004) Pharmacokinetic/pharmacodynamic relationships of antimicrobial drugs used in veterinary medicine. Journal of Veterinary Pharmacology and Therapeutics, 27 (6), 503-514.

Park, S. C. & Yun, H. I. (2003) Clinical pharmacokinetics of norfloxacin-glycine acetate after intravenous and intramuscular administration to horses. Research in Veterinary Science, 74 (1), 79-83.

Park, S. C., Yun, H. I. & Oh, T. K. (1996) Comparative pharmacokinetics and tissue distribution of norfloxacin-glycine acetate in flounder, (Paralichthys olivaceus) at two different temperatures. Journal Veterinary Medical Science, 58 (10), 1039-1040.

Park, S. C., Yun, H. I. & Oh, T. K. (1998) Comparative pharmacokinetic profiles of two norfloxacin formulations after oral administration in rabbits. Journal Veterinary Medical Science, 60 (5), 661-663.

Schentag, J. J. (1999) Antimicrobial action and pharmacokinetics/pharmacodynamics: the use of AUIC to improve efficacy and avoid resistance. Journal of chemotherapy, 11, 426-439

Stine, C. B., Nochetto, C. B., Evans, E. R., Gieseker, C. M., Mayer, T. D., Hasbrouck, N. R. & Reimschuessel, R. (2012) Depletion of melamine and cyanuric acid in serum from catfish Ictalurus punctatus and rainbow trout Onchorhynchus mykiss. Food Chemical Toxicology, 50 (10), 3426-3432.

Sun, M., Li, J., Gai, C. L., Chang, Z. Q., Li, J. T. & Zhao, F. Z. (2014) Pharmacokinetics of difloxacin in olive flounder Paralichthys olivaceus at two water temperatures. Journal of Veterinary Pharmacology and Therapeutics, 37 (2), 186-191.

Vancutsem, P. M., Babish, J. G. & Schwark, W. S. (1990) The fluoroquinolone antimicrobials: structure, antimicrobial activity, pharmacokinetics, clinical use in domestic animals and toxicity. Cornell Veterinary, 80 (2), 173-186.

Wajeeha, Khan, F. H., &Javed, I. (2006) Bioavailability and pharmacokinetics of norfloxacin after intramuscular administration in goats. Pakistan Veterinary Journal, 26 (1): 14-16

Wright, D. H., Brown, G. H., Peterson, M. L. & Rotschafer, J. C. (2000) Application of fluoroquinolone pharmacodynamics. Journal of Antimicrobial Chemotherapy, 46 (5), 669-683.

Yamaoka, K., Nakagawa, T. & Uno, T. (1978) Statistical moment in pharmacokinetics. Journal of Pharmacokinetic and Biopharmaceutics, 6, 547-558.

Pharmacokinetics and residue elimination of niacin norfloxacin premix in *Channaargus*

Meng-liZuo, Jia-yunYao, wen-linYin, Jin-yuShen

(Key Laboratory of Fish Health and Nutrition of Zhejiang Province, Zhejiang Institute of Freshwater Fisheries. Huzhou, China)

Abstract: To determine the parameters of the pharmacokinetics and residue elimination of niacin norfloxacin premix in Channaargus, the C. arguswere fed with niacin norfloxacin premix by oral gavages, the tissues muscle, blood, liver were purified by organic extraction and solid phase extraction. The concentrations of niacin norfloxacin in muscle, blood, liver were measured by ultra-performance liquid chromatography (UPLC). The results showed that the blood complied pharmacokinetic one-compartment model with first order absorption after C. arguswere oral gavages with niacin norfloxacin premix (20 mg/kg body weight). The main pharmacokinetic parameters were as follows, the peak time (T_{peak}) was 1.014 h, the maximum peak concentration (C_{max}) was 11.457 µg/mL, the absorption half-life ($T_{1/2Ka}$) was 0.532 h, the distribution half-life ($T_{1/2\alpha}$) was 0.955 h, the elimination half-life ($T_{1/2\beta}$) was 35.089 h. The drug residues time in muscle, blood, liver tissue were 5 d, 4 d, 12 d, respectively (less than 50 µg/kg), after continuous oral gavages with niacin norfloxacin premixes (20 mg/kg body weight) for 5 days.

Keywords: Channaargus; niacin norfloxacin premix; pharmacokinetics; residue

烟酸诺氟沙星预混剂在乌醴体内药代动力学及残留消除规律

摘要：研究烟酸诺氟沙星预混剂在乌醴体内的药代动力学及残留消除规律。采用混饲口灌方式给药乌醴，通过有机萃取及固相萃取等方法提纯净化各组织样本，利用超高效液相色谱（UPLC）法测定样本中药物的浓度，研究烟酸诺氟沙星预混剂在乌醴体内的药代动力学及残留规律。混饲口灌烟酸诺氟沙星预混剂（20.0 mg/kg body weight）后，乌醴血液符合药代动力学一级吸收一室开放模型。主要动力学参数如下：达峰时间 T_{peak} 为 1.104 h，最大峰浓度 C_{max} 为 11.457 µg/mL，分布半衰期 $T_{1/2\alpha}$ 为 0.995 h，消除半衰期 $T_{1/2\beta}$ 为 35.089 h；连续 5 d 混饲口灌烟酸诺氟沙星预混剂（20.0 mg/kg body weight）后，药物在肌肉、血液和肝脏等组织的残留低于 50 µg/mL 所需的时间分别为 5、4、12 d。

关键词：乌醴；烟酸诺氟沙星预混剂；药代动力学；残留

1　Introduction

Niacin norfloxacin is a Fluoroquinolone broad-spectrum antibiotic which is widely used for the treatment of systemic bacterial infections. Because of its fast absorption, highly concentration and widely distribution in the aquatic animal (Yang et al., 2004), it become the very popularly medicines in prevention and control of bacterial disease in aquatic animals (Zhang et al., 2010). But the niacin norfloxacin residues in animals can cause the immune suppressive effects and liver damage, and reduce the number of red blood cells, white blood cells in the body (Liao, 2011). Bacterial resistance has frequently been observed for niacin norfloxacin abuse under unknown condition of the drug pharmacokinetics in aquaculture. Therefore, it is important to study pharmacokinetics and residue elimination of niacin norfloxacin in farmed fish, in order to determine optimal dosage regimens and establish safe withdrawal periods, and to minimize the environmental effects of the drugs used in aquaculture.

So far, the pharmacokinetics and residue elimination of niacin norfloxacin have been reported in rainbow trout (Chen and Yin, 2011), litopenaeusvannamei (Cao et al., 2006), freshwater shrimp (He et al., 2012), hybrid tilapia (Fan et al., 2008), trionyxsinensis (Chen and Yin, 1997) andcarassiusauratus (Guo et al., 2008). All these studies showed that the pharmacokinetics of niacin norfloxacin may be affected by some factors such as species specificity (Gao, 2011), salinity (Fang and Zhen, 2006), route of drug administration (Koc F et al., 2009; Zhang et al., 2000), dose (Chen et al., 2010) and dosage form (Zhen et al., 2008), water temperature (Wu et al., 2008), and other experimental conditions. To our knowledge, there was little report on pharmacokinetics of niacin norfloxacin premix. Therefore, it is necessary to study pharmacokinetics and residue elimination of niacin norfloxacin in various farmed fish species, which was living under different conditions.

C. argus is an economically farmed fish in China. The goal of the research was to characterize the disposition of niacin norfloxacin premix in C. argus for improving disease control in aquaculture, and to minimize the potential risks for human and environmental health from residues in C. argus.

2 Materials and methods

2.1 Chromatographic conditions

Concentrations of niacin norfloxacin was determined by ultra-performance liquid chromatography (Waters ACQUITY UPLC, with an automatic injection performed on a Waters C8 reversed phase column.) using a method described by Gigosos et al. (2000). Monitoring wavelength was 278 nm. The mobile phase consisted of a phosphate buffer (0.05 M): acetonitrile (85: 15 v/v) solution (pH=2.5), and the flow rate was 0.1 mL/min.

2.2 Method validated

Mobile phase stock solution of niacin norfloxacin (purity 99%, IVDC, China) was prepared at a concentration of 1mg/mL. Working Standard calibration curve was prepared in a series of known concentrations of 0.001, 0.005, 0.01, 0.05, 0.1, 0.5, 1, 5, 10, 20, 50 and 100 μg/mL by dilution from the stock solution and was drawn by plotting, the known niacin norfloxacin concentrations against the average peak area. The niacin norfloxacin concentrations in unknown samples were read from the standard curve. The limit of detection (LOD) was defined as the UPLC-UV signal that was three times greater than the signal-to-noise ratio.

The recovery of niacin norfloxacin from the C. argus was determined using control samples spiked with 0.1, 0.5, 1, or 5 μg/mL niacin norfloxacin and extracted as previously described. The sample homogenate was allowed to interact with niacin norfloxacin at 4℃ for at least 2 h before the extraction. UPLC was then employed to quantify the niacin norfloxacin. Recovery was calculated by comparing the peak area of niacin norfloxacin from processed samples with that from the niacin norfloxacin standard in the mobile phase. The method precision was evaluated by analyzing one sample five times on the same day and five individually prepared samples over five separate runs on different days.

2.3 Dosing and sampling

The formulation of niacin norfloxacin premix for oral gavages was as follow, accurately weighing 6 g niacin norfloxacin premix raw materials (purity 10%, seasun, Beijing, China), the drug was mixed with crushed commercial fish feed and added water to 30 mL, the concentration of the slurry was 200 mg/mL.

Channaargus (200±20.0 g) were supplied with a drug-free feed to acclimatize for 1 week and starved for 24 h before drug administration, ten fish were randomly selected to acquisition muscles, blood and liver, make sure that the organization does not contain niacin norfloxacin. Fish were divided into three groups, the first group was treated as pharmacokinetics by oral gavages at a single dose of 20 mg/kg body weight; The second group was treated as residues by given a dose of 20 mg/kg body weight for 5 days (one dose each day) with the same slurry as that used for the pharmacokinetic study; The third group as control was fed with drug-free feed. The slurry was forced into the stomach of the C. argususing a 1 mL syringe fitted with a blunt needle. Only those individuals that did not regurgitate were used in the experiment. The water temperature was adjusted to 26±1℃ and pH 6.5. Ammonia nitrogen and nitrite concentration was 0.1 mg/L, 0.008mg/L respectively.

After oral gavages dosing in the study of pharmacokinetics and residues, ten fish were sampled at each time point. Blood (1 mL), liver (1 g) and muscle (1 g) of C. argus were immediately taken after the last dosing time at 0.25, 0.5, 1, 1.5, 2, 3, 4, 8, 12, 24, 48, 72 and 96 h for a single dose, and same samples tissues were taken at 1 h, 1 d, 2 d, 3 d, 4 d, 5 d, 7 d, 9 d, 11 d, 12 d, 15 d, 18 d and 21 d days after drug administration for multitude dosage. All samples were marked and immediately stored at -20℃ until analysis.

2.4 Niacin norfloxacin analysis protocol

The methodology for the analysis of the sample was performed by using the method reported by Mohan et al (2006). In brief, a sample of 1mL blood or 1g tissue was transferred to a 10-mL plastic centrifuge tube, the tissue sample was cut into small pieces by an operating scissors and add 1 mL of 0.05 M phosphate buffer (pH 2.5), homogenized on a homogenate machine, 5 mL of acetonitrile used as extraction was added to the tube, and the mixture was shaken vigorously for 5 min, ultrasonic extraction for 15 min, samples were then centrifuged at 6 000 rpm for 10 min. The supernatant was transferred to a clean centrifuge tube. And the residue was re-extracted twice using the same procedure as above to achieve a higher recovery. The combined supernatants were evaporated to dryness using a vortex evaporator. The resulting residue was dissolved in 1. 0 mL of mobile phase and then filtered through 0. 22 μm disposable syringe filters prior to analysis by the UPLC system.

2. 5　Pharmacokinetics analysis

The experimental data statistics was used by excel software. Pharmacokinetic analysis was performed by the computer program 3P97 (version 1. 0, edited by The Chinese Society of Mathematical Pharmacology, P. R. China).

3　Result

3. 1　Method validation

Standard curve was drawn by measured the niacin norfloxacin peak area (Ai) for each concentration (Ci). The standard curve equation and correlation coefficient was: Ai = 265172 Ci + 108870, r = 0. 9997 at 0. 01 - 100 μg/mL concentration range, the LOD was 0. 005 μg/mL. The norfloxacin mean recovery for blood ranged from 76. 09% to 86. 44%, for muscle ranged from 83. 33% to 90. 4%. The average inter-and intra-day precision (RSDs) were 6. 00% and 8. 20% for muscle and 6. 99% and 8. 23% for blood.

3. 3　The pharmacokinetics of niacin norfloxacin in C. argus

The C. argus were given a single dose of 20 mg/kg body weight of slurry containing niacin norfloxacin at 200 mg/mL, the concentration-time curves of the muscle, blood and liver in different sampling time points were showed in Fig. 1. As the results showed that the concentrations of niacin norfloxacin in the blood kept rising after the treatment, the maximum blood concentration (14. 996 μg/mL) was reached at 1. 5 h. Then the drug level declined rapidly and reached 50 μg/mL at 12 h. The drug concentration in the liver rising rapidly after 1 h, and the concentration in the liver was much higher than levels in blood and muscle, the drug concentration in the live reach peak concentration at 2 h that the niacin norfloxacin concentration was 91. 266 μg/mL, then drug concentration start falling. Niacin norfloxacin concentration of muscle showed bimodal phenomenon, each time point before 3 h drug concentrations were lower than the concentration of the corresponding points of liver and blood. The niacin norfloxacin levels were recorded at the first sample point (0. 5 h) after oral gavages, and the drug reached the first peak concentration (7. 962 μg/mL) at 1. 5 h, then it slowly decreased, another drug peak appeared at the 8h, the peak concentration was 5. 605 μg/mL.

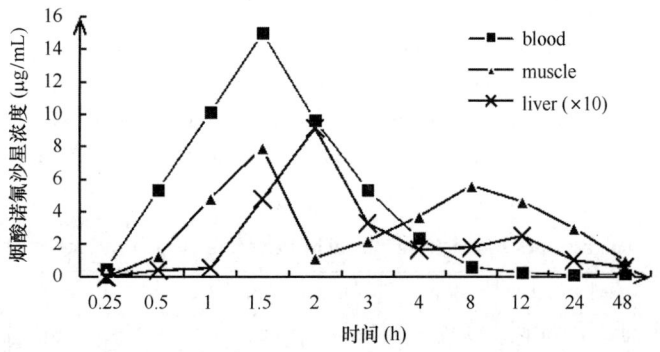

Fig. 1　The norfloxacin concentration-time curve in various organizations of Channa argus

The blood concentration-time curve of niacin norfloxacin in C. argus after oral gavages at a single dose of 20 mg/kg were shown in Figure 1. The data of pharmacokinetics fitted to one-compartment open pharmacokinetic model. The pharmacokinetic parameters calculated from this model were given in Table 1. The fitting had a weight of 1, the correlation R square was 0. 927, the regression coefficient was 0. 965, the goodness of fit was 1. 654. The absorption half-life ($T_{1/2ka}$), distribution half-life ($T_{1/2\alpha}$) and elimination half-life ($T_{1/2\beta}$) of niacin norfloxacin were 0. 532 h, 0. 955 h and 35. 089 h respectively.

Tab. 1　The pharmacokinetics parameters of niacin norfloxacin in blood from C. argus

Parameters	Unit	Parameters Values
A	(μg/mL)	53. 935
α	(1/h)	0. 726

Parameters	Unit	Parameters Values
B	(μg/mL)	0.076
β	(1/h)	0.020
K_a	(1/h)	1.304
Vd	(mg/kg)	0.837
$t_{1/2\alpha}$	(h)	0.955
$t_{1/2\beta}$	(h)	35.089
$t_{1/2Ka}$	(h)	0.532
K_{21}	(1/h)	0.022
K_{10}	(1/h)	0.589
K_{12}	(1/h)	0.069
AUC	(μg/mL $\cdot$ h)	32.942
CLs	(L/kg $\cdot$ h)	0.607
T_{peak}	(h)	1.014
C_{max}	(μg/mL)	11.457

Ka：absorption rate constant；α：distribution rate constant；β：elimination rate constant；$T_{1/2Ka}$：absorption half-life of the drug；$T_{1/2\alpha}$：distribution half-life of the drug；$T_{1/2\beta}$，elimination half-life of the drug；T_{peak}：the time point of maximum plasma concentration of the drug；C_{max}：the maximum plasma concentration；$V_{d/f}$：extensive apparent volume of the central compartment；K_{12}，K_{21}：first-order rate constants for drug distribution between the central and peripheral compartments；K_{10}：elimination rate constant from the central compartment；AUC，area under the concentration-time curve；CLs：total body clearance of the drug.

3.4　Residues of niacin norfloxacin in C. argus

The niacin norfloxacinresidue concentrations in each organisations after oral gavages were listed in Table 2. The elimination of niacin norfloxacin in muscle was rapid，the drug concentrations declined to 1.34 μg/g in muscle on 4 d and were undetectable in muscle on 5 d. The corresponding date after oral gavages was 0.60 μg/mL in blood on 3 d and was undetectable in blood on 4 d. The elimination of niacin norfloxacin in liver was slowly，the drug concentrations declined to 0.03 μg/mL in liver on 9 d and were undetectable in liver on 12 d.

Tab 2　The residues of niacin norfloxacin in various organizations of C. argus

Tissue	1 h	1 d	2 d	3 d	4 d	5 d	7 d	9 d	12 d	15 d	18 d
muscle	8.70	5.30	3.04	2.34	1.34	ND	ND	ND	ND	ND	ND
blood	21.21	6.00	2.72	0.60	ND	ND	ND	ND	ND	ND	ND
liver	85.16	37.69	15.76	4.58	2.56	1.80	0.83	0.03	ND	ND	ND

note：ND is undetectable.

4　Discuss

After C. argus was oral gavages withniacin norfloxacin premix（20 mg/kg），the peak time in blood occurred at 1.048 h，the absorption half-life was 0.514 h，the elimination half-life（$T_{1/2\beta}$）was 35.089 h，which indicated that the niacin norfloxacin in C. arguswas rapid absorption and slow elimination.

The concentration of niacin norfloxacin in the liver was the highest than that in other tissue. The residue time was also the longest，which suggests that the liver was the main detoxifying organs and the main part of drug residues in aquatic animal. In addition，the drug residues may also be associated with the enterohepatic circulation. Fang et al.（2013）reports this phenomenon，after oral gavages，the drugs in higher animals firstly through the gastric mucosa and capillary，and then pass through the hepatic portal vein into the blood

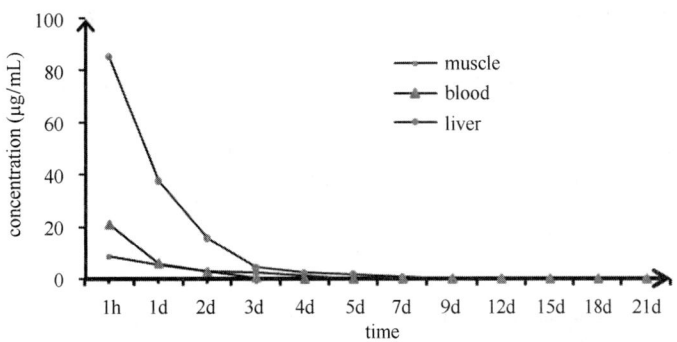

Fig. 2　The residues of niacin norfloxacin in various organizations of C. argus

vessels. the liver excrete drugs, at the same time the live bile into the duodenum, so some drugs can then through intestinal epithelial cells to absorb, the hepatic portal vein into the liver, to complete the cycle. It fit to pharmacokinetic of medicine-curve which appear bimodal phenomenon. The pharmacodynamics of blood drug concentration decline slowly, the effect time of the drug extended obviously which is reported in the American lobster blood sinus injection (Barron M G et al., 1988. James M O and Barron M G, 1988), American lobster (James M O and Herbert A H, 1988) and Litopenaeusvannamei (Park E D, 1995) that all achieved oral OMP, which show the concentration of the drug in liver pancreas and other organizations dozens of times on even one hundred times higher than that of phenomenon.

After feeding, the concentration of niacin norfloxacin in liver tissue was more than that in blood and muscle in 1.5 h. It suggest that the niacin norfloxacin absorption by liver was quicker than the blood. It may because that not all niacin norfloxacin reach the liver by circulate of blood system, according to reports (Wu et al., 2008), when the test dosage was greater than the actual demand, the drug in the body was transport by passive process, which was treated as external stimuli, it go beyond the body protection, the drug was degraded by the liver to alleviate the stimulation to the body. In addition, the reason may be that " the first effect", Wang et al (2004) report that liver tissue T_{peak} was 0.33 h after breeding bass with norfloxacin by oral gavages, and Liu et al (2007) feed yellow croaker norfloxacin by oral gavages, the maximum peak concentration in liver was 0.5 h, "the first effect" was that Gastrointestinal drug delivery, before which had not been absorbed into the blood circulation, the intestinal mucosa and hepatic start metabolizing it, and make it appear dose reduction phenomenon into the blood circulation. Further identified as " the first effect", which also require further experiment.

Bimodal phenomenon occurring in a muscle, it may be that the quinolones contains hydrophobic fluorine, which enhance it tissue penetration by its lipotropy, so the drugs can pass through the cell plasma membrane quickly, the quinolones can get into the muscle and other tissues by passive transportation, the niacin norfloxacin was fluoroquinolone drugs, the weight was 319.34, some studies have shown that the molecular weight between 250~500 of drugs can pass through the cell membrane easily (Sun et al., 2013), therefore, the first peak in muscles may be the result of the passive transportion that the niacin norfloxacin diffusion into muscle when through the circulating the blood, niacin norfloxacin also can be transported to muscle through the blood capillary, that may contribution to the second peak in the muscles.

In order to ensure consumer the C. argussafely. The european union, the united states, Japan and other regions rule that the concentration of niacin norfloxacin in the C. argus muscle and live must be under 50 μg/kg when it be consumption, so we suggest that withdrawal time of niacin norfloxacin was at least for the 9 d after the treatment when the C. argus feed by oral gavages with (20 mg/kg body weight) niacin norfloxacin premix in 26℃ temperature conditions.

Acknowledgements

The research was supported by Special Fund for Agro-scientific Research in the Public Interest (201203085) and the key project program of Huzhou City (GN10), as well as Special Fund for Agro-scientific Research in the Public Interest (No. 200803013).

References

B arron, M. G., Gedutis, C., James, M. O. 1988. Pharmacokinetics of sulphadimethoxine in the lobster, Homerusamericannus, following intraporicardial administration. Xenobiofiea. 18, 269-277.

Cao, L. M., Li, J., Liu, Q. 2006. Studies on pharmacokinetics of norfloxacin in Litopenaeusvannamei fed Artemia enriched with norfloxacin. Marine Sciences. 30, 45-51.

Chen, C., Lu, T. Y., Wuang, H. 2011. Pharmacokinetics of Norfloxacin in Rainbow Trout Oncorhynchusmykiss. chinese journal of fisheries. 24, 25－28.

Chen, L. L., Yang, H. S., Wu, G. H. 2010. Pharmacokinetics of Florfenicol in the Chinese Mitten－handed Crab (Eriocheirsinensis): Dose Effect. Chinese Journal of Zoology. 45, 102－109.

Chen, W. Y., Yin, C. H. 1997. Pharmacokinetics studies on Norfloxacin in Trionyxsinensis. Freshwater Fisheries. 21, 434－437.

Campbell, D. A., Pantazis, P., Kelly, M. S.. 2001. Impact and Residence Time of Oxytetracycline in the Sea Urchin, PsammechinusMiliaris, a Potential Aquaculture Spe－cies. Aquacult. 202, 73－87.

Fan, Z. Y., Zhang, Q. Z., Chen, G. Q. 2008. Residue Elimination of Norfloxacin Hydrochloride in Hybrid Tilapia (Oreochrom is aureus × O. niloticus). Freshwater Fisheries. 38, 55－58.

Fang, W. H., Zhen, G. X. 2006. pharmacokinetics of norfloxacin in hemolymph from whiteleg shrimp, penaeusvannamei following intramuscle injection and oral administration. Journal of the Acta hydro biologicasinica. 30, 541－546.

Fang, W. H., Li. G. L., Zhou, Sh., et al. 2013. Pharmacokinetics and Tissue Distribution of Thiamphenicol and Florfenicol in Pacific White Shrimp Litopenaeusvannamei in Freshwater following Oral Administration. Journal of Aquatic Animal Health. 23, 82－97

Fang, X. X., LI, J., Wang, Q. 2003. Residual characteristics of sulfamethoxazole and trimethoprim in lateolabraxjaponicus at different water temperature. Marine science. 27, 16－20.

GUO H. Y., Ma Y. G., Chen, Y. S. 2008. The Pharmacokinetics and Residues of Norfloxacin in Carassiusauratus. Freshwater Fisheries. 38, 46－50.

Guo, H. Y. 2011. Pharmacokinetics, Residues and Bioavailabilies of Norfloxacin in Carassiusauratus and Ctenopharyngodonidellus L. ChongQing: SouthwestUniversity,

Guo, T. 2005. Modern Pharmacokinetics. China Science and Technology Press, Beijing.

Gigosos, P. G., Reversado, P. R., Cadahia, O., Fente, C. A., Vazquez, B. I., Franco, C. M., Cepeda, A. 2000. Determination of quinolones in animal tissues and eggs by HPLC withphotodiode－array detection. Chromatogr. A. 871, 3－36.

HE, P., Yin, W. L., Shen, J. Y., et al. 2012. Pharmacokinetic Disposition of Streptomycin Sulfate in Japanese Eel (Anguilla japonica) after Oral and Intramuscular Administrations. scientific research. 3, 195－200.

Huang, W. H., Yang, X. L., Hua, K. Z. 2004. Disposition and elimination of norfloxacin in tissues from Litopenaeusvannamei. Journal of fisheries of china. 28, 19－24

James, M. O., Barron, M. G. 1988. Disposition of sulfadimethoxine in the lobster Oaomarmamerieanus). Vet Hum Toxieol. 30, 36－40.

James, M. O., Herbert, A. H. 1988. Disposition of ormetoprim in the lobster. Homarusamericanus K. Pharm Res. 5, 196－203.

Yamaoka, Y., Nakagawa, T. U. 1978. Application of Akaike's Information Criterion (AIC) in the Evaluation of Linear Pharmacokinetic Equations. Journal of Pharma－cokinetics and Biopharmaceutics. 6, 165－175.

Koc, F., Uney, K., Atamanalp, M. et al. 2009. Pharmacokinetic disposition of enrofloxacin in brown trout (Salmotruttafario) after oral and intravenous administrations. Aquaculture. 295 (2), 142－144.

Liao, B. C. 2011. Studies on pharmacokinetics and residues of norfloxacin in Hybrid Tilapia. Journal of the American Veterinary Medical Association. 26, 930－934.

Liu, Y. L., Wang, X. L., Yang, X. L. 2007. Studies on residues and depletion of norfloxacin in Bastard halibut (Paralichthysolivaceus) in vitro. Journal of the Marine Fisheries research. 31, 655－660.

Mohan, J., Sastry, K. V. H., Tyagi, J. S., Rao, G. S., Singh, R. V. 2006. Residues of fluoroquinolonedrugs in the cloacal gland and other tissues of Japanese quail. Brit. Poultry Sci. 47, 83－87.

Park, E. D., Lighmer, D. V., Milner, N., et al. 1995. Exploratoxy bioavailability and pharmacokinetic studier of sulphadimethoxine and ormetoprimin the penacidshrim, PenaeustannameiAquae. 130, 113－128.

Qu, X. R., Wang, Y. G., Li, S. Z. 2007. Studies on pharmacokinetics and the elimination regularity of norfloxacin residues in scopphthalmusmaximus. Journal of the Marine Fisheries research. 28, 25－29.

Sun, M., Li, J., Zhao, F. zh., Li, J. T., Chang, Zh. Q. 2013. Distribution and Elimination of Norfloxacin in FenneropenaeuschinensisLarvae, J. Ocean Univ. China (Oceanic and Coastal Sea Research). 12, 397－402.

Wang, Q., Liu, Q., Tang, X. L. 2004. Pharmacokinetics and residues of norfloxacin in Lateolabras japonicas. Journal of fisheries of china. 28, 13－18.

Wu, G. H., Zhang, J. B., Meng, Y. 2008. Pharmacokinetics of enrofloxacin in the Chinese mitten－handed crab, Eriocheirsinensis, at three water temperatures. Journal of Nanjing Agricultural University. 31, 105－110.

Yang, X. L., Wang, M. Q., Yang, Y., Ye, J. M. 2004. The method of RP－HPLC determination of norfloxacin in plasma and muscle of Pseudesciaenapolyactis. Journal of FisheriesofChina. 28, 1－6.

Zhang, H. Y., Zhao, M. J., Zhao, Z. 2010. Study Status in pharmacokinetics of Fishery Medicines In China. Progress in Veterinary Medicine. 31, 138－143.

Zhang, Y. F., Zhang, Z. X., Zhen, W. 2000. Study on pharmacokinetics of Norfloxacin in carp following different forms administration. Journal of the Journal Of fisheries Of China. 24, 559－563.

Zhen, Z. L., Liu, H. Y., Huang, H. 2008. Pharmacokinetics, Tissue Distribution of Sulfamethoxazole and Trimethoprim in Macrobranchiumrosenbergii：Dose Effect. Journal of Guang dong OceanUniversity. 28，54-59.

该文发表于《水产学报》【2004，28（增刊）：19-24】

诺氟沙星在凡纳滨对虾不同组织中处置和消除规律

Disposition and elimination of norfloxacin in tissues from *Litopenaeus vannamei*

房文红[1,2]，杨先乐[2]，周凯[1]

（1. 中国水产科学研究院东海水产研究所，农业部海洋与河口渔业重点开放实验室，上海 200090；2. 上海水产大学，上海，200090）

摘要：肌注和药饵给药两种方式下，诺氟沙星在凡纳滨对虾肌肉和肝胰脏等组织中药物的处置和消除差别较大。肌注给药后凡纳滨对虾肌肉和肝胰脏中药物达峰时间分别 1 min、12 min，药峰浓度分别为（5.14±1.25）μg·g^{-1}、（16.56±3.90）μg·g^{-1}，AUC 分别为 26.97 μg·h·g^{-1}、213.92 μg·h·g^{-1}，总体清除率分别为 370.78 g/kg·h、46.75 g/Kg·h。药饵给药后凡纳滨对虾肌肉、肝胰脏到达最高药峰的时间分别 4 h、2 h，它们的药峰浓度分别为 1.58±0.37 μg/g、404.36±109.07 μg/g；曲线下面积分别为 30.53 和 4694.9 μg·h/g，肝胰脏和肌肉中总体清除率分别为 6.39 g/kg·h 和 982.64 g/kg·h。药饵给药后肝胰脏中诺氟沙星浓度远高于血淋巴和肌肉，表现出诺氟沙星在肝胰脏中的"首过效应"。尽管如此，但药饵给药后肝胰脏中药物消除很快；根据诺氟沙星在水产品中推荐的残留限量标准（50 μg/kg），凡纳滨对虾肌肉和肝胰脏中诺氟沙星浓度分别在药饵给药后 144 h、192 h 低于此残留限量。

关键词：处置；消除；凡纳滨对虾；诺氟沙星

研究全血或血浆药物浓度-时间关系可以为抗生素的应用提供了重要依据，因为血液与组织无处不连。但抗微生物治疗的最终目的是根除特定感染部位的微生物致病菌，感染部位抗微生物药物的浓度影响着最终疗效[1]。然而，测定组织中药物浓度在人类医学难以实现，而在水产动物药动学研究中则能达到。在水产动物中测定不同组织中药物浓度有助于了解药物在不同组织间的转运过程，进一步为用药方案制定提供了理论依据。从食品安全角度，了解药物在组织中的消除规律可以为休药期的指定提供理论依据。国外有关水产动物组织中药物浓度变化的研究很多[2,3,4]，国内仅收集到土霉素在鳗鲡组织中残留的消除规律[5]、噁喹酸在鲈肌肉、肝脏、肾脏等组织中含量[6]等相关报道。本文主要探讨肌注和口服药饵两种给药方式下诺氟沙星在凡纳滨对虾肌肉和肝胰脏组织中转运和消除规律，为了便于比较同时也涉及到血药浓度。

1　材料与方法

1.1　实验药品

诺氟沙星原料药（含量98%）由上海三维制药厂提供，诺氟沙星标准品（含量99.5%，批号H040798），由中国兽药监察所提供。

1.2　试验动物

试验用凡纳滨对虾（*Litopenaeus vannamei*）取自上海市金山区漕泾对虾养殖场，经 7 d 暂养后用于试验。试验用海水盐度为 1.0 左右，接近淡水，水温为（25.0±1.5）℃。试验期间投喂适量螺蛳肉，并及时排出残饵和污物。

1.3　对虾给药

肌肉注射给药　肌肉注射诺氟沙星的剂量参考鱼类使用剂量[7]，给药剂量为 10 mg·kg^{-1}。凡纳滨对虾体重为 18～23 g，注射用诺氟沙星溶液浓度为 2%，根据体重计算每尾对虾给药量，一次性肌肉注射，给药部位从对虾第二腹节腹面稍斜插入，轻轻推进。试验期间对虾成活率为 100%。

药饵给药　凡纳滨对虾体重为 13～17 g，给药剂量为 30 mg·kg^{-1}，药饵含药量为 10 g·kg^{-1}，根据体重计算每尾对虾饵料投喂量，一次性投喂药饵。药饵投喂后 10 min 内摄食完全、没有残饵碎屑的对虾用于试验。10 min 后开始计时采样。试验期间对虾成活率为 100%。

药饵组成含有鱼粉、柔鱼粉和 α-淀粉等，添加诺氟沙星为 10 g·kg^{-1}，再加入 12%青蛤匀浆和 5%水，混合后搅拌至手捏成团，于电动搅肉机挤压成型。90℃下烘干 2 h，密封保存待用。

1.4 组织样品采集

血淋巴样品采集，用 1 mL 注射器插入对虾围心窦抽取，具体操作参见房文红等[8]。血淋巴样品采集后，立即将腹部与头胸甲分开。去除腹部甲壳、游泳足和肠即得肌肉样品；剥开头胸甲，用镊子将肝胰脏轻轻取出，去除食管等即得肝胰脏样品。

肌肉注射采样时间点为给药后 2、4、6、30 min，2、12、24、48、96、120、144、192、240 h，每个时间点采样 4 尾；药饵给药采样时间点为给药后 0.25、0.5、1、2、3、4、6、8、12、18、24、48、72、96、144、192、240 h，每个时间点采样 5 尾。

1.5 样品处理

血淋巴样品处理，参照房文红等[8]；肌肉和肝胰脏样品处理参照房文红等[8]。

1.6 组织样品中诺氟沙星含量分析

对虾血淋巴、肌肉和肝胰脏中诺氟沙星含量分析均采用反相高效液相色谱法。血淋巴分析采用的流动相为乙腈与磷酸盐缓冲液，流速为 0.5 mL/min，进样量为 10 μL，保留时间为 3.58 min 左右。肌肉和肝胰脏分析采用的流动相为甲醇与磷酸盐缓冲液，流速为 0.8 mL/min，进样量为 20 μL，保留时间分别为 7.5 min 和 8.7 min。

1.7 数据处理

血淋巴、肌肉和肝胰脏药物浓度-时间曲线下面积（AUC）采用梯形法计算：

$$AUC_n = \sum (t_{i+1} - t_i)(C_i + C_{i+1}),\ i = 0 \rightarrow (n-1)$$

式中，C_i、C_{i+1} 分别表示采样时间点 t_i、t_{i+1} 时组织中药物浓度；组织中药物总体清楚率 $CLs = dose/AUC_n$。

2 结果

2.1 肌注给药对虾组织中诺氟沙星处置过程

凡纳滨对虾肌肉注射诺氟沙星 10 mg/kg，其血淋巴、肝胰脏和肌肉中诺氟沙星浓度变化见图 1。肌注给药后在第一个采样点 1 min 时，肌肉中诺氟沙星浓度最高（5.141±1.251 μg/g），随后开始下降直到采样结束；血淋巴中诺氟沙星浓度在 2 min 为最高值（29.33±5.53 μg/mL），略迟于肌肉，随后同样开始下降，48 h 时血淋巴中诺氟沙星已低于检测限；而肝胰脏中诺氟沙星浓度在给药后逐渐上升，12 min 时达到最高浓度（16.56±3.90 μg/g），其后开始下降，直到采样结束。采用梯形法计算肌注给药后肝胰脏和肌肉中诺氟沙星曲线下面积（AUC）分别为 213.92 和 26.97 μg·h/g，由 AUC 推算所得的肝胰脏和肌肉中总体清除率分别为 46.75 g/kg·h 和 370.78 g/kg·h。见表 1。

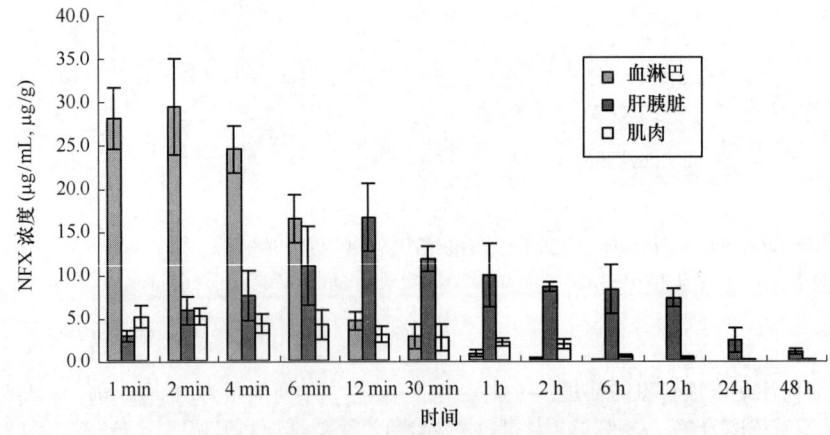

图 1　肌注给药南美白对是血淋巴、肝胰脏和肌肉中 NFX 浓度

Fig. 1　Concentrations of norfloxacin in the hemolymph, hepatopancreas and muscle of shrimp following intra-muscle injection

表1 肌注和药饵给药后凡纳滨对虾肝胰脏和肌肉中药物浓度参数

给药方式	参数	组织		
		肝胰脏	肌肉	血淋巴
肌注	C_{max}	(16.56±3.90) μg/g	(5.14±1.25) μg/g	(29.33±5.53) μg/mL
	T_{max}	12 min	1 min	2 min
	AUC	213.92 μg·h/g	26.97 μg·h/g	30.06 μg·h/mL
	CLs	46.75 g/kg·h	370.78 g/kg·h	332.67 g/kg·h
药饵	C_{max}	(404.36±109.07) μg/g	(1.58±0.37) μg/g	(2.86±1.15) μg/mL
	T_{max}	2 h	4 h	4 h
	AUC	4 694.9 μg·h/g	30.53 μg·h/g	43.50 μg·h/mL
	CLs	6.39 g/kg·h	982.64 g/kg·h	689.66 g/kg·h

2.2 药饵给药对虾组织中诺氟沙星处置过程

诺氟沙星以30 mg/kg的剂量对凡纳滨对虾进行药饵给药，给药后肝胰脏、血淋巴和肌肉中诺氟沙星浓度变化见图2。给药后，肝胰脏药物浓度首先表现为上升趋势，2 h到达药峰为404.36 μg/g，在随后的3~6 h肝胰脏中NFX浓度相差不大，表现出稳态水平，6 h后开始下降；肌肉中NFX浓度在给药后先表现为上升趋势，4 h到达最高值（C_{max} = 1.58 μg/g）后即开始下降；血淋巴中NFX表现出两个药峰，分别在4 h和12 h，药物浓度分别为2.86 μg/mL和2.037 μg/mL。同样以梯形法推算肝胰脏、肌肉中诺氟沙星药-时曲线下面积分别为4 694.9和30.53 μg·h/g，肝胰脏和肌肉中总体清除率分别为6.39 g/kg·h和982.64 g/kg·h。

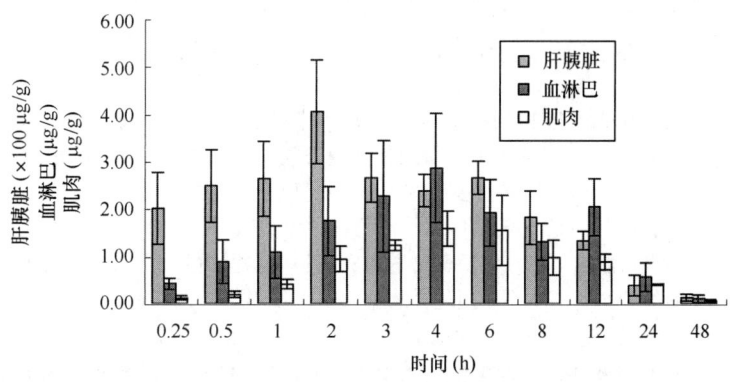

图2 药饵给药南美白对虾肝胰脏、血淋巴和肌肉中诺氟沙星浓度

Fig.2 Concentrations of norfloxacin in the hepatopancreas, hemolymph and muscle of shrimp following oral administration

2.3 肌注和药饵给药后诺氟沙星在凡纳滨对虾体内的消除

表2为肌注和药饵给药48 h以后凡纳滨对虾肌肉和肝胰脏中药物浓度。给药后48 h时，药饵给药对虾肌肉中诺氟沙星浓度略高于肌注给药，但在随后的各采样点肌肉中诺氟沙星含量均低于水产品诺氟沙星的残留限量0.050 μg/g。而肝胰脏中诺氟沙星在不同给药方式下相差较大，48 h时药饵给药肝胰脏中诺氟沙星含量高达12.89 μg/g，比肌注给药肝胰脏中诺氟沙星高12倍左右，且远远高于肌肉中含量；尽管如此，药饵给药组在随后的96 h内消除得很快，144 h时肝胰脏中诺氟沙星为0.068 μg/g，192 h时为0.036 μg/g，低于残留限量0.050 μg/g；肌注给药肝胰脏中诺氟沙星含量在144 h即低于残留限量标准，240 h几乎不能检出。

表2　凡纳滨对虾肌肉注射和药饵给药 48 h 后肝胰脏和肌肉组织中诺氟沙星浓度

Table 2　Norfloxacin concentrations in hepatopancreas and muscles from *P. vannamei* following intramuscle injection and oral administration

时间（h）Time	肌肉注射		药饵给药	
	肝胰脏（μg/g）	肌肉（μg/g）	肝胰脏（μg/g）	肌肉（μg/g）
48	0.969±0.352	0.047±0.023	12.894±7.568	0.052±0.032
72	——[a]	——[a]	5.620±2.673	0.028±0.005
96	0.067±0.043	0.036±0.021	2.025±0.796	0.023±0.007
144	0.010±0.008	0.023±0.005	0.068±0.024	0.024±0.030
192	0.006±0.005	0.015±0.015	0.036±0.017	0.020±0.003
240	ND[b]	0.012±0.002	0.030±0.017	0.019±0.001

注：[a]肌注给药 72 h 未设采样点；[b]未检出。

3　讨论

3.1　肌肉注射诺氟沙星在对虾体内的转运过程

对虾肌肉注射给药后，药物从给药部位沿着肌纤维散开，通过毛细血管（毛细管状的静脉）壁被吸收；在腹部的背面动脉和腹面动脉之间有一组血窦[8]，药物随着血淋巴通过开放的血窦流回心脏，同样随着血淋巴进入其他组织和器官。所以，药物浓度在对虾不同组织中表现为，肌肉中药物浓度自给药后一直呈下降趋势，血药浓度在给药后 2 min 达到最高，而肝胰脏中药物最高浓度是在 12 min。

3.2　药饵口服给药后对虾的"首过效应"

为了进一步分析，表3列出了凡纳滨对虾口服给药后三种组织中诺氟沙星浓度的比值。药饵给药后 0.25 h，肝胰脏中诺氟沙星含量是血淋巴中的 484 倍之多，2 h 时（肝胰脏药物达峰时间）其比为 232.1，即使在血药达峰时间（4h）其比仍为 83.3；与肌肉中药物含量相比，肝胰脏中诺氟沙星显得更高，肝胰脏/肌肉之比最高达 1 755.7，最低为 96.2；三种组织中，肌肉中药物含量最低。而且，以 30 mg/kg 剂量口服给药的血淋巴、肌肉组织中药物浓度均低于以 10 mg/kg 剂量肌注给药的相应浓度。由此可见，对虾口服给药后，大量药物进入肝胰脏后停留在肝胰脏，只有少量的药物进入血液循环和其他组织。对高等动物而言，口服给药时药物主要是通过胃肠粘膜及毛细血管，然后首先进入肝门静脉，通过肝脏后才能进入血液循环[9]。由此看来，药饵给药方式下对虾肝胰脏中药物浓度很高可能是诺氟沙星在对虾体内的"首过效应"。药饵给药方式下，肝胰脏很可能是对虾中央室的主要组成，然而肝胰脏中诺氟沙星消除较快，24 h 时药物浓度为 38.46 μg/g，约为药峰时（2 h）的 9.5%。在甲壳动物中肝胰脏既是主要的解毒器官又是脂肪贮存的场所。与本试验中药饵给药肝胰脏中药物浓度高的相关的报道有，美洲龙虾血窦内注射 SDM[10,4,11]、美洲龙虾[12]和凡纳滨对虾[13]口服 OMP 及克氏原螯虾暴露于 triclopyr[14]，均表现出肝胰脏中药物浓度高出其他组织的几十倍甚至上百倍。

表3　药饵给药对虾不同组织中 NFX 比率

时间（h）	0.25	0.5	1	2	3	4	6	8	12	24
肝胰脏/血淋巴	484.3	282.0	242.9	232.1	116.8	83.3	138.7	139.0	65.2	68.2
肌肉/血淋巴	0.28	0.22	0.38	0.54	0.54	0.55	0.81	0.74	0.43	0.71
肝胰脏/肌肉	1755.7	1306.6	638.9	428.2	214.6	150.3	171.6	187.3	150.7	96.2
血淋巴（μg/mL）	0.42	0.88	2.08	1.74	2.26	2.86	1.91	1.31	2.04	0.56

3.3　关于血药"再吸收"现象的解释

肝胰脏是消化系统的重要器官，在对虾科肝胰脏位于胃和中肠之间。药饵给药后 2 h 肝胰脏药物浓度达到最大（404.36 μg/g），随后开始下降，3 h 为 264.35 μg/g，3~6 h 肝胰脏中药物浓度呈现稳态水平，6 h 后又开始下降。药饵给药方式下，药物是通过肝胰脏后再进入血液循环系统。血淋巴第二个药峰的出现是在给药后 12 h，肝胰脏药物浓度在 3~6

390

h 的稳态水平可能为第二个血药峰出现提供了药的来源。这仅为对血药"再吸收"的一种现象解释，至于能否在理论上得到支持，留待以后进一步研究。

3.4　关于对虾肝胰脏疾病预防用药建议

在医学上有时利用药物在排泄器官中浓度高的特点来治疗排泄器官的疾病，如注射链霉素后，其尿液浓度为血浆浓度的 100 倍，可用于治疗泌尿系感染；红霉素等在胆道内的浓度较高，可用于治疗肝胆系统感染[9]。本文在药饵给药试验中，对虾肝胰脏中诺氟沙星浓度和曲线下面积（AUC）均远远大于血淋巴。而在四种弧菌对中国对虾注射感染试验中，对虾主要病变部位是肝胰脏、鳃和胃等，且肝胰脏又是最容易发生病变的器官[15]。作者认为，对于预防受细菌感染的肝胰脏疾病（如细菌性肠炎病[16]）采用口服诺氟沙星给药方式应是十分理想，给药剂量可以低于本试验给药剂量。

3.5　休药期的确定

诺氟沙星在水产品中的残留限量尚未确定，欧盟有关家禽及猪可食组织中诺氟沙星的允许残留为 50 μg/kg[17]，若参照此标准，将水产品中诺氟沙星的残留限量暂定为 50 μg/kg。

在水产养殖生产中，诺氟沙星给药主要是拌饲料口服，因此根据药饵口服给药方式下对虾可食组织中诺氟沙星残留消除规律来确定休药期。凡纳滨对虾药饵给药剂量为 30 mg/kg（药物/虾体重），肌肉中诺氟沙星含量在给药后 72 h 为 0.028±0.005 μg/g，肝胰脏在给药后 192 h 为（0.036±0.017）μg/g。根据上述水产品中诺氟沙星推荐用残留限量（MRL 为 50 μg/kg），可以初步确定诺氟沙星休药期为：去头虾为 7 d，有头虾为 10 d。

参考文献

1. Kuemmerle H P，Murakawa T，Nightingale C H.（北京协和医院药剂科译）.抗微生物药物药物动力学：原理·方法·应用。北京：中国医药科技出版社，1997，1-44。

2. Uno K，Aoki T，Ueno R. Pharmacokinetic study of oxytetracycline in cultured rainbow trout，amago salmon，and yellowtail. Nippon Suisan Gakkaishi Bull. 1992，58（6）：1151-1156.

3. Abedini S，Namdari R，Law F C P. Comparative pharmacokinetics and bioavailability of oxytetracycline in rainbow trout and chinook salmon. Aquaculture，1998，162（1）：23-32.

4. Barron M G，Gedutis C and James M O. Pharmacokinetics of sulphadimethoxine in the lobster. Homerus americannus，following intrapericardial administration. Xenobiotica，1988，18（3）：269-277.

5. 李美同，郭文林，仲峰等。土霉素在鳗鲡组织中残留的消除规律。水产学报，1997，21（1）：39-42。

6. Barron M G，Hansen S C，Ball T. Pharmacokinetics and metabolism of triclopyr in the crayfish（Procambarus clarki）. Drug Metab. Disposition. 1991，19（1）：163-167.

7. 陈文银，管国华。甲鱼血液中诺氟沙星浓度的反相高效液相色谱测定法。上海水产大学学报，1997，6（4）：301-303。

8. Dall W，Hill B J，Rothlisberg P C，et al.（陈楠生等译）.对虾生物学。青岛：青岛海洋大学出版社，1992，31-35。

9. 林志彬，金有豫主编. 医用药理学基础. 北京：世界图书出版公司，1998.4-18.

10. James M O，Barron M G. Disposition of sulfadimethoxine in the lobster（Homarus americanus）. Vet. Hum. Toxicol.，1988，30（1 suppl.）：36-40.

11. James M O，Herbert A H. Disposition of ormetoprim in the lobster，Homarus americanus. Pharm. Res.，1988，5：196-203.

12. Della-Rocca G，Zaghini A，Magni A. Pharmacokinetics of amoxicillin in sea bream（Sparus aurata）after oral and intravenous administrations. 1：comparison of serum concentrations evaluated with two different analytical methods. Boll. Soc. Ital. Patol. Ittica，1998，10（24）：45-50.

13. Park E D，Lightner D V，Milner N，et al. Exploratory bioavailability and pharmacokinetic studies of sulphadimethoxine and ormetoprim in the penaeid shrimp，Penaeus vannamei Aquaculture，1995，130（2，3）：113-128.

14. Kleinow K M，Lech J J. A review of the pharmacokinetics and metabolism of sulfadimethoxine in the rainbow trout（Salmo gairdneri）. Vet. Hum. Toxicol. 1988，30（1 suppl.）：26-30.

15. 李天道，于佳，俞开康。四种弧菌对中国对虾的致病性研究，海洋湖沼通报，1998，1：57~64。

16. 陈毕生编著。虾蟹养殖管理与病害防治技术。广州：广东科技出版社，1992，146-148。

17. Anadon A，Martinez-Larranaga M R，Velez C et al. Pharmacokinetics of norfloxacin and its N-desethyl-and oxo-metabolites in broiler chickens［J］. Am. J. Vet. Res. 1992.53（11）：2084-2088.

该文发表于《水产学报》【2007，31（5）：655-660】

诺氟沙星在大黄鱼体内的药代动力学及残留研究

Pharmacokinetics and residues of norfloxacin in Pseudoscia enapolyactis

刘玉林[1,2]，王翔凌[2]，杨先乐[2]，杨勇[2]

（1. 长江大学动物科学学院，湖北 荆州 434025；

2. 上海水产大学，上海高校水产养殖学 E-研究院，农业部渔业动植物病原库，上海 200090）

摘要： 在试验水温（22±2）℃时，按 10 mg·kg^{-1} 的剂量给大黄鱼单次口服诺氟沙星后，用高效液相色谱法测定血浆和组织中的药物浓度，研究了诺氟沙星在大黄鱼体内的代谢及消除。结果表明血药时间数据符合一级吸收二室开放模型，吸收分布迅速，但消除缓慢，半衰期（$T_{1/2Ka}$、$T_{1/2\alpha}$、$T_{1/2\beta}$）分别为 0.703 0、2.092 6、154.326 5 h，最大血药浓度为 0.886 4 μg·mL^{-1}，达峰时间为 2.091 4 h，药时曲线下面积（AUC）为 97.803 8 μg·h^{-1}·mL^{-1}。组织中肝脏的药物浓度最高，在测定的时间里各组织的药物浓度高于血浆。药物消除速度依次为：肾脏、肝脏、肌肉，消除半衰期分别为 135.88、173.25、223.55 h，肌肉作为可食性组织，且消除最慢，因此选取肌肉组织作为残留检测的靶组织，以 50 μg·kg^{-1} 为最高残留限量，因此在本试验条件下，建议休药期不低于 23 d；在治疗大黄鱼细菌性疾病时，以诺氟沙星 10 mg·kg^{-1} 剂量给药，一般 1 d 1 次，连用 2~3 d。

关键词： 大黄鱼；药代动力学；诺氟沙星；残留

诺氟沙星（norfloxacin，NFLX）是氟喹诺酮类药物之一，因其具有抗菌谱广、抗菌力强、无交叉耐药性[1-2]等特点，而广泛用于动物和人类多种感染性疾病的治疗[3]。近年来，该药主要应用于养殖鱼类细菌病的防治[4]，它对革兰氏阳性菌和革兰氏阴性菌在低浓度时即表现杀菌作用[5]。但此类药物在食品动物中的残留会引起人类病原菌对其产生耐药性，它产生的毒副作用还会对人体产生直接的危害，因而其在可食性动物组织中的残留问题已经日益引起人们的关注。

目前有关 NFLX 在大黄鱼体内的药动和残留的研究国内外尚未见报道，且同一种药物在不同动物种类、不同温度条件下其药动及残留特征也有所差别[7-8]，因此本试验模拟实际养殖条件，研究了 NFLX 在大黄鱼体内的药代动力学和残留状况，为制定合理的休药期以及临床用药提供了理论依据。

1 材料与方法

1.1 实验动物

大黄鱼（*Pseudoscia enapolyactis*）来自浙江象山黄皮岙育苗厂，健壮无病无伤，平均体重（210±10）g。试验前在海水小网箱中暂养 1 周，水温控制在（22±2）℃下，增氧泵充氧。饲喂不含任何药物的全价饲料。所有试验大黄鱼经抽查表明组织中均不含 NFLX。

1.2 药品及试剂

NFLX 标准品，含量 99.5%，由中国兽医药品监察所提供；NFLX 原料药，含量 97.8%，购自上海三维制药厂；甲醇和乙腈为 HPLC 级；其他试剂均为分析纯。

1.3 高效液相色谱仪及色谱条件

采用 Agilent1100 型高效液相色谱（HPLC）仪（色谱柱类型：ZORBAX SB-C$_{18}$，4.6 mm×150 mm，5 μm）；可变波长紫外检测器；流动相甲醇：磷酸缓冲液（0.05 mol·L^{-1}磷酸二氢钠）：四丁基溴化铵（0.05 mol·L^{-1}）= 24.5：73.5：2（$V/V/V$），用 H$_3$PO$_4$ 调 pH 值至 2.0，过滤脱气后现用；流速 1.0 mL·min^{-1}；柱温 40℃；紫外检测波长 280 nm；进样量 10 μL。

1.4 试验设计及采样

实验用大黄鱼 56 尾，随机分为 14 组，每组 4 尾，每一时间点取 1 组鱼，另取数尾未给药的大黄鱼作空白对照。将试验鱼用 75 mg·L^{-1} 的丁香酚稍麻醉后，用套有橡皮软管的卡介苗注射器以 10 mg·kg^{-1}给药剂量口灌 NFLX 溶液（2%的醋酸配制），无回吐者保留试验。给药前 4 h 禁食，给药后 4 h 投料，并于给药后 0.5、1、2、4、8、16、24、48、96、192、

288、432、576、720 h 分别采取血浆、肌肉、肝脏和肾脏样品，全部样品放入 -80℃ 超低温冰箱冷冻保存用于药物分析。

1.5　样品处理及测定

样品处理及测定基本参照杨先乐等[9] 的方法。血浆样品自然溶解后，取 1 mL 样品于 10 mL 玻璃离心管中并加入 1 mL 甲醇，组织样品加入 1 倍体积的 6% 高氯酸，各样品旋涡混合 2 min，10 000 r·m in⁻¹ 离心 10 min，上清 0.45 μm 微膜过滤后，进行高效液相色谱测定。

测得 NFLX 的标准曲线为 $Ai = 68.489 C_i + 2.3282$。在 0.1~2 μg·mL⁻¹ 范围内，线性关系良好（$r = 0.999\ 6$），本法的灵敏度为 0.01 μg·mL⁻¹。血浆平均回收率（100.4±4.61）%，肌肉平均回收率（95.57±1.97）%，肝脏平均回收率（89.12±2.43）%，肾脏的平均回收率为（88.73±3.31）%，所有样品批内变异系数小于 8.56%，批间变异系数小于 9.75%。

1.6　数据处理

血浆药代动力学模型拟合及参数计算采用 3P97 药动软件分析[10]；药物在不同时间内在血浆和组织中的浓度柱形图和药物消除的药时曲线图采用 Microsoft Excel 2000 软件绘制；消除方程、休药期计算采用 SAS 统计软件（8e）计算。消除方程采用 $C_i = C_0 e^{-kt}$ 公式经非线性过程处理，C_i 表示药物浓度，C_0 表示初始浓度，k 表示消除速率常数。

1.7　休药期的计算

NFLX 是按一级动力学过程从体内消除的，即在消除后期服从指数消除：$C_i = C_0 e^{-kt}$，可以根据消除后期测定的组织药物浓度及规定的最高残留限量（MRL），计算各组织药物浓度降至规定水平所需时间：

$$t = \frac{\ln(C_0 \text{MRL})}{k}$$

2　结果与分析

2.1　药代动力学

大黄鱼以 10 mg·kg⁻¹ 的剂量经单次口服 NFLX 后，药物在血浆中的动力学参数如表 1 所示。经房室模型分析，其血药浓度时间数据符合一级吸收二室开放模型：$C = 1.3837 e^{-0.331\ 2t} + 0.428\ 8 e^{-0.004\ 5t} - (1.8125)^{-0.985\ 9t}$。药物在口灌后吸收较迅速，在 0.5 h 即能达到 0.33 μg·mL⁻¹，然后 2 h 达到峰值 1.05 μg·mL⁻¹。药物在血浆中分布较快，其分布半衰期（$T_{1/2\alpha}$）为 2.092 6 h，但其消除非常缓慢，消除半衰期（$T_{1/2\beta}$）为 154.326 5 h，在给药后 288 h 仍能检测药物，其浓度为 0.080 1 μg·mL⁻¹，直到 432 h 未能检测到药物。

2.2　药物在组织中的分布及残留

大黄鱼单剂量口服 NFLX 后，药物在血浆和组织中的分布如图 1，可看出药物在血浆和组织中的分布较广泛，但在同时间点里，大部分组织中的药物浓度尤其是肝脏显著高于血浆。给药后 576 h 除肝脏和肾脏外，其他组织均未检测到药物。给药后 720 h 时各组织均未检测到药物。大黄鱼经口灌 NFLX 后，药物在血浆中的动力学参数如表 1 所示。

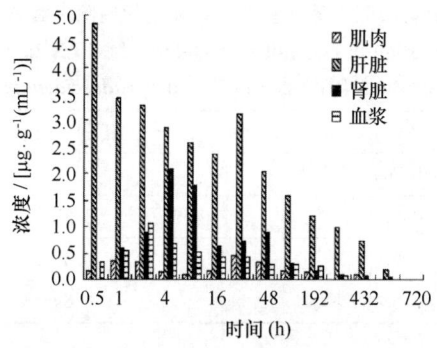

图 1　口灌给药后 NFLX 在大黄鱼血浆和组织中的药物浓度（10 mg·kg⁻¹）

Fig. 1　Concentrations of NFLX in plasma and tissues of Pseudosciaenapolyactisafter oral administration（10 mg/kg）

表1 大黄鱼单剂量口灌诺氟沙星（10 mg·kg⁻¹b. w）的动力学参数

Tab. 1 Pharmacokinetics parameters following single oral adiminnistration of
10 mg·kg⁻¹b. w of NFLX to *Pseudosciaenapolyactis*

参数 parameter	单位 unit	参数值 value	参数 parameter	单位 unit	参数值 value
A	$\mu g \cdot mL^{-1}$	1. 383 7	$T_{1/2\beta}$	h	154. 326 5
α	h^{-1}	0. 331 2	$T_{1/2Ka}$	h	0. 703 0
B	$\mu g \cdot mL^{-1}$	0. 428 8	C_{max}	$\mu g \cdot mL^{-1}$	0. 886 4
β	h^{-1}	0. 004 5	T_{max}	h	2. 091 4
Ka	h^{-1}	0. 985 9	AUC	$\mu g \cdot h \cdot mL^{-1}$	97. 803 8
$T_{1/2\alpha}$	h	2. 092 6	k_{12}	h^{-1}	0. 213 8
k_{21}	h^{-1}	0. 108 1	k_{10}	h^{-1}	0. 013 8

2.3 药物在组织中的消除规律

大黄鱼经口灌 NFLX 后，血浆及各组织的药物残留浓度随时间变化的规律见图2。

数据经回归处理得到血浆和4种组织中药物浓度（C）与时间（t）关系的消除曲线方程、相关指数（r^2）及消除半衰期（$T_{1/2}$）见表2。

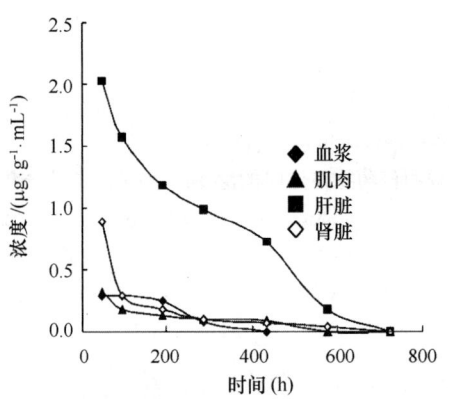

图2 大黄鱼口灌诺氟沙星（10 mg·kg⁻¹）后药物消除曲线

Fig. 2 Elimination curve ofNFLX in Pseudosciaenapolyactis after single oral administration at a dose of 10 mg·kg⁻¹

表2 大黄鱼口灌诺氟沙星后血浆和组织消除曲线方程及参数

Tab. 2 The equation of elimination curve and parameters of NFLX in plasma and tissues after oral
administration（10 mg·kg⁻¹）in *Pseudosciaenapolyactis*

组织 tissue	方程 equation	相关指数 r^2	消除半衰期（h）$T_{1/2}$
血浆 plasma	$C_{血} = 0.0442e^{-0.005\,1t}$	0. 78	135. 88
肌肉 muscle	$C_{肌} = 0.2720e^{-0.003\,1t}$	0. 82	223. 55
肝脏 liver	$C_{肝} = 2.6343e^{-0.004\,0t}$	0. 89	173. 25
肾脏 kidney	$C_{肾} = 0.6208e^{-0.005\,1t}$	0. 89	135. 88

2.4 休药期

欧盟对氟喹诺酮类药物（恩诺沙星、环丙沙星、单诺沙星）在食品动物中 MRL 的规定：在肌肉组织中为 100 μg·kg⁻¹，在肝脏和肾脏组织中为 200 μg·kg⁻¹，在有关家禽及猪可食组织 NFLX 的 MRL 为 50 μg·kg⁻¹[10]，NFLX 在水产品中的 MRL 尚未确定，因此本试验中 NFLX 在大黄鱼组织中的 MRL 暂以 50 μg·kg⁻¹ 为标准。按照 FDA 和欧盟推荐的回归计算方法[11-12]，通过休药期公式计算可知各组织的理论休药期为：肌肉 22.77 d，肝脏 41.30 d，肾脏 20.63 d。

根据休药期的定义，仅根据可食用组织中残留的浓度来确定休药期，因此在本试验条件下，以肌肉确定 NFLX 在大黄鱼体内的休药期为 23 d。

3　讨论

3.1　诺氟沙星在大黄鱼体内的药代动力学特征

大黄鱼以 10 mg·kg^{-1} 的剂量单次经口灌服 NFLX 后，药物在血浆中的达峰时间为 2.091 4 h，其最高药物浓度为 0.886 4 μg·mL^{-1}。吸收相半衰期 $T_{1/2Ka}$ 为 0.703 0 h，分布相半衰期 $T_{1/2\alpha}$ 为 2.092 6 h，消除相半衰期 $T_{1/2\beta}$ 为 154.326 5 h，由药动学参数可知药物在血浆中的吸收较迅速，分布快，但消除较缓慢。代表药物分布量的 AUC 值为 97.803 8 μg·h·mL^{-1}，说明大黄鱼口服 NFLX 的吸收效果较好。与相关研究相比较[13-14]，发现该药在大黄鱼血浆中的达峰时间比凡纳滨对虾肌注该药的达峰时间（2 min）长，与凡纳滨对虾药饵给药的达峰时间（2 h）相当，药物峰浓度也有较大差异，凡纳滨对虾中肌注给药的峰浓度最高，说明给药方式对药动学参数的差异影响较大。另外，药饵给药发现凡纳滨对虾存在血药再吸收现象，而本试验血浆中未发现此现象，可能是由于种属之间存在药物吸收特异性的原因。与斑节对虾单次肌注 NFLX 的研究相类似[15]，血药时间数据符合二室开放模型。王群等[16]研究发现 NFLX 口服给药后，在养殖鲈鱼内 0.33 h 内能达到峰浓度，吸收半衰期 $T_{1/2Ka}$ 为 0.73 h，消除半衰期 $T_{1/2\beta}$ 为 15.09 h。与本试验相比较，药物在大黄鱼体内的吸收与消除均较慢，能够在较长时间内维持血药浓度，可有效达到治疗疾病的目的。

从以上比较看出 NFLX 在大多数水产动物体内的房室模型符合一级吸收二室开放模型，但其药动学特征存在显著差异，此差异可能由于种属[15,17]、性别[18]、水温[19]、给药方式[14]、给药剂量[20-21]等原因造成。因此造成水产动物的药代动力学上差异的原因非常复杂，须慎重分析其差异才能制定出合理的给药方案。

3.2　诺氟沙星在大黄鱼组织中的残留及消除规律

在本试验条件下，大黄鱼单剂量口服 NFLX 后，药物在肝脏和肾脏中的浓度最高，这说明该药物在大黄鱼体内主要以肝脏的代谢和肾脏的排泄为主。并且从各点药物浓度可以看出在同一时间内组织中的药物浓度显著高于血浆，尤其是肝脏与血浆的药物浓度比最高值 14.58：1，说明该药物在体内的穿透力极强，广泛分布于各组织中。组织中的药物浓度较高，可用于全身及深部组织感染的治疗。所选取的组织和血浆中的消除半衰期 $T_{1/2}$ 在 135.88～223.55 h 间，给药后 0.5 h，其药物浓度肝脏为 4.825 1 μg·mL^{-1}，而肾脏在该时间点未检测到药物，可以看出各个组织的分布量以及消除速度差异较大。NFLX 在肝脏和肾脏的残留量较明显，该研究结果与所报道的养殖鲈鱼口服 NFLX 的结果相符[6]，但是与该药在牙鲆体内的残留特点有所区别[22]。造成这种区别的原因可能由于大黄鱼与鲈鱼同属于鲈形目，它们的亲缘关系较近，而牙鲆却是鲽形目的鱼类，大黄鱼与之亲缘关系较远的原因关于这一点，沙拉沙星在鲫和大西洋鲑体内的残留及消除研究也可以得到证实[7,21]。即使是属于同类但不同种的药物在同一种动物体内的残留消除规律也有差别，达氟沙星和盐酸沙拉沙星在鲫体内的残留的研究已经证实了这一点[7-8]。

另外在给药后，发现肝脏中的药物浓度在短时间内达到峰值的原因可能是"首过效应"引起的，对此也有相应的报道[14]。给药后肝脏在 0.5 h 达到最高浓度，然后药物浓度开始下降，到 24 h 药物又达到一个峰值，即肝脏中出现了药物重吸收现象，出现这种现象的原因可能由于"肝胆循环"所致。

药物在鱼体内的残留量，一般随药物给药的剂量的增高而增加，而消除速度随水温的降低而减慢[23]。因此不同温度及剂量对药物消除规律影响较大，方星星等[19]研究了在（22±3）℃、（16±2）℃两种水温条件下，复方新诺明在花鲈体内的残留及消除规律，研究表明，两种温度下其残留及消除情况差异较大，较高温度下药物吸收率高且消除速度快。Martinsen 等[21]采用的是多剂量给药方式，沙拉沙星在大西洋鲑中的药物消除规律明显的与单剂量给药方式有较大差异[20]。由于药物在体内的消除受许多因素的影响，因此关于休药期的制定应多方面考虑。本试验条件下所制定的休药期适应于本试验的条件，当条件有所区别时此休药期仅作参考。

3.3　临床用药建议

张雅斌等[24]研究了 NFLX 等 7 种抗菌药对鱼类细菌出血性败血病的致病菌的敏感性，发现 NFLX 的敏感性最高，其最小抑菌浓度（MIC）小于 0.2 μg·mL^{-1}，体内抗菌试验也表现出良好的抗菌作用，并且能提高感染鱼的存活率。采用本试验中的 10 mg·kg^{-1} 剂量口服给药时，在给药后 0.5 h，血药浓度已超过 NFLX 对大多数致病菌的最小抑菌浓度 0.2 μg·mL^{-1}，192 h 血药浓度为 0.243 0 μg·mL^{-1}，即给药 0.5～192 h 之间可有效抑菌。本试验的结果认为，以 NFLX 10 mg·kg^{-1} 对大黄鱼口服给药，达到此浓度的其间隔时间可长达 192 h。如果考虑到在生产实践中药物可能造成的损失、鱼类个体的差异、环境的差异，以及急性感染期间的情况，为了提高药物的治疗效果，可视具体情况将给药间隔相应缩短。据此，我们建议，治疗大黄鱼细菌性疾病，以 NFLX10 mg·kg⁻¹量给药，一般 1 d 1 次，连用 2~3 d。本试验对于保证诺氟沙星在水生动物疾病防治上的合理应用以及充分保证食用者的健康安全具有一定的理论价值。

参考文献

[1] 郭惠元. 我国诺酮类抗菌药物研究开发概况 [J]. 中国医药工业杂志, 1989, 20 (9)：421-424.

[2] Vancutsem P, Babish J G, ScwardW S. The fluoroquinolones antimicrobial：struettire, antimicrobial anmials and toxicity [J]. CornellVet, 1990, 80 (2)：173-186.

[3] 邓国东、杨桂香、陈杖榴. 氟喹诺酮类药物在动物性食品中的残留检测研究进展 [J]. 中国兽药杂志, 2000, 34 (3)：53-58.

[4] 张雅斌、王兆军. 养殖新技术 [M]. 长春：吉林科学技术出版社, 1999：134-137.

[5] 竺心影. 药理学 (第 3 版) [M]. 北京：人民卫生出版社, 1995：310.

[6] Sherri B T, Calvin C W, Jose E R, et al. Confirmation of fluoroquinolones in catfish muscle by electrospary liquid chromatography mass spectrometry [J]. J AOAC Int, 1998, 8 (3)：554-562.

[7] 王翔凌、方之平、操继跃、等. 盐酸沙拉沙星在鲫体内的残留及消除规律研究 [J]. 水生生物学报, 2006, (30) 2：198-203.

[8] 潘玉善、操继跃、方之平、等. 甲磺酸达氟沙星在鲫体内药物动力学及残留研究 [J]. 中国农业科学, 2006, 39 (2)：418-424.

[9] 杨先乐、王民权、杨勇、等. 诺氟沙星在大黄鱼血浆和肌肉中的 RP-HLLC 检测方法 [J]. 水产学报, 2004, 28 (增刊)：1-6.

[10] 魏树礼、张强. 生物药剂学与药物动力学 [M]. 北京：北京医科大学出版社, 1997：208-217.

[11] Anadon A, M artinez-LarmnagaM R, V elez C, et al. Pharmacokinefics of norfloxacin and its N-desethyl and oxometabolites in broiler chickens [J]. Am J V et Res, 1992, 53 (11)：2084-2088.

[12] Concordet D, Toutain P L. The withdraw altimeestimation of veterinary drugs revisited [J]. Journal of Veterinary Pharmacology and Therapeutics, 1997, 20：380-386.

[13] MartinezM, Friedlander P, Condon R, et al. Response to criticisms of the US FDA parametric approach for withdrawal time estimation：rebuttal and comparison to the nonparametric method proposed by concordet and toutain [J]. Journal of VeterinaryPharmacology and Therapeutics, 2000, 23：21-35.

[14] 房文红、杨先乐、周凯. 诺氟沙星在凡纳滨对虾不同组织中处置和消除规律 [J]. 水产学报, 2004, 28 (增刊)：19-24.

[15] 房文红、邵锦华、施兆鸿、等. 斑节对虾血淋巴中诺氟沙星含量测定及药代动力学 [J]. 水生生物学报, 2003, 27 (1)：13-17.

[16] 王群、刘淇、唐雪莲、等. 诺氟沙星在养殖鲈体内的代谢动力学和残留研究 [J]. 水产学报, 2004, 28：14-18.

[17] 张雅斌、张祚新、郑伟、等. 不同给药方式下鲤对诺氟沙星的药代动力学研究 [J]. 水产学报, 2000, 24：559-563.

[18] 杨先乐、刘至治、恒山雅仁. 盐酸环丙沙星在中华绒螯蟹体内药物代谢动力学研究 [J]. 水生生物学报, 2003, 27 (1)：18-21.

[19] 方星星、李健、王群、等. 复方新诺明在花鲈体内的残留及消除规律 [J]. 海洋科学, 2003, 27 (9)：16-20.

[20] H o S P, Cheng C F, W ang W S. Pharmacokinetic and depletion studies of sarafloxacin after oral adminstration to eel (Anguillaanguilla) [J]. Journal of Veterinary Medical Science, 1999, 61 (5)：459-463.

[21] Martinsen B, Horsberg T E, BurkeM. Multiple-dose pharmacokinetic and depletion studies of sarafloxacin in Atlantic salmon (Salmo salar) [J]. Journal of Fish Diseases, 1994, 17 (2)：111-121.

[22] 刘秀红、李健、王群. 诺氟沙星在牙鲆体内的残留及消除规律研究 [J]. 海洋水产研究, 2003, 24 (4)：13-18.

[23] Bjorklund H V, By lund G. Temperature related absorption and excretion of oxytetracycl-ine in rainbow trout [J]. Aquaculture, 1990, 84：363-372.

[24] 张雅斌、张祚新、郑伟、等. 诺氟沙星在鱼类细菌性疾病中的应用研究 [J]. 大连水产学院学报, 2006, 15 (2)：79-85.

该文发表于《水生生物学报》【2003, 27 (1)：13~17】

斑节对虾血淋巴中诺氟沙星含量测定方法和药代动力学初步研究

Analytical Method of Norfloxacin in the Giant Tiger Shrimp (*Penaeus Monodon*) Hemolymph and Brief Study on Pharmacokinetics

房文红，邵锦华，施兆鸿，杨宪时

(中国水产科学研究院东海水产研究所，上海 200090)

摘要：采用 Agilent 1100 液相色谱仪测定斑节对虾血淋巴中诺氟沙星含量，色谱柱为 ZORBAX SB-C18，流动相为：乙腈：磷酸盐缓冲液 (含 10 mmol/L 磷酸二氢钠和 5 mmol/L 溴化四丁基铵，用磷酸调节 pH2.0＝14.5：85.5

（V/V）；流速为 0.5 mL/min；柱温为 25℃；紫外检测器，检测波长为 280 nm；进样量为 10 μL。用外标法计算。上述条件下检测诺氟沙星浓度范围为 0.051~20.4 μg/mL，平均回收率为 96.49±1.60%，日内和日间精密度均小于 3%（RSD），适合于检测对虾血淋巴中 NFX 含量。使用该方法初步研究了斑节对虾一次性肌肉注射诺氟沙星的药代动力学，斑节对虾血淋巴药-时曲线可以用开放性二室模型（$C_0 = 11.73 \times e^{-1.133t} + 22.62 \times e^{-10.918t}$）来描述。由此推算诺氟沙星的药动学参数，分布和消除半衰期分别为 0.0635 和 0.612 h；曲线下面积（AUC）、总体消除率（CLs）和表观分布容积（V_d）分别为 12.425 μg·h/mL、804.83 mL/kg·h 和 710.35 mL/kg。

关键词：斑节对虾；HPLC 法；诺氟沙星；药代动力学

国外自 20 世纪 50 年代开展渔用药物在水产动物体内的药代动力学研究以来，涉及的水产动物和药物很多，水产动物主要是海水鱼类，如虹鳟[1]、大麻哈鱼[1,2]、大西洋鲑[2]和斑点鲴[3]等，药物主要有抗生素、磺胺和喹诺酮类等。而国内对渔药药代动力学研究远远落后，始于 90 年代。曾子建等研究了喹乙醇在鲤鱼体内的药代动力学[4]；李美同等采用 HPLC 法研究了日本鳗鲡组织中土霉素残留的消除规律[5]；台湾学者刘朝鑫用微生物法调查了硫酸新霉素在日本鳗鲡体内的吸收与消除[6]。在诺氟沙星（norfloxacin）方面，陈文银等用反相高效液相色谱法测定中华鳖口服诺氟沙星后血液中的浓度[7]，张雅斌等研究诺氟沙星在鲤鱼体内药代动力学[8]。在甲壳动物中，Park 等研究了南美白对虾对磺胺类药物的生物利用度及药代动力学[9]，James 和 Barron 报道了龙虾体内磺胺二甲氧嘧啶处置过程[10]。国内未见有关对虾药动学的报道。

诺氟沙星在对虾养殖中常用于红腿病、烂鳃病[11]和败血症[12]等疾病防治，因此很有必要了解诺氟沙星在对虾体内的药物代谢过程，正确指导虾病防治中如何使用药物，提供用药依据，制定用药方案，确定药物剂量与用药周期。本文主要探讨斑节对虾（*Penaeus monodon*）血淋巴中诺氟沙星的测定方法，初步探讨了一次性肌肉注射给药后斑节对虾体内诺氟沙星的药代动力学。

1　材料与方法

1.1　实验试剂

诺氟沙星原料药（含量 98%）由上海三维制药厂提供，诺氟沙星标准品（含量 99.5%，批号 H040798），由中国兽药监察所提供。乙腈，HPLC 级，上海化学试剂公司生产；溴化四丁基铵，分析纯，上海化学试剂公司生产；磷酸二氢钠，分析纯，上海新华化工厂生产。磷酸，分析纯，江苏教育试剂厂生产；醋酸，分析纯，上海试剂一厂生产；肝素钠，上海化学试剂公司生产。

2% 诺氟沙星注射液配制：称取 2.0 g 诺氟沙星原料药，用少量的 2% 醋酸溶解，然后转移到 100 mL 棕色容量瓶，再用 0.9% 生理盐水定容至刻度。

1.2　仪器高效液相色谱仪

美国 Agilent 公司的 Agilent 1100，色谱柱为 ZORBAX SB-C18，检测器为与其相配套的紫外检测器。低温冰箱，SANYO（日本），-80℃。移液器（1-10 μL，10-100 μL），eppendorf 公司（德国）。电子天平，METTLER TOLEDO AB204（瑞士）。TGL-16G 高速冷冻离心机（上海安亭科学仪器厂），以及震荡器和 0.45 μm 微孔滤膜等。

1.3　色谱条件

流动相：乙腈：磷酸盐缓冲液（含 10 mmol/L 磷酸氢二钠和 5 mmol/L 溴化四丁基铵，磷酸调节 pH 值到 2.0）= 14.5：85.5（V/V）。流速：0.5 mL/min。柱温：25℃。进样量：10 μL。检测波长：紫外，280 nm。

1.4　试验动物

试验用的斑节对虾取自金山县漕泾对虾养殖场，暂养在比重为 1.009 5、水温为 24~25℃海水中，暂养 7~10 d 后用于试验。试验在水体为 60 L 玻璃缸中进行，每只缸用 20 目筛绢分隔成 4 小格，每一小格放养体表无损、健康成虾 1 尾。试验期间不断充气和水体循环过滤，水温为（25±1）℃，试验期间投喂适量螺蛳肉，并及时排出残饵和污物。

1.5　斑节对虾给药和血淋巴采集

试验用斑节对虾体重为（17.8±1.9）g，给药剂量为 10 mg/kg，根据对虾体重计算给药量 15~20 μL，一次性肌肉注射，给药部位从对虾腹部第二腹节腹面稍斜插入，轻轻推进。试验期间对虾成活率为 100%。

将试验虾群体看作一个整体。给药前同批虾采一次空白血淋巴，于给药后 0.017、0.033、0.067、0.1、0.5、2、12、24、48、96、144、192 h 各点采集血淋巴样，每一时间点采集 4 尾对虾血淋巴，用于血淋巴诺氟沙星浓度分析。

血淋巴取样方法：先将取样虾用纱布擦干头胸甲，然后将 1 mL 注射器缓慢插入围心窦，抽取血淋巴 0.5 mL 左右，再转移到离心管中，在震荡器上震荡 1 min，于 5 000 rpm. 转速下离心 5 min，取出上层液贮存于-80℃低温冰箱中，直至药物浓度分析。注射器和离心管使用前用 1%肝素钠溶液均匀地涂布在壁上，吹干备用。

1.6 血样处理

从冰箱中取出血淋巴样品，于室温下自然融解后，加入等体积乙腈，旋涡快速混匀器上混合 2 min，于 5 000 rpm. 转速离心 10 min，吸取上清液经 0.45 μm 微孔滤膜过滤后，用于 HPLC 分析。

1.7 诺氟沙星标准曲线的制备

准确称取 0.025 5 g 干燥恒重的诺氟沙星标准品，置于 250 mL 烧杯中，用少许 2%醋酸溶解后，转移到 250 mL 容量瓶中，用磷酸盐缓冲液稀释至刻度，即成 0.102 mg/mL 诺氟沙星标准品母液，-20℃保存备用。取 1 mL 进样瓶 10 个，分别加入上述母液 0.5、1、2、5、20、40、60、80、100 μL，然后用流动相准确稀释至 500 μL，即成 0.102、0.204、0.408、1.02、2.04、4.08、8.16、12.24、16.32、20.40 μg/mL 诺氟沙星标准曲线溶液。HPLC 进样量 10 μL.

1.8 回收率测定

采用加样回收法。取 0.102 mg/mL 诺氟沙星标准母液 5、40 和 80 μL，分别加入到 245、210 和 170 μL 血淋巴中，混匀后按"血样处理"处理此血淋巴样品，HPLC 测定诺氟沙星浓度。每个浓度重复 5 次。根据诺氟沙星标准曲线计算出各血淋巴诺氟沙星实测浓度。回收率=实测浓度/理论浓度×100%。

1.9 精密度测定

日内精密度：在一日内的试验初、中、末，分别取 0.102 mg/mL 诺氟沙星标准母液 40 μL，加入到 210 μL 血淋巴中，混匀后按"血样处理"处理此血淋巴样品，进样测定其药物浓度。每次重复 3 次，记录诺氟沙星峰面积，计算日内精密度。日间精密度：一周内重复测定 2 次，每次重复 3 次，计算日间精密度。

1.10 数据处理

标准曲线线性回归采用 Microsoft Excel 2000 软件分析；药代动力学模型拟合及参数计算采用 SAS 统计分析软件（SAS© Proprietary Software Release 6.12）非线性回归分析部分 Marquardt 方法分析。

2 结果与讨论

2.1 诺氟沙星的色谱行为

HPLC 法的流动相主要参考 Nangia 等[13]应用反相离子对 HPLC 测定人血浆中氟喹诺酮类药物所使用的流动相，但其乙腈与磷酸盐缓冲液的比例经过数十次调整，定为 14.5：85.5（V/V）。其他色谱条件基本未改变，流速 0.5 mL/min，紫外检测波长为 280 nm，进样量 10 μL。分别用 0.40 μg/mL 诺氟沙星标准溶液、空白血淋巴添加诺氟沙星标准液连续多次进样，基线走动平稳，保留时间为（3.58±0.02）min。样品血淋巴中虽出现杂峰，但无干扰峰，且与药峰分离良好（见图 1）。

2.2 诺氟沙星线性范围和灵敏度

将 0.051、0.102、0.204、0.408、1.02、2.04、4.08、8.16、12.24、16.32、20.40 μg/mL 诺氟沙星标准曲线溶液进样后，用紫外检测器检测，记录其峰面积。以诺氟沙星浓度为横坐标（x），峰面积为纵坐标（y），将各点绘制到坐标图上（见图 2），诺氟沙星线性范围在 0.051~20.40 μg/mL 之间，两者间具有很好的相关性（$P < 0.01$）。机器噪声为 9.52×10^{-6} AU，以引起 2 倍基线噪声的药物浓度为最低检测限，本方法最低检测限为 0.02 μg/mL。

2.3 回收率

在空白血淋巴中添加诺氟沙星标准溶液，使其药物理论浓度分别为 2.04、16.32 和 32.64 μg/mL，然后按"血样处理"进行处理，测定诺氟沙星的含量。根据实测浓度和理论浓度计算出其回收率（见表 1），诺氟沙星在低、中、高三个浓度下平均回收率为（96.49±1.60）%。

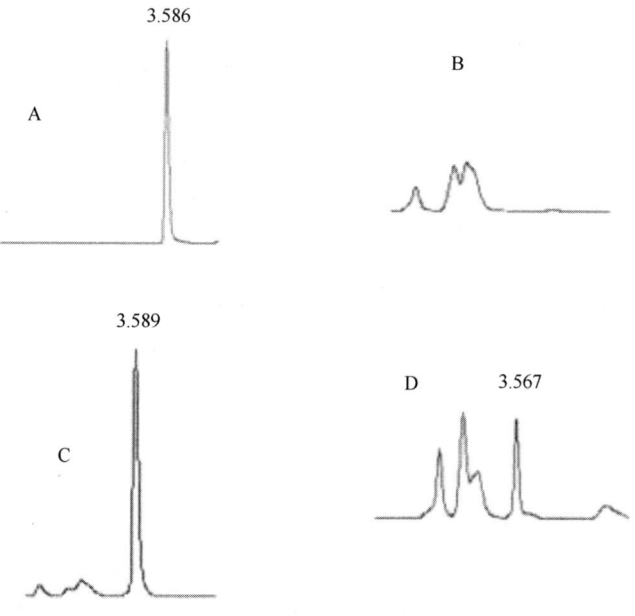

图 1 诺氟沙星色谱图

Fig. 1 HPLC chromatograms of norfloxacin

A 为诺氟沙星标准品色谱图，B 为对虾空白血浆色谱图，C 为对虾空白血浆加诺氟沙星标准品色谱图，
D 为对虾给药 2 h 后血浆色谱图；图中所标数值为诺氟沙星的保留时间

A，HPLC chromatogram from the standard sample of norfloxacin. B，HPLC chromatogram from shrimp plasma free of
norfloxacin. C，HPLC chromatogram from shrimp plasma supplied with norfloxacin. D，HPLC chromatogram
from shrimp plasma at 2 h after administration. The values represent the retention time.

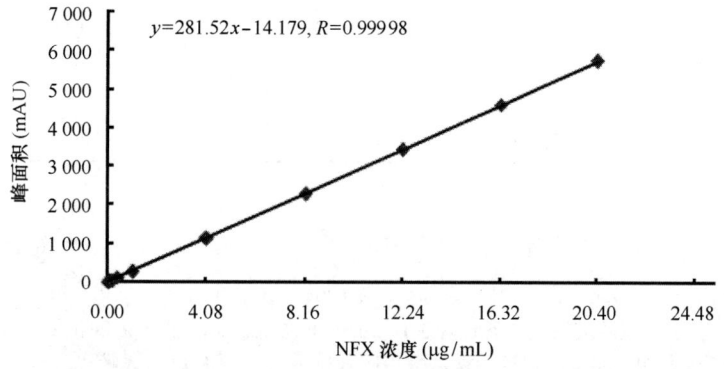

图 2 紫外检测器所检测的峰面积与 NFX 浓度的关系

Fig. 2 Realationship between peak area detected by UV detector and norfloxacin concentrations

表 1 诺氟沙星回归率

Table 1 The recoveries of RP–HPLC method for norfloxacin

组别 Group	理论浓度（μg/mL） Theoretic concentration	实测浓度（μg/mL） Determined concentration	回收率（%） Recovery	平均回收率（%） Mean recovery
1	2.04	1.982/1.992/1.956/1.942/1.960	96.39	
2	16.32	15.816/15.542/15.816/15.136/15.178	94.96	96.49±1.60
3	32.64	31.886/32.012/31.932/32.664/31.626	98.16	

2.4 精密度

日内精密度：在空白血淋巴中添加诺氟沙星标准溶液，血药浓度为 8.16 μg/mL，在试验刚开始时、试验中期、试验末期分别测定血药浓度，每次重复 3 次，记录其诺氟沙星峰面积，推算血药浓度。计算出相对标准偏差（RSD）为 2.34%。详见表 2。

表 2　诺氟沙星的日内精密度
Table 2　The intra-day precision of RP-HPLC method for norfloxacin

组别 Group	1	2	3
实测浓度（μg/mL） Determined concentration	8.117/7.719/7.740	7.963/7.926/8.066	7.551/7.985/7.799
平均浓度（μg/mL） Mean concentration	7.874		
标准偏差 SD	0.184		
相对标准偏差 RSD	2.34%		

日间精密度：在空白血淋巴中添加诺氟沙星标准溶液（血药浓度为 8.16 μg/mL），在一周内重复测定 2 次，计算其标准偏差和相对标准偏差。日间相对标准偏差（RSD）为 2.77%（表 3）。

表 3　诺氟沙星的日间精密度
Table 3　The inter-day precision of RP-HPLC method for norfloxacin

组别 Group	1	2	3
实测浓度（μg/mL） Determined concentration	8.117/7.719/7.740	7.963/7.926/8.066	7.551/7.985/7.799
平均浓度（μg/mL） Mean concentration	7.874		
标准偏差 SD	0.184		
相对标准偏差 RSD	2.34%		

2.5 斑节对虾一次性肌注给药后诺氟沙星药代动力学

诺氟沙星以 10 mg/kg 剂量、一次性肌肉注射对斑节对虾给药后，在不同时间点采集对虾血淋巴样品，测得其血淋巴中诺氟沙星浓度见表 4。血淋巴中诺氟沙星浓度上升迅速，1~2 min 的血淋巴诺氟沙星浓度相差不大，为最高值；随后血药浓度迅速下降，诺氟沙星处于分布相，30 min 以后血药浓度下降速度变缓，12 h 血药浓度比较低；48 h 以后几乎测不出诺氟沙星。

表 4　肌注诺氟沙星斑节对虾血药浓度的实测值与理论值
Table 4　The determined and theoretic concentrations of norfloxacin in hemolymph of Penaeus monodon

时间 Time	实测值（μg/mL） Determined concentration	理论值（μg/mL） Theoretic concentration
1 min	29.04±2.92	30.37±3.79
2 min	29.10±2.67	27.02±0.69
4 min	21.55±1.94	21.80±0.56
6 min	17.40±4.62	18.07±0.46
30 min	7.01±2.71	6.75±0.17
2 h	1.02±0.38	1.22±0.03

时间 Time	实测值（μg/mL） Determined concentration	理论值（μg/mL） Theoretic concentration
12 h	0.28±0.01	1.45E-5±3.7E-7
24 h	0.06±0.12	0.00
48 h	0.04±0.18	0.00

应用非线性最小二乘回归分析法对所测得数据进行一室、二室和三室的模型方程（$C_0 = C_1 e^{-L1t}$　$C_0 = C_1 e^{-L1t} + C_2 e^{-L2t}$　$C_0 = C_1 e^{-L1t} + C_2 e^{-L2t} + C_3 e^{-L3t}$）与数据拟合，使数据的加权平方和最小。拟合结果为应用二室模型方程与所测的数据拟合最好，这表明斑节对虾一次性肌肉注射给药后，诺氟沙星的处置可以用开放性二室模型来描述，其血淋巴药-时关系方程为：$C_0 = 22.62 \times e^{-10.918t} + 11.73 \times e^{-1.133t}$，该拟合曲线相关指数为 $R^2 = 0.9947$，并有此方程推算各采样时间点诺氟沙星血药浓度的理论值（见表4），与实测值十分接近。

根据血淋巴药-时关系方程计算药动学参数见表5。诺氟沙星经肌肉注射进入虾体后，在 1~2 min 内血药浓度达到最高值，随后进入分布和消除阶段。诺氟沙星的分布和消除半衰期分别为 0.063 5 h 和 0.612 h，曲线下面积（AUC）为 12.425 ug·h/mL，总体清除率（CLs）为 804.83 mL/kg·h，表观分布容积（V_d）为 710.35 mL/kg。

表5　斑节对虾肌肉注射诺氟沙星给药后的药代动力学参数
Table 5　Pharmacokinetic parameters of norfloxacin in Penaeus monodon following intrmuscular administration

参数 parameters	肌注给药 intrmuscular administration
A（μg/mL）	22.61
B（μg/mL）	11.73
α（h^{-1}）	10.918
β（h^{-1}）	1.133
$T_{\alpha 1/2}$（h）	0.0635
$T_{\beta 1/2}$（h）	0.612
k_{10}（h^{-1}）	0.765
k_{21}（h^{-1}）	4.474
k_{12}（h^{-1}）	4.812
AUC（μg·h/mL）	12.425
CLs（mL/kg·h）	804.83
V_d（mL/kg）	710.35

注：A、B 为浓度对时间曲线对数图上直线在零时的截距；α、β 分别为分布相、消除相的一级速率常数；$T_{\alpha 1/2}$、$T_{\beta 1/2}$ 分别为分布、消除半衰期；k_{10} 描述药物从中央室消除的一级速率常数；k_{12} 描述药物从第一室转运到第二室的一级速率常数；k_{21} 为描述药物从第二室转运到第一室的一级速率常数；AUC 为血淋巴药-时曲线下面积；CLs 为总体清除率；V_d 为表观分布容积。

3　讨论

3.1　采用该高效液相色谱法分析斑节对虾血淋巴中诺氟沙星浓度，药峰分离良好，检测浓度范围较广（0.051~20.4 μg/mL），且相关系数高达 $R^2 = 0.999\ 97$（$P < 0.01$）。诺氟沙星在低、中、高三个浓度下平均回收率为（96.49±1.60）%，这说明该血淋巴处理方法的回收率较高。该方法的日内精密度和日间精密度分别为 2.34% 和 2.77%，这说明该方法的精密度较高，即包括血淋巴样品的处理在内的高效液相色谱的全过程操作分析测定诺氟沙星方法具有较好的再现性。由此可见，该高效液相色谱法适合于分析斑节对虾血淋巴中诺氟沙星浓度。

3.2　鲤鱼肌注诺氟沙星给药后的药动学为开放性二室模型，与本研究的对虾肌注给药的药动学结果相一致[8]。在相同的给药剂量（10 mg/kg）下，肌注给药后对虾和鲤鱼血药浓度达到最高值的时间十分相近，分别为 2 min 和 2.5 min；但是对虾和鲤鱼最高血药浓度却相差较大，对虾为鲤鱼的 1.6 倍。对虾的消除半衰期 $T_{\beta 1/2}$ 为 0.612 h，而鲤鱼为 3.403 h；24 时对虾的血药浓度为 0.06 μg/mL，而鲤鱼 36 h 的血药浓度仍在 0.47 μg/mL，这说明诺氟沙星在对虾体内消除明显快于鲤鱼。

参考文献

[1]　Namdari R，Abedini S，Law F C P. A comparative tissue distribution study of oxytertracycline in rainbow trout，Oncorhynchus mykiss（Walbaum），and Chinook salmon，Oncorhynchus tshawytscha（Walbaum）[J]. Aquaculture Research. 1999，30（4）：279-286.

[2]　Namdari R，Abedini S，Albright L，et al. Tissue distribution and elimination of oxytertracycline in sea-pen cultured Chinook salmon，Oncorhynchus tshawytscha，and Atlantic salmon，Salmo salar，following medicated-feed treatment [J]. Journal of Applied Aquaculture，1998，8（1）：39-52.

[3]　Plakas S M，El-Said K R，Stehly G. R. Furazolidone disposition after intravascular and oral dosing in the channel catfish [J]. Xenobiotica，1994，24（1）：1095-1105.

[4]　曾子建，李逐波，吴绪田等. 喹乙醇在鲤鱼体内的药代动力学研究 [J]. 四川农业大学学报，1993，11（1）：109-112.

[5]　李美同，郭文林，仲峰，等. 土霉素在鳗鲡组织中残留的消除规律 [J]. 水产学报，1997，21（1）：39-42.

[6]　刘朝鑫. 硫酸新霉素在鳗鱼体内之吸收分布与排泄 [J]. 养鱼世界，1995，（5）：66-73.

[7]　陈文银，应春华. 诺氟沙星在中华鳖体内的药代动学研究 [J]. 水产学报，1997，21（4）：434-437.

[8]　张祚新，张雅斌，杨永胜，等. 诺氟沙星在鲤鱼体内的药代动力学 [J]. 中国兽医学报，2000，20（1）：66-69.

[9]　Park E D，Lightner D V，Milner N，et al. Exploratory bioavailability and pharmacokinetic studies of sulphadimethoxine and ormetoprim in the penaeid shrimp，Penaeus vannamei [J]. Aquaculture. 130；113-128.

[10]　Li Chung Li J，James M O. The oral bioavailability，pharmacokinetics and biotransformation of 9-Hydroxybenzo [a] pyrene in the American lobster，Homarus americanus [J]. Mar. Environ. Res. 1998. 46；505-508.

[11]　俞开康，战文斌，周丽. 海水养殖病害诊断与防治手册 [M]. 上海科学技术出版社. 2000，89-136.

[12]　樊海平，孟庆显，俞开康. 由二种气单胞菌引起的中国对虾败血病的研究 [J]. 海洋与湖沼，1995，26（3）：302-307.

[13]　Nangia A，Lam F，Hung C T. Reserved-phase ion-pair high-performance liquid chromatographic determination of fluroquinolines in human plasma [J]. J. Pharma. Sci. 1990，79（11）：988.

该文发表于《水生生物学报》【2006，30（5）：541-546】

肌注和药饵给药下诺氟沙星在南美白对虾血淋巴中药代动力学

Pharmacokinetics of Norfloxacin in Hemolymph from Whiteleg Shrimp，Penaeus Vannamei Following in Tramuscle Injection and oral Administration

房文红，郑国兴

（农业部海洋与河口渔业重点开放实验室中国水产科学研究院东海水产研究所，上海 200090）

摘要：本文分析了盐度 1‰ 和 15‰ 下诺氟沙星肌肉注射给药（剂量 10 mg/kg）和盐度 15‰ 下药饵口服给药（剂量 15 mg/kg 和 30 mg/kg）后南美白对虾血淋巴中药代动力学。在盐度 1‰ 和 15‰ 下，南美白对虾肌注给药后血药浓度的变化趋势基本相似，血药浓度与时间关系曲线适合用二室模型来描述，其药动学方程分别为 $C_0 = 35.422 \times e^{-9.778t} + 4.363 \times e^{-0.165t}$ 和 $C_0 = 35.144 \times e^{-13.335t} + 7.888 \times e^{-0.608t}$，但两盐度下部分药动学参数差别较大。药饵口服给药（30 mg/kg）后，血药浓度与时间关系曲线表现为双峰现象，出现的时间分别为给药后 4 h 和 12 h，峰浓度分别为 2.86 μg/mL 和 2.037 μg/mL。以 15 mg/kg 剂量药饵口服给药的血药浓度-时间变化与 30 mg/kg 剂量的结果相似，但第二峰浓度出现的时间为 8 h，且血药浓度与给药剂量没有明显的比例关系。

关键词：南美白对虾；诺氟沙星；药代动力学；血淋巴

在细菌性疾病的药物防治中，药代动力学研究具有十分重要的理论意义和实用价值，可为制定用药方案、确定药物剂量与用药周期提供科学依据。国外自上世纪 50 年代 Snieszko 等[1]报道了磺胺类药物在虹鳟体内药物浓度变化以来，涉及的水产动物和药物很多[2-7]，而国内对渔用药物药代动力学的研究远远落后。目前尚未见盐度、给药途径和给药剂量等因素对甲壳动物药动学影响的研究报道。作者曾对诺氟沙星在斑节对虾血淋巴中的药代动力学进行过初步研究[8]，本文比较研究了盐度、给药途径和给药剂量等对诺氟沙星在南美白对虾（*Penaeus vannamei*）血淋巴中药代动力学的影响。

1　材料与方法

1.1　实验试剂

诺氟沙星（NFX）原料药（含量98%）由上海三维制药厂提供，诺氟沙星标准品（含量99.5%，批号 H040798），由中国兽药监察所提供。2%诺氟沙星注射液配制：称取 2.00 g 诺氟沙星原料药，用少量的 2% 醋酸溶解，然后转移到 100 mL 棕色容量瓶，再用 0.9% 生理盐水定容至刻度。

1.2　试验动物

试验用南美白对虾（*P. vannamei*）取自上海市金山区漕泾对虾养殖场，经 7 d 暂养后用于试验。在盐度 1 和 15 下进行南美白对虾肌肉注射给药试验和在盐度 15‰ 下进行药饵口服给药试验，水温为（25.0±1.5）℃。试验期间投喂适量螺蛳肉，并及时排出残饵和污物。试验期间对虾成活率为 100%。

1.3　对虾给药

1.3.1　肌肉注射给药

肌肉注射试验用南美白对虾体重为 18~23 g，给药剂量为 10 mg/kg，根据对虾体重计算给药量在 10 μL 左右，一次性肌肉注射，给药部位从对虾腹部第二腹节腹面稍斜插入，轻轻推进。

1.3.2　药饵口服给药

药饵口服试验用南美白对虾体重为 13~17 g，给药剂量为 15 和 30 mg/kg，药饵含药量为 10 g/kg，根据对虾体重计算每尾对虾给饵量（约在 39~51 mg 之间），一次性投喂药饵。药饵投喂后 10 min 内摄食完全、没有残饵碎屑的对虾用于试验。10 min 后开始计时采样。

药饵由鱼粉、柔鱼粉和 α-淀粉等组成，添加诺氟沙星以达到药物含量为 10 g/kg。所有成分混合后，加入 12% 青蛤匀浆和 5% 水，于电动搅肉机挤压成型。90℃ 下烘干 2 h 左右，密封保存待用。

1.4　对虾血淋巴样品采集

先用纱布擦干取样虾的头胸甲，然后将 1 mL 注射器缓慢插入围心窦，抽取血淋巴 0.5 mL 左右，再转移到离心管中，在震荡器上震荡 1 min，于 5 000 rpm. 转速下离心 5 min，取出上层液贮存于 -70℃ 低温冰箱中，直至药物浓度分析。注射器和离心管使用前用 1% 肝素钠溶液均匀地涂布在壁上，吹干备用。

肌肉注射试验血淋巴采样时间点为给药后 1 min、2 min、4 min、6 min、30 min、2 h、12 h、24 h、48 h、96 h，每个时间点采样 4 尾。给药剂量为 30 mg/kg 的药饵口服试验组采样时间点为给药后 0.25、0.5、1、2、3、4、6、8、12、18、24、48、72、96、144、192、240 h；给药剂量为 15 mg/kg 的药饵口服试验组采样时间点为给药后 0.5、4、6、8、12 和 24 h，每个时间点采样 5 尾。

1.5　血淋巴样品处理

从冰箱中取出血淋巴样品，于室温下自然融解后，加入等体积乙腈，旋涡快速混匀器上混合 2 min，于 5 000 rpm. 转速离心 10 min，吸取上清液经 0.45 μm 微孔滤膜过滤后，用于高效液相色谱法（HPLC）分析。

1.6　血药浓度分析

对虾血淋巴中诺氟沙星浓度分析采用反相高效液相色谱法[8]。高效液相色谱仪采用美国艾捷伦公司的 Agilent 1100 系列，包括四元泵、自动进样器、柱温箱、荧光检测器及 HP 化学工作站。色谱柱为 ZORBAX SB-C18，4.6×150 mm。流动相为乙腈：磷酸盐缓冲液（含 10 mmol/L 磷酸氢二钠和 5 mmol/L 溴化四丁基铵，磷酸调节 pH 值至 2.0）= 14.5：85.5（*V/V*）。流速：0.5 mL/min。柱温：25℃。进样量：10 μL。检测波长：荧光检测器，Ex，280 nm，Em，418 nm。上述条件下，诺氟沙星的保留时间为（3.58±0.02）min；血淋巴中添加诺氟沙星浓度为 2.04、16.32 和 32.64 μg/mL 的回收率 96.39%、94.96% 和 98.16%；日内精密度和日间精密度相对标准偏差分别为 2.34% 和 2.77%。本方法的检测限为 0.01 μg/mL。

1.7　数据处理

肌肉注射试验的药-时曲线关系根据血药浓度-时间的半对数关系来确定药物的处置过程，采用房室模型分析药-时关系数据。模型方程的最初估计同样来自于血药浓度-时间半对数关系图形，然后采用非线性最小二乘方回归分析法对所测得的对虾血药数据进行拟合，使数据的加权平方和最小。肌肉注射给药的血药浓度-时间（*C-T*）数据采用二指数方程分

析，其拟合方程为

$$C_t = Ae^{-\alpha t} + Be^{-\beta t} \tag{1}$$

式中，C_t 为血药浓度，t 为时间，A 和 B 为系数，α 和 β 为混合速率常数。

始端（分布）和终端（消除）半衰期，即 $T_{\alpha 1/2}$、$T_{\beta 1/2}$ 分别由 $0.693/\alpha$ 和 $0.693/\beta$ 求得。曲线下面积 $AUC = A/\alpha + B/\beta$，总体清除率 $CLs = dose/AUC$，表观分布容积 $V_d = CLs/\beta$，稳态分布容积 $V_{ss} = (dose\,(A/\alpha^2 + B/\beta^2))/AUC^2$，中央室分布容积 $V_c = dose/(A+B)$，周边室分布容积 $V_p = V_{ss} - V_c$。

药饵口服给药试验南美白对虾血药浓度–时间关系曲线下面积（AUC）采用梯形法计算，

$$AUC_n = \sum (T_{i+1} - T_i)(C_i + C_{i+1}),\ i = 0 \rightarrow (n-1) \tag{2}$$

式中，C_i、C_{i+1} 分别表示采样时间点 T_i、T_{i+1} 的血药浓度，总体清楚率 $CLs = dose/AUC_n$。

药代动力学模型拟合及相关系数计算采用 SAS 软件（6.12 版本）中非线性回归分析部分 Marquardt 方法分析。

2 结果

2.1 肌注给药南美白对虾血淋巴中诺氟沙星浓度

诺氟沙星以 10 mg/kg 剂量、一次性肌肉注射南美白对虾给药后，在不同时间点采集对虾血淋巴样品，测得其血淋巴中诺氟沙星浓度见表 1，以时间为横坐标、血药浓度为纵坐标，绘制南美白对虾血淋巴药物浓度–时间曲线图（见图 1）。

表 1　肌注给药后南美白对虾血淋巴中诺氟沙星浓度
Table 1　Norfloxacin concentrations in hemolymph from *P. vannamei* following intramuscle injection

时间 Time	盐度 1	盐度 15
1 min	28.00±3.56	28.77±3.46
2 min	29.33±5.53	30.27±4.07
4 min	24.46±2.70	22.02±2.15
6 min	16.39±2.73	16.68±3.35
30 min	4.65±1.05	5.87±2.21
2 h	2.79±1.42	2.33±0.21
12 h	0.80±0.40	0.53±0.15
24 h	0.16±0.12	0.19±0.14
48h	0.04±0.02	0.05±0.04

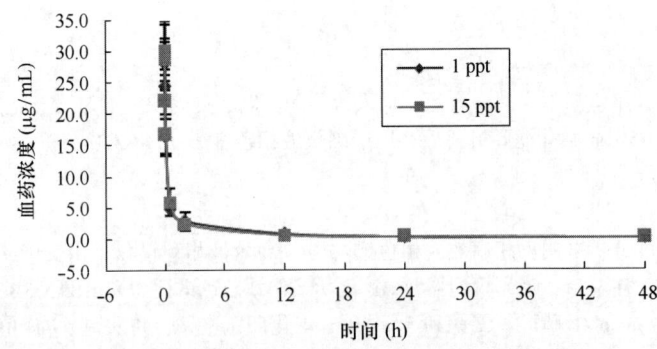

图 1　南美白对虾肌注给药后血药浓度–时间曲线
Fig. 1　The mean hemolymph concentration vs. time after intramuscular injection at a dose of 10 mg/kg

肌注给药后，在 1‰ 和 15‰ 盐度下南美白对虾血淋巴中诺氟沙星浓度首先均表现为迅速上升，给药后 1 min 血药浓度为次最高值［分别为（28.00±3.56）μg/mL 和（28.77±3.46）μg/mL］，2 min 时血淋巴诺氟沙星浓度为最高值（$T_{max} = 2$ min），C_{max} 分别为（29.33±5.53）μg/mL 和（30.27±4.07）μg/mL；随后血药浓度开始下降，下降速度较快，此时诺氟沙星处于分布相；30 min 以后血药浓度下降速度变缓，12 h 血药浓度比较低；48 h 以后几乎测不出诺氟沙星。从给药后血药浓度–时间变化曲线来看，在盐度 1 和 15 下，南美白对虾血药浓度的变化趋势相一致。

分别采用一室、二室、三室模型方程对南美白对虾血药浓度-时间关系进行数据拟合，结果为二室模型方程加权平方和最小，与根据血药浓度-时间半对数关系图形的最初估计结果（二室模型）相一致。南美白对虾在盐度 1 和 15 下，一次性肌肉注射给药后，诺氟沙星在对虾体内的处置过程用开放性二室模型来描述最为合适，其血淋巴药-时关系方程：盐度 1 时为 $C_0 = 35.422 \times e^{-9.778t} + 4.363 \times e^{-0.165t}$、拟合曲线相关指数为 $R^2 = 0.9949$（$P < 0.01$）；盐度 15 时为 $C_0 = 35.144 \times e^{-13.335t} + 7.888 \times e^{-0.608t}$、拟合曲线相关指数为 $R^2 = 0.9997$（$P < 0.01$）。以此药动学模型方程计算南美白对虾药动学参数见表 2。盐度 1 下，南美白对虾分布和消除半衰期（$T_{\alpha 1/2}$ 和 $T_{\beta 1/2}$）分别为 0.0709 h 和 4.208 h，曲线下面积（AUC）为 30.06 μg·h·mL^{-1}，总体清除率（CLs）为 332.7 mL·kg^{-1}·h^{-1}，表观分布容积（V_d）为 2016 mL·kg^{-1}。盐度 15 下，南美白对虾分布和消除半衰期（$T_{\alpha 1/2}$ 和 $T_{\beta 1/2}$）分别为 0.052 h 和 1.140 h，曲线下面积（AUC）为 15.61 μg·h·mL^{-1}，总体清除率（CLs）为 640.6 mL·kg^{-1}·h^{-1}，表观分布容积（V_d）为 1 054.3 mL·kg^{-1}。

表 2　南美白对虾肌肉注射诺氟沙星给药后的药代动力学参数

Table 2　Pharmacokinetic parameters of norfloxacin in P. vannamei following intramuscular administration at salinities of 1 and 15

药动学参数 Pharmacokinetic parameters	盐度 1	盐度 15
Dose 给药剂量（mg/kg）	10.0	10.0
A 分布相的零时截距（μg/mL）	35.422	35.144
B 消除相的零时截距（μg/mL）	4.363	7.888
α 分布相一级速率常数（h^{-1}）	9.778	13.335
β 消除相一级速率常数（h^{-1}）	0.165	0.608
$T_{\alpha 1/2}$ 分布半衰期（h）	0.0709	0.052
$T_{\beta 1/2}$ 消除半衰期（h）	4.208	1.140
V_{ss} 稳态分布容积（mL/kg）	1777.6	883.8
V_c 中央室分布容积（mL/kg）	251.4	232.4
V_p 周边室分布容积（mL/kg）	1526.2	651.4
AUC 血淋巴药-时曲线下面积（μg·h/mL）	30.06	15.61
CLs 总体清除率（mL/kg·h）	332.7	640.6
V_d 表观分布容积（mL/kg）	2016.4	1053.6

2.2　药饵口服给药南美白对虾血淋巴诺氟沙星浓度

南美白对虾以 30 mg/kg 单剂量药饵口服给药后血淋巴药物浓度-时间曲线见图 2。给药后 0.25 ~ 4 h，血淋巴诺氟沙星浓度几乎成直线上升，4 h 时血药浓度达到最高，C_{max} 为 2.86 μg/mL。随后血药浓度以较快的速度下降，8 h 后血药浓度又开始回升，12 h 时出现第二个药峰，血药浓度为 2.037 μg/mL；其后血药浓度再次下降，18 h 后血药浓度下降缓慢，处于消除阶段。

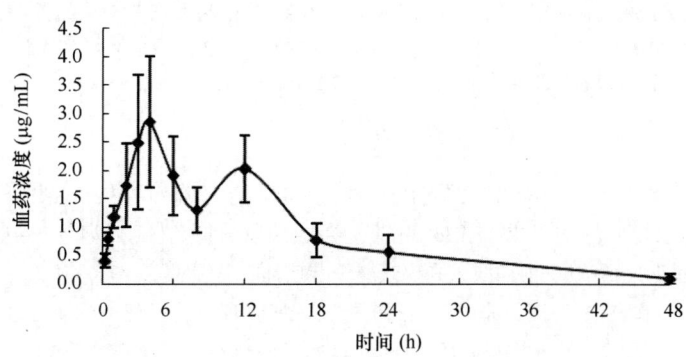

图 2　南美白对虾药饵给药后血药浓度-时间关系曲线

Fig. 2　Hemolymph concentrations of norfloxacin in *P. Vannamei* following aral administration

用梯形法计算药饵口服给药方式下诺氟沙星血药浓度-时间曲线下面积（AUC_{oral}）为 43.50 μg·h/mL，总体清除率为 689.66 mL/kg·h。

图 3 为 15 mg/kg 剂量药饵口服给药南美白对虾血淋巴中诺氟沙星浓度与 30 mg/kg 剂量下在几个时间采样点血药浓度的比较。与 30 mg/kg 剂量相同的是，15 mg/kg 剂量药饵口服给药后血药浓度同样出现两个吸收峰，第一峰出现在给药后 4 h，血药浓度为（1.44±0.27）μg/mL；不同的是 15 mg/kg 剂量药饵口服给药后第二个峰出现的时间是在 8 h，而 30 mg/kg 剂量出现的时间是在 12 h。两个剂量下的血药浓度与给药剂量没有明显的比例关系，尽管给药后 4 h、6 h，15 mg/kg 剂量下的血药浓度约为 30 mg/kg 剂量的一半，但其他各采样点的血药浓度差别较大，有时甚至低剂量的血药浓度大于高剂量组，如 15 mg/kg 和 30 mg/kg 剂量下在给药后 30 min 的血药浓度分别为 1.07±0.53 μg/mL 和 0.88±0.460 μg/mL。

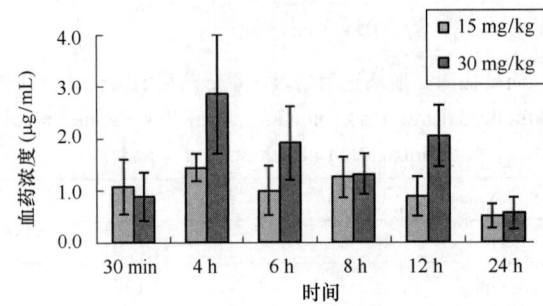

图 3 同人种给药剂量药饵口服给药对虾血药浓度比较

Fig. 3 Norfloxacin concentration in hemolymph following oral administration at dose of 15 mg/kg and 30 mg/kg

3 讨论

3.1 肌肉注射给药方式下盐度和种类对药动学的影响

南美白对虾在盐度 1 和 15 下，肌肉注射给药后血药浓度首先表现为迅速上升，在 2 min 都达到最高值，且到达药峰的 C_{max} 相近；其后开始下降，进入分布相和消除相，48 h 的血药浓度几乎相近。血药浓度的变化趋势与相近盐度下的斑节对虾肌注给药试验结果相似，且它们的血药浓度-时间关系曲线都属于开放性二室模型，但部分药动学参数在不同盐度和种类间存在一定差别，主要表现在 $T_{\beta1/2}$、AUC、CLs 和 V_d，南美白对虾在盐度 1‰ 时分别为 4.208 h、30.06 μg·h/mL、332.7 mL/kg·h、2016 mL/kg；南美白对虾在盐度 15‰ 时分别为 1.140 h、15.61 μg·h/mL、640.6 mL/kg·h、1 054 mL/kg；斑节对虾在此盐度下的相应值分别为 0.612 h、12.42 μg·h/mL、804.8 mL/kg·h、710 mL/kg[8]。仔细分析两种对虾的血药浓度差异，给药后 30 min 内的分布相各采样点的血药浓度差别并不大；30 min 以后血药浓度开始出现差异，此时血药浓度处于消除阶段，这可由消除半衰期加以验证，斑节对虾 $T_{\beta1/2}$ 为 0.612 h，南美白对虾在盐度 1 和 15 时 $T_{\beta1/2}$ 分别为 4.208 h 和 1.140 h。

导致对虾药动学差异的可能原因有二，一是种的差异，二是盐度不同。然而，种类和盐度的不同对药动学的影响，在不同的水生动物结果相异。Abedini 等[2]研究海水大鳞大麻哈鱼和淡水虹鳟 OTC 的比较药动学时，大鳞大麻哈鱼试验海水盐度高达 24，虹鳟为淡水，采用相同给药方式时（动脉给药或口服给药），其结果是种的差异和盐度对鲑鳟鱼的 OTC 吸收与消除并不起重要影响。而海水虹鳟和淡水虹鳟口服噁喹酸后，在给药后 24 h 内两种鱼的各组织内药物浓度差别不大，但 24 h 后海水虹鳟却表现出比淡水虹鳟高的代谢排除率（Ishida，1992）[3]。这一结果恰好与本试验的结果相似。因此，作者认为不可轻易将某一药物在一个动物的试验结果应用于其他动物。

3.2 关于诺氟沙星在不同动物间药动学的差异

与斑节对虾药动学结果[8]相似，在本试验中南美白对虾肌注诺氟沙星的血药浓度-时间曲线符合二室模型方程，且分布（平衡）和消除较快，$T_{\alpha1/2}$ 分别为 0.070 9 h（1 ppt）和 0.052 h（15 ppt），$T_{\beta1/2}$ 分别为 4.208 h（1 ppt）和 1.140 h（15 ppt）；鲤鱼肌肉注射诺氟沙星的血药浓度-时曲线关系模式属于开放性二室模型[9]，$T_{\alpha1/2}$ 和 $T_{\beta1/2}$ 分别为 0.127 9 h 和 3.403 2 h。南美白对虾药饵口服给药后血药浓度-时间曲线出现两个药物浓度峰现象；而鲤鱼单纯口服和混饲口服给药的药-时曲线关系不同，分别为二室和一室模型；中华鳖口服诺氟沙星的药-时曲线属一室模型[10]。由此可见，诺氟沙星在对虾和其他水生动间的药动学结果并不相同。

同一药物在不同水产动物体内所表现出的药动学差异明显。有研究者认为不同种类之间的药动学及参数差异可能是由于解剖学上的体积差异以及药物与血浆蛋白、组织结合的差异所致[5,11]。就甲壳动物与鱼类而言，其主要不同点是甲壳和血淋巴体积，甲壳被认为是甲壳动物药物处置场所[5,6]；甲壳动物的血淋巴体积与其他动物相比差别很大，甲壳动物血淋巴体积约占全体重的22%[5]，而鱼类血液体积约为5%左右[7]。血浆蛋白和组织与药物的结合程度主要取决于动物种类和所用药物，分布容积与组织结合直接相关、与血浆蛋白结合成反相关，即血浆蛋白结合低导致血管外分布增大。血浆蛋白与药物结合与分布容积的反相关性已在哺乳动物得到证实[11]。

3.3　关于肌注给药血药浓度达到峰值快

南美白对虾肌注给药后2 min血药浓度即达到最高值，到达药峰之快类似于南美白对虾血窦内注射[4]，但不能等同于血窦内/静脉注射给药。尽管对虾肌注给药达峰快，但仍有一吸收过程，而血窦内/静脉注射给药没有吸收，是直接进入分布相。肌注给药血药达峰时间快这一结果可以由对虾的开放式循环系统来解释。在对虾腹部的背面动脉和腹面动脉之间有一组血窦，这些血窦直接通心脏[12]。肌注给药后，药物沿肌纤维散开后通过毛细管状的静脉（或称毛细血管）汇入血窦，流回心脏。与某些动物实验中血液在血腔内流动慢的所不同的是，在显微镜下观察对虾尾肢上血窦内血液的循环，血液流动很快[12]。这又可以进一步解释对虾肌注后血药达峰快的原因。

3.4　药饵口服给药后血药浓度出现"双峰"现象

本研究在两个剂量（15 mg/kg和30 mg/kg）药饵口服给药试验中，南美白对虾的血药浓度-时间关系曲线均出现两个药物浓度峰，称为"双峰"现象。关于药物吸收出现双峰或多峰现象及其机制的研究在兔、鼠等其他动物已有报道[13,14]，这可能与肠肝循环、胃肠道吸收的非齐性（unhomogeneous）有关。药物药-时曲线第二峰的出现，相当于一次"自体给药"过程，势必影响药物作用强度及消除过程，同时使血药浓度-时间曲线和药动学模型变得更为复杂。本研究中"双峰"现象增大了药-时曲线下面积（AUC），同时也使消除半衰期延长。至于对虾血药浓度出现"双峰"现象的理论机制待今后继续研究。

参考文献

[1] Snieszko S F, Friddle S O, Griffin P J. Successful treatment of ulcer disease in Brook Trout (Salvelinus fontinalis) with Terramycin [J]. Science, 1951, 113: 717-718.

[2] Abedini S, Namdari R, Law F C P. Comparative pharmacokinetics and bioavailability of oxytetracycline in rainbow trout and chinook salmon [J]. Aquaculture, 1998, 162 (1): 23-32.

[3] Ishida N. Tissue levels of oxolinic acid after oral or intravascular administration to freshwater and seawater rainbow trout [J]. Aquaculture, 1992, 102: 9-15.

[4] Park E D, Lightner D V, Milner N, et al. Exploratory bioavailability and pharmacokinetic studies of sulphadimethoxine and ormetoprim in the penaeid shrimp, Penaeus vannamei [J]. Aquaculture, 1995, 130 (2, 3): 113-128.

[5] Barron M G, Gedutis C and James M O. Pharmacokinetics of sulphadimethoxine in the lobster. Homerus americannus, following intrapericardial administration [J]. Xenobiotica, 1988, 18 (3): 269-277.

[6] James M O, Barron M G. Disposition of sulfadimethoxine in the lobster (Homarus americanus) [J]. Vet. Hum. Toxicol., 1988, 30 (1 suppl.): 36-41.

[7] Plakas S M, Dickey R W, Barron M G, et al. Tissue distribution and renal excretion of ormetoprim after intravascular and oral administration in the channel catfish (Ictalurus punctatus) [J]. Can. J. Fish. Aquat. Sci., 1990, 47 (4): 766-782.

[8] 房文红，邵锦华，施兆鸿等．斑节对虾血淋巴中诺氟沙星含量测定及药代动力学 [J]．水生生物学报，2003，27（1）：13-17.

[9] 张祚新，张雅斌，杨永胜等．诺氟沙星在鲤鱼体内的药代动力学 [J]．中国兽医学报，2000，20（1）：66-69.

[10] 陈文银，印春华．诺氟沙星在中华鳖体内的药代动力学研究 [J]．水产学报，1997，21（4）：434-437.

[11] Oie S, Tozer T N. Effect of altered plasma protein binding on the apparent volume of distribution [J]. J. Pharm. Sci. 1979, 68: 1203-1208.

[12] Dall W, Hill B J, Rothlisberg P C, et al. (陈楠生等译)．对虾生物学 [M]．青岛：青岛海洋大学出版社，1992，31-35.

[13] 周怀梧，沈佳庆，吕明等．吡罗昔康在家兔体内的肠肝循环药物动力学分析 [J]．中国药理学报，1992，13（2）：180-182.

[14] 陈淑娟，杨毅梅，刘奕明等．蝙蝠葛碱大鼠体内药物代谢动力学研究 [J]．中国药理学通报，2001，17（2）：225-229.

该文发表于《动物学杂志》【2015，50（1）：103-111】

烟酸诺氟沙星在松浦镜鲤体内的药动学特征

Pharmacokinetics of Norfloxacin Nicotinate in Songpu Mirror Carp (*Cyprinus carpio specularis*)

韩冰[1,2]，王荻[1]，卢彤岩[1]

（1. 中国水科学研究院 黑龙江水产研究所，黑龙江 150070；2. 上海海洋大学 水产与生命学院，上海 201306）

摘要：结合单纯聚集法和二步法，应用高效液相色谱（HPLC）技术研究了分别以 10、30、60 mg/kg 剂量对松浦镜鲤（ *Cyprinus carpio specularis* ）口灌烟酸诺氟沙星后，药物在实验鱼血浆中的药动学特征。结果显示，3 种给药剂量下，诺氟沙星在松浦镜鲤血浆中的血药浓度和时间关系均可用一级吸收二室开放模型进行描述，吸收半衰期 $T_{1/2\,ka}$ 分别为 0.165、0.061、0.043 h，消除半衰期 $T_{1/2\beta}$ 分别为 18.282、29.969、42.051 h，达峰时间 T_{max} 分别为 0.333、0.327、0.302 h，达峰浓度 C_{max} 分别为 4.780、6.247、12.689 mg/L，药时曲线下面积（AUC）分别为 32.698、53.015、174.998 mg·h/L，表观分布容积 V_d 分别为 1.044、4.347、4.561 L/kg。说明随着给药剂量的增加，诺氟沙星的吸收和消除速率均加快，给药剂量对药动学特征有显著影响。

关键词：烟酸诺氟沙星；松浦镜鲤；药动学

该文发表于《广东农业科学》【2014，11：136-140】。

烟酸诺氟沙星预混剂在罗氏沼虾体内药代动力学及残留消除规律

The pharmacokinetics and eliminate residual of Niacin norfloxacin premix in *Macrobrachium rosenbergii*

左梦丽[1,2]，姚嘉赟[2]，沈锦玉[1,2*]，尹文林[2]，盛鹏程[2]

（1. 上海海洋大学，上海，201200；2. 浙江省淡水水产研究所，浙江湖州，313000）

摘要：研究烟酸诺氟沙星预混剂在罗氏沼虾体内的药代动力学及残留消除规律。采用混饲口灌方式给药罗氏沼虾，通过有机萃取及固相萃取等方法提纯净化各组织样本，利用超高效液相色谱（UPLC）法测定样本中药物的浓度，研究烟酸诺氟沙星预混剂在罗氏沼虾体内的药代动力学及残留规律。混饲口灌烟酸诺氟沙星预混剂（20.0 mg/kg body weight）后，罗氏沼虾肝胰腺符合药代动力学一级吸收二室开放模型。主要动力学参数如下：达峰时间 T_{peak} 为 2.905 h，最大峰浓度 C_{max} 为 275.3 μg/mL，分布半衰期 $T_{1/2\alpha}$ 为 2.178 h，消除半衰期 $T_{1/2\beta}$ 为 10.058 h；连续 3 d 混饲口灌烟酸诺氟沙星预混剂（20.0 mg/kg body weight）后，药物在肝胰腺、血液、肌肉组织的残留低于 50 μg/mL 所需的时间分别为 15、4、5 d。

关键词：罗氏沼虾；烟酸诺氟沙星预混剂；药代动力学；残留

该文发表于《浙江海洋学院学报（自然科学版）》【2008，27（2）：135-139】。

诺氟沙星在淡水青虾体内药物代谢动力学研究
Pharmacokinetic Study on Norfloxacin in Freshwater Shrimp

何平[1,2]，尹文林[2]，沈锦玉[2]

（1. 宁波大学，浙江 宁波 315000；2. 浙江省淡水水产研究所，浙江 湖州 313001）

摘要： 在 26±1℃ 的水温条件下，青虾一次性肌肉注射 25 mg/kg 诺氟沙星后，用反相高效液相色谱法测定青虾血淋巴和肌肉组织中诺氟沙星含量。青虾血淋巴药一时曲线和肌肉药一时曲线均可以用二室开放模型来描述，诺氟沙星在青虾血淋巴液中的主要药动学参数为：分布相半衰期 $T_{1/2\alpha}$ 为 1.66 h，消除相半衰期 $T_{1/2\beta}$ 为 1.69 h，达峰时间 T_{Peak} 为 1.82 h，峰浓度 C_{max} 为 6.008 1 μg/mL，曲下面积 AUC 为 30.75 $μg \cdot mL^{-1} \cdot h^{-1}$，吸收相半衰期 $T_{1/2ka}$ 为 1.66 h。肌肉中的主要药动学参数为：分布相半衰期 $T_{1/2\alpha}$ 为 0.08 h，消除相半衰期 $T_{1/2\beta}$ 为 4.42 h，达峰时间 T_{Peak} 为 0.03 h，峰浓度 C_{max} 为 16.72 μg/mL，曲下面积 AUC 为 12.34 $μg \cdot mL^{-1} \cdot h^{-1}$，吸收相半衰期 $T_{1/2ka}$ 为 0.08 h。结果表明青虾肌注诺氟沙星后，能比较迅速的被吸收，并且在组织中维持较高的药物浓度。

关键词： 青虾；诺氟沙星；药代动力学

该文发表于《中国渔业质量与标准》【2016，6（1）：22-28】（中文摘要）

诺氟沙星在日本鳗鲡体内的代谢动力学及残留消除规律

范红照[1]，林茂[1,2*]，鄢庆枇[1]，湛嘉[3]，李忠琴[2]

（1. 集美大学水产学院，福建厦门 361021；2. 鳗鲡现代产业技术教育部工程研究中心，福建厦门 361021；
3. 宁波出入境检验检疫局，浙江宁波 315010）

摘要： 为研究诺氟沙星（NFX）在鳗鲡体内的代谢和消除规律，以超高效液相色谱－串联质谱法测定日本鳗鲡在混饲口灌后血液和组织中 NFX 的含量变化，并进行药动学分析。结果表明，NFX 以 30 mg/kg 的剂量单次混饲口灌日本鳗鲡后，吸收分布迅速，达峰时间 T_{max}、吸收 $T_{1/2Ka}$ 和分布半衰期 $T_{1/2\alpha}$ 分别为 3.000、1.012 和 1.570 h；NFX 在鳗鲡体内消除较快，消除半衰期 $T_{1/2\beta}$ 为 15.267 h，总清除率（CL）为 1.315 L/（h·kg）。此外，峰浓度 C_{max} 为 1.273 mg/L，药时曲线下面积 $AUC_{(0-\infty)}$ 为 22.670 mg/（L·h）。NFX 以 30 mg/kg 的剂量连续 3 d 混饲口灌日本鳗鲡后，在肌肉、肝脏、肾脏和血浆中的消除速率常数分别为 0.144、0.125、0.102 和 0.093 1/d。根据 WT1.4 计算的理论休药期（WDT）分别为肌肉 22.97 d，肝脏 21.30 d，肾脏 33.40 d，血浆 18.29 d。本研究结果为诺氟沙星在水产动物中的实际应用提供理论依据。

关键词： 诺氟沙星；日本鳗鲡；药代动力学；残留；消除规律

该文发表于《水产学杂志》【2011，24（4）：25-28】

诺氟沙星在金鳟体内的药物代谢动力学研究
Pharmacokinetics of Norfloxacin in Rainbow Trout *Oncorhynchus mykiss*

陈琛，卢彤岩，王荻，李绍戊

（中国水产科学研究院 黑龙江水产研究所，黑龙江 150070）

摘要： 实验水温为（15±2）℃，金鳟（*Oncorhynchus mykiss*）［平均体质量（100±10）g］单剂量肌肉注射 30.0

mg/kg 诺氟沙星后，应用高效液相色谱（HPCL）法于 0.15、0.25、0.5、0.75、1、1.5、2、4、6、8、12、24、48、72 h 测定了鱼血浆、肝脏和肾脏组织中药物的浓度，研究了诺氟沙星在金鳟组织中的分布及药物动力学规律。结果表明，诺氟沙星在金鳟体内吸收分布迅速，符合药物动力学一级吸收二室开放模型，但消除缓慢。诺氟沙星在金鳟血浆、肝脏和肾脏中的主要动力学参数如下：分布半衰期 $T_{1/2\alpha}$ 分别为 0.866、1.985、0.388 h；消除半衰期 $T_{1/2\beta}$ 分别为 31.369、36.402、30.975 h；药时曲线下面积（AUC）分别为：308.005 μg/mL·h、622.721 μg/g·h、794.362 μg/g·h。

关键词：金鳟；诺氟沙星；药物代谢动力学

三、氟甲喹

氟甲喹化学名称是（9-氟-6,7-二氢-5-甲基-1-氧代-1,5-二氢-苯并（ü）喹啉-2-羧酸），属第二代氟喹诺酮类药物。氟甲喹以较快的速度分布到花鲈组织中，且从肌肉组织中的消除速度也快（Lin-linHu et al.，2011）。

该文发表于《JOURNAL OF VETERINARY PHARMACOLOGY AND THERAPEUTICS》【2016，39（2）：191-195】

Pharmacokinetics and bioavailability of flumequine in blunt snout bream（*Megalobrama amblycephala*）after intravascular and oral administrations

Ning Xu[*†], Jing Dong[*], Yibin Yang[*], Qiuhong Yang[*], Yongtao Liu[*†], Xiaohui Ai[*†]

（* Freshwater Fish Germplasm Quality Supervision and Testing Center, Ministry of Agriculture, Yangtze River Fisheries Research Institute, Chinese Academy of Fishery Sciences, Wuhan, China; † Hu Bei Freshwater Aquaculture Collaborative Innovation Center, Wuhan, China）

Abstract：In present study, the pharmacokinetic profile of flumequine (FMQ) was investigated in blunt snout bream (*Megalobrama amblycephala*) after intravascular (3 mg/kg body weight (b.w.)) and oral (50 mg/kg b.w.) administrations. The plasma samples were determined byultra-performance liquid chromatography (UPLC) with fluorescence detection. After intravascular administration, plasma concentration-time curves were best described by a two-compartment open model. The distribution half-life ($t_{1/2\alpha}$), elimination half-life ($t_{1/2\beta}$), and area under the concentration-time curve (AUC) of blunt snout bream were 0.6 h, 25.0 h, and 10612.7 h.μg/L, respectively. After oral administration, a two-compartment open model with first-order absorption was also best fit the data of plasma. The $t_{1/2\alpha}$, $t_{1/2\beta}$, peak concentration (C_{max}), time-to-peak concentration (T_{max}), and AUC of blunt snout bream were estimated to be 2.5 h, 19.7 h, 3946.5 μg/L, 1.4 h, and 56618.1 h.μg/L, respectively. The oral bioavailability (F) was 32.0%. The pharmacokinetics of FMQ in blunt snout bream displayed low bioavailability, rapid absorption, and rapid elimination.

Key words：pharmacokinetics; bioavailability; Flumequine; Megalobrama amblycephala

经静脉注射和口服给药后氟甲喹在团头鲂体内药动学和生物利用度研究

胥宁[*†]，董靖[*]，杨仪斌[*]，杨秋红[*]，刘永涛[*†]，艾晓辉[*†]

（* 农业部淡水鱼类种质资源监督检测中心，湖北 武汉 430223；* 中国水产科学研究院长江水产研究所 湖北 武汉 430223；† 淡水水产健康养殖湖北省协同创新中心 湖北 武汉 430223）

摘要：本研究主要考察了经口服和静脉给药后氟甲喹在团头鲂体内药动学特性。样品采用 UPLC 荧光检测器进行检测。经静脉注射之后，血浆药时曲线能够被二房室模型进行很好的拟合。分布半衰期、消除半衰期和药时曲线下面积分别为 0.6 h、25 h 和 10612.7 h·μg/L。经口服之后，一级吸收二房室模型也很好的拟合了血浆药时曲线。分布半衰期、消除半衰期、峰浓度、达峰时间和药时曲线下面积分别为 2.5 h、19.7 h、3946.5 μg/L、1.4h

和 10 612. 7 h·μg/L。口服生物利用度为 32. 0%。氟甲喹在团头鲂体内呈现出了低生物利用度、快速吸收和快速消除的特性。

关键词：药动学，生物利用度，氟甲喹，团头鲂

INTRODUCTION

Flumequine (FMQ) is a broad spectrum synthetic antimicrobial agent belonging to the second-generation quinolone group. It is used for treating certain bacterial diseases in aquatic animals, such as enteric infections, hemorrhage and gill rot disease. Nowadays, it is wildly used in aquaculture of China due to good antibacterial effect.

There is extensive information on the pharmacokinetics of FMQ in various aquatic species including Japanese sea perch (*Lateolabras janopicus*) (Hu et al., 2011), Muraenesox cinereus (Liang et al., 2006), European eel (*Anguilla anguilla*) (Hansen & Horsberg, 2000), rainbow trout (*Oncorhynchus mykiss*) (Sohlberg et al., 1990), Atlantic halibut (*Hippoglossus hippoglossus*), turbot (*Schophalmus maximus*) (Hansen & Horsberg, 1999), corkwing wrasse (*Symphodus melops*) (Samuelsen et al., 2000), Atlantic salmon (*Salmo salar*) (Elema et al., 1994), channel catfish (*Ictalurus punctatus*) (Plakas et al., 2000).

However, little information is available on the pharmacokinetic profile of FMQ in freshwater fish in China. Therefore, the aim of the present study is to investigate the pharmacokinetics and bioavailability of FMQ in blunt snout bream (*Megalobrama amblycephala*), a freshwater fish cultured widely in Southern China, after intravascular and oral administrations.

MATERIALS AND METHODS

Chemicals and regents

Analytical standard FMQ (purity grade 99. 0%) was purchased from Dr. Ehrenstorfer GmbH., Augsburg, Germany FMQ powder (purity grade 98. 0%) used for oral and intravenous administration was supplied by Zhongbo aquaculture Biotechnology Co. Ltd., Wuhan, China. The anesthetic of MS222 was purchased from Aibo Biotechnology Co. Ltd., Wuhan, China. The purity of other chemicals and reagents was analytical grade.

The FMQ solution (15 mg/mL) for oral administration was prepared by dissolving FMQ power in 0. 1 mol/L NaOH solution. The solution was adjusted to pH 10. 5 with HCl.

The FMQ solution (5 mg/mL) for intravenous administration was prepared by dissolving FMQ power in 0. 1 mol/L NaOH solution. The solution was adjusted to pH 10. 5 with HCl.

Animals

Blunt snout bream (300±10 g) were obtained from breeding base of Yangtze River Fisheries Research Institute. The fish were held in 30 plastic tanks (100×60×80cm, length×width×depth) receiving flowing well water (26 L/min) with six individuals per tank. The water quality parameters were measured once daily in each tank, and maintained at the following: total ammonia nitrogen levels ≤0. 73 mg/L; nitrite nitrogen concentrations at less than 0. 05 mg/L; dissolved oxygen levels at 6. 4-8. 2 mg/L; temperature at 25. 0±1. 2℃ and pH 7. 5±0. 2. All fish were fed a drug-free feed, which was detected by UPLC, and were acclimatized for 14 days prior to commencing the experiment. A total of 128 healthy fish were selected and divided into two groups, one group (48 fish for formal experiment with six individuals per tank and 10 fish for controlholding in one tank) for intravenous administration and one group (60 fish for formal experiment with six individuals per tank and 10 fish for control holding in one tank) for oral administration. Prior to drug administration, the fish fasted for 24 h. All the experimental procedures involving animals in the study were approved by the animal care center, Hubei Academy of Medical Sciences.

Intravascular administration and sample collection

Prior to intravascular administration, fish were anaesthetized with MS222 (50 mg/L) and weighted. The dose of FMQ was 3 mg/kg body weight (b. w.) for intravascular treatment. The control fish received corresponding volume of blank solution. The position of the needle in the caudal vein was confirmed by aspirating blood into the syringe prior to injection. If the fish was heavily bleeding after withdrawal of the needle or the needle had translocated during injection, the fish was excluded from the study and replaced. After intravascular treatment, six fishwere taken from tank for each sampling point at time intervals 0. 083, 0. 167, 0. 5, 1, 2, 4, 8, 16 h. At the end of the experiment, the control fish were allsampled. For each fish, approximately 2 mL of blood was drawn from the caudal vein away from the injection site. Blood samples were centrifuged at 4000 r/min for 10 min. The plasma was collected and stored at −20℃ until assayed. Following sampling, the fish were euthanized by an overdose in a solution of 300 mg/L MS 222.

Oral administration and sample collection

Prior to oral administration, fish were anaesthetized with MS222 (50 mg/L) and weighted. The fish were administered orally at a dose of 50 mg/kg b. w. by inserting a plastic hose with 1-mL micro-injector. The control fish received corresponding volume of blank solution. After treatment, each fish was observed in tank for regurgitation of FMQ. If FMQ solution was regurgitated, the fish was removed from tank and replaced.

After oral dose, approximately 2 mL blood was taken from each of 6 fish at each time point (0.083, 0.167, 0.5, 1, 2, 4, 8, 16, 24, 48 h). At the end of the experiment, the control fish were allsampled. Following sampling, the fish were euthanized by an overdose in a solution of 300 mg/L MS 222. Sample preparation and storage conditions were consistent with the protocol followed for the intravascular administration.

Analytical method

Protein of 0.5-mL plasma was precipitated by the same volume of acetonitrile. It was vortexed and centrifuged at 7 000 r/min for 5 min. The supernatant was filtered through a 0.22 μm disposable syringe filter for UPLC analysis (Hu et al, 2011). The samples were analyzed by a Waters ACQUITY™ UPLC with a fluorescence detector. The excitation was at 325 nm and emission was at 369nm. An ACQUITY UPLC BEH C18 column (1.7 μm, 2.1×100 mm) was used for separation. The mobile phase was tetrabutylammonium bromide buffer andacetonitrile (70:30, v/v). The buffer contained 0.01moL tetrabutylammonium bromide, and the pH was adjusted to 2.0 with phosphate acid. The column temperature was set to 30℃, and flow rate was 0.3mL/min.

The calibration curves of FMQ in plasma were prepared separately with five different concentrations between 10 and 8 000 μg/L using blank plasma. Precision and accuracy weredetermined by analyzing five replicates at a low, medium, andhigh concentration within the concentration range for the study.

Pharmacokinetic and statistical analyses

Pharmacokinetic parameters were calculated using Practical Pharmacokinetic Program 3P97 (Mathpharmacology Committee, Chinese Academy of Pharmacology, Beijing, China). The appropriate pharmacokinetic model was selected by the visual examination of concentration-time curve and by application of Akaike's Information Criterion (AIC) (Yamaoka et al., 1978). The oral bioavailability for FMQ was determined by calculating the area under concentration-time curve after oral and intravascular administration and using the following equation:

$$F = \frac{(\text{AUCoral administration}) \times (\text{dose intravascular injection})}{(\text{AUCintravascular injection}) \times (\text{dose oral administration})} \times 100\%$$

RESULTS

Method validation

The limit of detection for FMQ was 10 μg/L in blunt snout bream plasma by UPLC analysis. The good linearity of curves over range of 10 and 8 000 μg/L was exhibited by the coefficient of correlation $R^2 = 0.999$. The mean±SD recovery rates of FMQ in blunt snout bream plasma were 88.2±3.5% and its percentage relative standard deviations for inter-day and intra-day precision were 3.3±1.0% and 4.1±0.9%.

Pharmacokinetics of FMQ in blunt snout bream

The plasma concentration-time curve after intravascular and oral treatment was shown in Fig.1 and Fig.2. The pharmacokinetic parameters were listed in Table1. After oral dosing, the data were best described by a two-compartment open model with first-order absorption. FMQ was rapidly absorbed by gastro-intestinal tract of blunt snout bream. The drug concentration already reached 1924.8μg/L at first sampling time. Until 1.4 h after dosing, it increased to the peak plasma concentration (3 946.5 μg/L), and then the plasma concentration began to decline. The half-life of distribution ($t_{1/2\alpha}$), half-life elimination ($t_{1/2\beta}$), absorption rate constant (K_a), area under concentration-time curve (AUC) and bioavailability were estimated to 2.5 h, 19.7 h, 1.81/h, 56 618.1 h.μg/L and 32.0%, respectively. After intravascular treatment, the data best fit a two-compartment open model. The $t_{1/2\alpha}$, $t_{1/2\beta}$, and AUC were calculated to 0.6 h, 25.0 h, and 10612.7 h·μg/L, respectively.

Table 1 The pharmacokinetic parameters of flumequine for intravascular (3 mg/kg b. w.) andoral (50 mg/kg b. w.) administration in blunt snout bream (n=6)

Parameters	Units	i. v.	p. o.
A	μg/L	1 745.0	4 332.7
α	1/h	1.24	0.28
B	μg/L	255.7	1566.0

续表

Parameters	Units	i. v.	p. o.
β	1/h	0. 028	0. 035
K_a	1/h	–	1. 81
$t_{1/2\alpha}$	h	0. 6	2. 5
$t_{1/2\beta}$	h	25. 0	19. 7
$t_{1/2\,Ka}$	h	–	0. 4
k_{10}	1/h	0. 19	0. 091
k_{12}	1/h	0. 90	0. 12
k_{21}	1/h	0. 18	0. 11
AUC_{0-t}	h. μg/L	10 612. 7	56 618. 1
T_{max}	h	–	1. 4
C_{max}	μg/L	–	3 946. 5
V_d	L/kg	0. 45	2. 89

A：Zero-time blood drug concentration intercept of distribution phase；B：Zero-time blood drug concentration intercept of elimination phase；α：Distribution rate constant；β：Elimination rate constant；K_a：Absorption rate constant；$t_{1/2\alpha}$：Distribution half-life；$t_{1/2\beta}$：Elimination half-life；$t_{1/2\,Ka}$：Absorption half-life；k_{10}：Drug elimination rate constant from central compartment；k_{12}：First-order transport rate constant from central compartment to peripheral compartment；k_{21}：First-order transport rate constant from peripheral compartment to central compartment；AUC_{0-t}：Area under concentration-time curve；T_{max}：Time to peak concentration；C_{max}：Peak concentration；V_d：Apparent distribution volume.

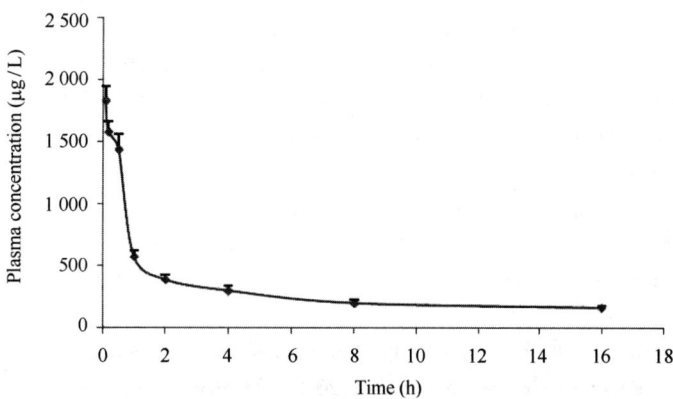

Fig. 1　Mean plasma concentration profiles of flumequine inblunt snout bream following intravascularadministration （3mg/kg. b. w. ）（n=6）

DISCUSSION

The present study firstly investigated the pharmacokinetics and oral bioavailability of FMQ in blunt snout bream. These data would be helpful to the dosage design of FMQ in clinical treatment.

A two-compartment open model proved to best fit the data of the two different administrations. After intravascular administration, the distribution half-life （$t_{1/2\alpha}$） and the terminal elimination half-life （$t_{1/2\beta}$） werecalculated to be 0. 6 h and 25. 0 h, respectively. Following identical drug administration, the $t_{1/2\alpha}$was longer than that in gilthead sea bream （0. 2 h）（Rigos et al. ，2003）but less than that in sea bass （1. 05 h）（Rigos et al. ，2002）. The $t_{1/2\beta}$was similar to that in sea bass （30 h），catfish （25 h）（Plakas et al. ，2000），halibut （32 h），turbot （34 h）（Hansen & Horsberg, 1999），wrasse （31 h）（Hansen & Horsberg, 2000），Atlantic salmon （30-40 h）（Elema et al. ，1995），but far less than that in eel, Anguillaanguilla （314 h）（Hansen & Horsberg, 2000）. The volume of apparent distribution （0. 45 L/kg）in this study was less than above-mentioned fish, such as gilthead sea bream （0. 57 L/kg），sea bass （1. 51 L/kg），halibut （2. 99 L/kg），turbot （3. 75 L/kg），cod （2. 41 L/kg），wrasse （2. 15 L/

413

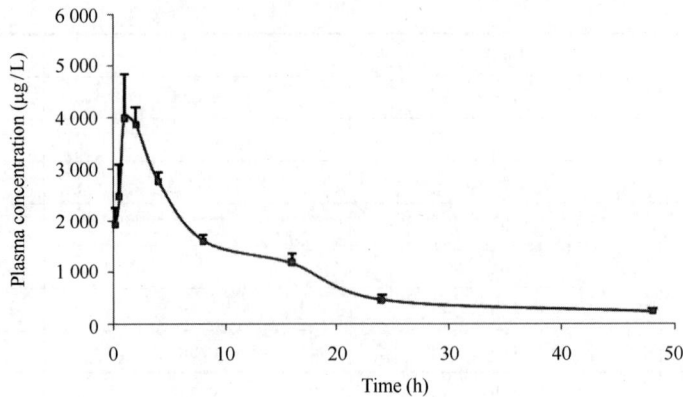

Fig. 2　Mean plasma concentration profiles of flumequine inblunt snout bream
following oral administration (50 mg/kg b. w.) (n=6)

kg), eel (3.4 L/kg), which suggested that tissue distribution of FMQ was low in blunt snout bream. The corresponding values of total body clearance (Cl) in halibut, turbot, eel, and wrasse were 0.12, 0.17, 0.012, and 0.14 L/h. kg, respectively. From those data, only value of eel was less than that of blunt snout bream. The reason for the slow elimination of FMQ in eels might be the lack of biotransformation and other possible physiological inter-species differences (Byczkowska-Smyk, 1958). Furthermore, only traces of 7-OH FMQ (one main metabolite of FMQ) and no glucuronide metabolites were found in eel (Van der Heijden et al., 1993).

After oral administration, the Cmax was 3 946.5 μg/kg at 1.4 h, which was equal to or slightly higher than that in previous study. The elimination half-life was estimated to be 19.7 h, which was less than that reported from above-mentioned study in halibut (43 h), turbot (42 h), cod (74 h), wrasse (41 h) after oral dose. Although many factors differed between our study and others, it might be accounted for by the different temperatures in studies. Our studies were performed at a temperature of 25.0±1.2℃. However, the temperatures in literatures were ranged from 8 to18℃. Previous studies demonstrated that a temperature increase of 1℃ corresponded to a 10% increase in metabolic and excretory rates of fish (Ellis, 1978). In addition, those references were obtained from sea water fish. Sohlberg, et al. (2002) reported that FMQ was eliminated at a substantially higher rate from fish in sea water than fresh water. This phenomenon might be explained by the influence of chelation on the compound's pharmacokinetic properties (Stein, 1996). Sea water fish drink actively, and the chelation of FMQ in the gut contents by Mg^{2+} and Ca^{2+} ions from sea water might have reduced the primary absorption and entero-hepatic cycling in seawater compared with freshwater conditions (Sohlberg et al., 2002). Rigos et al. (2003) also reported that the addition of 10 mM/L Ca^{2+} and 55 mM/L Mg^{2+} to the medium resulted in an 8-to > 120-fold reduction in FMQ activity. Moreover, the $t_{1/2\beta}$ values of FMQ in fish were longer than that in poultry and mammals. Previous study indicated that $t_{1/2\beta}$ value in chicken was 6.91 h (Anadon et al., 2008). A shorter $t_{1/2\beta}$ value was present in pigeons, only 1.57 h (Dorrestein et al., 1983). In mammals, $t_{1/2\beta}$ values were 6.35 h in pig (Villa et al., 2005), 11.5 h in sheep (Delmas et al., 1997), 3 h in calves (Ziv et al., 1986), respectively. The differences might be due to the metabolic capacity of species or different temperature.

The oral bioavailability (F) of FMQ was 32.0% in blunt snout bream after oral gavage of liquid dose, which was close to that in gilthead seabream (29.0%) following same oral dose with the present study but less than that reported from above-mentioned studies in halibut (56.0%), turbot (59.0%), cod (65.0%), wrasse (41.0%) following oral gavage of FMQ mixed in a fish-feed e-mulsion, Atlantic salmon (40.0%) following oral gavage of liquid dose. Elema et al. (1995) suggested that malabsorption, first-pass metabolism of FMQ might be possible causes. FMQ is a weak acid with a pKa value of 6.0, and its absorption might be reduced by the alkaline intestinal environment of fish. Otherwise, first-pass metabolism of drug might occur in intestinal mucosa and intestinal flora (Sousa et al., 2008), which needed further studies.

For fluoroquinolone antimicrobials, both C_{max}-to-MIC ratio and AUC-to-MIC ratio were used to predict antibacterial success. Most experts agreed that a C_{max} to MIC of 8-10or an AUC-to-MIC ratio greater than 100-125 prevented the overgrowth of resistant bacteria (Wright et al., 2000; McKellar et al., 2004). Minimal inhibition concentrations of FMQ against Vibrio anguillarum Serotype 1b, Photobacterium damsel ssp. piscicida, V. alginolyticus, V. damsel and V. fluvialis were 0.15, 0.3, 1.2, 0.019 and 0.15 μg/mL (Rigos et al., 2003). From our resultsafter given FMQ by oral gavage of liquid dose, the ratios of C_{max}/MIC were 26.09, 13.04, 3.26, 205.95 and 26.09 in this study, respectively, and that of AUC/MIC were 340.33, 170.17, 42.54,

2686. 84 and 340. 33, respectively. In contrast, ratios of C_{max}/MIC and AUC/MIC for V. alginolyticus were the lowest.

The pharmacokinetics of FMQ in blunt snout bream displayed low bioavailability, rapid absorption and rapid elimination. We showed that FMQ administered at a dose of 50 mg/kg b. w. by oral gavage of liquid dose offered an option for treatment of some diseases.

ACKNOWLEDGMENTS

This work is financially supported by the Special Fund for Agro-Scientific Research in the Public Interest of China (201203085) and National Nonprofit Institute Research Grant of Freshwater Fisheries Research Center, CAFS (2015JBFM29).

REFERENCES

Anadon, A., Martinez, M. A., Martinez, M., De La Cruz, C., Diaz, M. J. & Martinez-Larranaga, M. R. (2008) Oral bioavailability, tissue distribution and depletion of flumequine in the food producing animal, chicken for fattening. Food and Chemical Toxicology, 46 (2), 662-670.

Byczkowska-smyk, W. (1958) The respiratory surface of the gills in teleosts. 2. Acta Biologica Cracoviensia (Serie: Zoologie), 1, 831-837

Delmas, J. M., Chapel, A. M., Gaudin, V., &Sanders, P. (1997) Pharmacokinetics of flumequine in sheep after intraveous and intramuscular administration: bioavailability and tissue residue studies. Journal of Veterinary Pharmacology and Therapeutics, 20 (4): 249-257

Dorrestein, G. M., van Gogh, H., Buitelaar, M. N. & Nouws, J. F. (1983) Clinical pharmacology and pharmacokinetics of flumequine after intravenous, intramuscular and oral administration in pigeons (Columba livia). Journal of Veterinary Pharmacology and Therapeutics, 6 (4), 281-292.

Ellis, A. E. (1978) The anatomy and physiology of teleost. In Fish Pathology, 3 Edn. Eds Roberts, R., PP. 13-54. Bailliere Tindall, London.

Elema, M. O., Hoff, K. A. & Kristensen, H. G. (1994) Multiple-dose pharmacokinetic study of flumequine in Atlantic salmon (Salmo salar L.). Aquaculture, 128 (1-2), 1-11.

Elema, M. O., Hoff, K. A. & Kristensen, H. G. (1995) Bioavailability of flumequine after oral administration to Atlantic salmon (Salmo salar L.). Aquaculture, 136 (3-4), 209-219.

Hansen, M. K. & Horsberg, T. E. (1999) Single-dose pharmacokinetics of flumequine in halibut (Hippoglossus hippoglossus) and turbot (Scophthalmus maximus). Journal of Veterinary Pharmacology and Therapeutics, 22 (2), 122-126.

Hansen, M. K. & Horsberg, T. E. (2000) Single-dose pharmacokinetics of flumequine in cod (Gadus morhua) and goldsinny wrasse (Ctenolabrus rupestris). Journal of Veterinary Pharmacology and Therapeutics, 23 (3), 163-168.

Hansen, M. K. & Horsberg, T. E. (2000) Single-dose pharmacokinetics of flumequine in the eel (Anguilla anguilla) after intravascular, oral and bath administration. Journal of Veterinary Pharmacology and Therapeutics, 23 (3), 169-174.

Hu, L. L., Fang, W. H., Zhou, K., Xia, L. J., Zhou, S., Lou, B. & Shi, H. L. (2011) Pharmacokinetics, Tissue Distribution, and Elimination of Flumequine in Japanese Seabass (Lateolabrax japonicus) following Intraperitoneal Injection and Oral Gavage. Israeli Journal of Aquaculture-Bamidgeh, 63, 586.

Liang, Z. H., Lin, L. M., Liu, J. J., Jiang, Z. G. & Gong, Q. L. (2006) Pharmacokinetic study of oxolinic acid and flumequine in Muraenesox cinereus. Marine Fisheries Research, 27 (3), 6.

McKellar, Q. A., Sanchez Bruni, S. F. & Jones, D. G. (2004) Pharmacokinetic/pharmacodynamic relationships of antimicrobial drugs used in veterinary medicine. Journal of Veterinary Pharmacology and Therapeutics, 27 (6), 503-514.

Plakas, S. M., El Said, K. R. & Musser, S. M. (2000) Pharmacokinetics, tissue distribution, and metabolism of flumequine in channel catfish (Ictalurus punctatus). Aquaculture, 187 (1-2), 1-14.

Rigos, G., Tyrpenou, A., Nengas, I. & Alexis, M. (2002) A pharmacokinetic study of flumequine in sea bass, Dicentrarchus labrax (L.), after a single intravascular injection. Journal of Fish Diseases, 25 (2), 101-105.

Rigos, G., Tyrpenou, A. E., Nengas, I., Yiagnisis, M., Koutsodimou, M., Alexis, M. & Troisi, G. M. (2003) Pharmacokinetics of flumequine and in vitro activity against bacterial pathogens of gilthead sea bream Sparus aurata. Diseases of Aquatic Organisms, 54 (1), 35-41.

Samuelsen, O. B., Husgard, S., Torkildsen, L. & Bergh, O. (2000) The efficacy of a single intraperitoneal injection of flumequine in the treatment of systemic vibriosis in corkwing wrasse Symphodus melops. Journal of Aquatic AnimalHealth, 12 (4), 324-328.

Sohlberg, S., Czerwinska, K., Rasmussen, K. & Søli, N. E. (1990) Plasma concentrations of flumequine after intra-arterial and oral administration to rainbow trout (Salmo gairdneri) exposed to low water temperatures. Aquaculture, 84 (3-4), 355-361.

Sohlberg, S., Ingebrigtsen, K., Hansen, M. K., Hayton, W. L. & Horsberg, T. E. (2002) Flumequine in Atlantic salmon Salmo salar: disposition in fish held in sea water versus fresh water. Diseases of Aquatic Organisms, 49 (1), 39-44.

Sousa, T., Paterson, R., Moore, V., Carlsson, A., Abrahamsson, B. & Basit, A. W. (2008) The gastrointestinal microbiota as a site for the biotransformation of drugs. International Journal of Pharmaceutics, 363 (1-2), 1-25.

Stein, G. E. (1996) Pharmacokinetics and pharmacodynamics of newer fluoroquinolones. Clinical Infectious Diseases, 23 Suppl 1, S19-24.

Van der Heijden, M. H. T., Boon, J. H., Nouws, J. F. M., & Mengelers, M. J. B. (1993) residue depletion of flumequine in European eel. In Residues of Veterinary Drugs in Food. Proceedings of the EuroResidue II Conference, Eds Haagsma, N., Ruiter, A. Czedik-Eysenbergeds, P. B.,

Veldhoven, The Netherlands, May 3-5, 1993. pp. 357-361

Villa, R., Cagnardi, P., Acocella, F., Massi, P., Anfossi, P., Asta, F. & Carli, S. (2005) Pharmacodynamics and pharmacokinetics of flumequine in pigs after single intravenous and intramuscular administration. TheVeterinary Journal, 170 (1), 101-107.

Wright, D. H., Brown, G. H., Peterson, M. L. & Rotschafer, J. C. (2000) Application of fluoroquinolone pharmacodynamics. Journal of Antimicrobial Chemotherapy, 46 (5), 669-683.

Yamaoka, K., Nakagawa, T. & Uno, T. (1978) Statistical moment in pharmacokinetics. Journal of Pharmacokinetic and Biopharmaceutics, 6, 547-558.

Ziv, G., Soback, S., Bor, A. & Kurtz, B. (1986) Clinical pharmacokinetics of flumequine in calves. Journal of Veterinary Pharmacology and Therapeutics, 9 (2), 171-182.

该文发表于《ISRAELI JOURNAL OF AQUACULTURE-BAMIDGEH》【2011, 63 (1) 】

The pharmacokinetics, tissue distribution and elimination of flumequine in Japanese seabass (*Lateolabrax japonicus*) following intraperitoneal injection and oral gavage

Lin-lin Hu[a], Wen-hong Fang[a*], Kai Zhou[a], Lian-jun Xia[a],
Shuai Zhou[a], Bao Lou[b], Hui-lai Shi[b]

(a. East China Sea Fisheries Research Institute, Chinese Academy of Fisheries Sciences, Key Laboratory of Marine and Estuarine Fisheries Resources and Ecology, Ministry of Agriculture, Shanghai, 200090, China; b. Marine Fisheries Research Institute of Zhejiang Province, Zhoushan, 316100)

Abstract: The pharmacokinetics, tissue distribution and elimination of flumequine were investigated following intraperitoneal injection and oral gavage at a dose of 20 mg kg^{-1} in Japanese seabass (*Lateolabrax japonicus*) in seawater of a salinity of 26‰. The seawater temperature was at about 25℃. Following a single intraperitoneal injection, the profile of plasma concentrations vs. time was best described by a two-compartment open model with first-order absorption. The distribution and elimination half-lives ($t_{1/2\alpha}$ and $t_{1/2\beta}$) were found to be 0.71 h and 8.163 h, respectively. The total body clearance (CLs) and apparent volume of distribution (V_d) were estimated to be 256.0 mL h^{-1}kg^{-1} and 3.012 L kg^{-1}, respectively. The plasma concentration-time curve after oral gavage (force feeding) did not fit the compartment model. The parameters in the plasma were calculated by non-compartment model based on statistic moment theory. The area under the concentration-time curve from 0 to ∞ h (AUC$_{0-\infty}$), mean residue time from 0 to ∞ h (MRT$_{0-\infty}$), half-life ($t_{1/2z}$) and body clearance (CLz) were 26.40 μg·h mL^{-1}, 21.97 h, 14.90 h and 757.0 mL·h^{-1}·kg^{-1}, respectively. The plasma protein binding rates in vivo were determined to be 13.4% at 1 h and 31.3% at 12 h following intraperitoneal injection, and 19.3% at 4 h and 45.2% at 18 h after oral gavage. The parameters in the muscle and liver were calculated by statistic moment theory following intraperitoneal injection and oral gavage. After intraperitoneal injection, AUC$_{0-\infty}$, MRT$_{0-\infty}$ and body clearance (CLz) were 73.99 μg·h g^{-1}, 9.86 h and 270.0 g·h^{-1}·kg^{-1} for muscle and 140.54 μg·h·g^{-1}, 22.46 h and 142.0 g·h^{-1}·kg^{-1} for liver, respectively. After oral gavage, AUC$_{0-\infty}$, MRT$_{0-\infty}$ and body clearance (CLz) were 9.17 μg·h·g^{-1}, 37.62 h and 2 181.0 g·h^{-1}·kg^{-1} for muscle and 60.91 μg·h·g^{-1}, 45.80 h and 328.0 g·h^{-1}·kg^{-1} for liver, respectively. It was indicated that flumequine was rapidly distributed to tissues and eliminated fast from muscle when administrated to Japanese seabass.

Keywords: Pharmacokinetics; Lateolabrax japonicus; Flumequine; Tissue distribution; Elimination; Plasma protein binding

腹腔注射和口灌给药下氟甲喹在花鲈体内药动学、组织分布与消除

胡琳琳[a]，房文红[a*]，周凯[a]，夏连军[a]，周帅[a]，楼宝[b]，史会来[b]

(a. 中国水产科学院东海水产研究所，上海 200090；a. 农业部海洋与河口资源与生态重点实验室，上海 200090；
b. 浙江省海洋水产研究所，舟山 316100)

摘要：在海水盐度 26、水温 25℃左右，对花鲈进行腹腔注射和口灌氟甲喹给药（给药剂量 20 mg/kg），研究下其在花鲈体内药动学、组织分布与消除。单剂量腹腔注射，血药浓度–时间关系曲线适合采用一级吸收二室开放模型来描述。分布半衰期（$t_{1/2\alpha}$）和消除半衰期（$t_{1/2\beta}$）分别为 0.71 h 和 8.163 h，总体清除率（CLs）和表观分布容积（V_d）分别为 256.0 mL·h^{-1}·kg^{-1} 和 3.012 L·kg^{-1}。而口灌给药下药时曲线不适合采用房室模型来描述，其药动学参数采用统计矩原理来推算。曲线下面积（$AUC_{0-\infty}$）、平均驻留时间（$MRT_{0-\infty}$）、半衰期（$t_{1/2z}$）和总体清除率（CLz）分别是 26.40 μg·h·mL^{-1}、21.97 h、14.90 h 和 757.0 mL·h^{-1}·kg^{-1}。腹腔注射给药后 1 h 和 12 h 的体内蛋白结合率分别为 13.4% 和 31.3%，口灌给药后 4 h 和 18 h 时分别为 19.3% 和 45.2%。肌肉和肝脏中药动学参数采用统计矩原理推算。腹腔注射给药下，肌肉的 $AUC_{0-\infty}$、$MRT_{0-\infty}$ 和 CLz 分别是 73.99 μg·h·g^{-1}、9.86 h 和 270.0 g·h^{-1}·kg^{-1}，肝脏的分别为 140.54 μg·h·g^{-1}、22.46 h 和 142.0 g·h^{-1}·kg^{-1}。口灌给药下，肌肉的 $AUC_{0-\infty}$、$MRT_{0-\infty}$ 和 CLz 分别是 9.17 μg·h·g^{-1}、37.62 h 和 2 181.0 g·h^{-1}·kg^{-1}，肝脏的分别是 60.91 μg·h·g^{-1}、45.80 h 和 328.0 g·h^{-1}·kg^{-1}。研究结果表明，氟甲喹以较快的速度分布到花鲈组织中，且从肌肉组织中的消除速度也快。

关键词：药动学；花鲈；氟甲喹；组织分布与消除；血浆蛋白结合率

1 Introduction

Japanese seabass, *Lateolabrax japonicus*, is a euryhaline and eurythermal species widely distributed in China, Korea and Japan. Because of its economic value, it is widely cultivated in the coastal region of China. The Japanese seabass production is one of the highest in marine fish farming with an annual yield of about 90, 000 tonnes. Recently, diseases caused by Vibrio bacteria caused serious economic losses for Japanese seabass farming (Wang et al., 2002; Zhang et al., 2009). Some antimicrobials were used to try to solve such problems by fish farmers one if which was fumequine.

Flumeqine (FLU) is a synthetic antibiotic belonging to the group of "second generation" fluoroquinolines. It has a strong and rapid antibacterial activity against Gram–negative bacteria by inhiting DNA–gyrase (Drlica and Coughlin, 1989). It is used primarily for the treatment of enteric infections in human and domestic species (Rohifing et al., 1977). During the last decades, some studies showed that FLU had satisfactory kinetic properties in some farmed fish species (Rogstad et al., 1993; Martinsen & Horsberg, 1995) and low MICs against important bacterial fish pathogens (Ledo et al., 1987; Martinsen et al., 1992; Rigos et al., 2003), indicating that FLU has a potential efficacy for the treatment of diseases in fish. The pharmacokinetics of flumequine has previously been studied mainly in freshwater fishes and seawater fishes at low temperature. These freshwater fishes included rainbow trout (*Oncorhynchus mykiss*) (Sohlberg et al., 1994), European eel (*Anguilla anguilla*) (Boon et al., 1991; Van der Heijden et al., 1994), common carp (*Cyprinus carpio*) (Van der Heijden et al., 1994) and channel catfish (*Ictalurus punctatus*) (Plakas et al., 2000). Except for gilthead sea bream (*Sparus aurata*) (Malvisi et al., 1997; Rigos et al., 2003; Tyrpenou et al., 2003), the pharmacokinetics of many seawater fishes have been conducted at low temperature, which included Atlantic salmon (*Salmo salar*) (O'Grady et al., 1988; Rogstad et al., 1993; Elema et al., 1995; Martinsen & Horsberg, 1995), Atlantic halibut (*Hippoglossus hippoglossus*) (Samuelsen & Lunestad, 1996; Samuelsen & Ervik, 1997; Hansen & Horsberg, 1999), turbot (*Scophthalmus maximus*) (Hansen & Horsberg, 1999), Atlantic salmon (*Salmo salar*) (Ellingsen et al., 2002). These revealed that the kinetic profile of FLU had significant interspecies differences. Several studies have reported that the plasma protein binding of flumequine seemed to be dependent on the concentration of the drug in the plasma (Boon et al., 1991; Plakas et al., 2000).

The objective of the present study was to investigate the pharmacokinetics of flumequine after a single intraperitoneal injection and oral gavage in Japanese seabass held in warm seawater of a salinity of 26‰.

2　Materials and methods

2. 1　Experimental animals and aquaria conditions

Healthy juvenile Japanese seabass (*Lateolabrax japonicus*, 250–300 g) were obtained from a sea-cage farm situated in Zhoushan, Zhejiang Province, China. Fish were kept in 300-L fiberglass tanks (total 28 banks and 6 individuals per tank) with seawater at a salinity of 26‰. Water was filtered and recirculated, keeping aeration constant and water temperature at 25.0 ± 1.0 ℃. Fish were fed daily with artificial feed at 2%~3% of their average body weight and the remnants and excretions were removed in time. Fish were acclimated for 10 days prior to experiments with no history of FLU treatment.

2. 2　Chemicals

Flumequine standard was purchased from Sigma (St Louis, MO, USA). Flumequine raw material was supplied by Suzhou Hengyi Pharmaceutical Co., Ltd. The active ingredient was determined by high performance liquid chromatography and found to exceed 98.0%. The organic solvents used, such as dichloromethane, n-hexane, acetonitrile and tetrahydrofuran, were of High-performance liquid chromatography (HPLC) grade (Sigma-Aldrich, German). All other chemicals were of analytical grade.

2. 3　Intraperitoneal injection

Intraperitoneal dosing solution was prepared at a concentration of $10 \text{ mg} \cdot \text{mL}^{-1}$ by using flumequine raw material. To prepare 50 mL, 500 mg flumequine was dissolved with 0.03 M NaOH, and the solution was adjusted to pH 10.5 with 1 N HCl and brought to volume with ultra-pure water produced by Water Purification Systems (Milli-Q Advantage A10, Millipore Corporation, Billerica, MA 01821, USA). Dosing solutions were administrated at a single dose of 2 mL kg^{-1} body weight, corresponding to a final dose of 20 mg kg^{-1} body weight.

2. 4　Oral gavage

The mash feed was prepared at a concentration of $5 \text{ mg} \cdot \text{mL}^{-1}$ by using FLU raw material. A 500 mg FLU, which was dissolved with 0.03 M NaOH (pH was adjusted to 10.5 with 1 N HCl), was added to 15 g fish feed ingredients. They were mixed uniformly and then brought to volume of 100 mL with ultra-pure water. The oral single dose was 20 mg FLU per kg body weight. Thus, dosing feed was administered at a single dose of 4 mL kg^{-1} body weight, corresponding to a final dose of 20 mg kg^{-1} body weight. The mash feed was forced by catheter into the stomach of fish. Only those individuals which did not regurgitate were used in the experiment.

2. 5　Sample collection

Sampling times for intraperitoneal injection group were 10 min, 30 min, 1 h, 2 h, 4 h, 8 h, 12 h, 18 h, 1 d, 2 d, 3 d, 5 d, 8 d, 10 d, 15 d and 20 d post-dosing. Sampling times for oral gavage group were 30 min, 1 h, 4 h, 8 h, 12 h, 18 h, 1 d, 2 d, 4 d, 8 d, 15 d and 20 d post-dosing.

At each sampling time six animals were killed with a blow to the head. Blood was taken from the caudal vein using 2-mL syringes containing heparin to prevent clotting. The collected blood was centrifuged at 3000 r. p. m. for 10 min. Plasma was removed and placed in 1.5 mL Eppendorf vials. Muscle and liver were also collected and placed in individually marked plastic bags. All tissue samples were stored frozen at −70 ℃ until they were analyzed.

Plasma protein binding of the drugs was determined prior to freezing samples animals at 1 h and 12 h after intraperitoneal injection and at 4 h and 18 h after oral administration. Binding was determined by ultrafiltration using Microcon YM-10 Centrifugal Filter (Millipore Corporation, Billerica, MA 01821, USA). A 0.5 mL aliquot of plasma was used and centrifuged at 10000×g for 20 min at 4 ℃ (CF 16RXⅡ, Hitachi, Japan). The resulting 0.15 mL protein-free filtrate was frozen (−70 ℃) and later analyzed for parent compounds. The unbound fraction was calculated as the ratio of drug concentration in the ultrafiltrate to the concentration in the plasma.

2. 6　Extraction and purification

Plasma was thawed at room temperature. 400 μL of plasma were added with acetonitrile of the same volume. The mixture was vortexed for 2 min and centrifuged at 5000× g for 10 min under 4 ℃. The supernatant was filtered through 0.45 μm disposable syringe filter and then 10μl was injected into the HPLC.

Samples of muscle and liver were thawed and triturated in a mortar and pestle. 0.02 M H_3PO_4 (2.0 mL) was added to 2.0 g samples and vortexed for 2 min. Then 10 mL dichloromethane was added, vortexed for 5 min and centrifuged at 5000× g for 5 min under 4℃. The supernatant was transferred into a clean centrifuge tube, while the residue was re-extracted as above twice again. The combined supernatants were evaporated to dryness using the vortex evaporator. The resulting residue was dissolved in 1.0 mL of mobile phase and 2 mL of n-hexane. The mixture was shaken vigorously and then transferred into a 5 mL of centrifuge tube. After centrifugation at 5000× g for 5 min, the lower-layer liquid was filtered through 0.45 μm disposable syringe filters prior to analysis by HPLC system.

2.7　Chromatographic conditions

HPLC analysis was performed with Agilent 1100 consisted of double pump, auto-injector, column temperature tank and UV detector. The column was Zorbax SB C-18, 4.6×150 mm.

The mobile phase consisted of acetonitrile : tetrahydrofuran : 0.02 M H_3PO_4 (16：12：72, v/v/v). H_3PO_4 solution was filtered through 0.45 μm Millipore filter and sonicated for 10 min before use. The flow rate was 1.2 mL min^{-1} with analysis being performed at 40^0C. The detection wavelength was set at 324 nm and the injection volume was 10 μL.

The recoveries of the methods were (91.1±3.3)% for plasma, (87.0±2.1)% for muscle and (86.6±3.6)% for liver. The quantitative limits of FLU were 0.02 μg·mL^{-1} for plasma, 0.01 μg g^{-1} for muscle and liver.

2.8　Data analysis

The plasma and tissues concentration-time data following intraperitoneal injection and oral administration were analyzed by compartmental pharmacokinetic model and non-compartmental model (statistical moment theory), respectively. Pharmacokinetic compartmental and non-compartmental modeling of the drug concentration-time data was performed by DAS Version 2.0 (Drug and Statistics Software were developed by the Mathematical Pharmacology Professional Committee of China and authorized and supported by China State Drug Administration).

3　Results

3.1　Pharmacokinetics

Plasma concentrations of flumequine after intraperitoneal injection at a dose of 20 mg·kg^{-1} are showed in Fig. 1. The peak concentration of flumequine in plasma was 22.33 μg·mL^{-1} at the first sample point (10 min). The plasma concentrations of flumequine could be best described by a two-compartment open model with first-order absorption. The pharmacokinetic parameters derived from the compartment model are showed in Table 1. The distribution and elimination half-lives ($T_{1/2\alpha}$ and $T_{1/2\beta}$) were 0.71 h and 8.163 h, respectively. The total body clearance (CLs) and apparent volume of distribution were 256.0 mL·h^{-1}·kg^{-1} and 3.012 L·kg^{-1}, respectively.

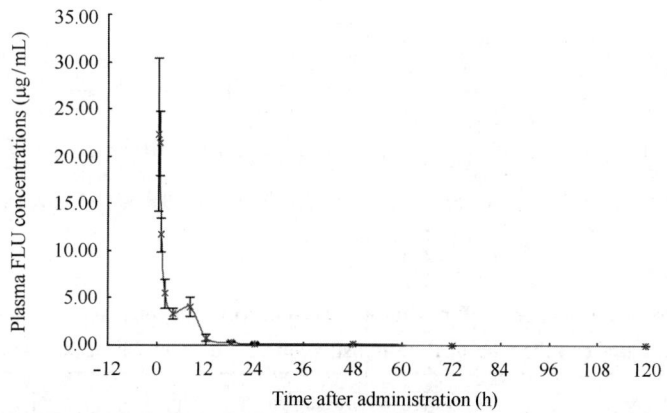

Fig. 1　The plasma concentrations of flumequine in Japanese seabass following intraperitoneal injection at a dose of 20 mg kg^{-1}. Standard deviation is given as vertical bars (n=6).

Table 1 **Pharmacokinetic parameters of flumequine in plasma fromLateolabras japonicus based on compartmental model following intraperitoneal injection at a dose of 20 mg kg^{-1} (n=6)**

Pharmacokinetic parameters	Unit	Values
C_{max}	$\mu g \cdot mL^{-1}$	22.33
T_{max}	h	0.167
α	h^{-1}	0.976
β	h^{-1}	0.085
$t_{1/2\alpha}$	h	0.71
$t_{1/2\beta}$	h	8.163
$t_{1/2ka}$	h	0.012
V_d	$L \cdot kg^{-1}$	3.012
CLs	$ml \cdot h^{-1} \cdot kg^{-1}$	256.0
AUC_{0-t}	$\mu g \cdot h \cdot mL^{-1}$	76.21
$AUC_{0-\infty}$	$\mu g \cdot h \cdot mL^{-1}$	78.10

C_{max}, maximum concentration in plasma; T_{max}, time when maximum concentration was obtained; α, distribution rate constant; β, elimination rate constant; $t_{1/2\alpha}$, distribution half-life of the drug; $t_{1/2\beta}$, elimination half-life of the drug; $t_{1/2ka}$, absorption half-life of the drug; V_d, apparent volume of distribution; CLs, total body clearance of the drug; AUC_{0-t}, area under the concentration-time curve from the zero to 120 h; $AUC_{0-\infty}$, area under the concentration-time curve from the zero to infinity.

Fig. 2 shows the plasma concentration-time profiles after oral administration of flumequine. Because of double-peak phenomenon in plasma, the concentration-time curve could not be fitted by the nonlinear least squares method using compartment model. Therefore, the plasma concentrations were estimated by non-compartmental analysis based on statistical moment theory. The obtained parameters are listed in Table 2.

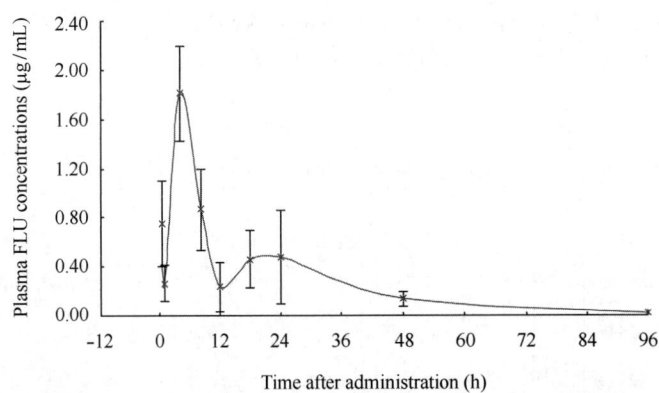

Fig. 2　The plasma concentrations of flumequine in Japanese seabass following oral gavage at a dose of 20 mg kg^{-1}. Standard deviation is given as vertical bars (n=6)

Table 2 **Pharmacokinetic parameters of flumequine inplasma from Lateolabras japonicus based on statistical moment theory following oral administration at a dose of 20 mg kg^{-1} (n=6)**

Pharmacokinetic parameters	Unit	Values
C_{max}	$\mu g \cdot mL^{-1}$	1.81
T_{max}	h	4.0
AUC_{0-t}	$\mu g \cdot h \cdot mL^{-1}$	23.35
$AUC_{0-\infty}$	$\mu g \cdot h \cdot mL^{-1}$	26.40
MRT_{0-t}	h	15.76

Pharmacokinetic parameters	Unit	Values
$MRT_{0-\infty}$	h	21.97
$t_{1/2z}$	h	14.90
CLz	$mL\ h^{-1}kg^{-1}$	757.0

C_{max}, maximum concentration in plasma; T_{max}, time when maximum concentration was obtained; AUC_{0-t}, area under the concentration−time curve from the zero to 120 h; $AUC_{0-\infty}$, area under the concentration−time curve from the zero to infinity; MRT_{0-t}, mean residue time of drug in body from zero to 120 h; MRT_{0-t}, mean residue time of drug in body from zero to infinity; $t_{1/2z}$, half-life of the drug; CLz, total body clearance.

Plasma protein binding values are 13.4% and 31.3% at 1 h and 12 h after intraperitoneal injection, respectively, and 19.3% and 45.2% at 4 h and 18 h after oral administration, respectively (Table 3).

Table 3　Plasma protein binding of flumequine in Japanese seabass, *Lateolabras japonicus* ($n=6$)

Route of administration	Sampling time	Plasma flumequine concentration ($\mu g \cdot mL^{-1}$)	Plasma protein binding (%)
Intraperitoneal injection	1 h	11.65±1.76	13.4±3.2
	12 h	0.79±0.30	31.3±3.9
Oral administration	4 h	1.81±0.39	19.3±2.7
	18 h	0.46±0.23	45.2±9.0

3.2　Tissue distribution and elimination

The tissue concentration−time curves of FLU after intraperitoneal injection and oral administration are showed in Fig. 3 and Fig. 4. Following intraperitoneal injection, the peak concentration in liver was 45.45 $\mu g\ g^{-1}$ at the first sampling time (10 min), followed by a rapid decrease until the fourth hour, afterwards which it reached the last data point. In the muscle, the peak concentration was 9.46 $\mu g\ g^{-1}$ at 1 h. Following oral administration, double−peak phenomenon was observed in liver, which the first and second peak concentrations were 1.71 $\mu g\ g^{-1}$ and 3.45 $\mu g\ g^{-1}$, at 30 min and 4 h post−dosing, respectively. The FLU concentrations in the muscle were lower than in the liver, as evidenced in Fig. 4. FLU was quickly cleared from muscle on day 4 where it reached a level of 0.020 $\mu g\ g^{-1}$.

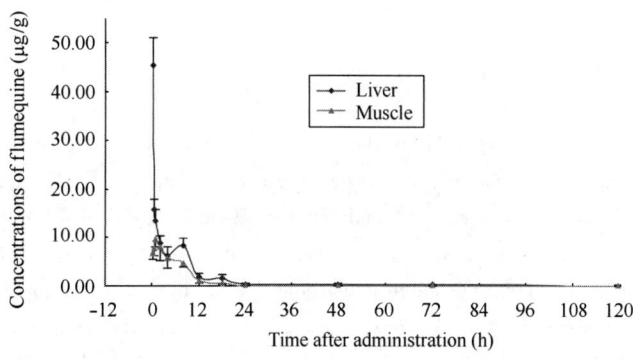

Fig. 3　The flumequine concentrations in muscle and liver from Japanese seabass following intraperitoneal injection at a dose of 20 mg kg^{-1}. Standard deviation is given as vertical bars (n=6)

The behaviors of FLU in tissues after intraperitoneal injection and oral administration were described by statistical moment theory. The pharmacokinetic parameters are listed in Table 4.

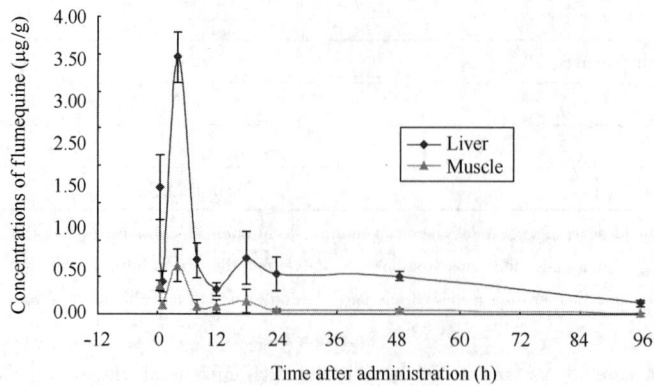

Fig. 4　The flumequine concentrations in muscle and liver from Japanese seabass following oral gavage at a dose of 20 mg kg^{-1}. Standard deviation is given as vertical bars （n=6）.

Table 4　Pharmacokinetic parameters of flumequine in tissues fromLateolabras japonicus based on statistical moment theory following intraperitoneal injection and oral administration at a dose of 20 mg · kg^{-1} （n=6）

pharmacokinetic parameters		Intraperitoneal injection		oral administration	
		muscle	liver	muscle	liver
C_{max}	μg g^{-1}	9. 46	45. 45	0. 64	3. 45
T_{max}	h	1. 0	0. 167	4. 0	4. 0
AUC$_{0-120}$	μg · h g^{-1}	73. 68	134. 22	8. 15	52. 93
AUC$_{0-\infty}$	μg · h g^{-1}	73. 99	140. 54	9. 17	60. 91
MRT$_{0-120}$	h	9. 24	14. 58	23. 70	30. 33
MRT$_{0-\infty}$	h	9. 86	22. 46	37. 62	45. 80
$t_{1/2z}$	h	24. 98	48. 29	36. 36	36. 25
CLz	g h^{-1}kg^{-1}	270. 0	142. 0	2181.0	328. 0

C_{max}, maximum concentration in plasma; T_{max}, time when maximum concentration was obtained; AUC$_{0-t}$, area under the concentration−time curve from the zero to 120 h; AUC$_{0-\infty}$, area under the concentration−time curve from the zero to infinity; MRT$_{0-t}$, mean residue time of drug in body from zero to 120 h; MRT$_{0-t}$, mean residue time of drug in body from zero to infinity; $t_{1/2z}$, half-life of the drug; CLz, total body clearance.

4　Discussion

In this study, the plasma profile in Japanese seabass following intraperitoneal injection could be evaluated by a two−compartment model, whereas FLU after oral administration could not be analyzed by any compartment model. In Atlantic halibut （*Hippoglossus hippoglossus*）, the pharmacokinetics of FLUafter intraperitoneal and oral administration have been described by a two−compartment model whereas after intravenous injection, it was described by a three−compartment model （Samuelsen & Ervik, 1997）. The pharmacokinetics of FLU in Atlantic Salmon （*Salmo salar*） after intravascular injection has been described by a two−compartment model whereas after oral administration, it was best described by a one−compartment model （Martinsen & Horsberg, 1995）. The disposition of FLU in freshwater rainbow was described by a three − compartmental model （Sohlberg et al., 1994）. We also attempted to use compartment model. However, the parameters calculated were not acceptable. The reason for this difference may be the variations between fish species, the route of administration or the experimental design （e. g., sampling time）.

Boon et al. （1991） found that the plasma protein binding of FLU seemed to be dependent on the concentration of the drug in the plasma in eel. In channel catfish, plasma protein binding of FLU in plasma from orally dosed animals was dependent on concentration or the time after dosing. The mean binding values were 67% at 24 h and 89% at 168 h （Plakas et al., 2000）. Although the binding values in the present work were much lower than that in channel catfish （Boon et al., 1991）, similar concentration−dependent or time−dependent changes were found in intraperitoneally and orally dosed Japanese seabass. Generally, if the drug bound to plasma proteins to form protein−drug complex, it could not transport across cell membrane and affected drug distribution. Oie and Tozer （1979） suggested that drugs with high protein binding showed low volumes of distribution, and, conversely, drugs with low protein binding showed high volumes of distribution. Comparing with channel catfish (Plakas et al., 2000), the pharmacokinetic values for FLU in Japanese seabass indicated high distribution volume (V_d, 3.012 L kg^{-1} for i. p.) and fast clearance (Cls, 256.0 mL h^{-1} kg^{-1} for i. p. and CLz, 757.0 mL h^{-1} kg^{-1} for oral administration). Hence, we also agreed that plasma protein binding might be an important variable influencing distribution and clearance of FLU, which was similar to Plakas et al. (2000). Plasma protein binding was determined to predict how much drug was in the free form and available for tissue distribution and therapeutic action. In this study, the low protein binding suggested that almost 60%~80% of FLU in plasma was available for treating systemic infections.

At peak time of 4 h, muscle: plasma and muscle: liver concentration ratios were 1: 2.8 and 1: 5.4, respectively. The AUC of muscle tissue was much lower than in plasma and liver tissue, with muscle tissue AUC being 35% of plasma AUC and 15% of liver AUC. This similarity that muscle had the lowest FLU concentration was also observed in channel catfish after oral administration of FLU (Plakas et al., 2000) and pike eel (*Muraenesox cinereus*) after intravascular injection of FLU (Liang et al., 2006). However, although the oral dosage (20 mg kg^{-1}) here was four times higher than 5 mg kg^{-1} by Plakas et al. (2000), the peak FLU concentrations in plasma, muscle and liver tissues from Japanese seabass were lower than those in channel catfish. In some fishes such as Atlantic salmon and sea bream, FLU was highly concentrated in skin and especially bone, relative to muscle (Steffenak et al., 1991, 1994; Malvisi et al., 1997). In channel catfish (Plakas et al., 2000), FLU concentrations in skin were a slightly higher than those in muscle, but a mass of FLU residue was concentrated in bile, relative to the other tissues. It was indicated that significant species differences occurred in tissue distribution of FLU.

In summary, flumequine was absorbed rapidly with the peak time of 4 h for muscle. And FLU in muscle tissue also eliminated fast. After 24 h, only small amount of flumequine was present in the muscle. After 120 h the drug could not be detected in muscle tissue (detection limit was 10 μg kg^{-1}). In accordance with Regulation (EC) No 1181/2002, maximum residue limit for FLU in finfish was 600 μg kg^{-1}. The results of this study indicated that flumequine residue represented a low potential hazard for the consumer in the light of the Manufactoring Readiness Level (MRL) value. Therefore, it was suggested that flumequine is a relatively safe antibacterial drug and can be used in prevention and treatment in the bacterial disease in Japanese seabass farming.

Acknowledgement

This work was supported financially by the National Support Programs of Eleventh Five-Year Plan (2006BAD03B04-03).

References

Boon, J. H., et al, 1991. Disposition of flumequine in plasma of European eel (Anguilla anguilla) after a single intramuscular injection. Aquaculture 99, 213-223.

Drlica, K. and Coughlin, S., 1989. Inhibitors of DNA gyrase. Pharmacol. Ther. 44, 107-122.

Elema, M. O., Hoff, K. A. and Kristensen, H. G., 1995. Bioavailability of flumequine after oral administration to Atlantic salmon (Salmo salar L). Aquaculture 136, 209-219.

Ellingsen, O. F., et al, 2002. Dosage regime experiments with oxolinic acid and flumequine in Atlantic salmon (Salmo salar) held in seawater. Aquaculture 209, 19-34.

Hansen, P. K. and Horsberg, T. E., 1999. Single-dose pharmacokinetics of flumequine in halibut (Hippoglossus hippoglossus) and turbot (Scophthalmus maximus). J. Vet. Pharmacol. Ther. 22, 122-129.

Ledo, A., et al, 1987. Effectiveness of different chemotherapeutic agents for controlling bacterial fish diseases. Bull. Eur. Assoc. Fish Pathol. 7, 20-22.

Liang Z. H., et al, 2006. Pharmacokinetic study of oxolinic acid and flumequine in Muraenesox cinereus. Marine Fisheries Research 27 (3), 86-92.

Malvisi, J., et al, 1997. Tissue distribution of flumequine after in-feed administration in sea bream (Sparus aurata). Aquaculture 157, 197-204.

Martinsen, B., et al, 1992. Temperature-dependent in vitro antimicrobial activity of four 4-quinolones and oxytetracycline against bacteria pathogenic to fish. Antimicrob. Agents Chemother 36, 738-743.

Martinsen, B. and Horsberg, T. E., 1995. Comparative single-dose pharmacokinetics of four quinolones, oxolinic acid, flumequine, sarafloxacin and enrofloxacin in Atlantic salmon (Salmo salar) held in seawater at 10 0C. Antimicrob. Agents Chemother 39, 1059-1064.

O'Grady, P., Moloney, M. and Smith, P. R., 1988. Bath administration of the quinolone antibiotic flumequine to brown trout Salmo trutta and Atlantic salmon S. salar. Dis. Aquat. Org. 4, 27-33.

Oie, S. and Tozer, T. N., 1979. Effect of altered plasma protein binding on the apparent volume of distribution. J. Pharm. Sci. 68, 1203-1205.

Plakas, S. M., Said, K. R. E. and Musser, S. M., 2000. Pharmacokinetics, tissue distribution, and metabolism of flumequine in channel catfish (Ictalurus punctatus). Aquaculture 187, 1-14.

Rigos, G., et al, 2003. Pharmacokinetics of flumequine and in vitro activity against bacterial pathogens of gilthead sea bream Sparus aurata. Dis. Aquat. Org. 54, 35-41.

423

Rogstad, A., Ellingsen, O. F. and Syvertsen, C., 1993. Pharmacokinetics and bioavailability of flumequine and oxolinic acid after various routes of administration to Atlantic salmon in seawater. Aquaculture 110, 207-220.

Rohifing, S. R., Gerster, J. F. and Kvam, D. C., 1977. Bioevaluation of the antibacterial flumequine for enteric use. J. Antimicrob. Chemother 3, 615 -620.

Samuelsen, O. B. and Ervik, A., 1997. Single dose pharmacokinetic study of flumequine after intravenous, intraperitoneal and oral administration to Atlantic halibut_ Hippoglossus hippoglossus. held in seawater at 9 0C. Aquaculture 158, 215-227.

Samuelsen, O. B. and Lunestad, B. T., 1996. Bath treatment, an alternative method for the administration of the quinolones flumequine and oxolinic acid to halibut Hippoglossus hippoglossus, and in vitro antibacterial activity of the drugs against some Vibrio sp. Dis. Aquat. Org. 27, 13-18.

Sohlberg, S., Aulie A. and Solie N. E., 1994. Temperature-dependent absorption and elimination of flumequine in rainbow trout (Oncorhynchus myciss Walbaum) in freshwater. Aquatic Organisms 9, 67-72.

Steffenak, I., Hormazabal, V. and Yndestad. M., 1991. Reservoir of quinolone residues in fish. Food Addit. Contam. 8, 777-780.

Tyrpenou, A. E., Kotzamanis, Y. P., and Alexis, M. N., 2003. Flumequine depletion from muscle plus skin tissue of gilthead seabream (Sparus aurata L.) fed flumequine medicated feed in seawater at 18 0C and 24 0C. Aquaculture 220, 633-642.

Van der Heijden, et al, 1994. Plasma disposition of flumequine in common carp (Cyprinus caprio L., 1758), African catfish (Clarias gariepinus Burchell, 1822) and European eel (Anguilla anguilla L., 1758) after a single per oral administration. Aquaculture 123, 21-30.

Wang, B. K., et al, 2002. Isolation and identification of pathogen (Vibrio harveyi) from sea perch, Lateolabrax japonicus. Journal of Fishery Sciences of China 9 (1), 52-55.

Wolfson, J.S., Hooper, D.C., Swartz, M.N., 1989. Mechanisms of action and resistance to quinolone antimicrobial agents. In: Wolfson, J. H., Hooper, D. C. (Eds.), Quinolone Antibacterial Agents. American Society for Microbiology, 5-34.

Zhang, J., et al, 2009. Isolation and identification of causative pathogen for visceral white spot in Lateolabrax japonicus stocked in sea-cage. Journal of Zhejiang Ocean University (Natural Science) 28 (2), 176-182.

四、双氟沙星和沙拉沙星

双氟沙星化学名称是（6-氟-1-对氟苯基-1，4-二氢-7-（4-甲基-1-哌嗪基）-4-氧代-3-喹啉羧酸），它是 1984 年合成的第三代氟喹诺酮类抗菌药物。

沙拉沙星化学名称是（1-（对-氟苯基）-6-氟-1，4-二氢-4-氧-7-（1-哌嗪基）-3-喹啉羧酸盐酸盐），也属于第三代氟喹诺酮类广谱抗菌药物。以原料药为碱基的盐酸盐是动物专用抗菌药，目前国内主要是盐酸盐，即盐酸沙拉沙星。沙拉沙星是双氟沙星在水产动物体内的主要代谢产物。

双氟沙星及其代谢产物沙拉沙星在中华绒螯蟹体内消除缓慢，且都呈现多峰现象（李海迪等，2009）。双氟沙星在异育银鲫血液及各组织中的中均出现重吸收峰现象，肝脏组织达峰浓度最大（章海鑫等，2012）。盐酸沙拉沙星在凡纳滨对虾体内达峰浓度高和生物利用度高，在血淋巴、肝胰腺和肌肉中的消除速率快慢不一，且在肌肉和肝胰腺组织中消除快（李国烈等，2012，2014）。双氟沙星能够在草鱼体内快速吸收，并且血液及各组织中均有分布（董亚萍等，2016）。

该文发表于《水生生物学报》【2017，（2）：379-383】

草鱼体内双氟沙星药代动力学研究

Pharmacokinetics of difloxacin in grass carp (*Ctenopharyngodon idellus*) after a single oral administration

董亚萍，孙晶，蒋新益，胡鲲，杨先乐

（上海海洋大学，国家水生动物病原库，上海 201306）

摘要：采用高效液相色谱法，经测定在 15℃水温状态下单次给草鱼灌喂 20 mg/kg 剂量的双氟沙星（DIF）后，草鱼体内双氟沙星的药代动力学以及各组织内双氟沙星的残留量。得出双氟沙星在各组织以及血液中的药时曲线均都符合二室开放性模型，双氟沙星能够在草鱼体内快速吸收，并且血液及各组织中均有分布，双氟沙星在草鱼体内的药物动力学方程为 $C = 5.056e^{-0.012t} + 19.041e^{-0.011t}$，其中双氟沙星在血液、肌肉、肝脏、肾脏中吸收半衰期 $T_{1/2\alpha}$ 分别为 0.176、0.562、4.562、1.477 h；消除半衰期 $T_{1/2\beta}$ 分别为 69.492、65.303、218.412 和 163.937 h；总体消除率 CL 分别为 0.495、11.181、10.789、7.102 L/h/kg；药时曲线下面积 AUC 分别为 81.550、1277.55、807.470、1432.150 μg/L/h。根据相关规定肌肉中双氟沙星最大残留量 300 μg/kg 为标准，建议休药期 26 天以上。

关键词：双氟沙星；药代动力学；草鱼

双氟沙星（difloxacin，DIF）为氟喹诺酮类第三代抗菌药物，对多种革兰阴性菌、革兰样性菌、球菌及支原体等均有较强的活性，以及具有能够广泛度分布在体内、与其他抗菌药物没有交叉耐药性等特点，因此被广泛应用于细菌性疾病的防治[1-2]。然而由于滥用抗菌药物会引起药物残留在动物体中，从而导致动物致病菌的耐药性，以及人类食用此类动物后，引起人类病原菌对抗菌药物产生耐药性，而其产生的毒副作用还会对人体产生直接的危害[3]。目前关于双氟沙星残留消除规律在人类、畜禽及水产动物类[4-8]均有报道。本研究探讨 15℃ 水温条件下单次口灌给药后，双氟沙星在草体内的残留消除规律，为药物的合理使用和水产品的安全提供参考。

1　材料和方法

1.1　实验动物

草鱼（Ctenopharynodon idellus）1 000 尾购于南通国营农场，健康无病，未用过任何药物。鱼平均体重为 60±7.8 g，试验前于 4 m×2 m×1 m 的网箱中暂养 7 d，并在给药前 3 d 停止投饲。试验期间投喂草鱼专用饲料。

1.2　实验药物

双氟沙星标准品（纯度≥99.5%）购于 Sigma 公司，盐酸双氟沙星原料药（纯度为 98%），由浙江新昌县永兴化工有限公司提供。乙腈、四丁基溴化铵为色谱级；二氯甲烷、正己烷、柠檬酸、柠檬酸三钠、氯化钠、磷酸氢二钠和磷酸二氢钠为分析纯。以上药品均购自国药集团化学有限公司。

1.3　给药、取样与样品前处理

给药双氟沙星以 20 mg/kg 鱼体重的剂量，用注射器灌入实验鱼前肠，无回吐现象的留作试验用。以 20 mg/kg 剂量口灌中华绒螯蟹给药后第 0.015 h、0.25 h、0.5 h、1 h、2 h、4 h、8 h、16 h、24 h、48 h、96 h、144 h、256 h，分别取取其血样、肌肉、肝脏、肠、肾等组织样品并分别合并成一个样品后于 -70℃ 保存备用。每组均设一个重复。给药前取空白组织样品作对照。

样品前处理血淋巴样品处理　取 1 mL 血液加入酸化乙腈（乙腈：盐酸：水 = 250：1：1）5 mL，漩涡混合 5 min，4℃ 条件下 8 000 r/min 离心 10 min，取上清液，45℃ 恒温氮气条件下吹干，加 1 mL 流动相溶解，经 0.45 μm 微孔滤膜过滤，HPLC 测定。

组织样品处理各组织样品准确称取 1.00 g 碾磨匀浆后加入 10 mL 离心管中，同时加入 1 mL NaCl（1 mol/L）和 PBS 缓冲溶液（pH 7.4、0.2 mol/L），以及 5 mL 二氯甲烷进行药物萃取，经旋涡混合振荡 10 min 后，4℃ 条件下 8 000 r/min 离心 10 min，取有机相；剩余残渣中再加入 5 mL 二氯甲烷，重复以上操作步骤进行第二次萃取，合并有机相，于 30℃ 恒温氮气中吹干，加 1 mL 流动相溶解后，再加入 5 mL 正己烷除脂；混匀、离心后将下层液体经 0.45 μm 微孔滤膜过滤留作 HPLC 分析。

1.4　色谱条件

样品分析仪为 Agilent 1100 高效液相色谱仪，分析柱为 RP-ODS C_{18} 柱。柱温 30℃；流动相为乙腈：四丁基溴化铵（0.030 mol/L 磷酸调节至 pH3.1）= 5：95（v/v）；检测波长 276 nm；流速 1.5 mL/min；柱温 40℃；自动进样量为 10 μL。

1.5　标准曲线绘制

以常规方法建立标准曲线，在空白组织中分别加入将双氟沙星标准品分别配制成 0.01、0.05、0.10、0.50、1.00、5.00 和 10.00 μg/mL 标准液，经 HPLC 分析后分别以双氟沙星的质量浓度为横坐标，峰面积为纵坐标绘制标准曲线，并求出回归方程。

1.6　回收率、日内及日间变异系数测定

取无药的空白血浆、肝脏、肌肉、肠道、肾脏分别加入 25 μg/mL，50 μg/mL 的双氟沙星标准溶液各 1 mL，每个组做 3 次平行试验每个浓度重复 3 次，测定回收率；随机取 3 个样品，每个样品每隔 3 h 平行测定一次，确定日内精密度；将 3 种不同浓度的 3 种标准溶液各测定 2 次，每隔 3 d 一次，以峰面积、标准差及相对标准偏差确定日间精密度。

标准曲线、药时曲线用 Excel 绘制；使用 DAS 2.1 软件对各组试验数据进行模型嵌合和参数计算。

2 结果

2.1 检测方法的可靠性

用 HPLC 法检测各个组织中的双氟沙星，其保留时间是 8.81 min，在保留时间内没有干扰峰的出现，药峰与各杂质峰分离较好，均为基线分离峰。双氟沙星标准溶液在 0.01-20.00 μg/mL 浓度范围内线性关系良好，其标准曲线方程分别为 $y = 6.428\,6x - 2.523\,8$（$R^2 = 0.999\,0$），计算出各个水平的平均回收率，血浆、肝脏、肌肉、肠道、肾脏中的回收率分别为 88.7%、74.2%、85.3%、77.5%、82.4%，测得的日内精密度系数不大于 2%；日间精密度系数不大于 8%，该实验方法是可靠的。

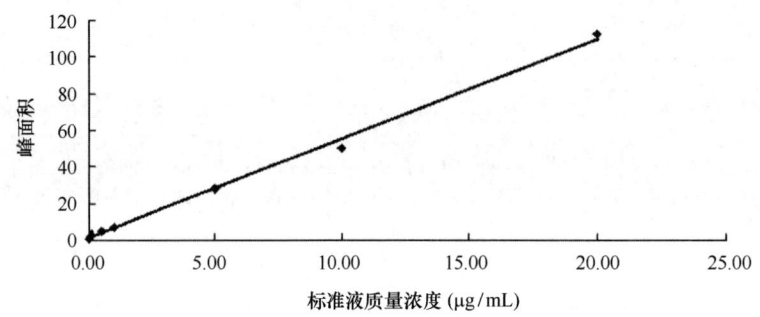

图 1　双氟沙星的标准曲线峰面积

Fig. 1　Standard curve of difloxacinrd liquid

2.2 回收率、日内及日间变异系数

计算出各个水平的平均回收率，血浆、肝脏、肌肉、肠道、肾脏中的回收率分别为 88.7%、74.2%、85.3%、77.5%、82.4%，测得的日内精密度系数不大于 2%；日间精密度系数不大于 8%，该实验方法是可靠的。

2.3 双氟沙星在草鱼体内的药代动力学参数和药时曲线

结果表明，双氟沙星在草鱼体内药动学特征均适合用一次式二室开放模型来描述。血浆药物动力学方程为 $C = 5.056e^{-0.012t} + 19.041e^{-0.011t}$。根据此模型，我们可获得一些药代动力学参数（表1）。图2显示的是在 15℃时以 20 mg/kg 体重的剂量单次给草鱼口灌 DIF 后，草鱼血液中的药时曲线。结果显示，口灌给药后 0.015 h 血液中就可检测到 DIF，0.25 h 达到最高峰，达峰浓度 C_{max} 为 32.620 μg/mL，0.015~0.25 h 时间段为 DIF 在草鱼体内的吸收过程。达峰后浓度持续下降，0.5~1 h DIF 有一个快速下降，1~48 h 药物浓度保持在一个较高的水平上；48~96 h 又是一个快速的下降期，256 h 血液浓度降到检测样品的最低值。组织中 DIF 药时曲线如图3所示：肌肉、肝脏、肾脏、肠道总体趋势一致，均有两个吸收峰；肝脏 DIF 药时曲线趋于平缓；达峰时间肾脏（4 h）= 肠道（4 h）>肌肉、肝脏（8 h）；C_{max} 肾脏（67.920 μg/mL）>肌肉（45.890 μg/mL）>肝脏（39.740 μg/mL）。

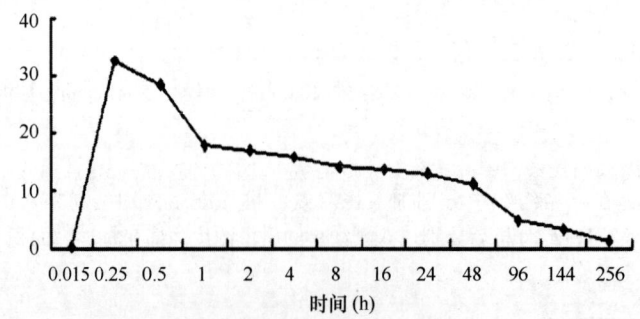

图 2　血液中单次口灌 20 mg/kg DIF 药时曲线（$n = 5$）

Fig. 2　The concentration—time curves of DIF in plasma following a single oral gavage of 20 mg/mL（$n = 5$）

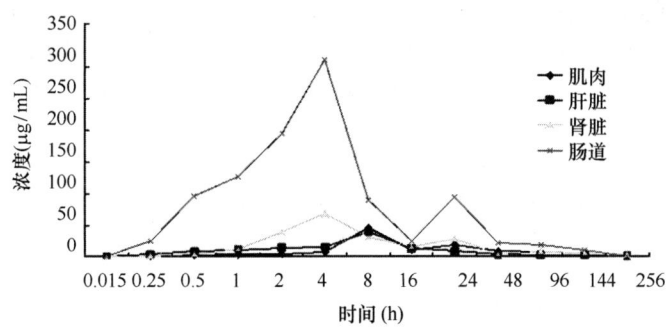

图 3　单次口灌 DIF 后组织中的药时曲线

Fig. 3　The concentration—time curves of DIF in tissue following a single oral gavage of 20 mg/kg

血液与组织中 DIF 药代动力学参数见表 1，从表 1 可以看出 DIF 消除最快和最慢的分别是血液和肌肉。DIF 消除半衰期（$T_{1/2\beta}$）变现最快的是肝脏、最缓慢的是肌肉组织；曲线下面积（AUC）变现最大的是肾脏，肌肉大于肝脏，最小的是血液。

表 1　双氟沙星在草鱼体内的药代动力学参数

Tab 1　Pharmacokinetics parameters of DIF in grass carps

模型参数	单位	血	肌肉	肝脏	肾脏
$T_{1/2\alpha}$	h	5.616	3.202	5.296	1.829
$T_{1/2\beta}$	h	69.472	65.303	218.412	163.937
$V_{1/F}$	L/kg	1.942	472.888	591.620	295.374
CL/F	L/h/kg	0.495	11.181	10.789	7.102
$AUC_{(0-\infty)}$	μg/L/h	81.550	1277.55	807.470	1432.150
K_{10}	1/h	2.525	0.024	0.018	0.024
K_{12}	1/h	2.524	0.173	0.109	0.300
K_{21}	1/h	0.011	0.020	0.003	0.066
K_a	1/h	0.018	1.233	0.152	0.469
$T_{1/2Ka}$	h	0.176	0.562	4.562	1.477
T_{max}	h	0.250	8.000	8.000	4.000
C_{max}	μg/L	32.620	45.890	39.740	67.920

注：$T_{1/2\alpha}$ 为分布相半衰期；$T_{1/2\beta}$ 为消除半衰期；$T_{1/2Ka}$ 为吸收半衰期；CL/F 为体内清除率；AUC 为药时曲线下面积；K_a 为吸收速率常数；T_{max} 为达峰时间；C_{max} 为达峰浓度。

3　讨论

3.1　双氟沙星在草鱼体内的代谢、残留以及消除

目前抗菌类药物在草鱼体内的药代动力学鲜有研究人员报道，其药时曲线大都可以用二室模型表示。秦改晓[10]等对单剂量口灌阿维菌素在草鱼体内的药动学及残留研究中得出，血药浓度与时间的关系符合二室开放性模型。在 28℃给草鱼口灌磺胺甲恶唑，发现在血浆以及肌肉中分布、吸收消除都相比于 18℃条件下的快，同时在两种不同温度条件下肌肉中的药代动力学参数都能够很好的用一级二室开放模型来模拟[15]。本实验中，15℃条件下，双氟杀星（DIF）在草鱼血液以及各组织中的药物-时间数据经药代动力学软件处理过之后也都可用开放性二室模型进行较好的拟合。也个别有研究表明，低温条件下呋喃唑酮在草鱼体内呈现的是一室模型[16]，经过以上研究更进一步说明了抗菌类药物在草鱼体内大都呈现的是二室模型。关于喹诺酮类在草鱼体内的代谢的研究比较少，一般都关注在畜禽类以及鲫、螃蟹体内的研究，同时通过本实验也为喹诺酮类在草鱼体内的药代动力学的研究提供参考。

C_{max}、T_{max} 和 $T_{1/2}$ 以及 K_a 是反映体内药物吸收速率以及分布速率的几个重要的指标。双氟沙星在不同水产动物以及不

同水温的情况下，其吸收速率会有很大的不同。如在 20℃ 条件下以 20 mg/kg 剂量口灌鲤，其 $T_{1/2ka}$ 为 0.54 h，最高血药浓度 C_{max} 为 10.28 μg/mL，达峰时间 T_{max} 为 2.21 h；以 20 mg/kg 剂量口灌中华绒螯蟹，其血淋巴中 $T_{1/2ka}$ 为 0.029 h，C_{max} 为 26.255 μg/mL，T_{max} 为 0.5 h；在 16℃ 和 25℃ 条件下以 20 mg/kg 口灌异育银鲫，其血淋巴中 $T_{1/2ka}$ 分别为 0.001 和 0.122 h，C_{max} 分别为 10.878 和 15.739 μg/mL，T_{max} 均为 0.25 h。本研究在 15℃ 水温条件下给草鱼以 20 mg/kg 剂量灌喂双氟沙星，其血淋巴中 $T_{1/2ka}$ 为 0.176 h，C_{max} 为 32.619 μg/mL，T_{max} 为 0.25 h。结果表明草鱼的 C_{max} 明显高于鲤和中华绒螯蟹，达峰时间 T_{max} 快于鲤和异育银鲫相同，说明 DIF 在草鱼体内能够更好、更快的吸收且达峰时间也较短，这可能与药物本身的性质以及实验动物不同有关。因此，在将双氟沙星运用在生产实践中时应充分了解不同水产动物的药代动力学规律，在使用上区别对待。

AUC 反映药物在动物机体中被利用的程度，双氟沙星（DIF）在草鱼肝脏、肌肉、肾脏组织中的 AUC 分别达到 807.74 μg/L/h、1277.55 μg/L/h、1432.15 μg/L/h，在肾组织中 AUC 明显高于其余组织，说明双氟沙星在肾组织的吸收要高于其余组织。$T_{1/2\beta}$ 是反映药物在体内消除的重要指标。本实验中双氟沙星在肝脏、肌肉与肾脏中，其 $T_{1/2\beta}$ 分别为 218.41 h、65.30 h 和 163.94 h。相比与猪、羊、鸡等畜禽类动物，双氟沙星在水产动物体内消除速率普遍比较慢，如以 5 mg/kg 的剂量给羊不同方式注射双氟沙星，其 DIF 的消除半衰期大小：静脉注射（11.43 h）<皮下注射（12.02 h）<肌肉注射（13.89 h）。给中华绒螯蟹以 20 mg/kg 剂量口灌双氟沙星，经测定：肌肉、精巢、肝胰腺、卵巢中的消除半衰期分别达到了 88.23 h、67.02 h、137.52 h、124.67 h[10]。由此看出双氟杀星（DIF）在水产动物体内的消除远远慢于畜禽动物。草鱼体内的磺胺类、阿维菌素的消除半衰期分别为 8.59 h 和 218.71 h，恩诺沙星、诺氟沙星的衰期分别为 24 h 和 44.75 h。由此看出，不同种类的药物在草鱼体内的消除半衰期会有很大差异，即便是同为喹诺酮类药物，不同衍生物其消除半衰期也不同。这些都能看出双氟沙星的残留消除不仅存在着组织和种属间的差异[11]，对于同一种水产动物不同的抗菌类药物也存在着差异。

总之，相比对畜禽动物来说，双氟沙星在不同水产动物体内均能够迅速的被吸收，达峰时间相对较短，峰浓度较高，广泛地分布于各组织中；而和畜禽类动物相比，最大的区别就是药物消除半衰期在水产动物体内更长。畜禽类动物可以通过肝、肾等器官的主动运转对药物加以消除，而水产动物的代谢器官功能较畜禽类动物弱，因此由于不同种属间代谢器官的差异，导致其对同一种药物代谢的机能也有所不同。

3.2 给药方案和休药期

有相关人员研究过盐酸双氟沙星对革兰氏阳性菌、假单胞菌、肠菌及厌氧菌的 MIC[12]，其中对嗜水气单胞菌的 MBC（最小杀菌浓度）和 MIC（最小抑菌浓度）均比较小[13]。因此双氟沙星在水产中常用来作为抗菌药物，并且在一定时间段内能达到较好效果。但是由于其在鱼体内消除半衰期长，消除缓慢，因此，在药物实际使用时，应相对延长给药间隔时间，避免耐药性的产生。本实验中草鱼以 10 mg/kg 的计量口服双氟沙星，0.015 h 之后就可以检测到该药，0.25 h 浓度达到最大为 32.620 μg/mL，之后血液中药物浓度持续下降，256 h 后降到最低值。考虑到药物在生产实践中会有环境以及水产动物个体等因素的差异，可根据具体情况来用药，由此我们建议，治疗草鱼的细菌性疾病，以 DIF 20 mg/kg 的剂量给药，适当缩短用药时长，建议一天 1 次，连续 3~4 天最佳。在由于草鱼的内脏没有食用价值，因此建议以双氟沙星在肌肉中的最大残留限量为标准。参考相关规定 DIF 在肌肉的残留限量为 300 μg/kg，根据上述药代动力学方程可计算出双氟沙星低于此残留限量的时间为 615 h（26 d）；由此建议在 15℃ 条件下双氟沙星在草鱼体内的休药期大于 26 d。

参考文献

[1] The Ministry of Agriculture《New Fishery Drugs Manual》Compilation Committee. New Fishery Drugs manual［M］. Beijing：China Agriculture Press，2005，211-212［农业部《新编渔药手册》编撰委员会. 新编渔药手册［M］. 北京：中国农业出版社. 2005，211-212］

[2] Liu Y K，Li Y L，Hu Y X. A general summary on the applications and studies of difloxacin［J］. Chinese Journal of Veterinary Drug，2004，38（6）：27-31［刘开永，李英伦，胡廷秀. 二氟沙星的研究概况. 中国兽药杂志，2004，38（6）：27-31］

[3] Granneman G R，Snyder K M，Shu V S. Difloxacin metabolism and pharmacokinetics in humans after single oral doses［J］. Antimicrobial Agents and Chemotherapeutics，1986，30（3）：689-693.

[4] Zeng Z L，Ding H Z，Huang X H，et al. Pharmacokinetics and bioavailability of difloxacin in Pigs［J］. Scientia Agricultura Sinica. 2003，36（7）：846-850［曾振灵，丁焕中，黄显会，等. 二氟沙星在猪体内的药物动力学及生物利用度研究. 中国农业科学，2003，36（7）：846-850］

[5] Garcia M A，Solans C，Aramayona J J. et al. Simultaneous det ermination of difloxacin and its primary metabolite sarafloxacin in rabbit plasma［J］. Chromatographia，2000，51：487-490

[6] Ding H Z，Zeng Z L，Huang X H，et al. Study on pharmacokinetics and bioavailability of difloxacin in chickens［J］. Chinese Journal of Veterinary Science and Technology，2004，34（6）：20-24［丁焕中，曾振灵，杨贵香，等. 二氟沙星在鸡体内的药物代谢动力学及生物利用度研究. 中国兽医科技，2004，34（6）：20-24］

[7] Marin P，Fernandez-Varon E，Escudero E，et al. Pharmacokinetics after intravenous，intramuscular and subcutaneous administration of diflox-

acin in sheep［J］. Research in Veterinary Science，2007，83：234-238

［8］ Chu D T W，Grannerman G. Difloxacin metabolian and pharmacokinetics in some animals after single oral doses［J］. Journal of Veterinary Pharmacology and Therapeutics，2001，24（suppl 2）：151-154

［9］ Ruan J M，Hu K，Zhang H X. Pharmacokinetics comparisons of difloxacin in crucian carp（Carassais auratus gibebio）at two different water temperatures［J］. Journal of Shanghai Ocean University，2011，20（6）：858-865［阮记明，胡鲲，章海鑫. 两种水温条件下异育银鲫体内双氟沙星药代动力学比较. 上海海洋大学学报. 2011，20（6）：858-865］

［10］ QinG X，Xu W Y，Ai X H，et al. Studies on the pharmacokinetics and residues of avermectin in grasscarp（Ctenopharyngodon idellus）after single oral administration［J］. Freshwater Fisheries，2012，42（4）：47-52［秦改晓，徐文彦，艾晓辉，等. 单剂量口灌阿维菌素在草鱼体内的药动学及残留研究. 淡水渔业. 2012，42（4）：47-52］

［11］ Liu Y K，Li Y L，Hu Y X. A general summary on the applications and studies of difloxacin［J］. Chinese Journal of Veterinary Drug，2004，38（6）：27-31［刘开永，李英伦，胡廷秀. 二氟沙星的研究概况. 中国兽药杂志. 2004，38（6）：27-31］

［12］ Jiang S C，Song K W. Quinolone Antimicrobial Agents Copyrigh［M］，Beijing：People's Military Medical Press. 1991，26-45［姜素椿，宋克王. 最新广谱喹诺酮类抗微生物药［M］. 北京：人民军医出版社. 1991，26-45］

［13］ Yang Y H，Tong H M，Lu T Y，et al. Pharmacodynamics of several fluoroquinolonesagainstA. hydrophila invitro［J］. Journal of Northeast Agricultural University，2003，3（4）：368-371［杨雨辉，佟恒敏，卢彤岩，等. 几种氟喹诺酮类药物对嗜水气单胞菌体外药效学研究. 东北农业大学学报，2003，3（4）：368-371］

［14］ Cao H P，Zhang H X，He S，Zheng，et al. Pharmacokinetics of Difloxacin in Healthy and Aeromonasis-Infected Gibel Carp Carassius auratus gibelio［J］. Israeli Journal of Aquaculture-Bamidgeh，2013，65：89

［15］ Ai X H，Liu C Z，Zhou Y T. A study on pharmacokinetic of sulphamethoxazole in grass carp at different temperatures and administration regimes［J］. Acta Hydrobiol Sinica 2005，3（2）：210-214［艾晓辉，刘长征，周运涛. 不同水温和给药方式下磺胺甲恶唑在草鱼体内的药动学研究. 水生生物学报，2005，3（2）：210-214］

［16］ Ding F K，Cao J Y，Ma L B，et al. Pharmacokinetics and tissue residues of difloxacin in crucian carp（Carassius auratus）after oral administration［J］. Elsevier 2006，256：121-128.

该文发表于《动物学杂志》【2012，47（6）：72~77】

双氟沙星在人工感染嗜水气单胞菌的异育银鲫体内的药代动力学

Pharmacokinetics of Difloxacin in the Aeromonas hydrophila-infected Carassiusaurutus gibelio

章海鑫[1,2]，曹海鹏[1]，阮记明[1]，胡鲲[1]，杨先乐[1]

（1. 上海海洋大学国家水生动物病原库，上海 201306；2. 江西省水产科学研究所，南昌 330039）

摘要： 为了阐明双氟沙星在健康与处于患病状态下的异育银鲫（Carassius aurutus gibelio）体内的药代动力学特征差异，为双氟沙星的正确合理用药提供参考，本研究通过人工创伤感染的方式采用最佳浓度的嗜水气单胞菌（Aeromonas hydrophila）感染异育银鲫，在此基础上进一步以双氟沙星在健康异育银鲫体内的药代动力学特征为对照，采用反相高效液相色谱法测定双氟沙星在感染嗜水气单胞菌的异育银鲫体内的药代动力学特征。实验结果表明，以鱼体重的 20 mg/kg 口灌给药后，双氟沙星在人工感染嗜水气单胞菌的异育银鲫与健康异育银鲫体内的总药时曲线均符合一级吸收开放性二室模型，其药动学方程分别为 $C = 6.227e^{-0.109t} - 8.074e^{-2.752t} + 1.847e^{-0.006t}$ 和 $C = 110.295e^{-0.331t} + 1.533e^{-0.01t} - 111.828e^{-0.412t}$，但与双氟沙星在健康异育银鲫体内的药代动力学参数相比，双氟沙星在人工感染嗜水气单胞菌的异育银鲫体内的吸收、分布、消除速度减慢，其在人工感染嗜水气单胞菌的异育银鲫体内的分布半衰期、消除半衰期、吸收速率常数、曲线下面积分别增加了 4.25 h、36.17 h、2.34 h 和 74.52 mg·h/L，达峰时间延长了 5.75 h，峰浓度降低了 61.16%，且未出现重吸收现象。本研究证实嗜水气单胞菌感染能够导致异育银鲫肝肾功能损伤，因而双氟沙星在人工感染嗜水气单胞菌的异育银鲫体内的吸收、分布、代谢和消除均会减慢。

关键词： 双氟沙星；嗜水气单胞菌；异育银鲫；药代动力学

该文发表于《动物学杂志》【2009，44（2）：12-20】

双氟沙星及其代谢产物在中华绒螯蟹体内药物代谢及残留消除规律

Pharmacokinetics and the Elimination Regularity of Difloxacin and its Metabolite Sarafloxacin in Eriocheir sinensis

李海迪，杨先乐，胡鲲，王翔凌

（上海海洋大学水产与生命学院农业部渔业动植物病原库，上海 201306）

摘要：采用高效液相色谱法，在口灌给药途径下，究双氟沙星及其代谢产物沙拉沙星在中华绒螯蟹（*Eriocheir sinensis*）体内的药代动力学。中华绒螯蟹以 20 mg/kg 剂量给药双氟沙星后，其血淋巴、肌肉、肝胰腺、精巢和卵巢中药物-时间曲线关系符合开放性二室模型。双氟沙星在中华绒螯蟹体内吸收迅速，在不同组织中分布较广，达峰时间短。血淋巴、肌肉、肝胰腺、精巢和卵巢中的 V_d 分别为 3.170 L/kg、2.122 L/kg、1.045 L/kg、1.051 L/kg 和 0.203 L/kg；双氟沙星在中华绒螯蟹体内消除缓慢，在血淋巴、肌肉、肝胰腺、精巢和卵巢中的消除半衰期 $T_{1/2\beta}$ 分别为 96.316 h、88.228 h、137.524 h、67.021 h 和 124.679 h；总体清除率 CL_s 分别为 0.783 L/h·kg、0.040 L/h·kg、0.013 L/h·kg、0.011 L/h·kg 和 0.008 L/h·kg。代谢产物沙拉沙星在中华绒螯蟹血淋巴、肌肉、肝胰腺、精巢和卵巢中药物水平的变化趋势与双氟沙星相似，都呈现多峰现象。但肌肉、肝胰腺、精巢和卵巢 4 种组织中代谢产物沙拉沙星出现药峰的时候恰好是这 4 种组织中双氟沙星下降缓慢时期。鉴于双氟沙星及其代谢产物沙拉沙星在中华绒螯蟹体内消除缓慢，而可食组织中脂肪含量较高，因此若规定可食组织中双氟沙星的最大残留限量以脂肪中最大残留限量（100 μg/kg）为标准，沙拉沙星的最大残留限量为 30 μg/kg，则建议双氟沙星在中华绒螯蟹中的休药期大于 24 d，才能保障食用者的安全。

关键词：中华绒螯蟹；双氟沙星；沙拉沙星；药代动力学

该文发表于《华中农业大学学报》【2012，31（4）：13-18】

盐酸沙拉沙星在凡纳滨对虾体内的残留消除规律

李国烈[1,2]，胡琳琳[1]，胡伟国[3]，房文红[1]，孙贝贝[1]，黄伟[3]，周帅[1]

（1. 中国水产科学研究院东海水产研究所/农业部海洋与河口渔业资源及生态重点开放试验室，上海 200090；
2. 四川农业大学动物医学院，雅安 625014；3. 上海市奉贤区水产技术推广站，上海 201400

摘要：在水温为（29.0±1.0）℃，盐度为 3.3% 条件下，以 30 mg/kg 的剂量单次药饵投喂盐酸沙拉沙星后，分别取凡纳滨对虾血淋巴、肝胰腺和肌肉样品，采用反相高效液相色谱法测定样品中的药物浓度。结果表明：盐酸沙拉沙星在肝胰腺中的药物残留浓度远高于血淋巴和肌肉，后两者的药物残留浓度比较低而且血淋巴中的残留浓度高于肌肉；给药后盐酸沙拉沙星在血淋巴、肝胰腺和肌肉中的消除速率快慢不一，在肌肉中消除速度最快，肝胰腺中的消除速度快于血淋巴。

关键词：盐酸沙拉沙星；凡纳滨对虾；残留；消除；休药期

该文发表于《南方水产科学》【2014，10（1）：50-56】

盐酸沙拉沙星在凡纳滨对虾体内药动学与生物利用度

Pharmacokinetics and bioavailability of sarafloxacin hydrochloride in *Litopenaeus vannamei*

李国烈[1,2]，王元[1]，房文红[1]，沈锦玉[3]，周俊芳[1]

（1. 中国水产科学研究院东海水产研究所农业部海洋与河口渔业资源及生态重点开放实验室，上海 200090；
2. 南充农产品质量监测检验中心，四川 南充 637000；3. 浙江省淡水水产研究所，浙江 湖州 313001）

摘要：采用 RP-HPLC 法，研究在盐度为 33、温度为（28.0±1.0）℃自然海水下盐酸沙拉沙星单剂量围心腔注射（剂量 10 mg·kg^{-1}）和单次药饵投喂（剂量 30 mg·kg^{-1}）给药后在凡纳滨对虾（*Litopenaeus vannamei*）体内药动学与生物利用度。围心腔注射给药后，血淋巴中药时曲线较适合用二室模型来拟合，而药饵投喂给药后血淋巴中药时曲线较适合采用一级吸收二室模型来拟合。药饵给药下盐酸沙拉沙星在凡纳滨对虾体内的生物利用度（*F*）为 61.6%。药饵投喂给药下大量药物分布到了肝胰腺，肝胰腺 C_{max} 和 AUC$_{0-t}$ 分别是血淋巴的 24.4 倍和 18.7 倍，分别是肌肉的 51.9 倍和 62.0 倍；药物在肝胰腺和肌肉中消除都很快，分别在给药后 5 d 和 36 h 即低于 0.1 mg·kg^{-1}。由此可见，盐酸沙拉沙星药饵给药下吸收好，达峰浓度高和生物利用度高，而且在肌肉和肝胰腺组织中消除快，是较为理想的防治对虾细菌性疾病的抗菌药物。

关键词：盐酸沙拉沙星；凡纳滨对虾；药动学；生物利用度

该文发表于《海洋渔业》【2013，35（3）：331-336】

盐酸沙拉沙星在中华绒螯蟹体内药动学及药效学研究

Pharmacokinetics and pharmacodynamics of sarafloxacin hydrochloride in *Eriocheir sinensis*

彭家红[1,2]，王元[1,2]，房文红[1]，姚小娟[1,2]，符贵红[1]

（1. 中国水产科学研究院东海水产研究所 农业部东海与远洋渔业资源开发利用重点实验室，上海 200090；
2. 上海海洋大学水产与生命学院，上海 201306）

摘要：本文研究了盐酸沙拉沙星（*Sarafloxacin hydrochloride*）在中华绒螯蟹（*Eriocheir sinensis*）雄蟹和雌蟹体内药动学，检测了盐酸沙拉沙星对致病性气单胞菌的最小抑菌浓度（MIC），建立了药动/药效（PK/PD）关系。盐酸沙拉沙星以 30 mg·kg^{-1} 单剂量口灌给药中华绒螯蟹，雌蟹和雄蟹血浆沙拉沙星浓度-时间关系曲线均符合一级吸收的二室开放动力学模型。雌蟹和雄蟹血药达峰时间 T_{max} 为 4 h，达峰浓度 C_{max} 分别为（5.619±1.192）mg·L^{-1} 和（7.775±1.426）mg·L^{-1}。沙拉沙星在雌蟹和雄蟹体内分布广泛，其表观分布容积 V_d 分别为 3.801 L·kg^{-1} 和 2.871 L·kg^{-1}；消除半衰期 $T_{1/2\beta}$ 分别为 51.982 h 和 55.956 h，总体清除率（CL$_s$）为 0.138 L/（kg·h）和 0.087 L/（kg·h），曲线下面积 AUC 分别为 209.1 mg·（L·h）$^{-1}$ 和 325.5 mg/（L·h）。中华绒螯蟹扣蟹致病菌嗜水气单胞菌 HX1b1、维氏气单胞菌 CMX313 和 CMX4 对盐酸沙拉沙星的最小抑菌浓度分别为 0.5、0.25 和 0.25 μg·mL^{-1}。以 30 mg·kg^{-1} 单剂量给药，嗜水气单胞菌 HX1B1 的 C_{max}/MIC=15.55（雄蟹）和 11.24（雌蟹）、AUC$_{0-24}$/MIC=197（雄蟹）和 145.4（雌蟹），维氏气单胞菌 CMX313 和 CMX4 的 C_{max}/MIC=31.1（雄蟹）和 22.48（雌蟹）、AUC$_{0-24}$/MIC=394（雄蟹）和 290.8（雌蟹）。由此可见，盐酸沙拉沙星在雌雄蟹体内的药动学数据有一定差异，但不是很大。以 30 mg·kg^{-1} 剂量拌料给药，即可以有效防治上述气单胞菌引起的细菌性疾病。

关键词：中华绒螯蟹；盐酸沙拉沙星；药动学；气单胞菌；药效学

该文发表于《上海海洋大学学报》【2011, 20 (6): 858-865】

两种水温条件下异育银鲫体内双氟沙星药代动力学比较

Pharmacokinetics comparisons of difloxacin in crucian carp (*Carassais auratus gibebio*) at two different water temperatures

阮记明[1,2], 胡鲲[1], 章海鑫[1], 王会聪[1], 付乔芳[1], 杨先乐[1]

(1. 上海海洋大学国家水生动物病原库, 上海 201306; 2. 江西农业大学动物科学技术学院, 江西 南昌 330045)

摘要: 运用反相高效液相色谱法 (RP-HPLC), 对比了单次口灌 20 mg/kg 鱼体重剂量条件下, 16℃和25℃两种水温时双氟沙星 (DIF) 在异育银鲫体内代谢动力学差异。数据经过药代动力学软件 (3P97) 处理, 结果显示: 两种温度条件下, 双氟沙星药时规律均符合开放性二室模型且存在明显的差异。其中, 吸收半衰期 $T_{1/2Ka}$ 与水温呈正相关, 在两种温度条件下分别可达 0.010 h 和 0.122 h; 消除半衰期 $T_{1/2\beta}$、表观分布容积 (V_d/F)、血浆和组织中药时曲线下总面积 (AUC) 等参数则与水温呈负相关, 16℃条件下血浆 $T_{1/2\beta}$、V_d/F 和 AUC 分别为 70.968 h、1.875/kg 和 763.761 μg/ (mL·h), 25℃则为 18.322 h、0.676/kg 和 243.244 μg/ (mL·h); 16℃条件下肝脏组织中 AUC 为 746.622 μg/ (mL·h), 肾脏组织中 AUC 为 1095.711 μg/ (mL·h), 肌肉组织中 AUC 为 1 222.750 μg/ (mL·h); 而25℃条件下肝脏组织中 AUC 则为 294 857 μg/ (mL·h), 肾脏组织中 AUC 为 258.587 μg/ (mL·h), 肌肉组织中 AUC 为 344.630 μg/ (mL·h)。另外, 在两种温度条件下, DIF 代谢产物沙拉沙星在血浆及肝脏、肾脏和肌肉组织中均能被检出, 且呈多峰现象。结果说明, 相对于25℃的水温条件, 16℃水温时双氟沙星能较好地分布于异育银鲫体内。另外, 16℃时 DIF $C_{max}/MIC_{90} \geq 10$ 的保持时间可达 72 h, 而 25℃ 时 $C_{max}/MIC_{90} \geq 10$ 的保持时间仅为 4~6 h; 因此, 双氟沙星在较低水温时 (16℃) 具有更高的药效。研究亮点: (1) 比较了不同水温条件下 DIF 在异育银鲫体内的药动学差异。在我国, 不同地域、不同季节的水温变化和差异均很大, 因此有必要对不同水温下的药动学进行比较研究, 以期为不同水温下的用药提供理论参考; (2) 从药物本身及其代谢产物两个层面整体考虑药效及药残等。

关键词: 水温; 异育银鲫; 双氟沙星; 沙拉沙星; 药代动力学

该文发表于《中国海洋大学学报》【2012, 42 (11): 81-86】

双氟沙星在异育银鲫体内药代动力学研究

ResearchsAbout Pharmacokinetics of DIF in *Carassais auratus gibebio*

章海鑫[1,2], 阮记明[1,3], 胡鲲[1], 郑卫东[4], 杨先乐[1], 王会聪[1]

(1. 上海海洋大学国家水生动物病原库, 上海 201306; 2. 江西省水产科学研究所, 江西 南昌 330039;

3. 江西农业大学动物科学技术学院, 江西 南昌 330045; 4. 中国水产科学研究院长江水产研究所, 湖北 武汉 430223)

摘要: 采用反相高效液相色谱法 (RP-HLPC), 研究了 (25±1)℃水温条件下单次经口灌注 20 mg·kg⁻¹双氟沙星 (DIF) 后, DIF 在异育银鲫体内的药代动力学特征及其代谢产物沙拉沙星 (SAR) 的残留规律。结果表明: DIF 在血液及各组织中的药时数据均符合二室开放模型。血液和各组织中均出现重吸收峰现象, 肝脏组织达峰浓度最大。其中 DIF 在血液、肝脏、肾脏、肌肉中吸收半衰期 $T_{1/2a}$ 分别为 0.144、0.274、0.137 和 8.636 h; 消除半衰期 $T_{1/2\beta}$ 分别为 18.322、26.013、33.066 和 52.504 h; 总体清除率 (CL) 分别为 0.082、0.068、0.077 和 0.058 L·h·kg⁻²; 药时曲线下面积 (AUC) 分别为 243.244 μg·mL·h⁻² 和 294.857、258.587、344.630 μg·g·h⁻²; 表观分布容积 V_d 分别为 0.676、0.220、0.226 和 1.258 L·kg⁻¹, 血浆蛋白结合率为 52.39%~64.10%。组织中均可检测到 DIF 的代谢产物 SAR, 其中血液中检测到的时间迟于其他组织; 根据肌肉组织的 DIF 和 SAR 最大残留限量为 100 和 30 μg·kg⁻¹, 建议 DIF 休药期为 18 d 以上。

关键词：药代动力学；双氟沙星（DIF）；反向高效液相色谱法；异育银鲫

五、氧氟沙星和达氟沙星

氧氟沙星化学名称是［（±）9氟-2，3-二氢-3-甲基-10-（4-甲基-1-哌嗪基）-7-氧-7-氢-吡啶并［1，2，3-de］-［1，4］-苯并恶嗪-6-羧酸]，属于第三代单氟喹诺酮类抗菌药物。达氟沙星（Danofloxain），又名单诺沙星，也是一种动物专用抗菌药。

该文发表于《水生生物学报》【2006，3（2）：54-57】

达氟沙星在史氏鲟体内药物代谢动力学比较研究

THE PHARMACOKINETIC OF DANOFLOXACIN IN HEALTHY AND DISEASED ACIPENSER SCHRENCKII INFECTED BY AEROMONAS HYDROPHILA

卢彤岩，杨雨辉，徐连伟，孙大江

（中国水产科学研究院黑龙江水产研究所，哈尔滨 150070）

摘要：采用高效液相色谱法测定以 10 mg/kg 体重剂量静脉注射和口服给药后史氏鲟血浆中达氟沙星的浓度。该法采用 C18 色谱柱，流动相为乙腈-水相（15：85），荧光激发波长和发射波长分别为 280 nm 和 450 nm，样品用甲醇沉淀蛋白，离心取上清液进样。达氟沙星在 0.005~1.0 μg/mL 范围内线性关系良好，本方法的最低检测限为 0.005 μg/mL。健康鱼单剂量静注达氟沙星（10 mg/kg），其药时数据符合无吸收的三室开放模型，方程为 $C = 5.830e^{-5.582t} + 4.162e^{-1.157t} + 0.852e^{-0.029t}$，主要动力学参数如下：$T_{1/2\alpha}$ 0.552 h；$T_{1/2\beta}$ 22.186 h；AUC34.226 mg/（L·h）；V 10.922 L/kg；V_b 10.144 L/kg；Ke 10.317 h。Ah 感染组的 V1 减小至 0.290 L/kg，静注感染组体内达氟沙星的消除没有显著的改变。健康口服组数据结果符合一级吸收二室开放模型，血药浓度和时间方程为 $C = 1.278e^{-0.073t} + 0.177e^{-0.089t} - 1.455e^{-0.329t}$。药动学常数分别为：$T_{1/2ka}$ 9.491 h，$T_{1/2\beta}$ 78.267 h，T_{max} 6.284 h，C_{max} 0.791 mg/mL；α0.073 h。但 Ah 感染改变达氟沙星口服给药后在史氏鲟体内的吸收、分布和消除。分布速率常数降低为 0.050/h。消除减慢，消除半衰期延长为 93.988 h，达峰时间延长至 9.060 h，峰浓度降低为 0.585 mg/mL。口服达氟沙星水溶液，健康及感染组史氏鲟对达氟沙星生物利用度分别为 96.503% 和 94.435%。本实验结果表明达氟沙星在健康史氏鲟体内分布广泛、吸收较完全。感染 Ah 对达氟沙星在史氏鲟体内的吸收、分布及消除规律均有不同程度的影响，其中口服给药的影响更为显著。达氟沙星可用于史氏鲟感染 Ah 的治疗。

关键词：达氟沙星；史氏鲟；药物代谢动力学

达氟沙星（Danofloxin）是继恩诺沙星后又一动物专用氟喹诺酮类药物，由于具有水溶性好、可多种途径给药、抗菌谱广、抗菌活性强、药动学特性优良、不良反应小及与其他抗菌药无交叉耐药性等优点，目前已被广泛用于兽医临床[1]。

嗜水气单胞菌（Aeromonas hydrophila 简称 Ah）是一种重要的致病菌，可以感染多种鱼类并造成大量死亡，随着鲟鱼养殖业的兴起，由 Ah 引起养殖鲟鱼的病害日趋严重[2]。在水产养殖生产中，抗菌药物的实际给药方案主要是根据健康鱼类的实验数据提出的，由此提出的方案假定药量和药效变化在患病和健康鱼类体内的反应是相同的，而未考虑疾病对药物量效关系的影响。事实上各种疾病均会导致动物机体的生理状况如器官组织的血流量和通透性、血液蛋白含量、药物的蛋白结合率、代谢能力等的变化，这些变化可能会影响药物在体内的吸收、分布、代谢和排泄，从而影响药物的效应[3]。

本研究探讨达氟沙星通过静注和口服在健康及实验室嗜水气单胞菌感染的史氏鲟体内代谢规律的影响。实验把疾病模型与药物动力学研究结合起来，为阐明达氟沙星在史氏鲟体内的吸收分布、消除规律和生物利用度以及 Ah 对达氟沙星在史氏鲟体内药物代谢动力学特征的影响提供重要的理论依据，从而为临床使用该药物提供参考。

1　材料与方法

1.1　试验动物　史氏鲟（Acipenser schrenckii）购自中国水产科学研究院鲟鱼繁育基地，体重范围 75~95 g，暂养于室内玻璃水族箱中 7 d 后进行实验。水族箱有效水体积为 180 L，试验期间以经过充分曝气的自来水作为水源，水温在（23±1)℃，持续以曝气的形式进行充氧。暂养期间每天投喂两次颗粒饵料。

1.2　药品和试剂　达氟沙星（甲磺酸达氟沙星）为河南郑州兽药厂生产，批号 990415，含量为 99.5%。乙腈为国产 HPLC 试剂，四丁基溴化铵、甲醇、磷酸、柠檬酸等均为国产分析纯试剂。

1.3　菌种　嗜水气单胞菌株由珠江水产研究所鱼病研究室馈赠，并经鉴定和保存。

1.4　仪器　分析仪器为自动进样的 Waters 2695 型配有 474 型荧光检测器的高效液相色谱仪，分离柱为 Waters C18 反相分离柱。BF2000 氮气吹干仪（BFC 八方世纪公司）。

1.5　疾病模型的复制　经过预实验确定嗜水气单胞菌对史氏鲟的半数致死剂量（LD50），从暂养池中选择健康的史氏鲟，腹腔注射经培养 24 h 的嗜水气单胞菌培养液，注射体积为 2 mL/kg（约 10^6 CFU/mL），选择具有明显发病特征的史氏鲟作为实验用感染鱼。

1.6　实验分组及给药　实验分为健康口服组、健康静脉注射组、感染口服组和感染静脉注射组。取 150 尾健康鲟用于健康口服和静注达氟沙星药动学研究。选择感染症状明显的史氏鲟用于疾病模型组口服和静注药动学试验。配制 5 mg/mL 达氟沙星灭菌水溶液供静注和口服给药，单剂量给药剂量均为 10 mg/kg。口服给药时用医用静滴软管一端固定于 1 mL 注射器上，一端轻插入实验鱼胃部，并按体重大小推入相应体积的药液。静注时注射入史氏鲟尾静脉，注射成功与否通过观察注射前回抽尾静脉血的方法进行确定。

1.7　样品采集　实验用史氏鲟，分别于静注给药后 0.017、0.125、0.25、0.5、0.75、1、2、4、8、12、24、48、72 和 120 h 及口服给药后 0.125、0.25、0.5、0.75、1、2、4、8、12、24、48、72 和 120 h 各时间点于尾部采血，1% 肝素抗凝后离心吸取血浆保存于 -24℃ 备用，每一时间点采集 5 尾史氏鲟血液用于药物浓度分析。

1.8　样品处理　取出血浆样品，于室温下自然解冻摇匀后取出 0.5 mL 血浆，加入 0.5 mL 甲醇，旋涡快速混匀器上混合 1 min，于 4 000 r/min 转速离心 8 min。取全部上清液在 BF2000 氮气吹干仪 70℃ 下吹干，再加入 0.5 mL 流动相混匀后经 0.45 μm 微孔滤膜过滤后，用于 HPLC 分析。

1.9　色谱条件　流动相中乙腈：水相为 15：85（V/V），每 1 000 mL 水相中含 12.4 mL 0.2 mol/L 磷酸氢二钠和 187.6 mL 0.1 mol/L 柠檬酸，每 1 000 mL 流动相中含 2.7 g 溴化四丁基铵，流动相 pH 为 3。柱温为 18℃，进样体积为 10 μL，荧光检测器激发和发射波长分别为 280 nm 和 450 nm。流速为 0.8 mL/min。

1.10　达氟沙星标准曲线的制备　采用外标法进行试验样品的测定，在 100 mL 棕色容量瓶中配制成 100 μg/mL 的达氟沙星标准品母液，在 4℃ 冰箱中避光保存备用，用流动相分别稀释成 0.005、0.01、0.025、0.05、0.1、0.25、0.5 和 1 μg/mL 达氟沙星水溶液。以峰面积为横坐标，药物浓度为纵坐标做标准曲线，用最小二乘法求出回归方程和相关系数。

1.11　回收率的测定　将 0.1、1、10 μg/mL 标准品分别加入 0.5 mL 的血浆中，使其终浓度分别为 0.01、0.1 和 1 μg/mL，按 1.8 方法进行处理后进行测定，各浓度重复测定 5 次，获得各浓度的峰面积，再按标准曲线回归方程计算达氟沙星的浓度，并与实验前加入量进行比较，回收率＝实际测定量/初始加入量×100%。

1.12　数据处理　药时数据用 MCPKP 药物代谢动力学参数程序计算，由程序自动拟合最佳模型。最后得出两种给药途径下健康及疾病模型史氏鲟体内药物代谢动力学参数、模型指数和系数，确定理论方程。

2　结果

2.1　感染嗜水气单胞菌疾病模型的复制

史氏鲟感染嗜水气单胞菌 8 h 后，出现典型的感染症状。感染鱼腹部、嘴四周、眼睛、腹部出血，肛门红肿，随机选择的感染鱼有淡红色腹水，无菌操作挑出患病鲟鱼肝脏，进行细菌分离，28℃ 培养 24 h 挑取单一菌落，在斜面上划线纯化，根据《发酵性革兰氏阴性杆菌分类鉴定编码手册》细菌鉴定方法，证实为嗜水气单胞菌。同时取未感染鱼作对照细菌培养，结果未感染鱼腹腔内无腹水，肝脏未分离培养出任何细菌。感染实验表明已成功复制出感染嗜水气单胞菌的疾病模型。

2.2　标准曲线及变异系数

标准曲线为 $y = 6\,521.906 + 59\,617\,810x$（$R = 0.999\,9$），达氟沙星在 $0.005 \sim 1.0$ μg/mL 范围内线性关系良好，当所测定的血浆的药物浓度大于线性范围的上限，则用流动相稀释后再进行测定，本方法的最低检测限为 0.005 μg/mL。高、中、低 3 个浓度的回收率均在 84% 以上，批内、批间、日内、日间变异系数均在 10% 以内，可满足达氟沙星药物代谢动力学的研究要求。

2.3　史氏鲟体内的血药浓度

史氏鲟健康及疾病状态下静注和口服达氟沙星水溶液后，不同时间理论及实际测定值见图 1~4。

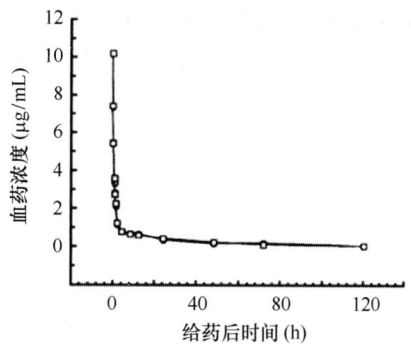

图1　健康静注达氟沙星后血药浓度的实测值和理论值

Fig. 1　Theoretic and observed concentration of danof loxacin in plasma of healthy Amur sturgeon after IV administration

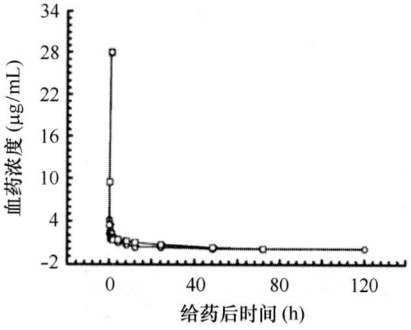

图2　感染后史氏鲟静注达氟沙星后血药浓度的实测值和理论值

Fig. 2　Theoretic and observed concentration of danofloxacin in plasma of infected Amur sturgeon after IV administration

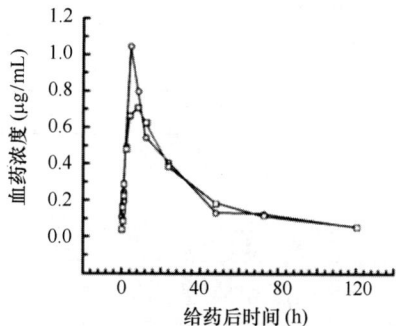

图3　康口服达氟沙星后血药浓度的实测值和理论值

Fig. 3　Theoretic and observed concentration of danofloxacin in plasma of healthy after oraladministration

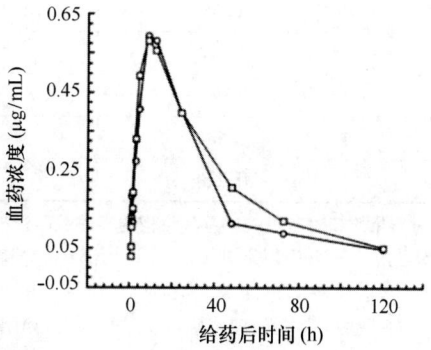

图4　感染 Ah 后口服达氟沙星后血药浓度的实侧值和理论值

Fig 4 Theoretic and observed concentration of danofloxacin in plasma of infected fish after oral administration

通过处理血药浓度-时间数据表明，史氏鲟口服达氟沙星的药时数据结果符合一级吸收二室开放模型，健康组的血药浓度和时间方程为 $C = 1.278e^{-0.073t} + 0.177e^{-0.089t} - 1.455e^{-0.329t}$。感染组口服给药的血药浓度和时间方程为 $C = 1.038e^{-0.050t} + 0.079e^{-0.074t} - 1.117e^{-0.221t}$，相关系数分别为 0.842 及 0.877。

史氏鲟静注达氟沙星药时数据结果符合无吸收三室开放模型，健康组的方程为 $C = 5.830e^{-5.582t} + 4.162e^{-1.157t} + 0.852e^{-0.0289t}$，疾病组的方程为 $C = 31.035e^{-13.4548t} + 2.830e^{-0.433t} + 0.648e^{-0.0336t}$，相关系数分别为 0.999 和 0.991。

2.4 生物利用度

达氟沙星在健康及感染 Ah 史氏鲟体内药物代谢动力学参数见表 1。口服达氟沙星水溶液，健康组生物利用度 96.5027%，感染组生物利用度为 94.4348%。

表 1 达氟沙星在健康和感染嗜水气单胞菌史氏鲟体内的药动学参数

Tab. 1 The pharmacokinetics parameters of danofloxacin lactate in Amur sturgeon by intravenous and oral admistration

参数 Parameters		口服 oral administration		静注 intravenous administration	
		健康组 Control	感染组 Infected group	健康组 Cntrol	感染组 Infected group
D	mg/kg	10.000	10.000	10.000	10.000
C_o	mg/mL	1.167	0.881	10.844	34.512
A	mg/mL	1.278	1.038	4.162	2.830
B	mg/mL	0.177	0.079	0.852	0.648
P	mg/mL	–	–	5.830	31.035
a	h^{-1}	0.073	0.050	1.157	0.433
β	h^{-1}	0.089	0.074	0.029	0.034
π	h^{-1}	–	–	5.582	13.455
$T_{1/2\pi}$	h			0.114	0.049
$T_{1/2a}$	h	9.491	13.959	0.552	1.479
$T_{1/2\beta}$	h	78.267	93.987	22.186	19.004
K_{12}	h^{-1}	0.028	0.013	1.628	8.826
K_{21}	h^{-1}	0.183	0.110	3.133	1.662
K_{13}	h^{-1}	–	–	1.502	1.814
K_{31}	h^{-1}	–	–	0.187	0.094
AUC	mg/L/h	33.029	26.591	34.226	28.157
V_1	L/kg			0.922	0.290
V_b	L/kg	–	–	10.144	10.562
CIb	L/kg/h	–	–	0.292	0.355
TCP	h	142.490	62.554	98.452	76.176
$T_{1/2ka}$	h	2.104	3.142		
T_{max}	h	6.284	9.060		
C_{max}	mg/mL	0.791	0.584		
K_a	h^{-1}	0.329	0.220		

注：D 为给药剂量；C_o 为初始浓度；A 为分布相的零时截距；B 为消除相的零时截距；P 为深室分布相的零时截距；α 为分布速率常数；β 为消除速率常数，π 为深室分布速度常数；Ka 为药物吸收速率常数；$t_{1/2\pi}$ 为深室分布相半衰期；$t_{1/2a}$ 为浅室分布相半衰期；$t_{1/2\beta}$ 为消除相半衰期；K_{12} 为药物由中央室到周边室内的一级运转速率常数；K_{21} 为药物自周边室到中央室的一级运转速率；K_{13} 为药物由中央室到周边室的一级运转速率常数；K_{31} 为药物由深室到中央室的一级运转速率常数；AUC 为血药浓度-时间曲线下面积；V_1 为中央室表观分布容积；V_b 为表观分布容积；Clb 为清除率；$t_{1/2ka}$ 为吸收相半衰期；t_{max} 为达峰时间；C_{max} 为峰浓度；K_a 为药物吸收速率常数。

3 讨论

3.1 静注给药的药动学特征

大量的药物代谢动力学试验表明，由于动物种属间的生理代谢特点不同，不同动物及同种动物不同年龄之间达氟沙星的药物代谢动力学特点均有所不同[3-7]。但有关达氟沙星在鱼类体内的药物代谢动力学规律尚未见报道。Shojaee Aliabadi F. 试验结果证明小牛静注达氟沙星有较高的分布容积和相对较快的清除率，半衰期为 2.65 h，清除半衰期分别为 10.2 h。山羊静注达氟沙星，血浆清除半衰期分别为 4.67 h，表观分布容积较高[4]。Atef M[5]等试验证实以 1.25 mg/kg 达氟沙星静注和健康的成年雌性山羊，血浆药物时间曲线符合二室开放模型，药物迅速分布和消除，半衰期分别为 17.71 h。药物在中央室分布率高，K_{12}/K_{21} 比率为 0.67，表观分布容积为 1.42 L/kg。研究证实以 1.25 mg/kg 在绵羊静注达氟沙星，表观分布容积为 2.766 L/kg，消除半衰期（$t_{1/2\beta}$）为 3.32 h，清除速率为 0.63 L/（kg·h）[6]。也有试验表明雏鸡静脉注射达氟沙星后，达氟沙星在鸡体内分布广泛、迅速，且消除较慢，因此达氟沙星适合于临床上治疗全身感染和深部组织感染[7]。本试验结果表明，健康及感染组史氏鲟以 10 mg/kg 静脉注射达氟沙星后，最佳药物代谢动力学模型符合无吸收三室开放模型，达氟沙星在浅室分布较快，在深室分布较慢，中央室表观分布较小，表观分布较大，表明达氟沙星在健康史氏鲟体内分布广泛，组织中的药物浓度较高。这与达氟沙星在其他动物体内的药物代谢动力学特征基本一致。

3.2 口服给药的药动学特征

史氏鲟口服达氟沙星的药时数据结果符合二室开放模型，健康史氏鲟以 10 mg/kg 剂量口服达氟沙星水溶液后表现为吸收速度较慢，吸收半衰期为 9.49 h，消除半衰期为 78.27 h，达峰时间为 6.28 h，峰浓度为 0.79 μg/mL。实验表明口服噁喹酸、氟甲喹、恩诺沙星和沙拉沙星的大西洋鲑，这 4 种药物的达峰时间均高于 6 h[8]，而口服诺氟沙星和环丙沙星的鲤鱼达峰时间分别为 0.44 h 和 0.73 h[8,9]。笔者认为由于大西洋鲑属于冷水性鱼类（适温 15~18℃），而鲤鱼属温水性鱼类（适温 23~28℃），二者的基础体温和环境温度差异很大，由于温度的作用而药物的代谢速度有明显的差别，因而药物在冷水性鱼体内的达峰时间均长于温水性鱼类。史氏鲟的生长适温为 21℃左右，介于温水性鱼类和冷水性鱼类之间[11]，本实验的试验水温为 23±1℃，达峰时间也介于温水性鱼和冷水性鱼类之间。

3.3 Ah 感染对静注给药的药动学的药动学的影响

动物机体的健康状况有时会对药物在体内的分布、代谢及吸收有一定的影响。有文献报道支原体与大肠杆菌合并感染能使司帕沙星在鸡体内的消除半衰期略有延长，机体消除率减少，药时曲线下面积增大，但影响不如口服给药明显[10]。认为是由于静注给药时药物直接入血，不需要吸收过程有关。研究表明猪大肠杆菌感染可使恩诺沙星的清除率显著下降、消除半衰落期延长，从而导致曲线下面积显著增大，表明大肠杆菌感染能显著改变恩诺沙星在猪体内的分布和消除，使中央室分布容积显著减小。本实验 Ah 感染明显改变达氟沙星在史氏鲟体内的中央室分布容积。但染组鱼体内达氟沙星的消除规律没有显著的改变。由于嗜水气单胞菌感染属全身性感染，会导致包括动物体肝脏、肾脏及其他器官组织生理状况及代谢功能的改变。而郑伟文等研究表明，嗜水气单胞菌感染后的中华鳖血清中肌酐的含量也由健康组的 21.8 显著升高至 39 μm/L。本试验结果表明 Ah 感染对达氟沙星在史氏鲟体内的分布规律有一定的影响。

3.4 Ah 感染对口服给药的药动学的影响

研究结果表明，由于不同药物的结构不同，药物代谢情况受实验动物感染情况的影响是不同的。研究表明雏鸡口服达氟沙星水溶液，支原体与大肠杆菌合并感染能改变药物在体内的动力学特征，使达氟沙星在感染鸡体内的吸收、分布和消除均减慢，达峰时间、有效浓度维持时间延长，药物在中央室与周边室的转运速度也减慢。刘雅红[11]则证明感染霉形体与大肠杆菌感染鸡内服给药氧氟沙星药动学参数没有显著改变。也有报道霉形体与大肠杆菌合并感染鸡内服司帕沙星，吸收减慢，达峰时间推迟，峰浓度增加，消除减慢。本实验中嗜水气单胞菌感染明显改变达氟沙星口服给药后在史氏鲟体内的吸收、分布和消除。分布速率常数降低，分布减慢。同时药物的消除也减慢，消除半衰期延长。达峰时间延长，峰浓度降低。可见在口服给药方式下嗜水气单胞菌感染对达氟沙星在史氏鲟体内的代谢规律有较大的影响。

3.5 达氟沙星在史氏鲟体内生物利用度的评价

生物利用度是确定药物剂量与作用强度之间关系的重要因素，也是进行药物动力学评价时的主要参数之一[14-17]。本实验结果表明，健康及感染组鱼生物利用度均较高，说明该药在史氏鲟体内吸收较完全。张秀英等证明达氟沙星内服给药的生物利用度为 90.08%。Aliabadi 证明山羊肌注达氟沙星的生物利用度为 100%。El-Gendi[12]实验表明火鸡肌注和口服达氟沙星的生物利用度分别为 96.56% 和 81.4%。Atef[5]则发现健康山羊肌注达氟沙星生物利用度为（65.70±10.28）%。Shem

–Tov[13]证明哺乳奶牛静注达氟沙星生物利用度大于100%。McKellar[6]等发现奶牛肌注达氟沙星的生物利用度为95.71%。表明达氟沙星在动物体内的吸收过程虽然存在着种属的差异，但生物利用度均较高。同时嗜水气单胞菌感染并未使达氟沙星在史氏鲟体内的生物利用度发生明显的改变。

体外抑菌实验结果表明达氟沙星对嗜水气单胞菌的最小抑菌浓度MIC为0.05 μg/mL，结合本实验结果达氟沙星可用于史氏鲟嗜水气单胞菌感染的治疗，建议给药方式以间隔2 d、给药剂量为10 mg/kg为宜。

参考文献

[1] 王建元，刘玉年. 氟喹诺酮类抗菌药在动物中的应用. 中国兽医学报，1996，16（3）：304-309.

[2] 储卫华，于勇. 鲟鱼嗜水气单胞菌的分离与鉴定. 淡水渔业，2003，33（2）：16-17.

[3] Shojaee Aliabadi F, Less P. Pharmacokinetic-pharmacodynamic integration of danof loxacin in the calf [J]. Res Vet Sci, 2003, 74 (3): 247-259.

[4] Aliabadi FS, Lees P. Pharmacokinetics and pharmacodynamics of danof loxacin in serum and tissue fluids of goats f ollowing intravenous and intramuscular administration [J]. Am J Vet, 62 (12): 1979-1989.

[5] Atef M, El-Gendi A Y, Aziza, et al. Some pharmacokineti c data for danof loxacin in healthy goats [J]. Vet Res Commun, 2001, 25 (5): 367-377.

[6] McKellar Q A, Gibson I F, McCormack R Z. Pharmacokinetics and pharmacodynamics of danofloxacin in serum and tissue fluids of goats following intravenous and intramuscular administration [J]. Biopharm Drug Dipos, 1998, 19 (2): 123-129.

[7] Martinsen B, Horsberg T E, Varma K J, et al. Single dose pharmacokinetic study of f lorfenicol in At lanti c salmon in seawater at 11℃ [J]. Aquculture, 1993, 112: 1-11.

[8] 张雅斌，张祚新，郑伟，等. 不同给药方式下鲤对诺氟沙星的药代动力学研究，水产学报，2000，24（6）：559-563.

[9] 杨雨辉，佟恒敏，卢彤岩，等. 环丙沙星在鲤鱼体内吸收、代谢和生物利用度的研究，水产学报，2003，27（6）：582-589.

[10] 孙大江，曲秋芝，马国军，等. 史氏鲟人工养殖研究现状与展望，中国水产科学，1998，5（3）：108-111.

[11] 刘雅红，冯淇辉. 氧氟沙星在健康和霉形体与大肠杆菌感染鸡的药动学研究，畜牧兽医学报，1998，29（6）：536-540.

[11] el-Gendi AY, el-Banna HA, Abo Norag M, et al. Disposition kinetics of danof loxacin and ciprofloxacin in broi ler chi ckens [J]. Dtsch Tierarztl Wochenschr. 2001, 108 (10): 429-434.

[12] Shem-Tov M, Rav-Hon O, Ziv G, et al. Pharmacokineti cs and penetration of danofloxacin from the blood into the milk of cows [J]. J Vet Pharmacol Ther, 1998, 21 (3): 209-213.

[13] 艾晓辉，陈正望. 喹乙醇在鲤鱼体内的药物代谢动力学及组织浓度. 水生生物学报，2003，27（3）：273-277.

[14] 杨先乐，刘至治. 盐酸环丙沙星在中华绒螯蟹体内药物代谢动力学研究. 水生生物学报，2003，27（1）：18-22.

[15] 王翔凌，方之平，操继跃，等. 盐酸沙拉沙星在鲫鱼体内的残留及消除规律研究. 水生生物学报，2006，30（2）：196-201.

该文发表于《大连海洋大学学报》【2011，26（2）：144-148】

氧氟沙星在吉富罗非鱼体内的药代动力学及残留的研究

Pharmacokinetics and residues of ofloxacin in GIFT tilapia strain Oreoch romisniloticus

王贤玉[1]，宋洁[1,2]，王伟利[1]，姜兰[1]，罗理[1]，柯剑[1,2]

（1. 中国水产科学研究院珠江水产研究所，广东 广州 510380；2. 上海海洋大学水产与生命学院，上海 201306）

摘要： 在水温为（25±2）℃下，按10 mg/kg的剂量给吉富罗非鱼 Oreochromis niloticus 单次口服氧氟沙星，用高效液相色谱法测定鱼血浆和组织中的药物浓度，研究氧氟沙星在吉富罗非鱼体内的代谢及消除规律。结果表明：血浆及组织药时数据均符合一级吸收二室开放模型，吸收分布迅速，但消除较为缓慢，血浆、肌肉、肝胰脏、肾脏中的达峰时间 T_{max} 分别为0.41、3.19、0.18、0.59 h；最大血药浓度分别为7.98 μg/mL、17.24、36.10、46.65 μg/g，组织中肝胰脏的药物浓度最高，在测定时间内各组织的药物浓度均高于血浆；药物消除速度依次为肾脏>肌肉>肝胰脏，消除半衰期 $T_{1/2B}$ 分别为12.90、19.45、28.27 h。以10 μg/kg为最高残留限量，在本试验条件下，建议休药期不低于8 d。在治疗罗非鱼疾病时，氧氟沙星的给药剂量为10 mg/kg，每天两次，连续使用2~3 d。

关键词： 吉富罗非鱼；药代动力学；氧氟沙星；残留

该文发表于《淡水渔业》【2015, 45 (2): 72-78】

氧氟沙星与左氧氟沙星在中华绒螯蟹体内的药代动力学和残留消除规律研究

Pharmacokinetics and elimination regularity of ofloxacin and levofloxacin in *Eriocheir sinensis*

吴冰醒，曹海鹏，阮记明，马荣荣，胡鲲，杨先乐

（上海海洋大学，国家水生动物病原库，上海 201306）

摘要：以 15 mg/kg 的剂量分别对中华绒螯蟹（*Eriocheir sinensis*）单次肌注氧氟沙星和左氧氟沙星，运用反相高效液相色谱法，测定了中华绒螯蟹血淋巴和肌肉、肝胰腺、精巢、卵巢中氧氟沙星和左氧氟沙星的浓度，分别阐明了它们在中华绒螯蟹各组织中的残留消除规律。结果显示，氧氟沙星和左氧氟沙星在中华绒螯蟹血淋巴内的药物-曲线时间关系符合开放性二室模型，其药代动力学方程分别为：$C_t = 21.31e^{-0.497t} + 9.779e^{-0.035t}$ 和 $C_t = 74.938e^{-1.822t} + 20.811e^{-0.038t}$。氧氟沙星在肌肉、肝胰腺、精巢、卵巢的消除半衰期分别为 88.89、65.02、83.26、114.76 h。左氧氟沙星在肌肉、肝胰腺、精巢、卵巢的消除半衰期分别为 86.03、344.83、106.19、154.73 h。结果表明，相同给药剂量下，左氧氟沙星在体内浓度更高，分布更广，对于中华绒螯蟹的细菌性疾病有着更好的预防与治疗作用。

关键词：氧氟沙星；左氧氟沙星；中华绒螯蟹（*Eriocheir sinens*）；药代动力学；残留消除

六、噁喹酸

噁喹酸属于第一代氟喹诺酮类药物，是一种动物专用的抗菌药物。凡纳滨对虾能很好地吸收噁喹酸，并有效地防治因弧菌引起的细菌性疾病（王元等，2016）。

该文发表于《水产学报》【2016, 40 (3): 512-519】

噁喹酸在凡纳滨对虾体内药动学和对弧菌的体外药效学

Pharmacokinetics of oxolinic acid in the Pacific white shrimp（*Litopenaeus vannamei*）and in vitro antimicrobial activity on Vibrios

王元[1]，殷桂芳[1,2]，符贵红[1]，周俊芳[1]，沈锦玉[3]，房文红[1]

（1. 中国水产科学研究院东海水产研究所农业部东海与远洋渔业资源开发利用重点实验室，上海 200090；
2. 上海海洋大学水产与生命学院，上海 201306；3. 浙江省淡水水产研究所，浙江 湖州 313001）

摘要：采用高效液相色谱法，研究了复方噁喹酸粉药饵投喂在凡纳滨对虾体内药动学和组织中消除规律，同时检测了噁喹酸对对虾源弧菌的最小抑菌浓度，建立了药动/药效（PK/PD）关系，提出了用药方案和休药期建议。结果显示，复方噁喹酸粉拌饵投喂给药，噁喹酸给药剂量为 30 mg/kg，凡纳滨对虾血浆噁喹酸浓度-时间关系曲线均符合一级吸收的二室开放动力学模型。血淋巴中噁喹酸达峰浓度 C_{max}、达峰时间 T_{max}、曲线下面积（AUC_{0-24}）和消除半衰期 $T_{1/2\epsilon}$ 分别为 14.70 mg/L、2 h、244.6 mg/L·h 和 18.56 h；肌肉、肝胰腺和鳃的峰浓度 C_{max} 分别为 4.11 mg/kg、17.20 mg/kg 和 7.01 mg/kg，消除半衰期 $T_{1/2\epsilon}$ 分别为 10.71 h、12.31 h 和 16.75 h。噁喹酸对 132 株弧菌的 MIC 主要分布在 0.15~1.25 μg/mL，MIC_{50} 和 MIC_{90} 分别为 0.62 μg/mL 和 1.25 μg/mL。PK/PD 相互关系参数 C_{max}/MIC_{90} 和 AUC_{0-24}/MIC_{90} 分别为 11.76 和 195.7。研究表明，噁喹酸以 30 mg/kg 剂量药饵给药，凡纳滨对虾能很好地吸收噁喹酸，可以有效地防治弧菌引起的细菌性疾病。

关键词：凡纳滨对虾；噁喹酸；药动学；体外药效；弧菌

噁喹酸（oxolinic acid，OA）是第二代喹诺酮类广谱杀菌药，通过干扰细菌 DNA 螺旋酶来影响细菌染色体超螺旋而达到杀菌作用[8]，广泛用于防治鱼类细菌性疾病[9]，尤其是海水弧菌病。药动学是制定用药方案的基础，有关噁喹酸在水产动物体内药动学研究在国外报道很多，主要以冷水鱼类为主[9]，且多为注射给药和口灌给药；在对虾中，仅 Uno 等报道了口灌给药下噁喹酸在日本对虾（Penaeus japonicus）[9]和斑节对虾（P. monodon）[10]体内药动学。国内关于噁喹酸的药动学研究仅在欧洲鳗鲡（Anguilla anguilla）[11]和大菱鲆（Scophthalmus maximus）[12]等鱼类。噁喹酸是我国兽药典中允许使用的抗菌药物之一[13]，但其用药方案和休药期十分笼统，缺乏针对鱼虾蟹等不同种属水产动物的具体用药方案。有关噁喹酸在水产动物体内药动学研究多是以噁喹酸原料药进行的，其结果对指导复方噁喹酸粉制剂的合理使用并不合适。本研究采用复方噁喹酸粉制剂以投喂药饵的给药方式，研究噁喹酸在对虾体内药动学、组织分布和消除规律；弧菌是危害对虾养殖的重要细菌性病原之一[14]，研究噁喹酸对分离自患病凡纳滨对虾（Litopenaeus vannamei）弧菌的体外抗菌活性，构建体内药动-体外药效联合作用关系，为我国对虾养殖中噁喹酸用药方案的制定提供理论依据。

1 材料与方法

1.1 试验材料

用于体外药效试验的 132 株弧菌，为本实验室分离自浙江、福建和海南等地沿海养殖发病凡纳滨对虾，分别为副溶血弧菌（Vibrio parahaemolyticus）71 株、哈维氏弧菌（V. harveyi）35 株、溶藻弧菌（V. alginolyticus）12 株和创伤弧菌（V. vulnificus）14 株。

用于药动学试验的凡纳滨对虾为东海水产研究所海南琼海研究中心室内池养殖，挑选规格均匀的用于试验，对虾体长（9.0±0.5）cm，试验前暂养一周；暂养和试验用水为盐度 33 的自然海水，保持 24 h 充气，水温为（30±1）℃。暂养期间投喂不含噁喹酸的人工配制的对虾全价饲料，试验期间对虾成活率为 100%。

1.2 药品和试剂

噁喹酸标准品（纯度≥98.0%）购于德国 Dr. Ehrenstorfer 公司，噁喹酸原粉（纯度≥98.0%）购于潍坊三江医药集团。复方噁喹酸粉[13]：7.5 g 噁喹酸加 2.5 g 维生素 C 混合均匀，然后再加 90 g 淀粉充分混匀。甲醇、乙腈和正己烷均为 HPLC 级，德国 Merk 公司产品；四丁基溴化铵和磷酸均为分析纯，购自国药集团上海化学试剂公司。营养肉汤和 M-H 细菌培养基购自青岛海博生物技术有限公司。

1.3 仪器和设备

高效液相色谱仪（Waters 2695，美国）；色谱柱为 Aglient Zorbax SB-C$_{18}$（4.6 mm×150 mm，5 μm）；荧光检测器（Waters 2475，美国）；超纯水仪（Millipore Milli-Q Advantage，美国）；均质器（Bertin precellys 240，法国）；生化培养箱（Sanyo MIR-254，日本）；微量移液器（Eppendorf，德国）；电子分析天平（Mettler 204，瑞士）；高速冷冻离心机（Hitachi CF16R Ⅱ，日本）；旋转蒸发仪（亚荣 RE-52A，中国上海）；0.22 μm 微孔滤膜（Rephile，中国上海）等。

1.4 最小抑菌浓度（MIC）的测定

最小抑菌浓度（MIC）采用试管二倍稀释法进行测定[15]，药物为噁喹酸原粉。

1.5 给药和取样

给药剂量为 30 mg/kg 体质量，配制含 0.6%噁喹酸的人工配合饲料，按 0.5%对虾体质量投喂药饵。投喂后 15 min 药饵全部吃完，分别在药饵吃完后 30 min、1 h、2 h、4 h、6 h、8 h、12 h、1 d、2 d、3 d、5 d、7 d、10 d、15 d 和 20 d 采集血淋巴、肝胰腺、肌肉和鳃等样品，置于-80℃冰箱保存备用。每个采样点采集 12 尾对虾，2 尾合并为一个样品，每个时间点共 6 个样品。对虾样品采集参照李国烈等方法[16]。

1.6 样品处理

1.6.1 血淋巴样品取出样品置室温融解后，1 000 r/min 离心 5 min，取 0.2 mL 上清于 1.5 mL 离心管中，加入等体积乙腈，涡旋振荡 2 min 后，12 000 r/min 离心 5 min，取上清液经微孔滤膜过滤，滤液用于 HPLC 分析。

1.6.2 肌肉、肝胰腺和鳃样品样品室温解冻后，取适量样品于均质器中均质；然后称取均质样品（肌肉 1.00 g、肝胰腺和鳃各 0.50 g）于 10 mL 离心管中，加入 1 mLPBS（0.1 mol/L，pH7.4），旋涡振荡器上振荡 2 min，加入 5 mL 乙腈，继续振荡 5 min，于 10 000 r/min 转速离心 5 min，收集上清液；残渣再重复提取两次，合并三次上清液于 100 mL 具塞茄形蒸发瓶中，于45℃、真空度0.075 Mpa下旋蒸至干；加入1 mL流动相和1 mL正己烷，盖上塞子置超声波清洗仪超声15 min，瓶底残留物完全溶解，将混合液体转移至2 mL离心管中，12 000 r/min离心5 min，用1 mL注射器吸取下层液体，经微孔滤膜过滤，滤液用于HPLC分析。考虑到对虾鳃较小，遂将2个样品合并在一起处理，因此鳃组织在每个时间点只有3个平行。

1.7 色谱条件

流动相：乙腈和0.01 mol/L四丁基溴化铵（磷酸调节pH值为2.75）；柱温40℃，样品温度25℃；进样量10 μL，流速1.0 mL/min；荧光检测器波长$Ex = 265$ nm、$Em = 380$ nm。

血浆和肌肉的流动相比例为乙腈：0.01M四丁基溴化铵（pH2.75）= 30：70（V/V），肝胰腺和鳃的流动相比例为18：82（V/V）。

上述样品前处理和色谱条件下，对虾血淋巴样品中噁喹酸检测限为0.02 μg/mL，肌肉、肝胰腺和鳃检测限为0.01 μg/g、0.02 μg/g和0.02 μg/g；血淋巴中噁喹酸回收率为86.0%～92.8%，肌肉、肝胰腺和鳃的回收率分别为78.9%～85.5%、82.1%～86.7%、85.2%～90.8%；血淋巴、肌肉、肝胰腺和鳃中噁喹酸日内和日间精密度均小于5%。

1.8 数据处理

凡纳滨对虾血药浓度-时间关系曲线方程采用房室模型进行拟合分析，血淋巴、肌肉、肝胰腺和鳃的药动学参数采用统计矩原理推算。模型的拟合和药动学参数的推算采用药物与统计软件（DAS3.1）进行分析，曲线下面积AUC_{0-24}则根据实际测定值采用梯形法算得。

2 结果

2.1 药饵给药后噁喹酸在凡纳滨对虾体内药动学

凡纳滨对虾投喂添加复方噁喹酸粉的颗粒饲料，噁喹酸剂量为30 mg/kg，给药后对虾血药浓度迅速升高，2 h即到达峰浓度（C_{max}为14.70 mg/L）；随后开始下降，且速度较快，48 h后下降速度缓慢；120 h时血药浓度低于检测限（图1）。血药浓度-时间关系曲线比较适合采用一级吸收二室模型来描述，拟合的药动学方程为：

$$C_t = 7.298e^{-0.075t} + 9.652e^{-0.037t} - 39.184e^{-2.607t}, \quad R^2 = 0.999, \quad AIC = -1.088.$$

药饵给药后，对虾组织中噁喹酸浓度变化趋势与血药相似（图2），但组织中噁喹酸峰浓度和达峰时间明显不同，肌肉、肝胰腺和鳃等组织中药物峰浓度分别为4.11 mg/kg、17.20 mg/kg和7.01 mg/kg，达峰时间分别为4 h、1 h和4 h；组织中噁喹酸消除较快，肌肉、肝胰腺和鳃在120 h低于或接近于检测限。

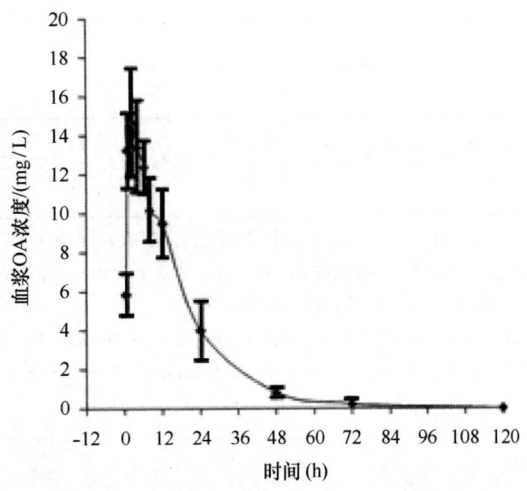

图1 药饵给药后凡纳滨对虾血淋巴噁喹酸浓度与时间关系曲线

Fig. 1 Profile of hemolymphoxolinic acid concentrations vs. time of pacific white shrimp after oral dosing via feed

采用统计矩原理求得对虾血淋巴、肌肉、肝胰腺和鳃等组织药动学参数，见表1。可以看出，各组织的曲线下面积（AUC）大小依次为血淋巴、肝胰腺、鳃和肌肉，各组织中噁喹酸消除快慢（$t_{1/2z}$）依次为肌肉、肝胰腺、鳃和血淋巴。

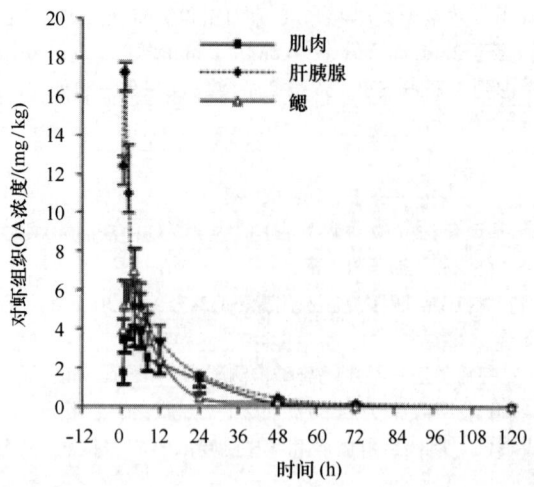

图 2　药饵给药后凡纳滨对虾组织中噁喹酸浓度与时间关系曲线

Fig. 2　Profile of tissueoxolinic acid concentrations vs. time of pacific white shrimp after oral dosing via feed

表 1　药饵给药后凡纳滨对虾各组织中噁喹酸药动学参数

Table 1　Pharmacokinetic parameters of oxolinic acid in pacific white shrimpafter oral dosing via feed

药动学参数 pharmacokinetic parameter		血淋巴 hemolymph	肌肉 muscle	肝胰腺 hepatopancreas	鳃 gill
C_{max}	mg/L［mg/kg］	14.70±2.78	4.11±1.01	17.20±0.47	7.01±1.05
T_{max}	h	2	4	1	4
AUC_{0-24}	mg/L·h［mg/kg·h］	244.6	58.42	98.82	71.73
AUC_{0-120}	mg/L·h［mg/kg·h］	403.2	79.69	110.8	83.00
$AUC_{0-\infty}$	mg/L·h［mg/kg·h］	403.7	80.34	118.6	83.22
MRT_{0-120}	h	23.50	14.93	9.64	11.73
$MRT_{0-\infty}$	h	23.71	15.52	13.34	12.09
Vz	L/kg	1.991	–	–	–
$T_{1/2z}$	H	18.56	10.71	12.31	16.75
CL_z	L/(kg·h)［kg/(kg·h)］	0.074	0.373	0.253	0.360

注：C_{max}表示峰浓度；T_{max}表示达峰时间；AUC_{0-24}、AUC_{0-120}和$AUC_{0-\infty}$分别表示 0~24 h、0~120 h 和 0~∞ 时间的曲线下面积；MRT_{0-120}和$MRT_{0-\infty}$分别表示 0~120 h 和 0~∞ 时间的平均驻留时间；$t_{1/2}$表示消除半衰期；V_z表示表观分布容积；CL_z表示总体清除率

Note：C_{max}-peak concentration；T_{max}-time to reach peak concentration；AUC_{0-24}、AUC_{0-120} and $AUC_{0-\infty}$ -area under the concentration-time curve from zero to 24 h, from zero to 120 h and from zero to infinity, respectively；$t_{1/2z}$-elimination half-life；MRT_{0-120} and $MRT_{0-\infty}$ -mean residue time from zero to 120 h and from zero to infinity, respectively；V_z-volume of distribution；CL_z-total body clearance

2.2　弧菌对噁喹酸的敏感性

噁喹酸对从患病对虾分离的 132 株弧菌的 MIC 值主要分布在 0.15~1.25 μg/mL，仅有 8 株弧菌的 MIC≥5 μg/mL，由此可见噁喹酸对弧菌较为敏感，132 株弧菌的 MIC_{50} 和 MIC_{90} 分别为 0.62 μg/mL 和 1.25 μg/mL（表 2）。

表 2　噁喹酸对对虾源弧菌的 MIC 值分布

Tab. 2　MIC distribtution values of oxolinic acid on vibrio from pacific white shrimp μg/mL

细菌种类 bacterial species	MIC 值分布 MIC distribution values									MIC_{50}/MIC_{90}
	0.08	0.15	0.31	0.62	1.25	2.5	5	10	>10	
副溶血弧菌（n=71） V. parahaemolyticus	0	6	26	25	10	1	0	1	2	0.62/1.25
哈维氏弧菌（n=35） V. harveyi	0	1	8	17	3	3	1	0	2	0.62/2.5
溶藻弧菌（n=12） V. alginolyticus	0	3	0	8	0	0	0	1	0	0.62/0.62
创伤弧菌（n=14） V. vulnificus	1	1	8	3	0	0	0	0	1	0.31/0.62
合计（n=132） Total number of strains	1	11	42	53	13	4	1	2	5	0.62/1.25

2.3　药动/药效（PK/PD）相互关系

C_{max}/MIC_{90} 和 AUC_{0-24}/MIC_{90} 值是评价 PK/PD 关系的重要参数，能显示体内药物抗菌效果。本研究中，噁喹酸对 35 株哈维氏弧菌的 C_{max}/MIC_{90} 和 AUC_{0-24}/MIC_{90} 值均为最小（分别为 5.88 和 97.8），对 132 株弧菌的两个 PK/PD 参数分别为 11.76 和 195.7，显示其具有较好的抗菌效应（表 3）。

表 3　对虾药动学和弧菌体外药效联合参数

Tab. 3　PK/PD parametersfrompharmacokinetics of oxolinic acid in shrimp and in vitro efficacy on vibrios

PK/PD 参数 PK/PD parameters	副溶血弧菌 V. parahaemolyticus	哈维氏弧菌 V. harveyi	溶藻弧菌 V. alginolyticus	创伤弧菌 V. vulnificus	弧菌 Vibrio
	（n=71）	（n=35）	（n=12）	（n=14）	（n=132）
C_{max}/MIC_{90}	11.76	5.88	23.71	23.71	11.76
AUC_{0-24}/MIC_{90}	195.7	97.8	394.5	394.5	195.7

3　讨论

3.1　噁喹酸在凡纳滨对虾体内药动学特征

采用房室模型拟合药时曲线关系，可推算得到较多的药动学参数，能很好地揭示药物在动物体内的动态变化规律。噁喹酸在水产动物体内药动学的报道较多，但在不同动物之间其房室模型差异较大。同为消化道给药（口灌或投喂药饵）下，大西洋鲑（Salmo salar）[0]、大西洋鳕（Gadus morhua）[0]、大西洋庸鲽（Hippoglossus hippoglossus）[0]和欧洲鳗鲡[0]的血浆药时曲线关系适合采用一室模型描述，大菱鲆[12]适合采用二室模型，而金头鲷（Sparus aurata）[8]则不能用房室模型拟合，日本对虾和斑节对虾在口灌给药下同样不适合采用房室模型描述。本研究中，拟合相关系数（R^2=0.999）和 AIC 值（AIC=-1.088）显示，噁喹酸药饵给药后凡纳滨对虾血药浓度与时间关系曲线关系适合采用一级吸收二室开放模型描述。造成房室模型差异的原因很多，例如动物种类和大小、水温、盐度、药物给予方式和房室的选择等，但是难以推断是哪种因子决定了最佳模型的选择。尽管本研究采用二室模型能很好地描述血药浓度–时间曲线关系，但还是采用了统计矩原理推算药动学参数，因为它们不受房室选择的影响[0]。

血药 C_{max}、T_{max} 和 AUC 是反映药物在体内吸收快慢和程度的重要参数[0]，同时也是评价体内药效的重要指标[0]。与大西洋鲑、大西洋鳕、金头鲷等低温养殖鱼类相比，凡纳滨对虾药饵给药后的血药峰浓度明显高于低温养殖鱼类（C_{max} 分别为 0.61 μg/mL、1.2 μg/mL 和 0.99 μg/mL），AUC 亦明显高于它们（分别为 26.8 μg/mL·h、140 μg/mL·h 和 26.75 μg/mL·h），达峰时间明显快于它们（分别为 12h、24 h 和 24 h）。在对虾属种类之间，凡纳滨对虾血药 C_{max} 和 AUC 与日本对虾

（C_{max} = 17.80 μg/mL、AUC = 573 μg/mL·h）较为接近[0]，但明显大于斑节对虾（C_{max} = 4.20 μg/mL、AUC = 43.6 μg/mL·h）[0]，其血药达峰均快于日本对虾[9]和斑节对虾[10]。由此分析，水温很可能是影响噁喹酸在水产动物体内吸收的主要因素，低温冷水性鱼类吸收很少；其次，种属间差异亦影响噁喹酸的吸收。与其他抗菌药相比，同是药饵给药下，噁喹酸在凡纳滨对虾的血药浓度略高于盐酸沙拉沙星[0]，远高于诺氟沙星[0]，可见噁喹酸能被凡纳滨对虾很好且快速地吸收。

消除半衰期（$T_{1/2z}$）和平均驻留时间（MRT）是评价药物消除快慢的重要参数[0]。尽管大西洋鳕和大西洋鲑都是低温养殖鱼类，但口灌给药后大西洋鳕血浆、肌肉和肝脏的 $T_{1/2\beta}$ 分别为 82 h、58 h 和 82 h[0]，而大西洋鲑血浆、肌肉和肝脏噁喹酸 $T_{1/2\beta}$ 分别为 18 h、20 h 和 10 h[0]。在对虾属中噁喹酸消除速率同样存在差异，如凡纳滨对虾血淋巴中噁喹酸消除速率与斑节对虾（MRT = 20.9 h，$T_{1/2\beta}$ = 19.8 h）相近[0]，快于日本对虾（MRT = 40.3 h，$T_{1/2\beta}$ = 34.3 h）[0]。由此可见，噁喹酸在不同动物体内的消除速率存在着种属差异，本研究中凡纳滨对虾血浆、肌肉和肝胰腺等组织均显示对噁喹酸具有较快的消除速率，这一结论在休药期计算中得到进一步证实。

3.2　关于凡纳滨对虾肝胰腺对噁喹酸的吸收

甲壳动物肝胰腺不仅具有代谢功能还具有吸收功能，经口摄入的化学药物和有机溶质等[0]，主要被肝胰腺吸收，其次是中肠[0]；Miyawaki 等已证明蛋白银、乳酸铁和橄榄油等物质在克氏螯虾肝胰腺进行主要吸收[0]。本研究中，凡纳滨对虾摄食噁喹酸药饵后，其肝胰腺中药物浓度达峰时间快于血淋巴、肌肉和鳃组织，这一现象同样出现在凡纳滨对虾口服氟苯尼考和甲砜霉素的结果中[0]。甲壳动物肝胰腺具有较强吸收作用的原因很可能与其消化系统的结构特征有关，例如对虾幽门胃和中肠前端包埋在肝胰腺中，肝胰腺和中肠同源均来自内胚层，对虾肝胰腺上皮细胞有整齐而密集的微绒毛，这种上皮细胞具有消化和吸收功能[0]。因此，过高的给药剂量将会导致对虾肝胰腺药物浓度过高而增加对肝胰腺的毒性危害，在甲壳动物给药时应避免过高给药剂量而引发对肝胰腺的毒性危害。

3.3　噁喹酸防治对虾弧菌病的有效性分析

结合血药浓度、作用时间和抗菌活性特征，依据 PK/PD 联合参数用于制定用药方案，对获得最佳临床疗效和指导合理用药有着重要意义[0]。噁喹酸属于浓度依赖型药物，因此，C_{max}/MIC 和 AUC_{0-24}/MIC 值是评价该类药物疗效的关键指标[0]。一般认为 C_{max}/MIC 之比为 8～10 倍时，临床有效率可达 90%；对于 AUC_{0-24}/MIC 比值，研究显示，AUC_{0-24}/MIC = 25 时可以应对不严重的细菌性感染，而对于严重的感染，该比值需要大于等于 100[0]，AUC_{0-24}/MIC 比值高于 250，可很快清除体内感染的细菌[0]。鉴于 AUC_{0-24}/MIC 比值在很大程度上不受给药间隔、药物种类、动物种类和感染部位的影响[0]，因此，研究者更多的采用 AUC_{0-24}/MIC 比值作为预测实际疗效的参数。

Bhavnani 等[0]认为，耐药菌株能提高 MIC_{90} 值，较大程度地影响 PK/PD，从而影响药物疗效。本研究采用 MIC_{90} 来评价弧菌对噁喹酸的敏感性，哈维氏弧菌的 C_{max}/MIC_{90} 和 AUC_{0-24}/MIC_{90} 比值明显低于其他种类弧菌。由于目前尚无噁喹酸对弧菌耐药折点的判别值，通过参考噁喹酸对气单胞菌的耐药折点（4 μg/mL）[0]，显示出本研究中的 132 株弧菌有 8 株为耐药菌株，其中 35 株哈维氏弧菌有 3 株为耐药菌株，由此升高了 MIC_{90} 值从而影响了 PK/PD 参数。总体而言，一日一次以 30 mg/kg 剂量药饵给药，针对 132 株弧菌的 MIC_{90}，其 PK/PD 参数 C_{max}/MIC_{90} 和 AUC_{0-24}/MIC_{90} 比值均显示能满足噁喹酸的有效治疗要求。

参考文献

[1]　Drlica K, Coughlin S. Inhibitors of DNA gyrase [J]. Pharmacology and Therapeutics, 1989（44）：107-122.

[2]　李健，王群，孙修涛，等. 噁喹酸对水生生物细菌病的防治效果及残留研究 [J]. 中国水产科学，2001，8（3）：45-49.
　　Li J, Wang Q, Sun X T, et al. Application of oxolinic acid against bacterial diseases in aquatic species and the body residues [J]. Journal of Fishery Sciences of China, 2001, 8（3）：45-49 (in Chinese).

[3]　Touraki M, Niopas I, Karagiannis V. Treatment of vibriosis in European sea bass larvae, Dicentrarchus labrax L., with oxolinic acid administered by bath or through medicated nauplii of Artemia franciscana（Kellogg）：efficacy and residual kinetics [J]. Journal of Fish Diseases, 2012, 35（7）：513-522.

[4]　Samuelsen O B, Bergh O I. Efficacy of orally administered florfenicol and oxolinic acid for the treatment of vibriosis in cod（Gadus morhua）[J]. Aquaculture, 2004, 235（1）：27-35.

[5]　Samuelsen O B, Ervik A, Pursell L, et al. Single-dose pharmacokinetic study of oxolinic acid and vetoquinol, an oxolinic acid ester, in Atlantic salmon（Salmo salar）held in seawater and in vitro antibacterial activity against Aeromonassalmonicida [J]. Aquaculture, 2000, 187（3）：213-224.

[6]　Samuelsen O B, Bergh O, Ervik A. A single-dose pharmacokinetic study of oxolinic acid and vetoquinol, an oxolinic acid ester, in cod, Gadus morhua L., held in sea water at 8°C and in vitro antibacterial activity of oxolinic acid against Vibrio anguillarum strains isolated from diseased cod [J]. Journal of Fish Diseases, 2003, 26（6）：339-347.

［7］　Samuelsen O B，Ervik A. A single-dose pharmacokinetic study of oxolinic acid and vetoquinol，an oxolinic acid ester，in Atlantic halibut，Hippoglossus hippoglossus L.，held in sea water at 9℃［J］. Journal of Fish Diseases，1999，22（1）：13-23.

［8］　Rigos G，Tyrpenou A，Nengas I，et al. A pharmacokinetic study of flumequine in sea bass，Dicentrarchus labrax（L.），after a single intravascular injection［J］. Journal of fish diseases，2002，25（2）：101-105.

［9］　Uno K. Pharmacokinetics of oxolinic acid and oxytetracycline in kuruma shrimp，Penaeus japonicus［J］. Aquaculture，2004，230（1）：1-11.

［10］　Uno K，Aoki T，Kleechaya W，et al. Pharmacokinetics of oxytetracycline in black tiger shrimp，Penaeus monodon，and the effect of cooking on the residues［J］. Aquaculture，2006，254（1）：24-31.

［11］　林丽聪，樊海平，廖碧钗，等. 噁喹酸在欧洲鳗鲡体内的药代动力学及残留研究［J］. 淡水渔业，2012，42（1）：24-29.
　　　　Lin L C，Fan H P，Liao B C，et al. Pharmacokinetics and residues of oxolinic acid inAnguilla anguilla［J］. Freshwater Fisheries，2012，42（1）：24-29（in Chinese）.

［12］　孙爱荣，李健，常志强，等. 噁喹酸在大菱鲆体内的药代动力学研究［J］. 中国海洋大学学报（自然科学版），2012，42（3）：75-79.
　　　　Sun A R，Li J，Chang Z Q，et al. Pharmacokinetics of oxolinic acid in Turbot（Scophthalmus maximus）after a single oral and intravenous administration［J］. Periodical of Ocean University of China，2012，42（3）：75-79（in Chinese）.

［13］　中国兽药典委员会. 中华人民共和国兽药典. 兽药使用指南. 化学药品卷. 2010 年版. 中国农业出版社. P357.
　　　　Chinese veterinary Pharmacopoeia Committee. Veterinary Pharmacopoeia of the People's Republic of China. Guid for Use of Veterinary Drug. Chemical Volume. 2010 Edition. China Agriculture Press. pp357（in Chinese）.

［14］　陈健舜，朱凝瑜，孔蕾，等. 凡纳滨对虾细菌性红体病病原的分子特征与耐药性［J］. 水产学报，2012，36（12）：1891-1900.
　　　　Chen J S，Zhu N Y，Kong L，et al. Molecular characteristics and antimicrobial sensitivity of bacterial pathogen from the outbreak of Litopenaeus vannamei red-body disease［J］. Journal of Fisheries of China，2012，36（12）：1891-1900（in Chinese）.

［15］　戴自英. 临床抗菌药物学［M］. 北京：人民卫生出版社，1985.
　　　　Dai Z Y. Clinical antibacterial pharmacy［M］. Beijing：People's Medical Publishing House，1985（in Chinese）.

［16］　李国烈，王元，房文红，等. 盐酸沙拉沙星在凡纳滨对虾体内药动学与生物利用度［J］. 南方水产科学，2014，10（1）：50-57.
　　　　Li G L，Wang Y，Fang W H，et al. Pharmacokinetics and bioavailability of sarafloxacin hydrochloride in Litopenaeus vannamei［J］. South China Fisheries Science，2014，10（1）：50-57（in Chinese）.

［17］　王广基，刘晓东，柳晓泉. 药物代谢动力学［M］. 北京：化学工业出版社，2005.
　　　　Wang G J，Liu X D，Liu X Q. Pharmacokinetics［M］. Beijing：Chemical Industry Press，2005（in Chinese）.

［18］　Wright D H，Brown G H，Peterson M L，et al. Application of fluoroquinolone pharmacodynamics［J］. Journal of Antimicrobial Chemotherapy，2000，46（5）：669-683.

［19］　房文红，郑国兴. 肌注和药饵给药下诺氟沙星在南美白对虾血淋巴中药代动力学［J］. 水生生物学报，2006，30（5）：541-546.
　　　　Fang W H，Zheng G X. Pharmacokinetics of norfloxacin in hemolymph from whiteleg shrimp，Penaeus vannamei following intramuscle injection and oral administration［J］. Acta Hydrobiologica Sinica，2006，30（5）：541-546（in Chinese）.

［20］　Loizzi，R. F.，Peterson，D. R.，1971. Lipolytic sites in crayfish hepatopancreas and correlation with fine structure. Comp. Biochem. Physiol. 39B，227-236.

［21］　Verri T，Mandal A，Zilli L，et al. d-Glucose transport in decapod crustacean hepatopancreas［J］. Comparative Biochemistry & Physiology Part A，2001，130（3）：585-606.

［22］　Miyawaki M，Taketomi Y，Tsuruda T. Absorption of experimentally administered materials by the hepatopancreas cells of the crayfish，Procambarus clarki［J］. Cell biology international reports，1984，8（10）：873-877.

［23］　Fang W，Li G，Zhou S，et al. Pharmacokinetics and Tissue Distribution of Thiamphenicol and Florfenicol in Pacific White Shrimp Litopenaeus vannamei in Freshwater following Oral Administration. Journal of Aquatic Animal Health. 2013，25：83-89.

［24］　陈宽智，鲍鹰，何伟宏. 东方对虾消化系统解剖和组织学的研究［J］. 山东海洋学院学报，1988，18（1）：43-53.
　　　　Chen K Z，Bao Y，He W H. Studies on the anatomy and histology of the digestive system of the Penaeus orientalis（Crustacea，Decapoda）［J］. Journal of Shandong College of Oceanology，1988，18（1）：43-53（in Chinese）.

［25］　Zhanel G G. Influence of pharmacokinetic and pharmacodynamic principles on antibiotic selection［J］. Current infectious Disease Reports，2001，3（1）：29-34.

［26］　曾苏. 药物代谢学［M］. 杭州：浙江大学出版社，2008.
　　　　Zeng S. Drug Metabolism［M］. Hangzhou：Zhejiang University Press，2008（in Chinese）.

［27］　Jacobs M R. Optimisation of antimicrobial therapy using pharmacokinetic and pharmacodynamic parameters［J］. Clinical microbiology and Infection，2001，7（11）：589-596.

［28］　Schentag J J，Meagher A K，Forrest A. Fluoroquinolone AUIC break points and the link to bacterial killing rates part 2：human trials［J］. Annals of Pharmacotherapy，2003，37（10）：1478-1488.

［29］　Owens R C，Bhavnani S M，Ambrose P G. Assessment of pharmacokinetic--pharmacodynamic target attainment of gemifloxacin against Streptococcus pneumonia［J］. Diagnostic microbiology and infectious disease，2005，51（1）：45-49.

445

［30］ Bhavnani S M, Ambrose P G, Jones R N. Pharmacodynamics in the evaluation of drug regimens. ［J］. The Annals of pharmacotherapy, 2002, 36（3）: 530–532.

［31］ Le Bris H E, Dhaouadi R, Naviner M, et al. Experimental approach on the selection and persistence of nti–microbial–resistant Aeromonads in faecal matter of rainbow trout during and after an oxolinic acid treatment ［J］. Aquaculture, 2007, 273（4）: 416–422.

第三节　磺胺类渔药在水产动物体内的药代动力学

磺胺类药物（sulfonamides）是人工合成的具有氨苯磺胺基本结构的广谱抗菌药物。它常作为动物抗感染治疗中重要药物之一。主要的磺胺药物有磺胺嘧啶、磺胺甲基异噁唑等，此外，作为外用的磺胺药物有磺胺嘧啶银等。

磺胺类药物在水产动物体内的代谢具有种属差异性。磺胺甲基异噁唑在中华绒螯蟹组织中消除较缓慢，尤其在肝胰腺中（郑宗林等，2005；唐俊等，2006）。磺胺甲噁唑在罗非鱼体内消除较快（袁科平等，2008）。投喂复方磺胺嘧啶后，磺胺嘧啶在吉富罗非（*Oreochromis niloticus*）胃肠道中有非齐性吸收现象；甲氧苄啶未见双峰现象（王伟利等，2016）。复方磺胺嘧啶在鳜（*Siniperca chuatsi*）体内消除缓慢，参考欧盟的最大残留限量规定，其休药期不少于36天（Weili Wang et al., 2015）。复方磺胺甲噁唑在松浦镜鲤（*Cyprinus carpio* Songpu）体内消除较快，建议休药期不低于15 d（韩冰等，2014）。采用尾静脉抽血和断尾取血两种取样方法对银鲫取血，测定磺胺二甲嘧啶，抽血采样组血药浓度明显高出断尾采样组。该药的休药期建议定为10 d（艾晓辉等，2001）。

复方磺胺制剂中的不同成分在鱼体内的代谢具有独立性。罗氏沼虾（*Macrobranchium rosenbergii*）口灌给药复方新诺明后，甲氧苄氨嘧啶（TMP）对磺胺甲基异噁唑（SMZ）的药代动力学并无显著影响（郑宗林等，2008）。

该文发表于《JVPT》【2015：39：309–314】

A pharmacokinetic and residual study of sulfadiazine/trimethoprim in mandarin fish（*Siniperca chuatsi*）with single–and multiple–dose oral administrations

Weili Wang, Li Luo, He Xiao, Ruiquan Zhang, Yuting Deng, Aiping Tan, Lan Jiang*

（Pearl River Fisheries Research Institute, Chinese academy of fishery science, Key Laboratory of Fishery Drug DevelopmentMinistry of Agriculture, Guangzhou, 510380, P. R. China）

Abstract: A pharmacokinetic and tissue residue study of sulfadiazine combined with trimethoprim（SDZ/TMP = 5/1）was conducted in Siniperca chuatsi after single–（120 mg/kg）or multiple–dose（an initial dose of 120 mg/kg followed by a 5-day consecutive dose of 60 mg/kg）oral administrations at 28℃. The absorption half–life（$t_{1/2\alpha}$）, elimination half–life（$t_{1/2\beta}$）, volume of distribution（V_d/F）, and the total body clearance（Cl_B/F）for SDZ and TMP were 4.3±1.7–6.3±1.8 h and 2.4±1.0–3.9±0.9 h, 25.9±4.5–53.0±5.6 h and 11.8±3.5–17.1±3.4 h, 2.34±0.78–3.67±0.99 L/kg and 0.39±0.01–1.33±0.57 L/kg, and 0.03±0.01–0.06±0.01 L/kg·h and 0.02±0.01–0.05±0.01 L/kg·h, respectively, after the single dose. The elimination half–life（$t_{1/2\beta}$）and mean residue time（MRT）for SDZ and TMP were 68.8±7.8–139.8±12.3 h and 34.0±5.5–56.1±6.8 h, and 99.3±6.1–201.7±11.5 h and 49.1±3.5–81.0±5.1 h, respectively, after the multiple–dose administration. The daily oral SDZ/TMP administration might cause a high tissue concentration and long $t_{1/2\beta}$, thereby affecting antibacterial activity. The withdrawal time for this oral SDZ/TMP formulation（according to the accepted guidelines in Europe for MRLs, <0.1 mg/kg of tissues for sulfonamides, and <0.05 mg/kg for TMP）should not be less than 36 days for fish.

Keywords: sulfadiazine, trimethoprim, pharmacokinetic, residual, fish

单次和连续给药下复方磺胺嘧啶在鳜鱼体内的药动学和残留消除

王伟利，罗理，肖贺，张瑞全，邓玉婷，谭爱萍，姜兰

（中国水产科学研究院珠江水产研究所农业部药理实验室，中国广州，510380）

摘要： 在28℃水温下，使用120 mg/kg剂量的复方磺胺嘧啶单次药饵饲喂鳜鱼，研究其在鳜鱼体内的药动学规律；相同条件下，使用首剂量120 mg/kg的复方磺胺嘧啶饲喂鳜鱼，之后按60 mg/kg的剂量连续五天饲喂实验鱼来研究药物在鳜鱼体内的残留消除规律。结果发现，单次给药情况下，磺胺嘧啶和甲氧苄啶的吸收半衰期$t_{1/2\alpha}$、消除半衰期$t_{1/2\beta}$、表观分布容积V_d/F和体清除率Cl_B/F分别为：（4.3±1.7）~（6.3±1.8）h 和（2.4±1.0）~（3.9±0.9）h，（25.9±4.5）~（53.0±5.6）h 和（11.8±3.5）~（17.1±3.4）h，（2.34±0.78）~（3.67±0.99）L/kg 和（0.39±0.01）~（1.33±0.57）L/kg，以及（0.03±0.01）~（0.06±0.01）L/kg·h 和（0.02±0.01）~（0.05±0.01）L/kg·h。连续给药情况下，磺胺嘧啶和甲氧苄啶的消除半衰期$T_{1/2\beta}$和平均滞留时间MRT分别为（68.8±7.8）~（139.8±12.3）h 和（34.0±5.5）~（56.1±6.8）h，以及（99.3±6.1）~（201.7±11.5）h 和（49.1±3.5）~（81.0±5.1）h。连续给药情况下，组织中药物浓度会显著提高，并且消除缓慢，具有较好的抗菌活性；参考欧盟对于磺胺类药物最大残留限量MRLs<0.1 mg/kg 和甲氧苄啶<0.05 mg/kg 的规定，休药期不少于36 d。

关键词： 磺胺嘧啶，甲氧苄啶，药动学，残留，鳜鱼

The combination of sulfonamides and trimethoprim（TMP）plays a significant role against bacterial diseases in different animal species and aquaculture（Prescott et al.，2000；Biswas et al.，2007；Bogialli et al.，2003）. The synergistic action of the combination of sulfonamides and TMP inhibits a different step in the bacterial folic acid biosynthesis pathway（Batzias et al.，2005），thereby lowering the minimum inhibitory concentration（MIC）of both drugs，broadening the bacterial spectrum and reducing bacterial resistance（Plumb，2002）. In vitro studies suggest that a 16：1 to 20：1 ratio of sulfonamides：TMP should be maintained to ensure synergisticaction against susceptible bacteria（Prescott et al.，2000）. To achieve the optimal treatment ratio in a veterinary clinical setting，a combination of sulfadiazine/trimethoprim（SDZ/TMP）was widely used at a 5：1 ratio（Prescott et al.，2000；Riviere & Spoo，2001；Batzias et al.，2005）.

In China，the mandarin fish（Siniperca chuatsi）has a relatively high market value，is widely cultured throughout the country，and is important in stocking fisheries in lakes and reservoirs（Liu et al.，1998）. To our knowledge，there is a lack of information at present regarding the pharmacokinetic behavior of SDZ and TMP in mandarin fish. Several pharmacokinetic studies of SDZ，TMP，or both have been conducted in other species such as chickens（Baert et al.，2003），cattle and calves（Shoaf et al.，1989；Kaartinen et al.，1999），ponies and horses（Van Duijkeren et al.，1994，2002），dogs（Sigel et al.，1981），pigs（Baert et al.，2001），and ostrich（Abu-Basha et al.，2008）. This study evaluates the pharmacokinetic and residual parameters of SDZ and TMP，and discusses their rational ratio in mandarin fish for well treatment of bacterial diseases.

The experiments were performed on 114 healthy mandarin fish（mean body weight：120±10 g），housed in glass tanks（six fish per tank）with a continuous flow of aerated tap water at 28±2℃. The pH，salinity，dissolved oxygen，nitrite and ammoniawere measured at the day before experiment，which was 7.0，0，4.95 mg/L，<0.005 mg/L and <0.01 mg/L respectively. The fish were fasted for 24 h before the start of the experimental treatments and fed approximately 1% of their body weight（bw）a day during the experiment. All of the procedures involving fish complied with local animal ethics regulations.

SDZ（200 mg，99%）and TMP standard（200 mg，99%）were purchased from China Institute of Veterinary Drugs Control，Beijing，P. R. China. SDZ（98%）and TMP（98%）raw powders were purchased from Beisha Pharmaceutical Co.，Foshan，P. R. China. Methanol，formic acidand acetonitrile（chromatography grade）were purchased from Dikma Technologies Inc.

The fish were randomly divided into two groups，and were gastric gavage by stomach tube（self made of fine silicone tube）. It was conveniently insert into the stomach through the throat when the swallow of the fish. Group A consisted of 66 fish and 60 fish were fitted with a stomach tube for a single oral administration of SDZ/TMP at 120 mg/kg bw to determine the pharmacokinetic parameters. Group B consisted of 48 fish and 42 fish were treated the same way；however，they also received a treatment of 60 mg/kg bw for 5 consecutive days to determine the tissue distribution and residue. The rest 6 fish in each group were untreated as blank controls. Samples of six treated fish were collected at the following time points：0.75，1.25，2，4，6，8，16，24，48，and 72 h after the single administration and 0.33，1，2，3，5，7，and 9 d after the multiple-dose administration. No dead fish were found during the experiment.

Control fish were killed and samples were collected before other fish were dosed. Two milliliters of blood were collected in heparinized syringes from the caudal vein of each fish, the fish were then euthanized with a blow to the head. Muscle (back), liver, andkidney samples were subsequently collected. There were no anesthesia been used. The experimental protocol was approved by the Committee on the Ethics of Animals of Pearl River Fisheries Research Institute, CAFS.

Blood samples were centrifuged at 1,000×g for 5 min to obtain clean plasma. The tissues samples were homogenized and stored with the plasma samples at−20℃ until analysis (The stability studies of SDZ and TMP were evaluated before the experiment. The maximum deviations of no more than 12% and 15% for SDZ and TMP, respectivelywere found at 30 d in the homogenates at−20℃. All of the samples were detected within 30 d).

The sample concentrations of SDZ and TMP were determined by a UPLC system and MS−MS detector (Waters Technologies, USA) performed in positive ion mode with multiple−reaction monitoring. The transitions of precursor/product ion pair m/z values were 250.8/155.7 and 250.8/107.7 for SDZ and 290.9/230.1 and 290.9/161.1 for TMP. The column (Acquity BEH C18 column, 50× 2.1 mm, 1.7 μm, Waters Technologies, USA) temperature was 40℃, and the gradient system was used with a mobile phase A (0.1% formic acid aqueous solution) and mobile phase B (acetonitrile) delivered at 0.25 mL/min. The linear gradient program was as follows: initial 90% A, 2 min 80% A, 2.5 min 70% A, and 3 min 90% A. The injection volume was 10 μL.

Approximately 1 mL of plasma or 1 g of homogenized tissue sample was transferred to a 15−mL polypropylene centrifuge tube, mixed with 10 mL ethyl acetate, and whirlpool concussed for 30 s. Then, the samples were sonicated for 10 min and centrifuged at 5, 000×g for 10 min. The supernatants were transferred to new 15−mL tubes. The residues of the tissue samples were re−extracted with 10 mL ethyl acetate. The combined organic phase was then evaporated at 40℃ under nitrogen and reconstituted in 2 mL methyl alcohol+4 mL 1% acetic acid. After being de−fatted via 6 mL n−hexane, the samples were cleaned using a solid phase extraction procedure performed with an Oasis HLB cartridge. Each residue was then redissolved with 1 mL of 50% mobile phase solution, filtered through a 0.22−μm nylon syringe filter, and finally transferred to an autosampler vial for analysis. The lower limit of quantification for SDZ and TMP was 5 ng/mL with a linear range of 0.01−2.5 ng/mL (g). The recoveries for SDZ and TMP at three concentrations (0.01, 0.1, 1.0 ng/mL [g]) were 93.7±6.2% and 95.6±7.5% (means±SD, n=5), respectively. The coefficients of variation (CV%) were all <8% for both intra−assay and inter−assay variability.

The pharmacokinetic parameters for SDZ and TMP in the plasma and tissues were analyzed using Data Analysis System 3.0 (Shanghai University of Traditional Chinese Medicine, China) by compartmental method. The SDZ/TMP residual analysis of the data was performed using the withdrawal time calculation program WT 1.4, which was developed in Germany and adopted by the Committee for Veterinary Medicinal Products of the European Union (Damte et al., 2012).

The mean concentration−time profiles for the SDZ/TMP combination at single−and multiple−dose oral administrations in mandarin fish are shown in Figures 1 & 2, respectively, whereas the pharmacokinetic variables are displayed in Tables 1 & 2.

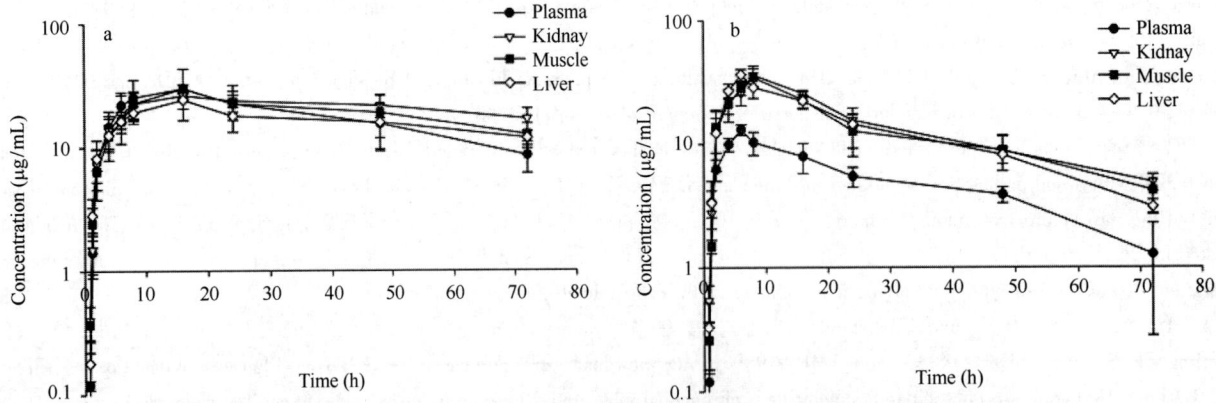

Fig. 1   Sulfadiazine (a) and trimethoprim (b) concentrations (mean±SD, n=6) vs time in mandarin fish (Siniperca chuatsi) at single−dose of 120 mg/kg (SDZ/TMP=5/1) bw oral administration at 28℃

Pharmacokinetic parameters—The plasma and all tissues drug concentration vs. time data aftersingle oral administration were best described by a one−compartment open model. The plasma mean absorption time for trimethoprim following oral administration was significantly faster than for sulfadiazine (2.4±1.0 h and 5.7±1.5 h, respectively). The plasma sulfadiazine concentration reached a peak (C_{max}) of 27.94±4.25 μg/mL at 16.0±3.1 h after a single oral administration. Whereas, plasma trimethoprim reached a peak

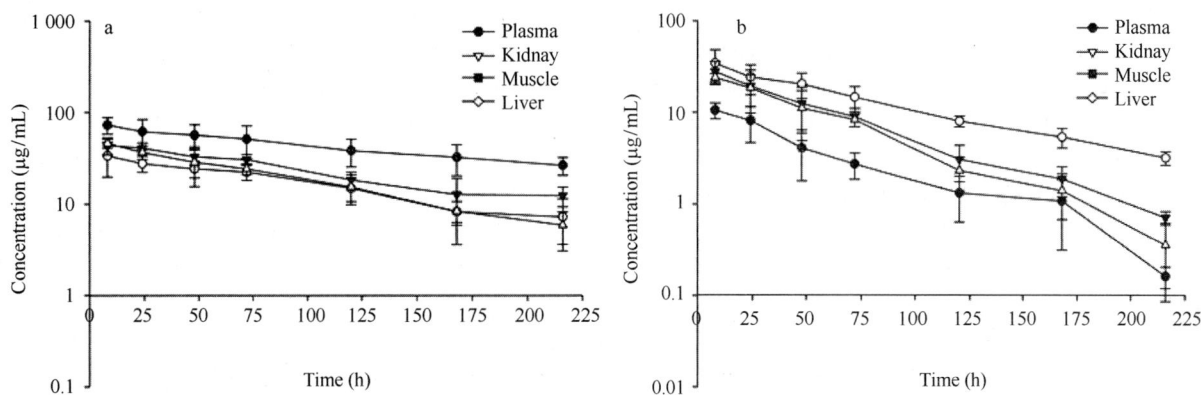

Fig. 2 Sulfadiazine (a) and trimethoprim (b) concentrations (mean±SD, n=6) vs time in mandarin fish (Siniperca chuatsi) at multiple–dose (an initial dose of 120 mg/kg followed by a 5 days consecutive dose of 60 mg/kg bw, SDZ/TMP=5/1) oral administration at 28℃

of 10. 89±2. 15 μg/mL at 7. 9±1. 8 h after a single oral administration. The plasma mean elimination half–life ($t_{1/2\beta}$) for trimethoprim following oral administration was significantly faster than for sulfadiazine (17. 1±3. 4 h and 25. 9±4. 5h, respectively). The plasma area under concentration–time curve (AUC) and volume of distribution (V_d/F) for sulfadiazine was significantly larger than for tri-methoprim (1601. 06±57. 03 μg · h/mLand 2. 34±0. 78 L/kg vs. 369. 37±27. 08 μg · h/mL and 1. 33±0. 57 L/kg, respectively). The plasma clearances (Cl_B/F) of the two drug were almost the same (0. 06±0. 01L · kg/h for sulfadiazine and 0. 05±0. 01L · kg/h for trimethoprim).

Residual kinetics analysis— After multiple–dose oral administration of the fish, the residual kinetic parameters of sulfadiazine and tri-methoprim in the plasma and all tissues were calculated using a non–compartmental model. Sulfadiazine reached a maximum concentra-tion (C_{max}) of 47. 23±4. 05 μg/mL, while its half–life ($t_{1/2\beta}$) was estimated to be 97. 9±12. 3 h in muscle. And trimethoprim reached a maximum concentration of 32. 02±3. 21 μg/mL, while its half–life ($t_{1/2\beta}$) was estimated to be 36. 5±4. 6 h in muscle. The mean residue time (MRT) of sulfadiazine was significant larger than of trimethoprim in muscle (141. 2±16. 5 h and 52. 7±2. 5 h, re-spectively). The muscle clearances (Cl/F) of the two drug were almost the same (0. 05±0. 02L · kg/hfor sulfadiazine and 0. 04± 0. 01L · kg/h for trimethoprim).

As we known, the elimination half–lives of sulfadiazine and trimethoprim in fish tissues have less been previously reported. In the present study it was shown that after 6 days of once daily administration of sulfadiazine/trimethoprim combination, sulfadiazine and tri-methoprim clearly demonstrated various kinetic patterns in fish tissues. The data from this study shown that the $t_{1/2\beta}$ and MRT in multi-ple administration were significant longer comparing to single oral administration, which was evident from the fact that sulfadiazine and trimethoprim reached a steady state after 6 days of treatment.

Establishing a withdrawal time guideline involves evaluations related to clinical veterinary practice, pharmaceutical formulations, equine medication pharmacokinetics, equine forensic science and relevant regulatory policy (Toutain, 2010; Tobin et al. , 2012). Based on the concentrations of SDZ and TMP MRLs (0. 1 and 0. 05 mg/kg) in the edible tissue of these fish (EU, EMEA/MRL/026/95), the withdrawal time (36 d for SDZ/TMP) was calculated in muscle.

Synergistic activity of the two drugs has been observed over a wide range of ratios which were from 1: 1 to 40: 1 and the optimal ratio was 20: 1 in vitro (Prescott et al. , 2000). It seem that the ratio of 5: 1 was the most popular in the livestock and poultry in-cluding aquaculture clinical use, such as ostriches (Abu–Basha et al. , 2008), rainbow trout (Lunden & Bylund, 2002). This may result from the much faster elimination characteristic of trimethoprim and much longer retaining time of sulfadiazine in vivo. TMP concentrations declined more rapidly than their respective values of SDZ for the shorter elimination half–life, regardless of the single– or multiple–dose condition. Therefore, the ratios increased with time after administration for all of the tissues. Similar result were showed in the hybrid striped bass (Morone chrysops × Morone saxitalis, Bakal et al. , 2004), with the $t_{1/2\beta}$ of 25. 9 h and 17. 1 h for sulfadimethoxine and ormetoprim, respectively. Given the unequal bactericidal effect of the two drugs and complicated internal drug metabolism environment, the ratio between SDZ/TMP is not always optimal in vivo, and a bactericidal action is not guaranteed. Therefore, the present pharmacokinetic study did not find the optimal 16: 1 to 20: 1 ratio of SDZ/TMP, nor was it de-tected in Oreochromis niloticus in unpublished work by our group and in the hybrid striped bass (Bakal et al. , 2004). Due to the nonconstant ratio of these drugs in the plasma of the animal, the actual drug ratio to use for determining minimum inhibitory concentration (MIC) is unclear. Bakal using the ratio of the AUC_{SDM}/AUC_{OMP} as a surrogate for the mean actual ratio, which was determined as 2.14：1 (Bakal et al. , 2004). The combination of sulfadimethoxine and ormetoprim appears to provide plasma concentrations high enough to inhibit the growth of Yersinia ruckeri, Edwardsiella tarda, and Escherichia coli. Though the optimal proportion of 20：1 was not acquired. The ratio of AUC_{SDZ}/AUC_{TMP} is 4.34：1 in mandarin fish by our research. While the ratio of SDZ：TMP was >45：1 in ostriches (Abu-Basha et al. , 2008) and 60-85：1 in broiler chickens (Baert et al. , 2003) throughout the treatment period after a single oral administration of SDZ/TMP at 5：1. Considering the different ratio of 2.14：1, 4.34：1 in fish, and >45：1, 60-85：1 in livestock and poultry, while in the same dosage ratio of 5：1, it is possiblly that the fish require a dosage of differs from those administered to mammals. More studies about the pharmacokinetics and pharmacodynamics of SDZ/TMP in fish is important to optimize dosing regimens in them.

Comparisons of the pharmacokinetic parameters in different literatures are extremely problematic. The $t_{1/2\beta}$ is 25.9 h and 17.1 h in plasma for SDZ and TMP in this study, respectively. The values for SDZ in European eels were 104.24 h and 46.21 h in plasma and muscle, respectively (Lin et al. , 2010); the values for sulfamethazine in Scophthalmus maximus were 8.8 h and 27.41 h in plasma and muscle, respectively (Zhang et al. , 2010); the value for sulfamethoxazole in grass carp was 31.94 h in plasma (Ai et al. , 2005); and the value for sulfadimethoxine and OMP in hybrid striped bass was 10.5 h and 3.9 h in plasma (Bakal et al. , 2004). There are significant difference. The $t_{1/2\beta}$ value was similar to that for grass carp, slower than that for striped bass and faster than that for European eels. Because all of the referenced studies refer to different fish species, sizes, water temperatures, experimental protocols and conditions, even different in drugs (SDZ vs SDM, TMP vs OMP), it is difficult to determine the main factor affecting the results. However, it is clear that the $t_{1/2\beta}$ value obtained from the fish in this study was much longer than has been reported for chickens (3.71 h for SDZ and 2.23 h for TMP; Baert et al. , 2003), and pigs (5.49 h for SDZ and 4.19 h for TMP; Baert et al. , 2001) after an oral administration of a SDZ/TMP (5/1). This result is most likely because of the higher Cl_B in chickens (0.09 L · kg/h for SDZ and 0.93 L · kg/h for TMP) and pigs (0.14 L · kg/h for SDZ and 0.54 L · kg/h for TMP).

In conclusion, the pharmacokinetics and elimination profile of SDZ/TMP were detected in Siniperca chuatsi. Significant differences were found with regard to the kinetic parameters of SDZ and TMP in fish compared with other species. The optimal SDZ/TMP ratio of 16：1 to 20：1 was not found in the present experiment after oral administration. Therefore, additional pharmacokinetic/pharmacodynamic studies should be performed to obtain an effective plasma level to extrapolate a suitable therapeutic dose of SDZ/TMP in fish.

Acknowledgment

This work was supported by the Special Fund for Agro-scientific Research in the Public Interest (Grant No. 201203085), Guangdong Province Science and Technology Plan Project (Grant No. 2011B02037001) and Guangdong Province Fish Disease Prevention Special Fund Project 2013 (ecological control of parasites in mandarin fish).

References

Abu-Basha, E. A. , Gehring, R. , Hantash, T. M. , Al-Shunnaq, A. F. & Idkaidek, N. M. (2008) Pharmacokinetics and bioavailability of sulfadiazine and trimethoprim following intravenous, intramuscular and oral administration in ostriches (Struthio camelus). Journal of Veterinary Pharmacology and Therapeutics, 32, 258-263.

Ai X H, Liu C Z, Zhou Y T. (2005). A study on pharmacokinetic of sulphamethoxazole in grass carp at different temperatures and administration regimes, Acta Hydrobiologica Sinica, 2, 210-214.

Baert, K. , De Baere, A. , Croubles, D. , Gasthuys, F. & De Backer, C. (2001) Pharmacokinetics and bioavailability of sulfadiazine and trimethoprim (trimazin 30%) after oral administration in non-fasted young pigs. Journal of Veterinary Pharmacology and Therapeutics, 24, 295-298.

Baert, K. , De Beare, S. , Croubles, S. & Backer, P. (2003) Pharmacokinetics and oral bioavailability of sulfadiazine and trimethoprim in broiler chickens. Veterinary Research Communications, 27, 301-309.

Bakal, R. S. , Bai. S. A. , Stoskopf, M. K. (2004) Pharmacokinetics of sulfadimethoxine and ormetoprim in a 5：1 ratio following intraperitoneal and oral administration, in the hybrid striped bass (Morone chrysops × Morone saxitalis). Journal of Veterinary Pharmacology and Therapeutics, 27, 1-6.

Batzias, G. , Delis, G. & Koutsoviti-Papadopoulou, M. (2005) Bioavailability and pharmacokinetics of sulfadiazine, N4-acetyl sulfadiazine and trimethoprim following intravenous and intramuscular administration of a sulfadiazine/trimethoprim combination in sheep. Veterinary Research Communications, 29, 111-117.

Biswas AK, Rao GS, Kondaiah N, AnjaneyuluA. S. R, MalikJ. K. (2007) Simple multiresidue method for monitoring of trimethoprim and sulfonamide residues in buffalo meat by high performance liquid chromatography. Journal of Agricultural and Food Chemistry, 55：8845-8850.

Bogialli S, Curini R, Corcia A D, Nazzari M, Sergi M. (2003) Confirmatory analysis of sulfonamide antibacterials in bovine liver and kidney：extraction with hot water and liquid chromatography coupled to a single-or triple-quadrupole mass spectrometer. Rapid Communications in Mass Spectrometry, 17 (11): 1146-1156.

Damte, D., Jeong, H. J., Lee, S. J., Cho, B. H., Kim, J. C. & Park, S. C. (2012) Evalution of linear regression statistical approaches for withdrawal time estimation of veterinary drugs. Food and Chemical Toxicology, 50, 773-778.

EU reference laboratory for residues of veterinary drugs. Committee for Veterinary Medicinal Products. No. EMEA/MRL/026/95.

Garwacki, S., Lewicki, J., Wiechetek, M., Grys, S., Rutkowski, J. & Zaremba, M. A. (1996) A study of the pharmacokinetics and tissue residues of an oral trimethoprim/sulfadiazine formulation in healthy pigs. Journal of Veterinary Pharmacology and Therapeutics, 19, 423-430.

Kaartinen, L., Löhönen, K., Wiese, B., Franklin, A. & Pyörälä, S. (1999) Pharmacokinetics of sulphadiazine-trimethoprim in lactating dairy cows. Acta Veterinaria Scandinavica, 40, 271-278.

Lai, H. T., Wang, T. Z. & Chou, C. C. (2011) Implication of light sources and microbial activities on degradation of sulfonamides in water and sediment from a marine shrimp pond. Bioresource Technoloy, 102, 5017-5023.

Lin L C, Fan H P, Liao B C, et al. (2010) Pharmacokinetics of sulfadiazine in European eels (Anguilla anguilla). Journal of inspection and quarantine, 4, 14-18.

Liu, J., Cui, Y. & Liu, J. (1998) Food consumption and growth of two piscivrous fishes, the mandarin fish and the Chinese snake head. Journal of Fish Biology, 53, 1071-1083.

Lunden, T. & Bylund, G. (2002) Effect of sulfadiazine and trimethoprim on the immune response of rainbow trout (Oncorhynchus mykiss). Veterinary Immunology and Immunopathology, 85, 99-108.

Plakas, S. M., Dickey, R. W., Barron, M. G. & Guarino, A. M. (1990) Tissue distribution and renal excretion of ormetoprim after intravascular and oral administration in the channel catfish (Ictalurus punctatus). Canadian Journal of Fisheries and Aquatic Sciences, 47, 766-771.

Plumb, D. C. (2002) Sulfadiazine/Trimethoprim. In Veterinary Drug Handbook. Ed. Plumb, D. C., pp. 762-766. Parma Vet Publishing, White Bear Lake, MN, USA.

Prescott, J. F., Baggott, J. D. & Walker, R. D. (2000) Antimicrobial Therapy in Veterinary Medicine, Sulfonomides, Diaminopyrimidines and Their Combinations, 3rd edn, pp. 290-314. Iowa State University Press, Ames, Iowa.

Riviere, J. E. & Spoo, J. W. (2001) Sulfonamides. In Veterimary Pharmacology and Therapeutics, 8th edn. Ed. Adams, H. R., pp. 796-797. Iowa state Iniversity Press, Ames, IA.

Roiha, I. S., Otterlei, E., Litlabø, A. & Samuelsen, O. B. (2010) Uptake and elimination of florenicol in Atlantic cod (Gadus morhua) larvae delivered orally through bioencapsulation in the brine shrimp Artemia franciscana. Aquaculture, 310, 27-31.

Samuelsen, O. B., Lunestad, B. T., Ervik, A. & Fjeld, S. (1994) Stability of antibacterial agents in an artificial marine aquaculture sediment studied under laboratory conditions. Aquaculture, 126, 283-290.

Samuelsen, O. B., Lunestad, B. T. & Jelmert, A. (1997) Pharmacokinetic and efficacy studies on bath-administering potentiated sulfonamides in Atlantic halibut, Hippoglossus hippoglossus L. Journal of Fish Diseases, 20, 287-296.

Shoaf, S. E., Schwark, S. W. & Guard, C. L. (1989) Pharmacokinetics of trimethoprim/sulfadiazine in neonatal male calves: effect of age and peneteration into cerebrospinal fluid. American Journal of Veterinary Research, 50, 396-403.

Sigel, C. W., Ling, G. V., Bushby, S. R. M., Woolley, J. L., DeAngelis, D. & Eure, S. (1981) Pharmacokinetics of trimethoprim and sulfadiazine in the dog: urine concentrations after oral administration. American Journal of Veterinary Research, 42, 996-1001.

The ministry of agriculture of the People's Republic of China. (2007) The ministry of agriculture announcement No. 958-12-2007. Determination of sulfonamides residues in aquatic products-liquid chromatography (HPLC). Beijing: China agriculture press.

The ministry of agriculture of the People's Republic of China. (2008) The ministry of agriculture announcement No. 1025-23-2008. Determination of sulfonamides residues in edible tissues of animal-liquid chromatography-tandem mass spectrmetry. Beijing: China agriculture press.

Tobin, T., Brewer, K., Stirling, K. H. (2012) World Rules for Equine Drug Testing and Therapeutic Medication Regulation. Wind Publications, Nicholasville, Kentucky.

Toutain, P. L. (2010) How to extrapolate a withdrawal time from an EHSLC published detection time: A Monte Carlo simulation appraisal. Equine Veterinary Journal, 42, 248-254.

Yamaoka, K., Nakagawa, T. & Uno, T. (1978) Application of Akaike's information criterion (AIC) in the evaluation of liner pharmacokinetic equations. Journal of Biochemistry, 100, 609-618.

Van Duijkeren, E., Vulto, A. & Van Miert, A. (1994) Trimethoprim/sulfonamide combinations in the hores: a review. Journal of Veterinary Pharmacology and Therapeutics, 17, 64-73.

Van Duijkeren, E., Ensink, J. M. & Meijer, L. A. (2002) Distribution of orally administraed trimethoprim and sulfadiazine into nonifected subcutaneous tissue chambers in adult ponies. Journal of Veterinary Pharmacology and Therapeutics, 25, 273-277.

Zhang C K, Wang M J, Gong X H, et al. (2010). Studies on pharmacokinetics of sulfamethazine residues in Scophthalmus maximus. Transactions of Oceanology and Limnology, 2, 86-90.

该文发表于《水生生物学报》【2005，29（2）：210-214】

不同水温和给药方式下磺胺甲噁唑在草鱼体内的药动学研究

A study on pharmacokinetic of sulphamethoxazole in grass carp at different temperatures and administration regimes

艾晓辉，刘长征，周运涛

（农业部淡水鱼类种质资源与生物技术重点开放实验室，中国水产科学研究院长江水产研究所，荆州 434000；
中国水产科学研究院淡水渔业研究中心，无锡 214081）

摘要：在 18℃和 28℃两个不同水温条件下，分别采用口灌和腹腔注射的给药方式，给予草鱼 100 mg/kg 体重单剂量的磺胺甲噁唑（SMZ），以 HPLC 法测定草鱼血浆和肌肉中药物浓度，用 MCPKP 药代动力学软件处理药时数据，结果表明，18℃条件下，草鱼口灌 SMZ 的分布半衰期 $T_{1/2\alpha}$、消除半衰期 $T_{1/2\beta}$、达峰时间 T_p 均显著长于 28℃（$P<0.01$），其峰浓度 C_{max} 显著低于 28℃下的值（$P<0.01$）；腹腔注射给药时，SMZ 在鱼体血浆和肌肉中的吸收与分布较口服快，消除较口服也快。在水温 18℃条件下，口灌 SMZ 在草鱼血浆中的药时数据符合带时滞的二室开放模型，腹腔注射的符合无时滞二室开放模型；水温 28℃条件下，口灌 SMZ 在草鱼血浆中的药时数据符合可忽略时滞的二室开放模型。根据 SMZ 的药代动力学规律、最低有效血药浓度和抗菌药物应用的一般原则，制定了 SMZ 的用药方案：治疗温水性鱼类细菌性疾病，在水温 18℃左右，口服或注射 100 mg/kg 剂量的 SMZ，以每天给药两次，3d 一个疗程为宜；在水温 28℃左右，以每天给药 3~4 次为宜。该项研究全面了解了在不同水温和不同给药方式下 SMZ 在鱼体内的药动学规律，为确定合理的临床用药方案以及无公害水产品中药物残留监测提供了可靠的理论依据。

关键词：磺胺甲噁唑；草鱼；药代动力学；水温；给药方式

　　磺胺类药物是水产养殖生产上广泛应用的抗微生物药，对鱼类多种革兰氏阴性细菌和一些革兰氏阳性细菌有较显著的抑菌效果，同时对某些真菌、衣原体等亦有抑制作用，是一种广谱的抑菌药物[1]。

　　近年来的研究资料表明，磺胺类药物在人体及不同动物体内的代谢情况存在明显的种属差异性[2-3]，同一种药物在同一种水生动物体内，即使性别不同，其代谢规律也存在差异性[4]。因此，只根据某一种实验动物或人的实验数据，为磺胺药在水产上的临床应用制订给药方案，很不合理。

　　磺胺甲噁唑（SMZ）在草鱼（*Tenopharyngodon idellus*）体内的药代动力学研究尚未见报道，本文就磺胺甲噁唑在不同水温和不同给药方式的条件下，在草鱼体内的药代动力学进行了研究，为制订磺胺甲噁唑在水产临床应用上的合理给药方案提供理论依据。

1　材料和方法

　　1.1　药品与试剂　磺胺甲噁唑（SMZ）粉，含量 99.7%，由昆山双鹤药业有限责任公司提供，批号 020709。二氯甲烷、无水醋酸钠为分析纯，85% 的磷酸为优级纯，配成 0.017 mol/L 磷酸缓冲液及 1 mol/L 的醋酸钠溶液备用；乙腈为色谱纯；蒸馏水为三蒸水。

　　1.2　仪器与色谱条件高效液相色谱仪（waters 600 型，由 717 自动进样器、515 高压泵，2487 双通道紫外检测器及 MILLEMNIUM^{32} 组成）；Wh^{-1} 微型漩涡混合仪；TDL-16B 高速离心机；电热恒温水浴锅；SCQ-250 超声波清洗器；IKA ALLbasic 组织捣碎机；0.45 μm 过滤器。ODS-C$_{18}$ 柱，150 mm×3.9 mm（id）；粒度 5 μm；流动相：乙晴：0.017 mol/L 磷酸缓冲液（40：60/V：V）；流速：1.0 mL/min；检测波长：287 nm；柱温：室温；进样量：20 μL。

　　1.3　试验鱼饲养及试验分组　草鱼，体重（76.14±5.27）g，健康无伤，由长江水产研究所鱼类育种试验场提供，试验前在 100 cm×68 cm×80 cm 的两个室内小水泥池中暂养 5~7 d，不投食，连续充氧，水温一个池控制在 18℃，另一池控制在 28℃，pH 值为 6.8~7.0。试验共分 3 组，每组鱼数 80 尾。（1）水温 18℃单次口腔灌药组；（2）水温 28℃单次口腔灌药组；（3）水温 18℃单次腹腔注射组。

　　1.4　给药方法　（1）口腔灌药组　称取 7.614 g SMZ 粉于 1 000 mL 烧杯中，加水约 800 mL，同时加入少量淀粉加热搅拌均匀至乳浊状，冷却后定容至 1 000 mL，每尾鱼灌 1 mL，即灌药剂量为 100 mg/kg 鱼体重，灌药采用文献[5]的方法进行。(2) 注射给药组用 0.2 mol/L 氢氧化钠溶液溶解 SMZ，使药液浓度为 0.761 4 g/100 mL 水，过滤、灭菌，进行鱼体安全试验后，每尾鱼腹腔注射该药液 1 mL，即注射剂量为 100 mg/kg 鱼体重，其余操作同单次灌药。

1.5 采样 每组分 16 个时间段采样，每次随机取草鱼 5 尾，从每尾鱼的尾动脉抽取血液 1.0 mL，分别置预先涂有肝素钠的塑料离心管中，做 5 个平行样，4 000 r/min 离心 5 min，分离出血浆；随后将取过血样的草鱼剖杀，迅速取出背脊两侧肌肉，分别装于白色塑料袋中，按取样时间顺序编号，保存于−20℃冰箱中。单次给药的采样时间为给药前取一次空白血样及肌肉样，给药后参照文献[6]的方法进行，分别于 0.25、0.5、1、2、3、4、6、8、12、24、48、72、96、120、144、168 h 采血样及组织样。

1.6 药物测定 鱼血浆及肌肉中 SMZ 的测定参照文献[3]的方法进行。该法的标准曲线的线性范围是在 0.02~10.0 μg/mL（$r=1.00$），回收率稳定在 75%~85%，日内精密度与日间精密度均小于 5%。

1.7 数据处理 采用 MCPKP 药物代谢动力学参数计算程序[7]，在微机上处理各给药组血浆、肌肉的血药浓度−时间数据，算出有关药动学参数。

2 结果

2.1 在不同水温条件和不同给药方式下，SMZ 在草鱼血浆中的药动学特征

以 100 mg/kg 鱼体重的单剂量给草鱼口灌 SMZ 后，在水温 18℃ 条件下，药物在（3.09±0.13）h 达到峰浓度，峰浓度均值为（152.67±3.21）μg/mL。在水温 28℃ 条件下，药物在（1.93±0.08）h 达到峰浓度，峰浓度均值为（414.05±7.31）μg/mL。SMZ 在两种不同水温条件下的血浆药动学最佳数学模型皆为带时滞的二室开放模型，28℃ 条件下的时滞可忽略不计。

以 100 mg/kg 鱼体重的单剂量给草鱼腹腔注射 SMZ 后，在水温 18℃ 条件下，药物在（0.888±0.251）h 达到峰浓度，峰浓度均值为（195.28±6.53）μg/mL。其血浆药动学最佳数学模型符合二室开放模型。

在 18℃ 和 28℃ 两种不同水温条件下给草鱼口灌 SMZ 以及在水温 18℃ 条件下给草鱼腹腔注射 SMZ，SMZ 在草鱼血浆中的主要药动学参数见表 1，给药后的时量曲线见图 1。

表 1 腹腔注射和口服 SMZ 在草鱼血浆中的主要药动学参数（$n=5$）

Tab. 1 Plasma pharmacokinetic parameters of SMZ after intraperitoneal and oral administration in grass carps（$n=5$）

参数		18℃（口灌）(Oral administration)	28℃（口灌）(Oral administration)	18℃（注射）(Intraperitoneal)
名称 Parameter	单位 Unit	参数值（X±SD）Value	参数值（X±SD）Value	参数值（X±SD）Value
D	mg/kg	100	100	100
C_o	μg/mL	225.34±0.19	978.10±0.03	351.37±5.89
A	μg/mL	187.41±0.30	4669.78±3.24	608.64±3.24
B	μg/mL	97.45±0.72	80.83±0.62	124.97±1.35
α	/h	0.203±0.013	0.488±0.051	1.16±0.05
β	/h	0.052±0.045	0.085±0.064	0.071±0.012
K_A	/h	0.724±0.131	0.605±0.028	1.87±0.20
K_{12}	/h	0.049±0.046	0.095±0.026	0.604±0.141
K_{21}	/h	0.112±0.112	0.113±0.028	0.444±0.12
K_{el}	/h	0.094±0.015	0.364±0.024	0.184±0.013
LAGTIME	h	0.210±0.042	0.0002±0.0007	0.00
$T_{1/2\alpha}$	h	3.41±0.15	1.14±0.13	0.596±0.117
$T_{1/2\beta}$	h	13.31±0.61	8.19±0.47	9.85±0.64
AUC	mg/L·h	2401.3±12.7	2684.7±11.5	1908±14.6
T_p	h	3.09±0.13	1.93±0.08	0.888±0.251

续表

参数 Parameter		18℃（口灌） （Oral administration）	28℃（口灌） （Oral administration）	18℃（注射） （Intraperitoneal）
名称 Parameter	单位 Unit	参数值（X±SD） Value	参数值（X±SD） Value	参数值（X±SD） Value
C_{max}	μg/mL	152.67±3.21	414.05±7.31	195.28±6.53
TCP	h	12.82±0.43	5.04±0.25	13.02±0.13

注：D 为给药剂量，C_0 为初始药物浓度，A 为分布相的零时截距，B 为消除相的零时截距，α 为分布相消除速率常数，β 为消除相消除速率常数，$T_{1/2\alpha}$ 为药物的分布相半衰期，$T_{1/2\beta}$ 为药物的消除相半衰期，K_{10} 为药物自中央室的消除速率常数，K_{12} 为药物自中央室到周边室的一级转运速率常数，K_{21} 为药物自周边室中央室到中央室的一级转运速率常数，AUC 为血药浓度-时间曲线下面积，T_p 为达峰时间，C_{max} 为达峰浓度，TCP 为有效血药浓度维持时间。

2.2 在不同水温条件和不同给药方式下，SMZ 在草鱼肌肉中的药动学特征

以 100 mg/kg 鱼体重的单剂量给草鱼口灌 SMZ 后，在水温 18℃条件下，草鱼肌肉中药物在（4.01±0.22）h 达到峰浓度，峰浓度均值为（57.22±2.27）μg/g。在水温 28℃条件下，药物在（2.06±0.12）h 达到峰浓度，峰浓度均值为（102.37±3.11）μg/g。在水温 18℃条件下，SMZ 在草鱼肌肉中药动学最佳数学模型皆为带时滞的二室开放模型，在水温 28℃条件下，SMZ 在草鱼肌肉中药动学最佳数学模型皆为无时滞的二室开放模型。

以 100 mg/kg 鱼体重的单剂量给草鱼腹腔注射 SMZ 后，在水温 18℃条件下，草鱼肌肉中药物在 2.37±0.36 h 达到峰浓度，峰浓度均值为 116.37±5.74 μg/g。其药动学最佳数学模型符合一室开放模型。

在 18℃ 和 28℃ 两种不同水温条件下给草鱼口灌 SMZ 以及在水温 18℃ 条件下给草鱼腹腔注射 SMZ，SMZ 在草鱼肌肉中的主要药动学参数见表 2，给药后的时量曲线见图 1。

由图 1 可见，在同等的温度条件下，以 100 mg/kg 鱼体重的单剂量给草鱼无论是口灌还是腹腔注射 SMZ 后，SMZ 在草鱼肌肉中的出峰时间（T_p）皆比血浆中长，而峰浓度则皆比血浆中低。SMZ 口灌给药后在草鱼肌肉中的消除速度比血浆中慢，而腹腔注射给药后，SMZ 在草鱼肌肉中的消除速度与血浆中差异不明显。

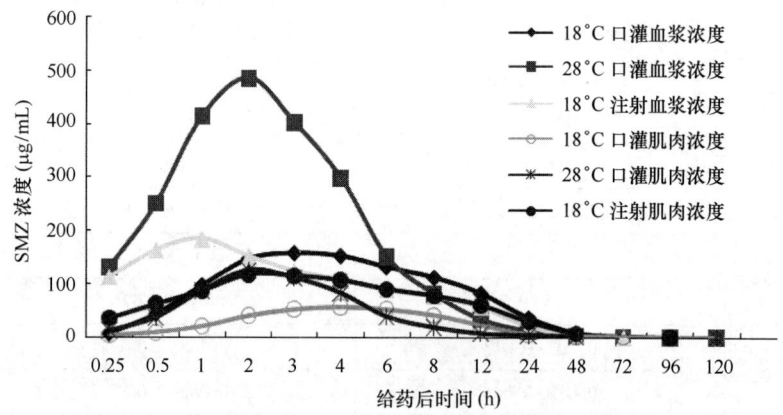

图 1　不同温度和给药方式下 SMZ 在草鱼血浆和肌肉中的浓度变化

Fig. 1　Variation of plasma and muscle SMZ concentration in grass carp after different administration at different temperatures

表 2　腹腔注射和口服 SMZ 在草鱼肌肉中的主要药动学参数（$n=5$）

Tab. 2　Muscles pharmacokinetic parameters of SMZ after intraperitoneal and oral administration of grass carp（$n=5$）

参数 Parameter		18℃（口灌） （Oral administration）	28℃（口灌） （Oral administration）	18℃（注射） （Intraperitoneal）
名称 Parameter	单位 Unit	参数值（X±SD） Value	参数值（X±SD） Value	参数值（X±SD） Value
D	mg/kg	100	100	100

参数		18℃（口灌） （Oral administration）	28℃（口灌） （Oral administration）	18℃（注射） （Intraperitoneal）
名称 Parameter	单位 Unit	参数值（X±SD） Value	参数值（X±SD） Value	参数值（X±SD） Value
C_0	μg/mL	151.54±4.46	258.13±8.13	137.76±11.48
A	μg/mL	8168.78±12.14	2173.33±16.14	—
B	μg/mL	4.24±0.34	6.21±0.31	—
α	/h	0.263±0.021	0.465±0.071	—
β	/h	0.014±0.009	0.022±0.006	—
K_A	/h	0.267±0.059	0.526±0.052	—
K_{12}	/h	0.076±0.058	0.139±0.079	—
K_{21}	/h	0.021±0.037	0.032±0.037	—
K_{el}	/h	0.179±0.056	0.316±0.077	—
LAGTIME	h	0.143±0.093	0.00	0.00
$T_{1/2\alpha}$（$T_{1/2ka}$）	h	2.64±0.33	1.49±0.11	0.535±0.421
$T_{1/2\beta}$（$T_{1/2k}$）	h	48.59±1.14	31.94±0.65	9.73±0.58
AUC	mg/L·h	843.66±6.9	816.27±4.82	1934.5±17.9
T_p	h	4.01±0.22	2.06±0.12	2.37±0.36
C_{max}	μg/mL	57.22±2.27	102.37±3.11	116.37±5.74

3　讨论

SMZ 在草鱼体内的药动学实验结果表明，在水温 18℃条件下，SMZ 在鱼体血浆和肌肉中的分布较慢，而消除也慢；在水温 28℃条件下，SMZ 在鱼体血浆和肌肉中的分布较快，而消除也相对较前者快。低温 18℃条件下，草鱼口服 SMZ 的分布半衰期 $T_{1/2\alpha}$、消除半衰期 $T_{1/2\beta}$、达峰时间 T_p 均显著长于 28℃高温（$P<0.01$），其峰浓度 C_{max} 显著低于 28℃高温（$P<0.01$）。该结论与 Herman and Degurse（1967）[8] 报道的在 7.7℃与 14℃两个不同水温条件下研究磺胺甲基嘧啶在虹鳟体内的代谢规律以及 Lee and Kou（1978）[9] 在 7℃与 14℃两个不同水温条件下研究 4-磺胺-6-甲氧嘧啶（SMM）在鳗鱼体内的代谢规律结论都是一致的。据 BjÖrklund 等（1990）[10] 报道，5℃、10℃、16℃下土霉素在虹鳟血清中达到峰值的时间分别为 24 h、12 h 和 1 h，消除时间高水温与低水温相差 108 d 之久。BjÖrklund 等（1990）[10] 认为，通常水温每升高 1℃，鱼类的代谢和消除速度将提高 10%。以上研究均表明，水环境的变化对水生动物药动学的规律影响很大。一般来说，在一定温度范围内，药物的代谢强度与水温成正比，水温越高，代谢速度加快。因此给药时水温是一个应予以考虑的重要因素，尤其是水温较低时停药期应予适当延长。

在不同的给药方式同等水温条件下，SMZ 在草鱼体内进行的药动学研究结果表明，腹腔注射给药时，SMZ 在鱼体血浆和肌肉中的吸收与分布较口服快，消除较口服也快，血浆中 SMZ 的分布半衰期 $T_{1/2\alpha}$ 为（0.596±0.117）h，消除半衰期 $T_{1/2\beta}$ 为（9.85±0.64）h，肌肉中 SMZ 的分布半衰期 $T_{1/2\alpha}$ 为（0.535±0.421）h，消除半衰期 $T_{1/2\beta}$ 为（9.73±0.58）h。注射给药吸收和消除速度比口灌给药快，这与张雅斌和张祚新（2000）[11] 报道的在不同给药方式下诺氟沙星在鲤体内的代谢规律是一致的。

SMZ 的消除相半衰期在不同的对象体内也有差异，其消除半衰期在人体内为 11~12 h，泰和鸡为 5.369 h，绿头鸭为 5.7 h，兔体内为 2.89 h[2]，在草鱼体内为 8.19~13.31 h（水温 18~28℃），比较而言，SMZ 在鱼体内为长效磺胺。

磺胺类药具有广谱抗菌作用，按磺胺类药在血中最低有效血药浓度为 50 μg/mL 计算，水温 18℃口灌给药、水温 28℃口灌给药及水温 18℃腹腔注射给药的有效血药浓度维持时间（TCP）分别为（12.82±0.43）h、（5.04±0.25）h、（13.02±0.13）h，再根据不同条件下 SMZ 的消除半衰期和抗菌药物应用的一般原则，治疗温水性鱼类细菌性疾病，在水温 18℃左右，口服或注射 100 mg/kg 剂量的 SMZ，以每天给药两次，3 d 一个疗程为宜；在水温 28℃左右，以每天给药 3~4 次为宜，考虑到水产养殖上投饲习惯一般是一天两次，投药时间为了与投饲习惯吻合，可将用药剂量增至 200 mg/kg。因而养殖者应根据实际情况调整用药剂量和次数。

　　关于 SMZ 的休药期，我们已对 SMZ 多次给药及与甲氧苄啶（TMP）混合给药的代谢动力学进行了研究，将结合本文研究结果另文报道。

参考文献

[1] Editorial board of "A handbook of medicines for aquatic animals" of Ministry of Agriculture. A Handbook of Medicines for Aquatic Animals [M]. Beijing: China Science and Technology Publishing House, 1998, 195~196. [农业部 "渔药手册" 编撰委员会. 渔药手册 [M]. 北京：中国科技出版社，1998：195~196]

[2] Ge Y H, Zhou J L. Studies on pharmacokinetics of SMZ and SMZ-TMP in domestic rabbits [J]. Journal of Hebei Agricultural University, 1989, 12 (1)：123~125. [葛印怀，周建玲. 磷胺甲基异噁唑（SMZ）与复方磷胺甲基异噁唑（SMZ-TMP）在家兔体内代谢动力学的研究 [J]. 河北农业大学学报，1989，12 (1)：123~125.]

[3] Liu C Z, Ai X H. A method for the detection of sulphamethoxazole and trimethoprim content in aquatic products with high performance liquid chromatography [J]. Freshwater Fisheries, 2004, 34 (2)：23~24. [刘长征，艾晓辉. 水产品中磷胺甲噁唑与甲氧苄啶含量的高效液相色谱测定法 [J]. 淡水渔业，2004，34 (2)：23~24.]

[4] Yang X L, Liu Z Z, Masahito YOKOYAMA. Pharmacokinetics of ciprofloxacinum in chinese mitten-handed crab, Eriocheir sinensis [J]. Acta Hydrobiologica Sinica, 2003, 27 (1)：18~22. [杨先乐，刘至治，横山雅仁. 盐酸环丙沙星在中华绒螯蟹体内药物代谢动力学研究 [J]. 水生生物学报，2003，27 (1)：18~22.]

[5] Ai X H, Chen Z W, Zhang C W, et al. Studies on pharmacokinetics and tissue concentration of olaquindox in carps [J]. Acta Hydrobiologica Sinica, 2003, 27 (3)：56~60. [艾晓辉，陈正望，张春光等. 喹乙醇在鲤体内的药物代谢动力学及组织浓度研究 [J]. 水生生物学报，2003，27 (3)：56~60.]

[6] Fang W H, Shao J H, Shi Z H. Analytical method of norfloxacin in the giant tiger shrimp (Penaeus monodon) hemolymph and brief study on pharmacokinetics [J]. Acta Hydrobiologica Sinica, 2003, 27 (1)：13~17. [房文红，邵锦华，施兆鸿等. 斑节对虾血淋巴中诺氟沙星含量测定及药代动力学 [J]. 水生生物学报，2003，27 (1)：13~17.]

[7] Xia W J, Cheng Z R. MCPKP-Amicrocomputer programme for kinetic analysis of drugs [J]. Journal of Pharmacology of China, 1988, 9：(2) 188~192. [夏文江，成章瑞. MCPKP—药物动力学分析的一种微机程序 [J]. 中国药理学报，1988，9 (2)：188~192.]

[8] Herman, R. L. and Degurse, P. E. Sulfamerazine residues in trout tissues [J]. Ichthyologica, 1967, 39：73~9.

[9] Lee. M. J. and Kou. G. H. Absorption of sulfamonomethoxine by eels in medicated bath. JCRR [J] Fish. Series, 1978, 34：69~75.

[10] Björklund, H and Bylund, G. Temperature-related absorption and excretion of oxytetracycline in rainbow trout [J]. Aquaculture, 1990, (84)：363~372

[11] Zhang Y B, Zhang Z X, Zheng W, et al. Study on pharmacokinetics of norfloxacin in carp following different forms administration [J]. Journal of Fisheries of China, 2000, 24 (6)：559~563. [张雅斌，张祚新，郑伟等. 不同给药方式鲤鱼对诺氟沙星的药代动力学研究 [J]. 水产学报，2000，24 (6)：559~563.]

该文发表于《上海水产大学学报》【2006，15 (4)：448-455】

磷胺甲基异噁唑在中华绒螯蟹体内的代谢和消除规律

Metabolism and elimination of sulfamethoxazole in tissues of Chinese mitten-handed crab（*Eriocheirsinensis*）

唐俊[1]，郑宗林[1,2]，杨先乐[1]，胡鲲[1]，喻文娟[1]

（1. 上海高校水产养殖 E-研究院，农业部水产种质资源与养殖生态重点开放实验室，
农业部渔业动植物病原库，上海水产大学，上海 200090；2. 西南大学水产学院，重庆，荣昌 402460）

摘要：在水温 25℃，以 100 mg/kg 蟹体重的磷胺甲基异噁唑（SMZ）对中华绒螯蟹单次口灌给药，采用 RPHPLC 方法研究了 SMZ 在中华绒螯蟹体内的代谢和消除规律。给药后 0.5~3 h，SMZ 在血淋巴和肝胰腺中的浓度迅速上升，至第 3 小时达到峰值（15.466±1.499）μg/mL 和（13.491±1.315）μg/g；而肌肉和性腺（卵巢和精巢）中 SMZ 却上升较慢，第 3 小时仅为（5.955±0.354）μg/g、（6.950±0.240）μg/g、（7.015±0.356）μg/g，第 6 小时才达到峰值（6.232±0.325）μg/g、（7.551±0.255）μg/g、（8.055±0.274）μg/g，峰值仅为血淋巴和肝胰腺的 1/2 左右。SMZ 在中华绒螯蟹性腺和肝胰腺中消除速度均较慢，但肌肉比肝胰腺稍快。给药后第 10 天，肌肉和卵巢中 SMZ 降至为（0.051±0.014）μg/g 和（0.099±0.003）μg/g，而肝脏和精巢中 SMZ 尚在 0.1 μg/g 以上，为（0.483±0.042）μg/g 和（0.123±0.006）μg/g。给药后第 20 天，仅肝胰腺中可检测到 SMZ，为 0.090 μg/g 左右。

肌肉、肝胰腺和性腺（卵巢和精巢）SMZ 浓度-时间消除曲线方程分别为 $C_0 = 18.537e^{-0.224t} + 4.775e^{-0.018t}$，$C_0 = 9.823e^{-0.021t} + 1.898e^{-0.006t}$，$C_0 = 4.405e^{-0.039t} + 3.894e^{-0.017t}$，$C_0 = 5.707e^{-0.033t} + 2.478e^{-0.013t}$，各组织中 $T_{1/2\beta}$ 分别为 1.6 d、4.74 d、1.7 d 和 2.2 d。若要使 SMZ 在中华绒螯蟹肌肉、肝胰腺和性腺中的浓度降至 0.1 μg/g 以下，则休药期分别需 13.473 d、24.61 d、10.33 d 和 14.52 d。试验表明，SMZ 在中华绒螯蟹组织中消除较缓慢，尤其在肝胰腺中，因此肝胰腺可以作为 SMZ 残留监测的首选组织。

关键词：磺胺甲基异噁唑；中华绒螯蟹；代谢；消除

该文发表于《淡水渔业》【2005，35（5）：8-11】

中华绒螯蟹血淋巴内磺胺甲基异噁唑的测定及药代动力学研究

Determination and pharmacokinetic of sulfamethoxazolein the haemolymphofmitten-handed crab （*Eriocheir sinensis*）

郑宗林[1]，叶金明[1,2]，杨先乐[1]，唐俊[1]，胡鲲[1]，喻文娟[1]

（1. 上海高校水产养殖 E-研究院，上海水产大学农业部水产种质资源与养殖生态重点开放实验室，农业部渔业动植物病原库，上海 200090；2. 江苏省扬州市水产生产技术指导站 225003）

摘要：以反相高效液相色谱法（RP-HPLC）测定中华绒螯蟹血淋巴内磺胺甲基异噁唑含量。血淋巴从心区抽取，经甲醇去蛋白，3 000 r/min 离心 10 min，取上清液进行 HPLC 检测。平均回收率，平均日间、日内精密度分别为 97.5%、1.20% 和 2.1%。本方法测定磺胺甲基异噁唑的最低检测限为 0.02 μg/mL。研究了中华绒螯蟹单次口灌磺胺甲基异噁唑（100 mg/kg 蟹体重）后的药代动力学。中华绒螯蟹血淋巴中药-时曲线可用开放式二室模型拟合，其主要药动学参数中分布半衰期（$T_{1/2\alpha}$）、消除半衰期（$T_{1/2\beta}$）分别为 20.190 h 和 79.586 h；曲线下面积（AUC）、总体清除率（CLs）和表观分布容积（V_d）分别为 431.268 mg·h/L、0.277 L/（kg·h）和 26.082 L/kg。本法适合于中华绒螯蟹血淋巴中磺胺甲基异噁唑含量分析。

关键词：中华绒螯蟹；反相高效液相色谱（RP-H PLC）；磺胺甲基异噁唑；药代动力学

该文发表于《华中农业大学学报》【2015：34（1）：103-107】

磺胺间甲氧嘧啶在罗非鱼体内的药物代谢动力学及休药期

Pharmacokinetics and withdrawal period of sulfamnomethoxine in tilapia

鞠晶[1,2]，王伟利[1]，姜兰[1]，肖贺[1,2]，罗理[1]，邓玉婷[1]，谭爱萍[1]

（1. 中国水产科学研究院珠江水产研究所/农业部渔药创制重点实验室/广东省免疫技术重点实验室，广州 510380；2. 上海海洋大学水产与生命学院，上海 201306）

摘要：28±2℃ 温度条件下，按 200 mg/kg 的剂量对罗非鱼单次药饵饲喂磺胺间甲氧嘧啶（SMM）后，采用 HPLC 法测定各组织中的药物浓度，研究 SMM 在罗非鱼体内的药代动力学及消除规律。结果显示，药物在各组织中（血液、肌肉、肝脏和肾脏）的最大峰浓度 C_{max} 分别为 22.66、7.13、45.50、22.77 μg/mL（μg/g）；达峰时间 T_{max} 分别为 7.52、7.02、1.00/8.00 和 2.00/10.00 h（T_{max1}/T_{max2}）；肝、肾组织中的药-时曲线有明显的双峰现象，并且第一个峰浓度高于第二个峰浓度，提示该药物在罗非鱼胃肠道中具有非齐性吸收现象；药物在各组织中的消除半衰期 $T_{1/2Ke}$ 分别为 5.21、4.84、14.12、6.80 h，显示 SMM 在罗非鱼体内代谢较快，属于较为短效的磺胺类药物；并且在肌肉和血液中的消除快于肝脏、肾脏。在相同的实验条件下，按 200 mg/kg 的剂量对罗非鱼连续 5 d 药饵饲喂 SMM，研究药物的消除规律；根据罗非鱼可食性组织肌肉的残留检测结果，参考中华人民共和国第 235 号公告中对动物源性食品中磺胺类药物总量的 MRL 规定，以 0.1 mg/kg 为残留限量，建议休药期不低于 5 d。

关键词：罗非鱼；磺胺间甲氧嘧啶；药代动力学；休药期；双峰现象；消除规律

该文发表于《华南农业大学学报》【2014，35（6）：13-18】

不同温度下复方磺胺嘧啶在罗非鱼体内的药代动力学比较

Pharmacokinetic of compound sulfadiazine in *Oreochromis niloticus* at different temperature

肖贺[1,2]，王伟利[1]，姜兰[1]，鞠晶[1,2]，罗理[1]，邓玉婷[1]，谭爱萍[1]

(1. 中国水产科学研究院 珠江水产研究所/农业部渔用药物创制重点实验室/广东省免疫技术重点实验室，广东 广州 510380；

2. 上海海洋大学，上海，201306)

摘要：【目的】探讨不同温度条件下复方磺胺嘧啶在罗非鱼血液中的药代动力学（简称药动学）特点及其变化。【方法】以罗非鱼为研究对象，在不同水温（18、23、28 和 33℃）饲养条件下按复方磺胺嘧啶（磺胺嘧啶：甲氧苄啶=5：1 {g/g}）120 mg/kg 的剂量单次饲喂给药，分别于给药后 0.5、1、2、4、6、8、10、24、48、72 h 采集血液样品，使用 HPLC 方法检测罗非鱼血浆中的药物浓度，研究复方磺胺嘧啶在罗非鱼血液中的吸收和消除变化规律。【结果与讨论】18、23、28 和 33℃时，磺胺嘧啶在血浆中的峰浓度分别为 12.40、19.60、22.48 和 30.75 μg/mL，甲氧苄啶在血浆中的峰浓度分别为 1.22、2.06、2.44 和 2.70 μg/mL，2 个药物的峰浓度均随着温度的升高而增大。磺胺嘧啶在血浆中的消除半衰期（$T_{1/2\,ke}$）分别为 18.22、17.89、16.90 和 12.99 h，甲氧苄啶在血浆中的消除半衰期（$T_{1/2\,ke}$）分别为 16.39、7.08、5.99 和 4.04 h，药物的消除随温度的升高而加快；各温度下药物在罗非鱼血液中的药动学均为 1 级动力学过程。给药后 10 h 内血浆中磺胺嘧啶和甲氧苄啶的比例分别为 9.57：1~11.01：1、6.30：1~8.40：1、5.4：1~9.15：1 和 4.2：1~20.6：1，均维持在 1：1~40：1 的理想抑菌配比范围内。研究结果表明温度对药物在罗非鱼体内的吸收及消除影响显著，可提高药物的最大血药浓度与消除速率，但对复方磺胺嘧啶在血浆中的比值影响不显著。

关键词：温度；血药浓度；复方磺胺嘧啶；药代动力学；罗非鱼

该文发表于《大连海洋大学学报》【2014，29（6）：618-623】

复方磺胺甲噁唑在松浦镜鲤体内药动学及残留研究

The Pharmacokinetics and Residues Study of Compound Sulfamethoxazole in *Cynipus carpio* L.

韩冰[1,2]，王荻[1]，卢彤岩[1]

(1. 中国水科学研究院 黑龙江水产研究所，黑龙江 150070；2. 上海海洋大学 水产与生命学院，上海 201306)

摘要：采用高效液相色谱法，研究了以 50 mg/kg 剂量在口灌给药条件下，复方磺胺甲噁唑（甲氧苄啶：磺胺甲噁唑=1：5）在松浦镜鲤 *Cyprinus carpio* Songpu 体内的药动学与残留消除规律。结果表明：单次给药后，甲氧苄啶在血浆和肌肉中药时关系符合一级吸收二室开放模型，在肝和肾中符合一级吸收一室开放模型，磺胺甲噁唑在血浆、肌肉、肝和肾中符合一级吸收二室开放模型；连续 5 d，每天一次给药后，甲氧苄啶于第 15 天开始低于 50 μg/kg，磺胺甲噁唑于第 9 天开始低于 100 μg/kg。在此实验条件下，建议休药期不低于 15 天。

关键词：复方磺胺甲噁唑；松浦镜鲤；药动学；残留

该文发表于《广东海洋大学学报》【2008，28（4）：54-58】

复方新诺明在罗氏沼虾中的药代动力学和组织分布

Pharmacokinetics，Tissue Distribution of Sulfamethoxazole and Trimethoprim in *Macrobranchium rosenbergii*

郑宗林[1,2]，刘鸿艳[1]，黄辉[1]，杨先乐[3]，周兴华[1]

（1. 西南大学荣昌校区水产系，重庆 402460；2. 四川农业大学动物医学院，四川 雅安 625014；
3. 农业部渔业动植物病原库，上海水产大学，上海 200090）

摘要： 研究了单剂量肌肉注射给药［甲氧苄氨嘧啶（SMZ）0.60 mg/kg］、口灌给药［磺胺甲基异噁唑（SMZ）100 mg/kg］和复方口灌给药（SMZ 100 mg/kg 和 TMP20 mg/kg）三种给药方式下，SMZ 和 TMP 在罗氏沼虾中的代谢动力学，分析了 SMZ 和 TMP 血淋巴、肌肉和肝胰腺中的分布特征。结果表明，SMZ 以肌肉注射方式给药后的血淋巴中，药物经时浓度曲线符合二室开放房室模型，而 SMZ&TMP 口灌给药后却不能用房室模型拟合。肌注给药后，SMZ 的消除半衰期（$T_{1/2\beta}$）为 24.92 h，单方和复方口灌给药后 SMZ 的消除半衰期（$T_{1/2\beta}$）分别为 40.70 和 47.197 h。复方口灌给药后 TMP 的消除半衰期（$T_{1/2\beta}$）为 28.77 h，TMP 对 SMZ 的药代动力学并无显著影响。口灌给药后，SMZ 和 TMP 在肝胰腺中的药物浓度高于同时间肌肉中的药物浓度，但肌肉和肝胰腺中的 SMZ 消除速度基本一致。

关键词： 罗氏沼虾；药代动力学；磺胺甲基异噁唑（SMZ）；甲氧苄氨嘧啶（TMP）；组织分布

该文发表于《中国渔业质量与标准》【2016：6（1）：29-35】

单次和连续药饵投喂方式下复方磺胺嘧啶在吉富罗非鱼体内的代谢消除规律

Pharmacokinetic and residues of compound sulfadiazine in GIFT (*Oreochromis niloticus*) at single and multiple dosage

王伟利[1]，肖贺[1,2]，姜兰[1]，罗理[1]，谭爱萍[1]，邓玉婷[1]，张瑞泉[1]

（1 中国水产科学研究院珠江水产研究所，农业部渔用药物创制重点实验室，广东省水产动物免疫技术重点实验室，
广东 广州 510380；2 上海海洋大学水产与生命学院，上海 201306）

摘要：【目的】研究复方磺胺嘧啶（磺胺嘧啶/甲氧苄啶=5/1）在罗非鱼体内的药动学及残留消除规律。【方法】在实验水温 28℃ 条件下，按复方磺胺嘧啶 120 mg/kg 的剂量对吉富罗非鱼（*Genetically improved farmed tilapia*，GIFT）单次药饵饲喂复方磺胺嘧啶后，采用 HPLC 法测定罗非鱼各组织中的药物浓度，分析复方磺胺嘧啶在罗非鱼体内的药动学规律。在相同条件下，按首剂量饲喂复方磺胺嘧啶 120 mg/kg 后再按 60 mg/kg 的剂量连续五天饲喂实验鱼来研究药物在罗非鱼肌肉组织中的残留消除规律。【结果与讨论】单次药饵饲喂罗非鱼后，磺胺嘧啶和甲氧苄啶的吸收半衰期（$T_{1/2ka}$），消除半衰期（$T_{1/2ke}$），表观曲线下面积（AUC）和清除率（CL/F）分别为（2.58~4.59 h 和 0.67~3.05 h），（13.43~16.97 h 和 5.99~13.49 h），（525.80~869.33 mg/L·h 和 42.37~214.25 mg/L·h）和（0.03~0.14 L/kg·h 和 0.09~0.39 L/kg·h）。磺胺嘧啶在各组织中（血液、肌肉、肝脏和肾脏）的最大峰浓度 C_{max} 分别为 22.49、14.79、41.3/39.7、34.9/25.46（C_{max1}/C_{max2}）mg/L（mg/kg）；达峰时间 T_{max} 分别为 10.82、8.73、1/8、2/10 h（T_{max1}/T_{max2}）。肝、肾组织中的药-时曲线有明显的双峰现象，提示该药物在罗非鱼胃肠道中有非齐性吸收现象；甲氧苄啶未见双峰现象，在各组织中（血液、肌肉、肝脏和肾脏）的最大峰浓度 C_{max} 分别为 2.45、5.04、13.44、12.85 mg/L（mg/kg）；达峰时间 T_{max} 分别为 6.04、12.03、3.45 和 3.64 h。根据

连续给药情况下罗非鱼可食性组织肌肉的残留检测结果，参考中华人民共和国第 235 号公告中对动物源性食品中磺胺类药物总量的 MRL 规定，以 0.1 mg/kg 为残留限量，建议休药期不低于 12 d。

关键词：吉富罗非鱼；复方磺胺嘧啶；药物动力学；休药期

该文发表于《淡水渔业》【2001，31（6）：52-54】

磺胺二甲嘧啶在银鲫体内的药动学及组织残留研究

艾晓辉，陈正望

（华中科技大学生命科学与技术学院，武汉 430070）

摘要：本文用比色法测定了磺胺二甲嘧啶（sulphadimine，SM_2）在银鲫血浆及肌肉、肝脏等组织中浓度并考察了采用尾静脉抽血和断尾取血两种取样方法对测定结果的影响情况。结果表明：抽血采样组血药浓度明显高出断尾采样组；以 200 mg/kg 鱼体重的剂量口灌给药，在水温 28.5？0.5℃ 的条件下，给银鲫口灌 200 mg/kg 的磺胺二甲嘧啶后，采用抽血方式采样，磺胺二甲嘧啶在银鲫体内的药动学最佳数学模型为一级吸收一室开放模型，其 $T_{1/2ka}$ 为 0.665 h，$T_{1/2k}$ 为 12.961 h，C_{max} 为 79.777 μg/mL。

关键词：磺胺二甲嘧啶；血浆及组织浓度；不同采样方法

该文发表于《水利渔业》【2008，28（3）：25-27】

磺胺甲噁唑在罗非鱼体内的药代动力学及组织浓度研究

袁科平，艾晓辉

（中国水产科学研究院长江水产研究所，湖北 荆州 434000）

摘要：研究了 (19±1)℃ 水温条件下，以 100 mg/kg 剂量单次口腔灌药后，磺胺甲噁唑（SMZ）在罗非鱼的肌肉、血液、肝脏组织中的残留和消除规律。各组织中药物浓度由高效液相色谱法测得。研究结果表明，罗非鱼血液中的药物浓度符合一级吸收一室开放模式；消除相半衰期 $T_{1/2k}$ 为 8.70 h，吸收相半衰期 $T_{1/2kα}$ 为 0.94 h，达峰时间 T_p 为 3.38 h，达峰浓度 C_{max} 为 18.41 μg/mL。建议 SMZ 在罗非鱼上的休药期为 10 d。

关键词：磺胺甲噁唑；罗非鱼；药代动力学

第四节 杀虫类渔药在水产动物体内的药代动力学

杀虫类渔药药动学的研究大部分集中于口服类药物。吡喹酮在黄鳝体内呈现出了快速吸收、快速消除和低生物利用度的特性（Ning Xu et al.，2016）。但盐度会影响其代谢过程。咸水中草鱼体内吡喹酮的代谢比淡水中的要迅速，咸水中鱼体组织内吡喹酮残留的浓度也更低（Xinyan Xie et al.，2015；谢欣燕等，2015）。阿维菌素在草鱼（*Ctenopharyngodon idellus*）体内消除较快（秦改晓等，2012；2012）。对虹鳟腹腔注射给药伊维菌素的吸收和消除速率均较口灌给药要快（康淑媛等，2015）。甲苯咪唑（MBZ）在银鲫体内吸收较慢（郭东方等，2009）。甲苯咪唑原型药在团头鲂体内所占的比例随着时间的延长逐渐降低，其主要代谢产物羟基甲苯咪唑（MBZ-OH）和氨基甲苯咪唑（MBZ-NH4）逐渐升高应作为残留检测的主要监控目标（胥宁等，2013）。

一、甲苯咪唑

该文发表于《中国水产科学》【2009，16（3）：434-441】

甲苯咪唑在银鲫体内的药代动力学研究

Pharmacokinetics of mebendazole in Johnny carp（*Carassius auratus gibelio*）

郭东方[1,2]，刘永涛[1]，艾晓辉[1]，袁科平[1]，王淼[1,2]，秦改晓[1,2]

(1. 中国水产科学研究院长江水产研究所，湖北 荆州 434000；2. 华中农业大学水产学院，湖北 武汉 430070)

摘要： 采用高效液相色谱法检测甲苯咪唑（MBZ）及其代谢物氨基甲苯咪唑（MBZ-NH2）和羟基甲苯咪唑（MBZ-OH）在银鲫（*Carassius auratus gibelio*）血浆及组织中的浓度，数据经 3P97 药代动力学程序分析。结果表明：(25±1)℃的水温条件下，银鲫单剂量口灌 MBZ 20 mg·kg^{-1}，血药经时过程符合二室开放式模型。主要药代动力学参数为：吸收速率常数 K_a 0.235 h^{-1}；消除半衰 $T_{1/2\beta}$ 为 52.26 h；药时曲线下面积 AUC 180.07 μg·h·mL^{-1}；表观容积分布 V_d/F 2.465 L·kg^{-1}；清除率 CLs 为 0.111 mL·h^{-1}·kg^{-1}；达峰时间 T_{peak} 为 6.20 h；质量浓度 C_{max} 为 4.14 μg·mL^{-1}。与哺乳类相比，MBZ 在银鲫体内吸收较慢，消除半衰期明显延长，药代动力学参数也有较大差异。建议在做到合理用药的同时加强药物检测力度。

关键词： 银鲫；甲苯咪唑；氨基甲苯咪唑；羟基甲苯咪唑；药代动力学；高效液相色谱

甲苯咪唑（MBZ）属于苯并咪唑类药物，也称甲苯哒唑、二苯酮咪胺酯、安乐士。化学名为（5-苯甲酰基-1H-苯并咪唑-2-基）氨基甲酸甲酯，是一种常见的广谱驱虫类药物。分子式是：$C_{16}H_{13}N_3O_3$，1971 年由 Van Geldev 首次合成，并获美国专利，比利时杨森公司同年投入临床使用。MBZ 作为一种广谱杀虫剂可用来治愈蛔虫、线虫、钩虫、鞭虫的感染[1]，在哺乳动物中用来预防山羊肠胃和肺中的线虫和蠕虫[2]，对肥育犬消化道寄生虫病具有较好的治疗效果[3]，可治疗小鼠脊形管圆线虫病。在水产上有报道甲苯咪唑可治疗欧洲鳗鲡指环虫病[4-6]。口服甲苯咪唑可治疗石斑鱼鳃的单殖吸虫病[8]。采用药溶可治疗红鲷多微小吸盘吸虫的感染[9]。药物的药代动力学研究是临床设计合理给药方案、发挥药物最佳疗效和减少副作用的理论基础。目前国内还未见甲苯咪唑在水产动物体内的药代学研究报道。本研究初步探讨该药在银鲫体内的药代学特征，为在鱼病害防治中科学合理地用药提供参考。

1　材料与方法

1.1　实验动物

健康银鲫（*Carassius auratus gibelio*）由中国水产科学研究院长江水产研究所窑湾试验场提供，体质量为（200±10）g。实验前在 1 m^3 水族箱内暂养 1 周，每天换水 1 次。实验用水为曝气 48 h 自来水，连续充氧，保持水中溶氧大于 5.0 mg·L^{-1}，实验水温为（25±1）℃，pH 值为 7.5~8.0，实验过程中停饲。

1.2　药品及试剂

MBZ 标准品（纯度≥99%，美国 Sigma 公司），羟基甲苯咪唑（MBZ-OH）和氨基甲苯咪唑（MBZ-NH$_2$）（纯度≥99%，Witega 实验室，Berlin-Adlershof GmbH）；MBZ 原粉（纯度≥99%，九州神龙有限公司）；乙腈、乙酸乙酯、正己烷等（均为色谱纯，Fisher ChemAlert 公司），二甲基亚砜和二甲基甲酰胺（分析纯，国药集团化学试剂有限公司）；三乙胺、磷酸二氢铵（分析纯，天津福晨化学试剂公司），磷酸氢二钠、磷酸二氢钠（分析纯，天津天大化学试剂公司）。

磷酸盐缓冲液：准确称取磷酸氢二钠（Na$_2$HPO$_4$·12H$_2$O）17.9 g 和磷酸二氢钠（NaH$_2$PO$_4$·2H$_2$O）4.6 g，分别溶于 1 L 蒸馏水中配成 0.05 mol·L^{-1} 的溶液，然后按 9：1 的体积比混合用氨水调节 pH 值到 9.5。

定容液的配制：A 液，称取 1.438 g 磷酸二氢铵溶解在 250 mL 双蒸水中配制成 0.05 mol·L^{-1} 的磷酸二氢铵溶液。B 液，二甲基甲酰胺。定容溶液为 A 液+B 液（体积比 7：3），用磷酸将溶液 pH 值调至 2.0，经抽滤倒入具塞玻璃瓶中置 4℃ 冰箱中备用。

标准混合液配制：取 MBZ 标准品 0.10 g 用二甲基亚砜溶解并定容到 100 mL，得到质量浓度 1 000 μg·mL^{-1} 的标准储备液。取 MBZ-OH、MBZ-NH$_2$两种药物标准品各 0.10 g 用二甲基甲酰胺溶解定容到 100 mL，得到质量浓度为 1 000 μg·mL^{-1}的标准储备液。分别从 3 种标准储备液中各取 10 mL 混合后用流动相稀释至 100 mL，得到质量浓度为 100.0 μg·mL^{-1}标准混合液。

1.3 仪器与设备

高效液相色谱仪（Waters 515 泵，717 自动进样器，2487 双通道紫外检测器及 Empower 色谱工作站）；自动高速冷冻离心机（日本 HITACHI 20PR-520 型）；Mettler-TOLEDO AE-240 型精密电子天平（梅特勒-托利多公司）；FS-1 高速匀浆机（华普达教学仪器有限公司）；调速混匀器（上海康华生化仪器制造厂）；恒温烘箱（上海浦东荣丰科学仪器有限公司）；HGC-12 氮吹仪（HENGAO T&D 公司）；Sartorius PB-10 型酸度计（德国赛多利斯公司）。

1.4 实验设计与采样

口灌给药：MBZ 原粉 200 mg 用少许二甲基亚砜溶解再用 100 mL 蒸馏水定容，此时药物质量浓度为 2 mg·mL^{-1}。按 20 mg·kg^{-1}单次给药，将药物用 2 mL 注射器接硅胶软管从银鲫口中插入，注意深度，无回吐鱼可用于实验。

采样：给药后的 0.25 h、0.5 h、1 h、2 h、4 h、8 h、12 h、24 h、48 h、72 h、120 h、168 h 尾静脉取血液，1%肝素钠抗凝后血液以 3 000 r·min^{-1}离心 10 min，取上清液，同时解剖分离出肌肉、皮、肝脏、肾脏；将血清和其他组织于-20℃保存备用。

每一时间点各取 5 尾鱼，作为 5 个平行样品分别处理测定。另取 5 尾未给药鱼作空白对照。

1.5 样品预处理

将冷冻保存的肌肉、皮、肝脏、肾脏和血液样品室温下自然解冻，剪取适量的肌肉组织，置高速匀浆机中匀浆至糜状，准确称取 5.0 g 于 50 mL 塑料离心管中，加 1 mL 的磷酸盐缓冲液和 20 mL 乙酸乙酯，置漩涡混合器上振荡 5 min，5 000 r/min 离心 10 min，取出提取液，再加入 15 mL 乙酸乙酯，重复操作一次，合并提取液于鸡心瓶中，置 40℃水温旋转蒸发至干。鱼皮先剪碎，准确称取 1.0 g；肾脏和肝脏都各取 0.5 g；血浆取 1.0 mL。分别加入 15 mL 具塞离心管中，加入 0.5 mL 的磷酸盐缓冲液，然后加 4 mL 乙酸乙酯，置调速混匀器上涡旋振荡混匀 1 min，以 4 000 r·min^{-1}离心 10 min，吸出上清液于另一支 15 mL 具塞离心管中，重复提取一次，合并上清液氮吹至干。旋转蒸发和氮吹至干的剩余物分别用 1 mL 定溶液定容，再加入 1 mL 的正己烷，置漩涡混合器上振荡 1 min，混合溶液于 4 000 r·min^{-1}离心 10 min，去除上层正己烷液，下层液经 0.22 μm 滤头过滤，用于 HPLC 测定。

1.6 HPLC 分析方法

1.6.1 色谱条件　Waters symmetry C$_{18}$反相色谱柱（250 mm×4.6 mm，5 μm）；流动相：乙腈+磷酸二氢铵溶液（0.05 mol·L^{-1}，0.1%三乙胺）（33：67，V/V）；流速：0.8 mL·min^{-1}；柱温：室温；紫外检测波长：298 nm；进样量：50 μL。

1.6.2 血液与组织标准曲线的制备　空白血液、组织（肌肉、肝脏、肾脏、皮）中加入 MBZ、MBZ-OH 和 MBZ-NH$_2$已知标准混合液，使其药物质量浓度范围分别为 0.01～100.00 μg·mL^{-1}和 0.01～100.00 μg·g^{-1}。按样品处理方法处理，作 HPLC 分析，以测得的各平均峰面积 y 为纵坐标，相应的质量浓度 x 为横坐标绘制标准工作曲线，求出回归方程和相关系数。用空白组织制成低质量浓度药物的含药组织，经预处理后测定，将引起三倍基线噪音的药物的质量浓度定义为最低检测限。

1.6.3 回收率与精密度测定　回收率=C_r/C_0×100%，其中 C_r 为用空白样品（血液或组织）加入一定量的 MBZ、MBZ-OH 和 MBZ-NH$_2$已知标准混合液，再按样品预处理方法进样后，测定 3 种药物的质量浓度；C_0 为加入一定量的 MBZ、MBZ-OH 和 MBZ-NH2 已知标准混合液，测得 3 种药物的质量浓度。在 4 种空白组织和血液中分别添加 4 个质量浓度水平的混合标准液，使组织中质量浓度分别为 0.01 μg·g^{-1}、0.10 μg·g^{-1}、1.00 μg·g^{-1}和 10.0 μg·g^{-1}，血液为 0.01 μg·mL^{-1}、0.10 μg·mL^{-1}、1.00 μg·mL^{-1}和 10.0 μg·mL^{-1}。每个浓度的样品，日内做 5 个重复，1 周内重复做 5 次，计算日内及日间精密度。

1.7 数据处理

药物动力学模型拟合及参数计算采用中国药理学会数学专业委员会编制的 3P97 药动软件分析；标准曲线，药物经时曲线图，消除方程及休药期计算和回归图，采用 Microsoft Excel 2003，SPSS（13.0）进行计算和绘制。

2 结果与分析

2.1 标准曲线方程与相关系数

空白血液、组织（肌肉、肝脏、肾脏、皮）中加入 MBZ、MBZ-OH 和 MBZ-NH$_2$ 已知标准混合液，使其药物在组织中的水平为 $0.01 \sim 100.00~\mu g \cdot g^{-1}$，血液中为 $0.01 \sim 100.00~\mu g \cdot mL^{-1}$。按样品处理方法处理，在 $0.01 \sim 100.00~\mu g \cdot g^{-1}$ 和 $0.01 \sim 100.00~\mu g \cdot mL^{-1}$ 范围内线性关系良好。以测得的各平均峰面积对相应质量浓度作线性回归，并制作标准曲线。5 种组织中 MBZ、MBZ-OH 和 MBZ-NH2 的标准曲线及相关系数分别见表 1。

表 1　MBZ 及其代谢物在银鲫血浆和组织中标准曲线和相关系数

Table. 1　Standard curve and correlation coefficient of MBZ and its metabolites in Johny carp plasma and tissues

标准曲线名称 Standard curve	分析物 Analyte	线性方程 Linear equation	相关系数 Correlation coefficient
血浆标准曲线 Standard stand of plasma	MBZ	$y = 14\ 231x - 10\ 985$	0.999 3
	MBZ-OH	$y = 6\ 392.2x - 4\ 556$	0.997 8
	MBZ-NH$_2$	$y = 17\ 363x - 15\ 116$	0.986 4
肌肉标准曲线 Standard curve of muscle	MBZ	$y = 25\ 574x - 8\ 920$	0.923 3
	MBZ-OH	$y = 17\ 221x - 6\ 756.3$	0.955 0
	MBZ-NH$_2$	$y = 26\ 269x - 18\ 943$	0.954 0
肝脏标准曲线 Standard curve of liver	MBZ	$y = 74\ 170x - 33\ 536$	0.936 3
	MBZ-OH	$y = 52\ 504x - 30\ 581$	0.946 4
	MBZ-NH$_2$	$y = 52\ 163x - 21\ 133$	0.988 0
肾脏标准曲线 Standard curve of kindey	MBZ	$y = 7\ 683.5x - 2\ 385$	0.994 3
	MBZ-OH	$y = 22\ 116x - 20\ 349$	0.999 1
	MBZ-NH$_2$	$y = 5\ 348x - 3\ 281.3$	0.952 5
皮肤标准曲线 Standard curve of skin	MBZ	$y = 29\ 389x - 10\ 828$	0.994 0
	MBZ-OH	$y = 28\ 312x - 17\ 262$	0.978 6
	MBZ-NH$_2$	$y = 23\ 094x - 8\ 939$	0.969 8

2.2 回收率与精密度

本实验条件下，各组织和血液中 4 个质量浓度水平的 MBZ 回收率为 81.2% ~ 86.3%，MBZ-OH 回收率为 86.5% ~ 93%，MBZ-NH 2 回收率为 70.6% ~75.0%。测得的日内精密度与日间精密度均小于 10%。

2.3 MBZ 在各组织中的浓度

将测到的 MBZ 的峰面积代入各自的标准曲线所绘制的回归方程（表 1），就可以求出药物在各组织中的浓度值（表 2）。

2.4 MBZ 在银鲫体内的药代动力学参数（表 3）

MBZ 以 $20~mg \cdot kg^{-1}$ 剂量单次口灌银鲫后，有关动力学数据经 3P97 实用药代动力学软件分析，血液中药物浓度与时间关系符合开放性二室模型，动力学方程为：$C = 17.38e^{-0.151t} + 1.944e^{-0.0133t}$。$K_{21} < K_{12}$，说明药物从中央室进入外周室的速率比外周室回到中央室的速率大。

2.5 色谱图

图 1（A-F）是标准液及银鲫经单次口灌 MBZ 后血浆和各个组织样品中 MBZ 水平达到峰值时的色谱图。由图 1 可见，甲苯咪唑经鱼体吸收后在其各个组织中出现了 MBZ-NH$_2$ 和 MBZ-OH 两种代谢物。原药和代谢物消除规律的同时研究，可更加合理地制定休药期。

表2 单剂量 20 mg·kg^{-1} (BW) 口灌 MBZ 在银鲫血液和组织中的残留

Tab. 2 MBZ concentration in plasma and different organized samples in Johnny carp after
oral administration at 20 mg·kg^{-1} (BW) n=5: $x-\pm SE$

时间 Time	MBZ 残留水平 MBZresidual level				
	肌肉/ (μg·g^{-1}) Muscle	血液/ (μg·g^{-1}) Blood	肝脏/ (μg·g^{-1}) Licer	皮/ (μg·g^{-1}) Skin	肾脏/ (μg·g^{-1}) Kidney
0.25	0.136±0.003	0.193±0.026	0.873±0.012	0.087±0.112	0.017±0.006
0.5	0.309±0.045	0.427±0.016	1.01±0.164	0.179±0.203	0.121±0.008
1	0.543±0.206	1.18±0.248	1.289±0.277	0.540±0.231	0.340±0.011
2	0.881±0.220	3.16±1.394	2.03±0.182	1.039±0.634	0.734±0.204
4	1.305±0.413	4.04±1.573	4.59±0.335	1.45±1.082	0.898±0.082
8	1.755±0.536	4.73±0.461	5.789±0.803	2.369±0.528	1.369±0.228
12	3.027±0.253	3.72±0.547	7.880±1.235	3.232±1.225	1.533±0.225
24	2.801±0.665	1.51±0.714	9.546±2.121	5.173±1.413	1.673±0.213
48	1.434±0.135	1.22±0.898	6.907±2.378	3.298±1.657	0.898±1.657
72	0.663±0.048	0.89±0.467	4.101±1.053	0.529±0.081	0.529±0.781
120	0.242±0.023	0.534±0.233	1.786±0.034	0.381±0.102	0.381±0.362
168	0.012±0.004	0.141±0.163	0.887±0.136	0.095±0.066	0.165±0.066

表3 单剂量 20 mg·kg^{-1} (BW) 口服 MBZ 在银鲫体内的药代动力学参数

Tab. 3 Pharmacokinetic parameters for MBZ in Johnny carp after single oral administration at 20 mg·kg^{-1} (BW)

参数 Parameter	单位 Unit	血液 Blood
A	μg·mL^{-1}	17.38
a	h^{-1}	0.151
B	μg·mL^{-1}	1.944
β	h^{-1}	0.0133
K_a	h^{-1}	0.235
Lag time	h	0.199
V_d/F	L·kg^{-1}	2.465
$T_{(1/2)\alpha}$	h	4.612
$T_{(1/2)\beta}$	h	52.26
$T_{(1/2)Ka}$	h	2.947
K_{21}	h^{-1}	0.044
K_{10}	h^{-1}	0.045
K_{12}	h^{-1}	0.074
AUC	μg·h·mL^{-1}	180.07
CL (s)	mL·h·kg^{-1}	0.111

参数 Parameter	单位 Unit	血液 Blood
T_{peak}	h	6. 20
C_{max}	$\mu g \cdot mL^{-1}$	4. 14

注：A、B 为药时曲线对数图上曲线在横轴和纵轴上的截距；α、β 分别为分布相、消除相的一级速率常数；K_{21} 由周边室向中央室转运的一级速率常数；K_{10} 由中央室消除的一级速率常数；K_{12} 由中央室向周边室转运的一级速率常数；V_d/F 观分布容积；AUC 药-时曲线下面积；Lag time 滞后时间；K_a 为一级吸收速率常数；$T_{(1/2)Ka}$ 为药物在中央室的吸收半衰期；$T_{(1/2)\alpha}$、$T_{(1/2)\beta}$ 分别为总的吸收和消除半衰期；T_{peak} 为单剂量给药后出现最高血药质量浓度的时间；C_{max} 为单剂量给药后的最高血药质量浓度；CL（s）为总体清除率.

Note：A，B-Drug concentration-time curve on logarithmic graph curve on the horizontal axis and vertical axis intercept；α，β-First-order rate constant for drug distribution phase and elimination phase；K21-First-order rate constant for drug distribution from peripheral compartment to central compartment；K10-First-order rate constant for drug elimination from central compartment；K12-First-order rate constant for drug distribution from central compartment to peripheral compartmenti；Vd/F-Apparent volume of distribution；AUC-Area under concentration-time curve；Ka-First-order rate constant for drug absorption；T(1/2)Ka-Absorption half-life in central compartment；T(1/2)α，T(1/2)β-The overall absorption half-life and Elimination half-life；Tpeak-The time at highest peak of concentration；Cmax-The biggest concentration of the drug in blood；CL（s）-The overall clearance rate.

3 讨论

3.1 MBZ 在银鲫体内的药动学特征

银鲫血浆中的 MBZ 药时数据符合二室开放模型。这与 Pandey 等[2] 报道的静脉注射山羊得到的药代动力学房室模型一致。王长虹等[10]、罗兰等[11]、李岩等[12] 报道了 MBZ 在兔血中的药代动力学模型为单室模型，KitzmanI 等[13] 报道了 MBZ 的同系物氟苯哒唑在斑点叉尾鮰体内药代动力学模型为二室模型。这些模型的差异可能与实验对象种属差异有关。T_{max}、K_a、$T_{(1/2)a}$ 是反应药物在体内吸收速率的重要指标。新西兰兔口服 MBZ 40 mg/kg[11]，其血药浓度达峰时间 T_{max} 为 4. 48 h，吸收半衰期 $T_{(1/2)a}$ 为 1. 74 h，吸收速率常数 K_a 为 0. 514 h^{-1}。Pandey 等[2] 报道的山羊单剂量口服 MBZ 40 mg/kg，其血浆中的药物吸收参数 T_{max} =（5. 8±0. 88）h，K_a =（0. 34±0. 06）h^{-1}，$T_{(1/2)a}$ =（3. 2±0. 06）h。本实验条件下得到 MBZ 在银鲫体内血浆药物浓度达峰时间 T_{max} 为 6. 2 h，吸收速率常数 K_a 0. 235 h^{-1}；吸收半衰期 $T_{(1/2)a}$ 4. 612 h。MBZ 在银鲫体内的达峰时间及吸收半衰期要慢于前两者，说明药物在银鲫体内的吸收速率低于哺乳动物。$T_{(1/2)\beta}$ 是反应药物在体内消除速率的重要参数。本实验得到的消除半衰期 $T_{(1/2)\beta}$ 52. 26 h 要大于 Krishnaiah 等[14] 报道的 2 种不同包衣的 MBZ 在人体内的消除半衰期，分别为（13. 2±5. 6）h 和（8. 2±2. 9）h。王长虹等[10] 报道 MBZ 乳剂和片剂在兔体内的消除半衰期分别为（4. 5±0. 7）h 和（6. 3±2. 6）h，消除时间都比在鱼体内短。与哺乳动物和人类相比，MBZ 在银鲫体内吸收较慢，消除半衰期明显延长，药代动力学参数也有较大的差异。

动物体内的组织蛋白结合能力和少量的细胞内渗透是造成个体间药代动力学参数差异的原因，药物制剂也是导致差异的因素之一。另外，药物也会因为在血液中的循环途径不同，影响到以后的分布和消除。

单次口服剂量为 20 mg·kg^{-1} 的 MBZ，药物在血浆和组织内的药-时浓度如表 2 所示；达峰浓度 C_{max} 从高到低依次为肝脏、皮、血液、肌肉、肾脏。MBZ 在肝脏中峰浓度最高，这与 Braithwaite 等[15] 报道的人口服 MBZ 主要被肝门静脉吸收，肝脏药物含量最高相符合。

3.2 MBZ 在银鲫体内代谢物

银鲫口服 MBZ 后，在其体内产生 MBZ-NH2 和 MBZ-OH 两种代谢物（图 1）。Iosifidou 等[16] 报道了用 1×10^{-6} 的 MBZ 浸泡欧洲鳗鲡 24 h，测得有 MBZ-OH 和 MBZ-NH2 两种代谢物，其中 MBZ-NH2 在体内的消除时间最长，需要 14 d。欧盟规定 MBZ 及其代谢物残留（总量）在羊、马可食性组织（肌肉、脂肪、肝脏、肾脏）中规定的最高残留限量（MRL）分别为 60 μg·kg^{-1}、60 μg·kg^{-1}、400 μg·kg^{-1}、60 μg·kg^{-1}[17]，而未规定水产品各组织中限量。本实验对（25±1）℃ 水温条件下银鲫血浆、肌肉、皮肤、肝脏、肾脏组织研究中发现了 MBZ 及其 2 种代谢物，进一步丰富了 MBZ 的在鱼类各组织中分布及消除资料，也为制定最高残留限量（MRL）及休药期（WDT）提供了参考资料。

3.3 MBZ 的用药方式及残留限量

MBZ 主要用作人和兽用驱虫药，中国近年在水产养殖中开始运用。李志青等[18] 报道了养殖欧洲鳗鲡在水温低于 28℃ 时，建议药浴水平为 0. 6~1. 0 mg/L，药浴时间不超过 24 h。水温超过 28℃，建议使用药浴水平为 0. 2~0. 3 mg/L，可长时

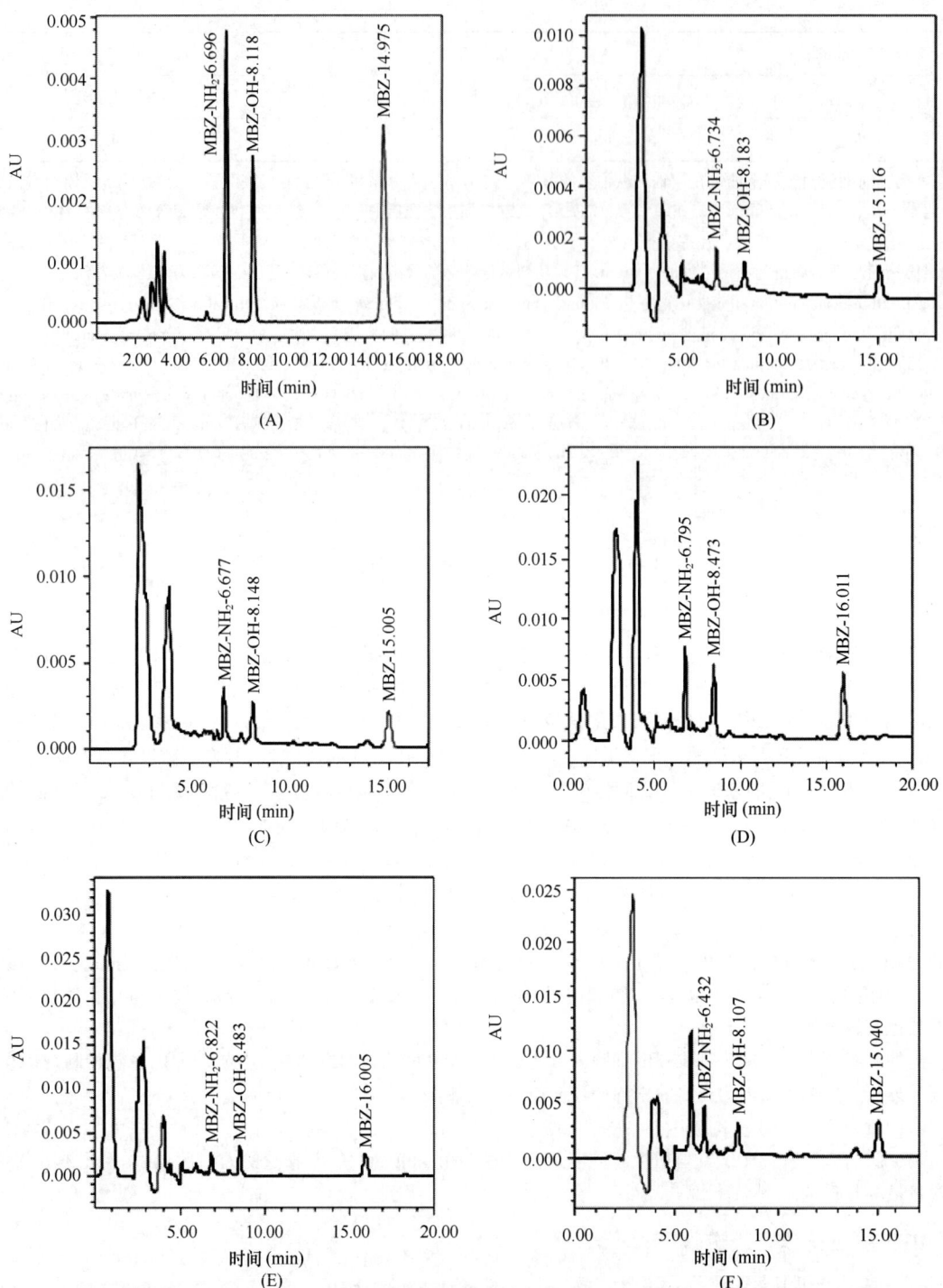

图 1　1 μg·mL^{-1}标准混合溶液（A）、给药后 8 h 银鲫血浆（B）、给药后 12 h 银鲫肌肉（C）、给药后 24 h 银鲫肝脏（D）、给药后 24 h 银鲫肾脏（E）、给药后 24 h 银鲫皮肤（F）的色谱图

Fig. 1　Chromatogram of 1 μg·mL^{-1} mixed standard solution（A），Johnny carp plasama sample at 8 h after oral MBZ（B），Johnny carp muscle sample at 12 h after oral MBZ（C），Johnny carp liver sample at 24 h after oral MBZ（D），Johnny carp kindey sample at 24 h after oral MBZ（E）and Johnny carp skin sample at 24 h after oral MBZ（F）

间药浴。李海燕等[19]研究建议生产上使用 0.8~1.0 mg/L 的 MBZ 治疗鳗的伪指环虫病。伪指环虫从产卵到孵化出幼虫一般需要 3~6 d，因此，建议使用 MBZ 后 7~8 d 再重复处理 1 次。此外，MBZ 还可用来治疗其他鱼类的单殖吸虫病，如鳜鳃上常寄生河鲈锚首虫。该药现用于水生动物多采用浸泡方式，药物直接进入水体对水环境会造成何种影响，国内外在这方面

尚无研究报道，药物在水体中的代谢情况还需进一步探讨。本实验采用灌胃方式给药以研究药物在银鲫体内的动力学，减少药物对水环境的影响，为药物拌饲投喂提供科学数据。

MBZ 的允许最高残留限量（MRL）在不同国家和地区是不同的。欧盟对羊和马属动物规定的 MRL 为 0.04 mg/kg[20]；美国 FDA 只批准用于马和狗[21]；日本肯定列表中暂时规定在水产品中 MBZ 的 MRL 为 0.02 mg/kg。由于 MBZ 可引起人全身不适，搔痒、皮疹、视神经视网膜病变，出现血尿现象等症状[22-24]，因此建议在做到合理用药的同时也要加大药物检测力度。

参考文献

[1] Swanepoela E, Liebenberga W, Melgardt M. Quality evaluation of generic drugs by dissolution test: changing the USP dissolution medium to distinguish between active and non-active mebendazole polymorphs [J]. Eur J Pharmaceutics Biopharm, 2003, 55: 345-349.

[2] Pandey S N, Roy B K. Disposition kinetics of mebendazole in plasma, milk and ruminal fluid of goats [J]. Small Ruminant Res, 1998, 27: 111-117.

[3] 吴忠，贺永健，周鸿海. 复方甲苯咪唑对肥育犬的影响 [J]. 动保药品，2006，31（2）：31-32.

[4] Terada M, Kino H, Akyol CV, Sano M. Effects of mebendazole on Angiostrongylus costaricensis in mice, with special reference to the timing of treatment [J]. Parasitnlngy Research, 1993, 79: 441-443.

[5] Buchmann K, Bjerregaard J. Mebendazole treatment of Pseudodactylogyrosis in an intensive eel-culture system [J]. Aquaculture, 1990, 86: 139-153.

[6] Buchmann K, Bjerregaard J. Comparative efficacies of commercially available benzimidazoles against Pseudodactylogyrus infestations in eels [J]. Dis Aquat Org, 1990, 9: 117-120.

[7] Mellergaard S. Mebendazole treatment against Pseudodactylogyrus infections in eel (Anguilla anguilla) [J]. Aquaculture, 1990, 91: 15-21.

[8] Kim K H, Choi E S. Treatment of Microcotyle sebastis (Monogenea) on the gills of cultured rockfish (Sebastes schelegeli) with oral administration of mebendazole and bithionol [J]. Aquaculture, 1998, 167: 115-121.

[9] Pantelis K, Nikos P, Pascal D. Treatment of Microcotyle sp. (Monogenea) on the gills of cage-cultured red porgy, Pagrus pagrus following baths with formalin and mebendazole [J]. Aquaculture, 2006, 252 (2-4): 167-171.

[10] 王长虹，孙殿甲. O/W 型甲苯咪唑口服乳剂兔体内药物动力学研究 [J]. 中国医院药学杂志，1999，19（10）：591-594.

[11] 罗兰，孙殿甲，苗爱东，等. 甲苯咪唑 β-CD 饱和物兔体内药物动力学研究 [J]. 西北药学杂志，2000，15（3）：118-119.

[12] 李岩，孙殿甲. 口服甲苯咪唑微丸的药物动力学及生物利用度 [J]. 中国医院药学杂志，1999，19（11）：649-651.

[13] KitzmanI J V, Holley J H, Huberi W G. Pharmacokinetics and metabolism of fenbendazole in channel catfish [J]. Veterin Res Commun, 1990, 14: 217-226.

[14] Krishnaiah Y S R, Rajua P V. Pharmacokinetic evaluation of guar gum-based colon-targeted drug delivery systems of mebendazole in healthy volunteers [J]. Controlled Release, 2003, 88: 95-103.

[15] Braithwaite P A, Roberts M S, Allan R J, et al. Clinical pharmacokinetics of high dose mebendazole in patients treated for cystic hydatid disease [J]. Eur J Clin Pharmacol, 1982, 22: 161-169.

[16] Iosifidou EG, Haagsma N, Olling M, et al. Residue study of mebendazole and its metabolites hydroxy-mebendazole and amino-mebendazole in eel (Anguilla anguilla) after bath treatment [J]. Drug Metabolism and Disposition, 1997, 25 (3): 317-320.

[17] EEC Council Regulation No. 2377/1990. Community procedure for the establishment of maximum residue limits of veterinary medicinal products in foodstuffs of animal origin [S].

[18] 李志青，樊海平，曾占状，等. 复方甲苯咪唑对指环虫的杀灭效果和欧洲鳗的毒性 [J]. 浙江农学院学报，2000，12（4）：221-223.

[19] 李海燕，李桂峰. 甲苯咪唑对鱼的急性毒性试验及驱虫效果研究 [J]. 广州大学学报，2005，4（1）：33-35.

[20] Commission regulation (EC) No 1680/2001 of 22 August 2001 amending Annexes I and II to Council Regulation (EEC) No 2377/90 laying down a Community [S].

[21] Code of Federal Regulations Title 21, Volume 6 Revised as of April 1, 2008, CITE: 21CFR520.132 [S].

[22] 席成亮. 复方甲苯咪唑致药物性荨麻疹 2 例报告 [J]. 中原医刊，1997，24（4）：30.

[23] 白玲，赵丕烈，陈东平. 甲苯咪唑的不良反应 [J]. 医药导报，1999，18（2）：128.

[24] 景宝洁. 甲苯咪唑引起肝损害 [J]. 药物不良反应杂志，2007，9（6）：433.

该文发表于《淡水渔业》【2013，43（6）：39-44】

甲苯咪唑在团头鲂体内主要代谢物及其变化规律研究

The study of main metabolites about mebendazole and its variation in *Megalobrama amblycephala*

胥宁[1,2]，刘永涛[1,2]，艾晓辉[1,2]，赵凤[1,4]，杨秋红[1]，吕思阳[1,3]

（1. 中国水产科学研究院长江水产研究所，武汉 430223；2. 淡水水产健康养殖湖北省协同创新中心，武汉 430223
3. 华中农业大学水产学院，武汉 430070；4. 南京农业大学渔业学院，无锡 214000）

摘要：为了确定甲苯咪唑在团头鲂（*Megalobrama amblycephala*）体内主要代谢产物，实验用 1mg/L 甲苯咪唑水溶液浸泡团头鲂，分别在浸泡后 1、2、4、8、12 和 24 h，分别采集肝脏、肌肉、胆汁和血液，采用高效液相串联质谱法（HPLC-MS-MS）定量和标准品比对法进行鉴定。结果显示，甲苯咪唑（MBZ）在团头鲂体内所占的比例随着时间的延长逐渐降低，羟基甲苯咪唑（MBZ-OH）和氨基甲苯咪唑（MBZ-NH$_4$）逐渐升高。MBZ-OH 和 MBZ-NH$_4$ 百分比都占到总残留的 10% 以上，二者为主要代谢产物，应作为残留检测的主要监控目标。

关键词：团头鲂（*Megalobrama amblycephala*）；甲苯咪唑；主要代谢物

二、阿维菌素和伊维菌素

该文发表于《淡水渔业》【2012，42（4）：47-52】

单剂量口灌阿维菌素在草鱼体内的药动学及残留研究

Studies on the pharmacokinetics and residues of avermectin in grass carp (*Ctenopharyngodon idellus*) after single oral administration

秦改晓[1]，徐文彦[1]，艾晓辉[2]，唐国盘[1]，崔锦[1]，李向辉[1]，袁科平[2]

（1. 郑州牧业工程高等专科学校，郑州 450011；2. 中国水产科学研究院长江水产研究所，武汉 430223）

摘要：按 0.1mg/kg 的剂量给草鱼（*Ctenopharyngodon idellus*）灌服阿维菌素，用高效液相色谱法检测用药后不同时间的血浆、肌肉和肝脏中的药物浓度，然后用 3P97 药代动力学软件处理药时数据，对药物在草鱼体内药动学及组织残留进行研究。结果表明：单剂量口灌阿维菌素在草鱼血浆中主要药动学参数为：AUC 1638.02 μg·h·L^{-1}，C_{max} 18.51 μg·L^{-1}，$T_{1/2\alpha}$ 21.77 h，$T_{1/2\beta}$ 218.71 h，T_{peak} 24.00 h，K_{10} 0.023 h^{-1}，K_{12} 0.007 3 h^{-1}，K_{21} 0.004 3 h^{-1}。给药后 14 d 血浆中检测不到药物，25 d 后肌肉和肝脏中药物消除。

关键词：阿维菌素；草鱼（*Ctenopharyngodon idellus*）；口灌；药动学；残留

该文发表于《西北农林科技大学学报：自然科学版》【2012，40（8）：13-20】

阿维菌素在草鱼体内的药物代谢动力学研究

Studies on pharmacokinetics of Avermectin in grass carp

(*Ctenopharyngodon idellus*)

秦改晓[1]，徐文彦[1]，艾晓辉[2]，唐国盘[1]，崔锦[1]，李向辉[1]，袁科平[2]

（1. 郑州牧业工程高等专科学校畜牧工程系，河南，郑州 450011；2. 中国水产科学研究院长江水产研究所，湖北，武汉 430223）

摘要：研究阿维菌素在草鱼体内的药物代谢动力学，为实际生产中阿维菌素的使用提供理论指导。【方法】用初始质量浓度为 0.3 μg/L 的阿维菌素水溶液药浴草鱼，于给药后 0.5，1，2，3，4，6，8，10，12，24，48，72，96，144，216，336，528 和 576 h 采取血浆、肌肉+皮、肝脏、肾脏、鳃等样品，采用高效液相色谱-荧光法测定阿维菌素在草鱼血浆中的质量浓度及在组织中的含量，数据经 3P97 药动学软件分析。在（26.0±1.0）℃的水温条件下，阿维菌素单剂量浸泡给药 0.3μg/L，血药经时过程符合二室开放式模型。主要药动学参数如下：分布半衰期（$T_{1/2\alpha}$）34.2 h，吸收半衰期（$T_{1/2(ka)}$）15.61 h，消除半衰期（$T_{1/2\beta}$）163.22 h，药时曲线下面积（AUC）2 486.02（μg·h）/L，达峰时间（T_{peak}）40.75 h，峰质量浓度（C_{max}）11.92 μg/L。药后 72 h 时草鱼肌肉、肝脏、肾脏和鳃中阿维菌素含量均达到最高值，其中肝脏中的含量最高，达到 17.8 μg/kg，其后依次为肾脏（12.1 μg/kg）、肌肉（10.7 μg/kg）和鳃组织（5.2 μg/kg），血浆中阿维菌素含量在 48 h 达到最高（11.2 μg/L）。肝脏、肾脏和鳃组织中阿维菌素含量均呈"双峰"曲线，前两者在 144 h 时都有第 2 次吸收高峰，分别为 15.0 和 8.4 μg/kg。草鱼血浆及各组织中阿维菌素在给药后 24 d 未检出，考虑到临床应用情况的复杂性及理论值与实测值之间的差距，建议对草鱼单剂量（0.3 μg/L）药浴阿维菌素后的休药期为 24 d。

关键词：草鱼；阿维菌素；药物代谢动力学；消除规律；高效液相色谱

该文发表于《中国农学通报》【2015，31（2）：101-106】

两种给药方式下伊维菌素在虹鳟体内的药物代谢动力学研究

Pharmacokinetics of ivermectin in rainbow trout by oral and injection

康淑媛[1,2]，韩冰[1,2]，王荻[1]，卢彤岩[1]

（1. 中国水科学研究院 黑龙江水产研究所，黑龙江 150070；2. 上海海洋大学 水产与生命学院，上海 201306）

摘要：研究旨在通过口灌和腹腔注射两种给药方式，研究伊维菌素在虹鳟体内的药学特征。采用高效液相色谱-紫外检测法（HPLC-UV）研究了以 0.3 mg/kg 剂量分别单次口灌和腹腔注射伊维菌素后，药物在虹鳟体内的药代动力学特征。结果表明：口灌给药方式下，伊维菌素在虹鳟血浆、肌肉中的药物浓度和时间关系符合一级吸收二室开放模型，而在肝脏、肾脏中符合一级吸收一室开放模型；腹腔注射给药方式下，伊维菌素在虹鳟血浆、肌肉、肝脏和肾脏中的药时关系均符合一级吸收一室开放模型。口灌给药后，伊维菌素在血浆中主要的药动学参数：AUC 为 123.709 mg/（L·h）、K_a 为 3.749 h^{-1}、$T_{1/2ke}$ 为 466.569 h，T_{max} 为 0.454 h，C_{max} 为 0.319 mg/L。腹腔注射给药后，伊维菌素在血浆中主要的药动学参数：AUC 为 52.560 mg/（L·h）、K_a 为 5.033 h^{-1}、$T_{1/2ke}$ 为 103.236 h、T_{max} 为 1.317 h、C_{max} 为 0.350 mg/L。说明在两种给药方式下，伊维菌素在虹鳟体内的药动学特征存在明显的差异，伊维菌素在腹腔注射给药方式下吸收和消除速率均较口灌给要快。

关键词：伊维菌素；虹鳟；药物代谢动力学

三、吡喹酮

该文发表于《Diseases of Aquatic Organisms》【2016, 119: 67-74】

Pharmacokinetics and residue depletionof praziquantel in ricefield eel (*Monopterus albus*)

Ning Xu[1,2,3], Jing Dong[1,2,3], Yibin Yang[1], Xiaohui Ai[1,3] *

(1. Freshwater Fish Germplasm Quality Supervision and Testing Center, Ministry of Agriculture, Yangtze River Fisheries Research Institute, Chinese Academy of Fishery Sciences, Wuhan 430223, China; 2. Freshwater Fisheries Research Center, Chinese Academy of Fishery Sciences, Wuxi, 214081, China; 3. Hu Bei Freshwater Aquaculture Collaborative Innovation Center, Wuhan 430070, China

Absteract: We investigated the pharmacokinetic characteristics of praziquantel (PZQ) in ricefield eels Monopterus albus. Pharmacokinetic parameters were determined following a singleintravenous administration (5 mg · kg^{-1} body weight [bw]) and a single oral administration (10 mg · kg^{-1} · bw) at 22.0±0.7℃. We also evaluated residue depletion in tissues following daily administrationof PZQ (10 mg kg^{-1} bw) that was given orally for 3 consecutive days at 22.0±0.7℃. Followingintravenous treatment, the plasma concentration-time curve was best described by a 3-compartment open model, with distribution half-life ($T_{1/2\alpha}$), elimination half-life ($T_{1/2\beta}$), and areaunder the concentration-time curve (AUC) of 0.54 h, 17.10 h, and 14505.12 h · μg/L, respectively. After oral administration, the plasma concentration-time curve was best described by a 1-compartmentopen model with first-order absorption, with absorption half-life (t1/2Ka), elimination halflife (t$_{1/2Ke}$), peak concentration (C_{max}), time-to-peak concentration (T_{max}), and AUC estimated to be2.28 h, 6.66 h, 361.29 μg/L, 5.36 h, and 6 065.46 h · μg^{-1}, respectively. The oral bioavailability (F) was 20.9%. With respect to residue depletion of PZQ, the t1/2βvalues of muscle, skin, liver, andkidney were 20.2, 28.4, 14.9, and 54.1 h, respectively. Our results indicated rapid absorption, rapidelimination, and low bioavailability of PZQ in rice field eels at the tested dosing conditions.

Key words: Pharmacokinetics, Residue depletion, Praziquantel, · Monopterus albus

吡喹酮在黄鳝体内药动学和残留消除研究

胥宁[1,2,3], 董晶[1,2,3], 杨移斌[1], 艾晓辉[1,3]

(1. 中国水产科学研究院长江水产研究所, 淡水鱼类种质资源质量监督检验中心, 中国武汉 430223;
2. 中国水产科学研究院淡水渔业研究中心, 中国无锡214081; 3. 湖北淡水养殖协同创新中心, 中国武汉430070)

摘要: 吡喹酮在黄鳝体内的药动学特性和残留消除规律。在水温为 (22.0±0.7)℃下, 经单次口服和静脉注射之后, 测定药动学参数。经连续灌胃3天之后, 评估吡喹酮在黄鳝体内残留消除。经静脉注射之后, 血浆药时曲线能够被3室模型进行很好的拟合。分布半衰期、消除半衰期和药时曲线下面积分别为0.54 h、17.1 h和14 505.12 h · μg/L。经口服之后, 血浆药时曲线能够被一级吸收1室模型进行很好的拟合。吸收半衰期、消除半衰期和峰浓度、达峰时间和药时曲线下面积分别为2.28 h、6.66 h、361.29 μg/L、5.36 h和6 065.46 h. μg/L。生物利用度为20.9%。对于吡喹酮的残留消除, 肌肉、皮肤、肝脏和肾脏的消除半衰期分别为20.2 h、28.4 h、14.9 h和54.1 h。本研究表明吡喹酮在黄鳝体内呈现出了快速吸收、快速消除和低生物利用度的特性。

关键词: 药动学、残留消除、吡喹酮、黄鳝

Introduction

Rice field eel (*Monopterus albus*), commonly known as "ginseng in water", is a fresh water fish species distributed in China,

Japan, and Southeast Asia. As its flesh contains large amounts of protein, vitamins and minerals, this species is valued as a food resource in addition to having medicinal values. The rice field eelhas become one of the most popular fish species in China and it is widely and intensively cultured in many provinces. For example, it is estimated that in excess of 30 million square meters is used for rice field eel cultured in the Hubei province, yielding an annual production of 65 thousand metric tons and 2.0 billion RMB by value in 2004 (Chen 2010).

With the development and expansion of rice field eel culture, a large number of disease issues have been identified. These diseases have impeded the development of the rice field eel aquaculture industry with serious economic consequences. Of these parasitic diseases are very important as they often cause reductions in feed conversion and growth, as well as causing direct and indirect mortalities of rice field eel. Previous studies have identified 19 species of parasite in rice field eels from China which include Neosentis celatus, Pallisentis (Neosentis) celatus, Trypanosoma monopter, Cryptobia agitate, Agamospirura sp., Eustrongylides sp., Philometrosis sp., Paraseuratoides acuminicauda, Polyuoncobothrium magnum, Polyoncobothrium sp., Azygia anguilae, Diplostomulum hupehensis, Diplostomulum hupehensis, Diplostomulum nieckashui, Phyllodistomum serrispatula, Cephalogonimidae gen et sp., Deretrema plaglorchis, Prosorchis tianjinensis, and Proterometra guangzhouensis (Wen 2003, Huang et al. 2014). Of these infections with Neosentis celatus, Pallisentis (Neosentis) celatusand Eustrongylides sp. are more common than others (Huang et al. 2014).

In order to reduce the economic losses caused by parasitesin rice field eels, it is important that an appropriate drug for therapy is chosen. Praziquantel (PZQ), 2-cyclohexylcarbonyl-1, 2, 3, 6, 7, 11b-hexahydro-4H-pyrazino [2, 1-a] iso-quinoline-4-one, is a broad-spectrum anthelmintic drug used in human and veterinary medicine. In aquaculture, PZQ is mainly usedfor controlling and killing acanthocephalans (Paterson et al. 2013), trematodes (Kim & Cho 2000, Montero et al. 2004, Mansell et al. 2005, Reimschuessel et al. 2011), cestodes (Pool 1985, Kline et al. 2009, Iles et al. 2012) and other parasites of fish. Although PZQ is applied widely in the treatment of aquatic animals, only a few studies have examined the pharmacokinetics and tissues residuesof PZQ in fish (Kim et al. 2001, Kim et al. 2003, Tubbs & Tingle 2006b, a, Osman et al. 2008, Xie et al. 2015). Furthermore, no information is available on the pharmacokinetic profile and residue depletion of PZQ in rice field eel (Monopterus albus). Therefore, the goal of this study was to investigate aspects of the pharmacokinetics of PZQ in the rice field eel including bioavailability and tissues residues.

Materials and methods

Chemicals and reagents. Analytical standard PZQ (purity grade 99.5%) and internal standard PZQ-d_{11} (purity grade 98.0%) wereobtained from Dr. Ehrenstorfer GmbH., Augsburg, Germany). ThePZQ powder (purity grade 98.0%) used for oral and intravenous administration was supplied by Zhongbo aquaculture Biotechnology Co. Ltd., Wuhan, China. The anesthetic tricaine methanesulfonate (MS222) was purchased from Aibo Biotechnology Co. Ltd., Wuhan, China. The purity of other chemicals and reagents was analytical grade.

Stock standard solutions of PZQ (100 μg/mL) and PZQ-d_{11} (100 μg/mL) wereprepared by dissolving these chemicalsin dimethyl sulfoxide. Stock standards were stored at-18℃ and were stable for six months. Intermediate solvent standard solutions (1 μg/mL) wereprepared by diluting PZQorPZQ-d_{11} stock solutions with acetonitrile. These solutions were stored at 4℃.

The PZQ solution for intravenous administration was prepared by dissolving PZQ power in a minute quantity of dimethyl sulfoxide (≤1% of total volume) and adjusting with physiological saline to the final concentration of 5 mg/mL.

The PZQ solution for oral administration was prepared by dissolving PZQ power in a minute quantity of dimethyl sulfoxide (≤1% of total volume) and adjusting with physiological saline to the final concentration of 10 mg/mL.

Animal Care and Husbandry. A total of 200ricefield eels (150±15 g) were obtained from culturing base of Yangtze River Fisheries Research Institute. The eels were held in plastic tanks (100×60×80cm, length×width×depth) that were supplied with flowing well water (26 L/min). Water quality parameters were measured daily in each tank, and water quality was maintained as follows: total ammonia nitrogen levels ≤ 0.75 mg/L; nitrite nitrogen levels 0.06 mg/L; dissolved oxygen levels at 6.5 - 7.9 mg/L; temperature at 22.0±0.7℃ and pH 7.3±0.2. Eels were fed a drug-free feed and were acclimatized for 14 days prior to commencing the experiment. The fish were divided into three groups, the pharmacokinetic study group (n=138), the residue studygroup (n=30) and controls (n=6). Prior to drug administration, the eels fasted for 24 h. The control samples were collected from six eels on the last sample date. All the experimental procedures involving animals in the study were approved by the Animal Care Center, Hubei Academy of Medical Sciences.

Pharmacokinetic study and sampling. Eels in pharmacokinetic studygroup were further divided into two groups, one group that received a single intravenous treatment and anotherreceiveda single oral treatment. The dose that was used for oral administration was the dose that is specified for treatment of fish bythe National Standards for Veterinary Medicine in China. Asthe solubility of PZQ is only 0. 04 g in water (100 g), then we used a low concentration (≤1% of total volume) of dimethyl sulfoxide to increase its solubility in physiological saline. The PZQ concentration for intravenous administration is determined to be as 5 mg/mL.

Eels in the intravenous treatment group were anaesthetized with MS222 (50 mg/L) andinjected at a dose of 5 mg/kg body weight (b. w.) by 1 mL micro-injector. The position of the needle in the caudal vein was confirmed by aspirating a small volume of blood into the syringe prior to injection. Fish that experienced heavy bleeding after withdrawal of the needle or in which the needle had translocated during injection, wereremoved from the study and replaced. Samples of 6eels were collected at 0. 083, 0. 167, 0. 5, 1, 2, 4, 8, 16, 24, 36, 48, 72 h after intravenous treatment. At sampling eels were anaesthetized with MS 222 (50 mg/L) and approximately1. 5mL of blood fromeach eel was drawn from the caudal vein away from the injection site. Blood samples were treated with heparin sodium solution (1% heparin sodium in water) andcentrifuged at 1 500 g for 10 min. The plasma was collected and stored at-20℃ until assayed. Following sampling, the fish were euthanized by an overdose in a solution of 300 mg/L MS 222.

Eels in the oral treatment group were anaesthetized with MS 222 (50 mg/L) prior to receiving an oral dose of 10 mg/kg b. w. PZQ. The oral dose was administered by inserting a plastic hose attached to a 1-mL micro-injector into the stomach. After treatment, each fish was observed in a separate tank for regurgitation of PZQ. If PZQ solution was regurgitated, the fish was removed from the study and replaced. Subsequently, six fish were taken from the tank at each sampling point (0. 083, 0. 167, 0. 5, 1, 2, 4, 8, 16, 24, 48, 72h) after the oral dose. At sampling fishwere anaesthetized with MS 222 (50 mg/L) and blood samples of approximately 1. 5 mL were collected and treated with 1% heparin sodium. Processing of the blood samples and storage of the plasma was conducted as described above. Following sampling, the fish were euthanized by an overdose in a solution of 300 mg/L MS 222.

Residue depletion and sampling. To examine PZQ tissues residues, eels were anaesthetized with MS 222 (50 mg/kg) prior to receiving an oral dose of 10 mg/kg b. w. PZQ, which was administered daily for 3 days. The oral dose was administered as described above. At 16, 24, 48, 72 and 96 h after the final dose 6 eels were anaesthetized with MS 222 (50 mg/L), then blood and tissue samples were taken. Approximately 1. 5 mL of blood was drawn from each fish, treated with 1% heparin sodium solution, and centrifuged at 1 500 g for 10 min. The resulting plasma collected and stored at 20℃ until assayed. Tissues (liver, kidney, muscle, and skin) were collected from standardized locations and transferred to 50 mL polypropylene tubes. These samples were stored frozen at-20℃ until assayed.

Analytical method. Levels of PZQ in plasma and tissues were determined by liquid chromatography triple quadrupole mass spectrometry (LC-MS-MS) (Thermo Scientific, San Jose, CA, USA). The LC-MS-MS systemconsisted of a Surveyor Llus HPLC and TSQ Quantum Access Max. Datawere obtained and processed using the Thermo Xcalibur software (Copyright 2. 1. 0). The chromatographic separation was performed on a Thermo Hypersil BDSoctadecylsilane column (100 mm× 2. 1 mm, 3 μm). The mobile phase was water containing 0. 1% formic acid andacetonitrile (70: 30, v/v). The total run time of a sample was 8. 0 min. The injection volume was 10 μL, and the column temperature was 30℃.

The instrument was operated using heated electrospray ionization (ESI) inpositive ion mode. The ion source parameters were optimized by monitoring thePZQ mass spectra. Selective reaction monitoring (SRM) was performed on PZQ protonated molecular ions using a spray voltage of 3500 V, vaporizertemperature of 300℃, ion transport tube temperature of 350℃, sheath gas (high puritynitrogen) of 40 arb, auxiliary gas (high purity nitrogen) of 10, collision gas (ultra-high purity argon) pressureof 1. 50 m Torr, scan width of 0. 010, and a scan time of 0. 200 s. Selective reaction monitoringfor PZQ was m/z 313. 3 to 174. 1 (qualitative) with a collision energy of 28 eV, and m/z 313. 3 to 203. 1 (quantitative) with a collision energy of 16 eV. Selective reaction monitoring for PZQ-d$_{11}$was m/z 324. 1 to 204. 1 (quantitative) withcollision energy of 17 eV.

Plasma and tissues samples were prepared using a modification of the methods given in Kim et al. (2001). Briefly, plasma samples were thawed at 25℃, vortexed for 2min and 1mL subsamples were transferred into 10 mL polypropylene tubes. To each tube 20 μL of internal standard (1 μg/mL) was added and the resulting plasma/internal solutions were vortexed for 1 min. 1 mL of acetonitrilewas added into the10 mL polypropylene tubes, mixed and incubated for 10 min at 4℃, after which they werevortexed and centrifuged at 5500 g for 5 min. The resulting supernatants were filtered through 0. 22 μm nylon filters prior toLC-MS-MS analysis.

Tissues were thawed at 25℃ and 2 g portions of the various tissues were weighed into 50mLpolypropylene tubes. To each tube10 μL of internal standard (1 μg/mL) was added and the tissue homogenized using ahomogenizer (PRO Scientific, USA). After homogenization, 5 mL of acetonitrile was added into the tube, and the samples were mixed and incubated for 10 min at 4℃, after which they were vortexed and centrifuged at 5 500 g for 5 min. The resulting supernatants were thentransferred tonew tubes. The tissues were extracted a second time as described above and the supernatants from the two extractions were pooled. Supernatants were evaporated to dryness with a gentle stream of nitrogen at 40℃ and the residue was reconstituted in 1mL of acetonitrile and water (30: 70). The resulting solutions were centrifuged at 5 500 g for 5 min, filtered through 0. 22 μm nylon filtersprior toLC-MS-MS analysis.

Calibration curves and recovery rates. Plasma and tissue samples were obtained from untreated fish following the methods described above. Plasma from untreated eels was spiked with a standard solution of PZQ and the internal standard to yield concentrations of 5, 50, 500, 3 000, and 6 000 μg/L and 10 μg/L internal standard. Tissues (muscle, liver, kidney and skin) from untreated fish were also spiked with a standard solution of PZQ and internal standard to yield concentrations of 5, 50, 500, 1000, and 2 000 μg/kg and 10 μg/kg internal standard. Samples were preparedas described above, and each concentration was assayed in triplicate. Precision and accuracy were determined by analyzing five replicates of spiked plasma and tissue samples at 5, 50, and 500 μg PZQ/L.

Pharmacokinetic and statistical analyses. Pharmacokinetic parameters were calculated using Practical Pharmacokinetic Program 3P97 (Mathpharmacology Committee, Chinese Academy of Pharmacology, Beijing, China). The appropriate pharmacokinetic model was selected by the visual examination of concentration-time curve and by application of Akaike's Information Criterion (AIC) (Yamaoka et al. 1978). The elimination half-life ($t_{1/2\beta}$) was calculated by $t_{1/2\beta} = 0.693/\beta$. The oral bioavailability for PZQ was determined by calculating the area under concentration-time curve after oral and intravenous administration and using the following equation:

$$F = \frac{(\text{AUC oral administration}) \times (\text{dose intravenous injection})}{(\text{AUC intravenous injection}) \times (\text{dose oral administration})} \times 100$$

Statistical analysis among groups was completedusing the SPSS 13.0 program.

Results

Method validation. The limit of detection for PZQ was 5 μg/Lin plasmaand 5 μg/kg in tissuesby LC-MS-MS analysis. The good linearity of curves over range of 5 and 6 000 μg/Lin plasma and 5 and 2 000 μg/kg in tissues was exhibited by the coefficient of correlation $R^2 = 0.999$ (Table 1). The mean±SD recovery rates of PZQ ranged from 95.28±1.23% to 99.80±2.62% in plasma, 93.70±2.92% to 99.93±3.32% in muscle, 95.61±4.55% to 103.89±3.89% in liver, 92.74±1.58% to 98.15±3.01% in kidney, and 94.52±3.52% to 100.92±2.89%in skin, respectively. Their percentage relative standard deviations for inter-day and intra-day precision were ≤ 10% (Table 2).

Table 1　The calibration equation and correlation coefficientof matrix-fortified in plasma and tissues

Tissues	Linear range (μg/L or μg/kg)	Calibration equation	Correlation coefficient (R2)
Plasma	5.0-6 000.0	y=0.059 9x-0.239 4	0.999 4
Muscle	5.0-6 000.0	y=0.075 3x-0.098 2	1
Liver	5.0-6 000.0	y=0.081 2x-0.133 2	0.999 7
Kidney	5.0-6 000.0	y=0.070 2x-0.457 8	0.999 1
Skin	5.0-6 000.0	y=0.075 2x-0.237 6	0.999 3

Table 2　Recovery efficiency of praziquantel in samples of plasma and tissues that were spiked with different concentrations (n=3)

Tissues	Amount added (μg/L or μg/kg)	Recovery (%)	Relative standerd deviation (%)
Plasma	5	95.28	1.23
	50	99.80	2.62
	500	99.71	2.43
Muscle	5	93.70	2.92
	50	96.71	2.87
	500	99.93	3.32
Liver	5	95.61	4.55
	50	98.21	3.57
	500	103.89	3.89

续表

Tissues	Amount added (μg/L or μg/kg)	Recovery (%)	Relative standerd deviation (%)
Kidney	5	92.74	1.58
	50	98.15	3.01
	500	96.45	2.77
Skin	5	94.52	3.52
	50	100.92	2.89
	500	97.28	2.57

Pharmacokinetics and depletion of PZQ in rice field eel. The plasma concentration-time curves after a single intravenoustreatment or a single oral treatment were shown in Fig. 1. The pharmacokinetic parameters were listed in Table 3. After intravenous administration at a dose of 5 mg/kg b. w. , the dataforplasma were best described by a three-compartment open model. The distribution half-life ($t_{1/2\alpha}$) and elimination half-life ($t_{1/2\beta}$) were 0.54h and 17.1 h, respectively. The area under plasma curve (AUC) and total body clearance (CL_b) were estimated to be 14 505.12 μg. h/L and 0.052 L/h, respectively. After oral administration at a dose of 10 mg/kg b. w. , the data were fit to a one-compartment open model with first order absorption. Theabsorption rate constant (K_a), elimination rate constant (K_e), absorption half-life ($t_{1/2Ka}$) and elimination half-life ($t_{1/2Ke}$) were 0.30/h, 0.10/h, 2.28/h and 6.66/h, respectively. The corresponding values of the apparent distributionvolume (V/F), AUC, peak concentration (C_{max}), time-to-peak concentration (T_{max}), were calculated to be 2.36 L, 6 065.46 μg. h/L, 361.29 μg/L and 5.36 h. The oral bioavailability (F) was 20.91%.

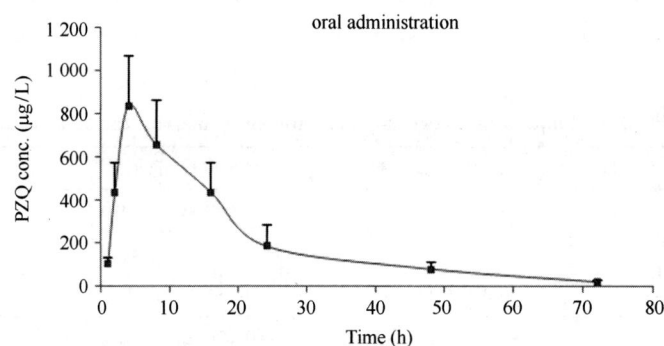

Fig. 1   Mean plasma concentration vs. time profiles of praziquantel in rice field eel following a single intravenous administration (5 mg/kg b. w.) and a single oral administration (10 mg/kg b. w.) (n=6 samples per time point)

P, A, B: Zero-time blood drug concentration intercept of triphasic disposition curve; π, α: Distribution rate constants; β: Elimination rate constant; K_a: Absorption rate constants; K_e: Elimination rate constants; $t_{1/2\pi, \alpha}$: Distribution half-lives; $t_{1/2\beta}$: Elimination half-life; $t_{1/2Ka}$: Absorption half-life; $t_{1/2Ke}$: Elimination half-life; K_{10}: Drug elimination rate constant from central compartment; K_{12}, K_{21}: First-order transport rate constant between the central compartment and the shallow peripheral compartment; K_{13}, K_{31}: First-order transport rate constant betweenthe central compartment and the deep peripheral compartment; AUC: Area under concentration-time curve; T_{max}: Time to peak concentration; C_{max}: Peak concentration; V/F: extensive apparent volume of the central compartment; CL_b: total body clearance of the drug.

Table 3   Pharmacokinetic parameters for praziquantelin rice field eel following a single intravenous administration (5 mg/kg b. w.) or a single oral administration (10 mg/kg b. w.) (n=6 for each treatment)

Intravenous administration		Oral administration	
Parameters	Values	Parameters	Values
P (μg/L)	20 227.29		
Π (1/h)	19.56		
A (μg/L)	1 499.86		

续表

Intravenous administration		Oral administration	
Parameters	Values	Parameters	Values
α（1/h）	1. 27		
B（μg/L）	498. 21	Ka（1/h）	0. 30
β（1/h）	0. 041	Ke（1/h）	0. 10
$t_{1/2\pi}$（h）	0. 035		
$t_{1/2\alpha}$（h）	0. 54	t1/2Ka（h）	2. 28
$t_{1/2\beta}$（h）	17. 10	t1/2Ke（h）	6. 66
$K_{12(}$1/h）	9. 74		
K_{21}（1/h）	2. 74		
$K_{13(}$1/h）	6. 62	V/F（L）	2. 36
K_{31}（1/h）	0. 24	Tpeak（h）	5. 36
K_{10}（1/h）	1. 53	Cmax（μg/L）	361. 29
AUC（μg. h/L）	14505. 12	AUG（μg. h/L）	6 065. 46
CLb（L/h）	0. 052	CLb（L/h）	0. 25

The mean concentrations of PZQ following multiple-dose treatment were given in Table 4. The concentrations of PZQ in muscle were significantly lower than that in other tissues and levels of PZQ were below the level of detection at 72 h after the final dose. Although the concentrations of PZQ in liver were significantly higher than that in muscle, kidney and skin from 16 h to 48 h post-treatment, it was also undetectable at 72 h. The residual time of PZQ in skin and kidney was longer than that of liver and muscle, and residues were still detected at 72 h after final dose in both of these tissues. The concentration of PZQ was highest in plasma at 16 hours, declining rapidly so that it was below the level of detection by 72 h after final dose.

Table 4　Praziquantel concentrations in plasma and tissues of rice field eel after oral administration ata dose of 10 mg/kg b. w. for 3 consecutive days（n=6 samples per time point）

Times（h）	PZQ Concentrations（μg/L or μg/Lkg）				
	Plasma	Liver	Kidney	Muscle	Skin
16	623. 8±23. 9	430. 7±30. 8	33. 1±4. 5	29. 3±3. 1	198. 1±20. 7
24	173. 5±18. 7	354. 7±12. 5	32. 2±3. 9	24. 2±4. 0	68. 9±11. 3
48	53. 2±5. 5	102. 4±8. 9	19. 9±2. 4	10. 0±2. 1	43. 7±5. 5
72	N. D.	N. D.	17. 4±1. 3	N. D.	21. 4±3. 2
96	N. D.	N. D.	N. D.	N. D.	N. D.

N. D. =not detected

Discussion

This study was designed to examine pharmacokinetic characteristics and depletion regularity of PZQ in the rice field eel (*Monopterus albus*). A three-compartment model best described the pharmacokinetics of PZQ in eelsafter intravenous treatment, whereas a one-compartment model with first-order absorption best fit the data from the oral treatment. Only a few studies have examined thepharmacokinetics of PZQ after a single dosein fish. The plasma concentration-time curves of grass carp (*Ctenopharyngodon idellus*) were best described by a two-compartmental model after a single oral administration at dose of 10 mg/kg b. w. in freshwater and brackish water at 22±1℃（Xie et al. 2015）. For kingfish (*Seriola lalandi*) following intravenous and oral administration at dose of 40 mg/kg b. w. at 18-19℃, a non-compartmentmodel best described the data（Tubbs & Tingle 2006a）. It is reported that these differences might result from fish species, sizes, temperatures, salinity of the environment and experimental protocols（Rigos et al. 2002）.

In this study, the time to peak concentration (T_{max}) and the peak concentration (C_{max}) following oral administration were estimated to be as 5.36 h and 361.29 μg/L, respecitvely. Tubbs and Tingle (2006a) reported a peak concentration of 12.73 μg/mL at 1 h after oral treatment at a dose of 40 mg/kg b.w.. The concentrationsthat they reported are much higher than achieved in this study and this difference might be due to their use of a much higher dose (Kim et al. 2003). The elimination half-life ($t_{1/2\beta}$) after intravenousand oral treatment in rice field eel were 17.10 h and 6.66 h, respectively. The $t_{1/2\beta}$ valueinking fishafter intravenous treatment was about half of that in rice field eel, and the $t_{1/2\beta}$ value after oral treatment was alsolower than that in rice field eel (Tubbs & Tingle 2006a). Hansen and Horsberg (2000) reported a much longer distribution half-life for the European eel (Anguilla) following intravenous and oral administrations of flumequine when compared to other fish species. These authors suggested that physiological and anatomical differences of eel gills may be responsible for this difference. The gills are regarded as one of important sites of elimination of xenobiotics in fish (Hansen et al. 2001), however, the gills and the circulation system of eels are characterized by an extremely low gill surface area, low blood hematocrit, small ventricular masses and relatively low cardiac output, which results in the gills being less perfused than in other fish (Byczkowska-Smyk 1958, Hansen & Horsberg 2000). In mammals, $t_{1/2\beta}$ values are significantly shorter than that in fish. For example, it was reported that$t_{1/2\beta}$ values in pig after intravenousat a dose of 10 mg/kg b.w. and oral treatment at a dose of 50 mg/kg b.w. were 1.5 h and 1.07 h, respectively (Zeng et al. 1993). In cattle, $t_{1/2\beta}$ values after intravenous treatment at a dose of 10 mg/kgb.w. were also shorter than that in fish, but itsvalue after oral treatment at a dose of 30 mg/kg b.w. was close to values obtained from fish (Cao et al. 2011). These differences between mammals and fishmight be due to differences in metabolic capacity. Similar differences have been observed between fish and mammalsfor the depletion of melamine and cyanuric acid (Stine et al. 2012).

The oral bioavailability (F) of PZQ in rice field eel was 20.9% in present study. Oral bioavailability of PZQ has been estimated for only a few fish species. Considerably higher F value (50.8%) was found in kingfish after oral administration (40 mg/kg b.w.) at 18-19℃ (Tubbs & Tingle 2006a). The large difference in F values between ours and their study is likely due to their use of a much higher (4X) dose of PZQ as other factors such as the route of drug administration and water temperatures. The bioavailability of PZQ has been reported to be relatively low or low in other animals including cattle (32.3%) (Cao et al. 2011) and pigs (3.2%) (Zeng et al. 1993). Low bioavailability may be caused by rapid and extensive metabolism of PZQ in the liver of animals to mono-and dihydroxylated derivatives. This view is supported by a lack of PZQ detection in the excretion products ofrats, dogs and monkeys that received intravenous and oral doses of^{14}C-PZQ (EMEA 1996). In addition, the metabolism of PZQ was found to be rapid; 15 minutes after oral and intravenousdosing, the serum radioactivity consisted of 99% and 54%metabolites in rats, 84% and 59%in dogs, and 99% and 50%in monkeys, respectively (EMEA 1996).

Factors effecting metabolism of PZQ in fish are required further investigation. It has been reported that PZQ is metabolized into hydroxylated derivatives in kingfish, however the predominant metabolite was different from that found in mammals (Tubbs et al. 2008). Some studies have demonstrated that the site of hydroxylation was in the cyclohexyl ring of mammals (Lerch & Blaschke 1998, Meier & Blaschke 2001, Schepmann & Blaschke 2001) however whether this is the case in fish is unknown.

The residue depletion of PZQ was determined in various tissues of rice field eel. At 16 hours post treatment PZQ concentrations were much higher in liver when compared to the other tissues, however, and these concentrations rapidly declined to below the level of detection by 72 hours. In kidney and skin levels were initially lower and PZQ was retained for longer periods of time. The $t_{1/2\beta}$ values of PZQ in liver, kidney, muscle and skin were 14.9 h, 54.1 h, 20.2 h and 28.4 h, respectively. Higher concentrations and longer $t_{1/2\beta}$ value were observed in muscle of rockfish (Sebastes schlegeli) after a single oral treatment of PZQ at dose of 400 mg/kg b.w. (Kim et al. 2001, Kim et al. 2003). A similar result was reported for rainbow trout (Oncorhynchus mykiss) following a single oral administration at dose of 500 mg/kg b.w. (Bjorklund & Bylund 1987). The differences between our results and those obtained for rockfish and rainbow trout are likely due to our use of a much lower concentrations of PZQ, as well as differences in water temperature and other conditions of culture. Regardless of the dose, rapid depletion of PZQ occurred in every study. In rainbow trout, 67%-96% of the maximum amounts had been excreted at 32 h after oral administration (Bjorklund & Bylund 1987). In mammals, PZQ is more rapidly eliminated (Zeng et al. 1993, Cao et al. 2011).

The present study demonstrated that PZQ was rapidly distributed and rapidly eliminated in rice field eel after oral and intravenous treatment. The residual time of PZQ was also short in rice field eel following daily administration over a period of 3 days. These data along with the known efficacy of PZQ against a wide range of parasites suggests that this drug may be suitable for the treatment of parasitic diseases of the rice field eel.

Acknowledgments

This work is financially supported by National Nonprofit Institute Research Grant of Freshwater Fisheries Research Center, CAFS

（2015JBFM29）and the Special Fund for Agro-Scientific Research in the Public Interest of China（201203085）.

References

Bjorklund H, Bylund G（1987）Absorption, distribution and excretion of the anthelmintic praziquantel（Droncit）in rainbow trout（Salmo gairdneri R.）. Parasitol Res 73：240-244

Byczkowska-Smyk W（1958）The respiratory surface of the gills in teleosts. Acta Biol Cracov Ser Zool 1：831-837

Cao JY, Liu EY, Zhao JL, Li KB, Dou SL（2011）Pharmacokinetics and bioavailability of praziquantel in cattle after oral, intramuscular and intravenous administration. Chinese Journal of Veterinary Science 21：612-614

Chen XL（2010）Study on preventive measures for Aeromonas hydrophila and Aeromonas sobria in rice field eel（Monopterus albus）. Master's thesis, Sichuan Agricultural University. Sichuan, China.

EMEA（1996）European Medicines Evaluation Agency. Praziquantel Summary Report by CVMP. EMEA/MRL/141/96 - Final September 1996 EMEA, London

Hansen MK, Horsberg TE（2000）Single-dose pharmacokinetics of flumequine in the eel（Anguilla anguilla）after intravascular, oral and bath administration. J Vet Pharmacol Ther 23：169-174

Hansen MK, Ingebrigtsen K, Hayton WL, Horsberg TE（2001）Disposition of 14C-flumequine in eel Anguilla anguilla, turbot Scophthalmus maximus and halibut Hippoglossus hippoglossus after oral and intravenous administration. Dis Aquat Organ 47：183-191

Huang JR, Chen XX, Huang BL, Chen S, Li D, Li L, Chen JP（2014）Parasites in rice field eels in mainland China and Related human infectious. Journal of Preventive Medicine Information 30：777-782

Iles AC, Archdeacon TP, Bonar SA（2012）Novel Praziquantel Treatment Regime for Controlling Asian Tapeworm Infections in Pond-Reared Fish. N Am J Aquacult 74：113-117

Kim CS, Cho JB, Ahn KJ, Lee JI, Kim KH（2003）Depletion of praziquantel in muscle tissue and skin of cultured rockfish（Sebastes schlegeli）under the commercial culture conditions. Aquaculture 219：1-7

Kim KH, Cho JB（2000）Treatment of Microcotyle sebastis（Monogenea：Polyopisthocotylea）infestation with praziquantel in an experimental cage simulating commercial rockfish Sebastes schlegeli culture conditions. Dis Aquat Organ 40：229-231

Kim KH, Kim CS, Kim JW（2001）Depletion of praziquantel in plasma and muscle tissue of cultured rockfish Sebastes schlegeli after oral and bath treatment. Dis Aquat Organ 45：203-207

Kline SJ, Archdeacon TP, Bonar SA（2009）Effects of Praziquantel on Eggs of the Asian Tapeworm Bothriocephalus acheilognathi. N Am J Aquacult 71：380-383

Lerch C, Blaschke G（1998）Investigation of the stereoselective metabolism of praziquantel after incubation with rat liver microsomes by capillary electrophoresis and liquid chromatography-mass spectrometry. J Chromatogr B Biomed Sci Appl 708：267-275

Mansell B, Powell MD, Ernst I, Nowak BF（2005）Effects of the gill monogenean Zeuxapta seriolae（Meserve, 1938）and treatment with hydrogen peroxide on pathophysiology of kingfish, Seriola lalandi Valenciennes, 1833. J Fish Dis 28：253-262

Meier H, Blaschke G（2001）Investigation of praziquantel metabolism in isolated rat hepatocytes. J Pharm Biomed Anal 26：409-415

Montero FE, Crespo S, Padros F, De la Gandara F, Garcia A, Raga JA（2004）Effects of the gill parasite Zeuxapta seriolae（Monogenea：Heteraxinidae）on the amberjack Seriola dumerili Risso（Teleostei：Carangidae）. Aquaculture 232：153-163

Osman HAM, Borhn T, El Deen AEN, El-Bana LF（2008）Estimation of skin and gill monogenea and musculature in O. niloticus treated by praziquantel using HPLC. Life Science Journal-Acta Zhengzhou University Overseas Edition 5：77-82

Paterson RA, Rauque CA, Valeria Fernandez M, Townsend CR, Poulin R, Tompkins DM（2013）Native fish avoid parasite spillback from multiple exotic hosts：consequences of host density and parasite competency. Biol Invasions 15：2205-2218

Pool DW（1985）The effect of praziquantel on the pseudophyllidean cestode Bothriocephalus acheilognathi in vitro. Z Parasitenkd 71：603-608

Reimschuessel R, Gieseker C, Poynton S（2011）In vitro effect of seven antiparasitics on Acolpenteron ureteroecetes（Dactylogyridae）from largemouth bass Micropterus salmoides（Centrarchidae）. Dis Aquat Organ 94：59-72

Rigos G, Tyrpenou A, Nengas I, Alexis M（2002）A pharmacokinetic study of flumequine in sea bass, Dicentrarchus labrax（L.）, after a single intravascular injection. J Fish Dis 25：101-105

Schepmann D, Blaschke G（2001）Isolation and identification of 8-hydroxypraziquantel as a metabolite of the antischistosomal drug praziquantel. J Pharmaceut Biomed 26：791-799

Stine CB, Nochetto CB, Evans ER, Gieseker CM, Mayer TD, Hasbrouck NR, Reimschuessel R（2012）Depletion of melamine and cyanuric acid in serum from catfish Ictalurus punctatus and rainbow trout Onchorhynchus mykiss. Food Chem Toxicol 50：3426-3432

Tubbs L, Mathieson T, Tingle M（2008）Metabolism of praziquantel in kingfish Seriola lalandi. Dis Aquat Organ 78：225-233

Tubbs LA, Tingle MD（2006a）Bioavailability and pharmacokinetics of a praziquantel bolus in kingfish Seriola lalandi. Dis Aquat Organ 69：233-238

Tubbs LA, Tingle MD（2006b）Effect of dose escalation on multiple dose pharmacokinetics of orally administered praziquantel in kingfish Seriola lalandi. Aquaculture 261：1168-1174

Wen AX（2003）Study on parasites in ricefield eels. Journal of Sichuan Agricultural University 21：43-46

Xie X, Zhao Y, Yang X, Hu K（2015）Comparison of praziquantel pharmacokinetics and tissue distribution in fresh and brackish water cultured grass carp（ctenopharyngodon idellus）after oral administration of single bolus. Bmc Veterinary Research 11

Yamaoka K，Nakagawa T，Uno T（1978）Application of Akaike's information criterion（AIC）in the evaluation of linear pharmacokinetic equations. J Pharmacokinet Biopharm 6：165-175

Zeng ZL，Chen ZL，Feng QH（1993）Pharmacokinetics and oral bioavailability of praziquantel in pigs following a single dose. Acta Vet Zootech Sin 24：170-174

渔药安全使用风险评估及其控制
（下册）

杨先乐　主　编
胡　鲲　副主编

海洋出版社

2020年·北京

目　录

第三章　新渔药及其制剂的创制

新渔药及其制剂的研究和创制是推动渔药发展的重要内容。渔药的研究与创制涉及到药效学研究，药动学与生物有效性研究，毒理学研究，临床试验研究以及药物相互作用的研究等内容。本章主要涉及到禁用药物孔雀石绿的替代药物制剂（美婷）和微生态制剂的创制等方面的内容。

第一节　禁用药物孔雀石绿替代制剂（美婷）的研究与创制

水霉病是水产养殖中一类由丝状真菌引起的疾病的统称，在全国各地均有流行并会可造成较大的经济损失。长期以来防治该病一直使用"孔雀石绿"，但由于该药具有高致癌，高毒性，高残留等安全性问题，2002 年，农业部就将其列入至《食品动物禁用的兽药及其他化合物清单》中。孔雀石绿的禁用对水霉病的防治带来了技术真空。近年来，孔雀石绿屡禁不止，严重威胁水产品质量安全面，引起了政府、民众和媒体的高度关注。为了从技术源头上彻底解决孔雀石绿的禁用问题，上海海洋大学科研团队聚焦水霉病及抗水霉制剂的研制，并取得了一些较有影响的成果。

水霉是一种真菌性病原（张书俊等，2009；夏文伟等，2011；孙琪等，2014；王浩等 2012）。水霉能感染鱼卵（Cao H. et al.，2014；Cao H. et al.，2013；Cao H. et al.，2012；夏文伟等，2011；欧仁建等，2012；李浩然等，2015）和成鱼（张书俊等，2009；夏文伟等，2011；孙琪等，2014）。水霉对温度、pH 和盐度适应较强，在我国主要淡水鱼类养殖区均有分布；可利用核糖体 rDNA 内部转录间隔区（Internal Transcribed Spacer，ITS）序列分析法对水霉菌进行分型研究（Siya Liu et al.，2017）。聚丙烯酰酸凝胶电泳法可用来分析水霉菌的蛋白（宋鹏鹏等，2013）。水霉菌细胞表面的生物膜的形成受到环境的影响（袁海兰等，2014）。水霉菌还存在某些毒力因子，如 *SpHtp*1，含有 *SpHtp*1 基因的菌株在分类地位上均隶属于寄生水霉。*SpHtp*1 基因可能在寄生水霉中特异性存在（Saprolegnia parasitica）（叶鑫等，2008）。

一、水霉病原生物学

该文发表于《Aquaculture and Fisheries》【2017，】

Sequence Analysis and Typing of *Saprolegnia* Strains Isolated from Freshwater Fish from Southern Chinese Regions

Siya Liu，Kun Hu，Peng-Peng Song，Ren-Jian Ou，Xian-Le Yang

（National Pathogen Collection Center for Aquatic Animals，Shanghai Ocean University，Shanghai 201306，China）

Abstract：Saprolegniasis, caused by Saprolegnia infection, is one of the most common diseases in freshwater fish. Our study aimed to determine the epidemiological characteristics of typical saprolegniasis in Chinese regions of high incidence. Saprolegnia were isolated and identified by morphological and molecular biological methods based on internal transcribed spacer（ITS）ribosomal DNA（rDNA）and phylogenetic trees of neighbor-joining（NJ）and maximum parsimony（MP）. The ITS sequences of eight strains were observed and compared with GenBank sequences, which indicated that all strains fell into three clades：CLADE1（02, LP, 04 and 14）, CLADE2（S1）, and CLADE3（CP, S2, L5 and the reference ATCC200013）. Isolates 02 and LP shared 80% sequence similarity with S. diclina, S. longicaulis, S. ferax, S. mixta, and S. anomalies. Further, isolates 04 and 14 shared 80% similarity with S. bulbosa and S. oliviae. Finally, extremely high ITS sequence similarities were identified between isolatesS1 and S. australis（100%）; CP and S. hypogyna（96%）; and S2, L5, ATCC200013 and S. salmonis（98%）. This research provides insights into the identification, prevention and control of saprolegniasis pathogens and the potential development of effective drugs.

Key words：freshwater fish；fish disease；genotyping；ITS sequence；Saprolegnia；saprolegniasis

分离自中国南方淡水鱼类主养区的水霉菌序列分析及分型研究

刘思雅，胡鲲，宋鹏鹏，欧仁建，杨先乐

（上海海洋大学国家水生动物病原库，中国上海，201306）

摘要：由水霉菌感染引起的水霉病是淡水鱼中最常见的疾病之一。我们的研究旨在确定典型水霉病在中国高发地区的流行病学特征。通过形态分类方法和 ITS rDNA 序列分析、以邻接法（NJ）和最大简约法（MP）构建的系统发育树等分子生物学方法对水霉菌进行了分类鉴定。观察 8 株菌株的 ITS 序列，并与 GenBank 序列比较，表明所有菌株分为三个进化枝：序列 1（02、LP、04 和 14），序列 2（S1）和序列 3（CP、S2、L5 和标准菌株ATCC200013）。分离株 02、LP 和菌种 S. diclina、S. longicaulis、S. ferax、S. mixta 以及 S. anomalies 有 80% 的序列相似性。分离株 04、14 与 S. bulbosa 和 S. oliviae 有 80% 相似性。在分离株 S1 和 S. australis；分离株 CP 和S. hypogyna；分离株S2、分离株L5、标准菌株 ATCC200013 和 S. salmonis 中鉴定出极高的 ITS 序列相似性，分别为100%、96%和98%。这项研究旨在为认识、预防和控制水霉病原菌以及有效药物的开发利用提供借鉴。

关键词：淡水鱼，鱼类疾病，基因分型，内转录间隔区序列，水霉属，水霉目

Introduction

Saprolegnia is one of the main aquatic diseases caused by filamentous fungi infections，which can cause severe losses of freshwater fish in both nature and commercial fish farms（Hatai and Hoshiai，1992；van et al.，2013）. These infections are usually termed saprolegniasis（Ke et al.，2009）and the main pathogens of Saprolegnia are Saproleniaceae saprolegnia and Saproleniaceae ambisexualis. These two pathogens are responsible for fungal infections of freshwater fish such as grass carp，crucian carp，silver carp，bighead carp，Megalobramaamblycephala，tilapia mossambica，pelteobagrusfulvidraco，channel catfish，and their eggs. There is a large range of adaptive temperature for Saprolegnia and cases of saprolegniasis taking place in the main freshwater fish cultured regions of China are reported all the year round. Saprolegnia infection spreads rapidly without a strict choice of hosts. It is hypothesized that dozens of fish species in different life stages including their eggs can be infected by Saprolegnia.

Currently，identification of Saprolegnia is mainly based on traditional morphological characteristics of the sexual reproductive structures such as oogonia，oospores and antheridia（Dick，1969；Leclerc et al.，2000；Eissa et al.，2013）. However，many species exhibit very similar or overlapping characters and their morphology is not stable and constant（Diéguez-Uribeondo et al.，1996）. Moreover，Saprolegnia from animals often fail to produce sexual structures in vitro（Dieguez-Uribeondo et al.，2007；Kozubíková-Balcarová et al.，2013）. Therefore，it is difficult and sometimes even impossible to identify，classify and name Saprolegnia using traditional morphological criteria（Seymour，1970；Dieguez-Uribeondo et al.，2007）.

Alternative approaches include polymerase chain reaction（PCR）amplification and sequencing of the internal transcribed spacer（ITS）region of ribosomal RNA（rRNA）genes. PCR amplification is easy even from small quantities of DNA due to the high copy number of rDNA genes，which makes it easy to analyze（Paul and Steciow，2004；Dieguez-Uribeondo et al.，2007）. Additionally，PCR has a high degree of variation even between closely related species，making it an ideal technique for identification and classification. In fact，molecular identification techniques based on ITS sequence analysis remedy the defects of traditional methods by morphology and has become a useful tool for the species separation，identification and determination（Huang et al.，1994；Paul et al.，1999；Paul，2001；Paul and Steciow，2004）.

Previous works on Saprolegnia has been carried out，but mostly based on morphological observations（Bangyeekhun et al.，2001；Fregeneda-Grandes et al.，2007）. The Eastern，Southern and Central regions of China，the most important freshwater fish farming areas，have a high incidence of Saprolegnia infections. Saprolegniasis brings significant economic losses of freshwater fish every year resulting from transportation，injuries，plummeted temperatures，and other factors（Sano，1998；Howe et al.，1999）. However，it has become increasingly difficult to prevent saprolegniasis in China's huge freshwater fish farming areas due to the unclear biological background information of the nature of the pathogens. On the other hand，additional variables must be taken into account including environmental conditions such as temperature，salinity，and pH，which differ from areas，climates，and seasons. The exact mechanism of these environmental conditions on Saprolegnia strains，in addition to the unclear taxonomic classifications of the strains， hinder the development of epidemiological investigation and Saprolegnia research.

The objective of this study was to investigate both phylogenetic and taxonomic aspects of eight representative isolates of pathogenic Saprolegnia obtained from different hosts and geographical origins including Guangzhou, Hubei and Shanghai. For this purpose, we have studied the morphological traits and sequenced the ITS to clarify their taxonomic positions. Moreover, the resistant abilities of strains on temperature, pH and salinity were evaluated and the phylogenetic background was analyzed. This work was designed to provide basic support to the epidemiological investigation and control on Saprolegnia infection and saprolegniasis in freshwater fish.

Materials and Methods

Saprolegnia strains

The eight samples of Saprolegnia strains were obtained from Hubei, Guangzhou, and Shanghai and their background information is presented in table 1. The standard strain ATCC 200013 was purchased from American Type Culture Collection (ATCC).

Table 1　Sources and background of the strains used in the present study

Isolate	Host	Origin	Acquisition time
02	Eggs of crucian	Shahu Town, Xiantao City, Hubei	May 10, 2010
04	Eggs of megalobramaamblycephala	Shahu Town, Xiantao City, Hubei	May 10, 2010
14	Water cultured crucian	Shahu Town, Xiantao City, Hubei	May 10, 2010
S1	Water cultured crucian	FoshanBairong hatchery, Guangdong	March 10, 2011
S2	Eggs of crucian	FoshanBairong hatchery, Guangdong	March 10, 2011
L5	Methylene blue treated eggs of crucian	FoshanBairong hatchery, Guangdong	March 10, 2011
LP	Caudal fin of weever	Xiceng Town, Qingpu District, Shanghai	April 13, 2012
CP	Skin of grass carp	Lingang New City River, Shanghai	August 20, 2012
ATCC200013	Rainbow trout	ATCC (Kanagawa, Japan)	March 23, 1990

Separation and purification

Separation and purification was performed as described by Oláh and Farkas with some modifications (Oláh and Farkas, 1978). The scrapings of isolates from the local lesions of diseased fish or their eggs were rinsed in 75% ethanol solution (chemically pure, Shanghai Anpel Scientific Instrument Co., Ltd., China) for 2-3 s and then dipped in sterile distilled water for 1-2 s before placed on potato dextrose agar (PDA, Qingdao Hope Biotechnology Co., Ltd., Shanghai, China) plates. After incubation at 20℃ (Mir-253, Sanyo, Japan) for 24 h, the fungi colony grew to the edge of the plate, and heat-sterilized rapeseeds were scattered on the plates and incubated at 20℃. The rapeseeds embedded in PDA were placed in sterilized and filtered river water and incubated at 20℃ until the zoospores ripened and released. Then 100 μL of Saprolegnia spore suspension was evenly coated on PDA plate and incubated at 20℃ until the fungi grow to cover the plate. A single colony was cut and put into another new PDA plate, in this way the strain was further purified by inoculations (Técher et al., 2010). If necessary, purification was repeated.

Saprolegnia growth in different environmental factors

To measure temperature dependency of growth, PDA plates with Saprolegnia strains were punched into disks (diameter of 6 mm), placed in the centers of new PDA plates and incubated at 4℃, 20℃ and 30℃, respectively. To measure the pH dependency of growth, disks punched from Saprolegnia PDA plates were placed in the center of new PDA plates that were previously adjusted to pH 7, pH 8, pH 9 and pH 10 using 1 M HCl and 1 M NaOH (Shanghai Anpel Scientific Instrument Co., Ltd., Shanghai, China), then incubated at 20℃. To measure the effect of NaCl concentration differences on Saprolegnia growth, PDA disks again were placed in the center of new PDA plates that included 0.5%, 1.0%, 1.5% and 2.0% NaCl, and incubated at 20℃ (Ali, 2005). All the PDA plates described above were incubated for 48 h and the radius (mm) of the developed fungus colonies were measured in triplicate at 12 h, 24 h, 36 h and 48 h, respectively.

Analysis of pathogenic strains

For each fish, both sides of the scales below the grass carp dorsal fin were removed with a scalpel blade. Wounds were scratched in the muscle (2 mm ~5 mm), the surface wiped of mucus, and the grass carp were added to the Saprolegnia spore concentration of 1×10^3 spores per mL. Each aquarium cylinder contained 10 grass carp. Saprolegnia infections were carried out when the temperature was between 25℃ to 15±1℃.

Saprolegnia Typing based on ITS sequence

Mycelium culture: After activation in slant tubes, the fungus was inoculated on PDA plates and incubated at 20℃ for 2-3 d. Then 4-5 blocks of agar (about 5 mm of diameter) were cut and evenly distributed on the surface of cellophane inside PDAY (PDA + 0.2% yeast extract) plates. After incubation at 20℃ for 36-48 h, the blocks with cellophane were carefully removed and mycelium on the surface was scraped with a blade. The scraped mycelium was placed in a small petri dish and dried in vacuum for 1-2 d. A portion of 0.05-0.10 g dried mycelium was placed in a centrifuge tube and preserved at -20℃ until required (Bockelmann et al., 2008).

DNA extraction: DNA extraction from strains was prepared according to the (cetyltrimethylammoniumbromide) CTAB method previously described (Jara et al., 2008). The extracted DNA was preserved at 4℃ until required. The strain DNA (1 μL) was diluted in 70 μL of diethylpyrocarbonate (DEPC) water and the mixed solution was measured at 260 nm (OD_{260}) and 280 nm (OD_{280}) by a spectrophotometer (DU800, Beckman Coulter, USA) to determine the DNA concentration.

PCR and agarose gel electrophoresis (AGE): PCR reactions were performed in 25 μL containing 17.2 μL ddH$_2$O, 2.5 μL 10X PCR buffer, 0.5 μL dNTPs (2.5 mM each), 1.5μL Mg^{2+} (25 mM), 1.0 μL of each primer (10 μM), 0.3 μL EX-Taq DNA polymerase (5 U/μL) and 1.0 μL cDNA template in a Bio-Rad iQ5 real-time PCR instrument (Hercules, CA, USA). The 10X buffer, dNTPs, Mg^{2+} and EX-Taq DNA polymerase were purchased from Shanghai Shengong Bioengineering Co. Ltd. Primers ITS1 (5'-TCCGTAGGTGAACCTGCGG-3') and ITS4 (5'-TCCTCCGCTTATTGATATGC-3') were synthesized by Shanghai Shengong bioengineering Co. Ltd (White et al., 1990; Dieguez-Uribeondo et al., 2007). The reaction profile included one initial cycle at 94℃ for 5 min; followed by 35 cycles at 94℃, 30 s; 58℃, 30 s and 72℃, 1 min. A final cycle at 72℃ for 5 min completed the amplification (Ke et al., 2009). The PCR products were then stored at 4℃. PCR fragment lengths were determined using 1.2% agarose gel electrophoresis (AGE) and image analysis after ethidium bromide staining (Shanghai Tiangen biochemistry technological Co., Ltd.) (Brito et al., 2009).

Sequence alignment and phylogenetic analysis

Sequences obtained were proofread using Sequencing Analysis 5.2 software (Applied Biosystems Inc., USA) and compared with BLAST in GenBank after the removal of ambiguous sequences. For the best alignment, the sequences were compared with homologous sequences of Achlyaconspicua (AF218144) as outgroup of phylogenetic analysis (Płoński and Radomski, 2010). The selected sequences were compared with Clustal X 1.83 and then manual corrections were made using Seaview software. Phylogenetic trees were constructed by the neighborjoining method (NJ) and the maximum parsimony method (MP), implemented with the MEGA4 computer package (Kumar et al., 2004) and PAUP4.0b10 computer package. To estimate the relative branch support of NJ and MP trees, bootstrap analysis with 1 000 replicates was conducted as default settings (Ke et al., 2009).

Results

Morphological strain characteristics

Optical microscopy was used to assess the morphological characteristics of all the isolated strains. Each strain (02, 04, 14, S1, S2, L5, LP and CP) had slight hyphae, zoospore and mature or immature archegonium, similar to our standard Saprolegnia strain ATCC200013 (Fig. 1). Clearly displayed in Figure 1, Saprolegnia is a multicore mycelium. The strain mycelium was prosperous with a transparent tubular structure while the sporangia was multi-cylindrical, stick-shaped or spike-shaped. Mature spores released from the tops of sporangia and into a group gathered in the sporangium mouth, into a group, or directly dispersed into the water. The secondary sporangia had typical lateral phenomenon. The archegonium was spherical or pear-shaped with multiple oospores. The mature spore had a large spherical or pear-shaped oil balls. Based on the morphology, all strains were considered to be legitimate members of the genus Saprolegnia.

All Saprolegnia strains were made into spore solution, and soak into the surface of the injured grass carp were when the temperature dropped. The surface of grass carp appear white cottony spots after 120 h infection, with typical characteristics of saprolegniasis. These strains have strong pathogenicity, and the average infection rate of grass carp can reach 93.3±4.25%.

Typing analysis of Saprolegnia strains

The OD_{260}/OD_{280} of extracted DNA for all strains was between 1.8 and 2.0, a high quality extraction with little protein contamination and low levels of DNA degradation. Each of the eight pathogenic Saprolegnia strains resolved clearly by AGE, indicating it was suitable for PCR analysis (Fig. 2A). Further, the electrophoretic mobility of each respective isolated PCR product and the referencestrain ATCC200013 was approximately 700 bp (Fig. 2).

All eight strains were classified into three clades that were named as CLADE1, CLADE2 and CLADE3. Strains 02, LP, 04 and

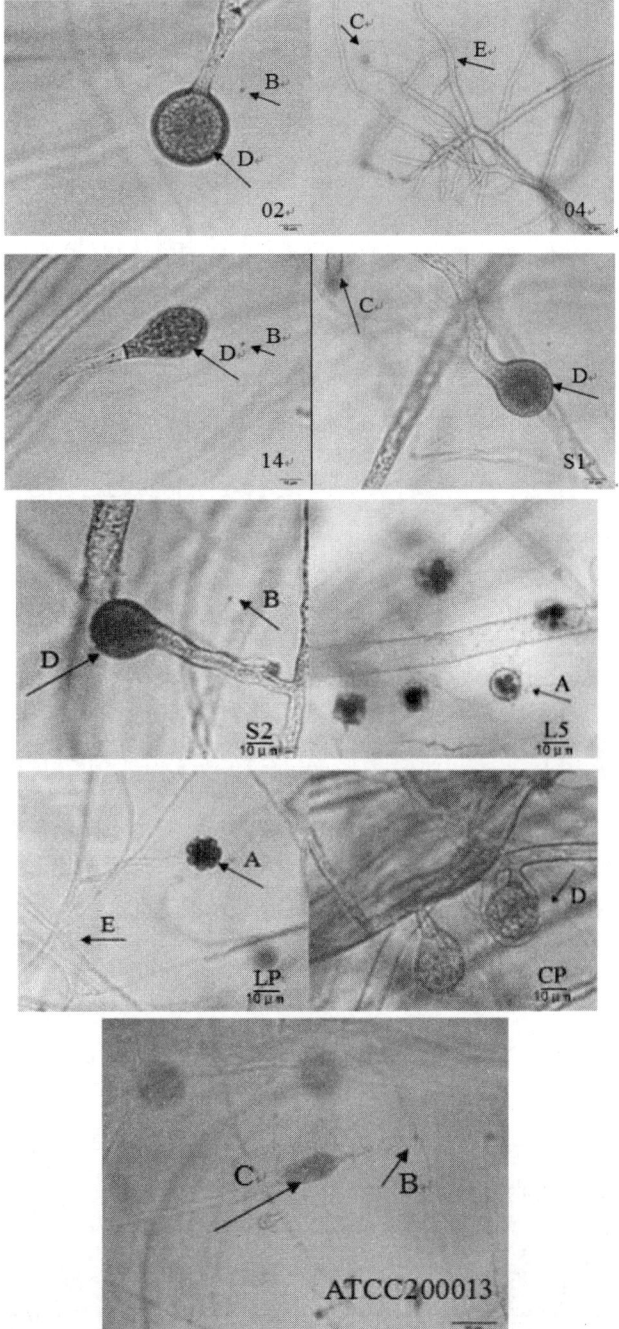

Figure 1   The morphological structures of nine strains were analyzed byoptical microscopy.
（A）mature archegonium；（B）zoospore；（C）swimming sporangium；（D）immature archegonium；
（E）hyphae. Size bars indicate magnification in each panel.

14 are clustered into CLADE1；CP, S2, L5 and the reference ATCC200013 are clustered into CLADE3；and only S1 belongs to CLADE2. Interestingly，there was a correlation between the isolated region and the Saprolegnia cluster features. Among those originated from Guangdong，S1 is clustered into CLADE2，while S2 and L5 are clustered into CLADE3. Among those from Shanghai，LP is clustered into CLADE1 and CP into CLADE3. However，the Saprolegnia from Hubei including 02，04 and 14 are clustered into CLADE1，but into different sub-branches（Fig. 3）. High similarity between the tested strains and known pathogenic sequences based on ITS rDNA was observed（Table 2）. There was 80% similarity 02 and LP with S. diclina, S. longicaulis, S. ferax, S. mixta, and S. anomalies. The similarities of ITS sequence between strains 04 and 14 and S. bulbosa, S. oliviae are also 80%. There is 100% simi-

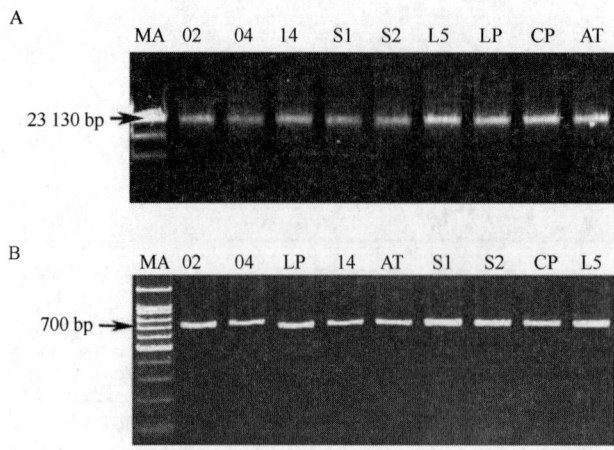

Fig. 2　Whole and PCR-amplified DNA analysis of pathogenic strains.

（A）Strains were isolated and total DNA was electrophoresed in 1. 2% agarose with the standard ATCC200013 strain and a size marker［MA：Product No. λ DNA HindIII（MD202）］.

（B）Isolated strains were amplified by PCR and electrophoresed in 1. 2% agarose with the standard ATCC200013 strain and a size marker［MA：Product No. 100bp DNA Ladder（MD109）］.

larity between S1 and S. australis；96% between CP and S. hypogyna；98% between S2, L5, ATCC200013 and S. salmonis.

The tested strains produced only one band following PCR amplification of ITS rDNA and they belong to three genotypes by phylogenetic analysis. The amplification plot of ATCC200013 was similar to that of ATCC provided by official description, which verifying that the genotyping method used in this study was reliable. The results also showed that some Saprolegnia strains with similar origins or backgrounds presented different clustering features. These data indicate thatprevention and control schemes should take into account strain-specific differences of Saprolegnia when attempting to control epidemics caused by these pathogens.

Table 2　Sequence analysis and results

Strain No.	ITS length	Similar strain	Similarity	Result
02	699bp	S. diclina, S. longicaulis, S. ferax, S. mixta, S. anomalies	80%	S. ferax
Lp	713bp	S. diclina, S. longicaulis, S. ferax, S. mixta, S. anomalies	80%	S. diclina
Cp	743bp	S. hypogyna	96%	S. hypogyna
04	722bp	S. bulbosa, S. oliviae	80%	S. bulbosa
14	737bp	S. bulbosa, S. oliviae	80%	S. oliviae
ATCC200013	733bp	S. salmonis	98%	S. salmonis
S1	742bp	S. australis	100%	S. australis
S2	758bp	S. salmonis	98%	S. salmonis
L5	727bp	S. salmonis	98%	S. salmonis

The infection experiment of pathogens

To test the infectivity and pathogenicity of these eight isolated strains, we infected crucians and analyzed the results. All eight strains and the control showed the similar saprolegniasis symptoms to those of spontaneous infections. Saprolegnia invades epidermal tissues, often beginning on the head or fins and spreading over the entire surface of the body（van West, 2006）. In this study, the injured tissue induced cotton-like appearance on the focal area. The zoospore of Saprolegnia invaded the wound and reproduced rapidly using the nutrition of tissues. Saprolegnia is able to cause cellular necrosis and epidermal damage, and tissue inflammation and necrosis often occurs from the winding and conglutination between the hyphae and cells on the wound. We observed these symptoms and also ob-

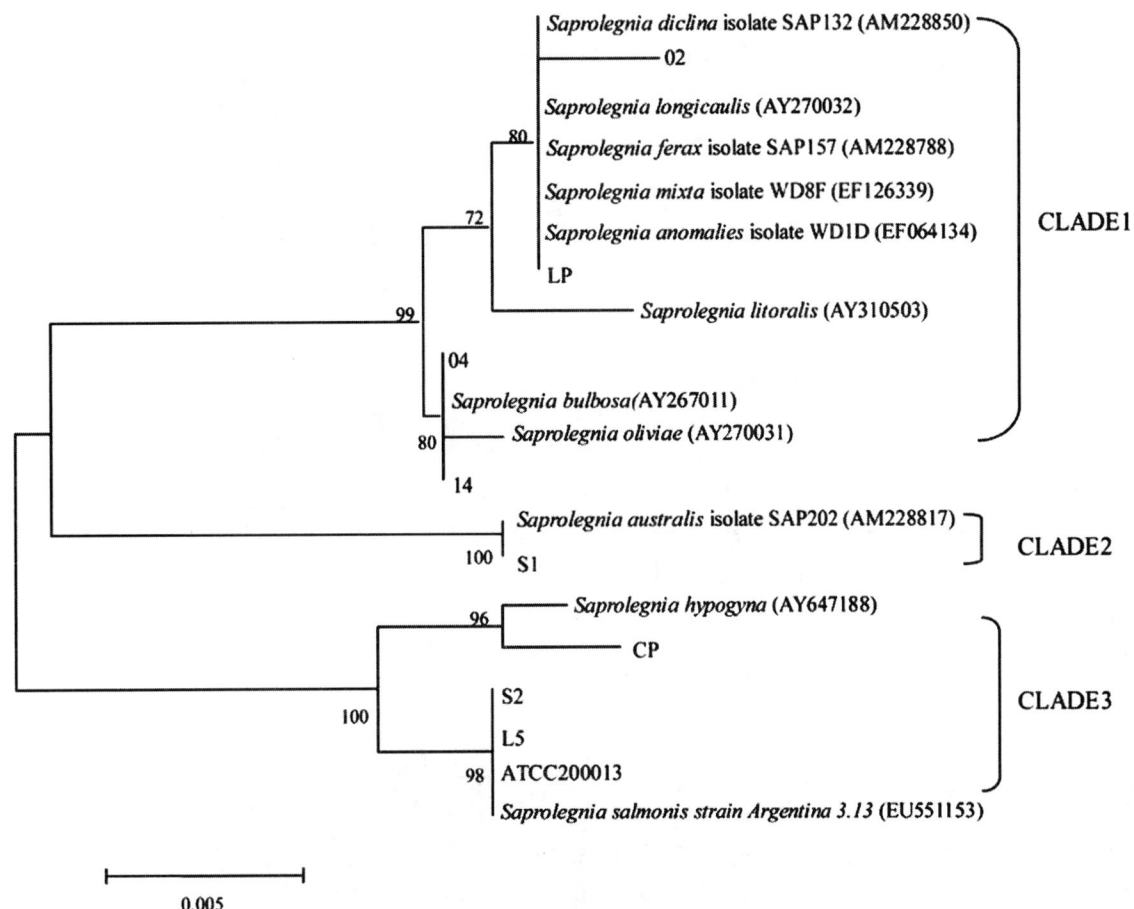

Fig. 3 Phylogenetic trees by NJ and MP methods.

Strains 02, 04, 14, LP, CP, ATCC200013, S1, S2 and L5 were classified based on ITS rDNA sequences.
All bootstrap values under the branches are indicated at 1000 repetitions and GenBank sequences are presented in
brackets behind the strain names.

served the crucians swimming slowly with anorexia. Those in especially bad conditions were often found hiding in deep mud. Severe Saprolegnia infections resulted in abnormal behavior and eventual death from exhaustion. Time to death by saprolegniosis is thought to depend on the initial infected site, type of tissue destroyed, growth rate of the pathogen, and the ability to withstand the invasion of the individuals (Noga, 1993). The crucians in control lived well without death during this experimental period. The isolation of pathogens in the challenge experiment and morphological identification suggested that all of these eight strains were pathogenic Saprolegnia.

The effects changing environmental conditions on Saprolegnia strains

We assessed the ability of Saprolegnia strains to grow in different environmental conditions by measuring the colony diameter on PDA plates following growth for 48 h. The effect of temperature on the growth of isolated strains was shown in Fig. 4A. Three strains, S2, ATCC, and CP, grew more efficiently at 20 °C, while the rest of the strains grew more efficiently at 30 °C (Fig. 4A). Further, the only strains that are able to grow at 4 °C were S2, 4, L5 and ATC200013. Therefore, the optimal temperature for Saprolegnia strains in this study ranged from 20℃ to 30℃.

While the separated pathogens could live and grow to some extent at pH ranging from 7.0 to 10.0, the growth of all the strains was severely inhibited at pH 9.0-pH 10 (Fig. 4B). All strains, especially strains 14, CP and ATCC200013, grew well at pH 7.0. Therefore, the optimal pH for growth of the tested strains is approximately 7.0.

We found an inverse correlation between NaCl concentration and strain growth in all of the isolated Saprolegnia strains (Fig. 4C). Further, some strains obviously grew more efficiently in more harsh (higher) NaCl concentrations than others, most notably, 4, 14, CP, and ATCC. These results indicated that the optimal NaCl content for all strains in this study was 0.05%.

In summary, these data indicate that the optimal growth conditions for all of the isolated pathogenic Saprolegnia strains in this study includes a low salt, neutral pH environment at 30 °C. The results were consistent with Koeypudsa's study (Koeypudsa et al.,

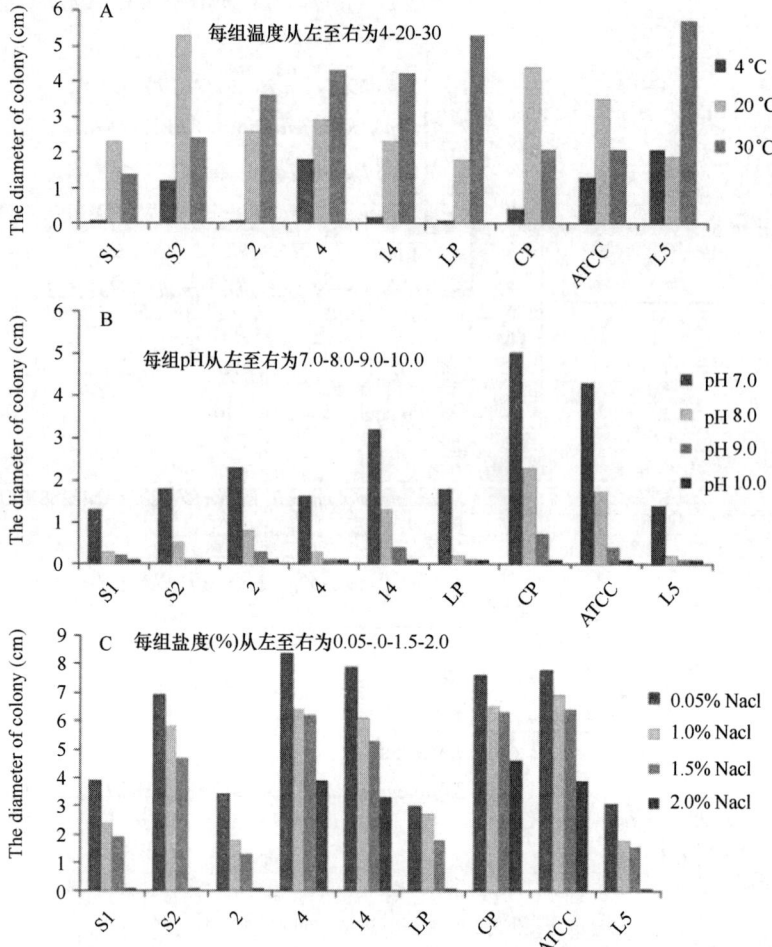

Fig. 4　Effect of environmental factors on the growth of isolated pathogenic Saprolegnia strains. Strains were tested for growth by analyzing colony diameter following growth on PDA plates.

（A）Growth was assessed at varying temperatures.　（B）Growth was assessed at varying pH levels.

（C）Growth was assessed at varying NaCl concentrations.

2005）.

Discussion

Shanghai, Guangdong and Hubei are three regions in China located at different latitude and longitude, which are very representative in perspective of temperature, salinity and pH value. For these reasons, we chose these regions as research subjects for typing of Saprolegnia strains causing saprolegniasis in freshwater fish. The results were concluded from the morphological features and genotyping method of ITS rRNA. By studying a number of sequences from isolates of different regions and hosts, wefound that the isolates of species distributed in three phylogenetically distinct clades that we named CLADE1（02, LP, 04 and 14）, CLADE2（CP, S2, L5 and the reference ATCC200013）and CLADE3（S1）.

Basic conditions for the pathogenesis of saprolegniasis

Saprolegniasis is one of the most common aquatic diseases in freshwater fish. The basic conditions for the incidence of saprolegniasisare are summarized in Table 3. Previous investigations suggested that drastic changes in temperature induced higher incidence of saprolegniasis, while stable temperature is beneficial to low incidence. In addition, aquatic organisms with tissue injuries are more likely to be infected by Saprolegnia than those without injuries. This is possibly due to the decreased immunity of the body resulting from physical damage which makes the spores geminate on the injure tissue. There is high incidence in water with considerable amount of pathogens because saprolegniasis of freshwater fish is not caused by primary disease, but rather from secondary diseases（Wardrip et al.,

1999; Edgerton et al., 2004).

Analysis on epidemic season/month and epidemic characters of saprolegnia

Based on data acquired from the National Monitoring on Aquaculture Disease and The National Fishery Technical Extension Center, China; several characteristics of Saprolegnia growth are known. A total of 265 saprolegniasis cases were investigated between January 2008 and July 2009 in the main Chinese regions of high Saprolegnia incidence (Table 4). The adaptive growth temperature for Saprolegnia ranges from 5–26℃ and propagated temperature ranges from 13–18℃. The months with high incidence are from February to May, while April is the peak with a proportion of 24.6%. Saprolegnia can infect a great number of aquatic organisms, from eggs to those in different ages, and spread rapidly. It will cause death of massive aquatic animals and bring an enormous loss to the economy and the environment.

A summary of a previous study on the prevalence of freshwater fish saprolegniasis in Guangdong, Hubei and Shanghai between 2008 and 2009 is shown in Fig. 5. Generally, the peak of incidence is in April. In Guangdong, saprolegniasis occurred between December and April, with the highest incidence in March. In Hubei, saprolegniasis occurred from December to May, with the highest incidence in April. This previous research provided a reliable method to control saprolegniasis through analysis and evaluation of the rate of Saprolegnia infection. In addition, our study took these findings one step further by developing a typing method to help identify and assess the frequency and incidence of saprolegniasis infection and eventually leading to potential methods of prevention and control.

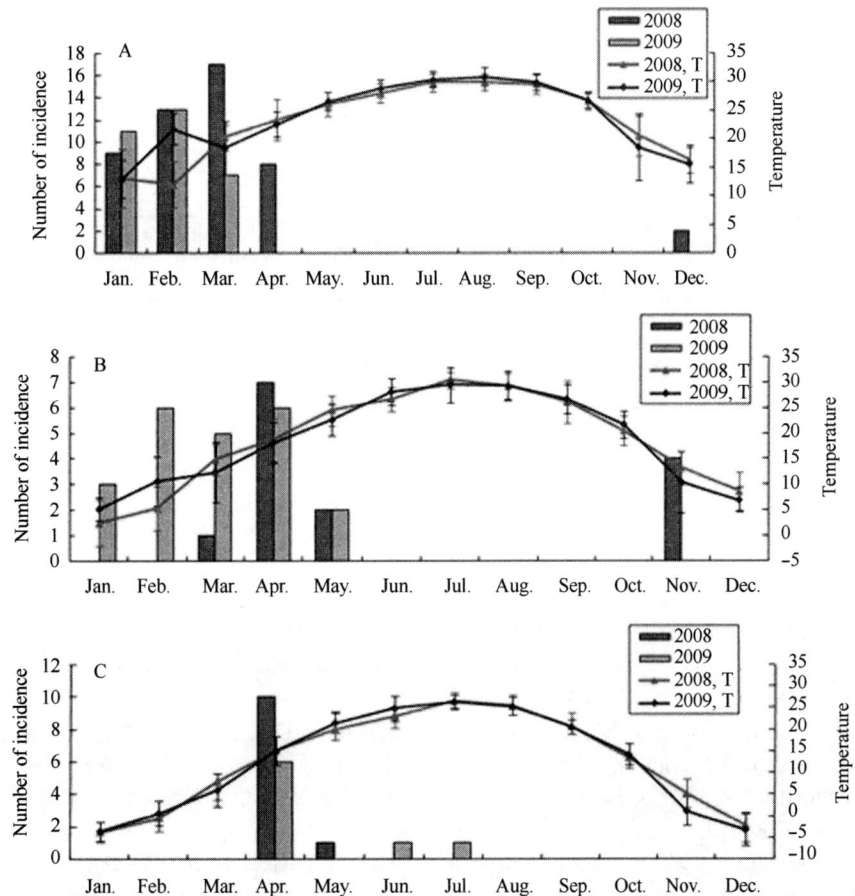

Fig. 5　The prevalence of freshwater fish saprolegniasis and temperature changes in Guangdong
(A) Hubei; (B) and Shanghai; (C) (years of 2008 and 2009).

Table 3　The prevalence of saprolegniasis under different environmental conditions

Environmental factor		Number of incidences	Percentage (%)
Temperature	> 25℃	3	3.0
	< 10℃	6	6.1

续表

Environmental factor		Number of incidences	Percentage （%）
	Upheaval	30	30.3
Tissue injure	Without injuries	5	5.1
	Injuries existed	24	24.2
Water condition	Almost no pathogens	4	4.0
	Several pathogens	27	27.3

Table 4　The prevalence of saprolegniasis and monthly temperature changes

Month 2008	Number of cases investigated	Percentage of total cases （%）	Month 2009	Cases investigated （n）	Percentage of cases （%）
Jan.	9	3.4	Jan.	16	6
Feb.	12	4.5	Feb.	24	9.1
Mar.	25	9.4	Mar.	21	7.9
Apr.	37	14	Apr.	28	10.6
May.	19	7.2	May.	12	4.5
Jun.	9	3.4	Jun.	5	1.9
Jul.	6	2.3	Jul.	5	1.9
Aug.	9	3.4			
Sep.	7	2.6			
Oct.	8	3			
Nov.	11	4.1			
Dec.	2	0.8			
Total	265	100			

References

［1］　Ali, E. 2005. Morphological and biochemical alterations of oomycete fish pathogen Saprolegnia parasitica as affected by salinity, ascorbic acid and their synergistic action. Mycopathologia, 159: 231-243. doi: 10.1007/s11046-004-6670-z

［2］　Bangyeekhun, E., Quiniou, S. M. A., Bly, J. E. and Cerenius, L. 2001. Characterisation of Saprolegnia sp. isolates from channel catfish. Dis Aquat Organ, 45: 53-59. doi: 10.3354/dao045053

［3］　Bockelmann, W., Heller, M. and Heller, K. J. 2008. Identification of yeasts of dairy origin by amplified ribosomal DNA restriction analysis （ARDRA）. International Dairy Journal, 18: 1066-1071. doi: http://dx.doi.org/10.1016/j.idairyj.2008.05.008

［4］　Brito, E. H. S., Brilhante, R. S. N., Cordeiro, R. A., Sidrim, J. J. C., Fontenelle, R. O. S., Melo, L. M., Albuquerque, E. S. and Rocha, M. F. G. 2009. PCR-AGE, automated and manual methods to identify Candida strains from veterinary sources: A comparative approach. Veterinary Microbiology, 139: 318-322. doi: http://dx.doi.org/10.1016/j.vetmic.2009.06.031

［5］　Dick, M. 1969. Morphology and taxonomy of the oomycetes, with special reference to Saprolegniaceae, Leptomitaceae, and pithyaceae. I. sexual reproduction. New Phytologist, 68: 751-775. doi:

［6］　Diéguez-Uribeondo, J., Cerenius, L. and Söderhäll, K. 1996. Physiological characterization of Saprolegnia parasitica isolates from brown trout. Aquaculture, 140: 247-257. doi: http://dx.doi.org/10.1016/0044-8486（95）01176-5

［7］　Dieguez-Uribeondo, J., Fregeneda-Grandes, J. M., Cerenius, L., Perez-Iniesta, E., Aller-Gancedo, J. M., Telleria, M. T., Soderhall, K. and Martin, M. P. 2007. Re-evaluation of the enigmatic species complex Saprolegnia diclina-Saprolegnia parasitica based on morphological, physiological and molecular data. Fungal Genet Biol, 44: 585-601. doi: 10.1016/j.fgb.2007.02.010

［8］　Edgerton, B. F., Henttonen, P., Jussila, J., Mannonen, A., Paasonen, P., Taugbøl, T., Edsman, L. and SOUTY-GROSSET, C. 2004. Understanding the causes of disease in European freshwater crayfish. Conservation Biology, 18: 1466-1474. doi:

［9］　Eissa A E, Abdelsalam M, Tharwat N, et al. Detection of Saprolegnia parasitica in eggs of angelfish Pterophyllum scalare （Cuvier-Valenciennes） with a history of decreased hatchability. International Journal of Veterinary Science and Medicine, 2013, 1 （1）: 7-14.

［10］ Fregeneda – Grandes, J. M., Rodríguez – Cadenas, F. and Aller – Gancedo, J. M. 2007. Fungi isolated from cultured eggs, alevins and broodfish of brown trout in a hatchery affected by saprolegniosis. Journal of Fish Biology, 71: 510 – 518. doi: 10. 1111/j. 1095 –8649. 2007. 01510. x

［11］ Hatai, K. and Hoshiai, G. 1992. Mass mortality in cultured coho salmon (Oncorhynchus kisutch) due to Saprolegnia parasitica coker. Journal of Wildlife Diseases, 28: 532–536. doi:

［12］ Howe, G. E., Gingerich, W. H., Dawson, V. K. and Olson, J. J. 1999. Efficacy of Hydrogen Peroxide for Treating Saprolegniasis in Channel Catfish. Journal of Aquatic Animal Health, 11: 222–230. doi: 10. 1577/1548–8667 (1999) 011<0222: EOHPFT>2. 0. CO; 2

［13］ Huang, T. –s., Cerenius, L. and Söderhäll, K. 1994. Analysis of genetic diversity in the crayfish plague fungus, Aphanomyces astaci, by random amplification of polymorphic DNA. Aquaculture, 126: 1–9. doi: http: //dx. doi. org/10. 1016/0044–8486 (94) 90243–7

［14］ Jara, C., Mateo, E., Guillamón, J. M., Torija, M. J. and Mas, A. 2008. Analysis of several methods for the extraction of high quality DNA from acetic acid bacteria in wine and vinegar for characterization by PCR–based methods. International Journal of Food Microbiology, 128: 336–341. doi: http: //dx. doi. org/10. 1016/j. ijfoodmicro. 2008. 09. 008

［15］ Ke, X. L., Wang, J. G., Gu, Z. M., Li, M. and Gong, X. N. 2009. Morphological and molecular phylogenetic analysis of two Saprolegnia sp. (Oomycetes) isolated from silver crucian carp and zebra fish. Mycol Res, 113: 637–644. doi: 10. 1016/j. mycres. 2009. 01. 008

［16］ Koeypudsa, W., Phadee, P., Tangtrongpiros, J. and Hatai, K. 2005. Influence of pH, temperature and sodium chloride concentration on growth rate of Saprolegnia sp. Journal of Scientific Research at Chulalongkorn University, 30: 123–130. doi:

［17］ Kozubíková–Balcarová E, Koukol O, Martín M P, et al. The diversity of oomycetes on crayfish: Morphological vs. molecular identification of cultures obtained while isolating the crayfish plague pathogen. Fungal biology, 2013, 117 (10): 682–691.

［18］ Kumar, S., Tamura, K. and Nei, M. 2004. MEGA3: Integrated software for Molecular Evolutionary Genetics Analysis and sequence alignment. Briefings in Bioinformatics, 5: 150–163. doi: 10. 1093/bib/5. 2. 150

［19］ Leclerc, M. C., Guillot, J. and Deville, M. 2000. Taxonomic and phylogenetic analysis of Saprolegniaceae (Oomycetes) inferred from LSU rDNA and ITS sequence comparisons. Antonie Van Leeuwenhoek, 77: 369–377. doi:

［20］ Noga, E. J. 1993. Water mold infections of freshwater fish: Recent advances. Annual Review of Fish Diseases, 3: 291–304. doi: http: // dx. doi. org/10. 1016/0959–8030 (93) 90040–I

［21］ Oláh, J. and Farkas, J. 1978. Effect of temperature, pH, antibiotics, formalin and malachite green on the growth and survival of Saprolegnia and Achlya parastic on fish. Aquaculture, 13: 273–288. doi: http: //dx. doi. org/10. 1016/0044–8486 (78) 90009–1

［22］ Paul, B. 2001. ITS region of the rDNA of Pythium longandrum, a new species; its taxonomy and its comparison with related species. FEMS Microbiol Lett, 202: 239–242. doi:

［23］ Paul, B., Galland, D. and Masih, I. 1999. Pythium prolatum isolated from soil in the Burgundy region: a new record for Europe. FEMS Microbiol Lett, 173: 69–75. doi:

［24］ Paul, B. and Steciow, M. M. 2004. Saprolegnia multispora, a new oomycete isolated from water samples taken in a river in the Burgundian region of France. FEMS Microbiol Lett, 237: 393–398. doi: 10. 1016/j. femsle. 2004. 07. 006

［25］ Płoński, P. and Radomski, J. P. 2010. Quick path finding—Quick algorithmic solution for unambiguous labeling of phylogenetic tree nodes. Computational Biology and Chemistry, 34: 300–307. doi: http: //dx. doi. org/10. 1016/j. compbiolchem. 2010. 10. 002

［26］ Sano, T. 1998. Control of fish disease, and the use of drugs and vaccines in Japan. Journal of Applied Ichthyology, 14: 131 – 137. doi: 10. 1111/j. 1439–0426. 1998. tb00630. x

［27］ Seymour, R. 1970. The genus Saprolegnia. Nova Hedwigia, 19: 1–124. doi:

［28］ Técher, D., Martinez–Chois, C., D' Innocenzo, M., Laval–Gilly, P., Bennasroune, A., Foucaud, L. and Falla, J. 2010. Novel perspectives to purify genomic DNA from high humic acid content and contaminated soils. Separation and Purification Technology, 75: 81–86. doi: http: //dx. doi. org/10. 1016/j. seppur. 2010. 07. 014

［29］ van West, P. 2006. Saprolegnia parasitica, an oomycete pathogen with a fishy appetite: new challenges for an old problem. Mycologist, 20: 99–104. doi: http: //dx. doi. org/10. 1016/j. mycol. 2006. 06. 004

［30］ van den Berg A H, McLaggan D, Diéguez–Uribeondo J, et al. The impact of the water moulds Saprolegnia diclina and Saprolegnia parasitica on natural ecosystems and the aquaculture industry. Fungal Biology Reviews, 2013, 27 (2): 33–42.

［31］ Wardrip, C. L., Seps, S. L., Skrocki, L., Nguyen, L., Waterstrat, P. R. and Li, X. 1999. Diagnostic Exercise: Fluffy, White, Cotton Candy–Like Growth on the Gills, Fins, and Skin of Salamanders (Ambystoma tigrinum). Contemp Top Lab Anim Sci, 38: 81 –83. doi:

［32］ White, T. J., Bruns, T., Lee, S. and Taylor, J. 1990. Amplification and direct sequencing of fungal ribosomal RNA genes for phylogenetics, in: Michael, A. I., David, H. G., John, J. S., Thomas J. WhiteA2 – Michael A. Innis, D. H. G. J. J. S. and Thomas, J. W. (Eds.), PCR Protocols. Academic Press, San Diego, pp. 315–322.

该文发表于《JOURNAL OF APPLIED ICHTHYOLOGY》【2014，30（1）：145-150. DOI：10. 1111/jai. 12316】

Saprolegnia australis R. F. Elliott 1968 infection in Prussian carp *Carassius gibelio*（Bloch，1782）eggs and its control with herb extracts

H. Cao[1,2,3]，R. Ou[1]，G. Li[2]，X. Yang[1]，W. Zheng[4] and L. Lu[1]

（1. Key Laboratory of Freshwater Fishery Germplasm Resources，Ministry of Agriculture of P. R. China，Shanghai Engineering Research Center of Aquaculture，National Pathogen Collection Center for Aquatic Animals，Shanghai University Knowledge Service Platform，Shanghai Ocean University Aquatic Animal Breeding Center（ZF1206），Shanghai Ocean University，Shanghai，China；2. Sihong Bureau of Aquatic Products，Sihong，China；3. Shanxi Veterinary Drug and Feed Engineering Technology Research Center，Yuncheng，China；4. Yangtze River Fisheries Research Institute，Chinese Academy of Fishery Sciences，Wuhan，China）

Summary：In order to control saprolegniosis in Prussian carp（Carassius gibelio（Bloch，1782）eggs，it is important to screen herb extracts as potential anti-Saprolegniadrugs in Prussian carp hatcheries. For this purpose，an oomycete water mould（strain SC）isolated from Prussian carp［Carassiusgibelio（Bloch，1782）］eggs suffering from saprolegniosis was characterised morphologically as well as from ITS rDNA sequence data. Initially identified as a Saprolegniasp. based on its morphological features，the constructed phylogenetic tree using the neighbour joining method further indicated that the SC strain was closely related to SaprolegniaaustralisR. F. Elliott 1968 strain VI05733（GenBank accession no. HE798564），and which could form biofilm communities as virulence factors. In addition，aqueous extracts from forty Chinese herbs were screened as possible anti-Saprolegniaagents. Among them，a 1 g·mL^{1}extract from Radix sanguisorbaewas the most efficacious anti-Saprolegniaagent，indicated by the minimum inhibitory concentration that was as low as 256 mg·L^1. Relative survival of 73 and 88% was obtained against the SC strain in fish eggs at concentrations of 256 and 1 280 mg·L^1，respectively. This is the first known report of SaprolegniaaustralisR. F. Elliott 1968 infection in C. gibelio（Bloch，1782）eggs involving the screening of R. sanguisorbaeextracts as potential anti-Saprolegniaagents.

水霉感染彭泽鲫鱼卵及利用中草药提取物控制其水霉病

曹海鹏[1,2,3]，欧仁建[1]，李贵[2]，杨先乐[1]，郑卫[4]，吕利群[1]

（1. 上海海洋大学；2. 泗洪渔业局；3. 山西兽药中心；4. 中国水产科学研究院长江水产研究所）

摘要：为了控制银鲫卵的水霉病，在银鲫孵化场筛选潜在的抗水霉草药提取物。将一个从患水霉病的银鲫卵中分离出的水霉（strain SC）根据其形态学特征和 rDNA 序列数据最初被鉴定为水霉。根据它的形态学特征和用邻接法构建的系统发生树进一步研究表明这个水霉种与南方水霉 R. F. Elliott 1968 种 VI05733（基因库登记号. HE798564）关系紧密，并且能建立生物膜团体作为毒力因子。另外，从 40 种中草药中提取的水提取物作为候选的抗水霉药物。在这些提取物中，1 g/mL 的地榆提取物是最有效的，研究表明其最小抑制浓度低至 256 mg/L。在防治鱼卵中水霉中，256 mg/L 和 1 280 mg/L 的地榆提取物浓度分别获得了 73%和 88%的相对存活率。这是第一个已知的关于地榆抵御异育银鲫卵感染水霉的报道。

Introduction

Prussian carp，Carassiusgibelio（Bloch，1782），is a popular freshwater fish species in China，Japan，Korea，East Europe，Siberia，the Russian Far East，Sakhalin and the basins of the largest Central Asian rivers Syr Darya and Amu Darya（Apalikova et al.，2011）. Especially in China，Prussian carp farming has become an emerging industry in the province of Hubei，Jiangsu，Hunan，Guangdong，Jiangxi，Anhui，Shandong，Sichuan，Guangxi，Liaoning，etc.，and has broughtgreat profits in recent years. Annual production was nearly 2. 2 million tonnes in 2010（Ge and Miao，2011）. However，underintensive culture conditions，

490

serious problems in Prussian carp hatcheries occur due to saprolegniosis caused by zoosporic oomycete fungi. During egg incubation, the oomycetes produce mycelia, which grow and spread from dead to healthy eggs thereby causing major economic losses (Noga, 1993; Cao et al., 2012a). More attention needs to be paid to control oomycete infections during incubation of the Prussian carp eggs.

Major pathogens of many fish species (Pickering and Willoughby, 1982) are species of the saprolegniaceous water mouldSaprolegnia, which can be found in most freshwater habitats and are responsible for significant infections of both living and dead fish as well as their eggs (Noga, 1993; Hussein and Hatai, 2002). A few studies have been conducted on Saprolegniaspecies of rainbow trout, zebra fish, silver crucian carp, Pengze crucian carp and yellow catfish eggs (Ke et al., 2009; Mousavi et al., 2009; Cao et al., 2012a, b). However, no specific report is available for S. australisR. F. Elliott 1968 infection in Prussian carp and its eggs. Additionally, the control of saprolegniosis remains a worldwide problem after the effective malachite green treatment was banned. Relevant studies need to be extensively conducted on the Saprolegniapathogens and potential anti-Saprolegniadrugs.

In the present study, the pathogenic S. australisR. F. Elliott 1968 of Prussian carp eggs suffering from saprolegniosis was isolated and its taxonomic position determined using phylogenetic analysis based on ITS rDNA sequence; its susceptibility to forty different Chinese herb extracts was further assayed to screen for potential anti-Saprolegniadrugs.

Materials and methods

Experimental fish eggs

A saprolegniosis outbreak occurred in June 2012 in a Prussian carp hatchery in Guangdong province, China. More than fifty infected Prussian carp eggs were examined externally and microscopically for the presence of oomycetes.

Another fifty affected fish eggs (as described by Evelyn et al., 1986) were sampled and transported to the laboratory. Healthy Prussian carp eggs used in the study were obtained from another hatchery with no history of any unknown mortalities or abnormalities (Cao et al., 2012b).

Isolation of fungi

The infected fish eggs were first disinfected for 2-3 s with 75% alcohol, then washed several times in sterile filtered water and plated on potato dextrose agar (PDA) (Sinopharm Chemical Reagent Co., Ltd) containing 100 mg·L^{-1} streptomycin and penicillin (Sinopharm Chemical Reagent Co., Ltd) to reduce bacterial contamination and incubated at 20℃ for 24 h. Autoclaved rape seeds were immediately placed at the edge of colonies grown on PDA plates and incubated until they were infested with hyphae. These were then transferred to sterile filtered river water and incubated until the zoospores were discharged. One hundred ll of zoospores were then spread on PDA plates and incubated at 4℃ for 48-72 h.

Infectivity experiment

Approximately 7 days prior to starting the test, rape seeds infested with the isolate were placed in sterile filtered river water at 20℃ for 72 h to induce zoospore formation; zoospore suspensions were then collected as described by Hussein and Hatai (2002). Then 720 healthy fish eggs were randomly distributed among six small glass aquaria (120 eggs per aquarium) containing aerated sterile filtered river water at 20℃. The fish eggs in the challenged aquarium were exposed to the isolate's zoospores at a concentration of 2 9 10^5 mL^1; in the control aquarium the eggs were held in aerated sterile filtered river water without the addition of zoospores. Each treatment was conducted in triplicate. The fish eggs were observed daily for 5 days. Eggs with hyphae were immediately removed for fungal isolation and identification according to Ghiasi et al. (2010). Infection of eggs by the isolate was recorded only if the challenge strain was reisolated and identified.

Morphological observation

The isolate was grown on PDA plates with several autoclaved rape seeds at 20℃ until the rape seeds were infested with the isolate as previously described. Rape seeds with hyphae were then transferred to 6-well cell culture plates containing sterile filtered river water and incubated at 20℃ for 14 days. Daily observations using an inverted microscope were carried out to check the emergence of zoosporangia, zoospore discharges, sporangial renewal and oogonia.

Molecular identification

Genomic DNA was extracted from pure cultures of the isolate using a Universal Genomic DNA Extraction Kit vers. 3 (Takara Biotechnology, Dalian). The internal transcribed spacer (ITS) gene was amplified by PCR using a pair of ITS gene primers (ITS1): 5′-TCCGTAGGTGAACCTGCGG-3′ as well as (ITS4): 5′-TCCTCCGCTTATTGATATGC-3′, and carried out according to the instructions of the fungi identification PCR kit (Takara Biotechnology, Dalian). Amplification was done after 35 cycles of denaturation at 94℃ for 0.5 min, annealing at 58℃ for 0.5 min and extension at 72℃ for 1.0 min followed by a final extension at 72℃ for 5 min using a PCR minicycler (Eppendorf Ltd., Germany). The PCR product was electrophoresed on 1% agarose gel and observed via ultraviolet trans-illumination. Sequencing was performed using a fluorescent labeled dideoxynucleotides termination method (with BigDye terminator) on an ABI 3730 automated DNA Sequencer.

The partial ITS rDNA sequence was assembled using MegAlign and Seqman software with a Macintosh computer. Searches were done with the National Centre for Biotechnology Information (NCBI) database using the Basic Local Alignment Search Tool (BLAST) program. The ITS rDNA gene sequence of the pathogenic isolate was constructed using the neighbour joining method (Chen and Liu, 2007).

Susceptibility assay to Chinese herb extracts

The susceptibility assay of the herb extracts against the isolate and the control Saprolegnia strains was conducted in triplicate. Forty Chinese herbs were obtained from Shanghai Fosun Industrial CO., LTD. (Table 1). A concentration of $1 \text{ g} \cdot \text{mL}^{1}$ of aqueous extract from each Chinese herb was prepared using the boiling method as described by Au et al. (2001). Herb extracts at concentrations of $0 -8192 \text{ mg} \cdot \text{L}^{-1}$ were used to assay the minimum inhibitory concentration (MIC). MIC of each herb extract was defined as the lowest concentration that did not allow any visible fungal colony growth, determined through the dilution plate method as described by Stueland et al. (2005a). S. parasitica strain JL and S. ferax strain HP, previously isolated and identified by Cao et al. (2012a, b), were used as controls. Data were presented as the mean the standard deviation (SD). $P < 0.05$ was considered statistically significant using oneway analysis of variance.

Protective efficacy assay

The protective efficacy assay was conducted in triplicate. Healthy fish eggs were obtained from Sand Lake Aquatic Technique Popularizing Station, Hubei China and maintained in twelve 10-L glass aquaria supplied with single-pass aerated sterile filtered river water at 20℃. Each aquarium was randomly stocked with 100 healthy eggs. The $1 \text{ g} \cdot \text{mL}^{1}$ of herb extracts were added directly to aerated sterile filtered river water in the treatment aquaria at final concentrations of 256 and 1280 mg $\cdot$ L^{1}; the control aquarium received no treatment. In the positive control treatment, aquaria malachite green was added at a final concentration of 0.2 mg $\cdot$ L^{1}, as recommended by Sudova et al. (2007). All test eggs were exposed to the isolate's zoospores at a concentration of $2\ 9\ 10^{5} \text{mL}^{1}$. The eggs were observed daily for 5 days; eggs with hyphae were immediately removed for fungal isolation and identification according to Ghiasi et al. (2010). Egg infections were recorded only if the challenge strain was re-isolated and identified. The cumulative infection rate and relative survival percentage were calculated using the following formulas: infected eggs/total eggs 9 100, cumulative infection rate in the control group—cumulative infection rate in the treatment group/cumulative infection rate in the control group 9 100 (Baulny et al., 1996; Khomvilai et al., 2006).

Table 1   MICs of extracts from Chinese herbs against the SC isolate and two control strains

Chinese herbs	MIC (mg · L^{-1})		
	Strain SC	Strain JL	Strain HP
Bupleurumchinense	ND	ND	ND
CatsiatoraLinn	ND	ND	ND
Cortex albiziae	ND	ND	ND
C. cinnamomi	ND	ND	ND
C. phellodendrichinensis	ND	ND	ND
Dryopterissetosa	ND	ND	ND
Eucommiaulmoides	ND	ND	ND
Exocarpiumbenincasae	ND	ND	ND
FibraureatinctoriaLour.	ND	ND	ND
Floscarthami	ND	ND	ND
Folium artemisiaeargyi	ND	ND	ND
F. nelumbinis	ND	ND	ND
F. isatidis Fructuscarpesii	4096$_{ND}$0	5461$_{ND}$2365	5461$_{ND}$2365
F. lyciibarbari	ND	ND	ND

续表

Chinese herbs	MIC (mg·L^{-1})		
	Strain SC	Strain JL	Strain HP
F. mume Gallachinensis Herbamenthae	54618192ND 23650	5461$_{ND}$2365 ND	68278192ND 23605
H. artemisiaeannuae Jasminumsambac (Linn.) Aiton	5461$_{ND}$2365	6827$_{ND}$2365	5461$_{ND}$2365
Melia azedarach Linn.	ND	ND	ND
OmphalialapidescensSchroet	ND	ND	ND
Pericarpiumcitrireticulataeviride	ND	ND	ND
P. granati PlantagoasiaticaL.	5461$_{ND}$2365	4096$_{ND}$0	5461$_{ND}$2365
Radix astragali	ND	ND	ND
R. aucklandiae	ND	ND	ND
R. et rhizomarhei R. glycyrrhizae	ND ND	ND ND	8192$_{ND}$0
R. ophiopogonis	ND	ND	ND
R. paeoniaerubra	ND	ND	ND
R. rehmanniaepreparata	427 149	427 149	512 0
R. sanguisorbae R. sophoraeflavescentis	$_{ND}$256 0	$_{ND}$256 0	$_{ND}$256 0
Rhizomaatractylodismacrocephalae	ND	ND	ND
R. phragmitis	ND	ND	ND
R. polygonatiodorati Salvia miltiorrhiza	ND ND	8192$_{ND}$0	ND ND
Sterculialychnophora	ND	ND	ND
Trachycarpusfortunei	ND	ND	ND

ND = herb extract MIC>8192 mg·L^{-1}.

Results

Morphological characterization

The SC strain was demonstrated as the causative agent of the Prussian carp egg saprolegniosis, exhibiting a colonizationrate of 73% and with apparent external signs similar to the naturally infected fish eggs (data not shown). No mortality or visible changes were observed in the control group. The SC strain exhibited identical morphological characteristics in asexual and sexual reproductions with Saprolegnia sp. as described by Johnson et al. (2003), e. g. hyphae aseptate; zoosporangia cylindrical, renewed internally with the secondary ones nesting inside discharged primary ones; zoospores dimorphic, discharge and behavioursaprolegnoid; oogonia lateral and spherical (Fig. 1). This initially identified the SC strain as a Saprolegnia isolate.

Phylogenetic analysis

The 750 bp ITS rDNA sequence of the SC strain was submitted to the GenBank database with the accession no. JN662488. Similarities between the ITS rDNA sequence of the SC strain and those of Saprolegniastrains in the GenBank database were 99. 0 to 100%, which confirmed the initial identification. The constructed phylogenetic tree using the neighbour joining method

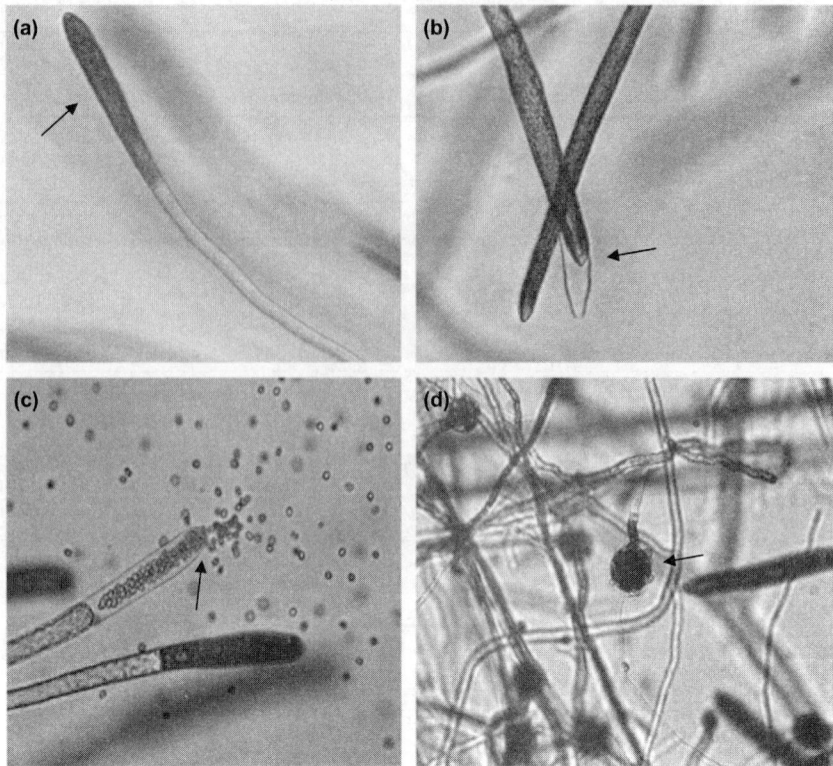

Fig. 1　Morphological characteristics of the SC strain. (a) cylindrical zoosporangium (arrow); (b) sporangial renewal by internal proliferation (arrow); (c) saprolegnoid discharge of zoospores (Arrow); (d) immature oogonium (arrow)

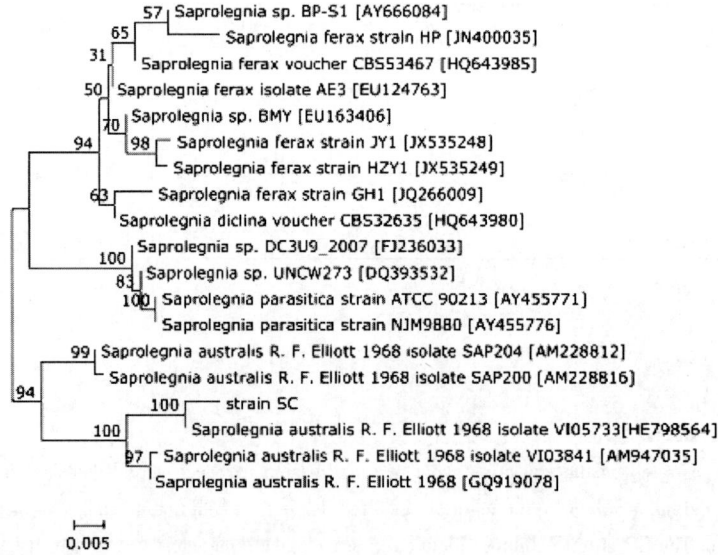

Fig. 2　Constructed ITS rDNA phylogenetic tree of 18 known Saprolegnia strains and the SC isolate using the neighbour-joining method. Bootstrap values (%) shown next to the clades, accession numbers indicated beside the strain names, and scale bars represent distance values

further demonstrated that the SC strain was closely related to SaprolegniaaustralisR. F. Elliott 1968 strain VI05733 (GenBank accession no. HE798564) (Fig. 2) that could form biofilm communities as the virulence factor (Ali et al., 2013). The molecular identification result from phylogenetic analysis was consistent with that found through morphological identification.

Susceptibility to the Chinese herb extracts

The susceptibility of the SC strain to 40 herb extracts is shown in Table 1. Extracts from R. sanguisorbaeand R. rehmanniae

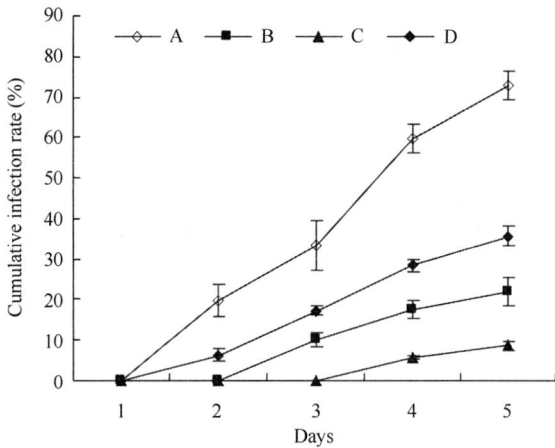

Fig. 3 Protective efficacy of Radix sanguisorbae extract on Saprolegniaaustralis infection in Prussian carp eggs. A, Cumulative infection rate in control group; B, Cumulative infection rate in treatment group with 256 mg · L^{-1}herb extract; C, Cumulative infection rate in treatment group with 1280 mg · L^{-1} herb extract; D, Cumulative infection rate in treatment group with 0. 2 mg · L^{-1} malachite green. Data points = mean values of three replicates; vertical bars on data points = standard deviation of the mean

preparata were demonstrated to show significant inhibition effects on the hyphae growth as indicated by their MICs ≤1 024 mg · L^{-1}for the SC isolate and the two controlstrains（P <0. 05）. In addition, the R. sanguisorbae extract had the best potential to control the SC isolate and the two control strains, because concentrations as low as 256 mg · L^{-1}of R. sanguisorbae was effective at inhibiting hyphae growth.

Protective efficacy

The protective efficacy of the R. sanguisorbaeextract on the Saprolegniainfection to Prussian carp eggs is shown in Fig. 3. Relative survival of 73% and 88% was obtained against the SC strain on eggs at concentrations of 256 mg · L^{1} and 1 280 mg · L^{1}, respectively, which was 22% and 37% higher than that achieved with a concentration of 0. 2 mg · L^{-1}of malachite green. The death of all the test eggs was caused by SaprolegniaaustralisR. F. Elliott 1968, as determined by fungal isolation and molecular identification（data not shown）.

Discussion

Saprolegnia australis R. F. Elliott 1968 has been implicated in significant oomycete infections of crayfish（Krugner-Higby et al., 2010）. However, little information is available on its infection of Prussian carp eggs. In the present study, the naturally occurring pathogen of saprolegniosis-infected Prussian carp eggs was identified as a SaprolegniaaustralisR. F. Elliott 1968 strain. The findings further confirmed that S. australisR. F. Elliott 1968 was a major cause of saprolegniosis in aquaculture production（Krugner-Higby et al., 2010）. In addition, the present results suggest that R. sanguisorbaeextract has the potential to control saprolegniosis during fish egg incubation. This herbal extract can be applied directly to fish hatching pools, and its use will relieve some of the environmental pressures in a fish hatchery.

Identification of the Saprolegniaspecies is complex and sometimes confusing. However, several typical morphological features involving asexual and sexual reproductive organs serve as classic indicators of Saprolegnia（Stueland et al., 2005b）. Thus, the SC strain in the present study was initially identified as a Saprolegniasp. based on its morphological characteristics that conform to the descriptions of Saprolegniasp.（Ke et al., 2009）（Fig. 1）. However, Saprolegniaspecies are usually difficult or even impossible to identify by traditional morphological criteria alone. Therefore, the ITS region of the SC strain is compared to those of Saprolegniaisolates（Whisler, 1996）. The phylogenetic analysis based on ITS rDNA region（Fig. 2）further clarified the taxonomic position of the SC strain and confirmed its initial identification as a Saprolegniasp.

Herb extracts as alternative saprolegniosis therapies attract a great deal of attention due to their sufficient quantities, low toxicity and costs（Cooper et al., 1997; Cao et al., 2006, 2012a; Cai et al., 2010）. R. sanguisorbae, an oriental herb medicine in China for hemostasis, has been used to treat hematemesis, hemoptysis, melena and hypermenorrhea（Chen et al., 2001）. This herb is found to show high antimicrobial activity against Staphylococcus aureus, Pseudomonas aeruginosa, Salmonella typhi, Streptococcus hemolyticus（Xia et al., 2009）, as well as against coronavirus through protein synthesis reduction（Kim et al., 2010）. However, little is known of its antimicrobial activity against oomycete pathogens. In our study, the growth of the SC isolate and the control strains was completely inhibited at as little as 256 mg · L^{-1} extract from R. sanguisorbae (Table 1), providing evidence of its potent antifungal property. This herb extract at concentrations of 256 and 1 280 mg · L^{-1} exhibited significant relative survival of 73% and 88%, respectively, against experimental SaprolegniaaustralisR. F. Elliott 1968 infection in fish eggs (Fig. 3). The patterns of susceptibility of the SC strain to herbal extracts are similar to those found among other Saprolegniaspecies and antifungal chemicals (Stueland et al., 2005a).

In conclusion, based on morphological features and phylogenetic analysis, the causative agent of saprolegniosis in Prussian carp eggs was identified as SaprolegniaaustralisR. F. Elliott 1968. This study reported the first SaprolegniaaustralisR. F. Elliott 1968 infection in Prussian carp eggs and demonstrated for the first time R. sanguisorbaeextract as a promising anti-Saprolegniadrug for the control of egg saprolegniosis.

References

A li, S. E.; Thoen, E.; Vralstad, T.; Kristensen, R.; Evensen, Ø.; Skaar, I., 2013: Development and reproduction of Saprolegniaspecies in biofilms. Vet. Microbiol. 163, 133-141.

Apalikova, O. V.; Podlesnykh, A. V.; Kukhlevsky, A. D.; Guohua, S.; Brykov, V. A., 2011: Phylogenetic relationships of silver crucian carp Carassiusauratusgibelio, C. auratuscuvieri, crucian carp Carassiuscarassiusand common carp Cyprinuscarpioas inferred from mitochondrial DNA variation. Russ. J. Genet. 3, 322-331.

Au, T. K.; Lam, T. L.; Ng, T. B.; Fong, W. P.; Wan, D. C. C., 2001: A comparison of HIV-1 integrase inhibition by aqueous and methanol extracts of Chinese medicinal herbs. Life Sci. 14, 1687-1694.

Baulny, M. O. D.; Quentel, C.; Fournier, V.; Lamour, F.; Gouvello, R. L., 1996: Effect of long-term oral administration of b-glucan as an immunostimulant or an adjuvant on some non-specific parameters of the immune response of turbot Scophthalmus maximus. Dis. Aquat. Org. 26, 139-147.

Cai, Y.; Lei, X.; Yang, D.; Lin, Y., 2010: Research on preventing and curing saprolegniasis of channel catfish hatching eggs by means of Chinese medical herbs. Hubei Agr. Sci. 5, 1170-1172.

Cao, H.; Huang, W.; Song, J., 2006: A review of the prevention and control of fish diseases by use of traditional Chinese herbal medicine. Mar. Sci. 4, 83-87.

Cao, H.; Xia, W.; Zhang, S.; He, S.; Wei, R.; Lu, L.; Yang, X., 2012a: Saprolegniapathogen from Pengze crucian carp (Carassiusauratusvar Pengze) eggs and its control with traditional Chinese herb. Isr. J. Aquacult. -Bamid. 64, 1-7.

Cao, H.; Zheng, W.; Xu, J.; Ou, R.; He, S.; Yang, X., 2012b: Identification of an isolate of Saprolegniaferaxas the causal agent of saprolegniosis of Yellow catfish (Pelteobagrusfulvidraco) eggs. Vet. Res. Commun. 36, 239-244.

Chen, J.; Liu, P., 2007: Tuber latisporumsp. nov. and related taxa, based on morphology and DNA sequence data. Mycologia3, 475-481.

Chen, C. P.; Yokozawa, T.; Sekiya, M.; Hattori, M.; Tanaka, T., 2001: Protective effect of Sanguisorbae radix against peroxynitrite-mediated renal injury. J. Tradit. Med. 18, 1-7.

Cooper, J. A.; Pillinger, J. M.; Ridge, I., 1997: Barley straw inhibits growth of some aquatic saprolegniaceous fungi. Aquaculture 156, 157-163.

Evelyn, T. P. T.; Prosperi-Porta, L.; Ketcheson, J. E., 1986: Experimental intra-ovum infection of salmonid eggs with Renibacteriumsalmoninarumand vertical transmission of the pathogen with such eggs despite their treatment with erythromycin. Dis. Aquat. Org. 1, 197-202.

Ge, X.; Miao, L., 2011: Current state and development suggestion on national conventional freshwater fishery industry. Chin. Fish. Qual. Stand. 3, 22-31.

Ghiasi, M.; Khosravi, A. R.; Soltani, M.; Binaii, M.; Shokri, H.; Tootian, Z.; Rostamibashman, M.; Ebrahimzademousavi, H., 2010: Characterization of Saprolegniaisolates from Persian sturgeon (Acipencerpersicus) eggs based on physiological and molecular data. J. Med. Mycol. 20, 1-7.

Hussein, M. M. A.; Hatai, K., 2002: Pathogenicity of Saprolegniaspecies associated with outbreaks of salmonid saprolegniosis in Japan. Fish. Sci. 5, 1067-1072.

Johnson, T. W.; Seymour, R. L.; Padgett, D. E., 2003: Biology and systematics of the Saprolegniaceae. University of North Carolina, Wilmington.

Ke, X.; Wang, J.; Gu, Z.; Li, M.; Gong, X., 2009: Morphological and molecular phylogenetic analysis of two Saprolegniasp. (Oomycetes) isolated from silver crucian carp and zebra fish. Mycol. Res. 113, 637-644.

Khomvilai, C.; Kashiwagi, M.; Sangrungruang, C.; Yoshioka, M., 2006: Preventive efficacy of sodium hypochlorite against water mold infection on eggs of chum salmon Oncorhynchus keta. Fish. Sci. 72, 28-32.

Kim, H. Y.; Eo, E. Y.; Park, H.; Kim, Y. C.; Park, S.; Shin, H. J.; Kim, K., 2010: Medicinal herbal extracts of Sophorae radix, Acanthopanacis cortex, Sanguisorbae radix and Torilisfructusinhibit coronavirus replication in vitro. Antivir. Ther. 15, 697-709.

Krugner-Higby, L.; Haak, D.; Johnson, P. T.; Shields, J. D.; Jones, W. M.; Reece, K. S.; Meinke, T.; Gendron, A.; Rusak, J. A., 2010: Ulcerative disease outbreak in crayfish Orconectespropin-quuslinked to Saprolegniaaustralisin big Muskellunge Lake, Wisconsin. Dis. Aquat. Org. 1, 57-66.

Mousavi, H. A. E.; Soltani, M.; Khosravi, A.; Mood, S. M.; Hosseinifard, M., 2009: Isolation and characterization of saprolegniaceae from rainbow trout (Oncorhynchus mykiss) eggs in Iran. J. Fish. Aquat. Sci. 6, 330-333.

Noga, E. J., 1993: Water mold infection of freshwater fish: recent advance. Ann. Rev. Fish. Dis. 3, 291-304.

Pickering, A. D.; Willoughby, L. G., 1982: Saprolegniainfections of salomonid fish. In: Microbial diseases of fish. R. J. Roberts (Ed.). Academic Press, London, pp. 271-297.

Stueland, S.; Heier, B. T.; Skaar, I., 2005a: A simple in vitro screening method to determine the effects of drugs against growth of Saprolegniaparasitica. Mycol. Prog. 4, 273-279.

Stueland, S.; Hatai, K.; Skaar, I., 2005b: Morphological and physiological characteristics of Saprolegniaspp. strains pathogenic to Atlantic salmon, Salmo salarL.. J. Fish Dis. 28, 445-453.

Sudova, E.; Machova, J.; Svobodova, Z.; Vesely, T., 2007: Negative effects of malachite green and possibilities of its replacement in the treatment of fish eggs and fish: a review. Vet. Med. 52, 527-539.

Whisler, H. C., 1996: Identification of Saprolegniaspp. pathogenic in chinook salmon. Final Report, DE-AC79-90BP02836, US Department of Energy, Washington.

Xia, H.; Sun, L.; Sun, J.; Zhong, Y., 2009: Progress on chemical ingredient and pharmacological activity of Sanguisorba officinalis L. Food Drug 7, 67-69.

该文发表于《 ISRAELI JOURNAL OF AQUACULTURE-BAMIDGEH》【2013, 65: 851】

Identification of an *Achlya klebsiana* Isolate as the Causal Agent of Saprolegniosis in Eggs of Yellow Catfish (*Pelteobagrus fuvidraco*) and Control with Herbal Extracts

Haipeng Cao[1], Renjian Ou[1], Shan He[2], Xianle Yang[1]

(1. National Pathogen Collection Center for Aquatic Animals, Key Laboratory of Exploration and Utilization of Aquatic Genetic Resources, Shanghai Ocean University, Shanghai 201306, P. R. China; 2. Shanghai Normal University, Shanghai 200235, P. R. China)

Abstract: Achlya species have been implicated in significant fungal infections of live fish and their eggs. In the present study, an oomycete water mold (strain YC) was isolated from yellow catfish (Peleobagrus fulvidraco) eggs suffering from saprolegniosis and characterized morphologically and by its internal transcribed spacer (ITS) gene sequence. Based on its morphological features, the mold was initially identified as an Achlya sp. isolate. Using the neighbor-joining method to construct a phylogenetic tree, the YC strain was closely related to A. klebsiana strain CBS101. 49 (GenBank accession no. AF119579), previously recorded as infecting plant roots as well as fish. Aqueous extracts from 31 Chinese herbs were screened as possible anti-Achlya agents. Of these, a 10 mg/mL extract from Rhizoma coptidis was the most efficacious. Significant protective efficacy of 66. 34% and 92. 20% was obtained against the YC strain in fish eggs by the R. coptidis extract provided at concentrations of 256 mg/L and 1 280 mg/L, respectively.

Key words: Achlya klebsiana, causal agent, yellow catfish eggs, saprolegniosis, Rhizoma coptidis

异丝绵霉菌种作为黄颡鱼卵水霉病病因识别和用草药提取物的控制

曹海鹏[1]，欧仁建[1]，何姗[2]，杨先乐[1]

(1. 国家水生动物病原库，水产种质资源发掘与利用教育部重点实验室，上海海洋大学上海201306 中国；
2. 上海师范大学上海200235 中国)

摘要：绵霉能造成鱼类及鱼卵的严重真菌性感染。本研究从黄颡鱼卵中分离得到一株水霉菌株 YC，根据形态特征和内部转录间隔区 (ITS) 基因序列，这株水霉菌被鉴定为绵霉属。用邻接法构建一个进化树，卵水霉菌与以前记录的感染植物根系和鱼的异丝绵霉菌 CBS101. 49 (基因库没有加入。AF119579) 亲缘关系相近。从31种中草药中水提取的物质作为可能抵抗绵霉菌种的药物被被筛选出来。其中，10 mg/mL 的黄连提取物是最有效的。分别以 256 mg/L 和 1 280 mg/L 的浓度提取黄连，获得了 66. 34% 和 92. 20% 的抵抗鱼卵中绵霉菌保护率。

关键词：异丝绵霉菌；因果代理；黄鲇鱼卵；水霉病；黄连

1 Introduction

The yellow catfish, Pelteobagrus fulvidraco, is a popular freshwater fish species in China, Japan, Korea, and southeast Asia due to its excellent meat (Lee and Lee, 2005; Chen et al., 2011). Yellow catfish farming is an emerging industry in Liaoning, Hubei, Sichuan, and Zhejiang Provinces of China, generating great profits. Annual production was nearly 100, 000 million tons in 2006 (Wu et al., 2010). However, under intensive cultural circumstances, saprolegniosis infections caused by zoosporic oomycete fungi are a serious problem in yellow catfish hatcheries. During egg incubation, the oomycete produces mycelia that grow and spread from dead to healthy eggs causing major economic losses (Noga, 1993; Cao et al., 2012). Thus, study of oomycete infections during the incubation of yellow catfish eggs is needed.

Species of the saprolegniaceous water mold Achlya are major pathogens of many fish species (Jeney and Jeney, 1995). They can be found in most freshwater habitats and are responsible for significant infections of living and dead fish as well as incubating eggs (Kales et al., 2007). Control of Achlya infections is a worldwide problem, since the traditional effective treatment by malachite green has been banned. Thus, relevant studies should be conducted on Achlya pathogens and potential anti-Achlya drugs.

Studies have been conducted on Achlya species in channel catfish and Nile tilapia (Khulbe et al., 1994; Osman et al., 2010). In the present study, a pathogenic A. klebsiana isolate from yellow catfish eggs suffering from saprolegniosis is described. Its taxonomic position was determined using phylogenetic analysis based on its internal transcribed spacer (ITS) rDNA sequence, and its susceptibility to Chinese herb extracts was assayed to screen for potential anti-Achlya drugs.

2 Materials and Methods

Fish eggs. A saprolegniosis outbreak occurred in June 2011 in a yellow catfish hatchery in Hubei Province, China. More than fifty infected yellow catfish eggs were examined externally and microscopically for the presence of oomycetes. Another fifty affected fish eggs were sampled into sterile plastic containers with aerated sterile filtered river water, and transported from the yellow catfish hatchery ponds to the laboratory. Healthy yellow catfish eggs were obtained from another hatchery that had no history of untoward mortalities or abnormalities (Xu et al., 2012).

Isolationof fungi. The infected fish eggs were disinfected for 2-3 s with 75% alcohol, washed several times in sterile filtered water, plated on potato dextrose agar (PDA) plates (Sinopharm Chemical Reagent Co., Ltd.) containing 100 ppm streptomycin and penicillin (Sinopharm Chemical Reagent Co., Ltd.) to reduce bacterial contamination, and incubated at 20℃ for 24 h. Autoclaved rape seeds were placed at the edges of colonies grown on the PDA plates, and the plates were incubated until they were infested with hyphae. The hyphae were transferred to sterile filtered river water and incubated until zoospores were discharged. Then, 100 μL zoospores were spread on PDA plates and incubated at 4℃ for 48-72 h.

Morphological observation. The isolatewasgrown on PDA plates with several autoclaved rape seeds at 20℃ until the seeds were infested with the isolate. Seeds with hyphae were transferred to 6-well cell culture plates containing sterile filtered river water and incubated at 20℃ for 14 days. Plates were observed under an inverted microscope every day to check for the emergence of primary cysts, zoospore discharges, and oogonium, and to examine antheridium morphology.

DNA extraction, PCR amplification, and sequencing. Genomic DNA was extracted from pure cultures of the isolate using a Universal Genomic DNA Extraction Kit ver. 3 (Takara Biotechnology, Dalian). The internal transcribed spacer (ITS) gene was amplified by PCR using a pair of ITS gene primers: ITS1 (5'-TCCGTAGGTGAACCTGCGG-3') and ITS4 (5'-TCCTCCGCTTATTGATATGC-3'), according to a Fungi Identification PCR Kit (Takara Biotechnology, Dalian). Amplification was done as follows: 35 cycles of denaturation at 94℃ for 0. 5 min, annealing at 58℃ for 0. 5 min, extension at 72℃ for 1. 0 min, and final extension at 72℃ for 5 min using a PCR minicycler (Eppendorf Ltd., Germany). The PCR product was electrophorised on 1% agarose gel and observed via ultraviolet trans-illumination. Sequencing was performed using the fluorescent labeled dideoxynucleotides termination method with a BigDye terminator on an ABI 3730 automated DNA sequencer.

Phylogenetic analysis. The partial ITS rDNA sequence was assembled using MegAlign, Editseq, and Seqman software on a powerful Macintosh computer. The National Center for Biotechnology Information (NCBI) database was searched with the Basic Local Alignment Search Tool (BLAST) program. The ITS rDNA gene sequence of the pathogenic isolate was constructed using the neighbor-joining method (Chen and Liu, 2007).

Infectivity experiment. Approximately 7 days prior to starting the test, rape seeds infested with the yellow catfish (YC) isolate were placed in sterile filtered river water at 20℃ for 72 h to induce zoospore formation. Zoospore suspensions were collected according to the method of Hussein and Hatai (2002). Briefly, after the zoospores were released, they were harvested by centrifugation at

498

5 000 r/min for 30 min, and diluted to a final concentration of 2×10^7/mL using sterile filtered river water. Healthy fish eggs were randomly distributed into six small glass aquaria (120 eggs/aquarium) containing aerated sterile filtered river water at 20℃. The eggs were exposed to zoospores of the isolate at a concentration of 2×10^5/mL. In the control aquarium, eggs were held in aerated sterile filtered river water without zoospores. Challenges were conducted in triplicate and eggs were observed daily for 5 days. Eggs with hyphae were immediately removed for fungal isolation (Ghiasi et al., 2010). Infection of eggs by the isolate was recorded only if the challenge strain was re-isolated and its identification morphologically confirmed (Xia et al., 2011).

Extracts of Chinese herbs. Thirty-one Chinese herbs were obtained from Shanghai Fosun Industrial Co., Ltd. (Table 1). Quantities (10 mg/mL) of each Chinese herb extract were prepared in triplicate using the boiling method (Chen et al., 2010). Briefly, 10 g herbs were boiled in 100 mL distilled water for 30 min. The boiled suspension was filtered through three-layer gauze. Concentration of the aqueous extract was adjusted by distilled water to a final concentration at 10 mg/mL and autoclaved at 121℃ for 20 min. Minimum inhibitory concentrations (MIC) of each extract (the lowest concentration that prevented visible fungal colony growth) were determined by the dilution plate method (Stueland et al., 2005).

Protective efficacy assay. Healthy fish eggs were obtained from Sand Lake Aquatic Technique Popularizing Station in Hubei, China, and maintained in three glass aquaria supplied with aerated sterile filtered river water at 20℃. Water did not circulate through the aquaria. The aquaria were stocked with 100 randomly selected healthy eggs and 10 mg/mL herb extracts were added to the aquaria at final concentrations of 256 mg/L and 1 280 mg/L. No extracts were added to the control aquaria. In the positive control aquaria, malachite green was supplemented at a final concentration of 0.2 mg/L (Sudova et al., 2007). Treatments were conducted in triplicate. Eggs were exposed to zoospores of the isolate at a concentration of 2×10^5/mL and observed daily for 5 days. Eggs with hyphae were immediately removed for fungal isolation (Ghiasi et al., 2010), and infection of eggs was recorded only if the challenge strain was re-isolated and its identification morphologically confirmed (Xia et al., 2011). The cumulative infection and protective efficacy rates were calculated according to the following formulae, respectively: 100 (infected eggs/total eggs) and 100 (cumulative infection rate in the control-cumulative infection rate in the treatment group) /cumulative infection rate in the control (Khomvilai et al., 2006).

3 Results

Morphological characterization. The YC strain was demonstrated as the causative agent of yellow catfish egg saprolegniosis, showing a colonization rate of 65% and apparent external signs similar to naturally-infected fish eggs (Fig. 1). No mortality or visible changes were observed in the control eggs. The YC strain exhibited morphological characteristics that were identical with those of asexual and sexual reproduction in Achlya sp. (Johnson et al., 2003): (a) hyphae of the strain were aseptate and transparent (data not shown), its sporangia were cylindrical, clavate, or fusiform, and it renewed in basipetalous succession, infrequently in a sympodial arrangement (Fig. 2); (b) it exhibited the typical "achlyoid" pattern of zoospore release (Steciow and Elíades, 2002), i. e., primary spores were discharged from the tip of the sporangium and clumped together to form a spore ball of primary cysts; from these, secondary swimming zoospores were released (Fig. 3); (c) oogonia were spherical or obpyriform with frequently diclinous, occasionally monoclinous, antheridial branches; they produced one to fifteen oospores that were eccentric and laterally generated a big oil globule (Fig. 4); (d) gemmae were globular or pyriform, formed mostly in chains (Fig. 5). Thus, the YCstrain was initially identified as an Achlya isolate.

Phylogenetic analysis. The 740 bp ITS rDNA sequence of the YC strain was submitted to the GenBank database with the accession no. JN627515. Similarities between the ITS rDNA sequence of the YC strain and those of Achlya strains in the GenBank atabase were 99.0% – 100%, confirming the initial identification. A phylogenetic tree, constructed using the neighbor-joining method, demonstrated that the YC strain was closely related to A. klebsiana strain CBS101.49 (GenBank accession no. AF119579; Fig. d6), a pathogen of plant mycosis (Riethmueller et al., 1999). The molecular identification resulting from the phylogenetic analysis was consistent with that found through morphological identification. Resulting from the phylogenetic analysis was consistent with that found through morphological identification.

Susceptibility to Chinese herb extracts. Extracts from Rhizoma coptidis and Bulbus allii showed good inhibition effects on hyphae growth of the YC strain as their MIC was less than 1 024 mg/L (Table 1). The R. coptidis extract had the best potential to control the YC strain because as little as 256 mg/L of the extract effectively inhibited hyphae growth.

Protective efficacy. The R. coptidis extract had good efficacy in controlling the Achlya infection in yellow catfish eggs (Fig. 7). Protective efficacy against the YC strain in eggs was 66.34% and 92.20% with extract concentrations of 256 mg/L and 1280 mg/L, respectively. This efficacy was 15.12% and 40.98% higher than that achieved with 0.2 mg/L malachite green. The death of all dead eggs was caused by the identified Achlya sp., as determined by fungal isolation and molecular identification (data not shown).

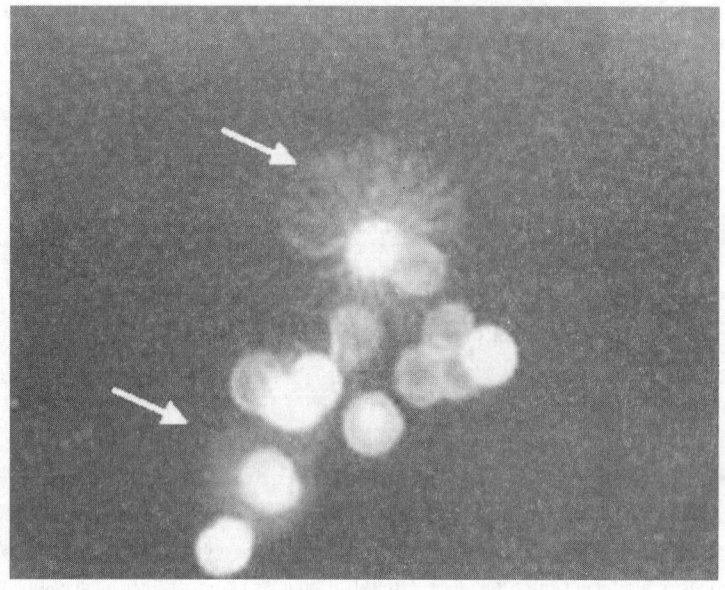

Fig. 1　Yellow catfish eggs suffering from saprolegniosis（arrows）.

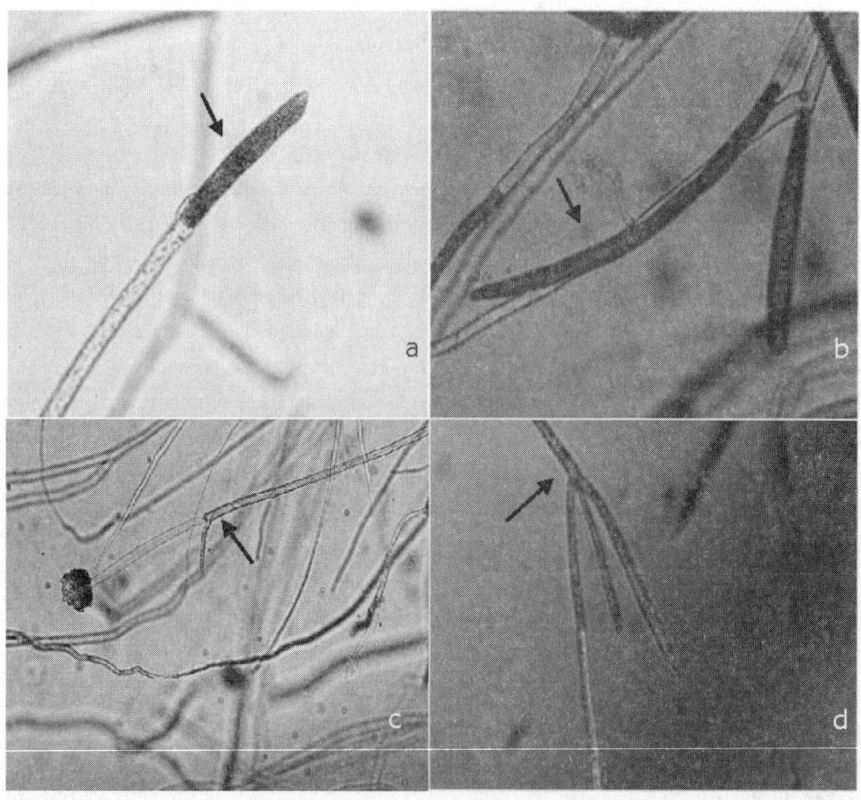

Fig. 2　Zoosporangia and their renewal in the YC strain of Achlya：（a）a cylindrical zoosporangium；
（b）a fusiform zoosporangium；（c）sporangial renewal in basipetalous succession；
（d）sporangial renewal in sympodial arrangement.

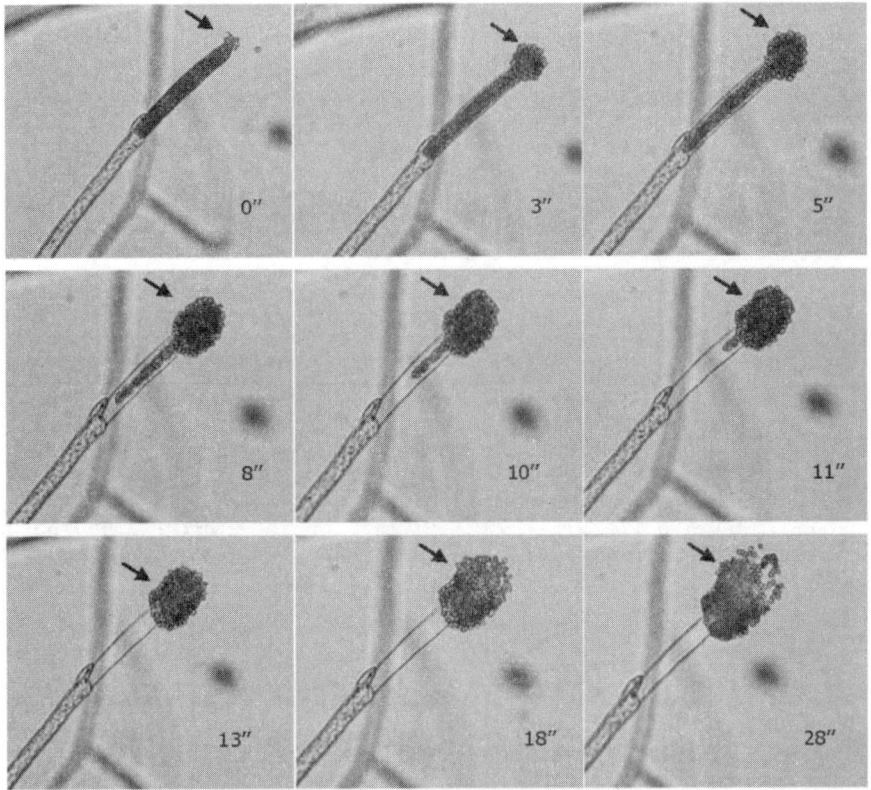

Fig. 3　Discharge of zoospores（arrows）in the YC strain of Achlya

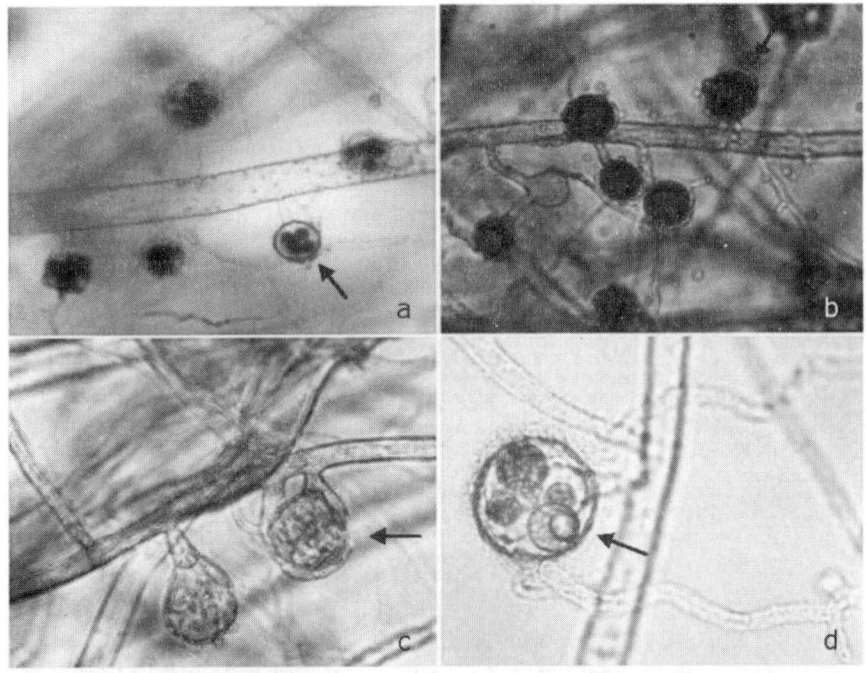

Fig. 4　Sexual reproduction characteristics of the YC strain of Achlya：（a）a mature oogonium，（b）an antheridial branch diclinous with the oogonium，（c）an antheridial branch monoclinous with the oogonium, and（d）a laterally formed oil globule in the oospore.

501

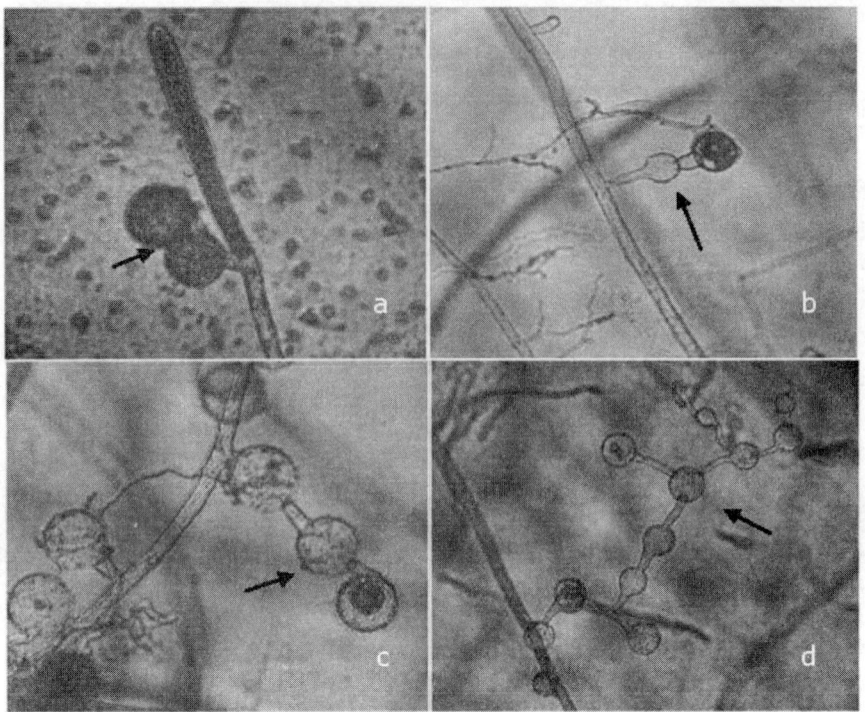

Fig. 5 Chained gemmae (arrows) of the YC strain Achlya, produced on branches of the main hyphea: (a) two catenulate gemmae, (b) agemma catenulate with an oogonium, (c) three catenulate gemmae, (d) thirteen catenulate gemmae.

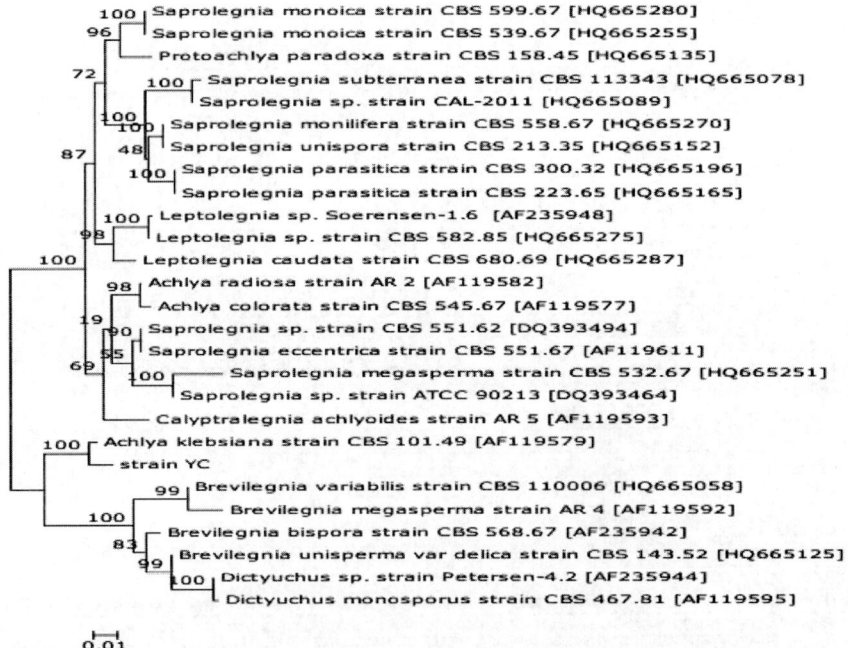

Fig. 6 phylogenetic tree showing 100% similarity between the YC strain and Achlya klebsiana, constructed using the neighbor-joining method.

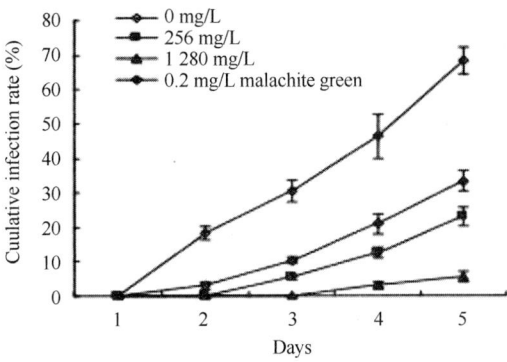

Fig. 7　Protective efficacy of a Rhizoma coptidis extract on Achlya klebsiana infection in yellow
catfish (Pelteobagrus fuvidraco) eggs.

Table 1　Minimum inhibitory concentrations (MIC) of Chinese herbs against the yellow catfish strain of Achlya.

Chinese herb extract	MIC (mg/L)	Chinese herb extract	MIC (mg/L)
Acorus calamus	10 923±4 730	Galla chinensis	16 384±14 189
Borneolum synthcticum	43 691±18 919	Herba artemisiae	65 536±0
Bulbus allii	512±0	Illicium verum	65 536±0
Cortex dictamni	65 536±0	Polygonatum sibiricum	43 691±18 919
Cortex moutan	43 691±18 919	Radix astragali	65 536±0
Cortex phellodendri	43 691±18 919	Radix llithospermi	65 536±0
Cortex pseudolaricis	1 365±591	Radix peucedani	43 691±18 919
Flos caryophylli	65 536±0	Radix polygoni	43 691±18 919
Flos lonicerae	32 768±0	Radix sophorae	43 691±18 919
Fructus cnidii	21 845±9 459	Rheum palmatum	16 384±14 189
Fructus foeniculi	32 768±28 378	Rhizoma coptidis	256±0
Fructus kochiae	43 691±18 919	Rhizoma curcumae	21 845±9 459
Fructus mume	5 461±2 365	Semen areca	43 961±18 919
Fructus perillae	16 384±0	Thallus laminariae	43 691±18 919
Fructus quisqualis	21 845±9 459	Zingiber officinale	8 192±0
Fructus toosendan	10 923±4 730		

4　Discussion

Achlya species are a major cause of saprolegniosis in aquaculture production (Steciow et al. , 2002) and have been implicated in significant oomycete infections of live Channa striatus and Cyprinus carpio and their eggs (Kitancharoen et al. , 1995; Chukanhom and Hatai, 2004). Based on morphological characteristics and molecular phylogenetic analysis, the naturally-occurring pathogen of the yellow catfish eggs in the present study was identified as an A. klebsiana strain.

Identification of Achlya species is complex and sometimes confusing. Several typical morphological features involving asexual and sexual reproductive organs serve as classic indicators of Achlya (Kitancharoen et al. , 1995). The YC strain in the present study was initially identified as an Achlya sp. based on morphological characteristics that conform to descriptions of Achlya sp. (Johnson et al. , 2003). However, Achlya species are usually difficult or even impossible to identify by traditional morphological criteria alone. Thus, the ITS region of the YC strain was compared tothat of Achlya isolates (Daugherty, 1998). Phylogenetic analysis based on ITS rDNA region further clarified the taxonomic position of the YC strain and confirmed its initial identification as an Achlya sp.

Hydrogen peroxide, salt, ozone, formaldehyde, and formalin formulations are used to control saprolegniosis in fish eggs (Forneris et al. , 2003; Rach et al. , 2004; Gieseker et al. , 2006), but these treatments cannot totally arrest the growth of Saprolegniaceous species（Van West, 2006）. Herb extracts as alternative saprolegniosis therapies attract a great deal of attention due to their low toxicity and costs（Cooper et al., 1997；Cai et al., 2010；Cao et al., 2012）. In our study, growth of the YC strain was completely inhibited by as little as 256 mg/L extract from R. coptidis, a traditional Chinese medicine with pharmacological properties（Ma et al., 2011）. Concentrations of 256 and 1280 mg/L provided significant protective efficacy of 66. 34% and 92. 20%, respectively, against experimental Achlya infections of fish eggs, possibly due to the ability of R. coptidis to produce coptisine that reportedly possesses antifungal activity（Chen et al., 2008；Seneviratne et al., 2008）. The patterns of susceptibility to herbal extracts of the YC strain were similar to those found between other saprolegniaceous species and antifungal chemicals（Stueland et al., 2005）.

In conclusion, based on morphological features and phylogenetic analysis, the causative agent of saprolegniosis in yellow catfish eggs is A. klebsiana and R. coptidis extract is a promising anti-Achlya drug for control of egg saprolegniasis.

Acknowledgements

This work was financially supported by the National High-Tech R&D Program, P. R. China（no. 2011AA10A216）, earmarked fund for Modern Agro-Industry Technology Research System, P. R. China（no. CARS-46）, Shangxi Provincial Science-Technology Innovation Program, P. R. China, and Lianyungang Key Technology Program for Agriculture, P. R. China.

References

C ai Y., Lei X., Yang D. and Y. Lin, 2010. Research on preventing and curing saprolegniasis of channel catfish hatching eggs by means of Chinese medical herbs. Hubei Agric. Sci., 5：1170-1172.

Cao H., Xia W., Zhang S., He S., Wei R., Lu L. and X. Yang, 2012. Saprolegnia pathogen from Pengze crucian carp（Carassius auratus var. Pengze）eggs and its control with traditional Chinese herb. Isr. J. Aquacult. -Bamidgeh, IJA_ 64. 2012. 685, 7 pages.

Chen J. and P. Liu, 2007. Tuber latisporum sp. nov. and related taxa, based on morphology and DNA sequence data. Mycologia, 3：475-481.

Chen J., Zhao H., Wang X., Lee F. S., Yang H. and I. Zheng, 2008. Analysis of major alkaloids in Rhizoma coptidis by capillary electrophoresis-electrospray-time of flight mass spectrometry with different background electrolytes. Electrophoresis, 29：2135-2147.

Chen X. Y., Xiao H. and X. Y. Fu, 2010. Antifungal activity of 55 Chinese herb medicines on Saprolegnia in Chinemys nigricans. J. Anhui Agric. Sci., 33：18855-18867.

Chen X., Xu J. and C. Ai, 2011. Study on nutrient requirement and formulated feed of yellow catfish, Pelteobagrus fulvidraco. Feed Ind., 10：48-51.

Chukanhom K. and K. Hatai, 2004. Freshwater fungi isolated from eggs of the common carp（Cyprinus carpio）in Thailand. Mycoscience, 45：42-48.

Cooper J. A., Pillinger J. M. and I. Ridge, 1997. Barley straw inhibits growth of some aquatic saprolegniaceous fungi. Aquaculture, 156：157-163.

Daugherty J., Evans T. M., Skillom T., Watson L. E. and N. P. Money, 1998. Evolution of spore release mechanisms in the Saprolegniaceae（Oomycetes）：evidence from a phylogenetic analysis of internal transcribed spacer sequences. Fungal Gen. Biol., 24：354-363.

Forneris G., Bellardi S., Palmegiano G. B., Saroglia M., Sicuro B., Gasco L. and I. Zoccarato, 2003. The use of ozone in trout hatchery to reduce saprolegniasis incidence. Aquaculture, 221：157-166.

Ghiasi M., Khosravi A. R., Soltani M., Binaii M., Shokri H., Tootian Z., Rostamibashman M. and H. Ebrahimzademousavi, 2010. Characterization of Saprolegnia isolates from Persian sturgeon（Acipencer persicus）eggs based on physiological and molecular data. J. Med. Mycol., 20：1-7.

Gieseker C. M., Serfling S. G. and R. Reimschuessel, 2006. Formalin treatment to reduce mortality associated with Saprolegnia parasitica in rainbow trout, Oncorhynchus mykiss. Aquaculture, 253：120-129.

Hussein M. M. A. and K. Hatai, 2002. Pathogenicity of Saprolegnia species associated with outbreaks of salmonid saprolegniosis in Japan. Fish. Sci., 68（5）：1067-1072.

Jeney Z. S. and G. Jeney, 1995. Recent achievements in studies on diseases of the common carp（Cyprinus carpio L.）. Aquaculture, 129：397-420.

Johnson T. W., Seymour R. L. and D. E. Padgett, 2003. Biology and Systematics of the Saprolegniaceae. Univ. North Carolina, Wilmington. pp. 452-481.

Kales S. C., DeWitte-Orr S. J., Bols N. C. and B. Dixon, 2007. Response of the rainbow trout monocyte/macrophage cell line, RTS11, to the water molds Achlya and Saprolegnia. Mol. Immunol., 44：2303-2314.

Khomvilai C., Kashiwagi M., Sangrungruang C. and M. Yoshioka, 2006. Preventive efficacy of sodium hypochlorite against water mold infection on eggs of chum salmon Oncorhynchus keta. Fish. Sci., 72：28-32.

Khulbe R. D., Bisht G. S. and C. Joshi, 1994. Epizootic infection due to Achlya debaryana in a catfish. Mycoses, 37：61-63.

Kitancharoen N., Hatai K., Ogihara R. and D. N. N. Aye, 1995. A new record of Achlya klebsiana from snakehead, Channa striatus, with fungal infection in Myanmar. Mycoscience, 36：235-238.

Lee O. K. and S. M. Lee, 2005. Effects of the dietary protein and lipid levels on growth and body composition of bagrid catfish, Pseudobagrus fulvidraco. Aquaculture, 243：323-329.

Ma B. L., Yao M. K., Han X. H., Ma Y. M., Wu J. S. and C. H. Wang, 2011. Influences of Fructus evodiae pretreatment on the pharmacokinetics of Rhizoma coptidis alkaloids. J. Ethnopharmacol., 137：1395-1401.

504

Noga E. J. , 1993. Water mold infection of freshwater fish: recent advance. Annu. Rev. Fish Dis. , 3: 291-304.

Osman A. , Ali E. , Hashem M. , Mostafa M. and I. Mekkawy, 2010. Genotoxicity of two pathogenic strains of zoosporic fungi (Achlya klebsiana and Aphanomyces laevis) on erythrocytes of Nile tilapia Oreochromis niloticus niloticus. Ecotox. Environ. Safe. , 73: 24-31.

Rach J. J. , Valentine J. J. , Schreier T. M. , Gaikowski M. P. and T. G. Crawford, 2004. Efficacy of hydrogen peroxide to control saprolegniasis on channel catfish (Ictalurus punctatus) eggs. Aquaculture, 238: 135-142.

Riethmueller A. , Weiss M. and F. Oberwinkler, 1999. Phylogenetic studies of Saprolegniomycetidae and related groups based on nuclear large subunit ribosomal DNA sequences. Can. J. Bot. , 12: 1790-1800.

Seneviratne C. J. , Wong R. W. and L. P. Samaranayake, 2008. Potent anti-microbial activity of traditional Chinese medicine herbs against Candida species. Mycoses, 51: 30-34.

Steciow M. M. and L. A. Elíades, 2002. A. robusta sp. nov. , a new species of Achlya (Saprolegniales, Straminipila) from a polluted Argentine channel. Microbiol. Res. , 157: 177-182.

Stueland S. , Heier B. T. and I. Skaar, 2005. A simple in vitro screening method to determine the effects of drugs against growth of Saprolegnia parasitica. Mycol. Prog. , 4: 273-279.

Sudova E. , Machova J. , Svobodova Z. and T. Vesely, 2007. Negative effects of malachite green and possibilities of its replacement in the treatment of fish eggs and fish: a review. Veterinarni Medicina, 52: 527-539.

Van West P. , 2006. Saprolegnia parasitica, an oomycete pathogen with a fishy appetite: new challenges for an old problem. Mycologist, 20: 99-104.

Wu S. , Gao T. , Zheng Y. , Wang W. , Cheng Y. and G. Wang, 2010. Microbial diversity of intestinal contents and mucus in yellow catfish (Pelteobagrus fulvidraco). Aquaculture, 303: 1-7.

Xia W. , Cao H. , Wang H. , Zhang S. and X. Yang, 2011. Identification and biological characteristics of a pathogenic Saprolegnia sp. from the egg of Pengze crucian carp (Carassius auratus pengzesis). Microbiol. China, 1: 57-62.

Xu J. , Cao H. , Ou R. and X. Yang, 2012. Identification and asexual reproduction characterization of a Saprolegnia ferax pathogen from the egg of yellow catfish (Pelteobagrus fulvidraco). Microbiol. China, 39 (4): 579-587.

该文发表于《VETERINARY RESEARCH COMMUNICATIONS》【2012，36（4）：239-244. DOI：10.1007/s11259-012-9536-8】

Identification of an isolate of Saprolegnia ferax as the causal agent of saprolegniosis of Yellow catfish (*Pelteobagrus fulvidraco*) egg

Haipeng Cao &Weidong Zheng &Jialu Xu &RenjianOu&Shan He &Xianle Yang

(Shanghai Ocean Univisity)

Abstract: Saprolegniaspecies have been implicated for significant fungal infections of both living and dead fish as well as their eggs. In the present study, an oomycete water mould (strain HP) isolated from yellow catfish (Peleobagrusfulvidraco) eggs suffering from saprolegniosis was characterized both morphologically and from ITS sequence data. It was initially identified as a Saprolegniasp. isolate based on its morphological features. The constructed phylogenetic tree using neighbour joining method indicated that the HP strain was closely related to Saprolegniaferaxstrain Arg4S (GenBank accession no. GQ119935), that had previously been isolated from farming water samples in Argentina. In addition, the zoospore numbers of strain HP were markedly influenced by a variety of environmental variables including temperature, pH, formalin and dithiocyano-methane. Its zoospore formation was optimal at 20℃ and pH7, could be well inhibited by formalin and dithiocyano-methane above 5 mg/L and 0.25 mg/L, respectively. To our knowledge, this is the first report on the S. ferax infection in the hatching yellow catfish eggs.

Keywords: Saprolegnia ferax. Causal agent. Phylogenetic analysis. Yellow catfish egg. Saprolegniosis

患水霉病的黄颡鱼卵中分离水霉菌的鉴定

曹海鹏，郑卫东，许佳露，欧仁建，何珊，杨先乐

（上海海洋大学）

摘要：水霉能使活鱼和死鱼以及卵感染真菌。本文从患有水霉病的黄颡鱼卵中分离出来一株水霉菌HP，并根据

形态学和序列数据进行了鉴定。基于其形态学特征它初步被鉴定为水霉。用邻接法构建的系统进化树表明该水霉株（*strain HP*）与之前从阿根廷养殖水体中分离出来的 *Saprolegniaferaxstrain Arg*4S（基因库编号 GQ119935）亲缘关系相近。另外，卵水霉（strain HP）游动孢子的数量明显的受各种环境环境变量的影响，包括温度、pH 值、福尔马林和二硫氰基甲烷。它的游动孢子最佳形成条件为20℃和pH7，分别在 5 mg/L 和 0.25 mg/L 以上浓度的福尔马林和二硫氰基甲烷能够很好地被抑制。据我们所知，这是第一个关于感染水霉孵化中的黄颡鱼卵的报道。

关键词： 水霉，病因，系统进化分析，黄颡鱼卵，水霉病

Introduction

Yellow catfish, Pelteobagrusfulvidraco, is a popular freshwater fish species in China, Japan, Korea, Southeast Asia and other countries due to its excellent meat quality (Chen et al. 2011; Lee and Lee 2005). Especially in China, the yellow catfish farming has become an emerging industry in the province of Liaoning, Hubei, Sichuan, Zhejiang, etc., and has brought a great profit in recent years (Ye et al. 2009). Its annual production was nearly 100, 000 million tons in 2006 (Wu et al. 2010). However, under intensive cultural circumstances, serious problems in yellow catfish hatcheries are saprolegniosis infections caused by zoosporic oomycete fungi. During the egg incubation, the oomycete produce mycelia which grow and spread from dead to healthy eggs causing major economic losses (Ke et al. 2010). Thus, more attention needs be paid to the oomycete infections in the incubation of yellow catfish eggs.

Species of Saprolegniaare major pathogens of many fish species (Pickering and Wiloughby 1982), can be found in most freshwater habitats and are responsible for significant infections of both living and dead fish as well as incubating egg (Noga1993; Hussein and Hatai 2002). A few studies have been conducted on Saprolegniaspecies in Rainbow trout eggs, Zebra fish, Silver crucian carp and others (Mousavi et al. 2009; Ke et al. 2009). However, no specific report is available on Saprolegniaspecies in yellow catfish eggs. Thus, relevant studies should be extensively conducted on the Saprolegnia pathogens and their controls.

In the present study, a pathogenic Saprolegniaferaxisolate from the yellow catfish eggs suffering from saprolegniosis was described, its taxonomic position was determined using phylogenetic analysis based on ITS rDNA sequence, and the influence of environmental factors and anti-Saprolegniaagents on its zoospores was also conducted. To our knowledge, this is the first report on the S. feraxinfection in the hatching yellow catfish eggs.

Materials and methods

Experimental fish eggs

A saprolegniosis outbreak occurred in June 2011 in a yellow catfish hatchery, Hubei province, China. More than fifty infected yellow catfish eggs were examined externally and microscopically for the presence of oomycete. Another fifty affected fish eggs were sampled and transported to the laboratory as described by Ruso (1987) from the yellow catfish hatchery ponds. Healthy yellow catfish eggs used in the study were obtained from another farm which had no history of untoward mortalities or abnormalities according to Ye et al. (2009).

Isolation of fungi

The sampled infected fish eggs were first disinfected for 2 to 3 s with 75% alcohol, then washed several times in sterile filtered water, placed on potato dextrose agar (PDA) plates (Sinopharm Chemical Reagent Co., Ltd) containing 100 mg/L of streptomycin and penicillin (Sinopharm Chemical Reagent Co., Ltd) to facilitate isolation of fungal strains and incubated at 20℃ for 24 h. The autoclaved rape seeds were immediately placed at the edge of colonies grown on PDA plates, and incubated until they were covered with hyphae. These were then transferred to sterile filtered river water and incubated until the zoospores were discharged. 100 μL of zoospores were then spread on PDA plates and incubated at 4℃ for 48 h to 72 h until used. The isolate was deposited in National Pathogen Collection Center for Aquatic Animals of China with the collection no. BYK003001.

Infectivity experiment

Approximately 7 days prior to starting the test, the rape seeds infested with the isolate were placed in sterile filtered river water (pH7.2) at 20℃ for 72 h to induce zoospore formation. Zoospore suspensions were then collected as described by Hussein and Hatai (2002). 720 healthy fish eggs were randomly distributed amongst 6 small glass aquaria (120 eggs/aquarium) containing aerated sterile filtered river water (pH7.2) at 20℃. The fish eggs were exposed to zoospores at a concentration of 2×10^5/mL, whereas in the control aquarium eggs were held in aerated sterile filtered river water (pH7.2). Each challenge was conducted in triplicate. The fish eggs were observed daily for 5 days. Eggs with hyphae were immediately removed for fungal isolation according to Ghiasi et al. (2010), and the infection of eggs was recorded and the infection rate was calculated as described by Hanjavanit et al. (2008).

Molecular identification

DNA extract, PCR amplification and sequencing Genomic DNA was extracted form pure cultures of the isolate using a universal genomic DNA extraction kit ver 3 (Takara Biotechnology, Dalian). The internal transcribed spacer (ITS) gene was amplified by PCR using a pair of ITS gene primers (ITS1): 5′-TCCGTAGGTGAACCTGCGG-3′ as well as (ITS4): 5′-TCCTCCGCTTATT-GATATGC-3′, and carried out according to the instructions of the fungi identification PCR kit (Takara Biotechnology, Dalian). Amplification was done after 35 cycles of denaturation at 94℃ for 0.5 min, annealing at 58℃ for 0.5 min and extension at 72℃ for 1.0 min followed by a final extension at 72℃ for 5 min using a PCR minicycler (Eppendorf Ltd., Germany). The PCR product was electrophorised on 1% agarose gel and observed via ultraviolet trans-illumination. Sequencing was performed using a fluorescent labeled dideoxynucleotides termination method (with BigDye terminator) on ABI 3730 automated DNA Sequencer.

Phylogenetic analysis

The partial ITS rDNA sequence was assembled using MegAlign, Editseq and Seqman software with a power Macintosh computer. Searches were done against the National Centre for Biotechnology Information (NCBI) database using the Basic Local Alignment Search Tool (BLAST) program. The ITS rDNA gene sequence of the pathogenic isolate was constructed using neighbour joining method as described by Cooke et al. (2000).

Environmental factor and antifungal drug influence assay

The influence of different factors on the zoospore formation of the pathogenic isolate was performed at a temperature of 5-35℃, pH of 4.0-10.0, a concentration of 0-25 mg/L formalin and 0-1.25 mg/L dithiocyano-methane as described by Smith et al. (1984). Briefly, sterile rape seeds were added to PDA plates containing isolate cultures to allow seeds to be inoculated with the isolate. For temperature influence assay, the seed with hyphae was respectively removed from PDA plates and added to 6-well plates containing 5 mL of sterile distilled water (pH7.0), then incubated at 5℃, 10℃, 15℃, 20℃, 25℃, 30℃, 35℃ for 72 h. For pH influence assay, the seed with hyphae was respectively removed from PDA plates, added to 6-well plates containing 5 mL of sterile distilled water with pH values of 4.0, 5.0, 6.0, 7.0, 8.0, 9.0, 10.0, and then incubated at 20℃ for 72 h. For formalin influence assay, the seed with hyphae was respectively removed from PDA plates, added to 6-well plates containing 5 mL of sterile distilled water (pH 7.0) filled with formalin at a final concentration of 0 mg/L, 5 mg/L, 10 mg/L, 15 mg/L, 20 mg/L, 25 mg/L, and then incubated at 20℃ for 72 h. For dithiocyano-methane influence assay, the seed with hyphae was respectively removed from PDA plates, added to 6-well plates containing 5 mL of sterile distilled water (pH7.0) filled with dithiocyanomethane at a final concentration of 0 mg/L, 0.25 mg/L, 0.50 mg/L, 0.75 mg/L, 1.00 mg/L, 1.25 mg/L, and then incubated at 20℃ for 72 h. The zoospore concentration was determined as described by Hussein and Hatai (2002). Each experiment was conducted in triplicate.

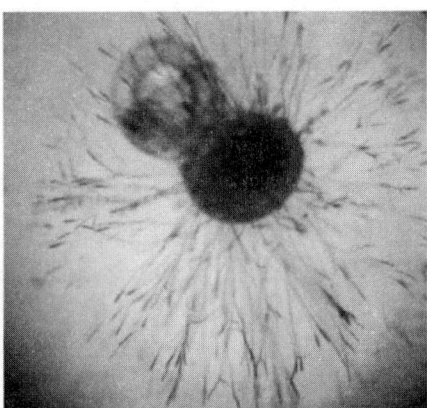

Fig. 1　The symptom of the yellow catfish egg suffering from saprolegniosis, arrowheads showed
that the mycelium penetrated egg (10×)

Results

Morphological characterization

The HP strain was demonstrated as the causative agent of the yellow catfish egg saprolegniosis, showing the infection rate of 45% and apparent external signs similar to the naturally affected fish eggs (Fig. 1). No mortality or visible changes were observed in the control and other isolates' challenge groups. The HP strain exhibited identical morphological characteristics in asexual and sexual reproductions with Saprolegniasp. as described by Johnson et al. (2003) (data not shown): hyphae aseptate; zoosporangia cylindrical or

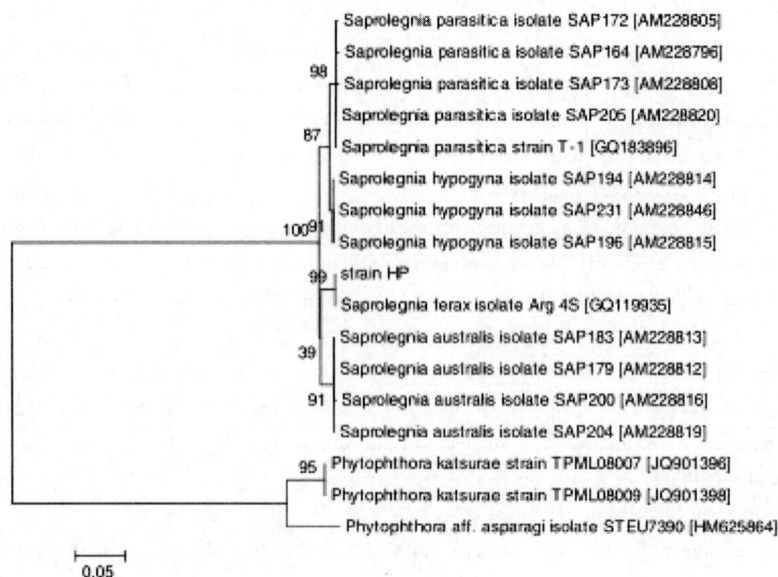

Fig. 2 Phylogenetic tree was constructed based on ITS rDNA sequences using neighbourjoining method. The bootstrap values (%) were shown besides the clades, accession numbers were indicated beside the name of strains, and scale bars represented distance values

clavate, renewed internally with secondary ones nesting inside discharged primary ones; zoospores dimorphic, discharge and behavior saprolegnoid; oogonia lateral and spherical; oospores centric or subcentric; antheridial branches predominantly monoclinous, rarely diclinous. Thus, the HP strain was initially identified as a Saprolegniaisolate.

Phylogenetic analysis

The 784 bp ITS rDNA sequence of the HP strain was submitted to GenBank database with the accession no. JN400035. Similarities between the ITS rDNA sequence of the HP strain and those of Saprolegniastrains in the GenBank database were 99.0%–100%, which confirmed the initial identification. The constructed phylogenetic tree using neighbour joining method further demonstrated that the HP strain was closely related to Saprolegniaferaxstrain Arg4S (GenBank accession no. GQ119935) (Fig. 2), that was isolated from farming water samples in Argentina. The molecular identification result from phylogenetic analysis was consistent with that found through morphological identification.

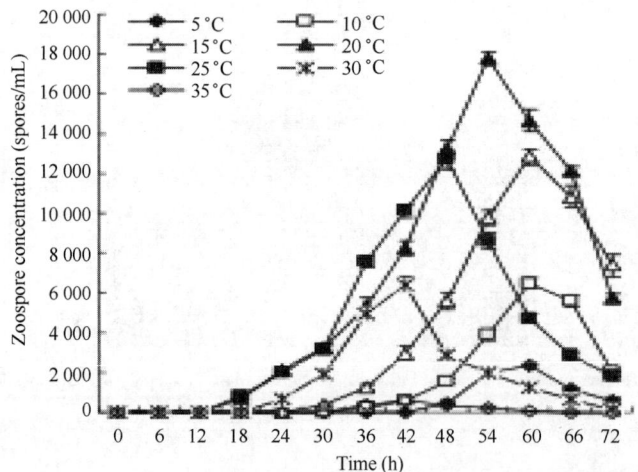

Fig. 3 Zoospore formation of the HP strain cultivated in the sterile distilled water at the temperature range of 5–35℃ for 72 h. All Data points are mean±SD from three independent experiments

Zoospore formation

The effects of temperature, pH, formalin and dithiocyanomethane on the zoospores of the HP strain were shown in Figs. 3, 4,

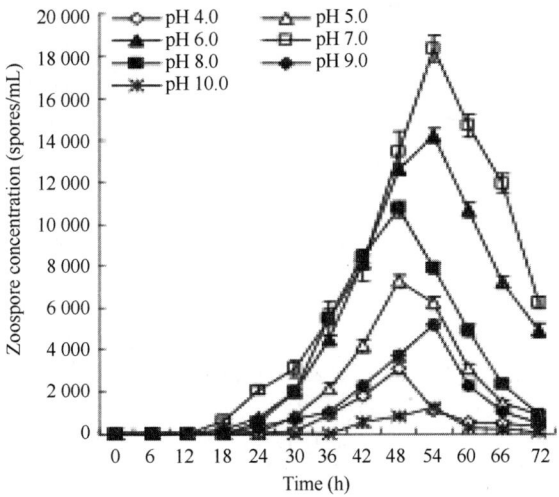

Fig. 4　Zoospore formation of the HP strain cultivated in the sterile distilled water at the pH range of 4. 0–10. 0 for 72 h. All Data points are mean±SD from three independent experiments

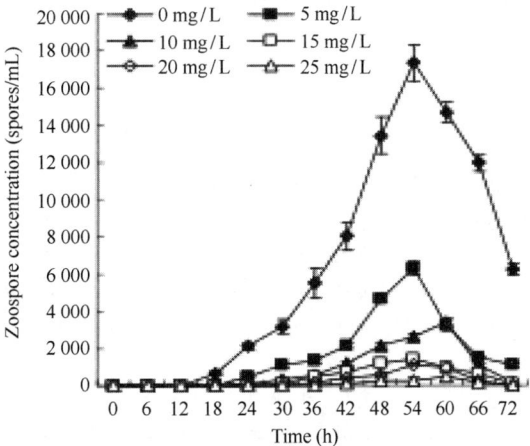

Fig. 5　Zoospore formation of the HP strain cultivated in the sterile distilled water containing 0–25 mg/L formalin for 72 h. All Data points are mean±SD from three independent experiments

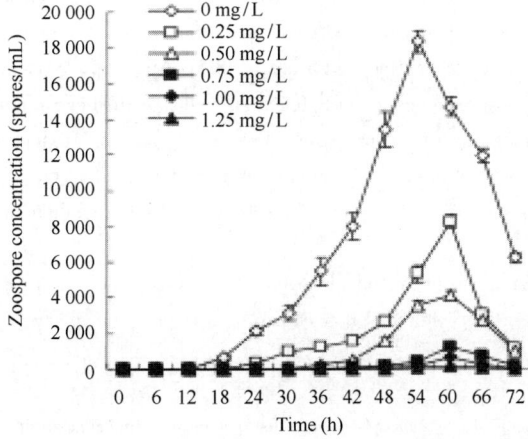

Fig. 6　Zoospore formation of the HP strain cultivated in the sterile distilled water containing 0–1. 25 mg/L dithiocyano–methane for 72 h. All Data points are mean±SD from three independent experiments

5, and 6. The numbers of zoospores produced by the HP isolate were markedly influenced by a variety of environmental variables including temperature, pH, formalin and dithiocyano-methane. The optimum temperature and pH for its zoospore formation were respectively 20℃ and 7. The use of formalin and dithiocyano-methane showed good inhibitory effects on the zoospore production, although variations were recorded at different concentrations. With a respective increase in the concentrations of formalin and dithiocyano-methane above 5 mg/L and 0.25 mg/L, the maximum zoospore numbers were significantly reduced by 63.78%-97.11% and 54.91%-98.91% in comparison with the control, respectively.

Discussion

Saprolegnia species have been implicated for significant fungal pathogens of both fish and their eggs. Thus, more attention should be paid to the Saprolegnia pathogens. In the present study, the naturally occurring pathogen was identified as S. feraxbased on morphological characteristics and phylogenetic analysis. The findings further confirmed that Saprolegnia species were the major cause of saprolegniosis in aquaculture production (Bangyeekhun et al. 2003). Furthermore, the influence of temperature, pH, formalin and dithiocyano-methane on its zoospore formation was also determined. As far as we know, this is the first report on the S. feraxinfection in incubated yellow catfish eggs.

The identification of Saprolegnia species was complex and sometimes confusing. However, several typical morphological features could be served as classical Saprolegniaidentification, involving asexual and sexual reproductive organs (Stueland et al. 2005). Thus, the HP strain in thepresent study was initially identified as Saprolegniasp. based on its typical morphological characteristics (data not shown), which were also in line with the precise descriptions on Saprolegniasp. by Ke et al. (2009). However, Saprolegniaspecies were usually difficult or even impossible to identify through traditional morphological criteria alone. Therefore, comparing ITS region method was further employed for identification of the HP strain, which was applied to Saprolegniaisolates from salmonid fish by Whisler (1996). The phylogenetic analysis based on ITS rDNA region (Fig. 2) further clarified the taxonomic position of the HP strain and confirmed its initial identification.

Zoospores of some moderately or highly pathogenic Saprolegniastrains were reported to have long hook hairs, that were believed to increase Saprolegniaattachment efficiency (Beakes1982). However, no long hooked hairs were observed on the zoospores of the pathogenic strain HP as well as other Saprolegniapathogens such as the Saprolegniaisolate SAP211 (GenBank accession no. AM228826) (Diéguez-Uribeondo et al. 2007). Thus, it was clear that long hooked hairs did not determine pathogenicity of Saprolegnia strains, other contributing factors such as the chymotrypsin-like activities were also likely to play a role in their pathogenicity (Peduzzi and Bizzozero1977).

The epidemic spread of saprolegniosis was primarily based on rapid dispersal from host to host by zoospores, because zoospores were able to exhibit chemotactic responses, electrotaxis, and autotaxis or autoaggregation to target new hosts for infection once released from the sporangium (Walker and Van West 2007). To date, there were a few chemicals used to control saprolegniosis in the fish eggs, such as formalin, ozone, hydrogen peroxide, dithiocyano-methane (Forneris et al. 2003; Rach et al. 2004; Gieseker et al. 2006), but these treatments couldn't totally control saprolegniosis (Van West 2006). This might be due to these antifungal agents unable to arrest the zoospore release from the sporangia. Thus, it is of great significance to study the zoospore formation under different factors. A few studies have been reported on the pH, temperature, salinity, formalin on the growth of Saprolegniastrains (Oláh and Farkas1978; Kitancharoen et al. 1996; Koeypudsa et al. 2005; Ali 2005), but little information is available about the effect of environmental variables on the zoospore formation of S. feraxfrom yellow catfish eggs. In our study, temperature, pH, formalin showed significant effects on the HP strain's zoospore formation (Figs. 3, 4, and 5), which was in accordance with the finding on S. diclinaobtained by Smith et al. (1984). Additionally, the obviously inhibitory effect on the zoospore formation was also shown by dithiocyano-methane (Fig. 6). This was possibly the result of electron transport inhibition in its respiratory chain (Zhang and Xu 2006).

In conclusion, based on morphological features and phylogenetic analysis, the causative agent was identified as S. ferax. This study reported the first S. feraxinfection in the cultured yellow catfish eggs from China.

References

Ali E. H. (2005) Morphological and biochemical alterations of oomycete fish pathogen Saprolegniaparasiticaas affected by salinity, ascorbic acide and their synergistic action. Mycopathologia 159: 231-243.

Bangyeekhun E., Pylkko P., Vennerstrom P., Kuronen H., Cerenius L. (2003) Prevalence of a single fish-pathogenic Saprolegniasp. colne in Finland and Sweden. Dis Aquat Org 53: 47-53.

Beakes G. (1982) A comparative account of cyst coat ontogeny in saprophytic and fish-lesion (pathogenic) isolates of the Saprolegniadeclina-parasiticacomplex. Can J Bot 61: 603-625.

Chen X., Xu J., Ai C. (2011) Study on nutrient requirement and formulated feed of Yellow catfish, Pelteobagrusfulvidraco. Feed Indus 32 (10): 48-51.

Cooke D. E. L., Drenth A., Duncan J. M., Wagels G., Brasier C. M. (2000) A molecular phylogeny of Phytophthoraand relelatedoomycetes. Fungal Genet Biol 30: 17-32.

Diéguez-Uribeondo J., Fregeneda-Grandes J. M., Cerenius L., PérezIniesta E., Aller-Gancedo J. M., Tellería M. T., Söderhäll K., Martín M. P. (2007) Re-evaluation of the enigmatic species complex Saprolegniadiclina-Saprolegniaparasiticabased on morphological, physiological and molecular data. Fungal Genet Biol 44: 585-601.

Forneris G., Bellardi S., Palmegiano G. B., Saroglia M., Sicuro B., Gasco L., Zoccarato I. (2003) The use of ozone in trout hatchery to reduce saprolegniosis incidence. Aquaculture 221: 157-166.

Ghiasi M., khosravi A. R., Soltani M., Binaii M., Shokri H., Tootian Z., Rostamibashman M., Ebrahimzademousavi H. (2010) Characterization of Saprolegniaisolates from Persian sturgeon (Acipencerpersicus) eggs based on physiological and molecular data. J Med Mycol 20: 1-7.

Gieseker C. M., Serfling S. G., Reimschuessel R. (2006) Formalin treatment to reduce mortality associated with Saprolegniaparasiticain rainbow trout, Oncorhynchus mykiss. Aquaculture 253 (1-4): 120-129.

Hanjavanit C., Kitancharoen N., Rakmanee C. (2008) Experimental infection of aquatic fungi on eggs of African catfish (ClariasgariepinusBurch). KKU Sci J 36: 36-43.

Hussein M. M. A., Hatai K. (2002) Pathogenicity of Saprolegniaspecies associated with outbreaks of salmonid saprolegniosis in Japan. Fish Sci 68: 1067-1072.

Johnson T. W., Seymour R. L., Padgett D. E. (2003) Biology and systematics of the saprolegniaceae. University of North Carolina, Wilmington.

Ke X., Wang J., Li M., Gu Z., Gong X. (2010) First report of Mucorcircinelloides occurring on yellow catfish (Pelteobagrusfulvidraco) from China. FEMS Microbiol Lett 302 (2): 144-150.

Ke X., Wang J., Gu Z., Li M., Gong X. (2009) Morphological and molecular phylogenetic analysis of two Saprolegniasp. (Oomycetes) isolated from silver crucian carp and zebra fish. Mycol Res 113: 637-644.

Kitancharoen N., Yuasa K., Hatai K. (1996) Effects of pH and temperature on growth of Saprolegniadeclinaand S. parasiticaisolated from various sources. Mycoscience 37: 385-390.

Koeypudsa W., Phadee P., Tangtrongpiros J., Hatai K. (2005) Influence of pH, temperature and sodium chloride concentration on growth rate of Saprolegniasp. J Sci Res Chula Univ 30 (2): 123-130.

Lee O. K., Lee S. M. (2005) Effects of the dietary protein and lipid levels on growth and body composition of bagrid catfish, Pseudobagrusfulvidraco. Aquaculture 243: 323-329.

Mousavi H. A. E., Soltani M., Khosravi A., Mood S. M., Hosseinifard M. (2009) Isolation and characterization of saprolegniaceae from Rainbow trout (Oncorhynchus mykiss) eggs in Iran. J Fish Aqua Sci 4 (6): 330-333.

Noga E. J. (1993) Water mold infection of freshwater fish: Recent advance. Annu Rev Fish Dis 3: 291-304.

Oláh J., Farkas J. (1978) Effect of temperature, pH, antibiotics, formalin and malachite green on the growth and survival of Saprolegniaand Achlyaparasitic on fish. Aquaculture 13 (3): 273-288.

Peduzzi R., Bizzozero S. (1977) Immunohistochemical investigation of four Saprolegniaspecies with parasitic activity in fish: serological and kinetic characterization of chymotrypsin-like activity. MicrobEcol 3: 107-119.

Pickering A. D., Wiloughby L. G. (1982) Saprolegniainfections of salomonid fish. In: Roberts RJ (ed) Microbial diseases of fish. Academic, London.

Rach J. J., Valentine J. J., Schreier T. M., Gaikowski M. P., Crawford T. G. (2004) Efficacy of hydrogen peroxide to control saprolegniosis on channel catfish (Ictalurus punctatus) eggs. Aquaculture 238: 135-142.

Ruso A. R. (1987) Role of habitat complexity in mediating predation by the gray damselfish Abudefdufsordiduson epiphytal amphipods. Mar EcolProgSer 36: 101-105.

Smith S. N., Armstrong R. A., Rimmer J. J. (1984) Influence of environmental factors on zoospores of Saprolegniadiclina. Trans Brit MycolSoc 82 (3): 413-421.

Stueland S., Hatai K., Skaar I. (2005) Morphological and physiological characteristics of Saprolegniaspp. strains pathogenic to Atlantic salmon, Salmo salarL. J Fish Dis 28: 445-453.

Van West P. (2006) Saprolegniaparasitica, an oomycete pathogen with a fishy appetite: New challenges for an old problem. Mycologist 20: 99-104.

Walker C. A., Van West P. (2007) Zoospore development in the oomycetes. Fungal Biol 21 (1): 10-18.

Whisler H. C. (1996) Identification of Saprolegniaspp. pathogenic in Chinook salmon. Final report, DE-AC79-90BP02836. US Department of Energy, Washington D. C.

Wu S., Gao T., Zheng Y., Wang W., Cheng Y., Wang G. (2010) Microbial diversity of intestinal contents and mucus in yellow catfish (Pelteobagrusfulvidraco). Aquaculture 303: 1-7.

Ye S., Li H., Qiao G., Li Z. (2009) First case of Edwardsiellaictaluriinfection in China farmed yellow catfish Pelteobagrusfulvidraco. Aquaculture 292: 6-10.

Zhang J., Xu Y. (2006) The use of dithiocyano-methane in fish diseases. Inl Fish 31 (5): 38-39.

该文发表于《ISRAELI JOURNAL OF AQUACULTURE-BAMIDGEH》【2012, 64, 685】

Saprolegnia Pathogen from Pengze Crucian Carp (*Carassiusauratus* var. Pengze) Eggs and its Control with Traditional Chinese Herb

Haipeng Cao[1], Wenwei Xia[1], Shiqi Zhang[1], Shan He[2],

Ruopeng Wei[3], Liqun Lu[1], Xianle Yang[1]

(1. Key Laboratory of Exploration and Utilization of Aquatic Genetic Resources, National Aquatic Pathogen Collection Center, Shanghai Ocean University, Shanghai 201306, China; 2. Shanghai Normal University, Shanghai 200235, China; 3. Shanxi Veterinary Drug and Feed Engineering Technology Research Center, Yuncheng 044000, China)

Abstract: In the present study, a pathogenic strain, JL, was isolated from Pengze crucian carp (Carassiusauratus) eggs suffering from saprolegniosis. It was initially determined as Saprolegnia sp. strain JL. Saprolegnia species have been implicated for significant fungal contamination, involving both living and dead fish and their eggs. A phylogenetic tree was constructed using the maximum parsimony method. The tree shows that the JL strain was closely related to Saprolegniaparasitica isolate SAP171, isolated from Salmo trutta suffering from saprolegniasis in Laukaa, Finland. The minimum inhibitory concentration (MIC) of 20 Chinese herbs was screened. PseudolarixkaempferiGord. (Pinaceae) was the most effective in inhibiting growth of the bacteria and was chosen for further trial. Significant protective efficacy of 52.63% and 73.68% was obtained against the JL strain in Pengze crucian carp eggs at P. kaempferi concentrations of 12.5 mg/mL and 25.0 mg/mL, respectively.

彭泽鲫鱼卵水霉病原及利用传统中药控制水霉病的研究

曹海鹏[1]，夏文伟[1]，张仕奇[1]，何珊[2]，魏若鹏[3]，吕利群[1]，杨先乐[1]

(1. 上海海洋大学；2. 上海师范大学；3. 山西兽药研究中心)

摘要：从患水霉病的彭泽鲫鱼卵中分离出一株病原菌 JL。初步确定为水霉属 JL 菌株。水霉种与真菌污染有很大的关系，包括活鱼和死鱼以及它们的鱼卵。系统发生树表明 JL 菌株与从芬兰的劳卡地区患水霉病的海鳟鱼中分离出来的寄生性水霉 SAP171 亲缘关系相近。20 种中草药的最小抑制浓度被筛选。金钱松抑制细菌的生长能，并且被选择做进一步实验。在对抗彭泽鲫鱼卵中的 JL 种的试验中，浓度为 12.5 mg/mL 和 25.0 mg/mL 的金钱松分别获得了 52.63% 和 73.68% 的显著保护效果。

Introduction

The crucian carp, Carassiusauratus, is a popular freshwater fish species with wide distribution in Asia, Europe, Africa, and North America. Crucian carp farming is an important industry, especially in China, with an annual production of 2 million tons. One of the most valuable crucian carps in China is the Pengze crucian carp, C. auratus var. Pengze. It is China's second largest export freshwater fish and 50,000 tons of its products are exported annually to Korea, Japan, Russia, and southeast Asia (Wang, 2009). However, one of the most serious problems in Pengze crucian carp hatcheries is oomycete infections caused by zoosporic fungi. During egg incubation, oomycete produce mycelia which grow and spread from dead to healthy eggs causing major financial losses. Fungal infections of freshwater fish often affect wild and farmed fish in freshwater environments (Pickering and Wiloughby, 1982). One of the most destructive is Saprolegnia sp., which is widespread in freshwater habitats around the world and responsible for significant contaminations involving living and dead fish as well as incubating fish eggs (Noga, 1993). Losses of millions of pounds in the salmon aquaculture business in Scotland, Chile, Japan, Canada, and the USA were attributed primarily to saprolegniosis (Hussein and Hatai, 2002). Control of saprolegniasis is a problem since the effective malachite green treatment has been banned worldwide. Saprolegnia species have been studied in rainbow trout eggs and zebra fish（Ke et al. , 2009; Mousavi et al. , 2009）. In the present study, morphological characteristics of a pathogenic Saprolegnia isolated from Pengze crucian carp eggs suffering from saprolegniasis is described, and its taxonomic position is determined by a nucleotide BLAST search in the NCBI website and phylogenetic analysis based on ITS rDNA sequence. PseudolarixkaempferiGord.（Pinaceae）was one of 20 Chinese herbs screened as a potential drug for controlling Saprolegnia infection.

Materials and Methods

Egg samples.

Fifty Pengze crucian carp eggs suffering from saprolegniosis were obtained as samples from the Sand Lake Aquatic Technique Popularizing Station in Hubei, China, in May 2010, where 6 million Pengze crucian carp eggs are hatched annually.

Isolation and purification of fungal strains.

The sampled eggs were disinfected for 2-3 seconds with 75% alcohol, then washed several times in sterile filtered water, placed on potato dextrose agar（PDA）plates（Sinopharm Chemical Reagent Co. , Ltd）containing 100 ppm streptomycin and penicillin（Sinopharm Chemical Reagent Co. , Ltd）to facilitate isolation of the fungus, and incubated at 25℃ for 24 h. Autoclaved rape seeds were placed at the edges of colonies that grew on the PDA plates, and were incubated until they were covered with hyphae. These were then transferred to sterile filtered river water and incubated until zoospores were discharged. 100 μL Zoospores were spread on PDA plates and incubated at 4℃ for 48-72 h until used.

Artificial challenge test.

Approximately seven days prior to the test onset, isolates were subcultured in sterile filtered river water containing several autoclaved rape seeds at 25℃ for 72 h. Zoospore suspensions were then collected. The test was carried out in nine glass petri dishes supplied with sterile filtered river water at 25℃. Each petri dish was randomly stocked with 40 healthy Pengze crucian carp eggs. These were challenged with the zoospore suspension at a concentration of 1×10^6/mL. Eggs in control dishes were held in sterile filtered river water only. The eggs were observed under a light microscope daily for five days. Eggs with hyphae were immediately removed for fungal isolation according to Ghiasi et al. （2010）, and mortalities were recorded.

Morphological observation.

Pathogenic isolates were grown on PDA plates with several autoclaved rape seeds at 25℃ until the rape seeds were covered with hyphae. The seeds with hyphae were transferred to six-well cell culture plates containing sterile filtered river water and incubated at 25℃ for 14 days. Observations under an inverted microscope were carried out every day to check the emergence of primary cysts, zoospore discharges, oogonia, antheridia, etc.

DNA extract, PCR, and sequencing.

Genomic DNA was extracted from pure cultures of pathogenic isolates using the Universal Genomic DNA Extraction Kit Ver 3. 0 （Takara Biotechnology（Dalian）Co. , Ltd. ）following the manufacturer's instructions. The 750 bp of the internal transcribed spacer（ITS）gene was amplified by PCR using two ITS gene primers: 5'-TCCGTAGGTGAACCTGCGG-3'（ITS1）and 5'-TCCTC-CGCTTATTGATATGC-3'（ITS4）, and carried out according to the instructions of the Fungi Identification PCR Kit（Takara Biotechnology（Dalian）Co. , Ltd. ）. Amplification was done after 35 cycles of denaturation at 94℃ for 0. 5 min, annealing at 58℃ for 0. 5 min, and extension at 72℃ for 1. 0 min, followed by a final extension at 72℃ for 5 min using a PCR minicycler（Eppendorf Ltd. , Germany）. The PCR product was electrophoresed on 1% agarose gel and observed via ultraviolet trans-illumination. Sequencing was performed by the fluorescent labeled dideoxynucleotides termination method（with a BigDye terminator）on an ABI 3730 automated DNA Sequencer.

Phylogenetic analysis.

The partial ITS rDNA sequence was assembled using MegAlign, Editseq, and Seqman software with a power Macintosh computer. Searches were done against the National Centere for Biotechnology Information（NCBI）database using the Basic Local Alignment Search Tool（BLAST）program. The ITS rDNA gene sequence of the pathogenic isolate was constructed using the maximum parsimony method as recommended by Chen and Liu（2007）.

Assay for susceptibility to Chinese herb extracts.

Prior to the susceptibility assay, twenty Chinese herbs were obtained from Shanghai Fosun Industrial Co. , Ltd. and 100 g of each herb was extracted according to Qiu et al. （2010）. The minimum inhibitory concentration（MIC）of each herb extract was determined by the dilution plate method described by Benger et al. （2004）. The MIC was the lowest concentration of each herb extract that prevented any visible fungal colony growth on the PDA plates.

Protective efficacy assay.

Healthy Pengze crucian carp eggs were obtained from Sand Lake Aquatic Technique Popularizing Station, Hubei, China, and maintained in three glass petri dishes supplied with sterile filtered river water at 25℃. Each dish was randomly stocked with 40 healthy eggs. Herb extracts were added to river water in the treatment dishes to final concentrations of 12.5 and 25.0 mg/mL. No extract was added to river water in the control dish. Eggs in the control and treatment dishes were then challenged with the zoospore suspension at a concentration of 1×10^6/mL. The tested eggs were observed under a light microscope daily for five days. Eggs with hyphae were immediately removed for fungal isolation, and dead eggs were recorded.

Results

Morphological characterization of the pathogenic Saprolegnia isolate.

Symptoms of Pengze crucian carp eggs suffering from saprolegniosis, covered with fungal hyphae, are shown in Fig. 1. Eight different fungal isolates were obtained from the infected eggs, but only one strain, named JL, was pathogenic, resulting in 45% mortality. The JL strain showed identical morphological characteristics of asexual and sexual reproduction as Saprolegnia sp. , such as aseptate and sparingly branched hyphae (not shown), clavate and straight or slightly bent zoosporangia (Fig. 2a), sporangial renewal by internal proliferation (Fig. 2b), sporangial discharges of zoospores (Fig. 2c−f), encysted zoospores (Fig. 2g), reniform secondary zoospores (Fig. 2h), germinating spores (Fig. 2i), terminal and intercalary oogonia with centric oospores (Fig. 3a), oogonia with monoclinous, androgynous, and diclinous antheridia (Fig. 3b − d). Thus, the JL strain was initially determined asSaprolegnia. Strain.

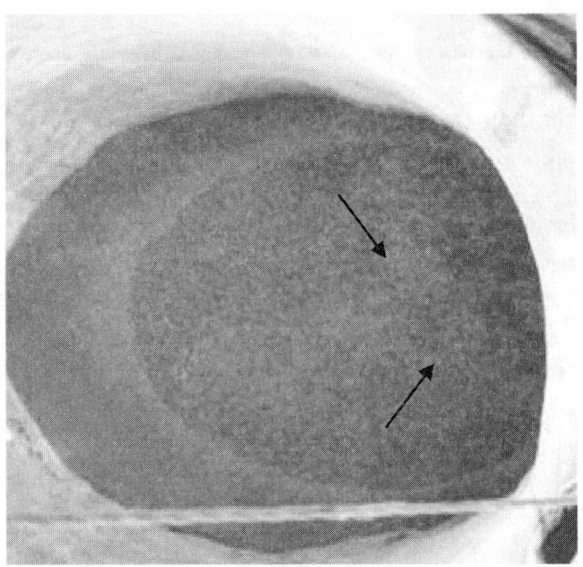

Fig. 1　Pengze crucian carp eggs suffering the JL strain was initially determined as from saprolegniosis, arrows show contaminated areas.

Molecular identification and phylogenetic analysis.

The 750 bp ITS rDNA sequence of the JL strain was submitted to the GenBank database with the accession no. HM637287. Similarities between the ITS rDNA sequence of the JL strain and those of Saprolegnia strains in the GenBank database were 99.0%, confirming the initial identification. The phylogenetic tree, constructed using the maximum parsimony method, further demonstrated that the JL strain was closely related to the Saprolegniaparasitica isolate SAP171 (GenBank accession no. AM228804; Fig. 4) that was isolated from Salmo trutta suffering from saprolegniasis in Laukaa, Finland (Diéguez−Uribeondo et al. , 2007). The molecular identification result of the phylogenetic analysis was consistent with that found through morphological identification.

Susceptibility to Chinese herb extracts.

Seven of the twenty herb extracts showed good inhibition effects on Saprolegniagrowth, i. e. , MIC was below 5.0 mg/mL (Table 1).

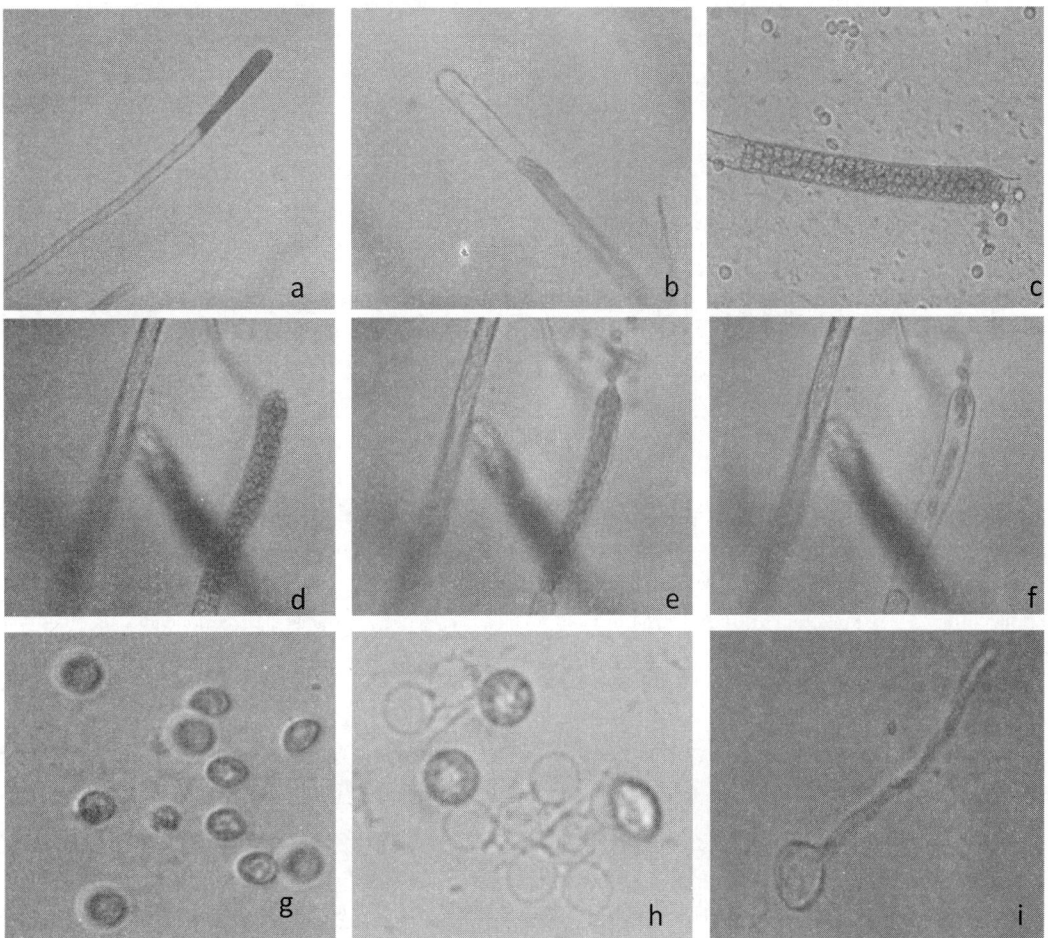

Fig. 2 Asexual reproduction of the JL strain of the Saprolegnia isolate: （a）immature zoosporangium, （b）sporangial renewal by internal proliferation, （c）saprolegnoid discharge of zoospores, （d-f）orderly release of primary zoospores from the apex of the zoosporangium, （g）primary and encysted zoospores, （h）empty cysts that underwent repeated zoospore e-mergence, and （i）germinating zoospores.

Table 1　Minimum inhibitory concentration （MIC） of the JL strain of Saprolegnia to 20 Chinese herb extracts （7×10^7 spores/mL）

Herb	MIC （mg/mL）
Syzygiumaromaticum	5.0-10.0
Fructuscnidll	>10.0
Thallus laminariae	>10.0
Fructusanisistellati	2.5-5.0
Perillafrutescens	>10.0
Sophoraflavescens	2.5-5.0
PseudolarixkaempferiGord.　（Pinaceae）	1.25-2.5
Polygonatumcanallculatum	2.5-5.0
Lithospermumerythrohizon	>10.0
Curcuma longa	>10.0
Bornelumsyntheticum	>10.0
Cortex phellodendri	2.5-5.0

续表

Herb	MIC（mg/mL）
Fructuskochiae	>10. 0
Herbaartemisiaescopariae	>10. 0
Foeniculum vulgare	>10. 0
Melaphischinensis	2. 5-5. 0
Melia azedarach L	>10. 0
Pericardium citrireticulatae	5. 0-10. 0
Cortex dictamni	>10. 0
Fragrant litsea	2. 5-5. 0

Protective efficacy

Since as little as 1. 25-2. 50 mg/mL of P. kaempferi was effective in inhibiting hyphae growth, this herb was further screened for its potential to inhibit Saprolegnia. Results suggest good protective effects of P. kaempferi for controlling Saprolegnia infection on crucian carp eggs（Fig. 5）. Significant protective efficacy against the JL strain in eggs（52. 63% and 73. 68%）was obtained at concentrations of 12. 5 mg/mL and 25. 0 mg/mL, respectively. The death of dead test eggs was caused by Saprolegnia sp., as determined by fungal isolation and molecular identification（data not shown）. Thus, P. kaempferi is a potential drug for the successful treatment of saprolegniasis of carp eggs.

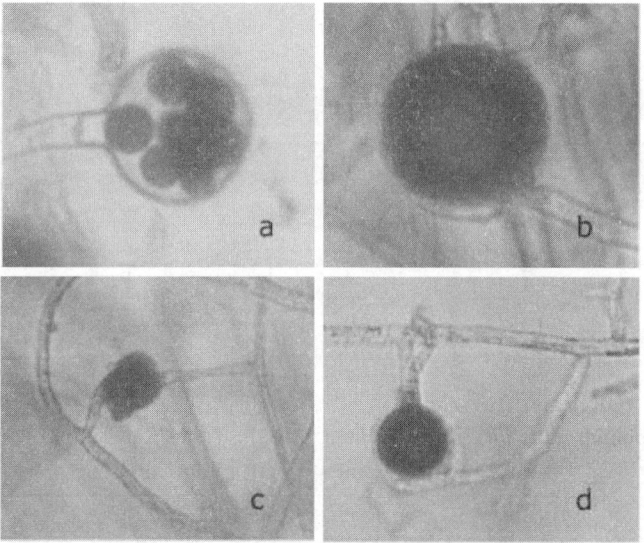

Fig. 3　Sexual reproduction of the JL strain of theSaprolegnia isolate: （a）mature oogonium with centricoospores, （b）oogonium with monoclinous antheridia, （c）oogonium with diclinous antheridia, and （d）oogonium with androgynous antheridia.

Discussion

In the present study, the naturally occurring pathogen was identified as Saprolegnia sp. based on morphological characteristics and phylogenetic analysis. Our findings confirm that Saprolegnia species are the major cause of saprolegniasis in aquaculture production（Bangyeekhun, 2003）. Determination of Saprolegniaspecies is complex and sometimes confusing. However, several typical morphological features involving asexual and sexual reproductive organs serve for classic Saprolegnia identification（Stueland et al., 2005a）. The JL strain in the present study was initially identified as Saprolegnia sp. based on typical morphological characteristics which comform with precise descriptions of Saprolegnia sp. by Van West（2009）. However, Saprolegnia species are usually difficult or even impossible to identify by traditional morphological criteria alone. Therefore, we compared ITS regions to further identify the JL strain, as done to Saprolegnia isolates from salmonid fish（Whisler, 1996）. The phylogenetic analysis based on the ITS rDNA region further clarified the taxonomic position of the JL strain 渔 and confirmed its initial identification as Saprolegnia sp.

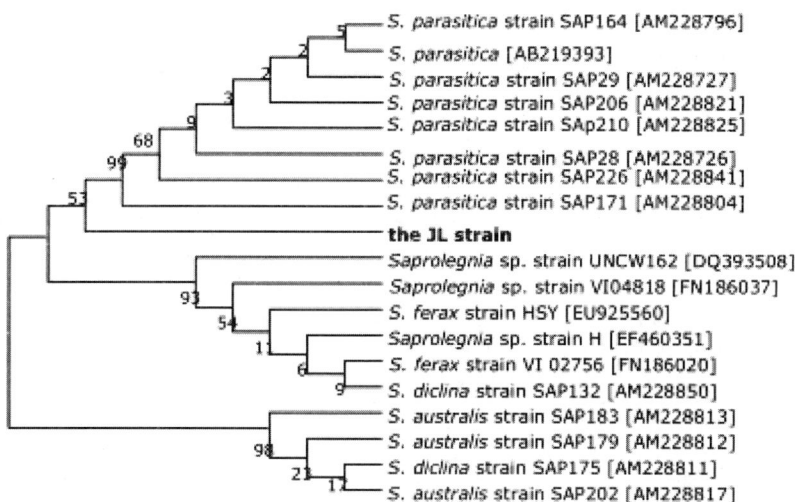

Fig. 4　Phylogenetic tree constructed using maximum parsimony method.

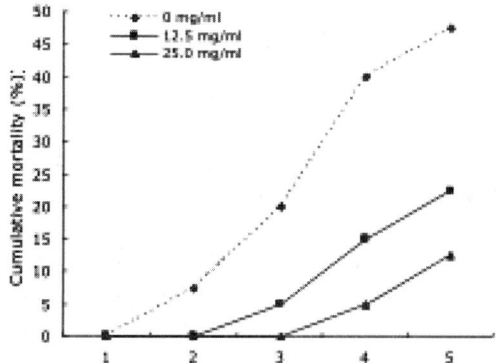

Fig. 5　Control by PseudolarixkaempferiGord. （Pinaceae） extract of artificial infection of
the JL strain of Saprolegnia on Pengze crucian carp eggs.

Zoospores of some moderately or highly pathogenic Saprolegnia strains have long hook cilia that are believed to increase Saprolegnia attachment efficiency （Beakes, 1982）. However, no such cilia were observed on the zoospores of pathogenic strain JL or on other Saprolegnia pathogens such as Saprolegnia sp. strain SAP211 （GenBank accession no. AM228826）（Diéguez-Uribeondo et al., 2007）. Other factors such as chymotrypsin-like activity could also contribute to their pathogenicity （Peduzzi and Bizzozero, 1977）.

No significant differences were found between the susceptibility of Saprolegniapathogens to antifungal chemicals （Stueland et al., 2005b）. Thus, only the JL strain was chosen for the susceptibility and protective efficacy assay. To date, the few chemicals used to control saprolegniasis include hydrogen peroxide, salt, ozone, formaldehyde, and formalin formulations （Forneris et al., 2003; Rach et al., 2004; Gieseker et al., 2006）, but these treatments do not totally arrest the growth of Saprolegnia species （Van West, 2006）. In our study, P. kaempferi completely inhibited the growth of the JL strain and, at concentrations of 12.5 mg/mL and 25.0 mg/mL, exhibited significant protective efficacy of 52.63% and 73.68%, respectively, against experimental Saprolegnia infections of eggs. This could be due to its ability to produce pseudolaric acid A and B, which possess antifungal activities （Yang and Yue, 2001）. In addition, field trials showed that when P. kaempferi was applied for five days, the incidence rates of saprolegniosis were reduced by up to 80% in crucian carp eggs and Megalobramaamlycephalaeggs at Sand Lake, Hubei, China （data not shown）. Thus, P. kaempferiis promising as an anti-Saprolegnia drug for controlling saprolegniasis.

References

B angyeekhun E., Pylkko P., Vennerstrom P., Kuronen H. and L. Cerenius, 2003. Prevalence of a single fish-pathogenic Saprolegnia sp. clone in Finland and Sweden. Dis.

Aquat. Organ. , 53：47-53.

Beakes G. , 1982. A comparative account of cyst coat ontogeny in saprophytic and fishlesion （pathogenic） isolates of the Saprolegniadeclina-parasitica complex. Can. J. Bot. , 61：603-625.

Benger S. , Townsend P. , Ashford R. L. and P. Lambert, 2004. An in vitro study to determine the minimum inhibitory concentration of Melaleuca alternifolia against the dermatophyte Trichophyton rubrum. The Foot, 14：86-91.

Chen J. and P. Liu, 2007. Tuber latisporum sp. nov. and related taxa, based on morphology and DNA sequence data. Mycologia, 99 （3）：475-481.

Diéguez-Uribeondo J. , Fregeneda-Grandes J. M. , Cerenius L. , Pérez-Iniesta E. , Aller-Gancedo J. M. , Tellería M. T. , Söderhäll K. and M. P. Martín, 2007. Reevaluation of the enigmatic species complex Saprolegniadiclina-Saprolegniaparasitica based on morphological, physiological and molecular data. Fungal Genet. Biol. , 44：585601.

Forneris G. , Bellardi S. , Palmegiano G. B. , Saroglia M. , Sicuro B. , Gasco L. and I. Zoccarato, 2003. The use of ozone in trout hatchery to reduce saprolegniasis incidence. Aquaculture, 221：157-166.

Ghiasi M. , Khosravi A. R. , Soltani M. , Binaii M. , Shokri H. , Tootian Z. , Rostamibashman M. and H. Ebrahimzademousavi, 2010. Characterization of Saprolegnia isolates from Persian sturgeon （Acipencerpersicus） eggs based on physiological and molecular data. J. Med. Mycol. , 20：1-7.

Gieseker C. M. , Serfling S. G. and R. Reimschuessel, 2006. Formalin treatment to reduce mortality associated with Saprolegniaparasitica in rainbow trout, Oncorhynchus mykiss. Aquaculture, 253：120-129.

Hussein M. M. A. and K. Hatai, 2002. Pathogenicity of Saprolegnia species associated with outbreaks of salmonid saprolegniosis in Japan. Fish. Sci. , 68：1067-1072.

Ke X. , Wang J. , Gu Z. , Li M. and X. Gong, 2009. Morphological and molecular phylogenetic analysis of two Saprolegnia sp. （oomycetes） isolated from silver crucian carp and zebra fish. Mycol. Res. , 113：637-644.

Mousavi H. A. E. , Soltani M. , Khosravi A. , Mood S. M. and M. Hosseinifard, 2009. Isolation and characterization of saprolegniaceae from rainbow trout （Oncorhynchus mykiss） eggs in Iran. J. Fish. Aquat. Sci. , 4：330-333.

Noga E. J. , 1993. Water mold infection of freshwater fish: recent advance. Annu. Rev. Fish Dis. , 3：291-304.

Peduzzi R. and S. Bizzozero, 1977. Immunohistochemical investigation of four Saprolegnia species with parasitic activity in fish: serological and kinetic characterization of chymotrypsin-like activity. Microb. Ecol. , 3：107-119.

Pickering A. D. and L. G. Wiloughby, 1982. Saprolegnia infections of salmonid fish. pp. 271-297. In: R. J. Roberts （ed.）. Microbial Diseases of Fish. Academic Press, London. Qiu Q. L. , Pan Q. Q. , Zhang Y. P. , Liu W. and D. Qian, 2010. Antibacterial effect of forty-five Chinese herb extractions to Vibrio isolates from mud crab, Scylla serrata （Forskǎl）. J. Zhejiang Ocean Univ. （Nat. Sci.）, 10 （1）：34-38.

Rach J. J. , Valentine J. J. , Schreier T. M. , Gaikowski M. P. and T. G. Crawford, 2004. Efficacy of hydrogen peroxide to control saprolegniasis on channel catfish （Ictalurus punctatus） eggs. Aquaculture, 238：135-142.

Stueland S. , Hatai K. and I. Skaar, 2005a. Morphological and physiological characteristics of Saprolegnia spp. strains pathogenic to Atlantic salmon, Salmo salar L. J. Fish Dis. , 28：445-453.

Stueland S. , Heier B. T. and I. Skaar, 2005b. A simple in vitro screening method to determine the effects of drugs against growth of Saprolegniaparasitica. Mycol. Progress, 4 （4）：273-279.

Van West P. , 2006. Saprolegniaparasitica, an oomycete pathogen with a fishy appetite: new challenges for an old problem. Mycologist, 20：99-104.

Wang H. , 2009. Development trends of Chinese crucian carp farming and development patterns of export-oriented Pengze crucian carps. Sci. Fish. , 8：1-3.

Whisler H. C., 1996. Identification of Saprolegnia spp. Pathogenic in Chinook Salmon. Final Report, DE-AC79-90BP02836. US Department of Energy, Washington D. C. 43 pp. Yang S. and J. Yue, 2001. Two novel cytotoxic and antimicrobial triterpenoids from Pseudolarixkaempferi. BioorganicMed. Chem. Lett., 11：3119-3122.

该文发表于《中国水产科学》【2009, 16 （1）：89-95】

施氏鲟水霉病病原的初步研究

Isolation of Saprolegnia parasitica as a pathogen from Acipenser schrenckii

张书俊[1,2]，杨先乐[1]，李聃[1]，高鹏[1]

（1. 上海海洋大学，农业部渔业动植物病原库，上海 201306；2. 上海出入境检验检疫局，上海 201202）

摘要： 从患水霉病的施氏鲟 （*Acipenser schrenckii*） 幼鱼肌肉中分离获得一株优势水霉菌株 XJ001。大麻籽上水霉 XJ001 培养物的形态观察表明，菌丝透明无间隔，分枝少；新孢子囊以层出方式产生或从旧孢子囊根部长出，游动孢子按水霉属方式释放；在7℃、15℃和25℃条件下不能产生有性器官，但能够产生大量的厚垣孢子。电镜观察发现第二孢子表面具有大量弯曲的长毛，其在25% GY培养液中呈现出间接萌发。根据形态及生理学特征将水霉XJ001鉴定为寄生水霉（*Saprolegnia parasitica*）。水霉XJ001的最适生长温度为25℃，对NaCl敏感，当PDA培养基中NaCl质量分数为1%和2%时，水霉生长抑制率分别为37.4%和75.6%；质量分数大于3%时，生长被完全抑制。用水霉XJ001对异育银鲫（*Carassius auratus gibelio*）进行人工感染，结果表明XJ001有较强的致病力，异育银鲫感染严重，体表出现棉花状白毛，感染率为83.3%，感染后的死亡率为100%，可见XJ001是施氏鲟水霉病的病原菌。

关键词：施氏鲟；寄生水霉；形态特征；致病力

水霉病是淡水养殖动物的常见病害，其病原菌主要为卵菌纲（Oomycetes）中水霉属（*Saprolegnia*）和绵霉属（*Achlya*）的一些种类[1]。近年来，水霉的分类学地位发生了改变，其所属卵菌纲由传统上认定的真菌界划出，同与其有相似特征的褐藻一起归入茸鞭生物界[2-3]。水霉分布范围很广，国内外的一些养殖场都有发生水霉病的报道。水霉对宿主没有严格的选择性，水产动物及其卵均可被感染，特别在高密度养殖条件下，鱼体间相互摩擦后造成的损伤、应激作用及恶劣水体环境，往往导致鱼体的抗感染能力低下，使得鱼体更易感染水霉，尤其是在气温变化较大的季节，水霉病的发生率更高。感染水霉病的鱼体特征明显，体表有棉花状的白毛。水霉病的发生会给水产养殖产业带来巨大的经济损失[4-5]，且由于不同水霉菌株间的致病性不同，危害亦不同[6-7]，若水霉病病原为高致病性菌株，则造成的危害将更大。因此，对分离得到的水霉菌株进行鉴定与致病性研究非常重要，并可据此判断其究竟是致病性菌株，还是腐生性菌株。

2006年12月中旬，上海市东海渔场施氏鲟（*Acipenser schrenckii*）幼鱼养殖池出现大量幼鱼死亡现象，患病鱼体体表有黄色丝状物质，与水霉病的症状相似。国外已有一些鲑鳟类和斑点叉尾鲴（*Ictalurus*）水霉病的报道[8-9]，而国内关于水产动物水霉病的研究则很少，仅见对花鲈（*Lateolabrax japonicus*）水霉病病原的研究报道[10]，而有关鲟类水霉病在国内则未见相关报道。本研究以发病施氏鲟为研究材料，对鲟水霉病病原菌进行分离纯化，研究该病原菌的形态、生理学特征及致病性，为水产动物水霉病病原研究及水霉病防治方法提供理论基础。

1　材料与方法

1.1　实验材料

1.1.1　实验鱼

施氏鲟采集于东海渔场水霉病暴发的施氏鲟幼鱼养殖池，体表有黄色菌丝，鱼体濒临死亡，体长（15.35±1.20）cm。异育银鲫（*Carassius auratus gibelio*）购于南通国营农场，鱼体规格均一，健康无伤病，体质量（35.00±1.12）g。实验前暂养于水箱，水箱规格70 cm×50 cm×40 cm。用加热棒控制水温至22℃，连续充气，按1%体质量投喂饵料，实验前24 h停止投喂。

1.1.2　培养基

马铃薯葡萄糖琼脂培养基（PDA），购自上海疾病预防控制中心，用于水霉的分离及保存；马铃薯葡萄糖液体培养基：新鲜去皮马铃薯200 g、葡萄糖20 g、蒸馏水1 000 mL，马铃薯先加热煮沸15 min，然后以纱布过滤去渣，加水补至原量，调pH至7.0，用于水霉菌丝的培养；GY液体培养基[11]：葡萄糖10 g，酵母提取物2.5 g，蒸馏水1 000 mL，调pH至7.0，用于孢子萌发方式的观察。

1.2　水霉分离纯化

用酒精棉球搽拭患病鱼体表，用解剖刀切取鱼背部、腹部、尾部肌肉（厚度3~4 mm），同时挑取鳃部菌丝，放置于PDA培养基平板中央。为了抑制细菌的生长，在培养基中加入硫酸链霉素（200 mg/mL），置于25℃培养箱培养3 d，在培养基表面形成水霉菌落。为了获得水霉的纯培养物，对Huang等[12]建立的单孢子分离方法进行了改良，取鱼体不同部位分离得到的菌丝琼脂块（直径3~4 mm）放入装有无菌水的培养皿（直径90 mm）中培养20 h，使菌丝释放出游动孢子，取100 μL该孢子悬液的稀释液至PDA平板，涂布均匀，置于25℃培养箱培养2 h后显微镜下观察，用解剖刀挑取含有单个萌发孢子的琼脂块并置于PDA平板中央，25℃下培养3 d，使水霉长满整个平板，置于4℃保存备用。、

1.3　优势水霉菌株的形态特征观察

1.3.1　一般形态显微观察

用解剖刀切取纯化后的水霉琼脂块（直径4 mm），加入含有马铃薯葡萄糖液体培养基的6孔板中，25℃培养24 h，待水霉菌落形成；挑取部分长出的菌丝，用无菌水洗涤3次，再放入含有无菌水的6孔板中，同1.2方法制备孢子悬液；取100 μL孢子悬液至含有灭菌大麻籽及无菌水的24孔板中，根据不同的观察目的，分别将24孔板置于不同温度条件下培

养。25℃下培养观察孢子囊形成、孢子释放及菌丝形态；分别在7℃、15℃、25℃下培养观察有性器官的形成，持续观察2个月。

1.3.2 第二孢子电镜观察

按照1.2方法制备第二游动孢子悬液，用无菌移液器吸取1 mL游动孢子悬液上层液体至24孔板中，然后加入电镜用铜网，25℃培养30 min，显微观察发现铜网小孔边缘有静止孢子形成（第二孢子），将铜网捞出后用磷钨酸对铜网上的第二孢子进行负染，用滤纸小心吸去染液，干燥后放入透射电镜下观察孢子表面的纹饰（Ornament）。

1.4 优势水霉菌株的生理特征

1.4.1 第二孢子萌发方式

按1.2方法制备游动孢子悬液，取上层悬液至玻璃试管中，采用漩涡振荡方法将游动孢子变为第二孢子，参照Yuasa等[13]的方法，取100 μL第二孢子悬液至含有25% GY培养液的24孔板中，20℃培养3 h，显微镜下观察孢子的萌发方式。

1.4.2 温度对水霉生长的影响

将水霉琼脂块（直径4 mm）接种至PDA平板中央，分别置于10℃、15℃、20℃、25℃、30℃下培养，每24 h测定不同温度下水霉菌落的直径大小，连续测定72 h。每组设3个平行。

1.4.3 NaCl对水霉生长的影响

将水霉琼脂块（直径4 mm）分别接种至含NaCl质量分数为0、1%、2%、3%、4%的PDA平板中央，25℃下培养72 h，每隔24 h观察测定各NaCl水平下菌落直径的大小。每组设3个平行。

1.5 优势水霉菌株的致病性检测

1.5.1 水霉孢子悬液的制备

按1.2方法，将6孔板中培养40 h的水霉菌丝用无菌水洗涤3次后加入到含有无菌水的三角烧瓶中，25℃培养20 h，用血球计数板测定孢子悬液的浓度为$2×10^5$ cell/mL，取280 mL该孢子悬液至水箱中（58 cm×28 cm×17cm），使水霉孢子浓度为$2×10^3$ cell/mL，实验水箱中水温为12℃。以不加水霉游动孢子组作为对照，每组3个平行。

1.5.2 攻毒感染

对文献[4,14]方法进行了改良，将暂养于22℃水温下的异育银鲫（由于未能获得施氏鲟幼鱼，因此选择异育银鲫进行攻毒感染）捞出，于自来水水流下冲洗，洗去体表黏液，用注射器针头在异育银鲫背部两侧肌肉各划一道伤口（长3 cm），然后放入加有水霉孢子的水箱中（水温12℃）。以未加水霉孢子的水箱作为对照，对照组中加入的异育银鲫与实验组中异育银鲫的处理方法相同，每箱各放入10条异育银鲫，每组3个平行。每天观察记录各组中异育银鲫的发病及死亡情况。

2 结果与分析

2.1 水霉病原菌株的分离

从发病施氏鲟的鳃丝、肌肉中共分离得到4株水霉，根据PDA平板上的菌落形态、菌丝的形态、直径、孢子囊形态和孢子排列、释放方式的显微观察比较，判断分离获得的4株水霉为同一种[15-17]，命名为XJ001。

2.2 水霉XJ001菌株形态特征

大麻籽上长出的水霉XJ001菌丝中等粗壮，随着菌丝长度的增加直径也呈增加趋势，直径范围为12.5~30 μm。菌丝透明无隔阂，分枝较少。随着培养时间的延长，菌丝内的胞质向其顶端注入，菌丝顶端逐渐变大，颜色变深，菌丝顶端部分与其下部产生明显的分界，形成动孢子囊。动孢子囊的形状呈棍棒状或纺锤状。随后囊内发生菌核分裂和细胞质分割，在动孢子囊的前端形成排放颈（Discharge neck）；最后囊内分割的细胞质形成许多大小一致的扁圆形动孢子，并充满整个动孢子囊，动孢子在囊中并排有序排列（图1）。随着动孢子在囊中出现晃动之后，一个个动孢子从排放颈中依次有序地"挤"出，向四面八方游去，剩下空的孢子囊（图2），此时产生的游动孢子称为第一动孢子。第一动孢子游动一段时间后停止游动，改为在原地旋转，借着旋转的力量使其缓慢移动，随后静止不动变成圆形，此时称之为第一孢子[18]；一段时间过后，原生质从第一孢子的细胞壁钻出，又成为游动孢子，游动速度比第一动孢子的速度快，称之为第二游动孢子（图3）。新孢子囊的再生方式有2种，一种是从空孢子囊中继续长出一个新孢子囊（图4）；另一种是从空孢子囊基部侧生出一个新孢子囊（图5）。孢子囊长115~195 μm，宽20~38 μm。根据动孢子的排列方式、释放方式及新孢子囊的生成方式，确定XJ001菌株为水霉属种类。在培养过程中，水霉XJ001菌株菌丝上出现了大量的厚垣孢子，较动孢子囊原生质聚集密度大且颜色深，厚垣孢子单独存在或若干个连接成节状（图6）。水霉XJ001菌株在7℃、15℃和25℃温度条件下均未产生有性器官。

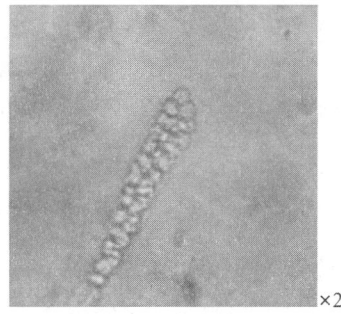

×200

图 1　水霉 XJ001 菌株成熟的孢子囊

Fig. 1　Mature zoosporangium of Saprolegnia XJ001

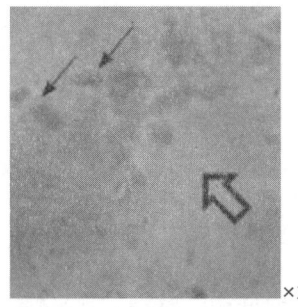

×200

图 2　水霉 XJ001 菌株第一游动孢子释放 "↑" 示游动孢子，"δ" 示空孢子囊

Fig. 2　Zoospore discharging from Saprolegnia XJ001

"↑" shows zoospore；"δ" shows empty zoosporangium

"↑"shows zoospore；"　"shows empty zoosporangium

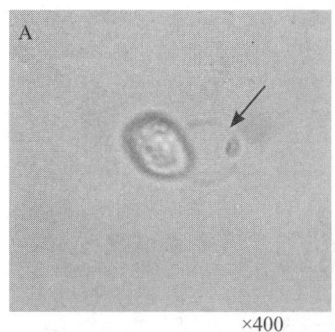
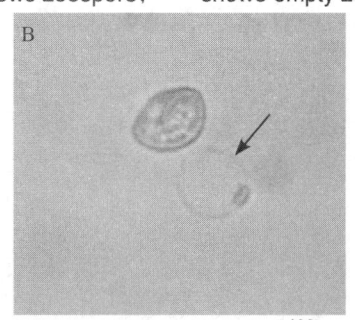

A　×400　　　　　B　×400

图 3　水霉 XJ001 菌株第二游动孢子的产生

A. 原生质从第一孢子的开口钻出（箭头）；B. 游动孢子游出后，留下空孢子壳（箭头）

Fig. 3　Zoospore emerging from primary cyst of Saprolegnia XJ001

A. Zoospore emerging through a small opening in the cyst wall（arrow）；

B. New zoospore released from cyst and the cyst become empty（arrow）

2.3　第二孢孢子电镜观察

电镜观察发现，水霉 XJ001 第二孢孢子表面有长毛样的纹饰，这些长毛在孢子表面以簇状的形式排列，每簇由若干根长毛聚成。不同孢子表面的长毛形态存在差异，有以下 2 种表面纹饰形态：第一种孢子表面的簇状长毛直立（图 7-A），且长毛的末端成钩状（图 7-B），这种孢子黏附于电镜铜网壁上；另一种孢子表面的长毛弯曲且扭成圆圈状（图 7-C），孢子独立存在，不与铜网黏附。

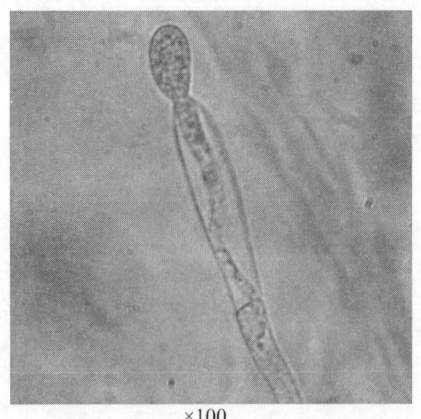

×100

图 4　新生孢子囊以层出方式长出

Fig. 4　Zoosporangium renewed by internal proliferation

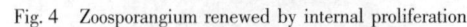

×100

图 5　新生孢子囊从旧的基部长出

Fig. 5　Zoosporangium renwed by basipetalous succession

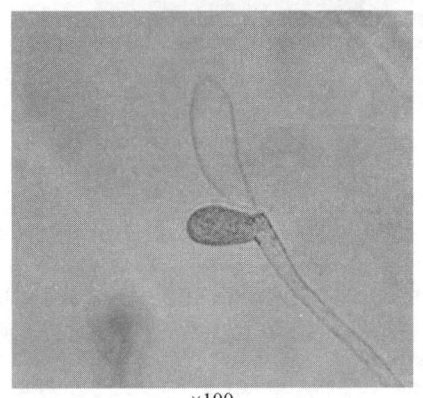

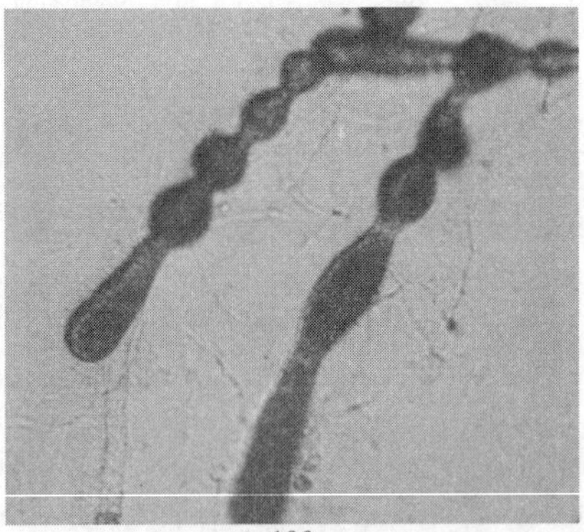

×100

图 6　水霉 XJ001 菌株成串的厚垣孢子

Fig. 6　Catenulated gemmae of Saprolegnia XJ001

2.4　第二孢孢子萌发方式

在 25% GY 培养液中，第二孢孢子存在直接萌发和间接萌发 2 种萌发方式，前者孢子及萌发出的菌丝均含有密实的细胞质，两者为一整体；后者则相反，孢子内没有细胞质，成空壳状，且近孢子端的一段菌丝也缺乏细胞质，并有隔膜，间接萌发（图 8）的比率为 52.8%。

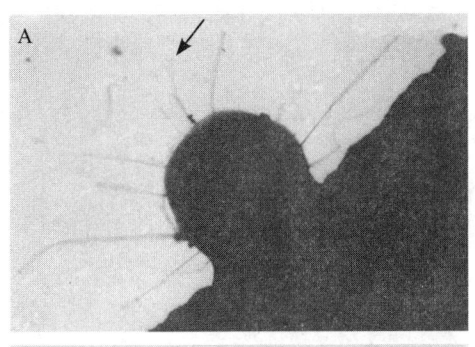
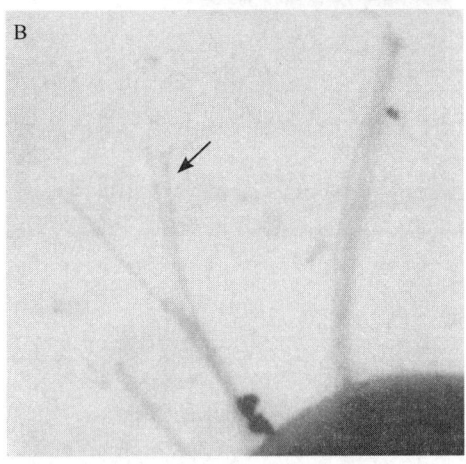
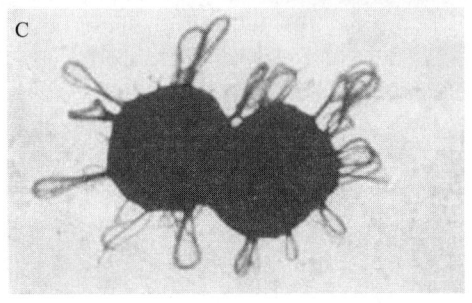

图7　水霉 XJ001 菌株第二孢孢子的形态

A. 孢子表面直立带钩的毛，×3 600；B. 孢子表面直立带钩长毛的末端结构放大图，×21 600；C. 孢子表面有弯曲成圈的毛，×2 900；箭头示毛末端钩状结构。

Fig. 7　Morphological characteristic of the secondary cyst of saprolegnia XJ001

A. The secondary cyst of saprolegnia XJ001 with long, hooked straight hairs in bundles (×3 600); B. the enlarge image of hooked hair end (×21 600); C. the secondary cyst of saprolegnia XJ001 with long, hooked loop hairs in bundles (×2 900). Arrow shows the hook at the hair end.

2.5　温度对水霉生长的影响

20℃与30℃温度条件下水霉生长无显著差异（P>0.05）；而在其他温度条件下水霉生长受温度影响显著（P<0.05），其中10℃和15℃温度条件下，培养时间不足24 h时，水霉生长差异不显著（P>0.05），但24 h之后，两种温度条件下水霉的生长出现显著差异（P<0.05），详见图9。水霉最适生长温度为25℃，该温度条件下水霉菌落在48 h后几乎铺满整个PDA平板。

2.6　NaCl 对水霉生长的影响

NaCl对水霉生长影响显著（P<0.05），水霉在NaCl质量分数为1%、2%时能够生长，但生长被显著抑制，当无NaCl组中的菌落在第3天铺满平板时，上述2个NaCl水平下的生长抑制率分别为37.4%和75.6%，1周后2% NaCl组中的水霉仍未能铺满平板；当NaCl质量分数大于3%时，水霉完全不能生长（图10）。

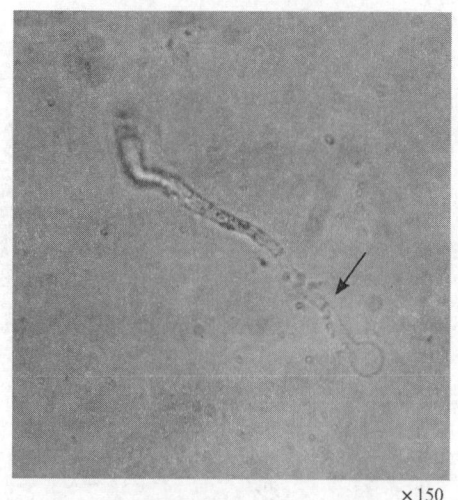

×150

图 8　水霉 XJ001 菌株第二孢子间接萌发箭头示中空无细胞质且有间隔的芽管

Fig. 8　Indirect germination of Saprolegnia XJ001 from secondary encysted spore Arrow
shows empty long germ tube with several septa.

2.7　水霉 XJ001 菌株的致病性

通过降低水温、刮伤、冲洗鱼体及充分暴露接触水霉 XJ001 的游动孢子（$2×10^3$ cell/mL），模拟了自然条件下水霉发生的条件因素。在感染后的第 2 天，鱼体感染水霉症状明显且发生大量死亡。感染初始阶段不易察觉，异育银鲫活动正常，过夜后，鱼体表的菌丝肉眼清晰可见，从头部至尾部几乎全被菌丝覆盖，表现出典型的水霉感染症状，鱼体活动能力几乎丧失，浮头于水面，最后衰竭死亡。水霉 XJ001 致病性强，鱼体从出现感染症状至死亡的发病周期短，一般为 3~4 d，且一旦发生感染均会引起死亡。水霉攻毒实验组中，异育银鲫的感染率为 83.3%，感染后的死亡率为 100%（图 11）。对照组中则没有鱼体出现水霉感染及死亡，感染率及死亡率与实验组差异显著（$P<0.05$）。

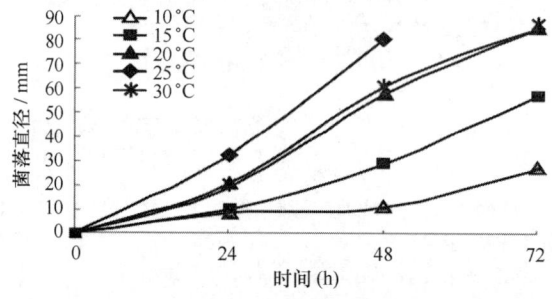

图 9　温度对水霉 XJ001 生长的影响

Fig. 9　Effect of temperature on hyphal growth of Saprolegnia XJ001

3　讨论

3.1　卵菌纲分类地位的改变

在中国水产动物病害学研究领域，一直将水霉归为真菌类[1,19-20]。然而随着对卵菌纲生物研究的深入，发现水霉与其他真菌存在很大不同，比如有不等鞭毛的游动孢子，细胞壁成分是纤维素而非几丁质以及线粒体是管状脊等，这些特征揭示了卵菌纲的系统发育与真菌界各纲明显不同；对真菌的分子系统发育学的研究也表明卵菌纲不在真菌单元论起源的进化路线上[21]。鉴于以上等原因，将卵菌纲从真菌界中划出成为独立的门，与硅藻、褐藻类等一并归为茸鞭生物界[2-3]。从而水霉也被描述为具有一些真菌特征，但非真菌的类真菌生物[22-23]，至此水霉的分类地位发生了改变。

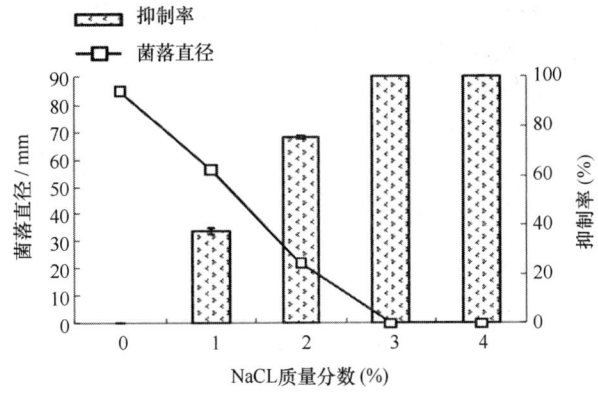

图 10　NaCl 对水霉 XJ001 生长的影响

Fig. 10　Effect of NaCl on hyphal growth of Saprolengia parasitica XJ001

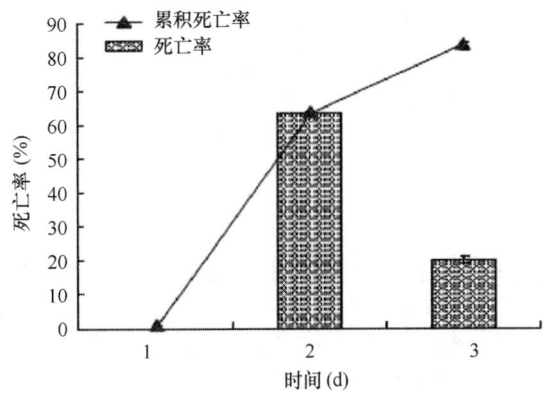

图 11　水霉 XJ001 对异育银鲫的毒力

Fig. 11　Pathogenicity of Saprolegnia paracitica XJ001 to Carassius auratus gibelio

3.2　关于水霉菌株 XJ001 的鉴定

水霉是淡水养殖种类的常见病害，对水霉的鉴定主要依据形态特征，本研究通过对从患病施氏鲟分离得到的水霉菌株 XJ001 形态结构观察，根据其在无性阶段的孢子排列及释放方式确定为水霉属种类；而属内具体种类的鉴定则主要依据有性结构，如藏卵器、卵球及雄器[15-16,24]。但一些从患病鱼体分离得到的种类在体外培养过程中不能产生有性结构[23]，Coker[15]首次将从患病鱼及鱼卵分离得到的只产生无性结构而未见有性结构的水霉命名为寄生水霉（Saprolegnia parasitica），此后有研究者根据此特征将分离于患病鱼体而不产生有性结构的水霉鉴定为寄生水霉[7-8]。本研究中水霉菌株 XJ001 在体外培养过程中未见形成有性结构，如据上述鉴定条件可鉴定其为寄生水霉，但又有学者报道了寄生水霉可产生有性结构[25]，因此只根据未产生有性结构这一特性就将本实验中水霉菌株 XJ001 鉴定为寄生水霉，依据不够充分。继 Coker 和 Kanouse 之后一些研究学者又发现寄生水霉与异丝水霉（Saprolegnia diclina）有一些共同特征，如藏卵器壁没有凹痕、有性结构推迟形成以及卵子类型和雄性结构菌枝起源相似[16]，这些共同特征增加了对上述 2 种水霉的鉴别难度[13,26]，使寄生水霉鉴定变得更加困难。出于方便，此类不易鉴别的水霉被称为异丝–寄生水霉（Saprolegnia diclina[24]-parasitica complex）[13]。Willoughby 认同了寄生水霉与异丝水霉是同一物种的提法，并根据藏卵器长度与宽度的比值，将异丝水霉分成了 3 类。其后的菌类学家认为，异丝水霉第一类型（Saprolegnia diclina type 1）与寄生水霉是同物异名[26]，两者具有相似的特征，即第二孢孢子表面有带钩的长毛、孢子萌发时表现出间接萌发方式[7]。Dieguez-Uribeondo 等[27]对异丝–寄生水霉种类的系统发育分析也证实了异丝水霉第一类型与寄生水霉的同源性，同时也发现这些同源菌株均具有上述 2 个特征，认为是此种类的特异性特征，并提出将从鱼体分离获得，具有上述 2 种特征，且有致病性的水霉菌株命名为寄生水霉。这与一些学者将水霉第二孢孢子表面的纹饰及孢子萌发方式也作为鉴定寄生水霉依据[27-30]的观点相同。这是目前为止作为寄生水霉鉴定最充分的依据，本研究从患病施氏鲟分离得到的水霉菌株 XJ001，在电镜下观察发现其第二孢孢子表面有一簇簇弯曲或直立带钩的长毛；孢子在 25% GY 培养基中表现出间接萌发方式，间接萌发率为 52.8%，这些特征与报道中寄生水霉的特征相似[23,31-32]，也符合上述寄生水霉的鉴定标准，因此可以确切地将 XJ001 鉴定为寄生水霉。在无性生殖阶段，XJ001

产生了大量的厚垣孢子，这可能与寄生水霉易产生厚垣孢子而不易形成藏卵器有关[7]。本实验电镜观察 XJ001 第二孢子过程中，发现了 2 种不同的孢子表面纹饰，第一种纹饰为一簇簇直立的长毛，每簇中的毛较少，可以看到长毛的末端有小钩；第二种纹饰为一簇簇弯曲的长毛，但每簇中的毛较多且弯曲，由于无法看到毛的末端，因此无法观察末端是否有小钩。虽然出现 2 种形态的孢子表面纹饰，但这两者纹饰的出现具有一定规律，第一种纹饰出现在黏附于电镜铜网边缘的孢子上，而第二种纹饰则出现在位于铜网孔中央、与铜网没有接触的孢子上。因此本研究推测孢子表面毛纹饰的不同形态可能与孢子是否黏附物体有关，因为直立的毛容易黏附到物体，且第二孢子表面纹饰可能有助于孢子黏附到鱼体表[29]，进而引发鱼体发生水霉感染。

3.3 水霉菌株的致病性及人工感染方法

采用人工感染方法研究了水霉菌株 XJ001 的致病性，结果表明人工感染后的异育银鲫表现出了典型的水霉发病症状，感染率为 83.3%，且感染后的死亡率高达 100%，说明分离得到的 XJ001 菌株是水霉病的病原菌。关于水霉人工感染方法报道已有较多[6,33-35]，主要是通过增加鱼体对水霉的敏感性。本研究通过降温、刮伤、洗去鱼体表黏液的方式降低鱼体的抵抗力，从而提高异育银鲫对水霉孢子的敏感性，感染率高、方法稳定、操作方便，基本不会因操作而造成鱼体死亡，适用于水霉人工感染研究。

参考文献

[1] 黄琪琰. 水产动物疾病学 [M]. 上海：上海科学技术出版社，1996：142.

[2] Hawksworth D. L., Kirk P. M., Sutton B. C., et al. Ainsworth & Bisby's dictionary of fungi [M]. 8th ed. Uk, Egham：International mycological Institute, 1995.

[3] 阿 历索保罗 C. J., 明斯 C. W., 布莱克韦尔 M.. 菌物学概论 [M]. 北京：中国农业出版社，2002.

[4] Hussein M. M. A., Hatai K.. Pathogenicity of Saprolegnia species associated with outbreaks of salmonid saprolegniosis in Japan [J]. Fish Sci, 2002, 68：1 067-1 072.

[5] Tampieri M. P., Galuppi R., Carelle M. S., et al. Effect of selected essential oils and pure compounds on Saprolegnia parasitica [J].

[6] Pharmaceut Biol, 2003, 41 (8)：584-591. Hatai K, Hoshiai G. Characteristics of two Saprolegnia species isolated from coho salmon with saprolegniasis [J]. J Aquat Anim Health, 1993, 5：115-118.

[7] Yuasa K., Hatai K.. Relationship between pathogenicity of Saprolegnia spp. isolates to rainbow trout and their biological characteristics [J]. Fish Pathol, 1995, 30 (2)：101-106.

[8] Hatai K., Hoshiai G.. Mass mortality in cultured coho salmon (Oncorhynchus kisutch) due to Saprolegnia parasitica Coker [J]. J Wildl Dis, 1992, 28 (4)：532-536.

[9] B angyeekhun E., Quiniou S. M. A., Bly J. E., et al. Characterisation of Saprolegnia sp. isolates from channel catfish [J]. Dis Aquat Organ, 2001, 45：53-59.

[10] 李爱华，聂品，卢全章. 花鲈水霉病及其病原的初步研究 [J]. 水生生物学报，1999, 23 (4)：388-390.

[11] Haitai K., Egusa S.. Studies on pathogenic fungus of mycotic granulomatosis. Ⅲ. Development of the medium for MGfungus [J]. Fish Pathol, 1979, 13：147-152.

[12] Huang T. S., Cerenius L., Soderhall K.. Analysis of the genetic diversity in crayfish plague fungus, Aphanomyces astaci, by random amplification of polymorphic DNA assay [J]. Aquaculture, 1994, 26：1-10.

[13] Yuasa K., Kitancharoen N., Hatai K.. Simple method to distinguish between Saprolegnia parasitica and S. diclina isolated from fishes with saprolegniasis [J]. Fish Pathol, 1997, 32 (3)：175-176.

[14] Gieseker C. M., Serfling S. G., Reimschuessel R.. Formalin treatment to reduce mortality associated with Saprolegnia parasitica in rainbow trout, Oncorhynchus mykiss [J]. Aquaculture, 2006, 253：120-129.

[15] Coker W. C.. The saprolegniaceae with notes on other water molds [M]. North Carolina, Chapel Hill：University of North Caroline Press, 1932.

[16] Seymour R.. The genus Saprolegnia [J]. Nova Hedwigia, 1970, 19：1-124.

[17] 魏景超. 真菌鉴定手册 [M]. 上海：上海科学技术出版社，1979. [18] 倪 达书. 鱼类水霉病的防治研究 [M]. 北京：中国农业出版社，1982.

[18] 战文斌. 水产动物病害学 [M]. 北京：中国农业出版社，2004.

[19] 朱越雄，杨斐飞，曹广力，等. 草鱼养殖水体水霉的分离培养及孢子的诱生 [J]. 水利渔业，2005, 25 (3)：70-72.

[20] 李明春，魏东盛，刑来君. 关于卵菌纲分类地位演变的教学体会 [J]. 菌物学研究，2006, 4 (3)：70-74.

[21] Pieter V. W.. Saprolegnia parasitica, an oomycete pathogen with a fishy appetite：new challenges for an old problem [J]. Mycologist, 2006, 20 (3)：99-104.

[22] Stueland S., Hatai K., Skaar I.. Morphological and physiological characteristics of Saprolegnia spp. strains pathogenic to Atlantic salmon, Salmo salar L [J]. J Fish Dis, 2005, 28：445-453.

[23] Willoughby L. G.. Saprolegniasis of salmonid fish in Windetmere: a critical analysis [J]. J Fish Dis, 1978, 1: 51−67.

[24] Kanouse B.. A physiological and morphological study of Saprolegnia parasitica [J]. Mycologia, 1932, 24: 431−452.

[25] K itancharoen N., Yuasa K., Hatai K.. Morphological aspects of Saprolegnia diclina type 1 isolated from pejerrey, Odonthetes bonariensisi [J]. Mycoscience, 1995, 36: 365−368.

[26] D ieguez-Uribeondo J., Fregeneda-Grandes J. M., Cerenius L., et al. Re-evaluation of the enigmatic species complex Saprolegnia diclinaSaprolegnia parasitica based on morphological, physiological and molecular data [J]. Fungal Genet Biol, 2007, 44: 585−601.

[27] Pickering A. D., Willoughby L. G., Mcgrogy C. B.. Fine structure of secondary cyst cases of Saprolegnia isolates from infected fish [J]. Trans Br Mycol Soc, 1979, 72: 427−436.

[28] Hatai K., Willoughby L. G., Beaks G. W. Some characteristics in Saprolegnia obtained from fish hatcheries in Japan [J]. Mycol Res, 1990, 94: 182−190.

[29] Beakes G. W.. A comparative account of cysts coat ontogeny in saprophytic and fish−lesion (pathogenic) isolates of the Saprolegnia diclina−parasitica complex [J]. Can J Bot, 1983, 61: 603−625.

[30] S oderhall K., Dick M. M., Clark G., et al. Isolation of Saprolegnia parasitica from the crayfish Astacus leptodactylus [J]. Aquaculture, 1991, 92: 121−125.

[31] Hallett I C., Dick M. W.. Fine structure of zoospore cyst ornamentation in the Saprolegiaceae and pythiaceae [J]. Trans Br Mycol Soc, 1986, 86 (3): 457−463.

[32] Pickerling A. D., Duston J.. Administration of cortisol to brown trout, Salmo trutta L., and its effects on the susceptibility to Saprolegnia infection and furunculosis [J]. J Fish Biol, 1983, 23: 163−175.

[33] Cross M. L., Willoughby L. G.. Enhanced vulnerability of rainbow trout (Salmo gairdneri) to Saprolegnia infection, following treatment of the fish with an androgen [J]. Mycol Res, 1989, 93: 379−402.

[34] B ly J. E, Lawson L. A., Dale D. J., et al. Winter Saprolegniosis in channel catfish [J]. Dis Aquat Organ, 1992, 13: 155−164.

该文发表于《微生物学通报》【2011, 38 (1): 57-62】

彭泽鲫卵源致病性水霉的鉴定及其生物学特性

Identification and biological characteristics of a pathogenic Saprolegnia sp. from the egg of Pengze crucian carp (*Carassius auratus* pengzesis)

夏文伟，曹海鹏，王浩，张世奇，杨先乐

（上海海洋大学，国家水生动物病原库，上海 201306）

摘要： 从患病的彭泽鲫卵上分离 3 株丝状真菌，经人工感染试验证实其中 1 株丝状真菌 JL1 对彭泽鲫卵具有致病性，并进一步研究了其形态与生长特性，开展了 ITS rDNA 序列分析。实验结果表明，菌株 JL1 菌丝为透明管状结构，中间无横隔，分枝较少；游动孢子囊多数呈棒状，游动孢子呈多排排列，发育成熟后从孢子囊中释放出来，并迅速游离；藏卵器呈球形，与雄器同枝或异枝。菌株 JL1 的 ITS rDNA 序列与 GenBank 基因库中水霉属菌株自然聚类，同源性高达 99%，与 *Saprolegnia* sp. H（登录号：EF460351）的亲缘关系最近。结合形态特征与 ITS 序列鉴定的结果，判定菌株 JL1 为水霉菌（*Saprolegnia* sp）。此外，菌株 JL1 在 5~30℃、pH 4~11 范围内均能生长，最适生长温度和 pH 范围分别为 25~30℃和 6~9。同时菌株 JL1 对 NaCl 敏感，质量分数为 2% 的 NaCl 即可抑制其生长，可以作为该病防治的依据。

关键词： 卵，水霉，鉴定，生物学特性

该文发表于《水生生物学报》【2014，1：180-183】

壳聚糖对草鱼人工感染水霉的影响

Effect of Chitosan on Artificial Infection of Grass Carp Saprolegniasis

孙琪，胡鲲，杨先乐

(上海海洋大学，国家水生动物病原库，上海 201306)

水霉病给水产养殖业造成巨大经济损失[1,2]，水霉是引起鱼类发生水霉病的主要病原，其适应温度范围广（5~26℃），分布范围亦很广[3]。水霉对宿主无严格的选择性[4,5]，除鱼卵外，鱼类水霉病均非原发性疾病，而是继发性疾病[6,7]。当鱼类皮肤或鳃受到机械损伤及其他病原体的伤害时，水霉孢子会在伤口处侵染萌发。目前国内市场上还没有一种药物可有效防治水霉病，防治水霉病成为水产养殖行业的一个重要难题。如何减少水霉病的发生成为水霉病研究的重要部分。由于水霉是腐生性的，未受伤的鱼不受感染[8,9]，所以促进鱼体伤口愈合成为减少水霉病发生的一条重要途径。

甲壳素（Chitin）是唯一一种在自然界中大量存在的氨基多糖，其化学命名为 β-（1→4）-2-乙酰氨基-2-脱氧-D-葡萄糖。壳聚糖（Chitosan）是甲壳素的脱乙酰基产物，也叫脱乙酰甲壳素，简称（CTS）[10]。甲壳素分布极其广泛，大量存在于海洋节肢动物（如虾、蟹）的甲壳中，也存在于昆虫、藻类细胞膜和高等植物的细胞壁中，在自然界的储量仅次于纤维素[11]，每年生物合成的甲壳素有 $10×10^{12}$ kg 之多。壳聚糖有促进伤口愈合的作用[12]，壳聚糖可促进内白细胞杀菌素和巨噬细胞的迁移，用壳聚糖处理的伤口，可以加速Ⅲ型胶原蛋白的分泌，从而促进了肉芽组织和上皮组织的形成[13]。Mori, et al[14]通过鼠体外实验证明壳聚糖对内皮细胞和表皮细胞有趋化性吸引作用，可以促进纤维母细胞及血管内皮细胞的迁移和增殖。Kawakami, et al.[15]通过对小鼠实验发现壳聚糖对伤口疼痛有很好的舒缓作用。Allan, et al.[16]发现壳聚糖与伤口接触时能起到清凉而舒服的润肤作用。

壳聚糖已在水处理、日用化学品、生物工程、农业、食品和医药等领域得到了广泛的应用[17-19]，其在水产养殖上的研究，主要以能增强鱼虾的免疫功能而逐渐引起国内外水产工作者的关注[20]，而有关壳聚糖对鱼类的其他作用，如对鱼类疾病有无预防、治疗效果的研究很少，就其在促进鱼类伤口愈合、预防水霉病等作用方面的研究还未涉及。因此，本研究将壳聚糖添加到受伤草鱼以及人工感染水霉草鱼的养殖水体中，研究壳聚糖对草鱼伤口愈合以及对人工感染水霉草鱼的影响。通过壳聚糖治疗鱼类体表外伤，减低鱼类感染水霉的概率，为壳聚糖应用于预防鱼类水霉病提供理论依据。

1 材料与方法

1.1 实验用鱼

草鱼（*Ctenopharyngodon idellus*），规格约为（50±10）g，体长（15±2）cm，购自江苏省南通国营农场，体表以及解剖观察健康。200 L 玻璃容器，24 h 增氧，水源为已曝气、消毒自来水，各项指标均符合渔业水质标准。草鱼实验前暂养10 d，每天定时投喂饲料 2 次，饲料配制参照 National Research Council（NRC）营养标准。

1.2 实验试剂

壳聚糖，纯度为 90%~95%，购自上海卡博工贸有限公司；马铃薯葡糖糖琼脂（PDA），购自国药集团化学试剂有限公司。

1.3 实验菌株

水霉菌株为寄生水霉（*Saprolegnia parasitica*），购自于 ATCC，菌株编号：200013。

1.4 实验方法

壳聚糖对草鱼伤口愈合的影响

取 5 个同样规格的鱼缸（65 cm×40 cm×50cm），本实验均为此规格鱼缸）加入等量已曝气水，每个鱼缸中分别加入相应量的壳聚糖，混匀，使其壳聚糖最终浓度分别为 0、12.5、25、50、100 mg/L。

取健康草鱼 100 条，在每条草鱼一侧背鳍下方的位置，用刀片划"Ⅱ"型伤口，伤口以不出血为标准，将带有伤口的草鱼随机地分配到 5 个鱼缸中，每缸 20 条。停止投喂，5 d 后观察伤口愈合情况。重复以上实验 3 次。

草鱼人工感染水霉实验

挑取保存于试管内的水霉菌丝，转接于马铃薯葡糖糖糖琼脂培养基（PDA，配方：葡萄糖 20 g/L，琼脂粉 20 g/L，马铃薯淀浸出粉 15 g/L）培养皿中，20℃培养 72 h，使菌丝长满整个培养基表面。取 20 个培养好的培养皿，将水霉菌连同培养基一起放入过滤网袋中，放入已曝气水体中，pH 6.5~6.8，在水温 20℃的条件下培养水霉后，用血球计数板每隔 2 h 计数一次水体中孢子数，直至水体中孢子浓度达到 1×10⁴个/mL，作为感染水体。

取养殖于 25℃水温中健康草鱼 20 条，用手术刀将其背鳍下方两侧的鳞片去掉，刮拭去掉鳞片部位的皮肤约 9 cm²，并用刀片轻微划伤皮肤，擦拭掉伤口表面的黏液。将 20 条划伤的草鱼置于感染水体中，并使水温在 1 h 内下降到 15℃，48 h 后开始观察记录实验结果。重复以上实验 10 次。

壳聚糖对水霉菌丝生长的影响

配置浓度分别为 0、12.5、25、50、100 mg/L 的壳聚糖悬浮液，分别取 5 mL 加入 6 孔板的相应各孔中，然后每孔中加入水霉菌块（5 mm×5 mm）。20℃培养 20 h 后，观察水霉菌丝生长情况。每个浓度分别作 3 组平行实验。

壳聚糖对草鱼人工感染水霉实验的影响

（1）感染水体中添加壳聚糖，对草鱼人工感染水霉实验的影响：实验组感染水体中含有壳聚糖，其浓度为 25 mg/L，对照组为已曝气自来水，pH 6.5~6.8，用上述草鱼人工感染水霉实验方法，将划伤的草鱼随机的放到上述两组鱼缸中，每缸各 20 条，48 h 后开始观察记录实验结果。重复以上实验 3 次。

（2）壳聚糖浸泡受伤草鱼后，对草鱼人工感染水霉实验的影响：按照草鱼人工感染水霉实验中划伤鱼体的方法处理草鱼 40 条，随机平均分配为两组，实验组水体中加入壳聚糖，使其浓度为 25 mg/L，对照组为已曝气自来水，pH 6.5~6.8，养殖 72 h 后，进行感染实验。

用上述草鱼人工感染水霉的实验方法，对以上两组草鱼进行人工感染，养殖 48 h 后，开始观察记录实验结果。重复以上实验 3 次。

1.5 数据分析

用 SPSS13.0 软件统计所有数据并进行方差分析（one-way-ANOVA），差异显著时采用最小显著性差异法（Least Significant Difference，LSD）对各组平均样之间进行多重比较，数据表示为平均值加减标准差（Mean±SD）。

2 结果

2.1 壳聚糖对草鱼伤口愈合的影响

壳聚糖浓度≥25 mg/L 时，对草鱼伤口愈合效果显著，且伤口表面附有黏液（图 1 a）；对照组草鱼伤口红肿，未愈合，甚至伤口有溃烂（图 1 b）。

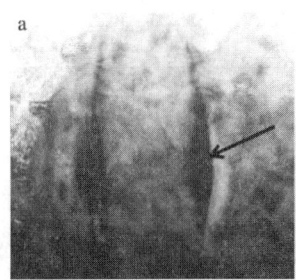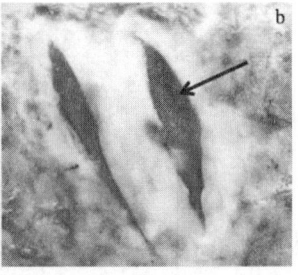

图 1 草鱼伤口愈合情况

Fig. 1 The wound healing situation of grass carp

a. 已愈合并附有黏液的伤口；b. 未愈合且已红肿、溃烂的伤口

a. The healing wound of grass carp；b. The swelling and fester wound of grass carp

对照组草鱼伤口愈合率为 1.67%。在水体中添加壳聚糖对草鱼伤口愈合率有显著影响（$P<0.05$）。在壳聚糖浓度为 50 mg/L 时，草鱼伤口愈合率与浓度为 100 和 25 mg/L 时的无显著差异（$P>0.05$）（表 1）。

2.2 草鱼人工感染水霉实验

10 次重复实验发现，本实验方法可成功地完成草鱼人工感染水霉，且有较高的感染率。从表 2 可看出，草鱼人工感染水霉平均感染率为 84.00%，且在感染过程中草鱼平均死亡率为 2.00%，各实验组间无显著差异（$P>0.05$）。

表 1 不同浓度壳聚糖对草鱼伤口愈合影响（n=3）

Tab. 1 Effect of different concentrations of chitosan on grass carp wound healing（n=3）

壳聚糖浓度 伤口愈合率 Chitosan concentration（mg/L）Healing rate（%）	死亡率 Mortality（%）
100 96.67±2.89ᵃ	1.67±2.89ᵃ
50 93.33±2.89ᵃ,ᵇ	0.00±0.00ᵃ
25 83.33±2.89ᵇ	3.33±2.89ᵃ
12.5 23.33±5.77ᶜ	0.00±0.00ᵃ
0 1.67±2.89ᵈ	1.67±2.89ᵃ

注：同列相同字母表示差异不显著（P>0.05），不同者表示差异显著（P<0.05）。

Note：Means in the same row with same superscripts indicate no significant（P>0.05），different superscripts indicate significant（P<0.05）

表 2 草鱼人工感染水霉实验结果

Tab. 2 The results of artificial infection of saprolegnia on grass carp

组别 感染率 Group Infection rate（%）	死亡率 Mortality（%）
1 80.00	0.00
2 85.00	5.00
3 90.00	0.00
4 90.00	0.00
5 90.00	0.00
6 85.00	0.00
7 75.00	15.00
8 80.00	0.00
9 80.00	0.00
10 85.00	0.00
平均值 Mean（%）84.00±5.16	2.00±4.83

2.3 壳聚糖对水霉菌丝生长的影响

20 h 后，对照组水霉菌丝平均长度为 7.67 mm。200 mg/L 壳聚糖悬浮液中水霉菌丝平均长度与对照组的显著差异（P<0.05），其他浓度组与对照组无显著差异（P>0.05）（表3）。从表3中可以看出：低浓度壳聚糖对水霉菌无直接的抑制或杀灭作用，在感染水体中添加壳聚糖不会对水霉菌产生直接影响。

2.4 壳聚糖对草鱼人工感染水霉实验的影响

感染水体中添加壳聚糖，对草鱼人工感染水霉实验的影响

在死亡率相同条件下，对照组草鱼人工感染水霉感染率为 86.67%。在感染水体中添加 25 mg/L 壳聚糖悬浮液时，其感染率为 61.67%与对照组有显著差异（P<0.05）（图2）。

表 3 20 h 后不同浓度的壳聚糖中水霉菌丝的长度（n=3）

Tab. 3 Hypha length of saprolegnia chitosan in different concentrations after 20 hours（n=3）

壳聚糖浓度 Chitosan concentration（mg/L）	菌丝长度 Hypha length（mm）
0	7.67±0.58
12.5	8.33±0.58

续表

壳聚糖浓度 Chitosan concentration（mg/L）	菌丝长度 Hypha length（mm）
25	8.33±1.15
50	8.00±1.00
100	8.00±1.73
200	4.33±0.58

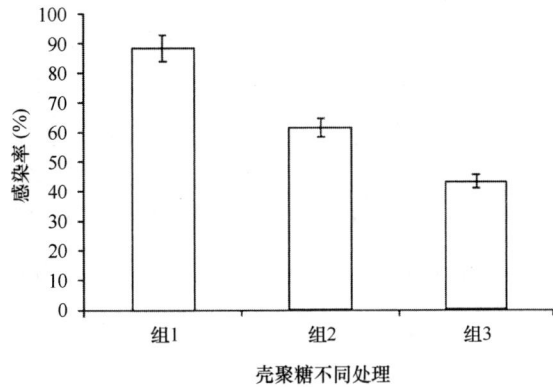

图 2　草鱼人工感染水霉感染率（$n=3$）

Fig. 2　The results of artificial infection of saprolegnia on grass carp with different chitosan treatments（$n=3$）

组1：对照组；组2：感染水体中添加壳聚糖，对草鱼人工感染水霉影响实验；组3：感染前，壳聚糖浸泡受伤草鱼后，对草鱼人工感染水霉的影响实验。

1 Control group；2 Artificial infection of saprolegnia on grass carp in 25 mg/L chitosan；3 Artificial infection of saprolegnia on grass carp after soaking in 25 mg/L chitosan for 72 h.

感染前，壳聚糖浸泡受伤草鱼后对草鱼人工感染水霉实验的影响

在死亡率相同条件下，对照组草鱼人工感染水霉感染率为86.67%。感染前，25 mg/L壳聚糖浸泡受伤草鱼3 d后，其感染率为43.33%与对照组有显著差异（$P<0.05$）（图2）。且未感染的草鱼伤口明显愈合，且伤口表面附有黏液。

在感染水体中添加25 mg/L壳聚糖悬浮液时，草鱼人工感染水霉感染率为61.67%；感染前，25 mg/L壳聚糖浸泡受伤草鱼3 d后，其感染率为43.33%。感染前，25 mg/L壳聚糖浸泡受伤草鱼3 d后，能极显著的降低其感染率（$P<0.05$）。

3　讨论

3.1　伤口愈合

本实验所用的壳聚糖利用虾、蟹壳生产，适量添加到受伤鱼体养殖水环境中5 d，壳聚糖在水体中的浓度达到25 mg/L以上时，对草鱼体表伤口有显著的愈合作用。

Usami，et al.[21]对牛、Okamoto，et al.[22]对狗、Kaoru，et al.[23]对鼠伤口研究均发现壳聚糖具有很好的促进伤口愈合作用，Graeme，et al.[24]对人体皮肤体外实验发现壳聚糖可促进纤维母细胞和角质细胞增殖。壳聚糖正被人医、兽医作为伤口愈合增效加速剂在医学界广泛使用[25]，本实验首次证实壳聚糖对非哺乳动物体表伤口有愈合作用，为进一步推广壳聚糖的应用提供理论依据。

3.2　人工感染

水霉病作为当今水产养殖业一个较重大难题，现今还不存在一种可用的特效药物防治水霉病。因此，预防水霉病的发生成为重中之重。通过10次草鱼人工感染水霉平行实验证明：本方法稳定可靠、可重复性高，可作为草鱼人工感染水霉基本方法用于科学研究。本实验验证了低浓度壳聚糖对水霉菌无直接的抑制或杀灭作用，在感染水体中添加壳聚糖对水霉菌无直接影响。在此基础上，分别采用在感染水体中添加25 mg/L的壳聚糖以及感染前用25 mg/L的壳聚糖浸泡受伤草鱼3 d，结果表明：草鱼人工感染水霉感染率均降低。而感染前用壳聚糖浸泡受伤草鱼，感染率相对更低，死亡率与对照组无

显著差异，说明 25 mg/L 壳聚糖对草鱼无毒害作用，在生产实践中可用于降低受伤鱼类水霉病发病率，减少经济损失。

参考文献

[1] Hussein M. M. A., Hatai K. Pathogenicity of Saprolegnia species associated with outbreaks of salmonid saprolegniosis in Japan [J]. Fisheries Science, 2002, 68 (5): 1067-1072.

[2] Tampieri M. P., Galuppi R., Carelle M. S., et al. Effect of selected essential oils and pure compounds on Saprolegnia parasitica [J]. Pharmaceutical Biology, 2003, 41 (8): 584-591.

[3] Monica M. S., Alan P., Kanak B.. Saprolegnia bulbosa sp. nov. isolated from an Argentine stream: taxonomy and comparison with related species [J]. FEMS Microbiology Letters, 2007, 268 (2): 225-230.

[4] Kanit C., Kishio H.. Freshwater fungi isolated from eggs of the common carp (Cyprinus carpio) in Thailand [J]. Mycoscience, 2004, 45 (1): 42-48.

[5] Kitancharoen N., Hatai K.. Some biological characteristics of fungi isolated from salmonid eggs [J]. Mycoscience, 1998, 39 (2): 249-255.

[6] Behrens P. W., Sicotte V. J., Delents J. Microalgae as a source of stable isotopically labeled compounds [J]. Journal of Applied Phycology, 1994, 6 (2): 113-121.

[7] Khulbe R. D., Bisht G. S., Chandra J. An ecological study on watermolds of some rivers of Kumaun Himalaya [J]. Journal of the Indian Botanical Society, 1995, 74 (3): 61-64.

[8] Behrens P. W., Sicotte V. J., Delents J.. Micro algae as a source of stable isotopically compounds [J]. Journal of Applied Physics, 1994, 6: 113-121.

[9] Song X. H., Chen K. W., Zhu C. F., et al. Effects of EM and deuterium sulfate treatment on the inhibition rate of Saprologenia parasitia during the hatching of Pseudobagrus fulvidraco fertilized eggs [J]. Freshwater Fisheries, 2007, 37 (1): 9-12, 18 [宋学宏，陈葵王，朱春峰，等. EM 及重氢硫酸盐对黄颡鱼卵孵化中水霉的抑制作用. 淡水渔业, 2007, 37 (1): 9-12, 18]

[10] Wang Y. T., Liu Y. H., Zhang S. Q. Advances in chemical modofication and application of chitin, chitosan and their derivatives [J]. Journal of Functional Polymers, 2002, 15 (3): 107-114 [汪玉庭，刘玉红，张淑琴. 甲壳素、壳聚糖的化学改性及其衍生物应用研究进展. 功能高分子学报, 2002, 15 (3): 107-114]

[11] Muzzarelli R., Jeuniaux C., Gooday G. W. Chitin in Nature and Technology [M]. NewYork: Plenum Press. 1986, 354.

[12] Chen Y., Dou G. F., Tan H. M., et al. Application of chitin and chitosan in the wound dressing. Journal of Functional Polymers, 2005, (1): 94-100 [陈煜，窦桂芳，谭惠民，等. 甲壳素和壳聚糖在伤口敷料中的应用. 高分子通报, 2005, (1): 94-100]

[13] Koch A. E., Polverini P. J., Kunkel S. L., et al. Interleukin-8 as a macrophage-derived mediator of angiogenesis [J]. Science, 1992, 258 (5089): 1798-1801.

[14] Mori T., Okumura M., Matsuura M., et al. Effects of chitin and its derivatives on the proliferation and cytokine production of fibroblasts in vitro [J]. Biomaterials, 1997, 18: 947-951.

[15] Kawakamia K., Miyatakea K., Okamotoa Y., et al. Analgesic effects of chitin and chitosan [J]. Carbohydrate Polymers, 2002, 49 (3): 249-252.

[16] Allan G. G., Altman L. C., Bensinger R. E., et al. In Chitin, Chitosan and Related Enzymes [M]. Amsterdam: Academic Press. 1984, 119-133.

[17] Ma N., Wang Q., Wang A. Q., et al. Progress in chemical modification of chitin and chitosan [J]. Progress in Chemistry, 2004, 16 (4): 643-653 [马宁，汪琴，王爱勤，等. 甲壳素和壳聚糖化学改性研究进展. 化学进展, 2004, 16 (4): 643-653]

[18] Chen J. Y., Le X. Y.. Applications of chitosan and its derivatives in agricultural production [J]. Chemical Research and Application, 2011, 23 (1): 1-8 [陈佳阳，乐学义. 壳聚糖及其衍生物在农业上的应用. 化学研究与应用, 2011, 23 (1): 1-8]

[19] Zhao X. R., Xia W. S.. Antimicrobial activities of chitosan and applications in food preservation [J]. Food Research and Development, 2006, 27 (2): 157-160 [赵希荣，夏文水. 壳聚糖的抗菌防腐活性及其在食品保藏中的应用. 食品研究与开发, 2006, 27 (2): 157-160]

[20] Hua X. M., Zhou H. Q., Zhou H., et al. Effect of dietary supplemental chitosan and probiotics on growth and some digestive enzyme activities in juvenile Fugu obscurus [J]. Acta Hydrobiologica Sinica, 2005, 29 (3): 299-305 [华雪铭，周洪琪，周辉，等. 饲料中添加壳聚糖和益生菌对暗纹东方鲀幼鱼生长及部分消化酶活性的影响. 水生生物学报, 2005, 29 (3): 299-305]

[21] Usami Y., Okamoto Y., Minami S., et al. Chitin and chitosan induce migration of bovine polymorphonuclear cells [J]. The Journal of Veterinary Medical Science, 1994, 56: 1215-1216.

[22] Okamoto Y., Minami S., Matsuhashi A., et al. Evaluation of chitin and chitosan on open wound healing in dogs [J]. The Journal of Veterinary Medical Science, 1993, 55: 743-747.

[23] Kaoru M., Hiroshi A., Masayuki I., et al. Hydrogel blends of chitin/chitosan, fucoidan and alginate as healing-impaired wound dressings [J]. Biomaterials, 2010, 31 (1): 83-90.

[24] Graeme I., Peter W., Edward J., et al. The effect of chitin and chitosan on the proliferation of human skin fibroblasts and keratinocytes in vitro [J]. Biomaterials, 2001, 22: 2959-2966.

[25] Takashi M., Toru F., Hiroshi U.. Topical formulations and wound healing applications of chitosan [J]. Advanced Drug Delivery Reviews,

2001, 52 (2): 105-115.

该文发表于《微生物学通报》【2012, 12:】

水霉菌环介导等温扩增检测方法的建立

Establishment of a loop-mediated amplification assay for the specific detection of Saprolegnia and its spores

王浩，张楠，杨先乐，吕利群

(上海海洋大学水产与生命学院，上海 201306)

摘要：【目的】建立一种快速简单检测水霉病病原菌的方法。【方法】针对水霉菌 ITS 区基因序列设计 4 条特异性引物，包括两条外引物和两条内引物，优化反应条件，观察检测结果。对该方法的特异性和敏感性进行研究。【结果】建立了环介导等温扩增技术检测水霉菌的方法，确定了其最适反应条件。该方法能够检测到浓度低至 10^3 个/mL 的水霉菌孢子，其灵敏度是普通 PCR 方法的 100 倍。【结论】建立的检测水霉菌的 LAMP 技术，具有操作简便快速等特点，可用做特异性水霉及其孢子的快速鉴定。

关键词：水霉，环介导等温扩增技术，检测

该文发表于《微生物学通报》【2012 (9): 1280-1289】

黄颡鱼卵致病性绵霉的分离鉴定与药敏特性

Identification and drug susceptibility of a pathogenic Achlya klebsiana strain from the eggs of yellow catfish (*Pelteobagrus fulvidrac*)

欧仁建[1]，曹海鹏[1]，郑卫东[2]，胡鲲[1]，许佳露[1]，杨先乐[1]

(1. 上海海洋大学国家水生动物病原库，上海 201306；2. 中国水产科学研究院长江水产研究所，湖北 武汉 430223)

摘要：【目的】对黄颡鱼水霉病病原进行分离鉴定，并对其药敏特性进行研究。【方法】参照传统方法对患水霉病的黄颡鱼卵上的丝状真菌进行分离，通过人工感染实验证实分离菌株的致病性，然后根据形态学特征和 ITS rDNA 序列分析对致病菌株进行鉴定，并进一步采用倍比稀释法研究其药敏特性。【结果】从患水霉病的黄颡鱼卵上分离到 8 株丝状真菌，经人工感染试验证实菌株 YC 对黄颡鱼卵具有致病性，并进一步研究了其形态与药敏特性，开展了 ITS rDNA 序列分析。结果表明，菌株 YC 为透明管状结构，中间无横隔；游动孢子囊多呈圆筒形、棍棒形或穗状，游动孢子发育成熟后不断从孢子囊顶端释放出来，成团聚集在游动孢子囊口，并经过一个时期的静休后，成团脱落或直接分散在水中游动；次生孢子囊具有典型的侧生现象；藏卵器呈球形或梨形，大多与雄器异枝，少数与雄器同枝，含 1~15 个卵孢子；成熟卵孢子中生或亚中生，偏生一个大油球。其 ITS rDNA 序列与 GenBank 基因库中绵霉属菌株自然聚类，同源性高达 99%，与异丝绵霉 (*Achlya klebsiana*) 菌株 CBS101.49 (GenBank 登录号 AF119579) 的亲缘关系最近。结合形态特征与 ITS 序列鉴定的结果，判定菌株 YC 为异丝绵霉 (*Achlya klebsiana*)。此外，在实验选用的中草药和消毒剂中，黄连和异噻唑啉酮分别对菌株 YC 的抑菌效果最好，其对菌株 YC 的最小抑菌浓度分别为 256 mg/L 和 2 mg/L。【结论】首次分离了黄颡鱼卵致病性异丝绵霉菌株 YC，并确定了其药敏特性，可以作为该病防治用药的依据。

关键词：黄颡鱼卵，绵霉，鉴定，最小抑菌浓度

该文发表于《四川动物》【2015，3：352-356】

鱼卵孵化酶在黄颡鱼鱼卵中的免疫组织化学定位和分布

Immunohistochemistry Localization of Hatching Enzyme in the Developing Embryos of *Pelteobagrus fulvidraco*

李浩然[1]，蒋石容[2]，欧仁建[3]，郭微微[1]，杨先乐[1]，胡鲲[1]

(1. 上海海洋大学，国家水生动物病原库，上海 201306；2. 湖南省华容县水产局，湖南 岳阳 414200；
3. 四川省成都市龙泉驿区水务局，成都 610100)

摘要： 孵化酶存在于各种水生动物的胚胎中，会影响水生生物胚胎发育，是造成鱼类人工繁殖过程中早脱膜现象的重要因素之一。本研究以黄颡鱼鱼卵为研究对象，采用免疫组织化学方法对不同孵化时间点的孵化酶进行定位分布研究，探讨孵化酶在黄颡鱼鱼卵中的时空表达规律。结果表明：在黄颡鱼鱼卵中，孵化酶主要分布在鱼卵外胚层及卵膜内层上；在鱼卵孵化的各个时期均有孵化酶存在，孵化酶在各时间点的相对表达量没有明显差异。研究结果为阐释孵化酶在鱼卵孵化过程中的作用奠定了基础。

关键词： 孵化酶；黄颡鱼；鱼卵；免疫组织化学；定位

该文发表于《微生物学通报》【2014，41（9）：1829-1836】

环境因子对水霉菌生物膜形成的影响

Effects of environmental factors on biofilm formation of Saprolegnia

袁海兰，苏建，胡鲲，曹海鹏，杨先乐

(上海海洋大学国家水生动物病原库，上海 201306)

摘要：【目的】体外构建水霉菌（Saprolegnia）生物膜（Biofilm，BF），研究环境因子对其生物膜形成的影响。【方法】采用改良的微孔板法研究静置培养条件下寄生水霉（*Saprolegnia parasitica*）ATCC200013 在 96 孔酶标板上的成膜情况，CCK-8 法（Cell Counting Kit-8）定量检测生物膜中水霉菌的活力。【结果】水霉菌的生物膜的 OD450 值在培养 24 h 达到峰值，48 h 后趋于稳定。随着初始孢子浓度升高，水霉菌生物膜 OD450 值升高，差异显著（$P<0.05$）。20~25℃ 生物膜形成量最多，OD_{450} 值显著高于其他温度组（$P<0.05$）。在起始 pH 值为 4~11 的沙氏葡萄糖液体培养基中，水霉菌均能形成生物膜。在培养基中加入 0.12 mmol/L 以上 $CaCl_2$，能促进生物膜形成；添加 0.03~2.00 mmol/L $MgCl_2$，水霉菌生物膜形成量与未添加 $MgCl_2$ 对照组无显著性差异；Cu^{2+} 对水霉菌生物膜的形成有显著影响，0.5 mmol/L 以上添加处理几乎不形成生物膜；NaCl 能明显抑制水霉菌生物膜的形成，当 NaCl 质量分数低于 0.12% 时，对生物膜形成的影响较小（$P>0.05$）。水霉菌在鲫皮和肌肉提取液包被后生物膜形成量与对照组相比无显著性差异，而鲫表皮黏液、鳃黏液包被后生物膜的形成量明显减少。【结论】研究首次采用微孔板法体外构建水霉菌生物膜，发现其生物膜的形成与多种环境因素有着密切的关系。这为了解水霉菌生物膜的形成规律提供了一定参考，为水霉菌生物膜的进一步研究奠定了基础。

关键词： 水霉菌，生物膜，环境因子

该文发表于《华中农业大学学报》【2014，33（5）：99-104】

*SpHtp*1 基因在不同种属水霉菌株中的分布

Distribution of *SpHtp*1 gene in different Saprolegnia strains

叶鑫，赵依妮，曹海鹏，胡鲲，杨先乐

（上海海洋大学国家水生动物病原库，上海 201306）

摘要：为研究 *SpHtp*1 基因在不同种属水霉菌株中的分布，本研究以寄生水霉 ATCC200013™ 的基因组 DNA 为模板，设计一对特异性引物，进行 PCR 扩增。将扩增片段回收，并克隆入 T 载体，转化大肠杆菌 DH5α，挑取阳性克隆测序。将测序结果与 GenBank 中登录的 *SpHtp*1 基因序列进行比较，同源性为 99%。以该方法调查从水霉病主要流行地区分离的 17 株水霉菌 *SpHtp*1 基因的分布情况。结果发现：在 17 株水霉菌株中，有 8 株水霉菌检测到 *SpHtp*1 基因，阳性率为 47%，且所有含 *SpHtp*1 基因的菌株在分类地位上均隶属于寄生水霉（*Saprolegnia parasitica*）。*SpHtp*1 基因可能在寄生水霉中特异性存在，可望用于寄生水霉的分离鉴定。

关键词：水霉；SpHtp1 基因；聚合酶链式反应；致病性

该文发表于《湖北农业科学》【2013，52（21）：5342-5344】

水霉菌蛋白制备方法的比较与分析

Comparison and Analysis of Preparation Methods of Saprolegnia Protein

宋鹏鹏，胡鲲，杨先乐

（上海海洋大学国家水生动物病原库，上海 201306）

摘要：比较了 3 种方法对 8 种水霉菌（*Saprolegnia*）破壁处理制备蛋白的效果，并进行聚丙烯酰胺凝胶电泳分析。结果表明，加石英砂和液氮混合研磨处理制备的蛋白效果最好，条带清晰，该方法操作简便、快捷、较为常用，可有效地区分水霉菌株分子遗传学背景的差异，适用于水霉菌分子流行病学研究。

关键词：水霉菌；聚丙烯凝胶电泳；处理方法

二、孔雀石绿的危害与禁用

孔雀石绿（Malachite Green，MG）又名碱性绿、盐基块绿、孔雀绿，其在水生生物体中的主要代谢产物为无色孔雀石绿（leucomalachite green）。从 20 世纪 90 年代开始，国内外学者陆续发现，孔雀绿及代谢产物无色孔雀绿具有一个共同的化学官能团，正因为这个三苯甲烷，使其具有高毒、高残留、高致癌和高致畸性。此外，它还能通过溶解足够的锌，引起水生动物急性锌中毒，并引起鱼类消化道、鳃和皮肤轻度发炎，从而影响鱼类的正常摄食和生长，它能阻碍肠道酶（如胰蛋白酶、α-淀粉酶的活性，影响动物的消化吸收功能。最近的研究也证明，孔雀石绿会给水产养殖业及其水产品带来严重的安隐患（柯江波等，2014；柯江波等，2015）。

鉴此，许多国家均将孔雀绿列为水产养殖禁用药物。1991 年美国药物管理局禁止使用该药。我国于 2002 年将孔雀绿列入《食品动物禁用的兽药及其化合物清单》及《无公害食品渔用药物使用准则》（中华人民共和国农业行业标准 NY5071-2002）中，2003 年英国等国的养鱼场也禁止使用。也有一些国家孔雀绿允许用于治疗观赏鱼的疾病，如水霉病、小瓜虫等疾病，但禁止在食用鱼中使用。

该文发表于《水产科技情报》【2005，32（5）：210-213】

对孔雀石绿的禁用及其思考

杨先乐，喻文娟，王民权，郑宗林

（农业部渔业动植物病原库，上海水产大学，上海市 200090）

1 孔雀石绿的性质及其危害

孔雀石绿又名碱性绿、盐基块绿、孔雀绿，英文名 Malachitegreen，分子式是 $C_{23}H_{25}ClN_2$，在水生生物体内的主要代谢产物为无色孔雀石绿（leucomalachitegreen，$C_{23}H_{26}N_2$）[1,2]，其化学结构如图 1 所示。

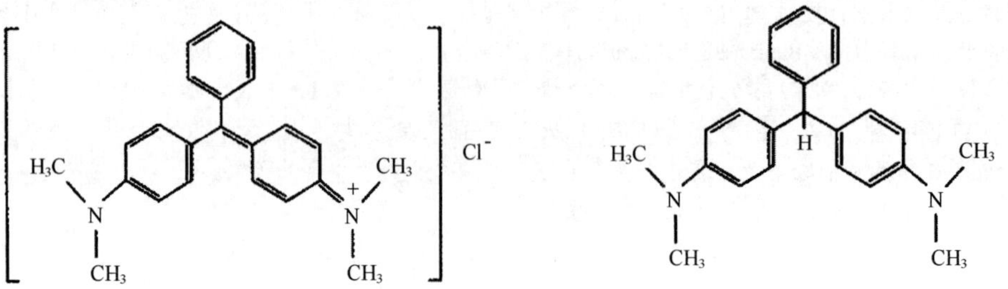

图 1　孔雀石绿结构式（左孔雀石绿；右无色孔雀石绿）

孔雀石绿是一种具有光泽的粉末状结晶，极易溶于水，水溶液呈蓝绿色，能溶于乙醇、甲醇或戊醇。它在浓硫酸中显黄色，稀释后显暗黄色，Ti^{2+} 使其水褪色，在氢氧化钠水溶液中由绿色变为白色沉淀，与 $AuBr^{4-}$、$GaCl_4$、$TaFb^-$、$TiBr^{4-}$、UO_2^{2+}、Zn^{2+} 等形成有色离子络合物，根据其这些特性，常采用光度测定法对之进行鉴别。

从 20 世纪 90 年代开始，国内外学者陆续发现，孔雀石绿及其在水生生物体内的代谢产物无色孔雀石绿具有高毒性、高残留、高致癌和高致畸性，因而各国陆续在渔业上禁止使用。

（1）高毒性　孔雀石绿对哺乳动物的细胞具有高毒性[3,4]，Culp 等（1999）[5]发现，孔雀石绿与无色孔雀石绿均能使大鼠的肝细胞空泡化，无色孔雀石绿还能使甲状腺滤泡上皮大量凋亡，减少公鼠甲状腺激素的释放；此外孔雀石绿还能抑制血浆胆碱脂酶产生作用，进而有可能造成乙酰胆碱的蓄积而出现神经症状[6]。水生动物对孔雀石绿均很敏感，其安全质量分数一般在 0.1 mg/L 以下（表1）。孔雀石绿还能通过溶解足够的锌，引起水生动物急性锌中毒，并引起消化道、鳃和皮肤轻度发炎，从而影响水生生物正常摄食和生长，阻碍肠道酶（如胰蛋白酶、α-淀粉酶等）的活性，影响动物的消化吸收功能。

表1　孔雀石绿对部分水生动物的安全质量分数

品种	安全质量分数（$\times10^{-2}$mg/L）	品种	安全质量分数（$\times10^{-2}$mg/L）
虹鳟鱼苗	2.5	淡水白鲳	3.1
加州鲈鱼	2.0	美国大口胭脂鱼	1.6
锦鲤	1.5	脊尾白虾幼体	1.0
异育银鲫	1.2	对虾苗	1.8
云斑鱼	1.0	中华大蟾蜍蝌蚪	1.5
翘嘴红鲌	3.1	中华鳖	2 630
蒙古裸腹溞	2.4		

（2）高残留　孔雀石绿及其代谢产物无色孔雀石绿能迅速在组织中蓄积，较多地在受精卵和鱼苗的血清、肝、肾、肌肉和其他组织中检测到[7,8]。无色孔雀石绿不溶于水，因而其残留毒性比孔雀石绿更强[9]。采用 SC/T3021-2004 标准《无公害食品水产品中孔雀石绿残留量的测定液相色谱法》可同时检测到孔雀石绿和无色孔雀石绿。

（3）高致癌　孔雀石绿的化学官能团是三苯甲烷（triphenylmethane），其分子中与苯基相连的亚甲基和次甲基受苯环影响有较高的反应活性，可生成自由基——三苯甲基，同时孔雀石绿也能抑制人类谷胱甘肽-S-转移酶的活性[10]，两者均能造成人类器官组织氧压的改变，使细胞凋亡出现异常，诱发肿瘤和脂质过氧化，而来源于上皮组织的恶性肿瘤即为"癌"。Panandiker 等（1992）[11]发现，孔雀石绿能使培养的仓鼠胚胎细胞产生过多的自由基造成脂质过氧化；RAO 等（1995）[12]认为，它能抑制培养的原始鼠肝脏细胞 DNA 和 EGF 的合成，诱使乳酸脱氢酶的释放，诱使肝脏肿瘤的发生；Sundarrajan 等（2000）[13]的研究表明，孔雀石绿能通过促进增殖细胞核抗原（PCNA）和 G1/S 期细胞周期蛋白（cyclins）的表达而诱发大鼠肝脏肿瘤。

关于孔雀石绿多苯环芳烃结构的致癌机理目前尚不完全清楚。Culp 等（1999）[14]研究发现，孔雀石绿代谢产物的衍生物及初级和次级的芳香胺在癌症发生中起着重要的作用，如小分子芳香胺，尤其是 1~4 个芳香环结构进入人体后，会穿透细胞膜，到达细胞核中的 DNA。芳香胺易产生活泼的亲电子"阳氮离子"，攻击 DNA 上的亲核位置，相互以共价键结合，从而破坏 DNA 引起癌变。除此之外，芳香胺分子的扁平部分还会"插入"正常细胞中 DNA 螺旋线结构的相邻碱基对之间，从而破坏 DNA，引起癌变。Meyer 等（1983）[15]与 Schnick 等（1978）[16]的试验也证实了孔雀石绿能使虹鳟产生癌病变。

（4）高致畸性 孔雀石绿能使淡水鱼鱼卵染色体发生异常[17]，即使孔雀石绿的质量分数低于 0.1 mg/mL，它仍能使兔和鱼繁殖致畸[18,19]；当孔雀石绿的质量分数高于 0.2 mg/L 时，对软体动物东风螺的受精卵或是幼体都会造成不同程度的发育畸形[20]。

（5）致突变 周立红等（1997）[21]认为，孔雀石绿可使鲻鱼和尼罗罗非鱼的红细胞产生微核，当质量分数为 0.5 mg/L 时，微核率分别达到 0.133% 和 0.071 7%，且微核率随药物质量分数的升高而提高。

2 孔雀石绿在渔业领域的使用情况

孔雀石绿原是一种三苯甲烷类染料，由苯甲醛和 N,N-二甲基苯胺在盐酸或硫酸中缩合生成四甲基代二氨基三苯甲烷的隐性碱体后，在酸介质中被二氧化铅秘氧化制得。因其氧化电位势与组成酶的某些氨基酸相近，在细胞分裂时发生竞争而阻碍蛋白肽的形成，从而产生抗菌杀虫作用，因此曾在渔业上广泛使用。2002 年，中华人民共和国农业部发布的 193 号公告以及 NY5072—2002 标准《无公害食品渔用药物使用准则》中明确规定在渔业领域禁用孔雀石绿。

近年来，我国在对孔雀石绿等违禁水产药物的监管方面取得了一些成绩，养殖户（场）纷纷建立了用药档案，较多省市和地区也构建了如"市-区县-乡镇"食品安全监管的网络，孔雀石绿等违禁水产药物的使用得到了基本控制。但是由于我国地域广阔，养殖面积大、养殖品种多，渔（农）民科学用药的意识薄弱，水产管理部门人员不足、经费有限，在对水产药物使用的监控、禁用水产药物的检测等方面存在着不足，致使孔雀石绿在不同程度上存在禁而不止的现象。某些不法商人为了谋取非法利润，仍在经销"孔雀石绿"，甚至在水产品保鲜和运输过程中添加孔雀石绿。

目前禁用药物孔雀石绿在我国渔业领域的使用量尚难确切地估计，若仅以泼洒给药方式防治淡水鱼类的水霉病，以最小化原则进行粗略估算，根据公式"孔雀石绿全年使用量（kg）= 防治该水产动物水霉病的用量（$2×10^{-6}$ kg/m³）×面积（667×10⁴ m²）×水深（1.5 m）"得出，孔雀石绿全年每 667 hm²（万亩）的使用量（kg）为 0.02 t；若全年有 670×10⁴ hm² 的养殖面积使用孔雀石绿（我国 2002 年的水产养殖面积为 681×10⁴ hm²），则其使用量可达 200 t 左右。如果考虑疗程及全年的发病次数，则应在此基础上再加一倍。这还不包括育苗场高频率、大剂量地使用，以及对其他水产养殖对象使用孔雀石绿的；不包括以浸浴给药方式使用孔雀石绿的，以及在防治水生动物鳃霉病、小瓜虫病、消毒、保鲜等方面使用孔雀石绿的。由此我们估计，目前在我国的渔业领域，孔雀石绿的使用量达到 600~800 t，令人触目惊心。

3 孔雀石绿屡禁不止的原因

3.1 价廉易得 孔雀石绿每千克仅 35 元左右，由于其使用剂量较低，且多采用浸泡、泼洒等给药方法，因此防治成本低。用该药防治水产养殖中的病害和进行水产品的运输保鲜，在有些地区是不足为怪的事情。

3.2 对水霉病等疗效好，缺少替代药 自从孔雀绿被禁用后，在水霉病的防治方面一直没有较好的替代药物，在对该病的控制上形成了一定的真空。而该病是我国水产养殖中的一种常见病，有时还可造成较大的损失，因此一些养殖户不得不用孔雀石绿来防治该病。

3.3 缺乏有力的监管制度 虽然国家已明文规定禁止使用孔雀石绿，但对孔雀石绿缺乏监管方法和监管网络，对违法经销和使用缺乏处罚措施，处罚力度也不够，以致不能从源头上阻断孔雀石绿在渔业上的非法使用。

3.4 孔雀石绿残留初筛方法缺乏 目前对孔雀石绿残留的检测主要是采取液相色谱法，但因该方法需有较昂贵的仪器，检测过程复杂，对检测技术的要求也较高，故液相色谱法不适用于普查。采用简易可行的初筛方法是控制孔雀石绿滥用、保证水产品安全的途径之一。

3.5 宣传和科学普及工作深度不够 养殖者对孔雀石绿的毒副作用缺乏深刻了解，虽一般都能知道孔雀石绿有毒，但对其能产生致畸、致癌、致突变的毒副作用缺乏认识；较多人对国家禁用孔雀石绿等药物的利害关系不了解，没有将其提高到一定的高度上去认识，因此在销售和使用时毫无顾忌，更无人对违法经营和使用者进行举报。

4 对在渔业领域彻底禁用孔雀石绿的思考

4.1 开展对渔业领域非法使用孔雀石绿的专项整治,强行制止孔雀石绿的使用 针对目前所存在的非法使用孔雀石绿的情况,对生产、销售与使用情况进行全面的检查和整改,对市场流通的水产品定期进行孔雀石绿残留抽检。

4.2 严格控制孔雀石绿的生产和销售,实行许可证制目前,在工业中的某些领域需要使用孔雀石绿,为了防止这部分孔雀石绿成为渔业非法使用的源头,必须严格规范孔雀石绿的生产、经营和销售活动。加强对市场上销售孔雀石绿企业、单位的监管力度,对违法销售孔雀石绿的经营单位或个人予以重罚,甚至追究法律责任。

4.3 建立与完善孔雀石绿检测和监管网络 百密难免一疏,仅凭执法部门对生产、销售企业和养殖户等的常规检查还不能彻底禁止孔雀石绿的使用,只有建立高效、健全的检测、监管制度和网络,才可能把住水产品质量安全的最后一道防线。

4.4 加强宣传和科学普及的力度 一方面要向广大渔(农)民宣传孔雀石绿的危害性,宣传国家有关禁止使用孔雀石绿的有关法律和法规,另一方面要加强科学用药技术的普及,使渔(农)民彻底改掉滥用药物的陋习。

4.5 加强对水产药物基础理论和应用的研究 目前所要解决的主要问题是:①孔雀石绿在主要水产动物体内的代谢和消除规律;②孔雀石绿防治水霉病替代品的研制;③简单、快速、灵敏的孔雀石绿检测方法的建立;④孔雀石绿毒性控制技术和分子结构改造的研究。

参考文献

[1] 郭德华, 叶长淋, 李波等. 高效液相色谱-质谱法测定水产品中孔雀石绿及其代谢物〔J〕. 分析测试学报, 2004, 23(21): 206~208.

[2] 《无公害食品水产品中孔雀石绿残留量的测定液相色谱法》(SC/T3021-2004).

[3] Clemmensen S., Jensen J.C., M eyer O. et al. Toxicological studies on malachite green: a triphenylmethane dye. A rch. Toxico. l 1984, (56): 43~45.

[4] Panadiker A., Fernandes C. and Rao K. V. K.. The cytotoxic properties of malachite green are associated with the increased demethylase, aryl-hydrocarbon hydroxylase and lipid peroxidation in prim aryculctures of Syrian hamsterembryocells. Cancer Let. t 1992, (67): 93~101.

[5] Culp S. J., Blankenship L. R., Kusewitt D. F. etal. Oxicity and metabolism of malachite green and leucomalachite green during short-term feeding to Fischer 344 rats and B6C3F1 mice. Chemico-Biological Interactions. 1999, 122(3): 153~170.

[6] Tuba Kücükkylync and Ynciözer. Inhibition of hum an plasmacholinesterase by m alachite green and related triarylmethanedyes: Mechanistic implications. Archives of Biochemistry andBiophysics. 2005. 7.

[7] Srivastava S, Sinha R. and Roy D.. Toxicological effects of malachite green. Aquatic Toxicology. 2004, 66(5): 319~329.

[8] M einertz J. R., Stehly G. R., G ingerichW. H. et al. Residues of〔14C〕malachite green in eggs and fry of rainbow trout (Oncorhynchusmykiss) (W albaum) after treatments of eggs. J. Fish D is. 1995, 18(3): 239~247.

[9] Mittelstaedt R. A., Mei N., Webb P. J. et al. Genotoxicity of malachite green and leucomalachite green in female Big Blue B6C3F1 m ice. M utation Research/Genetic Toxicology and Environmental Mutagenesis. 2004(561): 127~138.

[10] G lanville S. D. and Clark A. G.. Inhibition of hum an g lutathioneS-transferases by basic tripheny lm ethane dyes. Life Science. 1997, 1535 ~1544.

[11] PanandikerA., Fernandes C. and Rao K. V. K.. The cytotoxic properties of malachite g reen are associated with the increased demethylase, aryl hydrocarbon hydroxy lase and lipid peroxidation in primary cultures of Syrian ham sterembryo cells. Cancer Let. t 1992(67): 93~101.

[12] Rao K. V. K.. Inhibition of DNA synthesis in prim aryrat hepatocyte cultures by m alachite green: a new liver tum or prom oter. Toxicology Lette. r 1995(81): 107~113.

[13] Sundarrajan M., Fernandis A. Z. and Subrahmanyam G.. Overexpression of G1/S cyclins and PCNA and their relationship to ty rosine phosphorylation and dephosphorylation during tumor prom otion by m etanilyellowandmalachitegreen. Toxico logy Letters. 2000, 116(27): 119~130.

[14] Culp S. J., Blankenship L. R., Kusewitt D. F. et al. oxicity and metabolism of malachite green and leucomalachite green during short-temr feeding to Fischer 344 rats and B6C3F1 m ice. Chemico-Biological Interactions. 1999, 122(3): 153~170.

[15] Meyer F. P. and Jorgensen T. A.. Teratological and other effects of malachite green on the development of rainbow trout and rabbits. T rans. Am. F ish. Soc. 1983, 112(6): 818~824.

[16] Schnick R. A. and Meyer F. P.. Registration of thirty three fishery chemicals: status of research and estimated costs of required contract studies. Inves. t Fish. Contr. 1978(86): 19.

[17] Worle B.. 1995. Gene toxicological studies on fish eggs. Genotoxikologische Untersuchungen an Fischeiern, 92.

[18] Fernandes C., Lalitha V. S. and Rao K. V. K.. Enhancing effect of ma lachite green on the development of hepatic pre neoplastic lesions induced by N-nitrosodiethylamine in rats. Carcinogenesis 1991(12): 839~845.

[19] Rao K. V. K.. Inhibition of DNA synthesis in prim aryrat hepatocyte cultures by m alachite green: a new liver tum or promoter. Toxicology Letter 1995(81): 107~113.

［20］黄英，柯才焕，周时强．几种药物对波部东风螺早期发育的影响．厦门大学学报（自然科学版），2001，40（3）：821~826.
［21］周立红，徐长安．用微核技术研究孔雀石绿对鱼的诱变作用［J］．集美大学学报，1997，2（2）：55~57.

该文发表于《南方农业学报》【2014，45（12）：2274-2279】

孔雀石绿在养殖水和底泥中的残留消除规律

Elimination law of malachite green residue in water and sediment of aquaculture pond

柯江波，胡鲲，曹海鹏，杨先乐

（上海海洋大学/国家水生动物病原库/上海高校知识服务平台，上海 201306）

摘要：【目的】掌握养殖水体和底泥中孔雀石绿残留的消除规律，为其环境污染治理提供参考依据。【方法】通过人工模拟养殖生态系统，采用正交试验设计探讨不同光照强度（1 025、5 320、12 000 lx）、扰动强度（50、100、200 r/min）和 pH（6.0、8.0、10.0）对孔雀石绿在养殖水体和底泥中残留与消除规律的影响。【结果】随着时间的推移，养殖水体中孔雀石绿残留量呈逐渐减少的变化趋势，而底泥吸附的孔雀石绿残留量呈降低—回升—降低的变化趋势。孔雀石绿在养殖水体和底泥中的残留量均是在光照强度 12 000 lx、扰动强度 200 r/min、pH 8.0 的条件下消除最快。方差分析结果显示，光照强度是影响孔雀石绿在养殖水体中残留消除的主效环境因子，光照强度与扰动强度是影响孔雀石绿在底泥中残留量消除的主效环境因子，pH 变化对孔雀石绿在养殖水体和底泥中的残留量均无显著影响（$P>0.05$）。【结论】不同环境因子对养殖水体和底泥中孔雀石绿残留消除的影响作用表现为：光照强度>扰动强度>pH，因此清除水产养殖环境中的孔雀石绿残留应从底泥入手，通过暴晒和翻耕等方法制造高光照强度、高扰动强度条件以加速底泥中孔雀石绿的降解。
关键词：孔雀石绿；养殖水体；底泥；环境因子；残留；消除

三、抗水霉化学活性成分及药物的筛选

不同的药物对水霉的抑制能力和对鱼类的安全性各不相同（高鹏等，2007；张世奇等，2011）。Cu^{2+}不仅可有效抑制水霉菌丝的生长，还能影响水霉蛋白的合成（protein synthesis），能量的生物起源（energy biogenesis）和代谢（metabolism），从而抑制水霉的寄生和生长（Kun Hu et al.，2016），降低水霉菌对草鱼的感染率（Qi Sun et al.，2014）。此外，铜离子（Cu^{2+}）及金属络合剂（EDTA）会影响水霉菌生物膜的形成。EDTA 可促进抗菌药物的渗透（袁海兰等，2014；2015）。壳聚糖能在一定程度上抑制水霉的生长，防治水霉病（孙琪等，2014）。薯蓣素（Lei Liu et al.，2015）、嘧菌酯（Hu Xue-Gang et al.，2013）、大黄（Jia-yun YAO et al.，2016）及云木香/蛇床子/厚朴（Hu Xue-Gang et al.，2013）、五倍子/黄连/黄芩/乌梅（蔺凌云等，2015）等中草药均具有抑制水霉的活性。某些细菌，如黏质沙雷氏菌（*Serratia marcescens*）（张书俊等，2008）、芽孢杆菌（宋增福等，2012）能有效地抑制水霉。

该文发表于《水产科技情报》【2007，34（6）：247-250】

几种常用水产消毒剂对水霉的体外作用效果

高鹏，杨先乐，张书俊

（上海水产大学农业部渔业动植物病原库，上海 200090）

摘要：分别使用高锰酸钾、三氯异氰尿酸（强氯精）、苯扎溴铵、甲醛、聚维酮碘、戊二醛等 6 种水产常用消毒剂，对水霉游动孢子及菌丝进行了体外作用效果试验。结果表明：甲醛在浓度 18.8 mg/L 以上时对水霉游动孢子有明显的抑制作用；当浓度在 24 mg/L 浸泡 2 h 以上时开始对水霉菌丝有杀灭效果。其他 5 种消毒剂对水霉孢子

的抑制浓度、对水霉菌丝的杀灭浓度远高于实际生产安全浓度。

关键词：消毒剂；水霉；抑制

该文发表于《西南农业学报》【2011, 24 (4): 1568-1572】

防霉剂对水霉菌的抑菌效果研究
Antifungal Effects of Antimoulds on *Saprolegnia* sp.

张世奇，邱军强，曹海鹏，夏文伟，杨先乐

（上海海洋大学国家水生动物病原库，上海 201306）

摘要：本试验研究了丙酸钙、山梨酸钾、双乙酸钠、脱氢乙酸钠、异噻唑啉酮、尼泊金乙酯等 6 种常用的防霉剂对水霉游动孢子及菌丝的体外抑制效果。结果表明，异噻唑啉酮与尼泊金乙酯在浓度分别为 1.5 mg/L 和 6.3 mg/L 以上时可使 80% 的水霉游动孢子的活性被抑制，当浓度分别为 8 mg/L 和 32 mg/L 以上时对水霉菌丝具有杀灭作用。而其他四种防霉剂对水霉孢子的抑制浓度、对水霉菌丝的杀灭浓度远高于实际生产中可能的使用浓度。本试验还研究了异噻唑啉酮与尼泊金乙酯对水霉菌的后抑菌效应以及两者的联合抑菌效应，异噻唑啉酮在不低于 6.25 mg/L，尼泊金乙酯在不低于 25 mg/L 时能够显著减缓水霉菌丝的生长速率，12.5 mg/L 异噻唑啉酮和 25 mg/L 尼泊金乙酯还能显著改变菌丝的形态。异噻唑啉酮与尼泊金乙酯对水霉菌的抑制具有协同作用，其 FIC 指数为 0.75。

关键词：防霉剂；水霉菌；抑菌作用

该文发表于《PLOS ONE》【2016, 11 (2): · DOI: 10.1371/journal.pone.0147445】

Analysis of *Saprolegnia parasitica* transcriptome following treatment with copper sulfate

Kun Hu[a], Rong-Rong Ma[a,b], Jun-Ming Cheng[a], Xin Ye[a], Qi Sun[a], Hai-Lan Yuan[a], Nan Liang[a], Wen-Hong Fang[b], Hao-Ran Li[a] and Xian-Le Yang[a]

(a. National Pathogen Collection Center for Aquatic Animals, Shanghai Ocean University, 999 Hucheng Huan Road, Shanghai 201306, China;
b. East China Sea Fisheries Research Institute Chinese Academy of Fishery Sciences, Shanghai, 200090, China)

Abstract

Background：Massive infection caused by oomycete fungus Saprolegnia parasitica is detrimental to freshwater fish. Recently, we showed that copper sulfate demonstrated good efficacy for controlling S. parasitica infection in grass carp. In this study, we investigated the mechanism of inhibition of S. parasitica growth by copper sulfate by analyzing the transcriptome of copper sulfate-treated S. parasitica. To examine the mechanism of copper sulfate inhibiting S. parasitica, we utilized RNA-seq technology to compare differential gene expression in S. parasitica treated with or without copper sulfate.

Results：The total mapped rates of the reads with the reference genome were 90.50% in the control group and 73.50% in the experimental group. In the control group, annotated splice junctions, partial novel splice junctions and complete novel splice junctions were about 83%, 3% and 14%, respectively. In the treatment group, the corresponding values were about 75%, 6% and 19%. Following copper sulfate treatment, a total 310 genes were markedly upregulated and 556 genes were markedly downregulated in S. parasitica. Material metabolism related GO terms including cofactor binding (33 genes), 1, 3 -beta-D-glucan synthase complex (4 genes), carboxylic acid metabolic process (40 genes) were the most significantly enriched. KEGG pathway analysis also determined that the metabolism - related biological pathways were significantly enriched, including the metabolic pathways (98 genes), biosynthesis of secondary metabolites pathways (42 genes), fatty acid metabolism (13 genes), phenylalanine metabolism (7 genes), starch and sucrose metabolism pathway (12 genes). The qRT-PCR results were largely consistent with the RNA-Seq results.

Conclusion: Our results indicate that copper sulfate inhibits S. parasitica growth by affecting multiple biological functions, including protein synthesis, energy biogenesis, and metabolism.

Key words: Saprolegnia parasitica; transcriptome; copper sulfate

经硫酸铜处理后的寄生水霉转录组分析

胡鲲[a], 马荣荣[a,b], 程俊茗[a], 叶鑫[a], 孙琪[a], 袁海岚[a], 房文红[b],

梁楠[a], 房文红[b], 李浩然[a], 杨先乐[a]

a 国家水生动物病原库, 上海海洋大学, 上海沪城环路 999 号, 上海, 201306, 中国;

b 中国水产科学研究院东海水产研究所, 上海, 200090, 中国

摘要：背景：由卵菌寄生水霉引起的大量感染对淡水鱼是有害的。最近，我们发现，硫酸铜对于控制寄生水霉对于草鱼的影响显示出了较好的治疗效果。在这项研究中，我们通过分析硫酸铜影响寄生水霉的转录调查了硫酸铜对寄生水霉生长的抑制机制。为了观察硫酸铜抑制寄生水霉的机制，我们利用 RNA-seq 技术比较在寄生水霉使用硫酸铜和不使用硫酸铜两种情况的基因表达差异。

结果：总映射率和参考基因比分别为对照组 90.50% 和实验组 73.50%。在对照组中，注释的剪接联结点，部分异常的剪接联结点和全部异常的剪接联结点分别约为 83%、3% 和 14%。在治疗组中，相应的值分别为 19%、6% 和 75%。对于寄生水霉，硫酸铜处理后，共 310 个基因显著上调，共 556 个基因显著下调。物质代谢相关的 GO 术语包括辅因子结合（33 个基因），1,3-β-D-葡聚糖合酶复合物（4 个）、羧酸代谢过程（40 个基因）的显著富集。KEGG 途径的分析也确定了与代谢相关的生物制品的途径显著富集，包括代谢途径（98 个）、次生代谢产物的生物合成途径（42 个基因）、脂肪酸代谢（13 条基因）、苯丙氨酸代谢（7 个基因），淀粉和蔗糖代谢途径（12 个基因）。qRT-PCR 结果与 RNA 测序结果基本一致

结论：我们的研究结果表明，硫酸铜通过影响复杂的生物学功能来抑制寄生水霉的生长，包括蛋白质合成、能量代谢、生物合成。

关键词：寄生水霉；转录；硫酸铜

Introduction

Saprolegnia parasitica, a destructive oomycete pathogen that infects fish and fish eggs, is detrimental to freshwater hatcheries[1,2]. It often causes a severe disease, characterized by visible white or grey patches of filamentous mycelium on the freshwater fish or their eggs, which seriously affect their value, and can result in significant economic loss[3]. S. parasitica infection in aquaculture is prevalent, particularly under stress conditions such as rapid changing of temperatures or water quality[3]. The infection of S. parasitica can be controlled with effective substances. However, malachite green, an organic dye that is very efficient at controlling S. parasitica infection[4], was banned worldwide in 2002[3] because of its carcinogenic and toxicological effects on animal health[5,6]. Therefore, it is imperative to find an alternative.

Humic substances with higher molecular weights and aromaticity contain a high number of organic radicals that are efficient for reducing fungal growth. Humic substance development of internal oxidative stress could be the mechanism for the observed inhibition of S. parasitica growth[7]. Clotrimazole is another potent treatment agent of S. parasitica infection by inhibiting sterol 14 alpha-demethylase (CYP51)[8]. Meanwhile, Saprolmycin A-E are reported to have highly effective anti-S. parasitica selective activity as well[9]. Recently, our lab found copper sulfate could control S. parasitica infection in grass carp with fairly good efficacy, which suggested that copper sulfate might be used as a drug additive to control S. parasitica infection in the aquaculture industry[10]. However, the mechanism of inhibition of S. parasitica growth by copper sulfate is unclear and it is necessary to study.

In this study, we performed transcriptome analysis to investigate the mechanism of copper sulfate inhibition of S. parasitica growth. We found that copper sulfate might inhibit protein synthesis, energy biogenesis, and metabolism in S. parasitica. Our findings may describe the mechanism of action of copper sulfate for controlling S. parasitica growth in water.

Materials and Methods

The experimental protocol was established, according to the ethical guidelines of the Helsinki Declaration and was approved by the Human Ethics Committee of Shanghai Ocean University, China.

Maintenance and treatment of S. parasitica

Saprolegnia parasitica (ATCC 200013) was obtained from American Type Culture Collection (Manassas, USA) and cultured on potato dextrose agar (Sinopharm Chemical Reagent Co., Ltd., Shanghai, China) for 4 days at 20℃. A mold colony, composed of the countless S. parasitica, was treated with copper sulfate (0.5 mg/L, Shanghai Guoyao Chemical Reagent Co. Shanghai, China) for 30 min. Additional S. parasitica were treated with sterile H_2O as a control. Afterward, he S. parasitica were collected by centrifugation and stored in liquid nitrogen for further experiment.

RNA isolation, RNA sequencing (RNA-seq) library construction and sequencing

We extracted total RNA from samples using TRIzol reagent (Invitrogen, USA) according to the manufacturer's instruction. We removed DNA contaminants after treatment with RNase-free DNase I (Takara Biotechnology, Dalian, China), and confirmed RNA integrity with a minimum RNA integrated number value of 8 by an Agilent 2100 Bioanalyzer (Agilent, Santa Clara, USA). We isolated poly (A) mRNA using oligo-dT beads and then treated it with fragmentation buffer. We then transcribed the cleaved RNA fragments into first-strand complementary DNA (cDNA) using reverse transcriptase and random hexamer primers, followed by second-strand cDNA synthesis using DNA polymerase I and RNase H. The double-stranded cDNA was end-repaired using T_4 DNA polymerase, Klenow fragment, and T_4 polynucleotide kinase, followed by the addition of a single (A) base using Klenow 39-59 exo_2 polymerase, and ligation with an adapter or index adapter using T_4 quick DNA ligase. We selected adaptor-ligated fragments according to their size on agarose gel. The desired cDNA fragment size range was excised from the gel, and PCR was performed to selectively enrich and amplify the fragments. Following validation on an Agilent 2100 Bioanalyzer and ABI StepOnePlus Real-Time PCR System (ABI, California, USA), the cDNA library was sequenced on a flow cell using high-throughput 101-bp pair-end mode on an Illumina HiSeq 2500 (Illumina, San Diego, USA) unit.

Sequencing, data processing, and quality control

We filtered low-quality reads and removed 3′ adapter sequences using Trim Galore. The obtained reads were cleaned using FastQC software (http://www.bioinformatics.babraham.ac.uk/projects/fastqc/), and we evaluated the content and quality of the nucleotide bases in the sequencing data. Next, we conducted a comparative analysis with the reference genome (ASM15154v2). For each sample, sequence alignment with the reference genome sequences was carried out using Tophat[11].

Identification of differential expression genes

To investigate the expression level of each transcript of the two different treatment groups, we used Cuffnorm, a program to estimate the expression level (relative abundance) of a specific transcript expressed using FRKM (Fragments per kilobase of transcript per million fragments mapped)[12]. Generally, when FPKM>0.1, this indicates that a given transcript is expressed. The expression level of each transcript was transformed using base 2 $\log_2$ (FPKM+1). Meanwhile, we used the Cuffdiff program to calculate the fold change of a transcript and to screen all differentially expressed genes (DEGs). Two-fold changes with p-values<0.05 were considered significant.

GO functional annotation and enrichment analysis for differentially expressed genes

Initially, DEGs were annotated against the UniProt database (http://www.uniprot.org/). Next, we analyzed the functional annotation by gene ontology terms (GO; http://www.geneontology.org) with Blast 2 GO (https://www.blast2go.com/)[13].

All DEGs were mapped to GO terms in the GO database, and gene numbers were calculated for every term, followed by an ultrageometric test to find significantly enriched GO terms in DEGs compared to the transcriptome background. The formula used was as follows:

$$P = 1 - \sum_{i=0}^{m-1} \frac{\binom{M}{i}\binom{N-M}{n-i}}{\binom{N}{n}}$$

Where N is the number of all genes with GO annotation; n is the number of DEGs in N; M is the number of all genes annotated to specific GO terms; m is the number of DEGs in M. The calculated p-value was subjected to Bonferroni correction. A corrected p-value

<0. 05 was defined as a threshold, and then GO terms were considered significantly enriched in the DEGs.

DEGs pathway analysis

DEGs were annotated to the KEGG pathways database using the online KEGG Automatic Annotation Server (KAAS, available online: http: //www. genome. jp/keg/kaas/). Enriched DEG pathways were identified using the same formula as that in GO analysis. In KEGG pathway analysis, N is the number of all genes with KEGG annotation, n is the number of DEGs in N, M is the number of all genes annotated to specific pathways, and m is the number of DEGs in M.

DEG verification using qRT-PCR

The expression levels of DEGs identified in the RNA-seq analysis were verified with quantitative RT-PCR (qRT-PCR). Primers were designed using the Primer 5. 0 software (Premier company, Canada), and SpTub-b (S. parasitica Tub-b) was used as the reference gene[14,15].

The reactions were performed in a 25 μL volume, composed of 2 μL cDNA, 0. 5 μL of both the forward and reverse primer (10 μM), 12. 5 μL SYBR Premix Ex Taq (2×) and 9. 5 μL RNase-free H_2O. The thermal cycling program was 95℃ for 30 s, followed by 40 cycles of 95℃ for 5 s, 60℃ for 30s, and 72℃ for 30 s. Melting curve analysis was performed at the end of the qRT-PCR cycles to confirm the PCR specificity. Three replications were carried out and analysis of relative gene expression data using $2^{-\triangle\triangle CT}$ Method[16].

Results

Illumina sequencing and quality assessment

We performed RNA-seq using the Illumina sequencing platform to examine the effect of copper sulfate on the S. parasitica transcriptome. After filtering and quality checks of the raw reads (47, 614, 574) and (51, 163, 206) for the control and treatment groups, respectively), there were approximately 44 million (44, 676, 046) and 45 million (45, 325, 202) trimmed reads with trim rates of 95. 0% and 88. 6% for control and the treatment samples, respectively. Meanwhile, the respective average length of reads was 94. 5 and 93. 8 bp (Table 1), indicating successful sequencing of the S. parasitica transcriptome. And trimmed reads were used for the subsequent analysis.

Table 1　Summary of reads in S. parasitica transcriptome sequencing

Sample	Raw reads	Trimmed reads	Average length	Trim rate
Control	47, 614, 574	44, 676, 046	93. 8 bp	95. 0%
Copper sulfate	51, 163, 206	45, 325, 202	94. 5 bp	88. 6%

Comparative analysis with reference genome

The trimmed reads of S. parasitica transcriptome were compared with the reference genome sequence. The total mapped rates of the reads with the reference genome were 90. 50% in the control group and 73. 50% in the experimental group. Uniquely mapped reads were about 14 million (14, 952, 142) for the control and 29 million (29, 202, 690) for the experimental group, and accounted for 33. 47% and 64. 42% of total reads, respectively. Multiple mapped reads were about 25 million (25, 476, 517) for the control group and 4 million (4, 116, 175) of the experimental group, and make up 57. 03% and 9. 08% of total reads, respectively. Total unmapped reads accounted for 9. 50% in the control group and 26. 50% in the experimental group (Table 2).

Additionally, splice junctions were discovered using TopHat, which can identify splice variants of genes. In the control group, known splice junctions accounted for 83% of total splice junctions annotated in the reference genome sequence. Partial novel splice junctions were about 3% of total splice junctions, and complete novel splice junctions were about 14% of total splice junctions (Fig. 1 A). In the experimental group, known splice junctions, partial novel splice junctions and complete novel splice junctions were about 75%, 6% and 19%, respectively (Fig. 1 B). These results indicate that the transcriptome of S. parasitica treated by copper sulfate contained more partial novel and complete novel splice junctions than the S. parasitica of control group.

Table 2　Statistical results of trimmed reads mapping with reference genome

Map to genome	Control		Copper sulfate	
	Reads numbers	percentage	Reads numbers	percentage
Total reads	44, 676, 046	100. 00%	45, 325, 202	100. 00%
Total mapped	40, 428, 659	90. 50%	33, 318, 865	73. 50%

续表

Map to genome	Control		Copper sulfate	
	Reads numbers	percentage	Reads numbers	percentage
Uniquely mapped	14, 952, 142	33. 47%	29, 202, 690	64. 42%
Multiple mapped	25, 476, 517	57. 03%	4, 116, 175	9. 08%
Total unmapped	4, 247, 387	9. 50%	12, 006, 337	26. 50%

Analysis of DEGs

To identify DEGs in S. parasitica following copper sulfate treatment, we used Cuffdiff program to generate S. parasitica gene expression profiles (Fig. 2 and Fig. 3). The program identified 310 genes were markedly upregulated and 556 genes were markedly down regulated in S. parasitica following copper sulfate treatment, which indicates that copper sulfate affects S. parasitica gene expression.

GO annotation of DEGs

To investigate the biological functions in which DEGs are involved in S. parasitica, we classified the DEGs according to GO classification following copper sulfate treatment in S. parasitica. This analysis identifies the main biological functions DEGs exercise. The GO functional enrichment analysis also involved cluster analysis of expression patterns. Thus, the expression patterns of DEGs annotated with a given GO term were easily obtained. All annotated genes were classified into three GO domains: biological processes, cellular component, and molecular function. Dissimilar expression profiles were obtained from the DEGs in the treated and control samples, revealing the obvious effect of copper sulfate on S. parasitica metabolism and physiology. The expression profiles of the three GO domains were as follows (each domain showed the top 10 terms):

Molecular function: 1, 3-beta-D-glucan synthase activity (4 genes), transferase activity, transferring acyl groups other than amino-acyl groups (16 genes), coenzyme binding (22 genes), lipase activity (9 genes), ammonia-lyase activity (3 genes), amylase activity (3 genes), hydrolase activity, hydrolyzing O-glycosyl compounds (20 genes), hydrolase activity, acting on glycosyl bonds (21 genes), transferase activity, transferring acyl groups (25 genes), cofactor binding (33 genes) (Fig. 4).

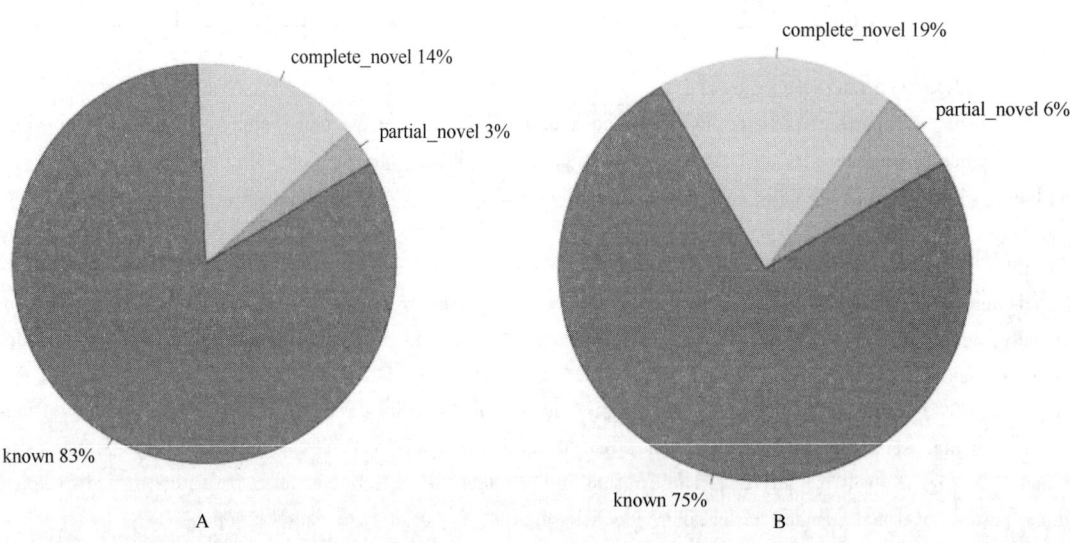

Fig. 1 Types of splice junctions in the S. parasitica transcriptome in the control (Fig 1A) and experimental group (Fig 1B) as compared to the reference genome. Pie charts break down the known, partial novel and novel splice junctions identified in this study. Known splice junctions: both ends of splice junction belong to the annotated splice junction in the known genetic model; partial novel splice junctions: one end of splice junction belongs to the annotated splice junction in the known genetic model and the other end belongs to the new splice junction; complete novel splice junctions: both ends of splice junction belong to the new splice junction.

Cellular component: organelle envelope lumen (1 gene), membrane part (65 genes), extracellular region (9 genes), endo-

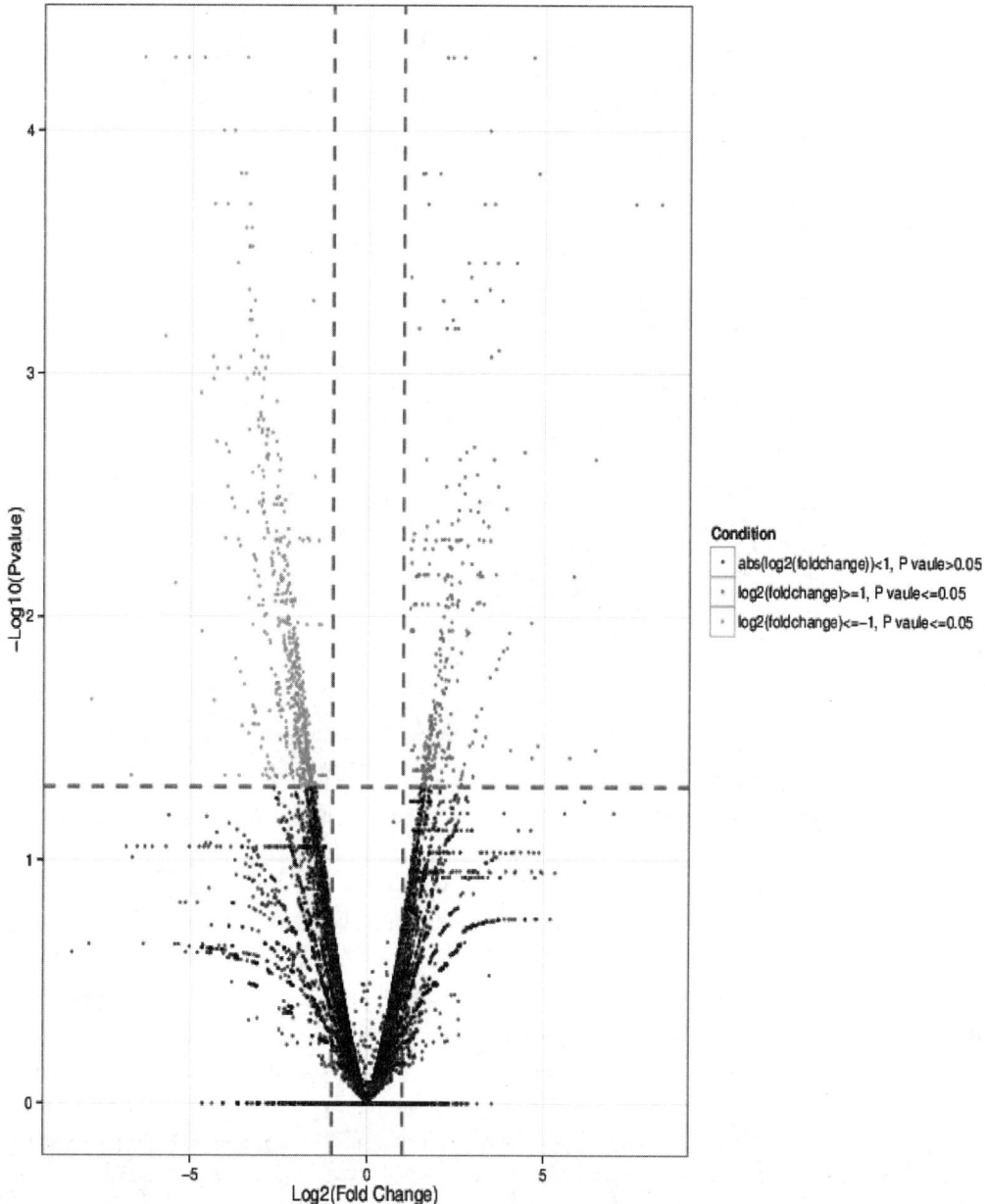

Fig. 2　Effect of copper sulfate treatment on the gene expression profile in S. parasitica gene expression profile following copper sulfate treatment. Volcanic plot of the degree of differences in the expression profile of S. parazitica after treatment with copper sulfate. X－axis, log2 (fold change); Y－axis, －log10 (Pvalue) . Red, the significantly upregulated genes, green, the significantly downregulated genes. Each dot represents one gene.

cytic vesicle (2 genes), intrinsic to membrane (62 genes), integral to membrane (62 genes), extracellular region part (4 genes), myosin complex (9 genes), actin cytoskeleton (11 genes), 1, 3－beta－D－glucan synthase complex (4 genes) (Fig. 4). Biological processes: single－organism catabolic process (9 genes), small molecule catabolic process (9 genes), small molecule metabolic process (90 genes), organic acid catabolic process (9 genes), carboxylic acid catabolic process (9 genes), cellular a-mino acid metabolic process (32 genes), single－organism metabolic process (141 genes), organic acid metabolic process (40 genes), oxoacid metabolic process (40 genes), carboxylic acid metabolic process (40 genes) (Fig. 4).

Cofactor binding (33 genes), 1, 3－beta－D－glucan synthase complex (4 genes), carboxylic acid metabolic process (40 genes) were the most significant GO terms. Globally, the 1515 GO terms were annotated, and the 262 GO terms underwent dramatic expression changes.

545

genes. Each dot represents on gene.

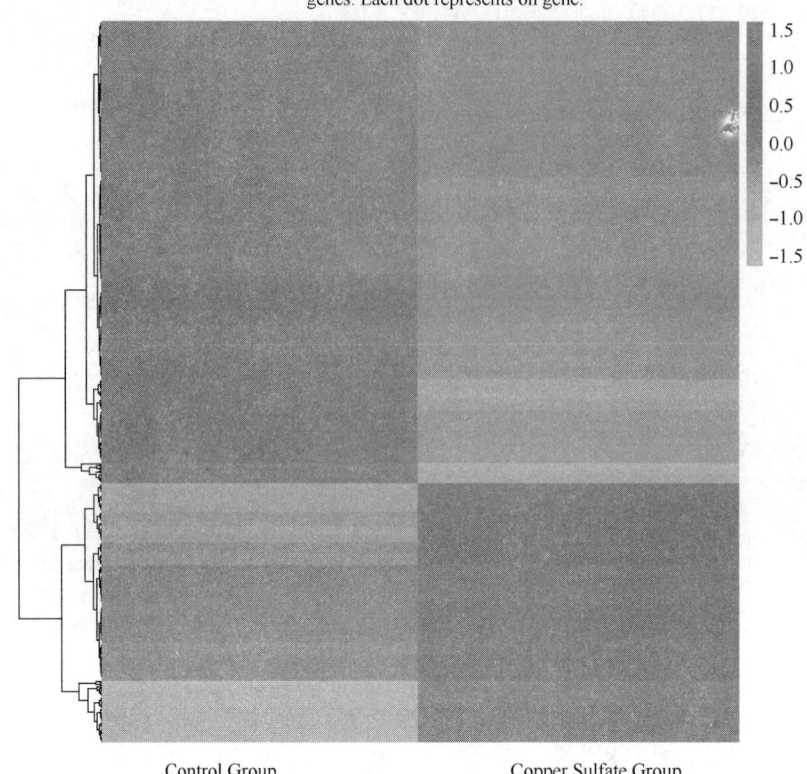

Control Group   Copper Sulfate Group

Fig. 3　Effect of copper sulfate treatment on gene expression profile using pattern clustering. Red, upregulated genes; green, downregulated genes. Each line represents one gene.

Following copper sulfate treatment, genes related to molecular function, including SPRG_ 10490 (ornithine-oxo-acid transaminase), SPRG_ 03427 (oxoglutarate dehydrogenase), SPRG_ 06771 (tyrosine aminotransferase), SPRG_ 15406 (trehalase), and biological process related genes such as SPRG_ 11730 (histidine ammonia-lyase), SPRG_ 05010 (tryptophanyl-tRNA synthetase), SPRG_ 01675 (homogentisate 1%2C2-dioxygenase), SPRG_ 11441 (uridine phosphorylase), SPRG_ 11440 (uridine phosphorylase) were repressed, while molecular function related genes like SPRG_ 07378 (threonine ammonia-lyase) and biological process related genes like SPRG_ 01336 (maleylacetoacetate isomerase), SPRG_ 00691 (imidazoleglycerol-phosphate dehydratase) were upregulated. Most of these genes are responsible for protein synthesis, suggesting that copper sulfate mostly affects protein synthesis in S. parasitica. The expression of genes related to cellular flow, signal transduction, and cellular transport was also altered in S. parasitica after treatment with copper sulfate (including downregulated genes such as SPRG_ 05020, SPRG_ 13728, SPRG_ 03925, SPRG_ 07166, SPRG_ 17366, SPRG_ 20259 and SPRG_ 00668, and upregulated genes, such as SPRG_ 19069, SPRG_ 06201 and SPRG_ 20593), which are closely linked with substance metabolism. Additionally, copper sulfate treatment specifically activated SPRG_ 08152 (hypothetical protein), SPRG_ 09968 (hypothetical protein) and SPRG_ 06695 (hypothetical protein), which only are only expressed in the copper sulfate treated samples. The functions of hypothetical proteins were not clear, but were closely related to the biological functions of organism.
Overall, the above analyses indicate that copper sulfate inhibits S. parasitica growth by affecting multiple biological functions, including protein synthesis and metabolism.

KEGG pathway analysis of differentially expressed genes

To further explore the DEG biological functions, we mapped DEGs to the KEGG database, and then enriched them to important pathways based on the whole transcriptome background. We mapped genes to 203 pathways. Many genes were found in multiple pathways; however, many genes were also restricted to a single pathway. There were 25 significantly enriched pathways, the most significantly enriched of which were metabolic pathways (Fig. 5). The metabolism-related biological pathways included metabolic pathways, biosynthesis of secondary metabolites and biosynthesis of amino acids. A total of 98 genes were mapped to the metabolic pathways, 42 genes were mapped to the biosynthesis of secondary metabolites pathways, and 18 genes were mapped to the biosynthesis

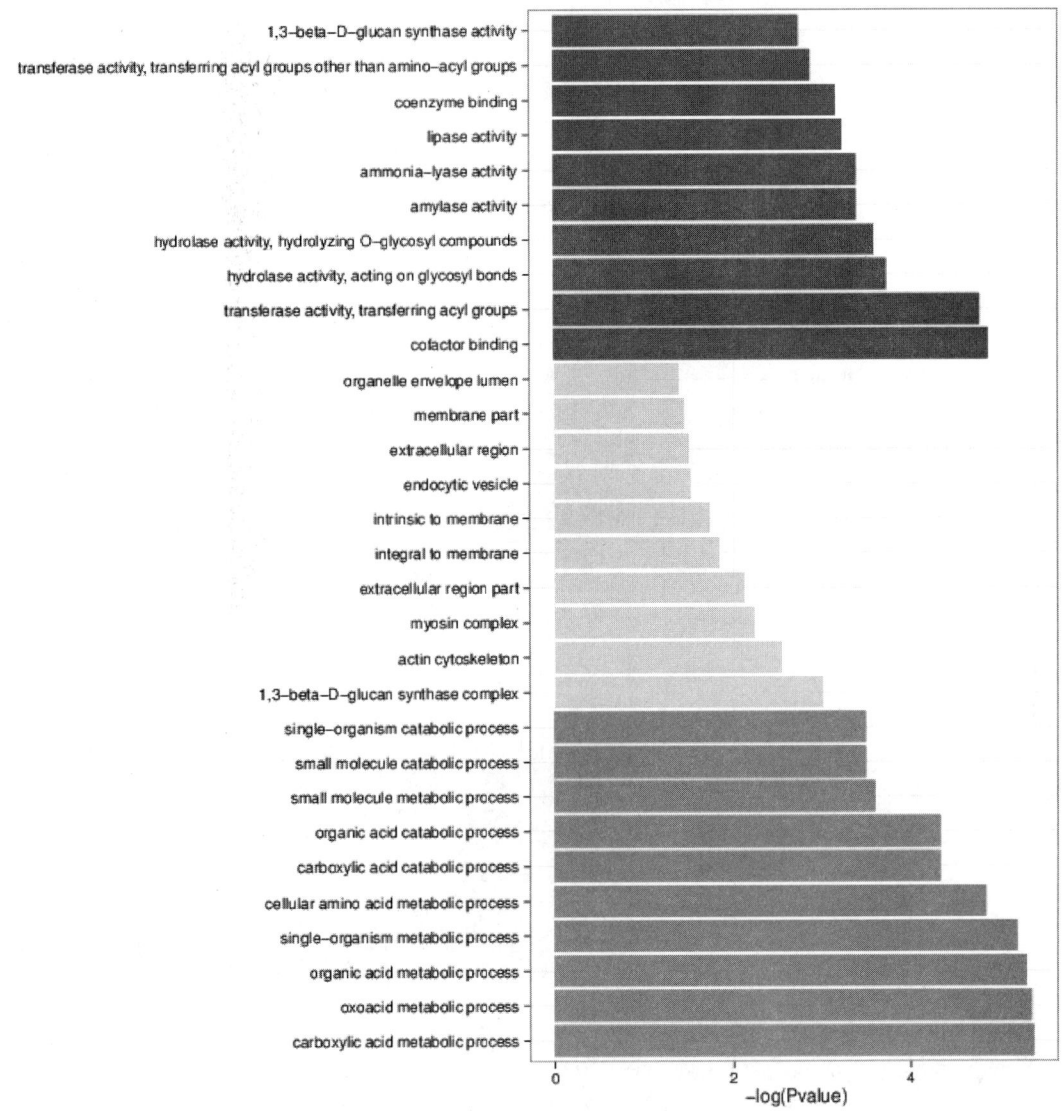

Fig. 4　Histogram representation of enriched category of GO annotation of DEGs in S. parasitica following copper sulfate treatment. GO categories (y-axis) were grouped into three main ontologies: biological process, cellular component, and molecular function. The x-axis indicates the statistical significance of the enrichment. All annotated genes were classified into three GO domains: biological processes, cellular component, and molecular function. Red histogram represents the biological processes, the yellow histogram represents cellular component, and the blue histogram represents molecular function. Dissimilar expression profiles were obtained from DEGs in the treated and control samples, revealing the obvious effect of copper sulfate on S. parasitica metabolism and physiology.

of amino acids. In addition, lipid metabolism-related biological pathways including fatty acid metabolism (13 genes), arachidonic acid metabolism (9 genes), and glycerolipid metabolism (6 genes), were also significantly enriched. Enriched amino acid metabolism-related biological pathways included phenylalanine metabolism (7 genes), tyrosine metabolism (7 genes), histidine metabolism (7 genes), valine, leucine and isoleucine biosynthesis (5 genes), and phenylalanine, tyrosine and tryptophan biosynthesis (5 genes).

　　The other significantly enriched pathways included carbohydrate metabolism-related starch and sucrose metabolic pathway (12 genes), metabolism of cofactors and vitamins related ubiquinone and other terpenoid-quinone biosynthesis pathway (4 genes), biosynthesis of other secondary metabolites related phenylpropanoid biosynthesis (4 genes), novobiocin biosynthesis (2 genes), tropane, piperidine and pyridine alkaloid biosynthesis (2 genes), immune diseases related primary immunodeficiency (4 genes), endocrine system related PPAR signaling pathway (8 genes), cyanoamino acid metabolism (3 genes), xenobiotic biodegradation

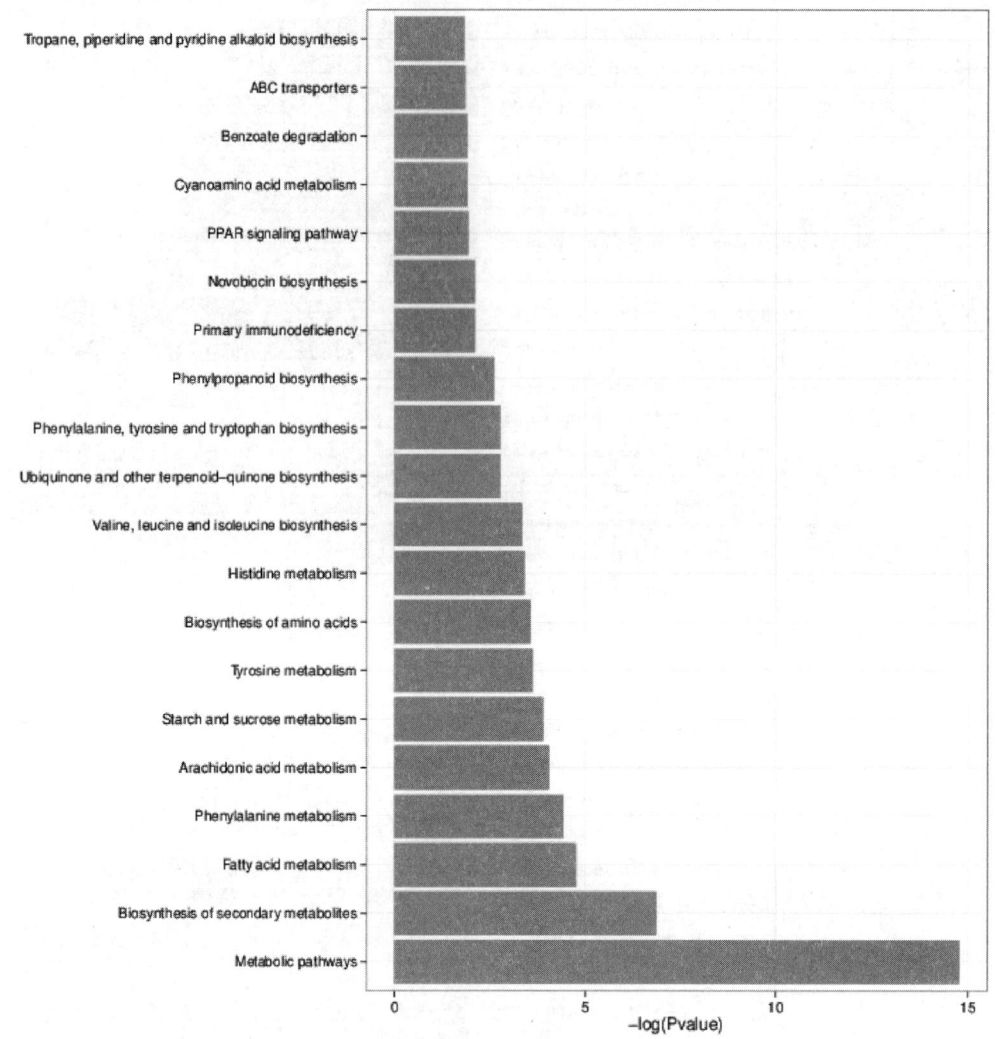

Fig. 5　Histogram representation of top twenty most enriched KEGG pathways of DEGs in S. parasitica following copper sulfate treatment.

Y-axis, KEGG pathway categories; x-axis, statistical significance of enrichment.

and metabolism related benzoate degradation (3 genes), styrene degradation (2 genes), nitrotoluene degradation (2 genes), membrane transport related ABC transporters (11 genes), digestive system related protein digestion and absorption (5 genes), cancers overview related chemical carcinogenesis (3 genes).

In the metabolism-related biological pathways, the expression of certain genes (such as SPRG_ 05378, SPRG_ 01336, SPRG_ 00691, SPRG_ 19437, SPRG_ 17560, SPRG_ 13611, SPRG_ 08603, SPRG_ 09764, and SPRG_ 01617) were upregulated, while SPRG_ 11730, SPRG_ 04491, SPRG_ 10490, SPRG_ 12259, SPRG_ 04186, SPRG_ 18377, SPRG_ 18063, SPRG_ 06771, SPRG_ 16261, and SPRG_ 02541 genes were downregulated after copper sulfate treatment. These findings were consistent with the GO enrichment analysis, which indicated that copper sulfate mostly prevented the growth of S. parasitica by affecting protein synthesis and metabolism.

In the lipid metabolism-related biological pathways, among the 13 genes mapped to fatty acid metabolism, 12 genes were downregulated (including SPRG_ 15243, SPRG_ 02257, SPRG_ 06557, and SPRG_ 01190), but SPRG_ 09764. In the arachidonic acid metabolism, three genes were upregulated in S. parasitica, (SPRG_ 12960, SPRG_ 16543 and SPRG_ 13907), six genes were downregleted, (SPRG_ 20516, SPRG_ 10696, SPRG_ 09110, SPRG_ 11344, SPRG_ 18867 and SPRG_ 11345) after copper sulfate treatment. These results showed that copper sulfate affected S. parasitica by influencing the energy biogenesis deregulation.

Overall, the results of the DEGs pathway analysis support the viewpoint that copper sulfate inhibits S. parasitica growth by affecting multiple biological functions, such as energy biogenesis, protein synthesis and metabolism.

Differential expression verification of differentially expressed genes

Nine genes with a clearly defined function were randomly selected from the differentially expressed genes identified by RNA-Seq to verify their expression using quantitative PCR. The primer sequences for all of the genes examined are listed in Table 3. The results verified that eight of the genes examined were consistent with the results of RNA-Seq (Fig. 6). However, one gene (SPRG_ 01336), was not consistent with the RNA-Seq data. Overall, these results suggest that the RNA-Seq results are generally reliable; however, further studies are still needed to be carried out to verify the results from this study.

Table 3　Oligonucleotide primers of qRT-PCR for DEGs validation

Gene name	Predict function	GO category	Pathway name	Primer name	Nucleotide sequence (5'-3')	Expected product
SpTub-b	–	–	–	SpTub-b-f	AGACGGGTGCTGGT AACAAC	136bp
				SpTub-b-r	AGCGAGTGCGTAATC TGGAAA	
SPRG_ 00691	Imidazoleglycerol-phosphate dehydratase	Carboxylic acid metabolic process (Biological process)	Metabolic pathways	Sp-1-f	GCGACGCTAACAAC GACTGG	123bp
				Sp-1-r	TTGTCGCCGTGCCC TGGTA	
SPRG_ 01336	Maleylacetoacetate isomerase	Carboxylic acid metabolic process (Biological process)	Metabolic pathways	Sp-2-f	TGTAGAAGCGGG CGTTGA	178bp
				Sp-2-r	CCGATGAGGCGG AAGAAG	
SPRG_ 03077	TKL protein kinase	–	Primary immunodeficiency	Sp-3-f	GGGCAGCATCTTC TTCACAG	195bp
				Sp-3-r	TCGCTTCAGAGTC AAGGGTC	
SPRG_ 07378	Threonine ammonia-lyase	Carboxylic acid metabolic process (Biological process)	Metabolic pathways	Sp-4-f	CAACTGGGTCAAGCA CTTTCG	126bp
				Sp-4-r	GCAAAGAGCCCGACT TGGT	
SPRG_ 11440	Uridine phosphorylase	Single-organism metabolic process (Biological process)	Metabolic pathways	Sp-5-f	TACAGCCTCTCGCA CATTGG	123bp
				Sp-5-r	CTTGGTGACTTCGTG GAGGAG	
SPRG_ 11441	Uridine phosphorylase	Single-organism metabolic process (Biological process)	Metabolic pathways	Sp-6-f	CCACCTCGGTCTC TCGTACT	142bp
				Sp-6-r	CTTGATCTCCAGC GTCTCGG	
SPRG_ 10490	Ornithine-oxo-acid transaminase	Cofactor binding (Molecular function)	Metabolic pathways	Sp-7-f	GCCCGTGAAGACAA GTATGGT	174bp
				Sp-7-r	CGTGAGTGCCGTTCA AGATGC	

续表

Gene name	Predict function	GO category	Pathway name	Primer name	Nuclectide sequence (5' −3')	Expected product
SPRG_ 11730	Histidine ammonia−lyase	Carboxylic acid metabolic process (Biological process)	Metabolic pathways	Sp−8−f	CAAGCCGTCCGAA CTCTTCA	179bp
				Sp−8−r	CCGAGCCCACAAA CACCAT	
SPRG_ 15406	Trehalase	Hydrolase activity, acting on glycosyl bonds (Molecular function)	Metabolic pathways	Sp−9−f	TGGGCACAAAGC CATAGTCA	98bp
				Sp−9−r	GATCATCCACGG TCTTCTCAA	

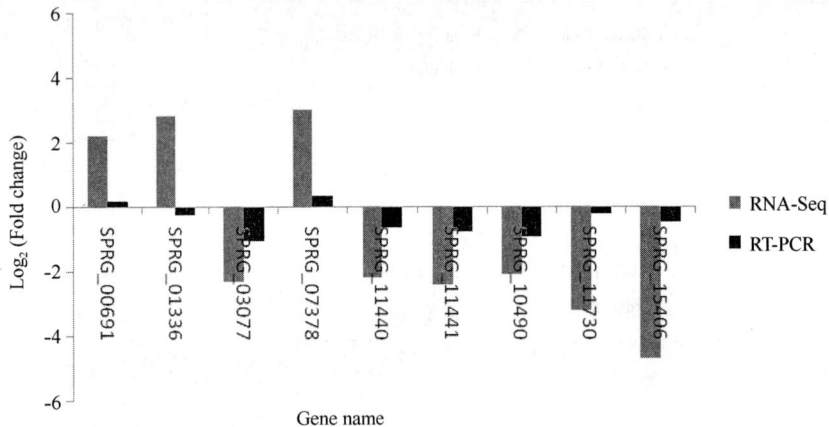

Fig. 6 Comparison of nine genes expression levels between RNA−Seq and RT−PCR. Negative values indicate that the gene expression of S. parasitica was downregulated following copper sulfate treatment; positive values indicate that the gene expression was upregulated.

Discussion

Transcriptome sequencing is a powerful technique for studying the mechanism of changes of biological characteristics of an organism, and has been used successfully in some species[17-19]. In our study, we examined the mechanism by which copper sulfate inhibits the growth of S. parasitica by examining changes in the transcriptome after treatment using the Illumina sequencing platform. The total mapped rates of reads were 90.50% in the S. parasitica transcriptome of control group as compared to the reference genome, which met the required quality of the sequencing data needed for follow−up studies. The total unmapped reads was 9.50% of the total reads in the controls; however, the unmapped reads was 26.50% in the experimental group. It was remarkable that unmapped reads rate of the transcriptome of S. parasitica following treatment with copper sulfate was higher. This discrepancy might be due to the fact that the existing annotation is incomplete, or because copper sulfate caused gene variation.

The transcriptome of S. parasitica treated by copper sulfate contained more partial novel and complete novel splice junctions than that of control group. These findings indicate that copper sulfate may change the alternative splicing of certain genes that affect the metabolism and activity of S. parasitica. Therefore, investigating splice junctions may help understand the mechanism by which copper sulfate inhibits S. parasitica.

In addition to differences in splicing, copper sulfate may also affect S. parasitica by directly altering the expression of genes involved in key cellular processes. In order to investigate this, we classified DGEs into 1515 GO terms, consisting of three domains: biological process, cellular component, and molecular function. Of these, 262 GO terms were found to have dramatic expression changes. We mapped the DGEs to 203 pathways, with 25 of these pathways significantly enriched. The most significantly enriched pathways in this analysis were the metabolism−related biological pathways, including metabolic pathways, biosynthesis of secondary metabolites and biosynthesis of amino acids, which are responsible for the main biological functions of S. parasitica. Our study supplements the previous studies by Jiang et al. , who investigated virulence genes in S. parasitica by sequencing its genome[20], and by Torto-Alalibo et al. , who investigated expressed sequence tags and disclosed S. parasitica putative virulence factors[21].

In GO annotation enrichment, most of DGEs were related to protein synthesis and metabolism. However, in KEGG pathway analysis, the results showed not only substance metabolism was affected, but the production of energy was also affected. Together, these results indicate that copper sulfate inhibit S. parasitica growth by affecting multiple biological functions, which differ from the mechanisms of other anti-S. parasitica reagents. Such reagents include humic substances with higher molecular weights and aromaticity[7], clotrimazole[8], and saprolmycin A-E[9]. These data indicate that copper sulfate represents a novel anti-S. parasitica treatment.

We found that copper sulfate treatment did not alter the expression of the host-targeting protein 1 (SpHtp1) gene, which is responsible for S. parasitica translocation into host cells, suggesting that copper sulfate may not interfere with S. parasitica-host interactions[22].

Most of the genes in S. parasitica were hypothetical proteins, and the functions were not clear in the existing data. In order to facilitate further research, nine of the DGEs with clear functions were chosen randomly from the RNA-Seq data for qPCR to verify the results of the RNA-Seq. The qPCR results were essentially consistent with the results of the analysis of the transcriptome. One of the genes analyzed had a discrepant result, however, but we still believe the quality of the transcriptome of S. parasitica meet the requirements for studies on functional genes.

Taken together, copper sulfate represents a novel anti-S. parasitica agent that likely functions of inhibiting S. parasitica by affecting energy biogenesis, protein synthesis and metabolism. Our findings provide the basis for further investigation of the potential application of copper sulfate for controlling S. parasitica growth in aquaculture in the future.

Conflict of Interest

The authors declare that they have no conflicts of interest concerning this article.

References

[1]　Meyer FP. Aquaculture disease and health management. J Anim Sci. 1991; 69: 4201-4208.

[2]　Rach JJ, S. Redman, D. Bast, and M. P. Gaikowski. Efficacy of hydrogen peroxide versus formalin treatments to control mortality associated with saprolegniasis on lake trout eggs. North American Journal of Aquaculture. 2005; 67: 148-154.

[3]　van West P. Saprolegnia parasitica, an oomycete pathogen with a fishy appetite: new challenges for an old problem. Mycologist. 2006; 20: 99-104.

[4]　Alderman DJ. Malachite green: a review. Journal of Fish Diseases. 1985; 8:

[5]　Srivastava S, Sinha R, Roy D. Toxicological effects of malachite green. Aquat Toxicol. 2004; 66: 319-329.

[6]　Stammati A, Nebbia C, Angelis ID, Albo AG, Carletti M, Rebecchi C, et al. Effects of malachite green (MG) and its major metabolite, leucomalachite green (LMG), in two human cell lines. Toxicol In Vitro. 2005; 19: 853-858.

[7]　Meinelt T, Paul A, Phan TM, Zwirnmann E, Kruger A, Wienke A, et al. Reduction in vegetative growth of the water mold Saprolegnia parasitica (Coker) by humic substance of different qualities. Aquat Toxicol. 2007; 83: 93-103.

[8]　Warrilow AG, Hull CM, Rolley NJ, Parker JE, Nes WD, Smith SN, et al. Clotrimazole as a potent agent for treating the oomycete fish pathogen Saprolegnia parasitica through inhibition of sterol 14alpha-demethylase (CYP51). Appl Environ Microbiol. 2014; 80: 6154-6166.

[9]　Nakagawa K, Hara C, Tokuyama S, Takada K, Imamura N. Saprolmycins A-E, new angucycline antibiotics active against Saprolegnia parasitica. J Antibiot (Tokyo). 2012; 65: 599-607.

[10]　Sun Q, K. Hu, X. L. Yang. The efficacy of copper sulfate in controlling infection of Saprolegnia parasitica. Journal of the World Aquaculture Society. 2014; 45: 220-225.

[11]　Trapnell C, Pachter L, Salzberg SL. TopHat: discovering splice junctions with RNA-Seq. Bioinformatics. 2009; 25: 1105-1111.

[12]　Trapnell C, Roberts A, Goff L, Pertea G, Kim D, Kelley DR, et al. Differential gene and transcript expression analysis of RNA-seq experiments with TopHat and Cufflinks. Nat Protoc. 2012; 7: 562-578.

[13]　Conesa A, Gotz S. Blast2GO: A comprehensive suite for functional analysis in plant genomics. Int J Plant Genomics. 2008; 2008: 619832.

[14]　Yan HZ, Liou RF. Selection of internal control genes for real-time quantitative RT-PCR assays in the oomycete plant pathogen Phytophthora parasitica. Fungal Genet Biol. 2006; 43: 430-438.

[15]　Van West P, de Bruijn I, Minor KL, Phillips AJ, Robertson EJ, Wawra S, et al. The putative RxLR effector protein SpHtp1 from the fish pathogenic oomycete Saprolegnia parasitica is translocated into fish cells. FEMS Microbiol Lett. 2010; 310: 127-137.

[16]　Livak KJ, Schmittgen TD. Analysis of relative gene expression data using real-time quantitative PCR and the 2 (-Delta Delta C (T)) Method. Methods. 2001; 25: 402-408.

[17]　Price DP, Nagarajan V, Churbanov A, Houde P, Milligan B, Drake LL, et al. The fat body transcriptomes of the yellow fever mosquito Aedes aegypti, pre-and post-blood meal. PLoS One. 2011; 6: e22573.

[18] Peng Y, Gao X, Li R, Cao G. Transcriptome sequencing and De Novo analysis of Youngia japonica using the illumina platform. PLoS One. 2014; 9: e90636.

[19] Li C, Weng S, Chen Y, Yu X, Lu L, Zhang H, et al. Analysis of Litopenaeus vannamei transcriptome using the next-generation DNA sequencing technique. PLoS One. 2012; 7: e47442.

[20] Jiang RH, de Bruijn I, Haas BJ, Belmonte R, Lobach L, Christie J, et al. Distinctive expansion of potential virulence genes in the genome of the oomycete fish pathogen Saprolegnia parasitica. PLoS Genet. 2013; 9: e1003272.

[21] Torto-Alalibo T, Tian M, Gajendran K, Waugh ME, van West P, Kamoun S. Expressed sequence tags from the oomycete fish pathogen Saprolegnia parasitica reveal putative virulence factors. BMC Microbiol. 2005; 5: 46.

[22] Wawra S, Bain J, Durward E, de Bruijn I, Minor KL, Matena A, et al. Host-targeting protein 1 (SpHtp1) from the oomycete Saprolegnia parasitica translocates specifically into fish cells in a tyrosine-O-sulphate-dependent manner. Proc Natl Acad Sci U S A. 2012; 109: 2096-2101.

该文发表于《Journal of the World Aquaculture Society》【2014, 45 (2): 220-225】

The Efficacy of Copper Sulfate in Controlling Infection of *Saprolegnia parasitica*

Qi Sun, Kun Hu and Xian-Le Yang

State Collection Center of Aquatic Pathogen, Shanghai Ocean University, 999 HuchengHuan Road, Shanghai 201306, China

Abstract: *Saprolegnia parasiticais* a severe fish pathogen that causes important economic losses worldwide. Copper is an important additive in the aquaculture industry for control of algal growth, ectoparasites, and fungal disease. However, at present no data is available on the specific interaction of copper sulfate with oomycete. In our study, the efficacy of copper sulfate on the mycelium and zoospore of S. parasitica was assessed in vitro, and S. parasitica infection experiment was conducted to assess its performance in vivo. The results indicated that copper sulfate at $\geqslant 0.5$ mg/L inhibited the growth of mycelium, no primary zoospores were released at $\geqslant 1.0$ mg/L. Additionally, 0.5 mg/L copper sulfate could reduce the infection rate of S. parasiticain the grass carp, Ctenopharyngodonidellus. This study demonstrates the good efficacy of copper sulfate on the control of S. parasitica infection in grass carps, and suggests that copper sulfate could be used as a drug additive to control the S. parasitica infection in the aquaculture industry.

硫酸铜控制寄生水霉感染的效果

孙琪，胡鲲，杨先乐

(国家水生动物病原库，上海海洋大学，沪城环路 999 号，中国上海，201306)

摘要：寄生类水霉菌是一种重要的造成全球性经济损失的病原。铜是一种在水产养殖产业中控制藻类、皮外寄生虫和真菌性疾病的制剂。但是，目前还没有可获得的数据表明关于硫酸铜和卵菌之间明确的相互作用关系。我们的研究在试管内评估了硫酸铜对寄生菌菌丝和游动孢子的功效，并且实施了寄生菌体外感染实验以评估它的性能。结果表明硫酸铜浓度在 $\geqslant 0.5$ mg/L 时抑制菌丝的生长，在浓度 $\geqslant 1.0$ mg/L 时没有初级孢子被释放出来。此外，0.5 mg/L 的硫酸铜能够降低寄生菌对草鱼的感染率。本文论证了硫酸铜在控制寄生菌对草鱼感染的良好效果，并且建议在水产养殖业上硫酸铜可以作为药物添加剂来控制寄生水霉在水产养殖中的感染。

Oomycete diseases are considered to be the most important fungi which afflict freshwater fish (Hatai et al., 1990; Bruno and Wood, 1999). Saprolegniaparasitica are the most important oomycete pathogens in fish (Van West, 2006). It is known to penetrate into epidermal tissues, usually colonizing the tail or head region and then proliferating to cover the entire body surface (Willoughby, 1994). The frequent occurrence of Saprolegnia sp. in freshwater sources may result in problems for freshwater hatcheries (Meyer 1991; Rach et al., 2005). Traditionally, S. parasitica infections were effectively controlled with malachite green (MG) (János and Farkas, 1978; Srivastava and Srivastava, 1978; Alderman, 1985). This compound, however, was banned worldwide in 2002

（Van West, 2006）due to its undesirable effects on animal health（Brock and Bullis, 2001; Srivastava et al. , 2004; Stammati et al. , 2005）. With the banning of MG, research has been directed toward the discovery, or the improvement in the application of suitable alternatives.

Copper sulfate has been documented to be effective for the control of parasitic infections（Ling et al. , 1993; Schlenk et al. , 1998; Noga, 2000; Burke and Miller 2008）, algal growth（Kiyoshi and Claude, 1993; James et al. 1998; Song et al. 2011）, and Saprolegniasis（Lio-Po et al. , 1982; Marking et al. , 1994; Lester, 2000; Straus et al. , 2012）, but in general the data concerning the comparison of antifungal activities against Saprolegniasis between copper sulfate and some other chemicals did not reveal how. The objective of this study was to make a specific investigation of the efficacy of copper sulfate on the mycelium and zoospore of S. parasitica, and to propose a method using copper sulfate to prevent S. parasitica infections in the aquaculture industry.

Materials and Methods

Saprolegnia parasitica Cultures

Saprolegnia parasitica ATCC 200013 was obtained from American Type Culture Collection（ATCC）. It was cultured on potato dextrose agar（PDA, Sinopharm Chemical Reagent Co. , Ltd, Shanghai, China）at 20℃. The cultures were at least 96-h old when used. Re-incubation of S. parasitica was performed every week. A suspension containing spores of S. parasitica was prepared by detaching hyphae from the surface of a culture grown on PDA and adding it to 100mL distilled water, which would be incubated at 20C for over 36h. The number of spores was quantified using a hemocytometer.

Fish and Rearing Conditions

Grass carps, Ctenopharyngodonidellus, （42. 7±4. 6 g, 15. 5±1. 7 cm）were obtained from Nantong fisheries farm（Yancheng, Jiangsu, China）. Before the experiment, all the healthy grass carps were acclimated in rearing tanks at 25℃ for 10 d. During the acclimation, they were fed to apparent satiation twice daily with a proper diet according to National Research Council（NRC）.

Effect of Copper Sulfate on S. Parasitica Mycelium Growth

The following dilutions of copper sulfate were prepared in water: 5. 0, 2. 0, 1. 0, 0. 5 and 0. 2 mg/L solutions and separately transfer 5 mL into six-well plates. Control was made by distilled water. Detach S. parasitica pieces from the culture grown on PDA and put them into the above solutions respectively and co-cultured in six-well plates at 20℃. Each treatment was performed in duplicate, and repeated for confirmation. The extent of S. parasitica mycelium growth in each test solution was then determined and compared after 24 h by microscopically observation（OLYMPUS SZX10, Olympus, Tokyo, Japan）.

Effect of Copper Sulfate on S. Parasitica Primary Zoospores Release Time

Detach S. parasitica pieces from the culture grown on PDA and put into the 1. 0, 0. 5, 0. 2, 0. 1, and 0. 05 mg/L copper sulfate solutions and incubate in six-well plates at 20℃. Control was made by distilled water. Each treatment was performed in duplicate, and repeated for confirmation. Mycelium that releases primary zoospores was observed by light microscope（OLYMPUS IX71, Olympus, Tokyo, Japan）at ×40 magnification and the first time when S. parasitica primary zoospores start to release and the peak time when S. parasitica primary zoospores reach its maximum were recorded. Quantify the primary zoospores count using a hemocytometer every hour thereafter the first time to 96th hour.

Effect of Copper Sulfate on S. Parasitica Spore Production by Spectrometry

One milliliter of the S. parasiticaspore solution, containing $5×10^5$ spores per mL, was added to each of the test vessels which contained a 50 mL potato dextrose broth（formulated as PDA but without the agar）diluted with an appropriate volume of the copper sulfate to make the 1, 0. 5, 0. 2, 0. 1, and 0. 05 mg/L solutions. Controls were made using undiluted potato dextrose broth. After 24 h, the absorbance of S. parasiticaspore were then taken for each test solution using a cell density meter set at 600 nm（WPA Biowave CO8000, Biochrom, Cambridge, UK）. （Caruana et al. 2012）and every 24 h thereafter for the following 3d. The number of S. parasiticaspores in each test solution was confirmed by making a count every 24 h by a hemocytometer to directly compare the number of spores against the turbidity readings. They are both based on duplicate counts, each test solution against a set of controls.

Treatment of Copper Sulfate for Wounded Grass Carp Infected with S. Parasitica

Saprolegnia parasitica infection was induced by wounded fish exposed to S. parasitica mycelium with a change in temperature from

25 to 15℃.

To assess the effect of copper sulfate on controlling infection of S. parasiticaon wounded grass carps, first, 20 replicate S. parasitica cultures grown on PDA were respectively transferred into the tanks which contained 20 L aerated and dechlorinated water. Then after 36 h when generally the primary spores have been released, an appropriate volume of copper sulfate was added into the above tanks respectively to make 1.0, 0.5, and 0.2 mg/L solutions (exposure tanks). Control was made by aerated and dechlorinated water. Finally, grass carps were removed from the holding tank (25℃), lightly wounded above the lateral line between the posterior margin of the dorsal fin, and then placed in the exposure tanks prepared above (15℃). The wound area was approximately 5 cm^2. Twenty grass carps were used for each trial. After 4 d, each tank was evaluated throughout the trial recording the number of infected and dead grass carps. Each treatment was performed in duplicate, and repeated for confirmation.

Statistics

Data in the table are presented as mean±SD. All statistical analyses were conducted using SYSTAT 12 (Systat Software Inc., Chicago, IL, USA).

Results

Effect of Copper Sulfate on S. Parasitica Mycelium Growth

Increasing copper sulfate exposure concentrations reduced the growth of S. parasiticamycelium after 24 h. Copper sulfate at ≥0.5 mg/L inhibited the growth of mycelium (Fig. 1).

Effect of Copper Sulfate on S. Parasitica Primary Zoospores Release Time

Increasing exposure concentrations inhibited the release of primary zoospores, reduced the primary zoospores count and copper sulfate at ≥1 mg/L prohibited the release of primary zoospores (Fig. 2).

Effect of Copper Sulfate on S. Parasitica Spore Production by Spectrometry

The comparison between the absorbance of S. parasiticaspore and the spore count at 24 h post-inoculation showed a good uniformity in all test solutions (Fig. 3), while the spore numbers indicated that the control behaved in an expected manner, with predicted increases in spore number throughout the duration of the experiment (Fig. 4). The results suggest that copper sulfate at ≥0.1mg/L was effective in either inhibiting (limiting the increase of spore number) or decreasing the number of spores in the culture.

Treatment of Copper Sulfate for Wounded Grass Carp Infected with S. Parasitica

Copper sulfate at three concentrations could reduce S. parasitica infection rate. There was a significant difference of infection rates between the experimental and control group (P<0.05). However, significant mortality difference existed between 1.0 mg/L group and the others (P<0.05, Table 1).

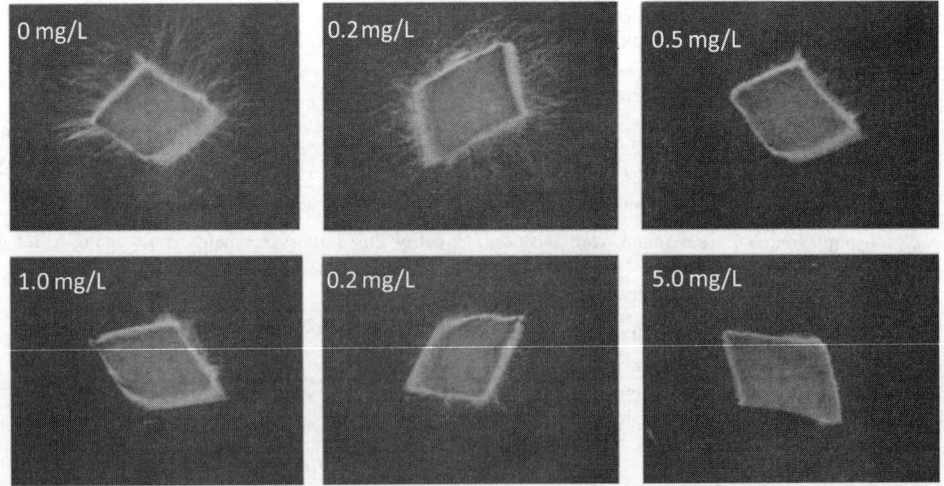

Fig. 1　Microscopic picture (OLYMPUS SZX10, Olympus, Tokyo, Japan) of Saprolegnia parasitica pieces hyphae extending in different copper sulfate concentrations after 24 h. All scale bars represent 10 mm.

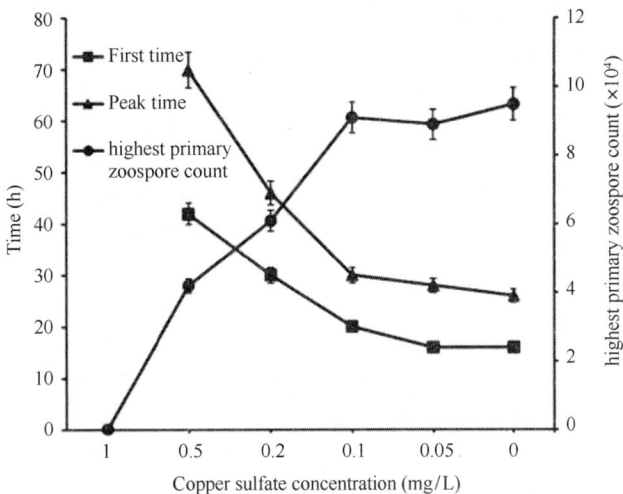

Fig. 2　The first time of mycelium releasing primary zoospores and its peak time when attained the highest primary zoospores count were raised along with the increasing exposure concentrations. No primary zoospores were released at 1mg/L of copper sulfate.

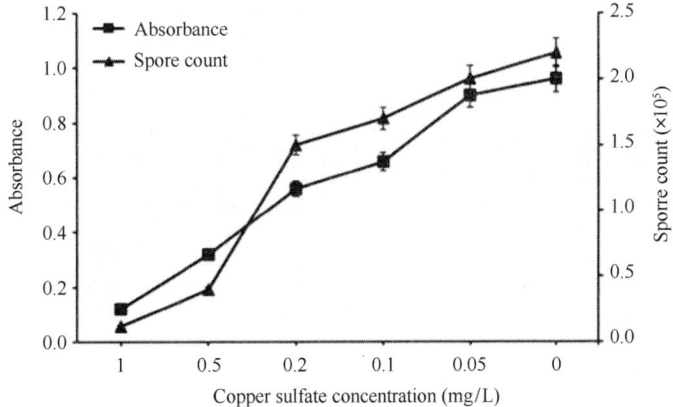

Fig. 3　The agreement between the Saprolegnia parasitica spore counts and the absorbance at 24 h.

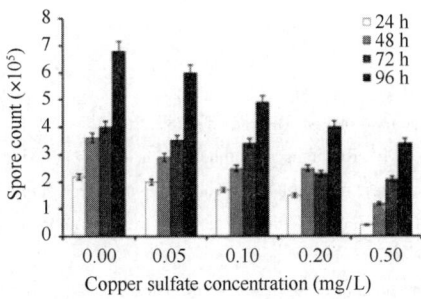

Fig. 4　Saprolegnia parasitica spore counts （mean±SE） made for each test solution of copper sulfate with time （hours） .

Discussion

The results indicated that copper sulfate could significantly influence the mycelium and zoospores of S. parasitica. Copper sulfate at $\geqslant 0.5$ mg/L inhibited the growth of mycelium; no primary zoospores were released at $\geqslant 1$ mg/L. Because copper sulfate at 0.5 mg/L has a significant difference of infection rates （$P < 0.05$） and no significant mortality difference （$P > 0.05$） compared with control group, it had the best efficacy on inhibiting the S. parasitica infections with grass carp.

Compared with the former researches concerning the effect of copper sulfate in Saprolegniasis which is cursory, we used particular methods both in vitro and in vivo to evaluate the specific efficacy of copper sulfate and revealed that it can inhibit the growth of the mycelium, inhibit the release of the primary zoospores, diminish the spores count, and prevent the S. parasitica infection.

Some researchers have showed that hydrogen peroxide, at concentrations of 500 ~ 1 000 mg/L, is effective against Saprolegniaspp. infections in salmonid eggs (Schreier et al., 1996; Gaikowski et al., 1998; Arndt et al., 2001), formalin at ≥50 mg/L treatment has a good efficacy in infection associated with Saprolegniasp. (Arndt et al., 2001; Barnes and Soupir, 2006; Gieseker et al., 2006), 25~500 mg/L carbon of humic substances reduced the growth of the fungal mycelium in vitro (Thomas et al., 2007), some selected plant extracts and bioflavonoids at 100 mg/L have efficacy in controlling infections of Saprolegniaaustralis (Caruana et al., 2012). Compared with copper sulfate, the effective concentration of these drugs mentioned above is much higher. However, due to the character of copper sulfate that some fish are highly sensitive to it (Chakoumakos et al., 1979; Dave and Xiu, 1991; Chen and Lin, 2001), it can only be used as a drug additive in low dose to reduce the S. parasitica infections.

In this study, we also developed a reliable and controlled method to induce infections from S. parasiticain grass carp to test whether copper sulfate can reduce infection rate associated with this disease. A precise infection was induced with temperature change, mechanical lesions, and S. parasitica exposure. The artificial infection was initially located on the lesions parts and then spread around. The infections in our model were rapidly induced compared with other methods that used a continuous spore challenge (Howe and Stehly 1998; Pottinger and Day 1999). The method induced severe uniform infections at the site of lesions in all fish in all trials and had a stably high infection rate which can be use as a Saprolegniasp. infection model for the further research of this disease.

Table 1 Copper sulfate concentration (mg/L), infection rate (%), and mortality (%) of grass carp infected with S. parasitica. [1]

Copper sulfate concentration (mg/L) Infection rate (%)	Mortality (%)
1.023.33±2.87[a]	43.33±5.77[a]
0.538.33±10.41[b]	3.33±2.87[b]
0.268.33±7.64[c]	3.33±5.77[b]
091.67±2.87[d]	1.67±2.89[b]

1 Infection rate and mortality are given as mean (±SD). Different letters denote significant differences between treatments.

Acknowledgments

The project has been financially supported by the 863 Program (grant no. 2011AA10A216), Special Fund for Agro-scientific Research in the Public Interest (grant no. 201203085), and the key project of prospering agriculture by science and technology of Shanghai agriculture committee (grant no. 2011-4-8).

Literature Cited

A lderman, D. J. 1985. Malachite green: a review. Journal of Fish Diseases 8: 289-298.

Arndt, R. E., E. J. Wagner, and M. E. Routledge. 2001. Reducing or withholding hydrogen peroxide treatment during a critical stage of rainbow trout development: effects on eyed eggs, hatch, deformities, and fungal control. North American Journal of Aquaculture 63: 161-166.

Barnes, M. E. and C. A. Soupir. 2006. Evaluation of formalin and hydrogen peroxide treatment regimeson rainbow trout eyed eggs. North American Journal of Aquaculture 69: 5-10.

Brock, J. A. and R. Bullis. 2001. Disease prevention and control for gametes and embryos of fish and marine shrimp. Aquaculture 197: 137-159.

Bruno, D. W. and B. P. Wood. 1999. Saprolegnia and other oomycetes. CABI Publishing, Oxon, UK.

Burke, J. M. and J. E. Miller. 2008. Dietary copper sulfate for control of gastrointestinal nematodes in goats. Veterinary Parasitology 154: 289-293.

Caruana, S., G. H. Yoon, M. A. Freeman, J. A. Mackie, and A. P. Shinn. 2012. The efficacy of selected plant extracts and bioflavonoids in controlling infections of Saprolegniaaustralis (Saprolegniales; Oomycetes). Aquaculture 358-359: 146-154.

Chakoumakos, C., R. C. Russo, and R. V. Thurston. 1979. Toxicity of copper to cutthroat trout (Salmo clarki) under different conditions of alkalinity, pH, and hardness. Environmental Science & Technology 13: 213-219.

Chen, J. C. and C. H. Lin. 2001. Toxicity of copper sulfate for survival, growth, molting and feeding of juveniles of the tiger shrimp, Penaeus monodon. Aquaculture 192: 55-65.

Dave, G. and R. Xiu. 1991. Toxicity of mercury, copper, nickel, lead and cobalt to embryos and larvae of zebrafish, Brachydaniorerio. Archives of Environmental Contamination and Toxicology 21: 126-134.

Gaikowski, M. P. , J. J. Rach, J. J. Olson, and R. T. Ramsay. 1998. Toxicity of hydrogen peroxide treatments to rainbow trout eggs. Journal of Aquatic Animal Health 10: 241-251.

Gieseker, C. M. , S. G. Serfling, and R. Reimschuessel. 2006. Formalin treatment to reduce mortality associated with Saprolegniparasiticain rainbow trout, Oncorhynchus mykiss. Aquaculture 253: 120-129.

Hatai, K. , L. G. Willoughby, and G. W. Beakes. 1990. Some characteristics of Saprolegnia obtained from fish hatcheries in Japan. Mycological Research 94: 182-190.

Howe, G. E. and G. R. Stehly. 1998. Experimental infection of rainbow trout with Saprolegniaparasitica. Journal of Aquatic Animal Health 10: 397-404.

James, P. M. , H. S. Thomas, L. S. Gordon, and B. T. Frieda. 1998. Copper tolerance by the freshwater algal species Oocystispusilla and its ability to alter free-ion copper. Aquatic Toxicology 44: 69-82.

János, O. and J. Farkas. 1978. Effect of temperature, pH, antibiotics, formalin and malachite green on the growth and survival of Saprolegniaand Achlyaparasiticon fish. Aquaculture 13: 273-288.

Kiyoshi, M. and E. B. Claude. 1993. Comparative evaluation of the solubility and algal toxicity of copper sulfate and chelated copper. Aquaculture 117: 287-302.

Lester, K. 2000. Fungal diseases in fish. Seminars in Avian and Exotic Pet Medicine 9: 102-111.

Ling, K. H. , Y. M. Sin, and T. J. Lam. 1993. Effect of copper sulphate on ichthyophthiriasis (white spot disease) in goldfish (Carassiusauratus). Aquaculture 118: 23-35.

Lio-Po, G. D. , M. E. Sanvictores, M. C. Baticados, and C. R. Lavilla. 1982. In vitro effect of fungicides on hyphal growth and Sporogenesisof Lagenidium spp. isolated from Penaeus monodon larvae and Scylla serrataeggs. Journal of Fish Diseases 5: 97-112.

Marking, L. L. , J. J. Rach, and T. M. Schreier. 1994. American fisheries society evaluation of antifungal agents for fish culture. The Progressive Fish-Culturist 56: 225-231.

Meyer, F. P. 1991. Aquaculture disease and health management. Journal of Animal Science 69: 4201-4208.

Noga, E. J. 2000. Fish disease: diagnosis and treatment. Wiley-Blackwell Publishers, Hoboken, New Jersey, USA.

Pottinger, T. G. and J. G. Day. 1999. A Saprolegniaparasiticachallenge system for rainbow trout: assessment of Pyceze as an anti-fungal agent for both fish and ova. Diseases of Aquatic Organisms 36: 129-141.

Rach, J. J. , S. Redman, D. Bast, and M. P. Gaikowski. 2005. Efficacy of hydrogen peroxide versus formalin treatments to control mortality associated with saprolegniasis on lake trout eggs. North American Journal of Aquaculture 67: 148-154.

Schlenk, D. , J. L. Gollon, and B. R. Griffin. 1998. Efficacy of copper sulfate for the treatment of ichthyophthiriasisin channel catfish. Journal of Aquatic Animal Health 10: 390-396.

Schreier, T. M. , J. J. Rach, and G. E. Howe. 1996. Efficacy of formalin, hydrogen peroxide, and sodium chloride on fungal-infected rainbow trout eggs. Aquaculture 140: 323-331.

Song, L. , T. L. Marsh, T. C. Voice, and D. T. Long. 2011. Loss of seasonal variability in a lake resulting from copper sulfate algaecide treatment. Physics and Chemistry of the Earth, Parts A/B/C 36: 430-435.

Srivastava, G. C. and R. C. Srivastava. 1978. A note on the potential applicability of malachite green oxalate in combating fish-mycoses. Mycopathologia 64: 169-171.

Srivastava, S. , R. Sinha, and D. Roy. 2004. Toxicological effects of malachite green. Aquatic Toxicology 66: 319-329.

Stammati, A. , C. Nebbia, I. De Angelis, A. A. Giuliano, A. Carletti, C. Rebecchi, F. Zampaglioni, and M. Dacasto. 2005. Effects of malachite green and its major metabolite, leucomalachite green, in two human cell lines. Toxicology In Vitro 19: 853-858.

Straus, D. L. , A. J. Mitchell, R. R. Carter, andJ. A. Steeby. 2012. Hatch rate of channel catfish Ictalurus punctatus (Rafinesque 1818) eggs treated with 100mg/L copper sulphate pentahydrate. Aquaculture Research 43: 14-18.

Thomas, M. , P. Andrea, M. P. Thuy, Z. Elke, K. Angela, W. Andreas, andS. Christian. 2007. Reductioninvegetativegrowth of the water mold Saprolegniaparasitica (Coker) by humic substance of different qualities. Aquatic Toxicology 83: 93-103.

van West, P. 2006. Saprolegniaparasitica, an oomycete pathogen with a fishy appetite: new challenges for an old problem. Mycologist 20: 99-104.

Willoughby, L. G. 1994. Fungi and fish diseases. Pisces Press, Stirling, Scotland.

该文发表于《FEMS MICROBIOLOGY LETTERS》【2015, 362 (24): DOI: 10.1093/femsle/fnv196】

Inhibition of dioscin on Saprolegnia *in vitro*

Lei Liu[1], Yu-Feng Shen[1], Guang-Lu Liu[2], Fei Ling[1],
Xin-Yang Liu[1], Kun Hu[3], Xian-Le Yang[3] and Gao-Xue Wang[1]

(1. College of Animal Science and Technology, Northwest A&F University, Xinong Road 22nd, Yangling, Shaanxi 712100, China; 2. College of Science, Northwest A&F University, Xinong Road 22nd, Yangling, Shaanxi 712100, China and

3. National Pathogen Collection Center for Aquatic Animals, College of Fisheries and Life Science, Shanghai Ocean University, 999 Hucheng Huan Road, Shanghai 201306, China)

薯蓣皂苷对水霉的体外抑制作用

刘镭[1], 沈毓峰[1], 刘广路[2], 凌飞[1], 刘鑫扬[1], 胡鲲[3], 杨先乐[3], 王高学[1]

(1. 西农大学动物科学技术学院, 西农路 22 号, 中国, 712100, 陕西杨凌; 2. 西农大学科学学院, 西农路 22 号, 中国, 712100, 陕西杨凌; 3. 国家水生动物病原库, 水产与生命学院, 上海海洋大学, 沪城环路 999 号, 201306 上海, 中国)

ABSTRACT: As one of the most serious pathogens in the freshwater aquatic environment, Saprolegnia can induce a high mortality rate during the fish egg incubation period. This study investigated the anti-Saprolegnia activity of a total of 108 plants on Saprolegnia parasitica in vitro and Dioscorea collettii was selected for further studies. By loading on an open silica gel column and eluting with petroleum ether-ethyl acetate-methanol, dioscin ($C_{45}H_{72}O_{16}$) was isolated from D. collettii. Saprolegnia parasitica growth was inhibited significantly when dioscin concentration was more than $2.0 \ mg \cdot L^{-1}$. When compared with formalin and hydrogen peroxide, dioscin showed a higher inhibitory effect. As potential inhibition mechanisms, dioscin could cause the S. parasitica mycelium morphologic damage, dense folds, or disheveled protuberances observed by field emission scanning electron microscopy and the influx of Propidium iodide. The structural changes in the treated mycelium were indicative of an efficient anti-Saprolegnia activity of dioscin. The oxidative stress results showed that dioscin also accumulated reactive oxygen species excessively and increased total antioxidant and superoxide dismutase activity. These situations could render S. parasitica more vulnerable to oxidative damage. Additionally, when dioscin concentration was less than $2.0 \ mg \cdot L^{-1}$, the survival rate of embryos was more than 70%. Therefore, the use of dioscin could be a viable way of preventing and controlling saprolegniasis.

Keywords: aquatic disease; dioscin; natural plant agent; membrane damage

摘要: 作为淡水水生环境最重要的病原菌之一, 水霉菌能在鱼卵孵化期导致高死亡率。本课题在体外通过寄生水霉一共研究了 108 种具有抗水霉活性的植物, 叉蕊薯蓣被筛选出来做进一步研究。通过一个开放式的硅胶柱层析并且用石油醚-乙酸乙酯-甲醇洗脱, 薯蓣素从叉蕊薯蓣中提取出来。当薯蓣素的浓度超过 $2.0 \ mg \cdot L^{-1}$ 时, 寄生水霉的生长被显著性抑制。和福尔马林与过氧化氢相比, 薯蓣素显示出更高的抑制效果。作为潜在的抑制机制, 通过扫描发射电镜和碘化丙啶的汇入能观察到薯蓣素可以引起水霉菌菌丝体形态损害, 密集的折叠或凌乱的突起物。菌丝结构的变化是薯蓣素具有有效的抗水霉活性的象征。氧化应激结果表明薯蓣皂苷还活性氧积累过多, 增强了总抗氧化性和超氧化物歧化酶活性, 这些情况会使寄生类水霉菌更易受到氧化类损伤。另外, 当薯蓣素浓度小于 $2.0 \ mg \cdot L^{-1}$ 时, 胚胎的成活率超过 70%。因此, 薯蓣素的使用可能是一个可行的防治水霉病的方式。

关键词: 水产疾病; 薯蓣素; 天然植物药剂; 膜损伤

INTRODUCTION

With the intensification and industrialization of aquaculture, Saprolegniales has become an important problem in freshwater ecosystems (Densmore and Green 2007; Fernandez-Beneitez et al. 2008; Ruthig 2009). Saprolegnia usually penetrates into epi-dermal tissues of fish and the zoospores could also colonize dead eggs which in turn will infect adjoining living eggs. A novel study showed that Saprolegnia diclina infection could result in egg chorion destruction, while Saprolegnia parasitica penetrated the chorion in a nondestructive manner (Songe et al. 2015).

Saprolegnia-infected eggs are easily recognized by fluffy, cotton-like white-to-greyish patches (Thoen, Evensen and Skaar, 2011).

As an effective fungicidal agent, malachite green was used widely in the treatment of saprolegniasis (Willoughby and Roberts, 1992; Alderman, 1994). However, its use has been banned around the world due to an undesirable effect on animal health (Srivastava, Sinha and Roy, 2004; Stammati et al., 2005). With the banning of malachite green, investigators tried to find some suitable alternatives, for instance hydrogen peroxide, sodium chloride, bronopol and formalin (37% formaldehyde v/v) (Marking, Rach and Schreier, 1994; Branson, 2002; Barnes, Stephenson and Gabel, 2003; Gieseker, Serfling and Reimschuessel, 2006).

However, hydrogen peroxide was found to harm the hatching of Oncorhynchus mykiss Walbaum eggs (Gaikowski et al. 1998), and sodium chloride seemed not to be as effective as hydrogen peroxide or formalin for controlling Saprolegnia infections (Schreier, Rach and Howe, 1996). Although formalin was considered to control Saprolegnia infections in eggs effectively (Waterstrat and Marking, 1995; Schreier, Rach and Howe, 1996; Rach et al., 2005; Barnes and Soupir, 2006), there were growing concerns regarding the risks to human health and the environment. In addition, alternative strategies using bronopol against S. parasitica were considered not very effective (Caruana et al., 2012). Although some cures are effective in a short period after application, their usage could be dangerous because of a lack of selectivity and small safety margins (Reverter et al., 2014). Therefore, natural plant agents are gradually receiving more attention as an alternative to chemotherapeutic agents (Khosravi et al., 2012). In this line of combat against saprolegnia-sis investigators have focused on the anti-Saprolegnia activity of plant essential oils and Chinese traditional medicines (Pattnaik, Subramanyam and Kole 1996; Keene et al. 1998; Rai, Kaushal and Acharya, 2002; Chukanhom, Borisuthpeth and Hatai 2005; Khosravi et al. 2012; Hu et al. 2013a).

An anti-Saprolegnia compound isolated from plants could enable us to develop new control strategies against Saproleg-nia infections. The present study aimed to evaluate the anti-Saprolegnia effect of 108 medicinal plants on S. parasitica in vitro, isolate an active compound from one selected plant, and determine acute toxicity on zebrafish embryos. Further, the way in which the compound inhibited Saprolegnia growth was investigated by testing mycelium damage and changes to the oxidative balance.

MATERIALS AND METHODS

Fungal strain

A pure strain of S. parasitica was received from the National Pathogen Collection Center for Aquatic Animals, Shanghai Ocean U-niversity, placed on potato dextrose agar (PDA) slants and stored at 4℃. This pure strain was isolated from Pengze cru-cian carp eggs which were obtained as samples from the Sand Lake Aquatic Technique Popularizing Station in 2010.

Preparation of plant materials

The plants (see Supplement 1 in the online supplementary material) were collected in September 2012 and the taxonomic identi-fication was made by Professor X. P. Song in Northwest A&F University (Shaanxi, China). The plant samples were washed, dried in an oven at 45℃ and then milled to a fine powder by using a strainer (30-40 mesh). The powdered samples were freeze-dried at −20℃ to ensure complete removal of water.

Screening experiment

The 108 resulting methanol extracts and minimum inhibitory concentrations (MICs) were performed as described previously (Hu et al. 2013a). The details are described in the Supplemental materials and methods available online.

Anti-Saprolegnia activity assays of the selected plants

Samples of 50 g of Dioscorea collettii and Dioscorea odorifera were extracted with petroleum ether, chloroform, ethyl acetate, methanol and water by sonication for 2 h in a water bath at 60-65℃, and the process was repeated three times. The ratio of sample to solvent was 1∶10 (m/v). Five solvent extracts were filtered, combined and evaporated under reduced pressure in a vacuum rotary e-vaporator. The extracts were dissolved in DMSO and then MICs of five extracts were performed as described previously (Hu et al. 2013a).

According to the five extract assay results, ethyl acetate extract of D. collettii and petroleum ether extract of D. odorifera were cho-sen, mixed with sterilized PDA at 50-60℃ and tested at 6.25, 12.5, 25.0, 50.0, 100.0 and 200.0 mg $\cdot$ L^{-1} in 7.5 cm diameter Petri dishes. PDA plugs (diameter: 0.5 cm) colonized with Saprolegnia were placed upside down in the center of extract-PDA plates. The mycelial growth diameter was measured after inoculation at 25 C for 48 h. The growth inhibition rate was calculated from mean values as: %IR = 100 (A − B) / (A − C), where IR = growth inhibition rate, A = mycelium growth diameter in control, B = mycelium growth in sample, and C = 0.5 cm.

Extraction and isolation

Dioscorea collettii was chosen for isolation of the active com-pounds. The details are described in the online Supplemental materi-als and methods. Eventually, a white needle-like crystal X was isolated from D. collettii.

Identification of the active compound

The active compound X was identified by the physicochemical properties, electron ionization mass spectrometry (ESI-MS), nu-clear magnetic resonance hydrogen spectrum (^{1}H NMR), and nuclear magnetic resonance carbon spectrum (^{13}C NMR). For ESI-MS a VG-ZAB-HS spectrometer (VG Company, Manchester, UK) was used at 70 eV; while a Bruker AM-400 spectrometer (Bruker, USA) was used for ^{1}H and ^{13}C NMR spectroscopy. Samples for NMR were dissolved in deuterated methanol (MeDO).

Bioassays

The anti-Saprolegnia efficacy was performed using the hemp seed microplate (HeMP) method, according to Stueland, Heier and Skaar (2005) with the modification that rape seeds were used instead of hemp seeds. A DMSO control group (1%, v/v, no signifi-cant anti-Saprolegnia activity, Hu et al. 2013b) and positive control groups (malachite green, formalin and hydrogen peroxide) were included. Each experiment included triplicate treatments. The details are described in the online Supplemental materials and methods.

Mycelium growth inhibition test

Dioscin was mixed with the sterilized PDA and tested at 0.5, 1.0, 2.0, 4.0, 8.0, 16.0 and 32.0 mg $\cdot$ L^{-1}. The Saprolegnia-colonized PDA (diameter: 0.5 cm) were inoculated in the center of dioscin-PDA plates. The growth inhibition rate was as described in the section 'Anti-Saprolegnia activity assays of the selected plants'. Liu et al.

Assay of oxidative stress

The S. parasitica mycelia exposed to dioscin (1, 2.5 and 5.0 mg $\cdot$ L^{-1}) for 24 h were harvested. The mycelium was centrifuged at 8.0×10^3 g for 15 min, added with 500 μL 0.2 M PBS (pH 7.2) and homogenized by an ultrasonic cell pulverizer (JY92-2D, Xinzhi Co., Ningbo, China) at 200 W for 10 min (ultrasonic: rest=2: 8 s) under ice-bath cooling. The homogenate was centrifuged at 1.0×10^4 g for 15 min at 4 C to obtain the supernatant. Following the instructions of commercial assay kits (Beyotime Institute of Biotechnology, Haimen, China), the fluorescence intensity of 2^7, 7^7-dichlorodihydrofluorescein diacetate (DCFH-DA) was measured using a microplate reader (M200: Tecan, Mannedorf, Switzerland), while total antioxidant capacity (T-"AOC) and superoxide dismutase (SOD) activities were measured under the microplate reader (ELX800, Gene, Hong Kong, China). In addition, DCFH-DA reacting with reactive oxygen species (ROS) to produce the fluorescence in mycelium was observed by using an upright fluorescence microscope (DM5000B, Leica, Germany).

Ultrastructural analysis

After exposure to 1.0, 2.5 and 5.0 mg $\cdot$ L^{-1} dioscin for 48 h, dioscin-treated mycelium was observed using a field emission scanning electron microscope (FE-SEM, S-3400N, Hitachi, Japan) and an upright fluorescence microscope (DM5000B, Leica). The details are in the online Supplemental materials and methods.

Acute toxicity evaluation

Zebrafish husbandry and embryo collection are described in the online Supplemental materials and methods. In this test, fertilized eggs (4 cell-protruding mouth stage) were exposed to 1.5, 2.0, 2.5, 3.0, 3.5, 4.0, 4.5 and 5.0 mg $\cdot$ L^{-1} dioscin for 72 h. Each group of 20 embryos were transferred to 6-well microtiter plates filled with 10 mL freshly prepared test solutions or controls per well and then incubated at (28±0.5) ℃ with a photoperiod of 16 h: 8 h light: dark. Five replicates were set for the tests. The survival rates of embryos were examined microscopically at regular intervals (12, 24, 36, 48, 60 and 72 h).

Statistical analysis

The data were analyzed by probit analysis which was used for calculating the 95% confidence interval of the EC_{50} using SPSS 18.0 for Windows (SPSS Inc., Chicago, IL, USA). The other data were analyzed using ANOVA followed by Tukey's post hoc tests using SPSS 18.0. A P value of <0.05 was considered statistically significant.

RESULTS

MIC test

Among 108 plants, D. collettii and D. odorifera were tested to have the lowest MIC values. Five solvent extracts of D. collettii and D. odorifera were evaluated against S. parasitica growth, and the data are listed in Supplement 2 in the online supplementary material. The ethyl acetate extract of D. collettii was most effective. The an-tifungal effects of the ethyl acetate extract of D. collettii and the petroleum ether extract of D. odorifera are shown in Supplement 5A and B in the online supplementary material. It was found that the ethyl acetate extract of D. collettii was the most effective, with a mean EC_{50} value of 35.20 mg $\cdot$ L^{-1} (the 95% confidence interval 21.23-46.80 mg $\cdot$ L^{-1}), followed by petroleum ether extracts of D. odorifera with a mean of 74.48 mg $\cdot$ L^{-1} (range 52.59-93.65 mg $\cdot$ L^{-1}). Therefore D. collettii was chosen as the target for further exploration.

Identification of the active compound

By analyzing the data of ^{1}H NMR (500 MHz, MeOD), ^{13}C NMR (126 MHz, MeOD), ^{13}C NMR DEPT 135 (Distortionless Enhancement by Polarization Transfer 135, 126 MHz, MeOD) and ESI-MS mass-to-charge ratio (m/z) (Supplement 3 in the online supplementary material) and comparing with the reported values in the literature (Yang, Ma and Liu 2005), X was identified as dioscin (C45H72O16) and its chemical structure is shown in Supplement 4 in the online supplementary material.

Anti-Saprolegnia effect of the extracted fractions and the active compound

The anti-Saprolegnia results of fractions and the active compound are shown in Table 1. Among the extracted fractions collected from the ethyl acetate extract, Fr. E exhibited the highest anti-Saprolegnia activity on S. parasitica mycelium at the concen-tration of $10-20$ mg · L^{-1}. Fr. E showed five subfractions, Fr. Ea-e. Fr. Ec possessed the highest anti-Saprolegnia activity at the concentration of $5-10$ mg · L^{-1}. A further separation of Fr. Ec gave two subfractions, Fr. Ec1 and 2, of which Fr. Ec2 was the most efficient at the con-centration of $5-10$ mg · L^{-1}.

For the active substance dioscin, the anti-Saprolegnia activity is evaluated as shown in Supplement 5C and D in the online sup-plementary material. After 48 h of exposure, the EC_{50} value of dioscin on S. parasitica was 4. 87 mg · L^{-1} (3. 61-6. 43 mg · L^{-1}). Based on the MIC and minimal lethal concentration (MLC) results, dioscin showed a relatively higher inhibitory effect on S. parasitica growth than formalin and hydrogen peroxide (Table 2).

Table 1　MICs of fractions and subfractions on Saprolegnia parasitica.

Fractions	MIC (mg · L^{-1}) [a]	Subfractions	MIC (mg · L^{-1}) [b]
Fr. A	>50	Fr. Ea	>20
Fr. B	>50	Fr. Eb	10—20
Fr. C	40—50	Fr. Ec	5—10
Fr. D	40—50	Fr. Ed	>20
Fr. E	20—30	Fr. Ee	>20
Fr. F	30—40	Sfr. Ec1	20—30
Fr. G	>50	Sfr. Ec2	5—10
Malachite green	0. 5—1	Malachite green	0. 5—1
Control (1% DMSO)	—	Control (1% DMSO)	—

a: HeMP Method.

b: Agar dilution assay.

Table 2　Results from the screening of substances against strains of Saprolegnia parasitica using the HeMP method. The MICs and MLCs are in between the ineffective (lowest) and the effective (highest) value.

Substance	MIC (mg · L^{-1}), 48 h exposure	MLC (mg · L^{-1}), 60 min exposure
Dioscin	5—10	20-30
Malachitegreen	<1	5-10
Formalin	40-80	>80
Hydrogen peroxide	>80	>80
Control (1% DMSO)	–	–

Effect of dioscin on ROS content

Figure 1 shows ROS production in dioscin treatment by the observed fluorescence of DCF. Weak fluorescence was present in the control of S. parasitica mycelium (Fig. 1A), and with increasing dioscin concentrations, fluorescence intensity was found slightly in-creased after exposure to 1. 0 mg · L^{-1} (Fig. 1B). Further, intracellular ROS generation was found to increase significantly, shown by the strong fluorescence in 2. 5 and 5. 0 mg · L^{-1} treatments (Fig. 1Cand D).

SOD, T-AOC and ROS

The results shown in Fig. 2 indicate that SOD and T-AOC activity in mycelium cells are significantly influenced by dioscin. The SOD activity was found to increase significantly in 2. 5 mg · L^{-1} (14. 60±0. 59 U (mg protein)$^{-1}$) and 5. 0mg · L^{-1} (16. 92±0. 47 U (mg protein)$^{-1}$) dioscin treatments, with 145% and 168% versus control group (Fig. 2A). Asshown in Fig. 2B, the T-AOC activity of mycelium cells was observed to increase by 60% and 90% in response to 2. 5and 5. 0 mg · L^{-1}dioscin treatments, respec-tively, compared with control group. A slight increase of ROS was observed in the 1. 0 mg · L^{-1} dioscin exposure group at 24 h (Fig. 2C). However, the ROS concentrations in the dioscintreated mycelia were significantly increased across the 2. 5 and 5. 0 mg · L^{-1}expo-sure groups, with 79% and 90% increases compared with control group.

Effect of dioscin on mycelium microstructure

The effect of dioscin on S. parasitica mycelium microstructure is shown in Fig. 3. Figure 3A shows normal mycelium with a cylindri-

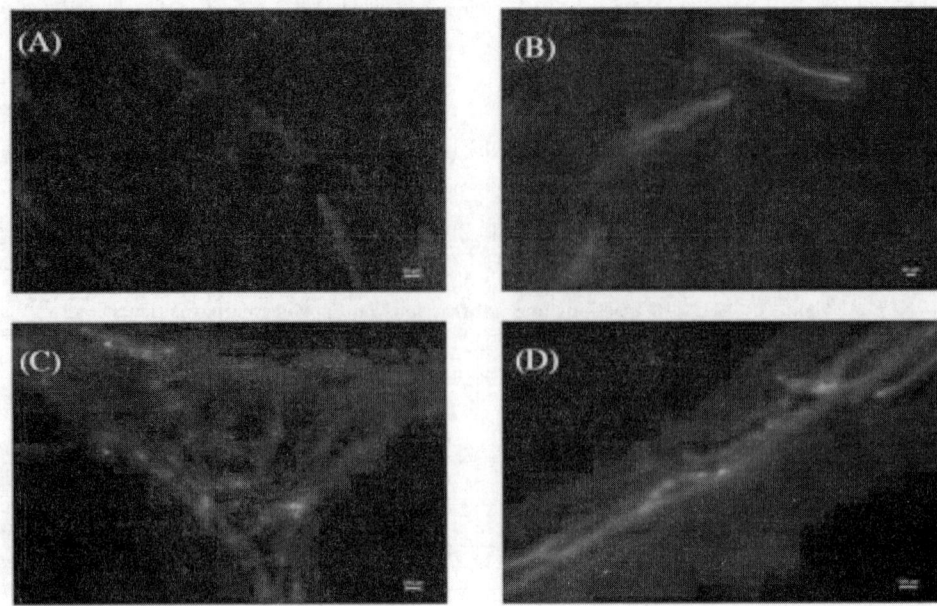

Fig. 1　The effect of dioscin on the appearance of intracellular ROS in Saprolegnia parasitica mycelium in 48 h. （A）control （50 μm）; （B）1.0 mg · L⁻¹ dioscin on mycelium （50 μm）; （C）2.5 mg · L⁻¹ dioscin on mycelium （100 μm）; （D）5.0 mg · L⁻¹ dioscin on mycelium （100 μm）.

cal shape and a smooth exterior. After 48 h exposure to 1.0 mg · L⁻¹ dioscin, the mycelium was obviously depressed or distorted （Fig. 3B）. As exposure concentrations exceeded 2.5 and 5.0 mg · L⁻¹, more severe damage of the mycelium was observed. Some mycelia showed ruptures or holes, or even forms of disruption of the membrane surface （Fig. 3B-D）.

Change in membrane permeability

For mycelium exposed to dioscin, the fluorescence intensity was observed to be significantly increased compared with controls （Fig. 4B-D）. By comparing the different exposure concentrations, the fluorescence intensity in the 5 mg · L⁻¹ group was found to be the strongest （Fig. 4D）.

Mortality rate

The mortality of zebrafish during the first 12 h of exposure to 4.5 and 5.0 mg · L⁻¹ dioscin is shown in Fig. 5. The embryo mortality gradually increased with increasing dioscin concentrations and extended exposure times. Dioscin could cause more than 50% mortality rate when the concentration exceeded 3.0 mg · L⁻¹; when dioscin concentration was lower than 2.0 mg · L⁻¹, the mortality rate of embryos was less than 30%. The half-maximal lethal concentration （LC50）at 72 h of dioscin was 2.74 μg · L⁻¹ （2.55-2.91 μg · L⁻¹）.

DISCUSSION

Although there is no previous research about anti-Saprolegnia activity of D. collettii extracts, some studies on the inhibitory effects of other plants on Saprolegnia spp. have been demonstrated such as ethanol extracts of Sophora flavescens, Hypericum perforatum, Chrysanthemum spp., Yucca spp., and Rumex （Cao et al., 2012; Caruana et al., 2012）. Compared with these plants, D. collettii was found to have a stronger effect against S. parasitica growth in our study. Furthermore, several investigators demonstrated the inhibitory effects of plant essential oils of Zataria multiflora, Geranium herbarium and Eucalyptus camaldolensis （Khosravi et al., 2012）. However, a variety of factors, such as solvent or dispersing agent, can influence the activity of essential oils （Kalemba and Kunicka, 2003）. By contrast, dioscin is dissolved and conserved more easily than essential oils. albicans （Sautour et al., 2004a, b）, protobioside from Dioscorea deltoida on Pyricularia oryzae （Shen et al., 2002）, and protodioscin from Dioscorea villosa on Candida tropicalis and Candida glabrata （Sautour, Miyamoto and Lacaille-Dubois, 2006）. Studies have shown that dioscin, a natural product, is associated with antifungal activity and antiinflammatory, lipid-lowering, anti-tumor and hepatoprotective properties （Liu et al.）, 2002; Sau-tour et al., 2004b; Sautour, Miyamoto and Lacaille-Dubois, 2006; Wang et al., 2007; Kaskiw et al., 2009）. However, to the best of our knowledge, no papers have been published in the open literature reporting the effects of dioscin therapy on the reduction of oomycetes.

562

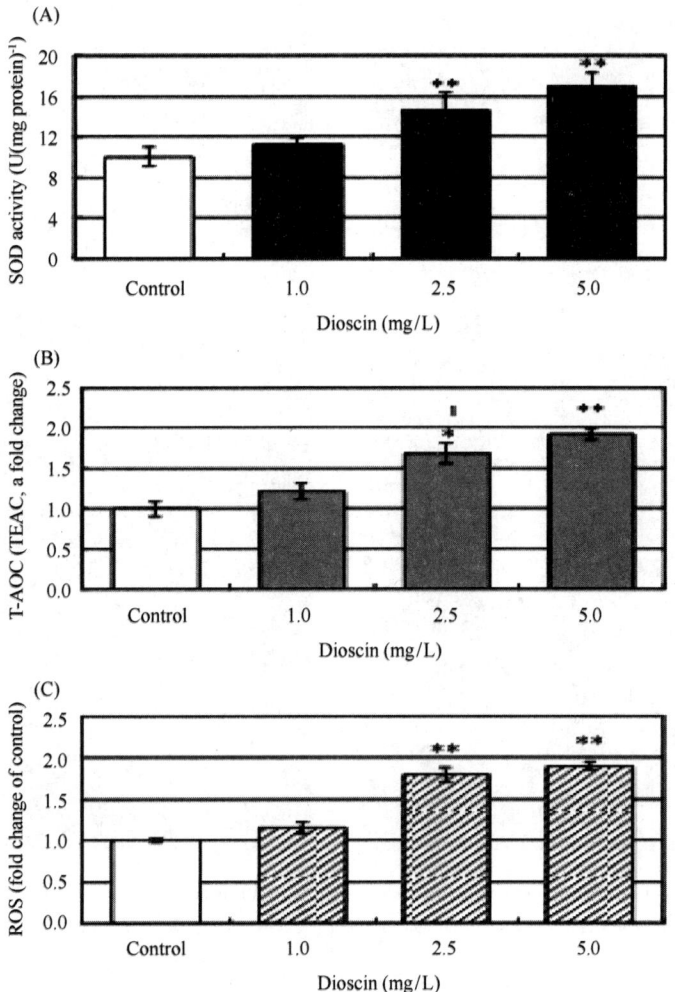

Fig. 2　The changes of SOD （A） and T-AOC （B） activities, and ROS concentration （C） in *Saprolegnia* parasitica after 1.0, 2.5 and 5.0 mg · L⁻¹ dioscin exposure. Values that are significantly different from the control are indicated by asterisks （one-way ANOVA, * P <0.05; ** P <0.01）.

Since the rhizome of D. collettii is rich in steroid saponins, D. collettii has been used as an anticancer agent in recent years （Sautour, Mitaine-Offer and Lacaille-Dubois, 2007）. Some steroid saponins from Dioscoreaceae have been isolated and the related biological activity was demonstrated, such as asperoside and measperoside from Dioscorea cayenensis on Candida albicans （Sautour et al. 2004a, b）, protobioside from Dioscorea deltoida on Pyricularia oryzae （Shen et al. , 2002）, and protodioscin from Dioscorea villosa on Candida tropicalis and Candida glabrata （Sautour, Miyamoto and Lacaille-Dubois 2006）. Studies have shown that dioscin, a natural product, is associated with antifungal activity and antiinflammatory, lipidlowering, antitumor and hepatoprotective properties （Liu et al. , 2002; Sautour et al. , 2004b; Sautour, Miyamoto and Lacaille-Dubois, 2006; Wang et al. , 2007; Kaskiw et al. , 2009）. However, to the best of our knowledge, no papers have been published in the open literature reporting the effects of dioscin therapy on the reduction of oomycetes.

In this study, dioscin exposure was found to induce pronounced membrane damage and disruption in the mycelium surface, increase the membrane permeability, and cause irreversible damage to the cellular membrane and even cell disintegration. This finding was in agreement with the conclusion of Cho et al. （2013） regarding the membranedisruptive action of dioscin. According to the report of Sautour, Mitaine-Offer and Lacaille-Dubois （2007）, dioscin was considered to damage biological membranes of some fungi, although *Saprolegnia* has different cell walls and general biology compared with fungi. Additionally, a significant enhancement of ROS in *Saprolegnia* after dioscin exposure was observed. It should be noted that the excessive ROS production can cause an oxidant-antioxidant imbalance making membrane lipids vulnerable to oxidative stress, and eventually release of cytochrome c leading to cell apoptosis （Xia et al. 1999）. Lv et al. （2013） found that dioscin could induce mitochondrial injury in glioblastoma multiforme, the collapse of

563

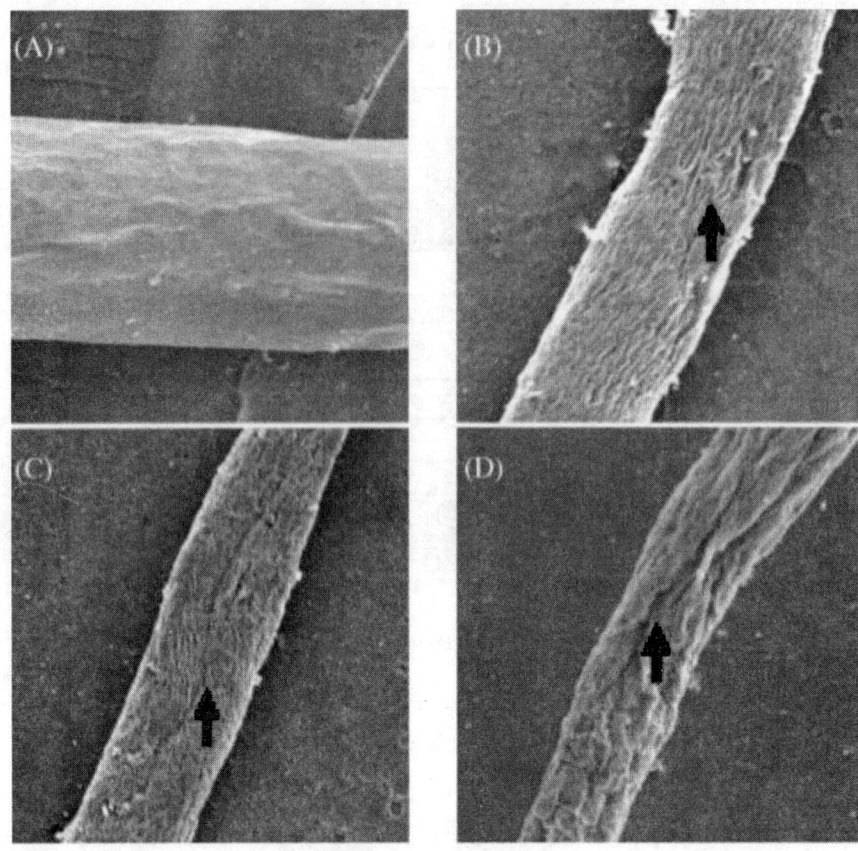

Fig. 3　Saprolegnia parasitica mycelium scanning electronmicrographs. (A) control; (B) 1.0 mg · L^{-1} dioscin on mycelium; (C) 2.5 mg · L^{-1} dioscin on mycelium; (D) 5.0 mg · L^{-1} dioscin on mycelium.

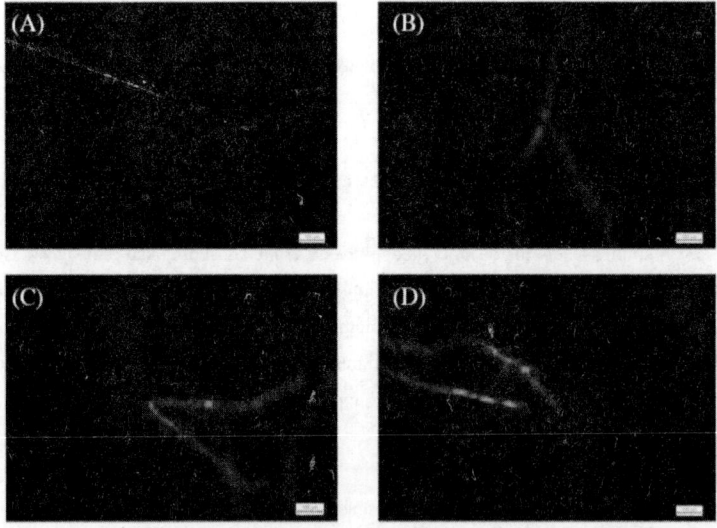

Fig. 4　The effect of dioscin on the membranes of Saprolegnia parasitica mycelium by analyzing the Propidium iodide (PI) influx. (A) control; (B) 1.0 mg · L^{-1} dioscin on mycelium; (C) 2.5 mg · L^{-1} dioscin on mycelium; (D) 5.0 mg · L^{-1} dioscin on mycelium. Scale bars represent 100 μm.

mitochondrial membrane potential and mitochondria permeability transition. In human myeloblastic leukemia HL-60 cells, biochemical study of dioscin indicated that a mitochondria dysfunction caused generation of ROS, leading to the changes in protein expression

（Wang et al. ，2007）. In the present study，T－AOC and SOD，as the integrated index，were also in－vestigated. Consistently，strong changes of antioxidant enzyme activities also showed an imbalance in antioxidant systems. Since SOD and T－AOC have the ability to counteract ROS（Van Rensburg et al. ，1995；Liu et al. ，2013；Liu，Zhu and Wang，2015），the increased SOD and T－AOC activities in the current study suggested that the protective system of S. parasitica，toacertain degree，was strengthened. Thereby，*S. parasitica* mycelium growth inhibited by dioscin could be attributed to oxidative damage.

Dioscin is considered as a lead compound for further preclinical studies in oncology and microbiology（Sautour，Mitaine－Offer and Lacaille－Dubois 2007），and it has been widely used as an important raw material for the synthesis of steroid hormone drugs such as cortisone（Brautbar and Williams 2002）. Even though dioscin is derived from many medicinal plants，the price is still high so that the potential treatment for saprolegniasis in aquaculture would be very expensive. Currently，the industrial production mainly involves diosgenin which can be converted into dioscin by combining with glucoses. Some investigators have determined that various combined extraction methods could be better for isolating diosgenin（Zhang et al. 2007；Li et al. 2009；Wang，Jin and Yu 2011）. When dioscin concentration is lower than 2. 0 mg · L^{-1}，the survival rate of embryos is more than 70%；simultaneously，dioscin could inhibit *S. parasitica* growth to some extent. Although there are some difficulties in using dioscin，further studies on the dioscin structure–activity relationship could result in the synthesis of a more highly active anti–Saprolegnia drug.

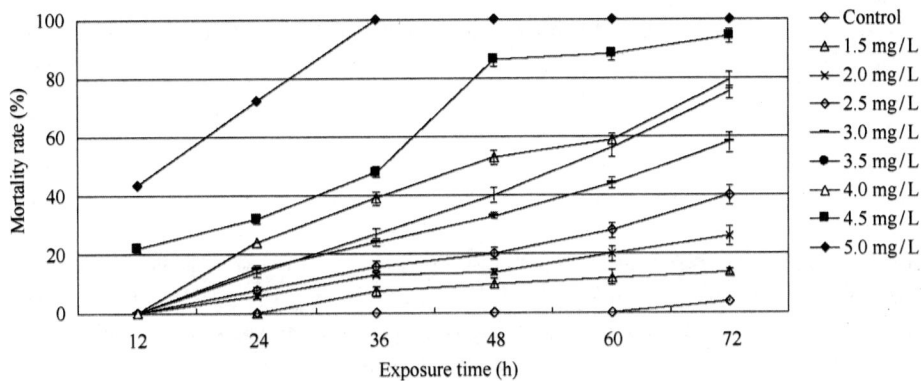

Fig. 5　The cumulative mortality rate at 12，24，36，48，60 and 72 h in zebrafish embryos exposed to dioscin.

Concluding remarks

In summary，this study investigated the anti–Saprolegnia activity of a total of 108 plants on *S. parasitica* in vitro by using liquid and solid dilution assay. Under bioassay–guided and multicolumn chromatography methods，dioscin，as an anti–Saprolegnia active compound isolated from *D. collettii*，exhibited a potent inhibiting effect on *S. parasitica* growth due to a membranedisruptive mechanism and oxidative damage. Further results showed that dioscin could inhibit *S. parasitica* significantly with low toxicity to zebrafish embryos when the concentration was 2. 0 mg · L^{-1}，and thus，dioscin could be used as a tool to combat saprolegniasis.

FUNDING

This work was supported by the National High Technology Research and Development Program of China（863 Program）（No. 2011AA10A216）.

SUPPLEMENTARY DATA

Supplementary data are available at FEMSLE online.

Conflict of interest：None declared.

REFERENCES

Alderman DJ. Control of oomycete pathogens in aquaculture. In：M"uller GJ.（ed. ）. Salmon Saprolegniasis. Portland，OR，USA：Bonneville Power Administration，1994，111 － 29. Barnes ME，Soupir CA. Evaluation of formalin and hydrogen peroxide treatment regimes on rainbow trout eyed eggs. NAmJ Aquacult，2006，69：5－10.

Barnes ME，Stephenson H，Gabel M. Use of hydrogen peroxide and formalin treatments during incubation of landlocked fall Chinook salmon eyed eggs. NAmJAquacult，2003，65：151－154.

Branson E. Efficacy of bronopol against infection of rainbow trout（Oncorhynchus mykiss）with the fungus Saprolegnia species. Vet Rec，2002，151：539–541.

Brautbar N, Williams J. Industrial solvents and liver toxicity: risk assessment, risk factor and mechanisms. Int J Hyg Envir Heal, 2002, 205: 479-491.

Cao HP, Xia WW, Zhang SQ, et al. Saprolegnia pathogen from Pengze crucian carp (Carassius auratus var. Pengze) eggs and its control with traditional Chinese herb. Isr J Aquacult-Bamid, 2012, 641-647.

Caruana S, Yoon GH, Freeman M, et al. The efficacy of selected plant extracts and bioflavonoids in controlling infections of Saprolegnia australis (Saprolegniales; Oomycetes). Aquaculture, 2012, 358: 146-154.

Cho J, Choi H, Lee J, et al. The antifungal activity and membrane-disruptive action of dioscin extracted from Dioscorea nippon-ica. Biochim Biophys Acta, 2013, 1828: 1153-1158.

Chukanhom K, Borisuthpeth P, Hatai K. Antifungal activities of aroma components from Alpinia galanga against water molds. Biocontrol Sci, 2005, 10: 105-109.

Densmore CL, Green DE. Diseases of amphibians. ILAR J, 2007, 48: 235-254.

Fernandez-Beneitez MJ, Ortiz-Santaliestra ME, Lizana M, et al. Saprolegnia diclina: another species responsible for the emer-gent disease 'Saprolegnia infections' in amphibians. FEMS Mi-crobiol Lett, 2008, 279: 23-29.

Gaikowski MP, Rach JJ, Olson JJ, et al. Toxicity of hydrogen peroxide treatments to rainbow trout eggs. J Aquat Anim Health, 1998, 10: 241-251.

Gieseker CM, Serfling SG, Reimschuessel R. Formalin treatment to reduce mortality associated with Saprolegnia par-asitica in rainbow trout, Oncorhynchus mykiss. Aquaculture, 2006, 253: 120-129.

Hu XG, Liu L, Chi C, et al. In vitro screening of Chinese medicinal plants for antifungal activity against Saprolegnia sp. and Achlya klebsiana. NAmJAquacult, 2013a, 75: 468-473.

Hu XG, Liu L, Hu K, et al. In vitro screening of fungicidal chemicals for antifungal activity against Saprolegnia. J World Aquacult Soc, 2013b, 44: 528-535.

Kalemba D, Kunicka A. Antibacterial and antifungal properties of essential oils. Curr Med Chem, 2003, 10: 813-829.

Kaskiw MJ, Tassotto ML, Mok M, et al. Structural analogues of diosgenyl saponins: synthesis and anticancer activity. Bioorg Med Chem, 2009, 17: 7670-7679.

Keene JL, Noakes DLG, Moccia RD, et al. The efficacy of clove oil as an anaesthetic for rainbow trout, Oncorhynchus mykiss (Wal-baum). Aquac Res, 1998, 29: 89-101.

Khosravi AR, ShokriH, SharifrohaniM, et al. Evaluation of the antifungal activity of Zataria multiflora, Geranium herbar-ium, and Eucalyptus camaldolensis essential oils on Saprolegnia parasitica-infected rainbow trout (Oncorhynchus mykiss) eggs. Foodborne Pathog Dis, 2012, 9: 674-679.

Li YS, Zhou Q, Zhang YZ, et al. Studies on the extraction process of diosgenin from Paris polyphylla from China Three Gorges. Chemical World, 2009, 2: 86-89.

Liu HW, Zhao QC, Cui CB, et al. Bioactive saponins from Dioscorea futschauensis. Pharmazie, 2002, 57: 570-572.

Liu L, Jiang C, Wu ZQ, et al. Toxic effects of three strobil-urins (trifloxystrobin, azoxystrobin and kresoxim-methyl) on mRNA expression and antioxidant enzymes in grass carp (Ctenopharyngodon idella) juveniles. Ecotoxicol Environ Saf, 2013, 98: 297-302.

Liu L, Zhu B, Wang GX. Azoxystrobin-induced excessive reactive oxygen species (ROS) production and inhibition of photosyn-thesis in the unicellular green algae Chlorella vulgaris. Environ Sci Pollut Res Int, 2015, 22: 7766-7775.

Lv L, Zheng L, Dong D, et al. Dioscin, a natural steroid saponin, induces apoptosis and DNA damage through reactive oxygen species: a potential new drug for treatment of glioblastoma multiforme. Food Chem Toxicol, 2013, 59: 657-669.

Marking LL, Rach JJ, Schreier TM. Evaluation of antifungal agents for fish culture. Prog Fish Cult, 1994, 56: 225-231.

Pattnaik S, Subramanyam VR, Kole C. Antibacterial and antifungal activity of ten essential oils in vitro. Microbios 1996; 86: 237-246.

Rach JJ, Redman S, Bast D, et al. Efficacy of hydrogen peroxide versus formalin treatments to control mortality associ-ated with saprolegniasis on lake trout eggs. NAmJAquacult, 2005, 67: 148-154.

Rai MK, Kaushal SK, Acharya D. In vitro effect of five Asteraceous essential oils against Saprolegnia ferax, a pathogenic fungus isolated from fish. Antiseptic, 2002, 99: 136-137.

Reverter M, Bontemps N, Lecchini D, et al. Use of plant extracts in fish aquaculture as an alternative to chemotherapy: current status and future perspectives. Aquaculture, 2014, 433: 50-61.

Ruthig GR. Water molds of the genera Saprolegnia and Leptoleg-nia are pathogenic to the North American frogs Rana cates-beiana and Pseudacris cruci-fer, respectively. Dis Aquat Organ, 2009, 84: 173-178.

Sautour M, Mitaine-Offer AC, Lacaille-Dubois MA. The Dioscorea genus: a review of bioactive steroid saponins. J Nat Med, 2007, 61: 91-101.

Sautour M, Mitaine-Offer AC, Miyamoto T, et al. Antifungal steroid saponins from Dioscorea cayenensis. Planta Med, 2004a, 70: 90-92.

Sautour M, Mitaine-Offer AC, Miyamoto T, et al. A new steroidal saponin from Dioscorea cayenensis. Chem Pharm Bull (Tokyo), 2004b, 52: 1353-1355.

Sautour M, Miyamoto T, Lacaille-Dubois MA. Steroidal saponins and flavan-3-ol glycosides from Dioscorea villosa. Biochem Syst Ecol, 2006, 34: 60-63.

Schreier TM, Rach JJ, Howe GE. Efficacy of formalin, hydrogen peroxide, and sodium chloride on fungal-infected rainbow trout eggs. Aquaculture, 1996, 140: 323-331.

Shen P, Wang SL, Liu XK, et al. A new steroidal saponin from Dioscorea deltoidea Wall var. orbiculata. Chinese Chem Lett, 2002, 13: 851-854.

Songe MM, Willems A, Wiik-Nielsen J, et al. Saprolegnia diclina IIIA and S. parasitica employ different infection strategies when colonizing eggs of Atlantic salmon, Salmo salar L. J Fish Dis, 2015, 10. 1111/jfd. 12368.

Srivastava S, Sinha R, Roy D. Toxicological effects of malachite green. Aquat Toxicol, 2004, 66: 319-329.

Stammati A, Nebbia C, Angelis ID, et al. Effects of malachite green (MG) and its major metabolite, leucomalachite green (LMG), in two human cell lines. Toxicol In Vitro, 2005, 19: 853-858.

Stueland S, Heier BT, Skaar I. A simple in vitro screening method to determine the effects of drugs against growth of Saproleg-nia parasitica. Mycol Prog, 2005, 4: 273-279.

Thoen E, Evensen Ø, Skaar I. Pathogenicity of Saprolegnia spp. to Atlantic salmon, Salmo salar L., eggs. J Fish Dis, 2011, 34: 601-608.

Van Rensburg S, Carstens M, Potocnik F, et al. Transferrin C2 and Alzheimer's disease: another piece of the puzzle found? Med Hypotheses, 1995, 44: 268-272.

Wang QY, Jin FX, Yu HS. Research progress of the preparation methods of diosgenin. J Anhui Agri Sci, 2011, 39: 2642-2644.

Wang Y, Che CM, Chiu JF, et al. Dioscin (saponin) -induced generation of reactive oxygen species through mitochon-dria dysfunction: a proteomic based study. JProteomeRes, 2007, 6: 4703-4710.

Waterstrat PR, Marking LL. Clinical evaluation of formalin, hy-drogen peroxide, and sodium chloride for the treatment of Saprolegnia parasitica on fall Chinook salmon eggs. Prog Fish Cult, 1995, 57: 287-291.

Willoughby LG, Roberts RJ. Towards strategic use of fungicides against Saprolegnia parasitica in salmonid fish hatcheries. J Fish Dis, 1992, 15: 1-13.

Xia Z, Lundgren B, Bergstrand A, et al. Changes in the genera-tion of reactive oxygen species and in mitochondrial mem-brane potential during apoptosis induced by the antidepres-sants imipramine, clomipramine, and citalopram and the ef-fects on these changes by Bcl-2 and Bcl-XL. Biochem Pharmacol, 1999, 57: 1199-1208.

Yang S, Ma Y, Liu XK. Steroidal constituents from Dioscorea parv-iflora (in Chinese). Yao Xue Xue Bao, 2005, 40: 145-149.

Zhang XC, Rong XD, Cheng DX, et al. Study on extraction of dios-genin from fresh Dioscorea zingiberensis C. H. Wright. Chem Bioeng, 2007, 6: 55-57.

该文发表于《NORTH AMERICAN JOURNAL OF AQUACULTURE》【2013, 75 (4): 468-473. DOI: 10. 1080/15222055. 2013. 808298】

In Vitro Screening of Chinese Medicinal Plants for Antifungal Activity *against Saprolegnia* sp. and *Achlyaklebsiana*

Hu Xue-Gang[1], Liu Lei[1], Chi Cheng[1], Hu Kun[2], Yang Xian-Le[2], Wang Gao-Xue[1]

(1. College of Animal Science and Technology, Northwest A&F University, Xinong Road 22nd, Yangling, Shaanxi 712100, China;

2. National Pathogen Collection Center for Aquatic Animals, College of Fisheries and Life Science,

Shanghai Ocean University, 999 HuchengHuan Road, Shanghai 201306, China)

Abstract: Saprolegniasis is a common fungal disease in aquaculture, causing severe damage to cultured fishes. To find natural agents for controlling and treating saprolegniasis, we investigated methanol extracts of 40 traditional Chinese medicinal plants. *Saprolegniasp. strain* JL and *Achlyaklebsianawere* used to evaluate the antifungal activity of the plants. Cnidiummonnieri, Magnolia officinalis, and Aucklandialappaat a concentration of 62. 5 mg/mL exhibited antifungal activity on Saprolegniaand Achlyaklebsianamycelium and were selected for further evaluation. The three plant species were extracted with four solvents (petroleum ether [PE], ethyl acetate, methanol, and water), and the extracts were evaluated with an in vitro bioassay using a rapeseed (Brassica napus) microplate method. Among the extracts tested, the PE extracts of the three plants exhibited the highest efficacy. The PE extract of A. lappaexhibited the best anti-Saprolegniaand anti-Achlyaactivities (50% effective concentrations = 11. 3 and 26. 1 mg/L, respectively), followed by C. monnieriand M. officinalis. Furthermore, the minimum fungicidal concentrations of PE extracts from the three herbs were identified as 25, 12. 5, and 25 mg/L, respectively, against Saprolegniaspores and 25, 25, and 12. 5 mg/L, respectively, against Achlyaklebsianaspores. These findings demonstrate that the three traditional Chinese medicinal plants—A. lappa, C. monnieri, and M. officinalis—have the potential for use in developing anovel therapy to control saprolegniasis in aquaculture.

体外筛选具有抗水霉和绵霉活性的中草药

胡学刚[1]，刘镭[1]，迟骋[1]，胡鲲[2]，杨先乐[2]，王高学[1]

(1. 西北农林科技大学动物科学技术学院，西农路 22 号，中国陕西，西安杨凌；

2. 上海海洋大学水产与生命学院国家水生动物病原库，沪城环路 999 号，中国上海)

摘要： 水霉病是水产养殖业中常见的真菌性疾病，对养殖鱼类产生严重的危害。为了找到控制和治疗水霉病的自然药物，我们研究了 40 种中草药的甲醇提取物。水霉菌 JL 种和异丝绵霉菌通常用来评估植物的抗真菌活性。浓度为的 62.5 mg/mL 蛇床子，厚朴和云木香对水霉和异丝绵霉展现出抗真菌活性，因此被选择出来做深入研究。这三种植物被四种化学溶剂萃取（石油醚，乙酸乙酯，甲醇和水），提取液用油菜籽微板法进行体外生物测定。在这些检测的提取液中，石油醚对三种植物的提取液表现出最高的功效。石油醚的云木香提取液能最好地抗水霉菌和异丝绵霉菌（半有效浓度分别为 11.3 mg/L 和 26.1 mg/L），其次是蛇床子和厚朴。此外，三种中草药的石油醚提取物对异丝绵霉菌的最小杀菌浓度分别被鉴定为 25 mg/L、25 mg/L 和 12.5 mg/L 这些结果表明这三种中草药——云木香，蛇床子，厚朴在农业上有发展一种控制水霉病的新型疗法的潜能。

In the aquaculture industry, fungal infection is one of the main factors responsible for mortality of cultured species and economic losses (Meyer 1991). Water molds (oomycetes) are widespread in freshwater and brackish water (Bangyeekhun et al. 2001; Hussein et al. 2001) and represent the most important fungal group affecting cultured fish and their eggs (Ke et al. 2009). Saprolegniasis causes severe damage in commercial aquaculture, including cultured salmon, trout, and anguillid eels, and in noncommercial aquariums. Fungal pathogens infect the surface of the fish's body after invading epidermal tissues, thus causing epidermal damage and cellular necrosis (Takada et al. 2010). The organic dye malachite green is extremely effective in the control and treatment of saprolegniasis and has been widely used as a therapeutic agent in aquaculture (Willoughby and Roberts 1992). However, malachite green has been banned for several years because it is a potential carcinogen and teratogen (Alderman 1985). Therefore, there is an urgent need to develop highly effective and environmentally safe fungicides for controlling *Saprolegniasis*.

People are becoming increasingly interested in traditional Chinese medicines because of their low toxicity and good therapeutic performance (Li and Chen 2005). Screening and proper evaluation of medicinal plants could reveal possible alternatives that may be both sustainable and environmentally acceptable (Eguale et al. 2007). However, there is little available information on the use of medicinal plants for the treatment of saprolegniasis in fish. We conducted the present study to find possible alternatives for protecting cultured fish. Our preliminary inspection of the literature indicated that 40 plant species had potential antifungal activity, and those plants were selected for use in antifungal activity testing. We obtained the methanol extracts of the 40 medicinal plants and evaluated their antifungal activity against *Saprolegniasp.* and *Achlyaklebsiana*; we also examined the in vitro bioactivity of different solvent extracts from three plant species: Cnidiummonnieri, Magnolia officinalis, and Aucklandialappa.

Methods

Fungal cultures. —Pure strains of *Saprolegniasp.* (strain JL; GenBank accession number HM637287; Cao et al. 2012) and Achlyaklebsianawere obtained from the National Pathogen Collection Center for Aquatic Animals at Shanghai Ocean University. The strains were cultured on potato dextrose agar (PDA) slants and were stored at 4℃ until use. To obtain long-term growth cultures, the water mold was subcultured on 7-cmdiameter PDA plates. A new subculture with fresh PDA medium was started 24 h before inoculating the rapeseed Brassica napuscultures.

To obtain the spore suspension, 20 fungal-colonized rapeseeds were placed with sterile forceps into a sterile, 300 mL Erlenmeyer flask (not shaken) and were incubated at 25℃ until abundant zoospores were produced. The spore suspension was filtered through carbasus, counted with a hemocytometer, and diluted with sterile water to an approximate concentration of 1×10^4 CFU/mL.

Plant materials. —Forty plant species (Table 1) that are traditionally used in Chinese folk medicine were collected in December 2011 and identified by X. P. Song at Northwest A&F University (Shaanxi, China). The voucher specimens were deposited at the Herbarium of the College of Life Science at Northwest A&F University. The plants were washed, cut into small pieces, and then placed in an oven at 45℃ until completely dried. The dried plants were separately crushed and reduced to a fine powder by using a strainer (30-40 mesh). The powdered samples were freeze dried at −20℃ to ensure complete desiccation.

Extraction of screened plants. —Dried powder (50.0 g) from each of the 40 plant species (Table 1) was separately extracted with methanol (500 mL, three times) for 2 h. The extracts were then filtered, combined, and evaporated under reduced pressure in a vacuum rotary evaporator (R-201; Shanghai Shen Sheng Biotech Co., Shanghai, China) at 50℃. The resulting extracts from each of the different plants were dissolved in dimethyl sulfoxide (DMSO) to obtain stock solutions of 200 mg/mL (sample/solvent), which were used to prepare the desired concentrations for the antifungal efficacy assay.

Extraction of antifungal plants. —The three plant species (Cnidiummonnieri, Magnoliaofficinalis, and Aucklandialappa) that had the highest antifungal activities were selected from amongthe40plantsinitiallyevaluated. Materialfromeachplant (50.0 g) was extracted with petroleum ether (PE), ethyl acetate, methanol, and water for 2 h (complete extraction), and the process was repeated three times. The ratio of sample to solvent was 1 : 10 (mass per volume). All of the extracts were filtered, combined, and evaporated under reduced pressure in a vacuum rotary evaporator. The extracts of the three plant species were dissolved in DMSO to obtain the 200 mg/mL (sample/solvent) stock solutions that were used in the assay.

Preliminary screening. —An agar dilution assay was used to test antifungal activity, and a modified fungal growth inhibition assay (with some modifications) was also performed (Stueland et al. 2005). To obtain effective drugs, a final test concentration of 500 mg/L for each drug was produced by dilution in PDA cooled at 60℃ in a sixwell, flatbottom tissue culture plate, and each concentration was carried out in triplicate. Three independent biological replicates were used. A DMSO control (2.5%) was included in the experiment.

Rapeseeds were used in this study, and Saprolegniawas used nthepreliminary screening via the rape seed microplate method. Whole rapeseeds were autoclaved at 121℃ for 20 min and cooled. The Saprolegniastrain was inoculated in the middle of a 7 cm-diameter Petri dish and incubated at 25℃ for 48 h. The sterilized rapeseeds were then placed in a circle around the Saprolegniacolony, and the agar plates were incubated at 25℃ for another 48 h. Subsequently, the *Saprolegnia*-colonized rapeseeds were transferred with sterile forceps to the surface of the cooled medium in the wells of a six-well, flat-bottom tissue culture plate. The plate was incubated at 25℃ for 48 h. The results were recorded visually by determining mycelium presence as positive ("+") or negative ("-").

Minimum inhibitory concentration evaluation. The medicinal plants for which mycelium presence was recorded as negative for at the concentration of 500 mg/L were tested for minimum inhibitory concentration (MIC) values ranging from 15.6 to 500 mg/L using the above method. Control plates were treated with DMSO (2.5‰) alone. The MIC was read visually at 48 h and was defined as the concentration of extract that inhibited growth by at least 80% or more relative to the control.

The MIC values of the four crude extracts from C. monnieri, M. officinalis, and A. lappawere tested at a different series of concentrations based on the initial tests, and the negative control groups containing no plant extract were set up under the same conditions as the test groups. A DMSO control was also included.

Mycelial growth inhibition test. In vitro antifungal activities of PE extracts from the three plants that had lower MIC values were assessed for mycelial growth inhibition rate. Each crude extract was added to assay flasks containing heat-sterilized PDA, and final concentrations were adjusted to the corresponding concentrations based on the initial tests. After mixing with a vortex mixer, aliquots (10 mL) of treated medium were poured into 7 cm-diameter Petri dishes. A DMSO control was included.

Table 1 Species names, medicinal names, antifungal activity of extracts at a test concentration of 500 mg/L, and the minimum inhibitory concentration (MIC) values of 40 plant species (+= no inhibition of the fungus; -= inhibition) as well as the chemicals malachite green and DMSO.

No	Species or chemical	Medical name	Family	Part used	Activity at 500 mg/L	MIC (mg/L) Saprolegnia sp.	MIC (mg/L) Achlya klebsiana
1	Scutellariabarbata	Herbascutellariabarbatae	Lamiaceae	Herb	+		
2	Curculigoorchioides	Rhizomacurculiginis	Amaryllidaceae	Rhizome	+		
3	Angelica pubescensf. biserrata	Radix angelicaepubescentis	Umbelliferae	Roots	+		
4	Cnidiummonnieri	Fructuscnidii	Umbelliferae		-	62.5	62.5
5	Areca catechu	Semen arecae	Palmaceae	Fruits	+		
6	Melia toosendan	Cortex meliae	Meliaceae	Seed Bark	+	125	250
7	Crataeguspinnatifida	Fructuscrataegi	Rosaceae	Fruits Roots	+		
8	Scutellariabaicalensis	Radix scutellariae	LabiataeLabiatae	Herb	-	125	250
	Menthahaplocalyx	Herbamenthae		Rhizome and roots	-		
9、10	Notopterygiumforbesii	Rhizoma et radix notopterygii	Umbelliferae		+		

No	Species or chemical	Medical name	Family	Part used	Activity at 500 mg/L	MIC (mg/L) Saprolegnia sp.	Achlya klebsiana
11	Phellodendronchinense	Cortex phellodendri	Rutaceae	Bark	+		
12	Lyciumbarbarum	Fructuslycii	Solanaceae	Fruits	+		
13	Acorustatarinowii	Rhizomaacoritatarinowii	Araceae	Rhizome	−	500	500
14	Polygonumcuspidatum	Rhizomapolygonicuspidati	Polygonaceae	Rhizome	+		
15	Acanthopanax (Eleutherococcus) senticosus	Radix et caulis acanthopanacissanticosi	Araliaceae	Roots	+		
16	Momordicacochinchinensis	Semen momordicae	Cucurbitaceae	Seeds	+		
17	Cistanchedeserticola	Herbavistanches	Orobanchaceae	Herb	+		
18	Sabina vulgaris	Sabina vulgaris	Cupressaceae	Stems	−	125	125
19	Houttuyniacordata	Herbahouttuyniae	Saururaceae	Herb	+		
20	Prunusmume	Fructusmume	Rosaceae	Fruits	+		
21	Zingiberofficinale	Rhizomazingiberis	Zingiberaceae	Rhizome	+		
22	Magnolia officinalis	Cortex magnoliae officinalis	Magnoliaceae	Bark	−	62.5	62.5
23	Rheum officinale	Radix et rhizomarhei	Polygonaceae	Rhizome and roots	+		
24	Dictamnusdasycarpus	Cortex dictamni	Rutaceae	Roots and bark	+		
25	Asparagus cochinchinensis	Radix asparagi	Liliaceae	Roots	+		
26	Liliumlancifolium	Bulbuslilii	Liliaceae	Herb	+		
27	Glycyrrhizauralensis	Radix glycyrrhizae	Leguminosae	Roots	+		
28	Foeniculum vulgare	Fructusfoeniculi	Umbelliferae	Fruits	+		
29	Tripterygium wilfordii	Radix et rhizomatripterygii	Celastraceae	Herb	−	125	250
30	Bupleurumchinense	Radix bupleuri	Apiaceae	Roots	+		
31	Aconitum carmichaelii	Radix aconitilateralispreparata	Ranunculaceae	Roots	+		
32	Alpiniaofficinarum	Rhizomaalpiniaeofficinarum	Zingiberaceae	Rhizome	+		
33	Salvia miltiorrhiza	Radix salviaemiltiorrhizae	Labiatae	Roots	−	250	500
34	Quisqualisindica	Fructusquisqualis	Combretaceae	Fruits	+		
35	Ophiopogonbodinieri	Radix ophiopogonis	Liliaceae	Roots	+		
36	Citrus reticulata	Semen citrireticulatae	Rutaceae	Seeds	+		
37	Lonicera japonica	Flosonicerae	Caprifoliaceae	Flower	+		
38	Aucklandialappa	Radix aucklandiae	Compositae	Roots	−	62.5	31.3
39	Pseudolarixkaempferi	Cortex pseudolaricis	Pinaceae	Roots and bark	−	500	500
40	Cinnamomum cassia	Cortex cinnamomicassiae	Lauraceae	Bark	+		
	Malachite green					1.0	1.0
	DMSO					35,000	35,000

Table 2 Minimum inhibitory concentration（mg/L）values of four crude extracts from
three plant species against Saprolegniasp. and Achlyaklebsianamycelium at 48 h, and
the minimum fungicidal concentration（mg/L）against spores at 72 h（-= an MIC was not detected）.

Species	Against	Petroleum ether		Ethyl acetate		Methanol		Water	
		Saprolegnia	Achlya	Saprolegnia	Achlya	Saprolegnia	Achlya	Saprolegnia	Achlya
Cnidiummonnieri	Mycelium	62.5	62.5	250	250	>500	>500	>500	>500
Magnolia officinalis	Spores Mycelium	2 562.5	2 562.5	- 250	- 250	- 500	- >500	- >500	- >500
Aucklandialappa	Spores Mycelium	12.5 62.5	2 531.3	- 250	- 125	- >500	- 500	- >500	- >500
	Spores	25	12.5	-	-	-	-	-	-

The fungal-colonized rapeseeds were inoculated in the center of the PDA in Petri dishes, and three replicates of each concentration were used. The mycelial growth diameter was measured after incubating at 25℃ for 48 h. This experiment was conducted in triplicate, and the growth inhibition rate（IR;%）was calculated from mean values as IR = 100 × [（A - B）/（A -C）], where A is mycelial growth in the control, B is mycelial growth in the sample, and C is the average diameter of the rapeseeds.

Spore germination test by theagar dilution method. —The PE extracts of the three plants（C. monnieri, M. officinalis, and A. lappa）were tested for activity against fungal spores. The spore germination test was carried out by an agar dilution method as outlined by the National Committee for Clinical Laboratory Standards（NCCLS 1997）. The plates were prepared by adding warm, liquid PDA with the required concentration of crude extract（100, 50, 25, 12.5, 6.3, or 3.2 mg/L）. Spores（10 μL; 1×10^4 CFU/mL）were spotted in triplicate, and the plates were incubated at 25℃ for 72 h. The minimum fungicidal concentration（MFC）was defined as the lowest concentration of extract that prevented visible growth or germination of spores; the presence of only one or two colonies was disregarded.

Statistical analysis. —The program SPSS（SPSS, Inc.）was used for all statistical analyses. The 50% effective concentration（EC_{50}）at the 95% confidence level was calculated by probit analysis（Finney 1971）.

Results

The anti-Saprolegniaactivities of the 40 plant species are presented in Table 1. Plants that showed reproducible activity at a concentration of 500 mg/L were considered for retesting. In this study, methanol extracts of 40 plants used in traditional Chinese medicine were screened. Of the 40 plants considered, 10 species（Cnidiummonnieri, Scutellariabaicalensis, Menthahaplocalyx, Acorustatarinowii, Sabina vulgaris, Magnolia officinalis, Tripterygium wilfordii, Salvia miltiorrhiza, Aucklandialappa, and Pseudolarixkaempferi）showed strong inhibition against Saprolegniaand Achlyaklebsiana. Three of those plants（C. monnieri, M. officinalis, and A. lappa）exhibited the most promising antifungal activities, as each had an MIC value of 62.5 mg/L.

Table 2 shows the MIC values of PE, Ethyl acetate, Methanol, and Water extracts from the three plant materials against Saprolegniasp and Achlyaklebsianamycelium at 48 h and the MFCs of PE extracts against spores at 72 h. The PE extracts exhibited the best antifungal activity against Saprolegniaand Achlyaklebsiana, with MIC values of 62.5 mg/L or less. The PE extracts of C. monnieriand A. lappashowed inhibition against Saprolegniaspores with MFC values of 25 mg/L, whereas the MFC for M. officinalis was 12.5 mg/L. The PE extracts of C. monnieriand M. officinalis showed inhibition against Achlyaklebsianaspores with MFC values of 25 mg/L, while the MFC value for A. lappawas 12.5 mg/L.

The antifungal efficacy of PE extracts from C. monnieri, M. officinalis, and A. lappai spresented in Table3. The PE extract of A. lappa was the most effective against Saprolegniaand Achlyaklebsianamycelium, demonstrating the lowest EC_{50} values of 11.3 and 26.1 mg/L, respectively.

Discussion

In previous studies, many in vitro screening methods have been described for the testing of new drugs against Saprolegnia. （Stueland et al.（2005））developed a simple, rapid in vitro screening method to test the fungistatic and fungicidal effects of chemical drugs against Saprolegnia. However, that method is not applicable to water-insoluble substances. Therefore, we used a rapeseed microplate method, which can quickly and conveniently evaluate the anti - Saprolegniaactivity of some waterinsoluble substances. The rapeseed microplate method may well be useful for other purposes and for other oomycetes and fungi as well, but this requires further investigation.

Table 3　Antifungal efficacy of petroleum ether extracts from Cnidiummonnieri, Magnolia officinalis, and Aucklandialappa (EC_{50} = 50% effective concentration, with 95% confidence interval in parentheses) against Saprolegniasp. and Achlyaklebsiana (y=probit (p); x=the $\log_{10}$ transformation of the concentration).

Species	Against	Regression equation	EC_{50} (mg/L)	X2
Cnidiummonnieri	Saprolegnia	y = −3.512 + 2.201x	39.422 (25.839−83.085)	2.216
	Achlyaklebsiana	y = −2.673 + 1.607x	46.117 (39.223−54.599)	2.659
Magnolia officinalis	Saprolegnia	y = −2.943 + 2.102x	25.136 (14.776−41.045)	0.097
	Achlyaklebsiana	y = −2.372 + 1.424x	46.317 (38.748−55.870)	1.474
Aucklandialappa	Saprolegnia	Y = −2.274 + 2.158x	11.314 (4.640−17.640)	1.526
	Achlyaklebsiana	Y = −2.356 + 1.662x	26.134 (22.128−30.609)	6.591

The roots of A. lappa (Saussureacostusin India; also known as "muxiang" in China), which is officially listed in the Chinese Pharmacopoeia (MOH, 2005), have been widely used as a traditional Chinese medicine for the treatment of various kinds of disorders, such as asthma (Shah, 1982; Sircar, 1984), cough, diarrhea, vomiting, indigestion, inflammation, and rheumatism (Shah 1982). The dried ripe fruit of C. monnieri (also known as fructuscnidii; "she chuangzi" in Chinese) has been used for the treatment of impotence, frigidity, and skin-related diseases, such assuppurative dermatitis and pudend alitching (Zhengetal., 1998; MOH, 2005). Anthelmintic activity against Dactylogyrus intermedius on the gills of Goldfish Carassiusauratuswas also observed (Wang et al., 2006, 2008; Cao et al. 2012) found that the water extract of C. monnierihad no inhibition against Saprolegnia, with an MIC value of over 10.0 mg/mL. However, we found that the PE extract of C. monnieriresulted in strong inhibition of Saprolegniaand an MIC value of 62.5 mg/L. The difference between the two studies may result from geographic and ecological differences that produce within-species variation in C. monnieriand in Saprolegnia. In addition, the bioactivity of the extracts can be affected by the solvent and extraction techniques used. The bark of M. officinalis (referred to as "houpo" in Chinese) has been used in traditional Chinese medicine for the treatment of abdominal distention and pains, dyspepsia, and asthmatic cough (MOH, 2005). Many studies have demonstrated that M. officinalis has a wide variety of pharmaceutical properties, such as antispasmodic, antioxidant, anticancer, and antidepressant activities (Lee et al., 2011). However, very little attention had been focused on the antimicrobial ability of M. officinalis, especially anti-Saprolegniaactivity, prior to our study.

The PE extracts of C. monnieri, M. officinalis, and A. lappas howed highly inhibitor yaction against Saprolegnia and Achlya, and this report provides the first documentation of antifungal activity from A. lappaand M. officinalis. Some compounds that are alternatives to malachite green (e.g., formalin, hydrogen peroxide, and sodium chloride; Schreier et al., 1996; Barnes et al., 2003; Gieseker et al. 2006) have been used in the treatment of superficial fungal infections. Although these chemicals can control the disease, they have some disadvantages in practice. Formalin was reported to be effective in treating Saprolegnia infections on eggs, but there are safety concerns for the user and the environment (Marking et al., 1994; Burka et al., 1997). Hydrogen peroxide, which is approved in the USA to control saprolegniasis on freshwater-reared finfish eggs, is strongly corrosive and combustible, with effective concentrations as high as 1 000 mg/L (Schreier et al., 1996). Sodium chloride, despite its safety, may be limited in applicability due to the high cost of acquiring effective concentrations as high as 30 000 mg/L.

To date, no treatments have been identified as wholly effective alternatives to malachite green.

In this study, solvent extracts were evaluated for antifungal activity. Additional research will be required to isolate the active compounds and to find methods for addressing water solubility. In addition, field studies of these extracts are required to determine their practical use in fish culture, and the in vivo activity of the extracts must be further investigated. In summary, we found that the PE extracts of A. lappa, C. monnieri, and M. officinalis have the potential for use in developing a novel therapy for the treatment of Saprolegniainfection.

References

Alderman, D. J. 1985. Malachite green: a review. Journal of Fish Diseases 8: 289-298.

Bangyeekhun, E., S. M. A. Quiniou, J. E. Bly, and L. Cerenius. 2001. Characterisation of Saprolegniasp. isolates from Channel Catfish. Diseases of Aquatic Organisms 45: 53-59.

Barnes, M. E., H. Stephenson, M. Gabel. 2003. Use of hydrogen peroxide and formalin treatments during incubation of landlocked fall Chinook Salmon eyed eggs. North American Journal of Aquaculture 65: 151–154.

Burka, J. F., K. L. Hammell, T. E. Horsberg, et al. 1997. Drugs in salmonid aquaculture—a review. Journal of Veterinary Pharmacology and Therapeutics 20: 333–349.

Cao, H., W. Xia, S. Zhang, et al. 2012. Saprolegniapathogen from Pengze Crucian Carp (Carassiusauratusvar. Pengze) eggs and its control with traditional Chinese herb. Israeli Journal of Aquaculture Bamidgeh [online serial] IJA 64. 2012. 685.

Eguale, T., G. Tilahun, A. Debella, et al. 2007. In vitro and in vivo anthelmintic activity of crude extracts of Coriandrumsativumagainst Haemonchuscontortus. Journal of Ethnopharmacology 110: 428–433.

Finney, D. J. 1971. Probit analysis, 3rd edition. Cambridge University Press, New York.

Gieseker, C. M., S. G. Serfling, R. Reimschuessel. 2006. Formalin treatment to reduce mortality associated with Saprolegniaparasiticain Rainbow Trout, Oncorhynchus mykiss. Aquaculture 253: 120–129.

Hussein, M. M. A., K. Hatai, T. Nomura. 2001. Saprolegniosis in salmonids and their eggs in Japan. Journal of Wildlife Diseases 37: 204–207.

Ke, X. L., J. G. Wang, Z. M. Gu, et al. 2009. Morphological and molecular phylogenetic analysis of two Saprolegniasp. (oomycetes) isolated from Silver Crucian Carp and Zebra Fish. Mycological Research 113: 637–644.

Lee, Y. J., Y. M. Lee, C. K. Lee, et al. 2011. Therapeutic applications of compounds in the Magnolia family. Pharmacology and Therapeutics 130: 157–176.

Li, H. B., F. Chen. 2005. Simultaneous separation and purification of five bioactive coumarins from the Chinese medicinal plant Cnidiummonnieriby high-speed counter-current chromatography. Journal of Separation Science 28: 268–272.

Marking, L. L., J. J. Rach, T. M. Schreier. 1994. Evaluation of antifungal agents for fish culture. Progressive Fish–Culturist 56: 225–231. Meyer, F. P. 1991. Aquaculture disease and health management. Journal of Animal Science 69: 4201–4208.

MOH (Ministry of Health). 2005. Pharmacopoeia Commission, People's Republic of China, Chemical Industry Press, Beijing. Pharmacopoeia of the People's Republic of China, English edition, volume 1. MOH.

NCCLS (National Committee for Clinical Laboratory Standards). 1997. Methods for dilution antimicrobial susceptibility tests for bacteria that grow aerobically; approved standard, 4th edition. NCCLS, M7-A4, Wayne, Pennsylvania.

Schreier, T. M., J. J. Rach. 1996. Efficacy of formalin, hydrogen peroxide, and sodium chloride on fungal-infected Rainbow Trout eggs. Aquaculture 140: 323–331.

Shah, N. C. 1982. HerbalfolkmedicinesinnorthernIndia. JournalofEthnopharmacology 6: 293–301.

Sircar, N. N. 1984. Pharmaco-therapeutics of dasemani drugs. Ancient Science of Life 3: 132–135.

Stueland, S., B. T. Heier, I. Skaar. 2005. A simple in vitro screening method to determine the effects of drugs against growth of Saprolegniaparasitica. Mycological Progress 4: 273–279.

Takada, K., H. Kajiwara, N. Imamura. 2010. Oridamycins A and B, antiSaprolegniaparasiticaindolosesquiterpenes isolated from Streptomyces sp. KS84. Journal of Natural Products 4: 698–701.

Wang, G. X., C. Chen, A. L. Cheng, et al. 2006. Dactylogyruskilling efficacies of 22 plant extracts and its 6 compounds. Acta Botanica BorealiOccidentaliaSinica 26: 2567–2573.

Wang, G. X., Z. Zhou, C. Cheng, et al. 2008. Osthol and isopimpinellin from Fructuscnidiifor the control of Dactylogyrus intermedius in Carassiusauratus. Veterinary Parasitology 158: 144–151.

Willoughby, L. G., R. J. Roberts. 1992. Towards strategic use of fungicides against Saprolegniaparasiticain salmonid fish hatcheries. Journal of Fish Diseases 15: 1–13.

Zheng, H. Z., Z. H. Dong, J. She, editors. 1998. Modern study of traditional Chinese medicine, volume 5. Xue Yuan Press, Beijing

该文发表于《JOURNAL OF THE WORLD AQUACULTURE SOCIETY》【2013, 44 (4): 528–535. DOI: 10.1111/jwas. 12052】

In Vitro Screening of Fungicidal Chemicals for Antifungal Activity against *Saprolegnia*

Xue-Gang Hu[1], Lei Liu[1], Kun Hu[2] and Xian-Le Yang[2], Gao-Xue Wang[1]

(1. College of Animal Science and Technology, Northwest A&F University, Xinong Road 22nd, Yangling, Shaanxi 712100, China;
2. National Pathogen Collection Center for Aquatic Animals, College of Fisheries and Life Science,
Shanghai Ocean University, 999 HuchengHuan Road, Shanghai 201306, China)

Abstract: Saprolegnia is an important fish fungal pathogen that often results in significant economic losses to freshwater aquaculture. To find effective drugs to control saprolegnias, 30 fungicidal chemicals used in agriculture were screened, in which kresoxim-methyl and azoxystrobin, with minimum inhibitory concentration (MIC) values of 1.0 and 0.5 mg/L, respectively, showed good in vitro antifungal activities against Saprolegnia. Azoxystrobin has the most promising anti-Saprolegnia activity with 50% effective concentration (EC_{50}) value of 0.212 mg/L against mycelial growth and minimum fungicidal concentration (MFC) value of 0.13 mg/L against spores, while EC_{50} and MFC values to kresoxim-methyl are 0.240 and 0.25 mg/L, respectively. Through the acute toxicity assay using goldfish, Carassiusauratus, azoxystrobin exhibited wider margin of safety with a safe concentration (SC) value of 0.553 mg/L than kresoxim-methyl with an SC value of 0.131 mg/L. These findings demonstrated that azoxystrobin has the potential for the development of therapy for the control of Saprolegniain in aquaculture. Both kresoxim-methyl and azoxystrobin were tested with a postantifungal effects (PAFE) assay and the results revealed that the two chemicals had no significant effect on fungal growth inhibition after a 1-hour exposure, indicating that the treatment needs to be carried out over an extended period.

体外筛选具有抗水霉活性的化学活性成分

胡学刚[1]，刘镭[1]，胡鲲[2]，杨先乐[2]，王高学[1]

（1. 西北农林科技大学动物科学技术学院，西农路22号，中国陕西，西安杨凌；
2. 上海海洋大学水产与生命学院国家水生动物病原库，沪城环路999号，中国上海）

摘要： 水霉菌是重要的鱼类真菌性病原体，经常导致淡水养殖业重大的经济损失。为了找到控制水霉菌的有效的药物，我们筛选了在农业上应用的30种抗真菌化学药物，其中醚菌酯和嘧菌酯的最小抑菌浓度分别为1.0 mg/L和0.5 mg/L，表现出很好的体外对抗水霉菌的抗真菌活性。嘧菌酯具有最有效的抗水霉活性，抑制菌丝生长的半有效浓度（EC_{50}）为0.212 mg/L，对孢子的最小抑菌浓度（MFC）为0.13 mg/L。醚菌酯的 EC_{50} 和MFC分别为0.240 mg/L and 0.25 mg/L。通过对金鱼和鲫鱼的急性毒性实验，嘧菌酯表现出比醚菌酯0.131 mg/L的安全浓度（SC）更广的安全系数，安全浓度（SC）为0.553 mg/L。结果表明嘧菌酯有发展成控制水产养殖业水霉菌的药物的潜力。醚菌酯和嘧菌酯都被进行了抗真菌效果试验，结果显示经过1 h曝光后这两个化学物质对抑制真菌生长没有显著影响，这表明治疗需要长期进行。

Water molds of the genus Saprolegniaspp. are responsible for significant fungal infections involving both dead and living fish as well as eggs in the aquaculture industry (Pickering and Willoughby, 1982; Noga, 1993; Ke et al., 2009). Saprolegniasis is the term used to describe infections caused by this genus of water molds (Roberts 1989; Beakes et al., 1994). This disease is quite harmful to intensively cultured freshwater fish especially salmonids (Hatai and Hoshiai 1992; Noga, 1993; DieguezUribeondo et al., 1996) and channel catfish, Ictalurus punctatus, (Bly et al., 1992) and is widespreading all stages of their life cycle from egg to adult. Saprolegniasis appears as cotton-wool-like tufts on the body surface causing destruction of the skin and/or fins due to cellular necrosis by hyphal penetration that is generally restricted to the epidermis and dermis (Fregeneda-Grandes et al., 2007).

In the past, malachite green, an organic dye, had been widely used as therapeutic agent to prevent Saprolegniainfections in aquaculture (Willoughby and Roberts, 1992). However, due to its teratogenic and carcinogenic potential (Alderman 1985; Doerge et al., 1998), the use of malachite green was prohibited in the USA in 1991 and in China in 2002. Current research for the alternative anti-Saprolegnia agents have focused mainly on formalin, hydrogen peroxide, and sodium chloride (Waterstrat and Marking 1995; Schreier et al. 1996; Barnes et al. 1998; Barnes et al., 2003). Although these chemicals can control the disease, they have some weak points in practice. Formalin was reported to be effective in treating Saprolegnia infections on eggs but raises safety concerns in relation to the user and the environment (Marking et al., 1994; Burka et al., 1997). Hydrogen peroxide, approved in the USA to control saprolegniasis on freshwater reared finfish eggs, is strongly corrosive and combustible with effective concentrations as high as 1000mg/L (Schreier et al., 1996), and sodium chloride, in spite of its safety, may be limited in its applicability due to its high cost in acquiring effective concentrations as high as 30 000mg/L. To date, no wholly effective alternative treatment to malachite green has been identified. Therefore, there is a continued need for therapeutants to control fungal infections (Rach et al., 1998). To find effective therapeutants to prevent and treat saprolegniasis, an attempt has been made under the present work to explore 30 fungicidal chemicals used in agriculture for their in vitro antifungal activities against Saprolegnia.

Materials and Methods

Fungal Culture

Pure strain, Saprolegnia sp. strain JL (accession No. HM637287) (Cao et al., 2012), was obtained (National Pathogen Collection Center for Aquatic Animals, Shanghai Ocean University), cultured on potato dextrose agar (PDA) slants and stored at 4C until used.

Tested Compounds

Thirty chemicals (Table 1) were purchased from Hubei KangBaoTai Fine-Chemicals Co., Ltd (Hubei, China). Each compound was dissolved in dimethyl sulfoxide (DMSO) (SigmaAldrich, St. Louis, MO, USA) to obtain 20.0 mg/mL (sample/solvent) of stock solutions, which were used for the preparations of the desired concentrations for antifungal efficacy assay.

Minimum Inhibitory Concentration Evaluation

The method used in this study for antifungal activity assay was performed with some modifications (Stueland et al. 2005b) using rapeseed, Brassica napus. Subcultures on fresh

PDA media were done 24 h prior and used to inoculate the rapeseeds. Final test concentrations of the drugs were produced by dilution in sterilized distilled water (SDW), and all final concentrations were prepared in 48 well flat bottom tissue culture plate. The compounds were tested at 100.0, 50.0, 25.0, 12.5, 6.3 and 3.1 mg/L to find a preliminary minimum inhibitory concentration (MIC) interval. On the basis of these encouraging preliminary MIC values (<3.1mg/L) for both kresoximmethyl and azoxystrobin, a smaller and more precise concentration gradient interval (4.0, 2.0, 1.0, 0.5, 0.25 and 0.13 mg/L) was then used to ascertain the exact MIC values for these compounds. To ensure good growth of Saprolegnia, 1% Sabouraud's medium (10 g neopeptone, 20 g glucose, 5g bacto yeast extract, 1 L distilled water, 1 mL 0.5% chloramphenicol, 400 000 IU penicillin and 200 000 IU polymyxin) was included in each test concentration. The plate was incubated at 25℃ for 48 h. The MIC values were recorded visually on the basis of mycelia growth.

All the independent experiments were conducted three times with triplicates at each test concentration. DMSO was the negative control and malachite green was the positive control.

Spores Germination Inhibition Test

The spore germination inhibition test was carried out by agar dilution method. Briefly, the Saprolegnia isolate was grown on PDA plates for 7-14 d, after which time spores were harvested from sporulating colonies and suspended in SDW. The concentration of spores in suspension were determined using a hemocytometer and adjusted to 1×10^4 CFU/mL approximately. The agar plates were prepared with the required concentration of the two chemicals (kresoxim-methyl and azoxystrobin), added to 10 mL of molten PDA (about 65℃). Ten microliters of the spore suspension were spotted in triplicate on these plates which were then incubated at 25℃ for 72 h. The minimum fungicidal concentration (MFC) was defined as the lowest concentration of the chemicals that prevented visible growth or germination of spores.

Table 1 The MIC values of 30 fungicidal chemicals against Saprolegnia.

No.	Chemicals	Purity (%)	CAS No.	MIC (mg/L)
1	Procymidone	99.0	32809-16-8	>100.0
2	Propamocarb	97.0	24579-73-5	>100.0
3	Thiram	95.0	137-26-8	>100.0
4	Carbendazim	96.5	10605-21-7	25.0
5	Thiophanate-methyl	96.0	23564-05-8	>100.0
6	Oxadixyl	97.0	77732-09-3	>100.0
7	Kresoxim-methyl	98.5	143390-89-0	>100.0
8	Azoxystrobin	98.0	131860-33-8	1.0 (<3.1)
9	Kasugamycin	97.7	19408-46-9	0.5 (<3.1)
10	Validamycin	96.0	37248-47-8	>100.0
11	Ethoxyquin	95.7	91-53-2	>100.0
12	Tridemorph	97.0	81412-43-3	100.0

No.	Chemicals	Purity (%)	CAS No.	MIC (mg/L)
13	Thiabendazole	96.5	148−79−8	50.0
14	Pyrimethanil	95.7	53112−28−0	>100.0
15	Cyprodinil	98.5	121552−61−2	>100.0
16	Chlorothalonil	98.7	1897−45−6	100.0
17	Triadimefon	98.5	43121−43−3	>100.0
18	Paclobutrazol	97.0	76738−62−0	25.0
19	Difenoconazole	98.0	119446−68−3	100
20	Flutriafol	96.0	76674−21−0	50.0
21	Diniconazole	95.5	83657−24−3	>100.0
22	Propiconazole	95.5	60207−90−1	>100.0
23	Dimethomorph	97.0	110488−70−5	100.0
24	Tecloftalam	97.5	76280−91−6	>100.0
25	Dimetachlone	99.5	24096−53−5	>100.0
26	Iprobenfos	98.5	26087−47−8	>100.0
27	Ethirimol	96.0	23947−60−6	>100.0
28	Sodium Dichloroisocyanurate	95.0	2893−78−9	>100.0
29	Mancozeb	98.0	8018−1−7	>100.0
30	Iprovalicarb	97.7	140923−17−7	>100.0
31	Malachite green	95.0	569−64−2	>100.0
32	DMSO	99.5	67−68−5	1.0

MIC = minimum inhibitory concentration; DMSO = dimethyl sulfoxide.

Mycelial Growth Inhibition Test

In vitro anti-Saprolegniaactivities of the two chemicals (kresoxim-methyl and azoxystrobin) were assessed on the basis of mycelial growth inhibition rate. Each chemical was added to assay flasks containing hot sterilized PDA (about 65℃) and final concentrations were adjusted to corresponding concentrations based on the initial tests. After mixing with a vortex, aliquots (10 mL) of treated medium were poured into 7-cm diameter Petri dishes. The Saprolegnia-colonized rapeseeds were inoculated in the center of the prepared media. The mycelial growth diameter was measured after inoculation at 25℃ for 48 h. The growth inhibition rate was calculated from mean values as:

$$\%IR = 100 (x - y) / (x - z)$$

where IR is the growth inhibition rate; x, the mycelial growth in control; y, the mycelial growth in sample; and z, the average diameter of the rapeseeds.

Post-antifungal Effects Test

Post-antifungal effects (PAFE) is defined as the delay of fungal regrowth persisting after exposure to an antifungal agent, which measures the growth recovery capacity after a limited exposure. The methods of Smith et al. (2011) and Zhang et al. (2011) with some modifications were used to determine the PAFE of the two chemicals. One Saprolegnia colonized rapeseed was added to each well of 48-well plate containing drug (1mL, dissolved in SDW) at concentrations ranging from 0.5 to 16×MIC. Following 1-h exposure, the drug was removed by washing five times. After the final wash, the rapeseed was transferred to the center of the PDA in a 7-cm diameter Petri dish. After incubating at 25℃ for 48h, the diameter of each dish was measured and the growth inhibition rate was calculated according to the above formula.

Acute Toxicity Test

Acute toxicities of the two chemicals were performed as described in our previous work (Wang et al., 2008). One-year-old healthy goldfish, Carassiusauratus, (mean weight 4.0±0.3g,) without any record of previous record of disease or parasitic infestation were obtained from Changxing fish farm (Xianyang, China). The fish were acclimatized for 30 d before experiment and maintained in several 100 L aquarium tanks under laboratory conditions (pH7. 2±0. 5, total ammonia nitrogen<0. 1 mg/L, nitrites<0. 1 mg/L, alkalinity 140 mg/L of $CaCO_3$, oxygen content higher than 85% saturation, 22±1℃, fed with commercial goldfish diet at 2% of body weight every morning). A 48 h exposure was performed in a 10 L aquarium tank with the same water quality parameters as the above conditions. It was conducted in triplicate using 10 individuals in each tank, containing 6 L of the test solution water. Each sample was assayed at a different series of concentrations ranged from 0. 5, 0. 6, to 1. 0mg/L for kresoxim-methyl and 2. 0, 2. 4, 2. 8, 3. 2, 3. 6, and 4. 0 mg/L for azoxystrobin (based on initial tests). Control groups were set under the same test conditions without chemicals. The fish were carefully observed for any signs of distress indicative of toxic insult such as increased respiration frequency and erratic behavior. Under these circumstances, the experiments were stopped and fish were transferred to fresh water. Mortalities were noted after 24 and 48h of exposure, during which no food was offered to the fish. The 50% lethal concentration (LC_{50}) and its confidence intervals were calculated for both compounds using Probit analysis. And the safe concentration (SC) was calculated from mean values according to Turubell formula (Zou et al. 2012):

$$SC = \frac{48hLC_{50} \times 0.3}{\left(\frac{24hLC_{50}}{48hLC_{50}}\right)^2}$$

Statistical Analysis

SPSS 17. 0 software (SPSS Inc. , Chicago, IL, USA) was used for all the statistical analysis. The EC_{50} and LC_{50} at the 95% confidence level were calculated by Probit analysis (Finney 1971).

Results

In this study, 30 compounds used in agriculture were screened for the MIC values (Table 1). Among them, 24 chemicals (80%) were considered to have no inhibition effect against Saprolegniawith MIC values ≥100. 0 mg/L while other six chemicals (thiram, kresoxim-methyl, azoxystrobin, tridemorph, triadimefon, and difenoconazole) showed antifungal activities with MIC ≤50. 0 mg/L, in which kresoxim-methyl and azoxystrobin have the most promising anti-Saprolegnia activities with the MIC values of 1. 0 and 0. 5 mg/L, respectively.

Table 2 lists the EC_{50} values of kresoximmethyl and azoxystrobin against Saprolegnia mycelium at 48h and MFC values against spores at 72h. The results reveal that azoxystrobin showed the best activity with the EC_{50} and MFC values of 0. 212 and 0. 13 mg/L, followed by kresoxim-methyl of 0. 240 and 0. 25 mg/L, respectively.

Figure 1 graphically showed the PAFE for kresoxim-methyl, azoxystrobin, and malachite green against Saprolegnia at concentrations of 0. 5, 1, 2, 4, 8, and 16×MIC. Compared to the PAFE curve for malachite green, the curve for kresoxim-methyl revealed that kresoximmethyl did not significantly inhibit the growth of Saprolegnia after 1 h exposure to the chemical. The growth inhibition rate of Saprolegnia after 1-h exposure to azoxystrobin appeared to be unaffected by the concentration of the chemical. The inhibition of Saprolegnia was 29% with concentrations of azoxystrobin ranging from 0. 25 to 8 mg/L.

Table 2　The EC_{50} values of kresoxim-methyl and azoxystrobin against mycelium at 48 h and MFC against spores at 72h

No.	Chemicals	EC_{50} (mg/L) (LCL-UCL)	MFC (mg/L)
7	Kresoxim-methyl	0. 240 (0. 149-0. 333)	0. 25
8	Azoxystrobin	0. 212 (0. 097-0. 992)	0. 13

EC_{50} = 50% effective concentration, LCL=lower confidence limit, UCL=upper confidence limit; MFC=minimum fungicidal concentration.

The results of acute toxicity assay for the two chemicals are shown in Table 3. The LC_{50} values at 24 and 48h of kresoxim-methyl are 0. 807 and 0. 658 mg/L, while that of azoxystrobin are 3. 290 and 2. 712 mg/L, respectively. Using the Turubell formula, the SC values of kresoxim-methyl and azoxystrobin are 0. 131 and 0. 553 mg/L, respectively.

Discussion

Fish infected with Saprolegnia are easily recognized by the cotton-like white to grayish patches on the skin and gills visible to the naked eye (Stueland et al. 2005a). It invades epidermal tissues, causing destruction and loss of epithelium integrity due to cellular necrosis by hyphal penetration of the basement membrane (Pickering and Willoughby 1982). The infection progresses very quickly and often results in mortality, causing huge losses of both fish and ova (Stueland et al. 2005a).

In this study, the method reported by Stueland et al. (2005b), which is a simple and rapid in vitro screening method for testing of the fungistatic and fungicidal effect of chemical drugs, was used to evaluate the antifungal activity against Saprolegnia with slight

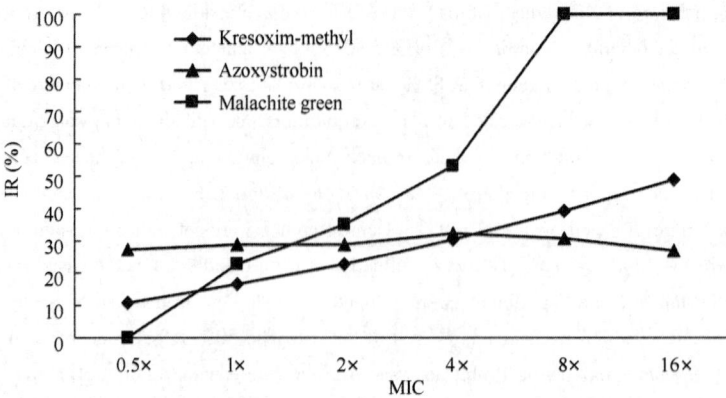

Fig. 1. Post-antifungal effects for kresoxim-methyl, azoxystrobin, and Malachite green against Saprolegnia

modifications. Rapeseed was used instead of hempseed as rapeseed is more abundant in Shaanxi province and is easy to obtain. More importantly, rapeseed was successfully used in an earlier Saprolegniastudy (Wen et al, 1996).

In intensive aquaculture, fish are more likely to suffer from fungal, bacterial, parasitic, viral infections and environmental, and nutritional diseases, many of which are similar to those concentration. of higher organisms. Therefore, the prevention and treatment of these diseases follow the same principles as diseases of other animals. Many of the drugs and chemicals used in chemotherapy of fishes are the same as for higher animals and plants. To find effective drugs to control saprolegniasis, 30 fungicidal chemicals used in agriculture were screened, in which two chemicals (kresoxim-methyl and azoxystrobin, with EC_{50} values of 0. 240 and 0. 212 mg/L, respectively) showed good in vitro antifungal activities against Saprolegnia. To the best of our knowledge, this is the first report on the anti-Saprolegniaactivity of kresoxim-methyl and azoxystrobin.

Both kresoxim-methyl and azoxystrobin, belonging to the group β-methoxyacrylatestrobilurins, are members of a class of fungicides derived from the fungal secondary metabolite strobilurin A produced by Strobilurustenacellus (Anke et al. , 1977). These fungicides are characterized by a broad spectrum of fungicidal activity against numerous foliar pathogens representing members of oomycetes (chromista), ascomycetes, basidiomycetes, and deuteromycetes (fungi), parasitizing different crop plants like pome fruits, grape vines, vegetables, and cereals (Grossmann and Retzlaff, 1997). The protective properties of kresoximmethyl were evaluated based on its effects on germination and sporulation of the rust fungi, Puccinia recondite and Uncinulaappendiculatus, and the powdery mildews Podosphaeraleucotrichaon apple, Uncinulanecatoron grape, Sphaerothecafuligineaand Erysiphecichoracearumon cucumber, and Erysiphegraminisf. sp. triticiand hordeion wheat and barley, using the detached leaves of the respective host plants (Stark-Urnau et al. , 1996). In a recent study, the mixture of kresoxim-methyl and boscalid was investigated for control efficacy against strawberry gray mold disease caused by Botrytis cinerea (Zhang et al. , 2008). Azoxystrobin was first marketed in 1996 and has since then been registered worldwide for use on a wide range of crops (Bartlett et al. , 2002). It was approved worldwide for disease control in most cereals, fruits, and vegetables, being currently registered for use in nearly 120 crops around 100 countries (Parra et al. , 2012). At present, this fungicide is being massively sold in a variety of formulations under different trade names. Nearly 4000 m. t. of azoxystrobin was used worldwide in 2009, with global annual sales over US $ 1 billion, which makes it the world's leading proprietary fungicide (Atkin 2010).

Kresoxim-methyl and azoxystrobin are widely applied to control crop disease in agriculture, and have not been banned for use in the aquaculture industry (MOA, 2002). According to Tomlin (2000), kresoximmethyl showed low toxicity to birds and mammals with 50% lethal dose (LD$_{50}$) values>2 150 and >5 000 mg/kg, while azoxystrobin with LD$_{50}$>2 000 and >5 000 mg/kg. In our study, the MIC of azoxystrobin was close to its SC while kresoxim-methyl showed higher toxicity to goldfish with MIC>SC. A 48 h exposure to kresoxim-methyl was more harmful to goldfish compared with azoxystrobin; however, behavior was normal when exposed to the minimum toxicity concentrations of 0. 5 mg/L kresoxim-methyl and 1. 0 mg/L azoxystrobin. On the basis of the acute toxicity, azoxystrobin has the potential for controlling saprolegniasis. Kresoxim-methyl may need chemical modifications to enhance its inhibitive effect or could be used in combination with other drugs.

In this study, kresoxim-methyl and azoxystrobin were tested the effects for in vitro antifungal activity against Saprolegniamycelium and spores. Both kresoxim-methyl and azoxystrobin showed no significant PAFE on Saprolegniagrowth inhibition. These results indicate that prolonged or multiple treatments will be required to treat against saprolegniasis using these chemicals. Cao et al. (2012) reported that a 5 d exposure of C. auratuseggs and Megalobramaamlycephalaeggs in Pseudolarixkaempferiextracts at 15 mg/mL reduced the inci-

578

dence of saprolegniasis by up to 80% in field trials. Considering the low toxicity of azoxystrobin, long-term exposure in low concentrations during egg stage might be a therapeutic strategy in practice.

In summary, azoxystrobin has the potential for the development of therapy for the treatment against Saprolegniainfection. However, the in vivo activity need to be further investigated and field studies are required before practical use of the chemical in fish aquaculture.

Table 3 The LC$_{50}$ values at 24 and 48 h and the SC values of kresoxim-methyl and azoxystrobin.

No.	Chemicals	LC$_{50}$ ((mg/L) (LCL-UCL)		MFC (mg/L)
		24 h	48 h	SC (mg/L)
7	Kresoxim-methyl	0.807 (0.713-0.930)	0.658 (0.537-0.747)	0.131
8	Azoxystrobin	3.290 (2.730-4.521)	2.712 (2.314-3.039)	0.553

LC$_{50}$ = 50% lethal concentration; LCL=lower confidence limit; UCL=upper confidence limit; SC= safe

Literature Cited

Alderman, D. 1985. Malachite green: a rewiew. Journal of Fish Diseases 8: 289-298.

Anke, T., F. Oberwinkler, W. Steglich, and G. Schramm. 1977. The strobilurins-new antifungal antibiotics from the basidiomycete Strobilurustenacellus. The Journal of Antibiotics 30: 806-810.

Atkin J. 2010. UBS Basic Materials Conference. Accessed April 14, 2011, at http://www2.syngenta.com/en/investor_relations/event_calendar.html.

Barnes, M.E., D.E. Ewing, R.J. Cordes, and G.L. Young. 1998. Observations on hydrogen peroxide control of Saprolegniaspp. during rainbow trout egg incubation. The Progressive Fish-Culturist 60: 67-70.

Barnes, M.E., H. Stephenson, and M. Gabel. 2003. Use of hydrogen peroxide and formalin treatments during incubation of landlocked fall chinook salmon eyed eggs. North American Journal of Aquaculture 65: 151-154.

Bartlett, D.W., J.M. Clough, J.R. Godwin, A.A. Hall, M. Hamer, and B. Parr-Dobrzanski. 2002.

The strobilurin fungicides. Pest Management Science 58: 649-662.

Beakes, G.W., S.E. Wood, and A.W. Burr. 1994. Features which characterize Saprolegniaisolates from salmonid fish lesions-a review. Pages 33-66 in G.J. Mueller, editor. Salmon Saprolegniasis. U.S. Department of Energy, Bonneville Power Administration, Portland, Oregon, USA.

Bly, J.E., L.A. Lawson, D.J. Dale, A.J. Szalai, R.M. Durborow, and L.W. Clem. 1992. Winter saprolegniosis in channel catfish. Diseases of Aquatic Organisms 13: 155-164.

Burka, J.F., K.L. Hammell, T.E. Horsberg, G.R. Johnson, D.J. Rainnie, and D.J. Speare. 1997. Drugs in salmonid aquaculture-a review. Journal of Veterinary Pharmacology and Therapeutics 20: 333-349.

Cao, H.P., W.W. Xia, S.Q. Zhang, S. He, R.P. Wei, L.Q. Lu, and X.L. Yang. 2012. Saprolegniapathogen from pengze crucian carp (Carassiusauratusvar. Pengze) eggs and its control with traditional chinese herb. The Israeli Journal of Aquaculture-Bamidgeh64: Special section, p1.

DieguezUribeondo, J., L. Cerenius, and K. Soderhall. 1996. Physiological characterization of Saprolegniaparasiticaisolates from brown trout. Aquaculture 140: 247-257.

Doerge, D.R., M.I. Churchwell, T.A. Gehring, Y.M. Pu, and S.M. Plakas. 1998. Analysis of malachite green and metabolites in fish using liquid chromatography atmospheric pressure chemical ionization mass spectrometry. Rapid Communications in Mass Spectrometry 12: 1625-1634.

Finney, D. 1971. Probit Analysis, 3rd edition. Cambridge University Press, Cambridge, UK. Fregeneda-Grandes, J.M., F. Rodriguez-Cadenas, M.T. Carbajal-Gonzalez, and J.M. Aller-Gancedo. 2007. Antibody response of brown trout Salmo truttainjected with pathogenic Saprolegniaparasiticaantigenic extracts. Diseases of Aquatic Organisms 74: 107-111.

Grossmann, K. and G. Retzlaff. 1997. Bioregulatory effects of the fungicidal strobilurinkresoximmethyl in wheat (Triticumaestivum). Pesticide Science 50: 11-20.

Hatai, K. and G. Hoshiai. 1992. Mass mortality in cultured coho salmon (Oncorhynchuskisutch) due to Saprolegniaparasiticacoker. Journal of Wildlife Diseases 28: 532-536.

Ke, X.L., J.G. Wang, Z.M. Gu, M. Li, and X.N. Gong. 2009. Morphological and molecular phylogenetic analysis of two Saprolegniasp. (Oomycetes) isolated from silver crucian carp and zebra fish. Mycological Research 113: 637-644.

Marking, L.L., J.J. Rach, and T.M. Schreier. 1994. American Fisheries Society evaluation of antifungal agents for fish culture. The Progressive Fish-Culturist 56: 225-231.

Ministry of Agriculture The People's Republic of China (MOA). 2002. Pollution-free food Guidelines for the use of fishery drugs. Agricultural industry standard of the People's Republic of China NY5071-2002.

Noga, E.J. 1993. Water mold infections of freshwater fish: recent advances. Annual Review of Fish Diseases 3: 291-304.

Parra, J., J.V. Mercader, C. Agullo, A. Abad-Somovilla, and A. Abad-Fuentes. 2012. Generation of antiazoxystrobin monoclonal antibodies from regioisomerichaptens functionalized at selected sites and development of indirect competitive immunoassays. AnalyticaChimicaActa 715: 105－112.

Pickering, A. D. and L. G. Willoughby. 1982. Saprolegniainfections of salmonid fish. Pages 38－48 in 50th Annual Report. Institute of Freshwater Ecology.

Rach, J. J., M. P. Gaikowski, G. E. Howe, and T. M. Schreier. 1998. Evaluation of the toxicity and efficacy of hydrogen peroxide treatments on eggs of warm－and coolwater fishes. Aquaculture 165: 11－25.

Roberts, R. J. 1989. Pages 467 Fish Pathology, 2nd edition. Bailliere Tindall Publishers, London, UK.

Schreier, T. M., J. J. Rach, and G. E. Howe. 1996. Efficacy of formalin, hydrogen peroxide, and sodium chloride on fungal－infected rainbow trout eggs. Aquaculture 140: 323-331.

Smith, R. P., A. Baltch, L. H. Bopp, W. J. Ritz, and P. P. Michelsen. 2011. Post－antifungal effects and time－kill studies of anidulafungin, caspofungin, and micafungin against Candida glabrataand Candida parapsilosis. Diagnostic Microbiology and Infectious Disease 71: 131－138.

Stark－Urnau, M., R. Gold, R. Guggenheim, and M. Duggelin. 1996. Sensitivity of different mildew and rust fungi against kresoxim－methyl. Pages 268－271 in G. H. J. Kema, R. E. Nike, and R. A. Daamen, editors. Proceedings of the 9th European and Mediterranean Cereal Rusts & Powdery Mildews Conference, Lunteren, the Netherlands.

Stueland, S., K. Hatai, and I. Skaar. 2005a. Morphological and physiological characteristics of Saprolegnia spp. strains pathogenic to Atlantic salmon, Salmo salarL. Journal of Fish Diseases 28: 445-453.

Stueland, S., B. T. Heier, and I. Skaar. 2005b. A simple in vitro screening method to determine the effects of drugs against growth of Saprolegniaparasitica. Mycological Progress 4: 273-279.

Tomlin, C. D. S. 2000. The Pesticide Manual, 12 edition. BCPC, Farnham, Surrey, UK.

Wang, G. X., Z. Zhou, C. Cheng, J. Yao, and Z. Yang. 2008. Osthol and isopimpinellin from Fructuscnidiifor the control of Dactylogyrus intermedius in Carassiusauratus. Veterinary Parasitology 158: 144-151.

Waterstrat, P. R. and L. L. Marking. 1995. Clinicalevaluation of formalin, hydrogen peroxide, and sodium chloride for the treatment of Saprolegniaparasiticaon fall chinook salmon eggs. The Progressive FishCulturist 57: 287-291.

Wen, B., Y. Zheng, C. L. Liu, and W. Chen. 1996. Techniques for isolating and cultivating Saprolegnia. ActaPhytopathologicaSinica 26: 277－281 (in Chinese with English abstract).

Willoughby, L. G. and R. J. Roberts. 1992. Towards strategic use of fungicides against Saprolegniaparasiticain salmonid fish hatcheries. Journal of Fish Diseases 15: 1-13.

Zhang, C. Q., Y. Zhang, and G. N. Zhu. 2008. The mixture of kresoxim－methyl and boscalid, an excellent alternative controlling grey mould caused by Botrytis cinerea. Annals of Applied Biology 153: 205-213.

Zhang, S. Q., J. Q. Qiu, H. P. Cao, W. W. Xia, and X. L. Yang. 2011. Antifungal effects of antimoulds on Saprolegniasp. Southwest China Journal of Agricultural Sciences 24: 1568-1572 (in Chinese with English abstract).

Zou, X., L. Z. Zhang, J. Y. Liu, Y. Wang, X. R. Huang, and P. Zhuang. 2012. Acute toxicity of the five fishery drugs on Siganusguttatus. Marine Fisheries 34: 189-194 (in Chinese with English abstract).

该文发表于《Journal of world fisheries society》【2016. DOI: 10. 1111/jwas. 12325】

Antifungal activity of rhein and aloe－emodin from Rheum palmatum on fish pathogenic *Saprolegnia* sp.

Jia-yun YAO[1], Ling-yun LIN[1], Xue-meiYUAN[1], Wen-lin YING[1], Yang XU[1], Xiao-yi PAN[1], Gui-jie HAO[1], Jin-yu SHEN[*1], Ting YE[2], Pei-hong GE[3]

(1. Zhejiang Institute of Freshwater Fisheries, Huzhou, Zhejiang, 313001, China;

2Quzhou Aquatic Product Technology Promotion Department, Quzhou, Zhejiang, 313001, China;

3. KaihuaAquatic Product Technology Promotion Department, Quzhou, Zhejiang, 324300, China)

Abstract: Saprolegnia infections cause severe economic losses among freshwater fish farming. In the present study, two known compounds rhein and aloe－emodin were isolated from Rheum palmatum, and the in vitro inhibitory activity of both compounds against mycelial growth and spore germination of Saprolegnia were tested. Both rhein and aloe－emodin were able to decrease Saprolegnia mycelial growth and sporeactivity in all tested concentrations after exposure for 48 h, while complete inhibition of mycelial growth was observed at 20 mg · L^{-1} for rhein and at 50 mg · L^{-1} for aloe－emodin, and spore germination was 100% prevented at 16 and 40 mg · L^{-1} for rhien and aloe－emodin, respectively. For the development of newly fungicidal agents, only rhein showing stronger in vitro anti－Saprolegnia activity were tested in vivo for the prevention and treatment efficacy on Saprolegnia infection of grass carp. Its acute activity to the fish（grass carp）was also evaluated. The results revealed that exposure to rhein at 20 mg·L^{-1} for 7 days could prevent 93.3% of abraded grass carp infecting with Saprolegnia, and 67.7% of infected fish could be recovered due to the exposure of rhein. The 48 h median lethal concentration（48h $-LC_{50}$）to grass carp was 148.5 mg·L^{-1}, which ensures the safety for the use of rhein. This study suggests that rhein has promising anti-Saprolegnia activity and may be an option in preventing and controlling of Saprolegnia infection.

Keywords：Antifungal activity, Saprolegnia, Grass carp, Hydroxyanthraquinone, Natural product

大黄抗水霉的活性成分的研究

姚嘉赟[1]，蔺凌云[1]，袁雪梅[1]，尹文林[1]，徐洋[1]，潘晓艺[1]，

郝贵杰[1]，沈锦玉[*1]，叶挺[2]，葛洪培[3]

（1. 浙江省淡水水产研究所 浙江 湖州 313001；2. 衢州水产技术推广站 浙江 衢州 313001；

3. 开化水产技术推广站 浙江 衢州 313001）

摘要：水霉病水产养殖过程中一类危害较大的真菌类疾病，经常引起鱼类的大量死亡给水产养殖业造成了巨大的经济损失。本研究利用活性追逐试验结合有机分离技术对大黄中的抗水霉的活性成分进行提取分离并获得 2 种抗菌活性较强的活性单体。经质谱、核磁氢谱、核磁碳谱、红外光谱等多种波谱分析技术鉴定这 2 种化合物分别为：大黄酸和芦荟大黄素。体外抗菌结果表明：大黄酸和芦荟大黄素在浓度为 20 mg·L^{-1} 和 50 mg·L^{-1} 可抑制水霉菌丝的生长，两者的浓度分别为 16 mg·L^{-1} 和 40 mg·L^{-1} 可抑制孢子的生长。体内试验结果表明：连续药浴浓度为 20 mg·L^{-1} 大黄酸 7 d，可有效减少感染水霉的数量（93.3%），而芦荟大黄素对应值为 67.7%。对草鱼的急性毒性实验结果表明大黄酸对草鱼的 48 h 半数致死浓度为 148.5 mg·L^{-1}。上述结果表明，大黄素有望开发为一种新型的抗水霉药物。

关键词：抗真菌活性，水霉，羟基蒽酮，天然产物

1　Introduction

Water molds of the genusSaprolegnia（order：Saprolegniales）is among the most important parasitic oomycete groups on fish, with species members being saprophytic and necrotrophic（Caruana et al., 2012）. Saprolegnia species are mostly occurred in freshwater environments, and they infect fish at all stages of their life cycle（van West, 2006；Caruana et al., 2012）. Infected fish show darkened coloration, listlessness, tend to be off their feed and are covered by white, cotton-like hyphae（Khosravi et al., 2012）. When the fungal hyphae are visible, mycelium may have already permeated into fish body tissues, and it unlikely to recovery naturally by the fish（Gieseker et al., 2006）.

Malachite green was considered as the most effective fungicidal agent against Saprolegnia infections in aquaculture（Li et al., 1996）. However, the use of malachite green has been banned due to its genotoxic carcinogenesis and residual toxicity. Alternative treatments are bronopol, sodium chloride, formalin, potassium permanganate and hydrogen peroxide at high concentrations（Waterstrat and Marking, 1995；Schreier et al., 1996；Barnes et al., 1998, 2003）, but none of these chemicals are effective relative to malachite green（Cortés et al., 2014）. Sanitary problems, environmental restrictions, and high cost have also limited the use of these synthetic antimicrobials. There is still an urgent need to develop new alternatives, being effective in combating mycotic infections and also safe for the fish and environment（Khosravi et al., 2012）.

Recently, the application of extracts or active compounds from medicinal plants to control Saprolegnia infection has attracted increasing attention because of demonstrable efficacy and low environmental hazard. Plant extracts, such as petroleum ether extracts of Aucklandialappa, Cnidiummonnieri, and Magnolia officinalis（Hu et al., 2013b）, dichloromethane extract from red seaweed Ceramiumrubrum（Hudson）（Cortés et al., 2014）, and pure compounds such as thymoquinone from Nigella sativa（Hussein et al. 2002）, all of them demonstrated notably antifungal activity against Saprolegnia. Rhubarb（Rheum palmatum）is a traditional Chinese medicinal（TCM）herb, and is widely distributed in Chinese, Japanese and European Pharmacopoeia（Wang et al., 2010）. Many bioactive components with antibacterial, anti-inflammation, and anti-agingproperties have been isolated from R. palmatum（You et al., 2013）. However, there is no available report on their antimicrobial efficacy against fish fungi. In a previous preliminary work, we investigated the anti-Saprolegnia activity of R. Palmatum and found that the n-butanol extracts of R. palmatum（NBu-E） was effective against Saprolegnia. The present study was conducted to assess the antifungal properties of NBu-E and isolate active constituents responsible for the activity using in vitro antifungal assay associated with bioassay-guided fractionation. Additionally, the acute toxicity of the active compounds against grass carp was evaluated.

2 Materials and methods

2.1 Isolation of the pure compounds

R. palmatum was purchased from Hangzhou, Zhejiang Province, China. The dried R. palmatumwere crushed by high-speed disintegrator (FW-100 Test InstrumentCo., Ltd., Tianjin, China) at 20, 000 rpm, and then the powders were passed through 40-mesh sieve. After dried in the cabinet drier at 50℃for 12 h, these powders were stored inthe desiccator.

Dried and powdered R. palmatum were exhaustively extracted with n-butanol at room temperature for 24 h. The extracts were subsequently filtered and the filtrates were evaporated to dryness under reduced pressure in a rotary evaporator (Shanghai Yarong, China) to yield the n-butanol extract. The dried n-butanol extract was subjected to open column chromatography on normal phase silica gel and sequentially eluted with petroleum ether, ethyl acetate and methanol with increasing polarity, eventually affording 538 fractions (300 mL each). TLC (thin layer chromatography) analysis was performed on silica gel using the same solvent system as the mobile phase. Spots on thin layer chromatography were visualized under ultraviolet (UV) light (254 and 365 nm), and fractions showing similar chromatograms were combined into five fractions (Fr. A: 1 ~ 90 fractions, Fr. B: 91 ~ 155fractions, Fr. C: 156 ~ 280 fractions, Fr. D: 281~390 fractions, Fr. E: 391~538 fractions). These five fractions were submitted to in vitro test, Fr. C and Fr. E showed the most active which were then applied to reversed-phase high performance liquid chromatography (RP-HPLC) with following chromatographic conditions: Alltima C18 (5μm, 10 mm×250 mm) column, MeOH-water-acetic acid (83:17:0.2, v/v) mobile phase, 5.0 mL min^{-1} flow rate, 30℃ column temperature and 500 μL capacity. Repetition of the chromatographic separations and recrystallization led to the isolation of two active compounds. The structures of these two compounds were elucidated by comparing their melting points (m.p.), mass spectrometry (MS) and nuclear magnetic resonance spectrum (1H NMR and 13C NMR). The compounds were dissolved in DMSO respectively to prepare a stock solution for tests.

2.2 Fungal culture and fish

Pure strain, Saprolegnia sp. strain JL (accession No. HM637287) (Cao et al. 2012), was obtained from National Pathogen Collection Centerfor Aquatic Animals (Shanghai Ocean University, China), and cultured on potato dextrose agar (PDA) slants and stored at 4℃ until used. To obtain long-growth cultures, the water mold was subcultured in 7 cm PDA-plate. A new subculture with fresh PDA medium was started 24 h before tests.

Healthy and uninfected grass carp (Ctenopharyngodon idella) with mean weight of 28.3±1.9 g were purchased from a local commercial supplier, and were housed and acclimated in a 1500L glass aquarium under laboratory conditions (22.0±1℃, pH 7.2±0.4, dissolved oxygen 6.3-7.1 mg·L^{-1}). The fish were fed commercial production feed (Tongwei, Sichuan, China) at approximately 3% cumulative body weight. Gross necropsy examinations and histopathology were conducted on 5 fish to assess their health status at the time of arrival.

2.3 Mycelial growth inhibition test

The in vitro inhibitory activity of rhein and aloe-emodin against Saprolegnia growth was tested according to the method described in Hu et al. (2013a, b). Rapeseeds (Brassica napus) were used to incubate with Saprolegnia strain instead of hempseeds described in Stueland et al. (2005). The rapeseeds were firstly autoclaved in at 121℃ for 20 min and cooled; and then the sterilized rapeseeds were placed in Saprolegnia PDA plates and incubated at 25℃ for 2 days. Meanwhile, each compound was added to assay flasks containing hot sterilized PDA (about 65℃) and final concentrations were adjusted to corresponding concentrations. After mixing with a vortex, aliquots (10 mL) of treated medium were poured into 7 cm diameter Petri dishes to produce PDA plates with compounds. The Saprolegnia-colonized rapeseeds were inoculated in the center of the prepared media. The mycelial growth diameter was measured after inoculation at 25℃ for 48 h. All the independent experiments were conducted in triplicates. Equivalent concentration of DMSO was set as the negative control. The growth inhibition rate was calculated from mean valuesas:

$$\%IR = 100 \times (A-B) / (A-C)$$

Where IR is the growth inhibition rate; Ais themycelial growth in control; B is the mycelia growth in sample; and C is the average diameter of the rapeseeds.

2.4　Spore germination inhibition test

The spore germination inhibition test was carried out according to Hu et al. (2013a, b). Briefly, theSaprolegnia isolate was grown on PDA plates for 7−14 d, after which time spores were harvested from sporulating colonies and suspended in sterilized distilled water (SDW). The concentration of spores in suspension was determined using a hemocyto meter and adjusted to 1×10^4 $CFUmL^{-1}$ approximately. The agar plates were prepared with the required concentration of the two compounds (rhein and aloe−emodin), added to 10 mL of molten PDA (about 65℃). 10 μL of the spore suspension were spotted in triplicate on these plates which were then incubated at 25℃ for 48 h. The minimum fungicidal concentration (MFC) was defined as the lowest concentration of the compounds that prevented visible growth or germination of spores.

2.5　In vivo tests of rhein against Saprolegnia infection

2.5.1　Basic procedure of disease induction.

Adequate amount of water mold was ground with mortar and poured into the exposure tank used for inoculation, where some boiled cracking corn kernels were put in there before hand. After a continuous observation for 72 h, if there is white fungal growth appeared on 90% corn kernels, the exposure tank can be used for the disease induction.

According to Gieseker et al. (2006), saprolegniasis was induced by abrading the fish, then exposing them to the fungal challenge with a changein temperature from 18 to 25℃. Fish were removed from the holding tank, anaesthetized individually (~20 ppt), and lightly abraded above the lateral line between the posterior margin of the dorsal fin and the anterior margin of the adipose fin, with a motorized tool equipped with a sanding drum (60−grit). The abrasion width was approximately 0.5 cm and approximately 2 to 3 cm in length. The scalesand overlying epidermis were removed, and the lesionappeared as a "close shave". To minimize variations in the procedure, only one person performed all of the abrasions. After the abrasion each fish was recovered from the anesthesia in fresh water, and then added to the exposure tank. After the bath exposure for 24 h, the infection phase was ended and the fish were randomly assigned into the experimental groups.

2.5.2　Prevention test of rhein against Saprolegnia infection.

This experiment was designed to test the efficacy of rhein in preventing grass carp from becoming infected by Saprolegnia. A total of 3 treatment groups with doses of 5.0, 10.0 and 20.0 $mg \cdot L^{-1}$ and a control group with no rhein were used. In this experiment, corresponding amount of compound were added into the exposure tanks (30 L) containing virulent Saprolegnia to obtain final solutions, before putting the abraded fish into tanks. The exposure was continued for 7 days. Each group contained of 3 parallel tanks and 10 fish in each tank. The incidence of fungal infection was recorded at day 7. A diagnosis of infection was based on the appearance of the fungal colony on the surface of the lesion.

2.5.3　Treatment test of rhein against Saprolegnia infection.

Ten fish successfully infected with Saprolegnia were put into a tank containing corresponding concentration of compound or blank water (with no test compound). A total of 3 treatment groups with doses of 5.0, 10.0 and 20.0 $mg \cdot L^{-1}$ and a control group with no test compounds were used. Each group contained of 3 parallel tanks. The number of recovery fish with no visible mycelium was recorded at day 7 after exposure, and the recovery rate was calculated.

2.6　Acute toxicity of rhein

Acute toxicity of rhein against fish (grass carp) was tested in 30 L glass tanks, each containing 15 L of test solution and 10 healthy grasscarp. Dilutions were prepared from the stock solution as the following concentrations: 60, 80, 100, 120, 140, 160, 180 and 200 $mg \cdot L^{-1}$. The tests were conducted in triplicate, as well as controls (under the same test conditions with no rhein). Throughout the tests, fish were not fed, and the water quality was in consistence with those in acclimatization period. Fish mortality in the treatment and control groups were recorded after 48 h of exposure; and then median lethal concentration (LC_{50}) was calculated.

2.7　Data analysis

The data in this study were analyzed by SPSS version 17.0, and presented as mean±SD. The homogeneity of the replicates of the samples was checked by the Mann−Whitney U test. The median effective concentration (EC_{50}) and median lethal concentration (LC_{50}) with their 95% confidence intervals (CI) were determined using the probit procedure.

3　Results

3.1　Structure identification of active compounds

The chemical structure confirmations of active compounds from R. palmatum were accomplished by comparing m p, MS, ^{1}H NMR and[13] C NMR data obtained to those published.

Compound 1 was obtained as yellow crystal, mp: 350-352℃. ^{1}H-NMR (CDCl3) δ13.68 (1H, s, COOH), 11.91 (2H, s, OH), 8.19 (1H, s, H-4), 7.85 (1H, t, J=8.22 Hz, H-2), 7.83 (1H, s, H-6), 7.76 (1H, d, J=7.54 Hz, H-6), 7.43 (1H, d, J=8.22Hz, H-7) . ^{13}C-NMR (CDCl3) δ191.2 (C-9), 180.9 (C-10), 165.4 (COOH), 161.5 (C-1), 160.9 (C-8), 138.1 (C-3), 137.7 (C-6), 133.9 (C-11), 133.0 (C-14), 124.7 (C-7), 124.2 (C-2), 119.6 (C-5), 118.8 (C-4), 118.7 (C-12), 116.1 (C-13); agree well with the data reported (Robert et al., 1993). So the compound was identified as rhein. Its molecular formula is $C_{15}H_8O_6$ (Zheng et al., 2008), and its structure is shown in Fig. 3. Compound 2 was obtained as pale yellow crystal; mp: 223-224. EI-MS m/z: 269 [M-H] -, 1H-NMR (CDCl3) δ: 11.98 (1H, s, 8-OH), 11.91 (1H, t, J=7.6 Hz, H-6), 7.81 (1H, t, J=7.6 Hz, H-6), 7.72 (1H, t, J=7.6 Hz, H-4), 7.38 (1H, d, J=8.4 Hz, H-7), 7.29 (1H, s, H-2), 5.65 (1H, s, CH2OH), 4.63 (2H, s, CH2OH), 13C-NMR (CDCl3) δ161.3 (C-1), 120.7 (C-2), 153.7 (C-3), 117.1 (C-4), 119.4 (C-5), 137.3 (C-6), 124.4 (C-7), 161.6 (C-8), 191.6 (C-9), 181.5 (C-10), 133.1 (C-4a), 115.9 (C-8a), 114.9 (C-9a), 133.3 (C-10a), 62.5 (CH2OH); Agree well with the data reported (Liu et al., 2005). So the compound was identified as aloe-emodin. Its molecular formula is $C_{15}H_{10}O_5$ (Liu et al., 2005), and its structure is shown in Fig. 1.

	R_1	R_2	Molecular formula
rhein	H	COOH	$C_{15}H_8O_6$
aloe-emodin	H	CH$_2$OH	$C_{15}H_{10}O_5$

3.2　Inhibitory effects of rhein and aloe-emodin on mycelial growth

The inhibitory effects of rhein and aloe-emodin against mycelial growth of Saprolegnia were shown in Table 1. Both rhein and aloe-emodin showed inhibitory activity at different extent. The EC_{50} value of rhein against Saprolegnia mycelium at 48 h was 4.7 mg · L^{-1} (95% confidence limit, 4.4-5.0 mg · L^{-1}), while mycelial growth was 100% inhibited by rhein at 20 mg · L^{-1}. For aloe-emodin, the EC_{50} value at 48 h was 15.6 mg · L^{-1} (95% CI, 13.3-17.8 mg · L^{-1}), and mycelial growth was 100% inhibited at 50 mg · L^{-1}.

	Concentration (mg · L^{-1})	0	1	2	3	4	5	6	7	10	20	40
Rhein	Mycelium expansion diameter (mm)	73.5±1.2	64.0±2.0	58.5±1.3	48.5±2.4	42.0±1.6	41.5±1.1	21.5±0.9	10.5±0.9	9.0±0.5	0	0
	Inhibition rate (%)		12.9	20.4	34	42.9	43.5	70.7	85.7	87.8	100	100
Aloe-emodin	Concentration (mg · L^{-1})	0	5	10	20	30	40	50	60			
	Mycelium expansion diameter (mm)	74.1±2.2	65.3±1.5	43.8±1.6	20.5±1.3	12.1±1.9	5.2±0.3	0	0			
	Inhibition rate (%)		11.8	40.9	72.3	83.7	93	100	100			

3.3　Inhibitory effects of rhein and aloe-emodin spore germination

As shown in Table 2, rhein showed stronger inhibitor activity against Saprolegnia spores than aloe-emodin did. The MFC values of rhein and aloe-emodin at 48 h were 16 and 40 mg·L⁻¹, respectively.

Table 2　Effects of rhein and aloe-emodin on spore germination of Saprolegnia after exposure for 48 h.

Rhein	Concentration (mg·L⁻¹)	0	1	2	4	8	16		
	Spore germination *	+++	+++	+++	++	+			
Aloe-emodin	Concentration (mg·L⁻¹)	+++	+++	+++	++	+	−	50	60
	Spore germination *	0	5	10	20	30	40	−	−

−: no germination, no growing mycelia on the plates.

+: germinating spores, no growing mycelia on the plates.

++: germinating spores, very few growing mycelia on the plates.

+++: germinating spores, profuse growing mycelia on the plates.

* Average of 3 replicates.

3.4　In vivo prevention and treatment effects of rhein against saprolegniasis

Only rhein which showed stronger anti-Saprolegnia activity were tested for its in vivo prevention and treatment effects on saprolegniasis of grass carp. Table 3 showed the prevention effects of rhein against saprolegniasis on grass carp. The results revealed that rhein have good capacity of preventing abraded grass carp infection with Saprolegnia. At concentration of 10.0 mg·L⁻¹, the prevention rate was 70%. When the concentration of rhein was up to 20.0 mg·L⁻¹, the prevention rate increased to 93.3%.

Table 4 showed the therapeutic effects of rhein against saprolegniasis on grass carp. The results demonstrated that rhein was effective in the treatment of saprolegniasis on grass carp. At concentration of 10.0 mg·L⁻¹, the 56.7% infected fish can be recovered from saprolegniasis. When the concentration of rhein was up to 20.0 mg·L⁻¹, the recovery rate increased to 67.7%.

Table 3　Prevention effects of rhein on grass carp against Saprolegnia infection.

Concentration (mg·L⁻¹)	Fish for tests (individual)	Fish infected (individual)	Protection rate (%)
0	10	10±0	0
5.0	10	4.3±1.2	57.0
10.0	10	3±1.7	70.0
20.0	10	0.67±0.6	93.3

Table 4　Treatment effects of rhein on grass carp infected with Saprolegnia.

Concentration (mg·L⁻¹)	Fish for tests (individual)	Fish recovered (individual)	Recovery rate (%)
0	10	0±0	0
5.0	10	4.3±0.6	43.0
10.0	10	5.67±1.2	56.7
20.0	10	6.67±1.2	67.7

3.5　Acute toxicity of rhein

The fish tolerated the rhein at concentration of 60 mg·L⁻¹ for 48 h without visible effects, but exposure to 200 mg·L⁻¹ resulted in an increased opercular movement and erratic behavior of fish within 4 h and the fish were all died after exposure for 48 h. The calculated 48 h-LC_{50} of rhein was 148.5 mg·L⁻¹, with the 95% CI of 140.1-156.9 mg·L⁻¹.

4. Discussion

Saprolegniosis caused by Saprolegnia spp. is the most common fungal disease affecting fish and their eggs in fresh and brackish water (Eli et al., 2011; Huang et al., 2015). This disease can be easily recognized by the cotton-like white to grayish patches on the skin and gills visible to the naked eye (Stueland et al., 2005). Saprolegnia species invade epidermal tissues, causing destruction and loss of epithelium integrity due to cellular necrosis by hyphal penetration of the basement membrane (Hu et al., 2013). The infection progresses very quickly and often results in mortality of both fish and ova. Due to the sanitary problems, environmental and governmental restrictions, solubility, toxicity and high cost of synthetic antifungal agents used currently, research on new alternatives is still necessary (Cortés et al., 2014).

The in vitro assay may be an excellent tool for the screening of the antimicrobial activity of potential drugs (Ling et al., 2013). A simple and rapid in vitro screening method to determine the effects of drugs against growth of Saprolegnia has been developed by Stuel and et al. (2005). In this study, this method was used with slight modifications. Rapeseed was used instead of hempseed since rapeseed is more abundant in China and is easy to obtain. More importantly, rapeseed was successfully used in earlier Saprolegnia studies (Hu et al., 2013a, b).

In present study, two known compounds rhein and aloe-emodin, were isolated from n-butanol extract R. palmatum. The in vitro antifungal activity of these two compounds against Saprolegnia in terms of their inhibitory effects on mycelial growth and spore germination was evaluated. Rhein at 20 mg · L^{-1} inhibited the mycelial growth completely and the germination of spores was prevented at 16 mg · L^{-1} when exposure for 48 h, while the values of aloe-emodin were 50 and 40 mg · L^{-1}, respectively. The 48 h-EC_{50} of rhein and aloe-emodin against Saprolegnia mycelium were 4.7 and 15.6 mg · L^{-1}, respectively. The efficacy of rhein and aloe-emodin in the inhibition of Saprolegnia mycelium might be a substantial improvement over many of the chemicals currently used (Stuel et al. 2005). Stuel and et al. (2005) reported that the minimum inhibitory concentrations of bronopol, formalin and hydrogen peroxide against mycelium of Saprolegnia parasitica were 10-50, 50-100 and 100-250 mg · L^{-1}, on the basis of a Hempseed MicroPlate (HeMP) -method for 48 h. Other fungicidal chemicals used in agriculture might possess higher antifungal activities against Saprolegnia. Based on the same method used in the present study, Hu et al. (2013a) reported that the EC_{50} values of kresoxim-methyl and azoxystrobin against Saprolegnia mycelium at 48 h were only 0.240 and 0.212 mg · L^{-1}, respectively; however, these chemicals were relatively toxic to fish with 48 h LC_{50} at 0.658 and 2.712 mg · L^{-1}. For the development of newly fungicidal agents, rhein showing stronger anti-Saprolegnia activity were tested for its prevention and treatment efficacy on saprolegniasis of grass carp. The results revealed that exposure to rhein at 20.0 mg · L^{-1} for 7 days could prevent 93.3% of abraded grass carp infecting with Saprolegnia, and 67.7% of infected fish could be recovered due to the exposure of rhein. There are few data in literatures could be used to compare, but these findings undoubtedly highlight that rhein showing promising anti-Saprolegniaproperty could be exploited as a fungicide in the prevention and control of Saprolegnia infection, while its 48 h-LC_{50} to the host (grass carp) was evaluated to be 148.5 mg · L^{-1} which ensures the safety for the use of rhein. Numerous early studies suggested that natural products (compounds and extracts) derived from some of the medicinal plants have the potential to be used as prevention and control against Saprolegnia (Hussein et al., 2009; Gormez and Diler, 2014; Huang et al., 2015). Huang et al. (2015) assessed the in vitro antifungal activity of 30 Chinese herb extracts to Saprolegnia sp., and found that Magnolia officinalis exhibiting enhanced growth inhibition appears to be valuable antifungal herbal species. Gormez and Diler (2014) evaluated the in vitro antifungal activity of essential oils from Tymbra, Origanum, Satureja species on the S. parasitica, and suggested that these essential oil has the potential to be used as health control of rainbow trout (Oncorhyncus mykiss) against S. parasitica infection.

To the best of our knowledge, the present study is the first study that has reported the antifungal activity of rhein and aloe-emodin derived from R. palmatum against fish pathogens, although Agarwal et al. (2000) has reported their antifungal activity against Candida albicans, Cryptococcus neoformans, Trichophyton mentagrophytes and Aspergillus fumigatus. Both rhein and aloe-emodin belong to hydroxyanthraquinone (HAQ) compounds, which HAQs have a same hydroxyanthraquinone nucleus composed of two ketone groups at C9 and C10, and twohydroxyl groups at C1 and C8. The different groups are substituted at C3 and C6 of phenyl ring (Fig. 1). It was reported that the aromatic hydroxyls on the anthraquinone ring were important moieties for the antimicrobial effects of HAQs (Tian et al., 2003). Wang et al. (2010) believed that the rigid planar structure of ketone and hydroxyl groups in neighborhood was also important. The antimicrobial potency of HAQs varied and might be related with the type and number of substituent groups on the molecular structure, and the stability and solubility of the particular compound (Wang et al., 2010). In the present study, rhein showed a stronger inhibitory activity against Saprolegnia than aloe-emodin did, which was consistent to the assay results on Staphylococcus aureus (Wu et al. 2005) and Bifidobacterium adolescentis (Wang et al., 2010). The reason was speculated to be that the polarity of carboxyl at C3 in rhein is larger than the hydroxylmethyl in aloe-emodin. The polar substituent groups might offer improvements on the antimicrobial activity.

In summary, the present study isolated two known compounds, rhein and aloe-emodin, from a traditional Chinese medicinal herb of R. palmatum, and demonstrated that both compounds showed in vitro antifungal activity against Saprolegnia. For the development of newly fungicidal agents, rhein showing stronger anti-Saprolegnia activity were tested for its in vivo efficacy on prevention and treatment of saprolegniasis on grass carp. Its acute toxicity to the host (grass carp) was also evaluated. The results revealed that rhein have potential in preventing and controlling of Saprolegnia infection at least for external uses. Further studies are required for field evaluation of rhein in the practical system.

Acknowledgements

The research was supported by the National Natural Science Foundation of China (31302211), the Commonness and Commonweal Technology Application Project Program of Zhejiang province (2014C32055).

Reference：

Agarwal, S. K.; Singh, S. S.; Verma, S.; Kumar, S.; 2000: Antifungal activity of anthraquinone derivatives from Rheum Emodi. J. Ethnopharmacol. 72, 43-46.

Barnes, M. E.; Ewing, D. E.; Cordes, R. J.; Young, G. L.; 1998: Observations on hydrogen peroxide control of Saprolegnia spp. during rainbow trout egg incubation. Prog. Fish-Cult. 60, 67-70.

Barnes, M. E.; Stephenson, H.; Gabel, M.; 2003: Use of hydrogen peroxide and formalin treatments during incubation of landlocked fall chinook salmon eyed eggs. N. Am. J. Aquac. 65, 151-154.

Cao, H. P.; Xia, W. W.; Zhang, S. Q.; He, S.; Wei, R. P.; Lu, L. Q.; Yang, X. L.; 2012: Saprolegniapathogen from pengzecrucian carp (Carassiusauratus var. Pengze) eggsand its control with traditional Chinese herb. Isr. J. Aquacult. -Bamid. 64, Special section, p1.

Caruana, S.; Yoon, G. H.; Freeman, M. A.; Mackie, J. A.; Shinn, A. P.; 2012: The efficacy of selected plant extracts and bioflavonoids in controlling infections of Saprolegniaaustralis (Saprolegniales; Oomycetes). Aquaculture 358-359, 146-154.

Cortés, Y.; Hormazábal, E.; Leal, H.; Urzúa, A.; Mutis, A.; Parra, L.; Quiroz, A.; 2014: Novel antimicrobial activity of a dichloromethane extract obtained from red seaweed Ceramiumrubrum (Hudson) (Rhodophyta: Florideophyceae) against Yersinia ruckeri and Saprolegniaparasitica, agents that cause diseases in salmonids. Electron. J. Biotechn. 17, 126-131.

Eli, A.; Briyai, O. F.; Abowei, J. F. N., 2011: A review of some fungi infection in African fish saprolegniasis, dermal mycoses; Branchiomyces infections, systemic mycoses and dermocystidium, Asian. J. Med. Sci. 3, 198-205.

Gieseker, C. M.; Serfling, S. G.; Reimschuessel, R.; 2006: Formalin treatment to reduce mortality associated with Saprolegniaparasitica in rainbow trout, Oncorhynchusmykiss. Aquaculture 253, 120-129.

Gormez, O.; Diler, O.; 2014: In vitro Antifungal activity of essential oils from Tymbra, Origanum, Satureja species and some pure compounds on the fish pathogenic fungus, Saprolegniaparasitica. Aquac. Res. 45, 1196-1201.

Hu X. G.; Liu, L.; Chi, C.; Hu, K.; Yang, X. L.; Wang, G. X.; 2013b: In vitro screening of Chinese medicinal plants for antifungal activity against Saprolegnia sp. and Achlyaklebsiana. N. Am. J. Aquac. 75, 468-473.

Huang, X. L.; Liu, R. J.; Whyte, S.; Du Z. J.; Chen, D. F.; Deng, Y. Q.; Wang K. Y.; Geng, Y.; 2015: The in vitro antifungal activity of 30 Chinese herb extracts to Saprolegnia sp. J. Appl. Ichthyol. doi: 10. 1111/jai. 12773

Hussein, M.; El-Feki, M.; Hatai, K.; Yamamoto, A.; 2002: Inhibitory effects of thymoquinone from Nigella sativa on pathogenic Saprolegnia in fish. Biocontrol. Sci. 7, 31-35.

Khosravi, A. R.; Shokri, H.; Sharifrohani, M.; Mousavi, H. E.; Moosavi, Z.; 2012: Evaluation of the antifungal activity of Zatariamultiflora, Geranium herbarium, and Eucalyptus camaldolensis essential oils on Saprolegniaparasitica - infected rainbow trout (Oncorhynchusmykiss) eggs. Foodborne Pathog. Dis. 9, 674-679.

Li, M. H.; Wise, D. J.; Robinson, E. H.; 1996: Chemical prevention and treatment of winter saprolegniosis ("winter kill") in channel catfish Ictaluruspunctatus. J. World Aquacult. Soc. 27, 1-6.

Ling, F.; Lu, C.; Tu, X.; Yi, Y.; Huang, A.; Zhang, Q; Wang, G.; 2013: Antiprotozoal screening of traditional medicinal plants: evaluation of crude extract of Psoraleacorylifolia against Ichthyophthiriusmultifiliis in goldfish. Parasitol. Res. 112, 2331-2340.

Liu Ying, Zhang Yuan hu, ShiRenbing. 2005. Studies on the chemical consitituents in herb ofMenthahaplocalyx. China journal of chinese material medica. 30 (14): 1086-1088.

Muhatadi, F. J.; Moss, M. J. R.; 1969: A synthesis of aloeemodin 1, 8-di-b-D-glucoside. Tetrahedron Lett. 4B, 3751-3752.

Nawa, H.; Uchibayashi, M.; Matsuoka, T.; 1961: Structure of rhein. J. Org. Chem. 26, 979-980.

Schreier, T. M.; Rach, J. J.; Howe G. E.; 1996: Efficacy of formalin, hydrogen peroxide, and sodium chloride on fungal-infected rainbow trout eggs. Aquaculture 140, 323-331.

Stueland, S.; Heier, B. T.; Skaar, I.; 2005: A simple in vitro screening method to determine the effects of drugs against growth of Saprolegniaparasitica. Mycol. Prog. 4, 273-279.

Tian, B.; Hua, Y. J.; Ma, X. Q.; Wang, G. L.; 2003: Relationship between antibacterial activity of aloe and its anthaquinone compounds. China J. Chinese Materia Med. 28, 1034-1037.

Van West, P.; 2006: Saprolegniaparasitica, an oomycete pathogen with a fishy appetite: new challenges for an old problem. Mycologist 20, 99-104.

Waterstrat, P. R.; Marking, L. L.; 1995: Clinical evaluation of formalin, hydrogen peroxide, and sodium chloride for the treatment of Saprolegniaparasitica on fall Chinook salmon eggs. Prog. Fish-Cult. 57，287-291.

Wu，Y. W.；Gao，W. Y.；Xiao，X. H.；Liu，Y.；2005：Calorimetric investigation of the effect of hydroxyanthraquinones in Rheum officinaleBaill on Staphylococcusaureus growth. Thermochim. Acta429，167-170.

You，X.；Feng，S.；Luo，S.；Cong，D.；Yu，Z.；Yang，Z.；Zhang J.；2013：Studies on a rhein-producing endophytic fungus isolated from Rheum palmatum L. Fitoterapia85，161-168.

Zheng X. T.，Shi P. Y.，Cheng Y. Y.；2008. Rapid analysis of a Chinese herbal prescription by liquid chromatography time-of-flight tandem mass spectrometry［J］. Journal of Chromatography A，1206（2）：140-146.

该文发表于《动物医学进展》【2015，8：55-58】

EDTA 联合抗菌药对水霉菌生物膜的影响

袁海兰[1]，欧仁建[2]，胡鲲[1]，杨先乐[1]

（1. 上海海洋大学国家水生动物病原库，上海 201306；2. 成都市龙泉驿区水务局，四川成都 610100）

摘要： 为探索金属离子螯合剂乙二胺四乙酸（EDTA）单独及联合抗菌药物对水霉菌生物膜的清除效应，采用微量稀释法测定 EDTA 对水霉菌游动孢子最小抑菌浓度（MIC）。用常用剂量和 2 倍剂量抗菌药亚甲基蓝、美婷及 1、4、16 倍 MIC 的 EDTA 联合二者常用剂量分别作用于水霉菌生物膜，CCK-8 法检测药物作用后生物膜活力变化。并用 Calcofluor White M2R 染色 EDTA 处理后水霉菌生物膜，荧光显微镜观察胞外基质变化。结果显示，亚甲基蓝、美婷在常用剂量下作用于水霉菌生物膜效果不佳，与对照组无显著性差异（$P>0.05$），亚甲基蓝在 2 倍剂量下仍不能发挥较好的效果，而亚甲基蓝、美婷与 EDTA 联合作用均表现出较好的作用。EDTA 作用后水霉菌生物膜结构破坏，胞外基质减少。研究结果表明，EDTA 可破坏水霉菌生物膜结构，与抗菌药联合使用对生物膜有显著的清除作用。

关键词： 水霉菌；生物膜；乙二胺四乙酸；抗菌药；联合用药

该文发表于《安徽农学通报》【2015，21（2）：11-12】

4 种中草药提取物对水霉的体外抑菌试验
Antifungi Effects of Extracts from 4 Kinds of Chinese Herb Medicines against Saprolegia

蔺凌云，袁雪梅，潘晓艺，姚嘉赟，徐洋，尹文林，郝贵杰，沈锦玉

（浙江省淡水水产研究所，浙江 湖州 313001）

摘要： 寻找鱼类水霉病害防治的新型安全高效药物。采用生长速率法和孢子萌发法测定五倍子、黄连、黄芩及乌梅等 4 种中草药的提取物对水霉菌丝及游动孢子的抑制效果。黄芩的乙酸乙酯提取物对水霉菌丝的生长抑制作用最显著，0.1 mg/mL 时抑制效率为 83%，0.5 mg/mL 时可完全抑制水霉菌丝的生长。黄芩的乙酸乙酯及无水乙醇提取物在浓度为 0.4 mg/mL 时，可使 80% 的水霉游动孢子萌发被抑制。4 种中草药的提取物对水霉菌丝生长及游动孢子的萌发均有一定程度的抑制作用。

关键词： 中草药；抑菌；水霉病

该文发表于《水生生物学学报》【2008，32（3）：301-306】

水霉拮抗菌的筛选及其拮抗作用的初步研究

SCREENINGOF ANTAGONISTIC BACTERIUM STRAIN AGAINST SAPROLEGNIOSIS AND THEPRELIMINARY STUDYOF *IN VITRO* ANTAGONISTIC ACTIVITY

张书俊，杨先乐，李聆，李怡，高鹏

(上海水产大学农业部渔业动植物病原库，上海 200090)

摘要： 为了获得水产动物水霉病病原菌的拮抗微生物，以水霉为靶标，对其拮抗菌进行了筛选，并研究了拮抗作用最强菌株的拮抗活性。实验结果表明：从发生过水霉病害的水体中分离得到了 130 株细菌，能抑制水霉菌落生长的有三株：LD038、LD057 和 LD106，其中以 LD038 的拮抗作用最强，经梅里埃 ATB 微生物自动鉴定系统鉴定为黏质沙雷氏菌（*Serratia marcescens*）。培养基对 LD038 的拮抗活性影响显著，LD038 在 PDA 平板上表现出的拮抗活性最强，大小依次为 PDA >BHI >GY >CA >Sabouraud。进一步对菌株 LD038 体外拮抗活性进行了研究。平板抑制实验表明，LD038 能抑制水霉菌丝生长和孢子萌发。琼脂扩散法中，LD038 菌落周围产生的抑菌圈直径为 20 mm，而无菌水对照组中水霉菌丝蔓过滤纸纸片并长至平板边缘；D38 菌线间孢子的萌发与孢子距菌线的距离有关，距 LD038 菌线 22 mm 范围内，孢子的萌发完全被抑制。无菌发酵上清液实验表明，与对照组相比，LD038 无菌发酵上清液对孢子的萌发和菌丝生长都表现出显著的抑制作用。其中孢子较菌丝对上清液更为敏感，5 倍稀释上清液下的孢子萌发率仅为 10%，且萌发后的菌丝生长也同样受到抑制，整个实验过程中菌丝均未形成网状，对照组中萌发形成的菌丝在 16 h 后呈网状。无菌上清原液和 2 倍稀释上清液中孢子均未萌发。对菌丝而言，仅无菌上清原液能彻底抑制菌丝生长，与对照组相比，2 倍和 5 倍稀释上清液显著延缓了菌丝的生长。当对照组中的菌丝铺满整孔时，2 倍和 5 倍稀释上清液中的菌丝长度分别为 0 和 2.6 mm。显微观察表明 LD038 的无菌发酵上清液导致水霉菌丝形态发生变化，较正常菌丝短而粗大，且出现了黑色原生质聚集。本研究为水霉病害的生物防治提供了一定的理论基础和依据。

关键词： 水霉；拮抗菌；生物防治；黏质沙雷氏菌

　　水霉病为淡水养殖种类的常见寄生性疾病，病原菌为卵菌纲的一些种类，分类学上，将此纲生物从原真菌界划分为茸鞭生物界[1]。目前引起水产动物水霉病的常见种类为水霉（*Saprolegnia*）和棉霉（*Achlya*）两个属中的一些种类[2]。水霉病流行很广[3]，一年四季均可发病[4]，晚冬初春季节尤为严重，对寄主无严格的选择性，水产动物及卵都可被感染，给养殖产业造成了巨大的经济损失[5,6]。对于水霉病的防治主要采用化学治疗方法，然而孔雀石绿作为经济、有效的水霉病防治药物，存在着免疫抑制[7]、三致作用（致癌、致畸、致突）、高残留等毒副作用[8]，从 1992 年起至今包括我国在内的许多国家都已将其列为水产养殖禁用药品[9]，由于缺乏能有效防治水霉病的药物，更加剧了水霉病害问题的严重性。到目前为止，可以防治水霉的主要化学药物有氯化钠、福尔马林、过氧化氢、臭氧等，但是仍缺乏高效的药物[10]。此外，化学药物的使用容易引起病原菌产生耐药性，对养殖环境的生态平衡也有一定的破坏作用，因此，采用生物控制水产病原菌得到了越来越多的重视。近年来，利用细菌防治水产病害有一些报道，Gibson, et al.[11] 利用中间气单胞菌（*Aeromonas media*）能够完全保护攻毒感染塔氏弧菌（*Vibrio tubiashii*）的牡蛎（*Crassostrea gigas*）；Lategan, et al.[12,13] 利用中间气单胞菌（*Aeromonas media*）A199 治疗人工感染水霉的澳洲鳗鲡（*Anguilla australis richardson*）及自发感染水霉的银锯眶鱼刺（*Bidyanus bidyanus*），取得了理想的治疗效果。然而在国内水产疾病研究领域，尚无水霉生物防治的相关报道。本文从水霉病害发生的养殖水体中分离筛选出一株对水霉有较强拮抗作用的拮抗株 LD038，并对其拮抗活性进行了初步的研究，为水霉病害的生物防治提供了一些理论基础和依据。

1 材料与方法

1.1 实验材料

1.1.1 样品采集

水样采集于东海渔场发生水霉病的养殖池塘；水霉病原菌，从患有水霉病鲟鱼肌肉组织中分离得到，属于水霉属，4℃保存。

1.1.2 培养基

普通营养琼脂（NA），用于水样中细菌的分离，购于上海疾病预防控制中心；马铃薯葡萄糖琼脂培养基（PDA），玉米粉琼脂培养基（CMA），用于水霉培养及平板抑制实验，购于上海疾病预防控制中心；脑心浸液琼脂培养基（BHI），用于水霉拮抗细菌的筛选，购于北京陆桥技术有限责任公司；GY 培养基：葡萄糖 10 g、酵母提取液 2.5 g、琼脂 16 g、蒸馏水 1 000 mL，pH 7.0，用于平板抑制实验；沙保劳（Sabouraud）琼脂培养基：蛋白胨 10 g，葡萄糖 10 g，琼脂 16 g，蒸馏水 1 000 mL，pH 7.0，用于平板抑制实验；马铃薯葡萄糖液体培养基：新鲜去皮马铃薯 200 g、葡萄糖 20 g、蒸馏水 1 000 mL，马铃薯先加热煮沸 15 min 后，经纱布过滤去渣，加水补足原量，pH 调至 7.0，此液体培养基用于拮抗细菌株的发酵培养。

1.2 拮抗菌株的筛选

1.2.1 细菌分离

按稀释涂布平板法，将水样稀释 10^3 倍后，取 0.1 mL 稀释液均匀涂布于普通培养基平板，置于 25℃下培养 3 d。挑取菌落形态不一样的细菌分离纯化后接种至斜面保存备用。

1.2.2 拮抗菌株的筛选

按照 Hussein, et al.[14] 的方法，将分离得到的细菌以划线的方式接种至 BHI 平板上。每个平板接种同一菌株的两条平行菌线，菌线之间的距离约为 3 cm，各自与平板边缘的距离也大约为 3 cm。将按此方法接有双条菌线的培养皿置于 25℃下培养 1 d，待菌落长出后，再在菌线之间接种水霉琼脂块（直接从长满水霉的 PDA 平板切取，直径为 3 mm）。25℃下培养 3 d 后，测定水霉菌落大小。以连续两天测定结果相同，即菌落停止生长后的值为最终水霉菌落大小，每组设三个平行。在对照 BHI 平板上（无接种细菌），水霉生长 3 d 后铺满整个平板。以水霉菌落边缘至细菌菌线距离作为细菌拮抗作用大小的判断标准，记录有拮抗作用菌株的编号。

1.3 不同培养基对拮抗作用的影响

将筛选出拮抗作用最强菌株 LD038 按 1.2.2 小节中的实验方法，接种至 GY、CA、BHI、PDA、沙保劳琼脂（Sabouraud）5 种培养基上，观察不同培养基上 LD038 对水霉抑制作用。以无划线接种 LD038 的培养基平板作为对照组。

1.4 LD038 无菌上清液的制备

将拮抗菌株 LD038 接种至马铃薯葡萄糖液体培养基中（50 mL），25℃，150 r/min 下摇床培养 24 h 后，按5%接种量将此活化菌液加入 150 mL 马铃薯葡萄糖液体培养基继续摇床培养 48 h，细菌发酵液经4℃，8 000 r/min 离心，取上清液，经 0.22 μm 孔径滤膜过滤后，收集无菌的发酵上清滤液。

1.5 水霉孢子悬液的制备

参照文献[15]，取 PDA 培养基上生长的水霉菌丝，在无菌水中漂洗三次后转移至盛有 20 mL 无菌水的培养皿中。20℃培养 24 h 后，弃去水霉菌丝，用血小球计数板在显微镜下计量孢子数目。用无菌水调整孢子浓度至 2×10^4 cyst/mL，4℃保存备用。

1.6 LD038 对水霉菌丝生长的抑制作用

采用琼脂扩散法，在 PDA 平板表面，距离培养皿边缘约 3 cm，沿四周等间距处，放入滤纸纸片。在滤纸纸片上滴加 20 μL LD038 菌液（1×10^8 cfu/mL），以无菌水作为对照，25℃下培养 1 d，LD038 以滤纸为中心在平板上形成菌落。在培养皿中央位置接种水霉琼脂块（直径为 3 mm），25℃下培养 3 d。测量 LD038 对水霉抑菌圈的大小，以 LD038 菌落边缘至受抑制的水霉菌落边缘为抑菌圈半径。

1.7 LD038 对水霉孢子萌发的抑制作用

按照 1.2.2 小节中的接种方法，将拮抗菌株 LD038 划线接种至 PDA 平板，25℃下培养 1 d。待菌落长出后，在拮抗菌菌线间滴加 100 μL 水霉孢子悬液（2×10^4 cyst/mL），涂布均匀。对照组中将 100 μL 水霉孢子悬液滴加至 PDA 平板上，涂布均匀，2 h 后观察孢子萌发的情况。

1.8 LD038 无菌发酵上清液对水霉菌丝及孢子萌发的抑制作用

将 1.4 节中制备的无菌上清液与无菌蒸馏水按照 1∶0、1∶1、1∶4 比例稀释后，加入 24 孔板中，每孔 1 mL，每个浓度组设三个平行；对照组中为 5 倍无菌蒸馏水稀释的马铃薯葡萄糖液体。每孔中加入新生水霉琼脂块（直径 2 mm）或 30 μL孢子悬浮液（2×10^4 cyst/mL），25℃下培养，于 2 h、4 h、16 h、28 h、40 h，显微镜下观察菌丝生长和孢子萌发情况，并测量菌丝长度。观察计算孢子萌发率时，每个平行组选择 5 个视野，取平均值。孢子萌发率以视野中萌发的孢子占总孢子数的比值计算。

1.9 LD038 细菌鉴定

采用法国生物-梅里埃公司生产的 ATB Expression 微生物自动鉴定系统，所用测试卡为 ID 32E，测定结果用随机软件进行分析。

2 结果

2.1 拮抗菌株的筛选

从水体中共分离到 130 株细菌，其中有拮抗作用的细菌共 3 株，编号为 LD038、LD057、LD106，但其中以 LD038 的拮抗作用最强（表 1）。

表 1 三株拮抗菌株对水霉生长的抑制作用

Tab. 1 The antagonistic effect of three antagonistic bacteriaisolated on saprolegnia hyphal growth

编号/Number	抑菌距离/Inhibition Zone（mm）
LD038	9
LD057	3
LD106	4

2.2 培养基对 LD038 拮抗作用的影响

结果表明，在不同培养基平板对照组中，水霉在 CA 上生长速度最慢，但 4 d 后铺满整个平板；在实验组中，除 Sabouraud 培养基平板上长满水霉外，其他 4 种培养基上的水霉都受到 LD038 不同程度的抑制作用，抑制活性大小依次为：PDA >BHI >GY >CA>Sabouraud（表 2、图 1）。因此选择 PDA 研究拮抗菌 LD038 对水霉的拮抗作用。

表 2 不同培养基对 LD038 拮抗作用的影响

Tab. 2 The effect of the culture media on inhibitory activity of the strain LD038

培养基/Culture media	抑菌距离/Inhibition Zone（mm）
GY	5
CA	3
BHI	9
PDA	12
Sabouraud	0

2.3 LD038 对水霉菌丝的抑制作用

PDA 平板表面上，拮抗菌 LD038 能显著抑制水霉菌丝的生长，接种水霉 3 d 后在滤纸周围形成清晰的抑菌圈，直径大小为 20 mm。

2.4 LD038 对孢子萌发的影响

在接种有 LD038 的 PDA 平板上水霉孢子萌发受到了抑制，孢子悬浮液涂布于平板 2 h 后，倒置显微镜下观察，对照组（无接种拮抗菌）孢子出现萌发，有芽管长出；而接种拮抗菌的 PDA 整个平板表面，孢子均没有萌发，随着培养时间的延长，在两条菌线间中央区域内，孢子开始萌发，并逐渐形成菌丝，最终在平板上形成肉眼可见的白色棉花状菌落；离拮抗菌落较近的区域，孢子虽有萌发，但抑制在芽管状态，没能形成菌丝；在拮抗菌落附近 22 mm 直径范围内，孢子没有萌发（图 2）。

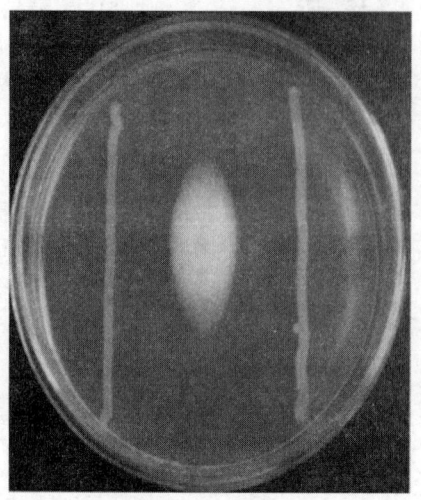

图 1　PDA 培养基上 LD038 对水霉生长的抑制

Fig. 1　*Saprolegnia inoculum*（center）inhibited by LD038 bacterial strain on PDA medium

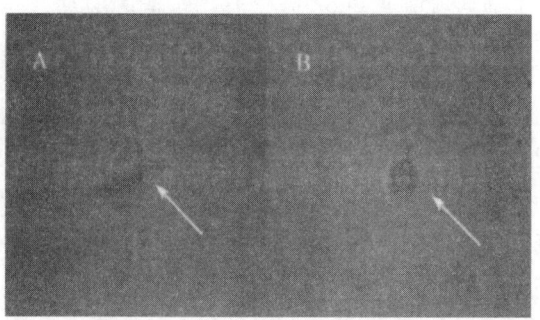

图 2　拮抗菌株 LD038 对孢子萌发的抑制作用（×400）

Fig. 2　Inhibition of cysts germination by the antagonistic bacteria strain LD038 after 3h incubation（×400）

A. 对照组中，PDA 平板上萌发后的孢子；B. LD038 存在时，孢子的萌发受到抑制

A. Germination proceeds on the PDA plate in the absent of the strain LD038；B. Germination is inhibited in the present of the strain LD038

2.5　LD038 无菌发酵上清液对水霉菌丝的抑制

各稀释浓度的 LD038 无菌上清均对菌丝生长具有一定的抑制作用，与对照组相比，菌丝长度呈现显著差异（$P<0.05$）。1∶0 实验组中，菌丝的生长被完全抑制，7 d 后观察菌丝也无生长；1∶1 和 1∶4 实验组中，无论是实验开始阶段还是后阶段，菌丝的生长速度明显慢于对照组，当对照组水霉菌丝铺满整个孔径时，上述两实验组的菌丝长度为 0 和 2.6 mm；随着培养时间的增加，20 h 后，1∶1 实验组中的菌丝也开始长出（图 3）。实验过程中，上述各实验组中的水霉琼脂块，在实验开始后的一段时间内，倒置显微镜（×400）下观察发现菌丝内有黑色的颗粒状物质，随着时间的延续，1∶1 和 1∶4 实验组中，逐渐有清晰透明的菌丝产生，并伸进上清液中，但菌丝分枝较多，且粗而短小，而对照组中的琼脂块的四周新生的菌丝在倒置显微镜下细长、清晰无杂质。

2.6　LD038 无菌发酵上清液对水霉孢子萌发的影响

在不同稀释倍数的拮抗菌 LD038 无菌上清液中，孢子的萌发均受到抑制，随着稀释倍数的增大，孢子萌发受抑制程度也逐渐减弱。对照组中，孢子在实验开始 1 h 后就开始出现芽管，1∶4 实验组在 2.5 h 后才开始萌发，而 1∶0 和 1∶1 实验组中水霉孢子则完全被抑制，直至实验结束后仍未萌发。1∶4 实验组中，虽有孢子萌发，但孢子萌发率显著低于对照组

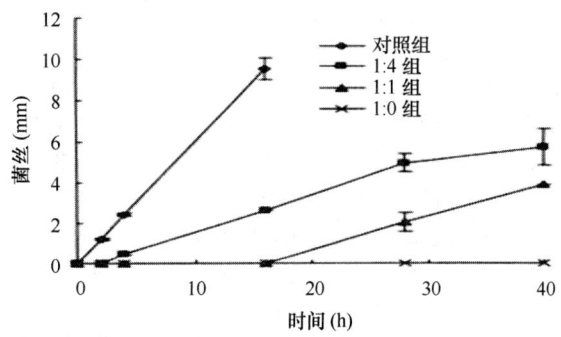

图 3　拮抗菌 LD038 无菌发酵液对水霉菌丝生长的影响

Fig. 3　The effect of cell-free culture supernatants of the strain LD038 on saprolegnia hyphal growth 8 h 后,
对照组菌丝铺满整孔 8 h later, wells were full of hypha in the control group

（$P<0.05$），萌发率为 10%（表 3）。萌发后菌丝的生长速度也显著低于对照组（$P<0.05$）。16 h 后,对照组中萌发后的菌丝结成网状,而 1∶4 实验组中孢子萌发后形成的菌丝还无分枝,至实验结束,也未形成网状（图 4）。

表 3　拮抗菌 LD038 无菌上清液对孢子萌发率的影响

Tab. 3　The effect of cell-free culture supernatants of the strain LD038 on saprolegnia cysts germination rate

组别 Group	萌发数 Numbers of cyst germinated	孢子数 Number of cysts	萌发率（%） Germination rate	均值（%） Average
对照组	9	9	100	
对照组	8	8	100	96.3
对照组	8	9	88.9	
1∶4 组	2	10	20	
1∶4 组	0	9	0	10
1∶4 组	1	10	10	

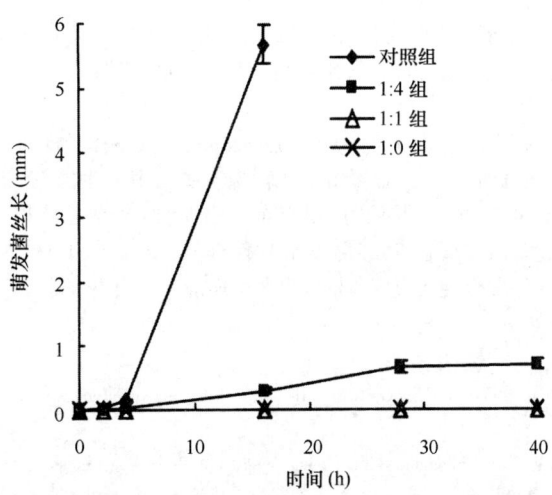

图 4　拮抗菌 LD038 无菌上清液对萌发后菌丝生长的影响 16 h 后,对照组的菌丝已经结成网状,无法测量

Fig. 4　The effect of cell-free culture supernatants of the strain LD038 on saprolegnia hyphal growth from germinated cysts 16 h later,
wells were confluent with hypha in the control group and it is unable to be measured

2.7　LD038 拮抗菌株的鉴定

LD038 经自动鉴定系统鉴定结果为黏质沙雷氏菌（Serratia marcescens），鉴定率（id）为 99.9%。

3 讨论

水产养殖动物生活的水环境中存在着微生态平衡问题，一旦打破这种平衡，就易导致水产养殖疾病的发生。当鱼体有创伤、养殖密度大、水环境恶化时最容易引起水霉病的发生，一方面是因为受伤后的鱼体自身免疫水平降低，抗感染能力下降；另一方面当水体中存在的微生态平衡被破坏后，可能导致水体中的部分微生物对水霉菌的抑制作用丧失。但也有些鱼在感染水霉病后能自行康复，可能是微生物间的相互关系在水霉病的控制中起着至关重要的作用。近年来，微生态制剂在水产养殖动物病害防治中受到越来越多的重视，并发挥着调节养殖环境，控制病害的作用，有望进一步取代化学治疗方法。目前研究发现一些细菌对水产病原菌有拮抗作用，如：Spanggaard et al.[16]从分离自鱼体的菌株中筛选出 45 株对弧菌病有拮抗作用的菌株；Das et al.[17]研究发现假单胞菌属的三株细菌在体外对嗜水气单胞菌（Aeromonas hydrophila）都有抑制作用；Hussein, et al.[14]从患水霉病鱼体的损伤处分离出假单胞菌属、气单胞菌属的细菌，发现均能抑制水霉的生长。由此可见，利用生物间的拮抗作用控制水产动物病害是有可能的。但是在分离、筛选拮抗菌的时候，如何增加筛选率是寻找拮抗微生物的一个重点和难点。本文从分离于水霉病害发生后的养殖水体的 130 株细菌株中筛选出 3 株对水霉有拮抗作用的细菌菌株，拮抗菌仅占分离细菌总数的 2.3%。这与筛选模型有很大的关系，建立一个合适的模型对增加筛选率有很大的帮助。同时筛选培养基的选择也是很重要的，既要适合水霉的生长也要满足细菌的营养，本文中选择以 BHI 作为筛选培养基，基本上适合所有细菌的生长，重要的是水霉的长势较好，优于 NA，特别是 BHI 含有天然的营养物质。在不同营养成分下，拮抗菌的代谢产物不同，也就会产生不同的拮抗效果。Lategan, et al.[18]发现拮抗菌 A199 在富含色氨酸的胰蛋白胨中和蛋白胨（TNPB）培养基上能产生具有拮抗作用的吲哚物质，但在色氨酸缺乏的 SE 培养基上则不能表现出拮抗作用。本文中，LD038 拮抗菌株同样在不同培养基上表现出不同的抑制作用，其中在 PDA 上表现出的抑制效果最强。可见培养基的选择也是导致拮抗菌筛选率低下的一个重要原因。

拮抗菌产生拮抗作用的方式有多种，如产生抗菌素[18]、抗菌蛋白[19]等拮抗物质；物种间的营养竞争[20]；生态位竞争，即拮抗菌比病原菌更易与生物机体结合，从而排斥病原菌而起到的保护作用[21]等。文中 LD038 菌株对水霉的拮抗作用，可以排除营养竞争因素。在实验过程中发现，将上清液以无菌水稀释 5 倍后，水霉菌丝及孢子都能生长、萌发，而无稀释及稀释浓度低的则不生长或生长受抑制。因此，应该是 LD038 代谢产物起到的拮抗作用。但 LD038 无菌发酵上清液中拮抗物质对水霉菌丝和孢子的作用机制，尚不清楚，显微镜下观察对水霉的作用表现为菌丝出现原生质聚集、菌丝肿大，孢子出现裂痕，与 Bae[22]、Yoshida, et al.[23]报道的拮抗菌代谢物质对陆栖真菌作用时的现象相似。

研究表明 LD038 及其无菌发酵上清液均对水霉有较强的拮抗作用。生长 1 d 的 LD038 对水霉产生的抑菌圈为 20 mm，而抗真菌药物咪康唑（10 μg/片）的抑菌圈 15 mm。低浓度无菌发酵上清液虽然没能彻底地抑制菌丝，但是显著地推迟了菌丝的初生及菌丝的生长速度。相比之下，孢子对上清液要敏感一些，5 倍稀释上清液中的孢子萌发率为 10%，且萌发后菌丝生长停止在菌丝分枝之前。孢子作为水霉生活史中的一个阶段，同时也是水霉病发生的传播因子[1]，因此对孢子清除也是防治水霉的有效途径。然而采用琼脂扩散法研究无菌发酵上清液在 PDA 平板上对水霉菌丝的抑制作用时，却没有表现出拮抗作用，这与 Hussein, et al.[14]报道的相似，估计与 LD038 发酵培养时的发酵条件和产生的拮抗物质的量有关，也有可能是与 LD038 产生的拮抗物质为几丁质酶有关。

LD038 经微生物鉴定系统鉴定为黏质沙雷氏菌（Serratia marcescens），其亦称灵杆菌（Bacillus prodigiosus），广泛存在于水、土壤和食品中[24]。由于能产生几丁质酶，从而被用于植物病害防治中的生防菌[25,26]。本文研究发现黏质沙雷氏菌（Serratia marcescens）对水霉也存在着拮抗作用，从而为水霉病的防治提供了新的思路和一定的理论基础。但是如何利用活菌还存在诸多困难，如水环境对生物防治的影响、细菌有无致病性等等。因此，LD038 的动物安全性试验及环境生态安全性还有待于进一步研究，以便对菌株的安全性及应用前景作出科学评价，从而更好地用于水霉的生物防治。

参考文献

[1] PieterV W. Saprolegnia parasitica, an oomycete pathogen with a fishy appetite: new challenges for an old problem [J]. Mycologist, 2006, 20 (3): 99-104.

[2] Huang Q Y. Aquatic animal diseases [M]. Shanghai: Shanghai Science and Technology Press. 1996, 142 [黄琪琰. 水产动物疾病学. 上海：上海科学技术出版社. 1996, 142].

[3] Monica M S, Alan P, Kanak B. Saprolegnia bulbosa sp. nov. isolated from an argentine stream: taxonomy and comparison with related species [J]. FEMS Microbiology Letters, 2007, 268 (2): 225-230.

[4] Ni D S. Study on control of saprolegniasis in fish [M]. Beijing: Agriculture Press. 1982, 8 [倪达书. 鱼类水霉病的防治研究. 北京：农业出版社. 1982, 8].

[5] Hussein M M A, Hatai K. Pathogenicity of saprolegnia species associated with outbreaks of salmonid saprolegniosis in Japan [J]. Fisheries Science, 2002, 68: 1067-1072.

[6] Tampieri M P, Galuppi R, Carelle M S, et al. Effect of selected essential oils and pure compounds on Saprolegnia parasitica [J]. Pharmaceu-

tical Biology, 2003, 41 (8)：584- 591.

[7] Prost H, Sopinska A. Evaluation of activity of the celluar protective process in carp with saprolegnia infection and treatment with malachite green and immunostimulant [J]. Medycyna Weterynaryjna, 1989, 45：603- 605.

[8] Zhai Y X, Guo Y Y, Geng X, et al. Advances in studies on metabolic mechanism and bio-toxicity of malachite green [J]. Periodical of Ocean University of China, 2007, 37 (1)：27— 32 [翟毓秀, 郭莹莹, 耿霞, 等. 孔雀石绿的代谢机理及生物毒性研究进展. 中国海洋大学学报, 2007, 37 (1)：27-32].

[9] Xie W, Ding H Y, Xi J Y, et al. Detection of residues of malachite green, crystal violet and their metabolites in aquatic animal samples [J]. Chinese Journal of Chromatography, 2006, 24 (5)：529- 530 [谢文, 丁慧瑛, 奚君阳, 等. 水产品中孔雀石绿、结晶紫及其代谢产物残留量的检测. 色谱, 2006, 24 (5)：529- 530].

[10] Fornerisa G, Bellardib S, Palmegianoc G B, et al. The use of ozone in trout hatchery to reduce saprolegniasis incidence [J]. Aquaculture, 2003, 221 (1)：157- 166.

[11] Gibson L F, Woodworth J, George A M. Probiotic activity of aeromonas media on the Pacific oyster, Crassostrea gigas, when challenged with Vibrio tubiashii [J]. Aquaculture, 1998, 169：111-120.

[12] Lategan M J, Torpy F R, Gibson L F. Control of saprolegniosis in the eel Anguilla australis Richardson, by Aeromonas media strain A199 [J]. Aquaculture, 2004, 240：19- 27.

[13] Lategan M J, Torpy F R, Gibson L F. Biocontrol of saprolegniosis in silver perch Bidyanus bidyanus (Mitchell) by Aeromonas media strain A199 [J]. Aquaculture, 2004, 235：77- 88.

[14] Hussein MM A, Hatai K. In vitro inhibitionof saprolegnia by Bacteria isolated from Lesions of Salmonids with Saprolegniasis [J]. Fish Pathology, 2001, 36 (2)：73- 78.

[15] Li A H, Nie P, Lu Q Z. A primary study of saprolegniasis and its pathogen of lateolabrax japonicas [J]. Acta Hydrobiologica Sinica, 1999, 23 (4)：388-390 [李爱华, 聂品, 卢全章. 花鲈水霉病及其病原的初步研究. 水生生物学报, 1999, 23 (4)：388- 390].

[16] Spanggaard B, Huber I, Nielsen J, et al. The probiotic potential against vibriosis of the indigenous microflora of rainbow trout [J]. Environmental Microbilogy, 2001, 3 (12)：755- 765.

[17] Das B K, Samal S K, Samantaray B R, et al. Antagonistic activity of cellular components of Pseudomonas species against Aeromonas hydrophila [J]. Aquaculture, 2006, 253：17-24.

[18] Lategan M J, Booth W, Shimmon R, et al. An inhibitory substance produced by Aeromonas media A199, an aquatic probiotic [J]. Aquaculture, 2006, 254：115- 124.

[19] Shen J Y, Yin W L, Cao Z, et al. Purification and partial characterization of antagonistic substance from Bacillus subtilis B115 [J]. Acta HydrobiologicaSinica, 2005, 29 (6)：689— 693 [沈锦玉, 尹文林, 曹铮, 等. 枯草芽孢杆菌 B115 抗菌蛋白的分离纯化及部分性质. 水生生物学报, 2005, 29 (6)：689-693].

[20] Smith P, Davey S. Evidence forthe competitive exclusion of Aeromonas salmonicida from fish with stress-inducible furunculosis by a fluorescent pseudomonad [J]. Journal of Fish Diseases, 1993, 16：521-524.

[21] Balcazar J L, Vendrell D, Blas I D, et al. In vitro competitive adhesion and production of antagonistic compounds by lactic acid bacteria against fish pathogens [J]. Veterinary Microbiology, 2007, 122 (3-4)：373-380.

[22] Bae D W, Lee J T, Son D Y, et al. Isolation of bacterial strain antagonistic to pyricularia oryzae and its mode of antifungal action [J]. Journal of Microbiology and Biotechnology, 2000, 10：811-816.

[23] Yoshida S, Hiradate S, Tsukamoto T, et al. Antimicrobial activity of culture filtrate of Bacillus amyloliquefaciens RC-2 isolated from mulberry leaves [J]. Phytopathology, 2001, 91：181-187.

[24] Yu H. Medical Microbiology [M]. Beijing：People's Hygienics Press. 1983, 299 [余贺. 医学微生物学. 北京：人民卫生出版社. 1983, 299].

[25] Chen X W, Fan H, Zhou K, et al. Study on the toxicity of serratia marcesens to the common vegetable insects [J]. Tianjin Agricultural Sciences, 2005, 11 (3)：5- 7 [陈秀为, 范寰, 周可, 等. 一株黏质沙雷氏菌对蔬菜常见害虫毒力的研究. 天津农业科学, 2005, 11 (3)：5— 7].

[26] Yin HX, Zhang J, Hou RT, et al. Isolation and identification of chitinase-producing bacterium and its synergistic effect on locust biocontrol [J]. Plant Protection, 2004, 30 (2)：37- 41.

[27] 尹鸿翔, 张杰, 侯若彤, 等. 一株几丁质酶产生菌的分离鉴定及其灭蝗增效作用. 植物保护, 2004, 30 (2)：37- 41].

该文发表于《淡水渔业》【2012，42（3）：28-31】

一株具水霉抑制特性的芽孢杆菌筛选及其分子生物学鉴定

Selection and molecular biological identification of a *Bacillus* sp. strain to inhibit the growth of *Saprolegnia ferax*

宋增福[1]，范斌[1]，佘林荣[2]，唐磊[1]，赵世林[1]，吕利群[1]，杨先乐[1]

（1. 上海海洋大学水产与生命学院、国家水生动物病原库，上海 201306；2. 广东海富药业有限公司，广东 潮州 515721）

摘要：以水霉菌（*Saprolegnia ferax*）的生物防治为切入点，通过对本实验室保存的 9 株芽孢杆菌（*Bacillus* spp.）与水霉菌 JL 菌株进行对峙抗抗试验，筛选出具有明显抑制水霉菌 JL 菌株的芽孢杆菌 BA1（$P<0.05$）。经进一步测定菌株 BA1 对不同来源水霉菌株试验，结果显示菌株 BA1 对水霉菌株 6#、10#、S2 也有明显的抑制作用（$P<0.05$）。通过测定菌株 BA1 的 16 S rDNA，利用 BLAST 软件进行同源性比对，发现与蜡样芽孢杆菌的同源性在 99% 以上。由此推测，菌株 BA1 为蜡样芽孢杆菌菌株，且具有明显抑制水霉菌作用，可以作为水产养殖上水霉病防治的生物制剂使用。

关键词：水霉菌（*Saprolegnia ferax*）；芽孢杆菌（*Bacillus* sp.）；抑菌试验；分子生物学鉴定

四、孔雀石绿的替代药物制剂"复方立达霉粉（复方甲霜灵粉）——美婷的研制"

孔雀石绿被禁用后，在防治水霉病方面一直没有较好的替代的药物，一定程度上形成了对水霉病防治的真空。为了从技术源头上彻底解决孔雀石绿的禁用，以上海海洋大学牵头的科研团队经过近 8 年的工作，研制出 1 种安全、有效的孔雀石绿替代制剂——"复方甲霜灵粉"（曾用名"复方立达霉粉"、商品名"美婷"）。"复方甲霜灵粉"对水霉病的有良好的预防和治疗效果（李浩然等，2015；肖国初等，2014），且对鱼、虾、蟹等水产动物和环境安全，使用成本基本上与孔雀石绿相当。

利用高效液相色谱法可以测定"美婷"的有效组分（马荣荣等，2013）和甲霜灵在水产品体内的残留。甲霜灵（立达霉）在水产品中的残留限量标准为 0.05 mg/kg。"复方甲霜灵粉"对鱼类等水生动物的休药期为 240 度日。据不完全统计，从 2009—2016 年期间，"复方甲霜灵粉"在我国东北、华北、华中、华南、华东、西北等计 22 个省、直辖市，累计试用面积达 2.5 万余亩；受试对象涵盖四大家鱼（青草鲢鳙）、大宗淡水鱼类（鲤、鲫、鳊、鲂）、主要出口鱼类（鳗等）和某些特种养殖鱼类（胭脂鱼、金鱼）等数十种。"复方甲霜灵粉"在大田生产实践中对鱼类水霉病具有良好的防治和治疗效果（夏文伟等，2010；曹海鹏等，2011；杨先乐等，2013）。其中对于鱼卵的效果尤为显著。"复方甲霜灵粉"的推广和使用获得了良好的经济效益、社会效益和生态效益。据不完全统计，取得直接经济效益达数十亿元以上（《复方甲霜灵粉在江苏、湖南、江西等地的生产性试验现场验收的意见》）。

该文发表于《食品科学技术学报》【2013，31（2）：11-14】

孔雀石绿的禁用及其替代药物美婷

Prohibition of Malachite Green and Its Alternative Drug Mei Ting

杨先乐，郭珺

（上海海洋大学 国家水生动物病原库，上海 201306）

摘要：水霉病是水产养殖中常见的真菌性疾病，且危害严重。过去防治水霉病最有效的药物是孔雀石绿，但孔雀石绿具有高毒性、致癌、致畸和致突变的副作用，我国于 2000 年将其列为禁用药物。然而，由于孔雀石绿抗水霉效果好、廉价易得、缺少相应代替药物等因素，偷用行为时有发生且屡禁不止，严重危害水产品的质量安全。由上海海洋大学研发的"美婷"制剂在全国多省市的实践应用中显示出较好的抗水霉效果和安全性，对彻底杜绝孔雀石绿的使用，保障水产品的质量安全具有重大意义。

关键词：孔雀石绿；禁用药物；美婷

水产品是一种优质、高效的动物蛋白，以其鲜美的味道和丰富的营养成为最受人们欢迎的食品之一。我国水产品的来源已完成了由捕捞到养殖的转变[1]，这对解决我国人均动物蛋白摄入不足、营养不良等社会发展现象具有重大意义。然而，水产养殖发展过程中药物的滥用、乱用现象屡见不鲜，尤其是禁用药物的使用，给水产品的质量安全带来了极大的隐患。孔雀石绿就是其典型代表。

1　水霉病与孔雀石绿

水霉病又称肤霉病、白毛病，属真菌性鱼病，一年四季都可发生，尤其是在早春和晚冬，水温 15～20℃ 时发病最严重。据统计，在 118 种生物源性疾病中，真菌性疾病占 4.24%，其中主要是水霉病。据报道青鱼、草鱼、鲢、鳙、花鲈、红点鲑、鲤、大银鱼、斑点叉尾鮰、大鲵、黄鳝、暗纹东方鲀、月鳢、鲑、黑鱼、中华绒螯蟹等大部分水产养殖动物都有可能发生水霉病。

水霉病主要症状是：病变部位长着大量白色的棉絮状菌丝，鱼体由于负担过重，游动失常，食欲减退，甚至会瘦弱而死。由于霉菌侵入机体，产生大量蛋白酶，病变部位在酶的作用下，分解鱼体组织中的蛋白，使鱼体受到刺激，分泌大量粘液。病鱼因此焦躁不安，食欲减退，最终死亡。感染水霉菌的卵，内菌丝侵入卵膜，卵膜外丛生大量外菌丝，因此常被称为"卵丝病"。而被寄生的卵因外菌丝呈放射球状，故又称为"太阳籽"。该病的病原主要是水霉（*Saprolegnia spp*），它对环境有很强的适应能力。水霉菌以孢子的形式繁殖，在菌丝的末端形成孢子囊，释放产生具有双鞭毛的游动孢子，在水体中自由游动，传播速度迅速。在环境条件不良的情况下，能产生厚垣孢子，大大提高其生存能力。水霉菌是腐生性的，极易感染。在换水或搬运时造成鱼体体表受伤，局部皮肤坏死。活细胞因能分泌一种抗霉性的物质，一般不会感染水霉．防治水霉病，曾用的有效药物是孔雀石绿（mal-achite green），它的代谢产物是无色孔雀石绿（leuco-malachite green，又称隐性孔雀石绿）。由于孔雀石绿的氧化电位与病原体组成酶的某些氨基酸相近，在细胞分裂时与其竞争而阻碍蛋白肽的形成，导致细胞分裂受到抑制，从而起到抗菌、杀虫作用[2]。孔雀石绿和无色孔雀石绿均是一种三苯甲烷类的染料，它们都具有三苯甲烷化学官能团。这个化学官能团不仅具有高毒性、高残留、高致癌和高致畸性，而且结构牢靠，不容易被破坏，因此能在水生动物体内长时间残留，还可通过食物链转移至哺乳动物和人类，对人类的健康和生态环境产生很大的危害[3-4]。据研究，孔雀石绿能使大鼠肝细胞空泡化，抑制血浆胆碱脂酶，造成乙酰胆碱的蓄积而出现神经症状[5]；同时也能抑制人类谷胱甘肽-S-转移酶的活性，器官组织氧压的改变，导致细胞凋亡出现异常，诱发肿瘤和脂质过氧化[6-7]。此外，孔雀石绿还能通过溶解足够的锌，引起水生动物急性锌中毒，导致鱼类消化道、鳃和皮肤轻度发炎，从而影响鱼类的正常摄食和生长，阻碍肠道酶（如胰蛋白酶、淀粉酶）的活性，影响动物的消化吸收功能[8-9]。鉴此，许多国家均将孔雀石绿列为水产养殖禁用药物。1991 年美国食品药品管理局禁止使用该药。我国农业部在 2000 年就下文禁止在渔业上使用孔雀石绿；2002 年将孔雀石绿列入《食品动物禁用的兽药及其化合物清单》及"水产养殖中禁止使用的药物"[10]。

2　孔雀石绿累禁不止及其原因

从苏丹红事件到三聚氰胺事件，食品的质量安全问题一直都是社会关注的焦点。孔雀石绿在过去很长一段时间被认为是水产养殖中防治水霉病和部分细菌感染的最有效和最廉价的药物，曾被广泛使用。由于它使用方便、效果明显、价格低廉，至今仍有人在水产养殖以及水产品运输过程中偷偷使用而屡禁不止。孔雀石绿的残留在国内外曾多次被检出。芬兰2003 年在监测的 227 个鳟样品中发现 5 个检出低残留的孔雀石绿代谢产。英国在 2004 年监测数据中，92 个样品中有 1 个检出无色孔雀石绿，其残留水平达 4.9 μg/kg，高于欧盟残留限值 2 μg/kg[11]。近年来我国水产品中发现孔雀石绿残留的现象也时有发生。2006 年 10 月，上海、北京等地相继发现多宝鱼存在孔雀石绿的残留超标事件，经查这些多宝鱼养殖过程中都违规使用了孔雀石绿等违禁药物，而其使用的"统一"牌多宝鱼饲料检测出含有大量孔雀石绿。这一轰动全国的多宝鱼事件，导致山东约 5 000 万尾多宝鱼囤积，经济损失近 20 亿元，青岛市场上 90% 的多宝鱼滞销，青岛多宝鱼养殖企业遭受重创。同年 11 月，香港对 15 个桂花鱼样本进行检验，结果发现 11 个样本含有孔雀石绿。受此事件影响，广东曾一度中断了对香港的活鱼供应，对港出口出现较大幅度的下滑，当地活鱼销量也只有原来的一成左右，价格也出现一定程度的缩水，给广东桂花鱼市场造成重大经济损失。2012 年 4 月湖南长沙市马王堆海鲜市场的 9 个批次的桂花鱼样本均被检出孔雀石绿，生产地均为广东等地。受该事件影响，广东省桂花鱼遭遇各地"封杀"，曾一度滞销，导致许多酒楼和餐馆无法供应桂花鱼，市场上的桂花鱼也突然消失。如同多米勒效应，孔雀石绿等禁用药物的屡禁不止引起了人们的恐慌心理，不仅影响了社会的和谐稳定，而且也阻碍了水产养殖的健康发展。

为何孔雀石绿屡禁不止？其原因主要是：

1）价廉易得。孔雀石绿仅 35 元/kg 左右，由于使用剂量较低，且多采用浸泡、泼洒等给药方法，防治成本低。因此，某些经销商在利益驱动下，仍在毫无顾忌地经销该药，某些养殖户和鱼贩也不顾国家的有关规定，购入和使用该药用于水产养殖的病害防治与水产品的保鲜和运输。

2）对水霉病等水产动物病害疗效较好，缺少有效的替代药物。自从孔雀石绿被禁用后，在水霉病的防治方面一直没有较好的替代药物，形成了一定的真空。而水霉病是我国水产养殖中的一个常见病，在一定程度上可造成较大的损失，因此一些养殖户不得不用孔雀石绿去防治该病。

3）缺乏有力的监管制度。虽然国家已明文规定禁止使用孔雀石绿，但对孔雀石绿缺乏有效的监管方法和监管网络，对违法经销和使用缺乏处罚措施与处罚力度，以至不能从源头上阻断孔雀石绿在渔业上的非法使用。

4）孔雀石绿残留初筛方法缺乏。目前孔雀石绿残留检测主要是采取液相色谱法或质谱法[12-13]，但因该方法需较昂贵的仪器，检测过程较复杂，技术水平要求也较高，不适用于普查。某些免疫试剂盒虽有研究，但技术尚不成熟，难以应用于生产。简易可行的初筛方法是保证水产品安全、控制孔雀石绿滥用的途径之一。

5）宣传和科学普及工作深度不够。养殖者对孔雀石绿副作用缺乏深刻的了解，虽一般都能知道孔雀石绿有毒，但对其能产生致畸、致癌、致突变的副作用缺乏认识；较多人对国家禁用孔雀石绿等药物的利害关系不了解，没有以较高的层面上去认识，因此在销售和使用时有人会毫无顾忌，更无人对违法经营和使用者进行举报。

3　孔雀石绿替代药物——美婷

"孔雀石绿"禁用之后，虽然有一些药物用于替代，如三氯异氰尿酸、甲醛、食盐、小苏打、亚甲基蓝、硫醚沙星、检科一号、水霉灵等，但不是效果不明显，就是存在着一定的毒副作用，或者成本太高，使得孔雀石绿的替代问题一直没有得到很好地解决。因此，开发和研制出安全、有效、经济、使用方便的替代孔雀石绿的新型药物，是从源头彻底解决孔雀石绿禁用的一个关键问题。

替代孔雀石绿的新型药物的研发和应用不仅具有深远的社会效益和生态效益，而且具有广阔的市场前景。经估算，目前我国淡水养殖面积为 380 万 hm²，每年水霉病的发病率一般是 5%~10%，在晚冬和早春该病的发病率更高，严重时可达 40% 左右。若仅以发病面积 40 万 hm²，以及用药 7.46 kg/hm² 计算，预计防治水霉病的药物年需求量最少是 300 万 kg 以上。

"美婷"是上海海洋大学国家水生动物病原库科研团队经过近 6 年多的努力研制出的"孔雀石绿"替代制剂，是"国家十一五科技支撑计划"、十二五"863"计划、"上海市农委科技兴农重点项目"资助所获得的最新科研成果。现已建立了美婷生产工艺，并已完成其作用机制、药理学、稳定性及安全性的研究。

"美婷"的作用机理是通过作用于菌丝体、孢子的细胞壁，以使其选择通透性，影响水霉菌对营养物质的吸收利用，使菌丝体细胞死亡、孢子失去萌发力，阻断其生活史。研究证明它对鱼体和鱼卵的水霉病具有良好的防治效果。当"美婷"的泼洒浓度为 5~10 g/m³，浸浴浓度为 20 g/m³ 时，可有效地防治鱼卵、鱼苗和成鱼的水霉病。试验表明，用"美婷"预防水霉病，用药浓度在 16 mg/L 时，它对草鱼、鲫的保护率分别可达到 96.7% 和 100%（而孔雀石绿分别只有 76.7% 和 73.3%），用 16 mg/L 的浓度治疗，对草鱼、鲫的有效率分别为 50% 和 53.3%（孔雀石绿分别为 56.7% 和 53.3%）[14]。

"美婷"的主要成分的组成是纯天然的植物生长调节剂、用于食品防腐剂和用于日用化妆品的防霉、防腐剂[15]，此外，还具有增强机体免疫的功能[16]，因此它对人类是安全的。经试验它不会对环境产生任何副面作用，一定程度上还可促进有益菌群和水体中的藻类、浮游生物的生长[17]。

在使用成本方面，防治鱼类的水霉病，泼洒的成本约为 0.4~0.47 元/hm²，浸浴约 0.7 元/m³ 水体，与孔雀石绿基本相当，基本可以为生产所接受。目前，"美婷"制剂已进入生产性扩大试验阶段。2008—2012 年期间，共生产应用于生产性试验的"美婷"13.5 t，在全国主要水产养殖区进行了中试推广验证。使用地点覆盖东北、华北、华中、华南、华东、西北等地区，北京、河北、河南、辽宁、吉林、上海、江苏、浙江、天津、黑龙江、安徽、福建、江西、湖北、湖南、山西、山东、广东、广西、重庆、四川、贵州等共 22 个省（直辖市），试验面积累计达 0.2 万 hm²。

针对鱼类（卵）的水霉病，"美婷"制剂的受试对象涵盖大宗淡水鱼类（青、草、鲢、鳙、鲤、鲫、鳊、鲂）、主要出口鱼类（鳗、罗非鱼、斑点叉尾鮰）、某些特种养殖鱼类（胭脂鱼、金鱼、鳜、黄颡鱼、江颡鱼、云斑鮰）等数十余种。此外还对运输途中受伤鳜、草鱼等进行了相应的试验。生产性试验反应，美婷能有效地降低水霉对鱼（或卵）的感染率，提高鱼的成活率、鱼卵的孵化率，减少经济损失，提高养殖效率。如对黄颡鱼卵、彭泽鲫卵，使用"美婷"后完全不会感染水霉病，对鱼苗、成鱼的水霉病预防作用也十分明显，可使水霉感染率平均降低 60%。

"美婷"应用产生的社会、生态与经济效益十分明显，不仅保障了水产品的质量安全，保护了水域生态环境，而且产生了巨大的经济效益。据估计，6 年来美婷生产性应用产生的直接经济效益达亿元以上。孔雀石绿的替代药物"美婷"已在生产应用中显示出良好的应用前景。随着"美婷"研究的深入，"美婷"生产文号的获得，将会从根本上解决孔雀石绿的禁用问题。

参考文献

[1]　范小建. 全面提高我国水产品质量安全水平的思考 [J]. 中国渔业经济, 2006 (3)：3-7.

[2]　万译文，洪波，曾春芳，等.水产品中孔雀石绿残留的研究进展 [J].水产科技情报，2010 (4)：191-194.

[3]　司徒建通.投入品对养殖水产品质量安全的影响 [J].农村养殖技术，2009 (22)：27-28.

[4]　宋怿，黄磊，穆迎春.我国水产品质量安全监管现状及对策 [J].农产品质量与安全，2010 (6)：19-21.

[5]　Mittelstaedt R A, Mei N, Webb P J, et al. Genotoxicityof malachite green and leucomalachite green in female BigBlue B6C3F1 mice [J]. Mutation Research/Genetic Toxi-cology and Environmental Mutagenesis, 2004, 561 (1-2)：127-138.

[6]　李永生，刘艳辉.禁用药物孔雀石绿的毒性、危害及控制措施 [J].吉林水利，2010 (5)：100-102.

[7]　Glanville S D, Clark A G. Inhibition of human glutathi-one S-transferases by basic triphenylmethane dyes [J]. Life Sciences, 1997, 60 (18)：1535-1544.

[8]　张佳艳，伍金娥，常超.孔雀石绿的毒理学研究进展 [J].粮食科技与经济，2011 (3)：43-47.

[9]　Srivastava S, Sinha R, Roy D. Toxicological effects ofmalachite green [J]. Aquatic Toxicology, 2004, 66 (3)：319-329.

[10]　中华人民共和国农业部.NY5071—2002 无公害食品渔用药物使用准则 [S].中国水产，2002 (12)：67-71.

[11]　李宁.孔雀石绿对健康的影响 [J].国外医学：卫生学分册，2005, 32 (5)：262-264.

[12]　郭德华，叶长淋，李波，等.高效液相色谱-质谱法测定水产品中孔雀石绿及其代谢物 [J].分析测试学报，2004, 23 （增）：206-208.

[13]　中国海洋大学，国家水产品质量监督检验中心.SC/T 3021—2004 无公害食品水产品中孔雀石绿残留量的测定 液相色谱法 [S].北京：中国标准出版社，2004：4-8.

[14]　高鹏.孔雀石绿替代药物的筛选及其对水霉病的防治效果研究 [D].上海：上海海洋大学，2008.

[15]　成绍鑫.生化黄腐酸与其他来源黄腐酸组成性质的初步比较 [J].腐植酸，2009 (2)：1-8.

[16]　薄芯，李京霞.生化黄腐酸对免疫系统的影响初探 [J].北京联合大学学报：自然科学版，1998, 12 (2)：44-47.

[17]　李文新，史清琪.生化黄腐酸BFA可持续发展的新支点 [J].科技与经济画报，2002 (3)：31-33.

该文发表于《科学养鱼》【2010，8：52-53】

美婷对鱼卵水霉病的防治试验

夏文伟，曹海鹏，杨先乐

（上海海洋大学国家水生动物病原库）

水霉病是由水霉属、绵霉属、丝囊霉属的多种丝状真菌引起的真菌性病害，在我国所有大宗淡水鱼类养殖区均有不同程度的暴发，对鱼卵，尤其是鲤、鲫、团头鲂等产的粘性卵危害极大。因而，有效防治水霉病是鱼类人工繁殖管理措施的重中之重。然而，自孔雀石绿在水产养殖中禁用后，水霉病防治成为鱼类人工繁殖过程中的一个难点，寻找既安全又能有效地防治鱼卵水霉病的药物具有积极的现实意义。国家大宗淡水鱼类产业技术体系渔药临床岗位（上海海洋大学）经过多年来不懈的研究，开发出一种专门防治水霉病的药物—美婷，已经证实在实验室条件下能有效预防和治疗斑马鱼卵水霉病发生。为进一步验证美婷对鱼卵水霉病的临床试验效果，本试验分别采用流水或静水孵化方式，以团头鲂卵和彭泽鲫卵为材料，观察了美婷防治水霉病的效果。

一、材料与方法

1. 试验地点和时间

试验地点在仙桃市沙湖水产技术推广站，时间为 2010 年 5 月。

2. 试验材料

鱼卵是仙桃市沙湖水产技术推广站人工繁殖的团头鲂卵和彭泽鲫卵；美婷，为上海海洋大学国家水生动物病原库自主研制生产；水霉净，主要成分苦参和地肤子，为四川某公司生产。

3. 试验方法

（1）美婷对流水孵化的团头鲂卵水霉病的防治试验试验分为阴性对照组、阳性对照组和试验组，每组各 2 个平行。将人工授精的团头鲂卵经黄泥脱粘后，分别转到 6 个容积为 200 L 的孵化箱中，孵化箱底部不断冲水使卵随水流不断上翻。试验组均匀泼洒终浓度为 100 mg/L 的美婷，每 12 h 用药 1 次，连用 3 次；阳性对照组均匀泼洒 100 mg/L 的水霉净，每 12 h 用药 1 次，连用 3 次；阴性对照组不使用任何药物。试验过程中观察鱼卵感染水霉的情况，并记录各组鱼卵的发霉率、孵化时间和出苗率。试验期间孵化用水水温为 20~23℃，pH 值为 7.5~8.5，溶氧 8~9 mg/L。

（2）美婷对静水孵化彭泽鲫卵水霉病的防治试验

试验分为阴性对照组、阳性对照组和试验组，每组各 2 个平行。将人工授精的彭泽鲫受精卵粘附在 6 块相同规格的网片上，24 h 后检查受精率，将网片置于 4 m×3 m×0.8 m 的 3 个相连的水泥池（水深 0.5 m）中，每池 2 块网片。试验组均匀泼洒 15 mg/L 的美婷，第一天用药 2 次，以后每天用药 1 次，连续用药 3 d，每次用药 30 min 后换水 1/6；阳性对照组均匀泼洒 15 mg/L 的水霉净，第一天用药 2 次，以后每天用药 1 次，连续用药 3 d，每次用药 30 min 后换水 1/6；阴性对照组不使用任何药物。试验过程中观察鱼卵感染水霉的情况，并记录各组鱼卵的发霉率、孵化时间和出苗率。试验期间孵化池保持微量的流水，进水口水温 19~24℃，pH 值为 7.5~8.5，溶氧 8~9 mg/L。

（3）孵化时间、发霉率和出苗率的测定

孵化时间（h）的测定通过记录从人工授精到孵化箱内有苗开始脱膜所需时间，发霉率和出苗率分别按照公式①和②测定：

①发霉率=发霉卵数/死卵数×100%；

②出苗率=出苗数/受精卵数×100%。

二、试验结果

1. 美婷对流水孵化的团头鲂卵水霉病的防治试验

试验结果见表 1。试验结果表明，美婷在使用后，虽然卵孵化时间较阴性对照组滞后 9 h，较阳性对照组滞后了 7 h，但鱼卵的发霉率较阴性对照组降低了 75.5%，较阳性对照组降低了 74.7%，而且鱼卵的出苗率较阴性对照组提高了 41.7%，较阳性对照组提高 7.6%。说明美婷可以明显提高鱼卵的出苗率，降低鱼卵的发霉率。

表 1　美婷对流水孵化团头鲂卵孵化时间、发霉率和出苗率的影响

组别	终浓度（mg/L）	受精率（%）	孵化时间（h）	发霉率（%）	出苗率（%）
实验组（美婷）	100	86	55	24	85
阳性对照（水霉净）	100	90	48	95	79
阴性对照（空白）	0	87	46	98	60

注：发霉率是在鱼苗脱膜期间测定，下同。

2. 美婷对静水孵化彭泽鲫卵水霉病的防治试验

试验结果见表 2。试验结果表明，美婷在使用后，虽然卵的孵化时间较阴性对照组滞后 11 h，较阳性对照组滞后 8 h，但鱼卵的发霉率较阴性对照组降低了 55.0%，较阳性对照组降低 55.0%，而且鱼卵的出苗率较阴性对照组提高了 244.4%，较阳性对照组提高 169.6%。说明美婷可以明显提高鱼卵的出苗率，降低鱼卵的发霉率。

表 2　美婷对流水孵化彭泽鲫卵孵化时间、发霉率和出苗率的影响

组别	终浓度（mg/L）	受精率（%）	孵化时间（h）	发霉率（%）	出苗率（%）
实验组（美婷）	15	33	136	45	62
阳性对照（水霉净）	15	30	127	100	23
阴性对照（空白）	0	35	124	100	18

三、小结

1. 美婷主要成分为生化防腐酸，能有效渗透霉菌组织的细胞壁，干扰细胞间酶的相互作用，使细胞内蛋白质变性，并改变霉菌细胞膜的选择通透性，影响其对营养物质的吸收和利用，使菌丝体细胞死亡、孢子失去萌发能力而起到杀灭作用。

2. 在流水孵化箱中均匀泼洒终浓度为 100 mg/L 的美婷，每 12 h 用药 1 次，连用 3 次，能明显降低团头鲂卵水霉病的发生率，提高出苗率，但在一定程度上会延长孵化时间。

3. 在静水孵化池中均匀泼洒 15 mg/L 的美婷，第一天用药 2 次，以后每天用药 1 次，连续用药 3 天，用药 30 min 后换水 1/6，能明显降低彭泽鲫卵水霉病的发生率，提高出苗率，但一定程度上会延长孵化时间。

4. 在相同用药方法和用量条件下，美婷对流水和静水孵化鱼卵的水霉病防治以及出苗效果均明显优于水霉净，可能由于水霉净为中草药制剂，直接使用在水体中起效慢。

该文发表于《科学养鱼》【2011，2：41】

大宗淡水鱼类黏性卵水霉病的发生与控制策略

曹海鹏，杨先乐

（上海海洋大学农业部渔业动植物病原库，上海 200090）

一、粘性鱼卵水霉病发生的条件

1. 鱼卵的损伤与死亡

研究表明，死卵、受伤的鱼卵容易被水霉感染，而活卵卵膜表面很少有水霉寄生。宋学宏等（2007）试验发现，绝大多数已死亡的黄颡鱼受精卵着生大量水霉，而活卵卵膜外面虽然有少量水霉菌丝，但菌丝不能通过卵周隙进入胚胎，胚胎可继续发育成稚鱼破膜而出。由此证明，鱼卵的损伤与死亡是引发水霉寄生的主要因素。

2. 水质条件

温度、pH 等水质环境因子是鱼卵孵化过程中重要的影响因素，同时也是水霉生长的重要理化因子。对水霉来说，在鱼卵孵化的正常温度与 pH 范围内，水霉同样可以生长繁殖，只是在不同的温度和 pH 条件下水霉生长与孢子萌发速度有快慢之分。水霉极强的生存能力是暴发水霉病的首要因素。对于鱼卵来说，过量的氨氮、亚硝酸盐等有害因子以及其他病原菌滋生对鱼卵也有极大的危害，在一定程度下可引起鱼卵的死亡，为水霉的寄生提供宿主条件。因此，水质条件的好坏是水霉病暴发的主要环境因素。

二、粘性鱼卵水霉病的预防与控制措施

从鱼卵水霉病的发病条件来看，鱼卵、水霉和水质是对其预防和控制的三大着手因素。如何在鱼卵孵化过程中避免鱼卵损伤和死亡；如何采取一些措施减少水霉的孢子萌发与繁殖速度，抑制其生长；如何保持良好的水质，保证鱼卵的健康，这些对防止鱼卵水霉病的发生均具有重大的意义。

1. 抑制水体中水霉菌的生长

据了解，目前大多数孵化池各有一个过滤池，过滤池中较多采用滤石过滤。然而，水霉在水体中是普遍存在的，因此，为了保证孵化池中水霉孢子数量少（保证水霉孢子含量在 100 cfu/mg 以下），我们建议：①过滤池中滤石定期清洗，再在过滤池中加入生石灰（挂袋），在进入孵化池的进水口再补充加入筛绢，采用物理过滤和化学消毒的方式对水体中的病原菌、水霉尽量清除；②在保证鱼卵孵化不受影响的前提下，孵化池中加入一定量的美婷、食盐等水霉预防药物，以浸浴的方式抑制水霉的生长，同时定期检测水体温度、pH 和水霉孢子（水体和死卵）的含量，以不断调节水体温度、pH 以及美婷、食盐等水霉预防药物的用量，抑制水霉的生长。尤其是食盐，研究表明：在一定高浓度的氯化钠存在的情况下，水霉的生长能够受到完全抑制，因此可以保持孵化池水体维持在一定的盐度。

2. 避免鱼卵的死亡，提高鱼卵的成活率

在鱼卵孵化的过程中，人为不正规操作、水质污染是鱼卵存活的主要威胁。因此，在鱼卵孵化的过程中，操作人员首先必须要专业，严格按照《鱼卵孵化技术规范》（DB43/T359-2007）避免操作过程中引起损伤，而且要防止闲杂人等进出；其次，操作人员要时刻观察鱼卵的健康状况，及时挑出死卵和受损伤的卵。如果操作不便，一旦发现出现大量死卵或受损伤的卵，也可以把鱼卵孵化的滤水绢取出用美婷等药物进行高浓度浸泡；再次，在鱼卵孵化的过程中要定期清洗滤水绢，调节流速等。

3. 改善水质

为了防止水体中氨氮、亚硝酸盐等有害因子富集与病原菌滋生对鱼卵产生危害，我们建议在孵化池水体中定期检测水

体 pH 以及氨氮、亚硝酸盐等含量以及细菌、弧菌总数，保持 pH 稳定，并使用适量的化学降氮制剂与消毒剂去除氨氮、亚硝酸盐、病原菌等。因为微生态制剂中的活菌在较低的水温（25℃以下）下作用效果不显著，故不提倡使用微生态制剂对水质进行处理。

三、结语

由于水霉菌只能抑制，无法彻底消除，因此我们建议在卵孵化过程中，重在预防，一旦发生水霉病，要及时将滤水池中的池水换掉，然后引入新水，并用高浓度的美婷等药物连续浸泡。

该文发表于《南方农业学报》【2015，4：697-701】

立达霉对七彩神仙鱼卵水霉病的防治效果

Control effect of Ridomil on saprolegniasis infected Discus Fish *Symphysodon aequifasciatus* Zygotes

李浩然，欧仁建，邱军强，郭微微，杨先乐，胡鲲

（上海海洋大学国家水生动物病原库，上海 201306）

摘要：【目的】 探讨分析立达霉对七彩神仙鱼（*Symphysodon aequfasciatus*）受精卵水霉病的防治效果及其安全性，为鱼类人工孵化过程中水霉病的防治提供参考依据。**【方法】** 采用倍比稀释法验证立达霉对水霉菌的体外抑菌效果，并利用水浴染毒法使七彩神仙鱼卵感染水霉菌，然后将其浸泡在不同浓度的立达霉溶液中，以孔雀石绿、无菌水为阳性对照和空白对照，观察不同药物处理对鱼卵发育的影响。**【结果】** 8.0 mg/L 立达霉和 0.10 mg/L 孔雀石绿即可明显抑制水霉菌丝生长。以 8.0 mg/L 立达霉分别浸泡受精 36 h 后和受精 3 h 内的七彩神仙鱼卵，鱼卵的水霉菌感染率和死亡率极显著（$P<0.01$）或显著（$P<0.05$）降低，而孵化率明显提高。鱼卵组织切片观察发现，空白对照组的鱼卵结构较紧密，0.10 mg/L 孔雀石绿组的鱼卵内部结构疏松、组织溶解破碎，8.0 mg/L 立达霉组的鱼卵内部结构未发生明显变化。**【结论】** 立达霉可有效防治观赏鱼类的水霉病，且不会对鱼卵产生明显毒害作用，生产推荐使用浓度为 8.0 mg/L。

关键词： 立达霉；七彩神仙鱼；鱼卵；水霉病；防治效果

该文发表于《华中农业大学学报》【2013，32（6）：1-5】

"美婷"原料药在草鱼肌肉组织中残留检测前处理方法

The Pretreatment Methods in the Determination of the raw materials of Mei Ting Residue in the grass carp muscle by High Performance Liquid Chromatography

马荣荣[1]，胡鲲[1]，王印庚[2]，吴冰醒[1]，杨先乐[1]

（1. 上海海洋大学 国家水生动物病原库，上海 201306；
2. 中国水产科学研究院黄海水产研究所 海水鱼类养殖与设施渔业研究室，山东 青岛 266071）

摘要： 研究了高效液相色谱法测定草鱼肌肉组织中"美婷"原料药残留的前处理方法。结果表明：以甲醇作抽提剂、二氯甲烷作净化剂、振荡离心，正已烷去脂、30℃旋转浓缩蒸发的前处理方法，快速、准确、经济、实用性强。该方法"美婷"原料药的最低检测限可达 1.0 μg/kg。在加标水平为 1.0~100 μg/kg 时，回收率为 67.72%~95.52%，相对标准偏差为 0.77%~32.17%。该前处理方法适合检测"美婷"原料药在鱼体内的残留。

关键词："美婷"原料药；残留；高效液相色谱法；前处理；鱼体肌肉

该文发表于《南方农业学报》【2014，45（4）：676-681】

复方立达霉粉的稳定性研究
Stability of compound preparation of metalaxyl

肖国初，胡鲲，邱军强，杨移斌，杨先乐

（上海海洋大学国家水生动物病原库，上海 201306）

摘要：【目的】探讨复方立达霉粉的稳定性，为其配方改善和保存条件提供理论依据。【方法】依据《中华人民共和国药典（二部）》的有关规定对复方立达霉粉进行高温、强光照、高湿等稳定性试验和破坏性试验，采用反相高效液相色谱法（RP-HPLC）定量分析复方立达霉粉有效成分。【结果】复方立达霉粉在高温条件（60℃）下放置5和10 d后，其有效成分含量分别下降1.88%和4.12%。样品在强光照条件 [（4 500±500 lx）] 下放置5和10 d后，其有效成分含量分别下降0.34%和1.27%。样品在相对湿度92.5%的条件下放置10 d，其性状由微黄色粉末变为深黄色且结块，有效成分含量下降4.49%，吸湿增重率达6.60%；在相对湿度75.0%的条件下放置10 d，复方立达霉粉样品性状无明显变化，各项指标均符合《中华人民共和国药典（二部）》的相关规定。【结论】复方立达霉粉对热较敏感，对光不敏感，容易吸潮，有一定的引湿性，因此在实际生产中，复方立达霉粉应密闭储存于阴凉通风干燥处，远离热源、火源，贮运过程中要防止受潮受热。

关键词：复方立达霉粉；高温；强光照；高湿；稳定性；反相高效液相色谱法（RP-HPLC）

该文摘自于《复方甲霜灵粉在江苏、湖南、江西等地的生产性试验现场验收的意见》

复方甲霜灵霉粉（美婷）生产性试验现场验收意见

[孔雀石绿替代药物复方甲霜灵霉粉（美婷）研究团队（上海海洋大学）]

上海海洋大学课题组在江苏省、湖南省和江西省等数十个县市开展了水霉病的流行病学调查，掌握了其流行特点，课题组研制的孔雀石绿替代制剂——"复方立达霉粉"（美婷）进行了田间应用试验及示范，累计受试面积达8.9万余亩；受试养殖对象包括草鱼、鲫、鲢、鳙、斑点叉尾鮰、青鱼、鳗、乌鳢等二十余个品种的成鱼、鱼苗和鱼卵。美婷对鱼类/鱼卵水霉病具有明显的防治和治疗效果，对鱼卵水霉的预防、治疗效果分别可达85%和60%，对成鱼水霉的预防、治疗效果分别可达75%和50%以上。减少了由于水霉造成的直接经济损失累积4.3亿元以上，减少间接经济损失5.4亿元以上。

"复方甲霜灵粉"（美婷）的规模性试验证明了该药的安全性和有效性，是替代孔雀石绿可行和有效的渔药制剂。

第二节　新型渔药的研究

新型渔药的研制是对传统渔药产品升级，对提高水产养殖效益、保障水产品安全起着重要的作用。中草药、增效剂、免疫增强剂及生长调节剂是新型渔药研究与创制的重要内容。

一、中草药

中草药（traditional Chinese herb medicine）具有毒副作用小、价格低廉、不易产生抗药性等优点，其中的某些药物成分不仅有抗病原微生物的作用，还具有免疫调节的作用，能提高机体自身的抗病防病能力。采用中草药作为部分替代药物或者是中西药合用，已经成为目前水产养殖上的一种趋势。

1. 促进水产动物生长、提高免疫机能的中草药

合理选择中草药添加到鱼类的饲料或药物中，可以调节水产动物的生理生化指标，保持机体代谢，促进鱼类生长。中

草药能改善水产养殖动物品质。复方中草药可显著提高罗非鱼肌肉的肌苷酸含量，降低肌肉脂肪、总胆固醇和丙二醛（汤菊芬，2009）；研究结果表明，它对提高凡纳滨对虾肌肉的常规营养成分和总氨基酸含量作用不显著，但能提高对虾肌肉中鲜味氨基酸的含量（汤菊芬等，2008）。

中草药能显著提高水产养殖动物的免疫力。复方中草药能提升罗非鱼血清中溶菌酶、SOD 及 POD 活性，还能诱导头肾和脾 TNF-α 和 IL-1β 的表达，降低血清 MDA 含量与嗜水气单胞菌感染后的死亡率（Jufen Tang et al.，2014）。黄芪多糖等中草药可诱导和促进 *Lysozyme-c*、*hsp*70 基因（汤菊芬，2011，李绍戊等，2012）的表达，提高吉富罗非鱼的机体免疫力，促进乳酸杆菌等有益菌群的增殖，避免免疫细胞应激损伤。中草药芽孢杆菌制剂可改善凡纳滨对虾的生长指标、改善养殖水体环境和改善其肠道菌群结构和增加抗病力（汤菊芬，2009；2015；2016）。复方中草药饲料添加剂能够显著促进红笛鲷（*Lutjanuss anguineus*）的生长，增强其血清的抗菌活力和溶菌酶活力（陆志款等，2008）。茯苓可以显著升高哲罗鱼 SOD 活性（刘红柏等，2013）。甘草、紫草等复方合剂显著提高鲫血清谷草转氨酶史（AST）和碱性磷酸酶（ALP）的活性，但明显降低了尿酸（URIC）含量（刘红柏等，2012）。玉屏风散方剂可明显提高氏鲟（*Acipenser schrencki*）鱼体内球蛋白含量及溶菌酶活性，并有效降低丙二醛含量（王荻等，2012）。中药方剂可对山女鳟（*Oncorhynchus masou* masou）脾及血细胞中 IL-1β 含量起到略微促进作用（王荻等，2013），提高鲫鱼肝脏中总抗氧化能力（T-AOC）（王瑞雪等，2012）、提升肝胰脏中溶菌酶活性（栾鹏等，2012）。黄芪多糖等中药能显著提高鲤 NO 和 NOS 的含量，而对鲤的生长和免疫有一定的促进作用（苏岭等，2010；2011）。中药方剂对鲟血清及肝脏中超氧化物歧化酶（SOD）的活性、总抗氧化能力（T-AOC）和谷胱苷肽过氧化物酶（GSH-PX）均有增强作用，可显著降低鲟血清中丙二醛（MDA）的含量（刘逊等，2012）；降低鱼体内丙二醛（MDA）含量，促进机体对热应激做出快速有效应答反应（李绍戊等，2012）。

该文发表于《Fish & Shellfish Immunology》【2014，39：401-406】

Immunostimulatory effects of artificial feed supplemented with a Chinese herbal mixture on *Oreochromis niloticus against Aeromonashy drophila*

Jufen Tang[1,a,b,c], JiaCai[1,a,b,c], Ran Liu[a,b,c], Jiamin Wang[a,b,c], Yishan Lu[a,b,c], Zaohe Wu[b,c,d] and Jichang Jian[a,b,c,*]

(a. College of Fishery, Guangdong Ocean University, Zhanjiang 524088, China;
b. Guangdong Provincial Key Laboratory of Pathogenic Biology and Epidemiology for Aquatic Economic Animals, Zhanjiang 524088, China;
c. Guangdong Key Laboratory of Control for Diseases of Aquatic Economic Animals, Zhanjiang 524088, China;
d. Zhongkai University of Agriculture and Engineering, Guangzhou 510225, China)

Abstract：The effects of a Chinese herbal mixture (CHM) composing astragalus, angelica, hawthorn, Licorice root and honeysuckleon immune responsesand disease resistant of Nile tilapia (*Oreochromis niloticus* GIFT strain) were investigated in present study. Fish were fed diets containing 0 (control), 0.5%, 1.0%, 1.5% or 2.0% CHM (w/w) for 4 weeks. And series of immune parametersincluding lysozyme, cytokine genes TNF-αand IL-1β, superoxide dismutase (SOD), peroxidase (POD), malondialdehyde (MDA) were measured during test period. After four weeks of feeding, fish were in fected with *Aeromonas hydrophila* and mortalities were recorded. Results of this study showed that feeding Nile tilapia with CHM-supplementation diet stimulated lysozyme activity, SOD activity and POD activity in serum, induced TNF-αand IL-1β mRNA expressionin head kidney and spleen, but decreased serum MDA content. All CHM-supplemental groups showed reduced mortalities following A. hydrophila infection compared with the group fed the control diet. These results suggested that this CHM can be applied as a tilapia feed supplement to elevatefish immunity and disease resistance against A. hydrophila.

Key words：*Oreochromis niloticus*；Chinese herbal mixture；Immune response；Disease resistance；*Aeromonas hydrophila*

复方中草药添加剂在尼罗罗非鱼抗嗜水气单胞菌
感染过程中的免疫刺激作用

汤菊芬[1a,b,c]，蔡佳[1a,b,c]，刘冉[a,b,c]，王家敏[a,b,c]，鲁义善[a,b,c]，吴灶和[b,c,d]，简纪常[a,b,c*]

（a. 广东海洋大学水产学院，中国湛江 524088；b. 广东省水产经济动物病原生物学及流行病学重点实验室，中国湛江 524088；
c. 广东水产经济动物病害控制重点实验室，中国湛江 524088；d. 仲恺农业工程学院，中国广东 510225）

摘要：本研究检测了含有黄芪，当归，山楂，甘草和金银花的复方中草药（CHM）在尼罗罗非鱼免疫反应及其在抗嗜水气单胞菌感染过程中的作用。投喂含 0（对照），含有 0.5%，1.0%，1.5% 或 2.0%CHM 的饲料 4 周，随后测定罗非鱼一系列免疫指标包括溶菌酶，细胞因子 TNF 和 IL-1β，超氧化物歧化酶（SOD），过氧化物酶（POD），丙二醛（MDA）的变化，以及感染嗜水气单胞菌后的死亡率。结果表明，饲喂 CHM 后血清中溶菌酶、SOD 及 POD 活性升高，还能诱导头肾和脾 TNF-α 和 IL-1β 的表达，降低血清 MDA 含量与嗜水气单胞菌感染后的死亡率。该复方中草药可以作为罗非鱼养殖中提高鱼体免疫力和抗嗜水气单胞菌感染的添加剂。

关键词：尼罗罗非鱼；复方中草药；免疫反应；抗病力；嗜水气单胞菌

1　Introduction

Tilapia（Oreochromis sp.）is one of the widely cultured fish species around the world. However, the diseases caused by bacterial pathogens in tilapia culture were becoming severe and resulted in significant morbidity and mortality[1-2]. In order to control the proliferation of these bacteria, antibiotics were used widely in intensive aquaculture. But prolonged use of antibiotics could lead to many negative side effects such as antibiotic resistant in bacteria, antibiotics residues in environment and fish products[3-4]. Therefore, exploring new methods for preventing infectious diseases have become very urgent in tilapia culture.

Immunostimulant can enhance non-specific immunity of fish and protectfish against infectious pathogens [5]. In general, using immunostimulant in combination with fish vaccine is a promising method for prevention of fish diseases[6]. Up to date, many substances including polysaccharide[7-9], polypeptide and protein[10-12] and various kindsof probiotics[13-15] have been used for fish immunostimulants. Herbs, which have been used as medicine and human immunity intensifier for thousand years in China[16], have attracted the attention of researchers in last decade. To date, it has been proved that administration of herb simproved the innate and adaptive immune response of different freshwater or marine fish and shellfish against bacterial, viral, and parasitic diseases[17]. In tilapia, herbal extracts or their products such as inosporacordifolia[2], Camellia sinensisL[18], Nyctanthesarbortristis[19], Toonasinensis[20] and Sophoraflavescens[1] have been documentedto enhance theimmune response and disease resistance of tilapia against bacterial pathogen[1,18-19]. However, to our knowledge, the immunostimulatory roles of Chinese herbal mixture in fish remain largely unknown.

In this study, a Chinese herbal mixture（CHM）consisted of astragalus, angelica, hawthorn, Licorice rootandhoneysuckle was prepared in our laboratory. And the effects of supplemental dietary CHM on immunity and disease resistance of Oreochromis niloticus against Aeromonas hydrophila were investigated.

2　Materials and methods

2.1　Fish and management

Tilapia（O. niloticus GIFT strain）weighing approximately（20±2）g, were obtained from a commercial aquaculture farm in Lianjiang, Guangdong province, China. The fish were acclimatized in tanks at（28±2）℃ aerated water for 2 weeks. The fish were fed twicedaily with control diet prepared in our laboratory. The control diet contained 5% fish meal, 9% soybean meal, 14% cottonseed meal, 25.5% rapeseed meal, 1.5% calcium hydrogen phosphate, 0.18% salt, 1.5% soybean oil. No other immunostimulant was added in the control diet. The water quality was maintained by canister external water filters during the experiment.

2.2　Immunostimulant diets preparation

The herbs used in this study were purchased from a local traditional Chinese medicine shop. These herbs were dried in an oven at 60℃ then smash sift out through 80 mesh sieves. The dried, powdered herbs were mixed with a certain proportion. The herbal mixture was incorporated into acontrol diet at different rates of 0.5%, 1.0%, 1.5%, 2% (w/w) then made into pellet feed. The experimental pellets were packed andstored in a freezer at 4℃ until further use.

2.3 Experimental design

To study the immune-related index, the tilapia were assigned into five groups of 24 each, intriplicate. The respective groups were fed CHM-supplemented diets at 0% (control group), 0.5%, 1%, 1.5%, 2.0% for 4 weeks. Each replicate consisted of sixrandomly sampled fish from four CHM-supplemented groups and the control group.

A. hydrophil used for challenge experiment was isolated from diseased O. niloticus and kept in our laboratory. To study the resistance of the tilapia to A. hydrophil, experimental and control groups consisting of 30 fish with three replicates weretested. All groups were fed the CHM containing at 0% (control group), 0.5%, 1.0%, 1.5%, 2.0% for 4 weeks. After fourth week of feeding, the fish were injected intraperitoneally (i.p) with 0.2 mL of a 1.0×10^8 cfu/mL A. hydrophilsuspended in phosphate-buffered saline (PBS), whereas the control group were injected intraperitoneally (i.p) with 0.2 mL PBS. Preliminary work determined that thisbacterial concentration causes death in approximately 75% of the fish in the control group. Mortalitieswere recorded daily for 2 weeks. A. hydrophil was re-isolated from the livers of the dead fish to confirm that the cause of mortality.

2.4 Sample collection

Blood samples were collected fromthe caudal vein of tilapia in each group every week. For serum separation, the blood was clotted for 30 minat room temperature and kept overnight at 4℃. The blood clots were then centrifuged at 500g for 10 min. The resulting supernatants were collected and stored at -80℃ until further use.

Tissue samples including head kidney and spleen for Real-time PCR (RT-qPCR) analysis were collected on 1st week and store at-80℃ before RNA isolation.

2.5 Lysozyme activity assay

Serum lysozyme activity was determined using commercial kits (Nanjing Jiancheng BioengineeringInstitute, Jiangsu, China). Briefly, 200 μL fish serum were added to 1.8mL of Micrococcus lysodeikticus suspension in phosphate buffer and followed by 15min reaction at 37℃. And 200 μL reacted suspension was removed into a 96-well plate and measured at 530 nm in a microplate reader (Bio-TEK, USA). Lysozyme activity was defined as μg per ml serum.

2.6 Real-time PCR analysis of cytokine genes

The mRNA expression of genes tumor necrosis factorα (TNF-α) and interleukin 1β (IL-1β) were examined in the head kidney and spleen. Total RNA of examine tissues were using TRIzol Reagent (Invitrogen) according to the manufacturer's protocol, the quality of total RNA was detected by electrophoresis on 1% agarose gel. The first-strand cDNA was synthesizedusing EasyScript First-Strand cDNA Synthesis SuperMix (Transgen, China) following DNase Ⅰ (NEB, USA) treatment. The specific primers used for qRT-PCR were listed in Table 1. Amplification of β-actin mRNA was used as internal controls. Primers used for β-actin amplification were listed in Table 1.

Table 1 Sequences of primers used in this study

Primers	Sequences (5'-3')
TNFα-S	CCTGGCTGTAGACGAAGT
TNFα-A	TAGAAGGCAGCGACTCAA
IL1β-S	GACAGCCAA AAG AGG AGC
IL1β-A	TCTCAGCGATGG GTGTAG
Actin-S	CCAGGCAGCTCGTAACTCT
Actin-A	GAAATCGTGCGTGACATCAA

The qRT-PCR assay was carried out through IQ5 Real-time PCR System (Bio-Rad laboratories). Dissociation curve analysis of amplification products was performed at the end of each PCR reaction. The amplification was carried out in a 25 μL reaction volume, containing 12.5 μL of 2×SYBR Premix Ex Taq (TaKara, Japan), 1 μL sense primer and 1 μL anti-sense primer (10 μm), 2 μL of 1 : 5 diluted cDNA and 8.5 μL of PCR-grade water. The thermal profile for qRT-PCR was 94℃ for 5 min followed by 40 cycles of 94℃ for 30 s, 55℃ for 30 s and 72℃ for 30 s. After the PCR program, qRT-PCR data were analyzed with IQ5 Software. The baseline was set automatically by the software. The relative expression levels of target genes wereanalyzed by $2^{-\triangle\triangle CT}$ method[21]. PCR efficiency was calculated according to the protocol in ref[22].

2.7 Superoxide dismutase (SOD) activity assay

Serum SOD activity was assayed through commercial test kits purchased from NanjingJiancheng Bioengineering Institute (Nanjing, China). One unit of SOD activity was defined as the amount of enzyme necessary toproduce a 50% inhibition of the nitrobluetetrazolium reductionrate measured at 550 nm. SOD activity was expressed as SOD units per ml serum.

2.8 Malondialdehyde (MDA) content assay

The MDA content of serum was detected using commercially available kit (Nanjing Jiancheng BioengineeringInstitute, Jiangsu, China) at 532 nm inmicroplate reader. MDA content was expressed as nmol/mL.

2.9 Total peroxidase (POD) activityassay

The total peroxidase activity of serum was measured according to method described in ref[23]. Briefly, 3, 3′, 5, 5′-tetramethylbenzidine hydrochloride (TMB, Sigma-Aldrich, Germany) andhydrogen peroxide (H_2O_2, Sigma-Aldrich) were applied as substrate for peroxidase activity. 15 μL serum was diluted to 50 μL in Hanks Balanced Salt Solution (HBSS) without Ca^{2+} and Mg^{2+} in flat-bottomed 96 well plates. Then 100 μL of 0.1 mm and 2.5 mm H_2O_2 were added and followed by 2 min incubation. 50μl of 2 m sulphuric acid was added to stopthe reaction after 2 min and the optical density (OD) was read at 450 nm in microplate reader. One unit of peroxidase activity was defined as the amount necessary to produce an absorbance change of 1 OD.

2.10 Statistical analysis

The data analyzed using one-way ANOVA were represented as mean ± standard error. Statistical analysis was performed using SPSS 11.0 software. Differences were analyzed viaconsidered statistically significant at P<0.05.

3 Results

3.1 Serum lysozyme activity

As shown in Fig. 1, significant increase in serum lysozyme activity was foundin 2.0% group compared to the control at 1st week and 3rd week, and in all CHM-supplemental groups at the end of this experiment (Fig. 1).

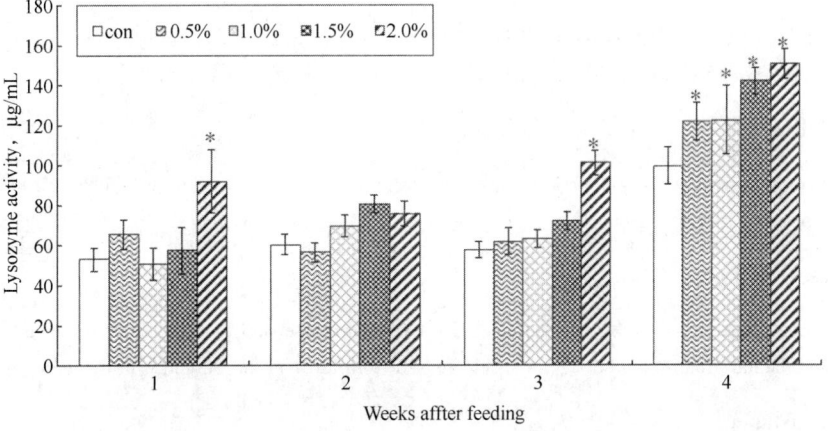

Fig. 1 Changes in serum lysozyme activity of O. niloticus fed with CHM-supplemental diet.
Data are expressed as means±SE. Significantly different (P<0.05) indicated by asterisks.

3.2　The effects of CHM supplementation on cytokine genes expression in headkidney and spleen

In spleen and head kidney, the CHM−supplemental diet significantly induced TNF−α gene transcription in 1.0%, 1.5% and 2.0% groups. And significant increase of IL−1β mRNA expression was observed in 1.5%, 2.0% groups in head kidney, in 2.0% group in spleen (Fig.2).

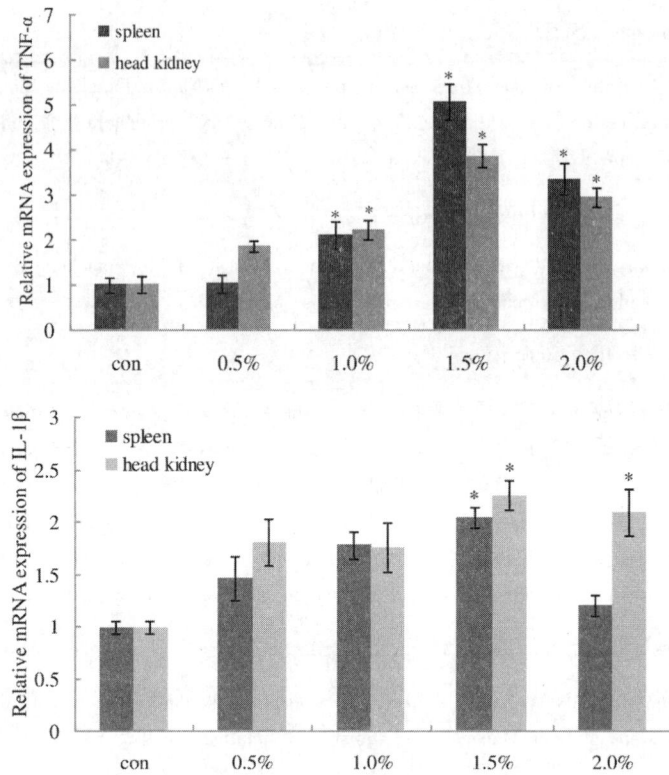

Fig.2　Effects of CHM−supplemental diets on transcription of TNF−a (A) and IL−1 b (B) genes in spleen and head kidney. Vertical bars represented the means±SE, and thesignificant differences of genes expressions between the CHM−supplemental andcontrol diet group were indicated with an asterisk (∗) at P<0.05.

3.3　Serum SOD activity

The effects of CHM−supplemental diet on serum SOD activity were shown in Fig. 3A. The SOD activity exhibited a statistically significant increase at 4th weekofthis experiment in all CHM−supplemental groups compared with the control group.

3.4　Serum POD activity

As shown in Fig. 3B, significant effects of CHM−supplemental diet onserum POD activity were detected in 2.0% group throughout the whole experimental period, in 1.0% group from 2nd week to the end of experiment.

3.5　Serum MDA content

MDA content decreased steadily when the supplemental CHM were increased up to 1.0% and further addition at 2nd week. And MDA contentsof all experimental groups showed significant decreased from 3rdweek to the end of this feeding experiment (Fig. 3C).

3.6　Disease resistance

In challenge experiment, the percentage mortality was significantly decreased in the fish fed CHM−supplemental diet when challenged with liveA. hydrophil (Fig. 4). The cumulative mortalities of the CHM−supplemental groups rangedfrom 26.7% to 66.7% compared with the control group. The 2.0% supplemental CHM conferred the best protection againstinfection, with a percentage mortality of 26.7% (P<0.005).

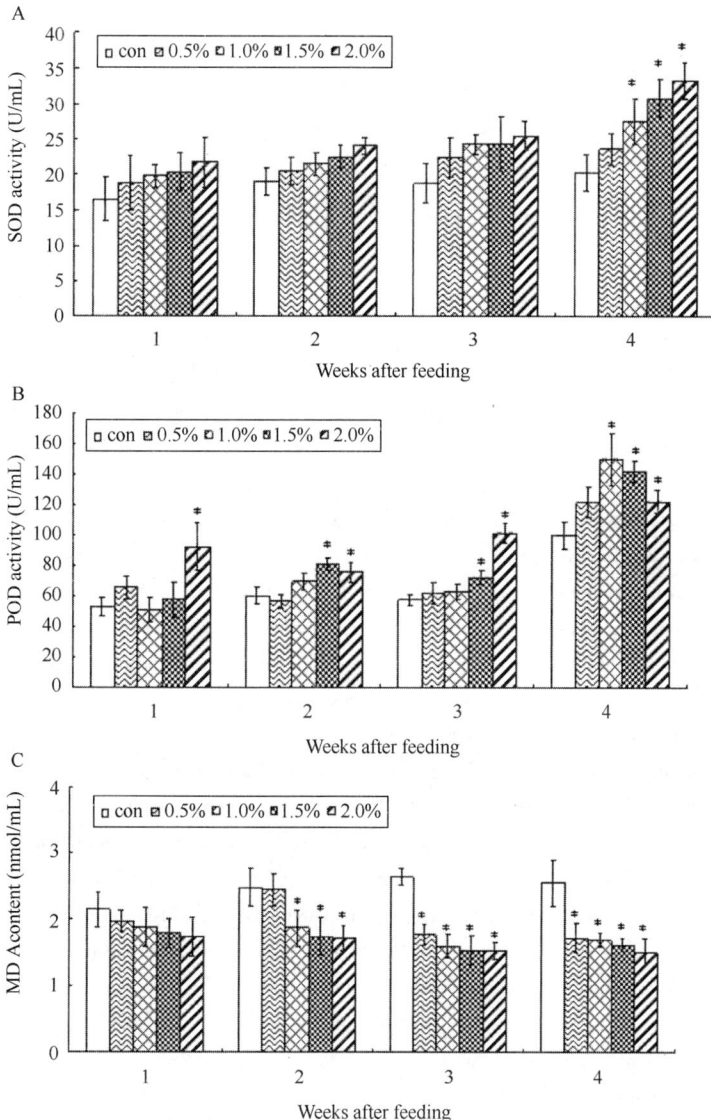

Fig. 3　Effects of CHM-supplemental diet on serum SOD activity （A）, POD activity （B） and MDA content （C）. Values are expressed as mean±SE. The P values less than 0.05 were considered as significant and indicated by asterisks.

4　Discussion

Though the immunostimulatory roles and disease resistance of herbs have been documented in many cultured fish[24-29], the effects of herbal mixture in fish are poorly known. In the present study, the roles of a Chinese herbal mixture on immunostimulatory response and pathogen resistance were investigated in O. niloticus.

As we known, lysozyme is an important antimicrobial effector in fish[30], which also serve as an opsonin in complement system and phagocytes activation[31]. In current study, total lysozyme activity of serum in CHM-dieted fish exhibited significant increase in all experimental groups at 4th week after feeding, indicating the crucial role of this CHM in inducing serumlysozyme activity. Similar findings were documented in tilapia fed with E. albaleaf extract[32], Sophora flavescensextract[1] or A. radixand S. radix[25]. Interesting, the increase of the lysozyme activity in this study was significant except at 2nd week. The similar pattern was also found on tilapia fed with Chinese herb medicines[33], Chinese sucker fed with Propolis and HerbaEpimedii extracts[34]. And the reason for this case could be attributed to the differences of composition, action mechanisms and action periodbetween different Chinese herbs. This hypothesis would be clarified in future research.

Tumor necrosis factor alpha （TNF-α） and interleukin-1β （IL-1β） are two well-studied cytokines in fish, which enhance va-

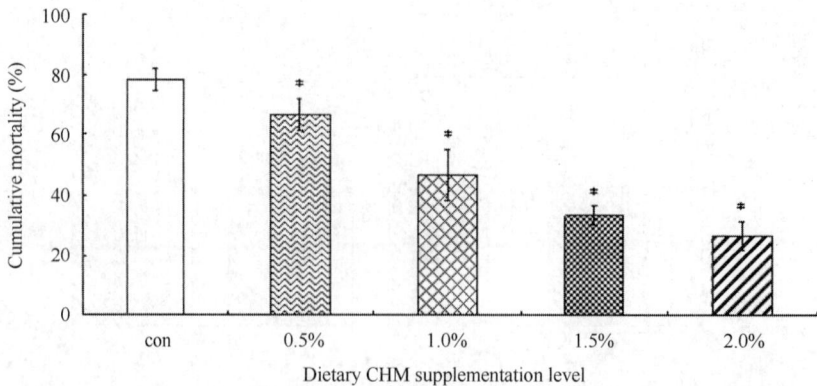

Fig. 4   Cumulative mortalities（%）of fish in control group and CHM-supplemental group. Each point represents mean±SE of triplicate groups（30 fish in each group）. Asterisks indicated significant differences（P<0. 05）.

rious cellular responses such as phagocytosis, chemotaxis, macrophage proliferation and lysozyme synthesis[35-36]. The result of qRT-PCR exhibited that the mRNA expressions of TNF-α and IL-1β in head kidney and spleen were significantly up-regulated when O. niloticus fed with relative high CHM supplementation levels（TNF-α in 1. 0%, 1. 5% and 2. 0%groups；IL-1β in 1. 5%, 2. 0% groups）, and no significant increases were found in 0. 5% group. The present results were not completely in agreement with the data recorded in Cyprinuscarpio L. When Cyprinuscarpio L injected with Astragalus polysaccharides（APS）, transcription of TNF-α was up regulated inthe spleen at high does group, while TNF-α mRNA level decreased significantly in the head kidney of lowdose of APS. And increased of IL-1β mRNA level was only found in the head kidney[37].

Pathogen infection and pollution lead to reactive oxygen species（ROS）accumulation, which may cause disruption of cellular functionand integrity[38-40]. Antioxidant can protect cells from ROS-induced damage[41]. The activities of two antioxidants SOD and POD activity were significantly induced by addition of Chinese herbal mixture in current research. The results of SOD activity were in line with C. carpio, L. rohit, O. mykiss, O. mossambicu fed with herbal mixture[42-43] or herb alone[2,19,44-46]. And the observation of POD activity in our study was consistent withC. gariepinus fed with Ficusbenghalensisand Leucaenaleucocephalaextract[47]. MDA is considered as an indicator of oxidative stress in fish[48], which decreased after feeding CHM-supplemented diets in this study. These data indicated that Chinese herbal mixture applied in this study could maintain fish healthy through protecting the tissue against peroxidation.

It's necessary to estimate the reflection of supplementaldiety on disease resistant to determine the efficacy of Chinese herbal mixture. In this study, all experimental groups showed a reduced mortality compared to control group after A. hydrophila infection. Similarly, decreased mortality on challenge with A. hydrophila were documented in O. niloticuswhen fed withA. membranaceusand L. japonica[49], and C. carpiogiven A. radix and G. lucidum[50].

To summarize, the Chinese herbal mixture prepared in this study could significantly enhanced immune responses in fish and provided disease resistance against A. hydrophil. Therefore, this CHM could be recommended as a potentialimmunostimulant to enhance immunity and disease resistant of cultured fish species.

Acknowledgments

The authors wish to thank all the laboratory members' suggestion on this manuscript. This work was supported byScience and Technology Planning Project of Guangdong Province（2007A020200006-10, 2008A020100014）and Special Fund for Agro-scientific Research in the Public Interest（201203085）.

References

［1］   Wu YR, Gong QF, Fang H, Liang WW, Chen M, et al. Effect of Sophoraflavescens on non-specific immune response of tilapia（GIFT Oreochromisniloticus）and disease resistance against Streptococcus agalactiae. Fish Shellfish Immunol 2013；34：220-227.

［2］   Alexander CP, Kirubakaran CJ, Michael RD. Water soluble fraction of Tinosporacordifolia leaves enhanced the non-specific immune mechanisms and disease resistance in Oreochromismossambicus. Fish Shellfish Immunol2010；29：765-772.

［3］   Jones OAH, Voulvoulis N, Lester JN. Potential ecological and human health risks associated with the presence of pharmaceutically active compounds in the aquatic environment. Environ Crit Rev Toxicol2004；34：335-350.

［4］   Cabello FC. Heavy use of prophylactic antibiotics in aquaculture：a growing problem for human and animal health and for the environment. Environ Microbiol2006；8：1137-1144.

［5］ Siwicki AK, Anderson DP, Rumsey GL. Dietary intake of immunostimulants by rainbow trout affects nonspecific immunity and protection against furunculosis. Vet ImmunolImmunopathol1994; 41; 125-139.

［6］ Raa J, Roerstad G, Engatad R, Robetsen B. The use of immunostimulants to increase resistance of aquatic organisms to microbial infections. DisAsianAquacult1992; 1; 39-50.

［7］ Sakai M, Kamiya H, Ishii S, Atsuta S, Koybayashi M. The immunostimulating effects of chitin in rainbow trout. Oncorhynchusmykiss. In; Shariff M, Subasinghe RP, Arthur JR, editors. Diseases in Asian aquaculture. Proceedings of the first symposium on diseases in Asian aquaculture. Manila; Asian Fisheries Society; 1992. p. 413-417.

［8］ El-Boshy ME, El-Ashram AM, AbdelHamid FM, Gadalla HA. Immunomodulatory effect of dietary Saccharomyces cerevisiae, b-glucan and laminaran in mercuric chloride treated Nile tilapia (Oreochromisniloticus) and experimentally infected with Aeromonashydrophila. Fish Shellfish Immunol 2010; 28; 802-808.

［9］ Geng X, Dong XH, Tan BP, Yang QH, Chi SY, Liu HY, et al. Effects of dietary chitosan and Bacillus subtilis on the growth performance, non-specific immunity and disease resistance of cobia, Rachycentroncanadum. Fish ShellfishImmunol 2011; 31; 400-406.

［10］ Sakai M, Otubo T, Atsuta S, Kobayashi M. Enhancement of resistance to bacterial infection in rainbow trout, Oncorhynchusmykiss (Walbaum) by oral administration of bovine lactoferrin. J Fish Dis 1993; 16; 239-247.

［11］ Villamil L, Figueras A, Novoa B. Immunomodulatory effects of Nisin in turbot (Scophthalmusmaximus L.). Fish Shellfish Immunol 2003; 14; 157-169.

［12］ Stafford JL, Wilson EC, Belosevic M. Recombinant transferring induces nitricoxide response in goldfish and murine macrophages. Fish Shellfish Immunol 2004; 17; 171-185.

［13］ Chiu CH, Cheng CH, Gua WR, Guu YK, Cheng W. Dietary administration of the probiotic, Saccharomyces cerevisiae P13, enhanced the growth, innate immune responses, and disease resistance of the grouper, Epinepheluscoioides. Fish ShellfishImmunol2010; 29; 1053-1059.

［14］ Harikrishnan R, Balasundaram C, Heo MS. Lactobacillus sakei BK19 enriched diet enhances the immunity status and disease resistance to streptococcosisinfection in kelp grouper, Epinephelusbruneus. Fish ShellfishImmunol 2010; 29; 1037-1043.

［15］ Harikrishnan R, Kim MC, Kim JS, Balasundaram C, Heo MS. Probiotics and herbal mixtures enhance the growth, blood constituents, and nonspecific immune response in Paralichthysolivaceus against Streptococcus parauberis. Fish ShellfishImmunol 2011; 31; 310-317.

［16］ Tan BK, Vanitha J. Immunomodulatory and antimicrobial effects of some traditional chinese medicinal herbs; a review. Curr Med Chem 2004; 11; 1423-1430.

［17］ Harikrishnan R, Balasundaram C, Heo MS. Impact of plant products on innate and adaptive immune system of cultured finfish and shellfish. Aquaculture 2011; 317; 1-15.

［18］ Abdel-Tawwab M, Ahmad MH, Seden MEA, Sakr SFM. Use of green tea, Camellia sinensis L., in practical diet for growth and protection of Nile tilapia, Oreochromisniloticus (L.), against Aeromonashydrophila infection. J World AquacSoc 2010; 41; 203-213.

［19］ Kirubakaran CJW, Alexander CP, Michael RD. Enhancement of non-specific immune responses and disease resistance on oral administration ofNyctanthesarbortristis seed extract in Oreochromismossambicus (Peters). Aquac Res 2010; 41; 1630-1639.

［20］ Wu CC, Liu CH, Chang YP, Hsieh SL. Effects of hot-water extract of Toonasinensis on immune response and resistance to Aeromonashydrophila in Oreochromismossambicus. Fish Shellfish Immunol 2010; 29; 258-263.

［21］ Livak KJ, Schmittgen TD. Analysis of relative gene expression data using real-time quantitative PCR and the $2^{-\triangle\triangle Ct}$ method. Methods 2001; 25, 402-408.

［22］ Schmittgen TD, Livak KJ. Analyzing real-time PCR data by the comparative C (T) method. Nat Protoc 2008; 3, 1101-1108.

［23］ Quade MJ, Roth JA. A rapid, direct assay to measure degranulation of bovine neutrophil primary granules. Vet ImmunolImmunopathol 1997; 58; 239-48.

［24］ Chakrabarti R, Rao YV. Achyranthesaspera stimulates the immunity and enhances the antigen clearance in Catlacatla. Int J Immunopharmacol 2006; 6; 782-790.

［25］ Yin G, Jeney G, Racz R, Xu P, Jun X, Jeney Z. Effect of two Chinese herbs (Astragalus radixandScutellaria radix) on non-specific immune response of tilapia, Oreochromisniloticus. Aquaculture 2006; 253; 39-47.

［26］ Divyagnaneswari M, Christybapita D, Michael RD. Enhancement of nonspecific immunity and disease resistance in Oreochromismossambicus by Solanumtrilobatum leaf fractions. Fish Shellfish Immunol 2007; 23; 249-259.

［27］ Thuy NTT, Mukherjee SC, Prasad KP. Studies on the immunostimulatory effect of certain plant extracts on fish. In; The sixth Indian fisheries forum. Mumbai, India; CIFE; 2002. Abstracts AH-13; 153.

［28］ Rao YV, Das BK, Jyotyrmayee P, Chakrabarti R. Effect of Achyranthesasperaon the immunity and survival of Labeorohitainfected with Aeromonashydrophila. Fish Shellfish Immunol 2006; 20; 263-273.

［29］ Punitha SMJ, Babu MM, Sivaram V, Shankar VS, Dhas SA, Mahesh TC, et al. Immunostimulating influence of herbal biomedicines on non-specific immunity in Grouper Epinephelustauvinajuvenile against Vibrio harveyiinfection. AquacultInt 2008; 16; 511-523.

［30］ Yeh SP, Chang CA, Chang CY, Liu CH, Cheng W. Dietary sodium alginate administration affects fingerling growth and resistance to Streptococcus sp. and iridovirus, and juvenile non-specific immune responses of the orange-spotted grouper, Epinepheluscoioides. Fish Shellfish Immunol 2008; 25; 19-27.

611

[31] Magnadottir B. Innate immunity of fish（overview）. Fish Shellfish Immunol2006；20：137-151.

[32] Christybapita, D, Divyagnaneswari, M, Michael, RD. Oral administration of Ecliptaalbaleaf aqueous extract enhances the non-specific immune responses and disease resistance of Oreochromismossambicus. Fish Shellfish Immunol2007；23：840-852.

[33] Wen ZQ, Pei JT, Wang WB, Wan SL, Liu HJ, Du JY, et al. The effects of Chinese herb medicine on non-specific immunity in tilapia Oreochromisniloticus. J Dalian Fish Univ2009；24（2）：136-140［China］.

[34] Zhang G, Gong S, Yu D, Yuan H. Propolis and HerbaEpimedii extracts enhance the non-specific immune response and disease resistance of Chinese sucker, Myxocyprinusasiaticus. Fish Shellfish Immunol2009；26（3）：467-472.

[35] Dinarello CA. Interleukin-1. Cytokine Growth Factor Rev1997；8：253-265.

[36] Goetz FW, Planas JV, MacKenzie S. Tumor necrosis factors. Dev Comp Immunol 2004；28：487-497.

[37] Yuan C, Pan X, Gong Y, Xia A, Wu G, Tang J, et al. Effects of Astragalus polysaccharides（APS）on the expression of immune response genes in head kidney, gill and spleen of the common carp, Cyprinuscarpio L. IntImmunopharmacol 2008；8：51-58.

[38] Di Giulio RT, Washburn PC, Wenning RJ, Winston GW, Jewell CS. Biochemical responses in aquatic animals：a review of determinants of oxidative stress. Environ ToxicolChem 1989；8：1103-1123.

[39] Han HJ, Kim DH, Lee DC, Kim SM, Park SI. Pathogenicity of Edwardsiellatardato olive flounder, Paralichthysolivaceus（Temminck& Schlegel）. J Fish Dis2006；29：601-609.

[40] Sturve J, Almroth BC, Förlin L. Oxidative stress in rainbow trout（Oncorhynchusmykiss）exposed to sewage treatment plant effluent. Ecotoxicol Environ Saf2008；70：446-452.

[41] Messaoudi I, Barhoumi S, Said K, KerkenA. Study on the sensitivity to cadmium of marine fish Salariabasilisca（Pisces, Blennidae）. J Environ Sci 2009；21：20-1624.

[42] Wu G, Yuan C, Shen M, Tang J, Gong Y, Li D, et al. Immunological and biochemical parameters in carp（Cyprinuscarpio）after Qompsell feed ingredients for long-term administration. Aquacult Res2007；38, 246-255.

[43] Yuan C, Li D, Chen W, Sun F, Wu G, Gong Y, et al. Administration of a herbal immunoregulation mixture enhances some immune parameters in carp（Cyprinuscarpuio）. Fish PhysiolBiochem 2007；33, 93-101.

[44] Rao, YV, Das BK, Jyotyrmayee P, Chakrabarti R. Effect of Achyranthesasperaon the immunity and survival of Labeorohita infected with Aeromonashydrophila. Fish Shellfish Immunol2006；20：263-273.

[45] Sahu S, Das BK, Mishra BK, Pradhan J, Sarangi N. Effect of Allium sativum on the immunity and survival of Labeorohitainfected with Aeromonashydrophila. J ApplIchthyol2007；23, 80-86.

[46] Nya, EJ, Austin B. Use of garlic, Allium sativum, to control Aeromonashydrophila infection in rainbow trout, Oncorhynchusmykiss（Walbaum）. J Fish Dis2009；32, 963-970.

[47] Verma VK, Rani KV, Sehgal N, Prakash O. Immunostimulatory effect of artificial feed supplemented with indigenous plants onClariasgariepinusagainst Aeromonashydrophila. Fish Shellfish Immunol2013；35：1924-1931.

[48] Modesto KA, Martinez CB. Roundup causes oxidative stress in liver and inhibits acetylcholinesterase in muscle and brain of the fish Prochiloduslineatus. Chemosphere2010；78：294-299.

[49] Ardo L, Yin G, Xu P, Varadi L, Szigeti G, Jeney Z, et al. Chinese herbs（AstragalusmembranaceusandLonicera japonica）and boron enhance the non-specific immune response of Nile tilapia（Oreochromisniloticus）and resistance against Aeromonashydrophila. Aquaculture 2008；275：26-33.

[50] Yin G, Ardo L, Thompson KD, Adams A, Jeney Z, Jeney G. Chinese herbs（Astragalus radix andGanodermalucidum）enhance immune response of carp, Cyprinuscarpio, and protection against Aeromonashydrophila. Fish Shellfish Immunol2009；26（1）：140-145.

该文发表于《西北农林科技大学学报：自然科学版》【2012，（3）：39-46】

玉屏风散对施氏鲟血液生化指标及非特异性免疫功能的影响

Effect of Jade screen power on the blood biochemical indicators and non-specific immunity of *Acipenser schrencki*

王荻，李绍戊，卢彤岩，刘红柏

（中国水产科学研究院黑龙江水产研究所，黑龙江 哈尔滨 150070）

摘要：【目的】研究玉屏风散方剂对施氏鲟（*Acipenser schrencki*）在适温及热应激条件下生长、生理生化指标、相关非特异性免疫功能的影响。【方法】随机选取体质量 0.08~0.10 kg 的 1+龄施氏鲟，分为试验组和对照组，每组

2个重复，每重复15尾鱼，在22℃下预饲2周后，试验组鱼按0.9 g/kg的剂量连续口灌给药14 d，分别于试验0，7 d，10 d，14 d给药前称量鱼体质量，计算鱼体质量增加率；给药结束后，将试验组和对照组半数鱼放入30℃的控温水族箱中热刺激2 h，然后分组采集所有供试鱼的血液及组织样本，进行各种酶活性及活性物质含量的测定。【结果】施氏鲟在玉屏风散口灌给药10~14 d期间体质量增加率快速提升。常温给药条件下，玉方总体能明显提高鱼体内球蛋白含量及溶菌酶活性，同时促进AKP含量的提升；而对白细胞吞噬率及吞噬指数无显著影响；且能对SOD活性及NO含量进行平衡调节，并有效降低丙二醛含量。另外，高温热刺激对玉方药效无显著不良影响。但玉方也在一定程度上导致鱼体血浆中乳酸脱氢酶、谷丙转氨酶活性及总胆红素和尿酸含量升高。【结论】玉屏风散对促进鱼类生长、增强其免疫及抗氧化能力有一定作用，可用作施氏鲟中药添加剂使用，但在使用中也应注意该方剂对鱼肝肾功能带来的负担和影响。

关键词：玉屏风散；施氏鲟；生理生化；非特异性免疫

该文发表于《水产学杂志》【2013：26（4）：38-41】

中草药添加剂对山女鳟体内 MHC 及 IL-1β 含量的影响

Effect of Chinese herb feed additives on MHC and IL-1β content in *Oncorhynchus masou*

王荻，穆桂强，刘红柏

（中国水产科学研究院/黑龙江水产研究所，哈尔滨 150070）

摘要：在（14.0±0.3）℃下，将体质量为（100.00±2.15）g的1+龄山女鳟（*Oncorhynchus masou*）饲养在循环水族箱（直径0.85 m，高1 m）中，每天按鱼体质量1.0%投喂对照（基础饲料，对照组）和含3种自组中药方剂（E1、E2和E3）水煎剂的饲料（给药剂量分别为53.4 g·kg⁻¹、18.6 g·kg⁻¹和2.4 g·kg⁻¹）30 d。停食1 d后测定脾、血浆、血细胞、肝4种组织中主要组织相容性复合体（MHC）及白细胞介素（IL-1β）的含量。结果表明：对照组中山女鳟体内MHC含量由高至低依次为：脾>血浆>血细胞>肝；IL-1β含量为：血浆>肝>脾>血细胞。方剂E2对山女鳟体内四种组织中MHC含量有促进作用，且在脾内作用最强；而对山女鳟脾及血细胞中IL-1β含量起到略微促进作用，但抑制血浆中IL-1β含量。本实验表明：方剂E2是最适用于山女鳟的免疫调节制剂。

关键词：中草药添加剂，山女鳟，MHC，IL-1β

该文发表于《东北农业大学学报》【2011，42（12）：107-113】

三种中药复方对鲤幼鱼生长及免疫性能的影响

The Effect of Four Kinds of Chinese Herbal Compound and APS on Growth and The Contents of NO And NOS in Organization of *Carassius auratus*

苏岭[1,2]，刘红柏[1]，王荻[1]，卢彤岩[1]，尹家胜[1,2]

（1. 中国水产科学研究院黑龙江水产研究所，哈尔滨 150070；2. 上海海洋大学水产与生命学院，上海 201306）

摘要：本试验研究了灌服三种复方中药水煎剂后鲤鱼生长、免疫以及血液生化指标的变化。选用健康1龄鲤鱼，按平均体重分四组，每组40尾，设置两个重复在相同条件下喂养。连续灌服中药28 d后采样检测各项指标。结果表明，三个灌服中药组鲤鱼的增重率均比对照组有显著的提高。方剂AH、AB在有细菌性抗原侵入时能提高血清中总蛋白（TP），白蛋白（ALB）及球蛋白（GLB）的含量。方剂AH能显著提高血清中碱性磷酸酶（AKP）的含量。反应肝功能的指标中，谷丙转氨酶（ALT）和谷草转氨酶（AST）的含量均能被三种中药显著降低。方剂CD组显著提高了血清中尿素氮（BUN）和肌酐（CREA）的含量。当有细菌性抗原侵入机体时，AH和AB组均能显著提高血清中铁的含量。血清内补体 C3、C4 研究结果表明，未注射细菌性抗原时三种中药复方均能使 C3、C4 含量显著高于对照组；注射细菌性抗原后，仅 C4 含量显著高于对照组。NO 和 NOS 的研究结果显示了较大的组织差异性。血清中，中药组鲤鱼 NO 和 NOS 的含量均显著高于对照组，而肝脏中仅在注射细菌性抗原后中药组 iNOS 的含量会显著高于对照组。综合结果显示，三种复合方剂均在不同程度上对鲤鱼的生长和免疫起到了促进作用

关键词：中药复方；鲤鱼；生化指标；免疫

该文发表于《华北农学报》【2012：27（增刊）：401-405】

中草药方剂对鲫鱼热应激蛋白 HSP 70 表达的影响

Effects of the Chinese herbal compounds on the heat shock protein 70 expression of Crucian Carp *carassius auratus*

李绍戊，王荻，苏岭，刘红柏

（中国水产科学研究院黑龙江水产研究所，哈尔滨 150070）

摘要：选用四种中草药方剂 AH、CD、IV 和 APS，每日按体重的 3% 混饲鲫鱼 30 天。采用实时荧光定量 RT-PCR 方法，监测适温（18℃）及热刺激（28℃）一小时后的鲫鱼脑及肝脏中 HSP 70 表达水平变化。实验结果表明，对照组鲫鱼体内两种组织中 HSP 70 的表达量在热刺激条件下均有不同程度的升高，且适温和热应激条件下，鲫鱼肝脏组织中 HSP 70 表达量均明显高于脑组织。而 AH 组、IV 组和 APS 组热刺激后 HSP70 mRNA 含量升高，但 CD 组热刺激后 HSP70 mRNA 转录水平降低，且在肝脏中表达水平降低极为显著。四种方剂中尤以补血养气为主要功能的 AH 方剂在鲫鱼肝脏中对 HSP 70 表达水平有极大的促进作用，在适温及热应激条件下，分别是对照组的 4.86 和 3.43 倍。因此，中草药方剂 AH 对促进鲫鱼 HSP 70 基因高水平表达有较好的作用，有助于鱼体抵抗热刺激防止热损伤，可作为有效的鲫鱼抗热应激备选药物。

关键词：中草药方剂；鲫鱼；荧光定量 PCR；HSP 70

该文发表于《水产学杂志》【2012，25（1）：46-48】

三种复方中药对鲫鱼血清、肝脏及肾脏中溶菌酶活性的影响

Three kinds of traditional Chinese medicine compounds effect of serum, liver and kidney of lysozyme activity on crucian carp *carassius curatus*

栾鹏[1,2]，王荻[1]，李绍戊[1]，卢彤岩[1]，刘红柏[1]

（1. 中国水产科学研究院黑龙江水产研究所，黑龙江 哈尔滨 150070；2. 东北农业大学动物医学学院，黑龙江 哈尔滨 150030）

摘要：在水温 18℃下研究了三种复方中药（AB、CD、AH）作为饲料免疫添加剂对体质量（44.6±0.2）g 1+龄健康鲫（*Carassius auratus*）血清、肝胰脏和脾脏中溶菌酶活性的影响。试验鱼分为 3 组，每天按体质量的 2.0% 分别投喂添加了 3 种中药复方的饲料，对照组投喂基础饲料，每组均设相同数量的鱼做平行，连续饲喂 28 d，停食 1 d 后采集血液及组织样品，测定溶菌酶活性。结果显示：3 种复方中药都能提高鲫体内溶菌酶活性。其中 AB 组提高血清中溶菌酶活性效果显著，AB、CD 组对肝胰脏中溶菌酶活性的提升效果显著，3 种复方中药对鲫脾脏中溶菌酶活性的提升均效果显著，说明 3 种复方中药均可增强鲫免疫力，有进一步研究和应用的价值，但对不同组织溶菌酶活性的影响不同，提示中药作用效果和机理的差异。

关键词：溶菌酶，中药，鲫鱼

该文发表于《水产学杂志》【2012：25（4）：24-28】

三种中草药方剂对鲫部分生化指标的影响

The Effect of Chinese Herbal Medicines on Biochemical Indices in Crucian Carp *Carassius auratus*

刘红柏[1]，宿斌[1,2]，王荻[1]

（1. 中国水产科学研究院黑龙江水产研究所，黑龙江 哈尔滨 150070；2. 南京农业大学渔业学院，江苏 南京 210095）

摘要：在水温20℃下研究了在饲料中分别添加3种中草药方剂（复方A由甘草、紫草等组成；复方B由生鸡内金、生牡蛎、山药等组成；复方C由茵陈、板蓝根、黄芩等组成），对平均体质量37.09±4.09 g鲫（Carassius auratus）肝胰脏和血液中部分生化指标的影响。每组15尾，每天按照鱼体质量的2.0%投喂饲料，对照组投喂不含中草药添加剂的基础饲料。实验结束后，取样、测定鱼体肝脏和血液部分生化指标。结果显示，复方A，B，C皆可显著提高血清谷草转氨酶（AST）和碱性磷酸酶（ALP）的活性，但却明显降低了尿酸（URIC）、总蛋白（TP）、总胆红素（TBIL）和球蛋白（GLB）的含量，其他几项生化指标升降不一；复方A，B，C对肝脏中Fe、ALP和肌酐（CREA）3项生化指标的提升效果比较明显，对Ca，P，AST、总胆固醇（CHOL）、乳酸脱氢酶（LDH），TP和GLB起相反作用。结果的差异可能与中草药配方组成成分、取样部位、测试的生化指标有关。

关键词：中草药方剂；鲫；生化指标

该文发表于《水产学杂志》【2010，23（3）：11-15】

四种复方中药和黄芪多糖对鲫鱼生长、组织中 NO 含量与 NOS 活性的影响

The Effect of Four Kinds of Chinese Herbal Compound and APS on Growth and The Contents of NO And NOS in Organization of *Carassius auratus*

苏岭[1,2]，刘红柏[1]，王荻[1]，卢彤岩[1]，尹家胜[1,2]

（1. 中国水产科学研究院黑龙江水产研究所，黑龙江 哈尔滨 150070；2. 上海海洋大学水产与生命学院，上海 201306）

摘要：将四种复方中药（分别编号为AB、AH、CD、IV）和黄芪多糖（APS）作为添加剂添加入基础料饲料中，连续饲喂1龄鲫鱼45 d，于第45 d及第60 d分别采样，测定鱼体重和肝脏、脾脏、肾脏中一氧化氮（NO）含量及一氧化氮合酶（NOS）活性，研究中药复方对鲫鱼生长及免疫功能的影响。结果显示：45 d时称量体重，复方AH、复方AB两组增重率显著高于对照组及其他三种中药复方组，饲喂60 d对照组及所有中药组增重率均无显著差距。投喂45 d，方剂AB显著降低鲫鱼肝脏、肾脏中NO的含量，五种中药方剂均能提高鲫鱼脾脏中NO含量。AH、AB、IV、APS四种方剂饲喂的鲫鱼肝脏中NOS含量明显高于对照组。投喂60天，五种中药方剂饲喂的鲫鱼脾脏中NO含量均高于对照组，除IV外，均差异显著。肝脏中NOS含量各组无显著差异，脾脏中IV组TNOS含量最高，APS组iNOS含量最高。肾脏中APS方剂组的iNOS含量显著高于对照组。

关键词：鲫鱼；复方中药；NO；NOS

该文发表于《中国农学通报》【2012，28（14）：134-138】

复方中药免疫添加剂对史氏鲟生长性能和抗氧化力的影响

The effects of Chinese herbal compound on growth and antioxidant defenses in Amur sturgeon as immune additive

刘逊[1,2]，王荻[1]，卢彤岩[1]，李绍戊[1]，刘红柏[1]

(1. 中国水产科学研究院黑龙江水产研究所，哈尔滨 150070；2. 东北农业大学动物医学院，哈尔滨 150030)

摘要： 探讨了复方中药对史鲟生长及抗氧化能的影响，为中药作为免疫添加剂在鲟鱼养殖生产上的应用提供基础。将 3 种复方中药（分别为方A、方B、方C）作为饲料添加剂加入基础料饲料中，连续饲喂 1[+] 龄史氏鲟 (Acipensers chrenckii Brandt) 35 天，测定增重率并检测血清、肝脏的超氧化物歧化酶 (SOD) 的活性、丙二醛 (MDA) 的含量、总抗氧化能力 (T-AOC) 和谷胱苷肽过氧化物酶 (GSH-PX) 活性，以研究中药复方对史氏鲟生长及抗氧化能力的影响。结果显示：中药方剂对鲟鱼血清及肝脏中超氧化物歧化酶 (SOD) 的活性、总抗氧化能力 (T-AOC) 和谷胱苷肽过氧化物酶 (GSH-PX) 均有增强作用，尤其 3 个方剂组血清中的 SOD 酶活性较对照组均有显著升高 ($P<0.05$)，A 组和 C 组鱼肝脏中的 SOD 酶活性较对照组均有显著升高 ($P<0.05$)；3 个方剂组中血清及肝脏中 T-AOC 活性均有所上升但与对照组相比无显著差异；只有 B 组和 C 组鱼血清中 GSH-PX 的活性与对照组有显著差异 ($P<0.05$)，3 个方剂组中肝脏的 GSH-PX 的活性与对照组均无显著差异；并且中药能显著降低鲟鱼血清中丙二醛 (MDA) 的含量，3 个方剂组中血清及肝脏中 MDA 含量均与对照组有显著差异 ($P<0.05$)，表明所选的 3 种复方中药添加剂有不同程度提高鲟鱼抗氧化功能的作用。

关键词： 中药复方；史氏鲟；生长性能；抗氧化

该文发表于《水产学杂志》【2012，25（3）：47-50】

三种中药方剂对施氏鲟抗氧化能力的影响

Effects of Three Chinese Medicinal Herb Complex Additives on Antioxidation in Amur Sturgeon（*Acipenser schrencki Brandt*）

李绍戊，王荻，卢彤岩，刘红柏

(中国水产科学研究院黑龙江水产研究所，黑龙江 哈尔滨 150070)

摘要： 在水温 22℃下，将平均体质量为 (0.08±0.10) kg 的施氏鲟 (Acipenser schrenckii) 在室内水族箱 (0.90 m×0.50 m×0.45 m) 中预饲两周后，每日口灌给药一次中药贯众、方剂一和方剂二水煎剂，剂量分别为 9、175 和 36 g·50 kg⁻¹ 体质量，口灌蒸馏水为对照组，连续给药 14 d。然后，取部分鱼高温 30℃ 热刺激 2 h 后，与常温 22℃ 给药的实验组共同采样，测定血液和肝脏内超氧化物歧化酶 (SOD)、一氧化氮合酶 (NOS) 的活性及丙二醛 (MDA) 和一氧化氮 (NO) 含量。结果表明，三种中药方剂均能提高施氏鲟抗氧化能力。贯众能有效提高施氏鲟组织中 SOD 活性，在热刺激情况下，促进鱼体快速应答，降低鱼体内 MDA 含量，但可能对肝脏造成一定负担，需慎重给药。方剂一能有效降低鱼体内 MDA 含量，促进机体对热应激做出快速有效应答反应；方剂二抗氧自由基效果较好，受高温热刺激影响不大。

关键词： 中药方剂；施氏鲟；超氧化物歧化酶 (SOD)；丙二醛 (MDA)；一氧化氮 (NO)；一氧化氮合酶 (NOS)

该文发表于《东北农业大学学报》【2012，43（6）：116-120】

三种复方中药对鲫鱼抗氧化能力的影响

Effect of three kinds of Chinese herbal compounds on antioxidation capability of liver in Crucian carp *carassius auratus*

王瑞雪[1,2]，王荻[1]，李绍戍[1]，卢彤岩[1]，刘红柏[1]

（1. 中国水产科学研究院黑龙江水产研究所，哈尔滨 150070；2. 东北农业大学动物医学学院，哈尔滨 150030）

摘要： 采用三种复方中药（编号分别为 AB、CD 和 AH）分别饲喂 1 龄鲫鱼 28 d，同时设置空白对照组，以探讨所用方剂对鲫鱼肝脏中抗氧化功能的影响。试验期间水温 20℃，每天按照体重的 2.0% 投喂制备的饲料。采集鲫鱼肝脏，检测超氧化物歧化酶（SOD）、丙二醛（MDA）、谷胱甘肽过氧化物酶（GSH-PX）、总抗氧化能力（T-AOC）四个代表性抗氧化指标。结果表明，三种复方中药组 SOD 活性均显著高于对照组。MDA 含量均低于对照组，其中 AB、CD 组与对照组差异显著（$P<0.05$）。GSH-PX 活性均高于对照组，其中 AH、CD 组与对照组差异显著（$P<0.05$）。三种复方中药均能提高 T-AOC，其中 AB、CD 组提高效果显著（$P<0.05$）。所选的三种复方中药均对鲫鱼肝脏抗氧化能力有促进作用，其中 CD 组方效果最好。

关键词： 中药；鲫鱼；抗氧化功能

该文发表于《东北农业大学学报》【2013，43（9）：127-134】

中草药方剂对哲罗鱼抗氧化能力的影响

Effect of Chinese herbal medicine on antioxidant of *Hucho taimen*

刘红柏[1]，宿斌[1,2]，王荻[1]

（1. 中国水产科学研究院黑龙江水产研究所，哈尔滨 150070；2. 南京农业大学渔业学院，南京 210095）

摘要： 哲罗鱼试验组每组 60 尾，每天按照表 3 分别投喂添加不同中草药（四种复方及 5 种单方）的饲料，对照组投喂基础饲料，连续投喂 28 d，停食 1 d 后采集血液（分离血清及红细胞）及肝脏等样品，进行 NO 含量、NOS 活性、SOD 活性、MDA 含量的测定。结果表明，血清中 NO 含量与对照组差异不显著，除当归组外 NOS 含量均有显著升高趋势；肝脏与血清相反，除当归组外 NO 含量均有上升，除黄芪高剂量组外 NOS 的活性均下降。血清中 SOD 酶活性没有明显变化，肝脏中，只有茯苓可以显著升高 SOD 活性，而红细胞中，黄芪、贯众和方二组可以使 SOD 活性有明显的升高；除方二、茯苓组，其余实验组均能明显降低血清中 MDA 的含量；除方五组外其他实验组均能显著降低 MDA 的含量。结果显示，中草药对哲罗鱼抗氧化能力的影响因药物种类、检测组织和指标而有所不同，但所选中草药均有一定程度增强机体抗氧化能力的作用。

关键词： 哲罗鱼；中药方剂；NO；NOS；SOD；MDA

该文发表于《饲料工业》【2008，29（20）：3-26】

复方中草药对凡纳滨对虾生长、肌肉营养成分和抗病力的影响

Effects of Chinese Herbal Medicine on the Growth，muscle nutritional compositionand Resistance to Diseases for *Litopenaeus vannamei*

汤菊芬[1,2]，陆志款[1,2]，彭卫正[3]，汤燕平[4]

（1. 广东海洋大学；2. 广东省水产经济动物病原生物学及流行病学重点实验室，广东 湛江 524005；

3. 湛江粤海饲料集团有限公司，广东 湛江 524017；4. 湛江粤华水产饲料有限公司，广东 湛江 524023）

摘要：试验在饲料中添加 0、0.5%、1%、1.5%、2%的自制复方中草药制剂，饲养体重为（1.21±0.08）g 的凡纳滨对虾 7 周，以研究不同添加量对凡纳滨对虾生长、肌肉营养组分和抗力的影响。试验结果表明，试验组凡纳滨对虾的增重率、成活率、饲料系数、蛋白质效率、抗病力等指标均优于对照组。本复方中草药制剂对凡纳滨对虾肌肉的常规营养成分和总氨基酸含量无显著影响，但能提高对虾肌肉中鲜味氨基酸的含量。从凡纳滨对虾的生长和抗病力等方面综合考虑，以添加 1.5%较为适宜。

关键词：中草药；凡纳滨对虾；生长；肌肉营养组成；抗病力

该文发表于《广东海洋大学学报》【2009：29（6）：46-49】

复方中草药对吉富罗非鱼生长及肠道菌群的影响

Effects of Chinese Herbal Medicine on Growth Performance and IntestinalMicrofora of GIFT Strain of Nile Tilapia（*Oreochromis niloticus*）

汤菊芬[1,2]，吴灶和[1,2]，简纪常[1,2]，鲁义善[1,2]，王蓓[1,2]

（1. 广东海洋大学水产学院，2. 广东省水产经济动物病原生物学及流行病学重点实验室，广东 湛江 524025）

摘要：将 A、B、C 三种复方中草药分别按质量分数 1.5%的比例添加到罗非鱼商品饲料中，喂养初始体重约 16.58±0.48 g 的吉富罗非罗 28 d，通过测定罗非鱼的增重率、饲料系数、肠道菌群等指标，研究三种复方中草药制剂对罗非鱼生长性能及肠道菌群的影响。结果显示：三种复方中草药制剂对罗非鱼的增重率、成活率、饲料系数等生长性能指标虽无显著影响（$P>0.05$），但与对照组相比，添加 1.5%的复方中草药制剂 C 可使吉富罗非鱼的增重率提高 10.82%，饲料系数降低 4.71%，蛋白质效率提高 15.68%；三种复方中草药制剂均可显著促进罗非鱼肠道细菌、双歧杆菌、乳酸杆菌的生长（$P<0.05$），C 方中草药制剂还可显著抑制大肠杆菌生长（$P<0.05$）。结果表明，C 方中草药制剂可显著促进罗非鱼肠道中乳酸杆菌等有益菌群的增殖，抑制大肠杆菌生长，从而有效促进罗非鱼生长。

关键词：中草药；吉富罗非鱼；生长；肠道菌群

该文发表于《饲料工业》【2009，30（16）：19-21】

复方中草药对罗非鱼生长及肌肉成分的影响

Effects of Chinese Herbal formula on the Growth Performance and meat quality of tilapia

汤菊芬[1,2]，吴灶和[1,2]，简纪常[1,2]，鲁义善[1,2]，汤燕平[3]

（1. 广东海洋大学水产学院，广东 湛江 524025；2. 广东省水产经济动物病原生物学及流行病学重点实验室，广东 湛江 524025
3. 湛江粤华水产饲料有限公司，广东 湛江 524023）

摘要：将 A、B、C 三种复方中草药分别按 1.5% 的比例添加到罗非鱼商品饲料中，喂养初始体重为（16.58±0.48）g 左右的吉富罗非鱼 28 d，通过测定罗非鱼的增重率、饲料系数、肌肉成分等指标，研究 A、B、C 三种复方中草药制剂对罗非鱼生长性能及肌肉成分的影响。结果显示：A、B、C 三种复方中草药制剂对罗非鱼的增重率、成活率、饲料系数及肌肉中的水分、蛋白质、灰分等指标无显著影响（$P>0.05$）。但均可显著提高罗非鱼肌肉的肌苷酸含量（$P<0.05$），降低肌肉脂肪、总胆固醇和丙二醛含量（$P<0.05$）。说明本试验所用三种复方中草药均能有效改善罗非鱼的肌肉品质。

关键词：复方中草药；罗非鱼；生长；肌肉成分

该文发表于《广东海洋大学学报》【2011，31（1）：58-61】

注射黄芪多糖对吉富罗非鱼 c 型溶菌酶基因表达量的影响

Effect of Astragalus Polysaccharides（APS）on the Expression ofLysozyme-c Gene in GIFT Strain of Nile Tilapia（*Oreochromis niloticus*）

汤菊芬，吴灶和，简纪常，鲁义善，王蓓，李楠

（广东海洋大学水产学院广东省水产经济动物病原生物学及流行病学重点实验室，广东 湛江 524025）

摘要：将黄芪多糖（APS）用无菌生理盐水配制成 2 mg/mL 和 20 mg/mL 针剂，腹腔注射吉富罗非鱼，以注射无菌生理盐水为对照。24 h 后分别提取吉富罗非鱼鳃、头肾、肝脏、脾脏等组织中的总 RNA 并反转录成 cDNA，利用 Real-time PCR 方法对不同组织中基因表达进行定量分析。结果表明：吉富罗非鱼腹腔注射 20 mg/mL 高剂量 APS 后，其鳃、头肾、肝脏等三个组织中的 *Lysozyme-c* 基因表达量显著高于对照组（$P<0.05$）；注射 2 mg/mL 低剂量 APS 后，*Lysozyme-c* 基因表达量仅在脾脏中出现显著上调（$P<0.05$）。APS 可通过诱导 *Lysozyme-c* 基因在鳃、头肾、肝脏和脾脏等组织在的表达量，来提高吉富罗非鱼的机体免疫力。

关键词：黄芪多糖；吉富罗非鱼；c 型溶菌酶基因

该文发表于《淡水渔业》【2011，41（6）：15-18】

注射黄芪多糖对吉富罗非鱼热应激蛋白基因表达的影响

Effects of astragalus polysaccharides（APS）injection onthe expression ofheat stress proteins 70（hsp70）gene in *Oreochromis niloticus*

汤菊芬[1,2]，吴灶和[2,3]，简纪常[1,2]，鲁义善[1,2]，王蓓[1,2]

（1. 广东海洋大学水产学院，广东 湛江 524025；2. 广东省水产经济动物病原生物学及流行病学重点实验室，广东 湛江 524025；3. 仲恺农业工程学院，广州 510225）

摘要：将黄芪多糖（APS）用无菌生理盐水配制成 2 mg/mL（低剂量）、20 mg/mL（高剂量）两种针剂，腹腔注射吉富罗非鱼（*Oreochromis niloticus*）。分别在注射后 0 h、24 h、48 h、72 h 和 96 h 采样提取吉富罗非鱼鳃、头肾、肝脏和脾脏组织中的总 RNA，并反转录成 cDNA，利用 Real-time PCR 方法对不同组织中的热应激蛋白（hsp70）基因表达进行定量分析。结果显示，在 24 h 时，高剂量组吉富罗非鱼鳃、头肾和肝脏三个组织中 hsp70 基因的表达量显著高于同期对照组（$P<0.05$），之后表达量快速降低，低剂量组 hsp70 基因表达量仅在脾脏中有上调。实验表明，APS 可通过诱导吉富罗非鱼 hsp70 基因表达来保护免疫细胞免受应激损伤，从而增强鱼体抵抗病原菌的能力。

关键词：黄芪多糖；吉富罗非鱼；热应激蛋白 70（hsp70）基因

该文发表于《广东海洋大学学报》【2008，28（4）：59-64】

一种复方中草药饲料添加剂在红笛鲷网箱养殖中的应用

The Application of Chinese Herbal Feed Additives in Net-cage Culture of *Lutjanus sanguineus*

陆志款，简纪常，吴灶和，鲁义善，汤菊芬

（广东海洋大学水产学院，广东省水产经济动物病原生物学及流行病学重点实验室暨广东省教育厅水产经济动物病害控制重点实验室，广东 湛江 524025）

摘要：将黄芪、当归、甘草和山楂等粉碎后按一定比例配伍制成复方中草药饲料添加剂后，按质量分数 0.5%、1.0%、1.5% 和 2.0% 添加到基础饲料中，连续投喂红笛鲷［体重（234.95±19.26）g，体长（24.76±1.42）cm］56 d，研究复合中草药制剂对红笛鲷生长、成活、免疫和抗病力的影响。结果表明，基础饲料中添加本复方中草药饲料添加剂能够显著促进红笛鲷的生长，增强其血清中的抗菌活力和溶菌酶活力，提高红笛鲷成活率及其对溶藻弧菌的免疫保护率，其中以添加 1.5%（质量分数）和连续饲喂 28 d 效果为最佳（$P<0.01$）。

关键词：红笛鲷；中草药饲料添加剂；抗菌活力；溶菌酶活力；成活率；免疫保护率

该文发表于《广东海洋大学学报》【2015，35（6）：47-52】

中草药复合益生菌制剂对凡纳滨对虾生长、抗病力及水质的影响

Effects of a Probiotics Combined with Chinese Herbal Medicine on Growth Performance, Water Quality and Resistance to Diseases for *Litopenaeus vannamei*

汤菊芬[1,2]，黄瑜[1,2]，蔡佳[1,2]，鲁义善[1,2]，吴灶和[2]，简纪常[1,2]*

（1. 广东海洋大学水产学院，广东 湛江 524088；2. 广东省水产经济动物病原生物学及流行病学重点实验室，广东 湛江 524088）

摘要：将枯草芽孢杆菌、纳豆芽孢杆菌、地衣芽孢杆菌等比例混合，制成芽孢杆菌制剂；将芽孢杆菌制剂与粪肠球菌、嗜酸乳杆菌等比例混合制成复合益生菌制剂；将分别用 3 种芽孢杆菌发酵的中草药等比例混合制成中草药芽孢杆菌制剂；将分别用 5 种益生菌发酵的中草药等比例混合，制成中草药复合益生菌制剂。在饲料中分别添加 4 种益生菌制剂（活菌为 2×10^7 cfu/g），研究 4 种制剂对凡纳滨对虾（*Litopenaeus vannamei*）生长、抗病力及水质的影响。结果表明：1）4 种益生菌制剂均可提高凡纳滨对虾的成活率、增重率和饲料利用率（$P<0.05$），中草药益生菌制剂组的促生长效果优于益生菌制剂组，以中草药复合益生菌制剂组促生长效果为最佳（$P<0.05$）；2）4 种益生菌制剂均可维持对虾养殖水体 pH 值、氨态氮和亚硝酸盐含量的稳定（$P<0.05$），中草药益生菌制剂对水质的改良效果优于益生菌制剂（$P<0.05$）；3）用 1×10^8 cfu/mL 的哈维氏弧菌（*Vibro harveyi*）菌液浴浸泡凡纳滨对虾 10 d，凡纳滨对虾的累计死亡率由大到小依次为中草药复合益生菌制剂组（31.11%）、复合益生菌制剂组（35.56%）、中草药芽孢杆菌制剂组（37.78%）、芽孢杆菌制剂组（44.44%）、对照组（93.33%）。在饲料中添加一定比例的中草药复合益生菌制剂可提高凡纳滨对虾的生长指标、改善养殖水体环境、提高对虾的抗病力。

关键词：凡纳滨对虾；饲料添加剂；益生菌制剂；中草药；复合制剂；生长；水质；抗病力

该文发表于《渔业科学进展》【2016，37（4）：104-109】

中草药复合微生态制剂对吉富罗非鱼生长、肠道菌群及抗病力的影响 *

汤菊芬[1,2]，黄瑜[1,2]，蔡佳[1,2]，丘金珠[1,2]，孙建华[3]，徐中文[3]，简纪常[1,2]*

（1. 广东海洋大学水产学院，湛江 524088；2. 广东省水产经济动物病原生物学及流行病学重点实验室，湛江 524088；3. 广东绿百多生物科技有限公司，湛江 524022）

摘要：通过在饲料中分别添加 2×10^7 CFU/g 的芽孢杆菌制剂、中草药芽孢杆菌制剂、复合微生态制剂和中草药复合微生态制剂，研究 4 种微生态制剂对吉富罗非鱼生长、肠道菌群及抗病力等的影响。结果显示：（1）饲料中添加 4 种微生态制剂均可以显著提高罗非鱼的增重率（$P<0.05$），对成活率和饲料利用率也有一定程度的提高（$P>0.05$），而中草药复合微生态制剂对罗非鱼促进生长效果最佳。（2）饲料中添加 4 种微生态制剂可以显著提高罗非鱼肠道中的细菌总数、芽孢杆菌、乳酸杆菌和双歧杆菌数量（$P<0.05$），大肠杆菌数量显著低于对照组（$P<0.05$），说明饲料中添加一定量的 4 种微生态制均可以改善罗非鱼的肠道菌群结构，以中草药复合微生态制剂的改善效果最佳。（3）经人工感染无乳链球菌后，罗非鱼对照组全部死亡，4 个实验组只有部分死亡。鉴定发现，吉富罗非鱼的死亡均由感染无乳链球菌所致，试验组罗非鱼的免疫保护率分别为 51.42%（B 组）、58.62%（C 组）、58.62%（D 组）和 68.93%（E 组），以中草药复合微生态制剂组的免疫保护率最高。综上所述，在饲料中添加一定比例的中草药复合微生态制剂可以提高吉富罗非鱼生长指标、改善其肠道菌群结构和增加抗病力。

关键词：中草药复合微生态制剂；吉富罗非鱼；生长；肠道菌群；抗病力

2. 防治水产动物细菌性疾病的中草药

中草药具有对细菌、真菌和寄生虫等病原较强的抑制和杀灭能力。大黄和公丁香对致病性哈氏弧菌及其生物膜的体外抑制作用，其中大黄的作用比公丁香更强（黎家勤等，2014）。合理运用中药的配伍，不仅可提高抗菌疗效，而且大大减少了药物在环境中的残留量。五倍子、石榴皮、大黄、虎杖、黄芩及黄连对鳗鲡致病性气单胞菌的具有较强抑制作用（李忠琴等，2011；2012），复方中草药可对鳗鲡主要致病菌呈现协同效应（李忠琴等，2011）。中草药配合氟苯尼考使用，可提高药效，还可减少药物对环境的污染和减缓耐药性问题（李忠琴等，2013）。虎杖、石榴皮、大黄以及五倍子等中药对 9 株养殖鳗鲡主要致病菌均有一定的协同抑制作用（关瑞章等，2011）。

本文发表于《水生生物学报》【2012，1：85-92】

六种中药及其复方对鳗鲡致病性气单胞菌的体外抑制作用

Antibacterial activity analysis for six kinds of Chinese herbs and their compounds against pathogenic *Aeromonas* from cultured eels *in vitro*

李忠琴，关瑞章，汪黎虹 刘宏伟

（集美大学水产学院，教育部鳗鱼产业技术工程研究中心，厦门 361021）

摘要：针对病鳗内脏中分离的 5 株致病性气单胞菌，采用琼脂稀释法测定五倍子、石榴皮、大黄、虎杖、黄芩及黄连各味中药的最低抑菌浓度（MIC）和最低杀菌浓度（MBC）；再根据棋盘法设计 15 种双联用药方和 4 种三联用药方，同样检测各组合配方的抑菌作用。实验结果表明，6 味中药对养殖鳗鲡 5 株致病性气单胞菌均有不同程度的抗菌效果，其中五倍子的抑菌作用最强，其次是大黄和石榴皮，而黄连的抑菌效果最差；15 种双联用药方较各味中药单用的抑菌活性绝大多数出现增强，抑菌浓度至少减低 39%，FIC<1 呈现协同效应的比例占 85.7%，其中 FIC≤0.5 表现显著增强抗菌活性的协同比例占 23.3%；4 种三联用药方对 5 株致病性气单胞菌均具有显著的协同抑制效应，复方中单味中药的抑菌浓度可以降低 80% 以上；而双联用 HC14 对 4 株致病菌出现 FIC≥2 的降低彼此抗菌活性的相互拮抗现象。由此说明合理运用不同中药的联用配伍，不仅可提高单味中药的抗菌疗效，而且大大减少了单一中药在实际养殖生产中的给药浓度，降低药物在环境中的残留量，防止残留药物造成环境污染，并且降低用药成本，提高水产养殖业的经济和社会效益。研究为中药复方防治细菌性鱼病提供科学理论参考。

关键词：中药；气单胞菌；抑菌作用；协同效应指数 FIC

鳗鲡肉质细嫩、营养丰富、药用价值高而备受人们青睐。气单胞菌感染鳗鱼可引起败血症、烂尾、烂鳃等细菌性疾病，是鳗鱼养殖的主要病害之一[1]。我国是世界上最大的鳗鲡生产国，随着高密度集约化养殖规模不断扩大，水体环境逐步恶化，化学性渔药长期使用，导致病菌耐药性不断增强，而鳗鲡自身抵抗力日益降低，病害问题日趋严重，给鳗鱼养殖业带来了巨大的损失[2-5]。本实验在我们前期研究基础[6,7]上，选择了 6 种中药组合成不同联用药方，采用超微药粉琼脂平板稀释法，针对 5 株养殖鳗鲡常见致病性气单胞菌进行体外中药单方和复方的抑菌实验，评价中药的体外抗菌效果，为鳗鲡细菌性疾病防治提供理论依据。

1 材料与方法

1.1 实验菌株

5 株致病菌均为本研究室近年从各地养殖场病鳗内脏中分离得到，经过人工感染实验证实为强致病菌。通过生理生化和基因鉴定，其中 B15、B18、B20 为豚鼠气单胞菌（Aeromonas caviae）；B10 为嗜水气单胞菌（A. hydrophila）；B09 为维罗纳气单胞菌（A. veronii）。

1.2 实验药材

虎杖（Rhizoma Polygoni Cuspidati，简称 RPC），产地福建；石榴皮（Punica Granatum Linn，简称 PGL），产地甘肃；大黄（Rheum Palmatum Linn，简称 RPL），产地甘肃；黄芩（Scutellaria Baicalensis Georgi，简称 SBG），产地河北；五倍子

（Rhus Chinensis Mill，简称 RCM），产地湖北；黄连（Coptis Chinensis Franch，简称 CCF），产地四川。

1.3 中药加工

将上述药材用中药粉碎机粉碎至 100 目左右，然后放入超微粉碎机进一步粉碎，收集超微药粉备用。在扫描电子显微镜下观察，90% 以上的颗粒粒径大小在 5~10 μm。

6 种中药两两组合成 15 个双联用药方（HC1-HC15）；选取其中 4 种高抗菌活性的中药通过三联用配伍，组成 4 个复方（TH1、TH2、TH3、TH4）。

1.4 菌悬液制备

从保种斜面上挑取菌苔划线接种 M-H 琼脂（Mueller-Hinton Agar）平板，置于 28℃恒温培养箱内培养 20h；挑取单个菌落划线至新鲜斜面上，28℃培养 24 h 后，用无菌生理盐水将菌体冲洗下来，将菌液浓度稀释至 10^7 CFU/mL，即可作为药敏实验接种的菌悬液。

1.5 体外药敏实验

实验组：以参与联用的 2 味或 3 味中药超微粉为溶质，MHB（Mueller-Hinton Broth）培养基为溶剂，采用琼脂稀释法，根据单味中药对致病菌的单方抑菌实验结果，分别以其单用的 1/4MIC、1/2MIC、MIC、2MIC 浓度，采用棋盘交叉法，每种药方配制出 16 个浓度配比的中药药敏培养基。

对照组：联用的 2 味或 3 味中药分别以单方的 1/4MIC、1/2MIC、MIC、2MIC 浓度的药敏平板作为单用药物对照组，以及不含药物平板作为空白对照组。

取已制备好的菌悬液 2 μL 点种于经过 121℃ 20 min 灭菌后的药敏培养基平板上，每个药物浓度梯度做两组平行。放置 28℃培养 24 h 和 48 h，观察药板上菌落的生长情况。以 24 h 无菌落形成的最低浓度为中药对该菌株的最低抑菌浓度（MIC），以 48 h 药板上无菌落形成的最低浓度为中药对该菌株的最低杀菌浓度（MBC）。若两个平行组的 MIC 值或 MBC 值不一致，实验需重做。

1.6 实验结果评价

$$双联用 FIC 指数 = MIC_{甲药联合}/MIC_{甲药单用} + MIC_{乙药联合}/MIC_{乙药单用}$$

FIC（Fractional Inhibitory Concentration）指数判读方法：FIC≤0.5 为显著协同作用，FIC 值居于 0.5~1 之间为协同作用，FIC=1 为相加作用，FIC 值居于 1~2 之间为无关作用，FIC≥2 为拮抗作用。

$$三联用 FIC 指数 = MIC_{甲药联合}/MIC_{甲药单用} + MIC_{乙药联合}/MIC_{乙药单用} + MIC_{丙药联合}/MIC_{丙药单用}$$

FIC 指数判读方法：FIC≤0.5 为显著协同作用，FIC 值居于 0.5~1 之间为协同作用，FIC=1 为相加作用，FIC 值居于 1~3 之间为无关作用，FIC≥3 为拮抗作用。

2 结果

2.1 中药对 5 株气单胞菌的抑制作用

本实验比较了 6 种中药单用分别对养殖鳗鲡 5 株致病性气单胞菌的体外抗菌活性（表1）。结果表明：五倍子、大黄、石榴皮、虎杖、黄芩和黄连对 5 株致病菌均有一定的抑制效果，但抑菌浓度差异明显，MIC 从 0.125 mg/mL 至 12.000 mg/mL；每种中药对 5 株致病菌的平均抑菌浓度依次为 0.175、1.200、0.825、3.075、1.500、6.300 mg/mL。

2.2 中药双联用对 5 株气单胞菌的抗菌活性

在中药单用抑菌实验的基础上，采用超微药粉琼脂稀释法检测 6 种中药两两联用组成 15 个双联用药方（HC1-HC15）的体外抗菌活性，实验结果（表2、表3），15 个双联用配方对 5 株致病菌的抑制作用各不同。从表中数据可知，五倍子、大黄、石榴皮、虎杖、黄芩和黄连分别与其余 5 味中药两两联用对 5 株致病菌的平均最低抑菌浓度范围分别为 0.063~0.075 mg/mL、0.300~0.600 mg/mL、0.188~0.413 mg/mL、0.225~1.875 mg/mL、0.375~2.175 mg/mL、0.900~2.250 mg/mL，与单用平均 MIC 相比，五倍子、大黄、石榴皮、虎杖和黄连的抑菌浓度分别至少降低 57%、50%、50%、39%、64%。而黄芩除了组方 HC14 外，在其余联用药方中的的抑菌浓度亦至少降低 40%。由此说明，达到相同抑菌效果，合理的联用药方可以大大减少其中单味中药的使用浓度。

2.3 中药三联用对 5 株气单胞菌的抑制效果

本试验研究了 4 个三联用中药复方对 5 株气单胞菌的抑制作用（表6、表7）。试验结果显示，复方中五倍子、大黄、石榴皮、虎杖的抑菌浓度与单用平均 MIC 相比，分别至少减低 82%、86%、79%、93%。4 个复方 TH1、TH2、TH3、TH4 平均 FIC 分别为 0.451、0.437、0.426、0.446，即复方中三味药的协同效应均为 FIC≤0.5，表明了此 4 个中药复方均具有非常好的协同抑菌作用；针对每株致病菌的协同效应指数 FIC 范围位于 0.375~0.561，其中复方 TH1、TH4、TH3 对 5 株致病菌都呈现显著协同效应，复方 TH2 的抑菌效果次之，其三味中药仅对致病菌 B15 菌株的协同效果相对较弱，FIC 为 0.561；而对其余致病菌株均表现显著协同作用。

表 1　六种中药对 5 株气单胞菌的体外抑菌浓度
Tab. 1　MIC and MBC of six Chinese herbs on the five strains of Aeromonas from eels

致病菌 Pathogens	五倍子 RCM (mg/mL)		大黄 RPL (mg/mL)		石榴皮 PGL (mg/mL)		虎杖 RPC (mg/mL)		黄芩 SBG (mg/mL)		黄连 CCF (mg/mL)	
	MIC	MBC	MIC	MBC	MIC	MBC	MIC	MBC	MIC	MBC	MIC	MBC
B09	0.250	0.250	1.500	3.000	0.375	0.750	3.000	3.000	1.500	3.000	12.000	12.000
B10	0.250	0.500	0.750	0.750	1.500	1.500	0.375	0.375	1.500	3.000	1.500	3.000
B15	0.125	0.125	0.750	1.500	0.750	0.750	3.000	6.000	1.500	1.500	6.000	6.000
B18	0.125	0.125	1.500	3.000	0.750	0.750	3.000	3.000	1.500	1.500	6.000	6.000
B20	0.125	0.125	1.500	3.000	0.750	0.750	6.000	6.000	1.500	1.500	6.000	6.000
平均值 Means	0.175	0.225	1.200	2.250	0.825	0.900	3.075	3.675	1.500	2.100	6.300	6.600

注：下划线"＿"标示 MBC 与 MIC 有差异的数值

Note：Underline "＿" indicated that there are differences between MBC and MIC values

表 2　中药双联用 15 个组方对 5 株气单胞菌的抑制效果
Tab. 2　Inhibitory effect of fifteen Chinese herbs combined with double on the five strains of Aeromonas from eels

致病菌 Pathogens	HC1 (mg/mL)		HC2 (mg/mL)		HC3 (mg/mL)		HC4 (mg/mL)		HC5 (mg/mL)		HC6 (mg/mL)		HC7 (mg/mL)	
	A	B	A	C	A	D	A	E	A	F	C	B	C	D
B09	0.063	0.375	0.063	0.375	0.063	0.094	0.063	0.375	0.063	3.000	0.188	0.188	0.375	0.188
B10	0.125	0.094	0.063	0.188	0.063	0.750	0.125	0.375	0.063	0.750	0.188	0.188	0.375	0.188
B15	0.063	0.375	0.063	0.188	0.063	0.375	0.063	0.375	0.063	1.500	0.375	0.188	0.375	0.094
B18	0.063	0.094	0.063	0.375	0.063	0.188	0.063	1.500	0.063	3.000	0.375	0.094	0.375	0.094
B20	0.063	0.188	0.063	0.375	0.063	0.188	0.063	0.375	0.063	1.500	0.375	1.500	0.375	0.375
平均值 Means	0.075	0.225	0.063	0.300	0.063	0.319	0.075	0.600	0.063	1.950	0.300	0.432	0.375	0.188

表 3（续）　中药双联用 15 个组方对 5 株气单胞菌的抑制效果
Tab. 3　Inhibitory effect of fifteen Chinese herbs combined with double on the five strains of Aeromonas from eels（Continued）

致病菌 Pathogens	HC8 (mg/mL)		HC9 (mg/mL)		HC10 (mg/mL)		HC11 (mg/mL)		HC12 (mg/mL)		HC13 (mg/mL)		HC14 (mg/mL)		HC15 (mg/mL)	
	D	E	D	B	C	F	C	E	B	F	D	F	B	E	E	F
B09	0.188	0.750	0.188	0.750	0.750	1.500	0.750	0.375	1.500	6.000	0.188	0.375	0.375	1.500	0.750	1.500
B10	0.750	1.500	0.750	0.188	0.375	0.375	0.375	0.375	0.375	0.750	0.375	0.750	0.750	0.375	0.750	1.500
B15	0.188	0.750	0.375	0.750	0.375	1.500	0.375	0.375	1.500	1.500	0.375	1.500	0.750	3.000	0.375	1.500
B18	0.188	0.750	0.375	0.750	0.375	1.500	0.750	0.375	3.000	1.500	0.375	1.500	3.000	3.000	0.750	1.500
B20	0.188	0.750	0.375	0.375	0.375	1.500	0.750	0.375	3.000	1.500	0.750	0.375	1.500	3.000	0.375	1.500
平均值 Means	0.300	0.900	0.413	0.563	0.450	1.275	0.600	0.375	1.875	2.250	0.413	0.900	1.275	2.175	0.600	1.500

表 4　中药双联用抑制 5 株致病菌的协同作用

Tab. 4　Synergistic antibacterial effect of Chinese herbs combined with double on five pathogenic bacteria from eels

致病菌 Pathogens	HC7		HC6		HC11		HC10		HC2		HC4		HC1	
	FIC	评价 Evaluation	FIC	评价 Evaluation	FIC	评价 Evaluation	FIC	评价 Evaluation	FIC	评价 Evaluation	FIC	评价 Evaluation	FIC	评价 Evaluation
B09	0.75	协同	0.19	显著协同	0.75	协同	0.63	协同	0.50	显著协同	0.50	显著协同	0.38	显著协同
B10	0.63	协同	0.75	协同	0.75	协同	0.75	协同	0.50	显著协同	0.75	协同	0.75	协同
B15	0.63	协同	0.56	协同	0.75	协同	0.75	协同	0.75	协同	0.75	协同	0.63	协同
B18	0.38	显著协同	0.28	显著协同	0.75	协同	0.50	显著协同	0.75	协同	1.50	无关	0.54	协同
B20	0.75	协同	0.50	显著协同	0.75	协同	0.50	显著协同	0.75	协同	0.75	协同	0.54	协同
协同率 Collaboration rate（%）	100		100		100		100		100		80		100	

注：协同率＝双联用药方对致病菌产生协同作用的菌株数量/5

Note：Collaboration rate is equal to the number of pathogen strains for synergistic antibacterial effect prescription divided by five

表 5（续）　中药双联用抑制 5 株致病菌的协同作用

Tab. 5　Synergistic antibacterial effect of Chinese Herbs combined with double on five pathogenic bacteria from eels（Continued）

致病菌 Pathogens	HC9		HC8		HC3		HC5		HC13		HC15		HC12		HC14	
	FIC	评价 Evaluation	FIC	评价 Evaluation	FIC	评价 Evaluation	FIC	评价 Evaluation	FIC	评价 Evaluation	FIC	评价 Evaluation	FIC	评价 Evaluation	FIC	评价 Evaluation
B09	0.75	协同	1.00	相加	0.50	显著协同	0.50	显著协同	0.53	协同	0.63	协同	1.00	相加	1.13	无关
B10	1.00	相加	1.50	无关	0.75	协同	0.75	协同	0.75	协同	1.50	无关	1.50	无关	2.25	拮抗
B15	0.75	协同	0.75	协同	1.00	相加	0.75	协同	0.75	协同	0.50	显著协同	0.75	协同	2.25	拮抗
B18	0.75	协同	0.75	协同	0.75	协同	1.00	相加	0.75	协同	0.75	协同	1.25	无关	3.00	拮抗
B20	0.56	协同	0.75	协同	0.75	协同	0.75	协同	1.06	无关	0.50	显著协同	0.75	协同	2.25	拮抗
协同率 Collaboration rate（%）	80		80		80		80		80		80		40		0	

注：协同率＝双联用药方对致病菌产生协同作用的菌株数量/5

Note：Collaboration rate is equal to the number of pathogen strains for synergistic antibacterial effect prescription divided by five

表 6　三联用中药复方对 5 株致病菌的抑制效果突出显示部分加注英文

Tab. 6　The inhibitory effect of Chinese triple compound on five strains of pathogenic bacteria

致病菌 Pathogens	TH1（mg/mL）			FIC	评价 Evaluation	TH2（mg/mL）			FIC	评价 Evaluation
	C	D	B			A	C	B		
B09	0.188	0.094	0.188	0.439	显著协同	0.031	0.188	0.188	0.375	显著协同
B10	0.094	0.188	0.094	0.501	显著协同	0.031	0.094	0.047	0.375	显著协同
B15	0.094	0.188	0.188	0.439	显著协同	0.031	0.188	0.188	0.561	协同
B18	0.188	0.188	0.188	0.439	显著协同	0.031	0.188	0.188	0.436	显著协同
B20	0.188	0.188	0.375	0.439	显著协同	0.031	0.188	0.375	0.436	显著协同
平均值 Means	0.150	0.169	0.207	0.451		0.031	0.169	0.197	0.437	

表 7（续） 三联用中药复方对 5 株致病菌的抑制效果

Tab. 7 The inhibitory effect of Chinese triple compound on five strains of pathogenic bacteria（Continued）

致病菌 Pathogens	TH3（mg/mL）			FIC	评价 Evaluation	TH4（mg/mL）			FIC	评价 Evaluation
	A	C	D			A	D	B		
B09	0.031	0.094	0.188	0.500	显著协同	0.031	0.094	0.188	0.437	显著协同
B10	0.031	0.094	0.094	0.375	显著协同	0.031	0.188	0.094	0.500	显著协同
B15	0.016	0.094	0.188	0.504	显著协同	0.016	0.188	0.188	0.441	显著协同
B18	0.016	0.094	0.188	0.375	显著协同	0.016	0.188	0.188	0.441	显著协同
B20	0.016	0.094	0.188	0.375	显著协同	0.016	0.188	0.188	0.410	显著协同
平均值 Means	0.022	0.094	0.169	0.426		0.022	0.169	0.169	0.446	

3 讨论

3.1 中药对 5 株鳗鲡致病性气单胞菌的抑制作用

本实验测试了 6 种中药对 5 株养殖鳗鲡致病性气单胞菌的 MIC 和 MBC，其平均抑菌强弱依次为五倍子>石榴皮>大黄>黄芩>虎杖>黄连；但对于菌株 B10 而言，6 种中药的抑菌效果略有差异，五倍子>虎杖>大黄>石榴皮>黄芩＝黄连。由此表明，因不同致病菌对不同中药所含有效抗菌成分的敏感度不同，在使用中药防治鱼病的实践中，应该先诊断致病的细菌种类，再根据致病菌的种类来区别用药。

6 味中药对 5 株试验菌的 MIC 与 MBC 比较结果显示，各中药试验组中均有部分菌株的 MIC 与 MBC 存在差异，其数量占实验菌株总数的比率范围为 33.3%。这说明同一味中药对不同种类致病菌的抑菌浓度和杀菌浓度存在一定程度的差异，为实际用药量提供参考。

3.2 中药联用的协同效应

根据方法 1.6 计算联用效应因子 FIC（表 4、表 5）。由计算结果可以看出，药方 HC14 联用（即虎杖与黄芩）出现负面效应，联用的两味中药对 5 株致病菌没有起到增强抗菌作用，甚至对其中 4 株菌反而产生拮抗作用；除此之外，多数联用配方对绝大多数致病菌起到了提高抗菌活性的协同作用，即 FIC<1 呈现协同效应的比例占 85.7%，其中 FIC≤0.5 表现显著增强抗菌活性的协同比例占 23.3%。而且在双联用基础上，所选 4 种中药的三联用药方对 5 株致病菌都起到显著协同增效的抑制作用，大大降低了复方中单味中药的抑菌浓度。

中药联用产生协同增效的可能原因是不同中药的抗菌有效化学成分不同，其抗菌机理也各不相同，将几味抑菌作用途径或靶点不同的中药联用，就可以多位点、多途径攻击致病细菌的细胞，提高药物的杀伤力。如此不仅可提高每味中药的抗菌活性，而且可以在较低浓度下就起到抑制或杀灭细菌的作用。据文献报道，如大黄的有效抗菌成分有大黄素和大黄酸，大黄素可以引起细菌 DNA 单链突变，造成 DNA 损伤从而抑制细菌生长[8,9]，而大黄酸可以抑制细菌菌体糖和糖代谢中间产物的氧化和脱氢，并与 DNA 结合，干扰蛋白质的合成[10,11]；五倍子和石榴皮的有效抗菌成分是鞣质和鞣酸，鞣质可与蛋白质结合，使细胞内的蛋白质凝固，抑制细菌代谢酶的活性发挥抗菌作用，而鞣酸具有较强的酸性，可以改变细菌膜的通透性，促进细胞内外物质的交换[12-14]。当五倍子或石榴皮与大黄联用时，鞣酸可促进大黄素和大黄酸等蒽醌类化合物进入致病菌细胞内，随后作用于相应靶点部位，故两者联用具有很好的协同抗菌作用。

3.3 中药联用的拮抗现象

本文实验结果表明，不同中药联用对不同致病菌产生抑菌活性高低存在差异，极少数出现拮抗现象。虽然 6 味中药组合之间大多数存在协同作用，少数配方中各味药的抗菌活性之间没有关联；但双联用药方 HC14 对致病菌 B10、B15、B18、B20 的协同指数 FIC>1，说明 HC14 药方的两味中药对这特定的 4 株致病菌没有起到相互增强抗菌活性，甚至反而出现相互削弱彼此的抗菌活性。由此表明不合理的中药联用配方会降低单味中药对致病菌的抑制作用。中药联用产生拮抗减效的原因可能是每味中药发挥药效的化学成分在联用时，可能由于多种有效成分相互结合、吸附等物理作用而降低药物的溶出、吸收，亦或甚至发生化学反应，失去原有的抗菌活性。本实验中药方 HC14 产生拮抗现象的机理有待后续的进一步研究。

不少研究者在检测中药复方体外抑菌作用的实验中，也发现药物联用后产生拮抗的现象。如李茜等[15]采用琼脂平板扩散法研究了大黄、连翘、黄柏、板蓝根等 23 种中药及 5 种复方制剂对由鲫鱼肠道分离得到的致病菌的抑制作用，实验结果显示：黄柏单用时对金黄色葡萄球菌的抑菌圈直径为 28 mm，而黄柏等复方对其抑菌圈直径减小为 7 mm；石榴皮单用时对嗜水气单胞菌的抑菌圈直径为 30 mm，而石榴皮等复方对其抑菌圈直径减小为 22 mm。曾惠芳等[16]研究了虎杖单独煎煮以及虎杖分别与甘草、五味子、山楂配伍共同煎煮后各种蒽醌的溶出量，结果显示虎杖单独煎煮时，总蒽醌的含量为 1 027 mg/L，虎杖与五味子合煎后总蒽醌的含量下降至 797 mg/L，虎杖与山楂合煎后总蒽醌的含量下降至 60.4 mg/L，推测其原因可能在于五味子、山楂中含有丰富的有机酸，可促进蒽醌类化合物水解，使得虎杖在与其配伍共同煎煮后总蒽醌的含量有所下降。

在实际生产应用上，中药往往以多种药物配伍的复方形式使用。多味中药的合理联用[17-20]，不仅可以通过主要化学成分间的增溶、助溶等物理作用，提高有效成分的溶出率，或者作用靶位的不同，促进药物的渗透吸收，提高生物利用度，增强药效；更可以在宏观上调节各药物的药性，提高机体的免疫力。本实验研究获得了高效抗鳗鲡致病菌的广谱性的中药复方，为进一步开发安全、高效、环境友好型的鱼病防治药物提供科学的参考数据。

参考文献

［1］ Dong C. F., Yu T. L., Yu F. S., et al. Isolation and serological survey of fish Aeromonas ［J］. Reservoir Fisheries, 2004, 6：10-12 ［董传甫，林天龙，俞伏松，等. 鱼源气单胞菌的分离鉴定及血清学调查. 水利渔业，2004，6：10-12］

［2］ Fan H. P.. Status and development strategy of eel aquaculture ［J］. Scientific Fish Aquaculture, 2006, 3：1-2 ［樊海平. 我国鳗鲡养殖业的现状与发展对策. 科学养鱼，2006，3：1-2］

［3］ Chen A. P.. Diseases, drug use and countermeasures of eel aquaculture ［J］. Scientific Fish Aquaculture, 2003, 8：38 ［陈爱萍. 鳗鲡养殖病害、用药情况及对策. 科学养鱼，2003，8：38］

［4］ Wu C. Y., Liu Z. J.. Problems and countermeasures of eel aquaculture industry in China ［J］. Fujian Fisheries, 2004, 4：16-19 ［吴成业，刘兆钧. 中国鳗业面临的问题与对策. 福建水产，2004，4：16-19］

［5］ Fan H. P., Zhong Q. F.. Overview and trends of cultured eel diseases in the first half of this year ［J］. Scientific Fish Aquaculture, 2006, 8：50 ［樊海平，钟全福. 今年上半年养殖鳗鲡病害发生概况与流行趋势. 科学养鱼，2006，8：50-51］

［6］ Liu H. W., Guan R. Z., Huang W. S.. Antibacterial activity analysis for Chinese herbal medicine in different processing against the major pathogens from eels ［J］. Journal of Jimei University, 2009, 14 (3)：229-233 ［刘宏伟，关瑞章，黄文树. 不同加工处理中草药对鳗鲡主要致病菌抑制作用的比较. 集美大学学报. 2009，14 (3)：229-233］

［7］ Wu L., Guan R. Z., Huang W. S.. The inhibition of florfenicol combined with four kinds of Chinese medicine against eel pathogens ［J］. Journal of Jimei University, 2008, 13 (1)：7-11 ［吴亮，关瑞章，黄文树. 氟苯尼考与4种中药联用对鳗鲡致病菌的抑制作用. 集美大学学报，2008，13 (1)：7-11］

［8］ Wang H. H., Chung J. G.. Emodin-induced inhibition of growth and DNA damage in the Helicobacter pylori ［J］. Current Microbiology, 1997, 35 (5)：262-266

［9］ Guo M. Z., Xu H. R., Li X. S.. Research of pharmacological effects of Rhein ［J］. Foreign Medical Sciences, 2002, 3 (24)：139-143 ［郭美姿，徐海荣，李孝生. 大黄酸药理作用的研究进展. 国外医学中医中药分册，2002，3 (24)：139-143］

［10］ Ni H., Xue X. P., Yang X. Z., et al. Mechanism of Rhein inhibited mouse peritoneal macrophage activation of inflammatory mediators ［J］. Tianjin Traditional Chinese Medicine, 2001, 18 (1)：35-36 ［倪弘，薛小平，杨秀竹，等. 大黄酸抑制小鼠腹腔巨噬细胞炎性介质活化的作用机理. 天津中医，2001，18 (1)：35-36］

［11］ Liu X. B., Du Z. Z., Wang D., et al. Antibacterial effect of Galla chinensisin vitro ［J］. Chinese Journal of Practical Medicine, 2008, 3 (7)：96-97 ［刘现兵，杜镇镇，王冬，等. 五倍子的体外抑菌作用研究. 中国实用医药，2008，3 (7)：96-97］

［12］ Qiao S. H., Jiang S. H., Zhang Y. N., et al. A Preliminary Study on antibacterial activity of Pomegranate ［J］. Pesticides, 2009, 48 (4)：299-303 ［乔树华，蒋红云，张燕宁，等. 石榴皮抑菌活性物质的初步研究. 农药，2009，48 (4)：299-303］

［13］ Yao J. M., Zhang L. J.. Extraction and Determination of tannin from Rhus chinensis and Granatum ［J］. Veterinary Science, 2004, 3：4-6 ［姚敬明，张李俊. 石榴皮、五倍子中鞣质的提取及测定试验. 中兽医学杂志，2004，3：4-6］

［14］ Sotohy S. A., Muller W., Ismail A. A.. In vitro effect of Egyptian tannin-containing plants and their extracts on the survival of pathogenic bacteria ［J］. Deutsche Tierarztliche Wochenschrift, 1995, 102 (9)：344-350

［15］ Li Q., Zhang Z. J., Hua R. Q., et al. Antibacterial test of 23 Chinese herbs and compounds against intestinal bacteria from carp in vitro ［J］. Freshwater Fisheries, 2007, 37 (4)：7-11 ［李茜，张懿瑾，华汝泉，等. 23种中草药及复方对鲫肠道3种细菌的体外抑菌试验. 淡水渔业，2007，37 (4)：7-11］

［16］ Zeng H. F., Su Z. R., Shi Q. R., et al. Study of physical and chemical changes in process of cooking Polygonum and its compounds ［J］. China Pharmacy, 1999, 10 (3)：112-113 ［曾惠芳，苏子仁，史俏蓉，等. 虎杖单煎、复方共煎过程中的物理化学变化初探. 中国药房，1999，10 (3)：112-113］

［17］ Liang Y. B.. Application of Chinese herbal medicine in the prevention of fish diseases ［J］. Rich Fishing Guide, 2004, 18：53-53 ［梁用本. 中草药在防治鱼病中的应用. 渔业致富指南，2004，18：53-53］

［18］ Hu X. Q., Hou Y. Q.. Effect of Chinese herbal extracts on carp growth and body composition ［J］. Food and Feed Industry, 2005, 5：40-41 ［胡先勤，侯永清. 中草药提取物对鲫鱼生长及体成分的影响. 粮食与饲料工业，2005，5：40-41］

［19］ Wang J. Q., Sun Y. X., Zhang J. C.. Effect of compound herbs such as honeysuckle on the growth, digestion and immunity of flounder ［J］.

Journal of Fisheries, 2006, 30 (1): 90-96 [王吉桥, 孙永新, 张剑诚. 金银花等复方草药对牙鲆生长、消化和免疫能力的影响. 水产学报, 2006, 30 (1): 90-96]

[20] Zheng S. M., Huang J. J., Wu Q., et al. Studies on separation and antibacterial activity of the effective components of Chinese herbal compounds gallnut for fishery antimicrobial agent [J]. Acta Hydrobiologica Sinica, 2010, 34 (1): 57-63 [郑曙明, 黄建军, 吴青, 等. 复方五倍子有效成分的分离鉴定及抑菌活性研究. 水生生物学报, 2010, 34 (1): 57-63]

本文发表于《湖南农业大学学报》【2011, 37 (3): 306-311】

中药对鳗鲡病原菌的体外抑制作用

Antibacterial activity of Chinese herbs against pathogenic bacteria from eels *in vitro*

李忠琴[1,2], 关瑞章[1,2]*, 汪黎虹[1,2], 刘宏伟[1,2]

(1. 集美大学水产学院, 福建 厦门 361021; 2. 教育部鳗鱼产业技术工程研究中心, 福建 厦门 361021)

摘要：研究虎杖、石榴皮、大黄、黄芩、五倍子、黄连、冬凌草、独活、贯众、木贼草等10种中药材单用和五倍子、大黄、石榴皮、虎杖、黄芩、黄连两两联用（15种配方）对从养殖病鳗中分离得到的12株常见病原菌的抑菌作用。结果表明：中药单用对病原菌均有不同程度的抑制作用，其中，五倍子的抑菌作用最强，其次是大黄和石榴皮，木贼草的抑菌作用最差；中药联用呈现协同效应（FIC<1）的占68.3%，其中有显著协同效应（FIC≤0.5）的占26.0%，与中药单用相比，中药联用对大多数病原菌的抑制作用增强，且可大大减少单一中药的给药浓度。

关键词：鳗鲡；病原菌；中药；抑菌活性

本文发表于《华中农业大学学报》【2011, 30 (6): 764-767】

4种中药三联用对鳗鲡致病菌的抑制作用

Inhibitory effect if four kinds of Chinese herbs and triple compounds absinest main pathogenic bacteria from cultivated eels

关瑞章[1,2] 李忠琴[1,2] 郭松林[1,2] 刘宏伟[1,2] 汪黎虹[1]

(1. 集美大学水产学院, 厦门; 2. 教育部鳗鱼产业技术工程研究中心, 厦门 361021)

摘要：应用超微粉碎法将虎杖、石榴皮、大黄以及五倍子4种中药进行加工处理至粒径为 $5\sim10\ \mu m$ 后, 组4个三联用的中药复方, 针对近年来从福建各地养殖场病鳗中常检测到的9株致病菌, 采用琼脂稀释法测定其抑菌最低浓度。试验结果表明：虎杖、石榴皮、大黄和五倍子组合的4个中药复方对9株养殖鳗鲡主要致病菌均有一定的协同抑制作用, 其协同效应因子FIC范围在0.38~0.76。其中, 联用效果最好的是复方1, 所联用的3种中药对9株致病菌均具有显著协同抗菌效应, FIC≤0.5。每个复方的联用3种中药的最低抑菌浓度均比单用浓度降低了79%以上, 不但可大大降低用药量, 而且用药成本比单用节省47%以上。

关键词：鳗鲡；致病菌；中药；抑菌协同效应

本文发表于《集美大学学报》【2011，16（4）：241-245】

中药双联用复方对养殖鳗鲡主要致病菌的抑制作用研究

Antibacterial Activity Analysis for Double-compounds of Chinese Herbs Against Pathogenic Bacteria from Cultured Eels

李忠琴[1,2]，关瑞章[1,2,3]，郭松林[1,2,3]，汪黎虹[1,2]，刘宏伟[1,2,3]

（1. 集美大学水产学院，福建 厦门 361021；2. 教育部鳗鱼产业技术工程研究中心，福建 厦门 361021；
3. 福建省高校水产科学技术与食品安全重点实验室，福建 厦门 361021）

摘要： 筛选单用抑菌作用较强的 6 种中药药材，依据棋盘法设计 15 种双联用复方，采用超微药粉琼脂稀释法测定各系列浓度组合的复方对养殖鳗鲡 18 株常见致病菌的抑制作用。实验结果表明，15 种双联用复方对 18 株致病菌中绝大多数菌株的抑制作用比 6 种中药单用的抑菌作用强，呈现协同效应（FIC≤1）的比例占 85.7%，其中显著协同（FIC≤0.5）占 20.4%；复方抑菌作用未发生变化（1<FIC<2）的占 11.1%；仅 3.2%出现拮抗作用（FIC≥2）。说明，合理的中药联用可提高中药单用对致病菌的抑制作用，降低复方中单个中药的抑菌浓度.

关键词： 鳗鲡；致病菌；中药双联用；协同抑菌效应

本文发表于《养殖与饲料》【2013，3：13-17】

氟苯尼考与中药联用对养殖鳗鲡主要病原菌的体外抗菌活性分析

The inhibitory effect of Florfenicol combined with Chinese herbs on the main pathogenic bacteria from cultivated eels

李忠琴 关瑞章 汪黎虹 刘宏伟 郭松林

（集美大学水产学院，教育部鳗鱼产业技术工程研究中心，厦门 361021）

摘要： 为了筛选防治养殖鳗鲡细菌性疾病的有效中西药联用药方，针对本研究室近年从养殖病鳗中分离得到的 18 株常见病原菌，进行药敏实验。利用超微粉碎机将黄连、黄芩、虎杖、石榴皮、大黄及五倍子 6 种中药材研磨成粒径为 5~10 μm 药粉后，分别与氟苯尼考配制成 6 种中西药双联用药方，采用超微药粉琼脂稀释法测定各药方的体外抑菌作用。实验结果表明，上述 6 种中西药双联用药方对 18 株常见病原菌均有不同程度的抑制效应，其联用 FIC 指数的范围为 0.38~3.00；6 种药方中氟苯尼考与黄芩联用效果最好，其次是氟苯尼考与黄连联用，再次是氟苯尼考与五倍子联用；而氟苯尼考与虎杖联用对 18 株病原菌的抑制效果差异显著，虽然仅对 7 株病原菌起到协同作用，但其中 FIC≤0.5 显著协同者占 85.7%，其余的 10 株病原菌抑制效应为无关，1 株为拮抗。联用 FIC≤1，表明联用药物的有效抑菌浓度比单用降低了 50%以上，氟苯尼考与 6 味中药两两联用对 18 株鳗鲡病原的协同指数 FIC≤1 所占比率为 74.1%，说明采用中西药双联用，不仅可以提高抗菌药效，还可以大大降低氟苯尼考在实际养殖生产中的使用浓度，从而减少对环境的污染和耐药菌株的产生。本研究为开发高效、环境友好型的渔药提供了科学参考数据。

关键词： 鳗鲡；病原菌；氟苯尼考；中西药联用

该文发表于《安徽农业科学》【2014，42（24）：8188-8190，8202】

大黄和公丁香对致病性哈氏弧菌及其生物膜的体外抑制作用

The Inhibitory Activity of Floscaryophylli and Rheumofficinaleon *Vibrio harveyi* and Its Biofilm

黎家勤[1,2,3]，庞欢瑛[1,2,3*]，简纪常[1,2,3]，汤菊芬[1,2,3]

（1. 广东海洋大学水产学院，广东 湛江 524088；2. 广东省水产经济动物病原生物学及流行病学重点实验室，广东 湛江 524088；3. 广东省教育厅水产经济动物病害控制重点实验室，广东 湛江 524088）

摘要：［目的］观察大黄（Rheumofficinale）和公丁香（Floscaryophylli）对哈氏弧菌（Vibrio harveyi）及其生物膜（Biofilm）的体外抑制作用。［方法］采用琼脂扩散法测定大黄和公丁香对哈氏弧菌的体外抑菌作用；采用倍比稀释法测定大黄和公丁香对哈氏弧菌最小抑菌浓度（MIC）和最小杀菌浓度（MBC）；采用 MTT 法评价 2 种中草药液对哈氏弧菌生物膜形成的影响。［结果］大黄对哈氏弧菌的抑菌圈直径为（15.31±0.4）mm，MIC 和 MBC 均为 7.813 mg/mL；公丁香对哈氏弧菌的抑菌圈直径为（11.53±0.6）mm，MIC 和 MBC 分别为 15.625 和 62.5 mg/mL；2 种中草药液浓度分别在 7.81 和 0.97 mg/mL 以上时对生物膜的形成有极显著抑制作用（$P<0.01$）。［结论］大黄和公丁香对哈氏弧菌及其生物膜均有明显抑制作用，其中大黄的作用比公丁香更强。

关键词：大黄；公丁香；哈氏弧菌；生物膜；体外抑制作用

3. 防治水产动物寄生虫疾病的中草药

利用固相萃取、高效液相色谱、中压快速制备等技术可从白屈菜、黄姜（姚嘉赟等，2010；Jia-yun YAO et al.，2011）中分离获得白屈菜碱（Xi-Lian LI et al.，2011）、白屈菜红碱（Jia-Yun YAO et al.，2011）等有效杀指环虫活性物质；从博落回中可分离获的血根碱（Jia-yun YAO et al.，2010）和从小果博落回中分离的二氢血根碱、二氢白屈菜红碱（Jia-yun YAO et al.，2011）均对车轮虫具有较好杀灭作用。黄姜对鱼类指环虫（姚嘉赟等，2010）、重楼对罗氏沼虾寄生纤毛虫（姚嘉赟等，2010；2012）均有较强的活性。灰色链霉菌 SDX-4 的代谢产物在体外、体内具有较强的抗小瓜虫活性（Jia-yun YAO et al.，2015），且对鱼体相对安全（Jia-yun YAO et al.，2014），有望开发为一种新型的杀虫药物。昆虫生长调节剂虫酰肼（蜕皮激素类似物）可影响典型海湾水蚤虫体在蜕皮期间的高死亡率（王伟利等，2011）。

该文发表于《Parasitology Rresearch》【2011，109：247-252】。

Activity of the chelerythrine，a quaternary benzo［c］phenanthridine alkaloid from *Chelidonium majus*，on *Dactylogyrus intermedius*

Xi-Lian LI[1]，Jia-Yun YAO[1,2*]，Zhi-Ming ZHOU[2]，Jin-Yu SHEN[2]，Xiao-Lin LIU[1*]

（1. College of Animal Science and Technology，Northwest A&F University，Yangling，
2. Institute of Freshwater Fisheries of Zhejiang，Huzhou，Zhejiang，313001，China Shaanxi，712100，China）

Abstract：Dactylogyrus intermedius is a significant monogenean parasite on the gills of cyprinid fishes and can cause severe economic losses in aquaculture and ornamental fish breeding. In the present study, bio-assay guided fractionation was employed to identify active compound from Chelidonium majus L. against D. intermedius. In vivo anthelmintic activity of petroleum ether, ethyl acetate, chloroform and n-butanol extracts of C. majus were tested. Among them, only then-butanolextracts exhibited promising anthelmintic efficacy, and therefore subjected to the further isolation and purification using various chromatographic techniques. A compound showing potent activity was obtained and identified by hydrogen, carbon-13 nuclear magnetic resonance spectrum and electron ionization mass spectrometry as chelerythrine. In vivo anthelmintic efficacy tests exhibited that chelerythrine was 100% effective against D. intermedius at a concentration of 1.60 mg · L^{-1}, with LC$_{50}$ values of 0.68 mg · L^{-1} after 48 h of exposure. The 48−h LC$_{50}$ value（acute toxicity tests）of chelerythrine was found to be 3.59 mg · L^{-1}for grass carp. These results provided evidence that chelerythrine can be selected as a lead compound for the development of new drugs against D. intermedius.

Keywords：Dactylogyrus intermedius；chelerythrine；anthelmintic；Chelidonium majus L. .

白屈菜中白屈菜红碱杀指环虫活性的研究

李喜莲[1]，姚嘉赟[1,2*]，周志明[2]，沈锦玉[2]，刘小林[1*]

（1. 西北农林科技大学动物科技学院，陕西 杨凌 712100；2. 浙江省淡水水产研究所，浙江 湖州 313001）

摘要：指环虫是寄生于鱼类鳃部的单殖亚纲类寄生虫，给水产养殖动物造成了巨大的经济损失。本研究利用活性追踪试验结合有机分离技术对白屈菜中的杀指环虫的活性成分进行提取分离。利用极性梯度法利用石油醚、乙酸乙酯、氯仿、正丁醇对白屈菜进行提取，体内药效结果表明正丁醇提取物杀虫活性最高。对正丁醇部位进行硅胶柱层析，并获得一种杀虫活性较强的活性单体。经质谱、核磁氢谱、核磁碳谱、红外光谱等多种波谱分析技术鉴定该化合物为白屈菜红碱。体内杀虫药效试验结果表明白屈菜红碱的100%杀指环虫浓度为 1.60 mg · L^{-1}，其半数致死浓度为 0.68 mg · L^{-1}。对草鱼的急性毒性实验结果表明白屈菜红碱对草鱼的 48 h 半数致死浓度为 3.59 mg · L^{-1}。上述结果表明，白屈菜红碱对指环虫具有较强的杀灭作用且对鱼体相对安全，有望开发为一种新型的杀虫药物。

关键词：指环虫，白屈菜红碱，抗寄生虫，白屈菜

1. Introduction

Monogenean parasite, Dactylogyrus intermedius is endemic ectoparasite in Asia, Central Europe, Middle East and North America（Paperna, 1964），cause serious economic damage in aquaculture industry in these regions. They are usually parasites on the gills of freshwater fish causing irritation, excessive mucus production, accelerated respiration and mixed infection with other pathogen（Dove and Ernst, 1998；Reed et al., 2009），leading to a serious damage to the host such as loss of appetite, lower growth performance and high mortalities（Topic et al., 2001； smail and Selda, 2007）.

To cope with the parasitism and deleterious consequences, various parasiticides such as formalin（Marshall, 1999），trichlorfon（Goven and Amend, 1982），a triazine derivative（Schmahl, 1993）and the twomost frequently used in practice, praziquantel（Schmahl and Mehlhorn, 1985）and mebendazole（Buchmann et al, 1993）against Dactylogyrus have been used to kill the parasite with varying levels of success. However, the frequent use of these chemical parasiticides has had limited efficacy in reducing monogenean infestations and is often accompanied by serious drawbacks, including the development of drug−resistant parasites, environmental contamination, and even toxicity to host, which have stimulated the search for new control strategies（Goven et al., 1980；Marshall, 1999）. Current research efforts are increasingly focused on developing alternative drug formulations including medical plants. Ekanem（Ekanem et al., 2004）found that the methanol extracts of the seeds of Piper guineense（Piperaceae）were active against Dactylogyrus. Crude extracts of Radix angelicae pubescentis, Fructus bruceae, Caulis spatholobi, Semen aesculi, and Semen pharbitidis were also found to exhibit a complete elimination of all Dactylogyrus intermedius in goldfish（Liu et al., 2010）. Five alkaloids from Macleaya microcarpa（Maxim）Fedde also showed effectiveness against D. intermedius（Wang et al., 2010）.

Chelidonium majus L.（Papaveraceae），which is widely distributed in Europe and Asia, is a plant of great interest for its wide use in various diseases in European countries and in Chinese herbal medicines（Pavão and Pinto, 1995；Colombo and Bosisio, 1996）. Extracts of C. majus are traditionally used in various complementary and alternative medicine（CAM）systems including homeopathy mainly in combating diseases of the liver（Taborska et al., 1995），stomach（Kim et al., 1997）and various skin disorders. The crude ethanolic extract has already been claimed to exhibit, anti−inflammatory（Lenfeld et al., 1981），anti−viral（Kery et al., 1987），anti−microbial（Colombo and Bosisio, 1996）and anti−tumor effects（Panzer et al., 2000）. The principal objective of this study was to assess the anthelmintic properties of C. majus and isolate the active constituents responsible for the activity using in vivo anthelmintic assay associated with bioassay−guided silica gel column chromatography isolation. Additionally, the acute toxicity for the grass carp of the active compound was evaluated.

2. Materials and methods

2. 1　Parasites and Hosts

Healthy grass carp (Ctenopharyngodon idella) (mean length±SD: 7. 8±1. 2 cm) were obtained from aquatic fry farm of Zhejiang Institute Freshwater Fisheries in China and maintained in a 1 m³ tank at 24±1℃ (controlled by automatic aquarium heater) with aeration for 7 days, then co-habitated with the ones infested with D. intermedius which were cultured in our laboratory. This procedure consisted of the collection of eggs, hatching of the eggs and re-infection with D. intermedius. The infected fish were prepared following the procedure described in a pervious study (Wang et al., 2008). After three weeks, 20 grass carp were then randomly selected, killed by spinal severance and examined for the presence of parasite under a light microscope (Olympus BX51, Tokyo, Japan) at 4×10 magnification prior to the experiment. Fish were chosen for the tests when the infection prevalence was 100% and the mean number of the parasite on the gills was 50~60.

For acute toxicity tests, parasite-free grass carp were obtained from commercial fish farm and maintained in a 1 m³ tank supplied with filtered groundwater under the same conditions as parasitized fish. On arrival, the absence of the parasites was carefully checked by examining 10 fish randomly selected.

2. 2　Preparation of crude extracts fractions and pure compound

The crude extracts, fractions and the pure compound isolated from C. majus were dissolved in 10 mL of dimethyl sulfoxide (DMSO) to get 1. 0 g/mL (sample/solvent) of stocking solutions which were used for the preparations of the desired concentrations for In vivo anthelmintic efficacy assay.

2. 3　In vivo anthelmintic efficacy test

In vivo anthelmintic efficacy tests were undertaken for the crude extracts (including petroleum ether, chloroform, ethyl acetate and n-butanol extracts), fractions isolated from n-butanol extract and the pure compound (crystal).

The In vivo tests were performed following the protocol described in a previous work (YAO et al., 2010). Briefly, the assays were conducted in 40 cm×30 cm×20 cm glass tanks, each containing 10 l of the test solution, and 10 parasitized fish. Initial tests were undertaken to determine the concentration boundaries for the efficacy tests, and the final tests were conducted with three replicates. The test samples were assayed at a different series of concentrations which were prepared from the stock solutions. Control groups with no chemical were set under the same experimental conditions as the test groups.

After 48 h treatment, grass carp in all treatments and control groups were killed by a spinal severance for biopsy. The lamella branchialis were placed on glass slides and the numbers of parasites on the gills were counted under a light microscope at 4×10 magnification to calculate the mean number of parasites per infected fish. The anthelmintic efficacy of tested samples was determined by comparison of the number of parasites in the treatments with those in the control groups, and calculated using the following equation described in our previous work (YAO et al., 2010).

$$A = (C - T) \times 100/C \tag{1}$$

Where A is the anthelmintic efficacy, C is the mean number of D. intermedius on the gills in the negative control, and T is the treatment groups.

2. 4　Bioactivity-guide fractionation

The chromatographic separation was monitored by the strategy of bioactivity-guided fractionation (invivo tests guided), only the extracts or fractions showed strong activity against D. intermedius were subjected to further separation and purification.

2. 5　Preparation of extracts

Dry and powdered C. majus (8. 6 kg) were extracted with 95% ethanol (20 l × 3 times) at room temperature for 24 h. The extract was evaporated to dryness under reduced pressure in a rotary evaporator to yield the ethanol extract (EE, 0. 74 kg). Part of the ethanol extract was reserved for activity assays whilst the rest of the extract was suspended in water, then, was extracted successively in a separating funnel with petroleum ether, ethyl acetate, chloroform and n-butanol. Each extract and the remaining aqueous part after solvent extraction were then evaporated to dryness under reduced pressure to give petroleum ether extract (38. 5 g), chloroform extract (130. 0 g), ethyl acetate extract (121. 7 g), n-butanol (258. 9 g) and remaining aqueous extract (109. 8 g). The in vivo tests showed that n-butanol extract was the highest in anthelmintic efficacy among all extracts. So, it was then subject to further separation.

2.6　Fractionation and isolation of pure compounds

The n-butanol extract (220.0 g) was subjected to open column chromatography on normal phase silica gel and eluted with a solvent mixture of chloroform-ethyl acetate solvent system to afford 627 fractions (300 mL each fraction). Fractions were monitored using thin-layer chromatography (TLC) on chloroform-ethyl acetate solvent system and fractions showing similar TLC chromatograms were combined into five fractions (Fr. A: 1~134 fractions, Fr. B: 135~318 fractions, Fr. C: 319~390 fractions, Fr. D: 391~510 fractions, Fr. E: 511~627 fractions). These five fractions were submitted to in vivo test, and Fr. D was the most active. Fr. D was then applied to reversed-phase high performance liquid chromatography (RP-HPLC) with following chromatographic conditions: Octadecyl Silane-A C_{18} (5 μm 120Å 10 mm×250 mm) column, acetonitrile-0.2% H_3PO_4 (22 : 78, v/v) mobile phase, 270nm wavelength, 5.0 mL min^{-1} flow rate, 30℃ column temperature and 500μl capacity. An active compound, chelerythrine, was obtained by RP-HPLC from Fr. D which was isolated by the above method. The structure of chelerythrine was established by EI-MS, ^{1}H NMR and ^{13}C NMR data by comparison with literature values (Wang et al., 2010; Radek et al., 1999; Pavlina et al., 2002). Its molecular formula is $C_{21}H_{18}NO_4$, and its structure was shown in Fig. 1.

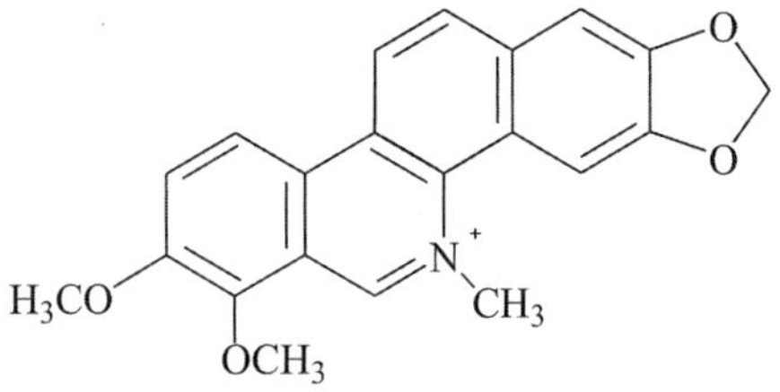

Fig. 1　Chemical structure of chelerythrine

2.7　Acute toxicity test of chelerythrine to grass carp

Acute toxicity of chelerythrine was performed using a standard lab protocol (Wang et al., 2009). The experiments were performed in an air-conditioned room in order to maintain a constant temperature at 24±1℃ during the experiment. Test grass carp (10 parasite free grass carp in each tank), were distributed in each tank, which was supplied with aeration-filtered groundwater. The experiment was carried out by adding chelerythrine to each tank in order to achieve 1.5, 2.0, 2.5, 3.0, 3.5, 4.0, 4.5, 5.0 mg·L^{-1}, control groups were set under the same test conditions without chemicals. Each concentration was considered to be one treatment group. In each treatment, three replicates were conducted. The deaths of fish were recorded when the opercula movement and tail beat stopped and the fish no longer responded to mechanical stimulus. The observed dead fish were removed from the water in time. Mortality of fish was recorded throughout the experiments, during which no food was offered to the fish. Under there circumstances, the experiments were stopped and fish were transferred to freshwater. Mortalities were registered at 48 h. The LC_{50} and its confidence intervals were calculated for chelerythrine using probit analysis.

3. Results and discussion

Currently, most countries have little effective and safe chemical treatments that can be used to control D. intermedius (Wang et al., 2010). More and more efforts have been spent on searching for more effective anti-D. intermedius drugs. Plants, microorganisms and marine organisms are potential sources of new drugs since they contain a countless quantity of natural products with a great variety of structures and pharmacological activities (Newman et al., 2003). Application of plants or their extracts to control monogenean parasites has been extensively studied during recent decades. Antiparasitic plant-derived compounds have been used as leads to develop semi-synthetic or synthetic drugs with better efficacy and safety (Tagboto and Townson, 2001). In the present study, an anti-D. intermedius compound was separated from C. majus by bioactivity-guided fractionation and identified as chelerythrine. In vivo test exhibited that chelerythrine was 100% effective against D. intermedius at a concentration of 1.60 mg·L^{-1}, with LC_{50} of 0.68 mg·L^{-1} after 48 h of exposure (Table. 1). Chelerythrine showed extraordinary activity when compared with mebendazole, which is widely used for the control of D. intermedius in practice (LC_{50} value=1.07 mg·L^{-1}) (Buchmann et al., 1993). To our best knowledge, this is the first report on in vivo anthelmintic investigation for C. majus.

Table 1 Anthelmintic activity and acute toxicity of crude extracts of C. majus and chelerythrineafter 48 h of exposure

Test samples	Anthelmintic efficacy		Acute toxicity LC$_{50}$ (mg · L^{-1})
	LC$_{50}$ (mg · L^{-1}) interval	95% confidence	
Ethanol extract	71.50	68.49–74.64	ND
petroleum ether extract	75.76	70.10–77.78	ND
chloroform extract	55.62	52.34–58.68	ND
ethyl acetate extract	25.50	24.38–27.90	ND
n−butanol extract	18.43	16.27–21.30	ND
aqueous extract	31.60	28.46–33.47	ND
chelerythrine	0.68	0.55–0.76	3.59
Mebendazole[a]	1.07	0.97–1.23	ND

ND not determined; [a] Positive control.

The different active ingredients in crude extract may have different polarity, so, the polar solvents with different polarity were applied to the extraction in order to avoid omitting the active ingredients (Wang et al, 2010). In this study, four extracts by different solvents (petroleum ether, ethyl acetate, chloroform and n−butanol) of increasing polarity were assayed for the anthelmintic activity. The n−butanol extract, exhibited the most significant activity with LC$_{50}$ value of 18.43 mg · L^{-1}, was subjected to bioassay−guided fractionation and purification, and a compound showing potent activity was obtained andidentified as chelerythrine. The other extract or fractions with lower anthelmintic activity were not further isolated, although they may contain compounds that have high activity, but present in low concentration. So, further phytochemical studies, towards the isolation and separation of the plant are recommended in our further research.

Chelerythrine, naturally occurring in a various plant belonging to the family poppy, is a representative quaternary benzo [c] phenanthridine alkaloid, shows antimicrobial, antifungal, and anti−inflammatory activity (Walterova et al., 1995; Vavreckova and Ulrichova, 1994; Simanek et al., 2003; Zdarilova et al., 2006; Malikova et al., 2006a). Besides, plant preparations containing a mixture of quaternary benzo [c] phenanthridine alkaloids including chelerythrine are used in toothpastes and mouthwashes as antiplaque agents (Zdarilova et al., 2006). Recently, chelerythrine was found to exhibit cytotoxic properties against several normal cells, such as human hepatocytes (Ulrichova et al., 2001) and rat cardiac myocytes (Yamamoto et al., 2001), as well as in cancer cells including, e.g. human primary uveal melanoma OCM−1 cells (Kemeny−Beke et al., 2006) and human promyelocytic leukemia HL−60 cells (Jarvis et al., 1994). The high cytotoxic potency of chelerythrine may responsible for the significant anthelmintic activity and high toxicity for the grass carp regarding chelerythrine with LC$_{50}$ values of 3.59 mg · L^{-1} (Table.1) in the present study. As recent researches indicated, mitochondria was the major cellular target of chelerythrine, in its induction of apoptosis invarious types of cells (Kemeny−Beke et al., 2006; Malikova et al., 2006b; Slaninova et al., 2001, 2007) and that chelerythrine exerted its cytotoxicity through multiple apoptosis−inducing pathways (Kaminskyy et al., 2008). As is well known, mitochondria are the key players that control and regulate apoptosis; mitochondrial dysfunction produced by toxic stress may lead to both apoptotic and necrotic cell death (Lemasters et al., 1999). A direct action on mitochondria might be involved in the eradication of the parasites. This may be the main mechanism responsible for the anti−D. intermedius activity of chelerythrine. However, the detailed mechanism of action regarding the anthelmintic activity of the chelerythrine should be further addressed.

In conclusion, using the strategy of bioactivity−guided isolation monitoring the chromatographic separation, a compound showing promising anti−D. intermedius activity was obtained from the ethanol extract of C. majus and identified as chelerythrine. The in vivo anthelmintic activity tests indicated that chelerythrine might be potential source of new anthelmintic drug for the control of D. intermedius. Also, further studies are required for field evaluations in the practical system and the mechanism of the antiparasitic activity remains to be performed.

Reference:

B uchmann K, Slotved HC, Dana D (1993) Epidemiology of gill parasite infections in Cyprinus carpio in Indonesia and possible control methods. Aquaculture 118: 9–21.

Colombo ML, Bosisio E (1996) Pharmacological activities of Chelidonium majusL.. Pharmacology Research 33: 127–134.

Dove A, Ernst I (1998) Concurrent invaders—four exotic species of Monogenea now established on exotic freshwater fishes in Australia. International journal for parasitology 28: 1755-1764.

Goven BA, Amend DF (1982) Mebendazole/trichlorfon combination: a new anthelmintic for removing monogenetic trematodes from fish. Journal of fish biology 20: 373-378.

smail KIR, Selda TÖ (2007) Helminth infections in common carp, Cyprinus Carpio L., 1758 (Cyprinidae) from Kovada Lake (Turkey). Türkiye. Parazitoloji. Dergisi 31: 232-236 (in German with English abstract).

Jarvis WD, Turner AJ, Povirk LF, Traylor RS, Grant S (1994) Induction of apoptotic DNA fragmentation and cell death in HL-60 human promyelocytic leukemia cells by pharmacological inhibitors of protein kinase C. Cancer Research 54: 1707-1714.

Kaminskyy V, Kulachkovskyy O, Stoika R (2008) A decisive role of mitochondria in defining rate and intensity of apoptosis induction by different alkaloids. Toxicology Letters 177: 168-181.

Kery RY, Horvath J, Nasz I, Verzar-Petri G., Kulcsar G, Dan P (1987) Antiviral alkaloid in Chelidonium majus L.. Acta Pharmaceutica Hungerica 57: 19-25.

Kemeny-Beke A, Aradi J, Damjanovich J, Beck Z, Facsko A, Berta A, Bodnar A (2006) Apoptotic response of uveal melanoma cells upon treatment with chelidonine, sanguinarine and chelerythrine. Cancer Letters 237: 67-75.

Kim DJ, Ahn B, Han BS, Tsuda H (1997) Potential preventive effects of Chelidonium majus L. (Papaveraceae) herb extract on glandular stomach tumor development in rats treated with N-methyl-N′-nitro-N-nitrosoguanidine (MNNG) and hypertonic sodium chloride. Cancer Letters 112: 203-208.

Lenfeld J, Kroutil M, Marsalek E, Slavik J, Preininger V, Simanek V (1981) Anti-inflammatory activity of quaternary benzophenanthridine alkaloids from Chelidonium majus. Planta Medica 43: 161-165.

Lemasters JJ, Qian T, Bradham CA, Brenner DA, Cascio WE, Trost LC, Nishimura Y, Nieminen AL, Herman B (1999) Mitochondrial dysfunction in the pathogenesis of necrotic and apoptotic cell death. Journal of Bioenergetics and Biomembranes 31: 305-319.

Liu YT, Wang F, Wang GX, Han J, Wang Y, Wang YH (2010) In vivo anthelmintic activity of crude extracts of Radix angelicae pubescentis, Fructus bruceae, Caulis spatholobi, Semen aesculi, and Semen pharbitidisagainst Dactylogyrus intermedius (Monogenea) in goldfish (Carassius auratus). Parasitology Research 106: 1233-1239.

Marshall CJ (1999) Use of Supaverm © for the treatment of monogenean infestation in koi carp (Cyprinus carpio). Fish Veterinary Journal 4 : 33-37.

Malikova J, Zdarilova A, Hlobilkova A (2006a) Effects of sanguinarine and chelerythrine on the cell cycle and apoptosis. Biomedical Papers of Medical Faculty University Palacky Olomouc Czech Rep 150: 5-12.

Malikova J, Zdarilova A, Hlobilkova A, Ulrichova J (2006b) The effect of chelerythrine on cell growth, apoptosis, and cell cycle in human normal and cancer cells in comparison with sanguinarine. Cell Biology and Toxicology 22: 439-453.

Newman DJ, Cragg GM, Snader KM (2003) Natural products as sources of new drugs over the period 1981-2002. Journal of natural products 66: 1022-1037.

Panzer A, Joubert AM, Bianchi PC, Seegers JC (2000) The antimitotic effects of Ukrain, a Chelidonium majus alkaloid derivative, are reversible in vitro. Cancer Letter 150: 85-92.

Pavlina S, Radek M, Jiri D, Roger D, Eddy LE (2002) Stuctural Studies of Benzophenanthridine Alkaloid free Bases by NMR Spectroscopy. Magn Reson Chem 40: 147-152.

Pavão ML, Pinto RE (1995) Densitometric assays for the evaluation of water soluble alkaloids from Chelidonium majus L. (Papaveraceae) roots in the Azores, along one year cycle. Arquipélago, Sér. Ciências Biol. Marinhas 13: 89-91.

Paperna I (1964) Adaptation of Dactylogyrus extensus (Muller and Van, 1932) to ecological conditions of artificial ponds in Israel. Journal of parasitology 50: 90-93.

Radek M, Jaromir T, Jiri D, Jiri S, Roger D, Vladimir S (1999) ^{1}H and[13]C NMR Study of Quaternary Benzo [c] phenanthridine Alkaloids. Magn Reson Chem 37: 781-787.

Reed PA, Francis-Floyd R, Klinger RC (2009) FA28/FA033: Monogenean parasites of fish. EDIS-Electronic Data Information Source-UF/IFAS Extension. University of Florida. http: //edis. ifas. ufl. edu/FA033. 17 May 2009.

Schmahl G., Mehlhorn H (1985) Treatment of fish parasites: 1. Praziquantel effective against Monogenea (Dactylogyrus vastator, Dactylogyrus extensus, Diplozoon paradoxum). Z. Parasitenkd 71: 727-737.

Schmahl G. (1993) Treatment of fish parasites 10. Effects of a new triazine derivative, HOE 092V, on Monogenea: a light and transmission electron microscopy study. Parasitology Research 79: 559-566.

Simanek V, Vespalec R, Sedo A, Ulrichova J, Vicar J (2003) Quaternary benzo (c) phenanthridine alkaloids - biological activities. In: Schneider, M. P. (Ed.), Chemical Probes in Biology. Kluwer Academic Publishers, Netherlands, pp. 245-254.

Slaninova I, Taborska E, Bochorakova H, Slanina J (2001) Interaction of benzo [c] phenanthridine and protoberberine alkaloids with animal and yeast cells. Cell Biology and Toxicology 17: 51-63.

Slaninova I, Slunska Z, Sinkora J, Vlkova M, Taborska E (2007) Screening of minor benzo (c) phenanthridine alkaloids for antiproliferative and apoptotic activities. Pharmaceutical Biology 45: 131-139.

Taborska E, Bochorakova H, Dostal J, Paulova H (1995) The greater celandine (Chelidonium majus L.) -A review of present knowledge. Ceska a Slovenska Farmacie 44 (2): 71-75.

Tagboto S，Townson S（2001）Antiparasitic properties of medicinal plants and other naturally occurring products. Advances in parasitology 50：199 -295.

Topi　PN，Hacmanjek M，Teskeredži　E（2001）Health status of rudd（Scardinius erythrophthalmus hesperidicus H.）in Lake Vrana on the Is- land of Cres，Croatia. Journal of applied ichthyology 17：43-45.

Ulrichova J，Dvorak Z，Vicar J，Lata J，Smrzova J，Sedo A，Simanek V（2001）Cytotoxicity of natural compounds in hepatocyte cell culture models -The case of quaternary benzo［c］phenanthridine alkaloids. Toxicology Letters 125：125-132.

Vavreckova C，Ulrichova J（1994）Biological activity of quaternary benzo［c］phenanthridine alkaloids--sanguinarine and chelerythrine. Chemicke Listy 88：238-248.

Wang GX，Zhou Z，Cheng C，Yao JY，Yang ZW（2008）Osthol and isopimpinellin from Fructus cnidii for the control of Dactylogyrus intermedius in Carassius auratus. Veterinary parasitology 158：144-151.

Wang GX，Han J，Feng TT，Li FY，Zhu B（2009）Bioassay-guided isolation and identification of active compounds from Fructus Arctii against Dac- tylogyrus intermedius（Monogenea）in goldfish（Carassius auratus）Parasitology Research 106：247-255.

Wang G. X，Zhou Z，Jiang DX，Han J，Wang JF，Zhao LW，Li J（2010）In vivo anthelmintic activity of five alkaloids from Macleaya microcarpa （Maxim）Fedde against Dactylogyrus intermedius in Carassius auratus. Veterinary parasitology171（3-4）：305-13.

Walterova D，Ulrichova J，Valka I，Vicar J，Vavreckova C，Taborska E，Harkrader RJ，Meyer DL，Cerna H，Simanek V（1995）Benzo（c） phenanthridine alkaloids sanguinarine and chelerythrine：biological activities and dental care applications. Acta University Palacki Olomouc Faculty of Medicine 139：7-14.

Yamamoto S，Seta K，Morisco C，Vatner SF，Sadoshima J（2001）Chelerythrine rapidly induces apoptosis through generation of reactive oxygen spe- cies in cardiac myocytes. Journal of Molecular and Cellular Cardiology 33：1829-1848.

Yao JY，Shen JY，Li XL，Xu Y，Hao G. J，Pan XY，Wang GX，Yin WL（2010）Effect of sanguinarine from the leaves of Macleaya cordataagainst Ichthyophthirius multifiliis in grass carp（Ctenopharyngodon idella）. Parasitology Research 107：1035-1042.

Zdarilova A，Malikova J，Dvorak Z，Ulrichova J，Simanek V（2006）Quaternary isoquinoline alkaloids sanguinarine and chelerythrine. In vitro and in vivo effects. Chemicke Listy 100：30-41.

该文发表于《Parasitology Rresearch》【2011，111：443-446】。

In vivo anthelmintic activity of chelidonine from Chelidonium majus L. against *Dactylogyrus intermedius* in *Carassius auratus*

Jia-Yun YAO[1]，Zhi-Ming ZHOU[1]，Xiao-yi PAN[1]，Gui-jie HAO[1]，Xi-Lian LI[2]，
Yang XU[1]，Jin-Yu SHEN[1*]，Hong-shun RU[1]，Wen-lin YIN[1]

（1. Institute of Freshwater Fisheries of Zhejiang，Huzhou，Zhejiang，313001，China；

2. College of Animal Science and Technology，Northwest A&F University，Yangling，Shaanxi，712100，China）

Abstract：Dactylogyrus intermedius is one of the most common and serious parasitic diseases of freshwater fish in aquacul- ture，and can cause morbidity and high mortality in most species of freshwater fish worldwide. To attempt controlling this par- asite and explore novel potential antiparasitic agents，the present study was designed to ascertain the anthelmintic activity of Chelidonium majus L. whole plant and to isolate and characterise the active constituents against D. intermedius. The ethanol extract from C. majus whole plant showed significant anthelmintic activity against D. intermedius（EC_{50}（median effective concentration）value=71. 5 mg · L^{-1}），and therefore subjected to the further isolation and purification using various chrom- atographic techniques. A quaternary benzo［c］phenanthridine alkaloid exhibited significant activity against D. intermedius was obtained and identified as chelidonine. In vivo anthelmintic efficacy tests exhibited that chelidonine was 100% effective a- gainst D. intermedius at a concentration of 0. 9 mg · L^{-1}，with EC_{50} values of 0. 48 mg · L^{-1} after 48 h of exposure，which is more effective than the positive control，mebendazole（EC_{50} value=1. 3 mg · L^{-1}）. In addition，the 48 h median lethal concentration（LC_{50}）for chelidonine against the host（Carassius auratus）was 4. 54 mg · L^{-1}. The resulting therapeutic in- dex for chelidonine was 9. 46. These results provided evidence that chelidonine might be potential sources of new anti-parasit- ic drugs for the control of Dactylogyrus.

key words：Dactylogyrus intermedius；chelidonine；anthelmintic；*Chelidonium majus L.*

白屈菜中白屈菜碱驱杀鲫体内指环的研究

姚嘉赟[1]，周志明[1]，潘晓艺[1]，郝贵杰[1]，李喜莲[2]，徐洋[1]，沈锦玉[1*]，茹洪顺[1]，尹文林[1]

(1. 浙江省淡水水产研究所，浙江 湖州 313001；2. 西北农林科技大学，陕西 杨凌 712100)

摘要： 指环虫是寄生于鱼类鳃部的单殖亚纲类寄生虫，给水产养殖动物造成了巨大的经济损失。本研究利用活性追逐试验结合有机分离技术对白屈菜中的杀指环虫的活性成分进行提取分离。前期结果表明乙醇提取物杀虫活性较高———其对指环虫的半数致死浓度为 71.5 mg·L^{-1}。对白屈菜的乙醇提取物进行硅胶柱层析，并获得一种杀虫活性较强的活性单体。经质谱、核磁氢谱、核磁碳谱、红外光谱等多种波谱分析技术鉴定该化合物为白屈菜碱。体内杀虫药效试验结果表明白屈菜碱的 100% 杀指环虫浓度为 0.9 mg·L^{-1}，其半数致死浓度（EC_{50}）为 0.48 mg·L^{-1}。其杀虫效果明显优于目前市场上的常用药物甲苯咪唑（$EC_{50}=1.3$ mg·L^{-1}）对草鱼的急性毒性实验结果表明白屈菜碱对草鱼的 48 h 半数致死浓度为 4.54 mg·L^{-1}。上述结果表明，白屈菜碱对指环虫具有较强的杀灭作用且对鱼体相对安全，有望开发为一种新型的杀虫药物。

关键词： 指环虫，白屈菜碱，抗寄生虫，白屈菜

1. Introduction

In recent years the gill monogean Dactylogyrus infestation of fish has increased in incidence and severity (Topi et al., 2001; smail and Selda, 2007). The infestation has traditionally been controlled by formalin; however, the use of formalin for the treatment of disease in dulfish has been discouraged due to its toxicity in small-scale trials (Diggles et al. 1993). Other chemicals, including toltrazuril, praziquantel, and mebendazole, have also been evaluated for chemotherapy of Dactylogyrus under laboratory conditions and also played a pivotal role in the treatment of Dactylogyrus infestation in aquaculture practice (Schmahl and Mehlhorn 1985; Schmahl et al. 1988; Treves-Brown 1999). However, the threats of anthelmintic resistance, risk of residue, environmental contamination, and toxicity to host caused by the frequently use of these drugs have led to the need of other alternative control methods (Goven et al. 1980; Klingerand Floyd, 2002).

Options like traditional medicinal plants are being examined in different parts of the world for the treatment of parasites infestation. Ekanem et al. 2004 found that the methanol extracts of the seeds of Piper guineense (Piperaceae) were active against Dactylogyrus. Two compounds, trillin and gracillin, from Dioscorea zingiberensis C. H. Wright have been reported to have the anthelmintic efficacy against Dactylogyrus intermedius (Wang et al. 2010). In our previous work, an alkaloid sanguinarine from Macleaya cordata was found to be effective against I. multifillis (Yao et al. 2010). Moreover, most of these natural ingredients exhibited higher efficacy than those widely used anthelmintics such as mebendazole (Wang et al. 2009). Therefore, traditional medicinal plants should be a reliable source for the discovery of novel and potential anthelmintic agents.

Chelidonium majus L. (Papaveraceae), a perennial traditional Chinese medicinal plant, has been widely used as herbal medicine to treat various infectious diseases like whooping cough, chronic tracheitis, beriberi and gastritis. Previous chemical and pharmacological studies found it mainly contained various isoquinoline alkaloids which exhibited antimicrobial, anticancer, anti-inflammatory, adrenolytic, sympatholytic and anti-acetylcholinesterase activities (Jiangsu New Medical College, 1977; Niu and He, 1994; Táborská et al., 1995; Colombo and Bosisio, 1996; Cho et al., 2006). Our previous study exhibited that ethanol extract from C. majus showed significant effect with an EC_{50} (median effective concentration) value of 71.5 mg·L^{-1} against D. intermedius (Li et al., 2011), it prompted us to conduct a further investigation on this plant, aiming to isolate and purify the active compounds responsible for its anthelmintic properties.

2 Materials and methods

2.1 Parasites and Hosts

Healthy goldfish, Carassius auratus, weighing 3.4±0.3 gwere obtained from aquatic fry farm of Zhejiang Institute Freshwater Fisheries in China and maintained in a 1-m^3 tank at 24±1℃ (controlled by automatic aquarium heater) with aeration for 7 days. The healthy goldfish were co-habitated at a ratio of 20% with the ones infected with D. intermedius which were reared in our laboratory. The infected goldfish were prepared according to our previous study (Li et al., 2011). After 15-18 days of co-habitation, 20 goldfish

637

were then randomly selected, killed by spinal severance and examined for the presence of parasite under a light microscope (Olympus BX51, Tokyo, Japan) at 4×10 magnification prior to the experiment. Fish were chosen for the tests when the infection prevalence was 100% and the mean number of the parasite on the gills was 50-60.

For acute toxicity tests, parasite-free goldfish were obtained from commercial fish farm and maintained in a $1-m^3$ tank supplied with filtered groundwater under the same conditions as parasitized fish.

2.2　Plant material

Chelidonium majus L. were collected in Zhejiang province, China, in March 2010 and identified by Prof. X L He in Northwest A&F University, China. A voucher specimen has been deposited in the Herbarium of College of Life Science of the university. They were cleaned and air dried for a week at 35-40℃ and pulverized in electric grinder. The powdered plant samples were stored at-20℃ until further use.

2.3　Activity-guided isolation of active compounds

Air-dried and powdered herbs of C. majus (3.2 kg) were exhaustively extracted with 50 L ethanol at room temperature by percolation, giving 405 g dry extract. The ethanol extract was subjected to column chromatography (120×10 cm) on a silica gel (3500 g, 100-200 mesh) and sequentially eluted with chloroform-ethyl acetate (1：0, 100：1, 50：1, 25：1, 10：1, 5：1, 2：1, 1：1, 0：1, 0：100, 0：50, 0：25, 0：10, 0：5, 0：2, 0：1, v/v, 500 mL each) with increasing polarity to provide 789 fractions. The fractions were combined on the basis of TLC monitoring, affording 7 fractions, which were submitted to anthelmintic assay. Fraction E, exhibiting high activity, was subjected to further preparative procedures.

Fr E (74.3 g) which displayed 100% effective against D. intermedius at the concentrations of 8.5 mg · L^{-1} was subjected to column chromatography (80×6 cm) on silica gel (1000 g, 100-200 mesh) using a mixture of petroleum ether-ethyl acetate-methanol (10：1：0, 5：1：0, 2：1：0, 1：1：0, 4：4：1, 2：2：1, 1：1：1, 2：2：3, 1：1：10, 0：0：1 v/v/v, 300 mL each) gradient as eluent to yield 136 fractions. The fractions showing similar TLC chromatograms were combined into 5 subfractions, and subfractions E3 displaying 100% effective against D. intermedius were subjected to column chromatography (40×2 cm) with Sephadex-LH20 (30g, 200-300mesh) using chloroform-methanol (1：0, 1：1, 0：1 v/v) as developing system, to afford a yellow lamellar crystal (288.5 mg).

2.4　In vivo anthelmintic efficacy test

In vivo tests were conducted in $40 \times 30 \times 20$ cm glass tanks, each containing 10 L of the test solution, and 10 parasitized fish (goldfish were placed on glass slides, and the presence of parasites on the paddle of the fish were determined microscopically, heavy infected fish were transferred into the glass tank immediately, fish infected with little parasites were returned to the tanks.). The test samples were assayed at a different series of concentrations based on initial tests. Negative control groups with no chemical were set up under the same experimental conditions, while mebendazole was used as the positive control. All treatments and control groups were conducted with three replicates.

After 48 h treatment, goldfish in all treatments and control groups were killed by a spinal severance for biopsy. The total number of D. intermedius in the collected gills and paddle from each fish was counted. D. intermedius were considered dead when appeared lysed or their motility (mouthpiece did not peristalsis in 2 min) was lost (Wang et al., 2006). The anthelmintic efficacy of tested samples was determined by comparison of the number of parasites in the treatments with those in the negative control groups, and calculated using the following equation described in our previous work (Yao et al., 2010).

$$E = (C - T) \times 100/C$$

Where E is the anthelmintic efficacy, C is the mean number of D. intermedius in the negative control, and T is the treatment groups.

2.5　Acute toxicity assay

The acute toxicity of the ethanol extracts and chelidonine from C. majus were assayed for the evaluation of their safety to the host. Tests were conducted in duplicate, using ten healthy goldfish in plastic pot with 10 L capacity, containing 5 L of test solution. The fish were not fed during the experiment. Dilutions were prepared from the stock solutions as the following concentrations: 3.0, 3.5, 4.0, 4.5, 5.0, 5.5, 6.0, 6.5 and 7.0 mg · L^{-1} for chelidonine; 170.0, 175.0, 180.0, 185.0, 190.0, 195.0, 200.0 and 205.0 mg · L^{-1} for ethanol extract; Control groups were set under the same test conditions with no chemical. The death of fish were recorded when the opercula movement and tail beat stopped and the fish no longer responded to mechanical stimulus. Any observed dead fish was removed in time from the water to avoid the deterioration of the water quality. Fish mortalities in the treatment and control groups were recorded after 48 h of exposure. The 48-h median lethal concentration (LC_{50}) and its confidence intervals were calculated using probit analysis.

2.6　Statistical analysis

The homogeneity of the replicates of the samples was checked by the Mann-Whitney U test. Probit analysis was used for calculating the LC_{50} and EC_{50} at the 95% confidence interval with upper confidence limit and lower confidence limit (Finney 1971). The therapeutic index (TI) was calculated by comparing LC_{50} versus the EC_{50}.

3　Results and discussion

Among all the tested extracts from C. majus whole plant, the ethanolic extract was found to be most effective when tested against the fish parasite D. intermedius (Li et al., 2011), indicating that the active constituents were mainly contained in this preparation. Therefore, bioassay-guided purification was conducted in the present study with this extract, yielding an active compound. The melting point of the active compound was determined to be 135-136℃. IR (KBr) cm^{-1}: 3640, 3371, 3000, 2897, 2783, 1652, 1485, 1354, 1266, 1233, 1078, 1042. UV$\lambda^{MeOHmax}$ nm: 289. EI-MS (70eV) m/z: 354.2 [M+1]$^{+}$, ^{1}H NMR (CD$_3$OD, TMS, 400MHz): δ 6.76 (1H, d, J=8.0 Hz, H-9), 6.735 (1H, d, J=8.0 Hz, H-10), 6.67 (1H, s, H-4), 6.648 (1H, s, H-21), 5.996 (1H, d, J=8 Hz, C$_{2,3}$-OCH$_2$O), 5.955 (1H, d, J=1.5Hz, C$_{2,3}$-OCH$_2$O), 5.937 (1H, d, J=1.5Hz, C$_{7,8}$-O-CH$_2$O), 4.241 (1H, d, J=1.8Hz, H$_{11}$), 4.10 (1H, d, J=15.7Hz, H$_6$), 3.45 (IH, d, J=15.7Hz, H$_6$), 3.57 (1H, brs, H$_{14}$), 3.23 (1H, d, J=17Hz, H$_{12}$), 3.10 (1H, dd, J=17.44Hz, H$_{5β}$), 3.00 (1H, brs, H$_{13}$), 2.29 (1H, s, NCH$_3$). ^{13}C NMR (CD$_3$OD, TMS, 400MHz) δppm: 148.03 (C$_8$), 145.50 (C$_3$), 145.16 (C$_2$), 142.94 (C$_7$), 131.2 (C$_{10a}$), 128.72 (C$_{4a}$), 125.54 (C$_{12a}$), 120.37 (C$_{10}$), 116.97 (C$_{6a}$), 111.84 (C$_4$), 109.46 (C$_9$), 107.42 (C$_1$), 101.2 (7, 8-OCH$_2$O-), 100.97 (2, 3-OCH$_2$O-), 72.33 (C$_{11}$), 62.83 (C$_{14}$), 53.93 (C$_6$), 42.42 (N-CH$_3$), 39.62 (C$_{21}$).

The data were in agreement with the reported literature values. Thus, the structure of compound was determined as chelidonine, its molecular formula is C$_{20}$H$_{19}$NO$_5$ (Zhang et al. 1995; Pandey et al., 1977; Swinehart et al., 1980).

The anthelmintic activity of chelidonine obtained in this work and ethanolic extract from C. majus were assayed on D. intermedius. As showed in Fig. 1, chelidonine was found to be 100% effective for the elimination of D. intermedius at the concentrations of 0.9 mg · L^{-1}, with the median effective concentration (EC_{50}) values of 0.48 mg · L^{-1}. In addition, chelidonine showed a much higher activity when compared with the original ethanolic extract (EC_{50} value=71.5 mg · L^{-1}, Table 1), indicating that it may or partly responsible for the anthelmintic activity of the ethanolic extract of C. majus. Chelidonine showed extraordinary activity when compared with mebendazole, which is frequently used for the control of Dactylogyrus (Buchmann et al., 1993) (EC_{50} value = 1.3 mg · L^{-1}, Table 1). The results obtained here for chelidonine were very promising and have provided scientific evidence for the development of chelidonine as an alternative anthelmintic agent. This is the first report on the anthelmintic activity of chelidonine.

In the acute toxicity test for chelidonine for C. auratus, most of the goldfish exhibited a slow reaction to external stimulation when the concentration exceeded 3.0 mg · L^{-1}. Mortality occurred when the concentration reached 3.5 mg · L^{-1} and the LC_{50} at 48 h was 4.54 mg · L^{-1} (Table 1) with the 95% fiducial limits of 4.17-5.01 mg · L^{-1}. In the case of parasiticides, the therapeutic index (TI) is estimated from the differential toxicity between the parasite and the fish. A chemical with a high index can be administered or bathed with greater safety than one with a low index, this allows for prolongation of the tretament time without affecting the fish. The TI of ethanolic extract and chelidonine is 2.58 and 9.46, respectively. This indicates that chelidonine is much safer than ethanolic extract when used in practice. Considering the low EC_{50} and high TI value of chelidonine, it has great potential for the development of a new parasiticide.

Chelidonine, naturally occurring in a various plant belonging to the family poppy, is a representative quaternary benzo [c] phenanthridine alkaloid. It was isolated from C. majus (Papaveraceae) as early as 1839, while it could be synthesized from 1971 onwards, and biosynthesis from dopamine was first reported in 1979 (summarised in (Cushman et al., 1980; Hanakoa et al., 1986; Simanek, 1985)). Chelidonine was shown to cause mitotic arrest by Lettre and Albrecht (1942) and has recently been shown to inhibit tubulin polymerization (Wolff and Knipling, 1993). The alkaloid possibly acts at the colchicine-binding site, as it is a weak competitive inhibitor of colchicines binding (Wolff and Knipling, 1993). As recent researches indicated, chelidonine was found to induce apoptosis in human acute tlymphoblastic leukaemia MT-4 cells. Apoptosis induction by chelidonine in these cells was accompanied by caspase-9 and -3 activation and an increase in the pro-apoptotic Bax protein (Philchenkov et al., 2008). As we known, the mitochondrial cell death pathway is known to be realized through the activation of caspase-9 (Hakem et al., 1998; Philchenkov, 2004). This may be the main mechanism responsible for the anti-D. intermedius activity of chelidonine. However, the detailed mechanism of action regarding the anthelmintic activity of the chelidonine should be further addressed.

In conclusion, the results of the present study provide a significant basis for use of the ethanolic extract from C. majus for the treatment of D. intermedius on goldfish. The extract as well as chelidonine isolated from C. majus found to be active in this study could be useful for the development of new anthelmintic agent. However, further studies are required for field evaluations in the practical system and the mechanism of the anthelmintic activity remains to be performed.

Reference:

Buchmann K., Slotved H. C., Dana D. (1993) Epidemiology of gill parasite infections in Cyprinus carpio in Indonesia and possible control methods. Aquaculture 118: 9-21.

Cho K. M., Yoo I. D., Kim W. G. (2006) 8-Hydroxydihydrochelerythrine and 8-hydroxydihydroSanguinarine with a potent acetylcholinesterase inhibitory activity from Chelidonium majus L. Biol Pharm Bull 29: 2317-2320.

Colombo M. L., Bosisio E. (1996) Pharmacological activities of Chelidonium majus L. (Papaveraceae). Pharmacol Res 33: 127-134.

Cushman M., Choong T., Valko J. T., Koleck M. P. (1980) Total synthesis of chelidonine. J. Org. Chem 46: 5067-5073.

Diggles B. K., Roubal F. R., Lester R. J. G. (1993) The influence of formalin, benzocaine and hyposalinity on the fecundity and viability of Polylabroides multispinosus (Monogenea: Microcotylidae) parasitic on the gills of Acanthopagrus australis (Pisces: Sparidiae). Int J Parasitol 23: 877-884.

Finney D. J. (1971) Probit analysis, 3rd edn. Cambridge University Press, CambridgeGoven BA, Gilbert J, Gratzek J (1980) Apparent drug resistance to the organophosphate dimethyl (2, 2, 2-trichloro-1-hydroxyethyl) phosphonate by monogenetic trematodes. J wildlife dis 16 (3): 343-346.

Hakem R., Hakem A., Duncan G. S., Henderson J. T., Woo M., Soengas M. S., Elia A., de la Pompa J. L., Kagi D., Khoo W., Potter J., Yoshida R., Kaufman S. A., Lowe S. W., Penninger J. M., Mak T. W. (1998) Differential requirement for caspase 9 in apoptotic pathways in vivo. Cell 94: 339-352.

Hanakoa M., Yoshida S., Annen M., Mukai C.. (1986) A novel and biomimetic synthesis of chelamins, chelidonine, sanguinarine, and dihydrosanguinarine from coptisine via a common intermediate. Chem. Lett 23: 739-742.

Huh J., Liepins A., Zielonka J., Andrekopoulos C., Kalyanaraman B., Sorokin A. (2006) Cyclooxygenase 2 rescues LNCaP prostate cancer cells from sanguinarine-induced apoptosis by a mechanism involving inhibition of nitric oxide synthase activity. Cancer Res 66: 3726-3736.

Smail K. I. R., Selda T. Ö. (2007) Helminth infections in common carp, Cyprinus Carpio L., 1758 (Cyprinidae) from Kovada Lake (Turkey). Türkiye. Parazitoloji. Dergisi 31: 232-236 (In German with English abstract).

Jiangsu New Medical College, 1977. Bai-qu-cai Chelidonium majus L. Dictionary of Chinese Materia Medica, vol. 1. Shanghai Scientific Technical Press, Shanghai, China, pp. 726-727.

Klinger R., Floyd R. F. (2002) Introduction to freshwater fish parasites. Document CIR716. Institute of Food and Agricultural Science, University of Florida.

Lettre H., Albrecht M. (1942) Narcotin, einMitosegift. Naturwiss 30: 184-185.

Li X. L., Yao J. Y., Zhou Z. M., Shen J. Y., Ru H. S., Liu X. L. (2011) Activity of the chelerythrine, a quaternary benzo [c] phenanthridine alkaloid from Chelidonium majus L., on Dactylogyrus intermedius. Parasitol Res (In press) DOI: 10. 1007/s00436-011-2320-9.

Niu C., He L. (1994) Review of the studies on Chelidonium majus L. Zhongguo Yaoxue Zazhi 29: 138-140 (C. A. 120: 279936z).

Pandey V., Ray A. B., Dasgupta G. (1977) Minor alkaloids of Fumaria indica seeds. Phytochemistry 18: 695.

Philchenkov A., Kaminskyy V., Zavelevich M., Stoika R. (2008) Apoptogenic activity of two benzophenanthridine alkaloids from Chelidonium majus L. does not correlate with their DNA damaging effects. Toxicol in Vitro 22: 287-295.

Schmahl G., Mehlhorn H. (1985) Treatment of fish parasites. 1. Praziquantel effective against Monogenea (Dactylogyrus vastator, Dactylogyrus extensus, Diplozoon paradoxum). Z Parasitenk 71: 727-737.

Schmahl G., Mehlhorn H., Haberkorn A. (1988) Sym. Triazinone (toltrazuril) effective against fish-parasitizing Monogenea. Parasitol Res 75 (1): 67-68.

Swinehart J. A., Stermitz F. R. (1980) Bishordeninyl terpene alkaloids and other constituents of Zanthoxylum culantrillo and Z. coriaceum. Phytochemistry. 19: 1219.

Táborská E., Bochoráková H., Dostál. J., Paulová H. (1995) The greater celandine (Chelidonium majus L.) -review of present knowledge. Ceska a Slovenska Farmacie 44: 71-75.

Topi P. N., Hacmanjek M., Teskeredži E. (2001) Health status of rudd (Scardinius erythrophthalmus hesperidicus H.) in Lake Vrana on the Island of Cres, Croatia. J. Appl. Ichthyol 17: 43-45.

Treves-Brown K. M. (1999) Availability of medicines for fish. Fish Vet J 4: 40-55.

Wang G. X,. Cheng C., Chen A. L., Shen Y. H., Zhang X. (2006) Dactylogyrus Killing Efficacies of 22 Plants Extracts and Its 6 Compounds.

Acta Bot. Boreal. −Occident. Sin 26（12）：2567−2573.

Wang G. X., Han J., Feng T. T., Li F. Y., Zhu B. (2009) Bioassay−guided isolation and identification of active compounds from Fructus Arctii against Dactylogyrus intermedius（Monogenea）in goldfish（Carassius auratus）. Parasito Res 106：247−255.

Wang G. X., Jiang D. X., Li . J, Han J., Liu Y. T., Liu X. L. (2010) Anthelmintic activity of steroidal saponins from Dioscorea zingiberensis C. H. Wright against Dactylogyrus intermedius（Monogenea）in goldfish（Carassius auratus）. Parasitol Res 107：1365−1371.

Wolff J., Knipling L. (1993) Antimicrotubule properties of benzophenanthridine alkaloids. Biochemistry 32, 13334−13339.

Yao J. Y., Shen J. Y., Li X. L., Xu Y., Hao G. J., Pan X. Y., Wang G. X., Yin W. L. (2010) Effect of sanguinarine from the leaves of Macleaya cordataagainst Ichthyophthirius multifiliis in grass carp（Ctenopharyngodon idella）. Parasitol Res 107：1035−1042.

Zhang G. L., Rucker G., Breitmaier E., Mayer R. (1995) Alkaloids from Hypecoum leptocarpum Phytochemistry 40：1813.

该文发表于《Aquaculture》【2011，318：235−238】。

Isolation of bioactive components from Chelidonium majus L. with activity *against Trichodina* sp.

Jia-yun YAO[1], Xi-lian LI[2], Jin-yu SHEN[1*], Xiao-yi PAN[1], Gui-jie HAO[1], Yang XU[1], Wen-lin YING[1], Hong-shun RU[1], Xiao-lin LIU[2]

（1. Zhejiang Institute of Freshwater Fisheries, Huzhou, Zhejiang, 313001, China；

2. College of Animal Science and Technology, Northwest A&F University, Yangling, 712100, China）

Abstract：Trichodinids are the most common ciliates parasite present on the skin of pond−reared fish and can cause severe economic losses in aquaculture and ornamental fish breeding. The present study aims to evaluate the antiparasitic activity of the active components from chelidonium majus L. against Trichodina sp.. Bioassay−guided fractionation and isolation of the compounds with antiparasitic activity were performed on the ethanolic extract of C. majus yielding three bioactive alkaloids namely：chelidonine（1）chelerythrine（2）and sanguinarine（3）Results from in vivo antiparasitic assays revealed that these compounds isolated could be 100% effective for the elimination of Trichodina sp. at the concentrations of 1.0, 0.8, and 0.7 mg · L^{-1}, with the median effective concentration（EC_{50}）values of 0.6, 0.33, 0.32 mg · L^{-1}, respectively. Furthermore, the promising chelidonine, chelerythrine and sanguinarine were subjected to acute toxicity tests for the evaluation of their safety to the host（Parabramis pekinensis）. It was found that the 48 h median lethal concentration （48h−LC$_{50}$）values determined by the acute toxicity tests on P. pekinensis were 2.5, 3.8 and 1.5 mg · L^{-1} respectively. These results provided evidence that the isolated compounds, especially chelerythrine, can be exploited as novel antiparasitic agents for the control of Trichodina sp..

Keywords：Chelidonium majus L. ; *Trichodina* sp. ; chelidonine; chelerythrine; sanguinarine

白屈菜杀车轮虫的活性成分的分离

姚嘉赟[1]，李喜莲[2]，沈锦玉[1*]，潘晓艺[1]，郝贵杰[1]，徐洋[1]，尹文林[1]，茹洪顺[1]，刘小林[2]

（1. 浙江省淡水水产研究所，浙江 湖州 313001；2. 西北农林科技大学，陕西 杨凌 712100）

摘要：车轮虫是寄生于鱼类鳃部的纤毛类寄生虫，经常引起鱼类的大量死亡给水产养殖业造成了巨大的经济损失。本研究利用活性追逐试验结合有机分离技术对白屈菜中的杀车轮虫的活性成分进行提取分离。对白屈菜的乙醇提取物进行硅胶柱层析，并获得3种杀车轮虫活性较强的活性单体。经质谱、核磁氢谱、核磁碳谱、红外光谱等多种波谱分析技术鉴定这3种化合物分别为：白屈菜碱，白屈菜红碱和血根碱。体内杀虫药效试验结果表明白屈菜碱，白屈菜红碱和血根碱对车轮虫的100%杀灭浓度分别为1.0，0.8和0.7 mg·L^{-1}，它们的半数致死浓度分别为0.6，0.33，0.32 mg·L^{-1}。对草鱼的急性毒性实验结果表明白屈菜碱，白屈菜红碱和血根碱对鳊鱼的48 h半数致死浓度为2.5，3.8和1.5 mg·L^{-1}。上述结果表明，白屈菜碱，白屈菜红碱和血根碱尤其是白屈菜碱有望开发为一种新型的杀车轮虫药物。

关键词： 白屈菜，车轮虫，白屈菜红碱，白屈菜碱，血根碱

1. Introduction

The continuing growth of fish aquaculture worldwide has been accompanied by the increasing importance of parasites as agents of disease. Trichodinids, which are common ectoparasites living on the gills of fish, present the largest group of metazoan fish parasites and are of major importance in the fish pathology (Lyholt and Buchmann, 1995). They can deteriorate the normal health condition of fish and lead to fish mortality by causing gill infestations and inhibiting oxygen exchange across gill lamella. Therefore, the control of this parasite has become an urgent need in fish farming.

Formaldehyde has traditionally been used for the control of trichodinids infection but appears to be inefficient in recent years (Madsen et al., 2000a). Other chemicals, including malachite green, potassium permanganate, acriflavin, bithionol have been e-valuated for chemotherapy of trichodinids (Madsen et al., 2000a). However, the frequent use of these chemical drugs has resulted in serious drawbacks such as development of parasites resistance, environmental contamination, drug residue and pressure to host, which has emerged the need of alternative strategy (Goven et al., 1980; Klingerand Floyd, 2002).

Recently, there have been increased research activities into the utilization of traditional plant-based medicines to control parasitic infections in human and animals (Orhan et al., 2006; Wang et al., 2010). Generally, the plant-based products are more non-re-sistible for frequent use as compared with the chemical drugs. Moreover, the novel products might be potential sources of new antiparasitic drug. Madsen et al. (2000b) reported that raw and squeezed garlic (Allium sativum) at 200 mg · L^{-1} had potential to treat trichodiniasis in eel. EI-Deen et al. (2010) found that the green tea extract (GTE) were active against Trichodina sp.. An alkaloid sanguinarine from Macleaya cordata has been reported to be active against I. multifillis (Yao et al., 2010).

Chelidonium majus L. (Papaveraceae), which is widely distributed in Europe and Asia, is a plant of great interest for its wide use in various diseases in European countries and in Chinese herbal medicines (Pavão and Pinto, 1995; Colombo and Bosisio, 1996). The principal objective of this study was to assess the antiparasitic properties of C. majus and isolate the active constituents responsible for the activity using in vivo antiparasitic assay associated with bioassay - guided silica gel column chromatography isolation. Additionally, the acute toxicities of active compounds from C. majus were evaluated.

2 Materials and methods

2.1 Parasites and hosts

Parabramis pekinensis (mean length±SD: 5.8±0.4 cm), naturally infected with Trichodina sp., were obtained fromaquatic fry farm of Zhejiang Institute Freshwater Fisheries in China and maintained in a 1 m^3 tank with an oxygen content higher than 85% saturation at 23±1℃. Fish were acclimatized under laboratory conditions for 7 days. On the seventh day, 10 fish were randomly selected, killed by spinal severance and examined for the prevalence and intensity of parasite under a light microscope (Olympus BX51, Tokyo, Japan) at 10×4 magnification prior to the experiment.

For acute toxicity tests, parasite-free P. pekinensis were obtained from commercial fish farm and maintained in a 1 m^3 tank supplied with filtered groundwater under the same conditions as parasitized fish. On arrival, the absence of the parasites was carefully checked by examining 10 fish randomly selected.

2.2 Plant material

Chelidonium majus L. were collected in Zhejiang province, China, in March 2010 and identified by Prof. X. L. He in Northwest A&F University, China. A voucher specimen has been deposited in the Herbarium of College of Life Science of the university. They were cleaned and air dried for a week at 35-40℃ and pulverized in electric grinder. The powdered plant samples were stored at-20℃ until further use.

2.3 Activity-guided isolation of active compounds

2.3.1 Selection of extraction solvent

Five powdered samples, each weighing 50.0 g, were respectively extracted under reflux with different solvent (petroleum ether, chloroform, ethyl acetate, ethanol, and water) of increasing polarity at 65℃ for 4 h. The process was performed in three replicates. Each extract was subsequently filtered, and the five filtrates were evaporated under reduced pressure in a vacuum rotary evaporator for the complete removal of solvents. The dried extracts were subjected to in vivo antiparasitic efficacy test. The bioassay showed that-the ethanol extract was the highest in antiparasitic efficacy among all extracts. So, it was then subject to further separation.

2. 3. 2 Extraction and isolation procedure

Air-dried and powdered herbs of C. majus (3. 2 kg) were exhaustively extracted with 50 L ethanol at room temperature by perco-lation, giving 405 g dry extract. The ethanol extract was subjected to column chromatography on a silica gel and sequentially eluted with petroleum ether, ethyl acetate andmethanol with increasing polarity. Repetition of the chromatographic separations and recrystallization led to the isolation of three known compounds (1-3) whose structures are presented in Fig. 1. The structures of these compounds were elucidated by comparing spectroscopic data with those reported as chelidonine (1) (Zhang et al., 1995; Pandey et al., 1977; Swinehart et al., 1980), chelerythrine (2) (Radek et al. 1999; Pavlina et al., 2002), sanguinarine (3) (Perez Gutierrez et al., 2002).

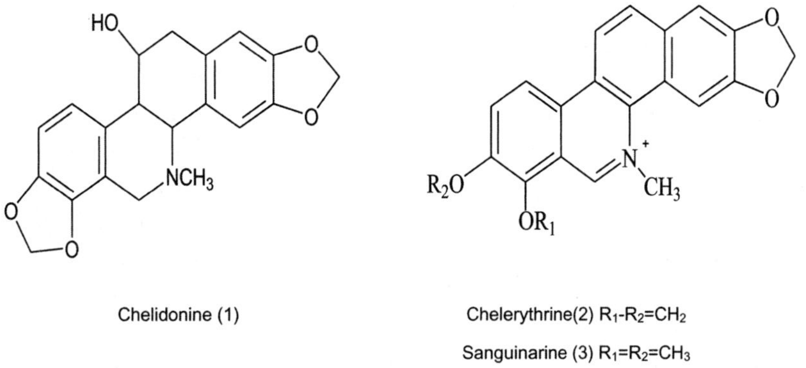

Chelidonine (1)

Chelerythrine(2) R₁-R₂=CH₂

Sanguinarine (3) R₁=R₂=CH₃

Fig. 1 Chemical structures of compounds isolated from chelidonium majus L.

1. 4 In vivo antiparasitic efficacy test

The crude extracts and the pure compounds isolated from C. majus were dissolved in 1 mL of dimethyl sulfoxide (DMSO) to get 0. 5 g · mL^{-1} (sample/solvent) of stocking solutions which were used for the preparations of the desired concentrations for in vivo an-tiparasiticefficacy assay. The highest concentration of DMSO in the treatment was less than 1% (Initial testswith 1% DMSO showed no antiparasitic activity).

In vivo tests were conducted in 40×30×20 cm glass tanks, each containing 10 L of the test solution, and 10 parasitized fish (P. pekinensis were placed on glass slides, and the presence of parasites on the paddle of the fish were determined microscopically, heavy infected fish were transferred into the glass tank immediately, fish infected with little parasites were returned to the tanks.). The test samples were assayed at a different series of concentrations based on initial tests. Negative control groups with no chemical were set up under the same experimental conditions, while formalin was used as the positive control. All treatments and control groups were con-ducted with three replicates.

After 48 h treatment, P. pekinensis in all treatments and control groups were killed by a spinal severance for biopsy. The total number of trichodinids in the collected gills and paddle from each fish was counted. The antiparasitic efficacy of tested samples was de-termined by comparison of the number of parasites in the treatments with those in the negative control groups, and calculated using the following equation described in our previous work (Yao et al., 2010).

$$E = (C - T) \times 100/C \qquad (1)$$

Where E is the antiparasitic efficacy, C is the mean number of Trichodina sp. in the negative control, and T is the treatment groups.

2. 5 Acute toxicity

Acute toxicity of chelidonine, chelerythrine and sanguinarine were carried out according to the method of Wang et al. (2010) . The tests were conducted in plastic pot of 5. 0 L capacity, each containing 2. 0 L of test solution, and 10 healthy P. pekinensis. The water temperature was 15-18℃, the pH ranged from 7. 0 to 7. 5 and the dissolved oxygen was approximately 6. 5-7. 8 mg · L^{-1} . Dilutions were prepared from the stock solutions as the following concentrations: 2. 0, 2. 3, 2. 6, 2. 9, 3. 1, 3. 4, 3. 7 mg · L^{-1}for chelidonine, 2. 5, 3. 0, 3. 5, 4. 0, 4. 5, 5. 0, 5. 5 mg · L^{-1} for chelerythrine, 1. 0, 1. 2, 1. 4, 1. 6, 1. 8, 2. 0, 2. 2 mg · L^{-1} for sanguinarine. The tests were conducted in triplicate, as well as controls (under the same test conditions with no chemicals) . The deaths of fish were recorded when the opercula movement and tail beat stopped and the fish no longer responded to mechanical stimulus. Any observed dead fish was removed from the medium in time to avoid deterioration of the water quality. Mortality of fish was recorded throughout the experiments, during which no food was offered to the fish. Under the circumstances, the experiments were stopped and fish were transferred to freshwater. Fish mortalities in the treatment and control groups were registered after 48 h of exposure.

2.6 Data analysis

The homogeneity of the replicates of the samples was checked by the Mann-Whitney U test. The LC_{50} and their 95% confidence intervals of test samples were calculated by log concentration-probit equation using the SPSS 16.0 probit procedure.

3 Results and discussion

There is growing interest in using plant extracts for the treatment of human and animal diseases, but little information is available on the use of the plant-derived products in the treatment of aquaculture diseases (Ekanem et al., 2004). In the present study, an attempt has therefore been made to exploit the active compounds from C. majus for their antiparasitic activity against Trichodina sp. by bioactivity-guided fractionation. In order to make full evaluation of the aerial parts of C. majus, five different solvents with increasing polarity (i. e., petroleum ether, chloroform, ethyl acetate, ethanol, and water) were applied for the extraction. All the resulted extracts were tested for antiparasitic activity against Trichodina sp.. As the ethanolic extract exhibited the highest activity, it was selected for further purification using various chromatograph techniques. Further fractionation of the ethanol extract led to the isolation of three isoquinoline alkaloids, which were identified as chelidonine, chelerythrine and sanguinarine. The antiparasitic activity of the three alkaloids obtained in this work and ethanolic extract were assayed on Trichodina sp.. As shown in Figs. 2, 3, and 4, chelidonine, chelerythrine and sanguinarine were found to be 100% effective for the elimination of Trichodina sp. at the concentrations of 1.0, 0.8, and 0.7 mg · L^{-1}, with the median effective concentration (EC_{50}) values of 0.6, 0.33, 0.32 mg · L^{-1}, respectively. All of the three isolated compounds showed much better activity than the original ethanolic extract (EC_{50} = 32.5 mg · L^{-1}), indicating that they may or partly responsible for the antiparasitic activity of the ethanolic extract of C. majus. And also, these alkaloids showed extraordinary activity when compared with formalin, which is commonly used for the control of Trichodina sp. in practice (EC_{50} value = 9.8 mg · L^{-1}, Table 1).

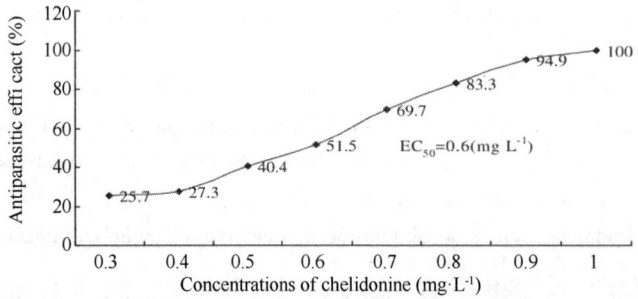

Fig. 2　Antiparasitic efficacy of chelidonine against Trichodina sp. after 48 h of exposure

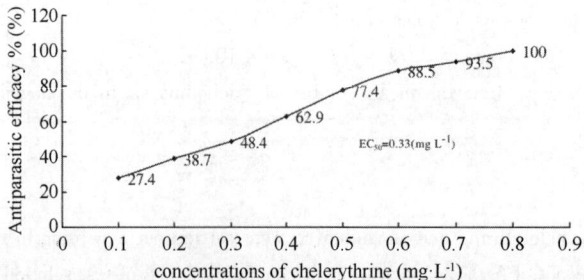

Fig. 3　Antiparasitic efficacy of chelerythrine against Trichodina sp. after 48 h of exposure

Chelidonine, chelerythrine and sanguinarine are quaternary benzo [c] phenanthridine alkaloids (QBAs) and display a wide spectrum of attractive biological activities. They are regularly used in folk medicine as antimicrobial, antifungal, and anti-inflammatory

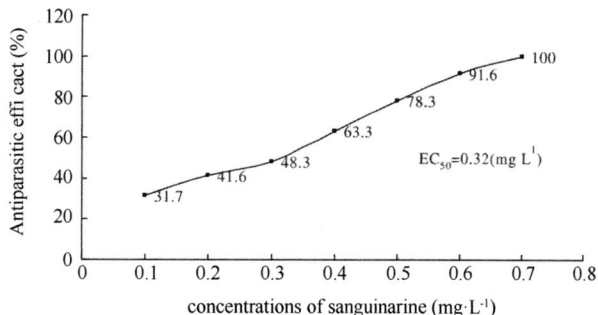

Fig. 4　Antiparasitic efficacy of sanguinarine against Trichodina sp. after 48 h of exposure

agents（Walterova et al. , 1995; Simanek et al. , 2003; Zdarilova et al. , 2006）. Recently, they were found to possess prominent cytotoxic（apoptotic）properties against several cells（Kemeny et al. , 2006; Vrba et al. , 2008, 2009）. The high cytotoxic potency of the three compounds may contribute to the significant antiparasitic activity in the present study. As recent researches indicated, mitochondrion was the major cellular target of isoquinoline alkaloids in its induction of apoptosis（Slaninova et al. , 2007; Kemeny et al. , 2006; Malikova et al. , 2006）. The mechanism of sanguinarine and chelerythrine towards mitochondria may due to induce reactive oxygen species（ROS）production leading to an impairment of plasma membrane integrity and quick development of necrotic processes（Chang et al. , 2007; Kaminskyy et al. , 2008; Yamamoto et al. , 2001）, which may be responsible for the antiparasitic activity of the active alkaloids. However, the exacts mechanism of action regarding their antiparasitic efficacy against Trichodina sp. remains to be further investigated.

The 48 h-LC_{50} values of three active alkaloids for P. pekinensis were presented in Table 1. As shown in Table 1, the LC_{50} values of chelidonine, chelerythrine and sanguinarine were 2. 5, 3. 8 and 1. 5 mg · L^{-1}, respectively. For chelerythrine, the toxic dose is almost 5 times the effective one. These findings ensure the safety for the use of this compound in controlling of Trichodina sp. infection, and suggest that chelerythrine has great potential for the development of a new parasiticide. In the case of chelidonine and sanguinarine, the median EC_{50} were nearly 2 times less than the toxic dose correspondingly, indicating that the two compounds are of high risk, which may limit the application in the control of Trichodina sp. .

Of great interest to note that sanguinarine showed toxicity 2 times more than chelerythrine. From a structural point of view, chelerythrine and sanguinarine molecules differ at positions 2 and 3 of the benzophenanthridine skeleton（Fig. 1）, sanguinarine has methoxy groups（$-OCH_3$）at both these positions, while chelerythrine has a methylendioxy group（$-OCH_2O-$）. It appears that alkaloids with methoxy groups at positions 2 and 3 are more toxic since this has also been observed with sanguirubine and chelirubine molecules, which also differ only at these positions, with the toxicity of sanguirubine, which contains methoxy groups, being greater（Slaninova et al. , 2007）.

Table 1　Antiparasitic activity and acute toxicity of ethanol extract and compounds from c. majus after 48 h of exposure

Test samples	Antiparasitic efficacy		Acute toxicity LC_{50}（mg · L^{-1}）
	EC_{50}（mg · L^{-1}）interval	95% confidence	
Ethanol extract	32. 5	31. 6-33. 4	ND
chelidonine	0. 5	0. 4-0. 6	2. 5
chelerythrine	0. 33	0. 21-0. 38	3. 8
sanguinarine	0. 32	0. 24-0. 43	1. 5
Formalin[a]	9. 8	8. 54-11. 35	ND

ND not determined; [a] Positive control.

EC_{50}: effective concentration 50%; LC_{50}: lethal concentration 50%.

In conclusion, the results obtained in the present study provided a scientific basis for use of the ethanolic extract from C. majus for the treatment of Trichodina sp. in P. pekinensis. This extract along with the active compounds isolated in this work, especially chelerythrine, could be useful for the development of new antiparasitic agents. However, further studies are required for field evaluations in the practical system and structure-activity of these active alkaloids remained to be further elucidated.

Reference：

Chang, M. C., Chan, C. P., Wang, Y. J., Lee, P. H., Chen, L. I., Tsai, Y. L., Lin, B. R., Wang, Y. L., Jeng, J. H., 2007. Induction of necrosis and apoptosis to KB cancer cells by sanguinarine is associated with reactive oxygen species production and mitochondrial membrane depolarization. Toxicol. Appl. Pharmacol. 218, 143−151.

Colombo, M. L., Bosisio, E., 1996. Pharmacological activities of Chelidonium majusL.. Pharm Res. 33, 127−134.

Ekanem, A., Wang, M., Simon, J., Obiekezie, A., Morah, F., 2004. In vivo and in vitro activities of the seed extract of Piper guineense Schum and Thonn against skin and gill monogenean parasites of goldfish (Carassius auratus auratus). Phytother res. 18, 793−797.

El−Deen, A. E. N., 2010. Green tea extract Role in removing the Trichodina sp. on Oreochromis niloticus fry in the Egyptian fish hatcheries. Report and Opinion. 2 (8), 77−81.

Goven, B., Gilbert, J., Gratzek, J., 1980. Apparent drug resistance to the organophosphate dimethyl (2, 2, 2−trichloro−1−hydroxyethyl) phosphonate by monogenetic trematodes. J Wildlife Dis. 16, 343−346.

Kaminskyy, V., Kulachkovskyy, O., Stoik, R., 2008. A decisive role of mitochondria in defining rate and intensity of apoptosis induction by different alkaloids. Toxicol Lett. 177, 168−181.

Kemeny, B. A., Aradi, J., Damjanovich, J., Beck, Z., Facsko, A., Berta, A., Bodnar, A., 2006. Apoptotic response of uveal melanoma cells upon treatment with chelidonine, sanguinarine and chelerythrine. Cancer Lett. 237, 67−75.

Klinger, R., Floyd, R. F., 2002. Introduction to freshwater fish parasites. Document CIR716. Institute of Food and Agricultural Science, University of Florida.

Lyholt, H. C. K., Buchmann, K., 1995. Infestations with the skin parasite Trichodina jadranica Raabe, 1958 (Ciliophira: Trichodinidac) in Danish ell farms. Bull. Scand. Soc. Parasit. 5 (2), 97.

Malikova, J., Zdarilova, A., Hlobilkova, A., Ulrichova, J., 2006. The effect of chelerythrine on cell growth, apoptosis, and cell cycle in human normal and cancer cells in comparison with sanguinarine. Cell Biol and Toxicol. 22, 439−453.

Madsen, H. C. K., Buchmann, K., Mellergaard, S., 2000a. Treatment of trichodiniasis in eel (Anguilla anguilla) reared in circulation. Aquacult. 221−231.

Madsen, H. C. K., Buchmann, K., Mellergaard, S., 2000b. Association between trichodiniasis in eel (Anguilla anguilla) and water quality in recirculation. Aquacult. 187, 275−281.

Orhan, I., Aslan, M., Sener, B., Kaiser, M., Tasdemir, D., 2006. In vitro antiprotozoal activity of the lipophilic extracts of different parts of Turkish Pistacia Vera L. Phytomedicine. 13, 735−739.

Pandey, V., Ray, A. B., Dasgupta, G., 1977. Minor alkaloids of Fumaria indica seeds. Phytochemistry. 18, 695.

Pavlina, S., Radek, M., Jiri, D., Roger, D., Eddy, L. E., 2002. Stuctural Studies of Benzophenanthridine Alkaloid free Bases by NMR Spectroscopy. Magn Reson Chem. 40, 147−152.

Pavao, M. L., Pinto, R. E., 1995. Densitometric assays for the evaluation of water soluble alkaloids from Chelidonium majus L. (Papaveraceae) roots in the Azores, along one year cycle. Arquipélago, Sér. Ciências Biol. Marinhas. 13, 89−91.

Radek, M., Jaromir, T., Jiri, D., Jiri, S., Roger, D., Vladimir, S., 1999. ^{1}H and ^{13}C NMR Study of Quaternary Benzo [c] phenanthridine Alkaloids. Magn Reson Chem. 37, 781−787.

Perez Gutierrez, R. M., Vargas Solis, R., Diaz Gutierrez, G., MartinezMartinez, F. J., 2002. Identification of benzophenanthridine alkaloids from Bocconia arborea by gas chromatography−mass spectrometry. Phytochem. Anal. 13, 177−180.

Simanek, V., Vespalec, R., Sedo, A., Ulrichova, J., Vicar, J., 2003. Quaternarbenzo (c) phenanthridine alkaloids−biological activities. In: Schneider, M. P (Ed.), Chemical Probes in Biology. Kluwer Academic Publishers, Netherlands, pp245−254.

Slaninova, I., Slunska, Z., Sinkora, J., Vlkova, M., Taborska, E., Screening of minor benzo (c) phenanthridine alkaloids for antiproliferative and apoptotic activities. Pharm Biol. 2007. 45, 131−139.

Swinehart, J. A., Stermitz, F. R., 1980. Bishordeninyl terpene alkaloids and other constituents of Zanthoxylum culantrillo and Z. coriaceum. Phytochemistry. 19, 1219.

Vrba, J., Petr, D., Jaroslav, V., Martin, M., Ulrichová, J., 2008. Chelerythrine and dihydrochelerythrine induce G1 phase arrest and bimodal cell death in human leukemia HL−60 cells. Toxicol in Vitro. 22, 1008−1017.

Vrba, J., Petr D., Jaroslav, V., Ulrichová, J., 2009. Cytotoxic activity of sanguinarine and dihydrosanguinarine in human promyelocytic leukemia HL−60 cells. Toxicol in Vitro. 23, 580−588.

Walterova, D., Ulrichova, J., Valka, I., Vicar, J., Vavreckova, C., Taborska, E., Harkrader, R. J., Meyer, D. L., Cerna, H., Simanek, V., 1995. Benzo (c) phenanthridine alkaloids sanguinarine and chelerythrine: biological activities and dental care applications. Acta University Palacki Olomouc Faculty of Medicine. 139, 7−14.

Wang, G. X., Han, J., Zhao, L. W., Liu, L. T., Jiang, D. X., Liu, X. L., 2010. Anthelmintic activity of steroidal saponins from Paris polyphylla. Phytomedicine. 17, 1102−1105.

Yamamoto, S., Seta, K., Morisco, C., Vatner, S. F., Sadoshima, J., 2001. Chelerythrine rapidly induces apoptosis through generation of reactive oxygen species in cardiac myocytes. J. Mol. Cell Cardiol. 33, 1829−1848.

Yao, J. Y. , Shen, J. Y. , Li, X. L. , Xu, Y. , Hao, G. J. , Pan, X. Y. , Wang, G. X. , Yin, W. L. , 2010. Effect of sanguinarine from the leaves of Macleaya cordataagainst Ichthyophthirius multifiliis in grass carp (Ctenopharyngodon idella) . Parasitol Res. 107, 1035-1042.

Zdarilova, A. , Malikova, J. , Dvorak, Z. , Ulrichova, J. , Simanek, V. , 2006. Quaternary isoquinoline alkaloids sanguinarine and chelerythrine. In vitro and in vivo effects. Chemicke Listy. 100, 30-41.

Zhang, G. L. , Rucker, G. , Breitmaier, E. , Mayer, R. , 1995. Alkaloids from Hypecoum leptocarpum Phytochemistry. 40, 1813.

该文发表于《Veterinary Parasitology》【2011, 183: 8-3】。

Antiparasitic efficacy of dihydrosanguinarine and dihydrochelerythrine from Macleaya Microcarpa against *Ichthyophthirius multifiliis* inrichadsin (*Squaliobarbus curriculus*)

Jia-yun YAO[1], Zhi-ming ZHOU[1], Xi-lian LI[2], Xiao-yi PAN[1], Gui-jie HAO[1], Yang XU[1], Hong-shun RU[1], Wen-lin YIN[1], Jin-yu SHEN[1]*

(1. Institute of Freshwater Fisheries of Zhejiang, Huzhou, Zhejiang, 313001, China
2. College of Animal Science and Technology, Northwest A&F University, Yangling, 712100, China)

Abstract: Ichthyophthirius multifiliis is a holotrichous protozoan that invades the gills and skin surfaces of fish and can cause morbidity and high mortality in most species of freshwater fish worldwide. The present study was undertaken to investigate the antiparasitic activity of crude extracts and pure compounds from the leaves of Macleaya Microcarpa. The chloroform extract showed a promising antiparasitic activity against I. multifiliis. Based on these finding, the chloroform extract was fractionated on silica gel column chromatography in a bioactivity-guided isolation affording two compounds showing potent activity. The structures of the two compounds were elucidated as dihydrosanguinarine (1) and dihydrochelerythrine (2) by hydrogen, carbon-13 nuclear magnetic resonance spectrum and electron ionization mass spectrometry. The in vivo tests revealed that dihydrosanguinarine and dihydrochelerythrine were significantly effective against I. multifiliis with LC_{50} values of 5. 18 and 9. 43 mg/L, respectively. The acute toxicities (LC_{50}) of dihydrosanguinarine and dihydrochelerythrine for richadsin were 13. 299 and 18. 231 mg/L, respectively. The overall results provided important information for the potential application of dihydrosanguinarine and dihydrochelerythrine in the therapy of serious infection caused by I. multifiliis.

Keyword: Ichthyophthirius multifiliis; Dihydrosanguinarine; Dihydrochelerythrine; Antiparasitic;

小果博落回杀小瓜虫活性成分的研究

姚嘉赟[1], 周志明[1], 李喜莲[2], 潘晓艺[1], 郝贵杰[1], 徐洋[1], 茹洪顺[1], 尹文林[1], 沈锦玉[1]*

(1. 浙江省淡水水产研究所, 浙江 湖州 313001; 2. 西北农林科技大学, 陕西 杨凌 712100)

摘要: 采用常压硅胶柱层析和 RP-HPLC 制备并结合体外杀虫活性追踪试验, 对小果博落回杀小瓜虫的活性成分进行追踪, 并分离获得两种具有较强杀虫活性的单体化合物, 波谱 (IR、MS、^{1}H NMR 和 ^{13}C NMR) 分析结果显示这两种活性物质分别为二氢白屈菜红碱和二氢血根碱。体内杀虫活性试验结果表明: 二氢血根碱和二氢白屈菜红碱对小瓜虫的半数致死浓度 (LC_{50}) 分别为 5. 18 和 9. 43 mg/L, 急性毒性试验结果表明, 两者对草鱼的半数致死浓度分别为: 13. 3 和 18. 2 mg/L。上述研究结果表明二氢血根碱和二氢白屈菜红碱均具有较好的杀虫效果, 可用于开发杀虫药物。

关键词: 小瓜虫; 二氢血根碱, 二氢白屈菜红碱, 抗寄生虫

1　Introduction

The ciliate Ichthyophthirius multifiliis is one of the most pathogenic parasites of fish, and it constitutes a major disease problem in aquaculture (Bisharyan et al 2003; Dickerson and Dawe, 1995; McCallum, 1985, 1986) . The morbidity rate due to this disease may reach up to 100%, causing great economic losses in aquaculture and ornamental fish breeding (Osman et al., 2009).

To cope with the parasitism and deleterious consequences, various parasiticides have been used in aquaculture. Formerly, ichthyophthiriasis was treated effectively with malachite green, a compound that has been banned on fish farms due to its confirmed carcinogenicity in most countries (Alderman, 1985; Wahli et al., 1993). Since the ban, formalin (Stoskopf, 1993; Rowland, 2009), copper sulfate (Straus, 2009; Rowland, 2009), permanganate (Straus and Griffin, 2001; Buchmann et al., 2003), hydrogen peroxide (Lahnsteiner and Weismann, 2007), potassium sodium percarbonate (Heinecke and Buchmann, 2009) have been used with varying levels of success. However, the treatments commonly used in commercial aquaculture as alternatives to malachite green are of low efficacy, cause environmental problems, or are unlikely to receive regulatory approval (Dickerson and Dawe 1995; Tieman and Goodwin 2001).

Therefore, an effective therapy has to be improved and new drugs with novel mechanisms of action are thus urgently required. Recently, there have been increased research activities into the utilization of traditional plant-based medicines to control I. multifiliis infection. Plant extracts from garlic (Buchmann et al., 2003), Mucuna pruriens and Carica papaya (Ekanem et al., 2004) have been assessed, showing some potential for killing the free-swimming stages of I. multifiliis. However, most of the compounds responsible for such activities have not been isolated and their structures determined so as to evaluate their potential for the development of novel antiparasitic drugs.

In our previous study, we investigated the anti-I. multifiliis efficacy of Macleaya cordata, a specie in the genus of Macleaya, and an active compound, sanguinarine, was isolated by using the strategy of bioactivity-guided isolation (Yao et al., 2010). It lead us to consider if M. microcarpa, as another specie of this genus, widely distributed in northwest and southwest of China, is also capable of controlling the parasite.

The principal objective of this study was to assess the antiparasitic properties of M. microcarpaand isolate active constituents responsible for the activity using in vitro antiparasitic assay associated with bioassay-guided fractionation. Additionally, the acute toxicity of the active compounds against richadsin of was evaluated.

2　Materials and methods

2.1　Fish

Richadsin (mean length: 6.3±0.3 cm), naturally infected with I. multifiliis, were obtained from aquatic fry farm of Zhejiang Institute Freshwater Fisheries in China and maintained in a 1 m^3 tank at 24±1℃ (controlled by automatic aquarium heater) with aeration for 7 days. On the seventh day, ten richadsin were randomly sampled, killed by spinal severance, and eight gill filaments of each fish were biopsied to determine the I. multifiliis infestation level and intensity under a light microscope (Olympus BX51, Tokyo, Japan) at 10×4 magnification before they were used for the assays.

For acute toxicity tests, parasite-free richadsin were obtained from commercial fish farm and maintained in a 1 m^3 tank supplied with filtered groundwater under the same conditions as parasitized fish. On arrival, the absence of the parasites was carefully checked by examining ten fish randomly selected.

2.2　Preparation of crude extracts, fractions and pure compounds

The crude extracts, fractions and the pure compounds isolated from M. microcarpa were dissolved in 10 mL of dimethyl sulfoxide (DMSO) to get 1.0 g/mL (sample/solvent) of stocking solutions which were used for the preparations of the desired concentrations for in vitro tests and in vivo test.

2.3　In vitro tests

Gill arches with mature I. multifiliis were cut into a 50 mL beaker containing 20 ml aerated groundwater to allow the trophonts to dislodge. About 30 min later, active moving trophonts were removed with the help of a pipette.

In vitro tests were performed as described in our previous work (Yao et al., 2010). Briefly, tests were conducted in each well of a 24-well tissue culture plate (Becton Dickinson Labware, NJ, USA), filled with 2 mL aerated groundwater, each containing the test samples and approximately 20 mature trophonts. Antiparasitic efficacy was determined by microscopic examination of each well at 1, 2, 3, and 4 h after treatment. The mortality of trophonts in groups with different test sample concentrations was calculated. Immobilized and lysed trophonts cells were considered dead. A negative control was included using aerated groundwater containing the same amount of DMSO as the test sample. All the tests were performed in duplicate.

2.4　Bioactivity-guide isolation

The chromatographic separation was monitored by the strategy of bioactivity-guided isolation (invitro tests guided), only the extracts or fractions showed strong activity against I. multifiliis were subjected to further separation and purification.

2.5　Preparation of extracts

Dry and powdered leaves of M. microcarpa (10 kg) were extracted with 95% ethanol (25 L × 3 times) at room temperature for 24 h. The extract was evaporated to dryness under reduced pressure in a rotary evaporator to yield the ethanol extract (EE, 1.36 kg). Part of the ethanol extract was reserved for activity assays whilst the rest of the extract was suspended in water and extracted with n-butanol. The yield of n-butanol extract (BE) and aqueous extract (AE) was 654.5 g and 126.6 g, respectively. The n-butanol extract was redissolved in chloroform/water (1:1, v/v) mixture, the pH value of the mixture was adjusted to 10-11 with 1% NaOH (m:v), then, extracted with chloroform for three times to give 256.0 g chloroform extract (CE) and 152.3 g remaining aqueous extract (RAE).

The dried extracts (BE, AE, CE, RAE) were dissolved in 10.0 mL DMSO to prepare at a final concentration of' 0.5 g/mL for in vitro tests. The bioassay showed that chloroform extract was the highest in antiparasitic efficacy among all extracts. So, it was then subject to further separation.

Fractionation and isolation of pure compounds

The chloroform extract (245.0 g) was subjected to open column chromatography on normal phase silica gel and eluted with a solvent mixture of petroleum ether/ethyl acetate (2:1, v/v) and finally eluted with methanol affording 6 major fractions (Fr.A: 1~115 fractions, Fr.B: 116~298 fractions, Fr.C: 299~365 fractions, Fr.D: 366~478 fractions, Fr.E: 479~535 fractions, Fr.F: 536~656 fractions). These six fractions were submitted to in vitro test, and Fr.B was the most active. Fr.B (22.5 g) was subjected to column chromatograph and successively eluted with petroleum ether/ethyl acetate gradients. Repetition of the chromatographic separations and recrystallization led to the isolation of two active compounds: Compound 1 (221.0 mg) and Compound 2 (144.0 mg).

2.6　Identification of active compounds

Based on the physico-chemical properties and electron ionization mass spectrometry (EI-MS, 70 eV VG-ZAB-HS, VG Company, England), nuclear magnetic resonance hydrogen spectrum (^{1}H-NMR) and nuclear magnetic resonance carbon spectrum (^{13}C-NMR) (Bruker Avance 400 spectrometer with SiMe$_4$ as internal standard, USA), the chemical structures of active compounds were identified.

2.7　In vivo test

The antiparasitic efficacies of the pure compounds were determined following the protocol described in a previous work (YAO et al., 2010). Briefly, the assays were conducted in 80 cm×60 cm×40 cm glass tanks, each containing 50 L of the test solution, and 10 parasitized fish. Initial tests were undertaken to determine the concentration boundaries for the efficacy tests, and the final tests were conducted with three replicates. The test samples were assayed at a different series of concentrations which were prepared from the stock solutions. Control groups with no chemical were set under the same experimental conditions as the test groups.

After 48 h treatment, richadsin in all treatments and control groups were killed by a spinal severance for biopsy. The lamella branchialis were placed on glass slides and the numbers of parasites on the gills were counted under a light microscope at 4×10 magnification to calculate the mean number of parasites per infected fish. The antiparasitic efficacy of tested samples was determined by comparison of the number of parasites in the treatments with those in the control groups, and calculated using the following equation described in our previous work (YAO et al., 2010).

$$E = (C - T) \times 100/C \tag{1}$$

Where E is the antiparasitic efficacy, C is the mean number of I. multifiliis on the gillsin the negative control, and T is the treatment groups.

2.8　Acute toxicity

Acute toxicities of active compounds were performed using a standard lab protocol (Wang et al., 2008). Each sample was assayed at a different series of concentrations ranged between 10.0-18.0 mg/L for compound 1, and 12.0-28.0 mg/L for compound 2 (based on initial tests). The fish were carefully observed for any signs of distress indicative of toxic insult such as increased respiration frequency and erratic behavior. Under there circumstances, the experiments were stopped and fish were transferred to freshwater. Mortality response was noted after 48 h of exposure, during which no food was offered to the fish.

2.9 Data analysis

The homogeneity of the replicates of the samples was checked by the Mann-Whitney U test. The LC_{50}, LC_{90} values and their 95% confidence intervals of test samples were calculated by log concentration-probit equation using the SPSS 16.0 probit procedure.

3 Results

3.1 In vitro antiparasitic efficacy of compounds

Two compounds (crystals) were separated from Fr. B by bioactivity-guide isolation and were then subjected to in vitro test to determine the active concertrations. The results showed that compound 1 was 100% effective against I. multifiliis, with LC_{50} and LC_{90} values of 4.14 and 6.29 mg/L respectively, at concentrations of 7.0 mg/L after 4 h of exposure (Fig. 1). Compound 2 was 100% effective against I. multifiliis, with LC_{50} and LC_{90} values of 7.99 and 9.25 mg/L respectively, at concentrations of 10.0 mg/L after 4 h of exposure (Fig. 2).

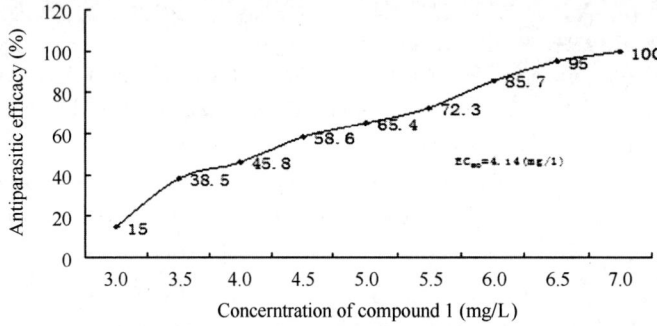

Fig. 1 Antiparasitic efficacy of compound 1 against I. multifiliisafter 4 h of exposure in vitro.

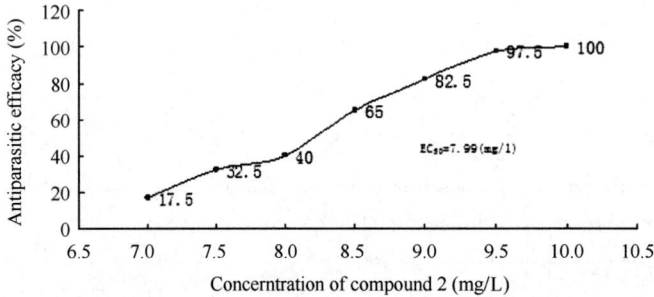

Fig. 2 Antiparasitic efficacy of compound 2 against I. multifiliisafter 4 h of exposure in vitro.

A noticeable effect on I. multifiliis trophonts was observed in treatment groups: ciliary movements stopped completely after starting the incubation at 6 mg/L or more compound 1 (9 mg/L or more compound 2), and the macronucleus was invisible, cilia could not be recognized by 4 h. In contrast, trophontsin the control groups were not affected during the observations in all trials and no abnormal morphology or loss of motility was seen after 4 h exposure.

3.2 Structure identification of active compounds

The chemical structure confirmations of active compounds from M. microcarpa were accomplished by comparing m p, MS, ^{1}H NMR and ^{13}C NMR data obtained to those published.

Compound 1 was obtained as colorless crystal, m p: 189-191℃. EI-MS m/z: 334 [M$^+$+1], 333 [M$^+$], 332 [M$^+$-1], 317 [M$^+$-H-CH3], 274; 1H-NMR (CDCl3) δ 7.10 (1H, s, H-1), 7.68 (1H, s, H-4), 4.18 (2H, s, H-6), 6.85 (1H, d, J=8.4 Hz, H-9), 7.32 (1H, d, J=8.4 Hz, H-10), 7.70 (1H, d, J=8.8 Hz, H-11), 7.48 (1H, d, J=8.8 Hz, H-12), 6.05 (2H, s, -OCH2O-2, 3), 6.04 (2H, s, -OCH2O-7, 8), 2.66 (3H, s, N-Me); ^{13}C-NMR

（CDCl3）δ 104.34（C-1），148.9（C-2），147.64（C-3），100.56（C-4），125.64（C-4a），142.42（C-4b），46.61（C-6），115.90（C-6a），144.31（C-7），147.21（C-8），107.60（C-9），116.51（C-10），127.39（C-10a），124.13（C-10b），119.98（C-11），123.46（C-12），131.04（C-12a），101.06（-OCH2O-2，3），101.52（-OCH2O-7，8），43.00（N-Me）；agree well with the data reported（Robert et al.，1993）. So the compound was identified as dihydrosanguinarine. Its molecular formula is $C_{20}H_{15}NO_4$（Robert et al.，1993），and its structure is shown in Fig.3.

Compound 2 was obtained aspale yellow crystalline powder；mp：163~165℃. EI-MS m/z：349［M⁺］，348［M⁺-1］，333［M⁺-H-CH₃］，318［M⁺-OMe］；¹H-NMR（CDCl₃）δ 7.10（1H，s，H-1），7.68（1H，s，H-4），4.30（2H，s，H-6），6.95（1H，d，J=8.4 Hz，H-9），7.53（1H，d，J=8.4Hz，H-10），7.72（1H，d，J=8.4Hz，H-11），7.48（1H，d，J=8.4Hz，H-12），6.09（2H，s，-OCH₂O-2，3），2.60（3H，s，N-Me），3.92（3H，s，OMe-7），3.96（3H，s，OMe-8）；¹³C-NMR（CDCl₃）δ 104.33（C-1），148.19（C-2），147.59（C-3），100.58（C-4），128.08（C-4a），142.63（C-4b），46.85（C-6），127.27（C-6a），145.58（C-7），152.15（C-8），111.63（C-9），118.78（C-10），124.76（C-10a），123.94（C-10b），119.75（C-11），123.29（C-12），131.07（C-12a），101.02（-OCH₂O-2，3），42.78（N-Me），60.94（OMe-7），55.79（OMe-8）；agree well with the data reported（Qin et al.，2004）. So the compound was identified as dihydrochelerythrine. Its molecular formula is $C_{21}H_{19}NO_4$（Qin et al.，2004），and its structure is shown in Fig.3.

Compound 1　　　　　　　　　　　　　　Compound 2

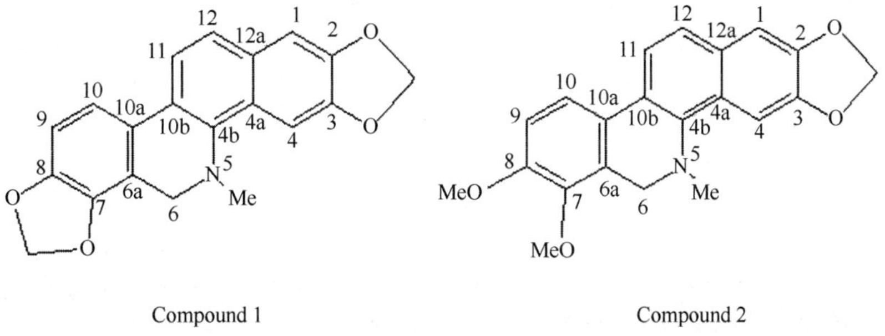

Fig.3　The chemical structures of dihydrosanguinarine（compound 1）and dihydrochelerythrine（compound 2）from M. microcarpa

3.3　In vivo test

The antiparasitic efficacies of dihydrosanguinarine and dihydrochelerythrine were depicted in Table 1. In vivo tests showed that both dihydrosanguinarine and dihydrochelerythrine were effective against I. multiffiliis with LC_{50} values of 5.18 and 9.43 mg/L, after 48 h of exposure, respectively. The solvent（DMSO）acted as a control showed no antiparasitic activity when treated at the highest concentration.

3.4　Acute toxicity of dihydrosanguinarine and dihydrochelerythrine

The LC_{50} values of dihydrosanguinarine and dihydrochelerythrine were determined from linear（$y = m + bx$）plots of the probit curves which were presented in Table 1. The probit mortality percentage of the richadsin was plotted against log-concentrations.

For dihydrosanguinarine, the linear equation $y = -15.139 + 13.471x$ which was derived from the regression analysis of probit mortality of richadsin in test solution bioassay. The calculated LC_{50} was 13.299 mg/L with the 95% confidence interval of 12.868-13.721 mg/L. For dihydrochelerythrine, the linear equation $y = -11.067 + 8.778x$ which was derived from the regression analysis of probit mortality with a LC_{50} of 18.231 mg/L and 95% confidence interval of 17.331-19.107 mg/L. There was no fish mortality occurred in the control groups during the experiments.

Table 1　Antiparasitic activity and acute toxicity of dihydrosanguinarine and dihydrochelerythrine after 48 h of exposure

Test samples	Antiparasitic efficacy		Acute toxicity
	LC_{50}（mg/L）	95%	LC_{50}（mg/L）
dihydrosanguinarine	5.18	4.969-5.375	13.299
dihydrochelerythrine	9.43	9.288-9.572	18.231

651

4　Discussion

Currently, most countries have no effective chemical treatments that can be used to control I. multifiliis (Eliška et al, 2010). More and more efforts have been spent on searching for more effective anti-I. multifiliis drugs. Medicinal plants are chemically complex and diverse. Botanical compounds provide a wide spectrum of biological and pharmacological properties (Patwardhan and Gautam, 2005). In the present study, two anti-I. multifiliis compounds were separated from M. microcarpa by bioactivity-guided isolation and identified as dihydrosanguinarine and dihydrochelerythrine. The invivo tests revealed that dihydrosanguinarine and dihydrochelerythrine were significantly effective against I. multifiliis with LC_{50} values of 5. 18 and 9. 43 mg/L, respectively. To our best knowledge, this study is the first report of the antiparasitic efficacy of dihydrosanguinarine and dihydrochelerythrine in fish.

Dihydrosanguinarine, naturally occuring in a various plant belonging to the family poppy, is a representative quaternary benzo [c] phenanthridine alkaloid, shows antimicrobial activity (Navarro and Delgado, 1999) and cytotoxicity against several cancer cell lines (Chen et al. , 1999). It was also cytotoxic to human promyelocytic leukemia HL-60 cells (Vrba et al. , 2009). Another active compound, dihydrochelerythrine is often found as major components in many traditional medicines. Various biological activities have been observed for it, including antimicrobial (Navarro and Delgado, 1999), antifungal (Meng et al. , 2009), anti-inflammatory activities (Hu et al. , 2006). Moreover, Antileishmanial (Fotie et al. , 2007) and anti-hepatitis B virus activities of dihydrochelerythrine were demonstrated (Wu et al. , 2007). Recently, cytotoxicity of dihydrochelerythrine to human promyelocytic leukemia HL-60 cells (Vrba et al. , 2008) and African green monkey kidney Vero cells (Fotie et al. , 2007) were reported. The high cytotoxic potency of the two alkaloids may responsible for the significant antiparasitic activity and high toxicity for richadsin regarding the two compounds (dihydrosanguinarine and dihydrochelerythrine) in the present study. As recent researches indicated, mitochondria was the major cellular target of benzo [c] phenanthridine alkaloid in its induction of apoptosis in human promyelocytic leukemia HL-60 cells and that benzo [c] phenanthridine alkaloid exerted its cytotoxicity through multiple apoptosis-inducing pathways (Vrba et al. , 2008, 2009). As is well known, mitochondria are the key players that control and regulate apoptosis; mitochondrial dysfunction produced by toxic stress may lead to both apoptotic and necrotic cell death (Lemasters et al. , 1999). A direct action on mitochondria might be involved in the eradication of the parasites. This may be the main mechanism responsible for the anti-I. multifiliis activity of benzo [c] phenanthridine alkaloids. However, the detailed mechanism of action regarding the antiparasitic activity of the two alkaloids should be further addressed.

Traditionally in drug discovery, scientists rely heavily on in vivo test in selecting the lead candidates. This involves various in-life studies that are expensive and time consuming (Chu et al. , 2002). In vitro testing has several advantages over in vivo testing, such as detailed mechanistic understanding, reduction, refinement, and replacement of animal experiments (ankveld et al. , 2010)。Moreover, the content of effective compounds contained in plants is often low, the in vitro test can be completed on a smaller scale (tests were completed in 24-well tissue plate in the present study) as compared to in vivo treatment, and so, small volumes of the active compound can meet the need of the experiment. So, it is suitabe for high throughput screeningin the early stage of the drug discovery process (Jarry et al. , 2006). In the present study, the strategy of in vitro bioactivity-guided isolation monitoring the chromatographic separation was used. By bioactivity-guided isolation, two active compounds were isolated from M. microcarpa. In vivo data showed that the two compounds had the same antiparasitic efficacy against I. multifiliis as the in vitro test had showed. These results imply that the in vitro bioactivity-guided isolation established in the present study can be useful as a monitor system for high-throughput drug candidate screening against I. multifiliis during the discovery phase, and it may become the modle of the new anti-I. multifiliis drug findings.

In conclusion, using the strategy of bioactivity-guided isolation monitoring the chromatographic separation, two compounds showing promising anti-I. multifiliis activity were obtained from M. microcarpa. Both compounds can be chosen as lead compounds for the development of new antiparasitic agents against I. multifiliis. Also, further studies are required for field evaluations in the practical system and the mechanism of the antiparasitic activity remains to be performed.

Reference：

A lderman, D. J. (1985). Malachite green: a review. Journal of fish diseases 8, 289-298.

Bisharyan, Y. , Chen, Q. , Hossain, M. M. , Papoyan, A. , Clark, T. G. (2003). Cadimun effect on Ichthyophthirius: evidence for metal-sequestration in fish tissues following administration of recombinant vaccines. Parasitology 126, 87-93.

Buchmann, K. , Jensen, P. B. , Kruse, K. D. (2003). Effects of sodium percarbonate and garlic extract on Ichthyophthirius multifiliis theronts and tomocysts: in vitro experiments. North American Journal of Aquaculture 65, 21-24.

Chen, J. J. , Duh, C. Y. , Chen, I. S. (1999). New tetrahydroprotoberberine N-oxide alkaloids and cytotoxic constituents of Corydalis tashiroi. Planta Medica 65, 643−647.

Chu, Cliff Chen, A. D. Soares, Chin−Chung Lin. (2002). The use of in vitro metabolic stability for rapid selection of compounds in early discovery based on their expected hepatic extraction ratios. Pharmaceutical Research 19 (11), 1606−1610.

Ekanem, A. P., Obiekezie, A., Kloas, W., Knopf, K. (2004). Effects of crude extracts of Mucuna pruriens (Fabaceae) and Carica papaya (Caricaceae) against the protozoan fish parasite Ichthyophthirius multifiliis. Parasitology Research 92, 361−366.

Eliška Sudová, David, L. S., Wienke, A., Meinelt, T. (2010). Evaluation of continuous 4−day exposure to peracetic acid as a treatment for Ichthyophthirius multifiliis. Parasitology Research 106, 539−542.

Fotie, J., Bohle, D. S., Olivie, M., Gomez, M. A., Nzimiro, S. (2007). Trypanocidal and Antileishmanial Dihydrochelerythrine Derivatives from Garcinia lucida. The Journal of Natural Products 70, 1650−1653.

Hu, J., Zhang, W. D., Liu, R. H., Zhang, C., Shen, Y. H., Li, H. L., Liang, M. J., Xu, X. K. (2006). Benzophenanthridine alkaloids from Zanthoxylum nitidum (Roxb) DC, and their analgesic and anti−inflammatory activities. Chemistry & Biodiversity 3, 990−995.

Heinecke, R. D., Buchmann, K. (2009). Control of Ichthyophthirius multifiliis using a combination of water filtration and sodium percarbonate: Dose−response studies. Aquaculture 288, 32−35.

Jarry, H., Spengler, B., Wuttke, W., Christoffel, V. (2006). In vitro assays for bioactivity−guided isolation of endocrine active compounds in Vitex agnus−castus. Maturitas 55, 26−36.

Lahnsteiner, F., Weismann, T. (2007). Treatment of Ichthyophthiriasis in Rainbow Trout and Common Carp with Common and Alternative Therapeutics. Journal of aquatic animal health 19, 186−194.

Lankveld D. P., Van Loveren H., Baken K. A., Vandebriel R. J. (2010). In Vitro Testing for Direct Immunotoxicity: State of the Art. Methods in molecular biology 598, 401−23.

Lemasters. J. J, Qian, T., Bradham, C. A., Brenner, D. A., Cascio, W. E., Trost, L. C., Nishimura, Y., Nieminen, A. L., Herman, B. (1999). Mitochondrial dysfunction in the pathogenesis of necrotic and apoptotic cell death. Journal of Bioenergetics and Biomembranes 31, 305−319.

Mccallum, H. I. (1985). Population effects of parasite survival of host death: experimental studies of the interaction of Ichthyophthirius multifiliis and its fish host. Parasitology 90, 529−547.

Mccallum, H. I. (1986). Acquired resistance of black mollies Poecilia latipinna to infection by Ichthyophthirius multifiliis. Parasitology 93, 251−261.

Meng, F. Y., Zuo, G. Y., Hao, X. Y., Wang, G. C., Xiao, H. T., Zhang, J. Q., Gui, L. X. (2009). Antifungal activity of the benzo [c] phenanthridine alkaloids from Chelidonium majusLinn against resistant clinical yeast isolates. Journal of Ethnopharmacology 125, 494−496.

Navarro, V., Delgado, G. (1999). Two antimicrobial alkaloids from Bocconia arborea. Journal of Ethnopharmacology 66, 223−226.

Osman H. A. M., El−Bana L. F., Noor El Deen A. E., Abd El−Hady O. K. (2009). Investigations on White Spots Disease (Ichthyophthriasis) in Catfish (Clarias gariepinus) with Special Reference to the Immune Response. Global Veterinaria 3 (2), 113−119,

Patwardhan, B., Gautam, M. (2005). Botanical immunodrugs: scope and opportunities. Drug Discovery Today 10, 495−502.

Qin, H. L., Wang, P., Li ZH. H., Liu, X., He W. Y. (2004). The establishment of the control substance and1H nuclear magnetic resonance fingerprint of Macleayamicrocarpa (Maxim.) Fedde. Chinese Journal of Analytical Chemistry 32 (9), 1165−1170 (in Chinese).

Robert, Williams, D. (1993). Alkaloids from agrobacterium cultures. Phytochemistry 32 (3), 719−729

Rowland, S. J., Mifsud, C., Nixon, M., Read, P., Landos, M. (2009) Use of formalin and copper to control ichthyophthiriosis in the Australian freshwater fish silver perch (Bidyanus bidyanus Mitchell). Aquaculture research 40, 44−54.

Stoskopf, M. K. (1993). Fish medicine. Saunders, London.

Straus, D. L., Griffin BR. (2001). Prevention of an initial infestation of Ichthyophthirius multifiliis in channel catfish and bluc tilapia by potassium permanganate treatment. North American Journal of Aquaculture 63, 11−16.

Straus, D. L., Hossain, M. M., Clark, T. G. (2009). Strain Differences in Ichthyophthirius multifiliis to Copper Toxicity. Diseases of aquatic organisms 83, 31−36.

Tieman, D. M., Goodwin, A. E. (2001). Treatments for Ich infestations in channel catfish evaluated under static and flow−through water conditions. North American Journal of Aquaculture 63, 293−299.

Vrba Ji í, Petr Doležel, Jaroslav Vi ar, Martin Modriansk, Jitka Ulrichová. (2008). Chelerythrine and dihydrochelerythrine induce G1 phase arrest and bimodal cell death in human leukemia HL−60 cells. Toxicology in Vitro 22, 1008−1017.

Vrba Ji í, Petr Doležel, Jaroslav Vi ar, Jitka Ulrichová. (2009). Cytotoxic activity of sanguinarine and dihydrosanguinarine in human promyelocytic leukemia HL−60 cells. Toxicology in Vitro 23, 580−588.

Wahli, T., Schmitt, M., Meier, W. (1993). Evaluation of alternatives to malachite green oxalate as a therapeutant for ichthyophthiriosis in rainbow−trout, Oncorhynchus mykiss. Journal of applied ichthyology 9, 237−249.

Wang, G. X., Zhou, Z., Cheng, C., Yao, J. Y., Yang, Z. W. (2008). Osthol and isopimpinellin from Fructus cnidii for the control of Dactylogyrus intermedius in Carassius auratus. Veterinary parasitology 158, 144−151.

Wu, Y. R., Ma, Y. B., Zhao, Y. X., Yao, S. Y., Zhou, J., Zhou, Y., Chen, J. J. (2007). Two new quaternary alkaloids and anti−hepatitis B virus active constituents from Corydalis saxicola. Planta Medica 73, 787−791.

Yao, J. Y., Shen, J. Y., Li, X. L., Xu, Y., Hao, G. J., Pan, X. Y., Wang, G. X., Yin, W. L. (2010). Effect of sanguinarine from the leaves of Macleaya cordataagainst Ichthyophthirius multifiliis in grass carp (Ctenopharyngodon idella). Parasitology Research 107, 1035−1042.

该文发表于《Parasitology Rresearch》【2010，107：1035-1042】。

Effect of sanguinarine from the leaves of *Macleaya cordataagainst Ichthyophthirius multifiliis* in grass carp（*Ctenopharyngodon idella*）

Jia-Yun YAO[1], Jin-Yu SHEN[1*], Xi-Lian LI[2], Yang XU[1], Gui-Jie HAO[1],
Xiao-Yi PAN[1], Gao-Xue WANG[2], Wen-Lin YIN[1]

（1. Institute of Freshwater Fisheries of Zhejiang, Huzhou, Zhejiang, 313001, China
2. College of Animal Science and Technology, Northwest A&F University, Yangling, 712100, China）

Abstract：The ciliate Ichthyophthirius multifiliis is one of the most pathogenic parasites of fish maintained in captivity. In this study, effects of crude extracts, fractions and compounds from the leaves of Macleaya cordata against I. multifiliis were investigated under in vitro conditions by bioactivity-guided isolation method. The dried ethanol extract of M. cordata was extracted successively in a separating funnel with petroleum ether, ethyl acetate, chloroform and n-butanol. Among them, only the chloroform extract showed promising activity and therefore, was subjected to further separation and purification using various chromatographic techniques. Four compounds were isolated from chloroform extract, but only one compound showed potent activity. The structure of the active compound was elucidated as sanguinarine byhydrogen, carbon-13 nuclear magnetic resonance spectrum and electron ionization mass spectrometry. In vitro antiparasitic efficacy tests exhibited that sanguinarine was 100% effective against I. multifiliis at a concentration of 0.7 mg · L^{-1}, with LC$_{50}$ and LC$_{90}$ values of 0.437 and 0.853 mg · L^{-1} after 4h of exposure. In vivo antiparasitic efficacy tests showed that the number of I. multifiliis on the gills in the treatment group（in 0.9 mg · L^{-1} sanguinarine）was reduced by 96.8%, in comparison to untreated group at 25℃ for 48h. Mortality of fish did not occur in the treatment group during the trail, although 40% of untreated fish died. Our results indicate that the studied plant extracts, as well as sanguinarine might be potential sources of new antiparasitic drug for the control of I. multifiliis.

key words：Ichthyophthirius multifiliis; sanguinarine; Macleaya cordata; antiparasitic

博落回中血根碱杀小瓜虫活性的研究

姚嘉赟[1]，沈锦玉[1*]，李喜莲[2]，徐洋[1]，郝贵杰[1]，潘晓艺[1]，王高学[2]，尹文林[1]

（1. 浙江省淡水水产研究所，浙江 湖州 313001；2. 西北农林科技大学陕，西杨 凌 712100）

摘要：多子小瓜虫是一种寄生于淡水养殖鱼类鳃部和表皮的纤毛虫类寄生虫，给世界淡水养殖业造成了具有很大危害，因此本研究拟从博落回中分离获取抗多子小瓜虫的药物。采用常压硅胶柱层析和 RP-HPLC 制备并结合体外杀虫活性追踪试验，对博落回杀小瓜虫的活性成分进行追踪，获得 4 种单体化合物，但只有一种具有较强杀虫活性的单体化合物，波谱（IR、MS、^{1}H NMR 和^{13}C NMR）分析结果显示该活性物质分别为血根碱。体外杀虫结果表明血根碱对小瓜虫幼虫的100%杀虫浓度为 0.7 mg/L，其 4h EC_{50} 和 EC_{50}分别为 0.437 和 0.853 mg/L。体内杀虫活性试验结果表明：经 0.9 mg/L 血根碱药浴作用 48 h 后，药物组小瓜虫体表数量减少了 96.8%，且其死亡率为 0，而对照组死亡率为40%。上述研究结果表明血根碱具有较好的杀虫效果，可用于开发杀虫药物。

关键词：小瓜虫，血根碱，博落回，抗寄生虫

1 Introduction

I. multifiliis is a holotrichous protozoan that invades the gills and skin surfaces of freshwater fish causing the disease ichthyophthiriosis, commonly referred to as ich or white spot. It is one of the most common and serious parasitic diseases of freshwater fish in aquaculture, and can causemorbidity and high mortality in most species of freshwater fish worldwide and can result in heavy economic losses for aquaculture （Paperna, 1972; Nigrelli et al. , 1976; Jessop, 1995; Traxler et al. , 1998）.

To cope with the parasitism and deleterious consequences, various parasiticides have been used in aquaculture. Formerly, ichthyophthiriasis was treated effectively with malachite green, a compound that has been banned on fish farms due to its confirmed carcinogenicity in most countries （Alderman, 1985; Wahli et al. , 1993; Committee on Mutagenicity of Chemicals in Food, Consumer Products and the Environment, 1999）. Since the ban, formalin （Bauer, 1970; Antychowicz, 1977; Stoskopf, 1993; Rowland, 2009）, hydrogen peroxide （Bauer, 1970; Rach et al. , 2000; lahnsteiner and weismann, 2007）, potassium permanganate （Straus and Griffin, 2001; Buchmann et al. , 2003）, sodium percarbonate （Buchmann et al. , 2003; Heinecke and Buchmann, 2009）, peracetic acid （Straus and Meinelt, 2009） and sodium chloride （Bauer, 1970; Stoskopf, 1993; lahnsteiner and weismann, 2007） have been used with varying levels of success. However, all compounds do not have widespread use because of its lower efficacy and problems associated with human handling （Dickerson and Dawe, 1995; Tieman and Goodwin, 2001）. Another agent used in the treatment of ichthyophthiriasis was copper sulfate （Ling et al. , 1993; Straus, 1993, 2009; Schlenk et al. , 1998; Rowland, 2009）. However, copper sulfate can be extremely toxic to fish in water of low alkalinity （Straus and Tucker, 1993; Wurts and Perschbacher, 1994; Perschbacher and Wurts, 1999）. Preventive method by increasing water flow in fish-holding systems has been suggested to control ichthyophthiriasis （Bodensteiner et al. , 2000）, but this method is difficult to use in many rearing facilities. Therefore, the urgent need for the development of affordable, effective and safe alternative agents to combat these diseases can not be overemphasized.

Recently, there have been increased research activities into the utilization of traditional plant – based medicines to control I. multifiliis infection. Plant extracts from garlic （Buchmann et al. , 2003）, Mucuna pruriens and Carica papaya （Ekanem et al. , 2004） have been assessed, showing some potential for killing the free-swimming stages. However, most of the compounds responsible for such activities have not been isolated and their structures have not been determined. So, it is hard to evaluate their potential for the development of novel antiparasitic drugs.

Macleaya cordata is a traditional Chinese medicine which is widely distributed in northwest and southwest of China. There were many reports about the antimicrobial, nematocidal, antitrichomonal activity of the crude extracts and active compounds of M. cordata （Navarro et al. , 1996; Navarro and Delgado, 1999; Satou et al. , 2002a; Calzada et al. , 2007）. However, there were few reports about the investigation of M. cordata in aquaculture. In this study, bioactivity-guided isolation was applied on ethanol extract of the leaf of M. cordata to provide active compounds, the antiparasitic efficacy of the compounds to eliminate I. multifiliis was also tested.

2 Materials and methods

2. 1 Fish

Heavy I. multifiliis-infected grass carp （Ctenopharyngodon idella）, mean length 8.7 ± 0.3 cm, were obtained fromaquatic fry farm of Zhejiang Institute Freshwater Fisheries in China and maintained in a 1 m³ tank with aerated groundwater at 25℃. Fish were acclimatized under laboratory conditions for 7 days, 10 fish were randomly selected, killed by spinal severance and examined for the prevalence and intensity of parasite under a light microscope （Olympus BX51, Tokyo, Japan） at 10×10 magnification prior to the experiment. The infection rate was 100% and the mean number of parasites on the gills was 60-80 per infected fish.

2. 2 Bioactivity-guide isolation

2. 2. 1 Preparation of extracts

Dried leaves of M. cordata （6 kg） were extracted using ethanol in a water bath at 65℃ for 2 h and then filtered. The residue was subjected to two successive extractions with ethanol. The extracts were combined and concentrated using a rotary vacuum evaporator at 70℃, resulting in the formation of a paste residue （1. 036 kg）.

The ethanol extract （1. 036 kg） was suspended in distilled water, then, was extracted successively in a separating funnel with petroleum ether, ethyl acetate, chloroform and n-butanol. Each extract and the remaining aqueous part after solvent extraction were then evaporated to dryness under reduced pressure to give petroleum ether extract （98. 5 g）, chloroform extract （130. 0 g）, ethyl acetate extract （221. 7 g）, n-butanol （258. 9 g） and remaining aqueous extract （259. 8 g） as seen in Fig. 1. The dried extract was dissolved in 10. 0 mL DMSO to prepare at a final concentration of' 0. 5 g · mL^{-1} for in vitro tests. The bioassay showed that chloroform extract was the highest in antiparasitic efficacy among all extracts. So, it was then subject to further separation.

2. 2. 2 Fractionation and isolation of pure compounds

Chloroform extract （130. 0 g） was dissolved in 200 mL ethanol, and the resulting solution was mixed with 300 g silica gel （300-400 mesh）, and then dried. The pasty residue was subject to column chromatography （150×12 cm） on a column of silica gel

(2500 g, 200-300 mesh) and eluted with a mixture of petroleum ether-ethyl acetate to afford 654 fractions (300 mL each fraction). Fractions were monitored using thin-layer chromatography (TLC) on petroleum ether-ethyl acetate solvent system and fractions showing similar TLC chromatograms were combined into seven fractions (Fr. A: 1~83 fractions, Fr. B: 84~134 fractions, Fr. C: 135~240 fractions, Fr. D: 241~314 fractions, Fr. E: 315~425 fractions, Fr. F: 426~535 fractions, Fr. G: 536~654 fractions). These seven fractions were submitted to in vitro test, and Fr. D (241~314 fractions, 10.6 g) was the most active. Fr. D was then applied to reverse-phase high performance liquid chromatography (RP-HPLC) with the following chromatographic conditions: Octadecyl Silane-A C_{18} (5 μm 120Å 10 mm×250 mm) column, MeOH-0.8% HAc (40∶60, v/v) mobile phase, 280 nm wavelength, 5.0 mL/min flow rate, 35℃ column temperature and 500 μL capacity. Four crystals were obtained by RP-HPLC from Fr. D which was isolated by the above method.

2.3 In vitro tests

In vitro tests were undertaken for the crude extracts (including petroleum ether, chloroform, ethyl acetate, n-butanol and water extract), fractions isolated from chloroform extract and the pure compounds (crystals). Tests were conducted in 24-well tissue culture plate (Becton Dickinson Labware, NJ, USA), filled with 2 mL aerated groundwater each containing the test samples and 20 trophonts. The crude extracts, fractions and the pure compound were dissolved in 1ml DMSO, respectively, and made up to 50 mL with distilled water which were used for the preparation of the different concentrations of the test solutions.

Gill arches with I. multifiliis were cut into a 100 mL beaker containing 30 mL aerated groundwater to allow the trophonts to dislodge. Active moving trophonts were removed with the help of a small brush. A total of 20 mature trophonts was distributed to each well of a 24-well tissue culture plate and exposed (at 25℃) to concentrations of crude extracts, fractions and pure compounds. The behavior of the trophonts was observed under a microscope (100×) every hour. Immobilized and lysed trophonts cells were considered dead. The mortality of trophonts in groups with different test sample concentrations was evaluated. A negative control was included using aerated groundwater containing the same amount of DMSO as the test sample. All the tests were performed in duplicate.

2.4 Identification of active compound

Based on the physico-chemical properties and electron ionization mass spectrometry (EI-MS, 70 eV VG-ZAB-HS, VG Company, England), nuclear magnetic resonance hydrogen spectrum (^{1}H-NMR) and nuclear magnetic resonance carbon spectrum (^{13}C-NMR) (Bruker Avance 400 spectrometer with $SiMe_4$ as internal standard, USA), the chemical structure of the active compound was identified.

2.5 In vivo tests

In vivo tests were undertaken for the active compound to determine the effective concentration. Tests were conducted in 80 cm×60 cm×40 cm glass tanks, each containing 50 L of the test solution, and 10 parasitized fish. The water pH ranged from 7.2 to 7.4. All tests were performed at 25℃.

Sanguinarine was used at a different series of concentrations based on the in vitro tests, 0.5 grams of sanguinarine were dissolved in 1ml DMSO and made up to 50 mL with distilled water as stock solution. Different concentrations of sanguinarine were prepared from the stock solution and dissolved in 50 l groundwater, and the final test presented here were conducted with sanguinarine concentrations of 0.2, 0.3, 0.4, 0.5, 0.6, 0.7, 0.8, and 0.9 mg · L^{-1}. Each concentration was considered to be one treatment group. In each treatment, two replicates were conducted. Control group without sanguinarine was set up under the same experimental conditions as the test groups. According to the solvent concentrations in the stock solutions of sanguinarine, control treatment was performed with 0.001 8 mL DMSO (the same amount of DMSO as the maximum concentration test group). The deaths of fish were recorded when the opercula movement and tail beat stopped and the fish no longer responded to mechanical stimulus. The observed dead fish were removed from the water in time. Mortality of fish was recorded throughout the experiments, during which no food was offered to the fish.

After 48 h treatment, the remaining grass carp in the treatment and control group were killed by a spinal severance. After that the lamella branchialis of each grass carpwere placed on glass slides and the surviving I. multifiliisin the gills were counted under a light microscope at 10×10 magnification, the mean number of parasites on each gill was calculated. The antiparasitic efficacy of each treatment was calculated according to the following formula:

$$E = (C - T) \times 100/C \tag{1}$$

Where E is the antiparasitic efficacy, C is the mean number of I. multifiliis on the gillsin the negative control, and T is the treatment groups.

2.6 Data analysis

The homogeneity of the replicates of the samples was checked by the Mann–Whitney U test. The LC_{50} and LC_{90} values and their 95% confidence intervals of sanguinarine were calculated by log concentration–probit equation using the SPSS 16.0 probit procedure.

3 Results

3.1 *In vitro* antiparasitic efficacy of the extracts and fractions

The results of antiparasitic efficacies for five extracts are depicted in Fig. 2, which indicated that the antiparasitic efficacy of the chloroform extract against I. multifiliis in both duplicated trials was 100%, at concentrations of 70.0 mg·L^{-1} after 4 h of exposure. Followed by the ethyl acetate and n-butanol extracts and the maximum antiparasitic efficacy were 80% (100.0 mg·L^{-1}) and 70% (120.0 mg·L^{-1}), respectively. The petroleum ether and water extracts exhibited the least activity with the maximum antiparasitic efficacy of 40% (120.0 mg·L^{-1}) and 27.5% (120.0 mg·L^{-1}), respectively. The solvent (DMSO) acted as a control showed no antiparasitic activity when treated at the highest concentration.

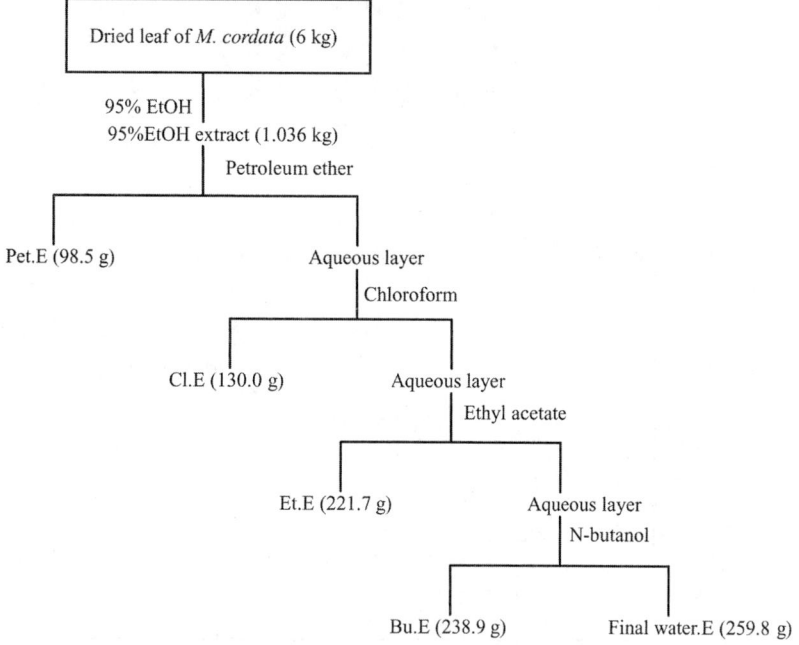

Fig. 2 Extraction scheme

The *in vitro* antiparasitic efficacy of seven fractions from the chloroform extract were depicted in Fig. 3, which showed that Fr. D had a 100% antiparasitic effective against I. multifiliis at concentrations of 9.0 mg·L^{-1}, after 4 h of exposure. Fr. E was 92.5% effective against I. multifiliis at a concentration of 30.0 mg·L^{-1} after 4 h of exposure, Fr. C was 87.5% effective against I. multifiliis at a concentration of 30.0 mg·L^{-1} after 4 h of exposure. The other 4 fractions (Fr. A, Fr. B, Fr. F, Fr. G) had low efficacy at 30.0 mg·L^{-1} with the efficacy of 22.5%, 25.0%, 25.0% and 15.0% respectively. Fr. D was considered to be the fraction that contained active compounds.

3.2 *In vitro* antiparasitic efficacy of Compounds

Four Compounds (crystals) were separated from Fr. D by RP–HPLC and were then subjected to in vitro test to determine the active compound. The results showed that Compound I (identified as sanguinarine) was 100% effective against I. multifiliis, with LC_{50} and LC_{90} values of 0.437 and 0.853 mg·L^{-1} respectively, at concentrations of 9.0 mg·L^{-1} after 4 h of exposure (Fig. 4 and Fig. 5), The other 3 compounds (Compound II, Compound III, Compound IV) had low efficacy at 9.0 mg·L^{-1} with the efficacy of 22.5%, 37.5% and 22.5% respectively (Fig. 5).

The *in vitro* treatment showed that sanguinarine had disastrous effect on the trophozoite stages of I. multifiliis. It led to severe alter-

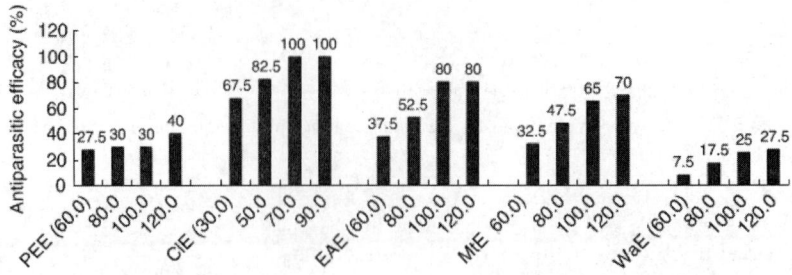

Fig. 3　Anthelmintic efficacy of five extracts from ethanol M. cordataagainst I. multifiliis at 4 h

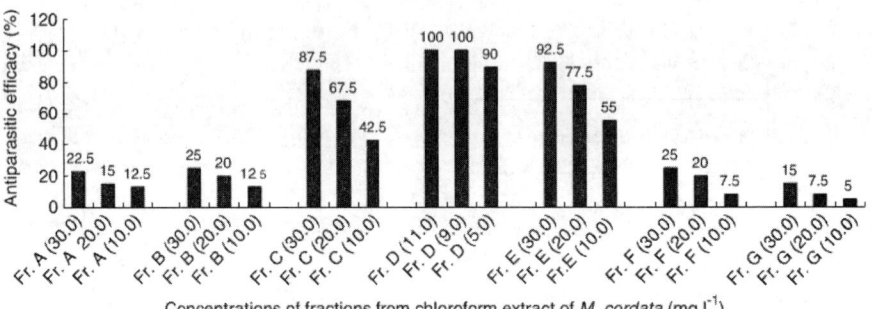

Fig. 4　Anthelmintic efficacy of the seven fractions against I. multifiliis at 4 h

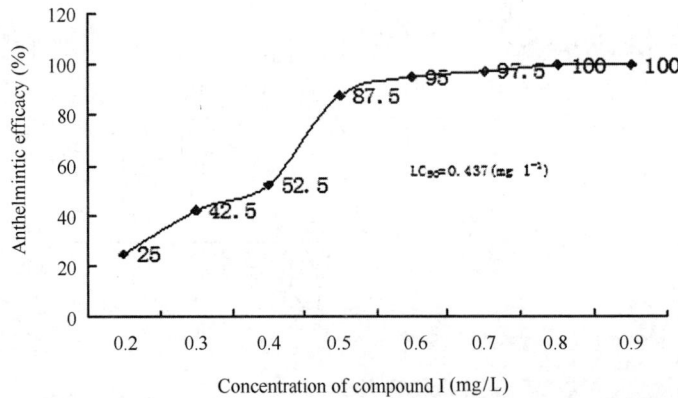

Fig. 5　Anthelmintic efficacy of compound I against I. multifiliis at 4 h

ations in the parasite's structure, after 4 h treatment, all of the parasites in treatment group were severely damaged: the outer cell membrane of I. multifiliis was destroyed, macronucleus was invisible, the cilia could not be recognized, and the cytoplasm of the trophozoites was characterized by vacuoles (Fig. 6, A). In contrast, all parasites' structure in untreated controls were intact with no deformities to the cell membrane. Macronucleus could be recognized obviously, movement of cilia could be easily observed (Fig. 6, B).

3.3　*In vivo* test

The treatment with the sanguinarine lead to a significant dose-dependent decrease in the number of I. multifiliison the gills compared to the controls. When the concentrations of sanguinarine were 0.2, 0.3, 0.4, 0.5, 0.6, 0.7, 0.8, and 0.9 mg · L^{-1}, the numbers of parasites on the gills were reduced by 16.1%, 17.3%, 32.9%, 53.9%, 75.3%, 82.3%, 89.4% and 96.8%, respectively (Fig. 7). 40% fish mortality was observed in the control group, while mortality of fish did not occur in the sanguinarine treatments.

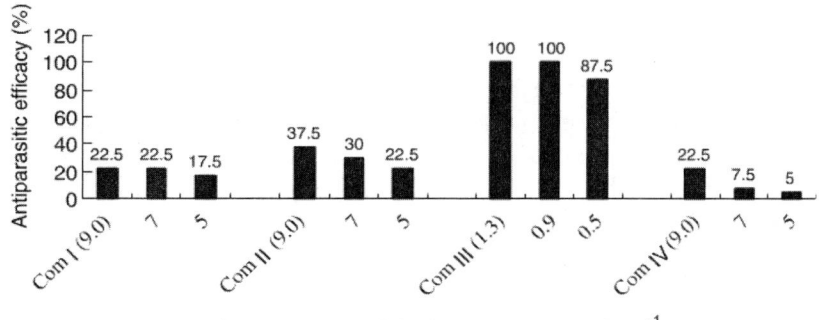

Fig. 6　Anthelmintic efficacy of four compounds from ethanol M. cordata against I. multifiliis at 4 h

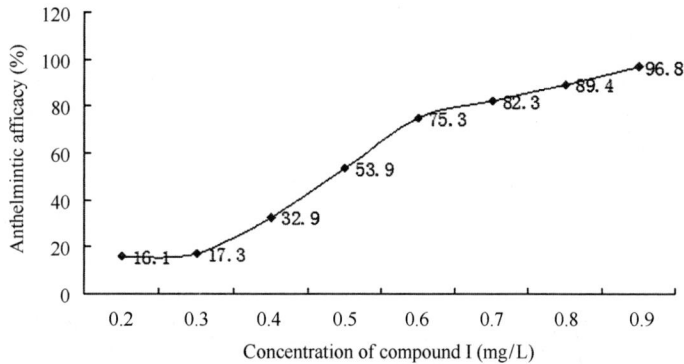

Fig. 7　In vivo anthelmintic efficacy of compound against I. multifiliis at 48 h

3. 4　Structure identification of the active compound

3. 3. 1　Compound I

The chemical structure confirmation of the compound from M. cordatawas accomplished by comparing m p, MS, ^{1}H NMR and[13] C NMR data obtained to those published.

Compound Ⅰ was obtained as reddish brown crystal; m p: 242–243℃; EI-MS (70eV) m/z: 334.3 (M+2), 333.3 (M+1), 332.3 (M$^+$), 331.3 (M−1), 315.3 (M−CH$_3$), 304.3, 274.3, 260.3;^{1}H NMR (CD$_3$OD, TMS, 400MHz): δ 8.6 (1H, d, H−4), 8.5 (1H, d, H−5), 7.9 (1H, d, H−8), 8.2 (1H, d, H−9), 7.56 (1H, d, H−11), 6.3 (2H, s, −OCH$_2$O−13), 8.1 (1H, d, H−15), 4.9 (3H, s, N−Mc), 6.3 (2H, s, −OCH$_2$O−20);^{13}C NMR (CD$_3$OD, TMS, 400MHz): δ 104.3 (C−1), 148.2 (C−2), 149.5 (C−3), 105.0 (C−4), 127.6 (C−5), 129.1 (C−6), 133.2 (C−7), 132.9 (C−8), 111.3 (C−9), 121.9 (C−10), 107.0 (C−11), 150.9 (C−12), 119.6 (C−13), 150.8 (C−14), 106.6 (C−15), 121.2 (C−16), 134.2 (C−17), 52.8 (N−Me), 150.8 (C−19), 118.3 (C−20); agree well with the data reported (Perez Gutierrez et al., 2002). So the compound was identified as sanguinarine. Its molecular formula is C$_{20}$H$_{14}$NO$_4$ (Perez Gutierrez et al., 2002), and its structure is shown in Fig. 8.

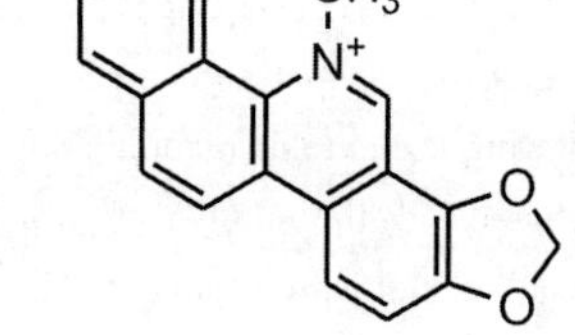

Fig. 8　Chemical structure of sanguinarine

4 Discussion

Plants, microorganisms and marine organisms are potential sources of new drugs since they contain a countless quantity of natural products with a great variety of structures and pharmacological activities (Newman et al., 2003). The diversity of natural products with antiparasitic activities has been reported and there are publications reportingon using plant extracts for the control of I. multifiliis, but little information is available on the isolation of natural compounds from plant responsible for the control of I. multifiliis. In the present study, an anti-I. multifiliis compound was separated from M. cordata by bioactivity-guided isolation and identified as sanguinarine. In vitro tests exhibited that sanguinarine was 100% effective against I. multifiliis at a concentration of 0.7 mg · L^{-1}. The in vivo test showed that bath treatments with 0.9 mg · L^{-1} of sanguinarine resulted in a significant reduction in the I. multifiliis burden of grass carp fish. Sanguinarine is an isoquinoline alkaloid that belongs to a group called benzo [c] Phenanthridine alkaloids. Many studies about the antimicrobial (Simanek, 1985; Facchini, 2001), anti-inflammatory (Lenfeld et al., 1981) and immunomodulatory (Chaturvedi et al., 1997) have been reported. Sanguinarine has also been reported to promote animal growth by increasing feed intake and decreasing aminoacid degradation from decarboxylation (Lenfeld et al., 1981; Drsata et al., 1996; Kosina et al., 2004). To our best knowledge, the effect of sanguinarine against parasitesin fish has not been investigated. Therefore, this study is the first report of the antiparasitic efficacy of sanguinarine assessed by both in vitro and in vivo experiments in fish. This result extended the general knowledge about the antiparasitic activity of the sanguinarine and the plants application to control fish parasite.

The crude extract fractions and sanguinarine isolated from M. cordata tested have demonstrated their potential in the control of I. multifiliis infection. Even if the parasites were not completely eliminated in the in vivo, a substantial reduction of the parasite burden, as shown in this study, led to the recovery of the fish and paved the way for the development of protective host immunity (Clark and Dickerson, 1997). This could be seen as a better alternative for aquaculture and the environment than the use of some chemicals.

The in vitro treatment showed that the types of damage were similar to those described when treatment has been conducted with immersion therapy in solutions of malachite green (Schmahl 1992, 1996). In both cases a degradation of the outer cell membrane and a complete breakdown of the macronucleus have been observed. This may be the main mechanism responsible for the anti-I. multifiliis activity of sanguinarine.

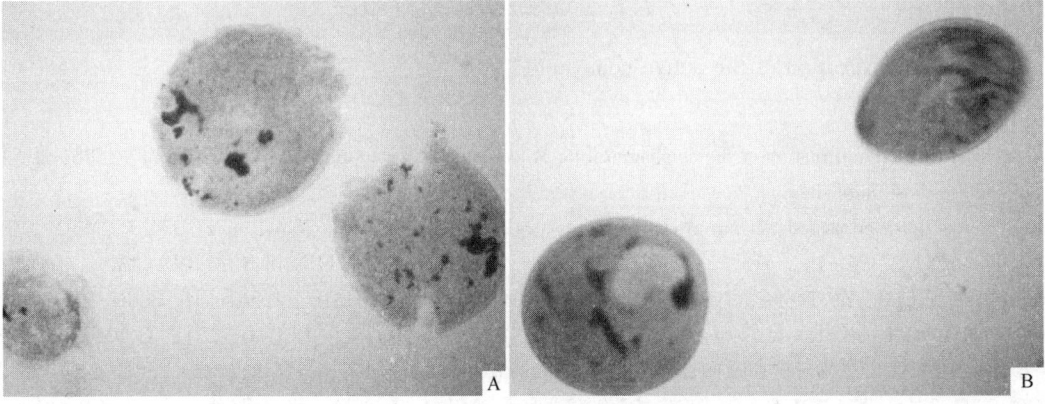

Fig. 9   Micrographs of I. multifiliis after 4 hours of treatment (10×10 magnification). (A) Treated with sanguinarine for 4 h at the concentration of 0.9 mg l-1 and (B) groundwater containing the same amount of DMSO as a control.

In vivo experiments with laboratory animals are an excellent experimental tool to describe the antiparasitic activity of an extract (fraction or compound); however, this approach has limitations if it is intended to be used as a monitoring system to isolate active compounds from herbal extracts. Animal studies are: (a) time consuming, (b) expensive, (c) ethical issue (such as: killing animals), (d) need more drugs, (e) insuitabe for high throughput screening (Jarry et al., 2006). So, in the present study, the strategy of in vitro bioactivity-guided isolation monitoring the chromatographic separation was used. Using in vitro assays as an alternative monitoring system can overcome these issues, moreover, the number of parasites can be recorded more precisely in in vitro tests, so the antiparasitic efficacy of active extracts can be calculated more precisely. The content of effective compounds contained in plants is often low, the in vitro test can be completed on a smaller scale as compared to in vivo treatment, and so, small volumes of the active compound can meet the need of the experiment. By bioactivity-guided isolation, four compounds were isolated from M. cordata and sanguinarine was determined as one of the active compounds.

In the present study, ethanol extracts of leaves of M. cordata was provided for bioactivity-guided isolation. The chloroform extract with the most promising activity was subjected to further separation and purification, and the active compound was identified as sanguinarine. The other extract or fractions with lower antiparasitic activity were not further isolated, although they may contain compounds that have high activity, but present in low concentration. So, further phytochemical studies, towards the isolation and separation of the plant are recommended in our future research.

For the lack of sanguinarine isolated from the leaf of M. cordata, the toxicity of sanguinarine was not investigated. However, Sanguinarine exhibited a low acute oral toxicity: for sanguinarine $LD_{50} = 1.7$ g kg^{-1} in rats; the acute i. v. toxicity for sanguinarine $LD_{50} = 29$ mg kg^{-1} in rats (Walterova et al., 1995). Rawling T. D et al. (2009) demonstrated that low levels (25−100 mg kg^{-1}) of Sangrovit © (The main component is sanguinarine) had a positive effect on tilapia growth performance with no apparent effects towards carcass composition, hepatic function or health status. During the in vivo test, all fish in the treatment group survived normally throughout the exposure period (48 h). The result indicated that sanguinarine is safe to fish at effective concentration.

In summary, using the strategy of bioactivity−guided isolation monitoring the chromatographic separation, a compound (sanguinarine) showing promising anti−I. multifiliis activity was obtained from the chloroform extract of M. cordata. The antiparasitic activity tests indicated that sanguinarine has the potential to the development for the new treatment against I. multifiliis. However, toxicity tests to different fish species must be determined, further studies are required for field evaluations in the practical system and the mechanism of the antiparasitic activity remains to be further investigated.

Reference：

Alderman D. J., 1985. Malachite green: a review. J Fish Dis. 8: 289−298.

Antychowicz J., 1977. The influence of some chemical compounds upon the development of Ichthyophthirius multifiliis. B Vet I Pulawy. 21: 35−43.

Asres K., Bucar F., Knauder E., Yardley V., Kendrick H., Croft S. L., 2001. In vitro antiprotozoal activity of extract and compounds from the stem bark of Combretum molle. Phytother Res. 15: 613−617.

Bauer O. N., 1970. Parasite diseases of cultured fishes and methods of their prevention and treatment. In: Dogiel VA, Petrushevski GK, and Polyanski YI, Parasitology of fishes. T. F. H. Publications, Hong Kong. pp: 265−298.

Bodensteiner L. R., Sheehan R. J., Wills P. S., Brandenburg A. M., Lewis W. M., 2000. Flowing water: an effective treatment for ichthyophthiriasis. J Aqua Ani Health. 12: 209−219.

Buchmann K., Jensen P. B., Kruse K. D., 2003. Effects of sodium percarbonate and garlic extract on Ichthyophthirius multifiliis theronts and tomocysts: in vitro experiments. N Am J Aquacult. 65: 21−24.

Calzada F., Yépez-Mulia L., Tapia-Contreras A., 2007. Effect of Mexican medicinal plant used to treat trichomoniasis on Trichomonas vaginalis trophozoites. J. Ethnopharmacol. 113: 248−251.

Chaturvedi M. M., Kumar A., Darnay B. G.., Chainy G. B. N., Agarwal S., Aggarwal B. B., 1997. Sanguinarine (pseudochelerythrine) is a potent inhibitor of NF-KB activation, IkBA phosphorylation, and degradation. J. Biol. Chem. 272: 30129−30134.

Clark T. G., Dickerson H. W., Gratzek J. B., Findly R. C., 1987. In vitro response of Ichthyophthirius multifiliis to sera from immune channel catfish. J Fish Biol. 31: 203−208.

Clark T. G., Dickerson H. W., 1997. Antibody-mediated effects on parasite behavior: evidence of a novel mechanism of immunity against a parasitic protist. Parasitol Today. 13: 477−480.

Committee on Mutagenicity of Chemicals in Food, Consumer Products and the Environment. 1999. Statement for COT: malachite green and leucomalachite green. Department of Health, London. http: //archive. food. gov. uk/dept_ health/pdf/mala. pdf. Cited 22 August 2003.

Cross D. G., 1972. A review of methods to control ichthyopthiriasis. Prog Fish Cult. 34, 165−170.

Dickerson H. W., Dawe D. L., 1995. Ichthyophthirius multifiliis and Cryptocaryon irritans (Phylum Ciliophora). in: Woo PTK (Ed) Fish diseases and disorders, vol 1. Protozoan and metazoan infections. CAB International, Wallingford.

Drsata J., Ulrichova J., Walterova D., 1996. Sanguinarine and chelerythrine as inhibitors of aromatic amino acid decarboxylase. J. Enzym. Inhib. 10: 231−237.

Ekanem A. P., Obiekezie A., Kloas W., Knopf K., 2004. Effects of crude extracts of Mucuna pruriens (Fabaceae) and Carica papaya (Caricaceae) against the protozoan fish parasite Ichthyophthirius multifiliis. Parasitol Res. 92: 361−366.

Ewing M. S., Kocan K. M., 1992. Invasion and development strategies of Ichthyophthirius multifiliis, a parasitic ciliate of fish. Parasitolo Today. 8: 204−208.

Facchini P. J., 2001. Alkaloid biosynthesis in plants: biochemistry, cell biology, molecular regulation, and metabolic engineering applications. Annu. Rev. Plant. Phys. 52: 29−66.

Heinecke R. D., Buchmann K., 2009. Control of Ichthyophthirius multifiliis using a combination of water filtration and sodium percarbonate: Dose-response studies. Aquaculture. 288: 32−35.

Ijah U. J. J., Oyebanji F. O., 2003. Effects of tannins and polyphenols of some medical plants on bacterial agents of urinary tract infections. Glob J Pure Appl Sci. 9: 193−198.

Jarry H., Spengler B., Wuttke W., Christoffel V., 2006. In vitro assays for bioactivity-guided isolation of endocrine active compounds in Vitex agnus-castus. Maturitas. 55: 26-36.

Jessop B. M., 1995. Ichthyophthirius multifiliis in elvers and small American eels from the East River, Nova Scotia. J Aqua Ani Health. 7: 54-57.

Kermanshai R., McCarry B. E., Rosenfeld J., Summers P. S., Weretilnyk E. A., Sorger G. J., 2001. Benzyl isothiocyanate is the chief or sole anthelmintic in papaya seed extracts. Phytochemistry. 57: 427-435.

Kosina P., Walterova D., Ulrichova J., Lichnovsky V., Stiborova M., Rydlova H., Vicar J., Krecman V., Brabec M. J., Simanek V., 2004. Sanguinarine and chelerythrine: Assessment of safety on pigs in ninety days feeding experiment. Food Chem Toxicol. 42: 85-91.

Lahnsteiner F., Weismann T., 2007. Treatment of Ichthyophthiriasis in Rainbow Trout and Common Carp with Common and Alternative Therapeutics. J Aqua Ani Health. 19: 186-194.

Lenfeld J., Kroutil M., Marsalek F., Slavik J., Preininger V., Simanek V., 1981. Antiinflammatory activity of quaternary benzophenanthridine alkaloids from Chelidonium majus. Planta Med. 43: 161-165.

Ling K. H., Sin Y. M., Lam T. J., 1993. Effect of copper sulphate on ichthyophthiriasis (white spot disease) in goldfish (Carassius auratus). Aquaculture. 118: 23-35.

MacLennan R. F., 1935. Observations on the life cycle of Ichthyophthirius, a ciliate parasitic on fish. Northwestern Scientist. 9: 12-14.

Rawling M. D., Merrifield D. L., Davies S. J., (2009). Preliminary assessment of dietary supplementation of Sangrovit © on red tilapia (Oreochromis niloticus) growth performance and health. Aquaculture. 294: 118-122.

Navarro V., Villarreal M. L., Rojas G., Lozoya X., 1996. Antimicrobial evaluation of some plants used in Mexican traditional medicine for the treatment of infectious diseases. J. Ethnopharmacol. 53: 143-147.

Navarro V., Delgado G., 1999. Two antimicrobial alkaloids from Bocconia arborea. J. Ethnopharmacol. 66: 223-226.

Newman D. J., Cragg G. M., Snader K. M., 2003. Natural products as sources of new drugs over the period 1981-2002. J. Nat. Prod. 66: 1022-1037.

Nigrelli R. F., Pokorny K. S., Ruggieri G. D., 1976. Notes on Ichthyophthirius multifiliis, a ciliate parasitic on freshwater fishes, with some remarks on possible physiological races and species. Trans Am Microsc Soc. 95: 607-613.

Paperna I., 1972. Infection by Ichthyophthirius multifiliis of fish in Uganda. Progressive Fish-Culturist. 34: 162-164.

Paolini V., Frayssines A., Farge D. L. F., Dorchies P., Hoste H., 2003. Effects of condensed tannins on established populations and on incoming larvae of Trichostrongylus colubriformis and Teladorsagia circumcincta in goats. Vet Res. 34: 331-339.

Perschbacher P. W., Wurts W. A., 1999. Effects of calcium and magnesium hardness on acute copper toxicity to juvenile channel catfish, Ictalurus punctatus. Aquaculture. 172: 275-280.

Gutierrez P. R. M., Vargas S. R., Gutierrez D. G., Martinez M. F. J., 2002. Identification of benzophenanthridine alkaloids from Bocconia arborea by gas chromatography-mass spectrometry. Phytochem. Anal. 13: 177-180.

Rach J., Gaikowski M. P., Ramsay R. T., 2000. Efficacy of hydrogen peroxide to control parasitic infestations on hatchery-reared fish. J Aqua Ani Health. 12: 267-273.

Rowland S. J., Mifsud C., Nixon M., Read P., Landos M., 2009. Use of formalin and copper to control ichthyophthiriosis in the Australian freshwater fish silver perch (Bidyanus bidyanus Mitchell) Aquac Res, 40: 44-54.

Satou T., Akao N., Matsuhashi R., Koike K., Fujita K., Nikaido T., 2002. Inhibitory effect of isoquinoline alkaloids on movement of second-stage larvae of Toxocara canis. Biol Pharm Bull. 25: 1651-1654.

Satrija F., Retnani E. B., Ridwan Y., Tiuria R., 2001. Potential use of herbal anthelmintics as alternative antiparasitic drugs for smallholder farms in developing countries. Livestock community and environment. Proceedings of the 10th Conference of the Association of Institutions for Tropical Veterinary Medicine, Copenhagen, Denmark.

Schäperclaus W., 1991. Diseases caused by ciliates. In: Schäperclaus W, Kulow H, Schreckenbach K (Eds) Fish diseases. Published for the U. S. Department of the Interior and the National ScienceFoundation, Washington, DC. Amerind, New Delhi, pp: 702-725.

Schlenk D., Gollon J., Griffin B. R., 1998. Efficacy of copper sulfate for the treatment of ichthyophthiriasis in channel catfish. J Aqua Ani Health. 10: 390-396.

Schmahl G., Ruider S., Mehlhorn H., Schmidt H., Ritter G., 1992. Treatment of fish parasites 9. Effects of a medicated food containing malachite green on Ichthyophthirius multifiliis Fouquet, 1876 (Hymenostomatida, Ciliophora) in ornamental fish. Parasitol Res. 78: 183-192.

Schmahl G., Schmidt H., Ritter G., 1996. The control of Ichthyophthiriasis by a medicated food containing quinine: efficacy tests and ultrastructure investigations. Parasitol Res. 82: 697-705.

Simanek V., 1985. Benzophenanthridine Alkaloids. Academic Press, New York

Stoskopf M. K., 1993. Fish medicine. Saunders, London.

Straus D. L., Tucker C. S., 1993. Acute toxicity of copper sulfate and chelated copper to channel catfish Ictalurus punctatus. J World Aquacult Soc. 24: 390-395.

Straus D. L., 1993. Prevention of Ichthyophthirius multifiliis Infestation in Channel Catfish Fingerlings by Copper Sulfate Treatment. J Aqua Ani Health. 5: 152-154.

Straus D. L., Griffin B. R., 2001. Prevention of an initial infestation of Ichthyophthirius multifiliis in channel catfish and blue tilapia by potassium per-

662

manganate treatment. North Am J Aquac. 63: 11-16.

Straus D. L., Hossain M. M., Clark T. G., 2009. Strain Differences in Ichthyophthirius multifiliis to Copper Toxicity. Dis Aquat Org. 83: 31-36.

Straus D. L., Meinelt T., 2009. Acute toxicity of peracetic acid (PAA) formulations to Ichthyophthirius multifiliis theronts. Parasitol Res. 104: 1237-1241.

Tieman D. M., Goodwin A. E., 2001. Treatments for Ich infestations in channel catfish evaluated under static and flow-through water conditions. N Am J Aquacult. 63: 293-299.

Tona L., Kambu K., Ngimbi N., Climanga K., Vlietinck A. J., 1998. Antiamoebic and phytochemical screening of some Congolese medicinal plants. J Ethnopharmacol. 61: 57-65.

Traxler G. S., Richard J., McDonald T. E., 1998. Ichthyophthirius multifiliis (Ich) epizootics in spawning sockeye salmon in British Columbia, Canada. J Aqua Ani Health. 10: 143-151.

Tucker C. S., Robinson E. H., 1990. Channel catfish farming handbook. Van Nostrand Reinhold, New York.

Wahli T., Schmitt M., Meier W., 1993. Evaluation of alternatives to malachite green oxalate as a therapeutant for ichthyophthiriosis in rainbow-trout, Oncorhynchus mykiss. J Appl Ichthyol. 9: 237-249.

Walterova D., Ulrichova J., Valka I., Vicar J., Vavreckova C., Taborska E., Harkrader R. J., Meyer D. L., Cerna H., Simanek V., (1995). Benzo [c] phenatridine alkaloids of sanguinarine and chelerythrine: biological activities and dental care applications. Acta Univ Palacki Olomuc Fac Med, 139: 7-16.

Willcox M. L., Bodeker G., 2000. Plant-based malaria control: research initiative on traditional antimalarial methods. Parasitol Today. 16: 220-221.

Wise D. J., Camus A. C., Schwedler T. E., Terhune J. S., 2004. Infectious diseases. In: Tucker CS, Hargreaves JA (Eds) Biology and culture of channel catfish. Elsevier, Amsterdam, 444-502.

Wurts W. A., Perschbacher P. W., 1994. Effects of bicarbonate alkalinity and calcium on the acute toxicity of copper to juvenile channel catfish (Ictaluruspunctatus). Aquaculture. 125: 73-79.

该文发表于《Parasitology Rresearch》【2015, 114 (4): 1425-1431】。

Evaluation of nystatin isolated from Streptomyces griseus SDX-4 against the ciliate, *Ichthyophthirius multifiliis*

JIA-Yun YAO[1], YANG XU[1], WEN-Lin YIN[1], XUE-Mei Yuan[1], LING-Yun LIN[1],
TING Xu[1], MENG-Li ZUO[1,2], XIAO-Yi PAN[1], JIN-Yu SHEN[1]*

(1. Zhejiang Institute of Freshwater Fisheries, HuZhou, Zhejiang, 313001, China;

2. ShangHai Ocean University, PuDong, ShangHai, 201306, China)

Abstract: The present study was conducted to evaluate the in vitro and in vivo anti-parasitic efficacy of active compounds from the bacterial extracellular products of Streptomyces griseus SDX-4 against Ichthyophthirius multifiliis. Bioassay-guided fractionation and isolation of compounds with antiparasitic activity were performed on n-butanol extract of S. griseus yielding a pure bioactive compound: Nystatin (Nys), identified by comparing spectral data (EI-MS, ^{1}H NMR and ^{13}C NMR) with literature values. Results from in vitro antiparasitic assays revealed that Nyscould be 100% effective against I. multifiliis theronts and encysted tomonts at the concentration of 6.0 mg $\cdot$ L^{-1}, with the median effective concentration (EC_{50}) values of 3.1 and 2.8 mg $\cdot$ L^{-1} for theronts and encysted tomonts (4 h), respectively. Results of in vivo test demonstrated that the number of I. multifiliis trophonts on the gold fish treated with Nys was markedly lower than the control group at 10 days after exposed to theronts (p<0.05). In the control group, 85.7% mortality was observed owing to heavy I. multifiliis infection at 10 days after the exposure. On the other hand, only 23.8% mortality owing to parasite infection was recorded in the groups treated with the Nys (4.0 and 6.0 mg $\cdot$ L^{-1}). In addition, our results showed that the survival and reproduction of I. multifiliis tomont exited from the fish were significantly reduced after treated with the 6.0 mg $\cdot$ L^{-1}Nys. The median lethal dose (LD$_{50}$) of Nys for goldfish was 16.8 mg $\cdot$ L^{-1}. This study firstly demonstrated that Nys has potent anti-parasitic efficacy against I. multifiliis, and it can be a good candidate drug for chemotherapy and control of I. multifiliis infections.

key words: Ichthyophthirius multifiliis; Nystatin; Streptomyces griseus; Antiparasitic

灰色链霉菌 SDX-4 中制霉菌素抗小瓜虫活性的研究

姚嘉赟[1]，徐洋[1]，尹文林[1]，袁雪梅[1]，蔺凌云[1]，许婷[1]，左梦丽[1,2]，潘晓艺[1]，沈锦玉[1]*

(1. 浙江省淡水水产研究所，浙江 湖州 313001；2. 上海海洋大学，上海 浦东 201306)

摘要：研究灰色链霉菌 SDX-4 代谢产物的体外、体内抗小瓜虫活性及成分。采用常压硅胶柱层析和 RP-HPLC 制备并结合体外杀虫活性追踪试验，对灰色链霉菌 SDX-4 代谢产物的活性成分进行追踪，并利用波谱（IR、MS、^{1}H NMR 和 ^{13}C NMR）分析确定其化学结构。获得 1 种活性较强的化合物，波谱分析结果表明其为制霉菌素。体外杀虫结果表明制霉菌素浓度为 6 mg·L^{-1} 时对小瓜虫包囊的杀灭率为 100%，其对包囊和成熟滋养体的 4 h 半数有效致死浓度（EC_{50}）分别为 3.1 和 2.8 mg·L^{-1}。体内杀虫活性试验结果表明：经 6.0 mg/L 制霉菌素药浴作用 10 d 后，药物组小瓜虫体表数量显著减少（$P<0.05$），且其死亡率为 23.8%，而对照组死亡率为 85.7%。急性毒性试验结果表明，制霉菌素对草鱼的半数致死浓度为 16.8 mg·L^{-1}。上述研究结果表明制霉菌素具有较好的杀虫效果，可用于开发杀虫药物。

关键词：小瓜虫，制霉菌素，灰色链霉菌，抗寄生虫

1 Introduction

Ichthyophthirius multifiliis（Ich）is a holotrichous protozoan that invades the gills and skin surfaces of freshwater fish causing the disease ichthyophthiriosis, commonly referred to as ich or white spot. It can cause sever morbidity and high mortality in most species of freshwater fish worldwide and can result in heavy economic losses for aquaculture（Traxler et al., 1998）. The life stages of the parasite include an infective theront, a parasitic trophont and a reproductive tomont（Dickerson, 2006；Matthews, 2005）.

Chemical treatment and vaccination have been used to control outbreaks of Ich. The most effective control of Ich infections has been achieved by use of malachite green, a compound that has been banned on fish farms due to its confirmed carcinogenicity in most countries（Alderman 1985；Wahli et al., 1993）. Other chemotherapeutants, such as formalin（Rowland et al., 2009），Chlorophyllin（Wohllebe et al., 2012），chloramine-T（Rintamäki-Kinnunen et al., 2005），copper sulphate（Rowland et al., 2009），potassium ferrate（VI）（Lin et al., 2011），potassium permanganate（Straus and Griffin, 2002），tricaine methanesulfonate（Xu et al., 2008）and bronopol（Shinn et al., 2012）have been evaluated for their antiparasitic efficacy. However, the threats of antiparasitic resistance, risk of residues, environmental contamination, and toxicity to fish caused by the frequently used of these chemicals have led to the need of other alternative control methods（Goven et al., 1980；Klinger and Floyd, 2002）. Vaccination against the parasites may provide an alternative to chemical treatment, but it costs highly and this cannot be ignored for fish farmer, restraining its widespread use. There is still an urgent need to develop newly antiparasite agents used in management of Ich disease.

Recently, there have been increased research activities into the utilization of environment-friendly products to control I. multifiliis infection（Yao et al., 2010, 2011；Yi et al., 2012）. Effective compounds of natural original expected to be more advantageous than synthetic antiparasitic agent, as they have generally a lower environmental impact and are easily biodegradable. Thus, a great number of biological control microorganisms have been screened and researched, including fungi（Crump et al., 1983；Chen and Dickson, 1996），bacteria（Becker et al., 1988），ray fungi, and protozoon（Chen and Liu, 2005；Sun et al., 2006）. Streptomycesare distributed widely in terrestrial and marine habitats（Pathom-aree et al., 2006）and are of commercial interest due to their unique capacity to produce novel metabolites. The genus Streptomyces was classified under the family Streptomycetaceae, which includes gram-positive aerobic members of the order Actinomycetales and suborder Streptomycineaewithin the new class Actinobacteria（Anderson and Wellington, 2001）. They produce approximately 75% of commercially and medically useful antibiotics and 60% of antibiotics used in agriculture（Watve et al., 2001）. In our previous study, we investigated the anti-I. multifiliis efficacy of extracellular products of Streptomyces griseus SDX-4 against I. multifiliis, and founded the n-butanol extracts of S. griseus（NBu-E）was effective agaisnt I. multifiliis（Yao et al., 2014）. The present study was conducted to assess the antiparasitic properties of NBu-E and isolate active constituents responsible for the activity using in vitro antiparasitic assay associated with bioassay-guided fractionation. Additionally, the acute toxicity of the active compounds against goldfish（Carassius auratus）of was evaluated.

2　Materials and methods

2. 1　Fish and parasites

Parasite-free goldfish were obtained from commercial fish farm and maintained in a 1 m³ tank supplied with filtered groundwater. The skin surface and gills of 10 randomly sampled fish were examined under a microscope to confirm that fish were not infected with gill parasites or skin parasites before the experiments. The fish were fed twice a day at a feeding rate of 1% body weight (BW) . All fish were acclimatized to laboratory conditions for 7 days before the experiment.

The sources of I. multifiliis, its propagation on grass carp and collection of the cysts have been described by Yao et al. (2014). Several heavy infected grass carp (Ctenopharyngodon idella) obtained from from aquatic fry farm of Zhejiang Institute Freshwater Fisheries in China were placed into filtered aquarium water for 30~60 min. Mature trophonts were allowed to dislodge from the host by body movements of the fish whilst in close proximity. Isolated trophonts were randomly divided into two batches, one was used to assay the activity of the fraction and active compound isolated from NBu-E for killing the tomonts, and the other placed in plastic beaker with aerated groundwater filtered and incubated for 24 h at 23℃. The water containing the hatched theronts in the plastic beaker was agitated and 50 μL of the water was withdrawn five times. The number of theronts in each 50 μL of water was counted, and the average number of parasites per milliliter was calculated to estimate the total number of hatched theronts.

2. 2　In vitro anti-parasitic efficacy against the theronts and tomonts

Tests against I. multifiliis theronts were performed as described in previous work (Yao et al., 2010, Lin et al., 2012). Briefly, approximate 100 viable theronts were distributed to each well of 24-well tissue culture plates (Becton Dickinson Labware, NJ, USA) and exposed to different concentrations frations and active compound from NBu-E. Mortality of theronts of each well was recorded by microscopic examination (×100 magnifications) at 10 min, and 1, 2, 3 and 4 h after exposure. A negative control was included using aerated groundwater containing the same amount of DMSO as the maximum concentration test group. The trial was repeated three times.

For the tomont trials, 30 tomonts were placed into each well of a 24-well tissue culture plate. One milliliter solutions with different concentrations of test sample were added to each well, respectively. A negative control was included using aerated seawater containing the same amount of DMSO as the maximum concentration test groups. The solutions were replaced by aerated groundwater with no test sample after 4-h exposure. Then the plates were incubated at 23℃ throughout the trial. The trial was allowed to stop until the parasites in the controls reached the theront stage. At the end of the trial, the dead tomonts was recorded and the mortality was counted. The parasites with the absence of internal cell motility or abnormal cell division and the ones cannot produce the theronts were considered dead. All treatments and control groups were conducted with three replicates.

2. 3　Bioactivity-guide isolation of pure compound from NBu-E

Strain SDX-4 was inoculated in Meat extract (D-glucose 1. 0%; peptone 0. 2%; yeast powder 0. 1%; beef extract 0. 1%; artificial sea water 50%; pH 7. 0~7. 2) for 7 days on a reciprocal shaker water bath at 30℃. The preparation of the n-butanol extract of S. griseus (NBu-E) was followed by our previous study (Yao et al., 2014) . The dried NBu-E were subjected to column chromatography on a silica gel and sequentially eluted with ethyl acetate and methanol with increasing polarity, eventually affording 425 fractions (400 mL each) . TLC (thin layer chromatography) analysis was performed on silica gel using the same solvent system as the mobile phase. Spots on thin layer chromatography were visualized under ultraviolet (UV) light (254 and 365 nm) or by spraying the plates with ethanol-sulphuric acid reagent, and fractions showing similar chromatograms were combined into five fractions (Fr. A: 1~65 fractions, Fr. B: 66~128 fractions, Fr. C: 129~230 fractions, Fr. D: 231~315 fractions, Fr. E: 316~425 fractions) . These five fractions were submitted to in vitro test, and Fr. D was the most active (EC_{50} for theronts and tomonts were 25. 6 and 32. 8 mg · L^{-1}) . Fr. D was then applied to macroporous resin HP20 (lvbaicao chemical Co., Ltd., Beijing province, China) as determined by preliminary experiments and eluted with 10% methanol (305nm wavelength, 5. 0 mL min^{-1} flow rate, 30℃ column temperature). A crystal was obtained from Fr. D which was isolated by the abovemethod. The structure of the compound was elucidated by comparing spectroscopic data with those reported as Nystatin (Thomas et al., 1982; Ling et al., 1986) .

2. 4　In vivo efficacy of Nys against Ich

In vivo test was conducted according to our previous study (Yao et al., 2014) with slight modifications. Briefly, 120 uninfected goldfish, weighing approximately 30 g each, were transferred to 1 000 L tanks and were acclimated to laboratory conditions for 7 days before the experiment. After acclimation, approximately 600, 000 I. multifiliis theronts were put into the 1 000 L tanks, the fish were held in the tank for 24 h with gentle aeration to promote infection. After exposure, the fish were divided into four groups: Nys-challenged (2.0, 4.0, 6.0 mg · L^{-1}) and control group (challenged with no chemical). Each group consisted of three replicates of 100 L groundwater and 10 infected goldfish. The Nys in each tank was replaced on day 3 and day 5 with a fresh solution at the same concentration.

On day 5 after exposure, three fish of each group were placed into 10 L tanks containing filtered aquarium water with no chemical for 45 min, mature parasites were quite freely dislodged from the host by body movements of the fish. Ten glass slides were placed in each tank when fish were transferred to the tanks. Glass slides were collected from the tank of each group and the number of trophonts was counted under the light microscope. For each treatment, 10 trophonts were distributed to each well (N = 3) of 24-well tissue culture plate and allowed to attach. The water in each well was removed carefully, and 2 mL of filtered aquarium water was added to each well. Until the parasites in the control groups reached the theront stage, the mortality and reproduction of tomonts in all wells were determined and calculated as described above.

Ten days after exposure to theronts, all remaining fish from each group were randomly sampled and the number of trophonts on the gills and fins was examined. The fish were carefully observed every an hour for any signs of distress indicative of erratic behavior, the fish was pick up to count the trophonts on the gills and fins as soon as dead. Fish mortality was recorded daily and the parasites on the gills and fins of dead fish were counted.

2.5 Acute toxicity of Nys on goldfish

The acute toxicity of Nyswas carried out according to the method of Yao et al. (2010) with slight modifications. The tests were conducted in 80-L glass tanks, each containing 50 L of test solution, and 10 healthy goldfish. Dilutions were prepared from the stock solution as the following concentrations: 12.0, 14.0, 16.0, 18.0, 20.0 and 22.0 mg · L^{-1} for Nys. The tests were conducted in triplicate, as well as controls (under the same test conditions with no chemicals). Fish mortalities in the treatment and control groups were recorded after 48 h of exposure, median lethal dose (LD$_{50}$) was calculated.

2.6 Data analysis

All data in this study were performed using the SPSS 16.0 probit procedure, the homogeneity of the replicates of the samples was checked by the Mann-Whitney U test. Tomont survival, tomont reproduction were compared with Student-Newman-Keuls test procedure for multiple comparisons (α = 0.05).

3 Results

3.1 In vitro antiparasitic efficacy of Nys against I. multifiliis theronts and tomonts

Nys was separated from Fr. D by bioactivity-guide isolation and was then subjected to in vitro test to determine the active concertrations. The results of Nys against I. multifiliis theronts were depicted in Fig. 1 As shown in Fig. 1 Nys concentrations vs I. multifiliis theront mortality demonstrated a distinct dose-response and a time-response relationship. The 10 min-, 1 h-, 2 h-, 3 h-and 4 h-EC_{50} (95% CI) of the compound against theronts of I. multifiliiswere 4.8 (4.4-5.1), 3.9 (3.6-4.1), 3.5 (3.3-3.7), 3.3 (3.1-3.5) and 3.1 (2.8-3.3) mg · L^{-1}, respectively.

The results of in vitro trial on the effect of Nys against tomonts were listed in Table 1. All tomonts were dead when exposed to Nys at 6.0 and 8.0 mg · L^{-1} concentrations, and no theronts were released. The mortality of tomonts exposed at 4.0, 2.0 and 1.0 mg · L^{-1} were 73.7%, 40.5% and 12.4%, respectively. Tested compound led to an obvious dose-dependent lethal effect against tomonts. The 4 h-EC_{50} (95% CI) against tomonts was 2.8 (2.5.0-3.1) mg · L^{-1}.

Table 1　I. multifiliis tomont survival and reproduction after 4-h exposure to Nystatin (mg · L^{-1})

Concentrations (mg · L^{-1})	Tomont mortality (%)	Reproduction
0 (control)	1.3±1.3a	488.3±25.7a
1.0	12.4±2.7b	463.7±34.3a
2.0	40.5±4.0c	325.0±26.5b

Concentrations（mg·L^{-1}）	Tomont mortality（%）	Reproduction
4.0	73.7±4.1d	107.0±18.1c
6.0	100±0.0e	0.0±0.0d
8.0	100±0.0e	0.0±0.0d

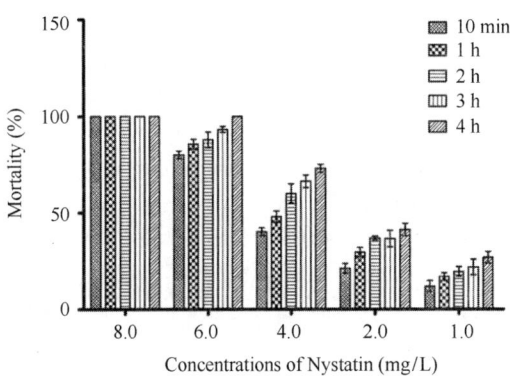

Fig. 1　Antiparasitic efficacy of nys against *I. multifiliis* theronts after 4 h exposure

The reproduction was represented as number of theronts releasedby each live tomont. Each value is expressed as mean±standard deviation of three replicates, and within a column, values followed by the different letters are significantly different（P<0.05, only compared with control）.

3.2　*In vivo* efficacy of Nys against Ich

In the control group, *I. multifiliis* on the body and fins of surviving fish were visible at 3 days and markedly increased at 6 days after starting the trial and increased over the duration of the trial. Mortality was recorded after the exposure due to *I. multifiliis* infection. 85.7% fish were died at 10 days in the control group after staring the trail（Table 2）. In contrast, in the groups treated with Nys, *I. multifiliis* were observed on the body and fins of surviving fish at 4 days after starting the trial, and the number of parasites was significantly lower than that of the control group（Table 2）. Furthermore, the number of visible *I. multifiliis* on the body surface of *I. multifiliis* in the group treated with the Nysdecreased over time. 23.8% mortality was recorded at 10 days after the exposure（6.0 mg·L^{-1}Nys）due to *I. multifiliis* infection, among the remaining fish, the average number of *I. multifiliis* trophonts on the gills and fins was 45.7±4.3（mean±standard deviation）. Furthermore, Nysat concentration of 6.0 mg·L^{-1} significantly decreased *I. multifiliis* tomont survival and reproduction when the tomonts were collected from infested fish bathed for 5 days, whereas there was no significant difference in tomont survival and reproduction between the fish in the control groups and the fish treated with 2.0 and 4.0 mg·L^{-1}Nys.

Table 2　In vivo efficacy of Nystatin against Ich

Concentration（mg·L^{-1}）	Tomonts obtained at day 5		Fish mortality（%）	Number of trophonts in each fish（gills and fins）
	Tomont mortality（%）	reproduction		
0（control）	0.0±0.0a	254.5±23.1a	85.7	332.5±43.3a（No. of fish=13）
Nystatin 2.0	75.2±6.3b	247.7±32.0a	57.1	206.0±38.5b（No. of fish=15）
4.0	56.8±4.2c	228.3±36.2a	23.8	58.3±21.2c（No. of fish=14）
6.0	45.7±4.3c	172.6±22.3b	23.8	43.0±12.5c（No. of fish=15）

The reproduction was represented as number of theronts released by each live tomont. Each value is expressed as mean±standard deviation of three replicates, and within a column, values followed by the different letters are significantly different（P<0.05）.

3.3　Acute toxicity of Nys

The fish tolerated Nys at concentration of 12.0 mg·L^{-1} for 48 h without visible effects, but exposure to 22.0 mg·L^{-1} resulted in an increased opercular movement and erratic behavior of fish with in 4 h, all fish died in tanks treated with 22. 0 mg · L^{-1} of Nys, the linear equation y =$-$10. 5+12. 6x which was derived from the regression analysis of probit mortality of goldfish in test solution bioassay, the calculated LD$_{50}$of Nys was 16. 8 mg · L^{-1} with the 95% confidence interval of 16. 1–17. 3 mg · L^{-1}. There was no fish mortality occurred in the control groups during the experiments.

4 Discussions

The disease ichthyophthiriasis, or 'white spot', caused by a parasitic ciliate, *I. multifiliis*, probably accounts for more damage to freshwater fish populations worldwide than any other eukaryote pathogen (Matthews, 2005). Currently there is no chemotherapeutant available to treat Ich effectively and economically. There is an urgent need to discover effective and safe parasiticides to control ichthyophthiriasis. In the present study, an anti-*I. multifiliis* compound was separated from then-butanol extracts of S. griseus by bioactivity-guided isolation and identified as Nys. In vitro tests exhibited that Nys were 100% effective against I. multifiliis theronts and encysted tomonts at concentrations of 6. 0 mg · L^{-1}. The in vivo test showed that bath treatments with Nys resulted in a significant reduction in the *I. multifiliis* burden of gold fish. Consequently, the parasite-induced mortality of the fish host was significantly reduced. To the best of our knowledge, it seems to be the first report demonstrating the protection efficacy of Nys from parasite infection for fish. This result extended the general knowledge about the antiparasitic activity of Nys to control fish parasite. It also highlights a new way to explore novel active compounds from bacterial extracellular products to combat fish disease.

Nys, produced by Streptomyces noursei, was the first polyene macrolide antifungal antibiotic to be discovered (Hazen and Brown, 1950). Nys possesses a broad spectrum with both antifungal and fungistatic activity (Recamier et al., 2010) being effective against azole-resistant strains of Candida and amphotericin B-resistant strains of Candidaalbicans (Ellepola & Samaranayake, 1999). Nys is indicated for treatment of cutaneous and mucocutaneous fungal infections caused by Candida species, the main yeast capable of infecting the oral mucosa, being C. albicans the most common species isolated (Campos et al., 2012). The antifungal activity of polyene antibiotics is related to their ability to form micelles with ergosterol molecules present in the cell membranes of fungal cells. This bonding causes structural damage and membrane permeabilisation leading to the loss of electrolytes and other cytoplasmatic components like proteins. As a consequence of which the fungal cells die (Coutinho & Prieto 2003). Thus, it is postulated that the antiparasitic mechanisms of action of Nys might be also attribute to these factors, however the exact mechanism of action regarding its antiparasitic efficacy remains to be further investigated.

Theront stage is external elements in the life cycle of *I. multifiliis* (Buchmann and Bresciani 2001). Killing the parasite at this stage will prevent its invasion of fish. It is therefore relevant to assess the susceptibility of free-living theronts to alternative substances. Most theronts (95. 3%) can survive in water for 48 h (Shinn et al., 2012) and possess an increased propensity to infectfish, especially when fish are raised at a high density. Rapidly eliminating theronts can prevent Ich infestationof host fish. This study showed that Nys could 100% killing the theront at the concentration of 6. 0 mg · L^{-1}. The exposure duration of Nys was much shorter than garlic extract which required 15 h to eradicate all theronts at a concentration of 62. 5 mg/L (Buchmann et al., 2003), and shorter than potassium ferrate, which took 30 min to eliminate all theronts at a concentration of 24 mg/L (Ling et al., 2010). The EC$_{50}$ of two active compounds were also lower than some plant compounds, such as dihydrosanguinarine (13. 299 mg · L^{-1}) and dihydrochelerythrine (18. 231mg · L^{-1}) (Yao et al. 2011). Thus Nys could effectively eradicate theronts in water.

Tomonts are another free-living stage of the *I. multifilliis* life cycle and each tomont reproduces hundred to thousands of infective theronts (Matthews, 2005; Dickerson, 2006), *I. multifilliis* infection can be easily amplified under practical fish farming conditions. So it is important to prevent infestation of Ich via termination of the reproduction of tomonts. Therefore, a 4 h exposure time for mortality of tomonts was employed in the present study. The results revealed that Nys could be 100% effective against *I. multifiliis* encysted tomonts at the concentration of 6. 0 mg · L^{-1}, with the median effective concentration (EC$_{50}$) values of 3. 1 and 2. 8 mg · L^{-1} for theronts and encysted tomonts (4 h). Buchmann et al. (2003) revealed that for I. multifiliis tomonts were more resistant to the parasiticides than theronts. This conclusion was supported by other studies (Yi et al., 2012; Ling et al., 2012). But interestingly, in contrast to their findings, we found that EC$_{50}$ value of the Nys against tomonts was closed to theronts. This may because the action site of Nys on tomonts and theronts is coordinat. However the exacts mechanism remains to be further investigated.

Under practical fish farming conditions, it would be necessary to apply drugs repeatedly because tomocysts release new infective theronts continuously over the course of several days. Li and Buchmann (2001) reported that the mean time between tomocyst formation and the release of theronts was found to be 18 h at 25℃, so Nys was administrated at every two days in the present study. In vivo test showed that fish treated with Nys at the concentrations of 6. 0 mg · L^{-1}carried significantly fewer parasites than the control. The observed reduction of parasites in the test groups could be attributed to the effects of the plant extracts because similar reduction in parasite burden was not observed in the control groups. These results suggest that continuous immerse with Nys at the concentration of 6.0 mg · L^{-1} at every two days for 3 times is suitable for controlling *I. multifiliis* infection.

In the present study, n-butanol extract of extracellular products of S. griseus was provided for bioactivity-guided isolation. The Fr. D from NBu-E with the most promising activity was subjected to further separation and purification, and the active compound was isolated and identified as Nys. The other extract or fractions with lower antiparasitic activity were not further isolated, although they may contain compounds that have high activity, but present in low concentration. So, further phytochemical studies, towards the isolation and separation of the plant are recommended in our future research.

In conclusion, using the strategy of bioactivity-guided isolation monitoring the chromatographic separation, a compound showing promising anti-*I. multifiliis* activity were obtained from NBu-E and elucidated as Nys. It can be chosen as a lead compound for the development of new antiparasitic agents against I. multifiliis. Also, further studies are required for field evaluations in the practical system and the mechanism of the antiparasitic activity remains to be performed.

Reference:

A lderman D J (1985) Malachite green: a review. J Fish Dis 8: 289-298.

Anderson A S, Wellington E M (2001) The taxonomy of Streptomyces and relatedgenera. Int J Syst Evol Microbiol 51: 797-814.

Becker J O, Zavaleta M E, Colbert S F, Schroth M N, Weinhold A R, Hancock J G, Vangundy S D (1988) Effects of rhizobacteria on root knot nematodes and gall formation. Phytopathology 78: 1466-1469.

Buchmann K, Sigh J, Nielsen C V, Dalgaard M (2001) Host responsesagainst the fish parasitizing ciliate Ichthyophthirius multifiliis. Vet Parasitol 100: 105-116.

Buchmann K, Jensen P B, Kruse K D (2003) Effects of sodium percarbonate and garlic extract on Ichthyophthirius multifiliis theronts and tomocysts: in vitro experiments. N Am J Aquacult 65: 21-24.

Campos FF, Calpena Campmany AC, Delgado GR, Serrano O L, Naveros BC (2012) Development and characterization of a novel nystatin-loaded nanoemulsion for the buccal treatment of candidosis: Ultrastructural effects and release studies. J Pharm Sci, 101, 739-3752.

Chen S Y, Dickson D W (1996) Fungal penetration of the cystwall of Heterodera glycines. Phytopathology 86: 319-327.

Chen SY, Liu X Z (2005) Control of the soybean cyst nematode by the fungi Hirsutella rhossiliensis and Hirsutella minnesotensisin greenhouse studies. Biol Control 32: 208-219.

Coutinho A, Prieto M (2003) Cooperative partition model of nystatin interaction with phospholipid vesicles. Biophys J 84, 3061-3078.

Crump D H, Sayre R M, Young LD (1983) Occurence of nematophagous fungi in cysts. Plant Dis 67: 63-64.

Dickerson H W (2006) Ichthyophthirius multifiliisand Cryptocaryon irritans (Phylum). In: Woo, P.T.K. (Ed.), Fish Diseases and Disorders. Protozoan and Metazoan Infections, vol.1, 2nd ed. CAB International, Wallingford, UK, pp. 116-153.

Ellepola AN, &Samaranayake LP (1999) The in vitro post-antifungal effect of nystatin on Candida species of oral origin. J Oral Pathol Med, 28, 112-116.

Goven B, Gilbert J, Gratzek J (1980) Apparent drug resistance to the organophosphate dimethyl (2, 2, 2-trichloro-1-hydroxyethyl) phosphonate by monogenetic trematodes. J Wildlife Dis 16 (3): 343-346.

Hazen EL, Brown R (1950) Two antifungal agents produced by a soil actinomycete. Science 112, 423.

Klinger R, Floyd RF (2002) Introduction to freshwater fish parasites. Document CIR716. Institute of Food and Agricultural Science, University of Florida, Florida.

Li A H, Buchmann K (2001) Temperature and salinity dependent development of a Nordic strain of Ichthyophthirius multifiliis from rainbow trout. J Appl Ichthyol 17: 273-276.

Ling D K, Chen S, Wang S W, SUN Z P, Ma J W (1986) Identification of two major components in Chinese nystatin. Acta Pharmaceutica Sinica. 21 (6): 454-457.

Ling F, Wang JG, Wang GX, Gong XN (2011) Effect of potassium ferrate (VI) on survival and reproduction of Ichthyophthirius multifiliis tomonts. Parasitol Res 109: 1423-1428.

Ling F, Wang J G, Lu C, Wang G X, Lui Y H, Gong X N (2012) Effects of aqueous extract of Capsicum frutescens (Solanaceae) against the fish ectoparasite Ichthyophthirius multifiliis. Parasitol Res 111: 841-848.

Matthews RA (2005) Ichthyophthirius multifiliis Fouquet and ichthyophthiriosis in freshwater teleosts. Adv Parasitol 59: 159-241.

Pathom-aree W, Stach J E M, Ward A C, Horikoshi K, Bull A T, Goodfellow M (2006) Diversity of actinomycetes isolated from challenger deep sediment (10, 898 m) from the Mariana Trench, Extremophiles 10: 181-189.

Recamier KS, Hernandez-Gomez A, Gonzalez-Damian J, Ortega-Blake I (2010) Effect of membrane structure on the action of polyenes: I. nystatin action in cholesterol and ergosterol-containing membranes. J Membr Biol, 237: 31-40.

Rintamäki-Kinnunen P, Rahkonen M, Mannermaa-Keränen A L, Suo-malainen L R, Mykrä H, Valtonen ET (2005) Treatment of ichthyophthiriasis after malachite green. I. Concrete tanks at salmonid farms. Dis. Aquat. Organ 64: 69-76.

Rowland SJ, Mifsud C, Nixon M, Read P, Landos M (2009) Use of formalin and copper to control ichthyophthiriosis in the Australian freshwater fish silver perch（Bidyanus bidyanus Mitchell）Aquacult Res 40：44-54.

Shinn P，Camacho A P，Bron S M，Conway J E，Yoon D，Guo G H，Taylor F C，Ni G H（2012）The anti-protozoal activity of bronopol onthe key life-stages of Ichthyophthirius multifiliisFouquet，1876（Cilio-phora）. Vet Parasitol 186：229-236.

Straus DL，Griffin BR（2002）Efficacy of potassium permanganate intreating ichthyophthiriasis in channel catfish. J. Aquat. Anim. Health14（2）：145-148.

Sun M H，Gao L，Shi Y X，Li B J，Liu X Z（2006）Fungi and actinomycetes associated with Meloidogyne spp. eggs and in China and their biocontrol potential. J Invertebr Pathol 93：22-28.

Thomas A H（1982）The heterogeneous composition of pharmaccutical grade nystatin. Analyst 20：294.

Traxler G S，Richard J，McDonald TE（1998）Ichthyophthirius multifiliis（Ich）epizootics in spawning sockeye salmon in British Columbia，Canada. J Aqua Ani Health 10：143-151.

Wahli T，Schmitt M，Meier W（1993）Evaluation of alternatives to malachite green oxalate as a therapeutant for ichthyophthiriosis in rainbow-trout，Oncorhynchus mykiss. J Appl Ichthyol 9：237-249.

Watve MG，Tickoo R，Jog MM，Bhole BD（2001）How many antibiotics are producedby the genus Streptomyces. Arch Microbiol 176：386-390.

Wohllebe S，Richter P，Häder DP（2012）Chlorophyllin for the control of Ichthyophthirius multifiliis（Fouquet）. Parasitol Res 111：729-733.

Xu D H，Shoemaker C A，Klesius P H（2008）Effect of tricaine methanesulfonate on survival and reproduction of the fish ectoparasite Ichthyophthirius multifiliis. Parasitol Res 103：979-982.

Yao JY，Shen JY，Li XL，Xu Y，Hao GJ，Pan XY，Wang GX，Yin WL（2010）Effect of sanguinarine from the leaves of Macleaya cordataagainst Ichthyophthirius multifiliis in grass carp（Ctenopharyngodon idella）. Parasitolo Res107：1035-1042.

Yao JY，Zhou ZM，Li XL，Yin WL，Ru HS，Pan XY，Hao GJ，XuY，Shen JY（2011）Antiparasitic efficacy of dihydrosanguinarine and dihydro-chelerythrine from Macleaya microcarpa against Ichthyophthirius multifiliisin richadsin（Squaliobarbus curriculus）. Vet Parasitol 183：8-13.

Yao JY，LiX C，Li G，Xu Y，Ai W M，Shen JY（2014）Anti-parasitic activities of specific bacterial extracellular products of Streptomyces griseus SDX-4 against Ichthyophthirius multifiliis，Parasitol Res 113（8）：3111-3117.

Yi YL，Lu C，Hu XG，Ling F，Wang GX（2012）Antiprotozoal activity of medicinal plants against Ichthyophthirius multifiliis in goldfish（Carassius auratus）Parasitol Res 111：1771-1778.

该文发表于《Parasitology Rresearch》【2014，113（8）：3111-3117】。

Antiparasitic activities of specific bacterial extracellular products of *Streptomyces griseus* SDX-4 against *Ichthyophthirius multifiliis*

Jia-Yun YAO[1]，Wen-Lin YIN[1]，Xin-cang LI[2]，Gang LI[3]，
Yang XU[1]，Wei-Ming AI[4]，Jin-Yu SHEN[1]*

（1. Zhejiang Institute of Freshwater Fisheries，Huzhou，Zhejiang，313001，China；2. East China Sea Fisheries Research Institute，Shanghai，200090，China；3. National Fisheries Technology Etension Center，Beijing，100125，China；4. Wenzhou Medical University，Wenzhou，Zhejiang，325035，China）

Abstract：The ciliate Ichthyophthirius multifiliis is one of the most pathogenic parasites of fish maintained in captivity. In this study，effects of bacterial extracellular products of Streptomyces griseus SDX - 4 against I. multifiliis were determined. The fermentation liquor of S. griseus was extracted successively in a separating funnel with petroleum ether，ethyl acetate，and n-butanol. In vitro assays revealed that the n-butanol extracts（NBu-E）and ethyl acetate（Eto-E）of S. griseus were observed to be more effictive against theronts than the other extracts with median effective concentration（EC_{50}）values of 0. 86 mg·L^{-1} and 12. 5 mg·L^{-1}，respectively，and significantly reduced the survival of the tomonts and the total number of theronts released by the tomonts（$P<0.05$）. All encysted tomonts were killed when the concentration of NBu-E was 30. 0 mg·L^{-1}. Results of in vivo test demonstrated that the number of I. multifiliis trophonts on the grasscarp treated with NBu-E was markedly lower compared to the control group at 11 days after exposed to theronts（$P<0.05$）. In the control group，100% mortality was observed owing to heavy I. multifiliis infection at 11 days after the exposure. On the other hand，only 9. 5% mortality owing to parasite infection was recorded in the groups treated with the NBu-E（30 mg·L^{-1}）. The median lethal dose（LD_{50}）of NBu-E for grasscarp was 152. 4 mg·L^{-1}. Our results indicate that n-butanol extract of S. griseus will be useful in aquaculture for controlling I. multifiliis infections.

Key words：*Ichthyophthirius multifiliis*；extracellular products；*Streptomyces griseus*；Antiparasitic

灰色链霉菌 SDX-4 代谢产物抗小瓜虫作用的研究

姚嘉赟[1]，尹文林[1]，李新苍[2]，李刚[3]，徐洋[1]，艾为明[4]，沈锦玉[1*]

(1. 浙江省淡水水产研究所，浙江 湖州 313001；2. 东海水产研究所，上海 200090；
3. 国家渔业技术推广中心，北京 100125；4. 温州医科大学，浙江 温州 325035)

摘要：多子小瓜虫是一种寄生于淡水养殖鱼类鳃部和表皮的纤毛虫类寄生虫，给世界淡水养殖业造成了具有很大危害。本研究利用不同极性大小的溶剂石油醚，乙酸乙酯和正丁醇回流提取灰色链霉菌 SDX-4 胞外代谢产物，进行杀灭多子小瓜虫幼虫和包囊的药效活性追踪。结果表明正丁醇和乙酸乙酯萃取部位的杀虫活性明显高于其他部位，其对小瓜虫幼虫的半数致死浓度（EC_{50}）分别为 0.86 mg·L^{-1}和 12.5 mg·L^{-1}。正丁醇萃取物的浓度为 30 mg·L^{-1}时可 100% 杀灭包囊并可抑制包囊释放幼虫的功能。体内感染实验结果表明，药物浸泡组（30 mg·L^{-1}正丁醇提取物）鱼体上小瓜虫的数量明显低于对照组，同时对照组鱼体的死亡率为 100% 而实验组的死亡率为 9.5%。正丁醇萃取物对草鱼的急性毒性实验结果表明其对草鱼的 48 h 半数致死浓度为 152.4 mg·L^{-1}。上述结果表明，灰色链霉菌 SDX-4 代谢产物正丁醇萃取物对小瓜虫具有较强的杀灭作用且对鱼体相对安全，有望开发为一种新型的杀虫药物。

关键词：小瓜虫；胞外产物；灰色链霉菌；抗寄生虫

1 Introduction

Ichthyophthirius multifiliis Fouquet is a serious protozoan parasite and epizootics have been reported in various freshwater fishes worldwide (Hines and Spira, 1974). The parasite damages fish gills and skin, results in high fish mortality, and leads to substantial economic losses for aquaculture (Traxler et al., 1998). The life stages of the parasite include an infective theront, a parasitic trophont and a reproductive tomont (Dickerson, 2006; Matthews, 2005).

To cope with the parasitism and deleterious consequences, various parasiticides have been used in aquaculture. Formerly, ichthyophthiriasis was treated effectively with malachite green, a compound that has been banned on fish farms due to its confirmed carcinogenicity in most countries (Alderman, 1985; Wahli et al., 1993). Since the ban, a number of variations on chemical bath treatments have been used to control ichthyophthiriasis, including formalin (Rowland, 2009), copper sulfate (Straus, 2009; Rowland, 2009), potassium ferrate (VI) (Lin et al., 2011), Chlorophyllin (Wohllebe et al., 2012), tricaine methanesulfonate (Xu et al., 2008). Static bath treatment is stressful to the fish (Kim and Choi, 1998) and the treatment chemicals are released into groundwater and soil which may lead to environmental pollution due to long degradation periods. Moreover, formaldehyde and copper sulfate may soon be banned from aquaculture usage owing to their adverse environmental and health effects (Jaafar and Buchmann, 2011).

In view of the limitation of chemical control, there is a growing incentive to develop environment-friendly products. Effective compounds of natural originare expected to be more advantageous than synthetic antiparasitic agent, as they have generally a lower environmental impact and are easily biodegradable. Thus, a great number of biological control microorganisms have been screened and researched, including fungi (Crump et al., 1983; Chen and Dickson, 1996), bacteria (Becker et al., 1988), ray fungi, and protozoon (Chen and Liu, 2005; Sun et al., 2006). Some biological control products like Paecilomyces lilacinus and Pochoniachlamydosporia have been commercially used in certain areas (Leij De et al., 1992). In our previous study, we isolated a Streptomyces griseusSDX-4, and founded its fermentation liquor was effective agaisnt I. multifiliis theronts. The present study was conducted to access the effects of extracellular products of S. griseus SDX-4 against I. multifiliis.

2 Materials and methods

2.1 Fish and parasites

Naive healthy grass carp, weighing approximately 30 g each, were obtained from special aquatic fry farm of Zhejiang Institute Freshwater Fisheries in China and were maintained in a 1500-L tank. The skin surface and gills of 10 randomly sampled fish were examined under a microscope to confirm that fish were not infected with gill parasites or skin parasites before the experiments. The fish were fed twice a day at a feeding rate of 3% body weight (BW). All fish were acclimatized to laboratory conditions (dissolved oxy-

671

gen: 5.37±0.76 mg · L^{-1}; pH: 7.63±0.39; nitrites: 0.019±0.008 mg · L^{-1}; ammonia: 0.131±0.068 mg · L^{-1}; Temperature: 23±1℃) for 7 days before the experiment.

The sources of *I. multifiliis*, its propagation on grass carp and collection of the cysts have been described by Yao et al. (2011). Several heavy infected grass carp (1530±234 g) obtained from a local fish pond were placed into filtered aquarium water for 30-60 min. Mature trophonts were allowed to dislodge from the host by body movements of the fish whilst in close proximity. Isolated trophonts were randomly divided into two batches, one was used to assay the activity of the extracellular products for killing the tomonts, and the other placed in plastic beaker with aerated groundwater filtered and incubated for 24 h at 23℃. The water containing the hatched theronts in the plastic beaker was agitated and 50μl of the water was withdrawn five times. The number of theronts in each 50 μL of water was counted, and the average number of parasites per milliliter was calculated to estimate the total number of hatched theronts.

2.2 Bacteria and extracellular products preparation

Strain SDX-4, isolated from a soil sample of East Sea in Zhoushan, Zhejinag province, China stored in our laboratory, and was inoculated in Meat extract (D-glucose 1.0%; peptone 0.2%; yeast powder 0.1%; beef extract 0.1%; artificial sea water 50%; pH 7.0-7.2) for 7 days on a reciprocal shaker water bath at 30℃. The combined fermentation liquor (250 L) was filtered and the supernatant was separated by centrifuging at 3000 rpm for 15 min, and then concentrated to dryness (96.8 g) by a vacuum freeze dryer (Chemical-free, Operon Co., Ltd., Korea). The dry powder was suspended in distilled water, then, was extracted successively in a separating funnel with petroleum ether, ethyl acetate and n-butanol. Each extract and the remaining aqueous part after solvent extraction were then evaporated to dryness under reduced pressure to give petroleum ether extract (PE-E, 4.3 g), ethyl acetate extract (Eto-E, 14.7 g), n-butanol (NBu-E 46.2 g) and remaining aqueous extract (RAE-E, 20.8 g). The dried extracts were dissolved in DMSO to prepare at a final concentration of 0.5 g · mL^{-1} for tests.

2.3 *In vitro* antiparasitic of extracellular products against *I. multifiliis* theronts

Tests were performed as described in previous work (Yao et al., 2010, Lin et al., 2012). Briefly, tests were conducted in each well of a 24-well tissue culture plate (Becton Dickinson Labware, NJ, USA), filled with 2 mL aerated groundwater, the theronts were placed into plates at a final concentration of about 100 theronts per well and exposed to different concentrations of test samples. Antiparasitic efficacy was determined by microscopic examination of each well at 1, 2, 3, and 4 h after treatment. The mortality of theronts in groups with different test sample concentrations was calculated. A negative control was included using aerated groundwater containing the same amount of DMSO as the maximum concentration test group. The trial was repeated three times.

2.4 *In vitro* antiparasitic of extracellular products against *I. multifiliis*tomonts

For thetomonts trial, 30 tomonts were placed into each well of a 6-well tissue culture plate. 5 mL NBu-Eat a concentration of 5, 10, 20 and 30 mg · L^{-1}, 5 mL Eto-E at a concentration of 10, 20, 40 and 60 mg · L^{-1}, were added to each well, respectively. A negative control was included using aerated groundwater containing the same amount of DMSO as the maximum concentration test group. The solutions were replaced by aerated groundwater with no extracts after 6-h exposure. Then the plates were incubated at 23℃ throughout the trial. The trial was allowed to stop until the parasites in the controls reached the theront stage. After counting dead tomonts (the parasites with the absence of internal cell motility or abnormal cell division and the ones can not produce the theronts, were considered dead), theronts in each well were enumerated as described above. The mortality and reproduction of tomonts were determined for each well. The tomont reproduction was expressed as number of theronts released by each tomont, calculated by total theronts/live tomonts. All treatments and control groups were conducted with three replicates.

2.5 In vivo efficacy of NBu-E and Eto-E against Ich

We set up seven experimental groups: NBu-E-challenged (10, 20 and 30 mg · L^{-1}), Eto-E-challenged (20, 40 and 60 mg · L^{-1}), and control group (challenged with no chemical). Each group consisted of three replicates of 50 L groundwater, each of which contained 10 grass carp.

210 uninfected grass carp, weighing approximately 30 g each, were transferred to 1 000 L tanks and were acclimated at 25℃ for 1 week. The tanks were aerated and supplied with groundwater. The temperature was maintained at 25±0.5℃ using a heating device. After acclimation, approximately 12, 000 theronts/fish were put into the 1 000 L tanks, the fish were held in the tank for 6 h with gentle aeration to promote infection. After exposure, the fish were divided into seven groups and each group was transferred to a 100 L tank with 10 infected grass carp. Each group was added with corresponding concentrations of the NBu-E and Eto-E of extracellu-

lar products, and the control group with no chemical. Triplicated tanks were used in each concentration. The extracellular products in each tank were replaced daily for 3 days with a fresh solution at the same concentration.

On day 5 after exposure, three fish of each group were placed into 20 L tanks containing filtered aquarium water without extracellular products, ten glass slides were placed in each tank when fish were transferred to the tanks. Glass slides were collected from the tank of each group and the number of trophonts was counted under the light microscope. For each treatment, 20 trophonts were distributed to each well (N=3) of 6-well tissue culture plate and allowed to attach. The water in each well was removed carefully, and 5 mL of filtered aquarium water was added to each well. Until the parasites in the control groups reached the theront stage, the mortality and reproduction of tomonts in all wells were determined and calculated as described above.

Eleven days after exposure to theronts, all remaining fish from each group except the control group were randomly sampled and the number of trophonts on the gills and fins was examined in the same way. In the control group, 100% mortality was recorded at 11 days after the exposure and the number of parasites on the gills and fins of seven dead fish was examined. Fish mortality was recorded daily and the parasites on the gills and fins of dead fish were counted.

2. 6　Acute toxicity

Only NBu-E showed greatest antiparasitic against I. multifiliis In vivo and in vitro, so the acute toxicity of NBu-E was carried out according to the method of Wang et al. (2010). The tests were conducted in 20-L glass tanks, each containing 10 L of test solution, and 10 healthy grass carp. Dilutions were prepared from the stock solution as the following concentrations: 100, 119, 140, 162, 185, 210, 237 and 250 mg $\cdot$ L^{-1} for NBu-E. The tests were conducted in triplicate, as well as controls (under the same test conditions with no chemicals). Fish mortalities in the treatment and control groups were recorded after 48 h of exposure, median lethal dose (LD$_{50}$) was calculated.

2. 7　Data analysis

All data in this study were performed using the SPSS 16. 0 probit procedure, the homogeneity of the replicates of the samples was checked by the Mann-Whitney U test. Tomont survival, tomont reproduction were compared with Student-Newman-Keuls procedure for multiple comparisons ($\alpha = 0.05$).

3　Results and discussions

The disease ichthyophthiriasis, or 'white spot', caused by a parasitic ciliate, I. multifiliis, probably accounts for more damage to freshwater fish populations worldwide than any other eukaryote pathogen (Matthews, 2005). Currently, most countries have few effective chemical treatments that can be used to control I. multifiliis (Sudová et al., 2010). More and more efforts have been spent on searching for more effectiveanti – I. multifiliisdrugs. Marine environment is the biggest reservoir of chemical and biological diversity. Therefore, research focus on marine environment has been gaining importance in recent years. However, still it has not been fully explored and there is tremendous potential to identify novel organisms with various biological properties. Streptomyces are distributed widely in terrestrial and marine habitats (Pathom-aree et al., 2006) and are of commercial interest due to their unique capacity to produce novel metabolites. The genus Streptomyceswas classified under the family Streptomycetaceae, which includes Gram-positive aerobic members of the order Actinomycetales and suborder Streptomycineaewithin the new class Actinobacteria (Stackebrandt et al., 1997; Anderson and Wellington, 2001). They produce approximately 75% of commercially and medically useful antibiotics and 60% of antibiotics used in agriculture (Watve et al., 2001). In this study, effects of specific bacterial extracellular products of a marine strain of S. griseus against I. multifiliis were determined. In vitro and in vivo tests results demonstrated that NBu-E and Eto-E of extracellular products were effective against I. multifiliis. As far as we know, it is the first study evaluating the extracellular products of microorganisms for the development of new pharmaceutical agents against fish pathogens, it also highlight a new way to explore novel active compounds to combat fish disease.

Theronts are the infective stage in the life cycle of I. multifilliis (Buchmann et al., 2001). Therefore it is important to kill theronts in order to control ichthyophthiriasis. It is therefore relevant to assess the susceptibility of free-living theronts to alternative substances. The results of antiparasitic efficacies against I. multifiliis theronts for four extracts are depicted in Fig. 1, which indicated that the antiparasitic efficacy of the NBu-E against I. multifiliis theronts in three duplicated trials was 100% at a concentration of 5. 0 mg $\cdot$ L^{-1} with a medical concentration (LC$_{50}$) of 0. 86 mg $\cdot$ L^{-1}. Followed by the Eto-E was 100% effective against I. multifiliis when the concentration was 30 mg $\cdot$ L^{-1}with LC$_{50}$ values of 12. 5 mg $\cdot$ L^{-1}, the PE-E and RAE-E exhibited the least activity with the maximum antiparasitic efficacy of 34% (40. 0 mg $\cdot$ L^{-1}) and 27. 6% (40. 0 mg $\cdot$ L^{-1}) (Fig. 1), respectively. The solvent (DMSO) acted as a control showed no antiparasitic activity when treated at the highest concentration. Results from this study suggest that NBu−E and Eto −E in the treatment of *I. multifiliis* theronts are more effective than many of the chemicals currently used in an attempt to control this parasite: Formaldehyde applied at 200 mg · L^{-1} for 1 h which kills 40% of theronts, hydrogen peroxide applied at 200 mg · L^{-1} for 1 h which kills 15% of theronts (Shinn et al. , 2012).

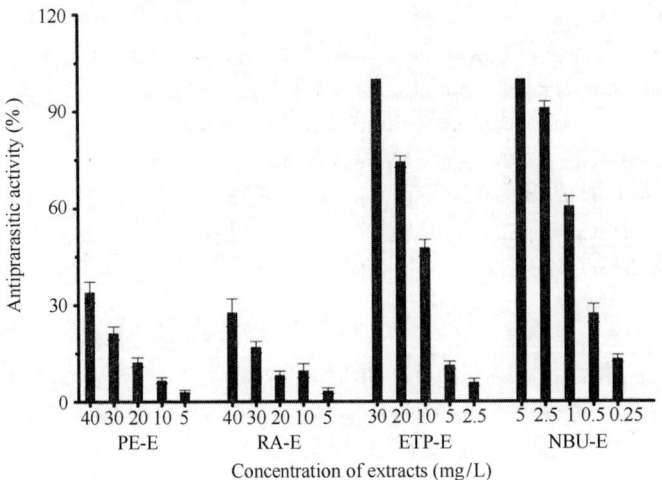

Fig. 1　Antiparasitic efficacy of the four extracts of *S. griseus* against *I. multifiliis* theronts after 4 h exposure

Buchmann et al. (2003) revealed that encysted tomonts resisted some substances better than theronts. This conclusion was supported by other studies (Yao et al. , 2011; Ling et al. , 2011; Yi et al. , 2012). It is therefore necessary and valuable to assess the susceptibility of tomonts to the four extracts of S. griseus in order to prevent the spread of ichthyophthiriasis. As Table 1 showed, NBu−E had a 100% antiparasitic effective against I. *multifiliis* tomonts at concentrations of 30.0 mg · L^{-1}, after 6 h of exposure. Exposure of *I. multifiliis* tomonts to 20.0 mg · L^{-1}NBu−E caused in 87.7% mortality, those remaining surviving continued to develop and release their theronts, the numbers of I. multifiliis released by the tomonts treated with NBu−E was smaller as compared to the controls (P<0.05). As for PE−E and RAE−E, at a concentration of 80.0 mg · L^{-1} exposed for 6 h showed little impact on the survival of encysted tomonts, only 15.5% and 25.5% of the encysted tomonts were killed 6 h post−start, respectively. No significant difference was noted on reproduction between treatments (PE − E and RAE − E concentration < 60 mg · L^{-1}) and the control. Intrestingly, tomonts exposed to 40 mg · L^{-1} and 60 mg · L^{-1} Eto−E shown an apparent delay in their development and subsequent release of theronts, but the reproduction of the tomonts treated with Eto−E showed no statistical difference as compared with control (Table 1). Therefore, it may contain active compounds in Eto−E that have the influence on the inhibition of *I. multifiliis* tomonts division. However, the active compounds response for the inhibition of divesion needed to be further isolated.

Table 1　*I. multifiliis* tomont survival and reproduction after 6−h exposure to the four extracts of S. griseus SDX−4 (mg · L^{-1})

Concentrations	Tomontmortality (%)	reproduction	Concentrations	Tomont mortality (%)	reproduction
0 (control)	1.5±1.1a	542.3±33.1a	0 (control)	1.5±1.1a	542.3±33.1a
NBu−E 5.0	33.3±2.2b	384.7±25.7b	RAE−E20.0	6.7±2.2b	522±43.3a
10.0	54.4±3.3c	217.1±33.5c	40.0	13.3±3.3c	518.7±38.7a
20.0	87.7±3.3d	105.3±15.4d	60.0	18.8±2.2d	523.3±39.1a
30.0	100±0.0e	0.0±0.0e	80.0	25.5±2.2e	441.7±32.1b
Eto−E10.0	27.7±2.2f	568.1±45.7a	PE−E20.0	3.3±1.1a	551.3±43.7a
20.0	44.5±2.2g	523.4±51.0a	40.0	7.8±2.2b	528.7±56.3a
40.0	62.2±4.4h	519.7±48.1a	60.0	10.0±1.1b	515.3±55.7a
60.0	64.4±3.3h	536.3±34.7a	80.0	15.5±4.4b	433.3±35.3b

The reproduction was representedas number of theronts releasedby each live tomont. Eachvalue is expressed as mean±standard deviation of three replicates, and within a column, values followed by the different letters are significantly different (P<0.05, only compared with control)

The *in vitro* assay may be excellent in measuring the intrinsic activity of a substance but cannot possibly emulate complex *in vivo* situations (Kirby, 1996). Therefore, an *in vivo* test was applied to determine the efficacy of NBu-E and Eto-E against I. multifiliis after exposure to theronts for 6 h. In the control group, *I. multifiliis* on the body and fins of surviving fish were visible at 4 days and markedly increased at 7 days after starting the trial and increased over the duration of the trial. Mortality was recorded after the exposure due to *I. multifiliis* infection. All fish were died at 11days in the control group after staring the trail (Table 2). In contrast, in the groups treated with the NBu-E, *I. multifiliis* were observed on the body and fins of surviving fish at 4 days after starting the trial, but the number of parasites was significantly lower than that of the control group (Table 2). Furthermore, the number of visible *I. multifiliis* on the body surface of *I. multifiliis* in the group treated with the NBu-E decreased over time. 9.5% mortality was recorded at 11 days after the exposure (60 mg · L^{-1} NBu-E) due to *I. multifiliis* infection, among the remaining fish, the average number of *I. multifiliis* trophonts on the gills and fins was 38.6±12.3 (mean±standard deviation). As for Eto-E (60 mg · L^{-1}), 57.1% mortality was recorded at 11 days, the average number of I. multifiliis trophonts on their gills and fins was 182.7±48.7 (mean±standard deviation). Furthermore the reproduction of tomonts obtained from the three fish transfer to freshwater at 5 days was remarkably reduced in the NBu-E groups. We did not collected enough theronts, therefore, we can not evaluated the infective ability of the theronts released from trophonts (from the 3 fish) after treated with NBu-E for 3 days, though, as shown in vivo test the number of released theronts significantly reduced. These results suggest that continuous immerse with NBu-E at the concentration of 30 mg · L^{-1} for 3 days is suitable for controlling *I. multifiliis* infection in these groups.

Table 2　*In vivo* efficacy of NBu-E and Eto-E against Ichat 11 days

Concentration (mg · L^{-1})	Tomonts obtained at day 3		Fish mortality (%)	Number of trophonts in each fish (gills and fins)
	Tomont mortality (%)	reproduction		
0 (control)	0.0±0.0a	511±43.3a	100	311±43.7a
NBu-E 10	36.7±3.3b	367±39.7b	42.8	189.7±35.0b
20	53.3±6.7c	271±21.0c	19.0	103.3±28.3c
30	76.7±3.3d	104±18.3d	9.5	38.6±12.3d
Eto-E 20	35.0±3.3e	489±52.7a	85.7	265±44.3e
40	43.3±6.7f	411±33.0b	71.4	222.0±35.7f
60	51.6±6.7g	366±13.3c	57.1	182.7±48.7g

The reproduction was represented as number of theronts released by each live tomont. Each value is expressed as mean±standard deviation of three replicates, and within a column, values followed by the different letters are significantly different (P<0.05).

In the acute toxicity test for grass carp, most of the fish exhibited fast respiratory frequency when the concentration of NBu-E exceeded 119 mg · L^{-1}. Mortality occurred when the concentration reached 119 mg · L^{-1}, All grasscarp died in tanks treated with 250 mg · L^{-1} of NBu-E, the linear equation $y = -26.38+12.06x$ which was derived from the regression analysis of probit mortality of grasscarp in test solution bioassay, the calculated LD$_{50}$ at 48 h was 152.4 mg · L^{-1} which is more 5 times than the effective one (30 mg · L^{-1}, Table 2). These findings ensure the safety or the use of NBu-E in the control of I. multifilliis infection, and suggest that it has great potential for the development of a new parasiticide.

The genus Streptomyceswas classified under the family Streptomycetaceae, are best known microorganism from soils where they contribute significantly to the turnover of complex biopolymers and antibiotics (Woese et al., 1987). Major types of antibiotics produced by Streptomycesare aminoglycosides, anthracyclins, glycopeptides, β – lactams, macrolides, nucleosides, peptides, polyenes, polyethers, and tetracyclines (Miyadoh, 1993). Different pharmacological activities of their metabolic products have been reported previously and relative mechanism of these activities has been demonstrated. Many researches revealed that production of extracellular cell-wall of Streptomyces spp. degrading enzymes (chitinase, β-1, 3-glucanase, cellulase, β-glucosidase, amylase, xylanase and mannase) inhibits pathogen growth significantly (Spear et al., 1993; Sepulveda and Crawford, 1998; Trejo-Estrada et al., 1998; Prapagdee et al., 2008; Gonzalez-Franco and Hernandez, 2009; Shafique et al., 2009). Additionally, some species such as Streptomyces pilosus, have ferrichrome siderophores that allows the sequesteration of iron as a growth limiting factor (Müller et al., 1984). Fathima and Adeline (2013) reported that both cells and cell-free extracts of Streptomyces griseus showed great antifungal activity and the enzymatic activities contribute to antifungal activities of S. griseus. All these contribute to effective biocontrol activity of the Streptomycesspp. For the present study, the antiparasiticmechanisms of action of extracellular products of S. griseus might be also attribute to these factors, however the exact mechanism of action regarding their antiparasitic efficacyr emains to be further investigated.

The results reported in this study have demonstrated that the NBu-E extract of extracellular products of S. griseus has potential for the control of I. multifilliis. However, the exact still has to be evaluated under field conditions, and toxicity tests to other fish species also need to be accessed. Future studies will investigate on the isolation and characterization of the active compounds in this NBu-E extract.

Reference:

Alderman D. J. (1985) Malachite green: a review. J Fish Dis 8: 289-298.

Anderson A. S., Wellington E. M. (2001) The taxonomy of Streptomyces and relatedgenera. Int J Syst Evol Microbiol 51: 797-814.

Becker J. O., Zavaleta M. E., Colbert S. F., Schroth M. N., Weinhold A. R., Hancock J. G., Vangundy S. D. (1988) Effects of rhizobacteria on root knot nematodes and gall formation. Phytopathology 78: 1466-1469.

Buchmann K., Sigh J., Nielsen C. V., Dalgaard M. (2001) Host responsesagainst the fish parasitizing ciliate Ichthyophthirius multifiliis. Vet Parasitol 100: 105-116.

Buchmann K., Jensen P. B., Kruse K. D. (2003) Effects of sodium percarbonate and garlic extract on Ichthyophthirius multifiliis theronts and tomocysts: in vitro experiments. N Am J Aquacult 65: 21-24.

Chen S. Y., Dickson D. W. (1996) Fungal penetration of the cystwall of Heterodera glycines. Phytopathology 86: 319-327.

Chen S. Y., Liu X. Z. (2005) Control of the soybean cyst nematode by the fungi Hirsutella rhossiliensis and Hirsutella minnesotensisin greenhouse studies. Biol Control 32: 208-219.

Crump D. H., Sayre R. M., Young L. D. (1983) Occurence of nematophagous fungi in cysts. Plant Dis 67: 63-64.

Dickerson H. W. (2006) Ichthyophthirius multifiliisand Cryptocaryon irritans (Phylum). In: Woo, P. T. K. (Ed.), Fish Diseases and Disorders. Protozoan and Metazoan Infections, vol. 1, 2nd ed. CAB International, Wallingford, UK, pp. 116-153.

Fathima A. Z., Adeline S. Y. T. (2013) Investigating the bioactivity of cells and cell-free extracts of Streptomyces griseus towards Fusarium oxysporum f. sp. cubense race. Biological Control 66: 204-208.

Gonzalez-Franco A. C., Hernandez L. R. (2009) Actinomycetes as biological control agents of phytopathogenic fungi. Techno Chih III (2): 64-73.

Hines R. S., Spira D. T. (1974) Ichthyophthiriasis in the mirror carp Cyprinus carpio (L.). III. Pathology. J Fish Biol 6: 189-196.

Jaafar R. M., Buchmann K. (2011) Toltrazuril (BAYCOX © VET.) in feed can reduce Ichthyophthirius multifiliis invasion of rainbow trout (salmonidae). Acta Ichthyologica Et Piscatoria 41: 63-66.

Kim K. H., Choi E. S. (1998) Treatment of Microcotyle sebastis (Monogenea) on the gills of cultured rockfish (Sebastes schelegeli) with oral administration of mebendazole and bithionol. Aquacult 167: 115-121.

Kirby G. C. (1996) Medicinal plants and the control of protozoal disease with particular reference to malaria. Trans R Soc Trop Med Hyg 90: 605-609.

Leij D. F. M., Kerry B. R., Dennehy J. A. (1992) The effect of fungal application rate and nematode density on the effectiveness of Pochonia chlamydosporia as a biological control agent for Meloidogyne incognita. Nematologica 38: 112-122.

Ling F., Wang J. G., Wang G. X., Gong X. N. (2011) Effect of potassium ferrate (VI) on survival and reproduction of Ichthyophthirius multifiliis tomonts. Parasitol Res 109: 1423-1428.

Ling F., Wang J. G., Lu C., Wang G. X., Lui Y. H., Gong X. N. (2012) Effects of aqueous extract of Capsicum frutescens (Solanaceae) against the fish ectoparasite Ichthyophthirius multifiliis. Parasitol Res 111: 841-848.

Matthews R. A. (2005) Ichthyophthirius multifiliis Fouquet and ichthyophthiriosis in freshwater teleosts. Adv Parasitol 59: 159-241.

Miyadoh S. (1993) Research on antibiotic screening in Japan over the last decade: aproducing microorganisms approach, Actinomycetologica 9: 100-106.

Müller G., Matzanke B. F., Raymond K. N. (1984) Iron transport in Streptomyces pilosus mediated by ferrichrome siderophores, rhodotorulic acid, and enantiorhodotorulicacid. J Bacteriol 160: 313-318.

Pathom-aree W., Stach J. E. M., Ward A. C., Horikoshi K., Bull A. T., Goodfellow M. (2006) Diversity of actinomycetes isolated from challenger deep sediment (10, 898 m) from the Mariana Trench, Extremophiles 10: 181-189.

Prapagdee B,. Kuekulvong C., Mongkolsuk S. (2008) Antifungal potential of extracellular extracts produced by Streptomyces hygroscopicusagainst phytopathogenic fungi. Int J Biol Sci 4: 330-337.

Rowland S. J., Mifsud C., Nixon M., Read P., Landos M. (2009) Use of formalin and copper to control ichthyophthiriosis in the Australian freshwater fish silver perch (Bidyanus bidyanus Mitchell) Aquacult Res 40: 44-54.

Shafique S., Bajwa R., Shafique S. (2009) Screening of Aspergillus nigerand A. flavus strains for extra-cellular amylase activity. Pak J Bot 41: 897-905.

Shinn P., Camacho A. P., Bron S. M., Conway J. E., Yoon D., Guo G. H., Taylor F. C., Ni G. H. (2012) The anti-protozoal activity of bronopol onthe key life-stages of Ichthyophthirius multifiliisFouquet, 1876 (Cilio-phora). Vet Parasitol 186: 229-236.

Spear L., Gallagher J., McHale L., McHale A. P. (1993) Production of cellulase and bglucosidase activities following growth of Streptomyces hygroscopicuson cellulose containing media. Biotechnol Lett 15: 1265-1268.

Stackebrandt E., Rainey F. A., Wardrainey N. L. (1997) Proposal for a new hierarchic classification system Actinobacteria classis Nov. Int J Syst Bacteriol 47: 479-491.

Straus D. L., Hossain M. M., Clark T. G. (2009) Strain Differences in Ichthyophthirius multifiliis to Copper Toxicity. Dis Aquat Org 83: 31-36.

Straus D. L., Meinelt T. (2009) Acute toxicity of peracetic acid (PAA) formulations to Ichthyophthirius multifiliis theronts. Parasitol Res 104: 1237-1241.

Sudová E., Straus D. L., Wienke A., Meinelt T. (2010) Evaluation of continuous 4-day exposure to peracetic acid as a treatment for Ichthyophthirius multifiliis. Parasitol Res 106: 539-542.

Sun M. H., Gao L., Shi Y. X., Li B. J., Liu X. Z. (2006) Fungi and actinomycetes associated with Meloidogyne spp. eggs and in China and their biocontrol potential. J Invertebr Pathol 93: 22-28.

Traxler G. S, Richard J., McDonald T. E. (1998) Ichthyophthirius multifiliis (Ich) epizootics in spawning sockeye salmon in British Columbia, Canada. J Aqua Ani Health 10: 143-151

Trejo-Estrada S. R., Paszczynski A., Crawford D. L. (1998) Antibiotics and enzymes produced by the biocontrol agent Streptomyces violaceusnigerYCED-9. J Ind Microbiol Biotechnol 21: 81-90

Trejo-Estrada S. R., Paszczynski A., Crawford D. L. (1998) In vitro and in vivo antagonism of Streptomyces violaceusniger YCED9 against fungal pathogens of turfgrass. World J Microbiol biot 14: 865-872

Wahli T., Schmitt M., Meier W. (1993) Evaluation of alternatives to malachite green oxalate as a therapeutant for ichthyophthiriosis in rainbow-trout, Oncorhynchus mykiss. J Appl Ichthyol 9: 237-249.

Watve M. G., Tickoo R., Jog M. M., Bhole B. D. (2001) How many antibiotics are producedby the genus Streptomyces, Arch Microbiol 176: 386-390

Woese C. R. (1987) Bacterial evolution. Microbiol Rev 51: 221-271

Wohllebe S., Richter P., Häder D. P. (2012) Chlorophyllin for the control of Ichthyophthirius multifiliis (Fouquet) Parasitol Res 111: 729-733

Xu D. H., Shoemaker C. A., Klesius P. H. (2008) Effect of tricaine methanesulfonate on survival and reproduction of the fish ectoparasite Ichthyophthirius multifiliis. Parasitol Res 103: 979-982

Wang G. X., Han J., Feng T. T., Li F. Y., Zhu B. (2009) Bioassay-guided isolation and identification of active compounds from Fructus Arctii against Dactylogyrus intermedius (Monogenea) in goldfish (Carassius auratus) Parasitol Res 106: 247-255

Yao J. Y., Shen J. Y., Li X. L., Xu Y., Hao G. J., Pan X. Y., Wang G. X., Yin W. L. (2010) Effect of sanguinarine from the leaves of Macleaya cordataagainst Ichthyophthirius multifiliis in grass carp (Ctenopharyngodon idella). Parasitolo Res107: 1035-1042

Yao J. Y., Zhou Z. M., Li X. L., Yin W. L., Ru H. S., Pan X. Y., Hao G. J., XuY., Shen J. Y. (2011) Antiparasitic efficacy of dihydrosanguinarine and dihydrochelerythrine from Macleaya microcarpa against Ichthyophthirius multifiliisin richadsin (Squaliobarbus curriculus). Vet Parasitol 183: 8-13

Yi Y. L., Lu C., Hu X. G., Ling F., Wang G. X. (2012) Antiprotozoal activity of medicinal plants against Ichthyophthirius multifiliis in goldfish (Carassius auratus) Parasitol Res 111: 1771-1778.

该文发表于《水产科学》【2010, 29 (5): 105-109】。

黄姜杀灭鱼类指环虫活性部位的初步研究

Study on active site of Dioscorea zingiberensis C. H. Wright to control the Dactylogyrus of fish

姚嘉赟，沈锦玉，尹文林，潘晓艺，郝贵杰，徐洋

（浙江省淡水水产研究所，浙江 湖州 313001）

摘要： 以寄生在金鱼鳃部的指环虫为指示寄生虫，采用活体感染、活体杀虫的方法，通过不同极性的溶剂回流提取黄姜，制备粗提物，进行杀灭金鱼指环虫的药效活性追踪试验，确定黄姜杀灭鱼类指环虫的活性部位，并对活性部位进行安全性评价。实验结果表明：黄姜的杀虫活性部位是70%乙醇部位，其浓度为20.0 mg/L时，平均最高杀虫率为100%。70%乙醇提取物经过进一步有机溶剂萃取后，杀虫药效试验表明，石油醚和最终水相萃取部位对指环虫具有杀灭作用，但石油醚萃取部位作用较佳，其浓度为5.0 mg/L时，平均最高杀虫率为100.0%。活性部位对草鱼急性毒性实验结果显示，石油醚萃取部位对草鱼的48 h半致死浓度（LC_{50}）为33.54 mg/L，其安全浓

度为 9.64 mg/L，表明黄姜是一种比较安全，杀虫效果良好的水产用中草药。

关键词：黄姜提取物；指环虫；活性部位；急性毒性试验

该文发表于《渔业科学进展》【2010，31（5）：105-109】。

重楼杀灭罗氏沼虾寄生纤毛虫的初步研究

Study on efficacy of Rhizorma Paridi against the ciliates on Giant Freshwater Shrimp（*Mactobrachium rosenbergii*）

姚嘉赟[1]，沈锦玉[1]，杨国梁[2]，王军毅[2]，尹文林[1]，徐洋[1]

（1. 浙江省淡水水产研究所，浙江，湖州 313001；2. 浙江南太湖淡水水产种业有限公司，313001）

摘要：利用系统溶剂极性法提取重楼，制备粗提物，以寄生罗氏沼虾上的纤毛虫（聚缩虫和靴纤虫）为指示寄生虫，进行杀灭罗氏沼虾寄生纤毛虫的药效活性追踪试验，初步确定重楼杀灭纤毛虫的活性部位，并对活性部位进行安全性评价。实验结果表明：不同极性的重楼提取物对纤毛虫均有一定的杀灭作用，其中重楼甲醇提物对纤毛虫的杀灭作用最强，其浓度为 45 mg/L 时，对纤毛虫杀灭率为 100%，确定其为杀灭纤毛虫的活性部位。活性部位对罗氏沼虾仔虾急性毒性实验结果显示，甲醇提取部位对仔虾的 96 h 半致死浓度（LC_{50}）为 272 mg/L，其安全浓度为 111.5 mg/L，表明重楼是一种比较安全，杀虫效果良好的水产用中草药。

关键词：重楼；纤毛虫；罗氏沼虾；活性部位

二、增效剂

某些化合物能提高抗菌素的杀菌效果。漆黄素（fisetin）和诺氟沙星对水生源粘质沙雷氏菌（*Serratia marcescens*）的体外协同效应（Jing Dong et al.，2016）。

本文发表在《FEMS MICROBIOLOGY LETTERS》【2016，363（1）】

In vitro synergistic effects of fisetin and norfloxacin against aquatic isolates of *Serratia marcescens*

Jing Dong[1,2]，Jing Ruan[3]，Ning Xu[1,2]，Yibin Yang[1,2]，Xiaohui Ai[1,2]

（1. Yangtze River Fisheries Research Institute，Chinese Academy of Fishery Sciences，Wuhan 430223，China；2. Hubei Freshwater Aquaculture Collaborative Innovation Center，Wuhan 430223，China；3. Institute of Hydrobiology，Chinese Academy of Sciences，Wuhan 430072，China；）

Abstract：*Serratia marcescens* is a common pathogenic bacterium which can cause infections both in human and animals. It can cause a range of diseases，from slight wound infections to life-threatened bacteraemia and pneumonia. The emergence of antimicrobial resistance has limited the treatment caused by the bacterium to a great extent. Consequently，there is an urgent need in developing novel antimicrobial strategies against this pathogen. Synergistic strategy is a new approach against the infections caused by drug resistant bacterium. In this paper，we isolated and identified the first multi-resistant pathogenic Serratia marcescens strain from diseased Soft-shelled turtle in China. Then we performed the checkerboard assay，the results showed that fisetin from 10 tested natural products has the synergistic effects against Serratia marcescens when combined with norfloxacin. The time-kill curve assay was further confirmed the results of checkerboard assay. We found that this novel synergistic effect could significantly reduce the dosage of norfloxacin against *Serratia marcescens*.

Key words：*Serratia marcescens*，resistance，fisetin，synergism

漆黄素和诺氟沙星对水生源粘质沙雷氏菌的体外协同效应

董靖[1,2]，阮晶[3]，胥宁[1,2]，杨移斌[1,2]，艾晓辉[1,2]

（1. 中国水产科学院长江水产研究所，中国 武汉 430223；2. 湖北淡水水产健康养殖湖北省协同创新中心，湖北 武汉 430223；
3. 中国科学院水生生物研究所，中国 武汉 430072）

摘要：粘质沙雷氏菌是一种常见的致病菌，可以导致动物和人的各种感染性疾病，包括从轻微的伤口感染到严重的可以威胁生命健康的菌血症和肺炎等。细菌耐药性的出现严重限制了抗生素在治疗细菌性感染中的应用。因此，迫切需要制订针对这种病原体的新抗菌战略。协同抗菌作用是针对由耐药细菌感染的新方法。在本文中，我们在患病甲鱼中分离和鉴定了一株重耐药致病粘质沙雷氏菌菌株。然后我们进行棋盘试验，结果表明，诺氟沙星与漆黄素具有协同抗菌作用。时杀菌曲线进一步证实了棋盘测定的结果。我们发现，这种新型的协同效应可能显著降低诺氟沙星对粘质沙雷氏菌用量。

关键词：粘质沙雷氏菌、耐药性、漆黄素、协同

Introduction

Serratia marcescens (*S. marcescens*), a gram-negative bacillus belonged to the Enterobacteriaceae family, can be widely isolated from the environment in the world. The organism was first considered as a nonpathogenic organism and widely used as a biological marker because it could be easily recognized by the red color (Lin et al., 2014). Furthermore, researchers have found that strains of S. marcescens can be used as a biodegradation of pentachlorophenol and pulp paper mill effluent (Wongsa et al., 2004, Singh et al., 2008, Singh et al., 2009). However, cases of infections were reported since 1951 and bacteraemia was first recorded in 1968. Therefore, it was known as an opportunistic pathogen which was related to several diseases both in human and animals (Hejazi & Falkiner, 1997). It could cause a variety of diseases including mastitis, urinary tract infection, wound infections and pneumonia which threatened human health (Cox, 1985, Gouin et al., 1993). Antibiotics play an important role in the battle against infections caused by S. marcescens. However, it was reported that S. marcescens strains was resistant to a large number of antibiotics, β-lactam, aminoglycoside and fluoroquinolone were included (Ishii et al., 2012, Yang et al., 2012).

Currently, fluoroquinolones with a broad-spectrum activity against infections caused by bacteria were widely used in aquaculture in China which had resulted in an increasing emergence of fluoroquinolones-resistance strains (Yang et al., 2012). Because of the appearance of resistance, the therapy using antibiotics were limited. Therefore, there was an urgent need to develop new strategies against this organism. Combination therapy is a novel strategy and maybe useful in the treatment of infections caused by drug-resistance strains of S. marcescens.

In this paper, we isolated a pathogenic strain of S. marcescens from Soft-shelled turtle in Hubei China. Biochemistry and phylogenetic analysis were performed to determine the taxonomic position of the strain. Furthermore, the sensitivity to antibiotics of the strain was examined using the Kirby-Bauer disk diffusion method and the synergistic effect of 10 natural products isolated from Chinese herb medicine with norfloxacin were tested. According to the results, we found that fisetin could be a potential drug in the control of drug-resistant S. marcescens infections.

Materials and methods

Micro-organism and reagents

Bacteria Strains were isolated from the livers of diseased Soft-shelled turtle in Hubei China with typical symptom of infection according to the previous report (Cao et al., 2010). Natural products and norfloxacin (purity>98%) were purchased from National Institutes for Food and Drug Control (Beijing, China). Stock solutions of the agents at a concentration of 40960μg/ml were prepared in dimethyl sulfoxide (Sigma-Aldrich, St. Louis, MO, USA).

Identification of S. marcescens

The strain was firstly determined by biochemical assay using the API 20E strips in the ATB 32GN system. Briefly, the strain was cultured in LB agar plates at 28℃ for 18-24 h, then single clone of the strain were re-suspended and loaded on an API 20E strip according to the manufacturer's instructions. The strips were incubated at 28℃ for 24 h. The biochemical characterization was determined according to the API identification index and database.

For phylogenetic analysis, 16S rRNA of the isolate was amplified using polymerase chain reaction. After sequencing, the 16S rRNA of the strain was identified using the Basic Local alignment Search Tool program (Altschul et al., 1990).

Susceptibility assay to antibiotics

The susceptibility of the strain to 16 antibiotics was determined using the Kirby-Bauer disk diffusion methods. After the incubation of 18-24 h, the diameters of inhibition zone were measured. Then the susceptibility was determined according to the method recommended by the Clinical and Laboratory Standards Institute (CLSI).

Determination of MICs

MICs of natural products and norfloxacin for S. marcescens were determined in triplicate using the broth microdilution method according to the CLSI. Briefly, serial 2-fold dilutions of the drugs at a starting concentration of 512 μgmL^{-1} were loaded into a 96-well microplate using Mueller-Hinton broth (MHB). Then the strain at a density of about 5×10^5cfu mL^{-1} in MHB was incubated with the drugs for further 18-20 h at 28℃. The MICs were defined as the lowest concentration inhibiting bacterial growth. S. marcescens ATCC 8100 was employed as control strain.

Statistical analysis

The MICs between norfloxacin alone and fisetin-norfloxacin combination were analyzed by Student's t-test with SPSS 14.0 statistical software (SPSS Inc., Chicago, IL), and a p value less than 0.05 was considered to be statistically significant.

Results

Identification of the bacterial strain

Only one bacterial strain was isolated from the liver of the diseased Soft-shelled turtle. When cultured the strain in LB medium at 37℃, the medium showed pink colonies. The biochemical characterization results showed that the strain was identified as a S. marcescens strain with 99.5% phenotypic identity (Table 1). Moreover, the result of the 16S rRNA sequence analysis showed that the sequence shared 99% homology with typical strain ATCC 8100.

Table 1 Biochemical characterization of Serratia marcescens isolate

Tested parameters	Reactions	Tested parameters	Reactions
D-melibiose	-	Lysine deaminase	+
Arabinose	-	D-amygdalin	-
Gelatin	+	Citrate	+
H$_2$S	-	D-glucose	+
Inositol	+	Arginine double hydrolase	-
D-mannitol	+	VP	+
Ornithine dehydrogenase	+	β-galactosidase	+
Oxidase	-	L-rhamnose	-
Sucrose	+	D-sorbitol	+
Tryptophan deaminase	-	Urease	-

+, positive; -, negative

All data represent of three independent experiments.

Antibiotic resistance

Most S. marcescens strains were reported as multi-resistance to several antibiotics (Yang et al., 2012). We found that the strain was resistant to most of the tested antibiotics (Table 2), intermediate to ciprofloxacin and norfloxacin (Table 2). No susceptible antibiotics were found among the tested agents (Table 2).

Table 2 Susceptibility of the isolated Serratia marcescens strain to antibiotics

Antibiotics	Classification	Resistance	Content (μg/disc)	Inhibition zone (mm)
Ciprofloxacin	Quinolones	I	5	18.5±1.82
Norfloxacin	Quinolones	I	10	13.15±1.24

Antibiotics	Classification	Resistance	Content（μg/disc）	Inhibition zone（mm）
Kanamycin	Aminoglycosides	R	30	10.26±0.87
Streptomycin	Aminoglycosides	R	10	9.83±0.25
Amikacin	Aminoglycosides	R	30	0
Neomycin	Aminoglycosides	R	30	0
Tobramycin	Aminoglycosides	R	10	0
Erythromycin	Macrolides	R	15	0
Azithromycin	Macrolides	R	15	0
Sulfisoxazole	Sulfa	R	300	0
Rifampicin	Rifamycins	R	5	0
Clindamycin	Lincosamides	R	2	0
Chloromycetin	Amphenicols	R	30	0
Amoxicillin	β-Lactams	R	20	0
Cefradine	Cephalosporins	R	30	0
Furazolidone	Nitrofurans	R	300	0

R, resistant; I, intermediate; S, susceptible.

All data represent of three independent experiments.

Antimicrobial activities of single agents

activities of norfloxacin and 10 natural products The *in vitro* antimicrobial were determined by MIC. The results of the agents alone and natural products combined with norfloxacin were listed in Table 3. From the results we found that the MIC value of norfloxacin could reach to 64 μg·mL^{-1} for the isolated strain, while the MIC values of natural products could reach the ranges from 64 μg·mL^{-1} to 1 024μg·mL^{-1}.

Table 3　The MIC values of natural products and norfloxacin（alone or in combination）against isolated Serratia marcescens strain

Natural product name	MIC of agents alone（μg·mL^{-1}）		MIC in combination（μg·mL^{-1}）		Results	
	MIC of natural product	MIC of norfloxacin	MIC of natural product	MIC of norfloxacin	FICI	INT
Apigenin	128		64	32	1	IND
Fisetin	256		32	1	0.14	SYN
Verbascoside	1024		512	16	0.75	IND
Rutin	1024		1024	16	1.25	IND
Chlorogenic acid	512	64	512	32	1.5	IND
Quercitrin	1024		256	32	0.75	IND
Resveratrol	64		16	32	0.75	IND
Dihydromyricetin	256		64	32	0.75	IND
Naringenin	512		64	32	0.625	IND
Daphnetin	128		64	32	1	IND

MIC, minimum inhibitory concentration; FICI, fractional inhibitory concentration index; INT, interpretation; SYN, synergism; IND, indifference; ANT, antagonism.

All data represent of three independent experiments. The fisetin-norfloxacin combination has a significant reduce when compared with norfloxacin alone by Student t-test. （p<0.05）

Checkerboard assay

To evaluate the synergistic effect of norfloxacin combined with natural products, checkerboard assays were performed and the FICI values were calculated for each agent. The FICI data were summarized in Table 3, the FICI value of fisetin combined with norfloxacin could reach to 0. 14. This result reveals that fisetin combined with norfloxacin showed a good synergistic effect under our experimental-conditions (Table 3). There was a significant decrease between norfloxacin alone and the fisetin-norfloxacin combination when analyzed by Student t-test (p<0. 05). No antagonistic effect was observed in all the tested agents.

Time-kill curves

To further approve the synergistic effect of fisetin, we performed time-kill curves with norfloxacin alone or combined with fisetin using the isolated *S. marcescens* strain. As shown in Figure 1, the combination of fisetin could increase the killing activity of norfloxacin from 12 h to 48 h. However, the synergy was only observed after 24 h. At the time point of 48 h, the yield of combination had a 2. 44 log10 cfu · mL^{-1} decrease compared with 32 μg · mL^{-1} of norfloxacin only. The killing activity of norfloxacin could be significantly increased when combined with fisetin.

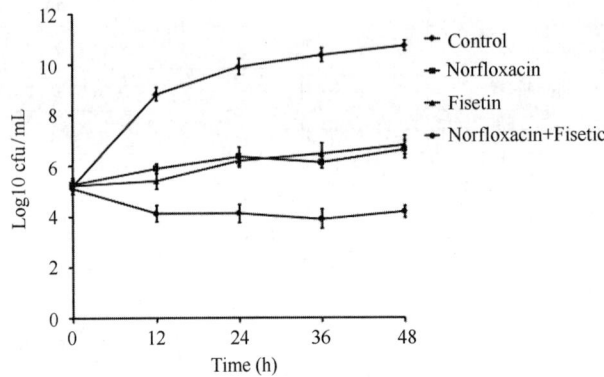

Fig. 1 Time-kill curve with norfloxacin and fisetin alone or in combination against *Serratia marcescens*.

Discussion

Ever since the first antibiotics were introduced for the treatment of bacterial infections, the development of novel antibiotics accompanied the spread of resistance in the world (Palumbi, 2001). Moreover, the decline in the identification of novel antibiotics leads us to the post-antibiotic era. There will be no available antibiotics in the treatment of bacterial infections in the future (Qiu et al., 2012). Particularly the emergence of multi-resistant bacteria makes the battle even tough, therefore new approaches against these pathogens have never suspended. Chinese traditional herbs have used for thousands years in the treatment of infectious diseases, natural products extracted from Chinese traditional herbs are included. Some of the natural products were earlier reported that had synergistic activity combined with antibiotics in the treatment of antibiotic resistant strains (Novy et al., 2011).

S. marcescens were first been identified as non-pathogenic bacteria and widely used in the world as a bio-marker. In recent years, S. marcescens has been reported could cause severe diseases both in human and animals (Mills & Drew, 1976, Korner et al., 1994). Because of the emergence of multi-resistant strains, the pursuit of new treatment overcoming the infections caused by S. marcescens is an important issue in clinic (Ishii et al., 2014). In this paper, we isolated a multi-resistant S. marcescens strain from the livers of diseased Soft-shelled turtle, which have never been reported in this animal. According to our result of susceptibility assay, we found the isolated strain was not sensitive to all the tested antibiotics. Aminoglycoside and quinolone antibiotics have shown good inhibitory activity against gram-negative strains. However, the strain in this study is resistant to all 5 tested Aminoglycoside antibiotics and intermediary to 2 quinolone antibiotics. These results are different from the strain isolated from diseased Siberian sturgeonsby Cao et al (Cao et al., 2014). Nevertheless, these results reveal that the resistance of S. marcescens strain becomes more severe. Although Cao et al had reported several herbal extract could inhibit the growth of S. marcescens, the concentration was high and not suitable for as a drug in aquaculture.

Synergistic strategies are one of the most important approaches for the treatment of infectious diseases caused by drug resistant strains. Several natural products have been previously reported that have synergistic activities when combined with antibiotics (Drago et al., 2011, Chen et al., 2014). Natural products combined with antibiotics can decrease the dosage of antibiotics, reduce the side-effects, enhance the antimicrobial activity and lower the drug resistance (Novy et al., 2011). Moreover, natural products have been identified have the activities against the expression and activity of virulence factors secreted by bacterial pathogens (Qiu et al., 2012, Dong et al., 2013, Dong et al., 2013). Fisetin was identified as an inhibitor of hemolysin secreted by Listeria monocytogenes (Wang et al., 2015). The synergistic activity against S. marcescens has never been conducted. In the paper, we analyzed 10 the interactions between natural products and norfloxacin using checkerboard method. According to the results, we found that fisetin had the activity of synergism when combined with norfloxacin. The results were further verified by the time-kill curve assay. These results indicated that fisetin-norfloxacin complex can be used as a candidate against multi-resistant S. marcescens infections.

Acknowledgements

This work is supported by National Nonprofit Institute Research Grant of Freshwater Fisheries Research Center, CAFS (No. 2015JBFM37).

Reference

Altschul S. F., Gish W., Miller W., Myers E. W. & Lipman D. J. (1990) Basic local alignment search tool. J Mol Biol215: 403-410.

Cao H. P., He S., Lu L. Q. & Hou L. D. (2010) Characterization and Phylogenetic Analysis of the Bitrichous Pathogenic *Aeromonas hydrophila* Isolated from Diseased Siberian sturgeon (Acipenser baerii). Isr J Aquacult-Bamid62: 181-188.

Cao H. P., Yang Y. B., He S., Zheng W. D., Liu K. & Yang X. L. (2014) Control by Herbal Extract of *Serratia marcescens* from Cultured Siberian Sturgeon Acipenser baerii Brandt. Isr J Aquacult-Bamid66.

Chen Z., Li X., Wu X., Wang W., Xin M., Shen F., Liu L., Liang J., Li L. & Yu L. (2014) Synergistic Activity of Econazole-Nitrate and Chelerythrine against Clinical Isolates of Candida albicans. Iran J Pharm Res13: 567-573.

Cox C. E. (1985) Aztreonam therapy for complicated urinary tract infections caused by multidrug-resistant bacteria. Rev Infect Dis7 Suppl 4: S767-771.

Dong J., Qiu J., Wang J., et al. (2013) Apigenin alleviates the symptoms of *Staphylococcus aureus pneumonia* by inhibiting the production of alpha-hemolysin. FEMS Microbiol Lett338: 124-131.

Dong J., Qiu J. Z., Zhang Y., et al. (2013) Oroxylin A Inhibits Hemolysis via Hindering the Self-Assembly of alpha-Hemolysin Heptameric Transmembrane Pore. Plos Computational Biology9.

Drago L., Nicola L., Rodighiero V., Larosa M., Mattina R. & De Vecchi E (2011) Comparative evaluation of synergy of combinations of beta-lactams with fluoroquinolones or a macrolide in *Streptococcus pneumoniae*. J Antimicrob Chemother66: 845-849.

Gouin F., Papazian L., Martin C., Albanese J., Durbec O., Domart Y., Veyssier P., Leroy J., Gres J. J. & Rollin C. (1993) A non-comparative study of the efficacy and tolerance of cefepime in combination with amikacin in the treatment of severe infections in patients in intensive care. J Antimicrob Chemother32 Suppl B: 205-214.

Hejazi A. & Falkiner F. R. (1997) *Serratia marcescens*. J Med Microbiol46: 903-912.

Ishii K., Adachi T., Hamamoto H. & Sekimizu K. (2014) *Serratia marcescens* suppresses host cellular immunity via the production of an adhesion-inhibitory factor against immunosurveillance cells. J Biol Chem289: 5876-5888.

Ishii K., Adachi T., Imamura K., Takano S., Usui K., Suzuki K., Hamamoto H., Watanabe T. & Sekimizu K. (2012) *Serratia marcescens* induces apoptotic cell death in host immune cells via a lipopolysaccharide-and flagella-dependent mechanism. The Journal of biological chemistry287: 36582-36592.

Korner R. J., Nicol A., Reeves D. S., MacGowan A. P. & Hows J. (1994) Ciprofloxacin resistant *Serratia marcescens* endocarditis as a complication of non-Hodgkin's lymphoma. J Infect29: 73-76.

Lin Q. Y., Tsai Y. L., Liu M. C., Lin W. C., Hsueh P. R. & Liaw S. J. (2014) Serratia marcescens arn, a PhoP-regulated locus necessary for polymyxin B resistance. Antimicrob Agents Chemother58: 5181-5190.

Mills J. & Drew D. (1976) Serratia marcescens endocarditis: a regional illness associated with intravenous drug abuse. Ann Intern Med84: 29-35.

Novy P., Urban J., Leuner O., Vadlejch J. & Kokoska L. (2011) In vitro synergistic effects of baicalin with oxytetracycline and tetracycline against *Staphylococcus aureus*. J Antimicrob Chemother66: 1298-1300.

Novy P., Urban J., Leuner O., Vadlejch J. & Kokoska L. (2011) In vitro synergistic effects of baicalin with oxytetracycline and tetracycline against *Staphylococcus aureus*. J Antimicrob Chemoth66: 1298-1300.

Palumbi S. R. (2001) Evolution-Humans as the world's greatest evolutionary force. Science293: 1786-1790.

Qiu J., Niu X., Dong J., Wang D., Wang J., Li H., Luo M., Li S., Feng H. & Deng X. (2012) Baicalin protects mice from *Staphylococcus aureus* pneumonia via inhibition of the cytolytic activity of alpha-hemolysin. J Infect Dis206: 292-301.

Singh S., Chandra R., Patel D. K., Reddy M. M. K & Rai V. (2008) Investigation of the biotransformation of pentachlorophenol and pulp paper mill effluent decolorisation by the bacterial strains in a mixed culture. Bioresource Technol99: 5703-5709.

Singh S., Singh B. B., Chandra R., Patel D. K. & Rai V. (2009) Synergistic biodegradation of pentachlorophenol by Bacillus cereus (DQ002384), Serratia marcescens (AY927692) and Serratia marcescens (DQ002385). World J Microb Biot25: 1821-1828.

Wang J. F., Qiu J. Z., Tan W., et al. (2015) Fisetin Inhibits Listeria monocytogenes Virulence by Interfering With the Oligomerization of Listeriolysin O. Journal of Infectious Diseases211: 1376-1387.

Wongsa P., Tanaka M., Ueno A., Hasanuzzaman M., Yumoto I. & Okuyama H. (2004) Isolation and characterization of novel strains of Pseudomonas aeruginosa and Serratia marcescens possessing high efficiency to degrade gasoline， kerosene， diesel oil， and lubricating oil. Curr Microbiol49：415-422.

Yang H. F.， Cheng J.， Hu L. F.， Ye Y. & Li J. B. (2012) Identification of a *Serratia marcescens* clinical isolate with multiple quinolone resistance mechanisms from China. Antimicrob Agents Chemother56：5426-5427.

三、免疫增强剂及生长调节剂

免疫增强剂可增强水产养殖动物的非特异性免疫能力，壳聚多糖等免疫增强剂能提高中华绒螯蟹血清中溶菌酶、超氧化物歧化酶（SOD）的活力，预防疾病的发生（沈锦玉等，2004）。

该文发表于《浙江农业学报》【2004，16（1）：25-29】。

免疫增强剂对中华绒螯蟹免疫功能的影响

Effects of immunostimulants on non-specific immunity of *Eriocheir sinensis*

沈锦玉，刘问，曹铮，尹文林，沈智华，钱冬，吴颖蕾

（浙江省淡水水产研究所，湖州 313001）

摘要：用三种不同免疫增强剂多糖（laminarin）、灭活细菌苗、壳聚多糖（Chitosan）注射或添加于饲料中饲喂中华绒螯蟹，在不同时间取样测定中华绒螯蟹血清中非特异性免疫指标的变化及中华绒螯蟹的免疫率。结果表明：经注射后48 h，A、B、C组中华绒螯蟹血清中溶菌酶活力分别为0.185、0.234、0.233，而对照组的平均值为0.094，经t检验，差异不显著；超氧化物歧化酶（SOD）的活力分别为283.2、263.5、289.8，而对照组平均为120.15，经t检验，A、B组血清中SOD活力差异极显著，C组差异显著。口服免疫增强剂后第九天，血清中溶菌酶、超氧化物歧化酶（SOD）的活力明显高于对照组，经t检验，各组血清中溶菌酶活力与对照相比，差异极显著；SOD的活力与对照相比，E组差异极显著，F、G组差异显著。注射免疫增强剂组蟹的免疫保护率达75%，而口服免疫增强剂组与对照组相比较，蟹的免疫保护率没有明显的差异。壳聚糖经生产性试用，能较好地预防病害的发生，中华绒螯蟹成活率比对照组提高10%以上。

关键词：中华绒螯蟹，免疫增强剂，非特异性免疫，溶菌酶，SOD

该文发表于《华南农业大学学报》【2011，32（3）：111-118】。

昆虫生长调节剂虫酰肼对典型海弯水虱的防控研究

The Effect of the Insect Growth Regulator Tebufenozide on *Alitropus typus*

王伟利[1]，陆晓苗[1]，彭华林[2]，邹为民[1]，姜兰[1]，赵飞[1]，谭爱萍[1]

（1. 中国水产科学研究院珠江水产研究所，广东 广州 510380；2. 广东省水生动物疫病预防控制中心，广东 广州 510222）

摘要：为了探索安全有效的典型海湾水虱控制方法，选用昆虫生长调节剂虫酰肼（蜕皮激素类似物）为试验药物，观察其对典型海湾水虱生长发育的影响。结果发现：按鱼体质量以剂量为 $0.02\ \mu L \cdot g^{-1} \cdot d^{-1}$ 的虫酰肼拌料连续饲喂宿主 5 d，宿主血液中的药物可影响典型海弯水虱的幼体、早期成体及较小型成体的正常蜕皮，从而引起虫体在蜕皮期间的高死亡率. 剂量减半、缩短给药时间或作用于该虫中等以上个体的情况下，该药效果不明显。

关键词：典型海弯水虱；蜕皮激素类似物；虫酰肼；蜕皮

第三节 微生态制剂的研究与应用

微生态制剂（Microecological preparation）是根据微生态学原理而制成的含有大量有益菌的活菌制剂，有的还含有它们的代谢产物或（和）添加有益的生长促进因子，具有维持动、植物和人类及其内、外环境的微生态平衡（或调整其微生态失调），提高其健康水平和保护环境的功能。我国微生态制剂在水产养殖中的应用始于20世纪80年代初期，微生态制剂因无毒副作用、无耐药性、无残留、成本低、效果显著、不污染环境，广泛地应用于水产养殖中。本章阐述了组成微生态制剂的功能性微生物研究与微生态制剂的研制与应用。

一、功能性微生物研究

对于微生态制剂的分类，各国有不同的规定。1989年，美国FDA公布了42种微生物菌种较为安全，可以直接饲喂。1999年，我国农业部允许使用的微生态制剂仅有12种，目前，我国市场上的微生态制剂产品很多，但主要成分都是益生菌群。水产养殖上应用较多的微生态制剂有光合细菌、放线菌、酵母菌、芽孢杆菌、硝化细菌、乳酸菌、弧菌等微生物以及EM菌群等。

1. 解磷降氮

肠球菌（*Enterococcus* sp）菌株对可溶性有机磷具有高效的去除效率（黄志华等，2010）。外源性碳源、硝酸盐对反硝化聚磷菌（RC11）磷酸盐代谢活动具有显著的影响，厌氧条件下菌株RC11具有利用硝酸盐作为电子受体进行反硝化除磷的功能（郑宗林等，2010）。

该文发表于《水产学报》【2010，34（12）：1901-1907】。

碳源对反硝化聚磷菌（RC11）磷酸盐代谢的影响

Influence of carbon source on phosphate metabolism characteristics of denitrifying phosphate-accumulating organisms RC11

郑宗林[1,2]，叶金明[3]，刘 波[1]，杨先乐[4]，周兴华[1]，向 枭[1]

(1. 西南大学（荣昌校区）鱼类繁殖与健康养殖中心，重庆 402460；
2. 中国水产科学研究院淡水渔业研究中心，农业部淡水鱼类遗传育种和养殖生物学重点开放实验室，江苏 无锡 214081；
3. 扬州水产生产技术指导站，江苏 扬州 225000；4. 上海海洋大学农业部渔业植物病原库，上海 201306)

摘要： 反硝化聚磷微菌由于具有同时脱氮和除磷的特点，能够最大程度的减少碳源需求，为解决生物脱氮除磷工艺的碳源竞争矛盾提供了新的思路和方法。以纯培养方式探讨了外源性碳源、硝酸盐对反硝化聚磷菌（RC11）磷酸盐代谢活动的影响。结果表明，好氧培养时，菌株RC11在外源碳源存在时发生了超量吸磷现象；缺氧培养时，菌株RC11可以利用硝酸盐而非亚硝酸盐作为电子受体进行反硝化聚磷。无外加碳源时，菌株RC11经历厌氧阶段后初期可以利用硝酸盐氧化到亚硝酸盐的过程中产生的能量进行摄磷；但当亚硝酸盐的积累达到高峰时，进入以亚硝酸盐为电子受体的反硝化阶段，由于亚硝酸盐氮不能作为氧分子的替代物进行反硝化除磷，菌株RC11实际上处于一个厌氧环境，会引发释磷；在厌氧条件下菌株RC11具有利用硝酸盐作为电子受体进行反硝化除磷的功能。

关键词： 碳源；反硝化聚磷菌；磷酸盐代谢

养殖污水处理再利用已是水产养殖业必须面临的事实。目前我国已有的养殖废水处理技术的研究，但考虑的重点是去除有机物及氨氮转化为硝酸盐，硝酸盐的毒性虽比氨氮低，但过度积累同样会影响水生动物的生长。当前对养殖废水处理还没有考虑氮磷的去除，只是形式上的转化。生物反硝化除磷技术因为同时具有去除有机物、氮和磷的优点而被广泛应用于生活污水的处理并取得了良好效益。生物反硝化除磷技术的本质是根据微生物的生理代谢特点为其创造适合的生长环境来达到废水处理的目的。按照理想的除磷理论，碳源（电子供体）和氧化剂（电子受体）不能同时存在，否则会影响脱氮和除磷效果[1]。在这种理论的指导下，水处理工艺都是按照厌氧-缺氧-好氧或者厌氧-好氧的运行来实现生物的脱氮除磷，并且认为聚 β-羟基丁酸（PHB）在实现生物除磷过程中起着关键作用，PHB 的合成量直接决定了生物除磷效率的高低[2]。业界也认为缺氧段硝酸盐的存在可能直接导致生物除磷的失败[3-5]，因为在缺氧条件下，常规的反硝化微生物可更快利用水体有机物进行反硝化作用，抑制了聚磷菌合成 PHB，导致好氧聚磷菌在厌氧段（严格上是缺氧环境）的无效释磷；好氧段缺乏碳源，无法获得过量吸磷的能量来源，致使生物除磷系统失败。后来研究人员又发现[6]，在一些脱氮工艺中存在反硝化除磷现象，即硝酸盐可以作为电子受体，它的存在可能更有利于简化生物脱氮除磷工艺，并被认为是一种可持续发展工艺。但关于聚磷微生物是否可以直接利用外源性碳源超量吸磷未见研究报道。本试验研究了外源性碳源对生物脱氮除磷的影响，以期为水产养殖的水质处理提供新的思路。

1　材料与方法

1.1　菌株来源

菌株 RC11 由重庆市荣昌县梅石坝渔场的池塘底泥中分离获得，经选择性富集培养和平板划线纯化，进行硝酸盐还原试验、多聚磷酸盐和聚 β-羟基丁酸累积试验筛选出菌体内有 PHB 和 Poly-P 积累的菌株 RC11，利用细菌自动分析鉴定仪等方法对其进行种属鉴定，菌株 RC11 隶属于假单胞菌属。

1.2　试验仪器

生物显微镜（OLYMPUS IX70），紫外分光光度计（THERMO SPECTRONIC HELIOS-γ），纯水器（LABCONCO 9000602），烘箱（SANYO MOV112F），高压蒸汽灭菌锅（SANYO MLS-3780），石英双重纯水蒸馏器（上海申生 1810D），台式离心机（太仓华美 TGLL-18K），超净工作台（上海淀山湖 SDJ-OS）。电子天平（METTLER TOLENDOAB104-N），磁力搅拌器。

1.3　培养基[7]

牛肉膏蛋白胨固体培养基（g/L）：牛肉膏 3；蛋白胨 10；蒸馏水 1 000 mL；琼脂 10；pH 7.0。

富磷肉汤培养基（g/L）：牛肉膏 3；蛋白胨 10；磷酸二氢钾 0.5；蒸馏水 1000 mL；pH 7.0。

合成废水培养基（g/L）：$(NH_4)_2SO_4$ 0.1；$Na_2S_2O_3$，0.1；$MgSO_4 \cdot 7H_2O$，0.2；$NaAc \cdot 3H_2O$，0.8；KH_2PO_4，0.06；微量元素溶液 1 mL。

硝酸盐培养基（g/L）：除 KNO_3 0.15 外，其余成分含量同合成废水。

无外加碳源时，菌株 RC11 的硝酸盐代谢特性测定，要将合成废水培养基中的 $NaAc \cdot 3H_2O$ 去除。

1.4　试剂

二苯胺试剂：二苯胺 0.5 g 溶于 100 mL 硫酸中，用 20 mL 蒸馏水稀释。

格里斯氏试剂：甲液是将对氨基苯磺酸 0.5 g 溶于 150 mL 醋酸（10%）；乙液是将 α-萘胺 0.1 g 溶于 20 mL 蒸馏水，然后加入稀醋酸（10%）150 mL。

苏丹黑 B 试剂：甲液是将苏丹黑 B 0.3 g 溶 100 mL 醇（70%），用力振荡，过夜过滤；乙液是二甲苯；丙液是 0.5% 番红水溶液。

阿尔培氏试剂：甲液是将甲苯胺兰 0.15 g，孔雀石绿 0.2 g 溶解于 2 mL 95% 乙醇中，再加入蒸馏水 100 mL 及 1 mL 乙酸，放置 24 h 过滤；乙液是将碘化钾 3 g 溶于 10 mL 蒸馏水中，再加碘 2 g，待完全溶解后加蒸馏水至 300 mL。

萘瑟氏试剂：甲液是将美兰 1 g 溶于 95% 乙醇 2 mL 加冰乙酸 5 mL 蒸馏水 95 mL，混匀过滤；乙液是将俾斯麦褐 0.2 g 溶于 100 mL，100℃ 蒸馏水中。

1.5　分析项目及检测方法

PHB：苏丹黑 B 染色法；细菌浓度：干燥恒重法；磷酸盐浓度：钼锑抗分光光度法；NO_3-N 浓度：酚二磺酸分光光度法；Poly-P：阿尔培氏法和萘瑟氏法；NO_2-N 浓度：N-（1-奈胺）-乙二胺光度法。

1.6　试验方法

标准曲线制定（1）硝酸盐标准曲线的制定[8]，是在一组 50 mL 比色管中，用分度吸管分别加入硝酸盐氮标准使用液 0、0.10、0.30、0.50、0.70、1.00、5.00、10.00 mL（含硝酸盐氮 0、0.001、0.003、0.005、0.007、0.010、0.030、0.050、0.070、0.100 mg），加水至约 40 mL，加 3 mL 氨水使成碱性，稀释至标线，混匀。在波长 410 nm 处，以水为参比，以 10 mm 0.01~0.10 mg 比色皿测量吸光度。由测得的吸光度值减去零浓度管的吸光度值，分别绘制不同比色皿光程长的吸光度对硝酸盐氮含量（mg）的校准曲线。

亚硝酸盐标准曲线的制定。配制方法同前，但于波长 540 nm 处，用光程长 10 mm 的比色皿，以水为参比，测量吸光度。

磷酸盐标准曲线的制定。配制方法同前，但于 700 nm 波长处，以零浓度溶液为参比，测量吸光度。

预培养 取菌株 RC11 斜面菌苔两环接种至 50 mL 肉膏蛋白胨液体培养基中活化 6 h 后转接到 150 mL 富磷肉汤，摇床培养（30℃，150 r/min）24 h 后，菌悬液于 10℃，5 000 r/min 条件下离心 5 min，去上清液，沉淀用无菌生理盐水洗涤两次后用少量灭菌生理盐水配成菌悬液备用。

外源性碳源对菌株 RC11 磷酸盐代谢的影响[9]（1）取预培养 24 h 的菌悬液 6 mL 加入到 150 mL 富磷肉汤中，摇床培养（30℃，150 r/min）36 h，按时取样分析培养基中磷酸盐浓度及不同时间点的单位体积的菌体干重，计算不同时间点细菌体内磷酸盐占菌体干重的比例。

以氧为电子受体时的磷酸盐代谢特性：取菌悬液加入到 150 mL 合成废水培养基中，置摇床振荡培养 6 h，定时取样分析培养基中磷酸盐浓度变化，并对菌体进行 poly-P 染色。

以硝酸盐作为电子受体时的磷酸盐代谢特性：取菌悬液接种至合成废水中，同时添加固体硝酸钾，使培养基的硝酸盐氮浓度达到 20 mg/L 左右，以液体石蜡油封闭培养基液面，并用胶塞密闭瓶口进行缺氧反应。定时取样分析培养基中磷酸盐及亚硝酸盐浓度变化，并对菌体进行 poly-P 染色。

无外加碳源时菌株 RC11 的磷酸盐代谢特性

以氧作为电子受体时的磷酸盐代谢特性：取预培养好的菌悬液接种至 2 瓶 250 mL 经除氧处理后的合成废水中，密封瓶塞，厌氧反应 3 h 后，进行充气培养，溶解氧浓度>4 mg/L，培养时间为 6 h，定时取样分析培养基中磷酸盐浓度变化，并对菌体进行 poly-P 染色。

以硝酸盐作为电子受体时的磷酸盐代谢特性：取预培养好的菌悬液接种至 2 瓶 250 mL 经除氧处理后的合成废水中，密封瓶口，厌氧反应 3 h 后，迅速添加固体硝酸钾，使其浓度达到 20 mg/L 左右，进入缺氧反应阶段，定时间取样测定培养基中亚硝酸盐和磷酸盐浓度，并对菌体进行 poly-P 染色。

2 结果与分析

2.1 标准曲线测定

磷酸盐测定的标准曲线为 $Y = 96.196X$，$R^2 = 0.999\,3$；硝酸盐测定的标准曲线为 $Y = 14.937X$，$R^2 = 0.998\,5$；亚硝酸盐氮测定的标准曲线为 $Y = 0.1177X$，$R^2 = 0.999\,2$。各曲线的相关性均良好。

2.2 外源性碳源对菌株 RC11 磷酸盐代谢的影响

菌株 RC11 在富磷肉汤培养基中好氧培养 36 h，菌体干重从 25 mg/L 增加到 700 mg/L，培养基中的磷酸盐浓度由试验开始时的 295 mg/L 下降到试验结束时的 253 mg/L，磷酸盐占菌体干重比例随培养培养时间的增加呈逐渐上升，最大值出现在培养时间为 24 h 时（图1），此时干菌体内的磷酸盐含量达 7%，而一般细菌体内磷酸盐含量为 1%~3%[10]，结合菌体内的多聚磷酸盐染色结果（图2，图3），提示菌株 RC11 在外源碳源存在时在好氧培养过程中时发生了超量吸磷现象。

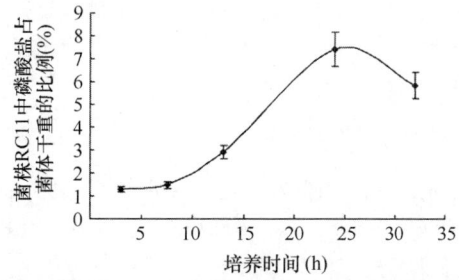

图 1　菌株 RC11 中磷酸盐占菌体干重的比例随培养时间的变化

Fig. 1　The ratio of phosphate in dry weight of strain RC11 at different incubation time

合成废水培养基接种菌株 RC11 好氧培养 8 h，培养基中磷酸盐的浓度显著降低（图4），磷酸盐浓度从 24.34 mg/L 降至 10.48 mg/L，即有 13.86 mg/L 磷酸盐从合成废水转移到了细菌内。

缺氧反应的整个反应周期可以分为两阶段，第一阶段为硝酸盐转化为亚硝酸盐阶段，发生在反应前 5 h，主要表现为

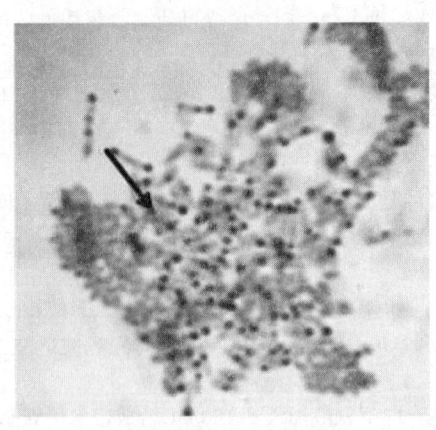

图2 菌株 RC11 内多聚磷酸盐颗粒×1 000 倍富磷肉汤培养基中好氧培养 36 h，阿尔培氏法，箭头所指为多聚磷颗粒.
Fig.2 The 1 000 times photo of poly-P in strain RC11In pohosphorus-rich broth medium with aerobic 36 h,
arrow indicates poly-P

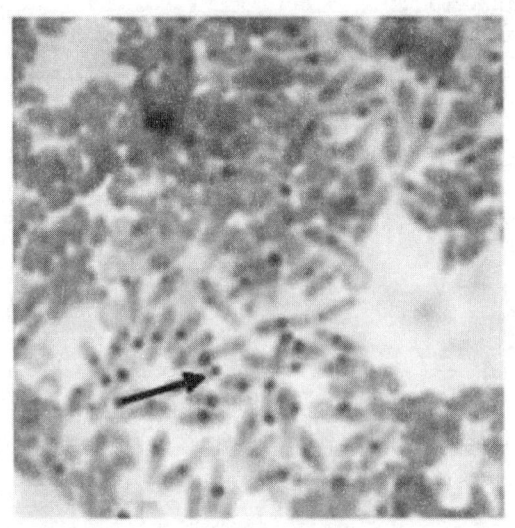

图3 菌株 RC11 内多聚磷酸盐颗粒 ×1 000 倍富磷肉汤培养基中好氧培养 36 h，萘瑟氏法，箭头所指为多聚磷颗粒.
Fig.3 The 1000 times photo of poly-P in strain RC11In pohosphorus-rich broth medium with aerobic 36 h,
Nessler's Reagent, arrow indicates poly-P

硝酸盐氮浓度的下降伴随着磷酸盐浓度的下降以及亚硝酸盐氮的积累，亚硝酸盐氮由 0 mg/L 上升到 20.61 mg/L，磷酸盐则由 24.53 mg/L 减少为 18.56 mg/L，下降了 5.97 mg/L，与同期进行的好氧组比较，其磷酸盐减少量仅为其的 60%；第二阶段为亚硝酸盐转化为氮气，培养基中磷酸盐浓度略为上升。

2.3　无外加碳源时菌株 RC11 磷酸盐代谢特性

图 5 是菌株 RC11 在经历厌氧阶段后的磷酸盐代谢情况。培养基中硝酸盐浓度在前 2 h 迅速从 20 mg/L 降为近 0，磷酸盐浓度也随之下降。随后数小时，随着硝酸盐浓度的下降，磷酸盐浓度却呈现上升趋势；试验结束时，培养基中磷酸盐含量又回到试验开始时的浓度。

同期进行的好氧培养组，菌株在经历厌氧释磷阶段后，在整个反应过程中培养基中的磷酸盐含量呈直线下降趋势，至试验结束时，培养基中的磷酸盐浓度只有开始时的近 9%，即培养基中 90% 的磷酸盐转移到了菌株 RC11 细胞内。

3　讨论

能量是生物进行一切生命代谢活动的基础。反硝化是微生物适应环境的结果，缺氧环境迫使微生物改变在好氧条件下的能量生成途径，即在电子传递链的终端利用硝酸盐或亚硝酸盐取代了分子氧作为电子受体，产生能量供生物各种反硝化

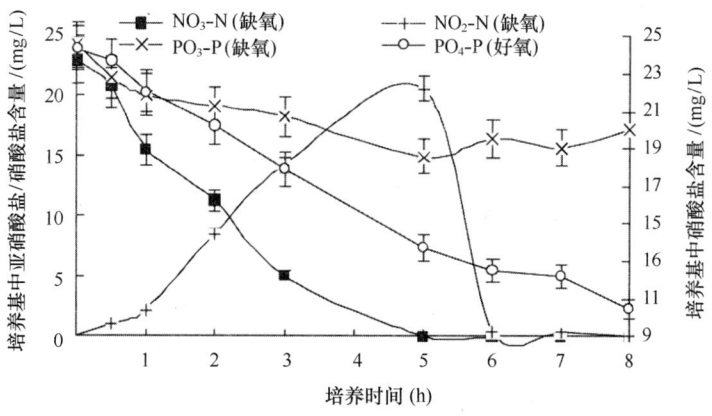

图 4 缺氧/好氧培养时菌株 RC11 培养基中亚硝酸盐/磷酸盐含量的变化情况（外加碳源）

Fig. 4 Changes in nitrite and phosphate concentration in the medium during aerobic or anoxic incubation of strain RC11 with the presence of external carbon source

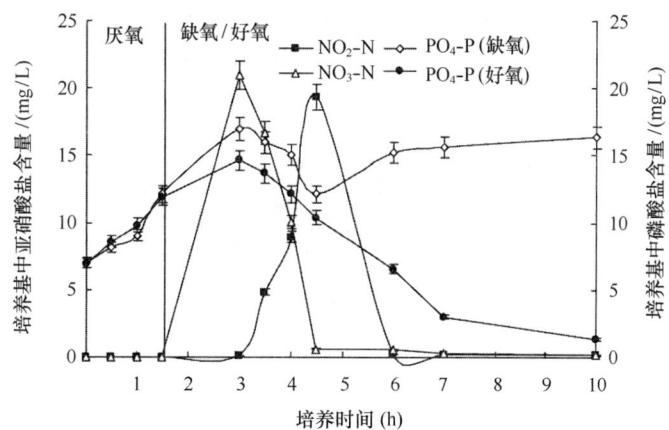

图 5 菌株 RC11 培养基中硝酸盐和亚硝酸盐/磷酸盐含量变化（无外加碳源，厌氧—缺氧—好氧培养）

Fig. 5 Changes in nitrate and nitrite/phosphate concentration in the medium during anaerobic–aerobic and anaerobic–anoxic incubation of strain RC11 with no external carbon source

是微生物适应缺氧环境的结果，缺氧环境代谢活动提供能量[10]；生物体内的多聚磷酸盐是一种能量储存库，将多余的能量以高能磷酸键的形式储存在体内，以供机体在特殊时期利用。由此可见，生物聚磷与生物反硝化其实是生物的两种不同生理活动过程，两者之间并没有直接的冲突[11]，一般的微生物在能量过剩时具有储存能量的能力，但是不同微生物储存能量所采取的方式不一样。有的微生物以糖原颗粒的形式将能量储存，有些微生物则在碳源充足而氮源不足时将能量以 PHB 的形式储存在菌体内[12]，而相当部分微生物则采用多聚磷酸盐的形式储存多余能量，这就是所谓的生物聚磷。从理论上讲，当能量过剩，并且微生物具有合成多聚磷酸盐的代谢途径的条件下，就可以发生过量摄磷行为。至于是利用外源碳源还是菌体内的碳源如 PHB，可能并不是生物是否过量摄磷的关键所在。

不同培养条件下的反硝化试验表明，菌株 RC11 具有利用硝酸盐作为电子受体进行反硝化除磷的功能。但在高浓度的亚硝酸盐存在时，细菌虽然可以将亚硝酸盐还原为氮气，但不能与聚磷作用相偶联。尤其值得注意的是，细菌 RC11 在有硝酸盐和乙酸钠同时存在的条件，也发生了磷酸盐的下降，这与目前大多数的研究结果相矛盾[13-15]，原因有待进一步研究。

从试验结果分析，这两种条件下，菌株 RC11 菌体内都积累了多聚磷酸盐颗粒。经过富磷培养后的菌株 RC11 其菌体内的磷酸盐含量达到其菌体干重的 7%，而一般的细菌其体内的磷酸盐含量只占菌体干重的 1.5%~2.0%[16]，这与 Suresh[17] 等的结果差别较大，通过对生物除磷系统中的假单胞菌的纯培养，发现这种细菌能够积累磷酸盐，其含量达到细胞干重的 31%；Deinema[18] 及 Buchan[19] 的研究也发现能够过量摄磷的不动杆菌体内磷酸盐含量也高达其细胞干重的 20% 左右。

关于聚磷菌能够利用外源性碳源进行过量摄磷的研究已有报道[19-22]，国内学者在进行聚磷菌磷酸盐代谢机理研究时，也是采用外源性碳源，而不是内源性碳如 PHB 等[22-23]。在污水除磷的工程当中，研究人员要实现生物除磷，活性污泥必须经过厌氧阶段以合成 PHB，且 PHB 的合成量决定了聚磷微生物的最终除磷效果，并据此提出了目前公认的生物除磷生化模型。但活性污泥需要经厌氧-好氧交替环境的运行方可实现生物除磷，这与活性污泥复杂的微生物组成密切相关，只有通过这种交替环境变化，才能使聚磷菌在竞争中取得优势，因为在聚磷菌体内有大量的多聚磷酸盐，可在厌氧条件下提供能量供机体进行新陈代谢，并具有合成 PHB 的代谢途径，能够将环境中的小分子有机酸转变成 PHB，以便在缺乏碳源的好氧环境下作为电子供体之用。其他微生物由于菌体内没有多聚磷酸盐，在厌氧环境不能提供能合成 PHB 的能量来源，在后续的好氧段有机物含量很低时，由于无碳源可用而被逐渐淘汰。这种厌氧-好氧交替运行的方法，其实质就是一种选择性的微生物培养方法。单一的好氧活性污泥法不能达到较高的生物除磷效果，可能是众多的好氧微生物共存消耗了大量的有机物质，使聚磷微生物没有足够的能量来实现超量吸磷而成为普通的好氧微生物。因此，要实现生物超量吸磷，微生物具备合成多聚磷酸盐的代谢途径是决定性因素，而厌氧-好氧交替运行的环境条件只是为聚磷微生物创造了一个有利环境而已。细菌合成 PHB 的能力可能只是微生物实现超量吸磷的非必要条件。也有研究者认为聚磷菌能否合成聚磷与基质类型有关[23]，在好氧条件下，向培养基中添加乙酸盐也能诱导聚磷菌释磷。这与本试验的结果不一致，造成这种差异的原因可能是研究对象不一致，前者以活性污泥为研究对象，后者则是利用纯种菌株进行试验。

早期的研究认为硝酸盐在厌氧段的存在能够影响微生物过量吸磷[24-26]，因为缺氧环境易于富集反硝化细菌，反硝化细菌能够优先利用环境中有机物进行反硝化而导致聚磷菌在缺氧段得不到足够的碳源用于后续好氧段的超量吸磷。但反硝化聚磷微生物的发现，使人们对生物除磷又有了新的认识，并提出了与好氧聚磷类似的聚磷机理。本文也对筛选自池塘底泥的一株反硝化细菌进行了磷酸盐代谢特性的验证性试验。分别考察了其在同时存在硝酸盐和乙酸钠的培养基 A 及按厌氧-缺氧方式运行时的磷酸盐代谢特性。发现该菌株能够利用硝酸盐进行反硝化聚磷，但在高浓度的亚硝酸盐存在时，细菌非但不能吸磷反而释磷，这与前人研究结果一致[27]。细菌 RC11 在硝酸盐和有机物同时存在的条件下，也一定程度表现出了吸磷现象，而对照反硝化细菌在硝酸盐和有机物共存的培养基中的磷酸盐浓度基本没有变化，说明菌株 RC11 能够利用外源性碳直接反硝化聚磷，这与目前公认的反硝化聚磷机理也不一致。虽然国外学者曾报道过脱氮副球菌在有外源性碳源存在的条下可以实现反硝化聚磷[28]。但有关菌株 RC11 是否具有聚磷特性还需作进一步的研究。

在本项目的研发过程和论文撰写过程中，得到美国建明工业（珠海）有限公司李正博士的帮助，在此表示感谢。

参考文献

[1] 林燕，杨永哲．硝酸盐浓度及缺氧好氧时段对反硝化聚磷诱导过程的影响 [J]．给水排水，2003，29（4）：3639.

[2] Comeau Y., Hall K. J., Oldham W. K. . Indirect polyphosphate quantification in activated sludge [J] . Water Pollut Res Canada, 1990, 25：161-174.

[3] Kuba T., Murnleitner E., van Loosdrecht M C M, et al.

[4] A metabolic model for biological phosphorus removal by denitrifying organisms [J] . Biotechnol Bioeng, 1996, 52：685-695.

[5] Shin H. S., Jun H. B., Park H. S. . Simultaneous removal of phosphorus and nitrogen in sequencing batch reactor [J] . Biodegradation, 1992, 3：105-111.

[6] Ng W. J., Ong S. L., Hu J. Y. . Denitrifying phosphorus removal by anaerobic/anoxic sequencing batch reactor [J] . Water Sci Tech, 2001, 43（3）：139-446.

[7] 李振林，微生物学及检验技术 [M]．广州：广东科技出版社，1996.

[8] 魏复盛，国家环境保护总局，水和废水检测分析方法编委会．水和废水检测分析方法 [M]．北京：中国环境科学出版社，2002.

[9] 周岳溪，钱易，顾夏声．假单胞菌磷代谢特性的研究 [J]．环境科学，1991，13（5）：41-43.

[10] 沈萍．微生物学 [M]．北京：高等教育出版社，2000.

[11] 罗宁，罗固源，许晚毅．从细菌的生化特性看生物脱氮与生物除磷的关系 [J]．重庆环境，2003，5：45-51.

[12] 周德庆．微生物学教程 [M]．北京：高等教育出版社，2001.

[13] Merzouki M., Beme N. Biologicaldenitrifing phosphorus removal in SBR：Effect of added nitrate concentration and sludge retention time [J] . Water Sci Tech, 2001, 43（3）：191-194.

[14] 李勇智，彭永臻，王淑滢．厌氧/缺氧 SBR 反硝化除磷效能的研究 [J]．环境污染治理技术与设备，2003，4（6）：9-12.

[15] 东秀珠，蔡妙英．常见细菌鉴定手册 [M]．北京：科学出版社，2001.

[16] Barnard J. L., Stevens G. M., Leslie P. L. . Design strategies for nutrient removal plants [J] . Water Sci Tech, 1985, 17（11/12）：233-242.

[17] 李军，杨秀山，彭永臻．微生物与水处理工程 [M]．北京：化学工业出版社，2002.

[18] Deinema M. H., Habets L. H. A., Scholten J., et al. The accumulation of polyphosphate in Acinetobacter spp. FEMS Microbiol Letters, 1980, 9：275-279

[19] Buchan L. Possiblebiologicalmechanism of phosphorus removal [J] . Water Sci Tech, 1983, 15（1）：8-103

[20] 习淑琪，吴迪．固定化污泥除磷的初步研究 [J]．污染防治技术，1999，12（4）：233-235.

[21] Kaltwasser H., Vogt G., Schlegel H. G. . Polyphosphatesynthese wathrend der nitrat-atmung von micrococcus denitrification stamm 11 Arch [J] . Mikroboil, 1961, 44：259-265.

［22］ 周岳溪，陈方荣．假单胞菌摄磷和释磷条件的研究．环境科学学报，1994，14（2）：312-315.

［23］ 周康群．反硝化聚磷一体化设备中的聚磷菌［J］．环境污染与防治，2002，24（3）：132-135.

［24］ Meinhold J.，Filipe C. D. M.，Daigger G. T.，et al. Characterization of the denitrifying fraction of phosphate accumulating organisms in biological phosphate removal ［J］. Water Sci Tech，1999，39（1）：31-42.

［25］ Jenkins D.，Tandoi V. . The applied microbiology of enhancedbiologicalphosphateremoval accomplishments and needs ［J］. Water Res，1991，25（12）：1471-1478.

［26］ Van S. W.，Rensink J. H.，Rijs G B. J. . Biological Premoval：state of the art in the Netherlands ［J］. Water Sci Tech，1993，27（2）：317-28.

［27］ Kerrn-Jespersen J. P.，Henze M.，Strube R. . Biological phosphorus release and uptake under alternating anaerobic and anoxic conditions in a fixed-film reacto. Water Res，1994，28（5）：1253-1255.

［28］ Yoram B.，Jaap V. R. . A typical polyphosphate accumulation by the denitrifying bacterium Paracoccus denitrificans ［J］. Applied and EvironmentalMicrobiology，2000，66（3）：1209-1212.

该文发表于《微生物学通报》【2010，37（7）：969-974】。

一株可溶性有机磷去除菌的分离及其生物学特性

Isolation of a Dissolved Organic Phosphorus-removing Strain and Its Biological Characteristics

黄志华[1△] 曹海鹏[1△] 杨先乐[1*] 李熙[2] 王炼[3] 安健[1] 龚雨州[1]

（1. 上海海洋大学 国家水生动物病原库，上海 201306；2. 长沙市动物防疫监督站，湖南 长沙 410006；

3. 四川省水产学校，重庆 401520）

摘要：以甘油磷酸钠（Sodium Glycerophosphate，以下简称 NaGly）作为外源可溶性有机磷，从富营养化的养殖池污泥中分离到 5 株可溶性有机磷去除菌株，通过除磷率比较，筛选出一株最为高效的菌株 D2，其对初始浓度为 5 mg/L 甘油磷酸盐磷（Phosphorus Glycerophosphate，以下简称 GP-P）的去除率可达 99.0%。此外，对其进行了 16S rRNA 基因序列测定，并进一步研究了其生长特性与除磷特性。试验结果表明，菌株 D2 为肠球菌（Enterococcus sp.），与屎肠球菌（Enterococcus faecium）菌株 KT4S13（登录号：AB481104）和 CICC6078（登录号：DQ672262）的 16S rRNA 基因序列相似性近 100% 其生长周期为：0~4 h 为生长迟缓期，4~8 h 为对数生长期，8~28 h 为稳定期，28 h 以后为衰亡期；且在 15~40℃、pH 4.0~9.0 以及 5~40 mg/L GP-P 条件下均能够生长，其中菌株 D2 最适生长的温度范围和 pH 范围分别为 30~35℃、6.0~7.0，而且 20~30 mg/L GP-P 能显著促进菌株 D2 生长。此外，菌株 D2 在进入衰亡期之前随着作用时间的延长，对 20 mg/L GP-P 的除磷率逐渐升高，在进入衰亡期后的 28~32 h 内对 20 mg/L GP-P 的除磷效果趋于稳定，其在 15~40℃、pH 4.0~9.0 以及 5~40 mg/L GP-P 条件下均具有除磷作用，其最适除磷温度范围、pH 范围和 GP-P 浓度范围分别为 25~35℃、6.0~7.0 和 5~10 mg/L。

关键词：可溶性有机磷，肠球菌，分离，生物学特性

2. 降解硝酸盐和亚硝酸盐

凝结芽孢杆菌（*Bacillus coagulans*）（安健等，2010）、蜡样芽孢杆菌（*Bacillus cereus*）（安健等，2012）、解淀粉芽孢杆菌（*Bacillus amyloliquefaciens*）（曹海鹏等，2013）和鲍曼氏不动杆菌（*Acinetobacter baumannii*）（王会聪等，2012）等均具有良好的好氧反硝化性能，且其反硝化活性主要发生在对数生长期。

该文发表于《中国水产科学》【2010，17（3）：561-569】。

好氧反硝化细菌 YX-6 特性及鉴定分析

Characteristics of aerobic denitrifying strain YX-6 and identification

安健[1,2]，宋增福[1]，杨先乐[1]，胡鲲[1]，路怀灯[2]，佘林荣[3]

(1. 上海海洋大学 水产与生命学院，国家水生动物病原库，上海 201306；2. 扬州绿科生物技术有限公司，江苏 扬州 225600；3. 广东海富药业有限公司，广东 潮州 515700)

摘要：从对虾池塘筛选得到 1 株高效的好氧反硝化细菌，命名为 YX-6。对该菌生长及反硝化性能间的关系进行研究；同时研究了不同温度、pH、盐度及碳源对该菌生长及反硝化性能的影响。结果表明，该菌反硝化作用主要发生在对数生长期，可将亚硝酸盐氮由 10 mg/L 降至 0；该菌 适生长及反硝化温度为 30℃；pH 值范围为 7~9 时适于该菌生长及反硝化性能的发挥。该菌 适盐度范围为 0~15；丁二酸钠、乙酸钠为该菌生长及反硝化的 适碳源。通过对 YX-6 菌株生理生化及 16S rRNA 分子鉴定，初步鉴定为凝结芽孢杆菌 (*Bacillus coagulans*)。对该菌株亚硝酸还原酶基因进行序列分析，结果表明，该菌含有亚硝酸还原酶 nirS 基因。

关键词：好氧反硝化；凝结芽孢杆菌；16S rRNA；亚硝酸还原酶基因

目前中国主要采取高密度集约化的养殖生产模式，水体中过量的饲料残渣及水产动物代谢物的排放，常常会导致水体亚硝酸盐氮的严重超标，并有可能进一步诱发细菌病、病毒病等病害，给水产养殖业造成巨大的经济损失。因此，养殖水体亚硝酸盐氮的有效去除是水产业亟需解决的问题之一。

好氧反硝化细菌能够在有氧条件下，将 NO_3^- 或 NO_2^- 还原为 N_2，是自然界氮素循环的重要途径之一。

当前，无论在养殖水体的生产应用还是基础研究方面，氮素消除技术仍然停留在厌氧除氮的初级水平上。由于好氧反硝化细菌的基础研究薄弱，基础数据积累少，制约了好氧反硝化除氮技术的发展。

自 Robertson 等[1]首次分离出好氧反硝化细菌以来，已有较多种类的细菌被报道具有好氧反硝化功能，其中部分菌种能够在高密度养殖条件下，有效地去除水体氮素的过度积累。目前已发现的好氧反硝化细菌主要存在于假单胞菌属 (*Pseudomonas*)[2-3]、产碱杆菌属 (*Alcaligenes*)[4-5]、副球菌属 (*Paracoccus*)[6-7] 和芽孢杆菌属 (*Bacillus*)[8-9] 等。研究表明，菌株的生长状况会影响反硝化作用效果[10-11]，因此，开展其生物学特性的研究对该类细菌的生产和应用具有重要的理论价值和实际意义。

本研究以对虾池塘分离的好氧反硝化细菌 YX-6 菌株为实验对象，拟在对其部分生物学特性及反硝化性能进行研究的基础上，确定该菌分类学和系统发育地位，旨在为丰富好氧反硝化细菌种质资源的同时，为该菌在养殖水体的实际应用提供理论基础和技术支撑。

1 材料与方法

1.1 材料

1.1.1 实验菌株
好氧反硝化细菌 YX-6 为本实验室从虾池中分离。

1.1.2 培养基
牛肉膏蛋白胨液体培养基 (LB)：蛋白胨 10 g/L，牛肉膏 5 g/L，NaCl 5 g/L，pH 7.0。反硝化培养基 (测定性能)：琥珀酸钠 4.72 g/L，$NaNO_2$ 0.05 g/L，KH_2PO_4 1.5 g/L，Na_2HPO_4 0.42 g/L，$MgSO_4 \cdot 7H_2O$ 1 g/L，pH 7.2。所有培养基经 0.11 MPa，121℃灭菌 20 min。

1.2 实验方法

1.2.1 细菌生长与好氧反硝化性能测定
实验菌株经纯培养 18 h，测其细菌光密度 A600 约为 0.605 后，按 1% (*V/V*) 菌液添加量接种到 200 mL 反硝化培养基中，30℃、200 r/min 摇床振荡培养 24 h，每隔 2 h 取样 1 次，每次取样 10 mL，先测定细菌光密度 A600，经 8 000 r/min 离心取其上清液测定培养液中亚硝酸盐氮的浓度。

1.2.2　温度对细菌生长及反硝化性能的影响

实验菌株经纯培养 18 h，测其细菌光密度 A600 约为 0.605 后，按 1%（V/V）菌液添加量接种到 100 mL 反硝化培养基中，分别在 10℃、15℃、20℃、25℃、30℃、35℃、40℃条件下经 200 r/min 摇床振荡培养 24 h 后，取样测定其 A600，同时测定其中亚硝酸盐氮的含量。

1.2.3　pH 对细菌生长及反硝化性能影响

实验菌株经纯培养 18 h，测其细菌光密度 A600 约为 0.605 后，按 1%（V/V）菌液添加量接种到 pH 值分别为 3、5、7、9、11 的 100 mL 反硝化培养基中，30℃、200 r/min 摇床振荡培养 24 h 后，取样测定其 A600，同时测定其中亚硝酸盐氮的含量。

1.2.4　盐度对细菌生长及反硝化性能的影响

配制盐度为 5、10、15、20、25、30 的人工海水，并添加反硝化培养基，将纯培养 18 h 后的菌株（光密度 A600 约为 0.605）按 1%（V/V）菌液添加量分别接种到 100 mL 反硝化培养基中，30℃、200 r/min 摇床振荡培养 24 h 后，取样测定其 A600，同时测定其中亚硝酸盐氮的含量。

1.2.5　不同碳源对细菌生长及反硝化性能的影响

将相同物质的 8 种碳源添加到反硝化培养基中，取经纯培养 18 h 后的试验菌株悬液，按 1%（V/V）菌液添加量分别接种于 100 mL 不同碳源的各反硝化液体培养基中，30℃、200 r/min 条件下振荡培养 24 h 后取样，测定不同碳源条件下细菌 A600 值，同时测定其中亚硝酸盐氮的含量。

1.2.6　细菌生理生化鉴定

采用 API 50CH 鉴定系统进行鉴定。

1.2.7　16S rRNA 序列分析及系统发育分析

用于 16S rRNA PCR 反应的引物为通用引物，PF：5′ AGAGATCCTGGCTCAG3′，PR：5′ GGTTACCTTGTTACG ACTT 3′，上海基康生物公司合成。PCR 反应体系（25 μL）：2.5 μL 10×PCR 缓冲液，3.5 μL MgCl₂，0.5 μL 模板 DNA，0.5 μL PF 和 PR，1 μL dNTP，0.5 μL Taq DNA 聚合酶，16 μL 超纯水。PCR 扩增条件为：94℃预变性 5 min；94℃变性 30 s，60℃退火 30 s，72℃延伸 1 min，循环 30 次；72℃延伸 10 min。PCR 产物连接 pMD18-T，转化 E.coli DH5α，在含氨苄青霉素的 LB 平板上筛选阳性克隆，提取质粒并测序。测序由 Invitrogen 公司完成。将 16S rRNA 所测序列通过 BLAST 检索程序与 GenBank 中已知 16S rRNA 序列进行分析，利用 Custal X1.8 进行序列比对，用 MEGA4.0 构建系统发育树。

1.2.8　亚硝酸盐还原酶基因序列分析

参照文献[12]，选取亚硝酸盐还原酶结构基因 nirS、nirK 特异性引物，对该菌亚硝酸盐还原酶进行 PCR 扩增，引物由上海基康生物公司合成。PCR 反应体系及条件同 1.2.7。PCR 产物通过电泳检测及纯化后，由 Invitrogen 公司测序并进行同源性分析。

1.2.9　分析方法

亚硝酸盐氮测定：采用 N-1 萘-乙二胺比色法[13]。A600：紫外分光光度计法。

1.3　数据统计与分析方法

实验结果用平均数±标准差（x±SD，n=3）表示，运用软件 SPSS.11.0，经 One-way ANOVA 分析，采用 Duncan's 多重检验分析实验结果的差异显著性，P<0.05 为差异显著。

2　结果与分析

2.1　细菌生长与反硝化性能测定结果

如图 1 所示，随着细菌的生长，4 h 以后细菌得以迅速增殖，此阶段为细菌对数生长期。同时，该菌能够在 14 h 内将初始量为 10 mg/L 的亚硝酸盐氮降解为 0，去除率达 100%。

2.2　温度对细菌生长及反硝化性能的影响

如图 2 所示，随着温度的升高，细菌 YX-6 生长 A600 值不断增大，当温度高于 20℃时，细菌均能正常生长，且可保持较高的亚硝酸盐氮去除率。温度为 30℃时，细菌生长 A600 值和亚硝酸盐氮去除率均为高；在 40℃高温时，该菌生长虽相对减缓，但仍然可保持 90% 以上的亚硝酸盐氮去除率；当温度低于 20℃时，细菌几乎无法正常生长，此时亚硝酸盐氮去除率也较低。

2.3　pH 对细菌生长及反硝化性能的影响

如图 3 所示，细菌在 pH 值为 3~9 时均可以正常生长，同时保持较高的反硝化效率。pH 值为 5~9 时，细菌生长 A600

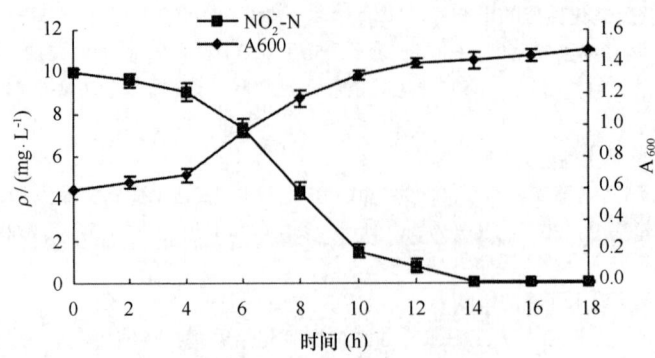

图 1 细菌 YX-6 生长与亚硝酸盐氮降解曲线

Fig. 1 Dynamics of YX-6 strain's growth and degradation curves of nitrite nitrogen

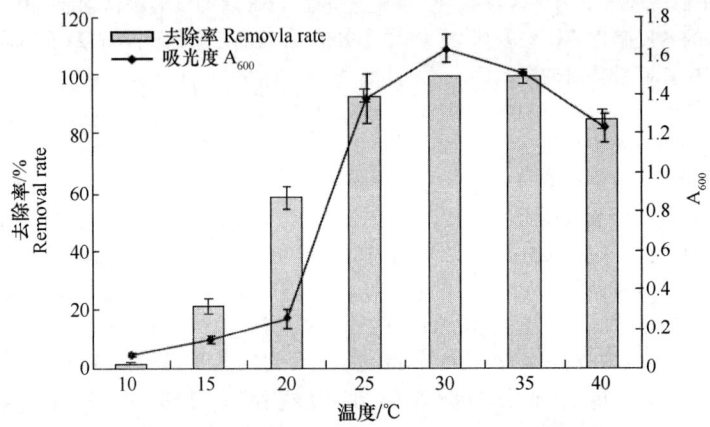

图 2 不同温度条件下细菌 YX-6 生长及亚硝酸盐氮降解曲线

Fig. 2 Dynamics of YX-6 strain's growth and degradation curves of nitrite nitrogen under different temperature

值仍不断增大，但此 pH 范围内，细菌对亚硝酸盐氮均可达到近 100% 的去除率；在 pH 值为 11 时，细菌生长极为缓慢，其亚硝酸盐氮去除率也较低。

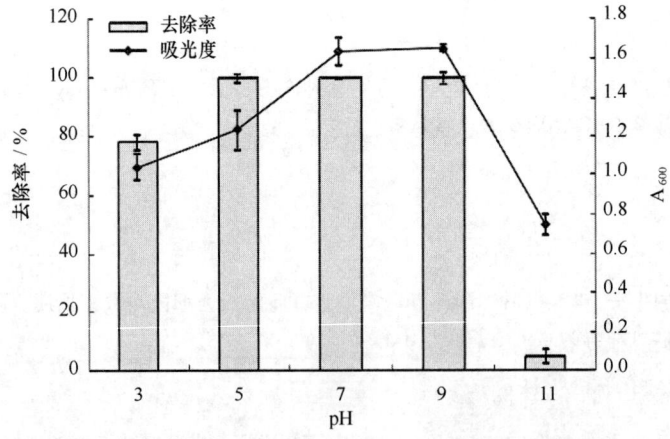

图 3 不同 pH 值条件下细菌 YX-6 生长及亚硝酸盐氮的降解曲线

Fig. 3 Dynamics of YX-6 strain's growth and degradation curves of nitrite nitrogen under different pH value

2.4 盐度对细菌生长及反硝化性能的影响

如图 4 所示，随着盐度的不断增大，细菌生长亚硝酸盐氮的去除率几乎为 100%，当盐度超过 15 A600 值依次降低。当盐度范围在 5~15 时，细菌对亚硝酸盐氮去除率逐渐降低。

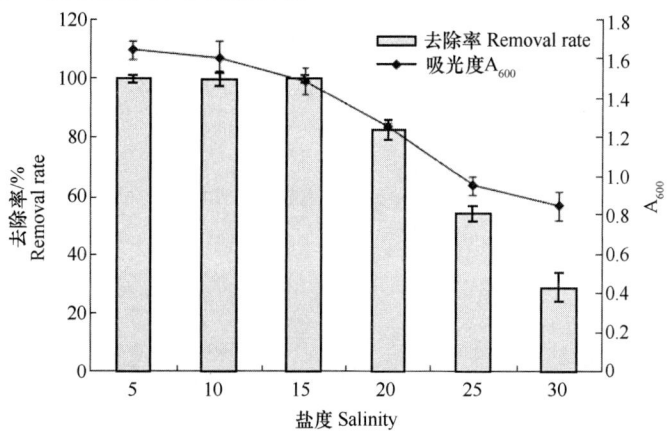

图4　不同盐度条件下细菌 YX-6 生长及亚硝酸盐氮的降解曲线

Fig. 4　Dynamics of YX-6 strain's growth and degradation curves of nitrite nitrogen under different salinity

2.5　不同碳源对细菌生长及反硝化性能的影响

细菌对亚硝酸盐氮的去除量也大，而该菌对蔗糖结果见表 1 所示，条件相同，在乙酸钠和丁二酸乳酸等碳源的利用率相对较低，菌体生长相对缓慢，钠为碳源的反硝化培养基中，细菌 A600 值大，同时，反硝化效率偏低。

表1　不同碳源对细菌 YX-6 生长及反硝化性能的影响（$n=3$；x±SD）

Tab. 1　Growth of YX-6 strain and degradation of nitrite nitrogen in different carbon sources

碳源名称 Carbon source	A600	起始 NO_2-N/（$mg \cdot L^{-1}$） Initial NO_2-N	终 NO_2-N/（$mg \cdot L^{-1}$） Final NO_2-N	去除率/% Removal rate
乙酸钠　Sodium acetate	1.149±0.036	10.011±0.026	0.068±0.034	99.3
酒石酸钾钠　Potassium sodium tartrate	0.371±0.027	10.011±0.034	4.833±0.141	51.7
乳酸　Lactic acid	0.220±0.003	10.005±0.054	7.011±0.153	30.0
柠檬酸钠　Sodium citrate	0.333±0.045	9.999±0.085	6.254±0.181	37.5
水杨酸钠　Sodium salicylate	0.179±0.006	10.000±0.017	7.862±0.192	21.4
蔗糖　Sucrose	0.111±0.006	10.011±0.026	8.415±0.156	15.9
葡萄糖　Glucose	0.340±0.007	10.006±0.009	5.640±0.079	43.6
丁二酸钠　Sodium succinate	1.067±0.064	10.017±0.017	0.085±0.017	99.2

2.6　细菌生理生化鉴定

对细菌 YX-6 进行了 50 项生理生化指标测试，初步鉴定该菌为凝结芽孢杆菌（Bacillus coagulans）（鉴定结果略）。

2.7　16S rRNA 分子鉴定及系统发育分析

该菌 16S rRNA 序列全长 1448 bp，GenBank 登录号 GU288875，同时该菌的 16S rRNA 序列经 BLAST 检索程序分析，并选取 GenBank 中已知好氧反硝化细菌及部分硝化细菌基因序列通过 Clustal X1.8 进行多重比对，并用 MEGA4.0 构建系统发育树，初步确定该菌株为凝结芽孢杆菌（Bacillus coagulans），同源性达 98%，结果如图 5 所示。

2.8　亚硝酸盐还原酶基因序列分析

细菌 YX-6 亚硝酸盐还原酶基因 nirS、nirK 基因扩增结果如图 6 所示，nirS 扩增结果为阳性，产物大小为 842 bp，而 nirK 扩增结果为阴性。同源检测结果表明，该菌 nirS 基因序列与 Pseudmonas aeruqinosa LESB58 菌株的 nirS 基因序列相似，同源性达 99%。

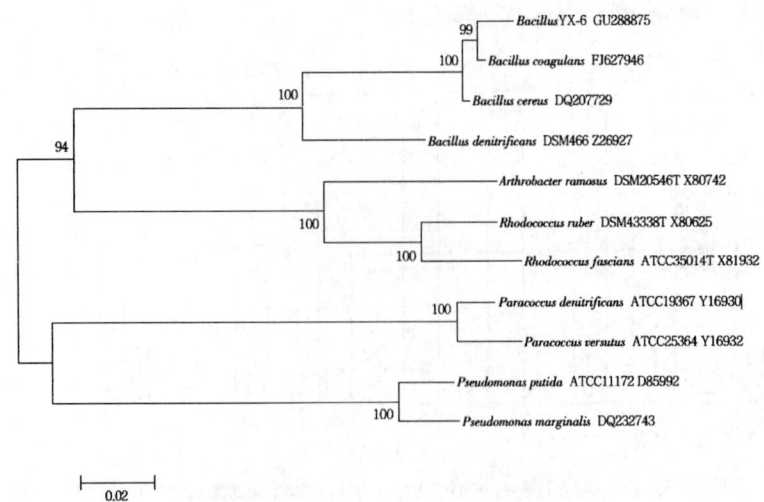

图 5　菌株 YX-6 与其他细菌间基于 16S rRNA 序列同源性的系统发育树分支节点的
数值表示自引导值（bootstrap），重复 1 000 次.

Fig. 5　Phylogenetic tree generated from an alignment of 16S rRNA of strain YX-6 and other bacteria
Node values represent bootstrap values repeating 1 000 times.

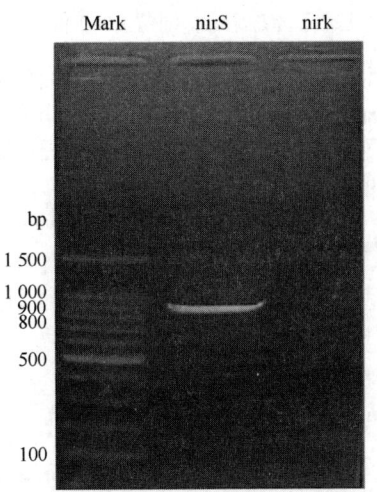

图 6　细菌 YX-6 亚硝酸盐还原酶基因 nirS、nirK 扩增结果
Fig. 6　Amplification of nirS and nirK gene from strain YX-6

3　讨论

好氧反硝化细菌的发现与应用，可有效地解决水体氮素污染。研究表明，反硝化作用主要发生在细菌对数生长期，这一阶段随着菌体的快速增殖，水体中硝态氮、亚硝态氮的量会迅速的下降[14]。该实验结果与马放等[15]的研究结果相符。

细菌生长及反硝化活性适温度范围是 25~35℃，当温度超过这一范围时，均会抑制细菌快速生长及反硝化性能的正常发挥[16]。本研究结果表明，水温在 20~40℃，该菌可以正常生长，并保持较高的反硝化活性，而水温 30℃ 是该菌生长及反硝化活性发挥的最适温度。此外，该菌在 40℃ 的高水温条件下，仍然生长良好，同时还可以保持高达 90% 以上的亚硝酸盐去除率，这一特性与曾报道过的好氧反硝化细菌特性不尽相同[4,10]。当水温低于 20℃ 时，细菌生长与反硝化性能发挥表现出了较明显的抑制作用。可见，温度对细菌生长及反硝化性能发挥有着重要的影响，而该菌更加适合高水温条件下的亚硝酸盐氮去除，这可能与该菌生长及反硝化酶的耐热机制有关，需进一步研究。

Gupta[17]、Timmermans 等[18]研究表明，细菌生长及反硝化酶活性的适 pH 值是中性或微碱性，pH 值过高或过低均会对菌株生长及反硝化性能的发挥产生影响。王宏宇等[19]、廖邵安等[20]对好氧反硝化菌株 C3（Pseudomonas sp.）、嗜麦芽寡养单胞菌 X0412（Stenotrophomonas maltophilia）研究也同样表明，pH 为中性条件有利于菌株生长及反硝化性能的发挥。本研究表明，该菌在 pH 值为 3~9 时，均可以较好生长，同时保持较高的亚硝酸盐氮去除率，其中，在 pH 值为 5~9 时，该菌亚硝酸盐氮去除率可始终保持近 100%。在 pH 值为 11 的高碱性环境中，该菌的生长及反硝化活性的发挥会受到一定的抑制。可见该菌对水体 pH 的改变显示出了较强的耐受能力，可在较宽的 pH 范围内保持正常生长及反硝化性能的稳定发挥，能应用于不同酸碱度的养殖水体。

养殖水体中，盐度也是影响好氧反硝化细菌生长与反硝化性能发挥的重要因素。Yoshie 等[21]指出，盐度对菌体生长及反硝化还原酶活性有一定的抑制作用。廖邵安等[20]对好氧反硝化菌株与盐度的影响关系也作了细致的研究，发现当水体盐度过高时，可对菌体反硝化活性产生抑制作用。本实验研究表明，在盐度 0~30，该菌株均可以正常生长并保持一定的反硝化活性，其中盐度为 0~15，是该细菌生长及反硝化活性发挥的 适盐度范围；当盐度超过 15 时，会对细菌的生长及反硝化性能产生一定的抑制作用。这一结果也与上述研究结果相同，有关盐度与该菌生长及反硝化酶的作用机制，仍需进行深入研究。

碳源不仅是细菌赖以生长的重要能源，而且对细菌反硝化作用有重要影响。张光亚等[22]人对好氧反硝化赤红红球菌（Rhodococcus rubber）HN 进行研究发现，细菌对酵母膏、乙酸钠、蛋白胨等碳源有较高利用率，其细菌生长 A600 值较大，而对衣康酸、甲酸钠等碳源则不可利用。王宏宇等[23]比较了不同碳源对好氧反硝化细菌 X31 生长性能的影响发现，细菌在丁二酸钠、乙酸钠和苹果酸钠为碳源的反硝化培养基中均可以保持良好生长，其中细菌 X31 在苹果酸钠为碳源时的菌体生长量为 大。本实验研究表明，该菌均可以利用上述 8 种实验碳源进行生长，并表现出一定的反硝化性能；其中该菌在乙酸钠和丁二酸钠为碳源时菌体生长量好，同时对亚硝酸盐氮的去除率也高，而对乳酸、蔗糖和水杨酸钠的碳源利用率较低，其菌体生长量及亚硝酸盐氮去除率相对较小，可见该菌对小分子碳源的利用率较高，而对大分子碳源的利用率较低，这一研究结果也与 Her 等[24]的研究结果一致。

本实验对好氧反硝化细菌 YX-6 生理生化指标进行了测定，并结合该菌 16S rRNA 序列比对分析结果，初步确定该菌株为凝结芽孢杆菌（Bacillus coagulans）。目前芽孢杆菌属的一些菌种如蜡样芽孢杆菌（Bacillus cereus）、枯草芽孢杆菌（Bacillus subtilis）、地衣芽孢杆菌（Bacillus licheniform）均有文章报道其具有好氧反硝化性能[8-9]。凝结芽孢杆菌作为益生功能菌已在较多领域中有所应用，本研究发现，凝结芽孢杆菌也同样具有好氧反硝化性能，并对该菌生物学特性及反硝化性能进行了较为细致的研究，但有关该菌反硝化相关作用机制仍需要进一步研究。

反硝化还原酶功能基因与细菌反硝化性能的强弱密切相关，有关反硝化还原酶的构成及相关基因的研究，已有较多报道[12,25]，其中亚硝酸还原酶作为亚硝酸盐氮去除过程中至关重要的酶类之一，其相关基因及作用机制的研究更是当前研究的热点。Stouthamer[26]证实了 T. pantotropha 中的亚硝酸还原酶是由 nirK 基因编码的含铜型的酶，而不是由 nirS 基因编码的细胞色素 cdl 型。廖绍安等[20]对菌株 S. maltophilia strain X0412 亚硝酸还原酶基因进行了扩增，发现该菌的亚硝酸还原酶是由 nirS 基因编码的细胞色素 cdl 型。可见，不同种属菌株之间的反硝化还原酶构成及酶的催化机理也不尽相同[27]。目前有关芽孢杆菌反硝化酶的相关研究国内外均未见报道。该实验对细菌 YX-6 亚硝酸还原酶基因进行扩增，结果表明，在该实验条件下，细菌 YX-6 亚硝酸还原酶中存在 nirS 基因，但有关该细菌中亚硝酸还原酶与亚硝酸盐氮去除之间的作用机制及影响因素方面的研究仍需要进一步探讨。

参考文献

[1]　Robertson L A, Kuenen J G. Aerobic denitrification: A controversy revived [J]. Arch of Microbiol, 1984, 139 (4): 351-354.

[2]　Takaya N, Maria A B, Yasushi S, et al. Aerobic denitrifying bacteria that produce low levels of nitrous oxide [J]. Appl Environ Microbiol, 2003, 69 (6): 3152-3157.

[3]　Su J J, Liu B Y, Liu C Y. Comparison of aerobic denitrification under high oxygen atmosphere by Thiosphaera pantotropha ATCC 35512 and Pseudomonas stutzeri SU2 newly isolated from the active sludge of a piggery wastewater treatment system [J]. J Appl Microbiol, 2001, 90 (3): 457-462.

[4]　Joo H S, Hirai M, Shoda M. Characteristics of ammonium removal by heterotrophic nitrification-aerobic denitrification by Alcaligenes faecalis No. 4 [J]. J Biosci Bioeng, 2005, 100 (2): 184-191.

[5]　Shwu L P, Chong N M, Chen C H. Potential applications of aerobic denitrifying bacteria as bioagents inwastewater treatment [J]. Biores Technol, 1999, 68 (2): 179-185.

[6]　Patureau D, Zumstein E, Delgenes J P, et al. Aerobic denitrifiers isolated from diverse natural and managed ecosystems [J]. Microbial Ecol, 2000, 39 (2): 145-152.

[7]　Robertson L A, Kuenen J G. Thiospaera pantotropha gen nov sp nov a facultatively anaerobic, facultatively autotrophic sulphur bacterium [J]. J Gen Microbiol, 1983, 129 (8): 2847-2855.

[8]　Joong K K, Kyoung J P, Kyoung S C, et al. Aerobic nitrificationdenitrification by heterotrophic Bacillus strains [J]. Biores Technol, 2005, 96 (17): 1897-1906.

[9]　杨希，刘德立，邓灵福，等. 蜡状芽孢杆菌好氧反硝化特性研究 [J]. 环境科学研究，2008，21 (3): 155-159.

[10]　马放，王弘宇，周丹丹，等. 好氧反硝化菌株 X31 的反硝化特性 [J]. 华南理工大学学报，2005，33 (7): 42-46.

[11] 周丹丹, 马放, 王宏宇, 等. 关于好氧反硝化菌筛选方法的研究 [J]. 微生物学报, 2004, 44 (6): 837-838.

[12] Braker G, Zhou J, Wu L, et al. Nitrite reductase genes (nirK and nirS) as functional markers to investigate diversity of denitrifying bacteria in pacific northwest marine sediment communities [J]. Appl Environ Microbiol, 2000, 66 (5): 2096-2104.

[13] 雷衍之. 养殖水环境化学实验 [M]. 北京: 中国农业出版社, 2006: 60-62.

[14] 刘晶晶, 汪苹, 王欢. 一株异养硝化-好氧反硝化菌的脱氮性能研究 [J]. 环境科学研究, 2009, 21 (3): 121-125

[15] 马放, 王弘宇, 周丹丹, 等. 好氧反硝化菌株 X31 的反硝化特性 [J]. 华南理工大学学报: 自然科学版, 2005, 33 (7): 42-46.

[16] 阎胜利, 温淑瑶. 土壤中反硝化作用的试验研究 [J]. 河海大学学报: 自然科学版, 1998, 26 (2): 90-94.

[17] Gupta A B. Thiosphaeera Pantotropha a sulphur bacterium capable of simultaneous heterotrophic nitrification and aerobic denitrification [J]. Enzyme Microb Technol, 1997, 21 (8): 589-595.

[18] Timmermans P, Van H A. Fundamental study of the growth and denitrification capacity of Hyphomicrobium sp [J]. Water Res, 1983, 17 (10): 1249-1255.

[19] 王宏宇, 马放, 苏俊峰, 等. 好氧反硝化菌株的鉴定及其反硝化特性研究 [J]. 环境科学, 2007, 28 (7): 1548-1552.

[20] 廖绍安, 郑桂丽, 王安利, 等. 养虾池好氧反硝化细菌新菌株的分离鉴定及特征 [J]. 生态学报, 2006, 26 (11): 3718-3724.

[21] Yoshie S, Noda N, Tsuneda S, et al. Salinity decreases nitrite reductase gene diversity in denitrifying bacteria of wastewater treatment systems [J]. Appl Environ Microbiol, 2004, 70 (5): 3152-3157.

[22] 张光亚, 陈培钦. 好氧反硝化菌的分离鉴定及特性研究 [J]. 微生物学杂志, 2005, 25 (6): 23-26.

[23] 王宏宇, 马放, 苏俊峰, 等. 不同碳源和碳氮比对一株好氧反硝化细菌脱氮性能的影响 [J]. 环境科学学报, 2007, 27 (6): 968-972.

[24] Her J J, Huang J S. Influences of carbon source and C/N ratio on nitrate/nitrate denitrification and carbon breakthrough [J]. Biores Technol, 1995, 54 (1): 45-51.

[25] Farver O, Brunori F, Cutruzzola S, et al. Intramolecular Electron Transfer in Pseudomonas aeruginosa cd1 nitrite reductase: Thermodynamics and kinetics [J]. Biophys J, 2009, 96 (7): 2849-2856.

[26] Stouthamer A H. Metabolic metabolic pathways in Paracoccus denitrificans and closely related bacteria in relation to the phylogeny of prokaryotes [J]. Antonie van Leeuwenhoek, 1992, 61 (1): 1-33.

[27] Tocheva E I, Rosell F, Mauk A G, et al. Side-on copper-nitrosyl coordination by nitrite reductase [J]. Science, 2004, 30 (4): 836-842.

该文发表于《微生物学通报》【2012, 39 (2): 162-171】

反硝化除磷菌筛选及其特性研究

Studies on the screening of denitrifying and phosphorus removal bacteria and its characteristics

安健[1△]　伏光辉[1△]　阮记明[2]　陈百尧[1]　龚琪本[1]　唐兴本[1]杨先乐[3]

(1. 江苏省连云港市海洋与水产科学研究所, 江苏 连云港 222044; 2. 江西农业大学 动物科学技术学院, 江西 南昌 330045; 3. 上海海洋大学 水产与生命学院 国家水生动物病原库, 上海 201306)

摘要:【目的】研究反硝化除磷菌特性。【方法】通过微生物筛选和生物学特性研究方法, 从对虾养殖池塘中筛选出多株可在有氧条件下同时具有反硝化除磷功能的菌种。【结果】菌株 LY-1 可在 18 h 内将初始量为 10 mg/L 的亚硝酸盐氮降低至 0.04 mg/L, PO_4^{3-}-P 降低至 0.05 mg/L。在 DO 浓度为 5.0~5.9 mg/L 时, 该菌反硝化除磷率近 100%。试验选取具有反硝化除磷功能的枯草芽孢杆菌为阳性对照菌, 大肠杆菌为阴性对照菌, 比较研究了菌株 LY-1 在不同 pH、温度、盐度、PO_4^{3-}-P 浓度、亚硝酸盐浓度时反硝化除磷的强弱, 在 pH 值为 5~9 范围时, 该菌亚硝酸盐氮去除率近 99%, PO_4^{3-}-P 去除率 86%; 温度为 30℃ 时, 该菌反硝化除磷率近 100%; 盐度为 5~15, PO_4^{3-}-P 浓度为 10 mg/L、亚硝酸盐氮浓度为 20 mg/L 时, 该菌亚硝酸盐氮和 PO_4^{3-}-P 去除率均可达 99%。【结论】菌株 LY-1 反硝化除磷性能显著高于对照菌 ($P<0.05$)。通过菌株 LY-1 形态学观察、生理生化及 16S rRNA 基因序列分析, 初步鉴定为蜡样芽孢杆菌 (Bacillus cereus)。

关键词: 反硝化除磷, 蜡样芽孢杆菌, 亚硝酸盐, 磷

该文发表于《环境污染与防治》【2013，35（6）：16-21】

具有降解亚硝酸盐活性的解淀粉芽孢杆菌的分离与安全性分析

Isolation and safety analysis of Bacillus amyloliquefaciens with nitrite removal activity

曹海鹏[1]　周呈祥[2]　何珊[1]　郑卫东[3#]　杨先乐[1]

（1. 上海海洋大学国家水生动物病原库，上海 301306；2. 安徽联大生物环保科技有限公司，安徽 滁州 239000；

3. 中国水产科学院长江水产研究所，湖北 武汉 430223）

摘要： 从养殖污泥中分离筛选了一株优良的亚硝酸盐降解菌 YX01，经 ARIS 2X 细菌自动鉴定系统以 16SRNA 系统发育分析，该菌株被鉴定为解淀粉芽孢杆菌（Bacillus amylolique faciens，GenBank 登录号为 JX649949），其 16S RNA 序列与基因库中芽孢杆菌属菌株的 16S RNA 序列有 99%～100% 的同源性，而且与解淀粉芽孢杆菌菌株 LD5（GenBank 登录号：GQ853414）的亲缘关系最近。该亚硝酸盐降解菌 YX01 在浓度为 $2.0×10^4$～$2.0×10^6$ cfu/L 时，对小球藻（Chlorella sp.）生长有促进作用；在浓度大于 $2.0×10^{11}$ cfu/L 时，对小球藻生长具有显著抑制作用。亚硝酸盐降解菌 YX01 对小球藻的半数抑制浓度（IC_{50}）大于 $2.0×10^{11}$ cfu/L，亚硝酸盐降解菌 YX01 对大型蚤（Daphniamagna）的 48 h 半数致死浓度（LC_{50}）大于 $2.0×10^{11}$ cfu/L，对草鱼（Ctenopharyngodon idellus）的 96 h 半数致死浓度（LC_{50}）大于 $2.0×10^{11}$ cfu/kg。经分析，亚硝酸盐降解菌 YX01 为无毒菌株。

关键词： 亚硝酸盐　解淀粉芽孢杆菌　鉴定　安全性

该文发表于《环境污染与防治》【2012，34（4）：32-36】

一株亚硝酸盐降解菌的分离鉴定与系统发育分析

Isolation, identification and phylogenetic analysis of a nitrite degrading bacterium

王会聪[1]　曹海鹏[1]　何珊[2]　杨先乐[1#]

（1. 上海海洋大学国家水生动物病原库，上海 201306；2. 上海师范大学教育系，上海 200234）

摘要： 从水产养殖污泥中分离筛选了一株优良的亚硝酸盐降解菌 AQ-3，其在 $1.0×10^7$ cfu/mL 时对 50 mg/L 亚硝酸盐的去除率高达 99.47%。通过细菌自动鉴定系统（API ID32GN 系统）以及 16S rDNA 系统发育分析，菌株 AQ-3 被鉴定为鲍曼氏不动杆菌（Acinetobacter baumannii）（JF751054），其 16S rDNA 序列与基因库中不动杆菌属菌株的 16S rDNA 序列有 99% 的同源性，而且与鲍曼氏不动杆菌菌株 14（FJ907197）的亲缘关系最近。此外，菌株 AQ-3 同时含有 nirS 基因、nirK 基因和 norB 基因，这些基因对 AQ-3 高效降解亚硝酸盐具有重要的意义。

关键词： 亚硝酸盐　鲍曼氏不动杆菌　鉴定系统　发育分析

二、微生态制剂的应用及其对鱼体或养殖生态环境的影响

微生态制剂不仅对水环境亚硝酸盐、氨氮、溶解氧、pH、总异养细菌数量、改善水体富营养化有较好的调节作用（李绍戊等，2012），而且也对水产养殖动物的生理功能具有偏好的影响，如可提高草鱼消化相关酶活性（王荻等，2011）、降低饲料系数（李绍戊等，2011）等。枯草芽孢杆菌（Bacillus substilis）在调水改水方面作用明显，对水体中亚硝酸盐含量、氨氮含量、溶解氧含量、pH 及总异养菌数量具有较好的改善作用（潘晓艺等，2007），并对中华鳖养具有其抗菌活性（沈锦玉等，2005）。此外，假单胞菌（Pseudomonas）M6 菌株还具有对孔雀石绿的脱色特性（曹海鹏等，2010）。

该文发表于《江苏农业科学》【2012，40（6）：219-222】

微生态制剂对东北三省淡水养殖池塘内浮游植物的影响

Influence of Microecologics on the Distribution of Phytoplankton in the Freshwater ponds in the Northeast China

李绍戊，王荻，尹家胜，卢彤岩

(中国水产科学研究院黑龙江水产研究所，黑龙江 哈尔滨 150070)

摘要： 随着人们对抗生素危害的日渐关注，绿色无残留的微生态制剂越来越广泛的被应用于水产动物养殖中。本实验选用2种常用微生物制剂 I 和 II，分别对黑龙江、吉林及辽宁淡水鱼养殖场进行投喂、泼洒及混用实验，并以池塘内浮游植物作为水质改良的标识物对2种微生态制剂在东北三省淡水养殖池塘内水质改良的作用进行评价。通过对东北三省淡水鱼类主要养殖季节7—9月份浮游植物的监测调查，分离鉴定的浮游植物分属7个门类，分别为绿藻门、硅藻门、蓝藻门、裸藻门、隐藻门、黄藻门和甲藻门。各省份淡水养殖池塘中浮游植物门类丰富度为：辽宁省（7门）>吉林省（6门）>黑龙江省（5门），均属于中-富营养型（β-α-ms）水体。两种微生态制剂在调节水质，改善水环境方面均有一定程度在作用，且 I 制剂在调节水体富营养化方面表现出良好的应用效果。

关键词： 微生态制剂；东北地区；养殖池塘；浮游植物

该文发表于《水产学杂志》【2011，24（4）：1-5】

微生态制剂对鲤、鲫和草鱼养殖池塘效益的影响

The Influence of Probiotics on Results of Common carp, Crucian Carp, and Grass Carp Culture

李绍戊，王荻，尹家胜，卢彤岩

(中国水产科学研究院黑龙江水产研究所，黑龙江 哈尔滨 150070)

摘要： 以枯草芽孢杆菌为主的主要用于促进鱼类消化生长的微生态制剂 I、多种混合微生物主要用于调节水质的微生态制剂 II，或是两者混合使用的方法，比较研究了黑龙江省及辽宁省池塘养殖的鲤（*Cyprinus carpio*）、鲫（*Carassius auratus*）和草鱼（*Ctenophyargodon idellus*）的增重率、饲料系数及药价等效益。结果表明，单种或者混合使用微生态制剂均对池塘养殖鲤及草鱼的增重率、饲料系数及用药价格有一定影响，同时使用两种微生态制剂提高鱼类增重率最显著，辽宁省和黑龙江省分别提高了423.57%和90%；而两省养殖池塘的药价也分别降低了29%和56.25%。微生态制剂 I 及混合同时使用两种微生态制剂均能显著降低饲料系数，辽宁及黑龙江分别降低0.35及0.05。结果可见，适当使用微生态制剂对辽宁草鱼及黑龙江池塘养鲤效益有较好影响，对减少用药、发展绿色养殖业具极大的推动作用。

关键词： 微生态制剂；养殖池塘；黑龙江；辽宁；效益

该文发表于《华北农学报》【2011，26（增刊）：219-223】

两种微生态制剂对草鱼消化相关酶活性的影响

Influence of Two Kinds of Microecologics on the Digestive Enzymes in Grass Carp

王荻，李绍戊，尹家胜，卢彤岩

（中国水产科学研究院黑龙江水产研究所，黑龙江 哈尔滨 150070）

摘要：选用以枯草芽孢杆菌为主要成分的 I 制剂及混合菌的 II 制剂两种常用、商品化微生物制剂，对辽宁省池塘养殖草鱼进行投喂、泼洒和混合使用试验，并比较测定鱼肝脏、肠道组织中淀粉酶、蛋白酶及脂肪酶活性值。结果表明，以调节水质为主要作用的泼洒制剂 II 对草鱼体内消化相关酶活性影响不大，但以促进消化生长为主要作用的投喂制剂 I 对酶活性有较大促进作用，而制剂 I/II 混用在调节水质、帮助消化的同时能最大程度的促进草鱼体内消化相关酶活性提升，有助于促进草鱼生长及繁殖发育。因此，微生态制剂在草鱼池塘养殖中，对鱼快速健康生长有较好的促进作用。

关键词：微生态制剂；草鱼；淀粉酶；蛋白酶；脂肪酶

该文发表于《水生生物学报》【2007，31（1）：139-141】

水产养殖中枯草芽孢杆菌的分子鉴定

Molecular Identification of Bacillus subtilis in Aquaculture

潘晓艺[1]，沈锦玉[1]，余旭平[2]，朱军莉[2]，尹文林[1]

（1. 浙江省淡水水产研究所，湖州 313001，2. 浙江大学，杭州 310029）

摘要：在测定细菌菌落形态及生理生化特性的基础上，应用 16S rRNA 作为分子指标直接对通过生理生化性状很难区别的枯草芽孢杆菌与蜡样芽孢杆菌进行分子分类鉴定。

关键词：枯草芽孢杆菌；生理生化性状；分子鉴定；16S rRNA

微生物种类繁多，以形态学和生理生化指标为基础的细菌鉴定较为复杂，是一项繁琐、费时的工作，对一种未知细菌的鉴定和分类，不仅需要分析多个生化指标，有时还要取决于工作人员的经验。因此迫切需要建立一些简单、方便、易于操作的分类鉴定方法对微生物进行分析，使人们在一定程度上更科学、更精确、更快速地找到分离物的分类地位。分子生物学为细菌等微生物的研究提供了新的理论和方法。目前，大分子 rRNA 及其基因 rDNA 已被作为一个分子指标[1]，广泛地应用于各种细菌的遗传特征和分子差异的研究[2]。大量已知细菌的 rDNA 数据都被测定并输入到国际基因数据库，使之成为对细菌鉴定和分类非常有用的参照系统。即人们可以通过对未知细菌 DNA 序列的测定和比较分析，达到对其进行快速、有效的鉴定和分类的目的。这一方法不仅为许多难以鉴定的微生物提供了一个新的手段；而且对于微生物资源和环境的研究也有重要的价值。本文在测定细菌菌落形态及生理生化特性的基础上，应用 16S rRNA 作为分子指标直接对通过生理生化性状很难区别的枯草芽孢杆菌与蜡样芽孢杆菌进行分子分类鉴定，并对所得数据进行分析和讨论。

1　材料和方法

1.1　供试菌株从养殖水域中分离得到的优势微生物经初筛、复筛之后，共筛选出 2 株具有不同生理功能的菌株，编号分别是 B110、B115。标准菌株枯草芽孢杆菌 63501，购自浙江省农业厅兽药检测所。

1.2 细菌的分类鉴定

1.2.1 常规生化鉴定 参照《常见细菌系统鉴定手册》[3]及《伯杰氏细菌鉴定手册》[4]，对纯化的菌株进行个体形态、菌落形态和培养特征的观察及主要生理生化试验。

1.2.2 细菌的分子生物学方法鉴定

1.2.2.1 细菌总 DNA 制备 细菌基因组 DNA 的提取参照文献[5]进行。抽提两株细菌的基因组 DNA 在 1% agarose 电泳观察检验，并作为 PCR 扩增的模板。

1.2.2.2 16S rDNA 的 PCR 扩增和产物纯化 用于 16S rDNA 的 PCR 反应的引物为一对通用引物[6]。PCR 反应体系（50 μL）为：10×Taqbuffer 5 μL，模板 DNA 1 μL，引物浓度为 10 μM 各 1 μL，10 μM dNTP 1 μL，5 U/μL Taq DNA 聚合酶 0.5 μL，超纯水 40.5 μL，混匀。反应条件：94℃预变性 5 min；94℃ 44 s，52℃ 60 s，72℃ 2 min，进行 30 个循环；最后 72℃延伸 5 min。取 PCR 产物 7 μL，1% 琼脂糖凝胶电泳检测 PCR 产物。PCR 产物用 PCR 产物纯化 kit 进行纯化。

1.2.2.3 PCR 产物的克隆与测序 PCR 产物经纯化后，克隆到 pMD 18-T 载体，转化 E. coliTG1，挑取白色菌落，碱法提取质粒，用 Pst I 和 EcoR I 双酶切及 PCR 鉴定确认阳性重组质粒。应用 MegaBACE 1000 自动测序仪（浙江大学分析测试中心）对阳性重组质粒进行测序，将所得序列运用 NCBI 的 BALST 程序进行同源序列比对。

2 结果与分析

2.1 菌株的形态特征 经初筛、复筛得到的 2 株芽孢杆菌的个体形态特征观察结果如下：这二株菌在牛肉膏蛋白胨斜面和平板培养基上，革兰氏阳性，杆状，能形成芽孢，芽孢呈椭圆形，芽孢囊不膨大，所以都属于芽孢杆菌属。

2.2 菌株的群体特征 在牛肉膏蛋白胨培养基倒成的平板上，接幼龄菌种，培养 24 h，观察菌株的菌落特征。在液体牛肉膏蛋白胨中培养，接幼龄菌种，静止培养 24 h，观察菌株的液体培养特征。群体特征观察结果列于表 1。

2.3 细菌的生理生化特征由于细菌个体微小，形态特征简单，在分类鉴定中单凭形态学特征不能达到鉴定的目的，故必须借助许多生理生化特征来加以鉴别，细菌对各种生理生化试剂的不同反应，显示出各类菌种的酶类不同。对试验中 2 株细菌的生理生化特征鉴定结果见表 2。

表 1 菌体的群体特征
Tab. 1 Population characterization of isolates

菌株 Isolate	菌落特征 Colony characteristics						静止液体培养 Still liquid culture	
	形状规则 Morphology	菌落直径（cm） Diameter	质地 Texture	边缘形状 Margin	光学透明度 Transparence	是否产色素 Pigment	菌膜 Pellicle	浑浊 Tubidity
63501	圆形 Circular	0.3	皱折 Wrinkle	Undulate	No	-	+	-
B110	圆形 Circular	0.5	皱折 Wrinkle	Undulate	No	-	+	-
B115	圆形 Circular	0.4	皱折 Wrinkle	Undulate	No	-	+	-

表 2 菌株的生理生化特征
Tab. 2 Physiological and biochemicalcharacterizations of isolates

测定项目 Characteristics	63501	B110	B115	测定项目 Characteristics	63501	B110	B115
VP 试验 VPtest	+	+	+	产气 Gas production	-	-	-
触酶 Catalase	+	+	+	水杨素 Sallcin	+	-	-
pH5.7	+	+	+	甲基红 MR test	+	-	-
硝酸盐 Nitrate	+	+	+	丙酸盐 Propionate	+	-	+
亚硝酸盐 Nitrite	+	+	+	无盐胨水 0% NaCl	+	+	+
柠檬酸盐 Citrate	+	+	+	7%NaCl	+	+	+

续表

测定项目 Characteristics	63501	B110	B115	测定项目 Characteristics	63501	B110	B115
葡萄糖 Glucose	+	+	+	10% NaCl	+	+	+
木糖 Xylose	+	+	+	5℃生长 Growth	−	−	−
淀粉 Amylase	+	+	+	10℃生长 Growth	−	−	−
明胶 Gelatinase	+	+	+	30℃生长 Growth	+	+	+
吲哚 Indole	−	−	−	40℃生长 Growth	+	+	+
尿素 Urease	+	+	+	50℃生长 Growth	±	+	−
苯丙氨酸 Phenylalanine	−	−	−	55℃生长 Growth	−	−	−

注：+表示阳性或生长；−表示阴性或不生长；±表示少量生长

note：+: positive or growth；−: negative or no growth；±: a little growth

根据表2各个菌株的生理生化特征，并结合个体形态和群体形态特征，参照东秀珠等《常见细菌系统鉴定手册》和布坎南等《伯杰氏细菌鉴定手册》，对此2株细菌的鉴定结果为枯草芽孢杆菌或蜡样芽孢杆菌。

2.4 细菌的16S rRNA基因序列鉴定菌株B110和B115的基因组经PCR扩增获得了预期的1.5 Kb的特异条带（见图1），PCR产物经回收，并克隆到pMD 18-T载体。阳性重组质粒经Pst I和EcoR I双酶切得到两条相应的条带，小片段约1.5 Kb，大片段约2.6 Kb（为pMD 18-T载体）；对阳性克隆进行PCR扩增，仍能获得预期大小约1.5Kb的片段，表明1.5Kb PCR扩增片段已克隆入pMD 18-T载体。

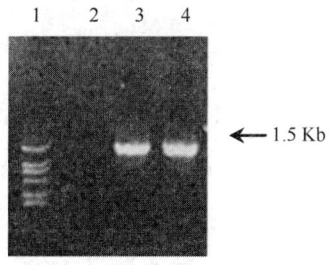

图1 细菌16S rDNA PCR产物

Fig. 1 PCR amplification of 16S rDNA of two isolates

1. Marker DL2000 2. 阴性对照 3. B110 16S rDNA 片段 4. B115 16S rDNA 片段

挑选其中一个阳性克隆进行测序，获得了细菌16S rRNA基因的部分序列，发现这两个菌株的序列基本相同，部分序列结果如下：

GTGTACAAGGCCCGGGAACGTATTCAC-
CGCGGCATGCTGATCCGCGATTAC-
TAGCGATTCCAGCTTCACGCAGTCGAGTT-
GCAGACTGCGATCCGAACTGAGAA-
CAGATTTGTGGGATTGGCTTAACCTCGCG-
GTTTCGCTGCCCTTTGTTCTGTCCATTG-
TAGCACGTGTGTAGCCCAGGTCATA-
AGGGGCATGATGATTTGACGTCATC-
CCCACCTTCCTCCGGTTTGTCACCG-
GCAGTCACCTTAGAGTGCCCAACTGAAT-
GCTGGCAACTAAGATCAAGGGTT-
GCGCTCGTTGCGGGACTTAACCCAA-
CATCTCACGACACGAGCTGACGACAAC-

703

CATGCACCACCTGTCACTCTGCCCCCGAAGGGGACGTCCTATCTCTAGGATTGTCAGAGGA

  测定的序列递交 NCBI 进行 BLAST 同源序列比对，与枯草芽孢杆菌的标准菌株的 16S rDNA 具有 99%～100%的同源性，结果表明 B110 和 B115 两株菌均属于枯草芽孢杆菌。

3 讨论

  由于常规生理生化方法很难正确地鉴定芽孢杆菌属中的各个种，特别是生理生化相似的种如枯草芽孢杆菌与蜡样芽孢杆菌。参照伯杰氏细菌鉴定手册（1990 年版）、东秀珠等《常见细菌系统鉴定手册》和国内其他参考文献都没有对这两类菌具有明显的区分特性，又由于细菌生理生化性状的可变性，因此用常规生化指标测定很难得到正确的鉴定。游春平等[7]用生理生化性状测定只能得到初步的鉴定结果。王雅平等[8]对其分离到的芽孢杆菌也只有鉴定到属（Bacillus spp.）。

  根据赵铭钦等[9]报道，蜡状芽孢杆菌与枯草芽孢杆菌的区别，前者细胞宽度大于0.9 μm，后者细胞宽度的小于 0.9 μm。作者认为仅凭细胞大小的差异来区分它们是不够充分的，况且在不同培养基上细胞大小有所变异。作者利用 16S rDNA 对其进行分子鉴定，不仅使鉴定结果迅速可靠，而且为今后可能的新种的鉴定打下了良好的基础。

  嗜水单胞菌引起的淡水鱼类细菌性败血症及其他多种水生动物病症，是我国在 90 年代乃至现在较严重的细菌性传染病，给水产养殖业造成巨大的损失[10]。该病发生后，主要方法是使用化学杀菌剂。长期使用化学杀菌剂不但会导致病菌对其产生抗药性[11]，而且水产品上残留农药也会对公众健康造成威胁和环境污染。本文中的二株菌 B110 和 B115 对鱼类养殖水环境具有明显的改善作用，其中 B115 对嗜水单胞菌具有强烈的抑杀作用，有关内容另文报道。通过对有益菌的研究，旨在利用有益生物来代替化学农药的防病新技术。这对开拓水产病害防治的新技术具有重要的意义。

参考文献

[1] Delong E. F., Wickham G. S., Pace N. R.. Phylogenetic stains：ribosomal RNA-based probes for the identification of single cells [J]. Science, 1989, 243：1360-1363

[2] Qu L. G.. General Molecular Genetic Study of Microbe Systematic Development [M]. In：(Ed：Fudan university), General Study of Microbe Genetics. Shanghai：FudanUniversity Press. 1993. [屈良鹄. 微生物系统发育的分子遗传学研究概述 [M]. 见：复旦大学主编. 微生物遗传学研究综述集. 上海：复旦大学出版社，1993]

[3] Dong X. Z., Cai M. Y.. General Manual of Systematic and Determinative Bacteriology [M]. Beijing：Science Press, 2001, 9-42 [东秀珠，蔡妙英等. 常见细菌系统鉴定手册 [M]. 北京：科学出版社，2001, 9-42]

[4] Holt J. G., Krieg N. R. Bergey's Manual of Systematic Bacteriology 9th ed [M]. BaltimoreLondon：Williams & Wilkins Co, 1994. 353-376

[5] Li D. B., Zhou X. P., Xu J. P., He Z. H.. Technique of Gene Engineering [M]. Shanghai：Science and Technology Press, 1996. [李德葆，周雪平，许建平，何祖华编. 基因工程操作技术 [M]. 上海科学技术出版社，1996]

[6] http：//silk. uic. ac. be/primer/database. html

[7] You C. P., Siao A. P., Li X. M., et al. Screening and Identification of Antagonistic Microorganisms against [J]. Pyricularia grisea. 2001, 23 (4)：519-521 [游春平，肖爱萍，李湘明等. 稻瘟病菌拮抗微生物的筛选及鉴定 [J]. 江西农业大学学报，2001, 23 (4)：519-521]

[8] Wang Y. P., Liu Y. Q., Pan N. S., et al. Screening of Antifungal Protein Producing Strain TG26 of Bacillus spp. and Studies on its Cultural Condition [J]. Acta Botanica Sinica, 1993, 35 (3)：222-228 [王雅平，刘伊强，潘乃穟等. 抗菌蛋白产生菌 TG26 的筛选及其培养条件 [J]. 植物学报，1993, 35 (3)：222-228]

[9] Zhao M. Q., Qiu L. Y., Liu W. C.. Identification and preliminary application study of tobacco flavoring strain [J]. Hei Long Jiang Tobacco, 1999, 3：17-20 [赵铭钦，邱立友，刘伟城等. 烤烟发酵增香菌株的鉴定和初步应用研究 [J]. 黑龙江烟草，1999, 3：17-20]

[10] Shen J. Y., Chen Y. Y., Shen Z. H., et al. Study on Pathogen of Outbreaks of Infectious Disease of Fishes in Zhejiang Province：Isolation, Pathogenicity, Physiological and Biochemical and Biochemical Characteristics of Aeromonas hydrophila [J]. Bulletin of Science and Technology, 1993, 9 (6)：397-401 [沈锦玉，陈月英，沈智华，等. 浙江养殖鱼类暴发性流行病病原的研究：嗜水气单胞菌的分离、致病性及生理生化特性 [J]. 科技通报，1993, 9 (6)：397-401]

[11] Boyd C. E., Massaaut L.. Risks associated with the use of chemicals in pond aquaculture [J]. Aqua Engin, 1999, 20 (2)：113-132.

该文发表于《水生生物学报》【2005, 29 (6): 689-693】。

枯草芽孢杆菌 B115 抗菌蛋白的分离纯化及部分性质

Purification and Partial Characterization of Antagonistic Substance from *Bacillus subtilus* B115

沈锦玉，尹文林，曹铮，潘晓艺，吴颖蕾

(浙江省淡水水产研究所, 湖州 313001)

摘要： 枯草芽孢杆菌 (*Bacillus subtilis*) B115 分离自水产养殖池塘，对鱼类细菌性败血症致病菌具有较强的拮抗能力。除去菌体培养液用浓盐酸沉淀，乙醇抽提所得的拮抗物粗提液对热不稳定，4℃保存不能超过两周，-18℃保存不能超过一个半月，对胰蛋白酶不敏感，对蛋白酶 K 部分敏感，对氯仿部分敏感，其作用活性 pH 范围较广，对 pH 值不敏感。粗提液经 CM 柱离子交换层析和 P-60 柱层析，得到一个拮抗活性峰 P2。粗提物经丙酮分级沉淀及高压液相色谱分离，得到较纯的 LP，经质谱仪测定分子量为 803.6D。1 L 发酵的细菌培养液可得到约 1 mg LP 纯品。

关键词： 枯草芽孢杆菌 B115；抗菌物质；分离纯化；理化特性

水产养殖的细菌性病害是造成农业损失的主要原因之一[1]，施用化学杀菌药及抗生素等尽管有一定效果，但长期使用会增加病原菌的抗性。另外，农药的大量使用还会造成环境污染及药物残留。利用微生物来抑制动物病原细菌为细菌病害的防治提供了新的可能。近年来，已筛选了许多防治植物病害的拮抗菌如芽孢杆菌 (*Bacillus* sp.)，并阐明了其抗菌物质对植物病原菌的抑制作用，抗菌机理及抗菌物质的理化特性等[2-5]。而在水产上枯草芽孢杆菌抗菌蛋白的性质未见报道。本室从池塘土壤中分离到一株强烈抑制鱼类细菌性败血症病原菌嗜水气单胞菌 (BSK-10)[1] 的拮抗菌 (*Bacillus subtilis*) B115，并从中分离到抗菌蛋白，这种物质除对鱼类出血性败血病病菌表现出强烈抑制作用外，对大黄鱼致病菌、中华绒螯蟹致病菌及金黄色葡萄球菌也有强烈的抑制作用，经同步培养，B115 对 BSK-10 的抑菌圈为 14 mm，对大黄鱼哈维氏弧菌的抑菌圈为 15 mm，对中华绒螯蟹致病菌 CL99920 的抑菌圈为 13 mm，对大黄鱼副溶血弧菌的抑菌圈为 13 mm，对金黄色葡萄球菌的抑菌圈为 10 mm（另文报道）。本文报道 B115 菌株产生的一种抗菌物质的分离纯化及其部分性质初步研究。

1 材料与方法

1.1 菌种

供试菌种为本室分离的一株枯草芽孢杆菌 (*Bacillus substilis*) B115。

1.2 主要仪器及试剂

高速低温离心机 (Sigma)；冷冻干燥机 (German)；层析系统 (Biologic LP, BIO-RAD)；离心真空干燥浓缩仪 (SPD 111V, Thermo Savant)；CM (Econo-Pac, Catalog No. 732-0001, BIO-RAD)；Bio-Gel P-60 Gel (BIO-RAD)；Waters 515；Bruker Daltonisos (esquire 3000)；蛋白酶 K (Merck, Germany)；胰蛋白酶为 Sigma 产品，其余试剂均为国产分析纯。

1.3 粗样品的制备

将 B115 菌株接种于 BPY（牛肉膏 5 g，蛋白胨 10 g，酵母浸膏 5 g，氯化钠 5 g，葡萄糖 5 g，水 1 000 mL）培养基，28℃摇床培养 24 h，摇床转速 180 r/min，4℃离心 30 min，6 000 r/min，收集上清液，以孔径为 0.45 μm 的细菌滤器过滤，得无细胞培养滤液。向上清液中加入浓盐酸，调节 pH 值至 3，8 000 r/min 离心 15 min，弃去上清。沉淀用 80% 乙醇抽提三次，抽提液真空抽干，将所得干粉悬浮在少量去离子水中，用 2 mol/L NaOH 调 pH 值至 7.0，使其全部溶解。再次用 1 mol/L的 HCl 调 pH 值至 3.0，8 000 r/min 离心 10 min，收集沉淀，向沉淀中加入少量 1% 的碳酸氢铵溶液，使其全部溶解。将此溶液冷冻干燥，得到淡黄色粉末状粗样品。每步实验中取少量进行抗菌活性检测。

1.4 粗提液的稳定性测定

1.4.1 热稳定性 粗提液在不同温度下处理 30 min，100℃水浴中处理 5 min、10 min、15 min 及样品在 4℃及-18℃冰箱放置 46 d（每隔 2 d 取样），各取 50 μL 进行抗菌活性测定。

1.4.2 对蛋白酶的稳定性 粗提液分别用浓度为 100 μg/mL 的胰蛋白酶和蛋白酶 K，在 37℃下处理 1 h，取 50 μL 进行抗菌活性检测。

1.4.3 对氯仿稳定性 粗提液与等量氯仿混合振荡抽提 60 mim，离心，分层后取上层水相，待残留氯仿挥发后，取 50 μL 进行拮抗活性检测。

1.4.4 对 pH 稳定性 粗提液的 pH 分别调至不同值，取 50 μL 进行抗菌活性检测。

1.5 抗菌成分的分离纯化

1.5.1 层析分离法

1.5.1.1 CM 离子交换层析 缓冲液 A 为 0.02 mol/L 的醋酸−醋酸钠缓冲液（pH5.0）加 0.05 mol/L 的氯化钠，缓冲液 B 为 0.02 mol/L 的醋酸−醋酸钠缓冲液（pH5.0）加 1.5 mol/L 的氯化钠。用缓冲液 A 充分平衡层析柱（20 cm×1.5 cm），流速 6~8 mL/min。取 1.3 制备的粗提液上样，流速 1 mL/min，用缓冲液 A 洗涤至出现基线，而后进行梯度洗脱，0~100%B 梯度洗脱，流速 1 mL/min，按每管 2 mL 进行收集，收集在 280 nm 处光密度值高的部分，同时测定各峰抑菌活性。保留有抑菌活性的蛋白峰。

1.5.1.2 Bio-Gel P-60 凝胶过滤层析 对 1.5.1.1 中有抑菌活性的部分进行纯化：Bio-Gel P-60 与 Sephadex G-100 相当，参照郑怡等[6]方法进行，用 0.02 mol/L 的磷酸钠缓冲液（pH7.2）平衡层析柱（100 cm×1.5 cm），取 1.5.1.1 中经抑菌检测有活性部分合并后上样，流速为 0.4 mL/min，使样品进入凝胶。最后用同样的缓冲液洗脱，流速 0.4 mL/min，自动部分收集器收集，1 mL/管。收集在 280 nm 处光密度值高的部分，同时测定各峰抑菌活性。

1.5.2 高压液相色谱分离法

1.5.2.1 丙酮分级沉淀 将粗样品溶解在 0.02 mol/L pH6.8 磷酸缓冲液中（0.1 g/mL），加入 6 倍体积冰浴的丙酮，12 000 r/min，4℃离心 10 min，弃去沉淀。在上清中继续加入原体积 10 倍的预冷的丙酮，12 000 r/min 4℃离心 10 min，收集沉淀，溶解于少量 0.02 mol/L pH6.8 磷酸缓冲液中。

1.5.2.2 高压液相色谱（HPLC）反相分析 将经丙酮沉淀得到的样品溶液用经 0.02 mol/L pH6.8 磷酸缓冲液平衡的 Bio Rad Hi pore 反相柱吸附，用 0~100% 的乙腈梯度洗脱，流速 1 mL/min，收集有抗菌活性的色谱峰，离心真空干燥浓缩，再将其冷冻干燥，即为 LP。

1.5.2.3 分子量的测定 采用 Bruker Daltonisos（esquire 3000）质谱仪。取冷冻干燥的样品 LP 少许，溶于甲醇中，用微量移液器吸取溶液，缓慢加入到质谱仪中；流动相为甲醇。

1.5.2.4 蛋白质浓度的测定 采用考马斯亮蓝染色法[7]，以牛血清白蛋白为标准蛋白。

1.6 抑菌实验

1.6.1 指示菌 选用鱼类细菌性败血症病原菌嗜水气单胞菌（BSK-10）为测定抑菌活性的指示菌，该菌由本实验室保存。

1.6.2 抑菌活性测定 将 BSK-10 培养于普通肉汤培养基，培养 24 h，取细菌浓度为 $1×10^7$ 个/mL 的菌悬液 0.1 mL，均匀地铺在装有 20 mL LB、直径为 90 mm 的培养皿上，用一个直径为 4.5 mm 的打孔器在培养皿的中心及边缘均匀打 5 孔，再分别接种 50 μL 处理液于孔内，每处理 3 个培养皿，用封口膜密封后培养在 30℃，1 d 后测定抑菌圈的直径。

2 结果

2.1 粗提物的稳定性

2.1.1 热稳定性 粗提液分别在不同温度下处理 30 min，取 50 μL 检测拮抗活性，结果如表 1。粗提物经 60~100℃处理 30 min 后仍保持部分抗菌活性，100℃处理时间越长，部分抗菌活性越弱，120℃处理 30 min 后丧失抗菌活性。样品在 4℃冰箱放置 15 d 后，−18℃冰箱放置 44 d 后，抗菌活性完全消失（见表 2），说明该抗菌物质不具有耐高温性，4℃保存不能超过两周，−18℃保存不能超过一个半月。

表 1　B115 粗提液经不同温度热处理后的抗菌活性
Tab. 1　Antagonistic activity of crude extract of B115 treated with various temperatures

	处理温度 t（℃）						煮沸时间（min）		
	40	60	70	80	100	120	5	10	15
抗菌圈直径（mm）	13	10	10	9	9	–	12	12	10

表2　B115粗提液不同时间保存后的抗菌活性

Tab. 2　Antagonistic activity of crude extract of B115 stored in various days

天数（d）		1	4	7	10	13	15	17	20	23	26	29	32	35	38	41	44	46
抗菌圈直径（mm）	4℃	14	12	11	9	8	6	–	–									
	-18℃	14	13	13	12	12	12	11	11	11	10	10	10	9	9	8	6	–

"–"表示没有抑菌圈

2.1.2　对蛋白酶的稳定性粗提液经不同蛋白酶处理后，取50 μL检测活性，结果如表3。粗提液对胰蛋白酶不敏感，对蛋白酶K部分敏感。

表3　处理后抗菌物质的抗菌活性

Tab. 3　Antagonistic activity of crude extract of B115 with various treatments

	pH			蛋白酶K	胰蛋白酶	氯仿	未处理
	4	5	9				
抗菌圈直径（mm）	14	14	14	11	14	–	14

2.1.3　对氯仿的稳定性　粗提液经氯仿处理后，用50 μL样品进行活性检测，结果发现抗菌圈直径为0；而未经氯仿处理的粗提液抗菌圈直径为14 mm。因此粗提液对氯仿敏感。

2.1.4　对pH值的稳定性粗提液在不同pH值下处理后，抗菌活性检测表明（见表3），B115粗提液活性pH值范围较宽，无论pH值为中性或偏酸性和偏碱性，滤液均对病菌有抑制活性，且三种pH值下的抑菌活性无差异，说明抗菌物质对pH值不敏感。

2.2　分离纯化

2.2.1　粗样品的制备　培养上清液用浓盐酸将pH值调至3时，测定样品上清及沉淀的抑菌圈直径，证明大部分抑菌物质在沉淀中，故用调节pH值的方法可初步浓缩抑菌物质。沉淀用80%乙醇抽提三次，测定每次抽提时上清及沉淀的抑菌圈直径，证明抑菌物质全溶在乙醇中，且乙醇容易挥发。利用抑菌物质的性质得到它的粗制样品。

2.2.2　凝胶层析的纯化结果　取1.3制备的粗提液经CM柱离子交换层析，洗脱曲线见图1，洗脱液经抑菌活性检测，第10~14管抑菌圈最大见图2。收集CM柱离子交换层析的活性部分（图1），上Bio-Gel P-60柱，得到两个主要吸收峰（图3）。经活性检测，约1 h后出现的是没有活性的小峰P1，约5.5 h出现的是有抑菌活性的峰P2。

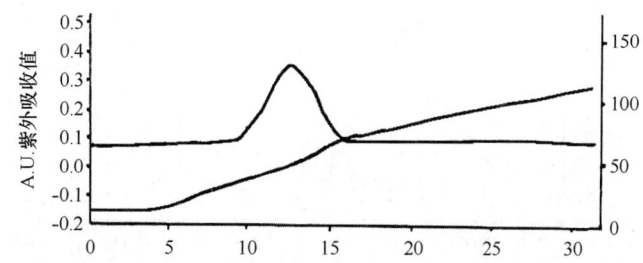

Fig. 1　Elution pattern of antibacterialproduction test of production through CM ion-exchange

2.2.3　高压液相色谱的纯化结果

2.2.3.1　抑菌物质的分离纯化经过两次酸沉淀后，冷冻干燥得到淡黄色粉末状粗样品，含有很多色素物质，丙酮分级沉淀能有效地去除这些杂质。反相柱层析时，得到的几个峰，只有峰3有抑菌活性（图4）。经分析峰3的含量大于80%。

2.2.3.2　抑菌物质的分子量测定　经质谱测定，抑菌物质LP的分子量为803.6D（图5）。由于LP的分子量较小，用常规的SDS-PAGE或者凝胶过滤色谱无法测定。最后得到的样品LP为白色粉末状，1 L发酵的细菌培养液可得到约1 mg LP纯品。

3　讨论

从拮抗细菌B115菌株中分离纯化到抗菌物质。其粗提液的pH值为中性或偏酸性和偏碱性时均对指示菌BSK-10有抑制活性。pH值为4、5、9时，滤液抑菌活性与对照组一致，而当pH值为3时，滤液中产生絮凝物质，离心后取沉淀，用pH值为7的生理盐水溶解，此溶液的抑菌活性并没有减弱，说明此抗菌物质对pH值不敏感。但经煮沸处理后，抗菌物质

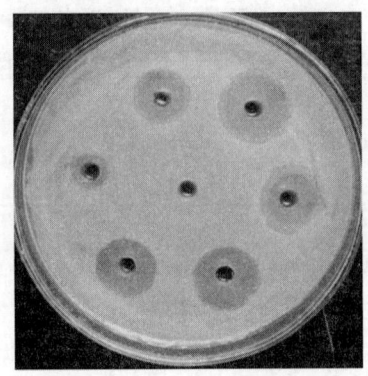

Fig. 2　The antibacterial activity after CM ion-exchange

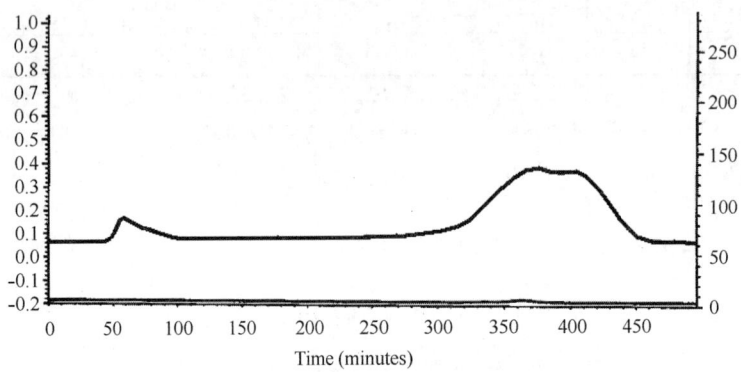

图 3　图 1 中活性峰经 P-60 柱的色谱图

Fig. 3　Column chromatography of fig. 1 peak on P-60

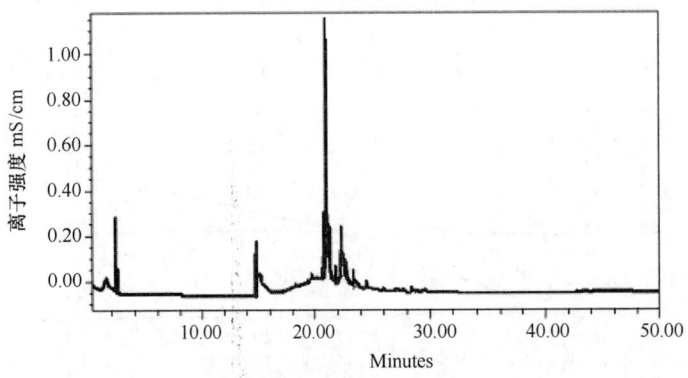

图 4　反相柱分离的色谱图

Fig. 4　Isolation of LP on reversed phase column

的抗菌活性随着煮沸时间的延长而减弱，经高压灭菌后，完全失去抑菌活性；在 4℃冰箱放置 15d 及-18℃冰箱放置 44 d 后，活性也完全消失，说明该抗菌物质不具有耐高温性，4℃保存不能超过两周，-18℃保存不能超过一个半月。

　　由于本文中抗菌物质的分离是按照蛋白质的特性进行分离纯化的，因此不排除该菌株同时产生其他非蛋白物质的可能。在进行分离物质 P2 的纯度及分子量的测定时，对 P2 进行 SDS-PAGE 电泳，结果 P2 在凝胶的最前缘，说明其分子量太小，不能用此电泳法测出。故改用 HPLC 及质谱法测定该物质的纯度及分子量。对 LP 样品用 Mr 截留为 $1×10^3$ 的超滤杯超滤浓缩后，上清液及浓缩液中均有部分抗菌活性，说明抑菌物质的分子量（Mr）较小，不到 $1×10^3$。用质谱测定其分子量仅为 803.6 的小肽。抗菌谱表明它是一种广谱的抗鱼类病原细菌的抗菌小肽，对许多危害严重的病原细菌都表现出很强的抗菌活性，在水产养殖中有很好的应用前景。由于抑菌物质的含量很低，因此，本文仅进行了抗菌活性测定、分离纯化及部分理化特性研究。至于它的组成成分、抗细菌机理等有待进一步研究。

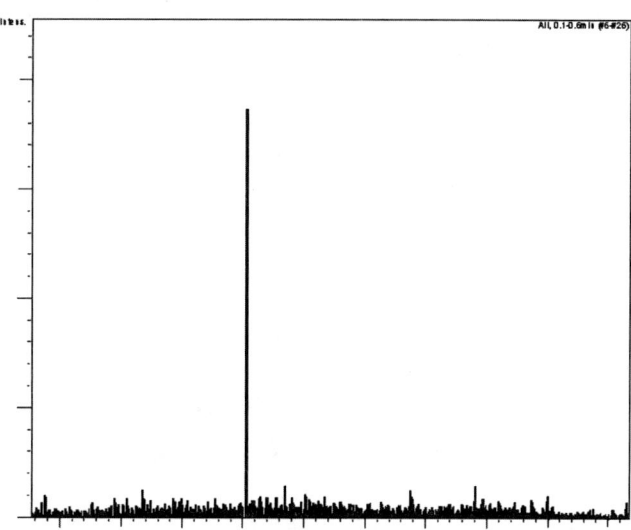

图 5　抗菌物质 LP 的质谱图谱

Fig. 5　MS profile of LP

自从 1952 年 J. Babad 等从枯草芽孢杆菌培养液中分离出抗真菌肽以来，又陆续分离出多种抗真菌肽[8-11]。这些小肽有许多共同特征：分子量很小，仅 1 000 D 左右。从已知的性质来看，LP 同文献报道的各种小肽类都有一定差别。芬枯草菌素 A （Fengycin A） 不溶于水，而且分子量为 1 447 D，同分离到的 LP 有很大区别[8]；但 LP 与刘颖等[12]从枯草芽孢杆菌培养液中分离出抗真菌肽性质很相似 （分子量 1057.3），但分子量相差 253；与童有仁等[12]从枯草芽孢杆菌培养液中分离出抗菌肽分子量 （50.3 kD） 相差悬殊，所以，LP 为不同于其他已发表的抗菌物质的一种抗菌小肽。

参考文献

［1］　Shen J. Y., Chen Y. Y., Shen Z. H., et al. Study on Pathogen of Outbreaks of Infectious Disease of Fishes in Zhejiang Province：Isolation, Pathogenicity, Physiological and Biochemical and Biochemical Characteristics of Aeromonas hydrophila. Bulletin of Science and Technology, 1993（6）：397-401.［沈锦玉，陈月英，钱冬，等. 浙江养殖鱼类暴发性流行病病原的研究：嗜水气单胞菌的分离、致病性及生理生化特性［J］. 科技通报，1993（6）：397-401］

［2］　Kong J., Wang W. X., Zhao B. G., et al. The studies on the antibiotic substances from Bacillus subtilis B-903 strain［J］. Acta Microbiologic Sinica, 1992, 32（6）：445~449.［孔建，王文夕，赵白鸽，等. 枯草芽孢杆菌 B-903 抗菌物质的研究［J］. 微生物学报，1992，32（6）：445~449］

［3］　Zhang X. J., Liu H. L., Pan X. M., et al. Condition on antagonistic substance production of B3 used for biocontrol of sharp eyespot of Wheat［J］. Journal of Nanjing Agricultural University, 1995, 18（1）：26~30.［张学君，刘焕利，潘小玫，等. 小麦纹枯病生防菌株 B3 产生抗菌物质条件的研究［J］. 南京农业大学学报，1995，18（1）：26~30］

［4］　Handelsman J., Stabb E. V.. Biocontrol of soilborne plant pathogens. The plant cell［J］, 1996, 8：1855-1869

［5］　WILSON C. L., CHALUTZ E.. Postharvest biological control of Penicillium rots of citrus with antagonist yeast sand bacteria［J］. Sci. Horticult., 1989, 40：105-112.

［6］　Zheng Y., YU P. and LIU Y. R.. Purification and Partial Characterization of Chlorella pyrenoidosa Lectin［J］. Acta Hydrobiologica Sinica. 2003, 27（1）：36-40.［郑怡，余萍，刘艳如. 蛋白核小球藻凝集素的分离纯化及部分性质研究［J］. 水生生物学报，2003，27（1）：36-40］

［7］　Li J. W., Xiao N. G., Yu R. Y.. Principle and Methods of Biochemical Experiment［M］. Beijing：Beijing University Press, 1994.［李建武，萧能庚，余瑞元，等. 生物化学实验原理和方法［M］. 北京：北京大学出版社，1994.］

［8］　Vanittanakom N., Loeffler W.. Fengycin-a novel antifungal lipopeptide antibiotic produced by Bacillus subtillis F-29-3. J antibiotics, 1986, 39（7）：888~901.

［9］　Peypoux F., Besson F., Michel G.. Characterzation of a new antibiotic of iturin group：Bacillomycin D. J Antibiotics, 1980, 33（10）：1146~1149.

［10］　Mhammedi A., Peypoux F., Besson F. et al. Bacillomycin F, a new antibiotic of Inurin group：isolation and characterization. J Antibiotics, 1982, 35（3）：306~311.

［11］　Kajimura Y., Sugiyama M., Kaneda M.. Bacillopetins, newcyclic lipopetide antibiotics from Bacillus subtillis FR-2. J Antibiotics, 1995, 48（10）：1095~1103.

[12] Liu Y., Xu Q., Chen Z. L.. Purification and Characterization of Antifungal Peptide LP-1 [J]. Acta Microbiologic Sinica. 1999, 39（5）：441-447. [刘颖，徐庆，陈章良. 抗真菌肽LP-1的分离纯化及特性分析 [J]. 微生物学报，1999, 39（5）：4 41-447]

[13] Tong Y. R., Ma Z. C., Chen W. L. et al. Purification and Partial Characterization of Antagonistic Proteins from Bacillus Subtilis B034 [J]. Acta Microbiologic Sinica. 1999, 39（4）：339-343. [童有仁，马志超，陈卫良，李德葆. 枯草芽孢杆菌B034拮抗蛋白的分离纯化及特性分析 [J]. 微生物学报，1999, 39（4）：339-343].

该文发表于《环境污染与防治》【2010, 32（12）：39-42】

孔雀石绿脱色菌 M6 的脱色特性与系统发育分析

Decolorization characterization of malachite green decolorized strain M6 and its phylogenetic analysis

曹海鹏[1]，葛秀卿[1]，何珊[2]，李怡[1]，杨先乐[1]

（1. 上海海洋大学水产与生命学院，国家水生动物病原库，上海 201306；2. 上海师范大学教育学院，上海 200234）

摘要：研究了温度、初始菌细胞数、孔雀石绿浓度、振荡速度对脱色菌 M6 脱色孔雀石绿的影响，并通过邻接法对脱色菌 M6 进行了系统发育分析。结果表明：（1）脱色菌 M6 对孔雀石绿脱色的最佳温度为 30℃，最佳振荡速度为 200 r/min。（2）随着初始菌细胞数的增大，脱色菌 M6 对孔雀石绿的脱色率明显升高。当初始菌细胞数为 $1.5 \times 10^2 \sim 1.5 \times 10^6$ cfu/mL 时，脱色菌 M6 对孔雀石绿的脱色率均不低于 86.92%；当初始菌细胞数大于 1.5×10^4 cfu/mL 时，脱色率均高于 90%。（3）当孔雀石绿质量浓度为 12.5~400.0 mg/L 时，脱色菌 M6 对孔雀石绿具有良好的脱色特性，脱色率均在 93% 以上。（4）脱色菌 M6 的 16S rDNA 序列与 GenBank 基因库中假单胞菌属（Pseudomonas）菌株的 16S rDNA 序列有 98%~99% 的高度同源性，而且与恶臭假单胞菌（Pseudomonas putida）菌株 JM9（登录号：FJ493139.1）的亲缘关系最近。

关键词：孔雀石绿 脱色 系统发育分析

三、致病微生物功能性研究

嗜水气单胞菌在传代过程中毒力基因分布相对较稳定。*aer*A、*ahp*A 为嗜水气单胞菌致病的主要毒力因子，与致病性存在相关性，*ahp*A 基因与毒力相关性最大，且在传代过程中极易发生变异，导致毒力逐渐减弱甚至消失（付乔芳等，2011）。*ast* 与致病力存在一定相关性，但不是主要毒力因子，hlyA、altA 与致病性不存在相关性（付乔芳等，2011）。

从患细菌性败血症的西伯利亚鲟（*Acipenser baerii*）的体内分离到一株致病嗜水气单胞菌（*Aeromonas hydrophila*）菌株 X1，将其制成灭活全菌苗。免疫结果表明，嗜水气单胞菌 X1 全菌苗能够明显提高西伯利亚鲟的血清抗体水平及总蛋白、免疫球蛋白、溶菌酶含量（李圆圆等，2008）。

该文发表于《生物技术通报》【2011, 9：130-135】

嗜水气单胞菌毒力基因在传代过程中的稳定性研究

Stability of Virulence Genes of Aeromonas hydrophila Strains During Subculture

付乔芳，邱军强，胡鲲，杨先乐，安健

（上海海洋大学水产与生命学院，国家水生动物病原库，上海 201306）

摘要：为了解嗜水气单胞菌毒力基因在传代过程中的稳定性，对传代前后嗜水气单胞菌（*Aeromonas hydrophila*）的毒力和毒力基因进行测定和检测。采用急性毒性试验方法测定 10 株嗜水气单胞菌原代、第 10 代及第 20 代的菌株对异育银鲫（*Carassais auratus gibebio*）的半数致死量，通过聚合酶链式反应（PCR）检测原代、第 10 代及第

20 代的菌株中 5 种毒力基因 aerA、hlyA、ahpA、altA 及 ast 的变化情况。试验结果表明，嗜水气单胞菌在传代过程中毒力基因分布相对较稳定。但是 ahpA 基因极易发生变异，在培养基上连续多次传代后，毒力逐渐减弱甚至消失。可推知，ahpA 基因与毒力相关性最大，并易受传代影响。

关键词：嗜水气单胞菌 传代 毒力基因 稳定性

该文发表于《生物学杂志》【2011, 28（6）：53-57】

嗜水气单胞菌国内分离株的毒力因子分布与致病性相关性分析

The Analyse of Virulence Factors–Pathogenicity Relationships of *Aeromonas hydrophila* Strains Isolated from China

付乔芳，邱军强，胡鲲，杨先乐，安健

（上海海洋大学水产与生命学院，国家水生动物病原库，上海 201306）

摘要：为了研究分析国内致病性嗜水气单胞菌的毒力因子的分布及与致病性的相关性及其防治，选取 5 种主要毒力因子气溶素（aerA）、溶血素（hlyA）、丝氨酸蛋白酶（ahpA）、抗金属蛋白酶（ast）、肠毒素（altA）设计 5 对特异性引物，用 PCR 方法进行检测，采用急性毒性实验方法测定菌株对异育银鲫的半数致死量（50% lethal dose，LD_{50}），细菌药物敏感试验纸片扩散（K-B）法检测对抗生素的敏感性，结果显示：除菌株 Hong12 的 LD_{50} 约为 10^9 cfu，毒力弱外，其他毒力均较强，LD_{50} 均在 10^6 cfu 左右。aerA、ahpA、hlyA、altA、ast 5 种毒力基因的携带率分别为 77.8%、88.9%、100%、100%、77.8%。通过比较嗜水气单胞菌毒力基因的分布情况与其对鲫鱼致病性试验结果可以发现，aerA、ahpA 为嗜水气单胞菌致病的主要毒力因子，与致病性存在相关性，ast 与致病力存在一定相关性，但不是主要毒力因子，hlyA、altA 与致病性不存在相关性。9 株嗜水气单胞菌对青霉素、羧苄青霉素、万古霉素不敏感，对复合磺胺、头孢噻肟、利福平中毒敏感，对呋喃妥因、氯霉素、四环素、氧氟沙星高度敏感。

关键词：嗜水气单胞菌；毒力因子；致病性；相关性

该文发表于《动物学杂志》【2008, 43（6）：1-9】

西伯利亚鲟源嗜水气单胞菌致病菌的分离及其全菌苗的免疫效果 *

Isolation of Pathogenic Bacteria *Aeromonas hydrophila* from *Acipenser baerii* and Immune Effect of Its whole Cell Vaccine

李圆圆，曹海鹏，邓璐，杨先乐

（上海海洋大学农业部渔业动植物病原库，上海 200090）

摘要：从患细菌性败血症的西伯利亚鲟（*Acipenser baerii*）的体内分离到一株致病菌株 X1，其对西伯利亚鲟的半数致死浓度（LC_{50}）为 5.62×10^5 cfu·mL，具有较强毒力；经 ATB 细菌鉴定仪生理生化鉴定和 16S rDNA 序列分析，菌株 X1 为嗜水气单胞菌（*Aeromonas hydrophila*）；其系统发育分析表明，菌株 X1 与嗜水气单胞菌 ATCC35654（登录号：X74676.1）的亲缘关系最近，其同源性为 99%。用 0.30% 福尔马林灭活，将菌株 X1 制成灭活全菌苗，对西伯利亚鲟进行注射免疫。研究结果表明，嗜水气单胞菌 X1 全菌苗能够明显提高西伯利亚鲟的血清抗体水平及总蛋白、免疫球蛋白、溶菌酶含量，而且在嗜水气单胞菌 X1 全菌苗中加入弗氏不完全佐剂，有利于进一步增强西伯利亚鲟血清抗体水平及总蛋白、免疫球蛋白、溶菌酶含量。此外，嗜水气单胞菌 X1 全菌苗对西伯利亚鲟抗嗜水气单胞菌 X1 人工感染也具有较好的免疫保护作用，其对西伯利亚鲟的免疫保护率为 50%，而且在嗜水气单胞菌 X1 全菌苗中加入弗氏不完全佐剂，嗜水气单胞菌 X1 全菌苗对西伯利亚鲟抗嗜水气单胞菌 X1 人工感染的免疫保护作用更好，其对西伯利亚鲟的免疫保护率为 70%。因此，将嗜水气单胞菌 X1 全菌苗用于西伯利亚鲟细菌性败血症的防治具有广阔的发展前景。

关键词： 西伯利亚鲟；嗜水气单胞菌；全菌苗；免疫效果

第四节　渔药剂型的研究

剂型（*Dosege forms*）指药物根据预防和治疗的要求经过加工制成适合于使用、保存和运输的一种制品形式，或是指药物制剂的类别，它是临床使用的最终形式，是药物的传递体，不同药物的剂型会产生不同的疗效。研究渔药的剂型，是充分发挥药物疗效的一个重要途径。

药物的剂型对药物的吸收代谢过程和药效有重要影响。目前，我国渔药的剂型还停留在以传统的粉剂、水剂等为主的阶段。研制新型渔药剂型可以有效提高渔药生物利用度、降低药物的毒副作用、增强药物的缓释和控释性能（胡鲲等，2010；沈丹怡等，2011）。

一、微胶囊缓释剂型

制备渔药的缓释制剂的方法有冷冻干燥法等（龚露旸等，2009；胡鲲等，2006；沈丹怡等，2011）。药物的缓释性能、稳定性与包埋壁材、工艺条件等因素紧密相关。其中，羧甲基纤维素的缓释性能优于淀粉，且冷冻温度的降低能促进微胶囊缓释性能（胡鲲等，2006）。以羧甲基纤维和淀粉为壁材的恩诺沙星微胶囊为例，羧甲基纤维素的抗紫外和热稳定性均优于淀粉（胡鲲等，2006）。

该文发表于《水生生物学报》【2014，38（4）：675-679】

诺氟沙星壳聚糖微胶囊缓释作用研究

THE ANALYSIS OF THE CONTROLLED-RELEASE PROPERTIES OF NORFLOXACIN MICROENCAPSULATION

赵依妮，李怡，胡鲲，阮记明，叶鑫，杨先乐

（上海海洋大学，国家水生动物病原库，上海 201306）

摘要： 利用冷冻干燥法制备了诺氟沙星-壳聚糖微囊制剂，并比较了诺氟沙星-壳聚糖微囊制剂与诺氟沙星原料药在草鱼组织中代谢动力学的差异性。结果表明，同诺氟沙星原料药相比，壳聚糖包埋对诺氟沙星在草鱼血浆中的吸收、代谢速率明显减缓，生物利用度明显增加。壳聚糖包埋条件下 $T_{1/2ka}$ 为单纯口灌诺氟沙星的 8 倍，达峰时间平均延长 2 倍有余，$T_{1/2\alpha}$ 和 $T_{1/2\beta}$ 也明显延长，消除率（CLs）是单纯口灌的 1/2；5 倍壳聚糖包埋条件下诺氟沙星在草鱼血浆中的吸收速度比 10 倍包埋条件下稍快，达峰时间短，消除速度明显加快，但分布半衰期和曲线下面积（AUC）大小二者比较接近。实验中诺氟沙星-壳聚糖微囊制剂同诺氟沙星原料药在草鱼体内代谢动力学差异表明壳聚糖包埋对诺氟沙星药效的释放及鱼体对药物的吸收代谢都具有明显的延缓效果，因此在实际的生产活动中可用壳聚糖作为缓释剂延长药效、延缓药物在养殖动物体内的吸收代谢，提高生物利用率，达到更好的预防和治疗效果。

关键词： 诺氟沙星；壳聚糖；微胶囊；草鱼；代谢规律

剂型作为渔药的应用形式，可改变渔药的作用性质，调节药物作用速度，决定药物的稳定性，降低药物的毒副作用，对渔药的疗效起着极其重要的作用[1]。当前渔药市场上主要的剂型有水剂、粉剂、颗粒剂、微囊剂等。近年来，随着一些用量大、残留多、毒性高的渔药旧剂型对环境污染日益严重，以及人们的环保意识的不断增强，市场对经济、安全、方便的渔药新剂型的需求量也与日俱增[2]。相关研究表明，不同剂型的渔药有不同的释放性，从而影响了药物的吸收和消除速率，进而影响了生物利用度[2]。近年来，诺氟沙星在水产动物细菌性疾病的防治上也逐渐成为重点研究对象和重要药品之一[3]。考虑到水产养殖动物简单的消化系统以及主要采用群体给药方式，因此大大降低了药物生物利用度。因此，如何有效提高渔药的生物利用度已成为目前渔药研发和创新的难题之一。高分子载药微球即微胶囊是近年来发展起来的新剂型，药物经其包囊后，不仅可以提高药物稳定性，控制其释放，延长在胃肠内的滞留时间，还可以大大提高药物的生物利用

度[4]。近年来对药物缓释载体的一系列研究表明，药物缓释载体在维持血药浓度，减少给药量和给药次数，通过不同方法延缓药物在体内的释放、吸收、代谢等方面具有重要作用[5]。其中，壳聚糖作为一种资源丰富的天然高分子化合物，因其不仅具有良好的生物相容性且毒副作用小、易降解吸收，同时还具有抗菌、消炎、止血等大多数聚合物所不具有的功能[6]，成为药物缓释剂的研究热点之一[7]。目前，有关各类常用渔药的药物动力学研究国内外皆有报道，但利用药物缓释载体，开发渔药新剂型，从而减少给药次数、提高药物稳定性和疗效、降低药物毒副作用等方面的研究尚属空白。

诺氟沙星（Norfloxacin, NFLX）是氟喹诺酮类代表药物之一[8,9]，被广泛用于动物和人类多种细菌性疾病的防治中[9]。本文利用诺氟沙星为模式药物，利用天然高分子化合物壳聚糖作为诺氟沙星缓释剂载体，采用冷冻干燥法制备诺氟沙星—壳聚糖微囊制剂，同时将其与诺氟沙星原料药在草鱼体内的代谢参数进行对比分析，据此评价壳聚糖作为药物缓释载体的缓释效果，对通过药物新剂型提高药物生物利用度、降低药物使用量、保障水产品质量安全具有实践价值。

1 材料与方法

1.1 实验动物和试剂

草鱼（*Ctenopharyngodon idellus*），购自江苏省南通国营农场，平均体重（100±10）g，体表以及解剖观察健康。运到实验室后，用2%NaCl溶液浸泡10 min，然后放置于高锰酸钾消毒后的水族箱（100 cm×80 cm×80 cm）中暂养1周，试验用水为充分曝气的自来水，水温控制在（23±2）℃下，自然光照，并使用增氧泵24 h充氧；暂养期间饲喂不含任何药物的全价饲料；每两天换1次水，每次换水量1/3；及时清除水底残饵和排泄物。所有实验草鱼经抽查表明组织中均不含NFLX。

NFLX标准品（纯度99.50%），由中国兽医监察所提供；NFLX原料药（含量97.8%），由军事兽医研究所提供；四丁基溴化铵和乙腈为HPLC级；其他试剂均为分析纯。

壳聚糖-诺氟沙星微囊制剂，由本实验室制备。按照质量比（w/w）分别为1∶5和1∶10两种比例，将NFLX与壳聚糖混合，加入少许2%盐酸溶解，并配制成40%的水溶液，振荡15 min，分装于冷冻瓶中，在-50℃下，抽真空，冷冻干燥，制备1/5和1/10的NFLX微胶囊样品，以下分别以1/5 NFLX和1/10 NFLX表示。

1.2 仪器及色谱条件

Aglient-1100型高效液相色谱仪配二极管阵列检测器、旋涡混合器、氮气吹干仪、精密电子天平（METTLER AB104-N）和高速冷冻离心机等。色谱柱为：ZORBAX SB2C-18分析住（150 mm×4.6 mm, 5 μm）。流动相为：乙腈∶四丁基溴化铵（0.030 mol/L, pH 3.1）＝5∶95（v/v）；检测波长276 nm；流速1.5 mL/min；柱温40℃；进样量为10 μL。以常规方法建立标准曲线。

1.3 实验设计及采样

实验用鱼162尾，随机分为3组，每组54尾，每一时间点取6条鱼，另取数尾未给药的草鱼作为空白对照组。所有实验用鱼均以40 mg/kg鱼体重的剂量口灌各剂型的诺氟沙星，确保各组诺氟沙星在鱼体内的含量相同，无回吐者保留试验。其中第一组仅为诺氟沙星；第二组为1∶5壳聚糖包埋的诺氟沙星；第三组口灌1∶10壳聚糖包埋的诺氟沙星。给药前5 h禁食，给药后5 h投料，并于给药后0.25、0.5、1、3、6、12、24、48、96、196 h围心腔取血，肝素钠抗凝；每个时间点6条鱼的血浆合并成一个样品后于-80℃保存备用。每组均设一个重复。

1.4 血液处理及HPLC测定

样品检测前，准确吸取1 mL样品加入5 mL酸化乙腈（乙腈∶盐酸∶水的体积比为250∶1∶1），旋涡混合10 min，4℃条件下8 000 r/min离心10 min，取上清液，45℃恒温氮气条件下吹干，加1 mL流动相溶解，经0.45 μm微孔滤膜过滤后取300 μL进行HPLC测定。以上样品处理及测定均参照阮记明等[10]的两种水温条件下异育银鲫体内双氟沙星药代动力学比较中使用的方法。

1.5 诺氟沙星标准曲线的制备

对照品标准曲线制备；以干燥恒重的诺氟沙星标准品制成0.00、0.025、0.10、0.20、0.50、1.00、2.00、5.00、10.00、15.00、20.00 μg/mL标准浓度系列，以HPLC仪分别测定其峰面积，然后作标准曲线，并求出回归方程和相关系数。

1.6 数据处理

各组织中药代动力学模型拟合及参数计算采用药动学程序软件DAS3.0口服给药模型对各组实验数据进行模型嵌合和

参数计算。药物在不同时间内在血浆中的浓度柱形图和药物消除的药时曲线图则采用 Microsoft Excel 2000 软件绘制。

2 结果

2.1 线性范围及标准曲线

在设定的色谱条件下，以高效液相色谱仪测定血液诺氟沙星含量，基线走动平稳，特异性强，重现性好。诺氟沙星的标准溶液在 0.025~20 μg/mL 浓度范围内呈线性关系，其回归方程及相关系数为 $A_i = 1444.3C_i - 212.25$。在 0.025~2 μg/mL 内线性关系良好（$r = 0.999$），以引起两倍极限噪音的药量为最低检测限，本法最低检测限为 0.01 μg/mL。诺氟沙星血浆平均回收率达（94.32±6.23）%，相对标准偏差为（6.28±0.83）%。

2.2 诺氟沙星血浆药代动力学规律

草鱼以 40 μg/mL 的剂量经单次口灌给药后，药物在鱼体血液中的吸收、分布与消除的动态如图 1 所示。有关动力学数据经房室模型分析，药物浓度与时间关系数据符合开放性二室模型（Two-compartment open model），其药动学参数详见表 1。

该模型药物以一级吸收过程进入体内，并按二室模型分布，即诺氟沙星口灌进入鱼体后，经消化道进入血液，其分布和消除情形是，血药浓度达峰时间（T_{max}）为 0.494 h，吸收速率常数（K_a）高达 15.513 h^{-1}，分布相较小，分布相半衰期（$T_{1/2\alpha}$）为 0.264 h，可见 NFLX 易被鱼体吸收。

表 1 草鱼单剂量口灌诺氟沙星动力学参数
Tab. 1 Pharm-acokinetics parameters following single oral administration of 40 μg/mL of
NFLX to Ctenopharyngodon idellus

参数 Parameter	单位 Unit	参数值 Value		
		NFLX	1/5 NFLX	1/10 NFLX
Dose	mg/kg	40.000	40.000	40.000
Cor.	—	0.999	0.987	0.987
C_0	μg/mL	24.623	25.470	23.621
K_a	h^{-1}	15.513	9.357	7.876
K_{12}	h^{-1}	2.224	1.655	1.778
K_{21}	h^{-1}	2.081	1.371	1.625
K_{10}	/h	0.333	0.073	0.043
$T_{1/2}$	h	0.264	2.240	2.761
$T_{1/2\beta}$	h	68.246	94.885	98.369
$T_{1/2Ka}$	h	0.126	1.061	1.035
AUC	μg/(L/h)	94.294	136.915	139.110
C_{max}	μg/mL	9.009	6.539	6.982
T_{max}	h	0.494	1.107	1.562
CL_s	mL/(kg/h)	59.262	23.751	23.064

2.3 NFLX 微胶囊药代动力学规律

将诺氟沙星与壳聚糖按 1:5 和 1:10 的质量比进行包埋（包埋率为 98.743%），并按每千克鱼体重口灌 40 mg 诺氟沙星原料药的剂量给药后，药物在鱼体血液中的吸收、分布与消除的动态如图 1 所示。有关动力学数据经 MCPKP 药动学程序软件处理的结果表明，无论何种情况，NFLX 在草鱼血浆中的药动学特征均适合用一次式二室开放模型来描述，相应的药动学参数见表 1。

由图 1 可知，在壳聚糖包埋条件下，NFLX 在血浆中的吸收、分布以及消除速度明显降低；在用 5 倍的壳聚糖包埋时，NFLX 在第 1.107 h 达到峰值 6.539 μg/mL，随后开始下降，但降幅较单纯口灌 NFLX 条件下延缓，在第 3、6、12 h 分别降到 3.341、2.493 和 2.107 μg/mL，第 48 h 为 1.325 μg/mL，约为第 1 h 的 1/6，第 96 h 降至 0.743 μg/mL，第 196 h 后仍能

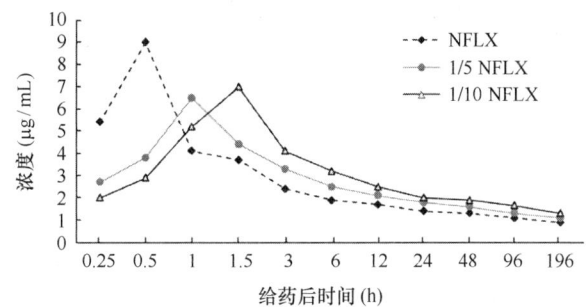

图 1　诺氟沙星及其在壳聚糖包埋条件下血浆药时曲线

Fig. 1　The concentration-time curves of NFLX and NFLX-embeded in plasma

检测到，浓度为 0.573 μg/mL；而在 10 倍壳聚糖包埋条件下，NFLX 在 1.562 h 达到最大值 6.982 μg/mL，此后开始下降，其下降幅度稍小于 5 倍壳聚糖包埋，第 24、48、96 h 分别为 2.184、1.752、1.084 μg/mL，第 196 h 为 0.815 μg/mL，仍高于 5 倍壳聚糖包埋第 96 h 的水平。

从表 1 可以看出，在壳聚糖包埋条件下，吸收相半衰期（$T_{1/2 ka}$）、分布相半衰期（$T_{1/2\alpha}$）和消除相半衰期（$T_{1/2\beta}$）均较低，达峰时间（T_p）延长，峰浓度降低（C_{max}）、消除率（CL_s）也明显降低；此条件下的药时曲线下面积（AUC）明显增加，5 倍壳聚糖包埋时为单纯口灌 NFLX 时的 1.45 倍。诺氟沙星 5 倍和 10 倍壳聚糖包埋相比，10 倍壳聚糖包埋条件下 NFLX 在草鱼血浆中的吸收相半衰期（$T_{1/2 ka}$）、分布相半衰期（$T_{1/2\alpha}$）、消除相半衰期（$T_{1/2\beta}$）都稍有延长；达峰时间（T_{max}）延长，峰值（C_{max}）却有增加；二者的清除速率（CL_s）和药时曲线下面积（AUC）则基本相同。

3　讨论

利用可生物降解的高分子材料作为药物载体在医药上的应用越来越引人注目，尤其是将药物包裹于高分子材料中制备的微胶囊。微胶囊其实质就是把分散的固体颗粒、液体或气体完全包封在一层膜中，形成球状的微球[11]。微胶囊可在体内特异性分布，利用对特定器官、组织的靶向性和药物的缓释性来降低药物的毒副作用，提高药物的生物利用度[7]。从 20 世纪 90 年代起，微胶囊的应用研究已经涉及药物控制释放、细胞和酶的固定化、动植物细胞培养以及生化物质分离等领域，成为医学、制药、材料等多学科的研究热点，并产生了广泛的应用前景[11]。目前，采用合成的或天然的高分子材料已制备出多种微胶囊，例如：壳聚糖、明胶、纤维蛋白原、乙基纤维素、蜂蜡等都是良好的微囊膜材料[12]。其中壳聚糖作为一种优良、价格便宜的天然高分子材料，其成膜性能是制作微囊理想材料，因而备受关注[13]。

目前，有关于壳聚糖应用于药物缓释、控释制剂方面的研究已经有了相当的研究深度和广度。诸多实验结果都验证了壳聚糖在药物缓释、控释上具有明显功效[12-15]。如恩诺沙星-壳聚糖纳米粒可以显著提高恩诺沙星的生物利用程度，参比恩诺沙星原料药的相对生物利用度为 262%[16]。He et al.[17] 采用喷雾干燥法制备的西米替丁和法莫替丁壳聚糖微囊，缓释作用明显，可明显提高药效。杨利芳等[18] 采用静电液滴工艺制备了胰岛素海藻酸钠-壳聚糖微囊，以四氧嘧啶为诱导剂建立糖尿病小鼠模型，对载药微囊的口服药效学进行评价，结果表明，以该微囊为基础的口服胰岛素制剂生物利用度高且缓释效果明显。何强芳等 [4] 以壳聚糖为载体，通过乳液化学交联法制备了 5-氟尿嘧啶（5-fu）壳聚糖微粒，实验结果表明，壳聚糖微囊对 5-fu 的缓释作用明显，释放周期比较长，壳聚糖可作为载 5-fu 的理想载体。

在本次实验中，口灌 40 μg/mL 诺氟沙星原料药后，由血液药动学参数可知诺氟沙星在草鱼血浆中的吸收较迅速，在各组织间的分布也较快，该结果同郭海燕等[19]、刘莉莉等[20] 的相关诺氟沙星在草鱼体内的代谢规律研究结果相一致；口灌 40 μg/mL NFLX-壳聚糖微囊制剂后，由血液药动学参数可知 NFLX 在草鱼体内代谢规律未发生变化，进而证明壳聚糖包埋对诺氟沙星在草鱼体内的代谢规律未发生影响。而 NFLX-壳聚糖微囊制剂同诺氟沙星原料药的药代动力学相比，在壳聚糖包埋条件下 NFLX 在草鱼体内的达峰时间（T_{max}）平均延长 2~3 倍，药时曲线下面积（AUC）明显增加，消除速率延缓，表明 NFLX-壳聚糖微胶囊的生物利用度明显增加。因此，在水产养殖动物疾病防治过程中，有必要根据临床应用实际需要，通过改变剂型以加快或减慢渔药作用时间，从而充分发挥渔药的疗效[2]。随着国内、外微囊技术研究的不断深入，壳聚糖在微囊体系中的应用将会有更加广阔的应用前景[11]。壳聚糖作为纯天然的高分子物质，因其良好的生物特性，且来源广、价格经济实惠，使其成为一种理想的载体材料[21]，在药物微囊化的过程中占有非常重要的地位。随着水产养殖业的迅猛发展，新型渔药制剂的研究开发日益得到广泛的关注，对于壳聚糖如何应用于渔药缓释、控释，利用壳聚糖制得的微胶囊或微球增加药物稳定性、降低毒副作用、延长药效、提高生物利用率是正在发展又非常有前景的课题，也是此领域的研究热点[13]。利用壳聚糖自身具有的优良特性且易被生物体吸收的特点及良好的控释、缓释性能，采用冷冻干燥技术制备了 NFLX-壳聚糖微胶囊，将其同 NFLX 原料药在健康草鱼血浆内的药物代谢规律进行比较分析，研究结果表明，壳聚糖可明显延长 NFLX 在鱼体的存留时间，提高生物利用度，进一步验证了壳聚糖用于渔药缓释、控释制剂研发的可行性。因此，发展以微胶囊为代表的控释缓释制剂等第三代制剂是未来渔药剂型的发展方向[2]。如果此类产品能够顺利问世，将成为临床预防治疗水产养殖动物细菌性疾病的又一有效方法，产生很好的社会效益和经济效益。

参考文献

[1] Wang Y. T.. New dosage form of fishery drug and the advantages of dosage form [J]. Fishery of China, 2012, (4)：56-57 [王玉堂. 渔药新剂型及剂型药使用优点. 中国水产, 2012, (4)：56-57].

[2] Hu K.. The dosage forms of fishery drug in China and their technical support to the safety use of fishery drug [J]. Journal of China Veterinary Medicine, 2011, 5 (45)：43-46.

[3] 胡鲲. 我国渔药剂型使用现状及其在渔药安全使用技术中的价值. 中国兽药杂志, 2011, 5 (45)：43-46.

[4] Ke X. L., Lu M. X., Cai J., et al. Studies on identification and pathogenicity of areococcus viridans isolated from tilapia [J]. Acta Hydrobiologica Sinica, 2011, 35 (5)：796-802 [可小丽, 卢迈新, 蔡炯, 等. 罗非鱼绿色气球菌的鉴定及致病性研究. 水生生物学报, 2011, 35 (5)：796-802].

[5] He Q. F., Li G. M., Wu H. Z., et al. Preparation and drug releasing property of 5-fluorouracil-loaded chitosan microsphere [J]. Application of Chemical, 2004, 21 (2)：192-195 [何强芳, 李国明, 巫海珍, 等. 5-氟尿嘧啶壳聚糖微球的制备及其释药性能. 应用化学, 2004, 21 (2)：192-195].

[7] Z J, Luo L. H.. The progress in chitosan-based drug release carriers research [J]. Polymer Bulletin, 2006, (7)：25-30.

[8] Li S. D., Quan W. Y., Cai J.. Progress in the applications of controlled release ability of chitosan drug agent [J]. Guangzhou Chemical Industry, 2007, 35 (4)：4-6 [李思东, 权维燕, 蔡鹰. 壳聚糖药物控释剂的应用研究进展. 广州化工, 2007, 35 (4)：4-6].

[9] Zhang Z. F., Zhou J. P., Huo M. R.. Chitosan microspheres drug delivery systems [J]. Pharmaceutical Development, 2006, 30 (6)：261-266 [张祖菲, 周建平, 霍美蓉. 壳聚糖微球给药系统. 药学进展, 2006, 30 (6)：261-266].

[10] Yang Y., Liu Y L., Wang X. L.. The pharmacokinetics and residues of Pseudosciaena crocea Norfloxacin [J]. Journal of Fisheries of China, 2007, 31 (5)：655-660 [杨勇, 刘玉林, 王翔凌. 诺氟沙星在大黄鱼体内的药代动力学及残留研究. 水产学报, 2007, 31 (5)：655-660].

[11] Ezejiofor N. A., Anusiem C. A., Orish C. N., et al. Acceleration of body clearance of isoniazid (Inh) by Norfloxacin and Ofloxacin [J]. Journal of US-China Medical Science, 2011, 1 (8)：19-23.

[12] Ruan J. M., Hu K., Zhang H. X., et al. Pharmacokinetics comparisons of difloxacin in crucian carp (Carassais auratus gibebio) at two different water temperatures [J]. Journal of Shanghai Ocean University, 2011, 20 (6)：858-865 [阮记明, 胡鲲, 张海鑫, 等. 两种水温条件下异育银鲫体内双氟沙星药代动力学比较. 上海海洋大学学报, 2011, 20 (6)：858-865].

[13] Jin L. Y., Bai Y., Zhang H. F., et al. The research progress of chitosan microcapsule and its application [J]. Journal of Henan University (Medical Science Edition), 2010, 29 (4)：242-245 [金邻豫, 白颖, 张海峰, 等. 壳聚糖在药物微囊化及应用中的研究进展. 河南大学学报 (医学版), 2010, 29 (4)：242-245].

[14] Li J. F., Zou Q., Lai X. F.. Review on progress of research on drug releasing mechanism of chitosan microspheres [J]. Journal of Biomedical Engineering, 2011, 28 (4)：843-846.

[15] 李峻峰, 邹琴, 赖雪飞. 壳聚糖微球药物释放机制研究进展. 生物医学工程学杂志, 2011, 28 (4)：843-846]

[16] Zhou H. Y., Chen X. G., Liu C. S., et al. Progress of chitosan on sustained drug delivery [J]. Marine Science, 2006, 30 (3)：85-89 [周惠云, 陈西广, 刘成圣, 等. 壳聚糖在药物缓释中的研究进展. 海洋科学, 2006, 30 (3)：85-89].

[17] Fang M., Gao J. L., Wang S., et al. Dielectric monitoring method of chitosan microspheres for sustained release of the drug effect [J]. Science Bulletin, 2010, (10)：867-874 [方明, 高金龙, 王胜, 等. 壳聚糖微球对药物缓释作用的介电监测方法. 科学通报, 2010, (10)：867-874].

[18] Zheng A. P., Liu H. H., Li H. B., et al. Preparation and invitro release of 5-fluorouracil-loaded chitosan microspheres for the intranasal administration [J]. Journal of China Medicine University, 2004, 35 (4)：318-323 [郑爱萍, 刘海宏, 李宏斌, 等. 5-氟脲嘧啶壳聚糖微球的制备及体外释放特性. 中国药科大学学报, 2004, 35 (4)：318-323].

[19] Gong L. Y., Hu K., Yang X. L.. Preparation and release characteristics of enrofloxacin chitosan nanoparticles in vitro [J]. Journal of Shanghai Ocean University, 2009, (3)：321-326.

[20] 龚露旸, 胡鲲, 杨先乐. 恩诺沙星壳聚糖纳米粒的制备及其体外释药特性研究. 上海海洋大学学报, 2009, (3)：321-326.

[21] He P., Davis S. S., Illum L.. Sustained release chitosan microspheres prepared by novel spray drying methods [J]. Journal of Microencapsul, 1999, 16 (3)：343-345.

[22] Yang L. F., Xue W. M., Gao Q., et al. Study on hypoglycemic effect of oral insulin-loaded alginate chitosan microspheres [J]. Chemical Engineering, 2008, 36 (1)：71-74 [杨利芳, 薛伟明, 高茜, 等. 载胰岛素壳聚糖微球口服降血糖作用的研究. 化学工程, 2008, 36 (1)：71-74].

[23] Guo H. Y., Chen Y. K., Zhu L., et al. The distribution and pharmacokinetics of Norfloxacin in tissues in grass carp [J]. Fisheries Science, 2009, 28 (1)：28-31 [郭海燕, 陈元坤, 朱林, 等. 诺氟沙星在草鱼体内的组织分布和药物动力学规律. 水产科学, 2009, 28 (1)：28-31].

[24] Liu L. L., Luo L., Luo Y. H., et al. Multi-residues determination of Norfloxacin, Enrofloxacin and Ciprofloxacin in grass carp musde tissue [J]. Journal of Southwest Agricultural University, 2010, (5): 171-176.

[25] 刘莉莉, 罗雷, 罗永煌, 等. 草鱼肌肉中诺氟沙星、恩诺沙星和环丙沙星的多残留检测. 西南大学学报 (自然科学版), 2010, (5): 171-176.

[26] Ji Z. X., An J., Wang D. S., et al, Preparation and application of chitosan microspheres [J]. Journal of Materials, 2010, 24 (13): 128-132 [姬振行, 安静, 王德松, 等, 壳聚糖微球的制备与应用. 材料导报, 2010, 24 (13): 128-132].

该文发表于《上海水产大学学报》【2006, 15 (1): 25-29】

恩诺沙星微胶囊的制备及其抗紫外、热稳定性的研究

Preparation of microencapsulation of enrofloxacin and analysis of anti-ultraviolet character & thermal stability

胡鲲, 杨先乐, 龚幼兰, 孙晶

(上海水产大学农业部渔业动植物病原库, 上海 200090)

摘要: 采用冷冻干燥法, 分别以羧甲基纤维素和淀粉为壁材, 制备恩诺沙星微胶囊制剂。抗紫外和热稳定性试验表明: 以羧甲基纤维素和淀粉为壁材的恩诺沙星微胶囊均有抗紫外和热稳定性能。其中, 羧甲基纤维素与恩诺沙星、淀粉与恩诺沙星包埋质量比为1:1、1:2时, 微胶囊表现出显著的抗紫外作用 (P <0.05); 羧甲基纤维素与恩诺沙星、淀粉与恩诺沙星包埋质量比为2:1、1:1、1:2时, 微胶囊表现出极显著的热稳定性 (P<0.01)。羧甲基纤维素的抗紫外和热稳定性均优于淀粉, 但差异不显著 。

关键词: 恩诺沙星; 微胶囊; 抗紫外; 热稳定性

该文发表于《上海水产大学学报》【2006, 15 (3): 375-379】

恩诺沙星微胶囊的制备及其缓释性能的研究

Preparation of microencapsulation of enrofloxacin by freezing-dry and determination of controlled-release properties

胡鲲[1], 朱泽闻[2], 杨先乐[1], 鲍莹[1], 宋晴[1], 王璐[1]

(1. 上海水产大学农业部渔业动植物病原库, 上海 200090; 2. 全国水产技术推广总站, 北京 100026)

摘要: 分别以羧甲基纤维素和淀粉为壁材, 利用冷冻干燥法在不同的工艺条件下制备得到恩诺沙星微胶囊制剂。缓释试验表明: 以羧甲基纤维素和淀粉为壁材的恩诺沙星微胶囊均表现出明显的缓释作用, 且羧甲基纤维素的缓释性能优于淀粉, 冷冻温度的降低能促进微胶囊缓释性能。其中, 恩诺沙星原料药在4.5 min 时的溶出率达到99.4%。-40℃和-50℃条件下制得的以羧甲基纤维素为壁材的恩诺沙星微胶囊在14 h 的溶出率分别为99.2%和99.7%; -40℃和-50℃条件下制得的以淀粉为壁材的恩诺沙星微胶囊在12 h 的溶出率为99.4%和99.6%。

关键词: 冷冻干燥; 恩诺沙星; 微胶囊; 缓释

二、壳聚糖纳米粒剂型

恩诺沙星壳聚糖纳米粒具有较强的缓释性能能力 (龚露旸等, 2009) 和良好的热稳定性、紫外稳定性 (龚露旸等, 2009)。壳聚糖包埋的诺氟沙星剂型能对药效的释放及其在鱼体内的代谢具有明显的缓释效果, 显著提高药物的生物利用率 (赵依妮等, 2014)。相比强力霉素原料药, 强力霉素壳聚糖纳米粒冻干粉该剂型在人工胃液、肠液和pH 7.4磷酸缓冲液中具有显著的缓释特性 (沈丹怡等, 2012)。

该文发表于《上海水产大学学报》【2009，18（3）：321-326】

恩诺沙星壳聚糖纳米粒的制备及其体外释药特性研究

Preparation and release characteristics of enrofloxacin chitosannano particles in vitro

龚露旸，胡鲲，杨先乐

(上海海洋大学国家水生动物病原库，上海 201306)

摘要：建立了离子交联法制备恩诺沙星壳聚糖纳米粒的方法，并优化了工艺条件，以低温超速离心法评价了恩诺沙星壳聚糖纳米粒包封率。利用透射电子显微镜分析了恩诺沙星壳聚糖纳米粒的显微形态，以透析法测定其体外释放度。结果表明：恩诺沙星壳聚糖纳米粒呈球形，外观形态饱满，表面光滑，分散性良好，平均粒径为 (131.1±10.2) nm，平均包封率 51.2%±2.9%。相比恩诺沙星原料药，恩诺沙星壳聚糖纳米粒在 1 h 内减少约 60% 药物的释放，24 h 总释放度为 79.9%。说明利用离子交联法制备恩诺沙星壳聚糖纳米粒快速、简便、操作性强，恩诺沙星壳聚糖纳米粒制剂缓释性能强，适于作为新剂型在水产药物中应用。

关键词：恩诺沙星；壳聚糖；纳米粒；体外释放

该文发表于《安徽农业科学》【2009，37（5）：2019-2021】

恩诺沙星纳米粒的制备及稳定性研究

Study on the Preparation and the Stability of Enrofloxacin Nanoparticles（ENR-NPs）

龚露旸，杨先乐，胡鲲

(上海海洋大学国家水生动植物病原库，上海 201306)

摘要：［目的］优化恩诺沙星纳米粒制备参数，考察其稳定性。［方法］以壳聚糖为辅料，采用离子交联法和冷冻干燥法制备恩诺沙星壳聚糖纳米粒，以包封率为参考指标设计正交试验，确定优化制备参数，以透射电镜观察其表观特征，并对其热稳定性及紫外稳定性进行考察。［结果］以优化参数制备的恩诺沙星壳聚糖纳米粒平均粒 237.4 nm，平均包封率（71.9±2.2）%，平均载药量（12.4±0.6）%。与原料药相比 4 h 热降解率和紫外降解率分别减少了 38.4% 和 13.3%。［结论］离子交联法和冷冻干燥法适用于制备恩诺沙星壳聚糖纳米粒，其产品具有良好的热稳定性和紫外稳定性。

关键词：恩诺沙星；壳聚糖；纳米粒；热稳定性；紫外稳定性

该文发表于《水产学报》【2012，36（6）：944-951】

强力霉素壳聚糖纳米粒冻干粉的体内外释药特性

Release characteristics of doxycycline chitosan nanoparticles freeze-dried powder *in vitro* and *in vivo*

沈丹怡[1,2]，艾晓辉[1]，刘永涛[1]，丁运敏[1]，余少梅[1,3]，索纹纹[1,3]

(1. 中国水产科学研究院长江水产研究所，湖北 武汉 430223；2. 浙江省慈溪市水产技术推广中心，浙江 慈溪 315300；
3. 华中农业大学水产学院，湖北 武汉 430070)

摘要：分别采用紫外分光光度法和超高效液相色谱法研究强力霉素原料药与强力霉素壳聚糖纳米粒冻干粉的体内外释药特性。结果表明：强力霉素原料药在人工胃液、肠液和 pH 7.4 磷酸缓冲液中的体外释放均可用零级动力学方程拟合，释放较快，在1h内能完全溶出；强力霉素壳聚糖纳米粒冻干粉具有显著的缓释特性，在各介质中的体外释放均可用双相动力学方程拟合，前期表现为快速释放，后期为缓慢释放，冻干粉在各介质中的释药速率从大到小依次为：pH2.0人工胃液>pH3.0人工胃液>pH4.0人工胃液>人工肠液>pH7.4磷酸盐缓冲液。强力霉素原料药包裹成强力霉素壳聚糖纳米粒冻干粉后，经（25±1）℃ 20 mg/kg单剂量口灌，在斑点叉尾的药—时曲线由双峰变为单峰，在血浆中的血药峰浓度（C_{max}）减小，峰值时间（T_{max}）和消除半衰期（$T_{1/2\beta}$）明显延长，药·时曲线下面积（AUC）变大，是一种理想的强力霉素新剂型。

关键词：斑点叉尾强力霉素；纳米粒；冻干粉；体外释药；药代动力学

强力霉素（doxycycline，DC）是四环素类广谱抗生素，对多种革兰氏阳性菌、阴性菌、立克次氏体、螺旋体等均有效[1]。DC抗逆性差，遇光、热、氧易降解变质[2]，在水产动物体内的药物释放也不稳定，有效血药浓度维持时间短，药物峰浓度时可能超出最低中毒浓度，对水产动物产生副作用，低浓度又不能显效。DC给药次数和频率大，既增加了劳力物力，又造成药物代谢缓慢，产生耐药性，引起肝肾毒性，导致药物残留[3]。壳聚糖纳米粒（chitosan nanopartiles，CS-NPs）可使药物有效隔绝空气中的光、热、氧，提高药物在运输及施药过程中的稳定性。由于其粒径小，表面积大，与生物膜的黏着性高，具有药物缓释和药物靶向能力[4-5]，能维持有效的血药浓度，延长半衰期，降低药物浓度过高引起的毒副作用，减少用药剂量和用药频率[6]，可用较少的剂量达到较高的药效。目前对于CS-NPs在模拟释放介质及动物体内的药代动力学研究较多，而在水产动物体内的释药特性研究鲜有报道。本实验以DC原料药为参比，研究DC-CS-NPs冻干粉在人工胃液、肠液、pH7.4磷酸盐缓冲液（PBS）的体外释药特性及其在斑点叉尾体内的药代动力学研究，可为DC新剂型的开发应用提供有用的理论依据。

1 材料与方法

1.1 实验动物

斑点叉尾（Ictalurus punctatus），由中国水产科学研究院长江水产研究所窑湾试验场提供，平均质量为（167±12）g，健康无病。实验前在水族箱内暂养一周，实验水温为（25±1）℃，pH值为7.5~8.0，饲喂不加抗菌药物的斑点叉尾全价配合饲料。实验前对斑点叉尾进行抽检表明血液中不含DC。

1.2 试剂

DC原料药（含量≥98%）；标准品（纯度≥99.0%）；DC-CS-NPs冻干粉（自制）；空白CS-NPs冻干粉（自制）；浓盐酸；高氯酸；氢氧化钠；磷酸二氢钾；磷酸二氢钠；胃蛋白酶；胰酶；乙腈；肝素钠；所用试剂浓盐酸为优级纯，乙腈为色谱纯，其他试剂为分析纯。

1.3 仪器

UV2802PC紫外分光光度；ACQUITY UPLC超高效液相色谱仪配TUV双通道紫外检测器及Empower 2色谱工作站；HITACHI 20PR-520型自动高速冷冻离心机；Sigma 3k15冷冻离心机；Met-tler-TOLEDO AE-240型精密电子天平；调速混匀器；Startorius PB-10酸度计；WIGGENS WH220加热磁力搅拌器；Alphai-4LD-plus冷冻干燥仪；AUCMA DW-86W150超低温冰箱；立式万用电炉。

1.4 DC原料药及DC-CS-NPs冻干粉的体外释药特性

标准曲线的制备配制10、20、30、50、70和90 μg/mL的DC系列标准溶液，分别测定270 nm处的吸光度（A），以A对浓度（C，μg/mL）作线性回归计算，绘制DC标准曲线，求出相关系数。

释放介质的配制不同pH人工胃液：配制pH为2.0、3.0、4.0的稀盐酸，加蒸馏水800 mL与胃蛋白酶10g，再加蒸馏水稀释1 000 mL。人工肠液：取KH₂PO₄ 6.8 g，加蒸馏水500 mL使溶解，用0.4%NaO溶液调节pH至6.8；另取胰蛋白酶10g，加蒸馏水适量使溶解，将两液混合后，加蒸馏水稀释至1 000 mL。pH7.4 PBS：取KH₂PO₄ 1.36 g，加0.1 mol/L NaOH溶液79 mL，加蒸馏水稀释至200 mL。

体外释药特性将透析袋用蒸馏水浸泡至沸，加入4 mL释放介质。称取DC原料药3.1 mg与DC-CS-NPs冻干粉20 mg（DC原料药与冻干粉的DC含量相等），溶于释放介质。将透析袋两端扎紧后，放入盛有100 mL 25℃同类释放介质的烧杯中，保持磁力搅拌速度为100 r/min，分别于0.083、0.167、0.25、0.5、1、2、4、6、8、12、16、24 h取外液1 mL，同时补充等量同类释放介质，维持释放介质体积不变。所取外液4℃13 000 r/min离心30 min，取上清液适当稀释后，以空白CS-NPs冻干粉为参比，用紫外分光光度（UV）法于270 nm处测定每批次外液吸光度，计算DC含量。每种释放介质平行测定3次，取平均值，绘制释放曲线，并按以下公式计算释放度：

透析外液累计强力霉素总量释放度（%）＝所投入强力霉素总量×100

释放模型拟合分别用零级、一级、Higuchi、Weibull 和双相动力学方程对 DC 原料药及 DC-CS-NPs 冻干粉体外释药的各时间点释放度 Q（%）对时间 t 进行拟合，拟合优度以 r 值进行判断。

1.5 体内血药浓度的测定及方法学考察

色谱条件　色谱柱：ACQUITY UPLC BEH C18（1.7 μm，2.1×50 mm）反相色谱柱；流动相：A 相为 0.01 mol/L 磷酸二氢钠（含 10%乙腈），B 相为乙腈（体积比 4∶1）；柱温：45℃；流速：0.300 mL/min；进样量：5 μL；紫外检测方式采用单波长运行模式，波长为 350 nm。

标准曲线的制备　取一定体积的标准混合液，用流动相稀释成 DC 浓度为 0.01、0.05、0.10、0.50、1.00、2.50、5.00 μg/mL 的标准溶液，分别进样测定，以 DC 峰面积（A，AU）对质量浓度（C，μg/mL）线性回归，求出回归方程和相关系数。

工作曲线的制备与最低检测限测定　取空白血浆加入 DC 已知系列标准溶液使之成为相当于血浆浓度 0.01、0.05、0.10、2.50、0.50、1.00、5.00 μg/mL 的样品，按照血浆样品处理方法处理，作 UPLC 分析。以测得的各平均峰面积 Y 为因变量，相应的质量浓度 X 为自变量绘制标准工作曲线，求出回归方程和相关系数。

用空白血浆制成低质量浓度药物的含药组织，经预处理后测定，将引起三倍基线噪音的药物质量浓度定义为最低检测限。

回收率、精密度测定　实验分两组，一组在空白血浆中分别添加 3 个浓度水平的 DC 标准液，使血液中药物浓度分别为 0.1、0.5、1.0 μg/mL，再按样品预处理方法处理后进样测定 DC 的质量浓度；另一组为加入 3 个浓度组的 DC 标准品测得的质量浓度。每个浓度组做 5 个平行，按以下公式计算回收率：

$$\text{回收率（\%）} = \frac{\text{加标样品测定值}}{\text{标准品测定值}} \times 100$$

每个浓度的样品，日内做 5 个重复，一周内重复做 5 次，计算日内及日间精密度。

1.6 DC 原料药及 DC-CS-NPs 冻干粉在斑点叉尾体内的药代动力学

试验设计与采样　将 DC 原料药和 DC-CS-NPs 冻干粉用蒸馏水配成混悬液，按 20 mg/kg 剂量单次口服灌胃给药，将无反胃回吐药的鱼用于实验。给药后 0.083、0.167、0.25、0.5、1、2、4、8、12、16、24、48、72、96、120、144、168 h 尾静脉取血，用 1%肝素钠抗凝。血液以 3 000 r/min 离心 10 min，取上清液，将血浆于-20℃保存，直至样品处理。每一时间点各取 5 尾鱼作为 5 个平行组，另取 5 尾鱼作为空白对照。

血浆样品处理将冷冻保存的血浆室温下自然解冻，摇匀后取 1.0 mL 置于 10 mL 具塞离心管中，加入 1.0 mL5%高氯酸提取液，置调速混匀器上涡旋振荡混匀 2 min，以 10 000 r/min 离心 10 min，取上清液于另一支 10 mL 具塞离心管中，经 0.22 μL 滤头过滤，UPLC 测定。

数据处理　药物动力学模型拟合及参数计算采用中国药理学会数学专业委员会编制的 3P97 药动软件分析。标准曲线、药物经时曲线图采用 Microsoft Excel 2003 计算绘制。

2　结果与分析

2.1 DC 原料药和 DC-CS-NPs 冻干粉的体外释药特性

标准曲线 DC 在 10～90 μg/mL 范围内吸光度（A）与浓度（C，μg/mL）线性相关，DC 标准曲线为 A＝0.0281C-0.0132，相关系数 r＝0.9998。

体外释药规律 DC 原料药和 DC-CS-NPs 冻干粉在人工胃液、肠液和 pH 7.4 PBS 的释放曲线如图 1 至图 3 所示。由图 1 至图 3 可知，DC 原料药在人工胃液、肠液和 pH 7.4PBS 释放较快，1 h 内释放度已接近 100%，均能完全溶出；DC 原料药在不同 pH 的人工胃液中，释放曲线相近，这是因为原料药没有被包裹，不会受胃蛋白活性影响其释放度。

与 DC 原料药相比，DC-CS-NPs 冻干粉则表现出明显的缓释特性，其在人工胃液、肠液和 pH 7.4PBS 的释放曲线分为突释和缓释两个阶段。纳米粒在 pH 2.0、pH 3.0、pH 4.0 的人工胃液和人工肠液中的突释阶段 2 h 内，释放率分别达 49.32%、45.29%、40.02%和 35.13%，突释效应依次减弱；纳米粒在 pH 7.4PBS 中 2h 的释放率仅为 24%，突释效应最弱。在缓释阶段 DC-CS-NPs 冻干粉中药物缓慢释放，24 h 在 pH 2.0、pH 3.0、pH 4.0 人工胃液的释放率分别为 91.11%、88.60%、85.36%，在人工肠液和 pH 7.4 PBS 的释放率分别为 78.12%和 72.91%，最终还有部分 DC 未被释出。

DC-CS-NPs 冻干粉的体外释药速度与释放介质的酶活性和体系 pH 相关。DC-CS-NPs 冻干粉在各介质的释药速率从大到小依次为 pH 2.0 人工胃液>pH 3.0 人工胃液>pH 4.0 人工胃液>工肠液>pH 7.4 PBS。纳米粒在人工胃液的释药速率较人工肠液大，这可能因为纳米粒在酸性介质中更易溶胀成为凝胶态，使得药物渗透出来；在 pH7.4 PBS 中由于没有酶类的作用，药物释放较慢；人工胃液在 pH 2.0～4.0 范围内，随着 pH 值的增大，DC-CS-NPs 冻干粉释放度减小，这是因为胃蛋白酶的最适 pH 值为 2.0，随着 pH 的增大，胃蛋白酶的活性减小。

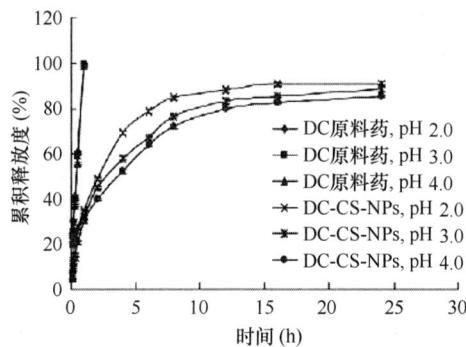

图 1　DC 原料药及 DC-CS-NPs 冻干粉在人工胃液的体外释放曲线

Fig. 1　The accumulation release curves of DC raw material drug and DC-CS-NPs freeze-dried powder in vitro in artificial gastric juice

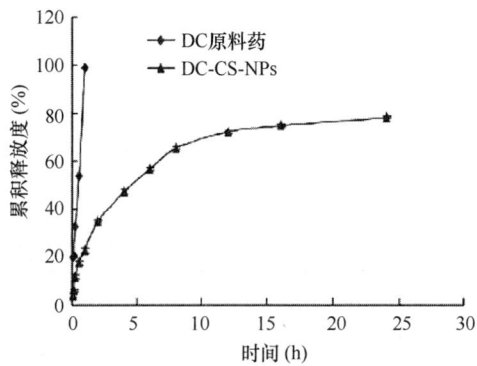

图 2　DC 原料药及 DC-CS-NPs 冻干粉在人工肠液的体外释放曲线

Fig. 2　The accumulation release curves of DC raw material drug and DC-CS-NPs freeze-dried powder in vitro in artificial intestinal juice

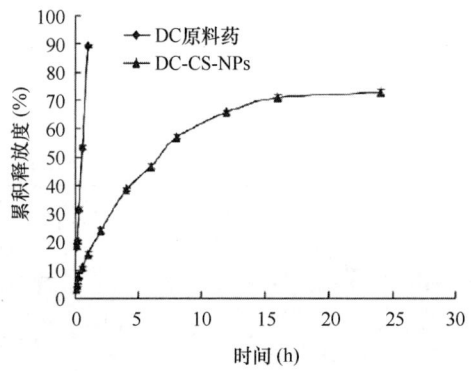

图 3　DC 原料药及 DC-CS-NPs 冻干粉在 pH 7.4 PBS 的体外释放曲线

Fig. 3　The accumulation release curves of DC raw material drug and DC-CS-NPs freeze-dried powder in vitro in pH 7.4 buffer phosphate

　　释放模型拟合分别用零级、一级、Higuchi、Weibull 和双相动力学方程对 DC 原料药及 DC-CS-NPs 冻干粉体外释药的各时间点进行拟合，结果见表 1、表 2。根据拟合方程的相关系数 r 可知，DC 原料药在人工胃液、肠液和 pH 7.4PBS 的体外释放均可用零级动力学方程拟合，释放较快，释放率与释放时间基本成正比，且 DC 原料药在 1 h 内能完全释出。而 DC-CS-NPs 冻干粉的体外释放具有明显的缓释作用，其在人工胃液、肠液和 pH 7.4 PBS 的体外释放均可用双相动力学方程描述，拟合情况较好。整个方程由冲击相和缓释相两相组成，冲击相表明纳米粒释放初期的突释效应，缓释相表示纳米粒在释放后期的缓释规律。

表 1 DC 原料药及 DC-CS-NPs 冻干粉在人工胃液的体外释药拟合结果

Tab. 1 The accumulation release simulation results of DC raw material drug and DC-CS-NPs freeze-dried powder in vitro in artificial gastric juice

数学模型 mathematical model	拟合模型及相关系数 fitted model and related coefficient		
	pH 2.0 人工胃液 pH 2.0 artificial gastric juice	pH 3.0 人工胃液 pH 3.0 artificial gastric juice	pH 4.0 人工胃液 pH 4.0 artificial gastric juice
DC			
零级 zero order	$Q=0.921\ 9t+0.106\ 2$ $r=0.983\ 9$	$Q=0.921\ 3t+0.104\ 9$, $r=0.983\ 4$	$Q=0.932\ 7t+0.080\ 2$, $r=0.989\ 3$
higuchi	$Q=0.984\ 0t^{1/2}-0.062\ 8$, $r=0.981\ 5$	$Q=0.985\ 6t^{1/2}-0.064\ 9$, $r=0.983\ 3$	$Q=0.983\ 9t^{1/2}-0.085\ 1$, $r=0.975\ 4$
weibull	$\ln\ln(1/(1-Q))=1.198\ 6$ $\ln t+1.230\ 4$, $r=0.916\ 1$	$\ln\ln(1/(1-Q))=1.180\ 0$ $\ln t+1.168\ 8$, $r=0.930\ 1$	$\ln\ln(1/(1-Q))=1.208\ 2$ $\ln t+1.086\ 5$, $r=0.934\ 9$
DC-CS-NPs			
一级 first order	$\ln(1-Q)=-5.982\ 3t+0.645\ 4$, $r=0.936\ 6$	$\ln(1-Q)=5.234\ 4t+0.525\ 1$, $r=0.942\ 1$	$\ln(1-Q)=-4.762\ 4t+0.483\ 8$, $r=0.942\ 9$
higuchi	$Q=0.207\ 5t^{1/2}+0.121\ 6$, $r=0.941\ 5$	$Q=0.199\ 5t^{1/2}+0.088\ 3$, $r=0.961\ 7$	$Q=0.195\ 3t^{1/2}+0.065\ 8$ $r=0.968\ 0$
weibull	$\ln\ln(1/(1-Q))=0.626\ 4\ln t$ $-0.812\ 5$, $r=0.990\ 4$	$\ln\ln(1/(1-Q))=0.631\ 8\ln t-$ $1.025\ 8$, $r=0.992\ 5$	$\ln\ln(1/(1-Q))=0.651\ 9\ln t-$ $1.184\ 7$, $r=0.991\ 6$
双相动力学 biphasickinetics	$Q=0.256\ 5e^{-0.056\ 9t}+0.681\ 5e^{-0.519\ 6t}$, $r=0.993\ 4$	$Q=0.614\ 9e^{-0.099\ 5t}+0.370\ 4e-1.370\ 4t$, $r=0.995\ 5$	$Q=0.370\ 3e^{-1.003\ 9}t+0.611\ 3e^{-0.081\ 3t}$, $r=0.994\ 7$

表 2 DC 原料药和 DC-CS-NPs 冻干粉在人工肠液、pH 7.4 PBS 的体外释药拟合结果

Tab. 2 The accumulation release simulation results of DC raw material drug and DC-CS-NPs freeze-dried powder in vitro in artificial gastric juice and pH 7.4 buffer phosphate

数学模型 mathematical model	拟合模型及相关系数 fitted model and related coefficient	
	人工肠液 artificial intestinal juice	pH=7.4 PBS pH=7.4 buffer phosphate
DC		
零级 zero order	$Q=0.938\ 9t+0.063\ 7$, $r=0.992\ 9$	$Q=0.850\ 0t+0.070\ 6$, $r=0.989\ 9$
一级 first order	$\ln(1-Q)=-4.544\ 3t+0.471\ 9$, $r=0.942\ 5$	$\ln(1-Q)=-2.196\ 4t+0.099\ 5$, $r=0.984\ 3$
Higuchi	$Q=0.979\ 3t^{1/2}-0.097\ 3$, $r=0.968\ 0$	$Q=0.897\ 5t^{1/2}-0.080\ 4$, $r=0.977\ 0$
Weibull	$\ln\ln(1/(1-Q))=1.213\ 9\ln t+1.017\ 8$, $r=0.929\ 2$	$\ln\ln(1/(1-Q))=0.995\ 9\ln t+0.558\ 3$, $r=0.964\ 4$
DC-CS-NPs		
一级 first order	$\ln(1-Q)=-0.069\ 1t-0.210\ 1$, $r=0.934\ 6$	$\ln(1-Q)=-0.061\ 2t-0.135\ 2$, $r=0.951\ 3$
Higuchi	$Q=0.180\ 5t^{1/2}+0.046\ 1$, $r=0.971\ 9$	$Q=0.172\ 5t^{1/2}+0.003\ 5$, $r=0.984\ 1$
Weibull	$\ln\ln(1/(1-Q))=0.643\ 2\ln t-1.380\ 4$, $r=0.992\ 4$	$\ln\ln(1/(1-Q))=0.682\ 9\ln t-1.708\ 9$, $r=0.997\ 1$
双相动力学 biphasickinetics	$1-Q=0.419\ 9e^{-0.030\ 4t}+0.541\ 7e^{-0.366\ 1t}$, $r=0.995\ 4$	$1-Q=0.715\ 9e^{-0.171\ 2t}+0.253\ 9e^{-0.000\ 1t}$, $r=0.998\ 1$

2.2　DC 原料药和 DC-CS-NPs 冻干粉的体内药代动力学

色谱图　DC 标准品、加样标准色谱图如图 4、5 所示。由图谱可见，本实验所选用的分析条件能较好地分离 DC，无杂质峰干扰，分离度较好。

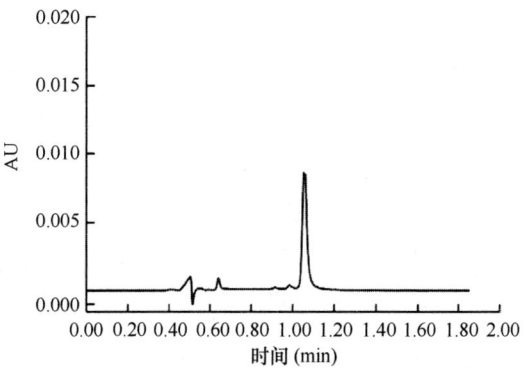

图 4　DC 标准溶液色谱图

Fig. 4　Chromatography of DC standard solution

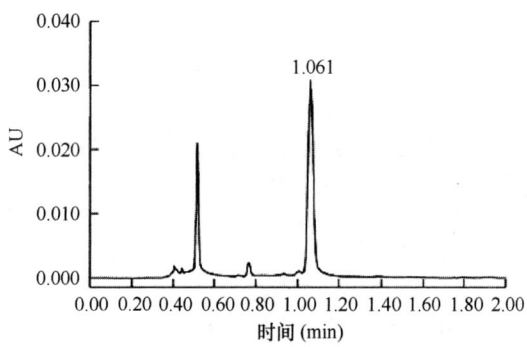

图 5　斑点叉尾血浆加标样品色谱图

Fig. 5　Chromatography of spiked plasma sample in channel catfish

标准曲线　以含 DC 浓度为 0.01、0.05、0.10、0.50、1.00、2.50、5.00 μg/mL 标准溶液进样，得标准曲线。结果表明，DC 在 0.01~5.00 μg/mL 范围内峰面积（A, AU）与质量浓度（C, μg/mL）线性相关，标准曲线方程为 A = 15 314 C +1 343.9，r=0.999 8。

血浆工作曲线方程与最低检测限的测定　在空白血浆中加入 DC 已知标准溶液，在 0.01~5.0 μg/mL 范围内线性关系良好。斑点叉尾血浆中 DC 的标准工作曲线为 Y=13 779X+1 284.9，r=0.999 7。样品中 DC 检测限为 3 μg/L。

回收率、精密度的测定　本试验条件下，测得血液中 3 个质量浓度水平的 DC 回收率为（91.45±2.75）%，日内精密度为（3.00±0.72）%，日间精密度为（4.82±0.97）%。

DC 原料药和 DC-CS-NPs 冻干粉在斑点叉尾体内的药动学　DC 原料药和 DC-CS-NPs 冻干粉分别按 20 mg/kg 剂量单次口灌斑点叉尾后，血液中的药物浓度随时间变化的规律如图 6 所示，药物在血液中的药动学参数如表 3 所示。口灌 DC 原料药后，药物在斑点叉尾血浆中呈现明显的双峰现象，其血药浓度时间数据符合一级吸收二室开放模型，动力学方程为 C=2.1689 $e^{0.1227t}$+1.2283 $e^{0.0160t}$+3.3972 $e^{0.8719t}$，血药达峰时间（T_{max}）为 3.039 1 h，血药峰浓度（C_{max}）为 2.4239 μg/mL，其分布半衰期（$T_{1/2\alpha}$）为 5.648 3 h，消除半衰期（$T_{1/2\beta}$）为 43.413 1 h，药-时曲线下面积（AUC）为 90.708 4 μg·h/mL。DC-CS-NPs 冻干粉经消化道吸收进入血液，在斑点叉尾体内血液中呈现单峰现象，经房室模型分析，DC-CS-NPs 冻干粉其血药浓度时间数据符合一级吸收二室开放模型，动力学方程为 C=6.1774 $e^{0.0884t}$+1.1566 $e^{0.0146t}$+7.3340 $e^{0.1673t}$，T_{max} 为 9.554 5 h，C_{max} 为 2.178 6 μg/mL，$T_{1/2\alpha}$ 为 7.8448 h，$T_{1/2\beta}$ 为 47.589 1 h，AUC 为 105.477 4 μg·h/mL。与 DC 原料药药代动力学参数相比，DC-CS-NPs 冻干粉的 T_{max} 明显延长，C_{max} 减小，$T_{1/2\beta}$ 延长，AUC 变大。DC-CS-NPs 冻干粉比原料药吸收缓慢，消除缓慢，具有一定的长效性。

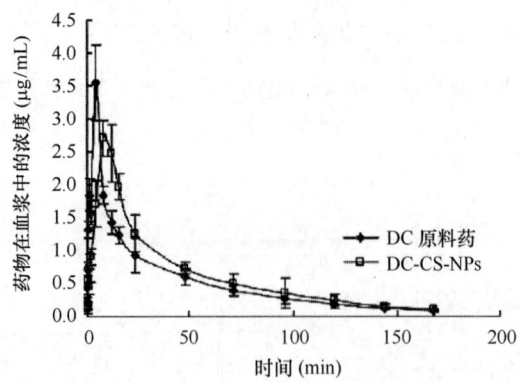

图 6　DC 原料药和 DC-CS-NPs 冻干粉在斑点叉尾鮰血浆中的药—时曲线

Fig. 6　Concentration-time curve of DC raw material drug and DC-CS-NPs freeze-dried powder in the plasma of channel catfish

表 3　20 mg/kg b·w 单次口灌 DC-CS-NPs 冻干粉在斑点叉尾鮰血液的药代动力学参数
Tab 3　Pharmacokinetic parameters for DC raw material drug and DC-CS-NPs freeze-dried powder in channel catfish after a single oral administration at 20 mg/kg b·w

参数 parameters	单位 unit	DC 原料药参数值 parameters of DC raw material drug	DC-CS-NPs 冻干粉参数值 parameters of DC-CS-NPs freeze-dried powder
A	μg/mL	2.168 9	6.177 4
α	h⁻¹	0.122 7	0.088 4
B	μg/mL	1.228 3	1.156 6
β	h⁻¹	0.016 0	0.014 6
K_α	h⁻¹	0.871 9	0.167 3
Lag time	h	0.000 9	0.005 5
V_d/F	L/kg	6.515 9	5.037 6
$T_{1/2\alpha}$	h	5.648 3	7.844 8
$T_{1/2\beta}$	h	43.413 1	47.589 1
$T_{1/2K\alpha}$	h	0.795 0	4.143 9
K_{21}	h⁻¹	0.057 9	0.034 2
K_{10}	1/h	0.033 8	0.037 6
K_{12}	h⁻¹	0.046 9	0.031 1
AUC	μg·h/mL	90.708 4	105.477 4
CLs	(mL/h)/kg	0.220 5	0.189 6
T_{max}	h	3.039 1	9.554 5
C_{max}	μg/mL	2.423 9	2.178 6

注：A, B 为药时曲线对数图上曲线在横轴和纵轴上的截距；α, β 分别为分布相、消除相的一级速率常数；K_{21} 由周边室向中央室转运的一级速率常数；K_{10} 由中央室消除的一级速率常数；K_{12} 由中央室向周边室转运的一级速率常数；AUC 药—时曲线下面积；Lag time 滞后时间；K_α 为一级吸收速率常数；$T_{1/2K\alpha}$ 为药物在中央室的吸收半衰期；$T_{1/2\alpha}$, $T_{1/2\beta}$ 分别为总的吸收和消除半衰期；T_{max} 出现最高血药质量浓度的时间；C_{max} 最高血药质量浓度。

Notes: A, B are drug concentration-time curve on logarithmic graph curve on the horizontal axis and vertical axis intercept; α, β are first-order rate constant for drug distribution phase and elimination phase; K_{21} is first-order rate constant for drug distribution from peripheral compartment to central compartment; K_{10} is first-order rate constant for drug elimination from central compartment; K_{12} is first-order rate constant for drug distribution from central compartment to peripheral compartment; AUC is area under concentration-time curve; Kα is first-order rate constant for drug absorption; $T_{1/2K\alpha}$ is absorption half-life in central compartment; $T_{1/2\alpha}$, $T_{1/2\beta}$ are the overall absorption half-life and Elimination half-life; T_{max} is the time at high-est peak of concentration; C_{max} is the biggest concentration of the drug in blood.

3　讨论

3.1　DC-CS-NPs 冻干粉的发展前景

现代水产养殖业日益趋向于规模化、集约化，渔药的使用为水产养殖业发展提供了有力的保障。然而目前我国渔药制剂技术发展滞后，部分药物如 DC 在水产上还是沿用兽药最传统的粉剂、注射剂等剂型，因此有必要研究出一种适用于水产养殖实际环境的新剂型药物，以改善药物在使用时的弊端。根据水生动物的种类、特点和规格及药物种类和性质等，有针对性地开发渔药剂型刻不容缓。

DC 味苦，抗逆性差，释放不稳定。因施药环境的特殊性，要求 DC 具有一定的稳定性。CS-NPs 是一种新型的药物缓释系统，具有缓释药物、靶向输送、减少胃肠道刺激、增加药物吸收、提高药物稳定性等优点[7]。将 DC 制成 DC-CS-NPs 冻干粉，不仅能较好地保持用药时药物的稳定性，便于药物在水环境中长期地发挥作用，也有利于药物储存、运输及保存；药物的包裹也能改善拌饲投喂时药物的适口性，有很好的应用价值；纳米粒有靶向性，其较小的粒径有利于延长局部用药时药物的滞留时间，增加药物与肠壁的接触时间，加大接触面积，延长药物作用时间，直达病灶部位，降低药物在其他正常组织的浓度；纳米粒的缓释功能可减少药物用药次数和用药剂量，降低耐药性，减少药物残留，有广阔的发展前景。

3.2　DC-CS-NPs 冻干粉的体外释放

DC-CS-NPs 纳米粒中的 DC 被包裹在纳米粒内或吸附在纳米粒表面，使得药物的释放速度具有选择性和可控性，从而能较好地发挥药物的治疗效果。DC-CS-NPs 冻干粉在人工胃液、肠液和 pH 7.4 PBS 的释药方式可分为突释和缓释两相：释放初期为突释现象，这是由吸附或镶嵌的纳米粒表面的 DC 快速释放所致，此外冷冻干燥技术使纳米粒的水分固化成结晶态的冰，体积增大，纳米粒内部形成更大的孔道，促进了药物的快速释放，增大了纳米粒的突释量[8]，突释有利于体内药物即刻到达有效的血药浓度。体外释药实验显示随着介质 pH 的增大，纳米粒的突释效应减弱，这可能因为游离的 DC 与纳米粒表面的电荷吸附作用加大，使得突释量降低。而释放后期 DC-CS-NPs 冻干粉呈缓释模式，此阶段以被包封的药物从纳米粒释放为主，释药速率平缓，释放均匀，表现出良好的缓释性能，缓释有利于维持有效的血药浓度[9]。体外释药试验显示在人工胃液、肠液和 pH 7.4 PBS 还有部分 DC 未被释放，但是由于鱼体内存在各种代谢酶和菌群，可逐步分解代谢壳聚糖，使药物最终完全释出[10]。

3.3　DC-CS-NPs 冻干粉的体内药代动力学特征

DC 进入水产动物体内后经胆汁排入肠道会再次被吸收，有显著的肝肠循环[1]，因此 DC 在水产动物体内的药物释放表现为双峰现象，再加上 DC 原料药一般每隔 24 小时给药一次，连续给药 3~5 d，多次给药后，药物在鱼体内会出现多个波峰，药效不稳定，有效血药浓度维持时间短，极易产生耐药性，从而影响药物的实际治疗效果，且重复给药易造成药物代谢缓慢，导致药物残留，严重影响无公害渔业的发展[11]。有学者开始探讨 DC 新剂型的改造，李四元等[12]对自制的 DC 脂质体在家兔体内进行药物动力学试验，一次静脉注射强力霉素脂质体制剂后，发现其 c—时曲线属一室开放模型，药代动力学及组织分布发生明显变化，DC 脂质体在家兔体内的 $T_{1/2\beta}$ 为 7.606 h，与 DC 原料药的 2.25 h 相比，药物代谢速度显著减慢，有效血药浓度时间明显延长。本研究自制的 DC-CS-NPs 冻干粉与 DC 原料药相比，T_{max} 明显延长，C_{max} 减小，$T_{1/2\beta}$ 延长，AUC 变大，在体内的药物的释放由双峰变为单峰，具有显著的缓释特性，明显地改变了吸收峰谷的差异，使吸收趋于平缓，能维持药物在体内长时间的稳态血药浓度，避免峰谷效应，克服了传统药物服用后在体内血药浓度波动大、副作用大的缺点。体内释药特性实验显示 DC-CS-NPs 冻干粉血药达峰浓度较 DC 原料药低，这也可间接表明 DC-CS-NPs 冻干粉具有被动靶向性，进入生物体内后更易被脾等富含网状内皮系统的组织摄取，优先分布于靶组织[13-14]，从而使得血浆中的药物浓度降低。DC 对斑点叉尾致病菌的最小抑菌浓度（MIC）为 0.10~1.25 μg/mL[15-16]。若 DC 原料药按 20 mg/kg 剂量口灌给药，则 16 h 药物浓度已小于 1.25 μg/mL。本研究中 DC 经包裹制成 DC-CS-NPs 冻干粉后，体内药代动力学发生改变，按 20 mg/kg 剂量口灌给药 24 h 药物浓度仍在 1.25 μg/mL 之上，延长了药物的作用时间。在临床应用上，可减少 DC-CS-NPs 冻干粉的用药剂量和用药频率，从而有效避免药物的滥用多用，提高药效发挥。

参考文献

[1]　刘辉旺，刘义明. 多西环素研究进展［J］. 兽医导刊，2008，（125）：27.28.

[2]　彭司勋. 药物化学［M］. 北京：化学工业出版社，1988：334.335.

[3]　熊喜华. 强力霉素脂质体的制备及药效学研究［D］. 长沙：湖南农业大学，2007.

[4]　汤爱国，张阳德.O-羧甲基乳糖酰化壳聚糖纳米粒-聚乳酸阿霉素纳米粒在大鼠体内的靶向性研究［J］. 中国现代医学杂志，2006，16（13）：1941.1943.

［5］ 潘卫三．新药制剂技术［M］．北京：化学工业出版社，2004：116.127.

［6］ 陈关平，操继跃．兽药控/缓释剂的研究进展［J］．中国兽药杂志，2003，37（8）：40.42.

［7］ Jeong Y. I.，Cheon J. B.，Kim S. H.，et al. Clonazepan re-lease from core-shell type nanoparticles in vitro［J］．Journal of Controlled Release，1998，51（2）：169.178.

［8］ 龙娜，吕竹芬．微球缓释系统的突释现象及其影响因素［J］．中国药师，2010，13（3）：421.423.

［9］ Sharma D.，Chelvi T. P.，Kaur J.，et al. Novel taxol formu-lation：polyvinylpyrrolidone nanoparticle-encapsulate taxol for drug delivery in cancer therapy［J］．Oncology Research，1996，8（7）：281.286.

［10］ 龚露旸，胡鲲，杨先乐．恩诺沙星壳聚糖纳米粒的制备及其体外释药特性研究［J］．上海海洋大学学报，2009，18（3）：321.326.

［11］ 丁昕颖，李平，丁得利．动物性食品中药物残留的现状分析［J］．中国禽业导刊，2008（21）：32.33.

［12］ 李四元．强力霉素脂质体的制备及药物代谢动力学研究［D］．长沙：湖南农业大学，2007.

［13］ 张阳德．纳米药物学［M］．北京：化学工业出版社，2005：17.46.

［14］ Brigger I.，Dubernet C.，Couvreur P.．Nanoparticles in cancer therapy and diagnosis［J］．Advanced Drug Deliv-ery Reviews，2002，54（5）：631.651.

［15］ 袁科平．网箱养殖斑点叉尾暴发性细菌性疾病防治研究［D］．武汉：华中农业大学，2008.

［16］ 孟小亮，陈昌福，吴志新，等．嗜水气单胞菌对盐酸多西环素的耐药性获得与消失速率研究［J］．长江大学学报：自然科学版，2009，6（1）：42.44.

该文发表于《淡水渔业》【2011，41（3）：55-60】

强力霉素壳聚糖纳米粒冻干粉的制备及其稳定性研究

Study on preparation and stability of doxycycline chitosan nanoparticles freeze-dried powder

沈丹怡[1,2]，丁运敏[1,2]，艾晓辉[2]，刘永涛[2]，余少梅[1,2]，索纹纹[1,2]

（1. 华中农业大学水产学院，武汉 430070；2. 中国水产科学研究院长江水产研究所，武汉 430223）

摘要： 采用离子交联法和冷冻干燥法制备强力霉素壳聚糖纳米粒冻干粉，并以包封率为评价指标，正交试验优化处方与工艺，用透射电镜观测其形态，纳米粒度及电位分析仪测定其粒径大小、分布和电位，并考察其热、光和高湿稳定性。结果显示，按壳聚糖浓度为 2.5 mg/mL、三聚磷酸钠浓度为 1.2 mg/mL、强力霉素用量为 8.0 mg、pH 值为 5.5 的处方制备的强力霉素壳聚糖纳米粒冻干粉表面蓬松，复溶性良好，电镜下观测发现纳米粒形态圆整、粒径大小均匀，平均粒径为（126.3±17.8）nm，平均包封率为（56.52±2.1）%，平均载药量为（16.83±0.27）%。与原料药相比，10 d 强力霉素壳聚糖纳米粒冻干粉的（4500±500）Lx 强光降解率减少 22.97%，40、60℃热降解率分别减少 18.72%和 31.41%，75%、92.5%高湿降解率分别减少 20.45%和 26.90%。强力霉素壳聚糖纳米粒冻干粉制备工艺切实可行，稳定性良好。

关键词： 强力霉素；壳聚糖；纳米粒；冻干粉；稳定性

三、水乳剂剂型

渔药的水乳剂型能降低药物的毒性，提高药效。阿维菌素水乳剂（EW）对白鲢（*Hypophthalmichthys molitrix*）、银鲫（*Carassius auratus*）、麦穗鱼（*Peudorasbora parva*）的安全浓度几近是其乳油制剂的两倍，而相同含量的阿维菌素水乳剂对锚头蚤的驱虫效果优于乳油（周帅等，2009）

该文发表于《上海海洋大学学报》【2009，18（3）：327-331】

渔用新剂型阿维菌素水乳剂的安全性和药效评价

Evaluation of safety and efficacy of abamectin emulsion in water for aquaculture

周帅[1,2]，房文红[2]，吴淑勤[3]

（1. 上海海洋大学生命学院，上海 200090；2. 中国水产科学研究院东海水产研究所，上海 20009；
3. 中国水产科学研究院珠江水产研究所，广州 510380）

摘要： 采用半静态法研究了渔用新制剂阿维菌素水乳剂（EW）对白鲢（*Hypophthalmichthys molitrix*）、银鲫（*Carassius auratus*）、麦穗鱼（*Peudorasbora parva*）的急性毒性实验及其对患病草鱼、银鲫体表锚头鳋的药效实验，且二者均以相同阿维菌素含量的乳油为参照进行对比，旨在评价水产养殖用阿维菌素水乳剂对养殖鱼类的安全性及其驱虫效果。结果显示，2%阿维菌素水乳剂对白鲢、银鲫、麦穗鱼的 96 h-LC_{50} 为 2.15 mg/L、1.15 mg/L、1.83 mg/L，安全浓度为 0.22 mg/L、0.12 mg/L、0.18 mg/L；而对应乳油 96 h-LC_{50} 为 0.98 mg/L、0.57 mg/L、1.08 mg/L，安全浓度为 0.09 mg/L、0.06 mg/L、0.1 mg/L。水乳剂的安全浓度几近乳油的两倍，因此使用起来更安全。对患锚头鳋的草鱼、银鲫施用 0.05 mg/L 2%阿维菌素水乳剂后 6 h 的校正防效为 96.27%、91.88%，24 h 即达到 100%、96.15%；施用 0.025 mg/L 2%阿维菌素水乳剂后 6 h 的校正防效为 43.65%、68.58%，24 h 达到 91.56%、90.80%；而施用 0.05 mg/L 2%阿维菌素乳油对应的校正防效分别为 47.31%、77.46% 和 95.69%、91.62%，可见相同含量的水乳剂驱虫效果优于乳油。

关键词： 阿维菌素、水乳剂、乳油、半致死浓度、药效、锚头鳋

第四章 渔药检测方法的研究与应用

第一节 仪器检测方法

常见的渔药检测方法有、高效液相色谱法（high performance liquid chromatography，HPLC）、液相色谱-串联质谱法（liquid chromatogramphy-mass spectrography，LC-MS）、免疫学方法微生物法等。

高效液相色谱法（HPLC）又称"高压液相色谱"、"高速液相色谱"、"高分离度液相色谱"、"近代柱色谱"等，它是以液体为流动相，采用高压输液系统，将具有不同极性的单一溶剂或不同比例的混合溶剂、缓冲液等流动相泵入装有固定相的色谱柱，在柱内各成分被分离后，进入检测器进行检测，从而实现对试样的分析。1903年俄国植物化学家茨维特（Tswett）首次提出"色谱法"（Chromatography）和"色谱图"（Chromatogram）的概念，1990年以后，生物工程和生命科学的迅速发展，为该方法的应用注入了新的活力。高效液相色谱法因其具有操作简单、方法灵活、灵敏度高等优点，是一种检测渔药残留的常用方法。

水产品及环境中的诺氟沙星（房文红等，2004；胡鲲等，2007；汤菊芬等，2014）、盐酸环丙沙星（杨先乐等，2001；郑宗林等，2006）、恩诺沙星（房文红等，2008；郑宗林等，2006）、双氟沙星及其代谢产物沙拉沙星（李海迪等，2009）、噁喹酸（殷桂芳等，2016）、氟甲喹（胡琳琳等，2011）等均能使用HPLC法进行检测，其中配荧光检测器可以取得比紫外检测器更好的检测灵敏度。

一、氟喹诺酮类药物残留检测

该文发表于《水产学报》【2001，25（4）：348-354】

中华绒螯蟹血淋巴内盐酸环丙沙星的反相高效液相色谱法的建立

The method of RP-HPLC determination of ciprofloxacinum in hemolymph of *Eriocheir sinenses*

杨先乐，刘至治，孙文钦，蔡完其

（上海水产大学农业部水产增养殖生态、生理重点开放实验室，上海 200090）

摘要： 应用反相高效液相色谱法（RP-HPLC）建立了盐酸环丙沙星在中华绒螯蟹血淋巴内含量的测定方法。本法采用肌肉注射的给药方法，给药剂量为（8.17 ± 0.56）mg·kg^{-1}和（6.25 ± 0.85）mg·kg^{-1}。血淋巴从心区抽取。经各种抗凝方法的比较，认为等量的ACD与等量的血淋巴混合，可得到持久的抗凝效果。血淋巴样经甲醇去蛋白，3 000 r/min离心10 min后，取上清液上HPLC。色谱柱，RP-ODS-5μ150×4.6 mm；流动相，甲醇-磷酸缓冲液-四丁基溴化铵=250：750：20（v/v/v）；流速，1 mL·min^{-1}；进样量，2 μL；检测波长，280 nm。研究表明，本法线性关系与重复性良好；雌、雄蟹平均回收率分别为（100.08 ± 3.43）%和（99.11 ± 8.20）%；日间精密度、平均提取精密度分别为（1.19 ± 0.92）%、（1.55 ± 0.01）%。用该方法对肌肉注射给药后48 h内中华绒螯蟹雄蟹血淋巴中的盐酸环丙沙星的血药浓度进行检测，揭示出由（37.13 ± 0.08）μg·mL^{-1}降至（5.96 ± 0.18）μg·mL^{-1}的代谢规律。

关键词： 中华绒螯蟹；盐酸环丙沙星；药物动力学；反相高效液相色谱法

盐酸环丙沙星（Ciprofloxacinum）属第三代喹诺酮类抗菌药，它通过与DNA回旋酶亚基A结合，抑制了酶的切割与连接功能，阻止了DNA的复制，而起到抗菌作用[1]。由于盐酸环丙沙星的抗菌活性高、抗菌谱广、不易产生耐药性，被广

728

泛地应用在水产动物细菌性疾病的防治上[2-4]。对由弧菌、嗜水气单胞菌等引起的中华绒螯蟹细菌性疾病，盐酸环丙沙星也是首选药物之一[5]。但由于对该药在中华绒螯蟹体内的药代动力学缺乏研究，因此，正确的使用剂量、疗程以及相应的停药期都是借鉴人、兽医的资料，存在较大的盲目性，这不仅影响了药物的使用效果，而且也给中华绒螯蟹的品质带来了一定的影响。

以往有关水产动物药动学方面的研究，大部分集中于鱼类，仅少数涉及到虾类、中华鳖等其他水产动物[6-16]。由于进行中华绒螯蟹药代动力学研究的给药、采样方法和其他水产动物有较大的区别，在某种程度上有一定难度，因此，用反向液相色谱测定法测定中华绒螯蟹体内药物的含量及药物动力学方面的研究鲜有报道。本文旨在摸索从中华绒螯蟹体内抽取血淋巴并建立血药浓度检测方法，为弄清楚盐酸环丙沙星在中华绒螯蟹体内的分布、代谢以及消除规律奠定基础，以便为中华绒螯蟹的合理用药指示方向。

1 材料与方法

1.1 中华绒螯蟹

由上海前卫特种水产养殖公司一场提供，雄蟹 130 只，体重（61.5±4.11）g；雌蟹 110 只，体重（47.81±7.62）g。试验前，在 4 m×5 m 的水泥池中分别暂养 50 d 和 15 d，并在试验所用水族箱（1.0 m ×0.5 m ×0.4 m）中驯养一周。试验水温（20.0±0.5）℃。

1.2 实验试剂

甲醇，HPLC 级，上海吴径化工厂生产；盐酸环丙沙星标准品（纯度为 99.54%），盐酸环丙沙星原粉（纯度为 98%），均由上海三维制药公司提供；抗凝剂肝素钠（1.45 μg·mL^{-1}）、血淋巴抗凝剂 AS（anticoagulant solution）及 ACD（acid citrate dextrose）参照蔡武城等的方法配制[17]。

1.3 高效液相色谱仪及色谱条件

高效液相色谱仪采用 SHIMADZU LC-6A 型系列，包括 SC-6A 系统控制仪，SIL-6A 自动进样器，SPD-6AV 紫外检测器，CTO-6A 柱温箱，LC-6A 高压泵，C-R5A 数据处理装置等。色谱条件：色谱柱为 RP-ODS C18 分析柱（15 cm×4.6 mm，5 μm），流动相按甲醇：磷酸缓冲液（0.05 mol·L^{-1}磷酸二氢甲+2.8 mL 磷酸）：四丁基溴化铵（0.05 mol·L^{-1}）= 250：750：20（v/v/v）进行配制，流速 1.0 mL·min^{-1}，检测波长 280 nm，柱温 40℃，检测灵敏度 0.01AUFS。

1.4 抗凝效果实验

分别采取血淋巴与抗凝剂直接混合（v/v =0.6：0.2）及抗凝剂浸润注射器与离心管后烘干的方式，比较三种抗凝剂肝素钠、AS 与 ACD 的抗凝效果。在此基础上，选择上述抗凝效果较好的抗凝剂，比较在血淋巴样品中加入不同体积抗凝剂的抗凝效果。

1.5 给药、血淋巴样的采集及血淋巴样前处理

盐酸环丙沙星原粉用双蒸水配成 5 和 3 mg·mL^{-1}的溶液（分别用于雄、雌蟹），通过中华绒螯蟹第四步足与体壁关节膜处肌肉注射给药，每只蟹 0.1 mL［即雄蟹：（8.17±0.56）mg·kg^{-1}；雌蟹：（6.25±0.85）mg·kg^{-1}］。从蟹背部心区开一小孔，用抗凝剂润湿管壁的卡介苗注射器抽取血淋巴，每只蟹 1 mL。给药前，先各抽取一组蟹（5 只）的血淋巴，作为空白对照；给药后，按所设时间点，定时抽取血淋巴，每个时间点取 5 只蟹，并做一个重复。用最佳抗凝方法 6 000 r·min^{-1} 15 min 分离血浆，样品置于-60℃冰箱备用。RP-HPLC 检测前，血浆自然溶解后摇匀，准确吸取 0.5 mL 10 mL 离心管中，加入甲醇 1 mL，旋涡混合 2 min，再加入甲醇 0.5 mL，同法混合 2 min，3 000 r·min^{-1}离心 10 min，吸取上清液于 1.5 mL 样品瓶中，自动进样器 2 μL 进样。

1.6 标准曲线的制备

用定容浓度依次为 1.015、2.030、3.045、4.060、5.075、10.15 μg·mL^{-1}的盐酸环丙沙星标准溶液系列，以 RP-HPLC 测得的峰面积为纵坐标，相应的浓度为横坐标，作出浓度范围为 1~10 μg·mL^{-1}的标准曲线，并通过 SYSTAT 软件，求出回归方程和相关系数。

1.7 回收率测定

采用加入法。将 0.5 mL 雄蟹空白血浆分别加入含盐酸环丙沙星标准溶液（浓度为 101.5 μg·mL^{-1}）15、20、25、50、

75 μL 的离心管中，处理与 RP-HPLC 测定同"血样"。同法进行雌蟹空白血浆的回收率测定。HPLC 测定后，按标准曲线回归方程计算回收率［回收率（%）= 样品实测浓度/样品实际浓度×100%］。

1.8 精密度测定

采用加法。

1.8.1 提取精密度（重复性）

在 5 只各盛有 0.5 mL 蟹空白血浆的离心管中分别加入浓度为 1011.33 μg·mL⁻¹ 的标准盐酸环丙沙星溶液 50 μL，其余同"血样"处理。平行测定后，计算出各管盐酸环丙沙星浓度及变异系数（RSD）。同法，对浓度为 101.5 μg·mL⁻¹ 的标准盐酸环丙沙星溶液再重复平行测定一次，计算 RSD。

1.8.2 日间精密度

将浓度为 2.03、4.06、10.15 μg·mL⁻¹ 的标准盐酸环丙沙星溶液 1 mL 加入 1.5 mL 样品瓶中，RP-HPLC 法连续测定 2 次，每隔 3 d 一次，计算平均峰面积及 RSD。

1.9 中华绒螯雄蟹血淋巴内盐酸环丙沙星药物浓度（以下简称血药浓度）的测定

按所建立的方法，测定肌肉注射给药 0.017 h（1 min）、0.083 h（5 min）、0.167 h（10 min）、0.5 h、1 h、1.5 h、2 h、4 h、8 h、12 h、24 h、48 h 后中华绒螯蟹雄蟹的血药浓度，以此检验所建立的方法。

2 结果

2.1 最佳抗凝剂和抗凝方法

通过在相同条件下，肝素钠、AS 与 ACD 三种抗凝剂的抗凝效果比较，结果表明在血淋巴中加入肝素钠者 1.5h 后，即出现 50% 以上的胶状凝集，而 AS 在 30 h 后才有少量絮状沉淀出现，ACD 36.5 h 后才有少量白色絮状物；采取润湿管壁后烘干方式，发现使用肝素钠者血淋巴样迅速凝固，而使用 ACD 者 72 h 后仍未见凝固现象。表 1 是 ACD 的不同处理方式与 ACD 与血淋巴不同体积比的抗凝效果。结果表明，ACD 采取润湿烘干方式不适于血淋巴较长时间保存，只有将等量的 ACD 与等量的血淋巴混合，才能取得较持久的抗凝效果。

2.2 中华绒螯蟹血淋巴中盐酸环丙沙星的色谱行为

在 1.3 所设计的色谱条件下，HPLC 基线走动平稳，盐酸环丙沙星保留时间为 10.3 min，样品血浆中虽有杂质峰出现，但无干扰峰，与药峰分离良好（图 1）。

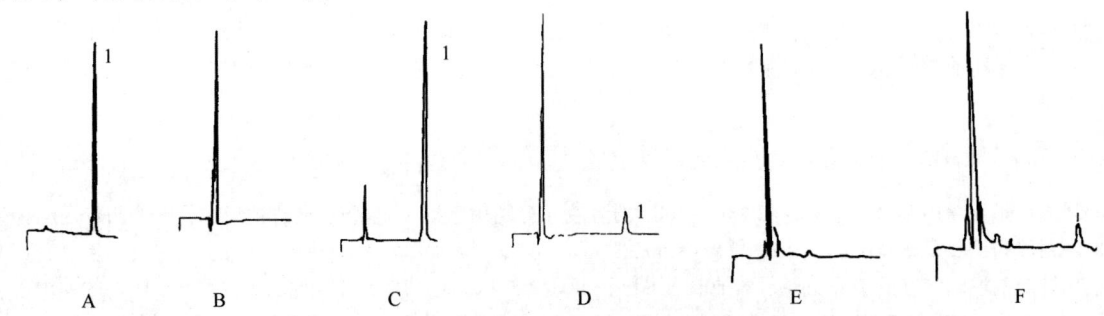

图 1 盐酸环丙沙星的色谱行为

Fig. 1 Liquid chromatogram of ciprofloxacinum

A. 标准盐酸环丙沙星；B. 雄蟹空白血浆；C. 加入标准盐酸环丙沙星后的雄蟹血浆；D. 给药 1 h 后的雄蟹血浆样品；
E. 雌蟹空白血浆；F. 给药 48 h 后的雌蟹血浆样品；1 所指示的色谱峰为盐酸环丙沙星峰

2.3 标准曲线的线性范围和灵敏度

在标准品盐酸环丙沙星浓度为 1~10 μg·mL⁻¹ 时，标准曲线为 $Y = -675.63 + 1204.723X$（Y-峰面积，X-盐酸环丙沙星浓度）（图 2）。相关系数为 $r = 0.9985$，相关性良好。根据标准曲线，以引起两倍基线噪音的药量为最低检测限，得出本法的灵敏度为 0.01 μg·mL⁻¹。

表 1 ACD 的处理方式及其不同体积的抗凝效果比较

Tab. 1 The comparison of anticoagulation of ACD by different treatments and various volume

处理方式	ACD 加入量（mL）	血淋巴量（mL）	ACD 与血淋巴的体积比	试验重复次数	抗凝效果
ACD 润湿管壁后烘干	-	0.6	-	2	+/+
	-	0.5	-	3	+/+/+
	-	0.2	-	2	+/+
ACD 与血淋巴混合	0	0.3	0	1	+
	0	0.5	0	1	+
	0.1	1.0	1/10	1	+
	0.2	1.0	1/5	1	+
	0.2	0.6	1/3	1	+
	0.2	0.5	2/5	2	+/+
	0.2	0.4	1/2	1	+
	0.3	0.5	3/5	4	+/+/+/+
	0.4	0.5	4/5	4	+/+/+/+

注：抗凝效果样品在低温（4℃）下放置 15 d 以上，"+"代表凝固，"-"代表不凝固。

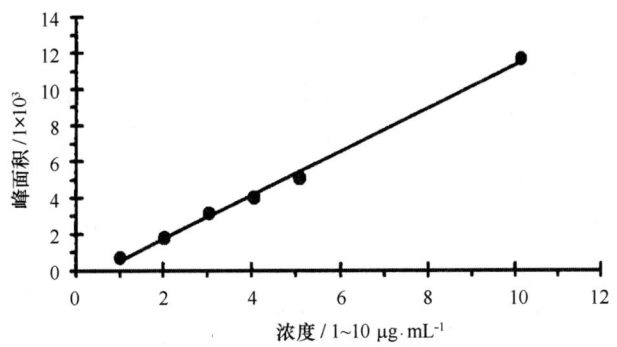

图 2 盐酸环丙沙星标准曲线

Fig. 2 The standard curve of ciprofloxacinum

2.4 回收率

分别对雌、雄蟹空白血淋巴中 5 种不同浓度的盐酸环丙沙星回收率测定结果表明，其范围在 88.71% ～ 105.76% 和 97.37% ～ 105.45% 间；平均回收率为（99.11±8.20)%和（100.08±3.43)%（表 2，表 3）。

表 2 中华绒螯蟹雌蟹血淋巴中盐酸环丙沙星回收率的测定结果

Tab. 2 The recovery of ciprofloxacinum in the hemolymph of female *Eriocheir sinensis*

项目	标准盐酸环丙沙星溶液（101.5 μg·mL⁻¹）加入量（μL）				
	15	20	25	50	75
样品实际浓度（μg·mL⁻¹）	7.53	10.01	12.49	24.67	36.55
样品实测浓度（μg·mL⁻¹）	6.68	9.19	13.21	26.07	37.86
回收率（%）	88.71	91.81	105.79	105.67	103.58
平均回收率（%）	99.11±8.20				

表3　中华绒螯蟹雄蟹血淋巴中盐酸环丙沙星回收率的测定结果

Tab. 3　The recovery of ciprofloxacinum in the hemolymph of male *Eriocheir sinensis*

项目	标准盐酸环丙沙星溶液（101.5 μg·mL⁻¹）加入量（μL）				
	15	20	25	50	75
样品实际浓度（μg·mL⁻¹）	0.76	1.01	1.25	2.48	3.67
样品实测浓度（μg·mL⁻¹）	0.74	0.99	1.27	2.43	3.87
回收率（%）	97.37	98.02	101.60	97.98	105.45
平均回收率（%）	100.08±3.43				

2.5　精密度

浓度为 1 011.33 μg·mL⁻¹ 与 101.5 μg·mL⁻¹ 的盐酸环丙沙星标准溶液的提取精密度（重复性）测定结果表明，其变异系数（RSD）分别为 1.55% 与 1.54%，平均为 1.55±0.01%（表4）。

表4　对中华绒螯蟹血淋巴中盐酸环丙沙星测定的提取精密度（重复性）

Tab. 4　The extracted-precision（or repeatability）of ciprofloxacinum in the hemolymph of *Eriocheir sinensis*

项　目	标准盐酸环丙沙星溶液（μg·mL⁻¹）	
	1011.33	101.5
加入量（μL）	50	50
各次实测浓度（μg·mL⁻¹）	25.82/25.63/26.24/25.22/26.09	2.58/2.62/2.63/2.53/2.61
平均实测浓度（μg·mL⁻¹）	25.80±0.40	2.59±0.04
RSD（%）	1.55	1.54
平均 RSD（%）	1.55±0.01	

表5是3种不同浓度的标准盐酸环丙沙星的日间精密度测定结果。由表5可知，RSD 最低为 0.21%，最高为 2.04%，平均为 1.19±0.92%。

表5　标准盐酸环丙沙星溶液测定的日间精密度

Tab. 5　The inter-day precision of standard ciproflxacinum

标准盐酸环丙沙星溶液浓度（μg·mL⁻¹）	峰面积		RSD（%）	平均 RSD（%）
	日间各次实测值	平均数±标准差		
2.030	2 214/2 151	2 182.5±44.55	2.04	
4.060	4 306/4 387	4 346.5±57.28	1.32	1.19±0.92
10.15	11 224/11 191	11 207.5±23.33	0.21	

2.6　注射给药后中华绒螯蟹雄蟹血淋巴内盐酸环丙沙星的代谢规律

应用上述所建立的 RP-HPLC 法，对肌肉注射盐酸环丙沙星的中华绒螯蟹雄蟹的血药浓度进行两次平行测定，药-时曲线如图3。结果表明，2 h 内，血药浓度迅速下降，由开始时的（37.13±0.18）μg·mL⁻¹ 降至 0.167 h（10 min）的（20.27±0.59）μg·mL⁻¹，1 h 的（15.16±0.78）μg·mL⁻¹ 以及 2 h 的（11.66±0.09）μg·mL⁻¹；2~24 h，血药浓度缓慢下降，维持在 10 μg·mL⁻¹ 左右；48 h 后，血药浓度降到 5.96±0.18 μg·mL⁻¹。

3　讨论

中华绒螯蟹属甲壳动物，循环系统为开放式，主要成分为血淋巴，它和鱼类有较大的区别。关于中华绒螯蟹血淋巴内药物检测方法的研究，至今尚无报道。动物体内药物浓度的检测方法有酶免疫法、薄层色谱、液相色谱、质谱和光谱等，而以 RP-HPLC 法较为常用，因为它具有灵敏度高、回收率高、精密度高及重现性好等特点，该法也常用于水产动物体内

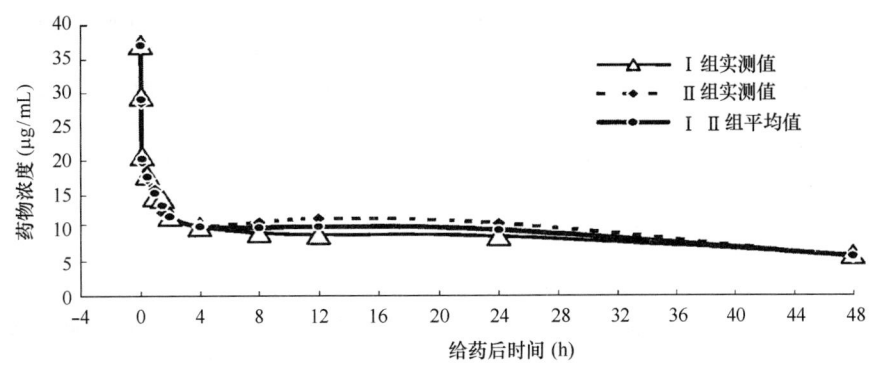

图3　肌肉注射盐酸环丙沙星后的药-时曲线

Fig. 3　The Concentration-Time curve of ciprofloxacinum after intramuscular administration

药物浓度的检测上[6,8-17]。对于甲壳动物中华绒螯蟹，是否亦能用该法检测其血淋巴内的药物含量，使用该法的具体条件是什么，这是本文所要解决的关键问题。本文的结果对此作出了肯定的回答：

（1）中华绒螯蟹的血淋巴与鱼类的血浆有较大的区别，本试验成功地摸索出了从中华绒螯蟹抽取血淋巴的方法，并筛选了合适的抗凝剂以及确定了抗凝剂的用量，为制备标准的分析样品奠定了基础。试验证明，ACD的抗凝作用最好，而且当其与血淋巴等量混合后，可取得较持久的抗凝效果；此种方法，完全可满足样品分析的要求。

（2）在试验所摸索的条件下，血淋巴内盐酸环丙沙星的色谱行为不仅基线走动平稳，保持时间较长，且无干扰峰，这就为药峰峰面积的测定提供了可能。在标准品盐酸环丙沙星浓度范围为 $1 \sim 10~\mu g \cdot mL^{-1}$ 时，标准曲线的相关系数 r = 0.9985，具有极好的相关性，由此表明了该方法为试验样品的准确定量提供了保障。

（3）回收率提示着检测方法的准确性。用HPLC法，陈会波等测定鳗康素的回收率为95.32%[19]，Uno等[20]测定虹鳟和五条鰤血浆中的磺胺间甲氧嘧啶（SMM）及其代谢物（AC-SMM）的回收率分别为96.6%和93.9%、98.0%和95.0%；用RP-HPLC法，陈文银和管国华[16]测定诺氟沙星在中华鳖血浆中的平均回收率为98.9%；而我们对雌、雄蟹空白血淋巴中5组不同加入量的标准盐酸环丙沙星检测的回收率范围分别为88.71%~105.76%和97.37%~105.45%，平均为（99.11±8.20）%和（100.08±3.43）%，表明该法具有较高的准确性。

（4）精密度是测定方法可靠性的保证。张祚新等[18]测定鲤鱼体内诺氟沙星的日内和日间精密度分别为4.72%和4.98%；Park等测定对虾体内的SDM和OMP（Ormemthoprim）在血淋巴中的含量时的精密度范围为 $0.05 \sim 50~\mu g \cdot mL^{-1}$[21]。而本试验的结果表明，日内精密度（或提取精密度）无论是在较高浓度还是在较低浓度样品均在1.54%~1.55%间，日间精密度为（1.19±0.92）%。由此可以看出，本测定方法具有较高的可靠性。

应用所建立的方法，我们检测了肌注给药后48 h内各时间点的中华绒螯蟹雄蟹血淋巴内的盐酸环丙沙星的浓度，结果基本表现出该药在其体内的代谢规律。由图3可以看出，给药后48 h内，血药浓度均在 $5~\mu g \cdot mL^{-1}$ 以上，预示着该药在中华绒螯蟹体内有很好的治疗疾病的前景。这说明，RP-HPLC测定法可完全用于中华绒螯蟹血淋巴内盐酸环丙沙星浓度的检测，由此可探讨中华绒螯蟹对环丙沙星的体内代谢状况，相关研究，我们将另有报道。

参考文献

［1］　竺心影. 药理学（第三版）［M］. 北京：人民卫生出版社. 1995. 307-311.

［2］　杨坚，渔药手册［M］. 北京：中国科学技术出版社. 1998. 179-180.

［3］　李太武，丁明进，宋协民，等. 皱纹盘鲍脓胞病病原菌——河流弧菌-Ⅱ的抗药机制的初步研究［J］. 海洋与湖沼，1996，27（6）：637-645.

［4］　林天然，余乃凤. 甲鱼急性肠胃炎鳃腺炎并发症的综合治疗［J］. 内陆水产，1997，7：25.

［5］　孟庆显，余开康. 鱼虾蟹贝疾病诊治和防治［M］. 北京：中国农业出版社. 1996. 255-281.

［6］　Nouws J. F. M., Crondel J. L., Schutte A. R., et al. Pharmacokinetics of ciprofloxacin in carp, African catfish and rainbow trout［J］. Vet Quart, 1988, 10：179-180.

［7］　Snieszko S. F., Friddle S. O., Gniffin. Successful treatment of ulcer disease in book trout（Saluelinus fontinalis）with terramycin［J］. Science, 1951, 113：179-182.

［8］　Bruno D. W.. An investigation into oxytetracycline residues in Atlantic salmon, Salmo. salar L［J］. J Fish Dis, 1989, 12：77-86.

［9］　Harry B., Gorran B.. Temprature-related absorption and excretion of oxytetracycline in rainbow trout（Salmo gairdneri L）［J］. Aquac, 1990, 84（3/4）：363-372.

［10］　李美同，郭文林，仲锋，等. 土霉素在鳗鲡组织中残留的消除规律［J］. 水产学报，1997，21（1）：39-42.

［11］ Sohlberg S., Czerwinska K., Rasmussen K., et al. Plasma concentrations of flumequine after intraarterial and oral administration torainbow trout (Salmo gairdneri) exposed to low water temperatures ［J］. Aquac, 1990, 84 (3/4)：355-361.

［12］ Boon J. H., Nouws J. M. F., VanderHeiden M. H. T., et al. Disposition offlumequine in plasma of European eel (Anguilla anguilla) after a single intra-muscular injection ［J］. Aquac, 1991, 99 (3/4)：213-223.

［13］ Hustredt S. O., Salte R. . Distribution and elimination of oxolinic acid in rainbow trout (Oncorhynchs mykiss Walbaum) after a single rapid intravascular injection ［J］. Aquac, 1991, 192：297-303.

［14］ Kleinow K. M., Jarboe H. H., Shoemaker K. E., et al. Comparative phamocokintics and bio-availability of oxolinic acid in channel catfish (Ictalurus punctatus) and rainbow trout (Oncorhynchus mykiss) ［J］. Can J Fish Aquat Sci, 1994, 51：1205-1211.

［15］ 李兰生、王勇强. 对虾体内氯霉素含量测定方法的研究 ［J］. 青岛海洋大学学报, 1995, 25 (3)：400-406.

［16］ 陈文银，管国华. 甲鱼血液中诺氟沙星浓度的反相高效液相色谱测定法 ［J］. 上海水产大学学报, 1997, 6 (4)：301-303.

［17］ 蔡武城、李碧羽、李玉民. 生化实验技术教程 ［M］. 上海：复旦大学出版社.1983：47.

［18］ 张祚新、张雅斌、杨勇胜、等. 诺氟沙星在鲤鱼体内的药代动力学 ［J］. 中国兽医学报, 2000, 20 (1)：66-69.

［19］ 陈会波、翁燕珍、郭少忠、等. 鳗康素的药代动力学研究 ［A］. 鱼病学研究论文集（二）C］. 北京：海洋出版社, 1995, 143-147.

［20］ Uno K., Aoki T., Ueno R., et al. Pharmacokinetics and metabolism of sulphamonomethoxine in rainbow trout (Oncorhynchus mykiss) andyellowtail (Seriola quinqueradiara) following bolusintra-vascular adminstration ［J］. Aquac, 1997, 153 (1-2)：1-8.

［21］ Park E. D., Lightner D. V., Milner N., et al. Exploratory bioavailability and pharmoacokinetic studies ofsulphadimethoxine and ormetoprim in the penaeid shrimp, Penaeus annamei ［J］. Aquac, 1995, 130：113-128.

该文发表于《水产学报》【2004，(B12)：7-12】

对虾肌肉、肝胰脏中诺氟沙星含量反相高效液相色谱法测定

房文红，于慧娟

（中国水产科学研究院东海水产研究所 农业部海洋与河口渔业重点实验室，上海 200090）

摘要： 采用反相高效液相色谱法建立了对虾肌肉、肝胰脏中诺氟沙星测定方法。分析用仪器为 Agilent 1100，色谱柱为 C18 柱，分析对虾组织样品中诺氟沙星用流动相为甲醇：磷酸盐缓冲液（含 0.044 mol·L^{-1}磷酸二氢钠和 2.8 mol·L^{-1}浓磷酸）：0.5 mol·L^{-1}溴化四丁基铵=20：76.2：3.8（分析肌肉用）或 19：77.2：3.8（分析肝胰脏用）（v/v/v），流速为 0.8 mL·min^{-1}，柱温为 32℃，进样量为 20 μL，荧光检测器检测波长 Ex 280 nm，Em 418 nm。肌肉中添加诺氟沙星线性范围为 0~7 920 μg·kg^{-1}，肌肉样品中平均回收率为 (66.19±2.72)%，日内和日间精密度相对标准偏差分别为 3.55 和 9.86%。肝胰脏中添加诺氟沙星线性范围为 0~23.76 mg·kg^{-1}，肝胰脏样品 NFX 回收率分别为 71.09%、86.97%和 89.05%，日内和日间精密度相对标准偏差分别为 2.53 和 4.90%。本方法测定肌肉样品的最低检测限为 10 μg·kg^{-1}，肝胰脏样品的最低检测限为 20 μg·kg^{-1}。该方法适合于对虾肌肉、肝胰脏中诺氟沙星含量分析。

关键词： 诺氟沙星；RP-HPLC 法；对虾；肌肉；肝胰脏

作者曾建立了对虾血淋巴中诺氟沙星测定方法[1]，而建立对虾肌肉和肝胰脏中诺氟沙星（norfloxacin，简称 NFX）含量的测定方法同等重要，有助于对虾药代动力学研究和药物残留检测。诺氟沙星的测定方法有非水滴定法、比色法、分光光度法，而生物样品中诺氟沙星的测定多采用高效液相色谱仪（HPLC），包括人血浆中诺氟沙星、畜禽等动物体内的血药浓度和组织中诺氟沙星含量的测定[2-4]；在水产动物中收集到采用 HPLC 法测定中华鳖血液[5]和鲤血浆[6]中诺氟沙星含量，但未见水产动物肌肉和肝胰脏等组织中诺氟沙星测定方法的报道。本试验采用反相高效液相色谱法建立对虾肌肉和肝胰脏组织中诺氟沙星含量测定方法。

1 材料与方法

1.1 实验试剂

诺氟沙星原料药（含量98%）由上海三维制药厂提供，诺氟沙星标准品（含量99.5%，批号 H040798）由中国兽药监察所提供。乙腈、甲醇，HPLC 级，Tedia 公司生产。

1.2　高效液相色谱仪及色谱条件

高效液相色谱仪：Agilent 1100 系列，包括四元泵、自动进样器、柱温箱、紫外检测器、荧光检测器及 HP 化学工作站。色谱柱为 ZORBAX SB-C18，4.6 mm×250 mm。流动相为甲醇、磷酸盐缓冲液（含 0.044 mol·L^{-1}磷酸二氢钠和 2.8 mol·L^{-1}浓磷酸）和 0.5 mol·L^{-1}溴化四丁基铵。流速：0.8 mL·min^{-1}。柱温：32℃。进样量：20 μL。检测波长：紫外检测器，280 nm；荧光检测器，Ex 280 nm，Em 418 nm。

1.3　试验动物和给药方法

试验用凡纳滨对虾（*Litopenaeus vannamei*）取自上海市金山区漕泾对虾养殖场，经 7 d 暂养后用于试验，暂养盐度为 1.0 左右，水温为（25.0±1.5）℃。试验期间投喂适量螺蛳肉，并及时排出残饵和污物。

试验用南美白对虾为 18～19 g，给药剂量为 10 mg·kg^{-1}，根据对虾体重计算给药量 10 μL 左右；给药为一次性肌肉注射，给药部位从对虾腹部第二腹节腹面稍斜插入，轻轻推进。

1.4　生物样品采集

肌肉采样　对虾血淋巴样品采集后立即将头胸甲和腹部分开，去除甲壳和肠，装入塑料密封袋，于-70℃保存直至使用。

肝胰脏采样　用镊子剥开头胸甲，同样用镊子夹住肝胰脏轻轻取出，去除食管等，将其放置在 2 mL 塑料离心管，于-70℃保存直至使用。

1.5　样品前处理

肌肉样品处理　肌肉样品在密封塑料袋内解冻，用剪刀将其剪碎。准确称取剪碎肌肉 2.00 g，加入等倍体积的 1：20（V/V）高氯酸，于研钵中研磨 8 min 左右，达到均质。然后，取出该均质 2.00 g 于玻璃离心管，5 000 r·min^{-1}离心 5 min，倒出上清液于另一刻度离心管；向残渣离心管中加入 1.0 mL 高氯酸（1：20，V/V），旋涡震荡器震荡 2 min，5 000 r·min^{-1}离心 5 min，将上清液合并于刻度离心管；再向残渣中加入 1.0 mL 高氯酸（1：20，V/V），重复一次，同样将上清液合并于刻度离心管。最后，用 1：20 高氯酸（V/V）定容至 5 mL 刻度，旋涡震荡混匀后经 0.45 μm 微孔滤膜过滤后上 HPLC 进样。

肝胰脏样品处理　将肝胰脏样品置于 15 mL 玻璃匀浆器中，其轴由马达带动，10 000 r·min^{-1}匀浆 2 min。称取肝胰脏匀浆约 0.40 g（精确至 0.01 g）于 10 mL 离心管中，加入 4 倍体积的高氯酸（1：20，V/V），在旋涡震荡器震荡 5 min，然后在 5 000 r·min^{-1}离心 5 min，取上清液经 0.45 μm 微孔滤膜过滤后上 HPLC 进样。

1.6　生物样品添加诺氟沙星标准曲线

肌肉中添加 NFX 标准曲线　称取 2.00 g 对虾空白肌肉 8 份，分别加入相应体积的 NFX 标准溶液，使之浓度为 0、0.025、0.05、0.20、0.60、1.00、4.00、8.00 μg·g^{-1}，然后按"1.5 样品前处理"处理此肌肉，HPLC 测定诺氟沙星浓度。

肝胰脏样品添加 NFX 标准曲线　称取 0.40 g 对虾空白肝胰脏 7 份，分别加入相应体积的 NFX 标准溶液，使之浓度为 0、0.1、0.4、2.0、8.0、12.0 和 24.0 μg·g^{-1}，然后按"1.5 样品前处理"处理此肝胰脏，HPLC 测定诺氟沙星浓度。

1.7　回收率测定

采用加样回收法。称取一定量的样品（空白肌肉 2.0 g，空白肝胰脏 0.40 g），加入相应体积的诺氟沙星标准母液，然后按"1.5 样品前处理"处理样品，HPLC 测定诺氟沙星浓度。每个浓度重复 3～5 次。根据诺氟沙星标准曲线计算出各肌肉或肝胰脏中诺氟沙星实测浓度。回收率=实测浓度/理论浓度×100。

1.8　精密度测定

日内精密度　在 1 日内的试验初、中、末，分别取一定量的样品（肌肉为 2.00 g，肝胰脏为 0.40 g）加入相应体积的诺氟沙星标准母液，分别按"1.5 样品前处理"处理肌肉或肝胰脏样品，HPLC 测定诺氟沙星浓度。每次重复 3 次，计算日内精密度 RSD。

日间精密度　在 1 周内重复测定 3 次，每次重复 3 次，计算日间精密度 RSD。

1.9　对虾肌肉和肝胰脏中诺氟沙星含量的测定

按所建立的方法，分别测定肌注给药后 1 min、4 min、6 min、12 min、30 min、6 h、24 h、48 h、96 h 和 144 h 时对虾

肌肉和肝胰脏中药物含量，每个采样时间点采样 4 个，以此检验所建立的方法。

2 结果

2.1 肌肉、肝胰脏中诺氟沙星色谱行为

本文分析对虾肌肉和肝胰脏中诺氟沙星的流动相组成为甲醇、磷酸盐缓冲液（含 0.044 mol·L^{-1}磷酸二氢钠和 2.8 mL·L^{-1}浓磷酸）和 0.5 mol·L^{-1}溴化四丁基铵，其比例经过数十次调整，测定肌肉中诺氟沙星时三者比例为 20：76.2：3.8（V/V/V）、肝胰脏时为 19：77.2：3.8（V/V/V）。测定肌肉中诺氟沙星时药物峰保留时间为 7.5 min、肝胰脏时为 8.7 min，基线走动平稳。肌肉、肝胰脏中虽然杂峰很多，但药物峰与杂峰仍分离良好，无干扰峰（见图 1、2）。

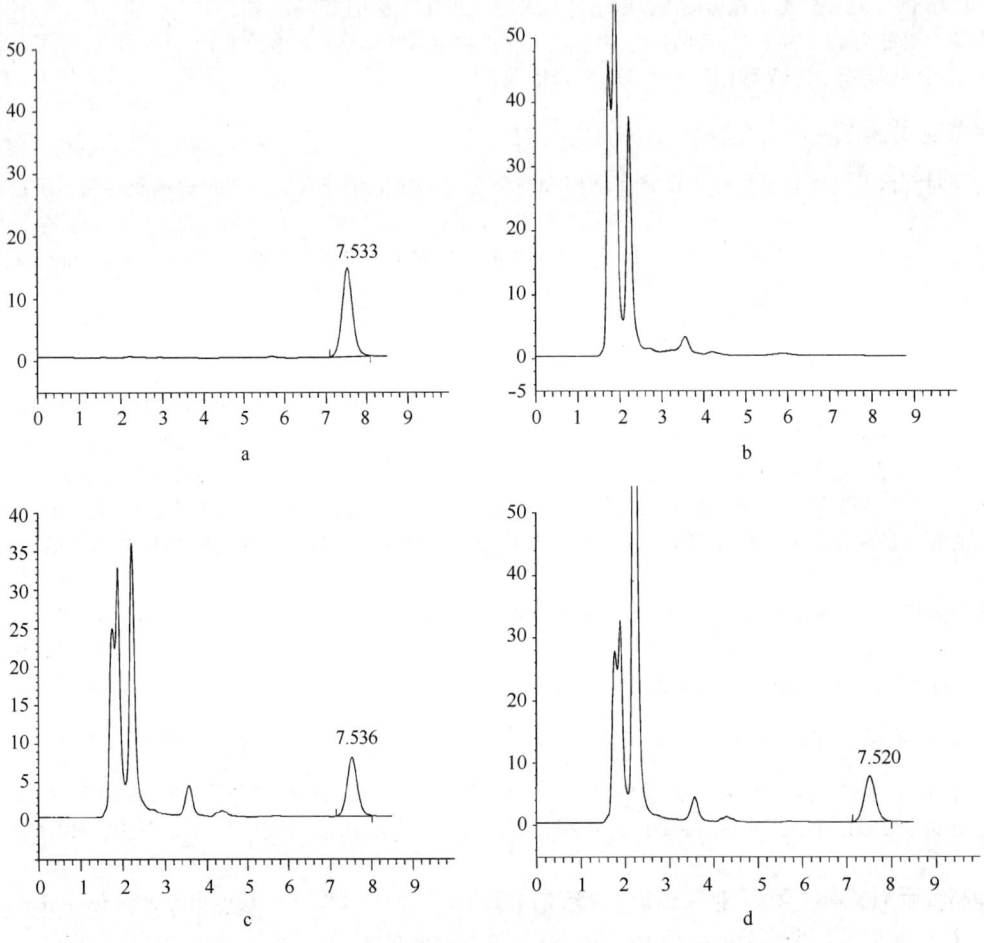

图 1　对虾肌肉中诺氟沙星色谱行为

Fig. 1　HPLC chromatograms of norfloxacin in muscle from shrimp

a. 诺氟沙星标准品色谱图；b. 对虾空白肌肉色谱图；c. 对虾空白肌肉加诺氟沙星标准品色谱图；d. 对虾肌注给药 2 h 后肌肉色谱图。图中所标数值为诺氟沙星的保留时间（min）。

a. HPLC chromatogram from the standard sample of norfloxacin；b. HPLC chromatogram from shrimp muscle free of norfloxacin；c. HPLC chromatogram from shrimp muscle supplied with norfloxacin；d. HPLC chromatogram from shrimp muscle at 2 h after intramuscle administration. The values represent the retention time（min）.

2.2 生物样品添加诺氟沙星的标准曲线

空白肌肉样品添加 NFX 标准曲线　以 NFX 浓度为横坐标（X），荧光检测器检测的峰面积（Y）为纵坐标，将此肌肉样品标准曲线分作两段回归，0～594 μg·kg^{-1}：$Y = 0.0302X + 0.2094$，$R^2 = 0.9993$；198～7920 μg·kg^{-1}：$Y = 0.0333X - 2.7376$，$R^2 = 0.9992$，两者间具有很好的相关性（$P < 0.01$）。

空白肝胰脏样品添加 NFX 标准曲线：同样以 NFX 浓度为横坐标（X），荧光检测器检测的峰面积（Y）为纵坐标，对

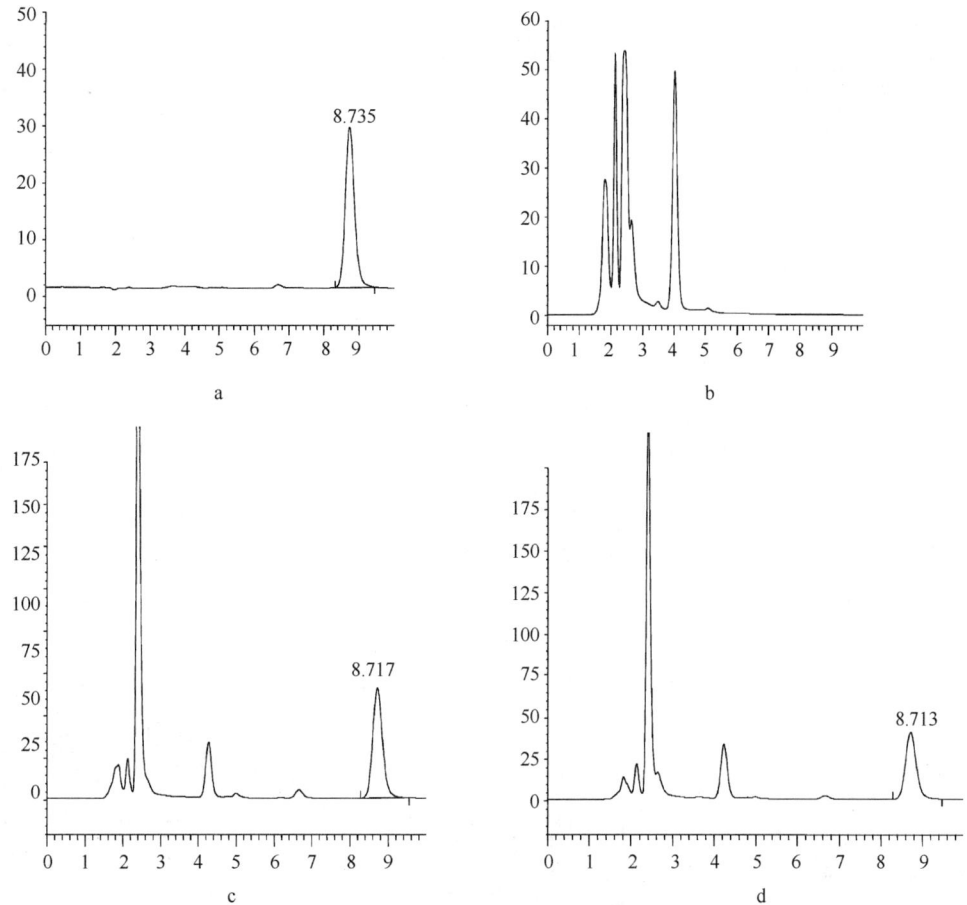

图 2　对虾肝胰脏中诺氟沙星色谱行为

Fig. 2　HPLC chromatograms of norfloxacin in hepatopancreas from shrimp

a. 诺氟沙星标准品色谱图；b. 对虾空白肝胰脏色谱图；c. 对虾空白肝胰脏添加诺氟沙星标准品色谱图；d. 对虾药饵给药后 6
h 肝胰脏色谱图。图中所标数值为诺氟沙星的保留时间（min）。

a. HPLC chromatogram from the standard sample of norfloxacin；b. HPLC chromatogram from shrimp hepatopancreas free of norfloxacin；
c. HPLC chromatogram from shrimp hepatopancreas supplied with norfloxacin；d. HPLC chromatogram from shrimp hepatopancreas at 6 h
after oral administration. The values represent the retention time（min）.

此肝胰脏样品标准曲线分作两段回归，$0 \sim 1.98$ mg·kg^{-1}：$Y = 35.728X + 0.0059$，$R^2 = 0.9995$；$1.98 \sim 23.76$ mg·kg^{-1}：$Y = 38.497X - 7.9231$，$R^2 = 0.9996$，两者间具有很好的相关性（$P < 0.01$）。

本方法测定肌肉样品的最低检测限为 10 μg·kg^{-1}，肝胰脏样品的最低检测限为 20 μg·kg^{-1}。

2.3　回收率

空白肌肉和肝胰脏中添加诺氟沙星回收率见表 1。肌肉添加 NFX 浓度为 0.0495、0.99 和 3.96 μg·g^{-1}时的回收率分别为 63.10、67.25 和 68.22%，平均回收率为（66.19±2.72）%。肝胰脏添加 NFX 浓度为 0.099、0.495 和 4.95μg·kg^{-1}时的回收率分别为 71.09%、86.97% 和 89.05%，平均回收率为（82.37±9.82）%。

表 1　肌肉和肝胰脏样品中诺氟沙星回归率

Tab. 1　The recoveries of RP-HPLC method for norfloxacin in muscles and hepatopancreas from shrimp

组别 group	理论浓度（mg·kg^{-1}） theoretic concentration	实测浓度（mg·kg^{-1}） determined concentration	回收率（%） recovery	平均回收率（%） mean recovery
肌肉 1	0.0495	0.03182/0.02814/0.03374	63.10±5.75	
肌肉 2	0.990	0.692/0.661/0.645	67.25±2.44	66.19±2.72

续表

组别 group	理论浓度（mg·kg⁻¹） theoretic concentration	实测浓度（mg·kg⁻¹） determined concentration	回收率（%） recovery	平均回收率（%） mean recovery
肌肉 3	3.960	2.662/2.718/2.725	68.22±0.87	
肝胰脏 1	0.099	0.0688/0.0722/0.0688/0.0713/0.0707	71.09±1.52	
肝胰脏 2	0.495	0.4236/0.4109/0.4411/0.4188/0.4581	86.97±3.83	82.37±9.82
肝胰脏 3	4.950	4.3595/4.3605/4.3243/4.5428/4.4541	89.05±1.05	

2.4 精密度

肌肉中添加诺氟沙星精密度试验 日内精密度：1 日内 3 次测定 9 个肌肉中添加诺氟沙星浓度，平均浓度为 2.639 mg·kg⁻¹，标准偏差为 0.094 mg·kg⁻¹，相对标准偏差为 3.55%。日间精密度：在 3 日内测定 9 个肌肉中添加诺氟沙星浓度，平均浓度为 2.352 mg·kg⁻¹，标准偏差为 0.232 mg·kg⁻¹，相对标准偏差为 9.86%。见表 2。

肝胰脏中添加诺氟沙星精密度试验 日内精密度：1 日内 3 次测定 9 个肝胰脏中添加诺氟沙星浓度，平均浓度为 4.494 mg·kg⁻¹，标准偏差为 0.114 mg·kg⁻¹，相对标准偏差为 2.53%。日间精密度：在 3 日内测定 9 个肝胰脏中添加诺氟沙星浓度，平均浓度为 4.747 mg·kg⁻¹，标准偏差为 0.232 mg·kg⁻¹，相对标准偏差为 4.90%。见表 3。

表 2 肌肉中添加诺氟沙星的日内精密度和日间精密度
Tab. 2 The intra-day and inter-day precision of RP-HPLC method for norfloxacin

精密度 precision	日内精密度 intra-day precision			日间精密度 inter-day precision		
	1	2	3	1	2	3
实测浓度（mg·kg⁻¹） determined concentration	2.662	2.709	2.595	2.702	2.308	2.120
	2.718	2.643	2.442	2.684	2.276	2.089
	2.720	2.700	2.557	2.531	2.282	2.179
平均值（mg·kg⁻¹）mean concentration	2.639			2.352		
标准偏差（μg·mL⁻¹）standard deviation	0.094			0.232		
相对标准偏差 relative standard deviation	3.55%			9.86%		

表 3 肝胰脏中添加诺氟沙星的日内精密度日间精密度
Tab. 3 The intra-day and inter-day precision of RP-HPLC method for norfloxacin

精密度 precision	日内精密度 intra-day precision			日间精密度 inter-day precision		
	1	2	3	1	2	3
实测浓度（mg·kg⁻¹） determined concentration	4.360	4.507	4.376	4.584	4.667	5.006
	4.324	4.598	4.523	4.448	4.738	5.030
	4.543	4.646	4.565	4.409	4.791	5.009
平均值（mg·kg⁻¹）mean concentration	4.494			4.747		
标准偏差（μg·mL⁻¹）standard deviation	0.114			0.232		
相对标准偏差 relative standard deviation	2.53%			4.90%		

2.5 对虾肌肉注射给药后肌肉和肝胰脏中诺氟沙星含量分析

应用上述所建立的反相高效液相色谱测定方法分析对虾肌注给药后肌肉和肝胰脏中诺氟沙星含量（见图 3），每个采样时间点为 4 个样品的平均值。肌注给药后 1 min，肌肉中诺氟沙星含量最高 ［（5.141±1.251）mg·kg⁻¹］，随后开始逐渐下降，144 h 时为（0.023±0.005）mg·kg⁻¹。而肝胰脏中诺氟沙星含量在给药后逐渐上升，12 min 时达到最高浓度 ［（16.557±3.900）mg·kg⁻¹］，随后开始逐渐下降，144 h 时为（0.031±0.008）mg·kg⁻¹。

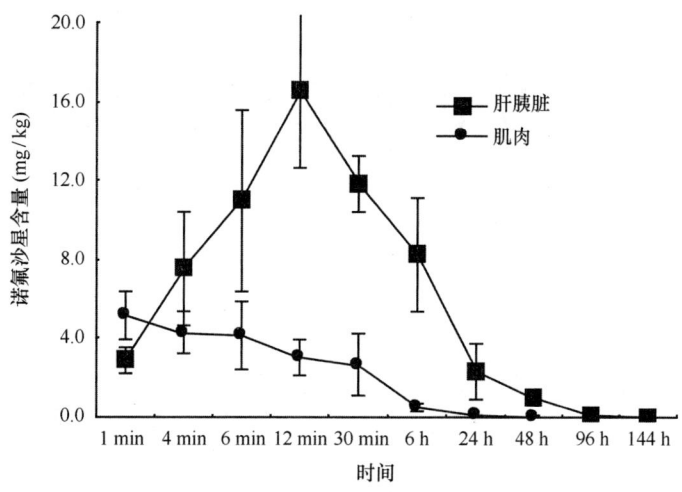

图3 肌注给药后对虾肌肉、肝胰脏中诺氟沙星含量

Fig. 3 Norfloxacin concetration of muscle and hepatopancrea from shrimp following intramuscular injection

3 讨论

3.1 关于流动相的选择

氟喹诺酮类药物的高效液相色谱分析，多数采用 RP-HPLC，且以甲醇（乙腈）-磷酸盐缓冲液（pH2.0~3.5）为流动相。作者在测定对虾血淋巴中诺氟沙星浓度所采用的流动相为乙腈：磷酸盐缓冲液（14.5：85.5，V/V）[1]，测定时血淋巴中药物峰与杂峰分离良好；在测定肌肉和肝胰脏中诺氟沙星时起初曾尝试采用测定血淋巴中诺氟沙星的流动相，但因肌肉和肝胰脏中干扰因子较多，调整流动相比例虽能改善分离效果，但达不到完全分离。后来采用甲醇-磷酸盐作为肌肉和肝胰脏分析用流动相。该流动相比乙腈-磷酸盐流动相更具优越性，甲醇比例下降，保留时间延长，峰形良好（BB 峰），有助于药物峰与杂峰的分离，而乙腈-磷酸盐流动相两者比例调整过大时，药物峰峰形质量下降，不成 BB 峰。

3.2 关于检测 NFX 的检测器

进行 HPLC 分析时，生物样品的基质、内源性物质及代谢产物相当复杂，对测定产生干扰不可避免。本试验分析肌肉、肝胰脏中诺氟沙星时，采用 1：20 高氯酸沉淀蛋白质及提取待测药物，曾同时采用紫外检测器和荧光检测器来比较检测结果，紫外检测器基线走动不如荧光检测器基线平稳，且干扰峰较多，尤其是肝胰脏样品更为复杂；而荧光检测器的选择性较好，干扰峰少，药物峰与杂峰分离良好，且在该流动相及色谱条件下，荧光检测器的灵敏度约高于紫外检测器 50%左右，所以诺氟沙星的检测采用荧光检测器更为适用，喹诺酮类其他药物也是如此。

3.3 关于回收率和精密度

分析方法的回收率试验和精密度试验应是包括样品处理在内的全过程操作[7]。生物样品含有大量的蛋白质，它们能结合药物，因此测定前须先将与蛋白质结合的药物解离出来。本文中沉淀肌肉和肝胰脏中蛋白质采用 1：20 高氯酸，既能沉淀蛋白质又能提取药物。分析生物样品时，回收率高反映了预处理过程中被测组分丢失少，但重要的是要重现性好，即使其回收率在 70%左右甚至更低，方法也仍然可用[7]。对虾肌肉中添加药物浓度在 0.0495~3.96 mg·kg⁻¹ 时，其回收率在 63.10%~68.22%之间。回收率虽偏低，但其日内和日间精密度分别为 3.55%和 9.86%，这说明具有较好的重现性。肝胰脏中添加诺氟沙星浓度在 0.099、0.495 和 4.95 mg·kg⁻¹ 时，回收率分别为 71.09%、86.97% 和 89.05%。在低浓度（0.099 mg·kg⁻¹）时回收率低于其他两个浓度的回收率，可能是因为肝胰脏中药物只提取 1 次，再加上浓度较低，有可能被离心管和肝胰脏吸附。但其日内精密度和日间精密度分别为 2.53%和 4.90%，同样说明了具有较好的重现性。

肌注给药后，在第一个采样点 1 min 时，肌肉中诺氟沙星含量最高，随后开始下降直到采样结束肌肉，而肝胰脏中诺氟沙星含量在给药后逐渐上升，12 min 时达到最高浓度，其后开始下降直到采样结束。该结果基本反映了肌注给药后诺氟沙星在肌肉和肝胰脏组织中吸收、分布和消除规律。诺氟沙星在家禽及猪所有可食组织中的允许残留为 50 μg·kg⁻¹[8]，若以此作为水产品的残留限量，本试验在给药 144 h 后肌肉和肝胰脏的检测结果均低于此值。再者，本方法的样品前处理无需对样品提取液进行浓缩，操作比较简单，适用于样品量较大时使用，可以节省时间。由此可见，在药代动力学研究和药残检测时，该方法完全可以应用于肌肉和肝胰脏中诺氟沙星含量分析。

参考文献

[1] 房文红, 邵锦华, 施兆鸿, 等. 斑节对虾血淋巴中诺氟沙星含量测定及药代动力学 [J]. 水生生物学报, 2003, 27 (1): 13-17.

[2] Nangia A., Lam F., Hung C T.. Reserved-phase ion-pair HPLC determination of fluroquinolines in human plasma [J]. J Pharma Sci, 1990, 79 (11): 988-991.

[3] 许丹科. 反相离子对色谱测定人血中氟哌酸 [J]. 药物分析杂志, 1990, 10 (5): 265-267.

[4] 邱银升, 朱式欧, 郝玉萍. 烟酸诺氟沙星在猪、鸡体内的组织残留研究 [J]. 中国兽药杂志, 1994, 28 (3): 10-12.

[5] 陈文银, 管国华. 甲鱼血液中诺氟沙星浓度的反相高效液相色谱测定法 [J]. 上海水产大学学报, 1997, 6 (4): 301-303.

[6] 张祚新, 张雅斌, 杨永胜, 等. 诺氟沙星在鲤鱼体内的药代动力学 [J]. 中国兽医学报, 2000, 20 (1): 66-69.

[7] 李发美. 医药高效液相色谱技术 [M]. 北京: 人民卫生出版社, 1999.

[8] Anadon A., Martinez-Larranaga M. R., Velez C., et al. Pharmacokinetics of norfloxacin and its N-desethyl-and oxo-metabolites in broiler chickens [J]. Am J Vet Res, 1992, 53 (11): 2084-2088.

该文发表于《分析科学学报》【2016, 32 (2): 183-187】

反相高效液相色谱法测定对虾组织中噁喹酸残留

Determination of Oxolinic Acid Residue in Shrimp, *Litopenaeus Vannamei* by Reversed-Phase High Performance Liquid Chromatogrphy

殷桂芳[1,2], 王元[1], 房文红[1], 沈锦玉[3], 陈进军[1,2]

(1. 中国水产科学研究院东海水产研究所农业部东海与远洋渔业资源开发利用重点实验室, 上海 200090;

2. 海海洋大学水产与生命学院, 上海 201306; 3 浙江省淡水水产研究所, 浙江 湖州 313001)

摘要: 在比较不同检测波长、流动相对噁喹酸色谱行为影响, 以及不同提取方法对噁喹酸回收率影响的基础上, 建立了测定对虾血淋巴、肌肉和肝胰腺等组织中噁喹酸含量的反相高效液相色谱法, 方法以乙腈和 0.01 mol·L^{-1} 四丁基溴化铵 (磷酸调节 pH 为 2.75) 为流动相, 色谱柱为 Aglient Zorbax SB-C$_{18}$ (4.6 mm×150mm, 5 μm), 柱温 40℃; 荧光检测器波长 Ex = 265 nm、Em = 380 nm; 进样量 10 μL, 流速 1.0 mL·min^{-1}。噁喹酸在 0.002~10 μg·mL^{-1} 范围内线性关系良好, 相关系数 R^2 = 0.9996~0.9999。采用乙腈提取对虾血淋巴、肌肉、肝胰腺和鳃组织中噁喹酸, 加标回收率为 84.70%~90.60%, 日内精密度为 1.60%~4.02%, 日间精密度为 2.91%~4.73%, 定量检测限分别为 0.02 μg·mL^{-1}、0.01μg·g^{-1}、0.02 μg·g^{-1} 和 0.02 μg·g^{-1}。该方法操作简单, 重现性好, 药峰无干扰, 适用于对虾生物样品中噁喹酸含量的分析。

关键词: 噁喹酸; 对虾组织; 残留; 检测; 反相高效液相色谱法

噁喹酸 (Oxolinic acid, 简写 OA) 是喹诺酮类的第二代广谱杀菌药, 通过干扰细菌 DNA 螺旋酶从而影响细菌染色体超螺旋而达到杀菌作用[1], 噁喹酸是我国国标渔药抗菌药物之一[2], 广泛用于防治鱼类系统性细菌性疾病[3], 尤其是海水养殖动物弧菌病[4,5]。不同国家和地区对水产品中噁喹酸最高残留限量的规定是不同的, 我国没有规定养殖虾中噁喹酸最高残留限量 (MRL), 仅规定了鱼 (肌肉和皮) 的 MRL 为 300 μg·kg^{-1}[6], 欧盟畜禽兽医残留限量标准也仅是规定了鱼 (肌肉和皮) 的 MRL 为 100 μg·kg^{-1}[7], 仅日本肯定列表中规定了虾产品中噁喹酸的 MRL 为 30 μg·kg^{-1}[8]。有关生物样品中噁喹酸的测定方法已有报道[9-14], 大多集中在鱼类。本文采用已报道的方法检测对虾组织中噁喹酸残留, 出现拖尾[9]、峰形不对称[9-10]、杂峰干扰[10-11]、峰宽增大[9,12]和回收率偏低[9,11,13-14]等现象, 均未能达到理想效果。我国水产养殖动物种类繁多, 不同水产动物采用同一种方法具有局限性。本研究优化了检测波长、流动相和生物样品提取方法, 建立了对虾血淋巴、肌肉、肝胰腺和鳃等组织中噁喹酸的反相高效液相色谱 (RP-HPLC) 测定方法, 为对虾药动学研究和药残检测奠定了可靠的技术基础。

1 实验部分

1.1 仪器

高效液相色谱仪 Waters 2695, 色谱柱为 Aglient Zorbax SB-C$_{18}$ (4.6 mm×150 mm, 5 μm); Waters 2475 荧光检测器;

Millipore Milli-Q Advantage 超纯水仪；Bertin precellys 240 均质器；Eppendorf 微量移液器；Hitachi CF16RⅡ高速冷冻离心机；亚荣 RE-52A 旋转蒸发仪；Rephile0.22 μm 微孔滤膜等。

1.2 试验材料和试剂

凡纳滨对虾（Litopenaeus vannamei）为东海水产研究所海南琼海研究中心室内池养殖，海水盐度 33，水温为 30℃±1℃，保持 24 h 充气，投喂不含噁喹酸的对虾全价饲料，无噁喹酸使用记录。挑选健康、体长 9.0 cm±0.5 cm 用于试验。

噁喹酸标准品（纯度≥98.0%）购于德国 Dr. Ehrenstorfer 公司，噁喹酸原粉（纯度≥98.0%）购于潍坊三江医药集团。甲醇、乙腈、四氢呋喃、乙酸乙酯、二氯甲烷和正己烷均为 HPLC 级，德国 Merck 公司产品；四丁基溴化铵、磷酸、甲酸和草酸均为分析纯，购自国药集团上海化学试剂公司。

噁喹酸母液：准确称取 0.010 0 g 噁喹酸标准品于 50 mL 烧杯中，加 0.5M NaOH 溶液约 200 μL 溶解，甲醇定容至 100 mL，即为 100 μg·mL^{-1} 的母液，4℃保存备用。

1.3 色谱条件

流动相为乙腈和 0.01 mol·L^{-1} 四丁基溴化铵（磷酸调节 pH 值为 2.75）；柱温 40℃，样品温度 25℃；进样量 10 μL，流速 1.0 mL·min^{-1}；荧光检测器波长 Ex=265 nm、Em=380 nm。

为使药峰和杂峰更好分开，不同组织样品的流动相比例有所不同，血浆和肌肉的流动相比例为乙腈：0.01M 四丁基溴化铵（pH 2.75）=25：75（V/V），肝胰腺和鳃的流动相比例为 18：82（V/V）。

1.4 样品前处理

1.4.1 血淋巴样品取出样品置室温融解后，1 000 r·min^{-1} 离心 5 min，取 0.2 mL 上清于 1.5 mL 离心管中，加入等体积乙腈，涡旋振荡 2 min 后，12 000 r·min^{-1} 离心 5 min，取上清液经微孔滤膜过滤，滤液用于 HPLC 分析。

1.4.2 肌肉、肝胰腺和鳃样品样品室温解冻后，取适量样品于均质器中均质；准确称取均质后样品（肌肉 1.00 g、肝胰腺和鳃 0.50 g）于 10 mL 离心管中，加入 1 mL PBS（0.1 mol·L^{-1}，pH 7.4），旋涡振荡器上振荡 2 min，加入 5 mL 乙腈，继续振荡 5 min，于 10 000 r·min^{-1} 转速离心 5 min，收集上清液；残渣再重复提取两次，合并三次上清液于 100 mL 具塞茄形蒸发瓶中，于 45℃旋蒸至干；加入 1 mL 流动相和 1 mL 正己烷，盖上塞子置超声波清洗仪超声 15 min，瓶底残留物完全溶解，将混合液体转移至 2 mL 离心管中，12 000 r·min^{-1} 离心 5 min，吸取下层液体，经 0.22 μm 微孔滤膜过滤，滤液用于 HPLC 分析。

2 结果与讨论

2.1 色谱条件的选择

2.1.1 检测波长的比较荧光检测器的灵敏度比紫外检测器的灵敏度高 10 倍以上[15]，噁喹酸具有光致发光特征可用荧光检测器检测。然而，选择合适的激发波长和发射波长，对检测的灵敏度和选择性都很重要，尤其是可以较大程度地提高检测灵敏度。表 1 为 3 组检测波长下的药峰峰宽、峰高和峰面积。3 组检测波长，Ex=265 nm、Em=380 nm 下的药峰峰高和峰面积最大，且药峰从基线开始到基线结束（BB 峰），在该波长下峰形和灵敏度都得到了提高。

表 1 不同检测波长下噁喹酸的峰宽、峰高和峰面积（n=3）

Table 1 Peak width, peak height and peak area of oxolinic acid under different detection wavelengths

Detection wavelength	Peak width（min）	Peak height（×10^{5}μv）	Peak area（×10^{6}μv·sec）
Ex=265 nm、Em=380 nm	0.96±0.11	1.228±0.029	2.143±0.064
Ex=312 nm、Em=369 nm	1.17±0.15	0.845±0.025	1.564±0.073
Ex=327 nm、Em=369 nm	1.03±0.05	1.076±0.023	1.991±0.063

2.1.2 流动相的比较

比较不同流动相下噁喹酸的色谱行为，5 种流动相分别为乙腈：四氢呋喃：0.02M 磷酸=16：15：69（V/V/V）[9]、乙腈：0.01M 草酸=35：65（V/V）[12]、乙腈：0.01M 草酸：甲醇=30：60：10（V/V/V）[11]、乙腈：0.1%甲酸溶液=60：40（V/V）[10]和乙腈：0.01M 四丁基溴化铵=25：75（V/V）[16]，对参考文献中的流动相比例做了适当调整。以 1 μg·g^{-1} 浓度添加噁喹酸的对虾肌肉样品在 5 种流动相下的色谱行为见图 1，其药峰峰宽、峰高和峰面积见表 2。综合分析，乙腈：四丁基溴化铵的流动相虽有轻微拖尾，但峰宽最小，峰形尖锐对称，且无杂峰干扰。

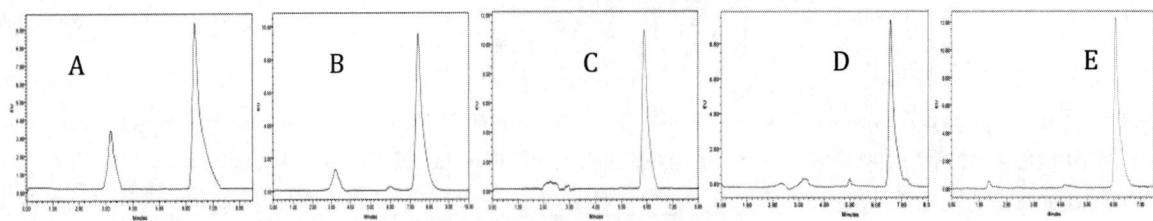

图1 不同流动相下噁喹酸色谱行为图

Fig. 1 Chromatographic behavior for oxolinic acid under different mobile phases

A：乙腈：四氢呋喃：0.02M 磷酸；B：乙腈：0.01M 草酸；C：乙腈：0.01M 草酸：甲醇；

D：乙腈：0.1%甲酸；E：乙腈：0.01M 四丁基溴化铵

表2 不同流动相下噁喹酸药峰峰宽、峰高和峰面积 （n=3）

Table 2 Peak width, peak height and peak area of oxolinic acid under different mobile phases

Mobile phases	Peak width （min）	Peak height （$\times 10^5 \mu v$）	Peak Area （$\times 10^6 \mu v \cdot sec$）
Acetonitrile：Tetrahydrofuran：0.02 M Phosphate	1.49±0.06	0.934±0.079	2.117±0.044
Acetonitrile：0.01M Oxalic acid	1.47±0.11	0.987±0.115	1.940±0.038
Acetonitrile：0.01M Oxalic acid：Methanol	1.11±0.06	1.045±0.045	1.907±0.083
Acetonitrile：0.1% Formic acid	1.34±0.13	0.891±0.034	1.892±0.061
Acetonitrile：0.01M Tetrabutyl ammonium bromide，pH2.75	0.98±0.06	1.143±0.029	1.926±0.054

通过以上比较分析，最终确定检测肌肉和血淋巴样品的流动相为乙腈和 0.01 mol·L^{-1}四丁基溴化铵（磷酸调节 pH 值为 2.75），两者比例为 25：75 （V/V），保留时间为 6.18 min（图2A，B）；因肝胰腺和鳃样品中内源性干扰，为使药峰和杂峰更好分开，将流动相比例调整为 18：82 （V/V），保留时间在 12.82 min（图2 C，D）。

2.2 提取液的选择

本试验采用加标法比较了不同提取液提取肌肉中噁喹酸的回收率（见表3），4 种提取液分别为乙腈[12]、酸化乙腈（1%醋酸乙腈）[10]、乙酸乙酯[9,11]和二氯甲烷[13-14]。从回收率来看，不同提取液对肌肉提取效果不同，乙腈提取效果最好，其次为酸化乙腈，再次为乙酸乙酯，二氯甲烷最差。二氯甲烷加入后，肌肉凝固变为较致密的一团难以分散，阻碍了噁喹酸的提取，从而导致回收率最低。

表3 不同提取液下肌肉中添加噁喹酸的回收率

Table 3 Recovery of standard addition of oxolinic acid in muscle under different extract solution

Extract solution	Added concentration （$\mu g \cdot g^{-1}$）	Measured concentration （$\mu g \cdot g^{-1}$）	Mean recovery （%）
Acetonitrile	1.00	0.902/0.856/0.848	86.87±2.91
Acidified acetonitrile	1.00	0.787/0.780/0.836	80.10±3.05
Ethyl acetate	1.00	0.676/0.683/0.718	69.23±2.25
Dichloromethane	1.00	0.612/0.598/0.656	62.19±3.03

2.3 标准曲线及检测限

应用优化的检测波长和流动相，建立噁喹酸标准曲线，在 0.002~10 μg·mL^{-1}范围内建立的标准曲线线性关系良好。检测血浆和肌肉组织样品采用的标准曲线为 $Y=242.51X-5.0293$ （$R^2=0.9996$），检测肝胰腺和鳃组织样品采用的标准曲线为 $Y=209.52X-4.0273$ （$R^2=0.9999$）。对虾血淋巴、肌肉、肝胰腺和鳃加标后噁喹酸定量限分别为 0.02 μg·mL^{-1}、0.01

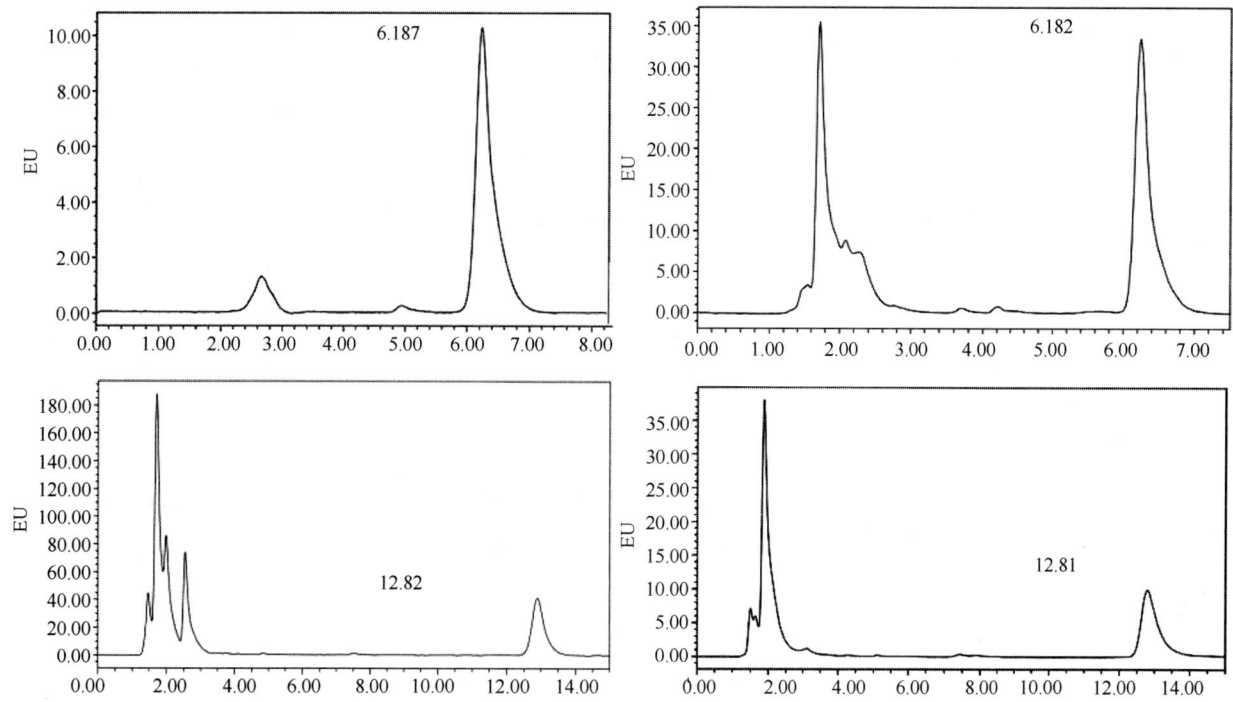

图 2　对虾组织样品中噁喹酸色谱行为图

Fig. 2　Chromatograms of oxolinic acid in shrimp tissues

A：药饵给药后 12 h 的血淋巴；B：药饵给药后 12 h 的肌肉；C：药饵给药后 1 h 的肝胰腺；D：药饵给药后 4 h 的鳃

A：Chromatogram of OA in shrimp hemolymph at 12 h after oral dosing；B：Chromatogram of OA in shrimp muscle at 12 h after oral dosing；

C：Chromatogram of OA in shrimp heptopancreas at 1 h after oral dosing；D：Chromatogram of OA in shrimp gill at 4 h after oral dosing.

$\mu g \cdot g^{-1}$、0.02 $\mu g \cdot g^{-1}$ 和 0.02 $\mu g \cdot g^{-1}$，加标后药峰峰形明显，无干扰峰影响。由此可见，该方法满足虾类产品中噁喹酸药残检测要求。

2.4　回收率及精密度

采用加标法测定对虾血淋巴、肌肉、肝胰腺和鳃中噁喹酸回收率，每个浓度 3 个平行，按"样品前处理"方法处理，采用 HPLC 测得其药物浓度，回收率=实测浓度/加标浓度×100% 计算。同时，采用加标法测定方法学的日内精密度和日间精密度，精密度由变异系数表示，变异系数 $C \cdot V$=标准偏差/平均值×100%。

加标法测得各组织回收率结果如表 4 所示，可以看出，各组织的噁喹酸回收率都较高，均在 80% 以上；通过精密度试验，得出对虾血淋巴、肌肉、肝胰腺和鳃中的日内精密度分别为：2.75%、1.60%、2.81% 和 4.02%，日间精密度分别为：2.91%、3.85%、4.73% 和 4.49%。

表 4　噁喹酸在对虾组织样品中的回收率（n=3）

Table 4　Recoveries of oxolinic acid fromshrimp tissues

Tissues	Added concentration（$\mu g \cdot mL^{-1}$，$\mu g \cdot g^{-1}$）	Determined concentration（$\mu g \cdot mL^{-1}$，$\mu g \cdot g^{-1}$）	Recovery（%）	Mean recovery（%）
Hemolymph	0.05	0.0432/0.0439/0.0449	88.02±1.68	90.60±2.49
	0.50	0.4569/0.4664/0.4679	92.75±1.19	
	5.00	4.5879/4.4793/4.5855	91.02±1.24	
Muscle	0.10	0.0860/0.0884/0.0893	87.94±1.71	86.67±1.39
	1.00	0.8971/0.8422/0.8688	86.87±2.75	
	10.00	8.8213/8.1926/8.5421	85.19±3.15	

续表

Tissues	Added concentration ($\mu g \cdot mL^{-1}$, $\mu g \cdot g^{-1}$)	Determined concentration ($\mu g \cdot mL^{-1}$, $\mu g \cdot g^{-1}$)	Recovery (%)	Mean recovery (%)
Hepatopancreas	0.10	0.0863/0.0811/0.0788	82.06±3.88	84.70±2.38
	1.00	0.8556/0.8775/0.8672	86.68±1.09	
	10.00	8.0641/8.8754/8.6720	85.37±4.22	
Gill	0.10	0.0802/0.0859/0.0877	84.60±3.89	86.84±3.49
	1.00	0.8650/0.8392/0.8521	85.21±1.29	
	10.00	9.0594/9.1104/9.0851	90.85±0.36	

3 结论

本文在比较检测波长、流动相和提取方法的基础上，建立了检测对虾组织样品中噁喹酸残留的高效液相色谱法，以乙腈为提取剂，方法操作简单，无杂峰干扰，重现性和稳定性高，方法定量检测限满足药残检测要求。

参考文献

［1］ Drlica K., Coughlin S.. Pharmacology and Therapeutics ［J］, 1989（44）：107-122.

［2］ Chinese veterinary Pharmacopoeia Committee（中国兽药典委员会）. Veterinary Pharmacopoeia of the People's Republic of China（中华人民共和国兽药典）. Veterinary drug use guid. Chemical volume. 2010 Edition（兽药使用指南. 化学药品卷. 2010 年版）. China Agriculture Press（中国农业出版社）. P357.

［3］ LI Jian（李健）, WANG Qun（王群）, SUN Xiu-tao（孙修涛）, MA Xiang-dong（马向东）, LIU De-yue（刘德月）, TONG Xiang（仝翔）. Journal of Fishery Science of China（中国水产科学）［J］, 2001,（3）：45-49.

［4］ Touraki M., Niopas I., Karagiannis V. Journal of fish diseases ［J］, 2012, 35（7）：513-522.

［5］ Samuelsen O. B., Bergh O. I.. Aquaculture ［J］, 2004, 235（1）：27-35.

［6］ The ministry of agriculture of the People's Republic of China（中华人民共和国农业部）. NO. 235 Bulletin of the ministry of agriculture of the People's Republic of China（中华人民共和国农业部公告第 235 号）［EB/OL］.（2002-12-24）［2012-01-30］. http：//www. moa. gov. cn/zwllm/tzgg/g..［EB/OL］.（2002-12-24）［2012-01-30］. http：//www. moa. gov. cn/zwllm/tzgg/g.

［7］ Union European. Commision Regulation（EU）No 37/2010 of 22December 2009 ［EB/OL］.（2009-12-22）［2012-01-30］. http：//ec. europa. eu/health/files/mrl/.

［8］ Electronic Code of Federal Regulations. Tolerances for residues of new animal drugs in food ［EB/OL］.（2011-07-28）［2011-09-30］. http：//ecfr. gpoaccess. gov/cgi/t/text/te.

［9］ GUO De-hua（郭德华）, WANG Min（王敏）, LI Fang（李芳）. Chemical Analysis and Meterage（化学分析计量）［J］, 2003, 12（1）：24-25.

［10］ LIAG Zeng-hui（梁增辉）, LIN Li-ming（林黎明）, LIU Jing-jing（刘靖靖）, JIANG Zhi-gang（江志刚）, GONG Qing-li（宫庆礼）. Marine Fisheries Research（海洋水产研究）［J］, 2006, 27（3）：86-92.

［11］ Karbiwnyk, C. M., Carr, L. E., Turnipseed, S. B., Andersen, W. C., Miller, K. E. *Analytica chimica acta* ［J］, 2007, 596（2）：257-263.

［12］ LIU Yan-ping（刘艳萍）, LENG Kai-liang（冷凯良）, WANG Qing-yin（王清印）, WANG Zhi-jie（王志杰）, SUN Wei-hong（孙伟红）, TAN Zhi-jun（谭志军）, GUO Meng-meng（郭萌萌）. Journal of Shanghai Ocean University（上海海洋大学学报）［J］, 2009,（3）：332-337

［13］ LI Yao-ping（李耀平）, LI Shou-song（李寿崧）, WU Wen-zhong（吴文忠）, HUANG Xiao-rong（黄晓蓉）, QIAN Jiang（钱疆）. Inspection and Quarantine Science（检验检疫科学）［J］, 2004, 14：37-39.

［14］ XU Bao-qing（许宝青）, LIN Qi-cun（林启存）, CAI Li-juan（蔡丽娟）, ZHANG Le（张乐）. Hangzhou Agricultural Science and Technology（杭州农业科技）［J］, 2006,（6）：15-16.

［15］ ZHANG Xiao-tong（张晓彤）, YUN Zi-hou（云自厚）. Performance liquid chromatographymethod（液相色谱检测方法）［M］. Beijing（北京）：Chemical Industry Press（化学工业出版社）, 2000：67-109.

［16］ Lin-tian（张林田）, HUANG Shao-yu（黄少玉）, CHEN Xiao-xue（陈小雪）, ZHANG Dong-hui（张冬辉）, CHEN Jian-wei（陈建伟）, SU Jian-hui（苏建晖）. Physical Testing and Chemical Analysis Part B（Chemical Analysis）（理化检验：化学分册）［J］, 2009：1086-1087.

该文发表于《分析科学学报》【2011，27（3）：375-378】

反相高效液相色谱法测定海水鱼组织中氟甲喹残留

ResDetermination of Flumequine in Marine Fish Tissues by Reversed Phase-High Performance Liquid Chromatography

胡琳琳，于慧娟，房文红，周帅

（中国水产科学研究院东海水产研究所，农业部海洋与河口渔业资源及生态重点开放实验室，上海 200090）

摘要： 采用反相高效液相色谱法（RP-HPLC）建立了美国红鱼和鲈鱼血浆、肌肉和肝脏组织中氟甲喹含量的测定方法。以乙腈：四氢呋喃：0.02 mol/L H_3PO_4 = 16：12：72（体积比）为流动相，色谱柱为 C18 柱，流速为 1.2 mL/min，紫外检测波长为 324 nm。氟甲喹在 0.05～10 µg/ml 范围内呈线性关系，相关系数为 0.999 9。采用二氯甲烷提取美国红鱼、鲈鱼肌肉和肝脏组织中氟甲喹，提取液的最佳蒸干温度为 40℃，蒸干后以流动相溶解残渣较优；血浆直接采用乙腈净化处理。结果表明，氟甲喹在美国红鱼和鲈鱼组织中加标平均回收率为分别 85.88%～92.78%和 84.05%～91.09%，精密度相对标准偏差分别为 3.85%～4.27%和 2.81%～3.97%。氟甲喹在血浆中的最低检测限为 0.05 µg/ml，肌肉和肝脏组织中最低检测限为 0.02 µg/g。该方法操作简单，重现性好，适合于海水鱼血浆、肌肉、肝脏等生物样品中氟甲喹含量分析。

关键词： 氟甲喹，美国红鱼，鲈鱼，血浆，肌肉，肝脏，RP-HPLC 法

氟甲喹（Flumequine）为第二代喹诺酮类合成抗菌药，其作用机理主要是通过抑制细菌 DNA 解旋酶 A 亚单位，干扰细菌脱氧核糖核酸、核糖核酸及蛋白质的合成，使细胞不能分裂而起到杀菌作用[1]。该药物对由大肠杆菌、沙门氏菌、假单胞菌、克雷伯氏杆菌、耶辛氏菌、鲑单胞菌以及鳗弧菌等引起的感染性疾病有效，广泛用于牛、猪、鸡和鱼等食品动物的细菌性疾病防治[2]。有关生物样品中氟甲喹的测定方法已有报道，仲锋等采用乙酸乙酯提取组织中氟甲喹，蒸馏后残渣采用 0.02 mol/L H_3PO_4 溶解，但该方法的变异系数较大[3]；Anadón 等采用乙酸乙酯提取鸡组织中氟甲喹，提取液蒸干时未提及具体蒸馏温度，流动相为有机溶剂与磷酸混合（乙腈：四氢呋喃：0.02 mol/L H_3PO_4）[4]；Samuelsen 和 Ervik 采用流动相梯度洗脱检测大西洋比目鱼组织中氟甲喹残留[5]，方法较为繁琐。本文采用上述方法均未能达到理想效果，于是在优化了海水鱼组织样品二氯甲烷提取液的蒸干温度和蒸干后残渣的溶解方法基础上，建立了海水鱼血浆、肝脏和肌肉组织中氟甲喹的高效液相色谱测定方法，为开展药动学研究和药残检测提供可靠的方法。

1 试剂与仪器

1.1 试验动物

试验用美国红鱼和鲈鱼采自浙江省舟山某网箱养殖公司，体重（180±10）g，无氟甲喹使用记录。暂养于水温为（25.0±1.5）℃、盐度为 25 左右的海水中。试验期间投喂适量饲料（不含氟甲喹），并及时排出残饵和污物，一周后用于试验。

1.2 药品与试剂

氟甲喹标准品（Flumequine），纯度高于 99.0%，购自 Sigma 公司；氟甲喹原料粉，纯度高于 98.0%，购自苏州恒益医药原料有限公司。二氯甲烷、正己烷、乙腈、四氢呋喃和甲醇均为 HPLC 级，为德国默克公司产品；磷酸、氢氧化钠、肝素钠等均为分析纯，购自国药集团上海化学试剂公司。

1.3 试验仪器

高效液相色谱（HPLC）仪为 Agilent 1100 系列，色谱柱为 Zorbax SB-C18（5µm，150×4.6 mm，I. D.）；超纯水仪，Millipore Advantage；电子分析天平，Mettler Toledoab 204，感量 0.000 1 g；高速冷冻离心机，Hitachi CF16RXII；旋涡混合器，IKA 18T。

2 实验方法

2.1 采样与样品前处理

2.1.1 采样

配制 10 mg/mL 氟甲喹注射液。腹腔注射给药剂量为 20 mg/kg 体重。于给药后 10 min、0.5 h、1 h、2 h、4 h、8 h、12 h、18 h 和 1 d 分别取血浆、肌肉和肝脏，分装后于 -40℃ 保存，待药物浓度分析。

2.1.2 血浆样品处理

血浆于室温下融解后，取 400 μL 血浆，加入等体积乙腈，旋涡快速混匀器上混合 2 min，于 10 000 rpm. 转速离心 10 min，吸取上清液经 0.45 μm 微孔滤膜过滤后，用于 HPLC 分析。

2.1.3 肌肉、肝脏样品处理

样品置室温解冻，用研钵研磨达到均质后精确称取样品 2.00 g，加入 2.0 mL 0.02 mol/L H_3PO_4，再加入 10.0 mL 二氯甲烷，振荡 5 min，于 5 000 rpm 下离心 5 min 后收集下清液。残渣中再加入 10 mL 二氯甲烷，振荡 5 min 后在 5 000 rpm 下离心 5 min，收集下清液；残渣再重复提取一次，合并三次所得下清液于试剂瓶中。水浴 40℃ 蒸干，加入 2 mL 的流动相，置于 20℃ 摇床中震荡至残留物完全溶解。加入 2 mL 正己烷脱脂，下清液经 0.45 μm 微孔滤膜过滤，用于 HPLC 分析。

2.2 色谱条件

流动相：乙腈：四氢呋喃：0.02 mol/L H_3PO_4 = 16：12：72（V/V/V）；流速：1.2 mL/min；检测波长：紫外检测波长 324 nm；进样量：10 μL。柱温：40℃。

3 结果与讨论

3.1 提取剂二氯甲烷蒸干温度对回收率的影响

取 100 μL 100 μg/mL 氟甲喹标准溶液于试剂瓶内，加入 20 mL 二氯甲烷，混匀后分别置于 30℃、35℃、40℃、45℃、50℃ 水浴中，每组 3 个平行，蒸干后用 1 mL 流动相溶解，10 000 rpm. 离心 10 min，上清液经 0.45 μm 微孔滤膜过滤后，用于 HPLC 分析。不同水浴蒸干温度对二氯甲烷提取剂中氟甲喹回收率的影响结果见表 1。

表 1 不同水浴温度对氟甲喹回收率的影响（n=3）

Tab. 1 Effect of recovery of flumequine in different water bath temperature（n=3）

蒸干温度（℃） Temperature	理论浓度（μg/mL） theoretic concentration	实测浓度（μg/mL） determined concentration	回收率（%） recovery
30	10	9.763±0.093	97.63±0.93
35	10	9.575±0.084	95.75±0.84
40	10	9.624±0.098	96.24±0.98
45	10	9.157±0.124	91.57±1.24
50	10	8.524±0.290	85.24±2.90

结果显示，30~40℃ 时，氟甲喹的回收率均达到 95% 以上，从 45℃ 开始，氟甲喹的回收率开始下降。随着蒸干温度的升高，二氯甲烷的蒸发速度随之加快，亦会带走部分药物，降低了氟甲喹的回收率。本试验中二氯甲烷提取液在水浴下也可达到蒸干目的，主要是因为其沸点较低（39.8℃），远低于乙酸乙酯（77.2℃）和乙腈（81.6℃），不需太高温度即可蒸馏。综合药物的回收率和蒸干速度，最终选取 40℃ 作为本试验的二氯甲烷提取液的蒸干温度，既能保证氟甲喹的高回收率，又能在最短的时间里提高试验效率。

3.2 不同溶解液对回收率的影响

生物样品提取液蒸干后，主要采用两种方法进行溶解，分别为 0.02 mol/L H_3PO_4 溶液[3,6] 和流动相（乙腈：四氢呋喃：0.02 mol/L H_3PO_4）[4]。本试验中分别采用流动相和 0.02 mol/L H_3PO_4 溶解蒸干后的残渣，肌肉组织中氟甲喹回收率见表 2。结果显示，以流动相作为溶解液，样品中氟甲喹的回收率达到 88.81%±0.55%，药物的提取效率得到较大提高，且重现性好。这可能与流动相中含一定比例的有机溶剂，对药物能起到一定助溶作用。所以，与单纯的 0.02 mol/L H_3PO_4 溶液相

比，选用流动相作为提取剂能提高氟甲喹的回收率。

表 2 不同溶解液对肌肉组织中氟甲喹回收率的影响 （$n=3$）
Tab. 2 Effect of recovery of flumequine in muscle by different dissolved solution （$n=3$）

溶解液 dissolved solution	理论浓度 theoretic concentration （μg/g）	实测浓度 determined concentration （μg/g）	回收率 recovery （%）	平均回收率 mean recovery （%）
0.02 mol/L H_3PO_4	10.0	7.565	75.65	74.11±1.80
	1.0	0.7213	72.13	
	0.5	0.2728	74.56	
流动相	10.0	8.847	88.47	88.81±0.55
	1.0	0.8945	89.45	
	0.5	0.4426	88.52	

3.3 氟甲喹标准品及样品色谱行为

喹诺酮类药物的高效液相色谱分析，多数采用甲醇（乙腈）-磷酸盐缓冲液（pH 2.0~3.5）为流动相，如 0.1%甲酸水：乙腈（40：60，V/V）[7]、4.5 mL/L 85%H_3PO_4水溶液（含 1.36 g/L KH_2PO_4）：乙腈：四氢呋喃（55：26.5：18.5，V/V/V）[8]、乙腈：四氢呋喃：0.02 mol/L H_3PO_4（16：12：72，V/V/V）[4]等。本试验在试用了三种流动相后，最终选用的乙腈：四氢呋喃：0.02 mol/L H_3PO_4=16：12：72（V/V/V）作为流动相，结果显示标准品和生物样品的基线平稳，氟甲喹药物峰分离良好，三种组织均不受杂质干扰，峰形对称，呈 BB 峰，氟甲喹标准品和样品的保留时间为 5.0 min 左右（图1），适合于大批量样品测试。

3.4 标准曲线及其检测限

本试验以氟甲喹药物浓度（X）为横坐标、以峰面积（Y）为纵坐标，建立了浓度范围为 0.05~10 μg/ml 内的药物标准曲线 $Y=13.299X-0.0851$，相关系数（R^2）为 0.9999，拟合结果显示标准曲线相关性较好。以三倍基线噪音的药物浓度为最低检测限，美国红鱼和鲈鱼血浆中氟甲喹最低检测限为 0.05 μg/mL，肌肉和肝脏组织中氟甲喹的最低检测限为 0.02 μg/g。本方法氟甲喹的最低检测限远低于我国农业部第 235 号公告中《动物性食品中兽药最高残留限量》中对氟甲喹在鱼体的最高残留限量 0.5 μg/mL 或 0.5 μg/g，符合药动学和残留检测的要求。

3.5 回收率和精密度

氟甲喹在美国红鱼和鲈鱼各组织中的回收率采用加标回收法，结果见 3。同样采用加标法进行日内精密度和日间精密度试验，结果见表 4。结果显示，氟甲喹在美国红鱼血浆、肌肉和肝脏组织中的平均回收率为分别（92.78±1.25）%、（88.27±0.60）%和（85.88±1.27）%，日内和日间精密度相对标准偏差分别为 3.85%、4.15%、3.90%和 4.16%、4.21%、4.27%。氟甲喹在鲈鱼血浆、肌肉和肝脏组织中的平均回收率分别（91.09±3.30）%、（87.05±1.13）%和（84.05±3.64）%，日内和日间精密度相对标准偏差分别为 3.28%、2.81%、3.64%和 3.82%、3.37%、3.97%。本方法在检测美国红鱼和鲈鱼等海水鱼不同组织中的氟甲喹时，具有很好的重现性和稳定性。

表 3 美国红鱼和鲈鱼各组织中氟甲喹回收率 （$n=3$）
Tab. 3 Recovery of flumequine in main tissues from the Sciaenops ocellatus and Lateolabras janopicus （$n=3$）

组织 tissue	添加浓度 added concentration （μg/mL, μg/g）	美国红鱼 S. ocellatus			鲈鱼 L. janopicus		
		实测浓度 determined concentration （μg/mL, μg/g）	回收率 Recovery （%）	平均回收率 mean recovery （%）	实测浓度 determined concentration （μg/mL, μg/g）	回收率 recovery （%）	平均回收率 mean recovery （%）
血浆 plasma	10.0	9.165	91.65	92.78±1.25	9.365	93.65	91.09±3.30
	1.0	0.9413	94.13		0.9226	92.26	
	0.5	0.4628	92.56		0.4368	87.36	

续表

组织 tissue	添加浓度 added concentration （μg/mL，μg/g）	美国红鱼 S. ocellatus			鲈鱼 L. janopicus		
		实测浓度 determined concentration （μg/mL，μg/g）	回收率 Recovery （%）	平均回收率 mean recovery （%）	实测浓度 determined concentration （μg/mL，μg/g）	回收率 recovery （%）	平均回收率 mean recovery （%）
肌肉 muscle	10.0	8.832	88.32	88.27±0.60	0.8834	88.34	87.05±1.13
	1.0	0.8765	87.65		0.8625	86.25	
	0.5	0.4442	88.84		0.4328	86.56	
肝脏 liver	10.0	8.472	84.72	85.88±1.27	8.824	88.24	84.05±3.64
	1.0	0.8723	87.23		0.8221	82.21	
	0.5	0.4285	85.70		0.4085	81.70	

表4　美国红鱼组织中氟甲喹 RP-HPLC 检测的日内、日间精密度（$n=3$）

Tab. 4　The intra-day and inter-day precision of RP-HPLC methods for flumequine（$n=3$）

组织 tissue	日内精密度 intra-day precision		日间精密度 inter-day precision	
	平均浓度±标准偏差 （AVE±SD）	相对标准偏差（RSD）	平均浓度±标准偏差 （AVE±SD）	相对标准偏差（RSD）
血浆 plasma	8.211±0.316	3.85%	8.189±0.340	4.16%
肌肉 muscle	8.268±0.343	4.15%	8.215±0.346	4.21%
肝脏 liver	8.153±0.318	3.90%	8.248±0.352	4.27%

表5　鲈鱼组织中氟甲喹 RP-HPLC 检测的日内、日间精密度（$n=3$）

Tab. 5　The intra-day and inter-day precision of RP-HPLC methods for flumequine（$n=3$）

组织 tissue	日内精密度 intra-day precision		日间精密度 inter-day precision	
	平均浓度±标准偏差 （AVE±SD）	相对标准偏差（RSD）	平均浓度±标准偏差 （AVE±SD）	相对标准偏差（RSD）
血浆 plasma	8.1770±0.269	3.28%	8.264±0.316	3.82%
肌肉 muscle	8.225±0.230	2.81%	8.157±0.275	3.37%
肝脏 liver	8.186±0.298	3.64%	8.181±0.325	3.97%

参考文献

［1］ 林志彬，金有豫主编. 医用药理学基础［M］. 北京：世界图书出版公司，1998：4-18.

［2］ 王祖兵. 国内研制开发的又一新型兽用抗菌药——氟甲喹［J］. 兽药市场指南，2005，9：14.

［3］ 仲峰，董琳琳，汪霞. 噁喹酸、氟甲喹在鱼组织中残留量的检测方法研究［J］. 中国兽药杂志，2002，36（11）：20-22.

［4］ Anadón A., Martínez M. A., Martínez M., et al. Oral bioavailability, tissue distribution and depletion of flumequine in the food producing animal, chicken for fattening［J］. Food and Chemical Toxicology, 2008, 46：662-670.

［5］ Samuelsen O. B., Ervik A.. Single dose pharmacokinetic study of flumequine after intravenous, intraperitoneal and oral administration to Atlantic halibut（Hippoglossus hippoglossus）held in seawater at 9℃［J］. Aquaculture, 1997（158）：215-27.

［6］ 农业部畜牧兽医局. 动物性食品中噁喹酸和氟甲喹残留检测方法（鱼）----高效液相色谱法［J］. 中国兽药杂志，2003，37（1）：16-17，8.

［7］ 梁增辉，林黎明，刘靖靖等. 噁喹酸、氟甲喹在鳗鱼体内的药代动力学研究［J］. 海洋水产研究，2006，27（3）：86-92.

［8］ Villa R., Cagnardi P., Acocella F., et al. Pharmacodynamics and pharmacokinetics of flumequine in pigs after single intravenous and intramuscular administration［J］. The Veterinary Journal, 2005, 170：101-107.

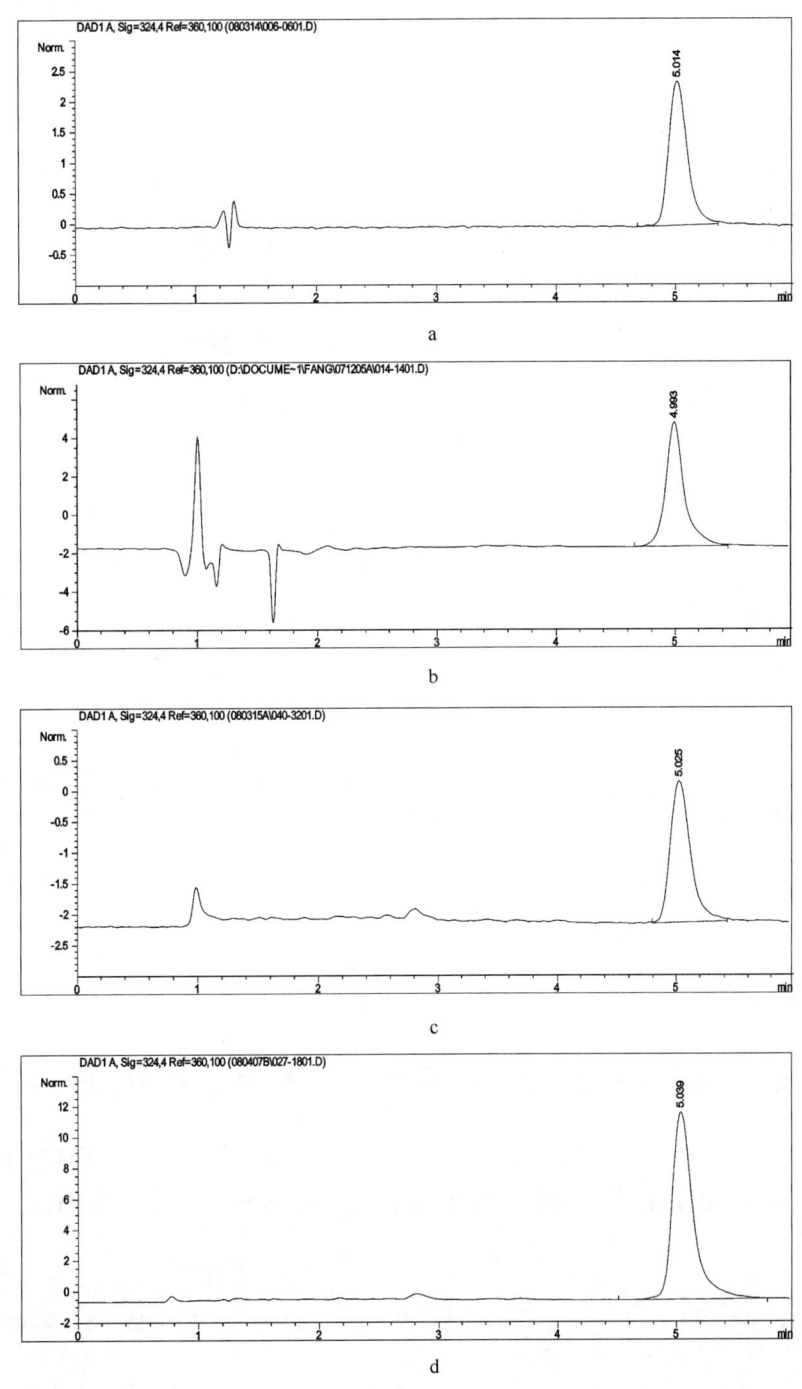

图 1　氟甲喹色谱行为图

a：标准溶液（1.98 μg/ml）色谱图；b：美国红鱼给药 1 h 后血浆样品的色谱分离谱图；c：美国红鱼给药 12 h 后肌肉组织样品的色谱分离谱图；d：美国红鱼给药 10 min 后肝脏组织样品的色谱分离谱图。

Fig. 1　The HPLC chromatogram of flumequine.

a：The HPLC chromatogram of standard solution of flumequine （1.98μg/ml）；b：The HPLC chromatogram of the plasma 1h after intraperitoneal injection of *Sciaenops ocellatus*；c：The HPLC chromatogram of the muscle 12h after intraperitoneal injection of Sciaenops ocellatus；d：The HPLC chromatogram of the liver 10min after intraperitoneal injection of *Sciaenops ocellatus*.

该文发表于《上海水产大学学报》【2006，15（2）：156-167】

RP-HPLC 法测定中华绒螯蟹主要组织中的恩诺沙星及其代谢产物

Development of RP-HPLC for determinationof enrofloxacin and itsMetabolitein tissues of Chinese mitten-handed crab (*Eriocheir sinensis*) by RP-HPLC

郑宗林，唐俊，喻文娟，胡鲲，叶金明，杨先乐

（上海高校水产养殖 E-研究院，上海水产大学农业部水产种质资源与养殖
生态重点开放实验室，农业部渔业动植物病原库，上海 200090；2. 扬州水产生产技术指导站，江苏 扬州 225003）

摘要：建立了反相高效液相色谱法（RP-HPLC）测定中华绒螯蟹肌肉、肝胰腺、精巢和卵巢组织内的恩诺沙星、环丙沙星含量。结果表明，恩诺沙星和环丙沙星在四种组织中的平均回收率分别为（70.5±12.61）%，（70.25±14.73）%；日内精密度分别为（3.27±0.84）%，（2.14±0.68）%；日间精密度分别为（5.38±2.46）%，（3.48±1.16）%。恩诺沙星、环丙沙星的最低检测限分别为 0.02 μg/g 和 0.004 μg/g。以 10 mg/kg 蟹体重剂量单次肌肉注射给药后，恩诺沙星在中华绒螯蟹肌肉、肝胰腺、精巢和卵巢内的 T_{max}，C_{max} 分别为：1 h，（4.323±0.56）μg/g；1 h，（6.042±0.72）μg/g；3 h，（2.381±0.43）μg/g；3 h，（2.101±0.29）μg/g，各组织的药物消除半衰期（$T_{1/2\beta}$）分别为：45.186 h，73.93 h，45.577 h，38.081 h。各组织中均能检测到环丙沙星，但含量均处较低水平，且代谢和消除起伏波动较大。本方法快速、灵敏、准确，适合于中华绒螯蟹组织中恩诺沙星、环丙沙星含量分析。

关键词：中华绒螯蟹；反相高效液相色谱法；恩诺沙星；环丙沙星；药代动力学

该文发表于《海洋渔业》【2008，30（4）：350-355】

采用 LC/M^{Sn} 法分析恩诺沙星在锯缘青蟹血浆中的代谢产物

Analysis on the metabolites of enrofloxacin in the hemolymph of mudcrab (*Scylla serrata*) bymeans of LC/MSn

房文红，胡琳琳，陈玉露，贾娴

（中国水产科学研究院东海水产研究所，农业部海洋与河口渔业重点开放实验室，上海 **200090**）

摘要：采用电喷雾离子阱质谱法在正离子检测方式下，对锯缘青蟹口灌恩诺沙星给药后血淋巴中恩诺沙星和代谢产物进行定性分析。通过对血浆中恩诺沙星及其代谢产物的质谱裂解行为进行碰撞诱导解离分析，获得了各化合物的多级质谱信息，通过比较各化合物结构和质谱裂解途径的异同点，发现有三个代谢产物，其中峰面积最大的为环丙沙星，其次为羟基化恩诺沙星，另一代谢产物可能是加氧恩诺沙星；推测恩诺沙星在锯缘青蟹体内的代谢反应主要是脱乙基反应、羟基化反应和氧化反应。通过定量分析，血浆中恩诺沙星和代谢产物环丙沙星的浓度分别为 19.1 μg/mL 和 0.526 μg/mL。

关键词：恩诺沙星；代谢产物；离子阱质谱；锯缘青蟹

该文发表于《药物分析杂志》【2009，29（7）：1142-1147】

高效液相色谱法测定中华绒螯蟹主要组织中双氟沙星及其代谢产物沙拉沙星方法

RP-HPLC determination of difloxacin and its metabolite in tissues of Chinese mitten-handed crab（*Eriocheir sinensis*）

李海迪，杨先乐，胡鲲，王翔凌

（上海海洋大学水产与生命科学院 农业部渔业动植物病原库，上海 201306）

摘要： 目的：建立高效液相色谱法（RP-HPLC）测定中华绒螯蟹（Eriocheirsinensis）肌肉、肝胰腺、精巢、卵巢组织中双氟沙星及其代谢产物沙拉沙星测定方法，并用于双氟沙星及其代谢产物在中华绒螯蟹体内的药代动力学研究。方法：色谱柱为 ZOR_ BAXSB_ C18（150 mm×4.6 mm，5 μm）；流动相：乙腈-0.030 mol·L^{-1}四丁基溴化铵（595），用磷酸调节至 pH 3.1；流速：1.5 mL·min^{-1}；紫外检测波长：276 nm；柱温：40℃。结果：双氟沙星在肌肉、肝胰腺、精巢和卵巢中回收率分别为（94.3±3.8）%、（78.9±4.5）%、（80.9±1.8）%、（75.5±2.3）%，沙拉沙星的回收率分别为（82.5±2.6）%、（92.8±4.7）%、（79.9±3.8）%、（80.6±4.4）%。日内精密度小于4.7%，日间精密度小于5.8%。双氟沙星的检测限为 0.01 μg·g^{-1}，沙拉沙星的检测限为 0.02 μg·g^{-1}。结论：本法灵敏度高，重复性好，能够快速可靠地测定双氟沙星及其代谢产物的浓度，适用于双氟沙星及其代谢产物沙拉沙星在中华绒螯蟹体内药代动力学研究。

关键词： 中华绒螯蟹；高效液相色谱法；双氟沙星；沙拉沙星

该文发表于《华中农业大学学报》【2007，26（5）：670-675】

绒螯蟹中 3 种氟喹诺酮类药物残留检测的前处理方法研究

Effects of Sample Pretreatment on Three Kinds of Quinolones Residues Determination in Fishery Products by High Performance Liquid Chromatography

胡鲲，杨先乐，唐俊，郑宗林，吕众

（1. 农业部渔业动植物病原库/上海水产大学水产品药物残留监控检测中心；

2. 上海市水产大学省部共建水产种质资源发掘与利用教育部重点实验室，上海 200090）

摘要： 研究了液相色谱法测定中华绒螯蟹组织中诺氟沙星、环丙沙星和恩诺沙星残留的前处理方法。结果表明：以提取前用无水硫酸钠作脱水剂、酸化乙腈作提取剂振荡提取、正己烷去脂、旋转蒸发并离心的前处理方法，灵敏、快速、准确、经济、实用性强。诺氟沙星、环丙沙星和恩诺沙星的最低检测限分别为 5.0 μg/kg、1.0 μg/kg、5.0 μg/kg；在加入标准品水平分别为 5.0~15.0 μg/kg、1.0~5.0 μg/kg 和 5.0~15.0 μg/kg 时，回收率分别为68.67%~82.89%、66.67%~74.83%和 72.67%~81.33%，相对标准偏差分别为 1.22%~7.32%、5.19%~8.18%和 2.52%~5.73%。该前处理方法适合检测诺氟沙星、环丙沙星和恩诺沙星残留。

关键词： 中华绒螯蟹；；诺氟沙星；盐酸环丙沙星；恩诺沙星；残留；高效液相色谱；前处理

该文发表于《中国兽药杂志》【2014，48（8）：55-60】

液质联用（HPLC-MS/MS）法同时测定水产品中的诺氟沙星、盐酸小檗碱、盐酸氯苯胍含量

Determination of Norfloxacin，Berberine hydrochloride and Robenidine hydrochlorideresidues in aquatic products by high performance liquid chromatography-mass spectrometry（HPLC-MS/MS）

汤菊芬[1,2]，蔡佳[1,2]，王蓓[1,2]，秦青英[1,2]，黄月雄[1,2]，廖建萌[3]，简纪常[1,2]

(1. 广东海洋大学水产学院，广东 湛江 524088；2. 广东省水产经济动物病原生物学及流行病学重点实验室，广东 湛江 524088；

3. 广东省湛江市质量计量监督检测所，广东 湛江 524022)

摘要：建立一种同时测定水产品中诺氟沙星、盐酸小檗碱和盐酸氯苯胍的残留量的 HPLC-MS/MS 法。样品经过 1%甲酸甲醇提取，正己烷脱脂，经冷冻、离心后，用高效液相色谱-串联质谱仪，选择反应监测（SRM）、正离子模式进行定性和定量分析。结果表明：诺氟沙星、盐酸小檗碱和盐酸氯苯胍在 1~100 μg/mL 范围内线性关系良好，线性相关系数≥0.999 6，检出限为 0.001 mg/kg。诺氟沙星、酸酸小檗碱、盐酸氯苯胍在 5、50 和 100 ng/g 三个添加水平下，三种水产品肌肉的平均加标回收率在 79.83%-104.06%之间，相对标准偏差在 2.97%-9.15%之间。本方法重现性好，灵敏度高，能够满足水产品中诺氟沙星、盐酸小檗碱和盐酸氯苯胍的残留检测和药代动力学的研究需要。

关键词：HPLC-MS/MS 法；诺氟沙星；盐酸小檗碱；盐酸氯苯胍；水产品

二、抗生素和磺胺类抗菌类药物残留的检测

由于难挥发及较强的极性等特点，HPLC 法也广泛地应用于氟苯尼考/氟苯尼考胺（刘永涛等，2008）、硫酸新霉素（刘永涛等，2011；宋洁等，2011）、磺胺甲噁唑/甲氧苄啶（刘长征等，2004）、盐酸小檗碱（汤菊芬等，2014）等抗生素、抗菌类药物在水产品中的残留检测中。

该文发表于《分析测试学报》【2008，27（3）：316-318】

水产品中氟苯尼考和氟苯尼考胺的高效液相色谱荧光检测法同时测定

Simultaneous Determination of Residues of Florfenicol and Florfenicol Amine in Fishery Products by HPLC with Fluorescence Detection

刘永涛[1,2]，艾晓辉[1]，李荣[1]，袁科平[1]，杨红[1]

(1. 中国水产科学研究院 长江水产研究所 农业部淡水鱼类种质监督检验测试中心，湖北 荆州 434000；

2. 华中农业大学 水产学院，武汉 430070)

摘要：建立了水产品中氟苯尼考和氟苯尼考胺同时检测的高效液相色谱荧光检测法，向样品中加入磷酸盐缓冲液，以丙酮为提取剂，加入二氯甲烷与丙酮形成共萃取剂与水相分层，有机相用氮气吹干，正己烷除脂肪。以乙腈-磷酸二氢钠溶液（0.01 mol/L，含 0.005 mol/L 十二烷基硫酸钠和 0.10%三乙胺）（2：3，V：V）为流动相，流速为 0.6 mL/min，荧光检测激发波长为 225 nm 发射波长为和 280 nm。本方法氟苯尼考在 10~10 000 μg/mL，氟苯尼考胺在 2~2 000 μg/mL 浓度范围内呈线性相关，相关系数分别为 r=1，r=0.999 9。当空白样品中氟本尼考添加水平为 20~200 μg/kg，氟苯尼考胺添加水平为 4~50 μg/kg 时，该方法的回收率为 79.26%~90.63%，相对标

准偏差为 3.66% ~ 6.21%，最低检测限分别为 5 μg/kg 和 1 μg/kg。结果表明：该方法具有灵敏度高，操作简便，准确，快速。适用于同时检测水产品中氟苯尼考和氟苯尼考胺的残留量。

关键词： 高效液相色谱；荧光检测法；氟苯尼考；氟苯尼考胺；水产品

氟苯尼考（Florfenicol, FF）又称氟甲砜霉素是甲砜霉素的单氟衍生物，属于氯霉素类广谱抗菌素，是由美国 Schering-Plough 公司首先研发的产品。其最大的优点是抗菌谱广、吸收良好、体内分布广泛。其抗菌活性也明显优于甲砜霉素和氯霉素而且氟苯尼考也无潜在致再生障碍性贫血作用。因此，氟苯尼考被广泛应用于治疗鱼类、牛和猪的细菌性疾病[1]。随着对氟苯尼考进一步研究发现氟苯尼考对动物有胚胎毒性，用药后还会出现厌食、腹泻等不良反应。氟苯尼考胺（Florfenicol Amine, FFA）是氟苯尼考的主要代谢产物，我国[2]和欧盟[3]对动物源性食品中的氟苯尼考的最高残留限量（MRL）做了规定，并以氟苯尼考及其代谢物氟苯尼考胺作为检出物。目前，报道的动物性食品中氟苯尼考和氟苯尼考胺同时检测的高效液相色谱法，除 Jeffery 等[4]采用 HPLC/MS 法外，其余均为 HPLC/UV 法[5,6]。国内外尚未见动物组织中氟苯尼考和氟苯尼考胺同时测定的高效液相色谱荧光检测法。本研究建立的高效液相色谱荧光检测法的检测限均优于国内外报道的 HPLC/UV 法，而且氟苯尼考胺的检测限与 Jeffery 等采用的 LC/MS 法相均为 1 μg/kg。本方法的建立为同时测定水产品中氟苯尼考和氟苯尼考胺的残留量提供了灵敏度高、简便、快速的分析手段。

1　材料与方法

1.1　仪器与设备

高效液相色谱仪（Waters 515 泵，717 自动进样器，2475 多通道荧光检测器及 Empower 色谱工作站）；自动高速冷冻离心机（日本 HITACHI 20PR-520 型）；Mettler-TOLEDO AE-240 型精密电子天平（梅特勒-托利多公司）；FS-1 高速匀浆机（华普达教学仪器有限公司）；调速混匀器（上海康华生化仪器制造厂）；HGC-12 氮吹仪（HENGAO T&D 公司）；恒温烘箱（上海浦东荣丰科学仪器有限公司）。

1.2　药品与试剂及配制方法

氟苯尼考标准品（纯度 99.0%，德国 Dr. Eherestorfer 公司）；氟甲砜霉素胺（纯度 ≥95%，sigma-aldrich 公司），乙腈（色谱纯，国药集团），二氯甲烷（分析纯，国药集团），磷酸氢二钠（$Na_2HPO_4 \cdot 12H_2O$），磷酸二氢钠（$NaH_2PO_4 \cdot 2H_2O$），正己烷（分析纯，国药集团），十二烷基硫酸钠（SDS）、丙酮（分析纯，武汉市江北化学试剂有限责任公司），三乙胺（分析纯，上海建北有机化工有限公司）。

1.2.1　标准溶液的配制

分别准确称取 0.01 g（精确至 0.0001 g）的氟苯尼考及氟苯尼考胺标准品，用乙腈分别溶解并定容至 100 mL，配制成 100 μg/mL 标准储备液，-20℃冰箱中冷藏保存。使用时分别用 5 mL 和 1 mL 移液管分别从氟苯尼考和氟苯尼考胺的标准储备液中移取 5 mL 和 1 mL 储备液，置于 50 mL 容量瓶中，用流动相定容至 50 mL，配制成含 10 μg/mL 和 2 μg/mL 氟苯尼考和氟苯尼考胺的混合标准溶液，然后用流动相依次稀释成含氟苯尼考浓度为 10000、5000、2500、1000、500、250、100、50、25、10 μg/mL，氟苯尼考胺浓度为 2000、1000、500、200、100、50、20、10、5、2 μg/mL 的混合标准溶液。

1.2.2　磷酸缓冲液的配制

磷酸缓冲液（pH 7.0）配制：准确称取磷酸氢二钠（$Na_2HPO_4 \cdot 12H_2O$）71.6 g 和磷酸二氢钠（$NaH_2PO_4 \cdot 2H_2O$）27.6 g，分别溶于 1 L 蒸馏水中配成 0.2 mol/L 的溶液，然后按 61:39 的比例混合即成。

1.2.3　定容溶液的配制

A 液：称取 0.895 g 磷酸氢二钠溶解在 250 mL 双蒸水中配制成 0.01 mol/L 的磷酸氢二钠溶液，用磷酸将溶液 pH 调至 2.8，经抽滤装置抽滤倒入具塞玻璃瓶中置4℃冰箱中备用；B 液：乙腈。定容溶液为 A 液：B 液（3:2，V:V）。

1.3　试验鱼

斑点叉尾鮰（Ictalurus punctatus）由中国水产科学研究院长江水产研究所窑湾试验场提供，体重（135±17.5）g。

1.4　实验方法

1.4.1　样品处理

1.4.1.2　组织样处理

试验鱼去皮沿背脊取肌肉，置高速匀浆机中匀浆 5 min，使肌肉组织呈糜状，放入小塑料袋中，-18℃冰箱中保存，备用。测定时，将匀浆冷冻保存的肌肉样品室温下自然解冻。准确称取 2.0 g，放于 10 mL 的具塞离心管中，加入 1 mL 的 pH 7.0 的磷酸盐缓冲液，置调速混匀器上涡旋振荡混匀 1 min，加入 4 mL 丙酮涡旋振荡混匀 1 min，以 5000 r/min 离心 5 min，将溶液转移到另一支 10 mL 离心管中，在向溶液中加入 3 mL 二氯甲烷，涡旋振荡混匀 1 min，5 000 r/min 离心 5 min，弃去上层水相，置 50℃氮吹仪上吹干，用 1 mL 定容溶液溶解残渣，涡旋振荡混匀 1 min，加入 3 mL 正己烷涡旋振荡混匀 1 min，5 000 r/min 离心 10 min，弃去正己烷，重复去脂一遍，水相经 0.22 μm 滤头过滤置自动进样瓶，HPLC 测定。

1.4.2 色谱条件

色谱柱：Waters symmetry C18 反相色谱柱（250 mm×4.6 mm，5 μm），流动相：乙腈–磷酸二氢钠溶液（0.01 mol/L，含 0.005 mol/LSDS 和 0.10%三乙胺）（2：3，v：v）用磷酸调 pH 至 4.5；流速：0.6 mL·min⁻¹；柱温：室温；激发波长为 225 nm，发射波长为 280 nm。

1.4.3 标准曲线的制备

将配制好的混合标准溶液，进行 HPLC 分析，进样量为 20 μL。分别以氟苯尼考和氟苯尼考胺的峰面积为横坐标（X），氟苯尼考和氟苯尼考胺的浓度为纵坐标（Y）绘制标准曲线，求出回归方程和相关系数。

2 结果

2.1 方法的线性范围和相关性

以上述氟苯尼考和氟苯尼考胺的混合标准溶液进样分析，得氟苯尼考和氟苯尼考胺的标准曲线。结果表明，氟苯尼考在 10~10 000 μg/mL，氟苯尼考胺在 2~2 000 μg/mL 浓度范围内成线性相关，线性方程和相关系数（见表 1），标准曲线（见图 1，图 2）。

表 1 FF、FFA 线性方程及限关系数（$n=5$）
Table 1 Linear equation and correlation coefficient of FF and FFA

分析物 Analyte	线性方程 Linear equation	相关系数 Correlation coefficient
FF	$Y=0.0009X+7.4874$	1
FFA	$Y=0.0002X+4.628$	0.9999

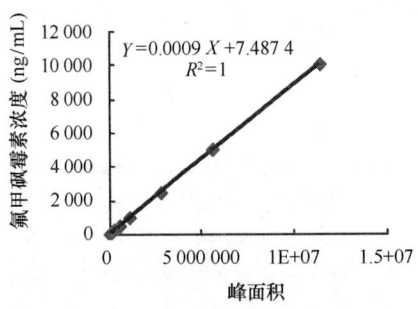

图 1 氟甲砜霉素标准曲线

Fig. 1 The standard curve of florfenicol

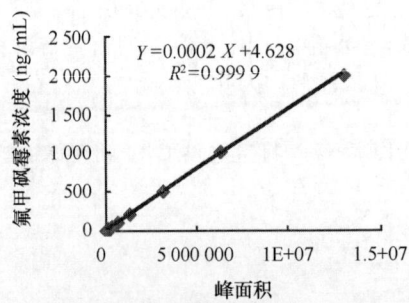

图 2 氟甲砜霉素胺标准曲线

Fig. 2 The standard curve of florfenicol amine

2.2　回收率测定

在斑点叉尾鮰空白肌肉组织样品中分别添加 4 个浓度水平的 FF 和 FFA 混合标准溶液：使样品含氟苯尼考和氟苯尼考胺分别为 20、50、100、200 μg/kg 和 4、10、20、40 μg/kg 每个浓度做 5 个平行，每个平行设一个空白对照，按样品前处理过程处理后测定回收率（见表 2），色谱图（见图 3~图 5）。

表 2　斑点叉尾鮰组织样品的 FF 和 FFA 回收率和相对标准偏差（n = 5）

Table 2　Recoveries and relative stand deviation of FF and FFA in Ictalurus punctatus spiked muscle samples

样品名称 Sample	分析物 Analyte	加标浓度 Spiked concentration（μg/kg）	平均回收率 Mean recovery（%）	相对标准偏差 RSD（%）
肌肉 Muscle	FF	20	79.26±0.03438	4.34
		50	88.64±0.04176	4.71
		100	89.91±0.03292	3.66
		200	89.99±0.04827	5.46
	FFA	4	81.02±0.05028	6.21
		10	90.63±0.04839	5.34
		20	86.18±0.04273	4.96
		40	89.98±0.05318	5.91

2.3　检测限

以 3 信噪比（S/N = 3）计算方法的最低检测限，鮰鱼肌肉中 FF 和 FFA 的最低检测限分别为 5 μg/kg 和 1 μg/kg。

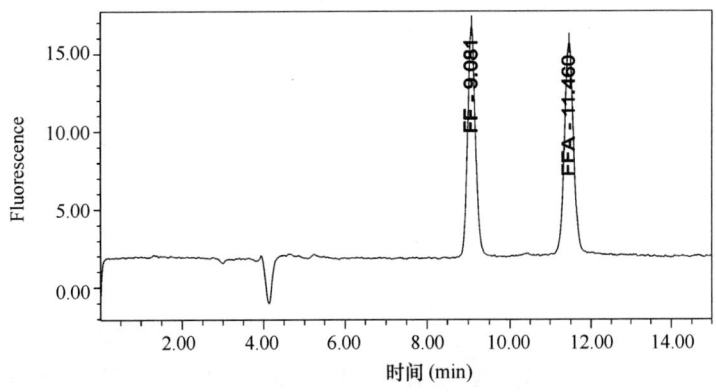

图 3　200 μg/mL FF 和 40 μg/mL FFA 混合标准溶液色谱图.

Fig. 3　Chromatography of 200 μg/mL FF and 40 μg/mL FFA mixed standard solution

2.4　色谱图

3　讨论

3.1　荧光检测波长的确定

本试验通过 Waters 2475 荧光检测器的扫描功能来确定最佳的激发波长和发射波长。通过扫描确定最佳激发波长和发射波长分别为 225 nm 和 300 nm，虽然氟苯尼考和氟苯尼考胺在此激发波长和发射波长下有强的荧光强度，但同时前面的杂质峰的荧光强度也会同时增大，本方法通过反复试验将激发波长和发射波长确定为 225 nm 和 280 nm，使氟苯尼考和氟苯尼考胺有较高的荧光强度，同时杂质的荧光强度也较小。本试验还发现在此激发和发射波长下同浓度的氟苯尼考胺的荧光强度是氟苯尼考荧光强度的 5 倍左右。

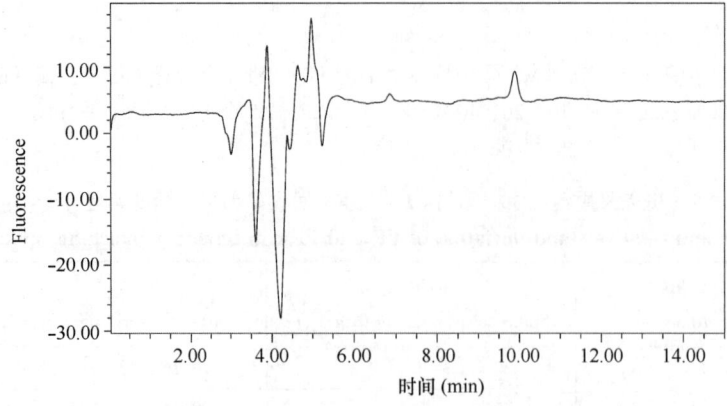

图 4　斑点叉尾鮰肌肉空白样品色谱图

Fig. 4　Chromatography of Ictalurus punctatus muscule blank sample

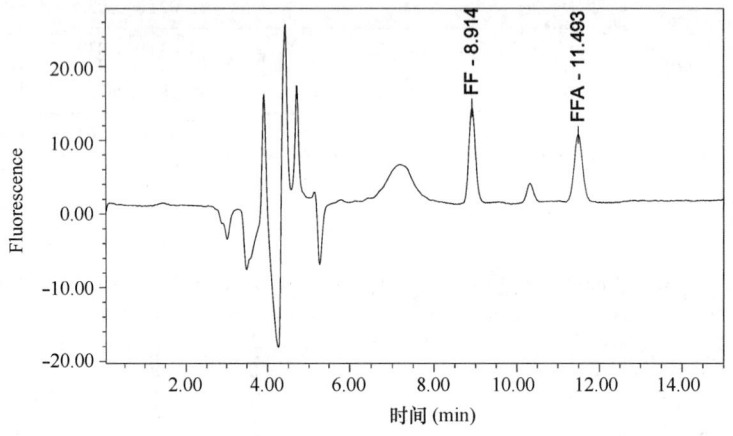

图 5　100 μg/kg FF 和 20 μg/kg 斑点叉尾鮰肌肉加标样品色谱图。

Fig. 5　Chromatography of 100 μg/kg FF and 20 μg/kg FFA spiked Ictalurus punctatus spiked muscule sample

3.2　最低检测限

在本试验条件下，斑点叉尾鮰肌肉组织中氟苯尼考和氟苯尼考胺的最低检测限分别为 5 μg/kg 和 1 μg/kg。国外 Hormazabal 等[5]用高效液相色谱紫外检测法同时测定鱼体肌肉中氟苯尼考和氟苯尼考胺的最低检测限均为 20 μg/kg，国内 郭霞等[6]报道用高效液相色谱紫外检测法测定鱼体肌肉中的氟苯尼考胺和氟苯尼考最低检测限分别为 5 μg/kg 和 10 μg/kg。 由此可以看用高效液相色谱荧光检测法灵敏度高更适合做为氟苯尼考和氟苯尼考胺的残留检测。

3.3　流动相对氟苯尼考和氟苯尼考胺保留时间和分析时间的影响

国外 Hormazabal 等[5]和国内郭霞等[6]报道的氟苯尼考和氟苯尼考胺同时检测的高效液相色谱紫外检测法所用流动相似。流动相的有机组分为甲醇，离子对试剂均为庚烷磺酸钠。本试验流动相中有机组分采用乙腈，离子对试剂采用十二烷基硫酸钠，采用乙腈使基线更加平稳而且缩短了氟苯尼考的分析时间，采用十二烷基硫酸钠作为离子对试剂同样可以起到增强氟苯尼考胺在反相色谱柱上保留的作用，而且十二烷基硫酸钠的价格要比庚烷磺酸钠便宜很多（注意：离子对试剂在 C18 柱上有较强的保留，对其他样品的测定有较大的影响，建议分析完样品后冲洗柱子或作为专用柱）。采用本试验条件下的流动相分析氟苯尼考和氟苯尼考胺出峰顺序和分析时间与相关报道不同。相关文献氟苯尼考胺出峰在氟苯尼考之前，本方法氟苯尼考胺出峰在氟苯尼考之后，而且同郭霞等的报道[6]缩短了分析时间。采用本方法的流动相可以节省分析时间提高分析效率。

4　结论

建立了高效液相色谱荧光检测法同时检测水产品中氟苯尼考和氟苯尼考胺的残留量。本方法首次采用高效液相色谱与荧光检测器联用同时检测水产品中氟苯尼考和氟苯尼考胺的残留量，而且该方法在色谱条件上较国内外相关文献有较大的改进，结果证明：该方法具有灵敏度高，操作简便，准确，分析时间短，效率高等优点。适用于同时检测水产品中氟苯尼考和氟苯尼考胺的残留量。

参考文献

[1]　张建民，钟超平．兽药广谱抗生素氟苯尼考的合成［J］．国外兽医药抗生素分册，1999，20（1）：13-16.

[2]　中华人民共和国农业部公告第 235 号《动物性食品中兽药最高残留限量》，2002.

[3]　EMEA/MRL/822/02-FINAL January 2002.

[4]　Jefery M. R., Ross A. P., Melissa C. F. et al. Simultaneous determination of residues of chloramphenicol, thiamphenicol, florfenicol, and florfenicol amine in farmed aquatic species by liquid chromatography/mass spectrometry［J］. journal of AOAC international, 2003, 86（3）：510-514.

[5]　Hormazabal V., Steffenak I., Yndestad M.. Simultaneous extraction and determination of florfenicol and the metabolite florfenicol amine in sediment by high-performance liquid chromatography［J］. Journal of Chromatography A, 1996, 724：364-366.

[6]　郭霞，张素霞，沈建忠，等．鱼肌肉中氟苯尼考和氟苯尼考胺残留的高效液相色谱检测法［J］．中国兽医科学，2006，36（09）：743-747.

该文发表于《淡水渔业》【2004，34（2）：23-24】

水产品中磺胺甲噁唑与甲氧苄啶含量的高效液相色谱测定法

刘长征[1,2]，艾晓辉[1,2]，伍刚[1,2]，袁科平[1,2]

（1. 农业部淡水鱼类种质资源与生物技术重点开放实验室，中国水产科学研究院长江水产研究所，湖北 荆州 434000；

2. 中国水产科学研究院淡水渔业研究中心，江苏 无锡 214081）

摘要： 本文研究了一种快速、方便的同时测定鱼类肌肉中磺胺甲噁唑与甲氧苄啶的 HPLC 法，用二氯甲烷和醋酸钠溶液提取样品中的磺胺甲噁唑及甲氧苄啶，浓缩提取液至干，用乙腈磷酸水溶液溶解残渣并去脂肪后过滤，清液供 HPLC 分析，紫外检测器在 287 nm 波长下检测定量。肌肉中杂峰能很好地与药物峰分离，空白样品添加药物，方法回收率稳定在 75%~85%，日内精密度与日间精密度均小于 5%，本方法可检出的样品中磺胺甲噁唑及甲氧苄啶低限为 0.008 μg/g。

关键词： 水产品，磺胺甲噁唑，甲氧苄啶，高效液相色谱

该文发表于《广东农业科学》【2011，（3）：142-147】

固相萃取-高效液相色谱串联质谱法测定水产品中硫酸新霉素残留量

Determination of Neomycinsulfate in AquaticProducts by High Performance Liquid Chromatography-Tandem Mass Spectrometry with Solid Phase Extraction

刘永涛[1,2]，李乐[3]，徐春娟[1,4]，胥宁[1,2]，董靖[1,2]，杨秋红[1]，杨移斌[1]，艾晓辉*[1,2]

（1. 中国水产科学研究院长江水产研究所，湖北 武汉 430223；2. 淡水水产健康养殖湖北省协同创新中心，湖北 武汉 430223；

3. 中国水产科学研究院，北京 100141；4. 上海海洋大学水产与生命学院，上海 201306）

摘要： 建立了一种分析水产品中硫酸新霉素残留量的固相萃取-高效液相色谱串联质谱法。样品经三氯乙酸水溶液提取，固相萃取小柱净化。用带加热电喷雾离子源的高效液相色谱串联质谱以选择反应离子监测（SRM）正离子模式检测，基质标准曲线定量。结果显示，在草鱼、斑点叉尾鮰、鳗鲡、南美白对虾和甲鱼空白肌肉中添加硫酸新霉素水平为 25~1 000 μg/kg 时，其加标回收率为 91.04%~114.57%，相对标准偏差为 1.91%~9.62%。硫酸新霉素的检测限为 10 μg/kg，定量限为 25 μg/kg。方法适用于水产品中硫酸新霉素残留量的测定，也适用于硫酸

新霉素在水产动物体内组织分布和消除研究。

关键词：水产品；硫酸新霉素；残留；高效液相色谱串联质谱

该文发表于《广东农业科学》【2011，3：142-147】

柱前衍生 HPLC 法检测罗非鱼肌肉中的硫酸新霉素

Determination of Neomycin sulfate in Tilapia muscle by high performance liquid chromatography with pre-column derivatization

宋洁[1,2]，王伟利[1]，姜兰[1]，王贤玉[1]

(1. 中国水产科学研究院珠江水产研究所，广东 广州 510380；2. 上海海洋大学水产与生命学院，上海 201306)

摘要：硫酸新霉素本身没有紫外吸收，必须通过柱前衍生来进行荧光检测，建立合适的衍生方法是提高检测灵敏度的关键。通过对硫酸新霉素衍生方法的优化，确定衍生化学试剂氯甲酸戊甲酯乙腈溶液的浓度为 8 mol/L，在 pH 值为 7.8 的硼酸缓冲液中衍生反应 15 min，10 h 内检测可得到理想的结果。对硫酸新霉素检测方法优化后，该药物在罗非鱼肌肉中检测限可达到 0.02 mg/kg，定量限为 0.05 mg/kg，在 0.05~5 mg/kg 浓度范围内有良好的线性关系（$r^2 = 0.9956$），在肌肉中分别按照 0.10、0.50、1.00 mg/kg 的标准添加硫酸新霉素，回收率分别为（76.43±1.46）%、（79.47±1.71）%、（86.82±1.13）%。

关键词：罗非鱼；硫酸新霉素；高效液相色谱法；氯甲酸芴甲酯；柱前衍生化

该文发表于《广东海洋大学学报》【2014，34（4）：62-66】

水产品中盐酸小檗碱的高效液相色谱-串联质谱（HPLC-MS/MS）法测定

Determination of Berberine Hydrochloride Residues in Aquatic Products by High Performance Liquid Chromatography-mass Spectrometry（HPLC-MS/MS）

汤菊芬[1,2]，王蓓[1,2]，蔡佳[1,2]，廖建萌[3]，秦青英[1,2]，黄月雄[1,2]，简纪常[1,2]

(1. 广东海洋大学水产学院；2. 广东省水产经济动物病原生物学及流行病学重点实验室，广东 湛江 524088；
3. 广东省湛江市质量计量监督检测所，广东 湛江 524022)

摘要：建立检测水产品中盐酸小檗碱残留量的高效液相色谱-串联质谱方法。样品经体积分数 1% 甲酸甲醇提取，正己烷脱脂，并经冷冻、离心后，用高效液相色谱-串联质谱仪分析。选择反应监测（SRM）正离子模式，定性离子对为 336/320 和 336/292，定量离子对为 336/320，盐酸小檗碱质量浓度在 1~100 ng/mL 范围内线性关系良好，线性相关系数为 0.999 7，检出限为 0.001 mg/kg。在添加浓度为 5~100 μg/kg 时，平均加标回收率为 81.29%~104.06%，相对标准偏差为 2.97%~7.87%。可满足水产品中盐酸小檗碱残留检测和药代动力学的研究需要。

关键词：盐酸小檗碱；残留；水产品；HPLC-MS/MS

三、其他药物残留的检测

甲苯咪唑及其代谢产物（郭东方等，2009）、螺旋霉素/替米考星/泰乐菌素/北里霉素（刘永涛等，2010）、三聚氰胺（刘永涛等，2009）、喹烯酮（杨莉等，2010）、喹乙醇（艾晓辉等，2003；HU Kun et al.，2003）等均能利用高效液相色谱法进行测定。

该文发表于《分析测试学报》【2009，28（4）：493-496】

HPLC/UV法对银鲫肌肉组织中甲苯咪唑及其代谢产物残留的同时测定

Simultaneous Determination of Mebendazole and Its Metabolites in Johnny Carp（*Carassius Auratus Gibelio*）Muscle by HPLC/UV Method

郭东方[1,2]，刘永涛[2]，艾晓辉[2]，袁科平[2]

（1. 华中农业大学水产学院，湖北 武汉 430070；2. 中国水产科学研究院长江水产研究所，湖北 荆州 434000）

摘要：建立了一种用高效液相色谱同时测定银鲫肌肉组织中甲苯咪唑（MBZ）及其代谢物氨基甲苯咪唑和羟基甲苯咪唑的方法。银鲫肌肉组织用乙酸乙酯提取，萃取物旋转蒸发至干后用1 mL二甲基甲酰胺-0105 mol/L磷酸盐缓冲液（体积比3：7）定容。色谱条件：WaterssymmetryC18反相色谱柱（250 mm×416 mm，5 μm）；流动相：乙腈-0105 mol/L磷酸二氢铵溶液（体积比33：67）；流速：018 mL/min；检测波长为298 nm；检测温度为室温。在10~120 μg/kg添加水平，MBZ、MBZ-NH2、MBZ-OH的回收率分别为81%~86%、71%~75%、86%~93%。MBZ的检出限为2 μg/kg，MBZ-OH和MBZ-NH2检出限均为3 μg/kg。

关键词：甲苯咪唑；羟基甲苯咪唑；氨基甲苯咪唑；银鲫；代谢物；HPLC/UV

甲苯咪唑是一种广谱杀虫剂，可用来治愈蛔虫、线虫、钩虫、鞭虫的感染[1]。在畜牧行业中该药物可用来预防绵羊肠胃和肺中的线虫和蠕虫[2-3]，在水产上被广泛用于治疗鳗鲡指环虫[4-5]，口服甲苯咪唑可治疗石斑鱼鳃的单殖吸虫[6]。甲苯咪唑在水产和畜牧行业中的应用越来越广泛，其大量运用可导致原药和代谢物在食物中的残留。为此欧盟定制了最高残留限量：甲苯咪唑（MBZ）、氨基甲苯咪唑（MBZ-NH2）和羟基甲苯咪唑（MBZ-OH）在动物肝脏、肌肉、肾脏和脂肪中的总含量分别为400、60、60和60 μg/kg[7]。国外文献中同时测定甲苯咪唑及其代谢物的方法很少，仅报道了鳗鲡组织[8]、羊的肝脏[9]和肌肉[10]以及牛肝脏中甲苯咪唑和羟基甲苯咪唑的检测[11]。目前国内尚未有文献报道甲苯咪唑及其代谢物同时检测的方法。本文以银鲫为研究对象，建立了一种能同时测定MBZ、MBZ-OH和MBZ-NH2的简单、有效、经济的检测方法。

1 实验部分

1.1 仪器与色谱条件

高效液相色谱仪（Waters515泵，717自动进样器，2487双通道紫外检测器，Empower色谱工作站）；自动高速冷冻离心机（日本Hitachi 20 PR-520型）；Mettler-TOLEDO AE-240型精密电子天平（梅特勒-托利多公司）；FS-1高速匀浆机（华普达教学仪器有限公司）；调速混匀器（上海康华生化仪器制造厂）；Sartorius PB-10型酸度计（德国赛多利斯公司）。

色谱条件：Waters symmetry C_{18}反相色谱柱（250 mm×4.6 mm，5 μm）；流动相：乙腈-磷酸二氢铵溶液（0.05 mol/L，0.1%三乙胺）（体积比33：67）；流速：0.8 mL/min；柱温：室温；紫外检测波长：298 nm；进样量：50 μL。

1.2 药品与试剂

MBZ标准品（纯度不小于99%，美国Sigma公司）；MBZ-OH，MBZ-NH2（纯度不小于99%，Witega Laboratorien，Berlin-Adlershof GmbH）；乙腈、乙酸乙酯、正己烷（均为色谱纯，Fisher Chem Alert公司）；三乙胺、磷酸二氢铵（均为分析纯，天津福晨化学试剂公司）；磷酸氢二钠、磷酸二氢钠（均为分析纯，天津天大化学试剂公司）。

1.3 磷酸盐缓冲液与定容液的配制

磷酸盐缓冲液：准确称取磷酸氢二钠（$Na_2HPO_4 \cdot 12 H_2O$）17.9g和磷酸二氢钠（$NaH2PO_4 \cdot 2 H_2O$）4.6g，分别溶于1 L蒸馏水中配成0.05 mol/L的溶液，然后按体积比9：1的比例混合，用氨水调节pH值至9.5。

定容液的配制：A液：称取1.438 g磷酸二氢铵溶解在250 mL双蒸水中配制成0.05 mol/L的磷酸二氢铵溶液；B液：二甲基甲酰胺。定容液为A液-B液（体积比7：3），用磷酸调溶液pH至2.0，经抽滤倒入具塞玻璃瓶，置4℃冰箱中备用。

1.4 实验方法

1.4.1 标准溶液的配制及标准曲线的制备 甲苯咪唑标准储备液的配制：称取甲苯咪唑 0.01 g 用二甲基亚砜溶解并定容到 100 mL，得到质量浓度 100 mg/L 的溶液。羟基甲苯咪唑和氨基甲苯咪唑标准储备溶液的配制：取两种药物各 0.01 g 分别用二甲基甲酰胺溶解定容到 100 mL，得到质量浓度皆为 100 mg/L 的溶液。测定时以储备液为基础，分别从 3 种储备液中取一定体积的标准溶液混合后稀释成含甲苯咪唑、羟基甲苯咪唑和氨基甲苯咪唑质量浓度各分别为 0.005、0.01、0.025、0.05、0.1、0.25、0.5、1.0、5.0、10.0 mg/L 的混合标准溶液。分别以甲苯咪唑、羟基甲苯咪唑和氨基甲苯咪唑的峰面积为纵坐标，质量浓度为横坐标绘制标准曲线，求出回归方程和相关系数。

1.4.2 样品处理 取 5 g 匀浆的银鲫肌肉样品于 50 mL 的塑料离心管中，加 1 mL 的磷酸盐缓冲液和 20 mL 乙酸乙酯，置旋涡混合器上振荡 5 min，5 000 r/min 离心 10 min，取出提取液，再加入 15 mL 乙酸乙酯，重复操作 1 次，合并提取液于鸡心瓶中，置 40℃ 水温旋转蒸发至干。剩余物用 1 mL 定容液溶解，再加入 1 mL 的正己烷，置旋涡混合器上振荡 1 min，混合溶液于 4 000 r/min 离心 10 min，除去上层正己烷液，下层液经 0.22 μm 滤头过滤，HPLC 测定。

2 结果与讨论

2.1 检测波长的选择

文献报道 MBZ 的吸收波长有 289[8]、298[11] 和 254[12] nm，本实验条件下得到 254 nm 处 MBZ 及其代谢物的峰面积最大，但干扰峰也最大。在 298 nm 和 289 nm 处干扰峰较小，且在 298 nm 处，MBZ 和 MBZ-NH2 的峰面积明显大于 289 nm 处的测得值，仅 MBZ-OH 的吸收峰面积略低于 289 nm 处的值。考虑要满足 3 种分析物的同时测定，故选择 298 nm 作为本实验的检测波长。

2.2 流动相的改进

本方法流动相与国外报道的相比有较大改进。Allan[12] 在流动相中加入了甲醇，作者通过实验发现采用该流动相会导致峰形变宽和拖尾现象；Steenbaar[13] 用乙腈和磷酸二氢铵溶液（体积比 3∶7）作流动相，发现 2 种代谢物的出峰时间不理想，同时有拖尾现象且峰形不对称。本实验通过反复调整，发现乙腈和磷酸二氢铵溶液的体积比为 33∶67 时效果令人满意，同时加入三乙胺以消除色谱柱填料表面硅羟基的作用，改善 MBZ、MBZ-OH、MBZ-NH2 的峰形。另外，本方法的色谱条件在保证待测物与杂质分开的前提下，出峰时间令人满意，在大批量的检测中节省了时间。

2.3 提取条件与定容液的优化

Steenbaar[13] 加入硫酸钠和碳酸钾溶液分别用来吸收水分和碱化样品，但由于硫酸钠吸水易结成块状，导致碳酸钾溶液因不能与样品充分接触而碱化作用减少，同时块状物又给样品前处理带来困难。Ruyck[9-10] 加入氢氧化钠，其碱性太强容易使检测物分解。这两种方法都容易导致大量的杂质峰出现，且回收率受到较大影响。作者用磷酸盐缓冲液除去样品中大分子蛋白质，氨水则可用来碱化样品，在液体环境中样品能够很好地碱化，处理起来方便快捷。实验中发现磷酸盐缓冲溶液的配制比例对实验结果有着较大的影响。经考察发现，0.05 mol/L 磷酸氢二钠-0.05 mol/L 磷酸二氢钠（体积比 9∶1）（氨水调节 pH 值至 9.5）得到的效果最佳。

本实验在固相萃取柱的选择与使用上也做了大量的比较研究。Hajee[8]、Steenbaar[13] 分别采用氨基柱和硅胶柱净化样品，Dowling[11] 用 C18 柱处理萃取液。通过实验发现这些柱子容易吸附药物，影响回收率。本实验不使用萃取柱，也可达到很高的回收率，同时节省了材料和时间。由于甲苯咪唑及其代谢物都不易溶于流动相，故向磷酸二氢铵溶液中加入二甲基甲酰胺作为定容液，解决了溶解问题。二甲基甲酰胺与磷酸二氢铵的体积比有 7∶3、6∶4、5∶5、4∶6、3∶7，实验表明体积比为 3∶7 时的效果最好，且没有乳化现象。用磷酸调节定容液 pH，以 pH 2.0 最为合适。

Ruyck[10] 报道用流动相和乙腈混合溶液定容残留物，易产生乳化现象。本实验用 1 mL 二甲基甲酰胺-0.05 mol/L 磷酸盐缓冲液定容旋转蒸发后的残留物，未出现乳化现象，杂质峰的分离较好，同时得到了很好的回收率。在优化条件下，可得到标样、空白样品和加标样品的色谱图（见图 1）。

2.4 方法的线性范围

以含甲苯咪唑、羟基甲苯咪唑和氨基甲苯咪唑质量浓度为 0.005、0.01、0.025、0.05、0.1、0.25、0.5、1.0、5.0、10.0 mg/L 混合标准溶液进样，得到标准曲线。结果表明，3 种分析物在 0.005~10 mg/L 范围内峰面积（A, AU）均与质量浓度（ρ, mg/L）呈线性，其中甲苯咪唑的线性方程：$A = 187.9\rho - 1542$（$r = 019\ 999$）；羟基甲苯咪唑的线性方程：$A =$

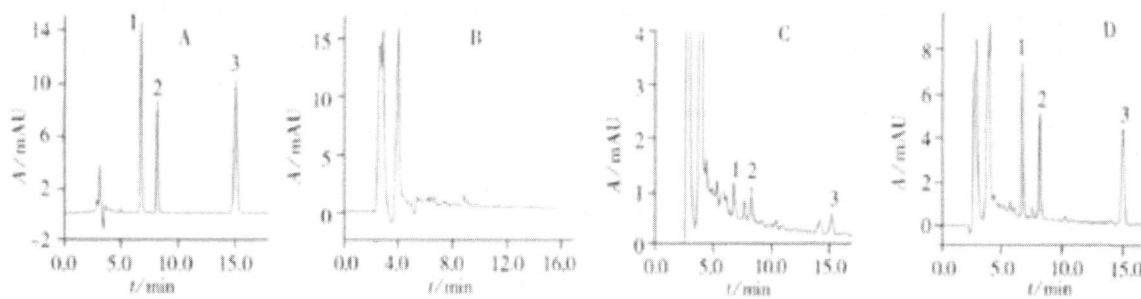

图1　1 mg/L混合标准溶液（A）、银鲫空白肌肉（B）、10 μg/kg银鲫肌肉加标样品（C）、
120 μg/kg银鲫肌肉加标样品（D）的色谱图

Fig. 1　Chromatogramsof1 mg/Lmixedstandardsolution（A），blankJohnnycarpmusclesample（B），Johnny carpmusclesamplespikedwith
10 μg/kgmixedstandard（C）and120 μg/kgmixedstandard（D）
1. MBZ-NH2；2. MBZ-OH；3. MBZ

98. 74p-1354（r=019 996）；氨基甲苯咪唑的线性方程：A=154. 2p-1346（r=019 999）。按上述的色谱条件测定加标样品，以3倍信噪比（3S/N）计算，样品中MBZ、MBZ-NH2和MBZ-OH检出限分别为2、3、3 μg/kg。

2.5　回收率与精密度

在银鲫空白肌肉样品中分别添加10、30、60、120 μg/kg的MBZ、MBZ-OH和MBZ-NH2混合标准溶液，每个含量做5次平行，设一个空白对照，按样品前处理过程处理后测定回收率，在银鲫空白肌肉样品中按1 d内重复测定5次，1周内重复5次测定，计算日内及日间精密度（见表1）。

表1　肌肉加标样的回收率和日内、日间相对标准偏差与精密度
Table1　Recoveries，intra-and inter-day accuracy and precision results for spiked muscle sample

Analyte	Added wA/（μg·kg^{-1}）	Meanrecovery R/%	RSD sr/%	ntra-dayRSD sr/%	nter-dayRSD sr/%
MBZ-NH2	10	71	8. 5	4. 5	6. 7
	30	71	7. 3	3. 5	7. 8
	60	73	5. 2	2. 5	4. 9
	120	75	3. 2	2. 8	3. 6
MBZ-OH	10	86	8. 3	9. 3	9. 9
	30	93	4. 0	5. 4	6. 6
	60	88	9. 9	5. 7	7. 9
	120	91	3. 0	4. 5	6. 6
MBZ	10	84	7. 2	8. 8	9. 5
	30	81	6. 8	4. 2	5. 6
	60	85	4. 6	7. 0	7. 4
	120	86	2. 8	1. 8	3. 2

3　结论

本文建立了高效液相色谱法同时测定银鲫（Carassius auratus gibelio）肌肉组织中甲苯咪唑及其代谢物氨基甲苯咪唑和羟基甲苯咪唑的分析方法。本方法样品制备简便快速，分离效果好，方法的线性范围、精密度、回收率和准确度均能满足鱼体组织中这3种分析物检测的要求。方法的建立为甲苯咪唑及其代谢物在鱼体内残留量的检测提供了技术保障，增加了我国动物性食品中检测方法的储备。

参考文献

［1］　Swanepoelae，Liebnbergaw，Melgardtmv. Quality evaluation of generic drugs by dissolutiontest：changing the USP dissolution medium to distinguish between active and nonactive mebendazole polymorphs [J] . Eur J Pharm Biopharm, 2003, 55 (3): 345-349.

[2]　CHUANG Chihcheng, CHEN Chiwu, FAN Chiakwung, et al. Angiostrongylus cantonensis: apoptosis of in xammatory cells induced by treatment with mebendazole or/and interleuk in 12 in mice [J] . Exp Parasitol, 2007, 115 (3): 226-232.

[3]　PANDEY S N, ROYBK. Disposition kinetics of mebendazole in plasma, milk and ruminal fluid of goats [J] . Small Ruminant Res, 1998, 27 (2): 111-115.

[4]　Bennett A., Guyatt H. . Reducing intestinal nematode infection: efficacy of albendazole and mebendazole [J] . Parasitol Today, 2000, 2 (16): 71-77.

[5]　Kathari O. S. P., PAPANDROULAKISN, DIVANACH P. Treatment of microcotyle sp. (monogenea) on the gills of cagecultured red porgy, pagrus following baths with formalin and mebendazole [J] . Aquaculture, 2006, 251 (3): 167-171.

[6]　Km K. H., Choie S. . Treatment of microcotyle sebastis (monogenea) on the gills of cultured rockfish (sebastes schelegeli) with oral adminis-tration of mebendazole and bithionol [J] . Aquaculture, 1998, 167 (2): 115-121.

[7]　EEC Council Regulation No. 2377/1990. Community procedure for the establishment of maximum residue limits of veterinary medicinal products in foodstuffs of animal origin [S] .

[8]　Hajee C. A. J., Haagsma N. . Liquid chromatographic determination of mebendazole and its metabolites, aminomebendazole and hydroxymeb-endazole in eel muscle tissue [J] . J AOAC Int, 1996, 79 (3): 645-651.

[9]　Ruyck H. D., Daeseleire E., R DDER H D. Development and validation of a liquid chromatography-electrospray tandem mass spectrometry method for mebendazole and its metabolites hydroxymebendazole and aminomebendazole in sheep liver [J] . Analyst, 2001, 126 (12): 2144 -2148.

[10]　RUYCK H D, DAESELEIRE E, R DDER H D, et al. Liquid chromatographic-electrospray tandem mass spectrometric method for the determi-nation of mebendazole and its hydrolysed and reduced metabolites in sheep muscle [J] . Anal Chim Acta, 2003, 483 (2): 111-123.

[11]　Dowling G., Cantwell H., Keeffem O., et al. Multi-residue method for the determination of benzimidazoles in bovine liver [J] . Anal Chim Acta, 2005, 529 (2): 285-292.

[12]　Allan R. J., Goodman H. T., Watson T. R. . Two high-performance liquid chromatographic determinations for mebendazole and its metabo-lites in human plasma using a rapid Sep PakC$_{18}$ extraction [J] . J Chromatogr, 1980, 183 (3): 311-319.

[13]　Steenbaar J. G., Hajee C. A. J., Haagsma N., et al. High - performance liquid chromatographic determination of the anthelmintic mebendazole in eel muscle tissue [J] . J Chromatogr, 1993, 615: 186-190.

该文发表于《分析测试学报》【2010, 29 (3): 316-320】

水产品中螺旋霉素、替米考星、泰乐菌素与北里霉素残留量的超高效液相色谱-紫外检测法同时测定

Simultaneous Determination of Spiramycin, Tilmicosin, Tylosin and Kitasamycin Residues in Fishery Products by UPLC-TUV Method

刘永涛[1,2*], 艾晓辉[1,2], 邹世平[1,2], 杨红[1,2]

(1. 中国水产科学研究院淡水渔业研究中心, 江苏 无锡 214081;

2. 中国水产科学研究院长江水产研究所/农业部淡水鱼类种质监督检验测试中心, 湖北 荆州 434000)

摘要: 建立了水产品肌肉组织中螺旋霉素、替米考星、泰乐菌素、北里霉素同时测定的超高效液相色谱-紫外检测法 (UPLC-TUV)。用乙腈作提取剂, 4%NaCl 水溶液防止样品乳化, 正己烷去除脂肪, 经固相萃取小柱净化等样品处理过程; 以乙腈-25 mol 磷酸二氢铵 (pH 2.5, 含10%乙腈) 为流动相, 以 ACQUITY UPLC BEH C18 为分离柱, 柱温为 45℃, 紫外检测启用波长事件模式或采用双波长 232 nm 和 287 nm 进行检测。方法在 0.100~20.0 mg/L 浓度范围内呈线性相关, 相关系数 $r=0.998\ 0$。平均回收率为 70.14%~95.78%, 相对标准偏差为 3.54%~11.35%, 螺旋霉素、替米考星、泰乐菌素和北里霉素的检测限分别为 25、25、50、75 μg/kg。本方法适用于水产品肌肉组织中螺旋霉素、替米考星、泰乐菌素和北里霉素的残留量测定。

关键词: 水产品; 螺旋霉素; 替米考星; 泰乐菌素; 北里霉素; 残留量; 超高效液相色谱

螺旋霉素、替米考星、泰乐菌素、北里霉素均属于大环内酯类抗生素, 大环内酯类药物主要用于畜禽, 对革兰氏阳性

菌和部分革兰氏阴性菌、霉形体、螺旋体等均有抑制作用。近年来，大环内酯类药物在水产养殖中的新应用日益受到关注。王秀珍等[1]报道了北里霉素与对照药物泰乐菌素防治草鱼"三病"的研究。随着研究的深入，替米考星等大环内酯类药物对人体的危害也逐渐被人们所认识，如大环内酯可产生交叉耐药性，高残留会对人类造成致敏性和毒副反应，尤其是敏感个体等[2]。1999年以来，欧盟禁止将泰乐菌素与螺旋霉素作为促生长剂添加到动物饲料中[3]。我国农业部颁布的NY 5071-2002《无公害食品·渔药使用准则》中规定禁止泰乐菌素在水产养殖中使用［4］。我国和欧盟还规定了动物组织中螺旋霉素、替米考星、北里霉素和泰乐菌素的最高残留量[5-6]。

目前，动物组织和牛奶中大环内酯类药物残留的测定多采用液相色谱-质谱法[7-12]，而测定水产品中该类药物的报道较少。Horie等[2]采用液相色谱电喷雾质谱法测定水产品中的大环内酯类药物，也有报道采用高效液相色谱-串联质谱法测定水产品中红霉素[13]、鳗鱼中大环内酯类残留[14]、鱼肉中林可酰胺类和大环内酯类抗生素残留[15]，以及液相色谱法同时测定水产品中螺旋霉素和泰乐菌素[3]。而水产品中螺旋霉素、替米考星、泰乐菌素和北里霉素同时测定的液相色谱和超高效液相色谱法尚未见报道。超高效液相色谱法由于具有分析速度快、灵敏度高等优点而得到广泛应用[16]。本文首次采用超高效液相色谱-紫外检测法（UPLC-TUV）建立了同时测定水产品中螺旋霉素、替米考星、泰乐菌素和北里霉素的分析方法。

1　实验部分

1.1　仪器、试剂与材料

ACQUITY UPLC超高效液相色谱（美国Waters公司）配TUV双通道紫外检测器及Empower 2色谱工作站；Mettler-TO-LEDO AE-240型精密电子天平（梅特勒-托利多公司）；HITACHI 20PR-520型自动高速冷冻离心机（日本日立公司）；FS-1高速匀浆机（华普达教学仪器有限公司）；调速混匀器（上海康华生化仪器制造厂）；HGC-12氮吹仪（天津市恒奥科技发展有限公司）；RE-2000E旋转蒸发仪（巩义市予华仪器有限责任公司）；固相萃取装置（德国CNW公司）；Oasis HLB固相萃取柱（3 mL/60 mg，Waters公司）。

替米考星标准品（纯度≥98%，Dr. Ehrenstorfer GmbH）；泰乐菌素标准品（纯度≥99% Dr. Ehrenstorfer GmbH）；红霉素标准品（纯度≥99%，Bomei）；北里霉素标准品（纯度≥72% Dr. Ehrenstorfer GmbH）；螺旋霉素标准品（纯度≥96% Dr. Ehrenstorfer GmbH）；乙酸乙酯、正己烷、乙腈、甲醇（液相色谱纯，美国J. T. Baker）；磷酸二氢铵（分析纯，天津市大茂化学试剂厂）；磷酸二氢钠（分析纯，天津市福晨化学试剂厂）；氯化钠（分析纯，天津市东丽区天大化学试剂厂）；酸性氧化铝（分析纯，上海市五四化学试剂有限公司）；乙醚（分析纯，上海市马陆制药厂）；磷酸（优级纯，天津市光复精细化工研究所）；鲫鱼从市场购买，南美白对虾肌肉样品为送检样品。

1.2　实验方法

1.2.1　样品处理

鱼，去鳞、去皮，沿脊背取肌肉；虾，去头、去壳，取肌肉部分；甲鱼等取可食部分；样品均质混匀-18℃冰箱中保存，备用。

测定时，将匀浆冷冻保存的肌肉样品室温下自然解冻。准确称取样品10.0 g置于50 mL离心管中，加入30 mL乙腈，涡旋振荡混匀1 min，4 500 r/min离心10 min，取上清液置于鸡心瓶中。向残渣中再加入15 mL乙腈重复提取一遍，合并提取液于鸡心瓶中，置旋转蒸发仪浓缩至近干，加入5 mL 4%NaCl溶液溶解残渣，涡旋振荡混匀30 s，再加入5 mL正己烷，涡旋振荡混匀1 min，静置5 min，弃去上层正己烷层，重复去脂一次。

1.2.2　样品净化

将HLB固相萃取小柱以5 mL乙腈、5 mL水活化后，将水相过固相萃取柱（柱流速保持1滴/s），用10 mL蒸馏水、10 mL 5%乙腈水溶液淋洗HLB固相萃取柱，用8 mL乙腈洗脱，洗脱液置50℃氮吹仪上吹干。向残渣中加入1 mL流动相，涡旋振荡混匀1 min，经0.22 μm滤头过滤置进样瓶，UPLC-TUV分析。

1.2.3　标准溶液的配制及标准曲线的制备

分别准确称取0.01 g（精确至0.000 1 g）的各标准品，用乙腈分别溶解并定容至100 mL，配制成100 μg/mL标准储液，-20℃冰箱中冷藏保存。测定时以储备液为基础，分别从四种储液中取一定体积的标准溶液混合后稀释成0.1、0.2、0.5、1、5、10、15、20 μg/mL的混合标准溶液。分别以替米考星、泰乐菌素、螺旋霉素、北里霉素面积为纵坐标，质量浓度为横坐标绘制标准曲线，求出回归方程和相关系数。

1.2.4　色谱条件

色谱柱：ACQUITY UPLC BEH C18（2.1×100 mm，1.7 μm）反相色谱柱；柱温：45℃；流速0.3 mL/min；进样量10 μL。梯度洗脱条件：A相为乙腈，B相为25 mM的磷酸二氢铵溶液（含10%乙腈），0~1.5 min，10% A；1.5~7.5 min 10% A~50% A；7.5~8.0 min 50% A~10% A；8.0~9.0 min，10% A；紫外检测方式采用单波长运行事件模式：0~

2.5 min，波长为287 nm，2.5~4.5 min 时，波长为232 nm，4.5 min~6.1 min 波长为287 nm，6.1 min~8.80 min 波长为232 nm，8.80 min~9.0 min 波长为287 nm。也可采用双波长模式：通道 A 232 nm，通道 B 287 nm。

2 结果与讨论

2.1 UPLC 法与 HPLC 法的比较

UPLC 系统较 HPLC 系统具有更细的管路，采用 1.7 μm 粒度的 C18 色谱柱，具有更小的死体积，更高的分辨率，更快的分析速度，消耗更少流动相。刘晔等[8]报道了猪肝中 4 种大环内酯类抗生素的高效液相色谱（二极管阵列检测器）检测法，其分析时间为 30 min，流速为 1 mL/min；采用 UPLC 测定 4 种大环内酯类抗生素只需 9 min 将其与杂质峰分离分开，而且采用 0.3 mL/min 的流速可以节省溶剂。因此，本文采用的 UPLC 法既节约了成本又缩短了分析时间。

2.2 样品处理的优化

由于实验采用 1.7 μm 粒度填料的 C18 色谱柱，该色谱柱对所分析的样品较高效液相色谱（5 μm 填料的反相色谱柱）有更大的峰容量和更好的分辨率，所以同等前处理条件下，超高效液相色谱图比高效液相色谱图有更多的杂质峰。考虑杂质峰对色谱基线及目标峰的影响，故采用 HLB 固相萃取柱净化样品。

分别以乙酸乙酯、乙醚、乙酸乙酯-乙醚（体积比 1:1）、乙腈、乙酸乙酯-乙腈（体积比 1:1）为提取剂进行提取，考察不同提取剂的影响。采用乙酸乙酯和乙醚为提取剂时，先加入 0.1 mol/L 的磷酸二氢钠（用氨水调节 pH 8.0），且不过固相萃取柱。结果表明，乙酸乙酯较乙醚更适合螺旋霉素和替米考星的提取，而乙醚对泰乐菌素和北里霉素提取回收率较高。分别用乙腈和乙酸乙酯-乙腈作为提取剂时，两者对于 4 种大环内酯类药物提取回收率高，但采用乙酸乙酯-乙腈作为提取剂在提取药物的同时也提取了样品的中极性和非极性杂质，而采用乙腈作提取剂提取杂质较单一，更有利于 HLB 固相萃取柱净化。采用乙腈提取，蒸干后加入磷酸二氢铵和 4% NaCl 溶液，结果发现采用 4% NaCl 溶液溶解的样品，溶液更清澈且更容易通过固相萃取柱。所以实验采用乙腈为提取剂，4% NaCl 溶液溶解，蒸发近干的样品用正己烷去脂，过 HLB 柱净化。

2.3 定溶溶液的选择

实验分别考察了二甲基甲酰胺-磷酸二氢铵、乙腈-磷酸二氢铵（体积比 1:1）和乙腈-磷酸二氢铵（1:1，用磷酸调至 pH 2.5）作为定容溶液的影响。结果表明，采用乙腈-磷酸二氢铵（1:1）作为定容溶液杂质峰较少。所以，实验采用乙腈-磷酸二氢铵（体积比 1:1）作为定容溶剂。

2.4 色谱条件的优化

分别考察了乙腈体积分数均为 10% 的乙腈-25 mol/L 磷酸二氢钠、乙腈-25 mol/L 磷酸二氢钠（用磷酸调 pH 2.5）、乙腈-25 mol/L 磷酸二氢铵、乙腈-25 mol/L 磷酸二氢铵（用磷酸调 pH 2.5）4 种流动相对螺旋霉素、替米考星、泰乐菌素、北里霉素的分离情况。结果表明，在此 4 种流动相条件下螺旋霉素和替米考星均有较好的峰形，但泰乐菌素和北里霉素峰形较差，以磷酸二氢铵（pH 2.5）为流动相时紫外检测灵敏度较以磷酸二氢钠为流动相的灵敏度高，同时用磷酸调节 pH 为 2.5 时色谱柱更易平衡，所以本实验选用乙腈-25 mmol/L 磷酸二氢铵（pH 2.5）为流动相，在该流动相条件下，分别研究了柱温在 30、40、45、50、60℃ 时对各药物分离的影响，结果发现随着柱温的升高各药物的保留时间提前，峰形变好，特别是泰乐菌素和北里霉素的峰形更好。但由于高温会缩短色谱柱的寿命，综合考虑各药物峰形、灵敏度和色谱柱寿命，最终选择柱温为 45℃。

2.5 方法的线性范围、回收率、精密度与检出限

按"1.2"进行测定，结果表明，4 种药物在 0.100~20.00 mg/L 范围内均呈线性相关。线性方程与相关系数见表 1。

在鲫鱼、南美白对虾空白肌肉样品中分别添加 0.10、0.50、1.0 mg/kg 3 个水平的混合标准溶液，每个水平平行测定 3 次，进行回收率实验。标准样品、空白样品及空白样品加标的色谱图见图 1。平均回收率为 70%~102%，相对标准偏差为 2.9%~11.2%（见表 1）。测定加标样品，以 3 倍信噪比分别计算螺旋霉素、替米考星、泰乐菌素和北里霉素在水产品肌肉组织中的检出限分别为 25、25、50、75 μg/kg。

表 1　4 种大环内酯类药物的线性方程、加标回收率与相对标准偏差

Table1　Linear equations，recoveries and relative standard deviations of 4 macrolide antibiotics（n=3）

Analyt	Linearequation *	orrelation coefficient	Added ρA/（mg · kg⁻¹）	Carassiusauratus（鲫鱼）RecoveryR/% RSDsr/%	Penaeusvannamei（南美白对虾）RecoveryR/% RSDsr/%
Spiramyci	$A = 10\ 630\ C+581.39$	0.998 7	0.10，0.50，1.0	83，74，77 10.2，3.2，7.4	75，91，96 4.4，2.9，3.8
Tilmicosi	$A = 7\ 209\ C+2\ 608.6$	0.999 3	0.10，0.50，1.0	90，94，92 11.2，3.4，8.4	81，102，86 8.5，4.1，4.5
Tylosi	$A = 6\ 642\ C-369.5$	0.999 4	0.10，0.50，1.0	84，75，99 10.6，3.9，6.7	72，80，86 9.4，8.8，6.9
Kitasamycin	$A = 5\ 037.43\ C+773.53$	0.998 0	0.10，0.50，1.0	77，70，77 9.3，3.1，5.4	75，81，84 10.2，6.5，4.6

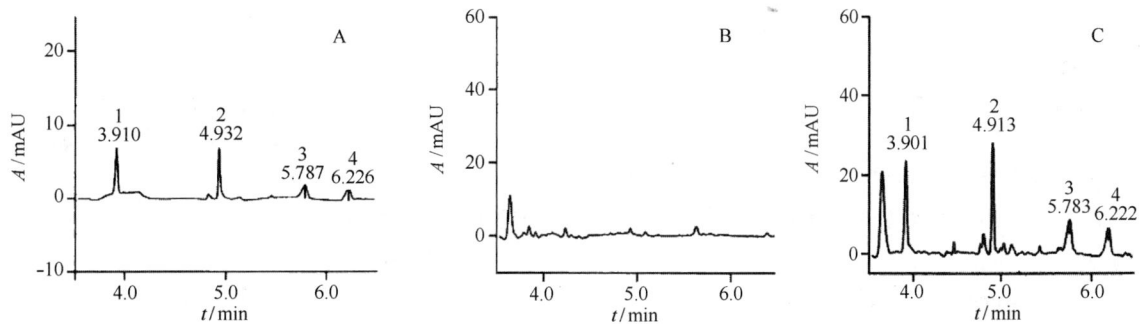

图 1　1.0 mg/L 混合标准 溶液（A）、鲫鱼空白肌肉（B）与 0.5 mg/kg 鲫鱼空白肌肉加标（C）的色谱图

Fig. 1　Chromatograms of 1.0 mg/L mixed standard solution（A），Carassius auratus muscle blank sample（B）and 0.5 mg/kg spiked Carassius auratus muscle blank sample（C）

peaks：1. Spiramycin，2. tilmicosin，3. Tylosin，4. kitasamycin

3　结论

本文采用 UPLC-TUV 建立了水产品中螺旋霉素、替米考星、泰乐菌素和北里霉素残留量同时测定的方法。该方法分析速度快、价格低廉、回收率高、检出限低、方法重复性好，可满足同时检测水 产品中螺旋霉素、替米考星、泰乐菌素和北里霉素的要求。

参考文献

［1］　王秀珍，刘国印 . 北里霉素防治草鱼"三病"试验［J］. 淡水渔业，1994，24（6）：18-19.

［2］　HORIEM，TAKEGAMIH，TOYA K，et al. Determination of macrolide antibiotics in meat and fish by liquid chromatography electrospray mass spectrometry［J］. Anal Ch im Acta，2003，492（1/2）：187-197.

［3］　杨方，李耀平，方宇，等 . 高效液相色谱法同时检测水产品中螺旋霉素与泰乐菌素药物残留［J］. 理化检验：化学分册，2007，43（4）：272-274.

［4］　中华人民共和国农业部 . NY 5071-2002. 中华人民共和国农业行业标准《无公害 食品渔用药物使用准则》［S］. 北京：中国标准出版社，2002.

［5］　中华人民共和国农业部 . 中华人民共和国农业部第 235 号公告［B］. 北京，2002.

［6］　Official Journal of the European Communities. Commission Regulation（EC）No 1181/2002［S］. 2002.

［7］　王凤美，陈军辉，林黎明，等 . UPLC-MS/MS 法对动物源性食品中 12 种大环内酯类抗生素残留的测定［J］. 分析测试学报，2009，28（7）：784-788.

[8] 赵东豪，贺利民，聂建荣，等. HPLC-MS/MS 检测猪肉中六种大环内酯类抗生素 [J]. 分析试验室，2009，28（1）：117-119.

[9] 孙雷，张骊，王树槐，等. 超高效液相色谱-串联质谱法对动物源食品中 13 种林可胺类及大环内酯类药物残留的检测 [J]. 分析测试学报，2009，28（9）：1058-1061.

[10] 王敏，林维宣，郭德华，等. 高效液相色谱-串联质谱法同时检测动物性食品中多种大环内酯类药物 [J]. 分析测试学报，2007，26（5）：675-678.

[11] 谢丽琪，岳振峰，唐少冰，等. 高效液相色谱串联质谱法测定牛奶中林可酰胺类和大环内酯类抗生素残留量的研究 [J]. 分析试验室，2008，27（3）：5-8.

[12] 李岩. 液相色谱-电喷雾质谱法测定动物源食品中残留大环内酯类抗生素 [D]. 大连：大连交通大学，2006：28-40.

[13] 于慧娟，马兵，惠芸华，等. 高效液相色谱-串联质谱法测定水产品中红霉素的残留 [J]. 分析试验室，2009，28（3）：51-54.

[14] 陈莹，陈辉，林谷园，等. 超高效液相色谱串联质谱法对鳗鱼中大环内酯类、喹诺酮类和磺胺类兽药残留量的同时测定 [J]. 分析测试学报，2008，27（5）：538-541.

[15] 岳振峰，陈小霞，谢丽琪，等. 高效液相色谱串联质谱法测定动物组织中林可酰胺类和大环内酯类抗生素残留 [J]. 分析化学，2007，35（9）：1290-1294.

[16] 陈佳，王刚力，姚佘文，等. 超高效液相色谱（UPLC）在药物分析领域中的应用 [J]. 药物分析，2008，（11）：1976-1981.

[17] 刘晔，王洪新，戴军，等. 固相萃取-高效液相色谱法测定猪肝中的大环内酯类抗生素 [J]. 食品与发酵工业，2008，34（5）：162-165.

该文发表于《分析试验室》【2009，28. Suppl：138-141】

水产品中三聚氰胺残留量的高效液相色谱检测法

Determination of Melamine residues in fishery products by HPLC

刘永涛，杨红，艾晓辉

（中国水产科学研究院长江水产研究所农业部淡水鱼类种质监督检验测试中心，湖北 荆州 434000）

摘要：建立了水产品肌肉组织中三聚氰胺的高效液相紫外检测法。肌肉组织中加入三氯乙酸作为提取剂，乙酸铅沉淀蛋白，过 PCX 混合阳离子交换柱净化等样品处理过程；乙腈-10 mmol 庚烷磺酸钠+10 mmol 柠檬酸溶液（8：92，V：V）为流动相，紫外检测波长为 240 nm。方法在 0.100~10.00 mg/L 浓度范围内呈线性相关，相关系数 r=0.999 9。平均回收率为 74.29%~89.04%，相对标准偏差为 0.44%~9.32%，三聚氰胺在水产品肌肉中的检测限为 0.1 mg/kg。本方法适用于水产品肌肉组织中三聚氰胺残留量的检测。

关键词：水产品；三聚氰胺；残留量；高效液相色谱

该文发表于《分析测试学报》【2010，29（4）：372-375】

高效液相色谱法对鱼体肌肉组织中喹烯酮及其代谢物残留量的同时检测

Simultaneous Determination of Residues of Quinocetone and Its Metabolin in Muscle Tissues of Fish by High Performance Liquid Chroma to graphy

杨莉[1,2]，艾晓辉[1]，袁科平[1]，汪开毓[2]

（1. 中国水产科学研究院长江水产研究所，湖北 荆州 434000；2. 四川农业大学动物医学院，四川 雅安 625014）

摘要：以斑点叉尾鮰（Ictaluruspunctatus）、鲤（CyprinuscarpioLinnaeus）和草鱼（Ctenopharyngodonidellus）为实验材料，建立了一种用高效液相色谱法同时检测鱼体肌肉组织中喹烯酮及其代谢物脱二氧喹烯酮残留量的方法。样品经乙腈、正己烷脱脂，旋转蒸发浓缩，乙酸乙酯洗脱，50℃氮气吹干及流动相定容后，采用 Waters515 高效液相色谱仪在 320 nm 波长处检测，流速为 110 mL/min，外标法定量。喹烯酮在 0108~2 150 mg/L 范围内呈线性，相关系数 r=019 997，加标回收率为 75%~84%，检出限为 3 148~7 135 μg/kg；脱二氧喹烯酮在 0102~0150 mg/L 范围内呈线性，相关系数 r=019 997，加标回收率为 74%~83%，检出限为 5 167~16 123 μg/kg。

关键词：斑点叉尾鲴；鲤；草鱼；喹烯酮；脱二氧喹烯酮；残留量；高效液相色谱

喹烯酮属喹啉 1，42 二氧类化合物，是中国农业科学院兰州畜牧与兽药研究所新研制的一种促生长饲料药物添加剂。该药可促进动物机体生长并提高饲料转化率，明显降低畜禽腹泻发生率，且毒性极小、无三致（致畸、致癌、致突变）作用、无残留、使用安全，是我国在国际上首创的一类新兽药[1-8]。研究表明，喹烯酮的代谢主要经脱单氧、脱双氧后生成 32 甲基喹啉 222 羧酸（又名脱二氧喹烯酮）。脱二氧喹烯酮因此被确定为最终存在于动物体内的主要残留物而被作为喹烯酮的残留标示物[9-11]。图 1 为喹烯酮及脱二氧喹烯酮的化学结构式。

目前，检测喹烯酮的方法有高效液相色谱法（HPLC）、反相高效液相色谱法（RP2HPLC）、液相色谱-串联质谱法（LC-MS/MS）和薄层层析法（TLC）等[12-18]。张丽芳等[9]采用高效液相色谱-串联质谱法测定鸡组织中的喹烯酮代谢物脱二氧喹烯酮。目前国内未见同时检测喹烯酮及其代谢物的研究报道。本文以斑点叉尾鲴、鲤、草鱼为实验材料，建立了同时测定鱼体肌肉组织中喹烯酮及其代谢物脱二氧喹烯酮残留量的高效液相色谱法，为完善我国水产品的质量保证体系提供了基础资料。

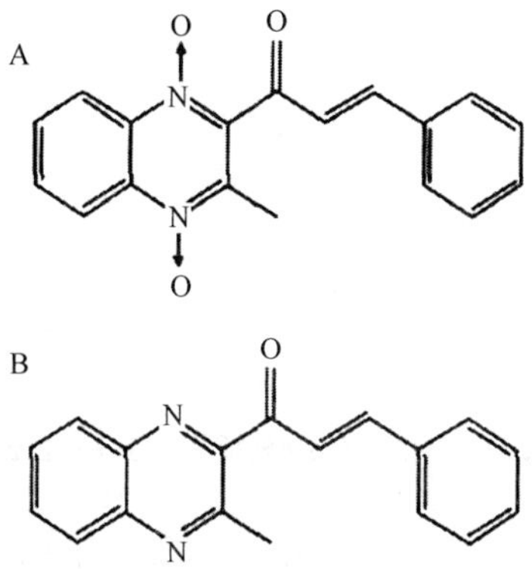

图 1　喹烯酮（A）与脱二氧喹烯酮（B）的化学结构式

Fig. 1　Chemical structures of quinocetone （A） and desoxyquinocetone （B）

1　实验部分

1.1　仪器与试剂

高效液相色谱仪（Waters515 泵，717 自动进样器，2487 双通道紫外检测器及 Empower 色谱工作站）；Mettler2ToledoAE2240 型精密电子天平（梅特勒-托利多公司）；Bushi2200 旋转蒸发仪（瑞士 Buchi 公司）；HS3120 型超声波清洗器（上海卓康生物科技有限公司）；自动高速冷冻离心机（日本 Hitachi20PR2520 型）；FS21 高速匀浆机（华普达教学仪器有限公司）；调速匀浆器（上海康华生化仪器制造厂）；HGC212 氮吹仪（HengaoT&D 公司）。

标准物质：喹烯酮、脱二氧喹烯酮（纯度98%以上）；喹烯酮、脱二氧喹烯酮标准溶液：用甲醇将喹烯酮和脱二氧喹烯酮分别配成 5 mg/L 喹烯酮储备液、1 mg/L 脱二氧喹烯酮储备液以及 500 μg/L 的储备混标，再根据检测要求用甲醇稀释成相应的工作液；乙腈、乙酸乙酯、正己烷均为色谱纯（Fisher 公司）；硫酸钠为国产分析纯，经 600~650℃灼烧 4 h 后，置密闭容器中备用；实验用水为二次蒸馏水。

1.2　样品预处理

准确称取 5 g 已匀浆好的鱼体肌肉样品（斑点叉尾鲴、鲤和草鱼）分别置于 50 mL 具塞离心管中。加入 15 mL 乙腈和 5 g无水硫酸钠在调速匀匀器上振荡混匀 2 min 后，超声提取 5 min，4 000 r/min 离心 10 min，将上清液转移至 250 mL 分液漏斗中。再向样品中加入 10 mL 乙腈，重复提取 1 次。合并乙腈提取液，在分液漏斗中加入 25 mL 正己烷（用乙腈饱和），盖塞振荡 2 min，静置分层，充分混合提取脂肪，将下层液体转移至 125 mL 梨形瓶中，于 40℃水浴旋转蒸发至干，用 4 mL 乙酸乙酯以每次 2 mL 淋洗梨形瓶中的提取物。合并淋洗液于 10 mL 具塞离心管中。将淋洗液在 50℃下氮气吹干，再用流动相定容至 2 mL，过 0145 μm 滤膜，供液相色谱检测。

1.3　实验方法

1.3.1　色谱条件　色谱柱：WaterssymmetryC18（250 mm×416 mm，5 μm）；流动相：乙腈-甲醇-水（体积比 5：2：3）；等度洗脱；流速：110 mL/min；柱温：室温；紫外检测波长：320 nm；进样量：20 μL。

1.3.2 测定方法　取样品溶液和标准溶液各 20 μL，注入高效液相色谱仪进行分析，以标准溶液峰的保留时间为依据进行定性，以标准溶液的峰面积计算样品中喹烯酮和脱二氧喹烯酮的含量。

2　结果与讨论

2.1　提取溶剂的选择

分别考察了乙腈、乙酸乙酯、甲醇、丙酮、二氯甲烷 5 种有机溶剂的提取效果。结果表明，选用乙酸乙酯作提取剂时，提取效果不及乙腈，且样品在乙酸乙酯中呈小块状散布；甲醇对药物提取率较低，提取率为 50%；丙酮挥干时间长且杂质很多，提取率仅 30%；而样品在二氯甲烷中固缩成团漂浮在溶剂上，高速离心后样品不能得到分散；选用乙腈为提取剂时，样品能够在溶剂中均匀分散，且提取液较为干净，提取率达 87%。因此实验选用乙腈为提取溶剂。

2.2　流动相的选择

用斑点叉尾鮰、鲤和草鱼肌肉作空白样品，有一干扰物的出峰时间与喹烯酮的保留时间很接近，影响对喹烯酮的准确定量。通过改变流动相，将干扰峰与喹烯酮色谱峰分离。在实验过程中，用 50% 乙腈作流动相时喹烯酮色谱峰与干扰峰得到很好地分离，但喹烯酮代谢物脱二氧喹烯酮 40 min 前未出峰；当用纯甲醇作流动相时喹烯酮与其代谢物的保留时间显著提前，但喹烯酮色谱峰与干扰峰重合严重。由此可见，乙腈有利于喹烯酮色谱峰与干扰峰的分离，而甲醇有利于喹烯酮及脱二氧喹烯酮提前出峰。对乙腈与甲醇两者比例进行调整，最终选用乙腈-甲醇-水（体积比 5∶2∶3）为流动相，既能使喹烯酮色谱峰避开干扰，又能使脱二氧喹烯酮在 20 min 前出峰。

2.3　方法的线性关系及检出限

将喹烯酮储备液及脱二氧喹烯酮储备液稀释成含喹烯酮 0108、0116、0131、0163、1125、2150 mg/L，脱二氧喹烯酮 0102、0103、0106、0113、0125、0150 mg/L 的混合标准溶液。分别取 1 mL 上述混合标准溶液过 0145 μm 滤膜，取 20 μL 供液相色谱检测。以测得的峰面积为横坐标，质量浓度为纵坐标绘制标准工作曲线，计算回归方程和相关系数。用空白组织制成低质量浓度药物的含药组织，经预处理后测定其检出限。结果表明，喹烯酮、脱二氧喹烯酮分别在 0108 ~ 2150、0102 ~ 0150 mg/L 范围内呈线性，相关系数均为 019 997，回归方程分别为 $Y = 87\,097X + 841\,323$、$Y = 961\,815X - 525\,192$。喹烯酮在斑点叉尾鮰、鲤和草鱼肌肉组织中的检出限（$S/N = 3$）分别为 3 148、6 110、7 135 μg/kg；脱二氧喹烯酮在以上 3 种鱼肌肉组织中的检出限分别为 5 167、13 179、16 123 μg/kg。

2.4　方法的回收率及精密度

分别以斑点叉尾鮰、鲤和草鱼为实验材料，进行加标回收率和精密度测定。在 3 种鱼的空白肌肉组织中分别添加 3 个水平的喹烯酮和脱二氧喹烯酮标准溶液，使其在组织中的含量分别为 40、150、300 μg/kg，每个水平做 5 次平行实验，按样品预处理过程处理后测定回收率，结果见表 1。

表 1　肌肉样品中喹烯酮与脱二氧喹烯酮的回收率及相对标准偏差（$n = 5$）
Table 1　Recoveries and RSDs of quinocetone and desoxyquinocetone in muscle spiked samples（$n = 5$）

Compound	Sample	Addedwa/（ug·kg^{-1}）	Recovery R/%	RSD S/%
Quinocetone（喹烯酮）	Ictalunis punctatus（斑点叉尾鮰）	40, 150, 300	83, 84, 80	8.4, 5.6, 4.0
	Cyprinus carpio Linnaeus（鲤）	40, 150, 300	76, 78, 79	5.5, 6.9, 5.0
	Ctenopharyngodon idellus（草鱼）	40, 150, 300	75, 80, 76	7.2, 6.2, 4.8
Desoxyquinocetone（脱二氧喹烯酮）	Ictalunis punctatus	40, 150, 300	78, 79, 83	8.4, 7.9, 5.4
	Cyprinus carpio Linnaens	40, 150, 300	80, 81, 78	9.2, 9.0, 4.7
	Ctenophanngodon idellus	40, 150, 300	77, 74, 79	8.7, 9.8, 5.9

在草鱼空白肌肉组织中分别添加 40、150、300 μg/kg 3 个水平的喹烯酮和脱二氧喹烯酮标准溶液，每个水平的样品，日内做 5 次重复，1 周内重复做 5 次，计算日内及日间精密度，结果见表 2。

表2　精密度实验结果

Table 2　The results of precision test

Compound	AddedwA/ ($\mu g \cdot kg^{-1}$)	Inter-RSD（$n=5$）		Intra-RSD（$n=5$）		RSD S/%
		X±SD（$\mu g \cdot kg^{-1}$）	RSDsr/%	X±SD（$\mu g \cdot kg^{-1}$）	RSDsr/%	
Quinocetone	40	30.16±2.16	7.2	30.45±2.83	9,3	9.3
	150	120.75±6.00	6.2	120.55±7.66	8.0	8.0
	300	229.50±11.75	4.8	233.72±16.38	6.6	6.6
Desoxyquino-cetone	40	30.64±2.66	8.7	32.44±2.94	9.0	9.0
	150	111.75±8.74	9.8	104.88±4.96	6.0	6.0
	300	237.30±14.88	5.9	236.26±15.51	6.2	6.2

2.5　样品测定

实验材料（斑点叉尾鮰、鲤和草鱼）来源于中国水产科学院长江水产研究所窑湾试验场，按照"112"方法进行处理，进液相色谱分析，均未检出喹烯酮和脱二氧喹烯酮。空白样品、空白样品加标的色谱图见图2。

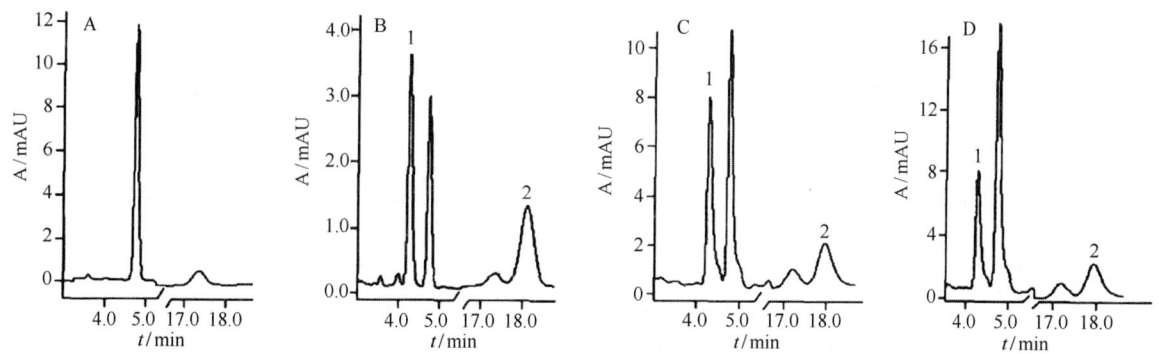

图2　草鱼肌肉组织空白样品（A）、草鱼肌肉加标150 μg/kg样品（B）、斑点叉尾鮰肌肉加标300 μg/kg样品（C）与鲤肌肉加标300 μg/kg样品（D）的色谱图

Fig. 2　Chromatograms of Ctenopharyngodon idellus musculature blank（A），150 μg/kg spiked Ctenopharyngodon idellus muscle sample（B），300 μg/kg spiked Ictalurus punctatus muscle sample（C）and 300 μg/kg spiked Cyprinus carpio Linnaeus muscle sample（D）1. quinocetone，2. desoxyquinocetone

3　结论

本文首次采用高效液相色谱法同时检测鱼体肌肉组织中喹烯酮及其代谢物脱二氧喹烯酮的残留量，通过乙腈提取，正己烷脱脂，优化流动相的配比，可有效分离干扰峰和喹烯酮及其代谢物的色谱峰。方法可操作性强、重复性好，适用于水产品中喹烯酮及脱二氧喹烯酮残留量的检测。

参考文献

[1]　张书松，王春秀，高春生.喹烯酮对早期断奶仔猪生长性能和腹泻的影响[J].河南农业科学，2009，9：112-113.

[2]　熊六凤，刘晓兰，伍海拔.喹烯酮对彭泽鲫生长性能的影响[J].水利渔业，2007，27（5）：55-56.

[3]　李剑勇，张继瑜，李金善，等.喹烯酮对肉仔鸡的促生长作用[J].畜牧与兽医，2007，39（9）：38-40.

[4]　许建宁，王全凯，崔涛，等.新兽药喹烯酮亚慢性经口毒性研究[J].中国兽药杂志，2005，39（3）：10-15.

[5]　王玉春，赵荣材，严相林，等.喹烯酮对小白鼠的致癌试验[J].中国兽药科技，1995，25（3）：24-25.

[6]　王玉春，赵荣材，严相林，等.喹烯酮对大白鼠的胚胎致畸性的研究[J].中国兽药科技，1993，23（8）：30-31.

[7]　严相林，李金善，王玉春.喹烯酮的Ames试验[J].中国兽药科技，1998，5：11-12.

[8]　李剑勇，李金善，徐忠赞，等.喹烯酮在猪、鸡体内的药代动力学研究[J].畜牧兽医学报，2002，34（1）：94-97.

[9]　张丽芳，薛飞群，刘元元，等.高效液相色谱-串联质谱法测定鸡组织中的喹烯酮标示残留物[J].分析测试学报，2006，25（5）：63-65.

[10] HUTCHINSON MJ, YOUNG P B, HEWITR S A, et al. Development and validation of improved method for confirmation of the carbadox metabolite, quinoxaline-2-carboxylic acid, in porcine liver using LC-electrospray MS/MS according to revised EU criteria for veterinary drug residue analysis [J]. Analyst, 2002, 127 (3): 342-346.

[11] European Commission. Commission Decision 2002/657/EC [S]. 2002, 8.

[12] 李剑勇，李金善，徐忠赞，等. 高效液相色谱仪检测鸡组织器官喹烯酮残留方法的建立 [J]. 动物医学进展，2004, 25 (5): 100-101.

[13] 金录胜. RP-HPLC 测定喹烯酮及其制剂的含量 [J]. 中国兽药杂志，2003, 37 (5): 25-27.

[14] 王霄旸，张丽芳，薛飞群，等. 采用高效液相色谱-串联质谱法检测鲫鱼组织 3-甲基喹啉-2-羧酸残留量 [J]. 中国兽医科学，2007, 37 (8): 718-720.

[15] 王艳春. 用 TLC 法检测饲料添加喹烯酮中的有关物质 [J]. 中国兽药杂志，1999, 33 (1): 25-26.

[16] 刘元元，张丽芳，邹思湘，等. 高效液相色谱法测定鱼饲料中的喹烯酮 [J]. 畜牧与兽医，2007, 39 (1): 35-37.

[17] 李剑勇，李金善，徐忠赞，等. 高效液相色谱仪检测猪组织器官喹烯酮残留方法的建立 [J]. 中国兽医科技，2003, 33 (8): 48-50.

[18] 李剑勇，赵荣材，李金善，等. 猪血液中 14C 标记喹烯酮的测定及药代动力学研究 [J]. 动物医学进展，2003, 24 (3): 78-79.

该文发表于《湖北农学院学报》【2003, 23 (4): 266-270】

鱼组织中喹乙醇残留量高效液相色谱检测方法研究

Determination of olaquindox residue in fish tissues by HPLC

艾晓辉，刘长征，文华

（农业部淡水鱼类种质资源与生物技术重点开放实验室，中国水产科学研究院长江水产研究所，湖北 荆州 434000）

摘要：通过比较文献中不同的色谱条件及样品处理过程，建立了测定鱼组织中喹乙醇残留量的高效液相色谱法。该法采用 C_{18} 色谱柱，选择甲醇-三蒸水（体积比 15∶85）为流动相，372 nm 为检测波长，样品用 150 g/L 的三氯乙酸沉淀蛋白，经 15 000 r/min 离心取上清液进样。喹乙醇在 0.2~12.8 μg/g 浓度范围内线性关系良好，检测限为 0.04 μg/g，平均回收率为 85.93%，不同浓度水平的日内和日间测定的相对标准差（RSD）均小于 10%。

关键词：鱼组织；喹乙醇；残留量；高效液相色谱法

该文发表于《J Shanghai Fish Univ》【2003, 12 (Suppl): 28-31】

Determining Olaquindox in fish tissue &feed by improved HPLC method（high performance liquid chromatography）

HU Kun, YANG Xian-le

（Fishery Pathogen Collection of The Ministry of Agriculture, Shangha iFisheries University, Shanghai 200090, China）

Abstract: The traditional determining method（high performance liquid chromatography, HPLC）to Olaquindox was improved in this thesis, including simplifying extracting steps and changing liquid system. The improved HPLC was used to determine the Olaquindox content in fish muscle tissue and feeds. The results showed that the minimum determination limits of this method is 0.100 mg/kg, the average recovery rates and relative standard deviations were 77.6%~90.1% and 0.495%~4.920% respectively. This method was simple, rapid, precise, stable and reliable and suitable to determine Olaquindox residues in aquatic products.

key words: Olaquindox; the improved HPCL method; muscle tissue of fish; feed

第二节　基于免疫学的快速检测方法

对渔药基于免疫学的快速检测方法有酶联免疫测定法（Enzyme-linked Immunosorbent Assay，ELISA）、胶体金免疫测定法（GICA）、荧光免疫测定法、免疫传感器法等，这些方法快速、灵敏、准确，适合于非实验室或半实验室条件下作为初筛方法使用。

一、环丙沙星

环丙沙星（ciprofloxacin，CIP）又称环丙氟哌酸，属于第三代喹诺酮类药物（quinolones，QNs）。由于安全隐患，我国政府将环丙沙星列为禁用渔药（中华人民共和国农业行业标准 NY 5071-2002《无公害食品 渔用药物使用准则》）；在方法检测限为 10 μg/kg 时，环丙沙星在水产品中的不得检出（中华人民共和国农业行业标准 NY5070-2002《无公害食品水产品中渔药限量》）。

1. 抗原的偶联和制备

环丙沙星（CIP）分子量为 331.13，不具备免疫原性，需要将其与一些大分子载体（如蛋白质、多肽或合成氨基酸）偶联，才能成为完全抗原。CIP 结构上 6-位碳原子上有一个氟原子，在 7-位碳原子上连接有亲水的哌嗪基，在 1-位氮原子上有环丙基。考虑到哌嗪基是抗菌活性基团，不易作为参与偶联反应的基团，因此选择 CIP 分子中的羧基作为与载体蛋白偶联的基团比较合适。通过将 CIP 与载体蛋白（如 BSA）偶联可以制备完全抗原。半抗原密度（半抗原与载体蛋白的比例）与免疫原性紧密相关，可以通过优化 CIP-BSA 偶联物提高免疫效果（Kun Hu et al., 2012；黄宣运等 2008）。

该文发表于《Bioscience Trends》【2012, 6 (2): 52-56】

Influence of hapten density on immunogenicity for anti-ciprofloxacin antibody production in mice

Kun Hu[1], Xuanyun Huang[2], Yousheng Jiang[1], Junqiang Qiu[1], Wei Fang[1], Xianle Yang[1], *

(1. Shanghai Ocean University, Shanghai, China;

2. East China Sea Fishery Research Institute, Chinese Academy of Fishery Sciences, Shanghai, China)

Summary：To generate antibodies against small molecules, it is necessary to couple them as haptens to large carriers such as proteins. However, the immunogenicity of the conjugates usually has no linear correlation with the hapten-protein ratio, which may lead to large variations in the character of the desired antibodies. In the present study, ciprofloxacin (CPFX) was coupled to bovine serum albumin (BSA) in five different proportions using a modified carbodiimide method. The conjugates were characterized qualitatively by spectrophotometric absorption and electrophoresis methods. Mass spectrometry and the trinitrobenzene sulfonic acid method were adopted to assay the density of conjugates quantitatively. As a result, CPFX-BSA conjugates with various hapten densities (21-30 molecules per carrier protein) were obtained. After immunization in mice, ELISA tests showed that the antisera titer increased gradually with the increase of hapten density. The antibody obtained from the mice showed high sensitivity toward CPFX. These results revealed the relationship between hapten density and immunogenicity as well as an optimized conjugation approach for immunization purposes.

key words：Hapten density, conjugation, immunogenicity, ciprofloxacin

半抗原密度对于小鼠中制备环丙沙星抗体免疫原性的影响

胡鲲[1]，黄宣运[2]，姜有声[1]，邱军强[1]，方伟[1]，杨先乐[1]

(1. 上海海洋大学，中国上海；2. 中国水产科学研究院东海水产研究所，中国上海)

摘要：从免疫原性的角度探讨了 CIP-BSA 偶联比率与偶联物半抗原密度之间的关系。偶联物免疫小鼠获得了抗血清，对抗体的分析结果表明：对于 CIP 而言，尽管随着半抗原-载体偶联比率的增加，偶联物半抗原密度会增加，但其所对应的抗体的效价并不是一直增加。当载体蛋白——半抗原的比例增加到 1：480，效价到偶联可以达到 $6.25×10^5$；当载体蛋白——半抗原的比例继续增加，相应的抗体效价并不随之增加。制备的抗体灵敏度良好，可达到 ng/mL 的水平，IC_{50} 值可达 4.93 ng/mL。研究结果显示半抗原密度与免疫原性存在联系，可通过优化偶联比例达到免疫的效果。

关键词：半抗原，偶联，免疫原性，环丙沙星

Introduction

A small hapten (molecular weight<1,000) is usually not immunogenic by itself[1,2]. It is well-known that immunogenicity can be acquired when a hapten is coupled with a macromolecule carrier, such as a protein, peptide or synthetic amino acid[3]. A hapten is generally coupled to a carrier protein through the ε-amino group. The carrier protein can increase both the strength and specificity of the antibody response efficiently. The coupling ratio of the hapten-protein is usually important for the properties of the antibody induced by the modified hapten. In most cases, increasing of the coupling ratio can enhance the strength and specificity of the immune responses[4]. On the other hand, the higher coupling ratio may decrease the activity of antibodies. The possible relationship between the immunogenicity and the hapten-protein ratio may cause large variations in the character of the desired antibodies. Screening an optimal hapten density for the conjugation is significant to improve the binding efficiency of hapten-protein conjugation[5]. Ciprofloxacin (CPFX) is an antibacterial agent applied as a veterinary medicine. The CPFX residues in consumed animal tissues raise potential risks for development of drug-resistance and chronic adverse effects in public health[6]. The maximum residue limits (MRLs) of CPFX in food stuffs of animal origin were established by the European Union[7]. Antibodyantigen reactions were extensively used to detect and quantify the CPFX residue in biological fluids. Such immunoassays were developed for determination of protein antigens as well as small molecule haptens. The central problem of fast detection by immunoassays was the availability of a specific antibody with high titer. Although there are reports about CPFX conjugation[1,8], the varying hapten density leads to the unstable character of the antibodies to a great extent. Hence, the optimized coupling ratio of CPFX-bovine serum albumin (BSA) has still not been attained. In our previous studies, a monoclonal antibody (mAb) against small hapten-CPFX with high affinity and specificity was produced and was used for rapid CPFX immunoassays in food stuff of animal origin[9]. Furthermore, with the CPFX-specific mAbs, an indirect competitive enzyme-linked immunosorbent assay (ELISA) was developed for the sensitive and specific detection of CPFX residues in fishery products[10]. In this study, CPFX as a hapten was covalently attached to BSA by a modified carbodiimide method using 1-ethl-3-carbodiimide methiodide (EDC)[1,8]. To generate the anti-CPFX antibody, the hapten density was optimized to improve the binding efficiency. It could be a benefit to the approach for preparing antibodies with good titre and sensitivity, which was significant for the CPFX immunoassay.

Materials and Methods

Chemicals and animals

CPFX (content $\geq$ 98.5%) was purchased from Zhejiang Guobang Pharmaceutical Co., Ltd. (Shangyu, Zhejiang, China). BSA was obtained from Dingguo Biotechnology Co., Ltd. (Beijing, China).

Trinitrobenzene sulfonic acid (TNBS) was from SigmaAldrich (St Louis, MO, USA). EDC (purity $\geq$ 99.3%) was obtained from Yanchang Confident Biochemical Technology Co., Ltd. (Shanghai, China). Chemical reagents such as NaCl and K2HPO4 were from Guoyao Chemical Reagent Co., Ltd. (Shanghai, China). All the chemical reagents and solutions used in this paper were analytical grade. The buffer was prepared with double distilled water.

Individual Balb/c mice were purchased from the Second Military Medical University (Shanghai, China). All mice were housed under controlled conditions and received food and water ad libitum. All animal experiments were performed in accordance with the guidelines of Regulation on Animal Experimentation and were approved by State Scientific and Technological Commission, State Council, China.

Coupling of the CPFX hapten and carrier protein

The conjugates of BSA and CPFX were synthesized by a modified carbodiimide method using EDC[11]. The conjugation reaction was performed with five different molecular ratios of BSA and CPFX (1:160, 1:320, 1:480, 1:640, and 1:800). CPFX was mixed with BSA (2 mg/mL) and EDC (60 mg/mL) while the reaction was carried out in phosphate buffer solution (pH 5.0) and incubated at 28℃ for 2 h. The mixture was dialyzed against the same buffer for 2 days and then freezedried. The CPFX−BSA conjugates C1, C2, C3, C4, and C5 (see Table 1 for the molecular ratios) obtained were stored at−20℃ until use.

Spectrophotometric analysis

The numbers of free Lys residue ε−amino groups in BSA conjugates were determined by the TNBS method[12]. Conjugate samples (0.2 mg each) were dissolved in 1 mL of 0.1 M sodium carbonate solution and then 0.5 mL of 0.01% TNBS solution was added. After incubation for 2 h at 37℃, 0.5 mL of 10% sodium dodecyl sulfate (SDS) solution and 0.25 mL of 1 M hydrochloric acid solution were added to terminate the reaction. The absorbance was measured at 335 nm. The number of amino groups left on the BSA molecule after the coupling reaction was determined by the difference of optical density (OD) values between the control group and the coupling group.

CPFX−BSA conjugates (C1−C5) were scanned from 250 nm to 350 nm using a Spectronic UV−Vis Spectrophotometer (Thermo Fisher Scientific Inc., Madison, WI, USA). The resolution length range and scanning speed were 1 nm and 1 nm/sec, respectively.

Gel electrophoresis analysis

Apparent molecular size of CPFX−BSA conjugates was analyzed by SDS−polyacrylamide gel electrophoresis (SDS−PAGE) using a PowerPac 300 type electrophoresis apparatus (Bio−Rad Laboratories, Hercules, CA, USA). The conjugates (C1−C5) were dissolved in sample buffer (10 mM Tris, 1 mM EDTA, 2.5% SDS, and 5% mercaptoethanol, pH 8.0) with a final concentration of 1 mg/mL. After heating at 100℃ for 5 min, 10 μL of each sample was loaded. The samples were separated at 80 V in the stacking gel and 100 V in a 12% separating gel. The protein samples were stained for 4 h using the Coomassie Brilliant Blue staining method and then de−stained by de−staining solution until the background was transparent.

Mass spectrometry analysis

The Agilent 1100 hp LCQ DECA Liquid Chromatography Mass Spectrum (Agilent Technologies, Santa Clara, CA, USA) was used to determine the molecular weights of the conjugates. The conjugates were dissolved in methanol and 10 μL of each was collected for injection. The chromatographic conditions were as follows. For gradient elution, ratios of mobile phases A and B (phase A, 20 mM sodium acetate + 0.018% triethylamine + 0.3% tetrahydrofuran, pH 7.2; phases B, 100 mM sodium acetate/acetonitrile/methyl alcohol (20:40:40, v/v/v, pH 7.2)) were 95:5 (v/v) during 0−5 min and gradually changed to 5:95 (v/v) during 5−17 min. The chromatography was performed at 25℃ with a flow rate of 0.2 mL/min using a Zorbax 300SB−C18 column (Agilent Technologies). The workstation was operated in positive−ion linear mode with the following parameters: 20 kV accelerating voltages, 100 nsec extraction delay, and 500 m/z low mass gate[13]. The molecular weight of the sample was analyzed with Data Explore™ software (Applied Biosystems, Carlsbad, CA, USA).

Table 1　Determination of hapten density of CPFX−BSA conjugates by chemical TNBS method and mass spectrometry analysis

conjugate	Protein−hapten mole ratio	Chemical TNBS method		Mass spectrometry analysis		ΔM/Mh[b] (hapten density)
		Percentage of amino group consumed	Approximate amount of amino group consumed	Observed quantity (Da)	Quantity variation (ΔM)	
Control	1:0	0	0	66210	0	0
C1	1:160	36.2	21.7 (22)	73015	6805	20.55 (21)
C2	1:320	41.5S	24.9 (25)	74202	7992	24.14 (25)
C3	1:480	41.5	24.9 (25)	74622	8412	25.41 (26)

conjugate	Protein-hapten mole ratio	Chemical TNBS method		Mass spectrometry analysis		$\Delta M/Mh^b$ (hapten density)
		Percentage of amino group consumed	Approximate amount of amino group consumed	Observed quantity (Da)	Quantity variation (ΔM)	
C4	1:640	47.1	28.2 (28)	75216	9006	27.20 (28)
C5	1:800	49.9	29.9 (30)	76100	9890	29.87 (30)

a. Mh indicates molecular weight of CPFX hapten, Mh = 331.1; b. Values in parentheses indicate deduced moles of CPFX binding on each BSA molecule.

Immunization

The female 4 weeks old Balb/c mice were kept in the sterile room with feeding every day. CPFX-BSA conjugates (150 μg each) and Freund's complete adjuvants were mixed in the ratio of 1:1 to immunize the Balb/c mice. The antigen sample with adjuvant was intraperitoneally injected into the mice and boosted after two weeks. One week later, the mice were immunized again in the same way without adjuvant. After 10 days, a part of the blood sample was collected to determine the antisera titer by ELISA, and then 150 μg each of CPFX-BSA conjugate was injected into the mice by tail intravenous injection. Three days later, the blood sample of the mice was collected and kept at 4℃ overnight. The sample was centrifuged at 7,000 rpm for 10 min. The supernatant was collected to determine the titer and lyophilized.

ELISA

A hundred μL of CPFX-conjugated ovalbumin, which was prepared in our laboratory (10), with a concentration of 400 ng/mL was added into wells of an ELISA plate (Dingguo Biotechnology Co., Ltd.) and incubated at 37℃ for 1 h. After coating, the wells were washed 3 times (3 min each) with phosphate buffer containing 0.5% Tween-20 (washing solution). The wells were blocked with 200 μL 5% milk and stored at 4℃ overnight. Two additional wells were filled with 100 μL buffer to serve as a control for nonspecific binding. After blocking, 100 μL of immunized mice sera (antisera) were added to the wells and incubated for 1 h at 37℃. After rinsing with washing solution, 100 μL of goat anti-mouse IgG conjugated with horseradish peroxidase (HRP) (1:6,000) was added and incubated for 0.5 h at 37℃. After washing 3 times, color development was initiated by adding 100 μL of 3, 3', 5, 5'-tetramethylbenzidine (TMB)/H2O2 and, 10 min later, stopped by adding 50 μL of 2 M H2SO4. The absorbance was determined at 450 nm with a BioTek EL311 micro-plate reader (BioTek, Winooski, VT, USA).

Detection of the sensitivity of antisera by indirect competitive ELISA

The procedure of indirect competitive ELISA was similar to the ELISA process with some modifications. After blocking, 50 μL of suitably diluted antisera was added to each well and 50 μL samples of various concentrations of CPFX solution (2, 4, 6, 8, and 10 ng/mL) were added. The subsequent steps were the same as described above.

Results and Discussion

CPFX-BSA conjugates with various hapten densities were synthesized as described in Materials and Methods. In UV scanning spectra (250-350 nm), the maximum absorption peaks of BSA, CPFX, and CPFX-BSA conjugates were at 280, 270, and 275 nm, respectively, and the higher binding molar ratios of the carrier hapten-protein (C1-C5) correlated with stronger UV absorbance (data not shown). SDS-PAGE analysis of the conjugates (C1-C5) revealed that the molecular weights of the conjugates were larger than that of BSA (data not shown). These results suggest the success of the coupling reaction between CPFX and BSA. The amount of free amino acid residues on the BSA molecule before and after coupling reaction was determined using the TNBS method and the amount of CPFX binding to the carrier BSA was calculated. With mass spectrometry analysis, the density of the hapten bound to the carrier protein was also determined by comparing the variations of molecular weight. As shown in Table 1, hapten densities of the conjugates increased gradually with increase of the hapten-protein molar ratios.

Next, mice were immunized with the conjugates C1-C5 and the antisera titer was determined by ELISA as described in Materials and Methods. As shown in Figure 1A, the titer rose according to the increase of the ratio between carrier protein and hapten and reached 5.76×10^5 at the ratio of 1:800 (C5). In contrast, the antisera titer with different hapten density did not show a significant difference (data not shown). The sensitivity of the antibody produced by the coupling reaction was determined by the ELISA method. The standard dilution curve analysis of the antibody prepared by the conjugate C5 exhibited good sensitivity up to a level of ng/mL

(Figure 1B) . By indirect competitive ELISA, the IC50 value was 10. 7 ng/mL (data not shown).

To acquire immunogenicity, the hapten needs to be coupled with a macromolecule (such as a carrier protein) and become a complete antigen. Generally, small molecules can covalently bind to a carrier protein. Different coupling methods are chosen and designed based on the different functional groups of the hapten. By the traditional EDC method, the conjugation efficiency for CPFX is not high enough as expected and the max value of hapten densities can reach 16[1]. In the traditional method, NHS is used as the activating group and can promote the carboxyl group to couple with the amino group (1, 8), while the modified one without NHS can lead to a higher coupling efficiency. Unlike other fluoroquinolones, CPFX is abundant in amino and carboxyl groups. To couple CPFX, NHS may not be necessary for this reaction. In this method, the extra hapten and EDC were removed by extensive dialysis to ensure the accuracy of the following analysis. BSA is widely used as a carrier protein, because it contains various amino residues on the surface. Compared to tyrosine, tryptophan and imidazole residues, the lysine residue epsilon amino group allows CPFX hapten to couple much more easily by a covalent bond[1].

The production of the conjugates can be confirmed by SDS-PAGE. The greater amount of protein molecule bound by the hapten and the larger molecular weight of the conjugate correspond with the shorter migration distance. Electrophoresis analysis can analyze the process of conjugate coupling. Compared with the molecular weight of BSA (66 kDa)[14], the differences between different conjugates are very slight (3 kDa) (data not shown) . As a result, the differences between the migration distances of the conjugates (C1-C5) were not significant. The hapten density of the conjugates can be calculated more precisely by comparing the molecular weight variations of the conjugates with mass spectrometry. There are 59 lysine residues in the BSA molecule[14], in which 26 residues exist on the surface[5]. The results show that the number of CPFX molecules did not increase linearly with the increase of the reaction ratio between CPFX and BSA; however, the number of CPFX bound to each BSA molecule can reach as high as 30 (Table 1) . It can be explained that the BSA molecule exists in an isomeric structure with high helicity in the environment at pH 5. 0[15]. This structure contains a part of lysine, which normally exists in the inner part of the protein, exposed on the surface of the carrier protein molecule. Another explanation could be that the hapten can couple with other amino residues besides lysine.

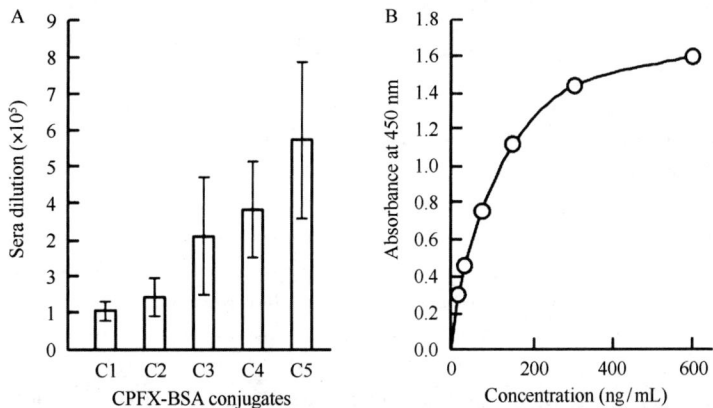

Figure 1　Qualitative analysis of the antibodies generated by various CPFX-BSA conjugates. (A) Titers of antiserum against CPFX obtained from mice immunized with CPFX-BSA conjugates with various hapten densities. Mice were immunized with C1, C2, C3, C4, and C5 having average hapten densities of 21, 25, 26, 28, and 30, respectively. (B) Standard dilution curve analysis of antibody generated by the conjugate C5.

The maximum absorption wavelength in the spectrogram can be used as the basis for the characteristic group analysis of the conjugates. Compared to the characteristic absorption peaks of the carrier protein and hapten, the ones of the conjugates (C1-C5) were shifted (data not shown) . The results, which indicate the presence of distinctive groups in the conjugates, could confirm the specific binding of BSA and CPFX. The UV spectra of the conjugates suggested that the increase of the hapten-protein ratio caused the absorbance of the conjugate to increase gradually. This also validated the conclusion that the hapten density increased with the increase of hapten-protein ratio. The amount of hapten bound to the carrier protein is an important factor which affects the quantity and quality of the antibody produced by the conjugate. However, the titer and affinity of the antibody did not increase linearly with the increase of the amount of the hapten bound to carrier protein[1]. Generally, the binding affinity was optimal when 15 to 30 hapten molecules had bound to the surface of the carrier protein[5]. The greater amount of hapten bound to the carrier protein, the higher, stronger and more specific antibody titer was produced by their conjugates. However, the immune reaction induced by the conjugate bound with less hapten was slow while the antibody produced by it had a higher affinity. The acquirement of optimal hapten density was important for the conjugate in the preparation of antibody. In addition, the antisera displayed better sensitivity than previously published[2]. These findings could contribute to improving immunoassay methods.

In conclusion, the protein-hapten mole ratio could increase the hapten density of conjugates, which has a great influence on the immunogenicity of CPFX. An optimum molar ratio of CPFX-BSA conjugates could induce antibodies with good sensitivity, which was suitable to be applied for immunization purposes.

Acknowledgements

This study was supported by the 863 Program (Grant No. 2011AA10A216) and Special Fund for Agroscientific Research in the Public Interest (Grant No. 201203085).

References

1　Erlanger B. F. The preparation of antigenic haptencarrier conjugates: A survey. Methods Enzymol. 1980; 70: 85-104.

2　Duan J. H., Yuan Z. H. . Development of an indirect competitive ELISA for ciprofloxacin residues in food animal edible tissues. J Agric Food Chem. 2001; 49: 1087-1089.

3　Fuentes M., Palomo J. M., Mateo C., Venteo A., Sanz A., Fernández-Lafuente R., Guisan J. M. . Optimization of the modification of carrier proteins with aminated haptens. J Immunol Methods. 2005; 307: 144-149.

4　Marco M. P., Gee S., Hammock B. D. . Immunochemical techniques for environmental analysis II. Antibody production and immunoassay development. Trends Analyt Chem. 1995; 14: 415-425.

5　Singh K. V., Kaur J., Varshney G. C., Raje M., Suri C. R. . Synthesis and characterization of hapten-protein conjugates for antibody production against small molecules. Bioconjug Chem. 2004; 15: 168-173.

6　Hernandez M., Aguilar C., Borrull F., Calull M. . Determination of ciprofloxacin, enrofloxacin and flumequine in pig plasma samples by capillary isotachophoresis-capillary zone electrophoresis. J Chromatogr B Analyt Technol Biomed Life Sci. 2002; 772: 163-172.

7　Diario Oficial de las Comunidades Europeas (DOCE), August 18, 1990. Council Regulation No. 2377/90 L224, 991, 2601.

8　Zhou Y., Li Y., Wang Z., Tan J., Liu Z. . Synthesis and identification of the antigens for ciprofloxacin. Chinese Journal of Veterinary Science. 2006; 26: 200-203. (in Chinese)

9　Jiang Y., Huang X., Hu K., Yu W., Yang X., Lv L. . Production and characterization of monoclonal antibodies against small hapten-ciprofloxacin. Afr J Biotechnol. 2011; 10: 14342-14347.

10　Hua K., Huang X., Jiang Y., Fang W., Yang X. . Monoclonal antibody based enzyme-linked immunosorbent assay for the specific detection of ciprofloxacin and enrofloxacin residues in fishery products. Aquaculture. 2010; 310: 8-12.

11　Huang X., Hu K., Fang W., Jin Y., Yang X. . Preparation of ciprofloxacin-carrier protein conjugation and identification of its products. Journal of Shanghai Fishery University. 2008; 17: 585-590. (in Chinese)

12　Habeeb A. F. . Determination of free amino groups in proteins by trinitrobenzenesulfonic acid. Anal Biochem. 1966; 14: 328-336.

13　Pingbo Z., Yamamoto K., Wang Y., Banno Y., Fujii H., Miake F., Kashige N., Aso Y. . Utility of dry gel from two-dimensional electrophoresis for peptide mass fingerprinting analysis of silkworm protein. Biosci Biotechnol Biochem. 2004; 68: 2148-2154.

14　Hirayma K., Akashi S., Furuya M., Fukuhara K. . Rapid confirmation and revision of the primary structure of bovine serum albumin by ESIMS and Frit-FAB LC/MS. Biochem Biophys Res Commun. 1990; 173: 639-646.

15　Rosenoer V. M., Oratz M., Rothschild M. A., eds. Albumin Structure, Function and Uses. Pergamon, New York, USA, 1977. (Received September 19, 2011; Revised November 29, 2011; Re-revised February 22, 2012; Accepted March 9, 2012)

该文发表于《上海水产大学学报》【2008, 17 (5): 585-590】

环丙沙星偶联物的制备及其产物的鉴定

Preparation of ciprofloxacin-carrier protein conjugation and identification of its products

黄宣运，胡鲲，方伟，金怡，杨先乐

(上海海洋大学农业部渔业动植物病原库, 上海 200090)

摘要：采用碳二亚胺法，制备了环丙沙星-小牛血清蛋白偶联物，评价了 pH、环丙沙星和碳二亚胺浓度对偶联结合比率的影响。同时采用红外光谱、紫外光谱、聚丙烯酰胺凝胶电泳和质谱方法对偶联产物的基团、分子量等特征进行分析鉴定。正交试验结果表明，pH 对环丙沙星-小牛血清蛋白的偶联结合比率具有显著性差异，碳二亚胺和环丙沙星浓度差异性不显著。在 pH 5.0，碳二亚胺和环丙沙星浓度分别为 60 mg/mL 和 5 mg/mL 的条件下，环丙沙星-小牛血清蛋白偶联结合比率为 30：1。该方法条件下的偶联反应成功。结论：该方法可用于环丙沙星-小牛血清蛋白的偶联，且简便易行。

关键词：环丙沙星；碳二亚胺法；偶联；鉴定

2. 单克隆抗体的制备及鉴定

相比多克隆抗体，单克隆抗体特异性更强，重现性好以及易于扩大化生产，以其为基础建立药物残留的免疫学分析方法是目前的主流发展方向。制备针对环丙沙星（Yousheng jiang et al.，2011；胡鲲等，2010；杨先乐等，2010）的单克隆抗体，可以建立 ELISA 检测方法能特异性地检测水产品中恩诺沙星和环丙沙星残留。

该文发表于《African J Biotechnol》【2011，10（65）：14342-14347】

Production and characterization of monoclonal antibodies against small hapten-ciprofloxacin

Yousheng Jiang，Xuanyun Huang，Kun Hu，Wenjuan Yu，Xianle Yang * and Liqun Lv

(College of Fisheries and Life Science, Shanghai Ocean University, Shanghai 201306, P. R. China.)

Abstract：High affinity and specificity monoclonal antibody against small hapten-ciprofloxaicn (CPFX) was produced and was used for rapid CPFX immunoassay in food stuffs of animal origin. Firstly, two kinds of antigens for CPFX were made using carrier proteins, bovine serum albumin (BSA) and ovalbumin (OVA), by a modified carbodiimide method and then the hapten-protein conjugates were characterized by ultraviolet spectrophotometry to detect hapten density before being used for the immunization and detection purposes. The production of monoclonal antibodies (Mabs) was sought following the generation of appropriate CPFX-BSA conjugate. Spleen cells from Balb/c mice immunized with CPFXBSA conjugate were fused with SP2/0 myeloma cells, and hybridomas secreting antibodies against CPFX were selected and cloned. One Mab against CPFX was produced and the average affinity of the Mab was 2.88×10^9 L/mol. The number of the hybridoma chromosome was 90 to 105. Except for enrofloxacin, the Mab had not any cross-reactivity with other fluoroquinolones. In the optimized ELISA, the Mab gave a 50% inhibition of 7.77 ng/mL with a detection limit of 1.56 ng/mL.

Key words：Ciprofloxacin, hapten, antibody, immunoassay.

单克隆抗体对抗小半抗原环丙沙星反应

姜有声，黄宣运，胡鲲，喻文娟，杨先乐，吕利群

(水产与生命科学学院，上海水产大学，中国 上海 201306)

摘要：高亲和力和特异性抗原的单克隆抗体对抗小半抗原环丙沙星（CPFX）确实有效同时，曾被用于动物来源食品中 CPFX 的快速检验。首先，对 CPFX 两种抗原中使用的载体蛋白，牛血清白蛋白（BSA）、卵清蛋白（OVA），用碳二亚胺法改进，然后用紫外分光光度法检测半抗原蛋白结合物中的抗原密度，再用于免疫与检测。生产单克隆抗体（单克隆抗体）要适当的遵循 BSA 结合物的产生。用 BSA 结合物免疫的 BALB/c 小鼠的脾细胞与 SP₂/0 骨髓瘤细胞发生融合，杂交瘤分泌经过选择和克隆的抗体对抗 CPFX。一个单克隆抗体因环丙沙星所产生单克隆抗体的平均亲和力为 2.88×10^9 L/mol，杂交瘤细胞染色体数为 90~105。除了恩诺沙星，单克隆抗体与其他氟喹诺酮类药物无交叉反应性。在经优化的酶联免疫吸附试验中，单克隆抗体在检出限为 1.56 ng/mL 中给了 7.77 ng/mL 的结果，抑制为 50%。

关键词：环丙沙星，半抗原，抗体，免疫

INTRODUCTION

Ciprofloxacin (CPFX) which was introduced in 1987 (Barry and Fuchs, 1991) is one of the third generation members of synthetic fluoroquinolone group. When compared with the developed fluorine-containing pyridinecarboxylic acid derivatives, CPFX has activity about four times greater against almost all of Gramnegative bacteria (Van Caekenberghe and Pattyn, 1984). The primary mechanism of CPFX activation is the inhibition of DNA gyrase and topoisomerase IV that control DNA topology and are vital for bacterial replication (Higgins et al., 2004). Due to its broad antibiotic activity, CPFX was widely used in the veterinary therapeutic treatment of several animal species (Margarita et al., 2002). However, the CPFX residues in food chain can cause a potential risk in human through the emergence of drugresistant bacteria (Zhou et al., 2008). Thus, the maximum residues limits (MRLs) 30μg/kg of CPFX and enrofloxacin in food stuffs of animal origin were established in European Union (Hammer and Heeschen, 1995).

At present, many chemical analytical techniques have been established for analyzing CPFX in samples, this include spectrophotometry (Nagarall et al., 2002), fluorimetry (Veiopoulou et al., 1997), high-performance liquid chromatography (Ramos et al., 2003) and capillary electrophoresis (Hernandez et al., 2000). However, they all have some limitations in term of time-consuming and require extensive sample cleanup (Ross and Larry, 2001). Antibody-antigen reaction was used to analyze chloramphenicol (Robert, 1996), tetrodotoxin (Raybould and Inouye, 1992) and enrofloxacin (Watanabe et al., 2002). It is advantageous, in that it is cheaper in cost and involves a less operation step which make it particularly useful in routine work (Nuria et al., 2007). When compared with other methods for CPFX detection, fewer studies about immunoassays have been made and this was mainly due to lack of specific and sensitive monoclonal antibody against CPFX which can support rapid CPFX immunoassay. In this study, we prepared specific monoclonal antibodies against CPFX and established a ci-ELISA method for CPFX detection.

ATERIALS AND METHODS

Ciprofloxacin, enrofloxacin, ofloxacin, pefloxacin, norfloxacin and sarafloxacin were purchased from Zhejiang Guobang pharmaceutical Co. Ltd. Bovine serum albumin (BSA), ovalbumin (OVA), tetramethylbenzidine (TMB) and peroxidase horseradish (HRP) were obtained from Dingguo biotechnology Co. Ltd. 1-Ethyl3carbodiimide methiodide (EDC, purity ≥ 99.3%) was obtained from Shanghai Yanchang biochemical technology Co. Ltd. RPMI 1640, HAT, polyethylene glycol-400 (PEG) and calf serum clarified were obtained from Sigma Chemical Co. Ltd. Myeloma cells SP2/0 were conserved by our laboratory. Balb/c mice were gotten from the Second Military Medical University. Chemical reagents such as NaCl, KCl and K2HPO4 were from Shanghai Guoyao Chemical Reagent Co. Ltd.

Preparation of CPFX-protein conjugates The conjugate of CPFX-BSA was synthesized by a modified carbodiimide method. 1 mL of CPFX (20 mg/ml in 0.01M PBS, pH 5.0) was mixed with 1 mL of BSA (8 mg/ml in 0.01M PBS, pH 5.0) and 1 mL of EDC (240 mg/ml in 0.01M PBS, pH 5.0). Then 1 mL of PBS (0.01 M, pH 5.0) was added and the reaction was carried out in the buffer and incubated at 28℃ for 2 h. The mixture was dialyzed for 2 days in PBS (pH 5.0), and the dialyzing buffer was changed every day. After freeze-drying, the conjugate CPFX-BSA was obtained and stored at-20℃. The same method was used to prepare CPFX-OVA conjugate.

Analysis of conjugates by ultraviolet spectrophotometry

The conjugates were diluted in PBS and were scanned by thermo spectronic ultravioletspectrophotometer, and the range of scanning wavelength was from 250 to 350 nm. Scanning step of the obtained spectrograms was 1 nm at scanning speed of 1 nm/s^{-1}.

Production of monoclonal antibodies

Balb/c mice of about 4 weeks old were injected with CPFX-BSA. A mixture of CPFX-BSA conjugate (2 mg/mL, 0.1 mL) and Freunds complete adjuvant (0.1 mL) was injected into mice intraperitoneally at the first time. The next week, a similar injection was administered using Freund's incomplete adjuvant and two booster injections were given by tail vein at 1 week intervals. Three days after the last injection, the mice were sacrificed for fusion. Spleen cells obtained from the final immunized mouse were fused with SP2/0 myeloma cells using polyethylene glycol. An indirect ELISA assay was used to screen hybridomas secreted CPFX positive antibodies. Selected hybridomas were subsequently cloned three or more times by limiting dilution, and one clone stable in culture was eventually chosen for further study. Hybridomas (1 to 5×106 cells) were injected into Balb/c mice (7 weeks old) through abdomens after liquid olefin injection for 7 days. The ascites can be obtained through the needle of a 20 mL injector about seven days later. The Mab was purified using Protein G Sepharose 4 Fast Flow.

Indirect ELISA

CPFX-OVA (1.6 μg/mL) were coated into a 96-well microplate (100 μL per well) for 1 h at 37℃. After three washes with PBST (0.5% tween), the wells were blocked with 1% BSA in PBS overnight at 4℃ and washed with PBST. Then the Mab (100 μL per well) were added for 1 h at 37℃. Some wells in the 96-well plate were incubated with PBS as negative control and block control. After washing with PBST, the plate was incubated in the horseradish peroxidase conjugated goat anti-mouse serum (1：6000) for 0.5 h at 37℃ and washed with PBST. The horseradish peroxidase reaction was developed in a substrate solution of TMB/H2O2 (100 μL per well) for 10 min, then the reaction was stopped by the addition of 2 M H2SO4 (50 μL per well) and the absorbance values were measured at 450 nm with the precise microplate reader. All experiments were made in triplicate.

Identify of hybridoma chromosome

Hybridoma chromosome was identified by the colchicines inhibition method. The number of the chromosome was counted and the photos were taken by an Olympus microscope.

Characterization of antibody

Subclass of antibody was determined by Invitrogen's Mouse MonoAb ID Kit (HRP). Mab affinity was measured by the method of a noncompetitive enzyme immunoassay according to Beatty et al. (1983). CPFX - OVA conjugates with the concentration of 200 ng/mL, 100 ng/mL, 50 ng/mL and 25 ng/mL was coated in a 96-well microplate (100 μL per well) for 1 h at 37℃. The subsequent steps were made as described in indirect ELISA.

A ci-ELISA was used to determine the sensitivity and specificity of Mab. CPFX - OVA conjugates with the concentration of 100 ng/mL was coated into a 96-well microplate, 50 μL of Mab (1：16000) were added to each well and 50 μL CPFX of different concentrations (50, 25, 12.5, 6.25, 3.125, 1.5625 ng/mL) were further added. The subsequent steps were the same as described in indirect ELISA. To assess the cross-reactivity of the Mab, tests were made using enrofloxacin, ofloxacin, pefloxacin, norfloxacin and sarafloxacin.

Samples preparation

Minced samples (5 g) were spiked with ciprofloxacin at concentrations of 10, 50 or 100 ng/mL. The spiked samples were homogenized with 10 mL ethyl acetate, vortexed for 10 min and centrifuged at 600 g for 10 min. The supernatants were added directly to the microtiter plates and analyzed by ci-ELISA.

RESULTS

Quantitative analysis of molecule conjugate ratio

The result of UV scanning spectrums is shown in Figure 1. The maximum absorption peaks of BSA, CPFX and CPFX-BSA conjugate were at 279, 270 and 274 nm, respectively. The maximum absorption peak of CPFXBSA was different from BSA and CPFX, which indicated the success of synthesis. The successful coupling reaction between CPFX and BSA could be speculated according to the methods of Yang et al. (1998), and the molecular conjugate ratio of CPFX to BSA was 30. CPFX-OVA was also analyzed by the UV scanning spectrums method and was found that the molecular conjugate ratio of CPFX to OVA was 12, so CPFX-OVA could be used in ELISA assay.

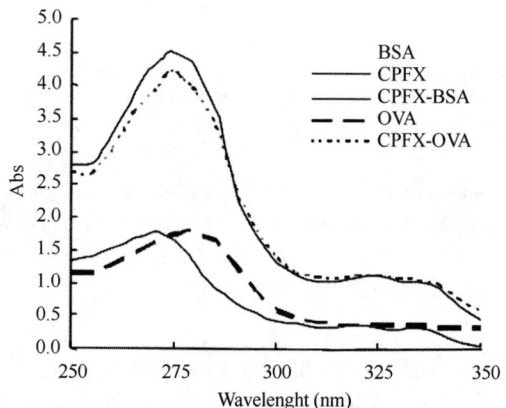

Fig. 1　The ultraviolet absorb spectral curve of BSA, OVA, CPFX-BSA and CPFXOVA

Chromosome identification of hybridomas

The average chromosome numbers of SP2/0 and spleen cells were 62 to 70 and 38 to 40, respectively. The chromosome number of hybridomas which secreted Mab (4F5) against CPFX were 90 to 105 (Figure 2), indicating the chromosome of hybridomas were from SP2/0 and spleen cells.

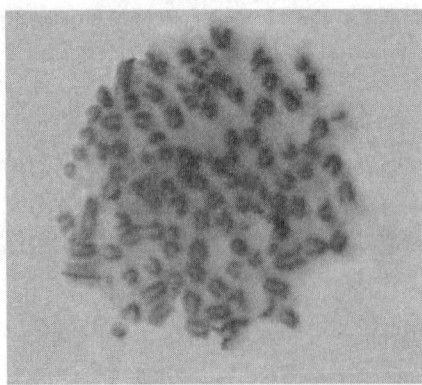

Fig. 2　The chromosome of hybridoma by the colchicines inhibition method

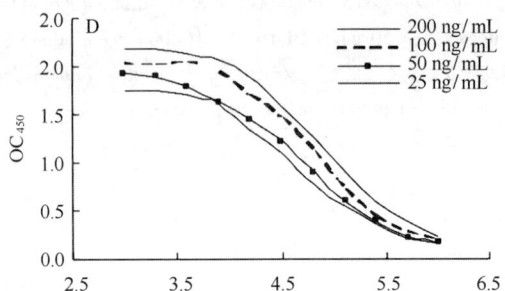

Fig. 3　Measurement of Mab affinity by non-competitive ELISA. The affinity of Mab was caculated by the equation as follow: Ka= (n-1) /2n ([Ab'] t- [Ab] t) . n= [Ag'] t/ [Ag] t. [Ag'] t and [Ag] t were different concentration of coated antigen. [Ab'] t and [Ab] t were the concentration of monoclonal antibody at 50% OD_{max} which was corresponding to the different coated antigen concentrations.

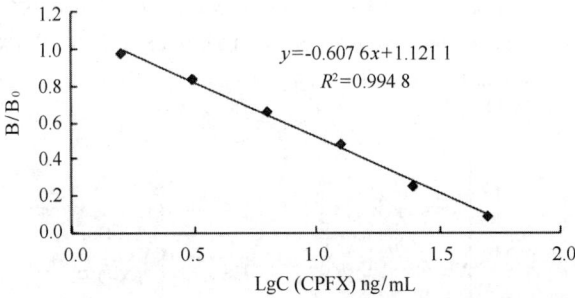

Figure 4　Standard curve of competitive direct ELISA for CPFX. Standard curves were prepared by plotting logarithm of CPFX concentration versus percent binding, which is calculated by the absorbance in the absence (B0) and presence (B) of CPFX in standard. In the range of 1. 56 to 50 ng/mL, the CPFX concentration showed the good linear relationship with B/B0.

The characterization of Mab

The antibody (4F5) was LgG2a with K light chain and theaverage affinity of 4F5 was 2. 88×10⁹ L/mol (Figure 3). Representative ci-ELISA curves for CPFX, obtained using monoclonal antibody are shown in Figure 4. The Mab exhibited a high sensitivity to CPFX with an IC_{50} value of 7. 77 ng/mL and a detection limit of 1. 56 ng/mL. The Mab 4F5 had cross-reactivity with enrofloxacin (Table 1) .

Detection of ciprofloxacin in various matrices

For used in the ELISA performance evaluation, three matrices were chosen. The recoveries are shown in Table 2. The recoveries in chicken muscle ranged from 95% to 122%, recoveries in chicken liver ranged from 98% to 115% and recoveries in aqua-production ranged from 80% to 95%.

Table 1 Cross-reactivity of monoclonal antibody against other drugs

Parameter	IC$_{50}$ (ng/mL)	CR (%)
Ciprofloxacin	7. 77	100
Enrofloxacin	8. 18	95
Ofloxacin	>100000	<0. 01
Pefloxacin	>100000	<0. 01
Norfloxacin	>100000	<0. 01
Sarafloxacin	>100000	<0. 01

Table 2 Recovery of ciprofloxacin from muscle, liver, aqua-production (n = 10)

Sample	Spiked lever (ng/mL)	Recovery (%)	CV (%)
Chicken Muscle	10	85. 8±8. 16	9. 51
	50	88. 8±7. 04	7. 92
	100	83. 3±11. 1	13. 3
Chicken Liver	10	82. 8±8. 68	10. 4
	50	85. 1±8. 14	9. 56
	100	92. 2±8. 7	9. 44
Aqua-production	10	81. 8±5. 71	6. 98
	50	81. 2±9. 72	12. 0
	100	89. 7±12. 3	13. 7

DISCUSSION

The CPFX is a small-molecular hapten (MW< 1000), which alone is not immunogenic, unless it is coupled withsome macromolecules such as proteins (Singh et al. , 2004). BSA was chosen as a carrier protein because it contained various amino residues on the surface and is easy to form a soluble conjugate compared with other carrier proteins. Moreover, the hapten-protein conjugate with high efficiency and this is important to the immunological analysis (Steffensa et al. , 2002). It is assumed that high antibody titers and moderate antibody affinities are usually obtained with hapten density of 15 to 30 molecules per carrier protein (Singh et al. , 2004). The coupling ratio of CPFX-BSA in this paper (30 : 1) was higher than CPFX-OVA (12 : 1), so CPFX-BSA can be used as immunogen.

Several synthesis methods of hapten with carrier protein were reported, such as mixed anhydride, carbodiimide method and glutaraldehyde method. However, few researches were reported about the conjugate of CPFX-protein by carbodiimide method which does not add the N-Hydroxysuccinimide. The production of monoclonal antibody also indirectly confirmed that the method of conjugating CPFX with carrier protein was successful. In ELISA assay, the conjugate CPFX-OVA were used as coating antigen to avoid the binding of Mabs with carrier protein BSA. The cross reactivity with other fluoroquinolones (FQs) of the Mabs was examined and the result showed that three Mabs had strong cross reactivity with other FQs in the four Mabs produced. One of the Mabs (4F5) against CPFX had cross-reactivity only with enrofloxacin, so the Mab (4F5) was chosen for further research. Our result shows that the Mab (4F5) had high cross reactivity with enroflxoacin. This corresponded to the work of Wang et al. (2007), who prepared monoclonal antibody against CPFX which also had high cross-reactivity with enroflxoacin. Duan and Yuan, (2001) prepared polyclonal antibodies against CPFX and the antibody exhibited a higher cross-reactivity to enroflxoacin (70%). This may be that the structure of CPFX and enrofloxacin was similar. Both CPFX and enrofloxacin had cyclopropyl group at position 1 in ring and the lack of an ethyl group on the piperazinyl ring was not detrimental to antibody binding.

Affinity is another important parameter of Mab, it is generally believed that affinity 10^9 to 10^{12} L/mol is adapted to immunodetection. The affinity of 4F5 in this paper is 2. 88×10^9 L/mol and the IC$_{50}$ is 9. 83 ng/mL, so it could be used for CPFX analyzing. An indirect ELISA for CPFX was reported (Duan and Yuan, 2001) and it was observed that the IC50 was 50 ng/mL, but the antibody was cross-reactive with enrofloxacin and norfloxacin.

In conclusion, we prepared a sensitive, specific monoclonal antibody against CPFX and it is useful for analyzing CPFX residues in the edible tissues of animals used as food through ci-ELISA.

ACKNOWLEDGEMENTS

This research was supported by Science and Technology of Shanghai Agriculture Committee (granted 6660106477) and key projects in the National Science and Technology Pillar Program during the eleventh fiveyear plan period (2006BAD03B04).

REFERENCES

1 Barry A. L., Fuchs P. C.　(1991). Anti-staphylococcal activity of temafloxacin, ciprofloxacin, ofloxacin and enoxacin. J. Antmicrob. Chemother. 28: 695-699.

2 Beatty J. D., Beatty B. G., Vlahos W. G.　(1983). Measurement of monoclonal antibody affinity by non-competitive enzyme immunoassay. J. Immunol. Methods, 100 (1-2): 173-179.

3 Duan J. H., Yuan Z. H.　(2001). Development of an Indirect Competitive ELISA for Ciprofloxacin residues in food animal edible tissues. J. Agric. Food Chem. 49: 1087-1089.

4 Hammer P., Heeschen W. (1995). Antibody-capture immunoassay for the detection of enrofloxacin. Milchwissenschaft, 50: 513-514.

5 Hernandez M., Borrull F., Calul M. (2000). Determination of quinolones in plasma samples by capillary electrophoresis using solid-phase extraction. J. Chromatogr. B. 742: 255-265.

6 Higgins P. G., Fluit A. C., Schmitz F. J. (2004). Fluoroquinolones: structure and target sites. Curr. Drug Targets, 4: 181-190.

7 Margarita H., Carme A., Francesc B., Marta C. (2002). Determination of ciprofloxacin, enrofloxacin and flumequine in pig plasma samples by capillary isotachophoresis-capillary zone electrophoresis. J. Chromatogr. B. 772: 163-172.

8 Nagarall B. S., Seetharamappa J., Melwanki M. B. (2002). Sensitive spectrophotometric methods for the determination of amoxycillin, ciprofloxacin and piroxicam in pure and pharmaceutical formulations. J. Pharmaceut. Biomed. 29: 859-864.

9 Nuria P. N., Ester G. I., Angel M., Rosa P. (2007). Development of a groupspecific immunoassay for sulfonamides Application to bee honey analysis. Talanta, 71: 923-933.

10 Ramos M., Aranda A., Garcia E., Reuvers T., Hoohuis H. (2003). Simple and sensitive determination of five quinolones in food by liquid chromatography with fluorescence detection. J. Chromatogr. B. 789: 373-381.

11 Raybould T. J. G., Inouye L. K.　(1992). A monoclonal antibody-based immunoassay for detecting tetrodotoxin in biological samples. J. Clin. Lab. Anal. 6 (2): 65-72.

12 Robert N. H. (1996). Chloramphenicol-specific antibody. Sci. Translational Med. 152: 203-205.

13 Ross C. B., Larry H. S. (2001). An antigen based on molecular modeling resulted in the development of a monoclonal antibody-based immunoassay for the coccidiostat nicarbazin. Analytica Chimica Acta. 444: 61-67.

14 Singh K. V., Kaur J., Varshney G. C., Raje M., Suri C. R. (2004). Synthesis and Characterization of Hapten-Protein Conjugates for Antibody Production against Small Molecules. Bioconjugate Chem. 15: 168173.

15 Steffensa G. C., Nothdurfta L., Buse G., Thissen H., Hocker H., Klee D. (2002). High density binding of proteins and peptides to poly (D, Llactide) grafted with polyacrylic acid. Biomaterials, 23: 3523-3531.

16 Van Caekenberghe D. L., Pattyn S. R. (1984). In vitro activity of ciprofloxacin compared with those of other new fluorinated piperazinylsubstituted quinoline derivatives. Antimicrob. Agents. Ch. 25: 518-521.

17 Veiopoulou C. J., Ioannou P. C., Lianidou E. S. (1997). Application of terbium sensitized fluorescence for the determination of fluoroquinolone antibiotics pefloxacin, ciprofloxacin and norfloxacin in serum. J. Pharmaceut. Biomed. 15: 1839-1844.

18 Wang Z. H., Zhu Y., Ding S. Y. (2007) Development of a monoclonal antibody-based broad-specificity ELISA for fluoroquinolone antibiotics in foods and molecular modeling studies of cross-reactive compounds. Anal. Chem. 79: 4471-4483.

19 Watanabe H., Satake A., Kido Y., Tsuji A. (2002). Monoclonal-based ELISA and immuno-chromatographic assay for enrofloxacin in biological matrices. Analyst. 127: 98-103

20 Yang I. G., Hu S. Y., Wei P. H. (1998). Enzyme immunoassaytechnic [M], Nanjing University Publishing Company. pp. 279-281.

21 Zhou X. J., Chen C. X., Yue L., Sun Y. X., Ding H. Z., Liu Y. H. (2008). Excretion of enrofloxacin in pigs and its effect on ecological environment. Environ. Toxicol. Pharm. 26: 272-277.

782

该文发表于《中国免疫学杂志》【2010，26（6）：538-543】

环丙沙星单克隆抗体的制备及其免疫学特性分析

Generation of the monoclonal antibody against Ciprofloxacin and analysis for its immunological traits

胡鲲，黄宣运，姜有声，方伟，杨先乐

（上海海洋大学水产与生命学院，上海 201306）

[摘要]　目的：制备环丙沙星单克隆抗体，并分析其免疫学特性，为建立环丙沙星残留的免疫学检测方法服务。方法：用人工抗原环丙沙星-小牛血清白蛋白偶联物免疫 BALB/c 小鼠，产生预期免疫应答后，将小鼠脾细胞与瘤细胞融合，经初筛、复筛和再克隆，筛选得到 1C9、3F6、6H2、6A7、6G11 和 8F56 株分泌环丙沙星单克隆抗体的杂交瘤细胞。对单克隆抗体的亚型、纯度、亲和力、灵敏度及特异性等免疫学特性进行鉴定和分析。结果：1C9、3F6 和 6A7 的抗体亚型为 IgG2a；6H2 和 8F5 的抗体亚型为 IgG1；6G11 的抗体亚型为 IgG3。SDS-PAGE 电泳结果显示单抗蛋白重链分子量约为 50kD，轻链分子量约为 25kD。6 株细胞产生的单抗均具有良好的特异性和灵敏度。其中，细胞株 1C9 细胞培养上清和腹水效价分别为 $1:6.4\times10^2$ 和 $1:5.6\times10^5$，亲和力常数可达 2.85×10^9 L/mol，IC_{50} 值可达 245.86 ng/mL，最低检测限为 45.25ng/mL，对氧氟沙星、双氟沙星、沙拉沙星、左氧氟沙星、孔雀石绿、氯霉素、呋喃西林等药物几乎不存在交叉反应，与恩诺沙星存在交叉反应，交叉反应率为 84.6%。结论：本方法制备的环丙沙星单克隆抗体具有较好的亲和力和特异性，可用于环丙沙星残留免疫学检测。

[关键词]　环丙沙星；单克隆抗体；亚型；亲和力；交叉反应

环丙沙星（Ciprofloxacin，CIP）属于第三代氟喹诺酮类药物[1]，是治疗人类疾病的重要抗生素。其在水产品等动物可食用组织中的残留，直接威胁了公共卫生安全和对外出口贸易[2]。鉴于 CIP 残留的巨大危害，美国和欧盟均将 CIP 列为食品中"严格监控的对象"。我国也将其列为"禁用渔药"[3]。利用免疫学方法作为初筛方法检测水产品中的药物残留，灵敏、快速、特异和简便[1,4,5]，具有十分广泛的应用前景和实用价值。相比多克隆抗体（PcAb），以单克隆抗体（Mab）为基础的免疫学检测方法特异性更强，重现性更好且更易于扩大化生产，是未来小分子药物残留快速检测的主流发展方向。目前，对于 CIP 残留的 ELISA 检测试剂盒大多是国外的产品，国内虽然开展了相关研究，但大多是建立在多克隆抗体（PcAb）的基础上[6,7]，或者处于单克隆抗体（Mab）的摸索阶段。

制备灵敏度高、强特异性的 CIP 单克隆抗体是建立免疫学分析方法的关键。本文以小牛血清蛋白（Bovine serum album，BSA）偶联 CIP，制备 CIP-BSA 完全抗原，并免疫 BALB/c 小鼠，将小鼠脾脏细胞与 SP$_2$/O 骨髓瘤细胞融合，采用间接 ELISA 法和竞争 ELISA 法对阳性克隆进行筛选，经过多次亚克隆，获得稳定分泌针对 CIP 的单克隆抗体，并对抗体进行分离、纯化和鉴定，为下一步试剂盒化组装奠定技术基础。

1　材料与方法

1.1　仪器和试剂

UNICAN 紫外分光光度计；FOR-MA-725 超低温冰箱；FORMA-3111 二氧化碳培养箱；BIO-TEK EL808IU 型酶标仪；OLIMPUS CK30 显微镜；Bio-Rad powerpac300 型电泳仪；UVP ca91786 型凝胶成像系统；SAVANT Novalyphe-NL150 型冷冻干燥机；Eppendorf centrifuge 5417R 型离心机。

CIP，含量≥98.5%，购自浙江国邦药业有限公司。牛血清蛋白（Bovine serum album，BSA）、Ovalbumin（OVA）、Tetramethylbenzidine（TMB）、Peroxidase（HRP）购于鼎国生物技术公司。乙基碳二亚胺（1Ethl-3carbodiimide methiodide，EDC），含量≥99.3%，购自上海延长生化科技发展有限公司。胎牛血清购自 HYCLONE 公司。RPMI1640 培养基、HAT、聚乙二醇（PEG-4000）购自 Gibco 公司。二甲亚砜购自 SIGMA 公司。小鼠单克隆抗体亚型快速鉴定试剂盒购自 Invitrogen 公司。NaCl、KCl、Na2HPO4 和 K2HPO4 等其他化学试剂购于上海国药有限公司。

BALB/c 小鼠购自第二军医大学。饲养环境：室温 20~26℃，湿度 20%~70%，饲养于无毒塑料网隔和不锈钢网中，内垫木屑，定期补给清洁水和饲料。

0.01 mol/L 的磷酸缓冲液（PBS）：称取 8.5 g 的氯化钠、2.85 g 磷酸氢二钠、0.2 g 氯化钾、0.27 g 磷酸二氢钾溶入 950 mL 的蒸馏水，滴加盐酸调节 pH 值至 7.4，加 50 mL 的蒸馏水，定容为 1 L。CIP-BSA 和 CIP-OVA 偶联物由本实验室制备[8]。

本文所涉及的化学试剂、溶液均为分析纯，缓冲溶液由双蒸水配制而成。

1.2 单克隆抗体的制备

1.2.1 免疫小鼠

选取 6 只 4 周龄雌性的 BALB/c 小鼠，取 2 mg/mL CIP-BSA 与弗氏完全佐剂等量混合，腹腔注射小鼠，每只小鼠的免疫剂量为 0.2 mg/次。首次免疫后第 14 d，抗原与弗氏不完全佐剂混合乳化，腹腔注射。以后免疫不加佐剂，分别在第 21 天、第 31 天免疫 1 次，采用尾部静脉注射免疫。第 31 天每只小鼠的免疫剂量为 0.1 mg。

1.2.2 细胞融合及培养

加强免疫后的第 3 天，处死小鼠，取脾脏，过 100 目网筛，用 RPMI1640（-）吹下形成单细胞悬液，用于细胞融合。无菌条件下取 BALB/c 小鼠胸腺，过 100 目网筛，用 RPMI1640（-）吹下形成单细胞悬液。将脾细胞悬液和胸腺细胞悬液分别以 1 000 r/min 离心 3 min，去上清液，脾细胞沉淀 37℃的 RPMI1640（-）重悬，胸腺细胞沉淀用 37℃的含 1%HAT 的 1640（含 10%胎牛血清）选择性细胞培养液重悬。取处于对数生长期的骨髓瘤细胞，1 000 r/min 离心 3 min，去上清液后，沉淀用 RPMI1640（-）重悬。将脾细胞悬液与瘤细胞悬液混合均匀后，1 000 r/min 离心 3 min，去上清液，轻弹离心管底，使两种细胞沉淀充分混匀，37℃水浴 5 min，吸取 37℃的 50%聚乙二醇溶液 1 mL，将管尖插入管底，轻轻搅动细胞沉淀，并缓缓添加 50%聚乙二醇，在 1 min 内加完。滴加 37℃的 RPMI1640（-）40 mL，前缓后快，1 000 r/min 离心 5 min，去上清液。将细胞沉淀用 3 mL 37℃的 1640（含 10%胎牛血清）细胞培养液重悬，取 2 mL 加入 2 mL 冻存液［9 份 1640（含 10%胎牛血清）+1 份二甲亚砜］冻存于-70℃，细胞悬液里补加含有 1%HAT 的 1640（含 10%胎牛血清）选择性细胞培养液和小鼠胸腺细胞 40 mL，混合均匀后每孔 100 μL 滴加到 10 个 96 孔板内。4.5%二氧化碳培养箱中 37℃培养。

1.2.3 检测

细胞融合 2 周后，杂交瘤细胞群落长到占 96 孔培养板的孔底 1/3 时开始检测。用间接酶联免疫法检测融合细胞的阳性孔：取 150 ng/mL 的 CIP-OVA 偶联物包被 96 孔酶标板，每孔 100 μL，37℃孵育 1 h，倒去多余液体，拍干后加入含有 5%牛奶的磷酸缓冲液（pH7.4，0.01 mol/L），200μL/孔，37℃孵育 1 h；倒去多余液体，用含 0.05%吐温 20 磷酸盐缓冲液（PBST）洗三次，每次 3 min（以下简称洗板）；拍干，加入细胞上清液，100μL/孔，37℃孵育 1 h；洗板，拍干后每孔加入 100 μL 过氧化物酶标羊抗鼠 IgG（1∶6000），37℃孵育 0.5 h；洗板，拍干后每孔加入 100 μLTMB 显色（1.5 mg/mL TMB：0.03%H$_2$O$_2$：0.2 mol/L NaAC =1∶5∶4，V/V/V），37℃孵育 10 min；每孔加入 2 mol/L 的硫酸 50 μL 终止反应；选取阳性值大于阴性值 2 倍，且阳性值大于 0.5 以上的细胞孔进行克隆。

阳性对照为免疫小鼠的高滴度血清稀释液（1000×）；阴性对照为骨髓瘤细胞上清；空白为 0.01 mol/L 的磷酸缓冲液。

1.2.4 克隆

采用有限稀释法对检测出的阳性杂交瘤细胞进行克隆：用乙醚麻醉小鼠，无菌取出胸腺细胞（作为饲养细胞），在 100 目网筛上研磨，用 RPMI1640（-）吹下形成单细胞悬液，1000 r/min 离心 3 min，去上清液，胸腺细胞沉淀用 30 mL 1640（含 10%胎牛血清）细胞完全培养液重悬。阳性细胞孔中的细胞用血球计数板计数，然后用 1640（含 10%胎牛血清）培养液以 10 的整次倍稀释，取 450 个杂交瘤细胞，放入 22.5 mL 的胸腺细胞 1640（含 10%胎牛血清）悬液中，用滴管吹打均匀后，滴一个 96 孔板，150 μL/孔。剩余的细胞悬液 7.5 mL 与预留出的 7.5 mL 胸腺细胞混合均匀，滴一个 96 孔板，平均每个孔滴加 150 μL。取阳性杂交瘤细胞转孔并扩增，收集细胞培养上清。

1.2.5 制备腹水

取 8 周的 BALB/c 小鼠，注射 0.5 mL 的降植烷。1 周后腹腔接种用无血清培养基稀释处于对数生长期的杂交瘤细胞，每只注射细胞数为 1×10^6~5×10^{6}个，间隔 4 d 后，观察小鼠腹水情况，待小鼠腹部膨大，精神变差，濒临死亡时，采集腹水。将腹水离心，取上清液，测定效价，分装保存在-70℃。

1.2.6 单抗的纯化

用硫酸铵沉淀法进行纯化[5]。

1.3 单抗的鉴定

1.3.1 间接 ELISA 方法的最佳工作条件的确定

1.3.1.1 确定包被抗原的最佳工作浓度

将抗原梯度稀释，以不同的浓度包被酶标板，4℃过夜；洗板，封闭，37℃孵育 1 h；洗板，加入过量的抗体，37℃孵育 1 h；洗板，加入过氧化物酶标羊抗鼠 IgG，37℃孵育 0.5 h；洗板，加入 TMB 显色液，37℃孵育 10 min；加入终止液，

OD$_{450}$读数。OD 值达到饱和值的最低的抗原包被浓度即为最佳工作浓度。

1.3.1.2 确定抗体稀释浓度

以 1.3.1.1 中确定的包被抗原浓度包被；测定步骤同上，OD 值为 1.0 时的抗体浓度为最佳工作浓度。

1.3.2 单抗抗体类型及亚类鉴定

采用 Invitrogen 公司小鼠单克隆抗体亚型快速鉴定试剂盒，按说明书方法操作。

1.3.3 效价测定

利用 ELISA 法测定杂交瘤细胞细胞培养上清效价及腹水效价。将细胞上清液及腹水倍比稀释。采取间接 ELISA 方法测定其效价，测定方法与 1.2.4 相同。倍比稀释的细胞上清液及腹水为一抗，辣根过氧化物酶标记的羊抗鼠 IgG 为二抗。同时设阴性、空白和阳性对照。

1.3.4 抗体的 SDS-PAGE 电泳纯度鉴定

分别取标准分子量蛋白、等量的经硫酸铵纯化的单克隆抗体进行 SDS-PAGE 分析[5]。

1.3.5 克隆抗体亲和力常数测定

利用非竞争性酶免疫法测定抗体的亲和力[9]。将包被抗原按 200、100、50、25ng/mL 包被酶标板，将单抗从 1000 倍开始倍比稀释，其余同间接 ELISA 检测方法。亲和力常数计算公式为 $Ka = (n-1)/2 (n [Ab']t - [Ab]t)$。其中 $n = [Ag']t/[Ag]t$，$[Ag']t$ 和 $[Ag]t$ 为不同包被抗原的浓度，$[Ab']t$、$[Ab]t$ 是对应不同包被抗原的浓度下，将抗体梯度稀释得到最大吸光度一半处的抗体浓度（mol/L）。

1.3.6 单克隆抗体的交叉反应率

用间接竞争 ELISA（Indirect-competitive ELISA，ic-ELISA）法测定单克隆抗体与其他药物之间的交叉反应率[5]。

1.3.7 竞争抑制曲线的绘制

CIP-OVA（150 ng/mL）包被酶标板，100 μL/孔，37℃孵育 1 h；倒去包被液，拍干后加入含有 5% 牛奶的磷酸缓冲液，200 μL/孔，37℃孵育 1 h；倒去多余液体，用 PBST 洗板；拍干，加入 50 μL 梯度浓度的 CIP 标准溶液，同时加入 50 μL 抗体（1:65 000），37℃孵育 1 h，设阴性对照和空白对照。其余同 1.2.3 方法。以竞争浓度（ng/mL）为横坐标，抑制率（B/B$_0$%）为纵坐标，绘制竞争抑制曲线，计算 IC$_{50}$（50% 抑制浓度）。最低检测限（LOD）为（B$_0$-3SD）/B$_0$ 所对应的浓度值。

2 结果

2.1 动物免疫

经过三次免疫后，6 只 BALB/c 小鼠尾部采血，用 ELISA 方法检测效价：2 和 4 号小鼠效价可达 1:1.25 ×10^5，1、3、5 和 6 号小鼠抗血清效价可到 1:6.25 ×10^5，可用于制备融合脾细胞。免疫双扩试验结果证明人工抗原刺激小鼠产生的抗血清对环丙沙星的具有特异性。

2.2 间接 ELISA 方法的最佳工作条件的确定

建立良好的 ELISA 筛选方法可以有效地避免假阳性结果，将阳性细胞株筛选出来。通过方阵法确定间接 ELISA 最佳工作条件为：CIP-OVA 包被原的包被浓度为 150 ng/mL。酶标二抗的工作浓度为 1:6 000。

2.3 细胞融合阳性杂交瘤细胞的筛选和克隆

融合后的细胞接种在 10 块 96 孔板上。其中，435 孔中出现体积略大，透亮并聚集生长的杂交瘤细胞，融合率约为 45.3%；第一次利用 ELISA 检测杂交瘤细胞分泌的上清，其中有 21 孔呈阳性，融合阳性率约为 4.83%。继续培养 2~3 d 后，从中筛选出 10 孔分泌阳性抗体的杂交瘤细胞。选取其中 OD 值比较大的阳性孔进行有限稀释克隆和扩大培养，进行一筛、二筛和三筛。选取阳性值大于阴性值 2 倍，且阳性值大 0.5 以上的细胞孔，分别为 C9、3F6、5H1、6A7、6G11、6H2 和 8F5。

利用竞争法对初筛的 7 株细胞进行常见的药物交叉反应分析，结果见表 1，其中 5H1 号阳性孔上清液针对氯霉素表现出较强的交叉反应，其余阳性孔交叉反应均较低。将 5H1 号阳性孔淘汰，利用其余 6 个阳性孔建株。

表1　ic-ELISA 测定单抗的交叉反应

Tab. 1　Determination of the cross-reaction of Mabs by ic-ELISA

Positive cell	Competitive drug									
	Ciprofloxacin	Enrofloxacin	Ofloxacin	Malachitegreen	Difloxacin	Chloramphenicol	Sarafloxacin	Furacilin	Levofloxacin	Negative control
1C9	0.079	0.067	2.472	2.396	2.46	1.862	2.454	2.46	2.587	2.627
3F6	0.086	0.075	2.497	2.485	2.548	2.23	2.456	2.522	2.583	2.555
5H1	0.075	0.05	2.708	2.588	2.671	0.272	2.671	2.699	2.86	2.847
6H2	0.068	0.227	3.017	2.998	3.081	2.815	3.058	3.037	3.058	2.998
6A7	0.078	0.062	2.7	2.788	2.872	1.191	2.788	2.757	2.718	2.692
6G11	0.081	0.089	2.645	2.486	2.759	1.878	2.453	2.498	2.548	2.475
8F5	0.071	0.061	2.828	2.755	2.828	1.158	2.755	2.802	2.915	2.766

2.4　单抗的鉴定及分析

2.4.1　单抗亚型的鉴定

利用小鼠单克隆抗体亚型快速鉴定试剂盒检测 6 株单克隆抗体蛋白亚型，结果显示：1C9、3F6 和 6A7 与 IgG2a，6H2 和 8F5 与 IgG1，6G11 与 IgG3 有明显的阳性反应。由此可以判定，1C9、3F6 和 6A7 的抗体亚型为 IgG2a，6H2 和 8F5 的抗体亚型为 IgG1、6G11 的抗体亚型为 IgG3。

2.4.2　单克隆抗体蛋白的纯度检测

纯化后的抗体电泳分析结果见图 1。1C9、3F6、6H2、6A7、6G11 和 8F5 6 株细胞分泌的单抗均具有较高的纯度。通过还原法，抗体蛋白变性解链，一条为重链，约在 50kD 处，另外一条为轻链约为抗体具有 25kD。5 株细胞分泌的单克隆抗体蛋白的总分子量均为 150 kD 左右。

2.4.3　单抗抗体亲和力测定

通过非竞争 ELISA 法测定细胞株的亲和力常数[9]。其中，细胞株 1C9 的亲和力常数最高，通过 1.3.5 的公式计算细胞株 1C9 的亲和力常数（图 2）分别为 4.34×10^9、3.03×10^9、2.52×10^9、2.63×10^9、2.35×10^9 和 2.23×10^9 L/mol，平均值为 2.85×10^9 L/mol。

2.4.4　细胞株交叉反应率的测定

恩诺沙星对 1C9、3F6、6H2、6A7、6G11 和 8F5 细胞株均对恩诺沙星表现出交叉反应（图 3）。其中，细胞株 1C9 对恩诺沙星的交叉反应率为 84.6%。

2.4.5　单抗效价的测定及竞争抑制曲线的绘制

杂交瘤细胞株 $1C_9$ 的细胞培养液上清和腹水效价分别为 1∶6.4×10^2 和 1∶5.6×10^5；其分泌的单抗蛋白竞争抑制曲线见图 4，曲线方程为 $Y = -54.896X + 18.24$，$R^2 = 0.9676$，IC_{50} 值为 245.86 ng/mL。以（B0-3SD）/B0 所对应的浓度值确定的最低检测限为 45.25 ng/mL。以细胞株 $1C_9$ 分泌的单抗蛋白建立间接竞争 ELISA 方法，检测 CIP，结果见表 2。在浓度为 0.265~1.261 μg/mL 时，ELISA 检测 CIP 的回收率在 82.64%~91.70% 之间，相对标准偏差在 5.04%~10.49% 之间。

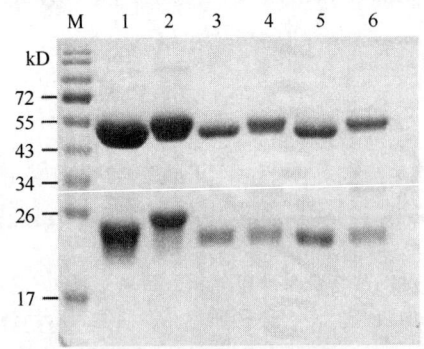

图 1　抗体的 SDS 电泳图

Fig. 1　SDS-PAGE for the purified Mabs

Note：M. Marker；1.1C9；2.3F6；3.6H2；4.6H2；5.6G11；6.8F5.

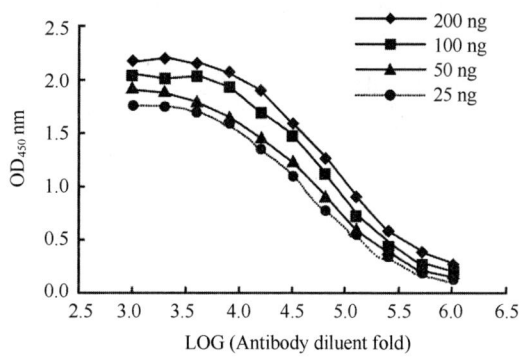

图 2　1C9 单抗的亲和力测定

Fig. 2　Affinity for 1C9 Mab

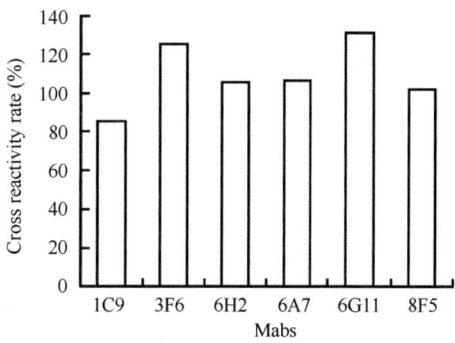

图 3　抗体与恩诺沙星的交叉反应

Fig. 3　Cross reactivity rates of Mabs for enrofloxacin

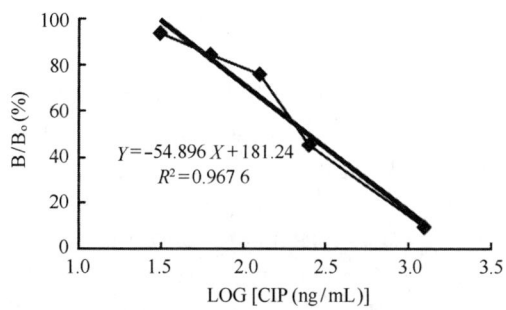

图 4　$1C_9$ 单抗的竞争抑制曲线

Fig. 4　Representative inhibition curve of 1C9

表 2　间接竞争 ELISA 方法测定 CIP（$n=5$）

Tab. 2　Determinate CIP by ic-ELISA method（$n=5$）

No.	Standard solution of CIP	iC-ELISA	
	Theoretical concentration（μg/mL）[a]	Recovery（%）[b]	RSD（%）[c]
1	0.265	82.64±8.67	10.49
2	0.532	88.61±8.98	10.13
3	1.261	91.70±4.62	5.04

Note：a. Theoretical concentration of CIP was determined by HPLC［2］；

b. Recovery. measured value/Theoretical concentration；c. Relative standard deviation，RSD. Standard deviation/Mean value.

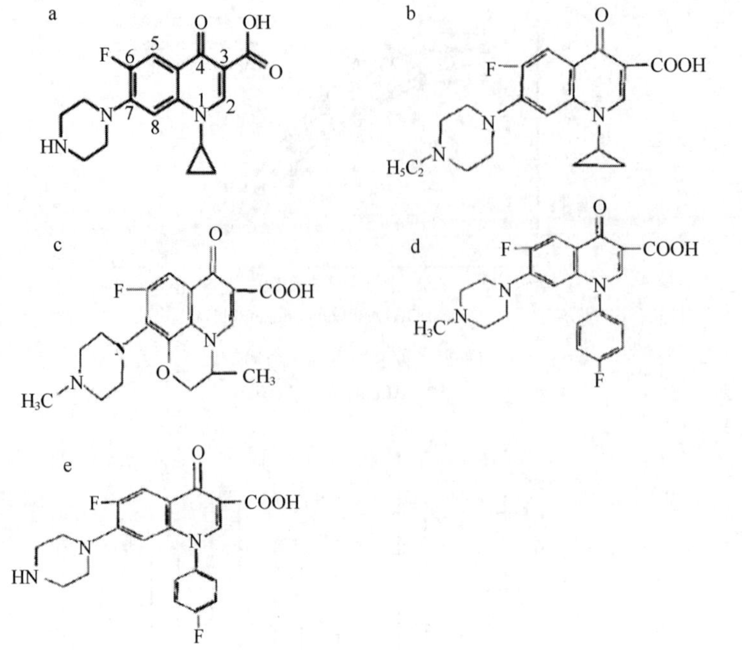

图 5 CIP（a）、恩诺沙星（b）、氧氟沙星（c）、双氟沙量（d）、沙拉沙星（e）结构图

Fig. 5 Molecular structure of Ciprofloxacin（a），Enrofloxacin（b），Ofloxacin（c），Difloxacin（d），Sarafloxacin（e）

3 讨论

制备高亲和力和强特异性的抗体是利用免疫检测药物残留的基础，由于 Mab 具有一系列的技术优势[5]，是药物残留初筛检测方法的主流发展方向，在食品安全领域具有重要的价值。

3.1 动物免疫方法

本文采用腹腔注射免疫法免疫 4 周龄的 BALB/c 小鼠。小鼠腹腔内血管密布，抗原的吸收迅速，并能诱发有效的免疫应答，且不会出现"免疫耐受现象"。对于免疫抗原的纯度对免疫效果的影响存在着一直存在着争议。一般认为：抗原纯度越高越好，但蔡勤仁[10]认为，"抗原是否经过纯化对免疫结果不产生影响，未经纯化的抗原或许更有利于机体的免疫"。本实验室采用改进的碳二乙胺法制备完全抗原（CIP-BSA 偶联物）[8]，尽管采取了透析等措施，但免疫原中或多或少地还是存在着少量的杂质（BSA-BSA、BSA-BSA-CIP 或 BSAEDC）。该免疫原免疫小鼠（使用佐剂），血清效价最高可达 $1：6.25×10^5$，符合细胞融合的要求。

3.2 Mab 的筛选

用间接 ELISA 对杂交的大量样本进行初筛，能有效地排除假阳性和纯化细胞。本研究在杂交过程中，探索发现时间非常重要：如克隆过早，会造成部分原本是阳性抗体由于尚分泌或量不足，造成假阴性结果。在本试验条件下，摸索出来的最佳检测时间为融合 14 d 左右。

3.3 Mab 的特异性

抗体的特异性是指与相应抗原或近似抗原物质的识别能力，也是衡量杂交瘤细胞分泌的单克隆抗体稳定性的重要指标[9]。抗体的特异性越强，发生交叉反应的机会越少。一般认为：半抗原末端的特征结构对特异性等免疫特性的影响最大[1]。目前，对于 QNs 类药物抗体的特异性研究较多，却存在着两种截然相反观点。一种观点认为：QNs 类药物 7 位上的取代基团是影响抗体特异性的主要因素，其对抗体特异性的影响远远大于 1 位的取代基[11]。另外一种观点正好相反：N-1 苯环取代基是影响抗体特异性的主要原因，而哌嗪环取代基的影响次之[12-14]。

CIP 分子结构中 N-1 位被环丙基取代，6 位被 F 取代，极大增强了药物活性和药效（图 5a），其中 C-6 位上的 F 原子、C-7 位上的哌嗪基和 C-3 位上羧基的上 O 原子是进行酶促反应的部位。恩诺沙星分子（图 5b）在 7 位上是 4-乙基-1 哌嗪基；氧氟沙星（图 5c）、左氧氟沙星和双氟沙星（图 5d）分子在 7 位上是 4-甲基-1 哌嗪基；沙拉沙星（图 5e）在 7 位上则是哌嗪基。

本文在阳性克隆孔的筛选阶段淘汰了对氯霉素表现出较强的交叉反应细胞株 5H1 株,利用 1C9 等 6 个阳性克隆建株。氧氟沙星、双氟沙星、沙拉沙星和左氧氟沙星对筛选的 6 株单克隆抗体细胞株几乎不存在交叉反应(表1),其原因在于这些药物分子的 1 位不是环丙基,而是空间结构相对庞大、复杂的 4 氟苯环,4 氟苯环产生的空间障碍,在一定程度上抵消了 CIP 分子中氟原子的电负性特征。

恩诺沙星对所有的单克隆抗体均表现出较高的交叉反应。从分子构成、空间结构和电子特性上看,CIP 与恩诺沙星非常相似:1 位上均为环丙基,7 位上分别为哌嗪基和 4-乙基-1 哌嗪基[12,13]。对于喹诺酮类药物,有许多类似的报道。Duan[6]发现恩诺沙星对其制备的 CIP 多克隆抗体也具有较高的反应率。王战辉[15]认为:CIP 在 1 位的取代基结构简单(为烷烃基)且靠近 3 位(羧基与载体蛋白偶联的部位),也是 CIP 抗体对恩诺沙星交叉反应率较大的原因之一。与之类似,邹明[14]在建立双氟沙星单克隆抗体时发现:沙拉沙星对于双氟沙星的抗体也表现出较高的交叉反应率,其原因在于沙拉沙星与双氟沙星结构在 1 位上均为氟苯环,差别只是在 7 位的取代基上分别为哌嗪基和 4-甲基-1 哌嗪基。

本文的研究结果也进一步证明:对于 CIP 半抗原,相比 7 位的取代基,1 位的取代基在决定抗体特异性方面发挥起更为重要作用的理论。该结果与 Holtzapple[13]、邹明[14]等人的观点非常一致。

此外,由于恩诺沙星在动物体内将可以转化为环丙沙星,中国、欧盟等国家或地区的法规中均有同时检测恩诺沙星和环丙沙星残留的要求,因此本文制备的抗体具有积极的现实意义。

参考文献

[1] 李俊锁,邱月明,王超.兽药残留分析[M].上海:上海科学技术出版社,2002:258-261.
[2] 胡鲲,杨先乐,唐俊,等.绒螯蟹中 3 种氟喹诺酮类药物残留检测的前处理方法研究[J].华中农业大学学报,2007;26(5):670-675.
[3] 中国水产科学研究院珠江水产研究所.NY5071—2002 无公害食品渔用药物使用准则[M].北京:中国标准出版社,2002.
[4] 胡鲲,杨先乐,张菊.酶联免疫法检测中华鳖肌肉中己烯雌酚[J].上海水产大学学报,2002;11(3):199-202.
[5] 冯仁青,郭振泉,宓捷波.现代抗体技术及其应用[M].北京:北京大学出版社,2006:47-50.
[6] Duan J H, Yuan Z H. Development of an indirect competitive ELISA for Ciprofloxacin residues in food animal edible tissues [J]. J Agric Food Chem, 2001; 49: 1087-1089.
[7] 曾华金,杨冉,屈凌波,等.抗环丙沙星抗血清制备及其 ELISA 检测[J].食品科学,2009;30(2):154-158.
[8] 黄宣运,胡鲲,方伟,等.环丙沙星偶联物的制备及其产物的鉴定[J].上海水产大学学报,2008;17(5):585-590.
[9] Beatty J D, Beatty B G, Wlanos W G. Measurement of monoclonal antibody affinity by non-competitive enzyme immunoassay [J]. J Immunol Methods, 1987; 100 (1-2): 173-179.
[10] 蔡勤仁,曾振灵,杨桂香,等.恩诺沙星单克隆抗体的制备及鉴定[J].中国农业科学,2004;37(7):1060-1064.
[11] Huet A C, Charlier C, Tittlemier S A et al. Simultaneous determination of (fiuoro) quinolone antibiotics in kidney, marine products, eggs and muscle by enzyme-linked immunosorbent assay (ELISA) [J]. J Agric Food Chem, 2006; 54: 2822-2827.
[12] Holtzapple C K, Carlin R J, Kubenal S L. Analysis of hapten-carrier protein conjugates by nondenaturing gel electrophoresis [J]. J Immunol Methods, 1993; 164: 245-253.
[13] Holtzapple C K, Buckley S A, Stanker L H. Production and characterization of monoclonal antibodies against sarafloxacin and cross-reactivity studies of related fluoroquinolones [J]. J Agric Food Chem, 1997; 45: 1984-1990.
[14] 邹明,陈杖榴.二氟沙星残留检测用单克隆抗体的制备及其鉴定[J].中国动物免疫,2009;26(2):35-37.
[15] 王战辉,沈建忠,张素霞.喹诺酮类药物抗体制备研究进展及策略分析[J].中国农业科学,2008;41(10):3311-3317.

该文发表于《食品科学》【2010,31(12):16-20】

环丙沙星单克隆抗体的制备及鉴定

Preparation and Identification of Monoclonal Antibody against Ciprofloxacin

杨先乐[1],黄宣运[1],胡鲲[1],方伟[1],王炼[2],姜有声[1]

(1. 上海海洋大学 国家水生动物病原库, 上海 201306; 2. 四川省水产学校, 四川 成都 401520)

摘要:通过改进的碳二亚胺法制备环丙沙星-牛血清白蛋白偶联物(CIP-BSA),用紫外扫描、十二烷基磺酸钠聚丙烯酰胺凝胶电泳(SDS-PAGE)方法鉴定偶联物。利用制备的 CIP-BSA 偶联物作为免疫原,免疫 Balb/c 小鼠,用杂交瘤技术制备环丙沙星单克隆抗体，并对其效价、亲和力和特异性进行鉴定。结果表明：CIP-BSA 的偶联比为 30：1。经间接酶联免疫吸附（ELISA）法筛选，共筛出两株敏感、特异的杂交瘤细胞（JY-1、JY-2）。这两株细胞的腹水经纯化后的抗体效价分别为 5.12×10^6、2.56×10^6，亲和力常数分别为 2.88×10^9、2.46×10^9 L/mol，两抗体均为 IgG1a，取亲和力高的 JY-1 进行竞争抑制测定，IC_{50} 值可以达到 9.83 ng/mL。该单抗仅与恩诺沙星存在交叉反应，交叉反应率为 95%。通过验证得到的 JY-1 细胞株分泌的单克隆抗体可用于水产品中环丙沙星残留的快速检测。

关键词：环丙沙星；半抗原；抗体；免疫

环丙沙星（ciprofloxacin，CIP）又称环丙氟哌酸，是第三代氟喹诺酮类药物[1]。该药杀菌能力强、抗菌谱广，曾普遍应用于水产动物的疾病治疗中。然而人类长期食用含环丙沙星残留的水产品会对健康造成损害，引发食源性疾病[2]。目前，环丙沙星已经被欧盟、中国等列为水产养殖禁用药物[3-4]。环丙沙星的检测方法主要有微生物法、高效液相色谱法、毛细管电泳法等[5]。这些方法不仅操作繁琐、仪器价格昂贵，而且不适合现场大规模的检测[6]。以免疫学方法为基础的快速检测方法，具有快速、灵敏、价格低廉、适合现场大规模检测等优点，是目前公认最适合作为药物残留检测的快速筛选方法之一。获得特异性强、灵敏度高的单克隆抗体是建立免疫学方法的关键。本实验通过改进的碳二亚胺法，提高免疫原——环丙沙星-牛血清白蛋白偶联物（CIP-BSA）的偶联密度，然后用杂交瘤技术制备高特异性的环丙沙星单克隆抗体，为环丙沙星免疫检测试剂盒的开发提供参考。

1 材料与方法

1.1 材料、试剂与仪器

CIP、恩诺沙星、氧氟沙星、双氟沙星、沙拉沙星、氯霉素、孔雀石绿（纯度 ≥ 98.5%）浙江国邦药业有限公司；牛血清白蛋白（BSA）、羊抗鼠 IgG 北京鼎国生物技术有限公司；环丙沙星-鸡卵血清白蛋白（CIP-OVA）实验室自制；乙基碳二亚胺（EDC，纯度 ≥ 99.3%）上海延长生化科技发展有限公司；弗氏完全、不完全佐剂 Sigma 公司；聚乙二醇（Polythylene Glycol-4000，PEG）、RPMI l640 培养基、HAT 选择性培养基 GIBCO 公司；其他化学试剂均购于上海国药有限公司。

UV-500 型紫外分光光度计北京优尼康生物科技有限公司；Bio-Rad powerpac300 型电泳仪 Bio-Rad 公司；UVP ca91786 型凝胶成像系北京百晶生物科技有限公司；Savant Novalyphe-NL150 型冷冻干燥机 Savant 公司；Bio-Tek EL311 酶标仪 Bio-Tek 公司。

1.2 实验动物

Balb/c 小鼠（雌性 4~5 周龄、7 周龄）购自第二军医大学动物实验中心。

1.3 方法

1.3.1 人工抗原的制备

采用改进的碳二亚胺法：将 8 mg 的 BSA 溶解在 4 mL PBS（0.01 mol/L，pH5.0）中，分别加入 240 mg EDC 和 20 mg CIP，在室温中反应 2 h 静置，装入透析袋中，在 PBS（0.01 mol/L，pH5.0）中透析 2d，每天换液一次。偶联物冷冻干燥后，-20℃保存备用。此偶联物记作 CIP-B SA。

1.3.2 人工抗原的鉴定

1.3.2.1 紫外扫描

将人工抗原 CIP-BSA 用 PBS（0.01 mol/L，pH5.0）稀释到 1.6 mg/mL，配制环丙沙星标准液和 BSA 标准液，在 250~350 nm 范围内进行扫描，扫描速度为 2 nm/s。以 CIP-BSA 的特征峰 274 nm，建立 CIP 和 BSA 的吸光度-质量浓度标准曲线，计算 CIP-BSA 中 CIP 的质量浓度。

$$偶联比 = \frac{CIP\ 质量浓度/CIP\ 相对分子质量偶联比}{BSA\ 质量浓度/BSA\ 相对分子质量}$$

1.3.2.2 SDS-PAGE 法

采用常规 SDS-PAGE 方法[7]，浓缩胶质量浓度为 5 g/100 mL，分离胶质量浓度为 10 g/100 mL，浓缩胶电压为 80 V，分离胶电压为 100 V。

1.3.3 单克隆抗体的制备

1.3.3.1 动物免疫

选取 4~5 周龄雌性的 Balb/c 小鼠，免疫前眼眶取血作为阴性血清。取 2 mg/mL CIP-BSA 与弗氏完全佐剂等量混合，腹腔注射小鼠，免疫剂量为 0.2 mg/次。首次免疫后两周，抗原与弗氏不完全佐剂混合乳化，腹腔注射。以后免疫不加佐剂，分别在第 3、4 周各免疫一次，采用尾部静脉注射免疫。最后一次免疫后进行细胞融合。

1.3.3.2　分泌抗环丙沙星单克隆抗体的杂交瘤细胞株的建立

取免疫小鼠脾细胞与骨髓瘤细胞 SP2/0，在 PEG-4000 的作用下进行融合，用 HAT 培养基培养于 96 孔板中，培养于体积分数 5% CO$_2$、37℃培养箱中。采用间接 ELISA 法筛选阳性杂交瘤细胞株，偶联物环丙沙星-鸡卵血清白蛋白（CIP-OVA）包被酶标板，100 μL/孔，4℃包被过夜；倒去包被液，洗板后加 5 g/100 mL 脱脂牛奶溶液封闭，200 μL/孔，37℃封闭 1 h，PBST 洗板 3 次，3min/次；加杂交瘤细胞上清液每孔 100μL，37℃孵育 1 h，同时设 PBS 空白对照和骨髓瘤细胞培养上清液阴性对照。洗板后加辣根过氧化酶（HRP）标记抗小鼠 IgG（稀释 6000 倍），37℃孵育 0.5 h 后洗板，加底物溶液显色。用酶标仪在 450 nm 处读取 OD 值。选取 OD 值最高的阳性孔进行有限稀释法亚克隆 3 次，测定上清液抗体效价，将抗体效价最高的单克隆细胞扩大培养并建株冻存。

1.3.3.3　单克隆抗体的腹水制备及纯化

取 7 周龄的 Balb/c 小鼠，每只注射 0.5 mL 的降植烷。一周后腹腔接种用无血清培养基稀释处于对数生长期的杂交瘤细胞，每只注射细胞数为 5×10^6 个，间隔 4d 后，观察小鼠腹水情况，待小鼠腹部膨大，精神变差，濒临死亡时，采集腹水。将腹水在 4℃、9 000 r/min 离心 30 min，收集上清液，取 10mL 上清液与等体积 PBS（0.01 mol/L，pH7.4）混合后，边搅拌边缓慢滴加 20 mL 饱和硫酸铵溶液，4℃沉淀 1h 后 9 000 r/min 离心 30 min，弃去上清液。加入 10 mL PBS（0.01 mol/L，pH7.4）溶解沉淀。将溶解后的液体装入透析袋，4℃条件下在 PBS（0.01 mol/L，pH7.4）中透析 2 d，每 8 h 换液一次。取纯化后的抗体，测定效价，分装保存在-70℃。

1.3.4　单克隆抗体的鉴定

1.3.4.1　亚型鉴定

采用间接 ELISA 法，Invitrogen 试剂盒测定单抗亚型。

1.3.4.2　抗体亲和力常数鉴定

利用非竞争性酶免疫法测定抗体的亲和力[8]。将包被抗原按 200、100、50、25 ng/mL 包被酶标板，将单抗从 1 000 倍开始倍比稀释，其余同间接 ELISA 检测方法。亲和力常数计算公式为 $K_a = (n-1)/2n([Ab']_t - [Ab]_t)$。其中 $n = [Ag']_t/[Ag]_t$，$[Ag']_t$ 和 $[Ag]_t$ 为不同包被抗原的浓度，$[Ab']_t$、$[Ab]_t$ 是对应不同包被抗原的浓度条件下，将抗体梯度稀释得到最大吸光度一半处的抗体浓度（mol/L）。

1.3.4.3　灵敏度鉴定

利用方阵法筛选抗原抗体最佳反应浓度，配制不同质量浓度 CIP 标准品（50、25、12.5、6.25、3.13、1.56 ng/mL），用直接竞争 ELISA 测定单抗对其抑制率。以抑制率 B/B0 为纵坐标（B 是 CIP 标准品不同质量浓度的 OD450 值，B0 是 CIP 质量浓度为 0 时的 OD450 值），以不同质量浓度 CIP 的对数值为横坐标，绘制标准抑制曲线，进行相关回归分析，计算单抗的 IC$_{50}$ 值。

1.3.4.4　特异性鉴定

以氟喹诺酮类药物以及氯霉素和孔雀石绿作为竞争物，用直接竞争 ELISA 法测定单抗对各竞争物的 IC$_{50}$ 值。以 CIP 的 IC$_{50}$ 值同各竞争物的 IC$_{50}$ 值的百分比为其交叉反应率。

2　结果与分析

2.1　偶联物的制备

对传统的 EDC 法[9-10]进行了改进，建立了不依赖于 N-羟基琥珀酰亚胺（N-hydroxysuccinimide，NHS）的 EDC 方法，成功制备得到 CIP-BSA 偶联物。整个偶联反应时间约为 2 h，极大提高了反应效率（碳二亚胺二步法[9]为 24 h）。偶联物经冷冻干燥后呈白色粉末，质地均匀，溶解性能良好。

2.2　偶联物分析

2.2.1　紫外扫描分析

BSA、CIP 和 CIP-BSA 的紫外扫描图谱见图 1。CIP 的最大吸收峰在 270 nm，BSA 的最大吸收峰在 278 nm，而 CIP-BSA 的最大吸收峰在 274 nm，CIP-BSA 的特征吸收峰发生了明显的偏移，证明偶联成功。经过紫外扫描得到 CIP 在 274 nm 处的吸光度-质量浓度标准曲线方程为 $Y = 14.724X + 0.033\ 2$（$R^2 = 0.997\ 6$），BSA 的吸光度-质量浓度标准曲线方程为 $Y = 0.548\ 1X - 0.003\ 9$（$R^2 = 0.999\ 7$）。将 CIP-BSA（1.6 mg/mL）的吸光度与 BSA（1.6 mg/mL）的吸光度的差值代入 CIP 标准曲线，即可得到 CIP 的浓度。根据公式计算得出 CIP-BSA 的偶联比可达 30∶1。

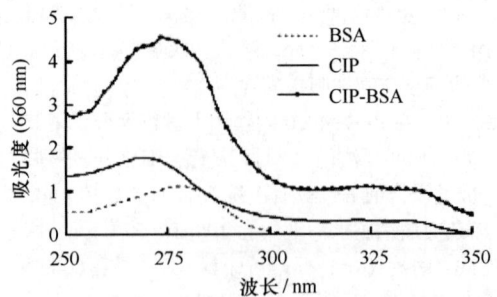

图 1 CIP、BSA and CIP-BSA 紫外扫描光谱

Fig. 1 Ultraviolet absorption spectra of CIP，BSA and CIP-BSA

2.2.2 SDS-PAGE 鉴定

BSA 和 CIP-BSA 的 SDS-PAGE 电泳结果见图 2，人工抗原 CIP-BSA 条带明显落后于 BSA，由此可以判定 CIP 与 BSA 偶联成功。

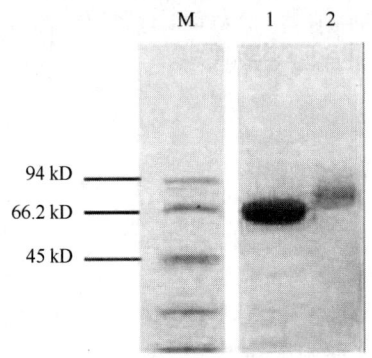

M. Mark; 1. BSA; 2. CIP-BSA

图 2 SDS-PAGE 鉴定 CIP-BSA

Fig. 2 SDS-PAGE analysis of CIP-BSA

2.2.3 细胞融合和阳性杂交瘤细胞的筛选用间接

ELISA 法筛选出的阳性杂交瘤细胞株，经过有限稀释法克隆，最终得到两株可分泌特异性抗体的细胞株，命名为 JY-1、JY-2。

2.2.4 单克隆抗体的亚型鉴定

通过 Invitrogen 试剂盒测定抗体亚型，表明 JY-1、JY-2 抗体均为 IgG1a 型单抗。

2.2.5 单克隆抗体的效价、蛋白含量及纯度鉴定

阳性值与阴性值的比值大于等于 2 时抗体的最大稀释度为抗体效价。经间接 ELISA 法测定，这两株细胞经纯化后的抗体效价分别为 5.12×10^6、2.56×10^6，根据 260 nm 和 280 nm 紫外吸收值，算得的蛋白含量为 1.348、1.583 mg/mL。纯化后的单抗经 SDS-PAGE 显示（图 3）：在变性解链状态下有两条带：一条为轻链，约在 25kD 处；一条为重链，约在 50kD 处，没有其他杂带，证明抗体纯化效果较好。

2.2.6 抗体亲和力鉴定

通过非竞争性酶免疫法测定抗体的亲和力，根据公式计算得出 JY-1 的亲和力为 2.88×10^9 L/mol，JY-2 的亲和力为 2.46×10^9 L/mol，选取亲和力高的 JY-1 抗体，进行后续实验。

2.2.7 抗体灵敏度测定

经方阵实验优化筛选，环丙沙星最佳间接竞争 ELISA 法测定最佳工作条件为包被抗原质量浓度为 100 ng/mL 包被酶标板；单克隆抗体稀释 50 000 倍；进行竞争反应前，抗体与待测样品预反应 0.5 h；酶标二抗稀释 6 000 倍，反应时间为 0.5 h。按上述标准实验条件测定抗体的灵敏度，由图 4 可知，曲线方程为 $Y = -0.6361X + 1.1313$，通过计算标准曲线的 IC_{50} 值为 9.83 ng/mL，检测限为 3.32 ng/mL。

2.2.8 JY-1 抗体的特异性

JY-1 与检测药物的交叉反应结果见表 1。结果显示，JY-1 除与恩诺沙星有较强的交叉反应外，与其他检测药物不存

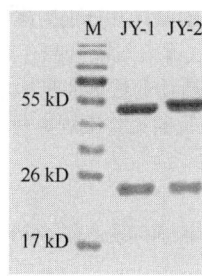

图 3　SDS 电泳测定单克隆抗体纯度
Fig. 3　SDS-PAGE analysis of purified Mabs from JY-1 and JY-2

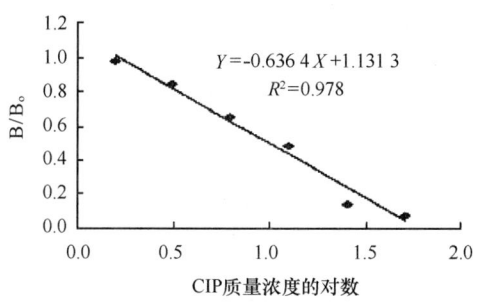

图 4　JY-1 抗体竞争抑制曲线
Fig. 4　Competitive inhibition curve of JY-1 Mab

在交叉反应，表明 JY-1 具有较好的特异性。

表 1　JY-1 抗体与其他药物的交叉反应率
Table 1　Cross-reactivity of JY-1 Mab with its analogues

竞争物	$IC_{50}/$（ng/mL）	交叉反应率/%
恩诺沙星	10.34	95
氧氟沙星	>100 000	< 0.01
双氟沙星	>100 000	< 0.01
沙拉沙星	>100 000	< 0.01
氯霉素	>100 000	< 0.01
孔雀石绿	>100 000	< 0.01

3　结论与讨论

环丙沙星属于小分子药物，其自身不具备免疫原性，因此需要对其偶联改造。目前，包括 CIP 在内的喹诺酮类药物半抗原的偶联方法，大都采用具有传统的两步 EDC 法[9-10]。其中，N-羟基琥珀酰亚胺（NHS）作为激活基团，起到促进羧基与氨基偶联的作用。对于 CIP 而言，具有特殊的一面：其分子结构中既有氨基又有羧基，可以直接与载体蛋白发生偶联反应，因此具有较好的选择余地。在前期工作的基础上[11]，改进的 EDC 法舍去了 NHS。改进的 EDC 法不仅简化了偶联条件，提高了偶联物的制备效率（将反应耗时由 24h 降低到 2h），而且相比传统 EDC 方法偶联 CIP[9]，能显著提高 CIP 人工抗原的偶联比（提高幅度达 130% 左右）。这对于制备特异性抗体具有重要的应用价值。

抗体的亲和力直接反映抗原抗体的结合能力，James[12]认为亲和常数 $10^7 \sim 10^{12}$ L/mol 为高亲和力，亲和常数 $10^5 \sim 10^6$ L/mol 为低亲和力。本研究中两抗体的亲和力常数分别为 2.88×10^9、2.46×10^9 L/mol，均属高亲和力的抗体，为建立环丙沙星残留的快速检测提供了必要的保证。

在实验中发现 CIP 与恩诺沙星存在较大的交叉反应，类似的结果也有报道，Duan 等[13]制备的 CIP 多克隆抗体，也对恩诺沙星具有较高的反应率。蔡勤仁等[14]以恩诺沙星-牛血清白蛋白为免疫原，筛选出两株单克隆抗体对环丙沙星的交叉反应率为 110%。一般而言，半抗原结构越相似，发生交叉反应的机率越大。从分子构成、能量构象[15]上看 CIP 与恩诺沙星非常相似：在 1 位的取代基均为环丙基，在 7 位上的取代基分别为哌嗪基和 4-乙基-1-哌嗪基，这可能是造成 CIP 单克隆抗体与恩诺沙星存在较强交叉反应的原因。JY-1 抗体与氧氟沙星、双氟沙星、沙拉沙星均不存在交叉反应。Holtzapple 等[16]利用 CAChe work System 软件合成了氟喹诺酮类药物的最低能量构象，尽管氟喹诺酮类药物的主要分子结构基本相似，但空间结构却不相同，这可能造成制备的 CIP 抗体与其他氟喹诺酮类药物存在较弱的交叉反应。

参考文献

[1] Barry A. L., Fuchs P. C.. Anti-staphylococcal activity of temafloxacin, ciprofloxacin, ofloxacin and enoxacin [J]. J Antmicrob Chemother, 1991, 28：695-699.

[2] Nelson J. M., Chiller T. M., Powers J. H., et al. Fluoroquinolone resistant Campylobacter species and the withdrawal of fluoroquinolones from use in poultry：a public health success story [J]. Clin Infect Dis, 2007, 44 (7)：977-980.

[3] Hammer P., Heeschen W. Antibody-capture immunoassay for the detection of enrofloxacin in raw milk [J]. Milchwissenschaft, 1995, 50：513-514.

[4] NY 5071- 2002 无公害食品鱼用药物使用准则 [S].

[5] 邓斌, 李琦华. 动物牲食品中环丙沙星残留检测方法的研究进展 [J]. 江西饲料, 2005 (2)：29-31.

[6] 王敏, 王权, 蔡小丽, 等. 无色孔雀石绿单克隆抗体的制备与鉴定 [J]. 畜牧与饲料科学, 2009, 30 (3)：1-3.

[7] 郭尧君. 蛋白质电泳实验技术 [M]. 北京：科学出版社, 1992：132-136.

[8] 万文徽. 单克隆抗体亲和常数的测定 [J]. 单克隆抗体通讯, 1993, 9 (2)：72-75.

[9] 周玉, 李岩松, 王哲, 等. 环丙沙星的偶联及鉴定 [J]. 中国兽医学报, 2006, 26 (2)：200-203.

[10] 郝贵增, 龙淼. 环丙沙星人工抗原的合成及鉴定 [J]. 现代食品科技, 2008, 24 (7)：658-660；673.

[11] 黄宣运, 胡鲲, 方伟, 等. 环丙沙星偶联物的合成及产物的鉴定 [J]. 上海水产大学学报, 2008, 17 (5)：585-590.

[12] James W. G.. Monoclonal antibodies：principles and practice [M]. London：Academic Press Inc Ltd, 1983：142-147.

[13] DUAN Jiuhua, YUAN Zonghui. Development of an indirect competitive ELISA for ciprofloxacin residues in food animal edible tissues [J]. Journal of Agricultural Food Chemistry, 2001, 49 (3)：1087-1089.

[14] 蔡勤仁, 曾振灵, 杨桂香, 等. 恩诺沙星单克隆抗体的制备及鉴定 [J]. 中国农业科学, 2004, 37 (7)：1060-1064.

[15] WANG Zhanhui, ZHU Yan, DING Shuangyang, et al. Development of a monoclonal antibody-based broad-specificity ELISA for fluoroquinolone antibiotics in foods and molecular modeling studies of cross-reactive compounds [J]. Anal Chem, 2007, 79 (12)：4471-4483.

[16] Holtzapple C. K., Buckley S. A., Stanker L. H.. Production and characterization of monoclonal antibodies against sarafloxacin and crossreactivity studies of related fluoroquinolones [J]. J Agric Food Chem, 1997, 45 (5)：1984-1990.

[17] Holtzapple C. K., Buckley S. A., Stanker L. H.. Production and characterization of monoclonal antibodies against sarafloxacin and crossreactivity studies of related fluoroquinolones [J]. J Agric Food Chem, 1997, 45 (5)：1984-1990.

3. 检测方法的建立与应用

基于单克隆抗体的酶联免疫法是一种快速、准确的初筛方法（Kun Hu et al., 2010；胡鲲等, 2010），能与作为高效液相色谱等复检方法相互补充，构建水产品药物残留检测技术平台。

该文发表于《Aquaculture》【2010, 310 (1-2)：8-12】

Monoclonal antibody based enzyme-linked immunosorbent assay for the specific detection of ciprofloxacin and enrofloxacin residues in fishery products

Kun Hu[a,1], Xuanyun Huang[b,1], Yousheng Jiang[a], Wei Fang[a], Xianle Yang[a,b,]

(a. Aquatic Pathogen Collection Centre of Ministry of Agriculture, Shanghai Ocean University, 999 Huchenghuanlu Road, Shanghai 201306, PR China b Laboratory of Fishery Environments and Aquatic Products Processing, East China Sea Fisheries Research Institute Chinese Academy of Fishery Sciences, Shanghai 200090, PR China)

Abstract：Ciprofloxacin (CPFX) residues in fishery products were dangerous for human health and still remain a bigchallenge in China. However, there are a few available assays to detect specifically CPFX residues in fishery This study aimed at generating CPFX-specific monoclonal antibodies (mAbs) and developing acompetitive indirect enzyme-linked immunosorbent assay (ci-ELISA) for the sensitive and specific detectionof CPFX residues in fishery products. Firstly, the CPFX was conjugated with BSA and the conjugates were immunized to Balb/c mice. After characterizing anti-CPFX sera, we generated hybridoma cell lines and found that one hybridoma clone, 1C9, produced anti-CPFX mAb. Characterization of IC9 mAb revealed that it belonged to IgG2a, with an affinity of 3.75×10^{10} mol/L for CPFX. Furthermore, we developed ic-ELISA using 150 ng/mL of CPFX-OVA for coating antigen and 0.22 μg of mAb. We found that this assay detected CPFX at 45.25 ng/mL and had 84.60% of cross-activity to enrofloxacin (ENR), but not to ofloxacin, levofloxacin, difloxacin, sarafloxacin, furacilin, chloramphenicol and nalachite green. Using this assay, we detected the CPFX in fishery products prepared with recovery rates from 80.57% to 94.61% and relative standard deviation (RSD) varying from 1.78% to 8.02%, similar to that by HPLC analysis. These data, together the protocol for the preparation of fishery products, indicate that the ic-ELISA is able to detect CPFX and ENR residues in fishery products sensitively and specifically within 2 h. Therefore, this assay can be used for monitoring CPFX and ENR in commercial fishery products, improving the safety of commercial meats in China.

key words: Ciprofloxacin, Monoclonal antibody, Enzyme-linked immunosorbent assay, Fishery products

基于酶联免疫吸附试验的单克隆抗体对水产品中的环丙沙星和恩诺沙星残留物的检测

胡鲲[a,1]，黄宣运[b,1]，姜有声[a]，方伟[a]，杨先乐[a,b]

(a. 上海海洋大学 农业部渔业动植物病原库，中国 上海 沪城环路 999 号，201306；
b. 中国水产科学院东海水产研究所 水产品质量安全与加工实验室，中国 上海 201306)

摘要：水产品中的环丙沙星（CPFX）残留物对人体健康有危害，其在中国仍然是一个重大挑战。然而，现有的一些的试验可以特异性检测水产品中的环丙沙星残留。本研究旨在生成环丙沙星特异性单克隆抗体（mAbs）并开发竞争性间接酶联免疫吸附测定（ci-ELISA）用于水产品中环丙沙星残留物的灵敏性和特异性检测。首先，将环丙沙星与 BSA 融合，并将融合物免疫接种 Balb/c 小鼠。表征抗环丙沙星血清后，生成杂交瘤细胞系，发现一个杂交瘤克隆-1C9，产生抗环丙沙星 mAb。IC9 mAb 的表征揭示它属于 IgG2a，对环丙沙星的亲和力为 3.75×10^{10} mol/L。此外，我们研发了一种方法：使用 150 ng/mL 的环丙沙星-OVAic-ELISA 包被抗原和 0.22 μg 的 mAb。研究发现本实验检测到 45.25 昨 ng/mL 的环丙沙星对恩氟沙星具有 84.60% 的交叉活性，但对氧氟沙星左氧氟沙星、二氟沙星、沙洛沙星、弗拉克林、氯霉素和孔雀石绿没有相似的活性。使用该测定，检测到水产品中的环丙沙星与 HPLC 分析类似：回收率 80.57%~94.61%，相对标准偏差（RSD）1.78%~8.02%。以上结论与水产品准备协议共同表明 ic-ELISA 能够在两小时内灵敏、特异性的检测到水产品中的环丙沙星和 ENR 残留。因此，该测定法可用于监测水产经济产品中的环丙沙星和恩诺沙星，提高中国经济肉类的安全性。

关键词：环丙沙星；单克隆抗体；酶联免疫吸附试验；水产品

1 Introduction

Ciprofloxacin (CPFX) is an antibacterial agent and used as aveterinary medicine in the world, particularly in developing countries, like China. The CPFX residues in the consumed animal tissues raise potential risks of the development of drug-resistance and chronic adverse effects in the public health (Hernandez et al., 2002). The maximum residues limits (MRLs) of CPFX in the food products of animal origin have been established in European Union (DOCE, 1990) and adapted by many other countries. However, CPFX-related fishery product-involved incidents often occur and have been a serious concern to human health. Therefore, the careful monitoring of CPFX residues in fishery products for human consumption by sensitive and specific methods is crucial for food safety and human health.

Currently, the available methods to determine the CPFX in commercial meats are mainly dependent on the complex high performance liquid chromatography (HPLC) (Garcia Ovando et al., 1999), capillary electrophoresis (Hernandez et al., 2002) andmicrobiological method (Montero et al., 2005). However, these methods usually require for the professional operators and expensive equipments. Enzyme-linked immunosorbent assay (ELISA) is a simple, sensitive, economical, and high throughput procedure for the specific detection of antigen. Furthermore, monoclonal antibody (mAb) based ELISA has greater advantage than that of polyclonal antibodies (pAbs), particularly for its specificity and stability. In addition, mAb can continue to supply the well-defined specific and stabilized antibodies with little batch variation. Several mAbs against fluoroquinolones (Holtzapple et al. , 1997; Watanabe et al. , 2002; Huet et al. , 2006; Wang et al. , 2007; Li et al. , 2008) have been generated. However, most commercial kits for the detection of CPFX residues are made by pAbs (Snitkoff et al. , 1998; Duan and Yuan, 2001). The few available mAbs against CPFX have been predominately applied to livestock and poultry (Huet et al. , 2006; Wang et al. , 2007; Li et al. , 2008). Currently, there is still no available of commercial kit for the detection of CPFX residues in the fishery products.

In this study, we aimed at generating mAbs against CPFX and developing mAb-based ci-ELISA for the detection of CPFX residues in commercial fishery foods. We first coupled CPFX into carrier proteins and immunized mice with the hapten – carrier complex. Subsequently, we generated hybridoma cell lines, which produced high affinity anti-CPFX mAb. Furthermore, we used the procedures for the preparation of fishery products and mAb-based ci-ELISA, with which we detected CPFX residues in the fishery products sensitively and specifically.

2　Materials and methods

2.1　Materials

Ciprofloxacin (Content ≥ 98.5%) was purchased from Zhejiang Guobang Pharmaceutical Co. , while bovine serum albumin (BSA), ovalbumin (OVA), and peroxidase horseradish (HPR) were obtained from Dingguo biotechnology. The 1 – Ethyl – 3 – (dimethylaminopropyl) carbodiimide (EDC, purity ≥ 99.3%) was from Shanghai Yanchang Biochemical Technology and 3, 3′.5, 5′-tetramethylbenzidine (TMB), HRP-conjugated goat anti-rabbit IgG, RPMI 1640, hypoxanthine aminopterin thymidine (HAT) medium, calf serum clarified, incomplete Freund's adjuvant (IFA) and complete Freund's adjuvant (CFA) were purchased from Sigma Chemical (St. Louis, MO, USA). Chemical reagents, such as sodium chloride, potassium chloride, sulfuric acid, and dipotassium hydrogen phosphate were from Shanghai Guoyao Chemical Reagent Co. The mouse monoclonal antibody isotyping reagents were purchased from Invitrogen (Carlsbad, CA, USA). The fish samples of Carassius auratus, Oreochromis niloticus, Ictalurus punctatus, Scophthalmus maximus, Eriocheir sinensis, Penaeus vannamei, Macrobrachium rosenbergii, Trionyx sinensis, and Oncorhynchus mykiss were obtained from a commercial farm in the suburbs of Shanghai, China.

2.2　Generation of mAbs against ciprofloxacin

The conjugates of CPFX-BSA and CPFX-OVA were prepared using a modified Carbodiimide method, as previously described (Huang et al. , 2008), and characterized by spectrophotometer absorption, electrophoresis, and mass spectra. Briefly, a total of 8 mg BSA, 20 mg CPFX and 240 mg EDC were mixed in 4 mL PBS (pH 5.0, 0.01 mol/L) at 28℃ for 2 h. The mixture was dialyzed against 200 volumes of PBS (pH 5.0, 0.01 mol/L) for two days. The conjugates were lyophilized and stored at −80℃. Similarly, the conjugate of CPFX-OVA was generated.

The mAbs were generated by a similar procedure described previously (Wang et al. , 2007). Briefly, individual Balb/c mice from the Second Military Medical University were immunized with 200 μg of CPFX-BSA (dissolved in 0.01 mol/L PBS pH 7.4) in 50% CFA and boosted with the same amount of antigen in 50% IFA for three times. Subsequently, blood samples were obtained from individual mice for measuring the CPFX-specific antibodies. After the antibodies titers reached 1 : 60, 000, the mice were sacrificed and their spleen cells were fused with SP2/0 myeloma cells, followed by screening the hybridomas using the indirect ELISA. Furthermore, the hybridoma cells were injected into Balb/c mice for the induction of ascites and mAb in the ascites was purified using a 5 mL protein G sepharose-4B fast flow column, according to the manufacturer's instruction (Beijing Search Biotech, Beijing). In addition, the subclass of the mAb was determined using the mouse monoclonal antibody isotyping reagents and its affinity was measured, as described previously (Zhou et al. , 2009). The cross-reactivity of anti-CPFX mAb toward quinolone analogues and antibiotics were assessed (Holtzapple et al. , 1997). The experimental protocols were approved by the Animal Research Protection Committee of Shanghai Ocean University.

2.3　Immunodiffusion

The glass-plates were poured with 3 mL 1% agar and punched with holes. Subsequently, the central holes were loaded with antiserum (1: 20) while the peripheral holes were added with single antigen of CPFX-OVA or OVA, respectively, followed by incubating at 37℃ for 24h. The precipitated antigen/antibody complex was considered as positive for immunodiffusion assay.

2.4 Sample preparation and extraction technique

The commercial C. auratus, O. niloticus, I. punctatus, S. maximus, E. sinensis, P. vannamei, M. rosenbergii, T. sinensis and O. mykiss (5.0 g/sample) were sampled in triplicate and placed into polypropylene centrifuge tubes. The fish samples were mixed with the extraction solution (30 mL) of 1 M HCl–acetonitrile (4/500, v/v). After being vortexed for 5 min and centrifuged at 495 g and 4℃ for 15 min, the supernatants were collected and the residue was re–extracted with the same procedures again, followed by centrifuging. Subsequently, the supernatants were harvested, loaded on a separatory funnel, and mixed with n–hexane (25 mL). After shaking vigorously for 5 min and then static settlement for 5 min, the aqueous layer was removed and the organic layer was evaporated for the dryness of the residues using a vortex evaporator. The resulting residues were dissolved in 1.0 mL of extraction solution and filtered through 0.45 μm disposable syringe filter equipped with cellulose acetate membranes. 2.5. Indirect competitive ELISA

The CPFX residues in the fish samples were detected using the mAb–based ic–ELISA, as described previously (Watanabe et al., 2002). Briefly, 150 ng of CPFX–OVA in 100 μL of PBS (0.01 mol/L, pH 7.4 were added into each well of the ELISA plates (Dingguo Biotechnology, Shanghai, China) and incubated at 37℃ for 1 h. After washing for three times with 250 μL washing solution (0.01 mol/L PBS, pH 7.4, containing 0.5% Tween 20, PBST), the wells were blocked with 1% BSA in PBST overnight at 4℃. After washing, 50 μL of standard CPFX at different concentrations was first mixed with 50 μL of anti–CPFX mAb (1 : 15, 000) and the mixture was then added in triplicate to each well, followed by incubation at 37℃ for 1 h. Individual wells with PBST alone or with anti–CPFX mAb alone were used as negative controls for the background reading in ic–ELISA. Subsequently, the plates were washed and the bound mAb was probed with HRP–goat anti–mouse sera (1/6000 in PBST) at 37℃ for 0.5 h. The bound HRP–second antibodies were detected with the substrate of 100 μL of TMB liquid substrate (1 mL TMB (1.5 mg): 5 mL of H2O2 (0.03%): 4 mL of 0.2 M NaAc pH 5.3) for 10 min, and the enzymatic reactions were terminated by the addition of 2 mol/L sulfuric acid (50 μL per well), followed by reading absorbance at 450 nm in a microplate reader. Standard curves were obtained by plotting absorbance against the logarithm of CPFX concentrations, which were fitted to a four–parameter logistic equation: $y = (D-A) / [1+ (x/C) B] +A$, where A is the minimum absorbance at infinite concentration, B is the curve slope at the inflection point, C is the concentration of IC_{50}, and D is the maximum absorbance in the absence of CPFX. The limit of detection (LOD) was calculated as the method, as described by Wang et al. (2007).

The CPFX residues in the prepared samples were also determined by HPLC using the procedure, similar to that of Tang et al. (2006).

2.6 Recovery rate and relative standard deviation of the assay

The CPFX standard solution was prepared by mixing CPFX with the metrices of individual fish, as described previously (Duan and Yuan, 2001). Briefly, mixing CPFX with the matrices of fish at different ratios (50 ng/g, 600 ng/g and 1250 ng/g) respectively, the CPFX residues in the mixture were extracted, as described above. The amount of CPFX residues in the filtrates was assayed by ic–ELISA and HPLC. The recovery rate and relative standard deviation (RSD) were calculated according to the following equations:

$$Recovery(\%) = \frac{Measured\ value}{The\ oretical\ value} \times 100\%;$$

$$RSD = \frac{S}{\bar{x}} \times 100\%; \quad S = \sqrt{\frac{\sum (x - \bar{x})^2}{n-1}}$$

RSD: Relative Standard Deviation; S: Standard Deviation; x: The average values of determined sample.

Recovery rates and RSD values in the given concentrations were analyzed by Student T–test. The correlation of the ic–ELISA with HPLC was determined by logistic regression analysis. A value of Pb0.05 was considered statistically significant.

3. Results

To generate mAb against CPFX, the CPFX was conjugated with BSA by a modified Carbodiimide method with a higher coupling efficiency. The conjugation rates of CPFX–BSA and CPFX–OVA reached 33: 1 and 12: 1, respectively, which were higher than 13: 1 and 6: 1 by conventional method (Zhou et al., 2006). The conjugates were characterized by spectrophotometric absorption, electrophoresis, and mass spectra (data not shown). The conjugates were immunized into Balb/c mice. The blood samples were collected from individual mice after each boosting and the titers of anti–CPFX sera were determined by ELISA. Following boosted for three times, the anti–CPFX sera in most mice reached a titer of 1: 62 500. The antibodies appeared to be specifically against the CPFX. The agglutination test (Table 1) indicated that the antibodies were only reacted with the CPFX–OVA, but not OVA. The high titer of anti–

CPFX specific sera in the mice provided the basis for the generation of mAbs against CPFX.

Table 1   Titers and specificity assaying results of antisera.

Mouse no.	Titers of antisera[a]	Immunodiffusion assay	
	CPFX–BSA[b]	CIP–OVA[c]	OVA[c]
1	1：62，500	Positive	Negative
2	1：12，500	Positive	Negative
3	1：62，500	Positive	Negative
4	1：12，500	Positive	Negative
5	1：62，500	Positive	Negative
6	1：62，500	Positive	Negative

a. The titrations of antibodies in mice after 3 boosting were determined by indirect ELISA.

b. Immunogen.

c. CIP-OVA or OVA were coated on the plates, respectively, and the speci CPFX sera was characterized by agglutination test.

Following fusion of SP2/0 myeloma cells with the splenic cells from the immunized mice with high titer of anti–CPFX antibodies, hybridoma cell lines were screened for antibodies against CPFX and subjected to cloning by the limited dilution. One of the hybridoma clones, 1C9, produced high concentrations of mAb and grew stably in vitro for at least 6 months. To generate large amounts of mAb, the 1C9 clone cells were injected into Balb/c mice for producing ascites and the mAb 1C9 was obtained from ascites, in which the protein concentration was 3.25 mg/ml. Characterization of 1C9 mAb revealed that the 1C9 mAb belonged to IgG2a subclass with an affinity of 3.75×10^{10} mol/L.

Next, we optimized the concentrations of antibodies and coating antigens for the ic–ELISA by checkerboard titration. We found that the experiment using 150 ng/mL of CPFX–OVA for coating antigen and 0.22 μg of mAb resulted in the best results. With this experimental condition, different concentrations of CPFX resulted in a standard curve in Fig. 1. The four–parameter logistic equation was $y = (0.92 + 0.07) / [1 + (x/2.43)^{10.18}] - 0.07$ while the value of R^2 was 0.997. The LOD detected by ic–ELISA was 45.25 ng/mL and the linear range was between 50 and 1250 ng/mL with a 50% inhibition at 245.86 ng/mL. The specificity of ic–ELISA to other similar antibiotics was assayed by the icELISA (Table 2). The ic–ELISA detected 84.6% cross–reactivity with ENR, a compound with high similarity of structure with CPFX (Fig. 2). In contrast, the ic–ELISA did not detect any of the other quinolone analogues and antibiotics tested, because there was no more than 0.01% cross–reactivity of the ic–ELISA to any of other antibiotics.

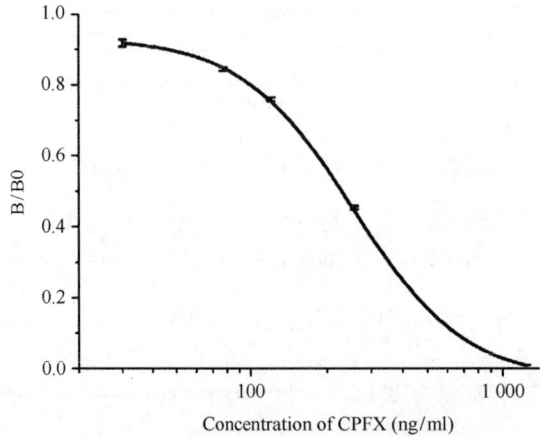

Fig. 1   Standard curve for detecting CPFX by ci–ELISA. The ci–ELISA was used for detection of different concentrations (0–1250 ng/mL) of CPFX and the resulting absorbance values were calculated as the value of individual concentrations of CPFX (b) /the value of control without CPFX (b0), respectively. The standard curve was established by different concentrations of CPFX versus the values of b/b0. Individual value represents the mean±SD of the concentration of CFPX from three independent experiments.

Table 2　Cross-reactivity of anti-CPFX monoclonal antibody towards other competitors

Competitor	IC$_{50}$（ng/mL）	Crossreactivity%	Competitor	IC$_{50}$（ng/mL）	Crossreactivity%
CPFX	245. 86	100	Sarafloxacin	N/A	<0. 01
ENR	290. 61	84. 6	Furacilin	N/A	<0. 01
Ofloxacin	N/A	<0. 01	Chloramphenicol	N/A	<0. 01
Levofloxacin	N/A	<0. 01	Malachite green	N/A	<0. 01
Difloxacin	N/A	<0. 01		N/A	

N/A means notavailable.

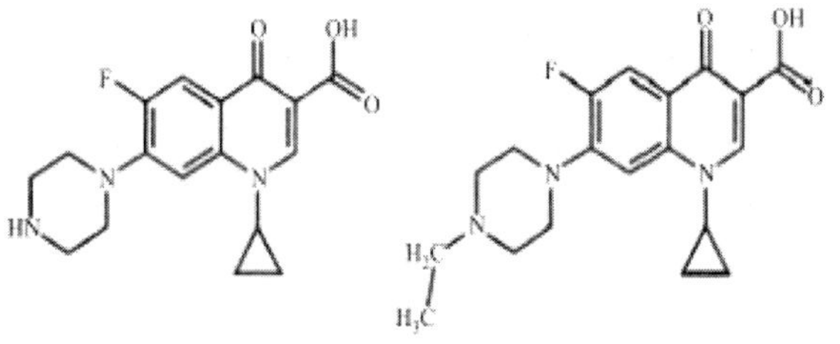

Fig. 2　Molecular structure of CPFX and ENR

To test the potential application of ic-ELISA, different fishery samples were mixed with different concentrations of CPFX, and the recovery rates of CPFX from Carassius auratus, Eriocheir sinensis, Trionyx sinensis, Oreochromis niloticus, Ictalurus punctatus, Scophthalmus maximus, Penaeus vannamei, Macrobrachium rosenbergii or Oncorhynchus mykiss were determined by ic-ELISA in Table 3. The recovery rates and RSD of CPFX from these fish samples were in the range of 80. 57%-94. 61% and 1. 78%-8. 02% respectively. Furthermore, the recovery rates and RSD of CPFX from different fish samples reached 84. 58%-96. 54% and 2. 47%-7. 01%, respectively, determined by HPLC. The recovery rates and RDS values of CPFX by ic-ELISA were similar to those by HPLC (PN0. 05), demonstrating the high sensitivity and accuracy of ic-ELISA for detecting CPFX in fishery products.

4　Discussions

Following the optimization of experimental conditions, we found that 150 ng/mL of CPFX-OVA for coating antigen and 0. 22 μg of mAbled to a specific detection of 45. 25 ng/mL CPFX with a 50% inhibition at 245. 86 ng/mL. Interestingly the ic-ELISA detected 84. 6% cross-reactivity with ENR, a compound with highly structural similarity to the CPFX, similar to that reported by Duan and Yuan（2001）and Wang et al.（2007）. The cross-reactivity of anti-CPFX with ENR suggested that the minor difference in the 7th position between CPFX and ENR（Fig. 2）may have little effect on the recognition and reactivity of antiCPFX. Indeed, the position 1 in both molecules contributes to the predominate antigenicity（Holtzapple et al. , 1993, 1997）. Importantly, the ic-ELISA did not detect the other seven quinolone analogues and antibiotics tested. In addition, the ic-ELISA detected the CPFX with recovery rates and RSD of 80. 57%-94. 61% and 1. 78%-8. 02%, respectively, which were similar to that determined by traditional HPLC. Moreover, the sensitivity of ic-ELISA for detection of CPFX was similar to that reported with mAb（Wang et al. , 2007）and higher than that using polyclonal antibodies（Snitkoff et al. , 1998; Duan and Yuan, 2001）. Apparently, the ic-ELISA is a highly sensitive and specific assay, and can be used for the detection of CPFX and ENR in fishery products with high reproducibility.

Table 3　RecoveriesofELISAMethodtoassayCPFXresiduesin fishery products（n=10）

Concentration（ng/g）	Assay method	Carassius auratus[a]		Eriocheir sinensis[b]		Trionyx sinensis[a]	
		Recovery（%）[c]	RSD（%）	Recovery（%）[c]	RSD（%）	Recovery（%）[c]	RSD（%）
Negative controls	HPLC	ND	---	ND	---	ND	---
	ELISA	ND	---	ND	---	ND	---
50	ELISA	87. 53±4. 08	4. 66	83. 67±6. 03	7. 21	86. 05±4. 59	5. 33

Concentration (ng/g)	Assay method	*Carassius auratus*[a]		*Eriocheir sinensis*[b]		*Trionyx sinensis*[a]	
		Recovery (%)[c]	RSD (%)	Recovery (%)[c]	RSD (%)	Recovery (%)[c]	RSD (%)
	HPLC	86.36±5.65	6.54	84.58±5.69	6.73	84.78±5.46	6.44
600	ELISA	92.54±6.34	6.85	85.25±6.51	7.64	88.67±3.90	4.40
	HPLC	90.98±5.53	6.08	84.65±5.01	5.92	88.14±4.53	5.14
1250	ELISA	92.38±2.86	3.10	92.20±5.16	6.08	92.65±1.65	1.78
	HPLC	93.24±6.54	7.01	96.54±4.12	4.27	94.35±2.33	2.47

Concentration (ng/g)	Assay method	*Oreochromis niloticus*[a]		*Ictalurus punctatus*[a]		*Scophthalmus maximus*[a]	
		Recovery (%)[c]	RSD (%)	Recovery (%)[c]	RSD (%)	Recovery (%)[c]	RSD (%)
Negative controls	HPLC	ND	---	ND	---	ND	---
	ELISA	ND	---	ND	---	ND	---
50	ELISA	86.09±6.08	7.06	93.09±6.52	7.00	85.04±5.22	6.14
	HPLC	89.65±5.31	5.92	88.47±3.25	3.67	87.54±3.21	3.67
600	ELISA	92.54±5.28	5.71	93.52±5.43	5.81	90.94±4.78	5.26
	HPLC	90.24±6.31	6.99	93.25±3.69	3.96	89.45±2.35	2.63
1250	ELISA	94.61±4.58	4.84	89.20±6.04	6.77	92.97±2.01	2.16
	HPLC	84.65±3.32	3.92	87.65±4.98	5.68	91.25±2.43	2.66

Concentration (ng/g)	Assay method	*Penaeus vannamei*[a]		*Macrobrachiumrosenbergii*[a]		*Oncorhynchus mykiss*[a]	
		Recovery (%)[c]	RSD (%)	Recovery (%)[c]	RSD (%)	Recovery (%)[c]	RSD (%)
Negative controls	HPLC	ND	---	ND	---	ND	---
	ELISA	ND	---	ND	---	ND	---
50	ELISA	86.29±3.46	4.00	82.37±6.61	8.02	80.57±2.64	3.27
	HPLC	87.35±5.96	6.82	87.54±5.40	6.17	81.31±2.58	3.17
600	ELISA	87.02±1.56	1.79	88.84±5.05	5.68	90.12±3.54	3.93
	HPLC	94.54±2.35	2.49	94.58±5.64	5.96	85.35±2.65	3.10
1250	ELISA	87.08±3.07	3.53	88.08±4.69	5.32	82.14±3,24	3.94
	HPLC	88.75±3.56	4.01	94.15±5.69	6.04	86.45±2.56	2.96

To prepare optimal fishery samples for detection of CPFX, we extracted the samples with the acidulated acetonitrile, which improved the extracting efficiency, making the extract easy to be concentrated and purified. Furthermore, we treated the extracts with n-hexane to maximally remove most insoluble components, such as fish fats, which usually interfere with the sensitivity and specificity of ic-ELISA in detecting small molecular compounds. These extraction strategies should improve the sensitivity and specificity of this assay, leading to high recovery rates from the fishery products.

Notably, ELISA has been demonstrated to be highly sensitive, specific, and suitable for simultaneous analysis of many samples. Furthermore, ELISA usually is relatively cheaper than other common methods and easily performed. In this study, we analyzed the CPFX residues in different fishery products, which were mainly exported from China, and detected small amount of CPFX in those fishery products within 2 h. Conceivably, the developed ic-ELISA, together with the extraction protocol, can be used for screening and characterization of CPFX-contained fishery products, thereby improving the safety of commercial fishery products.

In summary, we developed a rapid and sensitive ELISA for the detection of CPFX and ENR in aquaculture products' tissues with a sensitivity of 10 ng/g. The LOD was below current MRL established by the EU and China. Therefore, the method can be used as a convenient tool for the rapid detection of CPFX and ENR in aquaculture products.

5　Acknowledgements

This study was supported by the grants from the Key Project of Science and Technology of Shanghai Agriculture Committee (6660106477) and the earmarked fund for the Modern Agro-industry Technology Research System of China.

References

1　Diario Oficial de las Comunidades Europeas (DOCE), 1990. Council Regulation No. 2377/90L224, 991, 2601.

2　Duan J. H., Yuan Z. H., 2001. Development of an indirect competitive ELISA for ciprofloxacin residues in food animal edible tissues. J. Agric. Food Chem. 49, 1087–1089.

3　Garcia Ovando H., Gorla N., Luders C., Poloni G., Errecalde C., Prieto G., Puelles I., 1999. Comparative pharmacokineticsof enrofloxacin and ciprofloxacin in chickens. J. Vet. Pharmacol. Ther. 22, 209–212.

4　Hernandez M., Aguilar C., Borrull F., Calull M., 2002. Determination of ciprofloxacin, enrofloxacin and flumequine in pig plasma samples by capillary isotachophoresis— capillary zone electrophoresis. J. Chromatogr. B 772, 163–172.

5　Holtzapple C. K., Carlin R. J., Kubenal S. L., 1993. Analysis of hapten – carrier protein conjugates by nondenaturing gel electrophoresis. J. Immunol. Methods 164, 245–253.

6　Holtzapple C. K., Buckley S. A., Stanker L. H., 1997. Production and characterization of monoclonal antibodies against sarafloxacin and cross-reactivity studies of related fluoroquinolones. J. Agric. Food Chem. 45, 1984–1990. 12K. Hu et al./Aquaculture 310 (2010) 8–12.

7　Huang X. Y., Hu K., Fang W., Jin Y., Yang X. L., 2008. Preparation of ciprofloxacin–carrier protein conjugation and identification of its products. J. Shanghai Fish. Univ. 17 (5), 585–590 (In Chinese).

8　Huet A. C., Charlier C., Tittlemier S. A., Singh G., Benrejeb S., Delahaut P., 2006. Simultaneous determination of (Fluoro) quinolone antibiotics in kidney, marine products, eggs, and muscle by enzyme-linked immunosorbent assay (ELISA). J. Agric. Food Chem. 54, 2822–2827.

9　Li C., Wang Z. H., Cao X. Y., Beier R. C., Zhang S. X., Ding S. Y., Li X. W., Shen J. Z., 2008. Development of an immunoaffinity column method using broad-specificity monoclonal antibodies for simultaneous extraction and cleanup of quinolone and sulfonamide antibiotics in animal muscle tissues. J. Chromatog. A 1209, 1–9.

10　Montero A., Althaus R. L., Molina A., Berruga I., Molina M. P., 2005. Detection of antimicrobial agents by a specific microbiological method (Eclipse100©) for ewe milk. Small Rumin. Res. 57, 229–237.

11　Snitkoff G. G., Grabe D. W., Holt R., Bailie G. R., 1998. Development of an immunoassay for monitoring the levels of ciprofloxacin in patient samples. J. Immunoassay 19 (4), 227–238.

12　Tang J., Yang X. L., Zheng Z. L., Yu W. J., Hu K., Yu H. J., 2006. Pharmacokinetics and the active metabolite of enrofloxacin in Chinese mitten-handed crab (Eriocheir sinensis). Aquaculture 260, 69–76.

13　Wang Z. H., Zhu Y., Ding S. Y., He F. Y., Beier R. C., Li J. C., Jiang H. Y., Feng C. W., Wan Y. P., Zhang S. X., Kai Z. P., Yang X. L., Shen J. Z., 2007. Development of a monoclonal antibody-based broad-specificity ELISA for fluoroquinolone antibiotics in foods and molecular modeling studies of cross-reactive compounds. Anal. Chem. 79, 4471–4483.

14　Watanabe H., Satake A., Kido Y., 2002. Monoclonal-based enzyme-linked immunosorbent assays and immunochromatographic assay for enrofloxacin in biological matrices. Analyst 127, 98–103.

15　Zhou Y., Li Y. S., Wang Z., Tan J. H., Liu Z. S., 2006. Synthesis and Identification of the Antigens for ciprofloxacin. Chinese J. Vet. Sci. 26, 200–203 (In Chinese).

16　Zhou Y., Li Y. S., Pan F. G., Zhang Y. Y., Lu S. Y., Ren H. L., Li Z. H., Liu Z. H., Zhang J. H., 2009. Development of a new monoclonal antibody based direct competitive enzyme-linked immunosorbent assay for detection of brevetoxins in food samples. Food Chem. 118 (2), 467–471.

该文发表于《上海海洋大学学》【2010，19（6）：805-809】

基于单克隆抗体的酶联免疫法与高效液相色谱法检测
水产品中环丙沙星残留的比较

Comparing the competitive indirect enzyme-linked immunosorbent assay based onmonoclonal antibody and High PerformanceLiquid Chromatography methods to determinate the ciprofloxacin residues in fishery products

胡鲲[1]，黄宣运[1]，姜有声[1]，方伟[2]，朱泽闻[3]，杨先乐[1]

（1. 上海海洋大学农业部渔业动植物病原库，上海 201306；2. 广东省水生动物疫病预防控制中心，广东 广州 510222；
3. 全国水产技术推广总站，北京 100026）

摘要：为了快速、灵敏地检测禁用药物环丙沙星在水产品中的残留，建立了基于单克隆抗体的间接竞争酶联免疫法（competitive indirect ELISA, ci-ELISA），并比较了其与高效液相色谱法（High Performance Liquid Chromatography, HPLC）在灵敏度、准确性和重现性等方法学参数上的差异。实验结果显示，经过浓缩、净化等前处理过程，ci-ELISA 与 HPLC 的检测灵敏度分别可以达到 10 μg/kg 和 1 μg/kg，回收率分别为 79.28%～92.92% 和 86.27%～97.32%，相对标准偏差分别在 3.21%～8.88% 和 2.66%～5.52%。前者经济、高效，适合作为初筛方法大批量筛选样品；后者灵敏度、重现性更好，适于在实验室条件下作为验证方法。

关键词：环丙沙星；水产品；残留；间接酶联免疫法；高效液相色谱法

二、恩诺沙星

恩诺沙星（enrofloxacin，简称 ENR）是一种专用的氟喹诺酮类兽药。我国政府规定：恩诺沙星在水产品中的残留限量为 50 μg/kg（鳗鲡除外）。采用碳二亚胺法和活化酯法，可将 ENR 与鸡血清白蛋白（OVA）以及牛血清白蛋白（BSA）偶联，合成完全抗原，为建立其免疫学检测方法奠定基础（张庆华等，2010）。

该文发表于《中国免疫学杂志》【2010，26（6）：538-543】

恩诺沙星完全抗原的制备试验

Studies on synthesis of a complete antigen of enrofloxacin

张庆华，高建忠，熊清明，杨先乐

（上海海洋大学海洋种质资源发掘与利用重点实验室，上海 200090）

摘要：**目的：**制备恩诺沙星（EF）的完全抗原，进而为 EF 单克隆抗体检测试剂盒的研制奠定基础。**方法：**将小分子的药物 EF 和鸡血清白蛋白（OVA）以及牛血清白蛋白（BSA）分别采用碳二亚胺法和活化酯法两种不同的偶联方法合成完全抗原，通过紫外特征光谱和抗血清效价对完全抗原的合成效果进行分析。**结果**2 种不同的偶联方法均使 OVA 和 BSA 与 EF 偶联成功。偶联方法对于完全抗原的合成效果也存在差异，从紫外吸光率和偶联物结合比判断，OVA 采用活化酯法效果好，未经 EDA 处理的 OVA、活化酯法和碳二亚胺法合成的偶联物的结合比分别为 10.3、20.7、16.5；而 BSA 采用碳二亚胺法效果好，未经 EDA 处理的 BSA、活化酯法和碳二亚胺法合成的偶联物的结合比分别为 20.5、24.6 和 30.8。血清效价检测发现，以 BSA-EF 为完全抗原的效价为 25 600，以 OVA-EF 为完全抗原的效价为 6 400，这表明完全抗原的合成是成功的。**结论：**2 种不同方法合成的 OVA-EF 和 BSA-EF 完全抗原，可以作为免疫原用于制备抗体以及检测时的包被原进行 ELISA 检测。

关键词：恩诺沙星；鸡血清白蛋白；牛血清白蛋白；碳二亚胺法；活化酯法

三、孔雀石绿

通过改进的碳二亚胺法可以制备无色孔雀石绿-牛血清白蛋白偶联物（LMG-BSA）与副品红-牛血清白蛋白偶联物（PA-BSA）。以 LMG-BSA 与 PA-BSA 作为免疫原，可以制备针对 LMG 的多克隆抗体和单克隆抗体，其效价、亲和力、灵敏度和特异性均能满足孔雀石绿残留检测需要（陈力等，2011；陈力等，2011）。

该文发表于《西北农林科技大学学报：自然科学版》【2011，39（7）：28-34】

2 种不同无色孔雀石绿单克隆抗体的制备

Preparation of two different monoclonal antibodies against leuco malachite green

陈力[1]，胡鲲[1]，姜有声[1]，黄宣运[2]，安健[3]，杨先乐[1]

(1. 上海海洋大学，国家水生动物病原库，上海 201306；2. 中国水产科学研究院东海水产研究所，上海 200090；
3. 连云港市海洋与水产科学研究所，江苏 连云港 222044)

摘要：【目的】制备无色孔雀石绿（LMG）单克隆抗体，并分析其生物学特性，为建立 LMG 残留的免疫学检测方法奠定基础。【方法】通过改进的碳二亚胺法制备无色孔雀石绿-牛血清白蛋白偶联物（LMG-BSA）与副品红-牛血清白蛋白偶联物（PA-BSA），用紫外扫描、质谱法方法鉴定偶联物。利用制备的 LMG-BSA 与 PA-BSA 作为免疫原，免疫 4~5 周龄 Balb/c 小鼠，用杂交瘤技术制备 LMG 单克隆抗体，并对其效价、亲和力、灵敏度和特异性等进行分析。【结果】LMG-BSA 与 PA-BSA 的偶联比率分别为 7：1 和 37：1；经间接酶联免疫吸附（ELISA）法筛选，由 LMGBSA 和 PA-BSA 抗原各获得 1 株杂交瘤细胞，分别命名为 L-1、P-1。这 2 株细胞株分泌的单克隆抗体纯化后，经测定效价分别为 1.25×10^5 和 6.25×10^6，亲和力常数分别为 4.92×10^9 和 1.44×10^8 L/mol，半抑制质量浓度（IC_{50}）分别达 8.03 和 10.5μg/L。L-1 分泌的单克隆抗体仅与孔雀石绿存在交叉反应，交叉反应率为 100%；P-1 分泌的单克隆抗体与孔雀石绿及副品红均存在交叉反应，交叉反应率分别为 80% 与 100%。【结论】成功获得了 2 株 LMG 单克隆抗体，其均可用于水产品中 LMG 残留的快速检测。

关键词：无色孔雀石绿；副品红；半抗原；抗体；免疫

该文发表于《淡水渔业》【2011，41（6）：25-29】

应用改良的碳二亚胺法制备无色孔雀石绿完全抗原及鉴定

Complete antigen of leuco malachite green prepared byimproved carbodiimide method and identification

陈力[1]，胡鲲[1]，姜有声[1]，黄宣运[2]，安健[3]，杨先乐[1]

(1. 上海海洋大学，国家水生动物病原库，上海 201306；2. 中国水产科学研究院东海水产研究所，上海 200090；
3. 连云港市海洋与水产科学研究所，江苏 连云港 222044)

摘要：为完善无色孔雀石绿的免疫检测方法，通过改良的碳二亚胺法，改造无色孔雀石绿结构类似物副品红半抗原。经过紫外分光法、十二烷基磺酸钠-聚丙烯酰胺凝胶电泳（SDS-PAGE）、质谱法鉴定，偶联反应成功，且偶联比最高达 37：1。以副品红-牛血清白蛋白偶联物为完全抗原免疫 Balb/c 小鼠，产生针对无色孔雀石绿的多克隆抗体，效价可达 6.25×10^9 取抗体与无色孔雀石绿进行竞争抑制测定，IC_{50} 可以达到 11.8 mg/L。该抗体与孔雀石绿及副品红存在交叉反应，交叉反应率均为 100%。

关键词：无色孔雀石绿；副品红；碳二亚胺法；偶联；鉴定；多克隆抗体

四、氯霉素

采用碳二亚胺（EDC）方法制备氯霉素完全抗原 CAP-HS-BSA（CAP-HS，琥珀酸酯；BSA，牛血清白蛋白）（熊清明等，2004），可获得针对抗氯霉素的杂交瘤细胞株（方伟等，2010；方伟等，2009）。

该文发表于《中国海洋大学学报》【2010，40（3）：90-94】

氯霉素单克隆抗体的研制及其特性分析

Preparation and Characteristics of Monoclonals Antibodies Against Chloramphenicol

方伟[1,2]，姜有声[1]，黄宣运[1]，胡鲲[1]，杨先[1]

（1. 上海海洋大学国家水生动植物病原库，上海海洋大学水产品药物残留监控检测中心，上海 201306；

2. 广东省水生动物疫病预防控制中心，广东 广州 510222）

摘要：为了研制氯霉素单克隆抗体，建立氯霉素的简便有效的检测方法。采用碳二亚胺（EDC）方法制备氯霉素免疫抗原和包被抗原，经 SDS-PAGE 凝胶电泳，三硝基苯磺酸（TNBS）法鉴定表明偶联成功。采用制备的抗原 CAP-HS-BSA（CAP-HS，氯霉素琥珀酸酯；BSA，牛血清白蛋白）免疫 Balb/c 小鼠，通过杂交瘤技术获得了 2 株抗氯霉素的杂交瘤细胞株，细胞株体外传代和冻存复苏后抗体分泌稳定。经体内诱生法产生腹水，ELISA 检测纯化腹水的效价为 $1:10^6$。抗体的亲和常数分别为：$1.13\times10^{10} L/mol$，$1.41\times10^{10} L/mol$。ELISA 检测显示这 2 株单克隆抗体与氯霉素琥珀酸交叉反应率为 300%，与其他抗生素及结构类似物的交叉反应小，表明该单抗可满足建立免疫学检测方法的需要和开发应用的要求。

关键词：氯霉素；单克隆抗体；检测

该文发表于《华中农业大学学报》【2009，28（3）：334-338】

氯霉素人工抗原的合成与鉴定

Synthesis and Identification of Chloramphenicol Artificial Immunogen

方伟，胡鲲，姜有声，黄宣运，杨先乐

（上海海洋大学国家水生动植物病原库/上海海洋大学水产品药物残留监控检测中心/

上海市海洋大学省部共建水产种质资源发掘与利用教育部重点实验室，上海 201306）

摘要：采用重氮化法和碳二亚胺（EDC）法合成了氯霉素人工抗原。经紫外光谱法、电泳鉴定表明偶联成功；采用三硝基苯磺酸法测得载体蛋白 $\varepsilon-NH_3^+$ 残基利用率分别为 31.02%、22.27%，偶联物的结合比为 18.9、13.6；制备的抗原分别免疫小白鼠获得多克隆抗体，血清效价可达 10^{-6} 以上。经比较，碳二亚胺法更适于氯霉素人工抗原的制备。

关键词：氯霉素；人工抗原；合成；鉴定

五、己烯雌酚

利用针对兔 IgG（己烯雌酚抗体）的羊抗体与待测中华鳖肌肉中的己烯雌酚发生特异性的结合反应。采用酶联免疫法（ELISA）检测水产品中己烯雌酚残留，该方法灵敏度高、可靠性强，适用于大规模筛查水产养殖品中己烯雌酚残留（胡鲲等，2002）。

该文发表于《上海水产大学学报》【2002，11（3）：199-202】

酶联免疫法检测中华鳖肌肉中己烯雌酚

Enzyme linked immunosorbent assay（ELISA）for the quantitative analysis of diethylstilbestrol（DES）residues in the muscle of Trionyx sinensis

胡鲲，杨先乐，张菊

（农业部渔业动植物病原库，上海水产大学，上海 200090）

摘要：根据酶标免疫法原理，用含有针对兔 IgG（己烯雌酚抗体）的羊抗体与待测中华鳖肌肉中的己烯雌酚发生特异性的结合反应。洗涤后，加入的酶标底物连接未结合的己烯雌酚抗体，在酶基质的作用下与无色发色剂生成有色产物。反应后于 450 nm 处测量吸光度，吸光强度与样品中的己烯雌酚浓度成反比。该检测方法平均最低检测下限为 0.0125 μg/kg，平均回收率 74.8%，变异系数为 1.763%~3.084%。表明本方法快速、灵敏、可靠，适合对己烯雌酚残留限量检测的要求。

关键词：中华鳖；己烯雌酚；酶联免疫法

第三节 其他快速检测方法

通过药物敏感菌株和培养基显色剂，可以建立基于显色反应来判断结果的微生物检测方法。以腾黄八叠球菌（*Sarcinaureae*）为指示菌株，以氯化三苯四氮唑显色为显色剂，阳性样品呈肉眼可辨的红色，阴性样品呈无色，对恩诺沙星的最低检测限可达 50 μg/kg（胡鲲等，2009）。藤黄微球菌（*Micrococcus luteus*）还能检测水产品中杆菌肽残留（王正彬等，2015）

利用孔雀石绿溶于浓硫酸显黄色，稀释后显暗黄色，Ti（SO₄）₂使其褪色，加入 Na⁺变为白色沉淀，可检测孔雀石绿在水产品中的残留（钱科蕾等，2009；2010）。

该文发表于《上海水产大学学报》【2009，18（4）：472-478】

微生物显色法快速检测水产品中恩诺沙星残留

A rapid method to determine the residues of enrofloxacin in fishery products by microbiological chromotest

胡鲲，李怡，朱泽闻，黄宣运，王民权，方淑蓓，杨先乐

（1. 上海海洋大学农业部渔业动植物病原库，上海海洋大学水产品药物残留监控检测中心，
上海海洋大学省部共建水产种质资源发掘与利用教育部重点实验室，上海 201306；2. 全国水产技术推广总站，北京 100026）

摘要：针对水产品中恩诺沙星残留，通过筛选药物敏感菌株和显色剂，建立了一种基于显色反应来判断结果的微生物检测方法。以腾黄八叠球菌（*Sarcinaureae*）为指示菌株，以氯化三苯四氮唑显色为显色剂，阳性样品呈肉眼可辨的红色，阴性样品呈无色，最低检测限可达 50 μg/kg。经酶联免疫法（最低检测限可达 12 μg/kg，回收率在 74.25%~93.38%之间）和高效液相色谱法（最低检测限可达 5.0 μg/kg，回收率在 78.20%~91.27%之间）验证，该方法准确、可靠。不同于传统的方法，该方法更为快速、便捷、易于结果判定，是对现有检测方法的一种有益

的补充和完善，极其适合在生产、销售、流通等非实验室条件下作为初筛方法使用。

关键词：微生物显色法；恩诺沙星；残留；

该文发表于《华中农业大学学报》【2015，34（5）：105-110】

水产品中杆菌肽残留的微生物法检测

Microbiological inhibition method for determination of bacitracin residues in muscles of aquatic products

王正彬[1,2]，刘永涛[2]，董靖[2]，胥宁[2]，杨秋红[2,]，徐春娟[2]，艾晓辉[2]

(1. 华中农业大学水产学院，武汉 430070；2. 中国水产科学研究院长江水产研究所，武汉 430223)

摘要： 以藤黄微球菌（*Micrococcus luteus*）为检测菌种，样品经甲醇-0.1%甲酸水（体积比为3/7）超声提取，旋转蒸发浓缩，用杯碟法建立了在水产品可食性组织（草鱼肌肉、鳗鲡和对虾肌肉）中杆菌肽残留量的测定方法，结果显示：杆菌肽在草鱼肌肉中最低检测限为 0.25 μg/g，在对虾和鳗鲡肌肉中最低检测限为 0.5 μg/g，均达到了我国业部和欧盟规定的杆菌肽在肌肉组织中的最高残留限量检测要求。杆菌肽在各水产品肌肉组织中 1~10 μg/mL 范围内，药物质量浓度对数与抑菌圈直径的线性关系良好（$R^2 > 0.9900$），杆菌肽不同水产品肌肉组织添加质量浓度是 0.5~2.5 μg/g 时，各组织的回收率均 70.45%~93.4% 范围内，变异系数在 2.02%~12.62%。并采用3种不同细菌制作的检定平板确定出残留物为杆菌肽，方法有效。

关键词： 菌肽；杯碟法；残留；水产品

该文发表于《上海水产大学学报》【2010，19（1）：56-59】

Na^+-Ti $(SO_4)_2$法检测草鱼中孔雀石绿残留

Determination of malachitegreen residue in grass carp tissues by Na^+-Ti $(SO_4)_2$

钱科蕾[1,2]，胡鲲[1,2]，杨先乐[1,2]

(1. 上海海洋大学水产与生命学院，上海 201306；2. 上海海洋大学国家水生动物病原库，上海 201306)

摘要： 首次建立了 Na^+-Ti $(SO_4)_2$ 法检测草鱼中孔雀石绿残留的方法：孔雀石绿溶于浓硫酸显黄色，稀释后显暗黄色，Ti $(SO_4)_2$ 使其褪色，加入 Na^+ 变为白色沉淀，依据沉淀量对孔雀石绿残留进行定量分析。优化实验条件，采用 pH=7.4、硫酸钛和氯化钠的浓度均为 2.0 mol/L 时，可有效地检测孔雀石绿残留。方法学实验结果表明：该方法最低检测量为 0.08 mg/kg，在草鱼肌肉和肝脏中的回收率分别为 65.00%~75.63%，63.75%~71.35%，相对标准偏差分别为 4.82%~1.89%，4.93%~1.97%。该方法快速、简单、灵敏、准确，依据白色沉淀仅凭肉眼即可进行初筛，适合于快速检测孔雀石绿在水产品中的残留。

关键词： Na^+-Ti $(SO_4)_2$法；孔雀石绿；残留

该文发表于《安徽农业科学》【2009，37（12）：5347-5349】

Na$^+$-Ti（SO$_4$）$_2$法检测鲫鱼组织中孔雀石绿残留

Determination of Malachite Green Residue in *Crucian Carp* Tissues by Na$^+$-Ti（SO$_4$）$_2$

钱科蕾，胡鲲，杨先乐

（国家水生动物病原库，上海海洋大学，上海 201306）

摘要 ［目的］探究 Na$^+$-Ti（SO$_4$）$_2$法检测鲫鱼中孔雀石绿残留的方法。［方法］将孔雀石绿溶于浓硫酸显黄色，稀释后显暗黄色，用 Ti（SO$_4$）$_2$使其褪色，加入 Na$^+$变为白色沉淀，依据沉淀量对孔雀石绿残留进行定量测量并采用正交实验优化实验条件。［结果］采用 pH 值为 7.4、硫酸钛和氯化钠的浓度均为 2.0 mol/L 时，可有效地检测孔雀石绿残留。该方法最低检测限为 0.01 mg/kg，在鲫鱼肌肉和肝脏中的回收率分别为 86.63%～91.31%和76.63%～78.08%，相对标准偏差分别为 4.21%～2.19%和3.93%～1.86%。孔雀石绿残留在药浴 后鲫鱼肝脏中能检测到的含量最高（0.033 2 mg）；到 6 h 时再次出现检测峰值（0.031 0 mg），之后随鲫鱼暂养时间的延长检测量显著下降，到第 30 天肌肉和肝脏中均仅能检测到 0.000 1 mg。［结论］ 该方法实验成本低，操作简单，周期短，适用于检测水产品中孔雀石绿残留。

关键词：Na$^+$-Ti（SO$_4$）$_2$法；孔雀石绿；残留

第五章　病原微生物的耐药性及其控制耐药策略研究

渔药的不合理使用会导致病原微生物耐药性的产生。对主要水产动物病原菌开展耐药性调查、分析及其探寻耐药机制，制定延缓耐药的策略，有助于公共卫生安全和水产动物疾病的防治。

第一节　水产动物病原菌对药物的敏感性和耐药性

一、病原菌对药物的敏感性

病原菌对于药物的敏感性对临床用药具有重要的参考价值。病原菌对药物菌敏感性可用最低抑菌浓度（Minimal Inhibitory Concentration，MIC）或最低杀菌浓度（Minimal Bactericidal Concentration，MBC）来评价。从患病施氏鲟体内分离到的鲁氏耶尔森菌（*Yersinia ruckeri*）对恩诺沙星（enrofloxacin）、环丙沙星（ciprofloxacin）、四环素（tetracycline）等十余种药物敏感（Li Shaowu et al.，2013）。从西伯利亚鲟（*Acipenser baerii*）的分离到的气单胞菌（*Aeromonas*）、致伤弧菌（*Vibrio vulnificus*）等均对多粘菌素 B、妥布青霉素等药物较敏感（司力娜，2010）；来源于东北三省鲤致病性嗜水气单胞菌均对左氟沙星、多粘菌素 B、氧氟沙星较敏感（司力娜，2011）。三氯异氰尿酸和苯扎溴铵对海水养殖源弧菌的 MIC 值差异较大，且均呈现明显的浓度效应（姚小娟，2015）。即使是且来源于不同宿主动物的同种病原菌，其对药物的敏感性也差异很大（崔佳佳，2016）。溶藻弧菌、迟钝爱德华氏菌等病原菌还对诺氟沙星、恩诺沙星等药物具有明显的抗生素后效应（PAE）（房文红，2005）。

该文发表于《Journal of Aquatic Animal Health》【2013，25：9-14】

Isolation of Yersinia ruckeri strain H01 from farm raised Amur sturgeon, Acipenser schrenckii in China

Li Shaowu, Wang Di, Liu Hongbai, Lu Tongyan

(Department of Aquaculture, Heilongjiang River Fisheries Research Institute, Chinese Academy of Fishery Sciences, 43# Songfa Road, Daoli District, Harbin 150070, PR China)

Abstract: Yersinia ruckeri is the causative agent of enteric redmouth disease or yersiniosis, which affects salmonids and several other species of fish. However, there are no reports on the characteristics and pathogenicity of Y. ruckeri isolated from farm-raised Amur sturgeon, Acipenser schrenckii. Here, we isolated and characterized Y. ruckeri strain H01 from the diseased Amur sturgeon in China. The phenotypic and genotypic characteristics of Y. ruckeri were observed and its virulence was tested by examining experimentally infected sturgeons. Examination of the flagellar morphology of Y. ruckeri by transmission electron microscopy showed 5-8 peritrichous flagella located on the cell body. Actively dividing cells with an obvious cell membrane were approximately 0. 64 μm in diameter and between 1. 7 and 2. 5 μm in length. LD50 value was determined to be 7.2×10^6 CFU and Y. ruckeri could be re-isolated from the liver and kidneys of infected sturgeon. Antimicrobial susceptibility tests showed that H01 was susceptible to 10 antimicrobial agents. Part of the 16S rRNA sequences (563 bp) was amplified and sequenced to study the genotypic characterization in Y. ruckeri (GenBank accession number JQ657818). The phylogenetic tree revealed H01 was clustered together with Y. ruckeri strains. Together, this study describes the isolation, characterization and phenotypic/genotypic analysis of a Y. ruckeri strain isolated from farm-raised Amur sturgeon. The results discovered may provide some theoretical basis for the prevention and control of yersiniosis in Amur sturgeon.

key words: Yersinia ruckeri, Acipenser schrenckii, flagella, virulence, phylogenetic tree

人工养殖施氏鲟源鲁氏耶尔森菌的分离鉴定

李绍戊，王荻，刘红柏，卢彤岩

（中国水产科学研究院黑龙江水产研究所养殖室 松发街43号 黑龙江哈尔滨 150070 中国）

摘要： 鲁氏耶尔森菌（Yersinia ruckeri）是鱼类肠炎红嘴病或耶尔森菌病的主要病原，可感染鲑鳟鱼类及其他多种鱼类，但目前没有从养殖施氏鲟中分离到该菌。本研究从患病施氏鲟体内分离到一株鲁氏耶尔森菌，并对其表型、理化特性和致病性等进行了鉴定。通过透射电镜观察菌株形态可见，鲁氏菌有5~8条周生鞭毛并锚定在细胞上。菌株胞外有一层明显的细胞膜，细胞直径约0.64 μm，长度在1.7~2.5 μm之间，细胞膜的厚度为0.12 μm。人工感染试验结果表明，该菌株对施氏鲟的半致死浓度为7.2×10⁶ CFU，且从感染鱼的肝脏和肾脏中可再次分离到此菌。药敏试验结果显示，H01株对10种抗菌药物敏感。通过比对16S rRNA基因部分序列并构建系统进化树可见，H01株与鲁氏耶尔森菌菌株聚为一支。综上，本研究从人工养殖施氏鲟体内分离到一株鲁氏耶尔森菌并对其理化特性、基因特征等进行了分析，研究结果有助于为施氏鲟鲁氏耶尔森菌症的防控提供理论基础。

关键词： 鲁氏耶尔森氏菌；施氏鲟；鞭毛；毒力；系统进化树

1 Introduction

Yersinia ruckeri, a member of the family Enterobacteriaceae, is the causative agent of enteric redmouth (ERM) disease or yersiniosis, a general septicaemia affecting mainly salmonids (Fernandez et al., 2007; Tobback et al., 2007). Since its first description in the Hagerman Valley of Idaho (USA) in the 1950s, Y. ruckeri has been found to be widely distributed in many salmonid species in North America (Ross et al., 1966). More recently, the disease has been reported, and the bacterium has been isolated from fish hosts throughout the world, including Europe, North and South America, Australia, New Zealand and South Africa (Austin & Austin, 2007). Although salmonids are the main fish species susceptible to Y. ruckeri, susceptibility has also been reported in other fish species such as eels, goldfish, catfish, carp, sturgeon and tilapia (Vuillaume et al., 1987; Berc et al., 1999; Danley et al., 1999; Eissa et al., 2008). Infection with Y. ruckeri results in a bacterial septicaemia without disease specific signs but is most commonly detected due to exophthalmos and hemorrhages in the eye of fish. The enteric redmouth diseases exhibited obvious subcutaneous haemorrhage in the mouth and throat of rainbow trout (Furones et al., 1993). With respect to the clinical signs of the disease in other fish species, Vuillaume A et al. (1987) provided the first description of Y. ruckeri isolated from sturgeon, Acipenser baerii Brandt cultured in southwestern France. Haemorrhages were observed around the mouth, at the base of the rostrum and pectoral fins and around the urogenital pore. The belly was swollen and a bloody liquid exuded from the vent under finger pressure. However, there was no report on the Amur sturgeonuntil now. Accurate diagnosis of yersiniosis requires isolation and identification of Y. ruckeri from tissue samples. More commonly it is based on its phenotypic profile and can be readily differentiated from other taxa within the Enterobacteriaceae genera. In addition, more specific techniques such as the polymerase chain reaction (PCR) is available but is best used to confirm the identity of ambiguous isolates.

To our knowledge, few studies on Y. ruckeri in China have been performed. Xu et al. isolated a Y. ruckeri strain from diseased silver and bighead carps in 1991. Fan et al. (2010) reported the isolation and identification of Y. ruckeri FF003 from channel catfish, Ictalurus punctatus. FF003 had apparent pathogenicity to channel catfish, with symptoms of punctate haemorrhaging on the body surface, especially on the submaxilla, abdominal wall and the side of the body (Fan et al., 2010). In this study, a Y. ruckeri strain, H01 was isolated and identified from farm-cultured Amur sturgeon according to morphological, physiological and biochemical characteristics and phylogenetic analysis.

2 Materials and methods

2.1 Bacterial isolation

Y. ruckeri strain H01 was isolated from outbreaks which occurred in 2010 in a fish farm located in Hebei Province, China. Farms consist of concrete tanks and the freshwater is supplied from rivers, with water temperatures ranging from 15℃ to 17℃ and an average pH of 7.0. After the observation of clinical signs, we examined the gill and internal organs for parasitic and bacterial infections. For bacterial isolation, samples were aseptically collected from liver and kidney and directly streaked onto trypticase soy-agar (TSA, Difco) and incubated at 25℃ for 24~48 h. Pure cultures were kept frozen at −80℃ in tryptic soy broth (TSB, Difco) supplemented with 15% glycerol.

2.2 Biochemical characterization

Bacterial isolate was subjected to morphological, physiological and biochemical tests using classical tube and plate procedures. In parallel, the isolate was identified using Biolog, the microbial identification system (Biolog Ltd., Hayward, CA). The effect of temperature, salinity and pH value on the growth of the bacteria was detected. In addition, antimicrobial susceptibility of H01 strain to amikacin (AN: 30 μg), amoxicillin (AMC: 20 μg), carbenicillin (CB: 100 μg), chloramphenicol (CP: 30 μg), ciprofloxacin (CIP: 5 μg), enrofloxacin (EN: 5 μg), erythromycin (ETR: 15 μg), furazolidone (FZ: 300 μg), gentamicin (GM: 10 μg), kanamycin (KN: 30 μg), norfloxacin (NOR: 10 μg), novobiocin (NB: 5 μg), ofloxacin (OF: 5 μg), penicillineG (PG: 10 μg), tetracycline (TE: 30 μg), and sulfamethoxazole (SXT: 23.75 μg) was determined on the Mueller Hinton agar by the disc diffusion method (National Committee for Clinical Laboratory Standards. 2006).

2.3 Morphological characterization by negative-stain transmission electron microscopy (TEM)

The bacteria were streaked onto a TSA and incubated for 24 h at 25℃. The single colony was looped, suspended, and washed with phosphate buffered saline. One drop of the bacterial suspension was placed on parafilm for 5 min, then onto a formvar-coated, 300-mesh grid and allowed to dry. The grid was stained with 2% uranyl acetate (pH 6.5) for 30 s, dried at room temperature and then examined by transmission electron microscope (TEM; Hitachi-7650, Japan).

2.4 Virulence of Y. ruckeri H01 on A. schrenckii

Healthy juvenile sturgeons (body weight of 40~60 g) obtained from the fish farm (Hebei Province, China) were used for virulence assays. The health of the fish stock was checked upon arrival in the laboratory by collecting samples from the internal organs for microbiological analysis. Virulence assays were done by intraperitoneal (i.p.) injection with a dosage of 3×10^7 CFU. Fish (n=10) were maintained at 20℃ in a 0.5 m³ tanks containing aerated, static fresh water and equipped with a filtering system. All of the fish were examined microbiologically at 2 days post inoculation. Experimental control groups were challenged with sterile saline solution in a separate aquarium.

2.5 Determination of LD_{50}

Y. ruckeri H01 was grown on TSA agar overnight at 25℃, harvested, washed and was suspended in normal saline (pH 7.2). Serial 1.5-fold dilution of known number of bacteria/mL was made with normal saline and 0.2 mL of each dilution was inoculated intraperitonially into each group of fish (n=10). Fish were observed for 8 days and number of surviving fish was recorded. LD_{50} values were determined using the formula derived by Reed and Muench (Reed & Muench, 1938).

2.6 16S rRNA sequencing and analysis

Total bacterial DNA was extracted from pure bacterial cultures using Bacteria Genomic DNA Extraction Kit (Tiangen, China). The concentration of DNA was quantified spectrophotometrically and adjusted to a concentration of 100 ng/μL. 16S rRNA gene from DNA templates was amplified using the primer Yer-F: 5' -CGAGGAGGAAGGGTTAAGT-3' and Yer-R: 5' -AAGGCACCAAG-GCATCTCT-3' (Cunningham et al., 2010). The amplification mixture (50 μL) comprised 2 μL (10 pmol/μL) each of the Yer-F and Yer-R primers, 0.5 μL (2 U/μL) of Taq DNA polymerase, 5 μL of ×10 PCR reaction buffer, 1 μL of dNTP mixture (10mM each) and 0.5 μL of DNA template. The cycling program was as follows: initial denaturation at 95℃ for 5 min, followed by 30 cycles of denaturation at 94℃ for 30 sec, annealing at 55℃ for 30 sec, extension at 72℃ for 1 min, and a final extension at 72℃ for 10 min. Controls, without DNA, were simultaneously included in the amplification process. The integrity of the PCR products was assayed by the development of single bands following electrophoresis for 30 min at 120 V in 1% (w/v) agarose gels in TAE buffer. The PCR products were sequenced by Shanghai Shenggong Biotechnological Ltd (Shanghai, China). The sequence obtained was compared with known sequences by searching the National Center for Biotechnology Information (NCBI) databases with the BLAST program. A phylogenetic tree was constructed by comparing with the 16S rRNA sequence of the strain H01 and other relative bacteria species in the GenBank using MEGA 4.0 software based on the Neighbor-Joining method (Saitou & Nei, 1987).

3 Results

3.1 Bacterial isolation and biochemical characteristics

Diseased Amur sturgeon with variable size (from 15 g to 30 g) showed hemorrhages around the mouth, lower jaw, the base of the pectoral fins, abdomen and the urogenital pore as main external signals. Furthermore, they exhibited petechiae in some internal organs, including liver, hindgut and swim bladder. A dominant strain named H01 was isolated from the TSA plate after bacterial culture for 24 h at 25℃. The colonies were 1~2 mm in diameter, smooth, round, raised, and with entire edges. The optimum growth condition of Y. ruckeri H01 was pH=7.0, salinity=0%, and temperature=25℃ for 18 h, and their growth could be inhibited at the TSA medium when pH value varied from 0 to 4 and 10 to 14, or the concentration of NaCl was higher than 4%. H01 strain was identified as Y. ruckeri with Biolog similarity of 0.561. The biochemical characteristics of H01 strain was indicated in Table 1.

Table 1　Characteristics and biochemical reactions for H01 isolate in this study

Characteristics/Test	H01	Y. ruckeri (Danley ML et al., 1999)	FF003 (Fan FL et al., 2010)
Gram stain	−	−	−
Oxidation/fermentation	+	+	+
Oxidase	−	−	−
Motility	+	−	N/A
Urease	−	N/A	−
Lysine Decarboxylase	+	N/A	+
Ornithine Decarboxylase	+	+	+
Glutamyl transferase	+	+	N/A
Simmon's citrate	+	−	+
Gelatin	+	N/A	+
Indole production	−	−	−
VP reaction	−	−	+
Gas production	−	d	−
D−xylose	−	−	−
D (+) −Cellobiose	−	N/A	−
Melibiose	−	N/A	−
Raffinose	−	N/A	−
Sucrose	−	N/A	−
L−Rhamnose	−	−	−
L−arabinose	−	−	−
Trehalose	−	+	+
D (−) −Salicin	−	N/A	−
Sorbitolum	+	+	+
Inositol	−	N/A	−

Note: '+' means positive reaction; '−' means negative reaction; 'd' means variable reactions

TEM was used to confirm the ultrastructure of flagella that were attached to Y. ruckeri. As shown in Figure 1, there was an obvious cell membrane over the cell and actively growing cells were approximately 0.64 μm in diameter and between 1.7 and 2.5 μm in length. The thickness of the membrane was about 0.12 μm. Each bacterial cell body contained between 5~8 flagella in a peritrichous arrangement and was anchored firmly at the cell body. In addition, the cell membrane appeared to be shrunken and rough in shape.

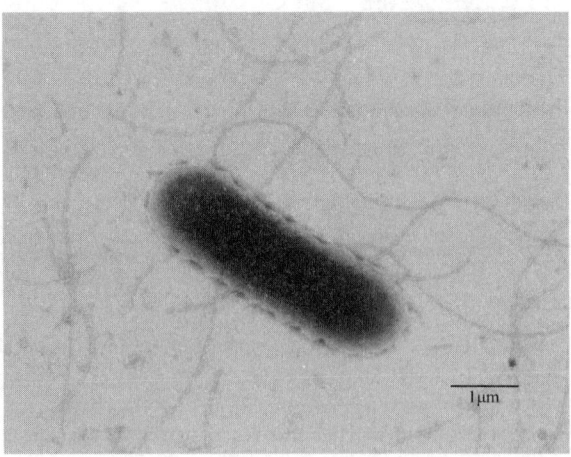

Fig. 1   Morphological characterization of Y. ruckeri H01 as shown by negative staining
transmission electron micrograph (25 000×)

Through testing antimicrobial susceptibility to 16 different types of antibiotics, it was found that the H01 strain was susceptible to 10 antibiotics, including amikacin, amoxicillin, carbenicilli, ciprofloxacin, enrofloxacin, gentamicin, kanamycin, norfloxacin, ofloxacin and tetracycline, while resistant to chloramphenicol, erythromycin, furazolidone, novobiocin, penicillin G, and sulfamethoxazole (Table 2).

Table 2   Antibiotic susceptibility pattern of Y. ruckeri strain H01

Antibiotics	inhibition zone diameter (N=3, mm)	Sensitivity (S/R *)	Antibiotics	inhibition zone diameter (N=3, mm)	Sensitivity (S/R *)
AN	18±2.64	S	GM	16.3±0.58	S
AMC	23.3±2.30	S	KN	20±2.64	S
CB	23±1.73	S	NOR	27.7±3.79	S
CP	0	R	NB	0	R
CIP	28.7±3.06	S	OF	28±4.00	S
EN	29.3±3.06	S	PG	9.33±0.58	R
ETR	0	R	TE	19±1.73	S
FZ	10.7±1.53	R	SXT	0	R

Note: *: S, sensitive; R, resistant

amikacin (AN), amoxicillin (AMC), carbenicillin (CB), chloramphenicol (CP), ciprofloxacin (CIP), enrofloxacin (EN), erythromycin (ETR), furazolidone (FZ), gentamicin (GM), kanamycin (KN), norfloxacin (NOR), novobiocin (NB), ofloxacin (OF), penicillineG (PG), tetracycline (TE), and sulfamethoxazole (SXT)

Artificial infection of Y. ruckeri was performed to determine the virulence of H01 strain. The results showed that 80% of experimental fish died in 2 days by intraperitoneal injection with a dosage of 3×10^7 CFU. The liver, kidney and spleen tissues were collected from diseased fish, which exhibited obvious haemorrhaging. In the microbiological examination, Y. ruckeri was re-isolated from those tissues, thus confirming infection. In addition, the number of death after i. p. injection was observed in 72 h and found to have a LD_{50} value of 7.2×10^6 CFU (Table 3).

Table 3 **The cumulative death number of fish injected with different concentrations of Y. ruckeri H01**

Time	Bacterial dosage (×10⁷ CFU)								
	3	2	1.33	0.89	0.59	0.39	0.26	0.17	0
	The cumulative number of death								
24 h	8/10*	4/10	3/10	2/10	1/10	0/10	0/10	0/10	0/10
48 h	9/10	7/10	5/10	4/10	3/10	1/10	0/10	0/10	0/10
72 h	10/10	9/10	7/10	6/10	4/10	3/10	1/10	0/10	0/10

* means the number of fish per group

The 563 bp region of 16S rRNA gene from H01 isolate revealed a 98%~99% sequence identity in a BLAST search. The sequence has been deposited with accession number JQ657818 in a genome database (i. e., GenBank) of the NCBI, and compared with that of different Yersinia genus and other several usual pathogens in aquaculture. A phylogenetic tree was constructed with a Neighboring joint method as shown in Figure 2. The phylogenetic tree analysis showed that H01 strain was clustered with Y. ruckeri isolates (the black arrow), while had a long distance with Aeromonas and Edwardsiella genus.

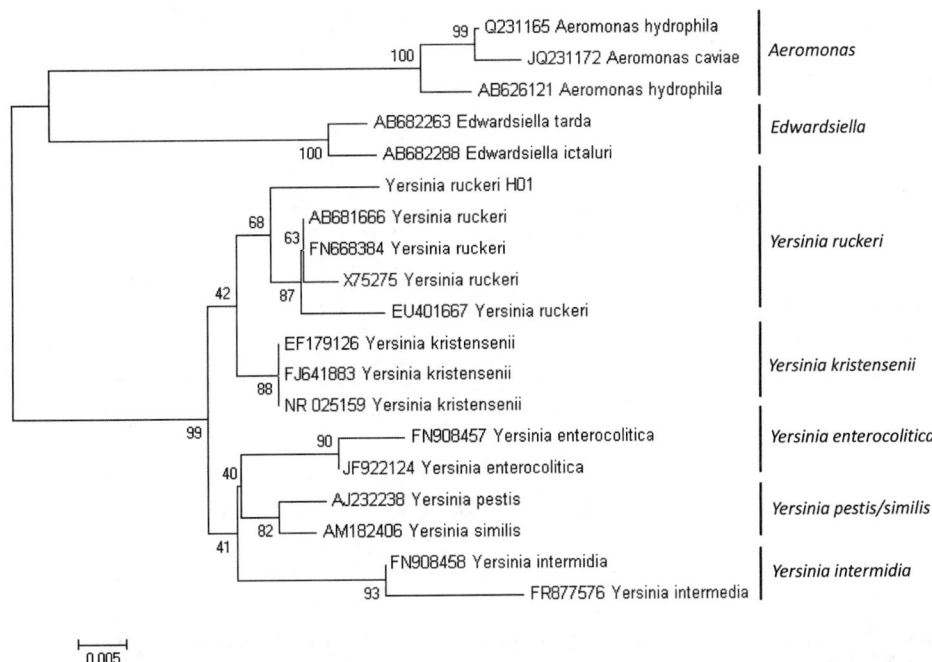

Fig. 2 Phylogenetic relationship among the Yersinia species based on 16S rRNA gene sequences

4 Discussion

After the first identification of Y. ruckeri as a pathogen of rainbow trout, Y. ruckeri has appeared in various fish species throughout the world (Buller, 2004). Outbreaks of yersiniosis are often associated with poor water quality, excessive stocking densities, and the occurrence of environmental stressors (Tobback et al., 2007). Y. ruckeri can also persist in an asymptomatic carrier state, where infection through carrier fish is especially important under stress conditions. However, yersiniosis was only found in Acipenser baerii Brandt but no other sturgeon species.

In this study, we described the isolation of a Y. ruckeri strain from cultured Amur sturgeon and the characteristics of the Y. ruckeri isolate. The diseased fish showed obvious haemorrhages around the mouth, lower jaw, the base of pectoral fins, abdomen and the urogenital pore as main external signals as well as petechiae in liver, hindgut and swim bladder in internal organs. Moreover, this clinical symptoms were very similar to diseased A. baerii Brandt reported by Vuillaume A et al. in 1987. The isolated strain H01 was mobile, flagellated, had a wrinkled cell membrane (Figure 1), and exhibited a high pathogenicity to this species, with LD$_{50}$ value of 7.2× 10⁶CFU, suggesting that Y. ruckeri H01 has morphological characteristics similar to those in strains isolated from salmon species and may be potentially pathogenic or virulent to sturgeons (Davies & Frerichs, 1989). Recently, the analysis of 16S rRNA gene sequence has been widely applied into the identification of pathogens isolated from diseased aquatic animals (LeJeune & Rurangirwa, 2000; Eddy et al., 2007; Pourahmad F et al., 2008). Here the 16S rRNA sequence of H01 strain was used to construct a phylogenetic tree and the results showed that H01 was close to Y. ruckeri with a homology of 99%. A bacterial challenge study indicated that H01 possessed a higher virulence to Amur sturgeonand could induce easily observable clinical symptoms. In the antimicrobial susceptibility tests, H01 was susceptible to 10 antimicrobial agents, indicating that the isolates might be unexposed or less exposed to many antimicrobial agents.

Although generally well controlled by means of vaccination and antibiotic treatment, there have been continuous yersiniosis outbreaks, especially in endemic areas. In some cases the losses due to this disease can be as high as 30%~70% of the stock (Horne et al., 1999). Over the past years, the use of molecular techniques has opened new possibilities to probe the mechanisms of pathogenicity of Y. ruckeri in salmonids (Fernández et al., 2004; Coquet et al., 2005). However, information about the pathogenicity or exact virulence of Y. ruckeri to the Amur sturgeon has not been available until now. Therefore, it is of future interest to investigate the pathogenicity or virulence along with corresponding immune responses inAmur sturgeon by using microarrays, silico experiments, 2D maps and so on.

In brief, here we described a Y. ruckeri strain isolated from diseased Amur sturgeon and investigated its basic characteristics, which may be helpful in providing a theoretical basis for prevention and control of yersiniosis. Furthermore, it would be important to isolate more bacteria strains of affected fish to understand the prevalence of Y. ruckeri inAmur sturgeon.

5 Acknowledgement

This work was supported by grants from The Central-Level Non-profit Scientific Research Institutes Special Funds (201003), National "Twelfth Five-Year" Plan for Science & Technology (2012BAD25B10) and Natural Science Foundation of Heilongjiang Province of China (C201140).

References

E issa, A. E., Moustafa, M., Abdelaziz, M., et al. 2008. Yersinia ruckeri infection in cultured Nile tilapia, Oreochromis niloticus, at a semi-intensive fish farm in lower Egypt. Afr. J. Aquat. Sci. 33, 283-286.

Austin, B. and Austin, D. A. 2007. Bacterial Fish Pathogens: Disease of Farmed and Wild Fish (4th ed.), Springer-Praxis.

Berc, A., Petrinec, Z., Matasin, Z., et al. 1999. Yersinia ruckeri septicaemia in experimentally infected carp (Cyprinus carpio L.) fingerlings. Acta. Vet. Hung. 47, 161-172.

Buller, N. B. 2004. Bacteria from Fish and Other Aquatic Animals: A Practical Identification Manual 74, CABI Publishing, Cambridge.

Danley, M. L., Goodwin, A. E., and Killian, H. S. 1999. Epizootics in farm-raised channel catfish, Ictalurus punctatus (Rafinesque), caused by the enteric redmouth bacterium Yersinia ruckeri. J. Fish. Dis. 22, 451-456.

Davies, R. L. and Frerichs, G. N. 1989. Morphological and biochemical differences among isolates of Yersinia ruckeri obtained from wide geographical areas. J. Fish. Dis. 12, 357-365.

Tobback, E., Decostere, A., Hermans, K., et al. 2007. Yersinia ruckeri infections in salmonid fish. J. Fish. Dis. 30, 257-268.

Eddy, F., Powell, A., Gregory, S., et al. 2007. A novel bacterial disease of the European shore crab, Carcinus maenas molecular pathology and epidemiology. Microbiology. 153, 2839-2849.

Fan, F. L., Wang, K. Y., Geng, Y., et al. 2010. Isolation, identification and phylogenetic analysis of Yersinia ruckeri in channel catfish Ictalunes Punctatus. Oceanlogia. Et. Limnologia. Sinica. 41, 862-868.

Coquet, L., Cosette, P., De, E., et al. 2005. Immobilization induces alterations in the outer membrane protein pattern of Yersinia ruckeri. J. Proteome. Res. 4, 1988-1998.

Fernández, L., Márquez, I., and Guijarro, J. A. 2004. Identification of specific in vivo-induced (ivi) genes in Yersinia ruckeri and analysis of ruckerbactin, a catecholate siderophore iron acquisition system. Appl. Environ. Microbiol. 70, 5199-5207.

Fernández, L., Mendez, J., and Guijarro, J. A. 2007. Molecular virulence mechanisms of the fish pathogen Yersinia ruckeri. Vet. Microbiol. 125, 1-10.

LeJeune, J. T. and Rurangirwa, F. R. 2000. Polymerase chain reaction for definitive identification of Yersinia ruckeri. J. Vet. Diagn. Invest. 12, 558-561.

Furones, M. D., Rodgers, C. J., and Munn, C. B. 1993. Yersinia ruckeri, the causal agent of enteric redmouth disease (ERM) in fish. Ann. Rev. Fish. Dis. 3, 105-125.

Horne, M. T. and Barnes, A. C. 1999. Enteric redmouth disease (Yersiniaruckeri) in: P. T. K. Woo, D. W. Bruno (Eds.), Fish diseases and disorders, Viral, bacterial and fungal infections, 3, 455-477, CABI Publishing.

Saitou, N. and Nei, M. 1987. The neighbor-joining method: a new method for reconstructing phylogenetic trees. Mol. Biol. Evol. 4, 406-425.

National Committee for Clinical Laboratory Standards. 1993. Performance Standards for Antimicrobial Disc Susceptibility Tests. Approved Standard NCCLS Publications M2-A5. Villanova, PA, USA.

Pourahmad, F., Cervellione, F., Thompson, K. D., et al. 2008. Mycobacterium stomatepiae sp. nov., a slowly growing, non-chromogenic species isolated from fish. Int. J. Syst. Evol. Micr. 58, 2821-2827.

Reed, L. J. and Muench, H. 1938. A simple method of estimating fifty percent end points. Am. J. Hyg. 27, 493-497.

Ross, A. J., Rucker, R. R., and Ewing, W. H. 1966. Description of a bacterium associated with redmouth disease of rainbow trout (Salmo gairdneri). Can. J. Microbiol. 12, 763-770.

Cunningham, S. A., Sloan, L. M., Nyre, L. M., et al. 2010. Three-Hour Molecular Detection of Campylobacter, Salmonella, Yersinia, and Shigella Species in Feces with Accuracy as High as That of Culture. J. Clin. Microbiol. 48, 2929-2933.

Vuillaume, A., Brun, R., Chene, P., et al. 1987. First isolation of Yersinia ruckeri from sturgeon, Acipenser baeri Brandt, in south west of France. Bull. Eur. Assn. Fish. P. 7, 18-19.

Xu, B. H. 1991. Yersinia ruckeri-a new kind of pathogens in silver and bighead carp. Chinese. Sci. Bull. 36, 620-620.

该文发表于《水产学杂志》【2010, 23 (4): 18-22】

养殖鲟鱼暴发病病原菌分离及药敏实验

The isolation and drug sensitive tests of pathogens found in sturgeon epizootic disease

司力娜[1,2], 李绍戊[2], 王荻[2], 刘红柏[2], 张颖[1], 卢彤岩[1]

(1. 中国水产科学研究院 黑龙江水产研究所, 黑龙江 哈尔滨 150070; 2. 东北农业大学 动物科学技术学院, 黑龙江 哈尔滨 150030)

摘要： 从濒死的西伯利亚鲟 (*Acipenser baerii*) 心、肝、肾、脾等组织中分离到细菌性病原 20 株, 并对分离的菌株进行了细菌生化编码鉴定管鉴定及药敏实验, 结果表明: 分离株分别包括致伤弧菌 (*Vibrio vulnificus*) 6 株、豚鼠气单胞菌 (*Aeromonas caviac*) 5 株、温和气单胞菌 (*Aeromonas sobria*) 5 株、亲水气单胞菌菌 (*Aeromonas hydrophila*) 4 株。16 种常见药物的药敏实验显示, 分离菌株均对多粘菌素 B、妥布青霉素等 6 种药物较敏感; 对红霉素、四环素等 6 种药物耐药, 仅有部分菌株表现出不同程度的耐药性及敏感度的变化趋势。该研究对于合理使用抗菌药物, 防治养殖鲟细菌性疾病及其他疾病引起的细菌继发性感染提供一定的数据基础。

关键词： 西伯利亚鲟; 细菌分离株; 生化鉴定; 药敏实验

该文发表于《江西农业大学学报》【2011, 33 (4): 786-790】

东北三省 15 株致病性嗜水气单胞菌分离株的药敏实验分析

Drug sensitive tests of fifteen strains of pathogenic Aeromonas hydrophila in Northeast China

司力娜[1,2], 李绍戊[1], 王荻[1], 尹海富[2], 卢彤岩[1]

(1. 中国水产科学研究院 黑龙江水产研究所, 黑龙江 哈尔滨 150070; 2. 东北农业大学 动物科学技术学院, 黑龙江 哈尔滨 150070)

摘要： 该研究对于合理使用抗菌药物并有效防治鲤鱼细菌性败血症提供一定的理论依据。采用细菌生化编码鉴定管与分子生物学方法 (嗜水气单胞菌特异性 16S rDNA 及气溶素 Aer 基因的部分片段进行扩增), 对从东北三省淡水鱼主养区患病鲤鱼体内分离并鉴定得到的 15 株致病性嗜水气单胞菌进行药敏分析试验。17 种常见药物的药敏实验结果表明, 东北三省的 15 个致病性嗜水气单胞菌分离株表现出较均一的耐药性及敏感度: 对左氟沙星、多粘菌素 B、氧氟沙星较敏感, 而对苯唑青霉素、替考拉丁、氨苄青霉素和林可霉素耐药性极强, 但仍有部分菌株表现出不同程度的耐药性及敏感度的变化趋势，值得进一步深入研究。

关键词：致病性嗜水气单胞菌；东北三省；分离鉴定；药敏实验

该文发表于《南方水产科学》【2015，11（1）：34-38】

三氯异氰尿酸和苯扎溴铵对海水养殖源弧菌的抑菌和杀菌效果

姚小娟[1,2]，王元[1]，赵姝[1]，胡梦华[1,2]，房文红[1]

(1. 中国水产科学研究院东海水产研究所，农业部东海与远洋渔业资源开发利用重点实验室，上海 200090；

2. 上海海洋大学水产与生命学院，上海 201306)

摘要：采用微量二倍稀释法进行三氯异氰尿酸和苯扎溴铵对 198 株海水养殖源弧菌的体外抑菌试验，在此基础上采用悬液定量杀菌法观察两种消毒剂对两株对虾致病性副溶血弧菌的杀菌效果，探讨海水养殖源弧菌是否对这两种消毒剂产生抗性。结果表明，三氯异氰尿酸对 198 株弧菌的最小抑菌浓度（MIC）分布在 31.13~498.0 mg·L^{-1}，苯扎溴铵对 198 株弧菌的 MIC 值分布在 20~320 mg·L^{-1}。在菌悬液杀菌试验中，随着三氯异氰尿酸和苯扎溴铵浓度降低，100% 杀灭弧菌的作用时间延长。当三氯异氰尿酸浓度分别为 7.78 mg·L^{-1} 和 15.56 mg·L^{-1} 时，100% 杀灭 FJ244 和 FJ246 菌株的作用时间均为 4 min；当苯扎溴铵浓度分别为 20 mg·L^{-1} 和 10 mg·L^{-1} 时，100% 杀灭 FJ244 和 FJ246 菌株的作用时间分别为 3 min 和 40 min。当消毒剂浓度分别递减为上述浓度一半时，即使延长消毒剂作用时间也未能杀灭弧菌。试验表明，海水养殖源弧菌对三氯异氰尿酸和苯扎溴铵产生明显的抗性。

关键词：三氯异氰尿酸；苯扎溴铵；弧菌；最小抑菌浓度；杀菌

该文发表于《江西农业大学学报》【2016，38（1）：152-159】

三北地区鱼源气单胞菌的分离鉴定与药敏试验

Isolation，Identification and Susceptibility testof Aeromonas from fish in three north areas of China

崔佳佳[1,2]，李绍戊[1]，王荻[1]，卢彤岩[1]

(1. 中国水产科学研究院 黑龙江水产研究所，哈尔滨 150070；2. 上海海洋大学 水产与生命学院，上海 201306)

摘要：选取三北地区鲤、鲫、草鱼、大西洋鲑等鱼体内分离到的气单胞菌共 16 株，结合革兰氏染色、氧化酶反应和分子生物学方法对其进行鉴定后，采用 K-B 药敏纸片法测定 16 株菌对 10 类 25 种药物的敏感性。结果表明：16 株菌均为革兰氏阴性、氧化酶阳性细菌；Blast 比对分析发现实验菌株的 16S rRNA 序列分别与 GenBank 数据库中维氏气单胞菌、嗜水气单胞菌、温和气单胞菌和杀鲑气单胞菌的不同菌株的 16SrRNA 基因序列同源性较高，相似性达 95% 以上；系统发育树显示 16 株菌株分别与维氏气单胞菌、嗜水气单胞菌、温和气单胞菌和杀鲑气单胞菌的不同菌株聚为一支，而与肠杆菌、志贺菌和弧菌分支较远，从而判定为维氏气单胞菌、嗜水气单胞菌、温和气单胞菌和杀鲑气单胞菌；16 株分离菌中有 3 株维氏气单胞菌、4 株嗜水气单胞菌、3 株温和气单胞菌和 6 株杀鲑气单胞菌；药敏试验中 16 株气单胞菌对氨苄西林、青霉素 G、阿莫西林、三甲氧苄氨啶、复方新诺明、磺胺异噁唑、氯霉素、利福平具有较高的耐药性，耐药率达 95% 以上，对氟喹诺酮类药物普遍较敏感，不同菌株之间的耐药谱存在很大差异。为气单胞菌的鉴定及三北地区淡水鱼细菌病的药物防治提供科学依据。

关键词：三北地区；气单胞菌；鉴定；耐药性

该文发表于《海洋渔业》【2005，27（1）：45-48】

诺氟沙星对溶藻弧菌和恩诺沙星对迟钝爱德华氏菌的抗生素后效应

房文红，周凯

（中国水产科学研究院东海水产研究所，农业部海洋与河口渔业重点开放实验室，上海，200090）

摘要： 采用试管二倍稀释法测定诺氟沙星对溶藻弧菌 s622、恩诺沙星对迟钝爱德华氏菌 751 的最小抑菌浓度（MIC）；采用菌落计数法测定诺氟沙星对溶藻弧菌、恩诺沙星对迟钝爱德华氏菌的抗生素后效应（PAE）。结果显示，诺氟沙星对溶藻弧菌 s622 在 1/2MIC、1MIC、2MIC 和 4MIC 浓度时的 PAE 分别为：0.47 h、0.76 h、1.26 h 和 2.05 h；恩诺沙星对迟钝爱德华氏菌 751 在 1/2MIC、1MIC、2MIC 和 4MIC 浓度时的 PAE 分别为：0.29 h、0.48 h、1.97 h、2.30 h。结果表明，诺氟沙星和恩诺沙星具有明显的 PAE；该结果提示，在制定给药方案时可以适当延长给药间隔时间，减少给药次数，仍能维持抗菌效果。

关键词： 诺氟沙星；恩诺沙星；溶藻弧菌 s622；迟钝爱德华氏菌 751；抗生素后效应

二、水产动物源病原菌耐药的基本状况

水产动物源病原菌对药物的耐药性广泛存在。中国南方九龙江流域表层水体环境中耐药细菌的种类较多，而且携带耐药基因的整合子分布较为广泛（Mao Lin et al.，2015）。鳗鲡及养殖水体普遍也存在多重耐药菌株，其中柠檬酸杆菌属和克雷伯菌属的多重耐药指数最高，而不动杆菌属则相对较低（吴小梅等，2015）。来源于鱼、虾、龟鳖等水产动物的气单胞菌均对氨苄西林、利福平、链霉素和萘啶酸等药物的耐药率相对较高（Yu-Ting Deng et al.，2014）。不同来源的气单胞菌的耐药率亦不尽相同：来源于爬行、两栖动物和观赏鱼的气单胞菌对氟喹诺酮类、头孢类等药物的耐药率比养殖鱼、虾类的高（吴雅丽等，2013）。此外，不同地区相同病原菌对抗生素的敏感程度也有一定程度的差异（连浩等淼，2015）。

该文发表于《Microb Drug Resist》【2014：20（4）：350-356】

Analysis of antimicrobial resistance genes in Aeromonas spp. isolated from cultured freshwater animals in China

Yu-Ting Deng[1], Ya-Li Wu[1,2], Ai-Ping Tan[1], Yu-ping Huang[1,2], Lan Jiang[1,*], Hui-Juan Xue[1,2], Wei-Li Wang[1], Li Luo[1], Fei Zhao[1]

（1. Key laboratory of fishery drug development，Ministry of Agriculture，P. R. China. Pearl River Fisheries Research Institute，Chinese Academy of Fishery Sciences，Guangzhou，China；2. Academy of Fishery and Life Science，Shanghai Ocean University，Shanghai，China）

Abstract： The development of resistance to antimicrobials used in aquatic animals is an increasing concern for aquaculture and public health. To monitor the occurrence of antimicrobial resistance and resistance genes in Aeromonas，a total of 106 isolates were collected from cultured freshwater animals in China from 1995 to 2012. Antimicrobial susceptibilities were determined by the disk diffusion method. The highest resistance percentage occurred with ampicillin，rifampin，streptomycin and nalidixic acid. Most strains were sensitive to fluoroquinolones，doxycycline，cefotaxime，chloramphenicol and amikacin. The isolates from turtle samples had the highest levels of resistance to 11 of the 12 tested antimicrobials when compared to those from fish or shrimp. PCR and DNA sequence results showed that all trimethoprim/sulfamethoxazole resistant strains contained sul1，and 37.0% were positive for tetA in tetracycline resistant strains. ant（3''）-Ia was identified in 13（24.5%）streptomycinresistant strains. Plasmid-borne quinolone resistance genes were detected in 5 A. hydrophila（4.7%），two of

817

which carried qnrS2 while the other three strains harbored aac（6'）-Ib-cr. Two cefotaximeresistant A. hydrophila were positive for bla[TEM-1] and bla[CTX-M-3]. To our knowledge, this is the first report characterizing antimicrobial resistance in Aeromonas isolated from cultured freshwater animals in China, and providing resistance information of pathogen in Chinese aquaculture.

Key words: Aquatic animal, Aeromonas, Antimicrobial resistance, Resistance gene

我国淡水水产动物源气单胞菌耐药基因流行情况分析

邓玉婷[1]，吴雅丽[1,2]，谭爱萍[1]，黄郁萍[1,2]，姜兰[1,*]，薛慧娟[1,2]，王伟利[1]，罗理[1]，赵斐[1]

（1. 中国水产科学研究院珠江水产研究所渔药药理学重点实验室，广州，中国；
2. 上海海洋大学水产与生命学院，上海，中国）

摘要：随着大量抗菌药物应用于水产动物细菌性病害的防治，细菌耐药菌逐渐增加，致使药物对气单胞菌疾病的防控能力减弱，还存在将耐药性传播给人类致病菌的潜在风险。为了解广东地区水产动物源气单胞菌的耐药情况，本研究采用 K-B 纸片法测定了 106 株于 1995—2012 年来源于不同种类患病水产动物的气单胞菌对 14 种抗菌药的耐药性，结果显示：气单胞菌对氨苄西林、利福平、链霉素和萘啶酸的耐药率相对较高，大部分菌株对氟喹诺酮类、多西环素、头孢喹肟、氯霉素和阿米卡星相对敏感。相对于鱼源和虾源菌株，龟鳖源分离菌株对 12 种测试药物的 11 种均表现出高水平耐药。采用 PCR 方法检测分离菌株的常见耐药基因，结果显示：所有的磺胺甲基异噁唑/甲氧苄啶耐药菌株均检测到 sul1 基因；37% 的四环素耐药菌株携带 tetA 基因；13 株（24.5%）链霉素耐药菌株检测到 ant（3''）-Ia 基因；5 株（4.7%）嗜水气单胞菌检测到质粒介导的喹诺酮类耐药基因，其中 2 株携带 qnrS 基因，3 株携带 aac（6'）-Ib-cr 基因。两株头孢喹肟耐药嗜水气单胞菌携带了 bla[TEM-1] 和 bla[CTX-M-3]。本研究首次报道了我国淡水水产养殖动物源气单胞菌的耐药及耐药基因流行情况，为我国水产动物源细菌耐药性的研究奠定基础。

关键词：水产动物，气单胞菌，耐药性，耐药基因

1　Introduction

Aeromonas is a Gram-negative, facultatively anaerobic, rod-shaped bacteria present ubiquitously in aquatic environments and frequently isolated from animals and humans[20]. It has also been isolated from different kinds of food, such as fish, vegetables, meat, and milk[20]. In 1984, the Food and Drug Administration (United States) labeled Aeromonas hydrophila as a "new" foodborne pathogen[19]. Since then, many clinical reports have indicated Aeromonas as capable of causing gastroenteritis and other complications, including wound infections, septicemia and endocarditis[20]. Contaminated drinking water and food are considered the major transmission routes for gastrointestinal infection with Aeromonas[9]. Thus, a high prevalence of Aeromonas species in the food chain should be considered a threat to public health[19].

China is currently the leader in aquaculture production, accounting for more than 70 percent of global production[16]. Rapid development of aquaculture has resulted in widespread use of antimicrobials. In China, nearly 50% of the antimicrobials produced was used for animal husbandry and aquaculture[36]. Several agents of fluoroquinolones, tetracyclines and sulfonamidesare approved and frequently used in veterinary medicines[29,36]. Resistance to quinolones and tetracycline, which are among the antimicrobials used to treat Aeromonas infections, has been widely documented[9,24]. Use of antimicrobials in animal feed can foster the development of antimicrobial resistance. Resistance genes may transfer to pathogenic bacteria and reduce efficacy of treatment for human and animal diseases caused by resistant pathogens[28]. Resistance to diverse groups of antimicrobials is a concerning characteristics of Aeormonas species.

A number of studies have shown a high prevalence of drug-resistant Aeromonas isolates from fish, environmental, and human clinical samples[1,5]. However, limited information is available about Aeromonas from cultured freshwater animals in China. In the present study, the occurrence and diversity of Aeromonas in cultured freshwater animal samples (fish, shrimp and turtles) were investigated. Antimicrobial resistance profiles and resistance genes were also characterized to evaluate the potential of aeromonads in these animals as a public health risk.

2 Materials and Methods

2.1 Sample collection, isolation and identification of Aeromonas

A total of 106 isolates were collected from diseased cultured freshwater animal samples, including 68 fish from 33 farms, 26 turtles from 15 farms and 12 shrimp from 8 farms. All samples were collected at fisheries hospital of Pearl River Fisheries Research Institute from November 1995 to February 2012. Gills, body surfaces, and livers were aseptically swabbed using sterile cotton buds, inoculated into LB broth for pre-enrichment at 28℃±2℃ for 18 h~24 h. The enriched cultures were streaked on Rimler-Shotts agar and incubated at 28℃±2℃ for 18~24 h. Yellow, oxidase-positive colonies were isolated and presumptively considered as Aeromonas species. Only one Aeromonas strain was selected from each sample. The presumptive Aeromonas colonies were further investigated by biochemical typing using ATB™ New System (BioMérieux, France). For further identification, polymerase chain reaction (PCR) amplification of 16S rRNA gene and gyrB genes was performed as described in previous studies[4,37]. Taxonomic identification of the sequences was performed using BLAST in GenBank (http://blast.ncbi.nlm.nih.gov/).

2.2 Antimicrobial susceptibility test

All strains were evaluated for resistance to 14 antimicrobials by the disk diffusion method, which comprised ampicillin (10 μg), cefotaxime (30 μg), sulfonamides (300 μg), trimethoprim/sulfamethoxazole (1.25/23.75 μg), rifampin (5 μg), nalidixic acid (30 μg), ciprofloxacin (5 μg), norfloxacin (10 μg), ofloxacin (5 μg), chloramphenicol (30 μg), tetracycline (30 μg), doxycycline (30 μg), streptomycin (10 μg), and amikacin (30 μg) discs (Oxoid). For quality control, Escherichia coli ATCC 25922 was used. The results were evaluated as susceptible (S), intermediate (I), and resistant (R) based on the interpretative criteria from the Clinical and Laboratory Standards Institute (CLSI)[7,8].

2.3 PCR amplification of resistance genes

DNA prepared by the whole cell boiled lysate protocol was used as PCR templates. All isolates were screened for PMQR genes (qnrA, qnrB, qnrS, qepA and aac (6') -Ib-cr). Isolates that showed relevant resistance profiles were screened for tetracyclines resistance genes (tetA, tetC, tetE), beta-lactamase resistance genes (CTX-M, TEM), aminoglycoside resistance genes [ant (3'')-I, strA] and sulfonamide resistance gene (sul1). The PCR primers (Table 1) were used according to previous studies[4,14,23,25,26,31,37]. PCR products were direct sequenced and the DNA sequences were analyzed using BLAST (http://blast.ncbi.nlm.nih.gov/).

Table 1 PRIMERS USED IN THIS STUDY AND EXPECTED SIZES OF PCR PRODUCTS

Primer	Nucleotide sequence (5′ to 3′)	Size (bp)	Reference
16S rRNA-F	AGA GTT TGA TCA TGG CTC AG	1501	4
16S rRNA-R	GGT TAC CTT GTT ACG ACT T		
gyrB-F	TCC GGC GGT CTG CAC GGC GT	1302	38
gyrB-R	TTG TCC GGG TTG TAC TCG TC		
qnrA-F	ATT TCT CAC GCC AGG ATT TG	519	26
qnrA-R	GAT CGG CAA AGG TCA GGT CA		
qnrB-F	GAT CGT GAA AGC CAG AAA GG	469	
qnrB-R	ACG ATG CCT GGT AGT TGT CC		
qnrS-F	ACG ACA TTC GTC AAC TGC AA	417	
qnrS-R	TAA ATT GGC ACC CTG TAG GC		
aac (6') -Ib-F	TTG CGA TGC TCT ATG AGT GGC TA	482	
aac (6') -Ib-R	CTC GAA TGC CTG GCG TGT TT		
qepA-F	GCA GGT CCA GCA GCG GGT AG	306	
qepA-R	CTT CCT GCC CGA GTA TCG TG		

819

续表

Primer	Nucleotide sequence (5′ to 3′)	Size (bp)	Reference
tet (A) -F	GTA ATT CTG AGC ACT GTC GC	956	31
tet (A) -R	CTG CCT GGA CAA CAT TGC TT		
tet (E) -F	GTG ATG ATG GCA CTG GTC AT	1198	
tet (E) -R	CTC TGC TGT ACA TCG CTC TT		
tet (C) -F	TCT AAC AAT GCG CTC ATC GT	588	
tet (C) -R	GGT TGA AGG CTC TCA AGG GC		
strA-strB-F	TTG AAT CGA ACT AAT AT	1640	
strA-strB-R	CTA GTA TGA CGT CTG TCG		
ant3-F	GTG GAT GGC GGC CTG AAG CC	526	25
ant3-R	ATT GCC CAG TCG GCA GCG		
sul1-F	CGG CGT GGG CTA CCT GAA CG	433	23
sul1-R	GCC GAT CGC GTG AAG TTC CG		
TEM-F	AAA GAT GCT GAA GAT CA	425	14
TEM-R	TTT GGT ATG GCT TCA TTC		
CTX-M-F	GTG CAG TAC CAG TAA AGT TAT GG	538	
CTX-M-R	CGC AAT ATC ATT GGT GGT GCC		

3 Results

3.1 Identification of Aeromonas spp.

A total of 106 Aeromonas isolates were identified to the species level by PCR amplification of 16S rRNA and gyrB genes. The dominantAeromonas species were A. hydrophila (54.7%) and A. veronii (19.8%). Other Aeromonas species included A. caviae (seven strains), A. sobria (six strains), A. trota (five strains), A. aquariorum (four strains), A. jandaei (three strains), and one strain of A. media (Table2). Two strains of Aeromonas were unidentified by genotyping methods.

Table 2 DISTRIBUTION OF AEROMONAS SPP. IN DIFFERENT CULTURED FRESHWATER ANIMAL SAMPLES

Aeromonas spp.	Shrimp	fish	Turtle
A. hydrophila (58)	4	35	19
A. veronii (21)	–	17	4
A. caviae (7)	2	4	1
A. sobria (6)	1	5	–
A. trota (4)	3	1	–
A. aquariorum (4)	1	2	1
A. jandaei (3)	–	3	–
A. media (1)	–	–	1
unidentified (2)	1	1	–
Total (106)	12	68	26

3.2 Antimicrobial susceptibility of Aeromonas

Although the overall resistance percentages were not very high, 41 (38.7%) isolates were resistant to three or more different classes of antimicrobials (Tables 3). Resistance was most prevalent forampicillin (84.9%), rifampin (56.6%), streptomycin (50.0%) and nalidixic acid (43.4%). Most isolates (> 80%) were sensitive tociprofloxacin, norfloxacin, cefotaxime, doxycyline, chloramphenicol, trimethoprim/sulfamethoxazole and amikacin.

The diversity ofantimicrobial resistance was compared among different animals (fish, shrimp, and turtles). Turtle samples had the highest diversity, with resistance to 13 of the 14 tested antimicrobials. Resistance to cefotaxime, ciprofloxacin and amikacin were only observed in the isolates from turtles. All the isolates from shrimp were susceptible to seven antimicrobials (cefotaxime, ciprofloxacin, norfloxacin, ofloxacin, doxycycline, chloramphenicol and amikacin), but rifampin resistance was prevalent (83.3% of the shrimp isolates, 10/12).

Table 3 ANTIMICROBIAL SUSCEPTIBILITIES OF 106 AEROMONAS ISOLATES FROM DIFFERENT CULTURED FRESHWATER ANIMALS

Antimicrobials	Breakpoints[a] (mm)		Percentage (no.) of strains resistant			
	S	R	All isolates (n=106)	Shrimp (n=12)	Fish (n=68)	Turtle (n=26)
Ampicillin	≧17	≦13	84.9 (90)	33.3 (4)	89.7 (61)	96.2 (25)
Cefotaxime	≧23	≦14	2.8 (3)	0	0	11.5 (3)
Sulfonamides	≧17	≦12	30.2 (32)	8.3 (1)	23.5 (16)	57.7 (15)
Trimethoprim/sulfamethoxazole	≧16	≦10	18.9 (20)	8.3 (1)	11.8 (8)	42.3 (11)
Rifampin	≧20	≦16	56.6 (60)	83.3 (10)	44.1 (30)	76.9 (20)
Nalidixic acid	≧20	≦14	43.4 (46)	41.7 (5)	33.8 (23)	69.2 (18)
Ciprofloxacin	≧21	≦15	4.7 (5)	0	0	19.2 (5)
Norfloxacin	≧17	≦12	9.4 (10)	0	2.9 (2)	30.8 (8)
Ofloxacin	≧25	≦21	12.3 (13)	0	5.9 (4)	34.6 (9)
Tetracycline	≧19	≦14	25.5 (27)	33.3 (4)	14.7 (10)	50.0 (13)
Doxycycline	≧14	≦10	11.3 (12)	0	4.4 (3)	34.6 (9)
Streptomycin	≧15	≦11	50.0 (53)	25.0 (3)	51.5 (35)	57.7 (15)
Amikacin	≧17	≦14	2.8 (3)	0	0	11.5 (3)
Chloramphenicol	≧18	≦12	13.2 (14)	0	7.4 (5)	34.6 (9)

[a]CLSI[7,8]; S, susceptible; R, resistant.

Comparison ofantimicrobial resistance profiles among Aeromonas species showed that A. caviae and A. sobria were susceptible to more antibiotics than A. hydrophila and A. veronii (Table 4). A. hydrophila isolatescollectivelyhad resistance to the widest range of antimicrobials, compared with the aggregate range of resistance in each of the other three species.

Table 4 ANTIMICROBIAL SUSCEPTIBILITIES OF THE MOST ISOLATED AEROMONAS SPP.

Antimicrobials	Percentage (no.) of strains resistant			
	A. hydrophila (n=58)	A. veronii (n=21)	A. caviae (n=7)	A. sobria (n=6)
Ampicillin	94.8 (55)	100.0 (21)	0	100.0 (6)
Cefotaxime	3.4 (2)	4.8 (1)	0	0
Sulfonamides	34.5 (20)	52.4 (11)	0	16.7 (1)
Trimethoprim/sulfamethoxazole	29.3 (17)	9.5 (2)	0	16.7 (1)
Rifampin	51.7 (13)	61.9 (13)	71.4 (5)	66.7 (4)
Nalidixic acid	39.7 (17)	81.0 (17)	28.6 (2)	33.3 (2)
Ciprofloxacin	8.6 (5)	0	0	0

续表

Antimicrobials	Percentage (no.) of strains resistant			
	A. hydrophila (n=58)	A. veronii (n=21)	A. caviae (n=7)	A. sobria (n=6)
Norfloxacin	8.6 (5)	0	0	0
Ofloxacin	19.0 (11)	9.5 (2)	0	0
Tetracycline	29.3 (17)	28.6 (6)	14.3 (1)	33.3 (2)
Doxycycline	17.2 (10)	9.5 (2)	0	0
Streptomycin	37.9 (22)	76.2 (16)	42.9 (3)	83.3 (5)
Amikacin	3.4 (2)	4.8 (1)	0	0
Chloramphenicol	22.4 (13)	4.8 (1)	0	0

3.3 PCR detection of resistance genes

The five PMQR genes were screened by PCR in all strains. qnrS and aac (6') −Ib−cr were found in five A. hydrophila isolates (4.7% of all Aeromonas isolates, 5/106). qnrS2was also identified in two isolates from fish and turtle samples (GenBank accession number KC542809 and KC542810 for the qnrS2 sequences). Interestingly, the qnrS2 positive strains were susceptible to all four quinolones tested in this study. On the other hand, aac (6') −Ib genes were found in eight nalidixic acid−resistant isolates. The cr variant of the aac (6') −Ib gene conferring resistance to ciprofloxacin was detected in three of the eight nalidixic acid−resistant isolates (GenBank accession number KC542812, KC554064 and KC554068).

Other resistance genes were also detected. Among 27 tetracycline−resistant Aeromonas isolates, only 10 isolates carried the tetA gene (detection frequency of 37.0%). The tetC and tetE genes were not found in these isolates. Of the 106 Aeromonas strains, 32 i-solates were resistant to sulfonamideand 20 were resistant to trimethoprim/sulfamethoxazole. All of the trimethoprim/sulfamethoxazole-resistant isolates contained the sul1 gene. The strA gene was not detected in any of 53 streptomycin−resistant strains, but ant (3′′)−Ia occurred in 13 isolates resistant to streptomycin (24.5%). Although cephalosporins are rarely used in aquaculture, three isolates from turtles were resistant to cefotaxime, suggesting the presence of an extended−spectrum β−lactamase (ESBL). Two types of ESBL genes were detected and confirmed as TEM−1 and CTX−M−3 in two of three cefotaxime resistantisolates. One strain of A. hydrophila contained both genes, and the other A. hydrophila only carried TEM−1 (GenBank accession number KC542811, KC542813 and KC542814).

All the resistance genes found in the isolates from different cultured freshwater animals and their resistance phenotypes were shown in Table5.

Table 5　PHENOTYPIC AND GENOTYPIC PROFILE OF RESISTANCE FROM AEROMONAS ISOLATES

Strain	Aeromonas spp.	Year	Source	Resistance profile[a]	Resistance genes
12C	A. trota	2006	shrimp	RIF \ TET \ STR	tet (A)
27F	A. veronii	2007	fish	AMP \ RIF \ NA \ TET \ STR	tet (A)
15C	A. caviae	2006	shrimp	RIF \ NA \ TET \ STR	tet (A)
S74	A. veronii bv sobria	1996	turtle	AMP \ RIF \ NA \ OFL \ TET \ DOX \ STR	tet (A)
43A	A. hydrophila	2007	turtle	AMP \ S3 \ SXT \ RIF \ NA \ CIP \ NOR \ OFL \ STR \ CHL	sul1
SH18	A. hydrophila	1998	fish	AMP \ S3 \ SXT \ RIF \ NA \ TET \ STR \ CHL	sul1 \ tet (A)
SH20	A. hydrophila	1998	fish	AMP \ S3 \ SXT \ RIF \ NA \ TET \ STR \ CHL	sul1 \ tet (A)

Strain	Aeromonas spp.	Year	Source	Resistance profile[a]	Resistance genes
G34	A. veronii	2001	fish	AMP \ S3 \ SXT \ NA \ STR	sul11 \ ant (3'') -Ia
2F	A. hydrophila	2006	fish	AMP \ S3 \ SXT \ NA \ TET \ STR	sul11 \ ant (3'') -Ia
12F	A. hydrophila	2007	fish	AMP \ S3 \ SXT \ RIF \ NA \ NOR \ OFL \ TET \ DOX \ STR \ CHL	sul11 \ ant (3'') -Ia
3B	A. hydrophila	2005	fish	AMP \ S3 \ SXT \ RIF \ NA \ NOR \ OFL \ TET \ DOX \ STR \ CHL	sul11 \ ant (3'') -Ia
26A	A. hydrophila	2003	turtle	AMP \ S3 \ SXT \ RIF \ NA \ NOR \ OFL \ TET \ DOX \ STR \ CHL	sul11 \ ant (3'') -Ia
29A	A. hydrophila	2003	turtle	AMP \ S3 \ SXT \ RIF \ NA \ NOR \ OFL \ TET \ DOX \ STR \ CHL	sul11 \ ant (3'') -Ia
E3	A. hydrophila	1996	fish	AMP \ S3 \ SXT \ RIF \ NA \ OFL \ TET \ DOX \ STR \ CHL	sul11 \ ant (3'') -Ia
34B	A. veronii	2004	fish	AMP \ S3 \ SXT \ RIF \ NA \ STR	sul1 \ ant (3'') -Ia
X21	A. sobria	1995	shrimp	AMP \ S3 \ SXT \ RIF \ NA \ TET \ STR	sul11 \ ant (3'') -Ia
SG2	A. hydrophila	2003	turtle	AMP \ S3 \ SXT \ RIF \ NA \ TET \ DOX \ STR \ CHL	sul11 \ ant (3'') -Ia
44A	A. hydrophila	2007	turtle	AMP \ S3 \ SXT \ RIF \ NA \ CIP \ NOR \ OFL \ STR \ CHL	sul1 \ aac6-Ib-cr
EG9	A. hydrophila	2003	turtle	AMP \ S3 \ SXT \ RIF \ NA \ CIP \ NOR \ OFL \ TET \ DOX \ STR \ CHL	sul11 \ ant (3'') -Ia, aac (6') -Ib
30A	A. hydrophila	2003	turtle	AMP \ S3 \ SXT \ RIF \ NA \ CIP \ NOR \ OFL \ TET \ DOX \ STR	sul11 \ ant (3'') -Ia, aac (6') -Ib
31A	A. hydrophila	2003	turtle	AMP \ S3 \ SXT \ RIF \ NA \ NOR \ OFL \ TET \ STR \ AMK \ CTX \ CHL	sul1 \ ant (3'') -Ia, aac (6') -Ib \ bla$_{CTX-M-3}$, bla$_{TEM-1}$
YD1	A. hydrophila	2004	turtle	AMP \ S3 \ SXT \ NA \ TET \ STR	sul1 \ aac (6') -Ib \ tet (A)
HJ2	A. hydrophila	2004	turtle	AMP \ S3 \ SXT \ RIF \ NA \ CIP \ NOR \ OFL \ TET \ DOX \ STR \ CHL	sul1 \ aac6-Ib-cr \ tet (A)
28A	A. hydrophila	2003	turtle	AMP \ S3 \ SXT \ RIF \ NA \ TET \ STR \ AMK \ CTX	sul1 \ aac6-Ib-cr \ aac (6') -Ib \ bla$_{TEM-1}$
17A	A. hydrophila	2005	turtle	AMP \ S3 \ TET \ DOX \ STR	qnrS2 \ tet (A)
T9	A. hydrophila	1998	fish	RIF \ STR	qnrS2
45A	A. veronii	2007	turtle	AMP \ RIF \ NA \ TET \ DOX \ STR \ AMK \ CTX \ CHL	aac (6') -Ib \ tet (A)

[a] AMP, ampicillin; S3, sulfonamides; SXT, trimethoprim/sulfamethoxazole; RIF, rifampin; NA, nalidixic acid; CIP, ciprofloxacin; NOR, norfloxacin; OFL, ofloxacin; TET, tetracycline; DOX, doxycycline; STR, streptomycin; AMK, amikacin; CTX, cefotaxim; CHL, chloramphenicol.

4 Discussion

As a general observation, majority of antimicrobial agents used in aquaculture are also used in human or veterinary medicine. With respect to Europe, no more than 2 or 3 antimicrobials agents are licensed for use in aquaculture in each country[12]. However, there are many countries with significant aquaculture industries where there is little effective regulation of access to, or use of, antimicrobials[35]. For example, there are a variety of agents have or are being used in Asia[3].

Fluoroquinolone, tetracyclines and sulfonamides have been commonly used for the last two decades to prevent and control motile Aeromonas septicemia or ulcerative infections in fish and shrimp, or red neck disease in soft shelled turtle[34]. Although only a few agents were licensed in Chinese aquaculture[29], imprudent and abuse use of antimicrobialshave lead to various antimicrobial resistance mechanisms encountered in different cultured species. In the current study, the results presented showed a detailed pattern of sensitivity of the various Aeromonas isolates to a variety of antimicrobials and provided useful information in the context of selective isolation and phenotypic identification of the aeromonads. In general, most of the isolates were susceptible for fluoroquinolones, doxycycline, cefotaxime, chloramphenicol and amikacin. These results in agreement to those published previously[1,18,30]. But the resistance to 14 antimicrobials tested was more prevalent in turtle-associated Aeromonas isolates than those from other animals. Turtles have high economic value and are a nutrient-rich food in China. The majority of turtles are raised in farms to meet the large demand in China. Their high market price encourages farmers to spend more money on effective antimicrobials to keep them alive. In addition, the cultured cycle of turtles is more than 4 years, which is longer than those of the cultured fishes and shrimps. Thus, the more antimicrobial are used in animals, the more selective pressure might impose on bacteria, and it may increase the resistance in aquaculture system.

Intensive use of antimicrobial in aquaculture provides a selective pressure creating reservoirs of drug-resistant bacteria and transferable resistance genes in fish pathogens and other bacteria in the aquatic environment. From these reservoirs, resistance genes may disseminate by horizontal gene transfer and reach human pathogens, or drug-resistant pathogens from the aquatic environment may reach humans directly[15]. Previous report has found that A. hydrophila, A. caviae and A. veronii biovar sobria might cause 85% of human gastrointestinal infections[32]. In this study, various species of Aeromonas linked to human disease have been observed, and have varying levels of susceptibilities to different antimicrobials. A. hydrophila and A. veronii isolates displayed greater levels of resistance as compared to other species. Multi-drug resistant Aeromonas species found in different cultured freshwater animals implied that difficulties might arise in the prophylaxis and therapy of Aeromonas infections and might increase the risks of infection to human.

In the present study, fluoroquinolones showed good activity against all species of Aeromonas, as reported in previous studies[1]. However, large amount of fluoroquinolones used in aquaculture in last decade has been associated with a trend of increasing prevalence of quinolone resistance. The occurrence of quinolone-resistant determinants on mobile genetic elements such as plasmids or transposons may accelerate the dissemination of resistance among bacteria[2,6,13]. Recently, plasmid-mediated transferable quinolone resistance (PMQR) determinants have been identified in Aeromonas strains[6,13]. In the current study, all Aeromonas isolates were screened for three types of PMQR genes [qnr, aac (6') -Ib-cr and qepA], but only qnrS2 and aac (6') -Ib-cr were detected. qnrS2 was first identified and confirmed in A. punctata from an environmental sample[6]. To date, only four qnr-determinantsincluding qnrS2, qnrS5, qnrB1 and qnrVC were identified in Aeromonas species[2,13,37]. Overall, qnrS2 seems to be the most commonly identified in aeromonads isolated from human clinical, diseased fish, and environmental samples[2,13]. We also found two qnrS2-positive strains of A. hydrophila isolated from turtle and fish samples. Another PMQR determinant, a cr variant of acc (6') -Ib that both conferred resistance to aminoglycoside and ciprofloxacin has been identified in A. allosaccharophila and A. media from environmental samples[9,33]. Eight nalidixic acid-resistant isolates were positive for aac (6') -Ib, but only three of them were cr variants. Despite the presence of the resistance gene, two isolates positive for qnrS2 were sensitive to all four quinolones tested and one isolate positive for acc (6')-Ib-cr was only resistant to nalidixic acid, because we tested the susceptibility of quinolones by disk diffusion method. Whether the PMQR positive strains reduced the susceptibility of fluoroquinolones, the MICs of fluoroquinolones should be further determined by double dilution method. To our knowledge, this is the first report of aac (6') -Ib-cr identified in A. hydrophila isolated from turtle. The prevalence of qnrS2 and aac (6') -Ib-cr in cultured freshwater animals may be due to the overuse of fluoroquinolones in Chinese aquaculture. Jiang demonstrated a high prevalence of PMQR determinants (25.2%, 55/218) in E. coli from fish gut samples in China[21]. Although fluoroquinolones have been reported as the treatment of choice for Aeromonas infection, the increasing prevalence of quinolone resistance in aquaculture makes their use on farms or in clinical therapy a concern.

Many reports have indicated that plasmid-located genes coding for tetracycline efflux proteins occur in Aeromonas species, contributing to the dissemination of tetracycline resistance in aquaculture systems[13,24,31]. The occurrence of various tetracycline resistance determinants has been described in aquaculture isolates, such as tetA, tetE, tetC and tetD, et al.[13,31]. Some authors also observed that two or more types of tet genes were present in the same isolates[13,31]. In contrast, we only observed tetA in the tetracycline-resist-

824

ant isolates. The difference in the distribution and diversity of tetgenes among studies is most likely due to differences in geographical locations and the status and history of antimicrobials used in each area.

A high prevalence of sulfonamide resistance and its determinants (sul1, sul2) in aeromonads has been described in many reports, and is most likely due to the overuse of sulfonamide drugs in animal feeding lots and fish ponds[10,17,22]. Resistance to trimethoprim is often associated with gene cassettes located in class 1 or class 2 integrons, and the sulfonamide resistance gene sul1 is part of the 3' -conserved segment of class 1 integron. Among our 32 sulfonamide-resistant isolates, 20 were resistant to trimethoprim/sulfamethoxazole and were positive for sul1. We also found that these 20 isolates carried a class 1 integron (unpublished).

Cephalosporins are used for human clinical cases, but are seldom used in aquaculture. Unexpectedly, we found three cefotaxime-resistant strains of A. hydrophila and A. veronii from turtle samples. Additionally, the ESBL genes (bla$_{TEM-1}$ and bla$_{CTX-M-3}$) were identified in two strains of A. hydrophila. The earliest report of ESBL-producing Aeromonas was described with a clinical isolate of A. caviae harboring bla$_{TEM-24}$[27]. Later, ESBLs-producing environmental isolates were also reported[38]. At least nine types of ESBLs (CTX-M, TEM, SHV, CMY, MOX, PER, VEB, TLA and GES) have been identified in Aeromonas species to date from clinical and environmental samples[11,38]. To our knowledge, this is the first report of ESBL genes (bla$_{CTX-M-3}$ and bla$_{TEM-1}$) detected in A. hydrophila from cultured freshwater animals. Both of these two genes were harbored in a single isolate. As we mentioned above, turtles have high economic value and have long cultured cycle. Farmers used much more effective antimicrobials such as cephalosporins and fluoroquinolones to treat the diseases. Interestingly, we also found that one strain harbored both bla$_{TEM-1}$ and aac (6') -Ib-cr. The occurrence of ESBLs and PMQR determinants in Enterobacteriaceae from fish samples has been demonstrated in previous study[21]. These results confirm aquaculture systems as a potential source of drug resistance bacteria that can transfer to humans. The incidence of drug abuse might also post a potential risk to public health.

In conclusion, multidrug-resistant Aeromonas were isolated from cultured freshwater animals and various types of antimicrobial resistance genes were identified. To the best of our knowledge, this is the first report of the PMQR gene aac (6') -Ib-cr as well as ESBL genes (bla$_{CTX-M-3}$,bla$_{TEM-1}$) in A. hydrophila from cultured freshwater animals. Therefore, our study underlined the potential for animals in aquaculture to act as a reservoir of resistance determinants, especially in China, where usage of antimicrobials should be more strictly regulated and antimicrobial resistance monitored in order to control the emergence of drug resistance.

5　Acknowledgements

This work was supported by Special Scientific Research Funds for Central Non-profit Institutes, Chinese Academy of Fishery Sciences (Grant No. 2012A0507), Special Fund for Agro-scientific Research in the Public Interest (Grant No. 201203085), and Guangdong province Science and Technology Plan Project (Grant No. 2011B020307001).

Disclosure Statement

No competing financial interests exist.

该文发表于《Environmental Science and Pollution Research》【2015, 22 (15): 11930-11939】

Genetic diversity of three classes of integrons in antibiotic-resistant bacteria isolated from Jiulong River in southern China

Mao Lin[1] & Jingjing Liang[1] & Xian Zhang[2] & Xiaomei Wu[1] & Qingpi Yan[1,3] & Zhuanxi Luo[2]

(1. Jimei University, Xiamen, Fujian, People's Republic of China; 2. Key Lab of Urban Environment and Health, Institute of Urban Environment, Chinese Academy of Sciences, Xiamen, People's Republic of China; 3. Key Laboratory of Healthy Mariculture for theEast China Sea, Ministry of Agriculture, Xiamen, People's Republic of China)

Abstract: We identified antibiotic-resistant bacterial isolates from the surface waters of Jiulong River basin in southern China and determinedtheir extent of resistance, as well as the prevalence and characterization of three classes of integrons. A phylogenetic analysis of 16S rDNA sequences showed that 20 genera were sampled from a total of 191 strains and the most common genus was Acinetobacter. Antimicrobial susceptibility testing revealed that the 191 isolates were all multi-resistant and there were high levels of resistance to 19 antimicrobials were tested, particularly the β-lactam, sulfonamide, amphenicol, macrolide and rifamycin classes. Moreover, class 1 integrons were ubiquitous while only five out of 191 strains harbored class 2 integrons and no class 3 integrons were detected. The variable region of the class 1 integrons contained 30 different gene cassette arrays. Nine novel arrays were found in 65 strains and seven strains had empty integrons. Among these 30 arrays, there were 34 different gene cassettes that included 25 resistance genes, six genes with unknown functions, two mutant transposase genes and a new gene. The unique array dfrA1-sat2-aadA1 was detected in all five isolates carrying the class 2 integron. We found that antibiotic-resistant bacterial isolates from Jiulong River were diverse and antibiotic resistance genes associated with integrons were widespread.

Key words: Integrons; Antibiotic-resistant bacteria; Jiulong River

中国南方九龙江分离耐药细菌中 3 类整合子的遗传多样性

林茂[1*]，梁晶晶[1]，张娴[2]，吴小梅[1]，鄢庆枇[1,3]，罗专溪[2]

(1. 集美大学水产学院，厦门 361021；2. 中国科学院城市环境与健康重点实验室，厦门 361021；
3. 农业部东海海水健康养殖重点实验室，厦门 361021)
集美大学水产学院；鳗鲡现代产业技术教育部工程研究中心，厦门 361021

摘要：本研究从中国南方的九龙江流域表面水样中分离耐药细菌，检测其耐药谱和 3 类整合子的流行特征。利用 16S rDNA 分子鉴定和系统发育树分析，将 191 株耐药细菌归属于 20 个菌属，其中不动杆菌属数量最多。药敏试验结果显示，这 191 株耐药细菌均具备多重耐药性，对 19 种抗生素具有较高的耐药率，特别是 β-内酰胺类、磺胺类、氯霉素类、大环内酯类和利福霉素类药物。在这 191 株细菌中，1 型整合子普遍存在，但只有 5 株检测到 2 型整合子，没有检测到 3 型整合子。这些 1 型整合子可变区共有 30 种基因盒阵列，在其中 65 株细菌中发现了 9 种新型基因盒阵列，7 株细菌为空整合子。这 30 种阵列共涉及 34 种基因盒，包括 25 种抗性基因、6 种编码未知功能蛋白的基因、2 种类似转座酶基因和 1 种尚未见报道的新型基因盒。5 株 2 型整合子的基因盒阵列都为 dfrA1-sat2-aadA1。研究的结果显示，九龙江流域表层水体环境中耐药细菌的种类较多，而且携带耐药基因的整合子分布较为广泛。

关键词：整合子，耐药细菌，九龙江

Introduction

Since penicillin was discovered in the early 20th century, antibiotics have been used to prevent and treat bacterial infections in humans and other animals with amazing efficiency. However, after widespread misapplication and failure to complete treatment regimens, resistant and multi-resistant bacteria have evolved under intense selective pressure. This prompted to state that the rise of antibiotic resistance, and not the antibiotics themselves, has been the real wonder of the medical use of antibiotics (Davies and Davies, 2010).

One aspect of antibiotic resistance is the genetic element integrons. Integrons are genetic elements that are site-specific enablers of recombination, with the ability to capture and excise one or more gene cassettes. Integrons play an important role in the multidrug-resistance of bacterial strains because their gene cassettes frequently contain a resistance gene (Hall and Collis, 1995) and these genes are mobilized by the integrons (Guerra et al., 2000). An integron has three core components: an integrase gene (*intI*) encoding a site-specific recombinase that catalyses the integration or excision of gene cassettes in the integron; the recombination site (*attI*) that is adjacent to intI and is recognized by the integrase as the insertion point for the gene cassettes; a promoter (Pc) that enables the genes in the cassettes to be expressed (Hall and Collis, 1998). A gene cassette is formed by a gene and a recombination site (attC or 59-be) that is different from attI. Integrons are classified into several groups according to the genetic correlation of the intI. Class 1, 2 and 3 integrons have received the most attention in recent studies (Domingues et al., 2012) and class 1 integrons were the most prevalent integrons in isolates from various resources, including clinical samples (Ibrahim et al., 2011) and non-clinical environments, such as animal (Hasan and Tinni, 2011), soil and water (Moura et al., 2007).

The Jiulong River is the second largest river in Fujian Province, China, following the Min River. The main channel is the North Stream and it converges with two main tributaries, the West Stream and South Stream, as well as many smaller tributaries. The drainage basin of Jiulong River is 14, 741 km² and the North Stream flows through 13 counties, including Xiamen City, one of the Special

Economy Zones in China. Naturally, Jiulong River is the central water resource for these counties. However, as livestock and aquatic farms have grown, the water of the Jiulong River basin has become contaminated with high levels of antibiotics (Zhang et al., 2011; Zhang et al., 2012). In this study, a total of 191 antibiotic-resistant bacteria were isolated from water samples from 21 selected sites in the Jiulong River. These strains were subjected to tests for 19 antimicrobials were tested and genomic DNA was used to determine the organismal composition of the strains and to assess the presence of three classes of integrons. The objectives of this study are: (1) to test the degree of resistance and multi-resistance of bacteria isolated from the Jiulong River; (2) to evaluate the diversity of these resistant strains; and (3) to investigate the prevalence and genetic diversity of class 1, 2 and 3 integrons.

Materials and methods

Sampling and bacteria screening

Water samples in sterile brown glass bottles were collected from 21 sampling points numbered J1-J21 along the Jiulong River basin (Fig. 1) in July 2012, and immediately carried back to the lab on ice. Each sample was diluted to 10^{-4} with normal saline solution.

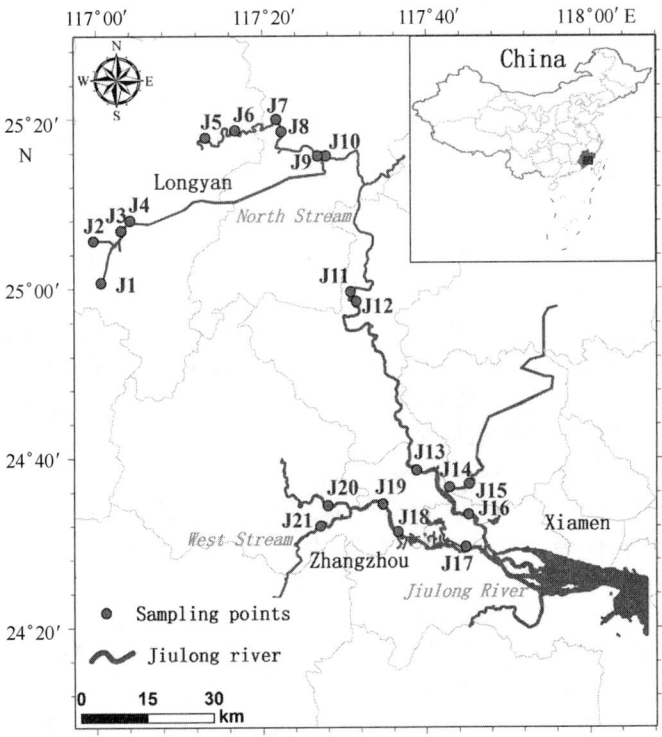

Fig. 1   Sampling sites along the Jiulong River basin (J1-J21)

Bacteria were cultured aerobically at 28℃ for 16 h to 24 h on LB nutrient agar medium (Haibo, Qingdao, China) with one of five antibiotics (ampicillin 100 mg/L, trimethoprim 30 mg/L, chloramphenicol 30 mg/L, streptomycin 30 mg/L and tetracycline 30 mg/L). For each site, individual colonies were screened on the basis of their morphology and then transferred to tubes with 5 mL LB broth medium (Haibo, Qingdao, China). Then they were incubated with shaking at 28℃ for 16 h to 24 h. The finalized unduplicated strains were confirmed based on the result of molecular identification and antimicrobial susceptibility.

Antimicrobial susceptibility

Isolates were subjected to susceptibility tests with 19 antibiotics: penicillin (PEN, 10 U), ampicillin (AMP, 10 μg), amoxicillin (AMX, 10 μg), cefotaxime (CTX, 30 μg) of β-lactams (β-Ls), streptomycin (STR, 10 μg), kanamycin (KAN, 30 μg), neomycin (NEO, 30 μg) of aminoglycosides (AGs), tetracycline (TCY, 30 μg), doxycycline (DOX, 30 μg) of tetracyclines (TCs), norfloxacin (NOR, 10 μg), ofloxacin (OFX, 5 μg), enrofloxacin (ENR, 5 μg) of quinolones (QNs), sulfamethoxazole (SMX, 300 μg), trimethoprim (TMP, 5 μg), sulfamethoxazole-trimethoprim (SXT, 23.7/1.25 μg) of sulfonamides (SAs), chloramphenicol (CHL, 30 μg), florfenicol (FLR, 30 μg) of amphenicols (APs), erythromycin (ERY, 15 μg), of macrolide (ML) and rifampicin (RIF, 5 μg) of rifamycin (RM) by standard disk diffusion method on Mueller-Hinton agar (Haibo, Qingdao, China). The results were interpreted using WHONET 5.6 according to the Clinical and Laboratory

Standards Institute guidelines (CLSI 2013).

DNA extraction, PCR, cloning, and sequencing of 16S rDNA genes and integrons

Genomic DNA was extracted with the TIANamp Bacteria DNA kit (Tiangen, Beijing, China) and stored at-20℃. The 16S rD-NA genes were amplified using bacterial universal primer sets 27F/1492R, while the primers intI1 F/R, intI2 F/R and intI3 F/R were used to amplify the integrases of three classes of integrons and sul1F/R were used to confirm the 3' -CS of integron 1. Primers hep58/59 and hep74/51 were used to amplify the variable regions of class 1 and class 2 integrons, respectively (Table 1) . Each 50 μL PCR mixture included 5 μL 10× Taq DNA buffer (with Mg^{2+}), 1 μLdNTPs (10 mM), 1 μL forward primer (10 μM), 1 μL reverse primer (10 μM), 0. 25 μL Taq polymerase (5 U/μL, Takara, Japan), 0. 5~1 μg template DNA and double distilled water. The reaction conditions were as follows: pre-denaturation at 95℃ for 10 min; 30 cycles of denaturation at 94℃ for 1 min, annealing at 55℃ for 1 min and extension at 72℃ for 1-5 mins (Table 1); and a final extension at 72℃ for 10 min. Amplified products were confirmed by electrophoresis using a 1% (w/v) agarose gel. Amplicons of the 16S rDNA genes and variable regions of the integrons were then purified with a Gel Extraction Kit (Axygen Scientific, USA) and cloned into pMD19-T vectors (Takara, Japan) for sequencing at GenScript Biotechnology (Nanjing, China) .

Sequences of 16S rDNA genes were clustered into several operational taxonomic units (OTUs, homology>97%) using Mothur and representative sequences of each OTU were aligned with reference sequences to construct a phylogenetic tree in MEGA 6. Sequences of the variable regions in integrons were assessed using a BLASTN search through NCBI (http: // blast. ncbi. nlm. nih. gov).

Results and discussion

Taxonomic composition of the antibiotic-resistant bacteria

Based on the morphology, 16S rDNA, phenotype of drug susceptibility, a total of 191 unique drug-resistant bacterial strains were isolated from the Jiulong River basin. The 16S rDNA gene sequences obtained from these strains were subsequently clustered into 28 OTUs using Mothur. A phylogenetic tree of the representative sequences and references suggests that these 191 organisms belong to 20 genera, including Comamonas, Delftia, Ralstonia, Vogesella in β-Proteobacteria, Chromobacterium, Acinetobacter, Pseudomonas, Aeromonas, Proteus, Providencia, Escherichia, Klebsiella, Enterobacter and Stenotrophomonas in γ-Proteobacteria, Bacillus in Bacilli, Corynebacterium, Microbacterium and Micrococcus in Actinobacteria, Soonwooa and Chryseobacterium in Flavobacteria (Fig. 2) .

As the largest group, members of *Acinetobacter* were more than a quarter of all bacteria sampled, followed by *Aeromonas*, *Comamonas* and *Pseudomonas* which accounted for 16. 2% (31/191), 14. 7% (28/191) and 12. 0% (23/191) of the sampled bacteria (Fig. 3) . *Acinetobacter spp.* have been found in many different types of opportunistic infections as nosocomial pathogens, including respiratory infections, bacteremia, meningitis, urinary tract infections, infective endocarditis and many other infections (Bergogne-Berezin and Towner 1996) . The genus Aeromonas has also been found in all kinds of clinical infections, such as gastroenteritis, hemolytic uremic syndrome, septicemia, new bacteremic syndromes, meningitis and others (Michael and Abbott 1998) . Comamonas and Pseudomonas have often been reported in terms of their pathogenicity (Isles et al. 1984; Oliver et al. 2000; Gul et al. 2007; Reddy et al. 2009; Opota et al. 2014) . These resistant bacteria were used to isolate from some waste water from hospitals and can cause considerable damage to humans and animals that suffer infections.

Prevalence of antibiotic-resistance in isolates

Antibiotic-resistant isolates from Jiulong River were tested for their susceptibilities to 19 antibiotics in eight classes.

As shown in Figure 4, the rate of resistance to 19 selected antimicrobial agents in 191 isolates varied from 8. 9% to 99. 5%. Except for the isolate S-B9B, belonging to Aeromonas, all strains were resistant to PEN. Over 80% of the bacteria were resistant to β-Ls and RMs, except for CTX (resistance in 49. 2% of the bacteria) . In addition, resistance to SAs and APs was also observed at high percentages ranging from 51. 3% to 79. 6%. The proportion of bacteria resistant to AGs and QNs was lower, between 42. 5% and 50. 8% and between 8. 9% and 36. 1%, respectively. OFX was the antibiotic with the lowest observed rate of resistance (8. 9%). Resistance to the TC antibiotics, TCY and DOX, was 58. 6% and 21. 0%. Our results were generally consistent with our hypothesis that within an antibiotic class, resistance rates correlated with the time the antibiotic has been used. For example, the resistance rates declined from 50. 8% in STR to 42. 5% in NEO in AG antibiotics, with the time in use decreasing from STR to NEO.

Figure 5 shows the extent of multi-resistance in isolates sampled in this study. All of the isolates were resistant to at least three kinds of antibiotics. 65. 4% of isolates were resistant to at least ten kinds of antibiotics. Two strains (A-X3B and T-X3D) were resistant to only three kinds of antibiotics and one strain (A-B11A) was resistant to four kinds of antibiotics. Strain A-X3B was resistant to PEN, AMP, and AMX of β-Ls, while strain T-X3D was resistant to PEN, SMX and TCY. Strain A-B11A was resistant to a similar group of antibiotics as strain A-X3B, but it was also resistant to ERY. It is worth noting that four strains (S-B8A, C-B10A, C-

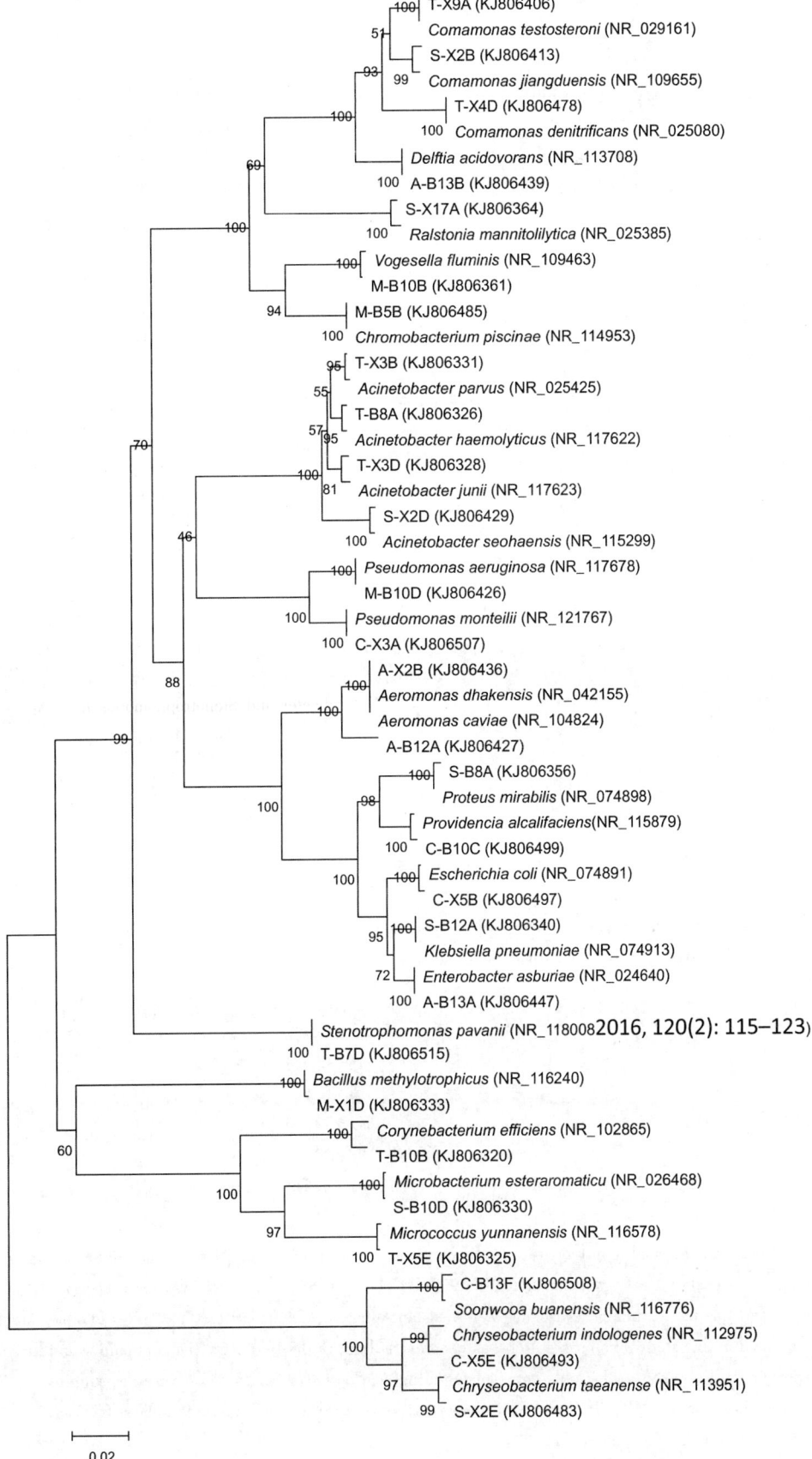

Fig. 2 Phylogenetic tree constructed with representative and their reference strains of 28 OTUs

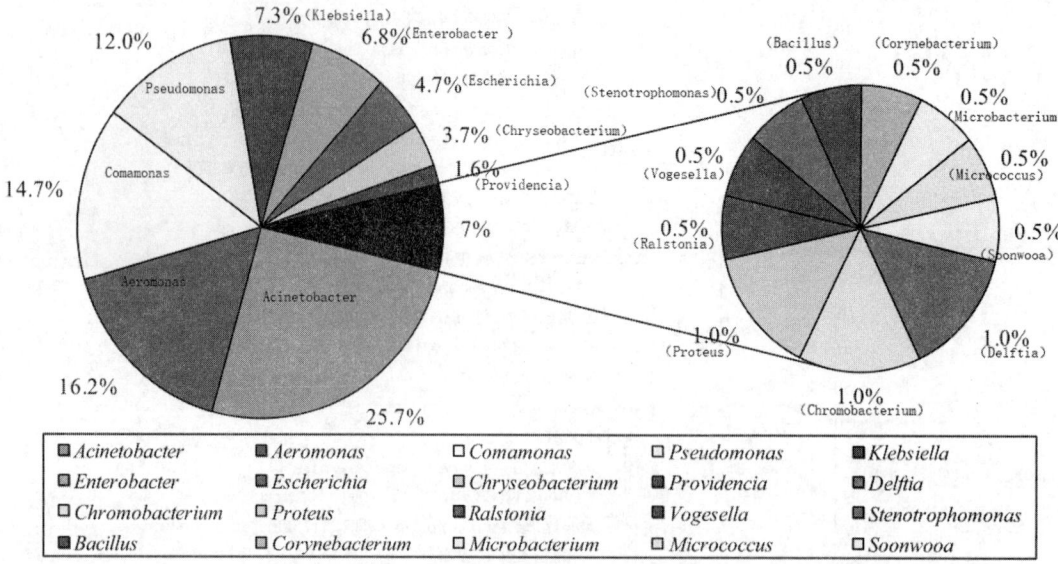

Fig 3 The taxonomic composition of the antibiotic-resistant bacteria isolated from the Jiulong River

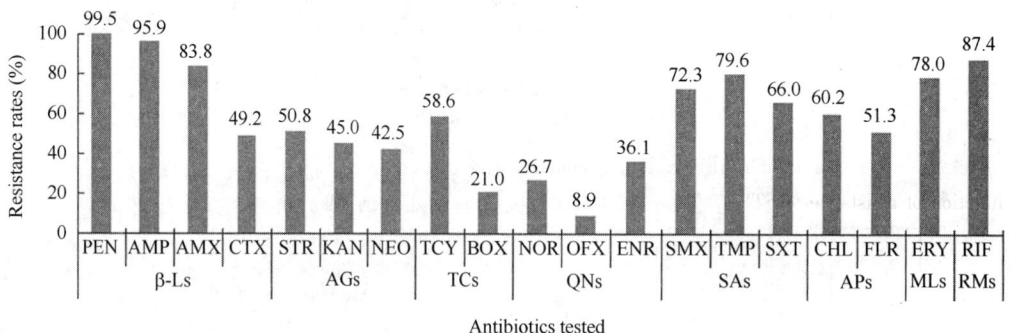

Fig. 4 Resistance rates to 19 antimicrobial agents in 191 isolates from the Jiulong River

X2B and S-X2E), belonging to different genera, were only susceptible to one kind of antibiotic and none of the 19 selected antibiotics could restrain the growth of C-B10C, a type of Providencia bacteria.

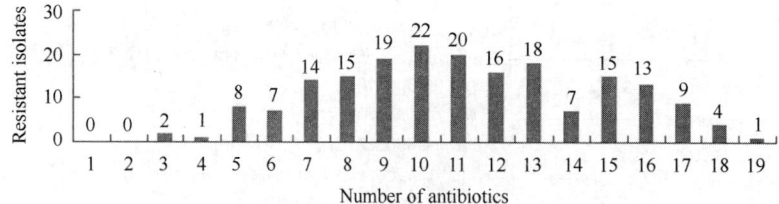

Fig. 5 Multiple antibiotic resistance of 191 resistant isolates from the Jiulong River

In a study of 183 coliforms isolated from ten rivers in the north-east coastal region of Turkey, the authors tested for antibiotic susceptibility (Ozgumus et al., 2009). They found that quite a few isolates were resistant to AGs such as STR (51.9%) and NEO (46.4%), similar to our findings, while fewer isolates were resistant to KAN (24%). Resistance rates to other antibiotics were lower than the rates we found in this study. Similar research has been carried out in the Bernesga River (Spain) and the results were generally consistent with our results: 289 Escherichia coli isolates sampled from river water and sediments exhibited multi-resistance to PEN and ERY, while resistance to AMP, TCY and STR along the river ranged from 65.5% to 94.7%, 16.7% to 75.9% and 34.5% to 65.5%, respectively (Sidrach-Cardona et al., 2014).

Integron detection and characterization

We detected three classes of integrons in 191 antibiotic-resistant strains isolated from Jiulong River by PCR. All isolates showed containing the intI1 gene, but only 42.4% (81/191) of them were detected the sul1 gene. The sul1 gene was a part of the 3'-CS of integron 1 and this might be due to the lost of the 3'-CS. Five out of 191 isolates contained the intI2 gene and none of the isolates had the intI3 gene. This pattern of intI was expected because class 1 integron (In1) is the most ubiquitous and has the central role in spreading resistant genes between bacteria (Stokes and Hall 1989; Cambray et al., 2010; Domingues et al., 2012), while the second class of integron (In2) is less common than In1 (Ozgumus et al. 2009; Koczura et al., 2014) and the third class (In3) has been found in just a few studies (Arakawa et al. 1995; Correia et al. 2003; Shibata et al. 2003; Xu et al. 2007; Poirel et al. 2010; Barraud et al., 2013).

Further studies were implemented to characterize the variable regions of In1 and In2. Out of 81sul1-positive isolates, 66 amplified a single copy and four isolates that were identified as Aeromonas (n=2) and Klebsiella (n=2), amplified two copies, with amplicon sizes varying from approximately 0.2 to 5.1 kb. 11 of the isolates (13.6%) failed to amplify the variable region. By contrast, all five intI2-positive isolatessuccessfully amplified the variable region and produced amplicons approximately 2.2 kb, which were identified as *Acinetobacter* (n=3), *Pseudomonas* (n=1) and *Proteus* (n=1).

30 gene cassette arrays and 34 cassettes were found in the In1. Seven isolates, including five isolates of Klebsiella sp., Comamonas sp. and Enterobacter sp., carried an empty In1 with no gene cassettes.

Among the recovered arrays, dfrA12-orfF-aadA2 was the most common and was found in fourteen isolates, ten of which were *Aeromonas*. Seven isolates contained arr-3-aacA4 and most of these isolates were identified as *Acinetobacter* (n=5). DfrA17-aadA5 was found in five isolates which were members of *Escherichia* (n=3) and *Enterobacter* (n=2). An array with the single cassette aadA2, was detected in four isolates of *Pseudomonas* (n=3) and Klebsiella (n=1). Other arrays commonly appeared in one or two strains (Fig. 6). Simultaneously, nine new arrays were discovered in this study, including aadA5-orfD, qacH-cmlA1-aadA2, aacA4-aadA1-tnaA', aacA4-ereA1-bla$_{OXA-2}$-aadA1, dfrA5-aacA4-nit1-nit2-catB3, dfrA17-aacA4-nit1-nit2-catB3, unknown-aacA4-ereA1-bla$_{OXA-2}$-dfrA1, dfrA16-bla$_{PSE-1}$-aadA2-cmlA1-aadA1 and dfrA16-bla$_{PSE-1}$-aadA2-cmlA1-orfA'-tna$_{IS4}$' (Fig. 7).

Of 34 different gene cassettes (Table S1), the most common were aadA2 (14.8%), encoding an aminoglycoside adenylyltransferase with a function of resistance to STR and spectinomycin. Other cassettes that encode this transferase also contained aadA1, aadA5, aadA8, aadB and aacA4 encodes an aminoglycoside acetyltransferase resistant to gentamicin and KAN. Like the AG resistant genes, various TMP resistant genes that encode dihyrofolate reductases also appeared in this study, including dfrA1, dfrA5, dfrA12, dfrA16, dfrA17, dfrA25, dfrA32 and dfrB4. Other resistant genes were present: β-lactamases (bla$_{OXA-2}$, bla$_{PSE-1}$), ADP-ribosyl transferases (arr-2, arr-3), chloramphenicol acetyltransferases (catB3, catB8), a chloramphenicol exporter (cmlA1), erythromycin esterases (ereA1, ereA2), a quaternary ammonium compound efflux (qacH), lincomycin nucleotidyltransferase (linF) and nitrilases (nit1, nit2). The emergence of nit1 and nit2 which encode nitrilases might be due to the existence of some waste plastic factories. In addition, four gene cassettes of unknown function (orfA', orfC, orfD and orfF) and two transposase-like genes (tnaA' and tna$_{IS4}$') were also detected. The orfA' gene and the tna$_{IS4}$' gene appear to be mutants that are 93% identical to the genes in Delftia acidovorans (CP000884). The tnaA' gene was also found to have 96% identity with that of Pseudomonas aeruginosa (JN790946, AY920928). It is worth noting that we found a completely novel gene cassette with unknown function in the array unknown-aacA4-ereA1-bla$_{OXA-2}$-dfrA1 (KJ522747).

In In2, the gene cassette array dfrA1-sat2-aadA1 (KP064397) was the only array we found that has also been reported in other research (Ramirez et al. 2012; Cao et al. 2014; Koczura et al. 2014). The strain A-X11A, one of the In2-positive isolates and a member of *Pseudomonas*, was also carrying an In1 with an aadA2 gene cassette. The cassette sat2 encodes a streptothricin acetyltransferase with a streptothricin resistance function. It has been reported that the intI2 gene contains a stop codon at the 179th amino acid, which may result in limited integration and excision of gene cassettes (Hansson et al., 2002) and this may be the reason that In2 is detected in fewer isolates and there are fewer gene cassette arrays in In2.

Integrons have been frequently detected in aquatic environments. For example, in the Turkish study mentioned above, 14 out of 183 coliforms carried In1 and five carried In2, while two E. coli strains carried both In1and In2. Six gene cassette arrays were found in In1, including dfrA7, dfrA16, dfr2d, dfrA1-aadA1, dfrA17-aadA5and bla$_{OXA-30}$-aadA1, which was quite different from the arrays identified in our study. However, the Turkish E. coli contained the same arraydfrA1-sat2-aadA1 in In2 as we found (Ozgumus et al., 2009). In the Portugese study (Tacão et al. 2014), intI1 was present in 35.1% of the 151 cefotaxime-resistant isolates sampled from 12 rivers and 41% of the intI1-positive strains successfully amplified the variable regions. Six of the isolates had empty integrons and there were nine other gene cassette arrays: aadA1, aadA2, aadA16/aacA4' fusion, aadA6-orfD, catB8-aadA1, dfrA1-aadA1, qnrVC4-aacA4'-17, bla$_{OXA-10}$-aacA4' and the most common array dfrA17-aadA5. Similarly, two co-existing In1 with disparate arrays were found in four isolates of Aeromonas hydrophila and E. coli. The study in Shandong Province, China, found 29

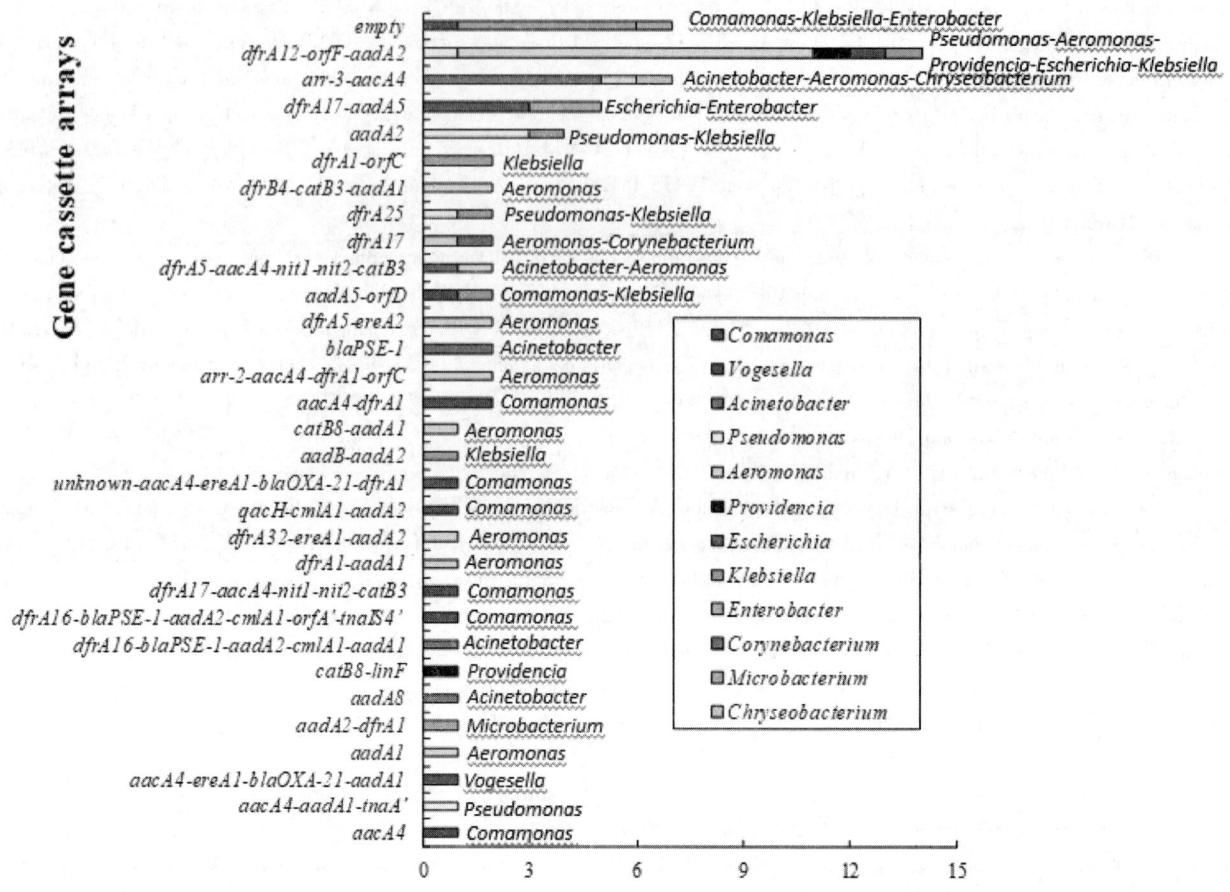

Fig. 6 31 different gene cassette arrays of class 1 integrons in 191 resistant isolates from Jiulong River

different gene cassette arrays in 34 different cassettes (Guo et al., 2011). Of all 29 cassette arrays, 22 were prevalent, especially dfrA17−aadA5 and dfrA12−orfF−aadA2. Seven novel arrays were also found. A total of 34 cassettes encoded dihyrofolate reductases, aminoglycoside adenylyltransferases, aminoglycoside acetyltransferases, β−lactamases, chloramphenicol acetyltransferases, erythromycin esterases, erythromycin esterases, nitrilases and putative proteins with unknown functions. A new cassette was also detected in that study. It seems that the prevalence and genetic diversity of integrons is fairly different in various locations, which may due to water usage in the region, the control of waste water emissions, as well as differing management practices of antibiotics.

Conclusions

There is a diverse community of antibiotic−resistant bacterial isolates in Jiulong River basin and some potentially pathogenic bacteria were identified, such as Acinetobacter, Aeromonas, Comamonas and Pseudomonas. Among these isolates, we found high rates of resistance to a wide range of antibiotics that are often used for prevention and therapy in humans and other animals. Many of these bacteria exhibited multi−resistance. Moreover, integrons were common components of the bacteria we sampled, and this means resistance genes can be easily shared among different strains. The discovery of novel gene cassettes suggests that additional antibiotic−resistant strains will emerge in the future. The severity of resistant isolates from Jiulong River might result from the waste water from the livestock and aquatic farms, as well as five waste water treatment plants and some waste plastic factories along the basin. We must treat the waste water more carefully, especially in the respect of the antibiotic−resistant bacteria. The government should also launch campaigns to illegal factories.

Acknowledgments This study is financially supported by the Special Fund for Agro−scientific Research in the Public Interest (NO. 201203085), National Natural Science Foundation of China (NO. 31202030) and Open Fund of Key Laboratory of Urban Environment and Health, Institute of Urban Environment, Chinese Academy of Sciences (NO. KLUEH201106).

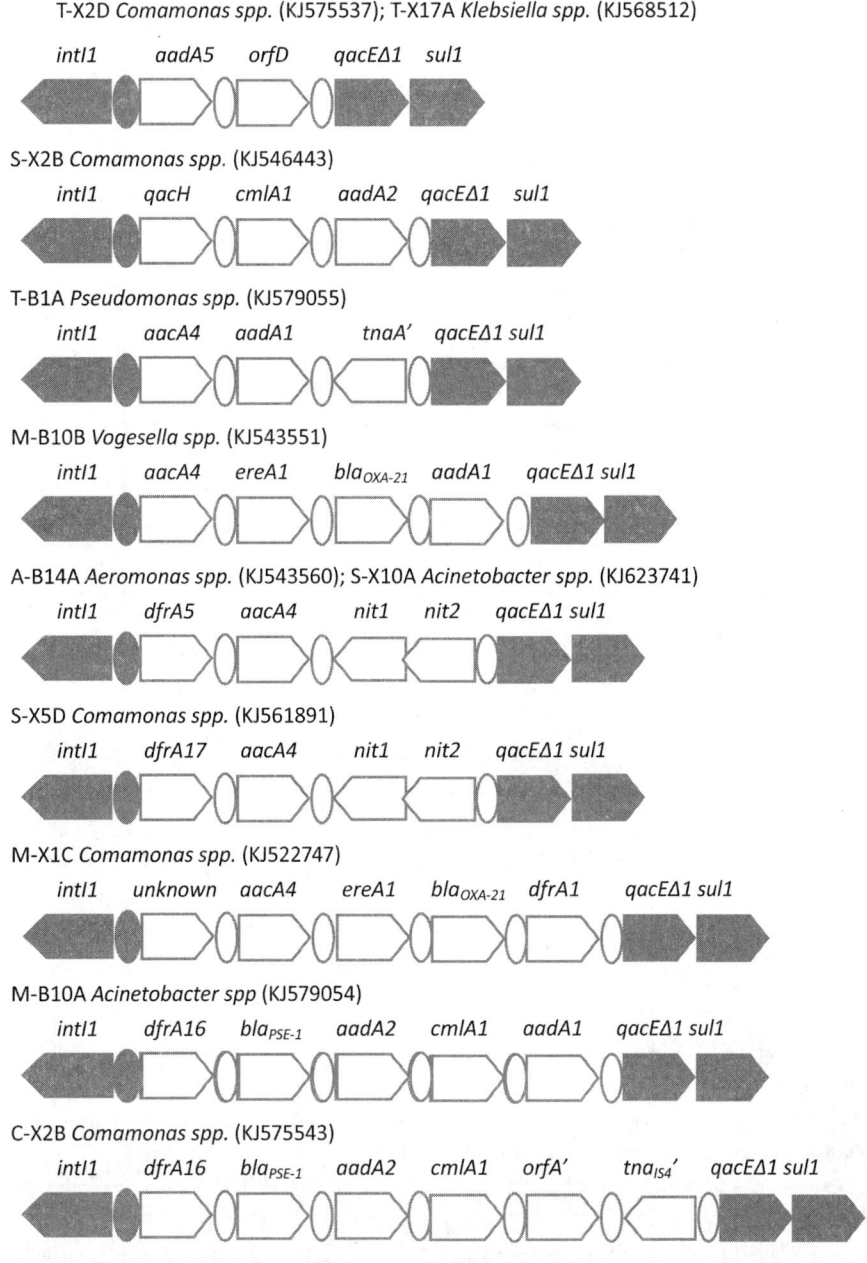

Fig. 7　Nine novel gene cassette arrays in class 1 integrons. The gray ovals represent attI and the blank ovals represent attC. The direction of gene transcription is denoted by the direction of the arrow.

References

Arakawa Y., Murakami M., Suzuki K., Ito H., Wacharotayankun R., Ohsuka S., Kato N., Ohta M. . (1995) A novel integron-like element carrying the metallo-beta-lactamase gene blaIMP. Antimicrob Agents Ch, 39: 1612-1615.

Barlow R. S., Gobius K. S. (2006) Diverse class 2 integrons in bacteria from beef cattle sources. J Antimicrob Chemother, 58: 1133-1138.

Barraud O., Casellas M., Dagot C., Ploy P. C. (2013) An antibiotic-resistant class 3 integron in an Enterobacter cloacae isolate from hospital effluent. Clin Microbiol Infec, 19: 306-308.

Bergogne-Berezin E., Towner K. J. (1996) Acinetobacter spp. as nosocomial pathogens: microbiological, clinical, and epidemiological features. Clin Microbiol rev, 9: 148-165.

Cambray G., Guerout A., Mazel D. (2010) Integrons. Annu Rev Genet, 44: 141-166.

Cao X., Zhang Z., Shen H., Ning M., Chen J., Wei H., Zhang K. . (2014) Genotypic characteristics of multidrug-resistant Escherichia coli isolates associated with urinary tract infections. APMIS, 122: 1088-1095.

Correia M., Boavida F., Grosso F., Salgado M. J., Lito L. M., Cristino J. M., Mendo S., Duarte A. (2003) Molecular characterization of a new class 3 integron in Klebsiella pneumoniae. Antimicrob Agents Chemoth, 47: 2838-2843.

Davies J., Davies D. (2010) Origins and evolution of antibiotic resistance. Microbiol Mol Biol Rev, 3: 417-433.

Domingues S., da Silva G. J., Nielsen K. M. (2012) Integron: Vehicles and pathways for horizontal dissemination in bacterial. Mob Genet Elements, 2: 211-223.

Guerra B., Soto S., Cal S., Mendoza M. C. (2000) Antimicrobial resistance and spread of class 1 integrons among Salmonella serotypes. Antimicrob Agents Chemoth, 44: 2166-2169.

Gul M., Ciragil P., Bulbuloglu E., Aral M., Alkis S., Ezberci F. (2007): Comamonas testosteroni bacteremia in a patient with perforated acute appendicitis. Short communication. Acta Microbiol Immunol Hung, 54: 317-321.

Guo X., Xia R., Han N., Xu H. (2011) Genetic diversity analyses of class 1 integrons and their associated antimicrobial resistance genes in Enterobacteriaceae strains recovered from aquatic habitats in China. Lett Appl Microbiol, 52: 667-675.

Hall R. M., Collis C. M. (1995) Mobile gene cassettes and integrons: capture and spread of genes by site-specific recombination. Mol Microbiol, 15: 593-600.

Hall R. M., Collis C. M. (1998) Antibiotic resistance in gram-negative bacteria: the role of gene cassettes and integrons. Drug Resist Updates, 1: 109-119.

Hansson K., Sundström L., Pelletier A., Roy P. H. (2002) IntI2 integron integrase in Tn7. J Bacteriol, 184: 1712-1721.

Hasan M. M., Tinni S. R. (2011) Detection of integrons from Escherichia coli isolates obtained from some selected animals in Republic of Korea. Indian J Microbiol, 51: 144-146.

Isles A., Maclusky I., Corey M., Gold R., Prober C., Fleming P., Levison H. (1984) Pseudomonas cepacia infection in cystic fibrosis: an emerging problem. JPediatr, 104: 206-210.

Koczura R., Semkowska A., Mokracka J. (2014) Integron-bearing Gram-negative bacteria in lake waters. Lett Appl Microbiol, 59: 514-519.

Li D., Qi R., Yang M., Zhang Y., Yu T. (2011) Bacterial community characteristics under long-term antibiotic selection pressures. Water Res, 45: 6063-6073.

Maynard C., Bekal S., Sanschagrin F., Levesque R. C., Brousseau R., Masson L., Larivière S., Harel J. (2004) Heterogeneity among virulence and antimicrobial resistance gene profiles of extraintestinal Escherichia coli isolates of animal and human origin. Journal of clinical microbiology, 42: 5444-5452.

Michael J., Abbott S. (1998) Evolving concepts regarding the genusAeromonas: an expanding panorama of species, disease presentations, and unanswered questions. Clin Infect Dis. 27: 332-344.

Moura A., Henriques I., Ribeiro R., Correia A. (2007) Prevalence and characterization of integrons from bacteria isolated from a slaughterhouse wastewater treatment plant. J Antimicrob Chemoth, 60: 1243-1250.

Oliver A., Canton R., Campo P., Baquero F., Blazquez J. (2000) High frequency of hypermutable Pseudomonas aeruginosa in cystic fibrosis lung infection. Science, 288: 1251-1254.

Opota O., Ney B., Zanetti G., Jaton K., Greub G., Prod'hom G. (2014) Bacteremia caused by Comamonas kerstersii in a patient with diverticulosis. J Clin Microbiol, 52: 1009-1012.

Ozgumus O. B., Sandalli C., Sevim A., Celik-Sevim E., Sivri N. (2009) Class 1 and class 2 integrons and plasmid-mediated antibiotic resistance in Coliforms isolated from ten rivers in northern Turkey. JMicrobiol, 47: 19-27.

Poirel L., Carattoli A., Bernabeu S., Bruderer T., Frei R., Nordmann P. (2010) A novel IncQ plasmid type harbouring a class 3 integron from Escherichia coli. J Antimicrob Chemoth, 65: 1594-1598.

Ramirez M. S., Morales A., Vilacoba E., Marquez C., Centron D. (2012) Class 2 integrons dissemination among multidrug resistance (MDR) clones of Acinetobacter baumannii, Curr Microbiol, 64: 290-293.

Reddy A. K., Murthy S. I., Jalali S., Gopinathan, U. (2009) Post-operative endophthalmitis due to an unusual pathogen, Comamonas testosteroni. J Med Microbiol, 58: 374-375.

Sidrach-Cardona R., Hijosa-Valsero M., Marti E., Balcazar J. L., Becares E. (2014) Prevalence of antibiotic-resistant fecal bacteria in a river impacted by both an antibiotic production plant and urban treated discharges. Sci Total Environ, 488: 220-227.

Shibata N., Doi Y., Yamane K., Yagi T., Kurokawa H., Shibayama K., Kato H., Kai K., Arakawa Y. (2003) PCR typing of genetic determinants for metallo-lactamases and integrases carried by gram-negative bacteria isolated in Japan, with focus on the class 3 integron. J Clin Microbiol, 41: 5407-5413.

Stokes H. W., Hall R. M. (1989) A novel family of potentially mobile DNA elements encoding site-specific gene-integration functions: integrons. Mol Microbiol, 3: 1669-1683.

Tacão M., Moura A., Correia A., Henriques I. (2014) Co-resistance to different classes of antibiotics among ESBL-producers from aquatic systems. Water Res, 48: 100-107.

White P. A., McIver C. J., Deng Y. M., Rawlinson W. D. (2000) Characterization of two new gene cassettes, aadA5 and dfrA17. FEMS Microbiol Lett, 182: 265-269.

Xu H., Davies J., Miao V. (2007) Molecular characterization of class 3 integrons from Delftia spp.. J Bacteriol, 189: 6276-6283.

Zhang D., Lin L., Luo Z., Yan C., Zhang X. (2011) Occurrence of selected antibiotics in Jiulongjiang River in various seasons, South China. J Environ Monit: JEM, 13: 1953-1960.

Zhang X., Zhang D., Zhang H., Luo Z., Yan C. (2012) Occurrence, distribution, and seasonal variation of estrogenic compounds and antibiotic residues in Jiulongjiang River, South China. Environ Sci Pollut Res Int, 19: 1392-1404.

该文发表于《水产学报》【2015，39（7）：1043-1053】

鳗鲡及其养殖水体分离耐药菌的多样性和耐药性分析

Diversity and antimicrobial susceptibility of drug-resistant bacteria isolated from Anguilla rostrata and the farming water

吴小梅[1]，林茂[1,2]*，鄢庆枇[1]，江兴龙[1,2]，张娴[3]

（1. 集美大学水产学院，厦门 361021；2. 鳗鲡现代产业技术教育部工程研究中心，厦门 361021；
3. 中国科学院城市环境与健康重点实验室，厦门 361021）

摘要： 为了更有针对性地防控水产动物细菌性病害的发生和流行，本研究对鳗鲡及其养殖水体耐药细菌的种属特征及耐药情况开展相关实验。首先采集鳗鲡不同部位（表皮、鳃、肠道）及其养殖水体的样品，经 5 种抗菌药物平板筛选耐药菌株，然后采用 K-B 纸片扩散法检测细菌对抗菌药物的耐药性，同时测定耐药菌株的 16S rDNA 序列，进而分析耐药菌的种属分布和多重耐药性。结果显示经耐药平板筛选分离纯化得到 108 株细菌，分别属于气单胞菌属、柠檬酸杆菌属、不动杆菌属等 20 个属；其中，93.5%的菌株对 3 种（含）以上的抗菌药物具有耐药性，86.1%的菌株对 3 类（含）以上的药物具有抗性。对阿莫西林的耐药率高达 90.7%，对四环素、利福平以及磺胺类和酰胺醇类药物类的耐药率在 60%~80%之间，对头孢噻肟、新霉素以及喹诺酮类的耐药性弱（低于20%）。鳗鲡肠道（0.40）、表皮（0.41）、鳃部（0.42）及水样（0.47）菌群的多重耐药指数显示各生态样品耐药程度较为严重，尤以水样为最。各菌属中，柠檬酸杆菌属（0.58）和克雷伯菌属（0.61）的多重耐药指数最高，而不动杆菌属（0.21）则相对较低。鳗鲡及养殖水体普遍存在多重耐药菌株，对此应引起足够的重视；水产动物及养殖环境耐药细菌对某些水产用药物如诺氟沙星、新霉素等耐药率低，可将其做为水产动物细菌性疾病治疗的首选药物。

关键词： 鳗鲡；16S rDNA；药敏试验；多重耐药性

　　细菌耐药性是全球关注的热门科研领域，病原菌多重耐药性的发展对人类健康和畜牧业生产都构成了严重的威胁。目前在金黄色葡萄球菌（*Staphylococcus aureus*）、肺炎克雷伯菌（*Klebsiella pneumoniae*）、铜绿假单胞菌（*Pseudomonas aeruginosa*）、鲍曼不动杆菌（*Acinetobacter baumannii*）、大肠杆菌（*Escherichia coli*）等医院感染的重要病原菌中均出现了多重耐药菌株，已引起了医学和微生物学者的忧虑和重视。而近年来在水产动物方面的研究报道也表明，水产动物源细菌的多重耐药性同样不容忽视。蔡俊鹏等研究发现九孔鲍（Haliotis diversicolor）养殖水体及其肠道菌群中有较多菌株对四环素（Tetracycline）、卡那霉素（Kanamycin）、青霉素 G（Penicillin G）、新霉素（Neomycin）和丁胺卡那霉素（Amikacin）等表现出多重耐药性。李绍戊等分离到的 28 株鱼源嗜水气单胞菌对 β-内酰胺类（β-Lactams）、大环内酯类（Macrolides）、氯霉素类（Chloramphenicols）和四环素类（Tetracyclines）药物的耐药率均超过 60%。Nguyen 等从 15 个鲶鱼养殖场分离的 116 株运动型假单胞菌（*Pseudomonas* spp.）和 92 株气单胞菌（*Aeromonas* spp.）具有多重耐药性的比例分别为 96.6%和 61.9%。动物源细菌的耐药性不仅削弱抗菌药物对动物细菌性疾病的控制效果，而且还存在向人类致病菌传播抗药性的潜在风险。因此，调查养殖动物及其环境菌群耐药性的变化趋势，据此分析可能带来的不利影响，具有非常重要的科研价值和现实意义。

　　2015 年 Cao 等在《Science》上的报道表明中国的水产养殖产量已超过了全球的 60%，而养殖鳗鲡（Anguilla spp.）是中国出口创汇的重要农产品之一。随着养殖集约化程度的增加，鳗鲡疾病时有发生，其中细菌性疾病的危害最为严重，常造成巨大的经济损失。据报道可引起鳗鲡疾病的致病菌有近 20 种，报道中分离得到的病原菌存在不同程度的多重耐药性。郑芳艳等错误！未找到引用源。由患溃烂病鳗鲡分离得到的创伤弧菌（*Vibrio vulnificus*）对去甲万古霉素（Norvancomycin）具有抗性。Lo 等从病鳗分离到的 94 株迟钝爱德华菌（*Edwardsiella tarda*）对多西环素（Doxycycline）和土霉素（Oxytetracycline）的耐药率均为 21.3%。雷燕等对养殖欧洲鳗鲡（Anguilla anguilla）体表溃疡病病原嗜水气单胞菌（*Aeromonas hydrophila*）的药敏试验显示，该菌株对 10 种抗菌药存在耐药性。谭爱萍等分离并鉴定的一株鳗源肺炎克雷伯菌，更是对 17

种药物具有耐药性。现代微生物学理论认为，细菌的耐药性除了源自基因突变和遗传，也可通过细菌间的水平传播获得，因此病原菌的多重耐药性在一定程度上受正常菌群和环境中耐药细菌的影响。此外，水产动物病原细菌一般为条件致病菌，在健康动物和养殖水体中长期存在。为此本研究从鳗鲡各个部位（肠道、鳃部和表皮）及其养殖水体分离耐药细菌，并分析耐药细菌的组成及耐药的程度，从而为评估鳗鲡养殖区的细菌耐药性风险，以及指导抗菌药物的合理使用提供必要的理论依据。

1 材料与方法

1.1 主要试剂

LB 培养基、MH 培养基购自青岛海博生物技术有限公司；抗菌药物购自北京楚和霞光生物技术发展中心；药敏纸片购自杭州滨和微生物试剂有限公司。Taq 酶、dNTP Mix、DNA ladder 购自大连 Takara 公司；细菌基因组 DNA 提取试剂盒购自天根生化科技（北京）有限公司；16S rDNA 通用引物（27F：5′-AGAGTTTGATCCTGGCTCAG-3′ 和 1492R：5′-ACGGC-TACCTTGTTACGACTT-3′）合成和 DNA 测序由南京金斯瑞生物公司完成；胶回收试剂盒购自美国 Axygen 公司。

1.2 样品采集与理化分析

分别自福建省 5 个养殖场采集美洲鳗鲡（Anguilla rostrata）及其养殖水样。水样以 HANA 水质仪（HI9804）检测温度、pH、溶氧、NH_3-N、NO_2-N 等理化指标，并利用 Waters 超高效液相系统（ACQUITY UPLC）-三重四极杆串联质谱（XEVO-TQ）联用仪检测抗菌药含量[15]。鳗鲡以 MS-222 麻醉后，用酒精棉擦拭体表，解剖，分别取鳃、肠道和表皮，剪碎加入生理盐水，匀浆待用。

1.3 耐药菌株的筛选、分离与纯化

鳗鲡各组织匀浆液和水样分别进行 10 倍梯度稀释后，分别涂布于含氨苄西林（Ampicillin，AMP，100 mg/L）、链霉素（Streptomycin，STR，100 mg/L）、甲氧苄啶（Trimethoprim，TMP，50 mg/L）、四环素（TET，100 mg/L）和氯霉素（Chloramphenicol，CHL，30 mg/L）等抗菌药物的 LB 平板上，30℃下培养 16~18 h；并根据菌落的形态特征挑出不同菌株进行纯培养和保种。

1.4 细菌基因组 DNA 的提取和 16S rDNA 的 PCR 扩增

各耐药菌株使用试剂盒提取总 DNA，以 1.2% 的琼脂糖凝胶电泳检测 DNA 后进行 PCR 扩增。PCR 反应体系：5 μL 10× Taq DNA Buffer，dNTPs（10 mmol/L）、引物 27F（10 μmol/L）和 1492R（10 μmol/L）各 1 μL，0.25 μL Taq 酶（5 U/μL），0.5~1 μg DNA 模板，补 ddH₂O 至 50 μL；反应条件：94℃预变性 5 min、94℃变性 1 min、55℃退火 1 min、72℃延伸 1.5 min，共 30 个循环，72℃终延伸 5 min。

1.5 16S rDNA 的测序及系统发育树的构建

16S rDNA 产物以 1.2% 的琼脂糖凝胶电泳检测并纯化后进行测序，所得序列在 GenBank 中进行 Blast 分析。耐药菌群 16S rDNA 序列用 MEGA5.2 软件进行比对和聚类分析，再用 Mothur 软件以阈值为 97% 进行 OTU（optional taxonomic unit，运算分类单元）分析，程序选出的代表菌株与 GenBank 中的参考菌株通过 MEGA5.2 软件采用邻接法（neighbour-joining method）构建系统进化树，选用 Kimura 2-parameter 距离模型，自举（Bootstrap）数据集设为 1 000 次。

1.6 药敏试验和耐药性分析

各菌株的药敏试验以大肠杆菌 ATCC25922 作为质控菌株，采用标准的 K-B 纸片琼脂扩散法。将活化好的耐药菌株用生理盐水稀释菌液，使稀释菌液达到 0.5 麦氏比浊管的浊度。无菌棉拭子蘸取上述菌液，在管壁上挤压去掉多余菌液，均匀涂布于 MH 平板上。待平板上的水分被琼脂完全吸收后，用无菌镊子取药敏纸片贴在平板表面压紧，于 30℃生化培养箱中倒置培养 16~18 h 后取出测量抑菌圈直径。参照美国临床实验室标准化委员会（Clinical and Laboratory Standards Institute，CLSI）标准判断结果。

根据药敏试验获得的耐药情况，参考 Krumperman 等的方法计算不同生态位菌群的多重耐药指数（=Σn/（D×S））。其中，"n"指某个菌株耐受抗菌药的数量，"Σn"即为该生态位所有菌株耐药的频数总和；"D"和"S"分别是受试的抗菌药物和菌株的数量。

2 结果与分析

2.1 养殖水样抗菌药含量检测

在 5 份养殖水样中，均检测到痕量的氯霉素（CHL）（7~22 ng/L）和氟苯尼考（Florfenicol，FFL）（3~1704 ng/L），

还有个别水样检测到甲氧苄啶（TMP）、磺胺甲恶唑（Sulfamethoxazole，SMZ）和红霉素（Erythromycin，ERY）（1~12 ng/L），而阿莫西林（Amoxicillin，AMX）、头孢噻肟（Cefotaxime，CTX）、链霉素（STR）、卡那霉素（KAN）、新霉素（NEO）、四环素（TET）、多西霉素（DOX）、诺氟沙星（Norfloxacin，NOR）、氧氟沙星（Ofloxacin，OFL）、恩诺沙星（Enrofloxacin，ENR）、利福平（Rifampicin，RIF）等均未检出（检测限为 0.1~10 ng/L）。

2.2　耐药细菌的分离及 16S rDNA 的序列分析

养殖水样及鳗鲡样品经五种抗菌药物（AMP、STR、TMP、TET 和 CHL）平板的筛选，分别获得 21、23、26、17 和 21 株耐药菌株，共计 108 株菌（后续检验未发现菌落特征、耐药谱和 16S rDNA 完全一致的重复菌株）。

以各菌株总 DNA 为模板，扩增获得 16S rDNA，经测序和同源性比对，鳗鲡不同体位及其养殖水体所分离的耐药菌株形成 23 个 OTU（表 1），归属于 4 纲 20 个属（图 1）。其中，γ-变形菌纲的气单胞菌属（24.1%）、柠檬酸杆菌属（Citrobacter）（17.6%）和不动杆菌属（Acinetobacter）（16.7%）的出现频率最高。在各样品中，鳃部耐药菌群多样性最为丰富，分布于 13 个属，又以气单胞菌属、柠檬酸杆菌属和不动杆菌属最为常见；其次，肠道分离耐药菌分属于 10 个菌属，其中气单胞菌属、柠檬酸杆菌属、不动杆菌属和希瓦菌属（Shewanella）居多；表皮分离耐药菌主要归属于气单胞菌属；水体耐药细菌种类最少，仅占 7 个属，主要分布于气单胞菌属、柠檬酸杆菌属和克雷伯菌属（Klebsiella），在鳗鲡样品中常见的不动杆菌属、希瓦菌属和变形杆菌属（Proteus）的耐药菌在水样中未见分离（图 2）。

2.3　药敏试验结果

108 株耐药细菌对 9 类 17 种抗菌药物敏感性试验结果（图 3）显示，耐药菌株对阿莫西林（AMX）的耐药率高达 90.7%，对四环素（TET）、磺胺甲恶唑（SMZ）、甲氧苄啶（TMP）、复方磺胺甲恶唑（Sulfamethoxazole/Trimethoprim，SXT）、氯霉素（CHL）、氟苯尼考（FFL）、利福平（RIF）的耐药率也在 60%~80% 之间。而对头孢噻肟（CTX）、新霉素（NEO）和诺氟沙星（NOR）、氧氟沙星（OFL）、恩诺沙星（ENR）的耐药率较低，分别为 11.1%、3.7%、6.5%、7.4% 和 15.7%。此外，对链霉素（STR）、卡那霉素（KAN）、红霉素（ERY）以及多西环素（DOX）的耐药率则介于 20%~40% 之间。对具体的菌属而言，柠檬酸杆菌（19 株）对四环素类药物 TET、DOX 和磺胺及其增效药物 SMZ、TMP、SXT 的耐药率（79%~100%）远高于总体耐药率，而不动杆菌（18 株）则恰恰相反，对上述两类药物的耐药率（0~17%）远低于总体耐药水平。

表 1　鳗鲡及其养殖水体中耐药细菌 16S rDNA 同源性比对和 OTU 分析结果
Tab. 1　16S rDNA homology alignment and OTU analysis of drug-resistant bacteria isolated from eel and the pond water

运算分类单元 OTU	菌株名 strain	代表菌株（登录号）representative（accession No.）	同源性最高菌株 strain of the highest homology		
			学名 scientific name	同源性（%）identity（%）	登录号 accession No.
1	M23, A5, S2, A11, S1, M3, C13, S11, S19, C24, M14, M25, S18, C22, A22, S9, C23, M27, M28, T15, A21, A20, A18, A13, S7, A14	C23（KM268975）	顶盖气单胞菌 *Aeromonas tecta*	100	NR_118043
2	T21, M12, M15, S20, S12, M20, T4, S21, T17, T16, S22, C20, C19, C18, T18, T12, C14, T23, M11	T21（KM269019）	弗氏柠檬酸杆菌 *Citrobacter freundii*	100	NR_028894
3	A8, C2, A3, C6, S14, A9, A15, C7, C16, C15, A16, C4, C1, A7, A2, S15	C16（KM268968）	医院不动杆菌 *Acinetobacter nosocomialis*	100	NR_117931
4	C5, C11, M2, T2, M1, C3, A1, T1, C12	C5（KM268927）	肺炎克雷伯菌 *Klebsiella pneumoniae*	99	NR_103936

运算分类单元 OTU	菌株名 strain	代表菌株（登录号）representative（accession No.）	同源性最高菌株 strain of the highest homology		
			学名 scientific name	同源性（%）identity（%）	登录号 accession No.
5	M21, M24, S23, M26, M18, T20	M18（KM268984）	新万景希瓦菌 *Shewanella seohaensis*	100	NR_ 025610
6	A17, M19, C21, A19, M22	C21（KM268973）	普通变形菌 *Proteus vulgaris*	100	NR_ 115878
7	M17, C17, S17	S17（KM269001）	木糖葡萄球菌 *Staphylococcus xylosus*	100	NR_ 036907
8	M5, M8, S4	M8（KM268937）	透明福格斯氏菌 *Vogesella perlucida*	99	NR_ 044326
9	A12, M13, S13	M13（KM268979）	类志贺邻单胞菌 *Plesiomonas shigelloides*	99	NR_ 044827
10	A4, T11	T11（KM269009）	粘质沙雷氏菌 *Serratia marcescens*	99	NR_ 114043
11	M7, T19	M7（KM268936）	成团泛菌 *Pantoea agglomerans*	99	NR_ 111998
12	A6, M4	A6（KM268919）	苏云金芽孢杆菌 *Bacillus thuringiensis*	99	NR_ 102506
13	T3, T13	T3（KM268949）	败血不动杆菌 *Acinetobacter septicus*	99	NR_ 116071
14	S8	S8（KM268945）	解脲金黄杆菌 *Chryseobacterium ureilyticum*	99	NR_ 042503
15	S16	S16（KM269000）	反硝化卓贝尔氏菌 *Zobellella denitrificans*	99	NR_ 043629
16	M6	M6（KM268935）	迟缓爱德华菌 *Edwardsiella tarda*	99	NR_ 024770
17	M16	M16（KM268982）	嗜麦芽寡养单胞菌 *Stenotrophomonas maltophilia*	99	NR_ 041577
18	S5	S5（KM268942）	香味类香味菌 *Myroides odoratus*	99	NR_ 112976
19	S24	S24（KM269008）	阿雅巴塔芽孢杆菌 *Bacillus aryabhattai*	100	NR_ 115953
20	T22	T22（KM269020）	摩氏假单胞菌 *Pseudomonas mosselii*	100	NR_ 024924
21	T14	T14（KM269012）	海藻希瓦菌 *Shewanella algae*	99	NR_ 117771
22	S3	S3（KM268940）	弗格森埃希菌 *Escherichia fergusonii*	99	NR_ 027549
23	S6	S6（KM268943）	双重氮植物杆菌 *Phytobacter diazotrophicus*	99	NR_ 115869

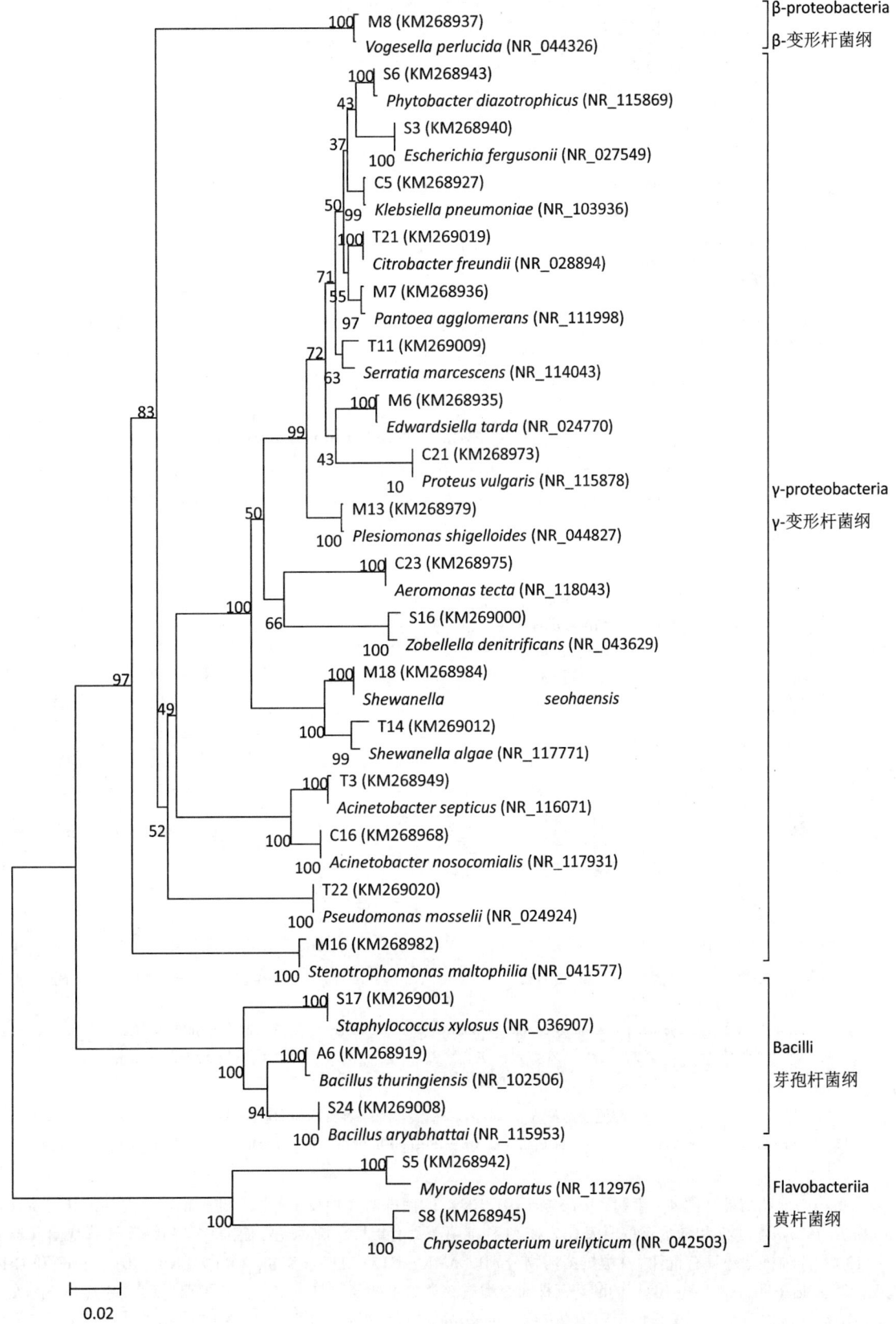

图 1　鳗鲡及其养殖水体中耐药细菌 16S rDNA 系统发育树

Fig. 1　Phylogenetic tree of 16S rDNA of drug-resistant bacteria isolated from eel and the pond water

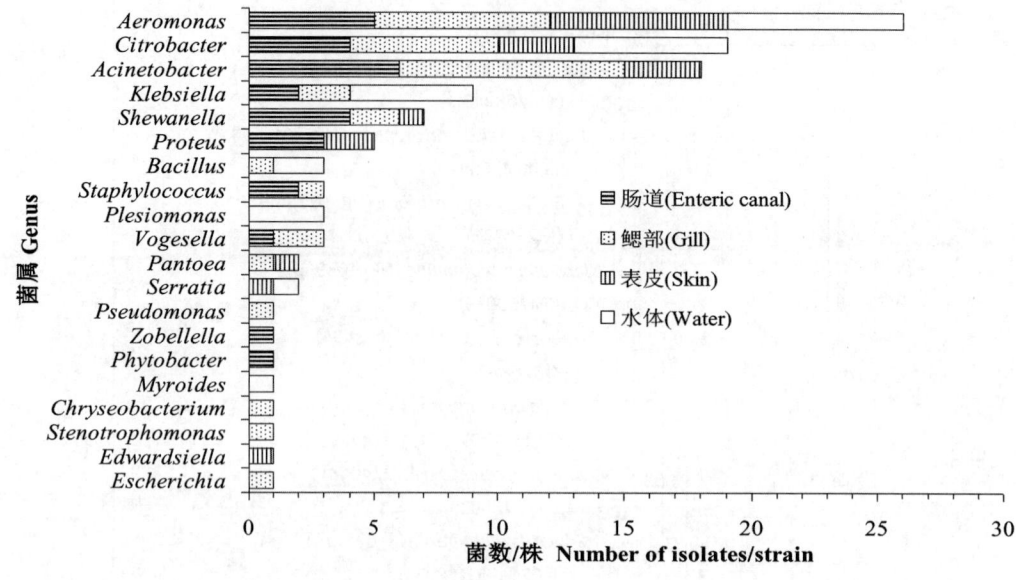

图 2　鳗鲡的鳃、肠道、表皮及其养殖水体中耐药细菌菌属分布

Fig. 2　Distribution of bacteria isolated from the gill, enteric canal and skin of Anguilla and the pond water

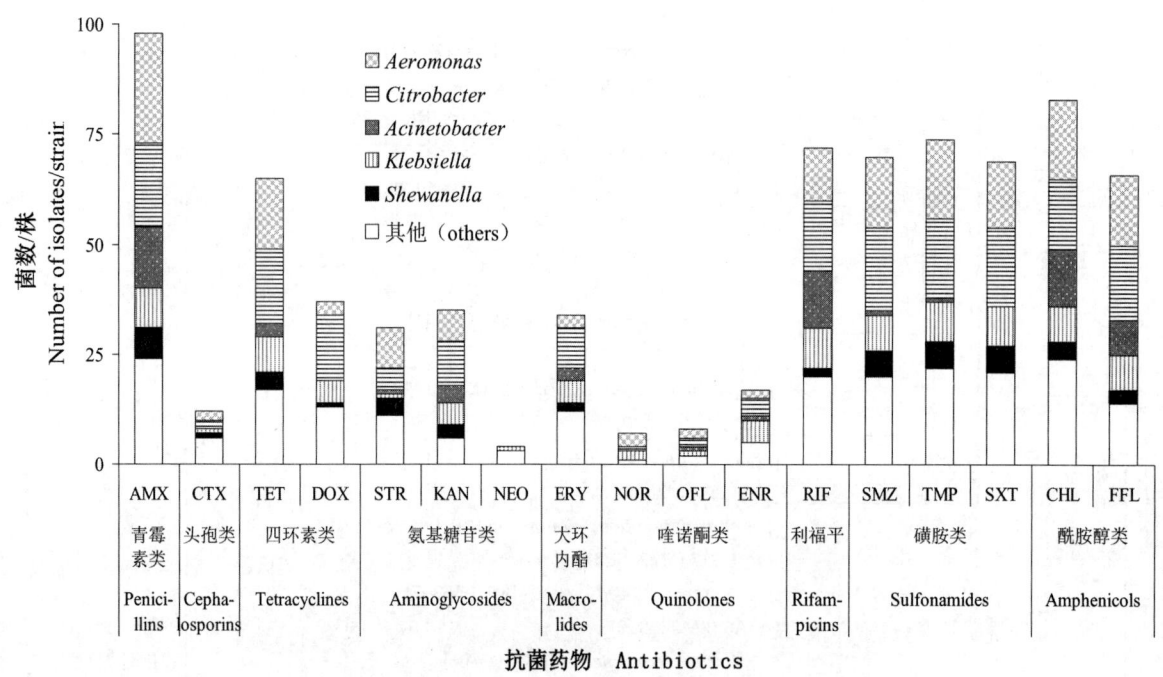

图 3　鳗鲡及其养殖水体分离耐药菌株对 17 种抗菌药的耐药率

Fig. 3　Frequencies of resistance to 17 antimicrobials of drug-resistant bacteria strains isolated from eel and the pond water

各菌株耐药谱统计结果（图 4）表明，93.5%（101/108）的菌株对 3 种或 3 种以上的药物具有抗性，其中费氏不动杆菌（*Citrobacter freundii*）T23 和肺炎克雷伯菌 C12 对 13 种抗菌药具有抗性，肺炎克雷伯菌 C11 和维氏气单胞菌（*Aeromonas veronii*）C13 对 14 种抗菌药具有抗性。9 类抗菌药物分别以 AMX、CTX、TET、STR、ERY、ENR、RIF、SMZ 和 CHL 等药物为代表计算，则有 86.1%（93/108）的耐药菌株对 3 类或 3 类以上的药物具有抗性，其中弗氏埃希菌（*Escherichia fergusonii*）S3、肺炎克雷伯菌 C11 和摩氏假单胞菌（*Pseudomonas mosselii*）T22 对 8 类抗菌药物具有抗性。进一步计算鳗鲡的肠道、表皮、鳃部及水样分离菌株的多重耐药指数，其数值分别为 0.40、0.41、0.42 和 0.47，结果也表明耐药程度较为严重，尤以水样为最。若以种属分别计算，分离菌株数最多的气单胞菌属、柠檬酸杆菌属、不动杆菌属、克雷伯菌属和希瓦菌属的多重耐药指数分别是 0.38、0.58、0.21、0.61 和 0.41，显示柠檬酸杆菌属和克雷伯菌属的耐药水平最高，而不动杆

菌属则相对较低。

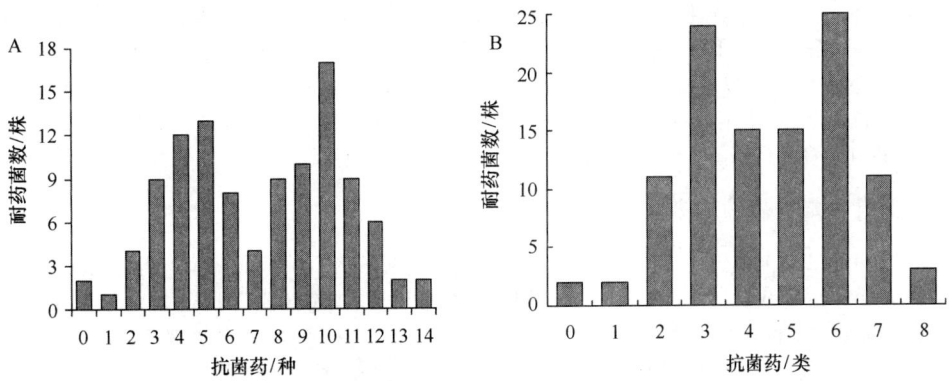

图 4 108 株细菌对 17 种（A）和 9 类（B）抗菌药的多重耐药性

Fig. 4 Multiple resistance to 17 varieties（A）and 9 classes（B）of antimicrobial for 108 bacteria strains

4 讨论

4.1 正常菌群与耐药菌群的种属分布

近年来，国内外学者对多种水产动物或其养殖水体的菌群组成进行了多方面的研究。赵庆新对鲤科鱼肠道菌群的分析表明，草鱼、白鲢、团头鲂和鲤鱼的肠道主要分布有哈夫尼亚菌属（*Hafnia*）、致病杆菌属（*Xenorhabdus*）、气单胞菌属、柠檬酸菌属、假单胞菌属、链球菌属（*Streptococcus*）、葡萄球菌属（*Staphylococcus*）等菌属。Salgado-Miranda 等分析 65 个养殖场虹鳟鱼不同部位菌群，共分离到 371 株菌，主要分布菌属有气单胞菌属、爱德华菌属（*Edwardsiella*）、肠杆菌属（*Enterobacter*）、埃希菌属（*Escherichia*）、克雷伯菌属、邻单胞菌属（*Plesiomonas*）、假单胞菌属和耶尔森菌属（*Yersinia*）。周金敏等报道黄颡鱼肠道和养殖水体菌群主要为气单胞菌属、棒杆菌属（*Corynebacterium*）、微球菌属（*Micrococcus*）、芽孢杆菌属（*Bacillus*）、葡萄球菌属、不动杆菌属、莫拉氏菌属（*Moraxella*）、黄杆菌属（*Flavobacterium*）、肠杆菌属和弧菌属（*Vibrio*）等。在这些水产动物及其养殖水体的菌群研究中，许多细菌种属在本研究所分离到的耐药细菌中也多有涉及。在鳗鲡菌群研究方面，Joh 等分析了日本鳗鲡不同体位（脾、肾、肝和鳃）及其水样菌群的组成，发现气单胞菌属、柠檬酸杆菌属分离到的频率最高。Ugur 等分离欧洲鳗鲡表皮细菌，发现其主要菌属是假单胞菌属、不动杆菌属和寡养单胞菌属（*Stenotrophomonas*）。而本研究中经抗菌药平板筛选得到的 108 株耐药菌株也以气单胞菌属、柠檬酸杆菌属和不动杆菌属为主，并包括假单胞菌属和寡养单胞菌属等共 20 个属，所得耐药菌株种属组成与正常菌群有极大相似之处，这表明耐药细菌的种属分布与正常菌群有较强的相关性。

4.2 生态菌群的多重耐药性分析

多重耐药菌株在分析样品中所占的比例，也就是多重耐药率，常用于耐药水平的分析。目前有较多报道在分析多重耐药性时，常依据是否对 3 种（含）以上抗菌药耐药来计算。如写腊月等研究的海水养殖源弧菌对 3 种（含）以上药物耐药的比例为 80.44%；王瑞旋等分离到的水产致病菌株该比例为 68.5%；Nguyen 等报道的假单胞菌（96.6%）和气单胞菌（61.9%）的多重耐药性比例亦据此分析。但是，根据严格定义，多重耐药性应是指某株细菌对 3 类（而不是 3 种）或 3 类以上的抗菌药物同时耐药，同一类化学构型的各种抗菌药存在交叉耐药的较大可能性，因此不应重复计算。Vincenti 等在研究医院分离菌株对 9 类 14 种抗菌药物的耐药性时，即以类为标准计算得到多重耐药率为 38.9%。抗菌药中，青霉素类和头孢菌素类多数情况下被合并在 β-内酰胺环类药物中，而实际上虽然它们都具有 β-内酰胺结构，但头孢菌素类的母核是 7-氨基头孢烷酸（7-ACA），而青霉素的母核则是 6-氨基青霉烷酸（6-APA），这一结构上的差异使头孢菌素能耐受青霉素酶，这使得它们在抗耐药菌的能力上表现迥异，文中结果（图 3）也验证了这一点。因此在计算多重耐药菌株数时，二者应作为两类药物进行判别。本研究中对 3 种或 3 种以上抗菌药物耐药的菌株为 93.5%，而对 3 类或 3 类以上的抗菌药物的多重耐药率（86.1%）则要低的多。

多重耐药指数也可用于评估多重耐药性风险，张博利用多重耐药指数大小分析不同采样环境的耐药性分布规律；王瑞旋等研究显示企鹅珍珠贝肠道细菌的多重耐药指数较水体低；Ghosh 等研究发现鱼的鳃部耐药率明显高于肠道细菌；Nguyen 等的研究则表明鲶鱼养殖场分离的假单胞菌的多重耐药指数（0.457）明显高于气单胞菌（0.293）。本文多重耐药指数的计算结果显示，水体菌群（0.47）的耐药性水平高于鳗鲡各组织（0.40~0.42），克雷伯菌属（0.61）和柠檬酸杆菌属（0.58）耐药水平高于希瓦菌属（0.41）、气单胞菌属（0.38）和不动杆菌属（0.21）。

许多学者认为耐药水平较高的细菌样品中受抗生素的污染可能较高，Krumperman 在其论文中指出多重耐药指数小于或等于 0.2 时样品中抗菌药物很少或从不使用抗菌药物，而当其高于 0.2 则说明其是常用抗菌药物的高风险污染来源。而本文鳗鲡养殖水样菌群的多重耐药指数（0.47）虽然较高，但在 5 份水样中，除了均检测到痕量的 CHL（7~22 ng/L）和 FFL（3~1704 ng/L），以及个别水样检测到 TMP、SMZ 和 ERY（1~12 ng/L），另外 11 种药物均未检出。这表明，抗菌药污染高的样品中可能得到更多的耐药菌，但抗菌药浓度低的样品中菌群的耐药水平不一定低，二者之间没有必然的关系。

4.3　生态菌群对特定药物的耐药率

水产养殖生态样品中分离得到的细菌对各种抗菌药物的耐药性情况虽然不尽相同，但有很多共通之处。最重要的特征之一是水环境分离细菌多为革兰氏阴性（G⁻）细菌，这使得它们对某些主要针对革兰氏阳性（G⁺）细菌的抗菌药物具有较强的抗药性，如 AMX 的作用机制是针对 G⁺细菌细胞壁的肽聚糖，而 G⁻细菌细胞壁的肽聚糖含量很低，因此细胞壁不易受损。本研究分离得到数量最多的气单胞菌属（26 株），在肠道、鳃、体表和水样中均匀分布，其对 AMX 高度耐药（耐药率大于 90%），对 TET 和磺胺类（SMZ、TMP、SXT），以及酰胺醇类（CHL、FFL）药物较耐药（60%~80%），而对头孢类（CTX）、NEO 以及氟喹诺酮类（NOR、OFL、ENR）药物耐药率低（小于 20%）。Nguyen 等分离得到的 92 株气单胞菌对 β-内酰胺类、氨基糖苷类、氟喹诺酮类和磺胺类等药物的耐药率与本研究结果相近，但其对 TET（34.2%）和 CHL（31.5%）的耐药率则要低的多。李绍戊等对 28 株鱼源嗜水气单胞菌的耐药性分析结果也大体相似，但其对四环素类、氨基糖苷类、喹诺酮类药物的耐药率比本研究高出 8.3%~26.7%，而对大环内酯类药物的耐药率（89.3%）更是高出 77.6%。水产养殖生态样品所分离细菌对特定药物的耐药率大体趋同，主要是因为水体的常见菌属比较相似，如气单胞菌、假单胞菌、不动杆菌和肠杆菌等。而不同来源样品的耐药率差异则可能与地区水文、宿主种、流行菌株和抗菌药物的使用等有关，因此研究特定养殖区域的耐药性情况，才能更全面的掌握该区域的耐药流行情况，提出更有针对性的用药指导。就本研究结果而言，取样区域的鳗鲡及养殖水体普遍存在多重耐药菌株（其中气单胞菌 C13 菌株耐 14 种药物），各养殖场样品的多重耐药指数（0.27~0.58）偏高，对此应引起足够的重视；而养殖生态菌群对某些水产用药物如诺氟沙星、新霉素、多西环素等耐药率低，可将其作为水产动物细菌性疾病治疗的首选药物。

参考文献

[1] Hu F. P., Zhu D. M., Wang F., et al. CHINET 2013 surveillance of bacterial resistance in China [J], Chinese Journal of Infection Chemotherapy, 2014, 14 (5): 365-374. ［胡付品，朱德妹，汪复，等. 2013 年中国 CHINET 细菌耐药性监测. 中国感染与化疗杂志，2014, 14 (5): 365-374.］

[2] Cai J. P., Zhou Y. P., Cai C. H. . Studies on antibiotic resistance of different groups of marine bacteria isolated from abalone (Haliotis diversicolor) farming waters and their digestion guts [J]. Microbiology China, 2004, 31 (6): 48-52. ［蔡俊鹏，周毅频，蔡创华. 九孔鲍养殖水体及其肠道不同菌群抗药性研究. 微生物学通报，2004, 31 (6): 48-52.］

[3] Li S. W., Wang D., Liu H. B., et al. Molecular characterization of integron-gene cassettes in multi-drug resistant Aeromonas hydrophila from fish [J]. Journal of Fishery Sciences of China, 2013, 20 (5): 1015-1022. ［李绍戊，王荻，刘红柏，等. 鱼源嗜水气单胞菌多重耐药菌株整合子的分子特征. 中国水产科学，2013, 20 (5): 1015-1022.］

[4] Nguyen H. N. K., Van T. T. H., Nguyen H. T., et al. Molecular characterization of antibiotic resistance in Pseudomonas and Aeromonas isolates from catfish of the Mekong Delta, Vietnam [J]. Veterinary Microbiology, 2014, 171: 397-405.

[5] Marshall B. M., Levy S. B. . Food animals and antimicrobials: impacts on human health [J]. Clinical Microbiology Reviews, 2011, 24 (4): 718-733.

[6] Cao L., Naylor R., Henriksson P., et al. China's aquaculture and the world's wild fisheries [J]. Science, 2015, 347 (6218): 133-135.

[7] Qiu J. X., Fang Z. S., Xie J., et al. Research advances on artificial breeding of Japanese eel (Anguilla japonica) [J]. Journal of Anhui Agriculture Sciences, 2013, 41 (34): 13269-13272. ［丘继新，方彰胜，谢骏，等. 日本鳗鲡人工繁殖的研究进展. 安徽农业科学，2013, 41 (34): 13269-13272.］

[8] Yang Q. H., Isolation and identification of the pathogenic Aeromonas spp. from farmed eels [D]. Xiamen: Dissertation for the Master Degree of Jimei University, 2012. ［杨求华. 养殖鳗鲡致病性气单胞菌的分离与鉴定. 厦门：集美大学硕士学位论文，2012.］

[9] Wu L. . Spot rapid diagnosis and medicine cure of pathogenic bacteria from the cultivated eels [D]. Xiamen: Dissertation for the Master Degree of Jimei University, 2007. ［吴亮. 鳗鲡病原菌的现场快速检测与药物防治. 厦门：集美大学硕士学位论文，2007.］

[10] Zheng F. Y., Shi C. B., Pan H. J., et al. Isolation and identification of pathogen from diseased Anguilla anguilla [J]. Journal of Shanghai Fisheries University, 2005, 14 (3): 242-247. ［郑芳艳，石存斌，潘厚军，等. 鳗鲡溃烂病病原的分离与鉴定. 上海水产大学学报，2005, 14 (3): 242-247.］

[11] Lo D. Y., Lee Y. J., Wang J. H., et al. Antimicrobial susceptibility and genetic characterisation of oxytetracycline-resistant Edwardsiella tarda isolated from diseased eels [J]. The Veterinary Record, 2014, 175 (8): 203-203.

[12] Lei Y., Xiao Y. . Identification and antibiotic sensitivity experiment of Aeromonas hydrophila isolated from skin ulcer of artificial breeding Anguilla anguilla [J] . Journal of Fujian Fisheries, 2012, 34（3）: 183-188. [雷燕, 肖洋. 养殖欧洲鳗鲡体表溃疡病病原菌的分离、鉴定及药敏试验. 福建水产, 2012, 34（3）: 183-188.]

[13] Tan A. P., Deng Y. T., Jiang L., et al. Isolation and identification and identificaton of a multiple-drug resistant Klebsiella pneumoniae from Anguilla marmorata [J] . Acta Hydrobiologica Sinica, 2013, 37（4）: 744-750. [谭爱萍, 邓玉婷, 姜兰, 等. 一株多重耐药鳗源肺炎克雷伯菌的分离鉴定. 水生生物学报, 2013, 37（4）: 744-750.]

[14] Bosshard P. P., Santini Y., Grüter D., et al. Bacterial diversity and community composition in the chemocline of the meromictic alpine Lake Cadagno as revealed by 16S rDNA analysis [J] . FEMS Microbiology Ecology, 2000, 31（2）: 173-182.

[15] Zhan J., Yu X., Zhong Y., et al. Generic and rapid determination of veterinary drug residues and other contaminants in raw milk by ultra performance liquid chromatography-tandem mass spectrometry [J] . Journal of Chromatography B, 2012, 906: 48-57.

[16] Clinical and Laboratory Standards Institute（CLSI）. M100-S19 Performance standards for antimicrobial susceptibility testing: Nineteenth informational supplement [S] . USA: Clinical and Laboratory Standards Institute, 2009.

[17] Krumperman P. H. . Multiple antibiotic resistance indexing of Escherichia coli to identify high-risk sources of fecal contamination of foods [J] . Applied and Environmental Microbiology, 1983, 46（1）: 165-170.

[18] Zhao Q. X. . An analysis of intestinal microflora of Cyprinidae [J] . Microbiology, 2001, 21（2）: 18-20. [赵庆新. 鲤科（Cyprinidae）鱼肠道菌群分析. 微生物学杂志, 2001, 21（2）: 18-20.]

[19] Salgado-Miranda C., Palomares E., Jurado M., et al. Isolation and distribution of bacterial flora in farmed rainbow trout from Mexico [J] . Journal of Aquatic Animal Health, 2010, 22（4）: 244-247.

[20] Zhou J. M., Wu Z. X., Zeng L. B., et al, Microflora in digestive tract of yellow catfish（Pseudobagrus fulvidraco）and in the water [J] . Journal of Huazhong Agricultural University, 2010, 29（5）: 613-617. [周金敏, 吴志新, 曾令兵, 等. 黄颡鱼肠道及养殖水体中菌群的分析, 华中农业大学学报, 2010, 29（5）: 613-617.]

[21] Joh S. J., Ahn E. H., Lee H. J., et al. Bacterial pathogens and flora isolated from farm-cultured eels（Anguilla japonica）and their environmental waters in Korean eel farms [J] . Veterinary Microbiology, 2013, 163（1）: 190-195.

[22] Ugur A., Yilmaz F., Sahin N., et al. Microflora on the skin of European eel（Anguilla Anguilla L., 1758）sampled from Creek Yuvarlakçay, Turkey [J] . The Israeli Journal of Aquaculture-Bamidgeh, 2002, 54（2）, 89-94.

[23] Xie L. Y., Hu L. L., Fang W. H., et al. Investigation and analysis of drug resistance of Vibrios from mariculture source [J] . Marine Fisheries, 2012, 33（4）: 442-446. [写腊月, 胡琳琳, 房文红, 等. 海水养殖源弧菌耐药性调查与分析. 海洋渔业, 2012, 33（4）: 442-446.]

[24] Wang R. X., Geng Y. J., Wang J. Y., et al. Antibiotic resistant genes in aquacultural bacteria [J] . Marine Environmental Science, 2012, 21（3）: 323-328. [王瑞旋, 耿玉静, 王江勇, 等. 水产致病菌耐药基因的研究. 海洋环境科学, 2012, 21（3）: 323-328.]

[25] Magiorakos A. P., Srinivasan A., Carey R. B., et al. Multidrug-resistant, extensively drug-resistant and pandrug-resistant bacteria: an international expert proposal for interim standard definitions for acquired resistance [J] . Clinical Microbiology and Infection, 2012, 18（3）: 268-281.

[26] Vincenti S., Quaranta G., De Meo C., et al. Non-fermentative gram-negative bacteria in hospital tap water and water used for haemodialysis and bronchoscope flushing: prevalence and distribution of antibiotic resistant strains [J] . Science of the Total Environment, 2014, 499: 47-54.

[27] Riaz S., Faisal M., Hasnain S. . Antibiotic susceptibility pattern and multiple antibiotic resistances（MAR）calculation of extended spectrum β-lactamase（ESBL）producing Escherichia coli and Klebsiella species in Pakistan [J] . African Journal of Biotechnology, 2013, 10（33）: 6325-6331.

[28] Zhang B. . Detection of antibiotic resistance and various resistance genes in Enterobacteriaceae isolated from the urban sewage [D] . Harbin: Dissertation for the Master Degree of Northeast Agriculture University, 2013. [张博. 城市污水中肠杆菌的抗生素耐药性和多种耐药基因的检测. 哈尔滨: 东北农业大学硕士学位论文, 2013.]

[29] Wang R. X., Lin Y. S., Guo Z. X., et al. Study on antibiotic-resistance of heterotophic bacteria from farming water and intestine of Pearl oyster（Pteria penguin）in Lingshui [J] . Journal of Tropical Oceanography, 2013, 32（6）: 96-100. [王瑞旋, 林韵锶, 郭志勋, 等. 海南陵水企鹅珍珠贝肠道及其养殖水体中异养细菌耐药性研究. 热带海洋学报, 2013, 32（6）: 96-100.]

[30] Ghosh K., Mandal S. . Antibiotic resistant bacteria in consumable fishes from Digha coast, West Bengal, India [J] . Proceedings of the Zoological Society, 2010, 63（1）: 13-20.

该文发表于《上海海洋大学学报》【2013，22（2）：219-224】

广东省水产动物源气单胞菌对抗菌药物的耐药分析

Antimicrobial susceptibilities of Aeromonas strains isolated from variousaquatic animals in Guangdong Province

吴雅丽[1,2]，邓玉婷[1]，姜兰[1]，谭爱萍[1]，薛慧娟[1,2]，王伟利[1]，罗理[1]，赵飞[1]

（1. 中国水产科学研究院珠江水产研究所 农业部渔药创制重点实验室，广东 广州 510380；

2. 上海海洋大学 水产与生命学院，上海 201306）

摘要： 为了解广东地区水产动物源气单胞菌的耐药情况，采用 K-B 纸片法测定了 112 株 1995—2012 年来源于不同种类患病水产动物的气单胞菌对 20 种抗菌药的耐药性，数据用 WHONET 5.6 耐药监测软件分析。结果显示，气单胞菌对氨苄西林和头孢噻吩的耐药率分别高达 85.7% 和 79.5%，其次对利福平、阿莫西林/克拉维酸、链霉素、萘啶酸、磺胺类、头孢西丁、四环素和磺胺甲基异恶唑/甲氧苄啶的耐药率分别达 57.1%、51.8%、49.1%、44.6%、31.2%、28.6%、28.6% 和 21.4%；对氟喹诺酮类（氧氟沙星、诺氟沙星、环丙沙星）、头孢噻肟、头孢曲松、亚胺培南、阿米卡星、呋喃妥因、氯霉素和多西环素相对敏感。比较不同来源气单胞菌的耐药情况，结果显示爬行、两栖动物和观赏鱼来源的分离菌株对氟喹诺酮类、头孢类等药物的耐药率比养殖鱼、虾类的高；气单胞菌对常用抗菌药呈现不同程度的耐药，不同来源的气单胞菌的耐药率亦不尽相同。水产动物源气单胞菌存在多重耐药菌株应引起重视，今后在气单胞菌疾病防治方面要慎重用药，并且有必要开展水产动物源的细菌耐药性监测，以指导水产养殖合理用药。

关键词： 气单胞菌；耐药性；水产动物源；检测

该文发表于《江西农业大学学报》【2015；37（2）：339-345】

三北地区冷水鱼常见病原菌的分布及耐药分析

Distribution and antimicrobial susceptibility analysis of common pathogenic bacteria isolated from cold-water fish in three north areas of China

连浩淼[1]，李绍戊[2]，张辉[1]，卢彤岩[2]

（1. 东北农业大学动物科技学院，哈尔滨 150030；2. 中国水产科学研究院黑龙江水产研究所，哈尔滨 150070）

摘要： 2013 年 5 月份到 2013 年 8 月期间对三北地区主要冷水鱼养殖的虹鳟、大西洋鲑、银鲑、西伯利亚鲟等发病鱼进行病原菌的分离鉴定及药敏试验。流行病学调查结果表明不同地区发病鱼表现出不同的临床症状，西北地区青海的虹鳟主要为烂鳃、出血等；东北地区辽宁的银鲑为体色发黑、内脏有出血点等；东北地区黑龙江的大西洋鲑主要表现为肠炎、烂鳃等；华北地区的北京西伯利亚鲟表现为体表严重出血、吻端周围红肿等。无菌条件下从患病鱼内部器官取样划线培养细菌，并进行氧化酶反应、革兰染色、API 20 生化鉴定、16S rRNA 保守序列分子鉴定等研究。结果表明从患病的辽宁银鲑中分离到 12 株杀鲑气单胞菌、1 株不动杆菌；从牡丹江大西洋鲑中分离 14 株杀鲑气单胞菌、6 株温和气单胞菌、2 株嗜水气单胞菌、6 株不动杆菌、2 株黄杆菌；青海虹鳟鱼中分离到 9 株温和气单胞菌、6 株维氏气单胞菌、6 株不动杆菌、2 株黄杆菌、2 株鲁氏耶尔森菌；从北京房山西伯利亚鲟鱼中分离到 3 株海豚链球菌、3 株不动杆菌、10 株停乳链球菌、2 株维氏气单胞菌、2 株嗜水气单胞菌、1 株黄杆菌。药敏试验结果表明：在 89 株菌中有 92.13% 和 96.63% 的菌株对氨苄西林和阿莫西林具有很强的耐药性，有 100%、93.26%、91.10%、92.13%、91.01% 和 94.38% 的菌株分别对左氧氟沙星、氧氟沙星、庆大霉素、卡那霉素、环丙沙星和恩诺沙星敏感，其中对左氧氟沙星无耐药菌株；另外，不同地区相同病原菌对抗生素的敏感程度也有一定程度的差异。本研究期望为冷水鱼细菌性疾病的防控提供参考。

关键词：冷水鱼；细菌性疾病；耐药性

三、病原菌对药物的耐药性

嗜水气单胞菌（A. hydrophila）基因型与耐药性存在一定的相关性，利用 ERIC-PCR 法可以对致病性嗜水气单胞菌的耐药模式进行基因分型（肖丹等，2011）。通过对气单胞菌耐药性克隆传播的追踪发现：复合水产养殖环境有可能有利于耐药菌从畜禽向水产养殖环境转移（黄玉萍等，2014）。广东地区龟鳖源气单胞菌对多种抗菌药物耐药的多重耐药现象普遍存在，且喹诺酮类耐药（PMQR）可能会在水产临床上更加快速而广泛地传播（谭爱萍等，2014）。从患病花鳗鲡（Anguilla marmorata）分离的肺炎克雷伯菌（Klebsiella pneumoniae）也表现出对氨苄西林、阿莫西林、磺胺甲基异恶唑等 17 种药物表现出多重耐药性（谭爱萍等，2013）。

该文发表于《中国水产科学》【2011，18（5）：1092-1099】

淡水养殖动物致病性嗜水气单胞菌 ERIC-PCR 分型与耐药性

ERIC-PCR genotyping and drug resistant analysis of pathogenic *Aeromonas hydrophila* from freshwater animals

肖丹，曹海鹏，胡鲲，杨先乐

（上海海洋大学 水产与生命学院，国家水生动物病原库，上海 201306）

摘要： 以质控菌株 ATCC7966 为对照，采用 ERIC-PCR 方法对 49 株分别采集于中国浙江、江苏、江西、广东、湖北、上海等主要淡水养殖区的致病性嗜水气单胞菌（*Aeromonas hydrophila*）进行了基因分型，并分析了嗜水气单胞菌对 12 种抗生素的耐药模式，探讨了嗜水气单胞菌的基因型与区域性分布及其耐药性的关联性。实验结果表明，50 株受试嗜水气单胞菌菌株可分为 Ⅰ、Ⅱ、Ⅲ、Ⅳ、Ⅴ、Ⅵ、Ⅶ、Ⅷ、Ⅸ、Ⅹ、Ⅺ、Ⅻ这 12 个基因型，其中Ⅷ型菌株最多，Ⅲ型和 X 型菌株最少。此外，50 株受试菌株对氨苄青霉素（AMP）的耐药率高达 100%，对头孢氨苄（CX）的耐药率高达 98%，94% 以上的菌株对氨基糖苷类及喹诺酮类抗生素未产生耐药性。受试菌株对 12 种抗生素呈现出 AMP、CX/AMP、CX/AMP/POL（多黏菌素）、CX/AMP/SMZ（复方新诺明）、CX/AMP/NM（新霉素）、CX/AMP/SMZ/POL、CX/AMP/SMZ/POL/GM（庆大霉素）、CX/AMP/SMZ/LOM（洛美沙星）/ENR（恩诺沙星）、CX/AMP/SMZ/LOM/ENR/OF（氧氟沙）/LEV（左氟沙星）这 9 种不同的耐药模式，其中Ⅻ型菌株的耐药谱均为 CX/AMP/POL 型，Ⅱ型和Ⅺ型菌株的耐药谱多为 CX/AMP/SMZ 型，Ⅳ型、Ⅵ型和Ⅷ型菌株的耐药谱多为 CX/AMP 型，X 型菌株的耐药谱为 CX/AMP/NM 型。据此推测，嗜水气单胞菌基因型与耐药性可能存在一定的相关性。

关键词： 嗜水气单胞菌；ERIC-PCR；耐药性

嗜水气单胞菌（*Aeromonas hydrophila*）是淡水养殖动物重要的病原菌之一，可引起淡水鱼类细菌性败血症、鳖 "红底板" 病等多种病害[1-2]，危害银鲫（*Carassius auratus gibelio*）、草鱼（*Ctenoph - aryngodon idella*）、花白鲢（*Hypophthalmichthys molitrix*）、团头鲂（*Megalobrama amblycephala*）、鳜（*Siniperca chuatsi*）、中华鳖（*Pelodiscus sinensis*）、三角帆蚌（*Hyriopsis cumingii*）等多类经济养殖品种。然而，近年来，随着抗生素的广泛使用，嗜水气单胞菌耐药问题日趋严重[3]，增加了嗜水气单胞菌病的防治难度。因此，开展水生动物嗜水气单胞菌的流行病学调查及其耐药性监测，对有效防治该病具有重要意义。

目前，嗜水气单胞菌流行病学调查的传统方法多以血清分型、生化分型等表型分析方法为主[4-5]，这些传统的分型方法常常存在误检率高等缺陷。分子分型是一种新兴的流行病学调查手段，它不仅克服了血清分型等传统方法的不足，还能结合菌株的耐药特性，从遗传学角度对耐药菌株的流行规律进行探讨[6]，以丰富目前细菌耐药机制的研究理论[7-8]。其中，ERIC-PCR 法作为一种半随机扩增多态 DNA 法，与其他分子分型方法相比，具有分型率较高、可比性强、准确性好、可重复性好、易于实现标准化、无需特殊仪器等优点，被广泛应用于大肠杆菌（*Escherichia coli*）、副猪嗜血杆菌（*Haemophilus parasuis*）、沙门菌（*Salmonella*）、

绿脓杆菌（*Pseudomonas aeruginosa*）、副溶血性弧菌（*Vibrio parahemolyticus*）等[9-13]人畜共患致病菌的分子流行病学调

查。鉴于此，本实验以质控菌株 ATCC7966 为对照，采用 ERIC-PCR 法对 49 株分别采集于中国浙江、江苏、江西、广东、湖北、上海等 6 个主要淡水养殖区的致病性嗜水气单胞菌进行了基因分型，分析了各基因型嗜水气单胞菌的耐药模式，并探讨了嗜水气单胞菌的基因型与区域性分布及其耐药性的关联性，以期为中国水生动物嗜水气单胞菌病的有效防控提供理论指导。

1 材料与方法

1.1 实验菌株

49 株嗜水气单胞菌分别分离于江苏、上海、浙江、广东、湖北、江西等地区患嗜水气单胞菌病的淡水养殖水生动物病变组织（表1）；质控菌株 ATCC7966 为国家水生动物病原库提供。

表1　菌株来源及背景

Tab. 1　Source and background of the strains

地域 region	来源 source	菌株数 number of strains	菌株名称（分离时间/分离地点） name（isolation time/location）of stains
江苏 Jiangsu	淡水鱼 freshwater fish	15	JS01（2007/武进），JS02（2007/武进），JS03（2007/武进），JS04（2007/武进），JS05（2007/武进），JS06（2007/武进），JS07（2007/武进），JS08（2007/南通），JS09（2007/南通），JS10（2007/南通），JS12（2009/盐城），JS13（2009/盐城），JS14（2009/盐城），JS15（2009/盐城），JS16（2009/盐城）
	中华鳖 *Pelodiscus sinensis*	1	JS11（2006 南通）
广东 Guangdong	淡水鱼 freshwater fish	9	GD01（2004/广州），GD02（1992/广州），GD03（2009/广州），GD04（2008/广州），GD05（2004/佛山），GD06（2004/佛山），GD07（2008/佛山），GD08（2008/中山），GD09（2009/中山）
	淡水鱼 freshwater fish	7	SH01（2004/金山），SH02（2005/金山），SH03（2006/青浦），SH04（1991/金山），SH06（2002/青浦），SH09（2010/南汇），SH10（2010/南汇）
上海 Shanghai	中华鳖 *Pelodiscus sinensis*	1	SH05（2003/青浦）
	三角帆蚌 *Hyriopsis cumingii*	2	SH07（2010/南汇），SH08（2010/南汇）
浙江 Zhejiang	淡水鱼 freshwater fish	5	ZJ01（2009/湖州），ZJ02（2009/杭州），ZJ03（2009/杭州），ZJ04（2009/宁波），ZJ05（2009/宁波）
江西 Jiangxi	淡水鱼 freshwater fish	4	JX01（2008 彭泽），JX02（2009 彭泽），JX03（2008 南昌），JX04（2009 南昌）
湖北 Hubei	淡水鱼 freshwater fish	5	HB01（2008/洪湖），HB02（2009/洪湖），HB03（2009/丹江口），HB04（2009/丹江口），HB05（2009/丹江口）

1.2 菌株鉴定

根据国家标准《致病性嗜水气单胞菌检测方法（GB/T 18652-2002）》[14]，采用细菌微量生化鉴定管（杭州天和）对 49 株嗜水气单胞菌和质控菌株 ATCC7966 进行生理生化鉴定，确认为致病性嗜水气单胞菌后，再进行 ERIC-PCR 分型及药敏试验。

1.3　嗜水气单胞菌基因组 DNA 的提取

将各嗜水气单胞菌菌株分别接种于 LB 液体培养基中，28℃恒温培养过夜，用细菌基因组 DNA 提取试剂盒（天根生化）分别提取各菌株基因组 DNA。

1.4　嗜水气单胞菌的 ERIC-PCR 分型

1.4.1　引物设计

ERIC-PCR 引物参照 Biswajit 等[15]的设计。其中 ERIC 正向引物（F）：5′-ATGTAAGCTCCTGGGGATTCAC-3′，ERIC 反向引物（R）：5′-AAGTAAGTGACTGGGGTGAGCG-3′，由上海捷瑞生物工程有限公司合成。

1.4.2　PCR 方法

PCR 反应采用 20 μL 反应体系，其中含 DNA 模板 2 μL（30 μg/L），ERIC 正、反向引物各 1 μL（10 mmol/L），Mix Buffer（TAKARA）2 μL，Taq 酶（TAKARA）2 U，用超纯水补足至 20 μL。PCR 程序如下：94℃预变性 7 min；94℃ 30 s，52℃ 1 min，65℃ 8 min，共循环 40 次；再 65℃维持 16 min。PCR 扩增产物在 2%琼脂糖凝胶 120V 电压电泳 60 min。

1.4.3　结果分析

应用 Quantity one 4.6 对电泳图谱进行分析。采用"1"和"0"的方式记录在 Excel 中，每个样品的扩增带存在时赋值为"1"，不存在时赋值为"0"。采用 NTSYSpc 2.1 计算软件，得到 Jaccard 的遗传相似性系数矩阵和遗传距离矩阵，采用非加权配对法（UPGMA 法）做遗传分析的树状图。凡相似度大于 0.8 者为同一亚型，小于 0.8 者为不同的基因型[16]。

1.5　药物敏感性实验

将各嗜水气单胞菌菌株制备成 10^5/mL 的菌悬液，均匀涂布于 M-H 培养基平板上后，用无菌镊子将药敏纸片（杭州天和）贴于平板上，于 28℃恒温培养 48 h 后测量抑菌圈直径，并根据 NCCLS[17]的判定标准进行药敏特性分析。

2　结果与分析

2.1　菌株鉴定

实验结果表明，49 株受试菌株和质控菌株 ATCC7966 均为革兰氏阴性杆菌，在 AHM 培养基上接种培养后，AHM 培养基顶部仍为紫色，底部为淡黄色且 49 株受试菌株和质控菌株 ATCC7966 均沿穿刺线呈刷状生长。此外，49 株受试菌株和质控菌株 ATCC7966 均可发酵葡萄糖、阿拉伯糖、蔗糖、七叶苷和水杨素，氧化酶实验、吲哚实验、脱脂奶平板实验结果均为阳性。根据《致病性嗜水气单胞菌检验方法（GB/T 18652-2002）》[14]中对致病性嗜水气单胞菌的判定标准，49 株受试菌株和质控菌株 ATCC7966 均为致病性嗜水气单胞菌。

2.2　嗜水气单胞菌的 ERIC-PCR 分型

实验结果表明，49 株嗜水气单胞菌和质控菌株 ATCC7966 经 ERIC-PCR 扩增，产生了 37 种不同的指纹图，且指纹图谱呈现出多态性（图 1）。PCR 产物为 2~14 条，主带为 2~8 条，大小介于 0.08~3.2 kb 之间。

2.3　嗜水气单胞菌 ERIC-PCR 指纹图谱聚类分析

实验结果表明，50 株嗜水气单胞菌可分为 12 个基因型，分别标记为Ⅰ、Ⅱ、Ⅲ、Ⅳ、Ⅴ、Ⅵ、Ⅶ、Ⅷ、Ⅸ、Ⅹ、Ⅺ及Ⅻ型（图 2）。其中Ⅷ型菌株最多，达 14 株；Ⅶ型和Ⅵ型菌株略少，分别为 8 株和 7 株；Ⅲ型和Ⅹ型菌株最少，分别为 1 株。根据 Hunter 等[18]介绍的计算方法，该分型方法的辨别指数值为 87%。此外，采集自不同地域的致病性嗜水气单胞菌呈现出不同的聚类特征。其中，除菌株 GD08、GD04 以外的广东分离株的基因型均为Ⅷ型，除菌株 JX04 以外的江西分离株均属Ⅵ型，而上海、浙江等地分离株的基因型均呈现多态性；而且不同基因型菌株的分布特点也有差异。例如，Ⅶ型菌株在广东、江苏、上海、浙江均有出现，而Ⅳ型、Ⅴ型、Ⅹ型菌株仅在上海出现（图 3）。

2.4　嗜水气单胞菌的耐药性分析

实验结果表明，50 株嗜水气单胞菌对氨苄青霉素（AMP）完全耐药，对头孢氨苄（CX）、复方新诺明（SMZ）、多黏菌素（POL）不同程度耐药，94%以上的菌株对左氟沙星（LEV）、氟哌酸（NOR）、氧氟沙星（OF）、阿米卡星（AMK）、恩诺沙星（ENR）、庆大霉素（GM）及新霉素（NM）敏感（图 4 和图 5）。此外，50 株嗜水气单胞菌对 12 种抗生素表现 AMP、CX/AMP、CX/AMP/POL、CX/AMP/SMZ、CX/AMP/NM、CX/AMP/SMZ/PO、CX/AMP/SMZ/POL/GM、CX/AMP/SMZ/LOM/ENR、CX/AMP/SMZ/LOM/ENR/OF/LEV 这 9 种耐药谱型。

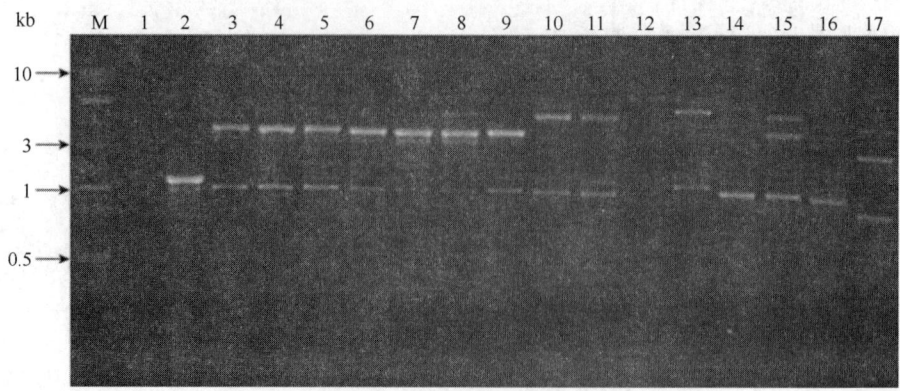

图 1　部分嗜水气单胞菌菌株 ERIC-PCR 指纹图

Fig. 1　ERIC-PCR fingerprints of partial Aeromonas hydrophila strains

M：marker；1. ATCC7966；2. SH01；3. JS01；4. JS02；5. JS03；6. JS04；7. JS05；8. JS06；
9. JS07；10. JS08；11. JS09；12. JS10；13. JX01；14. GD01；15. JX02；16. GD02；17. JS11.

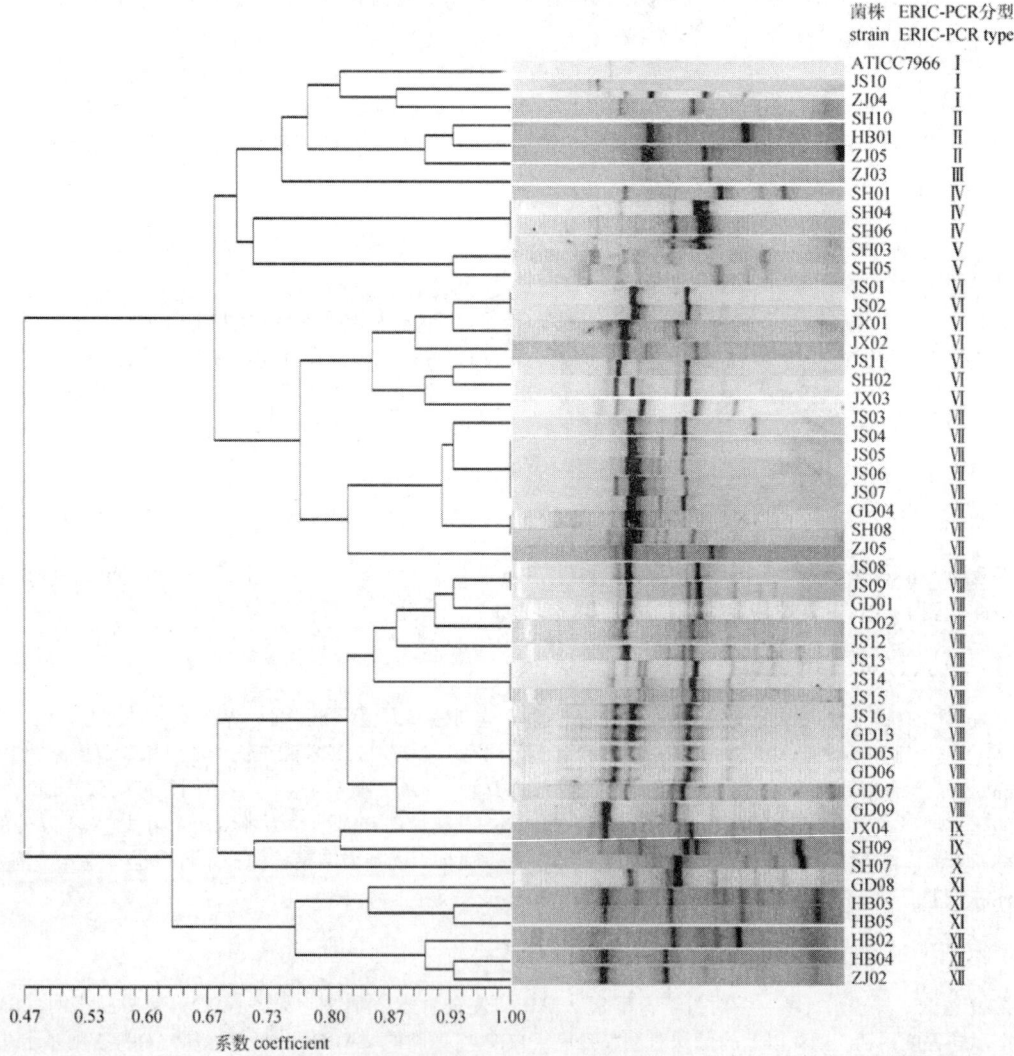

图 2　50 株嗜水气单胞菌指纹图谱聚类分析图

Fig. 2　Results of ERIC-PCR genotyping and UPGMA clustering of fifty strains

（图谱是根据条带在软件中识别处理用数字转换进行聚类分析后生成的，单纯从这图上肯定看不清）

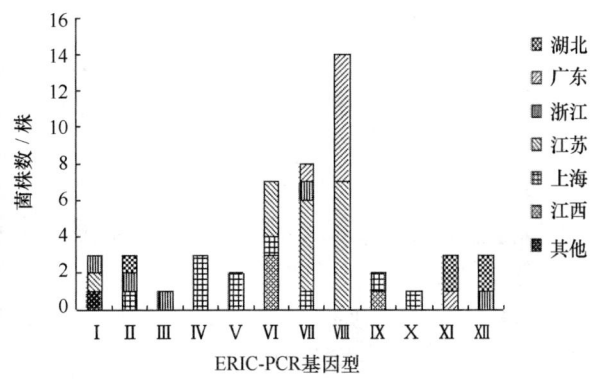

图3　嗜水气单胞菌 ERIC-PCR 基因型与分离地域间的关系

Fig. 3　Relationship between Aeromonas hydrophila strains' sources and ERIC-PCR genotypes

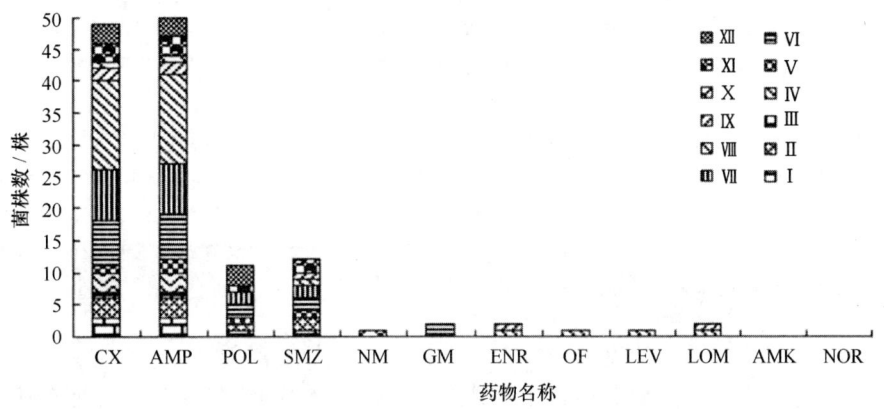

图4　50 株嗜水气单胞菌 ERIC-PCR 基因型与对 12 种抗生素耐药性间的关系

CX：头孢氨苄；AMP：氨苄青霉素；POL：多黏菌素；SMZ：复方新诺明；NM：新霉素；GM：庆大霉素；ENR：恩诺沙星；
OF：氧氟沙星；LEV：左氟沙星；NOR：氟哌酸；AMK：阿米卡星；LOM：洛美沙星.

Fig. 4　Relationship between fifty Aeromonas hydrophila strains' antibiotic susceptibility and ERIC-PCR genotypes

CX：Cephalexin；AMP：Ampicillin；POL：Polymyxin；SMZ：Cotrimoxazole；NM：Neomycin；GM：Gentamicin；ENR：
Enrofloxacin；OF：Ofloxacin；LEV：Levofloxacin；NOR：Norfloxacin；AMK：Amikacin；LOM：Lomefloxacin.

2.5　嗜水气单胞菌 ERIC-PCR 基因型和耐药性之间的关系

50 株嗜水气单胞菌 ERIC-PCR 基因型与其对 12 种抗生素耐药性之间的分布关系结果见图5。实验结果表明，耐 CX、AMP 的嗜水气单胞菌菌株在每种基因型中均有分布，耐 SMZ、POL 的嗜水气单胞菌菌株的基因型呈现多样性。此外，XII 型嗜水气单胞菌菌株的耐药谱均为 CX/AMP/POL，XI、II 型嗜水气单胞菌菌株的耐药谱主要为 CX/AMP/SMZ，IV 型、VI 型、VIII 型嗜水气单胞菌菌株的耐药谱主要为 CX/AMP，X 型嗜水气单胞菌菌株的耐药谱为 CX/AMP/NM 型。10%的菌株对 3 种以上抗生素耐药，它们分布在 V 型、VI 型、VIII 型、IX 型等基因型中。

3　讨论

3.1　关于嗜水气单胞菌 ERIC-PCR 基因分型

不同的嗜水气单胞菌菌株在地理分布、流行病学特征以及致病性、耐药性等方面均存在一定的差异，因此对其进行分子分型研究具有重要的意义[19]。目前，细菌基因分型的方法较多，如脉冲场凝胶电泳法（PFGE）、随机扩增多态性 DNA 法（RAPD）、重复序列 PCR 法（REP-PCR）、肠杆菌重复序列 PCR 法（ERIC-PCR）和限制性片段长度多态性分析法（RFLP）等。其中，PFGE 法准确性最高，但操作步骤繁杂，且需要特殊仪器，在普通实验室难以开展；而 ERIC-PCR 法操作简便，无需特殊仪器，且重复性和区分能力同 PFGE 非常相近，更易于在普通实验室中推广[20]。因此，本实验选用 ERIC-PCR 法对嗜水气单胞菌进行了基因分型。本实验结果表明，受试菌株通过 ERIC-PCR 法能够扩增出 2~14 条带，可

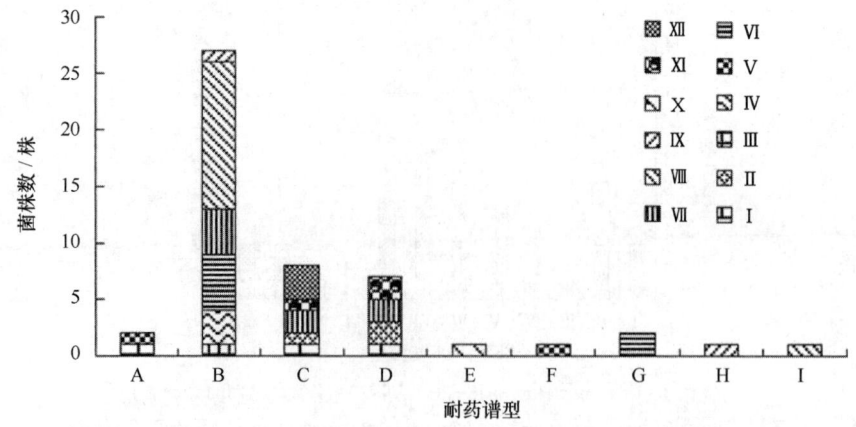

图 5   50 株嗜水气单胞菌菌株的耐药谱型与 ERIC-PCR 基因型关系

Fig. 5   Relationship between fifty Aeromonas hydrophila strains' resistance patterns and ERIC-PCR genotypes

A：AMP；B：CX/AMP；C：CX/AMP/POL；D：CX/AMP/SMZ；E：CX/AMP/NM；F：CX/AMP/SMZ/POL；
G：CX/AMP/SMZ/POL/GM；H：CX/AMP/SMZ/LOM/ENR；I：CX/AMP/SMZ/LOM/ENR/OF/LEV.

分为 12 个基因型，且质控菌株 ATCC7966 的扩增图谱与 Biswajit 等[15] 及 Ma 等[21] 对质控菌株 ATCC7966 的扩增图谱相似，说明本实验选用的基因分型方法适宜，分型能力也较强，分型结果可信。此外，Aguilera-Arreola 等[21] 通过 ERIC-PCR 法证实，背景来源相似的嗜水气单胞菌菌株有聚类趋势。然而，本实验结果表明，部分背景来源相似的致病性嗜水气单胞菌却呈现出不同的聚类特征，这可能与两项研究菌株来源不同有关。因此，在防控嗜水气单胞菌引起的地域性流行病时，必须有针对性地根据其不同的流行规律制定相应的防控方案。

3.2   关于嗜水气单胞菌的耐药性

继 2010 年"超级耐药细菌"对人类健康造成了巨大危害后，人畜共患致病菌的耐药性问题已经成为全球共同关注的卫生问题之一。嗜水气单胞菌作为人畜共患致病菌，除对水产动物有致病作用外，还能引起鸭败血症、人类腹泻及败血症等疾病[22]。然而，目前有研究表明，嗜水气单胞菌的耐药性问题已经相当严峻，对复方新诺明、阿莫西林、氟苯尼考等常用药物均表现出严重的耐药性[3,23-24]。林居纯等[3] 对 120 株嗜水气单胞菌的耐药性进行调查后发现，62% 的菌株对复方新诺明耐药，38%～90% 的菌株对 β-内酰胺类药物耐药。本实验结果也显示，约 30% 嗜水气单胞菌菌株对复方新诺明、多黏菌素等常用兽药也产生了耐药性，而且少数嗜水气单胞菌菌株对氧氟沙星、左氟沙星等也产生了耐药性。这除了与养殖户在养殖过程中滥用、乱用药物有关外，还与耐药菌株在全国各地的流通不无关系。因此，在解决嗜水气单胞菌的耐药问题时，不仅应在养殖过程中正确合理地选择和使用药物，而且还要进一步加强对耐药菌株的监测与流通控制。

3.3   关于嗜水气单胞菌基因型与耐药性之间的关系

据报道，发生在细菌基因组相关基因上的碱基突变、片段插入或缺失、基因转化、等位基因缺失等变化都有可能引起细菌的耐药性[25]。因此，如果能在细菌耐药表型和基因型之间找到某种客观的联系，或直接运用基因分型的方法找出导致耐药表型产生的单一多态性，就有可能对与该表型相对应的基因进行定位，从而可以在分子水平上对细菌的耐药性进行检测，为耐药细菌的分子流行病学提供科学依据[26]。例如，陈俭等[6] 对 47 株大肠杆菌的 ERIC-PCR 基因型与耐药表型进行了研究，认为菌株产生耐药性可能与特定的 ERIC 指纹图谱相关。目前，国内关于嗜水气单胞菌基因型与耐药表型之间的关系研究还鲜见报道。本实验发现，Ⅻ型嗜水气单胞菌菌株的耐药谱型为 CX/AMP/POL，74.1%～100% 的Ⅳ型、Ⅵ型和Ⅷ型嗜水气单胞菌菌株的耐药谱型为 CX/AMP，66.6% 的Ⅺ、Ⅱ型嗜水气单胞菌菌株的耐药谱型为 CX/AMP/SMZ，说明嗜水气单胞菌的基因型与耐药谱型间可能存在一定的联系，这与陈俭等[6] 的研究发现基本相同。此外，本实验结果还表明，少数基因型相同的菌株表现了不同的耐药性，少数耐药谱相同菌株的基因型也呈现出多态性，这可能是因为 ERIC-PCR 分型是一种以扩增条带多态性为主要依据的分型方法，相同大小的片段可能存在序列上的差异。而且，耐药性的发生机制极其复杂，受基因组靶基因突变、环境的选择压力、耐药质粒传递[27-28] 等多种因素的影响。

参考文献

[1]   沈锦玉. 嗜水气单胞菌的研究进展 [J]. 浙江海洋学院学报：自然科学版，2008，1 (27)：78-86.

[2]   文祝友，刘晓燕，金燮理. 三角帆蚌嗜水气单胞菌病的病理研究 [J]. 湖南农业大学学报，2001，1 (27)：57-59.

[3] 林居纯, 罗忠俊, 舒刚, 等. 嗜水气单胞菌临床分离菌对抗菌药物的耐药性调查 [J]. 安徽农业科学, 2009, 37 (15): 7024-7025.

[4] 钱冬, 陈月英, 沈锦玉, 等. 引起鱼类暴发性流行病的嗜水气单胞菌的血清型、毒力及溶血性 [J]. 微生物学报, 1995, 35 (6): 460-464.

[5] 朱芝秀, 曾志明, 何后军, 等. 江西地区嗜水气单胞菌流行株的生物学特性 [J]. 江西农业大学学报, 2007, 4 (25): 623-627.

[6] 陈俭, 姜中其, 唐一鸣. 仔猪黄白痢大肠杆菌耐药性检测及 ERIC 图谱分析 [J]. 浙江大学学报: 农业与生命科学版, 2010, 36 (2): 125-132.

[7] 卢强, 郑伟, 李莲瑞, 等. 嗜水气单胞菌喹诺酮类药物抗性决定区基因片段的克隆与突变分析 [J]. 吉林农业大学学报, 2003, 25 (1): 95-96, 101.

[8] 朱建萍, 叶展翔, 朱烨. 嗜水气单胞菌多重耐药性与质粒的关系分析 [J]. 社区医学杂志, 2009, 23 (7): 83-83.

[9] 张美玲, 周志华, 赵立平. 粪便样品中大肠杆菌多态性分子研究 [J]. 微生物学通报, 2005 (2): 5-9.

[10] 李鹏, 李军星, 李玉峰. 中国东南部地区副猪嗜血杆菌分离株 ERIC-PCR 指纹图谱分析 [J]. 中国兽医学报, 2009 (12): 1566-1570.

[11] 张宏梅, 石磊, 李琳. 沙门氏菌 ERIC 指纹分析 [J]. 云南农业大学学报, 2007 (7): 467-470.

[12] 周俊英, 周新, 付有荣, 等. 铜绿假单胞菌采用肠杆菌属重复基因间隔共有序列-PCR 的分析 [J]. 中华医院感染学杂志, 2006 (8): 846-849.

[13] 金莉莉, 董雪, 王秋雨, 等. 沈阳市副溶血弧菌重复序列 PCR 分型 [J]. 中国公共卫生, 2008 (3): 351-353.

[14] 南京农业大学动物医学院. GB/T18652-2002 致病性嗜水气单胞菌检验方法 [S]. 北京: 中国标准出版社, 2002.

[15] Biswajit M, Pendru R, Iddya K, et al. Typing of clinical and environmental strains of Aeromonas spp. using two PCR based methods and whole cell protein analysis [J]. J Micro-biol Methods, 2009, 78: 312-318.

[16] 伍晓锋, 黎毅敏, 卓超, 等. 18 株铜绿假单胞菌的耐药谱和 ERIC-PCR 分型 [J]. 中国抗生素杂志, 2007, 9 (32): 560-563.

[17] National Committee for Clinical Laboratory Standards. Performance standards for antimicrobial susceptibility testing [S]. Eighteenth informational supplement. NCCLS document M100-S18, 2008: 25-61.

[18] Hunter R P, Gaston A M. Numerical index of the discriminatory ability of typing systems: an application of Simpsons index of diversity [J]. J Clin Microbial, 1988, 6: 2465-2466.

[19] 卢强, 任瑞文, 胡岩, 等. 嗜水气单胞菌的随机扩增多态 DNA 分型 [J]. 中国兽医学报, 2002, 1 (22): 34-36.

[20] Liu P Y, Shi Z Y, Lau Y J, et al. Comparison of different PCR approaches for characterization of Burkholderia cepaciaisolates (Pseudomonas) [J]. J Clin Microbiol, 1995, 33 (12): 3304-3307.

[21] Aguilera-Arreola M G, Hernández-Rodríguez C, Zúñiga G, et al. Aeromonas hydrophila clinical and environmental ecotypes as revealed by genetic diversity and virulence genes [J]. FEMS Microbiol Lett, 2005, 2 (242): 112-118.

[22] 张运强, 李庆乐. 嗜水气单胞菌的致病性及其防治方法 [J]. 广西农业科学, 2007, 5 (38): 565-568.

[23] 童照威, 张龙琪, 王伟洪, 等. 嗜水气单胞菌感染现状及耐药分析 [J]. 中国微生态学杂志, 2008, 1 (20): 75-76.

[24] 李焕荣, 田丽英, 崔德凤, 等. 致病性嗜水气单胞菌对 31 种抗菌药物的敏感试验 [J]. 北京农学院学报, 2001, 3 (16): 34-37.

[25] 赵冰开. 细菌耐药性的产生机理与控制对策 [J]. 辽宁畜牧兽医, 2002 (3): 15-16.

[26] 黄静, 刘慧, 朱家馨, 等. 白念珠菌基因多态性与耐药性关系 [J]. 中山大学学报, 2007, 1 (28): 35-39.

[27] 杨向江. 嗜水气单胞菌对几种抗生素耐药性的形成及消失速率 [J]. 内陆水产, 1997 (6): 6-7.

[28] 宋铁英, 陈强, 郑在予, 等. 不同来源嗜水气单胞菌的抗菌素耐药性及耐药机制分析 [J]. 福建农业学报, 2007, 23 (2): 119-124.

该文发表于《中国水产科学》【2014: 21 (4): 777-785】

复合水产养殖环境中气单胞菌耐药性及其同源性分析

Antimicrobial resistance and homology analysis in Aeromonas isolated from integrated fish farms

黄玉萍[1,2], 邓玉婷[1], 姜兰[1], 谭爱萍[1], 吴雅丽[1], 王伟利[1], 罗理[1]

(1 中国水产科学研究院 珠江水产研究所, 农业部渔用药物创制重点实验室, 广东省免疫技术重点实验室, 广东 广州 510380; 2 上海海洋大学 水产与生命学院, 上海 201306)

摘要: 从广东省佛山市 4 个不同畜禽-鱼复合养殖场采集分离猪/鸭源、鱼源、水源、泥源气单胞菌 (*Aeromonas*) 共 57 株, 通过 K-B 药敏纸片法, 测定其对 8 类 24 种药物的敏感性; 提取基因组 DNA, 进行肠杆菌基因间重复序列 PCR（ERIC-PCR）及脉冲场凝胶电泳（PFGE）分子分型。57 株气单胞菌对氨苄西林、阿莫西林/克拉维酸、利福平具有较高耐药率；不同菌株间存在耐药谱差异。57 株气单胞菌通过 ERIC-PCR 分型，可分为 24 个基因型；采用 PFGE 分型可分为 46 个簇。两种分型方法均发现来源于同一养殖场的菌株存在相同或相似图谱的分离株，并且有相似的耐药谱，提示此为同一克隆株。实验结果表明，ERIC-PCR 及 PFGE 分子分型技术均适用于气单胞菌的相关性分析及其耐药性克隆传播追踪；复合水产养殖环境有可能有利于耐药菌从畜禽向水产养殖环境转移。本研究通过对不同来源气单胞菌进行耐药性分析及溯源追踪，旨在为规范畜禽-鱼复合养殖模式用药及建立健康的水产养殖模式提供参考。

关键词： 气单胞菌；复合养殖；耐药性；PFGE；ERIC-PCR；克隆传播

气单胞菌（Aeromonas）是广泛分布于自然界的人-畜-鱼条件致病菌，近年来由其引发的水产动物暴发性死亡各地多有报道。随着抗菌药物在畜禽及水产养殖中普遍性应用，气单胞菌耐药状况日趋严重[1-5]。在我国珠三角地区[6-7]和东南亚、南亚地区[8-9]存在一种对资源多重利用的复合水产养殖模式（立体养殖模式），较常见的是猪-鱼模式、鸭-鱼模式、猪-鸭-鱼模式。此种模式在鱼池边上修建鸭寮或猪棚，将鸭粪或猪粪直接排入鱼池，补充池塘中浮游生物生长所需的氮、磷等营养源。这种"以饲喂畜（禽），畜（禽）粪肥水，肥水养鱼"的新兴农业模式既减少了粪便对环境的污染，又节省了饲料。伴随着畜牧业的发展，虽然这种模式缓解了用地压力增加了经济效益，但是忽视了微生物带来的生态影响，对生态环境造成了极大的污染。畜禽粪便中携带的耐药菌或残留抗菌药有可能通过上述方式而影响水生动物微生物菌群的药物敏感性，从而造成耐药菌群在不同养殖群落间传播和扩散，加促了耐药菌的选择性演变，对抗菌药的研发使用发出了新的挑战。

基于 PCR 技术的肠杆菌科细菌基因间重复序列聚合酶链式反应（Enterobacterial repetitive intergenic consensus polymerase chain reaction，ERIC-PCR）分子分型方法由于其快速、简便、易操作等优点在细菌分型中得到普遍应用。脉冲场凝胶电泳（Pulsed field gel electrophoresis，PFGE）被誉为细菌分子分型技术的"金标准"，广泛应用于食源性疾病的溯源调查。

本研究通过采集畜禽-鱼复合养殖模式下不同来源的气单胞菌，采用药敏试验分析气单胞菌的耐药谱，了解其耐药情况，并将 ERIC-PCR 及 PFGE 应用于不同来源气单胞菌的相关性分析和耐药性追踪，以为规范畜禽-鱼养殖模式用药及设立健康水产养殖环境提供参考。

1 材料与方法

1.1 菌株来源

2012 年 9 月从广东省佛山市 4 个不同畜禽-鱼复合养殖场采集分离的 57 株气单胞菌，分别来源于猪肛门粪便（9 株）、鸭肛门粪便（1 株）、池塘底泥（6 株）、池塘水（7 株）以及鱼鳃（34 株），且每一样品来源仅分离一株气单胞菌。质控菌大肠埃希菌 ATCC25922 由华南农业大学兽医学院药理教研室馈赠。

1.2 菌株鉴定

1.2.1 生化鉴定

根据革兰氏染色及氧化酶试验结果采用 ATB 细菌鉴定仪及 ID 32 GN 革兰氏阴性杆菌鉴定试剂条（Biomerieux，法国）对分离细菌进行生化鉴定。

1.2.2 分子鉴定

根据 Omega 细菌基因组 DNA 提取试剂盒的使用说明（Omega，美国），提取分离细菌基因组 DNA。根据 Borrell 等[10]和 Yáñez 等[11]分别合成的 16SrRNA 基因序列和 gyrB 基因序列，并分别进行 PCR 扩增，PCR 产物直接送到上海英潍捷基贸易有限公司进行基因序列测定，并将测序结果上传 NCBI 的 Blast 检索系统进行序列同源性分析。扩增引物见表 1。

表 1　气单胞菌鉴定所用的引物信息
Tab. 1　PCR primers used in the identification of Aeromonas

引物 primer	序列（5'→3'） sequence（5'→3'）	基因长度/bp gene length/bp
16SrRNA-F	AGA GTT TGA TCA TGG CTC AG	1501
16SrRNA-R	GGT TAC CTT GTT ACG ACT T	

引物 primer	序列（5'→3'） sequence（5'→3'）	基因长度/bp gene length/bp
gyrB-F	TCC GGC GGT CTG CAC GGC GT	1130
gyrB-R	TTG TCC GGG TTG TAC TCG TC	

注：F，上游引物；R，下游引物。

Note：F，forward primer；R，reverse primer。

1.3 药物敏感性测定

药物敏感性测定采用纸片琼脂扩散法（K-B法），参照美国国家临床实验室标准委员会（Clinical and laboratory standards institute，CLSI）颁布的抑菌环直径诊断标准判定结果[12-13]，并以大肠埃希菌 ATCC25922 作为质控菌株。选取广谱抗菌药物共 8 类 24 种：氨苄西林（AMP，10 μg）、阿莫西林/克拉维酸（AMC，30 μg）、头孢噻吩（KF，30 μg）、头孢曲松（CRO，30 μg）、头孢西丁（FOX，30 μg）、头孢噻肟（CTX，30 μg）、亚胺培南（IPM，10 μg）、磺胺复合物（S3，300 μg）、磺胺甲基异噁唑/甲氧苄啶（SXT，25 μg）、利福平（RD，5 μg）、萘啶酸（NA，10 μg）、恩诺沙星（ENR，5 μg）、环丙沙星（CIP，5 μg）、诺氟沙星（NOR，10 μg）、氧氟沙星（OFX，5 μg）、四环素（TE，30 μg）、多西环素（DO，30 μg）、土霉素（OTC，30 μg）、氟苯尼考（FFC，30 μg）、氯霉素（C，30 μg）、呋喃妥因（F，300 μg）、阿米卡星（AK，30 μg）、庆大霉素（CN，10 μg）、新霉素（N，10 μg），均购自英国 OXOID 公司。

1.4 ERIC-PCR 分型

ERIC-PCR 参照 Versalov 等[14]的方法进行。采用 50 μL 的反应体系：ddH₂O 34 μL，PCR buffer5 μL，dNTP 4 μL，引物各 2 μL，模板 μL，rTaq1 μL。扩增程序为：94℃ 7 min；94℃ 30 s，52℃ 1 min，65℃ 8 min，30 个循环；65℃ 16 min。扩增引物参照 Millemann 等[15]，见表2。

表2 ERIC-PCR 引物序列 Tab. 2 ERIC-PCR primers

引物 primer	序列（5'→3'） sequence（5'→3'）
ERIC-F	ATGTAAGCTCCTGGGGATTCA
ERIC-R	AAGTAAGTGACTGGGGTGAGCG

注：F，上游引物；R，下游引物。

Note：F，forward primer；R，reverse primer。

1.5 PFGE 分型

将培养 14~18 h 的 57 株气单胞菌用低熔点琼脂糖凝胶包埋制胶，10%蛋白酶 K 54℃、175 r/min 消化 3 h，限制性内切酶 XbaI 酶（75U）37℃酶切 3 h。PFGE 条件为：1%PFGE 凝胶，0.5×TBE 缓冲液，14℃，电压 6V/cm、夹角 120°，脉冲参数 10~35 s，电泳 22 h。电泳后 0.5 μg/mL EB 染色 30 min，去离子水漂洗 2~3 次，脱色 1 h。Bio-Rad 公司凝胶成像系统读胶分析。

1.6 数据处理及同源性分析

应用 Quantity one 4.6 对 ERIC-PCR 图谱进行分析，采用 "1" "0" 记带法，同一位点有扩增带的记为 "1"，无扩增带记为 "0"，统计为 EXCEL 表格后，利用 NTSYSpc2.1 软件中的非加权组平均法（Unweighted pair group method using arithmetic averages，UPGMA）构建聚类分析树状图。ERIC-PCR 图谱经聚类分析，凡相似度大于 0.8 归为一个亚型，小于 0.8 为不同基因型[16]。

PFGE 图像经 Image Lab 3.0 处理分析后，将条带转化为分子量，并用 PASW Statistics 18.0 进行 R 型系统聚类分析。按照美国疾病预防和控制中心 Tenover 等[17]推荐的方法：PFGE 图谱条带大小和数量相近的为同一型别；3 个及以下条带出现差异的为同一型别的不同亚型，分别称为 1a、1b 和 1c；3 条以上差异者为不同型别。同一型的各亚型间认为在基因上有相关性，来源于同一亲代，而不同型的菌株认为在流行病学上无相关性。Pablos 等[18]研究认为，聚类相似度接近 100%的菌株可认为来源于同一菌株，参照以上标准，判读 PFGE 聚类结果。

2 结果与分析

2.1 细菌分离鉴定结果

结果如表 3 所示，生化鉴定结果和分子鉴定结果在种上存在一定差异。57 株气单胞菌生化鉴定上可分为 3 个种，分别为豚鼠气单胞菌（*Aeromonas caviae*）（34 株）、温和气单胞菌（*Aeromonas sobria*）（20 株）以及嗜水气单胞菌（*Aeromonas hydrophila*）（3 株）。而分子鉴定可鉴定到 5 个种，分别为维氏气单胞菌（*Aeromonas veronii*）（37 株）、豚鼠气单胞菌（*A. caviae*）（6 株）、温和气单胞菌（*A. sobria*）（9 株）、嗜水气单胞菌（*A. hydrophila*）（4 株）以及维氏气单胞菌温和亚种（*A. veronii bv. sobria*）（1 株）。

表 3　57 株气单胞菌基本信息、分型结果及耐药谱

Tab. 3　Basic information, genotyping and resistance profiles of 57 Aeromonas isolates

菌株编号 strains	养殖场 farms	来源 sources	生化鉴定 phenotypic identification	分子鉴定 molecular identification	ERIC-PCR 分型 ERIC-PCR genotyping	PFGE 分型 PFGE genotyping	耐药谱 resistance profiles
A184	IV	猪粪便	*A. sobria*	*A. veronii*	B1	25	AMP
A140	IV	鱼塘底泥	*A. sobria*	*A. veronii*	B1	25	AMP
A144	IV	鱼塘底泥	*A. sobria*	*A. veronii*	B2	25	AMP
A193	II	鱼	*A. caviae*	*A. veronii*	D7	44b	AMP/AMC/KF/TE/OTC/N
A197	II	鱼	*A. caviae*	*A. veronii*	F	44a	AMP/AMC/TE/OTC
A104	IV	鱼	*A. hydrophila*	*A. sobria*	A2	3a	AMP/AMC/RD
A110	IV	鱼	*A. caviae*	*A. veronii*	E1	3b	AMP/AMC/RD/NA/OFX/OTC
A183	IV	猪粪便	*A. sobria*	*A. veronii*	W2	37b	AMP/AMC/IPM/RD/NA/OFX
A99	IV	鱼	*A. sobria*	*A. sobria*	D3	37a	AMP/AMC/IPM/RD/NA/OFX/TE/OTC
A186	IV	猪粪便	*A. caviae*	*A. sobria*	D2	1a	AMP/AMC/IPM/RD/OFX/TE
A187	IV	猪粪便	*A. caviae*	*A. caviae*	V1	1b	AMP/AMC/KF/S3/SXT/RD/NA/OFX/TE/OTC/FFC/C
A122	I	鱼塘底泥	*A. caviae*	*A. veronii*	—	18	AMP/AMC/KF/FOX/S3/RD
A189	I	猪粪便	*A. hydrophila*	*A. hydrophila*	A3	23	AMP/AMC/KF/FOX/RD
A153	I	鱼	*A. sobria*	*A. veronii*	G	15	AMP/S3/RD/NA/N
A150	I	鱼	*A. caviae*	*A. sobria*	I	17	AMP/AMC/IPM/RD/CN/N
A149	I	鱼	*A. caviae*	*A. veronii*	U	31	AMP/AMC/S3
A126	I	鱼	*A. caviae*	*A. veronii*	J1	42	AMP/AMC/S3/NA/OFX/TE/OTC
A156	I	鱼	*A. caviae*	*A. hydrophila*	J2	10	AMP/IPM/S3/RD/NA/TE/OTC
A151	I	鱼	*A. sobria*	*A. veronii*	N1	4	AMP/AMC/IPM/NA/OFX/N
A157	I	鱼	*A. caviae*	*A. veronii*	N2	20	AMP/AMC/S3/TE/OTC/N
A163	I	鱼塘水	*A. caviae*	*A. veronii*	K2	24	AMP/AMC/S3/N
A148	I	鱼	*A. sobria*	*A. veronii*	K3	21	AMP/AMC/S3/RD/NA/N
A155	I	鱼	*A. sobria*	*A. sobria*	K4	—	AMP/AMC/IPM/S3/RD/TE/OTC
A162	I	鱼塘底泥	*A. caviae*	*A. veronii*	M1	30	AMP/AMC/S3/NA/TE/OTC/N
A191	I	鸭粪便	*A. caviae*	*A. sobria*	M3	26	AMP/AMC/S3/RD/N
A190	I	鱼塘水	*A. sobria*	*A. veronii*	M4	38	AMP/IPM/S3/RD
A152	I	鱼	*A. caviae*	*A. hydrophila*	V5	12	AMP/AMC/S3/RD/NA/TE/OTC
A188	I	猪粪便	*A. caviae*	*A. caviae*	V6	2	AMP/AMC/KF/FOX/RD/N
A164	I	鱼塘水	*A. sobria*	*A. veronii*	X1	32	AMP/KF/FOX/RD/NA/N
A154	I	鱼	*A. caviae*	*A. veronii*	X2	40	AMP/S3/RD/TE/OTC

续表

菌株编号 strains	养殖场 farms	来源 sources	生化鉴定 phenotypic identification	分子鉴定 molecular identification	ERIC-PCR 分型 ERIC-PCR genotyping	PFGE 分型 PFGE genotyping	耐药谱 resistance profiles
A194	II	鱼	A. caviae	A. veronii	H2	45	AMP/AMC/RD/NA/TE/OTC/N
A192	II	鱼	A. caviae	A. veronii	T2	46	AMP/OTC/N
A195	II	鱼	A. caviae	A. caviae	S1	29	AMP/AMC/KF/FOX/S3/RD/N
A196	II	鱼	A. sobria	A. veronii	S2	16	AMP/AMC/RD/NA/TE/OTC/N
A173	II	鱼塘底泥	A. sobria	A. veronii	Q	27	AMP/AMC/S3/RD/NA/N
A167	II	鱼塘水	A. caviae	A. veronii	R2	5	AMP/KF/FOX/S3/RD
A242	III	猪粪便	A. caviae	A. caviae	E2	9	AMP/AMC/KF/FOX/RD
A208	III	鱼	A. caviae	A. caviae	L	6	AMP/AMC/RD/NA/N
A209	III	鱼	A. caviae	A. veronii	V2	7	AMP/AMC/KF/FOX/S3/RD/N
A211	III	鱼	A. caviae	A. sobria	V3	35	AMP/AMC/S3/SXT/RD/NA/N
A210	III	鱼	A. caviae	A. caviae	V4	14	AMP/AMC/KF/FOX/RD
A185	IV	猪粪便	A. caviae	A. veronii	C	22	AMP/AMC/RD/TE/OTC
A100	IV	鱼	A. caviae	A. veronii	H1	28	AMP/S3/RD/NA/OFX/TE/OTC
A146	IV	鱼	A. caviae	A. veronii	A1	33	AMP/AMC/NA
A145	IV	鱼	A. sobria	A. veronii	E3	19	AMP/AMC/S3/RD/OFX
A103	IV	鱼	A. sobria	A. veronii bv. sobria	K1	—	AMP/AMC/S3/RD
A109	IV	鱼	A. caviae	A. sobria	T1	13	AMP/AMC/RD/NA
A141	IV	鱼塘底泥	A. caviae	A. veronii	O	—	AMP/AMC/NA
A142	IV	鱼塘水	A. sobria	A. veronii	R1	—	AMP/AMC/RD/TE/OTC
A147	IV	鱼	A. caviae	A. veronii	W1	34	AMP
A98	IV	鱼	A. caviae	A. veronii	D5	—	AMP/AMC/IPM/S3/RD
A112	IV	鱼	A. hydrophila	A. veronii	D6	39	AMP/S3/SXT/RD/NA/OFX/TE/OTC
A101	IV	鱼	A. sobria	A. sobria	D8	36	AMP/AMC/IPM/S3/RD
A139	IV	鱼塘水	A. caviae	A. veronii	D1	43	AMP/AMC/IPM/RD/NA/OFX/TE/OTC
A143	IV	鱼塘水	A. sobria	A. hydrophila	D4	11	AMP/AMC/RD/TE/OTC

2.2　药敏试验结果

实验结果显示，57株气单胞菌对氨苄西林的耐药率达100%，可能与气单胞菌对其天然耐药有关；对阿莫西林\克拉维酸及利福平具有较高耐药率（60%以上）（图1）；对萘啶酸、土霉素、四环素、新霉素和复合磺胺的耐药率在30%~40%；绝大部分菌株对β内酰胺类、磺胺甲基异噁唑/甲氧苄啶、氟喹诺酮类、氨基糖苷类、酰胺醇类、多西环素、呋喃妥因等16种药物敏感（耐药率在20%以下）。由表3的耐药谱可知，大部分菌株间存在耐药谱差异。

2.3　ERIC-PCR分型结果

57株气单胞菌除一株扩增无条带外，其余56株可扩出2~12条大小介于250~3000 bp的条带，且指纹图谱呈多态性分布（图2）。56株气单胞菌可分为24个基因型。其中A140、A184及A144具有相似的电泳图谱（图2），聚类分析后A140和A184鉴定为同一基因型（图3）。

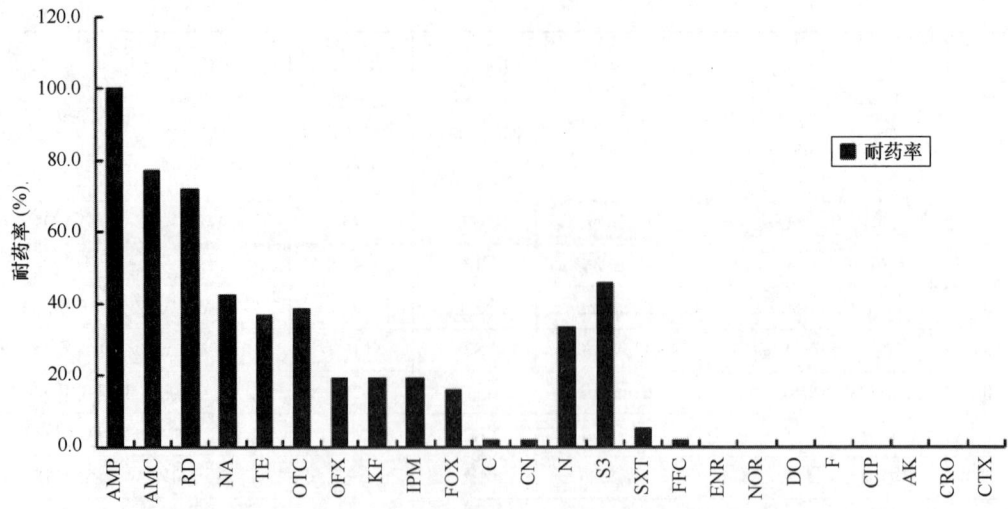

图 1　57 株气单胞菌对 24 种药物的耐药率

Fig. 1　Resistance rates of 24 antimicrobial among 57 Aeromonas isolates

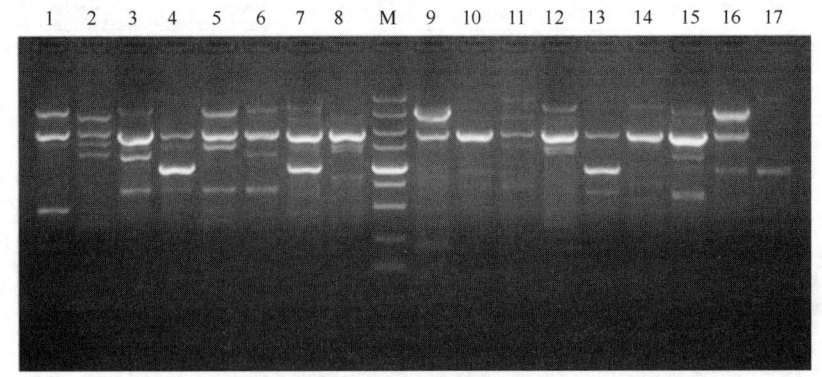

图 2　部分气单胞菌菌株 ERIC-PCR 指纹图谱

Fig. 2　ERIC-PCR patterns of partial Aerornonas isolates

M：DNA marker DL 5000；1：A146；2：A147；3：A140；4：A139；5：A183；6：A184；7：A141；8：A142；9：A185；
10：A186；11：A109；12：A111；13：A110；14：A112；15：A144；16：A143；17：A187

2.4　PFGE 分型结果

57 株气单胞菌经 XbaI 酶切后进行脉冲场凝胶电泳，除 5 株菌条带降解无法判读外，其他 52 株均可获得清晰的电泳条带（图 4）。依据 PFGE 指纹图谱所得聚类分析图（图 5），52 株气单胞菌可分为 46 个簇，其中分别来自于同一综合养殖场的猪源 A183 与鱼源 A99，猪源 A184、泥源 A140 与泥源 A144，猪源 A186 与猪源 A187，鱼源 A193 与鱼源 A197，鱼源 A104 与鱼源 A110 带型相似度均高于 99%，参照 PFGE 聚类分析判读结果，推测这些菌株来源于同一克隆株或具有克隆传播关系。

3　讨论

动物源细菌耐药性流行病学研究和耐药性监测等发现，耐药率与抗菌药的使用时间和普遍性密切相关，而且随着动物用药种类的不断增加，动物体内分离菌的耐药谱也迅速扩大[19]。2013 年 10 月最新颁布并于 2014 年 3 月 1 日实施的我国农业部公告 1997 号《兽用处方药品种目录（第一批）》中，抗生素类及合成抗菌药有 11 大类 51 种药物 138 个剂型，而水产批准使用的只有 5 大类 10 种药物 13 个剂型[20]。虽然大部分抗菌药的使用在水产养殖中未获得批准，但是本实验中从鱼体组织中分离得到的气单胞菌对部分畜禽常用抗菌药仍存在耐药现象，这可能与畜禽与鱼的复合养殖模式有关。畜禽养殖过程中，畜禽排泄物不经过处理直接排放，耐药菌株随粪便污水等直接进入池塘。池塘养殖水体交换缓慢而且营养丰富，不仅为耐药菌的增殖和富集提供场所，而且还为耐药基因在不同细菌之间交换提供便利。Su 等[21]在广东中山 4 个复合养

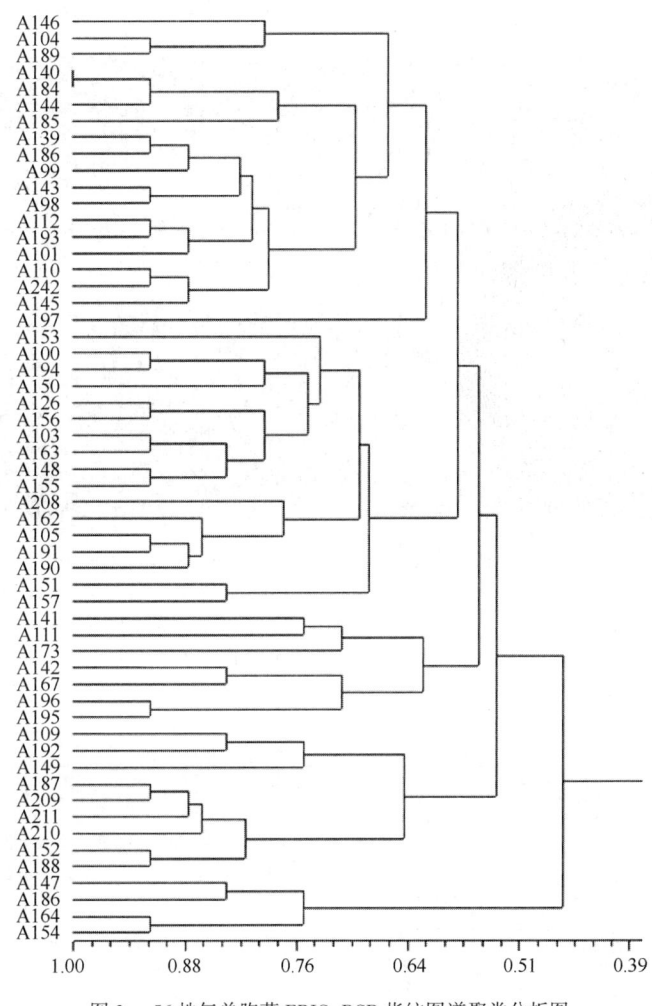

图3　56株气单胞菌ERIC-PCR指纹图谱聚类分析图

Fig. 3　Cluster analysis based on ERIC-PCR patterns of 56 *Aeromonas* isolates

殖场分离到的203株肠杆菌中，98.5%的菌株至少对12种检测药物中的一种药物耐药，超过50%的菌株含有四环素类和磺胺类耐药基因。Zhang 等[22]在广东珠三角复合养殖场分离的大肠埃希菌中检测到多种四环素类、磺胺类、β 内酰胺类耐药基因。张瑞泉[23]通过对传统水产养殖环境和复合水产养殖环境中的耐药菌和耐药基因的污染特征进行研究，发现复合水产养殖环境中的耐药菌及耐药基因污染水平显著高于传统水产养殖环境。复合水产养殖场俨然已成为耐药菌的存储库。本研究从 4 个不同复合水产养殖场分离的气单胞菌中，有来自于 2 个养殖场不同来源的 5 组多重耐药气单胞菌均发现了相同或相似基因分型，其中在Ⅳ号养殖场中，猪粪和底泥分离的气单胞菌之间、猪粪和鱼分离的气单胞菌之间存在相似或相同基因分型。由此可见，菌株有可能通过克隆增殖、耐药质粒的水平传递，或者两者共同作用使耐药菌或耐药基因在不同生物阶层转移。

动物养殖业（畜牧业、水产业）使用抗菌药物导致耐药菌株的出现和传播对人类健康的影响近年来已成为全世界关注的焦点。为明确复合养殖模式下气单胞菌耐药性的传播途径及其方式，探索耐药菌在畜禽与池塘环境、鱼之间垂直克隆传播的可能性，本研究采用 ERIC-PCR 和 PFGE 两种方法来对分离菌进行基因分型，比较不同来源菌株之间的同源相关性。结果发现同一 ERIC-PCR 聚类群的菌株，绝大部分分离自同一养殖场；同一 PFGE 聚类群内的菌株均分离自同一养殖场（表3）。另外，两种分型方法均发现同一养殖场不同来源的菌株具有相同或相似的基因分型，并且有相似的耐药谱，表明他们来源于同一耐药克隆株，也说明了同一养殖场动物与环境之间，不同动物（猪与鱼、鱼与鱼）之间可能存在着耐药气单胞菌的克隆传播。韩国学者 Kim 等[24]曾于 2006—2009 年在从不同患病鱼体及周边环境中分离到 16 株杀鲑气单胞菌，其不仅对四环素类及喹诺酮类药物高度耐药，并且经过 PFGE 分型，大部分菌株具有同源相关性，表明耐药菌在鱼体与环境间存在克隆传播关系。

对细菌进行基因分型的方法较多，ERIC-PCR 和 PFGE 是流行病学研究中最常用的方法，为病原菌微生物的致病性、流行性、变异性以及耐药性分析等方面提供了重要信息[25]。在本研究中，ERIC-PCR 及 PFGE 均可应用于气单胞菌的分子

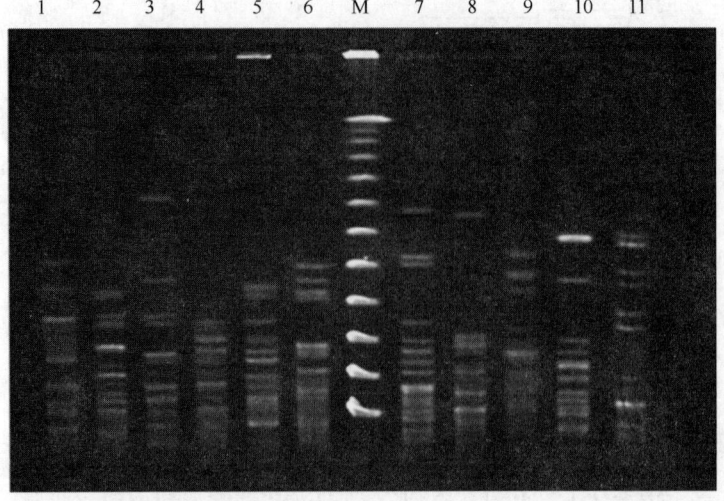

图 4　部分不同来源气单胞菌 XbaI 酶切 PFGE 指纹图谱

Fig. 4　PFGE patterns of partial Aerornonas isolates from different sources in integrated fish farms digested by XbaI

M：PFGE standard, lambda（λ）ladder（50~100 kb）；1：A148；2：A149；3：A150；4：A164；

5：A190；6：A191；7：A167；8：A194；9：A195；10：A196；11：A197；

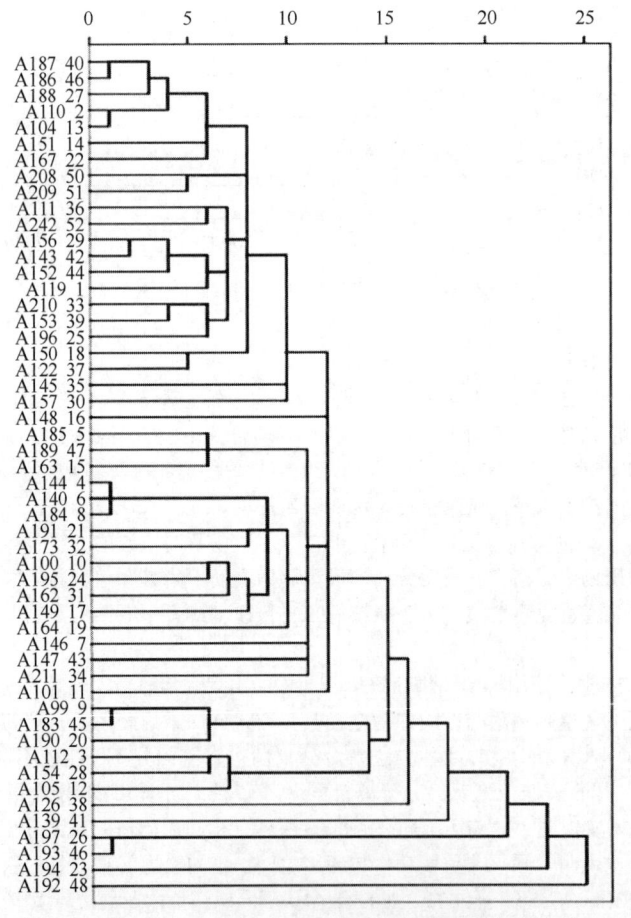

图 5　52 株气单胞菌 PFGE 指纹图谱聚类分析图

Fig. 5　Cluster analysis diagram for PFGE fingerprint about 52 Aeromonas isolate

分型，但两者分型结果存在一定差异。ERIC-PCR 分辨力较弱、重复性差、结果解析困难，但周期短、实验设备易用、操作更简便；而 PFGE 分辨率相对较高、能较好的反映出菌株间的进化关系，但试验流程长、对仪器和操作要求较高、不适于大样本量的应用。但本研究结果显示在溯源研究方面，由于 PFGE 具有较高的分辨力，该方法优于 ERIC-PCR。

对分离菌株的 ERIC-PCR 分型、PFGE 分型结果与耐药谱进行比较可以看出，相同基因分型的菌株其耐药谱不一定相同，而不同分型的菌株也可拥有完全相同的耐药谱型，提示同一克隆株的耐药菌在增殖散播的过程中存在着耐药质粒的获取或丢失，导致耐药性的改变，而不同克隆株之间还可能通过耐药质粒、转座子等耐药元件水平传播耐药性。澳洲学者 Ndi 等[26]发现同一虹鳟养殖场分离的携带相同 I 类整合子及耐药基因的环境细菌与鱼病原菌部分具有同源相关性，也有无相关性的，说明耐药性既可通过克隆传播，也可通过整合子将耐药基因在不同细菌之间进行水平传播。水平传播可增加耐药性在不同来源细菌快速传播和扩散的风险，因此，进一步开展畜禽-鱼复合养殖模式下细菌耐药性的传播机制及其传播方式研究，对水产养殖合理用药、阻断耐药性的扩散，甚至对人类的公共卫生等有着重要作用。

参考文献

［1］ 冷闯，邓舜洲，张文波，等. 中华鳖致病性嗜水气单胞菌的分离鉴定及药敏试验［J］. 动物医学进展，2012，33（2）：124-129.

［2］ 余波，徐景峨，谭诗文，等. 大鲵细菌性败血症病原的分离鉴定与药敏特性［J］. 中国兽医杂志，2011，47（1）：30-31.

［3］ 童桂香，黎小正，韦信贤，等. 斑点叉尾鲴套肠症的病原鉴定及其药敏特性［J］. 大连水产学院学报，2009，24（6）：475-481.

［4］ 司力娜，李绍戊，王荻，等. 东北三省15株致病性嗜水气单胞菌分离株的药敏实验分析［J］. 江西农业大学学报，2011，33（4）：786-790.

［5］ 李绍戊，王荻，刘红柏，等. 鱼源嗜水气单胞菌多重耐药菌株整合子的分子特征［J］. 中国水产科学，2013，5：1005-8737.

［6］ 张金宗. 浅议鱼、猪配套养殖和鱼病防治［J］. 齐鲁渔业，2008，4：14-16.

［7］ 谢绍河，邱镇洪. 珠、鱼、禽立体养殖模式探讨［J］. 水产科技，2008，2：24-26.

［8］ Shah S. Q., Colquhoun D. J., Nikuli H. L., et al. Prevalence of antibiotic resistance genes in the bacterial flora of integrated fish farming environments of Pakistan and Tanzania［J］. Environ Sci Technol, 2012, 46（16）：8672-8679.

［9］ Danq S. T., Petersen A., Van Truonq D., et al. Impact of medicated feed on the development of antimicrobial resistance in bacteria at integrated pig-fish farms in Vietnam［J］. Appl Environ Microbiol, 2011, 77（13）：4494-4498.

［10］ Borrell N., Acinas S. G., Figueras M. J., et al. Identification of Aeromonas clinical isolates by restriction fragment length polymorphism of PCR-amplified 16S rRNA genes［J］. J Clin Microbiol, 1997, 35（7）：1671-1674.

［11］ Yáñez M. A., Catalán V., Apráiz D., et al. Phylogenetic analysis of members of the genus Aeromonas based on gyrB gene sequences［J］. Int J Syst Evol Microbiol, 2003, 53（3）：875-883.

［12］ Clinical and Laboratory Standards Institute. Methods for antimicrobial dilution and disk susceptibility testing of infrequently isolated or fastidious bacteria［S］. Approved Guideline, CLSI document M45-A, Wayne, Pennsylvania, USA：2006.

［13］ Clinical and Laboratory Standards Institute. Methods for antimicrobial disk susceptibility testing of bacteria isolated from aquatic animals［S］. Approved Standards, CLSI document M42-A, Wayne, Pennsylvania, USA：2006.

［14］ Versalovic J., Koeuth T., Lupsaki J. R.. Distribution of repetitive DNA sequences in eubacteria and application to finger printing of bacterial genomes［J］. Nucleic Acids Res, 1991, 19（24）：6823-6831.

［15］ Millemann Y., Lesage-Descauses M. C., Lafont J. P., et al. Comparison of random amplified polymorphic DNA analysis and enterobacterial repetitive intergenic consensus-PCR for epidemiological studies of Salmonella［J］. FEMS Immunol Med Microbio, 1996, 9（2-3）：129-134.

［16］ 伍晓峰，黎毅敏，卓超. 18株铜绿假单胞菌的耐药谱和 ERIC-PCR 分型［J］. 中国抗生素杂志，2007，9（32）：560-563.

［17］ Tenover F. C., Arbeit R. D., Goering R. V., et al. Interpreting chromosomal DNA restriction patterns produced by pulsed-field gel electrophoresis：criteria for bacterial strain typing［J］. J Clin Microbiol, 1995, 33（9）：2233-2239.

［18］ Pablos M., Remacha M. A., Rodriquez-Calleja J. M., et al. Identity, virulence genes, and clonal relatedness of Aeromonas isolates from patients with diarrhea and drinking water［J］. Eur J Clin Microbiol Infect Dis, 2010, 29（9）：1163-1172.

［19］ 宁宜宝，宋立，张纯萍，等. 我国动物源细菌耐药状况、存在问题和对策［J］. 中国兽药杂志，2012，46（8）：50-53.

［20］ 中华人民共和国农业部. 兽用处方药品种目录（第一批）［Z］. 北京：2013.

［21］ Su H. C., Ying G. G., Tao R., et al. Occurrence of antibiotic resistance and characterization of resistance genes and integrons in Enterobacteriaceae isolated from integrated fish farms in South China.［J］. J Environ Monit, 2011, 13（11）：3229-3236.

［22］ Zhang R. Q., Ying G. G., Su H. C., et al. Antibiotic resistance and genetic diversity of Escherichia coli isolates from traditional and integrated aquaculture in South China［J］. J Environ Sci Health B, 2013, 48（11）：999-1013.

［23］ 张瑞泉. 畜牧、水产养殖环境及纳污环境中抗生素耐药菌和耐药基因污染特征研究［D］. 广州：中国科学院广州地球化学研究所，2013. 70-97.

［24］ Kim J. H., Hwang S. Y., Son J. S., et al. Molecular characterization of tetracycline-and quinolone-resistant Aeromonas salmonicida isolated in Korea［J］. J Vet Sc, 2011, 12（1）：41-48.

［25］ 夏季，邓少丽，陈鸣. 分子生物学技术应用于病原微生物分子分型的研究进展［J］. 国际检验医学杂志，2010，31（11）：1278

－1279.

[26] NdiOL, BartonMD. Incidence of class 1 integron and other antibiotic resistance determinants in Aeromonas spp. from rainbow trout farms in Australia [J]. J Fish Dis,

该文发表于《水产学报》【2014：38（7）：1018－1025】

养殖龟鳖源气单胞菌耐药性与质粒介导喹诺酮类耐药基因分析

Analysis of Antimicrobial susceptibility and Plasmid-mediated quinolone resistance genes in *Aeromonas isolated* from Turtles

谭爱萍[1]，邓玉婷[1]，姜兰[1]，吴雅丽[1]，冯永永[1,2]，黄玉萍[1,2]，罗理[1]，王伟利[1]

(1. 中国水产科学研究院珠江水产研究所，农业部渔用药物创制重点实验室，广东省水产动物免疫技术重点实验室，广东 广州 510380；2. 上海海洋大学水产与生命学院，上海 201306)

摘要：为了解养殖龟鳖源气单胞菌的耐药情况及质粒介导的喹诺酮类耐药（PMQR）、喹诺酮类耐药决定区（QRDR）与耐药表型之间的关系；实验采用K-B纸片法测定了1996—2013年从广东地区患病龟鳖分离的67株气单胞菌对23种常见抗菌药的耐药性，并检测5种PMQR基因qnrA、qnrB、qnrS、qepA和aac（6'）-Ib-cr，同时分析PMQR基因阳性菌株染色体上gyrA、parC基因QRDR的突变情况。结果显示，67株气单胞菌对氨苄西林、头孢噻吩和磺胺复合物的耐药率分别高达100%、92.54%和83.58%，对喹诺酮类药物呈现中等耐药，耐药率介于19.40%~64.18%，而对亚胺培南、呋喃妥因、阿米卡星、头孢噻肟敏感性较高，耐药率低于10%；79.10%（53/67）的菌株对3类或以上抗菌药物具有耐药性。19.40%（13/67）的菌株携带PMQR基因，其中，8.96%（6/67）携带qnrS1基因、5.97%（4/67）携带qnrS2基因、7.46%（5/67）携带aac（6'）-Ib-cr基因（其中两株同时携带qnrS2和aac（6'）-Ib-cr基因）。13株PMQR基因阳性菌株均分别携带1~4个质粒，大小介于0.8~15 kb；其中6株在gyrA基因及parC基因上均发生变异，3株仅在gyrA基因上发生变异，另外4株未发现QRDR的基因突变。研究表明，广东地区龟鳖源气单胞菌对多种抗菌药物耐药并存在多重耐药现象；而且PMQR机制的存在预示着喹诺酮类耐药性很可能会在水产临床上更加快速而广泛地传播，应引起重视。

关键词：龟鳖；气单胞菌；耐药；质粒介导；喹诺酮类

　　龟鳖作为我国主要的名特优水产养殖品种，其养殖业起步于20世纪70年代末，经过四十多年的迅猛发展，已逐步成为推动经济发展和农民致富的一项重要产业。但随着养殖密度的提高、养殖规模的扩大和养殖年份的增加，养殖龟鳖的各种疾病频发，特别是由气单胞菌引起的细菌性疾病，如龟鳖的肠道出血性败血症、穿孔病、红脖子病、红底板病、腐皮病等[1]，严重制约着龟鳖业的发展。

　　为控制病原菌对龟鳖养殖造成的损失，养殖者常使用各种抗菌药物对患病动物进行治疗；喹诺酮类药物具有抗菌谱广、高效低毒，副作用小等优点，在龟鳖病害防治过程中发挥了重要作用；但随着药物长期、大量、广泛的使用，亦不可避免地导致了细菌耐药性的出现。长期以来人们认为细菌对喹诺酮类耐药主要由染色体上的靶基因突变或者主动外排机制所致[2]；近期的研究陆续发现了各种质粒介导的喹诺酮类耐药机制（plasmid-mediated quinolone resistance，PMQR），其相关的基因包括qnr（qnrA、qnrB、qnrS、qnrC、qnrD、qnrR），aac（6'）-Ib-cr，qepA和oqxAB[3]，其中qnr的存在能促进染色体上喹诺酮类耐药决定区（quinolone resistance determining regions，QRDRs）的靶基因突变，导致高水平耐药[4]。PMQR耐药基因通常位于质粒上，这不仅增强了细菌对药物的适应性，而且耐药基因亦可在不同种属细菌之间水平传播；因此PMQR的研究已成为人类医学和畜牧兽医耐药研究的热点与焦点[5-6]。目前水产养殖中已出现喹诺酮类药物的临床耐药菌株[7-8]，开展水产动物源PMQR研究对于抑制喹诺类酮耐药性传播具有重要的意义。

　　本研究对1996—2013年从养殖患病龟鳖病灶或组织分离保存的气单胞菌进行药物敏感性分析、检测其PMQR基因携带情况及PMQR阳性菌QRDR的突变情况，旨在为龟鳖源气单胞菌喹诺酮类药物耐药性控制研究及临床合理用药提供理论依据。

1 材料和方法

1.1 试剂和材料

　　67株气单胞菌为本实验室1996—2013年从患病龟鳖病灶或组织分离、纯化和保存，依据参考文献[9]，通过生理生化特性和分子生物学鉴定为嗜水气单胞菌复合群（*Aeromonas hydrophila complex*）28 株、温和气单胞菌复合群（*Aeromonas sobria complex*）7 株和豚鼠气单胞菌复合群（*Aeromonas caviae complex*）32 株。嗜水气单胞菌标准菌株 ATCC 7966 由浙江省淡水水产研究所馈赠，大肠埃希菌 *Escherichia coli* ATCC 25922 由华南农业大学兽医学院药理教研室馈赠。

胰蛋白胨大豆（TSA）琼脂购自北京路桥技术有限责任公司，LB 肉汤、LB 琼脂、水解酪蛋白（MH）琼脂购自青岛海博生物技术有限公司。质粒提取试剂盒为 OMEGA 质粒小量提取试剂盒。PCR 所用试剂等均为康为世纪生物有限公司产品。

实验室所需药敏试纸为英国 Oxoid 公司产品，-20℃冷冻保存。纸片药物及含量（μg）分别为：氨苄西林（AMP）10、阿莫西林/克拉维酸（AMC）30、头孢噻吩（KF）30、头孢西丁（FOX）30、头孢噻肟（CTX）30、亚胺培南（IPM）10、磺胺复合物（S3）300、磺胺甲基异噁唑/甲氧苄啶（SXT）25、利福平（RD）5、萘啶酸（NA）10、恩诺沙星（ENR）5、环丙沙星（CIP）5、诺氟沙星（NOR）10、氧氟沙星（OFL）5、四环素（TE）30、多西环素（DO）30、土霉素（OTC）30、氟苯尼考（FFC）30、氯霉素（C）30、呋喃妥因（F）300、阿米卡星（AK）30、庆大霉素（CN）10、新霉素（N）30。

1.2　方法

药物敏感性测定采用 K-B 纸片扩散法，以 ATCC 25922、ATCC 7966 作为质控菌和对照菌进行药敏实验。将实验菌株用 LB 肉汤复苏后接种于 TSA 琼脂平板，28℃培养 18~20 h；挑取单菌落接种于 2 mL LB 营养肉汤，28℃振荡培养 3~4 h；用无菌生理盐水将菌液浓度稀释为 $1×10^6$ CFU/mL，用无菌棉拭子均匀涂抹至约 4 mm 厚度的 MH 培养基上，贴上药敏纸片，置 35℃培养 18~20 h 后观察结果。参照美国临床实验室标准化委员会（Clinical and Laboratory Standards Institute, CLSI）标准判断实验菌株的药物敏感性[10-11]，同时参照文献[12]判断多重耐药结果。数据结果经 WHONE T5.6 耐药监测软件和 SPSS18.0 软件开展分析。

细菌全基因组 DNA 提取采用水煮法制备细菌 DNA 模板。将实验菌株接种于 LB 固体培养基，28℃培养 12~16 h，挑取适量菌苔放入含 500 μL 1×TE 的 1.5 mL 灭菌离心管，混匀后煮沸 10 min，冰浴 5 min，1 200 r/min 离心 1 min，提取上清液即为 DNA 模版。

PMQR 基因的 PCR 扩增及目的基因片段序列分析以细菌全基因组 DNA 为模板，根据参考文献报道[13]设计相应的引物序列及 PCR 反应退火温度（见表 1）；对 PCR 扩增产物进行测序（引物合成及测序工作由上海博尚生物技术有限公司完成），将测序结果用 BLAST 软件（http：www.ncbi.nlm.nih.gov）在 GenBank 中对目的基因序列进行同源性检索分析。

表 1　PMQR 基因与 QRDR 基因的相应引物序列

Tab. 1　Primers for amplifying the PMQR and QRDR genes

引物 primers	序列（5'-3'） sequence	扩增产物大小/bp Size	退火温度/℃ annealing temperature
qnrA-F	ATTTCTCACGCCAGGATTTG	519	58
qnrA-R	GATCGGCAAAGGTCAGGTCA		
qnrB-F	GATCGTGAAAGCCAGAAAGG	469	58
qnrB-R	ACGATGCCTGGTAGTTGTCC		
qnrS-F	ACGACATTCGTCAACTGCAA	417	56
qnrS-R	TAAATTGGCACCCTGTAGGC		
qepA-F	GCAGGTCCAGCAGCGGGTAG	306	60
qepA-R	CTTCCTGCCCGAGTATCGTG		
aac（6'）-Ib-F	TTGCGATGCTCTATGAGTGGCTA	482	55
aac（6'）-Ib-R	CTCGAATGCCTGGCGTGTTT		
gyrA-F	CCATGAGCGTGATCGTAGGA	665	62
gyrA-R	CTTTGGCACGCACATAGACG		
parC-F	GTTCAGCGCCGCATCATCTAC	245	56
parC-R	TTCGGTGTAACGCATTGCCGC		

PMQR 阳性菌质粒的提取采用 OMEGA 质粒小量提取试剂盒提取。将实验菌株接种于 LB 营养肉汤，28℃振荡培养 12~16 h，按试剂盒使用说明进行质粒提取，检测各菌株携带质粒的情况。

PMQR 阳性菌 QRDR 基因突变分析参考文献报道[14]设计相应的 gyrA、parC 基因扩增引物序列及 PCR 退火温度（见表1），对 PCR 扩增产物进行测序（引物合成及测序工作由上海博尚生物技术有限公司完成），分析各实验菌株 gyrA、parC 基因的氨基酸突变情况。

2 结果与分析

2.1 药物敏感性测定

67 株受试菌株对氨苄西林、头孢噻吩和磺胺复合物的耐药率较高，分别为 100%、92.54% 和 83.58%；对亚胺培南、呋喃妥因、阿米卡星、头孢噻肟敏感性较高，耐药率在 10% 以下；对喹诺酮类药物呈现中等耐药，耐药率在 19.40% ~ 64.18%（表2）。此外，有 53 株菌株（79.10%）对 3 类或以上抗菌药物耐药，其中 15 株（23.88%）对 6 类抗菌药物耐药、14 株（20.90%）对 7 类抗菌药物耐药、3 株（4.48%）对 8 类抗菌药物耐药，表现出较强的多重耐药性（表3）。

表 2　67 株龟鳖源气单胞菌对 8 类 23 种抗菌药物的药敏试验

Tab. 2　usceptibility of 67 Aeromonas isolated from Chelonia isolates to 8 type （23） antimicrobial agents

类别 genre	药物 drug	菌株数（百分数）Number. of isolates （percentage）		
		耐药 R	中介 M	敏感 S
β-内酰胺类 β-lactans	氨苄西林（AMP）	67（100）	0（0）	0（0）
	阿莫西林/克拉维酸（AMC）	38（56.72）	25（37.31）	4（5.97）
	头孢噻吩（KF）	62（92.54）	0（0）	5（7.46）
	头孢西丁（FOX）	26（38.80）	10（14.93）	31（46.27）
	头孢噻肟（CTX）	3（4.48）	7（10.45）	57（85.07）
	亚胺培南（IPM）	6（8.96）	11（16.41）	50（74.63）
磺胺类 sulfonamides	磺胺复合物（S3）	56（83.58）	1（1.49）	10（14.93）
	磺胺甲基异噁唑/甲氧苄啶（SXT）	46（68.66）	4（5.97）	17（25.37）
喹诺酮类 quinolones	萘啶酸（NA）	43（64.18）	0（0）	24（35.82）
	恩诺沙星（ENR）	28（41.79）	10（14.93）	29（43.28）
	环丙沙星（CIP）	13（19.40）	16（23.88）	38（56.72）
	诺氟沙星（NOR）	15（22.39）	7（10.45）	45（67.16）
	氧氟沙星（OFL）	35（52.24）	3（4.48）	29（43.28）
四环素类 tetracyclines	四环素（TE）	37（55.22）	3（4.48）	27（40.30）
	多西环素（DO）	25（37.31）	1（1.49）	41（61.19）
	土霉素（OTC）	43（64.18）	0（0）	24（35.82）
氨基糖苷类 aminoglycosides	阿米卡星（AK）	4（5.97）	0（0）	63（94.03）
	庆大霉素（CN）	13（19.40）	2（2.99）	52（77.61）
	新霉素（N）	21（31.34）	2（2.99）	44（65.67）
酰胺醇类 amphenicols	氯霉素（C）	26（38.81）	4（5.97）	37（55.22）
	氟苯尼考（FFC）	18（26.87）	2（2.99）	47（70.15）
利福平类 rifampicins	利福平（RD）	47（70.15）	14（20.89）	6（8.96）
硝基呋喃类 nitrofurans	呋喃妥因（F）	5（7.46）	7（10.45）	55（82.09）

表3 67株龟鳖源气单胞菌对8类23种抗菌药物的多重耐药情况

Tab. 3 Multidrug resistance of 67 Aeromonas isolated from Chelonia isolates to 8 types（23）antimicrobial agents

耐药型 resistance types	药物种类 medicine types	菌株数 no. of isolates	百分比/% characteristic/total no. of isolates
3 耐 R	①②④；①②⑤；①③⑦	10	14.93
4 耐 R	①②④⑦；①③④⑦	4	5.97
5 耐 R	①②③④⑦；①②③⑥⑦；①②④⑥⑦	7	10.45
6 耐 R	①②③④⑤⑦；①②③④⑥⑦；①②③④⑦⑧；①②③⑤⑥⑦	15	23.88
7 耐 R	①②③④⑤⑥⑦；①②③④⑥⑦⑧	14	20.90
8 耐 R	①②③④⑤⑥⑦⑧	3	4.48

注：①β-内酰胺类 β-lactans；②磺胺类 sulfonamides；③喹诺酮类 quinolones；④四环素类 tetracyclines；⑤氨基糖苷类 aminoglycosides；⑥酰胺醇类 amphenicols；⑦利福平类 rifampicins；⑧硝基呋喃类 nitrofurans

2.2 PMQR 基因检测与序列分析

67 株受试菌株中，有 13 株（19.40%）携带了 PMQR 基因，部分菌株的 PCR 产物电泳结果见图 1 和图 2；所获得的 PMQR 基因序列与 BLAST 比对结果显示，有 6 株携带 qnrS1 基因（8.96%）、4 株携带 qnrS2 基因（5.97%）、5 株携带 aac（6′）-Ib-cr 基因（7.46%），其中 2 株同时携带 qnrS2 和 aac（6′）-Ib-cr 基因（2.99%）；获得的基因序列均提交至 Gen-Bank 并获取了序列号（见表 4）。在本研究实验中没有检测到 qnrA、qnrB 和 qepA 基因。

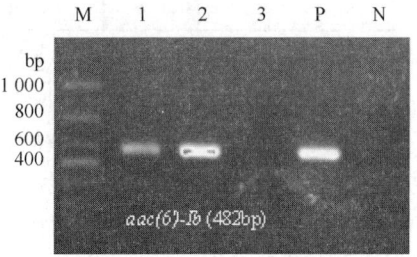

图 1 *aac*（6）-Ib 基因 PCR 电泳图

Fig. 1 The PCR products of *aac*（6）-Ib gene

M DL1000 分子量标准；1-3. 样品；P. 阳性对照；N. 空白对照

M Markers（DL1000）；1-3. sample；P. positive control；N. blank control

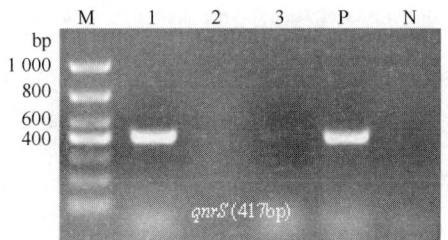

图 2 qnrS 基因 PCR 电泳图

Fig. 2 The PCR products of qnrS gene

M DL1000 分子量标准；1-3. 样品；P. 阳性对照；N. 空白对照

M Markers（DL1000）；1-3. sample；P. positive control；N. blank control

2.3 PMQR 阳性菌质粒的提取

携带有 PMQR 基因的 13 株菌株的质粒提取结果显示，13 株菌株分别携带了 1~4 个质粒，最大的约 15 kb，最小的约 0.8 kb（图 3）。

图 3 质粒图谱

Fig. 3 The restriction map of the plasmids

M. DL 15000 marks; 1–13. 28A, 13036, 12013, 12014, 12087, 13002,

44A, 12015, 12044, 1 肝 liver, 17A, 12045, 2 肾 kidney

2.4 PMQR 阳性菌株 QRDR 基因变异情况

携带了 PMQR 基因的 13 株菌株的 gyrA 和 parC 基因测序结果显示, 有 6 株同时存在 gyrA 基因编码的第 83 位丝氨酸 Ser→异亮氨酸 Ile 和 parC 基因编码的第 87 位丝氨酸 Ser→异亮氨酸 Ile 的突变, 有 3 株仅发生了 gyrA 基因靶位点的突变, 另外 4 株未发生 gyrA 或 parC 基因 QRDR 靶位点的突变。发生 QRDR 靶基因突变的菌株均表现出对不同喹诺酮类药物耐药, 详细结果见表 4。

表 4 PMQR 基因阳性菌的染色体 QRDR 变异情况及耐药谱

Tab. 4 Mtation of QRDR and qinolones resistance music in the PMQR positive strains

菌株 strains	分离年份 isolates year	PMQR	QRDR 突变 QRDR mutation		喹诺酮类耐药谱 quinolones resistance profiles
			gyrA	parC	
28A	2003	aac (6′) -Ib-cr [KC542812]	Ile83	Ile87	NA
17A	2005	qnrS2 [KJ162344]	/	/	/
44A	2007	aac (6′) -Ib-cr [KC554068]	Ile83	Ile87	NA \ ENR \ CIP \ NOR \ OFL
12013	2012	qnrS1 [KJ162351]	/	/	NA \ ENR \ OFL
12014	2012	qnrS1 [KJ162347]	Ile83	/	NA \ ENR \ OFL
12015	2012	aac (6′) -Ib-cr [KJ162339]	Ile83	Ile87	NA \ ENR \ OFL
12044	2012	qnrS1 [KJ162348]	Ile83	/	NA \ ENR \ OFL
12045	2012	qnrS1 [KJ162349]	Ile83	/	NA \ ENR \ CIP \ NOR \ OFL
12087	2012	qnrS2 [KJ162345]	/	/	/
2 肾 kidney	2013	qnrS1 [KJ162350]	Ile83	Ile87	NA \ ENR \ OFL
13002	2013	qnrS1 [KJ162346]	/	/	NA \ ENR \ OFL
13036	2013	aac(6′) -Ib-cr [KJ162341], qnrS2 [KJ162343]	Ile83	Ile87	NA \ ENR \ CIP \ NOR \ OFL
1 肝 liver	2013	aac(6′) -Ib-cr [KJ162340], qnrS2 [KJ162342]	Ilc83	Ile87	NA \ ENR \ CIP \ NOR \ OFL

3 讨论

3.1 养殖龟鳖源气单胞菌耐药性分析

龟鳖疾病具有潜伏期长、病程长、易导致并发症等特点, 严重阻碍了该类动物病害防治的开展。用于治疗龟鳖疾病的药物通常有消毒剂、抗菌药物、中草药等, 但这些药物的选用依据基本上是套用鱼类的使用标准, 效果不甚理想[1], 而临床上的不合理用药也使许多致病菌产生耐药性。吴雅丽等[15]发现临床分离的 112 株水产源气单胞菌对常用抗菌药呈现不同程度的耐药性, 爬行、两栖动物来源的分离菌株对氟喹诺酮类、头孢类等药物的耐药率比养殖鱼、虾类的高。蔡丽娟等[16]

发现甲鱼源的气单胞菌耐药率比鱼源的高。胡大胜等[17]发现广西龟鳖源分离菌的耐药率最强。本研究所测定的 67 株龟鳖源气单胞菌对不同种类的抗菌药耐药性不同，对氨苄西林、头孢噻吩耐药率高可能与其自身产 β 内酰胺酶天然耐药有关[18]；对头孢菌素类、氨基糖苷类、磺胺类、氟喹诺酮类药物等呈中等耐药或敏感。另外，本研究还发现 79.10% 的龟鳖源气单胞菌对 3 类或以上抗菌药物耐药，而且以耐 6 类和 7 类不同抗菌药物为主，显示龟鳖源气单胞菌对常见抗菌药物的耐药性和多重耐药性较为严重。临床上可考虑采取轮换用药方式以减轻药物的选择性压力，降低细菌耐药性产生的概率。

3.2　养殖龟鳖源气单胞菌喹诺酮类耐药性及其耐药基因分析

本研究结果显示，受试菌株对不同的喹诺酮类药物的耐药程度不一，对萘啶酸、氧氟沙星和恩诺沙星呈中等耐药，而对诺氟沙星和环丙沙星较敏感，这与 Han 等[19]、吴雅丽等[15]和薛慧娟等[20]报道的相似。这与细菌对不同喹诺酮类药物产生耐药的机制不同有关，萘啶酸的作用机制简单，靶位点单一，只作用于 DNA 回旋酶的 A 亚单位（gyrA），容易产生高水平耐药，而其他喹喏酮类药物（氟喹诺酮类）在基本结构上引进氟原子或其他不同基团后，与靶位点的结合能力增强，并介入了其他作用机制，不容易产生高水平耐药[2]。

1998 年 Martinez 等[21]首次报道了质粒上 qnr（quinolone resistance）基因介导的氟喹诺酮类药物耐药机制，并发现 qnr 基因可在不同细菌间迅速水平传播。随后，在世界多个国家和地区均发现携带 qnr 基因的不同种属的细菌。耐药基因 qnr 表达的产物可与 DNA 旋转酶及拓扑异构酶IV结合，对氟喹诺酮类抗菌药的作用靶位具有保护作用，现已发现有多种类似基因：qnrA、qnrB、qnrC、qnrD、qnrR 和 qnrS 等。目前气单胞菌中已报道的 qnr 基因有 qnrS2[7]，qnrS5［8］和 qnrB1[22]；本研究的 67 株龟鳖源气单胞菌，8 株携带 qnrS 基因（其中 6 株携带 qnrS1、2 株携带 qnrS2），且 qnrS1 还未见在气单胞菌中发现的报道。aac（6′）-Ib-cr 基因是一种由质粒携带的氨基糖苷乙酰化转移酶的变异基因，它可使细菌对环丙沙星及诺氟沙星的敏感性降低，也能引起对阿米卡星、卡那霉素等氨基糖苷类耐药[23]；在本研究检出的 5 株携带 aac（6′）-Ib-cr 基因的菌株中，有 4 株对环丙沙星或恩诺沙星耐药，由于环丙沙星是恩诺沙星的代谢产物，由此可见该基因的存在有可能影响恩诺沙星和诺氟沙星的作用效果。qepA 基因是由日本学者在大肠埃希菌株中发现的一种由质粒介导的耐喹诺酮类药物外排泵基因，已被证实位于可转移质粒上，包括一个 1 536 bp 的开放阅读框，含有丰富的 G~C 序列（72%），并可被两端带有 2 拷贝的 IS26 的转座子所介导，在其附近还连接着其他的一些耐药基因，如 rmtB、blaTEM-1 等[24]；本研究中未发现 qepA 基因，但由于其位于的质粒可携带多种耐药基因且可水平传播，应引起广大研究者的重视。

PMQR 是喹诺酮类耐药机制的重要组成部分，虽然 PMQR 基因的单独存在往往仅导致细菌对喹诺酮类药物的敏感性下降，表现为低水平耐药，但 PMQR 基因常与其他耐药机制如超广谱 β 内酰胺酶基因、16SrRNA 甲基化酶基因位于相同的质粒或转座子上，这些耐药元件使细菌耐药性传播得更快、范围更广。本研究结果显示 PMQR 基因已逐渐扩散在气单胞菌中，并且携带 PMQR 基因的耐药菌株携带了不同大小的质粒，因而有必要进一步深入开展质粒介导的喹诺酮类药物耐药性在气单胞菌中产生及传播机制研究。

喹诺酮类耐药决定区（QRDR）的靶基因突变是目前报道的气单胞菌对喹诺酮类耐药的主要机制[14,20]，其中 gyrA 基因 QRDR 的突变对喹诺酮类药物耐药的意义最大，其次是 parC、gyrB 和 parE。研究发现，耐药性与突变位点存在一定的关联，苑青艳等[25]通过耐药基因的扩增和测序发现低水平氟喹诺酮类耐药是由 gyrA 改引起，而高水平耐药是由 gyrA 和 parC 共同改变引起，且突变位点越多耐药性越强。本研究所检测的携带 PMQR 基因的 13 株龟鳖源气单胞菌中，发生 gyrA 或 gyrA 和 parC 靶位点突变的菌株对一种或多种喹诺酮类耐药，而未发生靶基因突变的 2 株菌株亦表现出喹诺酮类药物耐药表型，由此可见喹诺酮类耐药性的产生可能与多种耐药机制的共同作用有关。

喹诺酮药物耐药机制复杂，存在包括 PMQR 和 QRDR 突变在内的多种耐药机制；PMQR 往往仅导致喹诺酮类低水平的耐药，但 PMQR 基因的存在可以促进细菌染色体 QRDR 的变异，从而导致具有稳定遗传性的高水平耐药性产生。如何控制 PMQR 基因导致的耐药性产生及传播以及 PMQR 与 QRDR 变异之间的关系还有待进一步研究。

参考文献

[1]　HONG M L, FU L R, WANG R P, et al. The Advances in the Research of the Turtle's Disease [J]. Chinese journal of zoology, 2003, 38 (6)：115-119.［洪美玲，付丽容，王锐萍，等. 龟鳖动物疾病的研究进展 [J]. 动物学杂志，2003, 38 (6)：115-119.］

[2]　Fabrega A, Madurga S, Giralt E, et al. Mechanism of action of and resistance to quinolones [J]. Microbial Biotechnology, 2009, 2 (1)：40-61.

[3]　Strahilevitz J, Jacoby G A, Hooper D C, et al. Plasmid-mediated quinolone resistance：a multifaceted threat [J]. Clin Microbiol Rev, 2009, 22 (4)：664-689.

[4]　Martinez-Martinez L, Pascual A, Garcia I, et al. Interaction of plasmid and host quinolone resistance [J]. Journal of Antimicrobial Chemotherapy, 2003, 51 (4)：1037-1039.

[5]　QU F, LI J, ZHANG J L, et al. Occurrence of new resistance genes aac (6′) -Ib-Cr and qnrS in Aeromonas [J]. Infectious Disease Information, 2013, 26 (3)：175-177.［曲芬，李军，张鞠玲，等. 气单胞菌中检测到新的耐药基因 aac (6′) -Ib-cr 和 qnrS [J]. 传染

病信息，2013，26（3）：175-177.]

[6] YUE L, JIANG H X, LIU J H, et al. Detection of Plasmid-Mediated Quinolone Resistance in Clinical Isolates of Enterobacteriaceae from Avian [J]. Scientia Agricultura Sinica, 2009, 42（8）：2966-2971. [岳磊，蒋红霞，刘健华，等. 鸡源肠杆菌质粒介导喹诺酮类耐药基因检测 [J]. 中国农业科学，2009，42（8）：2966-2971.]

[7] Majumdar T, Das B, Bhadra R K, et al. Complete nucleotide sequence of a quinolone resistance gene（qnrS2）carrying plasmid of Aeromonas hydrophila isolated from fish [J]. Plasmid, 2011, 66（2）：79-84.

[8] Han J E, Kim J H, Choresca Jr C H, et al. First description of the qnrS-like（qnrS5）gene and analysis of quinolone resistance-determining regions in motile Aeromonas spp. from diseased fish and water [J]. Research in Microbiology, 2012, 163（1）：73-79.

[9] Abbott S L, Cheung W K, Janda J M. The genus Aeromonas：biochemical characteristics, atypical reactions, and phenotypic identification schemes [J]. Journal of Clincal Microbiology. 2003, 41（6）：2348-2357.

[10] Clinical and Laboratory Standards Institute. Performance Standards for Antimicrobial Susceptibility Testing：twenty-second Informational Supplement：M100-S22 [M]. Wayne, Pennsylvania, USA：Clinical and Laboratory Standards Institute, 2012：68-69.

[11] Clinical and Laboratory Standard Institute. Methods for antimicrobial dilution and disk susceptibility testing of infrequently isolated or fastidious bacteria：M45A-A2 [M]. Wayne, Pennsylvania, USA：Clinical and Laboratory Standards Institute, 2010：12-13.

[12] Magiorakos A P, Srinivasan A, Carey R B, et al. Multidrug-resistant, extensively drug-resistant and pandrug-resistant bacteria：an international expert proposal for interim standard definitions for acquired resistance [J]. Clinical Microbiology and Infection, 2012, 18（3）：268-281.

[13] Liu J H, Deng Y T, Zeng Z L, et al. Coprevalence of plasmid-mediated quinolone resistance determinants QepA, Qnr, and AAC（6'）-Ib-cr among 16S rRNA methylase RmtB-producing Escherichia coli isolates from pigs [J]. Antimicrobial Agents and Chemotherapy, 2008, 52（8）：2992-2993.

[14] Deng Y T, XUE H J, JIANG L, et al. Characterization of quinolone resistance mechanism of Aeromonas hydrophila selected in vitro [J]. Journal of South China Agricultural University, 2014, 35（1）：12-16. [邓玉婷，薛慧娟，姜兰，等. 体外诱导嗜水气单胞菌对喹诺酮类耐药及其耐药机制研究 [J]. 华南农业大学学报，2014，35（1）：12-16.]

[15] WU Y L, DENG Y T, JIANG L, et al. Antimicrobial susceptibilities of Aeromonas strains isolated from various aquatic animals in Guangdong Province [J]. Journal of Shanghai University, 2013, 22（2）：219-224. [吴雅丽，邓玉婷，姜兰，等. 广东省水产动物源气单胞菌对抗菌药物的耐药分析 [J]. 上海海洋大学学报，2013，22（2）：219-224.]

[16] CAI L J, XU B Q, LIN Q C, et al. Comparison and Analysis of Drug Resistance to Morbific Bacterium Aeromonas hydrophila Isolated from Aquatic Animals [J]. Fisheries Science, 2011, 30（1）：42-45. [蔡丽娟，许宝青，林启存. 水产致病性嗜水气单胞菌耐药性比较与分析 [J]. 水产科学，2011，30（1）：42-45.]

[17] HU D S, HUANG J, LUO Z H. et al. Bacterial disease monitoring and Antimicrobial susceptibilities investigation of Guangxi aquaculture [J]. China Science and Technology Achievements, 2010,（13）：49-51. [胡大胜，黄钧，罗振海，等. 广西水产养殖主要细菌病的监测与耐药性调查 [J]. 中国科技成果，2010，（13）：49-51.]

[18] Janda J M, Abbott S L. The genus Aeromonas：taxonomy, pathogenicity, and infection [J]. Clinical Microbiology Reviews, 2010, 23（1）：35-73.

[19] Han J E, Kim J H, Choresca Jr C H, et al. A small IncQ-type plasmid carrying the quinolone resistance（qnrS2）gene from Aeromonas hydrophila [J]. Letters in Applied Microbiology, 2012, 54（4）：374-376.

[20] XUE H J, DENG Y T, JIANG L, et al. Antimicrobial susceptibility and mutations of QRDR in Aeromonas hydrophila isolated from aquatic animals [J]. Guangdong Agricultural Sciences, 2012, 39（23）：149-153. [薛慧娟，邓玉婷，姜兰，等. 水产动物源嗜水气单胞菌药物敏感性及 QRDR 基因突变分析 [J]. 广东农业科学，2012，39（23）：149-153.]

[21] Martinez-Martinez L, Pascual A, Jacoby G A. Quinolone resistance from a transferable plasmid [J]. The Lancet, 1998, 351（9105）：797-799.

[22] Takasu H, Suzuki S, Reungsang A, et al. Fluoroquinolone（FQ）Contamination Does Not Correlate with Occurrence of FQ-Resistant Bacteria in Aquatic Environments of Vietnam and Thailand [J]. Microbes and Environments, 2011, 26（2）：135-143.

[23] Robicsek A, Strahilevita J, Jacoby G, et al. Fluoroquinolone-modifying enzyme：a new adaptation of a common aminoglycoside acetyhransferase [J]. Nature Medicine, 2006, 12（1）：83-88.

[24] Kunikazu Y, Junichi W, Satowa S, et al. New plasmid-mediated fluoroquinolone efflux pump, qepA, found in an Escherichia coli clinical isolate [J]. Antimicrob Agents Chemother, 2007, 51（9）：3354-3360.

[25] YUAN Q Y. Study on the Relationship between Fluoroquinolone Sensitivity and Gene Mutations of Streptococcus suis [D]. Harbin, northeast agricultural university, 2009. [苑青艳. 猪链球菌对氟喹诺酮类药物耐药性与耐药基因的研究 [D]. 哈尔滨：东北农业大学，2009.] 2011, 34, 589-599.

866

该文发表于《水生生物学报》【2013，37（4）：744-750】

一株多重耐药鳗源肺炎克雷伯菌的分离鉴定

Isolation and Identification of a multiple-drug resistant Klebsiella pneumoniae from Anguilla marmorata

谭爱萍[1]，邓玉婷[1]，姜兰[1]，吴雅丽[1,2]，薛慧娟[1,2]，王伟利[1]，罗理[1]，赵飞[1]

（1. 中国水产科学研究院珠江水产研究所，农业部渔药创制重点实验室，广州 510380；2. 上海海洋大学水产与生命学院，上海 201306）

摘要： 从患病花鳗鲡（Anguilla marmorata）的肝脏分离纯化到一株革兰氏阴性杆菌 HM2。人工感染试验结果显示，HM2 具有较强的致病力，96 h 的 LD50 为 $9.98×10^7$ CFU/mL。经细菌培养特性、生理生化特性和 ATB Expression 半自动细菌鉴定仪鉴定，结果符合肺炎克雷伯菌（Klebsiella pneumoniae）的特征。以细菌 16S rRNA 基因通用引物进行 PCR 扩增，获得大小为 1393 bp 的部分 16S rRNA 基因序列（Genbank 登录号为 JX282908），将所测序列与 GenBank 中的序列进行 BLAST 比对并构建系统进化树，结果表明其与肺炎克雷伯菌的同源性最高（100%），在系统发育树上与肺炎克雷伯菌聚为一簇，进一步确定菌株 HM2 为肺炎克雷伯菌。药物敏感性试验显示，HM2 对亚胺培南、链霉素、阿米卡星 3 种药物敏感，对氨苄西林、阿莫西林/克拉维酸、头孢噻吩、头孢曲松、头孢噻肟、头孢西丁、复合磺胺、磺胺甲基异噁唑/甲氧苄啶、利福平、萘啶酸、环丙沙星、诺氟沙星、氧氟沙星、呋喃妥因、四环素、多西环素、氯霉素 17 种药物具有耐药性。研究结果为指导临床合理用药提供了科学依据。

关键词： 花鳗鲡；肺炎克雷伯菌；致病性；鉴定；耐药

花鳗鲡（Anguilla marmorata）俗称鳝王、花锦鳝，属鳗鲡目鳗鲡科，是鳗鲡类中体型较大的一种，体长一般为 331～615 mm，体重 250 g 左右，最重的可达 30 kg 以上。由于其肉味鲜美，肉和肝的维生素 A 含量特别高，营养丰富，具有相当高的营养价值，且价格昂贵，历来被视为上等滋补食品，有"水中人参"之美誉。但是，由于工业有毒污水对河流的严重污染和捕捞过度，以及毒、电渔法对鱼资源的毁灭性破坏，拦河建坝修水库及水电站等阻断了花鳗鲡的正常洄游通道等原因，致使花鳗鲡的资源量急剧下降，因此，花鳗鲡成为濒危物种，是我国国家Ⅱ级保护野生动物。

肺炎克雷伯菌（Klebsiella pneumoniae），属于肠杆菌科（Enterobacteriaceae spp.），是人类呼吸道和肠道的常居菌，可引起人类肺炎、尿路感染、菌血症等感染[1]。除引起人类感染发病外，现已从禽类[2,3]、两栖类和鱼类[4-7]等多种动物体内分离到这种细菌，因此该菌作为人畜共同致病病原已被广泛重视。

近年来，花鳗鲡在广东的养殖规模逐渐增大。但随着养殖密度的不断提高，各种病害频发。除常见的寄生虫病外，由病原菌引起的鳗鲡烂鳃病、赤鳍病和爱德华氏菌病也尤为常见[8-10]，因此，不同种类的抗菌药物常用于各种细菌性疾病的防治。而目前，抗菌药物已在水产养殖中普遍试验，但由于不合理用药与滥用药物，导致细菌耐药性日渐增加，水产动物源耐药菌株的研究开始引起广大学者的关注。国内已报道从鱼、虾、龟等多种水产动物中分离到多重耐药的气单胞菌、弧菌、假单胞菌等水产病原菌[11-16]，这不仅给水产养殖动物疾病防治造成了困难，同时也严重影响了水产品质量乃至水环境的安全。本项研究针对从花鳗鲡分离的对多种抗菌药物耐药的菌株 HM2 的致病性和分类地位开展了研究。

1　材料与方法

1.1　试验材料

患病花鳗鲡及人工感染用花鳗鲡来源于广东顺德某花鳗鲡养殖场，体长 10～12 cm，体重 5～6 g。患病花鳗鲡其体色暗黑，红头，体表溃烂，腹部膨大，鳃丝充血肿胀，肝脏充血肿大。

胰蛋白胨大豆琼脂、MH 琼脂和肉汤、麦康凯琼脂、微量生化管及相关试剂购自广东环凯微生物科技有限公司，ID 32 E 肠杆菌细菌鉴定试剂条为法国生物-梅里埃公司出品，药敏试纸为英国 Oxoid 公司产品，PCR 所用试剂等均为康为世纪生物有限公司产品。

1.2　细菌分离培养

采用无菌操作取患病花鳗鲡肝脏，接种在胰蛋白胨大豆琼脂上，于 28℃恒温培养 18～20 h，挑取优势菌落划线纯化后，获得一株菌，编号为 HM2，并用脱脂牛奶作保护剂真空冻干，–20℃低温保存备用。

1.3 人工感染试验

采用静水式浸泡方法进行，实验期间水温（28±2）℃，采用 2 倍稀释法设定 6 个浓度梯度组，各组别细菌浓度分别为 $1.2×10^9$、$6×10^8$、$3×10^8$、$1.5×10^8$、$7.5×10^7$、$3.75×10^7$ CFU/mL，同时设不加菌液的阴性对照组，每组实验鱼 30 尾。所用 HM2 菌液于 MH 肉汤 28℃恒温震荡培养 20 h。人工感染后，连续饲养 7 d，观察、记录花鳗鲡的发病症状和死亡情况，并从濒死花鳗鲡的肝脏进行细菌再分离和鉴定。采用改良寇氏法计算 96 h 的 LD_{50}。

1.4 细菌生化鉴定及分子生物学鉴定

采用细菌常规鉴定方法，包括培养特性和生理生化试验，并同时进行 16S rRNA 基因序列分析。

1.4.1 培养特性

将 HM2 分别划线于胰蛋白胨大豆琼脂和麦康凯琼脂平皿上，于 28℃恒温培养 18~20 h，观察细菌的生长情况和菌落形态；同时从麦康凯琼脂平皿挑取单菌落制成涂片，革兰氏染色后，光学显微镜观察细菌的形态和染色特性。

1.4.2 生理生化试验

参照文献[17,18]将在麦康凯琼脂平皿上，于 28℃恒温培养 18~20 h 的 HM2 分别接种于吲哚、动力、VP、MR 等细菌微量生化管，28℃恒温培养 18~48 h，定时检查反应结果；同时将 HM2 接种于 ID 32 E 肠杆菌细菌鉴定试剂条，28℃恒温培养 24 h 后使用 ATB Expression 半自动细菌鉴定仪（法国生物–梅里埃 bio merieux 公司）进行生化指标的测定。

1.4.3 16S rRNA 基因序列分析

将 HM2 接种至 MH 营养肉汤中 28℃培养过夜，离心收集菌体，按照细菌基因组 DNA 提取试剂盒说明书提取细菌总 DNA。根据文献[19]设计合成 16S rRNA 引物，正向引物 P1：F-AGAGTTTGATCATGGCTCAG；反向引物 P2：R-GGTTACCT-TGTTACGACTT，由上海博尚生物技术有限公司合成。16S rRNA 基因 PCR 反应体系（50 μL）：2×Taq MasterMix 25 μL，P1 和 P2 引物（10 μmol/L）各 2 μL，DNA 模板 2 μL，最后以双蒸水补足，进行 PCR 扩增。反应条件为 95℃预变性 5 min，然后 95℃变性 30 s，55℃退火 30 s，72℃延伸 1 min，共 35 个循环，最后 72℃再延伸 10 min。PCR 扩增产物经 1%琼脂糖凝胶电泳后，阳性样品直接送博尚生物公司测序，测序结果与 GenBank 中相应序列进行同源性分析及系统进化树的构建。

1.5 药敏试验

采用纸片扩散法以大肠埃希菌（ATCC25922）作为质控菌进行药敏试验。在麦康凯琼脂平板上挑取单个菌落接种于 2 mL MH 肉汤中，28℃振荡培养 3~4 h，用无菌生理盐水稀释菌液，使稀释菌液达到 0.5 麦氏比浊管的浊度，再按 1：100 稀释菌液至含菌量约为 $1×10^6$ CFU/mL。用无菌棉拭子均匀涂抹至 MH 培养基上，贴上药敏纸片，正置 28℃培养 18~20 h 后观察，并根据美国临床实验室标准化委员会（Clinical and Laboratory Standards Institute，CLSI）标准判断结果[20,21]。

2 结果

2.1 人工感染试验

人工浸泡感染结果显示，在感染后 24 h，大部分组别花鳗鲡开始出现发病症状或死亡，表现为体色暗黑，体表充血，腹部膨大，鳃丝充血肿胀；其中 $1.2×10^9$ CFU/mL 组别的死亡率达 80%，$7.5×10^7$ CFU/mL 组别也出现死亡，死亡率为 10%；48 h 时 $3×10^8$ CFU/mL 及以上组别的死亡率达 100%，对照组和 $3.75×10^7$ CFU/mL 组别的死亡率为 0。根据改良寇氏法计算，96 h 的 LD_{50} 为 $9.98×10^7$ CFU/mL（表 1）。

表 1 人工感染试验结果

Tab. 1 Results of bath challenge trial

菌液浓度 Bacterial concentration（CFU/mL）	试验尾数 Number of tested fish	死亡率 Mortality（%）			
		24 h	48 h	72 h	96 h
		Mortality	Mortality	Mortality	Mortality
$1.2×10^9$	30	80	100	100	100
$6×10^8$	30	10	100	100	100
$3×10^8$	30	30	100	100	100
$1.5×10^8$	30	0	10	100	100

续表

菌液浓度 Bacterial concentration（CFU/mL）	试验尾数 Number of tested fish	死亡率　Mortality（%）			
		24 h Mortality	48 h Mortality	72 h Mortality	96 h Mortality
$7.5×10^7$	30	10	10	10	10
$3.75×10^7$	30	0	0	0	0
对照 control	30	0	0	0	0
LD$_{50}$（CFU/mL）	$9.98×10^7$				

2.2　细菌分类鉴定

2.2.1　细菌的形态与生化特性

HM2 为革兰氏阴性杆菌，在胰蛋白胨大豆琼脂平板上生长，菌落呈黄白色，在麦康凯琼脂平板上菌落呈浅粉色，边缘整齐，有黏性易成拉丝状；氧化酶、MR 阴性，VP 阳性；ATB Expression 半自动细菌鉴定仪鉴定结果显示，HM2 为肺炎克雷伯菌，ID 为 99.9%，T 值为 0.75，具体测定结果和其他生化特性测定结果（表 2）。常规鉴定结果显示，所分离的 HM2 菌株的培养特性和生理生化特性与文献[17,18]标准菌株基本一致，说明所分离的 HM2 菌株符合肺炎克雷伯菌的特征。从人工感染发病显症的华鳗鲡肝脏再分离的细菌也符合肺炎克雷伯菌的特征。

表 2　分离菌 HM2 的生理生化试验
Tab. 2　Physiological and biochemistrical characteristics of the strain HM2

测定项目 Test items	HM2	肺炎克雷伯菌 K. pneumoiae	测定项目 Test items	HM2	肺炎克雷伯菌 K. pneumoniae
氧化酶 Oxidase	−	−	α-葡萄糖苷酶 α-Glucosidase	−	−
动力 Mobility	−	−	L-阿拉伯糖 L-Arabinose	+	+
VP Voges-Proskauer	+	+	吲哚 Indol	−	−
MR Methyl red	−	−	α-甘露醇 α-Mannitol	+	+
酯酶 Lipase	−	−	酚红 Phenolrot	−	−
N-乙酰-β-葡萄糖甙酶 N-Acetyl-β-Glucosaminidase	−	−	5 酮基葡萄糖酸盐 5-Keto-Gluconat	−	−
D-麦芽糖 D-Maltose	+	+	L-阿拉伯糖醇 L-Arabitol	−	−
D-葡萄糖 D-Glucose	+	+	β-葡萄糖苷酶 β-Glucosidase	+	+
蔗糖 Saccharose	+	+	丙二酸盐 Malonate	+	+
D-半乳糖酸盐同化 D-Galacturonsaure	+	+	D-山梨醇 D-Sorbitol	+	+
尿素酶 Urease	+	+	β-葡萄糖醛酸酶 β-Glucuronidase	−	−
精氨酸双水解酶 Arginine dihydrolase	−	−	侧金盏花醇 Adonitol	−	+
鸟氨酸脱羧酶 Ornithine decarboxyla	−	−	L-鼠李糖 L-Rhamnose	+	+
赖氨酸脱羧酶 Lysin decarboxylase	+	+	α-半乳糖甙酶 α-Galactosidase	+	+
肌醇 Inositol	+	+	D-阿拉伯糖醇 D-Arabitol	+	+
α-麦芽糖 α-Maltosidase	−	+	古老糖 Palatinose	+	−
D-纤维二糖 D-Cellobiose	+	+	L-天冬氨酸芳胺酶 L-Aspartat-Arylamidase	−	−
D-海藻糖 D-Trehalose	+	+	β-半乳糖甙酶 β-Galactosidase	+	+
葡萄糖产气 Glucosegas	+	+			

注：+，表示阳性；−，表示阴性

Note：+，Positive：−，Negative

2.2.2　16S rRNA 基因序列分析

PCR 扩增出分离菌 HM2 的 16S rRNA 基因片段约为 1 500 bp，PCR 产物直接测序，获得的片段大小为 1 393 bp，Genbank 登录号为 JX282908。将获得的序列与 GenBank 数据库中已报道的 16S rRNA 基因序列进行 Blast、同源性分析及构建系统发育树。HM2 与肺炎克雷伯菌的同源性最高，同等片段长度同源性高达 100%，在系统发育树上与肺炎克雷伯菌聚为一簇（图 1），说明所分离 HM2 为肺炎克雷伯菌。

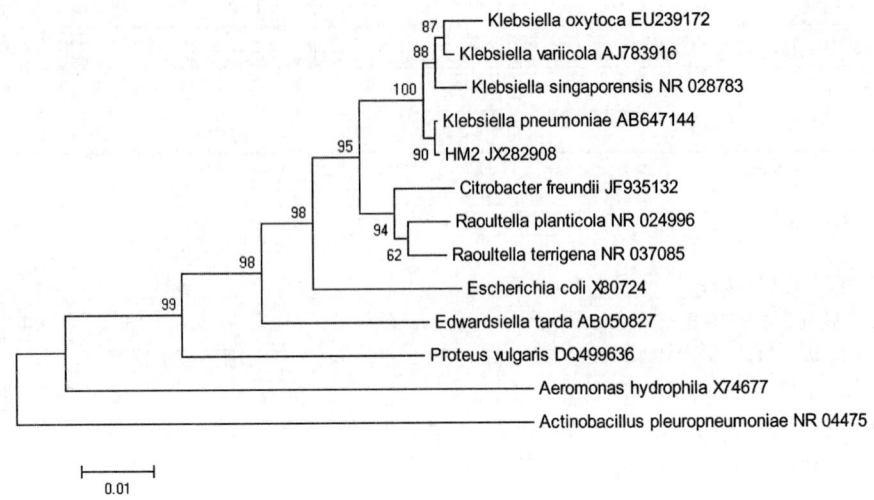

图 1　菌株 HM2 的聚类分析

Fig. 1　The cluster analysis of the strain HM2 based on the 16S rRNA

2.3　药敏试验

测定分离菌株 HM2 对 20 种常见抗菌药物的敏感性，结果显示，HM2 除对亚胺培南、链霉素、阿米卡星 3 种药物敏感外，对氨苄西林、阿莫西林/克拉维酸、头孢噻吩、头孢曲松、头孢噻肟、头孢西丁、复合磺胺、磺胺甲基异噁唑/甲氧苄啶、利福平、萘啶酸、环丙沙星、诺氟沙星、氧氟沙星、呋喃妥因、四环素、多西环素、氯霉素 17 种药物均耐药（表 3）。

表 3　HM2 株菌的药敏试验结果

Tab. 3　Susceptibility of HM2 strain toantimicrobial agents

抗菌药物 Antimicrobial agents	药物含量 Disk content（μg）	抑菌圈直径折点 Zone diameter Breakpoint（mm）		抑制圈大小 Inhibition zone size（mm）		HM2 的耐药判定结果 Resistance determination of HM2
		耐药 Resistance	敏感 Susceptible	HM2	ATCC25922	
氨苄西林 Ampicillin	10	≤13	≥17	0	0	R
阿莫西林/克拉维酸 Amoxicillin/Clavulanic acid	30	≤13	≥18	12	11	R
头孢噻吩 Cefalotin	30	≤14	≥18	0	9	R
头孢曲松 Ceftriaxone	30	≤19	≥23	0	33	R
头孢西丁 Cefoxitin	30	≤14	≥18	0	16	R
头孢噻肟 Cefotaxime	30	≤22	≥26	0	35	R
亚胺培南 Imipenem	10	≤13	≥16	22	19	S
复合磺胺 Sulfonamides	300	≤12	≥17	0	19	R
磺胺甲基异恶唑/甲氧苄啶 Sulfamethoxazole/Trimethoprim	23.75/1.25	≤10	≥16	0	23	R
利福平 Rifampin	5	≤16	≥20	0	16	R

续表

抗菌药物 Antimicrobial agents	药物含量 Disk content（μg）	抑菌圈直径折点 Zone diameter Breakpoint（mm）		抑制圈大小 Inhibition zone size（mm）		HM2 的耐药判定 结果 Resistance de- termination of HM2
		耐药 Resistance	敏感 Susceptible	HM2	ATCC25922	
萘啶酸 Nalidixic acid	30	≤13	≥19	0	31	R
环丙沙星 Ciprofloxacin	5	≤15	≥21	0	32	R
诺氟沙星 Norfloxacin	10	≤12	≥17	0	30	R
氧氟沙星 Ofloxacin	5	≤12	≥16	0	30	R
呋喃妥因 Nitrofurantoin	300	≤14	≥17	10	24	R
四环素 Tetracycline	30	≤14	≥19	0	28	R
多西环素 Doxycycline	30	≤10	≥14	0	27	R
链霉素 Streptomycin	10	≤11	≥15	16	19	S
阿米卡星 Amikacin	30	≤14	≥17	21	21	S
氯霉素 Chloramphenicol	30	≤12	≥18	0	30	R

注：R，耐药；S，敏感。

Note：R，Resistant；S，Susceptible.

3　讨论

3.1　肺炎克雷伯菌在水产养殖中的分布与致病性

Singh, et al[22]于 1992 年在鱼、虾、蟹等水产类食品中检测到肺炎克雷伯菌；唐毅等[5]发现在鲢鱼上分离的肺炎克雷伯菌能正常存在于养殖水体和鱼体的肠道内，并在鱼体受伤、水质恶化、气候变化等外在条件改变时才可能导致鱼体致病；此外，通过肺炎克雷伯菌回归感染实验发现被感染鱼的症状与自然发病一致，表现为鳃丝溃烂、肝胰脏充血、腹部肿大等。邓国成等[7]于 2008 年首次报道从鳗鲡成鱼分离到一株有致病性的肺炎克雷伯菌，被感染鱼也表现为体色暗黑、腹部膨大、肝脏出血、肠道腹水等。本实验从花鳗鲡小鱼苗所分离的 HM2 菌株经人工感染试验证实具有致病性，感染发病鱼的症状与自然发病的相似，表现为体色暗黑，体表充血，腹部膨大，鳃丝充血肿胀，与上述肺炎克雷伯菌感染的鲢鱼及鳗鲡的症状相似，表明肺炎克雷伯菌对花鳗鲡鱼苗和鳗鲡成鱼均具有致病性，但具体的致病机理还有待进一步研究。

本实验分离的菌株 HM2 经革兰氏染色、细菌培养特性及生理生化特性测定后，初步鉴定为革兰氏阴性杆菌；在胰蛋白胨大豆琼脂平板上生长，菌落呈黄白色，在麦康凯琼脂平板上菌落呈浅粉色，边缘整齐，有黏性易成拉丝状；氧化酶、MR、吲哚、鸟氨酸脱羧酶、精氨酸双水解酶阴性，VP、脲酶、赖氨酸脱羧酶、葡萄糖产气、蔗糖、L-阿拉伯糖阳性，具有肺炎克雷伯菌的特性，其结果与其他报道基本一致[4-7,22]。为进一步确定菌株 HM2 分类学地位，在传统生化鉴定的基础上，测定了该株 16S rRNA 基因的部分序列，并进行系统进化树的构建和同源性分析。从构建的系统进化树和 Blast 同源性分析结果看，菌株 HM2 与肺炎克雷伯菌聚为一类，与相应基因序列的同源性均大于 99%。综合细菌培养特性、生理生化特性和 16S rRNA 基因序列测定分析结果，将分离纯化的菌株 HM2 确定为肺炎克雷伯菌。

3.2　肺炎克雷伯菌的耐药性

徐海圣等[4]、陶锦华等[6]在浙江和广西从龟鳖分离的肺炎克雷伯菌对喹诺酮类、四环素类及磺胺类药物等耐药；而邓国成等[7]在广东从鳗鲡分离的致病菌对喹诺酮类、氨基糖苷类、四环素类及头孢菌素类药物中度敏感或高度敏感，而只对青霉素类耐药；本研究分离的 HM2 对亚胺培南及氨基糖苷类的链霉素、阿米卡星 3 种药物敏感，对头孢类、磺胺类、喹诺酮类以及四环素类等 17 种药物均耐药，表现为多重耐药，其耐药程度比上述报道的严重。肺炎克雷伯菌不仅是一种重要的病原菌，而且还是我国 CHINET 细菌耐药性监测系统定期监测的耐药菌之一[23,24]。本实验首次从养殖的花鳗鲡分离到对头孢类等多种药物耐药的肺炎克雷伯菌株，应引起广大养殖者和科研工作者的高度重视。

有研究表明[2-4]，肺炎克雷伯菌对氯霉素最敏感，而本实验结果却高度耐药；氯霉素已在水产养殖上禁用，但菌株 HM2 却表现出很强的耐药性，可能与水产养殖业的用药历史有关。头孢类药物是人医临床治疗细菌感染的一线抗感染药物，也常于畜牧养殖和小动物临床上使用，动物源肺炎克雷伯菌对头孢类药物的耐药率远高于人医临床分离株[25,26]；水产养殖上较少使用头孢类药物，但本实验所分离的 HM2 却对人医临床常用抗菌药头孢曲松、头孢西丁及头孢噻肟产生了耐药性。目前关于肺炎克雷伯菌耐药机制的研究在人医临床报道中较多，主要包括产超广谱β-内酰胺酶、生物被膜形成、外膜孔蛋白缺失、染色体基因突变、主动外排作用以及质粒介导的耐药机制等等[27]，而在水产动物源中报道相对较少。因此，水产源多重耐药菌株产生及其传播机制还有待进一步研究。

水产动物源多重耐药菌株随水产养殖污水流入江河后，其耐药因子是否会转入临床致病菌，经过食物、水等途径感染人体，使人源肺炎克雷伯菌的耐药现象更严重；或者耐药菌株是否会残留在水产动物体内通过食物链作用富集到人体内，在临床上产生耐药性，这将对人类生存健康造成巨大威胁。因此有必要开展对水产动物源细菌耐药性监测和研究工作，以指导水产养殖合理用药。

参考文献

[1] Wang L. F., Wang X. D.. Advances in the mechanism of drug resistance of Klebsiella pneumonia [J]. Chinese Journal of Antibiotics, 2004, 29 (6): 324-328 [王林峰, 王选锭. 肺炎克雷伯菌耐药机制研究进展. 中国抗生素杂志, 2004, 29 (6): 324-328].

[2] Hang Y. Y., Wan Y., Chen X. Z., et al. Study on the pathogenicity and biological character of chicken Klebsiellosis pneumonia [J]. Fujian Journal of Animal Science and Veterinary Medicine, 1996, 18 (2): 4-5 [黄印尧, 万沅, 陈信忠, 等. 鸡源肺炎克雷伯菌的致病性和生物学特性研究. 福建畜牧兽医, 1996, 18 (2): 4-5].

[3] Peng Y. Y., Liu H. Y., Wang H. J., et al. Study on the biological characteristics of Klebsiella pneumonia from chicken [J]. Journal of Southwest Agricultural University, 1995, 17 (4): 330-333 [彭远义, 刘华英, 王豪举, 等. 鸡肺炎克雷伯菌的生物学特性研究. 西南农业大学学报, 1995, 17 (4): 330-333].

[4] Xu H. S., Shu M. A.. Studies on the pathogens of the Klebsiella pneumoniae disease of Trionyx sinensis [J]. Journal of Zhejiang University (Science Edition), 2002, 29 (6): 702-706 [徐海圣, 舒妙安. 中华鳖肺炎克雷伯菌病的病原研究. 浙江大学学报 (理学版), 2002, 29 (6): 702-706].

[5] Tang Y., Zhang F., Xun H. C., et al. Isolation and identification of Klebsiella pneumoniae from silver carp [J]. Journal of Southwest University (Natural Science Edition). 2007, 29 (6): 73-76 [唐毅, 张芬, 孙翰昌, 等. 白鲢肺炎克雷伯菌的分离鉴定. 西南大学学报 (自然科学版), 2007, 29 (6): 73-76].

[6] Tao J. H., Li K. R., Wei P.. Diagnosis and metaphylaxis of Klebsiella pneumonia isolated from Mauremys mutica [J]. Guangxi Journal of Animal Husbandry and Veterinary Medicine, 2002, 18 (6): 20-21 [陶锦华, 李康然, 韦平. 石龟肺炎克雷伯菌感染的诊断与防治. 广西畜牧兽医, 2002, 18 (6): 20-21].

[7] Deng G. C., Luo X., Jiang X. Y., et al. Separation and identification of the eel Klebsiella pneumonia [J]. Journal of Shanghai Ocean University, 2009, 18 (2): 2193-2197 [邓国成, 罗霞, 江小燕, 等. 鳗肺炎克雷伯菌的分离与鉴定. 上海海洋大学学报, 2009, 18 (2): 2193-2197].

[8] Chen C. F., Wu Z. X., Gao H. J.. Isolation and identification of pathogenic bacteria causing edwardsiellosis in eel (Anguilla japonica) [J]. Journal of Huazhong Agricultural University, 1998, 17 (4): 382-388 [陈昌福, 吴志新, 高汉娇. 日本鳗鲡爱德华氏菌病病原菌的分离与鉴定. 华中农业大学学报, 1998, 17 (4): 382-388].

[9] Chen H. B., Lin Y. D., Weng Y. L., et al. Isolation, identification and drug resistance of the pathogens of red fin disease in eels [J]. Acta Hydrobiologica Sinica, 1992, 16 (1): 40-46 [陈会波, 林阳东, 翁燕玲, 等. 鳗鲡赤鳍病病原菌的分离鉴定和耐药性的研究. 水生生物学报, 1992, 16 (1): 40-46].

[10] Fan H. P., Xu J. E., Huang X. F.. Investigation and control for the bacterial diseases of cultured eels in Fu jian province [J]. Transactions of Oceanology and Limnology, 1996, 2: 67-70 [樊海平, 徐娟儿, 黄晓讽. 福建省养殖鳗鲡细菌性疾病的调查与防治. 海洋湖沼通报, 1996, 2: 67-70].

[11] Shen J. Y., Qian D., Liu W., et al. Studies on the pathogens of bacterial diseases of Macrobrachium nipponens [J]. Journal of Zhejiang Ocean University (Natural Science), 2000, 19 (3): 222-224 [沈锦玉, 钱冬, 刘问, 等. 养殖青虾"红鳃病"病原的研究. 浙江海洋学院学报 (自然科学版), 2000, 19 (3): 222-224].

[12] Chu W. H.. Studies on the pathogens, prevention and cure of Bacterial septicemia of Carassius auratus gibelio [J]. Reservoir Fisheries, 2000, 20 (1): 27-28 [储卫华. 异育银鲫细菌性败血症病原及防治研究. 水利渔业, 2001, 21 (1): 40].

[13] Shen J. Y., Yi W. L., Qian D., et al. Studies on the pathogens of bacterial diseases of Eriocheir sinensis [J]. Fishery Sciences of China, 2000, 7 (3): 89-92 [沈锦玉, 尹文林, 钱冬, 等. 中华绒螯蟹"腹水病"及"抖抖病"并发病病原的研究. 中国水产科学, 2000, 7 (3): 89-92].

[14] Tang X. C., Xue C. F., Gao Z. H.. Studies on the pathogens of Rana grylto Mejnegry [J]. Reservoir Fisheries, 2000, 20 (1): 27-28 [汤显春, 薛翠峰, 高智慧. 沼泽绿牛蛙致病细菌的研究. 水利渔业, 2000, 20 (1): 27-28].

[15] Tian T., Hu H. G., Chen C. F.. Identification and Pathogenicity of bacterial pathogens isolated in an outbreak of bacterial disease of bluntnose black bream, Megalobrama amblycephala [J]. Journal of Huazhong Agricultural University, 2010, 29 (3): 341-345 [田甜, 胡火庚, 陈昌福. 团头鲂细菌性败血症病原分离鉴定及致病力研究. 华中农业大学学报, 2010, 29 (3): 341-345].

[16] Li A. H.. Detection and properties of transferable R+ plasmid in fish-pathogenic bacteria [J]. Acta Hydrobiologica Sinica, 1998, 22 (4): 319-324 [李爱华. 鱼类病原菌中 R + 质粒的检出及其特性. 水生生物学报, 1998, 22 (4): 319-324].

[17] Dong X. Z., Cai M. Y.. Manual of Systematic and Determinative Bacteriology [M]. Beijing: Science Press. 2001, 91-92; 349-398 [东秀珠, 蔡妙英. 常见细菌系统鉴定手册. 北京: 科学出版社. 2001, 91-92; 349-398].

[18] Krieg N. R., Holt J. G.. Bergey's Manual of Systematic Bacteriology (volume 1) [M]. Bltimore: Williams & Wilkins. 1984, 461-465.

[19] Borrell N., Acinas S. G., Figueras M J, et al. Identification of Aeromonas clinical isolates by restriction fragment length polymorphism of PCR-amplified 16S rRNA genes [J]. Journal of Clinical Microbiology, 1997, 35: 1671-1674.

[20] Clinical and Laboratory Standards Institute. Performance standards for antimicrobial disk susceptibility tests for bacteria isolated from animals; Approved standard [M]. 3rd ed. Wayne, PA. CLSI, 2008a, M31-A3.

[21] Clinical and Laboratory Standards Institute. Performance Standards for Antimicrobial Susceptibility Testing: Eighteenth Informational Supplement [M]. Wayne, PA, CLSI. 2008, M100-S18.

[22] Singh B. R., Kulshreshtha S B. Preliminary examinations on the enterotoxigenieity of isolates of *Klebsiella pneumoniae* from seafoods [J]. International Journal of Food Microbiology, 1992, 16 (4): 349-352.

[23] Ye S. J., Yang Q., Yu Y. S.. CHINET 2005 surveillance of antimicrobial resistance in *Escherichia coli* and *Klebsiella pneumonia* [J]. Chinese Journal of Infection Chemotherapy, 2007, 7 (4): 283-286 [叶素娟, 杨青, 俞云松. 2005 年中国 CHINET 大肠埃希菌和肺炎克雷伯菌耐药性分析. 中国感染与化疗杂志, 2007, 7 (4): 283-286].

[24] Zhuo C., Su D. H., Ni Y. X., et al. CHINET 2009 surveillance of antimicrobial resistance in *E. coli* and Klebsiella spp in China [J]. Chinese Journal of Infection Chemotherapy, 2010, 10 (6): 430-435 [卓超, 苏丹虹, 倪语星, 等. 2009 年中国 CHINET 大肠埃希菌和克雷伯菌属细菌耐药性监测. 中国感染与化疗杂志, 2010, 10 (6): 430-435].

[25] Zou L. K., Wang H. N., Zeng B., et al. Phenotypic and genotypic characterization of β-lactam resistance in *Klebsiella pneumoniae* isolated from swine [J]. Veterinary Microbiology, 2011 149 (1-2): 139-146.

[26] Ma J., Zeng Z., Chen Z., et al. High prevalence of plasmid-mediated quinolone resistance determinants qnr, aac (6') -Ib-cr, and qepA among ceftiofur-resistant Enterobacteriaceae isolates from companion and food-producing animals [J]. Antimicrobial Agents and Chemotherapy, 2009, 53 (2): 519-524.

[27] Wang Y. H., Deng M., Zeng J.. Advances in the mechanism of drug resistanceof *Klebsiella pneumoniae* [J]. Chinese Journal of Nosocomiology, 2007, 17 (4): 478-480 [王玉红, 邓敏, 曾吉. 肺炎克雷伯菌耐药机制研究进展. 中华医院感染学杂志, 2007, 17 (4): 478-480].

第二节 病原菌耐药机制的研究

病原菌耐药性产生的机制较复杂, 但主要有固有耐药性和获得性耐药两种方式。本章主要分析了鱼源嗜水气单胞菌 (*Aeromonas hydrophila*)、溶藻弧菌 (*Vibrio alginolyticus*)、哈维氏弧菌 (*V. harveyi*) 耐药产生的机制。

一、鱼源嗜水气单胞菌耐药机制的研究

嗜水气单胞杆菌对恩诺沙星的耐药机制可能与控制细胞内药物蓄积的 ABC 转运蛋白的增加和拓扑异构酶 IV 减少密切相关 (Feng-Jiao Zhu et al., 2017)。水产养殖环境中耐药基因和整合子的分布广泛, bla$_{TEM}$, tetC, sulI, aadA, floR, qnrB 等耐药基因具有较高的出现频率 (Mao Lin et al., 2016)。鱼源嗜水气单胞菌呈多重耐药性, 且不同类型整合子的多重耐药率不同, 表明整合子系统在嗜水气单胞菌多重耐药性中发挥重要作用 (李绍戊, 2013)。嗜水气单胞菌对氟喹诺酮类耐药存在靶基因位点突变及主动外排等多种耐药机制; 在防治选择治疗药物需考虑交叉耐药情况 (崔佳佳, 2016)。tetE 基因可能是介导单胞菌对四环素类药物耐药的优势基因 (崔佳佳, 2016)。经药物的体外诱导, 嗜水气单胞菌对喹诺酮类药物的耐药基因 gyrA 易发生突变并引起对喹诺酮类药物低水平耐药; 随着耐药压力的逐步增大, 外排作用协同突变机制且细菌对喹诺酮类的耐药性增强 (邓玉婷, 2014)。分析嗜水气单胞菌对喹诺酮类药物的耐药基因 gyrA 和 parC 编码的喹诺酮类耐药决定区 (QRDR) 的突变, 结果表明, QRDR 区域的突变具有随机性, 且喹诺酮类药物耐药性的产生可能还与其他耐药机制存在关联 (薛慧娟, 2012)。

Transcriptome differences between enrofloxacin-resistant and enrofloxacin-susceptible strains of Aeromonas hydrophila

Feng-Jiao Zhu, Kun Hu, Zong-Ying Yangand Xian-Le Yang

(National Pathogen Collection Center for Aquatic Animals, Shanghai Ocean University, 999 Hucheng Huan Road, Shanghai 201306, China.)

Abstract: Background: Enrofloxacin is the most commonly used antibiotic to control diseases in aquatic animals caused by

A. hydrophila. A. hydrophila strains have begun to develop strong resistance to enrofloxacin; however, little is known about the molecular mechanisms of resistance. Aims：We conducted de novo transcriptome sequencing and compared the global transcriptomes of enrofloxacin-resistant and enrofloxacin-susceptible strains. Methods：*A. hydrophila* strains ATCC7966 was used as the enrofloxacin-sensitive strain, which was used to induce an enrofloxacin-resistant strain. After functional clustering analysis was performed to obtain a unigene database, the unigenes were annotated to the Swiss-Prot and National Center for Biotechnology Information (NCBI) non-redundant protein databases. Gene ontology (GO) and functional pathway analyses of differentially expressed genes (DEGs) were performed, and the DEGs were verified by qRT-PCR. Results：A total of 4,714 unigenes were assembled (2,353 unigenes were assembled from the resistant strains (7966QR) and 1,821 unigenes were assembled from the susceptible strains (ATCC7966); of these, 4,122 were annotated. A total of 3,280 unigenes were assigned to GO, 3,388 unigenes were classified into Cluster of Orthologous Groups of proteins (COG) using BLAST and BLAST2GO software, and 2,568 unigenes were mapped onto pathways using the Kyoto Encyclopedia of Gene and Genomes Pathway (KEGG) database. Furthermore, 218 unigenes were deemed to be DEGs. After enrofloxacin treatment, 135 genes were upregulated and 83 genes were downregulated in A. hydrophila. The GO terms biological process (126 genes) and metabolic process (136 genes) were the most enriched, and the terms for protein folding, response to stress, and SOS response were also significantly enriched. Metabolic pathways, genetic information processing, ABC transporters, and metabolism-related biological pathways were also significantly enriched. Conclusion：Our results indicate that enrofloxacin treatment affects multiple biological functions such as ABC transporters, cellular response to DNA damage stimulus, and SOS response of *A. hydrophila*. Our findings support that the *A. hydrophila resistance* mechanism of enrofloxacin is closely related to reduction of intracellular drug accumulation caused by ABC transporters and topoisomerase IV.

Key words：*Aeromonas hydrophila*; transcriptome; enrofloxacin; resistance mechanism

耐恩诺沙星嗜水气单胞菌菌株及敏感性菌株的转录组分析

朱凤娇，胡鲲，杨宗英，杨先乐

(上海海洋大学国家水生动物病原库，中国上海，沪城环路999号)

摘要：背景：在水产养殖中，恩诺沙星作为一种最常用的抗生素来控制由嗜水气单胞杆菌引起的水产动物疾病。目前已经发现了对恩诺沙星有耐药性的嗜水气单胞菌菌株。但是对其耐药性的分子机制还了解甚少。目的：我们利用从头转录组测序的方法比较分析对恩诺沙星敏感的嗜水气单胞菌和对恩诺沙星耐药的嗜水气单胞杆菌的转录组差异来寻找耐药基因。方法：选用标准菌株ATCC7966嗜水气单胞菌菌株作为恩诺沙星敏感菌株，用此菌株来诱导产生恩诺沙星耐药菌株。我们利用功能富集分析得到一个基因数据库，基因数据库中的利用Swiss-Prot和NCBI非冗余蛋白质序列数据库对该基因数据库中的序列进行注释。利用GO功能分析和pathway生物通路分析对差异表达基因进行分析，并通过实时荧光定量PCR验证差异表达基因。结果：研究组装得到了4714条基因序列其中耐药菌株组装得到2353条基因序列敏感菌株组装得到1821条基因序列。4122条组装得到基因序列得到了基因功能注释。其中GO功能注释3280条，利用BLAST和BLAST2GO软件分析COG注释3388条，利用KEGG数据库分析有2658条序列被映射到相应的生物通路。比较耐药菌株和敏感菌株的转录组差异分析，实验获得了218个明显差异基因，其中135个基因明显上调83个基因明显下调。GO功能分析显示大多数的差异基因分布在生物过程（126个基因）和代谢过程（136个基因）。其中蛋白质折叠，应激反应，和SOS反应明显富集。代谢途径，遗传信息处理和代谢相关的生物途径也有富集。结论：研究发现嗜水气单胞杆菌对恩诺沙星耐药性的产生主要是通过影响多种生理功能如ABC转运蛋白，DNA损伤修复，SOS反应等，同时研究结果表明嗜水气单胞杆菌对恩诺沙星的耐药机制可能与控制细胞内药物蓄积的ABC转运蛋白的增加和拓扑异构酶IV减少密切相关。

关键词：嗜水气单胞菌；转录组；恩诺沙星；耐药机制

Introduction

Following the decline in the capture fishing industry and diminishing wild fish stocks, the aquaculture industry has become an important source of food fish[1]. However, bacterial diseases hinder desirable production outputs. The gram-negative bacterium *Aeromonas hydrophila* is one of the major causative agents of disease and can cause serious damage in many animals[2], especially fish[3][4] as well as humans[5]. *A. hydrophila*, which is a representative of the Aeromonadaceae family, is an emerging aquatic pathogen that is distributed in a wide variety of aquatic systems[6] and[7]. It primarily inhabits freshwater and the intestines of freshwater animals. Farmers use a wide range of antibiotics or chemicals to control A. hydrophila infection[8]. Enrofloxacin is a third-generation fluoroquinolone with a broad antibacterial spectrum and high potency, which is commonly used to treat bacterial infections afflicting aquaculture[9]. Enrofloxacin has been mostly used to control A. hydrophila infections in aquaculture, and recently A. hydrophila has developed strong resistance to enrofloxacin among other drugs[10]. This resistance has rendered it increasingly difficult to treat diseases caused by A. hydrophila in aquaculture animal diseases. Moreover, as heavy antibiotic use is associated with negative effects such as antibiotic resistance in environment and fish[12]. Eventually, antibiotic use may be detrimental to environment and human health[11].

Based on previous reports quinolone-resistant bacteria adopt the following three main strategies of antibiotic resistance: First, chromosome-mediated changes in the topoisomerase target sites (changes in the amino acids in the quinolone resistance-determining region)[13]. Second, reduction in intracellular drug accumulation caused by efflux pump[14]. Third, bacterial protection conferred by plasmid-encoded qnr protein[15]. However, the mechanisms by which A. hydrophila is resistant to enrofloxacin remains unclear, and little is known about its molecular mechanisms of resistance. Here, we conducted de novo transcriptome sequencing for the comprehensive analysis of the global transcriptomes of enrofloxacin-susceptible and enrofloxacin-resistant strains. Transcriptomic profiling is used to analyze gene expression and signaling pathways in specific tissues or cells. The recent rapid development of next-generation sequencing technologies such as the Solexa/Illumina technology offers great advantages in analyzing the functional complexity of the entire transcriptome[16]. Next-generation sequencing techniques have been used for transcriptome analyses to simultaneously provide data on sequence polymorphisms and the levels of gene expression involved in cellular development, cancer, and the immune responses[17] and[18].

In the present study, we examined the genetic diversity of A. hydrophila using de novo transcriptome sequencing, and investigated the molecular mechanisms of enrofloxacin resistance in A. hydrophila.

Materials and Methods

Culture of susceptible strain and induction of an enrofloxacin-resistant strain

A. hydrophila ATCC7966, which was used in the present study, had been maintained at the National Pathogen Collection Center for Aquatic Animals, China. Because the genome background of the ATCC7966 standard strain is known, it is known to be sensitive to quinolones; therefore, it was selected as the sensitive strain. This strain was inoculated on LB agar and incubated at 28℃ for 24 h.

The type strain ATCC7966 was used to develop an enrofloxacin-resistant strain by culturing it in the presence of gradually increasing concentrations of enrofloxacin in vitro and judged by the critical concentration suggested by the Clinical and Laboratory Standards Institute (CLSI, 2011). The results of the drug sensitivity test were determined using the disc diffusion method. Resistance and sensitivity to enrofloxacin were determined on the basis of the drug sensitivity evaluation criteria issued by the American Association of Clinical Laboratory Standards in 2011[19]. Briefly, ATCC7966 cells were inoculated on LB agar containing 1/2 the minimum inhibitory concentration (MIC) of enrofloxacin (Shanghai Guoyao Chemical Reagent Co. Shanghai, China). Every two days, the cultured strains were inoculated into fresh LB agar containing 2× the previous concentration of enrofloxacin. The strains were inoculated to LB agar containing no enrofloxacin until the cultured strain of MIC was increased to the extent of drug resistance determination. The resultant resistant strains were screened on LB for 12 generations until the resultant strain (7966QR) could be deemed resistant.

Suspensions containing the susceptible (ATCC7966) and resistant (7966QR) strains were collected, centrifuged, and the final products were store at 4℃ until analysis.

RNA isolation, RNA-Seq library construction, and sequencing

Frozen samples were ground in a mortar with liquid nitrogen, and total RNA was extracted from approximately 1 mL of bacterial suspension with TRIzol reagent (Invitrogen, USA), according to the manufacturer's instructions. DNA contaminants were removed by treatment with RNase-free DNase I (Takara Biotechnology, Dalian, China). The resultant total RNA was dissolved in 200 μL RNase-free water. The concentration of total RNA was determined using a Nano-Drop2000 spectrophotometer (Thermo Scientific, USA), and its integrity was checked using an RNA 6000 Pico LabChip with the Agilent 2100 bioanalyzer (Agilent, USA) at 37℃ for 1 h, and the sample volume was diluted to 250 μL using nuclease-free water. Messenger RNA (mRNA) was further purified with a Micropoly (A) Purist kit (Ambion, USA) according to the manufacturer's protocol. mRNA was dissolved in 100 μL of RNA Storage Solution (Ambion), purified using oligo (dT) magnetic beads, fragmented by treatment with divalent cations and heat, and reverse transcribed into cDNA using reverse transcriptase and random hexamer primers. This was followed by second-strand cDNA synthesis using DNA polymerase I and RNaseH. The resultant double-stranded cDNA was end-repaired using T4 DNA polymerase, Klenow fragments, and T4 polynucleotide kinase followed by a single (A) base addition using Klenow 3′ to 5′ exopolymerase. This was then ligated with an adapter or index adapter using T4 Quick DNA ligase. The size range of the adapter-modified fragments was selected by gel purification and used as templates in PCR amplification. The cDNA library was validated with an Agilent 2100 Bioanalyzer and

ABI StepOnePlus Real-time PCR system, and sequenced on a flow cell using an Illumina HiSeq 2500 (Illumina, SanDiego, USA).

Sequencing, data processing, and quality control

We filtered low-quality reads and removed 3′-adapter sequences using Trim Galore. The obtained reads were cleaned using FastQC software (http: //www. bioinformatics. babraham. ac. uk/projects/fastqc/), and the content and quality of the nucleotide bases in the sequencing data were evaluated. Next, we conducted a comparative analysis with the reference genome (Aeromonas hydrophila subsp. Hydrophila ATCC 7966). For each sample, sequence alignment with the reference genome sequences was carried out using Tophat[20].

Assembly and functional annotation

High-quality reads were obtained after removing the adapter sequence, low-quality reads (reads with ambiguous bases N), and duplicate sequences using Trim Galore and FastQC, and then used FastQC software to clean reads and evaluate the performance of different k-mers. Next, the clean reads were combined by de Bruijn graphs and SOAPdenovo software based on sequence overlap to form longer fragments (without ambiguous 'N' reads), to create contigs[21]. Furthermore, the contigs were connected into transcript sequences and joined into scaffolds using paired-end reads. The paired-end reads were also used to fill the gaps in scaffolds, where the unigenes have the least Ns and cannot be extended on both ends. Based on the results of the assembly evaluation, the best results were selected and used for clustering analysis using TGI Clustering tools to achieve a unigene database[22]. The obtained unigenes were compared with the National Center for Biotechnology Information (NCBI), non-redundant protein (Nr), and UniProt databases using BLASTx (Basic Local Alignment Search Tool) search with an E value<0. 00001. Based on the results of the Nr annotation, we used Blast2GO software (https: //www. blast2go. com/) to analyze functional annotations by gene ontology terms (GO; http: // www. geneontology. org)[23]. The unigenes were also aligned to the Kyoto Encyclopedia of Genes and Genomes (KEGG), Clusters of Orthologous Group (COG), and Swiss-Prot databases to predict and classify gene functions to perform pathway annotation searching the unigenes that similarity>30% and E value<0. 00001, and merge all the information.

Analysis of differentially expressed unigenes

To estimate the expression level (relative abundance) of a specific transcript expressed as fragments per kilobase per million fragments mapped (FPKM), we used RSEM software with the default parameter settings[24]. The expression level of each transcript was transformed using base 2 $\log_2$ (FPKM+1). The fold change of a transcript and differentially expressed genes (DEGs) were estimated using DESeq software [25]. Twofold changes in expression level and differences with a p value of <0. 05 were considered significant.

GO functional and pathway enrichment analysis of differentially expressed genes (DEGs)

We annotated DEGs to analyze the transcriptome differences between enrofloxacin-resistant and enrofloxacin-susceptible strains of A. hydrophila. To this end, we used GO terms in accordance with previously published procedures[26]. This analysis first mapped all DEGs to GO terms in the database by calculating gene numbers for every term followed by an ultra-geometric test to identify significantly enriched GO terms in DEGs compared to the transcriptome background. The following formula was used:

$$P = 1 - \sum_{i=0}^{m-1} \frac{\binom{M}{i}\binom{N-M}{n-i}}{\binom{N}{n}}$$

Where N represents the number of all genes with GO annotation, n represents the number of DEGs in N, M represents the number of all genes annotated to specific GO terms, and m represents the number of DEGs in M. The calculated p value was subjected to Bonferroni correction. A corrected p value<0. 05 was defined as the "threshold." GO terms were considered significantly enriched in the DEGs.

Pathway analysis of DEGs

Pathways of DEGs were annotated against the KEGG database using the BLASTall program (http: //nebc. nox. ac. uk/bioinformatics/docs/blastall. html). Enriched DEG pathways were identified according to the same formula as that used in the GO analysis. In this case, N represented the number of all genes with KEGG annotations, n represented the number of DEGs in N, M was the number of all genes annotated to specific pathways, and m was the number of DEGs in M[26].

Verification of DEGs using qRT-PCR

Quantitative RT-PCR (qRT-PCR) was used to verify the expression level of DEGs that were identified by RNA-Seq analysis. Primers were designed using Primer 5 software and SpTub-b was used as the reference gene[27,28]. Reactions were performed in a 25 μL reaction volume composed of 2 μL cDNA, 0. 5 μL each of forward and reverse primers (10 μM), 12. 5 μL SYBR Premix Ex Taq (2×) and 9. 5 μL RNase-free H_2O. The thermal cycle protocol was as follows: 95℃ for 30 s followed by 40 cycles of 95℃ for 5 s, 60℃ for 30 s, and 72℃ for 30 s. Melting curve analysis was performed by the end of qRT-PCR to confirm PCR specificity.

Results

Illumina sequencing and quality assessment

Differences in gene expression between the enrofloxacin-susceptible and enrofloxacin-resistant strains of A. hydrophila were determined by sequencing the RNA-Seq data using the Illumina sequencing platform. After filtering and quality checks of the raw reads (26, 316, 850 and 26, 910, 746 reads for the 7966QR and ATCC 7966 strains, respectively), approximately 26 million (26, 123, 674) and 26 million (26, 730, 263) trimmed reads with trim rates of 99. 27% and 99. 33% were obtained for 7966QR and ATCC 7966, respectively. Meanwhile, the average lengths of reads for these two strains were 119. 66 and 120. 61 bp and their GC percentages were 55% and 54%, respectively (Table 1), indicating successful sequencing of the A. hydrophila transcriptome. Trimmed reads were used for the subsequent analysis.

Table 1　Summary of reads in A. hydrophila transcriptome sequencing

Sample	Raw reads	Trimmed reads	Average length	Trim rate	GC rate
ATCC7966	26, 316, 850	26, 123, 676	119. 66 bp	99. 27%	55%
7966QR	26, 910, 746	26, 730, 263	120. 61 bp	99. 33%	54%

Comparative analysis with reference genome

The trimmed reads of the A. hydrophila transcriptome were compared with the reference genome sequence. The total mapping rates of the reads with the reference genome were 94. 19% and 93. 29% in the ATCC 7966 and 7966QR groups, respectively. There were approximately 22 million (22, 717, 810) and 21 million (21, 997, 467) uniquely mapped reads for the ATCC 7966 and 7966QR groups, accounting for 85. 51% and 84. 78% of the total reads, respectively. There were approximately 2, 306, 148 and 2, 208, 047 multiple mapped reads in the ATCC 7966 and 7966QR groups, respectively, accounting for 8. 68% and 8. 51% of the total reads, respectively. The number of reads mapped in proper pairs accounted for 84. 04% and 83. 31% in the ATCC 7966 and 7966QR groups, respectively (Table 2).

Table 2　Statistical results of trimmed reads mapping with reference genome

Map to genome	7966QR		ATCC7966	
	Read numbers	Percentage	Read numbers	Percentage
Total reads	25, 946, 162	100. 00%	26, 566, 428	100. 00%
Total mapped	24, 205, 514	93. 29%	25, 023, 958	94. 19%
Uniquely mapped	21, 997, 467	84. 78%	22, 717, 810	85. 51%
Multiple mapped	2, 208, 047	8. 51%	2, 306, 148	8. 68%
Reads1 mapped	11, 003, 361	42. 41%	11, 370, 421	42. 80%
Reads2 mapped	10, 994, 106	42, 37%	11, 347, 389	42. 71%
Mapped to '+'	11, 004, 476	42. 41%	11, 368, 564	42, 79%
Mapped to '-'	10, 992, 991	42. 37%	11, 349, 246	42, 72%
Reads mapped in proper pairs	21, 614, 858	83. 31%	22, 327, 558	84, 04%

We then compared the unigenes of the sample species with the common data gene, and functional annotation was performed based on the similarity of the genes. The protein sequences were compared with the KOG, GO, and KEGG databases. The annotation of unigenes in the Swiss-Prot and TrEMBL databases accounted for 79. 77% and 99. 93% of the total unigenes (Table 3 and Fig 1). The unigene annotations in the COG, GO, and KEGG databases were about 82. 19%, 79. 57%, 62. 3%, respectively (Table 3). Transcripts were analyzed by COG classification. There were 3, 388 unigenes clustered into 25 functional categories (Fig. S1). The "amino acid transport and metabolism" and "signal transduction mechanisms" clusters were represented by the majority of transcripts (276 transcripts, 8. 15%; Fig S1). GO and KEGG database analysis of unigenes revealed that most unigenes were enriched in cellular processes, environmental information processing, genetic information processing, metabolism, and organismal systems (Fig S2 and Fig S3).

Table 3. Statistical results of the gene functional annotation

Database	Number of unigenes	Percentage（%）
Annotation in COG	3388	82. 19
Annotation in Swiss-Prot	3288	79. 77
Annotation in TrEMBL	4119	99. 93
Annotation in GO	3280	79. 57
Annotation in KEGG	2568	62. 3
Annotation in at least one database	4119	99. 93
Annotation in all databases	2313	56. 11
Total Unigenes	4122	100

Venn diagram for Database

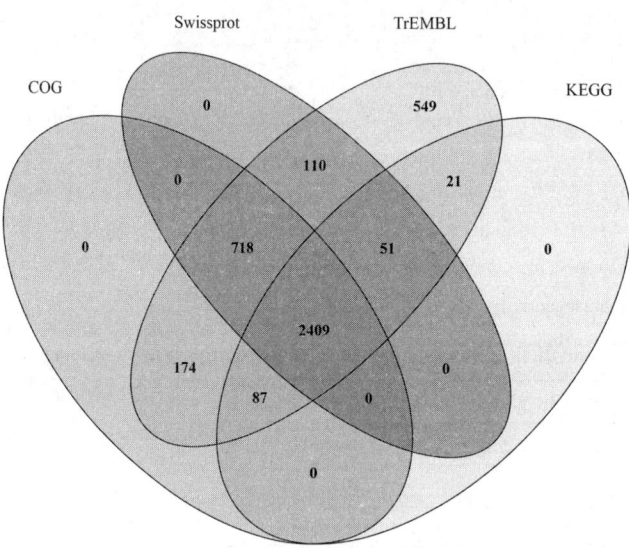

Fig. 1　Venn diagram representation of database annotations

Analysis of DEGs

The Cuffdiff program was used to generate A. hydrophila gene expression profiles to identify genes that are differentially expressed between the resistant and susceptible strains of A. hydrophila（Figs 2 and 3）. The program identified that among the DEGs, 135 genes were markedly upregulated and 83 were markedly downregulated, indicating that the gene expression had changed in the drug-resistant strains.

Gene Ontology（GO）annotation of DEGs

We used GO to determine the biological functions in which the DEGs are involved. GO functional enrichment analysis also involved cluster analysis of expression patterns. Thus, the expression patterns of DEGs annotated with a given GO term were easily obtained. All the annotated genes were classified into three GO domains: biological process, cellular component, and molecular function. Dissimilar expression profiles of the DEGs in the treated and control groups were used to determine the molecular mechanisms of enrofloxacin resistance in A. hydrophila.

The expression profiles of the three GO domains were as follows（Fig 4 and Table S1）.

Biological process: formate metabolic process（5 genes）, histidine catabolic process to glutamate and formate（4 genes）, amine metabolic process（7 genes）, histidine catabolic process to glutamate and formamide（4 genes）, formamide metabolic process（4 genes）, histidine catabolic process（4 genes）, imidazole-containing compound catabolic process（4 genes）, cellular biogenic amine metabolic process（6 genes）, cellular amine metabolic process（6 genes）, protein folding（6 genes）.

Cellular component: HslUV protease complex（2 genes）, proteasome complex（2 genes）, bacterial-type flagellum basal

878

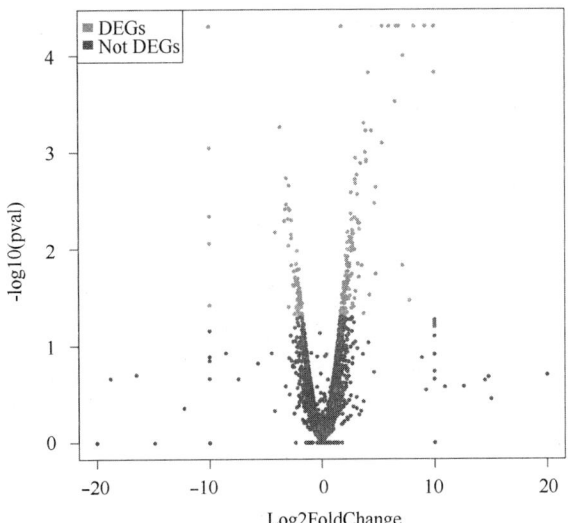

ATTCC7966 vs 7966QR

Volcanic plot of the degree of differences between the expression profiles of resistant and susceptible A. hydrophila strains. X-axis, $\log_2$ (fold change); Y-axis, $-\log_2$ (P value). Red, differentially expressed genes; blue, not differentially expressed genes. Each dot represents one gene.

Fig. 2　Effect of enrofloxacin treatment on the gene expression profile in resistant and susceptible strains of A. hydrophila

body, distal rod (2 genes), cytosolic proteasome complex (2 genes), bacterial-type flagellum basal body, and rod (2 genes).

Molecular function: oxidoreductase activity, acting on paired donors, with incorporation or reduction of molecular oxygen (3 genes); anion transmembrane-transporting ATPase activity (3 genes); oxidoreductase activity, acting on single donors with incorporation of molecular oxygen (3 genes); oxidoreductase activity, acting on single donors with incorporation of molecular oxygen, incorporation (3 genes); carbon-nitrogen lyase activity (3 genes); ATPase activity, coupled to transmembrane movement of ions (5 genes); and dioxygenase activity (3 genes).

The following DEGs related to the biological process were upregulated: AHA_ 0377 (formate metabolic process); AHA_ 0377; AHA_ 0378 (histidine catabolic process to glutamate and formate); AHA_ 0377; AHA_ 0378; AHA_ 0379; AHA_ 0380 (glutamate metabolic process); and cellular component relate genes such as AHA_ 4114, AHA_ 4115 (HslUV protease complex), AHA_ 4114, AHA_ 4115 (cytosolic proteasome complex), AHA_ 1948, AHA_ 3601 (oxidoreductase activity), AHA_ 0380, AHA_ 1413, and AHA_ 4201 (carbon-nitrogen lyase activity). The biological process-related genes such as AHA_ 1213, AHA_ 1652 (glycerol metabolic process), AHA_ 4006 (alditol metabolic process), AHA_ 1921, and AHA_ 2046 (endoplasmic reticulum) were downregulated. The functions of the hypothetical proteins were not clear, but were closely related to the biological functions of organisms.

Overall, the above analyses shed insights into the molecular mechanisms of enrofloxacin resistance in Aeromonas hydrophila.

KEGG pathway analysis of DEGs

To further explore the biological functions of the DEGs, DEGs were mapped to the KEGG database and enriched to important pathways based on the whole transcriptome. A total of 218 genes were mapped to 67 pathways. Many of the genes were found in multiple pathways, while many others were restricted to a single pathway. These pathways included metabolism, genetic information processing, cellular processes, organismal systems, and environmental information processing. The metabolism-related pathways were the most significantly (Fig 5 and Table S2). The metabolism-related biological pathways included metabolism of amino acids and their important derivatives, drug metabolism pathway, and carbohydrate metabolism. A total of 86 genes were mapped to metabolism-related biological pathways. In addition, lipid metabolism-related biological pathways, including fatty acid metabolism (3 genes), glycerolipid metabolism (2 genes), and glycerophospholipid metabolism (2 genes), were also significantly enriched. Enriched amino acid metabolism-related biological pathways included histidine metabolism (4 genes), d-alanine metabolism (2 genes), phenylalanine (4 genes), cysteine and methionine metabolism (2 genes), alanine metabolism (3 genes), glycine metabolism (3 genes), tyrosine metabolism (1 gene), lysine biosynthesis (1 gene), and arginine and proline metabolism (2 genes). Metabolism of xenobiotics by cytochrome P450 (1 gene) and drug metabolism-cytochrome P450 (1 gene) were the drug metabolism pathways that had been en-

879

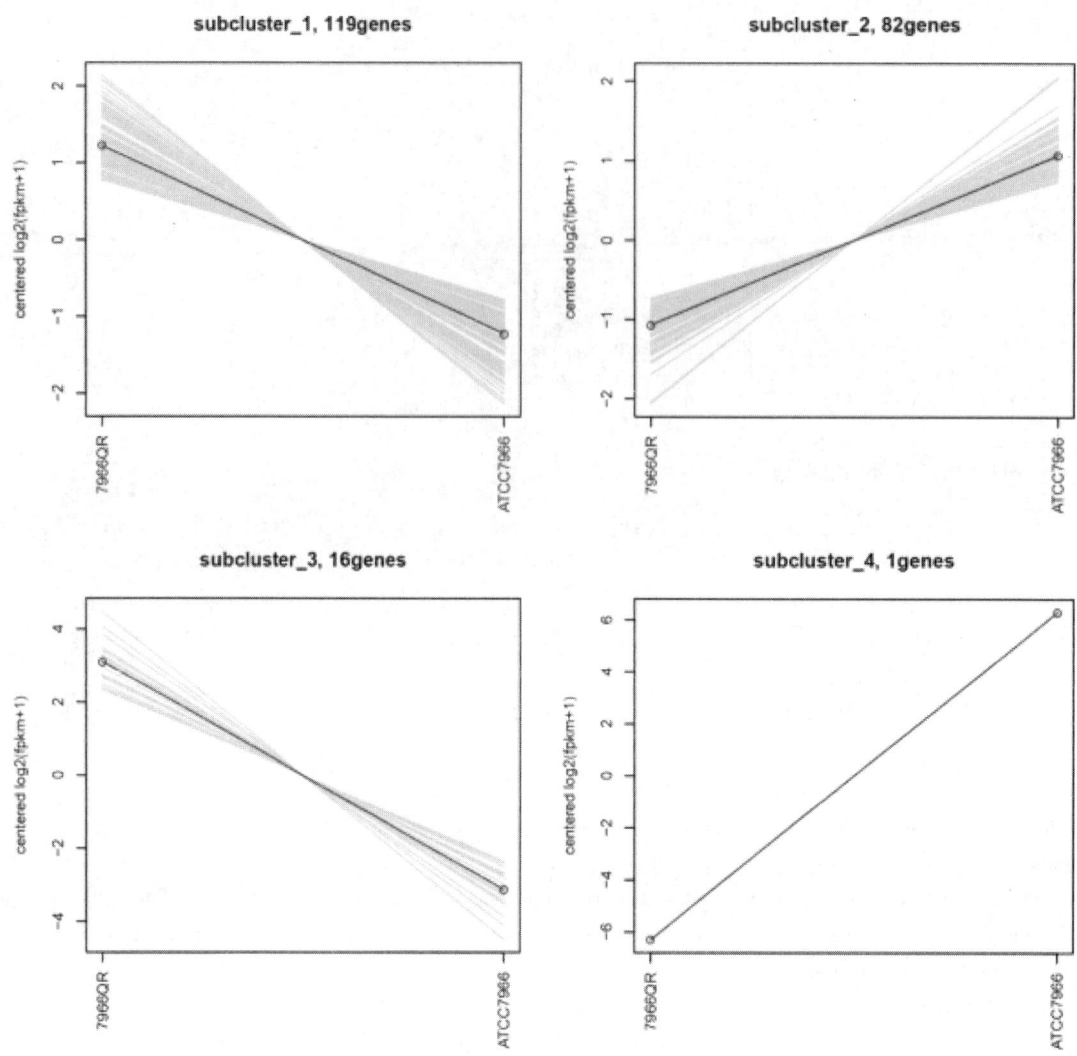

Red, upregulated genes; green, downregulated genes. Each list represent one sample; each line represents one gene.

Fig. 3 Heatmap representation of DEGs between resistant and susceptible strains of A. hydrophila.

riched. The carbohydrate metabolism pathways that were enriched included pentose phosphate pathway (2 genes), fructose and mannose metabolism (2 genes), glycolysis/gluconeogenesis (2 genes; Table S2).

The other significantly enriched pathways included environmental information processing (19 genes), cellular processes (3 genes), genetic information processing (9 genes), and organismal systems (5 genes). The environmental information processing pathways that were enriched included PI3K‐Akt signaling pathway (1 gene), ABC transporters (9 genes), bacterial secretion system (3 genes), phosphotransferase system (PTS) (1 gene), and two‐component system (5 genes). Enriched cellular processes included flagellar assembly (2 genes) and bacterial chemotaxis (1 gene). Enriched genetic information processing included protein processing in endoplasmic reticulum (1 gene), protein export (3 genes), RNA degradation (2 genes), sulfur relay system (1 gene), ribosome (2 genes). Additionally, the following pathways were enriched: antigen processing and presentation (1 gene), progesterone-mediated oocyte maturation (1 gene), estrogen signaling pathway (1 gene), NOD‐like receptor signaling pathway (1 gene) and plant-pathogen interaction (1 gene; Table S2).

The expression of certain genes such as AHA_ 2490 (PI3K‐Akt signaling pathway); AHA_ 2490 (NOD‐like receptor signaling pathway); AHA_ 0608, AHA_ 2812, AHA_ 0913, AHA_ 3728, AHA_ 1964, AHA_ 1687, AHA_ 4285, AHA_ 1595, and AHA_ 2813 (ABC transporters) were upregulated, while the AHA_ 1419 (environmental information processing), AHA_ 1331 (drug metabolism-cytochrome P450), AHA_ 2360, AHA_ 1331 (glycolysis/gluconeogenesis), AHA_ 1331 (metabolism of xenobiotics by cytochrome P450) genes were downregulated. The ABC transporter genes were expressed at higher levels

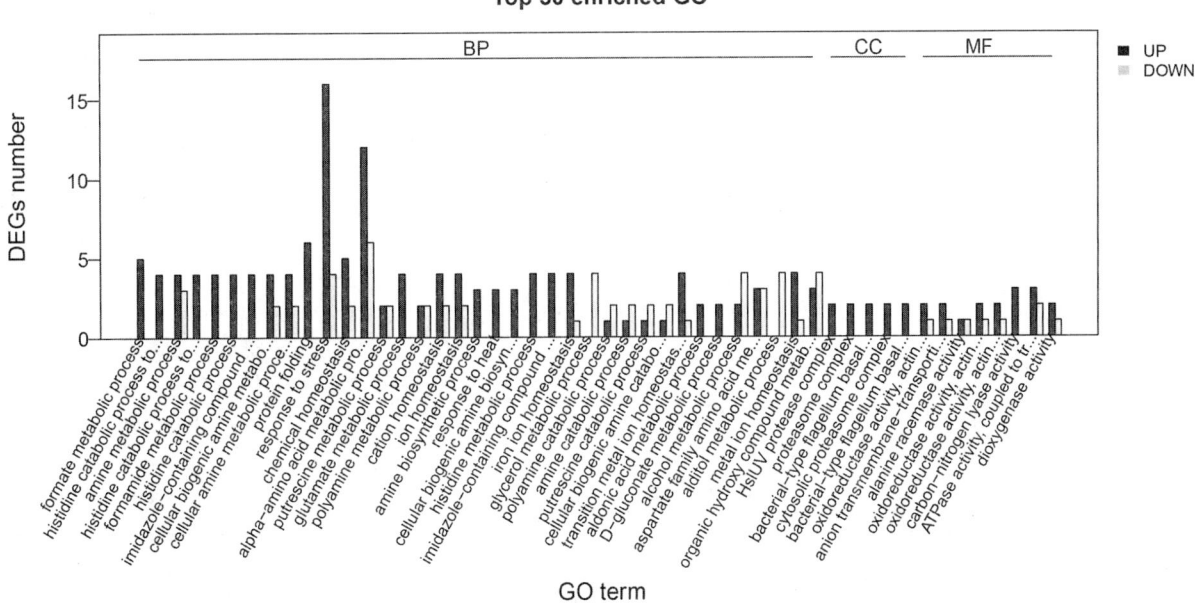

GO terms（X-axis）were grouped into three main ontologies：biological process, cellular component, and molecular function. The Y-axis indicates the number of DEGs. All annotated DEGs were classified into three GO domains：biological process, cellular component, and molecular function.

Fig. 4　Histogram representation of enriched pathway of KEGG annotation of DEGs between resistant and susceptible strains of A. hydrophila

than drug metabolism-cytochrome P450, indicating that A. hydrophila resistance to enrofloxacin may be mediated by a mechanism involving ABC transporters. These finding are consistent with the results of the GO enrichment analysis.

Overall, the results of the DEGs pathway analysis support the viewpoint that enrofloxacin inhibits A. hydrophila growth by affecting multiple biological functions, such as energy biogenesis, protein synthesis, and metabolism.

Verification of the differential expression of DEGs

Based on the results of the GO and KEGG analyses, the primers of eight genes significantly differed between the resistant and susceptible A. hydrophila strains and were therefore considered to be related to drug metabolism. With clear functional implication, these primers were designed to verify the expression of these DEGs identified in the RNA-Seq analysis. All primer sequences are listed in Table 4. Data revealed that the upregulation or downregulation of these six genes was consistent with the RNA-Seq results. Together, these results indicate that qRT-PCR and RNA-Seq results were reliable overall; however, further studies to determine the molecular mechanisms of resistance to enrofloxacin are required.（Fig 6）.

Table 4　Oligonucleotide primers of qRT-PCR for validation of DEGs

ACTIN	-	-	-	ACTIN-F	TGTGTAGCGGTGAAATGCG	140 bp
				ACTIN-R	CATCGTTTACGGCGTGGAC	
METL	Aspartokinase II	Aspartate family amino acid biosynthetic process（Biological process）	Metabolic pathways	METL-F	AAGGTGTAGTTGCTGGAGAGGT	130 bp
				METL-R	GCCGTGTGAAGAGACATCAAGGA	
METE	5 - methyltetra-hydropteroyl-triglutamate	Methylation（Biological process）	Metabolic pathways	METE-F	CTTACGAGGCGGGCATTCAG	151 bp
				METE-R	AAGCGGGTGATGGCAAAGC	
air-2	alanine race-mase	regulation of cell shape, peptidoglycan biosynthetic process（Biological process）	Metabolic pathways	air-2-F	AACGCTTTCTCTGGCTCCCTA	125 bp
				air-2-R	CGACATCAGCACGGCATTCA	
air	alanine race-mase	peptidoglycan biosynthetic process, alanine metabolic process（Biological process）	Metabolic pathways	air-F	ACCGCACCTTCACCCTCAA	209 bp
				air-R	GAACAGCACCACCTCGTCAC	

ACTIN	–	–	–	ACTIN-F	TGTGTAGCGGTGAAATGCG	140 bp
				ACTIN-R	CATCGTTTACGGCGTGGAC	
AHA2142	Acetyl – CoA acetyltransferase	signal transduction, metabolic process, cholesterol metabolic process (Biological process)	Metabolic pathways	AHA2142-F	GGAGACATTGCCGAAGTGACC	118 bp
				AHA2142-R	CTACCTCATAGTGCCGCTCAAC	
gyrB1	DNA gyrase subunit B	DNA topoisomerase type II (ATP-hydrolyzing) activity, ATP binding, metal ion binding, DNA replication origin binding, GTPase activity (Molecular process)	Metabolic pathways	gyrB1-F	GCGGAATGTTGTTGGTGAAGC	173 bp
				gyrB1-R	CTACGAAGGCGGCATCAAGG	
gyrA	DNA gyrase subunit A	DNA topoisomerase type II (ATP-hydrolyzing) activity, magnesium ion binding, protein heterodimerization activity (Molecular process)	Metabolic pathways	gyrA-F	GTCTTCTCGTCCACCTCCACT	222 bp
				gyrA-R	CAACATTCCGCCTCACAACCT	

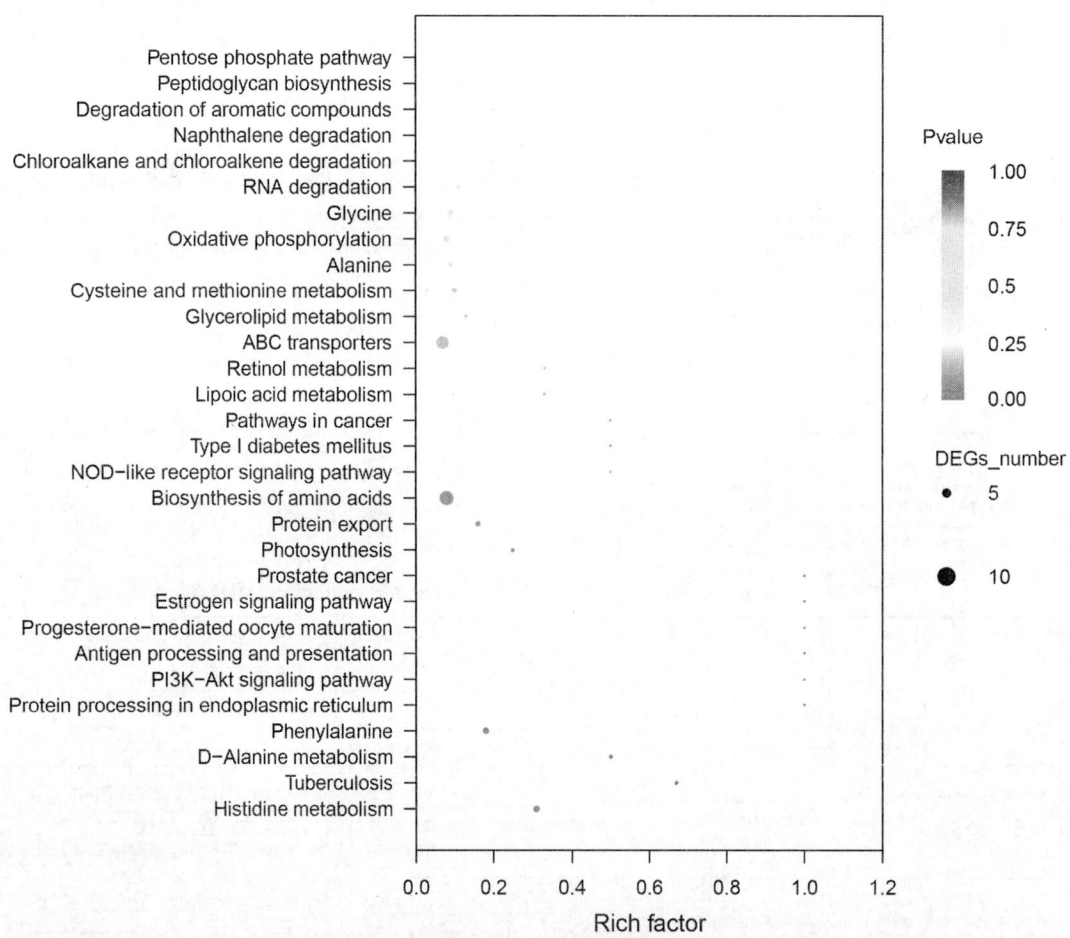

Scatter plot of the degree of differences in the expression profile of A. hydrophilia resistant and susceptible strains. X-axis, Rich factor; Y-axis, pathway name. The P value correlates to the color gradient; colors closer to red have a smaller P value. The size of the dots indicates the number of DEGs in each pathway.

Fig. 5　Scatter plot of enriched pathways of KEGG annotation of DEGs between resistant and susceptible strains of A. hydrophila.

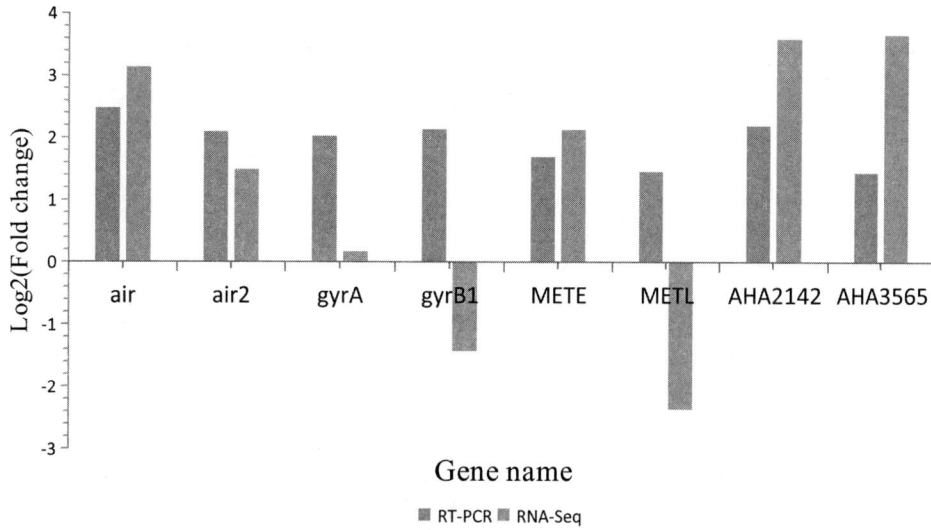

Negative values indicate that the gene expression of A. hydrophila was downregulated following enrofloxacin treatment; positive values indicate that the gene expression was upregulated.

Fig. 6　Comparison of the expression levels of eight genes determined by RNA-Seq and RT-PCR

Discussion

Transcriptome sequencing is a powerful technique for studying the mechanism of changes of biological characteristics of an organism, and has been used successfully in some species[29-31]. In our study, we examined the transcriptome of A. hydrophila using the Illumina sequencing platform and tried to explore the molecular mechanism of enrofloxacin resistance in A. hydrophilia. Compared with the reference genome, the total mapped rates of reads were 94. 19% and 93. 29% in the A. hydrophilia transcriptome of the enrofloxacin-susceptible (ATCC7966) and enrofloxacin-resistant (7966QR) groups, respectively, indicating that the quality of sequencing data met the demand for follow-up studies.

We obtained 218 DEGs and classified them into 1, 052 GO terms consisting of three domains: biological process, cellular component, and molecular function. Of these, 176 GO terms were found to have dramatic changes in expression. We mapped the DEGs to 68 pathways, of which 10 were significantly enriched. We divided the genes into five branches based on the KEGG metabolic pathway involved: cellular processes, environmental information processing, genetic information processing, metabolism, and organismal systems. The metabolism-related biological pathways and biosynthesis of amino acids were the most significantly enriched pathways in this analysis; these pathways are responsible for the main biological functions of A. hydrophila. Results of the KEGG pathway analysis revealed that a considerable percentage of genes were enriched in ABC transporters, metabolism of xenobiotics by cytochrome P450, and drug metabolism-cytochrome P450. All genes enriched in ABC transporters were upregulated, whereas all the genes enriched in metabolism of xenobiotics by cytochrome P450 and drug metabolism-cytochrome P450 were downregulated. We speculate that enrofloxacin promotes the protein expression of ABC transporters and inhibits the protein expression of cytochrome P450. This may be related to enrofloxacin resistance of A. hydrophila. Our study supplements the previous studies by Seshadri et al. who investigated the virulence genes in A. hydrophila by sequencing its genome[32]. Results of the GO annotation enrichment revealed that most DEGs were related to transmembrane transport and biosynthesis and degradation of amino acids. Furthermore, some of the genes were mapped to cellular response to DNA damage stimulus, and among the DEGs, gyrA was upregulated, which was verified by qRT-PCR. The qRT-PCR results were generally consistent with the results of the transcriptome analysis. Our results are in agreement with the results of Shakir et al. , who reported that changes in the topoisomerase target sites of chromosomes (amino acid changes in quinolone resistance-determining regions, QRDRs) could produce drug resistance and, A. hydrophila strains exhibiting high levels of resistance to antibiotics are common in the strains with gyrA and parD double mutations in QRDRs[5]. Together, these results indicate that A. hydrophilia resistance to enrofloxacin occurs primarily due to alterations in multiple biological functions, energy biogenesis, protein synthesis, and metabolism. Our findings support that the mechanism of enrofloxacin resistance in A. hydrophila is closely related to reduction of intracellular drug accumulation caused by ABC transporters and topoisomerase IV.

As most of the genes in A. hydrophilia encode putative proteins whose functions were not clear. In order to facilitate further research, we selected 8 of the DGEs whose functions were clearly known and subjected them to qPCR analysis to verify the RNA-Seq results of. The qPCR results were consistent with the results of the transcriptome analysis, except for two genes. Therefore, we believe that the qualities of the A. hydrophila transcriptome are adequate for further studies on its functional genes. These findings greatly extend the existing sequence resources relating to A. hydrophilia and provide abundant genetic information with which to further understand the molecular mechanisms of enrofloxacin-resistant A. hydrophilia in aquaculture.

Data archiving

All the sequencing reads performed in this study are available at NCBI SRA database (SRA1855752).

Acknowledgments

This study was supported by the Special Fund for Agro-scientific Research in the Public interest (Grant 201203085), the 863 Program (Grant 2011AA10A216), the National Natural Resources Platform and the Shanghai University Knowledge Service Platform.

References

[1] K. P. Plant, S. E. Lapatra. Advances in fish vaccine delivery. Dev. Comp. Immunol., 35 (2011), pp. 1256-1262. doi: 10. 1016/j. dci. 2011. 03. 007.

[2] L. Huang, Y. Qin, Q. Yan, G. Lin, L. Huang, B. Huang, W. Huang. MinD plays an important role in Aeromonas hydrophila adherence to Anguilla japonica mucus. Gene, 565 (2015), pp. 275-281. doi: 10. 1016/j. gene. 2015. 04. 031.

[3] P. Sahoo, K. Das Mahapatra, J. Saha, A. Barat, M. Sahoo, B. Mohanty, M. Sahoo, B. R. Mohanty, B. Gjerde, J. Odegård, M. Rye, R. Salte. Family association between immune parameters and resistance to Aeromonas hydrophila infection in the Indian major carp, Labeo rohita. Fish. Shellfish Immunol., 25 (2008), pp. 163-169. doi: 10. 1016/j. fsi. 2008. 04. 003.

[4] X. Mu, J. W. Pridgeon, P. H. Klesius. Comparative transcriptional analysis reveals distinct expression patterns of channel catfish genes after the first infection and re-infection with Aeromonas hydrophila. Fish. Shellfish Immunol., 35 (2013), pp. 1566-1576. doi: 10. 1016/j. fsi. 2013. 08. 027.

[5] D. Kregiel, K. Niedzielska. Effect of plasma processing and organosilane modifications of polyethylene on Aeromonas hydrophila biofilm formation Biomed Res. Int. (2014) 8 pages.

[6] R. Khushiramani, S. K. Girisha, P. P. Bhowmick, I. Karunasagar, I. Karunasagar. Prevalence of different outer membrane proteins in isolates of Aeromonas species World J. Microbiol. Biotechnol, 24 (2008), pp. 2263-2268. doi: 10. 1007/s11274-008-9740-4.

[7] I. I. Rodriguez, B. Novoa, A. Figureas. Immune response of zebrafish (Danio rerio) against a newly isolated bacterial pathogen Aeromonas hydrophila. Fish. Shellfish Immunol., 25 (2008), pp. 239-249. doi: 10. 1016/j. fsi. 2008. 05. 002.

[8] L. Liu, Y. X. Gong, B. Zhu, G. L. Liu, G. X. Wang, F. Ling. Effect of a new recombinant Aeromonas hydrophila vaccine on the grass carp intestinal microbiota and correlations with immunological responses. Fish. Shellfish Immunol., 45 (2015), pp. 175-183. doi: 10. 1016/j. fsi. 2015. 03. 043.

[9] Martinez, M., Mcdermott, P., and Walker, R. 2005. Pharmacology of the fluoroquinolones: A perspective for the use in domestic animals. Vet J. 172 (1): 10-28. doi: 10. 1016/j. tvjl. 2005. 07. 010.

[10] CS Del Castillo, J Hikima, HB Jang, SW Nho, TS Jung, Wongtavatchai J, Kondo H, Hirono I, Takeyama H, Aoki T. Comparative sequence analysis of a multidrug-resistant plasmid from Aeromonas hydrophila. 2013, 57 (1): 120-129. doi: 10. 1128/AAC. 01239-12.

[11] H. Cao, S. He, H. Wang, S. Hou, L. Lu, X. Yang. Bdellovibrios, potential biocontrol bacteria against pathogenic Aeromonas hydrophila. Vet. Microbiol., 154 (2012), pp. 413-418. doi: 10. 1016/j. vetmic. 2011. 07. 032.

[12] F. C. Cabello. Heavy use of prophylactic antibiotics in aquaculture: a growing problem for human and animal health and for the environment. Environ. Microbiol., 8 (2006), pp. 1137-1144. doi: 10. 1111/j. 1462-2920. 2006. 01054. x.

[13] Shakir Z, Khan S, Sung K, Khare S, Khan A, Steele R, Nawaz M. Molecular Characterization of Fluoroquinolone-Resistant Aeromonas spp. Isolated from Imported Shrimp. Applied and Environmental Microbiology. 2012. 78 (22): 8137-8141. doi: 10. 1128/AEM. 02081-12.

[14] Li Jl, Wang T, Shao B, Shen J, Wang S, Wu Y. 2012. Plasmid-mediated quinolone resistance genes and antibiotic residues in wastewater and soil adjacent to swine feedlots: potential transfer to agricultural lands. Environ Health Perspect. 2012 Aug; 120 (8): 1144-9. doi: 10. 1289/ehp. 1104776.

[15] Cattoir V, Poirel L, Aubert C, Soussy CJ, Nordmann P. 2008. Unexpected Occurrence of Plasmid-Mediated Quinolone Resistance Determinants in Environmental Aeromonas spp. 14 (2): 231-237. doi: 10. 3201/eid1402. 070677.

[16] Z. Wang, M. Gerstein, M. Snyder. RNA-Seq: a revolutionary tool for transcriptomics. Nat Rev Gent, 10 (1) (2009), pp. 57-63. doi: 10. 1038/nrg2484.

[17] L. X. Xiang, D. He, W. R. Dong, Y. W. Zhang, J. Z. Shao. Deep sequencing-based transcriptome profiling analysis of bacteria-challenged Lateolabrax japonicas reveals insight into the immune-relevant genes in marine fish. BMC Genomics, 11 (2010), p. 472. doi: 10. 1186/1471-2164-11-472.

[18] S. Czesny, J. Epifanio, P. Michalak. Genetic divergence between freshwater and marine morphs of Alewife (Alosa pseudoharengus): a

884

'next-generation' sequencing analysis. PLoS One, 7 (3) (2012), p. e31803. doi. org/10. 1371/journal. pone. 0031803.

[19] Clinical and Laboratory Standards Institute. Performance Standards for Antimicrobial Susceptibility Testing; Twenty-First Informational Supplement Wayne, PA: Clinical and Laboratory Standards Institute, 2011.

[20] Trapnell C, Pachter L, Salzberg SL. TopHat: discovering splice junctions with RNA-Seq. Bioinformatics. 2009; 25: 1105-1111. Doi: 10. 1093/bioinformatics/bip120 PMID: 19289445

[21] Luo R, Liu B, Xie Y, Li Z, Huang W, Yuan J, et al. (2012) SOAPdenovo2: an empirically improved memory-efficient short-read de novo assembler. Gigascience 1 (1): 18.

[22] Pertea G, Huang X, Liang F, Antonescu V, Sultana R, Karamycheva S, et al. (2003) TIGR Gene Indices clustering tools (TGICL): a software system for fast clustering of large EST datasets. Bioinformatics 19 (5): 651-652.

[23] Conesa A, Gotz S. (2008) Blast2GO: A comprehensive suite for functional analysis in plant genomics. Int J Plant Genomics 2008619832.

[24] Li B, Dewey CN. (2011) RSEM: accurate transcript quantification from RNA-Seq data with or without a reference genome. BMC Bioinformatics 12323.

[25] Anders S, Huber W. (2010) Differential expression analysis for sequence count data. Genome Biol 11 (10): R106.

[26] Liu, B., Jiang, G. F., Zhang, Y. F., Li, J. L., Li, X. J., and Yue, J. S. 2011. Analysis of transcriptome differences between resistant and susceptible strains of the citrus red mite Panonychus citri (Acari: Tetranychidae). PLoS One 6 (12): e28516. doi: 10. 1371/journal. pone. 0028516.

[27] Liu, B., Jiang, G. F., Zhang, Y. F., Li, J. L., Li, X. J., and Yue, J. S. 2011. Analysis of transcriptome differences between resistant and susceptible strains of the citrus red mite Panonychus citri (Acari: Tetranychidae). PLoS One 6 (12): e28516. doi: 10. 1371/journal. pone. 0028516.

[28] West, P. V., Bruijn, I. D., Minor, K. l., Phillips, A. I., Robertson, E. J., and Wawra, S. 2010. The putative RxLR effector protein SpHtp1 from the fish pathogenic oomycete Saprolegnia parasitica is translocated into fish cells. FEMS Microbiol. Lett. 310 (2): p. 127-137. doi: 10. 1111/j. 1574-6968. 2010. 02055x.

[29] Price, D. P., Nagarajan V, Churbanov A, Houke LL, Drade P, Milligan B. The fat body transcriptomes of the yellow fever mosquito Aedes aegypti, pre-and post-blood meal. PLoS One, 2011.; 6: e22573. doi: 10. 1371/journal. pone. 0022573.

[30] Peng, Y, Gao X, Li R, Cao G. Transcriptome sequencing and de novo analysis of Youngia japonica using the Illumina platform. PloS one. 2014; 9: e90636. doi: 10. 1371/journal. pone. 0090636

[31] Li, C, Weng S, Chen Y, Yu X, Lu L, Zhang H. Analysis of Litopenaeus vannamei transcriptome using the next-generation DNA sequencing technique. PLoS one. 2012; 7: e47442. doi: 10. 1371/journal. pone. 0047442.

[32] Rekha Seshadri1, Sam W. Joseph, Ashok K. Chopra, Jian Sha, Jonathan Shaw, Joerg Graf, Daniel Haft, Martin Wu, Qinghu Ren, M. J. Rosovitz, Ramana Madupu, Luke Tallon, Mary Kim, Shaohua Jin, Hue Vuong, O. Colin Stine, Afsar Ali, Amy J. Horneman, John F. Heidelberg. Genome Sequence of Aeromonas hydrophila ATCC 7966: Jack of All Trades. 2006; 188 (23): 8272-8282. doi: 10. 1128/JB. 00621-06.

该文发表于《Disease of Aquatic Organisms》【2016, 120 (2): 115-123】

Incidence of antimicrobial-resistance genes and integrons in antibiotic resistance bacteria isolated from eels and farming water

Mao Lin***, Xiaomei Wu*, Qingpi Yan, Ying Ma, Lixing Huang, Yingxue Qin, Xiaojin Xu

(Jimei University, Xiamen, Fujian, PR China and Engineering Research Center of the
Modern Technology for Eel Industry, Ministry of Education, PR China)

Abstract: The abundance of antimicrobial-resistance genes and integrons in 108 strains of antibiotic resistance bacteria isolated from eels and farming water in China were investigated in this study. Conventional PCR was implemented to examine common antibiotic-resistant genes, integrons, and their gene cassette arrays. The results showed that the antibiotic-resistance genes, bla_{TEM}, tetC, sulI, aadA, floR, qnrB, were detected at high percentages, as well as a number of other resistance genes. Class I integrons were present in 79. 63% of the strains; ten out of 108 isolates carried class II integrons; while class III integrons were not detected. Three strains carried both class I and class II integrons. Meanwhile, 73. 26% of the class I integron-positive isolates contained the $qacE\Delta1/sul1$ gene. Fourteen types of integron cassette arrays were found among class I integron-positive isolates. A new array, $dfrB4-catB3-bla_{OXA-10}-aadA1$, was discovered in this study. The gene cassette array dfrA12-orfF-aadA2 was found to be the most widely distributed. In summary, 23 different gene cassettes encoding resistance to 8 classes of antibiotics were identified in the class I integrons, and the main cassettes contained genes encoding resistance to aminoglycosides (aad) and trimethoprim (dfr). All class II integron-positive strains had only a single gene cassette array, dfrA1-catB2-sat2-aadA1. High levels of antimicrobial-resistance genes and integrons in eels and farming water suggest that the overuse of antimicrobials should be strictly controlled and that the levels of bacterial antimicrobial resistance genes in aquaculture should be monitored.

Key words: eels; antibiotic resistance bacteria; antimicrobial-resistant genes; integrons

鳗鲡及其养殖水体分离耐药细菌的抗性基因及整合子的流行特征

林茂*，吴小梅，鄢庆枇，马英，黄力行，覃映雪，徐晓津

(集美大学水产学院；鳗鲡现代产业技术教育部工程研究中心，厦门 361021)

摘要：本文研究了 108 株分离自中国鳗鲡及其养殖水体的耐药细菌中抗性基因及整合子的丰度情况。采用传统的 PCR 方法检测常见的抗生素耐药基因、整合子及其基因盒阵列，结果显示，bla_TEM、tetC、sulI、aadA、floR、qnrB 等耐药基因具有较高的出现频率；79.63%的菌株携带 I 型整合子；10 株携带 II 型整合子，其中 3 株同时携带 I 型整合子；未发现 III 型整合子。同时，73.26%的整合子阳性菌株含有 qacEΔ1/sul1 基因。在整合子阳性菌株中发现 14 种整合子基因盒阵列，其中的 dfrB4-catB3-bla_OXA-10-aadA1 是一种新的阵列类型，而 dfrA12-orfF-aadA2 则是分布最广泛的基因盒阵列。经鉴定，共有 23 种不同的基因盒编码 8 类抗生素的耐药基因，大多数基因盒为编码氨基糖苷类（aad）和甲氧苄啶（dfr）的耐药基因。全部 intIII 阳性菌株扩增出的基因盒阵列都是相同的 dfrA1-catB2-sat2-aadA1。水产养殖环境中耐药基因和整合子的分布广泛，可能意味着抗生素的过度使用，因此，必须通过对水产养殖环境中耐药基因的监控，严格管控耐药性风险。

关键词：鳗鲡；耐药细菌；耐药基因；整合子

1 Introduction

In recent years, aquaculture has developed rapidly in China, with more than 60% of the world's aquaculture volume contributed by China (Cao et al., 2015). Since 1989, China has been the largest producer of global aquaculture in the world 26 years in a row. With the continuing expansion of the aquaculture industry, many domestic fisheries are overexploited, leading to aquatic animal diseases becoming increasingly serious. Bacterial infections are an important type of aquatic animal disease. Antimicrobials are currently the main tools to effectively prevent and treat bacterial infections (Grave et al., 1999).

However, the overuse of antimicrobials has promoted the selection of antimicrobial-resistant bacteria. Bacteria have developed a series of mechanisms to resist the effects of antimicrobials, including changing the permeability of the cell membrane or cell wall, active efflux of antimicrobial agents, inactivation or modification of the antimicrobial to destroy its structure, and modification of antibiotic targets.

Bacterial resistance to antimicrobials is determined by specific resistance genes located on the chromosome or mobile genetic elements, such as plasmids, transposons and integrons, which facilitates their transfer between different bacterial species (Harnisz et al., 2015; Korzeniewska & Harnisz, 2013). Integrons are site-specific recombination systems, which can identify, capture, integrate, or shear extracellular free gene segments through a self-encoded integrase (Jacquier et al., 2009; Stokes & Hall, 1989). Integrons play an important role in the carriage and dissemination of antibiotic resistance genes (Chang et al., 2007; Fluit & Schmitz 1999). Generally speaking, on the basis of the sequence conservation, integrons contain three parts: the 5-conserved segment (5-CS), the 3-conserved segment (3-CS), and the variable region between in 5-CS and 3-CS, that may or may not contain one or more gene cassettes. Usually gene cassettes contain only a single gene and a recombination site (attC). More than 130 different gene cassettes carrying known antibiotic resistance genes have been identified in integrons (Partridge et al., 2009). Integrons consist of three essential elements: the integrase gene (intI) encoding a site-specific recombinase, the recombination site (attI) and a promoter (Pc). Three elements are present in the 5 -CS of integrons (Ochman et al., 2000; Rowe-Magnus & Mazel, 1999; Zhu et al., 2014). On the basis of integrase amino acid sequence similarity, integrons are divided into six classes. Most studies have focused on the class I, II, and III integrons that are commonly associated with antibiotic resistance, with the class I integrons being the most prevalent class (Labbate et al., 2009; Sarria-Guzmán et al., 2014; Uyaguari et al., 2013). The 3-CS of typical class I integrons consists of the genes qacEΔ1, sul1 and orf5 (Fig. S1). In recent years, many studies of integrons have been concerned with the carriage of antibiotic resistance markers by nosocomial or animal pathogens, while there is relatively limited research on their presence in aquatic animals (Dolejska et al., 2007; Nguyen et al., 2014; Spindler et al., 2012; Zhu et al., 2011).

Farmed eels are an important agricultural product exported by China. The recent development of this industry has led to a significant growth in the global demand for edible eels. The increase of aquaculture density frequently results in diseases of these animals, which are a major source of economic loss in the aquaculture industry. Above all, bacterial diseases are the most frequent and major cause of mass death in fish.

The aim of this study was to investigate: (i) the prevalence of selected antimicrobial-resistance genes in eels and farming water antibiotic resistance bacteria isolates in China; and (ii) the occurrence and distribution of class I, II and III integrons and their gene cassette arrays in these isolates.

2 Materials and methods

2.1 Sample collection, isolation and identification of antibiotic resistance bacteria, and antibiotic susceptibility testing

We previously isolated 108 strains from eels and farming water in Fujian province, China, and screened these on Luria-Bertani (LB) nutrient agar medium containing one of five antimicrobials. The phylogeny of these isolates was determined by 16S rDNA sequencing, and their antibiotic resistance was determined by the Kirby-Bauer disc diffusion method. These 108 strains have previously been shown to consist of many highly antibiotic resistant isolates (Wu et al. 2015).

2.2 DNA extraction

Total DNA of the 108 resistant strains was extracted using the TIANamp Bacteria DNA kit (Tiangen Biotech, China). Plasmid DNA from the 108 strains was extracted with the E. Z. N. A.™ Plasmid Mini Kit (OMEGA, USA). Template DNA was stored at −20℃.

2.3 Detection of antimicrobial resistance genes

A total of 108 resistant isolates were analyzed for the presence of 29 different antimicrobial resistance genes using polymerase chain reaction (PCR). The primers used in this study have been reported in previous studies and are shown in Table S1. Each 25 μL PCR mixture contained 2.5 μL 10 × Taq DNA buffer (with Mg^{2+}), 0.5 μL dNTPs (10 mM each), 0.5 μL forward primer and reverse primer (10 μM each), 0.125 μL Taq polymerase (5 U/μL), 0.5 ~ 1 μg total DNA, and double-distilled water (ddH₂O).

The conditions used for PCR amplification were as follows: pre-denaturation at 95℃ for 5 min; 30 cycles of 94℃ for 40 s, annealing for 40 s at 54 ~ 62 ℃ (Table S1) and extension at 72℃ for 50 s; followed by the final extension at 72℃ for 5 min. Amplification products were analyzed on an 0.8% ~ 1.2% agarose gel. The antimicrobial resistance genes that were detected were sequenced at GenScript Biotechnology (Nanjing, China) to verify the gene-specific primers. Sequences were compared with the Genbank database using NCBI Blast (http://www.ncbi.nlm.nih.gov/blast/).

2.4 Detection of three integrons and gene cassette array characterization

Primers used for the detection of integrons and their variable regions are shown in Table S2. Total DNA and plasmid DNA from all isolates were screened for class I, II, and III integrons using the primers intII F/R, intIII F/R, intIIII F/R, respectively. The variable regions of the integrons were amplified with primers hep58/59 (for the class I integrons) and hep74/51 (for the class II integrons). The gene cassette arrays were characterized for the corresponding positive isolates in the previous step. Each 50 μL PCR mixture included 5 μL 10 × Taq DNA buffer (with Mg^{2+}), 1 μL dNTPs (10 mM each), 1 μL of forward primer and reverse primer (10 μM each), 0.25 μL Taq polymerase (5 U/μL), 0.5 ~ 1 μg total DNA, and ddH₂O.

The conditions used for PCR amplification were as follows: pre-denaturation at 95℃ for 10 min; 30 cycles of 94℃ for 30 s, annealing 30 s at 55℃ and extension at 72℃ for 1 min (3 min for the variable regions); followed by the final extension at 72℃ for 10 min. Amplification products were then analyzed on an 0.8% ~ 1.2% agarose gel.

Amplification products of the variable regions were purified using the E. Z. N. A.™ Gel Extraction Kit (OMEGA, USA), cloned using the pMD™ 19-T vector Cloning Kit (TaKaRa, Japan), and sequenced. Using the primers hep58/59 or hep74/51, we sequenced the variable regions of approximately 1 500 bp. When sequences were longer than 1 500 bp, primer walking was used until the full sequence was obtained. All sequencing was performed by GenScript Biotechnology. More than 130 different gene cassettes carrying known antibiotic resistance genes have been identified in integrons and almost all antibiotic resistance genes have been identified. Therefore, the majority of integron gene cassette arrays could be identified directly with NCBI Blast. While a few new arrays could be identified with low sequence identity, we cut them into several segmentations to identify every segmentation of the gene cassettes, and then assembled gene cassettes according to the original sequence.

3 Results

3.1 Diversity and antimicrobial resistance of the drug-resistant bacteria

One hundred and eight resistant strains were successfully isolated and were classified into 20 genera, with high detection rates of bacteria in the genera Aeromonas, Citrobacter, and Acinetobacter. The percentage of bacteria with resistance to three or more antibiotics was 93.5%. The frequency of resistance to amoxicillin (90.7%) was high, as was resistance to tetracycline, rifampicin, sulfonamides, and amphenicols (60%~80%) (Fig. S2).

3.2 The abundance of antimicrobial resistance genes

Eight main classes of antimicrobial resistance genes were detected from total DNA of the 108 antibiotic-resistant isolates in this study (Fig. 1).

The aminoglycoside-resistance genes mainly included those that encode aminoglycoside acetyltransferases (aac), aminoglycoside adenyltransferases (aad), and aminoglycoside phosphotransferases (aph). Among all genes of this class that were investigated, the aadA gene was the most prevalent (74.07%), whilst aphA1 was found in 48 strains, the strA-B in 24 strains, and aac (3) -IV in only 7 strains. The aac (3) -I gene was not detected in any of the isolates.

The bla_{TEM} gene was found in all tested isolates. Other β-lactamase resistance genes were present at lower rates. The DHA gene was found in 39 strains. EBC, bla_{SHV}, and bla_{OXA} genes were found in 17, 16, and 5 strains, respectively, while the MOX gene was not detected in any of the 108 strains.

Among the resistance genes for antifolates, sulI had the highest detection rate (96.30%), followed by the dihydrofolate reductase genes dfrA12>dfrA5>dfrA1>dfrA7.

The tetracycline resistance genes tetA, tetB, tetC, tetD, and tetE were detected in all 108 strains in this study. Our results showed that only 13 strains carried tetE and 17 strains were positive for tetB, while tetA and tetD were found in 27 and 40 of the 108 strains, respectively. The tetC gene was found in almost all of the isolates (104 of 108 strains).

The quinolone resistance genes qnrA, qnrB, qnrS were relatively common in previous studies. Of the strains having quinolone resistance in our study, five were positive for qnrS (4.63% of the total). The qnrB gene was observed in over half of the isolates in this study (52.78%), while only one strain was found to possess the qnrA gene.

The amphenicol resistance genes (floR, cmlA, catI) encode resistance to florfenicol and chloramphenicol. The floR gene was present in large proportion of the isolates (75 out of 108, 69.44%), while cmlA was detected by PCR in 26 strains and only 4 strains carried the catI gene.

The arr2/3 gene encodes a rifampicin ADP-ribosylating transferase, conferring resistance to rifampicin. Almost half of the strains in the study were positive for arr2/3.

The ereA gene, which confers resistance to erythromycin, was only present in 8 strains.

The phenotype and genotype of the antibiotic-resistance markers were analyzed in each of the isolates (Fig. 2). The results showed that isolates with inconsistent antibiotic-resistance phenotype and genotype mainly contained resistance to rifampicins, quinolones and aminoglycosides.

3.3 Distribution of integrons and characterization of gene cassette arrays

PCR reactions using total DNA of the 108 strains as a template identified class I integrons in 86 of the antibiotic-resistant strains (79.63%), while 10 isolates (9.26%) carried class II integrons and class III integrons were not detected. Three strains carried both class I and class II integrons. The variable region from forty-one of the class I integron-positive isolates were successfully amplified and sequenced (Table S3). Of the class I integron-positive isolates, 33 had 'empty' class I integrons; that is, no gene cassette was present between the 5-CS and 3-CS. Additionally, gene cassette arrays were unable to be amplified from twelve strains with class I integrons. Meanwhile, 73.26% (63/86) of the class I integron-positive isolates contained the qacEΔ1/sulI gene as part of the 3'-CS of the class I integrons. Of the 10 isolates containing class II integrons, the variable regions were able to be amplified from all of these isolates. By testing for the presence of the integrons (both class I and class II) in plasmid DNA from each strain, we found that all in-

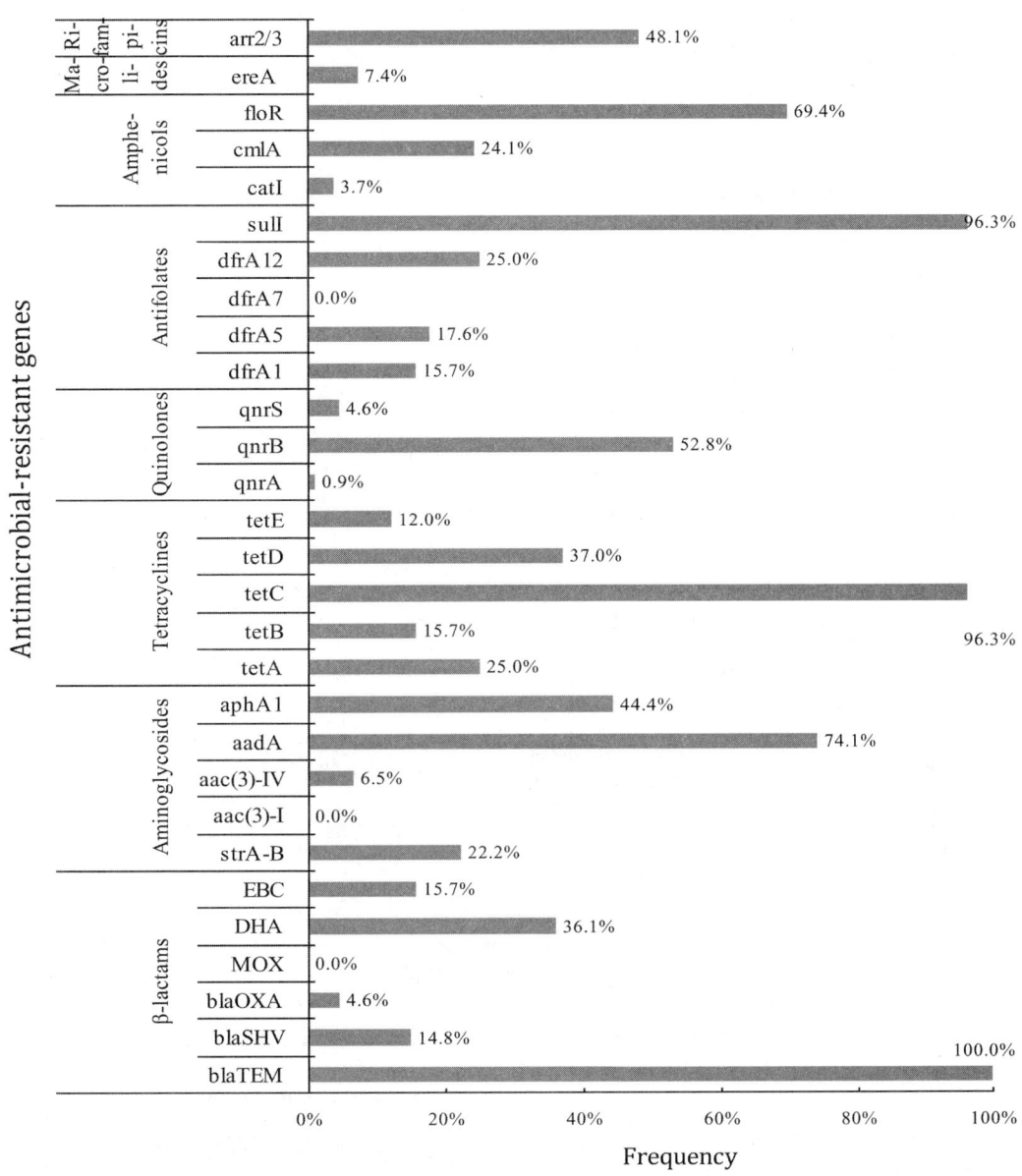

Fig. 1　The distribution of 29 different antimicrobial resistance genes in 108 strains of
antibiotic resistance bacteria isolated from eels and farming water

tegrons were located on plasmids in these strains.

Fourteen types of gene cassette arrays were found among class I integron-positive isolates（Fig. 3）. Furthermore, a new array dfrB4-catB3-bla$_{OXA-10}$-aadA1 was discovered in this study. The gene cassette array dfrA12-orfF-aadA2 was the most widely distributed, being present in 11 of the class I integron-positive isolates. Those isolates included Aeromonas（2）, Citrobacter（4）and She-wanella（5）. We found that 71.43%（5/7）of the Shewanella isolates contained the dfrA12-orfF-aadA2 array. While the aac（6′）-Ib-cr-arr-3-dfrA27 array was carried by nine strains. Other gene cassettes included dfrA17-aadA5, dfrA1-orfC, catB8, dfrB4-catB3-aadA1, arr2-aacA4-dfrA1-orfC, aacA4-ereA-bla$_{OXA-21}$-dfrA1, aadA1, aadA2, dfrB4-catB3-bla$_{OXA-10}$-aadA1, aac（6′）-IIc-ereA-tnpA-aac3-arr-ereA, catB3-aadA1, ereA. Among these gene cassette arrays, 82.93% of the arrays contained aminoglycoside resistance genes and trimethoprim resistance genes were also found in 82.93% of the arrays, with approximately 75% of the arrays containing both aminoglycoside and trimethoprim resistance genes in the same array. In summary, 23 different gene cassettes encoding resistance to 6 classes of antibiotics were identified in the class I integrons, including the following: genes encoding for resistance to aminoglycosides（aac（6′）-IIc, aac3, aacA4, aadA1, aadA2, aadA5）, trimethoprim（dfrA1, dfrA12, dfrA17, dfrA27, dfrB4）, β-lactams（bla$_{OXA-10}$, bla$_{OXA-21}$）, rifampicins（arr, arr2, arr3）, amphenicols（catB3, catB8）,

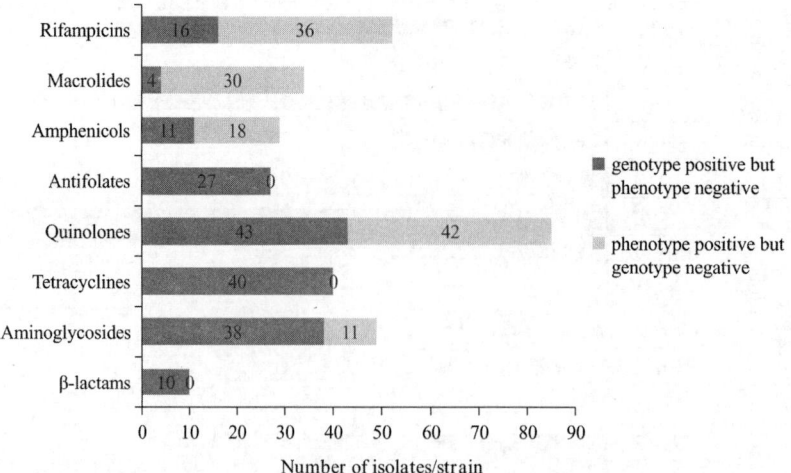

Fig. 2　The inconsistency of phenotype and genotype associated with antibiotic-resistance in the isolated strains.

quinolones（aac（6'）-Ib-cr），and macrolides（ereA）. Two open reading frames（orfC and orfF）of unknown function were detected and one transposase gene（tnpA）was also detected.

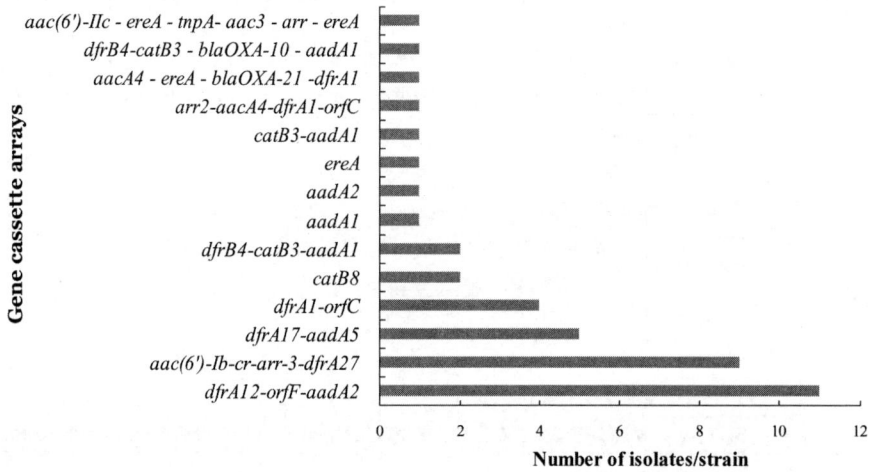

Fig. 3　The abundance of different class I integrin gene cassette arrays in resistant isolates from eels and farming water

In this study, class II integrons were identified by PCR in all five strains of Proteus. Nevertheless, other species that contained class II integrons were Aeromonas（2 strains），Staphylococcus（1 strain），Citrobacter（1 strain）and Shewanella（1 strain）. All 10 of these strains had the same gene cassette array, dfrA1-catB2-sat2-aadA1. The trimethoprim（dfrA1），amphenicols（catB2），and aminoglycoside（aadA1）resistance genes were also found as part of the gene cassettes in class II integrons. In addition, the sat2 gene cassette encodes a streptothricin acetyltransferase, which give rise to streptothricin resistance.

The frequency of antibiotic resistance in the integron-positive strains was higher than that of integron-negative strains（Fig. 4）. Only several integron-positive strains were resistant to streptomycin, norfloxacin, and ofloxacin.

4　Discussion

In this study, we reported the distribution of different classes of antibiotic-resistance genes in antibiotic resistance bacteria isolated from eels and farming water in China. We detected specific resistance genes based on primers designed during previous studies（Table S1）. Molecular characterization showed great diversity in the antimicrobial-resistance genes（Fig. 1）. Many previous studies have focused on one or two bacterial genera and/or strains to identify antimicrobial-resistance genes that correspond to antibiotic resistance bacteria. Dolejska et al.（2007）carried out an assessment of the occurrence of antimicrobial-resistance genes in Escherichia coli isolated from Black-headed gulls. The bla$_{TEM}$ gene was detected in 29 of 30 beta-lactam resistant isolates, while bla$_{SHV}$ and bla$_{OXA}$ genes were not found. Other resistance genes, such as cat, strA, aadA, tetA, tetB, sul1, sul2 were also detected in the resistant

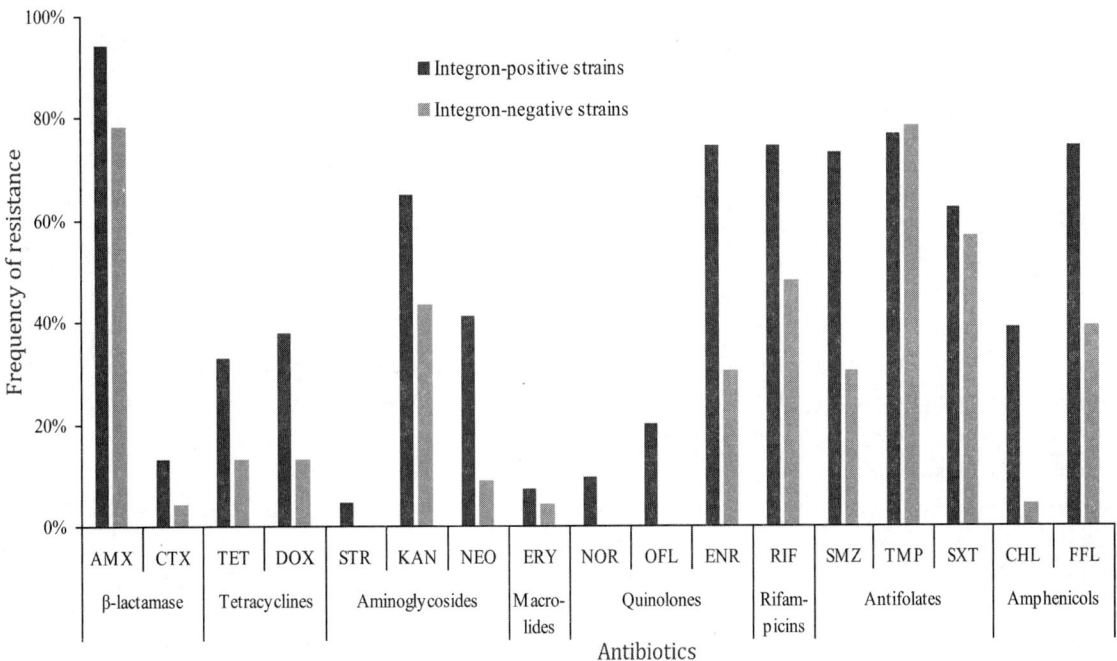

Fig. 4　Frequency of resistance to 17 antimicrobials among the integron-positive and integron-negative strains. Amoxicillin：AMX；Cefotaxime：CTX；Tetracycline：TET；Doxycycline：DOX；Streptomycin：STR；Kanamycin：KAN；Neomycin：NEO；Erythromycin：ERY；Norfloxacin：NOR；Ofloxacin：OFL；Enrofloxacin：ENR；Rifampicin：RIF；Sulfamethoxazole：SMZ；Trimethoprim：TMP；Sulfamethoxazole/Trimethoprim：SXT；Chloramphenicol：CHL；Florfenicol：FFL.

strains. These results are similar to our study, where bla_{TEM} was detected in all 108 antibiotic-resistant isolates, while few bla_{SHV}, bla_{OXA}, DHA and EBC genes were found, and the MOX gene was not detected. The bla_{TEM} gene has been shown to be the most prevalent gene detected in many antibiotic-resistant isolates from humans and animals (Ahmed et al. 2014；Bakour et al. 2013). Tetracycline resistance genes encoding active efflux pumps occur frequently among bacteria isolated from different environments. The most predominant gene is tetC, which was detected in all strains of Aeromonas spp. from rainbow trout farms in Australia (Ndi & Barton, 2011). Recently, a study on Escherichia coli from catfish showed that tetB was present in 75% of all isolates tested, which was the highest percentage of the tet genes tested (Nawaz et al. , 2009). While, Nawaz et al. (2006) also studied Aeromonas veronii from catfish, among these isolates tetE was the most predominant antibiotic-resistance gene, occurring in 73/81 (90.0%) strains. Therefore, it appears that it is not hard to find different tetracycline resistance genes in various ecosystems and the distribution of these genes differs with environment. Furthermore, in our study, the largest proportion of isolates tested was positive for tetC, followed by tetD and tetA. Resistance genes to antifolates include the genes encoding dihydrofolate reductase (dfrA1, dfrA5, dfrA7, dfrA12, dfrB4) and dihydropteroate synthase (sulI). The overall frequency of genes for sulfonamide and trimethoprim resistance in this study was sulI>dfrA12>dfrA5>dfrA1>dfrA7. Several studies have focused on dfr and sul genes together. Hu et al. (2011) showed that of 102 Stenotrophomonas maltophilia isolates analyzed in the study, 50.99% contained the sulI gene and 15.69% carried dfr genes. Therefore, the sul genes were more prevalent among this population of organisms than the dfr genes. Among the aminoglycoside-resistance genes, the aadA gene had the highest prevalence (74.07%) and the aphA1 gene was present in 44.44% of the isolates. This observation was consistent with another previous study (Glenn et al. , 2011), where the most prevalent aminoglycoside resistance genes among all isolates were aadA, aac (3), strA, strB, and aphA1. Among the other classes of antimicrobial-resistance genes (quinolones, amphenicols, rifampicin and macrolides) the most prevalent genes in the current study were qnrB, floR, arr2/3, and ereA, respectively. The distribution of antimicrobial-resistance genes depended on where the isolates came from (animal, water, sediment etc.), the detection method, and the genes that were detected. Zhang et al. (2013) studied resistance genes of aquaculture systems in Southern China and the results revealed that the most prevalent resistance genes from three different gene classes were tetA, sul2, bla_{TEM}, . Other resistance genes were detected in different degrees in these resistant isolates, including tetW, tetB, sul3, sul1, bla_{OXA} and bla_{CTX}. However, tetE, tetX and bla_{SHV} were not detected in any isolates. In addition, we found one strain of the 108 antibiotic-resistant isolates that carried eighteen resistance genes and all strains in this study contained at least two antibiotic-resistance genes.

The occurrence of the three integrons in different ecosystems have been reported in many studies, particularly on class I integrons and their gene cassette arrays. Ndi & Barton (2011) detected class I integrons in 31% of Aeromonas spp. isolated from trout farms in Australia, while they failed to detect any class II or class III integrons. In addition, they also found that class I integrons were present in 23% (30/129) of Pseudomonas spp, and again that class II and class III integrons were not found in any of the strains (Ndi & Barton 2012). Similarly, class I integrons have been reported from approximately 33% of Pseudomonas spp. and 28% of Aeromonas spp. isolates from catfish and the main gene cassette arrays were dfrA12-orfF-aadA2, aadA1, catB8-aadA1, dfrA1-orfX and dfrA21 (Nguyen et al. 2014). All isolates were negative for class II integrons. Due to the structure of the intIII gene, which contains an internal stop codon that renders intIII inactive, the occurrence of class II integrons is much lower than class I integrons. The most common gene cassette array found class II integron-positive isolates is dfrA1-sat2-aadA1 (Zhu et al. 2011). Class III integrons have only been detected in rare instances. In the present study, integrons were more widespread with class I integrons being found in 79.63% of strains and class II integrons found in 9.26% of the strains. Furthermore, 47.67% of the class I integron-positive strains harbored gene cassette arrays. The most frequently found array was dfrA12-orfF-aadA2, as has also been seen in previous studies. The second most commonly observed array was aac (6') -Ib-cr-arr-3-dfrA27, which was relatively uncommon in previous studies. Reasons for this might be that the sequence of the array was difficult to amplify and does not occur widely in most typical ecosystems. The qacEΔ1/ sul1 failed to amplify from several class I integron-positive isolates, which may be because they lack the non-essential structure 3'-CS. Nevertheless, the gene cassette arrays of the class II integrons all contained the array dfrA1-catB2-sat2-aadA1, which has only been reported in a few studies. However, the gene cassettes dfrA1, catB2, sat2, aadA1 were frequently detected in class II integrons. Three strains (one Shewanella and two Aeromonas strains) carried class I as well as class II integrons, and the class I integron gene cassette arrays of the three strains were all dfrA12-orfF-aadA2. Thus, we found that the distribution of integrons was connected with different gene cassettes and/or bacterial genera.

Many studies have investigated integrons in different bacterial genera that have been isolated from humans, such as Klebsiella and Acinetobacter. Research on bacterial integrons isolated from aquatic animals is comparatively small in number and most of these studies focused on the pathogens Aeromonas and Pseudomonas. Ndi & Barton have studied resistance genes and integrons in Aeromonas (2011) and Pseudomonas (2012) isolated from rainbow trout farms, and they found the frequency of resistance genes and integrons was higher in Aeromonas than Pseudomonas. In our study, we analyzed the high frequency strains from the genera Aeromonas, Citrobacter, and Acinetobacter. We found that 89.5% of Citrobacter isolates contained integrons and 58.8% integron-positive Citrobacter strains carried gene cassettes. The gene cassette arrays were the following three arrays: dfrA12-orfF-aadA2, aac (6') -Ib-cr-arr-3-dfrA27 and dfrA17-aadA5. Citrobacter isolates had high multiple antibiotic resistance indexes and Acinetobacter isolates were relatively lower, and the tendency towards antibiotic resistance indexes of Citrobacter and Acinetobacter was more obvious among the integron-positive isolates carrying gene cassettes. Of the Aeromonas strains 65.4% carried integrons, and 70.6% of these were detected as gene cassettes, which included six types of arrays. Only one strain (7.7%) carried a gene cassette array in the 72.2% of Acinetobacter strains that were positive for integrons. Overall, the frequency of Acinetobacter strains was far below average. These data indicated that integrons were closely associated with antibiotic resistance and across multiple bacterial genera.

Several earlier studies have found that some antimicrobial-resistance genes were closely associated with integrons. Ndi & Barton (2012) suggested that the presence of the aadA gene was commonly associated with integrons. The aadA1 gene was the most widely carried gene in strains isolated from seven countries (Abée-Lund & Rum 2001). The association of integrons with various resistance genes, such as sul and dfr genes, has been well documented (Ndi & Barton 2011; Petersen et al. 2000; Shah et al. 2014). The gene cassettes found in the present study demonstrate that the dominant cassettes belong to aminoglycoside (aac and aad) and trimethoprim (dfr) resistance genes, similar to other studies. The qacEΔ1/sul1 genes were found in 73.26% of class I integron-positive isolates, and were also found in 18.18% of class I integron-negative isolates. From the data generated in this study, it is clear that there is a strong relationship between antimicrobial-resistance genes and integrons. The genes identified with high prevalence in this study were bla_{TEM}, tetC, sul1, aadA, floR and qnrB, among others. Among these genes, sul1 and aadA were found in a high percentage of class I integrons. However, bla_{TEM}, tetC, floR, and qnrB were not often found within integrons. This may be related to specific conditions associated with horizontal gene transfer and the characteristics of different genes and integrons.

Several antimicrobials have been banned for use in aquatic animals, but antibiotic-resistance to these compounds still exists in several isolates. In earlier research, we identified the antimicrobail content in the same farming water from which we took samples for the present study. Trace amounts of chloramphenicol and florfenicol were detected and sulfamethoxazole, trimethoprim and erythromycin were detected in the water samples. The index of multiple-drug resistance was high. The results showed that high concentrations of antimicrobials might lead to large numbers of antibiotic-resistant isolates. In contrast, the antibiotic-resistance level of the isolates did not always correlate with the low concentration of antimicrobials in farming water. Horizontal transfer of antibiotic resistance genes might place emphasis on molecular mechanism to explain the frequency of antibiotic-resistance genes. Thus, research on the genes related to antimicrobial resistance is more specific as we can control the antibiotic-resistance of isolates. In addition, isolates with inconsistent antibiotic-resistance phenotype and genotype were mainly related to the rifampicins, quinolones and aminoglycosides. The strains resistant to quinolones were the most obvious in terms of the inconsistent antibiotic-resistance phenotype and genotype. The discordance for quinolones might be due to the fact that few qnr genes were tested that usually give low levels of resistance unless they are accompanied by other chromosomally-encoded resistance mechanisms. The inconsistencies might be connected with a lack of expression, repression of the drug-resistance genes, or genes which were not detected in our study. Bacteria may also have entered dormancy to escape the antimicrobial agents.

Antibiotic resistance genes and integrons are important contributors to the development of antibiotic resistance. Our study revealed that the frequency of antibiotic resistance in the integron-positive strains was higher than that of integron-negative strains (Fig. 3). Only several integron-positive strains were resistant to streptomycin, norfloxacin, and ofloxacin. Streptomycin is an aminoglycoside drug, to which resistance is conferred by aminoglycoside resistance genes, e. g. aac and aad genes. Thus, streptomycin resistance is usually associated with aminoglycoside resistance genes and integrons. Also, we found strains that were resistant to antifolate agents that were not associated with integrons. The genes may confer resistance to antifolates through other mechanisms. The data show that the higher frequency of antibiotic resistance of strains, the more resistance genes were detected. While some strains were found to be resistant to one class of antibiotics, we failed to detect a corresponding resistance gene. For example, we found 31.48% of strains were resistant to erythromycin, while only 7.41% strains contained the ere resistance gene. This means that the erythromycin-resistant strains contain other genes encoding resistance to erythromycin. In addition, some strains contained resistance genes, but did not display the corresponding drug-resistance. The bla_{TEM} gene was detected in all strains, but not all strains were resistant to β-lactamase drugs (amoxicillin and cefotaxime). This suggests that in some strains the genes were not expressed and did not play a role in resistance.

5　Conclusion

From this study we conclude that eels and farming water in China contain a large number of antibiotic resistant bacteria that carry a wide range of antimicrobial-resistance genes and integrons. These results show that these bacteria are potential threats to the eel farming industry, other animals and human health. Thus, we should constantly monitor the resistance genes of bacteria in aquaculture.

This study was financially supported by the Special Fund for Agro-scientific Research in the Public Interest (NO. 201203085), National Natural Science Foundation of China (NO. 31202030, NO. 31272669), and Natural Science Foundation of Fujian Province (NO. 2014J01131).

References

Abée-Lund TML, Rum HS (2001) Class 1 integrons mediate antibiotic resistance in the fish pathogen Aeromonas salmonicida worldwide. Microb Drug Resist 7: 263-272.

Ahmed A. M., Motoi Y., Sato M., Maruyama A., Watanabe H., Fukumoto Y., Shimamoto T. (2007) Zoo animals as reservoirs of gram-negative bacteria harboring integrons and antimicrobial resistance genes. Appl Environ Microb 73: 6686-6690.

Ahmed A. M., Shimamoto T., Shimamoto T. (2014) Characterization of integrons and resistance genes in multidrug-resistant Salmonella enterica isolated from meat and dairy products in Egypt. Int J Food Microbiol 189: 39-44.

Bakour S., Touati A., Sahli F., Ameur A. A., Haouchine D., Rolain J. (2013) Antibiotic resistance determinants of multidrug-resistant Acinetobacter baumannii clinical isolates in Algeria. Diagn Micr Infec Dis 76: 529-531.

Cao L., Naylor R., Henriksson P., Leadbitter D., Metian M., Troell M., Zhang W. (2015) China's aquaculture and the world's wild fisheries. Science 347: 133-135.

Chang Y., Shih D. Y., Wang J., Yang S. (2007) Molecular characterization of class 1 integrons and antimicrobial resistance in Aeromonas strains from foodborne outbreak-suspect samples and environmental sources in Taiwan. Diagn. Micr Infec Dis 59: 191-197.

Cong L. (2011) Analysis of antibiotic resistance and resistance genes of Salmonella from fish pond ecosystem. Jinan University (China).

Dolejska M., Cizek A., Literak I. (2007) High prevalence of antimicrobial-resistant genes and integrons in Escherichia coli isolates from Black-headed Gulls in the Czech Republic. J Appl Microbiol 103: 11-19.

Fluit A. C., Schmitz F. J. (1999) Class 1 integrons, gene cassettes, mobility, and epidemiology. Eur J Clin Microbiol 18: 761-770.

Glenn L. M., Lindsey R. L., Frank J. F., Meinersmann R. J., Englen M. D., Fedorka-Cray P. J., Frye J. G. (2011) Analysis of antimicrobial resistance genes detected in multidrug-resistant Salmonella enterica serovar Typhimurium isolated from food animals. Microb Drug Resist 17: 407-418.

Grave K., Lingaas E., Bangen M., Ronning M. (1999) Surveillance of the overall consumption of antibacterial drugs in humans, domestic animals and farmed fish in Norway in 1992 and 1996. J Antimicrob Chemoth 43: 243-252.

Guo M., Yuan Q., Yang J. (2013) Ultraviolet reduction of erythromycin and tetracycline resistant heterotrophic bacteria and their resistance genes in municipal wastewater. Chemosphere 93: 2864-2868.

Harnisz M., Korzeniewska E., Goła I. (2015) The impact of a freshwater fish farm on the community of tetracycline-resistant bacteria and the structure of tetracycline resistance genes in river water. Chemosphere 128: 134-141.

Hu L., Chang X., Ye Y., Wang Z., Shao Y., Shi W., Li X., Li J. (2011) Stenotrophomonas maltophilia resistance to trimethoprim/sulfamethoxazole mediated by acquisition of sul and dfrA genes in a plasmid-mediated class 1 integron. Int J Antimicrob Ag 37: 230-234.

Jacquier H., Zaoui C., Sanson-le Pors M., Mazel D., Berçot B. (2009) Translation regulation of integrons gene cassette expression by the attC sites. Mol Microbiol 72: 1475-1486.

Jun J. W., Kim J. H., Gomez D. K. (2010) Occurrence of tetracycline-resistant Aeromonas hydrophila infection in Korean cyprinid loach (Misgurnus anguillicaudatus). African Journal of Microbiology Research 4: 849-855.

Kehrenberg C., Friederichs S., De J. A., Michael G. B., Stefan S. (2006) Identification of the plasmid-borne quinolone resistance gene qnrS in Salmonella enterica serovar infantis. J Antimicrob Chemoth 58: 18-22.

Keyes K., Hudson C., Maurer J. J., Thayer S., White D. G., Lee M. D. (2000) Detection of florfenicol resistance genes in Escherichia coli isolated from sick chickens. Antimicrob Agents Ch 44: 421-424.

Korzeniewska E., Harnisz M. (2013) Extended-spectrum beta-lactamase (ESBL)-positive Enterobacteriaceae in municipal sewage and their emission to the environment. J Environ Manage 128: 904-911.

Labbate M., Case R. J., Stokes H. W. (2009) The integron/gene cassette system: an active player in bacterial adaptation. Methods in Molecular Biology 532: 103-125.

Maynard C., Bekal S., Sanschagrin F., Levesque R. C., Brousseau R., Masson L., Lariviere S., Harel J. (2004) Heterogeneity among virulence and antimicrobial resistance gene profiles of extraintestinal Escherichia coli isolates of animal and human origin. J Clin Microbiol 42: 5444-5452.

Nawaz M., Khan A. A., Khan S., Sung K., Kerdahi K., Steele R. (2009) Molecular characterization of tetracycline-resistant genes and integrons from avirulent strains of Escherichia coli isolated from catfish. Foodborne Pathog Dis 6: 553-559.

Nawaz M., Sung K., Khan S. A., Khan A. A., Steele R. (2006) Biochemical and molecular characterization of tetracycline-resistant Aeromonas veronii isolates from catfish. Appl Environ Microb 72: 6461-6466.

Ndi O. L., Barton M. D. (2011) Incidence of class 1 integron and other antibiotic resistance determinants in Aeromonas spp. from rainbow trout farms in Australia. J Fish Dis 34: 589-599.

Ndi O. L., Barton M. D. (2012) Resistance determinants of Pseudomonas species from aquaculture in Australia. Journal of Aquaculture Research & Development 03: 1-6.

Nguyen H. N. K., Van T. T. H., Nguyen H. T., Smooker P. M., Shimeta J., Coloe P. J. (2014) Molecular characterization of antibiotic resistance in Pseudomonas and Aeromonas isolates from catfish of the Mekong Delta, Vietnam. Vet Microbiol 171: 397-405.

Ochman H., Lawrence J. G., Groisman E. A. (2000) Lateral gene transfer and the nature of bacterial innovation. Nature 405: 299-304.

Partridge S. R., Tsafnat G., Coiera E., Iredell J. R. (2009) Gene cassettes and cassette arrays in mobile resistance integrons. FEMS Microbiol Rev 33: 757-784.

Perez-Perez F. J., Hanson N. D. (2002) Detection of plasmid-mediated AmpC-lactamase genes in clinical isolates by using multiplex PCR. J Clin Microbiol 40: 2153-2162.

Petersen A., Guardabassi L., Dalsgaard A., Olsen J. E. (2000) Class I integrons containing a dhfrI trimethoprim resistance gene cassette in aquatic Acinetobacter spp. FEMS Microbiol Lett 182: 73-76.

Ramos P. M., Velilla S. M. (2012) Aminoglycoside resistance by aph (3)-VIa and aac (3)-II genes in Acinetobacter baumannii isolated in Montería, Colombia. Salud Uninorte Barranquilla 28: 209-217.

Robicsek A., Strahilevitz J., Sahm D. F., Jacoby G. A., Hooper D. C. (2006) qnr prevalence in ceftazidime-resistant Enterobacteriaceae isolates from the United States. Antimicrob. Agents Ch 50: 2872-2874.

Rowe-Magnus D. A., Mazel D. (1999) Resistance gene capture. Curr Opin Microbiol 2: 483-488.

Sandalli C., Özgümüş O. B., Sevim A. (2010) Characterization of tetracycline resistance genes in tetracycline-resistant Enterobacteriaceae obtained from a coliform collection. World Journal of Microbiology and Biotechnology 26: 2099-2103.

Sarria-Guzmán Y., López-Ramírez M. P., Chávez-Romero Y., Ruiz-Romero E., Dendooven L., Bello-López J. M. (2014) Identification of antibiotic resistance cassettes in class 1 integrons in Aeromonas spp. strains isolated from fresh fish (Cyprinus carpio L.). Curr Microbiol 68: 581-586.

Schwaiger K., Harms K., Hölzel C., Meyer K., Karl M., Bauer J. (2009) Tetracycline in liquid manure selects for co-occurrence of the resistance genes tet (M) and tet (L) in Enterococcus faecalis. Vet Microbiol 139: 386-392.

Shah S. Q. A., Cabello F. C., L'Abée-Lund T. M., Tomova A., Godfrey H. P., Buschmann A. H., Sørum H. (2014) Antimicrobial resistance and antimicrobial resistance genes in marine bacteria from salmon aquaculture and non-aquaculture sites. Environ Microbiol 16: 1310-1320.

Spindler A., Otton LvM., Fuentefria D. B., Corção G. (2012) Beta-lactams resistance and presence of class 1 integron in Pseudomonas spp. isolated from untreated hospital effluents in Brazil. Antonie van Leeuwenhoek 102: 73-81.

Stokes H. W., Hall R. M. (1989) A novel family of potentially mobile DNA elements encoding site-specific gene-integration functions: integrons. Mol Microbiol 3: 1669-1683.

Sunde M., Norström M. (2006) The prevalence of, associations between and conjugal transfer of antibiotic resistance genes in Escherichia coli isolated from Norwegian meat and meat products. J Antimicrob Chemoth 58: 741-747.

Uyaguari M. I., Scott G. I., Norman R. S. (2013) Abundance of class 1-3 integrons in south Carolina estuarine ecosystems under high and low levels of anthropogenic influence. Mar Pollut Bull 76: 77-84.

Wu X., Lin M., Jiang X., Yan Q., Zhang X. (2015) Diversity and antimicrobial susceptibility of drug-resistant bacteria isolated from Anguilla rostrata and the farming water. Journal of fisheries of China 39: 1043-1053.

Xu H., Davies J., Miao V. (2007) Molecular characterization of class 3 integrons from Delftia spp. J Bacteriol 189: 6276-6283.

Zhang R., Ying G., Su H., Zhou L., Liu Y. (2013) Antibiotic resistance and genetic diversity of Escherichia coli isolates from traditional and integrated aquaculture in south China. Journal of Environmental Science and Health, Part B 48: 999-1013.

Zhu J., Jiang R., Wu K., Ma Z. (2012) Resistance mechanisms of multidrug-resistant strains of Klebsiella pneumoniae to rifampicin and sulfamethoxazole/trimethoprim compound. Chinese Journal of Nosocomiology 22: 452-456.

Zhu J. Y., Duan G. C., Yang H. Y., Fan QT, Xi Y. L. (2011) Atypical class 1 integron coexists with class 1 and class 2 integrons in multi-drug resistant Shigella flexneri isolates from China. Curr Microbiol 62: 802-806.

Zhu Y., Yi Y., Liu F., Lv N., Yang X., Li J., Hu Y., Zhu B. (2014) Distribution and molecular profiling of class 1 integrons in MDR Acinetobacter baumannii isolates and whole genome-based analysis of antibiotic resistance mechanisms in a representative strain. Microbiol Res 169: 811-816.

该文发表于《中国水产科学》【2013, 20 (5): 1015-1022】

鱼源嗜水气单胞菌多重耐药菌株整合子的分子特征

Molecular characterization of integron-gene cassettes in multi-drug resistant Aeromonas hydrophila from fish

李绍戊, 王荻, 刘红柏, 尹家胜, 卢彤岩

(中国水产科学研究院 黑龙江水产研究所, 哈尔滨 150070)

摘要: 为研究嗜水气单胞菌多重耐药菌株整合子-基因盒分布及分子特征, 首先采用K-B纸片扩散法检测28株鱼源嗜水气单胞菌对18种抗生素的耐药性, 然后利用PCR方法检测菌株中Ⅰ、Ⅱ、Ⅲ型整合酶基因并对其携带基因盒序列进行分析。结果表明, 分离到的鱼源嗜水气单胞菌呈多重耐药性, 对b-内酰胺类、大环内酯类、氯霉素类和四环素类药物的耐药率超过60%, 而对氟喹诺酮类药物较敏感, 且菌株间耐药谱差异较大。53.57%的菌株Ⅰ类整合子阳性, 21.43%的菌株Ⅱ类整合子阳性, 整合子阳性菌株对多种药物的耐药率均高于阴性株, 且Ⅰ类整合子阳性株多重耐药率明显高于Ⅱ类整合子阳性株, 表明整合子系统在嗜水气单胞菌多重耐药性中发挥重要作用。Ⅰ类整合子基因盒以aadA、dfrA、catB家族为主, 分别介导氨基糖苷类、甲氧磺胺嘧啶类和氯霉素类药物耐药; 基因盒的排列以aadA2+dfrA12类型为主。此外, Ⅰ类整合子阳性的嗜水气单胞菌多重耐药性在不同个体间也存在较大差异, 提示多重耐药菌株的耐药表型与基因盒的类型无直接相关性。

关键词: 嗜水气单胞菌; 多重耐药性; 整合子-基因盒系统

细菌耐药性以及多重耐药, 是当今国内外临床抗感染中遇到的重要难题和考验, 且呈现出耐药谱逐年上升和耐药性传播迅速的特点。耐药细菌的产生包括自然耐药和获得性耐药两种来源, 而后者主要通过细菌染色体基因突变、可移动遗传元件 (质粒、转座子、整合子等) 传递或捕获外源耐药基因从而使得细菌耐药[1]。其中, 整合子-基因盒系统是近年来发现的细菌耐药性传播的主要机制之一[2]。作为细菌的一种天然克隆与表达系统, 整合子通过对外源耐药基因的捕获、组合和排列, 借助于其强启动子使耐药基因表达; 同时, 一个整合子可整合多个不同的耐药基因盒, 从而使细菌具有多重耐药性[3-4]。迄今为止, 已发现80余种基因盒, 且基因盒的种类、数量及分布均呈现出差异性[5]。目前根据整合酶基因不同, 整合子至少可分为六类, 而以Ⅰ、Ⅱ、Ⅲ类为多, 其中Ⅰ类整合子主要存在于革兰氏阴性杆菌中, 介导细菌对大部分临床抗菌药物的耐药[6]。

嗜水气单胞菌 (Aeromonas hydrophila) 属弧菌科气单胞菌属, 是淡水养殖动物细菌性败血症的重要病原菌之一, 可导致多种水生动物表现出不同临床症状, 如肌肉和内脏出血、烂鳃、坏死、腹部膨胀、水肿等; 另外, 因其常与其他致病菌混合感染使得病情加重, 从而给水产养殖业造成严重危害[7]。近年来, 抗生素在鱼类细菌性疾病防治中得到广泛应用, 嗜水气单胞菌耐药问题日趋严重, 增加了嗜水气单胞菌病的防治难度[8-9]。因此, 开展鱼源嗜水气单胞菌的流行病学调查及

其耐药性监测，并对其携带耐药基因的可转移遗传因子整合子-基因盒系统进行研究，将具有重要的临床意义和流行病学价值。本研究采用 PCR 方法检测了 3 类整合子在 28 株鱼源嗜水气单胞菌分离株中的分布情况，旨在探索整合子及耐药基因盒在嗜水气单胞菌多重耐药株中的分布和构成差异，了解整合子与鱼源嗜水气单胞菌多重耐药性之间的关系，以正确评价整合子-耐药基因盒系统在嗜水气单胞菌多重耐药机制中的作用和地位。

1 材料与方法

1.1 实验菌株

28 株鱼源嗜水气单胞菌分别从黑龙江、吉林和辽宁等地区患淡水鱼出血病的鲤、鲫组织中分离。采用细菌微量生化鉴定管和 16S rRNA 基因序列对 28 株嗜水气单胞菌进行鉴定[10-12]，确认为嗜水气单胞菌后进行细菌药敏试验和整合子分布检测。

1.2 嗜水气单胞菌药敏实验

将嗜水气单胞菌分离株挑入 MH 肉汤培养基中，28℃震荡培养过夜。采用 K-B 纸片法，取 0.1 mL 培养菌液均匀涂板于 9 cm MH 琼脂平板上，然后等距贴上药敏纸片 4 片/平板，分别测定分离株对左氟沙星（Levofloxacin，LE）、氧氟沙星（Ofloxacin，OF）、环丙沙星（Ciprofloxacin，CP）、诺氟沙星（Norfloxacine，NF）、恩诺沙星（Enoxacin，EN）、庆大霉素（Gentamicin，GM）、卡那霉素（Kanamycin，KM）、链霉素（Streptomycin，SM）、阿莫西林（Amoxicillin，AMP）、氨苄西林（Ampicillin，AP）、羧苄青霉素（Carbenicillin，CB）、头孢克肟（Cefixime，CFX）、头孢噻呋（Ceftofur，CFT）、头孢吡肟（Cefepime，CP）、氯霉素（Chloramphenicol，CHL）、红霉素（Erythromycin，ER）、四环素（Tetracycline，TE）和复方新诺明（Compound Sulfamethoxazole，SMZ）共 18 种常见药物的敏感程度。将处理好的平板倒置于培养箱中，28℃培养 24 h，测定抑菌圈直径，并根据 CLSI 标准进行药敏结果判定[13]。实验菌株对抗生素的耐药率按照公式计算：耐药菌株数目（Rn）／总实验菌株数目（n）×100%。

1.3 整合子-基因盒系统检测及鉴定

采用细菌基因组 DNA 提取试剂盒及质粒提取试剂盒（Tiangen）分别提取 28 株嗜水气单胞菌基因组 DNA 和质粒 DNA，检测纯度后用于 PCR 扩增。

参考已发表文献，利用下列引物（表 1）分别对嗜水气单胞菌 I、II、III 类整合子进行 PCR 检测，并对 I 类整合子扩增片段进行序列测定及 BLAST 比对分析。PCR 反应采用 20 μL 反应体系，其中含 DNA 模板 0.5 μL，dNTP（10 mmol/L）0.5 μL，10× PCR buffer 2.5 μL，MgCl2（25 mmol/L）1.5 μL，整合子引物（10 μmol/L）各 1 μL，Taq DNA 聚合酶（MBI）2U，用 ddH2O 补足至 25 μL。PCR 反应程序为：95℃预变性 3 min；95℃变性 30 s，68℃退火 30 s，72℃延伸 1 min，40 个循环；72℃延伸 10 min；4℃。PCR 产物经 1% 琼脂糖凝胶电泳后用凝胶成像系统观察分析（Biorad，美国）。利用 BLAST 软件在线比对 I 类整合子可变区序列（http：//www.ncbi.nlm.nih.gov/blast/）。

表 1 整合子检测用引物信息

Tab. 1 Primers used in the detection of integrons in this study

引物名称 primer name	序列（5′-3′） sequence（5′-3′）	长度/bp length	温度/℃ temperature	参考文献 reference
Int1-F	CTGCGTTCGGTCAAGGTTCT	882	68	[14]
Int1-R	GCAATGGCCGAGCAGATCCT			
Int2-F	CACGGATATGCGACAAAAAGGT	746	59	[14]
Int2-R	GTAGCAAACGAGTGACGAAATG			
Int3-F	GCAGGGTGTGGACGAATACG	760	55	[14]
Int3-R	ACAGACCGAGAAGGCTTATG			
5′CS	GCCATCCAAGCAGCAAG	Variable	60	[15]
3′CS	AAGCAGACTTGACCTGA			

2　结果与分析

2.1　嗜水气单胞菌药敏试验结果

如表 2 所示，本实验中鱼源嗜水气单胞菌对 18 种抗生素中耐药率最高的是 β-内酰胺类（67.9%~100%）、大环内酯类（89.3%）、氯霉素类（64.3%）和四环素类（71.4%）药物，次之是氨基糖苷类（28.6%~53.6%）和磺胺类药物（42.8%），最低是氟喹诺酮类药物（21.4%~35.7%）。

耐药谱分析表明，28 株嗜水气单胞菌耐药谱差异较大，AH-43 分离株同时对 17 种抗生素耐药，其耐药 LE-OF-CIP-NF-EN-GEN-KN-STP-AMC-AMP-CB-CFX-CFT-CHL-EM-TE-SXT，AH-4 分离株仅对 2 种抗生素有耐药性（CFT-EM）。

表 2　鱼源嗜水气单胞菌分离株耐药率分析

Tab. 2　The Antibiotic resistance rate of Aeromonas hydrophila isolates from fish

抗生素类别 antibiotic type			药敏判定标准折点（breakpoint for antimicrobial susceptibility testing/mm）	菌株类别 strain type		
				R	I	S
氟喹诺酮类 quinolone		左氟沙星 levofloxacin，LF	16-18	9	0	19
		氧氟沙星 ofloxacin，OF	13-15	6	1	21
		环丙沙星 Ciprofloxacin，CIP	16-20	6	4	18
		诺氟沙星 norfloxacin，NF	13-16	10	9	9
		恩诺沙星 enrofloxacin，EN	16-18	8	0	20
氨基糖苷类 aminoglycoside		庆大霉素 gentamicin，GEN	13-14	8	7	13
		卡那霉素 kanamycin，KN	14-17	15	10	3
		链霉素 Streptomycin，STP	12-14	12	5	11
β-内酰胺类 β-lactam	青霉素类 penicillins	阿莫西林 amoxicillin，AMC	14-17	22	3	3
		氨苄西林 ampicillin，AMP	14-16	28	0	0
		羧苄青霉素 carbenicillin，CB	20-22	23	5	0
	头孢菌素类 cephalosporin	头孢克肟 cefixime，CFX	16-18	19	3	6
		头孢噻吩 cephalothin，CFT	15-17	27	0	1
		头孢吡肟 cefepime，CP	15-17	3	3	22
单一药物类 single drug class	氯霉素类 chloramphenicol	氯霉素 chloromycetin，CHL	13-17	18	1	9
大环内酯类 macrolides		红霉素 erythromycin，EM	14-22	25	3	0
四环素类 tetracyclines		四环素 tetracycline，TE	15-18	20	6	2
磺胺类 sulfonamides		复方新诺明 compound，SXT	13-16	12	0	16

注：R 表示耐受；I 表示中度敏感；S 表示敏感.

Note：R, resistant；I, intermediate；S, susceptible.

2.2　嗜水气单胞菌 3 类整合子的 PCR 扩增及检测结果

28 株嗜水气单胞菌中整合子的检出情况见表 3。Ⅰ类整合子检出率为 53.57%（15/28），Ⅱ类整合子检出率为 21.43%（6/28），另外 25% 菌株（7/28）整合子阴性。为初步研究整合子的定位，分别以嗜水气单胞菌菌株基因组 DNA 和质粒 DNA 为模板，使用 3 类整合子特异性引物扩增其整合酶基因。结果表明，Ⅰ型整合酶基因的 PCR 扩增条带大小为 882 bp

（图1），初步表明Ⅰ类整合子可定位于染色体、质粒或者两者都存在；Ⅱ型整合酶基因的PCR扩增条带大小为746 bp，表明Ⅱ类整合子均位于染色体上。

表3　嗜水气单胞菌整合子分布情况分析
Tab. 3　Distribution of integrons in Aeromonas hydrophila

菌株编号 strain number	整合子类型 integrons type	分布 distribution，C/P	菌株编号 strain number	整合子类型 integrons type	分布 distribution，C/P
AH-3	Ⅰ	C	AH-34	Ⅱ	C
AH-4	N		AH-38	Ⅱ	C
AH-5	Ⅰ	P/C	AH-39	Ⅰ	C
AH-8	Ⅰ	P/C	AH-43	Ⅰ	C
AH-9	N		AH-49	Ⅰ	C
AH-10	Ⅰ	P/C	AH-50	N	
AH-13	N		AH-52	N	
AH-15	Ⅰ	P/C	AH-53	Ⅰ	P/C
AH-17	Ⅰ	P/C	AH-56	Ⅱ	C
AH-18	N		AH-58	Ⅰ	P/C
AH-19	Ⅱ	C	AH-59	Ⅰ	P/C
AH-20	Ⅰ	P/C	AH-61	Ⅱ	C
AH-23	Ⅰ	P	AH-63	N	
AH-32	Ⅰ	C	AH-66	Ⅱ	C

注：N指整合子阴性；P指质粒；C指染色体.
Note：N，negative；P，plasmid；C，chromosome.

2.3　嗜水气单胞菌多重耐药与整合子相关性

研究结果表明，整合子阴性嗜水气单胞菌主要对β-内酰胺类、氯霉素类和大环内酯类药物耐药，而对氟喹诺酮类、氨基糖苷类以及磺胺类药物敏感（表4）。含有Ⅰ类和Ⅱ类整合子的嗜水气单胞菌其多重耐药性明显要高于阴性菌株，且整合子的有无对28株嗜水气单胞菌菌株的影响主要表现在对喹诺酮类、氨基糖苷类、β-内酰胺类和磺胺类药物的耐药性。由表4也可以看出，Ⅰ类整合子对嗜水气单胞菌多重耐药性的贡献要高于Ⅱ类整合子，显示出Ⅰ类整合子的存在与细菌多重耐药性之间具有十分密切的关系。另外，在含有Ⅰ类整合子的15株嗜水气单胞菌中，菌株之间的多重耐药谱也不尽相同。

表4　整合子-基因盒系统与嗜水气单胞菌多重耐药相关性
Tab. 4　Correlation between integron-gene cassette system and multi-drug resistance in Aeromonas hydrophila

抗生素类别 antibiotic type	名称 name	integron type								
		无整合子（n=7） no integron			Ⅰ类整合子（n=15） class Ⅰ integron			Ⅱ类整合子（n=6） class Ⅱ integron		
		%R	%I	%S	%R	%I	%S	%R	%I	%S
氟喹诺酮类 quinolone	LF	0	0	100	60	0	40	0	0	100
	OF	0	0	100	40	6.67	53.33	0	0	100
	CIP	0	0	100	40	26.67	33.33	0	0	100
	NF	0	57.14	42.86	60	20	20	16.67	33.33	50
	EN	0	0	100	53.33	0	46.67	0	0	100
氨基糖苷类 aminoglycoside	GEN	0	28.57	71.43	40	33.33	26.67	33.33	0	66.67
	KN	0	71.43	28.57	60	33.33	6.67	100	0	0
	STP	0	28.57	71.43	73.33	20	6.67	16.67	0	83.33

续表

抗生素类别 antibiotic type	名称 name	integron type								
		无整合子（n=7） no integron			Ⅰ类整合子（n=15） class Ⅰ integron			Ⅱ类整合子（n=6） class Ⅱ integron		
		%R	%I	%S	%R	%I	%S	%R	%I	%S
青霉素类 penicillins	AMC	28.57	42.86	28.57	100	0	0	83.33	0	16.67
	AMP	85.71	14.29	0	100	0	0	100	0	0
	CB	57.14	42.86	0	93.33	6.67	0	83.33	16.67	0
头孢菌素类 cephalosporin	CFX	57.14	14.29	28.57	60	13.33	26.67	83.33	0	16.67
	CFT	85.71	0	14.29	100	0	0	100	0	0
	CP	0	28.57	42.86	20	0	80	16.67	16.67	66.67
氯霉素类 chloramphenicol	CHL	57.14	0	42.86	73.33	6.67	20	50	0	50
大环内酯类 macrolides	EM	85.71	14.29	0	93.33	6.67	0	83.33	16.67	0
四环素类 tetracyclines	TE	28.57	57.14	14.29	86.66	6.67	6.67	83.33	16.67	0
磺胺类 sulfonamides	SXT	0	0	100	73.33	0	26.67	16.67	0	83.33

注：抗生素名称见表2.R 表示耐受；I 表示中间；S 表示敏感.

Note：Explanation for the name of antibiotic is shown in tab.2. R, resistant；I, intermediate；S, susceptible.

2.4　Ⅰ类整合子可变区序列分析

对含有Ⅰ类整合子的15株嗜水气单胞菌菌株使用5′CS 和3′CS 引物扩增其整合子可变区，并将 PCR 产物纯化后送测。序列分析结果表明，15株嗜水气单胞菌携带的基因盒以 aadA、dfrA、catB 家族为主，分别介导氨基糖苷类、甲氧磺胺嘧啶类和氯霉素类药物耐药；基因盒的排列以 aadA2+dfrA12 类型为主（表5）。

表5　28株嗜水气单胞菌Ⅰ类整合子可变区序列分析
Tab.5　Sequence analysis of variable region in class Ⅰ integrons in 28 Aeromonas hydrophila isolates

菌株编号 strain	耐药谱 drug resistant spectrum	基因盒 gene cassette
AH-3	STP-AMC-AMP-CB-CFT-TE-SXT	aadA2+dfrA12
AH-5	AMC-AMP-CB-CFT-EM	aadA2+dfrA27
AH-8	STP-AMC-AMP-CB-CFT-CHL-EM-TE-SXT	aadA2+dfrA12
AH-10	STP-AMC-AMP-CB-CFT-CHL-EM-TE-SXT	aadA2+aadA4a
AH-15	LF-NF-GEN-KN-STP-AMC-AMP-CB-CFX-CP-CHL-EM-TE-SXT	aadA2
AH-17	NF-KN-STP-AMC-AMP-CB-CFX-CP-CHL-EM-TE-SXT	aadB
AH-20	LF-CIP-EN-GEN-KN-AMC-AMP-CB-CFX-CFT-CHL-EM-TE	dfrA1+orfC
AH-23	LF-OF-CIP-EN-GEN-KN-STP-AMC-AMP-CB-CFX-CFT-CHL-EM-TE-SXT	catB3+aadA2
AH-32	LF-OF-CIP-NF-EN-GEN-KN-STP-AMC-AMP-CB-CFX-CHL-EM-TE-SXT	catB3+aadA2
AH-39	LF-NF-EN-GEN-KN-AMC-AMP-CB-CFX-CHL-EM	blaOXA21+catB3
AH-43	LF-OF-CIP-NF-EN-GEN-KN-STP-AMC-AMP-CB-CFX-CFT-CHL-EM-TE-SXT	aadA2+dfrA12
AH-49	NF-EN-AMC-AMP-CB-CFX-CFT-CHL-EM-TE	aadB+catB3
AH-53	LF-OF-CIP-NF-EN-KN-STP-AMC-AMP-CFX-CFT-EM-TE-SXT	dfrA12+orfC
AH-58	LF-OF-NF-STP-AMC-AMP-CB-CFT-EM-TE-SXT	aadA2
AH-59	LF-OF-CIP-NF-EN-KN-STP-AMC-AMP-CB-CFX-CFT-CHL-EM-TE-SXT	aadA2+dfrA12

　　I类整合子阳性的嗜水气单胞菌耐药性在个体间也存在较大差异。如表6所示，AH-3、5、8、10的耐药谱与其他菌株差异较大，尤其在对氟喹诺酮类药物的耐药性上，提示多重耐药菌株的耐药表型与基因盒类型不存在密切关系。

表6　I类整合子阳性嗜水气单胞菌耐药性分析

Tab. 6　Detection of drug resistance in Aeromonas hydrophila containing class I integron

抗生素 antibiotic	I类整合子 class I integron														
	3	5	8	10	15	17	20	23	32	39	43	49	53	58	59
LF	S	S	S	S	R	S	R	R	R	R	S	S	R	R	R
OF	S	S	S	S	I	S	S	R	R	S	R	S	R	R	R
CIP	S	S	S	S	I	I	R	R	R	I	R	S	R	I	R
NF	S	S	I	I	R	R	I	S	R	R	R	R	R	R	R
EN	S	S	S	S	S	S	S	R	R	R	R	R	R	S	R
GEN	S	S	S	I	S	S	R	R	R	R	R	I	I	I	I
KN	S	I	I	I	R	I	R	R	R	R	R	I	R	I	R
STP	R	S	R	R	R	R	I	R	R	I	R	I	R	R	R
AMC	R	R	R	R	R	R	R	R	R	R	R	R	R	R	R
AMP	R	R	R	R	R	R	R	R	R	R	R	R	R	R	R
CB	R	R	R	R	R	R	R	R	R	R	R	R	I	R	R
CFX	S	S	I	S	R	R	R	R	R	R	R	R	R	I	R
CFT	R	R	R	R	R	R	R	R	R	R	R	R	R	R	R
CP	S	S	S	S	R	R	S	S	S	R	S	S	S	S	S
CHL	S	S	R	R	R	R	R	R	R	R	R	R	R	I	R
EM	I	R	R	R	R	R	R	R	R	R	R	R	R	R	R
TE	R	I	R	R	R	R	R	R	R	S	R	R	S	R	R
SXT	R	S	R	R	R	R	S	S	R	S	R	S	R	R	R

注：抗生素名称见表2.

Note：Explanation for the name of antibiotic is shown in tab. 2.

3　讨论

　　作为一种人畜鱼共患致病菌，嗜水气单胞菌的耐药性问题已成为全球共同关注的卫生问题之一。近年来，鱼源嗜水气单胞菌临床分离株的多重耐药情况日益严峻，其对阿莫西林、四环素、复方新诺明、氟苯尼考等常用药物均表现出严重的耐药性。肖丹等[16]对49株嗜水气单胞菌耐药性调查表明，试验菌株对氨苄青霉素和头孢氨苄几乎100%耐药，而对氨基糖苷类和喹诺酮类药物未产生耐药性。林居纯等[17]研究发现62%的菌株对复方新诺明耐药，38%~90%菌株对β-内酰胺类药物耐药。本实验结果则表明，分离到的28株嗜水气单胞菌主要对β-内酰胺类、大环内脂类、氯霉素类和四环素类药物耐药严重，其耐药率均超过60%；其次是氨基糖苷类药物（如庆大霉素、卡那霉素和链霉素等）和磺胺类药物，耐药率最低的是氟喹诺酮类药物，包括恩诺沙星、诺氟沙星、左氟沙星、氧氟沙星和环内沙星等。另外，嗜水气单胞菌分离株对氨苄西林和头孢噻吩的耐药率达到96%以上。这些结果表明，从东北地区淡水鱼主养区分离到的嗜水气单胞菌已经逐步表现出对氟喹诺酮类药物的耐药性，推测其可能与该类药物的滥用以及耐药菌株的流通相关。因此，对细菌性疾病的防治要正确合理地选用抗生素类药物，严格把握用药剂量、方式等，同时加强对耐药菌株的监测。

　　目前，对嗜水气单胞菌的耐药机制研究不多，主要从耐药质粒、染色体基因突变方面进行。洪经等[18]通过分析质粒指纹图谱与耐药的相关性发现，嗜水气单胞菌的耐药性与所携带质粒的数量和大小无直接关系，来源相同、耐药类型相似的菌株质粒图谱及酶切质粒图谱相似。肖丹等[16]通过ERIC-PCR分型结合菌株耐药性分析研究表明，嗜水气单胞菌基因型与耐药性可能存在一定的相关性。王晓丰等[19]研究了嗜水气单胞菌对氟喹诺酮类药物的敏感性及耐药相关基因，结果表明喹诺酮抗性决定区QRDR的gyrA基因和parC基因发生氨基酸突变，但是同时指出菌株的耐药性与药物靶位基因QRDR区域是否发生突变并不存在线性关联，提示药物靶位的变化并非是氟喹诺酮类耐药菌株唯一的抗性机制。整合子作为一种可以

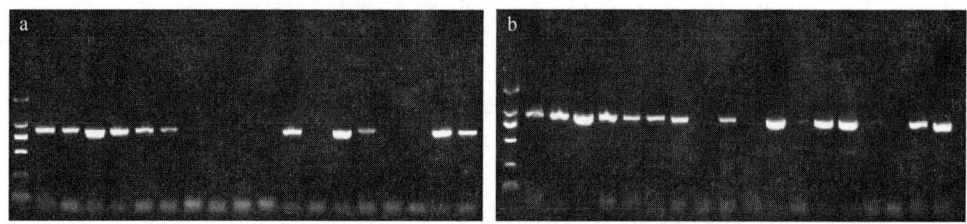

图 1 嗜水气单胞菌 I 类整合酶基因 Int 1 的 PCR 扩增结果

a：以细菌基因组 DNA 为模板进行 PCR 扩增；b：以细菌质粒为模板进行 PCR 扩增.

Fig. 1 PCR amplification of class Iintegronintegrase（int1）in Aeromonas hydrophilal.

a：PCR amplification using genome DNA as template；b：PCR amplification using plasmid DNA as template.

捕获外源基因的可移动遗传元件，在嗜水气单胞菌的多重耐药机制中也发挥着重要作用。由于整合子携带位点特异性重组系统，因而可以整合不同的耐药基因盒，使细菌具有多重耐药性。当耐药基因成为可移动单元的一部分时，其扩散机会大大增强。目前已发现了多个基因盒家族，包括 aad 基因家族、dfr 基因家族、cat 基因家族、oxa 基因家族等，可分别介导细菌对氨基糖苷类、甲氧嘧啶类、氯霉素类和苯唑西林类药物的耐药[20]。Mintra 等[21]研究指出，从尼罗罗非鱼中分离到的嗜水气单胞菌 46% 含有 I 类整合子，其基因盒类型以 dfrA12–aadA2 为主，而未检出 II、III 类整合子。本研究在 28 株嗜水气单胞菌中 I 类和 II 类整合子的检出率分别为 53.57% 和 21.43%，且 II 类整合子主要位于细菌染色体上，I 类整合子可同时位于染色体或质粒上。本研究通过对 I 类整合子可变区序列分析表明，基因盒以 aadA、dfrA、catB 家族为主，分别介导氨基糖苷类、甲氧磺胺嘧啶类和氯霉素类药物耐药；基因盒的组合以 aadA2+dfrA12 类型为主，这也与 Mintra 等[21]的研究结果相类似。

对整合子阳性和阴性菌株的耐药情况进行比较分析，可以发现，含 I 类和 II 类整合子的嗜水气单胞菌的耐药率明显高于阴性菌株，且整合子的有无对 28 株嗜水气单胞菌菌株的影响主要表现在氟喹诺酮类药物、氨基糖苷类药物、β–内酰胺类和磺胺类药物的耐药性。有研究指出，含整合子的菌株比不含有整合子的菌株更倾向于对各种不同抗菌药物产生耐药性[22]。在含有 I 类整合子的嗜水气单胞菌菌株中，个体间耐药谱差异也较大（表 6），AH–3、5、8、10 均对氟喹诺酮类药物敏感，而其他菌株则相对耐药。结合嗜水气单胞菌 I 类整合子携带基因盒的序列分析，可以看出多重耐药菌株的耐药表型与基因盒的类型无直接相关性。含有整合子的细菌并不意味着就一定耐药，其耐药性还与基因盒的表达强弱及其他一些尚不清楚的机制有关。此外，细菌耐药性的产生是多个调控机制共同发挥作用的，其模式非常复杂，且不符合简单的突变模型。

整合子–基因盒系统的研究为细菌耐药机制的探讨提供了新的研究方向，但仍需要结合其他研究如药物作用机理、细菌代谢途径等综合考虑细菌耐药产生机制。面对由整合子系统及其他耐药机制引起的细菌耐药日益严重的情况，必须积极探索有效的抗耐药途径，克服细菌耐药。

参考文献

［1］ Bennett P. M. Plasmid encoded antibiotic resistance：Acquisition and transfer of antibiotic resistance genes in bacteria［J］. Brit J Pharmacol，2008，153：347-357.

［2］ Hall R. M.，Brookes D. E.，Stokes H. W.. Site-specific insertionof genes into integrons：role of the 59-base element andde-termination of the recombination cross-over point［J］. Mol Microbiol，1991，5（8）：1941-1959.

［3］ Rowe-magnus D. A.，Mazel D.. Integrons：natural tools for-bacterial genome evolution［J］. Curr Opin Microbiol，2001，4（5）：565-569.

［4］ Mazel D.. Integrons：agents of bacterial evolution［J］. Nat RevMicrobiol，2006，4（8）：608-620.

［5］ 国宪虎，夏蕊蕊，徐海. 整合子基因盒系统及 β–内酰胺酶介导的细菌耐药［J］. 微生物学通报，2010，37（2）：289-294.

［6］ Partridge S. R.，Tsafnat G.，Coiera E.，et al. Gene cassettes and cassette arrays in mobile resistance integrons［J］. FEMS Mi-crob Rev，2009，33（4）：757-784.

［7］ 沈锦玉. 嗜水气单胞菌的研究进展［J］. 浙江海洋学院学报：自然科学版，2008，1（27）：78-86.

［8］ Vivekanandhan G.，Savithamani K.，Hatha A. A. M.，et al. Anti-biotic resistance of Aeromonas hydrophila isolated from marketed fish and prawn of South India［J］. Int J Food Mi-crobiol，2002，76：165-168.

［9］ 宋铁英，陈强，郑在予，等. 不同来源嗜水气单胞菌的抗生素耐药性及耐药机制分析［J］. 福建农业学报，2007，23（2）：119-124.

［10］ 司力娜，李绍戊，王荻，等. 东北三省 15 株致病性嗜水气单胞菌分离株的药敏实验分析［J］. 江西农业大学学报，2011，33（4）：786-790.

[11] 朱成科, 周晓扬, 张其中. 南方鲇幼鱼细菌性败血症病原与组织病理 [J]. 中国水产科学, 2011, 18 (2): 360-370.

[12] 樊海平, 吴斌, 曾占壮, 等. 日本鳗鲡体表溃疡病病原菌的分离、鉴定及单克隆抗体制备 [J]. 中国水产科学, 2009, 16 (2): 295-302.

[13] National Committee for Clinical Laboratory Standards. 1993. Performance Standards for Antimicrobial Disc Susceptibility Tests [S]. Approved Standard NCCLS Publications M2-A5. Villanova, PA, USA.

[14] Ryu S. H., Park S. G., Choi S. M., et al. Antimicrobial resistance and resistance genes in Escherichia coli strains isolated from commercial fish and seafood [J]. Int J Food Microbiol, 2012, 152 (1-2): 14-18.

[15] Levesque C., Piche L., Larose C., et al. PCR mapping of in-tegronsreveals several novel combinations of resistance genes [J]. Antimicrob Agents Chemother, 1995, 39: 185-191.

[16] 肖丹, 曹海鹏, 胡鲲, 等. 淡水养殖动物致病性嗜水气单胞菌 ERIC-PCR 分型与耐药性 [J]. 中国水产科学, 2011, 18 (5): 1092 -1099.

[17] 林居纯, 罗忠俊, 舒刚, 等. 嗜水气单胞菌临床分离菌对抗菌药物的耐药性调查 [J]. 安徽农业科学, 2009, 37 (15): 7024 -7025.

[18] 洪经, 潘连德. 鱼源嗜水气单胞菌质粒的指纹图谱及其与耐药性的关系 [J]. 水产学报, 2010, 34 (12): 1908-1915.

[19] 王晓丰, 薛晖, 丁正峰, 等. 嗜水气单胞菌对氟喹诺酮类药物的敏感性及耐药相关基因分析 [J]. 江苏农业学报, 2010, 26 (6): 1298-1303.

[20] 蒋宏, 刘继芬, 刘蓉. 整合子-基因盒系统与细菌多重耐药研究进展 [J]. 四川省卫生管理干部学院学报, 2008, 27 (2): 122-125.

[21] Lukkana M., Wongtavatchai J., Chuanchuen R. Class 1 in-tegrons in Aeromonas hydrophila isolates from farmed Nile tilapia (Oreochromis nilotica) [J]. J Vet Med Sci, 2012, 74 (4): 435-440.

[22] Rowe-Magnus D. A., Guerout A. M., Mazel D. Bacterialre-sistance evolution by recruitment of super-integron genecassettes [J]. Mol Microbiol, 2002, 43 (6): 1657-1669.

该文发表于《水产学报》【2016, 40 (3): 495-502】

养殖鱼源嗜水气单胞菌对氟喹诺酮类药物的耐药机制

In vitro study on fluoroquinoloneresistance mechanism of *Aeromonas hydrophila* from cultured fish

崔佳佳[1,2], 王荻[1], 卢彤岩[1], 李绍戊[1]

(1. 中国水产科学研究院 黑龙江水产研究所, 哈尔滨 150070; 2. 上海海洋大学 水产与生命学院, 上海 201306)

摘要: 为研究体外诱导敏感嗜水气单胞菌耐药后, 其敏感性变化与基因突变、外排作用的关系, 实验选取对氟喹诺酮类药物敏感的养殖鱼源嗜水气单胞菌临床分离菌株为研究对象, 分别在含亚抑菌浓度恩诺沙星 (EN) 和诺氟沙星 (NF) 的培养基上逐步诱导培养, 以获得高耐药菌株; 对诱导菌株 gryA 和 parC 基因进行扩增和测序分析; 测定诱导菌对诱导药物和 16 种非诱导药物的最小抑菌浓度 (MIC) 及添加外排泵抑制剂 N-甲基吡咯烷酮 (NMP) 后的 MIC 值变化; 并对诱导菌交叉耐药情况进行比较分析。结果发现, 诱导后菌株对 EN 和 NF 的 MIC 分别提高了 409.6 和 4096 倍, 对非诱导氟喹诺酮类药物和其他类药物的 MIC 也有较大变化; 药物诱导后各菌株 gyrA 基因和 parC 基因编码的氨基酸 QRDRs 区发生了典型的点突变: GyrA 发生 Ser83→Ile 变化, ParC 发生 Ser87→Ile/Arg 变化; 添加 NMP 后, 所有诱导菌株对两种药物的 MIC 值均有不同程度的下降; 诱导后菌株交叉耐药情况与菌株密切相关, 其中 3 和 8 号诱导菌株对 16 种非诱导药物均无交叉耐药反应, 而 EN 诱导菌株对氨基糖苷类和利福霉素类药物基本未产生交叉耐药反应, NF 诱导菌株对除庆大霉素以外的氨基糖苷类和利福霉素类药物基本未产生交叉耐药反应, 所有诱导后菌株均对四环素类和氯霉素类药物产生较严重的交叉耐药。研究表明, 嗜水气单胞菌对氟喹诺酮类耐药存在靶基因位点突变及主动外排作用等多种耐药机制; 且应慎重考虑在防治耐药菌株引发病害时, 交叉耐药情况对选择治疗药物的影响。

关键词: 嗜水气单胞菌; 氟喹诺酮类; 诱导耐药; 基因突变; 交叉耐药

嗜水气单胞菌 (*Aeromonashydrophila*, *Ah*) 属气单胞菌科 (Aeromonadaceae)、气单胞菌属 (*Aeromonas*), 是一种典型的人-兽-水生动物共患病的条件致病菌, 普遍存在于淡水、海水、淤泥以及土壤中, 可以感染多种水产动物[1-3]。近年来由其引发的养殖水产动物细菌性败血症暴发性死亡在各地多有报道，给水产养殖业造成巨大经济损失[4]的同时，也为产业健康持续发展造成严重干扰。

第三代喹诺酮类抗菌药在喹诺酮的 6 位引入氟原子，称氟喹诺酮类药物，诺氟沙星和恩诺沙星该类抗生素中的常用药物[5]，不仅具有良好的组织渗透性，且抗菌谱广、不良反应少、能快速杀灭细菌，在人医、兽医和渔业等方面均有广泛应用[6]。然而，由于氟喹诺酮类药物在水产临床细菌性疾病防治过程中的不规范、不合理使用，导致各地 Ah 耐药性的普遍产生，引起药物药效下降甚至失效，为临床治疗 Ah 引起的疾病造成极大困难。

目前，研究已知的细菌对氟喹诺酮类药物耐药机制共分 2 类，分别为特异性和非特异性耐药机制。特异性耐药机制包括靶点的改变和耐药性质粒的出现；非特异性耐药机制包括外排泵系统的表达、膜通透性的改变以及生物被膜的产生[7-8]。氟喹诺酮类药物的主要作用靶点是拓扑异构酶Ⅱ和拓扑异构酶Ⅳ。拓扑异构酶Ⅱ由 2 个 GyrA 亚基和 2 个 GyrB 亚基组成，拓扑异构酶Ⅳ由 2 个 ParC 亚基和 2 个 ParE 亚基组成。通过对耐药菌株与敏感菌株对比分析发现，耐药菌株的拓扑异构酶氨基酸序列中存在着不同的点突变，而这些点突变多集中在 GyrA、GyrB、ParC 和 ParE 亚单位的功能区（GyrA：67~106 aa，ParC：71~110 aa），通常称其为喹诺酮耐药决定区（The quinolone resistance-determining regions，QRDRs）[9]。目前研究认为，靶基因 gyrA 和 parC 发生突变，导致 GyrA 和 ParC 一个或多个结构发生改变，与 Ah 对氟喹诺酮类药物产生耐药性存在密切关系[10-12]。

氟喹诺酮类药物耐药菌株的快速增加及菌株耐药性的广泛传播引起了国内外研究者对该类药物耐药机制的强烈关注。Ah 耐药机制非常复杂，目前系统研究氟喹诺酮类药物体外诱导敏感菌，并对诱导后耐药菌株耐药机制进行研究的相关报道较少[13-14]。为探讨人工诱导引起 Ah 耐药的相关机制，本文拟通过恩诺沙星和诺氟沙星对养殖鱼源 Ah 体外诱导耐药，并对其 GyrA 和 ParC 变异情况及主动外排系统对菌株药物敏感性的影响及诱导后交叉耐药情况进行研究，用以分析 Ah 对氟喹诺酮类药物的耐药机制，为阐明 Ah 耐药的分子机理提供依据。

1　材料与方法

1.1　实验材料

1.1.1　实验菌株

诱导受试菌株 Ah1-9 号为中国水产科学研究院黑龙江水产研究所养殖室鱼病组分离保存，分离株来源于东北地区养殖鲤鱼，参考株 ATCC 7966 由珠江水产研究所邓玉婷博士馈赠。

1.1.2　实验试剂

诺氟沙星（Norfloxacin，NF，99.1%，C15648000）、恩诺沙星（Enrofloxacin，EN，98.5%，C13170000）和强力霉素（Doxycycline，DO，98.7%，C13084280）购自 Dr. Ehrenstorfer GmbH-Bgm.-Schlosser-Str. 6A-86199 Augsburg-Germany；左氧氟沙星（Levofloxacin，LE，98.5%，130455-201005）、环丙沙星（Ciprofloxacin，CP，99.0%，1134313-201405）与氧氟沙星（Ofloxacin，OFL，99.5%，130454-201206）购自中国药品检定研究院；庆大霉素（Gentamicin，GM，63.0%，130326-201015）购自中国食品药品检定研究院；链霉素（Streptomycin，SM，72.9%，130308-200713）、卡那霉素（Kanamycin，KM，66.5%，130556-200501）、阿米卡星（Amikacin，AM，65.5%，130335-200204）、新霉素（Neomycin，NEO，64.5%，130309-200811）、土霉素（Oxytetracycline，OTC，88.8%，130487-200703）、四环素（Tetracycline，TET，97.5%，130488-200403）、氯霉素（Chloramphenicol，99%，CH，130303-200614）、氟苯尼考（Florfenicol，98.2%，FFC，C13665000-200205）、甲砜霉素（Thiamphenicol，99.5%，THI，130433-200502）和利福平（Rifampicin，RIF，99.9%，130496-200702）均购自中国药品生物制品检定所；N-甲基吡咯烷酮（1-Methyl-2-pyrrolidinone，NMP）购自美国 Sigma 公司。

实验用药敏纸片购自杭州天和微生物试剂有限公司，纸片药物-含量（μg）分别为：LE-5、CP-5、NF-10、EN-5、OFL-5、SM-10、KM-30、AM-30、GM-120、NEO-30、OTC-30、TET-30、DO-30、CH-30、FFC-30、THI-30、RIF-5。

实验用胰蛋白胨大豆琼脂培养基（TSA）、胰蛋白胨大豆肉汤（TSB）、MH 肉汤、Mueller-Hinton 琼脂培养基均购自青岛科技园海博生物技术有限公司；细菌基因组 DNA 提取试剂盒购自天根生化科技有限公司；实验所用引物由哈尔滨博仕生物技术有限公司合成。

1.2　实验方法

1.2.1　体外诱导及遗传稳定性试验

将 EN、NF 和 NMP 分别于无菌条件下用超纯水配成 1.6、1.6 和 50 mg/mL 的储存液，分装后-80℃条件保存备用，使用时再稀释成所需浓度。

微量肉汤稀释法测 MIC：采用灭菌的 96 孔聚苯乙烯板，向 1~12 孔加 100 μL MH 肉汤，第 1 孔加入 100 μL 药液，倍比稀释至第 10 孔，第 12 孔不加药作为阳性对照，将培养的菌悬液经 MH 肉汤稀释后制备成相当于 0.5 麦氏比浊标准的菌悬液，向除第 11 孔外的每孔中加 5 μL 菌悬液，第 11 孔不加菌作为阴性对照，密封后置 28℃普通空气孵箱中，孵育 18~24 h 判断结果。

以受试菌株在不含药物的培养基中同步传代培养为对照，依据菌株药物 K-B 纸片扩散法测定的敏感性结果，将受试菌株分别接种于含有 1/4×MIC 的 EN 和 NF 的 TSA 固体培养基平板中，28℃培养传代，等比 2 倍提高诱导药物质量浓度对受试菌进行连续传代培养，至诱导药物质量浓度达到 128 μg/mL 为止，分别保存每代诱导菌株。

将诱导后的菌株在不含药物的 TSA 固体培养基中连续传代培养，并依次测定第 5、10、15 和 20 代传代菌株的 MIC，用以判断诱导后菌株耐药的遗传稳定性。

1.2.2 药敏试验

采用 K-B 纸片扩散法，以 ATCC 7966 为质控和对照菌株进行药敏试验，用以筛选氟喹诺酮类药物敏感的 Ah 菌株。

以无药物添加接种受试菌为阳性对照，无药物添加不接种受试菌为阴性对照，采用微量肉汤稀释法、结合 96 孔细菌培养板对受试菌、各代诱导菌株对 17 种受试药物的最小抑菌浓度（Minimum inhibitory concentration，MIC）。同时设置外排泵抑制剂实验组，在培养基中加入 NMP 至终浓度为 50 μg/mL，然后测定菌株 MIC 值变化。

1.2.3 交叉耐药试验

用微量稀释法测定诱导后的耐药菌株对非诱导药物的 MIC 值，并比较诱导前后的 MIC 值变化。菌株诱导后比诱导前对非诱导药物的 MIC 升高 4 倍或以上，定为交叉耐药[15]。

1.3 靶基因扩增

挑取纯化后单菌落接种于 3 mL TSB 营养肉汤的试管中，于气浴摇床中 28℃振荡过夜培养，无菌条件下 12 000 r/min 离心收集菌体于 1.5 mL 的 EP 管中，按照细菌基因组 DNA 提取试剂盒说明书提取细菌基因组 DNA。

根据参考文献[13,16]报道的相应引物，对 gyrA 和 parC 基因 QRDR 区及外排泵相关基因 qepA、oqxA 和 mdfA 进行扩增检测，并经 1%琼脂糖凝胶电泳检测目的基因扩增情况（表 1）。

表 1　扩增基因的相应引物序列

Tab. 1　Primers for amplifying genes

引物名称 primers	引物序列（5'-3'） sequences（5'-3'）	目的片段长度/bp fragment length	退火温度/℃ Tm
gyrA	F：CCTATCTTGATTACGCCATGAG R：CACATAGACGGAGCCACGAC	668	56
parC	F：TACCGAGCAGGCTTACTTGAA R：CCATCCTCCTCGTTCCACA	711	56
qepA	F：CGTGTTGCTGGAGTTCTTC R：CTGCAGGTACTGCGTCATG	403	52
oqxA	F：CTCGGCGCGATGATGCT R：CCACTCTTCACGGGAGACGA	392	52
mdfA	F：CATTGGCAGCGATCTCCTTT R：TTATAGTCACGACCGACTTCTTTCA	103	50

1.4 序列测定及分析

PCR 阳性扩增产物经哈尔滨博仕生物技术有限公司进行序列测定。采用 BLUST 方法，测序结果与 GenBank 上标准 Ah 序列进行比对分析，找出氨基酸序列突变位点。

2　结果

2.1 受试、诱导菌株 MIC 变化及稳定性

根据 K-B 药敏纸片法结果，共筛选鱼源氟喹诺酮类药物敏感 Ah 菌株 9 株为受试菌株，分别编号为 1-9。受试菌经不同浓度 EN 及 NF 体外连续诱导，获得高耐药诱导菌株，分别编号为 E1-9 和 N1-9。

受试菌株诱导前后 MIC 值见表 2。与诱导前相比，E1-9 对 EN 的 MIC 提高了 6.4~409.6 倍，对非诱导的氟喹诺酮类药

物 LE、CP、NF 和 OF 的 MIC 分别比诱导前增加了 32~512 倍、8~256 倍、1~64 倍和 8~256 倍；而 N1-9 对 NF 的 MIC 提高了 32~4096 倍，对非诱导的氟喹诺酮类药物 LE、CP、EN 和 OF 的 MIC 分别比诱导前增加了 2~128 倍、8~256 倍、40~1 280 倍和 4~128 倍；同时，个别菌株对氨基糖苷类 SM、GM、NEO，四环素类 TET 及利福霉素类 GIF 等其他类药物的 MIC 略有增加（2~4 倍），但多数菌株对氟喹诺酮类以外药物的 MIC 略有降低或无显著性变化。

加入外排泵抑制剂 NMP（50 μg/mL）后，诱导菌对 EN 和 NF 的 MIC 比未添加前均有所降低，降低幅度分别为 2~16 倍和 1~16 倍；而诱导菌株在无药平板上连续传代培养 20 代后，对 EN 和 NF 的 MIC 值基本不变，表明诱导菌株具有较好的遗传稳定性（表 2）。

表 2　诱导前后及加入 NMP 后和诱导后传代过程中菌株对 NF 与 EN 的 MIC 变化

Tab. 2　MICs of Norfloxacin and Enrofloxacin before and after selection as well as NMP added and after selection subcultured

	诱导前后 before and after selection				加 NMP 后 NMP added		5 代 5 generation		10 代 10 generation		15 代 15 generation		20 代 20 generation	
	诺氟 NF	恩诺 EN	诺氟 NF	恩诺 EN	诺氟 NF	恩诺 EN	诺氟 NF	恩诺 EN	诺氟 NF	恩诺 EN	诺氟 NF	恩诺 EN	诺氟 NF	恩诺 EN
1	0.5	1.25	128	128	32	64	256	128	128	128	128	128	256	128
2	0.5	2.5	256	64	64	32	256	64	256	64	128	64	128	64
3	4	5	128	64	8	4	128	32	128	64	64	64	256	64
4	1	1.25	128	64	64	64	256	16	256	32	128	16	128	32
5	0.125	0.3	64	32	16	32	128	32	128	32	32	32	32	32
6	0.0625	0.15	256	8	32	1	256	8	256	32	128	16	32	16
7	1	1.25	64	8	8	4	64	8	128	8	64	8	64	16
8	2	1.25	256	128	16	64	256	128	256	8	128	8	256	8
9	0.125	0.3	256	128	32	128	256	128	256	64	128	64	128	64

2.2　交叉耐药情况

经 EN 连续诱导至高耐药后的菌株 E1-9 分别对 GM、NEO、OTC、TET、DO、CHL、FFC 和 THI 中的一种或多种产生了交叉耐药，E1-9 对 GM，ENO、OTC、TET、DO、CHL、FFC 和 THI 产生了不同的交叉耐药情况，77.78% 的菌株对四环素类药物 DO 产生了交叉耐药；而仅有 12.5% 的菌株对氨基糖苷类药物 GM 和 NEO 产生了交叉耐药。E3 和 E8 未对 16 种非诱导药物中的任何 1 种产生交叉耐药，而 E6 和 E9 分别对 16 种非诱导药物中的 37.5% 和 31.25% 都产生了交叉耐药现象（表 3）。

表 3　E1-9 与 N1-9 交叉耐药情况

Tab. 3　Cross-resistance of E1-9 and N1-9

	KM	GM	NEO	OTC	TET	DO	CHL	FFC	THI	RIF
E1						+	+	+	+	
E2					+	+				
E3										
E4						+	+	+	+	

续表

	KM	GM	NEO	OTC	TET	DO	CHL	FFC	THI	RIF
E5		+		+	+	+				
E6			+		+	+	+	+	+	
E7					+	+				
E8										
E9				+	+	+		+	+	
N1		+								
N2					+	+				
N3										
N4		+			+		+		+	
N5		+			+	+		+		
N6		+			+	+	+	+	+	
N7	+	+			+	+				
N8										
N9		+			+	+	+	+	+	+

"+"：交叉耐药；"+"：Cross-resistance.

经 NF 连续诱导至高耐药后的菌株 N1-9 分别对 KM、GM、TET、DO、CHL、FFC、THI 和 RF 中的一种或多种产生了交叉耐药，N1-9 对表中 8 种药物产生了不同的交叉耐药情况，75% 的菌株对氨基糖苷类药物 GM 和四环素类药物 TET 产生了交叉耐药；而仅有 12.5% 的菌株对氨基糖苷类药物 KM 和利福霉素类药物 RF 产生了交叉耐药（表3）。N3 和 N8 未对 16 种非诱导药物中的任何 1 种产生交叉耐药，而 N9 对 16 种非诱导药物中的 43.75% 都产生了交叉耐药现象。

2.3 诱导菌株拓扑异构酶氨基酸序列突变情况

PCR 扩增所得 gyrA 和 parC 基因片段测序后比对分析，结果表明：E1-9 和 N1-9 菌株 gyrA 基因编码的 QRDR 区第 83 个氨基酸均发生 Ser83→Ile 的变化；且 E/N1-4 及 7-8 菌株 parC 基因编码的 QRDR 区第 87 个氨基酸发生 Ser87→Ile 的变化，而 E5 菌株的 Ser87→Arg。

另外，部分菌株 gyrA 和 parC 基因编码的非 QRDR 区氨基酸序列也有点突变的现象发生，包括 gyrA 基因编码区的 Ser116→Asn，Ala137→Ser，Asp202→Ala，Ile212→Val，Gly221→Ala，Val230→Ile；parC 基因编码区 His197→Leu/Tyr，Val200→Ala，Ala212→Val，Ala215→Thr/Qln，Val217→Ile，Pro231→Ser 的突变。

2.4 诱导菌株的外排泵相关基因的检测结果

诱导后 18 株耐药菌株的外排泵相关基因 oqxA、qepA 和 mdfA 的检测结果都为阴性。

3 讨论

本研究分别选取氟喹诺酮类渔用常用药物 EN 和 NF 对临床分离的养殖鱼源敏感 Ah 菌株从 1/4×MIC 逐级诱导至高耐药，表明 Ah 敏感菌株逐渐升高浓度的氟喹诺酮类药物的培养基中培养时，逐步筛选出了高度耐药菌，与马传玲[17]等报道的情况一致，应尽量避免抗菌药物低浓度与细菌长期接触产生耐药性而造成治疗失败[18]。

EN 和 NF 对受试菌诱导后，对非诱导的氟喹诺酮类药物的敏感性均出现不同程度的降低，可能与氟喹诺酮类药物有相似的结构和作用位点有关；而对其他类药物的敏感性除个别菌株降低外，大多数菌株均表现为增高趋势。

Hernould 等[19]首次报道了 Ah 存在引起内源性多重耐药的外排泵 AheABC，属于耐药结节化细胞分化超家族（resistance-nodulation-division superfamily，RND），其作用底物最少有 13 种，其中 9 种为抗生素，但对氟喹诺酮类无作用。诱导后菌株的药敏试验中添加外排泵抑制剂 NMP 后，诱导菌对所有被测药物的敏感性均存在不同程度的增高，提示外排泵系统可能在受试 Ah 菌株耐药机制中起到一定作用。但采用目前已报到的外排泵相关基因 oqxA、qepA 和 mdfA 的检测方法，其检测结果均为阴性，推测受试 Ah 菌株外排泵系统可能还存在其他作用机制，仍需进一步深入研究。

染色体拓扑异构酶基因点突变是引起氟喹诺酮类耐药性的主要原因，单基因自发突变可导致细菌低水平耐药性，而双突变则产生高水平耐药性[20]。有研究表明，细菌对氟喹诺酮类药物耐药性的演变多是由单点突变向多点突变进行[20-21]。

薛慧娟等[22]对临床分离的喹诺酮类耐药 Ah 的 QRDR 突变研究发现，萘啶酸耐药菌株的 GyrA 均发生突变，为 Ser83→Ile，而环丙沙星耐药菌株除了 GyrA 突变外，其 ParC 亦同时发生突变，为 Ser87→Ile 或 Ser87→Arg。国内外研究报道，临床分离的喹诺酮类耐药气单胞菌在 GyrA 第 83 位氨基酸多发生 Ser→Ile 突变，也有部分突变为 Arg、Val 或 Asn；而 ParC 在第 87 位变异最多，为 Ser→Ile，也有突变为 Thr 及 Arg[11-12]。本研究发现诱导后各菌株 GyrA 和 ParC 的氨基酸序列都有发生突变，虽然各菌株之间的突变位点和突变数目有差异，但与 Ah 对氟喹诺酮类药物产生耐药性密切相关的 QRDRs 区位点均出现了较一致的点突变，均表现为 Ser83→Ile 和 Ser87→Ile/Arg，与上述报道结果一致。而诱导后菌株 QRDRs 区以外的突变位点与突变数目的不同与 Ah 对氟喹诺酮类药物的 MIC 值高低无线性关系，推测 gyrA 基因和 parC 基因 QRDRs 区以外位点与 Ah 耐药性无直接联系，或受其他机制调节；且部分诱导菌株只在 GyrA 发生了点突变，而 ParC 未发生点突变，推测可能与氟喹诺酮类药物选择压力下，Ah 对氟喹诺酮类耐药性的增加，除了靶位点突变外，其他机制可能发挥了重要作用[14]相关。

同时，研究结果还表明诱导耐药后的 Ah 菌株产生了较严重的交叉耐药情况，其中 E9 和 N9 产生交叉耐药的药物种类最多，而 E3、E8、N3 和 N8 无交叉耐药，由此可见，不同菌株诱导耐药后对其他非诱导药物的敏感性变化较大，且与诱导前菌株自身情况密切相关，如 3、8 和 9 号菌株，值得注意的是，3 和 8 号菌株诱导前对 EN 和 NF 敏感性均较差，而 9 号菌株极敏感。另外，诱导后菌株对四环素类药物产生交叉耐药的最多，其次是氯霉素类药物。且 EN 诱导菌对氨基糖苷类和利福霉素类药物基本未产生交叉耐药反应，NF 诱导菌对除庆大霉素以外的氨基糖苷类和利福霉素类药物基本未产生交叉耐药反应，且 EN 和 NF 所有诱导后菌株均对四环素类和氯霉素类药物产生较严重的交叉耐药。

Ah 对氟喹诺酮类药物的耐药机制较复杂，本研究结果表明 Ah 诱导菌株的 GyrA 和 ParC 均在 QRDRs 区发生较一致的点突变，可能是引起菌株对氟喹诺酮类药物耐药的主要原因；而外排泵系统在耐药性变化中也起到了一定作用。且不同菌株耐药后易对四环素类和利福霉素类药物产生交叉耐药。因而，在指导氟喹诺酮类药物临床用药时，应严格按照科学用药指导足量用药以保证药效；同时应严格避免养殖水体中长期、低水平药物残留以避免耐药菌的产生，保证药物治疗效果，遇到氟喹诺酮类耐药菌时应注意耐药菌交叉耐药对后期选择药物治疗的影响。

参考文献

［1］　陆承平. 致病性嗜水气单胞菌及其所致鱼病［J］. 水产学报，1992，16（3）：282-288.
Lu C. P.. Pathogenic Aeromonashydrophila and the fish diseases caused by it［J］. Journal of Fisheries of China, 1992, 16（3）: 282-288（in Chinese）.

［2］　Cantas L., Midtlyng P. J., Sorum H.. Impact of antibiotic treatments on the expression of the R plasmidtra genes and on the host innate immune activity during pRAS1 bearing Aeromonashyrophila infection in zebrsfish（Daniorerio）［J］. BMC Microbiology, 2012, 19: 3-10.

［3］　Janda J. M., Abbott S. L.. The genus Aeromonas: Taxonomy, pathogenicity, and infection［J］. Clinical Microbiology Reviews, 2010, 23（1）: 35-73.

［4］　李绍戊，王荻，卢彤岩，等. 大西洋鲑杀鲑气单胞菌无色亚种的分离鉴定和致病性研究［J］. 水生生物学报，2015，39（1）：234-240.
Li S. W., Wang D., Lu T. Y., et al. Isolation, Identification and Pathogenicity of Aeromonas salmonicida subsp. Achromogenes from Atlantic Salmon（SalmoSalar）［J］. ActaHydrobiologicaSinica, 2015, 39（1）: 234-240（in Chinese）.

［5］　Xu H., Wang T., Zhao Q., et al. Analysis of fluoroquinolones in animal feed based on microwave-assisted extraction by LC-MS-MS determination［J］. Chromatographia, 2011, 74:（3/4）: 267-274.

［6］　杨先乐. 新编渔药手册［M］. 北京：中国农业出版社，2006，5：183-184.
Yang X. L.. New fishery medicinemanual［M］. Beijing: China Agriculture Press, 2006, 5: 183-184（in Chinese）.

［7］　Robicsek A., Strahilevitz J., Jacoby G. A., et al. Fluoroquinolone-modifying enzyme: a new adaptation of a common aminoglycoside acetyl-transferase［J］. Nature medicine, 2005, 12（1）: 83-88.

［8］　Seeger M. A., Diederichs K., Eicher T., et al. The AcrB efflux pump: conformational cycling and peristalsis lead to multidrug resistance［J］. Current drug targets, 2008, 9（9）: 729-749.

［9］　Hooper D. C.. New uses for new and old quinolones and the challenge of resistance［J］. Clinical infectious diseases, 2000, 30（2）: 243-254.

［10］　Alcaide E., Blasco M. D., Esteve C.. Mechanisms of quinolone resistance in Aeromonas species isolated from humans, water and eels［J］. Research in microbiology, 2010, 161（1）: 40-45.

［11］　Kim J. H., Hwang S. Y., Son J. S., et al. Molecular characterization of tetracycline-and quinolone-resistant Aeromonassalmonicida isolated in Korea［J］. Journal of veterinary science, 2011, 12（1）: 41-48.

［12］　Han J. E., Kim J. H., Cherescajr C. H., et al. First description of the qnrS-like qnrS5 gene and analysis of quinolone resistance-determining regions in motile Aeromonas spp. from diseased fish and water［J］. Research in microbiology, 2012, 163（1）: 73-79.

［13］　方一风，潘晓艺，蔺凌云，等. 嗜水气单胞菌对喹诺酮类药物耐药的分子机制［J］. 微生物学报，2014，（2）：174-182.
Fang Y. F., Pan X. Y., Lin L. Y., et al. Molecular mechanisms of quinolone resistance in Aeromonas hydrophila［J］. Acta Microbiologica Sinica, 2014,（2）: 174-182（in Chinese）.

[14] 邓玉婷, 薛慧娟, 姜兰, 等. 体外诱导嗜水气单胞菌对喹诺酮类耐药及其耐药机制研究 [J]. 华南农业大学学报, 2014, 35 (1): 12–16.

Deng Y. T., Xue H. J., Jiang L., et al. Characterization of quinolone resistance mechanism of Aeromonas hydrophila selected in vitro [J]. Journal of South China Agricultural University, 2014, 35 (1): 12–16 (in Chinese).

[15] 章喻军. 铜绿假单胞菌外膜耐药和保护原功能蛋白组学的研究 [D]. 厦门: 厦门大学图书馆, 2007: 43–44.

Zhang Y. J.. Functional proteomic on antibiotic resistance and protective immunogen of Pseudomonas aeruginosa outer membrane proteins [D]. Xiamen: Xiamen University Library, 2007: 43–44 (in Chinese).

[16] Swick M. C., Morgan–Linnell S. K., Carlson K. M., et al. Expression of multidrug efflux pump genes arcAB–tolC, mdfA, andnorE in Escherichia coli clinical isolates as a function of fluoroquinolone and multidrug resistance [J]. Antimicrobial Agents and Chemotherapy, 2011, 50 (2): 921–924.

[17] 马传玲, 温鸿, 尉骁, 等. 氟喹诺酮类药物对粪肠球菌的防耐药突变浓度研究 [J]. 中国抗生素杂志, 2015, 40 (1): 56–60.

Ma C. . L, Wen H., Yu X., et al. Mutant prevention concentration of fluoroquinolones for Enterococcusfaecalis [J]. Chinese Journal of Antibiotics, 2015, 40 (1): 56–60 (in Chinese).

[18] Paiiares R., Fenoll A., Linares J.. The epidemiology of antibiotic resistance in Streptococcus pneumoniae and the clinical relevance of resistance to cephalosporins, macrolides and quinolones [J]. International Journal of Antimicrobial Agents, 2003, 22: 15–24.

[19] Hernould M., Gagné S., Fournier M., et al. Role of the AheABC efflux pump in Aeromonashydrophila intrinsic multidrug resistance [J]. Antimicrobial agents and chemotherapy, 2008, 52 (4): 1559–1563.

[20] 夏蕊蕊, 国宪虎, 张玉臻, 等. 喹诺酮类药物及细菌对其耐药性机制研究进展 [J]. 中国抗生素杂志, 2010, 35 (4): 255–260.

Xia R. R., Guo X. H., Zhang Y. Z., et al. Quinolones and the mechanism of quinolone resistance in bacteria [J]. J Chin Antibiot, 2010, 35 (4): 255–260 (in Chinese).

[21] Fabrega A., Maduqa S., Giralt E., et al. Mechanism of action of and resistance to quinolones [J]. Microbial Biotechnology, 2009, 2 (1): 40–61.

[22] 薛慧娟, 邓玉婷, 姜兰, 等. 水产动物源嗜水气单胞菌药物敏感性及 QRDR 基因突变分析 [J]. 广东农业科学, 2012, 39 (23): 149–153.

[23] Xue H. J., Deng Y. T., Jiang L., et al. Antimicrobialsusceptibility and mutations of QROR inAeromonashydrophila isolated from aquatic animals [J]. Guangdong Agricultural Sciences, 2012, 39 (23): 149–153 (in Chinese).

该文发表于《微生物学报》【2016, 优先发表】

嗜水气单胞菌对四环素类药物诱导耐药表型及机理研究

Study on tetracyclines–induced phenotype and resistance mechanism in Aeromonas hydrophila

崔佳佳[1,2], 李绍戊[1], 王荻[1], 卢彤岩[1]

(1. 中国水产科学研究院 黑龙江水产研究所, 哈尔滨 150070; 2. 上海海洋大学 水产与生命学院, 上海 201306)

摘要: 初步探讨利用四环素类药物体外诱导嗜水气单胞菌耐药后, 嗜水气单胞菌对四环素类药物敏感性的变化及其耐药机制。筛选临床分离嗜水气单胞菌的四环素类敏感株, 从含有 $1/4 \times MIC$ 的强力霉素的 TSA 固体培养基开始, 等比 2 倍提高诱导药物质量浓度对受试菌进行连续传代培养, 以获得高耐诱导株; 测定诱导菌对强力霉素和 16 种非诱导药物的最小抑菌浓度 (MIC) 及添加外排泵抑制剂 N–甲基吡咯烷酮 (NMP) 后的 MIC, 分析其敏感性变化与外排作用的关系; 提取诱导菌的 DNA, PCR 扩增其 5 个 tet 基因并测序。诱导后菌株对强力霉素的 MIC 显著升高, 对非诱导四环素类药物也有不同程度提高, 对氟喹诺酮类药物的 MIC 比诱导前增加几十至上千倍; 对氨基糖苷类药物和利福平的 MIC 则有不同程度的降低; 添加 NMP 后, 所有诱导菌株对强力霉素的 MIC 值均有不同程度的下降; 四环素类耐药基因的检测结果表明, 在诱导后 7 号菌株中同时检测到 tetA 和 tetE; 在诱导前后的 2 号菌株中检测到 tetC; 在诱导前后的 1、3、4、5、6、7 号菌株中均检测到 tetE。本研究表明 tetE 基因可能是介导气单胞菌分离株对四环素类药物耐药的优势基因, 为阐明嗜水气单胞菌对四环素类药物耐药机制及耐药性与耐药基因之间的关系提供理论依据。

关键词: 嗜水气单胞菌; 四环素; 强力霉素; 体外诱导; 耐药基因

嗜水气单胞菌（*Aeromonas hydrophila*）属气单胞菌科（Aeromonadaceae）、气单胞菌属（*Aeromonas*），是一种典型的人-兽-水生动物条件致病菌，普遍存在于淡水、海水、淤泥以及土壤中，可以感染多种水生动物[1-3]。由嗜水气单胞菌引发的急性出血病，发病急、传播快、流行广、死亡率高，给淡水养殖业造成了巨大损失，成为制约淡水养殖业发展的重要因素之一[4]。

四环素类药物是一种广谱抗生素，通过干扰细菌蛋白质的合成而起到抑菌作用[5]，可用于防治鱼类的多种细菌性疾病，但因临床上对该类抗生素的不合理使用，使得嗜水气单胞菌对该类抗生素的耐药性日益增强[6-7]。

嗜水气单胞菌的耐药机制非常复杂，而单一条件变化即某一种抗生素诱导引起菌株耐药的分子机制易于分析，但对体外应用四环素类药物诱导使敏感菌耐药后的耐药机制的研究则未见报道。为探讨药物诱导引起嗜水气单胞菌耐药的反应机制，本文选用渔用常用药物——强力霉素作为四环素类药物的代表药物，以体外连续诱导的方式获得高度耐药菌，探索研究嗜水气单胞对四环素类药物的耐药机制，为阐明嗜水气单胞菌耐药的分子机制提供依据。

1　材料与方法

1.1　材料

1.1.1　菌株来源

诱导受试菌株为中国水产科学研究院黑龙江水产研究所养殖室鱼病组分离保存的鱼源嗜水气单胞菌，参考株由珠江水产研究所邓玉婷博士馈赠。

1.1.2　实验试剂：

实验用药物：强力霉素（Doxycycline，DO，98.7%，C13084280）、诺氟沙星（Norfloxacin，NF，99.1%，C15648000）和恩诺沙星（Enrofloxacin，EN，98.5%，C13170000）购自 Dr. Ehrenstorfer GmbH-Bgm. -Schlosser-Str. 6A-86199 Augsburg-Germany；左氧氟沙星（Levofloxacin，LE，98.5%，130455-201005）、环丙沙星（Ciprofloxacin，CP，99.0%，1134313-201405）与氧氟沙星（Ofloxacin，OFL，99.5%，130454-201206）购自中国药品检定研究院；庆大霉素（Gentamicin，GM，63.0%，130326-201015）购自中国食品药品检定研究院；链霉素（Streptomycin，SM，72.9%，130308-200713）、卡那霉素（Kanamycin，KM，66.5%，130556-200501）、阿米卡星（Amikacin，AM，65.5%，130335-200204）、新霉素（Neomycin，NEO，64.5%，130309-200811）、土霉素（Oxytetracycline，OTC，88.8%，130487-200703）、四环素（Tetracycline，TET，97.5%，130488-200403）、氯霉素（Chloramphenicol，99%，CH，130303-200614）、氟苯尼考（Florfenicol，98.2%，FFC，C13665000-200205）、甲砜霉素（Thiamphenicol，99.5%，THI，130433-200502）和利福平（Rifampicin，RIF，99.9%，130496-200702）均购自中国药品生物制品检定所；N-甲基吡咯烷酮（1-Methyl-2-pyrrolidinone，NMP）购自美国 Sigma 公司。

实验用药敏纸片购自杭州天和微生物试剂有限公司，纸片药物-含量（μg）分别为：LE-5、CP-5、NF-10、EN-5、OFL-5、SM-10、KM-30、AM-30、GM-120、NEO-30、OTC-30、TET-30、DO-30、CH-30、FFC-30、THI-30、RIF-5。

实验用胰蛋白胨大豆琼脂培养基（TSA）、胰蛋白胨大豆肉汤（TSB）、MH 肉汤、Mueller-Hinton 琼脂培养基均购自青岛科技园海博生物技术有限公司；细菌基因组 DNA 提取试剂盒购自天根生化科技有限公司；实验所用引物的合成及基因目的片段序列测定由哈尔滨博仕生物技术有限公司完成。

1.2　方法

1.2.1　药物敏感和体外诱导试验：

采用 K-B 纸片扩散法进行药敏试验，以 ATCC 7966 为质控和对照菌株进行药敏试验，筛选出 7 株四环素类药物敏感的嗜水气单胞菌菌株，分别编号为 1~7。

采用微量肉汤稀释法测定菌株对 17 种受试药物的最小抑菌浓度（Minimum inhibitory concentration，MIC）。依据菌株药物敏感性结果，将受试菌株分别接种在含有 1/4×MIC 的强力霉素的 TSA 固体培养基中，28℃ 培养传代，等比 2 倍提高诱导药物质量浓度对受试菌进行连续传代培养，以获得高耐诱导菌，分别编号为 D1~D7，并保存每代诱导菌株。

测试诱导试验中所有培养浓度下的诱导菌株的 MIC 值变化情况；测定在 MH 液体培养基中加入 50 μg/mL 的外排泵抑制剂 NMP 的诱导后菌株的 MIC。每次试验设置阳性对照（培养基加菌液）和阴性对照（仅加培养基）。

1.2.2　交叉耐药试验：

将诱导筛选出的耐药菌株在 TSA 固体培养基中培养后，用微量稀释法测定菌株对非诱导药物的 MIC，并比较诱导前后 MIC 的变化。若药物诱导后菌株对非诱导药物的 MIC 比诱导前升高 4 倍或以上，定为交叉耐药[8]。

1.2.3　遗传稳定性试验：

将诱导后的菌株在不含药物的 TSA 固体培养基上连续传代培养 20 次，每 1~2 d 接种传代 1 次，依次测定第 5、10、15、20 代传代菌株的 MIC，以此判断诱导后菌株耐药的遗传稳定性。

1.2.4 耐药相关基因扩增：

取诱导后菌株分离到的单菌落接种于 3 mL TSB 营养肉汤的试管中，28℃过夜振荡培养，无菌条件下取菌液在 1.5 mL 的 EP 管中 12 000 r/min 离心收集菌沉淀，按照细菌基因组 DNA 提取试剂盒说明书提取细菌基因组 DNA。

扩增四环素耐药相关基因所用引物[9-10]信息见表 1。PCR 反应体系为 DNA 模板 1.0 μL，dNTP（10 mmol/L）1.0μL，10×PCR buffer 5.0 μL，MgCl$_2$（25 mmol/L）3.0 μL，上下游引物（10 μmol/L）各 2 μL，TaqDNA 聚合酶（MBI）4U，用 ddH$_2$O 补足至 50 μL。将经 1%琼脂糖凝胶电泳检测为阳性的 PCR 产物切胶纯化后测序。测序结果通过 NCBI 中的 Blastn 系统进行序列相似性搜索，然后使用 DNAMAN 软件进行序列间相似性分析。

表 1 tet 基因的相应引物序列

Table 1 Primers for amplifying the tet genes

Primers	Sequences（5′-3′）	Product length/bp	Annealing Temperature/℃
tetA	F：GTAATTCTGAGCACTGTCGC R：CTGCCTGGACAACATTGCTT	956	55
tetB	F：TCATTGCCGATACCACCTCAG R：CCAACCATCATGCTATTCCATCC	391	55
tetC	F：CTGCTCGCTTCGCTACTTG R：GCCTACAATCCATGCCAACC	897	55
tetD	F：TGTGCTGTGGATGTTGTATCTC R：CAGTGCCGTGCCAATCAG	844	53
tetE	F：ATGAACCGCACTGTGATGATG R：ACCGACCATTACGCCATCC	744	57

2 结果

2.1 体外诱导菌株 MIC 的变化及交叉耐药情况

受试菌株用诱导药物 DO 从 1/4×MIC 开始，通过体外连续诱导培养方式获得了 DO 的高耐诱导菌。高耐诱导菌对 DO 的 MIC 与诱导前相比，提高了 6.4~102.4 倍，加入外排泵抑制剂 NMP（50 μg/mL）后，高耐诱导菌对 DO 的 MIC 比未添加前均有显著下降，降低幅度达 4~16 倍。而高耐诱导菌在无药平板上连续传代培养 20 代后，对 DO 的 MIC 值基本一致，表明具有良好的遗传稳定性（表 2）。

表 2 诱导前后及加入 NMP 后和诱导后传代过程中菌株对强力霉素的 MIC 变化

Table 2 MICs of Doxycycline before and after induction as well as NMP added and after induction sub-cultured

	Before selection	After selection	NMP added	5 generation	10 generation	15 generation	20 generation
1	2.5	64	16	64	64	64	32
2	5	128	32	128	128	128	64
3	1.25	64	32	64	64	64	64
4	1.25	128	16	128	128	128	128
5	2.5	128	8	64	64	64	64
6	5	128	16	128	128	128	64
7	5	32	4	32	16	16	16

高耐诱导菌对非诱导的 16 种药物的 MIC 值均有所变化，对氟喹诺酮类药物 LE、CP、NF、EN 和 OFL 的 MIC 分别比诱导前增加了 8~64 倍、1~64 倍、5~2560 倍、2~64 倍和 1~64 倍；对四环素类药物的 MIC 增加 1~32 倍；对氯霉素类药物的 MIC 增加 1~64 倍；对氨基糖苷类药物的 MIC 降低 1~16 倍；对 RIF 的 MIC 有小范围的降低（2~8 倍）（表 3）。

表3　诱导前后菌株对非诱导药物 MIC 的变化
Table 3　MICs of non-induced antibiotics before and after induction

	LE	CP	NF	EN	OFL	SM	KM	AM	GM	NEO	OTC	TET	CH	FFC	THI	RIF
1	0.625	2.5	0.5	1.25	2.5	>128	>128	32	16	>64	>64	40	2	2	8	40
D1	5	20	80	40	20	>128	>128	8	2	32	>64	>80	4	32	8	10
2	0.625	2.5	1	1.25	2.5	>128	16	32	2	2	64	20	0.5	1	2	10
D2	20	40	160	40	40	128	8	8	2	4	64	80	32	64	2	5
3	0.3	0.3	0.125	0.3	0.625	64	32	64	1	2	16	2.5	>64	1	>64	80
D3	10	10	80	20	20	32	8	16	2	4	64	80	4	64	8	40
4	0.15	0.3	0.0625	0.15	0.3	64	32	32	4	4	64	10	1	1	2	10
D4	10	20	160	10	20	8	4	4	2	2	>64	>80	2	8	2	10
5	0.625	2.5	1	1.25	2.5	>128	16	32	2	>64	>64	10	32	64	>64	10
D5	20	20	160	80	40	>128	>128	8	2	64	64	80	64	>64	>64	5
6	0.625	2.5	2	1.25	1.25	>128	>128	64	64	32	>64	80	>64	64	>64	10
D6	20	80	160	40	40	>128	128	8	8	4	>64	>80	>64	>64	>64	1.25
7	0.3	0.625	0.25	0.625	0.625	32	32	16	2	2	64	20	16	>64	>64	>80
D7	0.15	0.625	1.25	1.25	0.625	16	32	8	2	2	>64	40	16	>64	>64	>80

DO 诱导后菌株 D1-7 对 16 种非诱导药物中的部分药物产生了交叉耐药，具体情况见表 4。

表4　D1-7 交叉耐药情况
Table 4　Cross-resistance of D1-7

	LE	CP	NF	EN	OFL	OTC	TET	KM	CH	FFC
D1	+	+	+	+	+					+
D2	+	+	+	+	+		+		+	+
D3	+	+	+	+	+	+	+			+
D4							+			+
D5	+	+	+	+	+		+	+		
D6	+	+	+	+		+				
D7										

由表 4 可见，高耐诱导菌分别对氟喹诺酮类（LE、CP、NF、EN 和 OFL）、四环素类（OTC 和 TET）、氨基糖苷类（KM）和氯霉素类（CH 和 FFC）药物产生了不同程度的交叉耐药。其中，对 KM 和 CH 产生交叉耐药的菌株最少，仅为 12.49%；而对 5 中氟喹诺酮类药物产生交叉耐药的菌株最多，高达 71.43%。而 D7 未对任何一种药物产生交叉耐药现象，但 D2 和 D3 对测试的 16 种非诱导药物的 50% 均产生了交叉耐药。

2.2　耐药相关基因扩增结果

对诱导前菌株 1-7 和诱导后菌株 D1-7 进行了四环素类耐药基因 tetA、tetB、tetC、tetD 和 tetE 的检测，结果如图 1 所示：检测结果为阳性的菌株数量分别是 1、2、0、0 和 12；其中，tetA 仅在诱导后 D7 菌株中检出，tetC 仅在 2 和 D2 菌株中检出，而 tetE 在诱导前后的 1、3、4、5、6、7 号菌株中均可检测到，tetB 和 tetD 在所有受试菌株中均未检测到。

2.3　耐药相关基因序列分析

将扩增得到的阳性目的片段切胶回收测序后，使用 DNAMAN 软件分别对 2 个 tetC 和 12 个 tetE 基因片段进行多序列比对分析，结果显示其一致性均为 100%。将所得序列信息经 NCBI 中 Blastn 系统序列同源性分析，结果显示：tetA、tetC 和 tetE 基因片段分别与 GenBank 中登录的其他菌株 tet 基因序列相比较，构建系统发育树（图2）。

结果显示，tetA 基因与存在于气单胞菌（Aeromonas）（GenBank 登录号：EF471998.1），大肠杆菌（Escherichia coli）（GeneBank 登录号：KP294351.1）、沙门氏菌（Salmonella enterica subsp.）（GeneBank 登录号：KR091911.1）、鲍氏不动杆

911

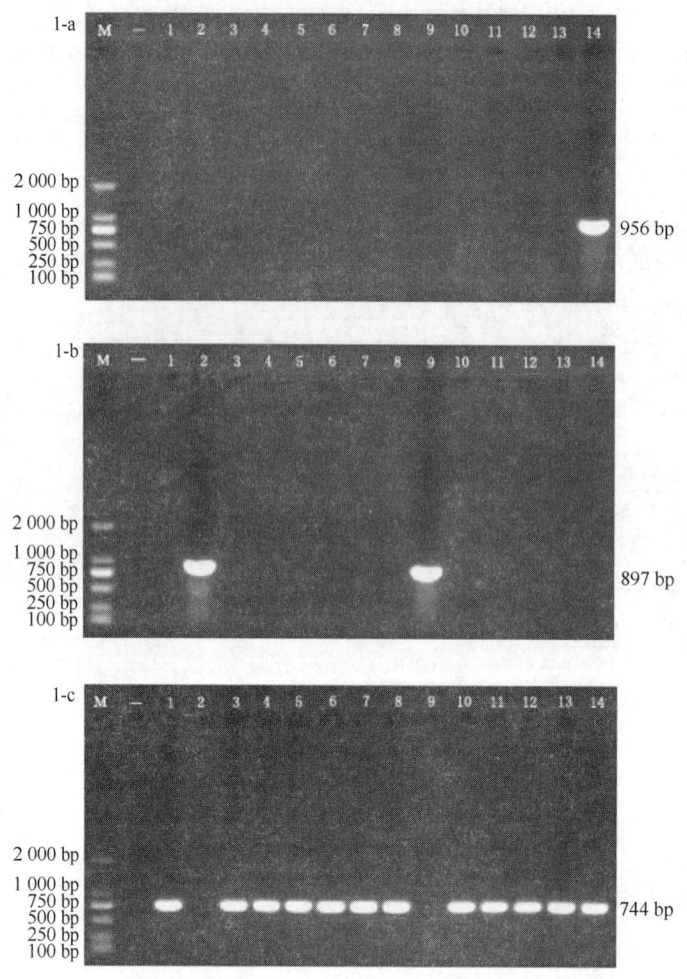

图 1 嗜水气单胞菌 tet 基因 PCR 扩增结果

Fig. 1 PCR amplification of tet genes in Aeromonas hydrophila

"1-a"，"1-b"，"1-c"：PCR amplification of tetA、tetC、tetE genes；"M"：2000bp DNA Ladder Marker；

"−"：Negative control；"1-7"：isolates before induction 1-7；"8-14"：isolates after induction D1-7

菌（Acinetobacter baumannii）（GenBank 登录号：KF483599.2）、肺炎克雷伯菌（Klebsiella pneumoniae）（GenBank 登录号：CP011314.1）、大肠杆菌（GenBank 登录号：KM023153.1）中的 tetA 基因聚为一簇；tetC 基因与来自嗜水气单胞菌（GenBank 登录号为：CP010947.1）、克隆载体 pLG338（GenBank 登录号为：KM604642.1）、大肠杆菌（GenBank 登录号为：EU751613.1）以及沙门氏菌（GenBank 登录号分别：AY046276.1、GU987054.1）中的 tetC 基因聚为一簇；tetE 基因则与来自中间气单胞菌（Aeromonas media）（GenBank 登录号为：CP007567.1）、嗜水气单胞菌 pAHH01 质粒（GenBank 登录号为：JN315882.1）、杀鲑气单胞菌（Aeromonas salmonicida subsp.）（GenBank 登录号为：CP000645.1）、杀鲑弧菌（Vibrio salmonicida）中质粒 pRVS1（GenBank 登录号为：NG_ 035496.1）中的 tetE 基因聚为一簇。

3 讨论

本研究选取渔用常用四环素类药物—DO 对临床分离的敏感嗜水气单胞菌分离株从 1/4×MIC 开始进行逐级诱导，诱导后的菌株对强力霉素耐药性显著增高，对非诱导的四环素类药物—四环素的耐药性也明显上升，四环素类药物 MIC 上升的幅度达 1~32 倍，可能与四环素类药物之间结构相似，作用位点相同有关；诱导后菌株对氟喹诺酮类药物的 MIC 增高几十到几千倍，与氟喹诺酮类药物存在严重的交叉耐药情况；诱导后菌株对氨基糖苷类和 RIF 的敏感性有不同程度的增强，推测其原因可能是菌株在外界环境改变时，通过改变自身对药物的耐受性以适应环境。但 7 号菌株诱导为高耐诱导菌 D7 后，未与任何非诱导药物产生交叉耐药，因而推断产生交叉耐药的情况可能与菌株自身特性相关，需要我们进一步进行研究验证。

诱导菌株药敏试验过程中添加 NMP 后，所有诱导菌对被测药物的敏感性均存在不同程度的增高，提示嗜水气单胞菌

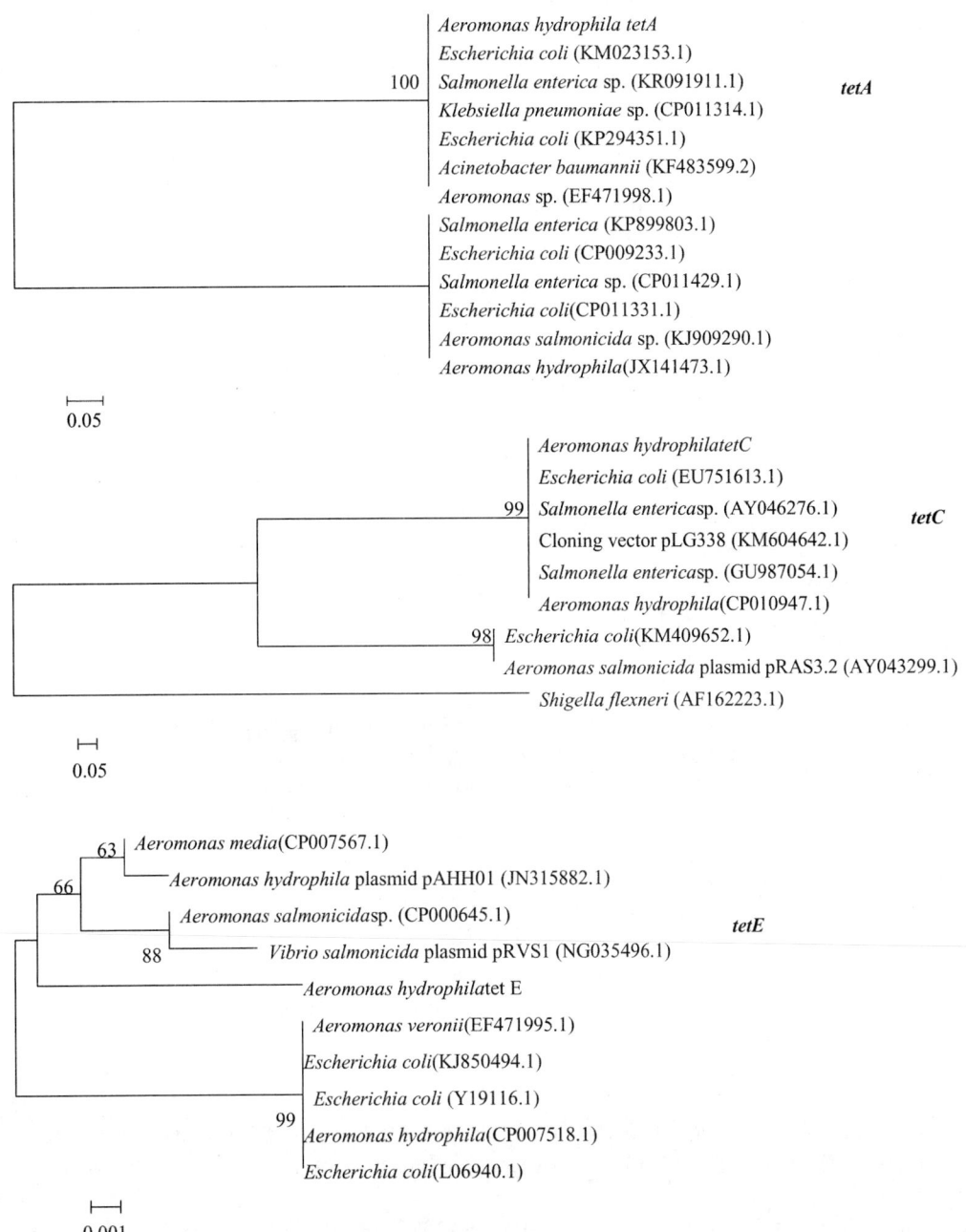

图 2　tetA、C、E 基因系统发育树

Fig. 2　Phylogenetic tree of tet A、C、E gene

The serial numbers in brackets mean GenBank accession Number; the numbers on branch point mean bootstrap support;

the scale means genetic distance.

中存在一种以四环素类为作用底物的外排泵；同时也存在其他类型的外排泵。Hernould 等[11]首次报道了嗜水气单胞菌存在引起内源性多重耐药的外排泵 AheABC，属于耐药节结化细胞分化（resistance nodulation division, RND）家族，其作用底物最少有 13 种，其中 9 种为抗生素，但对四环素类无作用。至于哪一类家族的外排泵对四环素类药物发挥了重要作用，目前国内外仍在探索之中，本试验为进一步对嗜水气单胞菌外排泵的耐药机制研究打下了基础。诱导后菌株对多种非四环素类耐药，且在外排泵抑制剂存在时降低 DO 的 MIC，说明诱导导致了染色体上外排泵基因的强表达，目前嗜水气单胞菌耐四环素类药物机制中，已发现 tetA、tetB、tetC、tetD、tetE、tetH、tetG 和 tetM 等，他们中的大多数编码外排泵蛋白，但就本实验所做的研究还无法说明那些基因可能出现突变，该问题也可作为未来研究的一个方向。

在革兰阴性菌中，两组 tet 基因已经在多种细菌中被检出，它们分别编码核糖体保护蛋白和外排泵蛋白[12]，而细菌对四环素类药物的耐药机制主要有两种：一种是通过核糖体保护蛋白保护核糖体免受四环素类药物的作用从而产生耐药[13]；另一种是最早被研究的一种耐药机制[14]，通过外排蛋白将四环素主动排出胞外，降低细胞内的药物浓度，从而保护了核糖体，使细菌产生了耐药性。本研究中诱导菌株在药敏试验中添加 NMP 后，诱导菌对所有被测药物的敏感性均存在不同程度的增高，证明嗜水气单胞菌中存在主动外排作用的一种耐药机制。

tetA、tetB、tetC、tetD 和 tetE 基因编码四环素的外排蛋白，外排蛋白是主要异化超家族（Major facilitato superfamily，MFS）中的一员[15]。研究已经证明，在气单胞菌中存在以上 5 个基因，且 tetE 基因在该 5 种耐药基因中检出率最高[16-19]；Nawaz 等[10]对分离自鲶鱼的 81 株耐四环素的维氏气单胞菌进行了相关研究，发现 tetA、tetB、tetC、tetD 和 tetE 基因检出率分别为 3.70%、28.40%、2.47%、2.47%和 90.12%，其中 tetE 基因的检出率显著高于其他 4 个基因，为其优势基因；以上结果均与本实验研究结果一致。然而，赵敏等[20]研究分离自斑点叉尾鲴的 42 株维氏气单胞菌，Jacobs[21]等在对 37 株分离自南非水产养殖环境中的气单胞菌的研究中，以及 Kim 等[22]对 8 株分离自韩国的耐四环素杀鲑气单胞菌的研究中都发现 tetA 基因在这些菌株中检出率最高。本研究中诱导前后共 14 株菌中，有 12 株检出 tetE 基因，占 86%；有 2 株检出 tetC 基因，占 14%；1 株检出有 tetA 基因，占 7%，验证了以上研究中 tetE 基因是介导气单胞菌分离株对四环素类药物耐药的优势基因的研究结论。

Nawaz[10]等和 Han[23]等研究表明 tetA 和 tetE 基因是主要的四环素耐药基因。一株菌中同时检测出 2 种或 2 种以上四环素类耐药基因的情况，在 Nawaz[10]等、Schmidt[17]等、Cížek[19]等以及赵敏等[20]各自的研究中都已有报道，但本研究选取的 7 个嗜水气单胞菌分离株及诱导后高耐诱导株仅 D7 出现了同时含有 2 个 tet 耐药基因的情况，也有研究发现，气单胞菌 tet 基因检出的多少与菌株 MIC 的高低没有对应关系[9]。此外，我们的研究还发现 1 号菌株在诱导前未检测到 tetA，诱导后的耐药株中检测到 tetA，是否是导致 1 号菌株耐药性升高的直接原因还需进一步深入研究，但同时说明了 tetA 基因在嗜水气单胞菌耐药机制中可能起到极其重要的作用。

参考文献

[1] 陆承平. 致病性嗜水气单胞菌及其所致鱼病 [J]. 水产学报，1992，16（3）：282-288.
Lu C. P. Pathogenic Aeromonashydrophila and the fish diseases caused by it [J]. Journal of Fisheries of China, 1992, 16 (3): 282-288. (in Chinese)

[2] Cantas L., Midtlyng P. J., Sorum H. Impact of antibiotic treatments on the expression of the R plasmidtra genes and on the host innate immune activity during pRAS1 bearing Aeromonashyrophila infection in zebrsfish (Daniorerio). BMC Microbiology, 2012, 12 (1): 1-10.

[3] Janda J. M., Abbott S. L.. The genus Aeromonas: Taxonomy, pathogenicity, and infection. Clinical Microbiology Reviews, 2010, 23 (1): 35-73.

[4] 李绍戊，王荻，刘红柏，等. 鱼源嗜水气单胞菌多重耐药菌株整合子的分子特征 [J]. 中国水产科学，2013，20（5）：1015-1022.
Li S. W., Wang D., Liu H. B.. Molecular characterization of integron-gene cassettes in multi-drug resistant Aeromonas hydrophila from fish [J]. Journal of Fishery Sciences of China, 2013, 20 (5): 1015-1022. (in Chinese)

[5] CLSI methods for broth dilution susceptibility testing of bacteria isolated from aquatic animal: Approved Guideline. CLSI document M49-A. CLSI. 2006.

[6] 李爱华，蔡桃真，吴玉深，等. 我国鱼类病原-嗜水气单胞菌的耐药性研究 [J]. 微生物学通报，2001，28（1）：58-63.
Li A. H., Cai T. Z., Wu Y. S., et al. Investigation on drug resistance of fish bacterial pathogen-Aeromonas hydrophila in China [J]. Journal of Microbiology, 2001, 28 (1): 58-63. (in Chinese)

[7] Chopra I., Roberts M.. Tetracycline antibiotics: mode of action, applications, molecular biology, and epidemiology of bacterial resistance. Microbiology and Molecular Biology Reviews, 2001, 65 (2): 232-260.

[8] 章喻军. 铜绿假单胞菌外膜耐药和保护原功能蛋白组学的研究 [D]. 厦门：厦门大学图书馆，2007：43-44.
Zhang Y. J.. Functional proteomic on antibiotic resistance and protective immunogen of Pseudomonas aeruginosa outer membrane proteins [D]. Xiamen: Xiamen University Library, 2007: 43-44. (in Chinese)

[9] Ndi O. L., Barton M. D.. Incidence of class 1 integron and other antibiotic resistance determinants in Aeromonas spp. from rainbow trout farms in Australia [J]. Journal of Fish Diseases, 2011, 34 (8): 589-599.

[10] Nawaz M., Sung K., Khan S. A., Khan A. A., Steele R.. Biochemical and molecular characterization of tetracycline-resistant Aeromonas-veronii isolates from catfish [J]. Applied and Environmental Microbiology, 2006, 72 (10): 6461-6466.

[11] Hernould M., Gagné S., Fournier M., Quentin C., Arpin C.. Role of the AheABC efflux pump in Aeromonashydrophila intrinsic multidrug resistance [J]. Antimicrobial agents and chemotherapy, 2008, 52 (4): 1559-1563.

[12] Zhang X. X., Zhang T., Fang H. H.. Antibiotic resistance genes in water environment [J]. Applied Microbiology and Biotechnology, 2009, 82 (3): 397-414.

[13] Kobayashi T., Suehiro F., CachTuyen B., Suzuki S.. Distribution and diversity of tetracycline resistance genes encoding ribosomal protection proteins in Mekong river sediments in Vietnam [J]. FEMS Microbiology Ecology, 2007, 59 (3): 729-737.

[14]　Alekshun M. N., Levy S. B.. Molecular mechanisms of antibacterial multidrug resistance [J]. Cell, 2007, 128 (6): 1037-1050.

[15]　Kumar A., Schweizer H. P.. Bacterial resistance to antibiotics: active efflux and reduced uptake [J]. Advanced Drug Delivery Reviews, 2005, 57 (10): 1486-1513.

[16]　Agersø Y., Bruun M. S., Dalsgaard I., Larsenb J. L.. The tetracycline resistance gene tet (E) is frequently occurring and present on large horizontally transferable plasmids in Aeromonas sppfrom fish farms [J]. Aquaculture, 2007, 266 (1): 47-52.

[17]　Schmidt A. S., Bruun M. S., Dalsgaard I., Larsen J. L.. Incidence, distribution, and spread of tetracycline resistance determinants and integron-associated antibiotic resistance genes among motile Aeromonads from a fish farming environment [J]. Applied and Environmental Microbiology, 2001, 67 (12): 5675-5682.

[18]　Jun J. W., Kim J. H., Gomez D. K., Choresca CHJr, Han J. E., Shin S. P., Park S. C.. Occurrence of tetracycline-resistant Aeromonashydrophila infection in Korean cyprinid loach (Misgurnus anguillicaudatus) [J]. African Journal of Microbiology Research, 2010, 4 (9): 849-855.

[19]　Cížek A., Dolejská M., Sochorová R., StrachotováK., Pia　ková V., Vesel　T.. Antimicrobial resistance and its genetic determinants in Aeromonadsisolated in ornamental (koi) carp (Cyprinuscarpio koi) and common carp (Cyprinuscarpio) [J]. Veterinary Microbiology, 2010, 1422 (3): 435-439.

[20]　赵敏, 汪开毓, 王均, 等. 斑点叉尾鮰源维氏气单胞菌对四环素类抗生素的耐药性及耐药基因的检测 [J]. 水生生物学报, 2014, 28 (2): 386-391.
　　　Zhao M., Wang K. Y., Wang J., Chen D. F., Huang L. Y., Wang H. C.. Tetracycline antibiotics resistance and its genetic determinants in Aeromonas veronii isolated from channel catfish (Ictalurus punctatus) [J]. Acta Hydrobiologica Sinica, 2014, 28 (2): 386-391. (in Chinese)

[21]　Jacobs L., Chenia H. Y.. Characterization of integrons and tetracycline resistance determinants in Aeromonasspp. isolated from South African aquaculture systems [J]. International Journal of Food Microbiology, 2007, 114 (3): 295-306.

[22]　Kim J. H., Hwang S. Y., Son J. S., Han J. E., Jun J. W., Shin S. P., Choresca CHJr, Choi Y. J., Park Y. H., Park S. C.. Molecular characterization of tetracycline- and quinolone-resistant Aeromonassalmonicida isolated in Korea [J]. Journal of Veterinary Science, 2011, 12 (1): 41-48.

[23]　Han J. E., Kim J. H., Choresca CHJr, Shin S. P., Jun J. W., Chai J. Y., Park S. C.. First description of ColE-type plasmid in Aeromonas spp. carrying quinolone resistance (qnrS2) gene [J]. Letters in Applied Microbiology, 2012, 55 (4): 290-294.

该文发表于《广东农业科学》【2012, (23): 149-153】

水产动物源嗜水气单胞菌药物敏感性及 QRDR 基因突变分析

Antimicrobial susceptibility and mutations of QRDR in Aeromonas hydrophila isolated from aquatic animals

薛慧娟[1,2], 邓玉婷[1], 姜兰[1], 吴雅丽[1,2], 谭爱萍[1], 王伟利[1], 罗理[1], 赵飞[1]

(1. 中国水产科学研究院珠江水产研究所/农业部渔药创制重点实验室, 广东 广州 510380;
2. 上海海洋大学水产与生命学院, 上海 200090)

摘要: 为了解临床分离的 59 株嗜水气单胞菌对喹诺酮类药物的耐药性及其靶基因 gyrA 和 parC 编码的喹诺酮类耐药决定区 (QRDR) 的突变情况。采用纸片琼脂扩散法测定 15 种抗菌药物对水产动物源嗜水气单胞菌的耐药情况, 采用 PCR 法扩增 gyrA 和 parC, 并通过测序分析其靶基因突变情况。结果表明: 24 株 (40.7%) 嗜水气单胞菌对喹诺酮类药物耐药, 其中对萘啶酸、诺氟沙星、氧氟沙星、环丙沙星的耐药率依次为 40.7%、16.9%、20.3%、8.5%。敏感菌株 QRDR 靶位点未发生突变, 24 株喹诺酮类耐药菌株中 gyrA 基因编码的第 83 位氨基酸均存在 Ser→Ile 突变; 19 株菌 parC 基因编码的第 87 位氨基酸也发生突变, 其中 18 株为 Ser→Ile 突变, 1 株为 Ser→Arg 突变, 有 5 株未检测到突变; ParC 亚基氨基酸突变的菌株其 GyrA 亚基均同时发生氨基酸突变。表明临床分离的嗜水气单胞菌对喹诺酮类药物耐药程度不一, 其中萘啶酸的耐药最为严重, 诺氟沙星、氧氟沙星和环丙沙星敏感性较高。喹诺酮类耐药菌株 GyrA 亚基 QRDR 氨基酸突变, 可能是引起萘啶酸耐药的主要原因。临床分离嗜水气单胞菌 QRDR 区域的突变具有随机性, 但靶位点 GyrA 的 Ser83→Ile 以及 ParC 的 Ser87→Ile 是最主要的突变方式, 另外喹诺酮类药物耐药性的产生可能还与其他耐药机制存在关联。

关键词: 嗜水气单胞菌; 喹诺酮类; gyrA 基因; parC 基因

该文发表于《华南农业大学学报》【2014，35（1）：12-16】

体外诱导嗜水气单胞菌对喹诺酮类耐药及其耐药机制研究

Characterization of Quinolone Resistance Mechanism of *Aeromonas hydrophila* Selected *In Vitro*

邓玉婷[1]，薛慧娟[1,2]，姜兰[1]，谭爱萍[1]，吴雅丽[1,2]，王伟利[1]，罗理[1]，赵飞[1]

（1. 中国水产科学研究院珠江水产研究所，农业部渔药创制重点实验室，广东省免疫技术重点实验室，广东 广州 510380；
2. 上海海洋大学水产与生命学院，上海 200090）

摘要：【目的】探讨在亚抑菌浓度喹诺酮类药物培养后，嗜水气单胞菌对喹诺酮类的药物敏感性变化及其耐药机制。【方法】以对喹诺酮类敏感的临床分离嗜水气单胞菌菌株和标准菌 ATCC7966 为研究对象，分别在含亚抑菌浓度萘啶酸（NAL）和环丙沙星（CIP）的培养基上逐步诱导培养。提取诱导菌的 DNA，PCR 扩增其 gryA 和 parC 基因，测序分析其 QRDR 突变情况；测定诱导菌对诱导药物和 11 种非诱导药物药物的最小抑菌浓度（MIC）及添加外排泵抑制剂羰基氰化氯苯腙（CCCP）后的 MIC，分析其敏感性变化与基因突变、外排作用的关系。【结果与结论】诱导后菌株对 NAL 和 CIP 的 MIC 分别提高了 1 024 和 64 000 倍，对非诱导药物也有不同程度提高；当 NAL 和 CIP 诱导浓度分别达到 16 和 32 μg/mL 或以上后，诱导菌株 gyrA 基因编码的氨基酸分别发生 Asp87→Tyr 和 Ser83→Arg 的变化，但两者 parC 基因编码的氨基酸均没有发生突变；添加 CCCP 后，只有氟喹诺酮类药物的 MIC 值略有下降，提示嗜水气单胞菌对喹诺酮类耐药存在靶基因突变及主动外排作用等多种耐药机制。

关键词：嗜水气单胞菌；喹诺酮类；体外诱导；基因突变；外排作用

二、鱼源溶藻弧菌和哈维氏弧菌耐药基因 qnr 的研究

qnr 基因是一种常见的耐药基因，溶藻弧菌（周维等，2015）、哈维氏弧菌（周维等，2016）均携带有喹诺酮类耐药基因 qnr。

该文发表于《生物技术》【2015，25（5）：414-419】

溶藻弧菌耐药基因 qnr 的克隆及生物信息学分析

Cloning and Bioinformatics Analysis of Quinolone Resistance Genefrom Vibrio alginolyticus

周维[1,2]，汤菊芬[1,2]，甘桢[1,2]，简纪常[1,2]，吴灶和[2,3]，丁燏[1,2]*

（1. 广东海洋大学水产学院，广东 湛江 524088；2. 广东省水产经济动物病原生物学及流行病学重点实验室，广东 湛江 524088；
3. 仲恺农业工程学院，广东 广州 510225）

摘要：[目的]：克隆溶藻弧菌（Vibrio alginolyticus）喹诺酮类耐药基因 qnr 并分析其蛋白结构，为研究该蛋白的生物学功能奠定基础。[方法]：根据 NCBI 上公布的相关序列设计 qnr 基因特异性引物，利用 PCR 方法扩增基因序列，进行 DNA 测序及生物信息学分析。[结果]：从溶藻弧菌染色体上获得 qnr 基因大小为 651 bp，编码 216 个氨基酸，进化分析可知其与副溶血弧菌有很近的亲缘关系，二级结构分析含有五肽重复序列，三维结构与大肠杆菌质粒上的 qnr 蛋白空间结构极为相似。[结论]：通过对蛋白质序列分析和结构预测，初步确定该基因为喹诺酮耐药基因，为水产细菌的耐药机制研究奠定基础。

关键词：溶藻弧菌，喹诺酮，耐药基因，五肽重复序列，生物信息学

该文发表于《广东海洋大学学报》【2016, 36 (1): 93-97】

哈维氏弧菌 qnr 基因的克隆及原核表达条件优化

Cloning and Optimization of Prokaryotic Expression of Quinolone Resistance Genein Vibrio harveyi

周维[1,2]，汤菊芬[1,2]，高增鸿[1]，甘桢[1,2]，简纪常[1,2]，吴灶和[2,3]，丁燏[1,2]*

(1. 广东海洋大学水产学院，广东 湛江 524088；2. 广东省水产经济动物病原生物学及流行病学重点实验室，广东 湛江 524088；3. 仲恺农业工程学院，广东 广州 510225)

摘要：根据 GenBank 中公布的哈维氏弧菌 qnr 基因序列设计引物扩增哈维氏弧菌 qnr 序列，将其插入 pET-32a 质粒，构建原核表达载体 pET32-qnr，并对诱导温度、时间、IPTG 浓度等条件进行优化。结果表明哈维氏弧菌 qnr 全长 651 bp，编码 216 个氨基酸；重组蛋白优化条件为 28℃、IPTG 浓度为 0.05 mmol/L 条件下诱导 6 h。

关键词：哈维氏弧菌；qnr 基因；原核表达；优化

第三节 控制病原菌耐药的策略

目前，渔药放耐药策略的研究主要集中在耐药突变选择窗理论、中草药防耐药应用及养殖条件对耐药的影响等方面 (潘浩，2016)。

该文发表于《生物技术通报》【2016, 32: (5): 34-39】

渔用药物防耐药策略研究进展

Progress in Anti-Drug Resistance of Fishery Drugs

潘浩[1,2]，王荻[1]，卢彤岩[1]

(1. 中国水产科学研究院黑龙江水产研究所，黑龙江 哈尔滨 150070；2. 上海海洋大学水产与生命学院，上海 201306)

摘要：目前，药物治疗仍然是防治水产动物病害的主要方法，但其不合理应用也诱发了大量耐药细菌的产生。如何制定合理的渔用常用药物防耐药用药策略已成为目前研究的重点。本文系统的综述了耐药突变选择窗理论、中草药防耐药应用及养殖条件影响在防耐药用药策略方向的研究进展。

关键词：联合用药；中草药；养殖环境；防耐药策略

近些年由于抗生素的过度和不合理利用，加速了细菌耐药性的产生[1-8]。究其原因，抗生素单从一方面对细菌进行抑制，一旦细菌适应并产生耐药，此类抗生素的效果就大打折扣。因此如何制定抑制、杀灭病菌的同时避免耐药性产生的科学合理的用药策略是目前渔用药物研究者研究的热点问题。

1 耐药突变选择窗理论及其防耐药用药策略

1.1 防耐药变异浓度和耐药突变选择窗

防耐药变异浓度 (mutant prevention concentration，MPC) 这一概念是由美国学者 Drlica[9] 于 2000 年提出的，指的是一个防止耐药的抗菌药物浓度阀值，即防止细菌产生耐药性的抑制细菌生长的药物浓度。它是评价抗菌药物抑制细菌生长效应的指标，它反映的是药物抑制细菌发生耐药突变的能力。Dong[10] 等研究发现，在琼脂平板上接种的细菌数量超过 1010CFU 的时候没有发现细菌变异，这个能使细菌不产生耐药性的药物浓度值就是 MPC。当抑菌药物浓度达到 MPC 以上的时候细菌必须同时发生两处耐药突变才能生长。而细菌同时发生两次耐药突变的概率仅为10~14，这是一个极低的概率，因而抗菌药物浓度在MPC之上时细菌大部分被杀死，基本不可能产生耐药突变菌株[10-11]。最小抑菌浓度（minimum inhibitory concentration，MIC），是测定抗菌药物抑制细菌活性大小的指标，其定义为体外培养18~24 h后能抑制培养基内细菌生长的最低药物浓度，当药物浓度低于MIC时，药物抑制效果消失但不会导致细菌耐药突变体产生。而当药物浓度介于MPC与MIC之间时，即便有很高的几率抑制甚至杀灭细菌，但也很容易出现耐药突变体，介于这两者之间的浓度范围就是耐药突变选择窗（mutant selection window，MSW）[12]。

1.2 基于MSW理论的防耐药用药策略

MSW理论认为当药物浓度在MPC之下并且在MIC之上的时候，才会导致细菌耐药突变体的选择

性富集并且产生抗性。因此可以通过选择使用低MPC、窄MSW的药物，调整用药方案或是通过药物的联合使用缩小或关闭MSW[13]。缩小或关闭MSW将会有极少甚至不会有耐药突变体产生，可以将选择性突变菌株扩增的几率降到最低。因此可以通过使药物浓度快速达到峰值并且在最短的时间内通过MSW，将血浆药物在MSW的滞留时间降到最低，使其处于MPC以上的时间升到最高，以达到最大限度地缩短突变选择时间的目的。方法二是缩小MIC与MPC的差距，缩小或者关闭MSW。其三是采取联合用药的方法，因为当两种或多种不同作用机制的药物同时存在于细菌生存的环境中时，细菌要继续生长必须同时发生两种及以上的耐药突变，而发生两种及以上的突变的概率极低，这样就可以达到最大限度关闭MSW以抑制细菌同时防止耐药菌株的产生。

联合用药是除了保证药物达到合适的药物浓度（MPC）之外的另一种经常采用的防耐药用药策略。单一药物治疗的关键是组织或血液药物浓度要高于药物用药安全剂量的MPC，并且在MSW以上的时间越长越好。然而在联合使用几种作用机制不同的药物时，由于临床安全剂量难以达到各自的MPC，因此可以通过匹配24 h用药时曲线下面积（AUC24）与最小抑菌浓度（MIC）的比值（AUC24/MIC）来关闭MSW，这样可以在达到理想的治疗效果的同时尽量避免耐药突变体的出现。

联合用药所引起的药物相互作用的改变是使药物效应改变[14]，这种改变可能是作用性质的变化，也可能是效应强度改变，从而对药物的有效性和安全性产生影响。药物联合使用所引起的药效包括：使原有的效应增强，即协同作用；使原有的效果减弱，即拮抗作用；产生毒性等不良反应，即毒副作用；以及可能出现的互不影响的结果。因此在用药的时候要充分了解药物的药效，避免拮抗作用与毒副作用的发生。

2 中草药防耐药用药策略

中草药由于其有效成分复杂、作用机制多样，实验发现体外抑菌试验很难解释中草药在体内的抗菌作用，这表明中草药除了其表现出来的抑菌作用外可能还存在其他作用机制协助进行抑菌，推测是通过其对细菌以及细菌所寄生的机体进行的多环节多途径的综合作用来实现[15]。近年来，中草药在抑菌机制及防耐药用药策略方面的应用也有了较系统的研究进展。

2.1 中草药抑菌成分与机制

近几年来，中草药抑菌成分的研究方向逐渐转向生物碱、黄酮类、多酚类、萜类等方面。随着医学技术的进步及新的仪器的研发，对含酚类成分的中草药，如低分子质量的有蒽醌及异戊烯黄酮，高分子质量的有多酚类及原花色素、水解单宁等中草药的研究也越来越多。

目前许多研究者认为，中草药的抗菌作用机制主要包括抑制菌体内能量的生成、抑制细菌生物膜的形成、改变细胞膜和离子通道的通透性、抑制细菌体内酶的活性、增强中性粒细胞的吞噬功能、代谢产物发挥抗菌作用、抑制细菌蛋白质的合成、减少细菌内毒素的释放。

2.2 中草药逆转细菌耐药性研究

在研究新型抗生素药物的同时，越来越多的学者开始把目光转向使抗性细菌恢复对抗生素药物的敏感性方面。由于成分复杂中草药在这方面有其特有的优势，并且一些相关研究也证实中草药在此方面具有较好的功效。

2.2.1 消除抗药性（R）质粒

在1970年之后，国内外许多研究学者在研究中试验通过消除细菌的R质粒使细菌恢复对抗菌药物的敏感性，但由于西药的毒副作用较强，一直未能取得较好的效果，因此开始把目光投向传统中草药。研究结果表明，中草药对消除细菌R质粒的效果良好[16]，而且中草药由于其作用机制对R质粒的消除作用体内明显强于体外，这可能是因为在体内细菌不仅受到药物的抑制作用与机体的免疫作用特别是先天性免疫系统的影响，而且同时还受到机体的内环境的影响有关。

目前，对中草药黄芩、黄连进行的相关研究较多，两者联用效果明显，可使R质粒消除率提高10倍以上，其效果远高于西药。不同组分的中草药对R质粒消除率也不同，如从艾叶得到的乙醇提取物对R质粒的消除率可达69.4%，艾叶提

取的挥发油对 R 质粒的消除率可达 16.67%[17]。中草药千里光对大肠杆菌 R 质粒消除作用显著，且含药血清消除作用达 14.9%[18]，明显强于其水浸液。从 R 质粒消除的表型来看，经千里光水浸液作用后细菌均表现为单一耐药性的丢失，而含千里光血清对其消除作用表现为多重耐药性的丢失，其中以四环素的耐药性消除最多。鞠洪涛[19]等研究表明，大蒜油等对大肠埃希氏菌氨苄青霉素的耐药性有明显的消除作用。

2.2.2　抑制细菌主动外排泵

主动外排机制是多数耐药菌耐药性产生的原因之一，也是细菌多重耐药性产生的原因之一[20]。诸多研究表明，中草药可以通过抑制多种外排泵的活性使耐药菌恢复对药物的敏感性。宋战昀[21]等构建了含有 norA 外排泵基因介导的金黄色葡萄球菌的多重耐药菌株，并筛选了 4 种中草药浙贝母、射干、穿心莲和菱角的提取物，发现这些中草药的提取物可以抑制外排泵介导的金黄色葡萄球菌的耐药性，并可以在不同程度上对金黄色葡萄球菌的耐药性产生逆转作用。adeABC 基因的过度表达导致鲍曼不动杆菌对环丙沙星产生耐药性[22]，从中草药萝芙木根中提取的生物碱利血平对外排泵基因 adeABC 的表达有抑制作用，从而使鲍曼不动杆菌部分恢复对环丙沙星的敏感性。

2.2.3　抑制超广谱 β-内酰胺酶（ESBLs）

β-ESBLs 产生抑菌作用是因为它可以水解抗菌药物的有效成分，降低抗菌药物的抑菌效果从而使细菌产生耐药性。目前研究表明，部分中草药可以抑制产 ESBLs 细菌菌株水解抗菌药物，使其恢复对抗菌药物的敏感性。刘平等[23]研究发现黄芩、黄柏、黄连、连翘、千里光这 5 种中草药的提取物可以抑制产 ESBLs 细菌的抗药性，与其他四种药物相比黄芩抑菌效果更好。三黄汤、黄连解毒汤、五味消毒饮可以逆转产酶大肠杆菌的抗药性，其机制也是通过抑制 ESBLs 的活性和表达以产生逆转效果的[24]。

2.2.4　抑制耐药基因的表达

余道军等[25]研究表明，提取蟾酥皮肤腺及耳后腺分泌的白色浆液，其水提液和醇提液在与含有耐药基因 TEM 和 CTX-M-9 的大肠杆菌作用 5d 后，可使细菌耐药基因的 mRNA 的表达丧失，失去翻译蛋白质的功能，恢复了对药物的敏感性。

2.3　中草药用药策略

目前对中草药对细菌的抑制作用机制和逆转细菌耐药性使其恢复对药物敏感性的研究还处于初级阶段，而且目前研究的大部分抑菌实验是针对孤立单一的已产生耐药性的菌株，还没有大量进行以菌群为单位的抑菌实验研究。中草药抗菌的理念与中医有相通之处，并非要将细菌全部杀灭，而是调整菌群与机体之间的"失衡状态"以实现菌群与所寄生的机体之间的"再平衡"为目的。这与现代医学提倡的抑制感染菌的同时又不伤害其他细菌维持菌群在机体内的平衡的理念相吻合。目前"突变选择窗"理论提倡的关闭突变选择窗策略与多种不同机制的药物同时作用殊途同归，这也与中草药的多途径、多靶点机制具有相似性，但是中草药抗菌的多途径、多靶点机制是一个很宽泛的概念，其各成分作用途径和靶点之间是否具有协同作用、抗菌作用机制与逆转细菌耐药性之间是否存在相关性等问题还有待深入研究。

3　养殖环境对防耐药策略的影响

有研究表明，温度、盐度、光照等环境条件和药物种类、给药方式、动物种类等因素可能会影响药物动力学参数及残留消除规律，如何合理调整养殖条件进而满足 MSW 防耐药用药策略要求的快速达峰，缩短血浆药物浓度在 MSW 的滞留时间以避免耐药性的产生，也逐渐受到渔药给药策略制定研究人员的关注。

3.1　水流对防耐药策略的影响

水流对鱼类生存、生长影响巨大。水流通过刺激鱼类的感觉器官，使其产生相应的行为反应和活动方式，进而影响鱼类的运动方式、摄食、生长、代谢等生命活动[26-30]。宋波澜[31]以多鳞四须鲃幼鱼为研究对象，研究逆流条件下多鳞四须鲃幼鱼体内代谢活动与药物残留，与静水对照组相比的结果表明，停药后各时间点（1、2、4 d）2.0 BL/s 流速组鱼肌肉盐酸诺氟沙星含量均显著减少（$P<0.05$），在停药第 6 d 药物浓度已低于检出限，而静水对照组则在停药 12 d 才低于检出限，说明逆流运动在一定程度促进鱼体内的代谢，加速体内残留药物的排除。药物在体内快速代谢减少了药物在 MSW 的滞留时间，可有效降低细菌耐药性的产生。刘海生[32]对宽鳍鱲幼鱼的研究也表明，水流刺激可以加速鱼的代谢活动。

3.2　温度对防耐药策略的影响

温度对鱼的生长具有很大的影响，通过控制水温来达到增强鱼体代谢的目的，从而加速药物在体内的代谢，减少滞留时间以降低细菌耐药性的产生。Lermen 等研究证实不同温度（15、23 和 31℃）对银鲶血液、肝脏和肌肉的代谢参数具有显著的影响作用[33]。杨贵强等[34]对溪红点鲑幼鱼的研究表明，在（5.5±0.5）℃～（17.5±0.5）℃温度范围内，随温度的上升耗氧率和排氨率均显著升高。汪文选等[35]对恩诺沙星在鲫鱼体内的药动学研究表明 20℃药时曲线下面积、消除半衰期明显大于 25℃，而清除率明显小于 25℃。在鲫鱼的肾脏和肝脏中，20℃药物的达峰时间、分布半衰期、消除半衰期、药时曲线下面积均明显大于 25℃，而清除率明显小于 25℃。对花尾胡椒鲷[36]、瓦氏黄颡鱼[37]、石斑鱼[38]、石鲷[39]、

919

鮸[40]、鲤[41]、鲶[42]等研究也表明温度通过影响鱼体组织的代谢率会对鱼体排泄率产生一定影响，并且氨氮排泄率也会随温度升高而增加。可见水温条件在一定程度上影响着药物在水产动物体内的代谢和消除，故而在实际养殖给药时应适当考虑不同水温条件以确定合理的用药策略，以达到降低细菌耐药性产生的目的，在确保食品安全的同时可以通过减少在体内的滞留时间以达到降低细菌耐药性的目的。

3.3 盐度对防耐药策略的影响

盐度是鱼类生理与生化反应的重要影响因子，主要通过对渗透压的调节来影响鱼类体内的能量代谢[43]。在高、低渗透压两种情况下鱼体会有不同的适应机制：高渗透压下通过增加内环境的渗透压；低渗透压下启用主动运输或离子转运蛋白的途经来转运离子[44]。而这两种机制都需要额外能量的消耗，这也在一定程度上加速鱼的机体代谢，从而会加速药物在体内的代谢，减少滞留时间以降低细菌耐药性产生的可能。李金兰等[45]对卵形鲳鲹的研究表明，在盐度15~30时，随盐度增大卵形鲳鲹的耗氨率、排氨率、代谢率、排泄率均呈U型抛物线变化，并在盐度20~25时有最小值。对中华圆田螺[46]、三疣梭子蟹[47]、中华绒螯蟹[48]、罗非鱼[42]的研究均表明盐度对机体的代谢有着显著的影响。因此可以考虑在用药后调节水体的盐度以加快药物在鱼体的代谢过程，减少细菌耐药性产生的可能。

3.4 光照对防耐药策略的影响

在水产养殖中，光照是一个重要而且复杂多变的生态因子。对生活在水体的生物而言，光照是一个具体而易变的因子[49]。王馨等[50]对三疣梭子蟹研究表明，在5种光照强度（0、200、800、1 500、3 500 Lx）下，三疣梭子蟹的耗氧率和排氨率均有显著差异，并且在1 500 Lx光照强度下，琥珀酸脱氢酶的活力处在较高水平，乳酸脱氢酶处在较低水平，表明蟹的有氧代谢能力高，整体代谢水平较高。另外有研究表明，光照可以通过影响牙鲆激素分泌而加速代谢[51-52]。制定最适的防耐药用药策略是否应考虑光照因素的影响尚待进一步深入研究。

4 展望

细菌对抗菌药物的耐药性问题是21世纪对人类威胁最大的健康问题之一。药物在使用的时候不仅要控制细菌感染，还应防止细菌耐药性的产生，更重要的是要将药物对目标细菌外的菌群的影响降到最低。在选择抗菌药物及用药方案的时候，不单要考虑药物的生物利用度、生物有效性、组织分布与药效参数，还应全面综合考虑细菌发生耐药突变的机制、MPC和MSW。对于耐药严重的细菌不能一味加大用药剂量或频频更换药物，要转换用药思路，采用相互拮抗的药物组合杀死耐药菌、协同型药物组合杀死敏感菌以控制感染，当然用药的前提是要筛选出合适的药物组合。目前中草药消除细菌耐药性的研究基本上还处于起步阶段，但中草药抑制剂作为纯天然制剂，同时作为一种新理论、新方法在临床中对解决日趋严重的耐药菌感染具有不可替代的作用和地位，同时也为遏制人源和动物源性细菌耐药性的交互传播、抑制细菌耐药性提供了新的思路与方法。

参考文献

[1] 刘荻萩, 李艳华, 蔡雪辉, 等. 链球菌对抗生素产生耐药性的作用机制 [J]. 国外医药（抗生素分册）, 2007, 03 (28): 103-107.
Liu D. D., Li Y. H., Cai X. H., et al. Antibiotic Resistance Mechanism of Streptococcus [J]. World Notes on Antibiotics, 2007, 03 (28): 103-107.

[2] 贾杰. 抗生素的耐药性与抗生素的应用 [J]. 中国热带医学, 2007, 09 (7): 1678-1680.
Jia J.. Resistance of microorganisms to antibiotics and use of antibiotics [J]. China tropical medicine, 2007, 09 (7): 1678-1680.

[3] 林居纯, 罗忠俊, 舒刚, 等. 嗜水气单胞菌临床分离菌对抗菌药物的耐药性调查 [J]. 安徽农业科学, 2009, 37 (15): 7024-7025.
Lin J. C., Luo Z. J., Shu G., et al. Investigation on Drug Resistance of Clinical Isolates fromAeromonashydrophilaon Antibiotics [J]. Journal of Anhui Agri, 2009, 37 (15): 7024-7025.

[4] 赵明秋, 沈海燕, 潘文, 等. 细菌耐药性产生的原因、机制及防治措施 [J]. 中国畜牧兽医, 2011, 05 (38): 177-181.
Zhao M. Q., Shen H. Y., Pan W., et al. The Causes, Mechanism and Preventive Measures of Bacterial Resistance [J]. China AnimalHusbandry&Veterinary Medicine, 2011, 05 (38): 177-181.

[5] 肖丹, 曹海鹏, 胡鲲, 等. 淡水养殖动物致病性嗜水气单胞菌ERIC-PCR分型与耐药性 [J]. 中国水产科学, 2011, 05 (18): 1092-1099.
Xiao D., Cao H. P., Hu K., et al. ERIC-PCR genotyping and drug resistant analysis of pathogenic Aeromonashydrophila from freshwater animals [J]. Journal of Fishery Sciences of China, 2011, 05 (18): 1092-1099.

[6] 高美英. 细菌对抗生素耐药性的起源分子基础与对策 [J]. 医药导报, 2003, 01 (22): 6-8.
Gao M. Y.. The molecular basis of bacterial origin and countermeasures of antibiotic resistance [J]. Herald of Medicine, 2003, 01 (22): 6-8.

[7] 高艳萍, 葛建新. 抗生素应用与细菌耐药性产生的辩证关系 [J]. 医学与哲学, 2008, 07 (29): 72-73.

Gao Y. P., Ge J. X.. The analysis and evaluation of the application of antibiotics and the bacterial drug resistance [J]. Medicine and Philosophy, 2008, 07 (29): 72-73.

[8] 吕昌彬. 关于抗生素滥用问题的思考 [D]. 山东中医药大学, 2014.
LV C. B.. Thinking about the problem of antibiotic abuse [D]. Shandong University of TraditionalChinese Medicin, 2014.

[9] Zhao X., Drlica K.. Restricting the selection of antibiotic-resistant mutants: a general strategy derived form fluoroquinolone studies [J]. Clini Infect Dis, 2001, 33: 147-156.

[10] Dong Y., Zhao X., Domagala J., Drlica K.. Effect of fluoroquinolone concentraton on selection of resistant mutants of Mycobacteriumbovis BCG and Staphylococcusaureus [J]. AntimicrobAgents Chemother, 1999, 43: 1756-1758.

[11] Joseph K., Zhao X., Domagala J., Drlica K.. Mutant prevention concentrationsoffluoroquinolones for clinical isolates of Streptococcus pneumoniae [J]. Antimicrobial Agents and Chemotherapy, Feb 2001; 433-438.

[12] 李燕玉, 裴斐, 赵西林, 等. 耐药突变选择窗与抗感染药物防突变浓度 [J]. 中国临床药理学与治疗学, 2004, 1 (9): 1-7.
Li Y. Y., Pei F., Zhao X. L., et al. Mutant selection window and mutant prevention concentration for antimicrobial agents [J]. Chinese Journal ofClinical Pharmacology and Therapeutics, 2004, 1 (9): 1-7.

[13] 吴彩霞, 刘朝明, 陈朝喜, 等. 基于药效和作用机制拮抗的防耐药性新策略研究 [J]. 中国兽医杂志, 2008, 44 (8): 73-74.
Wu C. X., Liu Z. M., Chen C. X., et al. The new antiresistance strategy of the efficacy and mechanism of action based on antagonism [J]. ChineseJournal of Veterinary Medicine, 2008, 44 (8): 73-74.

[14] 黄宇星, 刘二伟. 联合用药的药物相互作用及研究方法 [J]. 药物评价研究, 2014, 37 (3): 276-279.
Huang Y. X., Liu E. W.. Drug interactions and research methods of drug combination [J]. D rug Evaluation Research, 2014, 37 (3): 276-279.

[15] 刁人政. 中草药抗大肠杆菌及感染研究进展 [A]. 大连: 第九次全国中医药防治感染病学术交流大论文集 [C], 2009: 94.
Diao R. Z.. Chinese herbal anti-coli infection and Progress [A]. Dalian, 2009.

[16] 杨帆, 杨玉荣, 赵振生. 中草药消除细菌耐药性质粒研究进展 [J]. 动物医学进展, 2013, 34 (12): 160-164.
Yan F., Yang Y. R., Zhao Z. D. S.. Progress on Antiplasmid effect of China Herbal Medicine [J]. Progress in Veterinary Medicine, 2013, 34 (12): 160-164.

[17] 韩伟, 张铁, 钟秀会. 中草药对细菌R质粒消除作用的研究概况 [J]. 中国畜牧兽医, 2006, 33 (8): 52-54.
Han W., Zhang T., Zhong X. H.. Herbs on Bacteria eliminate the role of R plasmids Research Survey [J]. Chinese animal husbandry and veterinary medicine, 2006, 33 (8): 52-54.

[18] 张文平, 曹镐禄, 张文书, 等. 千里光对大肠埃希菌R质粒消除作用的血清药理学研究 [J]. 广东医学, 2007, 28 (8): 1238-1239.
Zhang W. P., Cao G. L., Zhang W. S., et al. Senecio serum pharmacological study of Escherichia coli plasmid to eliminate the role of R [J]. Guangdong Medical Journal, 2007, 28 (8): 1238-1239.

[19] 鞠洪涛, 韩文瑜, 王世若, 等. 中草药消除大肠埃希氏菌耐药性及耐药质粒的研究 [J]. 中国兽医科技, 2000, 30 (3): 27-29.
Ju H. T., Han W. Y., Wang S. R., et al. Research on Escherichia coli resistance and resistance plasmids herbal Elimination [J]. Chinese Journal of Veterinary Science and Technology, 2000, 30 (3): 27-29.

[20] 杭永付, 薛晓燕, 方芸, 等. 中药抗菌和逆转耐药作用机制研究进展 [J]. 中国药房, 2011, 47 (22): 4504-4507.
Hang Y. F., Xue X. Y., Fang Y., et al. Advances in medicine and antimicrobial resistance reversing mechanism [J]. China Pharmacy, 2011, 47 (22): 4504-4507.

[21] 宋战昀, 冯新, 韩文瑜, 等. 金黄色葡萄球菌norA外输泵中草药耐药抑制剂的筛选 [J]. 吉林农业大学学报, 2007, 29 (3): 329-333.
Song Z. Y., Feng X., Han W. Y., et al. Selection of Chinese Herbal Medicine Staphylococcus aureus norA Efflux Pump Inhibitor [J]. Journal of Jilin Agricultural University, 2007, 29 (3): 329-333.

[22] 范志茹. 耐环丙沙星鲍曼不动杆菌主动外排机制的研究及耐药逆转初步探讨 [D]. 中南大学, 2009.
Fan Z. R.. The mechanisms of active efflux transporters and reverse transcription test in ciprofloxacin-resistant Acinetobacter baumannii [D]. Central South University, 2009.

[23] 刘平, 叶惠芬, 陈惠玲, 等. 5种中草药对产酶菌的抑菌作用 [J]. 中国微生态学杂志, 2006, 18 (1): 39-40.
Liu P., Ye H. F., Chen H. L., et al. A study on the bacteriostatic action of5Chinese herbs against B-lactamases-producing bacteria [J]. Chinese Journal of Microecology, 2006, 18 (1): 39-40.

[24] 贾云鹏, 程宁. 中草药对大肠埃希菌抗生素耐药性逆转作用的实验研究 [J]. 陕西中医, 2009, 30 (3): 366-368.
Jia Y. P., Cheng N.. Experimental Study the influence of Escherichia coli antibiotic resistance reversal [J]. Shanxi Journal of Traditional Chinese Medicine, 2009, 30 (3): 366-368.

[25] 余道军, 徐叶艳. 蟾酥逆转大肠埃希菌耐药性的研究 [J]. 中华中医药学刊, 2010, 28 (5): 1033-1035.
Yu D. J., Xu Y. Y.. Reversal Effect of VenenumBufon is on Drug resistant Escherichia Coli [J]. Chinese Archives of Traditonal Chinese medicine, 2010, 28 (5): 1033-1035.

[26] Merino G. E., Piedrahita R. H., Conklin D. E.. Effect of water velocity on the growth of California halibut (Paralichthyscalifornicus) juveniles [J]. Aquaculture, 2007, 271: 206-215.

[27] 宋波澜, 林小涛, 王伟军, 等. 不同流速下红鳍银鲫趋流行为与耗氧率的变化 [J]. 动物学报, 2008, 54 (4): 686-694.

Song B. L., Lin X. T., Wang W. J., et al. Effects of water velocities on rheotaxisbehaviour and oxygen consumption rate of tinfoil barbs Barbodesschwanenfeldi [J]. ActaZoologicaSinica, 2008, 54 (4): 686-694.

[28] 李丹, 林小涛, 李想, 等. 水流对杂交鲟幼鱼游泳行为的影响 [J]. 淡水渔业, 2008, 38 (6): 46-51.

Li D., Lin X. T., L X, et al. Effects ofWater Current on Swimming Performance of Juvenile Hybrid Sturgeon [J]. Freshwater Fisheries, 2008, 38 (6): 46-51.

[29] 宋波澜, 林小涛, 许忠能. 逆流运动训练对多鳞四须鲃摄食、生长和体营养成分的影响 [J]. 水产学报, 2012, 36 (1): 134-142.

Song B. L., Lin X. T., Xu Z. N.. Effects of upstream locomotion training on metabolism and norfloxacin hydrochloride residues of juvenile tinfoilbarbs Barbodesschwanenfeldi [J]. Journal of Fisheries of China, 2012, 36 (1): 134-142.

[30] 李想, 林小涛, 宋波澜, 等. 流速对红鳍银鲫幼鱼游泳状态的影响 [J]. 动物学杂志, 2010, 45 (2): 126-133.

Li X., Lin X. T., Song B. L., et al. Effects of Water Velocities on Swimming Performances of Juvenile Tinfoil Barb Barbodesschwanenfeldi [J]. Chinese Journal of Zoology, 2010, 45 (2): 126-133.

[31] 宋波澜, 林小涛, 许忠能. 逆流运动对多鳞四须鲃幼鱼活动代谢及体内盐酸诺氟沙星残留的影响 [J]. 淡水渔业, 2012, 42 (6): 3-7.

Song B. L., Lin X. T., Xu Z. N.. Effects of upstream locomotion training on metabolism and norfloxacin hydrochloride residues of juvenile tinfoil barbs Barbodesschwanenfeldi [J]. Freshwater Fisheries, 2012, 42 (6): 3-7.

[32] 刘海生, 曹振东, 付世建. 水流刺激对宽鳍鱲幼鱼的游泳和代谢的影响 [J]. 重庆师范大学学报, 2015, 32 (1): 35-40.

Liu H. S., Cao Z. D., Fu S. J.. Effects of flow stimulation on the swimming performance and metabolism in Juvenile Zacco platypus [J]. Journal of Chongqing Normal University, 2015, 32 (1): 35-40.

[33] Lermen C. L., Lappe R., Crestani M., etal. Effect of differenttemperature regimes on metabolic and blood parameters of Silvercatfish (Rhamdiaquelen) [J]. Aquaculture, 2004, 239 (6): 497-507.

[34] 杨贵强, 徐绍刚, 王跃智, 等. 温度和摄食对溪红点鲑幼鱼呼吸代谢的影响 [J]. 应用生态学报, 2009, 20 (11): 2757-2762.

Yang G. Q., Xu S. G.,, Wang Y. Z., et al. Effects of water temperature and feeding on respiratory metabolism of juvenile Salvelinusfontinalus [J]. Chinese Journal ofApplied Ecology, 2009, 20 (11): 2757-2762.

[35] 汪文选, 卢彤岩, 王荻, 等. 两种水温条件下恩诺沙星在鲫鱼体内的药动学比较 [J]. 黑龙江畜牧兽医, 2010, 73 (3): 129-131.

Wang W. X., Lu T. Y., Wang D., et al. Both temperature conditions Pharmacokinetics of enrofloxacin in fish body [J]. Heilongjiang Animal Science And veterinary Medicine, 2010, 73 (3): 129-131.

[36] 陈锦云, 陈玉翠. 温度对瓦氏黄颡鱼幼鱼氨氮排泄的影响 [J]. 水产科学, 2006, 25 (5): 232-235.

Chen J. Y., Chen Y. C.. Effects of water Temperature on Ammonia – N Excretion and Body Compositions inPelteobagrusvachelli [J]. Fisheries Science, 2006, 25 (5): 232-235.

[37] Leung K. Y., Chu J cw, Wu R ss. Effects of body weight, water temperature ation size on ammonia excretion by the areolated grouper (Epinephelusareolatus) and mangrove snapper (Lutjanusargentimaculatus) [J]. Aquaculture, 1999, 170 (4): 215-227.

[38] 闫茂仓, 单乐州, 谢起浪, 等. 温度、盐度及体重对条石鲷幼鱼耗氧率和排氨率的影响 [J]. 海洋科学进展, 2008, 26 (4): 486-496.

Yan M. C., Shan L. Z., Xie Q. L., et al. Influence of Temperature, Salinity and Body Weight on Oxygen Consumption and Ammonia Excretion of Oplegnathusfasciatus Juvenile [J]. Advances in Marine Science, 2008, 26 (4): 486-496.

[39] 闫茂仓, 单乐州, 邵鑫斌, 等. 温度及体重对鮸鱼幼鱼耗氧率和排氨率的影响 [J]. 热带海洋学报, 2007, 26 (1): 44-49.

Yan M. C., Shan L. Z., Shao X. B., et al. Influences of temperature and weight on respiration and excretion of Miichthymiiuyjuvenile [J]. Journal of tropical Oceanography, 2007, 26 (1): 44-49.

[40] 徐绍刚, 杨贵强, 王跃智, 等. 温度对溪红点鲑耗氧率、排氨率和窒息点的影响 [J]. 大连水产学院学报, 2010, 25 (1): 93-96.

Xu S. G., Yang G. Q., Wang Y. Z., et al. Effects of water temperature on oxygen consumption, ammonia excretion and suffocation point in brook trout Salvelinusfontinalus [J]. Journal of DaLian Fisheries University, 2010, 25 (1): 93-96.

[41] 陈松波, 范兆廷, 陈伟兴. 不同温度下鲤鱼呼吸频率与耗氧率的关系 [J]. 东北农业大学学报, 2006, 37 (3): 352-356.

Chen S. B,. Fan Z. T., Chen W. X.. The relationship of respiratory rate and oxygen consumption rate in common carp (Cyprinus (C.) carpiohaematopterusTemminck et Schlegel) under different temperature [J]. Journal of Northeast Agricultural University2006, 37 (3): 352-356.

[42] Dalvi R. S., Pal A. K., Tiwari L. R., etal. Thermal tolerance and oxygenconsumption rates of the catfish Horabagrusbrachysoma (Günther) acclimated to different temperatures [J]. Aquaculture, 2009, 295 (8): 116-119.

[43] 王辉, 强俊, 王海贞, 等. 温度与盐度对吉富品系尼罗罗非鱼幼鱼能量代谢的联合效应 [J]. 中国水产科学, 2012, 19 (1): 51-61.

Wang H., Qiang J., Wang H. Z., et al. Combined effect of temperature and salinity on energy metabolism of GIFT Nile tilapia (Oreochromisniloticus) juveniles [J]. Journal of Fishery Sciences of China, 2012, 19 (1): 51-61.

[44] 薛泽. 温度、盐度、PH 和饵料对两种海洋桡足类摄食和代谢的影响 [D]. 中国海洋大学, 2013.

Xue Z.. Effect of temperature, salinity, PH and algal species on the ingestion and metabolism of two marine copepods [D]. China Ocean University, 2013.

［45］李金兰，陈刚，张健东，等. 温度、盐度对卵形鲳鲹呼吸代谢的影响［J］. 广东海洋大学学报，2014，01（34）：30-36.

Li J. L., Chen G., Zhang J. D., et al. Effects of Temperature and Salinity on the Respiratory Metabolism of Derbio［J］. Journal of Guangdong Ocean University, 2014, 01（34）：30-36.

［46］孙陆宇，温晓蔓，禹娜，等. 温度和盐度对中华圆田螺和铜锈环棱螺标准代谢的影响［J］. 中国水产科学，2012，19（2）：275-282.

Sun L. Y., Wen X. M., Yu N., et al. Influence of water temperature and salinity on standard metabolism of Cipangopaludinacathayensisand Bellamyaaeruginosa［J］. Journal of Fishery Sciences of China, 2012, 19（2）：275-282.

［47］庄平，贾小燕，冯广朋，等. 盐度对中华绒螯蟹雌性亲蟹代谢的影响［J］. 中国水产科学，2012，19（2）：217-222.

Zhuang P., Jia X. Y., Feng G. P., et al. Effect of salinity on the metabolism of female Chinese crabs, Eriocheirsinensis［J］. Journal of Fishery Sciences of China, 2012, 19（2）：217-222.

［48］路允良，王芳，高勤峰，等. 盐度对三疣梭子蟹成熟前后呼吸代谢的影响［J］. 水产学报，2012，09（36）：1392-1399.

Lu Y. L., Wang F., Gao Q. F., et al. Effects of salinity on the respiratory metabolism of pre-and post-maturity swimming crab（Portunustrituberculatus）［J］. Journal of Fisheries of China, 2012, 09（36）：1392-1399.

［49］Gilles Boeuf, Pierre-Yves Le Bail. Does light have an influence on fish growth［J］. Aquaculture, 1999, 177（1-4）：129-152.

［50］王馨，王芳，路允良，等. 光照强度对三疣梭子蟹呼吸代谢的影响［J］. 水产学报，2014，02（38）：237-243.

Wang X., Wang F., Lu Y. L., et al. Effects of light intensity on the respiratory metabolism of swimming crab（Portunustrituberculatus）［J］. Journal of fisheries of China, 2014, 02（38）：237-243.

［51］Nguyen N., Stellwag E. J., Zhu Y.. Prolactin-dependent modulation of organogenesis in the vertebrate：Recent discoveries in zebrafish［J］. Comparative Biochemistry and Physiology-PartC：Toxicology&Pharmacology, 2008, 148（4）：370-380.

［52］Falcón J., Gothilf Y., Coon S. L., et al. Genetic, temporal and developmental differences between melatonin rhythm generating systems in the teleost fish pineal organ and retina［J］. Journal of Neuroendocrinology, 2003, 15（4）：378-382.

第六章 渔药使用的风险评估

第一节 我国渔药生产、销售以及使用情况

渔药的安全使用与水产品质量安全有着密切的关系。它不仅影响水产品质量安全，还会影响到生态环境的安全，进而影响到人类的健康、行业的发展和社会的稳定（杨先乐等，2002；2013）。渔药在我国水产动物病害控制中起着举足轻重的作用，但我国在渔药基础理论研究、渔药市场规范以及渔药安全使用等方面还存在较多的问题，如乱用和滥用药物、缺少渔药评价的方法、法律和法规的缺失等（杨先乐等，2007）。目前我国渔药剂型还停留在以传统的粉剂、水剂等第一代剂型为主的阶段，发展渔药新剂型能提高药物生物利用度、降低毒副作用和提高药物稳定性（胡鲲等，2011）。某些商品性渔药制剂也存在一些问题，如渔用含氯消毒剂产品标示不完全、产品质量不稳定等。开发安全无公害的渔药制剂是渔药发展的一个方向（杨先乐等，2000）。

该文发表于《上海水产大学学报》【2007，16（4）：175-380】

我国渔药使用现状、存在的问题及对策

The application status and strategy of fishery medicine in China

（1. 上海高校水产养殖学 E-研究院，上海水产大学农业部渔业动植物病原库，上海 200090；
2. 西南大学水产学院，重庆 荣昌 402460）

摘要：概述了渔药在我国水产动物病害控制中的地位，渔药研发、管理和使用等方面的现状，同时指出了我国在渔药基础理论研究、渔药市场规范以及渔药安全使用等方面所存在的问题，并从加强渔药药理学等应用基础理论与高效低毒渔药及其制剂的研究，建立渔药的评价、研究与检测的基地，规范渔药的管理和使用等方面提出了相应的建议。
关键词：渔药；安全；管理；使用

药物防治是水产动物病害控制的三大措施之一，也是我国水产动物病害防治中是最直接、最有效和最经济的方式，因此在我国病害防治体系中受到普遍重视[1]。由于我国是世界上的水产养殖大国，养殖品种众多，养殖产量占全世界水产养殖总量的 70% 左右，因而我国也自然成为渔药生产、使用的大国[2]。我国渔药的种类较多，使用范围较广，由于使用不规范或渔药的滥用和错用，带来了诸多问题：某些渔药在水产品内的残留严重威胁了水产品安全、人民健康，影响了我国水产品的对外出口贸易；滥用渔药对环境的污染防碍了水产养殖的持续发展[3-4]。本文试图针对我国渔药使用现状提出一些粗浅的看法，旨在为水产养殖安全用药提供一些参考与借鉴。

1 我国渔药的使用现状

1.1 渔药在我国水产动物病害防治中占有重要地位

因为渔药来源广泛、生产简便、成本低廉、使用方便、疗效明显，是防治水产动物病害的首选途径，尤其对控制细菌性疾病，有较独特的效果[2]。宋晓玲等[5]研究认为至少有 15 种常用渔药对海洋弧菌具有杀灭作用，王雷等[6]通过体外抑菌试验，发现了 11 种天然渔药原料能抑制弧菌的繁殖和生长；吴后波等[7]发现磺胺类渔药能有效地控制哈氏弧菌（Vibio-Harveyi）引起的高体鰤弧菌病的发生，对该病的治愈率可达 75%。资料表明，全球兽用化学药物长期以来一直占据全部兽

药（包括疫苗）总销售额的85%以上[8]。据不完全统计，目前我国有专业性的渔药生产企业100多家，生产品种达500余种，渔药（指非生物性渔药）产量2.5万余吨，产值4亿元以上[9]。渔（农）民基本上靠渔药控制水产动物疾病。渔药在我国水产动物疾病的控制上有着不可替代的作用。

1.2　我国渔药的主要类别与特点

我国目前所使用的渔药主要有消毒剂、驱杀虫剂、水质（底质）改良剂、抗菌药、中草药5大类。以产量估算，其中消毒剂约占35%，抗菌药、中草药以及其他类渔药只占20%左右；以产值估算，消毒剂约占30%，驱杀虫剂、水质（底质）改良剂分别约占20%，其他渔药占30%左右[10]。

消毒剂的原料大部分是一些化学物质。生石灰是一种传统的消毒剂，使用较为普遍，除此之外，使用量较大的还有含氯消毒剂（如漂白粉、三氯异氰尿酸、二氧化氯等），含溴消毒剂（如溴氯海因、二溴海因等），含碘消毒剂（如聚维酮碘、双链季铵盐络合碘等）。其他类型的消毒剂，如醛类消毒剂（如甲醛、戊二醛等）、酚类消毒剂也有一定的应用。消毒剂可无选择地杀灭水体中的各种微生物，包括细菌繁殖体、病毒、真菌以及某些细菌的芽孢，但均会对水产动物产生一定的刺激，对环境造成一些不利影响[2]。

驱杀虫渔药一般具有较广的杀虫谱，对寄生于水产动物体表或体内的中华鳋、锚头鳋、鱼鲺、车轮虫、三代虫、指环虫、绦虫以及水中的松藻虫、水蜈蚣等均有较好的杀灭效果。这类渔药包括有机磷类、拟除虫菊酯类、咪唑类、重金属类以及某些氧化剂等，它们的驱杀虫方式主要是触杀和胃毒。其中敌百虫、硫酸铜、溴氰菊酯、氯氰菊酯、高效顺反氯氰菊酯、甲苯咪唑、苯硫咪唑等是我国常用的驱杀虫渔药[11]。

由于人们认识到水质和底质的优良对水产动物疾病的发生与否有着非常密切的关系，因此水质（底质）改良渔药的使用量逐年增加。这类渔药除了一些化学物质外（如沸石、过氧化钙等），较大部分是一些微生态制剂[12]。目前使用的微生态制剂主要是光合细菌类、芽孢杆菌类，乳酸菌类和酵母类的一些微生物制备的活性制剂。

抗菌类渔药是用来治疗细菌性传染病的一类药物，它对病原菌具有抑制或杀灭作用。从这类渔药的来源上可以分为天然抗生素（如土霉素、庆大霉素等），半合成抗生素（如氨苄西林、利福平等），以及人工合成的抗菌药（如喹诺酮类、磺胺类药物等）[11]。目前抗菌类渔药面临着产生负面效应、可能导致在水产品中的残留以及耐药性等问题[13-14]。

中草药渔药毒副作用小，不易产生耐药性，已作为渔药成份广泛使用。渔药中常用的中草药有大黄、黄柏、黄芩、黄连、乌桕、板蓝根、穿心莲、大蒜、楝树、铁苋菜、水辣蓼、五倍子、菖蒲等。从中草药提取有效杀虫、杀菌活性物质已取得了一定的进展[11]。

1.3　渔药使用安全是水产养殖的一个重要问题，已引起社会高度关注

水产品的药物残留问题已经引起了社会的普遍关注，要控制水产品的药物残留，保证水产品的安全，就必需要重视渔药的安全使用与科学管理。近几年先后出现的氯霉素、恩诺沙星、孔雀石绿等涉及到渔药使用安全的事件是长期不科学使用渔药所引起的矛盾的集中暴发[15]。为了控制渔药在水产品中的残留，保障水产品的安全，国家发布了一系列标准、法规和条例，并从2000年起开始对我国水产品中的渔药残留进行抽检，同时从源头抓起，加强对渔药的生产、销售和使用的管理[16]。

1.4　我国渔药使用管理的体系逐步完善，渔民规范用药的习惯正在形成

2004年新的《兽药管理条约》正式实施，推动了渔药规范使用和管理的进程[17]。根据该条约，各级渔业行政主管部门在一年多的时间内为规范用药做了大量的工作，如科学用药的宣传，水产养殖用药纪录的推广，药物残留在水产养殖动物体内的控制等，使我国渔药的使用已向规范的道路上迈出了坚实的一步。据"渔用药物代谢动力学及药物残留检测技术研究"课题组统计，由于向山东、河北、湖北、江苏、浙江、上海、海南等省市10余个渔区（渔场）的渔民宣传、指导水产养殖合理、规范用药，使这些渔区的养殖产品的无公害产品率达到100%，并取得了巨大的经济效益和社会效益。据10余个养殖示范区统计，其累计增收3.125亿元，取得间接经济效益41.284亿元。

2　我国渔药使用中所面临的问题

2.1　渔药药理学等基础理论滞后，导致渔药使用徘徊一个较低的水平

我国对渔药药动学的研究起始于20世纪90年代末，先后对氯霉素[18]、环丙沙星、诺氟沙星[19]、呋喃唑酮、土霉素、复方新诺明、噁喹酸、红霉素、喹乙醇和磺胺二甲嘧啶等[20]渔药在罗非鱼、中华绒螯蟹、南美白对虾、大黄鱼、鳗鲡、草鱼、鲤鱼及中华鳖等水生动物体内的代谢动力学和残留消除规律进行了研究，并比较研究了给药方式、给药剂量、种属差异、温度、盐度、性别、年龄等因子对药动学的影响，制定了一些渔药的最高残留限量及其相应的休药期。但离社会发展

的要求还存在较大差距，如研究的药物种类少，特别是对药物的代谢产物的研究尚未涉及；研究的水产动物种类有限；对病理模式下的药动学规律尚不清楚；对影响药动学和药效学的因素的研究还很薄弱；对中草药在水产动物机体内的药动学研究还鲜见报道；对药物代谢机制的研究，如南美白对虾、克氏螯虾中口服给药后存在的"首过效应"；南美白对虾口服诺氟沙星药饵、中华绒螯蟹口灌环丙沙星药液后，出现的"多药峰"现象均还停留在描述、推论阶段；在细胞和分子水平上建立水生动物药动学研究还在摸索中。这样使得我国在制定残留限量、休药期、给药剂量及用药规范等方面资料比较缺乏，导致渔药的使用存在着很大的盲目性。

2.2 有效安全的渔药及其剂型匮乏，制剂的工艺水平低

渔药企业是我国渔药开发的主体，由于企业强调急功近利，忽视基础研究，导致我国渔药的科技含量低，某些渔药虽然价格低、效果明显，但对人体和环境危害较大。目前我国大部分渔药来源于农药、人药或兽药，至今尚未形成自水产养殖专用的渔药系列[21]。据调查，其中驱杀虫渔药原料来于剧毒农药或农用化学药品的占80%以上，抗菌类渔药的原料与人、兽药成分同源的不低于70%。此外，对禁用渔药替代品的研究未能及时跟上，以致禁用渔药屡禁不止。剂型与制剂的多样性与其用途的专一性，是药物发展的重要标志之一。渔药的剂型和给予方式不同，机体吸收渔药的速率和数量不同，药效也发生差异，不少氧化剂由于没有合适剂型，极易失效，甚至发生爆炸造成意外[22]。

2.3 渔药市场无序竞争较为严重

2005年是我国渔药生产企业大调整的一年。渔药企业的GMP认证，渔药的地标升国标工作，虽促进了渔药生产企业的整合、调整和大洗牌，为渔药的规范生产创造了条件，但不同程度地使渔药市场出现了一些混乱[16]。某些无意认证的企业，不甘心"关、停、并"，纷纷借势转为经营企业，但仍以生产企业的模式运作；产品批准文号在交替之际，为某些渔利者带来了可乘之机；产品良莠不齐、以次充好的现象累见不鲜，破坏了诚实守信、守规经营的市场游戏规则；渔药产品的合格率得不到保证，渔药使用中出现的死鱼事件增加，水产品质量的保证也受到了一定的冲击[23-24]。据华东地区不完全调查，2005年因渔药本身原因所造成的死鱼案件就有60余起，所造成的直接经济损失达1 000万元以上[25]。

2.4 安全、科学用药的普及有限，乱用、滥用现象严重，威胁着水产品的安全

以专业户占主导地位的我国水产养殖行业，养殖技术水平还处在一个较低的水平，特别是在渔药使用上，一旦养殖品种患病，没有较好的对策，往往盲目用药，导致了严重的安全隐患。水产动物体内的渔药残留已经严重影响了水产品的质量安全，而渔药残留的产生无不与渔药的不规范使用紧密相关，据调查，我国渔药残留产生的原因主要是渔民规范用药、安全用药的意识差，主要表现在：①不遵守休药期有关规定或者缺乏休药期的意识；②不正确使用药物，使用渔药时，在用药剂量、给药途径、用药部位和用药动物的种类等方面不符合用药规定；③按错误的用药方法用药；④不做用药记录；⑤上市前使用渔药；⑥由于对水产动物疾病及其防治缺乏认识，疾病发生后乱用药、乱投药。渔药的滥用，也会对生态环境产生较大的毒副作用[26]。Samuelsen[27]的研究发现在海湾渔场使用土霉素后，淤泥中的土霉素在较低浓度下能存在很长时间（半衰期为87~144 d）；Vanloveren等[28]发现鱼类病毒性疾病的死亡率与污染物导致的免疫抑制有关，多氯联苯类化学物在很低浓度时会对鱼表现出免疫毒性作用。

2.5 渔药的滥用，导致病原体耐药性的增加

自1945年磺胺药成功地应用于治疗鳟鱼疖疮病以来，土霉素、卡那霉素、噁喹酸、氟甲喹等相继在鱼类中应用，化学药物治疗成为防治细菌性鱼病的重要手段[29]。但养殖生产中为了使治疗达到快速有效的结果，在药效不明显的情况下，往往过量用药，使用浓度是规定用量的3~5倍甚至更多[30]。由于渔药的使用范围和剂量加大，使用频繁，养殖水体病原体的耐药性问题日趋严重，国际上认可的鱼用氟尔康刚刚推出市场不久，其用量就已增加到常规用量的几倍；据对从欧洲鳗皮肤溃疡处分离出的嗜水气单胞菌的耐药性结果的测定，对诺氟沙星耐药的菌株率达60.0%，最小抑菌浓度（MIC）高达128 μg/mL，是使用浓度数十倍[31]。我国福建沿海危害鳗鲡的伪指环虫，已对敌百虫、甲苯咪唑等药物高度耐药，出现了无药可治的局面。随着耐药菌的大量出现，抗菌药的研制速度已无法解决日趋复杂的耐药性问题[32]。

2.6 渔药的给予方法所造成的负面影响尚缺乏有效的解决途径

渔药的给予大部分是要通过水媒体，其方式有口服、泼洒、浸浴和注射。当水产动物患病，食欲下降甚至不摄食时，口服法就不能实施，且该法不可避免在水中有部分的溶失；泼洒是渔药给予最常用的方法，但它不仅会对水产动物造成较大的应激，而且会对环境造成不利的影响；浸浴和注射因要捕捞水产动物进行处理，在应用时有较大的局限。给予方法所产生的负面效应，在一定程度上不仅不能有效地控制水产动物的疾病，而且还会对水产品的质量和环境的安全带来一些不利的影响[11]。值得提出的是，我国有较多养殖区还采用向水体里大量泼洒抗生素、泼洒农药等化学物质的方法治疗水产动物疾病，已对水域环境的安全造成了威胁[33]。

2.7　缺少渔药评价的方法和平台

正确、全面的药效评价体系不仅关系到渔药的研制与开发，而且关系到健康养殖和水产品安全。目前国内在评价药效时较片面，如对抗生素药效的影响因素研究较少，国内仅宋晓玲[34]等在体外测定了水中有机质含量对几种渔药最低抑菌浓度（MIC）和最低杀菌浓度（MBC）影响。大多数免疫刺激剂的评价标准、检测指标难以确定、检测手段还较落后。中草药由于所含成分较多，且许多化学结构不明，其不同的作用可由不同的成分产生，或某一作用是若干成分共同作用的结果，因此关于药效的评价受到了很大的局限，基本上停留在检测 MIC（MBC）和抑菌浓度指数（FIC）上[35]。国外对渔药毒理学的研究比较细致深入，涉及到毒理学各个研究范畴，如美国、加拿大、日本、以色列、英国等学者分别研究了硝酸银、林丹、马拉硫磷、敌敌畏等多种渔药，发现它们中许多渔药都具有潜在性毒害[8]，而我国很多渔药都缺乏严格而较全面的毒理学数据，如目前在水产上被批准使用的有机磷类、有机氯类、菊酯类渔药、重金属盐类化合物及中草药鲜有特殊毒理、水域生态毒性方面的研究，无法对其进行正确的安全性评价[36]。

3　我国渔药安全使用的建议

3.1　摆正药物防治的位置，发挥渔药在水产动物疾病控制中的积极作用

水产动物疾病的防治方法有药物防治、免疫防治和生态防治等，此3种方法各有其利弊。一般来说，药物防治无论在任何时期，都是疾病防治的一个重要和不可忽视的手段。我们强调水产养殖的安全，提倡健康养殖，并不是否认药物防治的作用。但我们所强调的药物防治，是要在保障水产品安全，环境安全的前提下，以控制用药、安全用药和高效用药的措施，提高水产动物疾病的防控水平[1]。

3.2　加强渔药安全使用的宣传、教育和培训工作，大力推行渔药的安全使用

《兽药管理条例》是渔药安全使用的基本法规，各级政府要统一组织，统一领导，互相配合，分工合作，围绕《兽药管理条例》及其配套的条例与规定进行广泛的宣贯，搞好执法人员、技术人员、养殖者的培训，使各项法规条例得以切实有效的落实；其次，兽医与渔业行政主管部门要对渔药的生产、销售和安全使用进行全过程的监督，加大执法力度，依法查处违规用药的案件，严格执行停药期规定，逐步建立完善水产养殖安全用药体系；第三，研究、开发和推广高效、速效、长效和对环境低污染、在鱼体内低残留的药物，将药物防治与水产动物的健康养殖、生态养殖和无公害养殖有机的结合[37]。

3.3　加强渔药药学的基础研究，为渔药的安全、合理使用提供依据

我国的渔药药学基础理论还相当薄弱，加强这方面的研究是渔药的安全使用的基础。当前应深化药动学、药效学研究，探讨渔药在水产动物体内的代谢规律，只有弄清渔药的疗效、毒性与渔药浓度之间的关系，渔药在体内的蓄积部位及蓄积程度，才能做到安全、合理地使用渔药，才能制定出合理的休药期，从而为临床安全和合理用药提供依据，为剂型的选择和新药的开发提供方向。此外，还应加强渔药对环境负面影响的研究，如渔药在水域环境中的蓄积、转移、转化；影响渔药降解的生物、物理、化学因素。加强环境修复技术的研究，1994年就有美国学者以枯草杆菌、地衣杆菌、多粘杆菌、假单孢菌等制成的系列制剂，用于水产废物分解；光合细菌能吸收分解水中的氨、氮、硫化氢等有害物质，具有很高的[38]水质净化能力。

3.4　建立渔药的评价方法

渔药的评价主要偏重于药效学和毒理学方面的评价。药效评价常以治愈率为检测指标，但易受到较多因素的影响，较难界定。离体测定是通常采用的方法。对抗生素药效的评价指标除 MIC、MBC 外，McDonzld 等[39]提出抗菌后效应（postantibioticeffect，PAE）是较确切地评价药效的一个指标。处于 PAE 期的细菌再与亚抑菌浓度（sub-MIC）渔药接触后，细菌的生长将受持续抑制，即产生抗菌后亚抑菌浓度效应（postantibioticsub-MiCeffect，PASME）。研究发现，处于 PAE 期的细菌对渔药敏感性提高，PAsME 的作用比相应的 PAE 及亚抑菌浓度作用（sub-MiCeffect，SME）大，甚至可以杀死细菌。刘涤洁等[40]关于恩诺沙星和环丙沙星对金葡菌的抗菌后效应及抗菌后亚抑菌浓度效应的研究，刘远飞等[41]关于单诺沙星和恩诺沙星对大肠杆菌和金葡球菌的抗菌后效应的研究均证实了这一点。除此之外，对渔药的评价还可参考 JECFA 的数据和相关资料。

3.5　加强对微生态制剂、免疫刺激剂、生物渔药及中草药等渔药的研究和开发

微生态制剂以其安全、低毒、有效正受到水产养殖者的重视。赵亮等[42]研究了对中华绒螯蟹水泥池全封闭育苗系统水

927

体添加光合细菌,实验组比对照组分别使氨氮下降41.5%、亚硝态氮下降41.4%,可使幼体成活率、单位产量和总出苗量提高20%以上。微生态制剂将会克服有益菌群"定植"、生产工艺提高等难题,利用生物工程技术,朝着高效、专一,益生菌和益生元相结合的方向发展。

免疫刺激剂是通过作用于非特异性免疫因子来提高水产动物的抗病能力的一种比化学药物安全性高,比疫苗应用范围广的特殊渔药。研究证实,一些富含多糖、生物碱、有机酸等多种成分的天然免疫物质,如蛋白质、氨基酸、高度不饱和脂肪酸、维生素和矿物质等都对水生动物的免疫功能具有显著的影响,如低聚糖、壳聚糖磺酸酯、几丁质、左旋咪唑等。随着对它们的使用方法、剂量、评价体系以及对水产动物免疫机制的研究进一步深入,免疫刺激剂将会在控制水产动物疾病上发挥出重要的作用。

生物渔药是根据生物之间相互依赖、相互抑制或相互竞争的关系,根据其生理特点或生态习性,抑制或消灭病原体的一种"生物制剂",如蛭弧菌[44]、噬锚头蚤的生物等。生物渔药将会推动我国渔药的研制的新思路和发展。

中草药具有来源广泛、使用方便、价廉效优、毒副作用小、无抗性、不易形成渔药残留等特点,具有广阔应用前景。中草药的发展方向是:①利用现代技术分离提取其有效成分及先导物,降低提取成本;②根据有效成分合成系列衍生物 42 或类似物,开发出人工合成的"中草药";③中草药作用靶点的研究,弄清中草药的作用机制;④中草药的细胞破壁技术;⑤中草药合理配伍的研究等。

3.6 加强新型渔药及其剂型、制剂的研究

研制窄谱性渔药,水产专用渔药,新型消毒剂,"三效三小"渔药(即"高效、速效、长效"与"毒性小、残留小、用量小")等是新型渔药的研制重点,新渔药的研究应该多来源、多途径、多方向、多思路。由于渔药药效受外界因子影响显著,应根据水产动物的种类和规格、发病类型及程度、渔药的性质研制出不同的渔药剂型。如运用新技术、新材料减少渔药在到达靶器官前的损耗,降低毒副作用;应用高分子材料制成的微胶囊剂将渔药包裹其中,避免渔药在环境中降解破坏,提高渔药的有效吸收;根据水产动物的食性和相应的诱食剂制成某种特殊的剂型进行给予;选用卤虫等活生物饵料作为载体将渔药直接输送到动物体内;研制出缓控、缓释、靶向制剂等。

3.7 建立渔药研究、检测的基地

从当前我国渔药安全使用的严峻形势和现状来看,建立高水平的渔药研究基地已经刻不容缓。目前我国关于渔药研究、检测基地在数量和质量上尚不能满足我国水产养殖发展的需要,建议相关职能部门在原有的基础上,建立相应的渔药检测研究基地,根据水产养殖生产实践中反映的问题开展针对性的工作,确保渔药的质量和使用安全。

参考文献

[1] 杨先乐,黄艳平.我国渔药管理工作刍议(上),(下)[J].科学养鱼,2002,(11):41-42.

[2] 郑宗林,向磊.我国水产药物使用现状分析[J].水产养殖,2002,(4):36-39.

[3] 黄艳平,杨先乐,湛嘉,等.水产动物疾病控制的研究和进展[J].上海水产大学学报,2004,13(1):60-66.

[4] 熊清明,杨先乐.氯霉素人工免疫原的合成与鉴定[J].上海水产大学学报,2004,13(3):279-282.

[5] 宋晓玲,张岩.常见水产药物对海洋弧菌的杀灭作用[J].海洋科学,1996(4):9-11.

[6] 王雷,李光友.11种天然药物对弧菌的体外抑制效果的观察[J].海洋与湖沼,1995,26(3):338-340.

[7] 吴后波,潘金培.海水网箱养殖高体弧菌病药物治疗的实验研究[J].热带海洋,2000,19(1):85-89.

[8] 张继瑜.欧盟和日本的兽药管理体制[J].新兽医,2005,(6):11-13.

[9] 王新春,郭键.水产养殖用药目前存在的问题及建议[J].黑龙江水产,2005,(2):13,15.

[10] 王俊菊.我国兽药产品中存在的问题及今后的发展方向[J].猪业在线,2005,(2):64-65.

[11] 农业部《新编渔药手册》编撰委员会.新编渔药手册[M].中国农业出版社,2005:318.

[12] 李健.我国新渔药研制开发与使用发展趋势[J].现代渔业信息,2000,15(8):6-8.

[13] 闫茂仓.当前鱼药中使用的问题[J].齐鲁渔业,2005,(22):2.

[14] 严天鹏.南美白对虾夏季养殖中常见病害及防治措施[J].渔业致富指南,2005,(10):49.

[15] 欧阳林山.浅析我国兽药监督抽检中存在的问题及建议[J].中国兽医杂志,2005,39(2):3-5.

[16] 殷文斌.改进渔用药物市场监管机制的几点思路[J].农业装备技术,2005,(31):3.

[17] 王泰健.我国兽药监管现状及改进建议[J].中国牧业通讯,2005,(8):10-15.

[18] 李爱华.氯霉素在草鱼和复合四倍体异育银鲫体内的比较药代动力学[J].中国兽医学报,1998,18(4):372-374.

[19] 陈文银,印春华.诺氟沙星在中华鳖体内的药代动力学研究[J].水产学报,1997,21(4):434-437.

[20] 王群,马向东,李绍伟.HPLC法研究磺胺类药物在水产动物体内的代谢和残留[J].海洋科学,1999,6:33-36.

[21] 战文斌,刘洪明,王越.水产养殖病害及其药物控制与水产品安全[J].中国海洋大学学报,2004,34(5):758-760.

[22] 包永胜，陈国兔. 对水产养殖用药监控的看法 [J]. 渔业致富指南，2005，（10）：48.

[23] 郭勇，刘东朴. 水产养殖用药现状及管理对策 [J]. 河北渔业，2003，（5）：14-15.

[24] 戚小伟，任浪，韦俊义. 兽药经营中存在的违法行为及对策 [J]. 四川畜牧兽医，2005，（5）：20.

[25] 李兆新，李晓川，付卫新，等. 我国渔药与渔药残留监控 [J]. 齐鲁渔业，2004，21（7）：23-25.

[26] 王奇欣. 新形势下对渔药管理工作的几点建议 [J]. 中国水产，2005，（5）：6-7.

[27] Samuelsen O. B. . The fate of antibiotics/chemotherapeutics in marine aquaculture sediments. In Chemotherapy in aquaculture from the orytoreality [J]. Aquaculture，152：13-24.

[28] VanloverenH，PSRoss，ADMEOsterhaus，et al. Contaminant-inducedimmunosuppressionandmassmortalitiesamongharborseals [J]. Toxicologyletters，2000，112：319-324.

[29] WollenbergerLB，Halling-Soerensen，KOKusk. Acute and ChronicToxicity of Veterinary Antibiotics to Daphniamagna [J]. Chemosphere，2000，40（7）：723-730.

[30] 赵长志. 浅析动物性食品药物残留危害及对策 [J]. 吉林畜牧兽医，2005，（8）：53.

[31] 樊海平，曾占壮. 欧洲鳗病原菌控菌药物的研究 [J]. 台湾海峡，2000，19（2）：228-232.

[32] 叶雪珠，赵燕申，王小丽. 水产品中药物残留的检测与控制 [J]. 水利渔业，2004，（24）4：19-20.

[33] 肖培弘. 水产养殖药物残留的危害及监控措施 [J]. 畜牧兽医科技信息，2005，（02）：9-11.

[34] 宋晓玲，张岩. 4株淡水鱼致病菌对常用渔用药物的药物敏感性研究 [J]. 海洋水产研究，2000，21（1）：47-51.

[35] 宋宏晓，林霖. 苦参的药理作用及在畜禽养殖业中的应用 [J]. 当代畜禽养殖业，2005，（5）：46-47.

[36] 崔效亮，薛克友，王玉莲，等. 我国兽药研究开发的现状及发展趋势 [J]. 中国兽药杂志，2005，39（7）：16-19.

[37] 高文学. 兽药、饲料经销商的路该怎样走———思考与对策 [J]. 饲料博览，2005，（3）：20-22.

[38] 陈世阳，席振东. 光合细菌的特性及其开发利用 [J]. 中国微生态学杂志，1990，2（3）：78-84.

[39] 赵香兰. 临床药代动力学基础与应用 [M]. 郑州大学出版社，2002：82-109.

[40] 刘涤洁，陈杖榴，冯淇辉. 恩诺沙星和环丙沙星对金葡菌的抗菌后效应及抗菌后亚抑菌浓度效应 [J]. 中国兽医学报，2002，22（3）：276-278.

[41] 刘远飞，佟恒敏，韩建春. 单诺沙星和恩诺沙星对大肠杆菌和金葡球菌的抗菌后效应 [J]. 中国兽医杂志，2003，39（7）：19-20.

[42] 赵亮，万全，孙德祥. 光合细菌对河蟹全封闭育苗系统水质的影响 [J]. 淡水渔业，2004，34（1）：28-29.

[43] 张海霞. 我省渔民转产转业培训工作取得新进展 [J]. 河北渔业，2005，（3）：54-55.

[44] 杨莉，王秀茹. 蛭弧菌对鲤感染嗜水气单胞菌预防效果的观察 [J]. 大连水产学院学报，2000，15（4）：288-292.

该文发表于《中国渔业质量与标准》【2013，3（4）：1-6】

水产品质量安全与渔药的规范使用

Food-safety and standard use of fishery drugs on aquatic products

杨先乐，郭微微，孙琪

（上海海洋大学国家水生动物病原库，上海 201306）

摘要： 中国是水产品生产和消费大国，水产品质量安全关系到国计民生。然而，近年水产品质量安全问题层出不穷，主要表现在药物残留超标、环境污染、有害微生物和重金属富集等方面。其中，渔药的规范使用与否与水产品质量安全有着密切的关系。它不仅影响水产品质量安全，还会影响到生态环境的安全，进而影响到人类的健康、行业的发展和社会的稳定。本文分析了中国渔药使用的现状，阐述了因渔药的滥用和乱用所导致的水产品质量安全的隐患，以及如何通过渔药规范使用和监管等一系列措施，最终确保水产品质量安全。

关键词： 渔药；水产品；质量；安全

中国是水产品生产和出口大国。自 1989 年以来，中国水产养殖得到了突飞猛进的发展，实现了由"捕捞增长型"向"养殖增长型"的转变[1]。2011 年水产养殖总产量达 4 023 万 t，占渔业总产量的 71.8%；出口总额达 177.9 亿美元，连续 11 年占农产品出口首位，且位居世界第一[2]。然而中国水产养殖在生产、运输、加工和监管过程中也暴露出大量问题，药物残留和重金属污染等问题逐渐成为影响水产品安全的重要因素[1]。随着生活水平的不断提高，人们对水产品的需求不断增加，对水产品质量要求也越来越高，水产品的质量安全问题已成为全社会普遍关注的焦点。

1　水产品质量安全牵动社会的神经

近年来，随着水产养殖业的深入发展，水产品质量安全越来越受到重视，各级政府在水产品质量安全管理方面采取较多措施，因此，总体上中国水产品的质量还是比较安全的。然而，伴随着渔业的快速发展，涉及水产品安全的诸多事件频繁发生，备受公众、媒体及政府部门等多方面的关注，如“出口欧盟对虾氯霉素事件”、“出口香港大闸蟹氯霉素事件”、“出口日本鳗鲡恩诺沙星事件”和“水产品中孔雀石绿事件”等。渔药的安全使用问题，是其矛盾集中爆发的原因之一。与水产品有关的食品安全事件严重威胁了人民身体健康，影响了公众对水产品安全的信心和水产品的对外出口贸易。提高水产品质量安全管理水平，建立完善的养殖水产品质量安全管理体系，已经成为促进中国渔业健康可持续发展的当务之急[3]。

1.1 中国水产品质量安全存在的问题

1.1.1 药物残留超标，违规使用禁用药品

为了片面追求眼前的经济利益，在养殖过程中使用添加含有激素类的饲料；滥用抗生素和违禁药物防治水产动物疾病；捕捞前不执行渔药的休药期制度；水产品生产、销售和加工过程中，存在使用孔雀石绿和氯霉素等禁用药物的违法行为；违规使用药物、水质改良剂、消毒剂、保鲜剂和防腐剂等；使用具有禁药成分而未列入禁药清单的渔药，导致药物残留超标[4]等。

1.1.2 养殖水环境污染

养殖水域的环境容易受到外界影响进而引发水产品质量安全问题。例如工矿企业废水和城镇生活污水及畜禽养殖用水未经处理或处理不彻底，导致养殖水体富营养化程度加剧；含有汞、镉、铬、铅及砷等生物毒性显著的重金属元素及其化合物通过食物链和生物富集作用，在水生生物体内积累，造成水产品重金属污染。如2007年太湖无锡水域暴发的蓝藻污染事件，不仅造成了养殖鱼类大量死亡，而且导致了饮用水危机；2012年广西龙江河的镉污染事件造成下游河水中的鱼大面积死亡，这些严重的水污染事件无一不给水产品安全带来了隐患[5]。

1.1.3 致病微生物与寄生虫等致病体对人体的感染

水产品中的生物危害主要是对人有危害的致病体，如病毒、细菌和寄生虫等。某些水产品中残存的寄生虫和能够引起甲肝、霍乱、副溶血性中毒等的致病原极易导致人体的感染。副溶血弧菌是水产品中引起食物中毒的主要致病菌[6]。2006年北京福寿螺事件是由于人们食用生的或加热不彻底的福寿螺后，感染了广州管圆线虫，从而引起食物中毒。

1.1.4 毒性物质中毒

水产品中的某些种类体内含有毒素，误食后会引起不同的中毒反应。主要有海参毒素甙中毒，淡水鱼中胆汁的氰甙、胆盐中毒，裸鲤、鲶、鳇和石斑鱼等鱼卵毒素中毒，食用鲨鱼和鳕的肝脏过量中毒，以及河豚毒素、肉毒鱼毒素、螺类毒素、海兔毒素和西加毒素中毒等[7]。此外，由贝类体内所蓄积的含毒赤潮藻类所引起的获得性毒素也是水产品毒素的一个重要方面，因其毒性大、反应快、无适宜解毒剂，严重时可引起死亡，在预防和治疗方面十分棘手。麻痹性贝毒（PSP）和记忆缺失性贝毒（ASP）是贝毒的常见种类[8]。2004年香港、广东等地发生多起因食用进口石斑鱼而引起的西加毒素中毒事件，使水产品毒性物质的危害得到广泛关注[9]。

1.1.5 转基因食品安全问题

目前在海洋生物领域研究的热点问题之一——转基因技术的使用备受争议。将外来基因植入水产动物，一方面使技术成熟的转基因产品拥有了生长速度快、抗病害能力强和繁殖水平高的优势，另一方面也给水产品的安全性带来隐患：外来基因的供体和受体两者的自身安全引起的产物安全问题；引入外来基因后产生预期外的性状或有毒产物[10]；载体中具有抗生素耐药性的选择标记基因通过生物筛选作用产生耐药菌株等[11]。

1.1.6 鱼类致敏原导致过敏现象

含致敏原的水生生物常见的有虾、龙虾、蟹和贝类，人类误食了含有这些过敏原的水产品之后，发生过敏反应表现为手脸发红、水肿、特异性皮肤炎症、横纹肌溶解和毒性反应等，但一般不会引起死亡。过敏原存在于水产品的肉、皮和骨中，鱼胶制品也包含一定的过敏原[12]。由过敏原致敏带来的水产品安全问题不可小觑。

1.2 水产品质量安全的重要性

近年来各类水产品质量安全事件不仅给养殖户和相关的企业带来了不可估量的经济损失，还给中国水产品的对外贸易产生了诸多负面影响。在影响国家和政府形象和渔业发展的同时，也侵害了消费者的利益，更使得全社会对整个水产行业的质量安全问题产生了信任危机。2001年9月底，欧盟因氯霉素残留问题自2002年2月1日起全面暂停从中国进口动物制品，导致2002年上半年中国水产品出口下降70%以上。2006年11月，上海市公布30件多宝鱼药残抽检结果，30件样品中全部被检出含硝基呋喃类代谢物，部分样品还被检出孔雀石绿、环丙沙星、氯霉素和红霉素等多种禁用渔药残留，事发后对于多宝鱼养殖行业几乎带来了毁灭性的打击。2007年6月，美国食品药品管理局（FDA）称，美国将暂停从中国进口鮰、鲮、虾和鳗鲡等4种水产品，原因是从部分中国进口的水产品中发现了微量抗生素、孔雀石绿和龙胆紫等禁止使用的药物或添加剂，这一决定直接导致安徽省滁州市的养殖鮰的网箱数量由2007年的5 000只减少到如今的1 000只，减少量达80%[13]。2006年2月，江苏省南通地区约40%水产品受重金属污染，其中文蛤的镉含量超标，属严重污染，由此导致

江苏省沿海贝类产品信誉整体受损，大量贝类产品滞港[14]。这些触目惊心的水产品质量安全事件无时不刻不在提醒大家，切实保证水产品的质量安全是促进水产养殖业健康发展迫在眉睫的问题。

1.3　中国政府对水产品质量安全的监管与人民的需求

近年来，食品及食用农产品质量安全事件频频发生，从三鹿奶粉的三聚氰胺事件开始，每一次食品安全事件都牵动着亿万国人的神经，仅2011年中国便发生包括瘦肉精、染色馒头、牛肉膏、毒豆芽、塑化剂和地沟油等令人闻之色变的数十起重大食品安全事件。当前防范质量安全事件发生，提高质量安全水平已成为中国社会发展面临的最重要课题之一。

相比而言，尽管也曾发生贝类毒素中毒、氯霉素等药物残留检出等事件，总体上，中国水产品质量安全水平趋稳向好，尤其是在养殖环节上，近几年并未发生重大质量安全事件。这既要归功于中国水产养殖业生产方式转变和产业素质整体提升，又与水产品质量安全监管工作不断加强和监管体系不断完善密不可分。然而，客观上由于受千家万户的生产方式和生产基础条件限制，中国水产品质量安全方面还存在一定的问题，集中表现在药物和重金属的残留，而且制约中国水产品质量安全水平提高的一些深层次的问题也还未得到解决。如从业人员水产品质量安全意识淡薄，水产品从生产到销售缺乏既可靠又切合实际的管理等。

随着人民生活水平的进一步提高，中国民众的食品安全消费意识越来越高。近几年，食品安全一直和房价、贫富差距和反腐倡廉等问题并列为两会热点，而据《中国食品安全报》调查，目前逾九成民众对食品质量安全问题感到不满。目前，随着中国水产品产量的进一步提高，民众对水产品的消费需求也发生很大变化，逐渐从"吃得起"向"吃得好、吃得安全"转变。人们的生活离不开健康饮食，保障水产品质量安全是食品安全的前提，是人们健康安全生活的迫切需求。

中国是水产品生产和消费大国，养殖为主、捕捞为辅是中国为了满足国内外供需和保护生态环境所实行的策略。水产养殖业的不断发展振兴了水产动物苗种培育与饲料营养等副业，但忽略了乱用药物的弊端。水产动物养殖过程中能否安全、合理、规范化地使用渔药，直接关系到水产品的质量安全。

2　渔药的规范使用对水产品质量安全的影响

2.1　药物防治是目前控制水产动物疾病的主要措施

在水产动物养殖过程中，药物防治是水产动物病害控制的三大措施之一，也是中国水产动物病害防治最直接、最有效和最经济的方式，因此在中国病害防治体系中受到普遍重视[15]。渔药具有来源广泛、生产简单、成本低廉、使用方便和疗效明显等特点，是防治水产动物病害的首选途径，尤其对控制细菌性疾病，有较独特的效果。在中国水产养殖业中，渔药的使用占有重要地位。

2.2　目前中国渔药的使用状况

渔药的规范使用是水产养殖中的一个重要环节。在养殖过程中，由于养殖户对规范使用渔药的意识淡薄，使得渔药乱用现象屡见不鲜，而水产品的药物残留则直接威胁人体健康。

2.2.1　安全科学用药的普及程度有限，乱用和滥用药物的现象依然存在

以农业为依托、以农民养殖为主体的中国水产养殖行业，在养殖及管理过程中存在着不可避免的缺陷。特别是在渔药使用上，一旦养殖水产动物患病，在没有准确诊断之前就盲目用药，致使存有安全隐患。中国渔药残留产生的主要原因是渔民规范和安全用药的意识淡薄；不遵守休药期有关规定或者缺乏休药期的意识；使用渔药时，在用药剂量、给药途径、用药部位和用药动物的种类等方面不符合用药规定；用药方法错误，不做用药记录；对水产动物疾病及其防治缺乏认识，疾病发生后乱用药、乱投药。值得注意的是，中国有少数养殖区还采用向水体里大量泼洒抗生素和农药等化学物质的方法治疗水产动物疾病[16]，这一陋习不仅影响水产品质量安全，也对生态环境产生严重的负面影响。

2.2.2　渔药的滥用，导致病原体耐药性的增加

药物防治是渔民防治水产动物疾病的主要手段。实际养殖生产中，在药效不明显的情况下，有的养殖户为了片面追求用药效果，药物的使用浓度往往达到规定用量的 3~5 倍甚至更多；也有一些养殖户低剂量反复用药来预防疾病发生。由于渔药的使用范围和剂量加大以及低剂量反复用药，养殖水体病原体的耐药性问题日趋严重。如福建沿海危害鳗鲡的伪指环虫，已对敌百虫和甲苯咪唑等药物产生高度耐药性，出现了难以根治的局面。随着耐药菌的大量出现，抗菌药的研制速度缓慢，已无法解决日趋复杂的耐药性问题。

2.2.3　缺少渔药评价的方法和指导原则

药效评价体系的全面、正确与否不仅关系到渔药的研制与开发，而且关系到健康养殖和水产品安全。在评价药效方面，目前国内缺少较为规范的药效评价指导原则，如对抗生素药效的影响因素研究等。大多数免疫调节剂的评价标准、检测指标难以确定，检测手段还较落后[17]。中草药由于所含成分较多，且许多化学结构不明，其药效评价有一定的局限性。此外，中国较多渔药都缺乏严格而较全面的毒理学数据，如目前在水产上被批准使用的有机磷类、有机氯类、菊酯类渔

药、重金属盐类化合物及中草药鲜有特殊毒理和水域生态毒性方面的研究，无法对其进行正确的安全性评价[18]。

2.3 不合理使用渔药产生的水产品质量安全隐患危及人类健康

随着水产品需求增加，水产品的药物残留问题已经成为社会普遍关注的热点问题。近几年先后出现的氯霉素、恩诺沙星等涉及到渔药使用安全的事件，是长期没有科学使用渔药而引起矛盾的集中暴发[20]。人体内不断蓄积药物残留不仅会破坏胃肠道内微生物平衡，使机体容易发生感染性疾病，而且还会产生毒性作用、过敏反应和致畸、致癌、致突变作用等。另外，耐药菌株也能够随着水产养殖动物传播给人类，危及人类的健康和生命。

2.4 法律和法规的建立是渔药规范使用和确保水产品质量安全的前提

为了控制渔药在水产品中的残留，保障水产品的安全，中国发布了一系列标准、法规和条例，并从2000年起开始对中国水产品中的渔药残留进行抽检，同时从源头抓起，加强对渔药的生产、销售和使用的管理[21]。2004年颁布并开始实施的《兽药管理条例》，推动了渔药规范使用和管理的进程[22]。《食品动物禁用的兽药及其他化合物清单》（农业部193号公告）禁止氯霉素、孔雀石绿等29种药物使用，限制8种渔药作为动物促生长剂使用；NY5071—2002《无公害食品渔用药物使用准则》规定了呋喃类、喹乙醇等32种禁用渔药；一大批药物残留检测方法标准（包括国家标准、行业标准和地方标准）的制定或修订等，使中国渔药的使用有法可依，有章可循；《水产养殖质量安全管理规定》（2003）和《农产品质量安全法》（2006）建立了渔药残留检测和监控体系，强调了"从鱼塘到餐桌"的药残全过程控制管理[23]。这些基础法规的制定，在一定程度上促使中国在渔药使用的法制轨道上向前迈进了一大步，为今后渔药使用的规范化、标准化提供了基础依据。

3 规范渔药的使用，确保水产品质量安全

影响中国水产品质量安全有着诸多因素，如从业人员技术水平低，质量安全意识淡薄；渔用饲料和渔药的生产和使用不规范；养殖水域污染严重；缺乏有效的检测方法和手段，监管体系不完善；质量保障体系不健全等。这些因素相互影响，相互制约。渔药的规范使用，所涉及的范围较广，内容较多，需要对以上诸方面进行统筹考虑。现从确保水产品质量安全的角度考虑，提出如下建议。

3.1 全行业强化渔药的规范、合理和安全使用意识

3.1.1 广泛开展安全用药的培训和普及工作

提高渔民素质，强化渔民规范、安全和合理用药意识，是保证水产品质量安全的前提和基础。中国水产养殖主要以粗放式的散户养殖为主，较多渔民安全用药意识淡薄，缺乏合理用药的知识。针对这一特点，各级政府部门特别是农业、水产相关部门应当积极宣传有关的渔药使用管理法规和条例，普及相关的科学知识。采取多种形式、多种手段，切实提高广大养殖户和渔药生产企业以及从业人员对渔药正确使用问题严重性的认识[24]。向渔民普及合理使用渔药知识，举办安全用药相关讲座，以提高养殖经营者用药知识和对休药期的认识。

自2007年以来，在农业部渔业局等有关司局的大力支持下，全国水产技术推广总站与中国水产学会联合开展的水产养殖规范用药科普下乡系列宣传活动已初步取得成效，影响作用明显。基层技术推广人员和渔业生产者的水产品质量安全意识明显提高，水产品质量得到了保证[25]。

3.1.2 加强规范用药的指导和监督工作

从基层抓起，组织水产养殖技术推广人员、病害防治人员深入水产养殖场或养殖户，对渔业生产过程中渔药的使用进行指导，督促养殖场（户）严格按照水产养殖的相关规定和标准用药，强化安全用药意识，切实保障水产品质量安全。同时，政府应加大安全用药的整治力度，对氯霉素、孔雀石绿和硝基呋喃等违禁药物开展专项整治工作，加大监控力度，对养殖和加工过程中的违法用药现象严肃查处，严厉打击。

3.2 加强渔药的科学研究

3.2.1 重视渔药的基础研究

渔药的基础研究为渔药的安全使用提供了保障。渔药的应用基础研究，如药效学、药动学及毒理学的研究，将为制定合理的用药方案提供依据；弄清药物在水产动物体内的代谢过程及代谢产物，渔药在环境中的蓄积、转移和转化规律[23]，将为控制渔药在鱼体内的残留限量，严格设定休药期和给药剂量，倡导健康无公害养殖，保证水产品质量安全提供依据。

3.2.2 增加科研投入，壮大渔药研究队伍

研究力量薄弱，渔药研发不够深入是导致渔药产业技术落后的关键因素之一。目前中国的渔药主要来源于兽药及人药，而专门用于水产动物的渔药种类还很缺乏。政府部门应加大科研资金投入，建立渔药的研发和评价基地。高校和水产科研院所应当建立适应渔药管理的人才培养制度，培养出能够服务于基层实际生产需要的专门人才。

3.2.3　积极研究和开发安全、高效、实用的新型渔药

研制具有科技含量高、疗效显著、安全无副作用的"三效"和"三小"的新型渔药，如新型的养殖环境改良剂、免疫增强剂、微生态制剂、疫苗、诊断试剂和抗病毒中草药等[21]。此外还需加大对制剂和剂型的研究。

3.3　强化渔药生产和使用的管理

渔药生产者和使用者是关系到渔药能否规范使用的直接因素。加强渔药行业企业的管理力度，严格执行从业资格人员的选拔是解决这一问题的重要途径。

3.3.1　加强渔药生产和经营企业的管理

渔药生产企业和渔药经营企业要严格执行《药品生产质量管理规范》（GMP）和《兽药经营质量管理规范》（GSP），要严格执行准入制度。对于目前已经开办或者今后想要从事渔药经营的企业，必须首先通过 GSP 认证才有资格获取《兽药经营许可证》，才能从事相关的渔药经营活动，从源头严把渔药质量关[23]。

3.3.2　对渔药从业人员的管理

渔药行业相关的从业人员主要包括生产者、经营者和使用者。渔药行业的从业人员要逐步推进执业兽医资格考试准入制度。中国《动物防疫法》和《国务院关于推进兽医管理体制改革的若干意见》都对推行执业兽医资格考试制度做出明确规定，提出了具体要求。在编在岗的执法人员，经过考试考核后可转为官方兽医；对新进入动物防疫机构的人员，要求取得执业兽医资格[21]。这些从业人员要有良好的规范使用渔药的素质和技术，他们在推广和销售渔药的同时，能够正确引导养殖生产者在安全的前提下正确使用渔药。

此外，在中国推行渔药处方制度是确保渔药安全使用的一种行之有效的方法。由专业人员根据实际情况开具正确的处方，对渔药的种类、使用方法、用量和休药期等进行指导和监督；而渔药处方药的销售也以处方作为唯一依据，以凭方购药的方式规范渔药的使用，杜绝药物滥用现象。同时要求水产养殖及相关人员严格按照有关规定用药，做到不向无《兽药经营许可证》的单位购买渔药；不使用国家已公布的禁用药；不使用无批准文号、无生产厂家、无生产批号的药物；严格按照药品使用说明按规定剂量给药；确保水产品在停药期满后才能进入市场，并自觉接受渔业行政部门以及质量检测部门的检查监督。

3.3.3　建立和健全渔药监管与检测机构

建立和健全渔药监管与检测机构，提高对水产品药物残留检测水平，使渔药的使用严格按照法律程序管理。制定渔药使用的行业准则，定期对水产养殖场或养殖户进行用药情况检查，调查用药记录，对没有用药记录者要指导和监督做好用药记录；定期对养殖场或养殖户的水产品进行药物残留量检测，做好水产品安全性评定，确保水产品从苗种投放开始到进入水产品市场的全程安全监管，让水产品安全进入市场，到达人们的餐桌上。

4　小结

中国是水产品生产和消费大国，水产品生产和加工在国民经济中占有重要地位。水产品药物残留及其他不安全因素将严重影响其质量，降低水产养殖生产的效益，危害人类的健康，进而影响全行业的整体形象。近年来所暴露出的问题，虽然付出了惨重的代价，但是也为中国水产行业的发展和繁荣指明了方向。前事不忘，后事之师。为了提高中国水产品质量，减少水产品药物残留量，应从源头抓起，强化渔药的规范和合理使用，使中国水产品质量安全提升到一个更新、更高的水平。

参考文献

[1]　宋亮，罗永康，沈慧星．水产品安全生产的现状和对策［J］．中国食品卫生杂志，2006，18（5）：445-449.
[2]　孙志敏．中国养殖水产品质量管理问题研究［D］．青岛：中国海洋大学，2007.
[3]　万建业，陈小桥，汪银焰，等．我国水产品质量安全存在的问题与对策［J］．现代农业科技，2011（10）：357-359.
[4]　张雨，黄桂英，刘自杰．中国食品安全的现状及其与国外的差距［J］．中国食物与营养，2003（3）：23-25.
[5]　NormanG，钱和．食品卫生原理［M］．华小娟，译．北京：中国轻工业出版社，2001.
[6]　谢从新．环境安全和水产品质量安全养殖模式［J］．养殖与饲料，2011（9）：4-5.
[7]　杨先乐，郑宗林．我国渔药使用现状、存在的问题及对策［J］．上海水产大学学报，2007，16（4）：374-380.
[8]　HallegraeffGM. Areviewof harmful algalbloomsandtheirapparentglobalincrease［J］．Phycologia，l993，32：79-99.
[9]　麦充志．我国食品安全与食源性疾病控制对策．中国医学研究杂志，2006，6（8）：940-942.
[10]　EwenSW PusztaiA. Effect of diets containing genetically modified potatoes expressing Galanthusnivalislectinonrats mallinte stine［J］．Lancet，1999，354：1353-1354.
[11]　张希春，杨晓光．转基因鱼的食用安全性评价及其分子生物学基础［J］．卫生研究，2004，33（2）：233-236.
[12]　李书谦．食品安全问题的思考与对策［J］．中国食物与营养，2002（4）：13-16.
[13]　任信林，凌武海．水产品质量安全存在问题及对策［J］．水产养殖，2011，12：48-50.

[14] 陈海仟, 吴光红, 张美琴, 等. 我国水产品重金属污染现状及其生物修复技术分析 [J]. 科学养鱼, 2010 (3): 3-5.
[15] 谢力, 相大鹏, 韦晓群, 等. 主要贸易国和地区食品中致敏原标识措施比较及其对我国的启示 [J]. 中国食品卫生杂志, 2011, 23 (5): 449-452.
[16] 郑宗林, 向磊. 我国水产药物使用现状分析 [J]. 水产养殖, 2002 (4): 36-39.
[17] 汪开毓, 阳涛. 我国水产品出口现状及质量安全问题 [J]. 中国渔业经济, 2007 (4): 48-51.
[18] 欧阳林山. 浅析我国兽药监督抽检中存在的问题及建议 [J]. 中国兽医杂志, 2005, 39 (2): 3-5.
[19] 杨先乐. 我国水产品的药物残留状况及控制对策 [J]. 水产科技情报, 2003 (2): 68-71.
[20] 王泰健. 我国兽药监管现状及改进建议 [J]. 中国牧业通讯, 2005 (8): 10-15.
[21] 马艳梅. 动物性食品兽药残留及其控制措施 [J]. 江西畜牧兽医杂志, 2005 (1): 6-10.
[22] 胡鲲, 杨先乐. 渔药基础理论及安全使用技术研究现状分析 [J]. 中国水产, 2011 (5): 28-29.
[23] 战文斌, 刘洪明, 王越. 水产养殖病害及其药物控制与水产品安全 [J]. 中国海洋大学学报, 2004, 34 (5): 758-760.
[24] 邵明洋. 兽药残留危害及监控对策 [J]. 中国兽药杂志, 2003 (2): 1-4.
[25] 李凯年, 逯德山. 水产品药物残留对中国水产品出口的影响与对策 [J]. 世界农业, 2007 (9): 53-56.

该文发表于《中国兽药杂志》【2011, 45 (5): 43-46, 50】

我国渔药剂型使用现状及其在渔药安全使用技术中的价值

The Dosage Forms of Fishery Drug in China and Their Technical Support to the Safety Use of Fishery Drug

胡鲲[1], 龚路旸[1], 朱泽闻[2], 杨先乐[1]

(1. 国家水生动物病原库, 上海海洋大学, 上海 201306; 2. 全国水产技术推广总站, 北京 100026)

[摘要]: 从渔药安全使用的角度, 对我国渔药剂型的现状和存在的问题进行了分析。通过总结渔药剂型在提高生物利用度、降低毒副作用、延长作用时间和提高稳定性等方面的作用, 阐述了渔药剂型在降低水产品药物残留风险方面的潜在应用价值, 期望为我国渔药剂型在渔药安全使用中提供参考。

[关键词]: 渔药剂型; 生物利用度; 毒性; 缓释控释; 稳定性; 安全使用技术

药物是防治水生动物病害的三大主要方法之一[1]。相比农药和兽药剂型 (dosageform) 已发展到以控释制剂为代表的第三代剂型, 目前我国渔药剂型种类尚较为单一: 还停留在以传统的粉剂、水剂等第一代剂型为主的阶段[1-3]。近年来发生的由渔药残留引起的水产品安全事件倍受公众、媒体和政府的关注。从建立渔药安全使用技术规范的角度而言, 剂型的单一、不合理 (如剂型不稳定、释放不均匀和抗逆性差等) 是导致渔药错用、滥用的重要因素, 也是间接引发渔药残留、危害水产品安全的隐患之一。

本文对我国渔药剂型的现状和存在的问题进行了分析, 并通过总结剂型在提高渔药的生物利用度、降低毒副作用、延长作用时间和提高稳定性等方面的作用, 独辟蹊径地阐述了渔药剂型在降低水产品药物残留风险方面的潜在应用价值, 期望以此达到保障水产品安全的目的。

1 我国渔药剂型的现状分析

1.1 剂型的概念、分类及发展

剂型是指根据预防和治疗病害的需要, 结合药物的理化性质, 应用加工技术将药物制备成适于直接使用、保藏和运输的制剂形式[1,4-5]。截至 2003 年, 世界各国药典收载的药物剂型达到 40 余种[5]。按照不同的分类标准, 可以把药物剂型分为不同的类别。常用的分类标准包括: 按形态分类 (可分为气体、液体、半固体和固体制剂等)、按分散系统分类 (可分为: 水剂、粉剂、颗粒剂、乳油剂、悬浮剂、片剂、微胶囊剂等)、按照给药途径和方法分类 (可分为注射剂、洗剂、硬/软膏剂、喷雾剂等)[1,5]。

Teorell (1937) 和 Wagner (1961) 最早提出药物剂型和生物因素一起对药物的吸收代谢过程和药效有重要影响, 并由此发展起一门新的学科———药剂学 (pharmaceutics)[6]。药剂学的发展经历了生物药剂时代、临床药剂时代和药物传输系统剂型三个阶段。人们逐渐认识到, 剂型与药物的效果有着密切的关系: 改变药物的性质、改变药物的作用速度、降低毒副作用、产生靶向作用和影响疗效等[4]。同时，人们对剂型的评价也从刚开始的仅从体外的化学标准和一般的性状评价药物剂型，过渡到考查剂型对药物在体内的吸收、分布、代谢、排泄（absorption, distribution, metabolism, excretion，简称 ADME）过程的影响。剂型的开发也树立了降低药物用量、提高疗效、减少毒副作用和方便使用的原则。

另外，值得特别指出的是药物剂型的发展与工艺技术水平的发展密不可分。传统的剂型技术基础为粉剂、片剂等第一、二代传统制剂工艺（包括液体输送、沉降、过滤和浸出等），而包括热分析（如冷冻制剂）、包合技术（如喷雾干燥法）、超微粉碎、生物酶降解、膜分离技术和超临界流体萃取技术等新技术的广泛应用为缓释制剂、靶向制剂等打下了坚实的基础[4-5]。

1.2　我国渔药剂型

目前，我国共有专业渔药生产厂家约有150多家，兼营渔药的厂家400多个，生产的渔药种类达100多种，产量2.5万吨，创产值3亿多元[2]。渔药剂型伴随着渔药防治水产动物病害技术的发展而发展。渔药一般是不允许或不提倡直接向水体泼洒使用的，常见的渔药（如环境改良/消毒药、抗微生物药、杀虫驱虫药、生长调节剂、防霉/抗氧化/麻醉剂、中草药、疫苗、微生态制剂等）都是以一定的剂型形式使用。

目前，我国渔药的剂型种类较为单一[3]。常见的渔药剂型包括气体剂型（如喷雾剂等）、液体剂型（如溶液剂等）、半固体剂型（如软膏剂等）和固体剂型（如粉剂等）。而固体剂型中的粉剂又在所有渔药剂型中所占的比例最大。各种渔药剂型的分类及其特点详见表1。

表1　常见渔药剂型的分类

种类	剂型形式	特点	适用范围	应用举例	备注
气体剂型	喷雾剂等	成本低，残留少，处理效率高，选择性差	主要为环境消毒类药物。	臭氧[7]；气态氯[1]等。	在特定环境下应用，范围较窄
液体剂型	溶液剂、注射剂、乳剂	使用方便，作用直接，药效迅速	主要为非挥发性或不易固体化药物	消毒剂（如福尔马林[8]、光合细菌制剂[9]、麻醉剂（如丁香酚[10]）、免疫增强剂（如脂多糖[11]）、疫苗（如嗜水气单胞菌疫苗[12]）、杀虫剂（如氯氰菊酯[13]）等	大多数为水剂
半固体剂型	软膏剂、糊剂	作用于水产动物特定的部位	主要为体表给药的药物或饲料添加剂	饲料添加剂（如脂溶性维生素[14]）	
固体剂型	粉剂、片剂、颗粒剂、微囊剂	运输、储藏和使用方便，作用时间长	适合各种类型的药物	消毒剂（如高锰酸钾[15]）、抗微生物制剂（如喹诺酮类药物[16]）、杀虫剂（如阿维菌素[1]）、抗氧化剂（如维生素E[17]）、中草药[1]、微生态制剂（如芽孢杆菌净水剂[18]）	粉剂在渔药剂型中所在比例最大

1.3　目前渔药剂型存在的主要问题

渔药大都由人药、兽药以及部分农药移植而来，而水产养殖的特殊性（如养殖对象、养殖模式和养殖环境等因素）对渔药剂型提出了独特的要求。目前现有的渔药剂型在已经远远不能满足生产实践的需求[2]，主要表现为：①生物利用度低，药效低，间接导致渔药剂量的过量使用，从而危害水产品安全，污染水域环境；②对水产动物毒副作用强，危害水产养殖动物的安全；③药效期短，释放速度快，容易导致药物的过量使用，造成水产品药物残留；④稳定性差，抗逆性不强，易受环境因素（如光、热等）影响。

2　剂型在保障水产品安全方面的作用

"安全、高效、环境友好型"是未来渔药的发展方向，这也对渔药剂型提出了一系列新的要求。合理的渔药剂型可以在水产品安全用药方面起到事半功倍的效果。

2.1　提高渔药生物利用度

生物利用度[19]（bioavailability），也称生物有效性，是指药物制剂中的主药成分被吸收进入动物血液内的程度和速度，包括血药浓度–时间曲线下面积（areaaunderthecurve，AUC）、血药浓度达峰时间（peaktime，T_{max}）和血浆药物峰浓度（peakconcentration，C_{max}）三个参数。生物利用度是衡量药物利用效率的重要指标，药物剂型是影响药物生物利用度的最主要因素之一[6,19-20]。剂型对药物在体内的 ADME 过程有着显著影响，其影响往往在药物代谢动力学的研究结果中得以体现中。虽然目前关于渔药剂型影响水产品体内生物利用度的报道极少，但关于人医药、兽药的报道极多，早已得到广泛的共识。这一点可以为我们充分地借鉴。

2.2　降低药物的毒副作用

适当的药物剂型能降低药物对动物的毒副作用[4]。凌昌全等比较了去甲斑蝥素缓释剂型与普通剂型对大鼠的急性毒性，试验结果表明缓释制剂对肝脏的刺激毒性低于普通制剂，缓释制剂安全性更高[21]。尚兰琴等发现相比水针剂，膦甲酸钠粉针剂对小鼠的毒性显著减小，尤其是对肾脏的损伤，并推断是由于粉针剂能够更好地水化药物，从而达到减毒的效果[22]。罗东在比较了干扰素（IFN）的不同种类剂型后发现，干扰素的脂质体、微球及微囊等新型制剂能有效降低干扰素的毒副作用，并降低用药剂量[23]。

2.3　增强药物的缓释和控释性能，延长药物作用时间

缓释（sustainedrelease）和控释（controlledrelease）分别是指在特定的介质中缓慢地以非恒定、恒定的速率释放药物，能大幅度减少用药次数或增长用药时间间隔[4,24-25]。缓释和控释具有一系列的优点：减少用药频率和剂量、血药浓度波动小、毒副作用小、延缓耐药性等。胡鲲等证明了：以羧甲基纤维素和淀粉为壁材的恩诺沙星微胶囊均表现出明显的缓释作用，且羧甲基纤维素的缓释性能优于淀粉，冷冻温度的降低能促进微胶囊缓释性能[26]。毕小平等利用聚乙烯醇（PVA）、甲壳素等辅料制备二氧化氯半固体新剂型，抑菌试验表明半固体新剂型除了维持杀菌效果，还延长了药效维持时间，具有缓释功能[27]。陈舒泛等以土霉素为试验药物，产业化生产生物微囊渔药，通过缓释作用达到防治鱼苗细菌性疾病的目的[28]。

2.4　提高渔药的稳定性

渔药从生产、运输、储藏到使用过程中需要经历一个较长的时期，往往由于化学（如光、热导致药物分解）、物理（如分散体系被破坏）、生物（如微生物污染）等因素导致药物疗效降低、失效甚至产生毒副作用[4,27]。药物的剂型与药物的稳定性紧密相关，研究药物的稳定性对规范和评价药物的质量、保障药效至关重要。水产养殖环境相对复杂，渔药面临的稳定性问题尤为突出，利用剂型提高渔药稳定性是一条有效的途径。以水产养殖常用消毒剂二氧化氯制剂为例，可以通过加入稳定剂、活化剂提高了二氧化氯产率，改善了稳定性[28]；蔡旭玲发现固体二氧化氯制剂对于大肠杆菌、金黄色葡萄球菌和白色念珠菌等病原菌的杀灭效果优于液体制剂[29]；而在消毒效果相当的情况下，气态二氧化氯制剂比液体制剂用量更小，对环境影响更小[30]。

2.5　其他

渔药剂型还在很大程度上影响着渔药使用方式便捷性、经济性、辅料的性质[31]以及动物生理状态[32]等。

3　加强渔药剂型研发的建议

加强渔药剂型的研究和开发，提升渔药剂型技术层次，优化渔药剂型构成，制定和颁布渔药剂型的技术规范是保障渔药安全使用的重要措施。

3.1　制定和颁布渔药剂型的技术标准和规范

目前，我国现行的法律、法规和标准均未对渔药剂型进行系统地分类和命名，渔药剂型生产、使用、销售和管理过程的混乱。制定和颁布一个实用性的渔药剂型及其代码的行业标准是未来渔药产业发展的大势所趋。2004 颁布的我国第一个农药剂型标准——中华人民共和国国家标准《农药剂型及其代码》[33-35]根据我国国情和国际化惯例，为国内 120 种农药剂型进行了分类和排列，规定了其代码，该标准的颁布和实施为农药剂型的规范化提供了技术规范，极大推动了和促进了产业的健康发展。渔药剂型可借鉴其思路。此外，渔药使用对象、环境和使用方式的特殊性也要求建立一整套适合渔药特点的渔药剂型评价技术规范，包括给药系统评价、稳定性评价等。

3.2　发展渔药专用辅料，建立适合产业特点的制剂生产工艺

药物辅料（Pharmaceutical Necessities）又称药物助剂（Pharmaceutical Aids），是指一类助于制剂成型、稳定、增溶、助溶、缓释、控释等不同功能和作用的各种辅料[36-38]，是制造、调配药物的必需品。常用的辅料包括载体材料[39]、缓释控释材料[40]、微囊材料等40余类。辅料是决定药物剂型的重要因素，是药剂学的发展的基础。发达国家辅料已经开始了专业化、系列化和规模化的应用，并成立了国际药用辅料协会（IPEC）[41]。目前我国专用渔药辅料相对几乎为空白。渔用专用辅料应该重点发展具有良好的吸附、包埋性能、空间拓展度高的材料[42]，并充分考虑其对水产动物的潜在安全性、材料的经济性和来源广泛性等因素。

3.3　发展适合水产养殖特点的渔药新剂型，优化渔药剂型结构

发展以微胶囊[4]为代表的控释缓释制剂等第三代制剂是未来渔药剂型的发展方向。改变渔药剂型过分单一的状况[8,38,42]，逐步由一元、二元制剂向多元制剂过渡；逐步提高新剂型的比例，提升渔药剂型技术层次，优化渔药剂型结构，达到增强药效、提高稳定性、减少用量，降低毒副作用的目的。

参考文献

[1]　杨先乐.新编渔药手册［M］.北京：中国农业出版社，2005.

[2]　邓永强，汪开毓，黄小丽.水产药物制剂生产存在的问题与对策［J］.科学养鱼，2004，7：42.

[3]　尹伦甫，陈昌福.绿色渔药新剂型浅谈［J］.科学养鱼，2009，（3）：76.

[4]　潘卫三.新药制剂技术［M］.北京：化学工业出版社，2004.

[5]　沈宝亨，李良铸，李明晔.药物制剂应用技术［M］.北京：中国医药科技出版社，2000.

[6]　魏树礼.生物药剂学与药物动力学［M］.北京：中国医科大学中国协和医科大学联合出版社，1997.

[7]　谭洪新，周琪，朱学宝.泡沫分离-臭氧消毒装置的水处理效果研究［J］.上海环境科学，2003，22（12）：987-990.2011，45（5）：43~46，50/胡鲲，等

[8]　章剑，赵淑梅.福尔马林在水产养殖中的应用［J］.渔业现代化，2001，5：18.

[9]　杨莺莺，李卓佳，贾晓平，等.水质净化作用菌光合细菌 PS2 的生物学特性及环境因子对其生长的影响［J］.上海水产大学学报，2003，12（4）：293-297.

[10]　赵艳丽，杨先乐，黄艳平，等.丁香酚对大黄鱼麻醉效果的研究［J］.水产科技情报，2002，29（4）：163-165.

[11]　石军，孙德文，陈安国.微生物多糖的应用研究进展［J］.畜禽业，2002，11：40-41.

[12]　孙玉华.鳖嗜水气单胞菌疫苗的试制［J］.上海水产大学学报，1998，7（1）：75-78.

[13]　王媛，杨康健，吴中，等.氯氰菊酯对鲫鱼血清中谷丙转氨酶及谷草转氨酶活力的影响［J］.水产科学，2005，24（9）：8-10.

[14]　周歧存，麦康森.水产动物对脂溶性维生素的营养需要［J］.水产科学中国饲料，1997，18：27-30.

[15]　杨莺莺，陈毕生，陈福华，等.5 种常用水产药物对网纹石斑鱼的急性毒性试验［J］.上海水产大学学报，2001，10（1）：93-95.

[16]　杨丽辉，佟恒敏.几种氟喹诺酮类药物对嗜水气单胞菌体外药效学研究［J］.东北农业大学学报，2003，34（4）：431-435.

[17]　王广军，谢骏.维生素 E 在水产养殖中的应用［J］.中国饲料，2003，4：22-24.

[18]　侯树宇，张清敏，多森，等.微生态复合菌制剂在对虾养殖中的应用研究［J］.农业环境科学学报，2004，23（5）：904-907.

[19]　李安良.生物利用度控制-药物化学原理、方法和应用［M］.北京：化学工业出版社，2004.

[20]　张玲玲，潘安，赵杰，等.2 种剂型麻保沙星在健康家犬体内药物动力学比较［J］.华中农业大学学报，2010，1：75-78.

[21]　凌昌全，郑晓梅，李柏.不同剂型去甲斑蝥素肝脏注射急性毒性比较［J］.中成药，2000，22（10）：715-717.

[22]　尚兰琴，魏雪涛，杨晓华，等.不同剂型膦甲酸钠针剂的肾脏毒性［J］.毒理学杂志，2006，20（2）：131-132.

[23]　罗东.干扰素剂型研究进展［J］.中国新医药，2003，2（9）：49-51.

[24]　杨勇博，孙庆申，车小琼，等.壳聚糖在药物剂型中的应用［J］.中国组织工程研究与临床康复，2008，12（41）：8131-8134.

[25]　李艳华，佟恒敏，闫清波，等.应用药物累积法研究复方中药Ⅱ的药代动力学研究［J］.中国兽医杂志，2007，43（7）：86-87.

[26]　胡鲲，朱泽闻，杨先乐，等.恩诺沙星微胶囊的制备及其缓释性能的研究［J］.上海水产大学学报，2006，15（3）：375-379.

[27]　毕小平，谢茵，胡向青，等.稳定性二氧化氯新剂型的研究［J］.山西医科大学学报，2003，3（3）：229-230.

[28]　吴礼龙，孙福招，肖明俐，等.一元固体稳定剂二氧化氯消毒剂的研究［J］.中国消毒学杂志，2010，27（3）：263-266.

[29]　蔡旭玲，汪保国，黄晓晖，等.二氧化氯消毒剂的剂型对杀菌效果的影响［J］.中国卫生检验杂志，2010，20（3）：551-552.

[30]　陈惠珍，王冰妹，黄美鲫，等.二种不同剂型二氧化氯对空气消毒效果研究［J］.中国卫生检验杂志，2007，17（6）：999-1058.

[31]　何华，陈颂仪，王宇新，等.药物制剂辅料β-环糊精的纯化及含量测定［J］.药学进展，2001，25（4）：243-245.

[32]　王吉桥，苏久旺，姜玉声，等.不同剂型和剂量的维生素 C 对幼刺参生长的影响［J］.海洋科学，2009，33（12）：56-63.

[33]　中华人民共和国国家标准.农药剂型名称及代码［S］.GB/T19378-2003.

[34]　王以燕，刘绍仁，宗伏霖，等.国家标准：《农药剂型名称与代码》释义（上）［J］.世界农药，2004，26（6）：23-26.

[35]　王以燕，刘绍仁，宗伏霖，等.国家标准：《农药剂型名称与代码》释义（下）［J］.世界农药，2005，27（1）：27-31.

[36] 张骞, 王冕, 谭晓伟. 我国农药剂型专利发展状况分析 [J]. 农药科学与管理, 2010, 7: 21-23.

[37] 邹杨, 杨在宾, 杨维仁, 等. 不同剂型丁酸钠与抗生素对肉仔鸡生产性能、肠道 pH 及挥发性脂 [J]. 动物营养学报, 2010, 22 (3): 675-681.

[38] 曾振灵, 刘义明, 黄显会. 兽药新剂型的研究与应用进展 [J]. 北方牧业, 2009, 2: 10-11.

[39] 贺云霞, 孙进, 程刚. 多药耐药性 P-糖蛋白在药物肠道吸收中的作用 [J]. 沈阳药科大学学报, 2004, 21 (5): 389-393.

[40] 陈关平, 操继跃. 兽药控/缓释剂的研究进展 [J]. 中国兽医杂志, 2003, 37 (8): 40-42.

[41] 张志芬. 从现代药剂学看药用辅料的发展 [J]. 药物分析杂志, 2005, 25 (12): 1576-1580.

[42] 李丽杰, 乔彦良, 闫祥华. 兽药新剂型的研究进展 [J]. 中国畜牧兽医, 2008, 35 (3): 85-87.

该文发表于《中国水产》【2002, 10: 74-75】

水产品药物残留与渔药的科学管理和使用 (一)

杨先乐

(农业部水产增养殖生态、生理重点开放实验室, 上海水产大学, 200090)

我国加入 WTO 后, 面临着全球经济一体化的形势, 面临着人们对安全、卫生、高质量的水产品日益增长的要求, 面临着我国水产养殖高效益、高水平持续发展的局面, 水产品的药物残留问题已经引起了社会的普遍关注。要控制水产品的药物残留, 保证水产品的安全, 势必与渔药的安全使用与科学管理密切相关。本文试图对此提出一些粗浅的看法, 以期为从事该方面工作的管理人员与业者提供一些借鉴。

一、冻虾仁氯霉素事件与欧盟的"禁令"

浙江舟山是我国冻虾仁主要出口基地之一, 舟山的冻虾仁因其个大味鲜而享誉海外。但是 2001 年底, 因抽检发现冻虾仁中氯霉素含量 0.2 $\mu g/kg$, 超过欧盟底限 (0.1 $\mu g/kg$, 而被拒货, 甚至被没收就地销毁; 今年 1 月 30 日欧盟正式做出禁止进口中国动物源性产品的决议。欧盟的这一禁令使中国蒙受了巨大损失: 近 5 万劳动力下岗, 十几万农户因企业无法履行收购合同而水产品被滞留; 94 家对欧盟出口水产品的企业损失达 6.23 亿美元。在此禁令下, 比利时、匈牙利、捷克等国家对中国水产品的进口均采取了十分苛刻的措施。

除了氯霉素外, 欧盟还规定了动物源性食品中严禁有以下 9 种无最大限量的药物活性物质, 它们是: 增噬力酸 (Arisolochiaspp. 又称马兜铃酸、木通甲素, 是一种免疫增强药)、氯仿 (Chloroform)、氯丙嗪 (Chlorpromazine, 又称冬眠灵, 氯善马秦, 为一种抗精神失常药)、秋水仙素 (Colochicine)、氨苯砜 (Dapsone)、二甲基硝咪唑 (Dimetridazole)、甲硝哒唑 (Metronidazole)、硝基呋喃 (Nitrofurans, 包括呋喃唑酮 (Furazolidone) 以及甲硝咪乙酰胺 (Ronidazole 又称洛硝哒唑, 一种抗原虫药) 等。

美国食品与药物管理局 (FDA) 也公布了禁止在进口动物源性食品中使用的 1 种药物名单, 它们是: 氯霉素、克伦特罗、己烯雌酚、二甲硝咪唑、其他硝基咪唑、异烟酰咪唑 (ipronidazole)、呋喃西林、呋喃唑酮、磺胺类药、氟乙酰苯醌 (fluoroquinolones)、糖肽 (glycopeptides)。

欧盟、美国等对水产品药物残留的苛求, 已使我国水产品的生产和出口蒙上了一层阴影, 同时也引起了我国业内外人士对水产品药残的高度重视

二、水产品药残及其危害

随着水产养殖面积的不断扩大, 养殖强度的不断增加, 养殖水环境的不断恶化, 水产养殖动物的病害也日趋严重, 人们为了控制水产动物病害的蔓延, 投入了较多的抗生素, 喹诺酮类、磺胺类、呋喃类等化学药物, 由于对用量、用药次数以及停药期的忽视, 造成了药物在水产品中残留, 进而对公众的健康和水域环境造成了潜在危害, 随着人们对动物性食品的要求越来越高, 食品 (包括水产品) 中药物的残留更引起了人们极大的关注。食品法典委员会 (CAC) 是药物残留监管的主要国际组织。早在 1984 年, 在 CAC 的倡导下, 由 FAO 和 WHO 联合发起, 开始组织食品中兽药残留立法委员会 (CCRVDF), 该组织于 1986 年正式成立。此后每年在美国华盛顿召开一次全体委员会, 制定和修改动物组织及产品中的兽药最高残留限量 (MRL) 以及休药期等法规。此外, 美国的食品药品管理局 (FDA) 以及下属的兽药中心 (CVM), 新加坡的农业食品和兽医机构 (AVA), 欧盟药物评估办公室 (AMEA) 以及下属的兽药产品委员会 (CVMP) 和下属各国的兽医药警戒机构等, 澳大利亚的国家农业及兽医化学注册管理部门 (NRA) 等在药物残留的监管中都起着很大的作用。

什么是药物残留? 根据食品兽药残留立法委员会 (CCRVDF) 的定义, 兽药残留是指动物产品的任何可食部分所含药物的母体化全物及/或其代谢物, 以及与药物有关的杂质的残留。所以药物残留既包括原药, 也包括药物在动物体内的代

谢产物。此外，药物或其代谢产物还能与内源大分子共价结合，形成结合残留，它们对靶动物具有潜在毒性作用。水产品中主要残留药物有喹诺酮类、抗生素类、磺胺类和呋喃类以及某些激素等。

一般来说，水产品中的药物残留大部分不会对人导致急性毒性作用，但是如果经常摄入含有低剂量药物残留的水产品，残留的药物即可在人体内慢慢蓄积而导致体内各器官的功能紊乱或病变，严重危害人类的健康。具体来说，药物具有以下的危害：

（1）毒性作用。人类长期摄入含有药物残留的水产品，药物会不断在体内蓄积，当浓度达到一定量时，就会对人体产生毒性作用。如磺胺类可引起肾脏损害，特别是乙酰化磺胺在酸性尿中溶解降低，析出结晶后损害肾脏；又如氯霉素，可以引起再生障碍性贫血，导致白血病的发生。

（2）产生过敏反应和变态反应。有些药物具有抗原性，它们能刺激机体形成抗体，产生过敏反应，严重者可引起休克，短时期内出现血压降低、皮疹、喉头水肿、呼吸困难等严重症状。如青霉素、四环素、磺胺类及某些氨基糖甙类抗生素等。呋喃类引起人体的过敏反应，表现在周围神经炎、药热、嗜酸性白细胞增多为特征的过敏反应。磺胺类药的过敏反应表现在皮炎、白细胞减少、溶血性贫血和药热。抗菌药物残留所致变态反应比起人对食物所受的其他不良反应所占的比例小得多。青霉素类药物引起的变态反应，轻者表现为接触性皮炎和皮肤反应，严重者表现为致死性过敏性休克。四环素的变态性反应比青霉素少，但四环素药物可引起过敏和荨麻疹。

（3）导致耐药菌株产生。由于药物在水产动物体内残留，并通过有药残的水产品在体内诱导某些耐药性菌株的产生，给临床上感染性疾病的治疗带来一定的困难，耐药菌株感染往往会延误正常的治疗过程。日本、美国、德国、法国和比利时学者研究证明，在乳、肉和动物脏器中都存在耐药菌株。当这些食品（如肉馅、牛肉调味酱等）被人食用后，耐药菌株就可能进入消费者消化道内。至今，具有耐药性的微生物通过动物性食品移生到人体内而对人体健康产生危害的问题尚未得解决。

（4）导致菌群失调。在正常情况下，人体肠道内的各种菌群是与人体的机能相互适应的，但是残留的影响会使这种平衡发生紊乱，造成一些非致病菌的死亡，使菌群的平衡失调，从而导致长期的腹泻或引起维生素的缺乏等反应，对人体产生危害。

（5）产生致畸、致癌、致突变作用。对人类会产生较强的三致作用的药物有孔雀石绿、双甲脒等。

（6）激素作用。激素类药物的残留会使人类生理功能造成紊乱，产生较大的影响。

三、水产品药残产生的主要原因

1970年美国FDA对造成兽药残留原因的调查结果表明：未遵守休药期的占76%，饲料加工或运输错误的占12%，盛过药物的贮藏器没有充分清洗干净的占6%，使用未经批准的药物占6%。

1985年美国兽医中心（CVM）的调查结果则是：不遵守休药期的占51%，使用未批准的药物占17%，未作用药记录的占12%。

我们认为造成水产品药物残留的主要原因是：

（1）不遵守休药期有关规定。休药期是指水产品允许上市前或允许食用前的停药时间。目前我国使用的较多水产药物均缺乏具体的休药期规定，在部分水产养殖业者中，休药期的意识还比较薄弱。

（2）不正确使用药物。使用渔药时，在用药剂量、给药途径、用药部位和用药动物的种类等方面不符合用药规定，因此造成药物在体内的残留。

（3）饲料加工、运送或使用过程中受到药物的污染。当将盛过抗菌药物的容器用于贮藏饲料，或盛过药物的贮藏器没有充分清洗干净而使用，都会造成饲料加工或使用过程中药物的污染。

（4）使用未经批准的药物。由于这些药物在水产动物体内的代谢情况缺乏研究，没有休药期的规定，如作为饲料添加剂来喂养水产动物，极易造成药物残留。

（5）养殖用水中含有药物。使用这种受污染的水，极易引起水产养殖动物的药物残留。

该文发表于《中国水产》【2002，11：74-75】

水产品药物残留与渔药的科学管理和使用（二）

杨先乐

（农业部水产增养殖生态、生理重点开放实验室，上海水产大学，200090）

四、水产品药残最高残留限量标准与我国水产品药物残留现状

由于水产动物病害日趋严重，渔药的使用不可避免，加上水环境中可能存在的渔药污染，水产品中不可能绝对不存在药物的残留，因此为了保护人类自身的健康，规定水产品药物的最高残留限量标准足保证水产品安全的基本条件水产品的最高残留限量标准应根据以下因素确定：①药物对人类健康的危害程度，如有三致作用的药物（禁用药物），残留限量要求就很低；一般性的药物，残留限量可适当高些；②残留药物（或其代谢产物）不会对人类产生较大危害和不会对有益菌群造成破坏的最高浓度；③不会导致耐药菌株产生的最高药物浓度；④残留药物可检测到的最高灵敏度；⑤国外有关国家和组织最高残留限量标准；⑥我国水产品生产和进出口的具体情况。目前，我国已经规定了水产品中渔药残留限量（见表1）。

表1 水产品中渔药残留限量

药物类别		药物名称		指标（MRL/（ug/kg））
		中文	英文	
抗生素类	四环素类	金霉素	Chlortetracycline	100
		土霉素	Oxytetracycline	100
		四环素	Tetracycline	100
	氯霉素类	氯霉素	Chloramphenicol	不得检出
磺胺类及增效剂		磺胺嘧啶	Sulfadiazine	100（以总量计）
		磺胺甲基嘧啶	Sulfamerazine	
		磺胺二甲基嘧啶	Sulfadimicline	
		磺胺甲噁唑	Sulfamethoxazole	
		甲氧苄啶	Trimethoprim	50
喹诺酮类		噁喹酸	Oxolinic acid	30
硝基呋喃类		呋喃唑酮	Furazolidone	不得检出
其他		乙烯雌酚	Diethylstilbestrol	不得检出
		喹乙醇	Olaquindox	不得检出

近年来国外一些组织和国家以设置贸易壁垒为目的，将药物残留限量的标准大幅度地提高，如氯霉素原来的标准是10 μg/kg，而现在降低到0.1 μg/kg，因此在抽查时，导致部分产品超标。此外国内也谣传一些水产品可能存在着某些激素，可能有较高的药物残留。从理论上来说，药物的残留越低（至接近零）越好，但是由于种种原因，要达到很低的药残比较困难。另一方面存在一定量的药残也不一定讨人类会造成很大的危害，这种有一定量药残但不会危害人类健康的水产品我们称为安全的水产品，而那种药残限量很低甚至没有药物残留的水产品则为高质量的水产品，高质量水产品与安全水产品应有很大的差别。

我国水产品药物残留状况究竟如何？通过我们对2000年和2001年水产养殖品药残抽检的情况，我们认为我国水产品在药物残留上虽然存在着一定的问题，但总的来说并非危言耸听，水产品质量总体上还是可以信赖的。2000年上海水产大学药物残留检测中心抽样检测了100个中华鳖肌肉样品、25个罗非鱼肌肉样品和50个鳗鲡肌肉样品，检测结果是100个中华鳖肌肉样品仅有1个检测样品已烯雌酚含量超标，占检测样品总数的1.0%；50个鳗鲡肌肉样品，土霉素含量超标的样品3个，占检测样品总数的6%；四环素和金霉素含量均低于残留限量标准；25个罗非鱼样品甲氧苄啶、磺胺、磺胺嘧啶、磺胺噻唑、碘胺吡啶、硫代异噁唑、磺胺甲噁唑、磺胺二甲嘧啶的含量的含量均低于残留限量，达标率为100%，2001年我们对来自江苏、浙江、湖北、湖南、河北、山东、大连、广东和福建9个省（市）的中华鳖、牙鲆、大菱鲆、鳗鲡、河蟹和鳜鱼等533份样品的已烯雌酚、氯霉素、土霉素类（土霉素/四环素/金霉素）和呋喃唑酮4大类渔药残留进行检测，合格样品达466份，检测合格率达87.4%。我们相信，通过我国目前实施的"无公害食品计划"，水产品的药残问题一定

会得到较大的控制。

五、水产品药残的监控

为了提供安全、高质量水产品，提高我国水产品的声誉，使我国的水产品稳固地占据国际市场有利地位，为使我国的水产养殖业在新时期持续发展，对水产品药物残留实施监控显得十分必要。我们认为水产品药物残留监控必需要采取以下措施：①建立有效的监控网络；②制定有针对性的监控计划；③树次可效仿和推广的先进典塑；④实施公正、有力的处罚手段。当然我们还必须采用高效、正确的分析、检测方法。目前在药残的检测上所采取的主要方法有：

微生物测定方法：微生物测定方法简单、快速、便宜，但较繁琐。Kusser［1990］、Pearson［1993］所介绍微生物测定法是利用对抗菌药物敏感的菌株 *Bacillus subtilis*，*R. cereus* 和 *Ehchenrichia coli* 作为指示生物体，将组织提取液对比稀释居与标准溶液对照。再根据最低抑菌浓度判断推算出组织提取液中的药物浓度。

分光光度法：分光光度法是应用分光光度计进行测定，比较简单、便宜、易操作，但主要缺点是精确度较低，特别是代谢物与原药结构相似，以至吸收光产生叠加不易区分，体液的内源性杂质干扰大时，则会出现测定结果偏高的现象。

气相色谱法及高效液相色谱法（HPLC）：气相色谱法使用有一定的局限，对那些不易气化的物质测定不准确，而且设备较昂贵，使用不普遍，高效液相色谱法以及由此改进所派生的反相高效液相色谱法是普遍使用的主要测定方法，虽然设备造价高，但对样品的分离鉴定不受挥发度、热稳定性及分子量的影响，具有分离效果好，测定精度高的优点。

为了更大面积和更快地对药残状况进行普查，现场快速测定是一个重要途径，在现场快速测定中发现有疑问的样品再送到实验室进行分析。目前有以下几种方法：①抗生素快速筛选法（FAST），该方法是在养殖场对水产动物肾组织中的抗生素或磺胺类进行快速筛选检验；②快速涂抹实验（STOP），是在加工厂内对可疑水产品中的抗生素残留进行检验的一种快速力法；③磺胺现场试验法（SOS），主要采集养殖动物排泄物，测定其中磺胺的残留。

在进行药残检测、分析时要注意以下四个问题：①执行官方采样程序，注意取样的科学性与代表性；②采取合理的样品前处理方法。③选择正确的药物分析方法；④作出准确的结果判断。要根据抽样、检测、养殖用药和国家的需要判断结果，做到客观、公正、正确。

六、控制水产品药残的途径—渔药的安全使用与科学管理

水产品中药物的残留，主要是因为渔药的不合理的使用引起的，要控制水产品的药残，从源头抓起，加强渔药的安全使用与科学管理是关键所在。

1. 渔药的科学管理

渔药管理工作的核心是重点解决好管理体系、法规和标准建设三个方面的问题。管理体系方面，要进行统一的渔药行政监督管理的可操作办法，改变渔药管理工作隶属于兽药而造成的渔药管而不实的现象；要逐步按照市场经济规律的要求，将渔药行政管理工作和渔药有偿性服务上作分离开来；加强渔民科学用药意识的教育和用药技术的培训。在管理法规方面，要制订相应的渔药管理条例，要有对新渔药正确评价、认证和检验的程序和机制，要加强渔药生产、销售的管理，要规定渔药的残留限量和休药期，要对养殖用药进行指导和监督，加强渔药滥用的处罚在管理标准方面，要加紧制定渔药最高残留限量标准，渔药检测标准，渔药研制与科学使用标准等，要加强相应的研究，为其标准的制定提供可靠的依据。

2. 渔药的科学使用

科学的使用渔药，就是要从药物、病原、环境、养殖动物本身和人类健康等方面的因素考虑，有目的、有计划和有效果地使用药物。科学用药是一个较复杂的问题，在用药时既要考虑提高水产动物疾病的防治效果，也要考虑药物对水产动物的品质和水域环境的影响科学用药要做到：①正确诊断，对症对方用药；②选药要有明确的指证，安全用药；③掌握影响药物疗效的一切因素，排除各种因素可能造成的对药物干扰，适宜用药；④适当加大或缩小用药的浓度、用药次数和用药的间隔时间，合理用药；⑤祛邪扶正并举，增强机体的抗病能力，控制用药；⑥认真观察、分析，根据情况采取停药、调整剂量和改换药物的措施，有效用药。

该文发表于《水产养殖》【2000，2：34-35】

上海市场渔用含氯消毒剂状况调查

杨先乐，蔡完其

（上海水产大学，200090）

近年来，渔用含氯消毒剂发展很快，各种产品不断涌现，作为一种主体消毒剂它在水产动物病害的防治上起到了较大

的作用。我们对上海市场上销售的渔用含氯消毒剂进行了调查与分析，并对其外观标示与有效氯含量进行了检测，以期为有关部门决策提供参考，为广大渔民对含氯消毒剂的选用提供借鉴。

1. 基本方法

1.1 市场调查与抽样 1995年5月，以走访形式对上海市郊的几个主要渔药销售商所经销的含氯清毒剂进行调查，了解1998年的销售情况；并对所销售的产品每种随机抽检2个样品进行检测。

1.2 样品的外观检查与标识 根据农业部1978颁布的《兽药规范》（简载华，1994）进行外观和标示的检查。

1.3 有效氯含量的测定 采取碘量法对各种含氯制剂的有效氯含量进行测定（郭振东，1986）

2 结果

2.1 上海市场渔用含氯消毒剂的组成 目前上海市场销售的主要含氯消毒剂有18种，可分以下几类：①次氯酸钙，产品主要有漂白粉与漂粉精等；②二氯异氰尿酸钠，产品有鱼虾安等；③三氯异氰尿酸，如鱼虾药王，君乐消毒剂等；④二氧化氯，如菌毒清、康丰III型ClO_2等；⑤其他，如溴氧海因等。在这些消毒剂中，漂白粉与漂粉精销售量占首位，达68.09%，其次是三氯异氰尿酸，达27.06%；其他占14.85%。

2.2 渔用含氯消毒剂的外观及其标示 对3类10个含氯消毒剂外观及其标示的检测结果显示：外包装不完好的占70%；生产许可证、产品批准文号和产品执行标准（即三证）不齐全的占60%；没有标示生产日期和有效期的分别占20%和40%。在所有10个产品中，包装及标示状况完好的仅有2个，占20%。

2.3 实际重量与标示重量的符合率 检测的10个样品中，超过标示重量的有4个，占40%，范围+0.2%～+2%，平均+0.86±0.83%；低于标示重量的4个，占40%，范围为-1.8%～-4%，平均-3.3±1.0%；产品无标示重量而无法检测的有2个，占20%。所检测的8个产品，检测的重量与标示重的平均误差-1.2±2.3%。

2.4 有效氯含量 采用碘量法对10个产品的有效氯含量测定的结果表明，漂白粉与漂白精的有效氯含量基本符合要求；而以三氯异氰尿酸为主要原料的产品有效氯含量最高者为87.87±0.11（2）%，最低者仅为27.21±0.11（2）%，平均为52.92±26.30（8）%，相差较悬殊；在检测的4个以ClO_2为主要原料的产品中，有1个产品性能不稳定，测定不到重复的结果，其余3个产品的有效氯含量分别为61.49±0.24（2）%、58.39±0.42（2）%和54.9±0.24（2）%，平均为58.09±3.20（6）%（表1）。

表1 上海市场含氯消毒剂有效氯含量

类别		有效氯含量（%）	
		范围	平均
次氧酸钙	漂白粉	27.73～28.11	27.92±2.70（2）*
	漂粉精	62.32～62.82	62.57±0.35（2）
三氯民氰尿酸		27.21～87.87	52.92±26.30（8）
二氧化氧		54.39～61.49	58.09±3.20（6）

*平均数 *标准差（测定样本数）

3 讨论

3.1 含氯消毒剂的发展经历了由漂白粉—漂粉精—有机氯消毒剂—二氧化氯—海因等的发展过程，按理来说，水产用含氯消毒剂应由海因类或二氧化氯类等取代漂白粉，但上海市场的情况却恰恰相反。造成这现象的原因可能是：①市场惯性。渔（农）民对一些新型含氯消毒剂还半信将疑，缺乏认识，市场的惯性是造成漂白粉等市场份额大的一个原因；②价格。新型消毒剂一般价格较贵，会导致养殖成本增加，而漂白粉相对较便宜；③质量。从有效氯含量的检测来看，漂白粉等次氯酸钙类无机氯消毒剂质量较好，这也是该类产品市场占有率较大的一个重要原因。针对这种情况我们应督促生产厂家对新型消毒剂进一步提高质量，降低成本，同时也应引导渔（农）民使用这些新型消毒剂。因为大多数新型含氯消毒剂具有用量少，无残留，性能稳定，作用期较长，使用安全和方便等特点。

3.2 上海渔用含氯消毒剂产品质量不容乐观。从调查结果来看，还存在产品标示不完全和产品质量不稳定等问题。鉴此我们建议有关部门进行适当的整改。

3.3 ClO_2是一种杀菌效能较强的新型消毒剂，它的质量除与次氯酸（盐）的质量相关之外，还与选择的酸（生产厂家俗称活化剂）的种类与质量密切相关，因为酸的好坏对所释放出的ClO_2量有较大的影响。本文对ClO_2采用碘量法测定，尚不能完全反应该类产品的质量，今后应采取比色法与吸收强度法对该类产品进行准确的质量检测。

第二节　渔药对生态环境的影响

渔药对生态环境影响是一个不容忽视的问题。水环境中残存的有机磷农药（杨先乐等，2002）和抗寄生虫渔药伊维菌素（江敏等，2008）等均存在一定的水生态风险，应引起重视。

该文发表于【*上海水产大学学报*，2002，11（4）：378-382】

有机磷农药对水生生物毒性影响的研究进展

Progress on research of toxic effect of organophosphorous pesticides on aquatic organism

杨先乐，湛嘉，黄艳平

（上海水产大学农业部渔业动植物病原库，上海 200090）

关键词：有机磷农药；毒性；水生生物

有机磷农药是人类最早合成而且仍在国内外农业生产中广泛使用的高效杀虫剂和植物生长调节剂。早期发展的大部分是高效高毒品种，如敌敌畏、甲拌磷乐果、对硫磷、甲胺磷等，而后逐步发展了许多高效低毒低残留品种，如乐果、马拉硫磷、二嗪磷、敌百虫和杀螟松等，成为农药的一大家族[1,2]。有机磷农药大多数是磷酸、磷酸酐或含硫类似物的中性酯或酰胺。除敌百虫、乐果为固体外，其余均为淡黄色或黄棕色液体，甚至是无色液体。绝大多数在水中不溶，而易溶于乙醇、乙醚、氯仿等有机溶剂，是典型的酶毒剂。由于其稳定性较低（半衰期大多数为几天至几十天），远不如有机氯农药在生物体内残留严重[3]，而替代了有机氯农药。在渔业生产中，也常用有机磷农药试剂来杀灭体外寄生虫等敌害生物。然而在其生产和使用过程中，大量成分复杂的有毒废水进人水环境，对水生生物造成危害，破杯水域生态环境。近十年来近岸水域受有机磷农药的污染不断导致了大批的鱼虾贝死亡事故。有机磷废水已开始成为人们普遍关注的污染物之一。关于有机磷农药对水生生物影响的研究已有大量的报道，本文就此项研究做一综述。

1　对水生植物影响及其毒作用机理

有机磷农药能抑制水生植物的生长和繁殖。Piska 等[4]对受乐果污染的印度莎母匹特湖初级生产力进行调查，结果表明，受乐果污染后，湖泊总产量及净产量下降，含氧量及生产量降低，这是由于自养生物受高浓度的乐果抑制，进而影响光合作用所致。唐学玺等[5]发现高浓度的久效磷对微藻细胞有严重的破坏作用，使叶绿素$_a$和类胡萝卜素降解，引起光合色素含量降低。汝少国等[6]报道了10种有机磷农药对扁藻生长的 EC_{50}（半数抑制浓度）值，并讨论了有机磷农药结构与微藻毒性大小的关系。

对水生植物毒作用机理研究相对较少，其机理尚不甚明了。水生植物不具备神经系统，其伤害机理有别于动物。王建华等[7]指出超氧化歧化酶和过氧化酶活性具有维持活性氧平衡，保护细胞膜的功能。McCord 和 Fridovich[8]指出植物细胞在污染胁迫下，往往打破活性氧产生和消除平衡，使其在细胞内过量积累。细胞内过量的活性氧可导致细胞膜结构的破坏和功能的丧失。唐学玺等[5]发现三角褐指藻在高浓度的久效磷的胁迫下，其超氧化歧化酶和过氧化酶活性降低，降低了细胞对活性氧的清除能力，活性氧于是在细胞内大量积累，膜脂过氧化作用加强，膜通透性增大，细胞内的电解质大量外漏，细胞严重受害，从而证实了活性氧也是参与有机磷农药对藻类伤害的主要因素之一。所以，有机磷对水生植物毒作用机理可能通过抑制植物体内重要酶的活性，导致活性氧的积累，对藻类产生主要毒害作用。

2　对水生动物的主要影响及其毒作用机理

2.1　急性中毒及判定

水生动物处于高浓度的有机磷中主要发生急性中毒。一般中毒症状表现为：开始可能出现急噪不安，有狂游冲撞等剧烈现象，然后游泳不稳定，呼吸困难，最后痉挛麻痹、失去平衡，直至昏迷致死。对于鱼类常见症状还有粘液增加，体色变黑。毒性的大小与生物种类、毒物的化学性质等因素有关[9-13]。在池塘中，敌害生物对有机磷的敏感性一般较鱼类强，故常用有机磷来防治寄生虫病。

2.2 慢性中毒

2.2.1 对生长、摄食、呼吸的影响

在亚致死浓度下的水生动物普遍表现为食欲减退，呼吸困难，食物转化率下降，随着新陈代谢水平的降低，生长减缓，甚至停止[14]。Shanmugavel 等[14]通过实验证明，在 10 mg/L 磷胺中的莫桑比克罗非鱼摄食率和食物转化率分别比正常情况下降低了 35% 和 47%，浓度越高，食物转化率越低。有机磷农药能使鳃丝溶解，导致呼吸障碍而引起死亡。一般认为在亚致死浓度中的水生动物耗氧能力下降，心跳减慢[15]。

2.2.2 对胚胎发育和繁殖的影响

有机磷农药能引起孵化率下降，对胚胎有致畸作用，可使幼体形体弯曲，身体瘦弱，眼睛色素沉淀，失去平衡，行为反常，身体上长出水泡，心包囊扩大，血液循环受阻[16-18]。Wong 等[16]发现大型蚤在 0.01 mg/L 的马拉硫磷中存活率下降，寿命缩短，幼体数量大大减少，但对繁殖第一代并无影响。

有机磷农药可抑制内分泌正常分泌水平，导致内分泌功能失调，影响性腺发育和分泌[17]。Kling 等[18]发现在杀螟硫磷中的性成熟罗非鱼的卵巢中无成熟的卵粒，平均卵径不到正常卵径的一半，且卵黄含量显著降低，实际繁殖力降低，但是卵敏感性一般较差。

2.2.3 组织中酶活性和血液参数的变化

有机磷最大的特点就是抑制各组织中的胆碱脂酶活性，尤其在致死浓度下乙酰胆碱酯酶和其他酯酶（如 ATP 酶）的活性下降更为明显，血液中的 AchE 比脑中的 AchE 更敏感。皮外注射阿托品或联苯，酯酶活性可以恢复正常。受害组织呈现碳氢化合，蛋白质，脂类的含量下降，糖原含量增加，酸性和碱性蛋白酶活性随暴露时间的延长而增加，氨基酸水平上升[19]，琥珀酸脱氢酶逐步被消耗，而乳酸脱氢酶和谷草转氨酶活性增加，酸性磷酸酶和碱性磷酸酶在最开始活性增加，其后活性下降[20]。

有机磷可引起血红素水平下降，红细胞数量下降，红细胞比容降低，血浆脂类总蛋白和碳氢化合物含量降低[20]。有机磷农药能引起机体的免疫反应，表现为白细胞组成和数量的改变，在一定的时间内数量上往往有明显的增加，其后减少[14]。此时，体内造血系统无疑已经出现障碍。

2.2.4 对内脏器官的损害作用

当有机磷农药被摄入后，对肝脏、胰脏、鳃、肠、肌肉等实质性脏器存在毒性效应。严重时，使这些组织的细胞坏死、破裂，致使其中的 RNA 和 DNA 含量降低。肝脏往往是影响最为严重的器官。有机磷在肝脏内转化为毒性更强的物质，如对硫磷转化为对氧磷，马拉硫磷转化为马拉氧磷，损害作用更大。Patil 等[14]指出久效磷引起大弹涂鱼肝脏细胞胞膜破裂，核偏移，脂肪降解，在 48~72 h 后便看不到正常的结构，盐分、铁离子、磷等含量降低，糖元含量升高，肝细胞数量呈减少趋势，核质更加密集。对肾脏的影响首先是近端小管受到损害接着肾小球造血组织及其他肾小管受害，形成空泡的细胞，胞质产生沉淀，随后坏死细胞出现。汝少国等[21,22]详尽研究了受久效磷影响的对虾内脏器官亚细胞的基本变化：细胞内质网严重水肿、扩张、囊泡化；高尔基膜囊水肿、扩张，高尔基小泡扩张，严重者破裂；线粒体内嵴局部瓦解；粘蛋白原和酶原颗粒明显增加；与久效磷对对虾的肝胰脏、肌肉的中毒症状相比，以中肠的中毒稍重。

2.2.5 生物积累效应

一般认为有机磷的生物富集问题不大。但据大量的研究表明，有机磷在体内的积蓄仍不可忽视。决定其在生物体内的积累量的关键是它的新陈代谢过程。Tsuda 等[23]指出鲤鱼在二嗪磷、马拉硫磷、杀螟硫磷中 12~48 h 时，其肌肉和内脏中的含量达到高峰。在一周后的二嗪磷在肝脏、肾脏、肌肉、胆囊的平均生物浓缩因子分别为 60.0、111.1、20.9 和 32.2。Dutta[24]曾研究了处于 6 mg/L 的马拉硫磷的囊鳃鲇 10 d 后的积蓄情况，以鳃残留量最大，其次为肝脏、胰脏，而肌肉中含量很低。残留量起初随染毒时间延长而逐步上升，达到一定量后，则急剧减少。

2.3 对水生动物毒作用机理

有机磷对具备神经系统的动物毒作用机理基本相同，均为抑制胆碱酯酶活性，引起乙酰胆碱代谢紊乱。在正常的神经传导过程中，酯酶和乙酰胆碱结合，形成酶-底物络合物，已酰基转移到胆碱酯酶上，形成酰化酯酶（该酶不稳定，半衰期仅为 0.1 s），迅速水解，又出现胆碱酯酶（即酶的复活），同时乙酰胆碱变成了无活性的乙酰和胆碱，等待下一个冲动的来临。但是当受到有机磷的危害时，胆碱酯酶受磷酸化作用，形成很稳定的磷酸化胆碱酯酶，该酶非常稳定（半衰期 10^4 s 以上），反应基本不可逆。这样乙酰胆碱代谢紊乱，大量蓄积而不能水解，致使后续神经元或效应器持续兴奋，引起痉挛麻痹，接着转入抑制。一般当组织胆碱酯酶抑制达 40%~60% 时，动物可在几秒钟内死亡[3]。

毒性与其化学结构极其相关，不同的化学结构，对胆碱酯酶分子上的阴离子和成酯部分亲和力不同，亲和力越强，毒

944

性越大。如（RO）$_2$-P≤结构中，R基含碳原子数多的毒性大，异丙基>乙基>甲基；硫联结构比硫结构毒性大。同种药物选择溶剂不同，毒性存在差别，强弱依次为乳油型>原油型>纯品型。

3　联合毒性效应

除了与生物种属品系，遗传特性、性别、年龄、环境、营养状况和疾病状态以及化合物性质，给毒剂量途径及给毒时间长短等外[25]，其他外来化合物的联合毒性应尤为注重，这些往往容易被忽视。

目前多数研究是针对单一因子对水生生物的毒性效应，而实际上水体中往往存在多种污染物，它们的作用无疑是综合的。要客观地反映污染物共存时对生物的危害程度，往往必须研究毒物的联合毒性效应。戴家银等[26]研究表明，铜离子与甲基异柳磷的联合毒性为拮抗作用、甲基异硫磷—甲胺磷的联合毒性为协同作用[27]。汝少国等[6]详尽地比较了久效磷、平硫磷、敌敌畏以及对硫磷的单一剂和混合剂对扁藻的急性毒性和联合毒性。

4　有机磷在生物体内的代谢转化及解毒

4.1　体内分解代谢

一个生物机体的存在依赖其内部补偿性机制来防止外来因素干扰。对进入体内的异物使之降解成水溶性的物质进而通过排泄排出体外。几乎所有有机磷农药分解的基本反应均是水解和氧化反应。缺氧条件下，由于微生物的影响，还可发生还原反应。化学结构不同，分解反应各异，如磷酸酯的主要分解反应是水解，硫代或二硫代磷酸酯发生氧化和水解反应。分解反应的快慢与较多的因素有关。一般地，植物体比动物体慢；在没有微生物参与的情况下，或处在低温、干燥暗淡的环境中，分解进行缓慢。吸入体内的有机磷分解一般较快，并生成无毒的产物。但在某些情况下，在代谢第一阶段也可生成比原来化合物更毒的物质。如乙拌磷、甲基乙拌磷的降解产物对乙酰胆碱酯酶的抑制作用比母本更强[28]。

4.2　有机磷农药的解毒

增效醚是目前用来治疗有机磷中毒的理想药物。在食物中拌喂增效醚，对虾在 0.01 mg/L 杀螟硫磷、0.2 mg/L 马拉硫磷、0.1 mg/L 二嗪磷溶液的存活率分别为对照组的（食物中不拌增效醚）11 倍、5 倍和 2.5 倍[29]。增效醚不影响机体对有机磷的吸收而是抑制其氧化，明显地降低有机磷的毒性。另据 Goel 等[30]报道维生素 B$_{12}$有抗甲基对硫磷毒性作用。

5　水域环境监控及含磷农药的废水处理

5.1　废水排放区域环境监控

有机磷农药污染给生物资源和水产养殖业构成严重的威胁，建立适当的有机磷农药毒性效应预测系统来监控水域生态环境显然极为重要。废水排放的生物学效应取决于其化学成分在水域中的持久性，被生物利用和累积的特性及其毒性效应。据 Capuzz[31]总结，废水排放的生物学效应可以在四个生物学组织水平上依次表现出来：①生化和细胞学水平；②综合了生理、生化和行为反应的生物个体水平；③种群水平，包括种群动态变化；④群落水平，可导致群落结构和动力学变化。值得注意的是这四个水平不是同时表现出来，只有当某一结构水平的补偿或适应机制开始失效时，才可能在下一个水平上表现出有害效应。在群落水平中，各种生物对有机磷农药的反应破感程度不一。据张甲耀等[32]报道，大型溞对有机磷农药非常敏感，作为生物监测材料有一定的可行性。

5.2　含磷农药的废水处理

有机磷农药废水的污染已受到国内外环保工作者的重视，各种处理方法层出不穷。文远高等[33]在反应池中投加粉末活性炭的采用间歇式活性污泥法（SBR 法）对有机磷农药废水的处理效果良好，对进水有机磷浓度的变化有较好的适应能力，且出水水质稳定。另外，利用微生物及其产生的降解酶进行水体中的有机磷农药去毒与净化是治理有机磷废水污染的有效新途径，有机磷农药促降解的微生物一般以荧光假单胞菌为主的混合菌，利用其提取的对硫磷水解酶来降解有机磷已显示出良好的应用前景[34]。

参考文献

[1] 范垂生，戚澄九. 农药与环境保护 [J]. 环境科学学报，1984，24（2）：50-55.

[2] 王梅林，郑家生，李永琪. 久效磷对僧帽牡蛎染色体毒性研究 [J]. 青岛海洋大学学报，1998，28（1）：75~81.

[3] 毛德寿，同宗灿. 环境生化毒理学 [M]. 沈阳：辽宁大学出版社，1986.

[4] Piska R. S., Waghray S., Toxic effects of dimethoate on primary production of lake ecosystem [J]. Indian J Environ Health，1991，33（1）：126-127.

[5] 唐学玺,李永祺,李春雁. 有机磷农药对海洋微藻致毒性的生物学研究 II. 久效磷胁迫下扁藻和三角褐指藻脂质过氧化伤害的研究 [J]. 海洋学报,1997,19(1):139-143.

[6] 汝少国,李永琪,敬永畅. 十种有机磷农药对扁藻的毒性 [J]. 环境科学学报,1996,16(3):337-341.

[7] 王建华,刘鸿先,徐同,等. 超氧化歧化酶在植物逆境和衰老中的作用 [J]. 植物生理通讯,1989,(1):1-7.

[8] McCord J. M., Fridovich 1. Superoxide dismutase:An enzyme function for erythrocuprein hemocuprein [J]. J boil Chem, 1969, 224:6049-6065.

[9] Jebakumar S. R. D., Jayaraman J.. Changes in protein, lipid and carbohydrate content in the freshwater fish Lepidocephalichthys thermalis during short-term sub-lethal exposure of malathion [J]. Ann Zool Agra, 1988, 26 (3):83-89.

[10] Natarajan E., Biradar R. S., George J. P., et al. Acute toxicity of pesticides to giant freshwater prawn Macrobrachium rosenbergii (De Man) [J]. J Aquaculture Tropics, 1992, 7 (2):183-188.

[12] Venugopal G., Swain D., George J. P., et al. Bioassay evaluation of toxicity of monocrotophos to a freshwater exotic carp, Cyprinus carpio communis (Linnaeus):Mortality and behaviour study [J]. J Environmental Biology, 1988, 9 (4):395-399.

[12] Govindan V. S., Jacob L., Devika R.. Toxicity and metabolic changes in Gambusia affinis exposed to phosphamidon [J]. J Ecotoxicol environmental Moniter, 1994, 4 (1):1-5.

[13] Bharathi C.. Toxicity of insecticides and effects on the behavior of the blood clam Anadara granosa [J]. Water air soil pollution, 1994, 75 (1-2):87-91.

[14] Shanmugavel S., Sampath K.. Sublethal effects of phosphamidon and methyl parathion on food intake, growth and conversion efficiency of the fish Oreochromis [J]. Environmental Ecology, 1988, 6 (2):257-261.

[15] Mohapatra B. C., Noble A.. Changes in oxygen uptake in mullet, Liza parsia (Hamilton-Buchanan) exposed to dichlorvos [J]. Fish technology society, 1993, 30 (2):112-114.

[16] Wong C. K., Chu hum F. Acute and chronic toxicity of malathion to the freshwater cladoceran Moina macrocopa [J]. Water air soil pollution, 1995, 84 (3-4):399-405.

[17] Singh P. B., Singh T. P.. Impact of malathion and gamma-BHC on steroidogenesis in the freshwater catfish, Heteropneustes fossilis [J]. Aquatic toxicology, 1992, 22, (. 1):69-80.

[18] Kling D. B.. Toxic effects of sublethal concentrations of Lebaycid upon the ovaries were examined in Sarotherodon leucostictus [J]. Z Angew zool, 1986, 73 (1):75-92.

[19] Sarbadhikary A., Sur R. K.. Effect of short duration exposure to methyl parathion followed by recovery of activities of some enzymes of the fish Oreochromis niloticus (Smith) [J]. Environmental ecology, 1992, 10 (2):333-340.

[20] Sarbadhikary A., Sur R. K., Effect of short duration exposure to methyl parathion on a target enzyme and some metabolic markers of the fish Oreochromis niloticus [J]. Environmental ecology, 1990, 8 (2):569-575.

[21] 汝少国,刘晓云,柳卫海,等. 久效磷对中国对虾细胞超微结构的影响 II. 对肠的毒性效应 [J]. 海洋水产研究,1997,18(1):1-8.

[22] 汝少国,李永祺,刘晓云,等. 久效磷对中国对虾细胞超微结构的影响 III. 对鳃的毒性效应 [J]. 应用生态学报,1997,8(6):655-658.

[23] Tsuda T., Aoki S., Kojima M., et al. Bioconcentration and excretion of diazinon, IBP, malathion and fenitrothion by carp [J]. Composition biochemistry physiology, 1990, 96 (1):108-119.

[24] Dutta G. R., Adhikari S.. Accumulation of malathion in different organs of Heteropneustes fossilis (Bloch) [J]. J Freshwater biology, 1994, 6 (2):183-186.

[25] Ravikumar S. S., Gupta T. R. C.. Toxicity of chlordane and malathion to silver carp and common carp [J]. First Indian Fisheries Forum, 1987, 4 (8):281-283.

[26] 戴家银,邓微云,王淑红. 重金属和有机磷农药对真鲷和平鲷幼体的联合毒性研究. 环境科学 [J]. 1997,18(5):44-46.

[27] 汝少国,李永琪,袁俊峰,等. 有机磷农药对扁藻的联合毒性研究. 青岛海洋大学学报 [J]. 1996,26(2):197-202.

[28] Gaelli R., Rich H. W., Scholtz R.. Toxicity of organophosphate insecticides and their metabolites to the water flea Daphnia magna, the Microtox test and an acetylcholinesterase inhibition test [J]. Aquatic toxicology, 1994, 30 (3):259-269.

[29] Kobayashi K., Wang Y.. Practical application of piperonyl butoxide for the reduction of organophosphorus toxicity [J]. Japanese society SCI fish, 1993, 59 (12):2053-2057.

[30] Goel S., Agrawal V. P.. Antitoxic effects of vitamin B sub (12) against methyl parathion poisoning in the fish Channa punctatus [J]. Environmental ecology, 1995, 13 (3):641-645.

[31] Capuzzo J. M.. Predicting pollution Effects in the Marine Environment [J]. Oceans, 1981, 24 (1):25-33.

[32] 张甲耀,肖化忠,张甫英,等. 三种有机磷农药萃取剂对水生生物的毒性效应 [J]. 中国环境科学,1996,16(5):382-385.

[33] 文远高,刘宏菊,房伟. SBR 法处理有机磷农药废水的研究 [J]. 工业水处理,2002,22(1):40-42.

[34] 虞云龙,陈鹤鑫,吕斌. 一硫代磷酸酯杀虫剂的酶促降解 [J]. 环境科学,1998,19(2):82-85.

该文发表于《渔业现代化，2008，35（4）：47-52》

伊维菌素在水产养殖中的应用及其水生态风险

江敏[1,2]，彭章晓[1]，吴昊[1]，胡鲲[1]，黄旭雄[1]

（1. 海洋大学，农业部水产种质资源与养殖生态重点开放实验室，上海 201306；
2. 水域环境生态上海高校工程研究中心，上海 201306）

摘要： 伊维菌素是一种广谱抗寄生虫渔药，在国外被广泛用于鲑鱼、海鲷养殖中的海虱防治，国内主要用于防治淡水鱼养殖中的各种寄生虫病，其在观赏鱼养殖中的应用也日益普遍。综述了目前国内外伊维菌素在水产养殖中的使用现状，并就其水生态风险进行评估。
关键词： 伊维菌素；水产养殖；水生态风险

伊维菌素（Ivermectin）是阿维菌素的衍生物，是目前世界上最优秀的广谱抗寄生虫药之一，具有广谱、高效、用量小、安全等优点，对生物体内外寄生虫，特别是线虫和节肢动物均有高效驱杀作用。自 1981 年投入市场作为兽用驱虫药后，在国内外临床兽医中已得到广泛应用[1]。随着研究的深入，人们发现伊维菌素也能作为渔药使用，且效果良好、价格低廉。但随着伊维菌素应用的日益广泛，其对水生生物尤其是底栖生物的毒性引起了人们的重视。对伊维菌素在水产养殖中的应用现状及其水生态风险进行了归纳与综述，为该药的后续研发、科学使用与管理提供参考。

1 伊维菌素的特点

1.1 理化性质和作用机理

伊维菌素是阿维菌素的氢化衍生物，为 22，23-二氢 B_{1a} 和 22，23-二氢 B_{1b} 的混合物，其中 B_{1a} 的活性最高，药物脂溶性高，其溶液在紫外 237 nm、245 nm 波长处有最大吸收峰[2]。

伊维菌素可与寄生虫体细胞上高特异性位点结合，增强虫体抑制性神经递质 r-氨基丁酸（GABA）的释放，以及打开谷氨酸控制的 Cl^- 通道。带负电的 Cl^- 引起神经元休止电位的超极化，使动作电位不能释放，神经传导受阻，最终导致寄生虫肌肉不能收缩，发生弛缓性麻痹而死[3]。

1.2 药代动力学

伊维菌素脂溶性高，被吸收后会广泛分布于生物体全身组织，并以肝脏和脂肪组织中含量最高，其在动物血液中消除时间较短，而在皮肤、肌肉和脂肪组织中消除时间较长。如伊维菌素在海鲷血液中的半衰期 $T_{1/2}=15.37$ h，而在虹鳟肌肉和胆汁中的半衰期分别达到 24 d 和 17 d[4-5]。数据显示[6]，伊维菌素在海鲷体内的药代动力学特征为吸收快（$T_{max}=2$ h）、生物利用率高（$AUC=10\ 700$ ng·h/mL）、血浆中药物半衰期短（$T_{1/2}=15.37$ h），对海鲷来说是一种优良的渔药。

2 伊维菌素在水产养殖中的应用

伊维菌素在养殖中的应用始于苏格兰的大西洋鲑鱼（*Salmo salar L.*）养殖。在过去 30 年中，大西洋鲑鱼养殖一直是苏格兰丘陵和岛屿地带最重要的经济发展产业。海虱是海生鲑鱼的体表寄生虫，会造成鲑鱼体重和品质的下降，给鲑鱼养殖造成严重影响，每年因海虱侵染造成的损失可达 1 500 万英镑；人们发现伊维菌素能驱杀海虱，且药剂成本相对较低，苏格兰环保局经过一系列科学调研后，于 1996 年作出了"一定条件下鲑鱼养殖场可以排放伊维菌素残留液"的规定[7-8]。随后，加拿大、英格兰、爱尔兰和其他一些北欧国家的鲑鱼养殖场也将其作为海虱防治的首选药。目前，伊维菌素的使用已推广至其他养殖品种，如金头鲷、虹鳟以及我国的四大家鱼等[9]。

谢卫华等[10]通过实际工作发现，伊维菌素防治斜管虫病效果很好，且用药方便、成本低廉。诸多优点使伊维菌素应用越来越广泛，水产专用的水体泼洒剂、混饲散剂已有生产，并用于虾蟹寄生虫病的防治[11]。在观赏鱼的寄生虫病防治中伊维菌素也越来越受人们的青睐[12]。

3 伊维菌素的水生态风险评估

3.1 伊维菌素对浮游和游泳生物的毒性

伊维菌素在水中溶解度很低，约为 10 μg/L。许多研究者进行了毒理试验，以探究水中低含量溶解态伊维菌素对浮游和游泳生物的毒性影响。表 1 显示，甲壳类浮游动物对伊维菌素非常敏感，而浮游藻类的敏感性却较低。

表 1　伊维菌素对浮游和游动生物的毒性数据[13-15]
Tab. 1　Toxicity of Ivermectin to plankton and nekton

	种类	测试时间	LC$_{50}$或 C$_{50}$（μg/L）
甲壳类	大型水溞（Daphnia magna）	48 h	0.005 7
	钩虾（Gammarus spp. *）	96 h	0.033
	普通新糠虾（Neomysis integer）	96 h	0.07
	无常小长臂虾（Palaemonetes varians）	96 h	54
	卤虫（Artemia Salina）	24 h	>300
	皱尾团水虱（Sphaemma rugicauda）	96 h	348
硬骨鱼类	虹鳟（Salmo gairdneri）	96 h	3.0
	蓝腮太阳鱼（Lepomismacmchims）	96 h	4.8
浮游藻类	绿藻（Pseudokirchneriella subcapitata）	48 h	> 4 000
	蛋白核小球藻（Chlorella pyrenoidose）	14 d	> 10 000

＊是迪氏钩虾（G. dueheni）和扎氏钩虾（G. zaddachi）的混合体

目前，国内外常用 K 值（K=PEC/PNEC，即预测浓度/预测无作用浓度）进行伊维菌素的环境风险评估，其中 PNEC 值通常由 LC$_{50}$/10 推算得到。在欧洲许多国家的鲑鱼养殖场中，伊维菌素的常用剂量为平均每千克鲑鱼施药 0.025 mg，每周 2 次，连续 4 周[16]。根据 Davies 等[17]的预测，在不考虑伊维菌素在海水中降解的情况下，渔场周围海水中伊维菌素的预计质量浓度 PEC 约为 4×10^{-5} μg/L，最大预计质量浓度为 15×10^{-5} μg/L。钩虾和糠虾是目前已知对伊维菌素最为敏感的海水生物，它们的 96 h LC$_{50}$值分别为 0.033 μg/L 和 0.07 μg/L，因此用这 2 种生物的 K 值对伊维菌素的环境风险进行评估具有代表性。计算可知，钩虾和糠虾的 K 值分别为 0.012 和 0.006，当 PEC 取最大值 15×10^{-5} μg/L 时，K 值分别为 0.05 和 0.02，远远小于 1，如果再把伊维菌素在水中的降解因素考虑进去，这个值会更小。这说明渔场周围溶解态的伊维菌素对海水中的浮游和游泳生物几乎没有影响。

目前已知的淡水浮游和游泳生物中，大型蚤对伊维菌素最为敏感，48 h LC$_{50}$仅为 0.005 7 μg/L，比钩虾还要敏感 6 倍，但是由于淡水养殖水体中伊维菌素的预测浓度迄今仍未有人报道，因此其环境风险就较难进行准确评估。

3.2 伊维菌素对底栖生物的毒性

伊维菌素的 K$_{oc}$为 12 600~15 700，这意味着进入水体的伊维菌素以溶解态形式存在的比例很低，绝大部分将会与含有机质丰富的底泥结合。伊维菌素在底泥中的降解十分缓慢，其降解半衰期>100 d，因此底泥中伊维菌素的含量会较高，这将可能对底栖生物，尤其是那些以底泥中有机碎屑为食的滤食性生物造成较大影响。表 2 总结了已发表的伊维菌素对部分底栖生物的毒性数据。

表 2　伊维菌素对水生底栖生物的毒性[18-19]
Tab. 2　Toxicity of Ivermectin to benthos

种　类	暴露时间	LC$_{50}$（μg/L）
沙蠋（Arenicola marina）	10 d	23＊
卷曲螺蠃蜚（Corophium volutator）	10 d	180＊
红海盘车（Asterias rubens）	10 d	23 600＊
沙蚕（Hediste diversicolor）	96 h	7.75

续表

种　类	暴露时间	LC_{50}（μg/L）
褐虾（Crangon septemspinosa）饲料暴露	96 h	>21.5
水暴露	96h	8 500
长牡蛎（Crassostreagigas）幼体	96 h	80~100
卵	96 h	460
贻贝（Mytilus edulis）	96 h	400
帽贝（Parella oulgata）	96 h	600
欧洲大扇贝（Pecten moximus）	96 h	300
单齿钟螺（Monodon talineata）	96 h	780
狗岩螺（Nucella lapillus）	96 h	390
欧洲玉黍螺（Littorina littorea）	96 h	580
婆罗囊螺（Hydrobia ulvae）	96 h	>10 000
新西兰泥螺（Potamopyrgus jenkinsii）	96 h	<9 000
普通滨蟹（Carcinus maenas）	96 h	957

＊标记的单位为 μg/kg 干底泥。

　　沙蹋是迄今为止发现的对伊维菌素最为敏感的底栖生物，其 10 d 的 LC_{50}=23 μg/kg，因此可将其作为代表进行伊维菌素对底栖生物的生态风险评价[16]。Davies 等[17] 对鲑鱼渔场及其周围水域底泥中伊维菌素含量进行了预测，认为施药中心区域底泥中其含量可达 66.823 μg/kg，周边将近 10 000~23 000 m² 的海域底泥中将超过 2 μg/kg。用 K 值法来评估伊维菌素对沙蹋的毒性效应，沙蹋的 PNEC=1.8 μg/kg，当 PEC 值取 2 μg/kg 时 K=1，当 PEC 值取 66.8 μg/kg 时 K=36，这说明渔场施药的中心区域底泥中伊维菌素会对许多底栖生物如多毛目环节动物沙蚕、沙蹋等造成急性毒性影响，而渔场周边区域底泥中的伊维菌素对底栖生物产生直接急性毒性的可能性较小，但很有可能具有长期的亚致死效应。

　　尽管底栖贻贝类、螺类等软体动物对伊维菌素的敏感性比多毛目环节动物（如沙蚕）低几百倍，但是这些软体动物对脂溶性的伊维菌素具有较大的生物富集能力。贻贝体内伊维菌素的质量浓度约为海水中的 750 倍[17]。因此，考虑生物富集因素时伊维菌素对这些底栖软体动物的生态风险也较大。同种生物不同的生活史阶段以及不同的药物接触方式都会造成底栖生物对药物的不同敏感性。生物的幼体阶段和蜕皮脱壳阶段一般对药物较为敏感，如伊维菌素对长牡蛎幼体 96 h 的 LC_{50} 为 80~100 μg/L，而对成体为 460 μg/L。褐虾喜欢摄食水体底部的残饵和有机碎屑，当以拌有伊维菌素的饲料对褐虾进行毒理试验时发现，褐虾对伊维菌素很敏感，96 h LC_{50}>21.5 μg/L，而当用水暴露的方式进行毒理试验时，发现它对伊维菌素不敏感，96 h 的 LC_{50}=8 500 μg/L[19]。

　　伊维菌素在亚致死浓度下对底栖生物的影响同样不能忽视。Allen 等[21] 研究发现，伊维菌素对欧洲玉黍螺的亚致死浓度仅为其 LC_{50} 的 1/1000，在此浓度下，欧洲玉黍螺不能爬行，仅能将身体缩至壳中，这使它很容易被其他肉食动物捕食。这种亚致死效应同样存在于其他底栖生物，如亚致死浓度下的伊维菌素会影响沙蹋和卷曲螺蠃蜚的脱壳率和存活率，进而影响到其种群生长情况。尽管要对伊维菌素的环境风险进行准确评估还需进行更多深入的研究，但就目前已知的信息来看，伊维菌素对底栖生物会有较大影响。

3.3　伊维菌素的原生态群落风险评估

　　Sanderson 等[22] 开创性地进行了伊维菌素的原生态环境风险评估，他们发现，伊维菌素在自然水体中的降解半衰期 DT_{50}=4 d，为实验室条件下测得的 2 倍；其在底泥沉积物中的含量几乎是不变的，这就意味着自然条件下伊维菌素很难降解。施药前期（10~68 d）高质量浓度伊维菌素处理组（1 μg/L）的 DO 和 pH 值分别比对照组高出 25%~30% 和 10%~12%，表明伊维菌素可能通过影响水中生物的活动来间接影响水体的理化性质。伊维菌素对水中浮游生物具有短期影响，长期影响效应不明显。其对水中叶绿素 a 的含量几乎无影响，对水溞类、桡足类和介形亚纲动物的影响很相似，短期内会使种群个体数量和物种丰度下降，而经过较长时间（229 d）后，这些浮游生物的种群规模和物种丰度几乎都能恢复，只不过桡足类生物恢复得更快一些，且不会出现优势种的变化，而水溞类恢复后会出现优势种的变化。生活在水体底层的水溞类（如圆形盘肠溞）则较难恢复，其优势种地位会被其他种取代。水中摇蚊幼虫、椿象、甲虫等也会受到较大影响，质量浓度为 1 μg/L 的处理组经过 265 d 后种群的数量和物种丰度仍未恢复。Sanderson 等进行的原生态下伊维菌素的环境风险评估更多侧重于对浮游生物的影响研究，几乎没有涉及到对底栖生物的影响研究，因此这方面的研究还有待进一步深入。

4　结语

伊维菌素从兽药转型为渔药并广泛应用于水产养殖中，表现出了一定的应用前景，但仍有许多问题有待深入研究：①伊维菌素在养殖鱼体内的药代动力学特征；②伊维菌素在水生生物体内的富集情况；③伊维菌素在养殖鱼体内的药物残留情况和保证水产品安全所必需的休药期；④伊维菌素的原生态群落风险评估。上述问题的解决会使人们更加深入地了解伊维菌素各方面的性质，有助于减小其水生态风险，使之更好地应用于水产养殖。

参考文献

[1] 符华林，吴蕾．伊维菌素脂质体的制备及质量控制研究 [J]．动物医学进展，2004，25（6）：102-104.

[2] 刘群，卢芳，艾青．紫外分光光度法测定伊维菌素片的含量 [J]．中国兽药杂志，2002，36（1）：21-22.

[3] DARRANM, YATES, ADRIAN J W. An ivermectin-sensitive glutamate-gated chloride channel subunit from Dirofilaria immitis [J]. International Journal for Parasitology，2004，34（9）：1075-1081.

[4] ROTH M, RAE G, MCGILL A S, et al. Ivermectin depuration in Atlantic salmon（Salmosalar）[J]. Journal of Agricultural and Food Chemistry，1993，41（12）：2434-2436.

[5] BADAR S, NATHAN R, CHARLES G, et al. Residue depletion of tritium-labeled ivermectin in rainbow trout following oral administration [J]. Aquaculture，2007，272（1-4）：192-198.

[6] KATHARIOSP, ILIOPOULOU-GEORGUDAKIJ, ANTIMISIARSS, et al. Pharmacokinetics of ivermectin in sea bream, Sparusaurata using a direct competitive ELISA [J]. Fish Physiology and Biochemistry，2002，26（2）：189-195.

[7] 张翼翾．苏格兰海虱防治的历史和现状 [J]．世界农药，2002，24（6）：41-45.

[8] Scottish Environment Protection Agency. The occurrence of the active ingredients of sea lice treatments in sediments adjacent to marine fish farms：results of monitoring surveys carried out by SEPA in 2001 &2002 [R]. SEPA Report TR-021009A_ G. 2004.

[9] HALLEY B A, JACOBTA, LUAYH. The environmental impact of the use of ivermectin：Environmental effects and fate [J]. Chemosphere，1989，18（7-8）：1543-1563.

[10] 谢卫华，张新强．伊维菌素治疗斜管虫鱼病效果 [J]．科学养鱼，2007（4）：59.

[11] 夏森，仇明生．提高养蟹经济效益的一些措施 [J]．渔业致富指南，2004（5）：43.

[12] 包海岩，张勤．庭院养殖观赏鱼鱼病防治一例 [J]．科学养鱼，2004（7）：51.

[13] JEANNE G, BERNARD V, KAREN D, et al. Effects of the parasiticide ivermectin on the cladoceran Daphnia magna and the green alga Pseudokirchneriella subcapitata [J]. Chemosphere，2007，69（6）：903-910.

[14] ALASTAIR G, REW D. Toxicity of ivermectin to estuarine and marine invertebrates [J]. Marine Pollution Bulletin，1998，36（7）：540-541.

[15] HALLEY B A, NESSEL R J, LUAYH. Environmental aspects of ivermectin usage inlivestock：general considerations [M] //Campbell W C. Ivermectin and abamectin. NewYork：Springer-Verlag，1989：162-172.

[16] DAVIES I M, GILLIBRAND P A, MCHENERY J G, et al. Environment risk of ivermectin to sediment dwelling organisms [J]. Aquaculture，1998，163（1-2）：29-46.

[17] DAVIES I M, MCHENERY J G, RAE G H. Environmental risk from dissolved ivermectin to marine organisms [J]. Aquaculture，1997，158（3-4）：263-275.

[18] THAIN J E, DAVIES I M, RAE G H, et al. Acute toxicity of ivermectin to the lugworm Arenicola marina [J]. Aquaculture，1997，159（1-2）：47-52.

[19] KILMARTIN J, CAZAHON D, SMITH P. Investigations of the toxicity of ivermectin to salmonids [J]. Bulletin of the European Association of Fish Pathologists，1997，17（21）：58-61.

[20] GRANT A, BRIGGS R D. Use of Ivermectin in marine fish farm：some concerns [J]. Marine Pollution Bulletin，1998，36（8）：566-568.

[21] ALLEN Y T, THAIN J E, HAWORTH S, et al. Development and application of long-term sublethal whole sediment tests with Arenicola marina and Corophium volutator using Ivermectin as the test compound [J]. Environmental pollution，2007，146（1）：92-99.

[22] SANDERSON H, LAIRD B, POPE L, et al. Assessment of the environmental fate and effects of ivermectin in aquatic mesocosms [J]. AquaticToxicology，2007，85（4）：229-240.

第三节　渔药的毒副作用

该文发表于《水生生物学报》【2005，29（1）：13-20】

喹乙醇在鱼体内蓄积及其对鱼类的影响

杨先乐[1]，胡鲲[1]，邱军强[1]，刁进宏[2]

（1. 农业部渔业动植物病原库，上海水产大学，上海 200090；2. 江苏省国营南通农场，南通 226000）

摘要：本文以含 350 mg/kg 和 400 mg/kg 喹乙醇的饲料经 99 d 分别连续投喂鲤和银鲫，以测定喹乙醇在鲤和银鲫体内的蓄积及其对鲤、银鲫的影响。试验结果表明，无论是银鲫单养还是鲤鲫混养，无论是鲤还是银鲫，试验组喹乙醇在肝脏组织中蓄积量最高，99 d 单养银鲫组可达 10.057 ± 0.015（3）mg/kg，鲤鲫混养组银鲫、鲤分别达 10.107 ± 0.226（3）mg/kg，9.883 ± 0.032（3）mg/kg，肾脏次之，99 d 单养鲫鱼组的银鲫，鲤鲫混养组的银鲫、鲤分别为 7.494 ± 0.064（3）mg/kg，7.777 ± 0.138（3）mg/kg 和 7.608 ± 0.086（3）mg/kg，肌肉中较低，分别为 0.1170 ± 0.003（3）mg/kg，0.160 ± 0.003（3）mg/kg，0.486 ± 0.006（3）mg/kg；经检验三种组织器官中的蓄积量差异显著（$P<0.05$）。由于喹乙醇在鱼体内的蓄积，虽鲤鲫混养组的银鲫、鲤的相对增重率达到 102.7% 和 110.7%，但是它们的抗应激反应率明显下降，而且随喹乙醇摄入量的增加下降趋势更为明显，当给予一定程度的应激刺激后，单养鲫鱼组的银鲫应激反应率由 58 d 的 44.8% 上升到 99 d 的 64.3%，鲤鲫混养组的银鲫和鲤分别由 58 d 的 66.7% 和 46.2% 分别上升到 99 d 的 88.3% 和 76.9%，无论是 58 d 还是 99 d，试验组与对照组间差异显著（$P<0.05$）。试验发现，摄入喹乙醇的银鲫肝脏炎症细胞浸润、细胞体积缩小、核固缩，甚至溶解；肾脏肾小球肿大，肾小管上皮细胞空泡变性；银鲫、鲤脾脏网状细胞肿胀变性、界限不明显等。研究认为，喹乙醇对养殖鱼类具有较大的毒副作用，应该禁止作为添加剂在鱼类饲料中使用。

关键词：鲤；银鲫；喹乙醇；蓄积；毒性

喹乙醇（Olaquindox），又称喹酰胺醇，由于它能提高饲料转化率，促进动物生长和具有广谱的抗菌作用，曾作为饲料添加剂在水产养殖中广泛使用[1-3]。然而由于喹乙醇大剂量、长期使用，导致养殖鱼类中毒，抗应激能力低下，出现"应激性出血"，并发生大量突发性死亡，给养殖生产造成了较大的损失[3-5]。

关于喹乙醇毒性的研究，在家畜、家禽上已有较多报道[6-8]。虽然近年来已发现了喹乙醇在水产养殖中的危害，但对它在水产动物机体内的蓄积研究及其影响至今鲜有报道，以至于人们仍存在侥幸心理，对喹乙醇的使用禁而不止，因喹乙醇添加而造成损失的事件仍时有发生。为了给水产养殖科学用药提供依据，本文对在常规养殖的鲤、鲫鱼中投喂添加喹乙醇的饲料后喹乙醇在鱼体的蓄积规律，以及其蓄积对鲤鲫生长、抗应激反应和组织病理的变化进行了研究。

1　材料和方法

1.1　试验鱼

银鲫（*Carassius auratusgibelio*，75~100 g/尾）、鲢（*Hypophthalmichthys molitrix*，约 200 g/尾）、鳙（*Aristichthysnobilis*，约 200 g/尾）、草鱼（*Ctenopharyngodon idella*，约 150 g/尾）及鳊（*Megalobramaamblycephala*，约 150 g/尾）均由江苏省国营南通农场提供；鲤（*Cyprinus carp*，140 g/尾）由连云港天利水产养殖公司提供，以上鱼种均健康无病，驯养一周后进行试验。

1.2　配合饲料

由上海曙光饲料厂提供，其中试验组银鲫配合料与鲤鲫混养鱼配合料喹乙醇添加量分别为 350 mg/kg 和 400 mg/kg，喹乙醇由江苏省南通天成保健品有限公司生产（生产批号 020109），含量为 99.4%。

1.3　试验方法

2002 年 7 月 31 日—11 月 6 日（共 99 d）在江苏省国营南通农场 4 个 6×4×1.2 m 的水泥池进行试验，水深 0.8 m，水温范围为 14℃～38℃，平均水温为 25.9℃。试验分为 4 组，其中一组主养鲫（主养银鲫 56 尾，并套养 5 尾鲢、1 尾鳙、3 尾草鱼、2 尾鳊）；另一组鲤鲫混养（除放养银鲫 15 尾，鲤 34 尾外，其余同单养鲫鱼组），每天均按体重的 2%～4% 投喂含有喹乙醇的银鲫配合料（单养鲫鱼组）或鲤鲫混养鱼配合料，并平行设立 2 个与试验对应的投喂不含喹乙醇配合料的对照组，按照常规方法进行饲养管理。试验开始后 58 d 和 99 d 分别对 4 个组约 1/3 的鱼给予同等强度的捕捞、捆箱等应激刺激（应激刺激的强度以对照组鱼不死亡为度），计算应激反应率

应激反应率 = （对照组成活率–试验组成活率）/对照组成活率×100%

同时每组随机抽取银鲫和鲤（各 3 尾）作喹乙醇蓄积量的检测。99 d 称量各组鱼的体重，计算其相对增重率：

相对增重率 = （试验组增重率–对照组增重率）/对照组增重率×100%

并对各组的鲤、银鲫的肝、肾、脾、肌肉等组织进行病理检查。喹乙醇的测定采用 HLPC 法，按中华人民共和国出口商品行业标准——出口肉中喹乙醇残留检验方法（SNO197-94）改进后进行（方法的建立将另文报道），病理检查基本参照龚志锦和詹榕洲的方法进行[9]。

2 结果

2.1 喹乙醇在鱼体内的蓄积

以甲醇和水（15：85）作流动相用 HLPC 法对试验鱼各组织中喹乙醇的含量进行检测，其结果表明：投喂喹乙醇的试验组，无论是单养鲫鱼组还是鲤鲫混养组，无论是银鲫还是鲤，无论是投喂后 58 d 还是 99 d，均表现出肝脏中喹乙醇蓄积量最高，肾脏次之，肌肉中稍低；58 d 单养鲫鱼组的银鲫肝、肾、肌肉喹乙醇的蓄积量分别为 6.4820±0.215（3）、4.659±0.193（3）、0.106±0.002（3）mg/kg，99 d 其蓄积量分别增加到 10.057±0.015（3）、7.494±0.064（3）、0.1170±0.003（3）mg/kg；鲤鲫混养组，银鲫喹乙醇的蓄积情况基本与单养鲫鱼组相同，58 d 和 99 d 肝、肾、肌肉分别为 6.517±0.120（3）、4.629±0.074（3）、0.125±0.004（3）mg/kg 和 10.107±0.226（3）、7.777±0.138（3）、0.160±0.03（3）mg/kg，而在鲤体内的蓄积分布似乎有所不同，在肌肉中的蓄积量要大于单养鲫鱼组，而在肝、肾中却稍略低于单养鲫鱼组，经 t 检验三种组织器官中的蓄积量差异显著（$P<0.05$）。58 d 和 99 d 肝、肾、肌肉中分别为 5.907±0.143（3）、4.147±0.052（3）、0.4360±0.018（3）和 9.833±0.032（3）、7.608±0.086（3）、0.486±0.006（3）mg/kg，各组织 58d 与 99d 无显著差异（以上对照均未检出，图 1、图 2）。

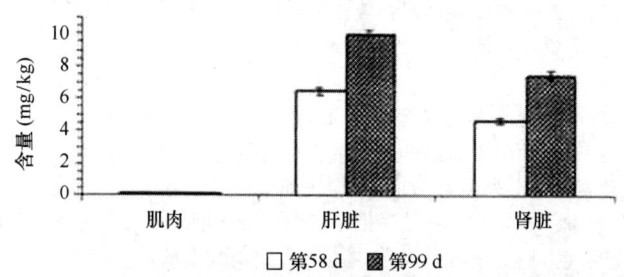

第58 d　第99 d

三种组织器官中的蓄积量差异显著 ($P<0.05$)，但58 d 与99 d
两个时间点间无显著差异

图 1　单养鲫鱼组喹乙醇在银鲫部分组织中的蓄积
Fig. 1　Accumulation of olaquindox in the tissues of erucian carp (*Carassius auratus gibelio*) in motocultural carp group
三种组织器官中的蓄积量差异显著 ($P<0.05$)，但 58 d 与 99 d 两个时间点间无显著差异

2.2 喹乙醇蓄积的试验鱼的抗应激反应

58 d 和 99 d 分别对试验鱼给予捕捞、捆箱等应激刺激，均会导致部分试验鱼（银鲫和鲤）眼球突出，上下颌、鳍基、鳞片下、鳃盖、腹部充血，出血，尤以腹部明显；试验鱼呼吸困难，不久便死亡；解剖可见腹腔充满红色腹水，鳔壁、肝、脾及肠系膜均有充血现象。表 1 结果表明随着喹乙醇在鱼体内的蓄积，试验鱼应激反应率明显上升，单养鲫鱼试验组银鲫 99 d 比 58 d 上升了 19.5%，鲤鲫混养试验组银鲫、鲤 99 d 的应激反应率比 58 d 分别上升了 16.7% 和 30.7%。

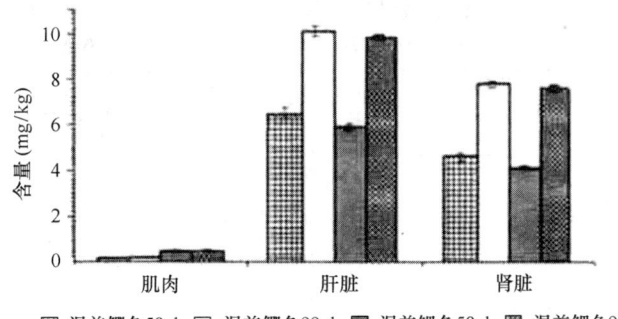

三种组织器官中的蓄积量差异均显著（$P<0.05$，$P<0.10$），58 d与

99 d两上时间点间无显著差异

图 2　喹乙酸在鲤鲫混养银鲫及鲤鱼部分组织中的蓄积

Fig. 2　Accumulation of olaquindox in the tissues of crucian carp (*Carassitu auratus gibelio*) and

common carp (*Cyprirus carpio*) in polycultural groups

三种组织器官中的蓄积量差异显著（$P<0.05$），但58 d与99 d两个时间点间无显著差异

表 1　喹乙醇蓄积对鱼类应激反应的影响

Tab. 1　The effect of olaquindox on stress of fishes

组别 Group		58 d The 58 th day			99 d The 99 th day		
		试验尾数 Total	存活尾数 Survival	存活率 S. R.（%）	试验尾数 Total	存活尾数 Survival	存活率 S. R.（%）
单养鲫鱼 Pru.	试验组 Test	18	10	55.2a	14	5	35.7c
	对照组 Control	18	18	100d	14	14	100d
	应激反应率 R. R.（%）[1]			44.8b			64.3e
鲤鲫混养 Pru. and Com.	试验组 Test　银鲫 Pru	3	1	33.3a	12	2	16.7c
	鲤 Com.	13	7	53.8a	13	3	23.1c
	对照组 Control　银鲫 Pru	3	3	100d	12	12	100d
	鲤 Com.	13	13	100d	13	13	100d
	应激反应率 R. R.（%）　银鲫 Pru			66.7b			83.3e
	鲤 Com.			46.2b			76.9e

S. R. –Survival rate；R. R. –Reaction rate；Pru. –Prussian carp；Com. –Common carp.

试验组与对照组间差异显著（$P<0.05$），应激反应率、存活率在58 d和99 d间存在显著性差异（$P<0.05$）。而在不同分组的鱼之间无显著性差异。注：同一列或同一行中上标非相同字母者，表示组间差异显著，上标相同字母者表示组间差异不显著。

2.3　喹乙醇蓄积对鱼体增重的影响

试验结果表明，喹乙醇对试验鱼有明显的增重作用，99 d单养鲫鱼组银鲫的相对增重率为22.1%，而鲤、鲫混养组银鲫为102.7%，鲤为110.7%（表2）。

2.4　喹乙醇蓄积的试验鱼肝、肾、脾、肌肉的组织病理变化

连续投喂含喹乙醇饲料99 d后，试验鱼的肝、肾、脾和肌肉均出现了不同程度的病变。银鲫肝脏炎症细胞浸润，部分肝细胞体积缩小，核发生固缩，甚至溶解消失，形成溶解灶，有时还可见严重脂肪变性和水泡变性（图版Ⅰ：1）；肾脏肾小球肿大，炎症细胞浸润，肾小管上皮细胞空泡变性（图版Ⅰ：4、6）；银鲫、鲤脾脏均表现炎症细胞增多，网状细胞肿胀变性，细胞界限不明显（图版Ⅰ：5、6），有时还可见少量细胞空泡变性（图版Ⅰ：3、5）；鲫肌肉肌纤维变形、折叠，肌间隙变宽，肌细胞横纹模糊不清，甚至溶解、消失（图版Ⅰ：2）。

表2 喹乙醇对鱼类生长的影响

Tab. 2 The effect of olaquindox on the growth of fishes

组别 Group		放养尾数 Total	每尾均重 Average body weight（g）		增重率 WGR.（%）
			开始时 Initial	结束时 Final	
单养鲫鱼 Pru.	试验组 Test	56	0.150	0.229	52.4
	对照组 Control	56	0.150	0.214	42.9
鲤鲫混养 Pru. and Com.	试验组 Test 银鲫 Pru	11	0.200	0.275	37.5
	鲤 Com.	34	0.279	0.323	15.8
	对照组 Control 银鲫 Pru	11	0.200	0.237	18.5
	鲤 Com.	34	0.279	0.300	7.5

WGR. -weighu gain rate；Pru. -Prussian；Com. -Common carp. 试验组与对照组间增重率差异极显著（$P<0.01$）。

增重率＝（试验结束时的重量-试验开始时的重量）/试验开始时的重量×100%。

3 讨论

3.1 喹乙醇的促生长作用

喹乙醇是以邻硝基苯胺为原料，通过化学方法而合成的一种喹啉类的化学药物，因为它对家畜家禽有明显的促生长作用，故又称为快育灵，曾广泛地添加于饲料中[10,11]。水产养殖中应用喹乙醇作饲料添加剂，对草鱼、建鲤等鱼类的生长均有明显的影响[1,2]，本试验也进一步证明了它对鲤和银鲫的促生长作用。从试验结果还可看出，喹乙醇对鲤与银鲫生长的影响基本上相同，而且喹乙醇的添加量加大（由350 mg/kg增加到400 mg/kg），它的促生长作用明显加强。这是因为喹乙醇能通过机体的内分泌系统，影响机体的代谢机能，促蛋白质合成而促进了鱼类的生长[12]。作者认为，鲤与银鲫为同一科同一属的鱼类，之所以喹乙醇对它们的影响基本一致，是因为喹乙醇对它们的作用机制基本相同。

3.2 喹乙醇在鱼体内的蓄积

试验揭示，喹乙醇均会在鲤、银鲫的肝脏、肾脏和肌肉中形成蓄积，而且蓄积会随着投喂时间的延长而不断增加；试验也证明，肝脏、肾脏是鱼类喹乙醇蓄积的主要器官，其中尤以肝脏为甚。这一结果与曾子建等、汪开毓等、孙永学等[4,13-14]报道的结果基本一致。肝脏、肾脏是喹乙醇运转和代谢的主要器官，在运转和代谢的过程中，这些器官也就成了喹乙醇的一个暂时储库，此外，喹乙醇也通过这些器官的运转，在鱼体其他组织中贮存起来。曾子建等、孙永学等[13,14]认为喹乙醇在体内吸收、分布比较快，消除比较慢，从而延长了喹乙醇在体内滞留的时间。如果长期使用，那么就会导致它在体内的蓄积，由此也导致了喹乙醇在鲤、银鲫体内的蓄积量随着投喂时间的延长而不断增加的现象。实验结果表明了喹乙醇在鲤、银鲫体内的蓄积情况基本相同，因此可以从一个方面说明喹乙醇在这两种鱼类中的运转与代谢规律基本一致。

3.3 喹乙醇的毒性

喹乙醇在体内的蓄积，使鱼类产生慢性中毒，表现在肝、肾、脾等实质性的器官损伤，肌肉组织不同程度的病变，其中尤以肝脏和脾脏较为突出。这不仅因为他们在肝脏等器官中蓄积量大，增加了对其损伤的可能性，而且还因为肝脏是喹乙醇的主要代谢器官[16,17]，由于其损伤，进一步造成了喹乙醇代谢障碍，延长了喹乙醇在体内滞留的时间而增加了喹乙醇中毒的可能性。鱼类对喹乙醇中毒的一个重要表现是鱼类的抗应激能力减弱，当给予相应的应激刺激后，鱼类就表现应激性出血。本试验揭示，鲤和银鲫的应激反应随着喹乙醇摄入时间的增加而增大，鲤和银鲫相比，银鲫较易产生应激，而鲤要产生相同强度的应激则要比银鲫摄入喹乙醇的时间要稍长些。这是因为喹乙醇较易在银鲫肝肾等组织中蓄积。Waldmann[16]认为，喹乙醇的毒性作用机理是与肾上腺的损伤有关，当肾上腺受损后而导致肾上腺皮质激素（醛固酮等）分泌减少，体内电解质平衡失调，一旦受到应激因子的刺激，即表现出应激性出血等中毒症状[17]。鱼类尚未完全证实具有肾上腺皮质激素，但作者认为鱼类因喹乙醇中毒引起的应激性出血症可能是因为肝肾等器官损害，引起内分泌失调而导致电解质平衡故障而造成。此外由于肝的损伤，肝合成Vc的功能减弱，使具有抗应激作用的Vc含量降低，而使鱼类的抗应激能力显著下降[14]。

董漓波采用定期递增染毒法测得喹乙醇的蓄积系数为3.24，认为它是中等蓄积的药物[18]，此外还有研究证明喹乙醇具有遗传毒性和诱变性[19-21]，试验进一步地证明了喹乙醇对养殖鱼类存在着明显的毒副作用。尽管喹乙醇具备较好的促生长性能，但因它具有较大的副作用而限制了它的应用。因此禁止喹乙醇作为饲料添加剂使用应该是不容置疑的。应加大对喹乙醇禁用的力度，以提高水产品品质和保证水产养殖业的持续稳步发展。

参考文献

[1] Ye J. Y., Chen Y. Y., Shen Z. H., The effects of HMQ added to feeds on growth and survival of one year old grass carp [J]. J. Zhejiang Coll. fish., 1992, 11 (1)：25-28 [叶金云，陈月英，沈智华. 喹乙醇（HMQ）添加量对一龄草鱼生长和成活率的影响. 浙江水产学院学报, 1992, 11 (1)：25-28].

[2] Wen L. Y., Li Y., Yang J. Q.. Theeffects of olaquindox added tofeeds on thegrowth of carp [J]. Freshwater Fisheries, 1995, 25 (1)：18-20 [文良印，李义，杨加琼. 饲料中添加喹乙醇对建鲤生长的影响. 淡水渔业, 1995, 25 (1)：18-20].

[3] Guo Q., Ren Z. L., Zeng H.. Utilization of olaquindox in aquaculture [J]. ChineseFeeds, 1997, 5：27-28 [郭庆，任泽林，曾虹. 喹乙醇在水产养殖中的应用. 中国饲料, 1997, 5：27-28] Animal Physiology and Animal Nutrition, 1998, 80 (2-5)：260-269.

[4] Geng Y., Wang K. Y.. The olaquindox should beusedwith care in aquaculture [J]. Reservoir Fisheries, 1999, 19 (6)：25-26 [耿毅，汪开毓. 喹乙醇在水产养殖上应慎用. 水利渔业, 1999, 19 (6)：25-26].

[5] Gao J.. Studies on security of olaquindox applied on carp breed [J]. Feeds Study, 2000, 2：36-37 [高俊. 喹乙醇在鲤养殖中安全性的研究. 饲料研究, 2000, 2：36-37].

[6] Waldmann K. H.. Clinical and hematological changes following intoxication with olaquindox in pigs [J]. Journal of veterinary medicine（series A）. 1989, 36 (9)：676-686.

[7] Zeng Z. L.. Thepharmalogy characters of olaquindox applied in broilers. Broilers' Culturing and Disease, 1995, 3 (3)：11-12 [曾振灵. 禽用喹乙醇的药物学特性. 养禽与禽病, 1995, 3 (3)：11-12].

[8] Tomoko Nagata. Determination of olaquindox residues in swine tissues by liquid chromatography [J]. J. Assoc Anal. chem. 1987, 4：706-707.

[9] Gong Z. J, Zan R. Z.. Pieces of pathology tissue and colourationtechnology [M]. shanghai：Shanghai Scientific and Technical Press, 1994, 140 [龚志锦，詹榕洲. 病理组织切片和染色技术 [M]. 上海：上海科学技术出版社 1994, 140].

[10] Kamphues J.. The discussionon growth-promoting feed additives-willing or not, veterinary nutritionists are especially involved [J]. Journal of

[11] Koehler Bernd, Karch Helge, Schmidt Herbert. Antibacterias that are used as growth promoters in animal husbandry can affect the release of Shiga-toxin-2-converting bacteriopphages and Shiga toxin 2 from Escherichia coli strans [J]. Microbiology, 2000, 146 (5)：1085-1090.

[12] Xu S. K.. Olaquindox [J]. Zhejiang Pasturage Technology. 1991. 16 (4)：27-29 [徐树宽. 喹乙醇. 浙江畜牧科技, 1991, 16 (4)：27-29].

[13] Zeng Z. J., Li Z. B., Wu X. T., et al. Studies on carp pharmacokinetics of olaquindox [J]. Journal of Sicuan Agriculture University, 1993, 11 (1)：109-112 [曾子建，李逐波，吴绪田，等. 喹乙醇在鲤体内的药代动力学研究 [J]. 四川农业大学学报, 1993, 11 (1)：109-112].

[14] Sun Y. X., Feng Q. H., Dong L. B.. Studies on biochemical and histopathological toxicity of olaquindox in broilers [J]. ActaVeterinaria et ZootechnicaSinica, 1998, 29 (6)：525-530 [孙永学，冯琪辉，董漓波. 喹乙醇在鸡体内的毒物动力学及生化毒性和病理学研究. 畜牧兽医学报, 1998, 29 (6)：525-530].

[15] Zhang W. Y.. Vc is used in the feeds of aquaculture [J]. Freshwater Fisheries. 1996. 16 (14)：18-20 [张维翯. 谈用于水产饲料中的维生素 C. 淡水渔业, 1996, 16 (4)：18-20].

[16] Waldmann K. H.. Clinical and haematological changes following intoxication with olaquindox [J]. J Veterinary Medicine（Serises A）, 1989, 36 (9)：676-686.

[17] Gu X. H.. The stress mechanism and diagnosis of domestic animal [J]. Current Pasturage, 1994, 4：2-5 [顾宪红. 家禽应激机理及其诊断. 当代畜牧, 1994, 4：2-5].

[18] Dong L. B.. Studiesontoxicityof chick and tissueremedyconsistencyofolaquindox [J], Journal of China South Agricultural University, 1993, 14 (4)：533-558 [董漓波. 喹乙醇对鸡的毒性及组织药物浓度的研究, 华南农业大学学报, 1993, 14 (4)：533-558].

[19] Wang K. Y., Geng Y.. Studies on haematological changes in common Carp by subacute toxicity test of olaquindox [J]. ActaHydrobiologica Sinica, 2003, 27 (1)：23-26 [汪开毓，耿毅. 鲤亚急性喹乙醇中毒的血液生化指标研究. 水生生物学报, 2003, 27 (1)：23-26].

[20] Wang K. Y., Geng Y.. Pathological Study on the AcuteOlaquindox Poisoning in Carp [J]. ActaVeterinaria et ZootechnicaSinica, 2002, 33 (6)：565-569 [汪开毓，耿毅. 鲤鱼急性喹乙醇中毒的病理学研究. 畜牧兽医学报, 2002, 33, (6)：565-569].

[21] Wang K. Y., Studies on pathologyand tissue concentration of the chronic olaquindox poisoninginCyprinus carp [J]. Journal of Fisheries of China, 2003, 27 (1)：75-82 [汪开毓，鲤慢性喹乙醇中毒的病理学和组织残留. 水产学报, 2003, 27 (1)：75-82].

该文发表于《动物学杂志》【2013，48（3）：446-450】

恩诺沙星对银鲫急性毒性及血液生化指标的影响

赵蕾[1,2]，曹海鹏[1]，陈辉[2]，侯三玲[1]，胡鲲[1]，杨先乐[1]

（1. 上海海洋大学国家水生动物病原库 上海 201306；2. 江苏世盛动物药业有限公司 大丰 224100）

摘要：为了解恩诺沙星对异育银鲫（Carassius auratus gibelio）的急性毒性及血液生化指标的影响，评价恩诺沙星对异育银鲫的安全性。本实验研究了恩诺沙星对异育银鲫的急性毒性，并观察了恩诺沙星在不同剂量下对异育银鲫血液生化指标的影响。实验结果表明，恩诺沙星对异育银鲫的半数致死剂量为 1 949.84 mg/kg，安全剂量为 194.98 mg/kg。当恩诺沙星以常规给药剂量（20 mg/kg）连续口灌异育银鲫 30 d 时，异育银鲫的血清总蛋白含量、谷丙转氨酶含量、谷草转氨酶、γ-谷氨酰转移酶含量与对照组没有显著变化；而当恩诺沙星以 40 mg/kg、80 mg/kg、160 mg/kg、320 mg/kg 的剂量连续口灌异育银鲫 30 d 时，异育银鲫的血清谷丙转氨酶含量较对照组分别增加了 23%、30%、46%、86%（$P<0.05$），血清谷草转氨酶含量较对照组分别增加了 37%、42%、69%、86%（$P<0.05$）。本研究确定了恩诺沙星对异育银鲫的安全剂量，证实了恩诺沙星引起异育银鲫出现肝功能失调的剂量范围，对恩诺沙星在异育银鲫养殖中的安全使用提供了科学依据，也证实常规给药剂量（20 mg/kg）的恩诺沙星不会导致异育银鲫出现肝损伤。

关键词：恩诺沙星；异育银鲫；急性毒性；血液生化指标

该文发表于《水产学杂志》【2011：24（2）：37-40】

恩诺沙星对小体鲟和史氏鲟体内 SOD 活力影响的比较研究

The effects of enrofloxacin on superoxide dismutase activities in two sturgeon species

李绍戊[1]，王荻[1]，马涛[2]，卢彤岩[1]

（1. 中国水产科学 黑龙江水产研究所，黑龙江 哈尔滨 150070；2. 哈尔滨工业大学 生命科学学院，黑龙江 哈尔滨 150006）

摘要：在水温 23℃、pH 值 6.5、溶氧 10 mg/L、氨氮含量小于 0.1 mg/L 和亚硝酸盐含量小于 0.001 mg/L 的条件下，将恩诺沙星按照 0、20、40、60、80 和 100 mg/kg 浓度，连续口服给平均体重 50.0±5.0 g 的小体鲟（Acipenser ruthenus Linnaeus）及史氏鲟（Acipenser schrenckii Brandt）5 d，停药 2 d 后测定其血浆及肝脏组织中过氧化物歧化酶 SOD 的活力，以期掌握不同恩诺沙星给药浓度下，两种鲟两种组织中 SOD 活力变化趋势，探讨分析该酶在药物代谢过程中的作用机制。结果表明：两种鲟两种组织内均含有一定量的 SOD 酶，且在对照组及所有给药组肝脏中酶活力均高于血浆中。不同给药浓度下，两种鲟两种组织中 SOD 活力均先受诱导升高，而后被抑制降低的变化趋势，且在 40 mg/kg 浓度组达到最大值。血浆中 SOD 活力受给药浓度影响较小，起伏较平稳，而小体鲟 SOD 活力始终高于史氏鲟。肝脏中 SOD 活力变化较剧烈，在低浓度组（<40 mg/kg）史氏鲟肝脏中 SOD 活力明显高于小体鲟，而在高给药浓度组（>40 mg/kg）则小体鲟略高于史氏鲟。

关键词：恩诺沙星；史氏鲟；小体鲟；SOD

该文发表于《中国畜牧兽医》【2011，38（7）：35-37】

诺氟沙星对两种鲟体内 SOD 活力影响的比较研究

A comparison study on the effects of Norfloxacin on the Superoxide Dismutase activities in two kinds of sturgeons

王荻[1]，李绍戊[1]，马涛[2]，卢彤岩[1]

（1. 中国水产科学 黑龙江水产研究所，黑龙江 哈尔滨 150070；2. 哈尔滨工业大学 生命科学学院，黑龙江 哈尔滨 150006）

摘要：将诺氟沙星按照 0、20、40、60、80 和 100 mg/kg 浓度，对小体鲟及史氏鲟连续口服给药 5 d，停药 2 d 后对其血浆及肝脏组织中过氧化物岐化酶 SOD 活力进行测定，以期掌握不同诺氟沙星给药浓度下，两种鲟两种组织中 SOD 活力变化趋势，并探讨评价该药在鲟养殖过程中施用的最适剂量，以及对鱼类肝脏的损伤情况。实验结果表明：两种鲟两种组织内均含有一定量的 SOD 酶，且在对照组及所有给药组肝脏中酶活力均高于血浆中。不同给药浓度下，两种鲟两种组织中 SOD 活力变化均呈现规律的先受诱导升高，而后被抑制降低的变化趋势，且在 40 mg/kg 给药浓度组达到最大值。血浆中 SOD 活力受给药浓度影响较小，起伏较平稳，在 40 mg/kg 给药浓度组史氏鲟血浆中 SOD 值高于小体鲟，而其他组别均为小体鲟较高。而肝脏中 SOD 活力变化较剧烈，且在对照组、诱导最高 SOD 活力的 40 mg/kg 给药浓度组和最高给药浓度组，小体鲟均高于史氏鲟，且小体鲟肝脏中 SOD 在 40 mg/kg 给药浓度组活力值极高，形成一个尖锐的峰值。结果表明，诺氟沙星对两种鲟给药浓度为 30~50 mg/kg 时，既能使药效最好而又不会对肝脏产生明显的损伤。

关键词：诺氟沙星；史氏鲟；小体鲟；超氧化物岐化

聚维酮碘对鱼类虽然毒性较低，但它能影响异育银鲫养殖水体中大型蚤等浮游动物的存在（高晓华等，2013）。克螨特作为一种广谱杀螨剂，对于鱼类毒性较强，其对金鱼 LC_{50} 为 1.716 mg/L，安全浓度为 0.172 mg/L（汤菊芬，2015）；对斑马鱼 LC_{50} 为 0.909 mg/L，安全浓度为 mg/L（黄瑜，2015）。

该文发表于《动物学杂志》【2013，48（2）：261-268】

水产用聚维酮碘对异育银鲫养殖的安全性评价

高晓华，曹海鹏，侯三玲，胡鲲，杨先乐

（上海海洋大学国家水生动物病原库，上海 201306）

摘要：评价水产用聚维酮碘对异育银鲫（*Carassius auratus gibelio*）养殖的安全性，为其在异育银鲫养殖中的安全应用提供了重要的科学依据，本研究参照国家标准及相关法规，在观察了聚维酮碘对小球藻（*Chlorella* sp.）生长抑制作用、对水产益生菌抑菌效果以及对大型蚤（*Daphnia magna straus*）、斑马鱼（*Brachydanio rerio*）和异育银鲫的急性毒性的基础上，分析其对异育银鲫及其养殖水体主要有害理化因子的影响。实验结果表明，聚维酮碘在终浓度为 6.00~14.00 mg/L 时对小球藻生长具有促进作用，对小球藻的半数抑制浓度大于 14.00 mg/L，对水产益生菌的最小抑菌浓度为 128~512 mg/L，对大型蚤、斑马鱼的半数致死浓度分别为 13.44 mg/L、17.63 mg/L。此外，聚维酮碘对异育银鲫的半数致死浓度为 74.77 mg/L，而且在养殖水体中加入聚维酮碘至终浓度为 0.20~1.40 mg/L后 14 d 内，随着聚维酮碘浓度的增加，各浓度组异育银鲫养殖水体的氨氮含量、亚硝酸盐含量均缓慢下降。本研究证实聚维酮碘低毒，但考虑到其可能对异育银鲫养殖水体中大型蚤等浮游动物存在潜在影响，建议其在异育银鲫养殖中的安全应用浓度应不高于 1.34 mg/L，在该安全应用浓度内不会引起养殖水中氨氮、亚硝酸盐等有害因子含量的增加。

关键词：聚维酮碘；大型蚤；斑马鱼；异育银鲫养殖；安全性评价

该文发表于《安徽农学通报》【2015，21（14）：140，142】

克螨特农药对金鱼急性毒性的初步研究

Preliminary study on acute toxicity of pesticide propargite in gold fish （*Carassius auratus*）

汤菊芬[1,2,3]，蔡佳[1,2,3]，王蓓[1,2,3]，鲁义善[1,2,3]，简纪常[1,2,3]，吴灶和[2,3,4]，黄瑜[1,2,3]

（1. 广东海洋大学水产学院，湛江 524088；2. 广东省水产经济动物病原生物学及流行病学重点实验室，湛江 524088；
3. 广东省水产经济动物病害控制重点实验室，湛江 524088；4. 仲恺农业工程学院，广州 510225）

摘要： 通过室内急性毒性测定试验来研究克螨特农药对金鱼的毒性，以期得出克螨特防治金鱼病虫害时的安全浓度。结果表明，克螨特对金鱼在用药96 h后的致死中浓度（LC_{50}）为1.716 mg/L，置信区间为1.565~1.882 mg/L。同时得出克螨特对金鱼的安全用药浓度（SC）为0.172 mg/L。

关键词： 金鱼；半致死浓度（LC_{50}）；克螨特

该文发表于《安徽农业科学》【2015，43（25）：138-139】

克螨特农药对斑马鱼急性毒性的初步研究

Preliminary study on acute toxicity of pesticidepropargite inzebrafish （*Daniorerio*）

黄瑜[1,2,3]，蔡佳[1,2,3]，王蓓[1,2,3]，鲁义善[1,2,3]，简纪常[1,2,3]，吴灶和[2,3,4]，汤菊芬[1,2,3]

（1. 广东海洋大学水产学院，湛江 524088；2. 广东省水产经济动物病原生物学及流行病学重点实验室，湛江 524088；
3. 广东省水产经济动物病害控制重点实验室，湛江 524088；4. 仲恺农业工程学院，广州 510225）

摘要： ［目的］寻找克螨特用于水生动物斑马鱼病虫害防控的安全浓度。［方法］通过室内急性毒性测定试验，设定7个克螨特浓度梯度来检测对斑马鱼的致死情况，从而计算出致死中浓度（LC_{50}）与安全用药浓度（SC）。［结果］克螨特对斑马鱼在用药96 h后的LC_{50}为0.909 ppm，置信区间分别为0.876~0.943 ppm，同时得出克螨特对斑马鱼的SC为0.091 ppm。［结论］该研究为克螨特杀灭斑马鱼寄生虫的用药量提供了重要的数据支持。

关键词： 斑马鱼；半致死浓度（LC_{50}）；安全浓度（SC）；克螨特

饲料中添加不同剂量的三聚氰胺，会降低罗非鱼的增重率，导致罗非鱼的肾脏细胞和肾脏组织受到严重损伤（汤菊芬，2011）。

该文发表于《饲料工业》【2011，32（20）：13-17】

饲料中添加三聚氰胺对吉富罗非鱼生长、免疫指标及肌肉残留量的影响

Effects of dietary melamine on growth，non-specific immunity and muscle residues in GIFT strain of Nile tilapia （*Oreochromis niloticus*）

汤菊芬，吴灶和，简纪常，鲁义善，王蓓

（广东海洋大学水产学院 广东省水产经济动物病原生物学及流行病学重点实验室，广东 湛江 524025）

摘要： 将三聚氰胺分别按0 mg/kg、500 mg/kg、2 000 mg/kg、5 000 mg/kg、10 000 mg/kg的比例添加于基础饲料中，喂养初始体重为（30.14±0.22）g的吉富罗非罗28 d。通过测定罗非鱼的增重率、血清免疫指标和三聚氰胺

958

在肌肉中残留量等指标，研究三聚氰胺对罗非鱼生长、免疫功能及肌肉残留量的影响。结果显示：饲料中添加不同剂量的三聚氰胺会降低吉富罗非鱼的增重率，在第21天和第28天，10 000 mg/kg组吉富罗非鱼的增重率显著低于同期对照组和其他实验组（$P<0.05$）；三聚氰胺对吉富罗非鱼血清中的LZM活力、SOD活力无显著影响（$P>0.05$），但AKP活力有显著影响（$P<0.05$），三聚氰胺对吉富罗非鱼血清中的AKP活力的影响呈先升高后降低趋势，在第14天，5 000 mg/kg和10 000 mg/kg组吉富罗非鱼血清中的AKP活性极显著高于同期对照组和其他实验组（$P<0.05$），在第21天和28天，所有实验组吉定罪罗非鱼的AKP活性显著低于同期对照组；从第7天开始，四个实验组吉富罗非鱼肌肉中均能检测到三聚氰胺残留，三聚氰胺在吉富罗非鱼肌肉中的残留量与三聚氰胺在饲料中的添加剂量成正相关关系。

关键词：三聚氰胺；吉富罗非鱼；生长；免疫指标；肌肉残留量

第四节　渔药风险评估模型的建立和应用

利用生理药动模型可以预测药物（如喹烯酮）在草鱼体内的残留消除规律，以及拟合药物60 d之后的药物残留消除曲线，所获得的结果与实测数据一致（胥宁等，2015）。

基于幂函数模型的比例化剂量反应关系的分析方法（假设检验法和可信区间法）可作为水产品组织中药物残留溯源分析的一种新的技术手段（胡鲲等，2013）。

该文发表于《水生生物学报》【2015，39（3）：517-523】

喹烯酮在草鱼体内生理药动模型的建立

The establishment of the physiological based pharmacokinetic model for quinocetone in grass carp (*Ctenopharyngodon idellus*)

胥宁[1,2]，刘永涛[1,2]，杨秋红[1]，艾晓辉[1,2]

(1. 中国水产科学研究院长江水产研究所，武汉 430223；2. 淡水水产健康养殖湖北省协同创新中心，武汉 430223)

摘要：为了预测喹烯酮在草鱼体内药物残留，建立其在草鱼体内生理药动学模型。通过搜集大量文献获得鱼的生理解剖参数，采用已有的喹烯酮试验数据拟合得到药物特异性参数。基于acslXtreme生理药动学软件，进行模型假设、血流图设计、质量平衡方程的建立和模型拟合。喹烯酮为小分子药物，其分布服从血流限速型，在肝脏代谢，从肾脏消除。喹烯酮通过口服进入肠道，然后经肝脏代谢进入血液循环，因此设定5个房室，即肝、肾、肌肉、肠和其他组织。经过一系列的计算和调试，最终建立喹烯酮在草体内5室生理药动模型，成功拟合连续饲喂药物60 d之后的药物残留消除曲线，其中肝脏中的预测结果比肾脏和肌肉高，与实测数据一致。因此，喹烯酮在鱼体内生理药动模型具有一定的应用价值，将是药物残留检测的新亮点。

关键词：喹烯酮；草鱼；生理药动学模型；药物残留；残留消除

近些年来，水产用药越来越广泛，由药物残留引发的食品中毒和水产品出口受阻事件越来越多，因此水产品中药物残留已经是社会关注的一个热点。传统的药物残留检测方法首先是鉴定水产动物体内的主要代谢产物，确定残留标示物和靶组织，然后经过残留消除试验进行仪器检测，然后确定最大残留限量和休药期[1,2]。传统检测方法耗时长，且需耗费大量的人力和物力。由于水产养殖动物种类繁多，品种、生理、疾病和外部环境的改变对药物残留都具有比较大的影响，因此传统方法已不能满足要求。生理药动学模型是基于动物生理解剖学、生物化学和药物代谢动力学等研究，利用房室模块和每个房室的质量平衡方程描述化合物体内处置的数学模型[3,4]。由于PBPK模型能够实现不同品种、生理特性、给药方式等情况下外推，因而此模型能够弥补残留分析方法的不足，更好的应用于药物残留分析中，具有很高的应用前景[5,6]。

喹烯酮是喹噁啉类二氧化合物的衍生物，其化学名称为3-甲基-2-苯乙烯酮基-喹噁啉-1，4-二氧化物，是我国批准的国家一类新兽药，允许作为饲料添加剂在动物饲料中使用。喹烯酮抗菌作用优良，是一种很好的抑菌剂，并可以提高饲料转化率，明显促进动物生长。毒理学结果也表明，喹烯酮毒副作用小，无三致作用，应用前景良好[7]。本文将建立喹烯酮在草鱼体内生理药动学模型，为喹烯酮在水产动物体内药物残留预测和种间类推打下基础。

1 材料与方法

1.1 材料

本次模型的建立，喹烯酮在草鱼体内药动学数据来自本实验室试验数据。药动学数据处理软件采用 3P97，生理药动学模型的建立、计算和拟合采用软件 acslXtremeV1.4（Aegis Technologies，美国）。

1.2 方法

模型的假设　喹烯酮分子量不超过 400D，为小分子物质，对于水产品药物残留我们关注的是可食性组织和重要器官的药物浓度，因此建立的模型时不包括包含特殊生理屏障的器官，因此可以假定它们在组织中分布的速度和程度主要取决于流经组织器官的血流量，即服从血流限速型分布；喹烯酮从胃肠道吸收进入血液循环为一级速率过程；喹烯酮在肝脏代谢，从肾脏消除，而且服从一级动力学过程。

模型血流图设计　血流图的设计重点关注肠道、肝脏、肾脏、肌肉和其他组织等四个房室，然后由血液循环链接（图1）。本模型建立喹烯酮（饲料添加药物，添加量为 75 μg/kg）连续给药 60 d 之后生理药动学预测模型，因此设定了连续口服给药模块。药物首先通过口服进入肠道，通过肠道的代谢吸收之后，进入肝脏，然后进入血液循环。

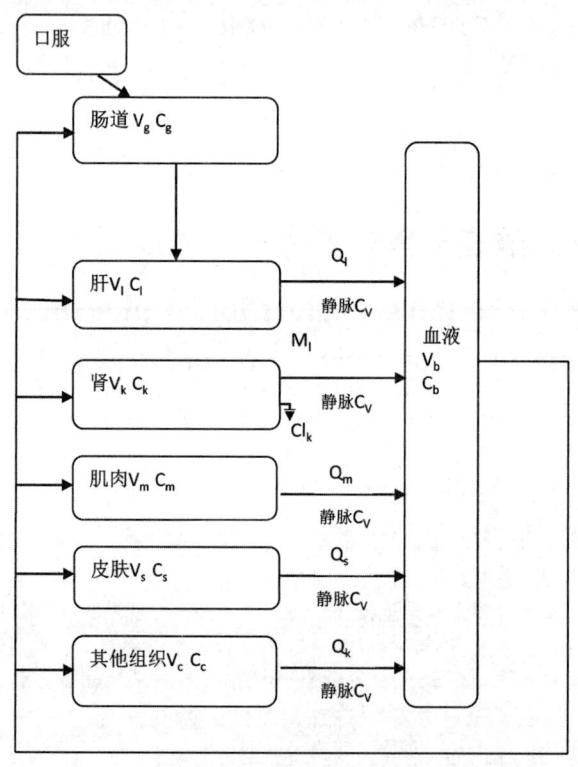

图 1　喹烯酮在草鱼的血流图

Fig. 1　The blood flow of quinocetone in grass carp

V_g、V_l、V_k、V_m、V_b 和 V_c 分别为肠道、肝、肾、肌肉、血液和其他组织的容积；C_g、C_l、C_k、C_m、C_b 和 C_c 分别为肠道、肝、肾、肌肉、血液和其他组织中药物浓度；Q_g、Q_l、Q_k、Q_m、Q_c 和 Q_t 分别为肠道、肝、肾、肌肉、其他组织的血流速率和心输出量；C_v 表示相对应的静脉血中药物浓度；M_l 表示肝脏的代谢率；Cl_k 表示肾脏的清除率

V_g、V_l、V_k、V_m、V_b and V_c were the volumes of intestine, liver, kidney, blood and other tissues respectively; C_g、C_l、C_k、C_m、C_b and C_c were the drug concentrations in the intestine, liver, kidney, blood and other tissues respectively; Q_g、Q_l、Q_k、Q_m and Q_c were the blood flow rate of the intestine, liver, kidney and other tissues respectively, Q_t was cardiac output; M_l was liver metabolism rate; Cl_k was kidney clearance

生理解剖参数和药物特异性参数的获得　鱼器官容积、器官血流量、心输出量等生理解剖参数全部来自于发表文献。药物特异性参数，如组织血浆分配系数、肾清除率和消除速率常数通过药动学数据（表1）拟合得到，参考方法来自于文

献[8,9]，血浆蛋白结合率来自于文献[10]。

表 1　草鱼单次口灌 50 mg/kg 喹烯酮之后各个组织中药物浓度[11]

Tab. 1　The concentration of quinocetone in various tissues of grass carp after a single oral administration with thedose of 50 mg/kg

时间 Time（h）	药物浓度 Drug concentration（μg/kg）			
	血浆 Plasma	肌肉 Muscle	肝脏 Liver	肾脏 Kidney
1	214.67±37.88	14.08±0.89	1489.47±368.76	774.63±278.57
2	116.53±23.74	22.97±0.41	1638.54±272.63	1271.19±356.21
4	109.62±13.98	23.76±2.46	1239.61±343.85	929.48±147.82
6	99.70±14.80	11.15±2.10	954.57±395.61	769.34±251.81
8	82.35±11.72	9.70±4.80	749.38±267.25	608.51±134.78
10	70.80±9.37	7.85±2.88	487.68±129.66	474.32±108.95
12	49.37±4.04	7.40±1.75	425.83±85.37	394.49±136.51
24	48.26±5.36	6.03±3.63	132.62±59.84	185.87±22.44
48	34.07±3.61	3.10±0.54	57.81±24.65	81.45±34.89
72	30.94±2.35	1.45±0.35	36.84±19.34	73.69±28.43
96	ND	2.63±0.83	10.10±4.18	59.69±13.54
120	ND	2.39±0.49	8.59±3.74	48.07±16.71
144	ND	ND	8.81±4.36	35.82±10.57
168	ND	ND	2.16±1.03	17.88±4.45
192	ND	ND	ND	5.58±1.32

ND：Not detected

质量平衡方程的建立肠道：$V_g \times \dfrac{\mathrm{d}c_g}{\mathrm{d}t} = \text{amount} + Q_g \times \left(C_b - \dfrac{C_g}{P_g}\right)$

肝脏：$V_l \times \dfrac{\mathrm{d}c_l}{\mathrm{d}t} = Q_g \times \left(\dfrac{C_g}{P_g}\right) + Q_l \times C_b - (Q_g + Q_l) \times \left(\dfrac{C_l}{P_l}\right)$

肾脏：$V_k \times \dfrac{\mathrm{d}c_k}{\mathrm{d}t} = Q_k \times \left(C_b - \dfrac{C_k}{P_k}\right) - Cl_k \times C_k$

肌肉：$V_m \times \dfrac{\mathrm{d}c_m}{\mathrm{d}t} = Qm \times \left(C_b - \dfrac{C_m}{P_m}\right)$

其他组织：$V_c \times \dfrac{\mathrm{d}c_c}{\mathrm{d}t} = Q_c \times \left(C_b - \dfrac{C_c}{P_c}\right)$

血液：

$$V_b \times \dfrac{\mathrm{d}c_b}{\mathrm{d}t} = Q_l \times \left(\dfrac{C_l}{P_l}\right) + Q_k \times \left(\dfrac{C_k}{P_k}\right) + Q_m \times \left(\dfrac{C_m}{P_m}\right) + Q_c \times \left(\dfrac{C_c}{P_c}\right) - Q_t \times C_b$$

以上各式中，Amount 表示通过口服进入肠道的药物量；Q_i 表示 i 器官的血流速率；P_i 表示 i 器官的组织血浆分配系数；C_i 表示 i 器官中的药物浓度；V_i 表示 i 器官的体积；Cl_k 表示肾清除率，参考图 1 注释。

模型的建立与参数灵敏性分析　模型的建立：在 acslXtreme 中，建立好各个模块，输入各个器官相应的参数，建立符合质量平衡方程的模型。通过预测值与实测值比对，预测值与实测值残差分析验证模型的准确性和特异性。

参数灵敏性分析：灵敏性分析的目的是评价各个参数对模型预测结果的影响。基本原理为：改变模型参数的值（Δx），代入模型计算出模型预测结果的改变量 [$f(x\pm\Delta x)$]；以预测结果的改变量除以模型参数值的改变量，得到灵敏系数；计算出的灵敏系数经过转换得到标准化的灵敏系数（Normalized sensitivity coefficient，NSC），NSC 可被用于判断参数的灵敏性。

$$SC = F'(x) = \frac{f(x + \Delta x) - f(x)}{\Delta x} \tag{1}$$

$$NSC = \frac{SC * x}{f(x)} \tag{2}$$

公式（1）所示，x：灵敏性分析的目标参数，Δx：该参数的改变量，$f(x)$：模型的输出结果，$f(x\pm\Delta x)$：模型预测结果的改变量。NSC 的计算如公式（2）所示，x 代表灵敏性分析的目标参数，$f(x)$ 代表模型的预测结果。灵敏性分析的具体运算通过专业模拟软件 ACSL xtreme 的优化模块（OptStat Module）来完成。

在本研究中，当 | NSC | <0.1 时，被认为该参数对模型影响较小，当 | NSC | >0.1 时，认为该参数灵敏，对模型影响较大。

不确定分析　模型预测结果的不确定性通过蒙特卡洛分析来评估。基本思路是，在设定的范围内随机抽取某个数值作为参数的初值，然后代入模型运算求得一个预测结果，如此反复多次运行之后就能生成一个预测结果的集合，这个集合与参数的分布相对应，反应了由参数变异引起的预测结果的变异。蒙特卡洛分析考察的对象为灵敏性参数，本研究评估了灵敏参数组织/血浆分配系数的变异性对模型预测结果的影响。

2 结果

2.1 生理参数

本文所用生理参数参考文献[12,13]（表2）。

表2　鱼类相关生理参数
Tab. 2　Physiological parameters of fish

组织 Tissue	组织重量（占体重的百分比%） Tissues weight（%）	器官血流量（占心输出量的百分比%） Blood flow of organs（%）
鳃 Gills	3.9	100
肝 Liver	1.16	18.14
肾 Kidney	0.8	10.23
肌肉 Muscle	46.5	39.77
其余组织 Carcass	31.11	31.86
肠 Intestine	8.52	15.39
静脉血 Venous blood	1.40	–
动脉血 Arterial blood	2.71	–

注：–不可估计
Note：–Not Estimation

2.2 药物特异性参数

药物特异性参数是通过本实验室已有的喹烯酮药动学数据拟合得到（表3）。

表3　喹烯酮在草鱼体内特异性参数
Tab. 3　Specific parameters of quinocetone in grass carp

组织 Tissues	组织血浆分配系数 Tissue/plasma partition coefficients	清除率（L/h.kg） Clearance（L/h.kg）	消除速率常数（1/h） Elimination rate constant（1/h）
肌肉 Muscle	0.148	–	0.0145
肝 Liver	5.505	0.156	0.0312
肾 Kidney	6.334	0.135	0.0387

注：–不可估计
Note：–Not Estimation

2.3 模型拟合

本模型运用数学和统计学的方法成功拟合了喹烯酮（给药方式为：饲料添加连续饲喂 60 d）在草鱼肠道、肝脏、肾脏

962

和其他组织中残留消除曲线。肠道中药物浓度最高预测值为 19 μg/kg，肌肉和肾脏为 24 μg/kg 和 27 μg/kg，肝脏为 70 μg/kg。比较 4 种组织，肠道最低，肌肉和肾脏相当，肝脏最高，这与文献报道的其他动物组织浓度趋势一致[14]。在本次研究中，所得到的预测结果与实测结果（表 4）也进行了比对，喹烯酮在饲料添加连续饲喂 60 d 之后，喹烯酮在草鱼体内实测浓度迅速降低，在停药 72 h 之后，肌肉中的浓度已经在 1 μg/kg 以下；216 h 之后，浓度已经不足 0.15 μg/kg，与预测趋势相符。

表 4 通过饲料添加 75 mg/kg 喹烯酮连续饲喂 60 天后草鱼肌肉中药物浓度[15]

Tab. 4 The concentrations of quinocetonein muscle of grass carp after feeding with 75 mg/kg quinocetone for 60 consecutive days

时间 Time（h）	肌肉 Muscle（μg/kg）
24	1.64±0.32
36	1.29±1.44
48	1.20±1.08
72	0.82±0.069
96	0.47±0.30
120	1.13±1.18
168	0.28±0.13
216	0.14±0.020
264	<0.1
360	<0.1

在本次研究中，对预测值和实测值进行了残差分析。如果残差值均偏向于 x 轴上侧，表明该模型低估了喹烯酮在组织中的浓度；如果残差值均偏向于 x 轴下侧，表明该模型高估了喹烯酮在组织中的浓度；如果残差值均匀的分布于 x 轴两侧，表明该模型很好的预测了喹烯酮在组织中的浓度。实测值只有肌肉测定的结果，所以对肌肉中的喹烯酮浓度进行了残差分析。图 2 表明，残差值均匀的分布于 x 轴两侧，表明模型很好的预测了喹烯酮在组织中的浓度。

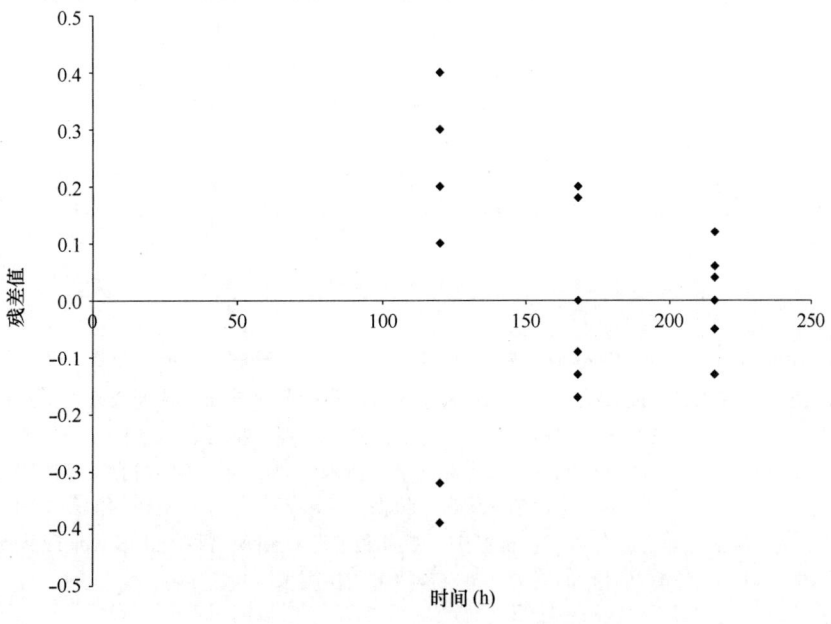

图 2 肌肉中喹烯酮预测浓度与实测值标准残差图

Fig. 2 Plots of the predicted concentration and the observed concentration ofquinocetone in the muscles

为了进一步对模型预测效果进行评价，本研究对模型预测值与实验测定的残留数据进行了相关性分析。通过计算，对于喹烯酮在肌肉中的动态变化，模型预测值与实验测定值的相关系数为0.9468。相关性分析结果表明，喹烯酮在肌肉中模型预测值和实验测定值之间相关性良好，模型较准确的预测了喹烯酮在肌肉中的残留消除。

2.4 灵敏性分析

通过灵敏性分析发现，以肝脏房室为例，生物利用度、肝脏-血浆分配系数、肾清除率以及一些生理参数（｜NSC｜＞0.1）对肝脏中的药物浓度影响较大。与肝脏中喹烯酮的浓度变化成正相关的参数有生物利用度、肝脏/血浆分配系数和肌肉容积，即肝脏中药物浓度随三个参数值的增大而增大。与肝脏中喹烯酮的浓度变化成负相关的参数有肾清除率、胃排空率、吸收速率常数、肝脏血流量、肾脏血流量、肌肉血流量，即肝脏中药物浓度随这些参数值的增大而减小。喹烯酮在其他组织房室中的灵敏参数与肝脏中的结果一致。

2.5 蒙特卡洛分析

图3表示组织/血浆分配系数在指定的范围内随机取值，代入模型运算1 000次后得到的喹烯酮在肌肉组织中残留水平的可能分布范围及频数。经过比较，大部分实测数值都落在其范围内，表明模型具有一定的预测群体动物中喹烯酮残留的能力。

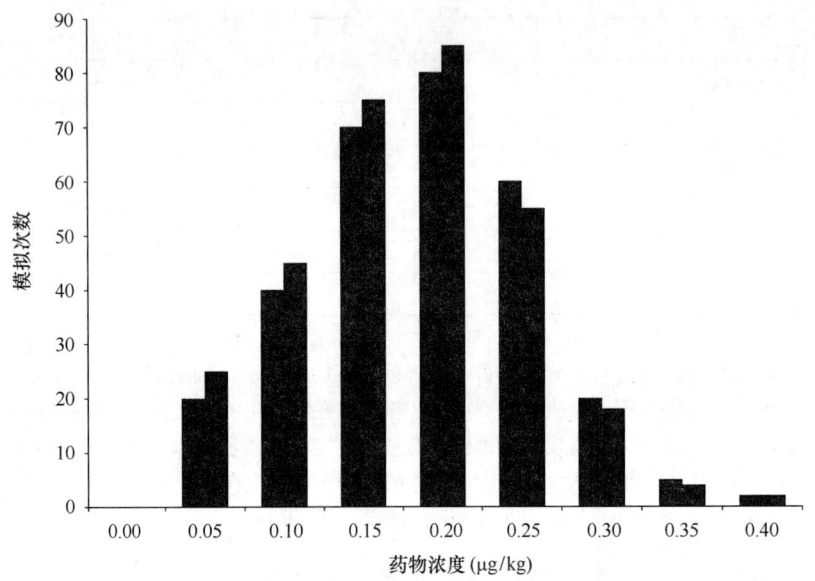

图3 组织/血浆分配系数对肌肉中药物浓度的不确定分析

Fig. 3 Monte Carlo simulations of tissue/plasma partition coefficients of quinocetone in the muscles

3 讨论

3.1 药动生理模型的优势

传统药代动力学存在的问题：动力学房室高度简化，生理学数据缺乏，间类推时会受限；无法清晰地给出药物在靶组织和其他组织中的浓度时间关系；不能反映生理变化对药物在机体内吸收、分布、代谢及排泄的影响。与传统药动学房室模型相比，生理模型可以减少繁杂的重复性试验，降低大量的经费。生理模型可以描述由于环境或病理因素引发参数的改变，准确模拟药物在不同情况下各器官的中的变化，更好地指导临床合理用药；预测可食性组织中的药物残留量、残留消除规律及休药期，为食品动物药物残留的风险评估提供科学依据；预测备选新药在机体内的药时曲线，简化新药筛选程序，降低新药研发风险[16]。在水产动物的药物残留检测中，涉及到的药物和动物种类繁多。生理药动学模型所具有种属间、化合物间、组织间和不同暴露方式间外推的能力，将发挥巨大的作用。

3.2 生理参数和药物特异性参数

生理药动学模型的建立，是基于具有详细的生理解剖参数、药物特异性参数和已有的药动学数据。通过目前的研究发现，鱼类的生理解剖参数不够丰富，我国鱼类的生参数的研究更是匮乏。有文献报道由其他鱼类的生理参数，代替我国鱼类的生理参数。Yang等[17]曾经采用虹鳟（*Oncorhynchusmykiss*）的生理参数代替鲫（*Carassiusauratus*）的生理参数，获得了氟苯尼考不同给药方式的类推。刘宇[6]从猪（Landrace）的参数外推到鱼，也实现了呋喃唑酮的残留预测。这种方法虽然具有一定的局限性，如果在不同种类的鱼之间进行类推，可能生理参数的差异的重要性就显现出来了。另外鱼类的代谢资料缺乏，许多药物在鱼类体内的代谢仍不完善，有些药物代谢和主要代谢物不明确；有些药物其代谢部位和代谢途径已经十分清楚，但由于缺乏足够的药物特异性参数，也难以建立模型。因此，关于我国鱼类的生理参数和药物代谢，很有必要进行深入研究。

3.3　喹烯酮生理药动学模型的建立

在模型的假设中，设定了5个器官模块，通过各个参数的代入拟合，对喹烯酮的残留消除成功拟合。目前，通过生理药动模型预测药物残留的文献较少，仅见在少数动物体进行研究。Burr等[18]建立了三聚氰胺在猪体内生理药动学模型，预测可食性组织中的休药期，其建立的是三房室模型，包括肝脏、肾脏和其他组织。药物是通过口服进入小肠，然后进入肝脏。Burr等[19]也建立了磺胺二甲嘧啶在猪体生理药动学模型，成功预测了在各个组织中的残留，其中包括5个房室，肝、肾、肌肉、脂肪和其他组织。Cortright等[20]建立的咪达唑仑在禽类体内生理药动学模型包括5个房室，肝、肾、肌肉、脂肪和其他组织。Law[21,22]建立了两个关于土霉素在鲑（Oncorhynehustshawytscha）体内的生理药动学模型，模型主要房室包括肝、肾、鳃、肌肉和其他组织，通过模型的建立预测了在各个组织中的休药期。Yang等[17]通过生理药动学模型的建立，成功完成了氟苯尼考对鲫口服和肌注不同给药方式的类推，其中有6个房室，包括鳃、肝、肾、肠、肌肉和其他组织。通过已有的研究发现，利用生理药动学模型预测药物在动物组织中的残留，只要包括主要的代谢器官、消除器官和可食性组织即可满足要求。

在灵敏性分析中发现，心输出量和消除率对药物体内的过程影响比分配系数的影响更为显著，这可能是因为前两者同时影响药物的分布和消除，而后者仅影响其中的某一方面，这也可能是血流限速型药物的特点。

本次研究通过预测值与实测值的残差分析和比对，蒙特卡洛不确定分析，对模型进行评价，成功建立喹烯酮在草鱼体内5室生理药动学模型。本模型能够对喹烯酮在草鱼体内的残留消除，较好地进行拟合和预测，在下一步的研究中我们将对模型的种内和种间外推进行详细的研究。

参考文献

［1］　Ruan J M, Hu K, Yang X L, et al. Blood-brain barrier permeability of DIF and its elimination comparative study between brain and peripheral tissues in Carassiusauratusgibelio［J］. ActaHydrobiologicaSinica, 2014, 38（2）: 272-278［阮记明, 胡鲲, 杨先乐, 等. 双氟沙星对异育银鲫血脑屏障渗透性及消除规律. 水生生物学报, 2014, 38（2）: 272-278］.

［2］　Zhang Y Z, Xu X L, Zhang H Q, et al. Simultaneous determination of twenty one sulfonamide residues in Trionyxsinensis by high performance liquid chromatography tandem mass spectrometry［J］. ActaHydrobiologicaSinica, 2008, 32（5）: 669-679［张永正, 徐晓琳, 张海琪, 等. HPLCMS/MS法检测中华鳖中磺胺类药物残留. 水生生物学报, 2008, 32（5）: 669-679］.

［3］　Luo X Y, Cai F Q, Yuan L G, et al. Rheogram of physiologically based pharmacokinetic model of ciprofloxacin in rats in vivo［J］. ScientiaAgriculturaSinica, 2010, 43（18）: 3857-3861［罗显阳, 蔡芳琴, 远立国, 等. 环丙沙星在大鼠体内的生理药动学模型的血流图设计. 中国农业科学, 2010, 43（18）: 3857-3861］.

［4］　Ding H Z, Zeng Z L. Physiologic-based pharmacokinetic model and its application in veterinary pharmacology［J］. Progress in Veterinary Medicine, 2007, 28（9）: 55-59［丁焕中, 曾振灵. 生理药动学模型及其在兽医药理学研究中的应用. 动物医学进展, 2007, 28（9）: 55-59］.

［5］　Yang B. Physiologically based pharmacokinetic model for the prediction of olaquindox residues in porecine edible tissues［D］. Thesis for Doctor of Science. Huazhong Agricultural University, Wuhan. 2010［杨波. 喹乙醇在猪可食性组织中残留的生理药代学模拟研究. 博士学位论文, 华中农业大学, 武汉. 2010］.

［6］　Liu Y. Physiologically based pharmacokinetic model for furazolidone in pigs［D］. Thesis for Doctor of Science. Huazhong Agricultural University, Wuhan. 2010［刘宇. 呋喃唑酮在猪体内生理药动学模型研究. 博士学位论文, 华中农业大学, 武汉. 2010］.

［7］　Jin X, Chen Q, Tang S S, et al. Investigation of quinocetone induced genotoxicity in HepG2 cells using the comet assay, cytokinesis block micronucleus test and RAPD analysis［J］. Toxicology in Vitro, 2009, 23（7）: 1209-1214.

［8］　Gallo J M, Lam F C, Perrier D G. Area method for the estimation of partition coefficients for physiological pharmacokinetic models［J］. Journal of Pharmacokinetics and Biopharmaceutics, 1987, 15（3）: 271-280.

［9］　Craigmill A L. A physiologically based pharmacokinetic model for oxytetracycline residues in sheep［J］. Journal of Veterinary Pharmacology and Therapeutics, 2003, 26（1）: 55-63.

［10］　Zhao F, Liu Y T, Xu N, et al. Determination of protein binding of quinocetone with grass carp（Ctenopharyngodonidellus）plasma［J］. Fresh Fisheries, 2014, 44（2）: 71-76［赵凤, 刘永涛, 胥宁, 等. 喹烯酮在草鱼血浆中蛋白结合率的测定. 淡水渔业, 2014, 44（2）: 71-76］.

［11］　Liu Y T, Ai X H, Wang F H, Yang H, Xu N, Yang Q H. Comparative pharmacokinetics and tissue distribution of quinocetone in crucian carp（Carassiusauratus）, common carp（Cyprinuscarpio L.）, and grass carp（Ctenopharyngodonidella）following the same experimental conditions. Jounal of Veterinary Pharmacology and Therapeutics, 2014, doi: 10.1111/jvp.12195. (Epub ahead of print).

[12] Law F C P, Abedini S, Kennedy C J. A biologically based toxicokinetic model for pyrene in rainbow trout [J]. Toxicology and Applied Pharmacology, 1991, 110 (3): 390-402

[13] Abbas R AND Hayton W. A physiologically based pharmacokinetic and pharmacodynamic model for paraoxon in rainbow trout [J]. Toxicology and Applied Pharmacology, 1997, 145 (1): 192-201.

[14] Li J Y, Li J S, Zhao R C, et al. Study on residues of quinocetone in edible chicken tissues [J]. Progress in Veterinary Medicine, 2008, 29 (4): 34-37 [李剑勇, 李金善, 赵荣材, 等. 喹烯酮在鸡食用组织中的残留研究. 动物医学进展, 2008, 29 (4): 34-37].

[15] Ai X H, Liu Y T, Xu N, et al. The study of establishment and application for the physiological based pharmacokinetic model of quinocetone in grass carp [A]. Report of Fundamental Research of Chinese Academy of Fishery Sciences, 2014, 1 (1): 1-10. [艾晓辉, 刘永涛, 胥宁, 等. 喹烯酮在草鱼体内生理模型的建立与应用研究. 中国水产科学研究院基本科研业务费报告, 2014, 1 (1): 1-10].

[16] Bao J, Luo X Y, Zhu L X, et al. Physiological compartment model of oxytetracycline in chicken [J]. China Poultry, 2010, 32 (22): 25-27 [鲍杰, 罗显阳, 朱理想, 等. 土霉素在鸡体内的生理房室模型研究. 中国家禽, 2010, 32 (22): 25-27].

[17] Yang F, Sun N, Sun Y X, et al. A physiologically based pharmacokinetics model for florfenicol in crucian carp and oral-to-intramuscular extrapolation [J]. Journal of Veterinary Pharmacology and Therapeutics, 2013, 36 (2): 192-200.

[18] Buur J L, Baynes R E, Riviere J E. Estimating meat withdrawal times in pigs exposed to melamine contaminated feed using a physiologically based pharmacokinetic model [J]. Regulatory Toxicology and Pharmacology, 2008, 51 (3): 324-331.

[19] Buur J, Baynes R, Smith G, et al. Use of probabilistic modeling within a physiologically based pharmacokinetic model to predict sulfamethazine residue withdrawal times in edible tissues in swine [J]. Antimicrobial Agents and Chemotherapy, 2006, 50 (7): 2344-2351.

[20] Cortright K A, Wetzlich S E, Graigmill A L. A PBPK model for midazolam in four avian species [J]. Journal of Veterinary Pharmacology and Therapeutics, 2009, 32 (6): 552-565.

[21] Smith D J, Gingerich W H, Beconi-Barker M G. Xenobiotics in Fish [A]. In: Law F C P (Eds.), A physiologically based pharmacokinetic model of oxytetracycline for salmonids [C]. New York: Kluwer Academic. 1992, 33-43.

[22] Smith D J, Gingerich W H, Beconi-Barker M G. Xenobiotics in Fish [A]. In: Law F C P (Eds.), A physiologically based pharmacokinetic model for predicting the withdrawal period of oxytetracycline in cultured Chinook salmon (oncorhynehustshawytscha) [C]. New York: Kluwer Academic. 1999, 105-121.

该文发表于《中国水产科学》【2013, 20 (4): 785-791】

利用比例化剂量反应关系溯源分析环丙沙星在尼罗罗非鱼血浆和组织体内残留风险

Research on evaluating the residual risk of ciprofloxacin in *Oreochomis niloticus* Linn by the Dose Proportionality method

胡鲲¹, 李浩然¹, 施建中², 阮记明¹, 章海鑫¹, 王会聪¹, 杨先乐¹

(1. 上海海洋大学 国家水生动物病原库, 上海 201306; 2. 江苏省南通农场渔业管理区, 江苏 南通 226017)

摘要: 为探讨应用比例化剂量反应关系溯源分析水产动物组织药物残留风险的可行性, 本研究在幂函数模型的基础上分别利用假设检验法和可信区间法对环丙沙星在尼罗罗非鱼 (*Oreochromis Niloticus Linn*) 血浆、肠道、肌肉和肝脏等组织中是否存在比例化剂量反应关系进行了判定和分析。实验结果表明: 在 20~80 mg·kg⁻¹ 的剂量范围内, 环丙沙星在尼罗罗非鱼肠道、血浆和肌肉组织中不呈现比例化剂量反应关系; 而在肝脏组织中呈现比例化剂量反应关系。对于肝脏组织, 曲线下面积-剂量 (AUC-D) 数据拟合结果理想, 幂函数方程为 $AUC = 0.162D^{1.409}$, 相关系数为 0.901。实验结果提示尼罗罗非鱼肝脏组织可作为环丙沙星残留溯源分析的候选靶器官; 通过定量推测偏差, 基于幂函数模型的比例化剂量反应关系研究方法可作为其组织中药物残留溯源分析的技术手段。该研究为水产品质量安全评价提供了一种新的思路。

关键词: 比例化剂量反应关系; 幂函数模型; 环丙沙星; 尼罗罗非鱼; 残留

比例化剂量反应关系 (Dose Proportionality, DP) 是指药物的最大达峰浓度 (Cmax) 或进入体循环的药量 (用血药浓度-时间曲线下面积 AUC 表示) 的变化与给药的剂量成正比例关系[1]。当药物具有 DP 时, 可准确地预测该药物在某一剂量范围内的残留消除过程，为药物的安全使用提供依据。对于非线性药物，DP能作为定量评价其偏离线性的程度的工具。DP的评价方法均有很多种，如：假设检验方法（Hypothesis Test）、可信区间法（Confidence Interval Criteria）等。评价方法不同得出的结果也不尽相同。目前，关于药物代谢的DP研究主要集中于人医药药物的安全使用评估中[2]。

对水产品中药物残留进行分析评估是建立水产品质量安全追溯体系、保障质量安全的技术关键。目前，针对水产动物开展的药物残留消除规律的研究报道[3~8]尽管很多，但涉及药物残留溯源分析的却极少。DP研究可以定量评价药代动力学参数和剂量之间的关系，据此可在一定的范围内由药物残留量逆向推算养殖过程中药物的摄入/给药剂量等参数。该方法在分析药代动力学参数的基础上通过数学模型分析，可充分挖掘原有参数的潜在应用价值；该方法为水产动物药物残留风险分析和评估提供了一种新工具和新探索，可为建立水产品药物安全追溯体系提供重要的技术支撑。目前，应用DP评价包括水产品在内的养殖用药安全尚未见相关报道。

环丙沙星（ciprofloxacin，CIP）属于第三代氟喹诺酮类药物，由于其在动物源食品中的残留严重威胁公共卫生安全，美国和欧盟均将其列为可食用动物组织中"严格监控的对象"[9]，我国也将其列为水产养殖过程中"禁止使用的药物"[10]。本文以CIP为模式药物，通过序贯单次给药设计（Sequential Single-dose Design，SSDD），在幂函数模型的基础上分别以假设检验法和可信区间法对CIP在尼罗罗非鱼（Oreochomis niloticus Linn）血浆、肠道、肌肉和肝脏组织中是否存在DP进行了判定，并对存在DP的组织中主要药代动力学参数（AUC或Cmax-给药剂量（D）进行了拟合，确定了可信区间；分析和探讨了水产动物不同组织作为药物残留溯源分析靶组织的可能性。该研究尝试建立一种基于DP的数学模型，为完善水产养殖品安全用药技术提供一种新思路。

1　材料与方法

1.1　仪器设备

Agilent-1100型高效液相色谱仪，配荧光检测器；Heidolph 4000型旋转蒸发仪；KINEMTICA AG型高速组织捣碎机；HitachiCR21G型高速离心机；IKA-C-MAG MS7型振荡器。

1.2　试剂与耗材

CIP，含量≥98.5%，购自浙江国邦药业有限公司。乙腈为色谱纯。无水硫酸钠、盐酸、正己烷、四丁基溴化铵为化学纯，购自上海安谱科学仪器有限公司。

CIP标准溶液的配制：准确称取CIP0.01 g，用少量盐酸溶解CIP，转移至100 mL容量瓶中，用磷酸盐缓冲液稀释至刻度，即成100 μg/mL的CIP母液，-20℃保存备用。

1.3　动物饲养、给药及取样

尼罗罗非鱼（Oreochromis Niloticus Linn）购于江苏省南通农场，健康无病，未用过任何药物。鱼平均体重50 ±10.4 g。试验前于4 m×2 m×1 m的网箱中暂养7 d，并在给药前3 d停止投饲。分别以20 mg·kg⁻¹、40 mg·kg⁻¹和80 mg·kg⁻¹的剂量，用套有橡皮软管的注射器对尼罗罗非鱼口灌CIP。给药后尼罗罗非鱼在25℃水温下饲养。在给药后0.017 h、0.25 h、0.5 h、1 h、2 h、3 h、6 h、12 h、24 h、48 h、72 h、96 h、120 h、144 h和168 h随机取试验尼罗罗非鱼3组，每组5尾。采取血浆、肠道、肌肉和肝脏组织，放入-80℃超低温冰箱保存备用。设立未给药的尼罗罗非鱼3组（每组5尾）为空白对照。所以实验用鱼的处置均遵循《动物实验规范（Regulation on Animal Experimentation）》[11]。

1.4　组织中CIP残留量分析

利用高效液相色谱法（High Performance Liquid Chromatography，HPLC）分析不同给药时间点的组织中CIP的浓度。紧密称取组织样品2.0 g或血浆2.0 mL，加入12 mL乙腈振荡提取，4 500 r/min离心5 min，取上清。对残渣重复提取一次，并合并上清。上清液与正己烷混合，充分振荡，静置，抛弃上层。将收集的下层在旋转蒸发器中蒸发至干，流动相溶解残渣并经微孔滤膜过滤后上HPLC。色谱条件：反相C₁₈柱色谱柱（4.6 mm ×150 mm）；流动相：乙腈：四丁基溴化铵溶液=5：95（V/V）；流速1.5 mL/min；激发波长：280 nm，发射波长450 nm；柱温40℃；进样量10 μL。将CIP母液稀释成浓度梯度为0.05~4 μg/mL系列标准溶液。以浓度（μg/mL）为横坐标，药物峰面积为纵坐标绘制标准曲线。利用外标法测定组织样品中药物的残留量。

药代动力学参数中的最大达峰浓度（C_{max}）和最大达峰时间（T_{max}）由样品直接测定得到。药时曲线下面积（AUC_{0-t}）等参数由DAS（版本3.0）药代动力学软件处理获得。实验数据由SPSS（16.0版本）软件进行处理和分析。

1.5　DP分析

利用DAS（Data Analysis System，3.0.5版本）软件，分别以假设检验法中的幂函数模型（Power Model）和相关分析法

（Unweighted Linear Regression）对 CIP 在尼罗罗非鱼肠道、肝脏、血浆和肌肉组织中 DP 进行判定：以 Y 代替 PK 参数（AUC 或 C_{max}），D 代表给药剂量，判定方法见表 1。

表 1　在不同评价方法中幂函数模型 a 的判定条件
Tab. 1　Condition for assessing dose proportionality with different assessment method in Power Model

评价方法/Assessment Method	参数及其判定依据条件/Parameters & Condition
假设检验方法/Hypothesis Tests	β；$\beta=1$
利用可信区间法/Confidence Interval Criteria	β；Lower Limit<90% Cl（L~H）<Upper Limit[b]

注：a：方程式为 $Y=\alpha \cdot D^{\beta}$

b：Upper Limit=1+ln（θ_H）/ln（h/l）；Lower Limit=1+ln（θ_L）/ln（h/l）；h、l 分别为最高给药剂量和最低给药剂量；θ_L、θ_H 应用平均生物等效性的标准为 0.8~1.25[12]。

2　结果

2.1　CIP 在尼罗罗非鱼组织中的残留分析方法

利用本文的 HPLC 法测定 CIP，在 0.05~4 μg/mL 的线性范围内，标准曲线线性关系良好，曲线方程为 y=73.377x+0.7064，R^2=0.9992。根据 3 倍噪音确定最低检测限，CIP 在尼罗罗非鱼肠道、肝脏、血浆和肌肉组织中的最低检测限可达 1.0 μg/kg，平均回收率为 70.0%~80.5%，相对标准偏差分别为 1.2%~7.9%。

2.2　不同给药剂量 CIP 在尼罗罗非鱼组织中的残留消除规律

25℃ 水温条件下，尼罗罗非鱼多剂量单次口灌给药 CIP 后，利用 HPLC 法测定不同时间点血浆中 CIP 残留量，CIP 在肠道、肝脏、血浆和肌肉组织中的代谢规律见图 1~4。CIP 在血浆中最大达峰时间最早，为 1 h；肝脏、肠道次之，肌肉组织中 CIP 达峰时间最晚，为 6 h。不同组织中 AUC、C_{max} 等药代动力学参数均与 CIP 的口灌剂量呈现正相关。应用 DAS 软件分析药物水平-时间关系进行二室模型数据拟合，利用房室模型推算所得的 CIP 在血浆中主要药动学参数见表 2。CIP 在肠道、肌肉、肝脏和血液组织中药物时量曲线方程见表 3。

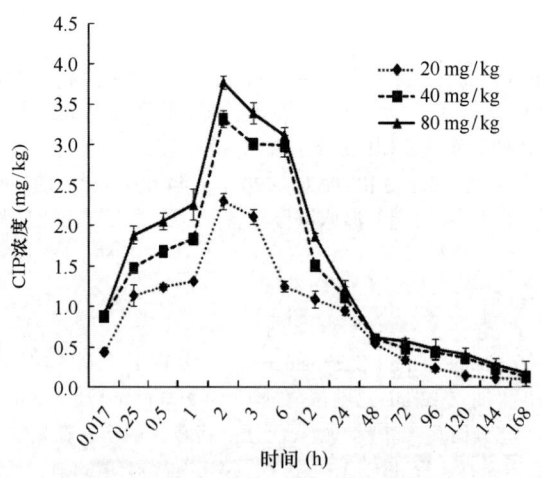

图 1　CIP 在尼罗罗非鱼肠道中的残留消除规律
Fig. 1　Pharmacokinetic of CIP after a single oral administration in intestine tissue of *Oreochomis niloticus* Linn

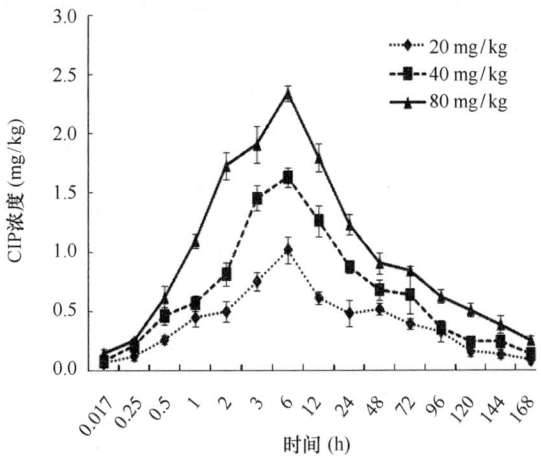

图 2　CIP 在尼罗罗非鱼肝脏中的残留消除规律

Fig. 2　Pharmacokinetic of CIP after a single oral administration in liver tissue of *Oreochomis niloticus* Linn

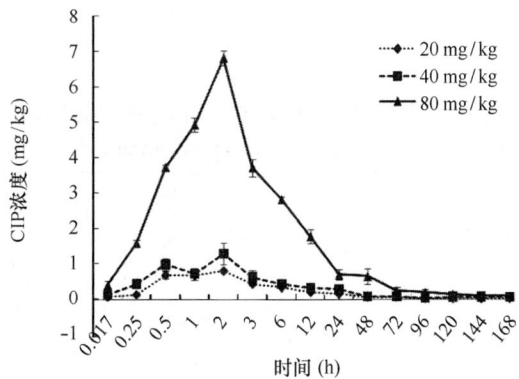

图 3　CIP 在尼罗罗非鱼血浆中的残留消除规律

Fig. 3　Pharmacokinetic of CIP after a single oral administration in plasma of *Oreochomis niloticus* Linn

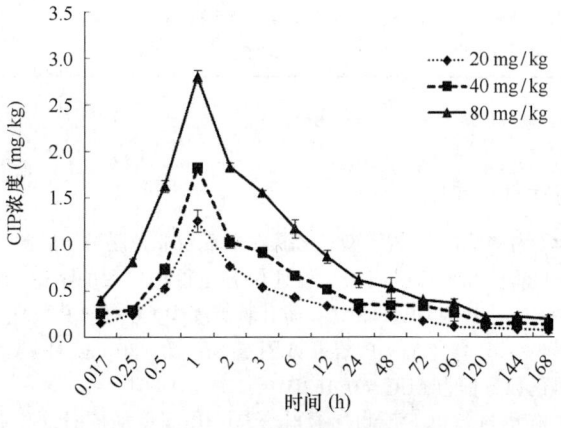

图 4　CIP 在尼罗罗非鱼肌肉中的残留消除规律

Fig. 4　Pharmacokinetic of CIP after a single oral administration in muscle tissue of *Oreochomis niloticus* Linn

表 2 单剂量口灌给药 CIP 尼罗罗非鱼不同组织中药物代谢的主要参数 ($\bar{x}\pm s$, $n=3$)

Tab. 2 Main pharmacokinetic parameters of CIP after a single oral administration in different dose to Oreochomis niloticus Linn ($\bar{x}\pm s$, $n=3$)

药物残留消除参数 Pharmacokinetic Parameters	不同给药剂量实验组 tests groups with different dosage											
	肠道 intestinal			肝 liver			血浆 plasma			肌肉 mucle		
	20 mg·kg⁻¹	40 mg·kg⁻¹	80 mg·kg⁻¹	20 mg·kg⁻¹	40 mg·kg⁻¹	80 mg·kg⁻¹	20 mg·kg⁻¹	40 mg·kg⁻¹	80 mg·kg⁻¹	20 mg·kg⁻¹	40 mg·kg⁻¹	80 mg·kg⁻¹
C_{max}/ ($\mu g \cdot g^{-1}$、$\mu g \cdot mL^{-1}$)	2.31	3.32	3.77	0.84	1.35	6.80	1.25	1.83	2.81	1.02	1.63	2.34
T_{max} (h)	2.00	2.00	2.00	1.50	1.50	2.00	1.00	1.00	1.00	6.00	5.00	6.00
$T_{1/2\alpha}$ (h)	25.24	37.63	33.16	17.41	17.41	17.77	18.91	14.77	17.39	34.79	104.67	105.91
$T_{1/2\beta}$ (h)	39.00	37.87	319450.18	20.46	21.12	18.38	99.95	87.32	158.03	83.08	104.75	107.81
V_1/ ($L \cdot kg^{-1}$)	14086.72	9312.33	32147.71	42536.26	29298.58	5680.82	34612.26	23819.60	13268.45	18916.29	24152.98	16038.25
CLs/ ($L \cdot h^{-1} \cdot kg^{-1}$)	151.14	103.25	368.85	1014.89	636.44	129.83	316.77	193.11	143.25	199.91	103.38	82.54
AUC$_{0-168}$/ ($\mu g \cdot h^{-1} \cdot mL^{-1}$)	154.30	228.18	254.27	25.26	38.05	193.83	64.37	104.35	149.75	114.45	164.52	256.16
AUC$_{0-\infty}$/ ($\mu g \cdot h^{-1} \cdot mL^{-1}$)	159.28	233.36	263.24	25.67	39.05	198.97	80.10	121.55	170.86	122.05	203.73	291.10

C_{max}：药物达峰浓度；T_{max}：药物达峰时间；$T_{1/2\alpha}$：吸收半衰期；$T_{1/2\beta}$：消除半衰期；V_1：中央室表观容积；CLs：清除率；AUC：药物浓度时间曲线下面积；

C_{max}: the maximum concentration; T_{max}: peak time; $T_{1/2\alpha}$: distribution half-life of the drug; $T_{1/2\beta}$: elimination half-life of the drug; V_1: volume of distribution CL$_s$: total body clearance of the drugs; AUC: area under the curve;

表 3 单剂量口灌给药 CIP 尼罗罗非鱼不同组织中药物时量曲线方程

Tab. 3 Drug-time curve formula of CIP after a single oral administration in different dose to Oreochomis niloticus Linn

组织/Tissue	剂量/Dosage		
	20 mg·kg⁻¹	40 mg·kg⁻¹	80 mg·kg⁻¹
肠道/intestinal	$C = 1.42e^{-0.02t} - 1.42$, $R^2 = 0.93$	$C = 0.82e^{-0.02t} + 1.33e^{-0.02t} - 2.14$, $R^2 = 0.94$	$C = 1.81e^{-0.02t} + 0.68e^{-0.01t} - 2.47$, $R^2 = 0.97$
肝脏/liver	$C = 1.44e^{-0.43t} + 0.27e^{-0.03t} - 1.70e^{-10.51t}$, $R^2 = 0.95$	$C = 5.16e^{-0.56t} + 0.35e^{-0.02t} - 5.50e^{-5.83t}$, $R^2 = 0.86$	$C = A4.70e^{-0.53t} + 1.78e^{-0.07t} - 26.47e^{-1.36t}$, $R^2 = 0.94$
血浆/plasma	$C = 11.29e^{-1.13t} + 0.39e^{-0.01t} - 11.67e^{-1.50t}$, $R^2 = 0.91$	$C = 8.22e^{-0.84t} + 0.50e^{-0.01t} - 8.70e^{-1.95t}$, $R^2 = 0.86$	$C = 11.33e^{-0.91t} + 1.02e^{-0.01t} - 12.35e^{-1.50t}$, $R^2 = 0.94$
肌肉/muscle	$C = 2.04e^{-0.08t} + 0.26e^{-0.01t} - 2.30e^{-0.51t}$, $R^2 = 0.87$	$C = 0.40e^{-0.01t} + 0.42e^{-0.01t} - 0.82$, $R^2 = 0.90$	$C = 8.82e^{-0.20t} + 1.47e^{-0.01t} - 10.29e^{-0.31t}$, $R^2 = 0.98$

R^2 为拟合度。

2.3 DP 不同假设方法的评价结果

分别采取假设检验法和可信区间法中的幂函数模型对 CIP 在尼罗罗非鱼肠道、肝脏、血浆和肌肉组织中 DP 进行判定，结果见表4。幂函数模型中，假设检验法和可信区间法二种评价方法得出一致的结论：在 $20\sim80$ mg·kg⁻¹ 的剂量范围内，尼罗罗非鱼肠道、血浆和肌肉组织中，CIP 的 DP 不成立；而肝脏组织中 CIP 的 DP 成立。

利用幂函数模型评价肝脏组织中 CIP 残留的 DP 结果见图5：在 $20\sim80$ mg·kg⁻¹ 的剂量范围内，曲线下面积-剂量（AUC-D）数据拟合结果理想，幂函数方程为 AUC $= 0.162D^{1.409}$，$R^2 = 0.901$。

假设检验结合可信区间方法，还能进行以下预测，①当给药 CIP 剂量加倍时，肝脏组织 AUC 的增加倍数 $2^\beta = 2.15$；②AUC加倍时，剂量需要增加倍数 $R = (2)^{1/\beta} = 1.87$；③按照 $0.80\sim1.25$ 的判别标准，DP 成立的最大剂量比为 $r = \theta_H \wedge [1/\max(1-L, U-1)] = 1.25 \wedge [1/\max(1-0.89, 1.04-1)] = 7.60$。

表4　利用不同评价方法中的幂函数模型判定 DP

Tab. 4　Results of dose proportionality assessed with different assessment method in Power Model

组织/Tissue	评价方法/Assessment Method	参数/Parameter（means±SD）	90%Cl（L~H）	下限/Lower Limit	上限/Upper Limit	DP 是否的成立/Dose Proportionality
肠道/intestinal	假设检验方法/Hypothesis Tests	β：0.37±0.05	0.28~0.45	—	—	NO
	利用可信区间法/Confidence Interval Criteria			0.83	1.16	NO
肝脏/liver	假设检验方法/Hypothesis Tests	β：1.11±0.18	0.89~1.04	—	—	YES
	利用可信区间法/Confidence Interval Criteria			0.83	1.16	YES
血浆/plasma	假设检验方法/Hypothesis Tests	β：0.63±0.05	0.54~0.72	—	—	NO
	利用可信区间法/Confidence Interval Criteria			0.83	1.16	NO
肌肉/muscle	假设检验方法/Hypothesis Tests	β：0.61±0.06	0.50~0.71	—	—	NO
	利用可信区间法/Confidence Interval Criteria			0.83	1.16	NO

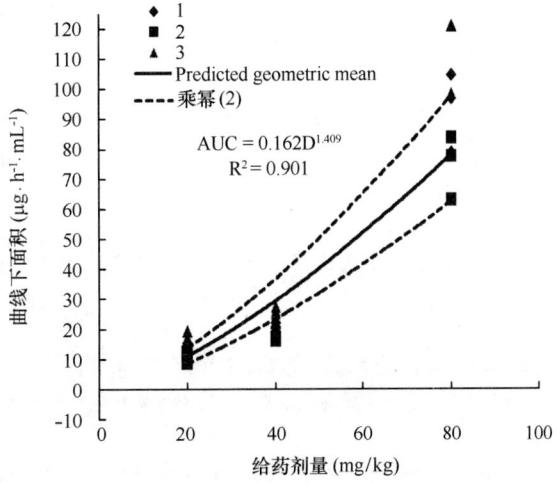

图5　利用幂函数模型评价肝脏组织 DP 的结果

Fig. 5　Results of DP assessed by power model for liver

3　讨论

3.1　药物残留消除规律的 DP 研究方法

药物在包括鱼类在内的动物体内的吸收、分布和消除过程涉及大量的酶及转运蛋白，也与动物自身的生理状态有关，是一个非常复杂的过程。这是不同的药物在不同的动物及不同的组织中可能呈现或不呈现 DP 的直接原因[2]。当药物具有 DP 特征时，很容易预测药物在该剂量范围内的代谢特征，为安全用药提供依据。而呈非线性特征的药物通常治疗窗比较小如饱和吸收的药物；同时剂量增加导致血药浓度升高过快，增加不良反应的风险。对其评价 DP，除了定性评价还应评价其偏离线性的程度。

DP 评价的方法多种多样，不同的评价方法得出的结论也可能不仅相同。常见的评价方法包括假设检验法、可行区间法等。前者以 DP 成立作为无效假设，不成立作为备择假设；后者将分类变量的剂量作为连续变量，不仅能判断 DP 是否成立；还可确定这种关系成立的最大剂量比和呈现非比例化的最小剂量比，因此可以预测研究剂量范围外的 PK 参数变化。在利用可信区间法中 θ_L、θ_H 应用平均生物等效性通常被定义为 0.8～1.25 的标准[12]，超过研究的剂量范围外推时，需慎重考虑。

幂函数模型（$Y = \alpha \cdot D^\beta$）是应用于假设检验法、可行区间法中的一种数学模型，通过对总体的回归参数进行检验，其结论是针对研究的整个剂量范围，并可估计此范围外的 PK 参数变化。与其他回归模型比较，幂函数剂量反应关系不成立时，仍可定量评价其偏离程度。幂函数模型较适合对 PK 模型进行判断和估计线性的偏离程度而被广泛地应用。

3.2 依据靶组织中药物残留开展药物溯源分析的可行性

药物的摄入方式、剂量、剂型及来源（内外源性）等因素与其在水产动物体内的代谢规律息息相关[13,14]，并在很大程度上影响药物代谢的各项参数（C_{max}、T_{max}、AUC 等）。从另一个角度而言，通过分析水产品组织中药物的分布特征和药物代谢动力学参数可以在一定程度上溯源分析药物的摄入剂量等信息。结合此理论，药物的 DP 研究依据统计学方法，通过分析不同组织中药代动力学参数（如 AUC）与给药剂量是否存在比例化剂量，可以在一定剂量范围内，可追溯测算养殖等环节中药物的摄入剂量等信息，并可评价其组织作为药物溯源分析靶器官的可行性。该方法对于水产品药物安全溯源分析具有重要的实践价值。

本文通过序贯单次给药设计，对尼罗罗非鱼单剂量口灌给药 CIP，其血浆、肠道、肌肉和肝脏等组织组织中的药物残留消除过程（图 1～图 4）和药物代谢动力学参数（表 2、表 3）与已有报道基本符合[8,15-17]。本研究采取最为常见的幂函数为数理模型，利用假设检验和可信区间 2 种方法分析了 CIP 在尼罗罗非鱼血浆、肠道、肌肉和肝脏组织中是否存在 DP，结果见表 4，4 种待测组织中肝脏组织 2 种评价方法的评价结果均符合 DP。

大量的药代动力学研究结果证明，恩诺沙星（enrofloxacin，ENR）及其代谢产物 CIP 在鱼类组织中代谢缓慢，尤以肝脏组织为代表性。房文红等（2007）发现 CIP 在欧洲鳗鲡肝脏中峰值最高、肌肉次之、血浆最低[8]。梁俊平等（2010）认为大菱鲆肝脏内 ENR 的峰浓度更大、达峰时间更短的原因可能为肝脏渗透性好[15]，且含丰富的血管。郭娇娇等（2011）发现 ENR 在杂交鲟体内肝脏消除半衰期分布达到最慢、血浆次之，肌肉最快[16]。余培建（2007）证实 ENR 及其代谢产物 CIP 在欧洲鳗鲡组织中残留消除较慢，并建议养成阶段不使用此药[17]。本文的比例化剂量评价结果与以上研究报道相一致。该结果提示，尼罗罗非鱼肝脏组织可作为 CIP 残留溯源分析的候选靶器官。

图 5 对 CIP 在 20～80 mg·kg⁻¹ 的剂量范围内曲线下面积-剂量（AUC-D）数据进行了拟合，取得较为理想的结果。此图作为 DP 常见的辅助方法，直观地显示出给药剂量与 PK 参数之间的关系。根据以上结果，基于平均生物等效性的原理，本文还对药剂量与 PK 参数之间的关系进行了外推预测。对于肝脏组织而言，即使在 DP 不成立时，以上计算公式仍可适用。其中①、②可以估计偏离程度；③可以提示偏离的原因（研究的剂量比过大或过小）。

由于 DP 研究中样本量不大（本文为 3 组），随着剂量的增加，PK 参数变异会加大，这会在一定程度上给研究的设计和验证带来困难。这一点可以通过积累基础数据和引入计算机模拟等辅助手段加以矫正解决。

参考文献

［1］ Yin Y., Chen C.. Optimizing first-time-in-human trial design for studying dose proportionality ［J］. Drug Informat J, 2001, 35: 1065-1078.

［2］ 盛玉成，何迎春，杨娟，郑青山. 药代动力学比例化剂量反应关系的研究方法及其线性评价 ［J］. 中国临床药理学杂志，2010，26（5）：376-381.

［3］ Intorre L., Cecchini S., Bertini S., et al. Pharmacokinetics of enrofloxacin in the soabass (Dicentrarchus labrax) ［J］. Aquaculture, 2000, 182 (1-2): 49-59.

［4］ Nouws J. F. M., Grendel J. L., Schutte A. R., et al. Pharmacokinetics of ciprefloxacin in carp, African catfish and rainbow trout ［J］. Vet Quart, 1988 (10): 211—216.

［5］ Della Rocca G., Di Salvo A., Malvisi J., et al. The disposition of enrofloxacin in seabrearn (Sparus aurata L.) after single intravenous injection or from medicated feed administration ［J］. Aquaculture, 2004, 232: 53-62.

［6］ Bowser P. R., Wooster G. A., St Leger J., et al. Pharmacokinetics of enfloxacin in fingerling rainbow trout (Oncorhynchus mykiss) ［J］. J Vet PharmacolTherap, 1992, 15 (1): 62-71.

［7］ Stoffregen D. A., Wooster G. A., Bustos P. S., et al. Multiple route and dose parmacokinetics of enrofloxacin in juvenile Atlantic salmon ［J］. J Vet Pharmacol Therap, 1997, 20: 111-123.

［8］ 房文红，于慧娟，蔡友琼，周凯，黄冬梅. 恩诺沙星及其代谢物环丙沙星在欧洲鳗鲡体内的代谢动力学 ［J］. 中国水产科学，2007，14（4）：622-629.

[9] Diario Oficial de las Comunidades Europeas (DOCE), 18th August 1990, Council Regulation No. 2377/90 L224, 991, 2601.

[10] 中华人民共和国农业行业标准. 无公害食品渔用药物使用准则 [s]. 2002, NY5071-2002.

[11] 中华人民共和国国家标准. 实验动物 环境及设施 [s]. GB 14925-2010. 中国质检出版社. 2011, 北京.

[12] Hummel Juergen, Mc Kendrick Sue, Brindley Charlie, French Raymond. Exploratory assessment of dose proportionality. Review of current approaches and proposal for a practical criterion [J]. Pharm Stat, 2009, 8: 38-49.

[13] 刘克辛, 韩国柱. 临床药物代谢动力学 [M]. 科学出版社, 2009, 北京.

[14] 陈西敬. 药物代谢动力学研究进展 [M]. 化学工业出版社, 2008, 北京.

[15] 梁俊平, 李健, 张喆, 王群, 刘德月, 王吉桥. 肌注和口服恩诺沙星在大菱鲆体内的药代动力学比较 [J]. 水生生物学学报, 2010, 34 (6): 1122-1129

[16] 郭娇娇, 潘红艳, 杨虎, 廖鑫, 宫智勇, 李谷. 恩诺沙星在杂交鲟体内的药物代谢动力学. 大连海洋大学学报, 2011, 26 (4): 362-366.

[17] 余培建. 药浴给药恩诺沙星积极代谢产物在欧洲鳗鲡体内的药代动力学研究 [J]. 福建水产, 2007, 1 (4): 38-43.

第五节　渔药安全使用的风险评估及其控制的建议

一、渔药在水产养殖疫病管理的地位

该文发表于《中国水产》【2006，5：14-15】

我国水生动物防疫实施中若干问题的探讨

朱泽闻[1]；胡鲲[2]

(1. 全国水产技术推广总站；2. 上海水产大学)

近年来我国水产养殖疫病问题日益突出，要求我们大力推动水生动物防疫工作的开展。但目前我国水生动物防疫的管理构架以及工作制度主要是按畜禽动物疫病防治思路进行设计，对水生动物疫病的特性针对不足。另外，由于我国渔业发展模式以水产养殖为主，致使我国水生动物疫病发生原因和传播途径有别于世界各国，因此，我国的水生动物疫病防治也不能照搬国际通常水生动物防疫的管理模式。这就要求我们必须在实践中不断总结和创新，发展符合我国国情的水生动物防疫理论体系。本文将重点对"十一五"期间我国水生动物防疫开展的外部环境以及实施中的若干理论问题进行探讨。

一、"十一五"期间我国水生动物防疫开展的外部环境分析

1. 水产养殖的发展将成为推动我国水生动物防疫工作的原动力

根据《21世纪初中国主要农产品需求和生产的预测》，到2010年我国水产品将达到5 250万吨，其中养殖水产品约3 700万吨，这就要求"十一五"期间我国水产养殖业在现有约3 000万吨的基础上，每年要完成约100万吨以上增量。但目前我国可利用的养殖水域日益减少，只有通过发展集约化养殖，提高单产，是实现水产养殖业稳步增长的可行途径。从疫病的流行的规律看，在集约化、高密度的养殖环境中疫病暴发的风险将大大增加，事实也证明集约化程度越高养殖疫病问题越突出，如对虾、鳗鱼、甲鱼、鲍鱼等集约化养殖以及高密度网箱养殖都是疫病最为严重的领域。因此，要实现我国水产养殖业持续发展，首先要解决好集约化养殖中的水生动物疫病问题，这无疑将成为推动我国水生动物防疫工作开展的强大的原动力。另一方面，集约化的现代养殖模式有利于水生动物防疫各项措施实施，有利于先进疫病防治科技的应用，有利于增强生产者的积极参与防疫的自觉性，有利于增加社会对防疫体系的投入，从而促进我国水生动物防疫工作的完善。

2. 水生动物防疫将成为保障我国食品卫生安全的重要措施之一

"十一五"是我国构建和谐社会的关键时期，城乡人民生活水平将进一步提高，生活健康和食品安全是和谐社会中人们追求的重要内容，而水产品作为广大消费者喜爱动物性产品，其质量安全上任何小的瑕疵，都可能引发人们对水产品的信任危机。20 世纪 80 年代，上海就因食用带有病毒性病原体的毛蚶而引起"甲肝"暴发，患病人数超过 30 万人。更重要的是严重水生动物疫病还会引发渔用药的滥用，造成药物残留，威胁人民健康。因此，水生动物防疫将视为保障我国食品卫生安全的重要措施之一。

3. 水产品出口贸易的发展要求我们高度关注水生动物疫病带来的风险

"十五"期间，我国水产品出口一直保持高速增长态势，年平均增长率超过 16%。据统计，2005 年 1—10 月，我国水产品出口量 202.8×10⁴ t，出口额 62.1 亿美元。预计 2005 年底，出口额将达 80 亿美元，约占渔业总产值的 15%，占我国农业产品出口总额的 30%左右，连续六年居大宗农产品首位。由于我国大量廉价的水产品出口，对日本、美国、欧盟等地的国内水产业造成巨大的冲击，并造成较大的贸易逆差，各国要求对中国水产品进行限制的呼声也不断高涨，并纷纷加强了动物疫病这一技术性贸易壁垒的应用。目前，日本、美国、欧盟等都单方面提高了检测标准，使我国优势水产品出口遭受巨大损失。可以预见，水生动物疫病将成为我国水产品出口发展中最不确定的因素，应高度关注其带来的风险。

4. 兽医管理体系改革为水生动物防疫工作的发展提供了有利的机遇

随着兽医体制的改革的深入，我国兽医管理体系将全面与国际接轨，加强统一兽医行政监督管理体系的建设，并逐步实行兽医资格认证制度和兽医行业准入制度。"十一五"期间，动物防疫的格局将可能进行重新的规划。目前，水生动物疫病作为动物疫病的一个重要组成部分已成广泛的共识，"十五"期间取得的成效也得到了广泛的认可，因此可以预见，兽医体系改革会充分考虑水生动物防疫的问题，"十一五"期间水生动物防疫工作实施步伐将加快。

二、我国水生动物防疫实施中的理论问题分析

1. 水生动物疫病的特点

与畜禽疫病相比，水生动物疫病有鲜明的特点：（1）水生动物疫病可通过水体传播，而且许多病原能在水环境中长期潜伏，并感染水生动物。因此，对于水生动物防疫，不仅要对染疫动物实施控制，而且还要对受污染的水体进行控制，控制难度和成本大，甚至有些病原在水体中存在就难以清除；（2）水生动物疫病，尤其是病毒性疫病的发病有特定的水温范围，在发病水温外即使携带病原也无症状临床，因此，临床症状只有在发病水温范围内才能作为疫病判断依据；（3）水生动物疫病的潜伏期受水温、气候等多方面影响，长短难以确定，因此，隔离检疫措施在水生动物中难以操作；（4）大多数水生动物免疫系统不发达，特异性免疫应答不明显。因此，水生动物疫病难以通过疫苗实施预防，也不能通过监测抗体来诊断病原，只能直接检测病原体，这就要求较高的实验室条件，而且往往需要杀死检测对象；（5）水生动物疫病防治是采用"群防群治"方式，即疫防对象往往是一定水体中全体水生动物，而禽畜动物以个体防治为主。

2. 水生动物防疫实施中的关键点

不规范的水产养殖行为是促使我国水生动物疫病发生、传播的主要原因，因此，我国水生动物疫病防治以水产养殖中疫病为主要控制对象，疫病防治思路应重点把握以下关键点：一是水生动物疫病预防主要通过健康养殖管理来实现。药物预防、生态预防、免疫预防是动物病害防治的主要方法，由于免疫预防在水生动物中操作困难、难以推广，而药物预防存在药残的危险，因此，只能通过生态预防来实现，即实施健康养殖，如制定科学的养殖模式、严格水产苗种和投入品的管理、实施休养轮养制、培育抗病品种等。二是区域监管是实施水生动物疫病防治的有效方式。由于水生动物一般个体小，且分散在水体中难以针对个体进行控制，因此，水生动物疫病防治只能通过区域监管来实现。即按疫病防治要求及水域特点，将水域划分为若干相对封闭性疫病防控区域，以疫病防控区域作为水生动物疫病防控的基本单位。三是水生动物疫病的控制和扑灭措施必须是在受染的水体和染疫水生动物中同时进行。

3. 防疫实施中的主要环节

根据我国水水生动物疫病发病的特点，防疫实施应重点突出"监、检、控、治"四个环节。"监"即监测预警，从源头对水生动物疫病进行监测，并提出预警措施，防止重大水生动物疫病的爆发或传播；"检"，即防疫检疫，在生产、运输、销售过程中加强水生动物以及产品的疫病的检疫，防止疫病的扩散；"控"和"治"即疫病控制扑灭和治疗，指在疫病的危害发生后，为防止危害进一步扩大，并消除影响而采取的措施。其中，对于二类以上重大疫病应通过控制扑灭系统进行隔离、控制、扑灭等，而对于三类疫病则通过病害防治进行治疗或无害化处理等。同时渔用药和药残监管可作为防治环节的延伸纳入其实施。从而，通过"监、检、控、治"四个环节有效的配合，实现从事前预防到事后处理的全过程监管的目标。

三、"十一五"期间水生动物防疫实施的建议

综上分析，我国"十一五"期间水生动物防疫管理实施中各环节建议重点开展如下工作：

1. 疫病的预防：实施《健康水产养殖行动计划》和水产养殖动植物病情测报。

2. 疫病的控制：启动并实施区域监管。国家疫病防治机构按水域特点科学划定防疫基本区域，由有关机构主动对基本区域内的水环境、疫病、投入品进行监测，以了解区域健康水平，并以此作为判断区域中水生动植物健康状况的依据。在检疫方面实施水产兽医官制度。

3. 疫病的扑灭：强化地方政府责任，以防疫区域为基本单位实施扑灭。

4. 疫病的诊疗：实施水产执业兽医制度。

5. 渔药的监管：实施水产处方药制度和药残的监控计划。

面对千万养殖生产者，仅依靠政府推动的水生动物防疫管理是无法达到最佳的效果，必须走"政府专业执法，全民防治监督"的发展道路。要充分发挥行业协会和龙头企业的核心作用，并通过宣传和引导提高从业者的防治意识，逐渐形成以政府专业执法为主，社会公共机构、相关企业协会以及广大养殖生产者共同参与监督的防疫机制。

二、我国渔药管理的对策

渔药管理具有特殊性，为解决由于渔药带来的水产品安全问题，当前渔药管理工作的核心——重点解决好管理体系、法规和标准建设三个方面的问题（杨先乐等，2002；杨先乐等，2002）。健全行政监督管理体系、制定技术标准、加强基础理论研究和新型渔药的研发、加强科普宣传培训是主要的措施（杨先乐，2003；王民权等，2006）。

该文发表于《科学养鱼》【2002，11：41】

我国渔药管理工作刍议（上）

杨先乐，黄艳平

（上海水产大学 农业部渔业动植物病原库，上海 200090）

我国渔药历来都是属于兽药的一部分，以兽药的方式进行管理，这是因为在 20 世纪 90 年代以前，我国的渔药基本上是空白，大部分借用人、兽药，没有专门的渔药，更没有渔药的生产企业和经营机构。但是近年来，这种现象发生了较大的变化。渔药的品种越来越多，由此引出的问题也越来越复杂。因此对渔药的管理进行专门研究已是十分迫切。我们在查阅国外的有关资料和参考国外的一些管理方式之后，结合我国的具体情况，对我国渔药的管理工作提出以下几点建议，供有关部门参考。

一、实行管理体制的改革——渔药管理的相对独立

我们认为，以现行的管理体制来管理渔药，存在着较多的弊端。渔药管理体制相对独立的改革方案，已经渐渐凸现。这是因为：

1. 渔药与兽药相比，有较大的特殊性。渔药虽然在一定程度上与人、兽药相似，它们在药物研制开发过程中对原料、安全性、分析方法的要求以及基本法规等方面的要求也有类似之处，但由于渔药作用的对象与人、兽有较大的区别（如温血性与变温性的区别），渔药的使用方式、作用过程与作用效果等与兽药有较大的不同，它们在代谢、残留等方面有着其特殊性。所以对渔药的管理与兽药相比应有较大的区别和不同。拿管理兽药的方式去管理渔药，不可能会达到完美的效果。因此建立专门的渔药管理机构，把渔药的管理从兽药中分离开来，实行渔药的专门管理已是势在必行。在这一方面，美国的经验已给了我们一些启示。

2. 渔药与兽药相比，还有其复杂性。渔药的使用除了要考虑对养殖的水产动物的影响、残留量对人类健康的影响外，还要考虑它对环境的影响。这一点在兽药中也是较少考虑的。

3. 渔药在兽药中所占的比例将越来越大。众所周知，我国是一个水产养殖大国，水产养殖的总产量占全世界养殖产量的近三分之二，而且还将会有所发展。目前养殖的品种从甲壳类、鱼类到爬行类有近百种，由于养殖高密度和集约化，由于病害控制的严峻性，渔药将会有更大的发展。在这一方面，国际上任何国家都是难以比拟的。国外一些发达的国家之所以将渔药归属于兽药管理，是因为它们的水产养殖在渔业中的比重很小，渔药也较少使用，而我国则不然。如果借用它们的管理模式，将会不利于我国渔药的正常、健康发展，进而会阻碍我国水产养殖业的发展。

4. 水产品的安全问题对渔药的管理提出了更高的要求。水产品中药物残留问题已经引起了多方面的高度重视，要解决这一问题，仅凭对上市的水产品进行药物残留检测，那将会是杯水车薪。问题的关键应该是加强渔药的研制、生产、销售和使用的管理，如果渔药还将隶属于兽药的范畴之内，水产品的安全问题不可能得到完美的解决。

5. 我国以往的管理模式已经显露出一些问题，出现了渔药管理方面的一些漏洞，形成了管理上的一些盲区（或盲点），有时也造成了一些管理上的错误。

6. 我国已经初步具备了渔药管理的条件。近一、二十年来，我们对渔药的研制、渔药对环境的影响、渔药药代动力学、渔药的监控与检测、渔药市场的管理、渔药的使用等做了较多的工作，并积累了较多的经验，我国已完全有能力将渔药从兽药中划分出来进行单独地管理。

二、渔药管理工作的核心——重点解决好管理体系、法规和标准建设三个方面的问题

关于管理体系，首先要认真研究解决进行统一的渔药行政监督管理的可操作办法，改变渔药管理工作隶属于兽药而造成的渔药管而不实的现象。同时，要逐步按照市场经济规律的要求，将渔药行政管理工作和渔药服务性工作分离开来。为体现渔药监督执法的公正性，应该逐步将渔药管理工作的各项经费纳入国家财政预算，从而建立起权、责统一，权、利分离的行政管理机制。此外，要逐步实行渔药行政的"垂直管理"，打破地方分割局面和消除地方保护主义现象，减少渔药管理部门的重复设置和各部门之间的矛盾，提高行政效能。其次，要推行官方渔药管理制度，逐步建立一支渔药管理的专业队伍。通过引导和监督企业建立自身质量管理体系和引导扶持企业的规模化、集约化生产，提高企业自主质量控制能力，从而提高渔药管理的行政人员检疫检验和监督的效率，为建立精简、高效的渔药管理队伍创造良好条件。

关于管理法规，要制订相应的渔药管理条例。渔药管理的法规建议，我们可参照各个国家对兽药（或渔药）的各项管理法令进行制订，囊括的内容应尽量全面，包括渔药的评价和认证，渔药的生产、销售、执照发放及管理，渔药的申报、注册、检验要求和程序，渔药标签的规范，渔药的残留限量和休药期的规定等。

关于管理标准，是我国一个比较薄弱的环节，现在已引起了政府等各方面的重视。当前渔药残留限量标准、渔药检测标准、渔药研制与科学使用标准的制定与实施等，是当务之急。当然，要制定与实施这些标准，必须要加强相应的研究，为其提供可靠的依据。这一点国外已有较多的借鉴。

该文发表于《科学养鱼》【2002，12：41】

我国渔药管理工作刍议（下）

杨先乐 黄艳平

（上海水产大学农业部渔业动植物病原库，上海 200090）

三、当前工作的重点———渔药管理中应强调的几个问题

1. 加强对新渔药的审批管理

加强对新渔药的审核管理，有效保证审批活动的科学、合理、规范，从而保证渔药使用的安全、有效，保证渔药管理法规的贯彻、实施。

一个好的渔药应效果好，毒副作用低，无公害，价格低廉，使用方便。因此对新渔药的审核应强调其在临床上的有效性及安全性。有效性是指产品持续地、不变地保持它所声称的作用。安全性是指药品对动物、动物源食品、用药的或者接触已用药动物的人和环境是否具有有害影响。

具体说明，我们认为我国渔药审批应要求申请方提供以下五方面的材料给渔药管理部门：①药物的有效性，提供新药适用的水生动物、用药方法、剂量和所治疗的疾病。②药物对水生动物的安全性，证明新药在数倍于使用浓度和数倍用药时间后，对动物无害。③新药物对人的安全性。提供用药后，水产品药物残留量降低至国家标准允许范围所需的时间，保障人食用药后的水产品不会发生安全事故。④药物对环境的影响。新药对水质、水生生物及人居住环境有否影响。⑤药厂生产新药的生化数据。

由于目前我国新渔药的开发力度还较弱，绝大部分渔药都是人、兽药用途的延伸，为了促进渔药的生产，建议水产部门成立专门的水产动物新药申请协调管理机构，任务是协调各渔药厂、各渔药研究机构进行新药试验和申请工作，以致最终通过审批。具体工作包括：①沟通水产新药申请单位和国家渔药管理中心的联系。②鼓励有关的科研机构和人员参与水

产新药的研制与试验。③寻找能给予新药试验资助的药厂。④指导新药试验单位进行年度总结，给渔药管理中心提交年度报告。⑤审查、记录和提供各种新药试验和申请的信息。⑥提供新药试验的申请和培训。

2. 对上市后渔药的管理与监测

对上市后渔药的管理与监测将有利于及时了解渔药使用过程中出现的问题并及时处理，保证渔药使用过程中的安全性。

建议渔业管理部门成立专门的部门对上市后的渔药进行管理与监测，搜集并评估渔药上市后的资料，主要包括：涉及动物或人的自发性不良反应报告，包括渔药缺乏预期药效或错误使用的情况；人体对渔药的可疑不良反应；耐药性的流行病学研究；对环境的潜在影响；违反渔药允许残留限量的事例；渔药风险、效益评估等。并以法规的形式规定上市渔药的许可证持有人必须指定专人负责该药的药物警戒工作内容，必须定期向有关职能部门提交报告

3. 抗生素耐药性产生的控制

对食用动物抗生素耐药性的控制问题已引起全世界的极大关注，世界卫生组织、世界粮农组织和国际流行病办公室等国际组织及许多国家都制定了相关政策，积极倡导谨慎使用抗生素，降低食用动物抗生素耐药性的产生。2000 年 6 月在日内瓦召开的谨慎使用抗生素的咨询会上，来自美国、丹麦、比利时、菲律宾、德国的专家分别介绍了各自国家在可食用动物控制抗生素耐药性方面的措施和经验，与会代表在充分讨论的基础上，形成了 "控制可食用动物抗生素耐药性的全球准则"。该准则将成为各国加强兽用抗生素管理、控制兽用抗生素耐药性的纲领性文件。这一纲领性文件对我国渔用抗生素的管理也将起到指导性的作用。

4. 动物源性食品残留监控体系的建立及实施

通过残留监控体系的建立及实施，了解诸如生产和销售过程中渔药的非法销售；大剂量药物的非法分销和使用；由于 GMP 实施不善，造成的动物饲料的交叉污染；未正确地遵循良好的养殖规范或停药要求；滥用药物作为饲料添加剂等各种违规要求，从而确保人类消费的动物源性食品的安全，确保所建立的法律、法规的正确实施，确保大众免于有害残留物的危害。

对动物源性食品残留的监控建议在国家残留监控计划确立的基础上，在国家监控组织的统一布局下实施。根据各渔药残留危害的历史检测信息及有关法规要求确定监控药物，建立一个全国性的、涉及多方管理当局的电脑信息系统。

对动物源性食品残留的监控，我们可参照美国的相关条文。

5. 加快 GMP 认证工作

加入 WTO 后我国渔药行业面临极大的挑战，要加快渔药 GMP 达标进程，提高渔药行业整体水平。

该文发表于《水产科技情报》【2003，30（2）：68-71】

我国水产品的药物残留状况及控制对策

杨先乐

（上海水产大学农业部渔业动植物病原库，上海 200090）

水产品药物残留通常指水产养殖动物的任何可食部分中药物的母体化合物及（或）其代谢物，以及与药物有关的杂质的残留。此外，药物或其代谢产物还能与内源大分子共价结合，形成结合残留，它们对靶动物具有潜在的毒性作用。因此，药物残留既包括原药，也包括药物在动物体内的代谢产物，还包括药物与内源分子的结合物。我国加入 WTO 后，面临着全球经济一体化的形势，面临着人们对安全、卫生水产品的要求日益增长及我国水产养殖高质量、高效益、持续发展的局面，我国水产品的药物残留问题已经引起了全社会的普遍关注。本文试图对我国水产品的药物残留状况作一个客观分析，并对其控制提出对策。

一、水产品药物残留引起国内外关注的由来

浙江舟山的冻虾仁因其个大味鲜而享誉海外，使舟山成为我国冻虾仁的主要出口基地之一。但 2001 年底欧盟官员到我国实地抽检时发现其药物残留量达到 0.2 ug/kg，超过欧盟规定的最低标准 0.1 ug/kg，2002 年 1 月 30 日，欧盟正式作出禁

止进口中国所有动物源性食品的决议。欧盟的这一禁令，是以氯霉素为把柄对中国水产品出口设置的贸易壁垒，它使我国蒙受了巨大损失：近 5 万劳动力下岗，十几万农户生产的水产品因企业无法履行收购合同而卖不出去；94 家对欧盟出口水产品的企业蒙受了高达 6.23 亿美元的损失。这一情况引起了我国社会各界对水产品药物残留状况的深思。

近 20 多年来，我国的渔业生产完成了从"捕捞增长型"到"养殖增长型"的转变，水产品的产量呈跨越式增长，解决了人们的"吃鱼难"问题。但是，我国水产养殖产量的增加是通过水产养殖面积的扩大，养殖强度的提高而获得的，是以消耗资源、牺牲养殖水域环境的代价换来的，也是养殖者和业内人士与日益增多、日益严重的水产养殖动物病害作斗争所取得的。养殖者为了控制水产动物病害的蔓延，减少病害造成的损失，投入了较多的抗生素类、喹诺酮类、磺胺类、呋喃类等化学药物，投入了较多的消毒剂。由于对鱼药认识的肤浅以及对药物的使用方法、用量和停药期的忽视，部分水产品中药物残留超标的问题越来越凸现出来。

在人们开始注重生活质量，对动物性食品的质量要求越来越高的时代，人们开始用怀疑的眼光去审视水产品："药蟹"的问题曾见之于报端，关于黄鳝、甲鱼是喂了避孕药的流言也一度盛传，有关鱼、虾、贝、藻药物残留超标的议论也时而可以听到。尤其是欧盟的禁令所引起的轩然大波更激起了社会各界对水产品药残问题的极大关注：行政管理者正在制定无公害食品行动计划，力争在 5 年之内使人吃上放心的水产品；科研人员正在加紧努力，除了对水产品监控提供相应的手段和技术外，还在开发新的低残留、高效药物，研究控制水产动物疾病发生和蔓延的有效方法；生产者也充分认识到了药物残留给自己和社会带来的损失和危害，正在为避免药物残留的出现而努力。

二、我国水产品药物残留的基本状况

目前，我国水产品的药物残留状况究竟如何？这个问题已引起了我国各级政府和广大人民的关注，社会上的流言蜚语也较多。如"鱼虾是药水里泡大的"、"蟹是吃药长大的"；水产品喂了禁药（指激素、环丙沙星等），不能吃等等，风风雨雨，危言耸听。

经过调查，我们认为，我国水产品的药物残留状况并不像有些人想象的那样十分严重，然而我们也应客观地承认，我国水产品的药物残留问题也并不乐观，确实存在着一定的问题，有时也比较严重。

（一）可能涉及药物残留的水产品和药物

我国的水产养殖对象品种很多，从甲壳类到爬行类共百余种，但重点养殖对象却只有 20 种左右。由于产品价格和养殖成本等原因，也由于养殖对象罹患暴发性疾病的可能性较小，因而出现药物残留的可能性不大。一般来说，淡水养殖鱼类药物残留的出现机率较小。原因是淡水鱼类价格较低，考虑到养殖成本，施药的频率和强度都较低；养殖的鲟鱼、海水螺类对水质的要求较高，疾病发生的频率也就较低。可能导致药物残留的都是那些价格相应较高，发病率较高，为了控制疾病发生而用药强度较大的一些品种。可能会有药物残留的养殖品种见表 1。

表 1　我国水产养殖对象中可能会出现药物残留的主要品种

类别	养殖品种
甲壳类	南美白对虾、中国对虾、刀额新对虾、罗氏沼虾、中华绒螯蟹
鱼类	大菱鲆、鲈鱼、大黄鱼、鳗鲡、黄鳝、罗非鱼
两栖类	牛蛙、美国青蛙
爬行类	中华鳖

我国水产动物的病害防治大多借用人兽原料药，或是这些药物用途的延伸。在水产养殖上，药物的种类较多，用药的范围也较广。除了消毒剂外，还有杀虫驱虫类、抗生素类、磺胺类、呋喃类和喹诺酮类等，个别情况下还使用某些激素。表 2 是我国目前使用较为普遍（或者曾经较为普遍），且有可能导致水产品出现药物残留的几种主要药物。

表 2　我国水产品中可能出现残留的主要药物

类别	药物
消毒剂类	硫酸铜、三氯异氰尿酸
杀虫驱虫类	菊脂类农药、甲苯（或苯硫）咪唑
抗生素类	土霉素、氯霉素
磺胺类	复方新诺明
呋喃类	呋喃唑酮、呋喃西林
喹诺酮类	盐酸环丙沙星、噁喹酸
其他	喹乙醇、已稀雌酚、甲基睾丸酮

（二）我国水产品药物残留的概况

2000—2002 年对某些地区水产养殖产品的药物残留情况作了相应的调查。管中窥豹，我们认为我国水产品在药物残留上虽然存在着一定的问题，但总的来说，水产品的质量总体上还是可以信赖的。2000 年从上海、江苏、浙江等省（市）的某些养殖区抽样检测了 100 个中华鳖肌肉样品、25 个罗非鱼肌肉样品和 50 个鳗鲡肌肉样品，检测结果是中华鳖肌肉样品中仅有 1 个样品的已烯雌酚含量超标，占检测样品总数的 1.0%；50 个鳗鲡肌肉样品中，土霉素含量超标的样品有 3 个，占检测样品总数的 6%；四环素和金霉素含量均低于残留限量标准；25 个罗非鱼样品，甲氧苄啶、磺胺、磺胺嘧啶、磺胺噻唑、磺胺吡啶、硫代异《唑、磺胺甲噁唑、磺胺二甲嘧啶的含量均低于残留限量，达标率 100%。2001 年又对来自江苏、浙江、湖北、湖南、河北、山东、大连、广东和福建等 9 个省（市）的中华鳖、牙鲆、大菱鲆、鳗鲡、中华绒螯蟹和鳜鱼等 533 份样品中已烯雌酚、氯霉素、土霉素类（土霉素/四环素/金霉素）和呋喃唑酮等 4 类鱼药的残留进行了检测，合格样品 466 份，检测合格率达 87.4%。同年 12 月抽查来自江苏、上海等地的中华绒螯蟹、鳗鲡样品各 30 个，其土霉素、四环素、氯霉素、呋喃唑酮、喹乙醇、乙稀雌酚等药物的残留都在最低残留限量以下。2002 年对来自上海、浙江、广东、广西、海南等地的南美白对虾 290 个样品中氯霉素、呋喃唑酮和乙烯雌酸的检测结果表明，仅有 3 个样品呋喃唑酮超标（标准是 10 ftg/kg 以下，而阳性检出结果在 12.53 μg/kg 以下，超标者占样品总数的 1%），氯霉素含量超标的有 49 个，最高为 7.51 ug/kg，大部分在 2~3 μg/kg（合格标准为 1 μg/kg），超标者占样品总数的 16.9%。

（三）我国水产品药物残留的主要表现

上述调查结果表明，我国水产品的药物残留主要体现在以下几个方面：（1）氯霉素是我国水产品药残监控的一个重点。这是因为：①氯霉素在防治水产动物疾病时具有效果较好、价格低的特点，曾在我国水产养殖上广泛使用。因为习惯，该药在水产养殖上还偶见使用，禁用尚不彻底，可能会在某些水产品中出现残留。②对氯霉素的检测限要求较高，欧盟规定是 0.1 ug/kg，我国目前制定的标准是 1 ug/kg。如此严格的检测限，一旦检测方法上有问题，则容易导致假阳性出现。③氯霉素在医治人类疾病上还被用作外用药物等，如果不注意，就可能在运输或加工途径中造成氯霉素污染。（2）对虾、鳗鲡、罗非鱼以及中华鳖等养殖对象，由于发病率较高，养殖中的用药强度较大，药物残留存在的可能较其他养殖对象要大。（3）有些渔（农）民缺乏科学用药的意识，盲目用药，有可能在某些水产动物上使用激素类药物，从而导致激素类药物残留。（4）一般来说，养殖强度较高且养殖历史不长的地区出现药物残留的可能性较大，如海南、广西等省（市）的有些地区。（5）分散的个体养殖户的养殖产品也很可能存在药物残留。

三、产生水产品药物残留的主要原因

水产品的药物残留，主要是养殖过程中滥用药物造成的。我们认为，根据我国目前的状况，我国水产品药物残留产生的主要原因有以下几个方面：

（一）法制、法规、标准等规章制度不健全，监管、督促、处罚等措施不力。

目前对滥用药物的管理尚无章可循，对药物使用、停药期、药物残留的检测等还缺少一定的标准，因而在药物残留的监控和管理上出现了较多的盲区。

（二）对鱼药及其检测方法的研究滞后。

我国对鱼药的研究起步较晚，对大部分鱼药，在药动学、药效学以及毒理学等方面都缺乏研究；当某些药物因毒性作用或与人药同源或同类而被禁用之后，不能及时推出新的药物，以致某些被禁用的鱼药被继续使用；或者使用一些未经批准的药物，由于对这些药物在水产动物体内的代谢情况缺乏研究，又没有休药期的规定，极易造成药物残留。

（三）渔民科学用药、安全用药的意识差。

具体表现在：（1）不遵守休药期的有关规定或者缺乏休药期的意识。（2）不能正确使用药物。在用药剂量、给药途径、用药部位和用药动物的种类等方面不遵守用药规定。（3）用药方法错误。（4）不做用药记录。（5）为了掩饰水产品上市前的临床症状，以获得较好的经济效益，在上市前使用鱼药。（6）对水产动物的疾病及其防治缺乏认识，片面理解防病与治病的关系，片面理解健康养殖和中草药的概念，认为健康养殖就是不用药，中草药就是绿色鱼药等，结果导致疾病发生后无法控制，出现乱用药、乱投药的混乱局面。

（四）其他。

如将盛过抗菌药物的容器用来贮藏饲料，或用未经清洗干净盛过药物的容器贮藏饲料原料或饲料，造成饲料在加工、运送或使用过程中的药物污染；利用受到药物污染的水，引起水产养殖动物体内的药物残留。

四、对水产品药物残留的主要对策

虽然水产品中的药物残留大多不会引起人体急性中毒，但是如果经常摄入含有低剂量药物残留的水产品，残留的药物可在人体内慢慢蓄积而产生毒性作用，使人体各器官的功能发生紊乱或病变，出现过敏反应和变态反应，导致耐药菌株产生，正常菌群失调，产生致畸、致癌、致突变作用和激素作用，严重危害人类的健康。因此，控制水产品的药物残留对维护社会稳定、保障经济发展均具有十分重要的意义。

控制水产品的药物残留，应该从源头抓起，要切实加强鱼药的安全使用和科学管理。

（一）强化对鱼药的科学管理

鱼药管理工作的核心是解决好管理体系、管理法规和管理标准的建设。

1. 管理体系建立和健全鱼药行政监督管理体系，改变鱼药管理工作隶属于兽药管理而造成的管而不实的现象；要逐步按照市场的经济规律，将鱼药的行政管理工作同鱼药的有偿性服务分离开来。

2. 管理法规要制订相应的鱼药管理条例，要建立对新鱼药进行正确评价、认证和检验的程序和机制，要加强对鱼药生产和销售的管理，要规定鱼药的残留限量和休药期，要对养殖用药进行指导和监督，加强对滥用鱼药的处罚。

3. 管理标准要加紧制定鱼药最高残留限量标准，鱼药检测标准，鱼药研制和扭学使用标准等。要加强相应的研究，为其标准的制定提供可靠的依据。

（二）将科学用药落到实处

科学用药可在很大程度上避免水产品的药物残留。科学使用鱼药就是要从药物、病原、环境、养殖动物本身和人类健康等方面的各种因素出发，有目的、有计划、有效地使用药物。科学用药是一个较复杂的问题，既要考虑提高对水产动物疾病的防治效果，又要考虑药物对水产动物品质和水域环境的影响。科学用药要做到：（1）正确诊断，对症、对方用药；（2）选药要有明确的指证，安全用药；（3）掌握影响药物疗效的一切因素，排除各种可能造成药物干扰的因素，适宜用药；（4）适当加大或缩小用药的浓度、用药次数和用药的间隔时间，合理用药；（5）祛邪扶正并举，增强机体的抗病能力，控制用药；（6）认真观察、分析，酌情采取停药、调整剂量和改换药物的措施，有效用药。

（三）加大鱼药研究和鱼药开发的力度。

大力开发"三效"（高效、速效、长效）、"三小"（剂量小、毒性小、副作用小）鱼药、水产专用药及生物鱼药，改变鱼药沿袭人药、兽药的现象。同时，制定出这些新型药物的合理休药期。

（四）加强对渔（农）民进行科学用药的教育和技术培训，提高渔（农）民科学用药、安全用药的意识和水平。

我们相信，经过切实努力，通过我国"无公害食品计划"的认真实施，水产品的药物残留问题一定会得到较好的解决，水产品的安全问题将不再是人们的忧患。

参考文献

1 田中良著. 刘世英、雍文岳译. 水产药详解. 378. 北京：农业出版社，1982.

2 贡玉清. 喹诺酮类药物的残留分析. 畜牧与兽医，2002，34（8）：31-33.

3 杨先乐，黄艳平. 我国渔药管理工作刍议. 科学养鱼，2002（）：41.

4 张合成. 采取有效措施保障水产品安全. 中国水产，2002（8）：14-17.

5 郑朝民. 水产动物药物残的控制对策. 福建水产，2002（9）：26.

6 酒井正博. 免疫赋活物质成为鱼病预防的可能性养殖，1994，3（2）：177-179.

7 周德庆，李晓川，李兆新等. NY 5070-2001 无公害食品水产品中渔药残留限量. 2001.

8 Ciprofloxacine，N. H. . An overview and prospective appraisal. AmJ. Med. ，1987，82（Suppl. 4A）：395.

9 Kusser W. C. ，Newman S. G. . Detection of OTC residues in fish tissues using a sensitive bioassay. Journal of Fish Disease，1990，13（6）：545-548.

该文发表于《科学养鱼》【2006，12：1-2】

我国渔药面临的困惑、对策及展望

王民权[1]；房文红[2]；杨先乐[1]

（1. 上海水产大学；2. 中国水产科学研究院东海水产研究所，200090）

渔药的发展与水产养殖业、水产动物疾病学、药学的发展密不可分，同时又受到现代生活与管理理念的影响，本文即对我国渔药面临的困惑、对策及展望作一阐述。

一、渔药发展所面临的困惑

1. 水产品安全

水产动物体内的渔药残留已经严重影响了水产品的质量安全，人们长期摄入含有渔药残留的水产品将影响身体健康，如磺胺类可引起肾脏损害；青霉素、四环素等药物会使敏感人群产生过敏反应，严重者可引起休克等症状。水产品安全问题已经引起社会普遍关注，使得各国对渔药残留限量的制定越来越严格，我国水产品出口因此屡遭绿色壁垒，近年出口的鳗鲡、小龙虾等水产品因药物残留超标而被退货。

2. 环境保护

渔药对生态环境的毒副作用不容忽视。国外对许多渔药在环境中的浓度、持续时间及在食物链中的富集做了研究。Samuelsen 等发现在海湾渔场使用土霉素后，淤泥中绝大部分土霉素可在第 1 周降解，但较低浓度残留则能存在很长时间（半衰期 87~144 d）；Wollenberger 等报道常用抗菌药如土霉素对大型溞有急性毒性作用，对水环境有潜在的不良作用；含氯消毒剂使用后水中含有大量的氯化消毒副产品，如三卤甲烷和卤乙酸，这些物质被证实与肿瘤发生有关。但国内研究涉及较少，对水体中泼洒大量的消毒剂及杀虫剂在水体中的蓄积、转移等尚缺乏研究。

3. 病原菌的耐药性

在当前水产养殖中，化学药物治疗鱼类疾病已成为一种普遍现象。但是，渔用抗菌药的使用范围和剂量日益加大，养殖水体病原菌耐药性问题日趋严重，如郑国兴等（1999）对从欧洲鳗皮肤溃疡处分离出的嗜水气单胞菌的耐药性测定结果显示，对诺氟沙星的耐药菌株率为 60.0%，最小抑菌浓度（MIC）高达 128 微克/毫升，高于使用浓度数十倍，加大了水产动物疾病控制的难度。

4. 有效安全渔药及其剂型贫乏

我国渔药自主研发能力比较薄弱，尤其是渔用化学药物主要为仿制、改进，真正从事实体化学品的研究极少，大部分渔药直接或间接地来源于人药、兽药或农药，至今尚未形成自主产品系列。专用渔药种类少，对禁用渔药替代品的研究未能及时跟上，导致禁用渔药继续使用的现象仍然存在，如孔雀石绿。随着耐药菌的大量出现，抗生素的研制速度已无法解决日趋复杂的耐药性问题。

5. 渔药药效评价的方法

正确、全面的药效评价体系不仅关系到渔药的研制与开发，而且关系到健康养殖和水产品安全。目前国内评价药效指标还不够完善。如对抗生素药效的影响因素研究较少；一些水产动物免疫刺激剂的评价标准、检测指标难以确定，检测手段比较落后；中草药的药效是由于多种成分共同作用的结果，因此对其药效的研究受到很大的局限，基本上停留在 MIC、最低杀菌浓度（MBC）、抑菌浓度指数（FIC）上。此外，国内很多渔药都缺乏严格而较全面的毒理学数据，如目前在水产上被批准使用的有机磷类、有机氯类、菊酯类渔药、重金属盐类化合物及中草药，鲜有特殊毒理学、水域生态毒性研究。

6. 渔药药学理论相对滞后

国内对渔药动力学的研究起始于 20 世纪 90 年代末，本世纪初，农业部渔业局组织部分水产院校、科研院所研究了氯霉素、环丙沙星、诺氟沙星、呋喃唑酮等在罗非鱼、中华绒螯蟹等水生动物体内的代谢动力学和残留消除规律，比较研究了给药方式、给药剂量、种属差异、温度、盐度、性别、年龄等因子对药物动力学的影响，并制定相应的最高渔药残留量的标准及其渔药的合理休药期。但研究的药物种类、药物的代谢产物、水产动物品种还远未能满足水产养殖的需要，对中草药在水生动物机体内的药动学研究鲜见报道，对"首过效应"、"多药峰"现象的药物代谢机制的研究目前还未能很好的解释，停留在描述、推导、分析阶段。

7. 渔药的剂型与给予方式

剂型与制剂的多样性与其用途的专一性，是药物发展的重要标志之一。但目前水产上应用的渔药剂型通常仅有溶液剂、散剂和片剂等，绝大多数药物的给予方式是口服和泼洒，浸浴和注射给药应用不多，而口服法不适于食欲下降甚至不摄食的患病动物，且不可避免地会在水中溶失一部分；浸浴或泼洒不利于通过皮肤、鳃和粘膜吸收的水产动物；肌肉或腹腔注射虽相对渔药吸收快，药效好，但仅使用于亲鱼和珍稀鱼类，对操作水平要求高，应用较少。此外，渔药的剂型和给予方式不同，机体吸收渔药的速率和数量也不同，药效最终也发生差异。

二、我国渔药发展的对策

1. 规范渔药使用，确保水产品、环境安全

渔药防治水产动物疾病，操作简单、使用方便、来源广泛、疗效明显，往往用药对症就能起到预期的防治效果，仍是广大养殖者首选的方式，尤其在细菌性疾病的控制上。但水产品中渔药的残留既不能危及人类健康，也不能破坏生态环境。因此，规范渔药的使用方法，限制与禁止将新近开发的人药作为渔药主要或次要成分，也限制与禁止直接向养殖水域泼洒抗生素；认真及时作好用药记录；严格遵守休药期，确保水产品与环境安全。

2. 控制病原菌的耐药性

耐药性是后天性的，经过一段时间的停药可以消失。国外在水产上实行 HACCP，使用的渔药均有记录，因此通过轮换给药、间隔给药等方法可达到抑制抗菌药的耐药性的目的。渔用抗菌药开发重点必须符合水产动物专用的要求，药物饲料添加剂应朝着无残留、抗菌、促生长、无交叉耐药性方向发展。另外，利用基因工程开发基因型渔药也是今后的发展趋势。

3. 加强渔药药效的理论研究

深化药动学、药效学研究，探讨同一药物在不同的水生动物体内代谢动力学种属差异性研究；研究环境因子（如温度、盐度、溶解氧等）对水生动物药动学影响；在细胞和分子水平上建立水生动物药代动力学研究方法或模型，开展中草药在水生动物体内的代谢、转化和消除规律研究；比较研究水生动物健康与非健康水平时的药物学，建立人工诱发疾病的水生动物药动学模型；运用群体药代动力学模型将药动学与药效学联合起来研究等等，最终掌握渔药在体内的代谢规律、渔药在体内的蓄积部位及蓄积程度，才能做到科学用药，制定细致、合理的休药期，为临床安全和合理用药提供依据，有利于指导剂型的选择和新药的开发。

4. 加强免疫增强剂、微生态制剂、生物渔药、中草药的研究和开发

免疫增强剂通过作用于非特异性免疫因子来提高水产动物的抗病能力，并减少使用抗生素等化学药物带来的负面影响，因此比化学药物安全性高，比疫苗应用范围广，如低聚糖、壳聚糖磺酸酯、几丁质等富含多糖、生物碱、有机酸等能显著提高水生动物的免疫功能。微生态制剂安全、低毒、有效，已经引起水产养殖者的重视，如反硝化聚磷菌。生物渔药是通过某些生物的生理特点或生态习性，去吞噬病原体或抑制病原体生长，如我们已从自然环境中筛选到一些能对海水弧菌、淡水气单胞菌等致病菌具有较强裂解作用的蛭弧菌，一周后能使致病菌浓度下降 4~5 个数量级。中草药具有来源广泛、使用方便、价廉效优、毒副作用小、无抗性、不易形成渔药残留等特点，在预防疾病中具有广阔应用前景，如三黄粉。

5. 新型渔药及其剂型、制剂的研究

通过多来源、多途径的方式改造药物化学结构，或研制开发窄谱性渔药、水产专用渔药、新型消毒剂、"三效三低"渔药（即高效、速效、长效与低毒、低残留、低抗药性）已刻不容缓。由于渔药药效受外界因子影响显著，应根据水产动物的种类和规格、发病类型及程度、渔药的性质等采用不同的剂型，如微胶囊剂、缓控释制剂。

三、展望

渔药的发展已进入调整前进期，水产品安全、生态环境安全是渔药发展的先决前提。渔药由国家标准化审批、渔药企业 GMP 规范化生产已成为不可逆转的趋势，专用渔药、生态环保型渔药、环境修复类渔药、生物渔药、疫苗类预防用药、诊断制剂、消毒剂及中草药的使用会越来越广泛。使用有毒害、有残留的化学药物，滥用抗生素将受到限制或禁止，渔药处方药将提上日程，与国际逐渐接轨。

三、关于禁用渔药和无公害渔药我国渔药管理的对策

由于具有"高毒性、高残留、高致癌、高致畸性和致突变"孔雀石绿被列为水产养殖禁用药物（杨先乐等，2005）。开发新型、无毒、低残留、高效的无公害渔药（杨先乐等，2003；杨先乐等，2003）可解决由渔药造成的水产品质量安全问题。

该文发表于《水产科技情报》【2005，32（5）：210-213】

对孔雀石绿的禁用及其思考

杨先乐，喻文娟，王民权，郑宗林

（农业部渔业动植物病原库，上海水产大学，上海 200090）

1　孔雀石绿的性质及其危害

孔雀石绿又名碱性绿、盐基块绿、孔雀绿，英文名 Malachitegreen，分子式是 $C_{23}H_{25}ClN_2$，在水生生物体内的主要代谢产物为无色孔雀石绿（leucomalachitegreen，$C_{23}H_{26}N_2$）[1,2]，其化学结构如图 1 所示。

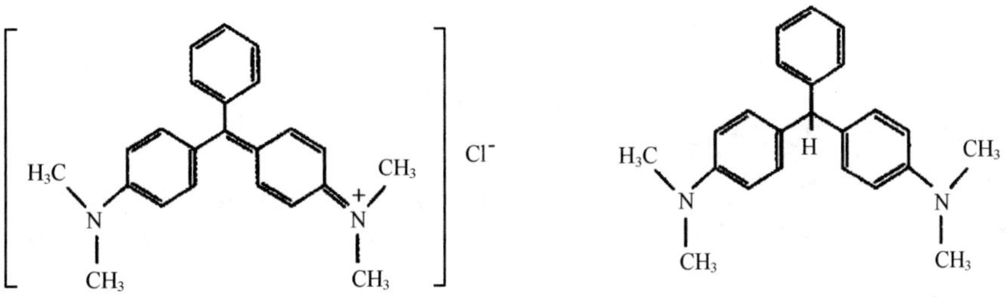

图 1　孔雀石绿结构式（左孔雀石绿；右无色孔雀石绿）

孔雀石绿是一种具有光泽的粉末状结晶，极易溶于水，水溶液呈蓝绿色，能溶于乙醇、甲醇或戊醇。它在浓硫酸中显黄色，稀释后显暗黄色，Ti^{2+} 使其水褪色，在氢氧化钠水溶液中由绿色变为白色沉淀，与 $AuBr^{4-}$、$GaCl_4$、$TaFb^-$、$TiBr^{4-}$、UO_2^{2+}、Zn^{2+} 等形成有色离子络合物，根据其这些特性，常采用光度测定法对之进行鉴别。

从 20 世纪 90 年代开始，国内外学者陆续发现，孔雀石绿及其在水生生物体内的代谢产物无色孔雀石绿具有高毒性、高残留、高致癌和高致畸性，因而各国陆续在渔业上禁止使用。

（1）高毒性孔雀石绿对哺乳动物的细胞具有高毒性[3,4]，Culp 等（1999）[5]发现，孔雀石绿与无色孔雀石绿均能使大鼠的肝细胞空泡化，无色孔雀石绿还能使甲状腺滤泡上皮大量凋亡，减少公鼠甲状腺激素的释放；此外孔雀石绿还能抑制血浆胆碱脂酶产生作用，进而有可能造成乙酰胆碱的蓄积而出现神经症状[6]。水生动物对孔雀石绿均很敏感，其安全质量分数一般在 0.1 mg/L 以下（表 1）。孔雀石绿还能通过溶解足够的锌，引起水生动物急性锌中毒，并引起消化道、鳃和皮肤轻度发炎，从而影响水生生物正常摄食和生长，阻碍肠道酶（如胰蛋白酶、α-淀粉酶等）的活性，影响动物的消化吸收功能。

表 1　孔雀石绿对部分水生动物的安全质量分数

品种	安全质量分数（$\times 10^{-2}$mg/L）	品种	安全质量分数（$\times 10^{-2}$mg/L）
虹鳟鱼苗	2.5	淡水白鲳	3.1
加州鲈鱼	2.0	美国大口胭脂鱼	1.6
锦鲤	1.5	脊尾白虾幼体	1.0
异育银鲫	1.2	对虾苗	1.8
云斑鱼	1.0	中华大蟾蜍蝌蚪	1.5
翘嘴红鲌	3.1	中华鳖	2630
蒙古裸腹溞	2.4		

（2）高残留孔雀石绿及其代谢产物无色孔雀石绿能迅速在组织中蓄积，较多地在受精卵和鱼苗的血清、肝、肾、肌肉和其他组织中检测到[7,8]。无色孔雀石绿不溶于水，因而其残留毒性比孔雀石绿更强[9]。采用 SC/T3021-2004 标准《无公害食品水产品中孔雀石绿残留量的测定液相色谱法》可同时检测到孔雀石绿和无色孔雀石绿。

（3）高致癌孔雀石绿的化学官能团是三苯甲烷（triphenylmethane），其分子中与苯基相连的亚甲基和次甲基受苯环影响有较高的反应活性，可生成自由基———三苯甲基，同时孔雀石绿也能抑制人类谷胱甘肽—S—转移酶的活性[10]，两者均能造成人类器官组织氧压的改变，使细胞凋亡出现异常，诱发肿瘤和脂质过氧化，而来源于上皮组织的恶性肿瘤即为"癌"。Panandiker 等（1992）[11]发现，孔雀石绿能使培养的仓鼠胚胎细胞产生过多的自由基造成脂质过氧化；RAO 等（1995）[12]认为，它能抑制培养的原始鼠肝脏细胞 DNA 和 EGF 的合成，诱使乳酸脱氢酶的释放，诱使肝脏肿瘤的发生；Sundarrajan 等（2000）[13]的研究表明，孔雀石绿能通过促进增殖细胞核抗原（PCNA）和 G1/S 期细胞周期蛋白（cyclins）的表达而诱发大鼠肝脏肿瘤。

关于孔雀石绿多苯环芳烃结构的致癌机理目前尚不完全清楚。Culp 等（1999）[14]研究发现，孔雀石绿代谢产物的衍生物及初级和次级的芳香胺在癌症发生中起着重要的作用，如小分子芳香胺，尤其是 1~4 个芳香环结构进入人体后，会穿透细胞膜，到达细胞核中的 DNA。芳香胺易产生活泼的亲电子"阳氮离子"，攻击 DNA 上的亲核位置，相互以共价键结合，从而破坏 DNA 引起癌变。除此之外，芳香胺分子的扁平部分还会"插入"正常细胞中 DNA 螺旋线结构的相邻碱基对之间，从而破坏 DNA，引起癌变。Meyer 等（1983）[15]与 Schnick 等（1978）[16]的试验也证实了孔雀石绿能使虹鳟产生癌病变。

（4）高致畸性孔雀石绿能使淡水鱼鱼卵染色体发生异常[17]，即使孔雀石绿的质量分数低于 0.1 mg/mL，它仍能使兔和鱼繁殖致畸[18,19]；当孔雀石绿的质量分数高于 0.2 mg/L 时，对软体动物东风螺的受精卵或是幼体都会造成不同程度的发育畸形[20]。

（5）致突变周立红等（1997）[21]认为，孔雀石绿可使鲻鱼和尼罗罗非鱼的红细胞产生微核，当质量分数为 0.5 mg/L 时，微核率分别达到 0.133% 和 0.0717%，且微核率随药物质量分数的升高而提高。

2 孔雀石绿在渔业领域的使用情况

孔雀石绿原是一种三苯甲烷类染料，由苯甲醛和 N，N-二甲基苯胺在盐酸或硫酸中缩合生成四甲基代二氨基三苯甲烷的隐性碱体后，在酸介质中被二氧化铅秘氧化制得。因其氧化电位势与组成酶的某些氨基酸相近，在细胞分裂时发生竞争而阻碍蛋白肽的形成，从而产生抗菌杀虫作用，因此曾在渔业上广泛使用。2002 年，中华人民共和国农业部发布的 193 号公告以及 NY5072-2002 标准《无公害食品渔用药物使用准则》中明确规定在渔业领域禁用孔雀石绿。

近年来，我国在对孔雀石绿等违禁水产药物的监管方面取得了一些成绩，养殖户（场）纷纷建立了用药档案，较多省市和地区也构建了如"市-区县-乡镇"食品安全监管的网络，孔雀石绿等违禁水产药物的使用得到了基本控制。但是由于我国地域广阔，养殖面积大、养殖品种多，渔（农）民科学用药的意识薄弱，水产管理部门人员不足、经费有限，在对水产药物使用的监控、禁用水产药物的检测等方面存在着不足，致使孔雀石绿在不同程度上存在禁而不止的现象。某些不法商人为了谋取非法利润，仍在经销"孔雀石绿"，甚至在水产品保鲜和运输过程中添加孔雀石绿。

目前禁用药物孔雀石绿在我国渔业领域的使用量尚难确切地估计，若仅以泼洒给药方式防治淡水鱼类的水霉病，以最小化原则进行粗略估算，根据公式"孔雀石绿全年使用量（kg）= 防治该水产动物水霉病的用量（$2×10^{-6}$ kg/m^3）×面积（$667×10^4$ m^2）×水深（1.5 m）"得出，孔雀石绿全年每 667 公顷（万亩）的使用量（kg）为 0.02 t；若全年有 $670×10^4$ hm^2 的养殖面积使用孔雀石绿（我国 2002 年的水产养殖面积为 $681×10^4$ hm^2），则其使用量可达 200 t 左右。如果考虑疗程及全年的发病次数，则应在此基础上再加一倍。这还不包括育苗场高频率、大剂量地使用，以及对其他水产养殖对象使用孔雀石绿的；不包括以浸浴给药方式使用孔雀石绿的，以及在防治水生动物鳃霉病、小瓜虫病、消毒、保鲜等方面使用孔雀石绿的。由此我们估计，目前在我国的渔业领域，孔雀石绿的使用量达到 600~800 t，令人触目惊心。

3 孔雀石绿屡禁不止的原因

3.1 价廉易得孔雀石绿每公斤仅 35 元左右，由于其使用剂量较低，且多采用浸泡、泼洒等给药方法，因此防治成本低。用该药防治水产养殖中的病害和进行水产品的运输保鲜，在有些地区是不足为怪的事情。

3.2 对水霉病等疗效好，缺少替代药自从孔雀绿被禁用后，在水霉病的防治方面一直没有较好的替代药物，在对该病的控制上形成了一定的真空。而该病是我国水产养殖中的一种常见病，有时还可造成较大的损失，因此一些养殖户不得不用孔雀石绿来防治该病。

3.3 缺乏有力的监管制度虽然国家已明文规定禁止使用孔雀石绿，但对孔雀石绿缺乏监管方法和监管网络，对违法经销和使用缺乏处罚措施，处罚力度也不够，以致不能从源头上阻断孔雀石绿在渔业上的非法使用。

3.4 孔雀石绿残留初筛方法缺乏目前对孔雀石绿残留的检测主要是采取液相色谱法，但因该方法需有较昂贵的仪器，检测过程复杂，对检测技术的要求也较高，故液相色谱法不适用于普查。采用简易可行的初筛方法是控制孔雀石绿滥用、保证水产品安全的途径之一。

3.5　宣传和科学普及工作深度不够养殖者对孔雀石绿的毒副作用缺乏深刻了解，虽一般都能知道孔雀石绿有毒，但对其能产生致畸、致癌、致突变的毒副作用缺乏认识；较多人对国家禁用孔雀石绿等药物的利害关系不了解，没有将其提高到一定的高度上去认识，因此在销售和使用时毫无顾忌，更无人对违法经营和使用者进行举报。

4　对在渔业领域彻底禁用孔雀石绿的思考

4.1　开展对渔业领域非法使用孔雀石绿的专项整治，强行制止孔雀石绿的使用　　针对目前所存在的非法使用孔雀石绿的情况，对生产、销售与使用情况进行全面的检查和整改，对市场流通的水产品定期进行孔雀石绿残留抽检。

4.2　严格控制孔雀石绿的生产和销售，实行许可证制。目前，在工业中的某些领域需要使用孔雀石绿，为了防止这部分孔雀石绿成为渔业非法使用的源头，必须严格规范孔雀石绿的生产、经营和销售活动。加强对市场上销售孔雀石绿企业、单位的监管力度，对违法销售孔雀石绿的经营单位或个人予以重罚，甚至追究法律责任。

4.3　建立与完善孔雀石绿检测和监管网络　　百密难免一疏，仅凭执法部门对生产、销售企业和养殖户等的常规检查还不能彻底禁止孔雀石绿的使用，只有建立高效、健全的检测、监管制度和网络，才可能把住水产品质量安全的最后一道防线。

4.4　加强宣传和科学普及的力度一方面要向广大渔（农）民宣传孔雀石绿的危害性，宣传国家有关禁止使用孔雀石绿的有关法律和法规，另一方面要加强科学用药技术的普及，使渔（农）民彻底改掉滥用药物的陋习。

4.5　加强对水产药物基础理论和应用的研究目前所要解决的主要问题是：①孔雀石绿在主要水产动物体内的代谢和消除规律；②孔雀石绿防治水霉病替代品的研制；③简单、快速、灵敏的孔雀石绿检测方法的建立；④孔雀石绿毒性控制技术和分子结构改造的研究。

参考文献

[1]　郭德华，叶长淋，李波等．高效液相色谱–质谱法测定水产品中孔雀石绿及其代谢物〔J〕．分析测试学报，2004，23（21）：206-208.

[2]　《无公害食品水产品中孔雀石绿残留量的测定液相色谱法》（SC/T3021-2004）.

[3]　Clemmensen S. , Jensen J. C. , M eyer O. et al.. Toxicological studies on malachite green：a triphenylmethane dye. A rch. Toxico. l 1984，（56）：43-45.

[4]　Panadiker A. , Fernandes C. and Rao K. V. K. . The cytotoxic properties of malachite green are associated with the increased demethylase，aryl-hydrocarbon hydroxylase and lipid peroxidation in prim arycultures of Syrian hamsterembryocells. Cancer Let. t 1992，（67）：93-101.

[5]　Culp S. J. , Blankenship L. R. , Kusewitt D. F. etal. Oxicity and metabolism of malachite green and leucomalachite green during short-term feeding to Fischer 344 rats and B6C3F1 mice. Chemico-Biological Interactions. 1999，122（3）：153-170.

[6]　Tuba Kücükkylync and Ynciözer. Inhibition of hum an plasmacholinesterase by m alachite green and related triarylmethanedyes：Mechanistic implications. Archives of Biochemistry andBiophysics. 2005. 7.

[7]　Srivastava S, Sinha R. and Roy D. . Toxicological effects of malachite green. Aquatic Toxicology. 2004，66（5）：319-329.

[8]　M einertz J. R. , Stehly G. R. , G ingerichW. H. et al. Residues of 〔14C〕 malachite green in eggs and fry of rainbow trout（Oncorhynchusmykiss）（W albaum）after treatments of eggs. J. Fish D is. 1995，18（3）：239-247.

[9]　Mittelstaedt R. A. , M ei N. , Webb P. J. et al.. Genotoxicity of malachite green and leucomalachite green in female Big Blue B6C3F1 m ice. M utation Research/Genetic Toxicology and EnvironmentalMutagenesis. 2004（561）：127-138.

[10]　G lanville S. D. andClarkA. G. . Inhibition of hum an g lutathioneS-transferases by basic tripheny lm ethane dyes. Life Science. 1997，1535-1544.

[11]　PanandikerA. , Fernandes C. and Rao K. V. K. . The cytotoxic properties of malachite g reen are associated with the increased demethylase，aryl hydrocarbon hydroxy lase and lipid peroxidation in primary cultures of Syrian ham sterembryo cells. Cancer Let. t 1992（67）：93-101.

[12]　Rao K. V. K. . Inhibition of DNA synthesis in prim aryrat hepatocyte cultures by m alachite green：a new liver tum or prom oter. Toxicology Lette. r 1995（81）：107-113.

[13]　Sundarrajan M. , Fernandis A. Z. and Subrahmanyam G. . Overexpression of G1/S cyclins and PCNA and their re lationship to ty rosine phosphorylation and dephosphorylation during tumor prom otion by m etanilyellowandmalachitegreen. Toxico logy Letters. 2000，116（27）：119-130.

[14]　Culp S. J. , Blankenship L. R. , Kusewitt D. F. et al. oxicity and metabolism of malachite green and leucomalachite green during short-temr feeding to Fischer 344 rats and B6C3F1 m ice. Chemico-Biological Interactions. 1999，122（3）：153-170.

[15]　Meyer F. P. and Jorgensen T. A. . Teratological and other effects of malachite green on the development of rainbow trout and rabbits. T rans. Am. F ish. Soc. 1983，112（6）：818-824.

[16]　Schnick R. A. and M eyer F. P. . Registration of thirty three fishery chemicals：status of research and estimated costs of required contract studies. Inves. t Fish. Contr. 1978（86）：19.

[17]　W orle B. . 1995. Gene toxicological studies on fish eggs. GenotoxikologischeUntersuchungen an Fischeiern，92.

[18] Fernandes C., Lalitha V. S. and Rao K. V. K.. Enhancing effect of ma lachite green on the development of hepatic pre neoplastic lesions induced by N-nitrosodiethylamine in rats. Carcinogenesis 1991 (12): 839-845.

[19] Rao K. V. K.. Inhibition of DNA synthesis in prim aryrat hepatocyte cultures by m alachite green: a new liver tum or promoter. Toxicology Letter 1995 (81): 107-113.

[20] 黄英，柯才焕，周时强. 几种药物对波部东风螺早期发育的影响. 厦门大学学报（自然科学版），2001，40（3）：821-826.

[21] 周立红，徐长安. 用微核技术研究孔雀石绿对鱼的诱变作用 [J]. 集美大学学报，1997，2（2）：55-57.

该文发表于《科学养鱼》【2003，5：38】

关于无公害渔药（上）

杨先乐

（农业部渔业动植物病原 库上海水产大学，上海 200090）

一、无公害渔药产生的背景

近年来，我国水产生产完成了从"捕捞增长型"到"养殖增长型"的转变，但由于环境污染加剧和养殖自身有机污染的不断积累，加上种苗流通及水产品贸易带来病原的扩散，我国水产养殖病害发生日趋严重。养殖者们为了控制水产动物病害的蔓延，为了挽回因病害发生可能造成的损失，投入了较多的抗生素类、喹诺酮类、磺胺类、呋喃类等化学药物，投入了较多的消毒剂。由于我国渔药的发展历史较短，大部分渔药是由兽药或者人药移植来的，缺乏对其药效学、药动学、毒理学及对养殖生态环境的影响等基础理论的研究，因此人们对渔药认识肤浅，对用法、用量和停药期忽视，部分水产品药物残留的问题越来越凸现出来。尤其是 2002 年初由于氯霉素事件，我国水产品对欧盟出口受阻，更引起了水产品药残的轩然大波。在这种情况下，我国政府制定了无公害食品行动计划，力争在五年之内使国人吃上放心的水产品，使我国水产品在国际贸易上占据有利的地位。因此，渔药的安全、合理、科学使用也就成了当今工作的焦点，而这一工作的基础就是要开发新的、无毒、低残留、高效的无公害渔药。

二、无公害渔药的基本要求

渔药是为提高水产养殖产量，用以预防、控制和治疗水产养殖动、植物病、虫、害，促进养殖对象健康生长以及改善养殖水体质量所使用的一切物质。无公害渔药是一个更高范畴上的渔药，它是我国实行"无公害食品计划"对渔药所提出的一种更高的要求，它不会对社会造成负面作用，不会对养殖对象、养殖环境以及对人类本身造成不良的影响。作为无公害渔药必须满足以下基本要求：

（1）它必须是有效的，甚至是高效的、速效的、长效的。如果是防病治病制剂，要求它能快速地、有选择性地杀灭病原体；如果是诊断制剂，它必须有较高的灵敏度、准确性和特异性；如果是水质改良制剂，施用后它应该对水产养殖环境有较明显的改善作用；如是营养和免疫增强剂，它能使养殖对象的生理机能和生长状态有明显的促进作用。总之，无公害渔药的药效应该是明显的。

（2）它的毒性较小。它必须容易分解或降解，其分解或降解的产物基本上是无害的或者很容易通过其他动物转换，从而在水产养殖对象的组织或水域环境中消失，避免在养殖对象组织中或环境中积累。另一方面无公害渔药必须要提供有关的毒性试验报告，其中包括急性毒性试验、胚胎毒性试验、行为反应测定、亚急（慢）性毒性试验和慢性毒性试验、特殊毒性试验，从而确定相关的毒理学指标和参数，确定它的毒性大小。任何毒理学指标不明了的药物，任何有致畸、致癌、致突变的药物均不可作为无公害渔药使用。

（3）它的副作用较小。药物在杀灭病原体或改变养殖环境或增强养殖对象免疫反应的同时，会产生一定的副作用，如给养殖对象带来较大的刺激，从而产生较大的应激反应；如影响养殖动物的正常摄食，从而使生长减缓；如对养殖对象的某些器官或组织带来不利的影响或负面效应；如使养殖环境发生一些变化而影响了养殖对象的正常生理活动或使环境的修复需要一定的条件和时间；等等。无公害渔药应使这种负面影响控制在最小的程度，对环境的影响能及时修复，对养殖对象的应激应控制在它们所能承受的范围内。

（4）它使用的剂量应尽量的小。只有较小的使用剂量，才会减少其毒副作用，也才会在使用成本上有较大的降低，获得较大的使用价值。

（5）必须制定出其合理的给药方法、给药剂量、给药间隔时间和休药期等参数。同时也应该对有可能引起的毒副作用提出警示。

（6）具有较好的稳定性。适宜在常温下保存，便于运输、销售和贮藏。

（7）剂型设计较合理，给药途径比较方便。

（8）有较大的价格优势。

三、无公害渔药研究与开发的基本原则

无公害渔药是一种科技含量较高、要求较高、开发研制难度较高的渔药。研究与开发无公害渔药既要考虑它应具有疗效显著、给予途径方便和价格便宜特点，而且更要注意它的安全性，它不会产生较大的危害和副作用。无公害渔药的研制应符合"渔业法"及"兽药管理条例"等相关法律法规的规定，也应符合国内外相关标准的要求。无公害渔药的报批应提供完整的、真实的、有效的研究报告，其中包括药物的设计原理和国内外研究现状的说明，包括药动学、药效学、毒理学以及药物的稳定性、临床试验等方面的报告，包括药物使用方法和安全性分析的报告。此外无公害渔药的生产应该严格按照 GMP 生产的要求进行。

该文发表于《科学养鱼》【2003，6：38】

关于无公害渔药（下）

杨先乐

（农业部渔业动植物病原库，上海水产大学，上海 200090）

四、怎样研究和开发无公害渔药

1. 筛选高效、低毒、低残留的化学原料或原料药作为渔药的母体

化学药物仍是防治水产动物病害最直接和最有效的手段之一。当前化学药物所存在的最主要问题是毒性和残留对水产动物、人类以及环境的危害和损害。如果化学药物克服了以上弊端，那么它在水产动物的病害防治上仍有着广阔的前景。有很多化学药物，通过其侧链的改变，或者某个化学基团的添加或减少，会使其性质发生很大的改变，有的也会使药物的毒性大大降低。因此在渔药原料上下工夫是开发和研究无公害渔药的一个重要途径。

2. 加强渔药药理学的研究

渔药药理学的研究是无公害渔药研究的基础，也是确定渔药是否无公害的前提。可以肯定地说，没有完整的药理学知识的渔药不能列为无公害渔药。此外还要针对水产动物的特点和特殊的生存环境，研究渔药的稳定性，研究渔药可能产生的降解产物的毒性，研究渔药可能对养殖对象产生的应激反应和控制应激反应的措施。

3. 采用新技术和新工艺，提高养殖对象对渔药的吸收，减少它对环境和水产动物的负面影响

新技术和新工艺等先进技术的引入，拓展渔药研究的思路，同时也为开发渔药新品种，增强渔药的药效，降低渔药的毒副作用提供了舞台。这些新技术有纳米技术、分子生物学技术及基因工程等。

4. 改变传统的用药理念，开发窄谱的抗生素和水产专用药物

窄谱的抗生素对目标致病菌有很强的杀灭作用，但它又不会对有益菌群产生影响，在施用这种抗生素后不会使水产动物体内外的微生态系统的平衡受到破坏。在研制窄谱性抗生素时一定要了解病原体的性质，病原体生长、繁殖的特点，然后针对它们生命过程中的某一环节进行抑制，如抑制细菌某种成分的合成，竞争细菌对某种酶的利用，阻断细菌在某个阶段的发育等。

渔药在以前一个很长的时间，都是借用人、兽药，是人、兽药用途的延伸。这种渔药研制方式除了简单和快捷之外，带来的负面效应也是较多的，有时甚至是严重的。耐药菌株的产生，会影响对人类疾病的药物治疗；药物的残留，会导致残留药物通过食物链在人体内积累，而最终危害人类自身的健康。开发水产专用药物能够避免以上弊端。

5. 研究与开发渔药的新剂型，增强渔药的药效和降低它的毒性

目前渔药的剂型以粉剂居多。这种单一的剂型，在一定程度上既影响了渔药的药效，也带来了较大的副作用。如氯制剂，当使用粉剂时会因它的瞬间作用而导致水产动物产生较大的应激，而用片剂时，就可克服它的这一缺点。渔药的剂型

应该朝多方向发展，如胶囊剂、缓释剂、长效制剂等的研制，通过剂型的开发，可以推动无公害渔药的发展。

6. 从控制与利用的抗病理念出发，研究与水产动物病原竞争的生物制剂或非生物制剂

对于水产动物疾病的防治，正确的观点是不一定要彻底消灭病原体，而只须控制它们的数量，使其处于劣势；所采取的主要措施就是利用有益菌群，用有益菌或中性菌甚至是一些处于不活跃的致病菌去与病原体竞争，抑制病原体的生长与繁殖。因此研制相应的微生态制剂或一些能与病原体发生竞争性抑制的非生物制剂就成了我们无公害渔药的一个发展方向，其中微生态制剂是其重点之一。

7. 研究与开发生物渔药

生物渔药是通过某些生物的生理特点或生态习性，去吞噬病原体或抑制病原体栖息的一种活的生物制剂。生物渔药发展前景十分诱人。人们已在这方面开始了一些探索，并取得了一些可人的成绩。如对噬锚头鳋生物的研究，噬菌体的探讨等。生物渔药的研制与应用会推动无公害渔药更大的发展。

五、无公害渔药研究与开发的前景

无公害渔药的研究与开发，必须要加强以下五个方面的工作：

1. 加强渔药基础理论和应用技术方面的研究，丰富和健全渔药的应用基础理论，建立完善渔药研究的技术手段。

2. 加强渔药复方制品、渔药制剂方面的研究，充分发挥药物组分的均衡与协同作用，发挥渔药剂型的作用，从而达到增强药效、降低毒性的目的。

3. 加强病原体对渔药耐药性的控制，解决某些寄生虫、致病菌对药物的耐受性问题，使水产动物寄生虫病的控制达到一个新的水平，同时也使困扰水产动物病害防治多年的某些寄生虫病，如孢子虫病、小瓜虫病等有所突破。

4. 加强中草药在水产动物病害防治上的研究，如在中草药有效成分的低成本提取方面；在中西药配伍、互相渗透方面；在中草药制剂的研制方面等。可以通过促进药物微核分子的扩展分散、乳化渗透、润湿粘合及增效等方法使其有所建树。

5. 扬长避短，合理地在无公害渔药的研制与开发中应用生物工程技术，改变渔药从自然界中筛选、化学合成，逐渐过渡到运用生物技术的方法、有目的地合成所需要的渔药。

6. 加大开发增强水产动物机体自身功能的渔药，如免疫增强剂、重要器官的保护剂等的研究和开发，使水产动物的抗病能力得以增强。

第七章　渔药安全使用技术的应用、示范和推广

一、渔药基础理论及发展

渔药的安全使用技术的推广应用是确保水产品质量安全、养殖生态环境安全的一个重要落脚点。而建立安全用药示范区是将研究的成果广泛推广应用的一个重要途径（孙琪等，2014）

渔药基础理论及安全使用技术的研究跟不上生产实践的需要，是导致渔药滥用、乱用的重要原因（杨先乐等，1999）。以"三效"（高效、速效、长效）、"三小"（毒性小、副作用小、剂量小）的渔药开发为目标，以禁用渔药替代药物研究为突破口是解决当前水产品质量安全问题的关键（胡鲲等，2011）。

该文发表于《中国水产》【2011，5：28-29】

渔药基础理论及安全使用技术研究现状分析

胡鲲，杨先乐

（上海海洋大学，上海 201306）

渔药基础理论及安全使用技术的研究跟不上生产实践的需要，是导致渔药滥用、错用的重要原因之一。近年来，相继发生的一系列水产品药物安全事件集中暴露出我国渔药基础理论研究薄弱的问题。加强渔药基础理论及安全使用技术研究对提升渔药产业水平、提高出口水产品竞争力、促进养殖业持续健康发展、保障公共卫生安全、增加农民收入具有重要的现实意义。

一、我国渔药基础理论及其安全使用技术研究的现状

我国渔药基础研究历史大致可以分为三个阶段：

第一阶段，20 世纪 50 年代。这一阶段重点是对主要病害进行有效药物筛选，初步形成治疗方案。渔药研究主要集中在病原筛选药物、药物有效浓度和安全浓度、药物应用范围及给药方法等方面，筛选出了一批针对性较强、药效显著的药物。如硫酸铜、硫酸铜和硫酸亚铁合剂、敌百虫、高锰酸钾和硝酸亚汞等治疗寄生虫病，磺胺药治疗细菌性肠炎病，食盐和小苏打治疗水霉病，漂白粉防治烂鳃病，石灰、茶饼和巴豆清塘等。

第二阶段，20 世纪 60—80 年代。这一阶段抗菌素和中草药研究呈现活跃态势。土霉素、金霉素、红霉素、链霉素等抗生素相继应用于细菌病防治；中草药防治鱼病主要是群众性经验，但这些工作仍停留在药效研究中。渔药的剂型、工艺都沿袭畜禽等兽药产品，未开发出适合水产动物特点的专用渔药。

第三阶段，20 世纪 90 年代开始至今。这一阶段开始比较系统地进行渔药基础理论（包括代谢动力学、药效学、毒理学等）研究，从机理上解决生产实践问题。取得了一系列成果：①建立了 20 余种渔药在水产动物体内残留检测方法；②建立了药物体外诱导细胞酶的模型，从细胞水平分析组织器官药物残留状况，为渔药的临床合理使用提供理论依据，为新药筛选设计、水产品药物残留检测及环境毒物的监测创建新的理论与技术平台；③建立了针对水产研制动物的药物安全性评价的技术方法。

在这些研究工作的基础上，渔业主管部门制定并颁布了一系列政策法规和技术标准：《食品动物禁用的兽药及其他化合物清单》（农业部 193 号公告）禁止氯霉素、孔雀石绿等 29 种药物使用，限制 8 种渔药作为动物促生长剂使用；《无公害食品渔用药物使用准则》（NY5071-2002）规定呋喃类、喹乙醇等 32 种为禁用渔药；一大批药物残留检测方法的标准（包括国家标准、行业标准和地方标准）被制定或修订，检测诺氟沙星等数十种药物；《兽药管理条约》（2004）规范了生产企业用药制度；《水产品养殖质量安全管理规定》（2003）和《农产品质量安全法》（2006）建立了渔药残留检测和监控体系，强调从"农场到餐桌"的药残全过程控制管理。

这些基础研究代表了我国渔药基础研究的主要研究成果，而且大部分成果已经在实际生产中应用，取得了良好的社会效益和经济效益，在一定程度上提升了公众对水产品安全的信心，提高了我国水产品在国际市场上的公信力。

二、我国渔药基础理论研究存在的问题

由于水产养殖品种众多、模式多样、环境复杂，我国渔药的基础理论研究还远远落后于水产养殖业发展需求，还存在着较多的盲点。大量的研究工作还有待于继续深入。

1. 渔药研究基础薄弱，技术理论体系不完善

我国水产养殖动物种类高达300余种，具有一定产业

规模的养殖品种也在25种以上，远超畜禽动物种类。我国现有国标渔药155种，已进行药动学研究并可指导其生产的渔药种类还不到现有常规使用渔药种类的10%。目前仅有的数据与资料还远远不能支撑渔药的安全使用技术体系的建立。此外由于药物代谢所存在的动物种属差异，同一动物存在的不同药物的代谢差异，更为渔药的合理使用带来了更大的难度。大量渔药的残留限量、休药期、给药剂量及用药规范等方面资料的匮乏，导致渔药的使用存在着很大的盲目性。

2. 渔药滥用现象普遍

我国水产养殖行业以家庭养殖为主，渔药使用的监管机制不完善。渔药滥用主要表现在：不遵守休药期规定，用药剂量、次数、给药途径随意，不遵守配伍禁忌，无用药记录等。渔药滥用是造成水产品药物残留的最主要原因。

3. 禁用药物替代制剂研究成果不显著

自2002年起，农业部先后颁布了一系列的法规和技术标准，规定了氯霉素、孔雀石绿等禁用渔药清单。由于对这些禁用药物的安全性危害认识不足或者是出于利益驱动，在基层水产养殖、运输等环节中这些禁用药物还屡屡被违法使用。

4. 渔药残留检测技术手段较落后

长期以来我国渔药残留检测方法一直是借用畜牧兽医的检测方法。我国现有的155种国标渔药中，仍有数十种渔药尚未建立相应的检测技术标准；对于已建立检测标准的渔药，还要不断提升技术水平以适应产业发展的需求；禁用药物缺乏有效的快速、检测手段，尤其缺乏非实验室条件下的快速、灵敏、高通量的水产品药物残留检测方法。

三、加强渔药基础理论及安全使用技术研究建议

当前，水产养殖业的迅猛发展及增长方式的转变，为我国渔药研究提出了新的要求，也提供了新的机遇。渔药研究也要转变观念，适应"健康、安全、绿色"产业发展要求，以"资源节约、环境友好、优质高效"为原则，以"三效"（高效、速效、长效）、"三小"（毒性小、副作用小、剂量小）的渔药开发为目标，以禁用渔药替代药物研究为突破口展开。我国渔药基础理论和安全使用技术的研究应重点加强以下方面的工作：

1. 加强渔药基础理论，特别是代谢动力学，为渔药的安全、合理使用提供依据。加强渔药药动学研究，深入探讨渔药在水产动物体内的代谢规律；提出渔药科学使用规范；研究渔药对环境的潜在安全性，弄清渔药在环境中的蓄积、转移、转化规律，为临床安全和合理用药提供依据。

2. 加强渔药残留检测监控体系的完善，推进标准化。根据水产养殖产业的发展变化趋势，整理现有的资源，重点针对渔药残留检测技术、渔药安全使用技术规范等空白领域制定和修订水产行业标准或国家标准。加强标准的宣传、推广和执行力度，进一步完善渔药残留检测监控体系。

3. 鼓励生物渔药等新型渔药的发展。生物渔药等新型渔药具有绿色、安全等特点，在水产养殖中具有广泛的应用前景。生物渔药的合理使用可以显著达到改良养殖水体环境、提高水产动物机体免疫力和抑制病原的目的。弄清不同养殖模式下生物渔药的作用机理、增强生物渔药在养殖环境中的定植能力和统一生物渔药技术标准能突破生物渔药研制和推广的瓶颈，使之起到部分替代药物的目的。

4. 整合技术资源，组织开展禁用渔药替代制剂的研发。加强禁用药物的替代制剂研究，从源头上杜绝违禁使用禁用药物事件的发生。以孔雀石绿替代制剂的研发为突破口，加大研发投入力度，提升公众对水产品安全的信心，保障产业的持续健康发展。

5. 建立和实行渔药处方制度，逐步健全可追溯制度。根据养殖生产中出现的病害进行正确的分析、诊断，制订合理的处置方案，开举药方，包括处方药和非处方药；建立《水产养殖用药记录》，大力推广健康养殖模式，规范渔药使用，最终提高产品质量和竞争力，增加渔业生产效益。

该文发表于《河南水产》【1999，4：10-11，14】

渔用药物的科学使用及其发展趋势

杨先乐

（上海水产大学，上海 200090）

改革开放以来，我国渔业取得令人瞩目的成就，水产品总产量持续大幅度增长。水产品从 1978 年的 465 万吨发展到 1997 年的 $3\,602\times10^4$ t，约占世界水产品总产量的 30%左右。养殖总产量的比重达到 56%，成为世界上唯一养殖产量超过捕捞产量的国家。1997 年与 1990 年相比，海淡水养殖产量增加 1 060 t，占新增产量的 75%，居世界首位。但是，随着水产养殖的发展和人工养殖方式集约化程度的提高，养殖环境的日趋恶化，病害的发生率也越来越高，虽然原有的疾病经过研究和成果推广多数得以控制，而新的病害却频频发生，并出现大规模流行。为了减少疾病给水产养殖业带来损失，药物防治已成为最主要和最直接的手段之一。因此科学使用渔药是提高水产动物疾病的防治效果与增强水产动物的品质的关键因素。根据我国目前水产动物疾病防治用药的现状及国内外有关文献，我们对渔用药物的现状和发展趋势提出一些见解。

该文涉及的主要内容是：1 渔用药物防治水产动物疾病的误区；2 科学用药的含义；3 怎样科学使用渔药；4 渔药的发展趋势。

二、渔药安全使用实用技术

1. 鳖用药物及其鳖病的防治

鳖用药物的功能包括：抑制或杀灭病原体、改良养殖环境和增强鳖自身的抗病能力（杨先乐，1996；杨先乐，1999）。鳖用药物的使用方式有挂篓（袋）、浸浴、涂抹、口服口灌和注射等（杨先乐，1995；杨先乐等，1995）。

该文发表于《渔业致富指南》【1999，12：49-51】

科学用药对鳖病的防治效果和鳖品质的作用

杨先乐

（中国水产科学研究院长江水产研究所，湖北省荆沙市 434000）

1　鳖用药物的基本作用

鳖用药物的功能与作用很多，但概述起来，主要是以下三个方面的作用：

（1）抑制或杀灭病原体

目前大部分鳖用药物都有灭杀和抑制病原体的作用，如抗细菌方面的药物如磺胺类、呋喃类、抗生素以及大蒜等，抗真菌方面的药物如食盐、福尔马林、中草药五倍子、菖蒲等，抗寄生虫方面的药物如硫酸铜、氨水、高锰酸钾等。这些药物通过间接的方式（如磺胺类药物、某些抗生素等）或直接的方式（如高锰酸钾、硫酸铜、福尔马林等）而达到杀灭病原体的作用。

（2）改良养殖环境

它们主要通过以下三个方面发挥其作用：（a）杀灭水体中的病原体，如漂白粉、优氯净、三氯异氰尿酸等；（b）改良鳖池的水质和底质，如生石灰、过氧化钙等，可提高鳖池的碱度，增加池底的通透性；（c）净化鳖池的水质，如沸石和光合细菌、硝化细菌等。这类药物，大部分间接或直接地都起到了这三个方面的作用。

（3）增强鳖自身的抗病能力

这类药物的功能表现在以下三个方面：（a）为鳖补充适当的营养物质，使鳖的新陈代谢处于最佳状态，如维生素等；（b）激发鳖的特异性免疫机能，如注射疫苗，使鳖对某种疾病具有特异性的抵抗力；（c）增强鳖的非特异性免疫机能，如某些免疫激活剂，以此调动鳖自身的补体、溶菌酶、C 一反应性蛋白、干扰素、铁传递蛋白等非特异性的防御因子抗御疾病。

鳖用药物的这三种作用并不是孤立的，某种药物的一种作用能影响其它两种作用，如环境改良药物对病原体的抵制杀灭和鳖自身抵抗力的增强都有影响；反过来，只有多种作用的药物协调作用，才能最终达到控制与消灭鳖病的目的。

2 鳖对外用药物反应的特点与用药（治疗）原则

（1）鳖对外用药物反应的特点

因为鳖的外部形态、内部构造以及生态习性等和鱼以及其它水生动物相比有很大的不同，因此它对药物反应与它们相比也有很大的区别。总的来看，它有如下的特点：

①对几种常用药物的敏感性低于鱼类。鳖对生石灰、硫酸铜、敌百虫、漂白粉等表现出极不敏感的现象，其安全浓度与加州鲈鱼苗相比相差几十倍至几百倍，与一般鱼类的常用量相比，也有比较悬殊的差别。出现这种差别的原因可能是因为鳖类革质的皮肤阻碍了体表对药物的吸收，也可能因为鳖常浮于水面，进行肺呼吸，减少了药物对呼吸器官的毒性作用。鳖对甲醛的安全浓度与加州鲈鱼的安全浓度和一般鱼类的常用量相比基本一致。出现这种现象的原因可能是：甲醛在气体状态下，可呈现强大的毒性作用，凝固蛋白质和溶解类脂，与蛋白质的氨基结合而使蛋白质变性，即使鳖离水进行肺呼吸，也很难逃离这种毒性作用。

②和其它几种常用药物相比，孔雀石绿对鳖的毒性较高，40 ppm 时可使鳖致死，安全浓度为 26.3 PPm，比一般鱼类常用量要低，如果使用不慎，可造成鳖的药物中毒。

③鳖对敌百虫虽不十分敏感，其致死浓度范围是 80—160 ppm，96 小时的 TLm（平均忍受限）为 160 ppm，安全浓度 28.5 ppm，但较低的浓度的敌百虫能导致它们的昏迷和活动能力减弱，这是因为敌百虫的水解产物抑制了胆碱脂酶的活性，使鳖水解破坏乙酰胆碱的能力减弱，引起神经失常。这种现象势必会影响鳖的生长、发育和繁殖。

④鳖是低耐盐的水生动物。24、48 及％小时的 TLm（中间忍受限＞分别为 28.3‰、14.1‰和 12.6‰，盐对它的安全浓度是 1.1‰，10‰的盐度就能使其致死，在 20‰盐度的水体中，最长存活时间为 36.7 小时，死亡速度（LP_{50}）为 33.9 小时；在 40‰盐度的水体中，最长存活时间只有 10.7 小时，LP_{50}下降到 5.7 小时。和鱼类相比，有较大的差异：草、鲢、鳙鱼苗可忍耐 4‰-5‰的盐度，成鱼可忍耐 10‰—12‰c 的盐度，并可在盐度为 5‰的水体中发育；鲤鱼的耐盐能力更强，可以生活在盐度高达 7‰的水体中。鳖的这种生理特征可能与鳖长期在含盐极低的溪河与淡水湖泊中生活的习性有关。⑤鳖对强酸性和强碱性水质的耐受力极强。96 小时内可以生存的 PH 值范围 2.0—11.5，这与鱼类的 PH 安全值范围 6.5—9.5 有很大的区别。

（2）药物治疗鳖病的原则

A 外用药物

根据鳖对一般常角药物的耐受性较强的特点，为了提高药物的治愈率，一般采取加强给药的原则。治疗时应注意以下几点：

①在使用该类外用药物时，除甲醛、高锰酸钾外，其用量可为鱼类常用量 1.5—2 倍，必要时，还可适当加大其使用浓度。

②不宜采用孔雀石绿治疗鳖病。由于孔雀石绿是三苯甲烷类染料，能溶解足够量的锌，会引起水生动物急性锌中毒；能阻碍肠道酶（如胰蛋白酶、α淀粉酶等）活性，影响水生动物的摄食与生长；更严重的是孔雀石绿可能是一种致癌物质；加上孔雀石绿对鳖的毒性较高，如处理不当，容易造成较大的损失。

③由于鳖对盐度的耐受力较低，在使用食盐治疗鳖病时，应严格控制其浓度和浸浴时间；长期浸浴，盐的浓度应控制在 1‰左右；高浓度的食盐药浴，最高浓度不应超过 40‰，浸浴时间应控制在 0.5 小时左右。

④利用鳖对强酸、强碱耐受力强的特点，在治疗鳖病时，可适量加大鳖池的酸碱度。最好的办法是泼洒生石灰，提高鳖池的碱度。鳖对生石灰的安全浓度是 239 ppm，因此，我们可将生石灰的泼洒量由原来的每亩（水深 1 米）的 15—25 公斤增加到 40—50 公斤（即 60—75 ppm），以提高疗效。

B 内服药（包括口服给予与注射给予）

用药的主要原则是选择效果高、吸收性能好、副作用小的药物，并要避免产生抗药性。应注意以下几点：

①应用磺胺类药物与抗生素时，应注意使用的剂量和次数。用量过大次数过多虽不会产生药害，但易产生耐药性，使治疗失败。

②轮换使用具有同一治疗效果的不同药物。

③由于鳖对食物的不挑剔性，药物添加较多时也不会影响其摄食，因此可采用中西药结合的方式口服给药，以发挥中、西药各自的长处，提高治疗效果。

④在使用内服药治疗时，可与外用药物配合使用以提高疗效。

3　鳖病药物的给药方法

A 鳖病药物给药方法的选择原则

①根据鳖病的病原体特性进行选择。

鳖类疾病的病原体有细菌、真菌、寄生虫等，寄生的部位有体表寄生和体内寄生，体表寄生有浅层体表寄生和深层体表寄生，我们应根据病原体的这些特性而选择不同的给药方式。浅层体表寄生的我们可采取遍洒和浸浴的方式，而深层体表寄生的只能采用浸浴与涂抹或深层给药的方式；有些细菌性病原体虽侵袭部位是体表，但因顽固，光靠外部给药尚难达到效果，往往也靠口服给药的方式，通过机体的吸收作用而发挥药效。

②根据鳖病的病程进行选择

对于病程较轻的病鳖，遍洒或口服方式给药即可达到药物的治疗效果，而病程较重的鳖病，则要采用浸浴给药的方式单独处理，有的时候，为了挽救个别病鳖，更快地发挥药效，则采取注射、涂抹或口灌的方法。

3 根据鳖的大小、鳖的年龄、鳖的体质等鳖的状况进行选择

稚鳖一般不采取注射的方法给药；鳖病情很重、鳖体质太弱时，摄食能力减弱或不摄食，口服给药就往往达不到治疗的目的；亲鳖在产卵期间，频繁地采取泼洒给药的方式，会影响其产卵率；商品鳖患病后，最适宜采取口服、遍洒或浅水浸浴的方式。

①根据药物的理化性质进行选择。

药物的药理、药效与其理化性质密切相关，因理化性质影响着机体对药物的吸收和分布，因此根据药物理化性质选择适当的给药方式是发挥药物药效的一个重要因素。有些药物，如磺胺嘧啶，浸浴给药则不易被吸收，无法发挥药效；而磺胺嘧啶钠在水中极易溶解，在浸浴给药时药效就会提高很多。

②根据鳖池和养殖条件进行选择。

鳖池和养殖条件也是选择给药方式的依据。当水源供给方便的时候，可采取浅水药浴的方式，并可清除病鳖专门用注射、涂抹的方法隔离治疗；小的鳖池，容易干池，鳖的清理也容易，当鳖病发生后，常将病鳖捞出，采取浸浴的方式；大的鳖池，则多采用遍洒与口服给药的方式。

4　给药量的确定

（一）外用药给药量的确定

1. 根据鳖对某种药物的安全浓度，药物对病原体的致死浓度而确定药物的使用浓度（ppm）c

2. 准确测量鳖池的体积或确定浸浴水体的体积。水体积的计算方法：水体积（米^3）＝面积（米^2）×平均水深（米）。

3. 计算出用药量：

用药量（克）＝需用药物的浓度（ppm，即克/米^3×水体积（米^3）

（二）内服药给药量的确定

我们常用给药的总量（毫克）和饲料药物的添加率（％）作为口服给药法的用药量指标，它们是根据用药标准量、鳖的总体重、给饵率来确定的。

1. 用药标准量：指每公斤鳖体重所用药物的毫克数（mg/kg），每种市售药物均有注明．：如果不知鳖的用药标准量，可参照人的用量和药物的水溶性进行推算，如磺胺类药物成人用量一般为 1 克/次，1 天 2 次，以成人 50 公斤体重计，则每公斤体重用药为 0.04 克；考虑药物在水中的损失，此用量扩大 2.5 倍（经验数字，一般为 2—5 倍），即每公斤鳖体重的标准用药量为 0.1 克（首次加倍应 0.2 克）。

2. 池中鳖的总体重（公斤）＝估计每只鳖的重量（公斤）×鳖的只数，或按投饵总重量｜公斤）/投饵率（％）进行推算。

3. 投饵率（％）：指每 100 公斤鳖体重投喂饲料的公斤数，根据鳖的不同养殖阶段进行确定。

4. 药物的添加率（％）：指每 100 公斤词料中所添加药物的毫克数。由下列公式得出：用药标准量（毫克/公斤）×投饵率（％）

5. 根据以上的数据，我们可从两个方面得到内服药的给药量：

（1）如果能估算出鳖的总体重，那么给药总量（毫克）＝用药标准量×鳖总体重。

（2）如果投饵量每日相应固定，且有一定的依据，那么给药总量（毫克）＝日投饵量（公斤）×药物添加率。

该文发表于《水产科技情报》【1996，23（4）：188-190】

鳖用药物

杨先乐，柯福恩

（中国水产科学研究院长江水产研究所）

鳖用药物主要用于预防、治疗鳖病，以达到控制疾病发生，使鳖更快、更好地生长、发育的目的。药物的种类和使用方法与鱼药有很多共同之处，但也有它的特殊性。

第一节　鳖用药物的基本作用

鳖用药物的功能与作用很多，概括起来主要有以下三个方面：

一、抑制或杀灭病原体病原体是鳖病发生的根本原因。病原体包括细菌、真菌、寄生虫及病毒。常用的抗菌药物有磺胺类、呋喃类、抗生素以及大蒜等；抗真菌药物有食盐、福尔马林、中草药五倍子、菖蒲以及新药鳖康宁浸浴剂等；抗寄生虫药物有硫酸铜、氨水、高锰酸钾等。这些药物通过内服或外用以直接或间接的方式起到杀灭病原体的作用。

二、改善养殖环境不良的栖息环境容易诱发鳖病。有些鳖用药物可以改善养鳖的生态环境，它们主要有以下三方面的作用：（1）杀灭水体中的病原体，如漂白粉、优氯净、三氯异氰脲酸等；（2）改良鳖池的水质和底质，如生石灰、过氧化钙等，可提高鳖池的碱度，增加池底的通透性；（3）净化鳖池的水质，如沸石、光合细菌、硝化细菌等。

三、增强鳖自身的抗病能力从增强鳖自身抗病力的目标出发，这类药物的功能大致有以下三方面：（1）为鳖补充适当的营养物质，如维生素等；（2）激发鳖的特异性免疫机能，使鳖对某种疾病具有特异性抵抗力，如注射疫苗；（3）增强鳖的非特异性免疫机能，调动鳖自身的补体、溶菌酶、C-反应性蛋白、干扰素、铁传递蛋白等非特异性防御因子，如某些免疫激活剂。

鳖用药物的这三种作用并不是孤立的，某种药物的一种作用能影响其它两种作用，如环境改良药物对病原体的抑制和杀灭、以及增强鳖自身的抵抗力等都有影响；反过来，也只有多种药物的协同作用，才能达到有效控制与预防鳖病的目的。

第二节　鳖对外用药物的反应特点和使用药物治疗鳖病的原则

一、鳖对外用药物的反应特点

鳖和鱼类相比，在体态、内部结构等方面有很大的不同，所以对水产上常用药物的反应也与鱼类有很大区别，大致有如下几个特点：

1. 鳖对生石灰、硫酸铜、敌百虫、漂白粉等药物的敏感性低于鱼类，其安全浓度与加州鲈鱼苗相比可高出几十倍乃至几百倍，与一般鱼类的常用量相比，也有显著差别（表5）。其原因可能是鳖的革质表皮阻碍了体表对药物的吸收；鳖用肺呼吸，常浮于水面，不存在药物对鳃组织的刺激。然而，鳖对甲醛的敏感反应几乎与加州鲈相似，这是因为甲醛挥发性气体有很大的刺激性，能凝固蛋白质和溶解类脂，使蛋白质变性。因而，用肺呼吸的鳖同样难逃药物对呼吸器官的影响。

2. 孔雀绿对鳖的毒性较高，浓度为 40 mg/L 时可使鳖致死，其安全浓度为 26.3 mg/L。如果使用不慎，易发生鳖的药物中毒。

3. 鳖对敌百虫不十分敏感，其致死浓度范围是 80~160 mg/L，96 h 的 TLm（平均忍受限）为 160 mg/L，安全浓度为 28.5 mg/L。但较低浓度的敌百虫能导致鳖的昏迷和活动能力减弱，从而影响它的生长、发育。

4. 鳖的耐盐力低。实验表明，24、48、96 h 的 TLm 分别为 28.3‰，14.1‰和 12.6‰；对盐的安全浓度为 1.1‰。在 20‰盐度的水体中，最长存活时间为 36.7 h，死亡速度（Lp_{50}）为 33.9 h；在 40‰盐度的水体中，最长存活时间只有 10.7 h，Lp_{50} 下降到 5.7 h。而草、鲢、鳙鱼苗可忍耐 4‰~5‰的盐度，成鱼可忍耐 10‰~12‰的盐度，并可在盐度为 5‰的水体中发育；鲤鱼的耐盐能力更强，可以生活在盐度高达 17‰的水体中。鳖的这种耐盐力低的特性与该物种长期适应于溪流、河川及淡水湖泊中有关。

5. 鳖对酸碱水质的耐受力强。由实验可知，鳖在 pH 值 2.0~11.5 之间 96 h 内无一死亡，这与鱼类的 pH 安全值范围 6.5~9.5 有很大区别（见表6）。

表 5　中华鳖稚鳖对几种药物的安全浓度与鱼类的比较

药物名称	中华鳖稚鳖的安全浓度 （mg/l）	加州鲈鱼苗的安全浓度 （mg/l）	一般鱼类的常用量（mg/l）	
			浸浴	遍洒
鳖康宁 *	18.3	–	–	–
高锰酸钾	19.5	0.74	10~20	1~2
孔雀绿	26.3	0.02	67	0.5~1.2
敌百虫	28.5	0.04	1~2	0.2~0.5
漂白粉	35.9	1.20	10	1~2
甲醛	45.9	42.0	250	25~30
最高可达80				
硫酸铜	94.9	1.42	8	0.7~0.9
生石灰	239.0	33.18	–	22.5~37.5

* 中国水产科学研究院长江水产研究所研制生产

表 6　鳖对极端 pH 水质的反应

pH 值	死亡率 （%）	死亡速度（h）		
		第一尾死亡时间	最后一尾死亡时间	Lp50（半数死亡时间）
13.00	100	3.3	7.1	4.8
12.75	100	3.8	11.1	5.7
12.50	100	17.0	24.8	19.0
12.25	50	19.5	36.3	40.8
12.00	25	23.3	23.3	–
11.50	0	–	–	–
11.00	0	–	–	–
10.50	0	–	–	–
10.00	0	–	–	–
3.00	0	–	–	–
2.50	0	–	–	–
2.00	0	–	–	–
1.50	75	13.3	42.0	24.0
1.00	100	8.0	15.5	10.0

二、药物使用的原则

（一）外用药物

根据鳖对一般常用药物耐受性的特点，为提高药物的治愈率，应注意以下几点：

1. 除甲醛、高锰酸钾外，使用"鳖康宁"等外用药物用量可为鱼类常用量的 1.5~2 倍，必要时还可适当加大。表 7 是几种药物的推荐使用浓度。

2. 不宜采用敌百虫和孔雀石绿治疗鳖病。孔雀绿是三苯甲烷类染料，能阻碍肠道酶（如胰蛋白酶、α 淀粉酶等）的活性，影响动物的消化吸收功能；对鳖的毒性较高；也有人认为，孔雀绿是一种致癌物质，所以不宜用它绿治疗鳖病。长江水产研究所研制生产的"鳖康宁"是一种疗效好、安全性能高的浸浴剂。

3. 使用食盐治疗鳖病时应严格控制其浓度和浸浴时间；长时间浸浴，盐度应控制在 1‰ 左右；用高浓度的食盐药浴，不应超过 40‰，浸浴时间控制在 0.5 h 左右。

4. 鳖对酸碱耐受力较强，因此常用石灰水消毒杀菌是安全可靠的。生石灰的安全浓度是 239 mg/L，将生石灰的泼洒量由原来的每千平方米（水深 1 m）的 22.5~37.5 kg 增加到 60~75 kg（即 60~75 mg/L），可提高疗效。

表7　几种常用药物的推荐使用浓度

药物名称	推荐使用浓度（mg/L）	
	药　浴	遍　洒
鳖康宁（泼洒剂）	4~8	0.6~1.0
高锰酸钾	10~15	2~4
漂白粉	15 左右	3~4
甲醛	50（1~2 h）	20
硫酸铜	10	1~1.5
生石灰	-	60~75

（二）内服药物（口服或注射给予）

选择疗效好，吸收性能强，副作用小的药物，并注意避免产生抗药性。

1. 使用磺胺类、抗生素类药物时，应掌握好使用剂量和次数。用量过大、次数过多，易产生耐药性。
2. 经诊断确定某种鳖病后，尽可能轮换使用具有同一治疗效果的不同药物。
3. 采用中西药结合的方式口服给药，发挥中、西药各自的长处，以提高疗效，节省费用。
4. 贯彻"防重于治"的原则，改善养殖环境条件；必要时内服药与外用药物配合使用。

该文发表于《农村实用技术与信息》【1995，12：18-19】

鳖鱼投药技术

杨先乐

（中国水产科学研究院长江水产研究所）

遍洒鳖对常用药物的敏感性较低，耐药性较强，使用时可适量加大浓度，或用常规浓度连续泼洒，在鳖病高峰期使鳖池几天内保持一定的药物浓度，温度较高、水质较肥时应适当降低药物浓度；反之应适当增大浓度。由于鳖水陆两栖，投药除了鳖池外，岸边、食台和晒台也要泼洒。为不打扰鳖的摄食和晒甲，使鳖与药物接触时间延长，下午 4~5 时施药最合适。防止过高的药物浓度使鳖致死，或使鳖逃离施药水域爬上岸边。微碱性水质能增强鳖的抗病力。而大部分药物都呈酸性，因此泼洒药物后 7 d 左右应用生石灰调节水质，用量为 50~85 g/m^2（水深 1 m）。

挂篓（袋）篓袋应挂在食场周围鳖出入处的水下 10 cm 左右的水层中，挂篓（袋）个数根据食场大小而定，要保证鳖能正常摄食，篓（袋）中的药物以鳖将食合上的饵料吃完而药物尚未溶完为度，但各篓（袋）总药量之和应低于全池泼洒该药的用量。挂篓（袋）一般持续 3~5 d。

浸浴病情轻的鳖可采取低浓度，长时间浸浴（8~20 h）、多次浸浴。浸浴时间过长易加速染病鳖鱼死亡。浸浴时，药水总量以淹没病鳖 10 cm 左右为度；一个容器浸浴病鳖的数量以鳖排满 1.5 层为宜；浸浴时要经常照看，防止鳖叠层，以免上层鳖受不到药浴，下层鳖呛药死亡；药浴水体的水温与病鳖生活的水温差异应不大于 2℃；浸浴后的药液不要倒入鳖池。可用一些收敛药物（如高锰酸钾、中草药金樱子、虎仗等）作为辅助浸浴剂，以促进病鳖表皮溃烂愈合。

涂抹涂抹前先将鳖用药水浸浴，清洗创面，然后用药膏或药液涂抹，涂抹药物的浓度不可太高，要防止药物流入鳖口，对洞穴病等深层感染，药物应抹到创伤的深层部位，涂抹后在无水状态下放置 0.5~2 h，不可太久。

口服针对病情选用敏感药物，同种（类）药物或有交叉伉药性的药物不宜长期使用。磺胺类药一般每公斤鳖每天用 0.1 g，第一天加倍；抗生素等每天 20 万国际单位，一个疗程 3~5 d。药量要足、次数不能少，否则达不到疗效还易导致病原体产生抗药性。药饵要选鳖喜食的饲料制作，避免使用含咸、酸和蒜味的原料。投喂氯霉素等较苦的药物应添加乳糖或蜂蜜。为保证鳖每天能将药饵吃完，投饲量应比平时少 10~20%。

口灌最好将药物做成较小的药丸，用适宜大小的木棒塞入鳖嘴，然后将药物送到喉部（药水用加套管的注射器送入）。为避免吐出，应稍等一段时间再将木棒抽出，操作时间不应过长。除治疗性口服药外，还可辅以一定的营养药物。

注射应注意药物配伍禁忌，如氯霉素不宜与盐酸四环素、卡那霉素混用；四环素、链霉素、硫酸卡那霉素忌与碱性药物同用。同时应补充一定的营养药物。注射部位有皮下、肌肉、和腹腔三处，以后肤皮下或肌肉注射效果为好。注射部位应作消毒处理。500 g 以上成鳖注射药液量为 0.5~0.8 mL，500 g 以下鳖注射药液量为 0.2~0.5 mL。注射动作要轻快，然后将鳖放入原饲养环境中，勿让其长时间离水。

该文发表于《水利渔业》【1995，5：36-37】

防治鳖病的给药方法

杨先乐，柯福恩

（中国水产科学研究院长江水产研究所，湖北省荆沙市 434000）

鳖病的药物防治，分为体外给药和体内给药。前者包括遍洒法、挂篓（袋）法，浸浴法和涂抹法；后者有口服法，口灌法及注射法。实践中经常体内外结合用药，以达到防治鳖病的目的。上述给药方法基本与鱼类用药相同，但应用于鳖又有其自身的特点。

1　遍洒法

鳖对一些常用药物敏感性较低，耐药性较强。如硫酸铜对鳖的安全浓度为 94.9×10^{-6}，是加州鲈鱼的 66.8 倍。因此使用时，可适当加大浓度，以提高效果。也可用常规浓度连续泼洒，在鳖病高峰期使鳖池几天内保持一定的药物浓度。另一方面，要避免过高的药物泼洒浓度，否则即使鳖不死亡，也会使其逃离施药水域爬到岸边，达不到预期目的。温度较高、水质较肥时应适当降低药物浓度；反之则增大浓度。另外，因为鳖水陆两栖，用药时除了泼入鳖池之外，还要在岸边，食台和晒台上泼洒。为不干扰鳖的摄食与晒甲，使鳖与药物接触时间延长，下午 4~5 时施药最合适。

微碱性的水质，能增强鳖的抗病力。而大部分药物都呈酸性，所以泼洒药物后 7 d 左右应用生石灰调节水质。用量为 50~85 g/m² （水深 1 m）。

2　挂篓（袋）法

篓（袋）宜挂在食场周围鳖出入处的水下 10 cm 左右的水层中。根据食场大小并要能保证鳖来正常摄食而确定篓（袋）的个数。篓（袋）中的药量应以鳖将食台上的饵料吃完而药物尚未溶完为适度，但各篓（袋）总药量之和应低于全池泼洒该药的用量。挂篓（袋）一般持续 3~5 d。

3　浸浴法

有两种方式：一是低浓度、长时间浸浴（8~20 h）；另一种是高浓度、短时间（0.5~2 h）、多次浸浴。前者适于病情较轻的鳖。后者适于病重的鳖。但浸浴时间都不宜过长（不超过 24 h），因为病鳖若长时间处在药物中，呼吸更困难，应激反应强烈，易加速其死亡。

浸浴操作要点：①浸浴药水总量以淹没病鳖 10 cm 左右为度；②一个容器浸浴病鳖的数量以鳖排满 1.5 层为宜；③浸浴时要轻常照看，避免鳖叠层，以至上层的鳖接受不到药浴，而下层的鳖呛药而死亡；④药浴水体的水温与病鳖原来的水温差异应不大于 2℃；⑤浸浴后药液不要倒入鳖池。

常用一些收敛药物，如高锰酸钾、中草药金樱子、虎仗等作为辅助浸浴剂，以促进病鳖表皮溃烂的愈合。

4　涂抹法

常用药膏或药液，并辅以一定量的收敛剂以促使创伤愈合。涂抹药物的浓度不可太高。涂抹前先将病鳖用药物浸浴，以清洗创面；涂抹时要防止药物流入鳖口，对洞穴病等深层感染，药物要涂抹在创伤的深层部位。涂抹后在无水状态下放置病鳖 0.5—2 h，不可太久。

5　口服法

针对病情选择敏感药物，同种（类）药物或有交叉抗药性的药物不宜长期使用。应交叉用药，避免发生抗药性。磺胺类药物一般每公斤鳖每天用药 0.1 g，第一天加倍；抗生素等每天 20 万国际单位；一个疗程 3~5 d。药量要足，次数也不能少，否则药浓度在鳖体内低于要求值，不但达不到防治目的，还可能导致病原体产生抗药性。

注意鳖的口味，药饵要选择鳖喜食饲料制作，避免使用含咸、酸和有蒜味的原料。若投喂氯霉素等较苦的药物，应添加乳糖或蜂蜜。为保证鳖每天能将药饵吃完，投饲量应比平时减少 10%～20%。

6 口灌法

除治疗性口服药外，还可辅以一定的营养药物，如葡萄糖等；为避免吐出，最好将药做成较小的药丸。用适宜大小的木棒塞入鳖嘴，然后将药物送到喉部（药水可用加套管的注射器送入）。为避免吐出，应稍等一段时间后再将木棒抽出。操作时间不应过长，不要使鳖产生强烈的应激反应。

7 注射法

应注意药物配伍禁忌，如氯霉素不宜与盐酸四环素、卡那霉素混用；四环素、链霉素、硫酸卡那霉素忌与碱性药物同用。同时应补充一定的营养药物。注射部位，有皮下、肌肉和腹腔三处。以后肢皮下或肌肉注射效果为好。注射部位应作消毒处理。500 g 以上成鳖注射药液量 0.5～0.8 mL，500 g 以下为 0.2～0.5 mL。注射动作要轻快，然后将鳖放入原饲养环境中，勿让其长时间离水。

8 特殊疗法

包括日晒疗法、深层给药法、加强药浴法、浅水泼洒法、剔除处理法等。对此作者在《淡水渔业》1995 年第 2 期上已有阐述。

2. 渔药及其鱼病的防治

南美白对虾与鱼、蟹等物种混养的模式中，需要查清水质，预判疾病，谨慎合理选择药物（章海鑫等，2012）。一氧化氮前体物既能提高鲫免疫力，对鲫大红鳃病具有一定的预防作用（杨先乐等，2013）。鲟养殖中常见疾病包括细菌性疾病和营养性疾病等（杨移斌等，2013）。定期消毒养殖池壁苔藓和注射基因工程疫苗是防治鲟鱼性腺溃烂病的有效途径（杨移斌等，2012）。高锰酸钾在水产上具有杀菌、收敛伤口、解毒和改良水质的作用（杨先乐，1995）。观赏鱼一般在进入水族箱后一周内较难养，疾病较多，这一时期很关键需合理用药，隔离外来病原生物，恢复其体质（陈少鹏等，2003）。作为一种新型消毒剂，聚六亚甲基胍（Polyhexamethyleneguanide, PHMG）能有效避免耐药性等问题，对水产养殖中一些棘手的疾病有较好作用（孙琪等，2014）。罗非鱼链球菌病的防治中需要认识到保肝护胆药物或环境解毒药物的重要性（胡利静等，2015）。鱼虾混养塘锚头鳋病的防治主要集中在中草药和化学药物防治两个方面（章海鑫等，2012）。

该文发表于《水产养殖》【2012，2: 39-40】

银鲫与南美白对虾混养模式下的几点用药建议

章海鑫[1]，阮记明[1,2]，杨先乐[1]，欧仁建[1]，王祎[1]

（1. 上海海洋大学国家水生动物病原库，上海 201306；2. 江西农业大学动物科学技术学院，江西 南昌 330045）

近年来，许多地区开始推广南美白对虾与鱼、蟹等物种混养的模式，并且具有了一定的经济和社会效益，目前这一模式在江苏北部等地区得到了大量的推广。但由于此种模式还处于发展阶段，很多技术都不是很成熟，特别是对于两种不同种属动物在病害防治中的药物运用问题不是很成熟。本文将对银鲫与南美白对虾混养模式下的药物使用应注意的问题进行一定的阐述。

由于银鲫与南美白对虾种属间差异较大，此种混养模式的病害防治和药物安全使用具有其独特的特点。养殖周期内，银鲫出现的病害主要有寄生虫性疾病（猫头蚤、中华蚤、车轮虫、指环虫、孢子虫、面条虫等）、细菌性疾病（肠炎、败血症、烂鳃病、赤皮病、肝胆综合症等）；而南美白对虾主要以细菌性的红体、红腿病、烂鳃红鳃病以及寄生性的纤毛虫病等为主。此外，两个养殖品种间流行病的病程和流行规律是不一样的，且银鲫和南美白对虾对各种水体和药物的耐受程度也不一样，所以对于疾病的预防和对症处理就显得特别的重要，既要有效的防治疾病，又要保证水体养殖动物的安全。对于此种混养模式下的病害防治原则是首先严格控制和管理水体水质，其次是要对可能暴发的疾病和疾病暴发的季节有个准确的预判，最后是根据具体病害情况合理选择和使用药物。

1 微生态制剂

养水是鱼虾疾病控制和预防的关键所在，对于银鲫和南美白对虾混养模式来说，养水就显得尤为重要。微生态制剂作为对养殖水质调控的手段在水产养殖实践中使用较多。生产实践上，微生态制剂在不同的养殖阶段选用不同的产品，并按照正确的使用方法使用，才能达到理想的效果：①微生态制剂一般在天气晴朗的上午，有风的时候使用效果较明显，此时，水体阳光充足，溶解氧高，微生物能够大量繁殖；②在高温低压的天气下不能随意的使用芽孢杆菌类及硝化细菌等好氧微生态制剂，若使用这类产品会与水体养殖动物竞争溶解氧资源，容易导致鱼、虾浮头；③在肥水、养水阶段可以选用一些含酵母菌、光合细菌类的产品，因为酵母菌和光合细菌能够为水体浮游动植物提供丰富的饵料来源；④蛭弧菌不能与其它的微生态产品同时使用，但在使用一周后可以使用 EM 菌，这样一般能达到理想的肥水状态；⑤微生态制剂不能与消毒剂同时使用，一般在消毒后一周左右使用微生态制剂效果较为理想；⑥酵母菌、乳酸菌、芽孢杆菌等不仅能调节水质，还能促进银鲫和南美白对虾的生长和提高饲料利用率。

2　杀虫药物

银鲫与南美白对虾之间寄生虫病害的差异较大，且南美白对虾对绝大多数的杀虫剂都敏感，所以对于杀虫剂的选择要慎重。南美白对虾对敌百虫、甲苯咪唑、菊酯类、硫酸铜等杀虫剂都非常敏感，对一些有机类杀虫剂也敏感，对于阿维菌素、伊维菌素等有一定的耐受浓度。而在此种养殖模式中，银鲫爆发寄生虫病的几率远大于南美白对虾，车轮虫、指环虫、猫头蚤等都极易的寄生在银鲫上，南美白对虾也常寄生有纤毛虫等。所以对于防治银鲫相关寄生虫疾病，应选用一些虾类不太敏感的药物（阿维菌素、伊维菌素等），也可以通过内服中草药抗虫剂来达到防治的目的。用药时应准确计算水体体积和药物浓度，药物应在晴天上午施用，之后开动增氧机增氧。一般用于防治银鲫寄生虫疾病的药物大多也可以防治南美白对虾纤毛虫病。另外，苏北地区是我国孢子虫病的高发地区，所以选择防治孢子虫病害的药物是非常必要的，预防可以通过内服一些药物进行，如氯苯胍，也可以选用一些中草药来进行预防，如青蒿和苦参等，治疗可以采用外用泼洒碘制剂或者环烷酸铜等药物。

3　消毒药物

目前，消毒仍然是防治水产疾病的主要手段之一。在银鲫与南美白对虾混养的模式中，消毒剂的使用量也很大，种类也有很多。但在使用消毒剂时，一定要考虑到银鲫与南美白对虾对其敏感性的高低，使用药物浓度不能大于敏感浓度。消毒时应选用一些刺激性较小的消毒剂例如二氧化氯、碘制剂等。而且使用时应选择在晴天的下午使用，消毒后加开增氧机等；特别是一些刺激性较大的消毒剂，必须慎重使用，如甲醛、苯扎溴铵等，一般是不用在此种养殖模式中，强氯精、溴氯海因等因对水体的伤害较大，刺激性较重，使用时应该慎重，特别是在鱼、虾苗较小时和低压闷热等天气状态下不能使用，还有就是单养南美白对虾中经常使用的茶籽饼不能用于此种混养池中。

4　内服药物

目前市场上内服药物有抗细菌、抗病毒以及免疫增强剂等几种。内服药物与消毒剂一样，是疾病治疗的首选，且使用量非常大。内服药物（抗菌药物）的选择首先是选择国标内服药物，对于其质量以及含量要严格保证。其次，内服药物的休药期一定要按照残留时间较长的动物来计算；不使用违禁药物，制定详细的药物使用计划。第三，不滥用药物，最好选用的内服药物能够是广谱抗菌药物，这样既能保证鱼、虾的绝对安全，又能在抗银鲫病菌的同时兼顾抗南美白对虾的病菌，常用的内服药物有氟苯尼考、甲砜霉素、磺胺嘧啶等，也可选用中草药进行预防，如大蒜素、板蓝根个、穿心莲等。

该文发表于《科学养鱼》【2013，7：57-58】

新型一氧化氮和高氯酸锶饲料添加剂在防治鲫大红鳃病中的应用

杨先乐[1]，王乙力[2]

(1. 上海海洋大学，上海 201306；2. 成都三阳科技实业有限公司，四川 成都 610000)

鲫大红鳃病是当地渔农的俗称，主要症状是鱼体鳃瓣组织发生严重充血、溢血。该病来势凶猛，流行面广，加之水体病菌、病毒及腐殖性有机物侵袭，死亡率一般在 60% 左右，有的几乎"全军覆灭"，对水产养殖业造成了严重威胁。2011—2012 年间琼、粤、江、浙四省精养鲫、罗非鱼的损失惨重，尤其在 2012 年江苏省大丰、射阳、兴化等地该病暴发严重。

一、发病原因分析

大红鳃病呈现出血性败血症状，出血点集中于体表鳃瓣中，此外在鱼鳃盖内表皮、腹肌层及眼眶等处也有红色小点，呈斑状出血（Ecchymoses）。引起该病的原因及致病途径有以下几种说法：

1. 生态环境恶化引起。由于水质败坏、缺氧等原因导致病原体孳生、感染鱼体，尤其以栖息于底层的鲫、罗非鱼为甚；加之水温适宜，导致病原体大量繁殖，从而破坏鱼体肝脏、肾、脾等组织。血液中红细胞被溶解，易在鳃丝部位渗透显露。据研究，其病原体在水底有机污染物中能存活15天，比一般肥水中水域的存活期要长1/3。加之缺氧，致鱼体血红蛋白的含量可由7.4克/升降至4.1克/升，因此贫血性鲫、罗非鱼极易感染该病。其病原菌有液化产气单胞菌、嗜水气单胞菌、点状气单胞菌、极毛菌、嗜水芽孢菌棕色亚种以及屈桡菌类黏细菌等等，其中尤其以一种具强毒的水肿型点状产气菌（*Aeromonepunctata*）危害最大，可致鱼体水肿、腹水等症状。

2. 病毒感染引起。有些学者将感染组织匀浆，再通过细菌滤器，负染后在电镜下观察，可见一种呈球形、直径0.1微米的病毒；病理研究在鱼神经系统及上皮细胞可见嗜伊红的颗粒。将含这种病毒的滤液感染健康鱼体，数日后可见鱼体肝脏肿大，鳃瓣充血，呈出血病症状。初步研究认为这种病毒是一种弹状病毒。

3. 液化产气单胞菌引起，环境仅是致病的重要条件之一。中科院微生物所的研究结果表明，所分离到的细菌属革兰氏阴性杆菌、兼厌气性，最宜生长温度28~30℃；丁二醇脱氢酶、甘油、葡萄糖（产气）呈阳性反应。台湾大学刘朝鑫认为，该菌能产生菌体外毒素，既损害肝脏、肾、脾组织，又破坏鳃丝结构，从而导致病鱼严重充血、溢血等症状。

4. 病毒与病菌共同感染所致，它们相互作用，导致感染加剧，直至鱼体死亡。

5. 生物媒体作用。该病的迅速暴发及蔓延，与生物媒介物传播有密切关联。据笔者试验证实，取病鱼池中藻类及鱼鳃上中华鳋等，进行培养分离出产气单胞杆菌，并接种于健康鱼体中，经24小时后可使鱼体发病。由此可知病鱼池内水中生物如轮虫、水蚤、藻类等有机生物通过换水、冲水至健康鱼池，导致了该病的传播。

6. 人为干扰影响。在养殖生产过程中，由于轮捕轮放等操作不慎，使鱼体受伤而导致病原体传播。在实验室，我们擦破健康鱼体皮肤鳞片，并将其移置于含病原菌1 000个细胞/毫升的器皿中片刻，再转入已脱落鳞片或受伤鱼池中，仅12小时，该批受伤鱼群便会感染该病。

7. 投喂过量含碳水化合物饲料，引发鱼体内糖、脂肪、蛋白质三者的代谢紊乱，使鱼体失去平衡，从而导致鱼出现高血糖、高肌糖及高肝糖症，奠定该病萌发的物质基础。据调查天然水域中健康鲫的血糖均值为96毫克/千克、鲤为59毫克/千克、罗非鱼70毫克/千克。在人工养殖应用专一的饲料条件下，其血糖值普遍上升，其中鲫、罗非鱼的血糖均值可增长2~3倍，明显呈高血糖病、尿毒症及出血性败血病。

二、防治对策

1. 彻底清塘

要彻底清除池塘中多年淤积的淤泥，消除池中病菌、病毒等病原体的附生基础。最适宜的清塘药物是生石灰，带水清塘用量是75~100千克/亩，干法清塘用量为30~40千克/亩。泼撒石灰时要全面均匀，决不留有死角。同时在池塘四周堤岸上，也须均匀泼洒消毒。

2. 加强鱼种、饲料及渔具的消毒

在放养前，鱼种用二氧化氯消毒，用量为0.3毫克/升，浸泡鱼15分钟；渔具用1毫克/升的浓度浸泡15~20分钟。

3. 预防

大量实践证明，应用新型免疫剂—氧化氮前体物既能提高免疫力，增强鱼的体质，又能加速细胞分裂，促进生长发育，达到"生物防治及促生长"的目的。研究表明，饲料中添加1.5‰的一氧化氮前体物，连续投喂85天后，鱼体血液指标可从初期的红细胞均值47万个/毫米^{3}上升到75万个/毫米^3，血小板均值由14万个/毫米^{3}上升到了22万个/毫米^3，血红蛋白均值由4.5克/升上升到7.1克/升，鱼种成活率可达85%~95%，增产率达9%~31%，饲料可节省14%左右。

4. 治疗

治疗可用超强氧化剂复合高氯酸锶饲料添加剂，它既可直接灭除病原体，又有促长增产作用。据美国农业部研究机构报告，每千克鱼体重内服高氯酸盐0.04克，可减少或灭除大肠杆菌、伤寒沙门菌、柯萨奇菌、肠道病毒、新城病毒及脊髓灰质类病毒等，同时还能转化成氯酸酶，促进吸收生长，增加体重3%~31%，饲料量减少7%~18%。即使增加含药量5~10倍也未曾发现有害作用。复合高氯酸锶还可外用，使用量为0.15~0.25克/米^3，全池均匀泼洒。在发病季节还可作预防使用。但须指出的是高氯酸盐类属易燃易爆的药品，必须经特殊处理后方可安全使用。

5. 加强饲养管理

必须重视池塘的饲养管理，遵循"四定"的投饲原则，并要严控饲料中碳水化合物含量。鲤、鲫、草、鲢、鳙等鲤科

鱼类，最佳含碳水化合物量为20%~40%，但实际使用量达到了80%~90%。由于习惯的原因，有些地区养鱼几乎全部用玉米、麸皮、米糠等碳水化合物饲料，从而易引发鱼类的代谢障碍，破坏鱼体内电介质平衡，导致体内糖分过量，是出血性败血病病原体及有些好糖性寄生虫（如蠕虫、甲壳类寄生虫）的良好媒质，因此必须加强饲料的质量管理和使用。

该文发表于《科学养鱼》【2012，10：64-65】

鲟鱼性腺溃烂病的防治

杨移斌[1]，夏永涛[2]，赵蕾[1]，邱军强[1]，胡鲲[1]，杨先乐[1]

（1. 上海海洋大学，国家水生动物病原库，上海 201306；2. 中国水产科学研究院，北京 100039）

鲟是大型经济鱼类，体内大部分是软骨，又体被硬鳞，故常被称作"软骨硬鳞"鱼类，其实鲟隶属于硬骨鱼纲、硬鳞总目、鲟形目（Acipenseriformes），因其皮是制作高档皮革制品的好原料，抗撕裂性、耐磨性、柔韧性可与鳄鱼皮媲美；鲟鱼是食用价值极好的大型经济鱼类，全身除体表骨板外其他部分（含骨骼）都可食用，营养价值极高，被列为高级滋补品，特别是素有"黑色黄金"之称的鲟鱼鱼子酱，更是驰名中外的高档食品。因为如此高的经济价值，随着鲟鱼养殖技术的提高，鲟鱼量变得越来越大，养殖区域不断增加，养殖规模及集约化程度不断上升。近年来，鲟鱼养殖中病害问题一直影响鲟鱼养殖的健康快速发展，如革兰氏阴性菌嗜水气单胞菌（Aeromonas hydrophila）、豚鼠气单胞菌（Aeromonascarvia）和类志贺邻单胞菌（Plesimonasshigelloides）感染鲟鱼致死，革兰氏阳性菌链球菌感染鲟鱼使其致病，这些疾病的发生使鲟鱼养殖受到了重大损失，对于鲟鱼疾病的研究刻不容缓。

2012年6月初在浙江省衢州市鲟鱼养殖基地发现养殖的鲟鱼出现一种急性传染疾病，死亡率高达80%，损失极为严重。本疾病研究中心对该养殖场发病的鲟鱼进行了取样，并展开流行病学调查，从发病的鲟体内分离到一株细菌，经实验室鉴定确认为嗜水气单胞菌，本文就此病的诊断及防治情况介绍如下，以期为鲟鱼养殖疾病防治工作提供参考。

一、疾病流行情况

2012年6月初浙江省衢州市鲟鱼养殖基地出现鲟鱼死亡，发病后死亡率高达80%，发病水温在24~26℃，发病个体30千克左右。水温在24℃及以上发病。发病鲟鱼种类集中在西伯利亚鲟、俄罗斯鲟、杂交鲟等，雌雄鲟鱼都可发病。发病鲟鱼个体目前发现最小为3龄，据流行病学调查发现其他养殖基地也有类似疾病。

二、临床诊断结果

发病初期病鲟萎蔫不振，或停留池底不动，或极度狂游，时而贴紧池壁游动，发病后期身体逐渐失去平衡形成腹部露出水面的现象，几乎不进食。出现症状后多数在一周内死亡，病鱼受到强烈刺激后数小时内急剧死亡。发病鲟鱼主要集中在西伯利亚鲟、俄罗斯鲟、杂交鲟等，基本症状表现为两种类型：一种是鲟鱼肛门一侧出现细线状红肿，有开始溃烂症状，解剖后看到肝脏充血肿大并出血，脾脏严重充血发黑，肾脏充血，腹腔有大量出血，导致倒提鲟鱼从生殖孔流出体外，体壁严重充血（也有体壁未充血的），性腺有大量出血点，产生溃烂，更严重的是生殖孔与腹腔相连的膈膜烂穿，肛门一侧溃烂开口，性腺掉出，直至鲟鱼死亡。另一种是肛门处一侧溃烂开口，生殖孔与腹腔相连的膈膜烂穿，性腺直接掉出、发生溃烂，未发现出血现象，解剖也未发现实质器官病变。有些鲟鱼也因急性感染，性腺并未掉出就已经死亡。

三、实验室诊断

1. 病原菌分离纯化

从10尾具有典型发病症状濒死的病鱼取性腺、肝脏等病灶部位少许以及少量血水划线于普通营养琼脂平板和TSA平板上，置于28℃恒温培养箱培养24小时后经纯化获得一株优势细菌。该菌在营养琼脂平板和TSA平板上生长良好，形成圆形、表面光滑、边缘整齐、中央菌落呈凸起状、直径0.8~1毫米的透明的菌落。

2. 生理生化鉴定和16S rDNA序列测定与系统发育分析

该菌落在兔鲜血琼脂平板上发生β-溶血，革兰氏染色为革兰氏阴性杆菌。菌株的生化鉴定结果为：除对D-甘露醇的反应与标准嗜水气单胞菌不同外，其余生化指标符合标准嗜水气单胞菌不同的特点。在生化鉴定的基础上，对该菌的16S rDNA基因进行PCR扩增，得到1 420 bp的扩增片段，测序后在国家生物技术信息中心（NCBI）进行比对，结果分离菌株与嗜水气单胞菌（*Aeromonashydrophila*）同源性最高，均为99%。因此，结合分离菌株形态、生化特性及16S rDNA鉴定结果将其鉴定为嗜水气单胞菌。

3. 人工感染试验

将分离到的菌株注射到健康的易发病品种西伯利亚鲟体内，12小时后试验鱼出现精神不振，或静止不动，或狂游不止，肛门出现红肿溃烂，14小时后出现死亡，24小时试验组全部死亡，而对照组无任何异常。解剖后看到肝脏充血肿大而且出血，脾脏严重充血发黑，肾脏充血，腹腔有大量出血导致倒提鲟鱼时从生殖孔流出体外，体壁严重充血，性腺有大量出血点，产生轻微溃烂，性腺未断肛门烂开而掉出体外，与自然发病鲟鱼症状一致。并从感染死亡鲟鱼肝脏、性腺及血水中分离到一株细菌，经形态、生理生化特性和16SrDNA基因序列分析，表明此菌株也是嗜水气单胞菌。综合发病情况、临床症状、病原学检测结果和回归感染试验，初步认为嗜水气单胞菌为该病的病原。

4. 药敏试验

对29种抗菌药物采用常规纸片法进行药敏试验，置于28℃培养24小时观察。结果表明嗜水气单胞菌对磺胺异噁唑、氯霉素、多黏霉素B、环丙氟哌酸、氟苯尼考、诺氟沙星、阿奇霉素、头孢噻肟、强力霉素、左氧氟沙星、妥布霉素、庆大霉素、丁胺卡那霉素等13种药物高度敏感，对新霉素等药物中度敏感。针对目前鲟鱼养殖中多发鲟鱼性腺溃烂病，病原为嗜水气单胞菌，是革兰氏阴性短杆菌，也是常见的顽固细菌性疾病。目前此鲟鱼疾病症状表现为全身性疾病，应该外用常规消毒剂，如碘制剂、漂白粉等，达到杀灭鲟鱼体表及池壁病原菌的效果。同时根据药敏试验结果注射庆大霉素进行体内杀菌，起到内外同时治疗的效果。由于嗜水气单胞菌属革兰氏阴性菌，会产生内毒素，在人工用药的情况下能杀死嗜水气单胞菌，但是留下的毒素却一样可以危害鲟鱼，产生溃烂，故需使用排毒药物，将毒素排出体外，另外适当投喂抗应激药物、保肝药物等，增加鲟鱼抵抗力。嗜水气单胞菌是水产上最常见的病原菌，为条件致病菌，主要发病原因是水温升高，鲟鱼本身受到刺激，抵抗力下降，而嗜水气单胞菌致病性却因水温升高而加强导致感染。还有水池中本来病原菌数量就不是很多，但是池壁生长的苔藓却成了病原菌滋生的良好场所，故要防止此类疾病再大规模暴发，第一是预防，第二是注射基因工程疫苗。

（1）预防：主要在于定期清理池壁苔藓，消毒，对池底淤泥进行改底处理，对食台进行消毒，保持卫生。春末夏初进行鲟鱼体内消毒，可以采取投喂庆大霉素或者氟苯尼考等，进行提前预防，添加中草药，提高鲟鱼抵抗力。

（2）注射基因工程疫苗：对于此类疫苗由于目前水产行业只有珠江水产研究所药厂有GMP认证资格生产，故市场基本没有出售，需专门研制自己使用，疫苗最大好处就是针对性强、对鱼体伤害小，如果成功那这个疾病也就算基本解决了。但是也很危险，有可能弄巧成拙。

该文发表于《水产养殖》【2013，2：46-48】

鲟鱼养殖常见疾病及防治

杨移斌[1]，夏永涛[2]，赵蕾[1]，曹海鹏[1]，胡鲲[1]，杨先乐[1]

（1. 上海海洋大学国家水生动物病原库，上海 201306；2. 中国水产科学研究院，北京 100039）

鲟鱼属于硬骨鱼纲、辐鳍亚纲、硬磷总目、鲟形目（Acipenseriformes）。我国的鲟类有2科3属8种，长江水系的有中华鲟（*Acipenser sinensis*）、白鲟（*Psephurus gladius*）和达氏鲟（*Acipenser dabryanus*）；黑龙江水系的有施氏鲟（*Acipenser schrenckii*）和达氏鳇（*Huso dauricus*）；新疆地区有裸腹鲟（*Acipenser nudiventris*）、小体鲟（*Acipenser ruthenus*）和西伯利亚鲟（*Acipenser baerii*）。鲟鱼是地球上最古老的鱼种之一，素有"活化石"之称，是我国的名特优珍品，肉味鲜美、骨软、营养价值高，鲟鱼肉和卵的蛋白含量高达18%和29%。鲟鱼全身都是宝，利用率极高。另外特别值得一提的是具有"黑色黄金"之称的鲟鱼籽酱，是由鲟鱼卵加工而成，更是驰名中外的高档食品，在国际市场十分走销。

我国第一个鲟鱼养殖场1992年在大连瓦房店成立，现今我国已开展人工养殖的品种有施氏鲟、达氏鳇、鳇鲟以及从国外引进的俄罗斯鲟、西伯利亚鲟等。另外鲟鱼具有食性广、生长速度快、成本低、价值高、个体大、易捕捞，以及抗病力强、容易饲养等特点，水产业者普遍看好鲟鱼的养殖前景，投入大量资金进行人工养殖。随着养殖品种及规模的不断增加，病害也正在逐步增加，已经严重影响了鲟鱼养殖产业的健康发展[1]。要解决鲟鱼养殖遇到的鱼病问题，已经刻不容缓。本文将就养殖中遇到的常见疾病加以概述，以期为鲟鱼病害防治提供一定的参考。

引起鲟鱼发病的原因有很多种，按照病原来分，可分为生物因素引起的疾病和非生物因素引起的疾病。

1　生物因素引起的疾病[2]

1.1　细菌性疾病

1.1.1　细菌性肠炎病

病因与症状：由点状产气单胞杆菌引起。病鱼肛门红肿，腹部、口腔出血，鱼体消瘦，轻压腹部有黄色液体从肛门流出，行动迟缓，不进食。解剖时，可以发现肠壁充血发炎，弹性差，无食物，内有很多淡黄色黏液。

治疗方法：用漂白粉挂袋消毒处理，浓度为 1 mg/L。口服大蒜素等抗菌药饵，大蒜素用量每千克饲料 2~3 g，疗程5~7 d，长期使用可以起到预防作用。

1.1.2　败血症[3-6]

病因与症状：由点状产气单胞菌（A. punotata f. instestinalis）引起的出血病。病鱼吃食量迅速下降，呆滞，呼吸困难，浮在岸边水面不动。病鱼体表充血，口腔有出血现象，肛门红肿。另外体表有溃烂产生。剖检腹腔内有淡红色混浊腹水，肝脏肿大土黄色，肝细胞发生了空泡变性，肠膜、脂肪组织、生殖腺及腹壁有出血斑点。肠内多无食物，肠壁及中肠以后部位螺旋瓣充血，后肠充满泡沫状黏液物质。

治疗方法：取肝脏、肠膜、生殖腺等组织，接种于 NA 培养基上，经一两天培养确定病优势菌，从而进行药敏试验，选出合适的抗生素加以控制疾病。从预防的角度来说，应定期在饲料中添加维生素 C、E 可加强鱼体的抗病力。

1.1.3　肿嘴病又称红嘴病

病因与症状：可能由嗜水气单胞菌（Aeromonas hydrophila）引起，嘴部肿大，四周充血，常伴有出血现象，口腔不能活动自如，进食能力差，肛门红肿，有时伴有水霉病发生，游动能力减弱。发病原因多是吃了发霉变质的饲料，也有的是因为养殖水域苔藓或者吃剩饲料等积累太多，导致病原菌滋生从而发病的。

治疗方法：此病属于细菌性疾病，抗生素治疗不失为一个好的方法，注射水产用青霉素钠效果良好。在日常管理中，应及时清除残饵，定期对食台进行清理，对发病的鱼应及时捞出以免交叉感染。

1.1.4　烂鳃病

病因与症状：由弧菌和假单胞菌或柱状嗜纤维菌引起。病鱼体表发黑，鳃盖内充血，鳃丝呈白色、浮肿、腐烂，并有黏液附着，形成团块状，严重时有些部分被腐蚀成不规则小洞，常伴随有浮头现象发生。

治疗方法：对于病鱼应当实行隔离治疗，及时降低养殖密度，减少发病数量。对发病鱼池每立方米 25~30 mL 的主要成分为福尔马林的药物。进行消毒，连续 2~3 d。

1.1.5　疖疮病[7]

病因与症状：由人工注射消炎不洁，点状产气单胞菌侵入机体引起。病鱼病灶部位肌肉组织长脓疮，隆起并出现红肿，抚摸有浮肿的感觉。脓疮内部充满脓汁，周围皮肤和肌肉都发炎充血，严重时肠道也充血。

治疗方法：此病以预防为主，人工注射采取腹腔、胸腔注射，操作要规范、细致。加强水质监测，增强鱼体抗病能力，发现病鱼时，可用注射器将脓疮内部的浓汁抽出，帮助排毒，可辅助病鱼康复。

1.2　真菌性疾病[8]

水霉病病因与症状：此病主要是由水霉属和绵霉属真菌引起的，鱼体受伤后经病原菌感染，伤处滋生大量絮状水霉呈絮状，严重的病鱼行动极为迟缓，鱼体消瘦，不摄食，直至死亡。另外在鲟鱼卵孵化时，卵表面长有白色或黄色絮状物形成"太阳仔"，是鲟鱼卵人工孵化中的常见疾病。

治疗方法：目前水霉病的治疗研究在全国各大水产疾病实验室都有涉及，由国家水生动物病原库研制的"美婷"可以很好的治疗水霉病，在发病的鱼池施药后一段时间能很明显的看到水霉脱落的现象。对于鱼卵孵化时可以采取提高鱼卵的受精率，减少死卵的产生，并且保持良好的孵化池用水，及时清除死卵的方法减少水霉病的发生。

1.3　寄生虫引起的疾病[9]

1.3.1　小瓜虫病（又称白点病）

病因与症状：肉眼观察病鱼鳃部、背鳍等部位有白点或片状，鳃丝和鳍条处较多。病鱼鱼体日益消瘦，游动迟缓且浮躁不安，食欲减退。显微镜下观察，白色小点球状，有的移动速度非常快，镜检分离白色病灶小点里的寄生虫即为小瓜虫。小瓜虫在组织里以组织细胞为营养，引起组织坏死，阻碍呼吸，导致鱼窒息死亡。

治疗方法：提高养殖用水温度至 25℃以上，小瓜虫病不治而好。此外用主要成分是福尔马林的药物浸泡。

1.3.2　车轮虫病

病因与症状：当鱼体和鳃耙上数量过多时直接影响生长，造成鱼体消瘦，游泳能力低下，不摄食，严重时造成苗种大量死亡。

治疗方法：加强管理，注意水质，提高鱼体抗病力。病鱼用 2%～3% 的食盐水浸泡后放入流水池进行养殖，一般能治愈。

1.3.3　斜管虫病

病因与症状：斜管虫大量寄生在病鱼体表、口腔、鳃部，病鱼在水中表现为急躁不安，体表呈蓝灰色薄膜样，口腔、眼周围黑色素增多。大量寄生时可引起皮肤及鳃上有大量黏液，鱼体与实物摩擦，形成擦伤发炎，坏死脱落，有时呼吸困难而死。

1.3.4　三代虫病

病因与症状：主要是投喂不干净或未消毒的水蚤造成的。患病鱼苗嘴部四周、鳃充血，体表有白色黏液，鱼体失去原有的光泽，游泳姿势不正常，并伴有缺氧浮头现象。

1.3.5　拟马颈鱼虱病

病因与症状：拟马颈鱼虱寄生在鱼鳍基部、肛门、鳃弓、口腔、鼻腔、咽部、食管等，尤以鳃弓、口腔等部位为最常见。

治疗方法：一般采用人工拨取虫体的方法、涂抗菌素软膏等或用 50 g/L 食盐水浸浴鱼体 1～2 h 有较好的疗效。

1.3.6　鲟锥虫病

病因与症状：病鱼行动迟缓，身体在水中呈 S 形或 L 形弯曲，常卧水底，不摄食，体呈黑色，无光泽。病鱼有时在水中上下急剧旋转，若不治疗，3～5 d 后死亡。病原体为锥虫。

2　非生物因素引起的[10]

2.1　营养性疾病

2.1.1　肝性脑病

病因与症状：患病鲟幼鱼体色、体表正常，偶有头部前端和吻端的腹面表皮脱落，背面粉红，患病初期有跳跃、乱窜等极度兴奋行为，离群独游，几乎不进食，后期处于昏迷状态，不摄食，然后就开始死亡。解剖时发现其肝脏呈紫色或者灰色，肝糜烂，胆囊正常，肠内无食。另外，还发现脑组织坏死、糜烂。

治疗方法：饲喂人工养殖的水蚯蚓，避免在饲料中添加药物，对疾病预防有很好的效果。

2.1.2　萎瘪病

病因与症状：病因主要是养殖密度过高，鱼摄食不够，或饵料成分不合理。在高密度养殖和大水面养殖中容易导致此类疾病，尤其是在越冬期间，投喂少，更容易大量发生此种情况。病鱼消瘦发黑，呈现出头大身细，病鱼游动缓慢，鳃丝发白，最终衰竭而死。

防治方法：加强饲养管理，控制养殖密度，投喂营养全面的饵料。捞出此类病鱼单独饲养，用添加促进消化和诱食剂的饵料进行喂养。

2.1.3　脂肪肝病

病因与症状：病鱼食欲不振，生长缓慢，抗病力下降。解剖后可发现肝、胰脏有大量脂肪淤积，肝、胰脏器官肿大，颜色不正常，触摸有较重油腻感。主要是饲料中的脂肪含量过高或营养搭配不合理造成的，还有就是可能养殖密度过大，养殖长期处于低氧状态。

治疗方法：以预防为主，搭配好饲料营养成分，定期添加维生素 C、E，加强鱼体脂肪代谢，提高抗病力。

2.2　由环境引起的疾病

气泡病病因与症状：多是因为水中氮气或氧气过饱和，使得鱼的体内形成微气泡，然后汇聚成大气泡。病鱼游动能力下降、上浮、贴边。解剖肉眼可见肠内有许多小气泡。镜检鳃发白、鳃丝上黏液增加，有许多小气泡。肠内有黄色黏液和小气泡，外观无其他病症。

治疗方法：主要是先改善水质，降低过饱和气体含量，再将病鱼放入 1% 食盐水中浸泡，消除鱼体内的微气泡，或直接将病鱼放入低温冷水中，饲养 1 周左右的时间，鱼体即可康复。

3　其他疾病

鲟鱼性腺溃烂病。病因与症状：发病鲟鱼多出现生殖孔红肿，继而生殖孔附近出现一条细小血丝，此时内部性腺已经开始溃烂，再后是生殖孔附近出现小孔，颜色发生变化，最后生殖孔烂开，性腺脱落出体外，同时生殖孔有粉色液体流出，直至病鱼死亡。病鱼解剖观察发现性腺从生殖孔开始布满小斑点，呈紫红色。发病鲟鱼集中在西伯利亚鲟和史氏鲟，发病规格目前发现最小是 2～3 龄。

治疗方法：目前尚未明确病原及发病机理，没有找到合适的治疗方法，一旦生殖孔附近溃烂出现小孔，死亡率将达到

80%左右，在此之前，注射水产用青霉素钠有一定的治疗效果。

4　小结

　　鲟鱼养殖经过这 10 多年的发展已经初具规模，发展速度也在不断加快。养殖规模不断扩大的同时，养殖环境也发生了巨大变化，高密度养殖使得养殖水体水质恶化，发生各类疾病，病原种类多，病因复杂，但是疾病防控却远远落后于其发展。因此鲟鱼疾病逐渐成为鲟鱼养殖的制约因素之一，迫切需要解决。另外针对养殖过程出现的疾病问题还是药物治疗为主，面对疾病，养殖户依赖于药物治疗，产品品质得不到保证，这也成了鲟鱼养殖的制约因素之一。如此形势下的鲟鱼养殖该怎样走出困境，笔者建议从以下三个方面入手：①在对鱼病病原、病因、病理、药物及防治方法深入研究的基础上，探索疾病发生的环境因素，总结出疾病发病机制，在疾病暴发前切断发病途径，使疾病得到有效的预防；②日常管理注重水质调节，保持良好的水质，调节水质应选用以微生态制剂为主的产品，不会对养殖鲟鱼造成太大刺激，也不会对养殖环境造成破坏；③同时政府应行使其职能加强行业监管力度，规范鲟鱼疾病防治技术，防止传染性疾病蔓延。

参考文献

[1]　崔禾，何建湘，郑维中. 我国鲟鱼产业现状分析及发展建议［J］. 中国水产，2006（6）：8-9；2006（7）：14-15.
[2]　何为，陈慧. 我国人工养殖鲟鱼的疾病防治［J］. 中国水产，2000（7）：26-27.
[3]　杨治国. 鲟鱼嗜水气单胞菌的分离及鉴定［J］. 淡水渔业，2001，31（5）：40-41.
[4]　孟彦，肖汉兵，张林，等. 施氏鲟出血性败血症病原菌的分离和鉴定［J］. 华中农业大学学报，2007，26（6）：822-826.
[5]　李圆圆，曹海鹏，何珊，等. 鲟鱼致病性嗜水气单胞菌 X1 的分离鉴定与药敏特性研究［J］. 微生物学通报，2008，35（8）：1186
　　　-1191.
[6]　曹海鹏，杨先乐，高鹏，等. 鲟细菌性败血综合征致病菌的初步研究［J］. 淡水渔业，2007，37（2）：53-56.
[7]　潘连德. 养殖鲟鱼非寄生性疾病的诊断与控制［J］. 淡水渔业，2000，30（6）：36-38.
[8]　王有基，胡梦红. 人工养殖鲟鱼的疾病与防治［J］. 北京农业，2004（12）：26-27.
[9]　王世权. 鲟鱼常见病的治疗［J］. 中国水产，1999（10）：32-33.
[10]　　朱欣，王存国. 鲟鱼常见病害的防治［J］. 中国水产，2002（4）：84-85.

该文发表于《中国水产》【1995，1：28】

高锰酸钾在水产病害防治中的应用

杨先乐

（长江水产研究所）

　　高锰酸钾是一种常规水产药物，广泛应用于鱼病防治之中。但近年来，由于新鱼药的问世，它在水产病害防治中的地位与作用已日渐低下，甚至被有些渔民朋友所遗忘。在各种新药蜂涌而至的今天，有必要总结近年来国内外应用高锰酸钾防治水产病害的成果，以提醒人们常规水产药物仍是水产病害防治中的一个不可忽视的品种。

一、高锰酸钾在水产上的作用及其机理

　　1. 杀菌、杀虫。高锰酸钾是一种强氧化剂，与有机物相遇即释放氧，并与之相结合，使病原菌（体）变性而失去感染性。

　　2. 收敛。在低浓度时，高锰酸钾还原后可形成一种二氧化锰与蛋白结合的蛋白盐类复合物，促使创伤愈合。

　　3. 解毒。对有机毒物，高锰酸钾能使之氧化而达到解毒效果。

　　4. 改良水质。在有机耗氧量高的池塘，施用高锰酸钾，可使有机物氧化降解，有机耗氧量减少，从而使池塘中溶氧消耗减少，水质改良。

二、高锰酸钾防治水产病害的优点

　　应用高锰酸钾防治水产病害，具有以下优越性：①对病原体无选择性杀菌、杀虫谱广；②不会产生耐药性；③对水体无污染；④不会形成药物的积累，影响食用鱼的品质；⑤使用方便；⑥来源广，价格较便宜。

三、高锰酸钾防治水产病害的范围及其使用方法

　　高锰酸钾一般采用浸浴的方法给药，但也有泼洒给药的方式。

1. 防治水生动物体外寄生虫，如鲺、鳋、指环虫、三代虫、嗜子宫线虫及不形成孢囊的原虫等。防治鲺病可用100 ppm的高锰酸钾浸浴病鱼5~10 min；防治鳋病可用2 ppm的高锰酸钾遍洒或以25 ppm的高锰酸钾浸浴15分钟，但该剂量很难杀灭雌虫和尚未在鱼体上寄生的幼虫，因此间隔2~3 d后应再用药一次；对于单殖吸虫和体外寄生原虫，可用4 ppm进行泼洒杀灭。也可用20 ppm浸浴15~30 min。

2. 防治污损性细菌所引起的疾病，如由屈挠菌所引起的细菌性鳃病，由杆状类细菌而导致的冷水性鱼类疾病等。使用方法是以4 ppm的浓度长时间的浸浴（或泼洒）防治鱼类鳃病；防治虾类鳃病所使用的浓度是5 ppm。

3. 防治真菌的感染，如水霉、鳃霉、毛霉所引起的水生动物的真菌病，使用的方法和剂量与防治体外寄生虫的方法基本相同。

4. 改善池塘水质。如果池塘有机物含量高，有机耗氧量大，可用2~3 ppm的高锰酸钾全池泼洒，可达到减少氧耗、净水的作用。

四、使用高锰酸钾的，注意事项

1. 在有机物含量高的污染水体中泼洒高锰酸钾时，水中的有机物能立即使其分解。因此在这种水体中应用该药时，应根据水质情况适当增加其用量。

2. 水温与高锰酸钾的氧化还原作用密切相关，水温高时，应适当降低其使用浓度；反之，应予增加。

3. 高锰酸钾反应后的最终产物二氧化锰在碱性条件下能形成沉淀，对鱼类等水生动物产生毒性。因此该药适宜在中性或微酸性的条件下使用；用药后，最好能予换水，以除去二氧化锰的毒性，特别是水由红色变为褐色时。

4. 高浓度泼洒高锰酸钾时，会大量杀死池塘中各种生物体，其尸体分解时将消耗池塘中的溶氧而造成鱼池缺氧。在这种情况下，应采取增氧措施。

5. 高锰酸钾对鱼类的致死浓度，随鱼的种类而有别，一般为22~62 ppm。超过这个浓度使用时，应严格控制药浴时间。

该文发表于《渔业现代化》【2003，3：20-21】

海水观赏鱼在运输与养殖过程中的用药

陈少鹏[1]，杨先乐[1]，黄艳平[1]，梁程超[2]

(1. 上海水产大学农业部渔业动植物病原库；2. 北京金龙牧海渔业技术发展有限公司)

近年来，各地兴起了海水观赏鱼的饲养热。但由于海水观赏鱼自身生活环境、生活习性及其捕捞、运输过程的特殊性，其饲养具有很大难度，养殖成活率较低。本文从用药着手，结合海水观赏鱼的捕捞、运输及养殖过程的特点，分析具体环境中的用药要求。

1 引种运输过程与鱼病的产生

海水观赏鱼的引种运输过程较为复杂，不仅是因为其至少包括7~8个过程，而且这么多过程必须在15~20 d完成，这对娇嫩珍贵的海水观赏鱼绝对是一场生死劫难。许多鱼都由于此阶段的处理不当而死亡，或是留下鱼病隐患（鱼病大多数与此有关），为以后养殖带来麻烦。

海水观赏鱼引种运输大致过程：

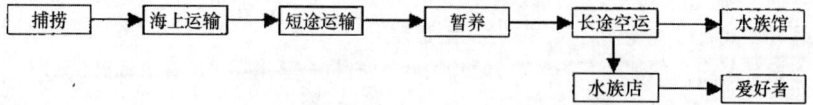

捕捞、运输和暂养过程在观赏鱼生产中占据很重要的地位，对于鱼病防治非常重要。这一阶段的特点是高密度，高密度产生的压迫感，使鱼群躁暴，惊恐不安，直接导致了鱼体表大量的外伤，并引起继发性感染。而处于应激状态的鱼群会大量分泌粘液，致使水体理化性质快速改变，pH迅速下降，氨氮急剧上升，引起鱼体不适，自身免疫力下降，同时也为有害微生物的繁衍创造了条件。因此，在这一阶段就应注意鱼病的防治。

2 用药特点

(1) 海水观赏鱼为观赏性非食用鱼类，一般不考虑药物残留问题，因此用药范围广，在考虑不污染水环境的前提下，大部分化学药品均可使用。

(2) 观赏鱼在捕捞运输中经过多次运输和暂养的过程，导致观赏鱼的抵抗力下降，病原体大量滋生，在其耐受范围内，消毒药的使用量较大。

(3) 生态缸中一般有多种生物因子，它们对药物具有不同的敏感性，在用药时应加以考虑。如水生植物（海草、海藻等）、腔肠动物（珊瑚、海葵等）、贝类（宝贝、日月贝等）、软体动物（海兔、海麒麟等）对大多数药物较敏感。正常治疗鱼病的药物剂量，往往会导致这几类生物大量死亡。

(4) 体色对于海水观赏鱼十分重要。因此，在对其用药时，应注意保持体色不受破坏。而长期单一使用较大剂量的强消毒药物，如高锰酸钾，氯制剂等，会使观赏鱼体色变淡，影响观赏性。

(5) 海水观赏鱼对温度、盐度、酸碱、溶氧及氨氮等的耐受性均较小，因此用药时应认真考虑这些理化因子同药物相互作用后所引起的变化，保障观赏鱼的安全。

3 用药原则

3.1 根据生活习性和环境准确用药

根据海水观赏鱼用药的特殊性，合理准确用药。如运输中理化因子不稳定，鱼的高密度导致水中 CO_2 增多、pH 下降，会对一些药物的作用产生影响。暂养池在用药过程中一般不停止过滤循环系统的使用，因此，要考虑过滤装置对药物的吸收、富集作用，避免再次用药导致观赏鱼的死亡。

3.2 慎用少用专用商品药制剂

观赏鱼店所卖的商品药制剂多是针对淡水观赏鱼所用，没有考虑海水观赏鱼的特殊性，缺乏海水中理化因子对药物影响的资料，所以其药效一般较差，有时还会产生副作用。如有些药对腔肠动物有极大的毒害作用。

3.3 全面考虑选择适宜药物

观赏鱼缸一般养有多种鱼类和水生生物，要考虑不同鱼和生物对药物毒性承受能力的不一致性，因此用药时对不同生物要分别对待。如狮子鱼对硫酸铜十分敏感，石头鱼对各种净水剂敏感。

3.4 确定合理的药饵给药位置

水族箱中光线的强弱分布一般错落有致，而不同的鱼对光的需求各不相同。如雀鲷科鱼一般喜光，因此投放药饵一般在强光区；而龙虾怕光，所以一般投放药饵于暗处。

3.5 注意药物的配伍禁忌

抗菌药和有益菌制剂的同时使用会导致有益菌制剂失效；多种药物的同时使用会造成活性碳等过滤装置失效；pH 调节剂和净水剂同时使用会同时影响两种药物的效果。

4 生产及饲养过程中的用药

4.1 运输

为减少外伤，运输中注意将不同类型的生物分舱分箱运输，可将体表粗壮、游泳迅速、凶猛的鱼类和体表柔软、游泳缓慢的鱼类分开。海上运为开放式活水船运输，始终能保持有清洁的海水，一般不用药。短途水车运输和长途空运均为封闭型运输，水体不能更新，且观赏鱼在长达 20 d 的捕捞和运输中一直处于应激态，鱼体免疫力下降。因此，麻醉药和抗菌药的使用尤显必要。麻醉药可降低鱼体的应激反应，减小粘液的分泌，同时减少鱼群的冲撞，避免外伤的产生。实验证明在适宜的水温下使用安定片可保证鱼在长途空运过程中的成活率达 98% 以上。抗菌药可抑制致病菌的滋生，减少伤口感染。一般用青霉素 20 mg/L 海水，配合麻醉剂的使用有很好的效果。

4.2 接鱼

经运输的鱼一般带有外伤，或携带有病菌、寄生虫等病原体。为减少外源性病原体带入到新的水体，同时为了防止伤口发生继发性感染，接货过程要求较为严格，一般要经过二至三道工序。首先，是水温和 pH 的调整；其次，淡水浴，方法是以淡水∶海水为 9∶1 的比例对鱼消毒 5~10 min；第三，若暂养池中的鱼短时间即恢复正常，则可用消毒剂进行二次消毒。

4.3 暂养

海水观赏鱼的暂养是运输过程的必要环节之一。暂养的观赏鱼在几经波折后体质较虚弱，因此在水池中高密度的暂养，极易发生暴发性流行病，并为以后鱼缸中养殖留下多种疾病隐患。在暂养中白点病、水绵病等最为常见。此阶段常用消毒剂是碘伏或硫酸铜、硫酸亚铁合剂，用量为碘伏 10 mg/L 海水，硫酸铜合剂 0.8 mg/L 海水。

4.4 净缸

缸体及缸中造景物在使用前必须先消毒，可使用高锰酸钾或氯制剂。海水消毒可用福尔马林，用于杀死小型甲壳类等；或用硫酸铜来净化天然海水。消毒过的缸体、造景、海水不能立即投入使用，必须经曝晒、曝气、试水等步骤后方可放鱼。

4.5 养殖

4.5.1 鱼病的治疗

观赏鱼缸的用药应尽量避免使用泼洒方法。一般来说，小型缸可将病鱼取出进行单独浸浴治疗。大型多生物造景缸中的鱼较难取出，在病情较轻的情况下，可采用药饵投喂的方法，如用盐酸四环素 50 mg/kg 鱼重制成药饵可防治弧菌、嗜水气单胞菌病；在病情较重的情况下可采用停止过滤，降低水位浸泡的方法，如降低水位全池泼洒 2~3 mg/L 海水的亚甲基兰治疗水霉病，但用此方法前应移去缸中所有对药物敏感的生物，且尽量选用在水中分解较快、残留较小的药品，使用后应尽量换去含药海水。对于水族馆等有条件的养殖单位，建议使用生物胶囊法。

4.5.2 有益菌制剂的使用

水质是海水观赏鱼养殖成功与否的关键。利用有益菌群抑制致病菌的繁衍是一种简便而有效的办法。因此，海水观赏鱼箱（缸）必备一套完善的过滤循环系统，为有益菌群的形成创造必要的条件。其使用过程必须注意以下几点：①有益菌制剂不可同抗生素等杀菌剂同时使用；②有益菌制剂加入后，几天内不能换水，在施用该制剂时，应注意水体中的有益菌浓度；③有益菌使用时应关闭缸顶的紫外杀菌灯。

5 小结

观赏鱼一般在进入水族箱后一周内较难养，疾病较多，这一时期很关键。只要在这一周为其营造较适的生活环境，根据上述用药原则合理用药，隔离外来病原生物，及时发现、治疗病鱼，恢复其体质，就会很顺利地养殖海水观赏鱼。

该文发表于《科学养鱼》【2014，11：54-55】

聚六亚甲基胍防治水产养殖"疑难杂症"试验研究

孙琪[1,2]；黄保华[1]；杨先乐[2]

（1. 上海黛龙生物工程科技有限公司，201412；2. 上海海洋大学，201306）

聚六亚甲基胍（*Polyhexamethyleneguanide*，PHMG）是进口原材料制成的新型多用途高分子聚合物，性能稳定，作用效果持久。其无毒、无味、无腐蚀、无刺激、无挥发，特别是无氯、碘、重金属等有害物质的残留，能确保鱼虾等水产品出口的质量要求。上海海洋大学国家水生动物病原库实验室已通过体外抑菌试验测定聚六亚甲基胍对淡水鱼类暴发病的嗜水气单胞菌、鳗鲡的细菌性疾病的致病鳗弧菌、对虾细菌性疾病的溶藻弧菌和鳗爱德华氏病的迟钝爱德华氏菌等四种致病菌的最小抑菌浓度（MIC），并与二溴海因、碘伏的杀菌效果作对照，结果表明聚六亚甲基胍比后二者抑菌效果高出数倍甚至数十倍。

当今水产养殖行业"疑难杂症"突出，市场上的药物鱼龙混杂、质量参差不齐，且一些传统药物往往因为其具有一定的刺激性而致使一些水产动物在发病时死亡量增加，同时，这些药物因多年长期使用，很多病原已对其产生了耐药性，进而最终导致现在对某些疾病无有效的防治手段或方法。聚六亚甲基胍是一种新兴且高效的杀菌物质，目前未在水产养殖业广泛应用。本试验室体外试验研究发现其对水产养殖中常见的一些病原有较好的抑杀作用，笔者通过跟踪走访调查，对其在水产养殖中一些棘手疾病试用试验效果进行了记录研究。

一、黄鳝"鳃出血"

湖北省荆州市公安县较大规模养殖黄鳝，当地养殖的黄鳝有一难治之症，其症状主要为体表有血点、血斑，肛门红肿、外翻，症状严重的黄鳝整个体表充血，鳃盖下出血，当地人把之称为黄鳝"鳃出血"。

养殖户曾先生有近百亩黄鳝养殖塘，从2013年冬天开始，其中一个塘口前半部分网箱中黄鳝陆续发生该病，用过当地常用消毒剂——聚维酮碘、戊二醛，同时内服过抗生素均无效果，最终大部分黄鳝发病死亡。而后，该病逐渐蔓延至另一半网箱，黄鳝开始陆续发病，不断有黄鳝死亡。该塘口整体水质较差，与旁边未发病塘口相比，水体颜色发黑，塘中四大家鱼每天也有死亡。

将病鱼随机平均置于含60升塘口水的水箱中，每组10尾，分别加入一定量的由上海某公司提供的聚六亚甲基胍（25%），使其最终浓度分别为0.5毫克/升、1毫克/升，塘口水为对照组，平行试验做2组。观察发现，1毫克/升组病鱼症状明显好转，逐渐治愈，保护率达90%以上；0.5毫克/升时，治疗效果相对较差，而对照组病鱼3天内全部死亡。

二、南美白对虾"倒芒症"

南美白对虾是当今世界养殖产量最高的三大虾类之一。目前广东、广西、福建、海南、浙江、江苏等地已逐步推广养殖。近年来，在江浙等地发现，养殖过程中虾会出现触角倒置、虾体无活力而逐渐死亡的现象，当地人把这种症状称之为"倒芒症"。南美白对虾的这一病症，传统消毒剂不能起到很好的防治效果，反而会刺激病虾的死亡量增加。

6—7月份，在江浙地区发现多例南美白对虾"倒芒症"。将病虾随机平均置于含30升塘口水的水箱中，每组10尾，分别加入一定量的聚六亚甲基胍（25%），使其最终浓度为0.5毫克/升，48小时后再加入一次0.5毫克/升的聚六亚甲基胍，塘口水为对照组，平行试验做2组。3天后发现：用药组病虾活力逐渐恢复，最终治愈，对照组病虾平均死亡率达85%以上。

三、异育银鲫"大红鳃"

大红鳃也是养殖异育银鲫过程中的常见病及难治病，发病的鲫鱼出现鳃部水肿，鳃丝呈"梳齿"状，有时伴随"花鳃"的症状，整个鳃呈西瓜红色，所以养殖户都称之为大红鳃，镜检可发现鳃充血严重，伴随有烂鳃及肝胆综合征，如若不及时治疗或可增加发生鳃出血病的概率。在治疗此病时，部分养殖户采用二硫氰基甲烷，而因二硫氰基甲烷自身毒性及刺激性较大，会导致病鱼死亡量激增。可有的养殖户无奈地称只能接受这种效果，因为其他消毒剂效果不行，只能通过毒性相对较大的药物将病鱼杀死，从而控制该病的进一步发展。

5月中旬，大丰斗龙港一养殖户张老板鱼塘刚进鱼种之后，发生大红鳃病。采用同样的试验方法，在60升水箱中配置1毫克/升的聚六亚甲基胍，试验发现3天后对照组病鱼逐渐死亡，镜检发现鳃丝充血且溃烂严重，解剖发现，腹水严重，内脏出血，更有甚者肝脏"完全溶化"；用药组病鱼康复，大红鳃的症状基本消失，鱼体相对正常。

四、异育银鲫"鳃出血"

苏北是我国淡水养殖的大区，养殖面积可达几百万亩，主要养殖对象为异育银鲫等，近年来异育银鲫的鳃出血病一直是养殖户最头痛的事情，被称为"不治之症"，且不能随意用药，一些传统消毒剂由于本身有一定的刺激性，在发生鳃出血时用这些药物反而会刺激鱼死亡量激增，该病给当地水产养殖业造成了巨大的经济损失。其主要症状为病鱼鳃部出血、身体泛红，内脏充血或出血。发病时水温一般在18~25℃，若不在该温度区间时也有可能发病。今年该病来势凶猛，4~7月，在大丰、东台、射阳等地区均有该病发生，发病鱼规格不等。

每个发病塘口均采用相同的试验方法，在60升水箱每次加入1.5毫克/升的聚六亚甲基胍处理病鱼，与对照组及不同发病塘口比较发现：聚六亚甲基胍对异育银鲫鳃出血病有一定的治疗效果，且该效果与发病时间有一定关系，病程越短，治疗效果越好，且用药量小；病程较长时，用了3次1.5毫克/升的聚六亚甲基胍可能都没法取得理想的效果。

当前研究已确定，该病由鲤疱疹病毒II型（*Cyprinidherpesvirus* II，CyHV-2）引起，属于病毒性疾病。纵观人、畜、禽、渔，这些行业中病毒性疾病最为棘手，即使最进步的人类医疗事业对人类一些病毒性疾病也束手无策，且人类病毒性疾病多以注射疫苗的形式预防为主。所以养殖户应该有此意识，病毒性疾病要靠预防，而不能完全指望依靠某一种药物的治疗效果。聚六亚甲基胍在众多药物中脱颖而出，对异育银鲫鳃出血病有一定的治疗效果，说明该药对引起鳃出血的病毒有一定的抑杀作用，此药物可用于预防鳃出血病的发生。

五、总结

聚六亚甲基胍对水产养殖水体中的有害微生物有很强的抑杀作用，能有效抑制水中细菌、病毒、真菌的滋生和繁衍，防止鱼、虾等水生动物因养殖水体中存在的细菌、病毒等病原体感染，能有效促进水产动物健康和快速生长。

如今水产养殖业种质差、鱼体弱、怪病多、药物杂，致使发病水产动物不能受药物刺激，一些疾病无药可治，养殖户对水产养殖业日渐失去信心。较传统消毒剂，聚六亚甲基胍具有无毒、无刺激、性质稳定等优点，应有良好的推广应用前景。面对尚无有效治疗药物且给水产养殖造成重创与巨大经济损失的各类"疑难杂症"，聚六亚甲基胍的治疗效果算是异军突起，可以给对养殖逐渐失去信心的养殖者带来希望和信心，对推进水产养殖病害防控和科学用药具有一定的指导意义。

该文发表于《科学养鱼》【2015，6：66】

罗非鱼链球菌病诊断及治疗方法

胡利静[1]；肖艳翼[1]；胡鲲[1]；杨先乐[1]；王雪琴[2]

(1. 上海海洋大学；2. 河北新世纪药业有限公司)

我国是养殖罗非鱼的主要国家，主要集中在广东、广西、福建及海南等省份，罗非鱼产业在全球占据了重要地位。但近几年来，罗非鱼的病害已趋向严重，尤其是罗非鱼的链球菌病，发病率、死亡率高，不仅给从事罗非鱼养殖者带来巨大的经济损失，并且严重制约着罗非鱼产业的持续发展。对一线养殖者来说，如何简单、快速、准确地判断罗非鱼链球菌病，是他们渴望掌握的一项技术。因此，了解链球菌的生长条件及掌握快速诊断链球菌病的方法对从事罗非鱼养殖者来说就显得尤为重要。现结合广东某养殖场罗非鱼链球菌病的诊断及治疗来进行简单的阐述。

一、病原

无乳链球菌为革兰氏阳性菌，拥有 B 群特异性抗原。水温 20~37℃、盐度 0、pH7.6 是链球菌的最佳生长条件。2014年 8—10 月广东省正值高温季节，该养殖场罗非鱼养殖密度较大，水质恶化，链球菌病暴发。

二、诊断方法

罗非鱼发病后，可以根据鱼的病症快速做出判断。罗非鱼链球菌病可分为急性型和慢性型，急性型发病鱼死亡前鱼体未表现出明显的症状，慢性型发病鱼可出现典型症状及病变。

1. 病鱼在鱼塘中的典型症状

病情较轻时，罗非鱼病鱼在池塘表面慢游、转圈，并逐渐失去平衡，有的病鱼在水中翻滚甚至跳跃；病情严重时，罗非鱼病鱼在水中打转挣扎，有的间或窜游。

2. 病鱼体表症状

眼球突出（明显症状），或在水中眼球不突出、离水后很快突出，严重时眼睛浑浊发白，甚至眼球脱落；眼围、鳃盖、鳍基部充血；腹部具有点状或斑点性出血。

3. 病鱼体内症状

鱼腹部有积水；肠道充血，肠管变粗，肠壁变薄；肝胆肿大，肝颜色变白；肝脾充血出血等。

4. 试验室病原检查

可以通过革兰氏染色法对病原菌进一步确认；还可以通过细菌的分离鉴定或免疫学检测方法来确认病原菌的种类。

三、治疗方法

链球菌对罗非鱼危害严重，首先要用药杀灭链球菌。对该广东发病养殖场用土霉素、四环素拌料投喂，每千克饲料拌药 1~2 克，严重时考虑到投饵量较正常减少，所以用药量有所增加，每千克饲料拌药 3~4 克。

其次，链球菌能产生外毒素，易导致罗非鱼器官功能紊乱。所以在治疗罗非鱼链球菌病时不仅要杀灭病原链球菌，还要考虑用药物消除外毒素或通过提高罗非鱼本身消解毒素的能力。对广东发病池塘采用投喂治肝的药：肝胆利康散，使用剂量为 1 包拌 40 千克（治疗）或 20 千克（预防）饲料；营养快线，使用剂量为 1 袋（500 克）拌 80 千克（治疗）或 40千克（预防）饲料。

最后还要考虑到鱼类生活的环境，并对水环境进行消毒，采用二氧化氯，使用剂量为（3~5）亩每米水深用500克，使用时注意勿使用金属桶盛装。

四、总结

1. 改善水体环境链球菌常存在于富营养化水体中或底泥中，水质不良、水温较高是该病的主要诱因。所以放养密度要合理，在养殖过程中要注意水质变化，尤其在高温季节要及时换水并进行底改，确保养殖水质符合要求。

2. 注重预防鱼类疾病以预防为主，链球菌为条件致病菌，在水体中很常见，鱼体较弱或鱼受伤时易感染链球菌，因此在日常管理中应保证罗非鱼的体质强壮，例如投喂配合饲料时拌复合维生素。

3. 养殖户心理战每位养殖者都希望养殖过程中投入少、回报多，在养殖过程中，尤其在预防与治疗用药方面，都希望药物既便宜又有效，减少药物方面的成本投入。在罗非鱼养殖过程中，有些养殖户为了节省成本，认为没有必要投喂复合维生素，他们觉得投喂与不投喂，鱼该得病时照样得。罗非鱼疾病的预防这一环节未做好，使罗非鱼发病机会增加。在链球菌病治疗过程中，养殖户没有认识到使用保肝护胆药物或环境解毒药物的重要性，往往只选择杀菌及消毒的药物，虽然也能控制病情，但停药后容易复发或后期因中毒而引发罗非鱼大量死亡，造成更严重的损失。这样的做法看似节约了药物成本，而实际上却让养殖户承担了更大的经济损失，同时也让养殖户对药物质量产生了怀疑。其中孰轻孰重，希望每位养殖户能客观理性看待。

该文发表于《水产科技情报》【2012，40（3）：156-177，162】

鱼虾混养塘锚头鳋病防治方法

章海鑫[1]；阮记明[2]；杨先乐[1]；许佳露[1]；刘攀[1]

（1. 上海海洋大学国家水生动物病原库，上海 201306；2. 江西农业大学动科院水产系，南昌 430223）

摘要：江苏东台市沿海地区为异育银鲫主要养殖地区，以异育银鲫与南美白对虾混养的模式为主。近年来，由于防虫药物使用受到一定限制以及使用的单一，导致异育银鲫锚头鳋病越来越严重。文章介绍了锚头鳋病的病原和流行情况，分析了该病在当地暴发的原因，介绍了锚头鳋病传统的防治方法，并通过实地考察锚头鳋病的防治情况，总结出了一些防治鱼虾混养塘锚头鳋病的新方法，推广使用后取得了良好的临床效果。

关键词：异育银鲫；南美白对虾；混养；锚头鳋病；防治

锚头鳋属于寄生桡足类，锚头鳋病为鱼类主要寄生性病害之一，对鱼类具有很大的危害性。锚头鳋主要有多态锚头鳋、鲤鱼锚头鳋、草鱼锚头鳋和四球锚头鳋等，寄主包括鲢、鳙、草鱼、鲤、鲫、乌鳢、泥鳅、鳜鱼、金鱼等[1]。锚头鳋在当年的幼鱼或成鱼上都能寄生，以头部钻入寄主的肌肉或者鳞片下（也有四球锚头鳋寄生于草鱼鳃弓上的报导），胸腹裸露于寄体外。在锚头鳋寄生的部位，肉眼可见针状的病原体[2]。

江苏省东台市沿海是我国异育银鲫的主养地区，养殖模式以异育银鲫与南美白对虾混养为主。虽然此种混养模式大大减少了异育银鲫及南美白对虾细菌性和病毒性疾病的发生[3]，但由于药物使用受到一定的限制，导致寄生虫疾病的危害越来越严重，锚头鳋病就是其中之一。特别是在春末与整个夏季，锚头鳋病对该地区养殖业的危害很大，且防治方法匮乏。本文对东台市沿海地区锚头鳋病害的流行情况和防治方法总结如下。

1. 锚头鳋病的流行情况

锚头鳋具有广温性，在2~33℃水温下均可繁殖，20~25℃为主要发病季节[4]。其次，锚头鳋具有寄主选择性，如多态锚头鳋寄生在鳙、鲢体表，草鱼锚头鳋寄生在草鱼体表，鲤锚头鳋寄生在鲤、鲫、鳙、鲢、乌鳢、青鱼等鱼种的体表、鳍和眼睛。东台市沿海地区混养塘内锚头鳋的种类以鲤锚头鳋为主，也有少量多态锚头鳋。锚头鳋雌性成虫营永久性寄生生活，虫卵孵化后3~4 d出现第一桡足幼体，经过6个桡足幼体期，13~21 d内完成从卵到成虫的生活周期；锚头鳋病的病程长，可以连续寄生；能够引起鱼种的死亡和影响成鱼的生长[4-5]。

鱼虾混养池锚头鳋病的发生时间为6—10月，8—9月为发病高峰期，锚头鳋主要寄生在银鲫的鳃、皮肤、鳍、眼、口腔等处，病鱼表现为急躁不安，食欲减退，继而逐渐消瘦，游动迟缓，甚至出现死亡。锚头鳋对银鲫鱼种的危害最大，当有4~5个锚头鳋寄生在鱼种上时即可引起死亡。成鱼寄生锚头鳋时，虽然不会直接导致死亡，但易引起细菌和病毒继发感染而致使组织发炎，进而影响摄食、生长和繁殖，严重时也会致鱼死亡。

2. 锚头鳋病的中草药和化学药物防治方法

对于锚头鳋病的危害很早就引起了关注，笔者也总结出了很多有效的防治方法，主要集中在中草药和化学药物防治两个方面。

中草药防治方法有[6]：（1）每亩（15亩＝1 hm²，下同）水面用苦楝树根 6 kg、桑叶 10 kg、麻饼或豆饼 11 kg、石菖 22 kg，研碎捣汁全池泼洒；（2）每亩水面用鲜五加 100 kg，分成小束，捆扎于竹杆上，使茎叶浸在水里，露出枝头，浸泡 10 d；（3）每亩水面投喂 150 kg 酒糟；（4）每亩水面用松树叶 15 kg，捣碎擂汁后全池泼洒等。

化学药物防治方法有：（1）用 90% 晶体敌百虫 1~4 mg/L 药浴杀死锚头鳋，每隔 7~10 d 药浴一次，经两次连续药浴治疗，能够较好地杀死猫头鳋成虫和幼虫[7]；泼洒敌百虫 150~350 g/亩，每 2 周使用一次，可以有效地防治锚头鳋病[8]。（2）水温在 15~20℃时，用浓度为 20 mg/L 的高锰酸钾溶液药浴 1~2 h；水温在 21~30℃时，用浓度为 10 mg/L 的高锰酸钾溶液药浴 1~2 h。（3）用 0.4% 的伊维菌素，以 25~30 mL/亩全池泼洒；或用 1% 阿维菌素 30 mL/亩，全池泼洒。（4）20 kg/亩全池泼洒生石灰，再全池泼洒 90% 敌百虫 133~266 g/亩[9]。

3. 鱼虾混养塘锚头鳋病的防治

东台市沿海地区的淡水养殖池塘以异育银鲫与南美白对虾混养为主，南美白对虾和锚头鳋同属于甲壳动物，它们对于药物的敏感性较为一致，而上述部分防治药物对南美白对虾敏感。因此，很多常规的防治药物都不能用来防治异育银鲫与南美白对虾混养塘内的锚头鳋病；另外，中草药对锚头鳋病的防治效果还有待进一步的观察。目前，东台市沿海地区防治锚头鳋的方法还是以使用伊维菌素和阿维菌素这两种药物为主，但由于使用的药物过于单一以及病害日益严重，导致这两种药物的使用频率过高，药量不断加大，锚头鳋已对之产生了抗药性，从而降低了防治效果。针对这种情况，笔者根据东台市沿海地区养殖的实际情况，结合当地养殖户的用药习惯，总结出了一些治疗锚头鳋病的方法，临床效果较为理想。

3.1 预防

无病先防，做好防病工作是保障养殖成功的必要条件。首先应该消除病源：用生石灰彻底清塘消毒，杀灭池水中的锚头鳋幼虫。鱼种在放养前，用 10~20 mg/L 的高锰酸钾溶液浸浴 15~20 min，保证鱼种不带虫；其次是调节水质，保证池鱼有良好的生活环境：定期用 0.2~0.5 mg/L 的二氧化氯或者 15~20 mg/L 生石灰全池泼洒，以改善和调节水质；通过生物菌群分解池底过多的有机质，可以有效地抑制锚头鳋排卵及其幼虫在无节幼体和桡足幼体阶段蜕皮，从而切断其生活史，最终达到消灭锚头鳋的目的[10]。

3.2 治疗

硫酸亚铁 100~150 g/亩，全池泼洒 2 h 后使用 1% 阿维菌素，用量为 30~40 mL/亩；用药后加注新水 3 d 后鱼体上虫体可减少 80%。用药 3~4 d 后，可以用二氧化氯或者聚维酮碘等消毒一次，以减少伤口感染病菌的机会。由于锚头鳋的生命周期一般为 13~21 d，因此在 2~3 周后再用药一次，能够杀死新变态的幼虫，就可基本达到控制锚头鳋的目的。

2011 年 6~10 月份，笔者在江苏东台市沿海地区的临床实践表明，以上方法对于防治鱼虾混养塘中的锚头鳋病具有较强的实用性，而且疗效非常不错。养殖户照此方法进行防治后，发病率下降了 60%，而治愈率达到 100%。

4. 体会

锚头鳋虫体一年四季均可在鱼体上发现，因此一年四季都有卵囊残留在水体。冬季锚头鳋不能繁殖，但能够过冬，等到春季来临，水温回升后，便会陆续排卵繁殖。当虫体达到一定的数量时会对鱼种造成危害。

目前很多人都使用敌百虫来防治锚头鳋病，但在异育银鲫与南美白对虾的混养塘中，南美白对虾对敌百虫较为敏感，因而不能随意全塘泼洒敌百虫，但可以在鱼种下塘前单独进行药浴。使用硫酸亚铁时不能盲目加量，用药后最好能够开启增氧机 1~2 h，这样既有利于药效的发挥，也能增强南美白对虾的抵抗力。

防治鱼虾混养塘内的锚头鳋病时，要先检查虫体是幼虫、成虫还是老虫。刚寄生到鱼体表的幼虫，状如细毛，白色，无卵囊；成虫身体透明，有韧性，身体后部有 1 对深绿色卵囊，显微镜下可见虫体内肠蠕动有力；老虫身体浑浊不清，变软，没有韧性，不久即会脱落死亡。掌握这些特点，可以更好地把握用药次数，避免盲目用药。

参考文献

[1] 王声瑜. 锚头鳋种类及防治 [J]. 齐鲁渔业，1999 (5)：21.
[2] 鱼类锚头鳋病及其防治方法 [J]. 科学养鱼，1999 (10)：41-42.
[3] 章海鑫，黄建珍，阮记明. 异育银鲫与南美白对虾混养技术要点 [J]. 水产养殖，2011 (11)：43-44.
[4] 孙全林，霍以群，李超美. 成鳗感染锚头鳋病的检疫和治疗 [J]. 动物检疫，1994 (2)：16.

[5]　谢申柱. 黄颡鱼常见病害防控技术 [J]. 农村百事通, 2009 [J]. 水产养殖, 2011 (11): 43-44.

[6]　林嵩海, 钟清莲, 林峰. 中草药防治鱼锚头鳋病 [J]. 中国兽医杂志, 1995 (1): 38.

[7]　姜鹰, 姜礼燔, 徐永福. 新敌百虫对鱼类指环虫病车轮虫病治理机制与毒理学评价 [J]. 内陆水产, 2007 (2): 44-45.

[8]　李南, 方平. 鱼类锚头鳋病的八种治疗方法 [J]. 北京水产, 2001 (2): 24-25.

[9]　谢志扬. 鲮鱼的锚头鳋病治疗 [J]. 鱼类病害研究, 1992, 14 (1): 36.

[10]　李永刚, 湘海燕, 金雪莲. 越冬期后鳜鱼锚头鳋病防治技术 [J]. 中国水产, 2007 (4): 58-59.

3. 清塘、杀（抑）藻类药物的应用

生石灰既使水中增加了钙离子浓度, 又能使被淤泥吸收固定的营养盐类释放出来, 增加水的肥度, 是一种良好的清塘剂（杨先乐等, 1984）。药物和生物防控相结合能较好地控制鲟养殖池的青苔（杨移斌等, 2013）。植物控藻化感物质具有突出的优势, 是一种新型的除藻方法（胡利静, 2015）。治理蓝藻仅使用单一的方法效果往往不理想, 需要多种手段综合防控（赵蕾等, 2013）。

该文发表于《中国水产》【1984, 4: 26】

用什么药物清塘好

杨先乐

清塘是清除敌害, 杀灭病原体, 改良池塘水质环境, 预防鱼病发生的一个有力手段, 也是预期夺得养鱼丰收的第一项技术措施。但是, 有些同志对清塘药物选择不当, 造成清塘不彻底, 收不到除害灭病的效果。

目前, 有些湖区的养殖渔民, 使用五氯酚钠清塘, 就是一例。五氯酚钠虽然也能杀死野杂鱼, 但它既不能彻底清除塘中的病原体, 也不能改良水质, 而对鱼类等水生动物有剧毒, 0.1~0.2 ppm 的浓度就能致死。放养工作中稍有疏忽, 很易使全池鱼类覆灭。五氯酚钠杀灭病原体的效果很低的原因是: ①只能在土表 3~5 分处形成药层。这些停留淤泥表层的药剂, 又常与淤泥中的铝、铁离子形成难溶性的盐类, 药性难于发挥。而大部分鱼类病菌、寄生虫, 潜藏和繁生于淤泥中, 药性渗透不到而杀不了它们; ②五氯酚钠的残效期短。只要经过 3~5 个晴天, 水中药剂就完全分解, 毒性基本消失。因此, 对耐药性稍强的病原体不起作用; ③它是一种触杀型药剂, 对某些病原微生物很难奏效。

我们提倡采用生石灰清塘。这是因为生石灰遇水能形成强碱性以及乳化时产生的高温等, 基本上能杀灭池塘中大部分危害鱼类的病原微生物和寄生虫。它的药效期也长, 能持续 7~8 d 天。它还能增加池水中的碱度和硬度, 改变腐殖质产生的酸性水质, 有利于鱼类生长而不利于病原体繁生。同时, 生石灰既使水中增加了钙离子浓度, 又能使被淤泥吸收固定的营养盐类释放出来, 增加水的肥度。两者, 都直接为水中生物提供了营养。所以, 生石灰是一举数得的清塘好药物。此外, 还有茶饼、漂白粉、氨水等也可用来清塘, 但比较起来, 都不及生石灰清塘效果好。

该文发表于《渔业致富指南》【2013, 3: 61-62】

鲟养殖池青苔危害及防控策略

杨移斌, 夏永涛, 邱军强, 杨先乐

鲟是一种大型淡水经济动物, 由于身体可利用度高, 经济价值高, 特别是鱼卵能制作价格昂贵的鱼子酱, 使得鲟养殖得到迅猛发展。随着养殖量增加、养殖品种日益广泛, 养殖环境变得更加恶劣, 鲟疾病频发, 逐渐成为鲟养殖一大瓶颈。

一、青苔发生的原因

目前鲟养殖主要包括网箱养殖, 陆地水泥池养殖等几种主要养殖方式, 其中陆地水泥池养殖采取的是流水形式养殖模式, 池子为圆形, 水流在池中旋转流动, 可以冲走残饵及其他污物, 使鲟养殖环境保持在一个相对较好的程度, 但是随着养殖年数增加, 水泥池老化等不可抗拒的因素造成污物残留于池壁、池底及下水管中, 为青苔的繁殖提供了必要的营养条件, 加上以下几个因素则很可能在鲟养殖高温时节发生青苔大面积滋生: 流水养殖, 水源经常带入青苔, 并在水池中大量繁殖; 鲟养殖为多年养殖模式, 不能做到每年彻底清塘杀死青苔; 鲟养殖投饵量很大, 而且经常不能很好的把握, 导致饵料残留; 流水养殖池中一般不进行施肥, 导致水质清瘦, 有益藻类得不到足够的营养元素生长, 故很容易被青苔取代。

二、青苔对鲟养殖的危害

严重消耗水体的氮、磷、钾、微量元素等营养元素，使水体清瘦，其他有益藻类如硅藻、绿藻等无法正常生长，同时也影响鲟鱼体内的水盐代谢平衡。

青苔大量繁殖，导致水体空间变得拥挤，鲟呼吸时常把青苔吸附在鳃表面，虽不致引起鲟死亡，但是也足以使鲟生活不适，吃食量下降，养殖效益下降。

鲟吃食张开大口，也能导致吃进很多青苔，常导致消化不良，影响生长，如青苔过度繁殖，还可能产生中毒现象。

池壁上青苔大量繁殖，在晚间大量消耗氧气，极易引起鲟浮头死亡。

青苔生长繁殖为病原菌生存及繁殖提供了很好的环境，这样导致鲟极易因碰伤，吃食等感染细菌，产生如性腺溃烂病、败血症、红嘴病等疾病，常呈暴发性流行，往往引起重大经济损失。故防控青苔在鲟陆地水泥池养殖中成了防治疾病的关键。

青苔大量繁殖后，水温升高，青苔在养殖池中泛起死亡，变黑腐烂，散发恶臭味，其在分解过程中产生大量有害物质，使水质变坏，提高了池中氨氮的含量，降低了水中的溶氧，严重时可能造成鲟中毒，结果导致养殖产量和经济效益降低。

三、药物防控

在鲟养殖水泥池空出后，排干池水，注意不要积水，使水泥池曝晒，并用生石灰按 50~100 kg/亩全池泼洒，达到彻底清塘的目的。

青苔暴发时，在池壁区域泼洒硫酸铜、硫酸亚铁合剂进行杀灭，并在 3~5 天后泼洒生石灰一次，用量为 5 kg/亩。

使用二氧化氯杀灭青苔一般适用于青苔多的水泥池。杀死青苔后，青苔腐烂，会造成水质恶化，应采用换水、调水、等方法进行调节。一般采用芽孢杆菌等进行调水。

四、生物防控

使用溶藻细菌进行防治。主要原理：直接溶藻，即溶藻菌与藻细胞直接接触，宿主细胞壁溶解；间接溶藻，细菌释放特异性或非特异性的胞外物质到环境中，非选择性的杀伤藻细胞，属于间接溶藻。

通过此细菌的溶藻作用，在青苔暴发前期进行控制，能起到一定的作用。

通过人工清理的方法，及时清除水体中青苔，经常洗刷池壁。

在进水口设置多层滤网，拦截可能进入水体的青苔源，从源头杜绝青苔暴发的可能性。

制定合理的饲料投喂量，减少残留。

五、结语

青苔对于鲟养殖是一个大的危害，能引起较大的经济损失，给养殖户带来了困扰。对于青苔的防治，目前防控方法比较多，但万变不离其宗，归根结底在于日常管理，管理得当，青苔问题自然可以很好的解决。鲟养殖是一个朝阳产业，养殖品种、产量正不断增加，但是我们也需要关注在养殖过程遇到的一些问题，从而妥善解决，为鲟健康养殖打下坚实基础，使得鲟养殖业能快速健康发展。

该文发表于《水产养殖》【2015，2：6-11】

淡水蓝藻控制方法的概述

胡利静，肖艳翼，蒋新益，杨先乐

(上海海洋大学，国家水生动物病原库，上海 201306)

水体富营养化曾经是世界各国所面临的重大环境问题之一。随着水体富营养化加剧，而海洋开发活动的日益增加，国际海运活动的愈加频繁，加上全球气候的变暖，导致赤潮发生频率逐年升高，已经成为一种全球性的海洋灾害[1,2]。而淡水中暴发蓝藻产生水华是水体富营养化的又一个突出表现。调研表明[3]，我国66%以上的湖泊、水库已处于富营养化的水平，其中重富营养和超富营养的占22%，使得富营养化已成为我国湖泊在一段相当长的时期内的重大水环境问题。近十几年来，太湖、滇池和巢湖等富营养化湖泊相继出现大面积的蓝藻暴发，引起蓝藻水华的种类主要有微囊藻（*Microcystis*）、鱼腥藻（*Anabaena*）、鞘颤藻（*Lyngbya*）、束丝藻（*Aphaniz omenon*）、颤藻（*Oscillatoria*）[4]，其中以产毒的微囊藻水华最为常见。有害蓝藻水华大规模的暴发不仅会降低水资源的利用效能，还会对海洋环境、水产养殖和渔业造成巨大的损失，而且部分蓝藻产生的毒素会对人类健康产生严重的危害[5]。现将蓝藻控制方法简介如下。

1 粗控制型

1.1 机械控制法

采用人工和机械打捞、生物曝气等多种方法清除蓝藻水华，能直接大量清除湖面蓝藻水华，且无明显负面影响。因此，在蓝藻水华大量发生季节，采用机械法清除湖面水华并加以综合利用，防止恶性增殖和二次污染，并由此降低水体营养水平污染负荷，从而可达到改善水环境之目的，是一项立竿见影的高效实用技术。沈银武等[6]在富营养化湖泊滇池蓝藻水华机械清除的研究中，利用重力振动、旋振和离心等方法收集富藻水，浓缩、脱水后得到藻泥分析其氮磷含量，表明在富营养湖泊中水华蓝藻大量暴发时，采用机械除藻，对控制蓝藻水华污染、有效降低内源氮、磷等污染物负荷具有十分良好的效果。

1.2 水生动物控制法

由于蓝藻颗粒较大，更容易被滤食性鱼类摄食到体内，从而在一定程度上延缓、阻碍蓝藻的生长。可供选择的鱼类有鲢、鳙、罗非鱼等。通过循环放养和重复养殖，可以调控生物之间的食物链关系，降低藻类现量，再通过成鱼捕获，获取水体中的营养物质，从而达到既有效控制水华，又取得一定渔获物的目的。刘建康[7]在实践研究中发现鲢鳙鱼能有效地遏制微囊藻水华，并且得出鲢鳙鱼遏制水华的有效放养密度为46~50 g/m³。虽然蓝藻被鲢鳙鱼摄食，一定程度上降低了蓝藻的密度，但蓝藻不易被鲢鳙鱼消化，致使含有蓝藻的鱼粪便漂浮在水面并不断解体，再次释放蓝藻，使蓝藻扩散而导致二次污染。为此，2006年研究人员利用"生态陷阱"技术治理临港新城滴水湖，利用水位控制技术，增加水体含泥量，使滤食性鱼类在摄入藻类的同时连带摄入更多的泥沙，使鱼粪因含泥沙而沉降到水底，从而让底栖生物再次摄食和降解。该技术优化了鲢鳙鱼控制蓝藻的成效，使蓝藻得到了有效的控制。

罗非鱼（*Oreochromis spp*）和鲢鳙鱼不同，它们兼有滤食捕食和啃食等多种摄食方式，能有效地摄食和利用水体中的浮游生物。陆开宏等[8]在藻华水体月湖和一些城镇公园水体中投放罗非鱼进行控藻研究，发现罗非鱼的牧食压力导致水体叶绿素的减少，蓝藻水华得到了有效控制。但Datta和Jana在印度一人工湖的小型围隔试验发现罗非鱼只有短期的控藻效应[9]。

1.3 植物控制法

植物控制法是将植物放入水体中，利用植物的代谢产物间接的抑制蓝藻生长的方法，具有化感作用的植物包含了漂浮水生植物、大型水生植物及陆生植物。

漂浮水生植物不但可以吸收水体的营养盐和有机物，减少形成水华的风险，还可以通过漂浮的特性，随蓝藻一起在水面漂浮盖住聚集的蓝藻颗粒层，阻碍其生长，间接促进其他藻类生长[10]。研究表明[11,12]，浮萍（*Lemnaminor*）、马来眼子菜（*Potanogetonmalaianus*）、苦草（*Vallisneriaspiralis*）、水花生（*Altemantheraphiloxeroides*）、金鱼藻（*Ceratophyllumoryzetorum*）等都是很好的选择种类。大型水生植物能够快速竞争性地吸收水体中的营养盐[13]，而且有遮光作用[14]，更能分泌产生化感物质抑制藻类生长[15,16]，大型水生植物与蓝藻之间具有相生相克的现象，因此水生高等植物被广泛应用于降低湖泊水体营养盐负荷，控制藻类生长，调节湖泊生态系统等[17,18]。而大麦秆、小麦秆等是天然的植物材料，对于水生高等植物、鱼和无脊椎动物等几乎不存在不利影响。大麦秆、小麦秆等不仅抑制藻类生长，还能有效改善水生生态系统结构，增加水域中的生物多样性[19]。

由于植物化感物质具有高效性、选择性和无污染性，应用在抑制藻类繁殖上不仅可以有效地抑制藻类的生长且不会污染水体造成二次污染，被认为是颇有前途的一种抑藻手段。植物化感作用抑制藻类主要有3种方式：①利用水体中活体水生植物分泌的化感物质抑制藻类[20]，例如穗花狐尾藻[21]（*Myriophyllumspicatum*L）；②利用植物秸秆，比如芦苇秆[22]、稻秆[23]、大麦秆[24,25]；③从植物中提取化感物质，例2-甲基乙酰乙酸乙酯[26]。

粗控制法所需的技术含量不大，而且不易引起水环境的二次污染，但是利用机械控制只能减缓蓝藻对水产养殖造成的危害，不能从根本上解决蓝藻问题。而对于植物控制法是直接往水体中投放秸秆会影响水体美观且见效慢，所以将收割干燥的植物体放入待处理水体，利用其腐败释放的化感物质抑制藻类，以及从各类植物中提取的化感物质施入水体抑制藻类的做法显得尤为重要，而且颇具有应用前景。

2　半粗半精控制型

2.1　絮凝剂控制法

对于水华的快速处理，使用无毒絮凝剂沉降是较好的选择之一。普通物理絮凝剂来源充足，天然无毒，使用方便，耗资少，尤其当水华大规模发生时。各类基于人工"取出"的技术或者人工生态工程由于成本问题而实施效果不理想时，絮凝除藻方法则更具优势。刘蕾等[27]发现絮凝剂的投加能明显降低水体的 pH 值和叶绿素 a 浓度，对改善水体的透明度有积极的作用，且絮凝剂浓度越高，降幅越大。肖利娟等[28]在研究中发现铁盐絮凝剂在应急处理方案中对蓝藻具有较好的去除效果。

无机絮凝剂除铁盐系外还有铝盐系，但有报道认为饮用水中铝可能对人体健康造成不良影响[29]。另外，还有黏土絮凝剂，而黏土除藻是通过阳离子交换以及凝集作用将藻细胞和颗粒凝聚沉降到水底，可以在一定程度上降低水面藻细胞密度。在采用黏土矿物絮凝沉降除藻方面，许多学者也进行了不少有益的尝试，张青等[30]在 2 种絮凝剂对水华蓝藻的去除能力的研究中发现在沉降时间上海泡石绒远快于高岭土。黏土除藻对于深水湖泊水华有一定的去除作用，但对于大多数的浅水型湖泊来说絮凝除藻效率低，而且黏土技术本身不能防止藻类的再次泛起和底泥二次污染，难以被广泛使用。

2.2　破坏藻自身控制法

破坏藻自身的方法主要采取化学药剂法、光化学降解法等措施。其中，最常用的化学药剂法主要是利用除草剂、杀藻剂等来控制水华藻类的繁殖，使用最广泛的当属硫酸铜，其中用铁盐、铝盐作增效剂可提高硫酸铜的除藻效果，但大量使用硫酸铜常会引发毒害水生生物及重金属在湖泊中的积累等二次污染，因此具有一定局限性。

2.2.1　化学药剂法———化学氧化法

预氧化剂的氧化性对藻类具有直接灭活作用，藻类同水中其他颗粒一样，可以通过絮凝沉淀过程除去，由于预氧化工艺具有一定的助凝作用，也可以提高除藻的效率。刘嵩等研究臭氧预氧化的除藻效果时发现臭氧预氧化可以明显地提高除藻效果，臭氧投量为 1.5 mg/L 时，除藻率达到最高值 88.3%，继续增加投入量，除藻率不再增高[1]。一些发达国家采用臭氧化除藻效果虽好，但运行成本高，投资太大，难以在发展中国家推广。当前使用较多的氧化除藻方法是预氯化工艺，其除藻机理可能主要是具有氧化成分物质使藻细胞膜系统发生脂质过氧化，造成膜系统和细胞器结构的破坏，引起膜内物质的外漏，最终导致藻细胞死亡[32]。但在氯化过程中会生成 THMs 等有害副产物，因此该工艺的使用受到很多国家的限制。

2.2.2　光化学降解法

蓝藻藻类的生长繁殖与其自身通过光照进行光合作用有着密不可分的关系，因此，利用遮光法可以使蓝藻藻类缺乏光合作用提供的各项代谢能量从而降低湖泊、养殖水体的蓝藻密度。陈雪初等[33]的研究表明遮光处理能够在营养盐浓度变化不大的状况下大幅度削减藻类生物量。万蕾等[34]在不同遮光方式的抑藻效果比较研究中也表明遮光是抑制藻类暴发的有效措施之一，不同水体可以根据具体情况选择不同的遮光方式。

光化学降解法治理蓝藻除了遮光法外还有紫外线法[35]、纳米半导体法[36,37]。紫外线是一种能量较高的短波，可用于消毒、抑制微生物生长。紫外线控藻法适合在水产养殖循环水等小型水体中使用，而对于大型水体中使用该法不太现实。纳米半导体材料具有光催化性能，在光照射下能形成氧化还原体系，这个氧化还原体系如果能对蓝藻光合作用中电子的输送过程起到抑制和阻断作用，那么就可能利用光催化来抑制蓝藻的生长。纳米半导体法应用在控藻上，还需要研究人员的进一步开发，研发其相应的除藻手段。

2.3　工程法控制型

工程法是以计算为理论，以技术为支撑的控藻方法，其中包括超声波控藻法、移动式加载絮凝磁分离水处理技术、粒子流复合除藻净化技术等。

超声波控藻是近年来受关注较多的一种物理控藻方法，可将蓝藻细胞内细胞器破碎或破坏其生理功能从而使蓝藻细胞死亡达到控藻效果。很多研究表明[38-40]超声波对蓝藻有明显的去除作用，而对水体中浮游动物、鱼类及沉水植物等其他主要水生生物不产生明显影响。移动式加载絮凝磁分离水处理技术，英文全称 Ballasted Flocculation Magnetic Separation（BFMS），是新一代水澄清和过滤技术。BFMS 技术是在传统的絮凝工艺中加入磁粉成为絮体的核体，增强了絮凝的效果，同时减少絮凝剂用量。张朝升等利用该技术发现其对水中藻类具有很好的去除效果，与传统工艺相比，有机物去除率平均提高 34.21%[41]。另外，粒子流复合除藻净化技术是将化学氧化法与电场法有机结合，采用二氧化氯氧化-电场法去除水中的藻类，该方法通过氧化作用破坏藻类的细胞结构和功能，影响藻细胞的光合作用，致使藻类逐渐凋亡。

半粗半精控制法中，化学药剂及絮凝剂对蓝藻的控制具有用量少、操作简单、见效快等优点，可用于应急的情况，但化学药剂及絮凝剂等容易引起二次污染，影响湖泊、池塘的生态平衡，威胁了鱼类等水产动物的生长。而工程控藻法不会引起环境的再次污染，并且效果较佳，但工程控藻法需要一定的技术做支撑，而且成本较高，不适用于小型湖泊和池塘。

3　精细控制型

国内外研究者在极力探究精细控制型的有效控藻方法，包括微生物控制法、营养物质控制法等，其中通过微生物控制蓝藻水华是近年来探索蓝藻控制的有效途径之一。

3.1　微生物控制法

微生物修复尤其是以菌治藻，主要是应用原位技术以及向水体投加有效微生物群从而改善污染水体的透明度、高锰酸盐指数（COD_{Mn}）、溶解氧（DO）、总氮（TN）、总磷（TP）、叶绿素 a 等以达到减轻水体富营养化程度的效果。在我国水产业应用光合细菌（Photolosyntheticbacteria，PSB）以来，已有许多文献报道了其在鱼、虾、贝类育苗和养殖方面的作用[42-46]。光合细菌能够分解水中的有害物质，起到净化水质、控制水体富营养化的目的[47]。丁成[48]研究固定化光合细菌对富营养化养殖水质的净化效果时发现固定化光合细菌在较短的时间内即可达最佳处理效果。水质指标不同其最佳净化条件也不尽相同，不能完全以哪个指标来单独评定和评价水质。常会庆等[49]进行了人工模拟富营养化水体处理试验，发现在富营养化水体中接种固定化光合细菌可明显降低水体中的养分，且对水体中藻类也起到一定的抑制作用。

近年来，部分学者相继报道了生物组合（光合细菌、硝化细菌及玉垒菌等）、有效微生物群（EM，以光合细菌、放线菌、酵母菌和乳酸菌为主）等在小型富营养化水体的应用方面的技术，这些技术均表现出较好的除藻效果，且工艺简单，成本低廉，是极具推广潜力的生物控藻新技术。李雪梅等[50]发现有效微生物群可有效抑制藻类的生长，防止水华的发生，改善景观。陈建等[51]在北京市延庆县妫水湖的现场围格水体中投加有效微生物菌群，结果表明投加 EM 菌不仅可以控制蓝藻水华的暴发，而且可以快速降低水体中氮磷浓度。

随着研究的深入，新的溶藻细菌被提出并引起研究者深厚的研究兴趣。溶藻细菌在水体及自然环境状态下广泛存在，是一种天然的物质，所以可在水体中筛选分离寡营养水体中的高效溶藻细菌，并研制高效溶藻菌剂，从而达到以菌控藻和原位治藻的目的。目前国内有关溶藻细菌的研究仅处于实验室研究方面，大量的研究集中于对溶藻细菌的分离和鉴定，已鉴定出的溶藻细菌有蜡状芽孢杆菌（Bacilluscereus）[52]、短小芽孢杆菌（Bacilluspumilus）[5]、荧光假单胞菌（Pseudomonas fluorescence）[54]等。以菌治藻作为水华生物防治的一个新对策，具有广泛的研究前景，何鉴尧等[55]在研究中发现 3 株溶藻细菌并非简单地杀灭水体中的全部或某些藻类，而是能明显地调节藻类群落结构，使其朝着较为稳定的方向发展，在修复藻型富营养化水体水生生态系统方面有很好的应用前景。

3.2　营养物质控制法

氨基酸是微生物生长的必需物质，但是某些氨基酸对特定蓝藻生长具有抑制作用，利用氨基酸来抑制蓝藻生长在国内外均有报道，Kaya 和 Sano 通过研究酵母提取液和溶藻细菌的胞外分泌物质，发现其共同有效成分——赖氨酸能专性抑制铜绿微囊藻的生长，而对其他藻类没有抑制作用[56]。随后 Kaya 在围隔试验表明[57]，在自然水体中赖氨酸可以有效抑制铜绿微囊藻的生长，同时发现赖氨酸和丙二酸联合处理时能有效抑制铜绿微囊藻的生长。但是对于使用氨基酸来抑制蓝藻生长的抑制效果的研究以及对于机理方面的研究还不深入，仅林必桂等[58]为了探讨其抑制机理，在光照条件下，研究了不同含量赖氨酸对铜绿微囊藻细胞的毒理效应。随后林必桂等[59]为了进一步探讨其抑制机理，采用人工光照培养箱培养方法，研究了不同浓度的赖氨酸对铜绿微囊藻各种生理生化特性的影响。氨基酸等营养物质作为抑制蓝藻的制剂具有良好的应用前景，有待于进一步研究开发利用。

精细控制蓝藻的方法具专一性控藻、易于自然降解不引起生态失衡的优点，但该法具药效开始时间长、成本高等缺点，需要进一步研发高效、低成本的药物制剂。

4　展望

近年来，在水产养殖环境日趋恶化，水体富营养化程度越来越高，蓝藻引发的藻类污染严重影响渔业的发展，必将引起水产行业的高度重视。处理蓝藻的方法很多，而且在以上处理蓝藻的方法中各个方法均有其各自的特点，其中以植物控藻中化感物质的控藻作用效果尤为突出。要让目前世界各国的水处理厂、湖泊以及池塘等水域的蓝藻得到控制，需要因地制宜，选择合适的除藻方法。同时必须加大科研步伐，研制开发高效低成本的新方法，因此，高效、低成本的除蓝藻新产品在今后的研究中仍需要开展较多的工作。

参考文献

[1]　孙冷，黄朝迎. 赤潮及其影响 [J]. 灾害学，1999，2：51-54.

[2]　李士虎，吴建新，李庭古，等. 赤潮的危害、成因及对策 [J]. 水利渔业，2003，(6)：38-54.

[3]　黄漪平. 太湖水环境及其污染控制 [M]. 北京：科学出版社，2001.

[4] 王扬才, 陆开宏. 蓝藻水华的危害及治理动态 [J]. 水产学杂志, 2004, 17 (1): 90-94.

[5] 齐雨藻. 赤潮 [M]. 广州: 广东科技出版社, 1999.

[6] 沈银武, 刘永定. 富营养湖泊滇池水华蓝藻的机械清除 [J]. 水生生物学报, 2004, 28 (2): 131-136.

[7] 刘建康. 用鲢鳙直接控制微囊藻水华的围隔试验和湖泊实践 [M]. 生态科学, 2003, 22 (3): 193-196.

[8] 陆开宏, 晏维金, 苏尚安. 富营养水体治理与修复的环境生态工程———利用明矾浆和鱼类控制桥墩水库蓝藻水华 [J]. 环境科学学报, 2002, 2 (6): 732-737.

[9] XIE P, LIU J K. Stuies on the influence of planktivorous fishes (silver carp and bighead carp) on the phytoplankton com-munity in a shallow eutrophic chinese lake (Donghu lake) using enclosure method [R]. //LIU J K (ed). Annual Report of FEBL for1990. International Academic Publishers, Beijing, 1991: 14-24.

[10] 张树林, 邢克智. 养殖水体水华发生及控制的研究 [J]. 水利渔业, 26 (2): 67-69.

[11] 顾林娣, 陈坚, 等. 苦草 (Vallisneria spiralis) 种植对藻类生长的影响 [J]. 上海师范大学学报 (自然科学版), 1994, 23 (1): 62-67.

[12] 马为民, 孙莉, 等. 三种高等水生植物对铜绿微囊藻生长的影响 [J]. 上海师范大学学报 (自然科学版), 2003, 32 (1): 101-102.

[13] 吴玉树, 余国莹. 根生沉水植物菹草 (Potamogeton crispus) 对滇池水体的净化作用 [J]. 环境科学学报, 1991, 11 (4): 411-416.

[14] 金送笛, 李永函, 倪彩虹, 等. 菹草 (Potamogeton crispus) 对水中氮磷的吸收及若干影响因素 [J]. 生态学报, 1994, 14 (2): 168-173.

[15] 潘继征, 李文朝, 陈开宁. 滇池东北岸生态修复区的环境效应———抑藻效应 [J]. 湖泊科学, 2004, 16 (2): 141-148.

[16] 孙文浩, 俞子文, 余叔文. 城市富营养化水域的生物治理和凤眼莲抑制藻类生长的机理 [J]. 环境科学学报, 1989, 9 (2): 188-195.

[17] 李锋民, 胡洪营. 大型水生植物浸出液对藻类的化感抑制作用 [J]. 中国给水排水, 2003, 12 (11): 18-21.

[18] 吴振斌, 邱东茹, 贺锋, 等. 水生植物对富营养化水体水质净化作用研究 [J]. 武汉植物学研究, 2001, 19 (4): 299-303.

[19] 吴为中, 芮克俭, 等. 大麦秆控藻研究进展 [J]. 生态环境, 2005, 14 (6): 972-975.

[20] 李磊, 侯文华. 荷花和睡莲种植水对铜绿微囊藻生长的抑制作用研究 [J]. 环境科学, 2007, 28 (10): 2180-2186.

[21] Nakai S, Inoue Y, Hosomi M, et al. Myriophyllum spicatum eleased allelopathic polyphenols inhibiting growth of microcys-tis aeruginosa [J]. Water Res. 2000, 34 (11): 3026-3032.

[22] 于淑池, 姜燕, 邓红英, 等. 芦苇秆浸出液对铜绿微囊藻抑制作用的研究 [J]. 淡水渔业, 2013, 43 (2): 66-70.

[23] 向丽, 邹华, 黄亚元, 等. 稻秆对铜绿微囊藻抑制作用的研究 [J]. 环境工程学报, 2011, 5 (2): 280-283.

[24] 张薛, 胡洪营, 门玉洁. 大麦秆提取液对铜绿微囊藻生长的影响研究 [J]. 环境科学学报, 2007, 27 (12): 1984-1987.

[25] 梁倩华, 章群, 范晓军, 等. 应用大麦秆控制原水蓝藻生长的初步研究 [J]. 生态科学, 2006, 25 (2): 122-123.

[26] 门玉洁, 胡洪营. 芦苇化感物质 EMA 对铜绿微囊藻生长及藻毒素产生和释放的影响 [J]. 环境科学, 200728 (9): 2058-2062.

[27] 刘蕾. 以铁盐为核心的絮凝剂在供水水库中对蓝藻水华的控制研究———围 隔实验 [J]. 环境科学, 2009

[28] 肖利娟, 韩博平. HA1 絮凝剂在供水水库水华应急处理中的应用 [J]. 环境科学, 2007, 28 (10): 2192-2197.

[29] 付军, 闫海, 王东升, 等. 聚铝及其加载黏土矿物高效絮凝沉降铜绿微囊藻的研究 [J]. 环境污染治理技术与设备, 2006, (6): 76-79.

[30] 张青, 邓金花. 2 种絮凝剂对水华蓝藻的去除能力 [J]. 江苏农业科学, 2009, 4: 316-317.

[31] 刘嵩. 臭氧预氧化的除藻效果及机理探讨 [J]. 山西建筑, 2010, 36 (27): 171-172.

[32] 刘洁生, 杨维东. 二氧化氯对球形棕囊藻叶绿素 a、蛋白质、DNA 含量的影响 [J]. 热带亚热带植物学报, 2006, 14 (5): 427-432.

[33] 陈雪初, 孙杨才, 张海春, 等. 遮光法控藻的中试研究 [J]. 环境科学学报, 2007, 27 (11): 1830-1834.

[34] 万蕾, 朱伟. 不同遮光方式的抑藻效果比较研究 [J]. 环境工程学报, 2009, 10 (3): 1749-1754.

[35] 袁侃, 毛献忠, 等. UV-C 辐照抑制铜绿微囊藻生长的动态实验研究 [J]. 环境科学, 2010, 31 (2): 311-317.

[36] 陆长梅, 张超英, 吴国荣, 等. 纳米级 TiO2 抑制微囊藻生长的实验研究 [J]. 城市环境与城市生态, 2002, 15 (4): 13-15.

[37] 尹海川, 林强, 涂学炎, 等. 纳米二氧化钛复合半导体光催化抑制蓝藻生长 [J]. 昆明理工大学学报 (理工版), 2005, 30 (1): 52-56.

[38] 储昭升, 庞燕. 超声波控藻及对水生生态安全的影响 [J]. 环境科学学报, 2008, 28 (7): 1335-1339.

[39] 邵路路, 陆开宏, 朱津永, 等. 低强度超声波抑制铜绿微囊藻生长的研究 [J]. 生态科学, 2012, 31 (4): 413-417.

[40] 胡冬雯. 超声波对 3 种水华爆发主因蓝藻的控制 [J]. 农业科学环境学报, 2013, 32 (7): 1432-1436.

[41] 张朝升, 张可方, 宋金璞, 等. 大梯度磁滤器处理微污染珠江源水 [J]. 中国给水排水, 2001, 17 (4): 70-73.

[42] 郭本华, 陆青艳, 宋志文. 光合细菌在环境污染防治中的应用 [J]. 青岛建筑工程学院学报, 2004, 25 (1): 61-65.

[43] ISAMUMAEDA, TADASHI MIZOGUCHI, OSHIHARU-MIURA, et al. Influence of sulfate reducing bacteria on outdoor hydrogenproduction by photosynthetic bacteriumwith seawater [J]. CurrentMicrobiology, 2001, 40 (3): 210-213.

[44] 周佳, 刘文睿, 黄遵锡. 高原湖泊光合细菌处理水产养殖污水的初步研究 [J]. 水产学报, 2006, 26 (2): 70-72.

[45] 张明, 史家樑. 光合细菌在水产养殖中的应用 [J]. 应用与环境生物学报, 1999 (5): 204-206.

[46] 张信娣，陈瑛. 光合细菌对养殖水体的调节作用 [J]. 水利渔业，2007，27 (1)：80-82.
[47] 李佐荣. 微生物在湖泊富营养化治理中的应用 [J]. 安徽农学通报，2007，13 (9)：63-64.
[48] 丁成. 固定化光合细菌对富营养化水质的净化效果 [J]. 安徽农业科学，2008，36 (14)：6028-6030.
[49] 常会庆，王世华，寇太记. 固定化光合细菌对水体富营养化的去除效果 [J]. 水资源保护，2010，26 (3)：64-67.
[50] 李雪梅，杨中艺，简曙光，等. 有效微生物群控制富营养化湖泊蓝藻的效应 [J]. 中山大学学报，2000，39 (1)：81-85.
[51] 陈建，丛军，等. 利用有效微生物菌群控制蓝藻水华研究 [J]. 环境工程学报，2010，4 (1)：101-104.
[52] 刘晶，潘伟斌，秦玉洁，等. 两株溶藻细菌的分离鉴定及其溶藻特性 [J]. 环境科学与技术，2007，30 (2)：17-20.
[53] 卢兰兰，李根保，沈银武，等. 溶藻细菌 DC2L5 的分离、鉴定及其溶藻特性 [J]. 水生生物学报，2009，33 (5)：860-865.
[54] 邓洁，李建宏，管章玲，等. 一株产碳酸酐酶附生菌对铜绿微囊藻 (Microcystis aeruginosa) 生长的影响 [J]. 湖泊科学，2012，(3)：429-435.
[55] 何鉴尧，潘伟斌，林敏，等. 溶藻细菌对富营养化水体藻类群落结构的影响 [J]. 环境污染与防治，2008，30 (11)：70-74.
[56] Kaya K, SanoT. Algicidal components inyeast extract as a componentss of microbial culture media [J]. Phycologia, 1996, 35：117-119.
[57] KAYAK, LIUYD, SHENYW, et al. Selective Control of Toxic Microcystis Water Blooms Using Lysine and Malonic Acid：An Enclosure Experiment [J]. Environmental Toxicology, 2005, 20 (2)：170-178.
[58] 林必桂，杨柳燕，肖林，等. 赖氨酸对铜绿微囊藻细胞的抑制机理 [J]. 生态与农村环境学报，2008，24 (4)：68-72.
[59] 林必桂，杨柳燕，肖林，等. 赖氨酸抑制铜绿微囊藻生长的机理研究 [J]. 农业环境科学学报，2008，27 (4)：1561-1565.

该文发表于《水产养殖》【2013，3：41-45】

养殖水体蓝藻水华的防治

赵蕾[1,2]，欧仁建[1]，陈辉[2]，杨移斌[1]，柯江波[1]，肖国初[1]，杨先乐[1]

(1. 上海海洋大学国家水生动物病原库，上海 201306；2. 江苏世盛动物药业有限公司，江苏 大丰 224100)

从本质上讲，水华是以蓝藻为载体的物质和能量转换的结果[1]。在含营养物质丰富的水体中，有些蓝藻常于夏季大量繁殖，并在水面形成一层蓝绿色而有腥臭味的浮沫，被称为"水华"。大规模的蓝藻爆发，被称为"绿潮"。中科院南京地理与湖泊研究所的孔繁翔研究员等在 2007 年对蓝藻水华的形成机制进行了研究，提出了 4 个阶段的理论假设：即蓝藻的生长与水华的形成可以分为休眠、复苏、生长、上浮及聚集 4 个阶段。每个阶段中，蓝藻的生理特性及主导环境影响因子有所不同。在冬季，水华蓝藻的休眠主要受低温及黑暗环境所影响。春季的复苏过程主要受湖泊沉积表面的温度和溶解氧控制，而光合作用和细胞分裂所需要的物质与能量，决定水华在春季和夏季的生长状况，一旦有合适的气象与水文条件，已在水体中积累的大量蓝藻群落将上浮到水体表面积聚，形成可见的水华[2]。水华的出现最根本的原因还是排入水体的污染物远远大于水体环境的自身容量[3]。

形成水华的蓝藻主要有微囊藻、鱼腥藻、色球藻、螺旋藻、拟项圈藻、腔球藻、尖头藻颤藻、裂面藻、胶鞘藻、束毛藻等十多个属，其中微囊藻属是分布最广、最为常见的蓝藻。当微囊藻之类具假空泡的蓝藻过量繁殖时，水的透明度极低，有光层变的很薄，蓝藻长时间处于低光照下，假空泡形成很快使细胞迅速上升，以致内压的升高尚来不及使假空泡破裂，藻体已升到光照过量的表层，形成斑状浮渣，浮渣分解时散发腥臭味，夜间大量消耗水中溶解氧，容易使鱼缺氧而死。而且蓝藻死亡后产生羟氨或硫化氢，对水生动物有毒，破坏水体，降低水体的利用价值[4]。

1　蓝藻水华的一般成因

1.1　内因——蓝藻生物学特性

蓝藻对高温、低光强和紫外线的适应，可以过量摄取无机碳和营养物质，低的氮磷比等因素都有利于蓝藻生长。具体描述如下。

在池塘水域经过 2 个多月的养殖，水体浮游生物大量生产和能量的转换，除硝酸盐较为丰富之外，另外两大营养盐类磷酸盐、硅酸盐已消耗殆尽。而蓝藻类及少数细菌具有能够利用空气中游离氮的能力，致使蓝绿藻体内积蓄较高量的蛋白氮物质，这一过程称为生物的固氮作用，使硝酸盐、氨氮的积累增添了来源，愈使喜欢高氮低磷的蓝藻类独具生长优势。在此季节蓝藻类水华往往是用不了三五天时间就会蔓延以至覆盖整个池塘水面[6]。

藻类群落中具有不被其他藻类可匹敌与抑制的能力，在环境适宜时往往会处于无节制的繁殖状态。

1.2　外因——蓝藻大量繁殖的主要环境条件

湖泊中营养盐含量增加，合适水文气象条件：高温、高光强、合适的小风速使蓝藻上浮到水面，漂移、合适的风向促成湖滨与港湾的静水堆积、水华形成。水华形成：大量蓝藻+水文气象条件。蓝藻生长：蓝藻+光照+温度+合适环境[5]。具体描述如下：一是连续晴热的高温天气，从藻相生态体系上看温度是其主要影响因素，大多数藻类能在很大温度范围内生活，但最适宜生长温度变幅较窄。一般来看，硅藻、金藻、黄藻适温较低在 14~18℃，绿藻较高为 20~23℃，而蓝藻更高，最喜欢生长在 20~32℃ 的温度阶段。其中危害最大的微囊藻类在 10~40℃ 都可生长，最适温度为 28~32℃。

二是水体呈强碱性（pH 值 8.8~10），在我国北方干旱与半干旱地区，盛夏季节往往是干旱高温持续数日，光照度极强，水体藻类光合作用极强，CO_2 消耗量最大，pH 值便急剧上升。

三是光照浮游植物生产的必要能源条件，故作为养殖用户要对光照的把握。

四是水体高氮低磷有机质含量较丰富。养殖水体富营养化的主要原因是营养物质、尤其是过量的溶解性营养盐（NH_3-N、NO_3-N、NO_2-N、PO_4-P 等）的积累。养殖水体中氮、磷等营养盐的来源有两类：一是外源，主要是养殖过程中向水体输入的废物（未食的饵料、养殖对象的粪便和排泄物等）以及随着水流进入养殖水体的农施化肥和家禽粪便等。二是内源，指养殖水体的底泥在一定条件下向水体释放的磷酸盐而增加的水体中氮、磷含量。在养殖过程中，残饵、粪便和排泄物不断进入水体，一部分有机物直接溶解于水中，一部分由于重力的作用而沉积水底，不仅增加养殖水体中的营养盐，而且增加底泥中的营养盐，这是一个连续渐进过程。另外，养殖水体中水生植物的排泄物、尸体以及其他腐殖质等，经过好气细菌的再次分解，可间接为藻类的繁殖提供营养。放养结构不合理，吃食性鱼与滤食性鱼比例失调，高温季节水质更新次数少，都会为浮游生物的大量繁殖提供方便。这些因素虽然不能直接造成养殖水体的富营养化，却可以使养殖水体的水质发生根本性的转变，为富营养化的形成提供了便利。

加之夏季干旱缺水，补充水量时常不足，强光、高温、高碱度的水体状态抑制了其他藻类的繁殖。而微囊藻正得其时，便得天独厚地繁衍开来[6]。

2 养殖水体预防管理措施

正确把住发病季节及时做好预防工作。蓝藻（微囊藻）喜生长在温度较高（28~32℃）、碱性较高（pH 值 8~12.5）的水中。所以由蓝藻大量繁殖引起的中毒发生在夏季及秋季。因此，预防工作主要措施要调节好水质，池水达到"肥、活、嫩、爽"[6]。

经常灌注清水，不使池水有机质含量过高，注意池水的 pH 值（定期泼洒生石灰），改变池水的酸碱度和水温，可抑制蓝藻过量繁殖。高温季节渔池每一个星期灌注清水 10~20 cm 深，每月、每 667 m^2、每米水深施用生石灰 15 kg。以有机肥为主的渔池，应与磷肥一起腐烂发酵连汁泼洒，可提高肥效，并且也可达到改良水质目的[6]。

2.1 降低水体 pH 值

可以采用有机酸和"降碱灵"降低池水 pH 值到 7.0~7.5 左右，降低 pH 值的过程应循序渐进，防止对水生动物造成较大的刺激。

2.2 调节池塘合适的 N、P 比

作为养殖水体，其他因素多不便于控制，所以对 N、P 比的调节很重要。1840 年 Liebig 提出最低量律和耐性定律。阐明了植物生长所需的元素中供给量最少（与需要量相关最大）的元素决定植物的产量。1925 年 Selford 提出，一种生物能否生存，要依赖一种综合的全部因子存在的环境，只要环境中的一项因子的量或质不足或过多，超过生物的忍耐程度，则该物种不能生存，甚至灭绝。许多实验研究提出，磷和氮等环境因子是制约蓝藻生长的重要因素[17-20]。蓝藻的繁殖以营养细胞分裂为主，适合蓝藻生长所需的氮磷比为 7.2∶1[21]。池塘富营养化，有益藻类生长适应的 N、P 比不一定合适，可以调节池塘合适的 N、P 比，有效控制蓝藻水华的发生。合理施用氮肥，加大磷肥比例，避免池水含氮过高引起蓝藻大量繁殖。过量施用氮肥，或氮、磷肥搭配不当，会造成水质老化，引起蓝藻过量繁殖，造成中毒死鱼。化肥养鱼，一般条件下，尿素和过磷酸钙的搭配比例：50 kg 尿素搭配 100~113 kg 过磷酸钙。但高温季节 7、8、9 三个月，由于浮游生物繁殖旺盛，磷需要量随之加大，加大磷肥用量，也对抑制蓝藻过量繁殖有好处，即 50 kg 尿素配 150 kg 过磷酸钙，特别是池塘和底层有机质比较多，加大磷肥用量尤为必要[7]。

2.3 降低池塘富营养化

可以通过施入有效溶解 P，利用池塘底质中的 N 源，为生物所利用，从而逐步降低池塘富营养化[7]。

3 治理措施

3.1 物理方法

3.1.1 彻底清塘消毒和加注不带蓝藻的新鲜水 由于蓝藻比其他藻类具有更强的竞争力，因此控制措施以预防为主、防

重于治。彻底清塘消毒可有效杀灭蓝藻，压低基数，减少大规模发生的可能。避免随加水带入蓝藻，对控制也有积极意义。

3.1.2 定期换鲜水对于含有较多蓝藻的池塘，经常、大量地换新鲜水，可稀释蓝藻的浓度，同时也稀释了蓝藻分泌的毒物浓度，促进其他藻类的生长和保持整个生态系统的动态平衡[11]，也可以带来其他藻类，减少蓝藻的种群优势。

3.2 化学方法

3.2.1 铜制剂蓝藻比其他藻类对铜离子更敏感，因此铜制剂常用作抑藻、杀藻剂。传统使用的铜制剂是晶体 $CuSO_4$。$CuSO_4$ 的药效持续时间短，受水质的碱度及水中的可溶性有机物、腐殖质及藻类自身释放的多肽的影响，使用时需要连续施加。高浓度的铜离子会造成游游植物的大量死亡引起水体严重缺氧，过多过量的使用还会引起鱼类的蓄积性中毒，造成肝、肾组织的损害影响鱼体的生长。故 $CuSO_4$ 不能经常使用，且浓度应严格控制。为了减轻铜离子对水生动物的影响，将铜离子制成铜基化合物，铜与三乙醇胺形成毒性更小的化合物。铜离子从铜基化合物中缓慢释放到水体中并维持一定浓度连续作用，抑制蓝藻的生长和大量繁殖。目前采用络合铜（络合铜溶液）其毒性小，安全；pH 值影响小；水溶液澄清，透明；且剂量准确。

3.2.2 除草剂可供选的有西玛三嗪、敌草隆、扑草净等，它们作用的主要特点是抑制光合作用。西玛三嗪能有效地抑制光合作用，能控制浮游生物而对鱼类无害的安全浓度是 0.5 mg/L，可有选择地杀死蓝藻。敌草隆、扑草净抑制光合作用的效果也较好，对鱼类的毒性较小。实验结果表明，除藻剂可能在短时间内除抑蓝藻，但不能从根本上解决湖泊富营养化问题[12]。

3.2.3 选择性施肥低氮磷比，有利于蓝藻进行固氮作用，高氮磷比则有利于绿藻繁殖。国外一些学者认为，氮磷比接近或等于 1：20 能有效控制固氮蓝藻的爆发。

3.2.4 二氧化氯微囊藻、球囊藻施用药后数量显著减少。二氧化氯制剂具有较强的杀菌作用，并可增加养殖水体中的溶解氧[8]。

3.3 生物方法

3.3.1 放养一定数量的滤食性鱼类虽然蓝藻不易被消化，但由于其颗粒较大，更容易被滤食性鱼类摄食到体内，在一定程度上延缓、阻碍了蓝藻的生长。可供选择的鱼类有白鲢、花鲢、白鲫等。实践表明，尾重 200 g 以上的白鲢对蓝藻有明显的抑制作用，每 667 m^2 总量达到 100 kg 时，基本不会爆发蓝藻。

3.3.2 投放漂浮水生植物如浮萍，不但可以吸收水体的营养盐和有机物，减少形成水华的风险，还可以通过漂浮的特性，随蓝藻一起在水面漂浮盖住聚集的蓝藻颗粒层，阻碍其生长，间接促进其他藻类生长[9]。

3.3.3 引种水生维管束植物维管束植物能有效吸收水体的营养盐类，还有较强的净化水质作用，但要防止植物大量死亡引起的"二次污染"。芦苇、水辣蓼都是很好的选择[9]。

3.3.4 施用对蓝藻有特异性侵染、裂解的病毒、细菌（益生菌）、真菌等微生物选择培养特异性的病毒、细菌、真菌[10]。在渔业水体中，微生物尤其是细菌在水体水生生态系统中起着重要作用。细菌不仅是有机物的主要分解者，在物质循环中起着重要的作用，而且是水生动物和鱼类的重要食物。在富营养化水体中，腐生菌极易繁殖，危害水产动物，在蓝藻暴发的水体，由于蓝藻毒素影响，细菌生长受到抑制，因此在富营养化水体和蓝藻水体，都不利于水生生物的生长。在水体中投放一定量的有益菌（其微生物组台以光合细菌、放线菌、酵母菌和乳酸菌为主），增加水体的益菌含量，能提高水体分解有机物的能力，促进水生生物的生长，形成细菌分解，生物吸收，水产动物生长的良性循环[11]。

3.3.5 引进或培养优良藻类引进某些对蓝藻有拮抗作用的优良藻类抑制蓝藻生长。调整水体的氮磷比也可以改善藻类的种群结构，磷氮比为 2 时，蓝藻可以大量发生，当磷氮比提高到 5 时，绿藻大量繁殖成为优势种群[3]。利用水生物利用氮、磷元素进行代谢活动以去处水体中氮、磷营养物质。日本科学家发现一种名叫"水网藻"的网片状或网带状形藻。此藻繁殖迅速大量吸收水中氮磷，从而抑制了其他藻的生长，达到以藻治藻的目的[10]。

3.3.6 引进食藻原生动物许多蓝藻是原生动物的良好食物源，蓝藻的许多属可为纤毛虫类、鞭毛虫类和变形虫类所捕食[3]。

原生动物作为控藻因子有以下优势：①原生动物取食范围广。已经分离到取食微囊藻、鱼腥藻、束丝藻等滇池优势藻种的原生动物。②原生动物食量大。实验室中观测到，只要原生动物数目达到某一阈值，体系中的藻细胞会被迅速消耗殆尽。③很多原生动物在食物耗尽时会形成包囊，过度食物缺乏期。当藻类重新增多时，包囊又会破壁复苏成为食藻营养体。④包囊结构具有很强的抗逆性，这种形式容易包装和运输。⑤容易繁殖，原生动物可以利用有机培养基大量发酵培养[11]。

3.4 其他方法变废为宝

通过对蓝藻进行资源化利用，化害为利，这是要重点考虑的问题。实际上，蓝藻是巨大的资源库。蓝藻含有丰富的维生素、多种微量元素、重要氨基酸、碳水化合物和酶素，含有 60%的植物蛋白，这些蛋白经过蓝藻的分解，更容易被人体吸收。蓝藻的蛋白质含量比任何一种食物都要高。蓝藻还含有脂肪酸、亚麻酸、脂质、核酸、维生素 B 群、维生素 C、E和植物元素，如葫萝卜素、叶绿素和藻蓝素（能抑制癌细胞增长）。可以利用蓝藻制作食物，营养十分丰富[12]。海藻提供了一个科技经济透视比任何农作物都要大的潜在天然生物柴油原材料。蓝藻经过光合作用从太阳、水和 CO_2 中获得能源，可以快速繁殖和每日采收。不同品种可以提供不同比例的油、碳水化合物和蛋白。蓝藻中提取的油可以转变为生物柴油，剩余的生物质可以转化为沼气和饲料等。目前，棕榈油生产生物柴油转化比例占其重量的 20%，而蓝藻转化比例能达到50%[7]。通过人工及时打捞迅速生长的藻类，进行人工水体净化，也是养殖水体富营养化防治的一种措施[1,13-15]。及时打捞换水法可以稀释蓝藻的浓度[23]。蓝藻抗病毒蛋白 N（Cyanovirin-N）简称 CV-N，相对分子质量 11kD，含有 101 个氨基酸残基，主要是由 β-叠片结构形成的链状蛋白，其中有 2 个内部二硫键。通过基因重组表达纯化的产物，其结构和活性均与天然 CV-N 相同。CV-N 与人类免疫缺陷病毒（HIV）包膜糖蛋白 120（gp120）具有高度亲和性，并能够阻断由包膜蛋白介导的细胞融合过程从而阻止病毒的扩散。CV-N 不仅能够有效地抑制多个亚型的人类免疫缺陷病毒 1（HIV-1），2（HIV-2），猴免疫缺陷病毒（SIV）。而且时单纯疱疹病毒、流行性感冒病毒及其他一些包膜病毒也有抑制作用。CV-N 的若干特性使其有可能成为一种非常有价值的新型抗病毒药物［12］。蓝藻抗病毒蛋白-N（cyanovirin-N，CV-N）是一种从椭孢念珠藻（*Nostocelliposporum*）中分离出的抗病毒活性蛋白，能够有效地抑制多个亚型的人类免疫缺陷病毒 I（HIV-Ⅰ），Ⅱ（HIV-Ⅱ）以及猴免疫缺陷病毒（SIV）。CV-N 的上述特性使它可能成为艾滋病（AIDS）的潜在高效治疗药物[16,22]。

总结：治理蓝藻仅使用一种方法有时效果不是很理想。如化学方法治理，虽然见效快，但往往会造成二次污染，而且蓝藻容易再次大量繁殖。如用生物方法治理，虽然不会造成二次污染，但生物治想获得理想的效果较慢。如果将上述方法综合利用的话，既能较快的杀灭蓝藻，又能使蓝藻不再大量繁殖，同时还不会造成二次污染，获得了比较满意的结果。综合治理的措施可以分为下列适当使用一些化学药物如铜盐、铁盐、铝盐等除藻，也可用黏土絮凝法杀灭蓝藻，即利用改性黏土对藻细胞的凝聚作用，吸附湖面上的蓝藻沉入湖底，而蓝藻是靠光合作用进行生长繁殖的浮游生物，一旦沉入湖底就无法继续生存，且死亡后的蓝藻并未对湖水造成二次污染。结果证明，用这种方法来应急治理蓝藻的污染，效果明显，治理后，微囊藻的数量平均减少了 28%。还可以用一种通过天然材料提取的生物酶，诱导蓝藻进行超常光合作用，从而加快其新陈代谢，超量消耗其自身养分，最后致其死亡，蓝藻祛除率达 13%以上。后经常定期地向蓝藻大量繁殖的水域注入不含蓝藻的水源，引进一些生物物种如鱼类等控制蓝藻再次大量繁殖[6]。2000—2003 年期间，中科院水生生物研究所在滇池进行了养殖白鲢和花鲢鱼治理蓝藻的实验，结果是 1 条 1 kg 左右的白鲢鱼 1 个月能吃掉 2 kg 蓝藻，同样大小的 1 条花鲢鱼，1 个月能消耗 3.66 kg 蓝藻。经 3 年治理，示范区内的水中蓝藻含量降低了 2/3[7]。蓝藻污染和防治问题已是世界性的问题。我国蓝藻的污染现象也很普遍，如滇池、太湖、巢湖等均受不同程度的污染，对蓝藻的治理越来越引起人们的关注[23]。研究蓝藻的防治方法正积极地进行。在治理过程中也有很多好的经验和方法，其中也不乏成功的先例。但是，最重要的是能从源头上减少对水体的污染。

参考文献

[1] 胡家文，姚维志. 养殖水体富营养化及其防治 [J]. 水利渔业，2005 (6)：74-76.

[2] 张哗. 太湖如何"减肥" [N]. 科技日报，2007，7 (19)：44-46.

[3] 王玲玲，沈熠. 水体富营养化的形成机理、危害及其防治对策探讨 [J]. 环境研究与监测，2007 (4)：33-35.

[4] 赵文. 水生生物学 [M]. 中国农业出版社. 2007：8，30-31.

[5] 孔繁翔. 国内外富营养化湖泊治理与蓝藻水华控制经验与进展 [J]. 百度，2007，07，16.

[6] 费久兴. 蓝藻引起的中毒症及防治技术 [J]. 渔业致富指南，2002 (14)：34.

[7] 王玉群，李志文. 蓝藻水华对鱼类的危害和蓝藻水华的控制（下）[J]. 科学养鱼，2006. 2：77.

[8] 艾晓辉，周剑光，毛爱民. 二氧化氯制剂对水化成分及浮游生物的影响 [J]. 淡水渔业，1999 (2)：11-14.

[9] 于虹漫. 浅谈蓝藻的危害与防治 [J]. 北京水产，2004 (1)：22.

[10] 胡家文，姚维志. 养殖水体富营养化及其防治 [J]. 水利渔业，2005 (6)：74-76.

[11] 何玉玲. 蓝藻爆发的诱因及治理方法 [J]. 黑龙江科技信息，2008 (1)：114.

[12] 胡敏，张健. 饮食业油烟快速检测—检气管法 [J]. 环境科学，2003，22 (11)：842-843.

[13] 胡川，康升云. 蓝藻发生与控制方法初探 [J]. 江西水产科技，2001 (4)：35-37.

[14] 施伟达. 池塘蓝绿藻的生态防治 [J]. 齐鲁渔业，2006 (9)：10.

[15] 李静会，高伟，张衡，等. 除藻剂应急治理玄武湖蓝藻水华实验研究 [J]. 环境污染与防治，2007 (1)：60-62.

[16] 庞峰，于红. 蓝藻抗病毒蛋白 N 的研究进展 [J]. 中国海洋药物，2006 (6)：51-55.

[17] SMITH. Low nitrogen phosphorus ratios favordominance by blue-green algae in lake phytoplankton [J]. Science, 1983, 221: 669-671.

[18] LEET J, MArSI MARA M. Ultrasonicirradiation for bluegralgae bloom control [J]. Envirol Technol, 2001, 22 (4): 383-390.

[19] GABAI, AISN N. Nitl in aquatic ecosystems [J]. Ambio, 2002, 31 (2): 102-112.

[20] KRATASYI Ⅱ. The use of bioluminescent biotests for study of natural and laboratory aquatic ecosystems [J]. Chemo-spher, 2001. 42（8）: 909-915.

[21] 陈善娜, 陈小兰, 刘开庆, 等. 水华蓝藻生消的生物数学理论及其应用 [J]. 云南大学学报（自然科学版）, 2005（2）: 161-165.

[22] 李奇渊, 章军. 抗人类免疫缺陷病毒蓝藻蛋白-N 研究进展 [J]. 中国新药与临床杂志, 2003, 22（6）. 371-375.

[23] 罗海南. 蓝藻的危害与防治 [J]. 科技信息（科学教研）, 2008（16）: 369-388.

4. 泛塘与中毒的解救

该文发表于《科学养鱼》【2014，9：62】

泛塘与中毒的辨别及解救措施

胡利静，杨先乐，王雪琴

在池塘养殖过程中，泛塘与中毒事件屡屡发生，严重时引起池塘大片死鱼现象，造成养殖户的经济损失，因此，准确辨别泛塘与中毒并及时采取相应的解救措施，才能够挽回局面，减少损失。

一、泛塘或中毒辨别方法

1. 从发生的时间上辨别

泛塘：按季节分，多发生在夏季高温季节，尤其在连续阴雨天及连续低气压闷热天气里最容易发生泛塘事件；按一天的时间段分，多发生在半夜到清早这段时间，尤其在水质较肥、放养密度较大的池塘泛塘的可能性较大。

中毒：一般没有季节、天气、白天黑夜时间段之分，也与放养密度大小无关。多与有毒污水进入池塘例如厂矿排污、暴发蓝藻的污水入池，或是治疗鱼病时用药过量或施用方法不当，或是农田施药时多次在池塘中清洗农药器械，抑或是个别不法分子因各种原因往池塘中投毒等有关。发生中毒死鱼的时间没有规律可言，比较随机。

2. 从鱼体的表现辨别

泛塘：因缺氧而浮头，塘鱼分散于池塘各处，露出身体，口一张一合平静地从水面空气中吸取氧气，严重缺氧时塘鱼白肚皮慢慢上翻，同时挣扎着保持平衡、反复几次，最终白肚皮朝天而死。

中毒：浮头现象不明显。其症状表现也因毒物的不同而不同。有的塘鱼表现为窜游、跳跃、不断挣扎，最终死亡；有的塘鱼表现为摄食不振、行动迟缓、体色慢慢变黑变硬，最终因丧失活动能力而死。

3. 从死亡鱼的种类上辨别

泛塘：泛塘死鱼一般以中上层鱼类较为常见，例如鲢、鳙鱼；底栖鱼类浮头死亡的较少，如鲤、鲫鱼。

中毒：中毒死亡鱼不分鱼的种类，甚至极耐缺氧的底栖鲤、鲫鱼或是泥鳅等都会发生死亡现象。

二、发生泛塘或中毒时要采取的解救措施

泛塘或中毒引起的鱼类死亡，如果发现及时，就有解救的可能，就有降低损失的机会。

对于泛塘的解救措施是及时大量注入新水并开动增氧机增氧。但是对于设施条件较差的养殖池塘在水源不便又缺乏增氧机的情况下，可及时施用高效增氧剂来救急增氧。

对于中毒池塘的解救措施是立即注入大量新水，放出池中老水，边排边注，直到鱼类恢复平静；但对于水源不足的池塘，应立即拉网捕鱼，迅速将鱼转移到无毒池塘或网箱中，待原池塘毒水妥善处理后转放回原池塘中。

三、结语

泛塘与中毒事件屡有发生，造成的损失或大或小，为减少不必要的经济损失，还应加强管理，防止疏忽大意。养殖人员应坚持每日巡塘，观察天气、水色，检查硬件设施正常运转等。另外，使用水源应保证其无污染及蓝藻等侵袭，避免在池塘中直接清洗农药器械及其他污物器材，尽量减少陌生人员等在养殖池塘周围徘徊等。

三、渔药安全使用的示范和推广

优良的产业资源需要先进的科技促进发展，实验室的成果需要在生产实践中推广和验证。渔药使用风险评估及其控制技术相关成果在江苏（孙琪等，2014；杨先乐等，2017）、湖南（杨先乐等，2017）、上海（房文红，2017）、浙江（沈锦

玉等，2017）、广东（简纪常等，2017）东北等地区开展了集成示范和推广应用，取得了良好的社会效益、经济效益和生态效益。

该文发表于《中国渔业质量与标准》【2014，4（5）：6-9】

苏北水产安全用药示范养殖区的实践与总结

The practice and summary of drug-used safety demonstration within aquaculture zones in northern Jiangsu province

孙琪[1]，赵依妮[1]，王会聪[2]，胡鲲[1]，杨先乐[1*]

（1. 上海海洋大学，上海 201306；2. 江苏农林职业技术学院，江苏 句容 212400）

摘要： 介绍了苏北水产安全用药示范养殖区建设情况，总结了示范区推进安全用药的经验，对加强中国水产养殖安全用药管理、保证水产品质量安全和提高水产品品质具有一定的指导意义，有利于提高中国水产品质量安全保障能力，推动水产养殖业的可持续健康发展。

关键词： 安全用药；质量；水产养殖

当前，中国水产养殖行业迅猛发展，但随之而来的安全用药及药物残留等问题也日益突出[1]。随着人们生活质量的不断提高以及安全意识的不断增强，人们要求水产品能安全、健康和放心食用。然而，由于一些养殖户没有合理的用药观念，只一味追求产量与效益，在养殖过程中使用不科学的养殖方法，乱用药、滥用药，使得当前水产养殖的生态环境不断恶化，病害及污染事故频繁发生，致使水产品质量无法得到保障[2]，特别是近年来发生的一系列质量安全事件，如大菱鲆（俗称"多宝鱼"）药残超标[3]、水产品中检测出孔雀石绿等禁用药物[4]，使水产品的质量安全备受人们关注。

渔药市场不规范，销售渠道混乱，产品标识不一，质量参差不齐，使得假、冒、伪、劣渔药混入市场[5-6]。而当前药物防治仍是中国水产动物病害防治中最简单、最直接、最有效和最经济的方式[7-9]。因此，在养殖过程中，安全用药在确保水产养殖业健康可持续发展、养殖水产品质量安全方面显得尤为重要。

要保证水产品的质量与安全，就要有科学的养殖模式、合理的养殖方法，安全用药。因此，上海海洋大学提出建立水产安全用药示范养殖区，引导养殖户学习正确的养殖观念，合理用药，安全养殖，带动周围养殖区域规范用药，进而推动整个水产养殖行业的健康发展，保障水产品的质量安全，让人们吃到放心的水产品。本文主要介绍了苏北水产安全用药示范养殖区的情况，对示范区的养殖模式进行了描述与探讨，总结多年来示范区内的一些养殖经验供养殖户参考，力求能带动周边水产养殖业的健康发展，最终推动中国水产养殖业的健康可持续发展。

1 苏北地区水产养殖概况

江苏是中国水产养殖大省，2005 年全省水产养殖面积达 82 万 hm²，养殖产量达 3 887 万 t，其中淡水养殖产量 2752 万 t[10]，2010 年全省水产养殖总面积为 7 亿多 m²[11]。苏北为全国沿海经济带重要组成部分，养殖水面占江苏养殖面积的 2/3 以上，池塘养鱼主要以青鱼、草鱼、白鲢、鳙、鲤、鲫以及团头鲂等常规鱼为主[10]。其中，盐城市射阳县盐场内有海淡水养殖面积约 2 000 多万 m²，淡水养殖精养面积 1 600 多万 m²，为全国最大的连片淡水鱼养殖基地；东台市淡水养殖面积约 800 万 m²，海水养殖面积 1 000 多万 m²[12]。

在建设安全用药示范养殖区之前，苏北地区大部分养殖户用药频繁且量大，费用至少可达 0.6 元/（m²·年）。其做法包括：为保证良好的水环境，每月用不同消毒药交替进行泼洒消毒，早春时使用聚维酮碘（含碘量 10%）消毒，温度上升后，使用二氧化氯、戊二醛+苯扎溴铵等消毒药物交替对水体进行消毒；清明过后，每月轮流使用不同抗寄生虫类药物预防相应的寄生虫病；初夏之后，每 1~2 个月使用不同抗生素与保肝护胆类药物内服，预防细菌性疾病等[12]。

通过频繁用药来预防各类疾病，虽短时间内在一定程度上减少了相应疾病的发生及死亡量，增加了收益，却激增了药物残留问题，降低了水产品的质量。从长远角度分析，滥用药物会引发病原产生耐药性[13]以及药物残留[14-16]等一系列问题。因此，滥用药物这种养殖方法不利于中国水产养殖业的可持续健康发展。

2　苏北水产安全用药示范养殖区实践与总结

2.1　苏北水产安全用药示范养殖区[11]

自2009年，上海海洋大学先后在江苏射阳、东台等地设立水产安全用药示范养殖点，逐渐扩大至目前300多万 m² 的安全用药示范养殖区。

苏北水产安全用药示范养殖区主要设在江苏省盐城市射阳县旺阳养殖公司与东台市（县/镇）三仓农场等地，各设14个示范点，每个示范点面积均为10 000 m²以上，总面积可达300多万 m²。主要养殖品种为鲫、草鱼、鲢、鲴和南美白对虾等。2013年示范区概况详见表1和表2。

表1　2013年射阳盐场安全用药示范区概况

Tab. 1　Details of drug-used safety demonstration area of 2013 in Sheyang salt field

区域 Area	养殖品种 Breeds	塘口面积/m² Pond area	塘口数/个 Pond number
示范区	鲤、草鱼和花白鲢	100 000	1
	黄金鲫和花白鲢		1
	异育银鲫鱼种		1
	鲴和花白鲢		1
	草鱼和异育银鲫		2
	草鱼和异育银鲫		2
	草鱼和异育银鲫		4
	草鱼和异育银鲫	200 000	2
对照区	草鱼和异育银鲫	100 000	6

表2　2013年东台市三仓农场安全用药示范区概况

Tab. 2　Details of drug-used safety demonstration area of 2013 in Sancang farm of Dongtai

区域 Area	养殖品种 Breeds	塘口面积/m² Pond area	塘口数/个 Pond number
示范区	异育银鲫、花白鲢和南美白对虾	106 667	4
	异育银鲫、花白鲢和南美白对虾	133 334	2
	异育银鲫、花白鲢和南美白对虾	100 000	6
对照区	异育银鲫、鲴和南美白对虾	116 667	1
		133 334	1
	异育银鲫、花白鲢和南美白对虾	100 000	4
	异育银鲫、花白鲢和南美白对虾	133 334	2

2.2　苏北水产安全用药示范养殖区建设方法

在设立示范区之前，首先了解了当地养殖条件，对该地区的整体养殖环境、养殖发病史、用药史进行调研，选择当地常规且发病较多的塘口作为示范点，然后对其之前的养殖模式进行深入分析总结。另外，在示范区周边选取一定数量、相同规格且养殖情况相似的区域作为对照区。安全用药示范区内实行统一日常养殖管理，按照全程监管和质量追溯的要求，示范区内设置明确标志牌（名称、面积、范围、塘口负责人和管理制度），落实责任单位和责任人。建立完整的生产档案，包括渔药使用情况，饲料投喂量。另外，需设立基础微生物实验室，设备包括无菌操作台、高压灭菌锅、培养箱和烘箱等，供实验人员操作细菌分离实验和药敏实验等，还需配置一台显微镜，用于镜检寄生虫等。

示范区内具体工作（4—10月）为每天定时巡塘，每10 d检测水质，每15 d随机取鱼做镜检、细菌分离实验，以检测鱼体寄生虫与细菌含量，进而决定是否需用药，以达到预防的目的。

使用专用记录册记录以上相应工作结果。内容包括巡塘日期、天气、检测水质时水体的各项指标、镜检寄生虫数量、种类、细菌分离实验菌落形态结构、数量（拍照）、药敏实验结果等；用药时，记录选用药物的名称、用量、用法及效果反馈等。同时，记录对照养殖区域内的用药情况、发病率以及死亡率。

为更好地建设安全用药示范区，提高养殖户规范用药的意识，组织多次安全用药讲座，邀请专家现场解答养殖户提出的各种问题。对示范区建设的意义及内容进行了宣传，并得到了积极的回应。

2.3 苏北水产安全用药示范养殖区建设结果与分析

通过安全用药示范区的建设，大大减少了水产养殖过程中常见疾病的发生，降低了发病率，与此同时，示范区内使用药物的种类与数量远远低于对照区，降低了用药成本，增加了养殖户的经济收益（表3和表4）。

表3 2013年射阳盐场安全用药示范区结果

Tab. 3 Results of drug-used safety demonstration area of 2013 in Sheyang salt field

区域 Area	主要病害 Disease	主要药物 Drugs	平均用药成本/（元·m^{-2}·年$^{-1}$） Average medical cost
示范区	鳃出血病	二氧化氯、二硫氰基甲烷、免疫增强剂和 Vc 等。	0.328
对照区	孢子虫病、车轮虫病、指环虫病、鳃出血病、大红鳃、出血病、肠炎烂鳃病	青蒿末+地克珠利、马杜霉素+磺胺二甲氧嘧啶、盐酸氯苯胍、甲苯咪唑、二硫氰基甲烷、二氧化氯、戊二醛+苯扎溴铵、大黄+黄岑+黄芪、恩诺沙星+氟苯尼考等	0.75

表4 2013年东台市三仓农场安全用药示范区结果

Tab. 4 Results of drug-used safety demonstration area of 2013 in Sancang farm of Dongtai

区域 Area	主要病害 Disease	主要药物 Drugs	平均用药成本/（元·年$^{-1}$） Average medical cost
示范区	—	二氧化氯、硫酸铜+复合磺酸盐、分裂素、底质改良剂和粒粒氧等。	0.34
对照区	孢子虫病、大红鳃和藻类爆发	灭藻肽、硫酸铜+复合磺酸盐、分裂素、氨基酸、硅藻种源、小分子肽+胨+营养激活素+藻种、底质改良剂、EM菌、芽孢杆菌、盐酸氯苯胍、环烷酸铜、蒲公英粉末、地克朱利和盐酸恩诺沙星等。	0.59

2.4 讨论

分析总结可知，相同地区的示范点间存在很大程度的相似性，不同地区的则有显著差异。总而言之，建设示范区需针对当地的养殖环境，设定不同的方案，不同地区的示范区需采取不同的措施。如：射阳盐场主要以一些常见疾病预防为主，而三仓农场则偏重于水质调控。为解决地域间的差异就需要工作人员在设定示范区（点）之前，对当地的养殖环境等做好充分的调研工作，并做出合理的规划。

建立完整的养殖生产档案，整个过程虽繁琐，却有重要借鉴意义。通过历年的记录总结，形成一些可借鉴的经验：1) 温度为20~28℃时，为鳃出血病的高发期，在25~27℃左右的时候，死亡率最高，该病常常会在这个时候暴发。这种季节时应注意在饲料中添加免疫增强剂以及保肝护胆类药物，增强鱼体机能，提高鱼体免疫力和抗应激能力；而非采用消毒剂对水体消毒，同时内服抗生素。后种做法会更加刺激鳃出血病的发生。2) 鳃出血病在温度超过30℃时，危害性逐渐减小，死亡率下降甚至停止；当温度低于10℃，其发病时间延长，死亡率低且呈慢性死亡，故该病常发于春末、初夏及初秋，高温季节几乎不发病，无需用高成本的药物预防该病，进而增加养殖成本。3) 五、六月份时，水温刚有回升，为大红鳃多发季节，此时使用二氧化氯 0.15~0.30 g/m^2 对水体进行消毒以预防该病的发生；如若发生该病，可及时用二硫氰基甲烷

0.075 mL/m²泼洒治疗。4）通过长期药敏实验总结出，大多数抗生素的抑菌效果不是特别明显，这可能是由于养殖户长期大量使用抗生素，导致病原菌产生耐药性的原因，故不提倡养殖户为达到预防不存在的病原的目的，而盲目地定期投喂抗生素。

5 年来，上海海洋大学通过建设苏北水产安全用药示范养殖区，运用定点、定时的监测与检测，杜绝了养殖户盲目地乱用药、滥用药的现象，为示范区的养殖户平均节省用药成本约 0.30 元/m²。利用前几年建设示范区积累的经验，2013 年在示范区的发展过程中，基本不用或很少用到抗生素类药物，极大程度降低用药成本，提高了鱼体本身品质，增加水产品质量安全，明显提高了养殖户的收益。与此同时，通过安全用药示范区的建设，规范了养殖户的用药观念，树立了正确的安全用药意识，大幅度降低了水产品中药物的残留量，促进了水产养殖业的健康可持续发展。

3　展望

苏北水产安全用药示范养殖区的成功建设不仅对该地区水产养殖行业的健康发展有极大的推动作用，同时对其他地区安全用药示范区的建设以及周边养殖业的发展具有指导意义。示范区内科学合理的用药提高了水产品的质量与安全性。该养殖方式可在其周边养殖区域内进行大规模推广，但由于它是针对其所在示范区内经验总结的一套养殖方式，各地区在地理环境、水环境、养殖品种和管理模式上均有一定的差异性，导致每种养殖方法就具有了针对性，所以在进行全国范围内的推广之前，还需要更大范围地推广安全用药示范区建设，更广范围地总结和验证示范区的养殖方式。但是，由于示范区内需要镜检及基础微生物实验操作，示范区建设前期需对相关设备进行投入及人才培养，同时，也需政府加大对安全用药示范养殖区推广的鼓励以及人才的引进。

参考文献

[1]　李振龙. 科学用药降低水产养殖风险 [J]. 中国水产，2007（6）：6-7.
[2]　张聪姜，启军. 我国水产品质量安全问题与对策建议 [J]. 山西农业科学，2010，38（3）：61-64.
[3]　宋迁红，卫红星. 从大菱鲆的药残谈水产养殖中的安全用药问题 [J]. 科学养鱼，2007（2）：5.
[4]　高平，陈昌福. 浅谈水产养殖动物疾病防治中的安全用药问题 [J]. 饲料工业，2007，28（6）：60-62.
[5]　刘锡胤，姜成嘉，张榭令. 关于推进水产养殖用药科学化管理的探讨 [J]. 现代渔业信息，2006，21（9）：16-17，23.
[6]　曹海鹏. 从水产药害事故谈渔药的安全使用 [J]. 当代水产，2011（8）：35-37.
[7]　王玉堂，吕永辉. 水产养殖用药与水产品质量安全 [J]. 农业工程，2011（3）：44-49.
[8]　叶金明，杨显祥，姜增华，等. 国内渔药使用现状、问题及合理化建议 [J]. 中国水产，2007（5）：65-69.
[9]　杨先乐，郑宗林. 我国渔药使用现状、存在的问题及对策 [J]. 上海水产大学学报，2007，16（4）：374-380.
[10]　罗永光. 苏北池塘养鱼技术调查与研究 [D]. 南京：南京农业大学，2006.
[11]　史志中. 江苏启动智能化低碳循环水养殖试验 [J]. 农机科技推广，2012（9）：45.
[12]　孙琪，胡鲲，杨先乐. 苏北水产安全用药示范养殖区建设 [J]. 渔业信息与战略，2013，28（2）：123-126.
[13]　孟思好，孟长明，陈昌福. 规范用药与致病菌药物敏感性检测 [J]. 渔业致富指南，2013（10）：67-68.
[14]　廉超，雒敏义，宫瑞. 浅析我国渔药研发管理现状及未来发展趋势 [J]. 水产学杂志，2012，25（1）：58-63.
[15]　陈洪大，LunestadBT. 水产养殖治疗性药物的使用、法规和药残控制 [J]. 水产科学，2009，28（7）：419-423.
[16]　孟思好，孟长明，陈昌福. 水产养殖用药与水产品质量安全问题（中）[J]. 科学养鱼，2009（6）：77.

湖南地区渔药安全使用风险评估及控制技术实践与总结

杨先乐，胡鲲，刘思雅

（上海海洋大学，上海 201306）

在"十五"、"十一五"和"十二五"期间，上海海洋大学科研团队较为系统地开展了渔药安全使用技术研究。积累了一大批符合我国水产养殖产业特点的技术成果和推广应用经验。"十二五"期间，上海海洋大学首次、全面地在全国范围内开展了渔药使用状况调查，并在主要水产养殖区域建立风险可控的渔药安全使用示范区，力争从技术上解决长期困扰水产养殖产业发展的瓶颈性难题。

1　渔药安全使用风险评估及控制技术主要成果

建立了一套从分子水平到组织器官水平的渔药药理学研究技术方法；较为系统地开展了鱼类药物代谢酶 CYP 和渔药受体 GABA 参与渔药代谢的功能研究；分析了多糖蛋白 Pgp 与药物残留量和剂型的关联性分析，并将之用于渔药的安全性评价。

针对主要养殖品种（如四大家鱼、黄颡鱼、小龙虾等），对主要渔药（如吡喹酮、沙拉沙星等）系统地开展了检测方法的研究，弄清了其代谢过程和残留消除规律，建立了其安全使用技术规范。

对主要禁用药物和抗生素（孔雀石绿、环丙沙星等）在水产品中的残留建立了快速检测试剂盒/试纸条；建立了能拟合拟合了禁用药物在鱼类组织中蓄积规律的数学模型。

研制了新型的渔药剂型，新型渔药制剂具有良好的缓释和抗逆（光、热等）性能。

建立了外源污染物对鱼类的安全性评价方法：利用 Pgp 等指示蛋白，对异噻唑啉酮、吡啶硫酮锌等常见日用/化工用消毒剂对鱼类的潜在安全性进行了评价。

对淡水鱼类主要致病菌——嗜水气单胞菌进行了分型，对其耐药性产生的机制进行了探讨，制定了延缓药物耐药性的新措施。

分析了聚维酮碘等消毒剂在养殖环境中的归趋性。

以速效降解水体中氨氮和维持高密度养殖环境下动物肠道微生态平衡、提高饵料利用效率为目的，研制出纳豆芽孢杆菌、复方植物乳杆菌、反硝化细菌等新型微生态制剂。

系统地收集整理了国内第一个渔药使用状况调查数据库。

建立并维护了国内唯一一个渔药安全领域的公共微信账号。

2　湖南示范区的建设情况

华容县位于湖南省北部边陲，岳阳市西境，北枕长江，南滨洞庭，境内水域众多，水资源十分丰富。拥有养殖水面 18.9 万亩（其中池塘养殖面积为 5.8 万亩；池塘网箱养鳝面积为 2.5 万亩，网箱 32 万口，网箱面积 140×10⁴ km²；湖泊水库大水面约为 10 万亩；稻虾连作面积 6 000 亩），滩涂面积 45 万亩，素以"鱼米之乡"著称。

华容养殖水域面积大、产量高，且养殖模式极具代表性。自 2004 年起，华容县水产品产量突破 10×10⁴ t，成为湖南省水产养殖区第一县。华容县水产养殖品种涵盖了鱼、虾、贝、龟、鳖、蛙、泥鳅等七大类 60 多个品种。其中包括高效特色水产品种小龙虾、黄颡鱼"全雄 1 号"、台湾泥鳅、黄鳝、淡水鲈鱼等。华容县的水产养殖模式包括：湖泊水库等大水面养殖、池塘精养、网箱养鳝、稻虾连作、虾莲共生等。2008 年水产品总产量 10.62×10⁴ t，渔业总产值为 11.15 万元，其中苗种占 240 万元。2009 年水产品总产量 10.12×10⁴ t，其中甲壳类占 0.1×10⁴ t，渔业总产值达到 11.77 万元，占农业总产值的 25.5%。2010 年水产品总产量 10.53×10⁴ t，渔业总产值迅速增长到 12.36 万元。2011 年水产品总产量 11.23×10⁴ t，甲壳类增产至 0.2×10⁴ t，渔业总产值达 12.9 万元。2012 年水产品总产量 11.99×10⁴ t，渔业总产值高达 18.9 万元。2013 年水产品总产量 12.35×10⁴ t，与前年同比增长 4.84%。2014 年水产品总产量 13.04×10⁴ t，同比增长 5.5%，渔业总产值 18.26 万元。2015 年渔获量 13.42×10⁴ t，渔业生产总值 18.46 万元，同比分别增长 2.9% 和 1.09%。根据往年的水产事业的增长趋势，预计 2016 年的水产品产量可达 13.91×10⁴ t，渔业生产总值约为 18.69 万元。逐年提高的产量和显著增加的收益实现了渔业致富。

渔药是水产动物病害防治最为直接和有效的方式。目前，华容县及周边地区几乎集中了全国所有的渔药生产、销售厂家和渔药品种。

3　示范区涉及的养殖户（场）的个数、人数、面积

上海海洋大学于 2012 年在湖南省华容县建立了《渔药使用风险评估及其控制技术研究与示范》"湖南省华容县级示范区"，并将以上技术在示范区进行了示范，2012—2016 年以湖南省田家湖渔业科技有限责任公司为示范区代表参与的养殖企业（户）与辐射个体户共 567 户（其中小龙虾 188 户、四大家鱼等 331 户、黄鳝 48 户），总面积达 107 819.53 亩（其中小龙虾 13 974.63 亩、四大家鱼等 91 498.41 亩、黄鳝 2 346.49 亩），详见图 1；涉及鱼类养殖品种包括四大家鱼、乌鳢、鳜鱼、黄颡鱼、鲈鱼等等，详见图 2；小龙虾的养殖模式有池塘养虾、稻虾连作和虾莲共生，详细分布见图 3；网箱养鳝的养殖面积分布见图 4。

4　所取得的经济、社会和生态效益

本项目的实施对保障水产品质量安全，保护水生态环境，促进水产养殖的可持续发展提供较大的技术支撑。

经济效益：2012 年至 2016 年期间，①减少因渔药不当使用造成的直接经济损失 1 456 万元以上，间接经济损失达 3 000万元以上。②水产养殖病害明显减少，病害的有效控制率明显提高，每年减少因病害造成的经济损失达 17% 以上，折合人民币 2 053 万元，直接经济效益近 1 亿元。

生态效益：目前可现场使用胶体金免疫层析法原理的试剂盒检测水产鱼虾、生产渔具与养殖用水中孔雀石绿的显隐性；可现场使用原理为间接竞争 ELISA 方法的快速检测试剂盒检测水产品鱼虾蟹组织中恩诺沙星的残留，可用免疫竞争法原理的试剂盒检测恩诺沙星代谢物环丙沙星的残留；可现场使用酶联免疫检测原理的检测盒快速检测氯霉素的残留等等。快速检测试剂盒/卡不仅操作简单可现场操作、检测时间短快速且特异性高、可准确定性，适合大批量水产品筛选。

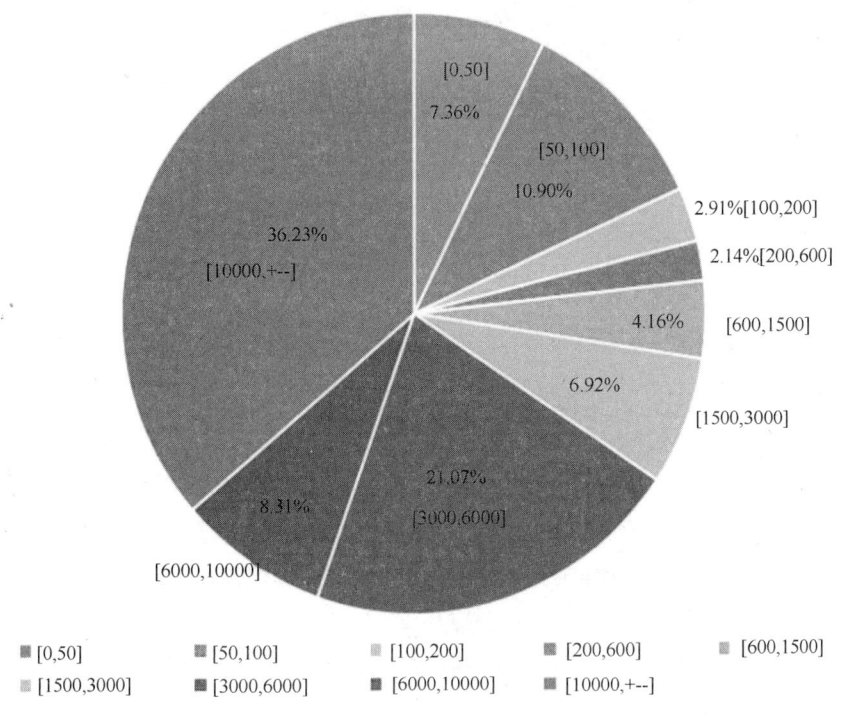

图1　"渔药使用风险评估及其控制技术"华容县示范区的养殖户面积比例分布（亩）

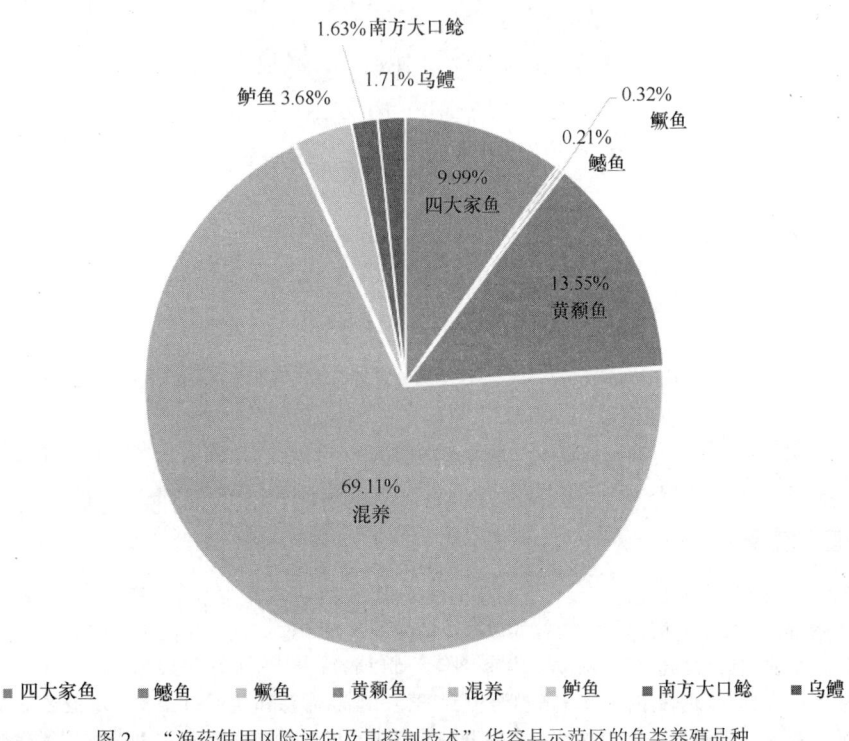

图2　"渔药使用风险评估及其控制技术"华容县示范区的鱼类养殖品种

　　以湖南省田家湖渔业科技有限责任公司、岳阳宏利渔业开发有限公司、华容县二郎湖渔场等为代表的示范区内所有养殖品均符合相关质量安全标准，水产品的质量合格率较2012年前（示范区未建立以前）提高20%以上，较2013年前（示范区中期）提高15%以上，示范区水产养殖外排水符合《淡水池塘养殖水排放要求》（SC/T9101-2007）和《海水养殖水排放要求》（SC/T9103-2007）的规定，为华容县生态环境的改善作出了较大的贡献。

　　社会效益：本研究成果，一方面可提高我国水产品在国际上的地位，打破欧盟等发达国家的"绿色贸易"壁垒，提高

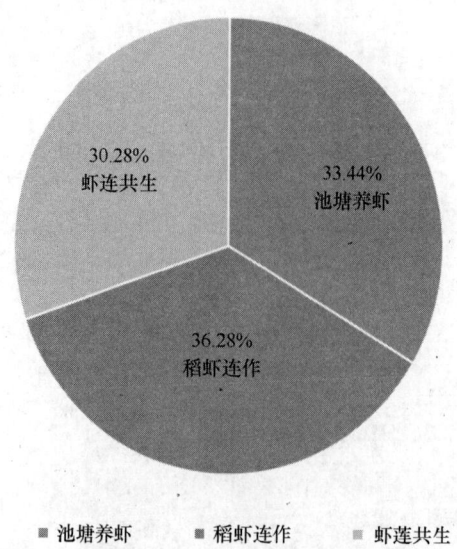

图3 "渔药使用风险评估及其控制技术"华容县示范区小龙虾的养殖模式

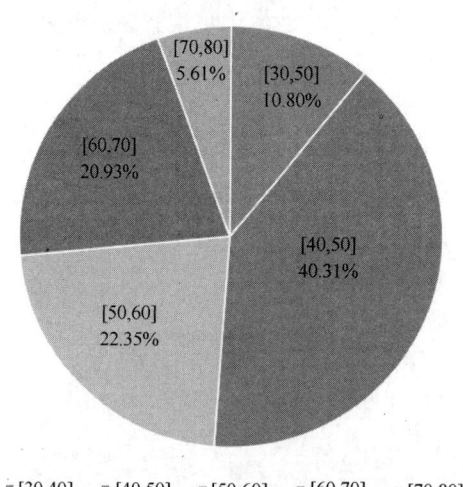

图4 "渔药使用风险评估及其控制技术"华容县示范区的网箱养鳝的养殖面积分布

我国水产品在国际上的竞争力；另一方面，通过改变渔药滥用的现象，保护人类健康。水产养殖业的健康可持续发展，可以为社会主义新农村建设增加农民收入、解决农村劳动力解决就业问题，构建和谐社会。

5 建设示范区的意义

"优良的产业资源需要先进的科技促进发展，实验室的成果需要在生产实践中推广和验证"。华容县水产局、水产技术推广站、渔政局等主管部门和以上海海洋大学为首的课题组清醒地认识到这一点，并很快达成共建"示范区"的共识。值得指出的是，尽快以往科研项目大都也有"示范"内容，但大都以养殖户、养殖场等零星单位为主体，随机性较强，不利于全程跟踪和技术服务；在具有影响力的水产主养区建立封闭性、全程跟踪和监管的县级"示范区"在全国尚属首次。

在"党和政府将包括水产养殖品在内的食品安全提升到前所未有的政治高度"和"渔药残留风险控制问题也成为影响我国水产养殖业可持续发展的主要瓶颈之一"的大背景下，针对华容县独特的水产养殖特点度身打造的"示范区"首创了一种"区域性封闭、全程跟踪示范"的新模式，既是新时期科技实践活动的一种新尝试，具有极强的现实意义和推广价值。

该"示范区"通过"渔药使用风险评估及其控制技术研究与示范"项目成果的实施，取得显著的经济效益、社会效益和生态效益。

上海奉贤"渔药使用安全风险评估及控制"示范区建设与示范

房文红

（中国水产科学研究院东海水产研究所）

1　地点及起始时间

技术示范：上海市奉贤区的对虾养殖企业（合作社），技术辐射：上海市奉贤区和青浦区的对虾养殖企业（合作社）

起始时间：2012—2016 年

2　实施单位、负责人

在上海奉贤对虾主养区，选择了 12 家水产养殖专业合作社建立了"渔药使用安全风险评估及控制示范区"，示范面积共计 1221 亩：上海礼和渔业养殖专业合作社（朱珺）、上海鸿庄水产养殖专业合作社（冯辉）、上海海锋水产养殖专业合作社（徐平）、上海景海水产养殖专业合作社（钱辉仁）、上海集贤虾业养殖专业合作社（胡忠）、上海康源虾业养殖专业合作社（项伟东）、上海锭顺水产养殖专业合作社（王雷杰）、上海金禾渔业专业合作社（唐建东）、上海荣盛虾业养殖专业合作社（潘银龙）、上海傲天水产养殖专业合作社（张进）、上海金贤水产养殖专业合作社（陈华强）和上海小宝虾业养殖专业合作社（林春浩）。

3　养殖品种和养殖方式

养殖品种：凡纳滨对虾；养殖方式：土池露天养殖。

4　主要工作内容

（1）根据抗菌药物在对虾体内吸收、分布和消除规律，结合对虾主要病原菌弧菌对抗菌药物的体外 MIC 结果，建立药动-药效联合模型，确定用药剂量和给药周期（见表 1）；根据抗菌药物在肌肉可食组织中的消除规律，结合抗菌药的最高残留，制定其休药期（见表 1）。

表 1　抗菌药在凡纳滨对虾体内用药方案和休药期

抗菌药物	用药方案【剂量，频率，周期】	休药期	最高残留限量
恩诺沙星	20~30 mg/kg，每日 2~3 次，连续使用 3~5 d	15 d	100 μg/kg
氟苯尼考	20~30 mg/kg，每日 3 次，连续使用 3~5 d	7 d	1 000 μg/kg
甲砜霉素	20~30 mg/kg，每日 2 次，连续使用 5 d	7 d	50 μg/kg
噁喹酸	30 mg/kg，每日 2 次，连续使用 5 d	5 d	300 μg/kg
复方新诺明	83.3 mg/kg SMZ +16.7 mg/kg TMP，每日 3 次，连续使用 3~5 d	7 d	100 μg/kg

（2）开展渔药使用情况调查，建立了区域性的《渔药使用状况数据库》

项目实施 5 年间，以上海郊区水产养殖为主，开展了渔药使用情况调查，共调查 250 余份，其中调查对虾养殖户 92 份，对示范区进行了长期跟踪调查。从调查结果来看，消毒剂、调水底改制剂和微生态制剂占据前三位，在调查养殖户中出现的频次分别为 72.8%、64.1% 和 51.1%；其次是抗应激药品，出现频次为 26.1%；抗菌药和杀虫药分别为 22.8% 和 9.8%，抗菌药和杀虫药呈现逐年减少的趋势，从养殖户出现的频次来看，5 年间分别为 42.1%（8/19）、40%（8/20）、35.3%（6/17）、22.2%（4/18）和 22.2%（4/18）。

（3）通过技术培训和现场指导，宣传渔药安全使用技术，规范了渔药投入品的管理和使用，加强档案管理制度的执行，实现了渔药安全使用，确保了水产品质量安全。累计培训 10 次，培训 720 余人次。

（4）对养殖产品进行药残抽检，发挥监督和监管作用。

随机抽取 24 家养殖企业（合作社）45 份养殖对虾产品进行药物残留检测，共检测了诺氟沙星、恩诺沙星、环丙沙星、沙拉沙星、氟甲喹、土霉素、甲砜霉素、氟苯尼考、复方新诺明和噁喹酸等抗菌药残留。45 个对虾肌肉样品中诺氟沙星、沙拉沙星、氟甲喹、甲砜霉素、复方新诺明和噁喹酸都没有检出；有 9 个样品检出恩诺沙星、3 个样品检出环丙沙星，但均低于最高残留限量（0.1 mg/kg）；有 3 个样品检测到土霉素，但低于最高残留限量（0.1 mg/kg）；有 3 个样品检测到氟苯尼考，但均低于最高残留限量（0.1 mg/kg）。

1031

5 效益

示范区内 10 家合作社、辐射区内 8 家企业（合作社）养殖生产的对虾先后获得农业部农产品质量安全中心"无公害农产品证书"。经初步统计，2016 年示范区养殖生产对虾累计总产值达 2 820 万元，平均亩产经济效益 2.31 万元；辐射区内累计总产值 3.459 亿元，平均亩产经济效益 1.814 万元。示范区抗菌药和杀虫药使用量降低了 20%，亩增效 10% 以上，养殖品均符合相关质量安全标准；18 家合作社养殖生产的对虾获得"无公害农产品证书"。

6 辐射情况

在上海奉贤和青浦对虾养殖产区进行了辐射应用，辐射面积共 19 070 亩（奉贤和青浦分别为 13 652 亩和 5 418 亩）。

7 经验、成果总结

（1）摸清药物在养殖动物体内药动学规律，建立药动-药效联合模型，为科学制定用药方案和休药期奠定坚实的理论基础。

（2）通过技术培训、现场指导、发放资料等方式，加强渔药安全使用技术宣传，确保技术能有效落地。

（3）对技术示范区和辐射区开展渔药使用现状调查，摸清渔药使用现状，为渔药和饲料等投入品的监管提供真实数据。

（4）建立完善的渔药、饲料等投入品的市场准入制度，杜绝禁用药物的使用，确保市场渔药、饲料等投入品安全。

（5）切实做好生产记录，规范渔药、饲料等投入品的使用与管理。

（6）宣传渔药安全使用技术，渔药使用者严格执行休药期。

（7）对养殖产品进行药残抽检，发挥监督和监管作用。

广东（高州市和湛江市）渔药安全使用风险评估及控制技术实践与总结

简纪常，汤菊芬

（广东海洋大学）

1 生产应用单位

（1）高州市朗业畜牧渔业科技养殖有限公司联系方式：广东省高州市石鼓镇 0668-6330978。

（2）湛江市水产技术推广站广东省湛江市赤坎区北桥路 6 号 0759-3106480。

2 面积

1.3 万余亩。

3 养殖品种和养殖方式

广东海洋大学于 2012—2016 年在湛江市和茂名高州市建立了 2 个《渔药使用风险评估及其控制技术研究与示范》示范区，区内共建立了等 10 个示范点，并将以上技术在示范区进行了示范，四年来参与的养殖场和企业共 156 户，面积达1.3 万余亩，养殖户面积比例见图 1、图 2。涉及的养殖品种包括对虾（凡纳滨对虾、斑节对虾、日本囊对虾）、罗非鱼、海水鱼（石斑鱼、卵形鲳鲹、美国红鱼、红笛鲷、黄鳍鲷等）。

4 主要工作内容

（1）氟苯尼考等 5 种渔药的检测方法和安全使用技术规范；

（2）中草药和中草药微生态制剂的应用；

（3）参与渔药使用状况调查数据库的建立和维护；

（4）参与建立并维护"渔药使用风险评估及其控制技术研究与示范"微信公众号；

（5）组织技术培训，参加培训人数达 1 000 人次；

（6）在"示范区"开展渔药使用状况调查和用药安全技术指导。

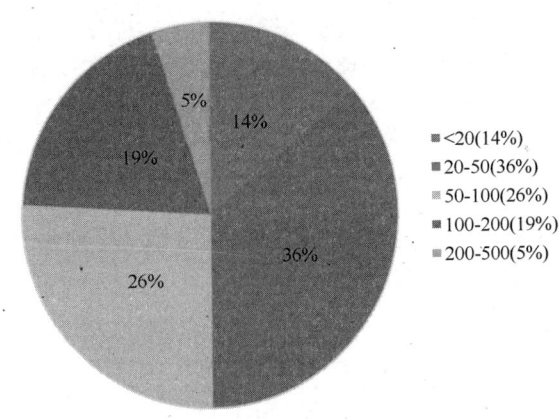

图 1　"渔药使用风险评估及其控制技术"示范区的养殖面积比例分布

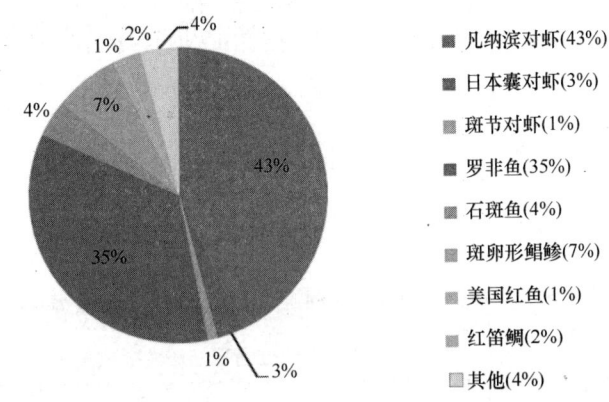

图 2　"渔药使用风险评估及其控制技术"示范区养殖品种比例分布

5　效益（包括社会，生态，经济效益，产品的安全性等）

本项目的实施对保障水产品质量安全，保护水生态环境，促进水产养殖的可持续发展提供强有力的技术支撑。

经济效益：2012—2016 年期间，①减少因渔药不当使用造成的的经济损失 1 000 万元以上，间接经济损失 2 500 万元以上。②水产养殖病害明显减少，病害的有效控制率明显提高，每年减少因病害造成的经济损失达 15% 以上，折合人民币 1 800 万元，养殖水产品产量提高 8% 以上，直接经济效益近 2 200 万元。

生态效益：化学药物的使用量降低 20% 以上，其中抗生素的使用量减低 30% 以上，有效保护了养殖生态环境。

示范区内所有养殖品种均符合相关质量安全标准，水产品的质量合格率较 2013 年前（示范区未建立以前）提高 15% 以上，示范区水产养殖外排海水符合《淡水池塘养殖水排放要求》（SC/T9101-2007）和《海水养殖水排放要求》（SC/T9103-2007）的规定，为生态环境的改善作出了较大的贡献。

社会效益：本研究成果，一方面可提高我国水产品的质量安全，从而提高在其国际贸易上的地位，打破欧美等发达国家的"绿色贸易"壁垒，提高我国水产品在国际上的竞争力；另一方面，科学合理使用渔药培养了渔民"科学合理用药"的意识，改变了渔药滥用的现象，保护生态环境和人类健康，促进水产养殖业的健康可持续发展，有利于增加农民收入、解决农村劳动力就业及社会主义新农村建设。

6　辐射情况

高州市示范区开展的技术应用示范主要辐射整个高州罗非鱼养殖。

湛江市示范区开展的技术应用示范主要辐射整个湛江市的对虾（凡纳滨对虾、斑节对虾、日本囊对虾）、罗非鱼、海水鱼（石斑鱼、卵形鲳鲹、美国红鱼、红笛鲷、黄鳍鲷等）养殖。

7　经验总结

与水产技术推广部门、渔业合作社及大型渔业生产企业联合建立示范区，利用他们的技术推广力量强、组织化程度高、辐射范围广的特点和平台示范效应，能有效调动渔民的积极性，同时可以让有关相关技术直接与渔民无缝对接，提高了技术推广和示范的效率。创新宣传手段，积极开展多种形式的活动，广泛宣传普及渔药安全知识，全力提升示范区内渔民科学用药的的意识和水平。一利用科研团队的实力，组织有关渔民参加渔药知识和安全使用的培训，贴宣传画和悬挂横幅，发放渔药安全知识很使用小册子和文化衫二让渔民加入"渔药使用风险评估及其控制技术研究与示范"微信公众号，通过手机和网络可以快速了解有关渔药的知识、药物使用风险、科学使用的知识以及最新的渔药鱼病咨询。

浙江德清县钟管镇"渔药使用安全风险评估及控制"示范区实践与总结

沈锦玉，姚嘉赟

（浙江省淡水水产研究所）

1 地点及起始时间

针对钟管镇水产养殖的发展及现状，2014—2016 年浙江省淡水水产研究所"渔药使用风险评估及其控制技术研究与示范（201203085-2-2）"课题组在德清县钟管镇建立"渔药安全使用风险评估及控制"示范区。

2 实施单位、负责人

针对钟管镇水产养殖的发展及现状，浙江省淡水水产研究所"渔药使用风险评估及其控制技术研究与示范（201203085-2-2）"课题组联合德清县水产技术推广站、钟管镇农办，在钟管镇建立"渔药安全使用风险评估及控制"示范区，该区以钟管镇的规模化养殖场、家庭农场、养殖公司以及具有能安全规范使用渔用药物的养殖户为示范点（表1）。

示范区承担单位：浙江省淡水水产研究所，负责人：沈锦玉，主管部门：浙江省海洋与渔业局。

表1 应用单位情况

序号	养殖场名称	负责人	联系电话	养殖场地址	示范面积（亩）
1	德清县小根养殖场	顾小根	13004258891	戈亭村	2100
2	浙江清溪鳖业股份有限公司	王根连	13362283831	曲溪新港园区	606
3	德清延炜龟鳖养殖场	姚志刚	13306825319	沈家墩村	330
4	德清腾跃家庭农场有限公司	姚胜	13626725389	钟管村溪口墩	35
5	德清钟管绿丰水产养殖场	沈金根	15967231299	干山村	422
6	德清钟管吴建荣水产养殖场	吴建荣	13059916899	东舍墩村	157
7	水产养殖户	朱法荣	13967281995	干山村	132
8	德清钟管月明家庭农场	汪月敏	13017901088	钟管村夏家湾	282
9	德清钟管三兴家庭农场	沈山春	13967265928	东坝头村	355
10	德清钟管阿祖养殖场	泮建祖	13867269976	干山村高桥头	143
11	德清陆氏水产养殖有限公司	房永根	13336832935	审唐村	330
12	德清钟管杭根法水产养殖场	杭根法	13059901468	北代舍村	122
13	水产养殖户	王福田	13587258162	青墩村	204
14	水产养殖户	蔡根芳	13665713036	干山村	113
15	德清钟管徐卧虎淡水水产养殖场	徐卧虎	13757231381	曲溪村	305
16	水产养殖户	潘益新	13567256578	曲溪村	116
17	水产养殖户	沈笑根	13754226222	干山村	122

3 示范面积、养殖品种和养殖方式

钟管镇的17家规模化养殖企业（户）作为示范点，示范面积5 874亩，养殖品种包括黄颡鱼、翘嘴红鲌、淡水青虾、

1034

龟鳖、鳜鱼、鲈鱼、太阳鱼、泥鳅等名优水产养殖品种，也包括青、草、鲢、鳙、鲫、鳊等常规水产养殖品种。

养殖模式也包括了高密度集约养殖模式、淡水青虾的两茬养殖模式、龟鳖温棚养殖模式、稻田生态养殖模式等等，示范区代表了的钟管镇水产养殖现状，而示范点对周边养殖户有相当强影响力，推动渔药安全使用技术的示范和推广。

4 主要工作内容

（1）现场指导和培训

为了提高各示范点的渔药安全，保证水产食品的质量和安全，课题组定期与德清县水产技术推广站、钟管镇农办的技术人员在苗种放养前、养殖生产过程中以及上市商品鱼上市前，到示范点进行现场指导和培训，检查水产养殖投入品的使用情况，发放课题组编写的《渔药安全使用技术手册》及其他技术资料。现场解答渔药安全使用的技术、水产动物病害防治技术等。并由钟管镇农办与示范点签定初级水产品质量安全承诺书，进行水产品安全监管记录。

（2）水产投入品的使用指导

在养殖生产过程中，严格禁止示范点使用"三无"饲料、饲料添加剂、变质及过期的饲料，预防和治疗的渔用药物严格选用按《渔药安全使用技术手册》规定的制剂、使用剂量和使用方法，并由具有 GMP 资质的兽药或渔药生产企业生产。禁止选用成分标注不明、抗生素原粉、超过有效期的渔药。使用后，严格执行休药期的规定。

（3）调控水质、减少化学药物的使用剂量和频度

课题组制备了枯草芽孢杆菌、干酪乳杆菌等有益微生物水质改良剂，分送于各示范点，并指导有益微生物水质改良剂、有益微生物底质改良剂的使用。

如吴建荣养殖场和小根养殖场，4—10 月期间，每月监测水质 1~2 次，每次监测 2~3 口池塘。2014—2016 年共采集水样 752 个水样，测定了养殖池塘中的 pH、溶氧、氨氮、亚硝酸盐氮、COD、总磷、总氮等 7 个指标，根据检测结果，指导各示范点应用枯草芽孢杆菌、干酪乳杆菌、复合 EM 菌等有益微生物水质和底质改良剂，使养殖池塘的水质指标符合渔业水质标准和淡水池塘养殖尾水排放标准。且各示范点的发病池塘数量、危害程度均大幅度下降，示范期间无大规模病害的暴发和流行。

（4）监测养殖池塘的病原微生物，做到以防为主

示范期间，监测养殖池塘的病原微生物特别是气单胞菌等的数量。根据监测结果，指导示范点根据养殖情况，及时使用消毒剂、抗菌药物。监测养殖池塘的病原微生物水样 37 个，通过早期的预防，减少了养殖池塘的发病率，使疾病得到了有效的控制。

（5）对症用药，严格执行休药期

在水产养殖生产过程中，示范点水产动物在发病后，课题组迅速到各示范点采集病样，或要求示范点、幅射区养殖户送检病样，目检和镜检体表、鳃后，实验室解剖观察内脏并进行细菌分离培养，并对病原菌进行药敏试验，根据药敏试验结果指导示范点安全用药；对疑似病毒病样品用 PCR 检测分析，提出隔离和防治措施。

5 效益（包括社会，生态，经济效益，产品的安全性等）

通过水质监测，指导有益微生物水质改良剂和底质改良剂的推广使用；养殖池塘病原微生物的监测，做到早防早治；对症下药；严格执行休药期等技术措施的实施，推动钟管镇渔用药物的使用向高效、安全、无公害方向提升，降低养殖成本，使示范区的水产动物发病率及因病导致的渔业经济损失大幅度的减少，同时通过严格执行休药期的规定，使初级水产品的抽查合格率达到 100%，保证了水产品的品质和质量。2014—2016 年示范区养殖总产值 31 756.7 万元，总利润 5 659.72 万元，化学药物的使用量减少了 51.96%，降低了因经济损失 58.18%。

6 幅射情况

通过培训、示范点的示范和带动，使渔药安全使用技术覆盖钟管镇的主要水产养殖村，初步解决了渔用药物的安全使用、水产动物病害防治、水产品质量监管等问题，使全镇的水产养殖业向可持续发展的健康养殖方向发展。2014—2016 年全镇的渔药安全使用技术的示范幅射区养殖户 524 户，养殖面积 13 346.6 亩，养殖品种主要有黄颡鱼、翘嘴红鲌、淡水青虾、鲈鱼、龟鳖、泥鳅、鳜鱼及常规的青、草、鲢、鳙、鲫、鳊等。

推广应用"水产益生菌固定化轻简化实用技术"、"恩诺沙星轻简化实用技术"、"诺氟沙星盐酸小檗碱预混剂轻简化实用技术"、"硫酸铜硫酸亚铁防治淡水虾纤毛虫病风险控制轻简化实用技术"、"溴氯海因在淡水青虾养殖中的安全使用轻简化实用技术"等新技术新方法。在养殖生产过程中，参照示范点的模式，选择饲料和饲料添加剂，杜绝了"三无"饲料、饲料添加剂、变质及过期的饲料的使用。在水质管理上，镇农办和示范点将课题组定期定点的水质监测数据、微生物制剂的选用种类、使用效果等及时通知养殖户，使用养殖户在养殖技术、经营管理上达到无公害养殖的要求。

在发病季节，及时通知养殖户，采取预防措施，做到早防早控。发病后，课题组、镇农办和示范点通过现场、电话、网络等多种途径对养殖户进行技术指导，及时采取治疗措施。预防和治疗的渔用药物严格选用按《渔药安全使用技术手

册》规定的制剂。禁止选用成分标注不明、抗生素原粉、超过有效期的渔药。使用后，严格执行休药期的规定。

"渔药使用风险评估及其控制技术研究与示范（201203085-2-2）"项目实施后，经统计抗生素、杀虫剂等化学药物的使用量从平均220元/亩减少到了150元/亩，亩减少化学药物成本70元，化学药物的使用减少31.82%。降低水产品的经济损失46.15%，2014—2016年利润5483.3万元。

7 经验、成果总结

"渔药使用风险评估及其控制技术研究与示范（201203085-2-2）"现场技术指导和培训，能极大地提高了示范点的渔药安全使用的意识，保证了渔用药物安全使用技术措施的实施。

通过定期水质的监测指导有益微生物制剂，改善养殖水质，使养殖池塘的pH、溶氧、氨氮、亚硝酸盐氮、COD、总磷、总氮等水质指标符合渔业水质标准和淡水池塘养殖尾水排放标准。减少了全年的换水次数，降低水产养殖对自然水域生态环境的影响，使示范点的水产养殖向生态环保、低碳节能方向发展。同时各示范点的发病池塘数量、危害程度均大幅度下降。

通过监测养殖池塘的病原微生物，做到早期的预防，减少了养殖池塘的发病率，使疾病得到有效的控制。通过细菌分离培养及病原菌药敏试验，指导示范点对症用药，降低了因病导致的渔业经济损失和用药成本，用药后严格执行休药期的规定，提高了水产品的质量。对疑似病毒病样品用PCR检测分析，提出隔离和防治措施，控制了病毒性疾病的蔓延和流行。

8 其他相关内容

2013—2016年浙江省淡水水产研究所"渔药使用风险评估及其控制技术研究和示范"课题组在德清县的钟管、乾元、新市、禹越、新安镇、下渚湖街道进行了水产动物流行病调查，通过查明水产动物病害的流行特点、分布区域、危害程度等，为示范区的水产动物病害的早防早控提供技术支撑，降低养殖生产过程中病害损失，最大程度地减少养殖生产中的抗菌药物、驱杀虫药物的使用剂量和频率，控制水产品中药物残留。